Asthma and COPD

Peter J. Barnes, DM, DSc, FRCP, FMedSci, FRS

Professor Imperial College, London, UK

Jeffrey M. Drazen, MD

Brigham and Women's Hospital, Boston, MA, USA

Stephen I. Rennard, MD

University of Nebraska Medical Center, Omaha, NE, USA

Neil C. Thomson, MD, FRCP

Professor Respiratory Medicine, University of Glasgow, Glasgow, UK

Asthma and COPD

Basic Mechanisms and Clinical Management

Second Edition

ELSEVIER

AMSTERDAM • BOSTON • HEIDELBERG • LONDON
NEW YORK • OXFORD • PARIS • SAN DIEGO
SAN FRANCISCO • SINGAPORE • SYDNEY • TOKYO

Academic Press is an imprint of Elsevier

Academic Press is an imprint of Elsevier

525 B Street, Suite 1900, San Diego, CA 92101-4495, USA
30 Corporate Drive, Suite 400, Burlington, MA 01803, USA
32 Jamestown Road, London, NW1 7BY, UK
Radarweg 29, PO Box 211, 1000 AE Amsterdam, The Netherlands

Second edition 2009

Notice

No responsibility is assumed by the publisher for any injury and/or damage to persons or property as a matter of products liability, negligence or otherwise, or from any use or operation of any methods, products, instructions or ideas contained in the material herein. Because of rapid advances in the medical sciences, in particular, independent verification of diagnoses and drug dosages should be made

Medicine is an ever-changing field. Standard safety precautions must be followed, but as new research and clinical experience broaden our knowledge, changes in treatment and drug therapy may become necessary or appropriate. Readers are advised to check the most current product information provided by the manufacturer of each drug to be administered to verify the recommended dose, the method and duration of administrations, and contraindications. It is the responsibility of the treating physician, relying on experience and knowledge of the patient, to determine dosages and the best treatment for each individual patient. Neither the publisher nor the authors assume any liability for any injury and/or damage to persons or property arising from this publication

British Library Cataloguing-in-Publication Data
A catalogue record for this book is available from the British Library

Library of Congress Cataloging-in-Publication Data
A catalog record for this book is available from the Library of Congress

ISBN: 978-0-12-374001-4

Printed and bound in USA

09 10 11 12 10 9 8 7 6 5 4 3 2 1

Contents

Preface to the 2nd Edition

In the 6 years since the publication of the first edition of *Asthma and COPD* there has been substantial progress in understanding the biological basis of both of these obstructive diseases. In contrast both our understanding of the physiological basis of these conditions and their treatment has improved very little. We firmly believe that the progress made in understanding the basic disease biology that has accrued since 2002 will lead, in the years to come to new and better treatments of these crippling obstructive airway diseases. As our knowledge advances, we now appreciate the areas of overlap and distinction between these two conditions which share many physiological similarities. Most of the chapters of this book highlight both the similarities and differences between these conditions.

In this era of widely available information on the internet it is reasonable to ask what value is served by a book of this type? The answer is simple. The editors have assembled the world's experts on these topics and commissioned them to write succinct reviews focusing on their particular area of expertise. For the student of the biology or treatment of these conditions, this compilation is an important resource of up-to-date information. For the researcher or clinician, the investigative or clinical problems that each of face on a daily basis are clearly illuminated by topic experts. We hope that you will find this summary of value to you in understanding these two highly variegated, closely similar but clearly distinct entities at the biological and clinical level.

Peter J. Barnes
Jeffery M. Drazen
Stephen I. Rennard
Neil C. Thomson

About the Editors

Peter J Barnes, DM, DSc, FRCP, FMedSci, FRS, is a Professor of Thoracic Medicine at the National Heart and Lung Institute, Head of Respiratory Medicine at Imperial College, and Honorary Consultant Physician at Royal Brompton Hospital, London. He qualified at Cambridge and Oxford Universities and was appointed to his present post in 1987. He has published over 1000 peer-review papers on asthma, COPD, and related topics and has edited over 40 books. He is also amongst the top 50 most highly cited researchers in the world and has been the most highly cited clinical scientist in the United Kingdom and the most highly cited respiratory researcher in the world over the last 20 years. He was elected a Fellow of the Royal Society in 2007, the first respiratory researcher for over 150 years. He is currently a Member of the Scientific Committee of the WHO/NIH global guidelines on asthma (GINA) and COPD (GOLD). He also serves on the Editorial Board of over 30 journals and is currently an Associate Editor of Chest and Respiratory Editor of PLoS Medicine. He has given several prestigious lectures, including the Amberson Lecture at the American Thoracic Society and the Sadoul Lecture at the European Respiratory Society.

Neil C Thomson, MD, FRCP, is a Professor of Respiratory Medicine at the University of Glasgow, Head of Respiratory Medicine within the Division of Immunology, Infection & Inflammation, and Honorary Consultant at Gartnavel General Hospital, Glasgow. He graduated from the University of Glasgow and undertook postgraduate training in Glasgow, London and McMaster University, Canada. He is a former Member of the Committee for Safety of Medicine and former Chair of the Scientific Committee of the British Lung Foundation. He has co-edited several textbooks on asthma and COPD and published over 150 peer-reviewed papers on asthma. His current research interests include corticosteroid insensitivity in smokers with asthma, biomarkers in asthma and COPD and assessment of novel treatments for asthma.

Jeffrey M Drazen, MD, was born in St. Louis, Missouri. He graduated from Tufts University with a major in Physics, and from Harvard Medical School. He served his medical internship and residency at Peter Bent Brigham Hospital in Boston and was a Clinical Fellow and Research Fellow at Harvard Medical School and Harvard School of Public Health. Thereafter, he joined the Pulmonary Divisions of the Harvard hospitals and served for many years as Chief of the Combined Pulmonary Divisions at Beth Israel and Brigham and Women's Hospitals.

Currently, he is a Senior Physician at the Brigham and Women's Hospital, and the Distinguished Parker B. Francis Professor of Medicine at Harvard Medical School, as well as Professor of Physiology, Harvard School of Public Health, and adjunct Professor of Medicine at Boston University School of Medicine.

He has served on the NIH Respiratory and Applied Physiology Study Section, the NIH Pulmonary Disease Advisory Council, the NIH Lung Biology and Pathology Study Section, and the NHLBI Advisory Council.

Through his research, he defined the role of novel endogenous chemical agents in asthma. This led to four new licensed pharmaceuticals for asthma used in the treatment of millions of people worldwide. He has published nearly 500 papers and edited 6 books.

He has been a member of the Editorial Boards of many prestigious journals, including the *Journal of Applied Physiology*, *American Journal of Physiology*, *Pulmonary Pharmacology*, *Experimental Lung Research*, *Journal of Clinical Investigation*, *American Journal of Respiratory Cell and Molecular Biology*, and the *American Journal of Medicine*. In addition, he has been an associate editor of the *Journal of Clinical Investigation* and the *American Review of Respiratory Disease*.

In 2000, he assumed the post of Editor-in-Chief of the *New England Journal of Medicine.* During his tenure, the *journal* has published major papers advancing the science of medicine, including the first descriptions of SARS and papers modifying the treatment of cancer, heart disease, and lung disease. The journal, which has more than half a million readers every week, has the highest impact factor of any medical journal publishing original research.

Stephen I Rennard, MD, is Larson Professor of Medicine in the Pulmonary and Critical Care Medicine Section of the Department of Internal Medicine at the University of Nebraska Medical Center in Omaha, Nebraska, and courtesy Professor of the Department of Pathology and Microbiology and the Department of Genetics, Cell Biology and Anatomy. He received an AB with honors in Folklore and Mythology from Harvard University and an MD with honors from the Baylor College of Medicine, Houston, Texas. He completed internal medicine training at Barnes Hospital, Washington University, St. Louis, Missouri and trained in Pulmonary Diseases at the National Institutes of Health where he remained for seven years, conducting research in the cell biology of lung disease.

He joined the University of Nebraska in 1984 as Chief of Pulmonary and Critical Care, a position he retained until 1997. He was the Director of the Nebraska Office of Tobacco Control and Research from 1997 until 2008.

Dr. Rennard currently serves on the Board of Directors of the COPD Foundation and the Alpha-1 Foundation. He is a member of the American Thoracic Society Committee on Corporate Relations and the National Heart Lung Education Program Executive Committee. He is an external advisor to the Thomas Petty Aspen Lung Conference and the University of California-Davis Pulmonary Training Grant.

Dr. Rennard is active in several professional societies and has previously served on the Board of Directors for the American Thoracic Society, the Council of the American Lung Association and was a Governor for the American College of Chest Physicians. He served on the American Board of Internal Medicine, Pulmonary Section and was a member of the expert panel which prepared the global GOLD guidelines for COPD for the WHO/NHLBI. He has served on several task force committees, including the ATS/ERS task force that prepared the ATS/ERS joint COPD Standards.

Professor Rennard maintains an active program of clinical investigation in COPD and smoking cessation and a program of basic research in the mechanisms of lung tissue repair and remodeling, including the role of stem cells in disease pathogenesis and repair.

List of Contributors

Ian M. Adcock (31)
National Heart and Lung Institute, Imperial College School of Medicine, London, UK

Alvar Agusti (44)
Hospital University Son Dureta, Servi Pneumologia, Palma de Majorca, Spain
Fundacion Caubet-CIMERA
CIBER Enfermedades Respiratorias

Jennifer A. Alison (38)
Discipline of Physiotherapy, Faculty of Health Sciences, University of Sydney, Lidcombe, NSW, Australia

Yassine Amrani (18)
Department of Infection, Immunity and Inflammation, University of Leicester, UK

Sandra D. Anderson (38)
Department of Respiratory Medicine, Royal Prince Alfred Hospital, Camperdown, NSW, Australia

Kjetil Ask (29)
Centre for Gene Therapeutics, Department of Pathology and Molecular Medicine, McMaster University, Hamilton, ON, Canada

Jon Ayres (39)
Department of Environmental and Occupational Medicine, Liberty Safe Work Research Centre, University of Aberdeen, Aberdeen, UK

Joan Albert Barberà (20)
Servei de Pneumologia, Hospital Clinic, IDIBAPS, Universitat de Barcelona, Barcelona, Spain

Peter J. Barnes (32, 33, 34, 40, 49, 50, 51, 52, 61)
National Heart and Lung Institute (NHLI), Clinical Studies Unit, Imperial College, London, UK

Bianca Beghè (41)
Department of Respiratory Diseases, University of Modena and Reggio Emilia, Modena, Italy

Richard C. Boucher (16)
Department of Medicine, The University of North Carolina at Chapel Hill, Chapel Hill, NC, USA

Andrew Briggs (62)
Section of Public Health and Health Policy, University of Glasgow, Glasgow, Scotland, UK

Guy G. Brusselle (10)
Department of Pulmonary Medicine, Erasmus University Medical Center, Rotterdam, The Netherlands

A. Sonia Buist (1)
Division of Pulmonary and Critical Care Medicine, Oregon Health Sciences University, Portland, OR, USA

Kimberlie Burns (16)
Department of Medicine, The University of North Carolina at Chapel Hill, Chapel Hill, NC, USA

Andrew Bush (65)
Department of Respiratory Pediatrics, Imperial College of Science, Technology and Medicine, Royal Brompton Hospital and National Heart and Lung Institute, London, UK

Carlos A. Camargo Jr. (64)
Department of Emergency Medicine, Massachusetts General Hospital, Harvard Medical School, Boston, MA, USA

Gaetano Caramori (31)
National Heart and Lung Institute, Imperial College School of Medicine, London, UK

George H. Caughey (9)
Cardiovascular Research Institute and Department of Medicine, University of California at San Francisco, USA

Bartolome R. Celli (54)
Division of Pulmonary and Critical care, Caritas -St. Elizabeth's Medical Center, Tufts University, Boston, MA, USA

Michael A. Chandler (47)
University of Nebraska Medical Center, Omaha, NE, USA

Richard N. Channick (56)
Division of Pulmonary and Critical Care Medicine, University of California, San Diego School of Medicine, La Jolla, CA, USA

Moira Chan-Yeung (36)
Department of Medicine, Respiratory Division, University of British Columbia, Vancouver, BC, Canada

Kian Fan Chung (27, 52)
National Heart and Lung Institute (NHLI), Imperial College, London, UK

Donald W. Cockcroft (35)
Division of Respirology, Critical Care and Sleep Medicine, Department of Medicine, University of Sasketchewan, Royal University Hospital, Saskatoon, Saskatchewan, Canada

Lorenzo Corbetta (41)
Department of Respiratory Diseases, University of Firenze, Firenze, Italy

Christopher J. Corrigan (55)
Department of Asthma, Allergy and Respiratory Science, Guy's Hospital, London, UK
MRC and Asthma UK Centre in Allergic Mechanisms of Asthma King's College London, UK

Manuel G. Cosio (13)
Meakins-Christie Laboratories and the Respiratory Division, Department of Medicine, McGill University, Montreal, Quebec, Canada

Harvey O. Coxson (43)
Department of Radiology, James Hogg iCAPTURE Centre for Cardiovascular and Pulmonary Research, University of British Columbia, Vancouver, BC, Canada

Adnan Custovic (46)
University of Manchester, Manchester, UK

Marc Decramer (58)
Respiratory Rehabilitation and Respiratory Division, University Hospital, Leuven, Belgium
Faculty of Kinesiology and Rehabilitation Sciences, Department of Rehabilitation Sciences, Katholieke Universiteit Leuven, Leuven, Belgium

Dawn L. DeMeo (2)
Channing Laboratory, Brigham and Women's Hospital, Harvard Medical School, Boston, MA, USA

Jeffrey M. Drazen (24, 33)
Department of Medicine, Pulmonary Division, Birgham and Women's Hospital, Boston, MA, USA
Harvard Medical School, Editorial Office New England Journal of Medicine, Boston, MA, USA

Leonardo M. Fabbri (41)
Department of Respiratory Diseases, University of Modena and Reggio Emilia, Modena, Italy

Jack Gauldie (29)
Centre for Gene Therapeutics, Department of Pathology and Molecular Medicine, McMaster University, Hamilton, ON, Canada

Maurice Godfrey (22)
Center for Human Molecular Genetics, Munroe Meyer Institute, University of Nebraska Medical Center, Omaha, NE, USA

Stefano Guerra (3)
Arizona Respiratory Center and Mel and Enid Zuckerman College of Public Health, University of Arizona, Tucson, AZ, USA

Ian P. Hall (48)
Division of Therapeutics and Molecular Medicine, University Hospital of Nottingham, Nottingham, UK

Trevor T. Hansel (49)
National Heart and Lung Institute (NHLI), Clinical Studies Unit, Imperial College, London, UK

James C. Hogg (6)
UBC McDonald Research Laboratories, St. Paul's Hospital, University of British Columbia, Vancouver, BC, Canada

Stephen T. Holgate (7)
School of Medicine, University of Southampton, Southampton, UK

Gabor Horvath (19)
Department of Pulmonology, Semmelweiss University, Budapest, Hungary

Charles G. Irvin (5)
Department of Medicine and Physiology, University of Vermont College of Medicine, Burlington, VT, USA

Kazuhiro Ito (30)
Section of Airway Disease, National Heart and Lung Institute, Imperial College, London, UK

Wim Janssens (58)
Respiratory Rehabilitation and Respiratory Division, University Hospital, Leuven, Belgium

Sebastian L. Johnston (37)
Department of Respiratory Medicine, National Heart and Lung Institute, Imperial College London, London, UK

Susan Kennedy (36)
School of Environmental Health, University of British Columbia, Vancouver, BC, Canada

Sergei A. Kharitonov (30, 42)
Section of Airway Disease, National Heart and Lung Institute, Imperial College, London, UK

Darryl A. Knight (15)
Department of Pharmacology and Therapeutics, University of British Columbia, Vancouver, BC, Canada

Martin Kolb (29)
Centre for Gene Therapeutics, Department of Pathology and Molecular Medicine, McMaster University, Hamilton, ON, Canada

Onn Min Kon (49)
National Heart and Lung Institute (NHLI), Clinical Studies Unit, Imperial College, London, UK

Samuel L. Krachman (57)
Section of Pulmonary and Critical Care Medicine, Temple University School of Medicine, Philadelphia, PA, USA

Vera P. Krymskaya (18)
Pulmonary, Allergy and Critical Care Division, Airways Biology Initiative, University of Pennsylvania, Philadelphia, PA, USA

Bart N. Lambrecht (10)
Department of Respiratory Medicine, University Hospital Ghent, Belgium

Department of Pulmonary Medicine, Erasmus University, Rotterdam, The Netherlands

Geoffrey J. Laurent (15)
UCLMS, Center for Respiratory Research, University St/Rayne Institute, London, UK

Bruce D. Levy (24)
Department of Medicine, Pulmonary Division, Brigham and Women's Hospital, Boston, MA, USA

Fabrizio Luppi (41)
Department of Respiratory Diseases, University of Modena and Reggio Emilia, Modena, Italy

Jean-Luc Malo (36)
Department of Chest Medicine, Sacré-Coeur Hospital, Montreal, Canada

James G. Martin (13)
Meakins-Christie Laboratories and the Respiratory Division, Department of Medicine, McGill University, Montreal, Quebec, Canada

Fernando D. Martinez (3)
Arizona Respiratory Center and Mel and Enid Zuckerman College of Public Health, University of Arizona, Tucson, AZ, USA

Simon D. Message (37)
Department of Respiratory Medicine, National Heart and Lung Institute, Imperial College London, London, UK

Lynne A. Murray (15)
Manager of Pharmocology Promedior, Inc.
Malvern, PA, USA

Paul M. O'Byrne (23)
Firestone Institute for Respiratory Health, St. Joseph's Healthcare and Department of Medicine, McMaster University, Hamilton, ON, Canada

Reynold A. Panettieri Jr. (18)
Pulmonary, Allergy and Critical Care Division, Airways Biology Initiative, University of Pennsylvania, Philadelphia, PA, USA

Martyn R. Partridge (68)
The Faculty of Medicine, Imperial College, London, UK

Rodolfo M. Pascual (63)
Section on Pulmonary, Critical Care, Allergy and Immunologic Diseases, Department of Internal Medicine, Wake Forest University School of Medicine, Medical Center Boulevard, Winston-Salem, NC, USA

Ian D. Pavord (42)
Department of Respiratory Medicine and Thoracic Surgery, Institute for Lung Health, Glenfield Hospital, Leicester, UK

James E. Pease (26)
Leukocyte Biology Section, National Heart and Lung Institute, Imperial College London, London, UK

Stephen P. Peters (63)
Section on Pulmonary, Critical Care, Allergy and Immunologic Diseases, Department of Internal Medicine, Wake Forest University School of Medicine, Medical Center Boulevard, Winston-Salem, NC, USA

Michael I. Polkey (59)
Royal Brompton Hospital and National Heart and Lung Institute, London, UK

Dirkje S. Postma (4)
Department of Pulmonology, University Medical Center Groningen, University of Groningen, Groningen, The Netherlands

Irfan Rahman (25)
Department of Environmental Medicine, Lung Biology and Disease Program, University of Rochester Medical Center, Rochester, NY, USA

Scott H. Randell (16)
Departments of Cell and Molecular Physiology and Medicine, The University of North Carolina at Chapel Hill, NC, USA

Stephen I. Rennard (34, 47, 66)
Pulmonary and Critical Care Medicine, University of Nebraska Medical Center, Omaha, NE, USA

Roberto Rodríguez-Roisin (20)
Servei de Pneumologia, Hospital Clinic, IDIBAPS, Universitat de Barcelona, Barcelona, Spain

Duncan F. Rogers (17)
Section of Airway Disease, National Heart and Lung Institute, Imperial College, London, UK

Brian H. Rowe (64)
Department of Emergency Medicine, University of Alberta, and Capital Health, Edmonton, AB, Canada

Lewis J. Rubin (56)
Division of Pulmonary and Critical Care Medicine, University of California, San Diego School of Medicine, La Jolla, CA, USA

Elizabeth Sapey (14)
Department of Medicine, Birmingham University, Birmingham, UK

Sanjay Sethi (53)
Division of Pulmonary, Critical Care and Sleep Medicine, Department of Medicine, University of Buffalo, State University of New York, and Veterans Affairs Western New York Health Care System, Buffalo, New York, USA

Pallav L. Shah (59)
Royal Brompton Hospital and National Heart and Lung Institute, London, UK

Steven D. Shapiro (8, 28)
Department of Medicine, University of Pittsburgh, Pittsburgh, PA, USA

Dean Sheppard (21)
Department of Medicine and Lung Biology Center, University of California, San Francisco, CA, USA

Stephanie A. Shore (8)
Molecular and Integrative Physiological Sciences Program, Harvard School of Public Health, Boston, MA, USA

Edwin K. Silverman (4)
Channing Laboratory and Division of Pulmonary and Critical Care Medicine, Brigham and Women's Hospital, Harvard Medical School, Boston, MA, USA

Dagmar Simon (12)
Department of Dermatology, Inselspital, University of Bern, Bern, Switzerland

Hans-Uwe Simon (12)
Department of Pharmacology, University of Bern, Bern, Switzerland

Michael W. Sims (18)
Pulmonary, Allergy and Critical Care Division, Airways Biology Initiative, University of Pennsylvania, Philadelphia, PA, USA

Don D. Sin (45)
The James Hogg iCAPTURE Center for Cardiovascular and Pulmonary Research, St. Paul's Hospital, Vancouver, BC, Canada

Helen Starkie (62)
Section of Public Health and Health Policy, University of Glasgow, Glasgow, Scotland, UK

Robert A. Stockley (14)
Lung Investigation Unit, University Hospital Birmingham, NHS Foundation Trust, Edgbaston, Birmingham, UK

Andrew J. Tan (49)
National Heart and Lung Institute (NHLI), Clinical Studies Unit, Imperial College, London, UK

Neil C. Thomson (32, 40, 60)
Department of Respiratory Medicine, Division of Immunology, Infection and Inflammation, University of Glasgow, Glasgow, UK,

Omar Tliba (18)
Pulmonary, Allergy and Critical Care Division, Airways Biology Initiative, University of Pennsylvania, Philadelphia, PA, USA

Martin J. Tobin (57)
Loyola University of Chicago, Stritch School of Medicine and Hines Veterans Administration Hospital, Maywood, IL, USA

Galen B. Toews (11)
Division of Pulmonary and Critical Care Medicine, University of Michigan Health System, Ann Arbor, MI, USA

Thierry Troosters (58)
Respiratory Rehabilitation and Respiratory Division, University Hospital, Leuven, Belgium
Faculty of Kinesiology and Rehabilitation Sciences, Department of Rehabilitation Sciences, Katholieke Universiteit Leuven, Leuven, Belgium

Adam Wanner (19)
Division of Pulmonary and Critical Care Medicine, University of Miami Miller School of Medicine, Miami, FL, USA

Jadwiga A. Wedzicha (67)
Department of Respiratory Medicine, University College of London, London, UK

Scott T. Weiss (2)
Channing Laboratory, Brigham and Women's Hospital, Harvard Medical School, Boston, MA, USA

Timothy J. Williams (26)
Leukocyte Biology Section, National Heart and Lung Institute, Imperial College London, London, UK

Ashley Woodcock (46)
University of Manchester, Manchester, UK

Olivia Wu (62)
Section of Public Health and Health Policy, University of Glasgow, Glasgow, Scotland, UK

Zhou Xing (29)
Centre for Gene Therapeutics, Department of Pathology and Molecular Medicine, McMaster University, Hamilton, ON, Canada

PART

1

Definitions, Epidemiology, and Genetics of Asthma and COPD

CHAPTER 1

Definitions

A. Sonia Buist

Division of Pulmonary and Critical Care Medicine, Oregon Health and Sciences University, Portland, OR, USA

Definitions of diseases evolve over time as our understanding of them changes. For example, until recently, the presence or absence of *reversibility* was considered to be the key distinction between asthma and chronic obstructive pulmonary disease (COPD) – with reversible airflow obstruction the hallmark of asthma, and irreversible airflow obstruction the hallmark of COPD. Better understanding of both diseases has brought new definitions that acknowledge the overlap and highlight the similarities and differences between them. The important change in our understanding is the recognition that chronic inflammation underlies both diseases. The nature of the inflammation differs, however, as does the response to anti-inflammatory medications, as described in detail in later chapters. This chapter draws heavily on the latest information on asthma and COPD that is included in the guidelines on the diagnosis and management of these diseases from two widely respected global initiatives, the Global Initiative for Asthma (GINA) [1] and the Global Initiative for Chronic Obstructive Lung Disease (GOLD) [2], as updated in 2006.

DEFINITIONS

Asthma

Most of the definitions for asthma have emphasized the characteristics of fluctuations over time in bronchoconstriction and the reversible nature of the disease [1, 3]. As the pathophysiological basis of asthma became clearer, definitions began to include a statement about the pathological characteristics. The 2006 revision of the GINA Guidelines [1] proposes an operational description of asthma as:

> *a chronic inflammatory disorder of the airways in which many cells and cellular elements play a role. The chronic inflammation is associated with airway hyperresponsiveness that leads to recurrent episodes of wheezing, breathlessness, chest tightness, and coughing, particularly at night or in the early morning. These episodes are usually associated with widespread, but variable, airflow obstruction within the lung that is often reversible either spontaneously or with treatment.*

This is very similar to the definition proposed by the National Asthma Education and Prevention Program in their 1997 guidelines [3]. Both definitions imply that asthma is one disorder, rather than multiple complex disorders and syndromes – a notion that is receiving increasing attention [4, 5].

COPD

Until quite recently, definitions of COPD used to include the terms "chronic bronchitis" and "emphysema". The GOLD Guidelines, first published in 2002 [6], and revised in 2006 [2], the American Thoracic Society/European Respiratory Society (ATS–ERS) Guidelines published in 2004 [7], and the NICE Guidelines [8, 9] published in 2004 deliberately omitted these terms and used just the umbrella term COPD. The main reason for this is that the use of many different terms for COPD has led to confusion on the part of health-care providers and the public. This in turn has stood in the way of COPD becoming widely recognized.

The 2006 revision of the GOLD guidelines [2]defines COPD as:

> *a preventable and treatable disease with some significant extrapulmonary effects that may contribute to severity in individual patients. Its pulmonary component is characterized by airflow limitation that is not fully reversible. The airflow limitation is usually progressive and associated with an abnormal inflammatory response of the lung to noxious particles of gases.*

This definition includes the phrase "preventable and treatable" to emphasize the importance of a positive attitude to outcome and encourage a more active approach to management. This definition also stresses that extrapulmonary effects [10, 11] are an integral part of COPD and need to be taken into consideration in the diagnosis and management.

SIMILARITIES AND DIFFERENCES

Over the past 30 years, the thinking about asthma and COPD has swung between the concept of asthma and COPD belonging to a spectrum of diseases that all cause airflow obstruction, to the concept of them as very different diseases, and most recently to them both being inflammatory diseases with important similarities and differences, a theme that is taken up more fully in subsequent chapters. The present thinking is illustrated in Fig. 1.1 from the 2006 GOLD guidelines, which shows both diseases involving an inflammatory response that causes airflow limitation, but through gene–environment interactions with different sensitizing agents, cell populations, and mediators. The airflow limitation ranges from completely reversible (the asthma end of the spectrum) to completely irreversible (the COPD end of the spectrum). It is important to emphasize that COPD and asthma often coexist, so the clinical picture may reflect both conditions which may complicate the diagnostic process and the pathological features. For example, some patients with COPD have features of asthma and a mixed population of inflammatory cells; some patients with longstanding asthma develop the pathological features of COPD.

Although the similarities between the diseases are striking, it is the differences in the inflammatory processes between the two diseases that define their natural histories, clinical presentations, and approaches to management. The major differences in the pulmonary inflammation between

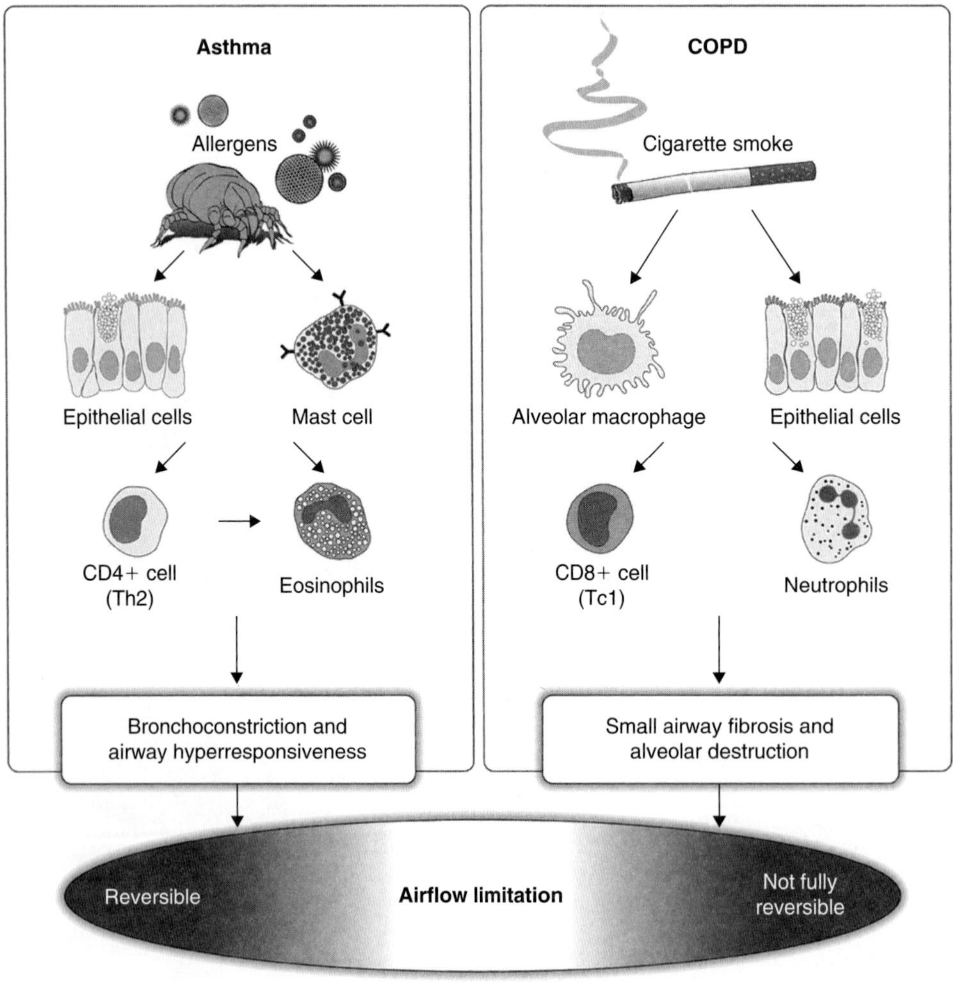

FIG. 1.1 Inflammatory cascade in COPD and asthma (adapted from Ref. [2] with permission).

TABLE 1.1 Differences in pulmonary inflammation in COPD and asthma.

	COPD	Asthma	Severe asthma
Cells	Neutrophils ++ Macrophages +++ $CD8^+$ T-cells (Tc1)	Eosinophils ++ Macrophages + $CD4^+$ T-cells (Th2)	Neutrophils + Macrophages $CD4^+$ T-cells (Th2), $CD8^+$ T-cells (Tc1)
Key mediators	IL-8 TNF-α, IL-1β, IL-6 NO +	Eotaxin IL-4, IL-5, IL-13 NO +++	IL-8 IL-5, IL-13 NO ++
Oxidative stress	+++	+	+++
Site of disease	Peripheral airways Lung parenchyma Pulmonary vessels	Proximal airways	Proximal airways Peripheral airways
Consequences	Squamous metaplasia Mucous metaplasia Small airway fibrosis Parenchymal destruction Pulmonary vascular remodeling	Fragile epithelium Mucous metaplasia ↑ Basement membrane Bronchoconstriction	
Response to therapy	Small b/d response Poor response to steroids	Large b/d response Good response to steroids	Smaller b/d response Reduced response to steroids

Source: From Ref. [2] with permission.
Notes: NO: nitric oxide; b/d: bronchodilator.

asthma and COPD are described in Table 1.1 from the 2006 GOLD report. These differences are further explored in later chapters.

Severity classification

Both asthma and COPD are often stratified by severity, but the reason for this stratification is different for the two conditions. In the 2006 GINA guidelines, the emphasis has changed from using a severity classification to guide management to using a classification based on asthma control to guide management. The GINA guidelines now recommend that asthma classification based on severity should now only be used for research purposes. Tables 1.2 and 1.3 show the GINA severity classifications based on lung function and asthma control, respectively.

For COPD, a severity classification based on spirometry is recommended for educational purposes, but not for guiding management (Table 1.4). The current recommendation is that treatment of COPD should be driven by the need to reduce and control symptoms, not by level of lung function.

Differentiating between asthma and COPD

It would be easy to differentiate between asthma and COPD if the latter occurred only in smokers and asthma in non-smokers. In fact, there is a clear diagnostic bias on the part of physicians, with COPD more likely to be diagnosed in men and asthma in women [12]. It is important to emphasize that both conditions may coexist in an individual, so many will have the clinical and pathophysiological features of both

TABLE 1.2 Classification of asthma severity by clinical features before treatment.

Intermittent
Symptoms less than once a week Brief exacerbations Nocturnal symptoms not more than twice a month ● FEV_1 or PEF ≥ 80% predicted ● PEF or FEV_1 variability < 20%
Mild persistent
Symptoms more than once a week but less than once a day Exacerbations may affect activity and sleep Nocturnal symptoms more than twice a month ● FEV_1 or PEF ≥ 80% predicted ● PEF or FEV_1 variability < 20–30%
Moderate persistent
Symptoms daily Exacerbations may affect activity and sleep Nocturnal symptoms more than once a week Daily use of inhaled short-acting β2-agonist ● FEV_1 or PEF 60–80% predicted ● PEF or FEV_1 variability > 30%
Severe persistent
Symptoms daily Frequent exacerbations Frequent nocturnal asthma symptoms Limitation of physical activities ● FEV_1 or PEFs 60% ≤ predicted ● PEF or FEV_1 variability > 30%

Source: From Ref. [1] with permission.

TABLE 1.3 Levels of asthma control.

Characteristic	Controlled (all of the following)	Partly controlled (any measure present in any week)	Uncontrolled
Daytime symptoms	None (twice or less/week)	More than twice/week	Three or more features of partly controlled asthma present in any week
Limitations of activities	None	Any	
Nocturnal symptoms/awakening	None	Any	
Meed for reliever/rescue treatment	None (twice or less/week)	More than twice/week	
Lung function (PEF or FEV_1)[a]	Normal	<80% predicted or personal best (if known)	
Exacerbations	None	One or more/year[b]	Once in any week[c]

Source: From Ref. [1] with permission.

[a] Lung function is not a reliable test for children 5 years and younger.

[b] Any exacerbation should prompt review of maintenance treatment to ensure that it is adequate.

[c] By definition, an exacerbation in any week makes that an uncontrolled asthma week.

TABLE 1.4 Spirometric classification of COPD severity based on post-bronchodilator FEV_1.

Stage 1: Mild	$FEV_1/FVC < 0.70$ $FEV_1 > 80\%$ predicted
Stage II: Moderate	$FEV_1/FVC < 0.70$ $50\% < FEV_1 < 80\%$ predicted
Stage III: Severe	$FEV_1/FVC < 0.70$ $30\% < FEV_1 < 50\%$ predicted
Stage IV: Very severe	$FEV_1/FVC < 0.70$ $FEV_1 < 30\%$ predicted or $FEV_1 <$ 50% predicted plus chronic respiratory failure

Source: From Ref. [2] with permission.

Notes: FEV_1: forced expiratory volume in 1 s; FVC: forced vital capacity: respiratory failure: arterial partial pressure of oxygen (PaO_2) less than 8.0 kPa (60 mmHg) with or without arterial partial pressure of CO_2 ($PaCO_2$) greater than 6.7 kPa (50 mmHg) while breathing air at sea level.

TABLE 1.5 Clinical features of COPD and asthma.

Diagnosis	Suggestive features
COPD	Onset in mid-life Symptoms slowly progressive Long history of tobacco smoking Dyspnea during exercise Largely irreversible airflow limitation
Asthma	Onset early in life (often childhood) Symptoms vary from day to day Symptoms at night/early morning Allergy, rhinitis, and/or eczema also present Family history of asthma Largely reversible airflow limitation

Source: From Ref. [2] with permission.

diseases. This makes differentiating the diseases sometimes challenging for the clinician, especially in older adults who are or have been smokers.

The clinician can be guided by information in the clinical history such as smoking history, age of onset of symptoms, history of atopic conditions, and description of acute episodes of shortness of breath (see Table 1.5).

Asthma usually has its onset in early childhood. However, adult-onset asthma does exist, and many are unable to remember childhood events that would provide a clue to the early stages of asthma. Therefore, unless symptoms are continuous from childhood, the onset of asthma symptoms in adult life may be hard to interpret, especially in the presence of other risk factors such as smoking. COPD typically becomes clinically apparent in the sixth and seventh decades of life. If an individual is physically active, he or she may notice reduced exercise tolerance earlier.

COPD in developed countries is mostly a disease of smokers. This is not necessarily true in developing countries where other risk factors, such as heavy outdoor and indoor/occupational air pollution, may be important risk factors that are causally related to COPD [2]. The relationship between asthma and smoking is complex. Individuals with asthma may be non-smokers, smokers, or ex-smokers. Since asthma genes and genes leading to the susceptibility to develop airflow obstruction with smoking are common in the population, the likelihood that an individual may have both is high. Likewise, COPD may occur in lifetime non-smokers [13–15]. In some, this may be longstanding asthma that may have been undiagnosed. In others, genes that are as yet unrecognized may be responsible. Alpha-1 antitrypsin deficiency is the model for such a genotype.

Pulmonary function tests can also provide guidance. Both diseases are characterized by airflow obstruction except in the early or mild stages. In asthma, lung function either remains normal or can be normalized with treatment in patients with mild intermittent or mild persistent disease [1]. COPD, in comparison with asthma, is defined by irreversible airflow limitation, and this becomes progressively greater as the disease advances. Lung function in asthma is characterized by reversibility and variability. *Reversibility* refers to the short-term response to an inhaled bronchodilating agent, and is preferably measured with spirometry or peak flow (acceptable but not as helpful as spirometry). *Variability* refers to the improvement and deterioration in symptoms and lung function over time (both short- and long-time periods) that is characteristic of asthma. The GINA guidelines [1] recommend that *reversibility* be defined as a ≥20% improvement post-bronchodilator in the 1-second forced expiratory volume (FEV_1) or peak flow (PEF). GINA further defines *variability* as a diurnal variation in PEF of more than 20% (with twice daily reading, more than 10%).

The GOLD COPD guidelines [2] recommend that spirometry be used to establish that the airflow limitation is not fully reversible (defined as a post-bronchodilator FEV_1/FVC ratio of <0.7 and $FEV_1 < 80\%$ predicted), and to stage the severity of the disease. However, GOLD emphasizes that bronchodilator reversibility testing does not predict a patient's response to treatment or predict disease progression.

OVERLAP BETWEEN ASTHMA AND COPD

Not acknowledged in the definitions is the fact that longstanding asthma can lead to airway remodeling and partly irreversible airflow obstruction [16]. So, in many (but not all) with longstanding asthma, there is an appreciable component of chronic irreversible airflow obstruction with reduced lung function and incomplete response (or at least, not complete reversibility) to a short-acting bronchodilator or to oral or inhaled corticosteroids [17]. This complicates the diagnosis of asthma in older adults, and requires that the goals of treatment be modified since maintenance of normal lung function can no longer be a realistic goal.

Whether longstanding asthma with remodeling can be called COPD is intensely controversial. In so far as there is irreversible or poorly reversible airflow obstruction in the remodeled lungs, the term seems appropriate. Conceptually and practically, the recognition that remodeling is a feature of longstanding asthma in many (but not all) reinforces the notion that these diseases constitute a spectrum of disease, as illustrated in Fig. 1.1, ranging from fully reversible to fully irreversible.

EXACERBATIONS

The definition of asthma highlights the importance of exacerbations as a feature of asthma, and emphasizes the fluctuations of the disease [1, 3]. The definition of COPD does not include any mention of exacerbations [2, 6–8]. Nevertheless, they may be as important in the natural history of COPD as they are in asthma [1] and account for approximately 70% of the COPD-related costs in the United States [2].

LIMITATIONS OF THE DEFINITIONS

Definitions for both asthma and COPD have limitations since they can reflect only our current understanding of the diseases, which is quite limited. Both diseases will continue to be redefined as our understanding of them deepens, and as new effective preventive strategies and treatments become available.

References

1. Global Strategy for the Diagnosis, Management, and Prevention of chronic Obstructive Pulmonary Disease. Global Initiative for Chronic Obstructive Lung Disease (GOLD), 2006; Available from: URL: www.goldcopd.org
2. Global Strategy for Asthma Management and Prevention. Global Initiative for Asthma (GOLD), 2006; Available from: URL: www.ginasthma.org
3. National Asthma Education and Prevention Program Expert Panel Report 2: *Guidelines for the Diagnosis and Management of Asthma*, National Institute of Health, National Heart, Lung, and Blood Institute. NIH Publication 97–4051, 1997.
4. Wenzel SE. Asthma: Defining of the persistent adult phenotypes. *Lancet* 368: 804–13, 2006.
5. A plea to abandon asthma as a disease concept. Editorial, *Lancet* 368:705, 2006.
6. Pauwels RA, Buist AS, Calverley MA, Jenkins CR, Hurd SS. Global strategy for the diagnosis, management and prevention of chronic obstructive pulmonary disease. NHLBI/WHO Global Initiative for Chronic Obstructive Lung Disease (GOLD) Workshop Summary. *Am J Respir Crit Care Med* 163: 1256–76, 2001.
7. Celli BR, MacNee W and committee members, Standards for the diagnosis and treatment of patients with COPD: A summary of the ATS/ERS position paper. *Eur Respir J* 23: 932–46, 2004.
8. National Institute for Clinical Excellence (NICE). Chronic obstructive pulmonary disease. National clinical guideline on management of chronic obstructive pulmonary disease in adults in primary and secondary care. *Thorax* 59(Suppl 1): 1–232, 2004.
9. National Institute for Clinical Excellence (NICE) Guideline available at URL: www.nice.org.uk/CG012niceguideline
10. Soriano JB, Visick GT, Muellerova H, Payvandi N, Hansell AL. Patterns of comorbidities in newly diagnosed COPD and asthma in primary care. *Chest* 128(4): 2099–107, 2005.
11. Agusti AG. Systemic effects of chronic obstructive pulmonary disease. *Proc Am Thorac Soc* 2(4): 367–70, 2005.
12. Dodge R, Cline MG, Burrows. B. Comparisons of asthma, emphysema, and chronic bronchitis diagnoses in a general population sample. *Am Rev Respir Dis* 133: 981–86, 1986.
13. Celli BR, Halbert RJ, Nordyke RJ, Schan. B. Airway obstruction in never smokers: Results from the Third National Health and Nutrition Examination Survey. *Am J Med* 118: 1364–72, 2005.
14. Menezes AM, Perez-Padilla R, Jardim JR, Muino A, Lopez MV, Valdivia G *et al.* Chronic obstructive pulmonary disease in five Latin American cities (the PLATINO study): A prevalence study. *Lancet* 366(9500): 1875–81, 2005.
15. Buist AS, McBurnie MA, Vollmer WM, Gillespie S, Burney P, Mannino DM, Menezes AMB, Sullivan SD, Lee TA, Weiss KB, Jensen RL, Marks GB, Gulsvik A, Nizankowska-Mogilnicka E. International Variation in the prevalence of chronic obstructive pulmonary disease (The BOLD Study): a population-based prevalence study. *Lancet* 2007; 370:741–50.
16. James AL, Wenzel SE. Clinical relevance of airway remodelling in airway diseases. Number 4 in Series "Airway Remodelling: from Basic Science to Clinical Practice". *Eur Respir J* 30: 134–55, 2007.
17. Fish JE, Peters SP. Airway remodeling and persistent airway obstruction in asthma. *J Allergy Clin Immunol* 104: 509–16, 1999.

CHAPTER 2

Epidemiology

Dawn L. DeMeo and Scott T. Weiss

Channing Laboratory, Brigham and Women's Hospital, Harvard Medical School, Boston, MA, USA

DEFINING THE DISEASES

Asthma

The study of asthma epidemiology has been plagued by lack of consensus regarding standards for diagnosis. Most definitions have included variable airflow obstruction, but asthma is a clinical syndrome, without a widely accepted case definition. Epidemiology studies have used questionnaires to assess for the presence of disease, but are limited by recall and misclassification bias. Some have suggested that symptoms should be assessed in conjunction with airway hyperresponsiveness [1]. Others argue that airway hyperresponsiveness and symptoms should be analyzed separately owing to the poor correlation between clinical asthma and hyperresponsiveness [2]. Population-based epidemiology studies have demonstrated a low sensitivity of airway hyperresponsiveness for detecting individuals with physician-diagnosed asthma versus a sensitivity of greater than 90% in clinic studies [3]. A standard definition of asthma is as a chronic inflammatory disease of the airways. A commonly referenced formal definition is that promulgated in the Global Initiative for Asthma (GINA) guideline, which defines asthma as

> *a chronic inflammatory disorder of the airways in which many cells and cellular elements play a role. The chronic inflammation causes an associated increase in airway hyper-responsiveness that leads to recurrent episodes of wheezing, breathlessness, chest tightness and coughing, particularly at night or in the early morning. These episodes are usually associated with widespread but variable airflow obstruction that is often reversible either spontaneously or with treatment [4].*

However, this definition is non-specific and does not provide absolute criteria to facilitate differentiating asthma from other airway diseases [5].

Beyond definitions, there are differences between languages for the words used to describe asthma symptoms. A novel approach to this problem has been used in the International Study of Asthma and Allergies in Children (ISAAC), which includes an asthma video questionnaire demonstrating clinical signs of asthma, as an attempt to improve uniformity in surveying for asthma [6]. This multi-modality approach represents an improvement in asthma surveillance and has captured more individuals with asthma. However, diagnostic heterogeneity remains, as even within the ISAAC project, prevalence estimates of current asthma may vary depending upon the assessment tool used (i.e. video versus questionnaire).

Chronic obstructive pulmonary disease

COPD includes chronic bronchitis, small airways disease and emphysema, and is characterized by airway obstruction that is fixed or only partially reversible. The diagnosis of COPD attached to a given patient varies depending on the degree of airflow obstruction considered diagnostic within a given set of guidelines. Thus the long-standing lack of international standardization of criteria for diagnosis of COPD makes understanding relative incidence and prevalence quite challenging. This is well illustrated by a study by Lindberg *et al.* [7], who compared COPD prevalence rates using European Respiratory Society (ERS), British Thoracic Society (BTS), Global Initiative for Chronic Obstructive Lung Disease

(GOLD), and clinical and spirometric American Thoracic Society (ATS) criteria. In this study, ERS criteria revealed a 14.0% prevalence of COPD, BTS criteria 7.6% prevalence, GOLD criteria 14.1% prevalence, clinical ATS criteria 12.2% prevalence, and spirometric ATS criteria 34.1%.

This example highlights the difficulty of comparison between international studies/standards and the efforts to understand COPD epidemiology on a global scale. Similar prevalence differences depending upon reference criteria have been observed by other authors [8, 9], further highlighting the challenges inherent in making comparisons between populations.

Recently, an international committee convened to refine and promote standardization of a COPD definition. The GOLD has defined COPD as

> *a preventable and treatable disease with some significant extra pulmonary effects that may contribute to the severity in individual patients. Its pulmonary component is characterized by airflow limitation that is not fully reversible. The airflow limitation is usually progressive and associated with an abnormal inflammatory response of the lungs to noxious particles or gases [10].*

GOLD stages are defined using post-bronchodilator spirometry, with an FEV_1/FVC ratio of less than 0.7 defining the presence of obstruction, and the FEV_1 reduction translating into the severity of COPD. A key element of this updated definition is that it includes capturing features of extra-pulmonary co-morbidities which have relevance for disease severity [11]. The GOLD initiative has aimed to promote studies to understand the increasing prevalence of COPD worldwide, as well as to standardize the collection of data for international comparison of research [12].

In summary, for many years both asthma and COPD have lacked widely accepted case definitions, but international efforts have been developed to rectify this. Definitions of both diseases include both symptoms and spirometric features. Refining the standardization of clinical diagnosis will facilitate comparability in epidemiological studies. However, given the wide variability in disease definitions for both diseases, comparisons of studies of asthma and COPD between populations and between countries must be viewed in the light of differences in criteria used for disease diagnosis. Although the GINA and GOLD guidelines have brought some standardization and consensus to diagnosing asthma and COPD respectively, diagnostic heterogeneity remains and further refinement of these guidelines is an important goal.

INCIDENCE

Incidence rates for asthma and COPD vary depending on the age of the population under consideration. For example, asthma is commonly diagnosed in early childhood and COPD is commonly diagnosed after age 60. Interestingly, cases of adult-onset asthma and early-onset COPD predominate in women, suggesting that there may be a common hormonal influence on the age-of-onset of these diseases [13, 14].

Global burden of disease estimates suggest that at least 300 million people worldwide have asthma, with the potential for an additional 100 million more cases by the year 2025 [15]. In childhood, incidence rates for asthma are highest among the youngest age groups and among male children until puberty [16–21]. In a study of an adult Swedish population, Toren and Hermansson [22] found the incidence rate for adult-onset asthma to be highest among females of all ages greater than 20, with an incidence of 1.3 per 1000 person-years; among women 16–20 years of age the rate was 3 per 1000 person-years. Analysis of data from a prospective cohort study in Finland demonstrated no increase in asthma incidence from 1982 to 1990 in adults aged 18–45 years [23]. Early investigation into the increasing prevalence of asthma in the United States was noted in a review of medical records from Olmsted County, Minnesota, where the annual incidence of asthma was found to increase from 183 per 100,000 in 1964 to 284 per 100,000 in 1983. The most significant increase was in children aged 1–14 years, suggesting a potential cohort effect early in life. Despite this increased incidence in asthma among children from 1964 to 1983, constant rates were observed among adults [16].

Although these data indicate that asthma incidence has been increasing, minimal information has been available for trends in COPD incidence. One challenge of studying COPD incidence is that the disease is usually moderately advanced at diagnosis, and thus true incidence rates have remained elusive. However, a recent study observed that cough and phlegm may identify a group at high risk for incident COPD [24]. This study by de Marco and colleagues focused on a cohort of 5002 subjects without asthma and with normal lung function. The observed incidence rate of COPD was 2.8 cases/1000/year; the incidence rate ratio for chronic cough and phlegm was 1.85 (95% CI 1.17–2.93) with a 10-year cumulative incidence of COPD in the overall cohort of 2.8%. Two other studies have reported on incidence rates of COPD using GOLD criteria, with a 10-year cumulative estimate of 13.5% noted by Lindberg *et al.* [25], and a 9-year cumulative incidence of 6.1% in a study by Johannessen *et al.* [26].

PREVALENCE

Trends in the prevalence of obstructive lung disease have been suggested by an analysis of the Third National Health and Nutrition Examination Survey (NHANES III) [27]. This survey included subjects with asthma, chronic bronchitis, and emphysema (Fig. 2.1). In this cohort, the case definitions were a physician diagnosis of chronic bronchitis, asthma or emphysema, respiratory symptoms, and low lung function. Of note, the investigators defined low lung function as present when the FEV_1/FVC ratio was <0.70 and the FEV_1 was less than 80% of predicted. Of the investigated population of 20,050 adults, 6.8% had low lung function as thus defined; 7.2% of the population had an FEV_1/FVC ratio less than 0.70 with an FEV_1 greater than 80% predicted, and were not included as having low lung function. Of the entire population, 8.5% reported obstructive lung disease. Importantly, 63.3% of those with documented low

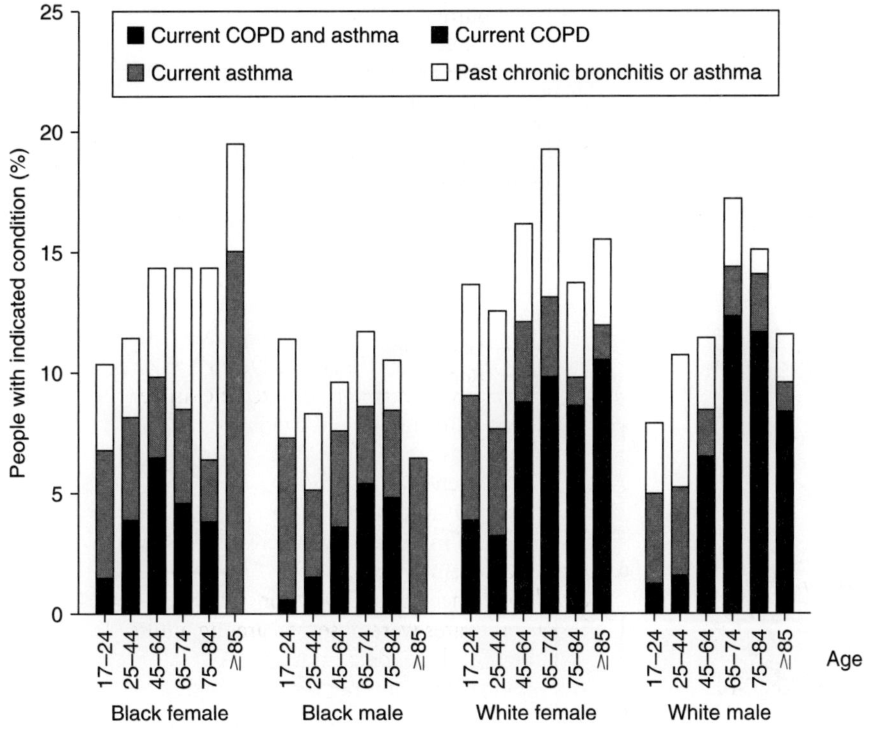

FIG. 2.1 Age-specific percentage of individuals, stratified by race and sex, with COPD and asthma, current COPD, current asthma, and past chronic bronchitis or asthma. Reproduced from National Center for Health Statistics, *Plan and Operation of the Third National Health and Nutrition Examination Survey, 1988–1994*, US Department of Health and Human Services publication PHS 94–1308, 1994, with permission.

lung function had no current or prior doctor diagnosis of obstructive lung disease. In addition to prevalence information regarding low lung function, NHANES III data suggest that there is still a significant proportion of disease that goes undiagnosed in the mild stages, leading to an underestimation of the true prevalence of obstructive lung disease.

Asthma

Data from the United States have suggested an increase in prevalence of asthma in children as well as in older adults (Fig. 2.2) [28]. During the last several decades studies have suggested an increase in prevalence worldwide of 5–6% per year. Data from the National Health Interview Survey (NHIS) reveal a 75% increase in self-reported asthma rates from 1980 to 1994. This trend was demonstrated in all age and race strata as well as in both genders. The most significant increase demonstrated by the NHIS was among children 0–4 years of age (increase of 160%) and persons 5–14 years of age (increase of 74%) [29].

Repeatedly, the prevalence of asthma amongst inner-city children has been demonstrated to be much higher than among similar aged children living outside the inner city [29–31]. It has been suggested that a doctor's diagnosis of asthma is made less frequently than asthma symptom reporting, raising concern that despite increasing prevalence there is still a tendency to under-diagnose asthma, and consequently underestimate true prevalence values [32].

The GINA project [15] has compiled asthma prevalence data for 20 regions of the world, with the main sources of data in children including the ISAAC project, and the main source of prevalence data in adults including the European Community Respiratory Health Survey (ECRHS). Higher prevalences of asthma (>10%) have been noted in developed countries including Canada, the United States, United Kingdom, New Zealand, and The Republic of Ireland [15]. In parallel with urbanization, the rates of asthma have increased in developing countries as well.

The ISAAC has as its aim to describe, across 155 centers, the prevalence and severity of asthma in children in 56 countries [6]. Phase 1 of this trial has demonstrated a large variation in the prevalence of asthma symptoms in children throughout the world, with the highest prevalence in centers from Australia, New Zealand, the United Kingdom, and Ireland [32–35]. While the prevalence of allergic rhinitis has been noted to be scattered in the groups with the highest prevalence of asthma, the lowest prevalence for rhinitis has been found in countries where the asthma prevalence was lowest, such as in Eastern Europe, Indonesia, Greece, and India. In addition to defining prevalence rates, the ISAAC study represents an effort to establish an international standard to facilitate comparability of data from epidemiological studies of asthma. The rising asthma prevalence rates in children heralds rising asthma rates for adults, as two-thirds of children with asthma still have symptoms by early adulthood [36].

Chronic obstructive pulmonary disease

COPD is best understood by understanding first the trends for smoking in populations. Although projected smoking rates throughout the world have increased, smoking prevalence in

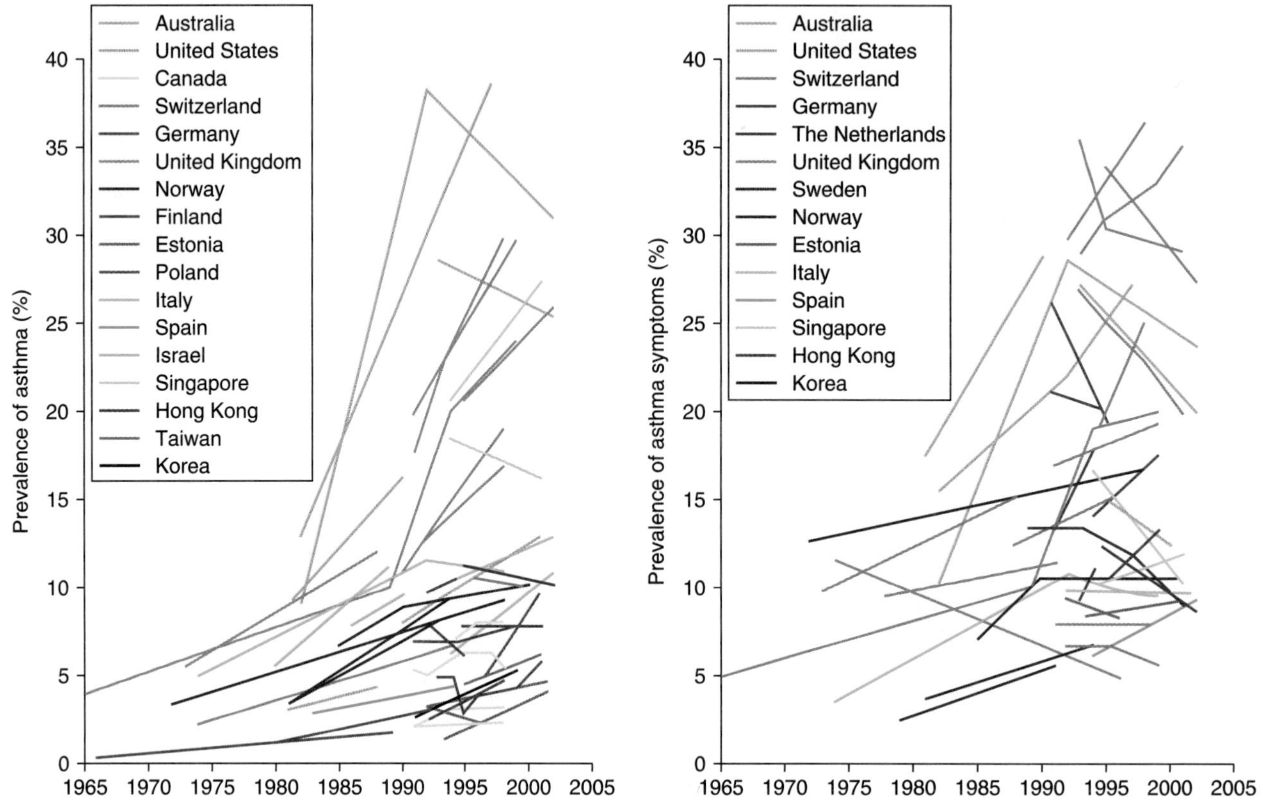

FIG. 2.2 Changes in the prevalence of asthma and asthma symptoms in children and young adults. Reproduced from Ref. [28], with permission.

the United States has been decreasing. Estimates from the year 2005 suggest that 21% of Americans smoke cigarettes, with cigarette smoking prevalence rates varying by both sex and race. Recent prevalence estimates [37] for smoking were highest for American Indian/Alaska Natives (37.5% men and 26.8% women). The prevalence of current smoking is 20.0% in white women, 17.3% in African American women, 24.0% in white men, and 26.7% in African American men.

However, in the context of COPD, susceptibility to cigarette smoke is not uniform and varies as a function of the amount of cigarette smoking [38]. Stang *et al.* [39] utilized smoking rates to create a mathematical model for estimating COPD prevalence. Using their model, they estimated that 15.3 million people in the United States aged 40 years or more have COPD; this was a reasonable estimate compared to the spirometric prevalence of 17.1 million as estimated by the NHANES III. Using this model, they also predicted the prevalence of COPD in Germany (2.7 million), the United Kingdom (3.0 million), Spain (1.5 million), Italy (2.6 million), and France (2.6 million), and suggested smoking rates as a useful surrogate for estimating COPD prevalence.

The World Health Organization prediction is that by 2020 COPD will rise from being the twelfth to the fifth most prevalent disease worldwide, and from being the sixth most common cause of death to the third most common [40]. Prevalence estimates of COPD in the United States in 1996, before widespread implementation of GOLD criteria, indicated approximately 15 million people have COPD with 14.1 million with chronic bronchitis and 1.8 with emphysema. By these measures, there was no change in the prevalence of emphysema from 1982 to 1996, although from 1983 to 1995 the prevalence of chronic bronchitis continued to increase. In a study of the Canadian population, prevalence rates of COPD were 4.6% in the 55–64 age group, 5.0% in the 65–74 age group, and 6.8% in the greater than 75 age group [41]. These data may be an underestimation, as there is a suggestion that COPD prevalence rates are underestimated in the elderly, especially in those with lower incomes [42]. By 2001, in distinction to the many asthma prevalence studies, only 32 prevalence surveys of COPD were reported (Fig. 2.3) [9] with comparability somewhat limited by COPD case definitions. The Burden of Obstructive Lung Disease (BOLD) project is a large-scale study of COPD, with one goal of measuring COPD prevalence using standard GOLD criteria. A recent report from the study in 1258 subjects reported a 26.1% prevalence of at least GOLD Stage 1 COPD, and 10.7% for GOLD Stage 2 or higher [44]. Pooled estimates based on GOLD criteria have suggested a prevalence of GOLD Stage 1 or higher COPD of 5.5% and 9.8% for GOLD Stage 2 or higher [9].

COPD is likely under-diagnosed in both North American and European populations. For example, the IBERPOC Project (*Estudio Epidemiologico de la EPOC en España*) was a population-based study of prevalence of COPD in Spain [45]. The prevalence of COPD in this population (26% current smokers, 24% ex-smokers, 76% men), defined according to ERS criteria, was 9.1%. Only

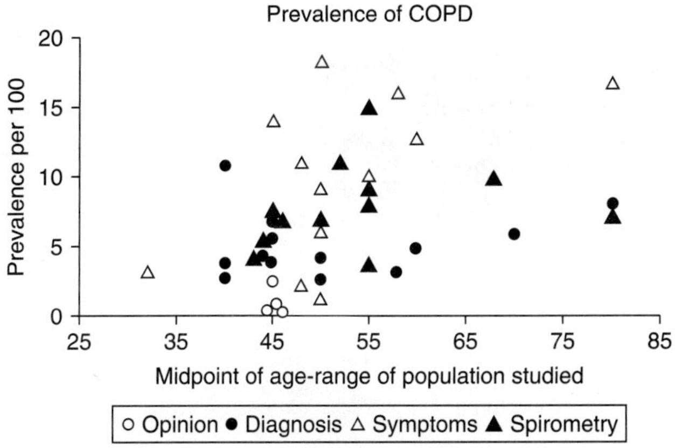

FIG. 2.3 Prevalence of COPD using estimates from expert opinion, physician diagnosis symptoms and spirometry. Data from Ref. [9] as reproduced in Ref. [43], with permission.

22% of those diagnosed had a prior diagnosis, while 48% had prior respiratory symptoms. More recently, despite prevalence estimates for GOLD Stage 1 or higher COPD in the BOLD study in Salzburg, only 5.6% of subjects reported ever receiving a doctor diagnosis of COPD [44]; the prevalence has yet to translate into widespread use of spirometry for COPD screening.

The high-prevalence estimates for both asthma and COPD have translated into significant morbidity associated with both diseases. In 2001, asthma was ranked as the 25th leading contributor to disability adjusted life years (DALYs) worldwide; COPD was ranked 12th. In this context, together with the high-prevalence estimates for both disease, the global burden of disease and economic costs for both diseases are high.

In summary, the prevalences of both asthma and COPD are increasing in both Western developed and developing countries. As both asthma and COPD are both likely under-diagnosed, the prevalence estimates underestimate the true burden of these diseases. Variability in definitions of both asthma and COPD contribute to inexact prevalence estimates and problems with comparisons of prevalence data. ISAAC (for childhood asthma), ECRHS (for adult asthma), and GOLD (for COPD) represent international efforts underway to standardize the definitions used in studies to enhance comparisons of incidence, prevalence, and burden of disease.

HEALTHCARE UTILIZATION AND HOSPITALIZATION TRENDS

In the United States, the estimated cost for year 2000 for asthma was projected to be 12.7 billion dollars (8.1 for direct cost, 2.6 related to morbidity, 2.0 related to mortality) and 30.4 billion dollars for COPD (14.7 for direct cost, 6.5 related to morbidity, 9.2 related to mortality). Utilization of health services has continued to increase for both diseases. An increase in health service use has been documented in many countries, including the United Kingdom, Canada, and the United States; the utilization increase has been concomitant with the documented increase in asthma prevalence [46–48].

Increased hospital visits have been documented worldwide, including in England, New Zealand, the United States, Greece, Australia, and Canada [49–56]. In capturing the epidemiology trend from the 1960s to the 1980s, Evans *et al.* reported on a 200% increase in rates of hospitalization of adults with asthma (and a 50% increase for children with asthma) [57]. Overall, hospitalizations with asthma as a primary diagnosis increased most steeply in the 1970s until the mid-1980s and then remained constant; this is in contrast to asthma as a secondary diagnosis which increased until 1997. From 1975 to 1995, office visits for asthma more than doubled, from 4.6 to 10.4 million [29].

Based on the National Ambulatory Medical Care Survey and the National Hospital Ambulatory Medical Care Survey, in 1998 more than 14 million visits were made to physicians for diagnoses related to COPD, and this was an increase from 5.5 million visits reported in 1980. In 2000, there were approximately 726,000 hospitalizations coded as COPD or allied conditions [58]. Care must be taken in interpretation of these data because of non-uniform case definitions; more than half of the discharge diagnoses were non-specifically coded as COPD or allied conditions.

In summary, increased health service utilization has occurred for asthma and COPD during the last two decades. Overall, hospital admissions and discharges have increased for both diseases.

MORBIDITY AND MORTALITY

Asthma

The New Zealand epidemic of asthma in the 1970s prompted a review of asthma deaths in Western countries; among such countries there was a notable increase of 1.5- to 2-fold in the asthma mortality rates between the mid-1970s and the mid-1980s [59]. The highest mortality rates in the United States have been in the inner-city regions, with particularly high-risk populations studied in East Harlem, New York City, and Cook County, Chicago [31, 60]. One study observed that socioeconomic and racial disparities were attributable to higher incidence of asthma exacerbations among inner-city children, with no excess utilization of medical resources [61].

International comparisons of mortality rates have been limited by differences in recording statistics of cause of death [62]. There is variability in the mortality rates ascribed to asthma, with significant variation by region of the world (Fig. 2.4), with the highest asthma case fatality rates in China (36.7/100,000 asthmatics) [15]. Mortality rates for asthma have decreased over the past 20 years, likely due to improvements in pharmacotherapy, although it has been suggested that most deaths from asthma now occur in adults, potentially consequent to medical non-compliance [63]. Comparison of mortality rates is difficult also because of the lack of standardized definitions for the disease, and because of environmental, genetic, socioeconomic, and occupational influences unique to a given population.

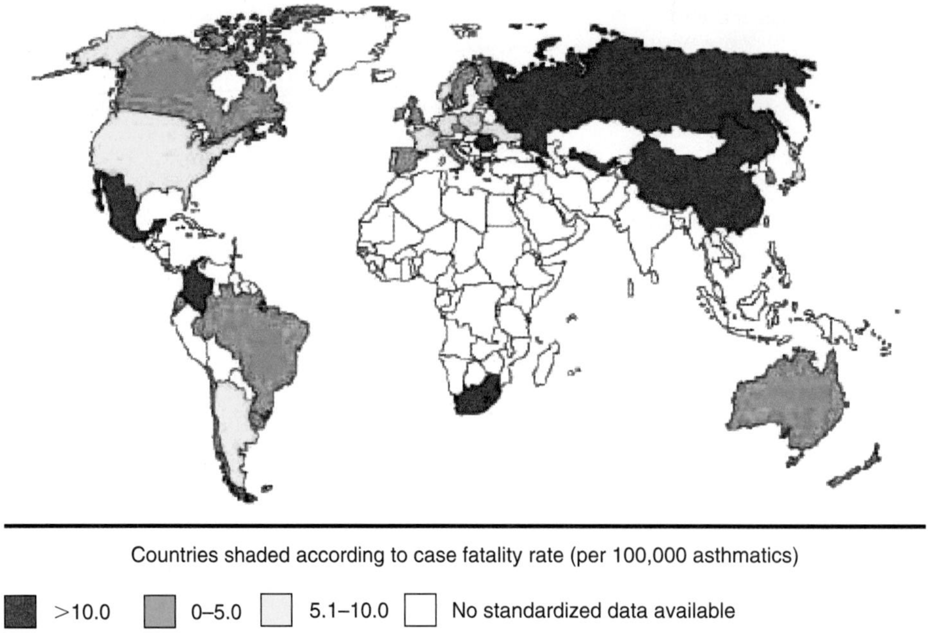

FIG. 2.4 Asthma case fatality rate in individuals 5–34 years of age calculated as the number of deaths per 100,000 cases of asthma. Reproduced from Ref. [15], with permission.

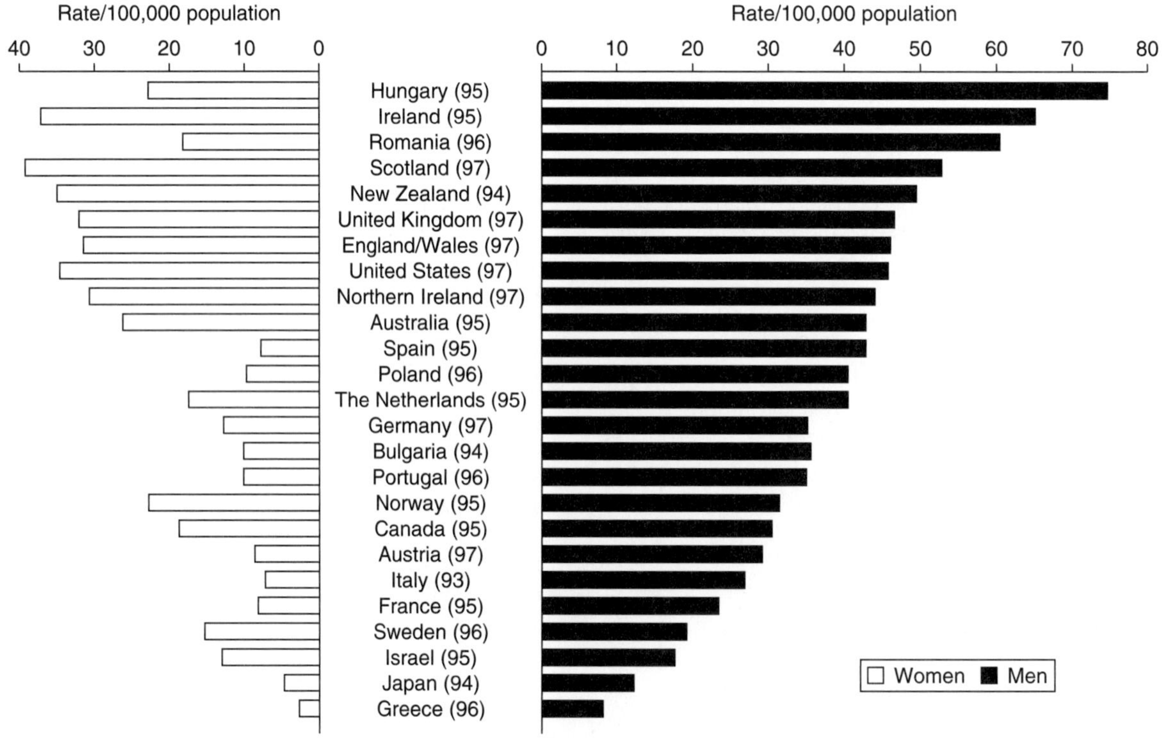

FIG. 2.5 Age-adjusted death rates for COPD by country and sex, for individuals aged 35–77. The year of data is shown in parentheses. Reproduced from Ref. [12], with permission.

Chronic obstructive pulmonary disease

Since 1960 mortality associated with COPD has continued to rise. In the year 2000, the number of women who died from COPD surpassed the number of men [58], and standardized mortality rates have been observed to be higher in women versus men with COPD (4.8 versus 2.7 respectively) [64]. In longitudinal follow-up of 13,756 individuals in the Atherosclerosis Risk in Communities (ARIC) study, rapid lung function decline over a 3-year period was associated with an increased risk for COPD-related hospitalization and death [65].

International mortality trends demonstrate high rates of deaths for COPD in many countries. These differences may be accounted for in part by different smoking behaviors, which include tobacco type, along with other environmental, infectious, and genetic factors. Differences among these death rates are striking (Fig. 2.5), but again lack of standardization in coding practices and death certification as well as practice differences and quality of care are relevant when comparing estimates [12].

Although overall asthma mortality remains low compared with COPD, mortality rates for both asthma and COPD have increased in the last decade. Differences in death rates for asthma and COPD between countries are multifactorial (genetic, environmental, occupational, socioeconomic), but differential coding of cause-of-death statistics hinders accuracy of estimates for both diseases.

SMOKING

Burrows and colleagues [38] have demonstrated that, for a given level of tobacco smoke exposure, FEV_1 varies substantially (Fig. 2.6). In addition, the dose–effect relationship between cigarette smoking and FEV_1 decline depends on the year of life when an individual is exposed. Dose and timing of tobacco smoke exposure have a differential effect on FEV_1 depending on the stage of the life cycle (Table 2.1). Cunningham *et al.* [66] observed that maternal smoking during pregnancy resulted in a 1.3% reduction in FEV_1 when children were 8–12 years old. Tager *et al.* [67] found that adolescents who smoke when aged 15–20 have an estimated 8% reduction in FEV_1. The Vlagtwedde/Vlaardingen study [68] demonstrated a large effect of cigarette smoking in decreasing maximal lung function in individuals less than age 20; this effect exceeded the effect of cigarette smoking on lung function decline seen in older subjects.

Smoking is a notable risk factor for both asthma and COPD in children and adults, and active smoking perpetuates pulmonary (and likely systemic) inflammation. Overall, smoking is associated with an increase in asthma incidence [69, 70], as well as an increased symptom severity, increased morbidity and lung function decline, and an increased risk of death [71–74]. Passive exposure to cigarette smoke increases the risk for the development of asthma and allergic sensitization [75–77]. There has also been a suggestion that non-specific airways responsiveness is increased by environmental and personal smoke exposure [78].

Maternal smoking is a risk factor for the development of asthma in children up to 1 year of age [79]. In a case-control study of children whose mothers were heavy smokers, one group demonstrated an odds ratio of 2.15 among 3–4-year olds for the development of asthma; these data were controlled for family history, past infections, gender, and other demographic variables [80]. In a 6-year follow-up, the odds ratio for asthma among those exposed to maternal smoking was 3.8 [81].

As noted above, the single most significant risk factor for COPD is tobacco smoking. Although an oft quoted estimate is that only 10–15% of smokers actually go on to develop obstructive lung disease, this is a marked underestimate, with many researchers suggesting that all individuals

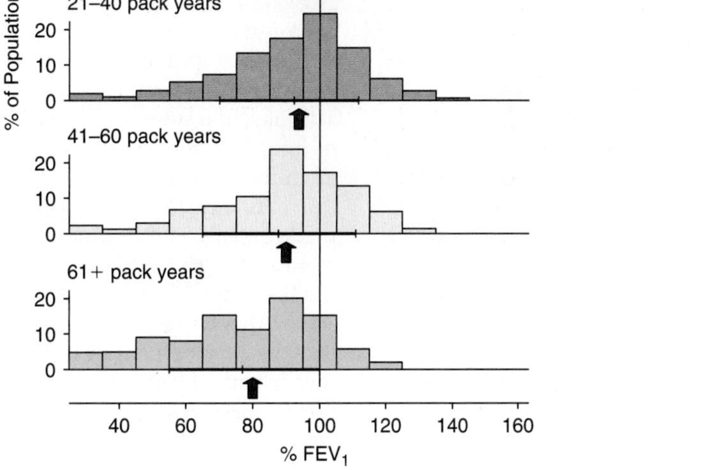

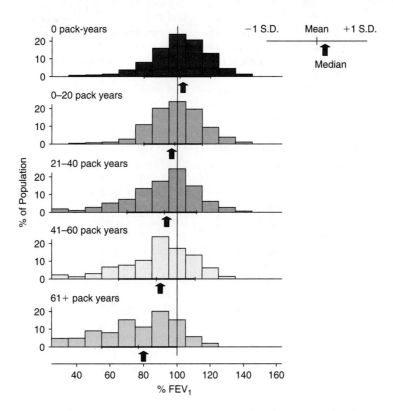

FIG. 2.6 Distribution of percentage predicted forced expiratory volume in 1 s (FEV_1) in adults with varying smoking histories as measured in pack-years. The proportion of smokers with normal flow decreased with increasing pack-year histories. Yet, many have near-normal FEV_1 with extensive smoking history. Subjects with "respiratory trouble" before age 16 were excluded. Medians and means ±1 SD are shown for each group in the abscissae. Note that among the 425 persons with 20+ pack-years, only 15% have an FEV_1 of 60% of predicted or less. Reproduced from Ref. [38], with permission.

TABLE 2.1 Effects of cigarette smoking at different stages of the life cycle.

Life phase (gender)	Cigarette dose	Total FEV_1 reduction	FEV_1 reduction (ml/year per packs/day)
In utero (M and F)	Variable exposure for 9 months	27.3 ml[a]	36
Adolescence (M)	15 cigarettes/day for 5 years[b]	390 ml	104
Adolescence (F)	10 cigarettes/day for 5 years[b]	340 ml	136
Adult (M)	Variable	N/A	13[c]
Adult (F)	Variable	N/A	7[c]

Source: Adapted from Weiss ST, Silverman EK. Risk factors for the development of chronic obstructive pulmonary disease. In: *Severe Asthma*, New York: Marcel Dekker, 2000.

Notes: M: male; F: female.

[a] Adjusted for gender and maternal smoking in the past year; based on 1.3% reduction and mean FEV_1 = 2.1 l, 1 pack/day in smoking mothers during pregnancy is assumed for relative FEV_1 reduction (Ref. [65]).

[b] Median values for cigarette smoking (Ref. [66]).

[c] Estimated values (Ref. [67]).

who smoke will go on to develop COPD. Among those smokers already with a decreased FEV_1, lung injury and subsequent decrements in lung function secondary to cigarette smoking are more dramatic. In the Lung Health Study, middle-aged smokers (with FEV_1 between 55% and 90%) who continued smoking for 5 years had further losses of several hundred milliliters of FEV_1 [82]. However, COPD has been identified in non-smokers as well. Prevalence has been noted to increase with age, to be higher in women than in men, to be particularly high in Hispanic individuals, and to be higher in low income versus affluent individuals [83]. Recent data from NHANES III suggests that 25% of cases of COPD in the United States occurs in lifelong non-smokers [84], a finding that is echoed by estimates of 23% in the United Kingdom [85] and 23% in Spain [86]. These observations reinforce that although cigarette smoking is the most important risk factor for COPD, it is not *necessary* for the development of irreversible airflow obstruction.

In summary, smoking has a lifetime influence on asthma and COPD. This starts *in utero* and continues into older age. Smoking is associated with enhanced airway responsiveness, both in asthma and COPD. Smoking is sufficient but not necessary for the development of COPD. Smoking is only one of several risk factors for asthma.

INTERMEDIATE PHENOTYPES

Allergy

Allergy represents immediate hypersensitivity to an antigen and is associated with an increased production of a specific immunoglobulin by sensitized lymphocytes. Elevations in specific IgE and/or total IgE, total eosinophil counts, and skin test reactivity to specific allergens have been used clinically to detect allergic individuals. As measured by skin test reactivity, allergy increases with age until about age 15, at which point it is maximal. The decline in skin test reactivity after age 35 confounds the measurement of this phenotype in older individuals. This reported association between skin test reactivity and decline in FEV_1 is not consistent in the literature. In retrospective studies, Taylor *et al.* [87] and Frew *et al.* [88] demonstrated no relation of skin test positivity to decline in FEV_1. However, Gottlieb *et al.* [89] investigated this prospectively in the Normative Aging Study and found that skin test positivity predicted increased annual rates of decline in both FEV_1 and FEV_1/FVC ratio.

Allergic inflammation is characteristic of asthma; 80–90% of childhood asthmatics are atopic, and the degree of atopy appears to be associated with prognosis in childhood asthma [90]. Studies have demonstrated that the asthmatic phenotype is associated with elevated serum IgE levels more so than skin test positivity [91], and that increased airways responsiveness is related to total serum IgE levels [92].

Weiss has suggested that immediate type I hypersensitivity is a risk factor for the development of chronic obstructive lung disease, and suggests that atopy may influence childhood asthma and limit maximal lung function, accelerate FEV_1 decline, and potentially enhance interaction with cigarette smoking to result in FEV_1 decline progressing to the development of COPD [93]. Hargreave and Leigh [94] demonstrated, in a subset of COPD patients, that eosinophilic inflammation is important in COPD exacerbations, and potentially leads to a decline in lung function. As well, an inverse association of serum IgE with the FEV_1/FVC ratio has been observed [95]. These data indicate that allergen sensitization may represent an intermediate phenotype which needs to be considered in understanding disease onset and progression in both asthma and COPD.

Airways responsiveness

Airways responsiveness to methacholine and histamine has been used in population-based studies to help define individuals susceptible to the development of obstructive lung disease. This intermediate phenotype is a feature of both asthma and a subset of patients with COPD. Baseline levels of lung function, allergy, age, and cigarette smoking history all influence airways responsiveness.

Airways hyperresponsiveness has been demonstrated to predict accelerated decline in lung function and the development of COPD [96]. More recently, airways responsiveness has been demonstrated to predict COPD mortality [97].

Airways responsiveness has been demonstrated to predict the development of asthma [98]. The prevalence of airway hyperresponsiveness exceeds the prevalence of asthma; the former is about 20% in the general population. Data from the Childhood Respiratory Disease Study demonstrate that increased airway responsiveness predicts the development of asthma in children and young adults with a 2- to 3-fold risk [99]. Some have observed this risk to be increased as much as 5-fold [100].

Airways hyperresponsiveness in COPD patients may be demonstrated in more than two-thirds of individuals in situations where it is actually measured [101]. Some individuals who develop COPD have an allergic asthma phenotype, as suggested by the Dutch hypothesis [102]. Alternatively the hyperresponsiveness may be a consequence of COPD. Results from a 25-year longitudinal study in the Netherlands revealed that increased airways responsiveness is an independent risk factor for FEV_1 decline [103]. Among individuals with early-onset COPD, the degree of baseline airways responsiveness determined the response to cigarette smoking; those with early-onset COPD who have increased airways responsiveness appeared more sensitive to the effects of cigarette smoke and had an accelerated decline in FEV_1 [104]. Thus, airway hyperresponsiveness, together with features of allergy and smoking, likely represent common risk factors for both asthma and COPD.

Gender/sex-related influences

The impact of gender and sex-biology-related features on airway disease are complex, and include both social paradigms (gender-related) as well as hormonal and genetic (sex-related) influences.

The epidemiology of asthma is characterized by age-related sex differences. Asthma and wheezing have been demonstrated to be more prevalent in young boys than young girls [20]. This trend disappears during puberty [105]. A recent analysis of the European Respiratory Health Survey [106] found that girls had a lower risk of developing asthma than did boys during childhood; about the time of puberty the risk was equal. After puberty, the risk in young women was higher than in young men, and this was a consistent trend in the 16 countries included in this study.

Women older than 20 years have higher prevalence and morbidity rates from asthma, and women are more like to present to the emergency department and be admitted with a diagnosis of asthma [107, 108]. In the multicenter Asthma Collaboration Study [109], women were more likely to be admitted to the hospital and report ongoing symptoms at follow-up, although overall men had less outpatient care and lower pulmonary function. Although it is unclear why women have more severe disease and higher mortality from asthma, hormonal, genetic, environmental, and social factors are likely contributory [110].

For many years COPD has traditionally been considered a disease of men, as men have been noted to have a higher prevalence of COPD at the population level. Yet, Prescott *et al.* [111] have suggested that women are more susceptible to the development of COPD, and observed that smoking was associated with a greater decrement in FEV_1 per pack-year of cigarette smoked, when compared to male smokers. Mannino *et al.* [112] analyzed data for deaths from obstructive lung disease from 1979 until 1993 and found that the mortality rates for men with COPD have started to stabilize but were continuing to increase among women, likely reflecting smoking trends. In the year 2000, the number of women dying from COPD has surpassed the number of men. Women have been observed to be more susceptible to the effects of tobacco smoking [113], more susceptible to develop severe early-onset COPD [14], have worse quality of life [114], and higher mortality from COPD [115] than men. These gender/sex differences most likely represent influences of both dose of tobacco exposure and underlying genetic and hormonal susceptibilities, and is a topic of active research.

In summary, in the adult years women have a higher prevalence of asthma. The prevalence of COPD in women is increasing, in keeping with the historical trends of cigarette smoking and disease. The gender and sex differences between asthma and COPD highlight the importance of hormonal and genetic influences relevant to disease expression in men versus women.

DEMOGRAPHICS

In the United States, morbidity from asthma has been demonstrated in multiple studies to be greater in children of African American descent. In the United States, physician-diagnosed asthma has been reported in 13.4% of African American children and 9.7% of white children [116]. African American children have also been reported to have greater limitation on activity due to asthma, with more hospital admissions and fewer doctors' visits when compared with white children [117]. Mortality from asthma has been higher for African American children when compared with children of other races since the mid-1980s [47, 57, 118–121]. Data from the NHANES II and the Hispanic Health and Nutrition Survey have noted that the prevalence of childhood asthma was higher in children of Puerto Rican versus African American descent (20.1% versus 9.1% respectively) [122]. An extensive review on the topic has noted that amongst the different ethnic groups living in the United States, the prevalence of asthma is highest in Puerto Ricans [123].

Studies in Chicago have demonstrated socioeconomic gradients and differing outcomes by race. In 1996, asthma hospitalization rates were more than twice as high as the United States' rates overall. Age-adjusted mortality was 4.7 times higher in non-Hispanic African Americans than in non-Hispanic whites [124]. An association with poverty has been suggested [125], and it has also been suggested that severe asthma may occur more frequently in poorer communities [126, 127]. The association of lower socioeconomic status with increased asthma prevalence is most likely multifactorial: the effects of indoor air pollution, passive cigarette smoke exposure, allergen exposure, and reduced access to medical care may all be relevant.

Some have suggested, based on the use of education as a surrogate for lower socioeconomic status, that there is an association of socioeconomic status with the development of obstructive lung disease. Bakke *et al.* [128] demonstrated that completion of only primary schooling was associated with an odds ratio of 2.9 for the development of obstructive lung disease when compared with those who achieved university level education. Exposure to smoking and occupational hazards decreased with increasing educational status.

AGE

Infants born prematurely, when compared to those born at term, have a risk for asthma that is increased approximately 4-fold [129]; reduced lung function measured with birth has been associated with incident asthma by age 10 [130]. There are data that breastfeeding is protective against asthma and, as previously noted, the risk for asthma increases in children exposed to cigarette smoke both *in utero* and during childhood. Asthma that begins after age 50 is thought to be more severe and less reversible than asthma that is incident in childhood [131]. In childhood, the remission of asthma before adulthood has been suggested to be about 50% [69, 132, 133]. Less information is available on the epidemiology of asthma in the middle-aged and elderly, yet some suggest that older patients are more severely affected than younger patients [134].

Some data support the proposition that individuals may "outgrow" asthma (with remission rates decreasing with increasing age) [135]. Other data suggest that remission of asthma and respiratory symptoms are uncommon [136]. Aging has been associated with increased airway obstruction overall [137] and asthma is a risk factor for COPD in adult lifetime never smokers [84]. The association of aging with the development of COPD most likely represents the cumulative insult of a lifetime of smoking and environmental exposure interacting with a susceptible host.

CONCLUSION

Ninety percent of all childhood asthma is diagnosed before the age of 6 years. Since there is a crude inverse relationship between respiratory symptoms and level of lung function, it is not surprising that as lung function increases in childhood respiratory symptoms decrease and often disappear. Thus, a large number of individuals are left with the intermediate phenotypes of increased airways responsiveness and/or allergy at the time that they reach their maximally attained level of lung function between the ages of 15 and 30. These intermediate phenotypes represent definable host characteristics that confer increased susceptibility to a variety of environmental exposures encountered in adult life, such as viral respiratory illness, occupation, allergen exposure, and perhaps most importantly, cigarette smoking.

The most clearly defined (non-genetic) susceptibility factors for premature or early-onset COPD are cigarette smoking, childhood asthma, increased airways responsiveness, and allergy. It is now clear that most airways hyperresponsiveness in adults antedates, precedes, and predicts the development of COPD. Obviously, this construct suggests that most genes that predict susceptibility to asthma may also be important genetic predictors of COPD susceptibility, and this is a topic of ongoing research.

Both asthma and COPD are defined as syndromes, and the definitions are loose; thus it is not surprising that there is substantial overlap between the conditions at any given age. Indeed, there are no substantive data to suggest that individuals cannot have both reversible and fixed airflow obstruction, and hence asthma and COPD at the same time.

The importance of early life and *in utero* events for the subsequent development of disease is a theme that is common to a variety of complex traits. These same issues also present themselves with disorders of the airways. The major barrier to applying a life-cycle approach to disease risk factors and natural history relates to the problem of recall bias and potentially missing or inadequate information about past events in both childhood and adult life, which may then distort the clinical picture. It is only through careful analysis of longitudinal cohort data that the true history of the relationship between the major environmental exposures and disease natural history can be deduced. We need to continue to gather such data, particularly data to link childhood asthma with adult COPD.

References

1. Toelle BG, Peat JK, Salome CM, Mellis CM, Woolcock AJ. Toward a definition of asthma for epidemiology. *Am Rev Respir Dis* 146(3): 633–37, 1992.
2. Pekkanen J, Pearce N. Defining asthma in epidemiological studies. *Eur Respir J* 14(4): 951–57, 1999.
3. Boushey HA, Holtzman MJ, Sheller JR, Nadel JA. Bronchial hyperreactivity. *Am Rev Respir Dis* 121(2): 389–413, 1980.
4. Global Initiative for Asthma. Global strategy for asthma management and prevention. Bethesda, MD: NHLBI/WHO Workshop Report, 2004.
5. Hargreave FE, Parameswaran K. Asthma, COPD and bronchitis are just components of airway disease. *Eur Respir J* 28(2): 264–67, 2006.
6. Asher MI, Keil U, Anderson HR *et al.* International Study of Asthma and Allergies in Childhood (ISAAC): Rationale and methods. *Eur Respir J* 8(3): 483–91, 1995.
7. Lindberg A, Jonsson AC, Ronmark E, Lundgren R, Larsson LG, Lundback B. Prevalence of chronic obstructive pulmonary disease according to BTS, ERS, GOLD and ATS criteria in relation to doctor's diagnosis, symptoms, age, gender, and smoking habits. *Respiration* 72(5): 471–79, 2005.
8. Viegi G, Pedreschi M, Pistelli F *et al.* Prevalence of airways obstruction in a general population: European Respiratory Society vs American Thoracic Society definition. *Chest* 117(5 Suppl 2): 339S–345S, 2000.
9. Halbert RJ, Natoli JL, Gano A, Badamgarav E, Buist AS, Mannino DM. Global burden of COPD: Systematic review and meta-analysis. *Eur Respir J* 28(3): 523–32, 2006.
10. Global Initiative for Chronic Obstructive Lung Disease. A collaborative project of the National Heart, Lung and Blood Institute and the World Health Organization. National Institutes of Health, National Heart, Lung and Blood Institute, 2006; Accessed June 30, 2007, at www.goldcopd.com.
11. Celli BR, Cote CG, Marin JM *et al.* The body-mass index, airflow obstruction, dyspnea, and exercise capacity index in chronic obstructive pulmonary disease. *N Engl J Med* 350(10): 1005–12, 2004.
12. Hurd SS. International efforts directed at attacking the problem of COPD. *Chest* 117(5 Suppl 2): 336S–338S, 2000.
13. Salam MT, Wenten M, Gilliland FD. Endogenous and exogenous sex steroid hormones and asthma and wheeze in young women. *J Allergy Clin Immunol* 117(5): 1001–7, 2006.
14. Silverman EK, Weiss ST, Drazen JM *et al.* Gender-related differences in severe, early-onset chronic obstructive pulmonary disease. *Am J Respir Crit Care Med* 162(6): 2152–58, 2000.
15. Masoli M, Fabian D, Holt S, Beasley R. The global burden of asthma: Executive summary of the GINA Dissemination Committee report. *Allergy* 59(5): 469–78, 2004.

16. Yunginger JW, Reed CE, O'Connell EJ, Melton LJ, O'Fallon WM, Silverstein MD. A community-based study of the epidemiology of asthma. Incidence rates, 1964–1983. *Am Rev Respir Dis* 146(4): 888–94, 1992.
17. McWhorter WP, Polis MA, Kaslow RA. Occurrence, predictors, and consequences of adult asthma in NHANESI and follow-up survey. *Am Rev Respir Dis* 139(3): 721–24, 1989.
18. Broder I, Higgins MW, Mathews KP, Keller JB. Epidemiology of asthma and allergic rhinitis in a total community, Tecumseh, Michigan. III. Second survey of the community. *J Allergy Clin Immunol* 53(3): 127–38, 1974.
19. Broder I, Higgins MW, Mathews KP, Keller JB. Epidemiology of asthma and allergic rhinitis in a total community, Tecumseh, Michigan. IV. Natural history. *J Allergy Clin Immunol* 54(2): 100–10, 1974.
20. Dodge RR, Burrows B. The prevalence and incidence of asthma and asthma-like symptoms in a general population sample. *Am Rev Respir Dis* 122(4): 567–75, 1980.
21. Martinez FD, Wright AL, Taussig LM, Holberg CJ, Halonen M, Morgan WJ. Asthma and wheezing in the first six years of life. The Group Health Medical Associates. *N Engl J Med* 332(3): 133–38, 1995.
22. Toren K, Hermansson BA. Incidence rate of adult-onset asthma in relation to age, sex, atopy and smoking: A Swedish population-based study of 15813 adults. *Int J Tuberc Lung Dis* 3(3): 192–97, 1999.
23. Huovinen E, Kaprio J, Laitinen LA, Koskenvuo M. Incidence and prevalence of asthma among adult Finnish men and women of the Finnish Twin Cohort from 1975 to 1990, and their relation to hay fever and chronic bronchitis. *Chest* 115(4): 928–36, 1999.
24. de Marco R, Accordini S, Cerveri I *et al.* Incidence of chronic obstructive pulmonary disease in a cohort of young adults according to the presence of chronic cough and phlegm. *Am J Respir Crit Care Med* 175(1): 32–39, 2007.
25. Lindberg A, Jonsson AC, Ronmark E, Lundgren R, Larsson LG, Lundback B. Ten-year cumulative incidence of COPD and risk factors for incident disease in a symptomatic cohort. *Chest* 127(5): 1544–52, 2005.
26. Johannessen A, Omenaas E, Bakke P, Gulsvik A. Incidence of GOLD-defined chronic obstructive pulmonary disease in a general adult population. *Int J Tuberc Lung Dis* 9(8): 926–32, 2005.
27. Mannino DM, Gagnon RC, Petty TL, Lydick E. Obstructive lung disease and low lung function in adults in the United States: Data from the National Health and Nutrition Examination Survey, 1988–1994. *Arch Intern Med* 160(11): 1683–89, 2000.
28. Eder W, Ege MJ, von Mutius E. The asthma epidemic. *N Engl J Med* 55(21): 2226–35, 2006.
29. Mannino DM, Homa DM, Pertowski CA *et al.* Surveillance for asthma – United States, 1960–1995. *Mor Mortal Wkly Rep CDC Surveill Summ* 47(1): 1–27, 1998.
30. Crain EF, Weiss KB, Bijur PE, Hersh M, Westbrook L, Stein RE. An estimate of the prevalence of asthma and wheezing among inner-city children. *Pediatrics* 94(3): 356–62, 1994.
31. Persky VW, Slezak J, Contreras A *et al.* Relationships of race and socioeconomic status with prevalence, severity, and symptoms of asthma in Chicago school children. *Ann Allergy Asthma Immunol* 81(3): 266–71, 1998.
32. Sole D, Yamada E, Vana AT, Costa-Carvalho BT, Naspitz CK. Prevalence of asthma and related symptoms in school-age children in Sao Paulo, Brazil – International Study of Asthma and Allergies in Children (ISAAC). *J Asthma* 36(2): 205–12, 1999.
33. Asher MI, Weiland SK. The International Study of Asthma and Allergies in Childhood (ISAAC). ISAAC Steering Committee. *Clin Exp Allergy* 28(Suppl 5): 52–66, 1998. discussion 90–1
34. Duhme H, Weiland SK, Rudolph P, Wienke A, Kramer A, Keil U. Asthma and allergies among children in West and East Germany: A comparison between Munster and Greifswald using the ISAAC phase I protocol. International Study of Asthma and Allergies in Childhood. *Eur Respir J* 11(4): 840–47, 1998.
35. Aguinaga Ontoso I, Arnedo Pena A, Bellido J, Guillen Grima F, Suarez Varela MM. [The prevalence of asthma-related symptoms in 13–14-year-old children from 9 Spanish populations. The Spanish Group of the ISAAC Study (International Study of Asthma and Allergies in Childhood). (published erratum appears in *Med Clin (Barc)* 112(13):494, April 17, 1999)]. *Med Clin (Barc)* 112(5): 171–75, 1999.
36. Gerritsen J, Koeter GH, Postma DS, Schouten JP, Knol K. Prognosis of asthma from childhood to adulthood. *Am Rev Respir Dis* 140(5): 1325–30, 1989.
37. Centers for Disease Control and Prevention. Tobacco use among adults – United States, 2005. *MMWR Morb Mortal Wkly Rep* 55(42): 1145–8, 2006.
38. Burrows B, Knudson RJ, Cline MG, Lebowitz MD. Quantitative relationships between cigarette smoking and ventilatory function. *Am Rev Respir Dis* 115(2): 195–205, 1977.
39. Stang P, Lydick E, Silberman C, Kempel A, Keating ET. The prevalence of COPD: Using smoking rates to estimate disease frequency in the general population. *Chest* 117(5 Suppl 2): 354S–359S, 2000.
40. Lopez AD, Murray CC. The global burden of disease, 1990–2020. *Nat Med* 4(11): 1241–43, 1998.
41. Lacasse Y, Brooks D, Goldstein RS. Trends in the epidemiology of COPD in Canada, 1980 to 1995. COPD and Rehabilitation Committee of the Canadian Thoracic Society. *Chest* 116(2): 306–13, 1999.
42. Enright PL, Kronmal RA, Higgins MW, Schenker MB, Haponik EF. Prevalence and correlates of respiratory symptoms and disease in the elderly. Cardiovascular Health Study. *Chest* 106(3): 827–34, 1994.
43. Mannino DM. Epidemiology and global impact of chronic obstructive pulmonary disease. *Semin Respir Crit Care Med* 26(2): 204–10, 2005.
44. Schirnhofer L, Lamprecht B, Vollmer WM *et al.* COPD prevalence in Salzburg, Austria: Results from the Burden of Obstructive Lung Disease (BOLD) Study. *Chest* 131(1): 29–36, 2007.
45. Sobradillo V, Miravitlles M, Jimenez CA *et al.* [Epidemiological study of chronic obstructive pulmonary disease in Spain (IBERPOC): Prevalence of chronic respiratory symptoms and airflow limitation]. *Arch Bronconeumol* 35(4): 159–66, 1999.
46. Fleming DM, Crombie DL. Prevalence of asthma and hay fever in England and Wales. *BMJ (Clin Res Ed)* 294(6567): 279–83, 1987.
47. Gerstman BB, Bosco LA, Tomita DK, Gross TP, Shaw MM. Prevalence and treatment of asthma in the Michigan Medicaid patient population younger than 45 years, 1980–1986. *J Allergy Clin Immunol* 83(6): 1032–39, 1989.
48. Manfreda J, Becker AB, Wang PZ, Roos LL, Anthonisen NR. Trends in physician-diagnosed asthma prevalence in Manitoba between 1980 and 1990. *Chest* 103(1): 151–57, 1993.
49. Mitchell EA. International trends in hospital admission rates for asthma. *Arch Dis Child* 60(4): 376–78, 1985.
50. Anderson HR, Bailey P, West S. Trends in the hospital care of acute childhood asthma 1970–8: A regional study. *BMJ* 281(6249): 1191–94, 1980.
51. Jackson RT, Mitchell EA. Trends in hospital admission rates and drug treatment of asthma in New Zealand. *N Z Med J* 96(740): 728–30, 1983.
52. Halfon N, Newacheck PW. Trends in the hospitalization for acute childhood asthma, 1970–84. *Am J Public Health* 76(11): 1308–11, 1986.
53. Priftis K, Anagnostakis J, Harokopos E, Orfanou I, Petraki M, Saxoni-Papageorgiou P. Time trends and seasonal variation in hospital admissions for childhood asthma in the Athens region of Greece: 1978–88. *Thorax* 48(11): 1168–69, 1993.
54. Carman PG, Landau LI. Increased paediatric admissions with asthma in Western Australia – a problem of diagnosis?. *Med J Aust* 152(1): 23–26, 1990.
55. Kun HY, Oates RK, Mellis CM. Hospital admissions and attendances for asthma – a true increase? *Med J Aust* 159(5): 312–13, 1993.
56. Wilkins K, Mao Y. Trends in rates of admission to hospital and death from asthma among children and young adults in Canada during the 1980s. *Can Med Assoc J* 148(2): 185–90, 1993.
57. Evans Rd, Mullally DI, Wilson RW *et al.* National trends in the morbidity and mortality of asthma in the US. Prevalence, hospitalization and death from asthma over two decades: 1965–1984. *Chest* 91(6 Suppl): 65S–74S, 1987.

58. Mannino DM, Homa DM, Akinbami LJ, Ford ES, Redd SC. Chronic obstructive pulmonary disease surveillance – United States, 1971–2000. *MMWR Surveill Summ* 51(6): 1–16, 2002.
59. Jackson R, Sears MR, Beaglehole R, Rea HH. International trends in asthma mortality: 1970 to 1985. *Chest* 94(5): 914–18, 1988.
60. Carr W, Zeitel L, Weiss K. Variations in asthma hospitalizations and deaths in New York City. *Am J Public Health* 82(1): 59–65, 1992.
61. McConnochie KM, Russo MJ, McBride JT, Szilagyi PG, Brooks AM, Roghmann KJ. Socioeconomic variation in asthma hospitalization: Excess utilization or greater need?. *Pediatrics* 103(6): e75, 1999.
62. Thom TJ. International comparisons in COPD mortality. *Am Rev Respir Dis* 140(3 Pt 2): S27–34, 1989.
63. Burr ML, Davies BH, Hoare A *et al.* A confidential inquiry into asthma deaths in Wales. *Thorax* 54(11): 985–89, 1999.
64. Ringbaek T, Seersholm N, Viskum K. Standardised mortality rates in females and males with COPD and asthma. *Eur Respir J* 25(5): 891–95, 2005.
65. Mannino DM, Reichert MM, Davis KJ. Lung function decline and outcomes in an adult population. *Am J Respir Crit Care Med* 173(9): 985–90, 2006.
66. Cunningham J, Dockery DW, Speizer FE. Maternal smoking during pregnancy as a predictor of lung function in children. *Am J Epidemiol* 139(12): 1139–52, 1994.
67. Tager IB, Munoz A, Rosner B, Weiss ST, Carey V, Speizer FE. Effect of cigarette smoking on the pulmonary function of children and adolescents. *Am Rev Respir Dis* 131(5): 752–59, 1985.
68. Xu X, Weiss ST, Rijcken B, Schouten JP. Smoking, changes in smoking habits, and rate of decline in FEV_1: New insight into gender differences. *Eur Respir J* 7(6): 1056–61, 1994.
69. Strachan DP, Butland BK, Anderson HR. Incidence and prognosis of asthma and wheezing illness from early childhood to age 33 in a national British cohort. *BMJ* 312(7040): 1195–99, 1996.
70. Kaplan BA, Mascie-Taylor CG. Smoking and asthma among 23-year-olds. *J Asthma* 34(3): 219–26, 1997.
71. Ulrik CS, Frederiksen J. Mortality and markers of risk of asthma death among 1,075 outpatients with asthma. *Chest* 108(1): 10–15, 1995.
72. Althuis MD, Sexton M, Prybylski D. Cigarette smoking and asthma symptom severity among adult asthmatics. *J Asthma* 36(3): 257–64, 1999.
73. Lange P, Parner J, Vestbo J, Schnohr P, Jensen G. A 15-year follow-up study of ventilatory function in adults with asthma. *N Engl J Med* 339(17): 1194–200, 1998.
74. Siroux V, Pin I, Oryszczyn MP, Le Moual N, Kauffmann F. Relationships of active smoking to asthma and asthma severity in the EGEA study. Epidemiological study on the Genetics and Environment of Asthma. *Eur Respir J* 15(3): 470–77, 2000.
75. Halken S, Host A, Nilsson L, Taudorf E. Passive smoking as a risk factor for development of obstructive respiratory disease and allergic sensitization. *Allergy* 50(2): 97–105, 1995.
76. Wartenberg D, Ehrlich R, Lilienfeld D. Environmental tobacco smoke and childhood asthma: Comparing exposure metrics using probability plots. *Environ Res* 64(2): 122–35, 1994.
77. Cook DG, Strachan DP. Health effects of passive smoking. 3. Parental smoking and prevalence of respiratory symptoms and asthma in school age children. *Thorax* 52(12): 1081–94, 1997.
78. Weiss ST, Utell MJ, Samet JM. Environmental tobacco smoke exposure and asthma in adults. *Environ Health Perspect* 107(Suppl 6): 891–95, 1999.
79. Weil CM, Wade SL, Bauman LJ, Lynn H, Mitchell H, Lavigne J. The relationship between psychosocial factors and asthma morbidity in inner-city children with asthma. *Pediatrics* 104(6): 1274–80, 1999.
80. Infante-Rivard C. Childhood asthma and indoor environmental risk factors. *Am J Epidemiol* 137(8): 834–44, 1993.
81. Infante-Rivard C, Gautrin D, Malo JL, Suissa S. Maternal smoking and childhood asthma. *Am J Epidemiol* 150(5): 528–31, 1999.
82. Anthonisen NR, Connett JE, Kiley JP *et al.* Effects of smoking intervention and the use of an inhaled anticholinergic bronchodilator on the rate of decline of FEV_1. The Lung Health Study. *JAMA* 272(19): 1497–505, 1994.
83. Whittemore AS, Perlin SA, DiCiccio Y. Chronic obstructive pulmonary disease in lifelong nonsmokers: Results from NHANES. *Am J Public Health* 85(5): 702–6, 1995.
84. Behrendt CE. Mild and moderate-to-severe COPD in nonsmokers: Distinct demographic profiles. *Chest* 128(3): 1239–44, 2005.
85. Birring SS, Brightling CE, Bradding P *et al.* Clinical, radiologic, and induced sputum features of chronic obstructive pulmonary disease in nonsmokers: A descriptive study. *Am J Respir Crit Care Med* 166(8): 1078–83, 2002.
86. Pena VS, Miravitlles M, Gabriel R *et al.* Geographic variations in prevalence and underdiagnosis of COPD: Results of the IBERPOC multicentre epidemiological study. *Chest* 118(4): 981–89, 2000.
87. Taylor RG, Joyce H, Gross E, Holland F, Pride NB. Bronchial reactivity to inhaled histamine and annual rate of decline in FEV_1 in male smokers and ex-smokers. *Thorax* 40(1): 9–16, 1985.
88. Frew AJ, Kennedy SM, Chan-Yeung M. Methacholine responsiveness, smoking, and atopy as risk factors for accelerated FEV_1 decline in male working populations. *Am Rev Respir Dis* 146(4): 878–83, 1992.
89. Gottlieb DJ, Sparrow D, O'Connor GT, Weiss ST. Skin test reactivity to common aeroallergens and decline of lung function. The Normative Aging Study. *Am J Respir Crit Care Med* 153(2): 561–66, 1996.
90. Nelson HS. The importance of allergens in the development of asthma and the persistence of symptoms. *J Allergy Clin Immunol* 105(6 Pt 2): S628–S632, 2000.
91. Burrows B, Martinez FD, Halonen M, Barbee RA, Cline MG. Association of asthma with serum IgE levels and skin-test reactivity to allergens. *N Engl J Med* 320(5): 271–77, 1989.
92. Sears MR, Burrows B, Flannery EM, Herbison GP, Hewitt CJ, Holdaway MD. Relation between airway responsiveness and serum IgE in children with asthma and in apparently normal children. *N Engl J Med* 325(15): 1067–71, 1991.
93. Weiss ST. Atopy as a risk factor for chronic obstructive pulmonary disease. Epidemiological evidence. *Am J Respir Crit Care Med* 162(3 Pt 2): S134–S136, 2000.
94. Hargreave FE, Leigh R. Induced sputum, eosinophilic bronchitis, and chronic obstructive pulmonary disease. *Am J Respir Crit Care Med* 160(5 Pt 2): S53–S57, 1999.
95. Sherrill DL, Lebowitz MD, Halonen M, Barbee RA, Burrows B. Longitudinal evaluation of the association between pulmonary function and total serum IgE. *Am J Respir Crit Care Med* 152(1): 98–102, 1995.
96. Rijcken B, Schouten JP, Weiss ST, Speizer FE, van der Lende R. The association of airways responsiveness to respiratory symptom prevalence and to pulmonary function in a random population sample. *Bull Eur Physiopathol Respir* 23(4): 391–94, 1987.
97. Hospers JJ, Postma DS, Rijcken B, Weiss ST, Schouten JP. Histamine airway hyper-responsiveness and mortality from chronic obstructive pulmonary disease: A cohort study. *Lancet* 356(9238): 1313–17, 2000.
98. Xu X, Rijcken B, Schouten JP, Weiss ST. Airways responsiveness and development and remission of chronic respiratory symptoms in adults. *Lancet* 350(9089): 1431–34, 1997.
99. Carey VJ, Weiss ST, Tager IB, Leeder SR, Speizer FE. Airways responsiveness, wheeze onset, and recurrent asthma episodes in young adolescents. The East Boston Childhood Respiratory Disease Cohort. *Am J Respir Crit Care Med* 153(1): 356–61, 1996.
100. Weiss ST, Tosteson TD, Segal MR, Tager IB, Redline S, Speizer FE. Effects of asthma on pulmonary function in children. A longitudinal population-based study. *Am Rev Respir Dis* 145(1): 58–64, 1992.
101. O'Connor GT, Sparrow D, Weiss ST. The role of allergy and non-specific airway hyperresponsiveness in the pathogenesis of chronic obstructive pulmonary disease. *Am Rev Respir Dis* 140(1): 225–52, 1989.
102. Sluiter HJ, Koeter GH, de Monchy JG, Postma DS, de Vries K, Orie NG. The Dutch hypothesis (chronic non-specific lung disease) revisited. *Eur Respir J* 4(4): 479–89, 1991.

103. Rijcken B, Schouten JP, Xu X, Rosner B, Weiss ST. Airway hyperresponsiveness to histamine associated with accelerated decline in FEV_1. *Am J Respir Crit Care Med* 151(5): 1377–82, 1995.
104. Tashkin DP, Altose MD, Connett JE, Kanner RE, Lee WW, Wise RA. Methacholine reactivity predicts changes in lung function over time in smokers with early chronic obstructive pulmonary disease. The Lung Health Study Research Group. *Am J Respir Crit Care Med* 153(6 Pt 1): 1802–11, 1996.
105. Venn A, Lewis S, Cooper M, Hill J, Britton J. Questionnaire study of effect of sex and age on the prevalence of wheeze and asthma in adolescence. *BMJ* 316(7149): 1945–46, 1998.
106. de Marco R, Locatelli F, Sunyer J, Burney P. Differences in incidence of reported asthma related to age in men and women. A retrospective analysis of the data of the European Respiratory Health Survey. *Am J Respir Crit Care Med* 162(1): 68–74, 2000.
107. Skobeloff EM, Spivey WH, St. Clair SS, Schoffstall JM. The influence of age and sex on asthma admissions. *JAMA* 268(24): 3437–40, 1992.
108. Baibergenova A, Thabane L, Akhtar-Danesh N, Levine M, Gafni A, Leeb K. Sex differences in hospital admissions from emergency departments in asthmatic adults: A population-based study. *Ann Allergy Asthma Immunol* 96(5): 666–72, 2006.
109. Singh AK, Cydulka RK, Stahmer SA, Woodruff PG, Camargo CA Jr. Sex differences among adults presenting to the emergency department with acute asthma. Multicenter Asthma Research Collaboration Investigators. *Arch Intern Med* 159(11): 1237–43, 1999.
110. Melgert BN, Ray A, Hylkema MN, Timens W, Postma DS. Are there reasons why adult asthma is more common in females? *Curr Allergy Asthma Rep* 7(2): 143–50, 2007.
111. Prescott E, Bjerg AM, Andersen PK, Lange P, Vestbo J. Gender difference in smoking effects on lung function and risk of hospitalization for COPD: Results from a Danish longitudinal population study. *Eur Respir J* 10(4): 822–27, 1997.
112. Mannino DM, Brown C, Giovino GA. Obstructive lung disease deaths in the United States from 1979 through 1993. An analysis using multiple-cause mortality data. *Am J Respir Crit Care Med* 156(3 Pt 1): 814–18, 1997.
113. Gan WQ, Man SF, Postma DS, Camp P, Sin DD. Female smokers beyond the perimenopausal period are at increased risk of chronic obstructive pulmonary disease: A systematic review and meta-analysis. *Respir Res* 7: 52, 2006.
114. Martinez FJ, Curtis JL, Sciurba F *et al.* Gender differences in severe pulmonary emphysema. *Am J Respir Crit Care Med* 176(3): 222–3, 2007.
115. Machado MC, Krishnan JA, Buist SA *et al.* Sex differences in survival of oxygen-dependent patients with chronic obstructive pulmonary disease. *Am J Respir Crit Care Med* 174(5): 524–29, 2006.
116. Taylor WR, Newacheck PW. Impact of childhood asthma on health. *Pediatrics* 90(5): 657–62, 1992.
117. Coultas DB, Gong H Jr., Grad R *et al.* Respiratory diseases in minorities of the United States. *Am J Respir Crit Care Med* 149(3 Pt 2): S93–131, 1994. [published erratum appears in *Am J Respir Crit Care Med* 150(1):290, July 1994]
118. Clark NM, Feldman CH, Evans D, Levison MJ, Wasilewski Y, Mellins RB. The impact of health education on frequency and cost of health care use by low income children with asthma. *J Allergy Clin Immunol* 78(1 Pt 1): 108–15, 1986.
119. Nelson DA, Johnson CC, Divine GW, Strauchman C, Joseph CL, Ownby DR. Ethnic differences in the prevalence of asthma in middle class children. *Ann Allergy Asthma Immunol* 78(1): 21–26, 1997.
120. Joseph CL, Foxman B, Leickly FE, Peterson E, Ownby D. Prevalence of possible undiagnosed asthma and associated morbidity among urban schoolchildren. *J Pediatr* 129(5): 735–42, 1996.
121. Weitzman M, Gortmaker SL, Sobol AM, Perrin JM. Recent trends in the prevalence and severity of childhood asthma. *JAMA* 268(19): 2673–77, 1992.
122. Carter-Pokras OD, Gergen PJ. Reported asthma among Puerto Rican, Mexican-American, and Cuban children, 1982 through 1984. *Am J Public Health* 83(4): 580–82, 1993.
123. Hunninghake GM, Weiss ST, Celedon JC. Asthma in Hispanics. *Am J Respir Crit Care Med* 173(2): 143–63, 2006.
124. Thomas SD, Whitman S. Asthma hospitalizations and mortality in Chicago: An epidemiologic overview. *Chest* 116(4 Suppl 1): 135S–141S, 1999.
125. Duran-Tauleria E, Rona RJ. Geographical and socioeconomic variation in the prevalence of asthma symptoms in English and Scottish children. *Thorax* 54(6): 476–81, 1999.
126. Mielck A, Reitmeir P, Wjst M. Severity of childhood asthma by socioeconomic status. *Int J Epidemiol* 25(2): 388–93, 1996.
127. Strachan DP, Anderson HR, Limb ES, O'Neill A, Wells N. A national survey of asthma prevalence, severity, and treatment in Great Britain. *Arch Dis Child* 70(3): 174–78, 1994.
128. Bakke PS, Hanoa R, Gulsvik A. Educational level and obstructive lung disease given smoking habits and occupational airborne exposure: A Norwegian community study. *Am J Epidemiol* 141(11): 1080–88, 1995.
129. von Mutius E, Nicolai T, Martinez FD. Prematurity as a risk factor for asthma in preadolescent children. *J Pediatr* 123(2): 223–29, 1993.
130. Haland G, Carlsen KC, Sandvik L *et al.* Reduced lung function at birth and the risk of asthma at 10 years of age. *N Engl J Med* 355(16): 1682–89, 2006.
131. Vergnenegre A, Antonini MT, Bonnaud F, Melloni B, Mignonat G, Bousquet J. Comparison between late onset and childhood asthma. *Allergol Immunopathol (Madr)* 20(5): 190–96, 1992.
132. Jonsson JA, Boe J, Berlin E. The long-term prognosis of childhood asthma in a predominantly rural Swedish county. *Acta Paediatr Scand* 76(6): 950–54, 1987.
133. Kelly WJ, Hudson I, Phelan PD, Pain MC, Olinsky A. Childhood asthma in adult life: A further study at 28 years of age. *BMJ (Clin Res Ed)* 294(6579): 1059–62, 1987.
134. Burrows B, Barbee RA, Cline MG, Knudson RJ, Lebowitz MD. Characteristics of asthma among elderly adults in a sample of the general population. *Chest* 100(4): 935–42, 1991.
135. Panhuysen CI, Vonk JM, Koeter GH *et al.* Adult patients may outgrow their asthma: A 25-year follow-up study. *Am J Respir Crit Care Med* 155(4): 1267–72, 1997. [published erratum appears in *Am J Respir Crit Care Med* 156(2 Pt 1): 674, August 1997]
136. Ronmark E, Jonsson E, Lundback B. Remission of asthma in the middle aged and elderly: Report from the Obstructive Lung Disease in Northern Sweden study. *Thorax* 54(7): 611–13, 1999.
137. Wise RA. Changing smoking patterns and mortality from chronic obstructive pulmonary disease. *Prev Med* 26(4): 418–21, 1997.

3

Natural History

CHAPTER

Stefano Guerra and Fernando D. Martinez

Arizona Respiratory Center and Mel and Enid Zuckerman College of Public Health, University of Arizona, Tucson, AZ, USA

INTRODUCTION

There is increasing interest in developing a better understanding of the natural history of the two most frequent chronic lower respiratory diseases, asthma and chronic obstructive pulmonary disease (COPD). The main reason for this interest is the shift in the therapeutic approach to both conditions. In the case of asthma, a growing body of evidence suggests that, although several current treatments for the disease are extremely effective in controlling symptoms, none can change its natural course. It is thus evident that new approaches to the primary and secondary prevention of asthma are needed. These approaches will require a thorough understanding of the factors that determine the inception of asthma and its progression with age. In the case of COPD, it is apparent that, although smoking is the main demonstrated cause of this condition, a simple strategy of discouraging tobacco consumption has proven insufficient to prevent the enormous impact of nicotine addiction on public health. A new understanding of the factors that increase the risk of COPD in smokers seems a more realistic approach, and could perhaps contribute to palliate the increasing social toll of COPD.

When discussing the natural history of asthma and COPD, an important issue is the overlap between these two conditions [1]. It has been conclusively shown that, at the population level, adults who report asthma are more likely to also report chronic bronchitis and/or emphysema as compared with subjects with no asthma [2] and recent epidemiological evidence shows that active asthma is a strong risk factor for acquiring a subsequent diagnosis of COPD [3]. This overlap may be explained by the fact that asthma and COPD share common risk factors or, alternatively, by the possible progression of persistent asthma into chronic airflow limitation, the clinical hallmark of COPD [4]. It is often not possible to determine whether persons with airflow limitation have asthma or COPD. However, understanding the natural history of the disease will have critical implications for identifying optimal prevention and treatment strategies.

METHODOLOGICAL APPROACH

Any discussion about the natural history of a chronic condition ought to address the different hypothetical phases that such condition undergoes during its lifetime course (Fig. 3.1).

There is first a pre-illness period during which subjects without overt disease have the susceptibility for the development of the condition, because of a genetic predisposition or because of injuries or developmental variations that have mutated one or more of the individual's disease-associated phenotypes. During this pre-illness phase, the susceptible individual is not yet destined to develop the disease. More likely, he or she may be exposed to environmental factors that, in the presence of genetic variants that predispose to the chronic condition, can further modify their phenotype. This model applies to both asthma and COPD, as they are both likely to represent heterogeneous diseases with multiple ways to reach final common pathways and an array of diverse pathogenetic mechanisms giving rise to more or less reversible airway obstruction.

It would obviously be of great help for any prevention strategy if markers of susceptibility were identified that could be measured during

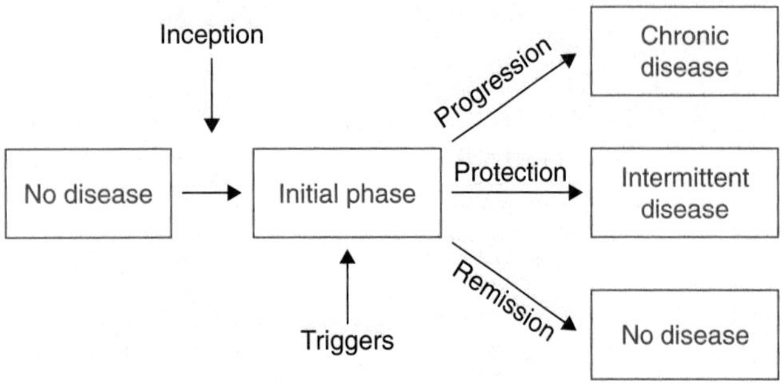

FIG. 3.1 Hypothetical representation of the natural history of a chronic condition. For explanation see text.

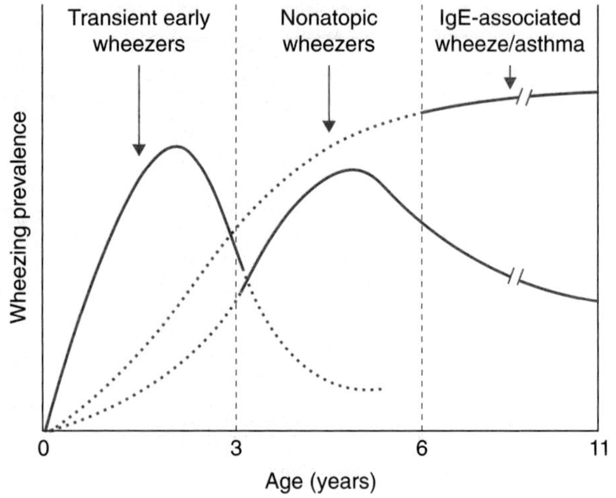

FIG. 3.2 Hypothetical yearly peak prevalence of wheezing for three different phenotypes in childhood. Prevalence for each age interval should be the sum of the areas under each curve. The dashed curves suggest wheezing can present different curve shapes due to many different factors, including overlap of groups. Adapted from Ref. [15].

this pre-illness phase. But the absence of a marker in a specific individual may not exclude susceptibility if that is not the pathway that may be active in that individual. Thus, because the marker *phenotype* may be influenced by determinants other than the factors that influence asthma or COPD risk, ascertainment of a marker may be fraught with difficulties. This "context-dependency" [5] of the phenotypes of markers of disease creates a further layer of complexity for the identification of susceptible individuals.

The above discussion points to the importance of determining the "true" time of incidence of the disease under study. Although the event that is easiest to ascertain is the age of initiation of symptoms, this event is modified by patients' perceptions. What is important is that time of onset ought to be ascertained before the disease has become chronic, that is before the individual has evidence of physiological or anatomical changes caused by the disease itself and that predispose to persistent or recurrent symptoms. During this *initial phase of the disease*, the affected individual has some clinical manifestations, but the condition has not yet fully developed. Moreover, some individuals may not go beyond this initial phase, because remission or stabilization may be fostered by the presence of environmental or genetic protective factors. During this initial phase secondary prevention is possible, but again knowledge of the natural history becomes essential.

NATURAL HISTORY OF ASTHMA PHENOTYPES

Asthma is a heterogeneous condition, and one of the main obstacles in understanding its natural history has been the lack of well-defined markers for the different disease phenotypes grouped under this common label. This hurdle has been addressed lately by a series of longitudinal studies that have assessed incidence and prevalence of asthma at different ages or defined the outcome of persons with asthma-like symptoms enrolled at different ages and especially during childhood [6–13].

Most cases of chronic, persistent asthma start in early life

Several longitudinal studies have confirmed that, in most cases of chronic, persistent asthma, the initial clinical manifestations of asthma occur during the first 5 years of life [6, 11, 13]. Studies have ascertained this age of onset of symptoms either by asking parents about the prevalence of wheezing or other asthma-like symptoms in their children at different ages, or by reviewing previous clinical charts in subjects who present to outpatient clinics with signs and symptoms of asthma at any age. Although these studies strongly suggest that asthma started at an early age in these individuals, a simple connection between these two phenomena is not warranted. Many children have asthma-like symptoms during viral infections in their preschool years, but most have transient conditions that subside with age, and only a minority will go on to have persistent asthma [14]. Thus it is difficult to distinguish who will go on to develop chronic asthma from those who will not.

Studies of this issue have suggested that there are at least three quite distinct groups of children with asthma-like symptoms coexisting up to the adolescent years (Fig. 3.2). The great majority of infants who wheeze during the first 1–2 years of life do so during viral infections, especially those caused by the respiratory syncytial virus (RSV) and by rhinovirus. Most of these children will have one or only a few episodes of wheezing, with no further symptoms beyond the age of 2–3 years. This condition, which has been identified as *transient wheezing of infancy* [16], is the most frequent form of recurrent airway obstruction in this age group, affecting over two-thirds of all infants with asthma-like symptoms. The main predisposing factors for transient wheezing of infancy are maternal smoking during pregnancy and lower levels of lung function, as assessed during the very first months of life and before any wheezing episode has occurred. Interestingly, infants (especially females) whose mothers smoke during pregnancy do have lower levels of lung function early in life than those whose mothers do not smoke [17]. This has suggested the possibility that both inherited and acquired characteristics of the lung, the airways, or both may create the structural conditions for the development of airway obstruction during lower respiratory illnesses [18].

The factors that determine the remission of early life wheezing in these children are not well understood, but it is likely that growth of the airways may outpace that of the lung parenchyma after age 2–5 months, the period of the highest incidence of wheezing episodes [19]. Since the airways are growing faster than the parenchyma, the conditions required for airflow to be dynamically obstructed may not be easily reached and noisy breathing is less likely [19]. It is also possible that changes in the regulation of airway tone may result in decreased bronchial hyperresponsiveness and decreased likelihood of respiratory symptoms with age [20, 21]. However, despite remission of wheezing with age in this group, children with early transient wheezing appear to have lower levels of lung function, in terms of FEF_{25-75}, FEV_1, and FEV_1/FVC ratio values, throughout childhood as compared with their peers who did not wheeze in the first 6 years of life [22]. It is established that individuals who enter adulthood with lung function deficits are more likely to develop COPD during the late adult years, suggesting that children with early transient wheezing may be particularly predisposed to COPD in adult life if exposed to cigarette smoking. Whether this is the case is at present unknown.

The factors that determine enduring or newly developed abnormalities in lung function during the first years of life are the subject of intense scrutiny, because they seem to be clearly associated with the risk of persistent asthma. Zeiger *et al.* [23] showed an inverse relationship between duration of symptoms and level of lung function in school-age children with asthma, and this suggests that the earlier onset of asthma symptoms is associated with greater losses in lung function. Sensitization to local aeroallergens is strongly associated with increased risk of chronic asthma-like symptoms into adult life. Using this knowledge, Halonen *et al.* [24] divided children who had a diagnosis of asthma by age 6 years into two main groups: those who at age 6 were skin-test positive for *Alternaria*, the main aeroallergen associated with asthma in the study area [25], and those who were not. They observed that, in the majority of children in both groups, symptoms had started before age 3 years. However, among children with asthma who were skin-test negative to *Alternaria*, inception of the disease had occurred mainly during the first year of life; while in those who were skin-test positive to *Alternaria*, peak incidence occurred during the second and third years of life (Fig. 3.3).

The fact that young children who will go on to develop chronic, atopy-related asthma by age 6 are diagnosed 1 or 2 years later than those with nonatopic asthma suggests that the mechanisms of disease are likely to be different in these two groups. Support for this contention comes from a slightly different analysis of the data from the same cohort on which Halonen and coworkers based their report quoted above. Stein *et al.* [14] assessed the outcome of children with confirmed lower respiratory tract illnesses due to RSV (RSV-LRI), after adjusting for all other known risk factors for subsequent asthma, including skin test reactivity, maternal history of asthma and birth weight, among others. They found that, as previously reported, risk of wheezing during the school years was higher in children

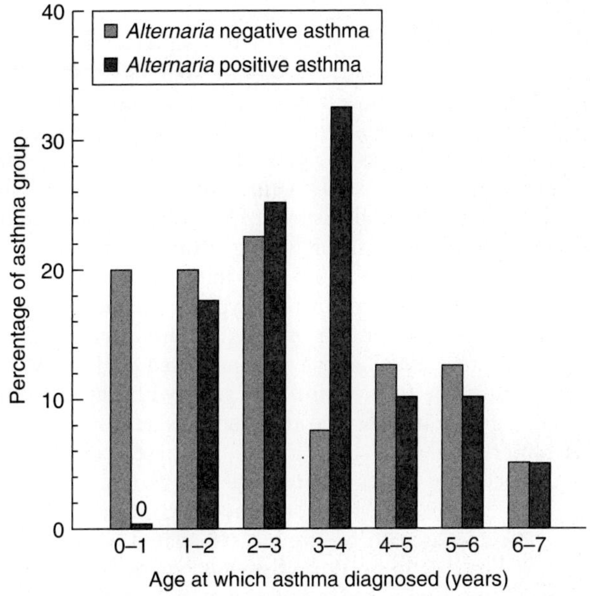

FIG. 3.3 Age of diagnosis of asthma for children who are skin-test positive for *Alternaria* or skin-test negative for *Alternaria* at age 6 in the Tucson Children's Respiratory Study. Adapted from Ref. [24].

with a history of RSV-LRI than in those with no such history. However, the risk decreased with age and was not statistically significant by early adolescence. There was also no association between RSV-LRI and subsequent risk of sensitization to local aeroallergens either at age 6 or at age 11 years. The only factor that was strongly associated with RSV-LRI in early life was diminished baseline levels of FEV_1, as measured at age 11. Interestingly, these deficits were reversed by use of a bronchodilator, suggesting that they were likely to be due to increased bronchomotor tone. Longitudinal studies performed in the early 1980s had also suggested that the outcome of RSV-LRIs is usually benign and unrelated to increased allergic sensitization [26].

It thus appears that at least three different forms of "asthma" coexist during infancy and early childhood (Fig. 3.2). Transient infant wheezing is quite frequent, is usually triggered by viruses (especially RSV), and, although it is confined to the first 3 years of life, may be associated with long-term lung function deficits. Other children who wheeze during viral infections in early life continue to have recurrent respiratory symptoms during the early school years. This form of nonatopic wheezing is associated with lower levels of lung function during the school years, but these lower levels seem to be reversible after use of a bronchodilator. Finally, children who will develop atopy-related asthma start having symptoms mainly during the second and third years of life.

Chronic asthma is most often related to atopy

There is now strong evidence indicating that, as a group, individuals with chronic asthma at any age between the elementary school years and mid-adult life are either sensitized to local aeroallergens, have elevated total levels of circulating IgE, or both [27]. Although the association between asthma and total and specific IgE is well established, the nature of the association is not well understood. For years it was thought that sensitization to specific allergens, especially in early life, was a cause of asthma [28], and that development of specific IgE against these allergens was the first step in the natural history of the disease. The strong association between risk of having asthma and sensitization to the allergens of house dust mites in coastal regions seemed to argue in favor of this hypothesis, and strategies for the primary prevention of asthma based on avoidance of exposure to these allergens were proposed [29]. However, several studies performed in desert areas with low exposure to house dust mites showed that the prevalence of asthma was either similar or even higher in these regions than that observed in zones where mite infestation rates were high [25, 30]. In desert regions, the allergens of the mold *Alternaria* appeared to be strongly associated with asthma. Moreover, studies in northern Sweden, where indoor exposure to either dust mites or molds is very low, showed that the prevalence of school-age asthma is very similar to that observed in southern Sweden, where exposure to house dust mites is high [31]. Interestingly, only 50% of all schoolchildren with asthma are sensitive to known aeroallergens in northern Sweden (especially cat and dog), compared with over 90% of children in more temperate regions, but skin-test negative children with asthma have significantly higher IgE levels than nonasthmatic children [31]. These observations suggest that, regardless of the nature of the association between elevated IgE and asthma, this link is not allergen-specific.

Recent epidemiological studies have challenged the view of a unidirectional link between atopy and asthma. Our group found that parental asthma is a strong predictor not only for asthma in the child (as one would expect), but also for the child's total IgE levels and skin test reactivity [32]. These relationships remained significant after adjusting for the intensity and other characteristics of parental atopy, supporting independent effects of parental asthma on the development of atopy in the child. Previous studies had also shown that, although the level of allergen exposure early in life predicts the development of allergic sensitization and allergic sensitization is associated with subsequent development of asthma, no direct link between allergen exposure early in life and subsequent development of asthma could be demonstrated.

All these studies suggest that the essential causal mechanism associated with asthma is not sensitization to any specific allergen, but more likely is an alteration in the regulation of responses to many different antigens with the potential of eliciting IgE responses, and especially to aeroallergens. IFNγ production might be among these altered immunological responses. In the Tucson cohorts of the Children's Respiratory Study [33, 34] and the Infant Immune Study [35], our group found strong associations between reduced IFNγ production from polyclonally stimulated peripheral blood mononuclear cells in the first year of life and the subsequent development of both allergic sensitization and asthma-related respiratory symptoms. Reduced IFNγ responses may be related to a delayed maturation of the immune system and, in turn, they may lead to susceptibility to viral infections, favor a Th2 skewed polarization of the immune system, and affect airway growth. Whether deficits of early IFNγ responses can explain the link between asthma and atopy or simply represent one of multiple pathways that are shared by these two phenotypes remains unknown.

Outcome of childhood asthma in adolescence and adult life: Remission of mild asthma and persistence of severe disease

Only a few population-based studies have addressed the outcome of childhood asthma in adolescence and adult life. These studies have used longitudinal assessment of asthma symptoms and lung function in children who were enrolled at birth [12] or during the early school years [36, 37] or who were part of birth cohorts for which assessment of asthma symptoms in early life was made through retrospective questionnaires [7, 8, 10, 38].

The transition of asthma from childhood into adolescence has been studied in the birth cohort of the Tucson Children's Respiratory Study [12]. In that study, most children who experienced only infrequent wheezing during childhood had remission of wheezing after puberty. However, almost 60% of children with frequent wheezing and/or a physician-confirmed diagnosis of asthma experienced wheezing at some point during adolescence. These results challenge the commonly held view that most children with asthma outgrow the disease in adolescence and suggest that complete remission of the disease may be the exception rather than the rule, at least among cases of moderate to severe asthma.

This link between severity and persistence of asthma has been confirmed by long-term prospective studies that have followed cohorts of children into mid-adult life. The longest such ongoing study is that initiated by Williams and McNicol in Melbourne, Australia [39]. Children with a history of recurrent episodes of wheezing were enrolled at the age of 7 years, together with a small group of controls without such a history. These individuals were then periodically reassessed, and the latest published data refer to information collected when their mean age was 42 years. Information about the age upon initiation of symptoms was obtained by use of questionnaires and, as a consequence, data on a crucial period of life for the development of asthma was possibly biased by preferential recall [36, 40]. The second study was the British 1958 birth cohort, in which over 18,000 subjects born between March 3 and March 9, 1958 were enrolled [8]. Of these persons, 31% contributed information for ages 7, 11, 16, 23, and 33 years. As with the Melbourne study, information for the first 7 years of life was obtained retrospectively at age 7. Similarly, in the Dunedin Multidisciplinary Health and Development Study (New Zealand) parents completed first respiratory questionnaires when children were already 9 years old. To date, participants have been followed up to age 26 with repeated questionnaires and lung function measurements [7]. Finally, the Tasmanian asthma survey enrolled over 8000 children born in 1961 who were also first contacted at age 7

[10]. Two thousand randomly chosen individuals were re-examined when they were 29–32 years of age.

The most important findings of all four studies were that asthma remits, at least transiently, in early adulthood in a sizable proportion of asthmatic children, and that the severity of asthma tracks significantly with age. This second point is important because it reflects a "stability" in the disease: most children with severe symptoms still have severe symptoms as adults, and asthmatic children with mild symptoms either have no asthma or have mild asthma as adults.

- Children enrolled in the Melbourne study were divided at the time of enrollment into five groups according to their previous history of wheezing: a control group (no wheezing), a group with mild wheezy bronchitis (less than five lifetime episodes of wheezing associated with colds), a group with wheezy bronchitis (five or more such episodes), a group with asthma (wheezing apart from colds), and a group with severe asthma (selected at age 10 based on severe impairment of lung function). At age 42 years, less than a quarter of children with either form of wheezy bronchitis showed frequent asthma episodes (wheezing during the previous 3 months, but less than once a week) or persistent asthma (once a week or more). However, 52% of children with asthma and 76% of those with severe asthma at the time of enrollment had frequent or persistent wheeze at age 42 [40].
- Similar results have been reported based on the British 1958 study: 27% of all children whose parents reported they had wheezed before age 7 reported wheezing during the previous year at age 33 [8].
- In the Dunedin study, 37% of children who ever wheezed had wheezing that persisted from childhood to adulthood or that relapsed after remission [7].
- For the Tasmanian study, 25.6% of subjects with "asthma or wheezy breathing" by age 7 reported current asthma at the age of 29–32 years, compared with only 10.8% of subjects without parental reports of childhood asthma [10]. As with the Melbourne study, those who had a history of more than 10 attacks of asthma by age 7 were almost twice as likely to have persistent wheezing as adults than those who did not.

A consistent factor associated with persistent asthma in all four cohorts was evidence of an allergic predisposition. In the Melbourne study, severe asthma in early life was associated with significantly higher prevalence of allergic rhinitis at age 35. In the British cohort, allergic symptoms (e.g. allergic rhinitis or eczema) were significantly associated with persistence of symptoms into adult life. In the Dunedin cohort, skin sensitization to house dust mites increased more than twice the odds for both persistence and relapse of wheezing, independent of other predisposing factors [7]. Finally, in the Tasmanian cohort, having a history of eczema in early life was also significantly associated with persistent wheezing at ages 29–32.

Childhood deficits in lung function are another factor that has been consistently associated with persistence of asthma into adult life across different cohorts. In the Tasmanian, the Melbourne, and the Dunedin studies, low lung function as measured in early school age was a significant predictor of subsequent persistent wheezing. In the Melbourne study, lung function for each subgroup of subjects classified according to their wheezing history by age 7 was repeatedly assessed up to the age of 42 years. Children with severe asthma at age 10 had very low initial levels of lung function, and this was to be expected because deficits in lung function were used to classify them as having severe asthma in the first place. When compared with their peers, these children did not show further deficits in lung function growth with age, and by age 42 their position relative to subjects with milder symptoms or with no asthma was substantially unchanged. A similar pattern of relatively stable lung function growth was observed for patients with mild asthma who had shown loss of lung function at enrollment, albeit much less pronounced than persons with severe asthma. Similarly, in the Dunedin study children with persistent wheezing had consistently lower values of the FEV_1/FVC ratio from age 9 up to age 26, as compared with controls. The mean deficit was almost 7%, but no significant differences in the slopes of change in the FEV_1/FVC ratio were found between persistent wheezers and controls. Thus, persistent wheezers had decreased values of FEV_1/FVC, but not steeper slopes of decline. These findings are very relevant for our understanding of the natural history of COPD, and will be discussed in more detail below.

The potential role of bronchial hyperresponsiveness in school age has been explored in two of the above referenced cohorts. In the Dunedin study, the presence at 9 years of age of either a value of methacholine $PC_{20} < 8$ mg/ml or an increase of $FEV_1 > 10\%$ from baseline in response to a bronchodilator was associated with fourfold increased odds for persistent wheezing and with almost sevenfold increased odds for relapse of wheezing [7]. In that study, bronchial hyperresponsiveness remained significantly associated with both outcomes after adjusting for other risk factors. In the Tucson Children's Respiratory Study, we found 68% of the children whose asthma persisted after the onset of puberty to have had a positive test for bronchial hyperresponsiveness at age 11, as compared with only 37% of children whose asthma remitted during adolescence [12]. Another interesting finding of that study was the strong association between childhood obesity and persistence of asthma after the onset of puberty. Multiple recent reports have consistently linked obesity not only to persistence but also to incidence of asthma [41–44]. The nature of this association remains at present not fully understood and warrants further research.

Relapse of asthma symptoms in patients whose asthma remitted in childhood

Little is known about prevalence and risk factors for relapses of asthma in adult life among subjects whose symptoms remitted during childhood. Taylor and colleagues found one-third of subjects who had childhood asthma that was in remission at age 18 to have relapses of the disease in young adulthood [45]. In the British cohort study, a group of over 1300 persons with a history of wheezing illnesses from birth to age 16 years had symptom remission by age 23. This group was 50% more likely to report wheezing at age 33 than those with no history of asthma or wheezy bronchitis during childhood [8]. These findings suggest that complete remission of asthma in adulthood may be less common than previously thought and are consistent with results from clinical studies supporting the presence of active, subclinical forms of the disease among asthmatics in clinical remission. In a Dutch cohort of 119 patients who were diagnosed with asthma during school age and re-examined 30 years later (at age 32–42), bronchial hyperresponsiveness and/or reduced lung function were present among 57% of those who were in clinical remission as adults [46]. Previous studies have shown that indicators of airway inflammation, such as eosinophils and IL-5 in bronchial biopsies, eosinophil percent in bronchoalveolar lavage, and exhaled nitric oxide levels, are higher among children and adolescents in clinical remission of asthma, as compared with their peer controls [47–49]. Bronchial hyperresponsiveness and other clinical abnormalities among these apparently remitting asthma cases may pose these subjects at increased risk for obstructive lung disease in adult life and their potential role as determinants of COPD will be discussed below.

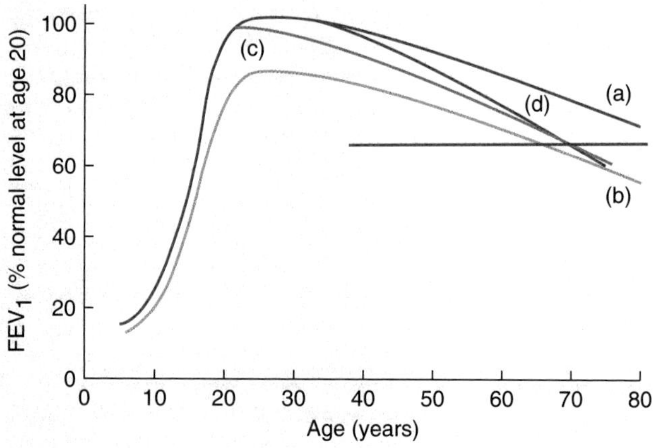

FIG. 3.4 Hypothetical mechanisms that may lead to a critically low level of lung function in adult life and to chronic airway obstruction (horizontal line): (a) normal growth and decline; (b) impaired lung growth with a lower plateau phase but a normal rate of decline compared to (a); (c) normal plateau with rapid initial decline in lung function and a subsequent normal rate of decline; (d) normal plateau with normal initial rate of decline but a subsequent accelerated loss in lung function. Adapted from Ref. [51].

NATURAL HISTORY OF COPD

According to the Global Initiative for Chronic Obstructive Lung Disease (GOLD) [50], the pulmonary component of COPD is characterized by airflow limitation that is not fully reversible, usually progressive, and associated with an abnormal inflammatory response of the lung to noxious particles or gases. Thus, the natural history of COPD overlaps extensively with that of the level of lung function (usually FEV_1, expressed as percent of predicted value in relation to a certain height) during adulthood. Three are the main factors that can determine the level of lung function achieved at any age during adult life (Fig. 3.4).

First, the individual may either start life with a low level of lung function or show a significant decline in lung function growth during the first years of life (line "b" in Fig. 3.4). The individual's level of lung function will fall after a certain age, at the same rate as normal peers, but at a lower level overall. The level of lung function attained by late adolescence or early adult life will thus be lower. Through the normal process of aging, decline in lung function will naturally occur, and this decline, although parallel to that occurring in all the population, will nevertheless predispose to the attainment of a level of lung function (expressed as percentage of normal level in early adulthood) that will be lower than the threshold for the expression of clinical symptoms of airway obstruction.

A second potential mechanism is that of an early decline in lung function occurring either at the end of adolescence or in early adult life (line "c" in Fig. 3.4). Again, the main result of such a mechanism would be the attainment of lower level of lung function in these individuals as compared with the rest of the population.

A third mechanism is that of a faster rate of decline of lung function during adult life (line "d" in Fig. 3.4). This may obviously occur both in individuals with a normal level of lung function at the beginning of adult life as in individuals who already have a diminished level of lung function to begin with.

The essential determinants of the natural history of COPD are the rate of growth of lung function up to late childhood, the timing and length of the plateau phase, and the rate of decline of lung function during adult life.

Growth of lung function during childhood

The three most important determinants of growth of lung function during childhood are:

- the level of lung function at birth,
- the incidence of lower respiratory illnesses during the first years of life,
- the development of persistent asthma-like symptoms during childhood.

Level of lung function at birth

The availability of lung function tests that can be performed shortly after birth, and which appear to show good correlation with more invasive tests of airway function [52], has allowed us to study the potential role of the development of the lung *in utero* on subsequent levels of lung and airway function attained by the individual. Since these methods have been available only for the last 25 years, it is not possible to determine the role of airway function in the postnatal period on events occurring beyond young adult life. Nevertheless, it is now apparent that there is quite significant tracking between the level of lung function measured shortly after birth and that measured during childhood [53, 54], adolescence [22], and up to young adulthood [55]. This may explain, at least in part, the strong association between lower respiratory illness in early life and level of lung function in late childhood and early adulthood (see the discussion below).

The factors that determine growth of lung and airways *in utero* are only partly understood. The two main determinants of the level of lung function measured shortly after birth are gender and exposure to tobacco smoke products *in utero* [56]. Boys have been found to have consistently lower maximal flows at functional residual capacity (V_{max}FRC) throughout the first year of life in comparison with girls. Indeed, females have larger airway size, relative to the size of lungs and to body size, as compared with males both at birth and during childhood, up to the plateau of lung growth [57]. However, the role of these gender differences in influencing airway function and predisposing to obstructive lung disease is not well understood. Maternal smoking during pregnancy is associated with lower levels of V_{max}FRC in both genders in comparison with unexposed infants [56]. Postnatal exposure to tobacco smoke appears to be less important than prenatal exposure, at least in most developed countries [58]. The effects of intrauterine exposure to tobacco smoke products persist for the first year of life [56], and even beyond the first year up to the early adolescent years [59].

Independent of what factors determine *in utero* lung growth, evidence has been accumulating that lung function at birth affects lung health far beyond infancy. In addition to the well-established association between reduced V_{max}FRC shortly after birth and wheezing in the first years of life [18, 56, 59, 60], recent studies have associated lung function measured in the first month of life to persistent wheezing and asthma at 10–11 years [54, 61]. Turner *et al.* found wheezing at 11 years to be associated with a reduced mean z score for V_{max}FRC at 1 month of age [54], whereas a reduced fraction of expiratory time to peak tidal expiratory flow to total expiratory time (t_{PTEF}/t_E) as assessed 3 days after birth predicted asthma and bronchial hyperresponsiveness at age 10 years in a Norwegian birth cohort [61]. Of note, in the latter study t_{PTEF}/t_E at birth did not correlate with FEV_1 or FVC at 10 years of age. These findings are complemented by results from the Tucson Children's Respiratory Study, in which we found V_{max}FRC measured at 2 month of age to correlate strongly with FEV_1 and the FEV_1/FVC ratio assessed at age 11, 16, and 22 years [55]. Participants in the lowest quartile for infant V_{max}FRC had a 5% mean deficit in the FEV_1/FVC ratio through age 22 compared to the upper three quartiles and this association was independent of wheezing and asthma status. Taken together, this evidence suggests that lung function at birth accounts for part of the lung function levels attained in young adult life and, in turn, may influence predisposition to subsequent obstructive lung disease.

Role of lower respiratory illnesses in early life

It has been known for decades that children who have lower respiratory illnesses during the first years of life have lower levels of lung function during childhood and into adult life [62]. One possible explanation for this association is that viral infections, which are the main etiological factor for lower respiratory illnesses in early life, may damage lung and airways, and this may predispose for lower levels of lung function [63]. This hypothesis seemed attractive, because it suggested a potential strategy for the prevention of early losses in lung function that could predispose to COPD. However, several studies in which the techniques to assess lung function in infants, described earlier, were used showed that children who developed lower respiratory symptoms during viral infections in early life had diminished pre-illness levels of lung function [18, 56, 59]. The hypothesis was thus suggested that lower levels of lung function observed after lower respiratory illnesses in early life could be explained by preexisting diminished lung function, the latter being therefore the link between early life episodes of airway obstruction and subsequent deficits in lung function.

Unfortunately, the number of infants in whom lung function has been ascertained in early life and who have been followed for a number of years after birth is rather small. Therefore, the possibility that lower respiratory illnesses by themselves may alter lung and airway growth in groups of susceptible individuals cannot be excluded. It is unlikely that pre-illness lung function may explain all forms of infection-associated wheezing during early childhood, and it is legitimate to surmise that immune responses to the viruses themselves may also play a significant role.

Evidence from studies in the United Kingdom would support the validity of both hypotheses [64–66]. In a cohort of children enrolled in 1964 and reassessed in 2001 for current lung function, smoking status, and respiratory symptoms, a significant linear trend was found between birth weight and FEV_1 at age 45–50 years after adjusting for confounders, including smoking [65]. This association was confirmed by a recent meta-analysis [66] of eight studies reporting a pooled increase of 48 ml in FEV_1 per 1 kg in birth weight. As modest as this association may appear, these findings support a direct influence of *in utero* growth on lung function in adulthood. However, another very long-term study in England provides evidence in support of additional effects of childhood respiratory infections on adult lung function [64]. In this study, men born between 1911 and 1930, whose birth weights, weights at 1 year, and childhood respiratory illnesses were recorded in early life, were studied at ages 59–70 years. Death from chronic obstructive disease, FEV_1, and respiratory symptoms were the main outcome variables. The main early life determinants of the level of FEV_1 in old age were birth weight and history of bronchitis or pneumonia in infancy, and these effects were independent of smoking habit and social class. These data would thus suggest that both intrauterine growth (and presumably lung development) and lower respiratory illnesses during the first years of life exert independent effects on the level of lung function attained late during adult life, and thus may be important determinants of the risk of COPD.

Persistent asthma-like symptoms during childhood

Epidemiological evidence supports that persistent asthma-like symptoms are a significant risk factor for the development of lower levels of lung function in childhood and young adult life [7, 36] and, in turn, may predispose to the development of COPD. One proposed explanation for these associations has been that asthma is a progressive disease [67]. It has thus been proposed that the presence of chronic airway inflammation is associated with significant lung remodeling and that the latter fosters a significant alteration in lung growth.

Although some studies were able to show impairment in the development of lung function in children with asthma [67, 68], the main outcome variable used in these studies was pre-bronchodilator lung function. The Childhood Asthma Management Program (CAMP) study was specifically designed to test the hypothesis that mild to moderate childhood asthma is associated with significant deterioration in airway growth and that treatment with inhaled anti-inflammatory therapy could reverse this deterioration in lung function [69]. Children aged 6–12 years were treated for 4–6 years with either an inhaled corticosteroid (budesonide), nedocromil, or a placebo. Post-bronchodilator FEV_1 was considered the outcome variable, because pre-bronchodilator FEV_1 could be affected by the degree of activity of the disease at the time of testing. The results of the study showed that, although children with asthma have levels of lung function that are significantly lower than those of children without the disease at any time between the ages of 6 and 15, these levels do not further deteriorate even among subjects who are not systematically treated with anti-inflammatory therapy. Moreover, systematic treatment with anti-inflammatory therapy improves bronchial hyperresponsiveness, but is not associated with a significant improvement in lung function with time. Based on these findings, more recently the Prevention of Early Asthma in Kids (PEAK) trial was designed to determine whether the lack of sustained effects of inhaled corticosteroids on the natural history of childhood asthma was attributable to the initiation of the intervention too late in life. In fact, the PEAK trial failed to detect any long-term effects of a 2-year treatment with inhaled corticosteroids on the natural history of asthma in a group of high-risk preschool children, despite the temporary beneficial effects of the treatment on disease symptoms and exacerbations [70].

An interpretation of these results is that, in children with a diagnosis of asthma, most of the deterioration in lung function observed during the school years occurs early in life. The results of studies by our group suggest that children who go on to have persistent asthma-like symptoms during childhood start life with levels of lung function that are slightly lower than those of children who will not have these persistent symptoms [71]. It is also possible that persistent childhood asthma affects lung function through mechanisms that are at least partly unrelated to airway inflammation and, therefore, insensitive to corticosteroid treatment. Finding the answer to these questions will have crucial implications for the prevention of long-term obstructive lung disease among these children.

Determinants of early losses in lung function

Very little is known about the factors that determine early losses in lung function at the age in which the plateau level of lung function is reached in early adult life. Xuan *et al.* [72] observed that, between the ages of 17 and 19 years, when growth in height had stopped, FEV_1 continued to grow in both males and females. However, children who had recent episodes of wheeze and those with evidence of bronchial hyperresponsiveness showed a reduced rate of growth in airway caliber. Unfortunately, lung function was not assessed after use of a bronchodilator, and it is thus not possible from these data to assess whether the changes are due to increased airway tone or to irreversible alterations in airway structure. In addition to active wheezing and bronchial hyperresponsiveness, early initiation to cigarette smoking has been associated with detrimental effects on the plateau level of lung function. Holberg *et al.* [73] showed that, at age 16, the level of lung function achieved by males who had started smoking was significantly lower than that of those who had not. No such effect was observed in females.

Determinants of increased slope of lung function decline

Several longitudinal studies have addressed the role of different intrinsic and extrinsic factors on the rate of decline of lung function after the plateau phase. The role of cigarette smoking has been clearly and consistently established. Camilli *et al.* [74] examined changes in FEV_1 in over 1700 adults enrolled in a prospective study of a general population sample. Individuals who smoked more than 10 cigarettes per day had excessive rates of decline in FEV_1 as compared with nonsmokers. The excess decline of smokers was age dependent, particularly in men: much of the excess loss of lung function occurred between 50 and 70 years of age. Interestingly, ex-smokers showed declines in FEV_1 values that were similar to those of nonsmokers. The authors also examined the effect of quitting smoking on the decline of FEV_1. In subjects younger than 35 years, quitting smoking during follow-up was associated with an actual increase in FEV_1. In men above 50 years, smoking cessation early in the study led to a return to normal rate of functional decline during follow-up. In addition to cigarette smoking, exposure to other toxic gases and particles – including environmental tobacco smoke, indoor and outdoor pollution, occupational dust, gases, and fumes – has been consistently associated with COPD and a steeper decline of lung function [75, 76].

However, the observation that only a more [77] or less [78] sizable fraction of smokers develop COPD strongly suggests the existence of intrinsic factors that can modify the effect of smoking and other deleterious exposures on the rate of decline of lung function. Bronchial hyperresponsiveness has been long hypothesized to be one of these factors [79, 80]. Indeed, bronchial hyperresponsiveness has been consistently shown to be a strong and independent risk factor for accelerated decline of lung function in prospective studies among smokers with early COPD as well as asymptomatic individuals [81–83]. Findings from these epidemiological studies have been confirmed in the clinical setting. For instance, Postma and coworkers [84] assessed the course of lung function after 2–21 years of follow-up in 81 nonallergic patients with considerable lung function impairment (<55% FEV_1/FVC ratio) at the beginning of the study. They reported that a more favorable rate of change in FEV_1 was not only associated with fewer pack-years of smoking but also with less nonspecific bronchial hyperreactivity and a higher degree of reversibility of airflow obstruction. These effects were independent of baseline FEV_1 value, both in smokers and in ex-smokers. Whether the association between bronchial hyperresponsiveness and decline in lung function is the consequence of an ongoing COPD-related inflammatory process and to what extent this inflammatory process differs from that associated with bronchial hyperresponsiveness in asthma remain at present open questions.

Airway inflammation and its structural *sequelae* may be also involved in the causal mechanisms linking accelerated decline of lung function in smokers with two other risk factors: chronic bronchitis (i.e. chronic cough and phlegm) and lower airway colonization/infection. Although the predictive value of chronic bronchitis for the development of COPD remains somewhat controversial [85, 86], this phenotype has been consistently associated with steeper rates of FEV_1 decline [87, 88]. It has been suggested that the clinical impact of chronic bronchitis may be dependent upon the stage of COPD, with this phenotype having stronger effects on the severe forms of COPD when lower airways become colonized and/or infected [89]. Acute exacerbations and lower airway colonization are strongly associated with decline of lung function among smokers and patients with COPD. In a prospective study [90], patients with frequent exacerbations (mean number of episodes per year >2.9) had a significantly faster decline of FEV_1 than infrequent exacerbators. Lower respiratory illnesses had a similar deleterious effect on decline of FEV_1 over 5 years among smokers in the Lung Health Study [91].

Of note, airway infections may also affect lung function decline through mechanisms that are independent of (and may indeed precede) the presence of COPD. Burrows and coworkers reported that smokers who recalled a history of "respiratory trouble" before 16 years of age had significantly steeper rates of decline in lung function as compared with those with no such history [92]. This effect was independent of a current or past diagnosis of asthma. It thus appears that intrinsic factors that modify the effects of smoking may be related to events occurring during the first years of life. This conclusion is compatible with the observation by Barker and coworkers quoted earlier [64], that elderly individuals with a history of lower respiratory illness early in life were more likely to have lower levels of lung function than those with no such history.

Although it is not well established how the effects of these intrinsic and extrinsic factors on the rate of decline of lung function are modulated, genetics are likely to play a major role. In the large cohort of the Tucson Epidemiological Study of Airway Obstructive Disease (TESAOD), our group observed significant intra-family correlation in the rate of decline of lung function within smokers [93], suggesting that these rates of decline have an important genetic component. However, the identification of the genetic components of COPD, as well as of those of asthma, is proving a challenging task and at present deficiency of α1-antitrypsin remains the only established genetic risk factor for this disease.

INTERSECTION OF THE NATURAL HISTORY OF ASTHMA AND COPD

It is not uncommon in the clinical setting to observe patients with asthma showing COPD-like phenotypes, and *vice versa*. Consistently, it has been long known that diagnoses of asthma, chronic bronchitis, and emphysema are frequently associated at the population level [2]. This evidence indicates that the natural history of asthma and COPD can converge in some cases during adult life. A possible explanation for these observations is that a significant proportion of cases with severe and/or persistent asthma can develop in the long-term COPD-like phenotypes. Support for this contention is provided by another report from the TESAOD study [19], in which adult subjects with active asthma were found to have a 12 times higher risk of acquiring a diagnosis of COPD over time than subjects with no asthma, after adjusting for covariates including smoking.

Since chronic airflow limitation is the central hallmark of COPD, much attention has been paid to understand the natural history of lung function of patients who develop COPD as a *sequela* of asthma. Much like for individuals from the general population (Fig. 3.4), in patients with asthma FEV_1 deficits can develop mainly as a result of either lower FEV_1 levels at the beginning of adult age or an accelerated FEV_1 decline during adulthood [51]. Interestingly, most birth cohort studies – including the Dunedin Multidisciplinary Health and Development Study [7], the Melbourne Asthma Study [36], and the British 1958 Birth Cohort [94] – have shown that, in a significant proportion of cases of persistent childhood asthma, lung function deficits are established before the early adult years, and track over time. These findings, together with the observation that, among children who start life with low levels of lung function, expiratory flows remain lower than those of their peers throughout childhood and adolescence [22], point toward the importance of early events in affecting lung growth and/or airway remodeling in childhood and, in turn, in influencing early predisposition to COPD. As for the second potential mechanism of lung function impairment (i.e. accelerated decline of lung function) in asthmatics, prospective cohort studies on adult populations have provided somewhat inconsistent findings. In the Busselton Health Study [95] and in the Copenhagen City Heart Study [96], asthmatics showed both initial FEV_1 deficits at age 20 years and an increased slope of FEV_1 decline in adulthood. However, the increase in FEV_1 decline associated with asthma was only 4 ml/year in the Busselton Study.

A possible reconciliation for these inconsistent results is provided by a recent study from the TESAOD cohort. We identified adult participants who had FEV_1/FVC ratio consistently lower than 70% (as a hallmark of persistent airflow limitation) and compared the natural history of their lung function based on the presence of asthma. Patients with asthma accounted for about one-third of all cases of persistent airflow limitation. Persistent airflow limitation was strongly associated with smoking among nonasthmatics and with eosinophilia among subjects who had asthma onset $\leq$ 25 years. Most importantly, the natural course of

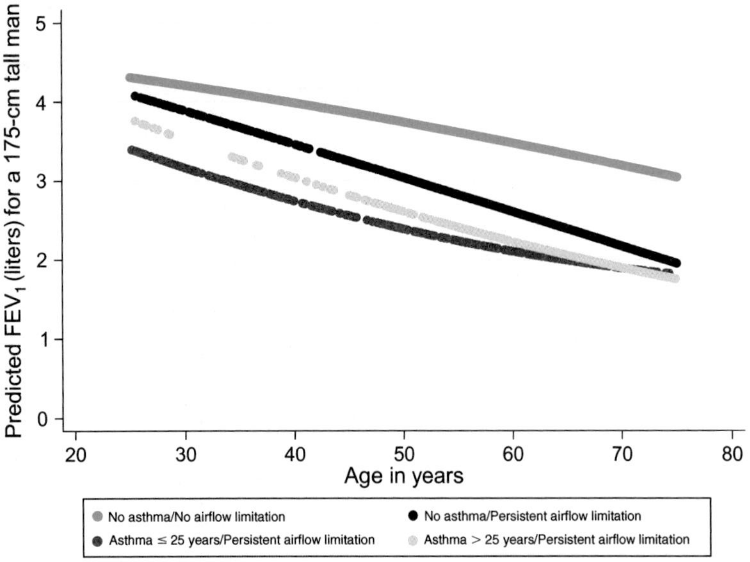

FIG. 3.5 Natural history of lung function among participants in the TESAOD study who had persistent airflow limitation with or without asthma. The black line refers to subjects with no asthma and persistent airflow limitation. The red line refers to subjects with asthma onset ≤ 25 years and persistent airflow limitation. The gold line refers to subjects with asthma onset >25 years and persistent airflow limitation. Predicted values for subjects with no asthma and no airflow limitation (healthy controls, green line) are also reported for comparison. Depicted values represent predicted values for a 175-cm tall male from the best-fitting random coefficients model. Adapted from Ref. [96].

lung function differed substantially between the two groups (Fig. 3.5), with subjects with asthma onset ≤ 25 years having lower FEV_1 levels at age 25 but not steeper FEV_1 decline in adulthood as compared with expected values from healthy controls. In contrast, subjects who developed persistent airflow limitation but had no asthma showed only moderate FEV_1 deficits at age 25 years but had greater than expected FEV_1 loss between age 25 and 75. Interestingly, the natural course of lung function of subjects who had adult-onset asthma (after age 25 years) included both moderate FEV_1 deficits in young adulthood and accelerated FEV_1 decline thereafter.

Taken together, findings from this and previous studies suggest that: (1) asthma accounts for a sizable fraction of COPD cases at the population level; (2) while development of traditional COPD is strongly associated with exposure to noxious agents (e.g. cigarette smoking), patients with asthma may develop persistent airflow limitation through pathways that are independent of smoking and related to allergic and eosinophilic inflammation; (3) among subjects who had severe and/or persistent asthma in childhood significant deficits in lung function are already present in early adult life; (4) during adulthood accelerated decline of FEV_1 might be present for subgroups of asthmatics, but this accelerated decline is likely to account for a limited proportion of the asthma-related deficits in lung function up to mid-adult life; (5) childhood versus adult-onset asthma may be associated with a different natural history of lung function that might explain part of the inconsistencies of results on FEV_1 decline in asthmatics across different studies.

The implications of these conclusions for prevention and treatment are crucial because they indicate that the timing and nature of prevention strategies for COPD-like phenotypes need to be different for subjects with and without asthma. While smoking prevention and cessation is vital against inception of traditional COPD, any intervention to prevent development of persistent airflow limitation among patients with childhood asthma will need to target these patients before they enter adult life because by then the bulk of their lung function impairment is already established. Similarly, the different trajectories by which patients with traditional COPD and patients with asthma may develop chronic airflow limitation in their adult life support the phenotypic heterogeneity that has emerged from clinical studies on these two groups [97]. Future research is required to determine whether multiple functional, morphological, and immunological assessments can be developed to capture the heterogeneity of these diseases and to identify optimal prevention and treatment strategies for these patients.

References

1. Burrows B, Bloom JW, Traver GA, Cline MG. The course and prognosis of different forms of chronic airways obstruction in a sample from the general population. *N Engl J Med* 317(21): 1309–14, 1987.
2. Soriano JB, Davis KJ, Coleman B, Visick G, Mannino D, Pride NB. The proportional Venn diagram of obstructive lung disease: Two approximations from the United States and the United Kingdom. *Chest* 124(2): 474–81, 2003.
3. Silva GE, Sherrill DL, Guerra S, Barbee RA. Asthma as a risk factor for COPD in a longitudinal study. *Chest* 126(1): 59–65, 2004.
4. Pauwels RA, Buist AS, Calverley PM, Jenkin CR, Hurd SS. Global strategy for the diagnosis, management, and prevention of chronic obstructive pulmonary disease. NHLBI/WHO Global Initiative for Chronic Obstructive Lung Disease (GOLD) Workshop summary. *Am J Respir Crit Care Med* 163(5): 1256–76, 2001.
5. Martinez FD. Context dependency of markers of disease. *Am J Respir Crit Care Med* 162(2 Pt 2): S56–S57, 2000.
6. Barbee RA, Dodge R, Lebowitz ML, Burrows B. The epidemiology of asthma. *Chest* 87(1 Suppl): 21S–25S, 1985.
7. Sears MR, Greene JM, Willan AR *et al.* A longitudinal, population-based, cohort study of childhood asthma followed to adulthood. *N Engl J Med* 349(15): 1414–22, 2003.
8. Strachan DP, Butland BK, Anderson HR. Incidence and prognosis of asthma and wheezing illness from early childhood to age 33 in a national British cohort. *BMJ* 312(7040): 1195–99, 1996.
9. Oswald H, Phelan PD, Lanigan A, Hibbert M, Bowes G, Olinsky A. Outcome of childhood asthma in mid-adult life. *BMJ* 309(6947): 95–96, 1994.

10. Jenkins MA, Hopper JL, Bowes G, Carlin JB, Flander LB, Giles GG. Factors in childhood as predictors of asthma in adult life. *BMJ* 309(6947): 90–93, 1994.
11. Anderson HR, Pottier AC, Strachan DP. Asthma from birth to age 23: Incidence and relation to prior and concurrent atopic disease. *Thorax* 47(7): 537–42, 1992.
12. Guerra S, Wright AL, Morgan WJ, Sherrill DL, Holberg CJ, Martinez FD. Persistence of asthma symptoms during adolescence: Role of obesity and age at the onset of puberty. *Am J Respir Crit Care Med* 170(1): 78–85, 2004.
13. Yunginger JW, Reed CE, O'Connell EJ, Melton LJ III, O'Fallon WM, Silverstein MD. A community-based study of the epidemiology of asthma. Incidence rates, 1964–1983. *Am Rev Respir Dis* 146(4): 888–94, 1992.
14. Stein RT, Sherrill D, Morgan WJ *et al.* Respiratory syncytial virus in early life and risk of wheeze and allergy by age 13 years. *Lancet* 354(9178): 541–45, 1999.
15. Stein RT, Holberg CJ, Morgan WJ *et al.* Peak flow variability, methacholine responsiveness and atopy as markers for detecting different wheezing phenotypes in childhood. *Thorax* 52(11): 946–52, 1997.
16. Martinez FD, Helms PJ. Types of asthma and wheezing. *Eur Respir J Suppl* 27: 3s–8s, 1998.
17. Tager IB, Ngo L, Hanrahan JP. Maternal smoking during pregnancy. Effects on lung function during the first 18 months of life. *Am J Respir Crit Care Med* 152(3): 977–83, 1995.
18. Martinez FD, Morgan WJ, Wright AL, Holberg CJ, Taussig LM. Diminished lung function as a predisposing factor for wheezing respiratory illness in infants. *N Engl J Med* 319(17): 1112–17, 1988.
19. Martinez FD. Sudden infant death syndrome and small airway occlusion: Facts and a hypothesis. *Pediatrics* 87(2): 190–98, 1991.
20. Montgomery GL, Tepper RS. Changes in airway reactivity with age in normal infants and young children. *Am Rev Respir Dis* 142(6 Pt 1): 1372–76, 1990.
21. Burrows B, Sears MR, Flannery EM, Herbison GP, Holdaway MD, Silva PA. Relation of the course of bronchial responsiveness from age 9 to age 15 to allergy. *Am J Respir Crit Care Med* 152(4 Pt 1): 1302–8, 1995.
22. Morgan WJ, Stern DA, Sherrill DL *et al.* Outcome of asthma and wheezing in the first 6 years of life: Follow-up through adolescence. *Am J Respir Crit Care Med* 172(10): 1253–58, 2005.
23. Zeiger RS, Dawson C, Weiss S. Relationships between duration of asthma and asthma severity among children in the Childhood Asthma Management Program (CAMP). *J Allergy Clin Immunol* 103(3 Pt 1): 376–87, 1999.
24. Halonen M, Stern DA, Lohman C, Wright AL, Brown MA, Martinez FD. Two subphenotypes of childhood asthma that differ in maternal and paternal influences on asthma risk. *Am J Respir Crit Care Med* 160(2): 564–70, 1999.
25. Halonen M, Stern DA, Wright AL, Taussig LM, Martinez FD. *Alternaria* as a major allergen for asthma in children raised in a desert environment. *Am J Respir Crit Care Med* 155(4): 1356–61, 1997.
26. Pullan CR, Hey EN. Wheezing, asthma, and pulmonary dysfunction 10 years after infection with respiratory syncytial virus in infancy. *BMJ (Clin Res Ed)* 284(6330): 1665–69, 1982.
27. Burrows B, Martinez FD, Halonen M, Barbee RA, Cline MG. Association of asthma with serum IgE levels and skin-test reactivity to allergens. *N Engl J Med* 320(5): 271–77, 1989.
28. Peat JK, Tovey E, Toelle BG *et al.* House dust mite allergens. A major risk factor for childhood asthma in Australia. *Am J Respir Crit Care Med* 153(1): 141–46, 1996.
29. Dust mite allergens and asthma—a worldwide problem. *J Allergy Clin Immunol* 83(2 Pt 1): 416–27, 1989.
30. Peat JK, Tovey E, Mellis CM, Leeder SR, Woolcock AJ. Importance of house dust mite and *Alternaria* allergens in childhood asthma: An epidemiological study in two climatic regions of Australia. *Clin Exp Allergy* 23(10): 812–20, 1993.
31. Perzanowski MS, Ronmark E, Nold B, Lundback B, Platts-Mills TA. Relevance of allergens from cats and dogs to asthma in the northernmost province of Sweden: Schools as a major site of exposure. *J Allergy Clin Immunol* 103(6): 1018–24, 1999.
32. Crestani E, Guerra S, Wright AL, Halonen M, Martinez FD. Parental asthma as a risk factor for the development of early skin test sensitization in children. *J Allergy Clin Immunol* 113(2): 284–90, 2004.
33. Martinez FD, Stern DA, Wright AL, Holberg CJ, Taussig LM, Halonen M. Association of interleukin-2 and interferon-gamma production by blood mononuclear cells in infancy with parental allergy skin tests and with subsequent development of atopy. *J Allergy Clin Immunol* 96(5 Pt 1): 652–60, 1995.
34. Stern DA, Guerra S, Halonen M, Wright AL, Martinez FD. Low IFN-gamma production in the first year of life as a predictor of wheeze during childhood. *J Allergy Clin Immunol* 120(4):835–41.
35. Guerra S, Lohman IC, Halonen M, Martinez FD, Wright AL. Reduced interferon gamma production and soluble CD14 levels in early life predict recurrent wheezing by 1 year of age. *Am J Respir Crit Care Med* 169(1): 70–76, 2004.
36. Phelan PD, Robertson CF, Olinsky A. The Melbourne Asthma Study: 1964–1999. *J Allergy Clin Immunol* 109(2): 189–94, 2002.
37. Horak E, Lanigan A, Roberts M *et al.* Longitudinal study of childhood wheezy bronchitis and asthma: Outcome at age 42. *BMJ* 326(7386): 422–23, 2003.
38. Marossy AE, Strachan DP, Rudnicka AR, Anderson HR. Childhood chest illness and the rate of decline of adult lung function between ages 35 and 45 years. *Am J Respir Crit Care Med* 175(4): 355–59, 2007.
39. Williams H, McNicol KN. Prevalence, natural history, and relationship of wheezy bronchitis and asthma in children. An epidemiological study. *BMJ* 4(5679): 321–25, 1969.
40. Phelan PD. Asthma in Children and Adolescents: An Overview. . London: Baillière Tindall, 1995.
41. Camargo CA Jr., Weiss ST, Zhang S, Willett WC, Speizer FE. Prospective study of body mass index, weight change, and risk of adult-onset asthma in women [see comments]. *Arch Intern Med* 159(21): 2582–88, 1999.
42. Castro-Rodriguez JA, Holberg CJ, Morgan WJ, Wright AL, Martinez FD. Increased incidence of asthmalike symptoms in girls who become overweight or obese during the school years. *Am J Respir Crit Care Med* 163(6): 1344–49, 2001.
43. Beuther DA, Sutherland ER. Overweight, obesity, and incident asthma: A meta-analysis of prospective epidemiologic studies. *Am J Respir Crit Care Med* 175(7): 661–66, 2007.
44. Guerra S, Sherrill DL, Bobadilla A, Martinez FD, Barbee RA. The relation of body mass index to asthma, chronic bronchitis, and emphysema. *Chest* 122(4): 1256–63, 2002.
45. Taylor DR, Cowan JO, Greene JM, Willan AR, Sears MR. Asthma in remission: can relapse in early adulthood be predicted at 18 years of age? *Chest* 127(3): 845–50, 2005.
46. Vonk JM, Postma DS, Boezen HM *et al.* Childhood factors associated with asthma remission after 30 year follow up. *Thorax* 59(11): 925–29, 2004.
47. van Den Toorn LM, Prins JB, Overbeek SE, Hoogsteden HC, de Jongste JC. Adolescents in clinical remission of atopic asthma have elevated exhaled nitric oxide levels and bronchial hyperresponsiveness. *Am J Respir Crit Care Med* 162(3 Pt 1): 953–57, 2000.
48. van den Toorn LM, Overbeek SE, de Jongste JC, Leman K, Hoogsteden HC, Prins JB. Airway inflammation is present during clinical remission of atopic asthma. *Am J Respir Crit Care Med* 164(11): 2107–13, 2001.
49. Warke TJ, Fitch PS, Brown V *et al.* Outgrown asthma does not mean no airways inflammation. *Eur Respir J* 19(2): 284–87, 2002.
50. Rabe KF, Hurd S, Anzueto A *et al.* Global strategy for the diagnosis, management, and prevention of COPD-2006 Update. *Am J Respir Crit Care Med*, 2007.
51. Weiss ST, Ware JH. Overview of issues in the longitudinal analysis of respiratory data. *Am J Respir Crit Care Med* 154(6 Pt 2): S208–S211, 1996.
52. Stocks J. Lung function testing in infants. *Pediatr Pulmonol Suppl* 18: 14–20, 1999.

53. Le Souef P, Turner S, Rye P *et al.* Pulmonary function at four weeks correlates with pulmonary function at 6 and 12 years. *Am J Respir Crit Care Med* 163: A541, 2001.
54. Turner SW, Palmer LJ, Rye PJ *et al.* The relationship between infant airway function, childhood airway responsiveness, and asthma. *Am J Respir Crit Care Med* 169(8): 921–27, 2004.
55. Stern DA, Morgan WJ, Wright AL, Guerra S, and Martinez FD. Poor aiway function in early infancy and lung function by age 22 years: a non-selective longitudinal cohort study. *Lancet* 370(9589): 758–64, 2007.
56. Young S, Arnott J, O'Keeffe PT, Le Souef PN, Landau LI. The association between early life lung function and wheezing during the first 2 yrs of life. *Eur Respir J* 15(1): 151–57, 2000.
57. Pagtakhan RD, Bjelland JC, Landau LI *et al.* Sex differences in growth patterns of the airways and lung parenchyma in children. *J Appl Physiol* 56(5): 1204–10, 1984.
58. Stein RT, Holberg CJ, Sherrill D *et al.* Influence of parental smoking on respiratory symptoms during the first decade of life: The Tucson Children's Respiratory Study. *Am J Epidemiol* 149(11): 1030–37, 1999.
59. Tager IB, Hanrahan JP, Tosteson TD *et al.* Lung function, pre- and post-natal smoke exposure, and wheezing in the first year of life. *Am Rev Respir Dis* 147(4): 811–17, 1993.
60. Murray CS, Pipis SD, McArdle EC, Lowe LA, Custovic A, Woodcock A. Lung function at one month of age as a risk factor for infant respiratory symptoms in a high risk population. *Thorax* 57(5): 388–92, 2002.
61. Haland G, Carlsen KC, Sandvik L *et al.* Reduced lung function at birth and the risk of asthma at 10 years of age. *N Engl J Med* 355(16): 1682–89, 2006.
62. Burrows B, Knudson RJ, Lebowitz MD. The relationship of childhood respiratory illness to adult obstructive airway disease. *Am Rev Respir Dis* 115(5): 751–60, 1977.
63. Samet JM, Tager IB, Speizer FE. The relationship between respiratory illness in childhood and chronic air-flow obstruction in adulthood. *Am Rev Respir Dis* 127(4): 508–23, 1983.
64. Barker DJ, Godfrey KM, Fall C, Osmond C, Winter PD, Shaheen SO. Relation of birth weight and childhood respiratory infection to adult lung function and death from chronic obstructive airways disease. *BMJ* 303(6804): 671–75, 1991.
65. Edwards CA, Osman LM, Godden DJ, Campbell DM, Douglas JG. Relationship between birth weight and adult lung function: Controlling for maternal factors. *Thorax* 58(12): 1061–65, 2003.
66. Lawlor DA, Ebrahim S, Davey Smith G. Association of birth weight with adult lung function: Findings from the British Women's Heart and Health Study and a meta-analysis. *Thorax* 60(10): 851–58, 2005.
67. Peat JK, Woolcock AJ, Cullen K. Rate of decline of lung function in subjects with asthma. *Eur J Respir Dis* 70(3): 171–79, 1987.
68. Agertoft L, Pedersen S. Effects of long-term treatment with an inhaled corticosteroid on growth and pulmonary function in asthmatic children. *Respir Med* 88(5): 373–81, 1994.
69. Long-term effects of budesonide or nedocromil in children with asthma. The Childhood Asthma Management Program Research Group. *N Engl J Med* 343(15): 1054–63, 2000.
70. Guilbert TW, Morgan WJ, Zeiger RS *et al.* Long-term inhaled corticosteroids in preschool children at high risk for asthma. *N Engl J Med* 354(19): 1985–97, 2006.
71. Martinez FD, Wright AL, Taussig LM, Holberg CJ, Halonen M, Morgan WJ. Asthma and wheezing in the first six years of life. The Group Health Medical Associates. *N Engl J Med* 332(3): 133–38, 1995.
72. Xuan W, Peat JK, Toelle BG, Marks GB, Berry G, Woolcock AJ. Lung function growth and its relation to airway hyperresponsiveness and recent wheeze. Results from a longitudinal population study. *Am J Respir Crit Care Med* 161(6): 1820–24, 2000.
73. Holberg CJ, Stern DA, Sherrill DL, Wright AL, Martinez FD, Morgan WJ. Effect of smoking on the development of lung function in adolescence. *Am J Respir Crit Care Med* 163: A260, 2001.
74. Camilli AE, Burrows B, Knudson RJ, Lyle SK, Lebowitz MD. Longitudinal changes in forced expiratory volume in one second in adults. Effects of smoking and smoking cessation. *Am Rev Respir Dis* 135(4): 794–99, 1987.
75. Pauwels RA, Rabe KF. Burden and clinical features of chronic obstructive pulmonary disease (COPD). *Lancet* 364(9434): 613–20, 2004.
76. Anto JM, Vermeire P, Vestbo J, Sunyer J. Epidemiology of chronic obstructive pulmonary disease. *Eur Respir J* 17(5): 982–94, 2001.
77. Rennard SI, Vestbo J. COPD: The dangerous underestimate of 15%. *Lancet* 367(9518): 1216–19, 2006.
78. Fletcher CM, Peto CM, Tinker CM, Spizer FE. The Natural History of Chronic Bronchitis and Emphysema. New York: Oxford University Press, 1976.
79. Orie NGM, Sluiter HJ, de Vries K, Tammeling GJ, Witkop J. The host factor in bronchitis. In: Orie NGM, Sluiter HJ (eds). *Bronchitis an International Symposium*. Assen, Netherlands: Royal vanGorcum, 1961.
80. Sluiter HJ, Koeter GH, de Monchy JG, Postma DS, de Vries K, Orie NG. The Dutch hypothesis (chronic non-specific lung disease) revisited. *Eur Respir J* 4(4): 479–89, 1991.
81. Tashkin DP, Altose MD, Connett JE, Kanner RE, Lee WW, Wise RA. Methacholine reactivity predicts changes in lung function over time in smokers with early chronic obstructive pulmonary disease. The Lung Health Study Research Group. *Am J Respir Crit Care Med* 153(6 Pt 1): 1802–11, 1996.
82. Brutsche MH, Downs SH, Schindler C *et al.* Bronchial hyperresponsiveness and the development of asthma and COPD in asymptomatic individuals: SAPALDIA cohort study. *Thorax* 61(8): 671–77, 2006.
83. Rijcken B, Schouten JP, Xu X, Rosner B, Weiss ST. Airway hyperresponsiveness to histamine associated with accelerated decline in FEV1. *Am J Respir Crit Care Med* 151(5): 1377–82, 1995.
84. Postma DS, de Vries K, Koeter GH, Sluiter HJ. Independent influence of reversibility of air-flow obstruction and nonspecific hyperreactivity on the long-term course of lung function in chronic air-flow obstruction. *Am Rev Respir Dis* 134(2): 276–80, 1986.
85. Vestbo J, Lange P. Can GOLD Stage 0 provide information of prognostic value in chronic obstructive pulmonary disease? *Am J Respir Crit Care Med* 166(3): 329–32, 2002.
86. de Marco R, Accordini S, Cerveri T *et al.* Incidence of chronic obstructive pulmonary disease in a cohort of young adults according to the presence of chronic cough and phlegm. *Am J Respir Crit Care Med* 175(1): 32–39, 2007.
87. Sherman CB, Xu X, Speizer FE, Ferris BG Jr, Weiss ST, Dockery DW. Longitudinal lung function decline in subjects with respiratory symptoms. *Am Rev Respir Dis* 146(4): 855–59, 1992.
88. Vestbo J, Prescott E, Lange P. Association of chronic mucus hypersecretion with FEV1 decline and chronic obstructive pulmonary disease morbidity. Copenhagen City Heart Study Group. *Am J Respir Crit Care Med* 153(5): 1530–35, 1996.
89. Vestbo J, Hogg JC. Convergence of the epidemiology and pathology of COPD. *Thorax* 61(1): 86–88, 2006.
90. Donaldson GC, Seemungal TA, Bhowmik A, Wedzicha JA. Relationship between exacerbation frequency and lung function decline in chronic obstructive pulmonary disease. *Thorax* 57(10): 847–52, 2002.
91. Kanner RE, Anthonisen NR, Connett JE. Lower respiratory illnesses promote FEV(1) decline in current smokers but not ex-smokers with mild chronic obstructive pulmonary disease: results from the lung health study. *Am J Respir Crit Care Med* 164(3): 358–64, 2001.
92. Burrows B, Knudson RJ, Cline MG, Lebowitz MD. A reexamination of risk factors for ventilatory impairment. *Am Rev Respir Dis* 138(4): 829–36, 1988.
93. Kurzius-Spencer M, Sherrill DL, Holberg CJ, Martinez FD, Lebowitz MD. Familial correlation in the decline of forced expiratory volume in one second. *Am J Respir Crit Care Med* 164(7): 1261–65, 2001.
94. Marossy AE, Strachan DP, Rudnicka AR, Anderson HR. Childhood Chest Illness and the Rate of Decline of Adult Lung Function Between Ages 35 and 45 Years. *Am J Respir Crit Care Med*, 2006.

95. James AL, Palmer LJ, Kicic E *et al.* Decline in lung function in the Busselton Health Study: The effects of asthma and cigarette smoking. *Am J Respir Crit Care Med* 171(2): 109–14, 2005.
96. Guerra S, Sherrill DL, Kurzius-Spencer M, Venker C, Halonen M, Quan SF, Martinez FD. The course of persistent airflow limitation in subjects with and without asthma. *Respir Med* (in press).
97. Fabbri LM, Romagnoli M, Corbetta L *et al.* Differences in airway inflammation in patients with fixed airflow obstruction due to asthma or chronic obstructive pulmonary disease. *Am J Respir Crit Care Med* 167(3): 418–24, 2003.

4

CHAPTER

Genetics of Asthma and COPD

Dirkje S. Postma[1] and Edwin K. Silverman[2]

[1]Department of Pulmonology, University Medical Center Groningen, University of Groningen, Groningen, The Netherlands

[2]Channing Laboratory and Division of Pulmonary and Critical Care Medicine, Brigham and Women's Hospital, Harvard Medical School, Boston, MA, USA

INTRODUCTION

It is now a well-established fact that genetic factors contribute to the development of asthma and chronic obstructive pulmonary disease (COPD). This chapter sets out the current knowledge focusing on the heritability of asthma and COPD and the genes found by positional cloning and association studies. It also focuses on particular genes to provide insight into the current knowledge on genetic approaches to understanding diseases like asthma and COPD.

HERITABILITY OF ASTHMA AND COPD

It has been known for centuries that asthma clusters in families (familial aggregation).

The first segregation studies in families with an asthmatic proband already showed that no single-gene accounts for a major part of the expression of the disease. A polygenic model with some evidence of oligogenic loci (i.e. a handful of loci being responsible for most of the genetic control) provided the best fit to the data [1].

For COPD, a role for genetic factors has been less obvious, because cigarette smoking is such a major environmental risk factor. The identification of alpha 1-antitrypsin (AAT) deficiency in the 1960s proved that genes could influence COPD susceptibility [2]. AAT deficiency is largely caused by homozygosity for the Z allele (PI ZZ) at the SERPINA1 locus, although multiple other rare deficiency alleles have been identified. PI ZZ subjects are at markedly increased risk for COPD, especially if they smoke cigarettes. However, the development of COPD among PI ZZ subjects is highly variable, and environmental, developmental, and genetic factors likely contribute to this variability. DeMeo and colleagues recently confirmed that male gender and asthma (especially in childhood), as well as cigarette smoking, are risk factors for COPD in PI ZZ subjects [3].

Evidence that COPD unrelated to AAT deficiency is also influenced by genetic factors has been provided by many familial aggregation studies. Pulmonary function levels in the general population cluster in families [4]. Moreover, studies of relatives of COPD subjects have confirmed familial aggregation of COPD, probably related to gene-by-smoking interactions. Silverman and colleagues found that first-degree relatives of severe, early-onset COPD subjects who smoke, had an approximately threefold increased risk of airflow obstruction and chronic bronchitis [5]; nonsmoking relatives had increased risk for abnormalities in sensitive spirometric measures such as FEF25-75 but not for severe airflow obstruction [6]. McCloskey and colleagues confirmed that COPD showed familial aggregation, which was only apparent in smokers [7]. Multiple genes likely contribute to the susceptibility for COPD and its phenotypic expression.

In the studies of the heritability of asthma and COPD, the absence of unambiguous criteria for their diagnosis and the fact that asthma and COPD are not one disease entity but are both heterogeneous diseases, for example COPD can encompass both chronic bronchitis and emphysema and the latter occurs homogeneously or locally in the lung, have been major obstacles to pinpointing the exact polymorphisms in genes contributing to disease development. Another explanation for the failure to identify loci may lie in the genetic component of the disease. For instance, different genes can be involved in the same phenotype (genetic heterogeneity) or the same genotype may result in different phenotypes (pleiotropy). Mutated genes do not always express the phenotype (incomplete penetrance) and vice versa. The specific phenotype under study can be expressed without the genetic mutation (phenocopy).

ASSESSING GENES FOR ASTHMA AND COPD

The human genome consists of approximately 3 billion base pairs. The sequence of the whole human genome has recently been published [8]. The genetic distances are expressed in centiMorgans (cM). One cM is about 100,000 base pairs on a physical map and corresponds to 1% recombination during meiosis. This means that one crossing-over event between two loci that are one cM apart occurs in every hundred meioses. The estimation of the total number of genes in the human genome is about 20,000 [9]. Most genes in the population have multiple locations where more than one variant is commonly found; the variants at these polymorphic locations are known as alleles. Only a minority of human DNA is responsible for the coding for a biological product.

The main strategies to identify susceptibility genes are positional cloning and the candidate gene approach.

Positional cloning starts with the investigation of families without a predetermined hypothesis regarding the location or identity of the underlying susceptibility gene or genes. Markers are randomly spaced throughout the entire genome and tested for linkage (i.e. coinheritance) with a disease phenotype. After the finding of linkage between a particular marker and a phenotype, further fine spaced typing of genetic markers (fine mapping) is required to pinpoint the exact gene causing the linkage. The approach is time consuming, as in-depth analysis of a particular region of linkage that still can cover a large part of a chromosome, requires considerable molecular analysis.

An alternative approach is to select candidate genes that putatively contribute to the underlying pathological process of the disease. The gene is screened for polymorphisms, which are tested for association with the disease or phenotype in question. The results can be interpreted in three ways, as shown in Box 4.1.

Replication of any genetic study is required to exclude spurious findings especially if multiple genes are involved in the disease process, like in asthma and COPD.

Candidate genes of unknown function or unknown role in disease pathogenesis can also be selected for genetic analysis based on their differential expression in diseased versus normal tissue [10, 11]. Another approach is to select a gene that is a proven cause of a monogenic syndrome that has the disease of interest as a component of its syndrome constellation. The hypothesis that can be tested is that mutations in the gene with a milder functional effect can contribute to the development of a complex genetic disorder in the general population. The gene *SPINK5* that encodes the serine protease inhibitor LEKTI on chromosome 5q32 as the cause of Netherton syndrome is such an example [12]. Netherton syndrome is a severe autosomal recessive disorder with a congenital skin disease associated with defective cornification and severe atopic manifestations. A common coding polymorphism of *SPINK5*, E420K, has been shown to be associated with atopy and atopic dermatitis in two independent family cohorts [13], although replication attempts have had varying success [14, 15]. In COPD, cutis laxa is a rare dermatological syndrome related to abnormal connective tissue elasticity; some cases are caused by mutations in the distal part of the elastin (ELN) gene. Emphysema often occurs in cutis laxa at a very early age. Kelleher and colleagues found a rare mutation in the first base of the last exon of ELN in an early-onset COPD subject [16]. This variant, which was not a private mutation in that pedigree, interfered with the assembly of the elastic fiber, changed the proteolytic pattern of the ELN protein, and altered cellular adhesion of the ELN molecule. Thus, in both asthma and COPD, the assessment of candidate genes from monogenic syndromes has led to some insights into the disease etiology.

Both the positional cloning and the candidate gene approaches have their own limitations. Population association between a disease and a genetic marker can arise as an artifact of the population structure. Linkage studies with modest numbers of affected sib pairs may be underpowered and fail to detect linkage, especially if there is genetic heterogeneity. Furthermore, although linkage analysis has been successful to identify genes underlying single-gene disorders, in complex diseases it is frequently very difficult to sufficiently narrow a region of linkage to just a single gene. Nevertheless positional cloning in asthma has identified some genes.

A truly comprehensive genetic association study must consider all putative causal alleles in a gene of interest or in the entire human genome if resources are available. Until recently, this was practically impossible. With the completion of the International HapMap Project [17], it is now

Box 4.1. POTENTIAL CAUSES OF GENOTYPE–PHENOTYPE ASSOCIATION

1. The trait of interest is due to a genotyped variant in the candidate gene.
2. The trait is determined by one or more genetic variants in linkage disequilibrium with a genotyped variant; that is, a genotyped variant is very close to the disease gene.
3. The association is the result of population admixture, that is, a certain trait has a higher prevalence in a specific ethnic subgroup within a mixed population. Any allele with a higher frequency within this subgroup will show association with the trait.

possible to target a large proportion of the genetic variation across the genome, either directly or indirectly (via LD). HapMap is a freely available reference panel of genotype data from different worldwide populations (http://www.hapmap.org). This resource can be used to guide the design of disease association studies and prioritization of single nucleotide polymorphism (SNP) genotyping assays. With this dataset, it is possible to study genetic variants for any locus of interest. The HapMap dataset has clearly demonstrated the existence of correlations between nearby variants. By taking advantage of these correlations, one can select informative SNPs (tagging SNPs) that provide information about neighboring variants that are not genotyped. Only a small fraction of SNPs need to be genotyped to capture the full information in a specific region. If a causal variant is not genotyped, its effect can be indirectly tested with the correlated tag SNP that has been genotyped.

The single-gene approach described earlier can also be extended to the entire genome. Genome-wide association scans are emerging as powerful tools to identify genes involved in complex diseases. On the basis of phase I HapMap data, it was shown that approximately 250,000–500,000 SNPs are required to capture all common SNPs in human populations. Although these numbers appear impressive, current technologies can evaluate 1,000,000 SNPs simultaneously [18, 19]. By selecting SNPs properly, we can now interrogate the entire genome in one assay. This allows us to undertake genome-wide association studies, combining the fine mapping and power of association analysis and the ease of case–control cohort recruitment with a genome-wide hypothesis–independent approach. These promising tools will certainly have a major impact on our understanding of the genetic basis of complex diseases such as asthma and COPD. Genome-wide association studies have already led to important and novel insights into the genetic architecture of a rapidly expanding list of complex diseases, including age-related macular degeneration and adult-onset diabetes mellitus [20, 21]. Although not yet reported in COPD, Moffatt and colleagues have performed a genome-wide association analysis in asthma [22]. They identified a region of highly significant association to asthma on chromosome 17q, which was replicated in several other study populations. Using microarray gene expression data, they observed that the expression of one of the genes in that region, ORMDL3, was strongly associated with the same SNPs that were associated with asthma – suggesting that asthma susceptibility is related to ORMDL3 gene expression.

LINKAGE ANALYSIS OF ASTHMA AND COPD

Linkage analysis includes a group of genetic epidemiological methods that are used to identify chromosomal regions that are likely to contain one or more genetic determinants influencing a phenotype of interest. A panel of genetic markers is genotyped within family units, which can be as small as sib pairs or as large as extended pedigrees, and the linkage between the markers and a hypothetical disease susceptibility locus is determined. The evidence for linkage is often expressed as an LOD score, which is the logarithm to the base 10 of the odds for linkage. Depending on the study design, LOD scores above 3.3 (for extended pedigrees) or 3.6 (for sib pairs) correspond to significant linkage [23]. Although the location of genes for classic Mendelian disorders can usually be determined quite accurately by linkage analysis, the locations of complex disease susceptibility loci are typically less precisely localized due to genetic heterogeneity, incomplete penetrance, and environmental phenocopies. Linkage can be assessed for quantitative phenotypes as well as disease affection status.

A large number of genome scan linkage studies have been performed in asthma, which were summarized by Celedon and colleagues [24]. Based on follow-up studies after linkage analyses, six potential susceptibility genes for asthma have been reported using positional cloning [25–30]. The only published genome scan linkage studies in COPD have been performed in the Boston Early-Onset COPD Study, implicating chromosomes 2q, 12p, and 19q as likely locations of COPD susceptibility genes [31, 32]. Of interest, a linkage study of spirometric phenotypes in a set of families unselected for respiratory disease identified chromosome 2q, a location of significant linkage to COPD, and chromosome 5q, a location of significant linkage to asthma, as likely locations for genetic influences on FEV_1/FVC in the general population [33]. In an adjacent region on chromosome 2q, Postma and colleagues found significant evidence for linkage to FEV_1/VC in families ascertained through asthmatic probands [34]. Thus, there may be genetic influences on asthma, COPD, and pulmonary function in the general population located within chromosome 2q.

ASSOCIATION STUDIES IN ASTHMA AND COPD

A large number of SNPs in the promoters and coding regions of a wide range of candidate genes have been examined for genetic association in asthma. There are now over 500 studies that have examined polymorphisms in over 200 genes for association with atopy and allergic disease phenotypes [35–37]. These studies have provided us with increasing insight into genetic susceptibility to asthma, the role of gene–environment interaction and the role of genetic variation in inter-individual response to treatment.

In this part we will focus on five of the candidate genes that have the best evidence for involvement in asthma susceptibility (Table 4.1); these genes have been identified in various ways, including positional cloning (*ADAM33*) and positional and/or biological candidates (*IL-13, IL-4, IL-4R, CD-14*). We will also review an example from asthma pharmacogenetics (*ADRB2*).

Subsequently, we will consider five of the COPD candidate genes with the strongest evidence for association to COPD susceptibility, including positional candidates (*TGFB1 and SERPINE2*) and biological candidates (*GSTP1, EPHX1, and SOD3*) (Table 4.2).

TABLE 4.1 Summary of positive and negative association studies in the genes *ADAM33*, *IL-3*, *IL-4*, and *IL-4Rα* with asthma and its related phenotypes.

Gene	Studies supporting association	Studies refuting association	Comments
A disintegrin and metalloprotease 33 *(ADAM33)*	[25] Caucasian	[38] Puerto Rican, Mexican	
	[39] African Americans, Caucasian, Hispanic, Dutch	[40]	Raby children
	[41] Japanese	[42] Chinese	Cheng allergic rhinitis
	[43]) Korean	[44] German	Schedel children. Positive and negative associations with various phenotypes
	[45] Caucasian		Jongepier association with FEV_1 decline
	[46] Caucasian		
	[47] Caucasian		Simpson low lung function in children
	[48]		
	[49] Japanese		Noguchi children
IL-13, IL-4, IL-4R	[50]	[51]	Celedon IL-13 83 trios children Deichmann IL-13+, IL-4R+
	[52]		IL-13+
	[53]		IL-13+
	[54]		IL-13 + Children
	[55]		IL-13 + Children
	[56]		IL-4R +/IL-4–Children
	[57]		IL-13 +/IL-4R–
	[56]		IL-4R+
	[53]		IL-13+
	[58]		IL-13+, IL-4+ Interaction+
	[59]		IL-13+ Children
	[60]		IL-13+
	[61]		IL-4R–
	[62]		IL-13+/IL-4R+ Interaction+
	[63]		IL-13+ Children
	[64]		IL-13+/IL-4+ Children
	[65]		IL-13+/Il-4/IL-4R Children, Interaction–
	[66]		IL-4+/IL-4R+ Interaction–
	[67]		IL-13+ Children
	[68)		IL-4+/ IL-4R+ Children
	[69]		Children
	[70]		IL-13+
	[71]		
	[72]		IL-13+/IL-4+/IL-4R–Interaction+ Children
	[73]		IL-4R/IL-13 Interaction + Children
	[74]		IL-13+ Children

SNPs in the IL13/IL4 pathway have been reported as single gene association. When interaction was present, this has been assigned as interaction+, and when interaction was absent, this has been assigned as interaction –.

ASTHMA GENETIC ASSOCIATION STUDIES

Gene found by positional cloning: ADAM33

ADAM33 was discovered by positional cloning in 460 Caucasian families from the US and UK on chromosome 20p. It was associated with asthma, specifically when bronchial hyperresponsiveness (BHR) was present, and there were variations in total and as well as specific IgE levels [25]. Multiple SNPs within *ADAM33* were significantly associated with asthma and BHR in a US or UK population, or with these two populations combined. The nomenclature typically used for ADAM33 SNPs is unusual; the exons have been assigned sequential letters, and the SNPs have been sequentially numbered within each exon (e.g. S_2 is the second SNP in exon S). SNPs occurring in the intron before an exon are designated with a minus sign (e.g. Q-1 is an intronic SNP before exon Q), and SNPs occurring in an intron immediately after an exon are designated with

TABLE 4.2 Summary genetic association evidence for five candidate genes in COPD.

Gene	Studies supporting association	Studies refuting association	Comments
Microsomal epoxide hydrolase (EPHX1)	[75]	[76]	Yoshikawa reported association with COPD severity
	[77]	[78]	
	[79]	[80]	
	[81]	[81]	Hersh found association in their case-control study but not their family study
		[82]	Cheng reported association with COPD severity
Glutathione *S*-transferase P1 (GSTP1)	[83]	[84]	
	[85]	[81]	He found increased risk for Val105 as opposed to Ile105
	[86]	[82]	Cheng found increased COPD risk for combination of Ile105 with GSTM1 and EPHX1 variants
Extracellular superoxide dismutase (SOD3)	[87]		Arg213Gly variant has low frequency
	[88]		
Transforming growth factor beta 1 (TGFB1)	[89]	[90]	
	[91]	[92]	
	[93]		
	[94]		Van Diemen found association with COPD affection status but not FEV_1 decline
Serpin peptidase inhibitor, Clade E, Member 2 (SERPINE2)	[95]	[96]	
	[97]		

a plus sign (e.g. F + 1 is an intronic SNP after exon F). The majority of associated SNPs were located in the 3′ half of the gene, extending from exon Q to the last exon, exon V. Subsequently, there have been a large number of case–control and family-based association studies focused on *ADAM33*, with the majority [39, 43, 45, 46, 48, 49, 82] but not all [38, 40, 42, 44], confirming the original finding (Table 4.1). For example, the Collaborative Study on the Genetics of Asthma, comprising eight US genetic centers, demonstrated positive association between SNPs in *ADAM33* and asthma as well as BHR in African-American, Hispanic, and white populations [39]. In a further study involving 1299 asthma cases, 1665 control subjects, and 4561 family members, Blakey and colleagues [48] applied a literature-based meta-analysis, which supplemented the database with new asthma cases from populations in Iceland and the United Kingdom, and demonstrated significant association for 4 of the 13 SNPs tested.

It is well recognized that in patients with chronic persistent asthma, baseline lung function declines more rapidly over time when compared with that of normal individuals [98]. Jongepier *et al.* showed in 200 patients with chronic asthma who were studied annually for 25 years [45] that the rare allele of the S_2 polymorphism (minor allele frequency = 0.25) was significantly associated with excess decline in FEV_1 over time and concluded that this variant of *ADAM33* was not only important in the development of asthma but also in disease progression, possibly related to enhanced airway remodeling. A further study by the same group investigated whether SNPs in *ADAM33* could also predict an accelerated FEV_1 decline at a population level [99]. A total of 1390 subjects from a Dutch general population cohort were genotyped for eight asthma-associated SNPs. Individuals homozygous for the minor alleles of SNPs S_2 and Q-1 and heterozygous for the SNP S_1 had a significantly accelerated FEV_1 decline over 25 years follow-up of 4.9, 9.6, and 3.6 ml/year, respectively, when compared with the wild-type allele. A further analysis demonstrated a higher prevalence of the SNPs F + 1, S_1, S_2, ST + 5, and T_2 in subjects with COPD at GOLD (Global Initiative for Chronic Obstructive Lung Disease) stage II or higher. Thus, in addition to asthma, it seems that polymorphic variation in *ADAM33* also influences the rate of decline of lung function at a population level, which may then lead to COPD. Subsequently, the authors confirmed the association with COPD in a second cohort [100] and interestingly found that the same ST_5 SNP was associated with the severity of BHR in COPD patients, as well as to the number of sputum cells and CD8 cells in bronchial biopsies. This links the increase in cells in COPD with hyperresponsiveness and *ADAM33*.

Finally, the Manchester Asthma and Allergy Study, a prospective cohort study of the development of asthma and allergies in children, investigated the relation between SNPs in *ADAM33* and lung function at ages 3 and 5 [47]. At the age of 5 years, four of the SNPs were associated with reduced FEV_1 (F + 1, N + 1, T1, and T2; $p < 0.04$). Linkage disequilibrium mapping of *ADAM33* pointed to functional SNPs lying between F + 1 and B + 1.

Taken together, these studies support the notion that *ADAM33* may be a gene involved in lung development and further remodeling over a lifetime (Fig. 4.1). This may then contribute to an accelerated loss of lung function and contribute as well to BHR. Given the associations with different SNPs, *ADAM33* may have differential function inducing different phenotypes in men.

Gene–gene interaction: IL-13, IL-4, and IL-4R

Il-13 and IL-4 are cytokines produced by Th2 cells that are capable of inducing isotype class switching of B-cells to produce IgE. They also share a receptor component, IL-4Rα, which is an important factor in the development or expression of atopy and asthma. The IL-13 receptor consists of one IL-4Rα subunit and either a low-affinity IL-13Rα1 or a high-affinity IL-13Rα2 subunit. The complete receptor for IL-4 is composed of one IL-4Rα subunit and an IL-4Rγ subunit, or IL-13α subunit (Fig. 4.2). IL-4 and IL-13 are located on chromosome 5q and share the IL-4 receptor α-chain, which is located at chromosome 16p12.

In *IL-13* seven frequent polymorphisms exist of which at least three could be of functional relevance: Polymorphism G2044A alters an amino acid in the protein domain involved with receptor binding (Arg 130Gln) [71]. This SNP and two other polymorphisms in the promoter (A-1512C and C-1112T) showed association with asthma and IgE regulation in different populations [54, 65]. Results from gene expression studies and transcription factor binding analysis on C-1112T strongly support a functional role of this polymorphism in gene expression [101]. Genetic variations in *IL-13* have been associated with asthma and related phenotypes in almost all studies that assess SNPs in *IL-13* (Table 4.1), and this is present in ethnically diverse populations living in variable environmental circumstances [50–55, 57–60, 62–65, 67, 70, 72–74, 102].

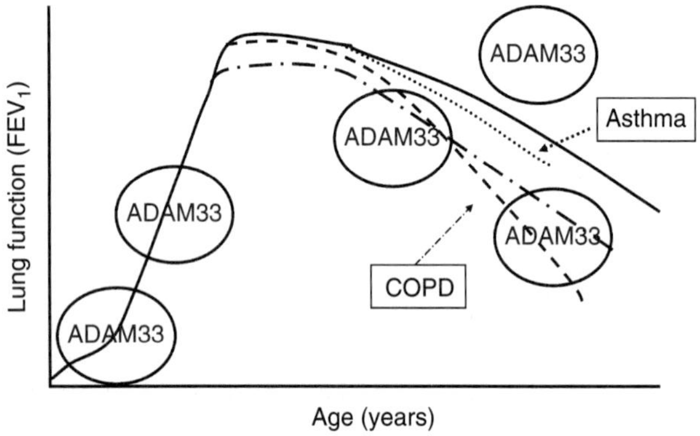

FIG. 4.1 The phases of lung function growth and decline: growth, plateau and decline phase. In all three phases *ADAM33* may contribute to the course of lung functions over a lifetime.

In the coding region of the IL-4Ralpha gene, at least 14 polymorphisms have been identified and for some of the more frequent genetic variations, functional data is available. SNP A148G alters an amino acid in the extracellular part of the receptor (I50V) which leads to increased IgE production in B cells [102]. Polymorphisms T1432C and A1652G lead to amino acid changes Ser478Pro and Gln551Arg in the intracellular domain of the IL-4Ra chain, respectively. Different studies have assessed the effect of these polymorphisms on atopic diseases [50, 56, 59, 61, 62, 64, 66, 68, 72, 73, 103]. However, overall, the associations between *IL-4Ralpha* polymorphisms and the diagnosis of asthma as well as serum IgE levels were only minor. In 2002, Howard *et al.* published a combined analysis of two polymorphisms in the IL-4/ IL-13 pathway [62]. In a longitudinal population of Dutch adults with asthma, one polymorphism in the promoter of *IL-13* (C-1112T) and one in the *IL-4Ra* gene leading to an amino acid change (Ser478Pro) were assessed. The SNP in *IL-13* (C-1112T) was previously found to be associated with BHR in this population. The SNP in *IL-4Ra* was associated with higher levels of IgE. As both traits are associated with asthma the authors studied interaction of these genes. When individuals with polymorphic alleles in both locations were compared to individuals with wild-type alleles in both locations the risk for asthma increased fivefold.

Also, the interaction between the polymorphism in the coding region of *IL-13* (G2044A) and the promoter polymorphism C-589T in *IL-4* has been studied [64]. The *IL-13* G2044A polymorphism and haplotypes consisting of *IL-13* G2044A and *IL-4* C-589T were associated with the development of atopy and atopic dermatitis. As children in this analysis were only followed up to the age of 24 months, no information on asthma was available.

Three polymorphisms in the *IL-4 receptor alpha* gene (Arg551Glu, Ile50Val, and Pro478Ser) have been investigated in combination with a promoter polymorphism in the *IL-4* gene (C-589T) [68]. The risk for asthma increased up to an

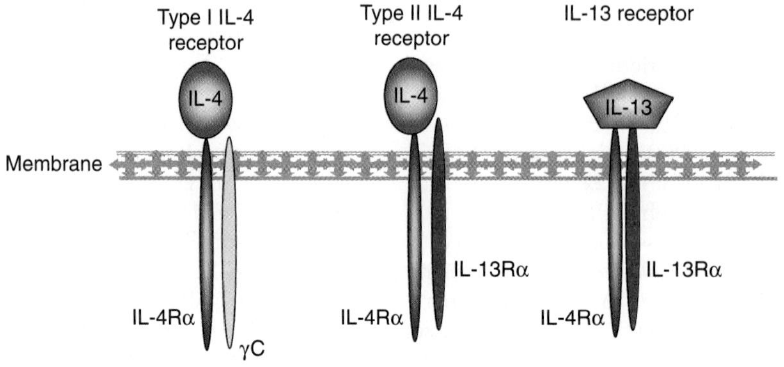

FIG. 4.2 Interaction between IL4 and IL13 and their respective receptors, that have some similarities.

odds ratio (OR) of 1.97 (95% CI 1.07–3.71) in individuals with the Arg allele at position 551 while the other two SNPs in the *IL-4Ra* gene did not show any significant effects. The combination of the *IL-4Ralpha* Arg allele with the IL-4 589T allele increased the OR to 3.70 (95% CI = 1.07 – 12.78). This study extends the initial findings by Howard in the sense that a second combination of genes seems to interact in the IL-4/ IL-13 pathway. In a similar British study, two polymorphisms in the *IL-4Ra* gene (Arg551Glu and Ile50Val, respectively) and two in the *IL-4* gene (the highly linked SNPs C-589T and C-33T, respectively) were analyzed concomitantly. While transmission disequilibrium tests and haplotype analysis showed significant associations with asthma for both genes individually, the authors state that no interaction between the two genes was observed. However, no data and no information on how interaction was tested, is provided [66].

Other researchers also failed to replicate interactions between polymorphisms in the IL-4/IL-13 pathway. Liu *et al.* investigated six polymorphisms in *IL-4*, *IL-13*, and *IL-4Ra* in almost 1000 children who were followed from birth. The effects of genetic variation in the IL-4/IL-13 pathway on specific and total serum IgE levels were analyzed [65]. The authors confirmed a role of *IL-13* polymorphisms in the regulation of total and specific IgE, but SNPs in *IL-4* and *IL4-Ra* were only associated with specific IgE responses in that study population. Although the interaction analysis is not shown, the authors state that no significant interactions between polymorphisms in the three genes were found. A further study on *IL-4, IL-4Ra, IL-13*, and *IL-13Ra1* provided association of the *IL-13* G2044A polymorphism with asthma in two populations, without significant interaction [53]. In contrast, recent studies by Kabesch *et al.* [72] and Chan *et al.* [73] found significant interactions between the genes in the IL-4/IL-13 pathway.

Differences in observations may result from small study population sizes and from characteristics of the population under study, for example the differences due to age, gender, prevalence of atopy, high- and low-risk population and so on. Thus the findings need more study. However, it is shown that gene–gene interaction may contribute to asthma and atopy development. Analyses of genes in pathways that are relevant to atopy, asthma and or its subphenotypes like BHR need further attention.

Gene–environment interaction: CD14

Innate immunity genes provide the interface between the immune system and microbial products. Many microbes have specific pattern molecules on their surfaces and these molecules interact with pattern-recognition receptors. CD14 is such a receptor. CD14 is part of the receptor complex for lipopolysaccharide (LPS, endotoxin) together with toll-like receptors (TLR) 2 and 4. CD14 does not have a transmembrane domain and thus does not signal itself but contributes to the affinity of the interaction between the microbial products and TLRs by formation of a receptor complex. Downstream effects include the upregulation of accessory molecules and release of cytokines, such as IL-10 and IL-12. These cytokines are potentially important in directing the adaptive immune response. The CD14 gene contains several SNPs, the most important one being the CD14/-159 (also called CD14/-260). Functional genomic studies showed that monocytic cells with the T allele are transcriptionally more active [103].

The initial study by Baldini *et al.* [104] demonstrated that the C-159T polymorphism was associated with elevated soluble CD14 levels, and reduced total serum IgE levels in children who were skin prick test positive to common aeroallergens. Not all studies have been consistent however, raising the possibility that the level of environmental exposure to endotoxin may alter the effect of CD14 polymorphisms [105]. A number of recent studies have supported this notion.

Endotoxin. The hygiene hypothesis proposes that microbial exposure during early life development reduces the development of asthma and allergic disease [106] and has been the subject of many genetic association studies [107]. The timing, dose, and route of microbial exposures, such as endotoxin, are likely to interact with genetic influences, thus altering the response to these exposures [106]. Exposure to endotoxin is known to occur indoors from contact with house dust. In the Barbados Asthma Genetics Study, subjects with low house dust endotoxin exposure, who had the CD14-159TT genotype had a reduced risk of asthma whereas high exposure increased the risk of asthma with this genotype [107]. Taken together, these studies provide support for an "endotoxin switch" in which there is a dose-dependent response to endotoxin exposure for specific risk genotypes [108]. Exposure to endotoxin is also encountered in occupational settings such as farming. Adult farmers with the CD14-159TT and -1691GG genotypes had significantly lower lung function and increased wheezing compared with other genotypes, possibly due to increased soluble CD14 levels interacting with inhaled endotoxin from the agricultural environment [103].

Animal exposure. The type of microbial exposure during immune system maturation may influence the development of atopy and asthma. In children with the CD14-159C allele who had regular contact with pets, serum IgE levels were higher than with the T allele [109]. The opposite occurred in children with regular contact with stable animals where the C allele was associated with lower IgE levels. In another study, early life farm environment and the CD14-159TT genotype combined to give the lowest risk of nasal allergies and atopy [110].

Environmental tobacco smoke. Exposure to environmental tobacco smoke may increase the risk of asthma in susceptible individuals. In a study of Puerto Rican and Mexican families, people with asthma who had the CD14 + 1437GG or GC genotypes and exposure to environmental tobacco smoke had a mean forced expiratory volume in 1 s that was lower by 8.6% of that predicted as compared with GG or GC subjects that were not exposed to environmental tobacco smoke [111]. In addition, people with asthma with the CD14-159TT genotype and exposure to environmental tobacco smoke had lower serum IgE levels. The mechanism for this interaction could involve exposure to endotoxin found in cigarettes. Sex differences may also exist in response to tobacco smoke. Girls whose mothers had smoked during pregnancy or whose parents had asthma had lower mean soluble CD14 levels [112].

Asthma pharmacogenetics: The $beta_2$-adrenergic receptor

The therapeutic response to $beta_2$-agonists is heterogeneous in asthmatic subjects [113], a finding suggested to be due in part to genetic variation in the $beta_2$-adrenergic receptor gene (*ADRB2*). A number of *ADRB2* polymorphisms have been described [114], with the greatest attention devoted to the SNPs causing amino acid substitutions at positions 16 and 27, namely the Gly16Arg and the Gln27Glu polymorphisms [115, 116]. Many of the early studies on the acute bronchodilator response to a beta-agonist differed substantially from one another with respect to study design, type of agonist utilized, and primary outcome assessed [117, 118]. Not surprisingly, there was marked inconsistency in results between trials. Subsequently, larger studies have demonstrated that bronchodilator responses were higher and more rapid among Arg16 homozygotes as compared to Gly16 homozygotes and heterozygotes [119, 120].

The second phenotypic outcome evaluated has been the potential for downregulation of beta-agonist receptor responsivity (tachyphylaxis) with the chronic administration of beta-agonists in association with *ADRB2* genetic variants. In a key study, 255 asthmatics of mild severity were randomized to either regular (180 μg QID) or as-needed albuterol use and were assessed over a 16-week period for evidence of tachyphylaxis and clinical deterioration (as measured by fall in AM peak expiratory flow rate (PEF)). No difference in PEF variation was observed between treatment groups [121]. However, Arg16 homozygotes significantly decreased their pre-albuterol PEF with regular utilization of albuterol therapy, compared to both Arg16 homozygotes receiving as-needed albuterol and Gly16 homozygotes in either treatment group [121]. The difference in evening PEF between these groups was 31.6 ± 10.2 l/min comparing Arg16 homozygous regular users versus Gly16 homozygous regular users ($p = 0.002$) and 31.1 ± 13.0 l/min versus the Arg16 homozygous as needed group ($p = 0.02$) [122]. Similar differences in morning PEF were noted (Fig. 4.3).

β_2AR desensitization was further evaluated in a prospective, genotype-stratified study of Arg16Gly on treatment-related changes in lung function. Israel *et al.* [123] matched asthmatic individuals homozygous Arg/Arg ($n = 37$) to those Gly/Gly ($n = 41$), by level of FEV_1. The genotype-stratified individuals were then randomized in a double-blind, cross-over study of regularly scheduled qid albuterol therapy versus placebo over two 16-week periods. Again, those with the Arg/Arg genotype had lower morning PEF versus placebo (-10 l/min, $p = 0.02$), while those with Gly/Gly had higher PEF (14 l/min, $p = 0.02$). The difference between Arg/Arg and Gly/Gly genotypes was significant for morning PEF (-24 l/min, $p = 0.0003$), evening PEF, FEV_1, morning symptom score, and need for rescue medication.

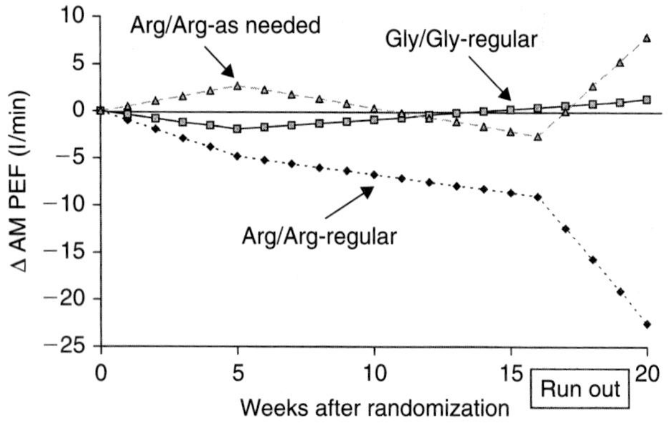

FIG. 4.3 Change in morning PEF (Peak Expiratory Flow) with beta-agonist (albuterol) treatment, according to *ADRB2* genetic variants.

Further studies have focused on asthma exacerbations as an outcome. With regular use of both short- and long-acting beta-agonists, presence of at least one Gly16 allele is protective against exacerbations [124]. Moreover, in a meta-analysis evaluating long-acting beta-agonist usage, subjects with at least one Arg allele demonstrated significant decrements in PC_{20} when compared to placebo, suggesting that the downregulation of the receptor accompanying tachyphylaxis may result in decreased bronchoprotection [125].

Several studies have assessed the effects of long-acting beta-agonists [126–128]. Wechsler *et al.* showed decreased responses to salmeterol in Arg/Arg subjects compared with Gly/Gly subjects [126]. In the Salmeterol or Corticosteroids (SOCS) trial, morning peak flow worsened in Arg/Arg asthmatic subjects ($n = 12$) who were treated with salmeterol alone after inhaled corticosteroids (ICS) withdrawal when compared with subjects who received placebo; no decrease in morning peak flow was observed in Gly/Gly subjects ($n = 13$). In the Salmeterol and/or Inhaled Corticosteroid (SLIC) trial, Arg/Arg ($n = 8$) asthmatic subjects did not show a sustained improvement in lung function as compared with Gly/Gly ($n = 22$) subjects on salmeterol, regardless of whether subjects were administered concomitant ICS. As these studies were all of small size that may confound the outcome of the analyses, larger studies are required to firmly establish whether polymorphisms in the *ADRB2* gene do affect the beneficial clinical effects of long-acting $beta_2$-antagonists (LABA) treatment when added to inhaled steroids. The size of the study population is of importance, particularly given the variability of individual responses to pharmacotherapy both between subjects and within subjects on a day-to-day basis.

Finally, previous studies have illustrated that relatively few *ABRB2* haplotypes account for the majority of the haplotypic diversity of this gene, and that the frequency of common haplotypes differs between population groups [115, 129]. An effect of *ADRB2* haplotype on acute bronchodilator responses to albuterol has been demonstrated in some studies [128] but not in others [115, 130].

Studies on *ADRB2* provide evidence that the gene is likely to be associated with bronchodilator response but that genetic prediction of this response may vary by ethnic group. Additionally, individuals homozygous for Arg16 demonstrate significant β_2AR desensitization and these individuals (approximately 15% of population) are at risk of clinical deterioration with the regular use of beta-agonists. Future studies have to assess whether this is still the case if patients are being treated with ICs.

COPD GENETIC ASSOCIATION STUDIES

Biological candidate gene: Microsomal epoxide hydrolase

Microsomal epoxide hydrolase is a xenobiotic enzyme that detoxifies highly reactive epoxides, which are created by cigarette smoking. Thus, variation in the activity of this enzyme has been hypothesized to contribute to variable COPD susceptibility. In an early COPD genetic association study, Smith and Harrison studied two genetic variants in the EPHX1 gene [75], one of which had been reported to cause reduced activity of this xenobiotic enzyme (Tyr113His) and the other which had been reported to cause increased enzyme activity (His139Arg). They analyzed two small case groups that included 68 subjects with spirometrically proven COPD and 94 patients with both pathologically proven emphysema and lung cancer. Their control group of 203 blood donors was not phenotyped. Both their COPD and emphysema/lung cancer case groups had significantly higher frequencies of homozygosity for the His113 "slow" allele associated with reduced EPHX1 enzyme activity. Somewhat paradoxically, a higher frequency of the Arg139 "fast" allele associated with increased EPHX1 enzyme activity was also found in the COPD cases compared to controls.

In a study of 180 former poison gas factor workers in Japan, followed to determine the impact of these toxic exposures on the development of lung disease, Yoshikawa and colleagues compared the distributions of the Tyr113His and His139Arg variants in 40 subjects with COPD ($FEV_1 < 70\%$ predicted) and the remaining 140 subjects [76]. No significant differences were observed in allele frequencies between these groups. However, when they subdivided their 40 COPD cases into 20 mild ($FEV_1 > 60\%$ predicted) and 20 moderate/severe ($FEV_1 < 60\%$ predicted) subjects, a higher frequency of the "slow" His113 allele was observed in the more severely affected subjects. Thus, they speculated that EPHX1 variants were associated with COPD severity rather than susceptibility.

Two other small case–control genetic association studies, performed in Asian populations and published in 2000, found no association between the Tyr113His and His139Arg variants and COPD [78, 80]. In a subsequent study of 184 COPD cases and 212 control smokers from Taiwan, the genotype frequencies for the Tyr113His and His139Arg SNPs did not differ between the cases and controls [41]. However, using the genotypes at both positions to assign two functional EPHX1 categories (slow/very slow versus normal/fast), did lead to a significant difference between cases and controls. Moreover, homozygosity for the His113 allele was more prevalent in severe versus mild COPD cases. More recently, Brogger and colleagues studied 244 subjects meeting spirometric criteria for COPD and 248 control subjects [77]. They found that the His113 variant was associated with COPD.

Sandford and colleagues studied the impact of haplotypes formed by the two widely studied EPHX1 SNPs on rate of decline in FEV_1 among continuing smokers in the Lung Health Study [79]. They selected 283 subjects with the most rapid FEV_1 decline and 308 subjects who had no measurable FEV_1 decline over 5 years. EPHX1 haplotypes were associated with rate of FEV_1 decline, and this association was more significant among individuals who reported a family history of COPD – potentially related to other COPD genetic risk factors, which act additively with EPHX1 or which interact with EPHX1 to increase COPD risk.

Hersh and colleagues studied a total of eight EPHX1 SNPs, which included the two widely studied nonsynonymous SNPs above, in a family-based study and a case–control study [81]. The family-based Boston Early-Onset COPD Study subjects included 127 extended pedigrees (949 total subjects) ascertained through a proband with severe, early-onset COPD; the case–control study included 304 severe COPD subjects from the National Emphysema Treatment Trial (NETT) and 441 control smokers from the Normative Aging Study. None of the EPHX1 SNPs were associated with quantitative or qualitative airway obstruction phenotypes in the early-onset COPD families under an additive model. However, the Arg139 allele showed modest evidence for association to COPD in the case–control study.

Overall, there is reasonable but not uniformly consistent evidence for association between EPHX1 SNPs and COPD. The small sample sizes of many of these studies could contribute to inconsistent replication. A further explanation might be differences in ethnicities; however, another potential contribution to these inconsistent results is phenotypic heterogeneity within the COPD cases. COPD includes both emphysema and airway disease in varying proportions between different subjects, and emphysema distribution also varies widely between subjects. DeMeo and colleagues studied emphysema distribution in 282 of the 304 NETT Genetics Ancillary Study participants studied by Hersh and colleagues who had chest CT scans with densitometric and radiologist assessment of emphysema severity and distribution [131]. The EPHX1 His139Arg SNP was significantly associated with emphysema distribution assessed by densitometry ($p = 0.005$) and by radiologist scoring ($p = 0.01$). To determine whether utilizing a more homogeneous set of COPD cases would improve the ability to detect genetic association, the association results using all NETT COPD cases were compared to the results using only 171 upper lobe predominant emphysema cases; the same set of 441 control subjects was used for both comparisons. Despite including a smaller number of COPD cases, the upper lobe predominant emphysema group association analysis indicated that the Arg139 "fast" allele was more protective (OR 0.60 with $p = 0.005$) compared to the association analysis using all COPD cases (OR 0.73 with $p = 0.02$). These results suggest that phenotypic heterogeneity could contribute to the inconsistent replication of EPHX1 (and other COPD candidate gene SNPs) associations with COPD; more comprehensive phenotyping of COPD cases, including chest CT scans, should be considered for future COPD genetic studies.

Another xenobiotic enzyme gene: GSTP1

Like EPHX1, glutathione *S*-transferase P1 (GSTP1) detoxifies oxygen radicals produced by cigarette smoke; it has been extensively studied as a COPD candidate gene. Ishii and colleagues studied two nonsynonymous SNPs in GSTP1 in 53 COPD cases and 50 control subjects from Japan [81]. A higher frequency of Ile105 homozygotes at the Ile105Val SNP was observed in COPD cases (79% versus 52%); some evidence suggests that the Ile105 variant confers reduced functional activity of the GSTP1 enzyme [132]. In a Korean population of 89 COPD cases and 94 smoking controls, Yim and colleagues found no association of the Ile105Val SNP with COPD [84]. In the Lung Health Study, He and his colleagues compared 544 subjects with the highest baseline FEV_1 values (mean $FEV_1 = 91.8\%$ predicted) to 554 subjects with the lowest baseline FEV_1 values (mean $FEV_1 = 62.6\%$ predicted); all of these subjects had COPD based on reduced FEV_1/FVC ratio [85]. They observed a higher frequency of Val105 homozygotes among the low lung function group (13.2% versus 9.3%), as well as a faster rate of FEV_1 decline in high lung function subjects who were Val105 homozygotes.

There are multiple potential explanations for the inconsistent results of these previous studies, including population differences (ethnicity, COPD status), failure to analyze the actual functional variant in GSTP1, and small samples in several of the studies. However, it is also possible that gene–gene interactions need to be considered to identify key COPD susceptibility genes. Studies of combinations of GSTP1 variants with other potential susceptibility genes have also been performed. Calikoglu and colleagues did find a higher frequency of Ile105 homozygotes in a male Turkish population of 149 COPD cases (61%) compared to 150 control subjects (38%) [86]. They also examined the impact of including variants in GSTT1 and GSTM1 in combination with GSTP1 and found a markedly increased risk of COPD when a combination of variants in these three genes was considered. In addition to EPHX1, Cheng and colleagues also studied the GSTP1 Ile105Val variant in their set of 184 COPD cases and 212 control subjects from Taiwan [41]. Although the Ile105Val SNP was not associated with COPD when analyzed individually, the combination of at least one EPHX1 "slow" allele (His113), homozygosity for the GSTM1 Null allele, and homozygosity for the Ile105 allele was significantly more common in COPD cases (36%) versus controls (8.5%). Formal tests of interaction in large samples will be required to replicate these findings and to determine whether these results represent the combination of susceptibility alleles acting independently or significant gene–gene interactions.

Association with a relatively uncommon allele: SOD3

Although genetic association studies typically analyze common genetic variants with allele frequencies above 10%, it is quite likely that less common variants contribute to the susceptibility, and resistance, to complex diseases like COPD. Extracellular superoxide dismutase (EC-SOD or SOD3) detoxifies oxygen radicals by scavenging superoxide anions. A variant in this gene, Arg213Gly, has been shown to have a functional effect on the processing of the EC-SOD protein, leading to substantially elevated plasma levels of the protein [133]. Young and colleagues genotyped the Arg213Gly variant in 230 COPD cases and 210 control smokers [87]. Importantly, as the Gly213 variant was hypothesized to confer resistance to COPD among smokers, their control subjects had normal spirometry with a heavy smoking history (mean pack-years = 42). The allele frequency of Gly213 was only 1% in the COPD cases but it was 5% in the control smokers ($p = 0.02$). Carriers of at least one Gly213 allele were significantly more common among controls than cases ($p = 0.005$).

In the Copenhagen City Heart Study, Juul and colleagues studied the Arg213Gly variant in 9258 Danish subjects [88]. They confirmed the marked effect of this variant on EC-SOD plasma levels, with mean values of 142 ng/ml in Arg213 homozygotes, 1278 ng/ml in Arg213/Gly213 heterozygotes, and 4147 ng/ml in Gly213 homozygotes. In a multivariate model adjusting for gender, age, smoking, and occupational dust exposure, smokers who were heterozygous (Arg213/Gly213) had substantially lower risk for COPD than Arg213 homozygotes (OR 0.4, with 95% CI 0.2–0.8). Among nonsmokers, no protective effect of the Gly213 variant for COPD was observed. Only two Gly213 homozygous subjects were identified in the entire cohort. This confirms the potentially protective effect of the Gly213 allele among smokers, and it emphasizes the utility of large sample sizes – especially when studying relatively uncommon genetic variants.

TGFB1: Convergence of biological and positional candidate gene approaches

Transforming growth factor beta 1 (TGFB1) is a cytokine that acts as a central regulator of the inflammatory response; thus, it is a logical biological candidate gene for COPD. A potentially functional variant has been identified at codon 10, which changes a leucine to a proline (Leu10Pro); individuals that carry at least one proline allele have higher TGFB1 mRNA in peripheral blood mononuclear cells and higher serum TGFB1 protein levels [134]. Wu and colleagues genotyped this nonsynonymous SNP in Caucasian subjects from New Zealand. They included 165 COPD cases and two control groups – 76 control smokers and 140 unphenotyped blood donors [89]. The Pro10 allele was less common in COPD cases (33%) than in either the blood donor controls (45%) or smoking controls (45%).

Celedon and colleagues identified TGFB1 as a positional candidate gene on chromosome 19q after performing linkage analysis in 72 extended pedigrees (585 individuals) ascertained through severe, early-onset COPD probands [91]. The initial evidence for linkage of FEV_1 to 19q was modest (LOD score = 1.40). However, genotyping additional short tandem repeat markers to increase the information available for linkage analysis, as well as performing a stratified analysis in smokers only (to identify genomic regions likely influenced by gene-by-smoking interactions), led to more substantial evidence for linkage (LOD score = 3.30). TGFB1 is located within the linkage region, and five SNPs were genotyped in both Boston Early-Onset COPD Study pedigrees for family-based association analysis and in a set of 304 COPD cases and 441 smoking control subjects. Three

SNPs were significantly associated with COPD phenotypes in each population, and one SNP in the promoter region of TGFB1 (rs2241718) replicated in both study populations. The Leu10Pro variant was only significantly associated in the case–control analysis, but a lower Pro10 allele frequency was noted in COPD cases – indicating the same directionality of effect as in the study by Wu and colleagues.

Following these initially supportive studies using the biological and positional candidate gene approaches, further replication efforts of TGFB1 in COPD susceptibility have had mixed results. In a small study comparing 84 COPD cases and 97 control subjects from China, significant differences in both allele and genotype frequencies were noted for two TGFB1 promoter SNPs [93]. However, association analysis of the Leu10Pro SNP and two promoter SNPs in 102 COPD cases and 159 control smokers from Korea showed no evidence for association [90]. Furthermore, in a comparison of 283 continuing smokers with rapid FEV_1 decline versus 307 continuing smokers with slow FEV_1 decline in the Lung Health Study, Ogawa and colleagues found no association between three TGFB1 SNPs (including the Leu10Pro SNP) and FEV_1 decline [92]. van Diemen and colleagues studied three TGFB1 SNPs including one promoter SNP (rs1800469, C-509T), one nonsynonymous SNP (rs1982073, Leu10Pro), and one 3′ untranslated region SNP (rs6957) in a general population sample of 1390 individuals in the Netherlands who were followed longitudinally for 25 years [94]. Although they found no association of TGFB1 SNPs with FEV_1 decline, rs6957 was significantly more common in 188 subjects meeting criteria for GOLD Stage II COPD than among the remainder of the cohort ($p = 0.001$). Additional studies of TGFB1 in COPD will be required, including larger numbers of SNPs in larger study populations, to definitively confirm or refute the role of TGFB1 genetic variants in COPD susceptibility.

SERPINE2: Integrating genomics and genetics in COPD

Although genetic linkage studies can point to chromosomal regions that are likely to include susceptibility genes for a complex disease, such regions typically contain many genes. In addition to the approaches of selecting a biologically plausible positional candidate gene (as for TGFB1) or dense genotyping of SNP markers (not yet reported for COPD) within linkage regions, assessment of gene expression can point to promising candidate genes that would not have been suspected based on our current knowledge of disease pathobiology. DeMeo and colleagues integrated gene expression microarray results from two data sets – one data set analyzed mouse lung development and the other data set compared gene expression in human lung tissue – for the genes located within their linkage region to airway obstruction on chromosome 2q [95]. Based on these gene expression results, SERPINE2 (Serpin Peptidase Inhibitor, Clade E, Member 2) was selected as a candidate gene, and 48 SNPs were genotyped in a family-based cohort (949 individuals from 127 Boston Early-Onset COPD Study pedigrees) and a case–control cohort (304 COPD cases and 441 control smokers). Eighteen SERPINE2 SNPs were associated with airflow obstruction phenotypes in the Boston Early-Onset COPD families; a SNP-by-smoking interaction term was included in those analyses. Five of these 18 SNPs demonstrated replicated associations in the case–control cohort. Although Chappell and colleagues wrote a Letter to the Editor indicating that they could not replicate these associations in their case–control cohort [96], Zhu and colleagues recently reported their analyses of 25 SERPINE2 SNPs in both a family-based cohort of 1910 subjects and a case–control cohort from Norway including 973 COPD cases and 956 control smokers [97]. The family-based association analysis demonstrated association of five SNPs to COPD, with p values as low as 0.002. The case–control analysis only found one SNP with $p < 0.05$ for COPD affection status; however, five SNPs were associated with FEV_1/FVC in COPD cases; these were the same five SNPs associated with COPD in the family-based association analysis. Thus, reasonable evidence for replication has been found, supporting the utility of future studies of SERPINE2 association and function to confirm whether it is a valid COPD susceptibility gene.

PERSPECTIVES ON GENETICS OF ASTHMA AND COPD

As demonstrated in this chapter, the research on genetics of asthma and COPD has become a promising new field. The main goal of the industries and the research institutes is to find the genes that increase susceptibility to develop asthma and COPD.

Both in asthma and in COPD the heterogeneity of the diseases with complex hereditary traits and many (sub)phenotypes are obstacles to find the right genetic information easily. Although a rapidly increasing number of genes are being implicated in asthma and COPD pathogenesis, the functional variants in these genes have not been definitively identified; this is a major area requiring investigation. In the pathogenesis of asthma and COPD it is reasonable to suspect that the interactions between multiple genes and multiple environmental triggers are important. This will require collaboration between many researchers combining their cohorts that have been carefully phenotyped for asthma and COPD. By combining large cohorts with systematic analyses of environmental risk factors, comprehensive assessment of genetic polymorphisms (including genome-wide SNP genotyping), and genetic analysis with advanced statistical techniques, we will better understand the genetic determinants and heterogeneity of asthma and COPD. Hopefully this will lead to new treatments, intervention strategies, and finally prevention of asthma and COPD.

ACKNOWLEDGEMENTS

Research on asthma and COPD in Groningen is supported by the Netherlands Asthma Foundation, ZonMW and KNAW. Dr. Silverman was supported by NIH grants R01 HL075478 and HL68926.

References

1. Holberg CJ, Elston RC, Halonen M, Wright AL, Taussig LM, Morgan WJ, Martinez FD. Segregation analysis of physician-diagnosed asthma in hispanic and non-hispanic white families. *Am J Respir Crit Care Med* 154: 144–50, 1996.
2. Laurell CB, Eriksson S. The electrophoretic α_1-globulin pattern of serum in α_1-antitrypsin deficiency. *Scand J Clin Lab Invest* 15: 132–40, 1963.
3. DeMeo DL, Sandhaus RA, Barker AF, Brantly ML, Eden E, McElvaney NG, Rennard S, Burchard EG, Stocks JM, Stoller JK, Strange C, Turino GM, Campbell EJ, Silverman EK. Determinants of airflow obstruction in severe alpha 1-antitrypsin deficiency. *Thorax* 62: 806–13. 2007.
4. Lewitter FI, Tager IB, McGue M, Tishler PV, Speizer FE. Genetic and environmental determinants of level of pulmonary function. *Am J Epidemiol* 120: 518–29, 1984.
5. Silverman EK, Chapman HA, Drazen JM, Weiss ST, Rosner B, Campbell EJ, O'Donnell WJ, Reilly JJ, Ginns L, Mentzer S, Wain J, Speizer FE. Genetic epidemiology of severe, early-onset chronic obstructive pulmonary disease: Risk to relatives for airflow obstruction and chronic bronchitis. *Am J Respir Crit Care Med* 157: 1770–78, 1998.
6. DeMeo DL, Carey VJ, Chapman HA, Reilly JJ, Ginns LC, Speizer FE, Weiss ST, Silverman EK. Familial aggregation of FEF(25-75) and FEF(25-75)/FVC in families with severe, early onset COPD. *Thorax* 59: 396–400, 2004.
7. McCloskey SC, Patel BD, Hinchliffe SJ, Reid ED, Wareham NJ, Lomas DA. Siblings of patients with severe chronic obstructive pulmonary disease have a significant risk of airflow obstruction. *Am J Respir Crit Care Med* 164: 1419–24, 2001.
8. Pennisi E. Human genome. Finally, the book of life and instructions for navigating it. *Science* 288: 2304–7, 2000.
9. Pennisi E. Genetics. Working the (gene count) numbers: Finally, a firm answer?. *Science* 316: 1113, 2007.
10. Schmidt-Weber CB. Gene expression profiling in allergy and asthma. *Chem Immunol Allergy* 91: 188–94, 2006.
11. Nomura I, Gao B, Boguniewicz M, Darst MA, Travers JB, Leung DY. Distinct patterns of gene expression in the skin lesions of atopic dermatitis and psoriasis: A gene microarray analysis. *J Allergy Clin Immunol* 112: 1195–202, 2003.
12. Chavanas S, Garner C, Bodemer C, Ali M, Teillac DH, Wilkinson J, Bonafe JL, Paradisi M, Kelsell DP, Ansai S, Mitsuhashi Y, Larregue M, Leigh IM, Harper JI, Taieb A, Prost Y, Cardon LR, Hovnanian A. Localization of the Netherton syndrome gene to chromosome 5q32, by linkage analysis and homozygosity mapping. *Am J Hum Genet* 66: 914–21, 2000.
13. Walley AJ, Chavanas S, Moffatt MF, Esnouf RM, Ubhi B, Lawrence R, Wong K, Abecasis GR, Jones EY, Harper JI, Hovnanian A, Cookson WO. Gene polymorphism in Netherton and common atopic disease. *Nat Genet* 29: 175–78, 2001.
14. Kabesch M, Carr D, Weiland SK, von Mutius E. Association between polymorphisms in serine protease inhibitor, kazal type 5 and asthma phenotypes in a large German population sample. *Clin Exp Allergy* 34: 340–45, 2004.
15. Jongepier H, Koppelman GH, Nolte IM, Bruinenberg M, Bleecker ER, Meyers DA, te Meerman GJ, Postma DS. Polymorphisms in SPINK5 are not associated with asthma in a Dutch population. *J Allergy Clin Immunol* 115: 486–92, 2005.
16. Kelleher CM, Silverman EK, Broekelmann T, Litonjua AA, Hernandez M, Sylvia JS, Stoler J, Reilly JJ, Chapman HA, Speizer FE, Weiss ST, Mecham RP, Raby BA. A functional mutation in the terminal exon of elastin in severe, early-onset chronic obstructive pulmonary disease. *Am J Respir Cell Mol Biol* 33: 355–62, 2005.
17. Altshuler D, Brooks LD, Chakravarti A, Collins FS, Daly MJ, Donnelly P. A haplotype map of the human genome. *Nature* 437: 1299–320, 2005.
18. Syvanen AC. Toward genome-wide SNP genotyping. *Nat Genet* 37(Suppl): S5–10, 2005.
19. Engle LJ, Simpson CL, Landers JE. Using high-throughput SNP technologies to study cancer. *Oncogene* 25: 1594–601, 2006.
20. Dewan A, Liu M, Hartman S, Zhang SS, Liu DT, Zhao C, Tam PO, Chan WM, Lam DS, Snyder M, Barnstable C, Pang CP, Hoh J. HTRA1 promoter polymorphism in wet age-related macular degeneration. *Science* 314: 989–92, 2006.
21. Sladek R, Rocheleau G, Rung J, Dina C, Shen L, Serre D, Boutin P, Vincent D, Belisle A, Hadjadj S, Balkau B, Heude B, Charpentier G, Hudson TJ, Montpetit A, Pshezhetsky AV, Prentki M, Posner BI, Balding DJ, Meyre D, Polychronakos C, Froguel P. A genome-wide association study identifies novel risk loci for type 2 diabetes. *Nature* 445: 881–85, 2007.
22. Moffatt MF, Kabesch M, Liang L, Dixon AL, Strachan D, Heath S, Depner M, von Berg A, Bufe A, Rietschel E, Heinzmann A, Simma B, Frischer T, Willis-Owen SA, Wong KC, Illig T, Vogelberg C, Weiland SK, von Mutius E, Abecasis GR, Farrall M, Gut IG, Lathrop GM, Cookson WO. Genetic variants regulating ORMDL3 expression contribute to the risk of childhood asthma. *Nature* 448: 470–73, 2007.
23. Lander E, Kruglyak L. Genetic dissection of complex traits: Guidelines for interpreting and reporting linkage results. *Nat Genet* 11: 241–47, 1995.
24. Celedon JC, Raby B, Weiss ST *et al.* (eds). Asthma genetics. In: Silverman EK, Shapiro SD, Lomas DA, and Weiss ST. *Respiratory Genetics*, pp. 219–96. London: Hodder Arnold, 2005.
25. Van Eerdewegh P *et al.* Association of the ADAM33 gene with asthma and bronchial hyperresponsiveness. *Nature* 418: 426–30, 2002.
26. Allen M *et al.* Positional cloning of a novel gene influencing asthma from chromosome 2q14. *Nat Genet* 35: 258–63, 2003.
27. Zhang Y, Leaves NI, Anderson GG, Ponting CP, Broxholme J, Holt R, Edser P, Bhattacharyya S, Dunham A, Adcock IM, Pulleyn L, Barnes PJ, Harper JI, Abecasis G, Cardon L, White M, Burton J, Matthews L, Mott R, Ross M, Cox R, Moffatt MF, Cookson WO. Positional cloning of a quantitative trait locus on chromosome 13q14 that influences immunoglobulin E levels and asthma. *Nat Genet* 34: 181–86, 2003.
28. Laitinen T, Polvi A, Rydman P, Vendelin J, Pulkkinen V, Salmikangas P, Makela S, Rehn M, Pirskanen A, Rautanen A, Zucchelli M, Gullsten H, Leino M, Alenius H, Petays T, Haahtela T, Laitinen A, Laprise C, Hudson TJ, Laitinen LA, Kere J. Characterization of a common susceptibility locus for asthma-related traits. *Science* 304: 300–4, 2004.
29. Nicolae D, Cox NJ, Lester LA, Schneider D, Tan Z, Billstrand C, Kuldanek S, Donfack J, Kogut P, Patel NM, Goodenbour J, Howard T, Wolf R, Koppelman GH, White SR, Parry R, Postma DS, Meyers D, Bleecker ER, Hunt JS, Solway J, Ober C. Fine mapping and positional candidate studies identify HLA-G as an asthma susceptibility gene on chromosome 6p21. *Am J Hum Genet* 76: 349–57, 2005.
30. Noguchi E, Yokouchi Y, Zhang J, Shibuya K, Shibuya A, Bannai M, Tokunaga K, Doi H, Tamari M, Shimizu M, Shirakawa T, Shibasaki M, Ichikawa K, Arinami T. Positional identification of an asthma susceptibility gene on human chromosome 5q33. *Am J Respir Crit Care Med* 172: 183–88, 2005.
31. Silverman EK, Palmer LJ, Mosley JD, Barth M, Senter JM, Brown A, Drazen JM, Kwiatkowski DJ, Chapman HA, Campbell EJ, Province MA, Rao DC, Reilly JJ, Ginns LC, Speizer FE, Weiss ST. Genomewide linkage analysis of quantitative spirometric phenotypes in severe early-onset chronic obstructive pulmonary disease. *Am J Hum Gen* 70: 1229–39, 2002.
32. DeMeo DL, Celedon JC, Lange C, Reilly JJ, Chapman HA, Sylvia JS, Speizer FE, Weiss ST, Silverman EK. Genome-wide linkage of forced mid-expiratory flow in chronic obstructive pulmonary disease. *Am J Respir Crit Care Med* 170: 1294–301, 2004.
33. Malhotra A, Peiffer AP, Ryujin DT, Elsner T, Kanner RE, Leppert MF, Hasstedt SJ. Further evidence for the role of genes on chromosome 2 and chromosome 5 in the inheritance of pulmonary function. *Am J Respir Crit Care Med* 168: 556–61, 2003.
34. Postma DS, Meyers DA, Jongepier H, Howard TD, Koppelman GH, Bleecker ER. Genomewide screen for pulmonary function in 200 families ascertained for asthma. *Am J Respir Crit Care Med* 172: 446–52, 2005.

35. Ober C, Hoffjan S. Asthma genetics 2006: The long and winding road to gene discovery. *Genes Immun* 7: 95–100, 2006.
36. Hoffjan S, Nicolae D, Ober C. Association studies for asthma and atopic diseases: A comprehensive review of the literature. *Respir Res* 4: 14, 2003.
37. Kere J, Laitinen T. Positionally cloned susceptibility genes in allergy and asthma. *Curr Opin Immunol* 16: 689–94, 2004.
38. Lind DL, Choudhry S, Ung N, Ziv E, Avila PC, Salari K, Ha C, Lovins EG, Coyle NE, Nazario S, Casal J, Torres A, Rodriguez-Santana JR, Matallana H, Lilly CM, Salas J, Selman M, Boushey HA, Weiss ST, Chapela R, Ford JG, Rodriguez-Cintron W, Silverman EK, Sheppard D, Kwok PY, Gonzalez Burchard E. ADAM33 is not associated with asthma in Puerto Rican or Mexican populations. *Am J Respir Crit Care Med* 168: 1312–16, 2003.
39. Howard TD, Postma DS, Jongepier H, Moore WC, Koppelman GH, Zheng SL, Xu J, Bleecker ER, Meyers DA. Association of a disintegrin and metalloprotease 33 (ADAM33) gene with asthma in ethnically diverse populations. *J Allergy Clin Immunol* 112: 717–22, 2003.
40. Raby BA, Silverman EK, Kwiatkowski DJ, Lange C, Lazarus R, Weiss ST. ADAM33 polymorphisms and phenotype associations in childhood asthma. *J Allergy Clin Immunol* 113: 1071–78, 2004.
41. Cheng SL, Yu CJ, Chen CJ, Yang PC. Genetic polymorphism of epoxide hydrolase and glutathione S-transferase in COPD. *Eur Respir J* 23: 818–24, 2004.
42. Wang P, Liu QJ, Li JS, Li HC, Wei CH, Guo CH, Gong YQ. Lack of association between ADAM33 gene and asthma in a Chinese population. *Int J Immunogenet* 33: 303–6, 2006.
43. Lee JH, Park HS, Park SW, Jang AS, Uh ST, Rhim T, Park CS, Hong SJ, Holgate ST, Holloway JW, Shin HD. ADAM33 polymorphism: Association with bronchial hyper-responsiveness in Korean asthmatics. *Clin Exp Allergy* 34: 860–65, 2004.
44. Schedel M, Depner M, Schoen C, Weiland SK, Vogelberg C, Niggemann B, Lau S, Illig T, Klopp N, Wahn U, von Mutius E, Nickel R, Kabesch M. The role of polymorphisms in ADAM33, a disintegrin and metalloprotease 33, in childhood asthma and lung function in two German populations. *Respir Res* 7: 91, 2006.
45. Jongepier H, Boezen HM, Dijkstra A, Howard TD, Vonk JM, Koppelman GH, Zheng SL, Meyers DA, Bleecker ER, Postma DS. Polymorphisms of the ADAM33 gene are associated with accelerated lung function decline in asthma. *Clin Exp Allergy* 34: 757–60, 2004.
46. Werner M, Herbon N, Gohlke H, Altmuller J, Knapp M, Heinrich J, Wjst M. Asthma is associated with single-nucleotide polymorphisms in ADAM33. *Clin Exp Allergy* 34: 26–31, 2004.
47. Simpson A, Maniatis N, Jury F, Cakebread JA, Lowe LA, Holgate ST, Woodcock A, Ollier WE, Collins A, Custovic A, Holloway JW, John SL. Polymorphisms in a disintegrin and metalloprotease 33 (ADAM33) predict impaired early-life lung function. *Am J Respir Crit Care Med* 172: 55–60, 2005.
48. Blakey J, Halapi E, Bjornsdottir US, Wheatley A, Kristinsson S, Upmanyu R, Stefansson K, Hakonarson H, Hall IP. Contribution of ADAM33 polymorphisms to the population risk of asthma. *Thorax* 60: 274–76, 2005.
49. Noguchi E, Ohtsuki Y, Tokunaga K, Yamaoka-Sageshima M, Ichikawa K, Aoki T, Shibasaki M, Arinami T. ADAM33 polymorphisms are associated with asthma susceptibility in a Japanese population. *Clin Exp Allergy* 36: 602–8, 2006.
50. Deichmann KA, Heinzmann A, Forster J, Dischinger S, Mehl C, Brueggenolte E, Hildebrandt F, Moseler M, Kuehr J. Linkage and allelic association of atopy and markers flanking the IL4-receptor gene. *Clin Exp Allergy* 28: 151–55, 1998.
51. Celedon JC, Soto-Quiros ME, Palmer LJ, Senter J, Mosley J, Silverman EK, Weiss ST. Lack of association between a polymorphism in the interleukin-13 gene and total serum immunoglobulin E level among nuclear families in Costa Rica. *Clin Exp Allergy* 32: 387–90, 2002.
52. van der Pouw Kraan TC, van Veen A, Boeije LC, van Tuyl SA, de Groot ER, Stapel SO, Bakker A, Verweij CL, Aarden LA, van der Zee JS. An IL-13 promoter polymorphism associated with increased risk of allergic asthma. *Genes Immun* 1: 61–65, 1999.
53. Heinzmann A *et al.* Genetic variants of IL-13 signalling and human asthma and atopy. *Hum Mol Genet* 9: 549–59, 2000.
54. Graves PE, Kabesch M, Halonen M, Holberg CJ, Baldini M, Fritzsch C, Weiland SK, Erickson RP, von Mutius E, Martinez FD. A cluster of seven tightly linked polymorphisms in the IL-13 gene is associated with total serum IgE levels in three populations of white children. *J Allergy Clin Immunol* 105: 506–13, 2000.
55. Liu X, Nickel R, Beyer K, Wahn U, Ehrlich E, Freidhoff LR, Bjorksten B, Beaty TH, Huang SK. An IL13 coding region variant is associated with a high total serum IgE level and atopic dermatitis in the German multicenter atopy study (MAS-90). *J Allergy Clin Immunol* 106: 167–70, 2000.
56. Takabayashi A, Ihara K, Sasaki Y, Suzuki Y, Nishima S, Izuhara K, Hamasaki N, Hara T. Childhood atopic asthma: Positive association with a polymorphism of IL-4 receptor alpha gene but not with that of IL-4 promoter or Fc epsilon receptor I beta gene. *Exp Clin Immunogenet* 17: 63–70, 2000.
57. Shirakawa I, Deichmann KA, Izuhara I, Mao I, Adra CN, Hopkin JM. Atopy and asthma: Genetic variants of IL-4 and IL-13 signalling. *Immunol Today* 21: 60–64, 2000.
58. Noguchi E, Nukaga-Nishio Y, Jian Z, Yokouchi Y, Kamioka M, Yamakawa-Kobayashi K, Hamaguchi H, Matsui A, Shibasaki M, Arinami T. Haplotypes of the 5' region of the IL-4 gene and SNPs in the intergene sequence between the IL-4 and IL-13 genes are associated with atopic asthma. *Hum Immunol* 62: 1251–57, 2001.
59. Leung TF, Tang NL, Chan IH, Li AM, Ha G, Lam CW. A polymorphism in the coding region of interleukin-13 gene is associated with atopy but not asthma in Chinese children. *Clin Exp Allergy* 31: 1515–21, 2001.
60. Howard TD, Whittaker PA, Zaiman AL, Koppelman GH, Xu J, Hanley MT, Meyers DA, Postma DS, Bleecker ER. Identification and association of polymorphisms in the interleukin-13 gene with asthma and atopy in a Dutch population. *Am J Respir Cell Mol Biol* 25: 377–84, 2001.
61. Wjst M, Kruse S, Illig T, Deichmann K. Asthma and IL-4 receptor alpha gene variants. *Eur J Immunogenet* 29: 263–68, 2002.
62. Howard TD, Koppelman GH, Xu J, Zheng SL, Postma DS, Meyers DA, Bleecker ER. Gene–gene interaction in asthma: IL4RA and IL13 in a Dutch population with asthma. *Am J Hum Genet* 70: 230–36, 2002.
63. DeMeo DL, Lange C, Silverman EK, Senter JM, Drazen JM, Barth MJ, Laird N, Weiss ST. Univariate and multivariate family-based association analysis of the IL- 13 ARG130GLN polymorphism in the Childhood Asthma Management program. *Genet Epidem* 23: 335–48, 2002.
64. He JQ, Chan-Yeung M, Becker AB, Dimich-Ward H, Ferguson AC, Manfreda J, Watson WT, Sandford AJ. Genetic variants of the IL13 and IL4 genes and atopic diseases in at-risk children. *Genes Immun* 4: 385–89, 2003.
65. Liu X, Beaty TH, Deindl P, Huang SK, Lau S, Sommerfeld C, Fallin MD, Kao WH, Wahn U, Nickel R. Associations between total serum IgE levels and the 6 potentially functional variants within the genes IL4, IL13, and IL4RA in German children: The German Multicenter Atopy Study. *J Allergy Clin Immunol* 112: 382–88, 2003.
66. Beghe B, Barton S, Rorke S, Peng Q, Sayers I, Gaunt T, Keith TP, Clough JB, Holgate ST, Holloway JW. Polymorphisms in the interleukin-4 and interleukin-4 receptor alpha chain genes confer susceptibility to asthma and atopy in a Caucasian population. *Clin Exp Allergy* 33: 1111–17, 2003.
67. Hummelshoj T, Bodtger U, Datta P, Malling HJ, Oturai A, Poulsen LK, Ryder LP, Sorensen PS, Svejgaard E, Svejgaard A. Association between an interleukin-13 promoter polymorphism and atopy. *Eur J Immunogenet* 30: 355–59, 2003.
68. Lee SG, Kim BS, Kim JH, Lee SY, Choi SO, Shim JY, Hong TJ, Hong SJ. Gene–gene interaction between interleukin-4 and interleukin-4 receptor alpha in Korean children with asthma. *Clin Exp Allergy* 34: 1202–8, 2004.
69. Hoffjan S, Ostrovnaja I, Nicolae D, Newman DL, Nicolae R, Gangnon R, Steiner L, Walker K, Reynolds R, Greene D, Mirel D,

Gern JE, Lemanske RF Jr., Ober C. Genetic variation in immunoregulatory pathways and atopic phenotypes in infancy. *J Allergy Clin Immunol* 113: 511–18, 2004.
70. Nieters A, Linseisen J, Becker N. Association of polymorphisms in Th1, Th2 cytokine genes with hayfever and atopy in a subsample of EPIC-Heidelberg. *Clin Exp Allergy* 34: 346–53, 2004.
71. Vladich FD, Brazille SM, Stern D, Peck ML, Ghittoni R, Vercelli D. IL-13 R130Q, a common variant associated with allergy and asthma, enhances effector mechanisms essential for human allergic inflammation. *J Clin Invest* 115: 747–54, 2005.
72. Kabesch M, Schedel M, Carr D, Woitsch B, Fritzsch C, Weiland SK, von Mutius E. IL-4/IL-13 pathway genetics strongly influence serum IgE levels and childhood asthma. *J Allergy Clin Immunol* 117: 269–74, 2006.
73. Chan IH, Leung TF, Tang NL, Li CY, Sung YM, Wong GW, Wong CK, Lam CW. Gene–gene interactions for asthma and plasma total IgE concentration in Chinese children. *J Allergy Clin Immunol* 117: 127–33, 2006.
74. Hunninghake GM, Soto-Quiros ME, Avila L, Su J, Murphy A, Demeo DL, Ly NP, Liang C, Sylvia JS, Klanderman BJ, Lange C, Raby BA, Silverman EK, Celedon JC. Polymorphisms in IL13, total IgE, eosinophilia, and asthma exacerbations in childhood. *J Allergy Clin Immunol* 120: 84–90, 2007.
75. Smith CAD, Harrison DJ. Association between polymorphism in gene for microsomal epoxide hydrolase and susceptibility to emphysema. *The Lancet* 350: 630–33, 1997.
76. Yoshikawa M, Hiyama K, Ishioka S, Maeda H, Maeda A, Yamakido M. Microsomal epoxide hydrolase genotypes and chronic obstructive pulmonary disease in Japanese. *Int J Mol Med* 5: 49–53, 2000.
77. Brogger J, Steen VM, Eiken HG, Gulsvik A, Bakke P. Genetic association between COPD and polymorphisms in TNF, ADRB2 and EPHX1. *Eur Respir J* 27: 682–88, 2006.
78. Yim JJ, Park GY, Lee CT, Kim YW, Han SK, Shim YS, Yoo CG. Genetic susceptibility to chronic obstructive pulmonary disease in Koreans: Combined analysis of polymorphic genotypes for microsomal epoxide hydrolase and glutathione S-transferase M1 and T1. *Thorax* 55: 121–25, 2000.
79. Sandford AJ, Chagani T, Weir TD, Connett JE, Anthonisen NR, Pare PD. Susceptibility genes for rapid decline of lung function in the Lung Health Study. *Am J Respir Crit Care Med* 163: 469–73, 2001.
80. Takeyabu K, Yamaguchi E, Suzuki I, Nishimura M, Hizawa N, Kamakami Y. Gene polymorphism for microsomal epoxide hydrolase and susceptibility to emphysema in a Japanese population. *Eur Respir J* 15: 891–94, 2000.
81. Hersh CP, Demeo DL, Lange C, Litonjua AA, Reilly JJ, Kwiatkowski D, Laird N, Sylvia JS, Sparrow D, Speizer FE, Weiss ST, Silverman EK. Attempted replication of reported chronic obstructive pulmonary disease candidate gene associations. *Am J Respir Cell Mol Biol* 33: 71–78, 2005.
82. Cheng L, Enomoto T, Hirota T, Shimizu M, Takahashi N, Akahoshi M, Matsuda A, Dake Y, Doi S, Enomoto K, Yamasaki A, Fukuda S, Mao XQ, Hopkin JM, Tamari M, Shirakawa T. Polymorphisms in ADAM33 are associated with allergic rhinitis due to Japanese cedar pollen. *Clin Exp Allergy* 34: 1192–201, 2004.
83. Ishii T, Matsuse T, Teramoto S, Matsui H, Miyao M, Hosoi T, Takahashi H, Fukuchi Y, Ouchi Y. Glutathione S-transferase P1 (GSTP1) polymorphism in patients with chronic obstructive pulmonary disease. *Thorax* 54: 693–96, 1999.
84. Yim JJ, Yoo CG, Lee CT, Kim YW, Han SK, Shim YS. Lack of association between glutathione S-transferase P1 polymorphism and COPD in Koreans. *Lung* 180: 119–25, 2002.
85. He JQ, Connett JE, Anthonisen NR, Pare PD, Sandford AJ. Glutathione S-transferase variants and their interaction with smoking on lung function. *Am J Respir Crit Care Med* 170: 388–94, 2004.
86. Calikoglu M, Tamer L, Ates Aras N, Karakas S, Ercan B. The association between polymorphic genotypes of glutathione S-transferases and COPD in the Turkish population. *Biochem Genet* 44: 307–19, 2006.
87. Young RP, Hopkins R, Black PN, Eddy C, Wu L, Gamble GD, Mills GD, Garrett JE, Eaton TE, Rees MI. Functional variants of antioxidant genes in smokers with COPD and in those with normal lung function. *Thorax* 61: 394–99, 2006.
88. Juul K, Tybjaerg-Hansen A, M S, Lange P, Nordestgaard BG. Genetically increased antioxidant protection and decreased chronic obstructive pulmonary disease. *Am J Respir Crit Care Med*, 173: 858–64, 2006.
89. Wu L, Chau J, Young RP, Pokorny V, Mills GD, Hopkins R, McLean L, Black PN. Transforming growth factor-beta1 genotype and susceptibility to chronic obstructive pulmonary disease. *Thorax* 59: 126–29, 2004.
90. Yoon HI, Silverman EK, Lee HW, Yoo CG, Lee CT, Chung HS, Kim YW, Han SK, Shim YS, Yim JJ. Lack of association between COPD and transforming growth factor-b1 (TGFB1) genetic polymorphisms in Koreans. *Int J Tuberc Lung Dis*, 10: 504–09, 2006.
91. Celedon JC, Lange C, Raby BA, Litonjua AA, Palmer LJ, DeMeo DL, Reilly JJ, Kwiatkowski DJ, Chapman HA, Laird N, Sylvia JS, Hernandez M, Speizer FE, Weiss ST, Silverman EK. The transforming growth factor-beta1 (TGFB1) gene is associated with chronic obstructive pulmonary disease (COPD). *Hum Mol Genet* 13: 1649–56, 2004.
92. Ogawa E, Ruan J, Connett JE, Anthonisen NR, Pare PD, Sandford AJ. Transforming growth factor-beta1 polymorphisms, airway responsiveness and lung function decline in smokers. *Respir Med* 101: 938–43, 2007.
93. Su ZG, Wen FQ, Feng YL, Xiao M, Wu XL. Transforming growth factor-beta1 gene polymorphisms associated with chronic obstructive pulmonary disease in Chinese population. *Acta Pharmacol Sin* 26: 714–20, 2005.
94. van Diemen CC, Postma DS, Vonk JM, Bruinenberg M, Nolte IM, Boezen HM. Decorin and TGF-beta1 polymorphisms and development of COPD in a general population. *Respir Res* 7: 89, 2006.
95. DeMeo D, Mariani TJ, Lange C, Srisuma SS, Litonjua AA, Celedon JC, Lake SL, Reilly JJ, Chapman HA, Mecham BH, Haley KJ, Sylvia JS, Sparrow D, Spira A, Beane J, Pinto-Plata V, Speizer FE, Shapiro SD, Weiss ST, Silverman EK. The SERPINE2 gene is associated with chronic obstructive pulmonary disease. *Am J Hum Genet* 178: 253–64, 2006.
96. Chappell S, Daly L, Morgan K, Baranes TG, Roca J, Rabinovich R, Millar A, Donnelly SC, Keatings V, MacNee W, Stolk J, Hiemstra PS, Miniati M, Monti S, O'Connor CM, Kalsheker N. The SERPINE2 gene and chronic obstructive pulmonary disease. *Am J Hum Genet* 79: 184–86, 2006, author reply 186–7.
97. Zhu G, Warren L, Aponte J, Gulsvik A, Bakke P, ICGN Investigators, Anderson WH, Lomas DA, Silverman EK, Pillai SG. The SERPINE2 gene is associated with chronic obstructive pulmonary disease in two large populations. *Am J Respir Crit Care Med* 176: 167–73, 2007.
98. Dijkstra A, Vonk JM, Jongepier H, Koppelman GH, Schouten JP, ten Hacken NH, Timens W, Postma DS. Lung function decline in asthma: Association with inhaled corticosteroids, smoking and sex. *Thorax* 61: 105–10, 2006.
99. van Diemen CC, Postma DS, Vonk JM, Bruinenberg M, Schouten JP, Boezen HM. A disintegrin and metalloprotease 33 polymorphisms and lung function decline in the general population. *Am J Respir Crit Care Med* 172: 329–33, 2005.
100. Gosman MM, Boezen HM, van Diemen CC, Snoeck-Stroband JB, Lapperre TS, Hiemstra PS, Ten Hacken NH, Stolk J, Postma DS. A disintegrin and metalloprotease 33 and chronic obstructive pulmonary disease pathophysiology. *Thorax* 62: 242–47, 2007.
101. Vercelli D. Genetics of IL-13 and functional relevance of IL-13 variants. *Curr Opin Allergy Clin Immunol* 2: 389–93, 2002.
102. Kruse S, Braun S, Deichmann KA. Distinct signal transduction processes by IL-4 and IL-13 and influences from the Q551R variant of the human IL-4 receptor alpha chain. *Respir Res* 3: 24, 2002.
103. LeVan TD, Von Essen S, Romberger DJ, Lambert GP, Martinez FD, Vasquez MM, Merchant JA. Polymorphisms in the

CD14 gene associated with pulmonary function in farmers. *Am J Respir Crit Care Med* 171: 773–79, 2005.
104. Baldini M, Lohman IC, Halonen M, Erickson RP, Holt PG, Martinez FD. A Polymorphism* in the 5' flanking region of the CD14 gene is associated with circulating soluble CD14 levels and with total serum immunoglobulin E. *Am J Respir Cell Mol Biol* 20: 976–83, 1999.
105. Martinez FD. Gene–environment interactions in asthma and allergies: A new paradigm to understand disease causation. *Immunol Allergy Clin North Am* 25: 709–21, 2005.
106. Liu AH, Leung DY. Renaissance of the hygiene hypothesis. *J Allergy Clin Immunol* 117: 1063–66, 2006.
107. Zambelli-Weiner A, Ehrlich E, Stockton ML, Grant AV, Zhang S, Levett PN, Beaty TH, Barnes KC. Evaluation of the CD14/-260 polymorphism and house dust endotoxin exposure in the Barbados Asthma Genetics Study. *J Allergy Clin Immunol* 115: 1203–9, 2005.
108. Yang IA, Fong KM, Holgate ST, Holloway JW. The role of Toll-like receptors and related receptors of the innate immune system in asthma. *Curr Opin Allergy Clin Immunol* 6: 23–28, 2006.
109. Eder W, Klimecki W, Yu L, von Mutius E, Riedler J, Braun-Fahrlander C, Nowak D, Martinez FD. Opposite effects of CD 14/-260 on serum IgE levels in children raised in different environments. *J Allergy Clin Immunol* 116: 601–7, 2005.
110. Leynaert B, Guilloud-Bataille M, Soussan D, Benessiano J, Guenegou A, Pin I, Neukirch F. Association between farm exposure and atopy, according to the CD14 C-159T polymorphism. *J Allergy Clin Immunol* 118: 658–65, 2006.
111. Choudhry S, Avila PC, Nazario S, Ung N, Kho J, Rodriguez-Santana JR, Casal J, Tsai HJ, Torres A, Ziv E, Toscano M, Sylvia JS, Alioto M, Salazar M, Gomez I, Fagan JK, Salas J, Lilly C, Matallana H, Castro RA, Selman M, Weiss ST, Ford JG, Drazen JM, Rodriguez-Cintron W, Chapela R, Silverman EK, Burchard EG. CD14 tobacco gene-environment interaction modifies asthma severity and immunoglobulin E levels in Latinos with asthma. *Am J Respir Crit Care Med* 172: 173–82, 2005.
112. Lodrup Carlsen KC, Lovik M, Granum B, Mowinckel P, Carlsen KH. Soluble CD14 at 2 yr of age: Gender-related effects of tobacco smoke exposure, recurrent infections and atopic diseases. *Pediatr Allergy Immunol* 17: 304–12, 2006.
113. Sin DD, Man J, Sharpe H, Gan WQ, Man SF. Pharmacological management to reduce exacerbations in adults with asthma: A systematic review and meta-analysis. *JAMA* 292: 367–76, 2004.
114. Hawkins GA, Tantisira K, Meyers DA, Ampleford EJ, Moore WC, Klanderman B, Liggett SB, Peters SP, Weiss ST, Bleecker ER. Sequence, haplotype, and association analysis of ADRbeta2 in a multiethnic asthma case–control study. *Am J Respir Crit Care Med* 174: 1101–9, 2006.
115. Contopoulos-Ioannidis DG, Manoli EN, Ioannidis JP. Meta-analysis of the association of beta2-adrenergic receptor polymorphisms with asthma phenotypes. *J Allergy Clin Immunol* 115: 963–72, 2005.
116. Green SA, Turki J, Innis M, Liggett SB. Amino-terminal polymorphisms of the human beta-2 adrenergic receptor impart distinct agonist-promoted regulatory properties. *Biochemistry* 33: 9414–19, 1994.
117. Ohe M, Munakata M, Hizawa N, Itoh A, Doi I, Yamaguchi E, Homma Y, Kawakami Y. Beta 2 adrenergic receptor gene restriction fragment length polymorphism and bronchial asthma. *Thorax* 50: 353–59, 1995.
118. Aziz I, McFarlane LC, Lipworth BJ. Comparative trough effects of formoterol and salmeterol on lymphocyte beta2-adrenoceptor – regulation and bronchodilatation. *Eur J Clin Pharmacol* 55: 431–36, 1999.
119. Martinez FD, Graves PE, Baldini M, Solomon S, Erickson R. Association between genetic polymorphisms of the beta-2 adrenoceptor and response to albuterol in children with and without a history of wheezing. *J Clin Invest* 100: 3184–88, 1997.
120. Lima JJ, Thomason DB, Mohamed MH, Eberle LV, Self TH, Johnson JA. Impact of genetic polymorphisms of the beta2-adrenergic receptor on albuterol bronchodilator pharmacodynamics. *Clin Pharmacol Ther* 65: 519–25, 1999.
121. Drazen JM, Israel E, Boushey HA, Chinchilli VM, Fahy JV, Fish JE, Lazarus SC, Lemanske RF, Martin RJ, Peters SP, Sorkness C, Szefler SJ. Comparison of regularly scheduled with as-needed use of albuterol in mild asthma. Asthma Clinical Research Network. *N Engl J Med* 33: 841–47, 1996.
122. Israel E, Drazen JM, Liggett SB, Boushey HA, Cherniack RM, Chinchilli VM, Cooper DM, Fahy JV, Fish JE, Ford JG, Kraft M, Kunselman S, Lazarus SC, Lemanske RF, Martin RJ, McLean DE, Peters SP, Silverman EK, Sorkness CA, Szefler SJ, Weiss ST, Yandava CN. The effect of polymorphisms of the beta(2)-adrenergic receptor on the response to regular use of albuterol in asthma. *Am J Respir Crit 3are Med* 162: 75–80, 2000.
122. Israel E, Chinchilli VM, Ford JG, Boushey HA, Cherniack R, Craig TJ, Deykin A, Fagan JK, Fahy JV, Fish J, Kraft M, Kunselman SJ, Lazarus SC, Lemanske RF Jr., Liggett SB, Martin RJ, Mitra N, Peters SP, Silverman E, Sorkness CA, Szefler SJ, Wechsler ME, Weiss ST, Drazen JM. Use of regularly scheduled albuterol treatment in asthma: Genotype-stratified, randomised, placebo-controlled cross-over trial. *The Lancet* 364: 1505–12, 2004.
124. Taylor DR, Drazen JM, Herbison GP, Yandava CN, Hancox RJ, Town GI. Asthma exacerbations during long term beta agonist use: Influence of beta(2) adrenoceptor polymorphism. *Thorax* 55: 762–67, 2000.
125. Lee DK, Currie GP, Hall IP, Lima JJ, Lipworth BJ. The arginine-16 beta2-adrenoceptor polymorphism predisposes to bronchoprotective subsensitivity in patients treated with formoterol and salmeterol. *Br J Clin Pharmacol* 57: 68–75, 2004.
126. Wechsler ME, Lehman E, Lazarus SC, Lemanske RF Jr., Boushey HA, Deykin A, Fahy JV, Sorkness CA, Chinchilli VM, Craig TJ, DiMango E, Kraft M, Leone F, Martin RJ, Peters SP, Szefler SJ, Liu W, Israel E. beta-Adrenergic receptor polymorphisms and response to salmeterol. *Am J Respir Crit Care Med* 173: 519–26, 2006.
127. Bleecker ER, Yancey SW, Baitinger LA, Edwards LD, Klotsman M, Anderson WH, Dorinsky PM. Salmeterol response is not affected by beta2-adrenergic receptor genotype in subjects with persistent asthma. *J Allergy Clin Immunol* 118: 809–16, 2006.
128. Nelson HS, Weiss ST, Bleecker ER, Yancey SW, Dorinsky PM. The Salmeterol Multicenter Asthma Research Trial: A comparison of usual pharmacotherapy for asthma or usual pharmacotherapy plus salmeterol. *Chest* 129: 15–26, 2006.
129. Drysdale CM, McGraw DW, Stack CB, Stephens JC, Judson RS, Nandabalan K, Arnold K, Ruano G, Liggett SB. Complex promoter and coding region beta 2-adrenergic receptor haplotypes alter receptor expression and predict in vivo responsiveness. *Proc Natl Acad Sci USA* 97: 10483–88, 2000.
130. Taylor DR, Epton MJ, Kennedy MA, Smith AD, Iles S, Miller AL, Littlejohn MD, Cowan JO, Hewitt T, Swanney MP, Brassett KP, Herbison GP. Bronchodilator response in relation to beta2-adrenoceptor haplotype in patients with asthma. *Am J Respir Crit Care Med* 172: 700–3, 2005.
131. DeMeo DL, Hersh CP, Hoffman EA, Litonjua AA, Lazarus R, Sparrow D, Benditt JO, Criner GJ, Make B, Martinez FJ, Scanlon PD, Sciurba FC, Utz JP, Reilly JJ, Silverman EK. Genetic determinants of emphysema distribution in the National Emphysema Treatment Trial. *Am J Respir Crit Care Med* 176: 42–48, 2007.
132. Sundberg K, Johansson AS, Stenberg G, Widersten M, Seidel A, Mannervik B, Jernstrom B. Differences in the catalytic efficiencies of allelic variants of glutathione transferase P1-1 towards carcinogenic diol epoxides of polycyclic aromatic hydrocarbons. *Carcinogenesis* 19: 433–36, 1998.
133. Olsen DA, Petersen SV, Oury TD, Valnickova Z, Thogersen IB, Kristensen T, Bowler RP, Crapo JD, Enghild JJ. The intracellular proteolytic processing of extracellular superoxide dismutase (EC-SOD) is a two-step event. *J Biol Chem* 279: 22152–57, 2004.
134. Suthanthiran M, Li B, Song JO, Ding R, Sharma VK, Schwartz JE, August P. Transforming growth factor-beta 1 hyperexpression in African-American hypertensives: A novel mediator of hypertension and/or target organ damage. *Proc Natl Acad Sci USA* 97: 3479–84, 2000.

Physiology and Pathology of Asthma and COPD

PART 2

Pulmonary Physiology

CHAPTER 5

Charles G. Irvin

Department of Medicine and Physiology, University of Vermont College of Medicine, Burlington, VT, USA

INTRODUCTION

Asthma is an inflammatory disease principally of the small airways but current understanding is that all the airways are involved [1]. The physiological manifestations of lung inflammation are reversible airflow limitation and airflow limitation that fluctuates widely with time; asthma can also result in persistent loss of lung function. In contrast, airflow obstruction in patients with chronic obstructive pulmonary disease (COPD) presents with both reversible and irreversible airflow limitation, the latter linked to loss of static elastic recoil due in part to the destruction of the architecture of the lung. The airflow limitation of COPD has much less temporal or periodic variation over the short term but patients with COPD usually exhibit steady consistent losses of lung function over a period of years. In reality the clinical presentation of both COPD and asthma is often quite variable and the pathophysiological data alone do not allow a definitive diagnosis. This confusing presentation of patients with features of both asthma and COPD is in keeping with the overlap observed in the epidemiology of these diseases [2, 3]. The focus of this chapter is on the pathophysiological presentation of COPD and asthma and the features that distinguish each disease as well as the similarities. This chapter will also examine the physiological processes that characterize airways disease.

LUNG VOLUMES AND ELASTIC RECOIL

Lung volumes and capacities

In diseases of the airway such as asthma and COPD there are characteristic increases in volume which distinguish these airway disorders from restrictive processes of the respiratory system that reduce lung volumes. Lung volumes have important bearing on the degree of disease since increases in the residual volume (RV) occur relatively early even with mild airway disease whereas increases in the total lung capacity (TLC) are usually an indication of more severe or long-standing disease (Fig. 5.1) [4–6].

The most useful lung volumes and capacities to measure are those that assess the physical limits of the lung and chest wall and define the extremes of the vital capacity (VC); TLC and RV. The volume in the lung at end expiratory position, the functional residual capacity (FRC), is important because of the influence that lung volume has on airway caliber and in turn on airflow resistance during eupnea. TLC, FRC, and RV are boundary conditions of respiration and accordingly provide the most information about the capacity and function of the respiratory system [6]. Assessing the changes in these lung volumes to evaluate disease severity and response to therapy allows one to distinguish between obstructive lung disease and restrictive lung disease (Table 5.1) [8, 9].

Obstructive lung diseases such as chronic bronchitis, asthma, and emphysema all result in increases in lung volume. An increase in TLC usually represents more chronic, severe disease, and in the case of COPD it is related to remodeling of the chest wall leading to the so-called barrel chest. The increase in TLC may have another more important role; that is the preservation of forced vital capacity (FVC) in face of a rise in RV that would then result in more pronounced falls in FEV_1 if the rise in TLC had not ameliorated the fall in FVC [4, 7]. The increase in TLC in parallel with each increase in RV can occur until the chest wall reaches some structural "limit" which probably varies from person to person [7, 10] and acutely

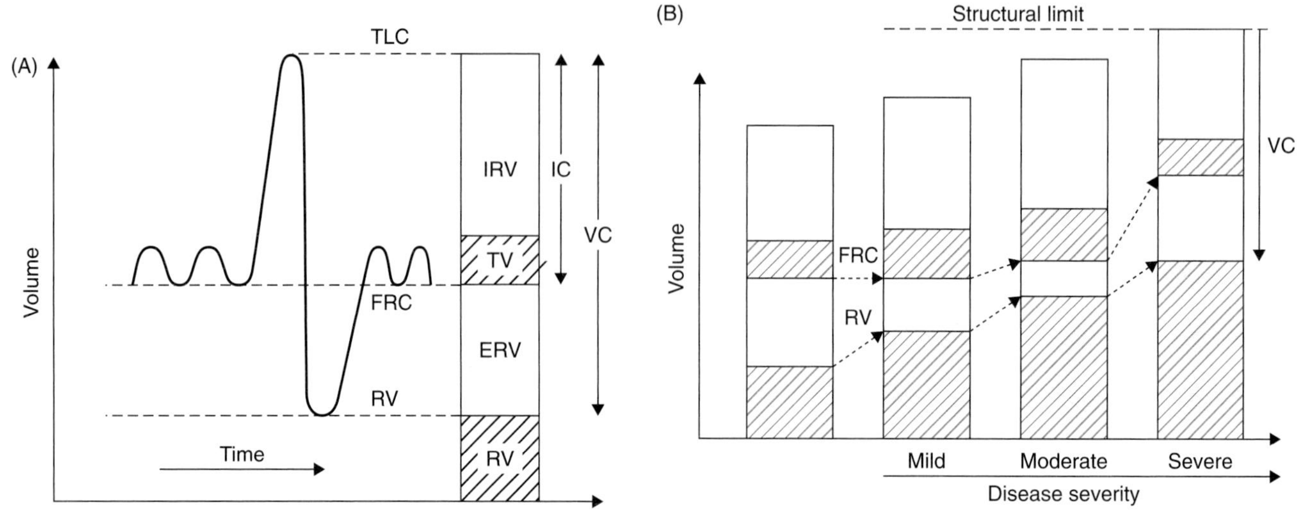

FIG. 5.1 (A) The figure presents a trace of volume displacement as a function of time during quiet breathing followed by a maximal inspiration to TLC, the subject then exhales the VC until no more air exits the lung. The volume of gas remaining in the lung at this point is RV. FRC is the volume left in the lung at the end of each normal breath. Expiratory reserve (ERV) and inspiratory reserve volume (IRV) are as noted. The inspiratory capacity (IC) is the volume of air inspired from FRC to TLC and is sometimes used as a surrogate for TLC. The three boundary conditions TLC, RV, FRC provide the most useful information as to the functioning of the lung (see text). (B) The figure present the changes in lung volumes and capacities that occur with increased airway disease severity as during an asthma attack or during the progression of COPD. The first (and last) change is usually an increase in RV (mild) as airway closure increases the RV continues to rise but FVC does not fall if the TLC rises in concert [4, 7]. However, at some point in the disease process (severe) or depending on the flexibility of the chest wall a structural "limit" is reached where the TLC cannot increase further. At this point the further rise in RV results in a fall in VC, FVC, and FEV_1 (see text).

TABLE 5.1 Boundaries of lung volume and physical determinants.

Volume or capacity	TLC (%)	Determinants
TLC: total lung capacity		• Respiratory muscle function • Structural limits of chest wall distortion • Compliance of chest wall and lung
FRC: function residual capacity	50	• Compliance of chest wall and lung • Expiratory time constants • Respiratory muscle activity
RV: residual volume	20–30	• Airway closure • Adequate expiratory effort • Children (<18 years): chest wall compliance

Note: TLC (%): total lung capacity for normal persons.

during an exacerbation of disease (Fig. 5.1B). The cause of the increase in TLC is unknown but the possibilities include either a loss of an inhibitory reflex that activates the muscles of the chest wall as TLC is reduced or simple structural overdistension [10–12]. For patients with COPD it would appear that the increase in TLC may also be due to remodeling of the chest wall and is commonly observed in X-ray or CT images. Assessing changes in FRC is of particular importance because resting lung volume has such an important bearing on airway caliber and airway smooth muscle (ASM) contractility [13–19]. Determination of FRC provides a reasonable alternative to the placement of an esophageal balloon to measure the balance of elastic forces of the respiratory system (see below). Increases in FRC observed in airways disease may represent a loss of recoil of the lung as occurs in emphysema or an increase in outward recoil of the chest wall. In the case of the latter, the inspiratory respiratory muscle in effect reset the equilibrium point of the respiratory system to a higher volume such as occurs during an asthma attack [6, 20]. For patients with asthma or COPD, the FRC increases by several mechanisms including persistent activity of the respiratory muscles [20], increased expiratory time constants [20], or alterations in recoil. Severe increases in FRC occur in patients with more severe forms of COPD because all these mechanisms occur in concert (Fig. 5.2).

Elastic recoil

Pressure–volume (PV) relationships are used to quantitatively display the elastic recoil of the components of the respiratory system: the chest wall and the lungs. Elastic recoil is very important in a number of respects. First, the strength of the chest wall muscle when coupled to the elastic recoil of the chest wall and lungs serve to determine the static lung volumes, specifically the maximal excursion of the chest wall as assessed by TLC and RV [6]. Second, the elastic recoil as influenced by changes in lung volume, alters intrapulmonary airway caliber [13–15, 21, 22]. Thirdly, elastic recoil is the most important load to smooth muscle contraction as changes in lung volumes have a profound effect on ASM performance [18, 22]. Lastly, elastic recoil is the major driving force for effort-independent maximal airflow, influencing peak flow and in particular FEV_1 [23, 24].

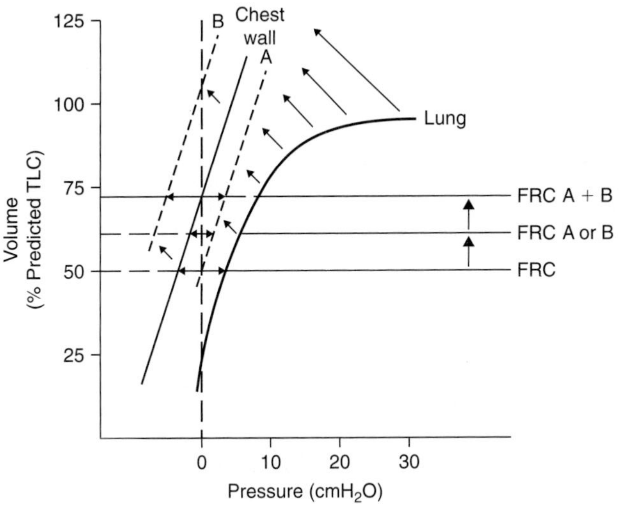

FIG. 5.2 Pressure–volume (PV) relationships in a normal subject (heavy solid lines), asthmatic patient (chestwall shifts to B) and in a patient with emphysema (where the lung PV shift to A and chest wall to B: A + B). The heavy solid line to the left is the PV curve of the chest wall showing outward recoil up until 66% of TLC and the line to the right is the PV curve of the lung. For the normal person the equilibrium point (FRC) where the outward recoil is equal but opposite to the inward recoil occurs at about 50% of the TLC. However, for the patient with loss of recoil of the lung (dashed line A) or chest (dashed line B) FRC rises because the inward recoil of the lung is less or the chest is more and a new equilibrium (FRC A or B) is reached. However, when both the chest wall (B) and lung PV (A) relationship change (both dashed lines) the FRC is increased further (FRC A + B).

Elastic recoil is diminished in all patients with emphysema whereas patients with asthma or the bronchitic form of COPD exhibit, on average, more normal recoil [25–27]. A caveat is that the loss of static elastic recoil in asthmatic patients has been reported by several workers [27–29]. In asthma the loss in elastic recoil can occur acutely [30], and can be reversed with bronchodilator treatment [28, 31]. Unlike emphysema where the loss of elastic recoil is thought to be due to microscopic destruction of parenchyma [32], these more acute losses in elastic recoil in some patients with asthma may relate to alterations in surface tension due to extravasation of plasma proteins from the inflammation or as a result of airway luminal fibrin formation within the airway lumen [33].

Determination of FRC coupled with an assessment of the chest wall (e.g. visual inspection) allows one to assess the likely position of the lung PV relationship without the issues associated with placement of an esophageal balloon. For example, finding a low FRC in a patient that was not overweight or obese would indicate a restrictive process; whereas an increase in measured FRC in a patient with normal appearing chest wall is suggestive of either persistent activity of the inspiratory muscles (especially if the patient has labored breathing) or a loss of inward elastic recoil of the lung or chest wall [6, 8].

Elastic recoil has a profound effect on airway caliber. As lung volume increases there is outward pull on intrapulmonary airways since these airways are embedded in the alveolar matrix of the parenchyma [34]. Hence, as lung volume increases or decreases the airway caliber increases or decreases [13–15]. Since, the contraction of ASM is dependent upon muscle fiber length and constitutes the load against which ASM must contract, lung volume sets the starting load. The observation explains why at low lung volumes the response to an inhaled bronchoconstricting agent is enhanced [16, 18, 19]. Hyperinflation diminishes luminal narrowing caused by ASM activation as shown by animal [16, 17] or human studies [18, 19].

DETERMINANTS OF AIRWAY CALIBER

The size of an intrapulmonary airway lumen at any given lung volume (e.g. FRC) is generally thought to depend on: (1) contents of the lumen (such as mucous plugs or liquid accumulation), (2) the structure of the airway both internal and external to the smooth muscle, (3) activation state or bronchomotor tone of the airways smooth muscle, (4) compliance or stiffness of the airway wall, and (5) tethering forces of the attached alveolar wall or lung volume. Transmural pressure (P_{tm}), which is the pressure inside minus pressure outside of the airway, represents the summation of forces working to open or close the airway and is resisted by the structure (or compliance) of the airway wall. In both asthma and COPD there is a considerable change referred to as remodeling of these structural components of the airway wall (Fig. 5.3).

Contents of the lumen

The biological processes that lead to obstructive lung disease leave the airway lumen filled with liquid, cells, and mucus; this accumulation is even more marked during acute exacerbations [35]. Indeed, this pathologic feature may be the most physiologically significant as obstruction of the airway causes much of the mismatch of ventilation to perfusion and in turn results in hypoxemia (see below). Moreover removal of such an obstruction is anything but straightforward as the contents change from being serous to more mucoid in nature. Acutely lumenal obstructions appear in the peripheral branches of the dependent portions of the tracheal–bronchial tree [36, 37]. These pathological changes in the distal lung are associated with 10-fold increase in peripheral resistance [38, 39] that increases further with disease severity [40–42]. Yet even with this 10-fold increase in peripheral resistance in the mild asthmatic, spirometry can still be within normal limits.

Structure of the airway wall

Inflammation of the airway wall characterizes all forms of obstructive airway disease. Structural alterations include increased extracellular deposits of the extracellular matrix that include fibronectin, collagen, vitronectin, and a host of other components [43]. The apparent thickening of the basement membrane below the airway epithelium is due

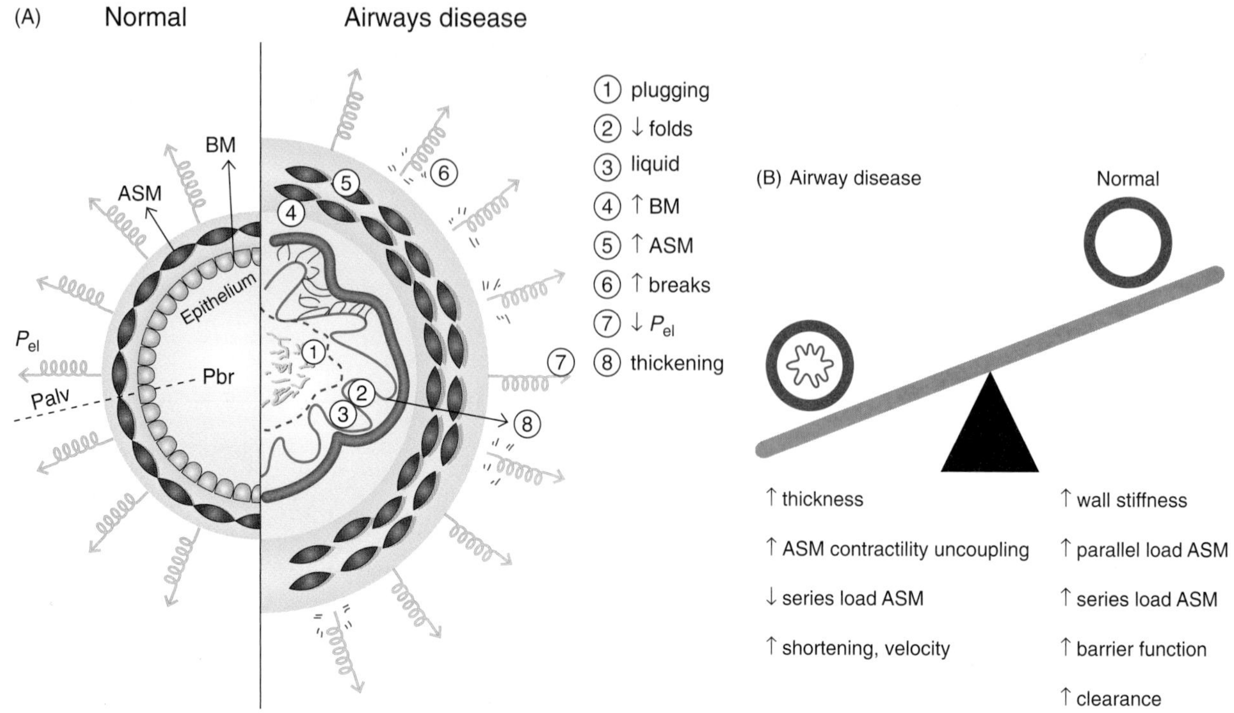

FIG. 5.3 (A) A schematic representation of the cross-section of an airway in a normal person (left-hand side of A) and a patient with obstructive airways disease (right-hand side of A). The changes that occur during airway remodeling in obstructive lung disease that could either cause airway narrowing and airway hyperresponsiveness or prevent airway hyperresponsiveness are presented. In asthma and COPD several factors are operational but the relationship of these structural changes to lung function is anything but clear. The mechanisms that cause or prevent airway narrowing are several and include: (1) Debris plugging of airway lumen, (2) decreased number of folds, (3) liquid in luminal folds of epithelium, (4) thickness of the basement membrane, (5) increased in contractility and/or size of ASM, (6) breaks in the alveolar tethers, (7) loss of elastic recoil, and (8) thickening of airway wall. P_{el}: elastic recoil forces; Palv: alveolar pressure; Pbr: pressure in the lumen of the bronchiole; Transmural pressure: the difference in Pbr Palv; ASM: airway smooth muscle. (B) Balance of forces favoring airway narrowing or airway opening.

to type I and III collagen deposition and is observed even in mild disease [44]. ASM is also increased both in volume (hypertrophy) and in number (hyperplasia) [43]. The functional consequence of remodeling of the airway is uncertain; indeed, there is little proof that there is a functional consequence. Arguments have been made that airway remodeling can both promote and protect the size of airway lumen [45] (Fig. 5.3B).

Activation state of the ASM

Bronchomotor tone has two consequences: first, depending on the compliance of the airway wall, smooth muscle tone narrows airways and reduces airflow but, secondly, smooth muscle tone also stiffens the airway wall as it contracts. A decrease of airway wall compliance would serve to prevent further collapse [45, 46]. This is best appreciated when ASM is activated at low lung volumes. Several studies have shown that when methacholine is inhaled in the supine posture, compared to the upright posture, more severe airway narrowing occurs [18, 19] and there is a marked increase in the maximal level of airway resistance. However, a decrease in bronchomotor tone will increase airway wall compliance and lead to dynamic airway narrowing during maximal expiratory flow [31, 47].

Compliance of the airway wall

The structural components of the airway wall represent the load to smooth muscle shortening, thus an increase in compliance would amplify ASM contraction and a decrease in compliance would be expected to antagonize ASM shortening. Cartilage, an important structural element of the airway wall, would serve to protect the airway lumen and has been shown to soften with disease [43, 48], an effect thought to be due to inflammatory mediators and the release of metalloprotineases. Folding or buckling of the mucosa has been speculated to either prevent airway narrowing or enhance airway luminal narrowing by changing wall compliance [46, 49]. In this situation as the wall is compressed by ASM contraction, the layer internal to the ASM folds or buckles up. The folds reinforce each other which serve to prevent further narrowing with an apparent increased load to the ASM. However, if the folds are few in number [49] and/or interstices of the folds become filled with liquid because of the small radius of curvature of an individual fold and the high surface tension found in the presence of inadequate surfactant [50], the airway lumen area would be narrowed or even obstructed.

Direct *in vivo* measurements of airway wall compliance yield conflicting results. Measurements from autopsy specimens of patients with COPD showed reduced compliance only in very small (<1 mm) airways [51]. Normal lungs show age-related decrease in wall compliance which complicates the interpretation of the losses observed in patients with COPD who are normally older [52, 53]. In asthma both increases [53, 54] and decreases [21] in wall stiffness have been reported. Decreased wall compliance or stiffening is likely the result of inflammation and/or wall thickening and hence would be related to asthma severity and disease status. Another potential mechanism of airway narrowing from the increase in wall compliance results from a decrease in forces tethering the airway open. These forces are decreased through two mechanisms. First, there are few alveolar attachments to the outer mucosa of the airway wall and second scar tissue has less tensile strength than normal tissue [55].

Tethering forces or the coupling of lung volume and resistance

The attachments of the alveolar wall to the external wall of the intrapulmonary airway are the anatomical link between the alveoli of the parenchyma and airway. These attachments form the physical linkage system that result in airways widening and the fall in airways resistance as lung volumes increase. This tethering force is the single most important factor stabilizing and influencing airway lumen integrity [13–15]. This fact is evidenced by the profound effect lung volume has on airflow. As noted above, for a given degree of smooth muscle activation greater airway narrowing occurs at low versus high lung volumes. For example, the simple act of assuming the supine position, which is not associated with a change in smooth muscle activation, markedly increases airway resistance [15, 18]. The effects of lung volume on airways resistance are very sensitive to even small losses (~1 cmH_2O) in lung elastic recoil pressure [16, 17, 35]. This process is not limited to changes within the parameter of the airway as denoted by ASM, since changes external to the ASM can also uncouple the lung volume–airway resistance relationship.

Changing from the upright to the supine position leads to airway hyperresponsiveness (AHR) in asthmatics but such changes also occur in normals [18, 56]. The mechanism for this phenomenon is most likely the loss of the tethering forces, as estimated by P_{el}. These forces pull open the airway wall as lung volume increases, thus a loss of outward recoil tethering the airway open is termed *uncoupling*. Airway resistance and lung volume uncoupling occurs rapidly with sleep onset in the patient with nocturnal asthma and further uncoupling during the night probably as a result of inflammatory events, resultant airway wall thickening, and mechanical uncoupling [21]. In COPD the loss of elastic recoil (P_{el}) is both greater and more persistent than the age-dependent loss of recoil that occurs in normal persons, and when considered in regards to loss of P_{el} with aging, accounts for the progressive loss of FEV_1 over the years.

The loss of P_{el}, together with breaks in alveolar attachments, results in the increased airway narrowing noted on forced expiration in patients with obstructive airway disease [47]. Smooth muscle contraction contributes to severe airway narrowing and loss of bronchodilation response typically seen in patients with emphysema [57, 58]. For patients with COPD or severe long-standing asthma this profound alteration in the function and structure of the airway accounts for the sustained and therapeutically resistant increase in airway resistance or decrease in FEV_1. The dilemma for the clinician is to ascertain what portion of the loss of structure and function is reversible and what portion is not.

AIRFLOW RESISTANCE

Chronic obstructive pulmonary disease

A common feature of the patient with COPD is airflow limitation; this is most often assessed by the fall in FEV_1. The current Global Initiative for Chronic Obstructive Lung Disease (GOLD) classification system for COPD uses post-bronchodilator FEV_1 as a major factor in classification of disease severity and in practice this variable makes up a significant part of the case definition for COPD [59]. Airflow limitation in COPD arises in the small (<2 mm diameter at FRC) airways which exhibit neutrophilic inflammation even in asymptomatic smokers [60, 61]. When patients have mild, asymptomatic disease there are subtle changes in airflow limitation [60] with changes in distribution and homogeneity of ventilation. As the severity of COPD increases there is increased involvement of more central airways but even when FEV_1 has fallen to 50–60% of predicted, more than half of the increased resistance can be attributed to narrowed intrapulmonary (<3 mm) airways [42]. Airway dimensions in COPD are reduced, even when the lungs are inflated to a standard inflation pressure (typically 30 cmH_2O) so one must conclude that the structural remodeling of collagen, mucous gland hyperplasia, general wall thickening, and loss of tethering elastic forces contributes to the airway narrowing [62].

These alterations in structure and function lead to profound ventilatory defects and a loss of collateral ventilation [63, 64]. In turn there is a trapping of gas due to both long expiratory time constants and check valve-like processes within the obstructed and emphysematous spaces. Re-establishment of collateral pathways with intrabronchial, one-way valves, much like the spiracles of insects, leads to deflation of these overextended spaces [65] and is currently being explored as a therapeutic option to the COPD patient to relieve this trapped gas [66].

In patients with severe COPD expiratory airflow limitation is observed at all lung volumes and often occurs even when there is a minimal expiratory effort; in these patients expiratory airflow limitation occurs on expiration with each breath [67]. In more mild cases, the airflow limitation is observed only on forced expiration because: (1) flow and volume during eupnea do not reach their structural limits due to the reserve in the respiratory system and (2) dynamic hyperinflation, which increases elastic recoil by increasing lung volume, further prevents reaching these limits during eupnea. The pathogenesis of airflow limitation in patents

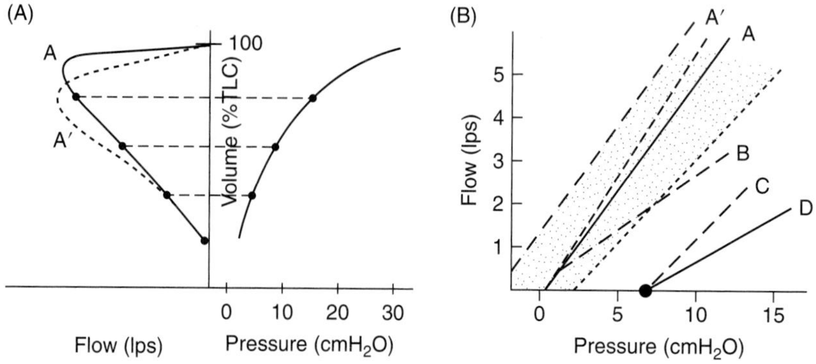

FIG. 5.4 (A) The role of the loss of lung recoil and airway wall collapse in affecting a fall in maximal expiratory flow (FEV_1) can be assessed by constructing a MFSR curve [73, 74]. (B) The conductance (resistance) of airways downstream of the point (equal pressure point) of airflow limitation is determined as the slope of V/P_{el} relationship. This V/P_{el} relationship or MFSR curve comes from selecting points from the both flow–volume and pressure–volume relationships at points of isovolume (horizontal dashed lines in A). The dotted line, flow–volume loop is from volume displacement of body plethysmograph whereas the solid line flow–volume loop is integrated volume from the mouth flow signal. The shaded area in (B) is from normal subjects [75]. MFSR A′ is the V/P_{el} without the effect of gas (airway wall compression) whereas MFSR A includes the effect of compression; the resistance is increased. MFSR B is a patient with increased resistance (e.g. airway narrowing) of the airways. MFSR C is a patient with emphysema where the sole abnormality is a loss of recoil, airway wall collapse and lower flow rates. MFSR D is a patient with increased resistance, loss of recoil, and increase pressure (volume) at which the airways close due to collapse and/or airway dynamic compression shifting the V/P_{el} relationship to the right.

with COPD or more severe forms of bronchitis and asthma accordingly is quite complex and is a reflection of the interplay between airway caliber, respiratory muscle function, lung volume, elastic recoil, and airway wall compliance [6, 14, 23, 24, 27, 62, 68].

The severity of COPD is defined as the post-bronchodilator FEV_1 [59], so it is useful to consider what a fall in FEV_1 signifies. The forced expiratory volume in 1 s – the FEV_1 – is a measurement taken from the volume–time relationship, known as a spirogram, during a forced exhalation from TLC. It is important to remember that the FEV_1 is a volume (liters) not a measure of airflow (liters per second) but it reflects the average airflow over the first second of a forced exhalation. A decrease in the FEV_1, as a percentage of its predicted value, indicates that the patient is unable to exhale a volume of air in 1 s similar in magnitude to that which could be exhaled by a person of the same sex, age, and height without lung disease. A considerable number of factors influence the magnitude of the FEV_1, but only one of these factors is airway luminal size. The magnitude of the FEV_1 is very much influenced by the size of the inspiration prior to beginning the forced expiration (TLC), elastic recoil, airway caliber, airway wall compliance, and size of the lung that is open. In the case of the latter as lung volume is lost by airway closure, filling or destruction, the VC, FVC, and hence FEV_1 decreases. It is important to realize that FEV_1 is a polyvalent index; where polyvalent indices are ones in which many factors contribute to the exact value of the index [69]; indeed most measures of lung function are polyvalent. Accordingly it is unclear what a low FEV_1 might signify, the cause of the decrease or even the best treatment needed to reverse the loss in FEV_1.

Empirically the FEV_1 is a very useful measure of overall lung function since it relates to morbidity and mortality of COPD, even death due to all causes [70]. The techniques used to measure FEV_1 have been well codified [71, 72]. The general performance and in particular the variability and reproducibility of the FEV_1 as an outcome measure is robust and for this reason the FEV_1 is a principal outcome variable for clinical research. Unfortunately a low FEV_1 is not specific for any single cause of airflow limitation, and other assessments of the loss in airflow limitation must be made.

Useful insight into the causes of airflow limitation can be gained by the construction of isovolume pressure–flow curves or maximum flow static recoil (MFSR) relationships which are determined from flow–volume and PV relationships (Fig. 5.4). To assess artifacts from compression of the gas in thorax, lung volume is measured from the volume displacement of the chest wall and is obtained by having the patient sit in a variable-volume, body plethysmograph. To measure static recoil an esophageal balloon attached to a pressure transducer is positioned in the lower esophagus to estimate pleural pressure. The subject then inhales to TLC and expires slowly to construct a deflation PV curve (Fig. 5.4A) and then the subject performs a maximal flow–volume curve. By taking a series of data pairs of flow and pressure from points of isovolume on the PV and flow–volume relationship, the MFSR curve can be constructed (Fig. 5.4B).

The slope ($\Delta V/\Delta P$) of the MFSR curve is resistance (or its inverse, conductance) of the segment of lung upstream of the point and is known as the equal pressure point, that is the point along the airway which P_{tm} is zero. Because there is still elastic recoil of the lung at RV, the MFSR relationship does not intersect with zero pressure. Patients with airflow limitation will exhibit different MFSR curves depending upon the cause of the airflow limitation. For patients where airflow limitation is solely the result of loss of static elastic recoil, the slope of the curve is normal; that is upstream resistance is normal and the loss of airflow is due solely to the loss of static elastic recoil. However, in a patient with airway narrowing due to inflammation, collagen deposition and/or collapse, the slope is decreased, that is resistance is elevated. By comparing flow–volume curves constructed from measuring both lung volumes by integrating the flow at the mouth to lung volume measured by the chest wall, the role of airway wall compression on the loss of airflow can be assessed [76]. The flow–volume curve derived from the flow and volume exiting the lung is displaced such that peak flow occurs at a higher volume but, after this initial peak, flow is lower at any given volume due to dynamic airway wall compression. In essence this analysis assesses the delay in gas exiting the lung; the more dynamic compression, the more temporal delay.

In most patients with COPD, a combination of factors contributes to airflow limitation [73–76]. Determining MSFR relationships can suggest which component might be amenable to treatment; bronchodilation for airway narrowing, pursed lip breathing for airway collapse and lung volume reduction surgery for improving static elastic recoil. In the case of the latter several studies using this analysis have shown that the effect of lung reduction surgery is a reduction of RV and the resultant increase in the VC [77] or a more complex response where both a change in volume and a change in resistance occurred [78].

BRONCHIAL RESPONSIVENESS

Bronchial responsiveness includes either the acute response to a one-time treatment with a bronchodilator – usually a β_2-receptor antagonist – or the response to a constrictor bronchial challenge. Determination of bronchial responsiveness better defines and classifies patients with airways disease. Bronchial responsiveness assessments are often used as major outcome variables in clinical investigations and treatment trials because of the precision of the measure, its association to underlying inflammation, the relevance to asthma triggers, and the changes in bronchial responsiveness as the result of treatment.

Bronchodilator responses

Bronchodilator responsiveness to inhaled bronchodilators is among the factors used to differentiate between whether a given patient is to be categorized as COPD or asthma. Professional Society guidelines and position statements for interpretation of spirometric results [72, 79] discuss the various consensus criteria used to derive the currently used bronchodilator criterion that identifies a significant response as one where a 12% or greater change in the post-bronchodilator FEV_1 or FVC from pre-bronchodilator values and a >200 ml absolute change occurs (Table 5.2).

Establishing a criterion that indicates a significant bronchodilator response has been controversial because of uncertainty as to just what constitutes a significant response [79, 80]. Various authors have suggested criteria that range from 10% to 15% change from baseline, some suggest using a change in percent of predicted or an exact volume change (e.g. 400 ml). Bronchodilator responses are reported as a percent change from the volumes obtained prior to use of a

TABLE 5.2 Criteria recommended for acute bronchodilator responses in adults [79, 81].

	FEV_1 (%)	FVC (%)	Comments
ACCP	15–25	15–25	% of baseline in 2 of 3 tests
Intermountain ATS	15	15	% of baseline
ATS 1991	12	12	% of baseline and 200 ml
ATS/ERS 2000	12	12	% of baseline and 200 ml

ACCP: American College of Chest Physicians; ATS: American Thoracic Society [79]; ERS: European Respiratory Society [81].

β-agonist; however, this enhances the response if the starting baseline of FEV_1 or FVC is low. The use of a definition that includes both an absolute change (e.g. >200 ml) or change expressed as a percent of predicted addresses this dilemma.

There are several approaches used to deliver bronchodilators to patients for assessing reversibility but two to four puffs of albuterol with or without a spacer is the most common. However, combinations of bronchodilators with or without an anticholinergic agent or the use of nebulizer treatments are sometimes used. In this case one administers the bronchodilator until side effects are observed or the total dose delivered is used [71, 72, 79–81]. A wait period is then imposed to allow for the drug to exhibit its physiological effects, for example 15–20 min in the case of albuterol. Unfortunately techniques to determine bronchodilator responsiveness are highly variable between studies and laboratories making direct comparison of study results difficult.

Bronchodilator responsiveness is commonly measured as a change in the FEV_1 but can also be assessed by expiratory flow–volume loops, specific airway conductance as determined with a body plethysmograph, or with the forced oscillation technique. Each technique measures different aspects of lung function and the result obtained may not be exactly equivalent. Moreover the pattern of the bronchodilator response can vary with some patients having a more central airways response (SGaw), whereas others have a more peripheral lung response as measured with FVC or the FEV_1 [4, 81, 82]. Such findings are of great interest in determining the inferred site of the functional defect such as central airways versus peripheral airways and have obvious ramifications to drug delivery.

Bronchoconstriction responses

AHR is either a bronchoconstricting response (fall in FEV_1) to a stimulus that normally does not cause a constriction or is a heightened bronchoconstriction to the stimulus. Examples of the former are exercise, cold air, or bradykinin where examples of the latter are histamine and methacholine. Airways hyperresponsiveness is often used to define the diagnosis of asthma, however, the presence of AHR is not specific since many lung diseases are associated with hyperresponsiveness including COPD [83].

Over the last 50 years the proposed mechanisms to explain AHR included anything from enhanced cholinergic tone of the airways to the current focus on pathologic alterations (remodeling) of the airway wall. While airways smooth muscle usually has a central role in most theories, recent experimental evidence suggests that hypercontractility of the ASM is not always required [84]. Bronchial thermoplasty in which radio frequency energy is applied to airways through the bronchoscope to abate the ASM, improves asthma symptoms with only modest improvements in AHR [85]. This interesting result suggests that the ASM may not be as essential to AHR as previously thought or may serve other purposes [86].

AHR is functionally defined as a shift of the dose–response relationship to an inhaled bronchoconstrictor agonist, for example, methacholine, such that lower amount of the agent yields equivalent biological response, that is a "leftward shift" of the dose–response curve [83]. The inhaled dose of bronchial constricting agent is plotted against the fall in lung function and from this relationship the PC_{20} is then measured. The PC_{20} is the interpolated dose of inhaled agonist that causes FEV_1 to fall exactly 20%. The PC_{20}, decreases as asthma severity increases [83, 87]. The relationship between the amount of inhaled bronchoconstrictor agent and FEV_1 often has a plateau in which further doses of drug illicit smaller and smaller physiological responses. In the case of severe asthmatics often a plateau of response cannot be demonstrated [88–91]. This loss of maximal response may signify the uncoupling, loss of protective mechanisms with resultant severe bronchoconstriction [90]. In clinical, and most research practice, dose–response relationships are usually not carried out until a plateau is revealed thus the PC_{20} only assesses the position (sensitivity) of the airways responsiveness relationship (Fig. 5.5).

Although there is considerable variability among subjects, AHR is considered present when the PC_{20}, using methacholine chloride as the bronchoconstrictor agonist, is less than 8 mg/ml [83, 87, 90]. Asthmatics will often demonstrate a reduced PC_{20} even though the FEV_1 is normal at the onset of the test; moreover, the PC_{20} in most studies does not correlate with the FEV_1 [88, 91]. As methacholine is well known to stimulate serous secretions from airway glands and cells, it seems reasonable to suggest the AHR of patients with mild to moderate asthma could be viewed as a measure of increased tendency for airway closure to occur.

In contrast among patients with COPD there is a correlation of FEV_1 to PC_{20} which suggests more of a structural mechanism for AHR consistent with the observed airway wall thickening [44, 91–93]. Moreover patients with α-1-antitrypsin deficiency demonstrated an elevated response in FEV_1 in concert with the loss of elastic recoil [94]. The mechanism of the fall in PC_{20} in COPD also seems unlikely to be the same as asthma since inhaled corticosteroids do not lead to an improvement in AHR for COPD patients [95] whereas inhaled steroids do improve AHR in asthmatic patients [96]. AHR has important prognostic value because as AHR increases (PC_{20} decreases) there is accelerated decline in FEV_1 in patients with COPD [97]. Taken together the data provide strong evidence that the physiological mechanisms that lead to AHR in COPD and asthma are probably quite different.

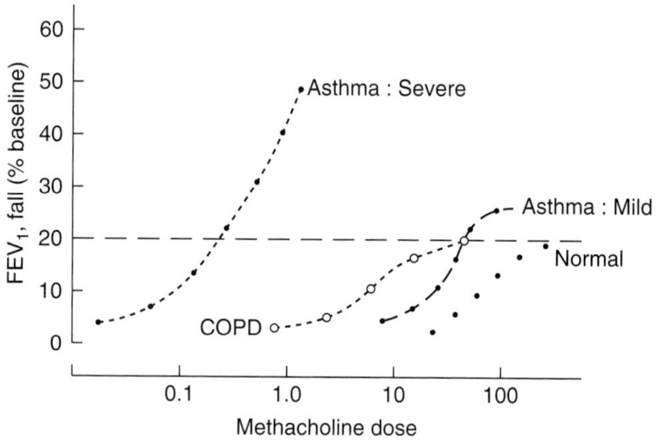

FIG. 5.5 Dose response curves for the dose of inhaled methacholine (mg/ml) versus FEV_1 expressed as a percent decrease from the starting baseline. Data points are individual values up until there is a plateau in the response or 50% fall in FEV_1 has occurred. The dashed line is the PC_{20} that is interpolated. Both mild asthma and COPD patients exhibit leftward (hyperresponsiveness) shifts of the curve without increase in the maximal response plateau. The patients with severe asthma show marked shift to the left without a plateau. Modified and redrawn from Ref. [91].

Lung volume also influences AHR based on recent studies on the effects of deep inspiration (DI). Large TLC breaths are known to be effective in some but not all asthmatics [98]. In some patients DI can even cause bronchoconstriction and such a response is associated with the degree of airway inflammation [99]. A DI can also provide protection to a subsequent bronchoconstricting response when the DI is performed prior to the inhalation of methacholine [100]. Loss of this bronchoprotective effect of DI occurs in asthmatic patients with mild disease whereas reversibility of an induced constriction by a DI is only lost as asthma becomes more severe [101, 102]. In patents with COPD the effects of DI are lost and this loss has been shown to be related to the loss of alveolar attachments [103].

GAS EXCHANGE

The respiratory system provides oxygen (O_2) and removes carbon dioxide (CO_2) from the body and failure of this critical role of gas exchange leads to CO_2 retention, hypercapnia, and/or hypoxemia. For ideal gas exchange to occur the ventilation must be matched with the vascular perfusion within the individual alveolar gas exchange units. The diffusion capacity (DL_{CO}) or transfer factor assesses the passive transfer of a test gas, carbon monoxide (CO), from alveolus to the blood. The DL_{CO} test is a frequently used measurement in the pulmonary function laboratory and is quite useful in categorizing the different types of airway disease.

Ventilation/perfusion relationship

The concept of ventilation to perfusion matching V_A/Q is commonly used in understanding how airway disease pathology affects gas exchange. There are four mechanisms used to explain gas exchange abnormalities and include: (1) alveolar hypoventilation, (2) impaired O_2/CO_2 diffusion, (3) shunt, and (4) V_A/Q mismatch, where the latter mechanism is by far and away the most common cause of gas exchange abnormalities in patients with obstructive lung disease.

Chronic obstructive pulmonary disease

The obstruction of airflow to specific regions of the lung parenchyma and widespread airway narrowing coupled with the well-described pulmonary vascular abnormalities leads to marked V_A/Q abnormalities and abnormal arterial blood gases. As respiratory failure ensues, alveolar hypoventilation further contributes to the observed hypercapnia and hypoxemia [104, 105]. Intrapulmonary shunt is not a major factor contributing to hypoxemia nor surprisingly is there a significant contribution of reduced gas diffusion probably because there is little ventilation of the areas that have emphysema. V_A/Q mismatching appears to fully explain the hypoxemia of patients with COPD [105, 106]. In the case of very mild disease in the peripheral airways as evidenced by FEV_1 of 80% predicted or greater, patients have an abnormal slope of phase III on the single breath N_2 test, a widened A-a O_2 gradient and mild V_A/Q mismatch but with a normal PaO_2 [107]. During an exacerbation of COPD the V_A/Q mismatch increases and accounts for about half of the observed hypoxemia whereas the remainder is low mixed venous O_2 due to increased O_2 consumption of the body presumably due to metabolic requirements of the disease [105, 107]. The A-a O_2 gradient is correlated to emphysema score [108]; hence, the abnormal gas exchange of COPD is explained by the pathology of the disease.

Asthma

Early studies with the multiple inert gas elimination technique (MIGET) in stable, mild asthmatics showed a bimodal distribution of pulmonary blood flow with about 25% of the blood flow going to units with low (<0.1) V_A/Q ratios. There was no evidence of areas of shunt or areas of high V_A/Q or increased dead space [109]. However, later studies showed less of this bimodal distribution and only some of the patients showed this bimodal pattern [110, 111]; whereas most patients showed widening of the distribution of blood flow. Patients in this latter study were receiving treatment with inhaled glucocorticosteroids which could explain the differences between studies. The gas exchange abnormalities of patients with mild disease were thought to be due to abnormalities in the periphery and consistent with abnormal peripheral lung function such as in RV or frequency dependence of compliance. Asthma severity is associated with widening of the V_A/Q relationship [112] but most patients still exhibit a near normal PaO_2 until they reach the point of seeking medical assistance (Fig. 5.6) [113].

Given that measurements of lung mechanisms show marked increase in peripheral resistance due to airway closure, one might expect to observe considerable shunting as blood flows by these closed airways and unventilated exchange units yet this is not the case [109–112]. The apparent explanation would include, preserved uploading of oxygen on the hemoglobin due to more than adequate capillary transit time, hypoxic vasoconstriction, redirecting blood away from affected areas, and adequate collateral ventilation thus bypassing the closed airway.

During an asthma exacerbation or "status asthmaticus" the presence of a bimodal pattern emerges but even in this situation little pure shunt is observed. Treatment of the asthma attack returns this pattern to normal [112] even at time of discharge from the emergency department [114]. The relative preservation of arterial blood gases up until respiratory failure is attributed to the mechanisms covered above and the high ventilatory and cardiac output that occurs during an attack. Bronchial challenge produces similar patterns but it has been noted that the disturbances in gas exchange lag behind the recovery of airflow rates [115, 116]. This pattern of recovery does not occur with leukotriene challenge, presumably because LTD_4 causes a more selective central airway constriction [117]. However, following exercise-induced bronchospasm airflow normalization occurs after the return of V_A/Q to normal, consistent with different mechanisms of action in contrast to a naturally occurring asthma exacerbation [118]. Taken together these findings suggest that our understanding of gas exchange in asthma, unlike COPD, is unclear.

Therapy has a marked effect on gas exchange. Paradoxical hypoxemia has long been noted as a result of acute β-agonist administration [119] and has been attributed to effects of β-agonists on pulmonary blood flow. V_A/Q measurement made after isoproterenol [110] showed transient (5 min) alterations in blood flow to areas low in V_A/Q; however, studies with a more selective β-agonist, salbuterol were without this effect [112, 120].

Diffusion capacity (transfer factor)

The diffusion capacity, DL_{CO}, is a measure of the disappearance of CO, from the alveolus over a 10 s breath hold and is taken as an index of the total area of potential gas exchange surface area. Because CO combines at the same site as oxygen on the hemoglobin molecule, blood flow, hematocrit, and cardiac output significantly influence the magnitude of DL_{CO}. The addition of helium to the test gas mixture inhaled allows for the calculation of the starting concentration of CO and the determination of the communicating gas volume by inert gas dilution. The standard single breath DL_{CO} yields two useful measurements; the DL_{CO} and a measurement of TLC (VA) [121].

Asthma & DL_{CO}

The DL_{CO} in patients with asthma is either within normal limits or high depending on several factors. In asthmatic patients with preserved lung function, the DL_{CO} is typically normal [122]. In moderate to severe asthma the DL_{CO} is usually elevated and will also increase with bronchodilator treatment. The high DL_{CO} values [123, 124] have been explained by hyperinflation, increased intrathoracic pressure, and a more likely cause, increases in pulmonary capillary

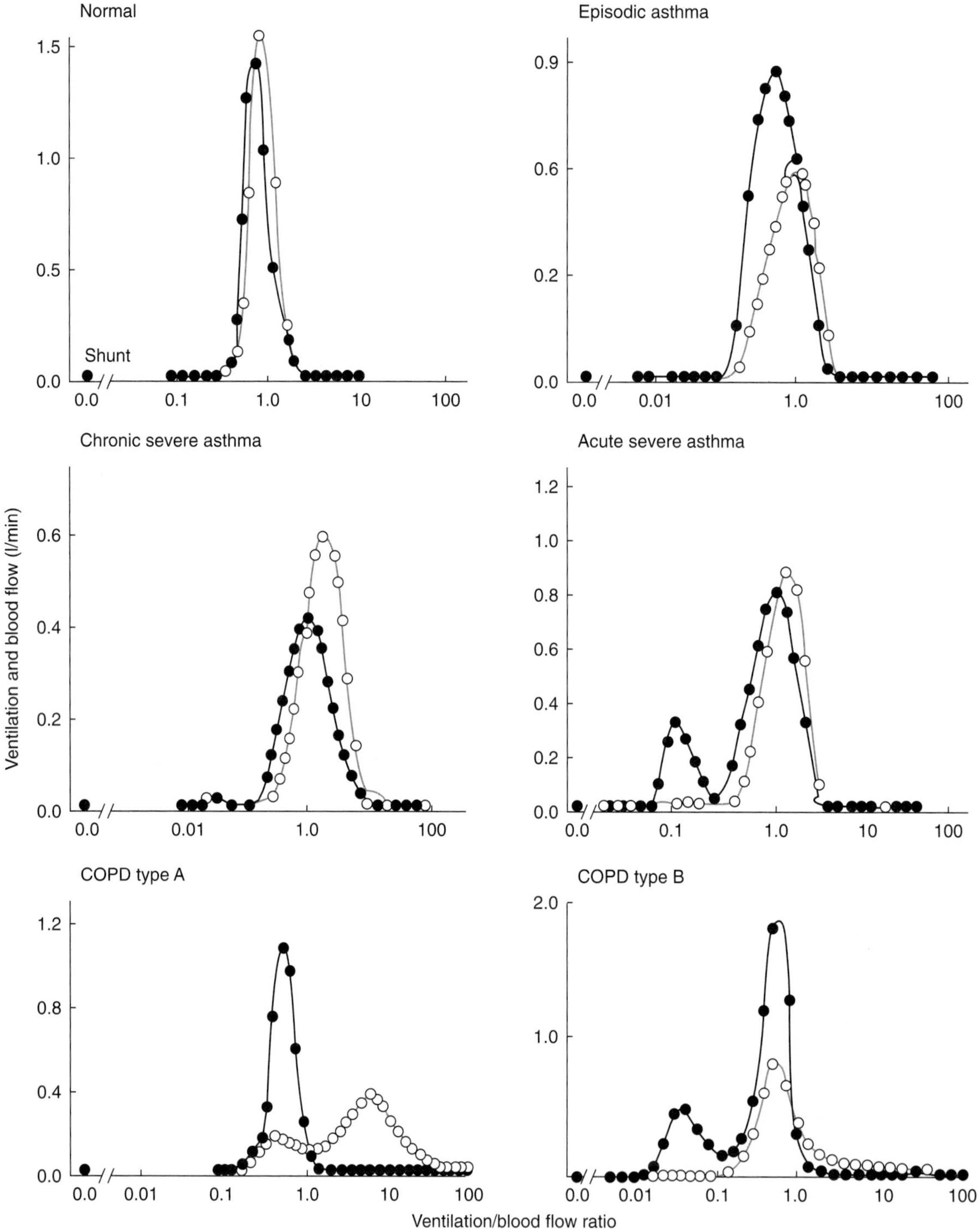

FIG. 5.6 The V_A/Q relationship as determined by the multiple inert gas elimination technique (MIGET). Note the different scaling for both axes in the various examples. The normal distribution of ventilation (open circles) and blood flow (solid circles) is over a narrow range of V_A/Q ratios and is unimodal. Patients with mild or chronic asthma have wider distributions but still exhibit unimodal distributions whereas acute severe asthma is bimodal. In emphysema (COPD, type A) there is a bimodal distribution in ventilation whereas chronic bronchitis (COPD, type B) exhibits bimodal blood flow. Shunt is not present in any of the examples. Reproduced from Refs [105, 111] with permission.

blood volume or extravasation of red blood cells into the alveolus. Although not frequently commented on or because current practice is to lyse the lavage sample, the red blood cell content is high in lavage samples of asthmatics, particularly those who are more severe [125]. Increased vascularization of the lung and airways might also increase the apparent pulmonary capillary blood volume and account for the elevated DL_{CO} [43].

Chronic obstructive pulmonary disease & DL_{CO}

In contrast to asthma patients with emphysema exhibit a reduction in DL_{CO} [121, 126]. This reduction in DL_{CO} has been shown to be closely related to evidence of pathologic emphysema [126] or emphysema as measured by CT [127, 128]. In asymptomatic smokers there is a detectable 10% decrease in DL_{CO} that reverses with smoking cessation likely due to reduced pulmonary capillary volume and CO back pressure [129, 130]. However, it should be noted that the correlation of the reduction in DL_{CO} to disease severity in the individual patients is poor, perhaps reflecting the heterogeneity of the pathology of a given patient. The DL_{CO} is typically within normal limits for COPD patients with primary chronic bronchitis. One study showed a significant relationship between the DL_{CO} and 3-year survival [131], although another study did not find this to be the case [132].

The measurement of alveolar volume (VA) provides additional insight into the physiological derangements present in the patient with airways obstruction. Because VA is a single breath dilution of an inspired gas, it is affected by both the presence of non-communicating lung volume, for example emphysematous bullae and the long-time constants caused by airways narrowing. In normal subjects, and when expressed as the volume of gas at body temperature, pressure and fully saturated, the measure of VA will agree (±200 ml) with TLC by body plethysmograph; however, patients with asthma or COPD will exhibit greater than 200 ml difference between TLC and VA [133]. The degree of difference between TLC and VA provides an approximate estimate of the total size of the non-communicating bullae and alveoli [126, 134] or gas maldistribution.

TABLE 5.3 Physiological features of asthma contrasted with COPD.

Physiological assessment	COPD	Asthma
Airflow limitation	Present	Sometimes present
Loss of elastic recoil	Frequently present	Sometimes present
Bronchodilator response	±	Usually present
AHR	+	+
Exercise-induced AHR	−	+
V_A/Q	Widened	Widened
DL_{CO}	↓*	WNL or ↑

Notes: AHR: airways hyperresponsiveness; V_A/Q: ventilation to perfusion ration; DL_{CO}: diffusion capacity of the lung.

*For emphysema, chronic bronchitis would be WNL.

TABLE 5.4 Pattern of changes in DL_{CO} and VA in patients with obstructive lung disease.

Clinical condition	FEV_1	VA	DL_{CO}
Emphysema	↓	↓	↓
Bronchitis	↓	↓	WNL
Asthma	↓	↓	WNL or ↑

Notes: DL_{CO}: diffusion capacity of carbon monoxide; FEV_1: forced expiratory volume in 1 s; WNL: within normal limit.

CONCLUSION

The clinical presentation of patients with obstructive lung disease often presents a confusing clinical picture (Table 5.3). Consider, for example, the patient who reports asthma-like symptoms to common triggers but who also smokes and has a greater than 20 pack per year smoking history. Such a patient might be expected to exhibit physiological features of both disorders. This is because the physiological mechanisms of asthma and COPD that differ in the details of the inflammatory processes, cells, mediators, and pathology may be very similar physiologically in terms of hyperinflation, loss of static recoil, and expiratory airflow limitation. Bronchodilator responsiveness is also not unique to asthma since patients with COPD often demonstrate reversibility of airflow limitation. Bronchial responsiveness is also present in both entities but the response to bronchoconstrictors is more related to lung function and structure in COPD and the patient with COPD generally does not respond to exercise by exhibiting bronchospasm. Gas exchange in general is similar between both diseases with increased V_A/Q mismatch but with little evidence of shunt or global hypoventilation. The DL_{CO} represents an important distinguishing feature between emphysema and other airways diseases and provides a means of discriminating the disease process when coupled with the measurement of spirometry and the response to a bronchodilator [135–137] (Table 5.4).

References

1. Bachert C, van Cauwenberge P, Khaltaev N. Allergic rhinitis and its impact on asthma. In collaboration with the world health organization. *Executive Summary of the Workshop Report*. December 7–10, 1999, Geneva, Switzerland. *Allergy* 57: 841–855, 2002.
2. Soriano JB, Davis KJ, Coleman B, Visick G, Mannino D, Pride NB. The proportional Venn diagram of obstructive lung disease: Two approximations from the United States and the United Kingdom. *Chest* 124: 474–81, 2003.
3. Viegi G, Matteelli G, Angino A, Scognamiglio A, Baldacci S, Soriano JB, Carrozzi L. The proportional Venn diagram of obstructive lung disease in the Italian general population. *Chest* 126: 1093–101, 2004.
4. Brown RH, Pearse DB, Pyrgos G, Liu MC, Togias A, Permutt S. The structural basis of airways hyperresponsiveness in asthma. *J Appl Physiol* 101: 30–39, 2006.
5. Kraft M, Cairns CB, Ellison MC, Pak J, Irvin C, Wenzel S. Improvements in distal lung function correlate with asthma symptoms after treatment with oral montelukast. *Chest* 130: 1726–32, 2006.
6. Irvin C, Cherniack R. Pathophysiology and physiologic assessment of the asthmatic patient. *Sem Resp Med* 8(3): 201–15, 1987.
7. Irvin CG. Lessons from structure-function studies in asthma: Myths and truths about what we teach. *J Appl Physiol* 101: 7–9, 2006.

8. Irvin C. Guide to the evaluation of pulmonary function. In: Hamid Q, Shannon J, Martin J, Hamilton J (eds). *Physiologic Basis of Respiratory Disease*. Ontario: BC Decker, 2005.
9. Wanger J, Clausen JL, Coates A, Pedersen OF, Brusasco V, Burgos F, Casaburi R, Crapo R, Enright P, van der Grinten CP *et al.* Standardisation of the measurement of lung volumes. *Eur Respir J* 26: 511–22, 2005.
10. Whittaker LA, Irvin CG. Going to extremes of lung volume. *J Appl Physiol* 102: 831–33, 2007.
11. Loring SH, O'Donnell CR, Butler JP, Lindholm P, Jacobson F, Ferrigno M. Transpulmonary pressures and lung mechanics with glossopharyngeal insufflation and exsufflation beyond normal lung volumes in competitive breath-hold divers. *J Appl Physiol* 102: 1–846, 2007.
12. Peress L, Sybrecht G, Macklem PT. The mechanism of increase in total lung capacity during acute asthma. *Am J Med* 61: 165–69, 1976.
13. Butler J, Caro CG, Alcala R, Dubois AB. Physiological factors affecting airway resistance in normal subjects and in patients with obstructive respiratory disease. *J Clin Invest* 39: 584–91, 1960.
14. Vincent NJ, Knudson R, Leith DE, Macklem PT, Mead J. Factors influencing pulmonary resistance. *J Appl Physiol* 29: 236–43, 1970.
15. Briscoe WA, Dubois AB. The relationship between airway resistance, airway conductance and lung volume in subjects of different age and body size. *J Clin Invest* 37: 1279–85, 1958.
16. Bates JH, Schuessler TF, Dolman C, Eidelman DH. Temporal dynamics of acute isovolume bronchoconstriction in the rat. *J Appl Physiol* 82: 55–62, 1997.
17. Nagase T, Martin JG, Ludwig MS. Comparative study of mechanical interdependence: Effect of lung volume on raw during induced constriction. *J Appl Physiol* 75: 2500–5, 1993.
18. Ding DJ, Martin JG, Macklem PT. Effects of lung volume on maximal methacholine-induced bronchoconstriction in normal humans. *J Appl Physiol* 62: 1324–30, 1987.
19. Meinero M, Coletta G, Dutto L, Milanese M, Nova G, Sciolla A, Pellegrino R, Brusasco V. Mechanical response to methacholine and deep inspiration in supine men. *J Appl Physiol* 102: 269–75, 2007.
20. Martin J, Powell E, Shore S, Emrich J, Engel LA. The role of respiratory muscles in the hyperinflation of bronchial asthma. *Am Rev Respir Dis* 121: 441–47, 1980.
21. Irvin CG, Pak J, Martin RJ. Airway-parenchyma uncoupling in nocturnal asthma. *Am J Respir Crit Care Med* 161: 50–56, 2000.
22. Irvin CG. Lung volume: A principle determinant of airway smooth muscle function. *Eur Respir J* 22: 3–5, 2003.
23. Mead J, Turner JM, Macklem PT, Little JB. Significance of the relationship between lung recoil and maximum expiratory flow. *J Appl Physiol* 22: 95–108, 1967.
24. Black LF, Hyatt RE, Stubbs SE. Mechanism of expiratory airflow limitation in chronic obstructive pulmonary disease associated with 1-antitrypsin deficiency. *Am Rev Respir Dis* 105: 891–99, 1972.
25. Gibson GJ, Pride NB, Davis J, Schroter RC. Exponential description of the static pressure–volume curve of normal and diseased lungs. *Am Rev Respir Dis* 120: 799–811, 1979.
26. Finucane KE, Colebatch HJ. Elastic behavior of the lung in patients with airway obstruction. *J Appl Physiol* 26: 330–38, 1969.
27. Schlueter DP, Immekus J, Stead WW. Relationship between maximal inspiratory pressure and total lung capacity (coefficient of retraction) in normal subjects and in patients with emphysema, asthma, and diffuse pulmonary infiltration. *Am Rev Respir Dis* 96: 656–65, 1967.
28. Gold WM, Kaufman HS, Wadd JA. Elastic recoil of the lungs in chronic asthmatic patients before and after therapy. *J Appl Physiol* 23: 433–38, 1967.
29. Woolcock AJ, Read J. The static elastic properties of the lungs in asthma. *Am Rev Respir Dis* 98: 788–94, 1968.
30. Mansell A, Dubrawsky C, Levison H, Bryan AC, Langer H, Collins-Williams C, Orange RP. Lung mechanics in antigen-induced asthma. *J Appl Physiol* 37: 297–301, 1974.
31. De Troyer A, Yernault JC, Rodenstein D. Influence of beta-2 agonist aerosols on pressure–volume characteristics of the lungs. *Am Rev Respir Dis* 118: 987–95, 1978.
32. Eidelman DH, Ghezzo H, Kim WD, Hyatt RE, Cosio MG. Pressure–volume curves in smokers. Comparison with alpha-1-antitrypsin deficiency. *Am Rev Respir Dis* 139: 1452–58, 1989.
33. Wagers SS, Norton RJ, Rinaldi LM, Bates JH, Sobel BE, Irvin CG. Extravascular fibrin, plasminogen activator, plasminogen activator inhibitors, and airway hyperresponsiveness. *J Clin Invest* 114: 104–11, 2004.
34. Mead J, Takishima T, Leith D. Stress distribution in lungs: A model of pulmonary elasticity. *J Appl Physiol* 28: 596–608, 1970.
35. Carroll NG, Mutavdzic S, James AL. Increased mast cells and neutrophils in submucosal mucous glands and mucus plugging in patients with asthma. *Thorax* 57: 677–82, 2002.
36. Lundblad LK, Thompson-Figueroa J, Allen GB, Rinaldi L, Norton RJ, Irvin CG, Bates JH. Airway hyperresponsiveness in allergically inflamed mice: The role of airway closure. *Am J Respir Crit Care Med* 175: 768–74, 2007.
37. Venegas JG, Winkler T, Musch G, Vidal Melo MF, Layfield D, Tgavalekos N, Fischman AJ, Callahan RJ, Bellani G, Harris RS. Self-organized patchiness in asthma as a prelude to catastrophic shifts. *Nature* 434: 777–82, 2005.
38. Wagner EM, Liu MC, Weinmann GG, Permutt S, Bleecker ER. Peripheral lung resistance in normal and asthmatic subjects. *Am Rev Respir Dis* 141: 584–88, 1990.
39. Kaminsky DA, Irvin CG, Gurka DA, Feldsien DC, Wagner EM, Liu MC, Wenzel SE. Peripheral airways responsiveness to cool, dry air in normal and asthmatic individuals. *Am J Respir Crit Care Med* 152: 1784–90, 1995.
40. Kraft M, Pak J, Martin RJ, Kaminsky D, Irvin CG. Distal lung dysfunction at night in nocturnal asthma. *Am J Respir Crit Care Med* 163: 1551–56, 2001.
41. Ohrui T, Yanai M, Sekizawa K, Morikawa M, Sasaki H, Takishima T. Effective site of bronchodilation by beta-adrenergic and anticholinergic agents in patients with chronic obstructive pulmonary disease: Direct measurement of intrabronchial pressure with a new catheter. *Am Rev Respir Dis* 146: 88–91, 1992.
42. Yanai M, Sekizawa K, Ohrui T, Sasaki H, Takishima T. Site of airway obstruction in pulmonary disease: Direct measurement of intrabronchial pressure. *J Appl Physiol* 72: 1016–23, 1992.
43. James AL, Wenzel S. Significance of airway remodeling in airway diseases. *Eur J Respir Dis* 30: 1420–41, 2007.
44. Roche WR, Beasley R, Williams JH, Holgate ST. Subepithelial fibrosis in the bronchi of asthmatics. *Lancet* 1: 520–24, 1989.
45. McParland BE, Macklem PT, Pare PD. Airway wall remodeling: Friend or foe? *J Appl Physiol* 95: 426–34, 2003.
46. Wiggs BR, Hrousis CA, Drazen JM, Kamm RD. On the mechanism of mucosal folding in normal and asthmatic airways. *J Appl Physiol* 83: 1814–21, 1997.
47. Krowka MJ, Enright PL, Rodarte JR, Hyatt RE. Effect of effort on measurement of forced expiratory volume in one second. *Am Rev Respir Dis* 136: 829–33, 1987.
48. Tandon MK, Campbell AH. Bronchial cartilage in chronic bronchitis. *Thorax* 24: 607–12, 1969.
49. Lambert RK, Codd SL, Alley MR, Pack RJ. Physical determinants of bronchial mucosal folding. *J Appl Physiol* 77: 1206–16, 1994.
50. Yager D, Butler JP, Bastacky J, Israel E, Smith G, Drazen JM. Amplification of airway constriction due to liquid filling of airway interstices. *J Appl Physiol* 66: 2873–84, 1989.
51. Wilson AG, Massarella GR, Pride NB. Elastic properties of airways in human lungs post mortem. *Am Rev Respir Dis* 110: 716–29, 1974.
52. Maisel JC, Silvers GW, Mitchell RS, Petty TL. Bronchial atrophy and dynamic expiratory collapse. *Am Rev Respir Dis* 98: 988–97, 1968.
53. Brackel HJ, Pedersen OF, Mulder PG, Overbeek SE, Kerrebijn KF, Bogaard JM. Central airways behave more stiffly during forced expiration

in patients with asthma. *Am J Respir Crit Care Med* 162: 896–904, 2000.
54. Wilson JW, Li X, Pain MC. The lack of distensibility of asthmatic airways. *Am Rev Respir Dis* 148: 806–9, 1993.
55. Saetta M, Ghezzo H, Kim WD, King M, Angus GE, Wang NS, Cosio MG. Loss of alveolar attachments in smokers. A morphometric correlate of lung function impairment. *Am Rev Respir Dis* 132: 894–900, 1985.
56. Skloot G, Permutt S, Togias A. Airway hyperresponsiveness in asthma: A problem of limited smooth muscle relaxation with inspiration. *J Clin Invest* 96: 2393–403, 1995.
57. Opazo Saez AM, Seow CY, Pare PD. Peripheral airway smooth muscle mechanics in obstructive airways disease. *Am J Respir Crit Care Med* 161: 910–17, 2000.
58. Pare PD, Wiggs BR, James A, Hogg JC, Bosken C. The comparative mechanics and morphology of airways in asthma and in chronic obstructive pulmonary disease. *Am Rev Respir Dis* 143: 1189–93, 1991.
59. Pauwels RA, Buist AS, Calverley PM, Jenkins CR, Hurd SS. Global strategy for the diagnosis, management, and prevention of chronic obstructive pulmonary disease. NHLBI/WHO global initiative for chronic obstructive lung disease (GOLD) workshop summary. *Am J Respir Crit Care Med* 163: 1256–76, 2001.
60. Cosio M, Ghezzo H, Hogg JC, Corbin R, Loveland M, Dosman J, Macklem PT. The relations between structural changes in small airways and pulmonary-function tests. *N Engl J Med* 298: 1277–81, 1978.
61. Niewoehner DE, Kleinerman J, Rice DB. Pathologic changes in the peripheral airways of young cigarette smokers. *N Engl J Med* 291: 755–58, 1974.
62. Hogg JC, Chu F, Utokaparch S, Woods R, Elliott WM, Buzatu L, Cherniack RM, Rogers RM, Sciurba FC, Coxson HO *et al.* The nature of small-airway obstruction in chronic obstructive pulmonary disease. *N Engl J Med* 350: 2645–53, 2004.
63. Hogg JC, Macklem PT, Thurlbeck WM. The resistance of collateral channels in excised human lungs. *J Clin Invest* 48: 421–31, 1969.
64. Terry PB, Traystman RJ, Newball HH, Batra G, Menkes HA. Collateral ventilation in man. *N Engl J Med* 298: 10–15, 1978.
65. Choong CK, Macklem PT, Pierce JA, Lefrak SS, Woods JC, Conradi MS, Yablonskiy DA, Hogg JC, Chino K, Cooper JD. Transpleural ventilation of explanted human lungs. *Thorax* 62: 623–30, 2007.
66. Wood DE, McKenna RJ Jr., Yusen RD, Sterman DH, Ost DE, Springmeyer SC, Gonzalez HX, Mulligan MS, Gildea T, Houck WV *et al.* A multicenter trial of an intrabronchial valve for treatment of severe emphysema. *J Thorac Cardiovasc Surg* 133: 65–73, 2007.
67. Eltayara L, Becklake MR, Volta CA, Milic-Emili J. Relationship between chronic dyspnea and expiratory flow limitation in patients with chronic obstructive pulmonary disease. *Am J Respir Crit Care Med* 154: 1726–34, 1996.
68. Hogg JC, Macklem PT, Thurlbeck WM. Site and nature of airway obstruction in chronic obstructive lung disease. *N Engl J Med* 278: 1355–60, 1968.
69. Mead J. Problems in interpreting common tests of pulmonary mechanical function. In: Macklem P, Permutt S (eds). *The Lung in Transition Between Death and Disease.* New York: Marcel Dekker Inc, 1979.
70. Ferguson GT, Enright PL, Buist AS, Higgins MW. Office spirometry for lung health assessment in adults: A consensus statement from the national lung health education program. *Chest* 117: 1146–61, 2000.
71. Enright PL, Johnson LR, Connett JE, Voelker H, Buist AS. Spirometry in the lung health study. I: Methods and quality control. *Am Rev Respir Dis* 143: 1215–23, 1991.
72. Miller MR, Hankinson J, Brusasco V, Burgos F, Casaburi R, Coates A, Crapo R, Enright P, van der Grinten CP, Gustafsson P *et al.* Standardisation of spirometry. *Eur Respir J* 26: 319–38, 2005.
73. Fry DL, Hyatt RE. Pulmonary mechanisms: A unified analysis of the relationship between pressure, volume and gas flow in the lungs of normal and diseased subjects. *Am J Physiol* 26: 672–89, 1960.
74. Pride NB, Permutt S, Riley RL, Bromberger-Barnea B. Determinants of maximal expiratory flow from the lungs. *J Appl Physiol* 23: 646–62, 1967.
75. Leaver DG, Tatterfield AE, Pride NB. Contributions of loss of lung recoil and of enhanced airways collapsibility to the airflow obstruction of chronic bronchitis and emphysema. *J Clin Invest* 52: 2117–28, 1973.
76. Ingram RH Jr., Schilder DP. Effect of gas compression on pulmonary pressure, flow, and volume relationship. *J Appl Physiol* 21: 1821–26, 1966.
77. Ingenito EP, Evans RB, Loring SH, Kaczka DW, Rodenhouse JD, Body SC, Sugarbaker DJ, Mentzer SJ, DeCamp MM, Reilly JJ Jr.. Relation between preoperative inspiratory lung resistance and the outcome of lung-volume-reduction surgery for emphysema. *N Engl J Med* 338: 1181–85, 1998.
78. Gelb AF, Brenner M, McKenna RJ Jr., Fischel R, Zamel N, Schein MJ. Serial lung function and elastic recoil 2 years after lung volume reduction surgery for emphysema. *Chest* 113: 1497–506, 1998.
79. ATS. Lung function testing: Selection of reference values and interpretive strategies. *Am Rev Respir Dis* 144: 1202–18, 1991.
80. Pellegrino R, Viegi G, Brusasco V, Crapo RO, Burgos F, Casaburi R, Coates A, van der Grinten CP, Gustafsson P, Hankinson J *et al.* Interpretative strategies for lung function tests. *Eur Respir J* 26: 948–68, 2005.
81. Smith HR, Irvin CG, Cherniack RM. The utility of spirometry in the diagnosis of reversible airways obstruction. *Chest* 101: 1577–81, 1992.
82. O'Donnell DE, Lam M, Webb KA. Spirometric correlates of improvement in exercise performance after anticholinergic therapy in chronic obstructive pulmonary disease. *Am J Respir Crit Care Med* 160: 542–49, 1999.
83. Crapo RO, Casaburi R, Coates AL, Enright PL, Hankinson JL, Irvin CG, MacIntyre NR, McKay RT, Wanger JS, Anderson SD *et al.* Guidelines for methacholine and exercise challenge testing-1999. This official statement of the American thoracic society was adopted by the ATS board of directors, July 1999. *Am J Respir Crit Care Med* 161: 309–29, 2000.
84. Wagers S, Lundblad LK, Ekman M, Irvin CG, Bates JH. The allergic mouse model of asthma: Normal smooth muscle in an abnormal lung? *J Appl Physiol* 96: 2019–27, 2003.
85. Cox G, Thomson NC, Rubin AS, Niven RM, Corris PA, Siersted HC, Olivenstein R, Pavord ID, McCormack D, Chaudhuri R *et al.* Asthma control during the year after bronchial thermoplasty. *N Engl J Med* 356: 1327–37, 2007.
86. Solway J, Irvin CG. Airway smooth muscle as a target for asthma therapy. *N Engl J Med* 356: 1367–69, 2007.
87. Hargreave FE, Ryan G, Thomson NC, O'Byrne PM, Latimer K, Juniper EF, Dolovich J. Bronchial responsiveness to histamine or methacholine in asthma: Measurement and clinical significance. *J Allergy Clin Immunol* 68: 347–55, 1981.
88. Ryan G, Latimer KM, Dolovich J, Hargreave FE. Bronchial responsiveness to histamine: Relationship to diurnal variation of peak flow rate, improvement after bronchodilator, and airway calibre. *Thorax* 37: 423–29, 1982.
89. Gibbons WJ, Sharma A, Lougheed D, Macklem PT. Detection of excessive bronchoconstriction in asthma. *Am J Respir Crit Care Med* 153: 582–89, 1996.
90. Sterk PJ, Bel EH. The shape of the dose–response curve to inhaled bronchoconstrictor agents in asthma and in chronic obstructive pulmonary disease. *Am Rev Respir Dis* 143: 1433–37, 1991.
91. Woolcock AJ, Anderson SD, Peat JK, Du Toit JI, Zhang YG, Smith CM, Salome CM. Characteristics of bronchial hyperresponsiveness in chronic obstructive pulmonary disease and in asthma. *Am Rev Respir Dis* 143: 1438–43, 1991.
92. Taylor RG, Joyce H, Gross E, Holland F, Pride NB. Bronchial reactivity to inhaled histamine and annual rate of decline in FEV_1 in male smokers and ex-smokers. *Thorax* 40: 9–16, 1985.

93. Tashkin DP, Altose MD, Bleecker ER, Connett JE, Kanner RE, Lee WW, Wise R. The lung health study: Airway responsiveness to inhaled methacholine in smokers with mild to moderate airflow limitation. The Lung Health Study Research Group. *Am Rev Respir Dis* 145: 301–10, 1992.
94. Cheung D, Schot R, Zwinderman AH, Zagers H, Dijkman JH, Sterk PJ. Relationship between loss in parenchymal elastic recoil pressure and maximal airway narrowing in subjects with alpha1-antitrypsin deficiency. *Am J Respir Crit Care Med* 155: 135–40, 1997.
95. Watson A, Lim TK, Joyce H, Pride NB. Failure of inhaled corticosteroids to modify bronchoconstrictor or bronchodilator responsiveness in middle-aged smokers with mild airflow obstruction. *Chest* 101: 350–55, 1992.
96. Sont JK, Willems LN, Bel EH, van Krieken JH, Vandenbroucke JP, Sterk PJ. Clinical control and histopathologic outcome of asthma when using airway hyperresponsiveness as an additional guide to long-term treatment. The AMPUL study group. *Am J Respir Crit Care Med* 159: 1043–51, 1999.
97. Tashkin DP, Altose MD, Connett JE, Kanner RE, Lee WW, Wise RA. Methacholine reactivity predicts changes in lung function over time in smokers with early chronic obstructive pulmonary disease. The Lung Health Study Research Group. *Am J Respir Crit Care Med* 153: 1802–11, 1996.
98. Orehek J, Charpin D, Velardocchio JM, Grimaud C. Bronchomotor effect of bronchoconstriction-induced deep inspirations in asthmatics. *Am Rev Respir Dis* 121: 297–305, 1980.
99. Lim TK, Pride NB, Ingram RH Jr.. Effects of volume history during spontaneous and acutely induced air-flow obstruction in asthma. *Am Rev Respir Dis* 135: 591–96, 1987.
100. Scichilone N, Kapsali T, Permutt S, Togias A. Deep inspiration-induced bronchoprotection is stronger than bronchodilation. *Am J Respir Crit Care Med* 162: 910–16, 2000.
101. Assefa D, Amin N, Dozor AJ. Effect of deep inspiration on airway caliber in children with asthma. *Pediatr Pulmonol* 38: 406–12, 2004.
102. Scichilone N, Marchese R, Soresi S, Interrante A, Togias A, Bellia V. Deep inspiration-induced changes in lung volume decrease with severity of asthma. *Respir Med* 101: 951–56, 2007.
103. Scichilone N, Bruno A, Marchese R, Vignola AM, Togias A, Bellia V. Association between reduced bronchodilatory effect of deep inspiration and loss of alveolar attachments. *Respir Res* 6: 55, 2005.
104. Begin P, Grassino A. Inspiratory muscle dysfunction and chronic hypercapnia in chronic obstructive pulmonary disease. *Am Rev Respir Dis* 143: 905–12, 1991.
105. Agusti AG, Barbera JA. Contribution of multiple inert gas elimination technique to pulmonary medicine. 2. Chronic pulmonary diseases: Chronic obstructive pulmonary disease and idiopathic pulmonary fibrosis. *Thorax* 49: 924–32, 1994.
106. Hlastala MP, Robertson HT. Inert gas elimination characteristics of the normal and abnormal lung. *J Appl Physiol* 44: 258–66, 1978.
107. Barbera JA, Roca J, Ferrer A, Felez MA, Diaz O, Roger N, Rodriguez-Roisin R. Mechanisms of worsening gas exchange during acute exacerbations of chronic obstructive pulmonary disease. *Eur Respir J* 10: 1285–91, 1997.
108. Barbera JA, Roca J, Ramirez J, Wagner PD, Ussetti P, Rodriguez-Roisin R. Gas exchange during exercise in mild chronic obstructive pulmonary disease. Correlation with lung structure. *Am Rev Respir Dis* 144: 520–25, 1991.
109. Wagner PD, Dantzker DR, Iacovoni VE, Tomlin WC, West JB. Ventilation–perfusion inequality in asymptomatic asthma. *Am Rev Respir Dis* 118: 511–24, 1978.
110. Wagner PD, Hedenstierna G, Bylin G. Ventilation–perfusion inequality in chronic asthma. *Am Rev Respir Dis* 136: 605–12, 1987.
111. Rodriguez-Roisin R, Roca J. Contributions of multiple inert gas elimination technique to pulmonary medicine. III: Bronchial asthma. *Thorax* 49: 1027–33, 1994.
112. Ballester E, Roca J, Ramis L, Wagner PD, Rodriguez-Roisin R. Pulmonary gas exchange in severe chronic asthma. Response to 100% oxygen and salbutamol. *Am Rev Respir Dis* 141: 558–62, 1990.
113. Mountain RD, Sahn SA. Clinical features and outcome in patients with acute asthma presenting with hypercapnia. *Am Rev Respir Dis* 138: 535–39, 1988.
114. Roca J, Ramis L, Rodriguez-Roisin R, Ballester E, Montserrat JM, Wagner PD. Serial relationships between ventilation–perfusion inequality and spirometry in acute severe asthma requiring hospitalization. *Am Rev Respir Dis* 137: 1055–61, 1988.
115. Rodriguez-Roisin R, Ferrer A, Navajas D, Agusti AG, Wagner PD, Roca J. Ventilation–perfusion mismatch after methacholine challenge in patients with mild bronchial asthma. *Am Rev Respir Dis* 144: 88–94, 1991.
116. Echazarreta AL, Gomez FP, Ribas J, Achaval M, Barbera JA, Roca J, Chung KF, Rodriguez-Roisin R. Effects of inhaled furosemide on platelet-activating factor challenge in mild asthma. *Eur Respir J* 14: 616–21, 1999.
117. Echazarreta AL, Dahlen B, Garcia G, Agusti C, Barbera JA, Roca J, Dahlen SE, Rodriguez-Roisin R. Pulmonary gas exchange and sputum cellular responses to inhaled leukotriene D-4 in asthma. *Am J Respir Crit Care Med* 164: 202–6, 2001.
118. Young IH, Corte P, Schoeffel RE. Pattern and time course of ventilation–perfusion inequality in exercise-induced asthma. *Am Rev Respir Dis* 125: 304–11, 1982.
119. Knudson RJ, Constantine HP. An effect of isoproterenol on ventilation–perfusion in asthmatic versus normal subjects. *J Appl Physiol* 22: 402–6, 1967.
120. Ballester E, Reyes A, Roca J, Guitart R, Wagner PD, Rodriguez-Roisin R. Ventilation–perfusion mismatching in acute severe asthma: Effects of salbutamol and 100% oxygen. *Thorax* 44: 258–67, 1989.
121. Cotton DJ, Graham BL. Single-breath carbon monoxide diffusion capacity or transfer factor. In: Hamid Q, Shannon J, Martin J (eds). *Physiologic Basis of Respiratory Disease*. Hamilton: BC Decker, 2005.
122. Saydain G, Beck KC, Decker PA, Cowl CT, Scanlon PD. Clinical significance of elevated diffusing capacity. *Chest* 125: 446–52, 2004.
123. Keens TG, Mansell A, Krastins IR, Levison H, Bryan AC, Hyland RH, Zamel N. Evaluation of the single-breath diffusing capacity in asthma and cystic fibrosis. *Chest* 76: 41–44, 1979.
124. Collard P, Njinou B, Nejadnik B, Keyeux A, Frans A. Single breath diffusing capacity for carbon monoxide in stable asthma. *Chest* 105: 1426–29, 1994.
125. Diaz P, Galleguillos FR, Gonzalez MC, Pantin CF, Kay AB. Bronchoalveolar lavage in asthma: The effect of disodium cromoglycate (cromolyn) on leukocyte counts, immunoglobulins, and complement. *J Allergy Clin Immunol* 74: 41–48, 1984.
126. McLean A, Warren PM, Gillooly M, MacNee W, Lamb D. Microscopic and macroscopic measurements of emphysema: Relation to carbon monoxide gas transfer. *Thorax* 47: 144–49, 1992.
127. Gould GA, MacNee W, McLean A, Warren PM, Redpath A, Best JJ, Lamb D, Flenley DC. CT measurements of lung density in life can quantitate distal airspace enlargement – An essential defining feature of human emphysema. *Am Rev Respir Dis* 137: 380–92, 1988.
128. Gevenois PA, De Vuyst P, de Maertelaer V, Zanen J, Jacobovitz D, Cosio MG, Yernault JC. Comparison of computed density and microscopic morphometry in pulmonary emphysema. *Am J Respir Crit Care Med* 154: 187–92, 1996.
129. Sansores RH, Pare P, Abboud RT. Effect of smoking cessation on pulmonary carbon monoxide diffusing capacity and capillary blood volume. *Am Rev Respir Dis* 146: 959–64, 1992.
130. Watson A, Joyce H, Hopper L, Pride NB. Influence of smoking habits on change in carbon monoxide transfer factor over 10 years in middle aged men. *Thorax* 48: 119–24, 1993.
131. Dubois P, Machiels J, Smeets F, Delwiche JP, Lulling J. CO transfer capacity as a determining factor of survival for severe hypoxaemic COPD patients under long-term oxygen therapy. *Eur Respir J* 3: 1042–47, 1990.

132. Anthonisen NR, Wright EC, Hodgkin JE. Prognosis in chronic obstructive pulmonary disease. *Am Rev Respir Dis* 133: 14–20, 1986.
133. Graham BL, Mink JT, Cotton DJ. Effect of breath-hold time on DL_{CO}(sb) in patients with airway obstruction. *J Appl Physiol* 58: 1319–25, 1985.
134. Wade JF 3rd, Mortenson R, Irvin CG. Physiologic evaluation of bullous emphysema. *Chest* 100: 1151–54, 1991.
135. Gelb AF, Gold WM, Wright RR, Bruch HR, Nadel JA. Physiologic diagnosis of subclinical emphysema. *Am Rev Respir Dis* 107: 50–63, 1973.
136. Klein JS, Gamsu G, Webb WR, Golden JA, Muller NL. High-resolution CT diagnosis of emphysema in symptomatic patients with normal chest radiographs and isolated low diffusing capacity. *Radiology* 182: 817–21, 1992.
137. Sciurba FC. Physiologic similarities and differences between COPD and asthma. *Chest* 126: 117S–124S, 2004, discussion 159S–161S.

6

CHAPTER

Airway Pathology

James C. Hogg

UBC McDonald Research Laboratories, St. Paul's Hospital, University of British Columbia, Vancouver, BC, Canada

NORMAL ANATOMY

The normal human bronchogram (Fig. 6.1) shows that the length of pathways from the trachea to the terminal airways differs depending on the pathway followed and it can take as few as 8 or as many as 24 divisions of airway branching to reach the gas-exchanging surface [1]. The small bronchi and bronchioles 2 mm in diameter are spread out from the fourth to the fourteenth generation of airway branching [2]. The total airway cross-sectional area expands rapidly beyond the 2 mm airways to provide for rapid diffusion of gas between the distal conducting airways and the gas-exchanging surface. The central conducting airways larger than 2 mm internal diameter are the major site of resistance to airflow in the normal lung because of their much smaller total cross-sectional area [3, 4].

The conducting airways are lined by epithelium and surrounded by an adventitial layer [5]. The submucosa between the epithelium and outer edge of the muscle layer is often referred to as the "lamina propria," but this term is technically incorrect because many airways are not completely surrounded by muscle [6]. Bronchi are defined by the presence of a layer of fibrocartilage external to the smooth muscle and tubuloalveolar glands, which communicate with the airway lumen via ducts [7]. Bronchioles lack both cartilage and glands and become respiratory bronchioles when alveoli open directly into their lumen [7]. The lining of the trachea and major bronchi consists of pseudostratified, ciliated columnar epithelium which gradually becomes more cuboidal

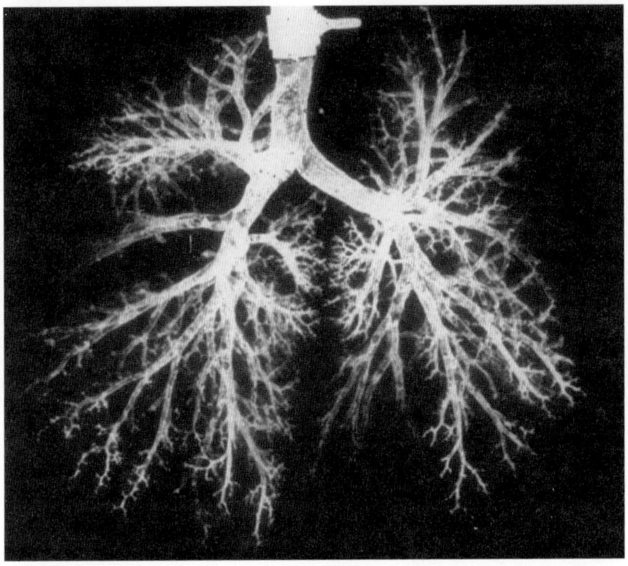

FIG. 6.1 Postmortem bronchogram from an otherwise normal 19-year-old man who died suddenly for reasons unrelated to the lung. Note the difference in airway length depending upon the pathway that is followed.

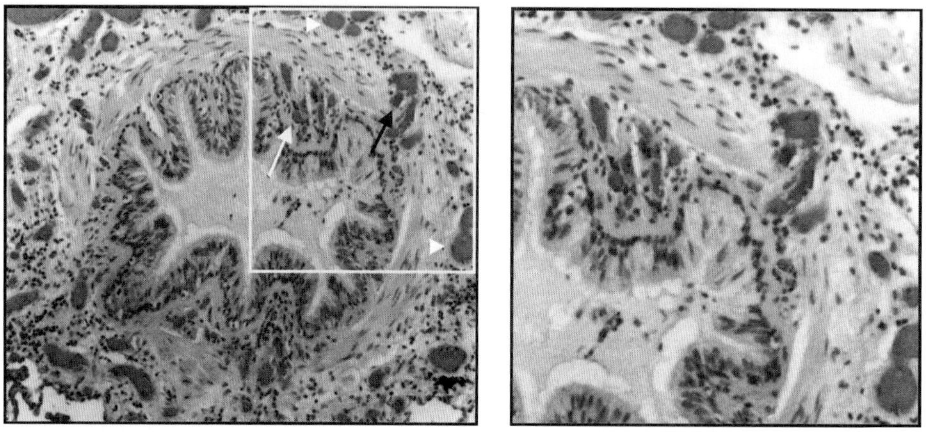

FIG. 6.2 Photomicrograph of a bronchiole from a patient who died of asthma. This shows the location of submucosal capillaries (white arrow), connecting vessels passing between muscle bundles (dark arrow) to larger postcapillary venules located outside the muscle layer. On the right, the same features are shown at a higher magnification.

with fewer ciliated cells as the alveolar surface is approached [6–8]. Light microscopy reveals that the basal aspect of airway epithelial cells is attached to a thin basement membrane (80–90 nm width) that contains primarily type IV collagen and elastin [9]. Transmission electron microscopy shows that a true basement membrane or basal lamina can be readily distinguished from the connective tissue observed with the light microscope [6, 7]. Quantitative studies [10–12] have shown that the trachea and mainstem bronchi of normal subjects consist of by volume:

- 30% cartilage;
- 15% mucous glands;
- 5% smooth muscle; and
- 50% connective tissue matrix containing the bronchial arterial, venous, and lymphatic vessels.

With progression toward the periphery of the bronchial tree, the amount of cartilage and glands decrease, and the percentage of smooth muscle increases to account for approximately 20% of the total wall thickness in the bronchioles.

The degree to which the smooth muscle surrounds the airway lumen varies according to site. In the trachea and mainstem bronchi, the airway smooth muscle is located within the posterior membranous sheath, whereas, in the bronchioles, it surrounds the entire lumen of the airway [7, 8, 13]. Consequently, the same degree of muscle shortening has a smaller effect on the caliber of the trachea and central airways than on the distal bronchi and bronchioles [14]. The adventitial layer consists of loose bundles of collagen admixed with blood vessels, lymphatics, and nerves. In the peripheral conducting airways, this layer interacts with the surrounding lung parenchyma through alveolar attachments that are distributed along the circumference of the adventitia. These alveolar attachments have the ability to limit the amount of airway narrowing produced by smooth muscle contraction, particularly at higher lung volumes [14].

The systemic arterial supply to the bronchial tree originates from the ventral side of the upper thoracic aorta in the left hemithorax, while on the right the origin of the bronchial vessels is more variable. They may originate from the first to the third intercostal artery, from the right internal mammary artery, or from the right subclavian artery [8, 15]. Miller's classic anatomical account [8] showed that two to three arterial branches accompany each of the larger bronchi and that anastomoses between these branches form an arterial plexus in the outer wall of the airway. Small branches of this plexus penetrate the smooth muscle layer to form a capillary network below the epithelium. Short connecting branches extend from this plexus through the muscle layer to form a second plexus of venules along the outer surface of the airway smooth muscle (Fig. 6.2). In some species, this outer plexus contains large venous sinuses that can extend into the submucosal layer of the major bronchi, where there is no smooth muscle between the cartilage and epithelium [16].

The venous blood from the first two or three subdivisions of the bronchial tree drains into the azygous and hemiazygous venous systems that empty into the vena cava. The remainder of the bronchial venous flow drains directly into the pulmonary circulation, although there is debate as to how much enters at the precapillary, capillary, and postcapillary levels [8, 15]. Airway disease increases the anastomotic flow between the pulmonary and the bronchial vascular systems; and in diseases such as bronchiectasis, injection of the bronchial arteries can result in rapid filling of the entire pulmonary vascular tree right back to the pulmonic valve [15].

The bronchial circulation accounts for about 1% of cardiac output and an average cardiac output of 5 l/min delivers approximately 72 l of blood to the conducting airways over 24 h. Each liter of blood contains between 4 and 11×10^9 white blood cells, made up of approximately 60% neutrophils, 30% lymphocytes, 5% monocytes, 3% eosinophils, and 2% basophils. The neutrophils do not divide after they leave the marrow have a relatively short half life within the circulation and remain within the vascular space unless an inflammatory response is present. Although much less is known about eosinophils and basophil kinetics they are predominately a tissue cell, indicating that their transit times through the tissue compartment is much longer than that for migrating neutrophils. Monocytes on the other hand leave the vascular space even in the absence of an inflammatory stimulus and divide in the tissue to form the alveolar macrophages

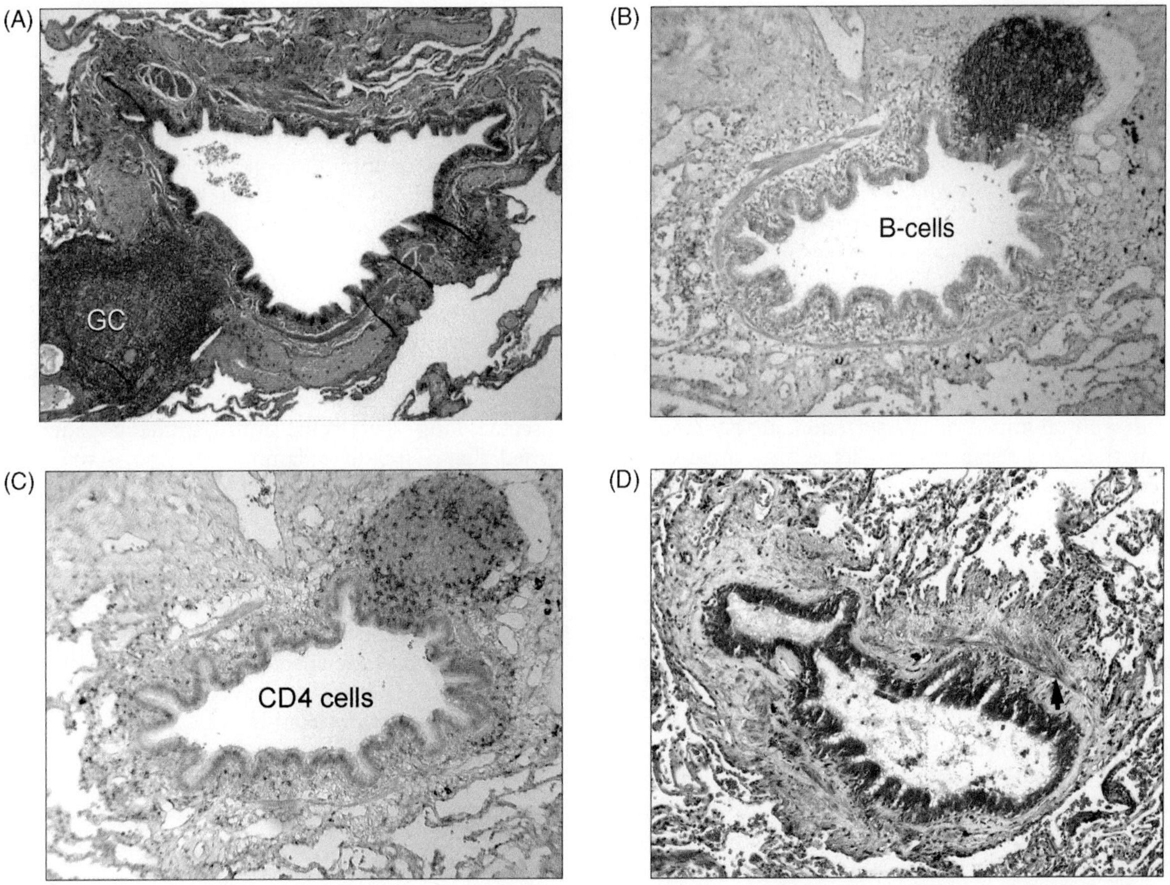

FIG. 6.3 (A) Shows a small bronchiole with a collection of lymphocytes containing a germinal center. (B) Shows another airway containing a lymphoid follicle where the germinal center stains strongly for B-cells. (C) Shows the same airway as in B where the CD4 cells appear at the edge of the follicle. (D) Shows yet another area where the wall is thickened and fibrosed and the arrow pints to the smooth muscle in the airway wall. Adapted from Ref. [17] with permission.

that are removed via the airways with very little if any re-entry into the tissue. On the other hand lymphocytes leave the circulation regularly with the assistance of specialized high endothelial cells and re-entering the circulating blood with the lymph as it drains into the venous system.

A study of lung tissue obtained from patients at all five Global Initiative for Chronic Obstructive Lung Disease (GOLD) categories of chronic obstructive pulmonary disease (COPD) severity has shown an association between the decrease in FEV_1 with both the extent of the accumulation of these cells (i.e. the percentage of airways containing cells) and severity of this accumulation (i.e. the total accumulated volume of cells) within small airway tissue [17]. The accumulating lymphocytes form follicles with germinal centers in both the small conducting airways [17] and the parenchyma [18] of the peripheral lung in both COPD [17, 18] and asthma [19]. These accumulations of lymphocytes (Fig. 6.3) are part of the bronchial associated lymphoid system or bronchial-alveolar lymphoid tissue (BALT) and differ from the regional lymph nodes in that they have no capsule and no afferent lymphatics. They are similar in structure to the tonsils and adenoids, the Peyers patches in the small bowel, and the appendix of the cecum which are also part of the mucosal immune system. They are organized in the same way as other lymph follicles (Fig. 6.3) in that the B-cells are found in the germinal center of the follicle and the CD4 and CD8 T-cells are located around the edges of the germinal center. Thus dendritic cells migrating from the epithelium and subepithelium circulate through the T- and B-cell rich regions of the follicle and can present antigen to uncommitted T and B lymphocytes moving through these regions in the lymph (Fig. 6.3). Moreover the recirculation of lymphocytes though the follicle and back into the circulation enhances the opportunity for T Helper cells and B-cells that have recognized the same antigen to interact with each other and initiate an adaptive immune response. The appearance of lymphoid follicles with germinal centers in the tissue provides histological evidence that an adaptive immune response has been mounted within the peripheral lung. Some investigators have begun to study this response in detail [18].

THE PATHOLOGY OF ASTHMA

Postmortem studies

At postmortem, the lungs of patients who have died in status asthmaticus remain markedly hyperinflated after the thorax is opened. This hyperinflation is due to air trapping caused by widespread plugging of the segmental, subsegmental, and the smaller conducting airways by mucus and cellular debris [20]. Although this luminal content may extend to the respiratory bronchioles, it usually stops short of these structures and does not fill the alveolar airspaces. Examination of the cut surface of the lung reveals the plugged airways, but – in contrast to parenchymal destruction seen in hyperinflated emphysematous lungs – the parenchyma of the asthmatic lung remains intact.

Huber and Koessler's [21] classic 1922 paper on the pathology of asthma reviewed 15 published cases and provided new data on 6 more. They noted that the pathology consisted of common features that allowed asthma to be distinguished from other conditions. They emphasized the presence of intraluminal mucus secretion, airway epithelial desquamation, and repair (e.g. goblet cell metaplasia), and airway inflammatory infiltrates consisting of an admixture of mononuclear cells and eosinophils and the presence of a thickened, "hyalinized" subepithelial basement membrane. Later studies based on electron microscopy and immunohistochemistry showed that this feature of the basement membrane was due to deposition of collagen fibrils and extracellular matrix below the true basement membrane, rather than thickening of the basal lamina. Huber and Koessler's report noted that the tenacious plugs that fill the airway lumen consist of an exudate of plasma containing inflammatory cells, particularly eosinophils, mixed with epithelial cells that had sloughed from the airway surface. Using an eyepiece micrometer to measure the external airway diameters, they concluded that the walls of bronchi and bronchioli of more than 2 mm outside diameter were thickened compared with non-asthmatic persons, and that this difference was due to an increased thickness of all of the components of the airway wall.

Over the next several decades, other reports confirmed and extended these findings [22–29]. Comparing Florey's [30] basic studies of the inflammatory process with these pathological findings show that the structural changes associated with asthma are consistent with an inflammatory process involving a mucus-secreting surface. However, this knowledge had relatively little impact on the allergists, pulmonary physicians, and physiologists until the 1970s [31], because they were preoccupied with the concept that asthma was due to IgE sensitized mast cells releasing mediators that caused excessive contraction of airway smooth muscle following specific antigen challenge.

Bronchoscopic studies

The nature of the airway pathology in persons with asthma was further revealed by studies of tissue obtained through the rigid and flexible bronchoscope. These techniques allowed investigators to obtain cells from living asthmatic patients by both bronchoalveolar lavage and bronchial biopsies [31–34]. This brought physiologists and clinicians into closer agreement with the pathologist's view that the inflammatory process was important to the pathogenesis of both bronchial hyperresponsiveness and reversible airflow obstruction.

A very important conceptual development based on the discovery that murine $CD4^+$ T-cell clones showed that these cells can be divided according to the cytokine messenger RNA (mRNA) and proteins that they produce [35]. These experiments established that one type of T-cell clone (Th-1) produced IL-2 and interferon-γ but no IL-4 or IL-5; whereas the second (Th-2) produced IL-4 and IL-5, but no IL-2 or interferon-γ. Both clones produced IL-3 and GMCSF, and interactions between Th-1 and Th-2 subtypes allowed one type of clone to inhibit the other. For example, IL-4 is a mast cell growth factor that also stimulates IgE production, and IL-5 promotes the differentiation and survival of eosinophils. Robinson and associates [36] put forward the hypothesis that asthma was the result of a "Th-2 response" based on bronchoalveolar lavage and bronchial biopsy findings. And subsequent studies of surgically resected lung specimens from asthmatic patients established that a similar inflammatory process was present in the smaller airways [37].

It has now become clear that the structural features of the airways from asthmatic patients result from an inflammatory process involving tissue with a mucus-secreting surface. This response appears to be driven by a subset of $CD4^+$ T lymphocytes producing cytokines that result in an excess of eosinophils and an overproduction of IgE. The end result is abnormal airway function, characterized by excessive airway narrowing in response to external stimuli, reversible airways obstruction, and gas trapping. Although the majority of the symptoms produced by the process can be rapidly reversed with appropriate treatment, the process can be life threatening and result in sudden death.

The relationship between airway structure and function

The concept that the same degree of smooth muscle shortening will cause greater reduction in airway caliber when the wall is thickened by disease has been suggested by several authors [38, 39]. Moreno and associates [39] calculated that the thickening of the airway wall observed in asthma would have only a minor effect on the caliber of the lumen of a fully dilated airway. However, when the smooth muscle in the airway shortens, the increased tissue between the muscle and lumen causes an excess reduction in airway caliber. Subsequent studies by James *et al.* [40] showed that the increase in wall thickness observed in the small airways of asthmatics was sufficient to close the lumen of these airways, even when smooth muscle shortening remained within the accepted normal range. This suggests that normal smooth muscle shortening may act in series with an abnormally thickened airway wall to narrow the airway lumen and it follows that the reduction in airway caliber produced by this mechanism would be rapidly reversed when the smooth muscle relaxed. This showed that an important feature of the pathology of asthma was the change in the structure of the airway wall produced by the remodeling of the tissue that occurs in relation to the inflammatory process. The important point is that these structural changes could result in excess airway narrowing with normal smooth muscle contraction and that excessive smooth muscle shortening will enhance the effect of these structural changes on the airway lumen.

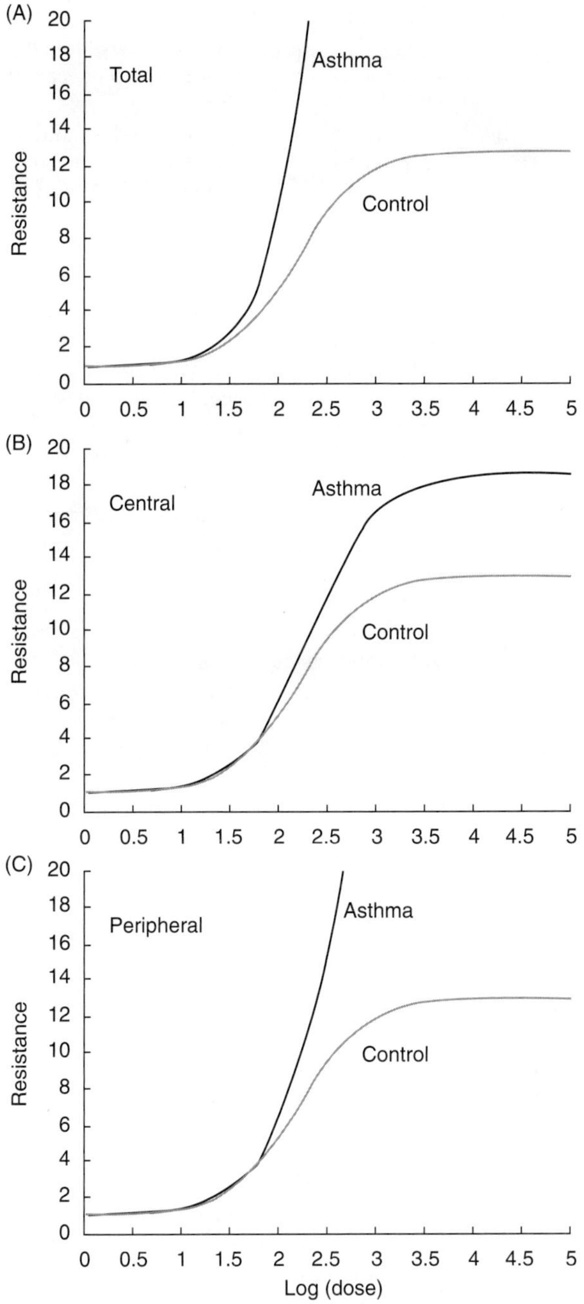

FIG. 6.4 Data from Wiggs *et al.* (Refs. [38, 39]) showing results from a computer model of the airways. The structural data from normal lungs were used to obtain the control measurements. These show that airway resistance increases from about 1 cmH_2O per liter per second to reach a plateau of 12 cmH_2O per liter per second, with maximum contraction of the airways smooth muscle. (A) When the structural features of asthmatic airways were used in the same computer model, the airway resistance continued to increase and did not reach a plateau at any physiologically meaningful value. (B) Data obtained from the model when the structural changes caused by asthma were limited to the central airway (i.e. those >2 mm diameter). Note that asthmatic changes in the central airways increased total resistance and resulted in a plateau slightly greater than the control lungs. (C) However, when the asthmatic changes were placed in the peripheral airway, the resistance increased without reaching a plateau. These data suggest that asthmatic changes in the peripheral airways have a much greater effect on overall airway function than changes that occur in the central airways. Reproduced from Ref. [38] with permission.

Wiggs *et al.* [41, 42] extended this concept using a computer model to test the effect of these structural changes on airway function. Their analysis (Fig. 6.4) showed that maximum stimulation of the smooth muscle caused airway resistance to increase and reach a plateau in the normal lung. This finding was consistent with previous observations by Woolcock *et al.* [43], who reported that the changes in the maximum volume of air that can be expired from the lung in 1 s (FEV_1) reached a plateau in normal subjects when maximally stimulated by inhaled bronchoconstrictors. However, when the data on airway structure were changed from normal values to those found in asthmatic lungs, a similar degree of airway smooth shortening resulted in a sustained rapid increase in airway resistance without a plateau. When the effect of disease on the central and peripheral airway function was examined separately, they found that the increase in airway resistance was primarily due to the effect of disease on the peripheral airways, where smooth muscle shortening produced widespread airway closure. The concept that the peripheral airways are the major site of obstruction in asthma has now been confirmed by direct measurements in living asthmatic patients reported by Yanai *et al.* [44].

The changes that are produced in the airway tissue also reduce the function of the conducting airways. Lambert [45] was the first to systematically study the normal folding pattern of the bronchial mucosa, and showed that in asthma the multiple mucosal folds that occur when normal airways narrow were replaced with fewer and larger folds that reduced airway caliber. Both Lambert and Wiggs *et al.* [46] suggested that the mucosal folding pattern was controlled by the stiffness of the subepithelial layer relative to that of the surrounding airway tissue. This analysis suggested that changes in the subepithelial connective tissue might play a key role in determining the pattern of mucosal folding. Lambert and Wiggs *et al.* argued that the formation of a large number of folds in the normal airway placed a load on the airway smooth muscle that tended to prevent airway closure at low lung volumes. It followed that a change in the mucosal folding pattern in asthma may be one way in which this disease causes peripheral airway dysfunction.

The caliber of the airway lumen is also influenced by airway surface tension that is normally low in the small airways because they are lined with surfactant [47]. Exudation of plasma and the secretion of mucus on to the small airway lumen surface should increase surface tension and cause the airways to narrow. The analysis of induced sputum has confirmed that plasma proteins, mucus, and inflammatory cells are present in the airway lumen even in mild asthma, suggesting that some of the abnormalities in airway function of asthmatics might result from the presence of this material in the lumen. Kuyper and colleagues [48] recently re-assessed the importance of the occlusion of the airway lumen by inflammatory exudates in deaths attributed to asthma. By examining the airway wall and luminal content of 275 airways from 93 patients with fatal asthma aged 10–49 years. Compared with control airways obtained from persons that died of causes unrelated to the lungs, the asthmatic airways showed much more extensive lumenal occlusion, by mucus and cells. They concluded that widespread airway occlusion was the major cause of death from asthma and death due to closure of empty airways by excessive bronchoconstriction must be a rare event. Moreover the extensive nature of the airway occlusion observed at autopsy in these cases suggests that bronchoconstriction superimposed on airways that have become partially occluded is the most probable cause of sudden death.

THE PATHOLOGY OF COPD

The inflammatory process contributes to the pathogenesis of chronic cough and sputum production [49], peripheral airways obstruction [3, 17, 50], and emphysematous destruction of the lung surface [51] that define COPD. As tobacco smoking produces lung inflammation in everyone, and only 15–20% of heavy smokers develop COPD, clinical disease from cigarette smoke-induced airway inflammation must develop in this minority of people because it amplified by either genetic or environmental risk factors [52].

Chronic bronchitis

Cough and sputum production are the features of airways disease that define chronic bronchitis [53] and these symptoms can be present either with or without airways obstruction [49]. Figure 6.5 shows the histology of a normal bronchus at low power where with the connection between the epithelial lining of the bronchial lumen to the mucus duct and gland is clearly seen. Reid [54] used the relative size of the mucus glands to the airway wall as a yardstick for measuring chronic bronchitis and downplayed the influence of inflammation in driving mucus production. However, a reevaluation of this problem some years later showed that chronic bronchitis was associated with inflammation of the airway mucosal surface, the submucosal glands, and gland ducts, particularly in the smaller bronchi between 2 and 4 mm in diameter [49]. The nature of the inflammatory process present in these airways is now quite well established with several studies reporting that $CD8^+$ lymphocytes are present in excess numbers in smokers with chronic bronchitis [55, 56].

Sputum production represents the clearance of a mucoid inflammatory exudate from the lumen of the bronchi. This exudate contains plasma proteins, inflammatory cells, and small amounts of mucus added from goblet cells on the surface epithelium and the epithelial glands. Although the size of the bronchial mucous glands tends to increase [49] in chronic bronchitis, Thurlbeck and Angus [57] showed that the Reid index was normally distributed with no clear separation between patients with chronic bronchitis and controls. Chronic bronchitis also results in an increase in airway smooth muscle, a generalized increase in the connective tissue in the airway wall, degenerative changes in the airway cartilage, and a shift in epithelial cell type that increases the number of goblet and squamous cells [58–62].

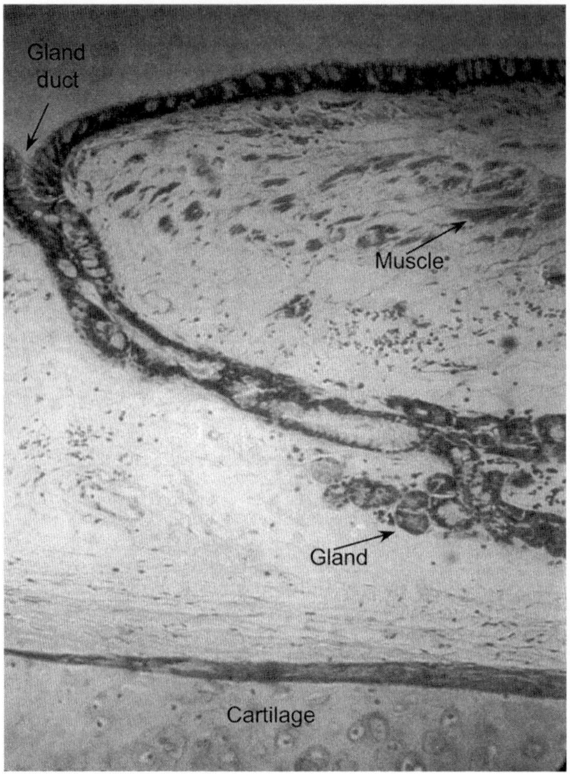

FIG. 6.5 Low-power photomicrograph of a normal bronchus showing the opening of a bronchial gland duct into the lumen and its connection with the submucosal gland. The bronchial muscle and cartilage are clearly labeled. The inflammatory process associated with chronic bronchitis involves the mucosa, gland ducts, and glands of the bronchi that are between 2 and 4 mm internal diameter. Original photograph taken by the late Dr. William Thurlbeck.

The site of airways obstruction

The site of airways obstruction in COPD is in the smaller conducting airways that include bronchi and bronchioles of less than 2 mm diameter [3]. Direct measurements of airways resistance in dogs [4] and in similar measurements in postmortem human lungs [3] established, in the normal lung, that the small peripheral airways offer very little resistance to air flow. Although van Brabant *et al.* [63] subsequently argued that these peripheral airways account for a larger proportion of the total airways resistance in the normal lung, there is general agreement that the peripheral airways are the major site of obstruction in COPD [3, 44, 63].

The causes of the reduced forced expiratory flow that define COPD include destruction of alveolar support of the peripheral airways [64], the loss of elastic recoil in the parenchyma supporting the airways [65], a decrease in the elastic force available to drive flow out of the lung [66], and structural narrowing of the airway lumen by a remodeling process [3, 17, 67]. Comparison of the histological appearance of peripheral airways at different degrees of severity of COPD shows a progressive increase in the magnitude of the inflammatory process with thickening of the airway walls accompanied by occlusion of the airway lumen by inflammatory mucous exudates [17]. However, multivariate analysis of

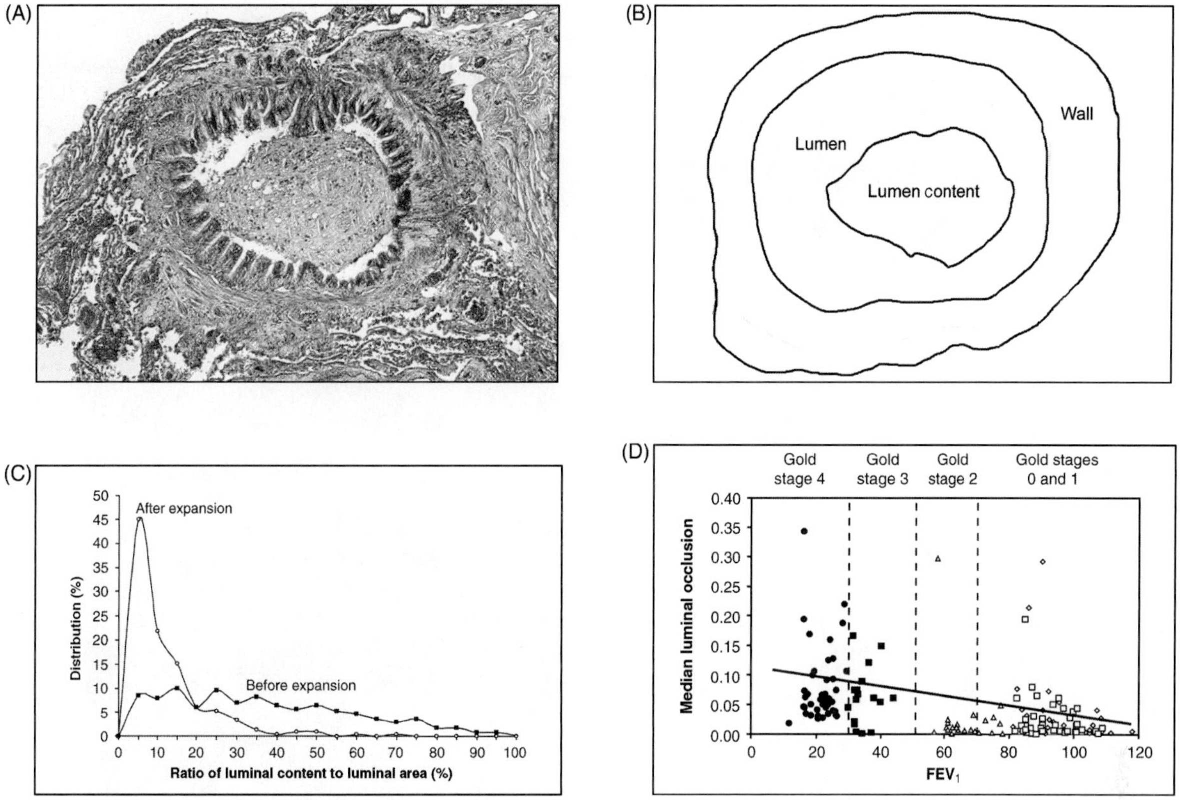

FIG. 6.6 (A) Shows a small bronchiole from a patient with severe (GOLD-4) COPD the lumen is nearly fully occluded and the mucosa is folded because it was fixed in a collapsed state. (B) Shows a redrawing of the same airway where the lumen occlusion is less severe because the airway has been fully inflated by removing the folds from the mucosa using a computer (C) Compares the frequency distribution of the ratio of luminal content area to the total area of the lumen for >562 airways from 42 patients with GOLD-4 COPD before and after the expansion of the lumen shown in A and B. As we do not breathe at either full expansion of the lung or at minimal volume many airways must have been partially or fully occluded in the normal breathing range. (D) Shows the relationship between the severity of the occlusion of the airway lumen with the level of FEV_1. Adapted from Ref. [17] with permission.

these data indicated that thickening of the airway walls and the occlusion of the lumen by inflammatory exudates containing mucus explained more of the variance in the decline in FEV_1 than the infiltration of the airway tissue by any of the inflammatory cell types examined [17]. Interestingly the severity of the occlusion of the airways lumen by inflammatory exudates containing mucus (Fig. 6.6) *has* also associated with premature death in persons with severe (GOLD-3) and very severe (GOLD-4) disease [68].

The lung inflammatory changes that are associated with cigarette smoking have been documented in autopsy studies [69–73], resected lung specimens [49–51, 74] lung biopsies [75], and indirectly by examining bronchoalveolar lavage fluid [76–78]. Collectively these data support the concept that cigarette smoke-induced lung inflammation is present in all smokers, including those with normal lung function and that it persists long after people have stopped smoking [17]. The reasons for both the amplification of the smoking-related inflammatory process in individual that develop COPD, the persistence of this response after the smoking has stopped and the precise relationship between the inflammatory response and the remodeling process needs further clarification.

Emphysema

Emphysema has been defined as "abnormal permanent enlargement of airspaces distal to terminal bronchioles, accompanied by destruction of their walls without obvious fibrosis" [79]. This definition emphasizes the destruction of the alveolar surface with a minimal reparative response in the lung matrix and the ability of this destructive process to reduce the gas-exchanging surface of the lung. The centrilobular form of emphysema (Fig. 6.7) resulting from dilatation and destruction of the respiratory bronchioles in the center of secondary lobule of Miller has been most closely associated with smoking [80, 82]. The panacinar form of emphysema, on the other hand, results from a uniform destruction of all of the acini in the entire secondary lobule and is characteristic of the lesions found in the lungs in α_1-antitrypsin deficiency [83]. The terms "distal acinar," "mantle," or "periseptal" emphysema are used to describe lesions that occur in the periphery of the lobule and along the lobular septae particularly in the subpleural region. These lesions have been associated with spontaneous pneumothorax in young adults and bullous lung disease in older individuals [84]. In far-advanced parenchymal destruction such as that frequently observed in end-stage

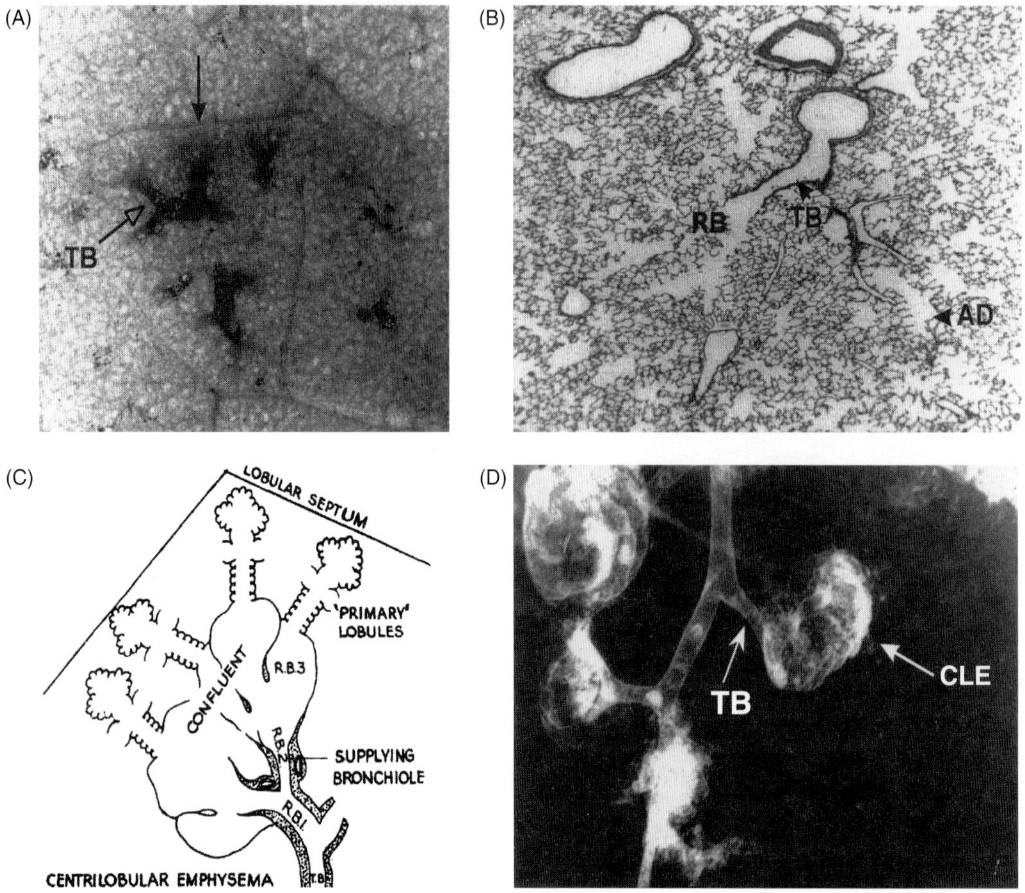

FIG. 6.7 (A) Normal lung photographed from the pleural surface after a postmortem bronchogram. The connective tissue outlining the secondary lobule is clearly seen (solid arrow) and a terminal bronchiole (TB) indicated by a clear arrow supplies a single acinus. (B) Photomicrograph of an acinus showing a terminal bronchiole (TB), respiratory bronchiole (RB), and alveolar ducts (AD). Their structure is represented by the fine spray of contrast at the end of the TB shown in the bronchogram in (A). (C) Diagram of the lesion in centrilobular emphysema, showing the dilation and destruction of the respiratory bronchioles. (D) Postmortem bronchogram showing a centrilobular emphysematous lesion (CLE) outlined by bronchographic material. Parts (C) and (D) reproduced from Refs [80, 81], with permission.

COPD, these descriptive terms are less helpful because centrilobular disease eventually destroys the entire lung lobule and destroyed lobules can coalesce to form much larger lesions.

The destruction of the lung by centrilobular emphysema results in a loss of elastic recoil with subsequent hyperinflation and a change in the pressure–volume relationship of the emphysematous lung tissue compared to the surrounding normal lung such that at any volume recoil pressure is less [81]. This decrease in lung elastic recoil diminishes expiratory flow by reducing the pressure available to drive air out of the lung [66], and interferes with gas exchange by reducing lung surface area [85].

Acute exacerbations of COPD

In a hospital-based study of 1205 admissions to five hospitals for acute exacerbations of COPD, infections accounted for 520 (43%), heart failure for 260 (22%), and 122 (10%) cases were attributed to a variety of etiologies that included arrhythmia, pulmonary embolism, pneumothorax, postoperative complications, and lung cancer [86], with no cause being established in 303 (25%) of these 1205 admissions. Early studies by Stuart-Harris and associates [87, 88] showed that COPD patients with acute upper respiratory infections are more likely to have signs of infection in the lower airways than normal controls. Probably because acute viral upper respiratory tract infections increase the risk of aspiration of mucoid exudate containing large numbers of bacteria from the upper airways, but also because viral infection reduces both the mucociliary clearance and bacterial killing in the lower airways [89–91]. *Streptococcus pneumonia*, *Staphylococcus aureus*, and *Hemophilus influenzae* are the most common bacteria infecting the lower airways during an acute viral infection [90, 91], and these infections account for the positive effect of antibiotics in some exacerbations of COPD [92]. Two important early studies established an association between acute illness and serologic laboratory evidence of viral infection in patients with COPD [93, 94]. The development of polymerized chain reaction (PCR)-based viral diagnostic techniques have improved the sensitivity of the diagnosis of viral infection and the use of real time PCR to measure viral copy numbers will likely improve its specificity. Early reports of studies based on these techniques indicate the importance of rhinoviral infection in causing acute exacerbations in an outpatient setting with influenza A and the corona virus OC-43 being more prominent causes of exacerbations that require admission to hospital [95–97].

The pathological features of the lung during acute exacerbations of COPD are incompletely defined. Primarily because postmortem studies of patients that die during exacerbations are complicated by terminal events and biopsy studies of patients with severe (GOLD-3 and very severe GOLD-4 disease during an exacerbation cannot be carried out without excess risk to the patient. One exception is an important study by Saetta *et al.* [98] in less severely ill patients with GOLD class 2 disease and a mean FEV_1 of 62 ± 7% predicted implicated the eosinophil as a cell of interest. Clearly more studies especially those that can make use of non-invasive or less invasive techniques than biopsy are needed to help clarify this important area of COPD.

SUMMARY

Asthma

The pathology of asthma is dominated by widespread plugging of the segmental, subsegmental, and smaller conducting airways that leads to hyperinflation but not destruction of the parenchyma. These airway plugs are a manifestation of the fluid and cellular exudative phase of an inflammatory process based in the airway tissue (Fig. 6.6). The cytokine and cellular content of this process is consistent with a Th-2 type of immune response [36, 37], and quantitative postmortem studies have established that the airway occlusion by plugs of this exudate is the major determinant of death from asthma [48].

COPD

The chronic bronchitis of COPD is defined by excess cough with sputum production and is associated with an inflammatory process located in the mucosa, gland ducts, and glands of intermediate-sized bronchi between 2 and 4 mm internal diameter. The airway obstruction in COPD is the result of a similar inflammatory response in the smaller bronchi and bronchioles under 2 mm internal diameter, where the repair process associated with chronic inflammation thickens the airway wall and narrows the lumen to cause fixed airway obstruction. Emphysematous destruction of the lung surface contributes to the decline in FEV_1 by reducing the elastic recoil force available to drive air out of the lung and is responsible for reduced gas exchange. The acute exacerbations of COPD that occur with increasing frequency as the disease progresses have several known causes including infection, right heart failure, and pulmonary embolism but it is discouraging that in up to 25% of cases admitted to hospital the cause is unknown.

References

1. Horsfield K, Cumming G. Morphology of the bronchial tree in man. *J Appl Physiol* 24: 373–83, 1968.
2. Weibel ER. Morphometry of the Human Lung. New York: Academic Press, 1963.
3. Hogg JC, Macklem PT, Thurlbeck WM. Site and nature of airways obstruction in chronic obstructive lung disease. *N Engl J Med* 48: 421–31, 1968.
4. Macklem PT, Mead J. Resistance of central and peripheral airways measured by a retrograde catheter. *J Appl Physiol* 22: 395–401, 1967.
5. Bai A, Eidelman DH, Hogg JC *et al.* Proposed nomenclature for quantifying subdivisions of the bronchial wall. *J Appl Physiol* 77: 1011–14, 1994.
6. Neutra M, Podykula H, Weiss L *Histology, Cell and Tissue Biology*, 5th edn, pp. 658–706. Amsterdam: Elsevier Biomedical, 1983.
7. Weiss L (ed.), *Histology, Cell and Tissue Biology*, 5th edn, pp. 788–868. Amsterdam: Elsevier Biomedical, 1983.
8. Miller WS. The blood vessels in the lung. *The Lung*, 3rd edn, pp. 73–84. Springfield, IL: Charles Thomas, 1943.
9. Roche WR, Beasley R, Williams JH, Holgate ST. Subepithelial fibrosis in the bronchi of asthmatics. *Lancet* i: 520–24, 1989.
10. Dunnill MS. The pathology of asthma with special reference to changes in the bronchial mucosa. *J Clin Pathol* 13: 27–33, 1960.
11. Dunnill MS. Quantitative methods in the study of pulmonary pathology. *Thorax* 17: 320–28, 1962.
12. Takizawa T, Thurlbeck WM. Muscle and mucous gland size in the major bronchi of patients with chronic bronchitis, asthma and asthmatic bronchitis. *Am Rev Respir Dis* 104: 331–36, 1971.
13. Plopper CG. Ten Have-Opbroek AAW Anatomical and histological classification of the bronchioles. In: Epler GR (ed.), *Diseases of the Bronchioles*, pp. 15–25. New York: Raven Press, 1994.
14. Macklem PT. Bronchial hyporesponsiveness. *Chest* 87: 158S–159S, 1985.
15. Cudkowicz L. The Human Bronchial Circulation in Health and Disease. . Baltimore: Williams & Wilkins, 1968.
16. Hill P, Goulding D, Webber SE, Widdicombe JG. Blood sinuses in the submucosa of the large airways of sheep. *J Anat* 162: 235–47, 1989.
17. Hogg JC, Chu F, Utokaparch S *et al.* The nature of small-airway obstruction in chronic obstructive pulmonary disease. *N Engl J Med* 350: 2645–53, 2004.
18. van der Strate BW, Postma DS, Brandsma CA *et al.* Cigarette smoke-induced emphysema: A role for the B cell?. *Am J Respir Crit Care Med* 173: 751–58, 2006, Epub 2006 Jan 6.
19. Elliot JG, Jensen CM, Mutavdzic S, Lamb JP, Carroll NG, James AL. Aggregations of lymphoid cells in the airways of nonsmokers, smokers, and subjects with asthma. *Am J Respir Crit Care Med* 169: 712–18, 2004.
20. Rigler LG, Koucky R. Roentgen studies of pathological physiology of bronchial asthma. *Am J Roentgenol* 39: 353–62, 1938.
21. Huber HL, Koessler KK. The pathology of bronchial asthma. *Arch Intern Med* 30: 689–760, 1922.
22. Cardell BS, Pearson RSB. Death in asthmatics. *Thorax* 14: 341–52, 1959.
23. Earle BV. Fatal bronchial asthma: A series of 15 cases with a review of the literature. *Thorax* 8: 195–206, 1953.
24. Houston JC, de Navasquez S, Trounce JR. A clinical and pathological study of fatal cases of status asthmaticus. *Thorax* 8: 207–13, 1953.
25. Dunnill MS, Massarella GR, Anderson JA. A comparison of the quantitative anatomy of the bronchi in normal subjects, in status asthmaticus, in chronic bronchitis and in emphysema. *Thorax* 24: 176–79, 1969.
26. MacDonald IG. The local and constitutional pathology of bronchial asthma. *Ann Intern Med* 6: 253–77, 1933.
27. Messer J, Peters GA, Bennet WA. Cause of death and pathological findings in 304 cases of bronchial asthma. *Dis Chest* 38: 616–24, 1960.
28. Richards W, Patrick JR. Death from asthma in children. *Am J Dis Child* 110: 4–21, 1965.

29. Saetta M, di Stefano A, Rosina C, Thiene G, Fabbri LM. Quantitative structural analysis of peripheral airways and arteries in sudden fatal asthma. *Am Rev Respir Dis* 143: 138–43, 1991.
30. Florey H. The secretion of mucus and inflammation in mucus membranes. In: Florey H (ed.), *General Pathology*, 3rd edn, pp. 167–96. London: Lloyd-Luke Medical Books, 1962.
31. Ciba Symposium on the Identification of Asthma. January 1971.
32. Djukanovic R, Wilson JW, Lai CKW, Holgate ST, Howarth PH. The safety aspects of fiberoptic bronchoscopy, bronchoalveolar lavage, and endobronchial biopsy in asthma. *Am Rev Respir Dis* 143: 772–77, 1991.
33. Djukanovic R, Roche WR, Wilson JW *et al.* Mucosal inflammation in asthma. *Am Rev Respir Dis* 142: 434–57, 1990.
34. Glynn AA, Michaels L. Bronchial biopsy in chronic bronchitis and asthma. *Thorax* 15: 142–53, 1960.
35. Mosmann TR, Coffman RL. TH1 and TH2 cells: Different patterns of lymphokine secretion lead to different functional properties. *Annu Rev Immunol* 7: 145–73, 1989.
36. Robinson DS, Hamid Q, Ying S *et al.* Predominant TH2-like bronchoalveolar T-lymphocyte population in atopic asthma. *N Engl J Med* 326: 298–304, 1992.
37. Hamid Q, Song Y, Kotsimbos TC *et al.* Small airways inflammation in asthma. *J Allergy Clin Immunol* 100: 44–51, 1997.
38. Friedman BJ. Functional anatomy of the bronchi. *Bull Pathophysiol Respir* 8: 545–51, 1972.
39. Moreno R, Hogg JC, Paré PD. Mechanics of airway narrowing. *Am Rev Respir Dis* 133: 1171–80, 1986.
40. James AL, Paré PD, Hogg JC. The mechanics of airway narrowing in asthma. *Am Rev Respir Dis* 139: 242–46, 1989.
41. Wiggs BR, Moreno R, Hogg JC, Hilliam C, Paré PD. A model of the mechanics of airway narrowing. *J Appl Physiol* 69: 849–60, 1990.
42. Wiggs BR, Bosken C, Paré PD, James A, Hogg JC. A model of airway narrowing in asthma and chronic obstructive pulmonary disease. *Am Rev Respir Dis* 145: 1251–58, 1992.
43. Woolcock AJ, Salome CM, Yan K. The shape of the dose-response curve to histamine in asthmatic and normal subjects. *Am Rev Respir Dis* 130: 71–75, 1984.
44. Yanai M, Sekizawa K, Ohrui T, Sasaki H, Takishima T. Site of airway obstruction in pulmonary disease: Direct measurements of intrabronchial pressure. *J Appl Physiol* 72: 1016–23, 1992.
45. Lambert R. The role of the bronchial basement membrane and airway collapse. *J Appl Physiol* 71: 666–73, 1991.
46. Wiggs BR, Hrousis CA, Drazen JM, Kamm RD. The implications of airway wall buckling in asthmatic airways. *Am J Respir Crit Care Med* 149: A585, 1994.
47. Macklem PT, Proctor DF, Hogg JC. The stability of peripheral airways. *Respir. Physiol.* 8: 191–203, 1970.
48. Kuyper LM, Paré PD, Hogg JC, Lambert RK, Ionescu D, Woods R, Bai TR. Characterization of airway plugging in fatal asthma. *Am J Med* 115: 6–11, 2003.
49. Mullen JBM, Wright JL, Wiggs B, Paré PD, Hogg JC. Reassessment of inflammation in the airways of chronic bronchitis. *BMJ* 291: 1235–39, 1985.
50. Cosio M, Ghezzo M, Hogg JC *et al.* The relation between structural changes in small airways and pulmonary function tests. *N Engl J Med* 298: 1277–81, 1978.
51. Retamales I, Elliott M, Meshi B *et al.* Amplification of inflammation in emphysema and its association with latent adenoviral infection. *Am J Respir Crit Care Med* 164: 469–73, 2001.
52. Speizer FE, Tager IB. Epidemiology of chronic mucus hypersecretion and obstructive airways disease. *Epidemiol Rev* 1: 124–42, 1979.
53. Ciba Guest Symposium Report. Terminology, definitions and classifications of chronic pulmonary emphysema and related conditions. *Thorax* 14: 286–99, 1959.
54. Reid L. Measurement of the bronchial mucous gland layer: A diagnostic yardstick in chronic bronchitis. *Thorax* 15: 132–41, 1960.
55. O'Shaughnessy TC, Ansari TW, Barnes NC, Jeffery PK. Inflammation in bronchial biopsies of subjects with chronic bronchitis: Inverse relationship of CD-8 + T lymphocytes with FEV_1. *Am J Respir Crit Care Med* 155: 382–87, 1997.
56. Saetta M, Di Stefano A, Turato G *et al.* T lymphocytes in the peripheral airways of smokers with chronic obstructive pulmonary disease. *Am J Respir Crit Care Med* 157: 822–26, 1998.
57. Thurlbeck WM, Angus GE. The distribution curve for chronic bronchitis. *Thorax* 19: 436–42, 1964.
58. Jamal K, Cooney TP, Fleetham JA, Thurlbeck WM. Chronic bronchitis: Correlation of morphological findings and sputum production and flow rates. *Am J Respir Dis* 129: 719–22, 1984.
59. Carlile A, Edwards C. Structural variations in the main bronchi of the left lung: a morphometric study. *Br J Dis Chest* 77: 344–48, 1983.
60. Mackenzie HI, Outhred KG. Chronic bronchitis in coal miners: Antimortem/postmortem comparisons. *Thorax* 24: 527–35, 1969.
61. Haraguchi M, Shemura S, Shirata K. Morphologic analysis of bronchial cartilage in chronic obstructive pulmonary disease and bronchial asthma. *Am J Respir Crit Care Med* 159: 1005–13, 1999.
62. Thurlbeck WM, Pun R, Toth J, Fraser RG. Bronchial cartilage in chronic obstructive lung disease. *Am Rev Respir Dis* 109: 73–80, 1974.
63. van Braband T, Cauberghs M, Verbeken E *et al J Appl Physiol: Respir. Environ. Exercise Physiol.* 55: 1733–42, 1983.
64. Dayman H. Mechanics of airflow in health and emphysema. *J Clin Invest* 3031: 1175–90, 1951.
65. Butler J, Caro C, Alkaler R, Dubois AB. Physiological factors affecting airway resistance in normal subjects and in patients with obstructive airways disease. *J Clin Invest* 39: 584–91, 1960.
66. Mead J, Turner JM, Macklem PT, Little J. Significance of the relationship between lung recoil and maximum expiratory flow. *J Appl Physiol* 22: 95–108, 1967.
67. Matsuba K, Thurlbeck WM. The number and dimensions of small airways in emphysematous lungs. *Am J Pathol* 67: 265–75, 1972.
68. Hogg JC, Chu FS, Tan WC *et al.* Survival following lung volume reduction in COPD: Insights from small airway pathology. *Am J Respir Crit Care Med* 16, 2007.
69. McLean KA. Pathogenesis of pulmonary emphysema. *Am J Med* 25: 62–74, 1958.
70. Niewoehner DE, Kleinerman J, Reisst DB. Pathologic changes in the peripheral airways of young cigarette smokers. *N Engl J Med* 291: 755–58, 1974.
71. Auerbach O, Garfinkle L, Hammond EC. Relation of smoking and age to findings in lung parenchyma: A microscopic study. *Chest* 65: 29–35, 1974.
72. Auerbach O, Hammond EC, Garfinkle L, Benante C. Relation of smoking and age to emphysema: Whole lung section study. *N Engl J Med* 286: 853–57, 1972.
73. Petty TL, Silverds GW, Stanford RE, Baird ME, Mitchell MS. Small airway pathology is related to increased closing capacity and abnormal slope of phase III in excised human lungs. *Am Rev Respir Dis* 121: 449–56, 1980.
74. Wright JL, Lawson LM, Paré *et al.* Morphology of peripheral airways in current smokers and ex-smokers. *Am Rev Respir Dis* 127: 474–77, 1983.
75. Ollerenshaw SL, Woolcock AJ. Characteristics of the inflammation in biopsies from large airways in subjects with asthma and chronic airflow limitation. *Am Rev Respir Dis* 145: 922–27, 1992.
76. Gadek JE, Fells JA, Crystal RG. Cigarette smoking induces a functional antiprotease deficiency in the lower respiratory tract of humans. *Science* 206: 315–16, 1979.
77. Hunninghake GW, Crystal RG. Cigarette smoking and lung destruction: Accumulation of neutrophils in the lungs of cigarette smokers. *Am Rev Respir Dis* 128: 833–38, 1983.
78. Stone DJ, Galor JD, McGowan SE *et al.* Functional 1-protease inhibitor in lower respiratory tract of cigarette smokers is not decreased. *Science* 221: 1187–89, 1983.
79. Snider GL, Kleinerman JL, Thurlbeck WM, Bengally ZH. Definition of emphysema. Report of a National Heart, Lung and Blood Institute, Division of Lung Disease Workshop. *Am Rev Respir Dis* 132: 182–85, 1985.

80. Leopold JG, Goeff J. Centrilobular form of hypertrophic emphysema and its relation to chronic bronchitis. *Thorax* 12: 219–35, 1957.
81. Hogg JC, Macklem PT, Thurlbeck WM. The elastic properties of the centrilobular emphysematous space. *J Clin Invest* 48: 1306–12, 1969.
82. Anderson AE, Hernandez JA, Holmes WL, Foraker AG. Pulmonary emphysema: Prevalence, severity and anatomical patterns with respect to smoking habits. *Arch Environ Health* 12: 569–77, 1966.
83. Pratt PC, Kilborn KH. A modern concept of emphysemas based on correlations of structure and function. *Hum Pathol* 1: 445–53, 1970.
84. Laurenzi GA, Toreno GM, Fishman AP. Bullous disease of the lung. *Am J Med* 36: 361–78, 1962.
85. McLean A, Warren PM, Gilooly M, Lamb D. Microscopic and macroscopic measurements of emphysema: Relation to carbon monoxide gas transfer. *Thorax* 47: 144–49, 1992.
86. Connors AF, Dawson NV, Thomas C *et al.* Outcomes following acute exacerbation of severe chronic obstructive lung disease. *Am J Respir Crit Care Med* 154: 959–67, 1996.
87. Stuart-Harris CH. The role of bacterial and viral infection in chronic bronchitis. *Arch Environ Health* 16: 586–95, 1968.
88. Stuart-Harris CH. Infection, the environment and chronic bronchitis. *J Roy Coll Physicians Lond* 5: 351–61, 1971.
89. Knipe DN. Virus–host interactions. In: Fields BM, Knipe DN, Howley PM (eds). *Field's Virology*, pp. 273–99. New York: Lippincott-Raven, 1996.
90. Hers JF, Masurel N, Mulder J. Bacteriology and histopathology of the respiratory tract in fatal Asian influenza. *Lancet* ii: 1141–43, 1958.
91. Lauria DB, Blumenfield HL, Ellis JT, Kilbourne ED, Rogers DE. Studies on influenza in the pandemic of 1957–58. II: Pulmonary complication of influenza. *J Clin Invest* 38: 213–65, 1959.
92. Anthonisen NR, Manfreda J, Warren CP *et al.* Antibiotic therapy in exacerbations of COPD. *Ann Intern Med* 106: 196–204, 1987.
93. Monto AS, Higgins MW, Ross HW. The Tecumseh study of respiratory illness. VIII: Acute infection in chronic respiratory disease and comparison groups. *Am Rev Respir Dis* 111: 27–36, 1975.
94. Smith CB, Golden CA, Canner RE, Renzetti AD. Association of viral and mycoplasmal pneumonia infections with acute respiratory illness in patients with COPD. *Am Rev Respir Dis* 121: 225–32, 1980.
95. Seemungal T, Harper-Owen R, Bhowmik A *et al.* Respiratory viruses, symptoms, and inflammatory markers in acute exacerbations and stable chronic obstructive pulmonary disease. *Am J Respir Crit Care Med* 164(9): 1618–23, 2001.
96. Ko FW, Ip M, Chan PK *et al.* Viral etiology of acute exacerbations of chronic obstructive pulmonary disease in Hong Kong. *Chest* 15, 2007. , Epub ahead of print
97. Tan WC, Xiang X, Qiu D, Ng TP, Lam SF, Hegele RG. Epidemiology of respiratory viruses in patients hospitalized with near-fatal asthma, acute exacerbations of asthma, or chronic obstructive pulmonary disease. *Am J Med* 115(4): 272–77, 2003.
98. Saetta M, Di Stefano A, Maestrelli P *et al.* Airway eosinophilia in chronic bronchitis during exacerbations. *Am J Respir Crit Care Med* 150: 1646–52, 1994.

Airway Remodeling

CHAPTER 7

Stephen T. Holgate

School of Medicine, University of Southampton, Southampton, UK

Airway remodeling may be defined as a process of sustained disruption and modification of structural cells and tissues leading to the development of a new airway-wall structure and as a consequence, new functions. Although this process was noted as long ago as 1922 [1], interest in the underlying in asthma has risen only in recent years, the debate about asthma causation largely being driven by concepts in immunology and inflammation. In chronic obstructive pulmonary disease (COPD), although airway structural changes are well recognized [2], the underlying pathogenesis has received much less attention, possibly due to the more peripheral location of pathology and the overshadowing interest in the adjacent emphysematous tissue destruction.

In both conditions, a number of "remodeling" processes occur; but in proportion, the changes observed are quite different. Principal among these changes are airway fibrosis, an elevation in smooth muscle mass, mucous metaplasia, and glandular hypertrophy, in addition to less well defined alterations of the bronchial vasculature and nerves to create an abnormal airway wall. In asthma, segmental and subsegmental bronchial walls are thickened over their entire size range [3]. In COPD, only the inner wall area of the large airways is convincingly thicker [4]. The peripheral airways (~2 mm diameter), normally devoid of either supporting cartilage or bronchial glands, are also conspicuously remodeled in COPD [5, 6].

NATURAL HISTORY AND CLINICAL IMPORTANCE

The natural history of remodeling is not well understood. Though it is surmised to be a consequence of long-term airways disease, studies have revealed early manifestations in asthmatics [7] and young smokers [8], suggesting that remodeling maybe as much part of the primary pathology of the disease rather than simply being a result of chronic inflammation. However, a considerable degree of variability in susceptibility to remodel the changes exists in both patient groups. Computerized tomography (CT) shows increased airway wall thickening in proportion to disease severity in both children and adults with asthma [9, 10], whereas in highly labile "brittle" asthma, this change is notably absent [11]. At a population level, asthma is associated with a greater than normal decay in lung function over time and associated loss of corticosteroid responsiveness [12]. In a small proportion of asthma patients with severe chronic disease the decline in lung function is more progressive and associated with persistence of symptoms despite corticosteroid therapy, thereby resembling a COPD-like state [13], although the pathology that underlies this in the two disorders differs [14], The accelerated decline in lung function in asthma may reflect not only the presence of airway structural modification, but also the inadequacy of currently available treatments to modify the remodeling processes.

Airway remodeling may have a number of clinical consequences for both asthma and COPD. For example, bronchial hyperresponsiveness has been postulated to be a function of the physical effects of overall airway wall thickening, in addition to airway inflammation and smooth muscle reactivity [15]. Of interest, however is the inverse relationship between airway wall thickening demonstrated by CT and by endobronchial ultrasound and airway hyperresponsiveness [16, 17]. Individual remodeling changes, such as an increased smooth muscle bulk or an enhanced mucus producing facility, have their own implications for both symptom severity and airway function.

MECHANISMS OF REMODELING

Epithelial damage, airway inflammation, and epithelial repair

Responses of the airway epithelial barrier to injury and manipulation and abnormalities in the ensuing processes of repair are the most likely causes of remodeling. In both adult and children with chronic asthma, there is a good evidence that following repeated epithelial injury by environmental factors (e.g. allergens, viruses, and pollutants) the repair processes involve intercellular signaling between the damaged and regenerating epithelium and the underlying subepithelial (myo)fibroblast sheet, to influence matrix synthesis and the mass and composition of structures lying beneath [18]. Under these conditions, the asthmatic airway is behaving like a chronic wound with release of proinflammatory cytokines and multiple growth factors leading to some permanent changes to airway wall morphology. These epithelial–mesenchymal interactions have invited parallels with the intercellular communication found during branching morphogenesis, and remodeling has been suggested to reflect a reactivation of these early life processes (discussed later).

In contrast to asthma, COPD is thought to be due almost exclusively to the toxic effects of tobacco smoke. Repetitive attempts by the epithelium to protect itself and repair the injury induced by this noxious agent leads to marked structural changes to the epithelium with thickening and squamous metaplasia accompanied by an enhanced mesenchymal response at some sites and alveolar destruction at others. The airway obstruction that accompanies these changes is resistant to corticosteroids and in this respect becomes permanent or "fixed."

The importance of airway inflammation, integral and intimately linked to the process of epithelial injury, is recognized in both diseases [19]. In asthma, much interest has focused on the $CD4^+$ T-lymphocyte that is skewed toward a Th2 type phenotype as the orchestrator of an immune response to allergens [20]. The predominant pattern of Th2 cytokine expression results in an inflammatory phenotype where eosinophils and mast cells are prominent, but macrophages, fibroblasts, and dendritic cells are also involved [21]. The inflammatory milieu differs considerably in COPD. Eosinophils may play a role in a subset that has features in common with asthma and also during acute exacerbations [22, 23], but their significance in the majority is outweighed by the contribution of macrophages and neutrophils with a $CD8^+$ T-cell preponderance [24–26]. Indeed $CD8^+$ T-cells are assuming increasing importance in orchestrating the neutrophil-mediated injury in COPD possibly linked to chronic infection. Known perpetrators of epithelial damage in COPD include tobacco smoke toxins, viruses, and bacteria, whereas in asthma house dust mite allergen [27], fungal and pollen enzymes [28], and toxic air pollutants are causally involved, with their effects being augmented by the airway inflammatory responses induced by these agents. Thus, the generation of reactive oxygen species and the products of infiltrating inflammatory cells including arginine-rich eosinophilic proteins [29], mast cell tryptase [30], neutrophil elastase, and metalloproteases derived from eosinophils and mast cells [31, 32], neutrophils and macrophages all induce epithelial injury. The result in asthma is a fragile bronchial epithelium with weakening of junctional adhesion structures, whereas in COPD the epithelium proliferates and thickens. Consequently, a feature of asthma is the shedding of columnar epithelial cells from their basal cell attachments [33–36]. In COPD, although atrophy and shedding of the epithelium may occur [37], focal squamous metaplasia is much more common [38, 39].

Altered airway adhesion molecule expression, a hallmark of an active repair process, has been demonstrated in asthma and COPD. This includes enhanced expression of CD44 [40], the integrins [41], and E cadherin [42] in asthma, while in both asthma and COPD there is augmented expression of ICAM-1 [43]. A recent important finding in asthma is the incomplete formation of tight junctions at the apex of the columnar epithelial cells which is preserved in tissue culture of differentiated asthmatic epithelium many weeks after removal from the patient's airways [44, 45] (Fig. 7.1).

Since tight junctions are crucial in controlling paracellular transport, their disruption in asthma could help to explain why inhaled allergens and other inhaled environmental agents initiate and maintain a chronic inflammatory response at this site. Epithelial expression of the epidermal growth factor receptor (EGFR) family is also increased in asthma, and reflects an injury and repair response seen in other epithelial tissues by interacting with six distinct ligands:

- epidermal growth factor (EGF)
- transforming growth factor α (TGF-α)
- amphiregulin (AR)
- heparin-binding EGF-like growth factor (HB-EGF)
- betacellulin (BTC)
- epiregulin.

Activation of the EGFR promotes both migration and proliferation of epithelial cells [46], and there is *in vitro* and *in vivo* evidence supporting a role for this receptor in bronchial epithelial repair [47, 48]. However, in asthma, increased EGFR expression does not appear to be coupled to an appropriate proliferative response by the repairing epithelium [49]. This impairment may explain why EGFR expression, as a marker of tissue injury, correlates positively with both asthma severity, neutrophil influx [50], and the extent of subepithelial fibrosis [51]. The impaired epithelial repair response in asthma may be well linked to the actions of the profibrogenic transforming growth factor beta (TGF-β) family of cytokines, which are known inhibitors of epithelial proliferation [52] and whose level is markedly increased in asthmatic airways [53].

Although there are substantial changes in epithelial structure in COPD, relatively little is known of the factors driving this. Cigarette smoke results in enhanced EGFR

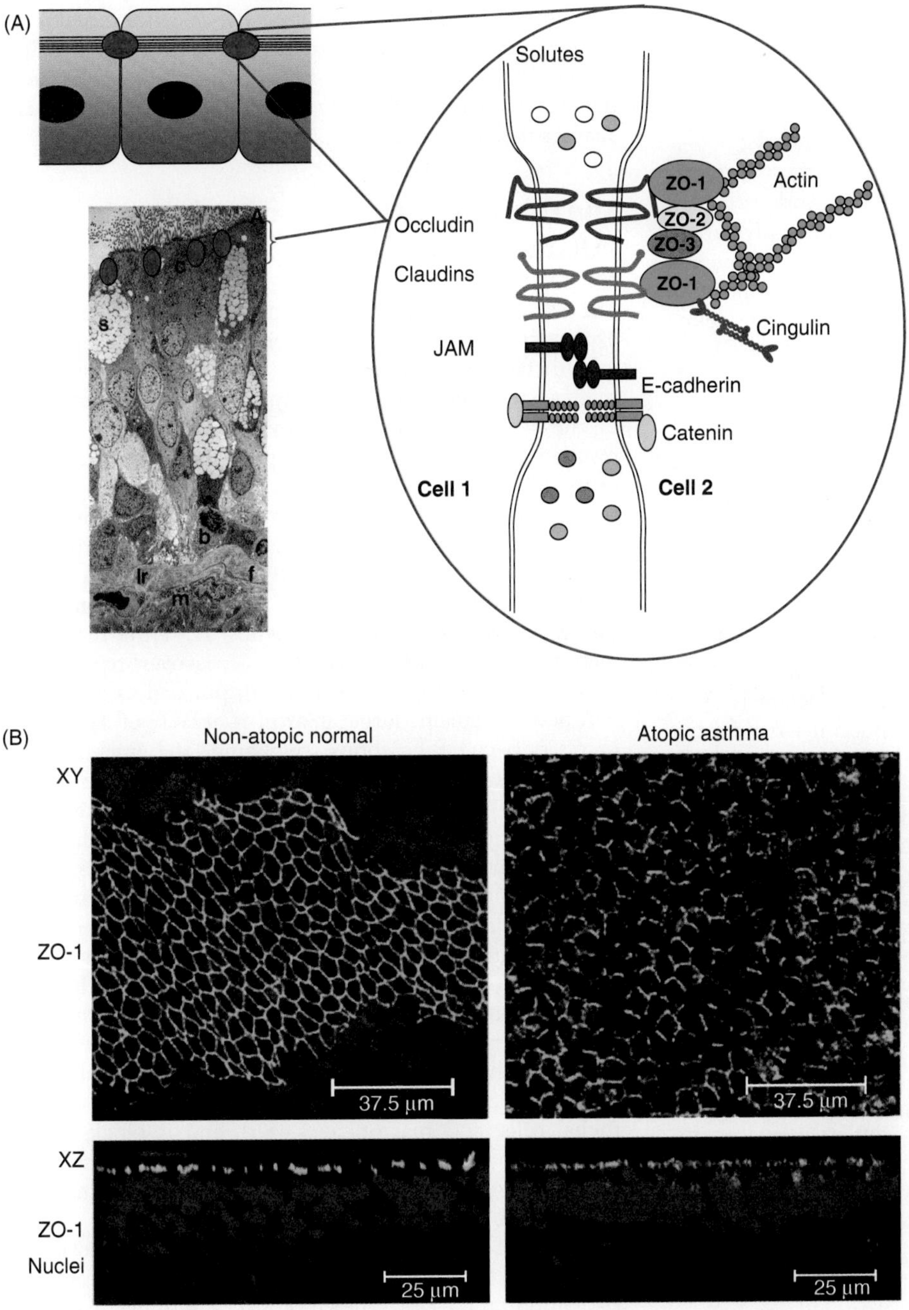

FIG. 7.1 Schematic representation of airway epithelial tight junction showing their localisation at the apex of epithelial cells (A). Confocal images of airway biopsies from atopic normal and asthmatic subjects stained by immunoflurescence for the tight junction protein ZO-1. Note the disrupted staining pattern in asthma (B).

expression in airways epithelium [54] and augmented expression of both EGF and TGF-β have been observed in subjects with chronic bronchitis [53]. Thus in both asthma and COPD, there occurs a potential imbalance between proliferative and anti-proliferative signaling in the airways in the former leading to a frustrated and incomplete response and in the latter an exuberant response.

The relationship between epithelial injury and fibrosis has been examined using co-cultures of bronchial epithelial cells and fibroblasts. In these studies, damage inflicted on the epithelial cells that resulted in enhanced myofibroblast proliferation and collagen gene expression secondary to the release of a number of growth factors including

- basic fibroblast growth factor (FGF1)
- acidic fibroblast growth factor (FGF2)
- insulin-like growth factor (IGF)
- platelet-derived growth factor (PDGF)
- transforming growth factor β (TGF-β)
- Endothelin-1 (ET-1) [55].

In addition to these, a range of other potentially important epithelial-derived products such as nerve growth factor (NGF), nitric oxide, metalloproteinases, fibronectin, (an extracellular matrix component with potent chemotactic properties for fibroblasts [56]) and a range of proinflammatory cytokines [57–60] are produced by epithelial cells during the repair process.

Remodeling of the extracellular matrix

The extracellular matrix is composed of a network of fibrous and structural proteins embedded in a hydrated polysaccharide gel to form the strong, resilient framework of the airway wall. Alteration in both the mass and the composition of this structure is one of the primary aspects of remodeling in both asthma and COPD, with enhanced fibroblast activity being most directly responsible.

The repair environment in the airways produces a number of changes in cytokine and growth factor expression to which the fibroblast is receptive. Studies in asthma have indicated that myofibroblasts, a cell population with morphological and biochemical features intermediate between those of fibroblasts and smooth muscle cells [61], are responsible for the particular pattern of matrix deposition in this disease. These cells, lying below and adjacent to the *lamina reticularis* in a layer referred to as the "attenuated fibroblast sheath" are major producers of collagen and proteoglycans implicated in matrix accumulation, as well as wound contraction [62]. In the absence of specific immunohistochemical markers for this cell type, myofibroblasts are identified by ultrastructural criteria and the expression of α-smooth muscle actin (α-SMA) which is also found in abundance in smooth muscle cells. Though myofibroblast hyperplasia has been demonstrated in the airway subepithelium in asthma [61, 63], with their numbers correlating with the extent of subepithelial collagen thickness [63, 64], the significance of this finding in relation to the remodeling response involving the full thickness of the airway wall is uncertain, however, their increased number is highly a characteristic of asthma. Myofibroblasts (or fibromyocyte as they have also been called) appear early after allergen challenge, perhaps implying a quiescent precursor cell that differentiates to acquire myofibroblast features [61]. Candidates for this include fibroblasts, smooth muscle cells, or perhaps other primitive mesenchymal or structural cells. Recent interest in the recruitment of $CD32^+$ bone marrow-derived fibrocytes from the circulation into the airways following allergen challenge is another possible source of these cells [65]. A further source of myofibroblasts is through epithelial–mesenchymal transition involving TGF-β as described in pulmonary fibrosis [66]. A number of factors have been identified which alter fibroblast α-SMA expression:

- TGF-β [67]
- Heparin [68]
- Interferon-γ
- granulocyte macrophage colony stimulating factor (GM-CSF).

Proliferation and migration of resident progenitors and/or myofibroblasts already present in the airways may contribute to the increase in these cells beneath the epithelium [63]. For example, smooth muscle mitogens and fibroblast growth factors have been detected in bronchoalveolar lavage (BAL) fluid after allergen challenge [69], and fibronectin has been observed to cause selective recruitment of myofibroblasts from a population of normal fetal lung fibroblasts [70]. In COPD, studies investigating the potential role of myofibroblasts suggest an increase in peripheral airways that show the greatest tendency to remodel [71].

The role of growth factors and inflammatory mediators

The use of transgenic animal models of mediator's over-expression has provided further evidence to link epithelial growth factor and inflammatory cytokine generation with matrix remodeling. Specifically, TGF-β, over-expressed in the rat airway epithelium by adenoviral gene transfer results in severe interstitial and pleural fibrosis [72], while over-expression of TGF-β, IL-6, or IL-11 as transgenes in the airway epithelium induces subepithelial fibrosis in the lungs of mouse models [73, 74]. Mouse models expressing the Th2 cytokines IL-5 [75], IL-9 [76], and IL-13 [77] all developed remodeling responses in their airways, although interestingly in mice over-expressing IL-4 the results are inconsistent [78, 79]. Interleukin-13 is over-expressed in asthmatic airways [80] and is being pursued as an important new therapeutic target in asthma because of its involvement in B-cell IgE isotype switching, Th2 inflammation, mucous metaplasia and remodeling effects [81]. The TGF-β superfamily of cytokines (that include the bone morphogenic proteins (BMPs)) are among the most studied with respect to airway fibrosis and remodeling. TGF-β is a multifunctional profibrotic growth factor and a key component in the regulation of tissue growth and differentiation during branching morphogenesis and wound repair as well as promoting smooth muscle differentiation [82]. It stimulates fibroblast growth and is responsible for the differentiation of fibroblasts into myofibroblasts [55, 83, 84]. TGF-β promotes synthesis of extracellular matrix components by a number of cells including fibroblasts, smooth muscle, epithelial cells, and macrophages [85], and it blocks matrix degradation by inhibiting proteolytic enzyme synthesis and augmenting the action of protease inhibitors with increased expression both in asthma and COPD [53, 86, 87]. Tissue macrophage TGF-β production is elevated in both diseases, while in asthma eosinophils and epithelial cells are also important sources [53, 88, 89], a significant correlation being demonstrated between the level of expression of TGF-β and the extent of subepithelial fibrosis and numbers of fibroblasts [53]. TGF-β levels in asthmatic airways are unaffected by corticosteroid treatment.

Another growth factor cluster is platelet-derived growth factor (PDGF). This is produced mainly in the airways by macrophages, epithelium, and most airway inflammatory cell types including mast cells and eosinophils as important sources of in asthma [90, 91]. PDGF promotes fibroblast chemotaxis, fibrosis, and smooth muscle mitogenesis [92] and would seem to be another very plausible mediator of airway remodeling with asthmatic bronchial fibroblasts showing enhanced responsiveness [93]. Nevertheless, it has been difficult to show increased expression of PDGF in the airways either in asthma or in COPD [94, 95], so that its relevance in these diseases has yet to be clearly demonstrated. A family of profibrotic mediators are the endothelins (1 and 2) whose levels are increased in both diseases [96–98]. These agents, in addition to their well described vasoconstrictor and bronchoconstrictor properties, the endothelins are potent activators of fibroblasts for enhanced matrix production [99].

Although in asthma epithelial EGFR expression has been shown to correlate with the extent of subepithelial fibrosis, the EGFR as an effector of matrix remodeling mesenchymal cells is also important, along with platelet-derived growth factor (PDGF), insulin-like growth factor, fibroblast growth factors and TGF-β. Stimulation of fibroblast-bound EGF and PDGF receptors works in concert with binding of β1 integrins also present on the surface of these cells by matrix components such as laminin or fibronectin to stimulate fibroblast chemotaxis and migration [100, 101]. Thus, alterations in growth factor and receptor expression, matrix composition, and adhesion molecule binding all have the potential to influence fibroblast activity in the airway repair response. Although EGF immunoreactivity has been demonstrated in the airway mucosa of both asthmatic and COPD patients, the number of EGF-expressing cells did not correlate with either basement membrane thickness or underlying fibroblast number [53]. Increased epithelial expression of TGF-α, another ligand of the EGF family results in lung fibrosis, but the effect is mediated via the epithelium to enhance other fibroblast growth factor secretion rather than through a direct action on fibroblasts [46].

The role of proteolytic enzymes

Alteration in matrix turnover is a critical factor in the remodeling process involving a range of proteolytic enzymes. Serine proteases and matrix metalloproteases that are produced by a variety of inflammatory and stromal cells can digest all the major components of the extracellular matrix [102]. Examples include MMP3 (stromolysin) and MMP9 in asthma [103] [104] and MMP1, MMP9, and MMP12 in COPD (with regard to their potential to induce emphysema) [105, 106]. In emphysema it is believed that a protease/anti-protease imbalance exists favoring the excessive proteolytic digestion of lung parenchyma and especially elastin, although there is now strong evidence that deposition of new matrix is an equally important component of COPD involving the more peripheral airways and contributing to increased airway stiffness [107]. Thus, although both MMP9 and neutrophil elastase levels are elevated in induced sputum of asthmatic and COPD patients, the augmented presence of these proteases is outweighed by a proportionally larger increase in their natural inhibitors TIMP1 and α1-antitrypsin [108, 109] which would favor fibrosis.

The release of growth factors that are encrypted in the extracellular matrix or from their cell membrane-bound precursors is another way that proteases may contribute to remodeling. For example, both FGF and TGF-β are bound as inactive forms to heparan sulfate and decorin respectively [110, 111] and can be released from these matrix stores by the proteolytic activity of plasmin and MMPs. In contrast, release of heparin-binding EGF (hb-EGF), a further member of the EGF family and a potent smooth muscle mitogen, is dependent on MMP3-induced cleavage of its transmembrane precursor [42]. Mast cell tryptase, acting via cellular protease activated receptor (PAR)2 acts directly on fibroblasts and smooth muscle cells to promote both mitogenesis and collagen secretion [112] and is also mitogenic for epithelialial cells and enhances profibrogenic growth factor release [113].

Effects of the altered matrix on remodeling

The capacity of the extracellular matrix to directly and indirectly influence cell migration, proliferation, and maturation is indicative of a highly dynamic function in addition to provide structural support [85, 114]. Matrix components such as decorin, versican, and fibronectin provide stimuli for inflammatory, epithelial, and stromal cells by serving as ligands for adhesion molecules [115] and by acting as a reservoir for release of cytokines, chemokines, and growth factors such as TGF-β [116]. Glycosaminoglycans, such as hyaluronic acid (HA), found to be increased in the BAL fluid of asthmatics [117], facilitate cell migration, and proliferation during injury and repair, [118] by interacting with receptors such as CD44 [119], while survival of eosinophils is prolonged by interaction with fibronectin and laminin, which inhibit apoptosis in part through the autocrine effect of GM-CSF [120]. Indeed, fibronectin is a matrix glycoprotein that has received considerable attention regarding its role in repair [85, 121]. It is produced by bronchial epithelial cells [122] fibroblasts, smooth muscle cells, and macrophages [123] and is upregulated by growth factors such as TGF-β [124] and integrin ligation [125]. Fibronectin is incorporated into the provisional matrix that forms after injury, where it acts as a chemoattractant for both epithelial cells and fibroblasts [126]. Elevated quantities of fibronectin have been detected in bronchial lavage fluid in COPD and in asthma where it is also found abnormally deposited in the subbasement membrane along with tenascin C, another matrix glycoprotein that is chemotactic for leukocytes and repair cells [63, 127, 128].

RESULTS OF MATRIX REMODELING IN ASTHMA AND COPD

The pattern and distribution of matrix deposition in the remodeled airway differs between asthma and COPD. Interest in asthma has centered mainly on the larger airways, whereas in COPD it is involvement of small airways where remodeling dominates the pathology. Matrix remodeling may be roughly subdivided into thickening of the subepithelial basement membrane (*lamina reticularis* or basal lamina), submucosal thickening, matrix deposition in airway smooth muscle and new matrix deposited in the airway adventitia outside the smooth muscle layer. In addition, the proliferation of new blood vessels, nerves, and muscle as well as mucous metaplasia and formation of larger and more numerous submucous glands are all part of the remodeling response.

Subepithelial basement membrane matrix deposition in asthma and COPD

Deposition of protein beneath the true epithelial basement membrane is characteristic of asthma. Electron microscopic and immunohistochemical analysis of bronchial biopsy specimens has shown that the true basement membrane, made up of the *lamina rara* and *lamina densa*, is normal in both size and composition and that the changes occur in the *lamina reticularis*. In this region, abnormal deposition of interstitial repair collagen subtypes I, III, and V takes place [129] in addition to non-collagenous matrix components that include fibronectin [129], laminin β_2 [130], and tenascin C [131]. Classical epithelial-derived membrane subtypes, such as collagens IV and VII, are absent from this abnormal subepithelial matrix layer. Recent studies in asthma using the blocking anti-IL-5 monoclonal antibody, mepolizumab, have shown that depletion of IL-5 and partial depletion of airway eosinophils results in reduced immunostaining for tenascin C, lumican, and collagen III in the *lamina reticularis* [132], but, interestingly, this is not paralleled by remission of asthma [133]. Similar effects on the matrix component of the late phase allergen response have also been reported [134]. It is possible that more prolonged anti-IL-5 treatment in more severe asthma might be more clinically effective since mepolizumab is highly effective in eosinophilic oesophagitis and nasal polyposis [135].

Studies of COPD have not revealed an equivalent to the subepithelial fibrosis as observed in asthma, although in post-transplant obliterative bronchiolitis, this change occurs secondary to epithelial damage, epithelial-mesenchymal transition, and proliferation of myofibroblasts [136, 137]. Studies of COPD show reveal normal basement membrane thickness [37, 138], although in a subset of patients with COPD who display overlap with the asthmatic phenotype thickening of the *lamina reticularis* is apparent especially in the presence of eosinophils [139]. These individuals, who have a BAL eosinophilia and significant corticosteroid reversibility, also show demonstrable basement membrane thickening, but the composition remains unknown [140]. Further studies are needed to properly define the clinical and pathological characteristics of this subgroup and any overlap with late onset "intrinsic" asthma.

Diffuse matrix deposition in asthma and COPD

Studies evaluating matrix deposition deep to the *lamina reticularis* in asthma have been few. An excess of collagen including types III and V have been found in large airway samples by some investigators [141, 142], but not by others [129, 143]. Other matrix proteins and proteoglycans found in excess in this region in asthma include decorin, lumican, biglycan, versican, and fibronectin [108]. Excess matrix deposition in COPD has been identified predominantly in the peripheral, noncartilaginous airways (<2 mm diameter) [39, 144]. However, one immunohistochemical investigation has demonstrated reduced decorin and biglycan in the peripheral airways, with staining patterns for type IV collagen and laminin similar to those observed in control lungs [145].

MUCOUS METAPLASIA

In normal lung, submucosal mucous glands and epithelial goblet cells are distributed throughout the cartilagenous airways in. In asthma and COPD, epithelial mucous metaplasia and hyperplasia occur with hypertrophy of the submucosal gland mass. An increase in both the number and the size of mucus-secreting cells leads to enlargement of these tracheobronchial glands in both diseases [146, 147]; however, a variable degree of replacement of serous with mucous acini occurs in COPD, but not in asthma [148].

Epithelial mucous cell metaplasia is observed in both central and peripheral airways in asthma [149, 150] but there exists a substantial degree of individual variation. In those who have died due to asthma goblet cells are found in the peripheral airways [151], compatible with most asthma deaths occurring in association with excessive mucous occlusion of airways [152]. In COPD, mucous metaplasia and hyperplasia are observed both centrally and peripherally [2, 153, 154] resulting in a more even distribution of secretary cells throughout the airways. Thus, the smaller (<400 μm diameter) airways which are normally populated with very few goblet cells become important contributors to the excess mucous which characterizes this disease [155, 156]. Indeed, the mucus produced in asthma and COPD is qualitatively and quantitatively abnormal with major alterations in its cellular and molecular composition. The elevated ratio of mucous/serous acini in COPD results in secretion of a more gel-like thicker mucus, which is also lacking in anti-proteases [157]. In severe asthma, bronchial obstruction from mucus plugging occurs in both central and peripheral airways [151, 158]. In COPD, partial or complete occlusion of the small (<2 mm diameter) airways with mucous plugs is common [5]. In both diseases, reduced levels of surfactant lining the small airways results in increased surface tension, airway collapse, and difficulty in re-expansion [159].

Expression of mucin genes, which encode the mucin glycoproteins, is the principal factor governing the differentiation of epithelial cells into goblet cells [160]. Both environmental and host factors, acting on the epithelium, have been shown to stimulate mucin gene upregulation and mucin secretion. Environmental factors include infectious agents [161] and environmental pollutants [162] while host factors include inflammatory mediators [163] and cell degranulation products [164, 165]. Acrolein, a low molecular weight component of cigarette smoke, induces epithelial MUC5AC gene expression, and mucous metaplasia in rats *in vivo* [166] ,while the Th2 cytokines IL-4, IL-9, and IL-13 have been closely linked to both augmented mucin gene expression, particularly the MUC5AC and MUC2 genes, and goblet cell differentiation both *in vitro* and *in vivo* [167–170].

A central role has been postulated for the neutrophil in stimulation of airway mucin production. This is based on the observation of augmented MUC5AC expression by cultured epithelium when exposed to neutrophils [171], possibly through neutrophil elastase and oxidative stress, each of which has been shown to augment both epithelial

MUC5AC mRNA and protein expression [171–174]. The EGFR is involved in regulation of airway mucin synthesis [175, 176] as mucin gene expression results from ligand-independent transactivation of the EGFR in response to oxidative stress [177] that can be blocked by selective inhibitors of the EGFR tyrosine kinase [131]. Both EGF and EGFR expression are elevated in bronchial glandular tissue in asthmatic subjects compared with controls [178] suggesting a similar role for the EGFR in asthma. IL-13 may cause induction of mucin gene expression via an EGFR-dependent pathway.

SMOOTH MUSCLE REMODELING

The presence of an augmented airway smooth muscle bulk in asthma and COPD is well established. In asthma, smooth muscle mass is increased in both large [179] and peripheral [180, 181] airways, although debate continues over whether hypertrophy or hyperplasia dominate. Autopsy studies suggest the existence of two distinct patterns. The type 1 pattern has increased muscle mass due to hyperplasia restricted to large central airways, while in type 2 there is smooth muscle thickening throughout the bronchial tree caused predominantly by hypertrophy, particularly in the small airways, with a mild degree of hyperplasia in the larger airways [182]. The situation is further complicated by the considerable degree of heterogeneity present. An increase in smooth muscle mass is not always demonstrated in mild asthma [183] or in post-mortem studies of asthmatics whom have died from other causes [184]. Why different patterns of smooth muscle enlargement should exist in asthma or what factors predispose some but not other asthmatics to the development of either one is unknown. Nor is it known whether smooth muscle changes occur before, in parallel with or as a consequence of the inflammation.

The peripheral airways are also the location of an increase in smooth muscle mass in COPD [183, 185–187], though it is suggested that the changes are less marked than those in asthma [183]. Studies often describe these changes under the heading of hypertrophy, but the extent to which hyperplasia plays a role is again unclear. Estimates of smooth muscle enlargement in the larger airways in COPD have varied [188]. Some investigators have observed no abnormality in smooth muscle area [179], while others report up to a twofold increase in bronchial smooth muscle thickness attributable to both hyperplasia and hypertrophy [189]. A clinical correlation with the presence of wheeze has also been suggested [190].

The mechanisms leading to the increase in smooth muscle in asthma (and COPD) are largely speculative and include the growth promoting potential of an enriched plasma environment due to microvascular leakage [191]. The mitogenic influences of mediators involved in the inflammatory response and an intrinsic abnormality of smooth muscle itself. However, most knowledge of factors that promote smooth muscle growth has come from *in vitro* cell culture studies [192].

An extensive list of smooth muscle mitogens has been identified:

- inflammatory mediators
- growth factors
- enzymes
- components of the extracellular matrix [192].

Group 1 are mediators, such as EGF and PDGF, which activate tyrosine kinase receptors. Group 2 includes those receptors coupled to GTP-binding proteins such as thrombin and mast cell tryptase [193]. Although ET-1 and LTD4 are smooth muscle mitogens in animal models, their effects on human smooth is controversial, although recent studies suggest an interaction with tyrosine kinase receptors possibly through the activation of metalloproteases and the mobilization of growth factors such as EGF ligands from their cell-bound precursors [193, 194]. TGF-β and PDGF isoforms are both differentiation factors for airway smooth muscle, but at the time of writing their role in smooth muscle development in asthmatic and COPD airways remains speculative.

A further association between inflammation and smooth muscle function has been suggested by the enhanced expression of adhesion molecules such as ICAM-1 and VCAM-1 on airway smooth muscle cells in response to TNF-α, IL-1, LPS, and IFNs [195]. Furthermore, the adherence of T-cells to smooth muscle cells results in stimulation of DNA synthesis [195]. The capacity of these ASM to express adhesion molecules, synthesize ECM components, and release inflammatory cytokines and chemokines including RANTES [196], eotaxin, and IL-8 [192] indicates that smooth muscle remodeling may itself be part of the inflammatory process. Towards this end, there is now good evidence that in asthma smooth muscle cells and their associated matrix support a distinct mast cell population that maybe highly relevant to the pathogenesis of hyperresponsiveness [197, 198].

In asthma and COPD, polymorphism of the disintegrin and metalloprotease 33 (ADAM33) molecule is strongly associated with BHR and an accelerated decline in lung function over time [199, 200]. ADAM33 is localized to airway smooth muscle and is implicated in vasculogenesis through release of its soluble form [201]. There is also much interest in the factors that are chemotractant for smooth muscle cells, their relationship to myofibroblasts and whether mechanisms for removal of muscle could be therapeutically beneficial. Towards this end, bronchoscopic thermoplasty is looking very promising as an entirely new approach to asthma treatment [202, 203].

VASCULAR AND NEURAL ALTERATIONS

Elevated airway-wall blood-vessel area has been demonstrated in adult and children's asthma and is greater than in controls or subjects with COPD [180, 183, 204]. However, it is unclear whether this vascular remodeling is primarily due to the formation of new blood vessels [205] or to enlargement of the existing microvasculature. A study of the membranous bronchioles of subjects with asthma and

COPD has suggested the latter [183]. Indeed fatal asthma is known to be associated with dilatation of bronchial mucosal blood vessels, congestion, and wall edema [206]. However, Li and Wilson [207] have discovered changes suggestive of new vessel formation in mild asthma, and the microenvironment in asthma has been shown to possess the potential for angiogenesis.

Mediators such as histamine, heparin, and tryptase all possess angiogenic properties, while expression of vascular endothelial growth factor (VEGF), a potent vascular growth factor, is upregulated by agents such as TGF-β, TNF-α, and TGF-α that are involved in the inflammatory milieu of the asthmatic airway. Of particular importance is VEGF which is released from epithelial cells as well as a broad range of inflammatory cells including mast cells. The pro-angiogenic function of soluble ADAM33 is also relevant here.

There is evidence in severe asthma of increased airway neural networks. This is most likely the consequence of nerve growth factor (and related neurotrophins) release from epithelial and inflammatory cells. Circulating levels of NGF relate [208] strongly to asthma severity and may prove to be a useful biomarker of this more severe phenotype [209]. In addition, in both asthma and COPD, the compromised epithelial barrier allows a greater exposure of nerve endings to environmental stimuli. Thus, the potential for release of neurotransmitters such as the tachykinins substance P and neurokinin A is increased. These agents, in addition to their effects on vascular and smooth muscle homeostasis, can contribute to local inflammation with the attraction and activation of inflammatory cells. Neurogenic inflammation is one way sensory information received from the inhaled environment could contribute to ongoing inflammation, although this has been difficult to prove in asthma or COPD but is easily shown in animal models of airway inflammation [210].

EFFECTS OF THERAPY

Whether any of the current anti-inflammatory strategies can significantly alter the course of remodeling is unknown. Long-term corticosteroid therapy has been shown to slow the annual rate of decline in lung function in adult asthmatics [211]. However, other studies have suggested that a negative relationship exists between response to treatment and duration of disease [212]. Added to this are reports showing persistent airflow obstruction in some patients, despite both oral and inhaled therapy [208]. The implication is that early therapy may, to a limited extent, prevent remodeling, but established structural change is steroid-insensitive. Several recent birth cohort studies [213, 214] in childhood asthma have shown that inhaled corticosteroids modify asthma symptoms while being continuously administered, but they have little or no effect on the natural history of asthma. Similar findings have been reported in the large CAMP and START inhaled corticosteroid intervention studies in older asthmatic children [215, 216]. This is somewhat surprising if it is thought that airway remodeling and the increase in smooth muscle is the direct consequence of chronic inflammation. Another possibility is that the structural changes either precede or occur in parallel rather than sequential to inflammation. While *in vitro* studies suggest possible inhibitory effects of bronchodilator drugs such as β2-adrenoceptor agonists and xanthines (e.g. theophylline) on smooth muscle and fibroblast proliferation [217, 218], similar effects on these measures *in vivo* are far from established.

It has been proposed that a more effective strategy to reduce airway smooth muscle in asthma would be to interfere with growth factor and other receptor mechanisms. Suggested approaches include blocking the actions of TGF-β in an attempt to limit smooth muscle differentiation as well as fibrosis. Monoclonal antibodies targeted to TGF-β are in clinical trials in diffuse fibrotic disorders such as scleroderma, but concerns over safety have so far prevented their exploration in asthma. The other functions of this growth factor such as the development of anti-inflammatory T-regulatory cells need to be taken into account as possible alternative on target side effects [219]. Whether targeting other mechanisms such as thrombin inhibitors, protease activated receptors themselves, endothelin and its receptors and other growth factors such as PDGF are very much in the early experimental stage. So far, bronchial thermoplasty seems the most promising approach although even with this, side effects of delayed airway fibrosis is a concern. One exciting possibility, however, is the inhibition of IL-13, a cytokine with multiple effects on inflammation, remodeling, and mucous metaplasia. After successfully pass in phase I safety studies, a number of human and humanized monoclonal antibodies targeting the IL-13 pathway are in clinical development in asthma. The clear advantage here is that blockade of this pathway may influence several interacting aspects of chronic asthma.

Attempts are in progress to reduce mucus secretion. While inhibiting EGFR signaling would be one route the downstream consequences of this on epithelial tissue repair responses would be serious, especially since epithelial repair in asthma is already known to be defective [220]. However, the chloride channel Gob 5 (hCLCA1) as an IL-13 sensitive mechanism that influences the secretion of mucus, is a promising target for small molecular weight inhibitors [221].

A protective role for basement membrane thickening has been postulated also in restricting inflammatory cell passage to the epithelial layer above, while smooth muscle hypertrophy may possibly help to maintain bronchiolar caliber in severe emphysema [164]. Thus if a new "anti-remodeling" agent were to emerge, the beneficial effects would have to be weighed against the consequences of interfering with what may, in some ways, be a valuable defence mechanism.

CONCLUDING COMMENTS

While airway wall remodeling undoubtedly occurs in both asthma and COPD, its physiological significance in health and disease remains speculative. Of concern is the lack

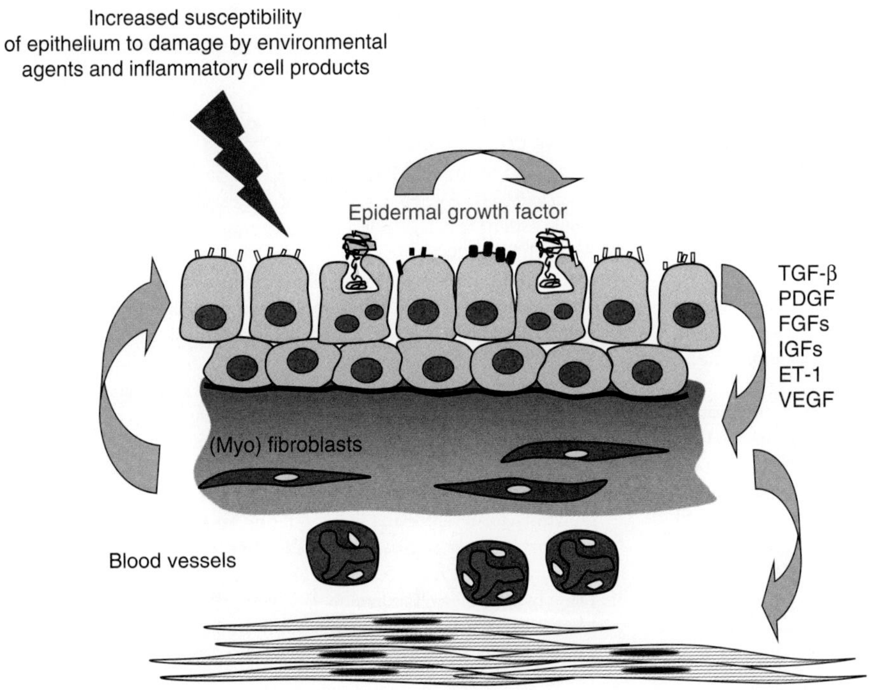

FIG. 7.2 Schematic representation of the epithelial mesenchymal trophic unit (EMTU) in asthma that becomes reactivated with the secretion of many growth factors by the epithelium and underlying myofibroblasts sheath.

of knowledge connecting inflammation to the structural changes. In asthma an overarching hypothesis is emerging that involves aberrant communication between a chronically damaged epithelium and the underlying mesenchyme reminiscent of reactivation of the epithelial mesenchymal trophic unit (EMTU) involved in fetal branching morphogenesis of the lung [222, 223] (Fig. 7.2). The inability of an efficient epithelial wound healing response is responsible for the production of a large number of growth factors implicated in fibrosis, vascuogenesis and new nerve formation, but whether this process is also involved in smooth muscle enhancement in asthma remains conjectural. The chronically damaged/repairing EMTU is also the source of a range of cytokines and growth factors to support chronic inflammation. Thus directing new therapies to increase resistance of the asthmatic airways to environmental insults may create a new approach to treatment.

References

1. Huber H. The pathology of bronchial asthma. *Arch Int Med* 30: 689–760, 1992.
2. Nagai A, West WW, Paul JL, Thurlbeck WM. The National Institutes of Health Intermittent Positive-Pressure Breathing trial: Pathology studies. I. Interrelationship between morphologic lesions. *Am Rev Respir Dis* 132(5): 937–45, 1985.
3. James AL, Pare PD, Hogg JC. The mechanics of airway narrowing in asthma. *Am Rev Respir Dis* 139(1): 242–46, 1989.
4. Tiddens HA, Pare PD, Hogg JC, Hop WC, Lambert R, de Jongste JC. Cartilaginous airway dimensions and airflow obstruction in human lungs. *Am J Respir Crit Care Med* 152(1): 260–66, 1995.
5. Hogg JC, Macklem PT, Thurlbeck WM. Site and nature of airway obstruction in chronic obstructive lung disease. *N Engl J Med* 278(25): 1355–60, 1968.
6. Macklem PT, Thurlbeck WM, Fraser RG. Chronic obstructive disease of small airways. *Ann Intern Med* 74(2): 167–77, 1971.
7. Pohunek P. Eosinophilic inflammation in the bronchial mucosa of children with bronchial asthma. *Eur Respir J* 19(25), 1997.
8. Niewoehner DE, Kleinerman J, Rice DB. Pathologic changes in the peripheral airways of young cigarette smokers. *N Engl J Med* 291(15): 755–58, 1974.
9. Vignola AM, Paganin F, Capieu L, Scichilone N, Bellia M, Maakel L *et al.* Airway remodelling assessed by sputum and high-resolution computed tomography in asthma and COPD. *Eur Respir J* 24(6): 910–17, 2004.
10. De BJ, Scheinmann P. The use of imaging techniques for assessing severe childhood asthma. *J Allergy Clin Immunol* 119(4): 808–10, 2007.
11. Boulet L, Belanger M, Carrier G. Airway responsiveness and bronchial-wall thickness in asthma with or without fixed airflow obstruction. *Am J Respir Crit Care Med* 152(3): 865–71, 1995.
12. Peat JK, Woolcock AJ, Cullen K. Rate of decline of lung function in subjects with asthma. *Eur J Respir Dis* 70(3): 171–79, 1987.
13. Brown PJ, Greville HW, Finucane KE. Asthma and irreversible airflow obstruction. *Thorax* 39(2): 131–36, 1984.
14. Mauad T, Dolhnikoff M. Pathologic similarities and differences between asthma and chronic obstructive pulmonary disease. *Curr Opin Pulm Med* 14(1): 31–38, 2008.
15. Moreno RH, Hogg JC, Pare PD. Mechanics of airway narrowing. *Am Rev Respir Dis* 133(6): 1171–80, 1986.
16. Park JW, Hong YK, Kim CW, Kim DK, Choe KO, Hong CS. High-resolution computed tomography in patients with bronchial asthma: Correlation with clinical features, pulmonary functions and bronchial hyperresponsiveness. *J Investig Allergol Clin Immunol* 7(3): 186–92, 1997.
17. Shaw TJ, Wakely SL, Peebles CR, Mehta RL, Turner JM, Wilson SJ *et al.* Endobronchial ultrasound to assess airway wall thickening: Validation *in vitro* and *in vivo*. *Eur Respir J* 23(6): 813–17, 2004.
18. Holgate ST, Holloway J, Wilson S, Bucchieri F, Puddicombe S, Davies DE. Epithelial-mesenchymal communication in the pathogenesis of chronic asthma. *Proc Am Thorac Soc* 1(2): 93–98, 2004.
19. Jeffery PK. Structural and inflammatory changes in COPD: A comparison with asthma. *Thorax* 53(2): 129–36, 1998.
20. Holgate ST, Wilson JR, Howarth PH. New insights into airway inflammation by endobronchial biopsy. *Am Rev Respir Dis* 145(2 Pt 2): S2–6, 1992.

21. Djukanovic R, Roche WR, Wilson JW, Beasley CR, Twentyman OP, Howarth RH *et al.* Mucosal inflammation in asthma. *Am Rev Respir Dis* 142(2): 434–57, 1990.
22. Saha S, Brightling CE. Eosinophilic airway inflammation in COPD. *Int J Chron Obstruct Pulmon Dis* 1(1): 39–47, 2006.
23. Chanez P, Vignola AM, O'Shaugnessy T, Enander I, Li D, Jeffery PK *et al.* Corticosteroid reversibility in COPD is related to features of asthma. *Am J Respir Crit Care Med* 155(5): 1529–34, 1997.
24. Saetta M, Di SA, Maestrelli P, Ferraresso A, Drigo R, Potena A *et al.* Activated T-lymphocytes and macrophages in bronchial mucosa of subjects with chronic bronchitis. *Am Rev Respir Dis* 147(2): 301–6, 1993.
25. Keatings VM, Collins PD, Scott DM, Barnes PJ. Differences in interleukin-8 and tumor necrosis factor-alpha in induced sputum from patients with chronic obstructive pulmonary disease or asthma. *Am J Respir Crit Care Med* 153(2): 530–34, 1996.
26. O'Shaughnessy TC, Ansari TW, Barnes NC, Jeffery PK. Inflammation in bronchial biopsies of subjects with chronic bronchitis: Inverse relationship of CD8+ T lymphocytes with FEV1. *Am J Respir Crit Care Med* 155(3): 852–57, 1997.
27. Herbert CA, King CM, Ring PC, Holgate ST, Stewart GA, Thompson PJ *et al.* Augmentation of permeability in the bronchial epithelium by the house dust mite allergen Der p1. *Am J Respir Cell Mol Biol* 12(4): 369–78, 1995.
28. Hassim Z, Maronese SE, Kumar RK. Injury to murine airway epithelial cells by pollen enzymes. *Thorax* 53(5): 368–71, 1998.
29. Frigas E, Loegering DA, Gleich GJ. Cytotoxic effects of the guinea pig eosinophil major basic protein on tracheal epithelium. *Lab Invest* 42(1): 35–43, 1980.
30. Redington AE, Polosa R, Walls AF, Howarth PH, Holgate ST. Role of mast cells and basophils in asthma. *Chem Immunol* 62: 22–59, 1995.
31. Shute JK, Parmar J, Holgate ST, Howarth PH. Urinary glycosaminoglycan levels are increased in acute severe asthma–a role for eosinophil-derived gelatinase B? *Int Arch Allergy Immunol* 113(1-3): 366–67, 1997.
32. Ohno I, Ohtani H, Nitta Y, Suzuki J, Hoshi H, Honma M *et al.* Eosinophils as a source of matrix metalloproteinase-9 in asthmatic airway inflammation. *Am J Respir Cell Mol Biol* 16(3): 212–19, 1997.
33. Lackie PM. The impact of allergen on the airway epithelium. *Clin Exp Allergy* 27(12): 1383–86, 1997.
34. Ollerenshaw SL, Woolcock AJ. Characteristics of the inflammation in biopsies from large airways of subjects with asthma and subjects with chronic airflow limitation. *Am Rev Respir Dis* 145(4 Pt 1): 922–27, 1992.
35. Laitinen LA, Heino M, Laitinen A, Kava T, Haahtela T. Damage of the airway epithelium and bronchial reactivity in patients with asthma. *Am Rev Respir Dis* 131(4): 599–606, 1985.
36. Jeffery PK, Wardlaw AJ, Nelson FC, Collins JV, Kay AB. Bronchial biopsies in asthma. An ultrastructural, quantitative study and correlation with hyperreactivity. *Am Rev Respir Dis* 140(6): 1745–53, 1989.
37. Ollerenshaw SL, Woolcock AJ. Characteristics of the inflammation in biopsies from large airways of subjects with asthma and subjects with chronic airflow limitation. *Am Rev Respir Dis* 145(4 Pt 1): 922–27, 1992.
38. Jeffery PK. Structural and inflammatory changes in COPD: A comparison with asthma. *Thorax* 53(2): 129–36, 1998.
39. Cosio M, Ghezzo H, Hogg JC, Corbin R, Loveland M, Dosman J *et al.* The relations between structural changes in small airways and pulmonary-function tests. *N Engl J Med* 298(23): 1277–81, 1978.
40. Lackie PM, Baker JE, Gunthert U, Holgate ST. Expression of CD44 isoforms is increased in the airway epithelium of asthmatic subjects. *Am J Respir Cell Mol Biol* 16(1): 14–22, 1997.
41. Peroni DG, Djukanovic R, Bradding P, Feather IH, Montefort S, Howarth PH *et al.* Expression of CD44 and integrins in bronchial mucosa of normal and mildly asthmatic subjects. *Eur Respir J* 9(11): 2236–42, 1996.
42. Bullock GR. Expression of E-cadherin and alpha, beta and gamma catenin is associated with epithelial shedding in human lung and nose. *J Clin Immunol* 101(1 part 2): 364–70, 1998.
43. Vignola AM, Campbell AM, Chanez P, Bousquet J, Paul-Lacoste P, Michel FB *et al.* HLA-DR and ICAM-1 expression on bronchial epithelial cells in asthma and chronic bronchitis. *Am Rev Respir Dis* 148(3): 689–94, 1993.
44. Holgate ST. Epithelium dysfunction in asthma. *J Allergy Clin Immunol* 120(6): 1233–44, 2007.
45. de Boer WI, Sharma HS, Baelemans SM, Hoogsteden HC, Lambrecht BN, Braunstahl GJ. Altered expression of epithelial junctional proteins in atopic asthma: Possible role in inflammation. *Can J Physiol Pharmacol* 86(3): :105–12, 2008.
46. Davies DE, Polosa R, Puddicombe SM, Richter A, Holgate ST. The epidermal growth factor receptor and its ligand family: Their potential role in repair and remodelling in asthma. *Allergy* 54(8): 771–83, 1999.
47. Madtes DK, Busby HK, Strandjord TP, Clark JG. Expression of transforming growth factor-alpha and epidermal growth factor receptor is increased following bleomycin-induced lung injury in rats. *Am J Respir Cell Mol Biol* 11(5): 540–51, 1994.
48. Van Winkle LS, Isaac JM, Plopper CG. Distribution of epidermal growth factor receptor and ligands during bronchiolar epithelial repair from naphthalene-induced Clara cell injury in the mouse. *Am J Pathol* 151(2): 443–59, 1997.
49. Demoly P, Simony-Lafontaine J, Chanez P, Pujol JL, Lequeux N, Michel FB *et al.* Cell proliferation in the bronchial mucosa of asthmatics and chronic bronchitics. *Am J Respir Crit Care Med* 150(1): 214–17, 1994.
50. Hamilton LM, Torres-Lozano C, Puddicombe SM, Richter A, Kimber I, Dearman RJ *et al.* The role of the epidermal growth factor receptor in sustaining neutrophil inflammation in severe asthma. *Clin Exp Allergy* 33(2): 233–40, 2003.
51. Puddicombe SM, Polosa R, Richter A, Krishna MT, Howarth PH, Holgate ST *et al.* Involvement of the epidermal growth factor receptor in epithelial repair in asthma. *FASEB J* 14(10): 1362–74, 2000.
52. Moses HL, Yang EY, Pietenpol JA. Regulation of epithelial proliferation by TGF-beta. *Ciba Found Symp* 157: 66–74, 1991.
53. Vignola AM, Chanez P, Chiappara G, Merendino A, Pace E, Rizzo A *et al.* Transforming growth factor-beta expression in mucosal biopsies in asthma and chronic bronchitis. *Am J Respir Crit Care Med* 156(2 Pt 1): 591–99, 1997.
54. Barsky SH, Roth MD, Kleerup EC, Simmons M, Tashkin DP. Histopathologic and molecular alterations in bronchial epithelium in habitual smokers of marijuana, cocaine, and/or tobacco. *J Natl Cancer Inst* 90(16): 1198–205, 1998.
55. Zhang S, Smartt H, Holgate ST, Roche WR. Growth factors secreted by bronchial epithelial cells control myofibroblast proliferation: An *in vitro* co-culture model of airway remodeling in asthma. *Lab Invest* 79(4): 395–405, 1999.
56. Shoji S, Rickard KA, Ertl RF, Robbins RA, Linder J, Rennard SI. Bronchial epithelial cells produce lung fibroblast chemotactic factor: Fibronectin. *Am J Respir Cell Mol Biol* 1(1): 13–20, 1989.
57. Cromwell O, Hamid Q, Corrigan CJ, Barkans J, Meng Q, Collins PD *et al.* Expression and generation of interleukin-8, IL-6 and granulocyte-macrophage colony-stimulating factor by bronchial epithelial cells and enhancement by IL-1 beta and tumour necrosis factor-alpha. *Immunology* 77(3): 330–37, 1992.
58. Laberge S, Ernst P, Ghaffar O, Cruikshank WW, Kornfeld H, Center DM *et al.* Increased expression of interleukin-16 in bronchial mucosa of subjects with atopic asthma. *Am J Respir Cell Mol Biol* 17(2): 193–202, 1997.
59. Kwon OJ, Jose PJ, Robbins RA, Schall TJ, Williams TJ, Barnes PJ. Glucocorticoid inhibition of RANTES expression in human lung epithelial cells. *Am J Respir Cell Mol Biol* 12(5): 488–96, 1995.
60. Aoki Y, Qiu D, Uyei A, Kao PN. Human airway epithelial cells express interleukin-2 *in vitro*. *Am J Physiol* 272(2 Pt 1): L276–86, 1997.
61. Gizycki MJ, Adelroth E, Rogers AV, O'Byrne PM, Jeffery PK. Myofibroblast involvement in the allergen-induced late response in mild atopic asthma. *Am J Respir Cell Mol Biol* 16(6): 664–73, 1997.

62. Gabbiani G, Le LM, Bailey AJ, Bazin S, Delaunay A. Collagen and myofibroblasts of granulation tissue. A chemical, ultrastructural and immunologic study. *Virchows Arch B Cell Pathol* 21(2): 133–45, 1976.
63. Brewster CE, Howarth PH, Djukanovic R, Wilson J, Holgate ST, Roche WR. Myofibroblasts and subepithelial fibrosis in bronchial asthma. *Am J Respir Cell Mol Biol* 3(5): 507–11, 1990.
64. Hoshino M, Nakamura Y, Sim J, Shimojo J, Isogai S. Bronchial subepithelial fibrosis and expression of matrix metalloproteinase-9 in asthmatic airway inflammation. *J Allergy Clin Immunol* 102(5): 783–88, 1998.
65. Schmidt M, Sun G, Stacey MA, Mori L, Mattoli S. Identification of circulating fibrocytes as precursors of bronchial myofibroblasts in asthma. *J Immunol* 171(1): 380–89, 2003.
66. Willis BC, duBois RM, Borok Z. Epithelial origin of myofibroblasts during fibrosis in the lung. *Proc Am Thorac Soc* 3(4): 377–82, 2006.
67. Ronnov-Jessen L, Petersen OW. Induction of alpha-smooth muscle actin by transforming growth factor-beta 1 in quiescent human breast gland fibroblasts. Implications for myofibroblast generation in breast neoplasia. *Lab Invest* 68(6): 696–707, 1993.
68. Desmouliere A, Rubbia-Brandt L, Grau G, Gabbiani G. Heparin induces alpha-smooth muscle actin expression in cultured fibroblasts and in granulation tissue myofibroblasts. *Lab Invest* 67(6): 716–26, 1992.
69. Naureckas ET, Ndukwu IM, Halayko AJ, Maxwell C, Hershenson MB, Solway J. Bronchoalveolar lavage fluid from asthmatic subjects is mitogenic for human airway smooth muscle. *Am J Respir Crit Care Med* 160(6): 2062–66, 1999.
70. Kawamoto M, Matsunami T, Ertl RF, Fukuda Y, Ogawa M, Spurzem JR *et al.* Selective migration of alpha-smooth muscle actin-positive myofibroblasts toward fibronectin in the Boyden's blindwell chamber. *Clin Sci (Lond)* 93(4): 355–62, 1997.
71. Hogg JC, Pierce RA. Remodelling of peripheral lung tissue in COPD. *Eur Respir J* 31(5): 913–14, 2008.
72. Sime PJ, Xing Z, Graham FL, Csaky KG, Gauldie J. Adenovector-mediated gene transfer of active transforming growth factor-beta1 induces prolonged severe fibrosis in rat lung. *J Clin Invest* 100(4): 768–76, 1997.
73. Tang W, Geba GP, Zheng T, Ray P, Homer RJ, Kuhn C III *et al.* Targeted expression of IL-11 in the murine airway causes lymphocytic inflammation, bronchial remodeling, and airways obstruction. *J Clin Invest* 98(12): 2845–53, 1996.
74. DiCosmo BF, Geba GP, Picarella D, Elias JA, Rankin JA, Stripp BR *et al.* Airway epithelial cell expression of interleukin-6 in transgenic mice. Uncoupling of airway inflammation and bronchial hyperreactivity. *J Clin Invest* 94(5): 2028–35, 1994.
75. Lee JJ, McGarry MP, Farmer SC, Denzler KL, Larson KA, Carrigan PE *et al.* Interleukin-5 expression in the lung epithelium of transgenic mice leads to pulmonary changes pathognomonic of asthma. *J Exp Med* 185(12): 2143–56, 1997.
76. Temann UA, Geba GP, Rankin JA, Flavell RA. Expression of interleukin 9 in the lungs of transgenic mice causes airway inflammation, mast cell hyperplasia, and bronchial hyperresponsiveness. *J Exp Med* 188(7): 1307–20, 1998.
77. Zhu Z, Homer RJ, Wang Z, Chen Q, Geba GP, Wang J *et al.* Pulmonary expression of interleukin-13 causes inflammation, mucus hypersecretion, subepithelial fibrosis, physiologic abnormalities, and eotaxin production. *J Clin Invest* 103(6): 779–88, 1999.
78. Jain-Vora S, Wert SE, Temann UA, Rankin JA, Whitsett JA. Interleukin-4 alters epithelial cell differentiation and surfactant homeostasis in the post-natal mouse lung. *Am J Respir Cell Mol Biol* 17(5): 541–51, 1997.
79. Rankin JA, Picarella DE, Geba GP, Temann UA, Prasad B, DiCosmo B *et al.* Phenotypic and physiologic characterization of transgenic mice expressing interleukin 4 in the lung: Lymphocytic and eosinophilic inflammation without airway hyperreactivity. *Proc Natl Acad Sci U S A* 93(15): 7821–25, 1996.
80. Saha SK, Berry MA, Parker D, Siddiqui S, Morgan A, May R *et al.* Increased sputum and bronchial biopsy IL-13 expression in severe asthma. *J Allergy Clin Immunol* 121(3): 685–91, 2008.
81. Izuhara K, Arima K, Kanaji S, Ohta S, Kanaji T. IL-13: A promising therapeutic target for bronchial asthma. *Curr Med Chem* 13(19): 2291–98, 2006.
82. Broide DH. Immunologic and inflammatory mechanisms that drive asthma progression to remodeling. *J Allergy Clin Immunol* 121(3): 560–70, 2008.
83. Desmouliere A, Geinoz A, Gabbiani F, Gabbiani G. Transforming growth factor-beta 1 induces alpha-smooth muscle actin expression in granulation tissue myofibroblasts and in quiescent and growing cultured fibroblasts. *J Cell Biol* 122(1): 103–11, 1993.
84. Zhang HY, Gharaee-Kermani M, Zhang K, Karmiol S, Phan SH. Lung fibroblast alpha-smooth muscle actin expression and contractile phenotype in bleomycin-induced pulmonary fibrosis. *Am J Pathol* 148(2): 527–37, 1996.
85. Rennard SI. Repair mechanisms in asthma. *J Allergy Clin Immunol* 98(6 Pt 2): S278–86, 1996.
86. Rennard SI. Repair mechanisms in asthma. *J Allergy Clin Immunol* 98(6 Pt 2): S278–86, 1996.
87. de Boer WI, van SA, Sont JK, Sharma HS, Stolk J, Hiemstra PS *et al.* Transforming growth factor beta1 and recruitment of macrophages and mast cells in airways in chronic obstructive pulmonary disease. *Am J Respir Crit Care Med* 158(6): 1951–57, 1998.
88. Minshall EM, Leung DY, Martin RJ, Song YL, Cameron L, Ernst P *et al.* Eosinophil-associated TGF-beta1 mRNA expression and airways fibrosis in bronchial asthma. *Am J Respir Cell Mol Biol* 17(3): 326–33, 1997.
89. Ohno I, Nitta Y, Yamauchi K, Hoshi H, Honma M, Woolley K *et al.* Transforming growth factor beta 1 (TGF beta 1) gene expression by eosinophils in asthmatic airway inflammation. *Am J Respir Cell Mol Biol* 15(3): 404–9, 1996.
90. Ohno I, Nitta Y, Yamauchi K, Hoshi H, Honma M, Woolley K *et al.* Eosinophils as a potential source of platelet-derived growth factor B-chain (PDGF-B) in nasal polyposis and bronchial asthma. *Am J Respir Cell Mol Biol* 13(6): 639–47, 1995.
91. Taylor IK, Sorooshian M, Wangoo A, Haynes AR, Kotecha S, Mitchell DM *et al.* Platelet-derived growth factor-beta mRNA in human alveolar macrophages *in vivo* in asthma. *Eur Respir J* 7(11): 1966–72, 1994.
92. Krymskaya VP, Hoffman R, Eszterhas A, Ciocca V, Panettieri RA Jr. TGF-beta 1 modulates EGF-stimulated phosphatidylinositol 3-kinase activity in human airway smooth muscle cells. *Am J Physiol* 273(6 Pt 1): L1220–27, 1997.
93. Chanez P, Vignola M, Stenger R, Vic P, Michel FB, Bousquet J. Platelet-derived growth factor in asthma. *Allergy* 50(11): 878–83, 1995.
94. Chanez P, Vignola M, Stenger R, Vic P, Michel FB, Bousquet J. Platelet-derived growth factor in asthma. *Allergy* 50(11): 878–83, 1995.
95. Aubert JD, Hayashi S, Hards J, Bai TR, Pare PD, Hogg JC. Platelet-derived growth factor and its receptor in lungs from patients with asthma and chronic airflow obstruction. *Am J Physiol* 266(6 Pt 1): L655–63, 1994.
96. Springall DR, Howarth PH, Counihan H, Djukanovic R, Holgate ST, Polak JM. Endothelin immunoreactivity of airway epithelium in asthmatic patients. *Lancet* 337(8743): 697–701, 1991.
97. Mattoli S, Soloperto M, Marini M, Fasoli A. Levels of endothelin in the bronchoalveolar lavage fluid of patients with symptomatic asthma and reversible airflow obstruction. *J Allergy Clin Immunol* 88(3 Pt 1): 376–84, 1991.
98. Chalmers GW, MacLeod KJ, Sriram S, Thomson LJ, McSharry C, Stack BH *et al.* Sputum endothelin-1 is increased in cystic fibrosis and chronic obstructive pulmonary disease. *Eur Respir J* 13(6): 1288–92, 1999.
99. Takuwa Y, Yanagisawa M, Takuwa N, Masaki T. Endothelin, its diverse biological activities and mechanisms of action. *Prog Growth Factor Res* 1(4): 195–206, 1989.

100. Li J, Lin ML, Wiepz GJ, Guadarrama AG, Bertics PJ. Integrin-mediated migration of murine B82L fibroblasts is dependent on the expression of an intact epidermal growth factor receptor. *J Biol Chem* 274(16): 11209–19, 1999.
101. Sakai T, de la Pena JM, Mosher DF. Synergism among lysophosphatidic acid, beta1A integrins, and epidermal growth factor or platelet-derived growth factor in mediation of cell migration. *J Biol Chem* 274(22): 15480–86, 1999.
102. O'Connor CM, FitzGerald MX. Matrix metalloproteases and lung disease. *Thorax* 49(6): 602–9, 1994.
103. Ohno I, Ohtani H, Nitta Y, Suzuki J, Hoshi H, Honma M *et al.* Eosinophils as a source of matrix metalloproteinase-9 in asthmatic airway inflammation. *Am J Respir Cell Mol Biol* 16(3): 212–19, 1997.
104. Dahlen B, Shute J, Howarth P. Immunohistochemical localisation of the matrix metalloproteinases MMP-3 and MMP-9 within the airways in asthma. *Thorax* 54(7): 590–96, 1999.
105. Finlay GA, O'Driscoll LR, Russell KJ, D'Arcy EM, Masterson JB, FitzGerald MX *et al.* Matrix metalloproteinase expression and production by alveolar macrophages in emphysema. *Am J Respir Crit Care Med* 156(1): 240–47, 1997.
106. Yoshioka A, Betsuyaku T, Nishimura M, Miyamoto K, Kondo T, Kawakami Y. Excessive neutrophil elastase in bronchoalveolar lavage fluid in subclinical emphysema. *Am J Respir Crit Care Med* 152 (6 Pt 1): 2127–32, 1995.
107. Kranenburg AR, Willems-Widyastuti A, Moori WJ, Sterk PJ, Alagappan VK, de Boer WI *et al.* Enhanced bronchial expression of extracellular matrix proteins in chronic obstructive pulmonary disease. *Am J Clin Pathol* 126(5): 725–35, 2006.
108. Vignola AM, Riccobono L, Mirabella A, Profita M, Chanez P, Bellia V *et al.* Sputum metalloproteinase-9/tissue inhibitor of metalloproteinase-1 ratio correlates with airflow obstruction in asthma and chronic bronchitis. *Am J Respir Crit Care Med* 158(6): 1945–50, 1998.
109. Vignola AM, Bonanno A, Mirabella A, Riccobono L, Mirabella F, Profita M *et al.* Increased levels of elastase and alpha1-antitrypsin in sputum of asthmatic patients. *Am J Respir Crit Care Med* 157(2): 505–11, 1998.
110. Redington AE, Roche WR, Holgate ST, Howarth PH. Co-localization of immunoreactive transforming growth factor-beta 1 and decorin in bronchial biopsies from asthmatic and normal subjects. *J Pathol* 186(4): 410–15, 1998.
111. Ornitz DM. FGFs, heparan sulfate and FGFRs: Complex interactions essential for development. *Bioessays* 22(2): 108–12, 2000.
112. Cairns JA, Walls AF. Mast cell tryptase stimulates the synthesis of type I collagen in human lung fibroblasts. *J Clin Invest* 99(6): 1313–21, 1997.
113. Cairns JA, Walls AF. Mast cell tryptase is a mitogen for epithelial cells. Stimulation of IL-8 production and intercellular adhesion molecule-1 expression. *J Immunol* 156(1): 275–83, 1996.
114. Raghow R. The role of extracellular matrix in postinflammatory wound healing and fibrosis. *FASEB J* 8(11): 823–31, 1994.
115. Miyake K, Underhill CB, Lesley J, Kincade PW. Hyaluronate can function as a cell adhesion molecule and CD44 participates in hyaluronate recognition. *J Exp Med* 172(1): 69–75, 1990.
116. Redington AE, Roche WR, Holgate ST, Howarth PH. Co-localization of immunoreactive transforming growth factor-beta 1 and decorin in bronchial biopsies from asthmatic and normal subjects. *J Pathol* 186(4): 410–15, 1998.
117. Bousquet J, Chanez P, Lacoste JY, Enander I, Venge P, Peterson C *et al.* Indirect evidence of bronchial inflammation assessed by titration of inflammatory mediators in BAL fluid of patients with asthma. *J Allergy Clin Immunol* 88(4): 649–60, 1991.
118. Hamann KJ, Dowling TL, Neeley SP, Grant JA, Leff AR. Hyaluronic acid enhances cell proliferation during eosinopoiesis through the CD44 surface antigen. *J Immunol* 154(8): 4073–80, 1995.
119. Rothenberg ME. CD44 – A sticky target for asthma. *J Clin Invest* 111(10): 1460–62, 2003.
120. Anwar AR, Moqbel R, Walsh GM, Kay AB, Wardlaw AJ. Adhesion to fibronectin prolongs eosinophil survival. *J Exp Med* 177(3): 839–43, 1993.
121. Rennard SI. Inflammation and repair processes in chronic obstructive pulmonary disease. *Am J Respir Crit Care Med* 160(5 Pt 2): S12–16, 1999.
122. Shoji S, Ertl RF, Linder J, Romberger DJ, Rennard SI. Bronchial epithelial cells produce chemotactic activity for bronchial epithelial cells. Possible role for fibronectin in airway repair. *Am Rev Respir Dis* 141(1): 218–25, 1990.
123. Vignola AM, Chanez P, Chiappara G, Merendino A, Zinnanti E, Bousquet J *et al.* Release of transforming growth factor-beta (TGF-beta) and fibronectin by alveolar macrophages in airway diseases. *Clin Exp Immunol* 106(1): 114–19, 1996.
124. Ignotz RA, Massague J. Transforming growth factor-beta stimulates the expression of fibronectin and collagen and their incorporation into the extracellular matrix. *J Biol Chem* 261(9): 4337–45, 1986.
125. Zhang Q, Sakai T, Nowlen J, Hayashi I, Fassler R, Mosher DF. Functional beta1-integrins release the suppression of fibronectin matrix assembly by vitronectin. *J Biol Chem* 274(1): 368–75, 1999.
126. Shoji S, Ertl RF, Linder J, Romberger DJ, Rennard SI. Bronchial epithelial cells produce chemotactic activity for bronchial epithelial cells. Possible role for fibronectin in airway repair. *Am Rev Respir Dis* 141(1): 218–25, 1990.
127. Roche WR, Beasley R, Williams JH, Holgate ST. Subepithelial fibrosis in the bronchi of asthmatics. *Lancet* 1(8637): 520–24, 1989.
128. Trebaul A, Chan EK, Midwood KS. Regulation of fibroblast migration by tenascin-C. *Biochem Soc Trans* 35(Pt 4): 695–97, 2007.
129. Roche WR, Beasley R, Williams JH, Holgate ST. Subepithelial fibrosis in the bronchi of asthmatics. *Lancet* 1(8637): 520–24, 1989.
130. Altraja A, Laitinen A, Virtanen I, Kampe M, Simonsson BG, Karlsson SE *et al.* Expression of laminins in the airways in various types of asthmatic patients: A morphometric study. *Am J Respir Cell Mol Biol* 15(4): 482–88, 1996.
131. Laitinen A, Altraja A, Kampe M, Linden M, Virtanen I, Laitinen LA. Tenascin is increased in airway basement membrane of asthmatics and decreased by an inhaled steroid. *Am J Respir Crit Care Med* 156(3 Pt 1): 951–58, 1997.
132. Flood-Page P, Menzies-Gow A, Phipps S, Ying S, Wangoo A, Ludwig MS *et al.* Anti-IL-5 treatment reduces deposition of ECM proteins in the bronchial subepithelial basement membrane of mild atopic asthmatics. *J Clin Invest* 112(7): 1029–36, 2003.
133. Flood-Page P, Swenson C, Faiferman I, Matthews J, Williams M, Brannick L *et al.* A study to evaluate safety and efficacy of mepolizumab in patients with moderate persistent asthma. *Am J Respir Crit Care Med* 176(11): 1062–71, 2007.
134. Phipps S, Flood-Page P, Menzies-Gow A, Ong YE, Kay AB. Intravenous anti-IL-5 monoclonal antibody reduces eosinophils and tenascin deposition in allergen-challenged human atopic skin. *J Invest Dermatol* 122(6): 1406–12, 2004.
135. Stein ML, Collins MH, Villanueva JM, Kushner JP, Putnam PE, Buckmeier BK *et al.* Anti-IL-5 (mepolizumab) therapy for eosinophilic esophagitis. *J Allergy Clin Immunol* 118(6): 1312–19, 2006.
136. Ward C, De SA, Fisher AJ, Pritchard G, Forrest I, Corris P. A descriptive study of small airway reticular basement membrane thickening in clinically stable lung transplant recipients. *J Heart Lung Transplant* 24(5): 533–37, 2005.
137. Sato M, Liu M, Anraku M, Ogura T, D'Cruz G, Alman BA *et al.* Allograft airway fibrosis in the pulmonary milieu: A disorder of tissue remodeling. *Am J Transplant* 8(3): 517–28, 2008.
138. Saetta M, Di SA, Maestrelli P, Ferraresso A, Drigo R, Potena A *et al.* Activated T-lymphocytes and macrophages in bronchial mucosa of subjects with chronic bronchitis. *Am Rev Respir Dis* 147(2): 301–6, 1993.
139. Jeffery PK. Remodeling and inflammation of bronchi in asthma and chronic obstructive pulmonary disease. *Proc Am Thorac Soc* 1(3): 176–83, 2004.

140. Chanez P, Vignola AM, O'Shaugnessy T, Enander I, Li D, Jeffery PK *et al.* Corticosteroid reversibility in COPD is related to features of asthma. *Am J Respir Crit Care Med* 155(5): 1529–34, 1997.
141. Wilson JW, Li X. The measurement of reticular basement membrane and submucosal collagen in the asthmatic airway. *Clin Exp Allergy* 27(4): 363–71, 1997.
142. Chu HW, Halliday JL, Martin RJ, Leung DY, Szefler SJ, Wenzel SE. Collagen deposition in large airways may not differentiate severe asthma from milder forms of the disease. *Am J Respir Crit Care Med* 158(6): 1936–44, 1998.
143. Godfrey RW, Lorimer S, Majumdar S, Adelroth E, Johnston PW, Rogers AV *et al.* Airway and lung elastic fibre is not reduced in asthma nor in asthmatics following corticosteroid treatment. *Eur Respir J* 8(6): 922–27, 1995.
144. Adesina AM, Vallyathan V, McQuillen EN, Weaver SO, Craighead JE. Bronchiolar inflammation and fibrosis associated with smoking. A morphologic cross-sectional population analysis. *Am Rev Respir Dis* 143(1): 144–49, 1991.
145. van Straaten JF, Coers W, Noordhoek JA, Huitema S, Flipsen JT, Kauffman HF *et al.* Proteoglycan changes in the extracellular matrix of lung tissue from patients with pulmonary emphysema. *Mod Pathol* 12(7): 697–705, 1999.
146. Mitchell RS, Stanford RE, Johnson JM, Silvers GW, Dart G, George MS. The morphologic features of the bronchi, bronchioles, and alveoli in chronic airway obstruction: A clinicopathologic study. *Am Rev Respir Dis* 114(1): 137–45, 1976.
147. Dunnill MS, Massarella GR, Anderson JA. A comparison of the quantitative anatomy of the bronchi in normal subjects, in status asthmaticus, in chronic bronchitis, and in emphysema. *Thorax* 24(2): 176–79, 1969.
148. Glynn AA, Michaels L. Bronchial biopsy in chronic bronchitis and asthma. *Istanbul Tip Fak Mecmuasi* 15: 142–53, 1960.
149. Aikawa T, Shimura S, Sasaki H, Ebina M, Takishima T. Marked goblet cell hyperplasia with mucus accumulation in the airways of patients who died of severe acute asthma attack. *Chest* 101(4): 916–21, 1992.
150. Shimura S, Andoh Y, Haraguchi M, Shirato K. Continuity of airway goblet cells and intraluminal mucus in the airways of patients with bronchial asthma. *Eur Respir J* 9(7): 1395–401, 1996.
151. Aikawa T, Shimura S, Sasaki H, Ebina M, Takishima T. Marked goblet cell hyperplasia with mucus accumulation in the airways of patients who died of severe acute asthma attack. *Chest* 101(4): 916–21, 1992.
152. Sheehan JK, Richardson PS, Fung DC, Howard M, Thornton DJ. Analysis of respiratory mucus glycoproteins in asthma: A detailed study from a patient who died in status asthmaticus. *Am J Respir Cell Mol Biol* 13(6): 748–56, 1995.
153. Adesina AM, Vallyathan V, McQuillen EN, Weaver SO, Craighead JE. Bronchiolar inflammation and fibrosis associated with smoking. A morphologic cross-sectional population analysis. *Am Rev Respir Dis* 143(1): 144–49, 1991.
154. Mitchell RS, Stanford RE, Johnson JM, Silvers GW, Dart G, George MS. The morphologic features of the bronchi, bronchioles, and alveoli in chronic airway obstruction: A clinicopathologic study. *Am Rev Respir Dis* 114(1): 137–45, 1976.
155. Niewoehner DE, Kleinerman J, Rice DB. Pathologic changes in the peripheral airways of young cigarette smokers. *N Engl J Med* 291(15): 755–58, 1974.
156. Cosio MG, Hale KA, Niewoehner DE. Morphologic and morphometric effects of prolonged cigarette smoking on the small airways. *Am Rev Respir Dis* 122(2): 221–65, 1980.
157. Kramps JA, Franken C, Meijer CJ, Dijkman JH. Localization of low molecular weight protease inhibitor in serous secretory cells of the respiratory tract. *J Histochem Cytochem* 29(6): 712–19, 1981.
158. Kim KC, McCracken K, Lee BC, Shin CY, Jo MJ, Lee CJ *et al.* Airway goblet cell mucin: Its structure and regulation of secretion. *Eur Respir J* 10(11): 2644–49, 1997.
159. Nagai A, West WW, Thurlbeck WM. The National Institutes of Health Intermittent Positive-Pressure Breathing trial: Pathology studies. II. Correlation between morphologic findings, clinical findings, and evidence of expiratory air-flow obstruction. *Am Rev Respir Dis* 132(5): 946–53, 1985.
160. Rogers DF. Airway goblet cells: Responsive and adaptable front-line defenders. *Eur Respir J* 7(9): 1690–706, 1994.
161. Li JD, Feng W, Gallup M, Kim JH, Gum J, Kim Y *et al.* Activation of NF-kappaB via a Src-dependent Ras-MAPK-pp90rsk pathway is required for Pseudomonas aeruginosa-induced mucin overproduction in epithelial cells. *Proc Natl Acad Sci U S A* 95(10): 5718–23, 1998.
162. Borchers MT, Wert SE, Leikauf GD. Acrolein-induced MUC5ac expression in rat airways. *Am J Physiol* 274(4 Pt 1): L573–81, 1998.
163. Howarth PH, Babu KS, Arshad HS, Lau L, Buckley M, McConnell W *et al.* Tumour necrosis factor (TNFalpha) as a novel therapeutic target in symptomatic corticosteroid dependent asthma. *Thorax* 60(12): 1012–18, 2005.
164. Breuer R, Christensen TG, Niles RM, Stone PJ, Snider GL. Human neutrophil elastase causes glycoconjugate release from the epithelial cell surface of hamster trachea in organ culture. *Am Rev Respir Dis* 139(3): 779–82, 1989.
165. Kim KC, Wasano K, Niles RM, Schuster JE, Stone PJ, Brody JS. Human neutrophil elastase releases cell surface mucins from primary cultures of hamster tracheal epithelial cells. *Proc Natl Acad Sci U S A* 84(24): 9304–8, 1987.
166. Borchers MT, Wert SE, Leikauf GD. Acrolein-induced MUC5ac expression in rat airways. *Am J Physiol* 274(4 Pt 1): L573–81, 1998.
167. Temann UA, Prasad B, Gallup MW, Basbaum C, Ho SB, Flavell RA *et al.* A novel role for murine IL-4 *in vivo*: Induction of MUC5AC gene expression and mucin hypersecretion. *Am J Respir Cell Mol Biol* 16(4): 471–78, 1997.
168. Dabbagh K, Takeyama K, Lee HM, Ueki IF, Lausier JA, Nadel JA. IL-4 induces mucin gene expression and goblet cell metaplasia *in vitro* and *in vivo*. *J Immunol* 162(10): 6233–37, 1999.
169. Louahed J, Toda M, Jen J, Hamid Q, Renauld JC, Levitt RC *et al.* Interleukin-9 upregulates mucus expression in the airways. *Am J Respir Cell Mol Biol* 22(6): 649–56, 2000.
170. Zuhdi AM, Piazza FM, Selby DM, Letwin N, Huang L, Rose MC. Muc-5/5ac mucin messenger RNA and protein expression is a marker of goblet cell metaplasia in murine airways. *Am J Respir Cell Mol Biol* 22(3): 253–60, 2000.
171. Takeyama K, Dabbagh K, Jeong SJ, Dao-Pick T, Ueki IF, Nadel JA. Oxidative stress causes mucin synthesis via transactivation of epidermal growth factor receptor: Role of neutrophils. *J Immunol* 164(3): 1546–52, 2000.
172. Kim KC, Wasano K, Niles RM, Schuster JE, Stone PJ, Brody JS. Human neutrophil elastase releases cell surface mucins from primary cultures of hamster tracheal epithelial cells. *Proc Natl Acad Sci U S A* 84(24): 9304–8, 1987.
173. Voynow JA, Young LR, Wang Y, Horger T, Rose MC, Fischer BM. Neutrophil elastase increases MUC5AC mRNA and protein expression in respiratory epithelial cells. *Am J Physiol* 276(5 Pt 1): L835–43, 1999.
174. Fischer B, Voynow J. Neutrophil elastase induces MUC5AC messenger RNA expression by an oxidant-dependent mechanism. *Chest* 117(5 Suppl 1): 317S–320S, 2000.
175. Takeyama K, Dabbagh K, Lee HM, Agusti C, Lausier JA, Ueki IF *et al.* Epidermal growth factor system regulates mucin production in airways. *Proc Natl Acad Sci U S A* 96(6): 3081–86, 1999.
176. Lee HM, Takeyama K, Dabbagh K, Lausier JA, Ueki IF, Nadel JA. Agarose plug instillation causes goblet cell metaplasia by activating EGF receptors in rat airways. *Am J Physiol Lung Cell Mol Physiol* 278(1): L185–92, 2000.
177. Goldkorn T, Balaban N, Matsukuma K, Chea V, Gould R, Last J *et al.* EGF-Receptor phosphorylation and signaling are targeted by H2O2 redox stress. *Am J Respir Cell Mol Biol* 19(5): 786–98, 1998.

178. Amishima M, Munakata M, Nasuhara Y, Sato A, Takahashi T, Homma Y *et al.* Expression of epidermal growth factor and epidermal growth factor receptor immunoreactivity in the asthmatic human airway. *Am J Respir Crit Care Med* 157(6 Pt 1): 1907–12, 1998.
179. Dunnill MS, Massarella GR, Anderson JA. A comparison of the quantitative anatomy of the bronchi in normal subjects, in status asthmaticus, in chronic bronchitis, and in emphysema. *Thorax* 24(2): 176–79, 1969.
180. Saetta M, Di SA, Rosina C, Thiene G, Fabbri LM. Quantitative structural analysis of peripheral airways and arteries in sudden fatal asthma. *Am Rev Respir Dis* 143(1): 138–43, 1991.
181. Carroll N, Elliot J, Morton A, James A. The structure of large and small airways in nonfatal and fatal asthma. *Am Rev Respir Dis* 147(2): 405–10, 1993.
182. Ebina M, Takahashi T, Chiba T, Motomiya M. Cellular hypertrophy and hyperplasia of airway smooth muscles underlying bronchial asthma. A 3-D morphometric study. *Am Rev Respir Dis* 148(3): 720–26, 1993.
183. Kuwano K, Bosken CH, Pare PD, Bai TR, Wiggs BR, Hogg JC. Small airways dimensions in asthma and in chronic obstructive pulmonary disease. *Am Rev Respir Dis* 148(5): 1220–25, 1993.
184. Sobonya RE. Quantitative structural alterations in long-standing allergic asthma. *Am Rev Respir Dis* 130(2): 289–92, 1984.
185. Mitchell RS, Stanford RE, Johnson JM, Silvers GW, Dart G, George MS. The morphologic features of the bronchi, bronchioles, and alveoli in chronic airway obstruction: A clinicopathologic study. *Am Rev Respir Dis* 114(1): 137–45, 1976.
186. Cosio MG, Hale KA, Niewoehner DE. Morphologic and morphometric effects of prolonged cigarette smoking on the small airways. *Am Rev Respir Dis* 122(2): 221–65, 1980.
187. Hale KA, Ewing SL, Gosnell BA, Niewoehner DE. Lung disease in long-term cigarette smokers with and without chronic air-flow obstruction. *Am Rev Respir Dis* 130(5): 716–21, 1984.
188. Thurlbeck WM. Pathology of chronic airflow obstruction. *Chest* 97 (2 Suppl): 6S–10S, 1990.
189. Hossain S, Heard BE. Hyperplasia of bronchial muscle in chronic bronchitis. *J Pathol* 101(2): 171–84, 1970.
190. Takizawa T, Thurlbeck WM. Muscle and mucous gland size in the major bronchi of patients with chronic bronchitis, asthma, and asthmatic bronchitis. *Am Rev Respir Dis* 104(3): 331–36, 1971.
191. Shiels IA, Bowler SD, Taylor SM. Airway smooth muscle proliferation in asthma: The potential of vascular leakage to contribute to pathogenesis. *Med Hypotheses* 45(1): 37–40, 1995.
192. Hirst SJ. Airway smooth muscle as a target in asthma. *Clin Exp Allergy* 30(Suppl 1): 54–59, 2000.
193. Panettieri RA Jr. Cellular and molecular mechanisms regulating airway smooth muscle proliferation and cell adhesion molecule expression. *Am J Respir Crit Care Med* 158(5 Pt 3): S133–40, 1998.
194. Cohen MD, Ciocca V, Panettieri RA Jr. TGF-beta 1 modulates human airway smooth-muscle cell proliferation induced by mitogens. *Am J Respir Cell Mol Biol* 16(1): 85–90, 1997.
195. Lazaar AL, Albelda SM, Pilewski JM, Brennan B, Pure E, Panettieri RA Jr. T lymphocytes adhere to airway smooth muscle cells via integrins and CD44 and induce smooth muscle cell DNA synthesis. *J Exp Med* 180(3): 807–16, 1994.
196. John M, Hirst SJ, Jose PJ, Robichaud A, Berkman N, Witt C *et al.* Human airway smooth muscle cells express and release RANTES in response to T helper 1 cytokines: Regulation by T helper 2 cytokines and corticosteroids. *J Immunol* 158(4): 1841–47, 1997.
197. Brightling CE, Bradding P, Symon FA, Holgate ST, Wardlaw AJ, Pavord ID. Mast-cell infiltration of airway smooth muscle in asthma. *N Engl J Med* 346(22): 1699–705, 2002.
198. Begueret H, Berger P, Vernejoux JM, Dubuisson L, Marthan R, Tunon-de-Lara JM. Inflammation of bronchial smooth muscle in allergic asthma. *Thorax* 62(1): 8–15, 2007.
199. Jongepier H, Boezen HM, Dijkstra A, Howard TD, Vonk JM, Koppelman GH *et al.* Polymorphisms of the ADAM33 gene are associated with accelerated lung function decline in asthma. *Clin Exp Allergy* 34(5): 757–60, 2004.
200. Gosman MM, Boezen HM, van Diemen CC, Snoeck-Stroband JB, Lapperre TS, Hiemstra PS *et al.* A disintegrin and metalloprotease 33 and chronic obstructive pulmonary disease pathophysiology. *Thorax* 62(3): 242–47, 2007.
201. Puxeddu I, Pang YY, Harvey A, Haitchi HM, Nicholas B, Yoshisue H *et al.* The soluble form of a disintegrin and metalloprotease 33 promotes angiogenesis: Implications for airway remodeling in asthma. *J Allergy Clin Immunol*, 2008.
202. Cox G, Thomson NC, Rubin AS, Niven RM, Corris PA, Siersted HC *et al.* Asthma control during the year after bronchial thermoplasty. *N Engl J Med* 356(13): 1327–37, 2007.
203. Pavord ID, Cox G, Thomson NC, Rubin AS, Corris PA, Niven RM *et al.* Safety and efficacy of bronchial thermoplasty in symptomatic, severe asthma. *Am J Respir Crit Care Med* 176(12): 1185–91, 2007.
204. Barbato A, Turato G, Baraldo S, Bazzan E, Calabrese F, Panizzolo C *et al.* Epithelial damage and angiogenesis in the airways of children with asthma. *Am J Respir Crit Care Med* 174(9): 975–81, 2006.
205. Battegay EJ. Angiogenesis: Mechanistic insights, neovascular diseases, and therapeutic prospects. *J Mol Med* 73(7): 333–46, 1995.
206. Dunnill MS, Massarella GR, Anderson JA. A comparison of the quantitative anatomy of the bronchi in normal subjects, in status asthmaticus, in chronic bronchitis, and in emphysema. *Thorax* 24(2): 176–79, 1969.
207. Li X, Wilson JW. Increased vascularity of the bronchial mucosa in mild asthma. *Am J Respir Crit Care Med* 156(1): 229–33, 1997.
208. Backman KS, Greenberger PA, Patterson R. Airways obstruction in patients with long-term asthma consistent with 'irreversible asthma'. *Chest* 112(5): 1234–40, 1997.
209. Olgart HC, de BF, Oster JP, Duvernelle C, Kassel O, Pauli G *et al.* Nerve growth factor levels and localisation in human asthmatic bronchi. *Eur Respir J* 20(5): 1110–16, 2002.
210. Groneberg DA, Quarcoo D, Frossard N, Fischer A. Neurogenic mechanisms in bronchial inflammatory diseases. *Allergy* 59(11): 1139–52, 2004.
211. Grol MH, Gerritsen J, Vonk JM, Schouten JP, Koeter GH, Rijcken B *et al.* Risk factors for growth and decline of lung function in asthmatic individuals up to age 42 years. A 30-year follow-up study. *Am J Respir Crit Care Med* 160(6): 1830–37, 1999.
212. Agertoft L, Pedersen S. Effects of long-term treatment with an inhaled corticosteroid on growth and pulmonary function in asthmatic children. *Respir Med* 88(5): 373–81, 1994.
213. Guilbert TW, Morgan WJ, Zeiger RS, Mauger DT, Boehmer SJ, Szefler SJ *et al.* Long-term inhaled corticosteroids in preschool children at high risk for asthma. *N Engl J Med* 354(19): 1985–97, 2006.
214. Murray CS, Woodcock A, Langley SJ, Morris J, Custovic A. Secondary prevention of asthma by the use of Inhaled Fluticasone propionate in Wheezy INfants (IFWIN): Double-blind, randomised, controlled study. *Lancet* 368(9537): 754–62, 2006.
215. The Childhood Asthma Management Program Research Group. Long-term effects of budesonide or nedocromil in children with asthma. *N Engl J Med* 343: 1054–63, 2000.
216. Busse WW, Pedersen S, Pauwels RA, Tan WC, Chen YZ, Lamm CJ *et al.* The Inhaled Steroid Treatment As Regular Therapy in Early Asthma (START) study 5-year follow-up: Effectiveness of early intervention with budesonide in mild persistent asthma. *J Allergy Clin Immunol* 121(5): 1167–74, 2008.
217. Tomlinson PR, Wilson JW, Stewart AG. Inhibition by salbutamol of the proliferation of human airway smooth muscle cells grown in culture. *Br J Pharmacol* 111(2): 641–47, 1994.
218. Levi-Schaffer F, Touitou E. Xanthines inhibit 3T3 fibroblast proliferation. *Skin Pharmacol* 4(4): 286–90, 1991.
219. Pyzik M, Piccirillo CA. TGF-beta1 modulates Foxp3 expression and regulatory activity in distinct CD4+ T cell subsets. *J Leukoc Biol* 82(2): 335–46, 2007.

220. Holgate ST. The airway epithelium is central to the pathogenesis of asthma. *Allergol Int* 57(1): 1–10, 2008.
221. Long AJ, Sypek JP, Askew R, Fish SC, Mason LE, Williams CM *et al.* Gob-5 contributes to goblet cell hyperplasia and modulates pulmonary tissue inflammation. *Am J Respir Cell Mol Biol* 35(3): 357–65, 2006.
222. Joad JP, Kott KS, Bric JM, Peake JL, Plopper CG, Schelegle ES *et al.* Structural and functional localization of airway effects from episodic exposure of infant monkeys to allergen and/or ozone. *Toxicol Appl Pharmacol* 214(3): 237–43, 2006.
223. Holgate ST, Davies DE, Powell RM, Howarth PH, Haitchi HM, Holloway JW. Local genetic and environmental factors in asthma disease pathogenesis: Chronicity and persistence mechanisms. *Eur Respir J* 29(4): 793–803, 2007.

Asthma and COPD: Animal Models

CHAPTER 8

Stephanie A. Shore[1] and Steven D. Shapiro[2]

[1]Molecular and Integrative Physiological Sciences Program, Harvard School of Public Health, Boston, MA, USA

[2]Department of Medicine, University of Pittsburgh, Pittsburgh, PA, USA

OVERVIEW

Animal models have been used extensively in the discovery of the biological processes that underlie asthma and COPD. For example, the demonstration by Gross *et al.* [1] that instillation of papain into the lungs of rats resulted in a syndrome similar to emphysema, contributed to the elastase/anti-elastase hypothesis regarding the pathogenesis of emphysema. Animal models have also been used to guide research into pharmacological interventions for the treatment of asthma and COPD. For example, the current use of leukotriene receptor antagonists for the treatment of asthma had its foundation partly in the demonstration of the marked bronchoconstrictor potency of cysteinyl leukotrienes in guinea pigs [2]. While many species, including dogs, cats, monkeys, horses, pigs, sheep, guinea pigs, and rats have been used to study asthma and COPD, most current investigations involve the use of mice. Consequently, this review will focus on models developed using this species. The advantages of using mice include cost of housing, a short breeding period, the availability of inbred species with known characteristics, a well characterized immune system and genome, and the ability to either overexpress or delete specific genes, sometimes conditionally, that is, at specific times or in particular cells. This ability to genetically manipulate mice is a powerful tool that permits very mechanistic experiments that cannot be accomplished in human subjects and has yielded some important advances in our understanding of lung disease. The disadvantages of using mice include differences in lung anatomy, lung development, pattern and route of breathing (mice are obligate nose breathers), the pathogen-free facilities in which most mice are housed which eliminate many environmental modifiers, and some differences in the immune system (mice have a more polarized T-lymphocyte response than humans). These disadvantages have been documented elsewhere [3–7] and may limit the utility of these models for studying human disease in some cases. Mouse models of asthma and COPD will be reviewed separately below.

MOUSE MODELS OF ASTHMA

Human asthma is characterized by intermittent reversible airway obstruction, AHR, and airway inflammation. Pathologically, airway remodeling including increased airway smooth muscle mass, subepithelial fibrosis, mucous hypersecretion, and increased airway vascularity are observed. Asthma does not typically develop in most non-human species. Consequently, many investigators have chosen to develop animal models in which an asthma-like phenotype is induced by some intervention. The etiology of asthma is complex, and still poorly understood, and it is clear that multiple factors including exposure to allergens, exercise, exposure to air pollution, certain viral infections, and the onset of obesity can also initiate symptoms. Because of the extensive nature of this topic, we have chosen to focus on only two of these factors: allergen and obesity.

Modeling allergic asthma

Many investigators have used models of allergic asthma in which mice are first sensitized to an allergen and then challenged with that allergen via the lungs. The sensitization and the challenge protocols vary widely among investigators, especially depending on whether AHR or airway remodeling are being targeted. The sensitization is usually i.p. injection of the allergen with an adjuvant. Allergens used have included ovalbumin (OVA), picryl chloride, sheep erythrocytes, short ragweed extracts, and house dust mite allergen. Although not environmentally very relevant, most investigators have used OVA as the allergen, perhaps because of its relatively low cost and easy availability. Mice are usually assessed 1–3 days after the last of several daily airway challenges and the characteristics monitored are AHR, airway inflammation, and serum IgE.

Consistent features of the model are increased OVA-specific IgE and IgG1, eosinophilia (intraluminal, peribronchial, and perivascular), and recruitment of lymphocytes to the airways. Increases in Th2-type cytokines either in bronchoalveolar lavage (BAL) fluid, in lung tissue, or in lymph nodes draining the lungs also occur. These changes are generally similar to those observed in the airways of asthmatics after allergen challenge, although the kinetics of the recruitment of eosinophils and lymphocytes differ: in humans, the peak in T-lymphocyte recruitment occurs about 24 h after challenge and precedes the peak of eosinophil recruitment, whereas in mice, T-cell recruitment continues long after eosinophils have peaked and returned to baseline [8].

AHR is often, though not always observed in allergen sensitized and challenged mice. There are few reports of comparisons of the various protocols in terms of their ability to induce AHR, but those available indicate that both systemic sensitization and pulmonary challenge are required to induce AHR [9]. For example, multiple airway challenges without systemic sensitization can result in pulmonary eosinophilia, but *in vivo* AHR is not observed under these conditions [10]. The route of airway challenge (nasal or intratracheal versus aerosol) may also be a factor [11]. Mouse strain also has an important impact on allergen-induced AHR. AHR is relatively easy to induce in Balb/c and A/J mice, but more difficult to induce in other strains, including C57BL/6 mice even though airway inflammation is robust in this strain [9, 12–14]. The relative resistance of C57BL/6 mice to allergen-induced AHR is unfortunate, since many of the knockout mice that could be used to study the mechanistic basis for allergen-induced AHR are originally developed on this background.

Recently, investigators have begun using repeated allergen challenges over several weeks to months to model the airway remodeling that characterizes human asthma [15]. In general, such challenges often result in tolerance, although in the A/J strain, eosinophilia and AHR are preserved over several months, and these mice develop subepithelial fibrosis and airway smooth muscle hypertrophy [12]. However, modifications to the protocols may improve the utility of mouse models for studies of airway remodeling. A recent report by Yu *et al.* [16] indicates that elimination of adjuvant during the sensitization phase results in preserved AHR, eosinophilia, and subepithelial fibrosis even after 9 weeks of repeated weekly intranasal challenges in mice on a C57BL/6 background. Others have observed subepithelial fibrosis and increases in airway smooth muscle mass after chronic challenge over several months with very low concentrations of OVA delivered intranasally, even in C57BL/6 mice [17].

Murine models of allergic airways disease have been used to examine the role of many cells, cytokines, chemokines, enzymes, and other factors in the pathology of asthma. These studies are too numerous to document here. Instead, we focus on three issues. One, the role of lymphocytes, can be seen as a major success of these models. Two others, the role of IgE and mast cells, and the role of eosinophils are instances where the models do not appear to reproduce the human findings.

Role of T-lymphocytes in allergen-induced airway responses: A major accomplishment of studies using these murine models was in establishing a role for lymphocytes in the development of allergic asthma. For example, following allergen sensitization and challenge, AHR does not develop in athymic mice or in mice with severe combined immunodeficiency, but adoptive transfer of lymphocytes from wild-type sensitized and challenged mice restores AHR in athymic mice [18, 19]. A particular role for $CD4^+$ T-lymphocytes was established by experiments demonstrating that MHC Class II deficient mice lacking mature $CD4^+$ T-lymphocytes and mice depleted of $CD4^+$ T-cells do not develop AHR or airway inflammation following allergen sensitization and challenge [20, 21]. The requirement for $CD4^+$ T-cells was also underscored by the observation that reconstitution of $CD4^+$ T-cells alone is sufficient to restore allergen-induced AHR to $RAG^{-/-}$ mice which lack both B- and T-lymphocytes [22]. It has since been established that other T-cell subsets including $\gamma\delta$-T-cells, iNKT cells, and regulatory T-cells may have roles in enhancing or attenuating allergic airways disease in these mouse models [23–26].

A major focus of research using these models over the last decade has been to establish how T-lymphocytes contribute to allergic airway responses. Early on, it was established that the ability of these cells to generate cytokines was likely to be important. Initial work focused on IL-4, IL-5, and IL-13 [19, 20, 22, 27–31]. More recently, other cytokines, including IL-17, have received increasing attention [32–34].

Role of IgE and mast cells in allergen-induced AHR: Mast cells express receptors (FcεRI) on their surface that bind the Fc portion of IgE with high affinity. Crosslinking FcεRI receptors upon binding of allergen to IgE results in the secretion of a panel of mediators including histamine, eicosanoids, and cytokines, many of which have the capacity to cause bronchoconstriction and to elicit AHR, though it is important to note that IgE independent activation of mast cells can occur in response to complement proteins, neuropeptides, and other moieties. In humans, activation of mast cells by IgE is important for both early and late responses to allergen inhalation: treatment with the humanized anti-IgE antibody, omalizumab, markedly reduces both the rapid and delayed declines in FEV_1 that occur following inhalation of allergen [35]. An important drawback of using mice is that such airway obstruction is not easily observed [36]. Small increases in pulmonary resistance (R_L) (less than 50% increase) can be observed following systemic administration of activating anti-IgE antibodies [37], but R_L changes little after airway challenge with allergen [36]. This is in marked contrast to many other species which develop marked increases in R_L after inhalation of allergen [38–40].

Mast cells and IgE also appear to play a role in the airway inflammation that occurs after allergen challenge in asthmatics [41], but the ability of mouse models to reproduce this phenomenon has not been consistent. Kung *et al.* [42] reported that aerosol OVA challenge of systemically sensitized mast cell deficient mice resulted in fewer eosinophils in BAL fluid and lung tissue compared to congenic controls, whereas others [10, 43–45] have failed to observe any effect of mast cell deficiency or IgE deficiency on the recruitment of eosinophils. At least part of this discrepancy is likely related to the precise protocol used to sensitize and challenge the mice. Kobayashi *et al.* used two distinct protocols to sensitize and challenge mast cell deficient mice and found reduced eosinophils compared to strain matched controls when relatively mild protocols for airway challenge that resulted in relatively few eosinophils were employed, but no difference between mast cell deficient and control mice when more frequent and robust airway challenges that produced marked increases in eosinophils were used [43].

Role of eosinophils in allergen-induced AHR: Eosinophils contain cationic proteins such as major basic protein (MBP) that have the capacity to induce AHR [46] and experiments in mice in the late 1990s provided strong evidence that eosinophils may be responsible for the AHR associated with allergic asthma. For example, in mice, transgenic overexpession of IL-5, a differentiation, chemoattractant, and survival factor for eosinophils, led to accumulations of eosinophils and AHR even in the absence of allergen challenge [47]. Following OVA sensitization and challenge, both accumulation of eosinophils in the airways and AHR were prevented in IL-5 deficient compared to wild-type mice and reconstitution of IL-5 in these knockout mice restored pulmonary eosinophilia and AHR [48, 49]. Similar results were obtained using anti-IL-5 antibodies [50]. The observation of a good correlation between airway eosinophils and AHR across multiple strains of mice, all sensitized and challenged in the same way, also argued for a relationship between these factors [13]. Data such as those cited above were strong enough to initiate clinical trials for anti-IL-5 in human asthma. However, the results of such trials have been disappointing. For example, a recent fairly large study showed that while anti-IL-5 antibody treatment results in marked suppression of both blood and sputum eosinophils, it does not impact pulmonary function, β-agonist use, symptom or quality of life scores in patients with persistent asthma symptoms despite inhaled corticosteroid use [51], although it has been argued that the study population (one of difficult to control asthma) could have biased the study [52]. Did our mouse models lead us astray? There are certainly data from mouse models arguing against a role for eosinophils in the AHR associated with allergen sensitization and challenge [19, 53, 54]. Mouse strain and the strength of the allergen challenge appear to impact whether a role for eosinophils are observed or not: experiments performed in Balb/c mice in which milder challenges are used indicate little role for eosinophils in OVA-induced AHR [5].

Summary: Mice are not useful for modeling the early or late responses to allergen, but models in which mice are sensitized and then challenged with allergen have been useful in establishing a role for T-lymphocytes in asthma and are currently being used to investigate the T-cell subsets and cytokines involved. Refinements in these models involving chronic challenges may ultimately allows us to use such models to study the airway remodeling that characterizes asthma, but are currently in their infancy. The protocol used to sensitize and challenge the mice, and the choice of mouse strain are key issues and can influence outcome.

Modeling obesity and asthma

Obesity is a risk factor for asthma. Both in adults and children, the prevalence and incidence of asthma increase with body mass index (BMI) (see recent reviews [55–58]). Obesity also appears to worsen asthma control and may increase asthma severity. In addition, both surgical and diet-induced weight loss improve many asthma outcomes. The efficacy of many standard asthma medications is altered in the obese [58, 59], suggesting that additional therapeutic strategies may be necessary in this population. Development of these strategies will require an understanding of the mechanistic underpinnings of the relationship between obesity and asthma. To that end, mice have been used to model the relationship between obesity and asthma. These studies have established that obese mice exhibit a pulmonary phenotype that includes both innate AHR, as well as increased responses to ozone (O_3) and allergen [60–65], and suggest a role for adipokines in the obesity–asthma relationship [66, 67].

Several types of obese mice have been used in these studies (see Ref. [68] for details). *Ob/ob* mice are genetically deficient in leptin, a satiety hormone synthesized by adipocytes and released into blood in proportion to adipose tissue mass. In the absence of leptin, *ob/ob* mice eat excessively, have a low resting metabolic rate, and by 8 weeks of age weigh at least twice as much as wild-type controls [69]. The increased body weight is entirely the result of an increase in fat mass. *Db/db* mice are genetically deficient in the long form of the leptin receptor, which is required for leptin's effect on satiety and metabolism [69]. Thus, *db/db* mice are similar to *ob/ob* mice in many respects: they overeat, are hypothermic and inactive, and massively obese. However, short forms of the leptin receptor with truncated or absent cytoplasmic domains are expressed in these mice, and are capable of some types of signaling. Hence, there can be subtle differences between *ob/ob* and *db/db* mice, including differences in their pulmonary phenotype [60]. Cpe^{fat} mice are genetically deficient in carboxypeptidase E (Cpe), an enzyme involved in processing of neuropeptides involved in eating behaviors. Obesity develops more slowly in these mice than in *ob/ob* or *db/db* mice, but can be quite marked [69]. Dietary obesity can also be induced by feeding weanling C57BL/6 mice a diet in which 45% or 60% of calories are derived from fat (predominantly in the form of lard). Diet-induced obesity (DIO) develops slowly and is much less marked than in the genetic obesities. All of these obese mice are hyperglycemic, hyperinsulinemic, and hyperlipidemic to some extent [68, 69].

Innate AHR: Airway responsiveness to intravenous methacholine is increased in *ob/ob*, *db/db*, and Cpe^{fat} mice, and in mice with DIO [60–65], indicating that it is a common feature of murine obesity. Consistent with the non-specific nature of human asthma, AHR is observed following both methacholine and serotonin challenges [60]. Both the magnitude and the duration of obesity appear to play some role in the development of AHR in obese mice. For example, AHR is more marked in *ob/ob* and *db/db* mice than in *Cpe^fat^* mice or mice with DIO. Mice do not exhibit AHR when they are raised on high-fat diets for approximately 16 weeks, but AHR is observed after 30 weeks on the diet, even though body weight averages about 40% more than low-fat fed controls at both time points [61].

The mechanistic basis for the AHR observed in obese mice is not known. However, the mice do not have any overt cellular inflammation in their lungs [60, 63]. All measurements were made in open-chested mice, mechanically ventilated at a fixed positive end-expiratory pressure (PEEP) and a fixed tidal volume, so changes in absolute or tidal volume likely do not play a role. In *ob/ob* and *db/db* mice, there are effects on lung development: despite their massive obesity, these mice have small lungs [60, 65]. However, lung size is not affected in Cpe^{fat} mice and mice with DIO, but these mice still exhibit AHR [61, 62]. Instead, it is likely that some aspect of the chronic low-grade systemic inflammation that characterizes the obese state [70] contributes to the AHR observed in obese mice. For example, preliminary data in both *db/db* and Cpe^{fat} mice indicate that treatment with anti-TNF-α antibodies reduces airway responsiveness [71].

Responses to O_3: Exposure to O_3, a common air pollutant, is a trigger for asthma. Hospital admissions for asthma are higher on days of high ambient O_3 concentrations [72] and in children, O_3 increases asthmatic symptoms even at concentrations below the US Environmental Protection Agency standard [73]. O_3 causes airway inflammation, as well as AHR, both of which may contribute to the ability of O_3 to exacerbate asthma.

Obesity impacts the effects of O_3 in the lung. Both the changes in pulmonary function and the AHR induced by O_3 exposure are increased in obese versus lean human subjects [74, 75]. Similar results are observed in obese mice regardless of the nature of their obesity. Acute exposure to O_3 (2 ppm for 3 h) increases R_L in obese but not lean mice and induces greater AHR in obese versus lean mice [60, 62, 65]. Compared to lean controls, O_3-induced injury and inflammation are also greater in *ob/ob* mice, *db/db* mice, Cpe^{fat} mice, and mice with DIO [60–62, 65]. However, the factors that contribute to the innate AHR of obese mice appear to be different from those that contribute to their increased response to acute O_3: increased O_3-induced inflammation is observed as early as 7 weeks of age in Cpe^{fat} mice [76], whereas AHR is not observed until they are 10 weeks old.

At least part of the increased response to O_3 observed in obese mice appears to derive from differences in IL-6 signaling. Following acute O_3 exposure, obese mice have increased IL-6 release into BAL fluid compared to lean mice [60, 62, 65]. Treatment with anti-IL-6 antibodies ablates the accelerated neutrophil influx that is observed following acute O_3 exposure in *ob/ob* mice and also attenuates the enhanced O_3-induced epithelial injury that is observed in these mice [77]. Pulmonary expression of signal transducer and activator of transcription-1 (STAT-1) is reduced in *ob/ob* mice [77] and may modify pulmonary responses to IL-6 in these mice.

Responses to allergen: Some, but not all, aspects of the response to allergen are augmented in obese mice. For example, following OVA sensitization and challenge, splenocyte proliferation, IL-2 production, and mast cell numbers are increased in mice with DIO compared to lean controls [78]. Increased OVA-induced AHR is also observed in obese mice [63]. However, this increase in airway responsiveness occurs in the absence of any differences in Th2 cytokines and in the face of *decreases* in BAL eosinophils compared to lean controls, suggesting that other factors must contribute. Interestingly, OVA-induced increases in serum IgE are also greater in *ob/ob* and *db/db* mice than in lean controls. As described above, IgE and mast cells do not appear to contribute prominently in most studies of allergen-induced airways disease in mice. However, the expression of 5-lipoxygenase activating protein is substantially increased in the adipose tissue of *ob/ob* mice [79]. This could lead to greater leukotriene synthesis and subsequent release from activated mast cells and thus enhance airway responsiveness. In this context, Peters-Golden *et al.* recently reported a more beneficial effect of the leukotriene antagonist montelukast in obese versus lean asthmatics [80].

Adipokines: Several hypotheses have been proposed to explain the association between obesity and asthma (Fig. 8.1). For example, obesity results in changes in a variety of adipose tissue derived factors (adipokines) that may contribute to the relationship between obesity and asthma. Serum leptin is increased in obesity and two cross-sectional studies indicate that leptin is also increased in asthmatics [81, 82]. Numerous reports indicate that leptin is pro-inflammatory (see recent review [83]), and these inflammatory effects could exacerbate asthma in the obese. Leptin treatment has been shown to augment allergen-induced AHR in lean mice, without affecting eosinophil influx or Th2 cytokine expression [66]. Indeed, rather than modifying adaptive immunity, leptin could be acting on the innate immune system: exogenous administration of leptin to lean mice increases their subsequent inflammatory response to acute O_3 exposure [65], a response characterized by release of acute phase cytokines and chemokines, and dependent to some extent on toll-like receptor activation [84]. However, there must be factors in addition to

Mechanisms proposed to explain the relationship between obesity and asthma

Common etiologies
- *In utero* conditions
- Genetics

Effects of obesity on lung mechanics
- ↓ FRC
- ↓ Tidal volume

Co-morbidities
- Gastroesophageal reflux
- Sleep disorder breathing
- Type-II diabetes
- Hypertension

Adipokines
- Cytokines
- Chemokines
- Energy regulating hormones
- Acute phase reactants
- Other factors

FIG. 8.1 Potential mechanisms whereby obesity may be associated with asthma. Reproduced from Ref. [68] with permission of the American Physiological Society.

leptin that contribute to the augmented O_3 responses of obese mice, since such responses are observed not only in *Cpe^fat^* and DIO mice [61, 62] that have marked increases in serum leptin, but also in *ob/ob* and *db/db* mice [60, 65], that lack either leptin or the leptin receptor.

In contrast to other adipokines, adiponectin, an insulin-sensitizing hormone, declines in obesity [85, 86]. In contrast to leptin, adiponectin has important anti-inflammatory effects in obesity [86]. In lean Balb/c mice, exogenous administration of adiponectin results in an almost complete suppression of OVA-induced AHR, airway inflammation, and Th2 cytokine expression in the lung [67]. Adiponectin has been shown to inhibit macrophage production of lymphocyte chemotactic factors [87]. An adiponectin-induced reduction in lymphocyte influx into the lungs would be expected to reduce pulmonary concentrations of Th2 cytokines and could thus be responsible for attenuating eosinophilia and AHR following OVA challenge. Three adiponectin receptors have been cloned: adipoR1, adipoR2, and T-cadherin [88, 89]. Lung tissue expresses all three adiponectin receptors. The expression of all three receptors declines in the lungs following allergen sensitization and challenge in mice [67], suggesting that asthma may be a state of adiponectin resistance. Coupled with obesity-related declines in adiponectin [85, 86] and additional declines in serum adiponectin that occur with allergen challenge [67], the data suggest that the obese asthmatic is likely to have defects in this important immunomodulatory pathway that augment the effects of allergen challenge.

MOUSE MODELS OF COPD

Modeling COPD

Animal models were critical to the development of the elastase:anti-elastase hypothesis generated 45 years ago. This hypothesis remains the cornerstone of our understanding of COPD pathogenesis today. The major finding at that time was that instillation of the proteinase papain into experimental animals caused airspace enlargement, the characteristic feature of emphysema [1]. Since then animals have been exposed to a variety of molecular, chemical, and environmental agents that lead to airspace enlargement [90]. In particular elastases [91–93], cigarette smoke [94, 95], and more recently inducers of apoptosis have been most informative [96, 97]. Mouse genetic mutants have also been critical in furthering our understanding of the pathogenesis of COPD [4, 95].

Elastases: Ever since the initial findings with papain, a variety of proteinases have been instilled into experimental animals. Of note, only those with elastolytic activity develop emphysema, collagenases do not [91]. Of most utility has been porcine pancreatic elastase (PPE) [91, 98, 99], a potent elastase that quickly generates airspace enlargement. This model has also been useful to test agents that have the capacity to restore lung structure such as retinoic acid [100]. A recent surprise has been the usefulness of PPE to model events upstream from elastolysis. This is likely due to the fact that elastin fragments themselves [101] are chemokines, and that the events following PPE instillation are very similar to the inflammatory cell cascade that one observes in COPD following cigarette smoke exposure. That is, initially PPE leads to acute airspace enlargement with elastin destruction and release of fragments of elastin and other matrix proteins. This results in subsequent neutrophilia followed within days by a predominant macrophage and T-cell influx. These activated inflammatory cells release endogenous elastases causing progressive airspace enlargement.

Environmental factors: Cigarette smoke is clearly the major etiologic factor for COPD. However, a variety of other environmental agents have been applied to determine their capacity to induce COPD. Nitrogen dioxide and ozone cause mild lung injury suggesting that they may be modifying environmental factors but unlikely primary environmental factors leading to COPD [90]. Cadmium chloride [102], a constituent of cigarette smoke, primarily causes interstitial fibrosis tethering open adjoining airspaces simulating emphysema. While this mechanism differs from airspace enlargement secondary to matrix destruction that characterizes emphysema, excess collagen deposition in the context of loss of elastin is a feature of human centrilobular emphysema.

Cigarette smoke [4, 94, 95]: The great strength of cigarette smoke-induced models of COPD is that we know that this is the causative agent for COPD. Hence, pathways leading to phenotypic changes of COPD mirror those of human COPD, limited only by differences in smoke exposure as well as biological differences between mice and humans. Over the years many species of mammals have been exposed to cigarette smoke using a variety of smoking chambers. Larger animals are limited by the long (years) duration of exposure required for development of COPD. Of small animals tested, the guinea pig is the most susceptible species and rat the least. Several inbred mouse strains are also susceptible to cigarette smoke-induced COPD, and hence they have dominated the literature due to our ability to manipulate gene expression as well as other advantages discussed above.

Following exposure to cigarette smoke, mice develop changes similar to humans including acute inflammation with neutrophils followed by subacute increases in macrophages, and $CD8^+ > CD4^+$ T-cells [103]. Proteinases are released and apoptosis of structural cells is observed. These processes ultimately lead to airspace enlargement that is easily detectable in most strains within 6 months. With respect to the airway, mice lose cilia upon cigarette smoke exposure, develop small airway fibrosis, and also develop goblet cell hypertophy [4, 104].

Apoptosis: Recently investigators have found that exposure to agents that initiate either endothelial cell death (via VEGFRII inhibition) [96] or epithelial cell death (via caspase 3 nodularin, ceramide delivery, and VEGF inhibition) [97, 105, 106] lead to non-inflammatory airspace enlargement. Clearly, to lose an acinar unit, one must destroy both extracellular matrix (ECM) and structural cells. Traditionally, we believed that inflammatory cell proteinases destroy ECM and cells unable to attach to the ECM float away and die. The apoptotic models suggest death of structural cells may be an initiating event, with subsequent release of matrix degrading proteinases.

Whether this occurs in human COPD as a primary event is uncertain, but does raise interesting testable possibilities.

Genetic models: Natural genetic mutants, transgenic mice, and gene-targeted mice, may develop spontaneous airspace enlargement of developmental origin. They may also develop progressive airspace enlargement acquired spontaneously over time that more closely reflects the destruction of mature lung tissue characteristic of emphysema [4, 99]. Overexpression or underexpression of a protein during development that leads to failed alveogenesis and enlarged airspaces is very informative regarding pathways required for normal lung development and hence repair in emphysema. However, this is very different from destruction and enlargement of normal mature alveoli that defines pulmonary emphysema.

If overexpression of a protein in transgenic mice leads to airspace enlargement in the adult, then *if* the protein is overexpressed in the endogenous process it is likely to play a role. For example, instillation of PPE mimics emphysema quite well, but of course PPE does not travel from the gut to the lung in true emphysema.

Of most utility in dissecting the pathogenesis of COPD are developmentally normal gene-targeted mice applied to cigarette smoke. Protection of a "knockout" with smoking suggests that the protein deleted is involved in a pathway that promotes COPD, while worsening COPD suggests a protective factor was deleted. There are many proteins that have been identified using these techniques as recently reviewed [99]. Knowledge gained from these studies will be synthesized below.

Limitations: The mouse lung structure is not identical to humans. Mice do not inhale cigarettes as humans would and these obligate nasal breathers bring the smoke through an extensive olfactory epithelium into the airway where there are few submucosal glands, much less airway branching, and no respiratory bronchioles – the initial site of centriacinar emphysema. Hence, mice do not develop "small airway disease" that is so important in human COPD. Nevertheless, the opportunity exists to infect mouse airways with relevant human pathogens to learn more about airway changes and exacerbations of COPD.

Mice require months of cigarette smoke exposure to develop mild emphysema. These are more likely to reflect Global Initiative for Chronic Obstructive Lung Disease (GOLD) stage I and II not the severe stages of COPD (III and IV). Moreover, there is a variety of smoking chambers now available that differ with respect to the mode, rate, and amount of smoke delivered. Particulate amounts, carboxyhemoglobin levels, and other surrogates of exposure are often measured to assess how well these mimic human exposure. The most important features are the resultant phenotypic changes in the exposed animals and their relationship to humans. One note of caution is that excessive acute exposure to smoke leads to acute lung injury with pronounced neutrophilia, which is not likely to reflect COPD.

Lessons from mouse models of COPD

Inflammatory cell network in COPD: Animal models have taught us that emphysema is not caused by a single cell type or proteinase. Rather, multiple inflammatory and immune cells and mediators are involved. Investigators are now focused on teasing out mechanisms that lead to this complex network and the interactions between these cells and their mediators that leads to COPD.

Upon acute exposure to cigarette smoke one observes activation of macrophages and structural cells that leads to a mild, acute, transient accumulation of neutrophils in the lung. Over days to weeks one also observes increased number of T-cells, particularly $CD8^+$ T-cells, and macrophages [107]. Neutrophils accumulate with each acute exposure, but due to the short-half life of the neutrophil do not accumulate to high levels. The longer-lived macrophages and T-cells progressively accumulate with chronic smoke exposure. In humans, one observes B-cell accumulation in bronchial–alveolar lymphoid tissue (BALT) late in the disease. While this could be secondary to bacterial colonization in humans, it has also been observed in mice exposed to cigarette smoke in the absence of infection [108].

An undoubtedly simplistic but unified pathway, supported by data from cigarette smoke-exposed mice, can account for much of the inflammation observed in response to cigarette smoke. Cigarette smoke, via its oxidant properties, can activate macrophages leading to TNF-α production with latent TNF-α translocating to the surface whereupon MMP-12 "sheds" active TNF-α leading to neutrophil recruitment and activation [109].

Release of MMP-12 from macrophages and collagenases MMP-8 and MMP-13 (MMP-1 not expressed in mice) from macrophages and likely epithelial cells leads to generation of elastin and collagen fragments respectively. Elastin fragments serve as chemokines for macrophages [101] and collagen fragments promote neutrophil recruitment [110]. Indeed, in humans CXC and CC chemokines may play a significant role in inflammatory cell migration, particularly when colonized with bacteria. However, in mice under sterile conditions, elastin fragments appear to be critical to the low-level macrophage accumulation one observes in response to cigarette smoke.

T-lymphocytes, particularly $CD8^+$ T-cells, are increased in lungs of patients with COPD [111]. Overexpression of the T-cell product IFNγ in transgenic mice, results in apoptosis, inflammation, and emphysema [112]. Moreover, IFNγ and IFNγ-IP-10 (interferon-gamma-inducible protein-10) have been shown to be increased in $CD8^+$ T-cells from patients with emphysema, and IP-10 was also shown to induce the production of MMP-12 in human alveolar macrophages [113]. In addition to this potentially indirect influence of $CD8^+$ T-cells on emphysema, they may also directly cause cytotoxicity contributing to emphysema. The importance of the $CD8^+$ T-cell in the pathogenesis of emphysema, was recently confirmed by subjecting wild-type (C57BL/6 J), $CD8^+$ T-cell deficient ($CD8^{-/-}$) mice, and $CD4^+$ T-cell deficient ($CD4^{-/-}$) mice to a model of cigarette smoke-induced emphysema [107]. In the absence of $CD8^+$ T-cells, but not $CD4^+$ T-cells, there was a marked inhibition in inflammation and emphysema. Integrating these results into the "network," one can conclude that $CD8^+$ T-cells via production of IP-10, signal through macrophage CXCR3 to activate the macrophage leading to proteinase and cytokine production. The macrophage then recruits neutrophils, and their proteinases then degrade ECM, particularly elastin and collagen, continuing the inflammatory cascade. Hence, $CD8^+$ T-cells may be the master regulator of inflammation in COPD. Of note, these results are entirely consistent with data from human COPD.

Defects in adaptive immunity are also likely to play an important role with respect to colonization of microorganisms in the airway and exacerbations. Prolonged cigarette smoke exposure (1 month) in mice results in a decrease in lung dendritic cell (DC) number, but DC migration to regional thoracic lymph nodes remains intact. Once in the lymph nodes, however, cigarette smoking suppresses maturation of DCs as demonstrated by the reduced expression of co-stimulatory molecules. Moreover, these cigarette smoke exposed DC cells have impaired capacity to activate $CD4^+$ T-cells with less IL-2 production and diminished T-cell proliferation [114]. Impaired $CD4^+$ T-cell activation would predispose patients to colonization with infectious microorganisms.

As discussed, B-cells and BALT are present in COPD lung tissue, however, their role in the disease process remains to be determined. Of note, recent animal [115] and human [116] studies have implicated autoimmune processes in the pathogenesis of COPD. In fact, elastin fragments themselves were shown to be an important autoimmunue target in COPD in human studies [116]. Other cell types including eosinophils, mast cells, and natural killer (NK) cells are also likely to be involved in this inflammatory cell network, but to date, they have not been carefully addressed.

Role of oxidative stress in COPD: Cigarette smoke, loaded with 10^{17} oxidant molecules per puff leads to increased oxidative stress in smokers and in patients with COPD. Oxidative stress, in part through inactivation of histone deacetylases, exposes DNA sequences and activates transcription factors for inflammatory genes, such as nuclear factor-κB (NF-κB) and activator protein (AP)-1 [117]. This finding, initially in humans, has been confirmed in rodents exposed to cigarette smoke [118]. Oxidants also have a significant role in promoting apoptosis, and although oxidants do not directly degrade ECM proteins, they might modify these proteins making them more susceptible to proteolytic cleavage. Oxidative stress also can alter the structure of pro-MMPs leading to autolytic cleavage and activation [119]. Gene deletion of Nrf-2 a master transcription factor for antioxidant genes also leads to enhanced emphysema in response to cigarette smoke exposure highlighting the importance of oxidative stress in COPD [120].

Role of proteinases in COPD: As mentioned at the beginning of this chapter and reviewed in depth in the "Matrix Degrading Proteinase" (Chapter 28), ECM destruction by proteinases is the cornerstone of emphysema. Animal models have played an important role in defining the contribution of different elastolytic enzymes in this process. In addition to the original instillation of papain by Gross [1], in one of the first applications of gene-targeted mice to lung disease, it was shown that macrophage elastase (MMP-12) deficient mice were protected from the development of emphysema [95]. As in several other models of disease, MMP-12 is pro-inflammatory, which in part is related to the generation of chemotactic elastin fragments [101].

Subsequently, several other animal models have demonstrated the importance of proteinases in COPD. Lung-specific transgenic mice expressing either the Th2 cytokine IL-13 ("Dutch mice") [121] or the Th1 cytokine IFN-γ ("British mice") [112] both develop proteinase-dependent emphysema. IFN-γ overexpressors develop a cysteine proteinase/apoptosis mediated form of emphysema, while IL-13 overexpression induces inflammation with MMP-12- and MMP-9-dependent airspace enlargement. IL-13 transgenic mice also develop small airway remodeling that is believed to be due to MMP-9-dependent activation of TGF-β with airway fibrosis [122]. The role of TGF-β in emphysema was also highlighted with $a_v b6^{-/-}$ mice, that are unable to activate TGF-β in the airspace, develop macrophage inflammation and MMP-12-dependent emphysema [123]. TGF-β regulation is complex as total absence of TGF-β releases the brake on inflammation and destruction in the airspace while too much TGF-β leads to fibrosis in the airways.

Other gene-targeted mice that develop emphysema include surfactant protein $D^{-/-}$ (SP-$D^{-/-}$) mice, which exhibit macrophage activation, MMP production, and consequent emphysema [124]. TIMP-3 deficiency leads to a combination of developmental airspace enlargement combined with progressive destructive emphysema in adults [125], supporting the role of MMPs in COPD. There is also evidence from animal models that neutrophil elastase (NE), a serine proteinase long thought to cause COPD, is also involved in experimental emphysema. $NE^{-/-}$ mice develop only 40% as much airspace enlargement as wild-type mice exposed to long-term cigarette smoke [126]. These studies have uncovered several interactions between NE and MMPs. MMPs degrade α1-AT and NE degrades TIMPs, each potentiating the other's proteinase activity. Moreover, NE mediates monocyte migration and as discussed, acute neutrophil inflammation secondary to smoking is related to MMP-12-dependent TNF-shedding [109].

Role of apoptosis in COPD: As discussed above, induction of structural cell apoptosis leads to airspace enlargement. However, there are some aspects of these models that do not mimic human disease. They are non-inflammatory and transient, which is likely related to their lack of matrix degradation. Most importantly, however, closer pathologic and physiologic examination of at least one model of epithelial cell apoptosis demonstrates that this model is not emphysema *per se*, but acute lung injury with collapsed alveoli tethering of open adjoining airspaces to appear enlarged (Mouded M and Shapiro SD, unpublished observations). Thus, like other traditional models of epithelial apoptosis, bleomycin being the classic, the result is acute lung injury and if severe, fibrosis. Endothelial apoptosis could have a different outcome, but it is likely that apoptosis does not initiate emphysema. Apoptosis, however, is likely to be a critically important modifier in the pathogenesis of COPD. For example, as in humans, apoptosis is observed in animal models following cigarette smoke exposure, likely due to loss of basement membrane attachment +/− oxidant effects of cigarette smoke. The fate of the apoptotic cell is critical to disease progression. Clearance by macrophages leads to anti-inflammatory TGF-β production and release of growth factors that promote repair [127]. Inefficient clearance leads to secondary necrosis, augmenting inflammation and destruction.

Role of repair in COPD: The ability of the adult lung to repair damaged alveoli appears limited. This is likely due to the structural and functional dependence upon the elastic fiber architecture combined with the difficulty in repairing elastic fibers. Whether one can temporally and spatially bring together the multiple microfibrillar components of an elastic fiber let alone coordinate matrix repair with endothelial and epithelial migration and differentiation over an injured matrix is unknown.

With the emergence of regenerative medicine and stem cell biology, there is hope that we can one day create new functional lung tissue. However, given the complex three-dimensional nature of the lung, its regeneration presents a special challenge relative to other organs. It is currently not known whether there is a master stem cell, separate progenitors for each cell type, nor what their reparative capacity is. More importantly, given the collapse of the elastic fiber cable network, it is unclear whether stem cells alone could regenerate a functional alveolar unit.

There has been some evidence that one can repair emphysematous lung tissue in animal models. The study by Massaro *et al.* [100] discussed above, showing that all-trans retinoic acid reversed PPE-mediated emphysema, has led to hope that we might be able to promote lung repair [100]. Use of retinoic acid did not have significant effects in human trials, but has opened the door to mechanisms that could be reparative.

There have been attempts to add growth factors, such as keratinocyte growth factor (KGF) to elastase-treated lungs. The results demonstrate the ability to protect from injury, but not repair damaged tissue [128]. The advent of lineage tagged and lineage ablated genetically engineered mice will be helpful to define progenitor cells and their reparative capacity in the future.

SUMMARY

It is important to keep in mind that there are important structural differences in the lungs of mice and humans and to interpret animal data with caution. In some instances, murine models of allergic asthma have not recapitulated the human findings, and have led to some costly failures in drug development. Nevertheless, animal models have been extremely useful in the evolution of current concepts regarding the etiology of asthma and COPD and genetically engineered mice continue to drive the generation and testing of new hypotheses.

ACKNOWLEDGEMENTS

The authors work is supported by National Institute of Environmental Health Sciences Grants ES-013307 and ES-00002, and National Heart, Lung, and Blood Institute Grants HL-084044 and HL-082541.

References

1. Gross P, Pfitzer E, Tolker M. Experimental emphysema: Its production with papain in normal and silicotic rats. *Arch Environ Health* 11: 50–58, 1965.
2. Drazen JM, Israel E, O'Byrne PM. Treatment of asthma with drugs modifying the leukotriene pathway. *N Engl J Med* 340: 197–206, 1999.
3. Shapiro SD. Animal models of asthma: Pro: Allergic avoidance of animal (model[s]) is not an option. *Am J Respir Crit Care Med* 174: 1171–73, 2006.
4. Shapiro SD. Transgenic and gene-targeted mice as models for chronic obstructive pulmonary disease. *Eur Respir J* 29: 375–78, 2007.
5. Boyce JA, Austen KF. No audible wheezing: Nuggets and conundrums from mouse asthma models. *J Exp Med* 201: 1869–73, 2005.
6. Wenzel S, Holgate ST. The mouse trap: It still yields few answers in asthma. *Am J Respir Crit Care Med* 174: 1173–76, 2006. Discussion 1176–78.
7. Zosky GR, Sly PD. Animal models of asthma. *Clin Exp Allergy* 37: 973–88, 2007.
8. Lommatzsch M, Julius P, Kuepper M, Garn H, Bratke K, Irmscher S, Luttmann W, Renz H, Braun A, Virchow JC. The course of allergen-induced leukocyte infiltration in human and experimental asthma. *J Allergy Clin Immunol* 118: 91–97, 2006.
9. Zhang Y, Lamm WJ, Albert RK, Chi EY, Henderson WR Jr., Lewis DB. Influence of the route of allergen administration and genetic background on the murine allergic pulmonary response. *Am J Respir Crit Care Med* 155: 661–69, 1997.
10. Hamelmann E, Tadeda K, Oshiba A, Gelfand EW. Role of IgE in the development of allergic airway inflammation and airway hyperresponsiveness – a murine model. *Allergy.* 54: 297–305, 1999.
11. Shore SA. Modeling airway remodeling: The winner by a nose? *Am J Respir Crit Care Med* 168: 910–11, 2003.
12. Shinagawa K, Kojima M. Mouse model of airway remodeling: Strain differences. *Am J Respir Crit Care Med* 168: 959–67, 2003.
13. Brewer JP, Kisselgof AB, Martin TR. Genetic variability in pulmonary physiological, cellular, and antibody responses to antigen in mice. *Am J Respir Crit Care Med* 160: 1150–56, 1999.
14. Ewart SL, Kuperman D, Schadt E, Tankersley C, Grupe A, Shubitowski DM, Peltz G, Wills-Karp M. Quantitative trait loci controlling allergen-induced airway hyperresponsiveness in inbred mice. *Am J Respir Cell Mol Biol* 23: 537–45, 2000.
15. Lloyd CM. Building better mouse models of asthma. *Curr Allergy Asthma Rep* 7: 231–36, 2007.
16. Yu M, Tsai M, Tam SY, Jones C, Zehnder J, Galli SJ. Mast cells can promote the development of multiple features of chronic asthma in mice. *J Clin Invest* 116: 1633–41, 2006.
17. Lim DH, Cho JY, Miller M, McElwain K, McElwain S, Broide DH. Reduced peribronchial fibrosis in allergen-challenged MMP-9-deficient mice. *Am J Physiol Lung Cell Mol Physiol* 291: L265–271, 2006.
18. Garssen J, Nijkamp FP, Van Der Vliet H, Van Loveren H. T-cell-mediated induction of airway hyperreactivity in mice. *Am Rev Respir Dis* 144: 931–38, 1991.
19. Corry DB, Folkesson HG, Warnock ML, Erle DJ, Matthay MA, Wiener-Kronish JP, Locksley RM. Interleukin 4, but not interleukin 5 or eosinophils, is required in a murine model of acute airway hyperreactivity. *J Exp Med* 183: 109–17, 1996.
20. Brusselle GG, Kips JC, Tavernier JH, van der Heyden JG, Cuvelier CA, Pauwels RA, Bluethmann H. Attenuation of allergic airway inflammation in IL-4 deficient mice. *Clin Exp Allergy* 24: 73–80, 1994.
21. Gavett SH, Chen X, Finkelman F, Wills-Karp M. Depletion of murine $CD4^+$ T lymphocytes prevents antigen-induced airway hyperreactivity and pulmonary eosinophilia. *Am J Respir Cell Mol Biol* 10: 587–93, 1994.
22. Corry DB, Grunig G, Hadeiba H, Kurup VP, Warnock ML, Sheppard D, Rennick DM, Locksley RM. Requirements for allergen-induced airway hyperreactivity in T and B cell-deficient mice. *Mol Med* 4: 344–55, 1998.

23. Zuany-Amorim C, Ruffie C, Haile S, Vargaftig BB, Pereira P, Pretolani M. Requirement for gammadelta T cells in allergic airway inflammation. *Science* 280: 1265–67, 1998.
24. Randolph DA, Stephens R, Carruthers CJ, Chaplin DD. Cooperation between Th1 and Th2 cells in a murine model of eosinophilic airway inflammation. *J Clin Invest* 104: 1021–29, 1999.
25. Meyer EH, DeKruyff RH, Umetsu DT. T cells and NKT cells in the pathogenesis of asthma. *Annu Rev Med* 59: 281–92, 2008.
26. Strickland DH, Stumbles PA, Zosky GR, Subrata LS, Thomas JA, Turner DJ, Sly PD, Holt PG. Reversal of airway hyperresponsiveness by induction of airway mucosal $CD4^+$ $CD25^+$ regulatory T cells. *J Exp Med* 203: 2649–60, 2006.
27. Lukacs NW, Strieter RM, Chensue SW, Kunkel SL. Interleukin-4-dependent pulmonary eosinophil infiltration in a murine model of asthma. *Am J Respir Cell Mol Biol* 10: 526–32, 1994.
28. Kips JC, Brusselle GG, Joos GF, Peleman RA, Devos RR, Tavernier JH, Pauwels RA. Importance of interleukin-4 and interleukin-12 in allergen-induced airway changes in mice. *Int Arch Allergy Immunol* 107: 115–18, 1995.
29. Grunig G, Warnock M, Wakil AE, Venkayya R, Brombacher F, Rennick DM, Sheppard D, Mohrs M, Donaldson DD, Locksley RM, Corry DB. Requirement for IL-13 independently of IL-4 in experimental asthma [see comments]. *Science* 282: 2261–63, 1998.
30. Hamelmann E, Wahn U, Gelfand EW. Role of the Th2 cytokines in the development of allergen-induced airway inflammation and hyperresponsiveness. *Int Arch Allergy Immunol* 118: 90–94, 1999.
31. Wills-Karp M, Luyimbazi J, Xu X, Schofield B, Neben TY, Karp CL, Donaldson DD. Interleukin-13: Central mediator of allergic asthma [see comments]. *Science* 282: 2258–61, 1998.
32. He R, Oyoshi MK, Jin H, Geha RS. Epicutaneous antigen exposure induces a Th17 response that drives airway inflammation after inhalation challenge. *Proc Natl Acad Sci USA* 104: 15817–22, 2007.
33. Karwot R, Maxeiner JH, Schmitt S, Scholtes P, Hausding M, Lehr HA, Glimcher LH, Finotto S. Protective role of nuclear factor of activated T cells 2 in CD8(+) long-lived memory T cells in an allergy model. *J Allergy Clin Immunol* 121(4): 992–99, 2008.
34. Pichavant M, Goya S, Meyer EH, Johnston RA, Kim HY, Matangkasombut P, Zhu M, Iwakura Y, Savage PB, DeKruyff RH, Shore SA, Umetsu DT. Ozone exposure in a mouse model induces airway hyperreactivity that requires the presence of natural killer T cells and IL-17. *J Exp Med* 205: 385–93, 2008.
35. Fahy JV, Fleming HE, Wong HH, Liu JT, Su JQ, Reimann J, Fick RB Jr., Boushey HA. The effect of an anti-IgE monoclonal antibody on the early- and late-phase responses to allergen inhalation in asthmatic subjects. *Am J Respir Crit Care Med* 155: 1828–34, 1997.
36. Zosky GR, Larcombe AN, White OJ, Burchell JT, Janosi TZ, Hantos Z, Holt PG, Sly PD, Turner DJ. Ovalbumin-sensitized mice are good models for airway hyperresponsiveness but not acute physiological responses to allergen inhalation. *Clin Exp Allergy* 38(6): 829–38, 2007.
37. Martin TR, Galli SJ, Katona IM, Drazen JM. Role of mast cells in anaphylaxis. Evidence for the importance of mast cells in the cardiopulmonary alterations and death induced by anti-IgE in mice. *J Clin Invest* 83: 1375–83, 1989.
38. Chand N, Nolan K, Sofia RD, Diamantis W. Changes in aeroallergen-induced pulmonary mechanics in actively sensitized guinea pig: Inhibition by azelastine. *Ann Allergy* 64: 151–54, 1990.
39. Eidelman DH, Bellofiore S, Martin JG. Late airway responses to antigen challenge in sensitized inbred rats. *Am Rev Respir Dis* 137: 1033–37, 1988.
40. Abraham WM, Oliver W Jr., King MM, Yerger L, Wanner A. Effect of pharmacologic agents on antigen-induced decreases in specific lung conductance in sheep. *Am Rev Respir Dis* 124: 554–58, 1981.
41. Djukanovic R, Wilson SJ, Kraft M, Jarjour NN, Steel M, Chung KF, Bao W, Fowler-Taylor A, Matthews J, Busse WW, Holgate ST, Fahy JV. Effects of treatment with anti-immunoglobulin E antibody omalizumab on airway inflammation in allergic asthma. *Am J Respir Crit Care Med* 170: 583–93, 2004.
42. Kung TT, Stelts D, Zurcher JA, Jones H, Umland SP, Kreutner W, Egan RW, Chapman RW. Mast cells modulate allergic pulmonary eosinophilia in mice. *Am J Respir Cell Mol Biol* 12: 404–9, 1995.
43. Kobayashi T, Miura T, Haba T, Sato M, Serizawa I, Nagai H, Ishizaka K. An essential role of mast cells in the development of airway hyperresponsiveness in a murine asthma model. *J Immunol* 164: 3855–61, 2000.
44. MacLean JA, Sauty A, Luster AD, Drazen JM, De Sanctis GT. Antigen-induced airway hyperresponsiveness, pulmonary eosinophilia, and chemokine expression in B cell-deficient mice. *Am J Respir Cell Mol Biol* 20: 379–87, 1999.
45. Takeda K, Hamelmann E, Joetham A, Shultz LD, Larsen GL, Irvin CG, Gelfand EW. Development of eosinophilic airway inflammation and airway hyperresponsiveness in mast cell-deficient mice. *J Exp Med* 186: 449–54, 1997.
46. Gleich GJ. The eosinophil and bronchial asthma: Current understanding. *J Allergy Clin Immunol* 85: 422–36, 1990.
47. Lee JJ, McGarry MP, Farmer SC, Denzler KL, Larson KA, Carrigan PE, Brenneise IE, Horton MA, Haczku A, Gelfand EW, Leikauf GD, Lee NA. Interleukin-5 expression in the lung epithelium of transgenic mice leads to pulmonary changes pathognomonic of asthma. *J Exp Med* 185: 2143–56, 1997.
48. Foster PS, Hogan SP, Ramsay AJ, Matthaei KI, Young IG. Interleukin 5 deficiency abolishes eosinophilia, airways hyperreactivity, and lung damage in a mouse asthma model [see comments]. *J Exp Med* 183: 195–201, 1996.
49. Hamelmann E, Takeda K, Haczku A, Cieslewicz G, Shultz L, Hamid Q, Xing Z, Gauldie J, Gelfand EW. Interleukin (IL)-5 but not immunoglobulin E reconstitutes airway inflammation and airway hyperresponsiveness in IL-4-deficient mice. *Am J Respir Cell Mol Biol* 23: 327–34, 2000.
50. Hogan SP, Mould A, Kikutani H, Ramsay AJ, Foster PS. Aeroallergen-induced eosinophilic inflammation, lung damage, and airways hyperreactivity in mice can occur independently of IL-4 and allergen-specific immunoglobulins. *J Clin Invest* 99: 1329–39, 1997.
51. Flood-Page P, Swenson C, Faiferman I, Matthews J, Williams M, Brannick L, Robinson D, Wenzel S, Busse W, Hansel TT, Barnes NC. A study to evaluate safety and efficacy of mepolizumab in patients with moderate persistent asthma. *Am J Respir Crit Care Med* 176: 1062–71, 2007.
52. O'Byrne PM. The demise of anti IL-5 for asthma, or not. *Am J Respir Crit Care Med* 176: 1059–60, 2007.
53. Hessel EM, Van Oosterhout AJ, Van Ark I, Van Esch B, Hofman G, Van Loveren H, Savelkoul HF, Nijkamp FP. Development of airway hyperresponsiveness is dependent on interferon-gamma and independent of eosinophil infiltration. *Am J Respir Cell Mol Biol* 16: 325–34, 1997.
54. Humbles AA, Lloyd CM, McMillan SJ, Friend DS, Xanthou G, McKenna EE, Ghiran S, Gerard NP, Yu C, Orkin SH, Gerard C. A critical role for eosinophils in allergic airways remodeling. *Science* 305: 1776–79, 2004.
55. Beuther DA, Weiss ST, Sutherland ER. Obesity and asthma. *Am J Respir Crit Care Med* 174: 112–19, 2006.
56. Ford ES. The epidemiology of obesity and asthma. *J Allergy Clin Immunol* 115: 897–909, 2005., quiz 910
57. Shore SA, Johnston RA. Obesity and asthma. *Pharmacol Ther* 110: 83–102, 2006.
58. Shore SA. Obesity and asthma: Implications for treatment. *Curr Opin Pulm Med* 13: 56–62, 2007.
59. Boulet LP, Franssen E. Influence of obesity on response to fluticasone with or without salmeterol in moderate asthma. *Respir Med* 101: 2240–47, 2007.
60. Lu FL, Johnston RA, Flynt L, Theman TA, Terry RD, Schwartzman IN, Lee A, Shore SA. Increased pulmonary responses to acute ozone exposure in obese db/db mice. *Am J Physiol Lung Cell Mol Physiol* 290: L856–l865, 2006.
61. Johnston RA, Theman TA, Lu FL, Terry RD, Williams ES Shore SA (in press) Diet-induced obesity causes innate airway hyperresponsiveness

to methacholine and enhances ozone induced pulmonary inflammation. *J Appl Physiol* 104: 1727–35, 2008.
62. Johnston RA, Theman TA, Shore SA. Augmented Responses to Ozone in Obese Carboxypeptidase E-Deficient Mice. *Am J Physiol Regul Integr Comp Physiol* 290: R126–R133, 2006.
63. Johnston RA, Zhu M, Rivera-Sanchez YM, Lu FL, Theman TA, Flynt L, Shore SA. Allergic airway responses in obese mice. *Am J Respir Crit Care Med* 176: 650–58, 2007.
64. Rivera-Sanchez YM, Johnston RA, Schwartzman IN, Valone J, Silverman ES, Fredberg JJ, Shore SA. Differential effects of ozone on airway and tissue mechanics in obese mice. *J Appl Physiol* 96: 2200–6, 2004.
65. Shore SA, Rivera-Sanchez YM, Schwartzman IN, Johnston RA. Responses to ozone are increased in obese mice. *J Appl Physiol* 95: 938–45, 2003.
66. Shore SA, Schwartzman IN, Mellema MS, Flynt L, Imrich A, Johnston RA. Effect of leptin on allergic airway responses in mice. *J Allergy Clin Immunol* 115: 103–9, 2005.
67. Shore SA, Terry RD, Flynt L, Xu A, Hug C. Adiponectin attenuates allergen-induced airway inflammation and hyperresponsiveness in mice. *J Allergy Clin Immunol* 118: 389–95, 2006.
68. Shore SA. Obesity and asthma: Lessons from animal models. *J Appl Physiol* 102: 516–28, 2007.
69. Leibel RL, Chung WK, Chua SC Jr. The molecular genetics of rodent single gene obesities. *J Biol Chem* 272: 31937–40, 1997.
70. Scherer PE. Adipose tissue: From lipid storage compartment to endocrine organ. *Diabetes* 55: 1537–45, 2006.
71. Lang JE, Williams ES, Shore SA. TNFα contributes to innate airway hyperresponsiveness in murine obesity. *Am J Respir Crit Care Med* 175: A912, 2007.
72. Tolbert PE, Mulholland JA, MacIntosh DL, Xu F, Daniels D, Devine OJ, Carlin BP, Klein M, Dorley J, Butler AJ, Nordenberg DF, Frumkin H, Ryan PB, White MC. Air quality and pediatric emergency room visits for asthma in Atlanta, Georgia, USA. *Am J Epidemiol* 151: 798–810, 2000.
73. Gent JF, Triche EW, Holford TR, Belanger K, Bracken MB, Beckett WS, Leaderer BP. Association of low-level ozone and fine particles with respiratory symptoms in children with asthma. *JAMA* 290: 1859–67, 2003.
74. Bennett WD, Hazucha MJ, Folinsbee LJ, Bromberg PA, Kissling GE, London SJ. Acute pulmonary function response to ozone in young adults as a function of body mass index. *Inhal Toxicol* 19: 1147–54, 2007.
75. Alexeeff SE, Litonjua AA, Suh H, Sparrow D, Vokonas PS, Schwartz J. Ozone Exposure and Lung Function: Effect Modified by Obesity and Airways Hyperresponsiveness in the VA Normative Aging Study. *Chest* 132: 1890–97, 2007.
76. Johnston RA, Williams ES, Shore SA. Airway responses to ozone during the development of obesity. *Proc Am Thorac Soc* 3: A528, 2006.
77. Lang JE, Williams ES, Mizgerd JP Shore SA (in press) Effect of obesity on pulmonary inflammation induced by acute ozone exposure: Role of interleukin-*Am J Physiol Lung Cell Mol Physiol* 294: L1013–20, 2008.
78. Mito N, Kitada C, Hosoda T, Sato K. Effect of diet-induced obesity on ovalbumin-specific immune response in a murine asthma model. *Metabolism* 51: 1241–46, 2002.
79. Back M, Sultan A, Ovchinnikova O, Hansson GK. 5-Lipoxygenase-activating protein: A potential link between innate and adaptive immunity in atherosclerosis and adipose tissue inflammation. *Circ Res* 100: 946–49, 2007.
80. Peters-Golden M, Swern A, Bird SS, Hustad CM, Grant E, Edelman JM. Influence of body mass index on the response to asthma controller agents. *Eur Respir J* 27: 495–503, 2006.
81. Guler N, Kirerleri E, Ones U, Tamay Z, Salmayenli N, Darendeliler F. Leptin: Does it have any role in childhood asthma? *J Allergy Clin Immunol* 114: 254–59, 2004.
82. Sood A, Ford ES, Camargo CA Jr. Association between leptin and asthma in adults. *Thorax* 61: 300–5, 2006.
83. Fantuzzi G. Adipose tissue, adipokines, and inflammation. *J Allergy Clin Immunol* 115: 911–19, 2005.
84. Williams AS, Leung SY, Nath P, Khorasani NM, Bhavsar P, Issa R, Mitchell JA, Adcock IM, Chung KF. Role of TLR2, TLR4, and MyD88 in murine ozone-induced airway hyperresponsiveness and neutrophilia. *J Appl Physiol* 103: 1189–95, 2007.
85. Kadowaki T, Yamauchi T, Kubota N. The physiological and pathophysiological role of adiponectin and adiponectin receptors in the peripheral tissues and CNS. *FEBS Lett* 582: 74–80, 2008.
86. Fantuzzi G. Adiponectin and inflammation: Consensus and controversy. *J Allergy Clin Immunol* 121: 326–30, 2008.
87. Okamoto Y, Folco EJ, Minami M, Wara AK, Feinberg MW, Sukhova GK, Colvin RA, Kihara S, Funahashi T, Luster AD, Libby P. Adiponectin Inhibits the Production of CXC Receptor Chemokine Ligands in Macrophages and Reduces T-Lymphocyte Recruitment in Atherogenesis. *Circ Res* 121, 2007.
88. Yamauchi T, Kamon J, Ito Y, Tsuchida A, Yokomizo T, Kita S, Sugiyama T, Miyagishi M, Hara K, Tsunoda M, Murakami K, Ohteki T, Uchida S, Takekawa S, Waki H, Tsuno NH, Shibata Y, Terauchi Y, Froguel P, Tobe K, Koyasu S, Taira K, Kitamura T, Shimizu T, Nagai R, Kadowaki T. Cloning of adiponectin receptors that mediate antidiabetic metabolic effects. *Nature* 423: 762–69, 2003.
89. Hug C, Wang J, Ahmad NS, Bogan JS, Tsao TS, Lodish HF. T-cadherin is a receptor for hexameric and high-molecular-weight forms of Acrp30/adiponectin. *Proc Natl Acad Sci USA* 101: 10308–13, 2004.
90. Snider GL, Lucey EC, Stone PJ. Animal models of emphysema. *Am Rev Respir Dis* 133: 149–69, 1986.
91. Kuhn C, Yu SY, Chraplyvy M, Linder HE, Senior RM. The induction of emphysema with elastase. II: Changes in connective tissue. *Lab Invest* 34: 372–80, 1976.
92. Janoff A, Sloan B, Weinbaum G, Damiano V, Sandhaus RA, Elias J, Kimbel P. Experimental emphysema induced with purified human neutrophil elastase: Tissue localization of the instilled protease. *Am Rev Respir Dis* 115: 461–78, 1977.
93. Kao RC, Wehner NG, Skubitz KM, Gray BH, Hoidal JR. Proteinase 3. A distinct human polymorphonuclear leukocyte proteinase that produces emphysema in hamsters. *J Clin Invest* 82: 1963–73, 1988.
94. Wright JL, Churg A. Cigarette smoke causes physiologic and morphologic changes of emphysema in the guinea pig. *Am Rev Respir Dis* 142: 1422–28, 1990.
95. Hautamaki RD, Kobayashi DK, Senior RM, Shapiro SD. Requirement for macrophage elastase for cigarette smoke-induced emphysema in mice. *Science* 277: 2002–4, 1997.
96. Kasahara Y, Tuder RM, Cool CD, Lynch DA, Flores SC, Voelkel NF. Endothelial cell death and decreased expression of vascular endothelial growth factor and vascular endothelial growth factor receptor 2 in emphysema. *Am J Respir Crit Care Med* 163: 737–44, 2001.
97. Aoshiba K, Yokohori N, Nagai A. Alveolar wall apoptosis causes lung destruction and emphysematous changes. *Am J Respir Cell Mol Biol* 28: 555–62, 2003.
98. Lucey EC, Keane J, Kuang PP, Snider GL, Goldstein RH. Severity of elastase-induced emphysema is decreased in tumor necrosis factor-alpha and interleukin-1beta receptor-deficient mice. *Lab Invest* 82: 79–85, 2002.
99. Mahadeva R, Shapiro SD. Animal models of pulmonary emphysema. *Curr Drug Targets Inflamm Allergy* 4: 665–73, 2005.
100. Massaro GD, Massaro D. Retinoic acid treatment abrogates elastase-induced pulmonary emphysema in rats. *Nat Med* 3: 675–77, 1997.
101. Houghton AM, Quintero PA, Perkins DL, Kobayashi DK, Kelley DG, Marconcini LA, Mecham RP, Senior RM, Shapiro SD. Elastin fragments drive disease progression in a murine model of emphysema. *J Clin Invest* 116: 753–59, 2006.
102. Snider GL, Lucey EC, Faris B, Jung-Legg Y, Stone PJ, Franzblau C. Cadmium-chloride-induced air-space enlargement with interstitial pulmonary fibrosis is not associated with destruction of lung elastin. Implications for the pathogenesis of human emphysema. *Am Rev Respir Dis* 137: 918–23, 1988.

103. Shapiro SD. COPD unwound. *N Engl J Med* 352: 2016–19, 2005.
104. Wright JL, Postma DS, Kerstjens HA, Timens W, Whittaker P, Churg A. Airway remodeling in the smoke exposed guinea pig model. *Inhal Toxicol* 19: 915–23, 2007.
105. Medler TR, Petrusca DN, Lee PJ, Hubbard WC, Berdyshev EV, Skirball J, Kamocki K, Schuchman E, Tuder RM, Petrache I. Apoptotic sphingolipid signaling by ceramides in lung endothelial cells. *Am J Respir Cell Mol Biol* 38(6): 639–46, 2008.
106. Tang K, Rossiter HB, Wagner PD, Breen EC. Lung-targeted VEGF inactivation leads to an emphysema phenotype in mice. *J Appl Physiol* 97: 1559–66, 2004. Discussion 1549.
107. Maeno T, Houghton AM, Quintero PA, Grumelli S, Owen CA, Shapiro SD. CD8+ T Cells are required for inflammation and destruction in cigarette smoke-induced emphysema in mice. *J Immunol* 178: 8090–96, 2007.
108. van der Strate BW, Postma DS, Brandsma CA, Melgert BN, Luinge MA, Geerlings M, Hylkema MN, van den Berg A, Timens W, Kerstjens HA. Cigarette smoke-induced emphysema: A role for the B cell? *Am J Respir Crit Care Med* 173: 751–58, 2006.
109. Churg A, Wang RD, Tai H, Wang X, Xie C, Dai J, Shapiro SD, Wright JL. Macrophage metalloelastase mediates acute cigarette smoke-induced inflammation via tumor necrosis factor-alpha release. *Am J Respir Crit Care Med* 167: 1083–89, 2003.
110. Weathington NM, van Houwelingen AH, Noerager BD, Jackson PL, Kraneveld AD, Galin FS, Folkerts G, Nijkamp FP, Blalock JE. A novel peptide CXCR ligand derived from extracellular matrix degradation during airway inflammation. *Nat Med* 12: 317–23, 2006.
111. Saetta M. Airway inflammation in chronic obstructive pulmonary disease. *Am J Respir Crit Care Med* 160: S17–20, 1999.
112. Wang Z, Zheng T, Zhu Z, Homer RJ, Riese RJ, Chapman HA Jr., Shapiro SD, Elias JA. Interferon gamma induction of pulmonary emphysema in the adult murine lung. *J Exp Med* 192: 1587–600, 2000.
113. Grumelli S, Corry DB, Song LZ, Song L, Green L, Huh J, Hacken J, Espada R, Bag R, Lewis DE, Kheradmand F. An immune basis for lung parenchymal destruction in chronic obstructive pulmonary disease and emphysema. *PLoS Med* 1: e8, 2004.
114. Robbins C, Franco F, Cernandes M Shapiro SD (in press) Cigarette smoke exposure impairs dendritic cell maturation and T cell proliferation in thoracic lymph nodes of mice. *J Immunol* 180: 6623–8, 2008.
115. Taraseviciene-Stewart L, Scerbavicius R, Choe KH, Moore M, Sullivan A, Nicolls MR, Fontenot AP, Tuder RM, Voelkel NF. An animal model of autoimmune emphysema. *Am J Respir Crit Care Med* 171: 734–42, 2005.
116. Lee SH, Goswami S, Grudo A, Song LZ, Bandi V, Goodnight-White S, Green L, Hacken-Bitar J, Huh J, Bakaeen F, Coxson HO, Cogswell S, Storness-Bliss C, Corry DB, Kheradmand F. Antielastin autoimmunity in tobacco smoking-induced emphysema. *Nat Med* 13: 567–69, 2007.
117. Ito K, Ito M, Elliott WM, Cosio B, Caramori G, Kon OM, Barczyk A, Hayashi S, Adcock IM, Hogg JC, Barnes PJ. Decreased histone deacetylase activity in chronic obstructive pulmonary disease. *N Engl J Med* 352: 1967–76, 2005.
118. Marwick JA, Kirkham PA, Stevenson CS, Danahay H, Giddings J, Butler K, Donaldson K, Macnee W, Rahman I. Cigarette smoke alters chromatin remodeling and induces proinflammatory genes in rat lungs. *Am J Respir Cell Mol Biol* 31: 633–42, 2004.
119. Parks WC, Shapiro SD. Matrix metalloproteinases in lung biology. *Respir Res* 2: 10–19, 2001.
120. Rangasamy T, Cho CY, Thimmulappa RK, Zhen L, Srisuma SS, Kensler TW, Yamamoto M, Petrache I, Tuder RM, Biswal S. Genetic ablation of Nrf2 enhances susceptibility to cigarette smoke-induced emphysema in mice. *J Clin Invest* 114: 1248–59, 2004.
121. Zheng T, Zhu Z, Wang Z, Homer RJ, Ma B, Riese RJ Jr., Chapman HA Jr., Shapiro SD, Elias JA. Inducible targeting of IL-13 to the adult lung causes matrix metalloproteinase- and cathepsin-dependent emphysema. *J Clin Invest* 106: 1081–93, 2000.
122. Lee CG, Homer RJ, Zhu Z, Lanone S, Wang X, Koteliansky V, Shipley JM, Gotwals P, Noble P, Chen Q, Senior RM, Elias JA. Interleukin-13 induces tissue fibrosis by selectively stimulating and activating transforming growth factor beta(1). *J Exp Med* 194: 809–21, 2001.
123. Morris DG, Huang X, Kaminski N, Wang Y, Shapiro SD, Dolganov G, Glick A, Sheppard D. Loss of integrin alpha(v)beta6-mediated TGF-beta activation causes Mmp12-dependent emphysema. *Nature* 422: 169–73, 2003.
124. Wert SE, Yoshida M, LeVine AM, Ikegami M, Jones T, Ross GF, Fisher JH, Korfhagen TR, Whitsett JA. Increased metalloproteinase activity, oxidant production, and emphysema in surfactant protein D gene-inactivated mice. *Proc Natl Acad Sci USA* 97: 5972–77, 2000.
125. Leco KJ, Waterhouse P, Sanchez OH, Gowing KL, Poole AR, Wakeham A, Mak TW, Khokha R. Spontaneous air space enlargement in the lungs of mice lacking tissue inhibitor of metalloproteinases-3 (TIMP-3). *J Clin Invest* 108: 817–29, 2001.
126. Shapiro SD, Goldstein NM, Houghton AM, Kobayashi DK, Kelley D, Belaaouaj A. Neutrophil elastase contributes to cigarette smoke-induced emphysema in mice. *Am J Pathol* 163: 2329–35, 2003.
127. Henson PM, Cosgrove GP, Vandivier RW. State of the art. Apoptosis and cell homeostasis in chronic obstructive pulmonary disease. *Proc Am Thorac Soc* 3: 512–16, 2006.
128. Plantier L, Marchand-Adam S, Antico VG, Boyer L, De Coster C, Marchal J, Bachoual R, Mailleux A, Boczkowski J, Crestani B. Keratinocyte growth factor protects against elastase-induced pulmonary emphysema in mice. *Am J Physiol Lung Cell Mol Physiol* 293: L1230–1239, 2007.

Inflammatory Cells and Extracellular Matrix

3

PART

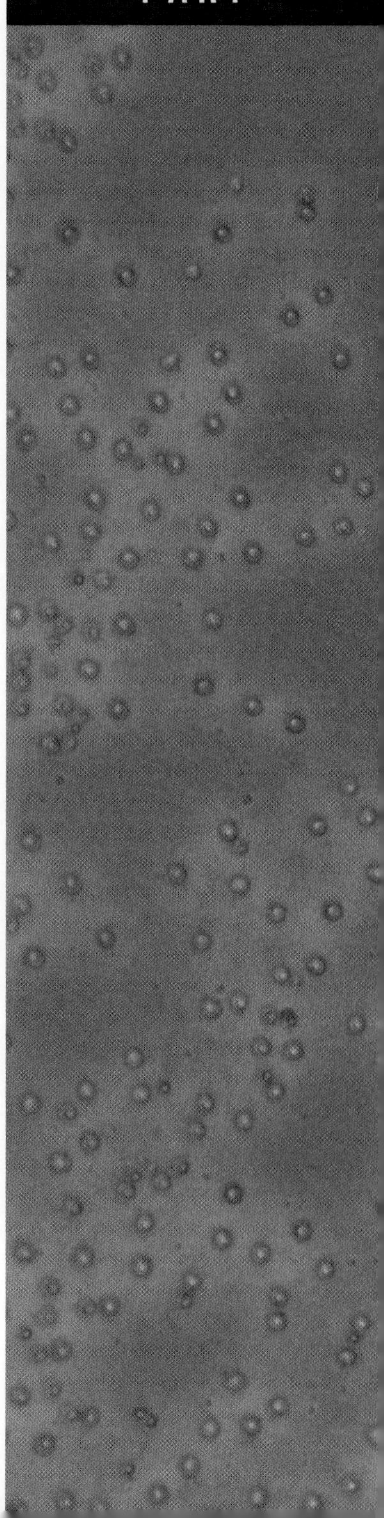

9

CHAPTER

Mast Cells and Basophils

George H. Caughey

Cardiovascular Research Institute and Department of Medicine, University of California at San Francisco, USA

Mast cells and basophils first came to attention over a century ago due to their ample stocks of intracellular granules with unusual staining characteristics. For many years their origins, normal functions, and roles in disease were obscure, and in some respects remain so. Nonetheless, knowledge of their biology increased tremendously in the past two decades [1–5].

Researchers once focused on mast cell and basophil release of histamine and eicosanoids in acute allergic events, which were considered to be a corruption of hypothesized normal function of defending against invasion by parasites such as worms and ticks. However, studies in mice now suggest that mast cells contribute to innate immune defense against airway bacteria like mycoplasma [6] and protection from immunologically nonspecific injury, as from venoms [7]. Mast cells and basophils also have the demonstrated potential to influence immune system development, regulation, and initiation of the immune response [8–11]. In some contexts, the overall contribution of mast cells is anti-inflammatory [6]. These roles deviate from traditional concepts of mast cell and basophil participation in adaptive responses involving antigen recognition by IgE.

In asthma investigations, attention is shifting from roles of these cells in acute responses to aeroallergen to roles in promoting persistent inflammation and remodeling in chronic disease. Their roles in other obstructive lung diseases have received less attention, but they may indeed contribute to conditions apart from asthma.

This chapter summarizes current thinking about roles of mast cells and basophils in obstructive airway disease.

ORIGIN AND FATE

Mast cells and basophils have shared origins but distinct distributions and fates. Mast cells, but not basophils, are normal residents of uninflamed airways. Mature mast cells rarely appear in blood, whereas mature basophils circulate and are recruited from blood to sites of allergic inflammation. Both cell types originate from shared progenitors in bone marrow; see Lee and Krilis [12] for a review. Immature mast cells released from marrow circulate briefly, exit the bloodstream to multiple tissue destinations, then differentiate, adopting a phenotype determined by their microenvironment. Basophils, on the other hand, mature in the marrow, circulate, then home to sites of inflammation if recruited to do so. Tissue mast cells are not fixed in location, for they migrate toward airway epithelium after antigen challenge and traffic to lymph nodes from sites of antigen exposure. Mast cells probably also proliferate in tissues. Conditions such as gut parasitosis lead to large local increases in mast cells.

Both types of cell survive degranulation and can restock their secretory granules with mediators. Life span *in vivo* has not been established, but *in vitro* studies predict that basophils are short-lived compared to mast cells, which survive for weeks in culture. In humans, the aggregate mast cell mass is much larger than that of basophils, which usually comprise $<1\%$ of circulating leukocytes. Among mammals, basophil numbers vary widely, ranging from numerous in guinea pigs, few in humans and mice, to nearly nonexistent in dogs, in which they are arguably unimportant. There are genetic variants of mice with almost no mast cells but none with an inherited, selective deficit of basophils; thus, it is easier to assess involvement of mast cells than of basophils in mouse disease models; see Galli [5] for a review.

DEVELOPMENT AND HETEROGENEITY

Paucigranular, mast cell-committed progenitors are released from marrow expressing surface receptor tyrosine kinase c-kit and low affinity IgG receptor ($F_c\gamma$RII) but not high affinity IgE receptor, $F_c\varepsilon$RI. *In vitro*, cells with mature characteristics, including $F_c\varepsilon$RI and protease-rich granules, differentiate from progenitors under the influence of IL-6 and kit ligand. Presumably, similar events occur *in vivo*, with kit ligand produced by endothelial, stromal, and epithelial cells being critical for mast cell survival. Mice with defective c-kit or its ligand lack mast cells. Their importance to mast cell development is underscored by gain-of-function c-kit mutations in systemic mastocytosis and mast cell malignancy, and development of generalized mastocytosis in response to exogenous kit ligand. The importance of local production of kit ligand is suggested by the finding in mice that intratracheal kit ligand provokes mast cell-dependent hyperreactivity [13].

In vitro, a variety of cytokines (especially IL-3, -6, -9 and -10, and TGF-β1) determines phenotype of maturing mouse mast cells. In the case of IL-9 (a candidate mouse "asthma gene"), airway overexpression in transgenic mice causes eosinophilic inflammation and hyperresponsiveness with mast cell hyperplasia [14].

Mast cells vary in features such as:

- proteoglycan and protease content;
- metachromasia;
- granule ultrastructure;
- responses to degranulating stimuli, such as substance P.

These phenotypic variations appear reversible in a given cell and changeable in cell populations in response to infection and injury. Classically, rodent mast cells are divided into "mucosal" and "connective tissue" groups, although the phenotype distribution is not strictly bimodal. Human mast cells are sorted into subsets based on content of granule proteases [15]. MC_T cells express tryptases but not chymase, whereas MC_{TC} express tryptases and chymase. Occasionally, chymase-only MC_C mast cells are seen. Bronchi contain a mixture of types, whereas alveolar interstitium contains predominantly MC_T. Because mast cells develop and are stimulated in tissues, some variation is due to differences in maturation, activation, or recovery from degranulation.

Less is known of factors influencing basophil differentiation. c-Kit appears less important than for mast cells because mature basophils normally express little or no c-kit and because levels are largely unaffected by defects in c-kit causing profound mast cell deficits in mice. Nonetheless, c-kit+ human basophil-like cells circulate in asthma and atopy and manifest phenotypic changes in which c-kit could play a role [16].

RECRUITMENT

In asthmatics, increased numbers of basophils and activated mast cells appear in sputum after allergen challenge [17], which may reflect migration from submucosal sites and bloodstream, respectively. As noted, mast cell precursors home to tissues even without inflammation. Constitutive homing and epithelial migration may involve responses to proteins such as:

- C5a;
- RANTES;
- IL-8;
- MIP-1α;
- MCP-1;
- VEGF;
- fractalkine;
- CXCL10;
- kit ligand.

These may orchestrate movement singly or in combination. Cultured human mast cells express a broad repertoire of chemokine receptors, which diminishes as cells mature, thus limiting mast cell movement after differentiation at a tissue site.

Kit ligand is notable in that it is specific for mast cells in comparison with other leukocytes and is produced by airway cells. Mast cell migration into tissues depends on integrins and other adhesion molecules [18]. Basophils express an array of chemokine receptors. Chemokines binding to CCR3 (such as RANTES and eotaxins) may be especially important [19, 20]. The multiplicity of chemoattractants predicts redundant recruiting pathways. Several chemoattractants also prime or activate one or both types of cells [20, 21], enhancing their role in asthma pathogenesis beyond that of recruitment alone.

ACTIVATION

IgE-dependent activation

Classic mast cell and basophil activation involves docking of allergens to IgE via $F_c\varepsilon$RI expressed as an assemblage of subunits ($\alpha\beta\gamma_2$) in the plasma membrane. $F_c\varepsilon$RIα expression is strongly influenced by the serum level of IgE itself [22]. Aeroallergens with repeating epitopes attach to receptor α chain-bound IgE, bridging receptors. Allergen-driven cross-linking initiates intracellular signaling characterized initially by phosphorylation of intracellular immunoreceptor tyrosine-based activation motif (ITAM) domains of receptor β and γ chains. In turn, these recruit and activate nonreceptor tyrosine kinases, especially lyn, syk, and btk, which access pathways leading to exocytosis and synthesis of eicosanoids and cytokines. Some of these activation pathway proteins have emerged as targets for new types of anti-allergic drugs

[23, 24]. For example, inhibition of syk blocks allergic airway inflammation in mice [25]. Intriguingly, syk protein is deficient in some humans with "nonreleaser" basophils [26], although the exact relationship between the nonreleaser phenotype and the asthma (or lack thereof) is not yet clear [27]. Furthermore, dexamethasone depresses syk activity, which may contribute corticosteroid efficacy in asthma [28]. The $F_c\varepsilon RI\beta$ chain is not essential in humans but amplifies the signal. $F_c\gamma RI$ signals are damped by inhibitory receptors, such as $F_c\gamma RII$ and gp49, which possess intracellular, immunoreceptor tyrosine-based inhibition motif (ITIM) domains [29].

Phosphorylated ITIMs attract tyrosine phosphatases, such as SHIP, which may inhibit $F_c\varepsilon RI$ signaling by dephosphorylating activated proteins in the signaling pathway.

$F_c\gamma RII$'s importance is particularly compelling [30]. When IgG and IgE antibodies are raised against polyvalent antigen, "heterotypic" cross-linking of $F_c\gamma RII$ and $F_c\varepsilon RI$ by allergen bound to IgG and IgE inhibits signaling by $F_c\varepsilon RI$. This may be a means by which "blocking antibodies" reduce atopic symptoms after allergen desensitization.

IgE-independent activation

Mast cells are activated by multiple nonimmunological inputs (Fig. 9.1). Physiological activators include:

- neuropeptides (e.g. substance P);
- purines (adenosine and ATP);
- byproducts of complement activation (e.g. C3a);
- eosinophil toxins;
- bacterial products (e.g. *E. coli* FimH);
- chemokines and lymphokines (e.g. IL-4);
- kit ligand.

These inputs are immunologically nonspecific and provide the means by which products of activation participate in neurogenic inflammation and innate host defense [31]. They also augment responses to allergen-specific mast cell activation (see below).

Thus, innate and adaptive responses are not mutually exclusive. C3a, an agent of innate immunity, is an example, for C3a receptor-null mice are protected from sequelae of airway allergen challenge [32].

It should be noted that human mast cell subpopulations do not respond uniformly to all stimuli. For example, lung mast cells tend to be less responsive than skin mast cells to substance P [33].

PRIMING AND INHIBITION

Priming describes the response to a substance that does not release mediators by itself but enhances the effect of another stimulus, such as cross-linked $F_c\varepsilon RI$. In cultured human mast cells, priming occurs with allergic cytokines, such as IL-4 and IL-5 [34], and also with kit ligand and adenosine [35]. Mechanisms of priming may be stimulus-specific and affect mediator synthesis and release in different ways. Interactions between primers are potentially complex, and, in the case of IL-4 and -5, may involve autocrine stimulation.

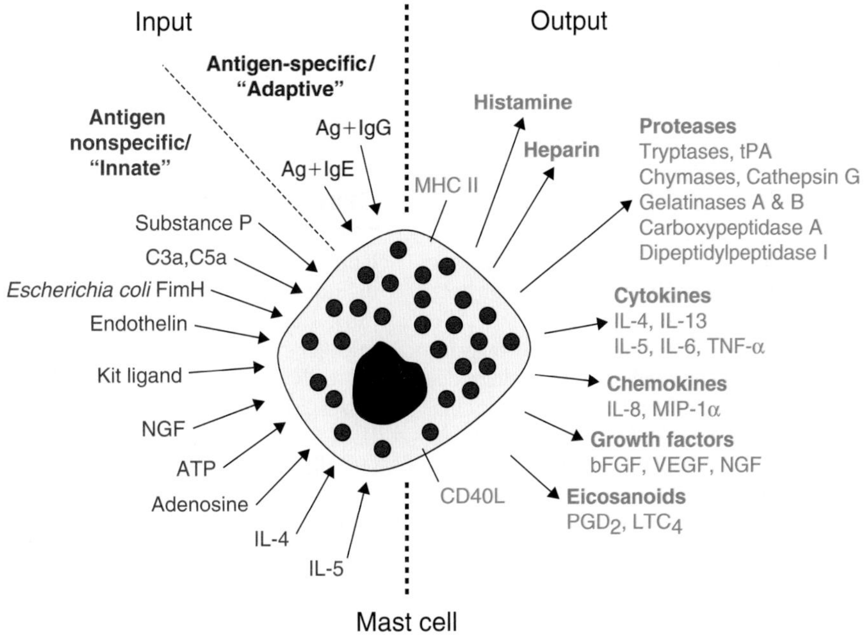

FIG. 9.1 Mast cell inputs and outputs. This contains a partial listing of factors influencing human mast cell production of mediators with postulated roles in acute and chronic airway inflammation in asthma. Inputs are divided into those directly involving immunoglobulin-mediated "adaptive" responses and those involving immunologically nonspecific "innate/natural" immunity.

In basophils, priming *in vitro* is especially impressive with IL-3, which augments release of histamine in response to MCP-4 and IL-4 in response to allergen, and is enhanced by eotaxin [21]. Basophils harvested from allergic asthmatics are primed [36], presumably by exposure to cytokines *in vivo*.

Priming may serve to activate cells when (and only when) necessary to protect the host; in asthmatics, dysregulated priming may contribute to the pathology of allergic inflammation. $F_c\varepsilon$RI-mediated activation can also be opposed by physiological influences, including adrenergic agonists and inhibitory receptors with ITIMs, as discussed above. Inhibition of activation is the basis of emerging strategies to combat allergic disease, including use of adenosine receptor antagonists, cytokine and chemokine antagonists, and activators of ITIM containing co-receptors [37].

MEDIATORS

Eicosanoids, histamine, and proteases

Stimulated mast cells and basophils release an astonishing variety of stored and newly synthesized "mediators". These include prostanoids (Chapter 23), leukotrienes (Chapter 24), proteases, chemokines (Chapter 26), lymphokines, growth factors (Chapter 29), and nitric oxide (Chapter 30), the properties and pharmacology of which are considered in other cited chapters and are not reviewed extensively here.

The principal eicosanoids synthesized after stimulation are PGD_2 (whose importance in allergic airway inflammation is demonstrated in PGD receptor-null mice [38]) and LTC_4, which is the major target of 5-lipoxygenase inhibitors and leukotriene receptor antagonists. The major granule-associated constituents of human mast cells are histamine, serine proteases, and proteoglycans (heparin and chondroitin). Histamine's importance is clearer in upper airway allergic disease, because antihistamines tend to be more effective in treating rhinitis than asthma. Proteases are the most abundant proteins in secretory granules [39], though this may not be true of normal basophils. As noted, human mast cells vary in expression of tryptases and chymase. In atopic and asthmatic individuals, circulating cells resembling hybrids of basophils and mast cells may express one or both types of protease [16].

Human tryptases are a polymorphic family of trypsin-like proteases implicated in asthma based on *in vitro* degradation of bronchodilating peptides, enhancement of bronchial contraction, promotion of airway smooth muscle and subendothelial fibroblast growth and collagen production, and pro-inflammatory properties [40]. Studies in sheep and guinea pigs suggest that inhibitors of tryptases reduce allergen-induced airway inflammation and bronchoconstriction [41].

Chymase is postulated to play roles in asthma by stimulating gland secretion and promoting airway remodeling via production of angiotensin II and activation of matrix metalloproteinases [39, 42, 43]. Apart from helping to package and stabilize proteases [44], the role of heparin and other proteoglycans is less clear. However, because heparin delivered into airways in pharmacological doses is anti-asthmatic, heparin released from stimulated mast cells may attenuate inflammatory effects of mediators released from mast cells and other effector cells [45].

There is little evidence that human basophils express mediators not also present in mast cells. However, proteins of unknown function (such as basogranulin) are recognized by monoclonal antibodies and may be basophil specific [16, 46].

Cytokines and chemokines

Activated human mast cells express cytokines similar to those of TH_2 cells, including IL-4, -5, -13, -16, and TNF-α. In airways in allergic rhinitis or asthma, mast cells are a substantial fraction of leukocytes expressing these "TH_2" cytokines [47].

The relative importance of mast cells and basophils versus lymphocytes as a source of these cytokines is unclear. However, mast cells differ from lymphocytes in storing cytokines in secretory granules. In the case of TNF-α, release from mast cell stores is a critical determinant of survival from peritonitis [48]. Asthmatic basophils produce prodigious amounts of IL-4 and IL-13 after antigen-specific activation [49]. In mice, basophils may be one of the IL-4-producing cell types that skew the immune environment in the lung toward allergic, Th2-type responses [50, 51]. The importance of IL-4 and IL-13 is strongly supported by studies in genetically modified mice, which suggest that airway overproduction causes epithelial hypertrophy, mucus metaplasia, eosinophilic inflammation, and hyperresponsiveness. Cytokine-dependent proliferation of mast cells is complex and is subject to autocrine regulation, as from self-stimulation by other types of mediators, such as leukotrienes [52]. Sustained overproduction of allergic cytokines by mast cell and basophils may heighten and perpetuate asthmatic inflammation. Chemokines expressed and secreted by mast cells include IL-8 and MIP-1α, which may recruit effector leukocytes to sites of inflammation [53].

Growth factors

Known cellular growth factors expressed by mast cells include bFGF [54], VEGF, [55] and nerve growth factor [56]. Less known growth factors are tryptases, which are mitogens for airway fibroblasts, smooth muscle, and epithelial cells [57] and may influence endothelial cells to form vessels, thereby promoting angiogenesis [58]. Mast cell IL-4 is fibrogenic when presented in the context of cell-to-cell contact [59]. These mediators suggest mechanisms by which mast cell and basophil mediators promote airway remodeling in the setting of persistent airway inflammation. Recent data from asthmatics suggest that infiltration of airway smooth muscle by mast cells (mastitis) is a phenomenon specific for asthma, which is to say that it is not seen in related inflammatory airway disorders such as eosinophilic bronchitis [(which unlike asthma is not associated with airway hyperresponsiveness (AHR)] and smoking-associated bronchitis [60]. Brightling and colleagues propose that airway smooth muscle mastocytosis in asthma is caused by a unique asthmatic smooth muscle phenotype, which includes production of mast cell chemoattractants like CXCL10 [61]. In turn, tryptase and other growth factors produced by mast cells recruited to airway smooth muscle may promote myocyte hyperplasia or hypertrophy [62, 63]. On the other hand, Wenzel and colleagues report a positive correlation between the chymase-positive subset of distal airway mast cells and lung function [64].

ROLES IN HOMEOSTASIS

Basophils and mast cells are presumed to exist for reasons other than to promote sneezing, itching, and wheezing. Lacking informative animal models or natural deficiency states, investigators have few direct clues regarding normal basophil function. However, compelling evidence of roles for mast cells has emerged from studies by using mast cell-deficient mice, which suggest critical roles in innate as well as adaptive immune responses, including defense against bacterial peritonitis and pneumonia [48, 65]. In similar mouse models (reviewed in [5]), mast cells promote a variety of immunologically nonspecific forms of inflammation, for example, ozone-inflicted lung injury [66], providing further evidence that they are activated by IgE-independent pathways.

ROLES IN ASTHMA

Animal models

Several experiments in mice support a role for mast cells in asthma-like allergic inflammation, including eosinophilia, hyperresponsiveness, and epithelial remodeling [14, 67–70]. Other studies suggest that IgE and mast cells do not make major contributions (reviewed in [71, 72]). Together, these studies reveal that IgE- and mast cell-dependence vary with strain of mouse, choice of antigen, mode of sensitization and challenge, and choice of physiological endpoints. Mast cell dependence is easier to detect in mice sensitized and challenged locally than systemically, using lower (more physiological) amounts of antigen, without adjuvants. Overall, murine studies suggest that mast cells and IgE are not essential for development of eosinophilic inflammation but do influence kinetics and magnitude of its expression.

Humans

Notwithstanding some studies questioning its importance in mice, IgE's long-suspected contributions to human asthma are supported by trials by using anti-IgE antibodies, which reduce circulating IgE to nearly undetectable levels and decrease symptoms and corticosteroid use in moderate, steroid-dependent asthmatics, [73] and rhinitics [74]. Anti-IgE joins a growing list of anti-asthmatic drugs influencing mast cells, basophils, and their products. These include:

- corticosteroids (which decrease mast cell numbers) [75];
- cromones;
- β-adrenergic agonists;
- theophylline;
- heparin (which inhibits degranulation);
- IL-4 antagonists (which inhibit priming) (58);
- leukotriene pathway inhibitors.

Others, effective in animal models, include tryptase inhibitors [76].

Interestingly, mast cell desensitization to chronic β-agonists used by asthmatics without corticosteroids may contribute to clinical deterioration [77]. Furthermore, corticosteroids may protect mast cells from desensitization [78].

Pharmacological evidence of roles for mast cells correlates with studies indicating activation in allergic airway disease [1, 79]. Mast cells release histamine and tryptase into bronchi following allergen challenge [80]. They appear degranulated in asthmatic airway, even in stable disease, [81, 82] and the percentage of mast cells expressing cytokines IL-4, IL-5, and TNF-α increases [47]. Mast cell and basophil numbers rise in asthmatic airways and correlate with hyperresponsiveness to acetylcholine [83]. However, fewer basophils are seen in asthmatic bronchial biopsies than mast cells or eosinophils, and they are more prominent in cutaneous than airway late-phase reactions [84]. Nonetheless, their arrival coincides with development of late-phase bronchoconstriction [85]. Airway basophils are thought to be a source of the late histamine release after allergen exposure, because this occurs without a corresponding peak in tryptase, which is more abundant in mast cells. On the other hand, other granulated leukocytes, including polymorphonuclear neutrophils, also can be sources of airway histamine [86].

More basophils appear in asthmatic sputum with late-phase responses to allergen than in those without such responses, and their numbers correlate with methacholine responsiveness [17]. This supports the hypothesis that mast cells and basophils are important in early- and late-phase responses, respectively. The further hypothesis that these cells promote chronic, persistent asthma is speculative. However, this notion is supported by the studies summarized above suggesting that proteases, cytokines, and growth factors from chronically activated mast cells and basophils promote airway remodeling and sustain TH_2-assisted IgE production and allergic inflammation.

ROLES IN CHRONIC OBSTRUCTIVE PULMONARY DISEASE

Several lines of evidence suggest connections between mast cells and chronic obstructive pulmonary disease (COPD) [87].

Elevated levels of histamine and tryptase in smokers' lavage fluid [88] imply that mast cells are activated by smoke. Studies in mice suggest that mast cells promote airway injury and epithelial remodeling in response to ozone [66], which could explain some human responses to smoke. Most mast cells close to human bronchial glands express chymase [89], which stimulates gland cell secretion [42] and may promote sputum production in bronchitis.

Conceivably, mast cells contribute to emphysematous lung destruction by secreting matrix altering proteases, for example chymase, tryptases, gelatinases, plasminogen activator, and dipeptidyl peptidase I [39]. Some investigators find increased mast cell numbers in proximal airways of asymptomatic smokers with mild COPD [90], although the relationship to lung function is less clear. Overall, evidence of mast cell involvement in asthma is more compelling, although further investigation of connections between mast cells and COPD is warranted.

SUMMARY

Mast cells protect from certain types of infection and injury by contributing to innate and adaptive immune responses. The homeostatic roles of basophils are less clear. Both cell types participate in the pathology of atopy and asthma by deploying an arsenal of inflammatory mediators, including proteases, growth factors, chemokines, and "TH_2" cytokines. Mast cells, being permanent airway residents, are more likely to encounter aeroallergens first and to participate in acute responses. Basophils, because of the time lag of recruitment, may be more important in late-phase responses.

Support is mounting for the hypothesis that both types of cells magnify the pathology of persistent inflammation in chronic asthma, including stimulation of IgE production, recruitment of eosinophils, and remodeling of epithelium.

References

1. Kaliner M. Asthma and mast cell activation. *J Allergy Clin Immunol* 83: 510–20, 1989.
2. Schulman ES. The role of mast cells in inflammatory responses in the lung. *Crit Rev Immunol* 13: 35–70, 1993.
3. Metcalfe DD, Baram D, Mekori YA. Mast cells. *Physiol Rev* 77: 1033–79, 1997.
4. Dvorak AM. Cell biology of the basophil. *Int Rev Cytol* 180: 87–236, 1998.
5. Galli SJ. Mast cells and basophils. *Curr Opin Hematol* 7: 32–39, 2000.
6. Xu X, Zhang D, Lyubynska N *et al.* Mast cells protect mice from mycoplasma pneumonia. *Am J Respir Crit Care Med* 173: 219–25, 2006.
7. Metz M, Piliponsky AM, Chen CC *et al.* Mast cells can enhance resistance to snake and honeybee venoms. *Science* 313: 526–30, 2006.
8. Nakae S, Suto H, Kakurai M *et al.* Mast cells enhance T cell activation: Importance of mast cell-derived TNF. *Proc Natl Acad Sci USA* 102: 6467–472, 2005.
9. Lu LF, Lind EF, Gondek DC *et al.* Mast cells are essential intermediaries in regulatory T-cell tolerance. *Nature* 442: 997–1002, 2006.
10. Jawdat DM, Rowden G, Marshall JS. Mast cells have a pivotal role in TNF-independent lymph node hypertrophy and the mobilization of Langerhans cells in response to bacterial peptidoglycan. *J Immunol* 177: 1755–62, 2006.
11. Min B, Le Gros G, Paul WE. Basophils: A potential liaison between innate and adaptive immunity. *Allergol Int* 55: 99–104, 2006.
12. Li L, Krilis SA. Mast-cell growth and differentiation. *Allergy* 54: 306–12, 1999.
13. Campbell E, Hogaboam C, Lincoln P, Lukacs NW. Stem cell factor-induced airway hyperreactivity in allergic and normal mice. *Am J Pathol* 154: 1259–65, 1999.
14. Temann UA, Geba GP, Rankin JA, Flavell RA. Expression of interleukin 9 in the lungs of transgenic mice causes airway inflammation, mast cell hyperplasia, and bronchial hyperresponsiveness. *J Exp Med* 188: 1307–20, 1998.
15. Irani AA, Schechter NM, Craig SS *et al.* Two types of human mast cells that have distinct neutral protease compositions. *Proc Natl Acad Sci USA* 83: 4464–68, 1986.
16. Li L, Li Y, Reddel SW *et al.* Identification of basophilic cells that express mast cell granule proteases in the peripheral blood of asthma, allergy, and drug-reactive patients. *J Immunol* 161: 5079–86, 1998.
17. Gauvreau GM, Lee JM, Watson RM *et al.* Increased numbers of both airway basophils and mast cells in sputum after allergen inhalation challenge of atopic asthmatics. *Am J Respir Crit Care Med* 161: 1473–78, 2000.
18. Hamawy MM, Mergenhagen SE, Siraganian RP. Adhesion molecules as regulators of mast-cell and basophil function. *Immunol Today* 15: 62–66, 1994.
19. Uguccioni M, Mackay CR, Ochensberger B *et al.* High expression of the chemokine receptor CCR3 in human blood basophils. Role in activation by eotaxin, MCP-4, and other chemokines. *J Clin Invest* 100: 1137–43, 1997.
20. Ochensberger B, Tassera L, Bifrare D *et al.* Regulation of cytokine expression and leukotriene formation in human basophils by growth factors, chemokines and chemotactic agonists. *Eur J Immunol* 29: 11–22, 1999.
21. Devouassoux G, Metcalfe DD, Prussin C. Eotaxin potentiates antigen-dependent basophil IL-4 production. *J Immunol* 163: 2877–82, 1999.
22. Saini SS, Klion AD, Holland SM *et al.* The relationship between serum IgE and surface levels of $F_c\varepsilon R$ on human leukocytes in various diseases: Correlation of expression with FcepsilonRI on basophils but not on monocytes or eosinophils. *J Allergy Clin Immunol* 106: 514–20, 2000.
23. Oliver JM, Kepley CL, Ortega E, Wilson BS. Immunologically mediated signaling in basophils and mast cells: finding therapeutic targets for allergic diseases in the human $F_c\varepsilon R1$ signaling pathway. *Immunopharmacology* 48: 269–81, 2000.
24. Barnes PJ. Novel signal transduction modulators for the treatment of airway diseases. *Pharmacol Ther* 109: 238–45, 2006.
25. Matsubara S, Li G, Takeda K *et al.* Inhibition of spleen tyrosine kinase prevents mast cell activation and airway hyperresponsiveness. *Am J Respir Crit Care Med* 173: 56–63, 2006.
26. Kepley CL. Antigen-induced reduction in mast cell and basophil functional responses due to reduced Syk protein levels. *Int Arch Allergy Immunol* 138: 29–39, 2005.
27. Youssef LA, Schuyler M, Gilmartin L *et al.* Histamine release from the basophils of control and asthmatic subjects and a comparison of gene expression between "releaser" and "nonreleaser" basophils. *J Immunol* 178: 4584–94, 2007.
28. Hiragun T, Peng Z, Beaven MA. Cutting edge: dexamethasone negatively regulates Syk in mast cells by up-regulating SRC-like adaptor protein. *J Immunol* 177: 2047–50, 2006.
29. Ott VL, Cambier JC. Activating and inhibitory signaling in mast cells: New opportunities for therapeutic intervention?. *J Allergy Clin Immunol* 106: 429–40, 2000.
30. Takai T, Ono M, Hikida M *et al.* Augmented humoral and anaphylactic responses in $F_c\gamma RII$-deficient mice. *Nature* 379: 346–49, 1996.
31. Galli SJ, Wershil BK. The two faces of the mast cell. *Nature* 381: 21–22, 1996.
32. Humbles AA, Lu B, Nilsson CA *et al.* A role for the C3a anaphylatoxin receptor in the effector phase of asthma. *Nature* 406: 998–1001, 2000.

33. Church MK, Lowman MA, Robinson C *et al.* Interaction of neuropeptides with human mast cells. *Int Arch Allergy Immunol* 88: 70–78, 1989.
34. Ochi H, De Jesus NH, Hsieh FH *et al.* IL-4 and -5 prime human mast cells for different profiles of IgE-dependent cytokine production. *Proc Natl Acad Sci U. S A* 97: 10509–13, 2000.
35. Marquardt DL. Adenosine. In: Barnes PJ, Grunstein MM, Leff AR, Woolcock AJ (eds). *Asthma*, Philadelphia: Lippincott-Raven, 1997.
36. Lie WJ, Knol EF, Mul FP *et al.* Basophils from patients with allergic asthma show a primed phenotype. *J Allergy Clin Immunol* 104: 1000–7, 1999.
37. Bingham CO III, Austen KF. Mast-cell responses in the development of asthma. *J Allergy Clin Immunol* 105: S527–S534, 2000.
38. Matsuoka T, Hirata M, Tanaka H *et al.* Prostaglandin D2 as a mediator of allergic asthma. *Science* 287: 2013–17, 2000.
39. Caughey GH (ed.), *Mast Cell Proteases in Immunology and Biology*. New York: Marcel Dekker, Inc, 1995.
40. Caughey GH. Of mites and men: Trypsin-like proteases in the lungs. *Am J Respir Cell Mol Biol* 16: 621–28, 1997.
41. Clark JM, Abraham WM, Fishman CE *et al.* Tryptase inhibitors block allergen-induced airway and inflammatory responses in allergic sheep. *Am J Resp Crit Care Med* 152: 2076–83, 1995.
42. Sommerhoff CP, Caughey GH, Finkbeiner WE *et al.* Mast cell chymase. A potent secretagogue for airway gland serous cells. *J Immunol* 142: 2450–56, 1989.
43. Fang KC, Raymond WW, Blount JL, Caughey GH. Dog mast cell α-chymase activates progelatinase B by cleaving the Phe88-Phe89 and Phe91-Glu92 bonds of the catalytic domain. *J Biol Chem* 272: 25628–35, 1997.
44. Humphries DE, Wong GW, Friend DS *et al.* Heparin is essential for the storage of specific granule proteases in mast cells. *Nature* 400: 769–72, 1999.
45. Ahmed T, Syriste T, Mendelssohn R *et al.* Heparin prevents antigen-induced airway hyperresponsiveness: Interference with IP3-mediated mast cell degranulation? *J Appl Physiol* 76: 893–901, 1994.
46. McEuen AR, Buckley MG, Compton SJ, Walls AF. Development and characterization of a monoclonal antibody specific for human basophils and the identification of a unique secretory product of basophil activation. *Lab Invest* 79: 27–38, 1999.
47. Bradding P, Roberts JA, Britten KM *et al.* Interleukin-4, -5, and -6 and tumor necrosis factor-α in normal and asthmatic airways: Evidence for the human mast cell as a source of these cytokines. *Am J Repir Cell Mol Biol* 10: 471–80, 1994.
48. Echtenacher B, Männel DN, Hültner L. Critical protective role of mast cells in a model of acute septic peritonitis. *Nature* 381: 75–77, 1996.
49. Devouassoux G, Foster B, Scott LM *et al.* Frequency and characterization of antigen-specific IL-4- and IL-13- producing basophils and T cells in peripheral blood of healthy and asthmatic subjects. *J Allergy Clin Immunol* 104: 811–19, 1999.
50. Min B, Prout M, Hu-Li J *et al.* Basophils produce IL-4 and accumulate in tissues after infection with a Th2-inducing parasite. *J Exp Med* 200: 507–17, 2004.
51. Voehringer D, Shinkai K, Locksley RM. Type 2 immunity reflects orchestrated recruitment of cells committed to IL-4 production. *Immunity* 20: 267–77, 2004.
52. Jiang Y, Kanaoka Y, Feng C, Nocka K, Rao S, Boyce JA. Cutting edge: Interleukin 4-dependent mast cell proliferation requires autocrine/intracrine cysteinyl leukotriene-induced signaling. *J Immunol* 177: 2755–59, 2006.
53. Moller A, Lippert U, Lissmann D *et al.* Human mast cell produce IL-8. *J Immunol* 151: 3261–66, 1993.
54. Powers MR, Qu Z, LaGesse PC *et al.* Expression of basic fibroblast growth factor in nasal polyps. *Ann Otol Rhinol Laryngol* 107: 891–97, 1998.
55. Boesiger J, Tsai M, Maurer M *et al.* Mast cells can secrete vascular permeability factor/vascular endothelial cell growth factor and exhibit enhanced release after IgE-dependent upregulation of F_c epsilon receptor I expression. *J Exp Med* 188: 1135–45, 1998.
56. Nilsson G, Forsberg-Nilsson K, Xiang Z *et al.* Human mast cells express functional TrkA and are a source of nerve growth factor. *Eur J Immunol* 27: 2295–301, 1997.
57. Cairns JA, Walls AF. Mast cell tryptase is a mitogen for epithelial cells-Stimulation of IL-8 production and intercellular adhesion molecule-1 expression. *J Immunol* 156: 275–83, 1996.
58. Coussens LM, Raymond WW, Bergers G *et al.* Inflammatory mast cells upregulate angiogenesis during squamous epithelial carcinogenesis. *Genes Dev* 13: 1382–97, 1999.
59. Trautmann A, Krohne G, Brocker EB, Klein CE. Human mast cells augment fibroblast proliferation by heterotypic cell-cell adhesion and action of IL-4. *J Immunol* 160: 5053–57, 1998.
60. Brightling CE, Bradding P, Symon FA *et al.* Mast-cell infiltration of airway smooth muscle in asthma. *N Engl J Med* 346: 1699–705, 2000.
61. Brightling CE, Ammit AJ, Kaur D *et al.* The CXCL10/CXCR3 axis mediates human lung mast cell migration to asthmatic airway smooth muscle. *Am J Respir Crit Care Med* 171: 1103–08, 2005.
62. Brown JK, Jones CA, Rooney LA, Caughey GH. Mast cell tryptase activates extracellular-regulated kinases (p44/p42) in airway smooth-muscle cells: Importance of proteolytic events, time course, and role in mediating mitogenesis. *Am J Respir Cell Mol Biol* 24: 146–54, 2001.
63. Caughey GH. Mast cell tryptases and chymases in inflammation and host defense. *Immunol Rev* 217: 141–54, 2007.
64. Balzar S, Chu HW, Strand M, Wenzel S. Relationship of small airway chymase-positive mast cells and lung function in severe asthma. *Am J Respir Crit Care Med* 171: 431–39, 2005.
65. Malaviya R, Ikeda T, Ross E, Abraham SN. Mast cell modulation of neutrophil influx and bacterial clearance at sites of infection through TNF-α. *Nature* 381: 77–80, 1996.
66. Longphre M, Zhang LY, Harkema JR, Kleeberger SR. Mast cells contribute to O_3-induced epithelial damage and proliferation in nasal and bronchial airways of mice. *J Appl Physiol* 80: 1322–30, 1996.
67. Martin TR, Takeishi T, Katz HR *et al.* Mast cell activation enhances airway responsiveness to methacholine in the mouse. *J Clin Invest* 91: 1176–82, 1993.
68. Kung TT, Stelts D, Zurcher JA *et al.* Mast cells modulate allergic pulmonary eosinophilia in mice. *Am J Respir Cell Mol Biol* 12: 404–9, 1995.
69. Kobayashi T, Miura T, Haba T *et al.* An essential role of mast cells in the development of airway hyperresponsiveness in a murine asthma model. *J Immunol* 164: 3855–61, 2000.
70. Williams CM, Galli SJ. Mast cells can amplify airway reactivity and features of chronic inflammation in an asthma model in mice. *J Exp Med* 192: 455–62, 2000.
71. Hamelmann E, Tadeda K, Oshiba A, Gelfand EW. Role of IgE in the development of allergic airway inflammation and airway hyperresponsiveness—a murine model. *Allergy* 54: 297–305, 1999.
72. Galli SJ. Commentary: Complexity and redundancy in the pathogenesis of asthma: Reassessing the roles of mast cells and T cells. *J Exp Med* 186: 343–47, 1997.
73. Milgrom H, Fick RB Jr., Su JQ *et al.* Treatment of allergic asthma with monoclonal anti-IgE antibody. *N Engl J Med* 341: 1966–73, 1999.
74. Adelroth E, Rak S, Haahtela T *et al.* Recombinant humanized mAb-E25, an anti-IgE mAb, in birch pollen-induced seasonal allergic rhinitis. *J Allergy Clin Immunol* 106: 253–59, 2000.
75. Djukanovic R, Wilson JW, Britten KM *et al.* Quantitation of mast cells and eosinophils in the bronchial mucosa of symptomatic atopic asthmatics and healthy control subjects using immunohistochemistry. *Am Rev Respir Dis* 142: 863–71, 1990.
76. Clark JM, Moore WR, Fishman CE *et al.* A novel tryptase inhibitor, APC 366, inhibits allergen-induced airway and inflammatory responses in allergic sheep. *Am J Respir Crit Care Med* 152: 2076–83, 1995.
77. Swystun VA, Gordon JR, Davis EB *et al.* Mast cell tryptase release and asthmatic responses to allergen increase with regular use of salbutamol. *J Allergy Clin Immunol* 106: 57–64, 2000.
78. Chong LK, Drury DEJ, Dummer JF *et al.* Protection by dexamethasone of the functional desensitization to β_2-adrenoceptor-mediated responses in human lung mast cells. *Br J Pharmacol* 121: 717–22, 1997.

79. Bousquet J, Jeffery PK, Busse WW *et al.* Asthma. From bronchoconstriction to airways inflammation and remodeling. *Am J Respir Crit Care Med* 161: 1720–45, 2000.
80. Wenzel SE, Fowler A, Schwartz LB. Activation of pulmonary mast cells by bronchoalveolar allergen challenge. *In vivo* release of histamine and tryptase in atopic subjects with and without asthma. *Am Rev Respir Dis* 137: 1002–8, 1998.
81. Beasley R, Roche WR, Roberts JA, Holgate ST. Cellular events in the bronchi in mild asthma and after bronchial provocation. *Am Rev Respir Dis* 139: 806–17, 1989.
82. Pesci A, Foresi A, Bertorelli G *et al.* Histochemical characteristics and degranulation of mast cells in epithelium and lamina propria of bronchial biopsies from asthmatic and normal subjects. *Am Rev Respir Dis* 147: 684–89, 1993.
83. Koshino T, Arai Y, Miyamoto Y *et al.* Airway basophil and mast cell density in patients with bronchial asthma: Relationship to bronchial hyperresponsiveness. *J Asthma* 33: 89–95, 1996.
84. Macfarlane AJ, Kon OM, Smith SJ *et al.* Basophils, eosinophils, and mast cells in atopic and nonatopic asthma and in late-phase allergic reactions in the lung and skin. *J Allergy Clin Immunol* 105: 99–107, 2000.
85. Guo CB, Liu MC, Galli SJ *et al.* Identification of IgE-bearing cells in the late-phase response to antigen in the lung as basophils. *Am J Respir Cell Mol Biol* 10: 384–90, 1994.
86. Xu X, Zhang D, Zhang H *et al.* Neutrophil histamine contributes to inflammation in mycoplasma pneumonia. *J Exp Med* 203: 2907–17, 2006.
87. Pesci A, Rossi GA, Bertorelli G *et al.* Mast cells in the airway lumen and bronchial mucosa of patients with chronic bronchitis. *Am J Respir Crit Care Med* 149: 1311–16, 1994.
88. Kalenderian R, Raju L, Roth W *et al.* Elevated histamine and tryptase levels in smokers' bronchoalveolar lavage fluid. Do lung mast cells contribute to smokers' emphysema? *Chest* 94: 113–19, 1988.
89. Matin R, Tam EK, Nadel JA, Caughey GH. Distribution of chymase-containing mast cells in human bronchi. *J Histochem Cytochem* 40: 781–86, 1992.
90. Ekberg-Jansson A, Amin K, Bake B *et al.* Bronchial mucosal mast cells in asymptomatic smokers relation to structure, lung function and emphysema. *Respir Med* 99: 75–83, 2005.

CHAPTER 10

Dendritic Cells in Asthma and COPD

Bart N. Lambrecht[1,2] and Guy G. Brusselle[2]

[1]Department of Respiratory Medicine, University Hospital Ghent, Belgium
[2]Department of Pulmonary Medicine, Erasmus University Medical Center, Rotterdam, The Netherlands

INTRODUCTION

The lung contains many subsets of dendritic cells (DCs) that are distributed in various anatomical compartments. In homeostatic conditions, a fine-tuned balance exists between plasmacytoid and myeloid DCs necessary for maintaining tolerance to inhaled antigen and for avoiding overt inflammation. These subsets of DCs also play important roles in establishment of airway inflammation seen in asthma and COPD. Based on these new insights on airway DC biology, several approaches that interfere with DC function show potential as new intervention strategies for these ever increasing diseases.

Obstructive airway disease, broadly divided clinically into asthma or COPD, is a significant cause of morbidity and mortality. In allergic asthma, allergen-specific T-helper type 2 (Th2) cells produce key cytokines like IL-4, IL-5, and IL-13 that regulate the synthesis of allergen-specific IgE and control tissue eosinophilic airway inflammation and remodeling of the airways. In the lungs of COPD patients, predominantly CD8 lymphocytes and neutrophilic airway inflammation are seen, concomitant with remodeling of small airways and the destruction of distal air spaces characteristic of emphysema. It is increasingly clear that DCs are essential for inducing activation and differentiation of not only naïve but also effector $CD4^+$ T- and $CD8^+$ T-cells in response to inhaled antigen, and it has been well established that these cells play a pivotal role in the initiation and maintenance phase of airway inflammation [1]. In this chapter, we will highlight the recent discoveries in airway DC biology with special emphasis on mouse models of asthma and COPD. The applicability to the human situation and the therapeutic potential of novel findings will be discussed where possible.

LUNG DC SUBSETS IN MOUSE AND MAN

It has long been established that the various lung compartments (conducting airways, lung parenchyma, alveolar compartment, pleura) contain numerous DCs, of which the precise lineage or origin have been poorly defined. Recently however, many groups have refined the ways in which lung DCs should be studied, both in mouse [2–5] and in human [6, 7]. It is clear that different DC subsets can be found in the lung, each with functional specialization. In the mouse, all of these express the integrin CD11c and the subsets are further defined on the basis of the expression of the myeloid marker CD11b, as well as anatomical location in the lung. The trachea and large conducting airways have a well-developed network of intraepithelial DCs, even in steady-state conditions. These cells in some way resemble skin Langerhans' cells, and have been shown to express langerin and CD103 while lacking expression of CD11b [4, 8]. In the submucosa of the conducting airways, $CD103^-CD11b^+CD11c^+$ myeloid DCs can be found, particularly under conditions of inflammation, and these cells are particularly suited for priming and restimulating effector CD4 cells in the lung [8, 9]. The lung interstitium that is accessible by enzymatic digestion also contains $CD11b^+$ and $CD11b^-$ DCs that access the alveolar lumen and migrate to the mediastinal lymph nodes (MLN) [2, 5, 10]. It should be noted that this population is contaminated with

DCs lining the small intrapulmonary bronchioles, as well as those lining the vessel walls. In the nearby alveolar lumen, $CD11c^{hi}$ alveolar macrophages control the function of these interstitial DCs. Plasmacytoid DCs are $CD11b^{-}CD11c^{int}$ cells expressing SiglecH and bone marrow stromal Ag-1 (recognized by the moAbs mPDCA-1 or 120G8). In the lungs, pDCs are predominantly found in the lung interstitium and produce large amounts of IFNα in response to triggering by CpG motifs or viral infection *ex vivo* [11].

The exact definition of the different DC subtypes in the human lung is incomplete and a subject of controversy. Two groups identified myeloid DC and plasmacytoid DC in single cell suspensions from digested human lung tissue by flow cytometry, using the blood dendritic cell antigen markers (BDCA) 1 through 3 [7, 12]. The Cluster of Differentiation (CD) nomenclature of these BDCA markers is as follows: BDCA1 = CD1c, BDCA2 = CD303, and BDCA3 = CD141. Within the low autofluorescent mononuclear cells of the lung digests, cells that were positive for the lineage markers, that is, CD3, CD19, CD20, and CD56, were excluded. Within the lineage negative pulmonary mononuclear cells, myeloid DC type 1 were characterized as $CD1c^{+}$/HLA-DR^{+} cells, myeloid DC type 2 were characterized as $CD141^{+}$/HLA-DR^{+} cells, and plasmacytoid DC were characterized as $CD303^{+}/CD123^{+}$ cells. *In vitro* studies suggest that Langerhans type DC (that are $Langerin^{+}$, $CD1a^{+}$, and Birbeck granule positive) derive from myeloid DC precursors that are $CD1c^{+}$ and $CD1a^{+}$ [13], and others have shown that Langerhans-type DC can be generated from monocytes under the influence of interleukin-15 [14]. It is unclear whether differences in terms of expression of surface markers reflect separate stages of differentiation of the DC, rather than different sublineages. It is tempting to speculate that, within these DC subsets, carrying different surface markers, functional specialization does occur. Indeed, pulmonary myeloid DC types 1 and 2 were shown to release proinflammatory cytokines (TNF-α, IL-1β, IL-6, and IL-8) in response to Toll-like receptor-2 (TLR-2) and TLR-4 ligands, whereas plasmacytoid DC released high amounts of interferon-α in response to the TLR-9 ligand CpG oligonucleotides [6]. Moreover, myeloid DC type 1 was shown to be strong inducers of T-cell proliferation in a mixed leukocyte reaction, while plasmacytoid DC induced little T-cell proliferation and myeloid DC type 2 had an intermediate T-cell-stimulatory capacity [6].

How, where, and by which DC subset inhaled antigen is sampled from the airway lumen has been a matter of debate. Jahnsen demonstrated that, analogous to that reported in the gut, a subset of rat airway intraepithelial DCs extend their processes into the airway lumen. This "periscope up" function is constitutively expressed within the airway mucosal DC population, providing a mechanism for continuous immune surveillance of the airway luminal surface in the absence of "danger" signals [15]. In the mouse, $CD103^{+}CD11b^{-}$ intraepithelial DCs express the tight junction proteins claudin-1, claudin-7, and zonula-2, allowing the sampling of airway luminal contents whereas keeping the epithelial barrier function intact [4]. This subset is also found in the alveolar septa, and DCs lining the alveolar wall can take up inhaled harmless ovalbumin (OVA) or bacterial anthrax spores by forming intra-alveolar extensions and then migrating to the MLN in a CCR7-dependent way [5, 8, 10]. It is still a matter of debate, however, whether the uptake and transport of inhaled antigen occurs exclusively by alveolar wall DCs, by intraepithelial DCs lining the large conducting airways or by both [2, 5]. Another controversial issue is the location and extent by which plasmacytoid DCs take up inhaled antigen. Two reports describe that within 24–48 h following exposure of inhaled fluorescently labeled Ag, almost 50–60% of pDCs are antigen-positive [11, 16], whereas another report saw only a minor percentage of Ag uptake in this subset [5]. It remains to be demonstrated if pDCs take up antigen in the periphery of the lung and subsequently migrate, whether they get their antigen from another migratory DC [17], or whether they take up free afferent lymph while residing in the lymph node. How much antigen crosses the epithelial barrier passively in the absence of DC uptake is unknown, but it heavily depends on the molecular weight of the Ag, its dose, as well as the potential to disrupt epithelial tight junctions. Control of epithelial barrier function could be under important genetic control as well, as many of the gene polymorphisms associated with atopy in humans control epithelial integrity (e.g. Spink5, S100 family). It is similarly possible that Ag uptake by lung pDCs would be facilitated by the presence of Ag-specific immunoglobulins acting on Fc receptors, thus enhancing endocytosis [18].

ROLE OF DCs IN ASTHMA

Outcome of antigen inhalation depends on the functional state of myeloid and plasmacytoid DCs

The usual outcome of inhalation of harmless protein antigen in the lungs is immunological tolerance (see Fig. 10.1 for a model depicting cellular interactions). As a result, when the antigen is subsequently given to mice in an adjuvant setting (e.g. in combination with the Th2 adjuvant alum) it no longer induces an immunological response that leads to effector cells causing inflammation [11, 19]. Inhalational tolerance is mediated in part by deletion of Ag reactive T-cells as well as induction and/or expansion of regulatory T-cells in the mediastinal nodes [17, 19–21]. The latter type of tolerance is dominant and can be transferred to other mice by adoptive transfer. Induction of tolerance to inhaled antigen is a function of lung DC subsets that migrate from the lung in a CCR7-dependent way [17]. It is often claimed that induction of tolerance is a function of "immature" DCs, meaning that these cells lack the expression of high levels of major histocompatibility complex (MHC), adhesion, and co-stimulatory molecules, However, Reis and Sousa recently argued that the term "mature DC" should be reserved for those DCs that have the potential to generate effector T-cells, and that expression of costimulatory molecules by DCs does not exclude the possibility that tolerance would be induced [22]. In the lungs, inhaled tolerance is dependent on signals delivered by CD86 and/or ICOS-L on DCs, supporting this view [20].

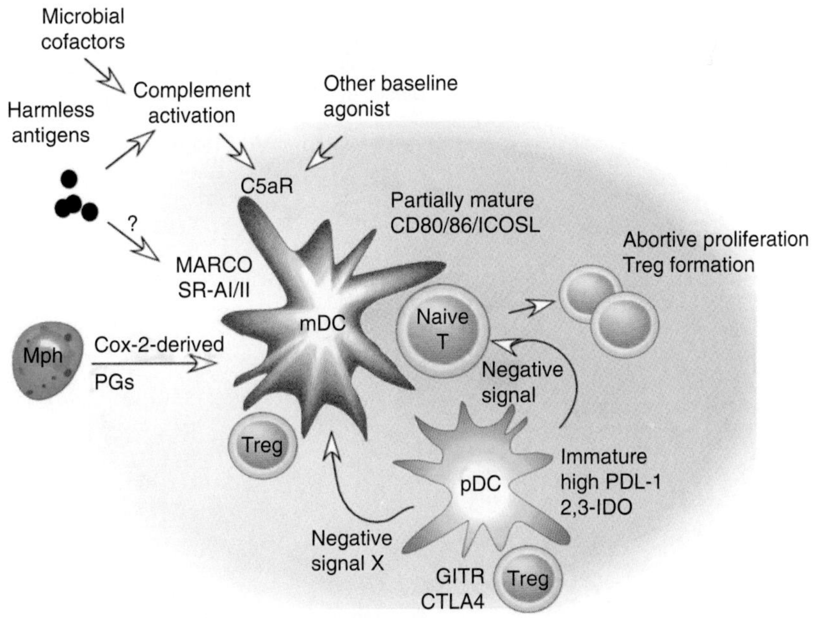

FIG. 10.1 Dendritic cell function following inhalation of harmless antigen in homeostatic conditions. When an inhaled antigen reaches the deeper airways and it is not cleared by mucociliary transport or by macrophages, it will be taken up by airway DCs. Both mDCs and pDCs take up antigen. mDCs seem to do this in the periphery, whereas pDCs only do this in the mediastinal nodes. In baseline conditions, the mDCs that reach the nodes are only partly "mature" and the T-cell response that they induce is characterized mainly by division but not by differentiation to effector cells. Eventually many dividing cells die. Additionally, mDCs induce Treg cells that suppress inflammation. At the same time, the pDCs control the level of activity of the mDCs so that these cells are kept in a quiescent state. This function of pDCs is controlled by Treg control occurring via GITR–GIRL ligand and CLTA4–CD80/86 interactions. The signals involved are not precisely known but could involve IDO, or some surface expressed ligand on pDCs (programmed death ligand-1, PDL1). The function of mDCs is constantly kept in check among others by tonic inhibition by cyclooxygenase-2 (Cox-2) derived prostaglandins (PGs), as well as by complement activation acting on the C5a receptor (C5a). The precise ligand for the class A scavenger receptors (MARCO and SR-AI/II) are not known.

Conventional lung DCs (either $CD11b^+$ or $CD11b^-$) are necessary for tolerance induction [17] but are also responsible for inducing Th2 sensitization, providing there is some form of activation (either LPS or TNFα) [23] leading to functional DC differentiation and their capacity to prime Th2 effector cells (see Fig. 10.2 for a model depicting cellular interactions leading to Th2 sensitization in the airways). In further support, Th2 sensitization can be induced by adoptive intratracheal transfer of GM-CSF-cultured bone marrow DCs, most closely resembling mature monocyte-derived $CD11b^+$ DCs, but not by Flt3L-cultured bone marrow-derived DCs that more resemble the immature steady-state DCs resident in the lymph nodes and spleen [11]. As activation of lung DCs is the common event leading to Th2 sensitization, it is likely that under homeostatic conditions, the degree of DC maturation is, therefore, constantly kept in check. One such pathway of tonic DC suppression seems to involve COX-2-derived prostaglandins or their metabolites, most likely derived from nearby alveolar macrophages [24]. Chimeric mice in which the PGD_2 receptor DP1 was selectively deleted on hematopoietic cells, demonstrated spontaneous maturation of lung mDCs and subsequently, response to harmless antigen was greatly enhanced, suggesting tonic inhibition of DC function by PGD_2 in the lung [25]. When exposed to selective PGD2 agonists, myeloid DCs induced the formation of $Foxp3^+$ Ag-specific Tregs that subsequently suppressed airway inflammation. A similar mechanism on DCs was found for stable PGI_2 analogs [26, 27].

When pDCs are depleted from the lungs, inhaled tolerance is abolished, and consequently pDC/mDC balance in the lung is tightly regulated among others by the cytokine osteopontin as well as complement C5a [11, 16, 28, 29]. How exactly pDC depletion leads to sensitization is still unsolved, but *in vitro* and *in vivo* data suggest that pDCs directly suppress the potential of mDCs to generate effector T-cells [11, 29]. Plasmacytoid DCs can also stimulate the formation of Treg cells, possibly in an ICOS-L-dependent way [11, 30]. Tregs expressing GITR could also induce the production of the tryptophan catabolizing enzyme indoleamine 2,3-dioxygenase (IDO) through reverse signaling in pDCs [31]. In mice depleted of pDCs, there was endogenous release of extracellular adenosine triphosphate (ATP) responsible for inducing mDC maturation and Th2 skewing potential. Th2 sensitization to inhaled OVA was abolished when ATP signaling was blocked using the broad spectrum P2X and P2Y receptor antagonist suramin. On the contrary, a non-degradable form of ATP was able to break inhalation tolerance to OVA [32]. At present, it is unclear how purinergic receptor triggering on DCs promotes Th2 development, but it could involve the formation of the inflammasome, a multi-protein complex that leads to activation of caspase 1 and processing and release of IL-1, IL-18, and possibly IL-33. One possibility is also that ATP indirectly influences DC function via the modification of mast cell and eosinophil function. Mast cells (e.g. through release of cytokines, PGD2 or sphingosine metabolites) [33]

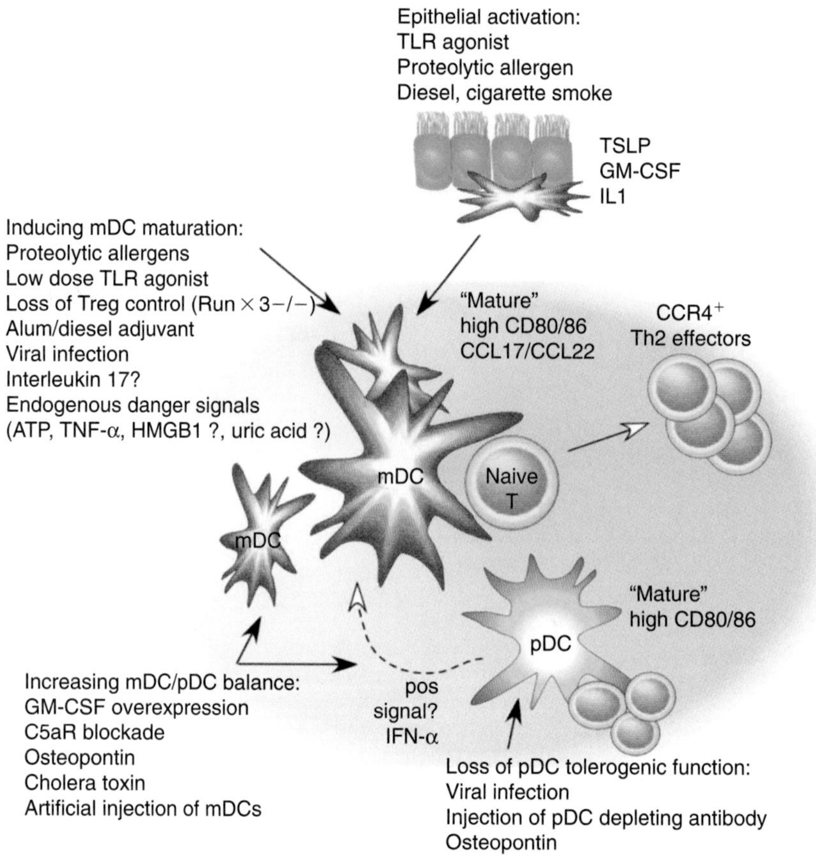

FIG. 10.2 Model of dendritic cell function during Th2 sensitization. Several known risk factors for atopy have been shown to interfere with DC function in the airways. Also, several experimental models have been developed in which sensitization occurs even after inhalation of harmless antigens to the lung, providing there is some form of DC activation. In these models, respiratory tolerance is broken. Some models have induced a shift in the pDC/mDC balance, and consequently mDCs induce priming because they are no longer suppressed by adequate numbers of pDCs. Activated mDCs also produce chemokines like CCL17 or CCL22 to further attract Th2 cells into the response. Some adjuvants induce proper activation of mDCs (yet not sufficient to induce IL-12) so that they now induce effector Th2 cells rather than regulatory T-cells. Some stimuli, like concomitant viral infection, might have an additional effect by inducing maturation of pDCs and their production of IFNα. This is a known maturation stimulus for mDCs and in this way, these cells might even contribute to sensitization upon viral infections. Activation of epithelial cells by proteolytic allergens, virus infection, TLR ligands, or air pollutants is an indirect way of activating and polarizing the DC network, through release of thymic stromal lymphopoeitin (TSLP) or granulocyte-macrophage colony stimulating factor (GM-CSF) or interleukin 1 (IL-1). The precise source and role of endogenous danger signals such as ATP, tumor necrosis factor α (TNF-α), high mobility group box 1 (HMGB1), or uric acid is currently being investigated.

and eosinophils (through release of leukotrienes and eosinophil-specific enzymes) also influence DC function [34]. The conditions regulating ATP release in the lungs will have to be studied more carefully before we can conclude how important the pathway of purinergic signaling is in sensitization to more common allergens, like house-dust mite.

Direct or indirect mechanisms of Th2 sensitization to inhaled antigen

Induction of tolerance or immunity to inhaled antigen by DCs is tightly controlled by signals from alveolar macrophages, Tregs, NKT cells, complement activation, nervous system interactions, and epithelial activation. While studying the literature on particular substances that can break inhalation tolerance and induce Th2 priming, one needs to wonder, therefore, whether a stimulus acts directly on mDCs or pDCs or whether its effects are mediated indirectly through modification of any of the above interactions. For systemically administered TLR agonists, like endotoxin, the activation of DCs occurs mainly through direct recognition by TLR4 expressed on the DC, but in epithelia, the response could be clearly different [35]. As an example, low dose endotoxin was able to break inhalation tolerance to inhaled OVA by inducing mDC maturation [23]. These effects could be mediated directly via TLR and Myd88-dependent pathways in DCs but could also be mediated via TLRs on bronchial epithelial cells [36]. Bronchial epithelial cells could produce chemokines as well as crucial growth and differentiation factors that subsequently attract, activate, and polarize lung DCs to prime Th2 responses. In this regard, the epithelial cytokines thymic stromal lymphopoeitin (TSLP) and GM-CSF might be crucial, as their overexpression in the lungs breaks inhalational tolerance [37, 38]. On the contrary, neutralization of these cytokines, during priming regimens, eliminates much of the adjuvant effects of diesel exhaust particles (DEP) [39, 40] or pro-allergic effects of house-dust mite [41]. Importantly, the production of these cytokines by bronchial epithelial cells in response to these triggers might be genetically regulated and this could be the explanation why some individuals become primed to inhaled antigen under the right environmental exposure [42]. Under some conditions, predictions about sensitizers or adjuvants can be made from *in vitro* experiments. Ambient particulate matter (APM) is ubiquitous in the environment and is associated with allergic diseases in inner cities. *In vitro*, inhaled APM can act directly on human DCs as a danger signal to direct a proallergic pattern of innate immune activation, thus explaining why it acts as an adjuvant [43]. Likewise, DEP induce maturation of human DCs indirectly, via promoting GM-CSF production in bronchial epithelial cells *in vitro* [40]. In mice, DEP and APM induce altered DC maturation directly, via nuclear factor-erythroid 2 (NF-E2)-related factor 2-mediated signaling, implicating oxidative stress in the activation of DCs [44]. Whether enhancement of inflammation in Nrf-2-deficient mice, which are hypersensitive to oxidative stress, is also the result of overzealous DC activation remains to be shown [45]. Another known sensitizer is cigarette smoke. When given concomitantly with harmless OVA, it induces Th2 responses, and this response is associated with enhanced DC maturation and migration [46]. DCs developed in a nicotinic environment (nicDCs) fail to support the terminal development of effector memory Th1 cells due to their differential expression of costimulatory molecules and lack of IL-12 production. In both human and mouse, nicDCs promoted the development of Th2 responses [47]. As maternal cigarette smoking is a solid risk factor for becoming sensitized in early life, it will be important to elucidate how it leads to DC activation (e.g. whether any indirect mechanisms acting via epithelial TLR4 contribute), as this might provide novel intervention strategies. Another unsolved question is how the nearby nervous endings react to viruses or air pollution and how this could affect the way DCs react to inhaled allergen. In this regard, the remodeling that occurs in the airway unmyelinated nerves following RSV virus could partly explain the subsequent enhanced risk of Th2 sensitization in mice.

Function of DCs in allergic inflammation and tissue remodeling

Not only do DCs play a role in the primary immune response to inhaled allergens, they are also crucial during the effector phase of asthma. The number of $CD11b^+$ DCs is increased in the conducting airways and lung interstitium of sensitized and challenged mice during the acute phase of the response [9, 48, 49]. However, during the chronic phase of the pulmonary response, induced by prolonged exposure to a large number of aerosols, respiratory tolerance develops through unclear mechanisms. During this regulatory phase, the number of mDCs as well as their costimulatory molecules in the lungs steadily decreased, and this was associated with a reduction of bronchial hyperresponsiveness (BHR), possibly mediated by the action of Treg cells [19, 50, 51]. Inflammation however reappears when mature inflammatory $CD11b^+$ DCs are given [50]. Increased levels of class A scavenger receptors (MARCO and SR-AI/II) were found in the lungs of asthmatic mice, possibly suppressing DC-driven inflammation. These receptors are expressed on lung macrophages, DCs, and basophils. Receptor-deficient mice had more eosinophilic airway inflammation, airway hyperresponsiveness (AHR), and increased migration of DCs to the MLN [52].

The role of mDCs in the secondary immune response was further supported by the fact that their depletion at the time of allergen challenge abrogated all the features of asthma, including airway inflammation, goblet cell hyperplasia, and BHR [9]. Again the defect was restored by intratracheal injection of GM-CSF cultured $CD11b^+CD11c^+$ mDCs, most closely resembling monocyte-derived "inflammatory DCs". The same effects were observed when DCs were depleted in the nose in an animal model for allergic rhinitis [53]. It therefore seems that "inflammatory" DCs are both necessary and sufficient for secondary immune responses to allergen. Upon allergen challenge, lung DCs upregulate the expression of CD40, CD80, CD86, ICOS-L, PD-L1, and PD-L2, particularly upon contact with Th2 cells [9, 11, 48, 49, 54]. Costimulatory molecules might be involved in activation of effector T-cells in the tissues or in regulation of Treg activity. In allergen-challenged mice, DCs might also be a prominent source of the inflammatory chemokines CCL17 and CCL22, involved in attracting $CCR4^+$ Th2 cells to the airways, and in producing eosinophil-selective chemokines [29, 49]. In helminth infections, recruitment of Th2 cells and eosinophils depends on an IL4/IL13-responsive bone marrow-derived cell, most likely a DC or alternatively an activated macrophage population [55]. A number of cytokines and innate immune response elements control the production of these chemokines. The pro-allergic cytokine TSLP induces the production of large amounts of CCL17 by mDCs, thus contributing to the recruitment of Th2 cells to the airways, explaining how it may act to enhance inflammation [37]. The complement factor C5a suppresses the production of CCL17 and CCL22 [29]. A similar effect was seen with the cytokine IL-17, explaining how it may suppress allergic inflammation when given during allergen challenge. *In vitro*, IL-17 reduced CCL17 production and antigen uptake by DCs and IL-5 and IL-13 production in regional lymph nodes *in vivo*. Furthermore, IL-17 is regulated in an IL-4-dependent manner as mice deficient for IL-4Rα signaling showed a marked increase in IL-17 concentration with inhibited eosinophil recruitment [56]. Emerging evidence suggests that IL4Rα expression on lung DCs is an important feedback mechanism through which IL-4 producing cells (effector Th2 cells, eosinophils, basophils) might promote further Th2 polarization in ongoing responses [57].

As the number and activation status of lung $CD11b^+$ DCs during the secondary challenge seems critical for controlling allergic inflammation, studying the factors that control recruitment, survival, or egress from the lung during allergic inflammation will be important, as this might reveal therapeutic targets. In an elegant study using mixed bone marrow chimeras in which half the hematopoietic cells were $CCR2^{-/-}$ and half were $CCR2^{+/+}$, it was shown by Robays *et al.* that CCR2 (and not CCR5 or CCR6) is crucial for releasing DC precursors from the bone marrow and attracting them into allergically inflamed lung. This was unexpected, as CCR6 is generally seen as the chemokine receptor attracting immature DCs into peripheral tissues [58]. Lung mDCs use CCR7 ligands and CCR8 for emigration to the draining lymph nodes but not the leukotriene C4 transporter multidrug-related protein-1 as they do in the skin [59]. Unexpectedly, disruption of CCR7-selective chemokines in paucity of lymphocyte T-cell (plt) mutant mice, deficient in CCL21 and CCL19, resulted in airway inflammation and Th2 activity that were enhanced [60]. Still, increased numbers of mDCs could be found in the draining lymph nodes of these mice. So, in addition to CCR7 ligands, there are other factors involved in the migration of DCs to the draining LN, including other chemokine receptors [59]. Eicosanoid lipid mediators like prostaglandins and leukotrienes can also influence the migration of lung DCs [61]. Leukotriene LTB4 promoted the migration of immature and mature skin DCs but these effects seem to be indirect [62]. It will be important to study if well-known inducers of LTB4 in the lungs, such as the environmental biopolymer chitin, derived from fungi, helminthes, and insects, also induce DC migration [63]. Additional "druggable" factors promoting the migration of DCs to the draining mediastinal nodes during inflammatory responses could be sphingosine-1-P and extracellular ATP [32, 64].

In humans, allergen challenge leads to an accumulation of myeloid, but not plasmacytoid DCs in the airways of asthmatics, concomitant with a reduction in circulating $CD11c^+$ cells, showing that these cells are recruited from the bloodstream in response to allergen challenge [65, 66]. A recent report suggests that pDCs are also recruited into the bronchoalveolar lavage (BAL) fluid but are poor antigen-presenting cells (APCs) [66]. The exact role of plasmacytoid DCs in ongoing allergen-specific responses in asthma is currently unknown. It was shown that pDCs accumulate in the nose, but not lungs, of allergen challenged atopics [67]. When pDCs were pulsed with pollen allergens, they were as efficient as mDCs in inducing Th2 proliferation and effector function [68]. Others have suggested, as in the mouse, that pDCs might also confer protection against allergic responses [16]. In children at high risk of developing atopic disease, the number of circulating pDCs was reduced.

ROLE OF DCs IN COPD

The inflmmatory basis of COPD

COPD is an inflammatory disease of the large and small airways and the lung parenchyma, which is caused most commonly due to the inhalation of noxious particles and gases and is associated with an abnormal systemic inflammatory response. The lungs of COPD patients are infiltrated with cells of the innate immune system such as neutrophils and macrophages [69], but there is also evidence for an activated adaptive immune response with accumulation of $CD8^+$ T-cells, B–cells, and the presence of lymphoid follicles [70, 71]. Four key elements appear to be crucial in the pathogenesis of COPD: increased *oxidative stress* (caused by cigarette smoke and by activated cells of the innate immune system, causing tissue damage), disturbance in the *protease–antiprotease balance* (with an elevated production of proteases and/or decreased levels of antiproteases such as α1-antitrypsin and tissue inhibitors of matrixmetalloproteïnases (TIMP)), increased *programmed cell death* and *profibrotic conditions* in the small airways. The exact pathogenetic mechanisms for ongoing pulmonary inflammation and damage in COPD, even after the initial inciting agent has disappeared (i.e. after smoking cessation), are poorly understood. Latent viral respiratory infections, chronic bacterial colonization of the lower airways, repetitive infectious exacerbations, autoimmune responses against changed epitopes in the lung, and genetic predisposition, are proposed as important driving mechanisms for the persistent inflammation in patients with COPD [72].

The DCs present in the human lung and lymphoid organs, linking innate and adaptive immune responses, could be a key element in the pathogenesis of COPD. In the remaining part of the chapter, we will discuss what is known about the effects of cigarette smoke on DCs and how DC populations are altered in "healthy" smokers (without airway obstruction) and patients with COPD. We will also formulate several hypotheses about the role of DC in the pathogenesis of COPD.

DC populations in lungs of smokers and COPD patients

There have been only a few studies addressing the number and distribution of DCs in the lungs of smokers and patients with COPD. Most of these descriptive studies have a cross-sectional design, so that the sequence of events cannot be established and any evidence for causality is (very) weak. At first sight, some data in the literature appear discrepant or even contradictory, but this may be due to differences in the area of interest (bronchial biopsies sampling large airways *versus* surgical resection specimens sampling small airways and parenchyma versus BAL sampling the alveolar lumen) (Fig. 10.3), differences in the examination techniques (electron microscopy, flowcytometry, or immunohistochemistry) or differences in the immunohistochemical markers used to identify and enumerate the DCs. It is also critical to discriminate between the effects of smoking *per se* on DC numbers, phenotypic markers, or functions versus the disease-specific effects of COPD on DC, irrespective of the current smoking status (see Fig. 10.3).

When studying pulmonary DC, it is important to take into account not only the different DC subsets mentioned earlier but also the different anatomic locations (i.e. distribution) of DC within the lung. Moreover, the different compartments of the lungs (large airways, small airways, lung parenchyma interstitium and alveolar lumen) are sampled by different methods (bronchoscopy, surgical lung resections, and BAL, respectively). In the large airways (trachea, bronchi), sampled by bronchoscopy with bronchial biopsies, the number of $CD1a^+$ DC was evaluated in healthy smoking controls and current smoking COPD patients, showing no significant differences between groups [73, 74]. Others evaluated the number of DC in large airways using electron

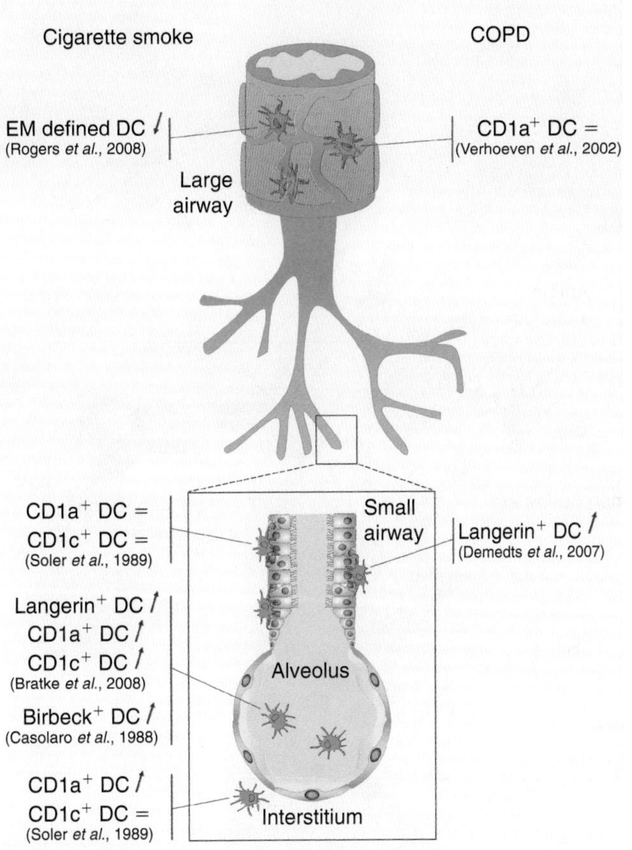

FIG. 10.3 Distribution of DCs in the lungs and the impact of smoking or COPD on different subsets of dendritic cells. Myeloid DCs can be retrieved in different compartments of the lung. Airway DCs are located as a network immediately above and beneath the basement membrane, either in the large airways (sampled by bronchial biopsies taken during bronchoscopy) or in the small airways (present in lung resection specimens). Alveolar DCs are present in the lumen of alveolar spaces and can be recovered by BAL, whereas interstitial DCs are located in the interstitium of the lung parenchyma between alveolar spaces (also present in lung resection specimens). The exact location of human plasmacytoid DCs has still to be elucidated. On the left side of the figure, studies evaluating the impact of cigarette smoking *per se* on the number of different myeloid DCs are indicated and summarized; on the right side, the studies comparing the number of $CD1a^+$ DC (in large airways) or Langerin+ DCs (in small airways) between patients with COPD and healthy-smokers – without airflow limitation – are indicated. COPD: chronic obstructive pulmonary disease; DCs: dendritic cells.

microscopy, based upon morphologic criteria derived from cultured $CD1c^+$ DC, and showed a significant decrease in the number of DCs in the large airways of smoking COPD patients versus ex-smoking COPD patients and never smoking controls [75].

In the small airways (bronchioli), which is the main location responsible for airway obstruction in COPD, the number of DC was evaluated using immunohistochemical staining for CD1a and CD1c (BDCA-1), showing no difference between smokers and non-smokers [76]. In patients with COPD, however, a significant increase in the number of CD207 (Langerin) positive cells was found compared to never-smokers and healthy-smokers (without COPD), suggesting an accumulation of Langerhans-type DC in COPD [77]. Moreover, the number of Langerin^+ DC further increased with the severity of the disease [77].

Studies sampling BAL fluid evaluated the number of DCs between never-smokers and smokers without COPD, showing a significant increase in the expression of Langerin and CD1a (markers of Langerhans cells) on myeloid DC [78] as well as an increase in Birbeck granule positive Langerhans-type DC in smokers versus non-smokers [79]. Finally, in the alveolar parenchyma, the number of $CD1a^+$ DC was increased in smokers versus never-smokers, whereas the number of $CD1c^+$ DC was not different [76].

Taken together, evidence points toward an accumulation of myeloid DC with Langerhans-type cell markers (CD207, CD1a, and Birbeck granules) in the small airways and alveoli of smokers and COPD patients. This is in agreement with data from experimental models of COPD, in which mice are chronically exposed to cigarette smoke and develop manifest pulmonary inflammation and emphysema. A clear accumulation of myeloid DC was seen in the BAL fluid and lungs of these mice [80], exposed to relatively high doses of cigarette smoke (with carboxyhemoglobin levels in serum comparable to the levels obtained in human smokers who smoke 20 cigarettes a day). However, exposing the mice to a lower dose of cigarette smoke appears to decrease the number of DCs in the lung in the absence of inflammation [81].

Functional differences in DCs exposed to cigarette smoke

Evidence from the mouse model of COPD and from human lung tissue suggests an activated macrophage inflammatory protein 3α (MIP3alpha)/CC-chemokine ligand (CCL20)–CC-chemokine receptor 6 (CCR6)-axis in COPD, responsible for the accumulation of myeloid DC in the lung [77, 82]. At the epithelial surface, DCs are capable of sensing danger signals and take up antigens to process them. However, little is known about the influence of cigarette smoke on the expression and function of innate receptors of DCs, including TLR and lectin-like receptors (such as Langerin and blood DC antigen-2 [83]). Once the DC has sampled the antigen, it will process the antigen, upregulate the expression of CC-chemokine receptor 7 (CCR7), and migrate toward the secondary lymphoid organs (i.e. MLN) (Fig. 10.4).

Several *in vitro* studies have evaluated the effect of nicotine and cigarette smoke extract (CSE) on the maturation process and T-cell stimulatory capacity of myeloid DC. Human monocyte-derived DCs, which have a predominant interstitial phenotype, show increased maturation and increased IL-10 and IL-12 production upon exposure to a high dose nicotine [84], whereas a low dose of nicotine results in a decreased immature DC-dependent T-cell proliferation and a decreased IL-12 production after LPS stimulation [85]. These nicotine-exposed DCs are indeed less able to induce differentiation of naïve T-cells into T-helper 1 cells [47, 86]. When CSE is added during the last 18 h of a monocyte-derived DC culture, mimicking the

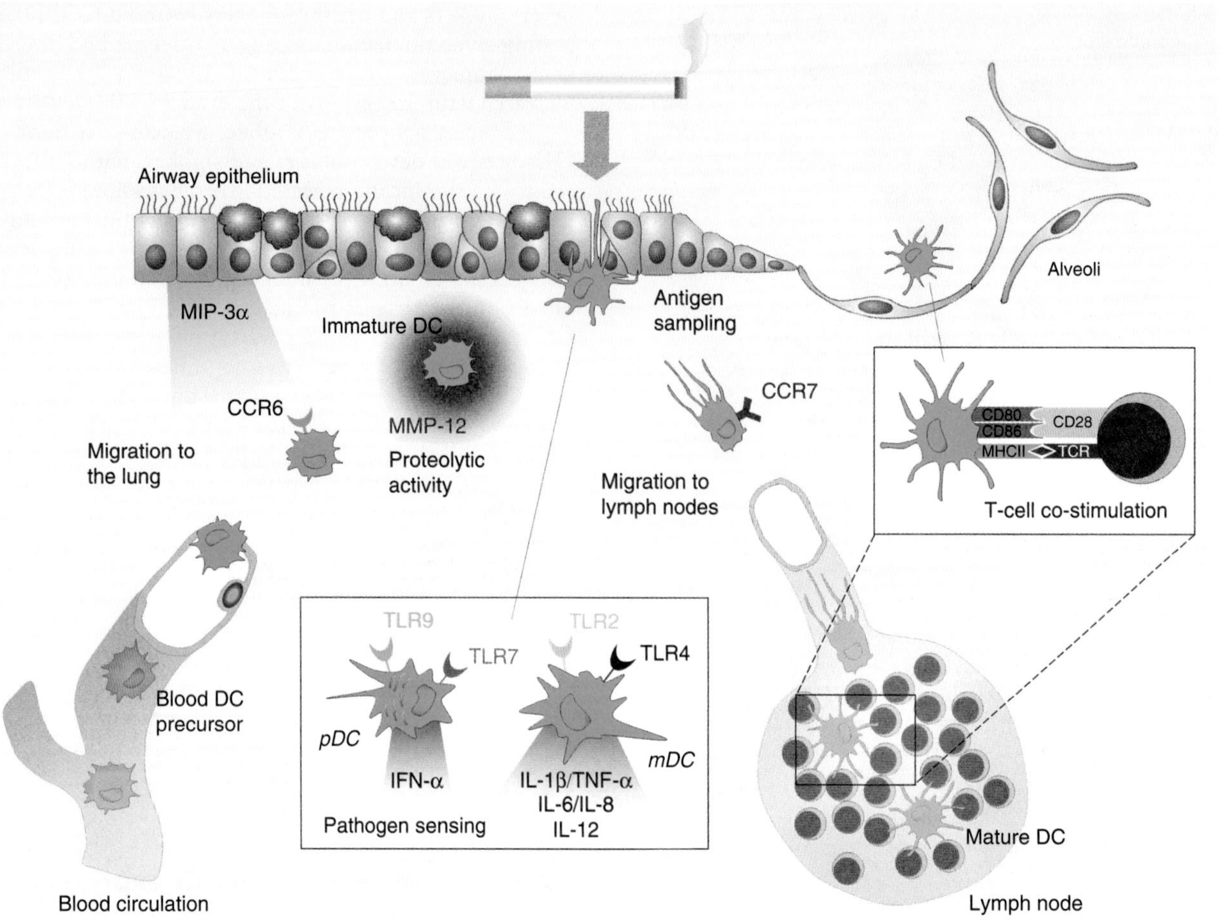

FIG. 10.4 Possible interrelationship between cigarette smoke, airway inflammation, and DCs in the pathogenesis of COPD. Injury to epithelial surfaces by cigarette smoke and/or oxidative stress induces the release of numerous DC-attracting mediators, including macrophage inflammatory protein-3α (MIP-3α)/CC-chemokine ligand 20 (CCL20), which is the only chemokine ligand of CC-chemokine receptor 6 (CCR6). Blood DC precursors are recruited out of the blood circulation, and they migrate to the site of antigen entry and tissue damage. Immature DCs display numerous TLR, including TLR2, TLR4, TLR7, and/or TLR9, which enable myeloid and plasmacytoid DCs to sense multiple pathogen-associated molecular patterns (PAMPs), which may precipitate infectious exacerbations in patients with COPD. After antigen uptake and activation, DCs lose all chemokine receptors except CC-chemokine receptor 7 (CCR7), which mediates the migration of DCs toward the secondary lymphoid organs in response to the chemokines CCL19 and CCL21. In the lymph nodes, mature DCs express high amounts of peptide-loaded MHC molecules and T-cell costimulatory molecules (such as CD40, CD80 and CD86). In BAL fluid, an increased expression of antigen presentation markers such as CD80 and CD86 was observed on myeloid DCs of smokers compared to never-smokers, whereas the expression of the lymph node homing receptor CCR7 was significantly decreased. CCL: CC-chemokine ligand; CCR: CC-chemokine receptor; DC(s): dendritic cell(s); MIP-3α: macrophage inflammatory protein-3α; MHC: major histocompatibility complex; PAMPs: pathogen associated molecular patterns; TLR: Toll-like receptors.

exposure to mainstream cigarette smoke, the DC-induced T-cell proliferation is significantly reduced, with a reduction of IL-12 production and an increase of IL-10 production by the DC. Moreover, the T-cell priming is skewed toward a T-helper 2 response and the LPS-induced maturation of CSE-exposed DC is inhibited, with a reduced upregulation of costimulatory molecules and an inhibition of chemotactic CCR7, which is involved in migration of the DC toward the secondary lymphoid organs. In summary, these *in vitro* data suggest that CSE impairs the maturation of interstitial type DC, skews the DC induced T-cell differentiation toward a Th2 response, and could possibly alter the migration of DC toward the lymph nodes. In contrast, *ex vivo* data from human myeloid DC obtained from BAL show an increased expression of maturation markers by myeloid DC combined with a decrease of CCR7 expression in smokers compared to non-smokers [78]. Although *in vitro* – for a specific stimulus – cigarette smoke exposed DC might mature less than non exposed DC, we hypothesize that *in vivo* the overall inflammatory response and danger signals due to the damaging effect of cigarette smoke in the lungs will still enhance maturation of DC in smokers and patients with COPD.

Apart from the integration of danger signals and antigens, DCs are also capable of producing substantial amounts of proteinases, contributing to the proteinase–antiproteinase imbalance in COPD. Mouse DCs show indeed an increase of MMP-12 production upon cigarette smoke exposure [87] and could thereby contribute to the pathogenesis of emphysema.

Further research evaluating the effects of CSE on the differentiation of DC and on the maturation of Langerhans-type DC compared to interstitial type DC is mandatory to understand the role of the DC in smoking-related diseases. Ideally, longitudinal studies sampling the large and small airways should be performed in never-smokers, smokers without COPD, and patients with COPD to enumerate the different DC subsets. Importantly, healthy-smokers and smoking COPD patients should be studied longitudinally before and after smoking cessation, to be able to discern the effects of smoking per se versus the disease-specific effects of COPD on DC numbers, phenotypic markers, and functions. Moreover, the role of the plasmacytoid DC in COPD has to be elucidated, both during stable disease and at exacerbations. As plasmacytoid DC have important antiviral and tolerogenic properties, they could play a crucial role in the pathogenesis of COPD. In addition, impaired maturational response of DC upon cigarette smoke exposure could lead to an increased susceptibility to viral and bacterial infections, contributing to the increased incidence of lower respiratory tract infections (including pneumonia and tuberculosis) in smokers and of infectious exacerbations in patients with COPD.

DCs AS DRUG TARGETS IN OBSTRUCTIVE AIRWAY DISEASES

If DCs are crucial in mounting immune responses during ongoing inflammation in the lung, then interfering with their function could constitute a novel form of treatment for obstructive airway diseases. Additionally, pharmacological modification of DCs might fundamentally reset the balance of the pro-inflammatory immune response in favor of regulatory T-cells and thus lead to a more long lasting effect on the natural course of allergic disease as well as progression in COPD. Glucocorticosteroids are currently the cornerstone of anti-inflammatory treatment in these diseases. Inhaled corticosteroids reduce the number of lung and nose DCs in patients with allergic asthma and allergic rhinitis [88]. Corticosteroids might also interfere with a GITRL-driven induction of the enzyme IDO in plasmacytoid DCs, thus broadly suppressing inflammation [31].

Recently, several other new molecules have surfaced that may alter DC function in allergic inflammation and thus treat disease. Many of these compounds were first discovered by their potential to interfere with DC-driven Th2 sensitization. The sphingosine-1-P analog FTY720 is currently used in clinical trials for multiple sclerosis and transplant rejection. When given to the lungs of mice with established inflammation, it strongly reduced inflammation by suppressing the T-cell stimulatory capacity and migratory behavior of lung DCs [64]. Also selective agonists of particular prostaglandin series receptors might suppress DC function. The DP1 agonist BW245C strongly suppressed airway inflammation and bronchial hyperreactivity when given to allergic mice by inhibiting the maturation of lung DCs [25]. A very similar mechanism was described for inhaled iloprost, a prostacyclin analog acting on the IP receptor expressed by lung DCs [27]. Extracellular ATP might be released by platelets upon allergen challenge. Neutralization of ATP via administration of the enzyme apyrase or the broad spectrum P2 receptor antagonist suramin reduced all the cardinal features of asthma by interfering with DC function [32]. A specific small molecule compound (VAF347) that blocks the function of B-cells and DCs was also shown to be effective in suppressing allergic airway inflammation in a mouse model of asthma [89]. Finally, specific inhibitors of *syk* kinase were shown to suppress DC function and eliminate established inflammation [90].

CONCLUSION

Our understanding of DC biology in the airways has grown considerably. The concept that different subtypes of DCs perform different functions not only during sensitization but also during established inflammation is a theme that will persist in the coming years. Slowly, therapeutic strategies are emerging from these basic studies in animal models of asthma and COPD, which can eventually reach clinical application. However, detailed knowledge of DC biology in human airways is still lacking.

References

1. Lambrecht BN, Hammad H. Taking our breath away: Dendritic cells in the pathogenesis of asthma. *Nat Rev Immunol* 3: 994–1003, 2003.
2. von Garnier C, Filgueira L, Wikstrom M, Smith M, Thomas JA, Strickland DH, Holt PG, Stumbles PA. Anatomical location determines the distribution and function of dendritic cells and other APCs in the respiratory tract. *J Immunol* 175: 1609–18, 2005.
3. de Heer HJ, Hammad H, Kool M, Lambrecht BN. Dendritic cell subsets and immune regulation in the lung. *Semin Immunol* 17: 295–303, 2005.
4. Sung SS, Fu SM, Rose CE Jr., Gaskin F, Ju ST, Beaty SR. A major lung CD103 (alphaE)-beta7 integrin-positive epithelial dendritic cell population expressing Langerin and tight junction proteins. *J Immunol* 176: 2161–72, 2006.
5. Wikstrom ME, Stumbles PA. Mouse respiratory tract dendritic cell subsets and the immunological fate of inhaled antigens. *Immunol Cell Biol* 85: 182–88, 2007.
6. Demedts IK, Bracke KR, Maes T, Joos GF, Brusselle GG. Different roles for human lung dendritic cell subsets in pulmonary immune defense mechanisms. *Am J Respir Cell Mol Biol* 35: 387–93, 2006.
7. Masten BJ, Olson GK, Tarleton CA, Rund C, Schuyler M, Mehran R, Archibeque T, Lipscomb MF. Characterization of myeloid and plasmacytoid dendritic cells in human lung. *J Immunol* 177: 7784–93, 2006.
8. del Rio M-L, Rodriguez-Barbosa J-I, Kremmer E, Forster R. CD103- and CD103+ bronchial lymph node dendritic cells are specialized in presenting and cross-presenting innocuous antigen to CD4+ and CD8+ T cells. *J Immunol* 178: 6861–66, 2007.
9. van Rijt LS, Jung S, Kleinjan A, Vos N, Willart M, Duez C, Hoogsteden HC, Lambrecht BN. *In vivo* depletion of lung CD11c+ dendritic cells during allergen challenge abrogates the characteristic features of asthma. *J Exp Med* 201: 981–91, 2005.
10. Cleret A, Quesnel-Hellmann A, Vallon-Eberhard A, Verrier B, Jung S, Vidal D, Mathieu J, Tournier JN. Lung dendritic cells rapidly mediate

anthrax spore entry through the pulmonary route. *J Immunol* 178: 7994–8001, 2007.

11. De Heer HJ, Hammad H, Soullie T, Hijdra D, Vos N, Willart MA, Hoogsteden HC, Lambrecht BN. Essential role of lung plasmacytoid dendritic cells in preventing asthmatic reactions to harmless inhaled antigen. *J Exp Med* 200: 89–98, 2004.
12. Demedts IK, Brusselle GG, Vermaelen KY, Pauwels RA. Identification and characterization of human pulmonary dendritic cells. *Am J Respir Cell Mol Biol* 32: 177–84, 2005.
13. Ito T, Inaba M, Inaba K, Toki J, Sogo S, Iguchi T, Adachi Y, Yamaguchi K, Amakawa R, Valladeau J *et al.* A CD1a+/CD11c+ subset of human blood dendritic cells is a direct precursor of Langerhans cells. *J Immunol* 163: 1409–19, 1999.
14. Mohamadzadeh M, Berard F, Essert G, Chalouni C, Pulendran B, Davoust J, Bridges G, Palucka AK, Banchereau J. Interleukin 15 skews monocyte differentiation into dendritic cells with features of Langerhans cells. *J Exp Med* 194: 1013–20, 200
15. Jahnsen FL, Strickland DH, Thomas JA, Tobagus IT, Napoli S, Zosky GR, Turner DJ, Sly PD, Stumbles PA, Holt PG. Accelerated antigen sampling and transport by airway mucosal dendritic cells following inhalation of a bacterial stimulus. *J Immunol* 177: 5861–67, 2006.
16. Xanthou G, Alissafi T, Semitekolou M, Simoes DC, Economidou E, Gaga M, Lambrecht BN, Lloyd CM, Panoutsakopoulou V. Osteopontin has a crucial role in allergic airway disease through regulation of dendritic cell subsets. *Nat Med* 13: 570–78, 2007.
17. Hintzen G, Ohl L, del Rio ML, Rodriguez-Barbosa JI, Pabst O, Kocks JR, Krege J, Hardtke S, Forster R. Induction of tolerance to innocuous inhaled antigen relies on a CCR7-dependent dendritic cell-mediated antigen transport to the bronchial lymph node. *J Immunol* 177: 7346–54, 2006.
18. Benitez-Ribas D, Adema GJ, Winkels G, Klasen IS, Punt CJ, Figdor CG, de Vries IJ. Plasmacytoid dendritic cells of melanoma patients present exogenous proteins to CD4+ T cells after Fc gamma RII-mediated uptake. *J Exp Med* 203: 1629–35, 2006.
19. Van Hove CL, Maes T, Joos GF, Tournoy KG. Prolonged inhaled allergen exposure can induce persistent tolerance. *Am J Respir Cell Mol Biol* 36: 573–84, 2007.
20. Akbari O, Freeman GJ, Meyer EH, Greenfield EA, Chang TT, Sharpe AH, Berry G, DeKruyff RH, Umetsu DT. Antigen-specific regulatory T cells develop via the ICOS-ICOS-ligand pathway and inhibit allergen-induced airway hyperreactivity. *Nat Med* 8: 1024–32, 2002.
21. Ostroukhova M, Seguin-Devaux C, Oriss TB, Dixon-McCarthy B, Yang L, Ameredes BT, Corcoran TE, Ray A. Tolerance induced by inhaled antigen involves CD4(+) T cells expressing membrane-bound TGF-beta and FOXP3. *J Clin Invest* 114: 28–38, 2004.
22. Reis e Sousa C. Dendritic cells in a mature age. *Nat Rev Immunol* 6: 476–83, 2006.
23. Piggott DA, Eisenbarth SC, Xu L, Constant SL, Huleatt JW, Herrick CA, Bottomly K. MyD88-dependent induction of allergic Th2 responses to intranasal antigen. *J Clin Invest* 115: 459–67, 2005.
24. Hashimoto K, Sheller JR, Morrow JD, Collins RD, Goleniewska K, O'Neal J, Zhou W, Ji S, Mitchell DB, Graham BS *et al.* Cyclooxygenase inhibition augments allergic inflammation through CD4-dependent, STAT6-independent mechanisms. *J Immunol* 174: 525–32, 2005.
25. Hammad H, Kool M, Soullie T, Narumiya S, Trottein F, Hoogsteden HC, Lambrecht BN. Activation of the D prostanoid 1 receptor suppresses asthma by modulation of lung dendritic cell function and induction of regulatory T cells. *J Exp Med* 204: 357–67, 2007.
26. Zhou W, Hashimoto K, Goleniewska K, O'Neal JF, Ji S, Blackwell TS, Fitzgerald GA, Egan KM, Geraci MW, Peebles RS Jr. Prostaglandin I2 analogs inhibit proinflammatory cytokine production and T cell stimulatory function of dendritic cells. *J Immunol* 178: 702–10, 2007.
27. Idzko M, Hammad H, van Nimwegen M, Kool M, Vos N, Hoogsteden HC, Lambrecht BN. Inhaled iloprost suppresses the cardinal features of asthma via inhibition of airway dendritic cell function. *J Clin Invest* 117: 464–72, 2007.
28. Oriss TB, Ostroukhova M, Seguin-Devaux C, Dixon-McCarthy B, Stolz DB, Watkins SC, Pillemer B, Ray P, Ray A. Dynamics of dendritic cell phenotype and interactions with CD4+ T cells in airway inflammation and tolerance. *J Immunol* 174: 854–63, 2005.
29. Kohl J, Baelder R, Lewkowich IP, Pandey MK, Hawlisch H, Wang L, Best J, Herman NS, Sproles AA, Zwirner J *et al.* A regulatory role for the C5a anaphylatoxin in type 2 immunity in asthma. *J Clin Invest* 116: 783–96, 2006.
30. Ito T, Yang M, Wang YH, Lande R, Gregorio J, Perng OA, Qin XF, Liu YJ, Gilliet M. Plasmacytoid dendritic cells prime IL-10-producing T regulatory cells by inducible costimulator ligand. *J Exp Med* 204: 105–15, 2007.
31. Grohmann U, Volpi C, Fallarino F, Bozza S, Bianchi R, Vacca C, Orabona C, Belladonna ML, Ayroldi E, Nocentini G *et al.* Reverse signaling through GITR ligand enables dexamethasone to activate IDO in allergy. *Nat Med* 13: 579–86, 2007.
32. Idzko M, Hammad H, van Nimwegen M, Kool M, Willart MA, Muskens F, Hoogsteden HC, Luttmann W, Ferrari D, Di Virgilio F *et al.* Extracellular ATP triggers and maintains asthmatic airway inflammation by activating dendritic cells. *Nat Med* 13: 913–19, 2007.
33. Bulanova E, Budagian V, Orinska Z, Hein M, Petersen F, Thon L, Adam D, Bulfone-Paus S. Extracellular ATP induces cytokine expression and apoptosis through P2 × 7 receptor in murine mast cells. *J Immunol* 174: 3880–90, 2005.
34. Idzko M, Panther E, Bremer HC, Sorichter S, Luttmann W, Virchow CJ Jr., Di Virgilio F, Herouy Y, Norgauer J, Ferrari D. Stimulation of P2 purinergic receptors induces the release of eosinophil cationic protein and interleukin-8 from human eosinophils. *Br J Pharmacol* 138: 1244–50, 2003.
35. Nolte MA, Leibundgut-Landmann S, Joffre O, Sousa CR. Dendritic cell quiescence during systemic inflammation driven by LPS stimulation of radioresistant cells in vivo. *J Exp Med* 204: 1487–501, 2007.
36. Noulin N, Quesniaux VF, Schnyder-Candrian S, Schnyder B, Maillet I, Robert T, Vargaftig BB, Ryffel B, Couillin I. Both hemopoietic and resident cells are required for MyD88-dependent pulmonary inflammatory response to inhaled endotoxin. *J Immunol* 175: 6861–69, 2005.
37. Zhou B, Comeau MR, De Smedt T, Liggitt HD, Dahl ME, Lewis DB, Gyarmati D, Aye T, Campbell DJ, Ziegler SF. Thymic stromal lymphopoietin as a key initiator of allergic airway inflammation in mice. *Nat Immunol* 6: 1047–53, 2005.
38. Lee HC, Ziegler SF. Inducible expression of the proallergic cytokine thymic stromal lymphopoietin in airway epithelial cells is controlled by NFkappaB. *Proc Natl Acad Sci USA* 104: 914–19, 2007.
39. Ohta K, Yamashita N, Tajima M, Miyasaka T, Nakano J, Nakajima M, Ishii A, Horiuchi T, Mano K, Miyamoto T. Diesel exhaust particulate induces airway hyperresponsiveness in a murine model: Essential role of GM-CSF. *J Allergy Clin Immunol* 104: 1024–30, 1999.
40. Bleck B, Tse DB, Jaspers I, Curotto de Lafaille MA, Reibman J. Diesel exhaust particle-exposed human bronchial epithelial cells induce dendritic cell maturation. *J Immunol* 176: 7431–37, 2006.
41. Cates EC, Fattouh R, Wattie J, Inman MD, Goncharova S, Coyle AJ, Gutierrez-Ramos JC, Jordana M. Intranasal exposure of mice to house dust mite elicits allergic airway inflammation via a GM-CSF-mediated mechanism. *J Immunol* 173: 6384–92, 2004.
42. Holgate ST. The epithelium takes centre stage in asthma and atopic dermatitis. *Trends Immunol* 28: 248–51, 2007.
43. Williams MA, Porter M, Horton M, Guo J, Roman J, Williams D, Breysse P, Georas SN. Ambient particulate matter directs nonclassic dendritic cell activation and a mixed TH1/TH2-like cytokine response by naive CD4+ T cells. *J Allergy Clin Immunol* 119: 488–97, 2007.
44. Chan RC, Wang M, Li N, Yanagawa Y, Onoe K, Lee JJ, Nel AE. Pro-oxidative diesel exhaust particle chemicals inhibit LPS-induced

dendritic cell responses involved in T-helper differentiation. *J Allergy Clin Immunol* 118: 455–65, 2006.
45. Rangasamy T, Guo J, Mitzner WA, Roman J, Singh A, Fryer AD, Yamamoto M, Kensler TW, Tuder RM, Georas SN *et al.* Disruption of Nrf2 enhances susceptibility to severe airway inflammation and asthma in mice. *J Exp Med* 202: 47–59, 2005.
46. Moerloose KB, Robays LJ, Maes T, Brusselle GG, Tournoy KG, Joos GF. Cigarette smoke exposure facilitates allergic sensitization in mice. *Respir Res* 7: 49, 2006.
47. Nouri-Shirazi M, Tinajero R, Guinet E. Nicotine alters the biological activities of developing mouse bone marrow-derived dendritic cells (DCs). *Immunol Lett* 109: 155–64, 2007.
48. Huh JC, Strickland DH, Jahnsen FL, Turner DJ, Thomas JA, Napoli S, Tobagus I, Stumbles PA, Sly PD, Holt PG. Bidirectional Interactions between antigen-bearing respiratory tract dendritic cells (DCs) and T cells precede the late phase reaction in experimental asthma: DC activation occurs in the airway mucosa but not in the lung parenchyma. *J Exp Med* 198: 19–30, 2003.
49. Beaty SR, Rose CE Jr., Sung SS. Diverse and potent chemokine production by lung CD11bhi dendritic cells in homeostasis and in allergic lung inflammation. *J Immunol* 178: 1882–95, 2007.
50. Koya T, Kodama T, Takeda K, Miyahara N, Yang ES, Taube C, Joetham A, Park JW, Dakhama A, Gelfand EW. Importance of myeloid dendritic cells in persistent airway disease after repeated allergen exposure. *Am J Respir Crit Care Med* 173: 42–55, 2006.
51. Lewkowich IP, Herman NS, Schleifer KW, Dance MP, Chen BL, Dienger KM, Sproles AA, Shah JS, Kohl J, Belkaid Y *et al.* CD4 + CD25 + T cells protect against experimentally induced asthma and alter pulmonary dendritic cell phenotype and function. *J Exp Med* 202: 1549–61, 2005.
52. Arredouani MS, Franco F, Imrich A, Fedulov A, Lu X, Perkins D, Soininen R, Tryggvason K, Shapiro SD, Kobzik L. Scavenger receptors SR-AI/II and MARCO limit pulmonary dendritic cell migration and allergic airway inflammation. *J Immunol* 178: 5912–20, 2007.
53. KleinJan A, Willart M, van Rijt LS, Braunstahl GJ, Leman K, Jung S, Hoogsteden HC, Lambrecht BN. An essential role for dendritic cells in human and experimental allergic rhinitis. *J Allergy Clin Immunol* 118: 1117–25, 2006.
54. Van Rijt LS, Vos N, Willart M, Kleinjan A, Coyle AJ, Hoogsteden HC, Lambrecht BN. Essential role of dendritic cell CD80/CD86 costimulation in the induction, but not reactivation, of TH2 effector responses in a mouse model of asthma. *J Allergy Clin Immunol* 114: 166–73, 2004.
55. Voehringer D, Reese TA, Huang X, Shinkai K, Locksley RM. Type 2 immunity is controlled by IL-4/IL-13 expression in hematopoietic non-eosinophil cells of the innate immune system. *J Exp Med* 203: 1435–46, 2006.
56. Schnyder-Candrian S, Togbe D, Couillin I, Mercier I, Brombacher F, Quesniaux V, Fossiez F, Ryffel B, Schnyder B. Interleukin-17 is a negative regulator of established allergic asthma. *J Exp Med* 203: 2715–25, 2006.
57. Webb DC, Cai Y, Matthaei KI, Foster PS. Comparative roles of IL-4, IL-13, and IL-4Ralpha in dendritic cell maturation and CD4+ Th2 cell function. *J Immunol* 178: 219–27, 2007.
58. Robays LJ, Maes T, Lebecque S, Lira SA, Kuziel WA, Brusselle GG, Joos GF, Vermaelen KV. Chemokine receptor CCR2 but not CCR5 or CCR6 mediates the increase in pulmonary dendritic cells during allergic airway inflammation. *J Immunol* 178: 5305–11, 2007.
59. Jakubzick C, Tacke F, Llodra J, van Rooijen N, Randolph GJ. Modulation of dendritic cell trafficking to and from the airways. *J Immunol* 176: 3578–84, 2006.
60. Grinnan D, Sung SS, Dougherty JA, Knowles AR, Allen MB, Rose CE 3rd., Nakano H, Gunn MD, Fu SM, Rose CE Jr.. Enhanced allergen-induced airway inflammation in paucity of lymph node T cell (plt) mutant mice. *J Allergy Clin Immunol* 118: 1234–41, 2006.
61. Hammad H, Lambrecht BN. Lung dendritic cell migration. *Adv Immunol* 93: 265–78, 2007.
62. Del Prete A, Shao WH, Mitola S, Santoro G, Sozzani S, Haribabu B. Regulation of dendritic cell migration and adaptive immune response by leukotriene B4 receptors: A role for LTB4 in up-regulation of CCR7 expression and function. *Blood* 109: 626–31, 2007.
63. Reese TA, Liang HE, Tager AM, Luster AD, Van Rooijen N, Voehringer D, Locksley RM. Chitin induces accumulation in tissue of innate immune cells associated with allergy. *Nature* 447: 92–96, 2007.
64. Idzko M, Hammad H, van Nimwegen M, Kool M, Muller T, Soullie T, Willart MA, Hijdra D, Hoogsteden HC, Lambrecht BN. Local application of FTY720 to the lung abrogates experimental asthma by altering dendritic cell function. *J Clin Invest* 116: 2935–44, 2006.
65. Upham JW, Denburg JA, O'Byrne PM. Rapid response of circulating myeloid dendritic cells to inhaled allergen in asthmatic subjects. *Clin Exp Allergy* 32: 818–23, 2002.
66. Bratke K, Lommatzsch M, Julius P, Kuepper M, Kleine HD, Luttmann W, Christian Virchow J. Dendritic cell subsets in human bronchoalveolar lavage fluid after segmental allergen challenge. *Thorax* 62: 168–75, 2007.
67. Jahnsen FL, Lund-Johansen F, Dunne JF, Farkas L, Haye R, Brandtzaeg P. Experimentally induced recruitment of plasmacytoid (CD123high) dendritic cells in human nasal allergy. *J Immunol* 165: 4062–68, 2000.
68. Farkas L, Kvale EO, Johansen FE, Jahnsen FL, Lund-Johansen F. Plasmacytoid dendritic cells activate allergen-specific TH2 memory cells: Modulation by CpG oligodeoxynucleotides. *J Allergy Clin Immunol* 114: 436–43, 2004.
69. Lapperre TS, Willems LN, Timens W, Rabe KF, Hiemstra PS, Postma DS, Sterk PJ. Small airways dysfunction and neutrophilic inflammation in bronchial biopsies and BAL in COPD. *Chest* 131: 53–59, 2007.
70. Hogg JC, Chu F, Utokaparch S, Woods R, Elliott WM, Buzatu L, Cherniack RM, Rogers RM, Sciurba FC, Coxson HO *et al.* The nature of small-airway obstruction in chronic obstructive pulmonary disease. *N Engl J Med* 350: 2645–53, 2004.
71. Gosman MM, Willemse BW, Jansen DF, Lapperre TS, van Schadewijk A, Hiemstra PS, Postma DS, Timens W, Kerstjens HA. Increased number of B-cells in bronchial biopsies in COPD. *Eur Respir J* 27: 60–64, 2006.
72. Curtis JL, Freeman CM, Hogg JC. The immunopathogenesis of chronic obstructive pulmonary disease: Insights from recent research. *Proc Am Thorac Soc* 4: 512–21, 2007.
73. Hoogsteden HC, Verhoeven GT, Lambrecht BN, Prins JB. Airway inflammation in asthma and chronic obstructive pulmonary disease with special emphasis on the antigen-presenting dendritic cell: Influence of treatment with fluticasone propionate. *Clin Exp Allergy* 29(Suppl 2): 116–24, 1999.
74. Verhoeven GT, Hegmans JP, Mulder PG, Bogaard JM, Hoogsteden HC, Prins JB. Effects of fluticasone propionate in COPD patients with bronchial hyperresponsiveness. *Thorax* 57: 694–700, 2002.
75. Rogers AV, Adelroth E, Hattotuwa K, Dewar A, Jeffery PK. Bronchial mucosal dendritic cells in smokers and ex-smokers with COPD: An electron microscopic study. *Thorax* 63: 108–14, 2008.
76. Soler P, Moreau F, Basset F, Hance AJ. Cigarette smoking-induced changes in the number and differentiated state of pulmonary dendritic cells/Langerhans cells. *Am Rev Respir Dis* 139: 1112, 1989.
77. Demedts IK, Bracke KR, Van Pottelberge G, Testelmans D, Verleden GM, Vermassen FE, Joos GF, Brusselle GG. Accumulation of dendritic cells and increased CCL20 levels in the airways of patients with chronic obstructive pulmonary disease. *Am J Respir Crit Care Med* 175: 998–1005, 2007.
78. Bratke K, Klug M, Bier A, Julius P, Kuepper M, Virchow JC, Lommatzsch M. Function-associated surface molecules on airway dendritic cells in cigarette smokers. *Am J Respir Cell Mol Biol*, 38: 655–60 2008.

79. Casolaro MA, Bernaudin J-F, Saltini C, Ferrans VJ, Crystal RG. Accumulation of Langerhans' cells on the epithelial surface of the lower respiratory tract in normal subjects in association with cigarette smoking. *Am Rev Respir Dis* 137: 406–11, 1988.
80. D'Hulst AI, Vermaelen KY, Brusselle GG, Joos GF, Pauwels RA. Time course of cigarette smoke-induced pulmonary inflammation in mice. *Eur Respir J* 26: 204–13, 2005.
81. Robbins CS, Dawe DE, Goncharova SI, Pouladi MA, Drannik AG, Swirski FK, Cox G, Stampfli MR. Cigarette smoke decreases pulmonary dendritic cells and impacts antiviral immune responsiveness. *Am J Respir Cell Mol Biol* 30: 202–11, 2004.
82. Bracke KR, D'Hulst A I, Maes T, Moerloose KB, Demedts IK, Lebecque S, Joos GF, Brusselle GG. Cigarette smoke-induced pulmonary inflammation and emphysema are attenuated in CCR6-deficient mice. *J Immunol* 177: 4350–59, 2006.
83. Arpinati M, Chirumbolo G, Urbini B, Martelli V, Stanzani M, Falcioni S, Bonifazi F, Bandini G, Tura S, Baccarani M *et al.* Use of anti-BDCA-2 antibody for detection of dendritic cells type-2 (DC2) in allogeneic hematopoietic stem cell transplantation. *Bone Marrow Transplant* 29: 887–91, 2002.
84. Aicher A, Heeschen C, Mohaupt M, Cooke JP, Zeiher AM, Dimmeler S. Nicotine strongly activates dendritic cell-mediated adaptive immunity: Potential role for progression of atherosclerotic lesions. *Circulation* 107: 604–11, 2003.
85. Nouri-Shirazi M, Guinet E. Evidence for the immunosuppressive role of nicotine on human dendritic cell functions. *Immunology* 109: 365–73, 2003.
86. Nouri-Shirazi M, Guinet E. A possible mechanism linking cigarette smoke to higher incidence of respiratory infection and asthma. *Immunol Lett* 103: 167–76, 2006.
87. Bracke K, Cataldo D, Maes T, Gueders M, Noel A, Foidart JM, Brusselle G, Pauwels RA. Matrix metalloproteinase-12 and cathepsin D expression in pulmonary macrophages and dendritic cells of cigarette smoke-exposed mice. *Int Arch Allergy Immunol* 138: 169–79, 2005.
88. Hammad H, Lambrecht BN. Recent progress in the biology of airway dendritic cells and implications for understanding the regulation of asthmatic inflammation. *J Allergy Clin Immunol* 118: 331–36, 2006.
89. Ettmayer P, Mayer P, Kalthoff F, Neruda W, Harrer N, Hartmann G, Epstein MM, Brinkmann V, Heusser C, Woisetschlager M. A novel low molecular weight inhibitor of dendritic cells and B cells blocks allergic inflammation. *Am J Respir Crit Care Med* 173: 599–606, 2006.
90. Matsubara S, Koya T, Takeda K, Joetham A, Miyahara N, Pine P, Masuda ES, Swasey CH, Gelfand EW. Syk activation in dendritic cells is essential for airway hyperresponsiveness and inflammation. *Am J Respir Cell Mol Biol* 34: 426–33, 2006.

11

Macrophages

CHAPTER

Galen B. Toews

Division of Pulmonary and Critical Care Medicine, University of Michigan Health System, Ann Arbor, MI, USA

Macrophages are a family of mononuclear leukocytes that are widely distributed throughout most tissues. They vary considerably in phenotype depending on the local microenvironment [1]. Macrophages are involved in the scavenging of dying cells, pathogens, and molecules through phagocytosis and endocytosis [2]. Macrophages secrete more than 100 substances. These secreted molecules

- induce cell movement;
- induce cell growth;
- induce cell death;
- influence cell differentiation;
- modify connective tissue structures;
- regulate blood vessel growth.

Macrophages within tissues make a vital contribution to immune and inflammatory responses [2]. Macrophages are crucial for

- early recognition of microbes, particulates, and immunogens;
- the initiation and regulation of inflammatory responses and adaptive immunity;
- the ingestion and killing of invading microbes.

ORIGIN AND DISTRIBUTION

The mononuclear phagocyte system (MPS) is composed of strikingly diverse differentiated cell types. Two subsets of murine monocytes arise from a common hematopoeitic progenitor, the macrophage, and dendritic cell precursor (MDP, or monoblast) [3]. "Inflammatory" monocytes express the cell surface protein Ly6c (Gr-1^+), the chemokine receptor CCR2 and the adhesion molecule L-selectin and are selectively recruited to inflamed tissues and lymph nodes. The second subset of monocytes ("resting") are found in resting and inflamed tissues. This subset is defined by high expression of the chemokine receptor CX_3CR and LFA-1 and by the lack of expression of Gr-1, CCR2, and L-selectin. Two monocyte subsets have also been identified in humans; $CD14^+$, $CD16^-$ monocytes resemble murine inflammatory monocytes and $CD14^{low}$ and $CD16^+$ monocytes resemble mouse resident monocytes [3]. Interleukin-3 (IL-3), granulocyte-macrophage colony stimulating factor (GM-CSF), and macrophage colony stimulating factor (M-CSF) stimulate a sequence of differentiation steps important in monocyte development by binding to specific receptors on progenitor cells. Cytokines released during acute inflammatory responses also regulate macrophage differentiation. Interleukin-1 (IL-1) and tumor necrosis factor (TNF-α) enhance M-CSF and GM-CSF production. Inhibitory inflammatory cytokines involved in macrophage differentiation include macrophage inflammatory protein-1α (MIP-1α) and transforming growth factor β (TGF-β) [4].

Monocytes/macrophages require continuous reconstitution. This can be achieved by self renewal of differentiated cells, by proliferation of bone marrow-derived precursors in peripheral tissues [5] or by the continuous extravasation and differentiation of circulating mononuclear phagocyte precursors [6]. Alveolar macrophages (AMs) are derived from blood monocytes and from proliferating macrophage precursors in the interstitium of the lung. Approximately 1% of the AM population in the normal lung is proliferating at any single time [7]. AMs have a life span of months and perhaps years [8].

RECOGNITION OF MICROBES/MICROBIAL PRODUCTS

Macrophages are a major cellular component of the innate immune system [9]. The molecular and cellular processes of the innate immune response defend the host in the first minutes or first hours after exposure to microbes. The recognition of microbes is problematic because of their molecular heterogeneity and their high mutation rates. The innate immune system uses a relatively small number of proteins encoded in the germ line to recognize a vast variety of molecular structures associated with microbes. The receptors recognize a few, highly conserved structures present in large groups of microorganisms. The receptors recognize molecular patterns rather than particular structures and accordingly have been termed pattern-recognition receptors (PRRs). The pathogen-associated molecular patterns (PAMPs) recognized by PRR are chemically quite distinct but share certain features. PAMPs are: (a) produced only by microbes and not by eukaryote hosts, (b) essential for survival or pathogenecity of the microbe, and (c) invariant structures shared by classes of pathogens. Characteristic PAMPs include lipopolysaccharides (LPSs) and teichoic acids, shared by gram-negative and gram-positive bacteria, respectively; mannans, conserved components of yeast cell walls; and the unmethylated CpG motif characteristic of bacterial but not mammalian DNA. PRRs expressed on macrophages include the mannose receptor, DEC205, CD14, scavenger receptors, integrins, and toll-like receptors (TLRs). Recognition of pathogens results in the activation of various types of innate immune receptors [10–15].

TLRs comprise a family of PRRs that recognize a variety of conserved PAMPs displayed by microbes. Ten TLRs have been described (an eleventh is functional in mice but not in man [16–22]). The principle role of TLRs appears to be induction of immune/inflammatory responses to a broad range of pathogens (viruses, helminths, bacteria, protozoa, and fungi). TLR4 is crucial for effective responses to gram-negative LPS [23]. LPS delivery to TLR4 requires the accessory proteins LPS-binding protein (LBP), CD14 and MD-2. Integrins such as CD11b/CD18 may facilitate responses to LPS [13, 24]. TLR4 also facilitates responses to *Streptococcus pneumoniae* pneumolysin and proteins of respiratory syncytial viruses [25, 26]. TLR2 mediates the responses to lypoproteins and lipoteichoic acids (LTA) from gram-positive bacteria (*S. pneumoniae*) and Mycobacteria [27, 28]. TLR2 plays a role in responses to *Borrelia burgdorferi*, *Aspergillus fumigatus*, and *Mycoplasma*. TLR2 responses are broadened by hetero-dimerization; TLR1/2 heterodimers respond to a different panel than TLR2/6 heterodimers. TLR2 and TLR4 respond to helminths.

TLR9 mediates the responses through recognition of CpG motifs in microbial DNA. TLR9 is activated by bacteria, Herpes viruses [29, 30], and *Aspergillus* [31]. TLR3 responds to double stranded viral RNA and TLR7 and TLR8 mediate responses to single stranded RNA [32, 33]. TLRs also respond to endogenous host molecules, including defensins [34], reactive oxygen species (ROS) [35], protein released from dying cells [36], surfactant protein A [37], fibrinogen [38], and breakdown products of tissue matrix, such as fibronectin fragments [39] and hyaluronic acid oligosaccharides [40].

Activation of TLRs initiates signal transduction pathways that initiate and amplify inflammatory responses in the lung and modulate adaptive immune responses. Molecules induced by PAMP-PRR interactions include

- signals that generate inflammatory responses, including TNFα, IL-1, IL-6, interferon (IFN)α/β and chemokines;
- signals that function as co-stimulators of T-cell activation, B7.1 and B7.2;
- signals that regulate the differentiation of lymphocytes, including IL-4, IL-5, IL-10, IL-12, TGF-β and IFN-γ [41].

Activation of TLRs can also contribute to the induction of programmed cell death.

HOMEOSTATIC REGULATION OF AM FUNCTION

Resident AMs are continually exposed to inhaled and aspirated particulates due to their position in the airway and alveolar lumen. In spite of this exposure to particulates, resident AMs produce only low levels of cytokines and are poorly phagocytic.

A mechanism by which the lung microenvironment instructs AM innate and adaptive immune functions has recently been defined [42]. Under homeostatic conditions, AMs closely adhere to alveolar epithelial cells (AECs). This cell–cell interaction induces TGF-β dependent expression of the integrin $\alpha v\beta 6$ on AEC. Localized activation of TGF-β in the vicinity of the macrophage suppresses macrophage phagocytosis and cytokine production. This inhibition of macrophage function by the $\alpha v\beta 6$–TGF-β complex is unique to the lung.

The mechanism whereby microbial infection overcomes suppression is also explained by AEC-macrophage-TLR responses. Infectious agents trigger TLRs which lead to a rapid loss of contact with AECs which in turn induces rapid loss of expression of $\alpha v\beta 6$ on AEC. Under these conditions TGF-β is no longer activated and macrophage activation and innate immune functions are no longer suppressed. Macrophages can now be primed to secrete pro-inflammatory cytokines and to ingest and kill microbes. AMs lack a feed forward mechanism of amplification mediated in other mononuclear phagocytes by autocrine secretion of IFN-β. AMs thus require exogenous IFN-β for a robust response [43]

GENERATION OF INFLAMMATORY RESPONSES

Resident AMs can effectively ingest and kill invading microbes if the bacterial burden is low and the microbe is minimally virulent [44, 45]. However, the recruitment of polymorphonuclear neutrophils (PMN) is essential for the effective containment of most virulent encapsulated bacteria within the lung and for the eventual clearance of these virulent microbes from the host. The generation of inflammation in the lower respiratory tract is a dynamic process that involves the coordinated expression of both pro- and anti-inflammatory cytokines. After the introduction of microbes or microbial products, TLR-mediated signals result in the production of TNF-α and IL-1β. TNF-α and IL-1β stimulate the expression of adhesion molecules on vascular endothelial cells. L-selectin on neutrophils interacts with its receptor/ligand (P- and E-selectin) on endothelial cells that lead to rolling. Intercellular adhesion molecule-1 (ICAM-1) expression is also induced on the surface of the endothelium; interactions between neutrophils and ICAM lead to firm adhesion [46, 47]. CXC chemokines and CC chemokines are also rapidly produced in macrophages following microbial stimuli. CXC chemokines include IL-8, (CXCL8), GROα, (CXCL1) ENA-78, and (CXCL5) are major PMN chemoattractants (Table 11.1). Thus, immediately after the recognition of bacterial products in the alveolar environment, PMNs begin to accumulate.

Macrophages also play a major role in amplifying the inflammatory response by stimulating cytokine production by cells that do not respond directly to bacterial products. Mononuclear phagocytes, neutrophils, and endothelial cells produce CXC chemokines in response to LPS. Alternatively, airway and AECs, pulmonary fibroblasts, and pleural mesothelial cells produce IL-8 in response to specific host-derived signals, such as TNF-α or IL-1 [48–50]. The importance of IL-1 and TNF-α as key cytokines in the initiation of this augmented inflammatory response is emphasized by the fact that all nucleated cells possess a functional receptor for IL-1 and TNF-α.

TABLE 11.1 The CXC chemokine/receptor family.

Systematic name	Human ligand	Mouse ligand	Chemokine receptor(s)
CXCL1	GROα/MGSA-α	GRO/KC?	CXCR2 > CXCR1
CXCL2	GROβ/MGSA-β	GRO/KC?	CXCR2
CXCL3	GRO-γ/MGSA-γ	GRO/KC?	CXCR2
CXCL4	PF4	PF4	Unknown
CXCL5	ENA-78	LIX?	CXCR2
CXCL6	GCP-2	Ckα-3	CXCR1, CXCR2
CXCL7	NAP-2	Unknown	CXCR2
CXCL8	IL-8	Unknown	CXCR1, CXCR2
CXCL9	Mig	Mig	CXCR3
CXCL10	IP-10	IP-10	CXCR3
CXCL11	I-TAC	Unknown	CXCR3
CXCL12	SDF-1α/β	SDF-1	CXCR4
CXCL13	BLC/BCA-1	BLC/BCA-1	CXCR5
CXCL14	BRAK/bolekine	BRAK	Unknown
(CXCL15)	Unknown	Lungkine	Unknown

A systematic name in parentheses means a human homologue has not been identified.
A question mark indicates that the mouse ligand homologue listed may not correspond to the human ligand.

Macrophages are also potent sources of bioactive lipids, which are important in inflammatory responses. Leukotriene (LT) synthesis is dependent on three sequential enzymes: cytosolic phospholipase A ($cPLA_2$), 5-lipoxygenase (5-LO), and 5-LO-activating protein (FLAP). These three proteins co-localize at a single membrane site to form a macromolecular complex termed a metabolon. There is now abundant evidence documenting that the nuclear envelope is the site at which this metabolon is assembled [51].

The LT synthetic pathway is initiated when $cPLA_2$ translocates from the cytosol to the nuclear envelope following activation signals. Arachidonate is released from nuclear envelope phospholipids and is bound by FLAP, an integral nuclear envelope protein. FLAP facilitates processing by 5-LO. On activation, 5-LO also translocates from its resting locale(s) in the cytosol and/or nucleoplasm to the nuclear envelope where it catalyzes the initial steps in LT synthesis [52, 53]. LTs, thus synthesized, are capable of either entering into the nucleus or being exported out of the cell [53]. Monocytes release LTB4 and LTC4 after being exposed to non-immunologic stimuli or immunologic stimuli. AMs produce a substantial excess of LTB4 compared with LTC4 after stimulation [54]. LTB4 is a potent chemotactic factor for PMNs and weaker chemotactic factor for eosinophils. LTB4 accounts for the majority of neutrophil chemotactic activity elaborated by AM immediately following stimulation [54]. LTB4 also promotes adherence of inflammatory cells to the endothelium. Leukotrienes also play a permissive role in inflammation by promoting the synthesis by macrophages of TNF-α, IL-8, (CXCL8), GROα, (CXCL1) ENA-78, (CXCL5), and IL-6.

An important implication of the metabolon concept is that the site of macromolecular assembly has evolved in a manner to best serve the needs of the cell. The observation that the LT metabolon is located within the nuclear envelope suggests that the autocrine actions of bioactive lipid mediators, including those potentially mediated within the nucleus, may be of more importance than those paracrine actions that have been classically recognized. 5-LO metabolites are important modulators of mitogenesis, apoptosis, and the activation of various transcription factors. A soluble nuclear receptor for LTB4 provides additional support for such interactions; interestingly, this receptor is a member of a super-family of transcription factors and its ligation induces gene transcription. Alternatively, reactive oxygen species, which are a byproduct of arachidonate 5-lipoxygenation could exert nuclear actions by activating transcription factors or otherwise modifying nuclear constituents.

Monocytes exit the blood in inflamed tissues in response to specific chemotaxins. Monocyte recruitment is

TABLE 11.2 The CC chemokine/receptor family.

Systematic name	Human ligand	Mouse ligand	Chemokine receptor(s)
CCL1	I-309	TCA-3/P500	CCR8
CCL2	MCP-1/MCAF	JE?	CCR2
CCL3	MIP-1α/LD78α	MIP-1α	CCR1/CCR5
CCL4	MIP-1β	MIP-1β	CCR5
CCL5	RANTES	RANTES	CCR1/CCR3/CCR5
(CCL6)	Unknown	C10/MRP-1	Unknown
CCL7	MCP-3	MARC?	CCR1/CCR2/CCR3
CCL8	MCP-2	MCP-2?	CCR3
(CCL9/10)	Unknown	MRP-2/CCF18/MIP-1γ	Unknown
CCL11	Eotaxin	Eotaxin	CCR3
(CCL12)	Unknown	MCP-5	CCR2
CCL13	MCP-4	Unknown	CCR2/CCR3
CCL14	HCC-1	Unknown	CCR1
CCL15	HCC-2/Lkn-1/MIP-1δ	Unknown	CCR1/CCR3
CCL16	HCC-4/LEC	LCC-1	CCR1
CCL17	TARC	TARC	CCR4
CCL18	DC-CK1/PARC AMAC-1	Unknown	Unknown
CCL19	MIP-3β/ELC/exodus-3	MIP-3β/ELC/exodus-3	CCR7
CCL20	MIP-3α/LARC/exodus-1	MIP-3α/LARC/exodus-1	CCR6
CCL21	6Ckine/SLC/exocus-2	6Ckine/SLC/exocus-2/TCA-4	CCR7
CCL22	MDC/STCP-1	ABCD-1	CCR4
CCL23	MPIF-1	Unknown	CCR1
CCL23	MPIF-2/Eotaxin-2	Unknown	CCR3
CCL25	TECK	TECK	CCR9
CCL26	Eotaxin	Unknown	CCR3
CCL27	CTACK/ILC	ALP/CTACK/ILC ESkine	CCR10

A systematic name in parentheses means a human homologue has not been identified. A question mark indicates that the mouse ligand homologue listed may not correspond to the human ligand.

critically dependent on CC chemokines [29]. Twenty-seven different CC chemokine ligands (CCL) have been described [55]. Chemokines that act mainly on monocytes are located on a cluster on human chromosome 17q11.2. Important monocyte chemoattractant CC chemokines include MCP-1 (CCL2), MIP-1α (CCL3), MIP-1β (CCL4), and RANTES (CCL5) (Table 11.2). CC chemokines are produced by monocytes, alveolar macrophages, lymphocytes, neutrophils, epithelial cells, fibroblasts, smooth muscle cells, and endothelial cells. These cells produce chemokines in response to a variety of factors including cigarette smoke, viruses, bacterial products, IL-1, TNF, C5a, LTB4, and INFs [56–66]. Bronchial epithelial cells release monocyte chemotaxins in response to cigarette smoke [63]. TNF-α and IL-1 are among the most potent stimuli for epithelial cell cytokine production. Macrophage stimulation of epithelial cells by TNF or IL-1 likely enhances and perpetuates the initial inflammatory response. Increased epithelial expression of MCP-1 has been observed by immunohistochemistry and biopsy specimens from patients with atopic asthma, and increased levels of this chemokine are also seen in bronchoalveolar lavage (BAL) fluid of allergic asthmatics when compared with normal subjects [65, 66].

The biological effects of chemokines are mediated by seven transmembrane domain receptors that are a subset of the G-protein-coupled receptor super-family. Sixteen receptors have been identified. Redundancy and binding promiscuity exists between ligands and receptors (Tables 11.1 and 11.2). A single chemokine may bind to several receptors and a single chemokine receptor can transduce signals for several chemokines [55].

Monocytes also respond to formyl peptides, C5a, and elastin fragments. In a murine model of cigarette smoke-induced emphysema, CC and CXC chemokines are undetectable but chemotactic extracellular matrix fragments, particularly elastin fragments, are present [67–69].

T-CELL INDEPENDENT MACROPHAGE ACTIVATION

The early nonantigen-specific activation of macrophages is one of the first events in the innate immune response and is often very effective in eliminating microbes. Innate immune macrophage activation fills a gap between microbial entry into the host and the development of antigen-specific immunity. Innate immune macrophage activation is a cytokine mediated macrophage-NK cell interaction [70]. Macrophages release IL-12 and TNF-α following microbial recognition; these cytokines induce NK cell IFN-γ which primes macrophages for microbicidal activity. IL-12 is regulated by both positive and negative feedback mechanisms. IFN-γ activated macrophages produce much higher levels of IL-12. IL-10, a product of macrophages and other cell types, is a potent inhibitor of IL-12 production [71]. Expression of IL-10 is delayed compared with the expression IL-12 *in vivo*. IL-10 is a critical component of the host's natural defense against excessive production of IL-12 and its pathological consequences.

ROLE OF MACROPHAGES IN INITIATION OF IMMUNE RESPONSES

The activation of T-lymphocytes is a complex biological function that requires the participation of an antigen-presenting cell (APC). Most cell types cannot perform these functions; "professional APCs" (macrophages, dendritic cells (DCs), and B-cells) are required. Antigen presentation involves the display of an antigenic epitope in association with a major histocompatibility complex (MHC) molecule. Three additional molecular interactions are crucial to the interaction between an APC and a T-lymphocyte:

1. adhesion molecules that promote the physical interaction between APC and T-cells;
2. co-stimulatory molecules which are membrane bound growth/differentiation molecules that produce T-cell activation [72];
3. soluble molecules such as TNF-α and IL-1.

AMs are ineffective in presenting antigen to T-cells [73–75]. AMs are less effective than monocytes in inducing proliferation of blood T-lymphocytes to soluble recall antigens. AMs can, however, restimulate recently activated T-cells effectively. AMs fail to activate $CD4^+$ T-cells because of defective expression of B7 co-stimulatory cell surface molecules. AMs activated with IFN-γ fail to express B7-1 or B7-2 antigens [76]. Resident pulmonary AMs also actively suppress T-cell proliferation induced by antigen [73].

The mechanisms whereby particulate antigens and microbial agents induce T-cell responses are being defined. DCs are present in the airways and in the interstitium of the lower respiratory tract [77]. DCs are clearly crucial for activating naive T-cells for proliferation and clonal expansion. Monocytes are also stimulatory to T-cell activation. Thus, inflammatory stimuli that recruit fresh monocytes to the lung might theoretically dilute the resident AM population with recruited monocytes and convert a normally immunosuppressive tissue milieu into one that is supportive of T-cell activation.

INNATE IMMUNE CONTROL OF ADAPTIVE IMMUNE RESPONSES

Naive T-lymphocytes can differentiate along different pathways to become distinct effector cells. The tissue microenvironment in which the specific immune response is generated crucially regulates this differentiation process via the secretion of specific cytokine signals (Fig. 11.1). IL-12 produced by macrophages during the early innate immune response and IFN-γ induced by IL-12 create an environment in which antigen-specific $CD4^+$ and $CD8^+$ T-cells are preferentially induced to differentiate into T1 cells that produce even higher levels of IFN-γ [78]. IL-4 is crucial to the development of T2 responses during the priming of naive T-cells. The crucial cellular source of early IL-4 production is uncertain. Basophils, mast cells, $\gamma\delta$ cells, T2 lymphocytes, and an NK 1.1^+ $CD4^+CD8^-$ T-lymphocyte have all been reported to produce IL-4 [79].

The type of APC that presents the antigen may also be crucial. DCs preferentially activate T2 cells in certain circumstances [80]. The mechanisms by which DCs favor the expression of T2 cells remains uncertain but may be related to the ability of DCs to secrete IL-1, a co-stimulator for T2 cells but not T1 cells, and to the absence of IFN-γ production by DCs which would inhibit the development of T2 cells. Differential expression of co-stimulatory molecules may also be crucial to this polarization process. AM fails to express B7.1 or B7.2 even when stimulated with IFN-γ. B7.2 is constitutively expressed on DCs and macrophages, whereas

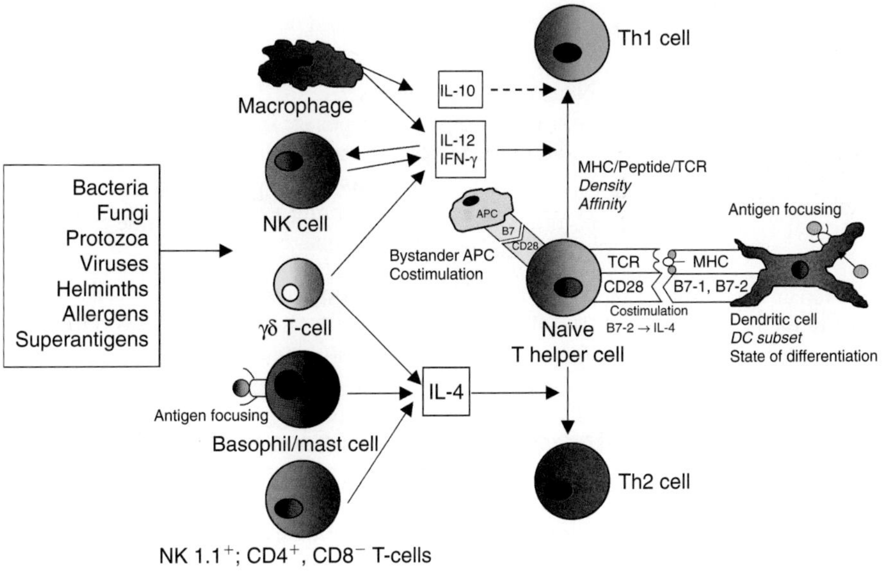

FIG. 11.1 Innate immune control of adaptive immune response. The tissue microenvironment regulates adaptive immune responses. DC migration and functional differentiation state are crucial determinants of immune priming. IgE on the surface of DC may focus uptake and processing of other antigens. Density and affinity of antigens regulates differential signals following T-cell receptor engagement. Co-stimulatory molecules influence T-cell differentiation. Innate immune cell cytokines are the major regulators of T-cell differentiation. (Solid arrows indicate stimulators; broken arrows indicate inhibitor.)

B7.1 is not. The outcome of B7.1 and B7.2 co-stimulation is different. B7.2 co-stimulates the production of IL-4 as well as IL-2 and IFN-γ. Thus, after B7.2 co-stimulation, an initial source of IL-4 is available. However, B7.2 co-stimulation provides only a moderate signal for T2 cell differentiation; additional signals are almost surely required to achieve high levels of IL-4 production [81]. B7 co-stimulatory signals can be delivered by bystander APC *in vitro* with the same efficiency as the APC engaging the TCR. Accordingly, T-cell activation *in vivo* may occur with MHC-TCR engagement being provided by one APC, whereas co-stimulation is delivered by a second bystander APC-type [82].

The tissue microenvironment in which the immune responses are generated is crucial in determining the type and intensity of T-lymphocyte responses. Cytokines secreted in the tissues, where the antigen is deposited, are also crucial for the maintenance and regulation of immune responses.

ROLE OF MACROPHAGE-DERIVED GROWTH FACTORS IN TISSUE REMODELING AND REPAIR

Fibrotic changes occur in airways and the lower respiratory tract following certain injuries. Macrophages produce numerous growth factors for fibroblasts including platelet-derived growth factor (PDGF), TGF-α and TGF-β, and insulin-like growth factor [83–85].

- PDGF induces fibroblast proliferation and collagen production.
- The wound healing effects of PDGF are macrophage dependent in most models [85].
- TGF-α stimulates the closure of wounds induced in cultures of type II AEC *in vitro*.
- TGF-α likely plays a role in epithelial cell repair of the lungs following injury. Macrophages also produce TGF-β, which has important effects on the turnover of matrix proteins and the proliferation of fibroblasts.
- TGF-β activates genes that favor the production of matrix proteins. TGF-β also down-regulates production of matrix metalloproteinases derived from PMNs and macrophages that digest matrix in the interstitium and alveolar spaces. In aggregate, these effects shift the balance in favor of matrix accumulation.

GM-CSF has a protective effect on repair processes following bleomycin. This protective effect is mediated, in part, by GM-CSF-induced production of PGE_2 by pulmonary macrophages [86]. PGE_2 has crucial down-regulatory effects on fibroblasts, including decreasing fibroblast proliferation and decreasing matrix produced by fibroblasts.

Macrophages may also play critical roles in regulating angiogenesis, which is a central biological event in repair and remodeling. CXC chemokines such as IL-8 or ENA-78 that contain the sequence GLU-LEU-ARG (ELR motif) are potent angiogenic factors. In contrast, CXC chemokines that lack the ELR motif (platelet factor 4, IFN-γ inducible protein 10) behave as potent angiostatic factors. Non-ELR-containing chemokines can inhibit angiogenic activity of both ELR-CXC chemokines and structurally unrelated macrophage-derived angiogenic factor bFGF [87].

Macrophages and chronic obstructive pulmonary disease

A marked increase in the number of macrophages and neutrophils in the airways of both humans and experimental animals is the most consistent, early effect of exposure to cigarette smoke [88]. Histologic studies of bronchial biopsies and lung parenchyma obtained from cigarette smokers demonstrate a predominance of macrophages and $CD8^+$ T-cells at sites of parenchymal destruction [89, 90]. The paucity of neutrophils in lung parenchyma and airway biopsies suggests neutrophils traffic rapidly from the blood into the airway lumen. The inflammatory process, once initiated, persists in ex-smokers [90, 91]. Surprisingly, the inflammatory responses persist in end-stage emphysema. Increased number of macrophages, CD8 and CD4 T–lymphocytes, and neutrophils are noted in lungs from patients undergoing lung volume reduction surgery.

Macrophages are likely to play an important role in initiating the neutrophilic inflammatory response in cigarette smokers. Macrophages may be activated by cigarette smoke and other inhaled particulates. Elevated levels of the CXC chemokines, IL-8, (CXCL8), GROα, and (CXCL1) are noted in cigarette smokers. IL-8 and (CXCL8) levels in sputum correlate with the magnitude of neutrophilic inflammation and with percent predicted FEV1 [92]. LTB4 is increased in the sputum of patients with chronic obstructive pulmonary disease (COPD); this potent neutrophilic chemoattractant likely participates in the generation of the neutrophilic inflammatory response.

The inflammatory response in COPD has several distinguishing features:

- Resident AMs are activated upon exposure to cigarette smoke.
- Neutrophils are rapidly recruited almost immediately after cigarette smoke exposure in response to macrophage and epithelial cell-derived chemokines and leukotrienes.
- Macrophages and $CD8^+$ and $CD4^+$ T-lymphocytes accumulate within days to weeks and continue to accumulate with time.
- The abnormal accumulation of inflammatory cells persists throughout the disease process, even when the inciting agent, cigarette smoke, is removed.

Macrophage proteinases and COPD

Early changes of emphysema include subtle disruption of elastic fibers, bronchiolar and alveolar distortion, and the appearance of fenestrae. Destruction of the elastic framework leads to loss of the intra-alveolar septae and macroscopic appearance of spaces of more than 1mm in diameter. This destructive process is accompanied by an increase in the mass of collagen, suggesting that active alveolar wall fibrosis occurs in the tissues which remain in otherwise emphysematous lungs [93]. The dominant, working hypothesis to explain the pathogenesis of emphysema has postulated an imbalance between proteolytic enzymes and proteinase inhibitors in the lung, favoring an excess of proteinases, particularly elastases. This hypothesis postulates that cigarette smoke, macrophages, chemoattractants, neutrophils, elastases, and proteinase inhibitors interact with lung connective tissue, primarily elastin, to cause repeated destruction and synthesis of pulmonary matrix.

Pulmonary macrophages may play an important role in this proteolytic process by releasing neutrophil chemotactic factors, which recruit neutrophils to the respiratory tract. Neutrophil elastase is believed to play an important role in COPD. Neutrophil elastase is one of the most potent elastases in the lung. Instillation of neutrophil elastase and proteinase 3 causes emphysema in animals [94–96]. Neutrophils are recruited into the lungs of animals and generate detectable elastin fragments following exposure to cigarette smoke. Neutrophil elastase is also a potent mucus secretagague.

Macrophages also secrete potent proteinases. Human AMs produce the cysteine (thiol) proteinases, cathepsins B, H, L, and S. Cathepsins L and S have relatively indiscriminate substrate specificities that include elastin and other matrix components. Cysteine proteinases are involved in lung destruction in IL-13 and IFN-γ transgenic mice [97, 98].

Pulmonary macrophages also produce matrix metalloproteinases (MMPs). Studies of human emphysematous lung tissue have demonstrated the presence of several MMPs. A correlation between MMP-1 and MMP-9 and emphysema was noted when smokers with emphysema were compared with smokers without emphysema [99]. Studies using transgenic and gene-targeted mice also lend support for the role of MMPs in emphysema. Transgenic mice over-expressing human MMP-1 in the lung develop airspace enlargement [100]. Studies of MMP-12 knockout mice provide specific loss-of-function data that MMP-12 is importantly involved in the development of emphysema in response to cigarette smoke. Exposure of MMP-12 +/+ mice to long-term cigarette smoke led to inflammatory cell recruitment followed by alveolar space enlargement similar to that seen in humans with emphysema. Mice deficient in MMP-12 (MMP-12 −/−) were protected from development of emphysema despite long-term smoke exposure [101]. Interestingly, MMP-12 −/− mice failed to recruit monocytes into the lung in response to cigarette smoke. MMP-12 −/− mice could egress from the pulmonary vasculature in response to MCP-1 instillation. However, even in the presence of MMP-12 −/− macrophages, the lungs of mice exposed to cigarette smoke had no changes in the mean linear intercept in alveolar duct areas when compared to mice not exposed to smoke.

These findings suggest that MMP-12 is expressed by resident AMs after exposure to cigarette smoke. MMP-12 generates monocyte chemotaxins, likely fragments of elastin, which are responsible for the monocyte recruitment in response to smoke. Macrophage-mediated lung destruction after exposure to cigarette smoke is, at least in part, directly related to the presence of MMP-12, which is most likely required for direct degradation of lung tissue. The generation of chemotactic elastin fragments provides a positive feedback loop perpetuating macrophage accumulation in lung destruction.

Cigarette smoke induces a distinct pattern of AM gene expression not seen in non-smokers or patients with asthma [102]. AM from smokers strongly express MMP-12, but most of the 72 genes upregulated in smokers were not identified in two transgenic murine models of emphysema.

MACROPHAGES AND ADAPTIVE IMMUNITY IN COPD

Cigarette smoking stimulates humoral and cellular components of the adaptive immune response. This response may provide exquisite specific memory for previous exposures to foreign materials. The histologic hallmark of an adaptive immune response is the presence of lymphoid follicules with germinal centers. Bronchial-alveolar lymphoid tissue (BALT) collections are found in 5% of smokers with normal lung function (GOLD0) and in those with mild (GOLD1) and moderate (GOLD2) severity. These collections of BALT increase sharply in severe (GOLD3) and very severe (GOLD4) COPD perhaps in response to colonization and infection of the lower airways [90].

MACROPHAGES AND APOPTOSIS IN COPD

There is growing evidence for a role for apoptosis in COPD [103]. AMs show a markedly reduced capacity to ingest apoptotic cells relative to their avid ingestion of inert particles [104]. This capacity is further reduced in COPD with attendant increased accumulation of apoptotic cells. Apoptotic cell recognition typically induces a unique anti-inflammatory state; defective clearance might be a factor encouraging lung inflammation [105].

MACROPHAGES AND INITIATION OF ANTIGEN-SPECIFIC T2 IMMUNE RESPONSES IN ASTHMA

Resident pulmonary AMs actively suppress T-cell proliferation induced by antigen or polyclonal stimuli [73]. Changes occur within the local inductive milieu of the lung in patients with asthma. AM suppression is reduced after exposure to allergens [106–108].

The tissue microenvironment is a crucial regulator of specific immune response generation (Fig. 11.1). The presence of IgE on APCs likely promotes the uptake and the processing of allergens and their eventual presentation to naïve T-cells. DCs express both $Fc_{\varepsilon}R$ I and $Fc_{\varepsilon}R$ II. These two receptors could function to capture allergen bound to allergen-specific IgE and thus focus the immune response through facilitated antigen presentation [109].

Antigens also deliver signals via quantitative variation in ligand density on APC. Peptide/MHC class II complexes that interact strongly with the TCR favor T1 responses, whereas weak interactions result in the priming of T2 responses. The overall binding affinity can be varied by modifying the peptide, which results in different signals. The mechanisms by which signals delivered via the TCR control differentiation is uncertain; differential TCR aggregation may result in differential intracellular signals that favor distinct cytokine gene expression or certain MHC–TCR interactions may favor differential co-receptor expression [110]. As noted above, co-stimulatory molecules may direct the polarization of T-cells into T1 or T2 cells; B7.2 provides only a moderate signal for T2 cell differentiation; and co-stimulatory signals may be delivered either by the APC that presents the antigen or by the bystander APC. Thus, macrophages may serve as bystander APC and influence DC-induced T-cell proliferation [111]. Soluble cytokines produced by cells of the innate immune response are likely the major regulators of T-cell differentiation (see "Innate Control of Adaptive Immune Responses" section).

MACROPHAGES IN THE EFFECTOR PHASES OF IMMUNE RESPONSES IN ASTHMA

Macrophages are likely the sources of cell-specific chemoattractants in patients with asthma. RANTES, (CCL5), MIP-1α, and (CCL3) are chemotactic for macrophages, eosinophils, and basophils. MCP-1 and (CCL1) are chemotactic for macrophages and basophils. IL-8 and (CXCL8) are chemotactic for neutrophils, basophils, and to lesser extent eosinophils. Additionally, macrophage-released early cytokines likely induce epithelial cells and fibroblasts to release chemoattractants and growth factors (GM-CSF, IL-8, (CXCL8), MCP-1, (CCL1), MCP-3, (CCL7), MCP4, (CCL13), MIP-1α, (CCL3), RANTES, and (CCL9)) in patients with asthma [112–115].

Leukotrienes are perhaps the best studied macrophage-produced mediators involved in the pathogenesis of asthma. The role of macrophage-derived leukotrienes remains uncertain. Cysteinyl leukotrienes (LTC4, D4, and E4) are minor products of AMs when compared to LTB4 production [51, 53, 113]. This metabolic profile argues against a role for AMs in LT overproduction in the lung. Alternatively, several features favor a role for AMs in LT production. The cells are located on the airway surface, express plasma membrane IgE receptors, and secrete substantially more LTC4 than do PMN. Cysteinyl leukotrienes are known to promote mucus hypersecretion, airway inflammation, and marked, prolonged contraction of smooth muscles in airways.

MACROPHAGES AND AIRWAY RESPONSES TO VIRAL INFECTION

Host innate immune responses to respiratory viral infections may play a crucial role in the pathogenesis of asthma. The site of initial viral replication is usually respiratory epithelial cells and macrophages. Host defense against virus depends on the targeted death of infected cells and the subsequent clearance of the cellular corpses by macrophages. Macrophages must avoid viral-induced death for this process to be effective. CCL5–CCR5 interactions are required to prevent apoptosis of virus infected macrophages *in vivo* and *in vitro*. Infected macrophages must resist cell death to efficiently clear virus infected apoptotic cells from the lung. If this clearance process is disrupted, pathogens are not cleared and residual apoptotic cells including activated macrophages can cause further damage [116].

References

1. Gordon S, Hughes DA. Macrophages and their origins: Heterogeneity in relation to tissue microenvironment. In: Lipscomb M, Russell S (eds). *Lung Macrophages and Dendritic Cells*, pp. 3–31. New York: Marcel Dekker, 1997.
2. Gordon S. Pattern recognition receptors: Doubling up for the innate immune response. *Cell* 111(7): 927–30, 2002.
3. Geissmann F, Jung Steffen Littman DR. Blood monocytes consist of two principal subsets with distinct migratory properties. *Immunity* 19: 71–82, 2003.
4. Sieff CA. Hematopoietic growth factors. *J Clin Invest* 79: 1549–57, 1987.
5. Takahashi K, Nalto M, Takeya M. Development and heterogeneity of macrophages and their related cells through their differentiation pathways. *Pathol Int* 46: 473–85, 1996.
6. Kennedy DW, Abkowitz JL. Mature monocytic cells enter tissues and engraft. *Proc Natl Acad Sci USA* 95: 14944–49, 1998.
7. Bitterman PB, Saltzman LE, Adelberg S *et al.* Alveolar macrophage replication: One mechanism for the expansion of the mononuclear phagocyte population in the chronically inflamed lung. *J Clin Invest* 74: 460–69, 1984.
8. Thomas ED, Rambergh RE, Sale GE *et al.* Direct evidence for bone marrow origin of the alveolar macrophage in man. *Science* 192: 1016–18, 1976.
9. Medzhitov R, Janeway CA. Jr. Innate immunity. *N Eng J Med* 343: 338–44, 1992.
10. Stahl PD. The mannose receptor and other macrophage lectins. *Curr Opin Immunol* 4: 49–52, 1992.
11. Jiang W, Swiggard WJ, Heufler C *et al.* The receptor DEC-2O5 expressed by dendritic cells and thymic epithelial cells is involved in antigen processing. *Nature* 375: 151–53, 1995.
12. Ulevitch RJ. Tobias PS. Receptor-dependent mechanisms of cell stimulation by bacterial endotoxin. *Annu Rev Immunol* 13: 437–57, 1995.
13. Wright SD, Ramos RA, Tobias PS *et al.* CD14, a receptor for complexes of lipopolysaccharide (LPS) and LPS binding Protein. *Science* 249: 1431–33, 1990.
14. Pugin J, Heumann D, Tomasz A *et al.* CD14 is a pattern recognition receptor. *Immunity* 1: 509–16, 1994.
15. Hampton RY, Golenbock DT, Penman M *et al.* Recognition and plasma clearance of endotoxin by scavenger receptors. *Nature* 352: 342–44, 1991.
16. Akira S, Takeda K. Toll-like receptor signaling. *Nat Rev Immunol* 4: 499–511, 2004.

17. Basu S, Fenton MJ. Toll-like receptors: Function and roles in lung disease. *Am J Physiol Lung Cell Mol Physiol* 286: L887–lL892, 2004.
18. Cook DN, Pisetsky DS, Schwartz DA. Toll-like receptors in the pathogenesis of human disease. *Nat Immunol* 5: 975–79, 2004.
19. Decker T. Sepsis: Avoiding its deadly toll. *J Clin Invest* 113: 1387–89, 2004.
20. Doherty TM, Arditi M. TB, or not TB: That is the question: Does TLR signaling hold the answer? *J Clin Invest* 114: 1699–703, 2004.
21. Iwasaki A, Medzhitov R. Toll-like receptor control of the adaptive immune responses. *Nat Immunol* 5: 987–95, 2004.
22. Ulevitch RJ. Therapeutics targeting the innate immune system. *Nature Rev Immunol* 4: 512–20, 2004.
23. Hirshfeld M, Ma Y, Weis JH *et al.* Cutting repurification of lipopolysaccharide eliminates signaling through both human and murine Toll-like receptor 2. *J Immunol* 166: 574–81, 2000.
24. Martin TR, Mathison JC, Tobias PS *et al.* Lipopolysaccharride binding protein enhances the responsiveness of alveolar macrophages to bacterial lipolysaccharide. Implications for cytokine production in normal and injured lungs. *J Clin Invest* 90: 2209–19, 1992.
25. Malley R, Henneke P, Morse SC *et al.* Recognition of pneumolysin by Toll-like receptor 4 confers resistance to pneumococcal infection. *Proc Natl Acad Sci USA* 100: 1966–71, 2003.
26. Kurt-Jones EA, Popova L, Kwinn L *et al.* Pattern recognition receptors TLR4 and CD14 mediate response to respiratory syncytial virus. *Nat Immunol* 1: 398–401, 2000.
27. Knapp S, Wieland CW, van't Veer C *et al.* Toll-like receptor 2 plays a role in the early inflammatory response to murine pneumococcal pneumonia but does not contribute to antibacterial defense. *J Immunol* 172: 3132–38, 2004.
28. Means TK, Jones BW, Schromm AB *et al.* Differential effects of a Toll-like receptor antagonist on *Mycobacterium tuberculosis*-induced macrophage responses. *J Immunol* 166: 4074–82, 2001.
29. Hochrein H, Schlatter B, O'Keeffe M *et al.* Herpes simplex virus type-1 induces IFN-production via Toll-like receptor 9-dependent and -independent pathways. *Proc Natl Acad Sci USA* 101: 11416–21, 2004.
30. Lund J, Sato A, Akira S *et al.* Toll-like receptor 9 mediated recognition of Herpes simplex virus-2 by plasmacytoid dendritic cells. *J Exp Med* 198: 513–20, 2003.
31. Bellocchio S, Moretti S, Perruccio K *et al.* TLRs govern neutrophil activity in aspergillosis. *J Immunol* 173: 7406–15, 2004.
32. Diebold SS, Kaisho T, Hemmi H *et al.* Innate antiviral responses by means of TLR7-mediated recognition of single-stranded RNA. *Science* 303: 1529–31, 2004.
33. Heil F, Hemmi H, Hochrein H *et al.* Species-specific recognition of single-stranded RNA via Toll-like receptor 7 and 8. *Science* 303: 1526–29, 2004.
34. Biragyn A, Ruffini PA, Leifer CA *et al.* Toll-like receptor 4-dependent activation of dendritic cells by β-defensin 2. *Science* 298: 1025–29, 2002.
35. Frantz S, Kelly RA, Bourcier T. Role of TLR-2 in the activation of nuclear factor-κB by oxidative stress in cardiac myocytes. *J Biol Chem* 276: 5197–203, 2001.
36. Park JS, Svetkauskaite D, He Q *et al.* Involvement of TLR2 and TLR4 in cellular activation by high mobility group box 1 protein (HMGB1). *J Biol Chem* 279: 7370–77, 2004.
37. Guillot L, Balloy V, McCormack FX, Golenbock DT, Chignard M, Si-Tahar M. Cutting edge: The immunostimulatory activity of the lung surfactant protein-A involves Toll-like receptor 4. *J Immunol* 168: 5989–92, 2002.
38. Smiley ST, King JA, Hancock WW. Fibrinogen stimulates macrophage chemokine secretion through Toll-like receptor 4. *J Immunol* 167: 2887–94, 2001.
39. Okamura Y, Watari M, Jerud ES *et al.* The extra domain A of fibronectin activates Toll-like receptor 4. *J Biol Chem* 276: 10229–33, 2001.
40. Termeer C, Benedix F, Sleeman J *et al.* Oligosaccharides of hyaluronan activate dendritic cells via Toll-like receptor 4. *J Exp Med* 195: 99–111, 2002.
41. Toews GB. Cytokines and pulmonary host defense against microbes. In: Nelson S, Martin TR (eds). *Cytokines in Pulmonary Disease: Infection and Inflammation*, pp. 1–17. New York: Marcel Dekker, 2000.
42. Takabayshi K, Corr M, Hayashi T *et al.* Induction of a homeostatic circuit in lung tissue by microbial compounds. *Immunity* 24: 475–87, 2006.
43. Punturieri T, Alvianani RS, Polack T *et al.* Specific engagement of TL4 or TLR3 does not lead to IFNβ mediated innate immune signal amplification and STAT-1 phosphorylation in resident murine alveolar macrophages. *J Immunol* 173: 1033–42, 2004.
44. Onofrio JM, Toews GB, Lipscomb MF *et al.* Granulocyte alveolar macrophage interactions in the pulmonary clearance of *Staphylococcus aureus*. *Am Rev Respir Dis* 136: 818–23, 1987.
45. Rehm SR, Gross GN, Pierce AK. Early bacterial clearance from murine lungs: Species-dependent phagocyte response. *J Clin Invest* 66: 194–99, 1980.
46. Albelda SM, Smith CW, Ward PA. Adhesion molecules and inflammatory injury. *FASEB J* 8: 504–12, 1994.
47. Springer TA. Traffic signals for lymphocyte re-circulation and leukocyte immigration: The multi-step paradigm. *Cell* 76: 301–14, 1994.
48. Strieter RM, Kunkel SL, Showell HJ *et al.* Endothelial cell gene expression of a neutrophil chemotactic factor by TNF, LPS, and IL-l. *Science* 243: 1467–69, 1989.
49. Strieter RM, Phan SH, Showell HJ *et al.* Monokine-induced neutrophil chemotactic factor gene expression in human fibroblasts. *J Biol Chem* 264: 10621–26, 1989.
50. Standiford TJ, Kunkel SL, Basha MA *et al.* Interleukin-8 gene expression by a pulmonary epithelial cell line: A model for cytokine networks in the lung. *J Clin Invest* 86: 1945–53, 1990.
51. Peters-Golden M, Brock TG. Intracellular compartmentalization of leukotriene biosynthesis. *Am J Respir Crit Care Med* 161: 536–40, 2000.
52. Woods J, Evans J, Ethier D *et al.* 5-lipoxygenase and 5-lipoxygenase activating protein are localized in the nuclear envelope of activated human leukocytes. *J Exp Med* 178: 1935–46, 1993.
53. Woods J, Coffey M, Brock TG *et al.* 5-lipoxygenase is located in the euchromatin of the nucleus in resting human alveolar macrophages and translocated to the nuclear envelope upon cell activation. *J Clin Invest* 95: 2035–40, 1995.
54. Martin TR, Pistorese B, Chi E *et al.* Effect of leukotriene B_4 in the human lung: Recruitment of neutrophils into the alveolar spaces without a change in protein permeability. *J Clin Invest* 84: 1609–19, 1989.
55. Zlotnik A, Yoshi O. Chemokines: A new classification system and their role in immunity. *Immunity* 12: 121–27, 2000.
56. Christensen PJ, Rolfe MW, Standiford *et al.* Characterization of the production of monocyte chemoattractant protein-1 (MCP-1) and interleukin-8 (IL-8) in an allogeneic immune response. *J Immunol* 151: 1205–13, 1993.
57. Liebler JM, Kunkel SL, Burdick MD *et al.* The production of IL-8 and MCP-1 by peripheral blood monocytes: Disparate responses to PHA and LPS. *J Immunol* 152: 241–49, 1994.
58. Standiford TJ, Kunkel SL, Lieber JM. The gene expression of MIP-1α from human blood monocytes and alveolar macrophages is inhibited by IL-4. *Am J Respir Cell Mol Biol* 9: 192–98, 1993.
59. Miller MD, Hata S, deWaal Malefyt R, Krangel MS. A novel polypeptide secreted by activated human T lymphocytes. *J Immunol* 143: 2907–16, 1989.
60. Kasama T, Strieter RM, Standiford TJ *et al.* Expression and regulation of human neutrophil-derived macrophages inflammatory protein-1α. *J Exp Med* 178: 63–72, 1993.
61. Paine R III, Rolfe MW, Standiford TJ *et al.* MCP-1 expression by rat type II alveolar epithelial cells in primary culture. *J Immunol* 150: 4561–70, 1993.
62. Standiford TJ, Kunkel SL, Phan SH *et al.* Alveolar macrophage-derived cytokines induce monocyte chemoattractant protein-1 expression from human pulmonary type II like epithelial cells. *J Biol Chem* 266: 9912–18, 1991.

63. Koyama S, Rennard SI, Leikauf GD *et al.* Bronchial epithelial cells release monocyte chemotactic activity in response to smoke and endotoxin. *J Immunol* 147: 972–79, 1991.
64. Rolfe MW, Kunkel SL, Standiford TJ *et al.* Expression and regulation of human pulmonary fibroblast-derived monocyte chemotactic peptide (MCP-1). *Am J Physiol* 263: L536–45, 1992.
65. Sousa AR, Lane SJ, Nakhosteen JA *et al.* Increased expression of the monocyte chemoattractant protein-1 in bronchial tissue from asthmatic subjects. *Am J Respir Cell Mol Biol* 10: 142–47, 1994.
66. Alan R, York T, Boyars M *et al.* Increased MCP-1, RANTES and MIP-1α in bronchoalveolar lavage fluid of asthmatic patients. *Am J Respir Crit Care Med* 153: 1398–404, 1996.
67. Hautamaki RD, Kobayashi DK, Senior RM *et al.* Macrophage elastase is required for cigarette smoke-induced emphysema in mice. *Science* 277: 2002–4, 1997.
68. Senior RM, Griffin GL, Mecham RP. Chemotactic activity of elastin-derived peptides. *J Clin Invest* 66: 859–62, 1980.
69. Hunninghake GW, Davidson JM, Rennard S *et al.* Elastin fragments attract macrophage precursors to diseased sites in pulmonary emphysema. *Science* 212: 925–27, 1981.
70. Scott P, Trinchieri G. The role of natural killer cells in host parasite interactions. *Curr Opin Immunol* 7: 34–40, 1995.
71. Macatonia SE, Hosken NA, Litton M *et al.* Dendritic cells produce IL-12 and direct the development of Th1 cells from naive CD4+ T cells. *J Immunol* 154: 5071–79, 1995.
72. Thompson CB. Distinct roles for the costimulatory ligands B7-1 and B7-2 in T helper cell differentiation. *Cell* 81: 979–82, 1995.
73. Toews GB, Vial WC, Dunn MM *et al.* The accessory cell function of human AMs in specific T cell proliferation. *J Immunol* 132: 181–86, 1984.
74. Lipscomb MF, Lyons CR, Nunez G *et al.* Human AMs: HLA-DR-positive macrophages that are poor stimulators of a primary mixed leukocyte reaction. *J Immunol* 136: 497–504, 1986.
75. Lyons CR, Ball EJ, Toews GB *et al.* Inability of human AMs to stimulate resting T cells correlates with decreased antigen-specific T cell-macrophage binding. *J Immunol* 137: 1173–80, 1986.
76. Chelen CJ, Fang Y, Freeman GJ *et al.* Human AMs present antigen ineffectively due to defective expression of B7 costimulatory cell surface molecule. *J Clin Invest* 95: 1415–21, 1995.
77. Toews GB. Pulmonary dendritic cells: Sentinels of lung-associated lymphoid tissues. *Am J Respir Cell Mol Bio1* 4: 204–5, 1991.
78. Trinchieri G, Gerosa F. Immunoregulation by interleukin-12. *J Leukocyte Biol* 59: 505–11, 1996.
79. Seder RA, Mosmann TA. Differentiation of effector phenotypes of CD4+ and CD8+ T cells. In: Paul WE (ed.), *Fundamental Immunology*, pp. 879–908. Philadelphia. New York: Lippencott. Raven, 1999.
80. Hauser C, Snapper CM, O'Hara J *et al.* T helper cells grown with hapten-modified cultured Langerhans cells produce interleukin-4 and stimulate IgE production by B cells. *Eur J Immunol* 10: 245–51, 1989.
81. Bluestone JA. New Perspective of CD28-B7-mediated T cell costimulation. *Immunity* 2: 555–59, 1995.
82. Ding L, Shevach EM. Activation of CD4+ T cells by delivery of the B7 costimulatory signal on bystander antigen-presenting cells (trans-costimulation). *Eur J Immunol* 24: 859–66, 1994.
83. Mornex JF, Martinet Y, Yamauchi K *et al.* Spontaneous expression of the c-sis gene and release of a platelet-derived growth factor-like molecule by human alveolar macrophages. *J Clin Invest* 78: 61–66, 1986.
84. Assoian RK, Fleudelys GE, Stevensen HC *et al.* Expression and secretion of type beta transforming growth factor by activated human macrophages. *Cell* 53: 285–93, 1988.
85. Rose R, Raines EW, Bowen-Pope DF. The biology of platelet-derived growth factor. *Cell* 46: 155–69, 1986.
86. Moore BB, Coffey MJ, Christensen PJ *et al.* GM-CSF regulates bleomycin-induced pulmonary fibrosis via a prostaglandin-dependent mechanism. *J Immunol* 165: 4032–39, 2000.
87. Strieter RM, Polverini PJ, Kunkel SL *et al.* The functional role of the ELR motif in CXC chemokine mediated angiogenesis. *J Biol Chem* 270: 27348–57, 1995.
88. Reynolds HY. Bronchoalveolar lavage. *Am Rev Respir Dis* 135: 250–63, 1987.
89. Seatta M, Di Stefano A, Turato G *et al.* CD8+ T lymphocytes in peripheral airways of smokers with chronic obstructive pulmonary disease. *Am J Respir Crit Care Med* 157: 822–26, 1998.
90. Hogg JC, Chu F, Utokaparch S *et al.* The nature of small-airway obstruction in chronic obstructive pulmonary disease. *N Engl J Med* 350: 2645–53, 2004.
91. Turato G, Di Stefano A, Maestrelli P *et al.* Effect of smoking cessation on airway inflammation in chronic bronchitis. *Am J Respir Crit Care Med* 152: 1666–72, 1995.
92. Keatings VM, Collins PD, Scott DM *et al.* Differences in interleukin-8 and tumor necrosis factor-α induced sputum from patients with chronic obstructive pulmonary disease or asthma. *Am J Respir Crit Care Med* 153: 530–34, 1996.
93. Lang MR, Fiaux GW, Gillooly, M *et al.* Collagen content of alveolar wall tissue in emphysematous and non-emphysematous lungs. *Thorax* 49: 319–26, 1994.
94. Damiano VV, Tsang A, Kucich U *et al.* Immunolocalization of elastase in human emphysematous lungs. *J Clin Invest* 78: 482–93, 1986.
95. Senior RM, Tegner H, Kuhn C *et al.* The induction of pulmonary emphysema induced with human leukocyte elastase. *Am Rev Respir Dis* 116: 469–77, 1977.
96. Kao RC, Wehner NG, Skubitz KM *et al.* Proteinase 3. A distinct human polymorphonuclear leukocyte proteinase that produces emphysema in hamsters. *J Clin Invest* 82: 1693, 1988.
97. Wang Z, Zheng T, Zhu Z *et al.* Interferon gamma induction of pulmonary emphysema in the adult murine lung. *J Exp Med* 192: 1587–600, 2000.
98. Zheng T, Zhu Z, Wang Z *et al.* Inducible targeting of IL-13 to the adult lung causes matrix metalloproteinases and cathepsin-dependent emphysema. *J Clin Invest* 106: 1081–93, 2000.
99. Finlay GA, O'Driscoll LR, Russel KJ *et al.* Matrix metalloproteinase expression and production by alveolar macrophages in emphysema. *Am J Respir Crit Care Med* 156: 240–47, 1997.
100. D'Armiento J, Dalal SS, Okada Y *et al.* Collagenase expression in the lung of transgenic mice causes pulmonary emphysema. *Cell* 71: 955–61, 1992.
101. Hantamaki RD, Kobayashi DC, Senior RM *et al.* Macrophage elastase required for cigarette smoke induced emphysema in mice. *Science* 277: 2002–4, 1997.
102. Woodruff PG, Koth LL, Yang YH *et al.* A distinctive alveolar macrophage activation state induced by cigarette smoking. *Am J Respir Crit Care Med* 172: 1383–92, 2005.
103. Kasahara Y, Tuder RM, Cool CD *et al.* Endothelial cell death and decreased expression of vascular endothelial growth factor and vascular endothelial growth factor receptor 2 in emphysema. *Am J Respir Crit Care Med* 163: 737–44, 2001.
104. Hu B, Sonstein J, Christensen PJ *et al.* Deficient *in vitro* and *in vivo* phagocytosis of apoptotic T cells by resident murine alveolar macrophages. *J Immunol* 165: 2124–33, 2000.
105. Fadok VA, Bratton DL, Konowal A *et al.* Macrophages that have ingested apoptotic cells *in vitro* inhibit proinflammatory cytokine production through autocrine/paracrine mechanisms involving TGF-β, PGE_2, and PAF. *J Clin Invest* 101: 809–98, 1998.
106. Aubus P, Cosso B, Godard P *et al.* Decreased suppressor cell activity of AMs in bronchial asthma. *Am Rev Respir Dis* 130: 875–78, 1984.
107. Spiteri MA, Knight RA, Jeremy JY *et al.* Alveolar macrophage-induced suppression of T cell hyperresponsiveness in asthma is reversed following allergen exposure *in vitro*. *Eur Respir J* 7: 1431–38, 1994.
108. Fischer HG, Frosch S, Reske K *et al.* Granulocyte-macrophage colony-stimulating factor activates macrophages derived from bone marrow cultures to synthesis of MHC class II molecules and to augment antigen presentation function. *J Immunol* 141: 3882–88, 1988.

109. Van der Heijden FL, Van Neerven RJI, Van Katwijk *et al.* Serum-IgE-facilitated allergen presentation in atopic disease. *J Immunol* 150: 3643–50, 1993.
110. Racioppi L, Ronchese F, Matis LA *et al.* Peptide-major histocompatibility complex class II complexes with mixed agonist/antagonist properties provide evidence for ligand-related differences in T cell receptor-dependent intracellular signaling. *J Exp Med* 177: 1047–60, 1993.
111. Ding L, Shevach EM. Activation of CD4+ T cells by delivery of the B7 costimulatory signal on bystander antigen-presenting cells (trans-costimulation). *Eur J Immunol* 24: 859–66, 1994.
112. Ying S *et al.* Eosinophil chemotactic chemokines (eotaxin, eotaxin-2, RANTES, monocyte chemoattractant protein-3 (MCP-3), and MCP-4), and C-C chemokines receptor 3 expression in bronchial biopsies from atopic and non-atopic (Intrinsic) asthmatics. *J Immunol* 163: 6321–29, 1999.
113. Gonzalo JA *et al.* The coordinated action of CC chemokines in the lung orchestrates allergic inflammation and airway hyperresponsiveness. *J Exp Med* 188: 157–67, 1998.
114. Elsner J *et al.* The CC chemokine antagonist Met-RANTES inhibits eosinophil effector functions through the chemokine receptors CCR1 and CCR3. *Eur J Immunol* 27: 2892–98, 1997.
115. Fels AO, Pawlowski NA, Cramer EB *et al.* Human AMs produce leukotriene B4. *Proc Natl Acad Sci USA* 79: 7866–70, 1982.
116. Tyner JW, Uchida O, Kajiwara N *et al.* CCL5-CCR5 interaction provides antiapoptotic signals for macrophage survival during viral infection. *Nature Med* 11: 1180–87, 2005.

12

CHAPTER

Eosinophils

Dagmar Simon[1] and Hans-Uwe Simon[2]

[1]Department of Dermatology, Inselspital, University of Bern, Bern, Switzerland
[2]Department of Pharmacology, University of Bern, Bern, Switzerland

Eosinophil infiltration of the airways is a characteristic feature of obstructive pulmonary diseases in particular of bronchial asthma. But eosinophils have also been connected with chronic obstructive pulmonary disease (COPD) in at least a subgroup of patients. Initially, the function of eosinophils was almost exclusively related to their destructive activity mediated by the toxic effects of the eosinophil granule proteins. In the last two decades research work has made it clear that eosinophils also play important roles in immunomodulation as well as in repair and remodeling processes. To describe the role of eosinophils in bronchial asthma or other eosinophilic diseases, we will first review their functional properties and the regulation of eosinophil numbers. We will then focus on the potential role of eosinophils in the pathogenesis of bronchial asthma and COPD. However, as the reader will realize our current knowledge on the physiologic and pathogenic role of eosinophils is still not complete.

MORPHOLOGY AND LOCATION

Among the leukocytes, eosinophils attract attention because of their characteristic appearance revealing a bilobed nucleus with highly condensed chromatin and a multitude of coarse granules. Eosinophils can easily be identified because of their strong affinity to the acidic dye eosin in blood and tissues even when they are paraformaldehyde fixed and paraffin embedded. This characteristic inspired Paul Ehrlich to name these granular leukocytes eosinophils after discovering them in 1879 [1]. Eosinophils are released from the bone marrow into the peripheral blood, where they represent 1–5% of the leukocytes with an upper limit of 0.4×10^9/l under normal conditions [2]. Some laboratories list higher upper values, in particular in children (as high as 0.75×10^9/l). Eosinophils circulate only a few hours in the peripheral blood before they enter the tissues, where they can survive for at least 2 weeks [3, 4].

The cytoplasmic granules are composed of four distinct populations, which can be identified under electron microscopy: primary and secondary granules, small granules and lipid bodies [5]. The primary granules contain galectin-10, the Charcot-Leyden crystal protein, which binds lysophospholipases [6]. The cytotoxic cationic proteins are stored in the secondary granules. These granules are formed by a core containing major basic protein (MBP) and a matrix composed of eosinophil cationic protein (ECP), eosinophil peroxidase (EPO), and eosinophil derived neurotoxin (EDN) [7]. Mature eosinophils also house small granules storing proteins, such as arylsulphatase B and acid phosphatase. The lipid bodies contain arachidonic acid, the basic component for eicosanoid production [8]. In addition, both the secondary granules and the lipid bodies store a number of cytokines, chemokines and growth factors [5].

In the absence of eosinophil-specific surface markers [9], MBP and ECP are widely used as targets for immunohistochemical studies. Some investigations revealed that eosinophil granule proteins are present not only in the cells, but also in the extracellular spaces suggesting eosinophil degranulation [10]. In addition, receptors, which are almost specifically expressed by eosinophils, for example, CC chemokine receptor (CCR)3 and interleukin (IL)-5 receptor alpha, or the transcription factor GATA-1 can be used as markers to identify eosinophils. Recent research using flow cytometry and microarray assay techniques demonstrated that eosinophils express typical surface molecule patterns consisting of CCR3, Siglec-8 (Siglec-F), FIRE (F4/80–like receptor), CD62 ligand, and paired Ig-like receptor A/B (Pir-A/B) at different stages of differentiation and activation [11].

Under physiologic conditions, eosinophils are found in the bone marrow, the lymphatic organs such as spleen, lymph nodes, and thymus, and throughout the gastrointestinal tract except the esophagus [12]. Recent studies reported the essential role of eosinophils in mammary gland branch formation [13] and maturation of the pubertal uterus in mice [14]. In the gastrointestinal tract, numbers of eosinophils depend on the location as well as on the exposure to pathogens in asymptomatic individuals [15]. Under normal conditions, the baseline levels of eosinophils are controlled by eotaxin-1 [16]. Moreover, in preinvolutional human thymi, eosinophils [17] and eosinophil precursors [18] have been identified, the latter suggesting differentiation of eosinophils in the thymus. Eosinophils have been associated with major histocompatibility complex (MHC) class I-restricted selection/deletion in the thymus, pointing to an immunomodulatory function under non-pathological conditions [19].

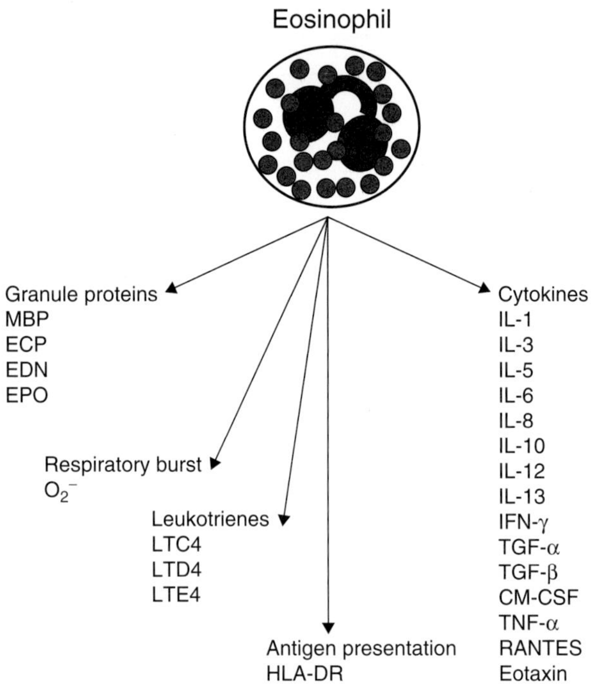

FIG. 12.1 Eosinophil effector molecules.

EOSINOPHIL FUNCTIONS

The view of eosinophil function has profoundly changed in the last two decades. Eosinophils have always been regarded as destructive effector cells that release toxic granule proteins. The primary function of eosinophils has been related to the protection against helminth parasites [20]. The observation that eosinophils surround parasites in the tissue as well as a number of in vitro and animal experiments revealed a direct role for eosinophils in killing parasites. Later it became clear that eosinophils also play a role in the pathogenesis of helminth infections by causing tissue damage and organ dysfunction [20]. Results of more recent research indicate that eosinophils play an important role in repair and remodeling processes after tissue damage as well as in immunomodulation [21]. Interestingly, eosinophils produce and release a broad spectrum of cytokines, chemokines, and leukotrienes as well as express cell surface receptors for these mediators, suggesting autocrine and paracrine mechanisms in regulating eosinophil functions [21]. The main effector molecules of eosinophils are summarized in Fig. 12.1.

Activation of eosinophils is required for their migration into the sites of inflammation in the tissue, for the production and release of their granule proteins and lipid mediators, as well as for the generation of reactive oxygen species [5]. Hematopoietins, such as IL-3, IL-5 and granulocyte-macrophage colony-stimulating factor (GM-CSF), increase functional responses of eosinophils to various agonists including lipid mediators, complement factors, or chemokines. This effect of hematopoietins is called priming.

Granule protein production and release

MBP is highly cytotoxic [22] and plays an essential role in host defense against helminth infection. Because of its cationic nature, it affects the charge of surface membranes resulting in disturbed permeability, disruption, and injury of cell membranes [23]. At low concentrations, MBP stimulates mediator production by other inflammatory cells (discussed later). ECP and EDN have been identified as ribonucleases [24] and possess antiviral activity [25]. ECP damages target cell membranes through the formation of pores or transmembrane channels [26]. Besides their ribonuclease activities, both EDN and ECP are potent neurotoxins [27]. Eosinophils store abundant amounts of ECP and may release it upon repetitive stimulation with the same agonist, implying that mature eosinophils do not require significant *de novo* ECP synthesis for secretion [28]. EDN was demonstrated to induce recruitment and cytokine release by dendritic cells [29]. EPO is involved in the generation of toxic oxygen species as part of the respiratory burst.

Upon activation, eosinophils specifically release their granule proteins by exocytosis or degranulation [30]. In classical exocytosis, single secretory granules are extruded to the cell exterior, whereas in compound exocytosis fusion of intracellular granules precedes their release through a single fusion pore to effectively target surfaces as it has been shown for helminth infection [31]. Piecemeal degranulation has been identified as mechanism by which cytoplasmic secretory vesicles containing cytoplasmic crystalloid granules with core components are released from eosinophils. In addition, granule deposition in the tissue may occur after cytolysis of the eosinophil [30]. A prerequisite for exocytosis is the docking of the vesicles/granules to the cell membrane mediated by membrane-associated proteins forming soluble NSF attachment protein (SNAP) receptors (SNAREs) [32].

Respiratory burst

The superoxide radical generation after activation of the NADPH oxidase complex presents a powerful weapon by which eosinophils dispose of pathogens; and this mechanism is even more powerful than that in neutrophils [33]. Activated NADPH oxidase catalyzes O_2 to O_2^-, which enters further redox pathways to generate hydrogen peroxide (H_2O_2) in the presence of superoxide dismutase, or hydroxyl and nitrogen dioxide radicals, after combining with nitric oxide. The granular protein EPO oxidizes bromide, nitrite, and thiocyanate in the presence of H_2O_2, which have the potency to disrupt cell membranes or cause intracellular oxidant stress [34].

Cytokine production and release

Eosinophils are able to produce, store, and secrete a wide spectrum of cytokines and chemokines and thus may function as immunomodulatory cells. Although eosinophils are terminally differentiated cells, their capacity to generate cytokines can be quite intriguing. Eosinophils from normal individuals constitutively express IL-4 and IL-10 with further upregulation in response to inflammatory signals [35]. Both the cytokines as well as IL-2, IL-5, and IL-13 are stored in the crystalloid core. Cytokines and chemokines that may activate cytokine production by eosinophils are IL-3, IL-5, GM-CSF, interferon (IFN)-γ, tumor necrosis factor (TNF)-α, and complement factor C5a [5, 36]. In response to stimulation, eosinophils may produce proinflammatory cytokines, which may overlap the typical spectrum of both T helper 1 and T helper 2 cytokines, regulatory cytokines, and chemokines: IL-1, IL-3, IL-5, IL-6, IL-8, IL-12, IL-13, IFN-γ, tumor growth factor (TGF)-α, TGF-β, GM-CSF, TNF-α, macrophage inflammatory protein (MIP)-1α, regulated upon activation, normal T-cell expressed and secreted (RANTES) and eotaxin [4, 5, 37–40]. Recently, eosinophils have been shown to be involved in innate immune responses by expressing Toll-like receptors [41]. By generating IL-10 and TGF-β, eosinophils are suggested to modulate T regulatory cell function and suppress inflammatory events [21].

Lipid mediator generation

In eosinophils, eicosanoid derivatives of arachidonic acid are generated by both cyclooxygenase and lipoxygenase pathways, which have been localized to the sites of the perinuclear membranes and the cytoplasmic lipid bodies [8]. Platelet-activating factor (PAF) as well as agonists of CCR3 can initiate the *de novo* formation of lipid bodies [8]. To synthesize leukotrienes (LT), eosinophils need to be both primed (e.g. by IL-3, IL-5, GM-CSF) and activated (e.g. by C5a, PAF) [42, 43]. To reach the threshold needed to phosphorylate cytosolic phospholipase A_2 (PLA_2), a concurrent stimulation of G-protein-coupled receptor and cytokine receptor resulting in an activation of the mitogen-activated protein kinase (MAPK) cascade is required [42]. The activation of PLA_2, which also requires calcium, is essential for leukotriene generation [42]. Because of their content of leukotriene C_4 synthetase, eosinophils are an important source of cysteinyl leukotrienes, predominatly LTC_4 and its active metabolites LTD_4 and LTE_4 [8]. Leukotrienes possess proinflammatory activities. They may act intracellularly as intracrine signaling molecules, for example, in regulating IL-4 release [8]. Leukotrienes may amplify the inflammatory cascade, for instance by acting as chemotactic factors or by triggering the release of cytotoxic proteins.

Antigen presentation

In vitro experiments have demonstrated that eosinophils are able to process antigens and express costimulatory molecules and MHC class II molecules, and therefore may function as antigen-presenting cells in stimulating T-cell responses [44]. Whereas blood eosinophils from normal and eosinophilic donors do not express MHC class II molecules, HLA-DR synthesis occurs following stimulation with specific cytokines including IL-3, IL-4, GM-CSF or IFN-γ [44, 45]. During transmigration into the tissues, HLA expression is further increased [46]. By expressing costimulatory molecules, such as CD80 and CD86, eosinophils provide additional secondary signals for lymphocytes [44]. However, the role of eosinophils as antigen-presenting cells to naïve T-cells is minor compared to dendritic cells [47].

REGULATION OF PRODUCTION, MIGRATION, AND SURVIVAL

Prerequisite for eosinophil infiltration and function in the tissues is a precise regulation of their production, migration, and accumulation as well as activation (Fig. 12.2).

Production

In the bone marrow, eosinophils are generated from CD34+ hematopoietic progenitor cells under regulation of critical transcription factors especially GATA-1 and the eosinophilopoietins IL-5, GM-CSF, and IL-3 [4]. Eosinophils share this progenitor with basophils [48]. IL-5 specifically regulates the production rate and differentiation of the eosinophil lineage [49]. The importance of IL-5 is evident from the studies of IL-5 deficient mice, which are unable to develop eosinophilia upon allergen sensitization and challenge [50]. In contrast, IL-5 transgenic mice exhibit extensive eosinophil production and tissue eosinophilia [51]. Anti-IL-5 therapy of asthma patients was shown to lead to an inhibition of eosinophil maturation in the bone marrow and a decrease of eosinophil progenitors in the bronchial mucosa [52]. Besides IL-5, IL-3 and GM-CSF have also been shown to increase eosinophil production [53]. Whereas the release of eosinophils into the peripheral blood circulation requires IL-5 [54]; the migration into the tissues under physiologic conditions is independent of IL-5 and mainly under the control of eotaxin-1 as shown for the

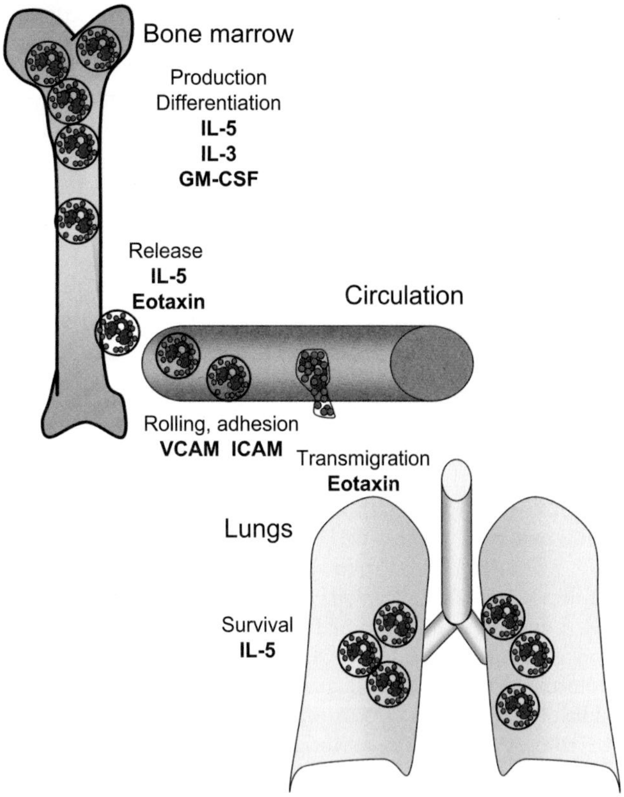

FIG. 12.2 Mechanisms and key players (bold) leading to eosinophilic inflammation in asthma.

gastrointestinal tract [55], the mammary glands [13], the uterus [14], and the thymus [16]. The receptor for the CC chemokine eotaxin is CCR3, which is expressed on both CD34+ progenitor cells and mature eosinophils [56, 57] and can be upregulated during allergic inflammation [57].

Migration into tissues

Priming of eosinophils circulating in the peripheral blood is a prerequisite for their transmigration through the endothelial cell layer, which has been demonstrated for eosinophils of asthmatic patients in contrast to healthy individuals [58] and in response to allergen challenge [59]. Migration comprises a series of sequential events including rolling, adhesion, and transmigration through the vascular endothelium. In the initiation and regulation of these processes a number of cytokines and chemokines are involved.

The initial steps of eosinophil recruitment are adhesion mediated by interaction of the P-selectin-glycoprotein ligand (PSGL)-1 on eosinophils with P-selectin on endothelial cells and rolling, a process in which PAF and eotaxin are involved [5]. While L-selectin is constantly expressed on eosinophils, P- and E-selectins on endothelial cells are upregulated by the inflammatory cytokines IL-1 and TNF-α [36]. The firm attachment of eosinophils to the endothelium is mediated by eosinophil surface very late activation antigen (VLA)-4 (integrin $\alpha4\beta1$) and CD11b/CD18 binding to vascular adhesion molecule (VCAM)-1 and intercellular adhesion molecule (ICAM)-1, respectively [36]. IL-4 was shown to induce the expression of the adhesion molecule VCAM-1, whereas IFN-γ stimulates ICAM-1 expression [60]. The expression of VLA-4 on eosinophils, which significantly contributes to eosinophil inflammation in allergic responses, can be upregulated by eotaxin [61]. After transmigration through the blood vessel, eosinophils enter the extracellular matrix where they bind to matrix proteins such as fibronectin. This binding is mediated by integrins and selectins on the eosinophils. Eosinophils migrate into the tissue along chemokine gradients. Their movement is characterized by adhesion/deadhesion to extracellular matrix proteins [36].

Eotaxin and RANTES are important chemokines for eosinophils and largely contribute to the movement of eosinophils from the peripheral blood to the sites of inflammation [62]. The principal receptor involved in eosinophil attraction is CCR3 [56]. In addition to eotaxin, RANTES and monocyte chemoattractant protein (MCP)-1, 2, 3, and 4 may also bind to CCR3 [63–65]. IL-5 alone has weak chemotactic activity, but increases the effect of eotaxin by enhancing CCR3-mediated responses [65]. Priming of eosinophils leads to an upregulation of additional chemokine receptors including CCR1, the receptor for MIP-1a, RANTES, and MCP-3 [66].

Complement factors such as the anaphylatoxins C3a and C5a [67] as well as PAF [68] have also been implicated in eosinophil recruitment. Moreover, leukotrienes (LTB_4) and prostaglandins (PGD_2) were found to induce eosinophil chemotaxis [69, 70], whereas lipoxin A_4 blocks eosinophil trafficking [71]. All these chemoattractants, C5a, PAF, LTB_4, and PGD_2 can be produced and released by macrophages that were shown to play a role in recruiting eosinophils to the lung and peritoneal cavity during helminth infection [11]. The T helper 2 cytokine IL-13 was shown to promote eosinophilia by increasing local IL-5 and/or eotaxin expression [72].

Survival

Tissue eosinophilia is regulated not only by recruitment of eosinophils from the bone marrow but also by eosinophil survival. Delayed apoptosis as a pathogenic mechanism has been shown in several eosinophilic diseases including asthma [73, 74]. However, resolution of inflammation is associated with disappearance of eosinophils [75]. Apoptotic eosinophils are taken up by macrophages [75] and epithelial cells [76]. In addition, eosinophils may be cleared by transepithelial migration [77]. In particular IL-5, but also IL-3 and GM-CSF delay apoptosis and prolong eosinophil survival [78, 3]. Recently, it has been shown that leptin may also block eosinophil apoptosis [79].

The signaling events important for cytokine-mediated antiapoptosis include activation of tyrosine kinases and signal transducers and activators of transcription (STAT) proteins resulting in the transcriptional activation of members of the Bcl-2 and inhibitor of apoptosis proteins (IAP) families [80–82]. IL-5 upregulates the antiapoptotic proteins Mcl-1 and Bcl-xL, and inhibits translocation of proapoptotic Bax, resulting in an inhibition of the caspase cascade and thus preventing apoptotic cell death [83, 84]. Moreover, following IL-5 stimulation, normal blood eosinophils express the antiapoptotic molecules survivin and cIAP-2 [82]. Although TNF-α and TNF-related apoptosis inducing ligand (TRAIL) are death receptor ligands, they do not induce eosinophil apoptosis under normal conditions [85]. On the contrary, both TNF-α and TRAIL were shown to prolong eosinophil survival [86, 87]. Stimulation of CD137, another member of the TNF-family, which is expressed on eosinophils of allergic patients, by a specific antibody together with IL-5 or GM-CSF blocks the inhibitory effect of the survival factors on eosinophil apoptosis [88]. Eosinophil death can be induced by interaction of the surface death receptors CD95 (Fas) and its ligand, CD95L (FasL), which is expressed by activated T-cells [89]. Therefore, T-cells, known as main producers of eosinophil survival factors, may also limit eosinophil expansion within inflammatory sites [90].

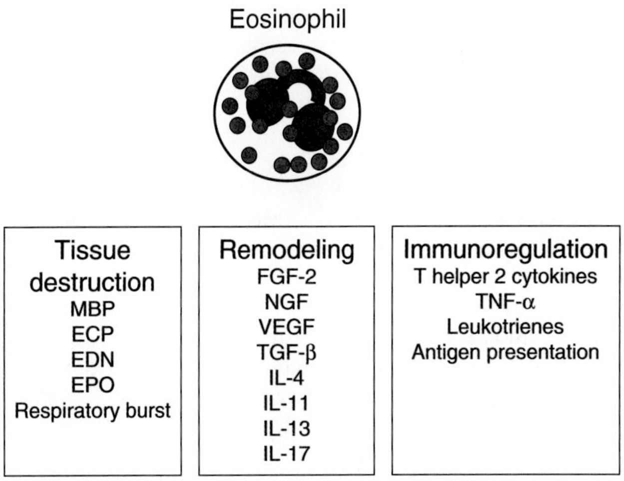

FIG. 12.3 Possible roles of eosinophils in the pathogenesis of asthma.

ROLE IN LUNG DISEASES

Asthma

Eosinophilia in the blood and/or tissues is a feature of multiple disorders. Although the initial cause and the organs vary, there are only the following two major pathways that mediate eosinophilia: (1) a cytokine-mediated increased differentiation and survival of eosinophils (extrinsic eosinophilic disorders) and (2) a mutation-mediated clonal expansion of eosinophils (intrinsic eosinophilic disorders) [2]. The most common cause of eosinophilia is the increased generation of IL-5-producing T-cells, as is observed in allergic diseases including bronchial asthma (Fig. 12.2).

There are a great number of reports in the literature demonstrating the correlation between asthma severity and levels of eosinophils and their products in blood, sputum, bronchoalveolar lavage fluid, or bronchial biopsies [91]. Eosinophils found in the airways of asthmatics are activated [92]. Since clinical trials investigating monoclonal anti-IL-5 antibody treatment in bronchial asthma provided rather disappointing results, a vehement controversy on the role of eosinophils in the pathogenesis of asthma started. Recent studies in genetically targeted mice showed that eosinophils are crucial for airway hyperresponsiveness and mucus accumulation [93] as well as for airway remodeling [94].

The different activities of eosinophils that contribute to the pathogenesis of bronchial asthma can be categorized as tissue destruction, remodeling, and immunoregulation (Fig. 12.3). During asthmatic inflammation, eosinophils interact with leukocytes including lymphocytes, mast cells, dendritic cells, macrophages and neutrophils, as well as with resident tissue cells, such as epithelial cells, endothelial cells, smooth muscle, fibroblasts, and nerve cells [21].

Tissue destruction

The release of eosinophil granule proteins, such as MBP, EPO, ECP, and EDN, has suggested that the main eosinophil effector function, causing desquamation and destruction of the epithelium, leads to airway damage and lung dysfunction [21, 95]. MBP has been localized to the sites of epithelial damage and has been detected in the sputum of asthmatic patients [40]. Eosinophils from asthmatic patients contain, and in response to IL-5 and GM-CSF release, significantly higher amounts of EDN compared to healthy controls [96]. Increased ECP levels in blood, sputum, and bronchoalveolar lavage fluid of patients with bronchial asthma have been reported to correlate with airway obstruction and disease activity [97]. Polymorphisms in the ECP gene are associated with the development of asthma [98].

Nitrotyrosine, a product of EPO, has been observed in the airways as well as in lung parenchyma of asthmatic patients [99, 100]. Eosinophils from patients with various types of asthma were reported to excessively produce reactive oxygen species [5]. These mechanisms may contribute to the damage of resident cells in the lungs. Patients with bronchial asthma have elevated levels of nitric oxide (NO) in their expired air [101], which is thought to reflect the extent of eosinophilic inflammation of the lower airways [102]. Expired NO is used as diagnostic tool in monitoring asthma [102]. Viral and non-viral infections are known to induce asthma exacerbations, in which an increase in blood and sputum eosinophil numbers often precedes deterioration of symptoms and lung function [103]. In response to respiratory syncytial virus infection, epithelial cells of the airways express chemokines leading to the recruitment and degranulation of eosinophils [104]. EDN and ECP are RNAses and were shown to function as anti-viral agents [25]. However, granule proteins released during viral infection may also lead to bronchial hyperreactivity and asthma exacerbation in susceptible individuals [25, 40].

It should be noted that basic proteins are believed to have additional function. For instance, MBP has been described to induce histamine release from mast cells and basophils [105], eosinophil degranulation and cytokine (IL-8) production [106], as well as to increase neutrophil [107, 108] and platelet functions [109]. In addition, MBP and EPO were reported to increase bronchial hyperresponsiveness by binding to the muscarinic M2 receptors of bronchial smooth muscles resulting in a vagal overstimulation and bronchoconstriction [110].

Remodeling

Eosinophils are involved in airway remodeling characterized by structural changes of the lung including epithelial hypertrophy, subepithelial deposition of extracellular matrix proteins in the lamina reticularis, mucus gland hypertrophy, airway smooth muscle hypertrophy, and vascular changes [111]. The matrix deposition (collagen, fibrin) in the lungs observed upon prolonged allergen challenge was significantly increased in wild-type mice compared to eosinophil-deficient [94]. Many of the factors implicated in remodeling are expressed by eosinophils and have been found to be increased in asthmatic airway including fibroblast growth factor (FGF)-2, IL-4, IL-11, IL-13, IL-17, nerve growth factor (NGF), and vascular endothelial growth factor (VEGF) [112]. In patients receiving anti-IL-5 antibody therapy, a reduced deposition of extracellular matrix proteins (tenascin, lumican, procollagen III) beneath the bronchial basement membrane has been observed, again suggesting an essential role of eosinophils in airway remodeling [113]. This decrease of extracellular matrix proteins was accompanied by a significant reduction of TGF-β expression [113]. TGF-β by autocrine–paracrine actions may stimulate eosinophils to generate IL-11, another cytokine with fibrogenic potential, which has been found to be significantly increased in the airways of asthmatics compared to healthy controls [114]. Furthermore, eosinophils express IL-6, whose function has been related to tissue fibrosis [114]. Since eosinophil granule proteins interact with several resident cell types in the lungs, they are suggested to play additional roles in tissue remodeling. MPB interacting with IL-1 and TGF-β stimulates lung fibroblasts [115], while EPO products affect endothelial cells [116]. By secreting metalloproteinases, especially matrix metalloproteinase (MMP)-9, eosinophils may directly affect airway remodeling [117]. MMP-9 has been found to be increased in severe persistent asthma and following allergen challenge [118]. That eosinophils play a role in the late-phase reaction and remodeling processes [112] is consistent with the observation that eosinophils are abundant in the submucosa of lung tissues from patients with slow-onset fatal asthma [119].

Eosinophils synthesize proangiogenic mediators such as VEGF and FGF-2 and thus are able to promote angiogenesis, which presents an additional important mechanism in chronic inflammation and tissue remodeling [120]. In bronchial biopsies of asthmatics, eosinophils are positively stained for VEGF and FGF-2 [121]. VEGF is released upon stimulation with GM-CSF and IL-5 [122]. Furthermore, eosinophils can promote angiogenesis directly by secreting extracellular matrix-degrading enzymes such as MMP-9 and heparanase [123] or indirectly by IL-8 secretion enhancing MMP-2 and MMP-9 mRNA expression by endothelial cells [120]. In contrast; VEGF produced by inflammatory and structural cells such as macrophages, neutrophils, epithelial cells, fibroblasts, and smooth muscle cells, induces migration and mediator release by eosinophils expressing VEGF receptors [124]. In a murine model of asthma, the administration of anti-VEGF receptor antibodies reduced eosinophil infiltration [125]. The mediators, expressed by eosinophils during inflammation, can also activate smooth muscle cells. For example, eotaxin-1 produced by eosinophils themselves can bind to CCR3, which is highly expressed on airway smooth muscle cells in asthmatic lungs [126]. Eosinophils have the capacity to generate IL-25, which has been detected in bronchial biopsies of patients with asthma, and is suggested to have remodeling functions by stimulating airway smooth muscle cells to express extracellular matrix components [114].

Immunoregulation

The inflammation in asthma is characterized by T helper 2 immune responses leading to increased production of cytokines such as IL-4, IL-5, IL-9, and IL-13. Since eosinophils have been demonstrated to express these cytokines, they are suggested to be involved in sustaining a local T helper 2 immune microenvironment [39]. This view is supported by the observation that in eosinophil-deficient mice the production of T helper 2 cytokines is reduced in the lungs after allergen challenge [93]. By producing cytokines associated with acute inflammatory responses such as TNF-α, eosinophils may enhance the inflammation in asthma [36]. Eosinophils are an important source of the cysteinyl leukotrienes [127]. In bronchial mucosa biopsies of asthmatics, eosinophils overexpress LTC_4 synthetase [128]. LTC_4 and its derivatives LTD_4 and LTE_4 were shown not only to increase mucus secretion [129] and bronchoconstriction [130], but also eosinophil survival [131]. In asthmatic patients, antigen challenge induces HLA-DR expression on recruited eosinophils, which enables them to function as antigen-presenting cells in stimulating T-cell responses [132, 133]. Moreover, eosinophils provide costimulatory signals for lymphocytes by expressing CD40, for which an increased surface expression has been found in atopic patients [134]. Endobronchial eosinophils are able to process inhaled antigens, traffic to the regional lymph nodes where they stimulate CD4+ T-cell responses [135]. Those lymph node eosinophils phenotypically resemble dendritic cells and express MHC class II, CD80, and CD86 [136].

Chronic obstructive pulmonary disease

In the pathogenesis of COPD, neutrophils, macrophages, and CD8+ T-cells are considered as key effector cells [137, 138]. The presence and role of eosinophils in COPD is controversial [138]. For instance, airway eosinophilia has been reported occurring during acute exacerbations of COPD [139] and has also been observed in patients with stable disease [140]. However, in one study, eosinophil numbers in induced sputum taken during a stable disease period were compared with those during an acute exacerbation in the same patients, and no differences have been observed [141]. Eosinophilia in COPD has been linked to smoking [142] as well as to an asthmatic component [143]. Moreover, airway obstruction was observed to correlate with an increase in the number of activated eosinophils in the airways [144].

Interestingly, eosinophilia in COPD does not seem to depend on IL-5 [145], but is linked to an upregulation of RANTES [146] as well as eotaxin and CCR3 [147]. The production of RANTES may be induced by TNF-α, which is increased in acute exacerbations [148]. Elevated levels of IL-6 and IL-8 were found in the sputum during exacerbations of COPD correlating with the number of inflammatory cells in particular eosinophils [141]. IL-8, although usually associated with neutrophil chemotaxis, has been suggested to recruit and activate eosinophils in COPD [139, 149].

The principal mechanisms by which eosinophils contribute to airway inflammation have been suggested as being similar to those in asthma [148]. In patients with COPD, higher numbers of eosinophils have been found in bronchial biopsies compared to healthy controls accompanied by increased eosinophil and ECP levels in bronchoalveolar lavage fluid and induced sputum [150]. In addition to ECP, EPO levels are raised in the sputum in patients with COPD [151]. Moreover, several studies reported eosinophil infiltration of the airway walls associated with increased amounts of ECP in bronchoalveolar lavage fluid and induced sputum during exacerbations of COPD [138, 148, 152]. Since COPD exacerbations may be associated with viral infections, eosinophils are suggested to have a role in antiviral host defense, for example, by releasing ECP and EDN acting as ribonucleases [138].

Eosinophilic airway inflammation is suggested to contribute to airflow obstruction and symptoms in patients with COPD [143]. The patients with sputum eosinophilia are more likely to respond to corticosteroid therapy [143]. Therefore, sputum eosinophilia may be used as predictive marker in identifying patients who will benefit from corticosteroid therapy [153].

ANTI-EOSINOPHILIC THERAPIES

Since eosinophils exhibit a multifunctional role in the pathogenesis of asthma, therapeutic approaches aiming to reduce their numbers and/or activity seem promising. There are several potential mechanisms for how to do so: reduce eosinophil production, stop eosinophil migration into inflamed tissues, activate eosinophil apoptosis, prevent eosinophil priming and activation, stop generation and release of mediators, and/or antagonize them. Here, we shortly discuss established as well as new therapeutic options which specifically attack eosinophils.

Immunosuppressive therapy

Corticosteroids are widely used in the treatment of bronchial asthma and COPD. Although corticosteroids increase the rate of eosinophil apoptosis [154], their main effect on eosinophils appears to be due to inhibition of cytokine and chemokine production by leukocytes, such as T-cells and resident airway cells [155]. The same mechanism is true for other immunosuppressive drugs such as cyclosporine.

Leukotriene antagonists

Leukotriene antagonists are well-established substances in the treatment of asthma that inhibit the effect of cysteinyl leukotrienes, which are generated in high amounts by eosinophils, mast cells, basophils, macrophages, and monocytes during asthmatic inflammation [156]. A second way to interfere with leukotrienes is the inhibition of their generation by blocking the 5-lipoxygenase pathway [157].

Interferons

The treatment with low-dose IFN-α was shown to be effective in the treatment of patients with severe corticosteroid resistant asthma with or without Churg–Strauss syndrome. The establishment of a correct T helper 1/T helper 2 balance as well as the induction of IL-10 are regarded as possible mechanisms [158]. IFN-γ therapy was shown to reduce blood eosinophil numbers in patients with severe steroid-dependent asthma. However, physiological asthma parameters, such as FEV_1 and peak expiratory flow rates, did not improve upon treatment with IFN-γ over 90 days [159].

Anti-IL-5 monoclonal antibodies

The first study using a monoclonal antibody to IL-5 investigated the effect and safety of a single dose of mepolizumab in patients with mild bronchial asthma [160]. Although peripheral blood and sputum eosinophil numbers significantly decreased, this treatment had no effect either on airway hyperresponsiveness before and after allergen challenge or on the allergen-induced late asthmatic response. A second study, in which the same anti-IL-5 antibody was given over 3 months, demonstrated that the expression of activation markers and T helper 2-cytokine receptors on eosinophils as well as the number of circulating T-cells and their capacity for generating cytokines were not affected [161]. Another antibody (SCH55700) was tested in patients with severe asthma. A single dose of this anti-IL-5 antibody resulted in a short improvement of baseline FEV1 and decrease of blood eosinophil numbers, but failed to improve asthma symptoms significantly [162]. Investigating eosinophil numbers at different sites revealed that anti-IL-5 therapy produced only a partial decrease of 52% in the bone marrow, and of 55% in the airways, but a 100% decrease in the peripheral blood [163]. The anti-IL-5 antibody seemed to exclusively reduce mature eosinophils in the bone marrow and in the peripheral blood, whereas CD34+/IL-5 receptor+ progenitor cells were not reduced [52]. However, the observed decrease of eosinophil progenitor cells in the bronchial mucosa suggested that local eosinophilopoiesis was impaired upon anti-IL-5 therapy. The reason why a single anti-IL-5 treatment failed to improve asthma might be the redundancy of different cells and cytokines contributing to eosinophilia and asthma symptoms.

Anti-TNF-α

The proinflammatory cytokine TNF-α plays a role in leukocyte recruitment and activation. Although anti-TNF-α therapy was shown to improve asthma symptoms, it did not affect the number of sputum eosinophils [164].

IL-12

IL-12 is a macrophage-derived cytokine that modulates T-cell responses and therefore might reduce eosinophilic inflammation. Recombinant human IL-12, administered to patients with mild asthma in three increasing dosages over 3 weeks, resulted in a decrease of eosinophil numbers in the peripheral blood and in the sputum after allergen challenge as well as in an improvement of the allergen induced airway hyperresponsiveness [165]. However, severe adverse events leading to withdrawal were seen in 4 out of 19 patients.

Selectin antagonists

Because recruitment of inflammatory cells from the peripheral blood into the airways is a characteristic feature of asthma, targeting selectins or their ligands is a promising approach in the treatment of asthma [166]. Bimosiamose, a pan-selectin antagonist, was studied in patients with mild allergic asthma revealing an inhibitory effect on late-phase responses but not on airway hyperresponsiveness or exhaled NO after allergen challenge [167].

CCR3 antagonists

To block the principal chemokine receptor of eosinophils, the effect of a low-molecular weight CCR3 antagonist has been investigated in a mouse asthma model. This revealed promising results as it markedly reduced eosinophil infiltration of the airways followed by a normalization of airway hyperresponsiveness, reduced mucus production and prevention of airway remodeling [168].

Integrin inhibitors

Since intergrins are involved in migration of inflammatory cells including eosinophils from the blood vessels into the tissues, antagonizing integrins offers another promising treatment option for asthma or allergic diseases. *In vitro* and *in vivo* experiments showed that a newly developed small molecule VLA-4 ($\alpha 4\beta 1$ integrin) inhibitor may prevent VLA-4/VCAM interactions resulting in reduced eosinophil migration and infiltration in the tissue [169].

ACKNOWLEDGMENTS

Work in the laboratory of Hans-Uwe Simon is supported by grants from the Swiss National Science Foundation (grant #310000-107526), the Stanley Thomas Johnson Foundation, Bern, and the OPO-Foundation, Zurich.

References

1. Ehrlich P. Beiträge zur Kenntnis der granulierenden Bindegewebszellen und der eosinophilen Leukozyten. *Arch Anat Physiol*: 166–69, 1879.
2. Simon D, Simon HU. Eosinophilic disorders. *J Allergy Clin Immunol* 119: 1291–300, 2007.
3. Simon HU, Yousefi S, Schranz C, Schapowal A, Bachert C, Blaser K. Direct demonstration of delayed eosinophil apoptosis as a mechanism causing tissue eosinophilia. *J Immunol* 158: 3902–8, 1997.
4. Rothenberg ME, Hogan SP. The eosinophil. *Annu Rev Immunol* 24: 147–74, 2006.
5. Kariyawasam HH, Robinson DS. The eosinophil: The cell and its weapons, the cytokines, its locations. *Semin Respir Crit Care Med* 27: 117–27, 2006.
6. Ackerman SJ, Liu L, Kwatia MA, Savage MP, Leonidas DD, Swaminathan GJ, Acharya KR. Charcot-Leyden crystal protein (galectin-10) is not a dual function galectin with lysophospholipase activity but binds a lysophospholipase inhibitor in a novel structural function. *J Biol Chem* 277: 14859–68, 2002.
7. Gleich GJ, Adolphson CR. The eosinophilic leukocyte: Structure and function. *Adv Immunol* 39: 177–253, 1986.
8. Bandeira-Melo C, Bozza PT, Weller PF. The cellular biology of eosinophil eicosanoid formation and function. *J Allergy Clin Immunol* 109: 393–400, 2002.
9. Prussin C, Metcalfe DD. IgE, mast cells, basophils and eosinophils. *J Allergy Clin Immunol* 111: S486–94, 2003.
10. Leiferman KM, Ackerman SJ, Sampson HA, Haugen HS, Venencie PY, Gleich GJ. Dermal deposition of eosinophil-granule major basic protein in atopic dermatitis: Comparison with onchocerciasis. *N Engl J Med* 313: 282–85, 1985.
11. Voehringer D, van Rooijen N, Locksley M. Eosinophils develop in distinct stages and are recruited to peripheral sites by alternatively activated macrophages. *J Leukoc Biol* 81: 1434–44, 2007.
12. Kato M, Kephart GM, Talley NJ, Wagner JM, Sarr MG, Bonno M, McGovern TW, Gleich GJ. Eosinophil infiltration and degranulation in normal human tissues. *Anat Rec* 252: 418–25, 1998.
13. Gouon-Evans V, Rothenberg ME, Pollard JW. Postnatal mammary gland development requires macrophages and eosinophils. *Development* 127: 2269–82, 2000.
14. Gouon-Evans V, Pollard JW. Eotaxin is required for eosinophil homing into the stroma of the pubertal and cycling uterus. *Endocrinology* 142: 4515–21, 2001.
15. Furuta G. Emerging questions regarding eosinophil's role in the esophago-gastrointestinal tract. *Curr Opin Gastroenterol* 22: 658–63, 2006.
16. Matthews AN, Friend DS, Zimmermann N, Sarafi MN, Luster AD, Pearlman E, Wert SE, Rothenberg ME. Eotaxin is required for the baseline level of tissue eosinophils. *Proc Natl Acad Sci USA* 95: 6273–78, 1988.
17. Müller E. Localization of eosinophils in the thymus by the peroxidase reaction. *Histochemistry* 52: 273–79, 1977.
18. Lee I, Yu E, Good RA, Ikehara S. Presence of eosinophilic precursors in the human thymus: Evidence for intra-thymic differentiation of cells in eosinophilic lineage. *Pathol Int* 45: 655–62, 1995.
19. Throsby M, Herbelin A, Pleau JM, Dardenne M. CD11c+ eosinophils in the murine thymus: Developmental regulation and recruitment upon MHC class I-restricted thymocyte deletion. *J Immunol* 165: 1965–75, 2000.
20. Klion AD, Nutman TB. The role of eosinophils in host defense against helminth parasites. *J Allergy Clin Immunol* 113: 30–37, 2004.
21. Jacobsen EA, Taranova AG, Lee NA, Lee JJ. Eosinophils: Singularly destructive effector cells or purveyors of immunoregulation. *J Allergy Clin Immunol* 119: 1313–20, 2007.
22. Gleich GJ, Frigas E, Loegering DA, Wassom DL, Steinmuller D. Cytotoxic properties of the eosinophil major basic protein. *J Immunol* 123: 2925–27, 1979.
23. Kroegel C, Costabel U, Matthys H. Mechanism of membrane damage mediated by eosinophil major basic protein. *Lancet* 1: 1380–81, 1987.

24. Gleich GJ, Loegering DA, Bell MP, Checkel JL, Ackerman SJ, McKean DJ. Biochemical and functional similarities between human eosinophil-derived neurotoxin and eosinophil cationic protein: Homology with ribonucleases. *Proc Natl Acad Sci USA* 83: 3146–50, 1986.
25. Rosenberg HF, Domachowske JB. Eosinophils, eosinophil ribonucleases, and their role in host defence against respiratory virus pathogens. *J Leukoc Biol* 70: 691–98, 2001.
26. Young JD, Peterson CG, Venge P, Cohn ZA. Mechanism of membrane damage mediated by human eosinophil cationic protein. *Nature* 321: 613–16, 1986.
27. Barker RL, Loegering DA, Ten RM, Hamann KJ, Pease LR, Gleich GJ. Eosinophil cationic protein cDNA. Comparison with other toxic cationic proteins and ribonucleases. *J Immunol* 143: 952–55, 1989.
28. Simon HU, Weber M, Becker E, Zilberman Y, Blaser K, Levi-Schaffer F. Eosinophils maintain their capacity to signal and release eosinophil cationic protein upon repetitive stimulation with the same agonist. *J Immunol* 165: 4069–75, 2000.
29. Yang D, Rosenberg HF, Chen Q, Dyer KD, Kurosaka K, Oppenheim JJ. Eosinophil-derived neurotoxin (EDN), an antimicrobial protein with chemotactic activities for dendritic cells. *Blood* 102: 3396–403, 2003.
30. Logan MR, Odemuyiwa SO, Moqbel R. Understanding exocytosis in immune and inflammatory cells: The molecular basis of mediator secretion. *J Allergy Clin Immunol* 111: 923–32, 2003.
31. Scepek S, Moqbel R, Lindau M. Compound exocytosis and cumulative degranulation by eosinophils and its role in parasite killing. *Parasitol Today* 10: 276–78, 1994.
32. Fasshauer D, Sutton RB, Brunger AT, Jahn R. Conserved structural features of the synaptic fusion complex: SNARE proteins reclassified as Q- and R-SNAREs. *Proc Natl Acad Sci USA* 95: 15781–86, 1998.
33. Shult PA, Graziano FM, Wallow IH, Busse WW. Comparison of superoxide generation and luminol-dependent chemiluminescence with eosinophils and neutrophils from normal individuals. *J Lab Clin Med* 106: 638–45, 1985.
34. Wang J, Slungaard A. Role of eosinophil peroxidase in host defense and disease pathology. *Arch Biochem Biophys* 445: 256–60, 2006.
35. Nakajima H, Gleich GJ, Kita H. Constitutive production of IL-4 and IL-10 and stimulation of IL-8 by normal peripheral blood eosinophils. *J Immunol* 156: 4859–66, 1996.
36. Lampinen M, Carlson M, Hakansson LD, Venge P. Cytokine-regulated accumulation of eosinophils in inflammatory disease. *Allergy* 59: 793–805, 2004.
37. Yousefi S, Hemmann S, Weber M, Hölzer C, Hartung K, Blaser K, Simon HU. IL-8 is expressed by human peripheral blood eosinophils. Evidence for increased secretion in asthma. *J Immunol* 154: 5481–90, 1995.
38. Lamkhioued B, Gounni AS, Aldebert D, Delaporte E, Prin L, Capron A, Capron M. Synthesis of type 1 (IFN-gamma) and type 2 (IL-4, IL-5, and IL-10) cytokines by human eosinophils. *Ann NY Acad Sci* 796: 203–8, 1996.
39. Schmid-Grendelmeier P, Altznauer F, Fischer B, Bizer C, Straumann A, Menz G, Blaser K, Wüthrich B, Simon HU. Eosinophils express functional IL-13 in eosinophilic inflammatory diseases. *J Immunol* 169: 1021–27, 2002.
40. Gleich GJ. Mechanisms of eosinophil-associated inflammation. *J Allergy Clin Immunol* 105: 651–63, 2000.
41. Nagase H, Okugawa S, Ota Y, Yamaguchi M, Tomizawa H, Matsushima K, Ohta K, Yamamoto K, Hirai K. Expression and function of Toll-like receptors in eosinophils: Activation by Toll-like receptor 7 ligand. *J Immunol* 171: 3977–82, 2003.
42. Thomet OAR, Wiesmann UN, Blaser K, Simon HU. Differential inhibition of inflammatory effector functions by petasin, isopetasin and neopetasin in human eosinophils. *Clin Exp Allergy* 31: 1310–20, 2001.
43. Takafuji S, Bischoff SC, De Weck AL, Dahinden CA. IL-3 and IL-5 prime normal human eosinophils to produce leukotriene C4 in response to soluble agonists. *J Immunol* 147: 3855–61, 1991.
44. Shi HZ. Eosinophil functions as antigen presenting cells. *J Leukoc Biol* 76: 520–27, 2004.
45. Lucey DR, Nicholson-Weller A, Weller PF. Mature human eosinophils have the capacity to express HLA-DR. *Proc Natl Acad Sci USA* 86: 1348–51, 1989.
46. Yamamoto H, Sedgwick J, Vrtis RF, Busse WW. The effect of transendothelial migration on eosinophil function. *Am J Respir Cell Mol Biol* 23: 379–88, 2000.
47. van Rijt LS, Vos N, Hijdra D, de Vries VC, Hoogsteden HC, Lambrecht BN. Airway eosinophils accumulate in the mediastinal lymph nodes but lack antigen-presenting potential for naïve T cells. *J Immunol* 171: 3372–78, 2003.
48. Denburg JA, Telizyn S, Messner H, Lim B, Jamal N, Ackerman SJ, Gleich GJ, Bienenstock J. Heterogeneity of human peripheral blood eosinophil-type colonies: Evidence for a common basophil–eosinophil progenitor. *Blood* 66: 312–18, 1985.
49. Sanderson CJ. Interleukin-5, eosinophils, and disease. *Blood* 79: 3101–9, 1992.
50. Foster P, Hogan P, Ramsay AJ, Matthaei KI, Young IG. Interleukin-5 deficiency abolishes eosinophilia, airway hyperreactivity and lung damage in a mouse asthma model. *J Exp Med* 83: 195–201, 1996.
51. Dent LA, Strath M, Mellor AL, Sanderson CJ. Eosinophilia in transgenic mice expressing interleukin-5. *J Exp Med* 172: 1425–31, 1990.
52. Menzies-Gow A, Flood-Page P, Sehmi R, Burman J, Hamid Q, Robinson DS, Kay AB, Denburg J. Anti-IL-5 (mepolizumab) therapy induces bone marrow eosinophil progenitors in bronchial mucosa of atopic asthmatics. *J Allergy Clin Immunol* 111: 714–19, 2003.
53. Nishinakamura R, Miyajima A, Mee PJ, Tybulewicz VL, Murray R. Hematopoiesis in mice lacking the entire granulocyte-macrophage colony-stimulating factor/interleukin3/interleukin-5 functions. *Blood* 88: 2458–64, 1996.
54. Palframan RT, Collins PD, Severs NJ, Rothery S, Williams TJ, Rankin SM. Mechanisms of acute eosinophil mobilization from the bone marrow stimulated by interleukin-5: The role of specific adhesion molecules and phosphatidylinositol 3-kinase. *J Exp Med* 188: 1621–32, 1998.
55. Mishra A, Hogan SP, Lee JJ, Foster PS, Rothenberg ME. Fundamental signals that regulate eosinophil homing to the gastrointestinal tract. *J Clin Invest* 103: 1719–27, 1999.
56. Ponath PD, Qin S, Post TW, Wang J, Wu L, Gerard NP, Newman W, Gerard C, Mackay CR. Molecular cloning and characterization of a human eotaxin receptor expressed selectively on eosinophils. *J Exp Med* 183: 2437–48, 1996.
57. Sehmi R, Dorman S, Baatjes A, Watson R, Foley R, Ying S, Robinson DS, Kay AB, O'Byrne PM, Denburg JA. Allergen-induced fluctuation in CC chemokine receptor 3 expression on bone marrow CD34+ cells from asthmatic subjects: Significance for mobilization of haemopoietic progenitor cells in allergic inflammation. *Immunology* 109: 536–46, 2003.
58. Bruijnzeel PL. Eosinophil tissue mobilization in allergic disorders. *Ann NY Acad Sci* 725: 259–67, 1994.
59. Luijk B, Lindemans CA, Kanters D, van der Heijde R, Bertics P, Lammers JW, Bates ME, Koenderman L. Gradual increase in priming of human eosinophils during extravasation from peripheral blood to the airways in response to allergen challenge. *J Allergy Clin Immunol* 115: 997–1003, 2005.
60. Spoelstra FM, Postma DS, Hovenga H, Noordhoek JA, Kauffman HF. Interferon-gamma and interleukin-4 differentially regulate ICAM-1 and VCAM-1 expression on human lung fibroblasts. *Eur Respir J* 14: 759–66, 1999.
61. Rosenberg HF, Phipps S, Foster PS. Eosinophil trafficking in allergy and asthma. *J Allergy Clin Immunol* 119: 1303–10, 2007.
62. Elsner J, Kapp A. Regulation and modulation of eosinophil effector functions. *Allergy* 54: 15–26, 1999.
63. Homey B, Zlotnik A. Chemokines in allergy. *Curr Opin Immunol* 11: 626–34, 1999.
64. Nagase H, Yamaguchi M, Jibidi S, Yamada H, Ohta K, Kawasaki H, Yoshie O, Yamamoto K, Morita Y, Hirai K. Eosinophil chemotaxis by chemokines: A study by a simple photometric assay. *Allergy* 54: 944–50, 1999.

65. Shahabuddin S, Ponath P, Schleimer RP. Migration of eosinophils across endothelial cell monolayers: Interactions among IL-5, endothelial-activating cytokines, and C-C chemokines. *J Immunol* 164: 3847–54, 2000.
66. Phillips RM, Stubbs VE, Henson MR, Williams TJ, Pease JE, Sabroe I. Variations in eosinophil chemokine responses: An investigation of CCR1 and CCR3 function, expression in atopy, and identification of a functional CCR1 promoter. *J Immunol* 170: 6190–201, 2003.
67. DiScipio RG, Daffern PJ, Jagels MA, Broide DH, Sriramarao P. A comparison of C3a and C5a-mediated stable adhesion of rolling eosinophils in postcapillary venules and transendothelial migration *in vitro* and *in vivo*. *J Immunol* 162: 1127–36, 1999.
68. Warringa RA, Mengelers HJ, Kuijper PH, Raaijmakers JA, Bruijnzel PL, Koenderman L. *In vivo* priming of platelet-activating factor-induced eosinophil chemotaxis in allergic asthmatic individuals. *Blood* 79: 1836–41, 1992.
69. Tager AM, Dufour JH, Goodarzi K, Bercury SD, von Adrian UH, Luster AD. BLTR mediates leukotriene (4)-induced chemotaxis and adhesion and plays a dominant role in eosinophil accumulation in a murine model of peritonitis. *J Exp Med* 192: 439–46, 2000.
70. Hirai H, Tanaka K, Yoshie O, Ogawa K, Kenmotsu K, Takamori Y, Ichimasa M, Sugamura K, Nakamura M, Takano S, Nagata K. Prostaglandin D2 selectively induces chemotaxis in T-helper type 2 cells, eosinophils, and basophils via seven-transmembrane receptor CRTH2. *J Exp Med* 193: 255–62, 2001.
71. Bandeira-Melo C, Bozza PT, Dias BL, Cordeiro RS, Jose PJ, Martins MA, Serhan CN. Lipoxin (LX) A4 and aspirin-triggered 15-epi-LXA4 block allergen-induced eosinophil trafficking. *J Immunol* 164: 2267–71, 2000.
72. Pope SM, Brandt EB, Mishra A, Hogan SP, Zimmermann N, Matthei KI, Foster PS, Rothenberg ME. IL-13 induces eosinophil recruitment into the lung by an IL-15- and eotaxin-dependent mechanism. *J Allergy Clin Immunol* 108: 594–601, 2001.
73. Simon HU, Blaser K. Inhibition of programmed eosinophil death: A key pathogenic event for eosinophilia?. *Immunology Today* 16: 53–55, 1995.
74. Kankaanranta H, Lindsay MA, Giembycz MA, Zhang X, Moilanan E, Barnes PJ. Delayed eosinophil apoptosis in asthma. *J Allergy Clin Immunol* 106: 77–83, 2000.
75. Woolley KL, Gibson PG, Carty K, Wilson AJ, Twaddell SH, Woolley MJ. Eosinophil apoptosis and the resolution of airway inflammation in asthma. *Am J Respir Crit Care Med* 154: 237–43, 1996.
76. Sexton DW, Blaylock MG, Walsh GM. Human alveolar epithelial cells engulf apoptotic eosinophils by means of integrin- and phosphatidylserine receptor-dependent mechanisms: A process upregulated by dexamethasone. *J Allergy Clin immunol* 108: 962–69, 2001.
77. Uller L, Andersson M, Greiff L, Persson CG, Erjefält JS. Occurrence of apoptosis, secondary necrosis, and cytolysis in eosinophil nasal polyps. *Am J Respir Crit Care Med* 170: 742–47, 2004.
78. Yamaguchi Y, Suda T, Ohta S, Tominaga K, Miura Y, Kasahara T. Analysis of the survival of mature eosinophils: Interleukin-5 prevents apoptosis in human eosinophils. *Blood* 78: 2542–47, 1991.
79. Conus S, Bruno A, Simon HU. Leptin is an eosinophil survival factor. *J Allergy Clin Immunol* 116: 1228–34, 2005.
80. Yousefi S, Hoessli DC, Blaser K, Mills GB, Simon HU. Requirement of Lyn and Syk tyrosine kinases for the prevention of apoptosis by cytokines in human eosinophils. *J Exp Med* 183: 1407–14, 1996.
81. Simon HU, Yousefi S, Dibbert B, Levi-Schaffer F, Blaser K. Anti-apoptotic signals of granulocyte-macrophage colony-stimulating factor are transduced via Jak2 tyrosine kinase in eosinophils. *Eur J Immunol* 27: 3536–39, 1997.
82. Vassina EM, Yousefi S, Simon D, Zwicky C, Conus S, Simon HU. cIAP-2 and survivin contribute to cytokine-mediated delayed eosinophil apoptosis. *Eur J Immunol* 36: 1975–84, 2006.
83. Huang HM, Huang CJ, Yen JJ. Mcl-1 is a common target of stem cell factor and interleukin-5 for apoptosis in prevention activity via MEK/MAPK and PI-3K/Akt pathways. *Blood* 96: 1764–71, 2000.
84. Dibbert B, Daigle I, Braun D, Schranz C, Weber M, Blaser K, Zangemeister-Wittke U, Akbar AN, Simon HU. Role for Bcl-xL in delayed eosinophil apoptosis mediated by granulocyte-macrophage colony-stimulating factor and interleukin-5. *Blood* 92: 778–83, 1998.
85. Simon HU. Molecules involved in the regulation of eosinophil apoptosis. *Chem Immunol Allergy* 91: 49–58, 2006.
86. Tsukahara K, Nakao A, Hiraguri M, Miike S, Mamura M, Saito Y, Iwamoto I. Tumor necrosis factor-alpha mediates antiapoptotic signals partially via p38MAP kinase activation in human eosinophils. *Int Arch Allergy Immunol* 120: S54–59, 1999.
87. Robertson NM, Zangrilli JG, Steplewski A, Hastie A, Lindemeyer RG, Planeta MA, Smith MK, Innocent N, Musani A, Pascual R, Peters S, Litwack G. Differential expression of TRAIL and TRAIL receptors in allergic asthmatics following segmental antigen challenge: Evidence for a role of TRAIL in eosinophil survival. *J Immunol* 169: 5986–96, 2002.
88. Heinisch IV, Bizer C, Volgger W, Simon HU. Functional CD137 receptors are expressed by eosinophils from patients with IgE-mediated allergic responses but not by eosinophils from patients with non-IgE-mediated eosinophilic disorders. *J Allergy Clin Immunol* 108: 21–28, 2001.
89. Hebestreit H, Yousefi S, Balatti I, Weber M, Crameri R, Simon D, Hartung K, Schapowal A, Blaser K, Simon HU. Expression and function of the Fas receptor on human blood and tissue eosinophils. *Eur J Immunol* 26: 1775–80, 1996.
90. Simon HU. Eosinophil apoptosis – pathophysiologic and therapeutic implications. *Allergy* 55: 910–15, 2000.
91. Wardlaw AJ, Brightling C, Green R, Woltmann G, Pavord I. Eosinophils in asthma and other allergic diseases. *Br Med Bull* 56: 985–1003, 2000.
92. Bousquet J, Chanez P, Lacoste JY, Barneon G, Ghavanian N, Enander I, Venge P, Ahlstedt S, Simony-Lafontaine J, Godard P. Eosinophilic inflammation in asthma. *N Engl J Med* 323: 1033–39, 1990.
93. Lee JJ, Dimina D, Macias MP, Ochkur SI, McGarry MP, O'Neill KR, Protheroe C, Pero R, Nguyen T, Cormier SA, Lenkiewicz E, Colbert D, Rinaldi L, Ackerman SJ, Irvin CG, Lee NA. Defining a link with asthma in mice congenitally deficient in eosinophils. *Science* 305: 1773–76, 2004.
94. Humbles AA, Lloyd CM, McMillan SJ, Friend DS, Xanthou G, McKenna EE, Ghiran S, Gerard NP, Yu C, Orkin SH, Gerard C. A critical role for eosinophils in allergic airways remodelling. *Science* 305: 1776–79, 2004.
95. Frigas E, Motojima S, Gleich GJ. The eosinophilic injury to the mucosa of the airways in the pathogenesis of bronchial asthma. *Eur Respir J* 13: S123–35, 1991.
96. Sedgwick JB, Vrtis RF, Jansen KJ, Kita H, Bartemes K, Busse WW. Peripheral blood eosinophils from patients with allergic asthma contain increased intracellular eosinophil-derived neurotoxin. *J Allergy Clin Immunol* 114: 568–74, 2004.
97. Venge P, Byström J, Carlson M, Hakansson L, Karawacjzyk M, Peterson C, Seveus L, Trulson A. Eosinophil cationic protein (ECP): Molecular and biological properties and the use of ECP as a marker of eosinophil activation in disease. *Clin Exp Allergy* 29: 1172–86, 1999.
98. Munthe-Kaas MC, Gerritsen J, Carlsen KH, Undlien D, Egeland T, Skinningsrud B, Torres T, Carlsen KL. Eosinophil cationic protein (ECP) polymorphisms and association with asthma, s-ECP levels and related phenotypes. *Allergy* 62: 429–36, 2007.
99. Saleh D, Ernst P, Lim S, Barnes PJ, Giaid A. Increased formation of the potent oxidant peroxynitrite in the airways of asthmatic patients is associated with induction of nitric oxide synthase: Effect of inhaled glucocorticoid. *FASEB J* 12: 929–37, 1998.
100. Kaminsky DA, Mitchell J, Carroll N, James A, Soultanakis R, Janssen Y. Nitrotyrosine formation in the airways and lung parenchyma of patients with asthma. *J Allergy Clin Immunol* 104: 747–54, 1999.
101. Alving K, Weitzberg E, Lundberg JM. Increased amount of nitric oxide in exhaled air of asthmatics. *Eur Respir J* 6: 1368–70, 1993.
102. Turner S. The role of exhaled nitric oxide in the diagnosis, management and treatment of asthma. *Mini Rev Med Chem* 7: 539–42, 2007.

103. Green RH, Brightling CE, McKenna S, Hargadon B, Parker D, Bradding P, Wardlaw AJ, Pavord ID. Asthma exacerbations and sputum eosinophil counts: A randomised controlled trial. *Lancet* 360: 1715–21, 2002.
104. Harrison AM, Bonville CA, Rosenberg HF, Domachowske JB. Respiratory syncytial viral-induced chemokine expression in the lower airways: Eosinophil recruitment and degranulation. *Am J Respir Crit Care Med* 159: 1918–24, 1999.
105. O'Donnell MC, Ackerman SJ, Gleich GJ, Thomas LL. Activation of basophil and mast cell histamine release by eosinophil granule major basic protein. *J Exp Med* 157: 1981–91, 1983.
106. Kita H, Abu-Ghazaleh RI, Sur S, Gleich GJ. Eosinophil major basic protein induces degranulation and IL-8 production by human eosinophils. *J Immunol* 154: 4749–58, 1995.
107. Moy JN, Gleich GJ, Thomas LL. Noncytotoxic activation of neutrophils by eosinophil granule major basic protein. Effect on superoxide anion generation and lysosomal enzyme release. *J Immunol* 145: 2626–32, 1990.
108. Page SM, Gleich GJ, Roebuck KA, Thomas LL. Stimulation of neutrophil interleukin-8 production by eosinophil granule major basic protein. *Am J Respir Cell Mol Biol* 21: 230–37, 1999.
109. Rohrbach MS, Wheatley CL, Slifman NR, Gleich GJ. Activation of platelets by eosinophil granule proteins. *J Exp Med* 172: 1271–74, 1990.
110. Jacoby DB, Gleich GJ, Fryer AD. Human eosinophil major basic protein is an endogenous allosteric antagonist at the inhibitory muscarine M2 receptor. *J Clin Invest* 91: 1314–18, 1993.
111. Pascual RM, Peters SP. Airway remodelling contributes to the progressive loss of lung function in asthma: An overview. *J Allergy Clin Immunol* 116: 477–86, 2005.
112. Kay AB, Phipps S, Robinson DS. A role for eosinophils in airway remodelling in asthma. *Trends Immunol* 25: 477–82, 2004.
113. Flood-Page P, Menzies Gow A, Phipps S, Ying S, Wangoo A, Ludwig MS, Barnes N, Robinson S, Kay AB. Anti-IL-5 treatment reduces deposition of ECM proteins in the bronchial subepithelial basement membrane of mild atopic asthmatics. *J Clin Invest* 112: 1029–36, 2003.
114. Foley SC, Prefontaine D, Hamid Q. Role of eosinophils in airway remodeling. *J Allergy Clin Immunol* 119: 1563–66, 2007.
115. Rochester CL, Ackerman SJ, Zheng T, Elias JA. Eosinophil-fibroblast interactions. Granule major basic protein interacts with IL-1 and transforming growth factor-beta in the stimulation of lung fibroblast IL-6-type cytokine production. *J Immunol* 156: 4449–56, 1996.
116. Wang JG, Mahmud SA, Thompson JA, Geng JG, Key NS, Slungaard A. The principal eosinophil peroxidase product, HOSNC, is a uniquely potent phagocyte oxidant inducer of endothelial cell tissue factor activity: A potential mechanism for thrombosis in eosinophilic inflammatory states. *Blood* 107: 558–65, 2006.
117. Wiehler S, Cuvelier SL, Chakrabarti S, Patel KD. p38 MAP kinase regulates rapid matrix metalloproteinase-9 releases from eosinophils. *Biochem Biophys Res Commun* 315: 463–67, 2004.
118. Mattos W, Lim S, Russell R, Jatakanon A, Chung KF, Barnes PJ. Matrix metalloproteinase-9 expression in asthma: Effect of asthma severity, allergen challenge, and inhaled corticosteroids. *Chest* 122: 1543–52, 2002.
119. Sur S, Crotty TB, Kephart GM, Hyma BA, Colby TV, Reed CE, Hunt LW, Gleich GJ. Sudden-onset fatal asthma. A distinct entity with few eosinophils and relatively more neutrophils in the airway submucosa?. *Am Rev Respir Dis* 148: 713–19, 1993.
120. Puxeddu I, Ribatti D, Crivellato E, Levi-Schaffer F. Mast cells and eosinophils: A novel link between inflammation and angiogenesis in allergic diseases. *J Allergy Clin Immunol* 116: 531–36, 2005.
121. Hoshino M, Takahashi M, Takai Y, Sim J, Aoike N. Inhaled corticosteroids decrease vascularity of the bronchial mucosa in patients with asthma. *Clin Exp Allergy* 31: 722–30, 2001.
122. Horiuchu T, Weller PF. Expression of vascular endothelial growth factor by human eosinophils: Upregulation by granulocyte macrophage colony-stimulating factor and interleukin-5. *Am J Respir Cell Mol Biol* 17: 70–77, 1997.
123. Temkin V, Aingorn H, Puxeddu I, Goldshmidt O, Zcharia E, Gleich GJ, Vlodavski I, Levi-Schaffer F. Eosinophil major basic protein: First identified natural heparanase-inhibiting protein. *J Allergy Clin Immunol* 113: 703–9, 2004.
124. Freisitzer C, Kaneider NC, Sturn DH, Mosheimer BA, Kähler CM, Wiedermann CJ. Expression and function of the vascular endothelial growth factor receptor FLT-1 in human eosinophils. *Am J Respir Cell Mol Biol* 30: 729–35, 2004.
125. Lee YC, Kwak YG, Song CH. Contribution of vascular endothelial growth factor to airway hyperresponsiveness and inflammation in a murine model of toluene diisocyanate-induced asthma. *J Immunol* 168: 3595–600, 2002.
126. Joubert P, Hamid Q. Role of airway smooth muscle in airway remodelling. *J Allergy Clin Immunol* 116: 713–16, 2005.
127. Weller PF, Lee CW, Foster DW, Corey EJ, Austen KF, Lewis RA. Generation and metabolism of 5-lipoxygenase pathway leukotrienes by human eosinophils: Predominant production of leukotriene C4. *Proc Natl Acad Sci USA* 80: 7626–30, 1983.
128. Cowburn AS, Sladek K, Soja J, Adamek L, Nizankowska E, Szczeklik A, Lam BK, Penrose JF, Austen KF, Holgate ST, Sampson AP. Overexpression of leukotriene C4 synthase in bronchial biopsies from patients with aspirin-tolerant asthma. *J Clin Invest* 101: 834–46, 1998.
129. Marom Z, Shelhamer JH, Bach MK, Morton DR, Kaliner M. Slow-reacting substances, leukotrienes C4 and D4, increase the release of mucus from human airways *in vitro*. *Am Rev Respir Dis* 126: 449–51, 1982.
130. Adelroth E, Morris MM, Hargreave FE, O'Byrne PM. Airway responsiveness to leukotrienes C4 and D4 and to methacholine in patients with asthma and normal controls. *N Engl J Med* 315: 480–84, 1986.
131. Lee E, Robertson T, Smith J, Kilfeather S. Leukitriene receptor antagonists and synthesis inhibitors reverse survival in eosinophils of asthmatic individuals. *Am J Respir Crit Care Med* 161: 1881–86, 2000.
132. Sedgwick JB, Calhoun WJ, Vrtis RF, Bates ME, McAllister PK, Busse WW. Comparison of airway and blood eosinophil function after *in vivo* allergen challenge. *J Immunol* 149: 3710–18, 1992.
133. Mengelers HJ, Maikoe T, Brinkman L, Hooibrink B, Lammers JW, Koenderman L. Immunophenotyping of eosinophils recovered from blood and BAL of allergic asthmatics. *Am J Respir Crit Care Med* 149: 345–51, 1994.
134. Ohkawara Y, Lim KG, Xing Z, Glibetic M, Nakano K, Dolovich J, Croitoru K, Weller PF, Jordana M. CD40 expression by human peripheral blood eosinophils. *J Clin Invest* 97: 1761–66, 1996.
135. Shi HZ, Humbles A, Gerard C, Jin Z, Weller PF. Lymph node trafficking and antigen presentation by endobronchial eosinophils. *J Clin Invest* 105: 945–53, 2000.
136. Duez C, Dakhama A, Tomkinson A, Marquillies P, Balhorn A, Tonnel AB, Bratton DL, Gelfand EW. Migration and accumulation of eosinophils toward regional lymph nodes after airway allergen challenge. *J Allergy Clin Immunol* 114: 820–25, 2004.
137. Watt AP, Schock BC, Ennis M. Neutrophils and eosinophils: Clinical implications of their appearance, presence and disappearance in asthma and COPD. *Curr Drug Targets Inflamm Allergy* 4: 415–23, 2005.
138. Tetley TD. Inflammatory cells and obstructive pulmonary disease. *Curr Drug Targets Inflamm Allergy* 4: 607–18, 2005.
139. Yamamoto C, Yoneda T, Yoshikawa M, Fu A, Tokuyama T, Tsukaguchi K, Narita N. Airway inflammation in COPD assessed by sputum levels of interleukin-8. *Chest* 112: 505–10, 1997.
140. Balzano G, Stefanelli F, Iorio C, De Felice A, Melillo EM, Martucci M, Melillo G. Eosinophilic inflammation in stable chronic obstructive pulmonary disease. Relationship with neutrophils and airway function. *Am J Respir Crit Care Med* 160: 1486–92, 1999.
141. Bhowmik A, Seemungal TAR, Sapsford RJ, Wedzicha JA. Relation of sputum inflammatory markers to symptoms and lung function changes in COPD exacerbations. *Thorax* 55: 114–20, 2000.

142. Lams BE, Sousa AR, Rees PJ, Lee TH. Immunopathology of the small-airway submucosa in smokers with and without chronic obstructive pulmonary disease. *Am J Respir Crit Care Med* 158: 1518–23, 1998.
143. Brightling CE, Monteiro W, Ward R, Parker D, Morgan MD, Wardlaw AJ, Pavord ID. Sputum eosinophilia and short-term response to prednisone in chronic obstructive pulmonary disease: A randomised controlled trial. *Lancet* 356: 1480–85, 2000.
144. Lams BE, Sousa AR, Rees PJ, Lee TH. Subepithelial immunopathology of the large airways in smokers with and without chronic obstructive pulmonary disease. *Eur Respir J* 15: 512–16, 2000.
145. Saetta M, Di Stefano A, Maestrelli P, Turato G, Mapp CE, Pieno M, Zanguochi G, Del Prete G, Fabbri LM. Airway eosinophilia and expression of interleukin-5 protein in asthma and in exacerbations of chronic bronchitis. *Clin Exp Allergy* 26: 766–74, 1996.
146. Zhu J, Qiu YS, Majumdar S, Gamble E, Matin D, Turato G, Fabbri LM, Bernes N, Saetta M, Jeffery PK. Exacerbations of bronchitis: Bronchial eosinophilia and gene expression for interleukin-4, interleukin-5, and eosinophil chemoattractants. *Am J Respir Crit Care Med* 164: 109–16, 2001.
147. Bocchino V, Bertorelli G, Bertrand CP, Ponath PD, Newman W, Franco C, Marruchella A, Merlini S, Del Donno M, Zhuo X, Olivieri D. Eotaxin and CCR3 are upregulated in exacerbations of chronic bronchitis. *Allergy* 57: 17–22, 2002.
148. Papi A, Luppi F, Franco F, Fabbri LM. Pathophysiology of exacerbations in chronic obstructive pulmonary disease. *Proc Am Thorac Soc* 3: 245–51, 2006.
149. Riise GC, Ahlstedt S, Larsson S, Enander I, Jones I, Larsson P, Andersson B. Bronchoal inflammation in chronic bronchitis assessed by measurement of cell products in bronchial lavage fluids. *Thorax* 50: 360–65, 1995.
150. Rutgers SR, Timens W, Kaufmann HF, van der Mark TW, Koeter GH, Postma DS. Comparison of induced sputum with bronchial wash, bronchoalveolar lavage and bronchial biopsies in COPD. *Eur Respir J* 15: 109–15, 2000.
151. Keatings VM, Barnes PJ. Granulocyte activation markers in induced sputum: Comparison between chronic obstructive pulmonary disease, asthma and normal subjects. *Am J Respir Crit Care Med* 155: 449–53, 1997.
152. O'Donnell R, Breen D, Wilson S, Djukanovic R. Inflammatory cells in the airways in COPD. *Thorax* 61: 448–54, 2006.
153. Pizzichini E, Pizzichini MM, Gibson P, Parameswaran K, Gleich GJ, Berman L, Dolovich J, Hargreave FE. Sputum eosinophilia predicts benefit from prednisone in smokers with chronic obstructive bronchitis. *Am J Respir Crit Care Med* 158: 1511–17, 1998.
154. Walsh GM, Sexton DW, Blaylock MG. Corticosteroids, eosinophils and bronchial epithelial cells: New insights into the resolution of inflammation in asthma. *J Endocrinol* 178: 37–43, 2003.
155. Bentley AM, Hamis Q, Robinson DS, Schotman E, Meng Q, Assoufi B, Kay AB, Durham SR. Prednisone treatment in asthma. Reduction in the numbers of eosinophils, T cells, tryptase-only positive mast cells, and modulation of IL-4, IL-5, and interferon-gamma cytokine gene expression within the bronchial mucosa. *Am J Crit Care Med* 153: 551–56, 1996.
156. Steinke JW, Culp JA. Leukotriene synthesis inhibitors versus antagonists: The pros and cons. *Curr Allergy Asthma Rep* 7: 126–33, 2007.
157. Berger W, De Chandt MT, Cairns CB. Zileuton: Clinical implications of 5-lipoxygenase inhibition in severe airway disease. *Int J Clin Pract* 61: 663–76, 2007.
158. Simon HU, Seelbach H, Ehmann R, Schmitz M. Clinical and immunological effects of low-dose IFN-alpha treatment in patients with corticosteroid-resistant asthma. *Allergy* 58: 1250–55, 2003.
159. Boguniewicz M, Schneider LC, Milgrom H, Newell D, Kelly N, Tam P, Izu AE, Jaffe HS, Bucalo LR, Leung DY. Treatment of steroid-dependent asthma with recombinant interferon-gamma. *Clin Exp Allergy* 23: 785–90, 1993.
160. Leckie MJ, ten Brinke A, Khan J, Diamant Z, O'Connor BJ, Walls CM, Mathur AK, Cowley HC, Chung KF, Djukanovic R, Hansel T, Holgate ST, Sterk PJ, Barnes PJ. Effects of an interleukin-5 blocking monoclonal antibody on eosinophils, airway hyper-responsiveness, and the late asthmatic response. *Lancet* 356: 2144–48, 2000.
161. Buttner C, Lun A, Splettstoesser T, Kunkel G, Renz H. Monoclonal anti-interleukin-5 treatment suppresses eosinophil but not T-cell functions. *Eur Respir J* 21: 799–803, 2003.
162. Kips JC, O'Connor BJ, Langley SJ, Woodcock A, Kerstjens HA, Postma DS, Danzig M, Cuss F, Pauwels RA. Effect of SCH55700, a humanized anti-human interleukin-5 antibody, in severe persistent asthma: A pilot study. *Am J Respir Crit Care Med* 167: 1655–59, 2003.
163. Flood-Page PT, Menzies-Gow AN, Kay AB, Robinson DS. Eosinophil's role remains uncertain as anti-IL-5 only partially depletes numbers in asthmatic airway. *Am J Respir Crit Care Med* 167: 199–204, 2003.
164. Berry M, Brightling C, Pavord I, Wardlaw AJ. TNF-a in asthma. *Curr Opin Pharmacol* 7: 279–82, 2007.
165. Bryan SA, O'Connor BJ, Matti S, Leckie MJ, Kanabar V, Khan J, Warrington SJ, Renzetti L, Rames A, Bock JA, Boyce MJ, Hansel TT, Holgate ST, Barnes PJ. Effects of recombinant human interleukin-12 on eosinophils, airway hyper-responsiveness, and the late asthmatic response. *Lancet* 356: 2114–16, 2114–153, 2000.
166. Romano SJ. Selectin antagonists: Therapeutic potential in asthma and COPD. *Treat Respir Med* 4: 85–94, 2005.
167. Beeh KM, Beier J, Meyer M, Buhl R, Zahlten R, Wolff G. Bimosiamose, an inhaled small-molecule pan-selectin antagonist, attenuates late asthmatic reactions following allergen challenge in mild asthmatics: A randomized, double-blind, placebo-controlled clinical cross-over-trial. *Pulm Pharmacol Ther* 19: 233–41, 2006.
168. Wegmann M, Göggel R, Sel S, Erb KJ, Kalkbrenner F, Renz H, Garn H. Effects of a low-molecular-weight CCR-3 antagonist on chronic experimental asthma. *Am J Respir Cell Mol Biol* 36: 61–67, 2007.
169. Okigami H, Takeshita K, Tajimi M, Komura H, Albers M, Lehmann TE, Rölle T, Bacon KB. Inhibition of eosinophilia *in vivo* by a small molecule inhibitor of very late antigen (VLA)-4. *Eur J Pharmacol* 559: 202–9, 2007.

13

CHAPTER

The Lymphocyte in Asthma and COPD

James G. Martin and Manuel G. Cosio

Meakins-Christie Laboratories and the Respiratory Division, Department of Medicine, McGill University, Montreal, Quebec, Canada

INTRODUCTION

Asthma and chronic obstructive pulmonary disease (COPD) are chronic inflammatory diseases of the airways that share a high prevalence and, when severe, impose a substantial burden on affected persons and society at large. Although controversy has surrounded the extent to which these conditions may be related, distinct patterns of airway inflammation have been described in the two conditions and evidence for the participation of different T-cell subsets in the pathogenesis of the two diseases is emerging. There is, however, some evidence of convergence of the inflammatory processes in advanced COPD and in severe asthma but, for the most part, there are clear and obvious differences. In particular, although the chronic inflammatory infiltrate in both diseases results in airway obstruction, obstruction is usually reversible in asthma, and is largely irreversible and progressive in COPD. These key differences in the pathophysiology of the two diseases are due to the type and consequences of the inflammation. Inflammation is largely limited to the airways in asthma whereas both the airways and parenchyma are affected in COPD and in such a way that irreversible fibrosis of the small airways and parenchymal destruction with emphysema may result. Both diseases when fully established show evidence of an adaptive immune response with an excess of activated T-cells, but of different phenotypes, secreting type 2 cytokines in asthma and type 1 cytokines in COPD. Descriptions of the immunopathology of asthma as well as data from animal models leave little doubt about the central place of the $CD4^+$ T-cell in virtually all of the key features of asthma, ranging from the characteristic eosinophil-rich inflammation to bronchoconstriction and tissue remodeling. T-cell cytokines recruit, activate, and enhance the survival of a range of inflammatory cells that are the more direct mediators of the pathologic processes. The participation of the T-cell in COPD is relatively less well understood but involves $CD8^+$ T-cells to a greater extent. Both diseases are now known to have features mediated by an adaptive immune response to antigens, attributable to inhaled soluble antigens in asthma, and to self-antigens either newly revealed or modified by tissue injury secondary to tobacco smoke in COPD. Why the mucosal immune system of the respiratory tract responds in such different manners to the antigenic stimuli, and results in the distinct disease phenotypes is still not very clear.

In this chapter we will review the evidence for the role of T-cells in the pathobiology of these two major causes of morbidity and mortality. We will also discuss the additional roles for certain T-cell subsets, such as the potential effect of $\gamma\delta$ T-cells in the maintenance of airway epithelial integrity, that extend the range of functions beyond the classical adaptive immune functions most commonly thought of in association with the T-cell.

HISTORICAL NOTE

Early preoccupations with the role of lymphocytes in asthma revolved around the role of B-cells in IgE production. It was observed as early as 1975 that human asthmatic subjects appeared to have a relative lymphopenia [1] and a defective T-cell response to stimulation by mitogens such as phytohemagglutinin and concanavalin A [2]. B-cells from atopic subjects were shown to spontaneously synthesize IgE *in vitro* and T-cells from the affected individuals were much less inhibitory of this synthesis compared to T-cells from unaffected persons [3]. Specific immunotherapy led to a correction of the reduced inhibitory T-cell subset [4]. Subsequently elevated $CD4^+$ to $CD8^+$ ratios in the blood of asthmatic subjects was also interpreted as suggesting a deficiency of suppressor T-cell function in asthmatics [5]. Differences in the ratio of $CD4^+$ to $CD8^+$ T-cells were noted following challenge of allergic asthmatic subjects as a function of whether the subjects demonstrated isolated early responses to challenge or had dual airway responses, with both early and late bronchoconstriction. The hypothesis proposed was that $CD8^+$ T-cells were recruited from the blood to the airways and prevented the development of the late airway response in single responders [6]. Activation of $CD4^+$ T-cells (surface expression of IL-2R, VLA-1, and HLA-DR) was also noted during acute asthma exacerbations [7]. HLA-DR expression on cells in bronchial biopsies of stable asthmatics was correlated with airway hyperresponsiveness [8].

A more obvious link of T-cells to airway inflammation emerged as evidence of the synthesis of high molecular weight chemotactic factors for eosinophils was obtained by injecting concanavalin A into the skin of a guinea pig [9]. However, without doubt, the single most important step forward in understanding the place of T-cells in allergic airway inflammation occurred with the description of Th1 and Th2 type $CD4^+$ T-cells in 1986 by Mosmann and colleagues [10]. For the subsequent 20 years research on the role of the T-cell in asthma has focused singularly on the exploration of this paradigm. Indeed much of the inflammatory process in allergic asthma is explained by this paradigm although added complexity has emerged as new subsets, such as regulatory T-cells, Th17 cells, $\gamma\delta$ T-cells, and *i*NKT cells that have been described, and have begun to be integrated into the pathogenetic schema of inflammatory airway disease.

Early observations on airway inflammation in subjects with COPD revealed typical innate immune inflammation. Young cigarette smokers were noted to have increased cellularity in the bronchoalveolar lavage (BAL) fluid that was attributable to macrophages but few neutrophils were present [11], while older smokers had increased numbers of both neutrophils and alveolar macrophages [12]. From these data originated the concept that the destruction of lung tissue resulted from the release of potent proteinases from these cells and the proteinase–anti-proteinase balance became the paradigm for the pathogenesis of COPD for almost 40 years. The nature and mechanisms of inflammation were reconsidered following the description by Finkelstein *et al.* of a prominent T-cell infiltration in the lungs of patients with COPD, the degree of which was strongly related to the extent of emphysema [13]. These findings suggested new mechanisms for the pathogenesis of COPD through T-cell involvement and prompted us to suggest that COPD might be an autoimmune disease triggered by cigarette smoking [14, 15].

LYMPHOCYTES IN THE RESPIRATORY TRACT

The lymphocytes in the respiratory tract are present in the form of localized bronchus-associated lymphoid tissue (BALT) and the nasal-associated lymphoid tissue (NALT) that are organized lymphoid aggregates, comprising T- and B-lymphocytes that may mediate adaptive immune responses to inhaled antigens. BALT is present at bronchial bifurcations and may be responsible for the synthesis and secretion of immunoglobulins in response to antigenic stimulation. Humoral immune responses elicited by BALT are primarily IgA secretion. The BALT is in many respects analogous to mucosal lymphoid aggregates in the intestine that has important functions in regulating intestinal mucosal immunity (reviewed in [16]). The epithelial cells overlying the BALT lack cilia and appear to be equivalent to M-cells in the intestinal epithelium. The M-cells are derived from basal cells in the respiratory epithelium and, as in the intestine overlying Peyer's patches, have the capacity to selectively phagocytose and pinocytose particles and molecules in their immediate environment. Lymphocytes are also distributed within the epithelium and more diffusely through the airway wall, invading all of the tissues in the setting of airway inflammation.

INCITING STIMULI

Both asthma and COPD are diseases that involve inflammation mediated by the respiratory mucosal immune system. In asthma aeroallergens are the best characterized inducers of disease although other causes such as strong irritants (often mediating damage through oxidant injury) and viral illness are also of importance. These same stimuli may trigger asthma attacks also. The place of the adaptive immune system in the pathogenesis of asthma caused by nonallergic mechanisms is not clear. It is possible that airway hyperresponsiveness may be induced as a result of airway remodeling from airway insults, bypassing the need for a major participation of adaptive immunity in the process. Model systems, based in recent years on murine and rodent models, have been heavily exploited to unravel the complex immunological reactions triggered by allergen. The literature is replete with references to murine asthma, for example, but it must be acknowledged that models based on a small number of allergen challenges model only a limited part of the disease. Growing literature on the role of viruses, such as respiratory syncytial virus and rhinovirus, will complement the results from allergen challenge to provide a more complete picture of the triggers and inducers of asthma. Likewise, the T-cell contribution to other forms of injury such as oxidant injury has received little attention. To date the mechanisms for activation of T-cells in asthma have dealt almost exclusively with allergic mechanisms.

The inflammatory reaction to cigarette smoke is central to COPD and the global initiative workshop summary for chronic obstructive lung disease (GOLD) definition of COPD states that *"the airflow limitation is usually progressive and associated with an abnormal inflammatory response of the lungs to noxious particles or gases"(GOLD Report update 2007, www.GOLDCOPD.com downloaded May 2008)* [17]. In both disease processes there is an important contribution of the innate immune response to the pathobiology and close links to lymphocyte function. Of particular importance is the dendritic cell, but macrophages and other cells may also play a role in conditioning the T-cell response. Cigarette smoke and other pollutants (ozone, NO_2, diesel particles) produce an acute innate immune reaction in the lung as do the triggers of asthma. The epithelium is exposed to all of these triggers over its surface and through damage, stress, or stimulation via receptors, such as Toll receptors, is important in initiating the response. The present evidence suggests that, by sending "danger" signals in response to cigarette smoke, the epithelium is responsible for the initiation and possibly maintenance of the innate immune response seen in smokers. Over 2000 different xenobiotic compounds have been identified in cigarette smoke, and it has been estimated that there are 10^{14} free radicals in each puff of cigarette smoke [18]. Not surprisingly, it has been shown that cigarette smoke is cytotoxic to epithelial cells and that the extent of injury produced is directly related to the concentration of smoke to which the cells were exposed [19].

The sustained innate immune response found in the lungs of cigarette smokers (neutrophils, macrophages, eosinophils, mast cells, $\gamma\delta$ T-cells, dendritic cells, and perhaps natural killer (NK) cells), and their products (cytokines, oxygen radicals, proteinases) are capable of producing matrix and cellular damage. Studies in mice have shown that after 24 h of cigarette smoke exposure, measurable increases in desmosine, a marker of elastin breakdown and hydroxyproline, a marker of collagen breakdown, are found in BAL and their levels are correlated with the number of neutrophils in the BAL [20, 21]. Evidence of connective tissue breakdown in human smokers also exists. Increased plasma and urine levels of elastin-derived peptides and desmosine [22, 23] have been found in COPD patients when compared with nonsmokers [24, 25]. However, the levels of elastin breakdown products are also elevated in smokers without COPD [24, 26–29]. Smokers with rapid decline in lung function, likely to develop COPD, were found to excrete 36% more desmosine than low decliners [30].

However the protease–anti-protease imbalance paradigm by itself cannot explain why only some smokers develop COPD. It might be closer to reality to consider the connective tissue breakdown as one of the many consequences of the innate inflammatory reaction triggered by smoking. Furthermore, these breakdown products or peptides once revealed to the immune system may be antigenic and be presented to T-cells, eventually triggering, in some smokers, a T-cell adaptive immune response against lung antigens (autoimmunity), as has been recently shown [31]. There is extensive literature showing that infectious and environmental agents have the potential to alter "self-proteins" that may then be recognized as antigens by the adaptive immune system. Among the important protein modifiers present in smokers are free radicals/oxidative stress and xenobiotic agents which abound in cigarette smoke. Nitric oxide (NO) *per se* or following its conversion to peroxynitrite by superoxide or other reactive oxygen species (ROS) may produce mitochondrial damage, DNA strand breaks, structural/functional modification of proteins, and ultimately cell death. Oxidative modification of proteins has been implicated in the immune mechanisms of rheumatoid arthritis, multiple sclerosis, and arteriosclerosis [32]. Induction of organ-specific autoimmune disease following tissue trauma, inflammation, and viral or bacterial infections has been frequently reported and likely occurs via tissue damage that results in the availability of previously hidden antigens or by molecular mimicry with viral or bacterial products [33]. Infections also provide abundant cytokines that may affect the expression of co-stimulatory molecules, an additional stimulus important in the perpetration of the immune response.

Airway or pulmonary parenchymal damage from cigarette smoke could easily follow the same mechanisms. The tissue injury and potential modification of self-proteins could generate antigens (e.g. release of cryptic antigens, modified proteins, necrotic cells, apoptotic cells, etc.) and the associated innate immune inflammation associated with smoking, could provide the necessary soluble mediators and co-stimulatory signals for the initiation or perpetuation of an adaptive immune response. That this is the case has been recently shown by Lee and coworkers who identified elastin as an antigen responsible for the T-cell response in COPD [31]. However in a disease like COPD in which multiple cells and tissues seem to be potentially injured, other cell products from necrotic, stressed, or apoptotic cells might also act as antigenic determinants. Whether a component of autoimmunity complicates asthma is not known but it would not be altogether surprising since cellular damage and oxidative stress is also a feature of the disease. This has been recently shown in smokers with COPD in which autoantibodies were found to bind at least three different autoantigens [34].

MECHANISMS OF T-CELL ACTIVATION AND DIFFERENTIATION: MODEL SYSTEMS

Dendritic cells, CD4$^+$, and CD8$^+$ T-cells

Adaptive immune responses are initiated by the dense network of dendritic cells in and under the epithelium [35]. These cells have projections that appear to sample the airway luminal contents and can take up antigen for presentation to T-cells (Fig. 13.1). This presentation occurs in draining lymph nodes following maturation of the dendritic cells under the influence of cytokines such as granulocyte-macrophage colony stimulating factor (GM-CSF). T-cells sample the dendritic cell surface for antigens in a more or less stochastic process. High levels of antigen on the cell surface shorten the time required for the T-cell to form stable contacts with the dendritic cell [36]. In addition to dendritic cell–T-cell contact in the lymph nodes it has also been observed that T-cells come into contact with dendritic cells in the airway wall [37]. Indeed following allergen challenge there is a marked increase in the number of dendritic cells in contact with T-cells. This interaction may short-circuit the longer process that is otherwise required

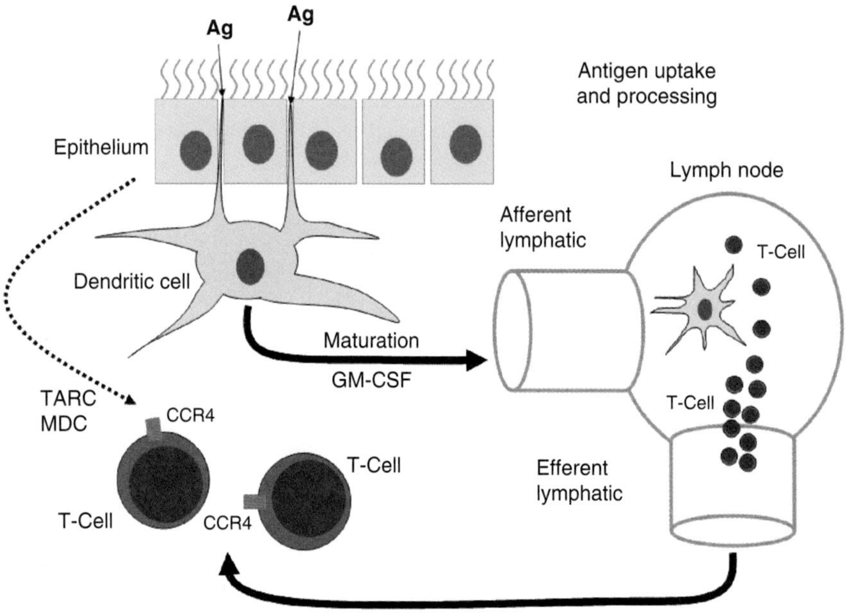

FIG. 13.1 Dendritic cells sample antigen material in the airway lumen through the cell processes that interdigitate between the epithelial cells. The dendritic cell then migrate to the draining lymph nodes and are sampled by T-cells that respond to the antigenic peptide on the surface of the dendritic cell. Maturation of the dendritic cell in response to the conditioning by cytokines such as granulocyte-macrophage colony stimulating factor (GM-CSF) enhances its antigen-presenting ability. T-cells with antigen specificity return to the mucosal surface following their egress from the lymph node through the efferent lymphatic. Chemokines such as TARC and MDC favor movement of Th2 cells into the tissues. These chemokines are produced by epithelial cells and dendritic cells, when appropriately stimulated.

and presumably is responsible for the rapid T-cell mediated response that follows allergen challenge *in vivo.*

Following engagement with the dendritic cell the T-cell is activated, secretes IL-2, and expresses the IL-2 receptor. Clonal expansion takes place and the T-cell phenotype is subsequently established in response to a variety of influences; cytokines, lipid mediators, and toll-like receptor (TLR) ligands all play a part in the latter phenomenon. The presentation of exogenous antigen in association with major histocompatibility complex (MHC) class II molecules leads to activation of $CD4^+$ T-cells with high-affinity antigen-specific T-cell receptors (Fig. 13.2). This is the predominant pathway activated by inhaled soluble antigens that commonly trigger allergic airway responses. Antigen produced intracellularly, such as may result from the infection or injury of pulmonary cells by virus or other agents like cigarette smoke, is processed by different intracellular pathways and presentation occurs in conjunction with MHC class I molecules. $CD8^+$ T-cell activation results. The processing and presentation of soluble exogenous antigen to $CD8^+$ T-cells in an MHC class I restricted manner is called cross presentation. Soluble antigens destined for cross presentation are taken up into early endosomes whereas those destined for classical MHC class II restricted presentation are taken up in lysosomes [38]. Heat shock proteins are released from cells undergoing lytic death, but not from cells dying of apoptosis [39] and necrotic cell lysates and also stressed cells have been reported to express cell surface HSP molecules [40] that could activate dendritic cells directly through receptors including CD91 [41] and TLR-4. Cell-associated proteins [42, 43] and particulate antigens [44] are much more effective at class MHC I restricted $CD8^+$ T-cells priming than are simple soluble molecules. Cross presentation is also essential for proper stimulation and functional development of $CD8^+$ T-cell effector and memory function since these depend on the encounter of both $CD4^+$ and $CD8^+$ T-cells with the same dendritic cell and the release of stimulatory factors from the $CD4^+$ T-cell. These encounters

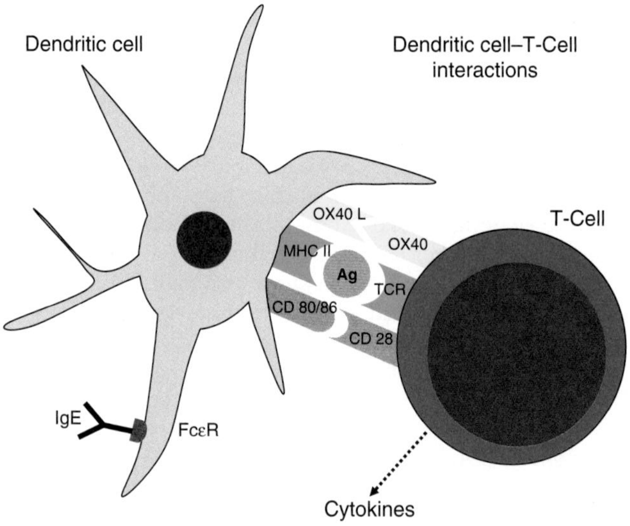

FIG. 13.2 The dendritic cell interacts with the T-cell through major histocompatibility molecules that harbor the antigenic peptide and the T-cell receptor. Various co-stimulatory molecules are required to allow effector functions to be established. OX40 and OX40L interactions are important for Th2 differentiation. CD28–CD80/86 interactions are important for both Th1 and Th2 development. The binding of IgE to the FcεR1 on the dendritic cells facilitates Th2 cell differentiation.

are not random but mediated by the upregulation of the chemokine receptor CCR5 on the $CD8^+$ T-cells, which directs these cells to sites of antigen-specific dendritic cell–$CD4^+$ interaction where the cognate chemokines CCL3 and CCL4 are produced [45].

Th1 and Th2 cells

The cytokine profile secreted by the $CD4^+$ T-cell is of critical importance in directing the inflammatory response,

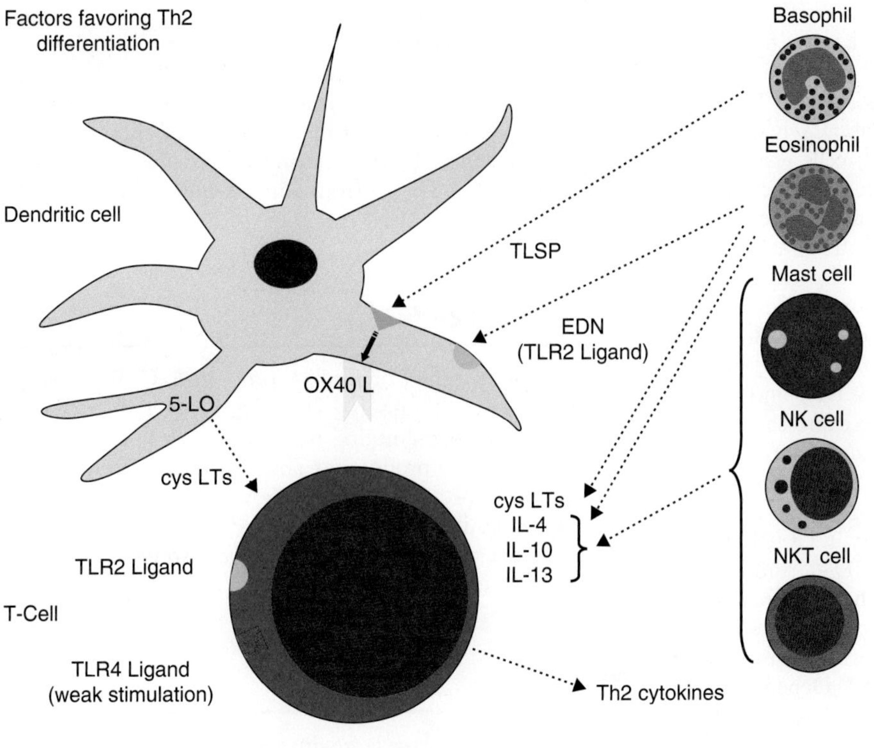

FIG. 13.3 The factors that favor the development of Th2 responses are several. Co-stimulatory molecules such as OX40 and OX40L, upregulation of OX40 by TSLP produced by basophils and structural tissue cells such as epithelial cells, IgE bound to the dendritic cell surface, TLR2 ligation by EDN, low lever TLR4 stimulation, cysteinyl leukotriene synthesis by dendritic cells and their subsequent interaction with the Cys-LT1R on the T-cell and Th2 cytokine production by a variety of cells such as mast cells, eosinophils, basophils and NKT cells all favor Th2 cell differentiation.

whether typical of allergic inflammation with a Th2 cytokine predominance or delayed type hypersensitivity typical of Th1-driven inflammation. The factors that determine the Th1 and Th2 type responses are several. The dendritic cell has an important role through the secretion of IL-12 in the case of the induction of Th1 responses and IL-4 and IL-10 in the case of Th2 and T-regulatory cell responses. Respiratory mucosal dendritic cells preferentially trigger Th2 responses, but after they have matured under the influence of cytokines such as GM-CSF they express high levels of MHC II, B7, and IL-12 and produce Th1 responses [46]. Concomitant pro-inflammatory stimuli condition the dendritic cell and its subsequent interaction with the T-cell (Fig. 13.3). Signals mediated via the TLRs are among the important stimuli of this sort. Endotoxin-poor antigen triggers Th2 responses whereas a strong concomitant stimulation of dendritic cells with lipopolysaccharide, for example, may convert the T-cell response into a Th1 response [47]. In contrast a TLR-2 agonist results in a Th2 biased T-cell response [48]. Recently it has been demonstrated that eosinophils, through eosinophil-derived neurotoxin (EDN), are involved in the process of T-cell differentiation [49]. EDN, a member of the RNase A superfamily, appears to be an endogenous ligand for TLR2. Cytokines also serve to condition the dendritic cell responses. Thymic stromal lymphopoietin (TSLP) induces OX40 ligand on dendritic cells that subsequently evokes Th2 cytokine production by $CD4^+$ T-cells. These cells produce TNF-α and do not express IL-10, differing from those Th2 cells that produce IL-10 as a part of their cytokine profile [50]. A problem with this paradigm for driving Th2 responses was that TSLP appeared to be produced in peripheral tissues but not in the lymph nodes. This issue has now been resolved by the demonstration that basophils may be triggered by allergen to release TSLP in lymph nodes and thereby promote Th2 cell differentiation in this site [51].

Signals that affect the T-cells are also important in directing the T-cell differentiation. It has been recognized for some time that IL-4 is important in creating a milieu that favors the development of Th2 responses. Mast cells and NK cells, as well as T-cells themselves, are sources of this key cytokine. Indeed mast cells express most of the cytokines that control T-cell differentiation pathways and may potentially regulate T-cell fate decisions (reviewed in [52]). Furthermore, signals to the dendritic cell as exemplified by TSLP and signals to the T-cell such as IL-4 synergize to produce, in this instance, a Th2 response [50]. It has long been known that the different T-cell phenotypes exert counter-regulatory effects, IFN-γ inhibits Th2 cell development whereas IL-10 inhibits Th1 cell development. Cysteinyl leukotrienes which are an important feature of allergic reactions are produced by the dendritic cells and appear to favors Th2 differentiation through an autocrine action that modulates IL-10 and IL-12 production by the dendritic cell [53]. Cys-LTs favors IL-10 production whereas inhibition of the Cys-LT1 receptor favors IL-12 production and IFN-γ synthesis by T-cells.

Other essential factors for Th1 differentiation are the transcription factors T-bet and IRF1. T-bet not only induces Th1 differentiation of naïve T-cells, but also "reprograms" polarized Th2 cells to produce the Th1 cytokine IFN-γ and negatively regulates the Th2-specific transcription factor GATA-3 thus promoting the Th1 polarization. A deficiency of T-bet is associated with an exaggerated asthmatic phenotype in mice [54].

Th1 and Th2 subsets are functionally separated based on their cytokine profiles but certain types of chemokine receptors are selectively expressed on either Th1 or Th2 cells and appear to have an important place in determining the patterns of accumulation of T-cell subsets as a function of lung pathology. Th1 cells express CCR5, CXCR3, and CXCR6 [55] whereas Th2 cells express CCR3, CCR4, and CCR8 [56]. TSLP acts on dendritic cells resulting in the synthesis of two chemokines, thymus and activation-regulated chemokine (TARC) and macrophage-derived chemokine (MDC) that serve as chemoattractants for Th2 cells through actions on CCR4. TARC is also expressed in other sites, such as the bronchial epithelium, and may contribute to the selective chemoattraction of Th2 cells to the airways in asthma [57] whereas interferon inducible protein 10 (IP-10/CXCL10) may contribute to $CD8^+$ and $CD4^+$ Th1 cell recruitment in COPD [58]. The expression of the chemokine receptors CXCR3, CCR5, and CXCR6 correlated with disease severity; moreover, the dendritic cells produced the respective ligands for these receptors, and the levels again correlated with disease severity in COPD [55].

The elucidation of the roles of $CD4^+$ and $CD8^+$ T-cell subsets in asthma has depended heavily on the use of small animal experimental models and either adoptive transfer or antibody depletion techniques. It is impossible to review the extensive literature in detail. Dependence of allergic airway responses on $CD4^+$ T-cells has been demonstrated by the technique of adoptive transfer. T-cells isolated from the drainage lymph nodes of the site of sensitization to allergen and administered to naïve recipients that undergo allergen challenge result in typical late allergic responses [59] and airway hyperresponsiveness [60, 61]. Conversely, $CD4^+$ T-cell depletion also abolishes airway responses to allergen challenge [62]. It is now clear that mixed T-cell responses may occur in response to exposure to soluble antigen. Adoptive transfer of antigen-sensitized $CD8^+$ T-cells to naïve recipient animals that are then allergen-challenged results in delayed airway narrowing and typical TR2-type cytokine expression [63]. $CD8^+$ T-cells are also responsible for virally induced airway hyperresponsiveness in the mouse and the accompanying IL-5-driven eosinophilia [64]. In these situations the $CD4^+$ and $CD8^+$ cells share some common properties such as the secretion of eosinophilia-promoting cytokines. $CD4^+$ T-cells have also been implicated in airway remodeling in the allergic rat. Again using adoptive transfer, it was shown that the ovalbumin-sensitized and challenged rat developed epithelial remodeling and airway smooth muscle remodeling through hyperplasia and reduction in apoptosis [65]. The capacity of ovalbumin stimulated $CD4^+$ T-cells to cause hyperplasia of isolated airway myocytes was demonstrated in culture, and was contact dependent.

The study of the roles of $CD4^+$ and $CD8^+$ T-cells in COPD lags years behind that of asthma. Until recently the evidence of the role of these cells in the immunity of COPD was mainly circumstantial. Recently it has been shown in mice exposed to cigarette smoke that the development of pathological and functional emphysema is dependent on the presence of $CD4^+$ and $CD8^+$ T-cells [66, 67] and only the animals with emphysema showed significant expression of Th1 cytokines. In another study it was shown that mice depleted of $CD8^+$ T-cells had a blunted inflammatory response and did not develop emphysema when exposed to long-term cigarette smoke, indicating an important role of these cells in the development of emphysema [68]. Proof of the concept that $CD4^+$ T-cells, and autoimmunity, may play an important role in COPD has been provided by demonstrating the induction of emphysema in naïve rats by the adoptive transfer of vascular endothelial-sensitized $CD4^+$ T-cells from antigen-induced emphysema animals [69].

Tc1 and Tc2 cells

In addition to their cytotoxic properties $CD8^+$ T-cells also show distinct subsets [Tc1 (Fig. 13.4) and Tc2] with cytokine profiles resembling the Th1 and Th2 patterns with IL-2 and IFN-γ production and IL-4, IL-5, and IL-10 production respectively [70]. These subsets termed Tc1 and Tc2 evoke similar degrees of delayed-type hypersensitivity reactions but the latter cells also cause a somewhat greater degree of eosinophilia [70].

Th17 cells

The Th17 cells produce large quantities of pro-inflammatory cytokines, most typically IL-17, and seem to be important in antibacterial responses and autoimmune diseases. Bacteria prime human dendritic cells to promote IL-17 production in memory Th cells through the actions of muramyldipeptide (MDP) on the innate pathogen recognition receptor nucleotide oligomerization domain 2 (NOD2). MDP enhanced TLR agonist induction of IL-23 and IL-1, promoting IL-17 expression in T-cells [71]. In mice, Th17 cells differentiate from naïve $CD4^+$ T-cells in response to transforming growth factor-β (TGF-β) and IL-6 and also require IL-23 for their maintenance or expansion. IL-23 signaling requires IL-12Rb1, a receptor component shared with IL-12. Differences in signaling involving IFN-γ-induced transcription factor appear to determine whether $CD4^+$ Th1 or Th17 cells are produced [72]. In a murine model, IL-17 is required during antigen sensitization to develop an allergic phenotype, as shown in IL-17R-deficient mice [73]. The increase in IL-17 expression with allergen challenge is inhibitory of the allergic response. Neutralization of IL-17 augments allergic responses whereas exogenous IL-17 reduces eosinophil recruitment and bronchial hyperreactivity. Eotaxin (CCL11) and TARC are downregulated by IL-17 as are IL-5 and IL-13 production in regional lymph nodes. IL-4 plays an inhibitory role in the magnitude of IL-17 expression. IL-17 is also responsible for a neutrophilic component to the bronchial inflammation induced by allergen challenge of sensitized mice [74]. Th17 cells have not yet been reported in COPD but it would be surprising if they did not play a role in smokers since they have a crucial role in the induction of autoimmune tissue injury. Furthermore IL-6 whose expression is elevated in smokers has the potential to prevent the TGF-β mediated differentiation of T-cells into regulatory T-cells (Tregs) by inhibiting the expression of FOXP3 and inducing the expression of IL-17 [75].

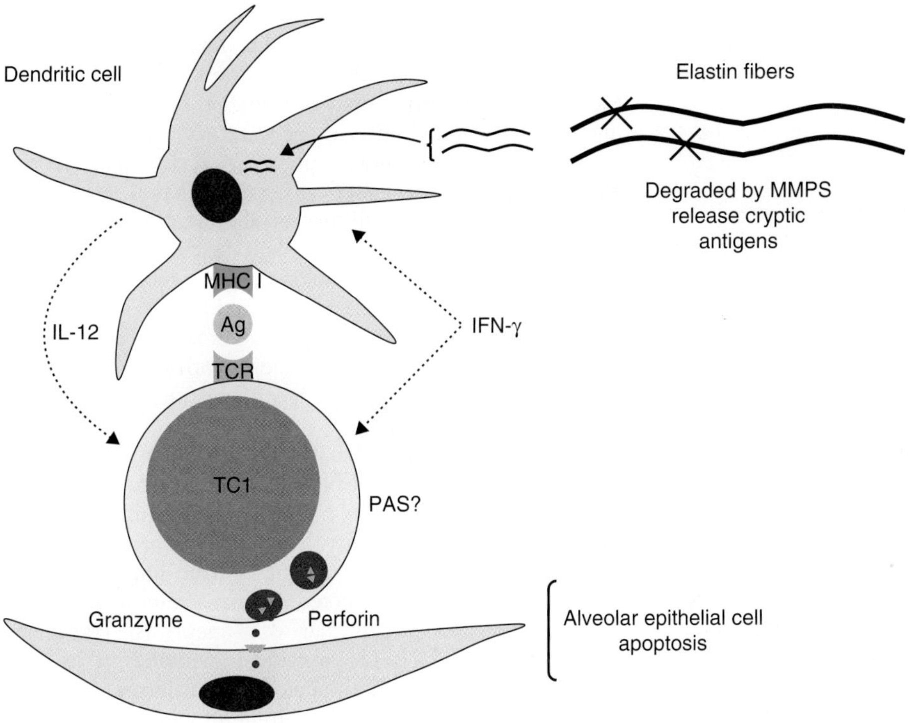

FIG. 13.4 Oxidative stress, matrix metalloproteinase release by inflammatory cells leads to degradation of the matrix with release of cryptic antigens from elastin fibers. These fibers elicit a cytotoxic T-cell response which induces apoptosis of alveolar epithelial cells by mechanisms that involve perforin, granzyme and the FAS–FASL pathways.

T-regulatory cells

There is growing evidence that a significant aspect of the inflammatory reaction in the airways is caused by a failure of suppression of the response. Active regulation appears to be important to peripheral tolerance to allergens in nonallergic individuals. In this regard Tregs are receiving substantial attention. A rapidly growing number of studies addresses the role of these cells in allergic processes and in models of asthma. Tolerance to allergens is associated with the synthesis of IL-10 and TGF-β by FOXP3 expressing $CD4^+CD25^+$ T-cells. The precise mechanisms by which T-cell inhibition is produced are still not well worked out. It appears that $CD4^+CD25^+$ Tregs are capable of inhibiting allergic responses even if they are harvested from IL-10-deficient mice and adoptively transferred into wild-type mice. The suppression of the Th2 responses is mediated by IL-10 producing $CD4^+$ T-cells that are activated by the transferred Tregs [76]. Intratracheal administration of $CD4^+CD25^+$ T-cells inhibits allergic airway responses (airway hyperresponsiveness and inflammation) through an IL-10-dependent mechanism acting through TGF-β expression. The T-cells required IL-10 treatment prior to transfer and even when unable to synthesize IL-10 themselves, they were still able to inhibit the allergic response [77]. The inflammatory process that follows allergen challenge in a murine model of allergic asthma and the ensuing airway hyperresponsiveness can be attenuated by Treg cells that are recruited into the airway mucosa [78]. These cells appear after the initial inflammation that is triggered by allergen inhalation and suggests that these cells may normally serve to limit the extent of the allergen-induced asthmatic reaction.

γδ T-cells

Important T-cell subsets that do not share the usual T-cell receptor, such as the γδ TCR-bearing cells and NK T-cells, have substantial importance for pulmonary defense and airway function. T-cells expressing the γδ T-cell receptor have a particular association with mucosal epithelial surfaces, but what they do is still largely enigmatic [1]. In contrast to αβ T-cells that recognize peptide antigens the γδ T-cells recognize small nonpeptidic molecules and in some cases nonclassical MHC molecules such as CD1. γδ T-cells are important in defense against intracellular organisms such as *Listeria, Pneumocystis carinii and Mycobacterium tuberculosis* [79]. However their functions extend beyond their innate immune defense properties. These cells produce large quantities of IFN-γ and have been shown to potently downregulate IgE responses to sensitization [80, 81]. As such these cells are important mediators of allergen tolerance. They also synthesize trophic factors such as keratinocyte growth factor (KGF) that is important in epithelial cell regeneration [82], at least in the small intestine and likely also in the airways. The administration of KGF is able to protect the epithelial barrier function in face of allergen challenge. IFN-γ production by $CD8^+$ γδ T-cells is potently inhibitory of late allergic airway responses in the rat [83]. However, in the mouse there are both pro-inflammatory and anti-inflammatory properties associated with γδ T-cells. In OVA-sensitized and challenged mice, γδ T-cells expressing Vγ1 enhance airways hyperresponsiveness [84], whereas cells expressing Vγ4 strongly suppress airways hyperresponsiveness [85]. The airways hyperresponsiveness-regulatory γδ T-cells had only minor effects on airway inflammation in these experiments in contrast to rodent models where they also inhibit Th2 inflammation. Cigarette smoking induces an increase of γδ T-cells both in human [15] and in mice [66] exposed to cigarette smoke, but their function in smokers is not clear.

*i*NKT-cells

*i*NKT cells express CD3 and an invariant T-cell receptor α-chain (reviewed in [86]). They respond to glycolipid antigens that are presented by CD1d, and produce large quantities of IFN-γ and IL-4 [87]. The role of *i*NKT cells in host defense in the lung has largely been inferred from experiments in knockout mice that lack *i*NKT cells, which show increased susceptibility to a wide range of bacterial and viral pathogens and parasites. It has recently been proposed that *i*NKT cells also have a role in asthma pathogenesis: experiments with *i*NKT-deficient mice show reduced responses to sensitization and challenge that can be largely reconstituted by the adoptive transfer of *i*NKT cells [87]. In a murine model, when stimulated by α-galactosylceramide *i*NKT cells produce IL-13 and mediate airway hyperresponsiveness and eosinophilic inflammation, providing an additional innate immune mechanism for an asthma phenotype [88].

T-CELL CYTOKINES IN MODELS OF ASTHMA AND COPD

The effects of the administration of cytokines to animals have been examined and provide some clues as to mechanisms of the links between the T-cells and airway responses. IL-2 administered to naïve rats leads to airway hyperresponsiveness to inhaled methacholine [89]. In high doses, it causes peribronchial edema and an eosinophilic airway inflammation. Low doses of IL-2 augment allergic responses, in terms of both the magnitude of bronchoconstriction and the degree of eosinophilic inflammation [90]. The increase in late response is attributable to an increase in sensitivity to leukotrienes rather than by enhanced synthesis [91, 92]. IL-4 is important in the sensitization phase to allergen in murine models and is necessary for allergen-induced airway hyperresponsiveness. However neutralization of its actions at the time of challenge does not abrogate hyperresponsiveness [93]. In contrast, IL-13 is of importance to airway hyperresponsiveness at the time of challenge to allergen [94–96], and its effects depend on intact signaling pathways for IL-13 in the epithelium [97]. How the epithelium transduces the effects of IL-13 is as yet unknown. Interestingly, the asthma phenotype caused by T-bet deficiency in mice is also attributable to IL-13, both airway hyperresponsiveness and airway remodeling were ameliorated by neutralization of IL-13 [98]. IL-5 has been shown in some models but not in others to mediate airway hyperresponsiveness [93, 99–101]. Recent studies help to clarify the issue of the controversial roles of IL-5 and eosinophilia in allergic asthma. When eosinophilia is absent, as in IL-5 and eotaxin-deficient mice, or in mice genetically engineered so as to lack eosinophils specifically, there is an abrogation of the allergic responses and airway hyperresponsiveness [102, 103]. However the mechanism appears to reflect an attenuated Th2 response and reduced expression of important cytokines such as IL-13. These findings are perhaps explained by the effects of cationic proteins, such as EDN on immune responses, as reviewed above. The adoptive transfer of genetically engineered T-cells overexpressing IL-10 is inhibitory of allergic airway responses and airways hyperresponsiveness [104], consistent with the emerging concept of this cytokine as an inhibitory cytokine when produced by T-cells in the airway. The induction of airway hyperresponsiveness by repeated ozone exposures has been recently demonstrated to require both IL-17 and NKT cells [105], expanding the range of pertinent stimuli that activate airway T-cells.

There is evidence that $CD8^+$ T-cells in the lungs of COPD patients express IFN-γ [106] which would enhance the inflammatory reaction in the lung in addition to that caused by their cytotoxicity. Transgenic overexpression of IFN-γ causes emphysema in the murine lung and IFN-γ-deficient mice are protected against cigarette smoking-induced increases in alveolar cell apoptosis. Transgenic overexpression of IL-13, a critical cytokine in asthma, causes emphysema with enhanced lung volumes and compliance, mucus metaplasia, and inflammation, in the adult murine lung. Matrix metalloproteinase (MMP)-2, -9, -12, -13, and -14 and cathepsins B, S, L, H, and K were induced by IL-13 in this setting. In addition, treatment with MMP or cysteine proteinase antagonists significantly decreased the emphysema and inflammation, but not the mucus in these animals. These studies demonstrate that IL-13 is a potent stimulator of MMP and cathepsin-based proteolytic pathways in the lung. They also demonstrate that IL-13 causes emphysema via an MMP- and a cathepsin-dependent mechanism(s) and highlight common mechanisms that may underlie COPD and asthma [107]. Transgenic overexpression of IL-18 also results in murine emphysema, and appears to result from the combined effects of both IL-13 and IFN-γ overproduction in the lung [108].

T-CELLS IN THE PATHOGENESIS OF ASTHMA: HUMAN STUDIES

T-cells are frequent in the airway wall. Most intraepithelial T-cells express CD8, whereas $CD4^+$ T-cells are more frequently found in the lamina propria. Both subsets mainly have an effector- and/or memory-cell phenotype, as defined by their expression of CD45RO (reviewed in [86]). In asthma the number of T-lymphocytes in the airway wall increases substantially. In agreement with the paradigm for T-cell subsets defined on murine T-cell cytokine profiles, a predominance of Th2 expressing $CD4^+$ T-cells has also been observed in the BAL cells from mild atopic asthmatics [109]. Furthermore, Th2 cytokine expressing cells increase in the BAL fluid of asthmatic subjects following allergen challenge [110]. In fact IL-4 and IL-5 mRNA expressing T-cells can be detected even in sputum samples from asthmatic subjects [111]. Consistent with various cellular and animal models there is also an upregulation of GATA-3, the principal transcription factor for Th2 cytokines in asthmatic airway tissues [112]. Lung lavage following segmental allergen challenge of the airways of asthmatic subjects has demonstrated that the CD4, CD8, and γδ T-cells all contribute to the Th2 response through the expression of IL-5 and IL-13 [113].

Bronchial biopsies from asthmatics are frequently positive for IL-5 mRNA expression that is related to the degree of eosinophilia and is consistent with a key role for this Th2 cytokine is causing eosinophilia in asthma [114]. In certain forms of occupational asthma involving low molecular weight sensitizers eosinophilia appears to be associated with IL-5 production by $CD8^+$ T-cells [115]. Anti-IL-5 antibody administration markedly reduces eosinophils in the sputum of asthmatics [116], and although tissue eosinophilia is affected also it is more resistant to the neutralizing antibody [117]. It has been difficult to demonstrate therapeutic benefits of anti-IL-5 treatments on surrogates for asthma such as airway hyperresponsiveness or late allergic airway responses. However, favorable effects on tissue matrix protein deposition were observed, likely through downstream effects on TGF-β, which is synthesized by eosinophils in asthma. Although it is widely believed that eosinophils are the major site of synthesis of cysteinyl leukotrienes in asthma there is no formal proof that they are indeed responsible for the cysteinyl leukotriene synthesis associated with late allergic responses.

Other Th2 cytokines, such as IL-4 and IL-13, are also expressed in the airway mucosa of asthmatic subjects, as is the Th1 cytokine IFN-γ and these cytokines are predominantly expressed in T-cells [118]. Clinical trials of a soluble IL-4 receptor have demonstrated only minor therapeutic effects [119]. A subsequent trial utilizing Pitrakinra, a recombinant mutant form of the wild-type human IL-4 that targets the effects of both IL-4 and IL-13, demonstrated that the late allergic airway response could be attenuated by inhalation of the biological [120]. There is increasing recognition of alternative mechanisms of inflammation mediated by T-cells. The presence of a neutrophil-rich infiltrate in some asthmatics is not well explained by the classical Th2 paradigm. IFN-γ is often associated with more severe asthma and the neutrophil chemoattractant CXCL8 (IL-8) is also found. The family of cytokines that includes IL-17 may help shed light on these alternative patterns of inflammation. IL-17 itself may also be a contributor to neutrophilic airway inflammation. IL-17 mRNA can be found in asthmatic sputum where it correlates with CXCL8 expression [121] and in bronchoalveolar lavage cells from asthmatic subjects ranging in severity from mild to severe [122].

Regulatory T-cells are being increasingly recognized to participate in inflammatory diseases and may suppress inflammation in asthma. It is a plausible hypothesis that defective inhibitory influences are as important in the genesis of inflammation as pro-inflammatory pathways. Indeed induction of regulatory T-cells by an appropriate education of the mucosal immune systems in early life is now proposed as an alternative hygiene hypothesis. Probiotics have been explored for their beneficial effects on eczema but clinical trials in asthma are lacking and data on modification of T-cell populations in the respiratory mucosa are also required before any definite statement about such a strategy can be recommended.

*i*NKT cells have been recently implicated in airway pathophysiology. The extent to which findings from murine models can be extrapolated to human asthma is unclear. Elevated numbers of $CD4^+$ *i*NKT cells have recently been reported in BAL from patients with asthma [123], but other studies have failed to confirm these findings and have concluded that these cells are relatively uncommon, representing less than 2% of the T-cells in the airways in either asthmatic or COPD patients [124]. Therefore it is not clear how pertinent current findings are to asthma in general. It is also not known what the appropriate stimulus for activation of *i*NKT cells is or how these cells may be activated by aeroallergens.

The final characteristic of asthma, namely remodeling has been linked to T-cell function in animals and human T-cells will also trigger proliferation of airway smooth muscle cells [125]. Recent studies describe T-cells among smooth muscle bundles in asthma [126], but their phenotype and possible effects on airway smooth muscle structure or function are as yet unknown.

T-CELLS IN THE PATHOGENESIS OF COPD: HUMAN STUDIES

We now recognize that antigens, specifically elastin or its by-products (probably among others) derived from the injury of the lung by the innate inflammatory response, are also an important trigger for the development of COPD. These antigens are taken up by dendritic cells not in the larger airways, like in asthma, but deeper inside the lung, small airways, and parenchyma, where the lung tissue suffers the most injury by cigarette smoke. There is evidence in the literature that cigarette smoking is associated with an expansion in the dendritic cell population in the lower respiratory tract [127] and with a marked increase in the number of mature cells in the lung parenchyma [128]. This is an indication that the lung response to cigarette-smoke exposure follows the established immune response design, including innate immunity and readiness for an adaptive immune response, if necessary. There is a similarity between asthma and COPD in that they are both initiated by the presentation of soluble antigenic products to dendritic cells. However, very dissimilar inflammatory patterns result from the dendritic cell–T-cell interaction. In asthma, the dendritic cell triggers a Th2 T-cell response, while in COPD it results in a Th1 and a cytotoxic $CD8^+$ T-cell response that is maintained and even progresses after the challenge by cigarette smoke ceases. These differences emphasize the importance of the dendritic cell, and its interaction with the antigen and the inflammatory milieu in determining the fate of the resulting inflammation and disease.

The study that initiated the present interest in the T-cell as a possible important cell in the pathogenesis of COPD was carried out by Finkelstein *et al.* in 1995 [13]. These authors used immunohistochemistry and morphometry to identify the inflammatory cells infiltrating the alveolar wall and to define the extent of emphysema in smokers and nonsmokers undergoing lung resection. Their findings were surprising at the time, as they reported that the most prominent inflammatory cell in the lung parenchyma of smokers was the $CD3^+$ T-lymphocyte that increased from a mean of 1546 cells/mm^3 in nonsmokers up to 10,000 cells/mm^3 in smokers. Furthermore, a clear correlation between the number of $CD3^+$ T-cells and the extent of emphysema was found, certainly suggestive of the protagonism of the T-lymphocytes in the pathogenesis of emphysema in smokers. Abundant but variable numbers of T-cells ($CD3^+$) together with other inflammatory cells were also found in the small airways of the same patient population [129, 130]. Interestingly the degree of airway reactivity, measured prior to surgery, correlated with the load of T-cells in the airways in smokers with centrilobular emphysema (CLE), but not in panlobular emphysema (PLE) or nonsmokers. Because similar total numbers of $CD3^+$ T-cells were present in the two forms of emphysema, we speculated that the T-cells in CLE were behaving differently, possibly indicating a mixed phenotype of Th1 and Th2 subsets. In support of this possibility is the finding that Th2 type cytokines IL-4 and IL-5 mRNA, typically found in asthma, are abundantly expressed by inflammatory cells in the wall of large airways in some smokers with chronic bronchitis and COPD [131] and it is possible that clones of T-cells in smokers with CLE may express a Th2 cytokine profile that might induce airway reactivity in these cases.

Following the report by Finkelstein *et al.* [130] proposing a role for the T-cell in COPD, several authors identified the $CD8^+$ T-cell as the predominant lymphocyte in the airways of smokers with COPD [132, 133]. Saetta showed that the only significant difference in the inflammatory cell infiltrate in asymptomatic smokers and smokers with COPD was the increase in $CD8^+$ T-cells in the small airways of patients with COPD. Furthermore, the number of $CD8^+$ T-cells were negatively correlated with the degree of airflow obstruction, as measured by the FEV_1, again suggesting a possible role for these cells in the pathogenesis of the disease. $CD4^+$ T-cells are also found, albeit in smaller numbers, in the airways of smokers with COPD and these cells express IFN-γ and activated signal transducer and activator of transcription 4 (STAT 4), a transcription factor that is essential for activation and commitment of the Th1 lineage. The number of T-cells expressing activated STAT 4 correlated with the degree of airflow obstruction [134]. These findings go along with the previously mentioned expression of CXCR3 and CCR5 by $CD4^+$ and $CD8^+$ T-cells and the ligand IP-10 and MIG in lung tissue and strongly support the idea that COPD is mediated by an active Th1 immune reaction in the lung comprising both $CD8^+$ and $CD4^+$ T-cells.

Majo while studying lungs from nonsmokers and smokers obtained at surgery found that the $CD8^+$ T-cells are also the predominant T-cells infiltrating the alveolar wall in smokers with COPD [15, 133], although $CD4^+$ T-cells are also increased [15, 135]. Furthermore, the number of both $CD8^+$ and $CD4^+$ increased with the amount smoked in smokers with COPD, but not in healthy smokers, indicating that the $CD4^+$ T-cell are also involved in the inflammatory process in COPD. Majo *et al.* also quantified and found an increased number of apoptotic cells in the lungs of smokers with COPD which correlated with the numbers of $CD8^+$ cytolytic T-cells in the alveolar wall [15]. This and other reports, showing an increased number of structural lung cells undergoing apoptosis in emphysematous lungs [136] support the idea that $CD8^+$ T-cells are inducing apoptosis of endothelial and epithelial cells in emphysema.

The predominance of $CD8^+$ T-cells over $CD4^+$ T-cells in the lungs of patients with COPD, resembles the inflammatory reaction seen in many viral diseases like Epstein-Barr and LCMV virus in which the numbers of $CD8^+$ T-cells are much greater than the $CD4^+$. Cell-associated proteins [42, 43] and particulate antigens [44] usually presented with HSP and apoptotic cells presented by phagosomes, are much more effective at MHC class I restricted $CD8^+$ T-cells priming than are simple soluble molecules and it can be speculated that these antigenic forms mimic the tissue debris associated with cell injury or viral infections [40] stimulating a larger $CD8^+$ proliferation.

The relatively mild infiltration with $CD8^+$ and $CD4^+$ T-cells found in patients with mild to moderate disease increases markedly in the lungs [135] and airways of severely diseased patients. All inflammatory cells, except B-lymphocytes, were found to be increased in the lungs of patients with severe emphysema, even though these patients had not smoked more than the control subjects. By far, the more numerous cells were the $CD4^+$ and the $CD8^+$ T-cells but neutrophils, macrophages, and even eosinophils were also increased. These studies are of importance showing that, in COPD, inflammation with an abundance of T-lymphocytes and other inflammatory cells continues late into the disease process.

An important finding in one of these studies was that the average duration of smoking cessation in the severe COPD subjects was 9.2 years [137] indicating that the adaptive immune system continues to be involved after smoking cessation, a phenomenon typically found in other autoimmune diseases. In favor of this possibility is the recent finding of oligoclonality of T-cells in the lungs of patients with COPD, a finding supportive of the idea that these cells are accumulating in the lung secondary to antigenic stimulation. If correct, COPD may be considered to have an autoimmune component to its pathogenesis that is triggered by smoking, as previously suggested [14, 15, 138].

Supporting the idea of COPD as an autoimmune disease is the key paper by Lee and colleagues characterizing antielastin antibodies and T-helper type Th1 type responses in human subjects, which correlate with emphysema severity [31]. In response to elastin peptides, but not collagen, T-cells from smokers with COPD produced IFN-γ and IL-10 but no IL-13 and these responses were decreased by blocking MHC class II molecules. A humoral response was also found in these patients that showed circulating antibodies against elastin but not collagen, findings that implicate not only the T-cell but the B-cell in the autoimmune reaction in COPD. T-regulatory cells have been well described in COPD and they may play an important role in the development of COPD in smokers. Lee [31] and Barcelo [139] have shown a lower number of Tregs in the lungs of smokers with COPD when compared with nonsmokers. Furthermore smokers with normal lung function had significantly higher Tregs in the lung than nonsmokers, suggesting an enhanced protective role by these T-cells, possibly controlling the development of autoimmunity and disease in smokers with normal lung function. It has been shown that $CD8^+$ expressing γδ T-cells have immune regulatory properties and might play a role in the development of COPD in some smokers. Pons described that, as seen with Tregs, smokers with normal lung function have increased numbers of γδ $CD8^+$ T-cells in the lungs while smokers with COPD do not [140]. These findings suggest a possible dysregulation of tolerance in smokers with COPD.

B-CELLS IN ASTHMA AND COPD

About one-third of adult patients with asthma are classified as nonatopic and they often suffer from more severe disease. The popular term intrinsic asthma became unpopular following the observation that IgE was elevated compared to control subjects in all asthmatics, irrespective of the skin test reactivity [141] which tends to wane with age. The systemic reaction to sensitization with the expression of specific IgE reflects an important B-cell contribution to this disease and there has been renewed interest in the contribution of IgE to asthma since the introduction of antibody treatment directed against IgE. Some further insights into the potential role of IgE have been obtained through recent observations examining IgE synthesis locally in airway tissues. IgE is characterized by its ε heavy-chain and it is produced after heavy-chain switching in B-cells from IgM, IgG, or IgA to IgE. The switch to IgE is initiated by the cytokines IL-4 or IL-13, produced principally by T_H2 cells, which drive ε germline gene transcription. Local IgE synthesis was first shown to occur within the nasal mucosa of patients with allergic rhinitis [142]. More recently immunopathological comparisons of bronchial biopsies from atopic and nonatopic patients with asthma have demonstrated expression of ε germline gene transcripts and expression of the high affinity IgE receptor, FcεRI mRNA. Evidence acquired by the examination of bronchial biopsies from atopic and nonatopic patients with asthma and appropriate controls suggested elevated synthesis of mature IgE in the asthmatic bronchial mucosa by local B-cells. It was argued that local IgE synthesis may well account for most, if not all, biologically significant IgE production, and circulating IgE may reflect spillover of IgE from synthesis at mucosal sites. It is therefore quite possible that synthesis of bronchial mucosal IgE contributes to both atopic and nonatopic asthma, making distinctions based on skin reactivity unnecessary and uninformative, except where it is necessary to document systemic sensitization as may be required in cases of occupational asthma. These observations also raise questions about current criteria for the restriction of monoclonal anti-IgE antibodies, such as omalizumab, to patients that are more obviously atopic. Clearly further studies are required in this area to clarify the situation.

B-cells have been shown to also play an important role in COPD. Lymphoid follicles containing monoclonal B-cells, dendritic cells, and T-cells, predominantly $CD4^+$, have been recently described in the airways [143] and parenchyma [144] of patients with COPD and also in mice with cigarette smoke-induced emphysema. These follicles function as inducible secondary lymphoid tissue for immune responses, where antigen presentation can be accomplished without lymphatic node migration. This could explain why dendritic cells found in the lungs of COPD patients express markers of maturation, such as CD80 and CD86 [145] but not CCR7 the homing receptor for lymph nodes [146]. It has been suggested that lymphoid follicles may develop in relation to microbial colonization and infection occurring in the later stages of COPD. However no bacterial or viral products were seen in the follicles suggesting that the B-cells in the follicles, which are monoclonal, proliferate in response to specific lung antigens. The recent demonstration of IgG autoantibodies with avidity for epithelial and endothelial cells along with the deposition of antigen–antibody immune complexes and complement in the lungs of patients with COPD further support this interpretation. Altogether these findings fulfill the conventional criteria that define the presence of antibody responses against self-antigens as autoimmunity [34]. Furthermore since the B-cells were producing IgG antibodies the participation of the $CD4^+$ T-cells primed for the same antigens as the B-cell is essential and suggest that a complex immunological process involving CD4 and CD8 T-cells along with the deposition of immune complexes and complement are involved in the mechanism of lung destruction in COPD.

CONCLUSIONS

The role of the T-cells in mediating the defining characteristics of asthma is still unclear. Airway inflammation is associated with findings that are quite consistent with the findings from animal models. Th2 type inflammation follows allergen challenge as expected. However, in certain subsets of asthma, such as severe asthma, there is often a neutrophilic inflammation and expression of cytokines and chemokines that are not associated with allergen-driven Th2 models. Airway hyperresponsiveness continues to defy elucidation and it likely has multiple causes. In acute murine models IL-13 is the best single cytokine candidate to explain allergen-induced changes in airway responsiveness. However, it is likely that complex cascades are involved in such reactions in human asthma and that airway hyperresponsiveness is the expression of these diverse pathways. While the understanding of the mechanisms of COPD has lagged behind the exploration of the inflammatory basis for asthma it is now clear that an antigen-driven adaptive immune reaction is also crucial in the production of the disease. It should be clearly understood that new emphasis on the adaptive immune response involving T-and B-cells does not detract from the importance of the innate immune inflammation, neutrophils, macrophages, γδ T-cells but rather completes the picture. These cells are responsible for the initial tissue injury and the production of antigens that result in the subsequent adaptive immune response. The autoimmune paradigm in COPD could explain why only a proportion of smokers develop COPD. Immunological tolerance based on Tregs is not an all or none phenomenon but may show various degrees of efficacy, so that disease may occur in the susceptible subject [147]. Different degrees of tolerance to self-antigens could explain the variable T-cell inflammation, wide range in function, and disease severity in response to similar degrees of cigarette exposure in smokers. Autoimmunity in COPD could also contribute to the widely recognized systemic manifestations of this disease. Hopefully the increased understanding of the mechanisms of these diseases will provide improved treatment options in time.

References

1. Gupta S, Frenkel R, Rosenstein M, Grieco MH. Lymphocyte subpopulations, serum IgE and total eosinophil counts in patients with bronchial asthma. *Clin Exp Immunol* 22: 438–45, 1975.
2. Strannegard IL, Lindholm L, Strannegard O. T lymphocytes in atopic children. *Int Arch Allergy Appl Immunol* 50: 684–92, 1976.
3. Saxon A, Morrow C, Stevens RH. Subpopulations of circulating B cells and regulatory T cells involved in *in vitro* immunoglobulin E production in atopic patients with elevted serum immunoglobulin E. *J Clin Invest* 65: 1457–68, 1980.
4. Canonica GW, Mingari MC, Melioli G, Colombatti M, Moretta L. Imbalances of T cell subpopulations in patients with atopic diseases and effect of specific immunotherapy. *J Immunol* 123: 2669–72, 1979.
5. Kus J, Tse KS, Vedal S, Chan-Yeung M. Lymphocyte sub-populations in patients with allergic and non-allergic asthma. *Clin Allergy* 15: 523–29, 1985.
6. Gonzalez MC, Diaz P, Galleguillos FR, Ancic P, Cromwell O, Kay AB. Allergen-induced recruitment of bronchoalveolar helper (OKT4) and suppressor (OKT8) T-cells in asthma. Relative increases in OKT8 cells in single early responders compared with those in late- phase responders. *Am Rev Respir Dis* 136: 600–4, 1987.
7. Corrigan CJ, Hartnell A, Kay AB. T-lymphocyte activation in acute severe asthma. *The Lancet* 1: 1129–32, 1988.
8. Poulter LW, Power C, Burke C. The relationship between bronchial immunopathology and hyperresponsiveness in asthma. *Eur Respir J* 3: 792–99, 1990.
9. Hirashima M, Tashiro K, Sakata K, Hirashima M. Isolation of an eosinophil chemotactic lymphokine as a natural mediator for eosinophil chemotaxis from concanavalin A-induced skin reaction sites in guinea-pigs. *Clin Exp Immunol* 57: 211–19, 1984.
10. Mosmann TR, Cherwinski H, Bond MW, Giedlin MA, Coffman RL. Two types of murine helper T cell clone. Definition according to profiles of lymphokine activities and secreted proteins.. *J Immunol* 136: 2348–57, 1986.
11. Hunninghake GW, Crystal RG. Cigarette smoking and lung destruction. Accumulation of neutrophils in the lungs of cigarette smokers. *Am Rev Respir Dis* 128: 833–38, 1983.
12. Martin TR, Raghu G, Maunder RJ, Springmeyer SC. The effects of chronic bronchitis and chronic air-flow obstruction on lung cell populations recovered by bronchoalveolar lavage. *Am Rev Respir Dis* 132: 254–60, 1985.
13. Finkelstein R, Fraser RS, Ghezzo H, Cosio MG. Alveolar inflammation and its relation to emphysema in smokers. *Am J Respir Crit Care Med* 152: 1666–72, 1995.
14. Cosio MG, Majo J, Cosio MG. Inflammation of the airways and lung parenchyma in COPD: Role of T cells. *Chest* 121: 160S–65S, 2002.
15. Majo J, Ghezzo H, Cosio MG. Lymphocyte population and apoptosis in the lungs of smokers and their relation to emphysema. *Eur Respir J* 17: 946–53, 2001.
16. Bienenstock J, McDermott MR. Bronchus- and nasal-associated lymphoid tissues. *Immunol Rev 2005* 206: 22–31, 2005.
17. Pauwels RA, Buist AS, Ma P, Jenkins CR, Hurd SS. Global strategy for the diagnosis, management, and prevention of chronic obstructive pulmonary disease: National Heart, Lung, and Blood Institute and World Health Organization Global Initiative for Chronic Obstructive Lung Disease (GOLD): Executive summary. *Respir Care* 46: 798–825, 2001.
18. Rahman I, Smith CA, Lawson MF, Harrison DJ, Macnee W. Induction of gamma-glutamylcysteine synthetase by cigarette smoke is associated with AP-1 in human alveolar epithelial cells. *FEBS Lett* 396: 21–25, 1996.
19. Sun W, Wu R, Last JA. Effects of exposure to environmental tobacco smoke on a human tracheobronchial epithelial cell line. *Toxicology* 100: 163–74, 1995.
20. Churg A, Zay K, Shay S, Xie C, Shapiro SD, Hendricks R, Wright JL. Acute cigarette smoke-induced connective tissue breakdown requires both neutrophils and macrophage metalloelastase in mice. *Am J Respir Cell Mol Biol* 27: 368–74, 2002.
21. Dhami R, Gilks B, Xie C, Zay K, Wright JL, Churg A. Acute cigarette smoke-induced connective tissue breakdown is mediated by neutrophils and prevented by alpha1-antitrypsin. *Am J Respir Cell Mol Biol* 22: 244–52, 2000.
22. Harel S, Janoff A, Yu SY, Hurewitz A, Bergofsky EH. Desmosine radioimmunoassay for measuring elastin degradation *in vivo*. *Am Rev Respir Dis* 122: 769–73, 1980.
23. Stone PJ, Gottlieb DJ, O'Connor GT, Ciccolella DE, Breuer R, Bryan-Rhadfi J, Shaw HA, Franzblau C, Snider GL. Elastin and collagen degradation products in urine of smokers with and without chronic obstructive pulmonary disease. *Am J Respir Crit Care Med* 151: 952–59, 1995.
24. Dillon TJ, Walsh RL, Scicchitano R, Eckert B, Cleary EG, McLennan G. Plasma elastin-derived peptide levels in normal adults, children, and emphysematous subjects. Physiologic and computed tomographic scan correlates. *Am Rev Respir Dis* 146: 1143–48, 1992.
25. Schriver EE, Davidson JM, Sutcliffe MC, Swindell BB, Bernard GR. Comparison of elastin peptide concentrations in body fluids from healthy volunteers, smokers, and patients with chronic obstructive pulmonary disease. *Am Rev Respir Dis* 145: 762–66, 1992.
26. Betsuyaku T, Nishimura M, Yoshioka A, Takeyabu K, Miyamoto K, Kawakami Y. Elastin-derived peptides and neutrophil elastase in bronchoalveolar lavage fluid. *Am J Respir Crit Care Med* 154: 720–24, 1996.
27. Davies SF, Offord KP, Brown MG, Campe H, Niewoehner D. Urine desmosine is unrelated to cigarette smoking or to spirometric function. *Am Rev Respir Dis* 128: 473–75, 1983.
28. Pelham F, Wewers M, Crystal R, Buist AS, Janoff A. Urinary excretion of desmosine (elastin cross-links) in subjects with PiZZ alpha-1-antitrypsin deficiency, a phenotype associated with hereditary predisposition to pulmonary emphysema. *Am Rev Respir Dis* 132: 821–23, 1985.
29. Viglio S, Iadarola P, Lupi A, Trisolini R, Tinelli C, Balbi B, Grassi V, Worlitzsch D, Doring G, Meloni F, Meyer KC, Dowson L, Hill SL, Stockley RA, Luisetti M. MEKC of desmosine and isodesmosine in urine of chronic destructive lung disease patients. *Eur Respir J* 15: 1039–45, 2000.
30. Gottlieb DJ, Stone PJ, Sparrow D, Gale ME, Weiss ST, Snider GL, O'Connor GT. Urinary desmosine excretion in smokers with and without rapid decline of lung function: The Normative Aging Study. *Am J Respir Crit Care Med* 154: 1290–95, 1996.
31. Lee SH, Goswami S, Grudo A, Song LZ, Bandi V, Goodnight-White S, Green L, Hacken-Bitar J, Huh J, Bakaeen F, Coxson HO, Cogswell S, Storness-Bliss C, Corry DB, Kheradmand F. Antielastin autoimmunity in tobacco smoking-induced emphysema. *Nat Med* 13: 567–69, 2007.
32. Rose N, Afanasyeva M. Autoimmunity: Busting the atherosclerotic plaque. *Nat Med* 9: 641–42, 2003.

33. Steinman L. State of the art. Four easy pieces: Interconnections between tissue injury, intermediary metabolism, autoimmunity, and chronic degeneration. *Proc Am Thorac Soc* 3: 484–86, 2006.
34. Feghali-Bostwick CA, Gadgil AS, Otterbein LE, Pilewski JM, Stoner MW, Csizmadia E, Zhang Y, Sciurba FC, Duncan SR. Autoantibodies in patients with chronic obstructive pulmonary disease. *Am J Respir Crit Care Med* 177: 156–63, 2008.
35. Lambrecht BN. Dendritic cells and the regulation of the allergic immune response. *Allergy* 60: 271–82, 2005.
36. Henrickson SE, Mempel TR, Mazo IB, Liu B, Artyomov MN, Zheng H, Peixoto A, Flynn MP, Senman B, Junt T, Wong HC, Chakraborty AK, von Andrian UH. T cell sensing of antigen dose governs interactive behavior with dendritic cells and sets a threshold for T cell activation. *Nat Immunol* 9: 282–91, 2008.
37. Huh JC, Strickland DH, Jahnsen FL, Turner DJ, Thomas JA, Napoli S, Tobagus I, Stumbles PA, Sly PD, Holt PG. Bidirectional interactions between antigen-bearing respiratory tract dendritic cells (DCs) and T cells precede the late phase reaction in experimental asthma: DC activation occurs in the airway mucosa but not in the lung parenchyma. *J Exp Med* 198: 19–30, 2003.
38. Burgdorf S, Kurts C. Endocytosis mechanisms and the cell biology of antigen presentation. *Curr Opin Immunol* 20: 89–95, 2008.
39. Srivastava PK, Menoret A, Basu S, Binder RJ, McQuade KL. Heat shock proteins come of age: Primitive functions acquire new roles in an adaptive world. *Immunity* 8: 657–65, 1998.
40. Srivastava P. Interaction of heat shock proteins with peptides and antigen presenting cells: Chaperoning of the innate and adaptive immune responses. *Ann Rev Immunol* 20: 395–425, 2002.
41. Porcelli SA, Modlin RL. The CD1 system: Antigen-presenting molecules for T cell recognition of lipids and glycolipids. *Ann Rev Immunol* 17: 297–329, 1999.
42. Bachmann MF, Kundig TM, Freer G, Li Y, Kang CY, Bishop DH, Hengartner H, Zinkernagel RM. Induction of protective cytotoxic T cells with viral proteins. *Eur J Immunol* 24: 2228–36, 1994.
43. Carbone FR, Bevan MJ. Class I-restricted processing and presentation of exogenous cell-associated antigen *in vivo*. *J Exp Med* 171: 377–87, 1990.
44. Kovacsovics-Bankowski M, Clark K, Benacerraf B, Rock KL. Efficient major histocompatibility complex class I presentation of exogenous antigen upon phagocytosis by macrophages. *Proc Natl Acad Sci USA* 90: 4942–46, 1993.
45. Castellino F, Huang AY, tan-Bonnet G, Stoll S, Scheinecker C, Germain RN. Chemokines enhance immunity by guiding naive $CD8^+$ T cells to sites of $CD4^+$ T cell–dendritic cell interaction. *Nature* 440: 890–95, 2006.
46. Stumbles PA, Thomas JA, Pimm CL, Lee PT, Venaille TJ, Proksch S, Holt PG. Resting respiratory tract dendritic cells preferentially stimulate T helper cell type 2 (Th2) responses and require obligatory cytokine signals for induction of Th1 immunity. *J Exp Med* 188: 2019–31, 1998.
47. Eisenbarth SC, Piggott DA, Huleatt JW, Visintin I, Herrick CA, Bottomly K. Lipopolysaccharide-enhanced, toll-like receptor 4-dependent T helper cell type 2 responses to inhaled antigen. *J Exp Med* 196: 1645–51, 2002.
48. Dillon S, Agrawal A, van DT, Landreth G, McCauley L, Koh A, Maliszewski C, Akira S, Pulendran B. A Toll-like receptor 2 ligand stimulates Th2 responses *in vivo*, via induction of extracellular signal-regulated kinase mitogen-activated protein kinase and c-Fos in dendritic cells. *J Immunol* 172: 4733–43, 2004.
49. Yang D, Chen Q, Su SB, Zhang P, Kurosaka K, Caspi RR, Michalek SM, Rosenberg HF, Zhang N, Oppenheim JJ. Eosinophil-derived neurotoxin acts as an alarmin to activate the TLR2-MyD88 signal pathway in dendritic cells and enhances Th2 immune responses. *J Exp Med* 205: 79–90, 2008.
50. Ito T, Wang YH, Duramad O, Hori T, Delespesse GJ, Watanabe N, Qin FX, Yao Z, Cao W, Liu YJ. TSLP-activated dendritic cells induce an inflammatory T helper type 2 cell response through OX40 ligand. *J Exp Med* 202: 1213–23, 2005.
51. Sokol CL, Barton GM, Farr AG, Medzhitov R. A mechanism for the initiation of allergen-induced T helper type 2 responses. *Nat Immunol*, 2007.
52. Sayed BA, Brown MA. Mast cells as modulators of T-cell responses. *Immunol Rev* 217: 53–64, 2007.
53. Machida I, Matsuse H, Kondo Y, Kawano T, Saeki S, Tomari S, Obase Y, Fukushima C, Kohno S. Cysteinyl leukotrienes regulate dendritic cell functions in a murine model of asthma. *J Immunol* 172: 1833–38, 2004.
54. Finotto S, Neurath MF, Glickman JN, Qin S, Lehr HA, Green FH, Ackerman K, Haley K, Galle PR, Szabo SJ, Drazen JM, De Sanctis GT, Glimcher LH. Development of spontaneous airway changes consistent with human asthma in mice lacking T-bet. *Science* 295: 336–38, 2002.
55. Freeman CM, Curtis JL, Chensue SW. CC chemokine receptor 5 and CXC chemokine receptor 6 expression by lung $CD8^+$ cells correlates with chronic obstructive pulmonary disease severity. *Am J Pathol* 171: 767–76, 2007.
56. Bonecchi R, Bianchi G, Bordignon PP, D'Ambrosio D, Lang R, Borsatti A, Sozzani S, Allavena P, Gray PA, Mantovani A, Sinigaglia F. Differential expression of chemokine receptors and chemotactic responsiveness of type 1 T helper cells (Th1s) and Th2s. *J Exp Med* 187: 129–34, 1998.
57. Sekiya T, Miyamasu M, Imanishi M, Yamada H, Nakajima T, Yamaguchi M, Fujisawa T, Pawankar R, Sano Y, Ohta K, Ishii A, Morita Y, Yamamoto K, Matsushima K, Yoshie O, Hirai K. Inducible expression of a Th2-type CC chemokine thymus- and activation-regulated chemokine by human bronchial epithelial cells. *J Immunol* 165: 2205–13, 2000.
58. Grumelli S, Corry DB, Song LZ, Song L, Green L, Huh J, Hacken J, Espada R, Bag R, Lewis DE, Kheradmand F. An immune basis for lung parenchymal destruction in chronic obstructive pulmonary disease and emphysema. *PLoS Med* 1: e8, 2004.
59. Watanabe A, Mishima H, Renzi PM, Xu LJ, Hamid Q, Martin JG. Transfer of allergic airway responses with antigen-primed $CD4^+$ but not $CD8^+$ T cells in Brown Norway rats. *J Clin Invest* 96: 1303–10, 1995.
60. Haczku A, Macary P, Huang TJ, Tsukagoshi H, Barnes PJ, Kay AB, Kemeny DM, Chung KF, Moqbel R. Adoptive transfer of allergen-specific $CD4^+$ T cells induces airway inflammation and hyperresponsiveness in brown-Norway rats. *Immunology* 91: 176–85, 1997.
61. Mishima H, Hojo M, Watanabe A, Hamid QA, Martin JG. $CD4^+$ T cells can induce airway hyperresponsiveness to allergen challenge in the brown norway rat. *Am J Respir Crit Care Med* 158: 1863–70, 1998.
62. Gavett SH, Chen X, Finkelman F, Wills-Karp M. Depletion of murine $CD4^+$ T lymphocytes prevents antigen-induced airway hyperreactivity and pulmonary eosinophilia. *Am J Respir Cell Mol Biol* 10: 587–93, 1994.
63. Isogai S, Taha R, Tamaoka M, Yoshizawa Y, Hamid Q, Martin JG. $CD8(^+)$ alpha beta T cells can mediate late airway responses and airway eosinophilia in rats. *J Allergy Clin Immunol* 114: 1345–52, 2004.
64. Schwarze J, Cieslewicz G, Joetham A, Ikemura T, Hamelmann E, Gelfand EW. CD8 T cells are essential in the development of respiratory syncytial virus-induced lung eosinophilia and airway hyperresponsiveness. *J Immunol* 162: 4207–11, 1999.
65. Ramos-Barbon D, Presley JF, Hamid QA, Fixman ED, Martin JG. Antigen-specific $CD4^+$ T cells drive airway smooth muscle remodeling in experimental asthma. *J Clin Invest* 115: 1580–89, 2005.
66. Guerassimov A, Hoshino Y, Takubo Y, Turcotte A, Yamamoto M, Ghezzo H, Triantafillopoulos A, Whittaker K, Hoidal JR, Cosio MG. The development of emphysema in cigarette smoke-exposed mice is strain dependent. *Am J Respir Crit Care Med* 170: 974–80, 2004.
67. Takubo Y, Guerassimov A, Ghezzo H, Triantafillopoulos A, Bates JH, Hoidal JR, Cosio MG. Alpha1-antitrypsin determines the pattern of emphysema and function in tobacco smoke-exposed mice: Parallels with human disease. *Am J Respir Crit Care Med* 166: 1596–603, 2002.
68. Maeno T, Houghton AM, Quintero PA, Grumelli S, Owen CA, Shapiro SD. $CD8^+$ T Cells are required for inflammation and

destruction in cigarette smoke-induced emphysema in mice. *J Immunol* 178: 8090–96, 2007.

69. Taraseviciene-Stewart L, Scerbavicius R, Choe KH, Moore M, Sullivan A, Nicolls MR, Fontenot AP, Tuder RM, Voelkel NF. An animal model of autoimmune emphysema. *Am J Respir Crit Care Med* 171: 734–42, 2005.
70. Li L, Sad S, Kagi D, Mosmann TR. CD8Tc1 and Tc2 cells secrete distinct cytokine patterns *in vitro* and *in vivo* but induce similar inflammatory reactions. *J Immunol* 158: 4152–61, 1997.
71. van Beelen AJ, Zelinkova Z, Taanman-Kueter EW, Muller FJ, Hommes DW, Zaat SA, Kapsenberg ML, de Jong EC. Stimulation of the intracellular bacterial sensor NOD2 programs dendritic cells to promote interleukin-17 production in human memory T cells. *Immunity* 27: 660–69, 2007.
72. Kano S, Sato K, Morishita Y, Vollstedt S, Kim S, Bishop K, Honda K, Kubo M, Taniguchi T. The contribution of transcription factor IRF1 to the interferon-gamma-interleukin 12 signaling axis and TH1 versus TH17 differentiation of $CD4^+$ T cells. *Nat Immunol* 9: 34–41, 2008.
73. Schnyder-Candrian S, Togbe D, Couillin I, Mercier I, Brombacher F, Quesniaux V, Fossiez F, Ryffel B, Schnyder B. Interleukin-17 is a negative regulator of established allergic asthma. *J Exp Med* 203: 2715–25, 2006.
74. Hellings PW, Kasran A, Liu Z, Vandekerckhove P, Wuyts A, Overbergh L, Mathieu C, Ceuppens JL. Interleukin-17 orchestrates the granulocyte influx into airways after allergen inhalation in a mouse model of allergic asthma. *Am J Respir Cell Mol Biol* 28: 42–50, 2003.
75. Bettelli E, Carrier Y, Gao W, Korn T, Strom TB, Oukka M, Weiner HL, Kuchroo VK. Reciprocal developmental pathways for the generation of pathogenic effector TH17 and regulatory T cells. *Nature* 441: 235–38, 2006.
76. Kearley J, Barker JE, Robinson DS, Lloyd CM. Resolution of airway inflammation and hyperreactivity after *in vivo* transfer of $CD4^+CD25^+$ regulatory T cells is interleukin 10 dependent. *J Exp Med* 202: 1539–47, 2005.
77. Joetham A, Takeda K, Taube C, Miyahara N, Matsubara S, Koya T, Rha YH, Dakhama A, Gelfand EW. Naturally occurring lung $CD4(^+)CD25(^+)$ T cell regulation of airway allergic responses depends on IL-10 induction of TGF-beta. *J Immunol* 178: 1433–42, 2007.
78. Strickland DH, Stumbles PA, Zosky GR, Subrata LS, Thomas JA, Turner DJ, Sly PD, Holt PG. Reversal of airway hyperresponsiveness by induction of airway mucosal $CD4^+CD25^+$ regulatory T cells. *J Exp Med* 203: 2649–60, 2006.
79. Boismenu R, Havran WL. Gammadelta T cells in host defense and epithelial cell biology. *Clin Immunol Immunopathol* 86: 121–33, 1998.
80. McMenamin C, Pimm C, McKersey M, Holt PG. Regulation of IgE responses to inhaled antigen in mice by antigen-specific gamma delta T cells. *Science* 265: 1869–71, 1994.
81. McMenamin C, McKersey M, Kuhnlein P, Hunig T, Holt PG. Gamma delta T cells down-regulate primary IgE responses in rats to inhaled soluble protein antigens. *J Immunol* 154: 4390–94, 1995.
82. Chen YP, Chou K, Fuchs E, Havran WL, Boismenu R. Protection of the intestinal mucosa by intraepithelial gamma delta T cells. *Proc Natl Acad Sci USA* 99: 14338–43, 2002.
83. Isogai S, Athiviraham A, Fraser RS, Taha R, Hamid Q, Martin JG. Interferon-gamma-dependent inhibition of late allergic airway responses and eosinophilia by $CD8^+$ gammadelta T cells. *Immunology* 122: 230–38, 2007.
84. Hahn YS, Taube C, Jin N, Takeda K, Park JW, Wands JM, Aydintug MK, Roark CL, Lahn M, O'Brien RL, Gelfand EW, Born WK. V gamma 4+ gamma delta T cells regulate airway hyperreactivity to methacholine in ovalbumin-sensitized and challenged mice. *J Immunol* 171: 3170–78, 2003.
85. Lahn M, Kanehiro A, Takeda K, Terry J, Hahn YS, Aydintug MK, Konowal A, Ikuta K, O'Brien RL, Gelfand EW, Born WK. MHC class I-dependent Vgamma4+ pulmonary T cells regulate alpha beta T cell-independent airway responsiveness. *Proc Natl Acad Sci USA* 99: 8850–55, 2002.
86. Holt PG, Strickland DH, Wikstrom ME, Jahnsen FL. Regulation of immunological homeostasis in the respiratory tract. *Nat Rev Immunol* 8: 142–52, 2008.
87. Akbari O, Stock P, Meyer E, Kronenberg M, Sidobre S, Nakayama T, Taniguchi M, Grusby MJ, Dekruyff RH, Umetsu DT. Essential role of NKT cells producing IL-4 and IL-13 in the development of allergen-induced airway hyperreactivity. *Nat Med* 9: 582–88, 2003.
88. Meyer EH, Goya S, Akbari O, Berry GJ, Savage PB, Kronenberg M, Nakayama T, Dekruyff RH, Umetsu DT. Glycolipid activation of invariant T cell receptor+ NK T cells is sufficient to induce airway hyperreactivity independent of conventional $CD4^+$ T cells. *Proc Natl Acad Sci USA* 103: 2782–87, 2006.
89. Renzi PM, Sapienza S, Du T, Wang NS, Martin JG. Lymphokine-induced airway hyperresponsiveness in the rat. *Am Rev Respir Dis* 143: 375–79, 1991.
90. Renzi PM, Sapienza S, Waserman S, Du T, Olivenstein R, Wang NS, Martin JG. Effect of interleukin-2 on the airway response to antigen in the rat. *Am Rev Respir Dis* 146: 163–69, 1992.
91. Nag S, Lamkhioued B, Renzi PM. Interleukin-2-induced increased airway responsiveness and lung Th2 cytokine expression occur after antigen challenge through the leukotriene pathway. *Am J Respir Crit Care Med* 165: 1540–45, 2002.
92. Renzi PM, Xu L, Yang XX, Powell WS, Martin JG. IL-2 enhances allergic airway responses in rats by increased inflammation but not through increased synthesis of cysteinyl leukotrienes. *J Allergy Clin Immunol* 104: 145–52, 1999.
93. Corry DB, Folkesson HG, Warnock ML, Erle DJ, Matthay MA, WienerKronish JP, Locksley RM. Interleukin 4, but not interleukin 5 or eosinophils, is required in a murine model of acute airway hyperreactivity. *J Exp Med* 183: 109–17, 1996.
94. Eum SY, Maghni K, Tolloczko B, Eidelman DH, Martin JG. IL-13 may mediate allergen-induced hyperresponsiveness independently of IL-5 or eotaxin by effects on airway smooth muscle. *Am J Physiol Lung Cell Mol Physiol* 288: L576–84, 2005.
95. Walter DM, McIntire JJ, Berry G, McKenzie AN, Donaldson DD, Dekruyff RH, Umetsu DT. Critical role for IL-13 in the development of allergen-induced airway hyperreactivity. *J Immunol* 167: 4668–75, 2001.
96. Wills-Karp M, Luyimbazi J, Xu X, Schofield B, Neben TY, Karp CL, Donaldson DD. Interleukin-13: Central mediator of allergic asthma. *Science* 282: 2258–61, 1998.
97. Kuperman DA, Huang X, Koth LL, Chang GH, Dolganov GM, Zhu Z, Elias JA, Sheppard D, Erle DJ. Direct effects of interleukin-13 on epithelial cells cause airway hyperreactivity and mucus overproduction in asthma. *Nat Med* 8: 885–89, 2002.
98. Finotto S, Hausding M, Doganci A, Maxeiner JH, Lehr HA, Luft C, Galle PR, Glimcher LH. Asthmatic changes in mice lacking T-bet are mediated by IL-13. *Int Immunol* 17: 993–1007, 2005.
99. Eum SY, Maghni K, Hamid G, Eidelman DH, Campbell H, Isogai S, Martin JG. Inhibition of allergic airways inflammation and airway hyperresponsiveness in mice by dexamethasone: Role of eosinophils, IL-5, eotaxin, and IL-13. *J Allergy Clin Immunol* 111: 1049–61, 2003.
100. Mould AW, Ramsay AJ, Matthaei KI, Young IG, Rothenberg ME, Foster PS. The effect of IL-5 and eotaxin expression in the lung on eosinophil trafficking and degranulation and the induction of bronchial hyperreactivity. *J Immunol* 164: 2142–50, 2000.
101. Nag SS, Xu LJ, Hamid Q, Renzi PM. The effects of IL-5 on airway physiology and inflammation in rats. *J Allergy Clin Immunol* 111: 558–66, 2003.
102. Lee JJ, Dimina D, Macias MP, Ochkur SI, McGarry MP, O'Neill KR, Protheroe C, Pero R, Nguyen T, Cormier SA, Lenkiewicz E, Colbert D, Rinaldi L, Ackerman SJ, Irvin CG, Lee NA. Defining a link with asthma in mice congenitally deficient in eosinophils. *Science* 305: 1773–76, 2004.
103. Mattes J, Yang M, Mahalingam S, Kuehr J, Webb DC, Simson L, Hogan SP, Koskinen A, McKenzie AN, Dent LA, Rothenberg ME, Matthaei KI, Young IG, Foster PS. Intrinsic defect in T-cell

production of interleukin (IL)-13 in the absence of both IL-5 and eotaxin precludes the development of eosinophilia and airways hyperreactivity in experimental asthma. *J Exp Med* 195: 1433–44, 2002.
104. Oh JW, Seroogy CM, Meyer EH, Akbari O, Berry G, Fathman CG, Dekruyff RH, Umetsu DT. CD4 T-helper cells engineered to produce IL-10 prevent allergen-induced airway hyperreactivity and inflammation. *J Allergy Clin Immunol* 110: 460–68, 2002.
105. Pichavant M, Goya S, Meyer EH, Johnston RA, Kim HY, Matangkasombut P, Zhu M, Iwakura Y, Savage PB, Dekruyff RH, Shore SA, Umetsu DT. Ozone exposure in a mouse model induces airway hyperreactivity that requires the presence of natural killer T cells and IL-17. *J Exp Med* 205: 385–93, 2008.
106. Saetta M, Mariani M, Panina-Bordignon P, Turato G, Buonsanti C, Baraldo S *et al.* Increased expression of the chemokine receptor CXCR3 and its ligand CXCL10 in peripheral airways of smokers with chronic obstructive pulmonary disease. *Am J Respir Crit Care Med* 165: 1404–9, 2002.
107. Zheng T, Zhu Z, Wang Z, Homer RJ, Ma B, Riese RJ Jr., Chapman HA Jr., Shapiro SD, Elias JA. Inducible targeting of IL-13 to the adult lung causes matrix metalloproteinase- and cathepsin-dependent emphysema. *J Clin Invest* 106: 1081–93, 2000.
108. Hoshino T, Kato S, Oka N, Imaoka H, Kinoshita T, Takei S, Kitasato Y, Kawayama T, Imaizumi T, Yamada K, Young HA, Aizawa H. Pulmonary inflammation and emphysema: Role of the cytokines IL-18 and IL-13. *Am J Respir Crit Care Med* 176: 49–62, 2007.
109. Robinson DS, Hamid Q, Ying S, Tsicopoulos A, Barkans J, Bentley AM, Corrigan C, Durham SR, Kay AB. Predominant TH2-like bronchoalveolar T-lymphocyte population in atopic asthma. *N Engl J Med* 326: 298–304, 1992.
110. Robinson D, Hamid Q, Bentley A, Ying S, Kay AB, Durham SR. Activation of $CD4^+$ T cells, increased TH2-type cytokine mRNA expression, and eosinophil recruitment in bronchoalveolar lavage after allergen inhalation challenge in patients with atopic asthma. *J Allergy Clin Immunol* 92: 313–24, 1993.
111. Olivenstein R, Taha R, Minshall EM, Hamid QA. IL-4 and IL-5 mRNA expression in induced sputum of asthmatic subjects: Comparison with bronchial wash. *J Allergy Clin Immunol* 103: 238–45, 1999.
112. Nakamura Y, Ghaffar O, Olivenstein R, Taha RA, Soussi-Gounni A, Zhang DH, Ray A, Hamid Q. Gene expression of the GATA-3 transcription factor is increased in atopic asthma. *J Allergy Clin Immunol* 103: 215–22, 1999.
113. Krug N, Erpenbeck VJ, Balke K, Petschallies J, Tschernig T, Hohlfeld JM, Fabel H. Cytokine profile of bronchoalveolar lavage-derived $CD4(^+)$, $CD8(^+)$, and gammadelta T cells in people with asthma after segmental allergen challenge. *Am J Respir Cell Mol Biol* 25: 125–31, 2001.
114. Hamid Q, Azzawi M, Ying S, Moqbel R, Wardlaw AJ, Corrigan CJ, Bradley B, Durham SR, Collins JV, Jeffery PK. Expression of mRNA for interleukin-5 in mucosal bronchial biopsies from asthma. *J Clin Invest* 87: 1541–46, 1991.
115. Maestrelli P, Del Prete GF, De CM, D'Elios MM, Saetta M, Di SA, Mapp CE, Romagnani S, Fabbri LM. CD8 T-cell clones producing interleukin-5 and interferon-gamma in bronchial mucosa of patients with asthma induced by toluene diisocyanate. *Scand J Work Environ Health* 20: 376–81, 1994.
116. Leckie MJ, ten Brinke A, Khan J, Diamant Z, O'Connor BJ, Walls CM, Mathur AK, Cowley HC, Chung KF, Djukanovic R, Hansel TT, Holgate ST, Sterk PJ, Barnes PJ. Effects of an interleukin-5 blocking monoclonal antibody on eosinophils, airway hyper-responsiveness, and the late asthmatic response. *Lancet* 356: 2144–48, 2000.
117. Flood-Page P, Menzies-Gow A, Phipps S, Ying S, Wangoo A, Ludwig MS, Barnes N, Robinson D, Kay AB. Anti-IL-5 treatment reduces deposition of ECM proteins in the bronchial subepithelial basement membrane of mild atopic asthmatics. *J Clin Invest* 112: 1029–36, 2003.
118. Ying S, Durham SR, Corrigan CJ, Hamid Q, Kay AB. Phenotype of cells expressing mRNA for TH2-type (interleukin 4 and interleukin 5) and TH1-type (interleukin 2 and interferon gamma) cytokines in bronchoalveolar lavage and bronchial biopsies from atopic asthmatic and normal control subjects. *Am J Respir Cell Mol Biol* 12: 477–87, 1995.
119. Borish LC, Nelson HS, Lanz MJ, Claussen L, Whitmore JB, Agosti JM, Garrison L. Interleukin-4 receptor in moderate atopic asthma-A phase I/II randomized, placebo-controlled trial. *Am J Respir Crit Care Med* 160: 1816–23, 1999.
120. Wenzel S, Wilbraham D, Fuller R, Getz EB, Longphre M. Effect of an interleukin-4 variant on late phase asthmatic response to allergen challenge in asthmatic patients: Results of two phase 2a studies. *Lancet* 370: 1422–31, 2007.
121. Bullens DM, Truyen E, Coteur L, Dilissen E, Hellings PW, Dupont LJ, Ceuppens JL. IL-17 mRNA in sputum of asthmatic patients: Linking T cell driven inflammation and granulocytic influx? *Respir Res* 7: 135, 2006.
122. Molet S, Hamid Q, Davoine F, Nutku E, Taha R, Page N, Olivenstein R, Elias J, Chakir J. IL-17 is increased in asthmatic airways and induces human bronchial fibroblasts to produce cytokines. *J Allergy Clin Immunol* 108: 430–38, 2001.
123. Akbari O, Faul JL, Hoyte EG, Berry GJ, Wahlstrom J, Kronenberg M, Dekruyff RH, Umetsu DT. $CD4^+$ invariant T-cell-receptor+ natural killer T cells in bronchial asthma. *N Engl J Med* 354: 1117–29, 2006.
124. Vijayanand P, Seumois G, Pickard C, Powell RM, Angco G, Sammut D, Gadola SD, Friedmann PS, Djukanovic R. Invariant natural killer T cells in asthma and chronic obstructive pulmonary disease. *N Engl J Med* 356: 1410–22, 2007.
125. Lazaar AL, Albelda SM, Pilewski JM, Brennan B, Pure E, Panettieri RA Jr., T lymphocytes adhere to airway smooth muscle cells via integrins and CD44 and induce smooth muscle cell DNA synthesis. *J Exp Med* 180: 807–16, 1994.
126. Begueret H, Berger P, Vernejoux JM, Dubuisson L, Marthan R, Tunon-de-Lara JM. Inflammation of bronchial smooth muscle in allergic asthma. *Thorax* 62: 8–15, 2007.
127. Casolaro MA, Bernaudin JF, Saltini C, Ferrans VJ, Crystal RG. Accumulation of Langerhans' cells on the epithelial surface of the lower respiratory tract in normal subjects in association with cigarette smoking. *Am Rev Respir Dis* 137: 406–11, 1988.
128. Soler P, Moreau A, Basset F, Hance AJ. Cigarette smoking-induced changes in the number and differentiated state of pulmonary dendritic cells/Langerhans cells. *Am Rev Respir Dis* 139: 1112–17, 1989.
129. Bosken CH, Hards J, Gatter K, Hogg JC. Characterization of the inflammatory reaction in the peripheral airways of cigarette smokers using immunocytochemistry. *Am Rev Respir Dis* 145: 911–17, 1992.
130. Finkelstein R, Ma HD, Ghezzo H, Whittaker K, Fraser RS, Cosio MG. Morphometry of small airways in smokers and its relationship to emphysema type and hyperresponsiveness. *Am J Respir Crit Care Med* 152: 267–76, 1995.
131. Zhu J, Majumdar S, Qiu Y, Ansari T, Oliva A, Kips JC, Pauwels RA, De RV, Jeffery PK. Interleukin-4 and interleukin-5 gene expression and inflammation in the mucus-secreting glands and subepithelial tissue of smokers with chronic bronchitis. Lack of relationship with $CD8(^+)$ cells. *Am J Respir Crit Care Med* 164: 2220–28, 2001.
132. O'Shaughnessy TC, Ansari TW, Barnes NC, Jeffery PK. Inflammation in bronchial biopsies of subjects with chronic bronchitis: Inverse relationship of $CD8^+$ T lymphocytes with FEV1. *Am J Respir Crit Care Med* 155: 852–57, 1997.
133. Saetta M, Di SA, Turato G, Facchini FM, Corbino L, Mapp CE, Maestrelli P, Ciaccia A, Fabbri LM. $CD8^+$ T-lymphocytes in peripheral airways of smokers with chronic obstructive pulmonary disease. *Am J Respir Crit Care Med* 157: 822–26, 1998.
134. Di SA, Caramori G, Capelli A, Gnemmi I, Ricciardolo FL, Oates T, Donner CF, Chung KF, Barnes PJ, Adcock IM. STAT4 activation in smokers and patients with chronic obstructive pulmonary disease. *Eur Respir J* 24: 78–85, 2004.
135. Retamales I, Elliott WM, Meshi B, Coxson HO, Pare PD, Sciurba FC, Rogers RM, Hayashi S, Hogg JC. Amplification of inflammation

in emphysema and its association with latent adenoviral infection. *Am J Respir Crit Care Med* 164: 469–73, 2001.
136. Kasahara Y, Tuder RM, Cool CD, Lynch DA, Flores SC, Voelkel NF. Endothelial cell death and decreased expression of vascular endothelial growth factor and vascular endothelial growth factor receptor 2 in emphysema. *Am J Respir Crit Care Med* 163: 737–44, 2001.
137. Shapiro SD. End-stage chronic obstructive pulmonary disease: The cigarette is burned out but inflammation rages on. *Am J Respir Crit Care Med* 164: 339–40, 2001.
138. Agusti A, Macnee W, Donaldson K, Cosio M. Hypothesis: Does COPD have an autoimmune component? *Thorax* 58: 832–34, 2003.
139. Barcelo B, Pons J, Ferrer JM, Sauleda J, Fuster A, Agusti AG. Phenotypic characterisation of T-lymphocytes in COPD: Abnormal $CD4^+CD25^+$ regulatory T-lymphocyte response to tobacco smoking. *Eur Respir J* 31: 555–62, 2008.
140. Pons J, Sauleda J, Ferrer JM, Barcelo B, Fuster A, Regueiro V, Julia MR, Agusti AG. Blunted gamma delta T-lymphocyte response in chronic obstructive pulmonary disease. *Eur Respir J* 25: 441–46, 2005.
141. Burrows B, Martinez FD, Halonen M, Barbee RA, Cline MG. Association of asthma with serum IgE levels and skin-test reactivity to allergens. *N Engl J Med* 320: 271–77, 1989.
142. Cameron L, Hamid Q, Wright E, Nakamura Y, Christodoulopoulos P, Muro S, Frenkiel S, Lavigne F, Durham S, Gould H. Local synthesis of epsilon germline gene transcripts, IL-4, and IL-13 in allergic nasal mucosa after *ex vivo* allergen exposure. *J Allergy Clin Immunol* 106: 46–52, 2000.
143. Hogg JC, Chu F, Utokaparch S, Woods R, Elliott WM, Buzatu L, Cherniack RM, Rogers RM, Sciurba FC, Coxson HO, Pare PD. The nature of small-airway obstruction in chronic obstructive pulmonary disease. *N Engl J Med* 350: 2645–53, 2004.
144. van der Strate BW, Postma DS, Brandsma CA, Melgert BN, Luinge MA, Geerlings M, Hylkema MN, van den BA, Timens W, Kerstjens HA. Cigarette smoke-induced emphysema: A role for the B cell? *Am J Respir Crit Care Med* 173: 751–58, 2006.
145. Demedts IK, Bracke KR, Van PG, Testelmans D, Verleden GM, Vermassen FE, Joos GF, Brusselle GG. Accumulation of dendritic cells and increased CCL20 levels in the airways of patients with chronic obstructive pulmonary disease. *Am J Respir Crit Care Med* 175: 998–1005, 2007.
146. Bratke K, Klug M, Bier A, Julius P, Kuepper M, Virchow JC, Lommatzsch M. Function-associated surface molecules on airway dendritic cells in cigarette smokers. *Am J Respir Cell Mol Biol*, 2008.
147. Brent L. The 50th anniversary of the discovery of immunologic tolerance. *N Engl J Med* 349: 1381–83, 2003.

14

CHAPTER

The Neutrophil and Its Special Role in Chronic Obstructive Pulmonary Disease

Elizabeth Sapey[1] and Robert A. Stockley[2]

[1]Department of Medicine, Birmingham University, Birmingham, UK
[2]Lung Investigation Unit, University Hospital Birmingham, NHS Foundation Trust, Edgbaston, Birmingham, UK

INTRODUCTION

There is a substantial body of evidence to support the hypothesis that the neutrophil is the primary effector cell in chronic obstructive pulmonary disease (COPD) and that neutrophil proteinases, especially neutrophil elastase (NE), are responsible for the main pathological features seen.

Studies have shown that patients with COPD have increased numbers of neutrophils in sputum [1, 2] and bronchoalveolar lavage fluid (BALF) [3] compared with asymptomatic smokers, and that the percentage of neutrophils in BALF is higher in patients with the greatest degree of airflow obstruction [4, 5]. Furthermore, in a recent study of bronchial biopsies, small airways intraepithelial neutrophil counts were greater in patients with COPD compared with both smoking and nonsmoking controls, and correlated with airway obstruction [6]. Increased neutrophil counts have been found in the bronchial walls and in the BALF samples from patients during exacerbations of COPD [7, 8], with increased neutrophil sequestration in the pulmonary microcirculation and such episodes relate to subsequent progression of airflow obstruction [9]. Resolution of the neutrophilic inflammation occurs approximately 5 days after appropriate treatment of the exacerbation, which coincides with clinical recovery [10]. In addition, a recent study by Donaldson *et al.* [11] reported that patients with higher numbers of neutrophils in sputum had a faster decline in FEV_1 compared with those with lower neutrophil counts, losing approximately 1% more than predicted each year.

Furthermore Parr *et al.* [12] demonstrated that baseline markers of neutrophilic inflammation relate to subsequent decline of lung function and CT quantification of emphysema over the subsequent 4 years. Gas transfer (the most direct physiological measure of emphysema) has been shown to be inversely proportional to the levels of neutrophil-associated markers, such as myeloperoxidase and human neutrophil lipocalin, in patients with COPD [13]. Both clinical and subclinical emphysema (noted on HRCT) are associated with an increase in NE and other neutrophil proteins in BALF [14, 15] and in established emphysema, severity is proportional to NE immunoreactivity in tissue [16] and enzyme activity in BALF [17]. Finally, both neutrophil counts and NE concentration appear to decline with smoking cessation [18] which is consistent with the benefits of this intervention.

COPD, while primarily a lung disease, is associated with increased co-morbidity including cardiovascular disease and systemic pathology such as muscle wasting and dysfunction. It has been hypothesized that persistent low grade inflammation may drive the co-morbidity and the systemic effects noted with this disease [19]. The systemic manifestations of COPD are important, as they are not only associated with increased morbidity, but are also predictive of disease outcome, especially body mass index (BMI) which forms part of the BODE index (body mass, airflow obstruction, dyspnea, and exercise capacity) [20]. There is controversy concerning whether increased and sustained pulmonary inflammation causes increased systemic inflammation. However C-reactive protein (CRP) (a generic marker of systemic inflammation) is raised in patients with COPD compared to healthy controls [21] irrespective of cigarette smoke exposure and ischemic heart disease (an independent cause of a raised CRP). It is unclear whether these systemic changes are incidental or driven by pulmonary inflammation. In a recent study of 16 patients with moderate to severe COPD, a relationship was present between absolute sputum neutrophil counts, CRP, and BMI less than 21 (see Fig. 14.1). These data indicate an association between the neutrophil load in the lungs and systemic manifestations of the disease, again highlighting the central role of this cell in many aspects of COPD.

The neutrophil was first implicated in the pathogenesis and progression of COPD in 1963, where it was noted that severe emphysema in early adulthood was linked with a deficiency of alpha 1 antitrypsin (α_1-AT), the serum inhibitor of the proteolytic enzyme NE [22]. Since then, the Proteinase/ Anti-proteinase theory has dominated research in COPD. This states that in health, proteinase activity is controlled by anti-proteinases, such as α_1-AT and secretory leukocyte proteinase inhibitor (SLPI), thereby limiting tissue damage. In COPD, an imbalance is believed to exist, either from the increased activity of proteinases or a real or functional deficiency in anti-proteinases, which leads to excessive tissue destruction [23] Tables 14.1 and 14.2 provides an overview of neutrophil proteinases and anti-proteinases in COPD.

This chapter provides an overview of the role of the neutrophil in COPD, starting with neutrophil maturation and structure. Cytokines and chemoattractants which are important in neutrophil activation and recruitment will be discussed and neutrophil migration into lung and neutrophil apoptosis will be outlined, with particular reference to COPD. The evidence for the actions of neutrophil proteinases in the pathogenesis of COPD will be reviewed, highlighting *in vitro* and *in vivo* work. Finally, the role that neutrophils may play in asthma, likely through similar pathogenic mechanisms is discussed briefly.

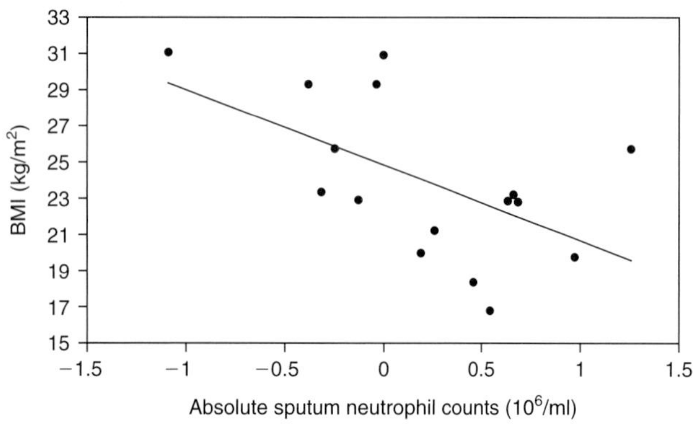

FIG. 14.1 The relationship between BMI and absolute sputum neutrophil counts in patients with moderate to severe stable COPD and chronic bronchitis. The relationship between absolute sputum neutrophil counts and BMI in 12 patients with stable moderate to severe COPD according to GOLD criteria and chronic bronchitis. Neutrophil counts were obtained from spontaneous sputum samples produced over 4 h from waking; $r = -0.8$, $p = 0.01$.

NEUTROPHIL STRUCTURE, FUNCTION, AND APOPTOSIS

Neutrophils have a characteristic multilobed nucleus and abundant storage granules in their cytoplasm. The mature neutrophil has three chemically distinct granule types, which

TABLE 14.1 Overview of neutrophil serine proteinases.

	Neutrophil elastase	Cathepsin G	Proteinase 3
Size	218 amino acids	235 amino acids	222 amino acids
Mass (kDa)	39	28.5	29
Substrate specificity	Val – Xaa > Ala – Xaa	Aromatic acids in P1 position	Same as NE, accepts basic amino acids in P1 position
Elastolytic activity		20% potency of NE	40% potency of NE

TABLE 14.2 Endogenous inhibitors of neutrophil proteinases.

Inhibitor	Source	Target proteinase	Inhibition
Alpha 1antitrypsin	Produced in the liver Found in serum	NE, CG, PR3	Irreversible
Alpha 1 antichymotrypsin	Liver and macrophages	CG	Reversible
Alpha 2 macroglobulin	Liver and macrophages	NE, CG, PR3, MMP-8, MMP-9	Irreversible
Secretory leukoproteinase inhibitor	Mucosa	NE, CG	Reversible
Elafin	Mucosa	NE, PR3	Reversible
Monocyte-neutrophil elastase inhibitor	Neutrophils/monocytes	NE, CG, PR3	Irreversible
Tissue inhibitors of metalloproteinases	Many cell types including epithelial cells, fibroblasts, neutrophils, and monocytes	MMP-8 and MMP-9	Irreversible

appear at different stages of maturation (see Table 14.3). The neutrophil differentiates and matures within bone marrow, developing from a bipotential progenitor cell, the granulocyte-macrophage colony forming unit. In the first developmental stage, the cell divides and differentiates from myeloblasts to promyelocytes. During this stage the azurophilic or primary granules are produced. These granules contain myeloperoxidase (MPO), antibacterial proteins (such as defensins, lysozyme, and azurocidin), and three serine proteinases, NE, cathepsin G (CG), and proteinase 3 (PR3). The proteinases are produced as preproenzymes, with gene expression stopping at the metamyelocyte stage [24] and are activated by a lysosomal cysteine proteinase, dipeptidyl peptidase [25]. The specific or secondary granules are formed as the cell enters the myelocytic stage. These granules contain lysozyme, lactoferrin, collagenase, and various membrane receptors. The small storage granules are formed last, at the metamyelocyte stage, and contain gelatinase and cathepsin B and D. At maturation (which takes approximately 2 weeks) each neutrophil contains a full complement of proteins which provide the mechanisms of cell migration, opsonophagocytosis, and a formidable arsenal against pathogens. However, the neutrophil proteinases (especially NE) also have the capacity to be intensely destructive, degrading structural lung proteins (including elastin, collagen, and gelatin) and appear to be involved in the posttranslational processing of enzymes, cytokines, and receptors [23]. Fully mature neutrophils leave the bone marrow in a nonactivated state and have a half-life of 4–10 h before marginating and entering tissue pools [26]. Once in tissue, neutrophils are usually removed by apoptosis leading to their recognition and phagocytosis by macrophages in the main and by other neutrophils when the macrophage clearance system is overwhelmed [27]. This mechanism prevents cell necrosis and the release of the remaining cellular content of proteinase and other mediators.

Apoptosis is regulated cell death, which allows the elimination of unwanted or damaged cells. At the present time, three caspase-dependent apoptosis pathways have been described. The first pathway is triggered in response to extracellular signals (the receptor-mediated extrinsic pathway), mediated by the binding of tumor necrosis factor related proteins (such as Fas ligand) to death receptors on the cell surface. This results in the formation of the death inducing signaling complex (DISC), which activates a series of caspases, leading to DNA fragmentation by DNAse [28–32]. A second pathway, the mitochondrial intrinsic pathway, responds to stress signals by the release of cytochrome c from mitochondria. Cytochrome c, apoptotic protease activating factor-1 (Apaf-1), and caspase-9 activate caspase-3 leading to apoptosis [33–35]. Finally, in the endoplasmic reticulum pathway, caspase-12 is activated in response to stress signals such as hypoxia [36, 37].

Clearance following apoptosis has no inflammatory sequelae. However, if phagocytosis fails, apoptotic cells undergo secondary necrosis with cell rupture, which is inflammatory in nature and increases both the proinflammatory proteins and proteinase burden in the lung [38]. Figure 14.2 describes the appearance of neutrophils during induced lung inflammation, showing both activated and apoptotic cells, apoptotic cells undergoing secondary necrosis, and neutrophils containing phagosomes with neutrophilic cell remnants.

The increased numbers of neutrophils seen in the lungs of patients with COPD may accumulate because of an increased influx from peripheral blood or because of prolonged neutrophil survival. Exposure to cigarette smoke

TABLE 14.3 Enzymes and other constituents of human neutrophil granules.

	Granules		
Constituents	**Azurophil**	**Specific**	**Small storage**
Antimicrobial	Myeloperoxidase Lysozyme Definsins Bacterial permeability-increasing protein (BPI)	Lysozyme Lactoferrin	
Neutral proteinases	Elastase Cathepsin G Proteinase 3	Collagenase Complement activator Phospholipase A2	Gelatinase Plasminogen activator
Acid hydrolases	Cathepsin D β-d-Glucuronidase α-Mannosidase Phospholipase A2		Cathepsin D β-d-Glucuronidase α-Mannosidase
Cytoplasmic membrane receptors		CR3, CR4 fMLP receptors Laminin receptors	
Others	Chondroitin-4-sulphate	Cytochrome b558 Monocyte Chemotactic factor Histaminase Vitamin B12 binding protein	Cytochrome b558

The enzymes and other constituents of human neutrophil granules.
fMLP: *N*-formylmethionyl-leucyl-phenylalanine

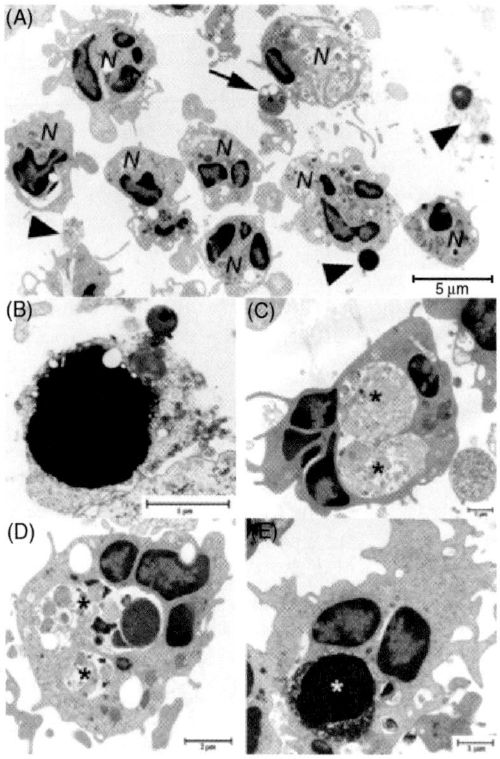

FIG. 14.2 Neutrophils during an LPS-induced lung inflammationElectron micrographs displaying neutrophils during an LPS-induced lung inflammation. (A) Highly activated neutrophils are shown (*N*) (characterized by phagosomes and/or cytoplasmatic protrusions), lying amongst apoptotic neutrophils (black arrow) and cell debris (black arrowhead). (B) Secondary necrosis (characterized by membrane rupture of cells with an otherwise apoptotic morphology) of neutrophils is shown. Neutrophils containing large phagosomes (asterisks) enclosing neutrophilic cell remnants, such as apoptotic nuclei and neutrophil granulae, are shown in images (C–E). Used with kind permission from Rydell-Tormanen *et al.*, Respir Res 7: 143, 2006.

appears to stimulate neutrophil differentiation and maturation, causing peripheral leukocytosis, [39, 40] which has been found to correlate with the severity of airflow obstruction [41, 42]. This suggests that a peripheral neutrophilia facilitates an increased neutrophil transmigration into the lung, leading to a higher proteinase burden with subsequent tissue damage. Platz and colleagues [43] described a reduction of peripheral neutrophil apoptosis (based on Annexin V-PE binding and nuclear morphology) during exacerbations of COPD compared with healthy controls, which returned to control levels upon resolution of symptoms. This increased-cell longevity would facilitate increased neutrophil egression into lung, again predisposing to enhanced lung damage. Once in the lungs, it has been suggested that reduced neutrophil apoptosis may account for the high numbers seen in the lung and the increased inflammation described in COPD although recent studies present conflicting results [44]. Rytila *et al.* [45] found no differences in the proportion of apoptotic neutrophils in induced sputum, or in the *in vitro* anti-apoptotic activity detected in the sputum fluid phase in patients with COPD, healthy smokers or healthy controls despite patients with COPD having a significantly higher percentage of neutrophils in the samples. The authors suggested that the increased neutrophilic presence in COPD was due to an increased influx rather than a reduction in apoptosis. Yoshikawa *et al.* [46] also found higher numbers of neutrophils in induced sputum from patients with COPD compared with healthy controls, but described a reduced neutrophilic chemotactic response to IL-8 and *N*-formyl methionyl-leucyl phenylalanine (fMLP) in the COPD patients which worsened with increasing disease severity. In this study, the authors postulated that an increased influx could not explain the higher numbers of cells seen in COPD, and that increased survival should be considered. This issue has yet to be resolved.

PROMIGRATORY STIMULI

Neutrophils migrate into the lung in response to soluble mediators. Promigratory stimuli can be classified as nonchemotactic cytokines, chemotactic cytokines, or chemoattractants.

Cytokines comprise families of molecules including interleukins, lymphokines, monokines, growth factors, interferons, and chemokines. Two of the most important proadhesive cytokines in COPD are tumor necrosis factor alpha (TNF-α) and interleukin-1β (IL-1β).

TNF-α is produced by activated monocytes and macrophages (but also epithelium, endothelium, and probably smooth muscle cells) and has been implicated in the pathogenesis of COPD [47, 48]. Increased levels of TNF-α have been measured in serum, sputum [49], and in bronchoalveolar lavage samples from patients with COPD, and in smokers demonstrating a dose-dependent relationship with cigarette exposure, [50] with further increases during exacerbations [51]. TNF-α has also been associated with the systemic manifestations of COPD, including a low BMI [52, 53] (perhaps via leptin, [54]) and abnormal resting energy expenditure [55]. TNF-α overexpression (due to genetic polymorphism) has been linked to early COPD development or rapid progression [56]. In support of this, mouse models with an inducible TNF-α gene construct have shown that overexpression of TNF-α is associated with the development of emphysema associated with a general increase in lung inflammation [57] probably by inducing MMP production [58]. Conversely TNF-α receptor knock-out mice demonstrate reduced smoking or elastase-induced emphysema in comparison with the wild type [59, 60] suggesting a central role for this cytokine.

IL-1β is also produced by macrophages (although neutrophils and epithelial cells produce the cytokine) and increased levels have been found in sputum of patients with stable COPD, which increase further during exacerbations [61]. Furthermore, IL-1β production is enhanced by cells cultured from smokers with COPD following cigarette smoke exposure compared with controls [62] and there is some evidence that overexpression of IL-1β (caused by polymorphisms such as the –511 SNP with a cytosine/thymine transition) may increase susceptibility to COPD [63, 64]. In a study of patients with severe COPD, data from our

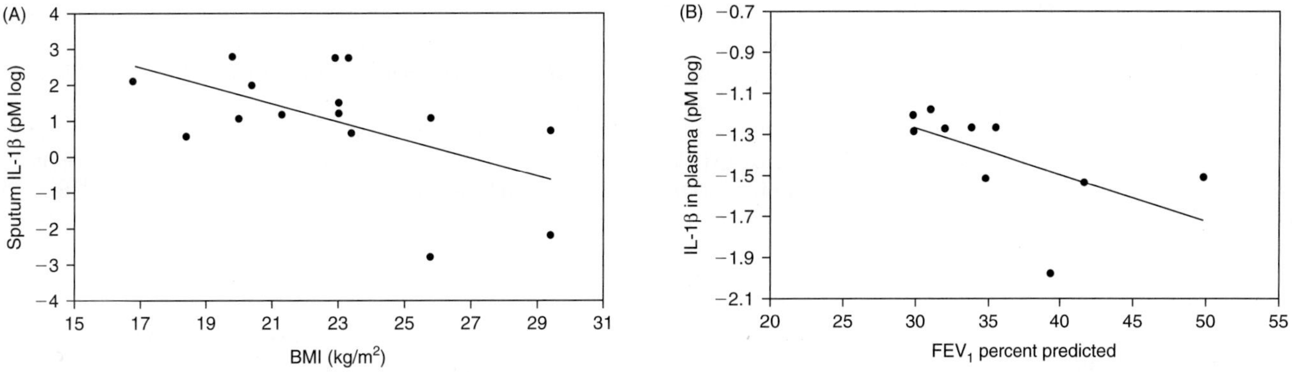

FIG. 14.3 The relationship between IL-1β and disease severity in COPD. Patients had moderate to severe COPD (as defined by GOLD) and chronic bronchitis. Each data point represents the average of 11 measurements of IL-1β and a single BMI (A) or (B) FEV_1 for one patient over 1 month. Spontaneous sputum samples were collected over a 4 h period from waking. The regression lines are drawn and the correlation co-efficient (*r*) and significance (*p*) are as follows: IL-1β and BMI $r = -0.7, p = 0.01$; IL-1β and FEV_1 $r = -0.56, p = 0.02$.

group have shown a negative correlation between plasma IL-1β and FEV_1 and sputum sol IL-1β and BMI, clearly linking IL-1β with COPD disease severity (see Fig. 14.3). Furthermore, animal studies using inducible IL-1β systems and complex knock-out mice have described a neutrophilic infiltrate, distal airspace enlargement, increased thickness of the conducting airways, and enhanced mucin production when IL-1β was overexpressed [65].

TNF-α and IL-1β are not directly chemotactic, but they act by increasing expression of integrins on the neutrophil surface and increasing endothelial expression of selectins and intercellular adhesion molecule 1 (ICAM-1), aiding capture and firm adhesion of neutrophils to the vascular endothelium (see later). They also increase the production of directly chemotactic mediators from endothelial cells to enhance the migration process.

Neutrophils possess at least five receptors for chemotactic stimuli and a number of chemotactic mediators have been implicated in the pathogenesis of COPD, especially the chemokine CXCL8 and the neutrophil chemoattractant LTB4.

CXCL8 is primarily produced by leukocytes (monocytes, T-cells, neutrophils, and natural killer cells) and airway epithelial cells. Production is not constitutive but is induced by proinflammatory cytokines, such as IL-1β and TNF-α [66], bacteria and bacterial products, [67, 68] viruses, such as adenovirus and rhinovirus [69, 70], and oxidants, for example, from cigarette smoke [71]. Once secreted, CXCL8 binds to CXC receptors (1 and 2) on leukocytes resulting in activation of protein kinase B and GTPases, which lead to enhanced neutrophil adherence to endothelial cells (by increasing expression of β_2-integrins) and directed cell migration. CXCL8 also activates Ras and "eventually" mitogen-activated protein kinases and extracellular signal-related kinases in neutrophils, causing degranulation.

In COPD, sputum CXCL8 correlates with levels of neutrophil activation markers, such as MPO and NE [72], and relates to airflow obstruction, [73] oxygen saturation, cigarette exposure [49], and progression of emphysema quantified by CT scan [12]. It has been suggested that oxidative stress (caused by cigarette smoke and bacterial and viral infections) induces CXCL8 production in both airway epithelial and endothelial cells, leading to neutrophil adhesion, chemotaxis, and degranulation.

LTB4 is mainly produced by monocytes, alveolar macrophages, and activated neutrophils and its production is upregulated by a number of inflammatory mediators including C5a, IL-1β, TNF-α, granulocyte-macrophage colony stimulating factor (GM-CSF), PAF, NE, and even LTB4 itself [74, 75]. LTB4 enhances neutrophil aggregation and chemotaxis via two neutrophil surface receptors. A low affinity receptor induces degranulation and increases oxidative metabolism whereas a high affinity receptor induces aggregation, chemokinesis, and adhesion via the integrin Mac-1 [76]. LTB4 may also activate endothelial cell monolayers *in vitro* enhancing neutrophil emigration [77]. LTB4 concentrations are elevated in sputum [78] and exhaled breath condensate [79] from patients with COPD. Concentrations correlate with the degree of airway neutrophilia [80] and increase further during bacterial exacerbations [10] returning to baseline once bacteria have been successfully eradicated [81]. In addition, LTB4 concentrations also correlate with the subsequent decline in gas transfer implicating neutrophilic infiltration [12].

In vitro chemotaxis studies, using functional antibodies against CXCL8 and an LTB4 receptor antagonist, have demonstrated that 30% of the chemotactic activity of sputum can be accounted for by CXCL8, and up to 50% by LTB4 [82, 83]. Furthermore, Woolhouse and colleagues [82] found that LTB4 concentrations correlated with the overall chemotactic activity of sputum, suggesting that LTB4 in particular may be central to the neutrophilic influx seen in COPD, at least in the stable state.

It is unclear whether other inflammatory mediators are important in the pathogenesis of COPD. At least five other chemokines mediate neutrophil responses in humans; three forms of growth related oncogenes (GRO) (α, β, and γ), epithelial cell derived neutrophil activating peptide (ENA)-78, and neutrophil-activated peptide (NAP)-2. GROα is detectable in bronchial secretions [84] and both GROα and ENA-78 have been measured in BALF [85]. One study has suggested that chemotaxis toward GROα and NAP2 is increased in COPD compared with healthy controls and smokers, [86] but their contribution to the pathogenesis of COPD is unknown. A number of other inflammatory mediators including C5a, fragments of fibrin, elastin and collagen, α_1-AT polymers, and bacterial products such as fMLP, are also neutrophil chemoattractants found in the lung, but their importance in neutrophil recruitment in COPD has yet to be established. The numbers of fMLP receptors is elevated in both healthy smokers and subjects with COPD who smoked, but not nonsmoking patients with COPD [87] suggesting that fMLP may play a role in the recruitment of neutrophils leading to the development of disease and its progression whilst smoking and especially during bacterial exacerbations (where bacterial formyl peptides may be most important).

The individual importance of each proinflammatory or promigratory protein in COPD has yet to be unraveled, certainly many of these proteins appear to have similar actions and a degree of overlap or redundancy is likely to exist in the inflammatory cascade. However, there is increasing interest in identifying the key inflammatory mediators in COPD as potential targets for therapeutic interventions. To date, trials which have aimed to block a specific mediator have not proved helpful [88] including a recent trial of anti-TNF-α [89]. It is likely that a greater understanding of each mediator, its naturally occurring antagonists and their relation to COPD is required to develop and clarify more successful treatment strategies. However, Gompertz *et al.* were able to show in a short phase-2 study that antagonism of five lipoxygenase activating proteins reduced LTB4 in the airway secretions and subsequent markers of neutrophilic inflammation [90]. Although not dramatically effective, this may provide a model for future studies.

NEUTROPHIL MIGRATION

Neutrophils are present at both the bronchial and alveolar level in COPD, and therefore it is likely that neutrophil migration occurs from the bronchial and pulmonary circulation. Most of our understanding of neutrophil migration has been derived from *in vitro* studies using systemic vessels in the mesentery and dermis. Studies using samples of lung tissue suggests that while migration in the bronchial circulation occurs in a similar fashion to migration in other tissues, migration in the pulmonary capillaries may depend on distinct mechanisms.

Neutrophil migration in the bronchial circulation

Initiation of migration

In the bronchial circulation, neutrophils migrate from vessel to tissue in a step-like process, dictated by the sequential activation of adhesive proteins and their ligands on neutrophils and endothelial cells.

Migration begins with the capture of neutrophils from flowing blood which causes the cell to roll along the endothelial surface. Tethering and rolling of the neutrophil along the vessel wall is a normal feature of circulating neutrophils and is due to reversible binding of transmembrane glycoprotein adhesive molecules called "selectins," which are found both on neutrophils and endothelial cells [91].

Leukocyte Selectin (L-Selectin) is constitutively expressed on projecting microvilli on the surface of neutrophils and binds an endothelial ligand which is believed to be a sialomucin oligosaccharide (potentially a fucosylated variant of CD34). L-Selectin-induced neutrophil capture can be transient in noninflamed tissue ("stick and release"), however, during both nonpathogenic and pathogenic rolling interactions, L-Selectin, once bound, is shed from neutrophils [92]. Both L-Selectin binding and shedding is enhanced in the presence of inflammatory products such as TNF-α, CXCL8, fMLP, and LPS [93]. Once bound and cleaved, L-Selectin molecules cannot be replaced and low expression has been associated with neutrophil apoptosis [94].

In the presence of inflammation, at least two further endothelial bound selectins are expressed,

Platelet Selectin (P-Selectin) and Endothelial Selectin (E-Selectin): P-Selectin is stored intracellularly in Weibel-Palade bodies in endothelial cells [95] and can be mobilized rapidly in response to various inflammatory mediators including oxygen-free radicals, components of the complement cascade, and many cytokines [96]. Its neutrophilic counterligand is P-Selectin glycoprotein ligand-1 (PSGL1) which is uniformly expressed on the surface of neutrophils. P-Selectin–PSGL1 interactions occur after L-Selectin–ligand interactions, and have greater longevity. In the presence of other adhesion molecules, P-Selectin/PSGL1 binding slows neutrophil rolling velocities and eventually causes cell tethering to the endothelium but in the absence of other adhesive events, binding is also transient [97].

E-Selectin is not stored, and peak expression is seen 4–6 h after endothelial exposure to inflammatory mediators [98]. It binds to E-Selectin ligand 1 [99] and it is thought to maintain neutrophil tethering after P-Selectin has been downregulated. E-Selectin expression is increased in both serum and BAL in patients with COPD and relates to lung function [100] suggesting a common causality.

Firm adhesion

The next step in neutrophil migration is the transition from reversible rolling to firm adhesion with the endothelium. This is achieved by the sequential activation of neutrophil receptors called integrins [101, 102].

The integrins are heterodimeric transmembrane glycoproteins that comprise an α and a β subunit, which together form an extracellular binding site. Integrins are found on many haematopoietic cells, with differing α and β subunits. The two most important integrins in neutrophils share a β_2 subunit (CD18), and are called macrophage antigen 1 (MAC-1; CD11b/CD18) and lymphocyte-associated function antigen 1 (LFA-1; CD11a/CD18). A third CD18 integrin, p150,95 can also promote neutrophil trafficking, but MAC-1 appears to be the most important integrin in models of neutrophil migration [103, 104].

MAC-1 is stored in secretory granules [105, 106] and is rapidly mobilized to the cell surface after exposure to inflammatory stimuli (including fMLP, TNF-α and LPS).

Inflammatory stimuli also promote transcription and translation of the MAC-1 gene via a G-protein, Rho, [107] further increasing its surface expression [96]. Some MAC-1 is expressed constitutively on the neutrophil, but these proteins are incapable of binding ligands unless activated by intra or extracellular signals, where conformational changes occur exposing a requisite binding epitope [108]. Neutrophils from patients with COPD have increased baseline surface expression of MAC-1, which increases further following activation, [109] suggesting that this adhesion molecule may play a role in neutrophil migration in this disease.

MAC-1 has a high affinity for ICAM-1, an immunoglobulin-like protein that is expressed in low numbers on the endothelial cell surface, but is rapidly induced in inflammation [110]. Vascular cell adhesion molecule-1 (VCAM-1) is also an immunoglobulin-like molecule expressed by endothelial cells which it binds selectively to $\alpha_4\beta_1$-integrin [called very late antigen-4 (VLA-4)] on neutrophils [111]. MAC-1/ICAM-1 interactions cause enhanced expression of both ICAM-1 and VCAM-1 on endothelial cells, suggesting that both may be important in inflammatory driven neutrophil migration [112] and levels of ICAM-1 are raised significantly in COPD in contrast to asthma or healthy controls and correlate with overall pulmonary neutrophil infiltration [113].

Migration

The final step of neutrophil recruitment from the bronchial circulation to the lungs is transendothelial migration. This occurs preferentially at tricellular junctions [114] and depends upon activation of platelet endothelial cell adhesion molecule (PECAM1) [115] which is distributed evenly around the neutrophil and at intercellular junctions of endothelial cells. PECAM1 is thought to act as a homing beacon that directs migration toward cellular junctions and blocking PECAM1 on either neutrophils or endothelial cells using antibodies does not prevent adhesion but does prevent migration through the basement membrane both *in vitro* and *in vivo* [116].

Once through the endothelial cell layer, leukocytes bind to matrix components such as collagen and laminin via β_1 integrins, with VLA-6 and -9 being perhaps the most important in allowing neutrophils to move through venule basement membrane and lung tissue [117–119]. Endothelial/neutrophil PECAM1 interactions lead to increased neutrophil surface expression of VLA-6 ($\alpha_6\beta_1$) and VLA-6 facilitates passage thorough the basement membrane and beyond. To support this, neutrophils from PECAM1 knock-out mice do not have the associated rise in VLA-6 which is seen in the wild type [120]. Migration is accompanied by release of neutrophil proteinases especially NE, [121] which may facilitate passage by matrix degradation, exposing laminin for VLA-6 binding.

Direction of migration and neutrophil movement

Neutrophils migrating within the lung encounter multiple chemoattractant signals in complex spatial and temporal patterns as endothelial, epithelial cells, and immune cells respond to infection or injury. Individual chemoattractants can vary in their ability to affect neutrophils and *in vitro* models have demonstrated that neutrophils can migrate up and down chemical gradients, responding to one signal, migrating to its concentration peak and then migrating up a novel, more distant chemoattractant gradient, from endothelium to tissue.

Neutrophils can ignore a high concentration source (due to receptor saturation, desensitization, and/or receptor sequestration) and preferentially respond to novel chemoattractants [122]. Evaluations of chemotaxis by the under-agarose method has suggested that neutrophils are able to migrate in the direction of the vector sum of two or more differing signals [123]. If two sources of the same agonist are used, migration toward the second attractant is poor, [122] so two distinct agonists are required for precise targeting. These findings may account for the observation that activated cells characteristically secrete multiple induced chemoattractants concurrently. Cells can regain their prior sensitivity, but this process takes time, requiring recycling of receptors. For example neutrophils preincubated with low chemotactic levels of LTB4 showed a significant reduction in chemotaxis to LTB4 which improved after 10 min [122].

Neutrophil movement itself is achieved by the formation of locally protruding actin-rich pseudopods while retracting other regions of the cell body [124]. In order to achieve this, directed neutrophil chemotaxis begins with cell polarization. Extracellular chemoattractant gradients causes the localized accumulation of PtdIns(3,4,5)P_3 on the side of the cell facing the highest chemoattractant concentration. PtdIns(3,4,5)P_3 signaling guides the localized polymerization of F-actin (filamentous actin), which in turn leads to pseudopod extension, which confers cellular motility. The persistent presence of PtdIns(3,4,5)P_3 at a particular site on the plasma membrane causes cells to acquire an elongated shape in which one F-actin enriched pseudopod becomes the leading edge of the migrating cell, whereas retraction of pseudopods at the rear and sides of the cell is mediated by cortical myosin II [125]. Although there may be many proximal signal pathways by which polarization is regulated, the key event appears to be activation of the RhoGTPases, Rho and Rac via integrins [126]. The $\alpha_4\beta_1$ integrin forms a complex with paxillin and GIT1 which inhibits Rac activation at the sides and rear of the cell, but this inhibitory complex is impeded at the leading edge of the cell by phosphorylation of the α_4 integrin cytoplasmic tail, which allows Rac activation [127]. At the front of the cell Rac activation leads to the F-actin polymerization while at the rear of the cell Rho activation leads to assembly of myosin, with both GTPases working in a co-operative manner to establish and stabilize cell polarity [128].

Recent studies have elucidated how the neutrophil "steers" from one direction to another. Pseudopods are made in spatially restricted sites by splitting of the leading edge of the cell, and the generation of these protrusions appears to be random in both their direction and timing. In a study where cells were exposed to a promigratory stimulus, which was then relocated once an accurate trajectory was established, pseudopod generation occurred randomly, but pseudopods that extended toward the stimulus were more likely to be retained, while those which extended in inaccurate directions were retracted (see Fig. 14.4). Therefore it appears that neutrophils migrate up chemoattractant gradients by

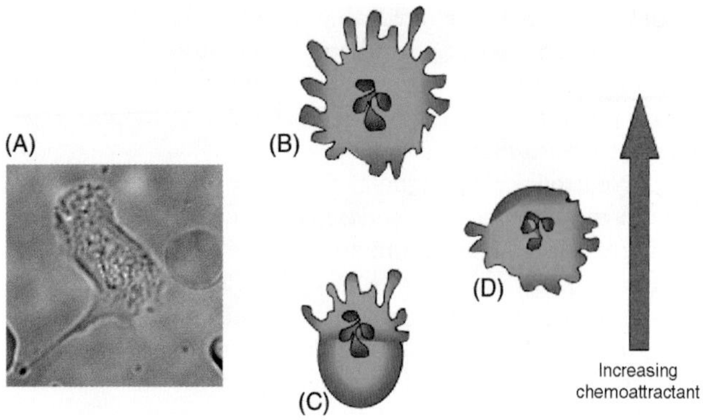

FIG. 14.4 Neutrophil migration by pseudopod generation toward a promigratory stimulus. (A) A polarized (elongated) neutrophil migrating toward IL-8 in a Zigmond chamber. The direction of the chemoattractant gradient is shown by the large arrow. (B) A representation of pseudopods generated over the time course. Here, pseudopod generation was noted across the breadth of the leading edge of the cell in a random fashion. (C) Pseudopods generated toward the stimulus were more likely to be retained (the sustained pseudopods over the time course is shown schematically). (D) Pseudopods that extended in inaccurate directions were more likely to be retracted (the retracted pseudpods over the time course are shown schematically). It appears that neutrophils migrate up chemoattractant gradients by choosing the best aligned of competing randomly generated pseudopods.

choosing the best aligned of competing randomly generated pseudopods [129].

Transmigration through extracellular matrix

Neutrophil proteinases are released during migration through extracellular matrix [121] but it has been difficult to ascertain whether proteinases are necessary for neutrophil migration. Chemotaxis through artificial substrates in response to fMLP can be inhibited by 50% by α_1-AT [130] and CG antibodies; synthetic inhibitors of CG and α_1-antichymotrypsin (α_1-ACT) also reduce neutrophil migration [131]. Furthermore, fMLP-stimulated migration across an artificial basement membrane is also reduced by inhibitors of both NE and MMP-9 [132]. However, *in vitro* studies of endothelial monolayers and basement membrane matrices have shown consistently that proteinase inhibitors are ineffective at stopping neutrophil migration [133, 134] although degradation of basement membrane components is reduced [135]. In animal studies, neutrophils from mice whose genes for NE and CG had been "knocked out" showed normal migration both *in vitro* and *in vivo* when exposed to LPS although pathogen clearance was impaired [136] and mice deficient in gelatinase B had normal neutrophilic migration into the lungs [137]. However, animal studies of cigarette smoke inhalation suggest that neutrophil influx into the lung is reduced in the presence of proteinase inhibitors [138, 139].

Therefore the exact mechanisms remain unresolved, but the conflicting results may well reflect the study model used, including the density of the ECM studied and in particular the presence or absence of cross-linked collagen (compare, e.g. [140, 141]). The majority of recent studies have used tumor cells, but there is evidence to suggest that migrating cells can change their behavior depending on the matrix and stimuli they are exposed to. For example, in a noncross-linked collagen 3D matrix, cells migrate along fibers, but can squeeze between fibers in the presence of anti-proteinases, therefore migration is not inhibited but redirected. On the other hand in a dense cross-linked collagen 3D matrix, cells require proteinases to migrate and migration is inhibited in the presence of anti-proteinases. Cell-derived matrices are cross-linked but have gaps and migrating cells appear to preferentially seek out these gaps to migrate, which may or may not therefore require proteinases, depending on the size of the gap [142]. If neutrophil migration *in vivo* is partially proteinase dependent, *in vitro* migration may not require degradation of "loose" extracellular matrix by proteinases to allow cell passage, but instead the generation/activation of inflammatory chemokines and cytokines or modulation of adhesion molecules (enhancing expression, increasing their activation, or exposing binding sites [117]) by proteinases may be sufficient, and this may be more easily decreased or prevented by inhibitors, explaining the apparent experimental contradictions.

Neutrophil migration in the pulmonary circulation

The majority of neutrophils appear to enter the lung from the pulmonary capillary network rather than the postcapillary venules [143, 144]. Neutrophil emigration into the lung from the pulmonary circulation is less well understood than from the bronchial circulation, however there seem to be important differences in mechanism, probably due to the size of capillary networks. 40–60% of pulmonary capillary segments are narrower than a spherical neutrophil, which on average is 7–8 μm in diameter [145, 146]), so neutrophil rolling is unlikely because of size constraints. Neutrophils entering the pulmonary capillary network have to undergo a shape change, from a sphere to an oblong, to allow passage through these narrow vessels. This slows their transit time, and indeed, radiolabeled neutrophils from patients with stable COPD have been shown to have a slower transit time in the pulmonary circulation compared with red blood cells [147]. When activated, neutrophils become less deformable due to actin polymerization, and this slows their progression through the capillary network further [148]. In such circumstances there is no need for transient adhesion via the selectins to initiate neutrophil rolling along the endothelium, as the leukocytes are already in close contact with the endothelium, and indeed there is evidence to suggest that some migration can occur without adhesion molecules [149, 150].

In the bronchial circulation, adhesion and migration appear to be primarily dependent upon ICAM-1 interactions, however, in the pulmonary circulation both CD18-dependent and -independent adhesion pathways have been described and the path utilized appears to be stimulus-specific. For example, in animal models, bacteria such as *Streptococcus pneumoniae* and *Staphylococcus aureus* and hydrochloric acid have induced CD18-independent neutrophil migration, while human studies have demonstrated CD18-independent migration toward host-derived chemoattractants such as LTB4, CXCL8, and sputum [150,151]. In contrast, IL-1, phorbol myristate acetate, and gram-negative bacterial stimuli including LPS elicit migration via pathways predominantly mediated by CD18 [152–155] CD18-dependent migration may not only be stimulus-driven but also inflammatory mediator driven. Rabbits produce both CXCL8 and TNF-α during a bacterial pneumonia but CXCL8 and TNF-α production during a gram-negative pneumonia are 2- and 10-fold greater respectively than seen during a gram-positive pneumonia [156]. It may be the varying concentrations of these proinflammatory stimuli which elicit a CD18-dependent response, perhaps by inducing nuclear factor-κβ [152] rather than the bacterial insult itself. Animal models have suggested that selectins are required where CD18-dependent migration occurs, but not for CD18-independent migration. For example, L-Selectin knock-out mice had significantly less neutrophil recruitment into the lungs in response to LPS (a CD18-dependent stimulus) while recruitment in response to Streptococcus pneumoniae (a CD18-independent stimulus) was unaffected [157].

Migration in COPD

Current evidence suggests that the increase in neutrophils in the lungs is not due to delayed apoptosis and it has been hypothesized that neutrophils in COPD reflect an enhanced chemotactic response. Certainly, neutrophils from patients with chronic bronchitis and emphysema demonstrate enhanced migration toward common chemoattractants *in vitro* compared with healthy controls [158].

Unfortunately, studies examining the expression of adhesion molecules have produced conflicting results. Noguera [109] measured MAC-1, LFA-1, and L-Selectin expression on neutrophils from controls and patients with COPD, prior and poststimulation with TNF-α. Neutrophils from patients with COPD had enhanced expression of these adhesion molecules compared with controls, and differences were even more pronounced following stimulation with chemoattractants. Woolhouse *et al.* [159] demonstrated upregulation of CD11b (a component of MAC-1) on neutrophils from smokers with COPD compared with controls. However, Gonzalez [160] found no differences between levels of adhesion molecules in smokers with and without airflow obstruction and Yoshikawa *et al.* [161] found reduced chemotaxis toward sputum sol in patients with COPD compared with healthy controls.

The increased neutrophil migration and degranulation seen in COPD has led to a belief that neutrophils may be different from those in health. The differences may be accounted for by genetic polymorphisms and be apparent during maturation in bone marrow. Alternatively neutrophils could be "primed" following their release into the circulation, perhaps by inflammatory cytokines, so that they are more responsive, with an increased ability to degranulate compared with those of healthy individuals. This priming may also occur during transmigration and studies have confirmed differences between neutrophils prior to and following migration even in healthy controls with increased expression of proteinases on the cell surface, increased adhesion molecule expression, and enhancement of the respiratory burst [162, 163]. However, none of these options have been studied in COPD. Whatever the differences in neutrophils, they cannot reflect environmental factors (such as prolonged cigarette smoke exposure) as only a proportion of smokers develop significant airflow obstruction indicating a genetic susceptibility [23]. In susceptible individuals, environmental exposure may lead to increased neutrophil pooling in the lung, increased degranulation and hence increased tissue damage; although it remains unclear whether current observations reflect a predisposing factor for COPD or a consequence of the disease.

THE EFFECTS OF NEUTROPHIL PROTEINASES IN COPD

The actions of neutrophil elastase

A significant physiological role of NE is bacterial killing. This is achieved when opsonized bacteria are ingested in phagosomes which fuse with lysozymes containing proteinases and oxidants. NE is also intensely destructive and this is believed to be the primary cause of lung damage seen both in emphysema and in chronic bronchitis. NE was the first of the serine proteinases to be shown to produce emphysema in animal models. Intratracheal instillation of purified NE-induced emphysema in dogs and hamsters [164–166] and intratracheal instillation of other elastases have given similar results in various animal models–Papain (plant elastase) in rats, [167] neutrophil lysates in dogs [168], and porcine pancreatic elastase in rats and hamsters [169, 170]. In fact the development of emphysema is specific to elastase activity with emphysema severity relating proportionally to the elastolytic potency of the elastase used [169, 170] and in these models emphysema can be prevented by specific elastase inhibitors [171, 172].

NE acts upon a wide range of proteins and can degrade elastin, fibronectin, and collagen, [173, 174] and can also decrease the activity of immunoglobulins and activate components of the complement cascade [175, 176]. NE may also affect wound healing, by its actions on transforming growth factor β and the epithelins [177,178]. During activation, azurophil granule proteinases (including NE) are expressed on the neutrophil membrane [179] and *in vitro*, over 95% remains associated with the cell by a charge-dependent mechanism, while less than 5% is released into the liquid mileu or directly onto tissue. Thus NE most commonly causes damage by close contact between cells and matrix [180, 181]. Recent elegant studies have shown that NE polarizes toward the leading edge of the neutrophil as it migrates. Some is then left behind as the cell moves on where it may cause collateral damage for instance to the connective tissue [182].

Free NE activity has been detected in secretions of patients with COPD [72] and this is felt to be fundamental in the development of the condition *in vivo*. Free NE may accumulate from degranulating neutrophils, or (in contrast with apoptotic cells) may be freely released during cell necrosis [183]. Also the process of phagocytosis may cause the release of significant quantities of proteinases into the media ("sloppy eating"), especially during "frustrated phagocytosis," when cells attempt to ingest large particles [184]. Free NE can also be released from activated macrophages, which scavenge the proteinase from apoptotic neutrophils via endocytosis and subsequently release it during the first 24h of their own inflammatory response [185]. This is important, as although cell-associated proteinases have partial resistance to native inhibitors such as α_1-AT, [180] free NE is more readily inactivated by both serum- and tissue-based inhibitors [186] if sufficient quantities are present, which means that in health, free NE should be completely inactivated at a short distance from the activated cell.

Indeed, it has been shown that the concentration of free NE released from neutrophils falls exponentially away from the cell [187]. In healthy subjects, the serum concentration of α_1-AT is 30μMol, which is at least two orders of magnitude lower than NE concentrations in the neutrophil granule, and inhibits NE on a one to one molar basis. Therefore following degranulation NE cannot be completely inhibited until it has diffused far enough to reduce its concentration to 30μMol. This phenomenon is called quantum proteolysis [187] and was clearly demonstrated in a series of experiments using serum from patients with normal or deficient α_1-AT where an area of obligate enzyme activity existed even with serum from healthy subjects but was far greater using serum from patients with α_1-AT deficiency [188]. Membrane bound NE is less susceptible to anti-proteinases [180] and the combination of increased free and membrane bound NE present in COPD may be sufficient to overcome local inhibitors, causing the tissue destruction which is characteristic of the disease.

NE interacts with matrix proteins and cells, and this affects not only its own activity, but also the efficacy of its inhibitors and other proteinases. For example, although free NE is irreversibly inhibited by α_1-AT, elastin bound NE is poorly inhibited by α_1-AT, while the inhibitory effect of SLPI is unaffected [189]. Adhesion to goblet cells appears to alter the neutrophil membrane, enhancing the release of membrane bound NE into the intercellular space [190] potentially causing mucus secretion. Furthermore, animal models have demonstrated that NE can induce secretory cell metaplasia which is prevented by NE-specific inhibitors [190–192]. It has been suggested that mucous gland epidermal growth factor receptor interacts with NE as part of a signaling cascade [193]. In human studies, there is a clear relationship between the amount of mucus production and the concentration of active NE in the lung secretions.

NE can damage the respiratory epithelium *in vitro* [194], reduce ciliary beating [195], and trigger a state of oxidative stress in cells [196] all of which are abrogated by NE inhibitors. NE can also induce apoptosis of epithelial cells [197] and detachment of bronchial epithelial cells from the extracellular matrix [198] and both PR3 and NE induce detachment and apoptosis of endothelial cells [199] which has also been implicated in the pathogenesis of COPD [200].

NE stimulates the release of LTB4 by macrophages [201] and may cause release of CXCL8 from bronchial epithelial cells, which enhances neutrophil migration into the lung. This inflammatory response is greater in patients with α_1-AT deficiency who have higher levels of LTB4 and elastase providing an amplification to neutrophilic recruitment and subsequent increase in the potential for tissue destruction compared with patients with normal anti-proteinase function thereby explaining the more severe and rapidly progressive disease of α_1-AT deficiency patients [78].

The most important risk factor for COPD is cigarette smoke exposure and the relationships between smoke inhalation and NE have been studied. In animal models NE knock-out mice are partially protected (45%) against the development of emphysema, [202] however, these models are limited, as CG and PR3 persist, and CG has been shown to cause secretory cell metaplasia [166] whereas PR3 causes both emphysema and secretory cell metaplasia [203]. Furthermore, animal models have suggested that both synthetic [171] and natural NE inhibitors [172, 204, 205] can limit emphysema when delivered simultaneously with the elastase insult. Most animal models suggest that neutrophil influx is greatest during the early stages of lung damage with macrophage influx, and their metalloproteinases accumulating at a later stage [206, 207]. However, even when given in established disease, NE inhibitors still limit inflammation and connective tissue breakdown [138].

COPD is characterized by periods of stability of symptoms punctuated with exacerbations [which are intermittent worsening of symptoms and (most probably) of the inflammatory load [208]]. NE activity relates to sputum purulence, which is due to MPO which can be graded visually [10]. During bacterial exacerbations of COPD, sputum purulence, neutrophil influx, NE activity, and tissue degradation products increase [10, 209, 210] and therefore it is likely that proteinase-induced tissue damage also increases. More frequent exacerbations are associated with a faster decline in lung function as well as increased morbidity and mortality and it may be that the increases in NE load and activity during these periods causes progressive lung damage which is reflected in markers of disease severity [208]. Table 14.4 summarizes the actions of NE in COPD.

Matrix metalloproteinases

There have been significant advances in our understanding of the role of matrix metalloproteinases (MMPs) in COPD. Both neutrophils and macrophages produce large amounts of MMPs and their inhibitors, the tissue inhibitors of metalloproteinases (TIMPs). The MMPs are proteolytic enzymes that are secreted as proenzymes (activated by other MMPs and NE) and remain bound to cell membranes. MMPs not only degrade matrix proteins, but also inactivate anti-proteinases such as α_1-AT and α_1-ACT, activate enzymes involved in the clotting cascade, and interact with cytokines and adhesion molecules, so their potential role in the pathogenesis of COPD is complex and wide ranging.

Over 24 mammalian MMPs have been described, and although they are primarily grouped according to the proteins they degrade (collagenases, gelatinases, etc.), most can degrade many substrates and together they can degrade all extracellular substrates. MMPs are inhibited by α_2-macroglobulin and by the four TIMPs described to date

TABLE 14.4 The actions of neutrophil elastase in COPD

Bacterial killing	Intracellular: engulfed organisms in phagosome Extracellular: targeting and cleaving bacterial virulence factors in released granule proteins
Degradation	ECM components Cystatin C TIMPs T-Lymphocyte surface antigen
Activation	MMP-2, MMP-3, MMP-9, Cathepsin B
Modification of inflammatory mediators	Enhances epithelial secretion of CXCL8 Enhances macrophage secretion of LTB4 Inhibits cellular response to TNFRII prolonging half-life of TNF Increases SLPI expression but reduces secretion of SLPI Increases elafin expression Increases alpha 1 antitrypsin expression by monocytes and alveolar macrophages
Cell migration	NE/alpha 1 antitrypsin complexes are chemotactic for neutrophils Modification of ICAM-1 expression enhancing adhesion
Cell apoptosis	Increases epithelial and endothelial cell apoptosis
Cell function	Disruption and detachment of epithelial cells Reduces ciliary beating of columnar epithelium Enhances oxidative stress Increases mucin MUC5AC protein content Increases bacterial adherence and colonization

This table is not exhaustive and references are not included for reasons of space. For an excellent review of the cell signaling by neutrophil proteinases see Ref. [211].

[212]. The main MMPs secreted by neutrophils are MMP-9 (Gelatinase B), which degrades collagen, elastin and gelatin; and MMP-8 (neutrophil collagenase) which degrades collagen type's I–III.

MMP-9 is increased in lung tissue, BALF, and plasma taken from patients with COPD, [213, 214] and levels are negatively correlated with airflow obstruction and relate to the number of sputum neutrophils [215, 216]. MMP-9 not only acts as a proteinase, but also modifies cellular functions by regulating cytokines and matrix bound growth factors and therefore may have a role in lung remodeling after the inflammatory insult resolves [217, 218]. MMP-9 deficient mice exposed to intratracheal LPS showed no differences in histological tissue damage, neutrophil migration, or infiltration compared to the wild-type mouse, suggesting a limited role for this MMP in the pathogenesis of COPD [137]. In a further study, MMP-9 knock-out mice displayed greater neutrophil influx than the wild type, perhaps because MMP-9 can degrade neutrophil chemoattractants [191]. However smoke-exposed guinea pigs displayed a reduction in the severity of emphysema and MMP-9 activity in BALF after introduction of a broad spectrum MMP inhibitor, although this may also have reflected inhibition of other metalloproteinases [219]. Genetic polymorphisms of MMP-9 have, however, been identified which cause enhanced protein expression, and in a Japanese study, polymorphism −1562C/T has been associated with an increased risk of smoking-induced emphysema [220] although this has yet to be replicated in other populations [221].

MMP-8 has been less studied in COPD compared with MMP-9, however one study described both raised concentrations and increased activity of MMP-8 in the induced sputum of patients with COPD compared with smokers without evidence of COPD. Furthermore both the levels and activity of MMP-8 correlated inversely with FEV_1 and positively with sputum neutrophil counts [222]. There are currently no animal studies to elucidate the importance of MMP-8 in COPD.

INTERACTIONS BETWEEN MMPS AND NE IN COPD

It is likely that the serine proteinases and MMPs act synergistically in lung disease. NE degrades TIMPs [223] facilitating MMP activity, and can activate several MMPs converting proenzymes, including MMP-9, to the active form [224]. Conversely MMP-12, for instance, inactivates α_1-AT thereby enhancing NE activity [225]. The majority of animal "knock out" models of emphysema support a multifaceted pathogenic process in the disease, as inhibition of serine, cysteine, or MMPs all show partial protection from the development of emphysema [138, 191]. The noticeable exception is MMP-12 and there have been several conflicting animal studies comparing MMP-12 activity, macrophage and neutrophil influx, and emphysema. MMP-12 knock-out mice (Line 129 mice) exposed to cigarette smoke did not develop emphysema and lung inflammation was reduced, although macrophage recruitment to the lungs was normal in response to monocyte chemoattractant protein-1 (MCP-1). All elastolytic activities in the wild-type mice were derived from MMP-12 and emphysema severity was increased following instillation of MCP-1 suggesting a very specific mechanistic pathway [226]. In a study of Sprague-Dawley rats, neutrophil numbers increased in the first month following smoke exposure, but breakdown of connective tissue and the development of emphysema occurred later and related to a subsequent increase in macrophages [206]. Furthermore, neutrophil depletion (with a significant reduction in NE activity) did not protect against smoke-induced emphysema, while macrophage depletion (with normal neutrophil activity) did. These results have questioned the importance of the neutrophil in the pathogenesis of COPD (at least in rodents) and have suggested that the macrophage was the only effector cell.

However, in a further study BALF from cigarette smoke-exposed mice showed an acute influx of neutrophils with evidence of both elastin and collagen degradation in a dose-dependent manner at 6 and 24h. This had resolved by 48h and was reduced by neutrophil depletion or the administration of the serine proteinase inhibitor α_1-AT [207] although macrophage numbers were not affected. Other animal models have also found an early rise in neutrophil number and activity, and a corresponding rise in matrix breakdown products after exposure to cigarette smoke [138]. In guinea pigs, NE inhibitors reduced TNF-α and emphysema by approximately 30%, but only if given at the start of exposure and when inhibitors were given after 4 months of exposure, these effects were not seen.

TNF-α receptor knock-out mice were also protected from the acute neutrophil infiltrate and connective tissue breakdown seen following cigarette smoke exposure in wild-type mice [211] and mice that produce lower levels of TNF-α following an inflammatory insult (Line 129 mice) also appeared to have less neutrophils and less emphysema compared with the wild-type mice [197, 211]. In fact TNF-α appears to be central to about 70% of the tissue destruction seen following cigarette smoke in mouse models, probably by enhancing neutrophil migration into the lung [227].

In order to understand the relationship between macrophages and neutrophils in COPD, MMP-12 knock-out mice were studied further. Cigarette smoke exposed MMP-12 knock-out mice did not display the early neutrophilic infiltrate or the release of desmosine and hydroxyproline (matrix breakdown products) that is characteristic of wild-type mice, although low levels of macrophage infiltration did occur. When the MMP-12 knock-out mice underwent intratracheal instillation of normal macrophages, a neutrophil influx was seen and the use of an MMP inhibitor also prevented a neutrophilic infiltrate and subsequent matrix breakdown [211]. Following these observations, studies demonstrated that cigarette smoke-induced production of TNF-α from alveolar macrophages in wild-type mice, but not in MMP-12 knock-out mice while levels of TNF-α mRNA were the same in both groups. It was surmised that MMP-12 processes TNF-α after secretion and it is likely, therefore, that MMP-12 is needed to activate TNF-α which in turn initiates neutrophil recruitment, leading to degranulation and tissue damage [228]. Figure 14.5

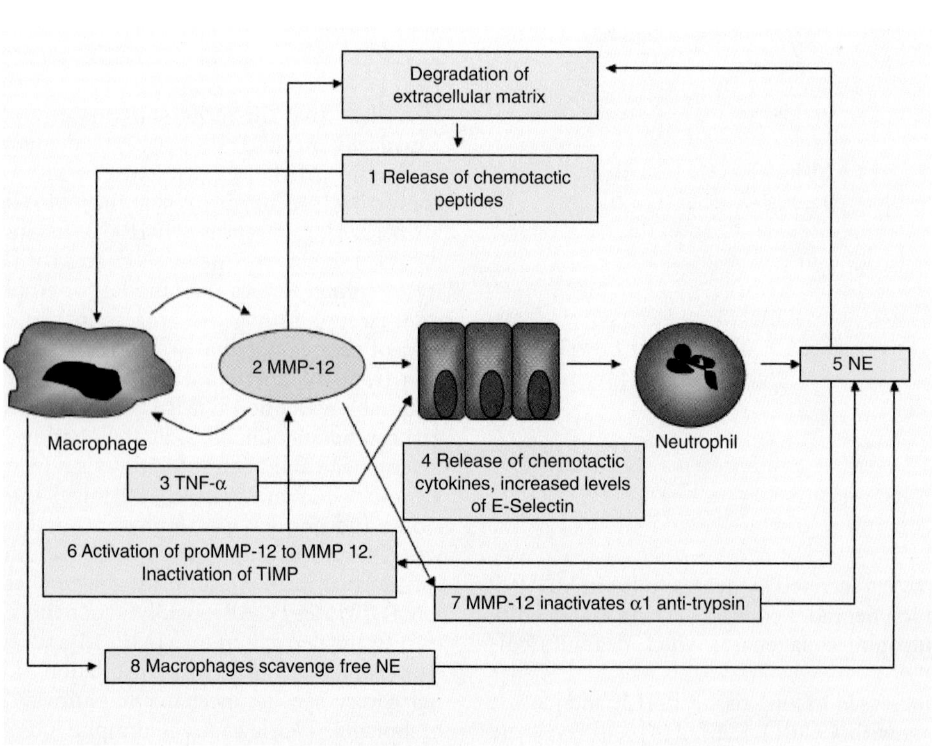

FIG. 14.5 The proinflammatory actions of MMP-12 and the interactions between neutrophils and macrophages
1 Cigarette smoke causes the release of chemotactic peptides which recruit macrophages and neutrophils to airways.
2 Activated macrophages secrete MMP-12 which activates epithelial cells and causes degradation of extracellular matrix components.
3 MMP-12 facilitates the release of active TNF-a which activates epithelial cells further.
4 Activated epithelial cells release chemotactic cytokines and increase the expression of proadhesive proteins such as E-Selectin.
5 Neutrophils migrate to areas of activated epithelium and release free NE during frustrated phagocytosis or sloppy eating. NE promotes epithelial cell apoptosis and increases expression of proinflammatory mediators such as LTB4 and CXCL8.
6 NE proteolytically converts proMMP-12 to active MMP-12 and proteolytically inactivates TIMP, both of which enhance MMP-12 activity.
7 MMP-12 proteolytic inactivation of a1-AT causes the further release of promigratory proteins and allows increased NE activity.
8 Macrophages scavenge free NE which is then reused to further degrade extracellular matrix components.

summarizes the proinflammatory properties of MMP-12 and macrophage/neutrophil interactions.

When interpreting results from *in vitro* work and animal models, it is crucial to consider that there may be important variations in the pathogenesis of COPD between cell types and differing species, and indeed, conflicting results are common. However it seems likely that neutrophil infiltration is an early event and that both neutrophil numbers and elastolytic activity relate well to lung damage in the early stages of disease. Macrophages seem central to neutrophil recruitment, probably via activation of TNF-α by metalloproteinases, and may be needed to sustain the inflammatory process and hence development of emphysema.

THE NEUTROPHIL IN ASTHMA

Although current concepts of asthma pathogenesis focus primarily on $CD4^+$ T cells and eosinophilic inflammation, there are robust data suggesting that the neutrophil may also be important in the pathophysiology of asthma. Absolute neutrophil counts appear to be raised in a subset of patients with asthma, and indeed, in one study up to 60% of patients studied had non-eosinophilic airway inflammation present in induced sputum samples [229]. Measures of chronic asthma severity, such as FEV_1, correlate with the degree of airway neutrophilia in stable disease [230–233]. Furthermore, neutrophilic inflammation has been clearly documented in acute exacerbations of asthma, in status asthmaticus, and in nocturnal asthma [234–236] and has been reported as the predominant cell type in fatal asthma [237, 238]. These findings cannot be explained by patient's smoking status, co-morbidity, or the presence of acute infection alone, suggesting that neutrophils may well have an important role in this disease [239]. Finally, although eosinophilic inflammation is sensitive to corticosteroid treatment, neutrophils respond poorly to this treatment modality, and may be the underlying cause of steroid resistant cases of asthma [240, 241].

It has been proposed that neutrophils enter the airways in response to infection. In murine models, pulmonary infection with mycoplasma or Sendai virus is associated with a neutrophilic infiltrate. This leads to a chronic (and in most cases a permanent) airway hyperresponsiveness that persists long after the infection has resolved [242, 243]. *In vivo* there is serologic and polymerase chain reaction derived evidence associating chronic asthma with airway infection with both viruses and atypical bacteria (which have in turn been associated with increased neutrophil counts) [244, 245]. However, even if infection initiates neutrophil influx into the lungs, treatment of the infection has not been associated with a reduction of symptoms, suggesting that on-going infection may not be sustaining the neutrophilic features of the disease process.

Asthmatic epithelial cells constitutively hypersecrete both IL-8 and IL-17 (both of which are chemotactic for neutrophils), [246, 247] and it has been suggested that these cells are abnormal in asthma, both when quiescent and active, driving inflammation by the release of proinflammatory mediators [248]. This theory has yet to be fully explored.

Whatever the stimulus, once present, neutrophils have the potential to cause the pathological features seen by the release of proteinases and reactive oxygen species, as described previously. In particular there is destruction of the epithelial cell layer [194], which is often a feature of the airways in asthma [249]. Indeed recent studies have confirmed the importance of proteinase function in both the acute inflammatory response and the subsequent mucous cell metaplasia that accompanies postinfectious murine models of asthma, [24] although these findings have yet to be confirmed in humans. Further work is required, but current evidence seems to support the premise that while eosinophils may be dominant in the early stages of disease in atopic individuals, neutrophils form a key feature of more severe, nonatopic asthma [250].

CONCLUSIONS

There is strong evidence supporting the belief that the neutrophil is central to the pathogenesis of COPD. The neutrophil is the only cell that has been shown to be able to cause all of the pathological changes of the disease including emphysema, mucus hypersecretion, epithelial destruction, and reduced ciliary beating. *In vitro* models, animal models, and human studies have led to conflicting results regarding the importance of various components of the inflammatory response. However, based upon current evidence it appears likely that the neutrophil and in particular, NE is associated with at least the initial lung damage seen in COPD although macrophages appear necessary either to initiate the acute neutrophilic response (via TNF-α) and/or sustain the subsequent inflammatory response causing initiation and progression of emphysema. Studies so far have not provided a means to study the effects of other complex processes, such as the recurrent exacerbations in COPD, or the effect of smoking cessation and to clarify our understanding of the overlapping actions of proteinases. Detailed studies of cells and chemoattractants in man will remain crucial in dissecting the mechanisms. Although studied in much less detail, it is likely that neutrophils also play a role in asthma, with the involvement of similar pathways.

References

1. Rutgers SR, Timens W, Kaufmann HF, van der Mark TW, Koeter GH, Postma DS. Comparison of induced sputum with bronchial wash, bronchoalveolar lavage and bronchial biopsies in COPD. *Eur Respir J* 15: 109–15, 2000.
2. Stanescu D, Sanna A, Veriter C, Kostianev S, Calcagni PG, Fabbri LM *et al.* Airways obstruction, chronic expectoration and rapid decline in FEV1 in smokers are associated with increased levels of sputum neutrophils. *Thorax* 51: 267–71, 1996.
3. Martin TR, Raghu G, Maunder RJ, Springmeyer SC. The effects of chronic bronchitis and chronic air flow obstruction on lung cell populations recovered by bronchoalveolar lavage. *Am Rev Respir Dis* 132: 260, 1985.
4. Thompson AB, Daughton D, Robbins RA *et al.* Intraluminal airway inflammation in chronic bronchitis: Characterisation and correlation with clinical parameters. *Am Rev Respir Dis* 140: 1527–37, 1989.

5. Lacoste JY, Bousquet J, Chanez P *et al.* Eosinophilic and neutrophilic inflammation in asthma, chronic bronchitis and chronic obstructive pulmonary disease. *J Allergy Clin Immunol* 92: 537–48, 1993.
6. Pilette C, Colinet B, Kiss R, Andre S, Kaltner H, Gabius HJ, Delos M, Vaerman JP, Decramer M, Sibelle Y. Increased galectin 3 expression and intra-epithelial neutrophils in small airways in severe chronic obstructive pulmonary disease. *Eur Respir J* 45, 2007, epub ahead of print.
7. Balbi B, Bason C, Balleari E *et al.* Increased bronchoalveolar granulocytes and granulocyte/macrophage colony stimulating factor during exacerbations of chronic bronchitis. *Eur Respir J* 10: 846–50, 1997.
8. Tsoumakidou M, Tzanakis N, Chrysofakis G, Kyriakou D, Siafakas NM. Changes in sputum T-lymphocyte subpopulations at the onset of severe exacerbations of chronic obstructive pulmonary disease. *Respir Med* 99(5): 572–79, 2005.
9. Selby C, Drost E, Lannan S *et al.* Neutrophil retention in the lungs of patients with chronic obstructive pulmonary disease. *Am Rev Respir Dis* 143: 1359–64, 1991.
10. Gompertz S, O'Brien C, Bayley D, Hill SL, Stockley RA. Changes in bronchial inflammation during acute exacerbations of chronic bronchitis. *Eur Respir J* 17(6): 1112–19, 2001.
11. Donaldson GC, Seemungal TA, Patel IS, Bhowmik A, Wilkinson TM, Hurst JR, Maccallum PK, Wedzicha JA. Airway and systemic inflammation and decline in lung function in patients with COPD. *Chest* 128(4): 1995–2004, 2005.
12. Parr DG, White AJ, Bayley DL, Guest PJ, Stockley RA. Inflammation in sputum relates to progression of disease in COPD: A prospective study. *Respir Res* 7: 136, 2006.
13. Ekberg-Jansson A, Andersson B, Bake B, Boijsen M, Enanden I, Rosengren A *et al.* Neutrophil associated activation markers in healthy smokers relates to a fall in DLCO and to emphysematous change on high resolution CT. *Respir Med* 95(5): 363–73, 2001.
14. Yoshioka A, Betsuyaku T, Nishimura M, Miyamoto K, Kondo T, Kawakami Y. Excessive neutrophil elastase in bronchoalveolar lavage fluid in subclinical emphysema. *Am J Respir Crit Care Med* 152: 2127–32, 2005.
15. Betsuyaku T, Nishimura M, Takeyabu K, Tanino M, Venge P, Xu S *et al.* Neutrophil granule proteins in bronchoalveolar lavage fluid from patients with subclinical emphysema. *Am J Respir Crit Care Med* 159: 1985–91, 1999.
16. Ge YM, Zhu YJ, Luo WC, Gong YH, Zhang XQ. Damaging role of neutrophil elastase in the elastic fiber and basement membrane in human emphysematous lung. *Chin Med J* 103(7): 588–94, 1990.
17. Fujita J, Nelson NL, Daughton DM, Dobry CA, Spurzem JR, Irino S *et al.* Evaluation of elastase and antielastase balance in patients with chronic bronchitis and pulmonary emphysema. *Am Rev Respir Dis* 142: 57–62, 1990.
18. Rennard SI, Daughton D, Fujita J, Oehlerking MB *et al.* Short term smoking reduction is associated with reduction in measures of lower respiratory tract inflammation in heavy smokers. *Eur Respir J* 3(7): 752–59, 1990.
19. Sin DD, Anthonisen NR, Soriano JB, Agusti AG. Mortality in COPD: Role of co-morbidities. *Eur Respir J* 28(6): 1245–57, 2006.
20. Celli BR, Cote CG, Marin JM, Caanova C, Montes de Oca M, Mendez RA, Pinto-plata V, Cabral HJ. The body mass index, airflow obstruction, dyspnoea and exercise capacity index in chronic obstructive pulmonary disease. *N Engl J Med* 350(10): 1005–12, 2004.
21. Pinto-Plata VM, Mullerova H, Toso JF, Feudjo-Tepie M, Soriano JB, Vessey RS, Celli BR. C-Reactive protein in patients with COPD, control smokers and non-smokers. *Thorax.* 61(1): 23–28, 2006.
22. Laurell CBaSE. The electrophoretic alpha-globulin pattern of serum in alpha antitrypsin deficiency. *Scand J Clin Invest* 15: 132–40, 1963.
23. Stockley RA. Neutrophils and the pathogenesis of COPD. *Chest* 121: 151S–55S, 2002.
24. Fouret P, du Bois RM, Bernaudin JF, Takahashi H, Ferrans VJ, Crystal RG. Expression of the neutrophil elastase gene during human bone marrow cell differentiation. *J Exp Med* 169: 833–45, 1989.
25. Adkison AM, Raptis SZ, Kelley DG, Pham CTN. Dipeptidyl peptidase I activates neutrophil-derived serine proteases and regulates the development of acute experimental arthritis. *J Clin Invest* 109: 363–71, 2002.
26. Burnett D. Neutrophils. In: Stockley RA (ed.), Pulmonary Defenses, pp. 113–26. WileyEurope Chichester, 1997.
27. Rydell-Tormanen K, Uller L, Erjefalt JS. Neutrophil cannibalism – a back up when the macrophage clearance system is insufficient. *Respir Res* 14(7): 143, 2006.
28. Muzio M, Stockwell BR, Stennicke HR, Salvesen GS, Dixit VM. An induced proximity model for caspase-8 activation. *J Biol Chem* 273: 2922–26, 1998.
29. Muzio M, Chinnaiyan AM, Kischkel FC, O'Rourke K, Shevchenko A, Ni J, Scaffidi C, Bretz JD, Zhang M, Gentz R, Mann M, Krammer PH, Peter ME, Dixit VM. FLICE, a novel FADD-homologous ICE/CED-3-like protease, is recruited to the CD95 (Fas/APO-1) death-inducing signaling complex. *Cell* 85: 817–27, 1996.
30. Scaffidi C, Fulda S, Srinivasan A, Friesen C, Li F, Tomaselli KJ, Debatin KM, Krammer PH, Peter ME. Two CD95 (APO-1/Fas) signaling pathways. *EMBO J* 17: 1675–87, 1998.
31. Hirata H, Takahashi A, Kobayashi S, Yonehara S, Sawai H, Okazaki T, Yamamoto K, Sasada M. Caspases are activated in a branched protease cascade and control distinct downstream processes in Fas-induced apoptosis. *J Exp Med* 187: 587–600, 1998.
32. Tang D, Kidd VJ. Cleavage of DFF-45/ICAD by multiple caspases is essential for its function during apoptosis. *J Biol Chem* 273: 28549–52, 1998.
33. Liu X, Kim CN, Yang J, Jemmerson R, Wang X. Induction of apoptotic program in cell-free extracts: Requirement for dATP and cytochrome c. *Cell* 86: 147–57, 1996.
34. Zou H, Henzel WJ, Liu X, Lutschg A, Wang X. Apaf-1, a human protein homologous to C. elegans CED-4, participates in cytochrome c-dependent activation of caspase-3. *Cell* 90: 405–13, 1997.
35. Slee EA, Harte MT, Kluck RM, Wolf BB, Casiano CA, Newmeyer DD, Wang HG, Reed JC, Nicholson DW, Alnemri ES, Green DR, Martin SJ. Ordering the cytochrome c-initiated caspase cascade: Hierarchical activation of caspases-2, -3, -6, -7, -8, and -10 in a caspase-9-dependent manner. *J Cell Biol* 144: 281–92, 1999.
36. Rao RV, Hermel E, Castro-Obregon S, del Rio G, Ellerby LM, Ellerby HM, Bredesen DE. Coupling endoplasmic reticulum stress to the cell death program. Mechanism of caspase activation. *J Biol Chem* 276: 33869–74, 2001.
37. Szegezdi E, Fitzgerald U, Samali A. Caspase-12 and ER-stress-mediated apoptosis: The story so far. *Ann NY Acad Sci* 1010: 186–94, 2003.
38. Rydell-Tormanen K, Uller L, Erjefait JS. Direct evidence of secondary necrosis of neutrophils during intense lung inflammation. *Eur Respir J* 28(2): 268–74, 2006.
39. Corre F, Lellouch J, Schwartz D. Smoking and leucocyte counts: results of an epidemiological study. *Lancet* 2: 632–34, 1971.
40. Van Eeden SF, Hogg JC. The response of human bone marrow to chronic cigarette smoking. *Eur Respir J* 15: 915–21, 2000.
41. Yeung MC, Buncio AD. Leukocyte count, smoking and lung function. *Am J Med* 76: 31–37, 1984.
42. Sparrow D, Glynn RJ, Cohen M, Weiss ST. The relationship of the peripheral leukocyte count and cigarette smoking to pulmonary function among adult young men. *Chest* 86: 383–86, 1984.
43. Pletz MW, Ioanas M, de Roux A, Burkhardt O, Lode H. Reduced spontaneous apoptosis in peripheral blood neutrophils during exacerbations of COPD. *Eur Respir J* 23(4): 532–37, 2004.
44. Uller L, Persson CG, Erjefalt JS. Resolution of airway disease: Removal of inflammatory cells through apoptosis, egression or both. *Trends Pharmscol Sci* 27(9): 461–66, 2006.
45. Rytila P, Plataki M, Bucchieri F, Uddin M, Nong G, Kinnula VL. Djukanovic. Airway neutrophilia in COPD is not associated with increased neutrophil survival. *Eur Respir J* 28: 1163–69, 2006.
46. Yoshikawa T, Dent G, Ward J, Angco G, Nong G, Nomura N, Hirata K, Djukanovic R. Impaired neutrophil chemotaxis in chronic obstructive pulmonary disease. *Am J Respir crit Care Med* 175(5): 473–79, 2007.

47. Churg A, Dai J, Tai H, Xie C, Wright JL. Tumour necrosis factor alpha is central to acute cigarette smoke induced inflammation and connective tissue breakdown. *Am J Respir Crit Care Med* 166: 849–54, 2002.
48. Sakao S, Tatsumi K, Igari H, Watanebe R, Shino Y, Shirasawa H, Kuriyama T. Association of tumour necrosis factor alpha gene promoter polymorphism with low attenuation areas on high resolution CT in patients with COPD. *Chest* 122: 416–20, 2002.
49. Hacievliyagil SS, Gunen H, Mutlu LC, Karabulut AB, Temel I. Association between cytokines in induced sputum and severity of chronic obstructive pulmonary disease. *Respir Med* 100(5): 846–54, 2006.
50. Kuschner WG, D'Alessandro A, Wong H, Blanc PD. Dose dependent cigarette smoking- related inflammatory responses in healthy adults. *Eur Respir J* 9: 1989–94, 1996.
51. Aaron SD, Angel JB, Lunau M *et al.* Granulocyte inflammatory markers and airway infection during acute exacerbations of chronic obstructive pulmonary disease. *Am J Respir Crit Care Med* 163: 349–55, 2001.
52. Di Francia M, Barbier D, Mege JL, Orehek J. Tumor necrosis factor alpha levels and weight loss in chronic obstructive pulmonary disease. *Am J Respir Crit Care Med* 150: 1453–55, 1994.
53. de Godoy I, Donahoe M, Calhoun WJ, Mancino J, Rogers RM. Elevated TNF-alpha production by peripheral blood monocytes of weight-losing COPD patients. *Am J Respir Crit Care Med* 153: 633–37, 1996.
54. Calikoglu M, Sahin G, Unlu A, Ozturk C, Tamer L, Ercan B, Kanik A, Atik U. Lepitn and TNFα levels in patients with chronic obstructive pulmonary disease and their relationship to nutritional parameters. *Respiration* 71(1): 45–50, 2004.
55. Nguyen LT, Bedu M, Caillaud D, Beaufrere B, Beaujon G, Vasson M, Coudert J, Ritz P. Increased resting energy expenditure is related to plasma TNF alpha concentration in stable COPD patients. *Clin Nutr* 18(5): 269–74, 1999.
56. Brogger J, Steen VM, Eiken HG, Gulsvik A, Bakke P. Genetic association between COPD polymorphisms in TNF, ADRB2 and EPHX1. *Eur Respir J* 27(4): 682–88, 2006.
57. Vuillemenot BR, Rodriguez JF, Hoyle GW. Lymphoid tissue and emphysema in the lungs of transgenic mice inducibly expressing TNFα. *Am J Respir Crit Care Med* 30: 438–48, 2004.
58. Wright JL, Tai H, Wang R, Wang X, Churg A. Cigaretter smoke up-regulates pulmonary vascular matrix metalloproteinases via TNF alpha signaling. *Am J Physiol Lung Cell Mol Physiol* 292(1): L125–33, 2007.
59. Churg A, Wang RD, Tai H, Wang X, Xie C, Wright JL. Tumour necrosis factor alpha drives 70% of cigarette smoke-induced emphysema in the mouse. *Am J Respir Crit Care Med* 170: 492–98, 2004.
60. Lucey EC, Keane J, Kuang PP, Snider GL, Goldstein RH. Severity of elastase-induced emphysema is decreased in TNFα and IL-1β receptor deficient mice. *Lab Invest* 82: 79–85, 2002.
61. Chung KF. Cytokines in chronic obstructive pulmonary disease. *Eur Respir J* 34: 50s–59s, 2001.
62. Rusznak C, Mills PR, Devalia JL, Sapsford RJ, Davies RJ, Lozewicz S. Effect of cigarette smoking on the permeability and IL-1beta and sICAM-1 release from cultured human bronchial epithelial cells of never smokers, smokers and patients with chronic obstructive pulmonary disease. *Am J Respir Cell Mol Biol* 23(4): 530–36, 2000.
63. Hegab EA, Sakamoto T, Saitoh W *et al.* Polymorphisms of IL-4, IL-13 and ADRB2 genes in COPD. *Chest* 126: 1832–39, 2004.
64. Asada M, Yamaya M, Ebihara S, Yasufa H, Tomita N, Kubo H, Sasaki H, Hegab A, Sakamoto T, Sekizawa K. Interleukin-1β gene polymrphisms associated with COPD. *Chest* 128: 1072–73, 2005.
65. Lappalainen U, Whitsett JA, Wert SE, Tichelaar JW, Bry K. Interleukin-1beta causes pulmonary inflammation, emyphysema and airway remodelling in the adult murine lung. *Am J Respir Cell Mol Biol* 32(4): 311–18, 2005.
66. Matsushima K, Morishita K, Yoshimura T, Lavu S, Kobayashi Y, Lew W *et al.* Molecular cloning of a human monocyte-derived neutrophil chemotactic factor (MDNCF) and the induction of MDNCF mRNA by interleukin-1 and tumour necrosis factor. *J Exp Med* 169: 1485–90, 1988.
67. DiMango E, Zar HJ, Bryan R, Prince A. Diverse *Pseudomonas aeruginosa* gene products stimulate respiratory epithelial cells to produce interleukin-8. *J Clin Invest* 96: 2204–10, 1995.
68. Khair OA, Davies RJ, Devalia JL. Bacterial-induced release of inflammatory mediators by bronchial epithelial cells. *Eur Respir J* 9(9): 1913–22, 1996.
69. Alcorn MJ, Booth JL, Coggeshall KM, Metcalf JP. Adenovirus type 7 induces interleukin-8 production via activation of extracellular regulated kinase. *J Virol* 75: 6450–59, 2001.
70. Johnston SL, Papi A, Monick MM, Hunninghake GW. Rhinoviruses induce interleukin 8 mRNA and protein production in human monocytes. *J Infect Dis* 175: 323–29, 1997.
71. Deforge LE, Preston AM, Takeuchi E, Kenney J, Boxer LA, Remick DG. Regulation of interleukin 8 gene expression by oxidant stress. *J Biol Chem* 268: 25568–76, 1993.
72. Hill AT, Bayley D, Stockley RA. The interrelationship of sputum inflammatory markers in patients with chronic bronchitis. *Am J Respir Crit Care Med* 160: 893–98, 1999.
73. Yamamoto C, Yoneda T, Yoshikawa M, Fu A, Tokuyama T, Tsukaguchi K *et al.* Airway inflammation in COPD assessed by sputum levels of Interleukin-8. *Chest* 112(2): 505–10, 1997.
74. Borgeat PE, Krump E, Palmanteir R, Picard S, Dallaire N, McDonald P *et al.* The synthesis of leukotrienes by the human neutrophil. In: Holgate S (ed.), *SRS-A to leukotrienes; the dawning of a new treatment*, pp. 69–84. Blackwell Oxford, 2003.
75. Crooks SW, Stockley RA. Leukotriene B4. *Int J Biochem Cell Biol* 30: 173–78, 1998.
76. Tonnesen MG, Anderson DC, Springer TA, Knedler A, Avdi N, Henson PM. Adherence of neutrophils to cultured human microvascular endothelial cells. Stimulation by chemotactic peptides and lipid mediators and dependence upon the Mac-1, LFA-1, p150,95 glycoprotein family. *J Clin Invest* 83: 637, 1989.
77. Nohgawa M, Sasada M, Maeda A, Asagoe K, Harakawa N, Takano K *et al.* Leukotriene B4 activated endothelial cells promote transendothelial neutrophil migration. *J Leuk Biol* 62: 203–9, 1997.
78. Hill AT, Bayley D, Campbell EJ, Hill SL, Stockley RA. Airways inflammation in chronic bronchitis; the effects of smoking and alpha1-antitrypsin deficiency. *Eur Respir J* 15: 886–90, 2000.
79. Montuschi P, Kharitonov S, Ciabattoni G, Barnes PJ. Exhaled leukotrienes and prostaglandins in COPD. *Thorax* 58(7): 585–88, 2003.
80. Bhowmik A, Seemungal TAR, Sapsford RJ. Relation of sputum inflammatory markers to symptoms and lung function changes in COPD. *Thorax* 55: 114–20, 2000.
81. White AJ, Gompertz S, Bayley DL, Hill SL, O'Brien C, Unsal I *et al.* Resolution of bronchial inflammation is related to bacterial eradication following treatment of exacerbations of chronic bronchitis. *Thorax* 58(8): 680–85, 2003.
82. Woolhouse IS, Bayley DL, Stockley RA. Sputum chemotactic activity in chronic obstructive pulmonary disease: Effect of α1-anti-trypsin deficiency and the role of leukotriene B4 and interleukin 8. *Thorax.* 57: 709–14, 2002.
83. Beech KM, Kornmann O, Buhl R *et al.* Neutrophil chemotactic activity of sputum from patients with COPD: Role of interleukin 8 and Leukotriene B4. *Chest* 123: 1240–47, 2003.
84. Traves SL, Culpitt SV, Russell RE, Barnes PJ, Donnelly LE. Increased levels of the chemokine GROalpha and MCP-1 in sputum samples from patients with COPD. *Thorax* 57: 590–95, 2002.
85. Morrison D, Strieter RM, Donnelly SC, Burdick MD, Kunkel SL, MacNee W. Neutrophil chemokines in bronchoalveolar lavage fluid and leukocyte-conditioned medium from smokers and non-smokers. *Eur Respir J* 12: 1067–72, 1998.
86. Traves SL, Smith SJ, Barnes PJ, Dinelly LE. Specific CXC but not CC chemokines cause elevated monocyte migration in COPD. A role for CXCR2. *J Leuk Biol* 76(2): 441–50, 2004.
87. Matheson M, Rynell AC, McClean M, Berend N. Cigarette smoking increases neutrophil formyl methionyl leucyl phenylalanine receptor numbers. *Chest* 123(5): 1642–46, 2003.

88. de Boer W. Perspectives for cytokine antagonist therapy in COPD. *DDT* 10(2): 93–106, 2005.
89. Rennard SI, Fogarty C, Kelsen S, Long W, Ramsdell J, Allison J, Mahler D, Saadeh C, Siler T, Snell P, Korenblat P, Smith W, Kaye M, Mandel M, Andrews C, Prabhu R, Donohue JF, Watt R, Lo KH, Schlenker-Herceg R, Barnathan ES, Murray J. The safety and efficacy of infliximab in moderate to severe chronic obstructive pulmonary disease. *Am J Respir Crit Care Med* 175(9): 926–34, 2007.
90. Gompertz S, Stockley RA. A randomised controlled trial of a Leukotriene synthesis inhibitor in patients with COPD. *Chest* 122: 289–94, 2002.
91. Wagner JG, Roth RA. Neutrophil migration mechanisms with an emphasis on pulmonary vasculature. *Pharmacol Rev* 52(3): 349–74, 2000.
92. Walchek B, Kahn J, Fisher JM, Wang BB, Fisk RS, Payan DG *et al.* Neutrophil rolling altered by inhibition of L-Selectin shedding in vitro. *Nature* 380: 720–23, 1996.
93. Drost EM, MacNee W. Potential role of IL8, platelet activating factor and TNF-α in the sequestration of neutrophils in the lung; effects on neutrophil deformability, adhesion receptor expression and chemotaxis. *Eur J Immunol* 32: 393–403, 2002.
94. Matsuba KT, Van Eeden SF, Bicknell SG, Walker BAM, Hayashi S, Hogg JC. Apoptosis in circulating PMNs: Increasing susceptibility in L-selectin deficient PMN. *Am J Physiol* 272: H2852–hH2858, 1997.
95. Malik AR, Lo SK. Vascular endothelial adhesion molecules and tissue inflammation. *Pharmacol Rev* 48: 213–29, 1996.
96. Spertini O, Luscinskas FW, Munro JM, Griffin JD, Gimbrone MAJ, Tedder TF. Leukocyte adhesion molecule-1 (LAM1, L-selectin) interacts with an inducible endothelial cell ligand to support leukocyte adhesion. *J Immunol* 147: 2565–78, 1991.
97. Alon R, Chen S, Puri KD, Finger EB, Springer TA. The kinetics of L-selectin tethers and the mechanisms of selectin mediated rolling. *J Cell Biol* 138: 1169–80, 1997.
98. Klein CL, Bittenger F, Kohler H, Wagner M, Otto M, Hermanns I *et al.* Comparative studies on vascular endothelium in vitro. Effects of cytokines on the expression of E-selectin. *Pathobiology* 65: 83–92, 1995.
99. Steegmaier M, Levinovitz A, Isenmann S *et al.* The E-selectin ligand ESL-1 is a variant of a receptor for fibroblast growth factor. *Nature* 373: 615–20, 1995.
100. Rise GC, Larsen S, Lofdahl CG, Andersson BA. Circulating cell adhesion molecules in bronchial lavage and serum in COPD patients with chronic bronchitis. *Eur Respir J* 7(9): 1673–77, 1994.
101. Crockett-Torabi A, Fantone JC. The Selectins: Insights into selectin induced intracellular signalling in leukocytes. *Immunol Res* 14: 237–51, 1995.
102. Williams MA, Solomkin JS. Integrin mediated signalling in human neutrophil functioning. *J Leuk Biol* 65: 725–36, 1999.
103. Diamond MS, Staunton DE, de Fougerolles AR *et al.* ICAM-1 (CD-54): a counter receptor for MAC-1 (CD11b/CD18). *J Cell Biol* 111: 3129–39, 1990.
104. Rainger GE, Buckley C, Simmons DL, Nash GB. Cross-talk between cell adhesion molecules regulates the migration velocity of neutrophils. *Curr Biol* 17: 316–25, 1997.
105. Borregaard N, Cowland JB. Granules of the human neutrophilic polymorphonuclear leukocyte. *Blood* 89: 3503–21, 1997.
106. Sengelov H. Secretory vesicles of human neutrophils. *Eur J Haem* 57(Suppl 58): 6–24, 1996.
107. Laudanna C, Campbell JJ, Butcher EC. Role of Rho in chemoattractant activated leukocyte adhesion through integrins. *Science* 271: 981–83, 1996.
108. Michishita M, Videm V, Arnaout MA. A novel divalent cation-binding site in the A domain of the beta 2 integrin CR3 (CD11b/CD18) is essential for ligand binding. *Cell* 72: 857–67, 2003.
109. Noguera A, Batle S, Miralles C, Iglesias J, Busquets X, MacNee W *et al.* Enhanced neutrophil response in chronic obstructive pulmonary disease. *Thorax* 56: 432–37, 2001.
110. Hashimoto M, Shingu M, Ezaki I, Nobunaga M, Minamihara M, Kato K *et al.* Production of soluble ICAM1 from human endothelial cells induced by IL-1β and TNFα. *Inflammation* 18: 163–73, 1994.
111. Davenpeck KL, Steeber DA, Tedder TF, Bochner BS. Rat neutrophils express α4 and β1 integrins and bind to vascular adhesion molecule 1 and mucosal adressin cell adhesion molecule. *Blood* 91: 2341–46, 1988.
112. Clayton A, Evans RA, Pettit E, Hallett M, Williams JD, Steadman R. Cellular activation through the ligation of intercellular adhesion molecule1. *J Cell Sci* 111: 443–53, 1998.
113. Qu Q, Zhoa M, Gu G. ICAM-1 and HLA-DR expression in bronchial epithelial cells from patients with chronic obstructive pulmonary disease. *Zhonghua Jie He He Hu Xi Za Zhi* 21(7): 411–14, 1998.
114. Burns MJ, Walker DC, Brown ES, Thurmon LT, Bowden RA, Keese CR *et al.* Neutrophil transendothelial migration is independent of tight junctions and occurs preferentially at tricellular corners. *J Immunol* 159: 2893–903, 1997.
115. Newman PJ. The biology of PECAM-1. *J Clin Invest* 99: 3, 1997.
116. Muller WA. The role of PECAM1 in leukocyte emigration. Studies in vitro and in vivo. *J Leuk Biol* 66: 698–704, 1995.
117. Yadav R, Larbi KY, Young RE, Nourshargh S. Migration of leukocytes through the vessel wall and beyond. *Thromb Haemost* 90(4): 598–606, 2002.
118. Shang XZ, Issekutz AC. β2 (CD18) and β1 (CD29) Integrin mechanisms in migration of human polymorphonuclear leucocytes and monocytes through lung fibroblast barriers. Shared and distinct mechanisms. *Immunol* 92: 527–35, 1997.
119. Shang XZ, Yednock T, Issekutz AC. α9β Integrin is expressed on human neutrophils and contributes to neutrophil migration through human lung and synovial fibroblast barriers. *J Leuk Biol* 66: 809–16, 1999.
120. Thompson RD, Noble KE, Larbi KY, Dewar A, Duncan GS, Mak TW *et al.* Platelet-endothelial cell adhesion molecule-1 (PECAM-1) deficient mice demonstrate a transient and cytokine-specific role for PECAM1 in leukocyte migration through perivascular basement membrane. *Blood* 97(6): 1854–60, 2001.
121. Wright DC, Gallin JL. Secretory responses of human neutrophils: Exocytosis of specific (secondary) granules by human neutrophils during adherence in vitro and during exudation in vivo. *J Immunol* 123: 258–94, 1979.
122. Foxman EF, Campbell JJ, Butcher EC. Multistep navigation and the combinatorial control of leukocyte chemotaxis. *J Cell Boil* 139: 1349–59, 1997.
123. Foxman EF, Kunkel EJ, Butcher EC. Integrating conflicting chemotactic signals: the role of memory in leukocyte navigation. *J Cell Boil* 147(3): 577–87, 1999.
124. Devreotes P, Janetopoulos C. Eukaryote chemotaxis: Distinctions between directional sensing and polarisation. *J Biol Chem*: 20445–48, 1988.
125. Charest PG, Firtel RA. Big roles for small GTPases in the control of directed cell movement. *Biochem J* 401: 377–90, 2007.
126. Huttunlocher A. Cell polarisation mechanisms during directed cell migration. *Nature Cell Biol* 7(4): 336–37, 2005.
127. Nishiya N, Kiosses WB, Han J, Ginsberg GH. An α_4 integrin-paxillin-Arf-GAP complex restricts Rac activation to the leading edge of migrating cells. *Nature Cell Biol* 7: 343–45, 2005.
128. Ohta Y, Hartwig JH, Stossel TP. A Rho and ROCK regulated GAP for Rac binds filamin A to control actin remodelling. *Nature Cell Biol* 28: 803–14, 2006.
129. Andrew N, Insall RH. Chemotaxis in shallow gradients is mediated independently of PtdIns 3-kinase by biased choices between random protrusions. *Nature Cell Biol* 9(2): 193–200, 2007.
130. Stockley RA, Shaw J, Afford SC, Morrison HM, Burnett D. Effect of alpha one proteinase inhibitor on neutrophil chemotaxis. *Am J Respir Cell Mol Biol* 2: 163–70, 1990.
131. Lomas DA, Stone SR, Llewellyn-Jones C, Keogan MT, Wang ZM, Rubin H *et al.* The control of neutrophil chemotaxis by inhibitors of cathepsin G and chymotrypsin. *J Biol Chem* 270: 437–23443, 1995.
132. Delclaux C, Delacourt C, D'Ortho MP, Boyer V, Lafuma C, Harf A. Role of gelatinase B and elastase in human polymorphonuclear neutrophil migration across basement membrane. *Am J Respir Cell Mol Biol* 14: 288–95, 1996.

133. Allport JR, Ding HT, Ager A, Steeber DA, Tedder TF, Luscinskas FW. L-selectin shedding does not regulate human neutrophil attachment, rolling, or transmigration across human vascular endothelium in vitro. *J Immunol* 158: 4365–72, 1997.
134. Mackeral AJ, Cottell DC, Russell KJ, Fitxgerald MX, O'Conner CM. Migration of neutrophils across human pulmonary endothelial cells is not blocked by matrix metalloproteinase or serine proteinase inhibitors. *Am J Respir Cell Mol Biol* 20: 1209–19, 1999.
135. Steadman R, St John PL, Evans RA, Thomas GL, Davies M, Heck LW *et al.* Human neutrophils do not degrade major basement membrane components during chemotactic migration. *Int J Biochem Cell Biol* 29: 993–1004, 1997.
136. Tkalcevic J, Novelli M, Phylactides M, Iredale JP, Segal AW, Roes J. Impaired immunity and enhanced resistance to endotoxin in the absence of neutrophil elastase and cathepsin G. *Immunity* 12(201): 210, 2000.
137. Betsuyaku T, Shipley JM, Liu Z, Senior RM. Neutrophil emigration in the lungs, peritoneum, and skin does not require gelatinase B. *Am J Respir Cell Mol Biol* 20: 1303–9, 1999.
138. Wright JL, Farmer SG, Churg A. Synthetic serine elastase inhibitor reduces cigarette smoke induced emphysema in guinea pigs. *Am J Respir Crit Care Med* 166: 954–60, 2002.
139. Delacourt C, Herigault S, Delclaux C, Poncin A, Levame M, Harf A *et al.* Protection against acute lung injury by intravenous or intratracheal pretreatment with EPI-HNE-4; a new potent neutrophil elastase inhibitor. *Am J Respir Cell Mol Biol* 26: 290–97, 2002.
140. Chun TH, Sabeh F, Ota I, Murphy H, McDonagh KT, Holmbeck K *et al.* MTI-MMP dependent neovessel formation with the controls of a 3 dimension extra cellular matrix. *J Cell Biol* 167: 757–67, 2004.
141. Sabeh F, Ota I, Holmbeck K, Birkedal-Hansen H *et al.* Tumor cell traffic through the extra cellular matrix is controlled by the membrance anchored collagenases MTI-MMP. *J Cell Biol* 167: 769–81, 2004.
142. Even-Ram S, Yamada K. Cell migration in a 3D matrix. *Curr Opin Cell Biol* 17: 524–32, 2005.
143. Doerschuk CM, Allard MF, Hogg JC. Neutrophil kinetics in rabbits during infusion of zymosan-activated plasma. *J Appl Physiol* 67: 88–95, 1989.
144. Downey GP, Worthen GS, Hyde DM. Neutrophil sequestration and migration in localised pulmonary inflammation. capillary localisation and migration across the interalveolar septum. *Am Rev Respir Dis* 147(1): 168–76, 1993.
145. Hogg JC. Neutrophil kinetics and lung injury. *Physiolog Rev* 67(1249): 1295, 1987.
146. Doerschuck CM, Beyers N, Coxson HO, Wiggs B, Hogg JC. Comparison of neutrophil and capillary diameters and their relation to neutrophil sequestration in the lung. *J Appl Physiol* 74: 3040–45, 1993.
147. Selby C, Drost E, Gillooly M *et al.* Neutrophil sequestration in lungs removed at surgery: The effect of microscopic emphysema. *Am J Respir Crit Care Med* 149: 1526–33, 1994.
148. Erzurum SC, Downey GP, Doherty DE, Schwabb B, Elson EL, Worthen GS. Mechanisms of lipopolysaccharide retention. *J Immunol* 149: 154–62, 1992.
149. Doerschuk CM. The role of CD18 mediated adhesion in neutrophil sequestration induced by infusion of activated plasma in rabbits. *Am J Respir Cell Mol Biol* 7: 140–48, 1992.
150. Morland CM, Morland BJ, Darbyshire PJ, Stockley RA. Migration of CD18 deficient neutrophils in vitro; evidence for a CD18 independent pathway induced by IL8. *Biochem Biophys Acta* 1500(1): 70–76, 2000.
151. Mackerel AJ, Russel KJ, Brady CS, FitzGerald MX, O'Connor CM. Interleukin 8 and leukotriene B4 but not formylmethionyl leucylphenylalanine stimulate CD 18 independent migration of neutrophils across human pulmonary endothelial cells in vitro. *Am J Respir Cell Mol Biol* 23: 154–61, 2000.
152. Doerschuk CM, Winn RK, Coxson HO, Harlan JM. CD18 dependent and independent mechanisms of neutrophil emigration in pulmonary and systemic microcirculation of rabbits. *J Immunol* 144: 2327–33, 1990.
153. Ramamoorthy C, Sasaki SS, Su DL, Sharar SR, Harlan JM, Winn RK. CD18 adhesion blockade decreases bacterial clearance and neutrophil recruitment after intrapulmonary E Coli but not after Staph Aureus. *J Leuk Biol* 61: 167–72, 1997.
154. Qin L, Quinlan WM, Doyle NA, Graham L, Sligh JE, Takei F *et al.* The role of CD11/CD18 and ICAM1 in acute pseudomonas aeruginosa induced pneumonia. *J Immunol* 157: 5016–21, 1996.
155. Hellewell PG, Young SK, Henson PM, Worthen GS. Disparate role of the β2 integrin CD18 in the local accumulation of neutrophils in pulmonary and cutaneous inflammation in rabbit. *Am J Respir Cell Mol Biol* 10: 391–98, 1994.
156. Shoburg DS, Quinlan WM, Hebert CA, Ashkenazi A, Doerschuk CM. Cytokine production in response to stimuli that elicit either CD18 dependent or CD18 independent neutrophil emigration. *Am J Respir Crit Care Med* 147: A1088, 1994.
157. Doyle NA, Bhagwan SD, Meek BB, Kutkoshi GJ, Steeber DA, Tedder TF *et al.* Neutrophil margination, sequestration and emigration in the lungs of L-selectin deficient mice. *J Clin Invest* 99: 526–33, 1997.
158. Burnett D, Chamba A, Hill SL, Stockley RA. Neutrophils from subjects with chronic obstructive disease show enhanced chemotaxis and extracellular proteolysis. *Lancet* 2: 1043–46, 1987.
159. Woolhouse IS, Bayley DL, Lalor P, Adams DH, Stockley RA. Endothelial interactions of neutrophils under flow in chronic obstructive pulmonary disease. *Eur Respir J* 25(4): 612–17, 2005.
160. Gonalez S, Hards J, van Eeden S, Hogg JC. The expression of adhesion molecules in cigarette smoke induced airways obstruction. *Eur Respir J* 9: 1995–2001, 1996.
161. Yoshikawi T, Dent G, Ward J, Angco G, nong G, Nomura N, Hirata K, Djukanovic R. Impaired neutrophil chemotaxis in chronic obstructive pulmonary disease. *Am J Respir Crit Care Med* 175(5): 473–79, 2007.
162. Yadav R, Larbi KY, Young RE, Nourshargh S. Migration of leukocytes through the vessel wall and beyond. *Thromb Haemost* 90(4): 598–606, 2002.
163. Dangerfield J, Larbi KY, Huang MT, Dewar A, Nourshargh S. PECAM-1 (CD£!) homophillc interaction up-regulates α6β1 on transmigrated neutrophils in vivo and plays a functional role in the ability of α6 integrins to mediate leukocyte migration through the perivascular basement membrane. *J Exp Med* 196(9): 1201–11, 2002.
164. Janoff A, Sloan B, Weinbaum G, Damiano V, Sandhaus RA, Elias J *et al.* Experimental emphysema induced with purified human neutrophil elastase; tissue localisation of the instilled protease. *Am Rev Respir Dis* 115(3): 461–78, 1977.
165. Senior RM, Tegner H, Kuhn C, Ohlsson K, Starcher BC, Pierce JA. The induction of pulmonary emphysema with human leukocyte elastase. *Am Rev Respir Dis* 116(3): 469–75, 1977.
166. Lucey EC, Stone PJ, Breuer R, Christensen TG, Calore JD, Catanse A *et al.* Effect of combined human neutrophil cathepsin G and elastase on induction of secretory cell metaplasia and emphysema in hamsters, with in vitro observations on elastolysis by these enzymes. *Am Rev Respir Dis* 132: 362–66, 1985.
167. Gross P, Pfizer EH, Tolker B, Babyok MA, Kaschak M. Experimental emphysema; its production with papain in normal and silicoctic rats. *Arch Environ Health* 11: 50–58, 1964.
168. Marco V, Mass B, Meranze DR, Weinbaum G, Kimbel P. Induction of experiemental emphysema in dogs using leukocyte homogenates. *Am Rev Respir Dis* 104: 595–98, 1971.
169. Blackwood CE, Hosannah Y, Perman E, Keller S, Mandl L. Elasteolytic titre of inducing enzyme as determinant of the response. *Proc Soc Exp Biol Med* 144: 450–54, 1973.
170. Snider GL, Hayes JA, Franzblau C, Kagan HM, Stone PJ, Korthy A. Relationship between elasteolytic activity and experimental emphysema inducing properties of papain preparations. *Am Rev Respir Dis* 110: 254–57, 1974.

171. Lucey EC, Stone PJ, Powers JC, Snider GL. Amelioration of human neutrophil elastase-induced emphysema in hamsters by pretreatment with an oligopeptide chlormethyl ketone. *Eur Respir J* 2: 421–27, 1989.
172. Rudolphus A, Kramps JA, Mauve I, Dijkman JH. Intratracheally instilled antileukoprotease and alpha 1 proteinase inhibitor; effect on human neutrophil elastase-induced experimental emphysema and pulmonary localisation. *Histochem J* 26: 817–27, 1994.
173. Beith JG. Elastases: catalytic and biological properties. In: Mecham RP (ed.), *Biology of extracellular matrix: Regulation of Matrix Accumulation*, pp. 217–320. Orlando FL: Academic, 1986.
174. Kafienah W, Buttle DJ, Burnett D, Hollander AP. Cleavage of native type 1 collgane by human neutrophil elastase. *Biochem J* 330: 897–902, 1998.
175. Niederman MS, Merrill WW, Polomski LM, Reynolds HY, Gee JB. Influence of IgA and elastase on trachea cell bacterial adherence. *Am Rev Respir Dis* 133: 255–60, 1986.
176. Vogt W. Cleavage of the fifth component of complement and generation of a functionally active C5b6-like complex by human leukocyte elastase. *Immunolbiology* 201: 470–77, 2000.
177. Fick RBJ, Robbins RA, Squier SU, Schoderbeck WE, Russ WD. Complement activation in cystic fibrosis respiratory fluids: In vivo and in vitro generation of a C5a and chemotactic activity. *Paediatr Res* 20: 1258–68, 1986.
178. Ashcroft GS, Lei K, Jin W *et al.* Secretory leukocyte protease inhibitor mediates non redundant functions necessary for normal wound healing. *Nat Med* 6: 1147–53, 2000.
179. Owen CA, Campbell MA, Sannes PL, Boukedes SS, Campbell EJ. Cell-surface bound elastase and cathespin G on human neutrophils; a novel, non-oxidative mechanism by which neutrophils focus and preserve catalytic activity of serine proteinases. *J Cell Biol* 131: 775–89, 1995.
180. Owen CA, Campbell MA, Boukedes SS, Campbell EJ. Cytokines regulate membrane-bound leukocyte elastase on neutrophils; a novel mechanism for effector activity. *Am J Physiol Lung Cell Mol Physiol* 272: L385–lL393, 1997.
181. Campbell EJ, Campbell MA, Owen CA. Bioactive proteinase three on the cell surface of human neutrophils: Quantification, catalyic activity and susceptibility to inhibition. *J Immunol* 165: 3366–74, 2000.
182. Cepinskas G, Sandig M, Kvietys PR. PAF induced elastase dependent neutrophil transendothelial migration is associated with the mobilisation of elastase to the neutrophil surface and localised to the migrating front. *J Cell Sci* 112(Pt 12): 1937–45, 1999.
183. Fadok VA, Bratton DL, Guthrie L, Henson PM. Differential effects of apoptotic versus lysed cells on macrophage production of cytokines Role of proteinases. *J Immunol* 166: 6847–54, 2001.
184. Ohlsson K, Linder C, Lundberg E, Axelsson L. Release of cytokines and proteases from human peripheral blood mononuclear and polymorphonuclear cells following phagocytosis and LPS stimulation. *Scand J Clin Lab Invest* 56: 461–70, 1996.
185. Russell REK, Thorley A, Culpitt SV, Dodd S, Donnelly LE, Dermattos C *et al.* Alveolar macrophage-mediated elastolysis; roles of matrix metalloproteinases, cysteine and serine proteases. *Am J Physiol Lung Cell Mol Physiol* 283: L867–73, 2002.
186. Weitz JL, Huang AJ, Landman SL, Nicholson SC, Silverstein SC. Elastase-mediated fibrinogenolysis by chemoattractant stimulated neutrophils occurs in the presence of physiologic concentrations of antiproteinases. *J Exp Med* 166: 1836–50, 1987.
187. Liou TG, Campbell EJ. Quantum proteolysis resulting from release of single granules by human neutrophils; a novel, nonoxidative mechanism of extracellular proteolytic activity. *J Immunol* 157: 2624–31, 1996.
188. Campbell EJ, Campbell MA, Boukedes SS, Owen CA. Quantum proteolysis by neutrophils; implications for pulmonary emphysema in a1antitrypsin deficiency. *J Clin Invest* 104: 337–44, 1999.
189. Rice WG, Weiss SJ. Regulation of proteolysis at the neutrophil–substrate interface by secretory leukoprotease inhibitor. *Science* 249: 178–81, 1990.
190. Takeyama K, Agusti C, Ueki IF, Lausier J, Cardell LO, Nadel JA. Neutrophil-dependent goblet cell degranulation; role of membrane bound elastase and adhesion molecules. *Am J Physiol* 19: L294–302, 1998.
191. Lanone S, Zheng T, Zhu Z, Liu W, Lee CG, Ma B *et al.* Overlapping and enzyme specific contributions of matrix metalloproteinases-9 and -12 in IL-13 induced inflammation and remodelling. *J Clin Invest* 110(463): 474, 2002.
192. Zheng T, Zhu Z, Wang Z, Homer RJ, Ma B, Riese RJJ *et al.* Inducible targeting of IL-13 to the adult lung causes matrix metalloproteinase and cathepsin dependent emphysema. *J Clin Invest* 106: 1445–46, 2000.
193. Shim JJ, Dabbagh K, Ueki IF, Dao-Pick T, Takeyama K, Tam DCW *et al.* IL-13 induces mucin production by stimulating epidermal growth factor receptors and by activating neutrophils. *Am J Physiol Lung Cell Mol Physiol* 280: L134–40, 2001.
194. Smallman LA, Hill SL, Stockley RA. Reduction of ciliary beat frequency in vitro by sputum from patients with bronchiectasis; a serine proteinase effect. *Thorax* 39: 663–67, 1984.
195. Amitani R, Wilson R, Rutman A, Read R, Ward C, Burnett D *et al.* Effect of human neutrophil elastase and pseudomonas aeruginosa proteinases on human respiratory epithelium. *Am J Respir Cell Mol Biol* 4: 26–32, 1991.
196. Aoshiba K, Yasuda K, Yasui S, Tamaoki J, Nagai A. Serine proteinases increase oxidative stress in lung cells. *Am J Physiol Lung Cell Mol Physiol* 281: L556–64, 2001.
197. Nakajoh M, Fukushima T, Suzuki K, Yamaya M, Nakayama K, Sekizawa K *et al.* Retinoic acid inhibits elastase-induced injury in human lung epithelial cell lines. *Am J Respir Cell Mol Biol* 28: 296–304, 2002.
198. Rickard K, Rennard S. Neutrophil elastase causes detachment of bronchial epithelial cells from extracellular matrix. *Am Rev Respir Dis* 139: 406, 1989.
199. Ballieux BE, Hiemstra PS, Klar-Mohamad N, Hagen EC, Van Der Woude FJ, Daha MR. Detachment and cytolysis of human endothelial cells by proteinase 3. *Eur J Immunol* 24: 3211–15, 1994.
200. Tuder RM, Zhen L, Cho CY, Taraseviciene-Stewart L, Kasahara Y, Salvemini D *et al.* Oxidative stress and apoptosis interact and cause emphysema due to vascular endothelial growth factor receptor blockade. *Am J Respir Cell Mol Biol* 29(1): 88–97, 2003.
201. Hubbard RC, Fells G, Gadek J, Pacholok S, Humes J, Crystal RG. Neutrophil accumulation in the lung in a1-antitrypsin deficiency; spontaneous release of leukotriene B4 by alveolar macrophages. *J Clin Invest* 88: 891–97, 1991.
202. Shapiro SD. Animal models for COPD. *Chest* 117: 223S–27S, 2000.
203. Kao RC, Wehner NG, Skubitz KM, Gray BH, Hoidal JR. A distinct human polymorphonuclear leukocyte proteinase that produces emphysema in hamsters. *J Clin Invest* 82: 1963–73, 1988.
204. Stone PJ, Lucey EC, Virca GD, Christensen TG, Breuer R, Snider GL. Alpha-1 protease inhibitor moderates human neutrophil elastase-induced emphysema and secretory cell metaplasia in hamsters. *Eur Respir J* 3: 673–78, 1990.
205. Lucey EC, Stone PJ, Ciccolella DE, Breuer R, Christensen TG, Thompson RC *et al.* Recombinant human secretory leukocyte-protease inhibitor; in vitro properties and amelioration of human neutrophil elastase induced emphysema and secretory cell metaplasia. *J Lab Clin Med* 115: 224–32, 1990.
206. Ofulue A, Ko M, Abboud R. Time course of neutrophil and macrophage elastinolytic activities in cigarette induced emphysema. *Am J Physiol Lung Cell Mol Physiol* 275: L1134–44, 1998.
207. Zay K, Loo S, Xie C, Devine DV, Wright JL, Churg A. Role of neutrophils and a1 antitrypsin in coal and silica-induced connective tissue breakdown. *Am J Physiol Lung Cell Mol Physiol* 276: L269–79, 1999.
208. Sapey E, Stockley RA. COPD exacerbations: Aetiology. *Thorax* 61(3): 250–58, 2006.
209. Dowson LJ, Guest PJ, Stockley RA. Longitudinal changes in physiological, radiological and health status measurements in

alpha(1)-antitrypsin deficiency and factors associated with decline. *Am J Respir Crit Care Med* 164: 1805–9, 2001.
210. Donaldson GC, Seemungal TA, Bhowmik A, Wedzicha JA. Relationship between exacerbation frequency and lung function decline in chronic obstructive pulmonary disease. *Thorax* 57: 847–52, 2002.
211. Churg A, Dai J, Tai H, Xie C, Wright JL. Tumour necrosis factor-a is central to acute cigarette smoke-induced inflammation and connective tissue breakdown. *Am J Respir Crit Care Med* 166: 849–54, 2002.
212. Baker AH, Edwards DR, Murphy G. Metalloproteinase inhibitors: Biological actions and therapeutic opportunities. *J Cell Sci* 115: 3719–27, 2002.
213. Segura-Valdez L, Pardo A, Gaxiola M, Uhal BD, Beceril C, Selman M. Upregulation of gelatinases A and B, Collagenases 1 and 2 and increased parenchymal cell death in COPD. *Chest* 117: 684–94, 2000.
214. Mao JT, Tashkin DP, Belloni PN, Baratelli F, Roth MD. All-trans retinoic acid modulates the balance of matrix metalloproteinase-9 and tissue inhibitor of metalloproteinase-1 in patients with emphysema. *Chest* 124(5): 1724–32, 2003.
215. Beeh KM, Kornmann O, Buhl R. Sputum matrix metalloproteinase-9, tissue inhibitor of metalloproteinase-1 and their molar ratio in patients with COPD, idiopathic pulmonary fibrosis and healthy controls. *Respir Med* 97(6): 634–39, 2003.
216. Vignola AM, Bonanno A, Mirabella A, Riccobono L, Mirabella F, Profita M *et al.* Increased levels of elastase and alpha-1 antitrypsin in sputum of asthmatic patients. *Am J Respir Crit Care Med* 157: 505–11, 1998.
217. Atkinson JJ, Senior RM. Matrix metalloproteinase-9 in lung remodelling. *Am J Respir Cell Mol Biol* 28(1): 12–24, 2003.
218. Li H, Cui D, Tong X, Ma N, Cui X, Lu L *et al.* The role of matrix metalloproteinases in extracellular matrix remodelling in chronic obstructive pulmonary disease rat models. *Zhonghua Nei Ke Za Zhi* 41(6): 393–98, 2002.
219. Selman M, Cisneros-Lira J, Gaxiola M, Ramirez R, Kudlacz EM, Mitchell PG *et al.* Matrix metalloproteinases inhibition attenuates tobacco smoke-induced emphysema in guinea pigs. *Chest* 123: 1633–41, 2003.
220. Minematsu N, Nakamura H, Tateno H, Nakajima T, Yagaguchi K. Genetic polymorphisms in matrix metalloproteinase-9 and pulmonary emphysema. *Biochem Biophys Res Commun* 289(1): 116–19, 2001.
221. Wallace AM, Sandford AJ. Genetic polymorphisms of matrix metalloproteinases: Functional importance in the development of chronic obstructive pulmonary disease? *Am J Pharmacogenom* 2(3): 167–75, 2002.
222. Vernooy JH, Lindeman JH, Jacobs JA, Hanemaaijer R, Wouters EF. Increased activity of matrix metalloproteinase-8 and matrix metalloproteinase-9 in induced sputum from patients with COPD. *Chest* 126(6): 1802–10, 2004.
223. Itoh Y, Nagase H. Preferential inactivation of tissue inhibitor of metalloproteinase-1 that is bound to the precursor of metalloproteinase-9 (progelatinase B) by human neutrophil elastase. *J Biol Chem* 270: 16518–21, 1995.
224. Ferry G, Lonchamp M, Pennel L, de Nanteuil G, Canet E, Tucker GC. Activation of MMP-9 by neutrophil elastase in an in vivo model of acute lung injury. *FEBS* 402: 111–15, 1997.
225. Desrochers PE, Jeffrey JJ, Weiss SJ. Interstitial collagenase (matrix metalloproteinase-1) expresses serpinase activity. *J Clin Invest* 87(6): 2258–65, 1991.
226. Hautamaki RD, Kobayashi DM, Senior RM, Shapiro SD. Requirement for macrophage elastase for acute cigarette smoke-induced emphysema in mice. *Science* 277: 2002–4, 1997.
227. Churg A, Wang RD, Tai H, Wang X, Xie C, Wright JL. Tumour necrosis factor-alpha drives 70% of cigarette smoke induced emphysema in the mouse. *Am J Respir Crit Care Med* 170(5): 492–98, 2004.
228. Churg A, Wang RD, Tai H, Wang X, Xie C, Dai J, Wright JL. Macrophage metalloelastase mediates acute cigarette smoke induced inflammation via tumor necrosis factor-alpha release. *Am J Respir Crit Care Med* 167(8): 1083–89, 2003.
229. Gibson PG, Simpson JL, Saltos N. Heterogeneity of airway inflammation in persistent asthma. *Chest* 119: 1329–36, 2001.
230. Wenzel SE, Szefler SJ, Leung DYM *et al.* Bronchoscopic evaluation of severe asthma. *Am J Respir Crit Care Med* 156: 737–43, 1997.
231. Jatakanon A, Uasuf C, Maziak W *et al.* Neutrophilic inflammation in severe persistent asthma. *Am J Respir Crit Care Med* 160: 1532–37, 1999.
232. Louis R, Lau LCK, Bron AO *et al.* The relationship between airway inflammation and asthma severity. *Am J Respir Crit Care Med* 161: 9–16, 2000.
233. Woodruff PG, Khashayar R, Lazarus SC *et al.* Relationship between airway inflammation; hyperresponsiveness and obstruction in asthma. *J Allergy Clin Immunol* 108: 753–58, 2001.
234. Ordonez CL, Shaughnessy TE, Matthay MA, Fahy JV. Increased neutrophil numbers and IL8 levels in secretions in acute severe asthma: clinical and biological significance. *Am J Respir Crit Care Med* 161: 1185–90, 2000.
235. Martin RJ, Cicutto LC, Smith HR, Ballard RD, Szefler SJ. Airways inflammation in nocturnal asthma. *Am Rev Respir Dis* 143: 351–57, 1991.
236. Carroll N, Carello S, Cooke C, James A. Airway structure and inflammatory cells in fatal attacks of asthma. *Eur Respir J* 19: 709–15, 1996.
237. Sur S, Crotty TB, Kephart GM, Hyma BA, Colby TV, Reed CE, Hunt LW, Gleich GJ. Sudden onset fatal asthma: distinct entity with few eosinophils and relatively more neutrophils in the airway submocosa? *Am Rev Respir Dis* 148: 712–19, 1993.
238. Woodruff PG, Fahy JV. A role for neutrophils in asthma? *Am J Med* 112: 498–500, 2002.
239. Keatings VM, Jatakanon A, Worsdell YM, Barnes PJ. Effects of inhaled and oral glucocorticoids on inflammatory indices in asthma and COPD. *Am J Respir Crit Care Med* 155: 542–48, 1997.
240. Tillie-Leblond I, Gosset P, Tonnel AB. Inflammatory events in severe acute asthma. *Allergy* 60: 23–29, 2005.
241. Martin RJ, Chu HW, Honour JM, Harbeck RJ. Airway inflammation and bronchial hyper-responsiveness after mycoplasma pneumonia infection in a murine model. *Am J Respir Crit Care Med* 24: 577–82, 2001.
242. Castleman WL, Sorkness RL, Lemanske RF, McAllister PK. Viral bronchiolitis during early life induces increased numbers of bronchiolar mast cells and airway hyper-responsiveness. *Am J Pathol* 137: 821–31, 1990.
243. ten Brinke A, van Dissel JT, Sterk PJ *et al.* Persistent airflow limitation in adult-onset non-atopic asthma is associated with serological evidence of *Chylamydia pneumoniae* infection. *J Allergy Clin Immunol* 107: 449–54, 2001.
244. Black PN, Scicchitano R, Jenkins CR *et al.* Serological evidence of infection with *Chylamydia pneumoniae* is related to the severity of asthma. *Eur Respir J* 15: 254–59, 2000.
245. Kraft M, Cassell GH, Henson JE *et al.* Detection of *Mycoplasma pneumoniae* in the airways of adults with chronic asthma. *Am J Respir Crit Care Med* 158: 998–1001, 1998.
246. Bayram H, Devalia JL, Khair OA *et al.* Comparison of ciliary activity and inflammatory mediator release from bronchial epithelial cels of non atopic nonasthmatic subjects and atopic asthmatic patients and the effects of diesel exhaust particles in vitro. *J Allergy Clin Immunol* 102: 771–82, 1998.
247. Linden A. Role of interleukin 17 and the neutrophil in asthma. *Int Arch Allergy Immunol* 126: 179–84, 2001.
248. Akk AM, Simmons PM, Chan HW, Agapov E, Holtzman MJ, Grayson MH, Pham CTN. Dipeptidyl peptidase I- dependent neutrophil recruitment modulates the inflammatory response to Sendai virus infection. *J Immunol* 180: 3535–42, 2008.
249. Jeffrey PK, Wardlow AJ, Nelson FC, Collins JV, Kay AB. Bronchial biopsies in ssthma: An ultrastructural quantitative study and correlation with hyperactivity. *Am Rev Respir Dis* 140: 1745–53, 1989.
250. Kamath AV, Pavord ID, Ruparelia PR, Chilvers ER. Is the neutrophil the key effector cell in severe asthma? *Thorax* 60: 529–30, 2005.

15

CHAPTER

Fibroblasts

Lynne A. Murray[1], Darryl A. Knight[2] and Geoffrey J. Laurent[3]

[1]Manager of Pharmocology Promedior, Inc, Malvern, PA, USA
[2]Department of Pharmacology and Therapeutics, University of British Columbia, Vancouver, BC, Canada
[3]UCLMS, Center for Respiratory Research, University St/Rayne Institute, London, UK

INTRODUCTION

Fibroblasts are typically spindle-shaped cells with an oval flat nucleus found in the interstitial spaces of organs. In the lung, they reside in highly complex multicellular environments, usually closely apposed to the epithelium or endothelium. They are the primary source of extracellular matrix (ECM) proteins, which, in addition to providing a scaffold for cells, play key roles in determining cell phenotype and function. In these contexts, fibroblasts contribute to injury responses in both the initiation and the resolution phases.

In chronic lung diseases including asthma, chronic obstructive pulmonary disease (COPD), and idiopathic pulmonary fibrosis (IPF), there are changes in the number and phenotype of fibroblasts. These changes play a critical role in the loss of normal tissue architecture and function associated with these diseases. Although considerable research effort has focused on modulating leukocyte function and inflammation, relatively few studies have investigated the effects of existing or novel therapies on fibroblast function. Moreover, there have not been sufficient efforts to develop new approaches to regulate fibroblast function in these diseases. This chapter explores the current knowledge about the role of fibroblasts in lung homeostasis and pathological disorders associated with chronic remodeling. A better understanding of the phenotypes of disease-associated fibroblasts may highlight pathways specific to disease pathologies, giving rise to targeted therapeutics. Finally, we review the potential of opportunities arising for better therapeutic intervention strategies targeting fibroblasts that will either halt or potentially reverse fibrosis.

FIBROBLAST FUNCTION

Fibroblasts play a myriad of important roles in normal tissue function. In the lung, they coordinate organogenesis and budding of the lung from the foregut, through intimate bi-directional communication with adjacent epithelial cells. They are also key cells in the production and homeostasis of the ECM. In the lung, the greatest number of fibroblasts are found in the subepithelial layer of the conducting airways and the interstitium of the lung parenchyma. Here they are in a prime location to interact with the epithelial and endothelial cells. These interactions are likely important in disease settings and will be further discussed in this chapter. Fibroblasts are metabolically active cells – capable of synthesizing, secreting, and degrading ECM components – including collagens, proteoglycans, tenascin, laminin, and fibronectin. These cells continually synthesize ECM proteins although the amount they secrete is tightly regulated. For example, up to 90% of all procollagen molecules are degraded intracellularly prior to secretion, depending on tissue and age. Further, fibroblasts generate matrix metalloproteinases (MMPs) and their inhibitors, tissue inhibitor of metalloproteinases (TIMPs), thus controlling tissue architecture and matrix turnover rates.

MMPs exert proteolytic activities on various proteins including many ECM components and are thus central to ECM formation [1]. They have been shown to be elevated in asthma and COPD (reviewed in [1]); as well as in IPF where MMP1, 2, and 9 were colocalized to the epithelium surrounding fibrotic lesions, whereas increased TIMP2 was also observed suggesting that the MMP activity may be inhibited and that the fibrotic region not degraded [2].

Another function of MMPs is to activate growth factors and chemokines, thus potentially promoting the fibrotic and inflammatory milieu [3–5].

MYOFIBROBLASTS

Myofibroblasts express α-smooth muscle actin (α-SMA) and have contractile and secretory properties that are central to controlling tissue architecture [6]. They express a panel of markers that have been correlated with the site of origin. For example, myofibroblasts found in the peripheral and subpleural regions of fibrosis express α-SMA, vimentin, and desmin, whereas cells found in other regions of the lung do not express desmin [7]. *In vitro*, fibroblast-to-myofibroblast *trans*-differentiation can be induced by transforming growth factor-β_1 (TGF-β1), and it has been hypothesized that TGF-β1 found locally at sites of fibrosis *trans*-differentiate resident fibroblasts into myofibroblasts [8, 9].

It has been proposed that the contractile properties of myofibroblasts are central to wound healing by limiting the amount of exposed wound area [10]. However, the sustained presence of contractile myofibroblasts in the interstitium of the lung may cause a retraction of parenchymal tissue mediating alveolar collapse and resulting in the characteristic honeycombing seen in the lungs of IPF patients, or add to the increase in alveolar size, which is characteristic of COPD [11]. There is also a significant increase in myofibroblast numbers in the airways of asthmatic patients following allergen challenge [12], which correlate with subepithelial collagen deposition [13].

In normal wound healing, myofibroblasts sequentially perpetuate and then dampen inflammation via the secretion of chemokines, cytokines, arachidonic acid metabolites, and protease inhibitors [14]. When activated, they express cell surface adhesion molecules allowing specific interactions with immune and inflammatory cells, including lymphocytes, mast cells, and neutrophils. If these processes become dysregulated, fibrosis may ensue with catastrophic consequences for lung function.

Given the importance of myofibroblasts to wound healing and fibrosis, defining both the origin and the mechanisms leading to their clearance will greatly add to our understanding of the role of this cell in fibrotic diseases. At sites of normal wound healing, once sufficient ECM has been deposited and remodeled, fibroblasts and myofibroblasts undergo apoptosis [15, 16]. This serves to limit the excessive deposition of ECM and also dampen the pro-inflammatory and pro-fibrotic milieu. However, for reasons still unclear, myofibroblasts persist in fibrotic conditions.

FIBROBLASTS AND FIBROSIS

Most insight into the potential role of fibroblasts at driving pulmonary remodeling, as well as phenotypic differences in fibroblasts found in fibrotic regions versus those located in normal tissue, has been garnered from *in vitro* studies by using fibroblasts isolated from IPF lungs. Fibroblasts isolated from fibrotic environments are phenotypically different from non-fibrotic fibroblasts [17–19]. Furthermore, fibroblasts from a pro-fibrotic environment exhibit both altered responsiveness to growth factors and enhanced chemokine receptor expression, which has also been observed in murine models of pulmonary remodeling [20]. Taken together, these studies suggest a distinct heterogeneity in fibroblast function and phenotype in the fibrotic lung.

The progression and the severity of many lung diseases, notably IPF, are tightly associated with regions of fibroblast accumulation and proliferation; the extent of these regions, termed fibroblastic foci, has become a reliable indicator of survival. The increased number of (myo)fibroblasts seen in these diseases implies that they are hyperproliferative and/or resistant to apoptosis. However, whether they proliferate faster than normal fibroblasts is still controversial [17, 21, 22]. We and other investigators have reported that fibroblasts derived from IPF lungs proliferate faster than the cells derived from normal lung tissue [17]. In contrast, few others have also shown that the growth rate of IPF fibroblasts was significantly slower than that of normal fibroblasts. However, the specific techniques used to isolate primary fibroblasts differ between laboratories; therefore direct comparisons may not be suitable.

Moreover, discrepancies may be due to the site in the lung from which fibroblasts are harvested, since the magnitude of inflammation and fibrosis are heterogeneous in distribution. Thus, areas of active fibrosis may yield hyperproliferative fibroblasts, compared to the areas of established fibrosis where cells may be hypoproliferative.

To begin to address this diversity, recent studies have used microarray technologies to profile global gene expression in pulmonary fibrosis in man and mouse models. These studies have showed that the expression of almost 500 genes are increased more than twofold in fibrotic lungs, including many genes related to cytoskeletal reorganization, ECM, cellular metabolism and protein biosynthesis, signaling, proliferation, and survival [8]. Studies examining human lung fibroblast global gene expression in response to TGF-β_1 have shown almost 150 genes upregulated, representing several functional categories described earlier. These included 80 genes that were not previously known to be TGF-β_1-responsive [8]. There was excellent concordance between gene expression in human and experimental models, giving us some confidence in the value of our efforts to model human disease.

ORIGIN OF FIBROBLASTS AND MYOFIBROBLASTS: PLASTICITY OF RESIDENT FIBROBLASTS

As remodeling of the lung is associated with the accumulation of fibroblasts and myofibroblasts, understanding the derivation of these cells is critical to our understanding of disease processes. Current thinking suggests that there may be multiple pathways through which fibroblasts and myofibroblasts are derived [14]. These include proliferation and plasticity of resident cells, such as fibroblasts and epithelial cells. Further, emerging data highlight a significant role of circulating, bone marrow-derived cells at remodeling lung architecture [23].

FACTORS INVOLVED IN FIBROBLAST–MYOFIBROBLAST DIFFERENTIATION

As has been described, fibroblasts can be induced to differentiate into myofibroblasts. It is becoming increasingly accepted that the fibroblast–myofibroblast transition begins with the appearance of the protomyofibroblasts that are hyperproliferative and migratory but do not synthesize significant amounts of ECM proteins [24]. Under specific conditions, the protomyofibroblast evolves and may result in a transition into a differentiated myofibroblast, characterized by the presence of organized stress fibers containing α-SMA. Myofibroblasts can, according to the experimental or clinical situation, express other smooth muscle cell contractile proteins, such as myosin heavy chains or desmin. Generally, these cells are thought to be hypoproliferative but responsible for secretion of the bulk of collagens I and III.

Although the processes involved in the appearance of protomyofibroblasts are at present not well explored, the transition from the protomyofibroblast to a differentiated myofibroblast has been related to the production of TGF-β1 by inflammatory cells, and possibly by fibroblasts themselves. The action of TGF-β1 also depends on the local presence of the cellular fibronectin ED-A splice variant. Thus, myofibroblast differentiation is regulated by both a cell product and the ECM. Moreover, it is becoming more accepted that mechanical factors play an important role in both transitions through either TGF-β generation [25] or TGF-β activation [26].

EPITHELIAL CELLS: EPITHELIAL–MESENCHYMAL TRANSITION

Another potential pool of fibroblasts may arise by a process called epithelial–mesenchymal transition (EMT). EMT is a dynamic process by which epithelial cells undergo phenotypic transition to fully differentiated and motile mesenchymal cells, such as fibroblasts and myofibroblasts [27, 28]. This process occurs normally during early fetal development where there is seamless plasticity between epithelial and mesenchymal cells. The differentiation between airway epithelial cells of one type and another, for example type I pneumocytes transitioning into goblet cells, has been previously described [29–32]. However, the switching of an epithelial cell into a phenotype that moves beyond the original cell's embryonic lineage has recently been hypothesized as a driving factor in fibrosis [33–35].

Epithelial cells exposed to TGF-β1, alone or in combination with other growth factors such as epidermal growth factor (EGF), begin the process of EMT by the increased expression of MMPs that enable basement membrane degradation and cell detachment. The cells also undergo cytoskeletal changes as well as altered expression of surface molecules. For example, a downregulation of E-cadherin and zona occludens 1 (ZO-1) with a concomitant upregulation of vimentin as well as the ED-A fibronectin is needed for migration and transition to a mesenchymal phenotype [36, 37]. The majority of the work evaluating EMT has been performed *in vitro*; however, the full extent of this pathogenic pathway *in vivo* is currently being evaluated. In animal models of kidney fibrosis, it has been estimated that up to 20% of the fibroblasts found in the fibrotic lesions were derived from the epithelium through EMT [33]. The idea of EMT promoting the fibrosis observed in asthma and IPF is rapidly beginning to evolve [32]. It is still not known whether EMT contributes to the excess ECM deposition. Future work correlating the timecourse of EMT induction with disease staging will also be very insightful. This pathogenic process may provide novel therapeutic targets such that inhibiting or reversing EMT may provide clinical benefit to patients with fibrosis-associated diseases.

CIRCULATING PROGENITOR MESENCHYMAL CELLS

Along with the epithelium, recent studies have also highlighted a role for bone marrow-derived circulating cells, or fibrocytes, in promoting lung fibrosis by differentiating into fibroblasts or myofibroblasts [23, 38].

Fibrocytes appear to be pleiotropic in function and a variety of extracellular and intracellular markers have been used to characterize them, including CD45, indicating haematopoietic origin; CD34 or CD13; and various chemokine receptors which have been demonstrated to modulate migration *in vitro* and *in vivo* [39–43]. However, the underlying features of fibrocytes are that they are derived from the bone marrow and are positive for type I collagen. Moreover, it has recently been postulated that an overexuberant recruitment of these cells to sites of pulmonary injury contributes to the aberrant deposition of collagen, which ultimately induces pathologic fibrosis [44–48].

Fibrocytes are pleuripotent in their ability to differentiate into other cell lineages, as has been demonstrated with fibrocyte-derived adipocytes [49]. Furthermore, these cells are extremely plastic *in vitro*, making both the derivation and the characterization of these cells unclear. Exposure of fibrocytes to TGF-β1 *in vitro* results in the cells transitioning into a myofibroblast phenotype and producing fibronectin and type III collagen [48]. Using an adoptive transfer model of bone marrow cells from green fluorescent protein (GFP) transgenic mice into recipient mice challenged with intratracheal bleomycin (to initiate lung injury), recruited GFP+ fibrocytes were shown to differentiate into fibroblasts while resident lung fibroblasts differentiated into myofibroblasts [39]. These data would indicate that the fibrocytes may serve to replenish the pool of fibroblasts from which myofibroblasts are derived.

There is increasing evidence that fibrocytes may also have a pathogenic role in asthma [41, 48]. A correlation in the number of fibrocytes in the basement membrane of asthmatics and the extent of subepithelial fibrosis has been recently described [50]. Increased numbers of these cells have been reported in the airways of asthmatics and allergen challenge further increased cell recruitment to the subepithelial region [48]. Interestingly, CD34 colocalized with α-SMA, suggesting that CD34+ fibrocytes traffic to sites of active remodeling and differentiate into myofibroblast-like cells [48]. Further studies are necessary to delineate the fate of fibrocytes and the total impact of this pathway to other diseases such as COPD.

FIBROBLAST ACTIVATION

Fibroblasts are activated by numerous signals such as mechanical forces imposed during bronchoconstriction, ECM interactions, and hypoxia. Furthermore, a large number of mediators, produced by various cell types, and proteases of the coagulation cascade are known to promote fibroblast proliferation, collagen synthesis, migration, and differentiation.

Transforming growth factor β

TGF-β is one of the most potent pro-fibrotic mediators *in vitro* and a strong candidate as a central player in remodeling diseases including asthma and IPF. The emergence of therapeutics directed against the TGF-β pathway will shed light on the role of this growth factor in diseases. Although antagonism of TGF-β by a number of strategies ameliorates experimental fibrosis, inhibition of this growth factor family as a valid therapeutic target requires rigorous interpretation, since it plays an important role as an inhibitor of immune responses and normal cell differentiation.

TGF-β1 is upregulated in the lungs of IPF patients and asthmatics [51–56] (reviewed in [57]). Interestingly, expression of TGF-β1 is nearly absent in the bronchial epithelial cells but is highly expressed in inflammatory cells beneath the basement membrane where subepithelial fibrosis predominates [58]. Transient over-expression of TGF-β1 or pulmonary delivery of this cytokine to mouse lungs induces a pronounced interstitial fibrosis mediated by aberrant ECM generation and deposition, as well as the presence of myofibroblasts [59]. Further, over-expression of TGF-β1 in the lungs of mice also induces a fibrotic response through modulation of apoptotic and inflammatory pathways, as well as aberrant expression of MMP12 [60, 61]. Epithelial cells and eosinophils also produce large amounts of TGF-β2, especially following allergen challenge [62–65]. However, the role in remodeling and fibrosis is not as well studied as that of TGF-β1, although it has been shown to induce epithelial cell mucus production [63] and exert pro-fibrogenic effects [66].

TGF-β1 regulates numerous biological activities, such as proliferation, apoptosis, and differentiation via Smad-dependant and Smad-independent mechanisms [67, 68] (Fig. 15.1). TGF-β is produced in an inactive form tethered to a latency-associated peptide (LAP) by covalent bonding. Activation of, and signaling through, its receptors requires exposure of the active site of the ligand either through conformational change (as induced by integrin-mediated activation) or through cleavage of the LAP.

Using a transgenic mouse model of Smad-3 deficiency, TGF-β/Smad-3 signaling was shown to be required for alveolar integrity and ECM homeostasis, and this pathway is involved in pathogenic mechanisms mediating tissue destruction and fibrogenesis [69]. TGF-β signaling pathway involves the phosphorylation of downstream Smad proteins, comprising the receptor-regulated (R)-Smad (Smad 2, 3), the

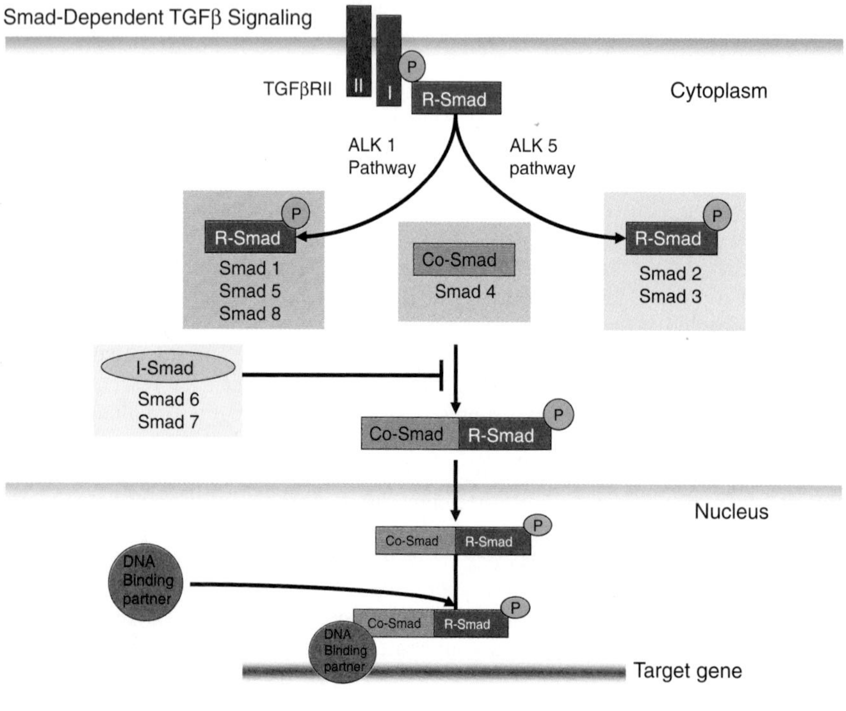

FIG. 15.1 TGF-β signaling pathways. TGF-β is believed to be a major signaling pathway regulating fibroblast functions. All three TGF-β isoforms bind to the TGF-βRII which then phosphorylates the TGFRI chain. This activates the kinase activity of TGF-RI, which then activates Smad proteins. These, in turn, bind Smad 4, translocate to the nucleus, and modulate gene expression. Inhibitory Smads, including Smad 6 and 7, can inhbibit Smad signaling. TGF-β can also signal through a variety of Smad independent pathways. Activation of these pathways may occur through the kinase activity of either TGF-βRI or TGF-βRII.

co-Smad, Smad 4). These activated Smad complexes translocate to the nucleus, bind to specific consensus sequences on target DNA to upregulate the transcription of many genes. Regulation of this pathway occurs through the synthesis of the inhibitory Smad-7. Although Smad signaling is generally considered the predominant pathway, Smad-independent signaling through P38, Akt/PI-3 K, and extracellular signal-regulated kinase (ERK) pathways is also observed. These pathways generally lead to different cellular responses [70, 71]. In this regard, while Smad signaling generally arises through activation of the type I TGF-β receptor, P38 signaling occurs through activation of the type II receptor, implying different mechanisms of activation and regulation [70].

IL-13 and IL-4

Interleukin (IL)-13 and IL-4 are pleiotropic, Th2-associated cytokines,with numerous distinct and overlapping functions. They share overlapping but not redundant roles due to the shared IL13Rα1 receptor subunit. IL-13 is elevated in the lungs of IPF patients and is associated with fibrotic pathologies and aberrant remodeling at various tissue sites [72, 73]. The pro-fibrotic role for IL-13 in asthma and pulmonary fibrosis has been recently reviewed [74, 75]. There are differences in the downstream events following IL-4 or IL-13 signaling. IL-13 activates epithelial cells and goblet cells causing mucous production, goblet cell hyperplasia and EMT [76, 77]. Various animal models of pulmonary fibrosis have indicated a more pro-fibrotic role for IL-13 than IL-4. Indeed it has been hypothesized that IL-4 is involved in the initiation of fibrosis whereas IL-13 is central to the maintenance of the fibrotic response, reviewed in Chapter 8 [78].

Both IL-13 and IL-4 induce pro-fibrotic responses in fibroblasts, responses that are comparable to those induced by TGF-β1 [79, 80]. Both IL-4 and IL-13 induce fibroblast proliferation [81]. Interestingly, IL-4 and IL-13 induced greater fibroblast proliferation from patients with mild asthma compared to responses elicited on fibroblasts derived from severe asthma patients [82]. This suggests an altered fibroblast phenotype which is dependent on disease staging. IL-13 and IL-4 have also been demonstrated to play pro-fibrotic roles on airway epithelial cells by promoting epithelial cell proliferation and also by inducing mitogenic TGF-β2 production from these cells [83]. Interestingly, IL-13 and not IL-4 promotes the differentiation of fibroblasts to myofibroblasts [81, 82]. The other mechanism by which IL-13 can mediate fibrosis is via the second IL-13 receptor, IL-13Rα2 [84]. This receptor was, until very recently, thought to be a decoy receptor as it has a very short cytoplasmic tail and is frequently shed from the surface of cells [85, 86]. However IL-13 binds to IL-13Rα2 with a much higher affinity than IL-13Rα1 [87, 88]. Recent data suggests that IL-13Rα2 has signaling capabilities resulting in TGF-β1 secretion from macrophages [84].

Thrombin and coagulation pathway-associated proteases

Thrombin and several other proteases of the coagulation pathway have been implicated in a number of pulmonary diseases where fibrosis is often a feature. These include acute respiratory distress syndrome (ARDS), and the interstitial lung diseases including IPF [89–93]. In one of these studies, thrombin was shown to represent a large proportion of the fibroblast proliferation capacity of bronchoalveolar lavage (BAL) fluid in patients with lung fibrosis, suggesting that it might be a major player [91]. Animal models have strengthened the connection between these proteases and the occurrence of fibrosis. For example, increased thrombin is found in the lungs of mice challenged with bleomycin and pharmacological inhibition of thrombin significantly reduced the collagen deposition [90, 94].

It is now clear that proteases of the coagulation cascade, including factor VIIa, factor Xa, and thrombin, exert pro-inflammatory and pro-fibrotic effects via the protease-activated receptors (PAR). The PAR comprise four members, PAR1 to PAR4 and between them the coagulation proteases of the extrinsic pathway can target all four receptors. In terms of influencing fibroblast function, PAR1, the high affinity thrombin receptor, is the major receptor by which thrombin and factor Xa exert their potent pro-fibrotic effects. This receptor has emerged as a promising new target to prevent fibrosis in the setting of both IPF and ARDS, based on studies demonstrating that thrombin inhibition partially blocks experimental fibrosis [95] and that mice deficient for PAR1 are protected from lung inflammation, pulmonary edema, and lung collagen accumulation following bleomycin injury [96]. Recently, the role of PAR has been linked to epithelial cell activation of TGF-β. It has been known for some time that mice deficient for the β_6 subunit are protected from pulmonary fibrosis due to a defective ability to activate TGF-β [97] but recent studies suggest that activation of PAR1 leads to the activation of latent TGF-β via this $\alpha_v\beta_6$-dependent mechanism [94]. Together these data suggest thrombin may be upstream to the activation of TGF-β.

The PAR have also been implicated in airways diseases (reviewed in [98–100]). The mechanism for this effect is still being explored but it may be due to pro-inflammatory effects of PAR activation. For example, PAR-2-deficient mice had decreased eotaxin/CCL11 influx along with reduced eosinophilia following antigen challenge in an allergen sensitization and challenge model of asthma [101, 102]. In relation to COPD, Churg and coworkers recently suggested that α_1-antitrypsin inhibition of TNF-α and MMP12 production in animals exposed to cigarette smoke may in part be due to inhibition of serine proteases activating PAR1 on macrophages [103].

CLINICAL TARGETING OF FIBROSIS IN LUNG PATHOLOGIES

Currently there are no therapeutic molecules in use in clinical respiratory medicine that specifically target fibroblasts or fibroblast function. Further, myofibroblasts are likely to be a key to the chronicity of IPF and other related diseases and therefore strategies that modulate the functions of these cells may have a significant impact on the treatment of fibrotic remodeling.

Data emerging suggests that targeting-specific mediators during asthma, such as IL-13, may impact disease through modulating fibroblast function. Moreover, increased pathogenic functions ascribed to fibroblasts indicate that this cell type may be attractive to target therapeutically. Aberrant activation of fibroblasts during fibrosis results in excessive collagen deposition, which ultimately results in a loss of tissue function. As more is learnt about fibroblasts and cells capable of generating excess ECM, specific cell markers may be identified that could allow for a selective targeting with minimal detriment to normal tissue. Therefore, targeting fibroblasts directly may result in a decrease in ECM synthesis and a potential beneficial clinical outcome, but unlike immunosuppressants and non-specific anti-inflammatory agents, novel therapeutics would leave cellular defense mechanisms necessary for host defense in the lung intact.

References

1. Demedts IK *et al.* Matrix metalloproteinases in asthma and COPD. *Curr Opin Pharmacol* 5(3): 257–63, 2005.
2. Fukuda Y *et al.* Localization of matrix metalloproteinases-1, -2, and -9 and tissue inhibitor of metalloproteinase-2 in interstitial lung diseases. *Lab Invest* 78(6): 687–98, 1998.
3. Li Q *et al.* Matrilysin shedding of syndecan-1 regulates chemokine mobilization and transepithelial efflux of neutrophils in acute lung injury. *Cell* 111(5): 635–46, 2002.
4. Ludbrook SB *et al.* The integrin alphavbeta3 is a receptor for the latency-associated peptides of transforming growth factors beta1 and beta3. *Biochem J* 369(Pt 2): 311–18, 2003.
5. Sheppard D. Roles of alphav integrins in vascular biology and pulmonary pathology. *Curr Opin Cell Biol* 16(5): 552–57, 2004.
6. Ohta K *et al.* Immunohistochemical identification and characterization of smooth muscle-like cells in idiopathic pulmonary fibrosis. *Am J Respir Crit Care Med* 152(5 Pt 1): 1659–65, 1995.
7. Zhang K *et al.* Myofibroblasts and their role in lung collagen gene expression during pulmonary fibrosis. A combined immunohistochemical and in situ hybridization study. *Am J Pathol* 145(1): 114–25, 1994.
8. Chambers RC *et al.* Global expression profiling of fibroblast responses to transforming growth factor-beta1 reveals the induction of inhibitor of differentiation-1 and provides evidence of smooth muscle cell phenotypic switching. *Am J Pathol* 162(2): 533–46, 2003.
9. Desmouliere A *et al.* Transforming growth factor-beta 1 induces alpha-smooth muscle actin expression in granulation tissue myofibroblasts and in quiescent and growing cultured fibroblasts. *J Cell Biol* 122(1): 103–11, 1993.
10. Gabbiani G, Badonnel MC. Contractile events during inflammation. *Agents Actions* 6(1–3): 277–80, 1976.
11. Dacic S, Yousem SA. Histologic classification of idiopathic chronic interstitial pneumonias. *Am J Respir Cell Mol Biol* 29(3 Suppl): S5–9, 2003.
12. Gizycki MJ *et al.* Myofibroblast involvement in the allergen-induced late response in mild atopic asthma. *Am J Respir Cell Mol Biol* 16(6): 664–73, 1997.
13. Brewster CE *et al.* Myofibroblasts and subepithelial fibrosis in bronchial asthma. *Am J Respir Cell Mol Biol* 3(5): 507–11, 1990.
14. Hinz B *et al.* The myofibroblast: One function, multiple origins. *Am J Pathol* 170(6): 1807–16, 2007.
15. Desmouliere A *et al.* Apoptosis mediates the decrease in cellularity during the transition between granulation tissue and scar. *Am J Pathol* 146(1): 56–66, 1995.
16. Darby I, Skalli O, Gabbiani G. Alpha-smooth muscle actin is transiently expressed by myofibroblasts during experimental wound healing. *Lab Invest* 63(1): 21–29, 1990.
17. Moodley YP *et al.* Fibroblasts isolated from normal lungs and those with idiopathic pulmonary fibrosis differ in interleukin-6/gp130-mediated cell signaling and proliferation. *Am J Pathol* 163(1): 345–54, 2003.
18. Scaffidi AK *et al.* Oncostatin M stimulates proliferation, induces collagen production and inhibits apoptosis of human lung fibroblasts. *Br J Pharmacol* 136(5): 793–801, 2002.
19. Keerthisingam CB *et al.* Cyclooxygenase-2 deficiency results in a loss of the anti-proliferative response to transforming growth factor-beta in human fibrotic lung fibroblasts and promotes bleomycin-induced pulmonary fibrosis in mice. *Am J Pathol* 158(4): 1411–22, 2001.
20. Hogaboam CM *et al.* Differential monocyte chemoattractant protein-1 and chemokine receptor 2 expression by murine lung fibroblasts derived from Th1- and Th2-type pulmonary granuloma models. *J Immunol* 163(4): 2193–201, 1999.
21. Jacoby DB, Gleich GJ, Fryer AD. Human eosinophil major basic protein is an endogenous allosteric antagonist at the inhibitory muscarinic M2 receptor. *J Clin Invest* 91(4): 1314–18, 1993.
22. Ramos C *et al.* Fibroblasts from idiopathic pulmonary fibrosis and normal lungs differ in growth rate, apoptosis, and tissue inhibitor of metalloproteinases expression. *Am J Respir Cell Mol Biol* 24(5): 591–98, 2001.
23. Quan TE, Cowper SE, Bucala R. The role of circulating fibrocytes in fibrosis. *Curr Rheumatol Rep* 8(2): 145–50, 2006.
24. Desmouliere A, Darby IA, Gabbiani G. Normal and pathologic soft tissue remodeling: Role of the myofibroblast, with special emphasis on liver and kidney fibrosis. *Lab Invest* 83(12): 1689–707, 2003.
25. Lindahl GE *et al.* Activation of fibroblast procollagen alpha 1(I) transcription by mechanical strain is transforming growth factor-beta-dependent and involves increased binding of CCAAT-binding factor (CBF/NF-Y) at the proximal promoter. *J Biol Chem* 277(8): 6153–61, 2002.
26. Sheppard D. Transforming growth factor beta: A central modulator of pulmonary and airway inflammation and fibrosis. *Proc Am Thorac Soc* 3(5): 413–17, 2006.
27. Zavadil J, Bottinger EP. TGF-beta and epithelial-to-mesenchymal transitions. *Oncogene* 24(37): 5764–74, 2005.
28. Zavadil J *et al.* Integration of TGF-beta/Smad and Jagged1/Notch signalling in epithelial-to-mesenchymal transition. *EMBO J* 23(5): 1155–65, 2004.
29. Borok Z *et al.* Modulation of t1alpha expression with alveolar epithelial cell phenotype in vitro. *Am J Physiol* 275(1 Pt 1): L155–L164, 1998.
30. Danto SI *et al.* Reversible transdifferentiation of alveolar epithelial cells. *Am J Respir Cell Mol Biol* 12(5): 497–502, 1995.
31. Torday JS, Torres E, Rehan VK. The role of fibroblast transdifferentiation in lung epithelial cell proliferation, differentiation, and repair in vitro. *Pediatr Pathol Mol Med* 22(3): 189–207, 2003.
32. Kim KK *et al.* Alveolar epithelial cell mesenchymal transition develops in vivo during pulmonary fibrosis and is regulated by the extracellular matrix. *Proc Natl Acad Sci USA* 103(35): 13180–85, 2006.
33. Kalluri R, Neilson EG. Epithelial-mesenchymal transition and its implications for fibrosis. *J Clin Invest* 112(12): 1776–84, 2003.
34. Selgas R, Jimenez-Heffernan J, Lopez-Cabrera M. On the epithelial–mesenchymal transition of mesothelial cells. *Kidney Int* 66(2): 866–67, 2004.
35. Valcourt U *et al.* TGF-beta and the Smad signaling pathway support transcriptomic reprogramming during epithelial–mesenchymal cell transition. *Mol Biol Cell* 16(4): 1987–2002, 2005.
36. Grunert S, Jechlinger M, Beug H. Diverse cellular and molecular mechanisms contribute to epithelial plasticity and metastasis. *Nat Rev Mol Cell Biol* 4(8): 657–65, 2003.
37. Iwano M *et al.* Evidence that fibroblasts derive from epithelium during tissue fibrosis. *J Clin Invest* 110(3): 341–50, 2002.
38. Quan TE *et al.* Circulating fibrocytes: Collagen-secreting cells of the peripheral blood. *Int J Biochem Cell Biol* 36(4): 598–606, 2004.

39. Hashimoto N *et al.* Bone marrow-derived progenitor cells in pulmonary fibrosis. *J Clin Invest* 113(2): 243–52, 2004.
40. Moore BB *et al.* The role of CCL12 in the recruitment of fibrocytes and lung fibrosis. *Am J Respir Cell Mol Biol* 35(2): 175–81, 2006.
41. Phillips RJ *et al.* Circulating fibrocytes traffic to the lungs in response to CXCL12 and mediate fibrosis. *J Clin Invest* 114(3): 438–46, 2004.
42. Moore BB *et al.* The role of CCL12 in the recruitment of fibrocytes and lung fibrosis. *Am J Respir Cell Mol Biol*, 2006.
43. Moore BB *et al.* CCR2-mediated recruitment of fibrocytes to the alveolar space after fibrotic injury. *Am J Pathol* 166(3): 675–84, 2005.
44. Abe R *et al.* Peripheral blood fibrocytes: Differentiation pathway and migration to wound sites. *J Immunol* 166(12): 7556–62, 2001.
45. Chesney J, Bucala R. Peripheral blood fibrocytes: Mesenchymal precursor cells and the pathogenesis of fibrosis. *Curr Rheumatol Rep* 2(6): 501–5, 2000.
46. Chesney J *et al.* Regulated production of type I collagen and inflammatory cytokines by peripheral blood fibrocytes. *J Immunol* 160(1): 419–25, 1998.
47. Bucala R *et al.* Circulating fibrocytes define a new leukocyte subpopulation that mediates tissue repair. *Mol Med* 1(1): 71–81, 1994.
48. Schmidt M *et al.* Identification of circulating fibrocytes as precursors of bronchial myofibroblasts in asthma. *J Immunol* 171(1): 380–89, 2003.
49. Hong KM *et al.* Characterization of human fibrocytes as circulating adipocyte progenitors and the formation of human adipose tissue in SCID mice. *FASEB J* 19(14): 2029–31, 2005.
50. Nihlberg K *et al.* Tissue fibrocytes in patients with mild asthma: A possible link to thickness of reticular basement membrane? *Respir Res* 7: 50, 2006.
51. Coker RK *et al.* Diverse cellular TGF-beta 1 and TGF-beta 3 gene expression in normal human and murine lung. *Eur Respir J* 9(12): 2501–7, 1996.
52. Khalil N *et al.* TGF-beta 1, but not TGF-beta 2 or TGF-beta 3, is differentially present in epithelial cells of advanced pulmonary fibrosis: An immunohistochemical study. *Am J Respir Cell Mol Biol* 14(2): 131–38, 1996.
53. Khalil N *et al.* Increased production and immunohistochemical localization of transforming growth factor-beta in idiopathic pulmonary fibrosis. *Am J Respir Cell Mol Biol* 5(2): 155–62, 1991.
54. Yoshida K, Gage FH. Cooperative regulation of nerve growth factor synthesis and secretion in fibroblasts and astrocytes by fibroblast growth factor and other cytokines. *Brain Res* 569(1): 14–25, 1992.
55. Okumura KK *et al.* Cortisol and TGF-beta inhibit secretion of platelet-activating factor-acetylhydrolase in a monocyte-macrophage model system [corrected]. *Mol Hum Reprod* 3(11): 927–32, 1997.
56. Vignola AM *et al.* Transforming growth factor-beta expression in mucosal biopsies in asthma and chronic bronchitis. *Am J Respir Crit Care Med* 156(2 Pt 1): 591–99, 1997.
57. Howell JE, McAnulty RJ. TGF-beta: Its role in asthma and therapeutic potential. *Curr Drug Targets* 7(5): 547–65, 2006.
58. Magnan A *et al.* Altered compartmentalization of transforming growth factor-beta in asthmatic airways. *Clin Exp Allergy* 27(4): 389–95, 1997.
59. Sime PJ *et al.* Adenovector-mediated gene transfer of active transforming growth factor-beta1 induces prolonged severe fibrosis in rat lung. *J Clin Invest* 100(4): 768–76, 1997.
60. Kang HR *et al.* TGF-beta 1 stimulates pulmonary fibrosis and inflammation via a bax-dependent, bid-activated pathway that involves matrix metalloproteinase-12. *J Biol Chem*, 2007.
61. Lee CG *et al.* Early growth response gene 1-mediated apoptosis is essential for transforming growth factor {beta}1-induced pulmonary fibrosis. *J Exp Med* 200(3): 377–89, 2004.
62. Torrego A *et al.* Expression and activation of TGF-beta isoforms in acute allergen-induced remodelling in asthma. *Thorax* 62(4): 307–13, 2007.
63. Chu HW *et al.* Transforming growth factor-beta2 induces bronchial epithelial mucin expression in asthma. *Am J Pathol* 165(4): 1097–106, 2004.
64. Tschumperlin DJ *et al.* Mechanical stress triggers selective release of fibrotic mediators from bronchial epithelium. *Am J Respir Cell Mol Biol* 28(2): 142–49, 2003.
65. Thompson HG *et al.* Epithelial-derived TGF-beta2 modulates basal and wound-healing subepithelial matrix homeostasis. *Am J Physiol Lung Cell Mol Physiol* 291(6): L1277–85, 2006.
66. Puddicombe SM *et al.* Involvement of the epidermal growth factor receptor in epithelial repair in asthma. *FASEB J* 14(10): 1362–74, 2000.
67. Pinkas J, Teicher BA. TGF-beta in cancer and as a therapeutic target. *Biochem Pharmacol* 72(5): 523–29, 2006.
68. Sanders YY, Kumbla P, Hagood JS. Enhanced myofibroblastic differentiation and survival in Thy-1(-) lung fibroblasts. *Am J Respir Cell Mol Biol*, 2006.
69. Bonniaud P *et al.* Smad3 null mice develop airspace enlargement and are resistant to TGF-beta-mediated pulmonary fibrosis. *J Immunol* 173(3): 2099–108, 2004.
70. Shi Y, Massague J. Mechanisms of TGF-beta signaling from cell membrane to the nucleus. *Cell* 113(6): 685–700, 2003.
71. Scaffidi AK *et al.* Regulation of human lung fibroblast phenotype and function by vitronectin and vitronectin integrins. *J Cell Sci* 114(Pt 19): 3507–16, 2001.
72. Wynn TA. Fibrotic disease and the T(H)1/T(H)2 paradigm. *Nat Rev Immunol* 4(8): 583–94, 2004.
73. Hancock A *et al.* Production of interleukin 13 by alveolar macrophages from normal and fibrotic lung. *Am J Respir Cell Mol Biol* 18(1): 60–65, 1998.
74. Wills-Karp M. Interleukin-13 in asthma pathogenesis. *Immunol Rev* 202: 175–90, 2004.
75. Finkelman FD *et al.* Interleukin-4- and interleukin-13-mediated host protection against intestinal nematode parasites. *Immunol Rev* 201: 139–55, 2004.
76. Yasuo M *et al.* Relationship between calcium-activated chloride channel 1 and MUC5AC in goblet cell hyperplasia induced by interleukin-13 in human bronchial epithelial cells. *Respiration* 73(3): 347–59, 2006.
77. Atherton HC, Jones G, Danahay H. IL-13-induced changes in the goblet cell density of human bronchial epithelial cell cultures: MAP kinase and phosphatidylinositol 3-kinase regulation. *Am J Physiol Lung Cell Mol Physiol* 285(3): L730–39, 2003.
78. Murray LA *et al.* Fibroblasts and the Extracellular Matrix, in Allergy and Allergic Diseases. In: Kay AB, Kaplan AP, Bousquet J, Holt PG, (eds). 2nd edition. Blackwell Publishing Limited, 412–435, 2008.
79. Oriente A *et al.* Interleukin-13 modulates collagen homeostasis in human skin and keloid fibroblasts. *J Pharmacol Exp Ther* 292(3): 988–94, 2000.
80. Hashimoto S *et al.* IL-4 and IL-13 induce myofibroblastic phenotype of human lung fibroblasts through c-Jun NH2-terminal kinase-dependent pathway. *J Allergy Clin Immunol* 107(6): 1001–8, 2001.
81. Saito A *et al.* Potential action of IL-4 and IL-13 as fibrogenic factors on lung fibroblasts in vitro. *Int Arch Allergy Immunol* 132(2): 168–76, 2003.
82. Kraft M *et al.* IL-4, IL-13, and dexamethasone augment fibroblast proliferation in asthma. *J Allergy Clin Immunol* 107(4): 602–6, 2001.
83. Richter A *et al.* The contribution of interleukin (IL)-4 and IL-13 to the epithelial-mesenchymal trophic unit in asthma. *Am J Respir Cell Mol Biol* 25(3): 385–91, 2001.
84. Fichtner-Feigl S *et al.* IL-13 signaling through the IL-13alpha2 receptor is involved in induction of TGF-beta1 production and fibrosis. *Nat Med* 12(1): 99–106, 2006.
85. Gauchat JF *et al.* A novel 4-kb interleukin-13 receptor alpha mRNA expressed in human B, T, and endothelial cells encoding an alternate type-II interleukin-4/interleukin-13 receptor. *Eur J Immunol* 27(4): 971–78, 1997.
86. Orchansky PL *et al.* An interleukin (IL)-13 receptor lacking the cytoplasmic domain fails to transduce IL-13-induced signals and inhibits responses to IL-4. *J Biol Chem* 272(36): 22940–47, 1997.
87. Rahaman SO *et al.* IL-13R(alpha)2, a decoy receptor for IL-13 acts as an inhibitor of IL-4-dependent signal transduction in glioblastoma cells. *Cancer Res* 62(4): 1103–9, 2002.
88. Wu AH, Low WC. Molecular cloning and identification of the human interleukin 13 alpha 2 receptor (IL-13Ra2) promoter. *Neuro-oncol* 5(3): 179–87, 2003.

89. Ludwicka-Bradley A, Bogatkevich G, Silver RM. Thrombin-mediated cellular events in pulmonary fibrosis associated with systemic sclerosis (scleroderma). *Clin Exp Rheumatol* 22(3 Suppl 33): S38–46, 2004.
90. Howell DC, Laurent GJ, Chambers RC. Role of thrombin and its major cellular receptor, protease-activated receptor-1, in pulmonary fibrosis. *Biochem Soc Trans* 30(2): 211–16, 2002.
91. Hernandez-Rodriguez NA *et al.* Role of thrombin in pulmonary fibrosis. *Lancet* 346(8982): 1071–73, 1995.
92. Ohba T *et al.* Scleroderma bronchoalveolar lavage fluid contains thrombin, a mediator of human lung fibroblast proliferation via induction of platelet-derived growth factor alpha-receptor. *Am J Respir Cell Mol Biol* 10(4): 405–12, 1994.
93. Kimura M *et al.* The significance of cathepsins, thrombin and aminopeptidase in diffuse interstitial lung diseases. *J Med Invest* 52(1–2): 93–100, 2005.
94. Jenkins RG *et al.* Ligation of protease-activated receptor 1 enhances alpha(v)beta6 integrin-dependent TGF-beta activation and promotes acute lung injury. *J Clin Invest* 116(6): 1606–14, 2006.
95. Howell DC *et al.* Direct thrombin inhibition reduces lung collagen, accumulation, and connective tissue growth factor mRNA levels in bleomycin-induced pulmonary fibrosis. *Am J Pathol* 159(4): 1383–95, 2001.
96. Howell DC *et al.* Absence of proteinase-activated receptor-1 signaling affords protection from bleomycin-induced lung inflammation and fibrosis. *Am J Pathol* 166(5): 1353–65, 2005.
97. Munger JS *et al.* The integrin alpha v beta 6 binds and activates latent TGF beta 1: A mechanism for regulating pulmonary inflammation and fibrosis. *Cell* 96(3): 319–28, 1999.
98. Lan RS, Stewart GA, Henry PJ. Role of protease-activated receptors in airway function: A target for therapeutic intervention? *Pharmacol Ther* 95(3): 239–57, 2002.
99. Moffatt JD, Page CP, Laurent GJ. Shooting for PARs in lung diseases. *Curr Opin Pharmacol* 4(3): 221–29, 2004.
100. Laurent GJ. No bit PARt for PAR-1. *Am J Respir Cell Mol Biol* 33(3): 213–15, 2005.
101. Takizawa T *et al.* Abrogation of bronchial eosinophilic inflammation and attenuated eotaxin content in protease-activated receptor 2-deficient mice. *J Pharmacol Sci* 98(1): 99–102, 2005.
102. Schmidlin F *et al.* Protease-activated receptor 2 mediates eosinophil infiltration and hyperreactivity in allergic inflammation of the airway. *J Immunol* 169(9): 5315–21, 2002.
103. Churg A *et al.* Alpha1-antitrypsin suppresses TNF-alpha and MMP-12 production by cigarette smoke-stimulated macrophages. *Am J Respir Cell Mol Biol* 37(2): 144–51, 2007.

16

CHAPTER

Epithelial Cells

Scott H. Randell[1], Kimberlie Burns[2] and Richard C. Boucher[2]

[1]Departments of Cell and Molecular Physiology and Medicine, The University of North Carolina at Chapel Hill, NC, USA

[2]Department of Medicine, The University of North Carolina at Chapel Hill, NC, USA

INTRODUCTION

Respiratory tract epithelial cells constitute a front-line physical barrier between the organism and the environment. They serve vital functions in host physiology and mucosal defense and are central to the pathogenesis of asthma and chronic obstructive pulmonary disease (COPD). This chapter concisely addresses epithelial cell development, structure, cell lineages, integrated function in mucus clearance, and presents an overview of their role in immune responses relevant to asthma and COPD.

LUNG DEVELOPMENT AND EPITHELIAL CELL DIFFERENTIATION

The tracheo-bronchial and lung epithelium are endoderm-derived, originating as a foregut evagination at 3.5 to 7 weeks of gestation in humans [1]. Complex, reciprocal molecular interactions between epithelium and the underlying mesenchyme govern development, resulting in elaborate branching morphogenesis that forms the airways and lungs [2]. There is a centrifugal pattern of cyto-differentiation and structural development that culminates in the normally formed bronchial tree and lungs at end gestation, including maturation of the pulmonary surfactant system. Detailed discussion of molecular regulation of lung epithelial development and abnormalities surrounding premature birth is beyond the scope of this chapter.

Survival of premature birth and its complications has increased dramatically in recent years, creating a growing population of individuals born at early gestational age that are exposed to variable treatment regimes, including differing preparations of exogenous surfactant [3], conventional versus high-frequency oscillatory ventilation [4], nitric oxide [5], and inhaled versus systemic glucocorticoids [6]. The spectrum of lung disease resulting from premature birth has also evolved and now presents a histopathologic picture characterized by diffuse developmental abnormalities in lung structure [7]. Wheezing is more common after premature birth, especially in susceptible populations and in the presence of chorioamnionitis [8]. Lung function may normalize somewhat during later childhood and despite a higher prevalence of respiratory symptoms, airway hyperactivity was not present in one cohort of adults born prematurely [9, 10]. Time will tell if those born prematurely are predisposed to COPD, especially if they smoke.

Prenatal airway and lung development sets the stage for the transition to air breathing at birth, including the essential function of lung liquid absorption, which depends on presence of epithelial ion channels [11]. Available evidence also indicates a period of dynamic postnatal adaptation and maturation of the epithelium. For example, in both animals and humans, mucous secretory cells are induced in late gestation and are highly abundant in the neonatal period but normally regress by adulthood [12–17]. Although a few studies exist [16, 17], there is a relative paucity of quantitative data on airway epithelial cell kinetics and differentiation during the late prenatal and early postnatal stages of normal human airway development. Environmental exposures and infections during infancy and early childhood mold airway and lung structure and function and likely have a lasting influence on the ultimate development of disease. There is intense interest and controversy regarding the hygiene hypothesis, which proposes that inadequate or inappropriate early postnatal stimulation skews the immune system, promoting the development of allergy, asthma, and autoimmunity [18, 19]. Environmental exposures and infections occur in the context of a dynamic airway epithelial proliferation and cytodifferentiation program during which responsiveness to various stimuli (pollution, second-hand tobacco smoke, infection) is likely altered. However, the basic cellular composition of the developing and maturing postnatal human airway epithelium is still inadequately described. Mechanisms regulating changes in epithelial proliferation and maturation, differential sensitivity to relevant stimuli and the impact of any differences on interactions with the developing immune system during this potentially critical period are still not well understood.

STRUCTURE OF THE EPITHELIUM AND "TROPHIC UNITS"

In the normal adult human, epithelial cell populations vary systematically as a function of airway level. Portions of the nasal passages and the trachea to approximately the 6–10th generation bronchi are lined by a tall columnar pseudostratified mucociliary epithelium consisting principally of basal, intermediate, ciliated, and mucous secretory (goblet) cells (Fig. 16.1). In this region, submucosal glands are abundant. Proceeding distally, cartilage support and glands diminish and become absent in the bronchioles, which are initially also lined by a columnar pseudostratified epithelium. In terminal and respiratory bronchioles, the epithelium transitions to a simple columnar and eventually cuboidal epithelium, where Clara cells replace mucous secretory cells [20]. Individual pulmonary neuroendocrine cells (PNECs) and highly innervated collections of PNECs (neuroendocrine bodies) are distributed within the pseudostratified, columnar, and cuboidal epithelial zones and extend into the proximal alveolar region [21]. PNECs have been implicated in lung development, in supporting epithelial stem cell niches, and possibly as peripheral chemosensors that alter postnatal growth and adaptation [22]. Finally, the alveolar region epithelium is composed mainly of thin type I and cuboidal type II alveolar epithelial cells with critical roles to create the thin diffusion barrier and produce surfactant, respectively. Just as the relative abundance of surface cell types changes along the respiratory tract axis, each segment is characteristically vascularized, innervated, and variably populated by mesenchymal and bone marrow-derived cells.

The epithelium, its underlying matrix, submucosal cells, and tissue structures constitute the "epithelial-mesenchymal trophic unit" [23]. Reciprocal cellular communication within the trophic unit underlies lung development and likely regulates steady-state cell behavior and repair after injury, for example by the activation of TGF-β [24]. Studies in a primate model suggest that neonatal environmental pollutant and allergen exposure alter the epithelial-mesenchymal trophic unit, resulting in a permanent compromise of airway growth and development [25]. Pathologic epithelial-mesenchymal interactions occurring via IL-1β and TGF-β have been recently implicated in the development of airway

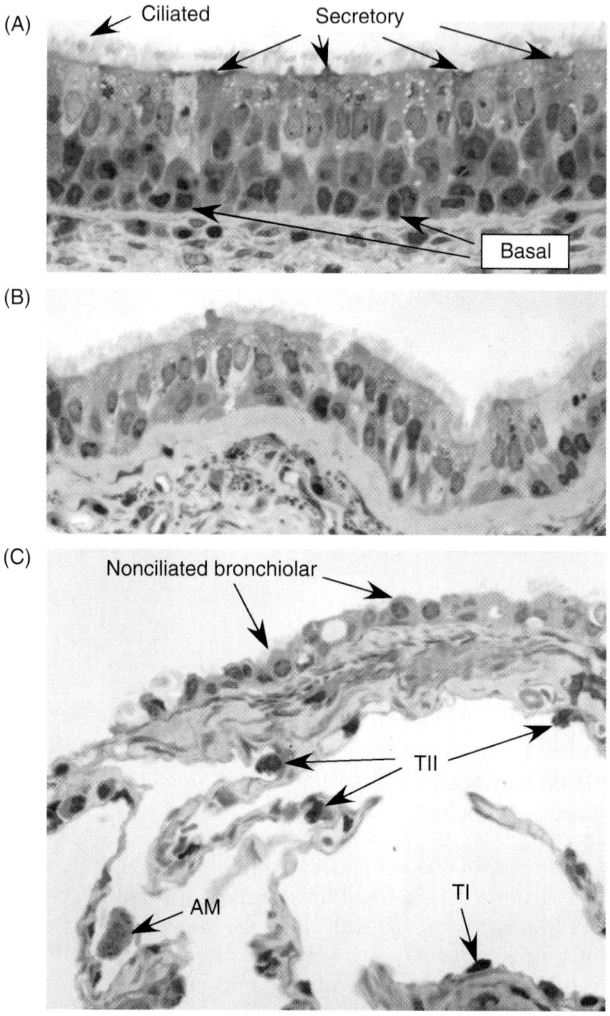

FIG. 16.1 Lung epithelial cell morphology. (A) Normal human bronchus, (B) bronchiole, and (C) terminal bronchiole and alveoli. One micrometer-thick plastic sections, Richardsons stain, original magnification = 400X, AM = alveolar macrophage, TI and TII = Type I and II alveolar epithelial cells, respectively.

epithelilal squamous metaplasia and sub-epithelial fibrosis that is characteristic of COPD [26]. Sub-epithelial remodeling, including basal lamina thickening, airway wall fibrosis, increased vascularity, and smooth muscle hypertrophy/hyperplasia, similarly contribute to the pathophysiology of asthma [27, 28]. However, the detailed cellular composition and molecular mechanisms of communication within epithelial-mesenchymal trophic units at each level in the human airway in health and disease are not well understood. Airway epithelial cell renewal during the steady state and after injury occurs within the context of the trophic unit.

LUNG EPITHELIAL STEM AND PROGENITOR CELLS

In thoroughly studied epithelia such as the epidermis and intestine, temporal and spatial patterns of cell renewal, and the regulation of progenitor–progeny relationships are becoming well understood [29, 30]. Stem cells reside in specific niches and proliferation, migration, and cell differentiation are highly orchestrated within and between the epithelium and mesenchyme. Stem cell niches in the airway and lung epithelium are less understood than in many other organs, but there has been recent progress. The subject of lung epithelial stem cells has been recently reviewed [31]. It is important to note that most studies of cell lineages have been performed in rodent models and strict translation to humans remains unknown. Briefly, metabolic pulse labeling with DNA precursors has long shown that both basal and columnar secretory cell types of the normal adult pseudostratified airway epithelium divide, albeit relatively infrequently [32, 33]. However, repair of epithelial damage is brisk. Many cell types rearrange their cytoskeleton and migrate to cover denuded basal lamina. Subpopulations of cells proliferate and ultimately re-differentiate [34–36]. There has been controversy, but it appears that while ciliated cells actively participate in repair by migration and "dedifferentiation" [37], they do not enter the pool of dividing cells [38]. Subsets of cells within both the basal and columnar cell populations can reconstitute a full mucociliary epithelium in an *in vivo* transplantation model [39, 40]. Thus, many cell types can serve as progenitors during repair of injury. This high degree of "plasticity" in the airway epithelium is likely driven by the necessity for rapid epithelial repair to prevent fibrotic ingrowth and airway obliteration.

Classically, stem cells are defined by their capacity to both self-renew and differentiate, whereas progenitor cells have limited self-renewal capacity. The applicability of classical definitions from well-known model systems to the highly "plastic" airway epithelium is questionable, but there does appear to be a hierarchy of "stemness" within the pool of identified progenitor cells. Colony formation on plastic dishes *in vitro* [41], clonal growth [42], lineage tracing [43], and DNA label retention [44] suggest that although many cells can contribute to repair of injury, basal cells likely represent a stem cell compartment in the adult pseudostratified epithelium (Fig. 16.2).

The simple columnar to cuboidal bronchiolar epithelium of rodents is composed of Clara, ciliated, PNEC, and brush cells. Secretoglobin (SCGB) 1a1, also known as CC10 or CCSP, commonly serves as a marker for Clara cells. However, SCGB1a1 is expressed in a spectrum of cells, present in multiple airway levels, with varying differentiation potentials and behaviors. Following oxidant gas exposure of rodents which damages ciliated cells, surviving bronchiolar Clara cells proliferate to restore the epithelium [45, 46]. When the toxin naphthalene is administered to mice, almost all Clara cells are killed due to selective metabolic activation in this cell type. Ciliated cells shed their cilia and migrate to cover the denuded bronchiolar basal lamina, but do not re-enter the cell cycle [38], whereas surviving Clara cells proliferate [47]. A population of naphthalene-resistant Clara cells resides within, or close to, PNEC clusters and DNA label retention studies suggest that this unit constitutes a stem cell niche [48]. Following naphthalene injury, PNECs also proliferate, but they are apparently a distinct lineage system not requiring, nor generating, Clara cells [49]. A second stem cell niche has been proposed in

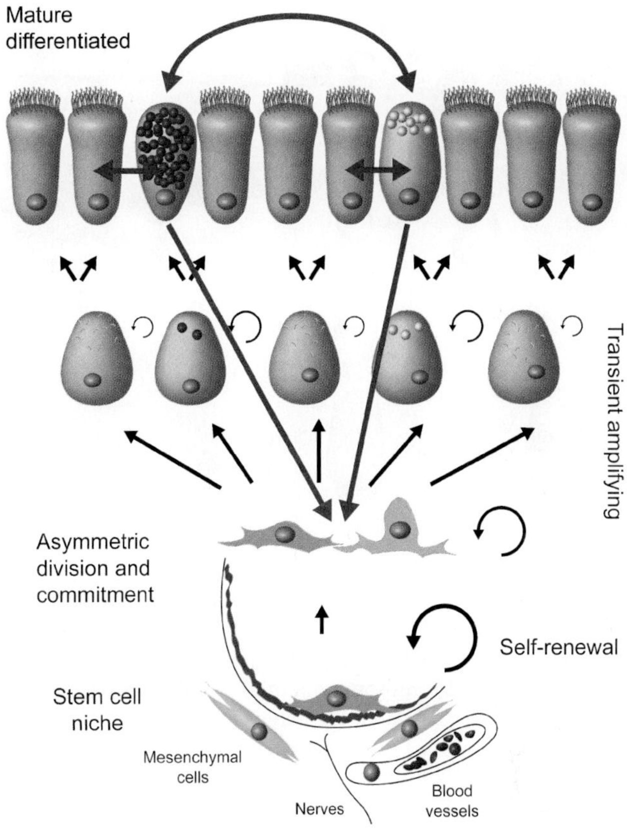

FIG. 16.2 Cell hierarchy in the pseudostratified airway epithelium. In this model, a subset of basal cells gives rise to transiently amplifying cells with basal and intermediate cell morphologies that generate differentiated cells (straight black arrows). The circular arrows represent growth capacity. The red arrows indicate plasticity-differentiated cells are capable of "dedifferentiation" (backwards arrows), "redifferentiaton," and phenotypic conversion during repair of injury. Reproduced, with permission, from Randell SH. Airway epithelial stem cells and the pathophysiology of chronic obstructive pulmonary disease. *Proc Am Thorac Soc* 8: 718–25, 2006.

the broncho-alveolar junction zone [50, 51]. Here, specific Clara-like cells co-express SCGB1a1, SP-C, SP-A, CD-34 and Sca-1, and proliferate in response to naphthalene or bleomycin injury [51]. The cells reportedly grow clonally *in vitro*, differentiate into both Clara and distal lung type I and type II epithelial cells, and generate adenocarcinomas when expressing an active K-ras oncogene *in vivo* [51]. Epithelial cell lineages in the bronchioles and proximal alveolar region are illustrated in Fig. 16.3. It is important to confirm, in humans, the dual airway and alveolar potential of the putative broncho-alveolar stem cells found in mice.

It has been known for many years that alveolar type II cells proliferate to repair injuries that damage the large, thin, and putatively fragile type I alveolar epithelial cells [52]. Palisades of reactive, cuboidal type II cells are frequently present when there is chronic distal lung injury. Although there is evidence for a spectrum of proliferative potential amongst the type II cell population in the distal alveolar epithelium [53], epithelial stem cell hierarchies in the distal alveoli are still poorly understood. The importance of effective epithelial repair in the alveolus is underscored by the

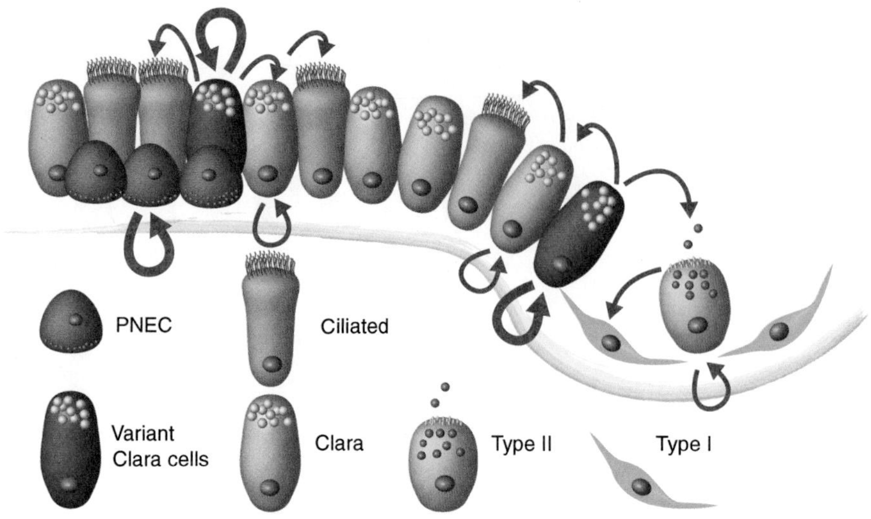

FIG. 16.3 Cell hierarchy in the bronchiolar epithelium. The circular arrows represent growth capacity and differentiation potential. The pulmonary neuroendocrine cell (PNEC) lineage is apparently separate from other epithelial cell types in the adult. Variant Clara cells present in association with PNEC bodies serve as progenitor cells of the bronchiolar epithelium. Variant Clara cells found near the bronchiole-alveolar junction may generate both bronchiolar and alveolar epithelium. Reproduced, with permission, from Randell SH. Airway epithelial stem cells and the pathophysiology of chronic obstructive pulmonary disease. *Proc Am Thorac Soc* 8: 718–25, 2006.

recent discovery that mutations in telomerase reverse transcriptase or the RNA component of the telomerase complex are a cause of idiopathic pulmonary fibrosis [54, 55]. In these cases, deficient chromosome maintenance due to telomerase deficiency likely impairs epithelial regeneration, resulting in aberrant epithelial–mesenchymal interactions that ultimately cause progressive fibrosis. Altered airway epithelial–mesenchymal interactions, whether due to deficient epithelial repair or not, may similarly result in airway sub-epithelial fibrosis and remodeling in asthma and COPD. The relationship between epithelial repair capacity and the development of pulmonary emphysema has not yet been defined.

Two somewhat controversial topics have entered the lung epithelial stem cell dialogue. The first is whether circulating cells become epithelial cells by undergoing a process called "trans-differentiation." In the original report it was claimed that, following lethal doses of ionizing radiation and bone marrow transplantation, circulating cells of bone marrow origin reconstituted a substantial percentage of airway and alveolar epithelial cells [56]. Subsequent studies suggest that apparent replacement of the epithelium from circulating cells may be due to fusion events between myeloid cells and epithelial cells and that transdifferentiation is unlikely [57]. A recent analysis of cord blood stem cell conversion to epithelium is illustrative, and suggests that if transdifferentiation occurs, circulating cells may generate only a very small percentage of the epithelium [58]. The growing consensus is that most epithelial replacement occurs via lung tissue-specific resident or so-called somatic, stem cells. The presence of regional-specific stem cells as the target cells for transformation and cancer initiation may underlie the heterogeneity of lung cancer types [59]. A second area of recent interest is epithelial to mesenchymal transition (EMT), which is a normal developmental process, for example in kidney development, where epithelial cells loose polarity and gain features of mesenchymal cell types. Recent studies suggest that lung epithelial cells exposed to TGF-β *in vitro* may convert into myofibroblast-like cells and that EMT occurs in lung fibrosis models in animals and in human idiopathic pulmonary fibrosis [60, 61] (reviewed in Ref. [62]). The significance of EMT for airway remodeling in asthma and COPD remains unknown. The metastasis of lung cancer cells is thought to involve a process similar to EMT that is regulated by TGF-β and the extracellular matrix [63].

Due to the promise of regenerative medicine for chronic lung disease after neonatal respiratory disorders or for more typical adult forms of pulmonary emphysema, there is intense interest in lung stem cells. Furthermore, the important issue of lung cancer initiating cells and stem cells within lung cancers is highly related to stem cell biology in the normal lung. Recent advances point towards additional studies needed to characterize lung stem cells and their niches, and to elucidate steady state and repairing cell lineages. The identification of unique gene and protein markers indicative of cell position in the lung epithelial stem cell hierarchy will enable generation of novel transgenic animals that will foster more extensive and precise lineage mapping, which is claimed to be the gold standard by some stem cell authorities [64]. Furthermore, such markers may enable purification and functional testing of candidate stem and progenitor cell populations from both animals and humans. A greater understanding of the molecular pathways regulating normal cell behavior and disease phenotypes underlies rational approaches to cellular therapy for lung diseases, whether it involves augmenting epithelial repair by stimulating growth or by inhibiting growth as in the case of epithelial hyperplasia/metaplasia and lung cancer.

INTEGRATED EPITHELIAL FUNCTION

A continuous epithelial layer, composed of characteristic cell types along the respiratory tract axis, forms a physical barrier between the outside environment and the underlying matrix and circulatory system. The airways deliver warmed and humidified inhaled air to the alveoli while protecting gas exchange structures from potentially harmful airborne chemicals, particles, and pathogens. The epithelium provides both general and regionally specialized functions that are key to normal physiology and host defense. The airway epithelium is a principal generator of nitric oxide [65] and coordinates fluid and mucous secretion to foster effective mucociliary transport and productive cough [66]. The distal lung alveolar epithelium plays a key role in lung fluid balance, the prevention and resolution of alveolar edema after injury [67] and is the source of pulmonary surfactant. Furthermore, the epithelium in both the conducting and respiratory zones plays a key role during "innate" and "adaptive" immune responses.

MUCOCILIARY AND MUCUS CLEARANCE

A layer of fluid and mucus normally flows over conducting airway surfaces toward the pharynx. This cleansing action is a critical, physiologically regulated, protective mechanism of the airways and lungs, whose failure causes recurrent infections characteristic of the genetic diseases cystic fibrosis and primary ciliary dyskinesia [68]. Effective mucus clearance results from coordinated secretion of mucins and other proteins along with sufficient water for hydration, which is locally regulated by airway epithelial ion transport. Adequately hydrated mucus is propelled by beating cilia and, when necessary, by cough. The combination of mucociliary and cough clearance equals mucus clearance. As measured by movement of inhaled and deposited radioactive tracer particles, mucus clearance declines in proportion to airflow obstruction in acute exacerbations of asthma, and returns toward normal after effective therapy [69–71]. Several studies indicate that mucus clearance is decreased by tobacco smoking and in COPD [72–75]. The accumulation of inflammatory mucoid secretions in the small airways in COPD, which likely reflects both increased local mucin production and decreased clearance [76], is correlated with the degree of physiologic impairment [77]. Drug therapies such as β-adrenergic agonists, cholinergic antagonists and corticosteroids used to treat asthma and COPD may have beneficial effects on mucus transport, but consistent positive effects on mucus clearance have been difficult to substantiate [78, 79]. Additional specific therapies directed toward balancing mucin production and ion and water transport to improve mucus clearance will likely be useful.

MUCINS AND MUCUS

Mucus is composed of mucins, proteins, lipids, ions and water present on mucosal surfaces. Mucins are highly glycosylated proteins that can be released from mucous secretory cell granules and/or by shedding from the surface of many cell types. Intermolecular disulfide bonds within the principal gel-forming lung mucins, MUC5AC and MUC5B, enable multimerization and extensive interdigitation creating a visco-elastic network capable of trapping a wide variety of inhaled particles [80]. Increased mucous secretory cell number and increased mucin gene expression and glycoprotein production characterize both asthma and COPD. Regulations of airway epithelial secretory cell differentiation, mucin gene and glycoprotein expression and the secretory pathways that mediate mucin release have been reviewed [81–83]. In addition to the surface epithelium, submucosal glands in the cartilaginous airways are a significant source of fluid, mucins, and other proteins such as anti-microbial factors [84–86]. Gland hypertrophy is a pathologic hallmark of both asthma and COPD and likely contributes to mucin hypersecretion [27, 28].

CILIATED CELLS

Ciliated cells have long and branched microvilli plus approximately 200 cilia per cell that occupy a large percentage of the normal airway epithelial luminal surface (Fig. 16.4). Cilia in the trachea and bronchi are denser and longer than in the bronchioles, but cilia average approximately 250 nm in diameter and 6 μm in length. The shaft, or axoneme, of motile cilia consists of nine outer microtubule doublets, a central microtubule pair, inner and outer dynein arms, nexin links, and radial spokes. The axoneme enters the ciliated cell and is anchored by a cytoplasmic basal body similar to the centriole of mitotic cells. Greater than 200 unique proteins constitute the cilia and basal body [87]. In normal airways, cilia beat at 10–20 Hz in a coordinated manner that generates directional, metachronal waves to assist in the propulsion of fluid, mucus and trapped particles towards the pharynx.

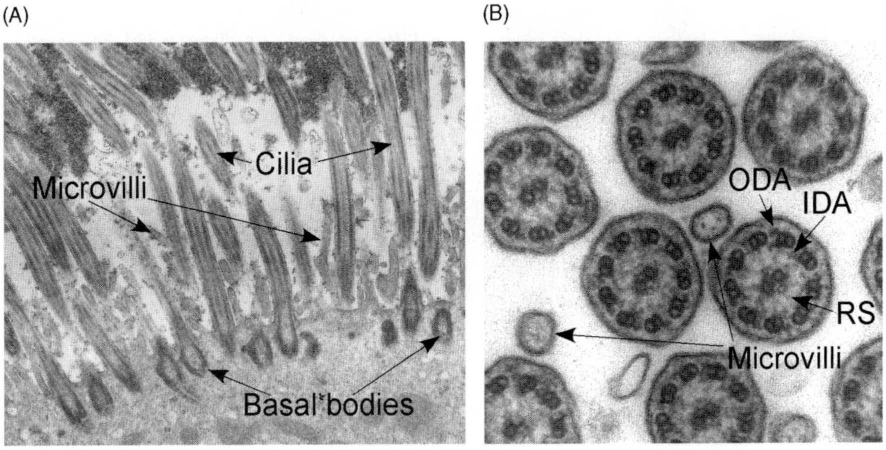

FIG. 16.4 Ciliated cell ultrastructure. Conventional transmission electron microscopy of a normal human nasal biopsy. (A) Longitudinal section across the apical border of a ciliated cell showing a clear area near the cell surface with cilia tips enmeshed in viscous material. Original magnification = 12,000 X. (B) Cross sections of cilia axonemes, illustrating the 9 + 2 arrangement of microtubules in motile cilia. ODA: outer dynein arm; IDA: inner dynein arm; RS: radial spoke, original magnification = 50,000 X.

ION AND WATER TRANSPORT AND INTEGRATED FUNCTION FOR EFFECTIVE MUCUS CLEARANCE

Mucous secretory and ciliated cells distinctly contribute to mucociliary clearance which, combined with cough, constitutes mucus clearance, a key airway defense mechanism [88]. Collectively, the system is organized and physiologically regulated to balance mucin production with ion and water transport to maintain surface hydration at levels sufficient to foster mucus transportability. Airway surface liquid is normally present in two layers, an overlying viscous mucus layer and a periciliary layer (PCL) in which cilia can freely beat. A "slip" layer provides "lubrication" at the interface of these two layers [66]. The mucus layer and the PCL can be visualized after special fixation techniques of tissues or in culture preparations of live, differentiated airway epithelial cells [89–91] (Fig. 16.5). By balancing Cl^- secretion and Na^+ absorption at the epithelial luminal membrane, the airway epithelium regulates PCL volume to fully hydrate the PCL and maintain its depth at the height of the outstretched cilia [90, 92, 93]. There are two principal apical membrane channels, the cystic fibrosis transmembrane conductance regulator (CFTR) and the Ca^{2+}-activated Cl^- channel (CaCC), that mediate Cl^- transport. Na^+ transport occurs through the epithelial Na^+ channel (ENaC). CFTR, CaCC, and ENaC activity is locally and coordinately regulated by changes in the PCL concentration of nucleotides and nucleosides [94–97]. In this system, ATP released by cells activates $P2Y_2$ purinergic receptors, which stimulates IP_3 formation, causing Ca^{2+} release that activates CaCC. Decreased PIP_2 is thought to coordinately inhibit ENaC. ATP can also be metabolized to adenosine, which will activate A_{2B} adenosine receptors, generating cAMP and activating protein kinase C to ultimately open CFTR. Active CFTR is also reported to directly inhibit ENaC. Apical membrane Cl^- secretion through CFTR and/or CaCC increases airway surface hydration whereas Na^+ entry through ENaC decreases airway surface hydration. Water follows its osmotic gradient transcellularly through aquaporins and, a small part, via para-cellular pathways. In the distal lung, similar processes, mainly involving Na^+ absorption through ENaC but also involving CFTR, are thought to regulate normal alveolar fluid balance and are important in the resolution of pulmonary edema after acute lung injury [67].

Endogenous mediators active in regulation of airway and lung surface liquid have additional roles, which may pose conundrums relevant to potential therapies for asthma and COPD. For example, although ATP promotes airway surface hydration via activation of epithelial $P2Y_2$ purinergic receptors, ATP is also a "danger signal" involved in regulation of the immune system [98]. Excess airway ATP may contribute to an asthma-like phenotype via pro inflammatory effects, including the recruitment and activation of dendritic cells (DCs) [99]. Similarly, adenosine activation of epithelial A_{2B} receptors is important to activate CFTR and hydrate the epithelial surface, but high adenosine levels in adenosine deaminase deficient mice promotes inflammation, and inhaled adenosine causes broncho-constriction in asthmatic humans [100]. Pharmacologic therapies directed at reducing adenosine or antagonizing adenosine receptors may have the unwanted effect of creating a cystic fibrosis-like

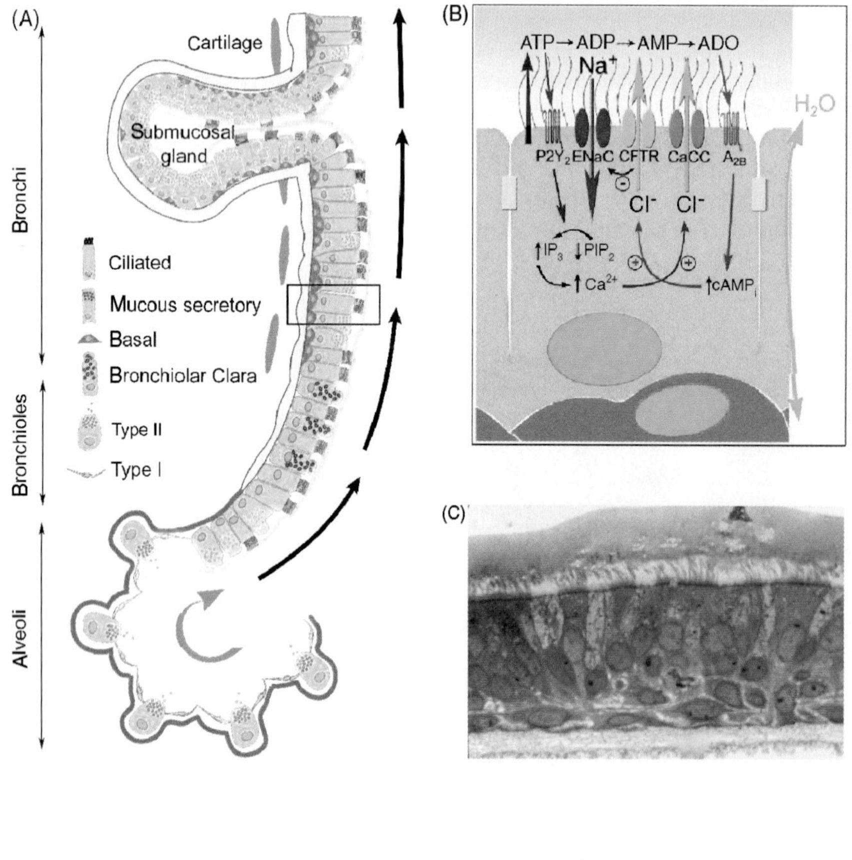

FIG. 16.5 Respiratory tract ion and water transport and mucus clearance. (A) The airway epithelium coordinates fluid and mucous secretion to foster effective mucus clearance. It is still uncertain how much alveolar lining fluid leaves the distal lung via the airways (gray arrow). (B) Ion channels in the apical plasma membrane, regulated by local concentrations of nucleotides and nucleosides, maintain adequate airway surface hydration. (C) Airway surface liquid consists of two layers, a clear periciliary layer and overlying mucus. Perflurocarbon-OsO_4 fixed, well-differentiated human bronchial epithelial cell culture, thin plastic section, original magnification = 500X, Richardson's stain. (A) and (B) are adapted, with permission, from Ref. [68].

phenotype [97]. Furthermore, although elevated adenosine produces "whole animal" pro-inflammatory effects, adenosine acting via A_{2A} receptors on macrophages is anti-inflammatory [101]. The pleiotropic effects of mediators such as ATP and adenosine are likely concentration-, compartment-, receptor-, and cell type-specific and must be considered when proposing therapies based on manipulation of their levels or receptors.

LUNG EPITHELIAL FUNCTION IN IMMUNITY

The mammalian immune system is broadly divided into "innate" and "adaptive" branches. Innate immunity is the first line of defense that is constantly on guard and requires no prior exposure to harmful substances or pathogens to induce a protective response. Adaptive immunity takes time to develop via specific, antigen triggered cellular interactions between DCs, T-, and B-cells and has profound memory. The two branches are highly intertwined. For example, the cell and chemical milieu produced by the innate immune system is critical for the development of adaptive immunity, and conversely, products of adaptive immune cells can alter the function of cells involved in innate immunity. The airway and lung epithelium is integral to innate and adaptive immunity and is both an effector and target in the altered immune function characteristic of asthma and COPD.

ROLE IN INNATE IMMUNITY

The airway and lung epithelium is a selective physical barrier that coordinates removal of potentially harmful substances by mucus clearance. It also plays important roles in host defense by secreting a wide spectrum of protective, antimicrobial, homeostatic, and immune regulatory factors including antioxidants, defensins, collectins, complement components, lactoferrin, lysozyme, and protease inhibitors (reviewed in Refs [102, 103]). Normally protective innate immune functions can be enhanced when pathogen/microbial associated patterns or other chemical danger signals are recognized by innate immune system receptors expressed by epithelial cells. The receptors include transmembrane molecules such as the toll-like receptors (TLRs) [104], protease-activated receptors (PARs) [105] and purinergic receptors [98] as well as cytosolic proteins including the nucleotide-binding, leucine-rich repeat receptors (NLRs) [106] and the retinoic acid-inducible gene I-like helicases (RLHs) [107]. The substances detected by these multiple classes of receptors include endogenous homeostatic factors such as ATP as well as molecules derived from microbes and pathogens including bacterial lipopolysaccharide and viral RNA.

Receptor engagement induces downstream signaling that modifies cellular behavior in many ways to eliminate the perceived danger. Airway epithelial ion channels can be activated to induce fluid secretion accompanied by mucin release and increased cilia beating to "flush away" the offending stimulus. Preformed substrates such as arachidonic acid can be metabolized and released and/or gene expression/stability can be modified to change patterns of epithelial cell mediator production. Receptor engagement and downstream signaling may induce pro-inflammatory cytokines and chemokines that selectively attract neutrophils, monocytes, eosinophils, and basophils to ostensibly eliminate pathogens. The process is dynamic, and repeat or chronic exposure to substances that activate innate epithelial cell immune system receptors will modulate epithelial cell behavior upon subsequent exposure [108]. The epithelium thus plays a vital innate immune role to perceive and respond to danger, with mechanisms in place to modulate and orchestrate inflammation depending on the type, degree, and chronicity of stimulation. Functional polymorphisms in genes involved in epithelial innate immunity are likely important modifiers related to the development of lung disease.

ROLE IN ADAPTIVE IMMUNITY

Beyond regulating luminal dwell time and permeability of allergens and pathogens, the airway and lung epithelium play key regulatory roles in adaptive immunity. Cytokines, chemokines, and other molecules produced by the epithelium at steady state, and especially after innate immune stimulation, modulate adaptive immunity by attracting DCs, T-, and B-cells and altering their function (reviewed in Ref. [109]). In turn, the epithelium responds to products of adaptive immune cells, establishing complex feedback loops that ultimately determine tissue inflammation and lesion development. The epithelium also plays a functional role by facilitating the transport of secretory immunoglobulin.

The conducting airway epithelium and alveolar interstitium contain resident DCs that sample the luminal microenvironment and present antigens to T-cells locally, or after migrating to regional lymph nodes. The numbers of DCs in the airways and lungs, and their functional state, are altered by numerous chemokines, cytokines, and other molecules involved in DC attraction and activation that are secreted by epithelial cells [110]. Chronic inflammatory diseases are regulated by T-cell activation and polarization and recent discoveries have added new T-cell subsets to the landscape. Organ parenchymal cells participate in the differentiation of Th_1, Th_2, Th_{17}, and T-regulatory (T_{reg}) cells from naïve T-cells and create feedback loops that regulate and sustain local responses. For example, epithelial IL-6 production positively reinforces Th_{17} cell polarization [111]. Conversely, T-cell products affect the epithelium, such as the well-known induction of mucin by the Th_2 cytokine IL-13 [112, 113]. Th_{17} cell production of IL-22 and its action on the epithelium appears to play an important role in epithelial repair and mucosal defense against gram-negative bacteria [114]. The airway and lung are populated by immunoglobulin secreting plasma cells. B-cell stimulation and class switching is strongly influenced by other adaptive immune cells but there is mounting evidence that epithelium-derived factors participate in regulation of B-cell class switching and immunoglobulin production [109]. The epithelium plays an integral role in adaptive immunity by transporting secretory immunoglobulin from the lamina propria where it is produced into the lumen, a process that may be impaired in asthma and COPD [115].

To summarize, complex reciprocal interactions between the epithelium and the adaptive immune system regulate the integrated response to allergens and infectious agents and play important roles in asthma and COPD.

References

1. Wert SE. Normal and abnormal structural development of the lung. In: Polin RA, Fox WW, Abman SH (eds). *Fetal and Neonatal Physiology*, pp. 783–94. Philadelphia: Saunders, 2004.
2. van Tuyl M, Post M. Molecular mechanisms of lung development and lung branching morphogenesis. In: Polin RA, Fox WW, Abman SH (eds). *Fetal and Neonatal Physiology*, pp. 812–21. Philadelphia: Saunders, 2004.
3. Pfister RH, Soll RF, Wiswell T. Protein containing synthetic surfactant versus animal derived surfactant extract for the prevention and treatment of respiratory distress syndrome. *Cochrane Database Syst Rev*, 2007, CD006069.
4. Henderson-Smart DJ, Cools F, Bhuta T, Offringa M. Elective high frequency oscillatory ventilation versus conventional ventilation for acute pulmonary dysfunction in preterm infants. *Cochrane Database Syst Rev*, 2007, CD000104.
5. Barrington KJ, Finer NN. Inhaled nitric oxide for respiratory failure in preterm infants. *Cochrane Database Syst Rev*, 2007, CD000509.
6. Shah SS, Ohlsson A, Halliday H, Shah VS. Inhaled versus systemic corticosteroids for the treatment of chronic lung disease in ventilated very low birth weight preterm infants. *Cochrane Database Syst Rev*, 2007, CD002057.
7. Baraldi E, Filippone M. Chronic lung disease after premature birth. *N Engl J Med* 357: 1946–55, 2007.
8. Kumar R, Yu Y, Story RE, Pongracic JA, Gupta R, Pearson C *et al.* Prematurity, chorioamnionitis, and the development of recurrent wheezing: A prospective birth cohort study. *J Allergy Clin Immunol* 121: 878–84, 2008.
9. Narang I, Baraldi E, Silverman M, Bush A. Airway function measurements and the long-term follow-up of survivors of preterm birth with and without chronic lung disease. *Pediatr Pulmonol* 41: 497–508, 2006.
10. Narang I, Rosenthal M, Cremonesini D, Silverman M, Bush A. Longitudinal evaluation of airway function 21 years after preterm birth. *Am J Respir Crit Care Med* 178: 74–80, 2008.
11. Barker PM, Southern KW. Regulation of liquid secretion and absorption by the fetal and neonatal lung. In: Polin RA, Fox WW, Abman SH (eds). *Fetal and Neonatal Physiology*, pp. 822–34. Philadelphia: Saunders, 2004.
12. Randell SH, Shimizu T, Bakewell W, Ramaekers FC, Nettesheim P. Phenotypic marker expression during fetal and neonatal differentiation of rat tracheal epithelial cells. *Am J Respir Cell Mol Biol* 8: 546–55, 2006.
13. Otani EM, Newkirk C, McDowell EM. Development of hamster tracheal epithelium: IV. Cell proliferation and cytodifferentiation in the neonate. *Anat Rec* 214: 183–92, 1986.
14. Leigh MW, Gambling TM, Carson JL, Collier AM, Wood RE, Boat TF. Postnatal development of tracheal surface epithelium and submucosal glands in the ferret. *Exp Lung Res* 10: 153–69, 1986.
15. Van Winkle LS, Fanucchi MV, Miller LA, Baker GL, Gershwin LJ, Schelegle ES *et al.* Epithelial cell distribution and abundance in rhesus monkey airways during postnatal growth and development. *J Appl Physiol* 97: 2355–63, 2004.
16. Gaillard DA, Lallement AV, Petit AF, Puchelle ES. *In vivo* ciliogenesis in human fetal tracheal epithelium. *Am J Anat* 185: 415–28, 1989.
17. Lamb D, Reid L. Acid glycoproteins produced by the mucous cells of the bronchial submucosal glands in the fetus and child: A histochemical autoradiographic study. *Br J Dis Chest* 66: 248–53, 1972.
18. Racila DM, Kline JN. Perspectives in asthma: Molecular use of microbial products in asthma prevention and treatment. *J Allergy Clin Immunol* 116: 1202–5, 2005.
19. Schaub B, Lauener R, von Mutius E. The many faces of the hygiene hypothesis. *J Allergy Clin Immunol* 117: 969–77, 2006.
20. Mercer RR, Russell ML, Roggli VL, Crapo JD. Cell number and distribution in human and rat airways. *Am J Respir Cell Mol Biol* 10: 613–24, 1994.
21. De Proost I, Pintelon I, Brouns I, Kroese AB, Riccardi D, Kemp PJ, Timmermans JP, Adriaensen D. Functional Live Cell Imaging of the Pulmonary Neuroepithelial Body Microenvironment. *Am J Respir Cell Mol Biol* 2008 doi:10. 1165/rcmb. 2008-0011OC.
22. Cutz E, Yeger H, Pan J. Pulmonary neuroendocrine cell system in pediatric lung disease-recent advances. *Pediatr Dev Pathol* 10: 419–35, 2007.
23. Evans MJ, Van Winkle LS, Fanucchi MV, Plopper CG. The attenuated fibroblast sheath of the respiratory tract epithelial-mesenchymal trophic unit. *Am J Respir Cell Mol Biol* 21: 655–57, 1999.
24. Araya J, Cambier S, Morris A, Finkbeiner W, Nishimura SL. Integrin-mediated transforming growth factor-beta activation regulates homeostasis of the pulmonary epithelial-mesenchymal trophic unit. *Am J Pathol* 169: 405–15, 2006.
25. Plopper CG, Smiley-Jewell SM, Miller LA, Fanucchi MV, Evans MJ, Buckpitt AR *et al.* Asthma/allergic airways disease: Does postnatal exposure to environmental toxicants promote airway pathobiology? *Toxicol Pathol* 35: 97–110, 2007.
26. Araya J, Cambier S, Markovics JA, Wolters P, Jablons D, Hill A *et al.* Squamous metaplasia amplifies pathologic epithelial-mesenchymal interactions in COPD patients. *J Clin Invest* 117: 3551–62, 2007.
27. Holgate ST. Epithelium dysfunction in asthma. *J Allergy Clin Immunol* 120: 1233–44, 2007.
28. Jeffery PK. Remodeling and inflammation of bronchi in asthma and chronic obstructive pulmonary disease. *Proc Am Thorac Soc* 1: 176–83, 2004.
29. Fuchs E. Skin stem cells: Rising to the surface. *J Cell Biol* 180: 273–84, 2008.
30. Barker N, Clevers H. Tracking down the stem cells of the intestine: Strategies to identify adult stem cells. *Gastroenterology* 133: 1755–60, 2007.
31. Kotton DN, Fine A. Lung stem cells. *Cell Tissue Res* 331: 145–56, 2008.
32. Donnelly GM, Haack DG, Heird CS. Tracheal epithelium: Cell kinetics and differentiation in normal rat tissue. *Cell Tissue Kinet* 15: 119–30, 1982.
33. Breuer R, Zajicek G, Christensen TG, Lucey EC, Snider GL. Cell kinetics of normal adult hamster bronchial epithelium in the steady state. *Am J Respir Cell Mol Biol* 2: 51–58, 1990.
34. Desai LP, Aryal AM, Ceacareanu B, Hassid A, Waters CM. RhoA and Rac1 are both required for efficient wound closure of airway epithelial cells. *Am J Physiol* 287: L1134–44, 2004.
35. Shimizu T, Nishihara M, Kawaguchi S, Sakakura Y. Expression of phenotypic markers during regeneration of rat tracheal epithelium following mechanical injury. *Am J Respir Cell Mol Biol* 11: 85–94, 1994.
36. Erjefalt JS, Persson CG. Airway epithelial repair: Breathtakingly quick and multipotentially pathogenic. *Thorax* 52: 1010–12, 1997.
37. Park KS, Wells JM, Zorn AM, Wert SE, Laubach VE, Fernandez LG *et al.* Transdifferentiation of ciliated cells during repair of the respiratory epithelium. *Am J Respir Cell Mol Biol* 34: 151–57, 2006.
38. Rawlins EL, Ostrowski LE, Randell SH, Hogan BLM. Lung development and repair: Contribution of the ciliated lineage. *Proc Natl Acad Sci USA* 104: 410–17, 2007.
39. Liu JY, Nettesheim P, Randell SH. Growth and differentiation of tracheal epithelial progenitor cells. *Am J Physiol Lung Cell Mol Physiol* 266: L296–307, 1994.

40. Avril-Delplanque A, Casal I, Castillon N, Hinnrasky J, Puchelle E, Peault B. Aquaporin-3 expression in human fetal airway epithelial progenitor cells. *Stem Cells* 23: 992–1001, 2005.
41. Randell SH, Comment CE, Ramaekers FCS, Nettesheim P. Properties of rat tracheal epithelial cells separated based on expression of cell surface alpha-galactosyl end groups. *Am J Respir Cell Mol Biol* 4: 544–54, 1991.
42. Schoch KG, Lori A, Burns KA, Eldred T, Olsen JC, Randell SH. A subset of mouse tracheal epithelial basal cells generates large colonies *in vitro*. *Am J Physiol* 286: L631–42, 2004.
43. Engelhardt JF, Schlossberg H, Yankaskas JR, Dudus L. Progenitor cells of the adult human airway involved in submucosal gland development. *Development* 121: 2031–46, 1995.
44. Borthwick DW, Shahbazian M, Todd KQ, Dorin JR, Randell SH. Evidence for stem-cell niches in the tracheal epithelium. *Am J Respir Cell Mol Biol* 24: 662–70, 2001.
45. Evans MJ, Johnson LV, Stephens RJ, Freeman G. Renewal of the terminal bronchiolar epithelium in the rat following exposure to NO_2 or O_3. *Lab Invest* 35: 246–57, 1976.
46. Evans MJ, Shami SG, Cabral-Anderson LJ, Dekker NP. Role of nonciliated cells in renewal of the bronchial epithelium of rats exposed to NO_2. *Am J Pathol* 123: 126–33, 1986.
47. Van Winkle LS, Buckpitt AR, Nishio SJ, Isaac JM, Plopper CG. Cellular response in naphthalene-induced Clara cell injury and bronchiolar epithelial repair in mice. *Am J Physiol* 269: L800–18, 1995.
48. Hong KU, Reynolds SD, Giangreco A, Hurley CM, Stripp BR. Clara cell secretory protein-expressing cells of the airway neuroepithelial body microenvironment include a label-retaining subset and are critical for epithelial renewal after progenitor cell depletion. *Am J Respir Cell Mol Biol* 24: 671–81, 2001.
49. Reynolds SD, Hong KU, Giangreco A, Mango GW, Guron C, Morimoto Y *et al.* Conditional Clara cell ablation reveals a self-renewing progenitor function of pulmonary neuroendocrine cells. *Am J Physiol* 278: L1256–63, 2000.
50. Giangreco A, Reynolds SD, Stripp BR. Terminal bronchioles harbor a unique airway stem cell population that localizes to the bronchoalveolar duct junction. *Am J Pathol* 161: 173–82, 2002.
51. Kim CF, Jackson EL, Woolfenden AE, Lawrence S, Babar I, Vogel S *et al.* Identification of bronchoalveolar stem cells in normal lung and lung cancer. *Cell* 121: 823–35, 2005.
52. Adamson IY, Bowden DH. The type 2 cell as progenitor of alveolar epithelial regeneration. A cytodynamic study in mice after exposure to oxygen. *Lab Invest* 30: 35–42, 1974.
53. Reddy R, Buckley S, Doerken M, Barsky L, Weinberg K, Anderson KD *et al.* Isolation of a putative progenitor subpopulation of alveolar epithelial type 2 cells. *Am J Physiol* 286: L658–67, 2004.
54. Tsakiri KD, Cronkhite JT, Kuan PJ, Xing C, Raghu G, Weissler JC *et al.* Adult-onset pulmonary fibrosis caused by mutations in telomerase. *Proc Natl Acad Sci USA* 104: 7552–57, 2007.
55. Armanios MY, Chen JJ, Cogan JD, Alder JK, Ingersoll RG, Markin C *et al.* Telomerase mutations in families with idiopathic pulmonary fibrosis. *N Engl J Med* 356: 1317–26, 2007.
56. Krause DS, Theise ND, Collector MI, Henegariu O, Hwang S, Gardner R *et al.* Multi-organ, multi-lineage engraftment by a single bone marrow-derived stem cell. *Cell* 105: 369–77, 2001.
57. Kotton DN, Fabian AJ, Mulligan RC. Failure of bone marrow to reconstitute lung epithelium. *Am J Respir Cell Mol Biol* 33: 328–34, 2005.
58. Sueblinvong V, Loi R, Eisenhauer PL, Bernstein IM, Suratt BT, Spees JL *et al.* Derivation of lung epithelium from human cord blood-derived mesenchymal stem cells. *Am J Respir Crit Care Med* 177: 701–11, 2008.
59. Giangreco A, Groot KR, Janes SM. Lung cancer and lung stem cells: Strange bedfellows? *Am J Respir Crit Care Med* 175: 547–53, 2007.
60. Kim KK, Kugler MC, Wolters PJ, Robillard L, Galvez MG, Brumwell AN *et al.* Alveolar epithelial cell mesenchymal transition develops *in vivo* during pulmonary fibrosis and is regulated by the extracellular matrix. *Proc Natl Acad Sci USA* 103: 13180–85, 2006.
61. Willis BC, Liebler JM, Luby-Phelps K, Nicholson AG, Crandall ED, du Bois RM *et al.* Induction of epithelial-mesenchymal transition in alveolar epithelial cells by transforming growth factor-beta1: Potential role in idiopathic pulmonary fibrosis. *Am J Pathol* 166: 1321–32, 2005.
62. Willis BC, Borok Z. TGF-beta-induced EMT: Mechanisms and implications for fibrotic lung disease. *Am J Physiol Lung Cell Mol Physiol* 293: L525–34, 2007.
63. Shintani Y, Maeda M, Chaika N, Johnson KR, Wheelock MJ. Collagen I promotes epithelial-to-mesenchymal transition in lung cancer cells via transforming growth factor-beta signaling. *Am J Respir Cell Mol Biol* 38: 95–104, 2008.
64. Nystul TG, Spradling AC. Breaking out of the mold: Diversity within adult stem cells and their niches. *Curr Opin Genet Dev* 16: 463–68, 2006.
65. Bove PF, van der Vilet A. Nitric oxide and reactive nitrogen species in airway epithelial signaling and inflammation. *Free Radic Biol Med* 41: 515–27, 2006.
66. Randell SH, Boucher RC, for the University of North Carolina Virtual Lung Group. Effective mucus clearance is essential for respiratory health. *Am J Respir Cell Mol Biol* 35:20–8, 2006.
67. Berthiaume Y, Matthay MA. Alveolar edema fluid clearance and acute lung injury. *Respir Physiol Neurobiol* 159: 350–59, 2007.
68. Livraghi A, Randell SH. Cystic fibrosis and other respiratory diseases of impaired mucus clearance. *Toxicol Pathol* 35: 116–29, 2007.
69. Messina MS, O'Riordan TG, Smaldone GC. Changes in mucociliary clearance during acute exacerbations of asthma. *Am Rev Respir Dis* 143: 993–97, 1991.
70. O'Riordan TG, Zwang J, Smaldone GC. Mucociliary clearance in adult asthma. *Am Rev Respir Dis* 146: 598–603, 1992.
71. Shah RV, Amin M, Sangwan S, Smaldone GC. Steroid effects on mucociliary clearance in outpatient asthma. *J Aerosol Med* 19: 208–20, 2006.
72. Goodman RM, Yergin BM, Landa JF, Golinvaux MH, Sackner MA. Relationship of smoking history and pulmonary function tests to tracheal mucous velocity in nonsmokers, young smokers, ex-smokers, and patients with chronic bronchitis. *Am Rev Respir Dis* 117: 205–14, 1978.
73. Vastag E, Matthys H, Koehler D, Gronbeck L, Daikeler G. Mucociliary clearance and airways obstruction in smokers, ex-smokers and normal subjects who never smoked. *Eur J Respir Dis* 66(Suppl. 139): 93–100, 1985.
74. Camner P, Mossberg B, Philipson K. Tracheobronchial clearance and chronic obstructive lung disease. *Scand J Respir Dis* 54: 272–81, 1973.
75. Smaldone GC, Foster WM, O'Riordan TG, Messina MS, Perry RJ, Langenback EG. Regional impairment of mucociliary clearance in chronic obstructive pulmonary disease. *Chest* 103: 1390–96, 1993.
76. Hogg JC. Pathophysiology of airflow limitation in chronic obstructive pulmonary disease. *Lancet* 364: 709–21, 2004.
77. Hogg JC, Chu F, Utokaparch S, Woods R, Elliott WM, Buzatu L *et al.* The nature of small-airway obstruction in chronic obstructive pulmonary disease. *N Engl J Med* 350: 2645–53, 2004.
78. Restrepo RD. Inhaled adrenergics and anticholinergics in obstructive lung disease: Do they enhance mucociliary clearance? *Respir Care* 52: 1159–73, 2007.
79. Bennett WD. Effect of beta-adrenergic agonists on mucociliary clearance. *J Allergy Clin Immunol* 110(6 Suppl.): S291–97, 2002.
80. Sheehan JK, Kesimer M, Pickles R. Innate immunity and mucus structure and function. *Novartis Found Symp* 279: 155–66, 2006.
81. Rose MC, Voynow JA. Respiratory tract mucin genes and mucin glycoproteins in health and disease. *Physiol Rev* 86: 245–78, 2006.
82. Thornton DJ, Sheehan JK. From mucins to mucus: Toward a more coherent understanding of this essential barrier. *Proc Am Thorac Soc* 1: 54–61, 2004.
83. Williams OW, Sharafkhaneh A, Kim V, Dickey BF, Evans CM. Airway mucus: From production to secretion. *Am J Respir Cell Mol Biol* 34: 527–36, 2006.
84. Inglis SK, Wilson SM. Cystic fibrosis and airway submucosal glands. *Pediatr Pulmonol* 40: 279–84, 2005.

85. Wine JJ, Joo NS. Submucosal glands and airway defense. *Proc Am Thorac Soc* 1: 47–53, 2004.
86. Ballard ST, Spadafora D. Fluid secretion by submucosal glands of the tracheobronchial airways. *Respir Physiol Neurobiol* 159: 271–77, 2007.
87. Fliegauf M, Omran H. Novel tools to unravel molecular mechanisms in cilia-related disorders. *Trends Genet* 22: 241–45, 2006.
88. Knowles MR, Boucher RC. Mucus clearance as a primary innate defense mechanism for mammalian airways ("Perspective"). *J Clin Invest* 109: 571–77, 2002.
89. Sims DE, Westfall JA, Kiorpes AL, Horne MM. Preservation of tracheal mucus by nonaqueous fixative. *Biotech Histochem* 66: 173–80, 1991.
90. Matsui H, Grubb BR, Tarran R, Randell SH, Gatzy JT, Davis CW *et al.* Evidence for periciliary liquid layer depletion, not abnormal ion composition, in the pathogenesis of cystic fibrosis airways disease. *Cell* 95: 1005–15, 1998.
91. Tarran R, Boucher RC. Thin-film measurements of airway surface liquid volume/composition and mucus transport rates *in vitro*. *Meth Mol Med* 70: 479–92, 2002.
92. Tarran R, Button B, Boucher RC. Regulation of normal and cystic fibrosis airway surface liquid volume by phasic shear stress. *Annu Rev Physiol* 68: 543–61, 2005.
93. Tarran R, Grubb BR, Gatzy JT, Davis CW, Boucher RC. The relative roles of passive surface forces and active ion transport in the modulation of airway surface liquid volume and composition. *J Gen Physiol* 118: 223–36, 2001.
94. Lazarowski ER, Tarran R, Grubb BR, van Heusden CA, Okada S, Boucher RC. Nucleotide release provides a mechanism for airway surface liquid homeostasis. *J Biol Chem* 279: 36855–64, 2004.
95. Tarran R. Regulation of airway surface liquid volume and mucus transport by active ion transport. *Proc Am Thorac Soc* 1: 42–46, 2004.
96. Tarran R, Trout L, Donaldson SH, Boucher RC. Soluble mediators, not cilia, determine airway surface liquid volume in normal and cystic fibrosis superficial airway epithelia. *J Gen Physiol* 127: 591–604, 2006.
97. Rollins BM, Burn M, Coakley RD, Chambers LA, Hirsh AJ, Clunes MT, Lethem MI, Donaldson SH, Tarran R. A2B Adenosine Receptors Regulate the Mucus Clearance Component of the Lungs Innate Defense System. *Am J Respir Cell Mol Biol* 2008. doi:10.1165/rcmb.2007-0450OC.
98. Bours MJ, Swennen EL, Di Virgilio F, Cronstein BN, Dagnelie PC. Adenosine 5′-triphosphate and adenosine as endogenous signaling molecules in immunity and inflammation. *Pharmacol Ther* 112: 358–404, 2006.
99. Idzko M, Hammad H, van Nimwegen M, Kool M, Willart MA, Muskens F *et al.* Extracellular ATP triggers and maintains asthmatic airway inflammation by activating dendritic cells. *Nat Med* 13: 913–19, 2007.
100. Mohsenin A, Blackburn MR. Adenosine signaling in asthma and chronic obstructive pulmonary disease. *Curr Opin Pulm Med* 12: 54–59, 2006.
101. Mohsenin A, Mi T, Xia Y, Kellems RE, Chen JF, Blackburn MR. Genetic removal of the A_{2A} adenosine receptor enhances pulmonary inflammation, mucin production, and angiogenesis in adenosine deaminase-deficient mice. *Am J Physiol Lung Cell Mol Physiol* 293: L753–L761, 2007.
102. Martin TR, Frevert CW. Innate immunity in the lungs. *Proc Am Thorac Soc* 2: 403–11, 2005.
103. Zaas AK, Schwartz DA. Innate immunity and the lung: Defense at the interface between host and environment. *Trends Cardiovasc Med* 15: 195–202, 2005.
104. Akira S, Uematsu S, Takeuchi O. Pathogen recognition and innate immunity. *Cell* 124: 783–801, 2006.
105. Shpacovitch V, Feld M, Hollenberg MD, Luger TA, Steinhoff, M. Role of protease-activated receptors in inflammatory responses, innate and adaptive immunity. *J Leukoc Biol* 83: 1309–22, 2008.
106. Wilmanski JM, Petnicki-Ocwieja T, Kobayashi KS. NLR proteins: Integral members of innate immunity and mediators of inflammatory diseases. *J Leuk Biol* 83: 13–30, 2008.
107. Takeuchi O, Akira S. MDA5/RIG-I and virus recognition. *Curr Opin Immunol* 20: 17–22, 2008.
108. Wu Q, Lu Z, Verghese MW, Randell SH. Airway epithelial cell tolerance to Pseudomonas aeruginosa. *Respir Res* 6: 26, 2005.
109. Kato A, Schleimer RP. Beyond inflammation: Airway epithelial cells are at the interface of innate and adaptive immunity. *Curr Opin Immunol* 19: 711–20, 2007.
110. Hammad H, Lambrecht BN. Dendritic cells and epithelial cells: Linking innate and adaptive immunity in asthma. *Nat Rev Immunol* 8: 193–204, 2008.
111. Schmidt-Weber CB, Akdis M, Akdis CA. TH17 cells in the big picture of immunology. *J Allergy Clin Immunol* 120: 247–54, 2007.
112. Cohn L, Homer RJ, MacLeod H, Mohrs M, Brombacher F, Bottomly K. Th2-induced airway mucus production is dependent on IL4Ralpha, but not on eosinophils. *J Immunol* 162: 6178–83, 1999.
113. Gavett SH, O'Hearn DJ, Karp CL, Patel EA, Schofield BH, Finkelman FD *et al.* Interleukin-4 receptor blockade prevents airway responses induced by antigen challenge in mice. *Am J Physiol Lung Cell Mol Physiol* 272: L253–L261, 1997.
114. Aujla SJ, Chan YR, Zheng M, Fei M, Askew DJ, Pociask DA *et al.* IL-22 mediates mucosal host defense against Gram-negative bacterial pneumonia. *Nat Med* 14: 275–81, 2008.
115. Pilette C, Durham SR, Vaerman JP, Sibille Y. Mucosal immunity in asthma and chronic obstructive pulmonary disease: A role for immunoglobulin A? *Proc Am Thorac Soc* 1: 125–35, 2004.

17 CHAPTER

Airway Mucus Hypersecretion in Asthma and COPD: Not the Same?

Duncan F. Rogers

Section of Airway Disease, National Heart & Lung Institute, Imperial College, London, UK

INTRODUCTION

Patients with asthma or chronic obstructive pulmonary disease (COPD) invariably exhibit characteristics of airway mucus hypersecretion, comprising sputum production [1, 2], excessive mucus in the airway lumen [3] (Figs. 17.1 and 17.2), goblet cell hyperplasia [3–6] and submucosal gland hypertrophy [3, 4] (Fig. 17.3). The pathophysiological sequelae of mucus hypersecretion are airway obstruction, airflow limitation, ventilation-perfusion mismatch, and impairment of gas exchange. In addition, compromised mucociliary function, with reduced mucus clearance, can encourage bacterial colonization leading to repeated chest infections and exacerbations, particularly in COPD. In COPD, a city-wide epidemiological study in Copenhagen showed that a number of clinical parameters, including deterioration in lung function, risk of hospitalization and death, were increased in patients with COPD who had chronic mucus hypersecretion compared with patients with either no sputum production or with only a small amount of intermittent sputum production [7] (Fig. 17.4). In addition, patients who had chronic airway mucus hypersecretion and were also susceptible to chest infections showed increased mortality and morbidity compared with patients with mucus hypersecretion but without infections (Fig. 17.4). In contrast, there is much less information on the contribution of mucus to pathophysiology and clinical symptoms in asthma. However, excess mucus may not only obstruct the airway lumen, but also contribute to the development of airway hyperresponsiveness [8]. The latter possibilities led to the suggestion that chronic mucus hypersecretion reflects lack of asthma control, leading in turn to accelerated loss of lung function and increased mortality [7].

The association of airway mucus hypersecretion with asthma and COPD is not necessarily cause and effect. The excess mucus may merely be a result of the inflammatory processes causing the disease. Consequently, treating the mucus problem will not inevitably treat the disease, especially if the changes that have led to increased mucus production are irreversible. Nevertheless, as discussed above, mucus hypersecretion does contribute to clinical symptoms in certain groups of patients with asthma and COPD. This suggests that it is important to develop drugs that address the airway mucus problem in these patients. The first issue pertaining to the rational development of drugs to treat airway mucus hypersecretion is to understand the nature of the underlying disease processes in asthma and COPD and how these relate to the mucus hypersecretory phenotype in each condition. Secondly, it is important to know which aspect(s) of mucus hypersecretion should be targeted.

The present chapter addresses the above issues by discussion of similarities and differences in the pulmonary inflammatory "profile" between asthma and COPD and how these may relate to the generation of the airway mucus hypersecretory state in the two conditions. To set these discussions in context, the chapter begins with consideration of sputum, airway mucus, and mucins, followed by an outline of the characteristics of mucus hypersecretion in asthma and COPD, emphasizing differences between the two conditions and the theoretical requirements for drugs aimed at effectively inhibiting airway mucus hypersecretion.

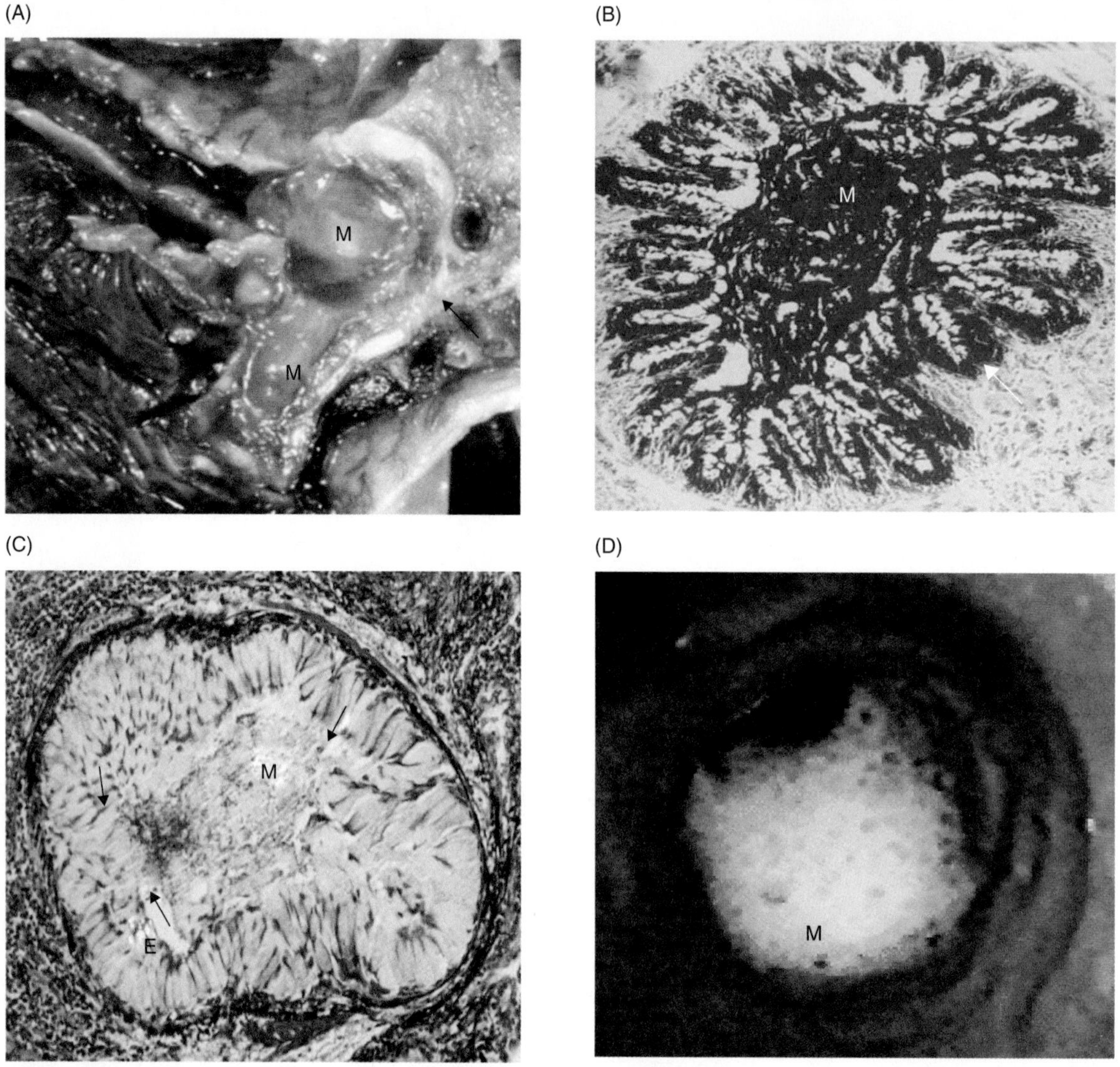

FIG. 17.1 Mucus obstruction of the airways in asthma and COPD. (A) Mucus plugging in asthma. Complete occlusion by mucus plugs (M) of an intrapulmonary bronchus (arrow), cut in longitudinal section, in a patient who died of an acute-severe asthma attack. (B) Bronchoconstriction and luminal mucus in fatal asthma. Intrapulmonary airway (transverse section) of a patient who died of an acute-severe asthma attack showing airway epithelium (arrow) thrown into folds by smooth muscle contraction, and occlusion by mucus (M) of remaining luminal space. This relatively small amount of mucus would not be expected to significantly reduce airflow in the relaxed, non-constricted airway. (C) Airway mucus "tethering" in asthma. Transverse section through a bronchiole of a patient with asthma showing incomplete release, or "continuity," of secreted mucin from airway epithelial goblet cells (arrows). Similar appearances of mucus tethering are not observed in the airways in COPD. (D) Mucopurulent secretions (M) in an intrapulmonary bronchus in a patient with COPD.

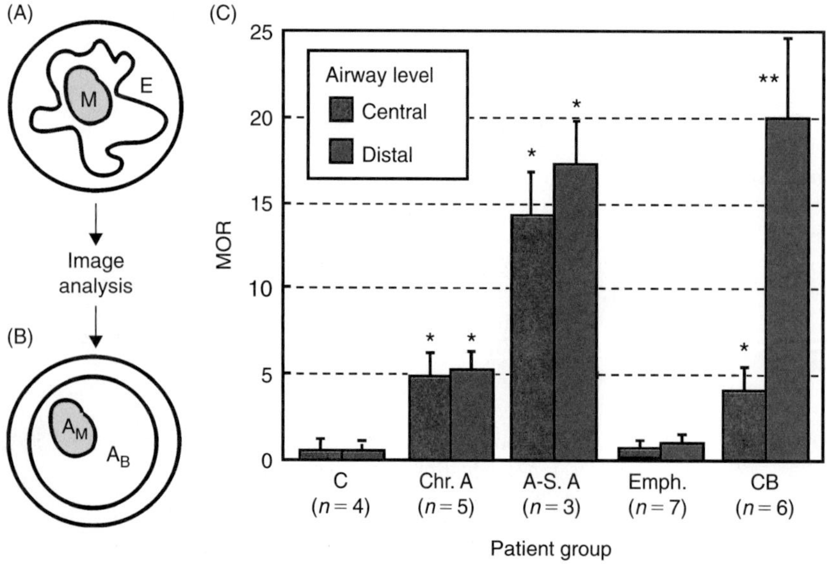

FIG. 17.2 Relative amounts of airway luminal mucus in asthma and COPD. (A) Lungs from patients dying either of causes other than lung disease (controls, C), or from chronic asthma (Chr. A), acute-severe asthma (A-S. A), emphysema (Emph.), or chronic bronchitis (CB), were cut in histological section and stained with hematoxylin–eosin and elastica–Goldner stains to visualize mucin (M). The stained sections were viewed by light microscopy and intrapulmonary airways assessed for amount of luminal mucus. E: bronchial epithelium. (B) Airways of interest were subjected to computer-based image analysis. The irregular perimeter of the bronchial epithelium was digitally converted to a circle, from which the area of mucus (A_M) was expressed as a ratio of the area of the bronchus (A_B), to give a mucus-occupying ratio, MOR (A_M/A_B), as shown in panel C. $^*p < 0.01$, $^{**}p < 0.01$ compared with controls. Redrawn from data in Refs [3, 4].

SPUTUM

Coughing or "hawking" to bring up "phlegm" is a sign of respiratory disease [2, 9]. It indicates airway mucus hypersecretion, coupled with impaired mucus clearance and mucus retention, and is a feature of many patients with asthma [10] or COPD [11]. The expectorated excessive secretions are referred to as sputum. The characteristics of sputum change with infection and inflammation. Inflammation leads to mucus hypersecretion, ciliary dysfunction, and changes in the composition and biophysical properties of airway secretions [12]. Inflammatory cells, particularly neutrophils, which are recruited to the airway to combat infection, disappear from the airway either through programmed cell death (apoptosis) or by necrosis. Necrotic neutrophils release proinflammatory mediators that damage the epithelium and recruit more inflammatory cells. They also release DNA [13] and filamentous actin (F-actin) from the cytoskeleton. DNA and F-actin copolymerize to form a second rigid network within airway secretions [14, 15]. Neutrophil-derived myeloperoxidase imparts a characteristic green color to inflamed airway secretions, which are termed mucopurulent (17.1). Excessive infection turns the secretions dark yellow, green, or brown, and these are termed purulent.

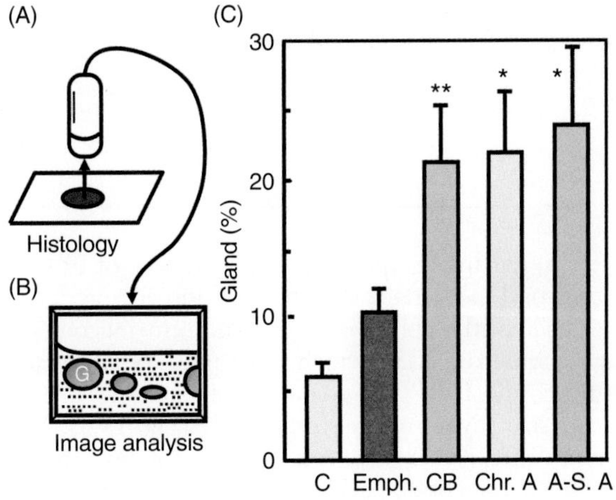

FIG. 17.3 Relative size of airway submucosal glands in asthma and COPD. (A) Lungs from patients dying either of causes other than lung disease (controls, C), or from chronic asthma (Chr. A), acute-severe asthma (A-S. A), emphysema (Emph.), or chronic bronchitis (CB), were cut in histological section and stained with hematoxylin–eosin and elastica–Goldner stains to visualize the submucosal glands. The stained sections were viewed by light microscopy and intrapulmonary airways assessed for size of glands. (B) Airways of interest were subjected to computer-based image analysis, and the area proportion of glands (G) expressed as a percentage of the wall area, as shown in panel C. $^{*}p < 0.01$, $^{**}p < 0.01$ compared with controls. Redrawn from data in Refs [3, 4].

AIRWAY "MUCUS"

Airway mucus is a complex dilute aqueous solution of lipids, glycoconjugates, and proteins. It comprises salts, enzymes and anti-enzymes, oxidants and antioxidants, exogenous bacterial products, endogenous antibacterial secretions, cell-derived mediators and proteins, plasma-derived mediators and proteins, and cell debris such as DNA. Airway mucus is considered to form a liquid bi-layer whereby an upper gel layer floats above a lower, more watery sol, or periciliary liquid, layer [16] (Fig. 17.5). There may be a thin layer of surfactant lying in between, and separating, the gel and sol layers [17]. The functions of the sol layer are debated, but are presumed to include "lubrication" of the beating cilia. The surfactant layer might facilitate spreading of the gel layer over the epithelial surface. The gel layer traps particles and is moved on the tips of the beating cilia. The inhaled particles are trapped in the sticky gel layer

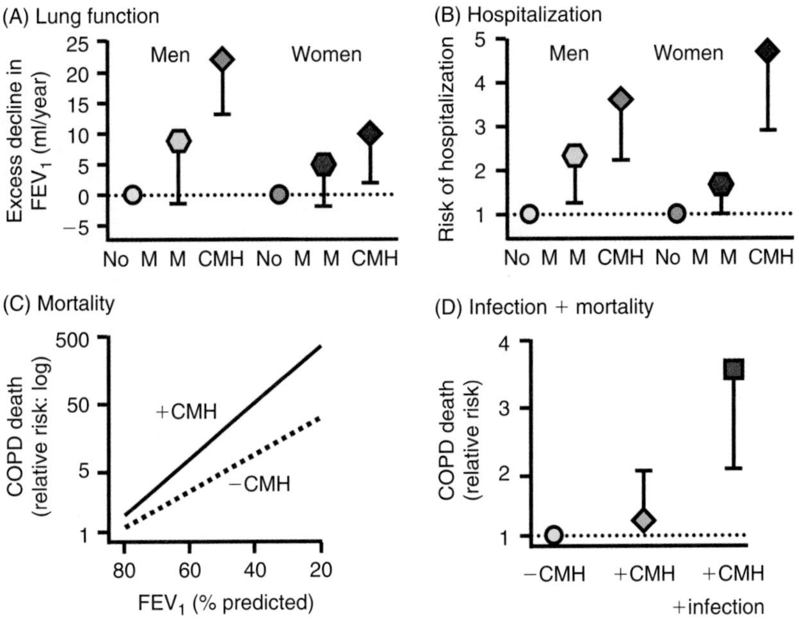

FIG. 17.4 Clinical impact of airway mucus hypersecretion in COPD. The Copenhagen Heart Study [7] showed that patients with chronic mucus hypersecretion (CMH), as defined by long-term phlegm production, had an accelerated decline in lung function (A), increased risk of hospitalization (B), increased risk of death (C), and a further increased risk of death when linked with pulmonary infection (D), when compared with subjects without mucus production (−CMH) or patients with some sputum production but without infection. No M: no sputum (mucus) production; M: non-chronic sputum (mucus) production. Redrawn after data in Ref. [7].

and are removed from the airways by mucociliary clearance. When the mucus reaches the throat, it is either swallowed and delivered to the gastrointestinal tract for degradation or, if excessive, as in respiratory disease, it is coughed out [18].

Respiratory tract mucus requires the correct combination of viscosity and elasticity for optimal efficiency of ciliary interaction [19, 20]. Viscosity is a liquid-like characteristic and is the resistance to flow and the capacity to absorb energy when moving. Elasticity is a solid-like property and is the capacity to store the energy used to move or deform it. Viscoelasticity confers a number of properties upon the mucus that allow effective interaction with cilia. These properties have been variously described in terms of spinnability, adhesiveness, and wettability [21]. An important characteristic of mucus is that it is non-Newtonian: its viscosity decreases as the applied force increases [22]. Consequently, the ratio of stress to rate of strain is non-linear, with the result that the more forcefully the cilia beat, the more easily the mucus moves. Viscoelasticity is conferred on the mucus primarily by high molecular weight mucous glycoproteins, termed mucins.

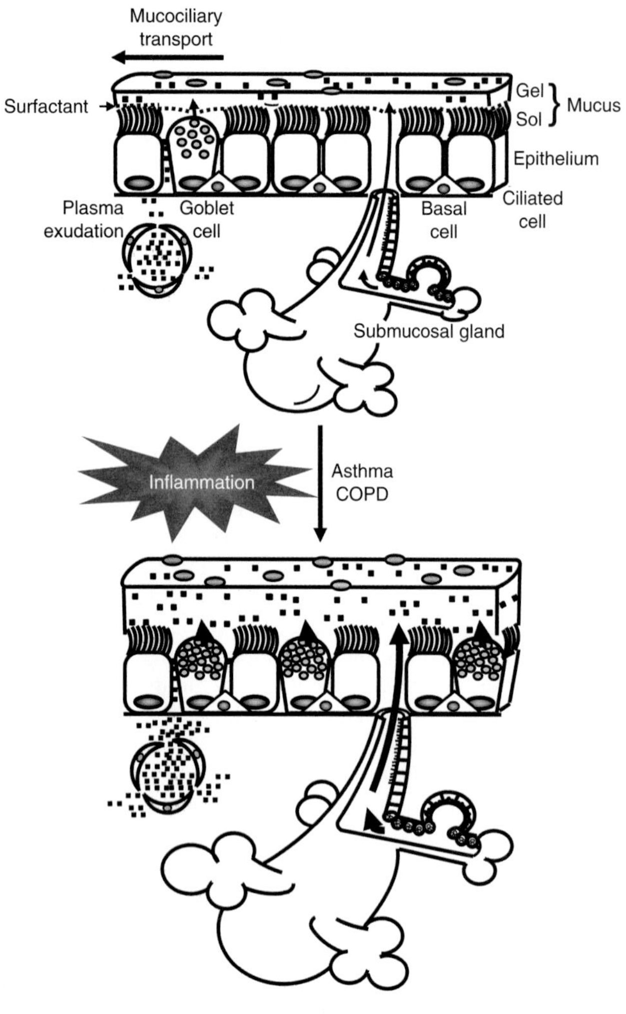

FIG. 17.5 Airway mucus secretion and hypersecretion. Upper panel: In healthy airways, mucus forms a bi-layer over the epithelium, with surfactant (dotted line) separating the gel and sol layers. Mucins secreted by goblet cells and submucosal glands confer viscoelasticity on the mucus which facilitates mucociliary clearance of inhaled particles and irritants. Mucus hydration is regulated by salt (and, hence water) flux across the epithelium: the glands also secrete water. Plasma proteins exuded from the tracheobronchial microvasculature bathe the submucosa and contribute to the formation of the mucus. The above processes are under the control of nerves and regulatory mediators. Lower panel: Airway inflammation (in asthma and COPD) induces changes associated with a mucus hypersecretory phenotype, including increased plasma exudation (more prominent in asthma than COPD), goblet cell hyperplasia, via differentiation from basal cells, and associated increased mucus synthesis and secretion, and submucosal gland hypertrophy (with associated increased mucus production), leading to increased luminal mucus (and airway obstruction).

RESPIRATORY TRACT MUCINS

In health, mucins comprise up to 2% by weight of the airway mucus [23]. In the airways, mucins are produced by goblet cells in the epithelium [24] and sero-mucous glands in the submucosa [25], which in respiratory diseases such as asthma and COPD may increase in number (hyperplasia) and size (hypertrophy) respectively (Fig. 17.5). Mucins are long, thread-like, complex glycoconjugates. They consist of a linear peptide backbone (termed apomucin), which is encoded by specific mucin (MUC) genes (see below), to which hundreds of carbohydrate side-chains are O-linked, but also with additional N-linked glycans [26]. The glycosylation pattern is complex and extremely diverse [27], and is associated with complimentary motifs on bacterial cell walls, thereby facilitating broad-spectrum bacterial attachment and subsequent clearance [28, 29]. Within the main protein core are variable numbers of tandemly repeated serine- and/or threonine-rich regions which are unique in size and sequence for each mucin [18], and represent sites for mucin glycosylation. These complex glycoproteins are polydisperse, linear polymers that can be fragmented by reduction to give monomers termed "reduced subunits" [30–33]. There are at least two structurally and functionally distinct classes of mucin, namely the membrane-associated mucins and the secreted (gel-forming or non-gel-forming) mucins (Table 17.1). Membrane-tethered mucins, which have a hydrophobic domain that anchors the mucin in the plasma membrane, contribute to the composition of the cell surface [18]. Secretory mucins are stored intracellularly in secretory granules and are released at the apical surface of the cell in

TABLE 17.1 Human mucin genes expressed in the airways and lung.

Classification of gene product	Genes
Membrane associated	MUC1, MUC4, MUC11, MUC13, MUC20
Secreted, cysteine rich (gel forming)	MUC2, MUC5AC, MUC5B, MUC19
Secreted, cysteine poor	MUC7

response to stimuli. It would appear that mucus production is such a fundamental homeostatic process that virtually all acute interventions examined trigger airway mucin secretion (Table 17.2). In addition, many of these same mediators when administered more chronically not only induce mucin secretion but also upregulate mucin gene expression, with concomitant increases in mucin synthesis and goblet cell hyperplasia (Table 17.2).

TABLE 17.2 Inducers of airway mucus secretion, goblet cell hyperplasia and mucin (MUC) gene expression/mucin synthesis.

Stimulation	Secretion	Hyperplasia	MUC
Cytokines			
IL-1β	+	NP	NP
IL-6	+	NP	Yes
IL-9	NP	NP	Yes
IL-13 (IL-4)	+	Yes	Yes
TNF-α	++	Yes*	Yes*
Gases			
Cigarette smoke, ozone	++	Yes	Yes
Nitric oxide	−/+	NP	NP
Reactive oxygen species	0/+	NP	NP
Inflammatory mediators			
Bradykinin	+	NP	NP
Cysteinyl leukotrienes	++	NP	NP
Endothelin	0/+	NP	NP
Histamine	+	NP	NP
PAF	+	Yes*	Yes*
Prostaglandins	0/+	NP	NP
Proteinases	+++	Yes	NP
Purine nucleotides	++	NP	NP
Neuronal pathways			
Cholinergic nerves	++	NP	NP
Cholinoceptor agonists	++	Yes	NP
Nicotine	++	Yes	NP
Tachykininergic nerves	+	NP	NP
Substance P	++	NP	NP
Neurokinin A	+	NP	NP
Miscellaneous			
EGF (+TNF-α)	NP	Yes	Yes
Sensitization followed by challenge	+	Yes	Yes

Notes: +++: highly potent; ++: marked effect; +: lesser effect; 0: minimal effect; EGF: epidermal growth factor; IL: interleukin; NP: effect not published.
*Effect only observed with PAF (platelet activating factor) and TNF-α (tumor necrosis factor-α) in combination.

MUCIN GENES AND GENE PRODUCTS

Twenty human mucin (MUC) genes have so far been identified. Of these, only nine, namely MUC1, MUC2, MUC4, MUC5AC, MUC5B, MUC7, MUC11, MUC13, and MUC20 are expressed in the human respiratory tract [18] (Table 17.1). Of these, only MUC2, MUC5AC, and MUC5B, the classic gel-forming mucins, are found in airway secretions. However, only MUC5AC and MUC5B glycoproteins, localized adjacent to each other on chromosome 11p15.5, are considered the major gel-forming mucins in both normal respiratory tract secretions as well as in airway secretions from patients with respiratory diseases [34–39]. Small amounts of MUC2 may, however, be found in secretions from "irritated" airways (see below).

Advances in biochemistry, molecular biology, and biophysics mean that the structure and functions of the airway MUC gene products are becoming better characterized [23, 26]. The predicted sequences of the MUC1, MUC 4, MUC11, MUC13, and MUC20 mucins suggest they are membrane bound, with an extracellular mucin domain and a hydrophobic membrane-spanning domain (Table 17.1). In contrast, MUC2, MUC5AC, MUC5B, MUC6, and MUC7 gene products are secreted mucins (Table 17.1). The technology for studying the contribution to physiology and pathophysiology of the individual MUC gene products lags well behind that of investigation of gene expression [18]. MUC1, MUC2, and MUC8 genes are expressed in both the epithelium and submucosal glands, whereas MUC4, MUC5AC, and MUC13 are expressed primarily in the epithelium. In contrast, MUC5B and MUC7 genes are expressed primarily in the glands. Use of currently available antibodies confirms that the MUC5AC gene product is a goblet cell mucin, whilst MUC5B predominates in the glands, albeit that some MUC5AC and MUC7 is also usually present [23]. Interestingly, MUC4 mucin localizes to the ciliated cells. The mucin content of secretions from patients with hypersecretory respiratory diseases may differ from normal (see below).

MUC5AC

MUC5AC mucin, initially isolated as a tracheobronchial mucin [40], is found in airway secretions pooled from healthy individuals [35, 37]. Increased levels of MUC5AC protein have also been shown to be present in the airways of patients with asthma [5], which suggests that this mucin may contribute to the pathophysiology of asthma. MUC5AC is the main mucin produced by the goblet cells in the tracheobronchial surface epithelium. However, MUC5AC can be found highly expressed not only in human bronchial epithelium, but also in bronchial submucosal glands, nasal mucosa, gastric epithelium, endocervix epithelium, and submucosal glands. This mucin has been found to be highly oligomerized, which makes it an ideal gel-forming molecule. The expression of many genes, such as MUC5AC, in airway epithelial cells is regulated by various neurohumoral factors and inflammatory mediators (Table 17.2).

MUC5B

MUC5B mucins are also a major component of tenacious mucus plug from the lungs of a patient who died in *status asthmaticus* [38, 41] and in sputum from patients with chronic bronchitis [39], which suggests that MUC5B is a major component of lung mucus from patients with obstructive lung diseases [42]. MUC5B mucin exists as differently charged glycoforms (termed the low-charge and high-charge glycoforms) and is secreted primarily by the mucous cells in the bronchial submucosal glands [34, 38, 43, 44]. However, it has been shown that MUC5B mucins are also synthesized by goblet cells [39], and are expressed in the tracheal and bronchial glands, salivary glands, endocervix, gall bladder, and pancreas. MUC5B is unique in that it does not appear to be polymorphic.

From the above, it appears that, in healthy individuals, MUC5B is mainly expressed in the airway submucosal glands, which are restricted to the more proximal, cartilaginous airways. In contrast, MUC5AC expression is generally restricted to goblet cells in the upper and lower respiratory tracts [45, 46]. Thus, the composition of normal mucus can be altered depending on the relative contribution to the secretions of these different cellular sources [47]. In respiratory diseases associated with airway mucus hypersecretion, such as asthma and COPD, further changes in the composition of the mucus, and in the mucus secretory phenotype in general, are observed, as discussed below.

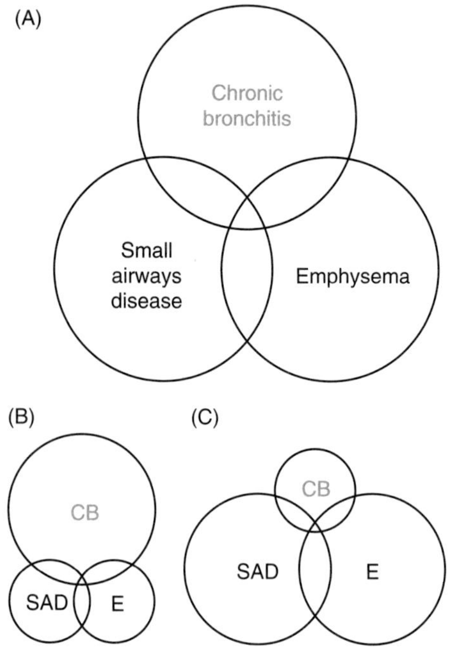

FIG. 17.6 The pathophysiological components of COPD. (A) COPD comprises three interlinked conditions, namely chronic bronchitis (CB: long-term airway mucus hypersecretion), small airways disease (SAD: also known as chronic bronchiolitis), and emphysema (E: alveolar destruction). The relative contribution to airway obstruction in any one patient of each component is invariably unclear. (B) Patient in whom mucus hypersecretion predominates (may be identifiable by excessive sputum production). (C) Patient in whom mucus hypersecretion contributes proportionally less to airflow limitation than SAD and E.

AIRWAY MUCUS HYPERSECRETORY PHENOTYPE IN COPD

COPD comprises three overlapping conditions, namely chronic bronchitis (airway mucus hypersecretion), chronic bronchiolitis (small airways disease), and emphysema (airspace enlargement due to alveolar destruction) [10] (Fig. 17.6). The following discussion considers the "bronchitic" component of COPD. The airways of patients with COPD contain excessive amounts of mucus [6], which is markedly increased above that in control subjects [3, 48]. The excessive luminal mucus is associated with increased amounts of mucus-secreting tissue. Goblet cell hyperplasia is a cardinal feature of chronic bronchitis [6], with increased numbers of goblet cells in the airways of cigarette smokers either with chronic bronchitis and chronic airflow limitation [49] or with or without productive cough [50]. Submucosal gland hypertrophy also characterizes chronic bronchitis [3, 6, 51, 52], and the amount of gland correlates with amount of luminal mucus [3].

The number of ciliated cells and the length of individual cilia is decreased in patients with chronic bronchitis [53]. Ciliary aberrations include compound cilia, cilia with an abnormal axoneme or intracytoplasmic microtubule doublets, and cilia enclosed within periciliary sheaths [54]. These abnormalities coupled with mucous hypersecretion are presumably associated with reduced mucus clearance and airway mucus obstruction in the bronchitic component of COPD.

AIRWAY MUCUS HYPERSECRETORY PHENOTYPE IN ASTHMA

Asthma is a chronic inflammatory condition of the airways characterized by variable airflow limitation that is at least partially reversible, either spontaneously or with treatment [55, 56]. It has specific clinical and pathophysiological features [57], including mucus obstruction of the airways [8]. The latter is particularly evident in a proportion of patients who die in *status asthmaticus*, where many airways are occluded by mucus plugs [58–60]. The plugs are highly viscous and comprise plasma proteins, DNA, cells, proteoglycans [61], and mucins [41, 58, 61]. Incomplete plugs are found in the airways of asthmatics who have died from causes other than asthma [62], which indicates that plug formation is a chronic, progressive process. There is also more mucus in the central and peripheral airways of both chronic and severe asthmatics compared with control subjects [4]. Analysis of asthmatic sputum indicates that the mucus comprises DNA, lactoferrin, eosinophil cationic protein, and plasma proteins such as albumin and fibrinogen [63–65], as well as mucins [5, 64, 66]. The increased amount of lumenal mucus reflects an increase in amount of airway secretory tissue, due to both goblet cell hyperplasia [4, 5] and submucosal gland hypertrophy [58], although the latter is not characteristic of all patients with asthma [4].

Airway epithelial fragility, with epithelial shedding in extreme cases, is a significant feature of asthma [67]. Shedding includes loss of ciliated cells, with presumably a concomitant reduction in mucus clearing capacity. As in COPD above, abnormalities in airway ciliated cells and cilia have been described in asthma. The ciliated cells themselves may be damaged, with loss of cilia, vacuolization of the endoplasmic reticulum and mitochondria, and microtubule damage [68–70]. These abnormalities could be caused by some of the inflammatory mediators generated in the airways of asthmatic patients, for example, eosinophil major basic protein [71]. A variety of other ciliostatic and ciliotoxic compounds are also present in asthmatic airway secretions [72].

In summary of this section, the combination of an increased amount of mucus-secreting tissue, with associated mucus hypersecretion, the production of viscid mucus, and abnormal ciliary function leads to reduced mucus clearance and the development of airway mucus obstruction in asthma.

MUCOCILIARY CLEARANCE IN ASTHMA AND COPD

Clearance of mucus from the airways is impaired in patients with a variety of respiratory diseases, including asthma and COPD [73]. However, it should be noted that there are often discrepancies in results between studies that are invariably due to differences in methodology [74, 75], but may also be due to observations made at different stages of disease.

COPD

Mucus clearance is generally considered to be impaired in patients with COPD [73]. However, the validity of these studies is dependent upon patient selection and the exclusion of patients with asthma. For example, patients classified as having obstructive chronic bronchitis and with a bronchial reversibility of less than 15% had slower lung mucus clearance than patients with reversibility greater than 15% and who, therefore, were likely to be asthmatic [76]. Nevertheless, mucus clearance is significantly reduced in heavy smokers [77] and in patients with chronic bronchitis [78]. Lung mucus clearance differs between patients with chronic airway obstruction, with or without emphysema [79]. Both groups of patients were smokers or ex-smokers and had productive cough, but lung elastic recoil pressure was reduced in the emphysema group. Mucus clearance from central airways was similar. In contrast, clearance from the peripheral lung was faster in the emphysema group than in the patients without emphysema. Importantly, forced expirations and cough markedly increased peripheral clearance in the non-emphysema group but not in the emphysema group. Comparable findings in a subsequent study led to the suggestion that cough compensates relatively effectively for decreased mucus clearance in patients with chronic bronchitis [80]. Conversely, cough is not so effective in COPD patients with impaired lung elastic recoil.

Asthma

Airway mucociliary clearance is well documented as being impaired in asthma [72]. Clearance is impaired even in patients in remission [81] and in those with mild stable disease [82]. Mucus clearance is proportionally reduced in symptomatic asthmatics [83] and during exacerbations [84]. In addition, the normal slowing of mucus clearance during sleep is more pronounced in asthmatic patients [85, 86], and this could be a contributory factor in nocturnal asthma. The mechanisms underlying the reduced mucus clearance in asthma are not clearly defined, although airway inflammation is considered to be a major contributor [72].

MECHANISMS OF DEVELOPMENT OF AIRWAY GOBLET CELL HYPERPLASIA

Airway goblet cell hyperplasia is a predominant feature of asthma and COPD (see above), and is an often-used end point in animal models of respiratory disease [87]. The cellular composition of the airway epithelium can alter both by cell division and by differentiation of one cell into another [88]. There are at least eight cell types in the airway epithelium of the conducting airways. Of these, the basal, serous, and Clara cells are considered progenitor cells with the capacity to undergo division followed by differentiation into "mature" ciliated or goblet cells. In specific experimental conditions, for example, exposure to cigarette smoke, goblet cell division contributes in part to the hyperplasia. However, differentiation of non-granulated airway epithelial cells is a major route for production of new goblet cells [88–90]. In experimental animals, production of goblet cells is usually at the "expense" of the progenitor cells, most notably serous and Clara cells, which decrease in number as goblet cell numbers increase. Serous-like cells and Clara cells are found in macroscopically normal bronchioles in human lung [91]. Whether there is a reduction in number in respiratory disease is not reported, but merits investigation. Reduction in the relative proportion of serous and Clara cells has pathophysiological significance because they produce a number of antiinflammatory, immunomodulatory, and antibacterial molecules vital to host defence [92, 93]. For example, serous cells produce lysozyme, lactoferrin, the secretory component of IgA, peroxidases, and at least two protease inhibitors. Clara cells produce Clara cell 10-kDa protein, also known as uteroglobulin, Clara cell 55-kDa protein, Clara cell tryptase, β-galactoside-binding lectin, possibly a specific phospholipase, and surfactant proteins A, B, and D. Thus, in respiratory diseases associated with airway mucus hypersecretion it seems that not only is there goblet cell hyperplasia, with associated mucus hypersecretion, but also a reduction in serous and Clara cells, with concomitant potential for impaired host defence.

DIFFERENCES IN MUCUS HYPERSECRETORY PHENOTYPE BETWEEN ASTHMA AND COPD

In order to develop appropriate models of airway mucus obstruction and devise drugs to aid mucus clearance, it is necessary to understand the similarities and differences in the features of mucus obstruction for different hypersecretory conditions. There are a number of differences in the pathophysiology of airway mucus hypersecretion between asthma and COPD (Fig. 17.7).

Airway inflammation in asthma and COPD

Any differences in airway mucus hypersecretory phenotype are presumably related, either directly (genetic predisposition) or indirectly (environmental exposure), or a combination of the two, to differences in the airway inflammatory phenotype. Thus, although the underlying pulmonary inflammation of asthma and COPD shares many common features, there are specific characteristics unique to each condition [94–96]. These differences in turn may contribute to differences in pathophysiology of airway mucus hypersecretion between the two conditions.

Asthma is invariably an allergic disease that affects the airways, rather than the lung parenchyma, and is characterized by Th2-lymphocyte orchestration of pulmonary eosinophilia. The reticular layer beneath the basement membrane is markedly thickened and the airway epithelium is fragile, features not usually associated with COPD. The bronchial inflammatory infiltrate comprises activated T-cells (predominantly $CD4^+$ cells) and eosinophils. Neutrophils are generally sparse in stable disease. In contrast, COPD is currently perceived as predominantly a neutrophilic disorder governed largely by macrophages and epithelial cells. It is associated primarily with cigarette smoking. Three conditions comprise COPD, namely mucous hypersecretion, bronchiolitis, and emphysema. The latter two features are not associated with asthma. In addition, and in contrast to asthma, $CD8^+$ T-lymphocytes predominate and pulmonary eosinophilia is generally associated with exacerbations.

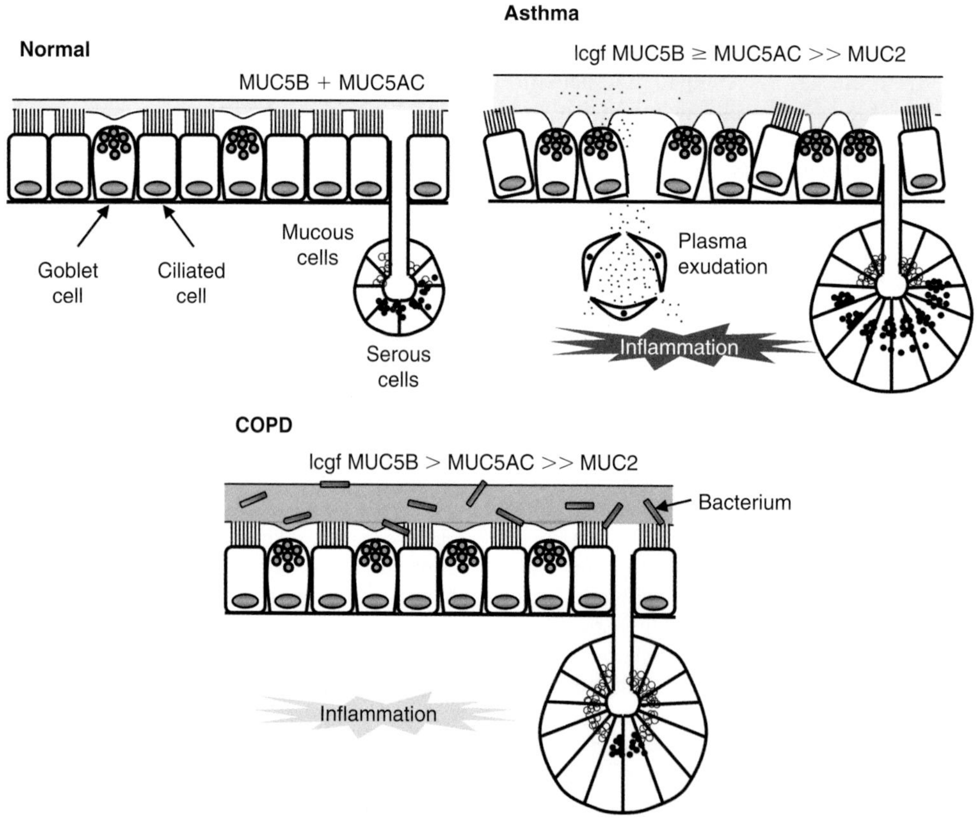

FIG. 17.7 Putative differences in the airway mucus hypersecretory phenotype between asthma and COPD. Compared with normal, in asthma, there is airway inflammation (predominantly eosinophils and Th2 lymphocytes), an increased amount of luminal mucus with an increased content of MUC5AC and MUC5B mucins, with the probability of an increased ratio of the low-charge glycoform (lcgf) of MUC5B to MUC5AC, the appearance of small amounts of MUC2 in the secretions, epithelial "fragility" with loss of ciliated cells, marked goblet cell hyperplasia, submucosal gland hypertrophy (although, in contrast to COPD – see below – without a marked increase in mucous cell to serous cell ratio), "tethering" of mucus to goblet cells, and plasma exudation. In COPD, there is airway inflammation (predominantly macrophages and neutrophils), increased luminal mucus, increased amounts of MUC5AC and MUC5B mucins, an increased ratio of lcgf MUC5B to MUC5AC above that in asthma, small amounts of MUC2, goblet cell hyperplasia, submucosal gland hypertrophy (with an increased proportion of mucous to serous cells), and a susceptibility to bacterial infection. Many of these differences require further investigation in greater number of subjects.

Both asthma and COPD have a characteristic "portfolio" of inflammatory mediators and enzymes, many of which differ between the two conditions [97, 98]. At a simplified level, histamine, interleukin (IL)-4, and eotaxin are associated with asthma, whilst IL-8, neutrophil elastase, and matrix metalloproteinases are associated with COPD. Thus, there are specific differences between asthma and COPD in their airway inflammation and remodeling. These differences may in turn exert different influences on the development of airway mucus obstruction in the two conditions (Fig. 17.7).

Airway mucus in asthma and COPD

Airway mucus in asthma is more viscous than in COPD, with the airways of asthmatic patients tending to develop, and subsequently become blocked by, gelatinous "mucus" plugs [8]. Whether or not mucus in asthma has an intrinsic biochemical abnormality is unclear. In general terms, sputum from patients with asthma is more viscous than that from patients with chronic bronchitis [65, 99, 100]. Mucus plugs in asthma differ from airway mucus gels in chronic bronchitis in that they are stabilized by non-covalent interactions between extremely large mucins assembled from conventional sized subunits [41]. This suggests an intrinsic abnormality in the mucus due to a defect in assembly of the mucin molecules, and could account for the increased viscosity of the mucus plugs in asthma. Plug formation may also be due, at least in part, to increased airway plasma exudation in asthma compared with COPD [101, 102]. In addition, in direct contrast to COPD, exocytosed mucins in asthma are not released fully from the goblet cells, leading to "tethering" of luminal mucins to the airway epithelium [103]. This tethering may also contribute to plug formation. One explanation of mucus tethering is that neutrophil proteases, the predominant inflammatory cell in COPD [10], cleave goblet cell-attached mucins. In asthma, the inflammatory cell profile, predominantly airway eosinophilia [57], does not generate the appropriate proteases to facilitate mucin release.

Mucin species in asthma and COPD

Different mucin species, or at least different proportions of these species, appear to be present in respiratory tract secretions in COPD and asthma. MUC5AC and a low-charge glycoform of MUC5B are the major mucin species in airway secretions from patients with COPD and asthma [35–37, 39, 47]. There is significantly more of the low-charge glycoform of MUC5B in the respiratory diseases than in normal control secretions [47]. An interesting difference between the disease conditions is that there is a proportional increase in the MUC5B mucin over the MUC5AC mucin in airway secretions from patients with COPD compared with secretions from patients with asthma [37]. The above data require confirmation in more samples. The significance of the change in MUC5B glycoforms between the different diseases is unclear. However, it may relate to differences in propensity of bacterial colonization of the lungs. It is noteworthy that it is COPD, a disease in which patients are prone to infection [11], that has the proportional reduction in serous cells, rather than asthma, a condition in which patients are not so notably prone to chest infection.

Goblet cells in asthma and COPD

In contrast to normal airways, goblet cells in the airways from patients with COPD contain not only MUC5AC but also MUC5B [39, 104] and MUC2 [23, 105]. This distribution is different to that in the airways of patients with asthma, where MUC5AC and MUC5B show a similar histological pattern to normal controls [106, 107]. It is noteworthy that although MUC2 is located in goblet cells in irritated airways, and MUC2 mRNA is found in the airways of smokers [48], MUC2 mucin is either not found in airway secretions from normal subjects or patients with chronic bronchitis [35], or is found only in very small amounts in asthma and COPD [47, 108]. The significance of the above combined observations is unclear, but suggests that there are differences in goblet cell phenotype between asthma and COPD.

Submucosal glands in asthma and COPD

A notable difference between asthma and COPD is in the bronchial submucosal glands [109]. In asthma, although hypertrophied, the glands are morphologically normal with an even distribution of mucous and serous cells. In contrast, in chronic bronchitis, gland hypertrophy is characterized by a markedly increased number of mucous cells relative to serous cells, particularly in severe bronchitis. The reduction in number of gland serous cells may have clinical significance. The serous cells are a rich source of antibacterial enzymes such as lysozyme and lactoferrin [25]. Thus, the airway mucus layer in COPD patients may have a reduced antibacterial capacity compared with that in asthma. This reduction, coupled with the change in MUC5B glycoforms in COPD (see above), could further explain, at least in part, the much higher incidence of bacterial chest infections in COPD compared with asthma.

Ciliated cells in asthma and COPD

It is not clear whether or not there are differences in the airway ciliary abnormalities between COPD and asthma (see above). However, epithelial fragility and shedding are features of asthma rather than COPD [95], which suggests that there may be greater loss of, and damage to, the ciliated cells in asthma compared with COPD.

From the above, it may be seen that there are theoretical and actual differences in the nature of airway mucus obstruction between COPD and asthma. How these relate to pathophysiology and clinical symptoms in the two conditions is, for the most part, unclear. However, these dissimilarities indicate that different treatments are required for effective treatment of airway mucus obstruction in different respiratory diseases.

POTENTIAL THERAPIES FOR AIRWAY MUCUS HYPERSECRETION

Current and potential therapies for asthma and COPD are discussed elsewhere in this volume. However, there are a number of avenues for specific treatment of airway mucus hypersecretion that are worthy of brief mention. Some of the mechanisms for development of airway goblet cell hyperplasia and the associated mucus hypersecretory phenotype in asthma and COPD are becoming clearer [8, 110]. Many regulatory and inflammatory mediators and enzymes increase mucus secretion and induce MUC gene expression, mucin synthesis, and goblet cell hyperplasia in experimental systems (Table 17.2). These mediators are intermediates in a cascade of pathophysiological events leading from initiating factors (such as allergen exposure in asthma or cigarette smoking in COPD) setting up a chronic inflammatory/repair response, which in turn leads to mucus hypersecretion and associated airway obstruction and clinical symptoms (Figs 17.5–17.7). A small number of key molecules may be involved in translating the actions of the different inflammatory mediators into airway mucus hypersecretion, namely epidermal growth factor (EGF) and its receptor tyrosine kinase signaling pathway [111], the mitogen-activated kinase kinase and extracellular signal-regulated kinase (MEK/ERK) pathway [112], calcium-activated chloride (CLCA) channels [113, 114], and the retinoic acid receptor (RAR)-α signaling pathway [115]. A wide variety of small molecule antagonists and inhibitors of these pathways are currently in pharmacotherapeutic development [12]. In addition, inhibition of the process of mucin exocytosis, with concomitant inhibition of secretion and, thereby, mucus hypersecretion, is currently receiving attention from two different directions. Firstly, MARCKS protein is a chaperone molecule essential for movement of newly synthesized mucin-containing granules to the apical surface of mucin-secreting cells [116]. Inhibition of MARCKS by a MARCKS-related peptide inhibits mucin secretion *in vitro* in human airway epithelial secretory cells [116] and mucus hypersecretion in a mouse model of asthma [117]. Secondly, SNARE complex proteins are essential for the "docking" of mucin granules to the internal apical surface of mucin-secreting cells. We have developed a fully recombinant fusion protein comprising the translocation domain and endopeptidase of *Clostridium botulinum*, which normally enters nerves and cleaves SNARE proteins mediating neurotransmission [118], with EGF to target EGF receptors on airway mucus-secreting cells. This molecule now cleaves SNARE proteins linking mucin granules with mucous cell apical membranes, and thereby inhibits mucin secretion from human respiratory epithelial cells [119] and mucus hypersecretion in a rat model of COPD [120].

CONCLUSIONS

Production of airway mucus is a vital homeostatic mechanism that protects the respiratory tract from a barrage of inhaled insult. However, abnormal production of mucus can contribute to respiratory disease. Airway obstruction by mucus is a common feature of a number of severe respiratory conditions, including asthma and COPD. These diseases share pulmonary inflammation and remodeling as a pathophysiological characteristic, although with differing profiles of inflammatory cells and mediators. One aspect of the remodeling is airway mucus hypersecretion, associated with the shared features of goblet cell hyperplasia and submucosal gland hypertrophy. They also each have a number of unique features that characterize their airway mucus obstruction. For example, plasma exudation, mucus plug formation, and mucus tethering are features of asthma, whereas submucosal gland hypertrophy with a disproportionate increase in the ratio of mucous to serous cells is a significant feature in COPD. Understanding of the relative importance of the differences and similarities in the pathophysiology of the different mucus hypersecretory phenotypes between these two respiratory diseases should lead to rational development of pharmacotherapeutic interventions. At present, there is considerable interest in inhibitors of EGF receptor tyrosine kinases, MEK/ERK and CLCA channels to suppress goblet cell hyperplasia, and in molecules to selectively interfere with mucin exocytosis (such as MARCKS-related peptide and retargeted clostridial endopeptidases) and thereby inhibit airway mucus hypersecretion.

References

1. Medical Research Council. Definition and classification of chronic bronchitis for clinical and epidemiological purposes. *Lancet* i: 775–79, 1965.
2. Openshaw PJ, Turner-Warwick M. Observations on sputum production in patients with variable airflow obstruction; implications for the diagnosis of asthma and chronic bronchitis. *Respir Med* 83: 25–31, 1989.
3. Aikawa T, Shimura S, Sasaki H, Takishima T, Yaegashi H, Takahashi T. Morphometric analysis of intraluminal mucus in airways in chronic obstructive pulmonary disease. *Am Rev Respir Dis* 140: 477–82, 1989.
4. Aikawa T, Shimura S, Sasaki H, Ebina M, Takishima T. Marked goblet cell hyperplasia with mucus accumulation in the airways of patients who died of severe acute asthma attack. *Chest* 101: 916–21, 1992.
5. Ordonez CL, Khashayar R, Wong HH, Ferrando R, Wu R, Hyde DM, Hotchkiss JA, Zhang Y, Novikov A, Dolganov G, Fahy JV. Mild and moderate asthma is associated with airway goblet cell hyperplasia and abnormalities in mucin gene expression. *Am J Respir Crit Care Med* 163: 517–23, 2001.
6. Reid L. Pathology of chronic bronchitis. *Lancet* i: 275–78, 1954.
7. Vestbo J. Epidemiological studies in mucus hypersecretion. *Novartis Found Symp* 248: 3–12, 2002.
8. Rogers DF. Airway mucus hypersecretion in asthma: An undervalued pathology?. *Curr Opin Pharmacol* 4: 241–50, 2004.
9. Turner-Warwick M, Openshaw P. Sputum in asthma. *Postgrad Med J* 63(Suppl 1): 79–82, 1987.
10. Global Initiative for Asthma. Global strategy for asthma management and prevention. www.ginasthma.org, 2006.
11. Global Initiative for Chronic Obstructive Lung Disease. Global strategy for the diagnosis, management, and prevention of chronic obstructive pulmonary disease. www.goldcopd.org, 2006.
12. Rogers DF, Barnes PJ. Treatment of airway mucus hypersecretion. *Ann Med* 38: 116–25, 2006.
13. Lethem MI, James SL, Marriott C, Burke JF. The origin of DNA associated with mucus glycoproteins in cystic fibrosis sputum. *Eur Respir J* 3: 19–23, 1990.

14. Shah SA, Santago P, Rubin BK. Quantification of biopolymer filament structure. *Ultramicroscopy* 104: 244–54, 2005.
15. Tomkiewicz RP, Kishioka C, Freeman J, Rubin BK. DNA and actin filament ultrastructure in cystic fibrosis sputum. In: Baum G (ed.), *Cila, Mucus and Mucociliary Interactions*, pp. 333–41. New York: Marcel Dekker, 1998.
16. Knowles MR, Boucher RC. Mucus clearance as a primary innate defense mechanism for mammalian airways. *J Clin Invest* 109: 571–77, 2002.
17. Morgenroth K, Bolz J. Morphological features of the interaction between mucus and surfactant on the bronchial mucosa. *Respiration* 47: 225–31, 1985.
18. Rose MC, Voynow JA. Respiratory tract mucin genes and mucin glycoproteins in health and disease. *Physiol Rev* 86: 245–78, 2006.
19. King M. Experimental models for studying mucociliary clearance. *Eur Respir J* 11: 222–28, 1998.
20. King M. Physiology of mucus clearance. *Paediatr Respir Rev* 7(Suppl 1): S212–S214, 2006.
21. Houtmeyers E, Gosselink R, Gayan-Ramirez G, Decramer M. Regulation of mucociliary clearance in health and disease. *Eur Respir J* 13: 1177–88, 1999.
22. Sleigh MA, Blake JR, Liron N. The propulsion of mucus by cilia. *Am Rev Respir Dis* 137: 726–41, 1988.
23. Davies JR, Herrmann A, Russell W, Svitacheva N, Wickström C, Carlstedt I. Respiratory tract mucins: Structure and expression patterns. *Mucus Hypersecretion in Respiratory Disease*, pp. 76–88. Chichester: John Wiley & Sons, 2002.
24. Rogers DF. Airway goblet cell hyperplasia in asthma: Hypersecretory and anti-inflammatory?. *Clin Exp Allergy* 32: 1124–27, 2002.
25. Finkbeiner WE. Physiology and pathology of tracheobronchial glands. *Respir Physiol* 118: 77–83, 1999.
26. Thornton DJ, Rousseau K, McGuckin MA. Structure and function of the polymeric mucins in airways mucus. *Annu Rev Physiol*, 70: 459–486, 2008.
27. Hanisch FG. O-glycosylation of the mucin type. *Biol Chem* 382: 143–49, 2001.
28. Dell A, Morris HR. Glycoprotein structure determination by mass spectrometry. *Science* 291: 2351–56, 2001.
29. Moniaux N, Escande F, Porchet N, Aubert JP, Batra SK. Structural organization and classification of the human mucin genes. *Front Biosci* 6: D1192–206, 2001.
30. Sheehan JK, Thornton DJ, Somerville M, Carlstedt I. Mucin structure. The structure and heterogeneity of respiratory mucus glycoproteins. *Am Rev Respir Dis* 144: S4–9, 1991.
31. Thornton DJ, Davies JR, Kraayenbrink M, Richardson PS, Sheehan JK, Carlstedt I. Mucus glycoproteins from "normal" human tracheobronchial secretion. *Biochem J* 265: 179–86, 1990.
32. Thornton DJ, Devine PL, Hanski C, Howard M, Sheehan JK. Identification of two major populations of mucins in respiratory secretions. *Am J Respir Crit Care Med* 150: 823–32, 1994.
33. Thornton DJ, Sheehan JK, Carlstedt I. Heterogeneity of mucus glycoproteins from cystic fibrotic sputum. Are there different families of mucins?. *Biochem J* 276: 677–82, 1991.
34. Hovenberg HW, Davies JR, Carlstedt I. Different mucins are produced by the surface epithelium and the submucosa in human trachea: Identification of MUC5AC as a major mucin from the goblet cells. *Biochem J* 318: 319–24, 1996.
35. Hovenberg HW, Davies JR, Herrmann A, Linden CJ, Carlstedt I. MUC5AC, but not MUC2, is a prominent mucin in respiratory secretions. *Glycoconj J* 13: 839–47, 1996.
36. Sheehan JK, Howard M, Richardson PS, Longwill T, Thornton DJ. Physical characterization of a low-charge glycoform of the MUC5B mucin comprising the gel-phase of an asthmatic respiratory mucous plug. *Biochem J* 338: 507–13, 1999.
37. Thornton DJ, Carlstedt I, Howard M, Devine PL, Price MR, Sheehan JK. Respiratory mucins: Identification of core proteins and glycoforms. *Biochem J* 316: 967–75, 1996.
38. Thornton DJ, Howard M, Khan N, Sheehan JK. Identification of two glycoforms of the MUC5B mucin in human respiratory mucus. Evidence for a cysteine-rich sequence repeated within the molecule. *J Biol Chem* 272: 9561–66, 1997.
39. Wickstrom C, Davies JR, Eriksen GV, Veerman EC, Carlstedt I. MUC5B is a major gel-forming, oligomeric mucin from human salivary gland, respiratory tract and endocervix: Identification of glycoforms and C-terminal cleavage. *Biochem J* 334: 685–93, 1998.
40. Guyonnet DV, Audie JP, Debailleul V, Laine A, Buisine MP, Galiegue-Zouitina S, Pigny P, Degand P, Aubert JP, Porchet N. Characterization of the human mucin gene MUC5AC: A consensus cysteine-rich domain for 11p15 mucin genes?. *Biochem J* 305: 211–19, 1995.
41. Sheehan JK, Richardson PS, Fung DC, Howard M, Thornton DJ. Analysis of respiratory mucus glycoproteins in asthma: A detailed study from a patient who died in status asthmaticus. *Am J Respir Cell Mol Biol* 13: 748–56, 1995.
42. Rose MC, Nickola TJ, Voynow JA. Airway mucus obstruction: Mucin glycoproteins, MUC gene regulation and goblet cell hyperplasia. *Am J Respir Cell Mol Biol* 25: 533–37, 2001.
43. Desseyn JL, Buisine MP, Porchet N, Aubert JP, Laine A. Genomic organization of the human mucin gene MUC5B. cDNA and genomic sequences upstream of the large central exon. *J Biol Chem* 273: 30157–64, 1998.
44. Sharma P, Dudus L, Nielsen PA, Clausen H, Yankaskas JR, Hollingsworth MA, Engelhardt JF. MUC5B and MUC7 are differentially expressed in mucous and serous cells of submucosal glands in human bronchial airways. *Am J Respir Cell Mol Biol* 19: 30–37, 1998.
45. Audie JP, Janin A, Porchet N, Copin MC, Gosselin B, Aubert JP. Expression of human mucin genes in respiratory, digestive, and reproductive tracts ascertained by *in situ* hybridization. *J Histochem Cytochem* 41: 1479–85, 1993.
46. Reid CJ, Gould S, Harris A. Developmental expression of mucin genes in the human respiratory tract. *Am J Respir Cell Mol Biol* 17: 592–98, 1997.
47. Kirkham S, Sheehan JK, Knight D, Richardson PS, Thornton DJ. Heterogeneity of airways mucus: Variations in the amounts and glycoforms of the major oligomeric mucins MUC5AC and MUC5B. *Biochem J* 361: 537–46, 2002.
48. Steiger D, Fahy J, Boushey H, Finkbeiner WE, Basbaum C. Use of mucin antibodies and cDNA probes to quantify hypersecretion *in vivo* in human airways. *Am J Respir Cell Mol Biol* 10: 538–45, 1994.
49. Saetta M, Turato G, Baraldo S, Zanin A, Braccioni F, Mapp CE, Maestrelli P, Cavallesco G, Papi A, Fabbri LM. Goblet cell hyperplasia and epithelial inflammation in peripheral airways of smokers with both symptoms of chronic bronchitis and chronic airflow limitation. *Am J Respir Crit Care Med* 161: 1016–21, 2000.
50. Mullen JB, Wright JL, Wiggs BR, Pare PD, Hogg JC. Structure of central airways in current smokers and ex-smokers with and without mucus hypersecretion: Relationship to lung function. *Thorax* 42: 843–48, 1987.
51. Reid L. Measurement of the bronchial mucous gland layer: A diagnostic yardstick in chronic bronchitis. *Thorax* 15: 132–41, 1960.
52. Restrepo G, Heard BE. The size of the bronchial glands in chronic bronchitis. *J Pathol Bacteriol* 85: 305–10, 1963.
53. Wanner A. Clinical aspects of mucociliary transport. *Am Rev Respir Dis* 116: 73–125, 1977.
54. McDowell EM, Barrett LA, Harris CC, Trump BF. Abnormal cilia in human bronchial epithelium. *Arch Pathol Lab Med* 100: 429–36, 1976.
55. American Thoracic Society. Standards for the diagnosis and care of patients with chronic obstructive pulmonary disease (COPD) and asthma. *Am Rev Respir Dis* 136: 225–44, 1987.
56. British Thoracic Society. The British guidelines on asthma management. *Thorax* 52(Suppl 1): S1–21, 1997.
57. Eapen SS, Busse WW. Asthma. *Clin Allergy Immunol* 16: 325–53, 2002.
58. Dunnill MS. The pathology of asthma with special reference to changes in the bronchial mucosa. *J Clin Pathol* 13: 27–33, 1960.

59. Houston JC, De Navasquez S, Trounce JR. A clinical and pathological study of fatal cases of status asthmaticus. *Thorax* 8: 207–13, 1953.
60. Saetta M, Di Stefano A, Rosina C, Thiene G, Fabbri LM. Quantitative structural analysis of peripheral airways and arteries in sudden fatal asthma. *Am Rev Respir Dis* 143: 138–43, 1991.
61. Bhaskar KR, O'Sullivan DD, Coles SJ, Kozakewich H, Vawter GP, Reid LM. Characterization of airway mucus from a fatal case of status asthmaticus. *Pediatr Pulmonol* 5: 176–82, 1988.
62. Dunnill MS. The morphology of the airways in bronchial asthma. In: Stein M (ed.), *New Directions in Asthma*, pp. 213–21. Park Ridge: American College of Physicians, 1975.
63. Fahy JV, Liu J, Wong H, Boushey HA. Cellular and biochemical analysis of induced sputum from asthmatic and from healthy subjects. *Am Rev Respir Dis* 147: 1126–31, 1993.
64. Fahy JV, Steiger DJ, Liu J, Basbaum CB, Finkbeiner WE, Boushey HA. Markers of mucus secretion and DNA levels in induced sputum from asthmatic and from healthy subjects. *Am Rev Respir Dis* 147: 1132–37, 1993.
65. Lopez-Vidriero MT, Reid L. Chemical markers of mucous and serum glycoproteins and their relation to viscosity in mucoid and purulent sputum from various hypersecretory diseases. *Am Rev Respir Dis* 117: 465–77, 1978.
66. Lopez-Vidriero MT, Reid L. Bronchial mucus in health and disease. *Br Med Bull* 34: 63–74, 1978.
67. Bousquet J, Jeffery PK, Busse WW, Johnson M, Vignola AM. Asthma. From bronchoconstriction to airways inflammation and remodeling. *Am J Respir Crit Care Med* 161: 1720–45, 2000.
68. Beasley R, Roche WR, Roberts JA, Holgate ST. Cellular events in the bronchi in mild asthma and after bronchial provocation. *Am Rev Respir Dis* 139: 806–17, 1989.
69. Carson JL, Collier AM, Fernald GW, Hu SC. Microtubular discontinuities as acquired ciliary defects in airway epithelium of patients with chronic respiratory diseases. *Ultrastruct Pathol* 18: 327–32, 1994.
70. Laitinen LA, Heino M, Laitinen A, Kava T, Haahtela T. Damage of the airway epithelium and bronchial reactivity in patients with asthma. *Am Rev Respir Dis* 131: 599–606, 1985.
71. Gleich GJ, Loegering DA, Frigas E, Filley WV. The eosinophil granule major basic protein: Biological activities and relationship to bronchial asthma. *Monogr Allergy* 18: 277–83, 1983.
72. Del Donno M, Bittesnich D, Chetta A, Olivieri D, Lopez-Vidriero MT. The effect of inflammation on mucociliary clearance in asthma: An overview. *Chest* 118: 1142–49, 2000.
73. Wanner A, Salathe M, O'Riordan TG. Mucociliary clearance in the airways. *Am J Respir Crit Care Med* 154: 1868–902, 1996.
74. Clarke SW, Pavia D. Lung mucus production and mucociliary clearance: Methods of assessment. *Br J Clin Pharmacol* 9: 537–46, 1980.
75. Pavia D, Sutton PP, Agnew JE, Lopez-Vidriero MT, Newman SP, Clarke SW. Measurement of bronchial mucociliary clearance. *Eur J Respir Dis Suppl* 127: 41–56, 1983.
76. Moretti M, Lopez-Vidriero MT, Pavia D, Clarke SW. Relationship between bronchial reversibility and tracheobronchial clearance in patients with chronic bronchitis. *Thorax* 52: 176–80, 1997.
77. Goodman RM, Yergin BM, Landa JF, Golivanux MH, Sackner MA. Relationship of smoking history and pulmonary function tests to tracheal mucous velocity in nonsmokers, young smokers, ex-smokers, and patients with chronic bronchitis. *Am Rev Respir Dis* 117: 205–14, 1978.
78. Agnew JE, Little F, Pavia D, Clarke SW. Mucus clearance from the airways in chronic bronchitis – Smokers and ex-smokers. *Bull Eur Physiopathol Respir* 18: 473–84, 1982.
79. van der Schans CP, Piers DA, Beekhuis H, Koeter GH, van der Mark TW, Postma DS. Effect of forced expirations on mucus clearance in patients with chronic airflow obstruction: Effect of lung recoil pressure. *Thorax* 45: 623–27, 1990.
80. Ericsson CH, Svartengren K, Svartengren M, Mossberg B, Philipson K, Blomquist M, Camner P. Repeatability of airway deposition and tracheobronchial clearance rate over three days in chronic bronchitis. *Eur Respir J* 8: 1886–93, 1995.
81. Pavia D, Bateman JR, Sheahan NF, Agnew JE, Clarke SW. Tracheobronchial mucociliary clearance in asthma: Impairment during remission. *Thorax* 40: 171–75, 1985.
82. Bateman JR, Pavia D, Sheahan NF, Agnew JE, Clarke SW. Impaired tracheobronchial clearance in patients with mild stable asthma. *Thorax* 38: 463–67, 1983.
83. Foster WM, Langenback EG, Bergofsky EH. Lung mucociliary function in man: Interdependence of bronchial and tracheal mucus transport velocities with lung clearance in bronchial asthma and healthy subjects. *Ann Occup Hyg* 26: 227–44, 1982.
84. Messina MS, O'Riordan TG, Smaldone GC. Changes in mucociliary clearance during acute exacerbations of asthma. *Am Rev Respir Dis* 143: 993–97, 1991.
85. Bateman JR, Pavia D, Clarke SW. The retention of lung secretions during the night in normal subjects. *Clin Sci Mol Med Suppl* 55: 523–27, 1978.
86. Pavia D, Lopez-Vidriero MT, Clarke SW. Mediators and mucociliary clearance in asthma. *Bull Eur Physiopathol Respir* 23(Suppl 10): 89s–94s, 1987.
87. Rogers DF. *In vivo* preclinical test models for studying airway mucus secretion. *Pulm Pharmacol Ther* 10: 121–28, 1997.
88. Ayers MM, Jeffery PK. Proliferation and differentiation in mammalian airway epithelium. *Eur Respir J* 1: 58–80, 1988.
89. Nadel JA, Burgel PR. The role of epidermal growth factor in mucus production. *Curr Opin Pharmacol* 1: 254–58, 2001.
90. Rogers DF. Airway goblet cells: Responsive and adaptable front-line defenders. *Eur Respir J* 7: 1690–706, 1994.
91. Rogers AV, Dewar A, Corrin B, Jeffery PK. Identification of serous-like cells in the surface epithelium of human bronchioles. *Eur Respir J* 6: 498–504, 1993.
92. Basbaum CB, Jany B, Finkbeiner WE. The serous cell. *Annu Rev Physiol* 52: 97–113, 1990.
93. Singh G, Katyal SL. Clara cell proteins. *Ann NY Acad Sci* 923: 43–58, 2000.
94. Djukanovic R. Airway inflammation in asthma and its consequences: Implications for treatment in children and adults. *J Allergy Clin Immunol* 109: S539–S548, 2002.
95. Jeffery PK. Differences and similarities between chronic obstructive pulmonary disease and asthma. *Clin Exp Allergy* 29(Suppl 2): 14–26, 1999.
96. Saetta M, Turato G, Maestrelli P, Mapp CE, Fabbri LM. Cellular and structural bases of chronic obstructive pulmonary disease. *Am J Respir Crit Care Med* 163: 1304–9, 2001.
97. Barnes PJ. New treatments for COPD. *Nat Rev Drug Discov* 1: 437–46, 2002.
98. Barnes PJ, Chung KF, Page CP. Inflammatory mediators of asthma: An update. *Pharmacol Rev* 50: 515–96, 1998.
99. Charman J, Reid L. Sputum viscosity in chronic bronchitis, bronchiectasis, asthma and cystic fibrosis. *Biorheology* 9: 185–99, 1972.
100. Shimura S, Sasaki T, Sasaki H, Takishima T, Umeya K. Viscoelastic properties of bronchorrhoea sputum in bronchial asthmatics. *Biorheology* 25: 173–79, 1988.
101. Rogers DF. Physiology of airway mucus secretion and pathophysiology of hypersecretion. *Respir Care* 52: 1134–46, 2007.
102. Rogers DF, Evans TW. Plasma exudation and oedema in asthma. *Br Med Bull* 48: 120–34, 1992.
103. Shimura S, Andoh Y, Haraguchi M, Shirato K. Continuity of airway goblet cells and intraluminal mucus in the airways of patients with bronchial asthma. *Eur Respir J* 9: 1395–401, 1996.
104. Chen Y, Zhao YH, Di YP, Wu R. Characterization of human mucin 5B gene expression in airway epithelium and the genomic clone of the amino-terminal and 5'-flanking region. *Am J Respir Cell Mol Biol* 25: 542–53, 2001.

105. Davies JR, Carlstedt I. Respiratory tract mucins. In: Salathe M (ed.), Cilia and Mucus: From Development to Respiratory Defense, pp. 167–78. New York: Marcel Dekker Inc., 2001.
106. Groneberg DA, Eynott PR, Lim S, Oates T, Wu R, Carlstedt I, Roberts P, McCann B, Nicholson AG, Harrison BD, Chung KF. Expression of respiratory mucins in fatal status asthmaticus and mild asthma. *Histopathology* 40: 367–73, 2002.
107. Groneberg DA, Eynott PR, Oates T, Lim S, Wu R, Carlstedt I, Nicholson AG, Chung KF. Expression of MUC5AC and MUC5B mucins in normal and cystic fibrosis lung. *Respir Med* 96: 81–86, 2002.
108. Davies JR, Svitacheva N, Lannefors L, Kornfalt R, Carlstedt I. Identification of MUC5B, MUC5AC and small amounts of MUC2 mucins in cystic fibrosis airway secretions. *Biochem J* 344(Pt 2): 321–30, 1999.
109. Glynn AA, Michaels L. Bronchial biopsy in chronic bronchitis and asthma. *Thorax* 15: 142–53, 1960.
110. Rogers DF. The role of airway secretions in COPD: Pathophysiology, epidemiology and pharmacotherapeutic options. *COPD* 2: 341–53, 2005.
111. Burgel PR, Nadel JA. Roles of epidermal growth factor receptor activation in epithelial cell repair and mucin production in airway epithelium. *Thorax* 59: 992–96, 2004.
112. Hewson CA, Edbrooke MR, Johnston SL. PMA induces the MUC5AC respiratory mucin in human bronchial epithelial cells, via PKC, EGF/TGF-alpha, Ras/Raf, MEK, ERK and Sp1-dependent mechanisms. *J Mol Biol* 344: 683–95, 2004.
113. Toda M, Tulic MK, Levitt RC, Hamid Q. A calcium-activated chloride channel (HCLCA1) is strongly related to IL-9 expression and mucus production in bronchial epithelium of patients with asthma. *J Allergy Clin Immunol* 109: 246–50, 2002.
114. Zhou Y, Shapiro M, Dong Q, Louahed J, Weiss C, Wan S, Chen Q, Dragwa C, Savio D, Huang M, Fuller C, Tomer Y, Nicolaides NC, McLane M, Levitt R. A calcium-activated chloride channel blocker inhibits goblet cell metaplasia and mucus overproduction. Mucus Hypersecretion in Respiratory Disease, pp. 150–65. Chichester: Wiley, 2002.
115. Donnelly LE, Rogers DF. Antiproteases and retinoids for treatment of chronic obstructive pulmonary disease. *Expert Opin Ther Patents* 13: 1345–72, 2003.
116. Li Y, Martin LD, Spizz G, Adler KB. MARCKS protein is a key molecule regulating mucin secretion by human airway epithelial cells *in vitro*. *J Biol Chem* 276: 40982–90, 2001.
117. Singer M, Martin LD, Vargaftig BB, Park J, Gruber AD, Li Y, Adler KB. A MARCKS-related peptide blocks mucus hypersecretion in a mouse model of asthma. *Nat Med* 10: 193–96, 2004.
118. Foster KA. A new wrinkle on pain relief: Re-engineering clostridial neurotoxins for analgesics. Drug Discov Today 10: 563–69, 2005.
119. Foster KA, Adams EJ, Durose L, Cruttwell CJ, Marks E, Shone CC, Chaddock JA, Cox CL, Heaton C, Sutton JM, Wayne J, Alexander FC, Rogers DF. Re-engineering the target specificity of clostridial neurotoxins – a route to novel therapeutics. *Neurotox Res* 9: 101–7, 2006.
120. Adams EJ, Foster K, Barnes PJ, Rogers DF. Inhibition of airway mucin secretion by a retargeted clostridial endopeptidase in a rat model of COPD. *Am J Respir Crit Care Med*, 175, (abstracts issue): A995, 2008.

18

CHAPTER

Airway Smooth Muscle

Yassine Amrani[1], Omar Tliba[2], Vera P. Krymskaya[2], Michael W. Sims[2] and Reynold A. Panettieri Jr.[2]

[1]Department of Infection, Immunity and Inflammation, University of Leicester, UK

[2]Pulmonary, Allergy and Critical Care Division, Airways Biology Initiative, University of Pennsylvania, Philadelphia, PA, USA

INTRODUCTION

Airway smooth muscle (ASM) functions as the primary effector cell that regulates bronchomotor tone. In asthma and chronic obstructive pulmonary disease (COPD), bronchoconstriction evoked by smooth muscle shortening promotes airway obstruction, a hallmark of asthma and COPD. Recent evidence also suggests that ASM may undergo hypertrophy and/or hyperplasia and modulate inflammatory responses by secreting chemokines and cytokines. This chapter reviews current studies focusing on excitation–contraction coupling, signaling pathways modulating ASM growth, and cytokine-induced effects on ASM synthetic responses.

REGULATION OF CALCIUM SIGNALING IN ASM

Because increases in cytosolic calcium directly regulate the initiation and the development of ASM contraction, changes in calcium homeostasis may promote bronchial hyperresponsiveness in asthma. Evidence suggests that various mediators and cytokines important in the pathogenesis of asthma, which augment agonist-induced contractile function, also directly alter calcium signaling. This section reviews (1) the critical calcium-dependent mechanisms involved in the regulation of ASM contraction, and (2) the potential molecular mechanisms that regulate calcium signaling.

Calcium responses induced by bronchoconstrictor agents

Using the fluorescent dye Fura-2, studies performed on ASM cells have shown that agonist-induced elevation of the cytosolic free calcium ($[Ca^{2+}]_i$) concentration is biphasic. The initial rapid and transient phase of $[Ca^{2+}]_i$ elevation results from the depletion of inositol-1,4,5 trisphosphates (IP_3)-sensitive calcium stores; see the review by Amrani and Panettieri [1]. This transient increase in intracellular calcium initiates the contraction process through activation of a calcium–calmodulin-dependent myosin light chain kinase and subsequent phosphorylation of the 20 kDa myosin light chain, the central regulatory mechanism of ASM contraction; see the review by Giembycz and Raeburn [2]. The subsequent sustained elevation of intracellular $[Ca^{2+}]_i$ is regulated by the activation of a calcium influx across the plasma membrane since depletion of extracellular calcium or the use of calcium channel inhibitors prevents this sustained phase [3, 4]. The role of the calcium influx in ASM cells after agonist stimulation is not completely understood, but evidence suggests that calcium entry plays an important role in maintaining the plateau phase of ASM contraction via a PKC-dependent mechanism [2].

Pathways regulating $[Ca^{2+}]_i$ influx

Although agonist-mediated increases in $[Ca^{2+}]_i$ have been extensively characterized, much less is known about the cellular mechanisms linking the transient and the sustained phase. In 1986, Putney proposed the concept of store-dependent calcium entry (also called the capacitative model) where depletion of intracellular stores directly regulates calcium entry at the plasma membrane via store-operated calcium channels (SOCC). This calcium influx contributes to the refilling of the internal stores [5]. The existence of capacitative calcium entry in excitable cells, such as smooth muscle cells, was demonstrated using drugs, such as thapsigargin, that cause depletion of sarco-endoplasmic reticulum (SER) by directly inhibiting the activity of the SER-associated calcium-ATPases (SERCA) [6].

In human ASM cells, although thapsigargin-sensitive SERCA2a and 2b isoforms exist, SERCA2b is the predominant isoform. In these cells, depletion of intracellular calcium stores in response to thapsigargin activates a calcium influx with a magnitude that is dependent upon the duration of stimulation with thapsigargin [1]. These data suggest that pathways that activate calcium influx in ASM are linked to the filling state of SERCA2b-associated internal calcium stores. Release of calcium from SER also regulates calcium-dependent chloride and nonselective cation channels, leading to membrane depolarization and opening of voltage-dependent channels; see the review by Janssen [7]. Pretreatment of human ASM cells with thapsigargin also abrogates the calcium responses induced by bradykinin, histamine, or carbachol, suggesting that thapsigargin-sensitive calcium stores involve, at least in part, those activated by the contractile agonists [3].

By mobilizing the same calcium stores, it is plausible that both thapsigargin and agonists activate similar SOCC-dependent calcium influx pathways in ASM. This is supported by the fact that the amplitude of the sustained increase in $[Ca^{2+}]_i$ is dependent on the amplitude of the initial transient phase [3]. Whether additional mechanisms are involved, such as concomitant activation of receptor-operated calcium channels (ROCC) previously described in ASM [3, 4] or involvement of a released soluble mediator controlling calcium influx [8], remains unknown.

Collectively, these data show that the thapsigargin-sensitive intracellular calcium stores in ASM not only serve as a source of calcium to initiate the transient response to agonists, but also participate in the regulation of calcium influx in ASM cells as proposed by the capacitative model. Figure 18.1 summarizes the potential sources of calcium mobilized by contractile agonists to regulate ASM contraction.

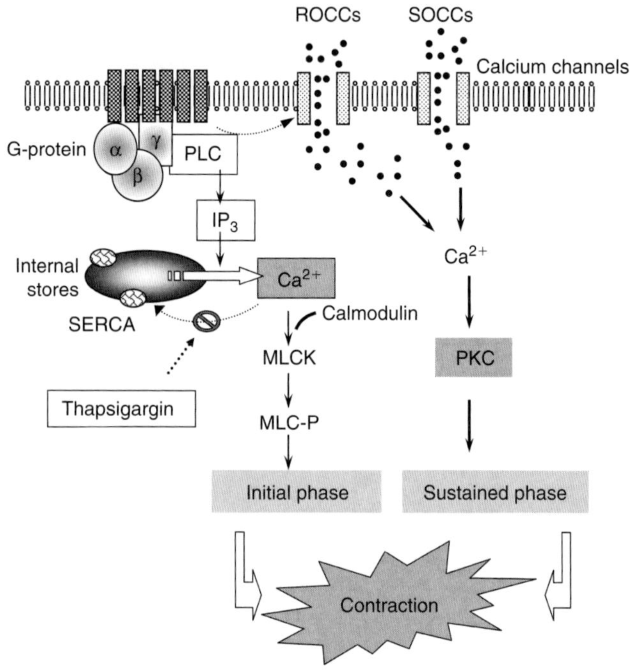

FIG. 18.1 Sources of calcium involved in the regulation of agonist-induced ASM contraction. Contractile agonists activate G-protein-coupled receptors that stimulate PLC and evoke contraction via the mobilization of calcium from IP_3-sensitive internal stores. The increase in $[Ca^{2+}]_i$ activates calcium-/calmodulin-sensitive myosin light chain kinase (MLCK), with subsequent phosphorylation of the 20-kDa myosin light chain (MLC) and initiation of the cross-bridge cycling between actin and myosin. The sustained phase of contraction is thought to involve calcium entry through both ROCCs and SOCCs. In contrast to ROCCs that remain open as long as the contractile agonist is present, SOCCs are triggered by the filling state of sarco-endoplasmic reticulum calcium ATPases (SERCA)-associated internal stores.

Calcium homeostasis in ASM: A possible target for proinflammatory agents

Amplification of contractile receptor-coupled calcium signaling

The observation that cultured ASM cells derived from hyperresponsive Fisher rats have an enhanced calcium signal to serotonin suggests the existence of altered calcium metabolism in a model of airway hyperresponsiveness [9]. The underlying mechanism is unknown, but evidence shows that proinflammatory factors may play an important role in regulating calcium metabolism in ASM. Recent studies demonstrate that tumor necrosis factor (TNF)-α and IL-1β enhance bovine smooth muscle contractility to acetylcholine and other contractile agonists by involving an increased mobilization of intracellular $[Ca^{2+}]_i$ [10]. Accordingly, proinflammatory cytokines may "prime" ASM cells for a nonspecific increase in calcium responsiveness, an effect that appears not to involve a change in the receptor affinity for its ligand [11, 12]. The fact that TNF-α potentiates increases in calcium induced by thapsigargin, which depletes SER calcium stores, suggests a possible modulatory effect of TNF-α on SER-associated regulatory proteins such as SERCA. In addition, TNF-α and IL-1β also augment agonist-evoked phosphoinositide turnover, suggesting that cytokines may modulate receptor-coupled phospholipase C (PLC) activity; see the review by Amrani *et al.* [11]. In support of this hypothesis, TNF-α also enhanced calcium signals in response to NaF [1], an agent that directly activates G-proteins in ASM cells, and upregulated expression of G_q and G_i proteins in human ASM cells [13].

Together, these data suggest that TNF-α can induce a "hyperresponsive" phenotype by enhancing signaling pathways downstream from G-protein-coupled receptor (GPCR) activation.

TABLE 18.1 Potential modulators of calcium metabolism in ASM cells.

	Factors	Effect	Species	References
Modulation of receptor-coupled signal transduction	IL-1β, TNF-α	↑ Ca^{2+} transient, IP_3 turnover	Human	[11, 12]
	Acrolein	↑ Ca^{2+} transient and frequency oscillations	Rat	[14]
	Major basic protein	↑ Ca^{2+} transient and plateau	Bovine	[15]
	Mechanical strain	↑ IP_3 turnover	Bovine	[16]
	PLA_2	↑ Ca^{2+} signals	Bovine	[17]
	Allergen challenge	↑ Ca^{2+} signals, ↓ IP_3 5-phosphatase	Rat	[9]
	Chronic hypoxia	↑ Ca^{2+} frequency oscillations, receptor affinity for agonist	Rat	[18]
Modulation of receptor density	Fenoterol	↑ Histamine H1 receptor	Bovine	[19]
	IL-1β	↑ Bradykinin B2 receptor	Human	[12]
	TNF-α	↓ Muscarinic receptor	Human	[1,13]

(↑) increase; (↓) decrease; (IP_3) inositol 1,4,5 trisphosphate.

Other reports have also described that proinflammatory stimuli modulate GPCR/PLC signaling pathways in cultured ASM cells (Table 18.1). Pretreatment of bovine ASM cells with either eosinophil-derived polycationic proteins or myelin basic protein increased bradykinin-induced transients as well as sustained elevations of $[Ca^{2+}]_i$ [15]. Incubation with either the aldehyde pollutant acrolein or the proinflammatory enzyme PLA_2 significantly increased the intensity as well as the frequency of calcium transients in response to acetylcholine [14, 17]. Finally, other experimental conditions, such as chronic hypoxia or mechanical strain, also modulate calcium signaling induced by agonists in ASM cells. Belouchi *et al.* [18] reported that the amplitude of calcium transients in response to low concentrations of acetylcholine was significantly higher in freshly isolated tracheal smooth muscle cells from hypoxic rats than in those obtained from normoxic animals. Investigators also showed that mechanical strain augmented carbachol-induced IP_3 turnover through the regulation of G-protein and/or PLC activities [16].

Alteration in the density of contractile receptors

Exposure of cultured ASM to various proinflammatory mediators also modulates the density of contractile agonist receptors. ASM cells exposed to TNF-α have a dramatic decrease in muscarinic receptor density [3, 13]. In contrast, levels of the bradykinin B2 receptor are rapidly increased in human bronchial smooth muscle exposed to IL-1β via a prostanoid-dependent regulation of gene transcription [12]. Surprisingly, fenoterol, a β_2-agonist, as well as other cAMP-elevating agents, significantly upregulates the expression of histamine H1 receptor in bovine ASM cells, an effect that involves both increased gene expression and mRNA stability [19]. This increase in H1 receptor expression was associated with an increase in ASM responsiveness to histamine in contraction studies. This may be important since fenoterol may lead to the worsening of asthma by modulating, at least in part, bronchial hyperresponsiveness; see the review by Beasley *et al.* [20]. Figure 18.2 summarizes the potential mechanisms involved in the modulation of agonist-evoked calcium signaling in response to various stimuli.

Increase in the calcium sensitivity of the contractile apparatus

Increase in the sensitivity of myofilaments to calcium also represents a mechanism by which the contractile function of ASM can be enhanced. Abnormal calcium sensitivity has been described in ASM derived from allergen-sensitized animals or from passively sensitized tissues; see the review by Schmidt and Rabe [21]. The role played by proinflammatory agents in the impairment of calcium sensitivity has been suggested *in vitro*. Nakatani *et al.* [22], by simultaneously measuring $[Ca^{2+}]_i$ and isometric tensions in response to acetylcholine, showed that brief exposure to TNF-α enhanced the calcium sensitivity of contractile elements in bovine tracheal smooth muscle. Surprisingly, TNF-α did not affect the calcium signals induced by acetylcholine. In guinea pig tracheal smooth muscle, short-term treatment with TNF-α also increased calcium sensitivity of the contractile apparatus [23].

Together, these studies suggest that, in addition to the modulation of receptor-coupled signal transduction, increase in calcium sensitization of contractile elements represents another downstream target potentially modulated by proinflammatory agents.

Because ASM is an essential effector cell modulating bronchoconstriction, changes in ASM properties can be regarded as a potential mechanism contributing to the increased ASM contractility associated with asthma. A variety of stimuli present in asthmatic airways may induce hyperresponsiveness by modulating calcium signaling in

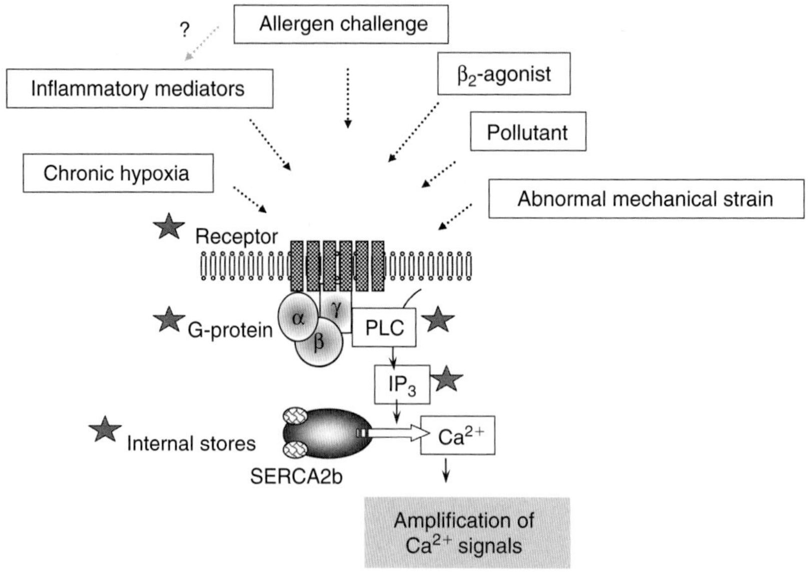

FIG. 18.2 Potential modulators of contractile receptor-coupled calcium signaling in ASM. A variety of stimuli may induce a "hyperresponsive" phenotype, which modulates G-protein-coupled receptor activation in response to agonists. (?) indicates the possible role played by inflammatory mediators in allergen-induced changes in calcium responsiveness to contractile agonist. The potential intracellular targets underlying the increase in calcium responsiveness induced by these various stimuli are shown with a bold star. IP_3: inositol-1,3,5 trisphosphate; PLC: phospholipase C; $[Ca^{2+}]_i$: intracellular calcium.

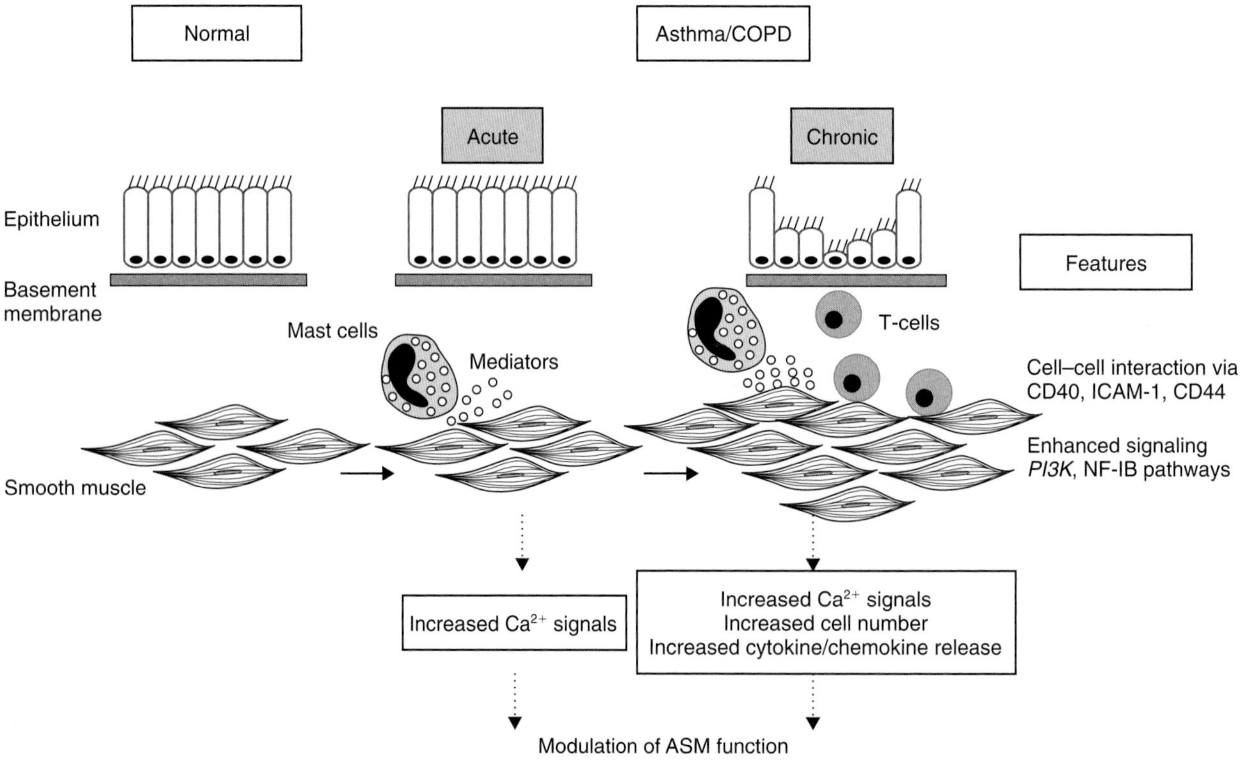

FIG. 18.3 Factors affecting ASM function in acute and chronic state of asthma/COPD. During the acute inflammation, a variety of mediators, such as cytokines, can modulate ASM contractile function by enhancing calcium signaling to agonists. These mediators regulate the recruitment and activation of eosinophils and T-lymphocytes in the airway mucosa, a characteristic histopathological feature of the chronic disease. The persistence of airway inflammation via the production of cytokines and chemokines by both inflammatory as well as structural cells has the potential to directly stimulate ASM proliferation or indirectly as a result of T-cell-ASM interaction mediated by cell surface expression of various CAM proteins such as ICAM-1, CD40, and CD44. Enhanced activation of PI3K or the transcription factor NF-κB in ASM may also stimulate mitogenic and synthetic functions.

response to contractile agonists as summarized in Fig. 18.3. Whether this amplification of calcium signaling contributes to the increased contractility in asthmatic patients remains to be investigated.

ASM CELL PROLIFERATION

Increased ASM mass is a characteristic histopathological finding in the airways of chronic severe asthmatics and of COPD patients. Although increased ASM mass is, in part, due to ASM cell proliferation, the mechanisms that regulate ASM cell growth remain unclear. Many studies have characterized the stimulation of ASM growth in response to mitogenic agents such as polypeptide growth factors, inflammatory mediators, and cytokines. The observation that contractile agonists induce smooth muscle cell proliferation may be a critical link between the chronic stimulation of muscle contraction and the myocyte proliferation [24]. Although the specific cellular and the molecular mechanisms by which agonists induce cell proliferation remain to be elucidated, similarities exist between signal transduction processes activated by these agents and those of known growth factors, which can also stimulate smooth muscle contraction. The complex interaction between signaling pathways that induce myocyte proliferation and those that inhibit cell growth by stimulation of apoptosis may promote airway remodeling as seen in the bronchi of patients with asthma, bronchiolitis obliterans, or chronic bronchitis.

Smooth muscle cell proliferation is stimulated by mitogens that fall into two broad categories:

1. those that activate receptors with intrinsic tyrosine kinase activity (RTK);
2. those that mediate their effects through receptors coupled to heterotrimeric GTP binding proteins (G-proteins) and activate nonreceptor-linked tyrosine kinases found in the cytoplasm.

Although both pathways increase cytosolic calcium through activation of PLC, different PLC isoenzymes appear to be involved. Activated PLC hydrolyzes phosphatidylinositol bisphosphate (PIP_2) to inositol trisphosphate (IP_3) and diacylglycerol (DAG). These second messengers activate other cytosolic tyrosine kinases as well as serine and threonine kinases (protein kinase C, G, and N) that have pleiotropic effects including the activation of proto-oncogenes, which are a family of cellular genes (*c-onc*), that control normal cellular growth and differentiation. Recent reviews attempted to summarize current progress in our understanding of cellular mechanisms leading to smooth muscle cell proliferation [25–28]. This section will focus on the role of PLC activation and the phosphatidylinositol 3-kinase (PI3K) signaling pathway in smooth muscle cell mitogenesis.

PLC activation

Receptors with intrinsic tyrosine kinase activity and those coupled to G-proteins both activate specific PLC isoforms. These PLCs are the critical regulatory enzymes in the activation of the PI pathway. The γ family of PLC contains src-homology SH2 and SH3 domains and is regulated by tyrosine phosphorylation. In ASM cells, some growth factors which activate receptors with intrinsic tyrosine kinases have been identified. Platelet-derived growth factor (PDGF) and epidermal growth factor (EGF) in human ASM cells [29–31] and IGF-1 in bovine and rabbit ASM cells [32–34] have been shown to induce myocyte proliferation. However, the role of PLC-γ_1 activation in modulating ASM cell growth remains unknown.

Other PLC isoforms are controlled by G-proteins and/or calcium. Although the role of PLC activation in mediating G-protein-dependent cell growth is complex, G-protein activation appears critically important in transducing contractile agonist-induced cell growth. G-proteins are composed of three distinct subunits, α, β, and γ, the latter two existing as a tightly associated complex [35]. Although α-subunits were considered the functional components important in downstream signaling events, recent evidence suggests that $\beta\gamma$-subunits also play a critical role in modulating cell function [36].

Advances in single-cell microinjection techniques in combination with the development of neutralizing antibodies to specific Gα subunits have enabled investigators to characterize the role of G-protein activation in cell proliferation. Using these techniques, studies with 3T3 fibroblasts have determined that, while both thrombin and bradykinin required G_q activation to mobilize cytosolic calcium, to generate IP_3 and to induce mitogenesis, thrombin, but not bradykinin, induces cell growth by stimulating G_{i2} [37]. These studies determined that a single mitogen may require functional coupling to distinct subtypes of G-proteins in order to stimulate cell growth. Collectively, these data also provide a mechanism to explain why some, but not all, agonists induce cell proliferation while mobilizing comparable levels of cytosolic calcium.

Recently, the role of PLC activation and IP_3 in mediating contractile agonist-induced ASM cell growth has been explored. Several contractile agonists, which mediate their effects through GPCRs, induce ASM cell proliferation. Studies have determined that histamine [38] and serotonin induce canine and porcine ASM cell proliferation. Endothelin-1, leukotriene D_4 and U-46619, a thromboxane A_2 mimetic, induce rabbit ASM cell growth [39], and thrombin induces mitogenesis in human ASM cells [30]. Although the mechanisms that mediate these effects are unknown, agonist-induced cell growth probably is modulated by activation of G-proteins in a manner similar to that described in vascular smooth muscle (VSM).

Using human ASM cells, Panettieri *et al.* [30] examined whether contractile agonist-induced human ASM cell growth was dependent on PLC activation and IP_3 formation. These investigators examined the relative effects of bradykinin and thrombin on myocyte proliferation and PI turnover. Thrombin, but not bradykinin, stimulated ASM cell proliferation despite a fivefold greater increase in [^{3}H]-inositol phosphate formation in cells treated with bradykinin as compared with those treated with thrombin. Inhibition of PLC activation with U-73122 had no effect on thrombin- or

EGF-induced myocyte proliferation. In addition, pertussis toxin completely inhibited thrombin-induced ASM cell growth but had no effect on PI turnover induced by either thrombin or bradykinin [30]. Taken together, these studies suggest that thrombin induced human ASM cell growth by activation of a pathway that was pertussis toxin-sensitive and independent of PLC activation or PI turnover.

Compared to RTK-dependent growth factors, contractile agonists, with the exception of thrombin and sphingosine-1-phosphate, appear to be less effective human ASM mitogens [40]. In cultured human ASM cells, 100 μmol/l histamine or serotonin induces two- to threefold increases in [^{3}H]-thymidine incorporation as compared with that obtained from unstimulated cells. EGF, serum, or phorbol esters, which directly activate protein kinase C, induce 20–30 fold increases in [^{3}H]-thymidine incorporation [41]. In rabbit ASM cells, endothelin-1 induces cell proliferation by activating phospholipase A_2, and by generating thromboxane A_2 and LTD_4 [39, 42]. In human ASM cells, however, endothelin-1, thromboxane A_2 and LTD_4 appear to have little effect on ASM cell proliferation despite these agonists inducing increases in cytosolic calcium [30, 43, 44].

Clearly, interspecies variability exists with regard to contractile agonist-induced cell proliferation. These models, however, may prove useful in dissecting downstream signaling events that modulate the differential effects of contractile agonists on ASM cell proliferation.

PI3K signaling pathway

The PI3K signaling pathway is a highly conserved signal transduction network regulating a variety of cellular functions including cell proliferation, differentiation, transformation, cell motility, and apoptosis [45, 46]. PI3Ks are a subfamily of lipid kinases that catalyze the addition of phosphate molecules specifically to the 3-position of the inositol ring of phosphoinositides [47]. Tumor suppressor PTEN (phosphatases and tensin homolog) negatively regulates PI3K signaling by specifically dephosphorylating the 3′-OH position of the inositol ring of the lipid second messengers phosphatidylinositol-3,4,5-trisphosphate. [48–50], PI3K lipid products are not substrates for the PI-specific PLC enzymes that cleave inositol phospholipids into membrane-bound DAG and soluble inositol phosphates. The 3-phosphoinositides function as second messengers and activate downstream effector molecules such as PDK1, p70s6 kinase (S6K1), protein kinase Cζ, and Akt [47, 51]. The ability of PI3K to regulate diverse functions may be due to the existence of multiple isoforms that have specific substrate specificities and that reside in unique cytoplasmic locations within the cell [52].

PI3K isoforms can be divided into three classes based on their structure and substrate specificity. Class IA PI3Ks are cytoplasmic heterodimers composed of a 110-kDa (p110α, β, or δ) catalytic subunit and an 85-kDa (p85, p55, or p50) adaptor protein. Catalytic subunits p110α and p110β are ubiquitously expressed in mammalian cells. Catalytic subunit p110δ is expressed predominantly in lymphocytes and lymphoid tissues and, therefore, may play a role in PI3K-mediated signaling of immune responses [53]. Class IA isoforms are mainly activated by receptor and nonreceptor tyrosine kinases while Class IB p110γ is activated by Gβγ subunits of GPCRs [54]. Class II isoforms are mainly associated with the phospholipid membranes and are present in the endoplasmic reticulum and Golgi apparatus [55]. Class III isoforms, structurally related to yeast vesicular sorting protein Vps34p [56], are mammalian homologs that use only membrane phosphatidylinositol as a substrate and generate phosphatidylinositol-3-monophosphate.

PI3K activation by multiple inputs, such as growth factors, insulin, cytokines, cell–cell, and cell–matrix adhesion, provided by both receptor and nonreceptor tyrosine kinases leads to activation of serine–threonine kinase Akt (Fig. 18.4). Downstream of PI3K the tumor suppressor complex tuberous sclerosis complex (TSC)1/TSC2 is subject to direct inhibitory phosphorylation by Akt [57–59]. TSC2 forms a complex with tumor suppressor TSC1 and regulates mTOR/S6K1 signaling [60] by directly controlling the activity of the small GTPase Rheb via the GTPase activating protein (GAP) domain of TSC2 [61]. Rheb directly binds to Raptor [62, 63] and controls the activity of the mTOR/Raptor complex (mTORC1), which, in turn, directly phosphorylates and activates S6K1 (Fig. 18.4).

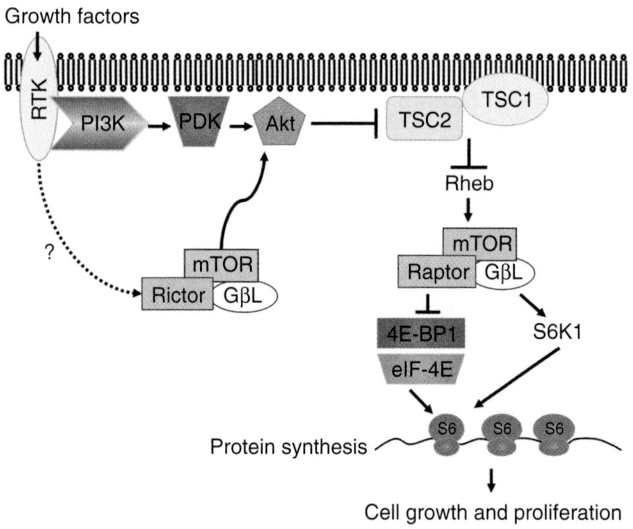

FIG. 18.4 The PI3K signaling pathway is critical for regulating cell growth and proliferation. Growth-factor (GF)-induced activation of PI3K initiates a signaling cascade involving sequential activation of phosphoinositide-dependent kinase (PDK) and Akt/PKB. Akt/PKB-dependent phosphorylation of TSC2 results in dissociation of TSC1/TSC2 membrane complex and activation of Rheb. Next, Rheb activates mTOR complex 1 (mTORC1) which consists of mTOR, Raptor, and GβL. This is followed by activation of S6K1 with subsequent phosphorylation of ribosomal protein S6. In parallel, mTOR-dependent phosphorylation of the binding protein 4E releases from the inhibitory complex eukaryotic initiation factor 4E (eIF-4E), which initiates protein synthesis, resulting in cell growth and proliferation. Rapamycin-insensitive complex mTORC2, which consists of mTOR, Rictor, and GβL, regulates Akt/PKB phosphorylation; however, the precise mechanism of mTORC2 regulation by Rheb and TSC1/TSC2 remains to be elucidated. 4E-BP1: eIF-4E binding protein; Rheb: Ras homolog enriched in brain; RTK: receptor tyrosine kinase.

mTOR is a part of both complexes: the rapamycin-sensitive mTOR/Raptor (TORC1) phosphorylating S6K1, and the rapamycin-insensitive mTOR/Rictor (TORC2) phosphorylating Akt [64] and regulating Rac1 activation [65]. The PI3K-TSC1/TSC2-Rheb-mTOR-S6K1 signaling pathway is highly conserved across species, and controls cell growth and proliferation [45, 46, 61].

Studies demonstrate that PI3K signaling cascade plays a critical role in regulating human airway and pulmonary arterial VSM cell proliferation [66, 67]. Wortmannin and LY294002, two potent inhibitors of PI3K, inhibit DNA synthesis in bovine ASM, porcine, and rat VSM cells stimulated with PDGF, basic fibroblast growth factor, angiotensin II, or serum [68–71]. Stimulation of α_1 adrenergic receptors with noradrenaline activated mitogenesis, Ras, MAPK, and PI3K in human VSM cells in a wortmannin-sensitive manner [72]. In rat thoracic aorta VSM cells, wortmannin completely blocked angiotensin II-induced Ras activation but had no effect on mitogen-activated protein kinase (MAPK) activation or protein synthesis [73]. Thrombin, which induces human ASM cell growth by activating a receptor presumably coupled to both G_i and G_q proteins [30], requires PI3K activation to mediate its growth effects [74]. In bovine ASM, the mitogenic effects of PDGF or endothelin-1 (ET-1) have been attributed to their ability to stimulate PI3K or S6K1. These data suggest that in both airway and VSM cells PI3K is involved in mitogenic signaling induced by numerous agents.

In order to determine whether PI3K activation is necessary or sufficient to stimulate human ASM DNA synthesis, Krymskaya *et al.* [67] used transfected cells with a chimeric model of Class IA PI3K in which the inter-SH2 region of p85 regulatory subunit was covalently linked to its binding site at the p110 N-terminal region of the catalytic subunit [75]. Transient expression of constitutively active p110* was sufficient to induce DNA synthesis in ASM cells in the absence of mitogens [67]. Interestingly, in human ASM cells the level of p110*-induced DNA synthesis was markedly lower than that induced by EGF, thrombin, or serum; and thus, although PI3K activation is sufficient to induce DNA synthesis in ASM cells, other signaling pathways that act in parallel or that are more effective inducers of PI3K may play a role in modulating mitogen-induced DNA synthesis in these cells.

In numerous cell types, PI3K has been shown to be an important mediator of S6K1 activation in response to serum and growth factors [76]. S6K1 is a critical enzyme for mitogen-induced cell cycle progression through the G_1 phase and translational control of mRNA transcripts that contain a polypyrimidine tract at their transcriptional start site [77].

EGF and thrombin significantly stimulate S6K1, and wortmannin, LY294002, and rapamycin completely block this activation in ASM cells [74]. Moreover, transient expression of constitutively active p110* PI3K activates S6K1 in the absence of stimulation with mitogens, while overexpression of a dominant-negative Δp85 PI3K abolished EGF- and thrombin-induced S6K1 activation [74]. Thus, EGF and thrombin induce activation of S6K1 in human ASM cells and mitogen-induced activation of S6K1 appears PI3K dependent.

A recent study demonstrated that GPCR activation by inflammatory and contractile agents can synergize with RTK activation to augment human ASM growth. In EGF-stimulated cells, GPCR-mediated potentiation does not appear mechanistically linked to increased EGFR (epidermal growth factor receptor) or p42/p44 MAPK activation but is associated with sustained activation of S6K1 for several hours after the initial early phase of activation. These findings not only provide insight into mechanisms by which inflammation contributes to ASM hyperplasia/hypertrophy in diseases such as asthma and COPD, but also suggest a general mechanism by which GPCRs and RTKs interact to promote cell growth [78].

PDGF, EGF, and thrombin, which transduce their signals by activating distinct pathways through RTKs and GPCR, induce PI3K activity in human ASM cells [67, 74, 79, 80]. Importantly, cross talk between GPCR and RTK signaling pathways occurs at the PI3K levels and promotes a synergy among growth factors and contractile agonists [78, 79, 81]. Src protein tyrosine kinase integrates GPCR- and RTK-induced signaling pathways that ultimately regulate cell proliferation, migration, differentiation, and gene transcription [82, 83]. Src kinases have a well-established role in the regulation of cell proliferation. The Src tyrosine kinase is the first tyrosine kinase to be identified and linked to the regulation of cell proliferation [82, 83]. Because more than one mechanism is involved in Src activation, this enzyme is critical in transducing signals, which originate through GPCRs and RTKs and result in integrated cellular responses, such as DNA synthesis and migration. The cellular functions regulated by Src were identified by using predominantly overexpressed Src and its mutants in established immortalized cell lines and *in vitro* assays. Using selective cell-permeable Src inhibitors PP1 and PP2, Src inhibition of DNA synthesis was demonstrated in guinea pig ASM [84] and rat aortic smooth muscle cells [85]. In our study using pharmacological and molecular approaches, we demonstrate that Src is necessary and sufficient for human ASM DNA synthesis. Interestingly, PP2 had differential effects on EGF- versus PDGF- and thrombin-induced DNA synthesis: IC_{50} for EGF was $\sim 5.5\,\mu M$, which is a magnitude higher compared to the IC_{50} for PDGF and thrombin, which was $\sim 0.3\,\mu M$ for both agonists. The high sensitivity of thrombin-induced DNA synthesis to Src inhibition is potentially due to a critical role of Src in modulating mitogenic signaling from activated GPCRs to its downstream targets [86, 87]. We also found that the differential Src regulation of cell proliferation correlated with differential activation of Src-associated PI3K. Thus, upon stimulation of human ASM cells with PDGF or thrombin, PI3K activity was associated with Src but not in cells stimulated with EGF. Since PI3K activation is critical for cell proliferation, the lack of Src-associated PI3K activation may explain the insensitivity of EGF-induced ASM DNA synthesis to blocking Src. Importantly, our previously published study identified the signaling pathway of EGF-dependent PI3K activation in HASM cells, which involved activation of the ErbB2 receptor of the EGF receptor family [88]. Collectively, these data suggest a differential role of Src in growth promoting signaling stimulated by EGF, PDGF, and thrombin.

ASM: AN IMMUNOMODULATORY CELL

Asthma is a disease characterized, in part, by reversible airflow obstruction, hyperresponsiveness, and inflammation. COPD, which includes chronic bronchitis, emphysema, and bronchiectasis, is characterized by predominantly irreversible airflow obstruction associated with neutrophilic airway inflammation. Traditional concepts concerning airway inflammation have focused on trafficking leukocytes and on the effects of inflammatory mediators, cytokines, and chemokines secreted by these cells. ASM, the major effector cell responsible for bronchomotor tone, has been viewed as a passive tissue responding to neurohumoral control and inflammatory mediators. New evidence, however, suggests that ASM may secrete cytokines and chemokines and express cell adhesion molecules (CAMs) that are important in modulating submucosal airway inflammation. The cellular and molecular mechanisms that regulate the immunomodulatory functions of ASM may offer new and important therapeutic targets in treating these common lung diseases.

Chemokine and cytokine release by ASM cells

The various cell types that infiltrate the inflamed submucosa present the potential for many important cell–cell interactions. Eosinophils, macrophages, neutrophils, and lymphocytes initiate and perpetuate airway inflammation by producing proinflammatory mediators. Evidence also suggests that exposure of ASM to cytokines or growth factors alters contractility and calcium homeostasis [1] and induces SMC hypertrophy and hyperplasia; see the review by Lazaar *et al.* [89].

Studies now show that ASM cells secrete a number of cytokines and chemoattractants. Studies of bronchial biopsies in mild asthmatics reveal constitutive staining for RANTES, a CC chemokine, in ASM [90]; *in vitro*, RANTES secretion is induced by TNF-α and IFN-γ [91–93]. Other chemokines that are secreted by ASM cells include eotaxin, an eosinophil chemoattractant, and monocyte chemotactic proteins (MCP)-1, MCP-2, MCP-3 [93–95].

Similarly, the CXC chemokine IL-8 is also secreted by ASM in response to TNF-α, IL-1β, and bradykinin, a contractile agonist [96, 97]. IL-8 has been described to regulate both proliferation and migratory function of ASM cells [98]. Understanding the factors that stimulate IL-8 production may be of therapeutic benefit for the treatment of asthma. Studies found that IL-8 is induced by several factors including β tryptase, the most abundant mast cell product [99] as well as IL-17/IL-1β via both transcription factors, AP-1 and NF-κB [100]. Others found that the antimicrobial peptide LL-37 also induces the production of IL-8 [101] providing a novel mechanism through which inflammatory cell products could participate in the pathogenesis of asthma. Additional studies showed that IL-4 also stimulates eotaxin/CCL-11 expression in ASM cells, in part via the activation of STAT6 transcription factor [102] while Rahman *et al.* found that (p38, p42/p44 ERK, JNK) signaling pathways play a critical role in IL-17A-induced eotaxin-1/CC chemokine ligand 11 [103]. Extracellular matrix (ECM) proteins seem to play an important role in the regulation of eotaxin induced by different interleukins, IL-13, IL-1β, or TNF-α in ASM cells cultured from subjects with asthma compared with healthy donors [104]. Interestingly, increased expression of eotaxin is observed during the co-culture of ASM with mast cells [105], suggesting that cell–cell interaction is important to consider in the regulation of causes of airway inflammation.

In asthma and COPD, β-adrenergic agonists, by elevating $[cAMP]_i$, promote bronchodilation. Studies have investigated whether cAMP pathways also modulate chemokine secretion by ASM cells. In TNF-α-stimulated ASM cells, eotaxin and RANTES expression was potently and effectively inhibited by isoproterenol, PGE_2, dibutyl-cAMP, or phosphodiesterase inhibitors, rolipram and cilomilast [91, 106]. TNF-α-induced IL-8 secretion was inhibited by the combination of cAMP-mobilizing agents and corticosteroids [107]. Similarly, sphingosine-1-phosphate, which activates a Gs-protein-coupled receptor and increases $[cAMP]_i$, abrogated TNF-α-induced RANTES secretion in ASM cells [108].

Current evidence suggests that chemokine secretion induced by inflammatory mediators is inhibited by dexamethasone in human ASM cells. Cytokine-induced secretion of RANTES [91–93], MCP [93], eotaxin [106], and GM-CSF [109] was abrogated with corticosteroids. In most of these studies, corticosteroid and cAMP-mobilizing agents also acted additively to inhibit chemokine secretion.

IL-6, a pleiotropic cytokine, induces smooth muscle cell hyperplasia [110], but also modulates B- and T-cell proliferation and immunoglobulin secretion. IL-6 secretion by ASM cells is inducible by multiple stimuli, including IL-1β, TNF-α, TGF-β, and sphingosine-1-phosphate, a recently described mediator in asthma [91, 108, 111, 112]. Although corticosteroids are potent inhibitors of IL-6 secretion [112], agents that elevate $[cAMP]_i$ actually increase IL-6 secretion [91]. This suggests an intriguing role for IL-6 in regulating airway hyperreactivity and is in agreement with murine models in which IL-6 overexpression promotes airway hyporesponsiveness [113].

Finally, additional cytokines that are secreted by human ASM cells include IL-1β and other IL-6 family cytokines, such as leukemia inhibitory factor and IL-11 [111, 114, 115]. ASM cells may also play a role in promoting both the recruitment and the survival of eosinophils by secretion of GM-CSF and IL-5 [109, 114, 116].

Receptors involved in cell adhesion

CAMs mediate leukocyte–endothelial cell interactions during the process of cell recruitment and homing [117] (Chapter 21). The expression and activation of a cascade of CAMs that include selectins, integrins, and members of the immunoglobulin superfamily, as well as the local production of chemoattractants, leads to leukocyte adhesion and transmigration into lymph nodes and sites of inflammation involving nonlymphoid tissues.

In addition to mediating leukocyte extravasation and transendothelial migration, CAMs promote submucosal or subendothelial contact with cellular and ECM components. New evidence suggests that CAMs mediate inflammatory cell–stromal cell interactions that may contribute to airway inflammation. ASM cells express intercellular adhesion molecule (ICAM)-1 and vascular cell adhesion molecule (VCAM)-1, which are inducible by a wide range of inflammatory mediators. Contractile agonists, such as bradykinin and histamine, in contrast, have little effect on ASM CAM expression [118]. ASM cells also constitutively express CD44, the primary receptor of the matrix protein hyaluronan [118]. Activated T-lymphocytes adhere via LFA-1 and VLA-4 to cytokine-induced ICAM-1 and VCAM-1 on cultured human ASM cells. Moreover, an integrin-independent component of lymphocyte–smooth muscle cell adhesion appeared to be mediated by CD44–hyaluronan interactions [118].

Current hypotheses suggest that steroids may directly inhibit gene expression, such as CAMs, by altering gene promoter activity and/or by abrogating critical signaling events such as NF-κB or AP-1 activation that then modulate gene expression as shown in Fig. 18.5. Interestingly, dexamethasone had no effect on TNF-α- or IL-1β-induced NF-κB activation in human ASM cells [119]. Further, cytokine-induced ICAM-1 expression in ASM cells, which is completely dependent on NF-κB activation, was not affected by dexamethasone, whereas IL-1β-induced cyclooxygenase-2 expression was abrogated [119–121]. In contrast, cytokine-induced CAM expression is sensitive to cAMP-mobilizing agents [122].

Receptors involved in leukocyte activation and immune modulation

CAMs can function as accessory molecules for leukocyte activation [117, 123, 124]. Whether CAMs expressed on smooth muscle serve this function remains controversial. ASM cells do express major histocompatibility complex (MHC) class II and CD40 following stimulation with IFN-γ [125, 126]. Recent studies also suggest that human ASM cells express low levels of CD80 (B7.1) and CD86 (B7.2) [127]. The physiological relevance of these findings remains unknown since ASM cells cannot present alloantigen to CD4 T-cells, despite the expression of MHC class II and costimulatory molecules [125].

Functionally, however, adhesion of stimulated CD4 T-cells can induce smooth muscle cell DNA synthesis [118]. This appears to require direct cell–cell contact and cannot be mimicked by treatment of the cells with T-cell conditioned

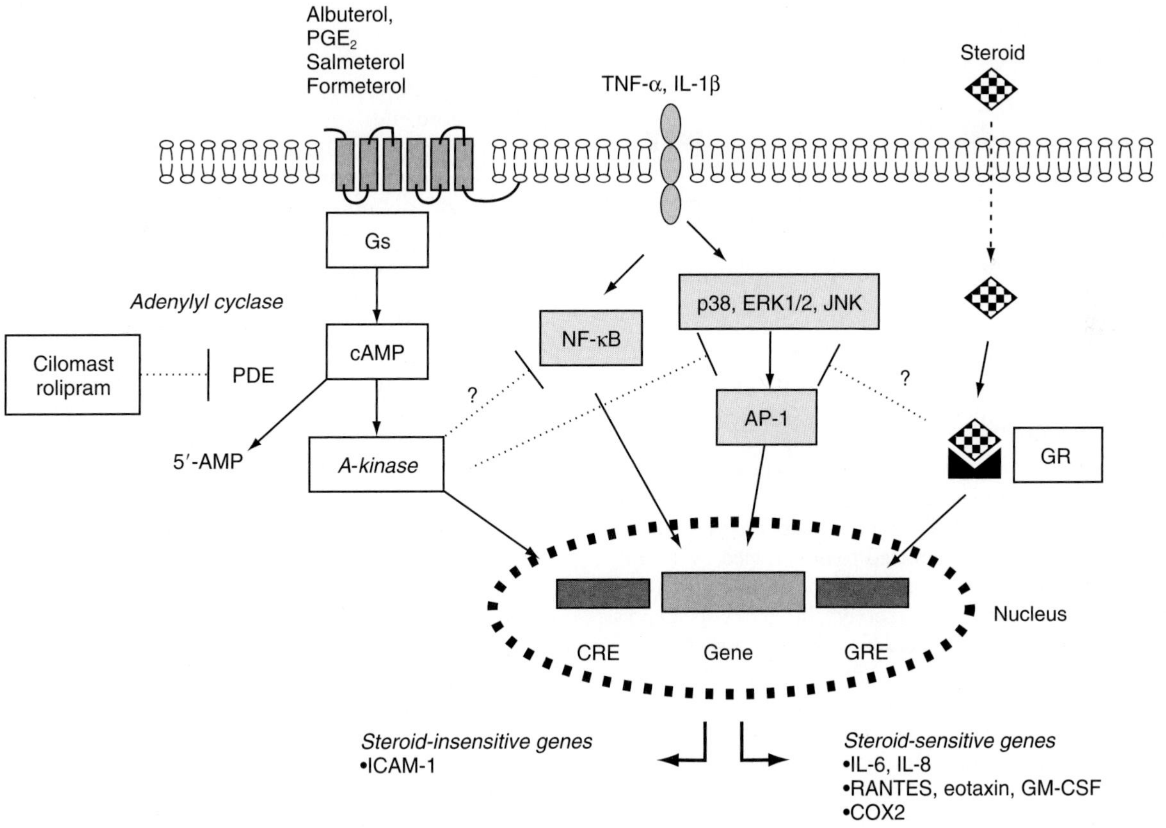

FIG. 18.5 Putative signaling pathways that regulate cytokine-induced synthetic responses in ASM cells. $[cAMP]_i$-mobilizing agents inhibit cytokine-induced chemokine and CAM expression. IL-6 secretion, however, is augmented by agents that increase $[cAMP]_i$. Cytokine-induced NF-κB activation in human ASM is corticosteroid-insensitive. A-kinase: cAMP-dependent protein kinase A; CRE: cAMP response element; ERK: extracellular signal-regulated kinase; GR: glucocorticoid receptor; GRE: glucocorticoid response element; Gs: guanine-nucleotide binding protein; JNK: jun kinase; NF-κB: nuclear factor κB; PDE: phosphodiesterase.

medium. In addition, ligation of CD40 on ASM increases intracellular calcium as well as IL-6 secretion, while cross-linking VCAM-1 cells activates PI3K and augments growth factor-induced ASM cell proliferation [126, 128]. These studies highlight the fact that direct interactions between leukocytes and smooth muscle cells via immune receptors, such as CD40, or adhesion receptors, such as ICAM-1 or VCAM-1, contribute to the modulation of the local milieu resulting in smooth muscle cell activation.

Increased amounts of nitric oxide (NO) have been detected in patients with asthma; see the review by Gaston *et al.* [129] and Chapter 32. NO appears to have a selective-suppressive effect on the Th1 subset of T-helper cells, suggesting that increased levels of NO may therefore lead to the predominantly Th2-type response associated with asthma. NO synthase has been demonstrated in ASM cells where it results in an inhibition of ASM cell proliferation [130, 131]. Thus the role of ASM-derived NO needs to be further defined, as it may have both beneficial and deleterious effects in the airway.

ASM cells produce large amounts of PGE_2, and to a lesser extent other prostanoids, following stimulation with proinflammatory cytokines [120]. Although PGE_2 is a potent bronchodilator, it also has significant immunological effects. For example, PGE_2 can decrease expression of CD23 (FcγRII), which has been shown to be expressed on human ASM cells [132], and synergize with IL-4 to induce IgE synthesis [133]. Release of PGE_2 by ASM cells may also be important in modulating the inflammatory milieu by inhibiting smooth muscle cell secretion of chemokines [134].

Therapeutic modulation of inflammatory gene expression

Issa and colleagues uncovered a novel mechanism by which steroids suppress cytokine-induced inflammatory genes in human ASM cells [135]. MPK-1 was found to be essential in mediating the inhibitory action of dexamethasone on growth-related oncogene protein (GRO)-α induced by IL-1β and TNF-α, thus confirming the role of MAPK in inflammatory gene regulation. Activation of MAPK pathways by IL-1β and TNF-α seem to be important in the regulation of thymic stromal lymphopoietin, a novel factor that promotes dendritic cell-mediated T-helper (Th)-2 inflammatory responses [136]. Other steroid-sensitive genes include TLR2, TLR3, and TLR4 reported to be induced by TNF-α in ASM cells [137]. CD38 was among the steroid-sensitive genes as dexamethasone completely abrogated its expression and increased enzymatic activity following treatment with TNF-α [138]. The observation that IFN can act as an effective inhibitor on some, but not all, TNF-α-inducible inflammatory genes [139] suggests that a better understanding of the IFN-dependent pathways could provide new insight into the design of more potent anti-inflammatory drugs. Macrolides, such as erythromycin and azithromycin, have been recently shown to suppress IL-17-induced production of IL-8 and 8-isoprostane in ASM cells while immunosuppressive agents had no effect [140]. Targeting IKKβ was recently found to be an alternative option for the suppression of a number of inflammatory genes including ICAM-1, IL-6, IL-8, GM-CSF, RANTES, MCP-1, GRO-α, and NAP-2 [141].

Whether ASM is a therapeutic target for steroids in asthma is yet unclear. Several *in vitro* studies in cultured ASM cells confirmed that glucocorticoids (GCs) are effective in suppressing induction of inflammatory genes including IL-6, eotaxin, RANTES, GM-CSF, ICAM-1, and CD38 8 to name just a few [106, 119, 142–145]. Further, *in situ* hybridization and immunohistochemistry studies revealed the *in vivo* expression of glucocorticoid receptor (GR) in ASM of both normal and asthmatic subjects [146]. We previously showed that cultured ASM cells express both GR isoforms, GR-α, and under certain inflammatory conditions, GR-β [143]. Interestingly, Roth *et al.* [147] showed GC treatment induced GR nuclear translocation from cytosol to the nucleus in bronchial smooth muscle cell lines from healthy individuals or subjects with asthma and those with emphysema. Similar results were obtained in cultured ASM obtained from Roth *et al.* in 2002 [148, 149]. The activation of GR by GC through its binding to a glucocorticoid response element (GRE) DNA sequence has also been demonstrated in cultured ASM cells [143, 147, 149].

Interestingly, although steroid effects on human ASM cells have been investigated [150], their modulatory effects on gene expression are complex and poorly characterized. First, the effects of steroids are gene specific. For example, reports showed that dexamethasone effectively inhibits cytokine-induced IL-6, RANTES or eotaxin expression, while the same steroid has little effect on cytokine-induced ICAM-1 expression [106, 119, 142]. Second, steroid-suppressive effects are time dependent since dexamethasone partially abrogates cytokine-mediated ICAM-1 expression at early time points, but has no effect at later time points [119]. Third, steroid inhibitory action is also stimulus specific. While dexamethasone significantly inhibits IL-1β-induced GM-GSF secretion, it has only a partial effect on GM-GSF secretion induced by thrombin [144].

Recently, Miller *et al.* [151], using an animal model of asthma, showed that GC treatment significantly reduced allergen-induced increases in peribronchial collagen deposition and levels of total lung collagen but failed to reduce allergen-induced increases in the thickness of the peribronchial smooth muscle. In line with this, Roth *et al.* [147] showed that GC failed to inhibit the proliferative response in bronchial ASM from patients with asthma. Further *in vitro* studies showed that the anti-proliferative and anti-migratory effects of the GC were impaired by contact of ASM with collagen. GC also failed to inhibit ASM cells cultured on collagen, but not on laminin [148], suggesting that in the collagen-rich microenvironment of the inflamed and fibrotic asthmatic airway, ASM may contribute to steroid resistance. In addition, we also recently showed that steroid-suppressive effects are dramatically impaired in ASM cells treated with multiple cytokines [143]. Indeed, the specific combination of TNF-α with IFNs, but not with IL-1β or IL-13, impairs the actions of multiple classes of GCs, fluticasone, dexamethasone, and budesonide [143] by a mechanism involving the upregulation of GR-β isoform. Recently, Tliba and colleagues [152] found for the first time that an increased level of IRF-1 by "pro-asthmatic" cytokines is insensitive to GCs and can dramatically influence the anti-inflammatory action of steroids via mechanisms that remain to be determined. This study supports the hypothesis that inflammation is a key modulator of steroid responsiveness in structural airway cells.

Collectively, the fact that ASM cells can secrete and express different immunomodulatory proteins, combined with the functional expression of GR, give ASM cells a particular importance as cellular targets for inhaled GC. However, further *in vivo* studies are needed to confirm the aforementioned *in vitro* observations to determine the role of ASM in modulating the anti-inflammatory effects of GC, notably in steroid-resistant patients.

While ASM has long been established as a key effector cell in the pathogenesis of asthma, its role in COPD has only recently been appreciated. Several studies have demonstrated increased ASM mass in the small airways of smokers with airflow obstruction compared to smokers with normal lung function [153–156]. In contrast to asthma, however, where large, central airways are the predominant site of ASM hypertrophy and hyperplasia, ASM hypertrophy in COPD is uniquely observed in small airways with a diameter of approximately 2 mm [155, 157, 158]. The extent of ASM hypertrophy observed in small airways correlates with the severity of airflow obstruction [155, 156], though associated increases in matrix deposition in the walls of these airways may also contribute to airflow obstruction by limiting the extent to which ASM relaxation results in increased airway luminal diameter [159]. Airway inflammation appears to be one important cause of ASM hypertrophy in COPD, as activated T-lymphocytes have been demonstrated to bind VCAM-1 and CD44, inducing DNA synthesis [118].

While ASM has historically been viewed as a target tissue, regulating bronchomotor tone and responding to neurohumoral control and inflammatory mediators, emerging evidence suggests that ASM may also play an important immunomodulatory role in the small airway inflammation characteristic of COPD [160, 161]. ASM cells are capable of expressing numerous proinflammatory and chemotactic cytokines and chemokines germane to COPD, in addition to several relevant growth factors and proteases, as outlined above. Further, the pattern of ASM gene expression depends highly on the inflammatory environment [93, 96, 97, 162, 163], underscoring the importance of interactions between ASM and the nearby epithelium. A great deal of work remains to be done to elucidate the role of ASM in the pathogenesis of COPD, but all available evidence indicates that it is likely an important component of the pathologic processes leading to airway remodeling and irreversible airflow obstruction.

SUMMARY

New evidence suggests that ASM, the major effector cell regulating bronchomotor tone, also plays an important role as an immunomodulatory cell. ASM, exposed to a variety of mediators and cytokines, can undergo phenotypic changes that may be important in the pathogenesis of asthma and COPD. ASM, which secretes chemokines and cytokines, may participate in or even perpetuate the mucosal inflammatory changes via the activation and the recruitment of inflammatory cells. Further, cytokine-induced alterations in calcium homeostasis render the ASM hyperresponsive to contractile agonists. In chronic severe asthma and COPD, ASM hypertrophy and hyperplasia may potentially render the asthmatic airway irreversibly obstructed.

Further elucidation of the cellular and the molecular mechanisms that regulate ASM function in asthma and COPD will offer new therapeutic targets in the treatment of asthma, chronic bronchitis, and emphysema.

References

1. Amrani Y, Panettieri RA Jr. Cytokines induce airway smooth muscle cell hyperresponsiveness to contractile agonists. *Thorax* 53: 713–16, 1998.
2. Giembycz MA, Raeburn D. Current concepts on mechanisms of force generation and maintenance in airway smooth muscle. *Pulm Pharmacol* 5: 279–97, 1992.
3. Amrani Y, Magnier C, Wuytack F, Enouf J, Bronner C. Ca^{2+} increase and Ca^{2+} influx in human tracheal smooth muscle cells: Role of Ca^{2+} pools controlled by sarco-endoplasmic reticulum Ca^{2+}-ATPase 2 isoforms. *Br J Pharmacol* 115: 1204–10, 1995.
4. Murray RK, Kotlikoff MI. Receptor-activated calcium influx in human airway smooth muscle cells. *J Physiol (Lond)* 435: 123–44, 1991.
5. Putney JW. Excitement about calcium signaling in inexcitable cells. *Science* 262, 676–78, 1993.
6. Gibson A, McFadzean I, Wallace P, Wayman CP. Capacitative Ca^{2+} entry and the regulation of smooth muscle tone. *Trends Pharmacol Sci* 19: 266–69, 1998.
7. Janssen LJ. Calcium handling in airway smooth muscle: Mechanisms and therapeutic implications. *Can Respir J* 5: 491–98, 1998.
8. Randriamampita C, Tsien RY. Emptying of intracellular Ca^{2+} stores releases a novel small messenger that stimulates Ca^{2+} influx. *Nature* 364: 809–14, 1994.
9. Tao FC, Tolloczko B, Eidelman DH, Martin JG. Enhanced $Ca^{(2+)}$ mobilization in airway smooth muscle contributes to airway hyperresponsiveness in an inbred strain of rat. *Am J Respir Crit Care Med* 160: 446–53, 1999.
10. Reynolds AM, Holmes MD, Scicchitano R. Cytokines enhance airway smooth muscle contractility in response to acetylcholine and neurokinin A. *Respirology* 5: 153–60, 2000.
11. Amrani Y, Chen H, Panettieri RA Jr.. Activation of tumor necrosis factor receptor 1 in airway smooth muscle: A potential pathway that modulates bronchial hyper-responsiveness in asthma? *Respir Res* 1: 49–53, 2000.
12. Schmidlin F, Scherrer D, Daeffler L, Bertrand C, Landry Y, Gies JP. Interleukin-1β induces bradykinin B2 receptor gene expression through a prostanoid cyclic AMP dependent pathway in human bronchial smooth muscle cells. *Mol Pharmacol* 53: 1009–15, 1998.
13. Hotta K, Emala CW, Hirshman CA. TNF-α upregulates $G_i\alpha$ and $G_q\alpha$ protein expression and function in human airway smooth muscle cells. *Am J Physiol (Lung Cell Mol Physiol)* 276/20: L405–11, 1999.
14. Roux E, Hyvelin JM, Savineau JP, Marthan R. Calcium signaling in airway smooth muscle cells is altered by *in vitro* exposure to the aldehyde acrolein. *Am J Respir Cell Mol Biol* 19: 437–44, 1998.
15. Wylam ME, Gungor N, Mitchell RW, Umans JG. Eosinophils, major basic protein, and polycationic peptides augment bovine airway myocyte Ca^{2+} mobilization. *Am J Physiol. (Lung Cell Mol Physiol)* 274: L997–1005, 1998.
16. An SS, Hai CM. Mechanical strain modulates maximal phosphatidylinositol turnover in airway smooth muscle. *Am J Physiol (Lung Cell Mol Physiol)* 277: L968–74, 1999.
17. Takata Y, Nishimura Y, Maeda H, Yokoyama M. Phospholipase A2 augments contraction and intracellular calcium mobilization through

thromboxane A2 in bovine tracheal smooth muscle. *Eur Respir J* 14: 396–404, 1999.
18. Belouchi NE, Roux E, Savineau JP, Marthan R. Effect of chronic hypoxia on calcium signalling in airway smooth muscle cells. *Eur Respir J* 14: 74–79, 1999.
19. Mak JC, Roffel AF, Katsunuma T, Elzinga CR, Zaagsma J, Barnes PJ. Up-regulation of airway smooth muscle histamine H(1) receptor mRNA, protein, and function by beta(2)-adrenoceptor activation. *Mol Pharmacol* 57: 857–64, 2000.
20. Beasley R, Pearce N, Crane J, Burgess C. Beta-agonists: What is the evidence that their use increases the risk of asthma morbidity and mortality? *J Allergy Clin Immunol* 104: S18–S30, 1999.
21. Schmidt D, Rabe KF. Immune mechanisms of smooth muscle hyperreactivity in asthma. *Allergy Clin Immunol* 105: 673–82, 2000.
22. Nakatani Y, Nishimura Y, Nishiumua T, Maeda H, Yokoyama M. Tumor necrosis factor-α augments contraction and cytosolic Ca^{2+} through phospholipase A2 in bovine tracheal smooth muscle. *Eur J Pharm* 392: 175–82, 2000.
23. Parris JR, Cobban HJ, Littlejohn AF, MacEwan DJ, Nixon GF. Tumor necrosis factor-α activates a calcium sensitization pathway in guinea-pig bronchial smooth muscle. *J Physiol* 518: 561–69, 1999.
24. Panettieri RA Jr., Kotlikoff MI. Cellular and molecular mechanisms regulating airway smooth muscle cell physiology and pharmacology. In: Fishman AP, Elias JA, Fishman JA, Grippi MA, Kaiser LR, Senior RM (eds). *Pulmonary Diseases and Disorders*, 3rd edn, pp. 107–17. New York: McGraw-Hill, 1998.
25. Condliffe aM, Cadwallader KA, Walker TR, Rintoul RC, Cowburn AS, Chilvers ER. Phosphoinositide 3-kinase: A critical signaling event in pulmonary cells. *Respir Res* 1: 24–29, 2000.
26. Hershenson MB, Abe MK. Mitogen-activated signaling in airway smooth muscle. A central role for Ras. *Am J Respir Cell Mol Biol* 21: 651–54, 1999.
27. Page K, Hershenson MB. Mitogen-activated signaling and cell cycle regulation in airway smooth muscle. *Front Biosci* 5: d258–67, 2000.
28. Thomas CF Jr., Limper AH. Phosphatidylinositol kinase regulation of airway smooth muscle cell proliferation. *Am J Respir Cell Mol Biol* 23: 429–30, 2000.
29. Hirst SJ, Barnes PJ, Twort CHC. Quantifying proliferation of cultured human and rabbit airway smooth muscle cells in response to serum and platelet-derived growth factor. *Am J Respir Cell Mol Biol* 7: 574–81, 1992.
30. Panettieri RA Jr., Hall IP, Maki CS, Murray RK. α-Thrombin increases cytosolic calcium and induces human airway smooth muscle cell proliferation. *Am J Respir Cell Mol Biol* 13: 205–16, 1995.
31. Stewart AG, Grigoriadis G, Harris T. Mitogenic actions of endothelin-1 and epidermal growth factor in cultured airway smooth muscle. *Clin Exp Pharmacol Physiol* 21: 277–85, 1994.
32. Cohen P, Noveral JP, Bhala A, Nunn SE, Herrick DJ, Grunstein MM. Leukotriene D_4 facilitates airway smooth muscle cell proliferation via modulation of the IGF axis. *Am J Physiol (Lung Cell Mol Physiol)* 269/13: L151–57, 1995.
33. Kelleher MD, Abe MK, Chao T-SO, Jain M, Green JM, Solway J, Rosner MR, Hershenson MB. Role of MAP kinase activation in bovine tracheal smooth muscle mitogenesis. *Am J Physiol (Lung Cell Mol Physiol)* 268/12: L894–901, 1995.
34. Noveral JP, Bhala A, Hintz RL, Grunstein MM, Cohen P. The insulin-like growth factor axis in airway smooth muscle cells. *Am J Physiol (Lung Cell Mol Physiol)* 267/11: L761–65, 1994.
35. Hepler JR, Gilman AG. G proteins. *Trends Biochem Sci* 17: 383–87, 1992.
36. Sternweis PC. The active role of βγ in signal transduction. *Curr Opin Cell Biol* 6: 198–203, 1994.
37. Lamorte VJ, Harootunian AT, Spiegel AM, Tsien RY, Feransisco JR. Mediation of growth factor induced DNA synthesis and calcium mobilization by G_q and G_{i2}. *J Cell Biol* 121: 91–99, 1993.
38. Panettieri RA Jr., Yadvish PA, Kelly AM, Rubinstein NA, Kotlikoff MI. Histamine stimulates proliferation of airway smooth muscle and induces c-fos expression. *Am J Physiol (Lung Cell Mol Physiol)* 259/3: L365–71, 1990.
39. Noveral JP, Grunstein MM. Role and mechanism of thromboxane-induced proliferation of cultured airway smooth muscle cells. *Am J Physiol (Lung Cell Mol Physiol)* 263/7: L555–61, 1992.
40. Panettieri RA Jr.. Airways smooth muscle cell growth and proliferation. In: Raeburn D, Giembycz MA (eds). *Airways Smooth Muscle: Development and Regulation of Contractility*, pp. 41–68. Basel, Switzerland: Birkhauser Verlag, 1994.
41. Panettieri RA Jr., Cohen MD, Bilgen G. Airway smooth muscle cell proliferation is inhibited by microinjection of the catalytic subunit of cAMP dependent kinase. *Am Rev Respir Dis* 147: A252, 1993.
42. Noveral JP, Rosenberg SM, Anbar RA, Pawlowski NA, Grunstein MM. Role of endothelin-1 in regulating proliferation of cultured rabbit airway smooth muscle cells. *Am J Physiol (Lung Cell Mol Physiol)* 263/7: L317–24, 1992.
43. Panettieri RA Jr., Murray RK, DePalo LR, Yadvish PA, Kotlikoff MI. A human airway smooth muscle cell line that retains physiological responsiveness. *Am J Physiol: Cell Physiol* 256/25: C329–35, 1989.
44. Panettieri RA Jr., Tan EML, Ciocca V, Luttmann MA, Leonard TB, Hay DWP. Effects of LTD_4 on human airways smooth muscle cell proliferation, matrix expression, and contraction *in vitro*: Differential sensitivity to cysteinyl leukotriene receptor antagonists. *Am J Respir Cell Mol Biol* 19: 453–61, 1998.
45. Engelman JA, Ji L, Cantley LC. The evolution of phosphatidylinositol 3-kinases as regulators of growth and metabolism. *Nat Rev Genet* 7: 606–19, 2006.
46. Krymskaya VP. Targeting phosphatidylinositol 3-kinase pathway in airway smooth muscle: Rationale and promise. *BioDrugs* 21: 85–95, 2007.
47. Rameh LE, Cantley LC. The role of phosphoinositide 3-kinase lipid products in cell function. *J Biol Chem* 274: 8347–50, 1999.
48. Li J, Yen C, Liaw D, Podsypanina K, Bose S, Wang SI, Puc J, Miliaresis C, Rodgers L, McCombie R, Bigner SH, Giovanella BC, Ittmann M, Tycko B, Hibshoosh H, Wigler MH, Parsons R. PTEN, a putative protein tyrosine phosphatase gene mutated in human brain, breast, and prostate cancer. *Science* 275: 1943–47, 1997.
49. Maehama T, Taylor GS, Dixon JE. PTEN and myotubularin: Novel phosphoinositide phosphatases. *Annu Rev Biochem* 70: 247–79, 2001.
50. Steck PA, Pershouse MA, Jasser SA, Yung WKA, Lin HK, Ligon AH, Langford LA, Baumgard ML, Hattier T, Davis T, Frye C, Hu R, Swedlund B, Teng DHR, Tavtigian SV. Identification of a candidate tumour suppressor gene, MMAC1, at chromosome 10q23.3 that is mutated in multiple advanced cancers. *Nat Genet* 15: 356–62, 1997.
51. Toker A. Protein kinases as mediators of phosphoinositide 3-kinase signaling. *Mol Pharmacol* 57: 652–58, 2000.
52. Vanhaesebroeck B, Waterfield MD. Signaling by distinct classes of phosphoinositide 3-kinases. *Exp Cell Res* 253: 239–54, 1999.
53. Chantry D, Vojtek A, Kashishian A, Holtzman D, Wood C, Gray PW, Cooper JA, Hoekstra MF. p110δ, a novel phosphatidylinositol 3-kinase catalytic subunit that associates with p85 and is expressed predominantly in leukocytes. *J Biol Chem* 272: 19236–41, 1997.
54. Stoyanov B, Volonia S, Hanck T, Rubio I, Loubtchenkov M, Malek D, Stoyanova S, Vanhaesebroeck B, Dhand R, Nurnberg B, Gierschik P, Seedorf K, Hsuan JJ, Waterfield MD, Wetzker R. Cloning and characterization of a G protein-activated human phosphoinositide 3-kinase. *Science* 269: 690–93, 1995.
55. Domin J, Gaidarov I, Smith MEK, Keen JH, Waterfield MD. The class II phosphoinositide 3-kinase PI3K-C2α is concentrated in the trans-Golgi network and present in clathrin-coated vesicles. *J Biol Chem* 275: 11943–50, 2000.
56. Volinia S, Dhand R, Vanhaesebroeck B, MacDougall L, Stein R, Zvelebil MJ, Domin J, Panaretou C, Waterfield MD. A human phosphatidylinositol 3-kinase complex related to the yeast Vps34p-Vps15p protein sorting system. *EMBO J* 14: 3339–48, 1995.
57. Dan HC, Sun M, Yang L, Feldman RI, Sui XM, Yeung RS, Halley DJJ, Nicosia SV, Pledger WJ, Cheng JQ. PI3K/AKT pathway regulates

TSC tumor suppressor complex by phosphorylation of tuberin. *J Biol Chem* 277: 35364–70, 2002.

58. Manning BD, Tee AR, Longdon MN, Blenis J, Cantley LC. Identification of the tuberous sclerosis complex-2 tumor suppressor gene product tuberin as a target of the phosphoinositide 3-kinase/Akt pathway. *Mol Cell* 10: 151–62, 2002.
59. Potter CJ, Pedraza LG, Xu T. Akt regulates growth by directly phosphorylating Tsc2. *Nature Cell Biol* 4: 658–65, 2002.
60. Goncharova EA, Goncharov DA, Eszterhas A, Hunter DS, Glassberg MK, Yeung RS, Walker CL, Noonan D, Kwiatkowski DJ, Chou MM, Panettieri Jr RA, Krymskaya VP. Tuberin regulates p70s6 kinase activation and ribosomal protein S6 phosphorylation. A role for the TSC2 tumor suppressor gene in pulmonary lymphangioleiomyomatosis (LAM). *J Bio chem* 277: 30958–67, 2002.
61. Goncharova EA, Krymskaya VP. Pulmonary lymphangioleiomyomatosis (LAM): Progress and current challenges. *J Cell Biochem* 103: 369–82, 2008.
62. Long X, Ortiz-Vega S, Lin Y, Avruch J. Rheb binding to mammalian target of rapamycin (mTOR) is regulated by amino acid sufficiency. *J Biol Chem* 280: 23433–36, 2005.
63. Long X, Lin Y, Ortiz-Vega S, Yonezawa K, Avruch J. Rheb binds and regulates the mTOR kinase. *Curr Biol* 15: 702–13, 2005.
64. Sarbassov DD, Guertin DA, Ali SM, Sabatini DM. Phosphorylation and regulation of Akt/PKB by the rictor-mTOR complex. *Science* 307: 1098–101, 2005.
65. Jacinto E, Loewith R, Schmidt A, Lin S, Ruegg MA, Hall A, Hall MN. Mammalian TOR complex 2 controls the actin cytoskeleton and is rapamycin insensitive. *Nature Cell Biol* 6: 1122–28, 2004.
66. Goncharova EA, Ammit AJ, Irani C, Carroll RG, Eszterhas AJ, Panettieri RA Jr., Krymskaya VP. Phosphatidylinositol 3-kinase is required for proliferation and migration of human pulmonary vascular smooth muscle cells. *Am J Physiol (Lung Cell Mol Physiol)* 283: L354–L363, 2002.
67. Krymskaya VP, Ammit AJ, Hoffman RK, Eszterhas AJ, Panettieri RA Jr.. Activation of class IA phosphatidylinositol 3-kinase stimulates DNA synthesis in human airway smooth muscle cells. *Am J Physiol (Lung Cell Mol Physiol)* 280: L1009–L1018, 2001.
68. Bacqueville D, Casagrande F, Perret B, Chap H, Darbon JM, Breton-Douillon M. Phosphatidylinositol 3-kinase inhibitors block aortic smooth muscle cell proliferation in mid-late G1 phase: Effect on cyclin-dependent kinase 2 and the inhibitory protein p27KIP1. *Biochem Biophys Res Commun* 244: 630–36, 1998.
69. Saward L, Zahradka P. Angiotensin II activates phosphatidylinositol 3-kinase in vascular smooth muscle cells. *Circ Res* 81: 249–57, 1997.
70. Scott PH, Belham CM, Al-Hafidh J, Chilvers ER, Peacock AJ, Gould GW, Plevin R. A regulatory role for cAMP in phosphatidylinositol 3-kinase/p70 ribosomal S6 kinase-mediated DNA synthesis in platelet-derived-growth-factor-stimulated bovine airway smooth-muscle cells. *Biochem J* 318: 965–71, 1996.
71. Weiss RH, Apostolids A. Dissociation of phosphatidylinositol 3-kinase activity and mitogenic inhibition in vascular smooth muscle cells. *Cell Signal* 7: 113–22, 1995.
72. Hu ZW, Shi XY, Hoffman BB. α1 Adrenergic receptors activate phosphatidylinositol 3-kinase in human vascular muscle cells. *J Biol Chem* 271: 8977–82, 1996.
73. Takahashi T, Kawahara Y, Okuda M, Ueno H, Takeshita A, Yokoyama M. Angiotensin II stimulates mitogen-activated protein kinases and protein synthesis by a Ras-independent pathway in vascular smooth muscle cells. *J Biol Chem* 272: 16018–22, 1997.
74. Krymskaya VP, Penn RB, Orsini MJ, Scott PH, Plevin RJ, Walker TR, Eszterhas AJ, Amrani Y, Chilvers ER, Panettieri RA Jr. Phosphatidylinositol 3-kinase mediates mitogen-induced human airways smooth muscle cell proliferation. *Am J Physiol (Lung Cell Mol Physiol)* 277/21: L65–L78, 1999.
75. Hu Q, Klippel A, Muslin AJ, Fantl WJ, Williams LT. Ras-dependent induction of cellular response by constitutively active phosphatidylinositol-3 kinase. *Science* 268: 100–2, 1995.
76. Chung J, Grammer TC, Lemon KP, Kazlauskas A, Blenis J. PDGF- and insulin-dependent pp70^{S6k} activation mediated by phosphatidylinositol-3-OH kinase. *Nature* 370: 71–75, 1994.
77. Jefferies HBJ, Fumagalli S, Dennis PB, Reinhard C, Pearson RB, Thomas G. Rapamycin suppresses 5′TOP mRNA translation through inhibition of p70s6k. *EMBO J* 16: 3693–704, 1997.
78. Krymskaya VP, Orsini MJ, Eszterhas AJ, Brodbeck KC, Benovic JL, Panettieri RA Jr., Penn RB. Mechanisms of proliferation synergy by receptor tyrosine kinase and G protein-coupled receptor activation in human airway smooth muscle. *Am J Respir Cell Mol Biol* 23: 546–54, 2000.
79. Billington CK, Kong KC, Bhattacharyya R, Wedegaertner PB, Panettieri RA, Chan TO, Penn RB. Cooperative regulation of p70S6 kinase by receptor tyrosine kinases and G protein-coupled receptors augments airway smooth muscle growth. *Biochemistry* 44: 14595–605, 2005.
80. Krymskaya VP, Goncharova EA, Ammit AJ, Lim PN, Goncharov DA, Eszterhas A, Panettieri RA Jr. Src is necessary and sufficient for human airway smooth muscle cell proliferation and migration. *FASEB J* 19: 428–30, 2005.
81. Billington CK, Penn RB. Signaling and regulation of G protein-coupled receptors in airway smooth muscle. *Respir Res* 4: 1–23, 2003.
82. Brown MT, Cooper JA. Regulation, substrates and functions of src. *Biochim Biophys Acta* 1287: 121–49, 1996.
83. Thomas SM, Brugge JS. Cellular functions regulated by Src family kinases. *Annu Rev Cell Dev Biol* 13: 513–609, 1997.
84. Tsang F, Hwa Choo H, Dawe GS, Fred Wong WS. Inhibitors of the tyrosine kinase signaling cascade attenuated thrombin-induced guinea pig airway smooth muscle cell proliferation. *Biochem Biophys Res Commun* 293: 72–78, 2002.
85. Sayeski PP, Ali MS. The critical role of c-Src and the Shc/Grb2/ERK2 signaling pathway in angiotensin II-dependent VSMC proliferation. *Exp Cell Res* 287: 339–49, 2003.
86. Luttrell LM, Ferguson SSG, Daaka Y, Miller WE, Maudsley S, Della Rocca GJ, Lin FT, Kawakatsu H, Owada K, Luttrell DK, Caron MG, Lefkowitz RJ. β-Arrestin-dependent formation of β_2 adrenergic receptor-Src protein kinase complexes. *Science* 283: 655–61, 1999.
87. Marinissen MJ, Gutkind JS. G-protein-coupled receptors and signaling networks: Emerging paradigms. *Trends Pharmacol Sci* 22: 368–76, 2001.
88. Krymskaya VP, Hoffman R, Eszterhas A, Kane S, Ciocca V, Panettieri RA Jr. EGF activates ErbB-2 and stimulates phosphatidylinositol 3-kinase in human airway smooth muscle cells. *Am J Physiol (Lung Cell Mol Physiol)* 276/20: L246–55, 1999.
89. Lazaar AL, Amrani Y, Panettieri RA Jr.. The role of inflammation in the regulation of airway smooth muscle cell function and growth. In: Busse W, Holgate S (eds). *Asthma and Rhinitis*, 2nd edn, pp. 1402–13. Oxford, UK: Blackwell Science, Ltd., 2000.
90. Berkman N, Krishnan VL, Gilbey T, Newton R, O'Connor B, Barnes PJ, Chung KF. Expression of RANTES mRNA and protein in airways of patients with mild asthma. *Am J Respir Crit Care Med* 154: 1804–11, 1996.
91. Ammit AJ, Hoffman RK, Amrani Y, Lazaar AL, Hay DWP, Torphy TJ, Penn RB, Panettieri RA Jr. TNFα-induced secretion of RANTES and IL-6 from human airway smooth muscle cells: Modulation by cAMP. *Am J Respir Cell Mol Biol* 23: 794–802, 2000.
92. John M, Hirst SJ, Jose PJ, Robichaud A, Berkman N, Witt C, Twort CHC, Barnes PJ, Chung KF. Human airway smooth muscle cells express and release RANTES in response to T helper 1 cytokines. Regulation by T helper 2 cytokines and corticosteroids. *J Immunol* 158: 1841–47, 1997.
93. Pype JL, Dupont LJ, Menten P, Van Coillie E, Opdenakker G, Van Damme J, Chung KF, Demedts MG, Verleden GM. Expression of monocyte chemotactic protein (MCP)-1, MCP-2, and MCP-3 by human airway smooth-muscle cells. Modulation by corticosteroids and T-helper 2 cytokines. *Am J Respir Cell Mol Biol* 21: 528–36, 1999.
94. Chung KF, Patel HJ, Fadlon EJ, Rousell J, Haddad E-B, Jose PJ, Mitchell J, Belvisi M. Induction of eotaxin expression and release from

human airway smooth muscle cells by IL-1β and TNFα: Effects of IL-10 and corticosteroids. *Br J Pharmacol* 127: 1145–50, 1999.
95. Ghaffar O, Hamid Q, Renzi PM, Allahkverdi Z, Molet S, Hogg JC, Shore SA, Luster AD, Lamkhioued B. Constitutive and cytokine-stimulated expression of eotaxin by human airway smooth muscle cells. *Am J Respir Crit Care Med* 159: 1933–42, 1999.
96. John M, Au B-T, Jose PJ, Lim S, Saunders M, Barnes PJ, Mitchell JA, Belvisi MG, Chung KF. Expression and release of interleukin-8 by human airway smooth muscle cells: Inhibition by Th-2 cytokines and corticosteroids. *Am J Respir Cell Mol Biol* 18: 84–90, 1998.
97. Pang L, Knox AJ. Bradykinin stimulates IL-8 production in cultured human airway smooth muscle cells: Role of cyclooxygenase products. *J Immunol* 161: 2509–15, 1998.
98. Govindaraju V, Michoud MC, Al-Chalabi M, Ferraro P, Powell WS, Martin JG. Interleukin-8: Novel roles in human airway smooth muscle cell contraction and migration. *Am J Physiol: Cell Physiol* 291: C957–65, 2006.
99. Mullan CS, Riley M, Clarke D, Tatler A, Sutcliffe A, Knox AJ, Pang L. Beta-tryptase regulates IL-8 expression in airway smooth muscle cells by a PAR-2-independent mechanism. *Am J Respir Cell Mol Biol* 38(5):600–8, 2008.
100. Dragon S, Rahman MS, Yang J, Unruh H, Halayko AJ, Gounni AS. IL-17 enhances IL-1beta-mediated CXCL-8 release from human airway smooth muscle cells. *Am J Physiol (Lung Cell Mol Physiol)* 292: L1023–29, 2007.
101. Zuyderduyn S, Ninaber DK, Hiemstra PS, Rabe KF. The antimicrobial peptide LL-37 enhances IL-8 release by human airway smooth muscle cells. *J Allergy Clin Immunol* 117: 1328–35, 2006.
102. Odaka M, Matsukura S, Kuga H, Kokubu F, Kasama T, Kurokawa M, Kawaguchi M, Ieki K, Suzuki S, Watanabe S, Homma T, Takeuchi H, Nohtomi K, Schleimer RP, Adachi M. Differential regulation of chemokine expression by Th1 and Th2 cytokines and mechanisms of eotaxin/CCL-11 expression in human airway smooth muscle cells. *Int Arch Allergy Immunol* 143(Suppl 1): 84–88, 2007.
103. Rahman MS, Yamasaki A, Yang J, Shan L, Halayko AJ, Gounni AS. IL-17A induces eotaxin-1/CC chemokine ligand 11 expression in human airway smooth muscle cells: Role of MAPK (Erk1/2, JNK, and p38) pathways. *J Immunol* 177: 4064–71, 2006.
104. Chan V, Burgess JK, Ratoff JC, O'Connor B J, Greenough A, Lee TH, Hirst SJ. Extracellular matrix regulates enhanced eotaxin expression in asthmatic airway smooth muscle cells. *Am J Respir Crit Care Med* 174: 379–85, 2006.
105. Liu L, Yang J, Huang Y. Human airway smooth muscle cells express eotaxin in response to signaling following mast cell contact. *Respiration* 73: 227–35, 2006.
106. Pang L, Knox AJ. Regulation of TNF-α-induced eotaxin release from cultured human airway smooth muscle cells by β_2-agonists and corticosteroids. *FASEB J* 115: 261–69, 2001.
107. Pang L, Knox AJ. Synergistic inhibition by β_2-agonists and corticosteroids on tumor necrosis factor-α-induced interleukin-8 release from cultured human airway smooth-muscle cells. *Am J Respir Cell Mol Biol* 23: 79–85, 2000.
108. Ammit AJ, Hastie AT, Edsall LC, Hoffman RK, Amrani Y, Krymskaya VP, Kane SA, Peters SP, Penn RB, Spiegel S, Panettieri RA Jr.. Sphingosine 1-phosphate modulates human airway smooth muscle cell functions that promote inflammation and airway remodeling in asthma. *FASEB J* 15: 1212–14, 2001.
109. Saunders MA, Mitchell JA, Seldon PM, Yacoub MH, Barnes PJ, Giembycz MA, Belvisi MG. Release of granulocyte-macrophage colony stimulating factor by human cultured airway smooth muscle cells: Suppression by dexamethasone. *Br J Pharmacol* 120: 545–46, 1997.
110. De S, Zelazny ET, Souhrada JF, Souhrada M. IL-1β and IL-6 induce hyperplasia and hypertrophy of cultured guinea pig airway smooth muscle cells. *J Appl Physiol* 78: 1555–63, 1995.
111. Elias JA, Wu Y, Zheng T, Panettieri RA Jr. Cytokine- and virus-stimulated airway smooth muscle cells produce IL-11 and other IL-6-type cytokines. *Am J Physiol (Lung Cell Mol Physiol)* 273/17: L648–55, 1997.
112. McKay S, Hirst SJ, Bertrand-de Haas M, de Jonste JC, Hoogsteden HC, Saxena PR, Sharma HS. Tumor necrosis factor-α enhances mRNA expression and secretion of interleukin-6 in cultured human airway smooth muscle cells. *Am J Respir Cell Mol Biol* 23: 103–11, 2000.
113. DiCosmo BF, Geba GP, Picarella D, Elias JA, Rankin JA, Stripp BR, Whitsett JA, Flavell RA. Airway epithelial cell expression of interleukin-6 in transgenic mice. Uncoupling of airway inflammation and bronchial hyperreactivity. *J Clin Invest* 94: 2028–35, 1994.
114. Hakonarson H, Maskeri N, Carter C, Chuang S, Grunstein MM. Autocrine interaction between IL-5 and IL-1β mediates altered responsiveness of atopic asthmatic sensitized airway smooth muscle. *J Clin Invest* 104: 657–67, 1999.
115. Knight DA, Lydell CP, Zhou D, Weir TD, Schellenberg RR, Bai TR. Leukemia inhibitory factor (LIF) and LIF receptor in human lung: Distribution and regulation of LIF release. *Am J Respir Cell Mol Biol* 20: 834–41, 1999.
116. Hallsworth MP, Soh CPC, Twort CHC, Lee TH, Hirst SJ. Cultured human airway smooth muscle cells stimulated by interleukin-1β enhance eosinophil survival. *Am J Respir Cell Mol Biol* 19: 910–19, 1998.
117. Springer TA. Adhesion receptors of the immune system. *Nature* 346: 425–34, 1990.
118. Lazaar AL, Albelda SM, Pilewski JM, Brennan B, Puré E, Panettieri RA Jr.. T lymphocytes adhere to airway smooth muscle cells via integrins and CD44 and induce smooth muscle cell DNA synthesis. *J Exp Med* 180: 807–16, 1994.
119. Amrani Y, Lazaar AL, Panettieri RA Jr. Up-regulation of ICAM-1 by cytokines in human tracheal smooth muscle cells involves an NF-κB-dependent signaling pathway that is only partially sensitive to dexamethasone. *J Immunol* 163: 2128–34, 1999.
120. Belvisi MG, Saunders MA, Haddad E-B, Hirst SJ, Yacoub MH, Barnes PJ, Mitchell JA. Induction of cyclo-oxygenase-2 by cytokines in human cultured airway smooth muscle cells: Novel inflammatory role of this cell type. *Br J Pharmacol* 120: 910–16, 1997.
121. Pang L, Knox AJ. Effect of interleukin-1β, tumour necrosis factor-α and interferon-γ on the induction of cyclo-oxygenase-2 in cultured human airway smooth muscle cells. *Br J Pharmacol* 121: 579–87, 1997.
122. Panettieri RA Jr., Lazaar AL, Puré E, Albelda SM. Activation of cAMP-dependent pathways in human airway smooth muscle cells inhibits TNF-α-induced ICAM-1 and VCAM-1 expression and T lymphocyte adhesion. *J Immunol* 154: 2358–65, 1995.
123. Dustin ML, Springer TA. Role of lymphocyte adhesion receptors in transient interactions and cell locomotion. *Annu Rev Immunol* 9: 27–66, 1991.
124. van Seventer GA, Newman W, Shimuzu Y, Nutman TB, Tanaka Y, Horgan KJ, Gopal TV, Ennis E, O'Sullivan D, Grey H. Analysis of T cell stimulation by superantigen plus major histocompatibility complex class II molecules or by CD3 monoclonal antibody: Costimulation by purified adhesion ligands VCAM-1, ICAM-1, but not ELAM-1. *J Exp Med* 174: 901–13, 1991.
125. Lazaar AL, Reitz HE, Panettieri RA Jr., Peters SP, Puré E. Antigen receptor-stimulated peripheral blood and bronchoalveolar lavage-derived T cells induce MHC class II and ICAM-1 expression on human airway smooth muscle. *Am J Respir Cell Mol Biol* 16: 38–45, 1997.
126. Lazaar AL, Amrani Y, Hsu J, Panettieri RA Jr., Fanslow WC, Albelda SM, Puré E. CD40-mediated signal transduction in human airway smooth muscle. *J Immunol* 161: 3120–27, 1998.
127. Hakonarson H, Kim C, Whelan R, Campbell D, Grunstein MM. Bi-directional activation between human airway smooth muscle cells and T lymphocytes: Role in induction of altered airway responsiveness. *J Immunol* 166: 293–303, 2001.
128. Lazaar AL, Krymskaya VP, Das SK. VCAM-1 activates phosphatidylinositol 3-kinase and induces p120(Cbl) phosphorylation in human airway smooth muscle cells. *J Immunol* 166: 155–61, 2001.
129. Gaston B, Drazen JM, Loscalzo J, Stamler JS. The biology of nitrogen oxides in the airways. *Am J Respir Crit Care Med* 149: 538–51, 1994.

130. Hamad AM, Johnson SR, Knox AJ. Antiproliferative effects of NO and ANP in cultured human airway smooth muscle. *Am J Physiol* 277: L910–18, 1999.
131. Patel HJ, Belvisi MG, Donnelly LE, Yacoub MH, Chung KF, Mitchell JA. Constitutive expressions of type I NOS in human airway smooth muscle cells: Evidence for an antiproliferative role. *FASEB J* 13: 1810–16, 1999.
132. Hakonarson H, Grunstein MM. Autologously up-regulated Fc receptor expression and action in airway smooth muscle mediates its altered responsiveness in the atopic asthmatic sensitized state. *Proc Natl Acad Sci USA* 95: 5257–62, 1998.
133. Roper RL, Conrad DH, Brown DM, Warner GL, Phipps RP. Prostaglandin E2 promotes IL-4-induced IgE and IgG1 synthesis. *J Immunol* 145: 2644–51, 1990.
134. Lazzeri N, Belvisi MG, Patel HJ, Yacoub MH, Fan Chung K, Mitchell JA. Effects of prostaglandin E(2) and cAMP elevating drugs on GM-CSF release by cultured human airway smooth muscle cells. Relevance to asthma therapy. *Am J Respir Cell Mol Biol* 24: 44–48, 2001.
135. Issa R, Xie S, Khorasani N, Sukkar M, Adcock IM, Lee KY, Chung KF. Corticosteroid inhibition of growth-related oncogene protein-alpha via mitogen-activated kinase phosphatase-1 in airway smooth muscle cells. *J Immunol* 178: 7366–75, 2007.
136. Zhang K, Shan L, Rahman MS, Unruh H, Halayko AJ, Gounni AS. Constitutive and inducible thymic stromal lymphopoietin expression in human airway smooth muscle cells: Role in chronic obstructive pulmonary disease. *Am J Physiol (Lung Cell Mol Physiol)* 293: L375–82, 2007.
137. Sukkar MB, Xie S, Khorasani NM, Kon OM, Stanbridge R, Issa R, Chung KF. Toll-like receptor 2, 3, and 4 expression and function in human airway smooth muscle. *J Allergy Clin Immunol* 118: 641–48, 2006.
138. Kang BN, Tirumurugaan KG, Deshpande DA, Amrani Y, Panettieri RA, Walseth TF, Kannan MS. Transcriptional regulation of CD38 expression by tumor necrosis factor-alpha in human airway smooth muscle cells: Role of NF-kappaB and sensitivity to glucocorticoids. *FASEB J* 20: 1000–2, 2006.
139. Keslacy S, Tliba O, Baidouri H, Amrani Y. Inhibition of TNFα-inducible inflammatory genes by IFNγ is associated with altered NF-κB transactivation and enhanced HDAC activity. *Mol Pharmacol* 71: 609–18, 2007.
140. Vanaudenaerde BM, Wuyts WA, Geudens N, Dupont LJ, Schoofs K, Smeets S, Van Raemdonck DE, Verleden GM. Macrolides inhibit IL17-induced IL8 and 8-isoprostane release from human airway smooth muscle cells. *Am J Transplant* 7: 76–82, 2007.
141. Catley MC, Sukkar MB, Chung KF, Jaffee B, Liao SM, Coyle AJ, Haddad el B, Barnes PJ, Newton R. Validation of the anti-inflammatory properties of small-molecule IkappaB kinase (IKK)-2 inhibitors by comparison with adenoviral-mediated delivery of dominant-negative IKK1 and IKK2 in human airways smooth muscle. *Mol Pharmacol* 70: 697–705, 2006.
142. Ammit AJ, Lazaar AL, Irani C, O'Neill GM, Gordon ND, Amrani Y, Penn RB, Panettieri RA Jr. Tumor necrosis factor-α-induced secretion of RANTES and interleukin-6 from human airway smooth muscle cells: Modulation by glucocorticoids and β-agonists. *Am J Respir Cell Mol Biol* 26: 465–74, 2002.
143. Tliba O, Cidlowski JA, Amrani Y. CD38 expression is insensitive to steroid action in cells treated with tumor necrosis factor-alpha and interferon-gamma by a mechanism involving the up-regulation of the glucocorticoid receptor beta isoform. *Mol Pharmacol* 69: 588–96, 2006.
144. Tran T, Fernandes DJ, Schuliga M, Harris T, Landells L, Stewart AG. Stimulus-dependent glucocorticoid-resistance of GM-CSF production in human cultured airway smooth muscle. *Br J Pharmacol* 145: 123–31, 2005.
145. Vlahos R, Stewart AG. Interleukin-1alpha and tumour necrosis factor-alpha modulate airway smooth muscle DNA synthesis by induction of cyclo-oxygenase-2: Inhibition by dexamethasone and fluticasone propionate. *Br J Pharmacol* 126: 1315–24, 1999.
146. Adcock IM, Gilbey T, Gelder CM, Chung KF, Barnes PJ. Glucocorticoid receptor localization in normal and asthmatic lung. *Am J Respir Crit Care Med* 154: 771–82, 1996.
147. Roth M, Johnson PR, Borger P, Bihl MP, Rudiger JJ, King GG, Ge Q, Hostettler K, Burgess JK, Black JL, Tamm M. Dysfunctional interaction of C/EBPalpha and the glucocorticoid receptor in asthmatic bronchial smooth-muscle cells. *N Engl J Med* 351: 560–74, 2004.
148. Bonacci JV, Schuliga M, Harris T, Stewart AG. Collagen impairs glucocorticoid actions in airway smooth muscle through integrin signalling. *Br J Pharmacol* 149: 365–73, 2006.
149. Roth M, Johnson PR, Rudiger JJ, King GG, Ge Q, Burgess JK, Anderson G, Tamm M, Black JL. Interaction between glucocorticoids and beta2 agonists on bronchial airway smooth muscle cells through synchronised cellular signalling. *Lancet* 360: 1293–99, 2002.
150. Hirst SJ, Lee TH. Airway smooth muscle as a target of glucocorticoid action in the treatment of asthma. *Am J Respir Crit Care Med* 158: S201–6, 1998.
151. Miller M, Cho JY, McElwain K, McElwain S, Shim JY, Manni M, Baek JS, Broide DH. Corticosteroids prevent myofibroblast accumulation and airway remodeling in mice. *Am J Physiol (Lung Cell Mol Physiol)* 290: L162–69, 2006.
152. Tliba O, Damera G, Banerjee A, Gu S, Baidouri H, Keslacy S, Amrani Y. Cytokines induce an early steroid resistance in airway smooth muscle cells: novel role of interferon regulatory factor-1. *Am J Respir Cell Mol Biol* 38(4): 463–72, 2008.
153. Bosken CH, Wiggs BR, Pare PD, Hogg JC. Small airway dimensions in smokers with obstruction to airflow. *Am Rev Respir Dis* 142: 563–70, 1990.
154. Cosio MG, Hale KA, Niewoehner DE. Morphologic and morphometric effects of prolonged cigarette smoking on the small airways. *Am Rev Respir Dis* 122: 265–321, 1980.
155. Hogg JC, Chu F, Utokaparch S, Woods R, Elliott WM, Buzatu L, Cherniack RM, Rogers RM, Sciurba FC, Coxson HO, Pare PD. The nature of small-airway obstruction in chronic obstructive pulmonary disease. *N Engl J Med* 350: 2645–53, 2004.
156. Saetta M, Di Stefano A, Turato G, Facchini FM, Corbino L, Mapp CE, Maestrelli P, Ciaccia A, Fabbri LM. CD8+ T-lymphocytes in peripheral airways of smokers with chronic obstructive pulmonary disease. *Am J Respir Crit Care Med* 157: 822–26, 1998.
157. Kuwano K, Bosken CH, Pare PD, Bai TR, Wiggs BR, Hogg JC. Small airways dimensions in asthma and in chronic obstructive pulmonary disease. *Am Rev Respir Dis* 148: 1220–25, 1993.
158. Tiddens HA, Pare PD, Hogg JC, Hop WC, Lambert R, de Jongste JC. Cartilaginous airway dimensions and airflow obstruction in human lungs. *Am J Respir Crit Care Med* 152: 260–66, 1995.
159. Cosio M, Ghezzo H, Hogg JC, Corbin R, Loveland M, Dosman J, Macklem PT. The relations between structural changes in small airways and pulmonary-function tests. *N Engl J Med* 298: 1277–81, 1978.
160. Chung KF. The role of airway smooth muscle in the pathogenesis of airway wall remodeling in chronic obstructive pulmonary disease. *Proc Am Thoracic Soc* 2: 347–54, discussion 371–72, 2005.
161. Panettieri RA Jr. Airway smooth muscle: An immunomodulatory cell. *J Allergy Clin Immunol* 110: S269–74, 2002.
162. Hardaker EL, Bacon AM, Carlson K, Roshak AK, Foley JJ, Schmidt DB, Buckley PT, Comegys M, Panettieri JRA, Sarau HM, Belmonte KE. Regulation of TNF-α- and IFN-γ-induced CXCL10 expression: Participation of the airway smooth muscle in the pulmonary inflammatory response in chronic obstructive pulmonary disease. *FASEB J* 18: 191–93, 2004.
163. Jarai G, Sukkar M, Garrett S, Duroudier N, Westwick J, Adcock I, Chung KF. Effects of interleukin-1beta, interleukin-13 and transforming growth factor-beta on gene expression in human airway smooth muscle using gene microarrays. *Eur J Pharmacol* 497: 255–65, 2004.

19

CHAPTER

Tracheobronchial Circulation

Adam Wanner[1] and Gabor Horvath[2]

[1]Division of Pulmonary and Critical Care Medicine, University of Miami Miller School of Medicine, Miami, FL, USA

[2]Department of Pulmonology, Semmelweiss University, Budapest, Hungary

INTRODUCTION

Since the blood circulation typically participates in inflammatory processes at the tissue level, the vasculature of the tracheobronchial tree can be expected to undergo structural and functional changes in asthma and chronic obstructive pulmonary disease (COPD), conditions that are associated with airway inflammation. As will be shown in this chapter, there are distinct differences in the vascular abnormalities between asthma and COPD. The purpose of this chapter is to review the pathophysiological role of the tracheobronchial circulation in asthma and COPD and the vascular effects of pharmacologic interventions. The *in vivo* measurement of tracheobronchial blood flow in man usually captures subepithelial blood flow in the conducting airways, also termed airway blood flow (Qaw); therefore, this review will focus on Qaw in asthma and COPD, with a brief discussion of the normal airway circulation as background.

NORMAL AIRWAY CIRCULATION

The airway circulation, which derives its blood from the systemic circulation, is the principal vascular supply to the airway wall (Fig. 19.1) [1]. Bronchial arteries usually arise from the aorta or intercostal arteries and form a peribronchial plexus surrounding the bronchial wall. Branches penetrate the muscular layer to form a subepithelial plexus. The two interconnected plexuses follow airways as far as the terminal bronchiole. Although bronchial capillaries anastomose freely with the pulmonary circulation along the airways and anastomoses have been demonstrated at pre-, post-, and capillary-levels [2], venous blood from the intraparenchymal bronchial vasculature mainly drains through post-capillary pulmonary vessels [3] to the left heart [4]. Venous blood from the trachea and major bronchi drains through the vena azygos and superior vena cava to the right heart.

Under physiological conditions, total bronchial blood flow comprises 0.5–1% of cardiac output. The major part of blood flow is distributed to the subepithelial tissues where the microvasculature comprises 10–20% of tissue volume [5]. Subepithelial blood flow (Qaw) has been reported to range between 30 and 95 ml min^{-1} 100 g wet $tissue^{-1}$ in different species including in human [6, 7]. Its main function presumably is to nourish the epithelium that has one of the highest metabolic rates in the body. In healthy human adults, Qaw in the tracheobronchial tree to a 200 μm depth from the epithelial surface amounts to between 25 and 40 μl min^{-1} ml^{-1}, where ml reflects the anatomical dead space lined by the epithelium [8–10].

Qaw is influenced by a variety of factors (Table 19.1). Although adrenergic and cholinergic nerves have been demonstrated in the airway wall [11, 12], physiological and pharmacological studies suggest that the main nervous control of the airway circulation is by the sympathetic nervous system [11]. In different animal models, sympathetic nerve stimulation [13], close-arterial injection of adrenergic agonists [14], and inhalation of selective α-adrenergic agonists [15] have been shown to induce vasoconstriction. Selective β_2-adrenergic agonists increase bronchial blood flow in sheep [6, 8], a finding subsequently confirmed in humans [16] (Table 19.1). Parasympathetic nerve stimulation and intravascular administration of cholinergic

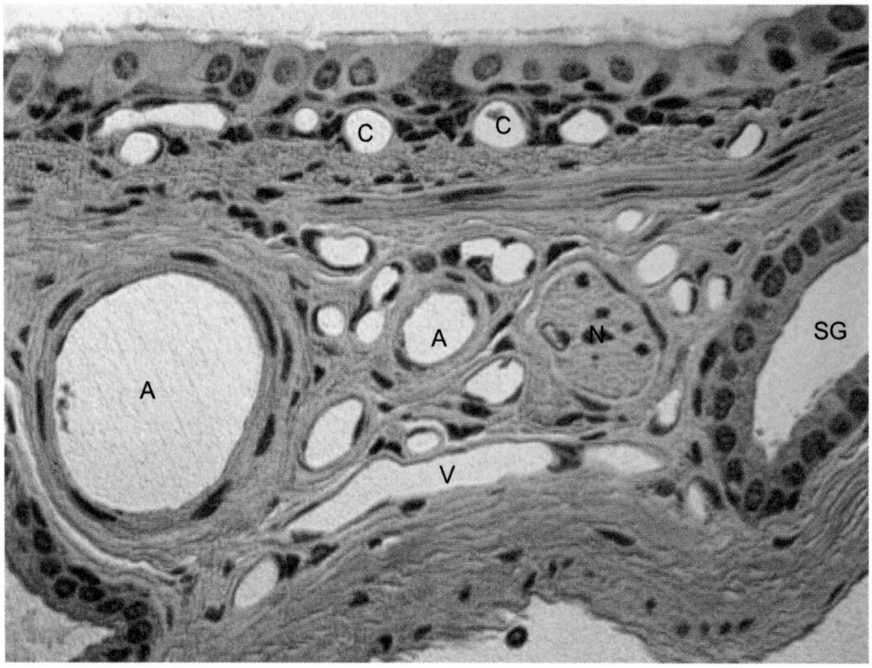

FIG. 19.1 Section of sheep bronchus (glutaraldehyde-fixed at 100 mmHg aortic pressure, paraffin embedded, HE-stained, ×200). A: arterioles; C: capillaries; V: venule; SG: submucosal gland; N: autonomic neuron. Section kindly provided by A. Mariassy.

TABLE 19.1 Vasoactive substances that potentially could participate in asthma-and COPD associated airway inflammation.

Vasodilators	Vasoconstrictors
β-adrenergic agonists	α-adrenergic agonists
Cholinergic agonists	NPY
VIP, PHI, PHM	LTC_4, LTD_4
NO	TNF-α
SP, NKA, CGRP	ET-1
Histamine, bradykinin, PAF	GCS (indirect)
5-HT	PGE_1, $FGF_{2\alpha}$, PGD_2
VEGF	

VIP: vasoactive intestinal polypeptide; PHI: polypeptide histidine isoleucine; PHM: polypeptide histidine methionine; NO: nitric oxide; SP: substance P; NKA: neurokinin A; CGRP: calcitonin gene related peptide; PAF: platelet-activating factor; 5-HT: serotonine; PG: prostaglandin; VEGF: vascular endothelial growth factor; NPY: neuropeptide Y; LT: leukotriene; TNF-α: tumor necrosis factor-α; ET-1: endothelin-1; GCS: glucocorticosteroids.

agonists cause vasodilatation in several animal models [11], whereas aerosolized acetylcholine has only minimal effects [17]. Nonadrenergic, noncholinergic mechanisms have also been described. Vasoactive intestinal polypeptide (VIP), polypeptide histidine isoleucine (PHI), or polypeptide histidine methionine (PHM) released by parasympathetic nerves are vasodilators, whereas neuropeptide Y (NPY) released by sympathetic nerves is a vasoconstrictor in the tracheobronchial circulation [13, 14]. Sensory neuropeptides, as substance P (SP), neurokinin A (NKA), and calcitonin gene-related peptide (CGRP) released from unmyelinated sensory afferents induce vasodilation [11]. Nitrergic mechanisms have recently been reported to regulate bronchovascular tone and partially mediate β-adrenergic vasodilator responses [15, 18, 19]. Finally, physicochemical stresses including hydrostatic pressure, temperature, humidity, deviations from normal osmolarity, hypoxia, and acid–base relations have all been reported to influence airway blood flow [20]. These factors may be of relevance in asthma and COPD where hydrostatic airway and pleural pressures are altered by the abnormal lung mechanics; there may be disturbances in gas exchange; there is airway inflammation; and there may be hemodynamic changes with elevated pulmonary arterial, right atrial, and left atrial pressures due to secondary pulmonary hypertension or cardiovascular comorbidity.

AIRWAY CIRCULATION IN ASTHMA

Allergic and non-allergic inflammation is considered a major factor in the vascular changes that have been associated with asthma [82]. The inflammatory mechanisms include the complex actions of inflammatory cells and mediators, neurotransmitters, and neuropeptides on vascular endothelial and smooth muscle cells. The main vascular manifestations are hyperemia, hyperpermeability, and edema formation.

Structure

It has been recognized for many years that asthmatics have an increased cross-sectional submucosal vascular area, an

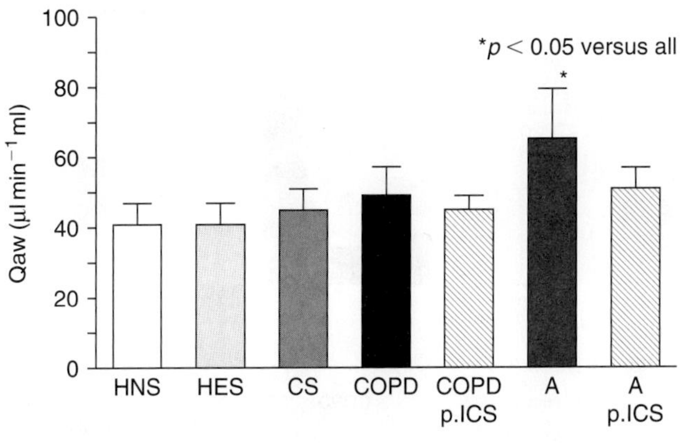

FIG. 19.2 Airway blood flow (Qaw) in healthy non-smokers (HNS), healthy ex-smokers (HES), current smokers (CS), COPD, COPD after long-term treatment with an inhaled corticosteroid (COPD p.ICS), asthma (A), and asthma after long-term treatment with an inhaled corticosteroid (A p.ICS). Mean values with SE in brackets. Adapted from Refs [9] and [29].

increased size and number of vessels in their airway wall, and intimal thickening [21–24]. This can be considered the vascular component of airway wall remodeling. The stimulus for new vessel formation in asthma is unclear, although various endothelial growth factors, inflammatory cytokines, and other putative angiogenic factors have been proposed [25–27]. It appears that the structural changes are already present during childhood and may even be present in atopic children without clinical asthma [28].

Function

Airway blood flow

Allergic and non-allergic airway inflammation has been shown to be associated with an increase in Qaw (Fig. 19.2) [6, 14]. This is also the case in subjects with stable asthma [8–10, 30]. An increase in Qaw is generally thought to result from dilatation of resistance arteries. Numerous inflammatory mediators have a vasodilator effect [8]. Histamine has a triphasic effect characterized by an initial vasodilation, followed by vasoconstriction and then a long-lasting vasodilation [8]. Sensory neuropeptides released from afferent nerves are also strong vasodilators. Reflex bronchial vasodilation is largely mediated by cholinergic and noncholinergic parasympathetic vagal pathways [31]. Not all hemodynamically active inflammatory mediators are vasodilators. For example, endothelin-1, a potent vasoconstrictor is increased in the airway of asthmatics [32]. Its release may be induced by pro-inflammatory cytokines such as tumor necrosis factor-α [33]. However, the net effect of inflammatory mediators is vasodilation.

While baseline Qaw is increased in asthma, vascular reactivity is markedly deranged, with blunted β_2-adrenergic vasodilator responsiveness and enhanced α-adrenergic vasoconstrictor responsiveness (Fig. 19.3) [8, 9, 30, 34]. The former observation suggests abnormal endothelium-dependent vasodilation in the airway circulation.

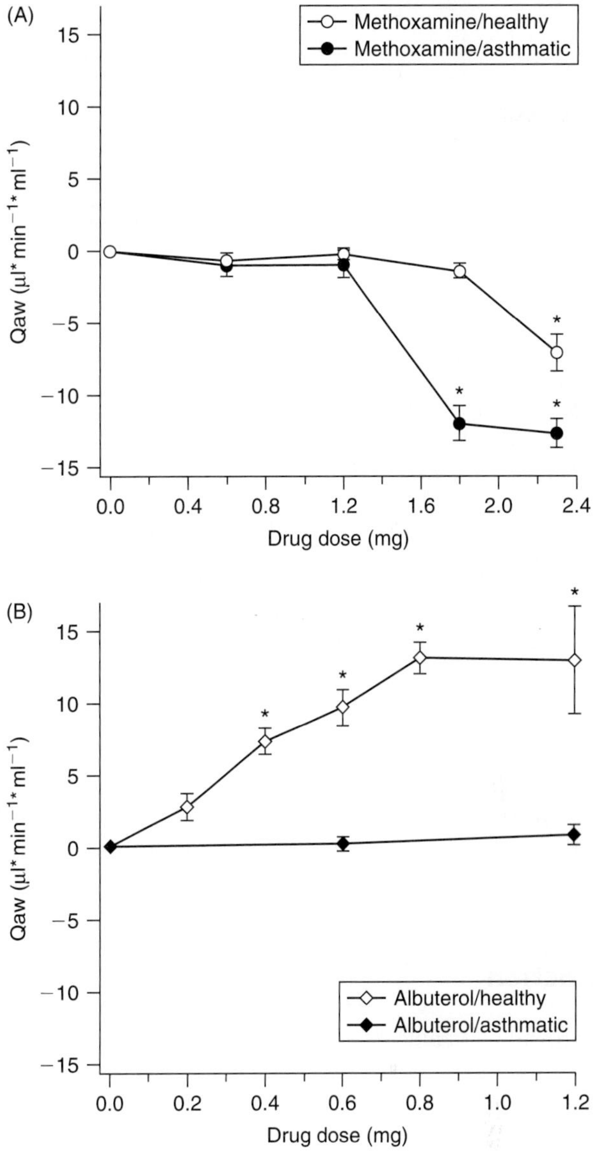

FIG. 19.3 Effects of inhaled methoxamine and albuterol on airway blood flow (Qaw) in 11 healthy (A) and 11 asthmatic (B) subjects. Mean values ±SE. *$p < 0.05$ versus baseline. Reproduced from Ref. [8].

Microvascular hyperpermeability

Asthma-associated leakage of macromolecules from the microvasculature occurs through endothelial pores and vesicles or through the formation of intercellular gaps in post-capillary venules. Inflammatory mediators and sensory autonomic nerve stimulation have been shown to increase microvascular permeability by producing intercellular gaps [11, 27] and induce interstitial edema formation. In addition, plasma components can collect in the airway lumen by passing through paracellular gaps between epithelial cells thereby contributing to excessive airway secretions. Vasodilation and hyperperfusion-related microvascular congestion have been shown to potentiate microvascular hyperpemeability [35]. In this regard, blood flow is related to edema formation.

AIRWAY CIRCULATION IN COPD

Structure

Early reports on the structure of the airway circulation in COPD suggested that intrapulmonary bronchial artery branches are obliterated or narrowed in emphysematous regions of the lungs and that the bronchial venous system is greatly expanded [36, 37]. These findings have not been supported by recent studies that showed increased airway wall vascularity in stable COPD with a slight increase in the number of vessels in small airways, but to a much lesser degree than in asthma, where there is an increased vascularity throughout the bronchial tree [29, 38]. Thus, in contrast to asthma, angiogenesis in the airway wall does not appear to be a typical feature of COPD.

However, in patients with COPD accompanied by bronchiectasis, there is an extensive remodeling of the airway circulation with a marked proliferation and enlargement of the bronchial arteries and numerous pre-capillary anastomoses between the pulmonary and bronchial circulation [39–41]. This vascular remodeling is a likely cause of hemoptysis encountered by patients with bronchiectasis. It is also likely that the primary stimuli for the development of bronchiectasis and airway vascular hypertrophy are suppurative airway infections that typically are seen in these patients. In support of this assumption, chronic lung abscesses also have been shown to lead to a localized hypertrophy and hyperplasia of the airway circulation [41].

Function

We found no significant difference in Qaw among ex-smokers with COPD ($51 \pm 3\,\mu l\,min^{-1}\,ml^{-1}$ anatomical deadspace), healthy ex-smokers (HES) ($41 \pm 3\,\mu l\,min^{-1}\,ml^{-1}$), healthy current smokers ($46 \pm 2\,\mu l\,min^{-1}\,ml^{-1}$) and healthy lifetime non-smokers ($41 \pm 4\,\mu l\,min^{-1}\,ml^{-1}$), although the values tended to be higher in COPD patients (Fig. 19.2) [29]. Similarly, Paredi *et al.* [38] reported normal Qaw values in COPD. These findings are consistent with the morphological observations that do not support the presence of significant new vessel formation in stable COPD in contrast to asthma. However, airway blood flow is increased in patients with COPD and bronchiectasis, reflecting the above described hyperplasia of the airway circulation in bronchiectasis [37].

Albuterol-induced vasodilation is blunted in ex-smokers with COPD (Fig. 19.4) [29], similarly to patients with asthma [8, 9]. Albuterol-induced vasodilation has been used as an index of endothelium-dependent vasodilation and endothelial function, especially in current smokers [43]. Since the COPD patients were ex-smokers and the asthmatics lifetime non-smokers, it cannot be assumed that the abnormal albuterol responsiveness was due to endothelial dysfunction. However, inasmuch as oxidative stress has been closely associated with endothelial dysfunction [43, 44] and endogenously generated oxidants have been implicated in the pathogenesis of both COPD and asthma,

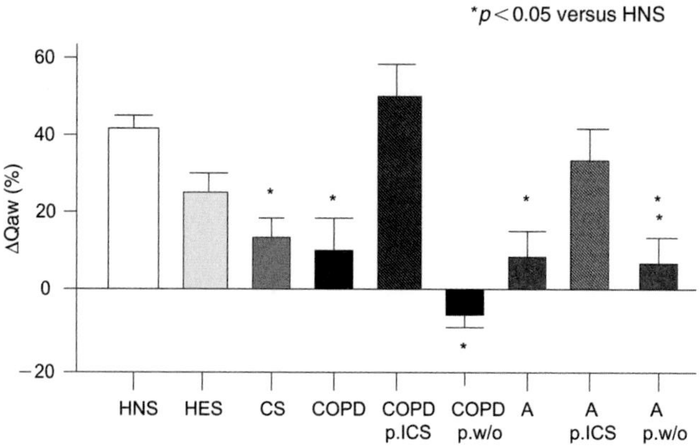

FIG. 19.4 Airway blood flow reactivity to inhaled albuterol (180 μg) in healthy non-smokers (HNS), healthy ex-smokers (HES), current smokers (CS), COPD, COPD after long-term treatment with an inhaled corticosteroid (COPD p.ICS) and after washout of the medication (COPD p.w/0), asthma (A), and asthma after long-term treatment with an inhaled corticosteroid (A p.ICS) and after the washout of the medication (A p. w/o). Vertical axis: change in airway blood flow. Mean values with SE in brackets. Adapted from Refs [29] and [42].

it is conceivable that the blunted albuterol responsiveness of Qaw reflects an abnormality in endothelium-dependent relaxation. To what extent the defective vascular reactivity contributes to the development of COPD is unknown. It could be speculated that the blunted vasodilator response in the airways reflects systemic vascular disease that often co-exists with COPD [45, 46].

Cigarette Smoking

Acute effects

In several extrapulmonary vascular beds, smoking a cigarette has been shown to decrease blood flow transiently [47–49]. The airway circulation is the first and only systemic vascular bed to be exposed to cigarette smoke directly and is exposed to the highest concentration of cigarette smoke constituents. Their effects on airway blood flow therefore may be different from their systemic vascular effects. In anesthetized pigs, cigarette smoke increases airway blood flow transiently during the duration of inhalation, an effect that resembles inhalation challenge with NO [50, 51]. This could be interpreted as indicating a vasodilator effect of NO contained in cigarette smoke. Inasmuch as cigarette smoke is an irritant, more prolonged vascular effects could be expected in conscious humans. Furthermore, cigarette smoke contains nicotine, a sympathomimetic that has variable hemodynamic actions in different vascular beds [52, 53].

We found that Qaw increased by a mean of 81% at 5 min after smoking a cigarette and returned to baseline by 30 min in healthy smokers [54], whereas nicotine administered by nasal spray (systemic action) or oral inhalation (local and systemic action) had no effect on Qaw, suggesting a pharmacologic or irritant effect of other cigarette smoke constituents. Thus, smoking a cigarette seems to cause transient vasodilation in the airway and vasoconstriction in extrapulmonary systemic vascular beds. Whether or not repetitive vasodilator responses to cigarette smoking have a role in the pathogenesis of COPD is not known.

Long-term effects

While healthy current and ex-smokers have a near-normal baseline Qaw as mentioned above, they have an abnormal vasodilator response to albuterol, presumably as a manifestation of endothelial dysfunction as shown by Mendes *et al.* [29] (Fig. 19.4). There was a partial recovery of albuterol responsiveness in healthy ex-smokers while albuterol responsiveness remained blunted in the ex-smokers with COPD. The partial reversibility of albuterol responsiveness after smoking cessation in healthy ex-smokers but not in ex-smokers with COPD suggests that the abnormal vascular reactivity is sustained in COPD, possibly due to ongoing airway inflammation.

It may well be that the abnormal albuterol responsiveness in smokers reflects systemic endothelial dysfunction and that there is a correlation between systemic endothelial dysfunction and albuterol-induced vascular relaxation in the airway among healthy smokers. This remains to be investigated.

FUNCTIONAL CONSEQUENCES

The vascular changes described above could participate in the physiological and clinical manifestations of asthma and COPD; however, this is still a matter of debate.

Mucosal thickness

As vessels are known to occupy a significant portion of the inner airway wall, it is possible that the vascular component of inflammation increases mucosal thickness enough to increase airflow resistance. Airway vascular engorgement, submucosal edema, and luminal fluid accumulation have all been proposed to contribute to the excessive airway narrowing and enhanced airway responsiveness in asthma. Several animal and human studies have examined this experimentally; the results have been conflicting with respect to the effects of vascular congestion on airway caliber [55–59]. The observation that rapid infusion of intravenous fluids causes airway obstruction in humans supports this theory [57]. However, another study has not been able to confirm this finding [58, 59]. There is less controversy on the effect of airway edema on airway caliber: edema of the inner wall internal to the smooth muscle contributes to airway hyperresponsiveness as the same degree of muscle shortening causes greater luminal narrowing than in the normal airway [60].

Drug, mediator, and cell transport

Inflammation-associated blood flow, permeability, and interstitial barrier changes could alter transport functions of the airway vasculature as suggested by previous studies. Desmopressin uptake by the nasal mucosa was enhanced when blood flow was increased [61], whereas inhaled histamine [62] and antigen challenge-induced [63] bronchoconstriction were prolonged when bronchial blood flow was decreased. Therefore one might speculate that the inflammatory increase in airway blood flow would enhance the clearance of locally released inflammatory substances (e.g. spasmogens) and decrease the magnitude and duration of the effect of inhaled bronchoactive drugs. In addition, increased airway blood flow could favor the distribution of systemically administered drugs to the airways and the accumulation of inflammatory cells in the airway wall. Thus, the inflammatory increase in airway blood flow appears to have clinically beneficial as well as undesirable effects on inflammation and its drug treatment.

Heat and water exchange

Exercise-induced bronchoconstriction is a common condition in asthmatics. Reactive hyperemia in response to airway cooling and increased airway liquid and interstitial osmolarity caused by water loss have been proposed to have a critical role [63, 64]. The former may narrow the airway while the latter may promote mediator release from inflammatory cells. However, while hyperventilation with frigid air, a substitute for exercise-related hyperventilation, has been shown to cause a marked increase in Qaw in healthy subjects [81], this may not be the case in asthmatics, who have a blunted vasodilator responsiveness. This remains to be studied.

Mucociliary clearance

The role of airway blood flow in supporting the mucociliary apparatus is incompletely understood. Mucociliary clearance was impaired in the immediate postoperative period after lung transplantation [65], and clearance of inhaled particles was significantly impaired when bronchial blood flow was stopped in sheep [66]. The effect of increased airway blood flow on mucociliary clearance in asthma has not been examined.

PHARMACOLOGIC INTERVENTIONS

In asthma and COPD, the dearth of physiological data on the pathogenetic role of the airway circulation precludes the formulation of meaningful therapeutic recommendations. An exception may be the reversal of microvascular hyperpermeability, an intervention that is likely to increase airway patency and decrease airflow resistance. Although such an effect is difficult to demonstrate experimentally, anti-inflammatory agents, especially inhaled glucocorticosteroids can be expected to reduce airway edema by inhibiting microvascular hyperpermeability based on observation in extrapulmonary vascular beds [36, 37, 67]. In contrast, the physiological and clinical consequences of the increased Qaw and the blunted vasodilator responsiveness (e.g. to β_2-a-drenergic agonists) of the airway circulation in asthma and COPD are not known. These vascular changes could contribute to the pathogenesis of the disease or constitute a vascular adaptation that may counteract the effects of airway inflammation on airway function. Likewise, the blunted vasodilator response in the airways of patients with COPD may be contributing to or mitigate the airway disease of COPD.

It therefore might be more useful to focus on the effects of adrenergic agonists and glucocorticosteroids on Qaw because these agents remain the mainstay of pharmacotherapy in obstructive lung disease.

Adrenergic agonists

Inhaled methoxamine, an α-adrenergic agonist, causes a dose-dependent transient decrease in Qaw, with an enhanced responsiveness in asthmatics compared to healthy subjects (Fig. 19.3) [8]. Higher doses of methoxamine cause bronchoconstriction in asthmatics. However, the vasoconstrictor effect is already seen at methoxamine doses that do not cause bronchoconstriction or have systemic side effects. Thus combining an inhaled β_2-adrenergic agonist or glucocorticosteroid with a low dose of an α-adrenergic agonist might increase residence of the former drugs in the airway tissue and enhance and prolong their biological effects.

The vasodilator effect of inhaled albuterol, a β_2-adrenergic agonist, is blunted in asthmatics and patients with COPD (Fig. 19.3) [8, 9, 30, 34]. In as much as vasodilation may be an undesirable side effect of inhaled β_2-adrenergics when administered for bronchodilation, the vascular hyporesponsiveness to this drug in asthma and COPD could be considered a therapeutic advantage. Unfortunately, inhaled glucocorticosteroids that are typically co-administered with β_2-adrenergic agonists restore vasodilator responsiveness to albuterol [42]. This effect is seen after long-term glucocorticosteroid treatment and after pretreatment with a single dose of a glucocorticosteroid, suggesting that both genomic and non-genomic glucocorticosteroid actions are involved [34].

Glucocorticosteroids

There is increasing evidence that glucocorticosteroids have profound effects on the airway vasculature. Among other actions, such as reduction of microvascular hyperpermeability, glucocorticosteroids regulate Qaw.

Acute vasoconstriction

Inhaled glucocorticosteroids decrease Qaw rapidly and transiently by a non-genomic action that is mediated by the noradrenergic nervous system as the effect can be blocked with an oral α_1-adrenoceptor blocking agent. Thus a single dose of inhaled fluticasone (880 and 1760 mg) causes a decrease in Qaw within 30 min, with a return toward baseline by 90 min in asthmatics and healthy subjects [68]. Other inhaled glucocorticosteroids such as bleclomethasone and budesonide have similar effects (Fig. 19.5) [10].

Reversal of increased Qaw in asthma

Treatment with inhaled fluticasone (440 mg daily) for 2 weeks reverses the asthma-associated increase of Qaw independent

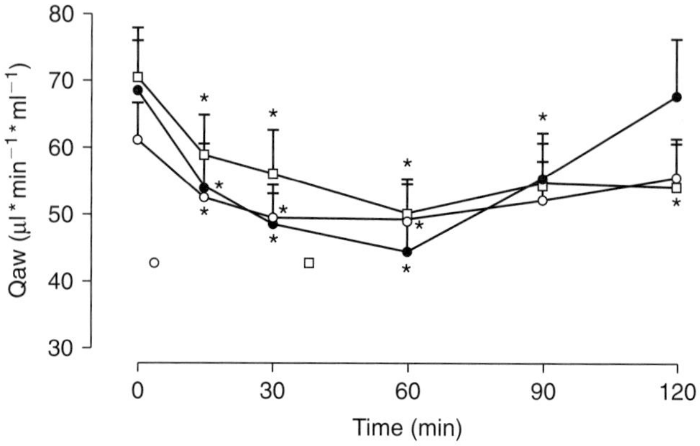

FIG. 19.5 Airway blood flow (Qaw) before and after the inhalation of fluticasone (•), beclomethasone (o) and budesonide (Δ) in asthmatic subjects ($n = 10$). Data are presented as mean ±SD. * $p < 0.05$ versus baseline value. Reproduced with from Ref. [10].

of the acute vasoconstrictor action of the drug, presumably through a genomic action (Fig. 19.4) [42]. A similar effect is seen with oral montelukast, suggesting that Qaw is a marker of airway inflammation in asthma [69]. The observed restoration of a nearly normal Qaw by long-term glucocorticosteroid treatment may reflect a reversal of inflammatory vasodilation and/or an effect on vascular remodeling. For instance, inhaled beclomethasone (200–1500 mg daily) has been reported to decrease the subepithelial area occupied by vessels in asthmatics [70]. The underlying mechanism is not known, but glucocorticosteroids have been reported to reduce vascular endothelial growth factor expression in pulmonary vascular smooth muscle cells [71].

Restoration of β2-adrenergic vasodilation

In patients with asthma and COPD in whom inhaled albuterol fails to increase Qaw, inhaled glucocorticosteroid therapy restores normal albuterol responsiveness (Fig. 19.4) [29, 42]. Inasmuch as β_2-adrenergic vasodilation is mediated by the endothelium, these conditions may be associated with a defect in endothelial function, and glucocorticosteroids may turn out to have a role in the treatment of endothelial dysfunction. Given the prevalence of cardiovascular disease in COPD and its relation to endothelial dysfunction [45, 46], there is a need to examine the effects of glucocorticosteroids on endothelial function in the extrapulmonary systemic circulation as well. Data base studies indeed have suggested that inhaled glucocorticoteroids have a beneficial effect on cardiovascular morbidity and mortality in COPD [72–74].

Glucocorticosteroid—β_2-adrenergic interaction

These stress hormones have been shown to potentiate each other's effect in target tissues including the airway. It has been known for some time that through a genomic action, glucocorticosteroids can upregulate the synthesis and G-protein coupling of β_2-adrenergic receptors, thereby enhancing the tissue responsiveness to β_2-adrenergic agonists [75, 76].

In addition, glucocorticosteroids have acute, non-genomic actions in the airway that can potentiate the effects of β_2-adrenergic agonists as shown above. Inhaled β_2-adrenergic agonists are cleared from the airway tissue by local metabolism and vascular clearance. Glucocorticosteroids inhibit within minutes the uptake of organic cations including adrenergic agonists into non-neuronal cells containing the agonists' metabolizing enzymes [77–80]. This could increase the concentration of adrenergic agonists at adrenergic receptor sites in airway tissue, thereby potentiating the agonists' effects on airway and airway vascular smooth muscle. The latter effect has now been demonstrated *in vivo* in patients with asthma in whom pretreatment with a single glucocorticosteroid dose potentiated the vasodilator response to albuterol within minutes [34]. In addition, the acute vasoconstrictor action of inhaled glucocorticosteroids could reduce the vascular clearance thereby further enhancing the biological effect of inhaled β_2-adrenergic agonists. Although these effects thus far have not been studied with long-acting β_2-adrenergic agonists, the current practice of treating obstructive lung disease with long-acting β_2-adrenergic agonist-glucocorticosteroid combination formulations is supported by the above shown rapid interaction between the two classes of drugs.

References

1. Bernard SL, Glenny RW, Polissar NL *et al.* Distribution of pulmonary and bronchial blood supply to airways measured by fluorescent microspheres. *J Appl Physiol* 80(2): 430–36, 1996.
2. Charan NB, Carvalho PG. Anatomy of the normal bronchial circulatory system in humans and animals. In: Butler J (ed.), *The bronchial circulation*, pp. 45–77. New York: Marcel Dekker, 1992.
3. Baile EM, Paré PD, Ernest D . Distribution of blood flow and neutrophil kinetics in bronchial vasculature of sheep. *J Appl Physiol* 82(5): 1466–71, 1997.
4. Wagner EM, Mitzner W, Brown RH. Site of functional bronchopulmonary anastomoses in sheep. *Anat Rec* 254: 360–66, 1999.
5. Mariassy AT, Gazeroglu H, Wanner A. Morphometry of the subepithelial circulation in sheep airways. *Am Rev Respir Dis* 143: 162–66, 1991.
6. Wanner A, Chediak AD, Csete ME. Airway mucosal blood flow: Response to autonomic and inflammatory stimuli. *Eur Respir J* 12: 618s–623s, 1990.
7. Baile EM, Pare PD. Methods of measuring bronchial blood flow. In: Butler J (ed.), *The Bronchial Circulation*, pp. 101–96. New York: Marcel Dekker, 1992.
8. Brieva J, Wanner A. Adrenergic airway vascular smooth muscle responsiveness in healthy and asthmatic subjects. *J Appl Physiol* 90: 665–69, 2001.
9. Kumar SD, Emery MJ, Atkins ND *et al.* Airway mucosal blood flow in bronchial asthma. *Am J Respir Crit Care Med* 158: 153–56, 1998.
10. Mendes ES, Pereira A, Danta I *et al.* Comparative bronchial vasoconstrictive efficacy of inhaled glucocorticosteroids. *Eur Respir J* 21: 989–93, 2003.
11. Widdicombe JG, Webber SE. Neuroregulation and pharmacology of the tracheobronchial circulation. In: Butler J (ed.), *The bronchial circulation*, pp. 249–89. New York: Marcel Dekker, 1992.
12. Canning BJ, Fischer A. Localization of cholinergic nerves in lower airways of guinea pigs using antisera to choline acetyltransferase. *Am J Physiol* 272: L731–38, 1997.
13. Franco-Cereceda A, Matran R, Alving K *et al.* Sympathetic vascular control of the laryngeo-tracheal, bronchial and pulmonary circulation in the pig: Evidence for non-adrenergic mechanisms involving neuropeptide Y. *Acta Physiol Scand* 155(2): 193–204, 1995.
14. Laitinen LA, Laitinen MA, Widdicombe JG. Dose-related effects of pharmacological mediators on tracheal vascular resistance in dogs. *Br J Pharmacol* 92: 703–9, 1987.
15. Barker JA, Chediak AD, Baier HJ *et al.* Tracheal mucosal blood flow responses to autonomic agonists. *J Appl Physiol* 65(2): 829–34, 1998.
16. Onorato DJ, Demirozu MC, Breitenbücher A *et al.* Airway mucosal blood flow in humans. Response to adrenergic agonists. *Am J Respir Crit Care Med* 149(5): 1132–37, 1994.
17. Charan NB, Carvalho P, Johnson SR *et al.* Effect of aerosolized acetylcholine on bronchial blood flow. *J Appl Physiol* 85(2): 432–36, 1998.
18. Carvalho P, Johnson SR, Charan NB. Non-cAMP-mediated bronchial arterial vasodilation in response to inhaled β-agonists. *J Appl Physiol* 84(1): 215–21, 1998.
19. Carvalho P, Thompson WH, Charan NB. Comparative effects of α-receptor stimulation and nitrergic inhibition on bronchovascular tone. *J Appl Physiol* 88: 1685–89, 2000.
20. Wanner A. Circulation of the airway mucosa. *J Appl Physiol* 67(3): 917–25, 1989.
21. Erjefalt I, Persson CGA. Effects of adrenaline and terbutaline on mediator-increased vascular permeability in the cat trachea. *Br J Pharmacol* 77: 399, 1982.
22. Kuwano K, Bosken CH, Paré PD. Small airways dimensions in asthma and in chronic obstructive pulmonary disease. *Am Rev Respir Dis* 148: 1220–25, 1993.
23. Green FHY, Butt JC, James AL, Carroll NG. Abnormalities of the bronchial arteries in asthma. *Chest* 130: 1025–33, 2006.
24. Li X, Wilson JW. Increased vascularity of the bronchial mucosa in mild asthma. *Am J Respir Crit Care Med* 156(1): 229–33, 1997.
25. Bousquet J, Jeffery PK, Busse WW *et al.* Asthma. From bronchoconstriction to airways inflammation and remodeling. *Am J Respir Crit Care Med* 161: 1720–45, 2000.
26. Simcock DE, Kanabar V, Clarke GW, O'Connor BJ, Lee TH, Hirst SJ. Proangiogenic activity in bronchoalveolar lavage fluid from patients with asthma. *Am J Respir Crit Care Med* 176: 146–53, 2007.
27. Kanazawa H, Nomura S, Asai K. Roles of angiopoetin-1 and angiopoetin-2 on airway vascular permeability in asthmatic patients. *Chest* 131: 1035–41, 2007.
28. Barbato A, Turato G, Baraldo S, Bazzan E, Calabrese F, Panizzolo C, Zanin MA, Zuin R, Maestrelli P, Fabbri LM, Saetta M. Epithelial damage and angiogenesis in the airways of children with asthma. *Am J Respir Crit Care Med* 174: 975–81, 2006.
29. Mendes ES, Campos M, Wanner A. Airway blood flow reactivity in healthy smokers and in ex-smokers with or without COPD. *Chest* 129: 893–98, 2006.
30. Paredi P, Kharitonov SA, Barnes PJ. Correlation of exhaled breath temperature with bronchial blood flow in asthma. *Respir Res* 6: 15–21, 2005.
31. Pisarri TE, Zimmerman MP, Adrian TE *et al.* Bronchial vasodilator pathways in the vagus nerve of dogs. *J Appl Physiol* 86(1): 105–13, 1999.
32. Mattoli S, Soloperto M, Marini M, Fasoli A. Levels of endothelin in the bronchoalveolar lavage fluid of patients with symptomatic asthma and reversible airflow obstruction. *J Allergy Clin Immunol* 88(3 Pt 1): 376–84, 1991.
33. Wagner EM. TNF-α induced bronchial vasoconstriction. *Am J Physiol Heart Circ Physiol* 279: H946–H951, 2000.
34. Mendes ES, Campos MA, Wanner A. Acute effect of a glucocorticosteroid on beta-adrenergic airway and airway smooth muscle reactivity in patients with mild asthma. *J Allergy Clin Immunol* 121: 700–4, 2008.
35. Erjefalt I, Persson CGA. Effects of adrenaline and terbutaline on mediator-increased vascular permeability in the cat trachea. *Br J Pharmacol* 77: 399–408, 1982.
36. Cudkowicz L, Armstrong JB. The Bronchial arteries in pulmonary emphysema. *Thorax* 8: 46–58, 1953.
37. Liebow AA. The bronchopulmonary venous collateral circulation with special reference to emphysema. *Am J Pathol* 29: 251–63, 1953.

38. Paredi P, Ward S, Cramer D, Barnes P, Kharitonov SA. Normal bronchial blood flow in COPD is unaffected by inhaled corticosteroids and correlates with exhaled nitric oxide. *Chest* 131: 1075–81, 2007.
39. Liebow AA, Hales MR, Lindskog GE. Enlargement of the bronchial arteries, and their anastomoses with the pulmonary arteries in bronchiectasis. *Am J Pathol* 25: 211–31, 1949.
40. Fishman AP. The clinical significance of the pulmonary collateral circulation. *Circulation* 24: 677–90, 1961.
41. Charan NB, Turk GM, Dhand R. The role of bronchial circulation in lung abscess. *Am Rev Respir Dis* 131: 121–24, 1985.
42. Brieva JL, Danta I, Wanner A. Effect of inhaled an glucocorticosteroid on airway mucosal blood flow in mild asthma. *Am J Respir Crit Care Med* 161: 293–96, 2000.
43. Schindler C *et al.* Mechanisms of beta-adrenergic receptor-mediated vasodilation. *Clin Pharmacol Ther* 75: 49–59, 2004.
44. Zeiher AM *et al.* Long-term cigarette smoking impairs endothelium-dependent coronary arterial vasodilator function. *Circulation* 92: 1094–100, 1995.
45. Barr RG, Meisa-Vela S, Austin JHM, Basner RC, Keller BM, Reeves AP, Shimbo D, Stevenson L. Impaired flow-mediated dilation is associated with low pulmonary function and emphysema in ex-smokers. *Am J Respir Crit Care Med* 176: 1200–7, 2007.
46. McAllister DA, Maclay JD, Mills NL, Mair G, Miller J, Anderson D, Newby DE, Murchison JT, McNee W. Arteial stiffness is independently associated with emphysema severity in patients with chronic obstructive pulmonary disease. *Am J Respir Crit Care Med* 176: 1208–14, 2007.
47. Czernin J, Waldherr C. Cigarette smoking and coronary blood flow. *Prog Cardiovasc Dis* 45: 395–404, 2003.
48. Yamamoto Y, Nishiyama Y, Monden T, Satoh K, Ohkawa M. A study of the acute effect of smoking on cerebral blood flow using 99mTc-ECD SPET. *Eur J Nucl Med Mol Imaging* 30: 612–14, 2003.
49. Monfrecola G, Riccio G, Savarese C, Posteraro G, Procaccini EM. The acute effect of smoking on cutaneous microcirculation blood flow in habitual smokers and nonsmokers. *Dermatology* 197: 115–18, 1998.
50. Alving K, Fornhem C, Weitzberg E, Lundberg JM. Nitric oxide mediates cigarette smoke-induced vasodilatory responses in the lung. *Acta Physiol Scand* 146: 407–8, 1992.
51. Alving K, Fornhem C, Lundberg JM. Pulmonary effects of endogeneous and exogeneous nitric oxide in the pig: relation to cigarette smoke inhalation. *Br J Pharmacol* 110: 739–46, 1993.
52. Benowitz NL. Cigarette smoking and cardiovascular disease: Pathophysiology and implications for treatment. *Prog Cardiovasc Dis* 46: 91–111, 2003.
53. Vleeming W, Rambali B, Opperhuizen A. The role of nitric oxide in cigarette smoking and nicotine addiction. *Nicotine Tob Res* 4: 341–48, 2002.
54. Kanwaldeep R, Mendes E, Wanner A. Acute effect of cigarette smoke and nicotine on airway blood flow and airflow in healthy smokers. *Lung* 184: 363–68, 2006.
55. Cabanes LR, Weber SN, Matran R *et al.* Bronchial hyperresponsiveness to methacholine in patients with impaired left ventricular function. *N Engl J Med* 320(20): 1317–48, 1989.
56. Csete ME, Abraham WM, Wanner A. Vasomotion influences airflow in peripheral airways. *Am Rev Respir Dis* 141(6): 1409–13, 1990.
57. Gilbert IA, Winslow CJ, Lenner KA *et al.* Vascular volume expansion and thermally induced asthma. *Eur Respir J* 6(2): 189–97, 1993.
58. Blosser S, Mitzner W, Wagner EM. Effects of increased bronchial blood flow on airway morphometry, resistance, and reactivity. *J Appl Physiol* 76(4): 1624–29, 1994.
59. Tang GJ, Freed AN. The role of submucosal oedema in increased peripherial airway resistance by intravenous volume loading in dogs. *Eur Respir J* 7: 311–17, 1994.
60. James AL, Paré PD, Hogg JC. The mechanics of airway narrowing in asthma. *Am Rev Respir Dis* 139: 242–46, 1989.
61. Olanoff LS, Titus CR, Shea MS *et al.* Effect of intranasal histamine on nasal mucosal blood flow and the antidiuretic activity of desmopressin. *J Clin Invest* 80(3): 890–95, 1987.
62. Kelly L, Kolbe J, Mitzner W *et al.* Bronchial blood flow affects recovery from constriction in dog lung periphery. *J Appl Physiol* 60(6): 1954–59, 1986.
63. Csete ME, Chediak AD, Abraham WM *et al.* Airway blood flow modifies allergic airway smooth muscle contraction. *Am Rev Respir Dis* 144(1): 59–63, 1991.
64. McFadden ER. Hypothesis: Exercise-induced asthma as a vascular phenomenom. *Lancet* 335: 880–83, 1990.
65. Paul A, Marelli D, Shennib H *et al.* Mucociliary function in autotransplanted, allotransplanted, and sleve resected lungs. *J Thorac Cardiovasc Surg* 98: 523–28, 1989.
66. Wagner EM, Foster WM. Importance of airway blood flow on particle clearance from the lung. *J Appl Physiol* 81(5): 1878–83, 1996.
67. Hashimoto M, Tanaka H, Shoshaku A. Quantitative analysis of bronchial wall vascularity in the medium and small airways of patients with asthma and COPD. *Chest* 127: 965–72, 2005.
68. Kumar SD, Brieva JL, Danta I *et al.* Transient effect of inhaled fluticasone on airway mucosal blood flow in subjects with and without asthma. *Am J Respir Crit Care Med* 161: 918–21, 2000.
69. Mendes ES, Campos MA, Hurtado A, Wanner A. Effect of montelukast and fluticasone propionate on airway mucosal blood flow in asthma. *Am J Respir Crit Care Med* 169: 1131–34, 2004.
70. Orsida BE, Li X, Hickey B *et al.* Vascularity in asthmatic airways: Relation to inhaled steroid dose. *Thorax* 54: 289–95, 1999.
71. Nauck M, Roth M, Tamm M *et al.* Induction of vascular endothelial growth factor by platelet-activating factor and platelet-derived growth factor is down-regulated by corticosteroids. *Am J Respir Cell Mol Biol* 16: 398–406, 1997.
72. Huiart L, Ernst P, Ranouil X, Suissa S. Low-dose inhaled corticosteroids and the risk of acute myocardial infarction in COPD. *Eur Respir J* 25: 634–39, 2005.
73. Macie C, Wooldrage K, Manfreda J, Anthonisen NR. Inhaled corticosteroids and mortality in COPD. *Chest* 130: 640–46, 2006.
74. Man SFP, Sin DD. Effects of corticosteroids on systemic inflammation in chronic obstructive pulmonary disease. *Proc Am Thor Soc* 2: 78–82, 2005.
75. Mak JC, Nishikawa M, Barnes PJ. Glucocorticosteroids increase beta-2 adrenergic receptor transcription in human lung. *Am J Physiol* 268: L41–146, 1995.
76. Adcock IM, Stevens DA, Barnes PJ. Interactions of glucocorticoids and β2- agonists. *Eur Respir J* 9: 160–68, 1996.
77. Horvath G, Wanner A. Inhaled corticosteroids: Effects on the airway vasculature in bronchial asthma. *Eur J Respir Dis* 27: 172–87, 2006.
78. Horvath G, Sutto Z, Torbati A, Conner GE, Salathe M, Wanner A. Norepinephrine transport by the extraneuronal monoamine transporter in human bronchial arterial smooth muscle cells. *Am J Physiol Lung Cell Mol Physiol* 285: L829–37, 2003.
79. Horvath G, Lieb T, Conner GE, Salathé M, Wanner A. Steroid sensitivity of norepinephrine uptake by human bronchial arterial and rabbit aortic smooth muscle cells. *Am J Respir Cell Mol Biol* 25(4): 500–6, 2001.
80. Horvath G, Mendes ES, Schmid N, Schmid A, Conner GE, Salathe M, Wanner A. The effect of corticosteroids on the disposal of long-acting beta2-agonists by airway smooth muscle cells. *J Allergy Clin Immunol* 120: 1103–9, 2007.
81. Kim HH, LeMerre C, Demirozu CM *et al.* Effect of hyperventilation on airway mucosal blood flow in normal subjects. *Am J Respir Crit Care Med* 154(5): 1563–66, 1996.
82. Jeffrey P. Remodeling and inflammation of bronchi in asthma and chronic obstructive pulmonary disease. *Proc Am Thor Soc* 1: 176–83, 2004.

20 CHAPTER

Pulmonary Vessels

Roberto Rodríguez-Roisin and Joan Albert Barberà

Servei de Pneumologia, Hospital Clínic, IDIBAPS, Universitat de Barcelona, Barcelona, Spain

CHRONIC OBSTRUCTIVE PULMONARY DISEASE

Introduction

Pulmonary vascular circulation abnormalities are common in chronic obstructive pulmonary disease (COPD). It is likely that, pulmonary hypertension associated with COPD is the most frequent form of pulmonary hypertension. Concepts gathered in the systemic circulation have contributed to a better understanding of changes occurring in pulmonary vessels. This has promoted a change in the notion of the pathogenesis of pulmonary hypertension from a vasoconstrictive phenomenon to a cell proliferative disorder. Some of the concepts used to explain the pathogenesis of the most severe forms of pulmonary hypertension have been extrapolated to COPD, yielding to new approaches to its pathobiology. In this section, we will review the current knowledge on the mechanisms of pulmonary vascular changes associated with COPD.

Lung structure

Pulmonary vascular remodeling is broadly defined as an active process of profound structural changes reflecting an abnormal tissue repair response to an acute and/or chronic insult. In COPD, the latter process results from profound fibrocellular changes that lead to an enlargement of the vessel wall – an active process that affects preferentially small and pre-capillary arteries of the pulmonary vasculature identified at different degrees of COPD severity.

In patients with end-stage COPD and pulmonary hypertension or *cor pulmonale*, post-mortem studies have shown deposition of longitudinal muscle, fibrosis, and elastosis that cause intimal enlargement in pulmonary muscular arteries [1, 2]. In the arterioles there is a development of a medial coat of circular smooth muscle, bounded by a new elastic lamina, with deposition of longitudinal muscle and fibrosis of the intima. As compared with patients with mild COPD and healthy nonsmokers, Santos *et al.* [3] showed substantial enlargement of the intima in lung tissue specimens obtained from lung volume reduction surgery. Conversely, medial thickness was slightly reduced in patients with severe emphysema when compared with patients with milder form of COPD.

In patients with mild-to-moderate COPD, morphometric studies of pulmonary muscular arteries have shown enlargement of the intimal layer with luminal reduction [4–7] (Fig. 20.1). Intimal enlargement occurs in muscular arteries of different sizes, although it is more pronounced in small arteries with an external diameter less than 500 μm [5]. In addition, the number of small pulmonary muscular arteries, with a diameter $<$200 μm, is increased [8], indicating muscularization of arterioles. Intimal hyperplasia results from proliferation of poorly differentiated smooth muscle cells and deposition of elastic and collagen fibers [9]. Changes in the media of muscular arteries are less conspicuous and the majority of morphometric studies have failed to show differences in the thickness of the muscular layer when comparing patients with mild-to-moderate COPD with control subjects [5]. Moreover, the magnitude of this intimal thickening, not different from that shown by mild

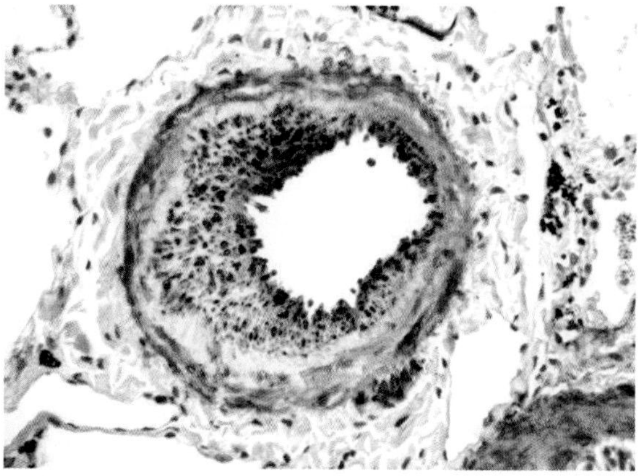

FIG. 20.1 Photomicrograph of a pulmonary muscular artery from a patient with COPD showing prominent intimal hyperplasia and lumenal narrowing. Immunostaining with antibody against α-smooth muscle actin reveals marked proliferation of smooth muscle cells in the intima.

COPD patients, is also present in heavy smokers with normal spirometry [6, 7].

Clinical background

In COPD pulmonary hypertension is considered to be present when mean pulmonary artery pressure (PAP) exceeds the upper normal limit of 20 mmHg [10]. In general, the degree of pulmonary hypertension in COPD is of low-to-moderate magnitude and rarely exceeds 35–40 mmHg. Both right atrial pressure and pulmonary capillary wedge pressure tend to be normal, as well as the cardiac output [10], a pattern at variance with other pulmonary hypertensive conditions (i.e. idiopathic pulmonary arterial hypertension, congenital heart disease, thromboembolic disease) in which PAP can reach extremely high levels, close to those of the systemic circulation, usually in the context of a reduced cardiac output.

Pulmonary hypertension in COPD evolves over time and its severity correlates with the degree of airflow obstruction and the impairment of pulmonary gas exchange [11, 12]. However, the rate of progression of pulmonary hypertension in COPD is slow and usually PAP is only moderately elevated, even in patients with advanced stages [11]. Weitzenblum *et al.* [11] studied the natural history of PAP in a cohort of over 100 COPD patients followed up to 5 years with PAP increases on an average of 0.6 mmHg/year. Although the rate of increase of PAP was slightly higher in patients without pulmonary hypertension at the onset of the follow-up compared with those who already had pulmonary hypertension, the PAP increase was more closely related to the progress of gas exchange disturbances than to the initial PAP values [11].

The beginning of pulmonary hypertension in the natural history of COPD has intrigued investigators. Kessler *et al.* [13] assessed the outcome of pulmonary hemodynamics in patients with moderate COPD without pulmonary hypertension at rest and observed 7 years later that PAP had increased by 2.6 mmHg (annual rise, 0.4 mmHg/year), thus 25% of the patients already having pulmonary hypertension at rest. The incidence of resting pulmonary hypertension was greater in the subset of patients who initially developed pulmonary hypertension during exercise (32%), as compared with those without exercise-induced pulmonary hypertension (16%), such that increased PAP during exercise was an independent predictor for the subsequent development of pulmonary hypertension [13]. These results indicate that in COPD pulmonary circulation, derangement can be present years before pulmonary hypertension is apparent at rest. Further, this study suggests that exercise tests might be of interest to demonstrate pulmonary circulation disturbances, which are consistent with morphometric studies that show conspicuous structural changes in pulmonary muscular arteries in mild COPD patients [4, 6].

Pulmonary hypertension is a common complication of COPD. Its presence is associated with increased mortality, more frequent exacerbations, and greater use of health-care resources. At present, there is no specific and effective treatment for this condition in COPD. However, recent advances in the understanding of the pathobiology of pulmonary hypertensive states, along with the development of effective treatments for pulmonary arterial hypertension, have opened a new perspective that might be relevant in COPD. Nevertheless, the actual prevalence of pulmonary hypertension in COPD remains elusive, because it has not been assessed systematically using right heart catheterization across the wide clinical spectrum of COPD. Hemodynamic data currently available are limited to patients with advanced disease.

Three studies have provided data in a large series of patients. The first study is of patients with severe emphysema screened for lung volume reduction surgery and the incidence of pulmonary hypertension (PAP > 20 mmHg) was very high (91%), although of mild-to-moderate magnitude (20–35 mmHg); [13] in 5% of patients, however, PAP values exceeded 35 mmHg. Although there were significant but modest correlations between PAP and FEV_1 and arterial PO_2, PAP was more closely related to pulmonary artery occlusion pressure, slightly increased in most of the patients, suggesting that gas trapping may contribute to this increased capillary pressure. The second study is a retrospective analysis of over 1000 COPD patients. The prevalence of severe pulmonary hypertension (PAP > 40 mmHg) was 1.1% [14]. This cohort of patients with disproportionate pulmonary hypertension had moderate airflow limitation, severe hypoxemia, hypercapnia, and very low diffusing capacity, indicating that these patients may share some characteristics of idiopathic pulmonary arterial hypertension. In a third study with approximately 200 COPD patients who were candidates for lung volume reduction surgery or lung transplantation, [15] pulmonary hypertension (i.e. PAP > 25 mmHg) was observed in 50% of the patients, although of mild severity (26–35 mmHg). In 10% of the patients, pulmonary hypertension was moderate (36–45 mmHg) and, in 4% of the patients, severe (>45 mmHg). A cluster analysis identified a subset of patients characterized by moderate impairment of airway function and high PAP values, along with severe hypoxemia, supporting the concept of the existence of a cohort of COPD patients with "disproportionate" pulmonary hypertension.

Pathophysiology

Hypoxic pulmonary vasoconstriction, right ventricular function, peripheral edema, and exercise are four key factors influencing the development of pulmonary hypertension.

Pulmonary vasoconstriction in response to hypoxia reduces perfusion in poorly ventilated or nonventilated lung units and diverts blood flow to better-ventilated units, thereby restoring ventilation–perfusion (V_A/Q) imbalance toward normal and increasing PaO_2. Teleologically, hypoxic vasoconstriction plays a fundamental role in matching pulmonary blood flow to alveolar ventilation. Failure to maintain this balance, and the consequent V_A/Q mismatching, is the major cause for both hypoxemia and hypercapnia in COPD [16]. Studies using inert gases [17] have shown that hypoxic pulmonary vasoconstriction release while breathing high oxygen concentration worsens the underlying V_A/Q imbalance in patients with different degrees of COPD severity [16]. In general terms, the contribution of hypoxic vasoconstriction to correct or restore V_A/Q matching tends to be greater in patients with less severe COPD [18]; conversely, hypoxic vasoconstriction remains less influential in patients with severe structural derangement of pulmonary muscular arteries [4]. Furthermore, the magnitude of contraction in isolated pulmonary artery rings induced by a hypoxic stimulus is negatively correlated to endothelial function and directly related to the arterial PO_2 [19], suggesting that the vascular dysfunction is associated with an altered response to hypoxia that further worsens gas exchange. Accordingly, the contribution of hypoxic vasoconstriction to gas exchange in COPD should be taken into account when administering vasodilators that might potentially inhibit such a response since they may also impair gas exchange [17].

In patients with COPD and pulmonary hypertension, the PAP is not markedly elevated and the rate of progression of pulmonary hypertension is slow. Therefore, the right ventricle has time to adapt to the modest increase in afterload. When PAP is chronically elevated the right ventricle dilates and both end-diastolic and end-systolic volumes increase. The stroke volume of right ventricle is usually maintained, whereas the ejection fraction decreases. Subsequent hypertrophy of the right ventricular wall in persistent pulmonary hypertension reduces its tension and hence the afterload. The reduction in right ventricular ejection fraction (RVEF) is inversely related to PAP; nonetheless, a decrease in RVEF does not imply that there is a true ventricular dysfunction [20]. Measurements of end-systolic pressure–volume relationships have shown that a stable COPD contractility of the right ventricle lies within normal limits, irrespective of the PAP values [21, 22]. However, during COPD exacerbations, when PAP increases markedly, the contractility of the right ventricle is reduced in patients presenting clinical signs of right heart failure [23, 24]. Cardiac output in COPD is usually preserved and might rise during exacerbations [25, 26], even when right heart failure is apparent. Therefore, the usual definition of heart failure as a reduction in cardiac output does not apply to COPD. These questions relating to the nature of right heart failure in COPD have led to a reassessment of the concept of *cor pulmonale*. The suggestion has been made that this term should be abandoned in favor of a more precise definition based on objective evidence of right ventricular hypertrophy, enlargement, functional abnormality, or failure.

Although peripheral edema may be a sign of venous congestion due to upstream transmission of right ventricular filling pressures, in advanced COPD edema is more related to hypercapnia rather than to raised jugular pressures [27, 28]. Some patients may present peripheral edema without hemodynamic signs of right heart failure or PAP changes [29, 30]. This has lead to reconsideration of the mechanisms for peripheral edema formation in COPD [29, 30–32].

In COPD, peripheral edema results from a complex interaction between the hemodynamic changes and the balance between edema promoting and protective factors. In COPD patients with pulmonary hypertension and coexisting chronic respiratory failure, hypoxemia and hypercapnia aggravate venous congestion by further activating the sympathetic nervous system, which is already stimulated by right atrial distension. Sympathetic activation decreases renal plasma flow, stimulates the renin–angiotensin–aldosterone system and promotes tubular reabsorption of bicarbonate, sodium, and water. Vasopressin, which also contributes to edema formation, is released when patients become hyponatremic and its plasma levels rise in patients with hypoxemia and hypercapnia [31].

Atrial natriuretic peptide is released from distended atrial walls and may act as an edema-protective mechanism since it has vasodilator, diuretic, and natriuretic properties. Nevertheless, these effects are usually insufficient to counterbalance the edema-promoting mechanisms. Peripheral edema may develop or worsen during episodes of COPD exacerbation. Weitzenblum *et al.* [26] identified a subgroup of patients with more marked peripheral edema that was attributed to hemodynamic signs of right heart failure (increase in end-diastolic pressure). Compared with patients with normal end-diastolic pressure, patients with right heart failure had more marked increase in PAP, and more severe hypoxemia and hypercapnia. This suggests that worsening of pulmonary hypertension during exacerbations contributes to edema formation.

Physical exercise produces abnormal increases in PAP, especially in patients who have pulmonary hypertension at rest [33]. As noted above, patients that appear to be more prone to the development of pulmonary hypertension may show an abnormal increase in PAP during exercise, years before pulmonary hypertension is apparent at rest [13]. Different studies have identified a number of mechanisms for exercise-induced pulmonary hypertension in COPD, including hypoxic vasoconstriction, reduction of the capillary bed by emphysema, extramural compression by increased alveolar pressure, or impaired release of endothelium-derived relaxing factors [34–36]. These mechanisms may coexist hence several factors may contribute to the development of exercise-induced pulmonary hypertension in a given individual. In COPD, PAP during exercise is greater than predicted by the PVR equation, suggesting active pulmonary vasoconstriction on exertion [34]. The latter may be due to an enhancement of hypoxic vasoconstriction by decreased mixed venous PO_2, increased tone of the

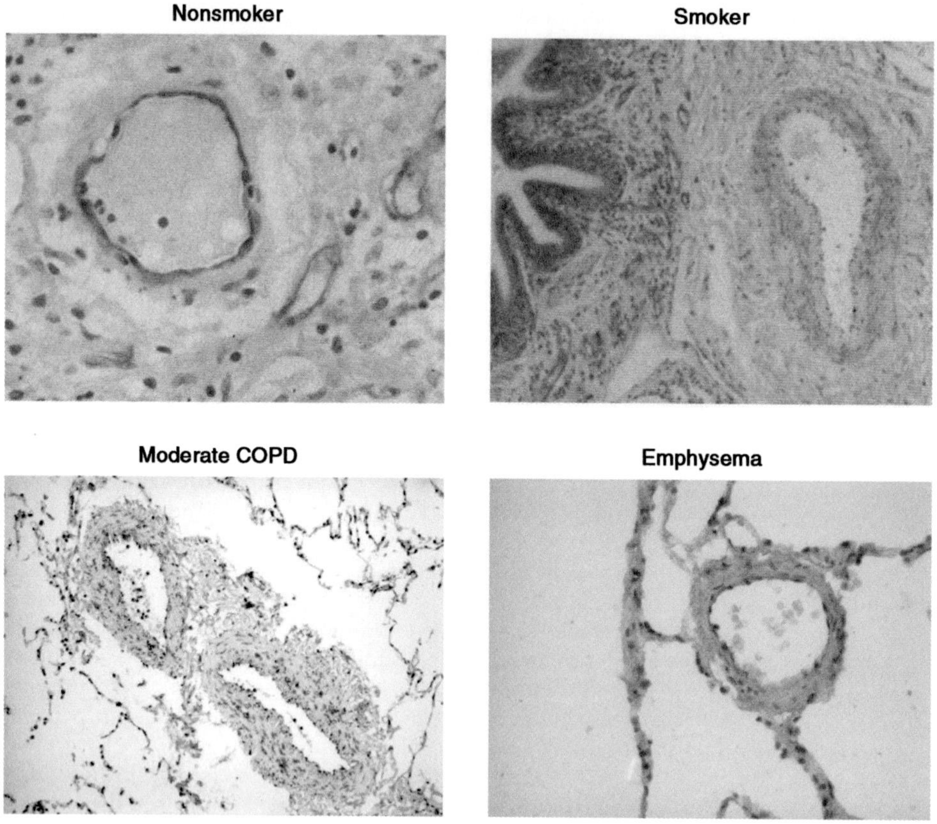

FIG. 20.2 Immunodetection of endothelial nitric oxide synthase (eNOS) in pulmonary muscular arteries from a nonsmoker control subject, a smoker with normal lung function, a patient with moderate COPD, and a patient with severe emphysema. Whereas endothelial cells of the nonsmoker subject show strong immunoreactivity to eNOS, the intensity of the signal is weak or absent in pulmonary arteries of the smoker subject and patients with COPD.

sympathetic nervous system, and/or decreased arterial pH [37]. During exercise, patients may develop dynamic hyperinflation due to expiratory flow limitation that results in increased alveolar pressure which is transmitted to the capillary wedge pressure [38]. Furthermore, increased ventilation during exercise, in the presence of airflow limitation, results in significant swings of intrathoracic pressure that may reduce cardiac output by altering systemic venous return or by increasing left ventricular afterload [39]. Impairment of endothelial release of vasorelaxing agents such as nitric oxide (NO) may also contribute to an impaired dilator response to increases in flow [40]. However, in a subset of COPD patients who developed pulmonary hypertension during exercise, external administration of NO did not block the abnormal increase in PAP, suggesting that the defective endogenous NO release plays a marginal role in exercise-induced pulmonary hypertension [41]. Since pulmonary hypertension may develop at moderate levels of exercise, it has been suggested that repeated episodes of pulmonary hypertension during daily activities, such as stairs climbing or walking, could contribute to the development of right ventricular hypertrophy [42].

Pathobiology

The field of vascular biology has made a major breakthrough over the last decades since the seminal studies identified the key role of endothelium in the regulation of vascular homeostasis [43]. Endothelial dysfunction is a common disturbance in hypertensive states of both the systemic and pulmonary circulations. Vascular actions of endothelium are mediated through the balanced release of potent vasoactive mediators, such as NO, prostacyclin, endothelin-1 (ET-1), and angiotensin. Some of these mediators are also potent modulators of proliferative and inflammatory responses.

Endothelial dysfunction has been shown in pulmonary arteries of both end-stage [44] and mild [7] COPD patients. Impairment of endothelial function results from changes in the expression and release of vasoactive mediators. Endothelium-derived NO is a potent endogenous vasodilator with anti-proliferative effects in the vessel walls. In pulmonary arteries of COPD patients with pulmonary hypertension, the expression of endothelial NO synthase (eNOS) is reduced [45]. Interestingly, in smokers, eNOS expression is also reduced [46] (Fig. 20.2), likely explaining the endothelial dysfunction present in mild disease states [7]. Prostacyclin, which is also synthesized by endothelial cells, exerts similar actions to NO and we know that the expression of prostacyclin synthase is reduced in pulmonary arteries of patients with severe emphysema [47]. ET-1 is a potent vasoconstrictor that also exerts a mitogenic effect on arterial smooth muscle cells, its expression being augmented in pulmonary arteries of patients with COPD and pulmonary hypertension, [48] but not in patients with mild-to-moderate disease [46, 49]. These findings indicate that endothelial dysfunction, with changes in the expression and release of endothelium-derived vasoactive mediators that regulate cell growth, is a common characteristic finding of COPD. It may appear early in the natural history of the disease and could contribute to further changes in vascular structure and function. Hypoxia, inflammation, and cigarette smoke might all contribute to the origin of pulmonary endothelial damage in COPD.

Hypoxia is a potential mechanism that may explain endothelial cell damage and vascular remodeling in COPD. Hypoxia may downregulate eNOS expression and induce smooth muscle cell and adventitial fibroblast proliferation [50, 51]. In addition, hypoxia elicits the contraction of pulmonary arteries. However, the role of hypoxia as the major etiological factor for pulmonary vascular impairment in COPD is now being re-assessed. First, COPD patients show a wide variation in the individual response of the pulmonary circulation to changes in inspired oxygen concentration, [52] and the correlation between PaO_2 and PAP remains weak [12]. Second, long-term oxygen therapy cannot reverse pulmonary hypertension completely [53]. Third, structural abnormalities and endothelial dysfunction in pulmonary arteries can be observed in nonhypoxemic patients with mild COPD and also in smokers with normal spirometry [4, 6, 7]. Therefore, hypoxia does not completely explain the pulmonary vascular changes in COPD.

Inflammation may play a role in the pathogenesis of pulmonary vascular abnormalities associated with COPD, although its precise effects have not been established yet. Inflammatory cell infiltration of small airways is associated with intimal hyperplasia of pulmonary muscular arteries [4]. Further, there is an increased number of $CD8^+$ T-lymphocytes infiltrating the adventitia of pulmonary arteries in patients with mild-to-moderate COPD [7] (Fig. 20.3). Inflammatory cells are a source of cytokines and growth factors that may target the endothelial cells and contribute to the development of structural and functional abnormalities of the vessel wall. Indeed, the number of inflammatory cells infiltrating the pulmonary arteries is directly related to both endothelial dysfunction and to intimal hyperplasia [7]. Smokers with normal spirometry also exhibit an increased number of $CD8^+$ T cells with a reduction of the $CD4^+/CD8^+$ ratio in the arterial adventitia, as compared with nonsmokers [7] (Fig. 20.3). This suggests that cigarette smoking might induce inflammatory changes in pulmonary arteries before the development of lung functional disturbances.

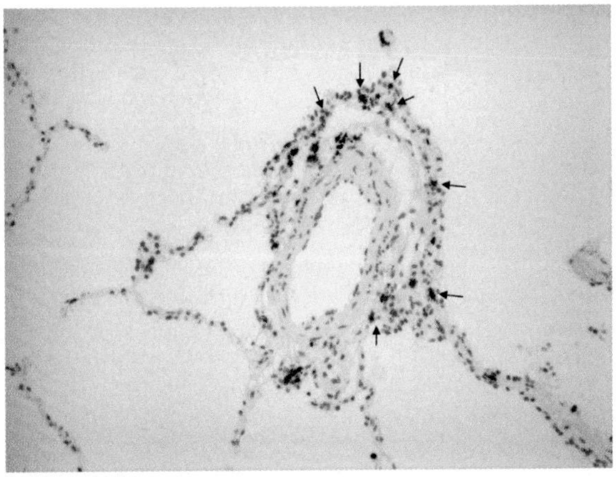

FIG. 20.3 Small pulmonary muscular artery with several $CD8^+$ lymphocytes infiltrating the adventitia (arrows).

Cigarette smoke products may directly influence pulmonary vessel structure. Smokers with normal spirometry show prominent changes in pulmonary arteries, such as smooth muscle cell proliferation [9], impairment of endothelial function [7], reduced expression of eNOS [46], increased expression of vascular endothelial growth factor (VEGF) [3], and $CD8^+$ T-cell infiltrate [7]. Most of these changes cannot be differentiated from those seen in COPD patients. These changes are clearly absent in nonsmokers, and have been replicated in part in a guinea pig model using chronic exposure to cigarette smoke [54]. VEGF protein content in lung tissue is reduced in severe emphysema such that the expression of VEGF varies according to the severity of COPD (Fig. 20.4). In the animal model, cigarette smoke exposure induces muscularization of pre-capillary vessels and increases PAP [55]. It is of note that vascular abnormalities are clearly shown after 2 months of cigarette smoke exposure without any evidence of emphysema [55], suggesting that cigarette smoke-induced vascular abnormalities may precede the development of pulmonary emphysema [56]. Moreover, rapid changes in gene expression of VEGF, VEGF receptor-1, ET1, and inducible NOS [57], mediators that regulate vascular cell growth and vessel contraction, likely involved in the pathogenesis of pulmonary vascular changes of COPD, are induced in this animal model.

Cigarette smoking is a well-known risk factor for the development of systemic vascular disease. Active and passive

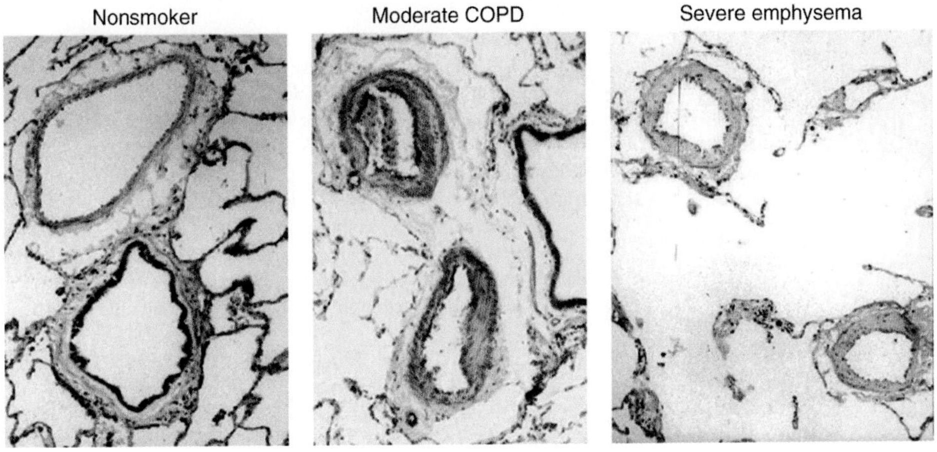

FIG. 20.4 Photomicrographs of pulmonary muscular arteries from a nonsmoker, a patient with moderate COPD, and a patient with severe emphysema, immunostained with antibody against vascular endothelial growth factor (VEGF). Positive cells (brown) were located in the intima. Patients with moderate COPD showed the greatest immunoreactivity to VEGF.

exposure to tobacco smoke produces endothelial dysfunction in both coronary and systemic arteries [58]. Exposure of pulmonary artery endothelial cells to cigarette smoke extract causes the inhibition of both eNOS and prostacyclin synthase at the level of protein content and mRNA expression [59, 60]. Cigarette smoke contains a number of products that have the potential to produce endothelial injury; among these, the aldehyde acrolein seems to play a prominent role since it reduces the expression of prostacyclin synthase in endothelial cells [60].

In summary, several evidences suggest that the initial event in the natural history of pulmonary hypertension in COPD could be pulmonary endothelial injury induced by cigarette-smoke products with the subsequent downregulation of eNOS and prostacyclin synthase expression, and impairment of endothelial function. When the disease progresses, sustained exposure to hypoxemia and inflammation may induce further pulmonary vascular remodeling, thus amplifying the initial effects of cigarette smoke.

BRONCHIAL ASTHMA

Structural background

In patients who died of sudden fatal asthma, there is an inflammatory response in the peripheral airways characterized by lumenal occlusion, resulting from airway narrowing due to increased smooth muscle thickness, and, importantly, an inflammatory infiltrate, including both mononuclear cells and eosinophils together with airway secretions [61]. A salient finding is that the muscular pulmonary arteries neighboring inflamed and narrowed airways show an adventitial inflammatory infiltrate, mainly consisting of increased numbers of eosinophils. Increased intimal fibrous thickening with bands of longitudinal muscle within the intima and also externally to the elastica lamina, characteristic features of chronic alveolar hypoxia, were not exhibited by these vessels. That both the intimal and medial thickness of muscular pulmonary arteries was within normal limits suggested that the effects of regional hypoxia produced by mucus plugging were transient and short enough in duration to fail to induce the structural derangement usually shown in patients with chronic hypoxic lung diseases [61]. These pulmonary vasculature findings are of interest for understanding the behavior of the pulmonary circulation and its interplay with the V_A/Q matching, particularly while breathing 100% oxygen. In addition to an increased smooth muscle thickness in the peripheral airways, this study [61] demonstrated an increased thickness of the non-muscular airway wall components, most likely resulting from inflammatory cell infiltrate, plasma exudate, and vascular congestion, a combination of changes possibly occurring during an acute eruption of the inflammatory cascade. However, not all the airways were occluded by mucus plugging, as almost half of the airways were without secretions, whereas the other half showed some degree of luminal occlusion.

Gas exchange pathophysiology

Two studies have investigated the status of pulmonary gas exchange and the behavior of the pulmonary circulation in patients with stable moderate-to-severe persistent asthma. Patients with unstable, persistent asthma were characterized by longstanding severe airflow limitation ($FEV_1 < 40\%$ predicted).

One of the studies showed that the V_A/Q distributions demonstrated a broad unimodal blood flow V_A/Q pattern [62]. Accordingly, very little perfusion was associated with alveolar units with low V_A/Q ratios and intrapulmonary shunt was conspicuously absent; areas of high V_A/Q regions were also negligible and dead space was within the normal range. The overall amount of V_A/Q imbalance, as assessed by the dispersion of pulmonary blood flow and that of alveolar ventilation, was slightly increased. The correlation between V_A/Q markers and the degree of airflow obstruction, expressed as either FEV_1 or peak expiratory flow or mid-expiratory airflow rates, was very modest. Another study [63] showed similar V_A/Q findings. As in the previous study [62], there was no correlation between the V_A/Q mismatching and PaO_2, pointing to the vital role played by the extra-pulmonary determinants of pulmonary gas exchange, such as cardiac output and minute ventilation. In this study [62], pulmonary hemodynamics were within normal limits. Upon 100% oxygen breathing, it was shown that V_A/Q disturbances further deteriorated. Since FEV_1 did not change, this V_A/Q impairment may suggest, other things being equal, release of hypoxic pulmonary vasoconstriction. Despite the changes in the amount of the dispersion of blood flow distribution, it is of note that there were no changes in PAP, indicating that the V_A/Q descriptors are more sensitive to hyperoxic mixtures changes than conventional measures of pulmonary vascular pressures.

References

1. Wilkinson M, Langhorne CA, Heath D, Barer GR, Howard P. A pathophysiological study of 10 cases of hypoxic *cor pulmonale*. *Q J Med* 249: 65–85, 1988.
2. Wright JL, Petty T, Thurlbeck WM. Analysis of the structure of the muscular pulmonary arteries in patients with pulmonary hypertension and COPD: National Institutes of Health Nocturnal Oxygen Therapy Trial. *Lung* 170: 109–24, 1992.
3. Santos S, Peinado VI, Ramirez J, Morales-Blanhir J, Bastos R, Roca J, Rodriguez-Roisin R, Barbera JA. Enhanced expression of vascular endothelial growth factor in pulmonary arteries of smokers and patients with moderate chronic obstructive pulmonary disease. *Am J Respir Crit Care Med* 167: 1250–56, 2003.
4. Barberà JA, Riverola A, Roca J, Ramirez J, Wagner PD, Ros D, Wiggs BR, Rodriguez-Roisin R. Pulmonary vascular abnormalities and ventilation–perfusion relationships in mild chronic obstructive pulmonary disease. *Am J Respir Crit Care Med* 149: 423–29, 1994.
5. Magee F, Wright JL, Wiggs BR, Paré PD, Hogg JC. Pulmonary vascular structure and function in chronic obstructive pulmonary disease. *Thorax* 43: 183–89, 1988.
6. Peinado VI, Barberà JA, Ramirez J, Gomez FP, Roca J, Jover L, Gimferrer JM, Rodriguez-Roisin R. Endothelial dysfunction in pulmonary arteries of patients with mild COPD. *Am J Physiol* 274: L908–L913, 1998.

7. Peinado VI, Barbera JA, Abate P, Ramirez J, Roca J, Santos S, Rodriguez-Roisin R. Inflammatory reaction in pulmonary muscular arteries of patients with mild chronic obstructive pulmonary disease. *Am J Respir Crit Care Med* 159: 1605–11, 1999.
8. Hale KA, Ewing SL, Gosnell BA, Niewoehner DE. Lung disease in long-term cigarette smokers with and without chronic airflow obstruction. *Am Rev Respir Dis* 130: 716–21, 1984.
9. Santos S, Peinado VI, Ramirez J, Melgosa T, Roca J, Rodriguez-Roisin R, Barberà JA. Characterization of pulmonary vascular remodelling in smokers and patients with mild COPD. *Eur Respir J* 19: 632–38, 2002.
10. Naeije R. Should pulmonary hypertension be treated in chronic obstructive pulmonary disease? In: Weir EK, Archer SL, Reeves JT (eds). *The diagnosis and treatment of pulmonary hypertension*, pp. 209–39. Mount Kisco: Futura Publishing, 1992.
11. Weitzenblum E, Sautegeau A, Ehrhart M, Mammosser M, Hirth C, Roegel E. Long-term course of pulmonary arterial pressure in chronic obstructive pulmonary disease. *Am Rev Respir Dis* 130: 993–98, 1984.
12. Scharf SM, Iqbal M, Keller C, Criner G, Lee S, Fessler HE. Hemodynamic characterization of patients with severe emphysema. *Am J Respir Crit Care Med* 166: 314–22, 2002.
13. Kessler R, Faller M, Weitzenblum E, Chaouat A, Aykut A, Ducolone A, Ehrhart M, Oswald-Mammosser M. Natural history of pulmonary hypertension in a series of 131 patients with chronic obstructive lung disease. *Am J Respir Crit Care Med* 164: 219–24, 2001.
14. Chaouat A, Bugnet AS, Kadaoui N, Schott R, Enache I, Ducolone A, Ehrhart M, Kessler R, Weitzenblum E. Severe pulmonary hypertension and chronic obstructive pulmonary disease. *Am J Respir Crit Care Med*, 2005.
15. Thabut G, Dauriat G, Stern JB, Logeart D, Levy A, Marrash-Chahla R, Mal H. Pulmonary hemodynamics in advanced COPD candidates for lung volume reduction surgery or lung transplantation. *Chest* 127: 1531–36, 2005.
16. Barberà JA. Chronic obstructive pulmonary disease. In: Roca J, Rodriguez-Roisin R, Wagner PD (eds). *Pulmonary and Peripheral Gas Exchange in Health and Disease*, pp. 229–61. New York: Marcel Dekker, Inc, 2000.
17. Roca J, Wagner PD. Contribution of multiple inert gas elimination technique to pulmonary medicine. 1. Principles and information content of the multiple inert gas elimination technique. *Thorax* 49: 815–24, 1994.
18. Barberà JA, Roger N, Roca J, Rovira I, Higenbottam TW, Rodriguez-Roisin R. Worsening of pulmonary gas exchange with nitric oxide inhalation in chronic obstructive pulmonary disease. *Lancet* 347: 436–40, 1996.
19. Peinado VI, Santos S, Ramirez J, Roca J, Rodriguez-Roisin R, Barberà JA. Response to hypoxia of pulmonary arteries in COPD: An *in vitro* study. *Eur Respir J* 20: 332–38, 2002.
20. Weitzenblum E, Chaouat A. Right ventricular function in COPD: Can it be assessed reliably by the measurement of right ventricular ejection fraction? *Chest* 113: 567–69, 1998.
21. Crottogini AJ, Willshaw P. Calculating the end-systolic pressure–volume relation. *Circulation* 83: 1121–23, 1991.
22. Biernacki W, Flenley DC, Muir AL, MacNee W. Pulmonary hypertension and right ventricular function in patients with COPD. *Chest* 94: 1169–75, 1988.
23. MacNee W. Pathophysiology of cor pulmonale in chronic obstructive pulmonary disease. Part One. *Am J Respir Crit Care Med* 150: 833–52, 1994.
24. MacNee W. Pathophysiology of cor pulmonale in chronic obstructive pulmonary disease. Part Two. *Am J Respir Crit Care Med* 150: 1158–68, 1994.
25. Barberà JA, Roca J, Ferrer A, Felez MA, Diaz O, Roger N, Rodriguez-Roisin R. Mechanisms of worsening gas exchange during acute exacerbations of chronic obstructive pulmonary disease. *Eur Respir J* 10: 1285–91, 1997.
26. Weitzenblum E, Apprill M, Oswald M, Chaouat A, Imbs JL. Pulmonary hemodynamics in patients with chronic obstructive pulmonary disease before and during an episode of peripheral edema. *Chest* 105: 1377–82, 1994.
27. Campbell EJM, Short DS. The cause of oedema in "cor pulmonale". *Lancet* I: 1184–86, 1960.
28. Baudouin SV. Oedema and cor pulmonale revisited. *Thorax* 52: 401–2, 1997.
29. MacNee W, Wathen CG, Flenley DC, Muir AD. The effects of controlled oxygen therapy on ventricular function in patients with stable and decompensated cor pulmonale. *Am Rev Respir Dis* 137: 1289–95, 1988.
30. MacNee W, Wathen CG, Hannan WJ, Flenley DC, Muir AL. Effects of pirbuterol and sodium nitroprusside on pulmonary haemodynamics in hypoxic cor pulmonale. *Br Med J (Clin Res Ed)* 287: 1169–72, 1983.
31. Lee-Chiong TL, Matthay RA. The heart in the stable COPD patient. In: Similowski T, Whitelaw WA, Derenne JP (eds). *Clinical management of chronic obstructive pulmonary disease*, pp. 475–532. New York: Marcel Dekker, Inc, 2002.
32. Palange P. Renal and hormonal abnormalities in chronic obstructive pulmonary disease (COPD). *Thorax* 53: 989–91, 1998.
33. Burrows B, Kettel LJ, Niden AH, Rabinowitz M, Diener CF. Patterns of cardiovascular dysfunction in chronic obstructive lung disease. *N Engl J Med* 286: 912–18, 1972.
34. Agustí AGN, Barberà JA, Roca J, Wagner PD, Guitart R, Rodriguez-Roisin R. Hypoxic pulmonary vasoconstriction and gas exchange during exercise in chronic obstructive pulmonary disease. *Chest* 97: 268–75, 1990.
35. Wright JL, Lawson L, Paré PD, Hooper RO, Peretz DI, Nelems JM, Schulzer M, Hogg JC. The structure and function of the pulmonary vasculature in mild chronic obstructive pulmonary disease. *Am Rev Respir Dis* 128: 702–7, 1983.
36. Harris P, Segal N, Bishop JM. The relation between pressure and flow in the pulmonary circulation in normal subjects and in chronic bronchitis. *Cardiovasc Res* 2: 73–83, 1968.
37. Naeije R, Barberà JA. Pulmonary hypertension associated with COPD. *Crit Care* 5: 286–89, 2001.
38. Butler J, Schrijen F, Henriquez A, Polu JM, Albert RK. Cause of the raised wedge pressure on exercise in chronic obstructive pulmonary disease. *Am Rev Respir Dis* 138: 350–54, 1988.
39. Montes de Oca M, Rassulo J, Celli BR. Respiratory muscle and cardiopulmonary function during exercise in very severe COPD. *Am J Respir Crit Care Med* 154: 1284–89, 1996.
40. Rubanyi GM, Romero JC, Vanhoutte PM. Flow-induced release of endothelium-derived relaxing factor. *Am J Physiol* 250: H1145–49, 1986.
41. Roger N, Barberà JA, Roca J, Rovira I, Gomez FP, Rodriguez-Roisin R. Nitric oxide inhalation during exercise in chronic obstructive pulmonary disease. *Am J Respir Crit Care Med* 156: 800–6, 1997.
42. Weitzenblum E. The pulmonary circulation and the heart in chronic lung disease. *Monaldi Arch Chest Dis* 49: 231–34, 1994.
43. Furchgott RF, Zawadzki JV. The obligatory role of endothelial cells in the relaxation of arterial smooth muscle by acetylcholine. *Nature* 288: 373–76, 1980.
44. Dinh-Xuan AT, Higenbottam T, Clelland C, Pepke-Zaba J, Cremona G, Butt AY, Large S, Wells FC, Wallwork J. Impairment of endothelium-dependent pulmonary-artery relaxation in chronic obstructive pulmonary disease. *N Engl J Med* 324: 1539–47, 1991.
45. Giaid A, Saleh D. Reduced expression of endothelial nitric oxide synthase in the lungs of patients with pulmonary hypertension. *N Engl J Med* 333: 214–21, 1995.
46. Barberà JA, Peinado VI, Santos S, Ramirez J, Roca J, Rodriguez-Roisin R. Reduced expression of endothelial nitric oxide synthase in pulmonary arteries of smokers. *Am J Respir Crit Care Med* 164: 709–13, 2001.
47. Nana-Sinkam SP, Lee JD, Sotto-Santiago S, Stearman RS, Keith RL, Choudhury Q, Cool C, Parr J, Moore MD, Bull TM, Voelkel NF, Geraci MW. Prostacyclin prevents pulmonary endothelial cell apoptosis induced by cigarette smoke. *Am J Respir Crit Care Med* 175: 676–85, 2007.
48. Giaid A, Yanagisawa M, Langblen D, Michel RP, Levy R, Shennib H, Kimura S, Masaki T, Duguid WP, Path FRC, Stewart DJ. Expression

of endothelin-1 in the lungs of patients with pulmonary hypertension. *N Engl J Med* 328: 1732–39, 1993.
49. Melgosa M, Peinado VI, Santos S, Morales J, Ramirez J, Roca J, Rodriguez-Roisin R, Barberà JA. Expression of endothelial nitric oxide synthase (eNOS) and endothelin-1 (ET-1) in pulmonary arteries of patients with severe COPD. *Eur Respir J* 22: 20s, 2003.
50. Rabinovitch M, Gamble W, Nadas AS, Miettinen OS, Reid L. Rat pulmonary circulation after chronic hypoxia: Hemodynamic and structural features. *Am J Physiol* 236: H818–27, 1979.
51. Stenmark KR, Fasules J, Hyde DM, Voelkel NF, Henson J, Tucker A, Wilson H, Reeves JT. Severe pulmonary hypertension and arterial adventitial changes in newborn calves at 4,300 m. *J Appl Physiol* 62: 821–30, 1987.
52. Ashutosh K, Mead G, Dunsky M. Early effects of oxygen administration and prognosis in chronic obstructive pulmonary disease and cor pulmonale. *Am Rev Respir Dis* 127: 399–404, 1983.
53. Weitzenblum E, Sautegeau A, Ehrhart M, Mammosser M, Pelletier A. Long-term oxygen therapy can reverse the progression of pulmonary hypertension in patients with chronic obstructive pulmonary disease. *Am Rev Respir Dis* 131: 493–98, 1985.
54. Wright JL, Churg A. A model of tobacco smoke-induced airflow obstruction in the guinea pig. *Chest* 121: 188S–91S, 2002.
55. Wright JL, Churg A. Effect of long-term cigarette smoke exposure on pulmonary vascular structure and function in the guinea pig. *Exp Lung Res* 17: 997–1009, 1991.
56. Yamato H, Churg A, Wright JL. Guinea pig pulmonary hypertension caused by cigarette smoke cannot be explained by capillary bed destruction. *J Appl Physiol* 82: 1644–53, 1997.
57. Wright JL, Tai H, Dai J, Churg A. Cigarette smoke induces rapid changes in gene expression in pulmonary arteries. *Lab Invest* 82: 1391–98, 2002.
58. Celermajer DS, Adams MR, Clarkson P, Robinson J, McCredie R, Donald A, Deanfield JE. Passive smoking and impaired endothelium-dependent arterial dilatation in healthy young adults. *N Engl J Med* 334: 150–54, 1996.
59. Su Y, Han W, Giraldo C, Li YD, Block ER. Effect of cigarette smoke extract on nitric oxide synthase in pulmonary artery endothelial cells. *Am J Respir Cell Mol Biol* 19: 819–25, 1998.
60. Nana-Sinkam SP, Lee JD, Sotto-Santiago S, Stearman RS, Keith RL, Choudhury Q, Cool C, Parr J, Moore MD, Bull TM, Voelkel NF, Geraci MW. Prostacyclin prevents pulmonary endothelial cell apoptosis induced by cigarette smoke. *Am J Respir Crit Care Med* 175: 676–85, 2007.
61. Saetta M, Di Stefano A, Rosina C, Thiene G, Fabbri LM. Quantitative structural analysis of peripheral airways and arteries in sudden fatal asthma. *Am Rev Respir Dis* 143: 138–43, 1991.
62. Ballester E, Roca J, Ramis L, Wagner PD, Rodriguez-Roisin R. Pulmonary gas exchange in severe chronic asthma. Response to 100% oxygen and salbutamol. *Am Rev Respir Dis* 141: 558–62, 1990.
63. Corte P, Young IH. Ventilation–perfusion relationships in symptomatic asthma. Response to oxygen and clemastine. *Chest* 88: 167–75, 1985.

Adhesion Molecules in Asthma and Airway Diseases

CHAPTER 21

Dean Sheppard

Department of Medicine and Lung Biology Center, University of California, San Francisco, CA, USA

All cells express a number of transmembrane proteins that could be characterized as "adhesion" receptors. Most of these proteins have been thus classified based on their role in facilitating attachment of cells to either components of the extracellular matrix or other cells. However, many adhesion receptors also transduce biological signals that modulate cell behavior that are either loosely connected or unrelated to such adhesion events. In this chapter, I will review the structures of some of the most important families of adhesion receptors, will summarize what is known about the patterns of adhesion receptor expression and function on cells relevant to asthma and other chronic airway diseases, and will discuss examples of *in vivo* evidence implicating some of these proteins in pathways which contribute to phenotypic features of airway diseases.

ADHESION MOLECULE FAMILIES

Adhesion molecules can be divided into families based on their primary amino acid sequence similarity. Some adhesion molecule families principally mediate cell–cell adhesion, others are principally involved in cell adhesion to the extracellular matrix, and some are involved in both types of adhesion events. Families involved in cell–cell adhesion include claudins, occludins, cadherins, and immunoglobulin superfamily members. Claudins, occludins, and cadherins are all monomeric proteins that principally mediate homotypic cell–cell adhesion, usually through direct binding to members of the same family expressed at the lateral margins of adjacent cells (Table 21.1). These protein families play critical roles in establishing polarity of epithelia and endothelia and in maintaining and regulating endothelial and epithelial permeability barriers. Cadherins are the core proteins of cell–cell junctions termed "adherans junctions" which serve as central organizers of epithelial and endothelial cell–cell adhesion by homotypic interactions of cadherin extracellular domains [1, 2]. Claudins are critical components of tight junctions that determine the degree of paracellular permeability of endothelia and epithelia to large macromolecules and even small ions [3]. These proteins also bind directly to family members on adjacent cells, but, in contrast to cadherins, claudins can associate with either identical family members or nonidentical members in tight junctions. Multiple claudin isoforms are expressed on airway epithelial cells [4]. Occudin [5], the product of a single gene with multiple spice variants, can serve as a central organizer of other tight junction proteins [6].

Immunoglobulin superfamily members play more complex roles in cell adhesion. Some family members, such as the endothelial adhesion protein, platelet endothelial cell adhesion molecule (PECAM), or the small subfamily of junctional adhesion molecules (JAMs), also play roles in homotypic cell–cell adhesion, but JAMs have also been reported to bind to members of other adhesion protein families (e.g., integrins) and to play roles in heterotypic cell–cell adhesion (e.g., between endothelial or epithelial cells and migrating leukocytes). Other immunoglobulin superfamily members (e.g., intercellular adhesion molecules, ICAMs and vascular adhesion molecule, VCAM) are restricted to roles in heterotypic cell–cell adhesion, principally through their roles as integrin ligands. Selectins, another small family of adhesion molecules, also principally mediate heterotypic cell–cell

TABLE 21.1 Selected adhesion molecules of potential importance in asthma and COPD

Name	Structure	Binding partners	Location	Comments
Claudins	Multiple membrane spanning (tetraspanin)	Homotypic and heterotypic bindings to Claudins	Tight junctions	Regulate ion permeability at tight junctions
JAMs	Single membrane spanning Ig family members	Homotypic and/or heterotypic binding to JAMS. Also bind selected integrins	Tight junction	Mediate leukocyte *trans*-endothelial migration, endothelial and epithelial cell–cell adhesion
Selectins	Single membrane spanning monomers with lectin domain	*Trans*-membrane proteoglycans with sialyl-Lewis X motif	Platelets, leukocytes, and endothelial cells	Mediate leukocyte rolling, can initiate intracellular signals. Potential drug target for asthma
VCAM1	Single membrane spanning Ig family member	Integrins α4β1 and α9β1	Endothelial cells, tumor cells, smooth muscle, pericytes	Counter-receptor for integrins α41 and α9β1. Important for recruitment of leukocytes
ICAM1	Single membrane spanning Ig family member	β2 Integrins	Endothelial and epithelial cells	Counter-receptor for β2 integrins. Possible drug target for asthma
α4β1 Integrin	Transmembrane heterodimer	VCAM1, fibronectin, ADAMs proteins	Endothelium, pericytes, smooth muscle, lymphocytes, eosinophils, and mast cells	Potential drug target for asthma
αvβ6 Integrin	Transmembrane heterodimer	TGFβ1 and -3, fibronectin, tenascin C	Epithelial cells	Activates TGF-β. Negatively regulates macrophage protease expression and emphysema
αvβ8 Integrin	Transmembrane heterodimer	TGFβ1 and -3, vitronectin	Airway epithelial cells, fibroblasts, dendritic cells, and T-cells	Activates TGF-β. Involved in cross talk between airway fibroblasts and epithelial cells

interactions, in that case through interactions with carbohydrate components of proteoglycan counter-receptors.

Receptors for components of the extracellular matrix can also play complex roles in cell biology. Syndecans are transmembrane proteoglycans that appear to function solely as receptors for components of the extracellular matrix. However, integrins, initially identified as extracellular matrix receptors, are now known to recognize a wide variety of spatially restricted proteins as ligands, including cellular counter-receptors, growth factors, and proteases [7].

Integrins were initially identified based on their ability to mediate stable adhesion to fixed extracellular matrix ligands. Integrin-mediated cell adhesion depends on the ability of β subunit cytoplasmic domains to bind to a group of adaptor proteins that link integrins to the actin cytoskeleton (Fig. 21.1). By connecting the relatively rigid extracellular matrix to the relatively rigid actin cytoskeleton, integrins play important roles in maintaining cohesion of multicellular organs and organisms. These factors also position integrins as essential components of cellular force transducers which allow cells to detect and respond to mechanical deformation. The proteins involved in linking integrins and actin are highly conserved from *Caenorhabditis elegans* to humans.

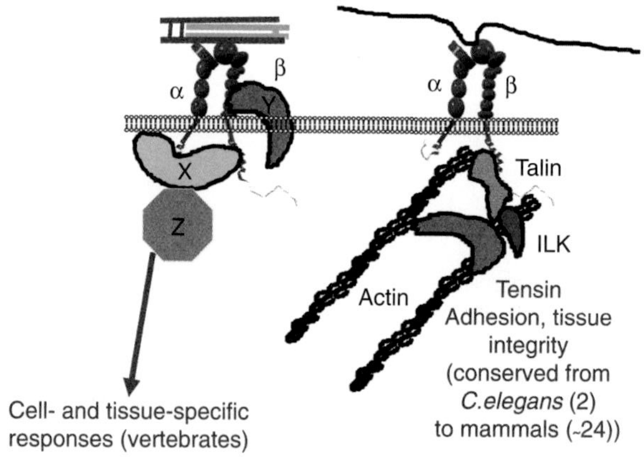

FIG. 21.1 Integrin functions. Integrins evolved to link the actin cytoskeleton and the extracellular matrix, a function that is highly conserved from *C. elegans* to humans (shown on right). This linkage is accomplished through formation of a highly conserved multiprotein complex that includes talin, tensin, and the integrin-linked kinase to the integrin β subunit cytoplasmic domain. In vertebrates, integrins have evolved to recruit a multitude of additional cytoplasmic and transmembrane adaptor and signaling proteins to these complexes, allowing these receptors to modulate a wide variety of cellular behaviors (shown on left).

During the evolutionary transition from invertebrates to vertebrates, there was a rapid increase in the size of the integrin family, and integrins have been adapted to bind to a number of additional ligands and to modulate a wide variety of cellular behaviors, including cell proliferation, differentiation, and survival [8]. These new functions have been accomplished by recruitment of cytoplasmic kinases (such as src and focal adhesion kinase) to the multiprotein complexes assembled on integrin cytoplasmic domains and by lateral associations with a number of additional signaling molecules and adaptor proteins (e.g., tetraspanins, growth factor receptors, urokinase receptor, u-PAR, and the integrin-associated protein).

ROLES OF ADHESION RECEPTORS ON LEUKOCYTES AND ENDOTHELIAL CELLS IN REGULATING AIRWAY INFLAMMATION

Airway and lung parenchymal inflammation are important features of asthma and chronic obstructive pulmonary disease (COPD). Inhibition of recruitment of specific populations of inflammatory cells into the airways has thus been considered as a potential therapeutic intervention in these diseases. Recruitment of inflammatory cells from the bloodstream into tissues is a multistep process [9], as outlined in Fig. 21.2. Key steps in this process include initial slowing of rapidly moving cells, a process called rolling [10]. These slowed cells must then undergo firm adhesion to completely arrest their movement in the bloodstream and then need to resume a different form of slow migration to exit the vasculature either in induced gaps between adjacent endothelial cells or by passing directly through the induced holes within individual endothelial cells. Once they have exited the bloodstream, leukocytes then migrate through the interstitial space where some arrest in the vicinity of nerves and airway smooth muscle, others arrest within the airway epithelium, and some traverse the epithelium and enter the airspaces. Each of these steps of adhesion and migration involves a suite of adhesion receptors expressed on the leukocytes, endothelial cells, or other target extravascular cells.

The first step in this process, leukocyte rolling, is principally mediated by interactions of a selectin on leukocytes (L selectin) with specific carbohydrate ligands on endothelial cells and selectins on endothelial cells (P and E selectins) with additional carbohydrate ligands on leukocytes [11]. This process is principally regulated by the induction of expression of L selectin ligands and P and E selectins only on endothelial cells that have been activated by pro-inflammatory cytokines, such as interleukin-1 and TNF-α. In this process, selectins play a critical role in facilitating the rolling of neutrophils along the vessel wall. However, it is now clear that certain members of the integrin family, for example the integrins α4β1 and α4β7, can mediate rolling even in the absence of input from selectins [12, 13]. Selectin-mediated rolling may play an important role in asthma, since inhibition of selectins or their ligands has been shown to reduce allergen-induced airway hyperresponsiveness in models of allergic asthma [14, 15].

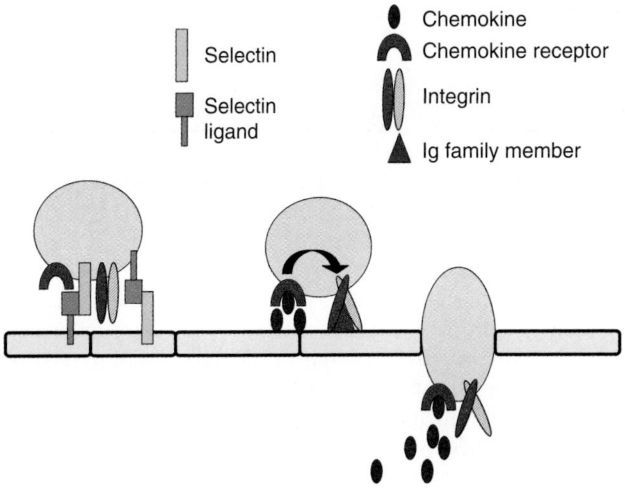

FIG. 21.2 Multistep process of leukocyte rolling arrest and *trans*-endothelial migration.Initial slowing of leukocytes from rapidly moving blood is usually mediated by interactions of selectins on leukocytes and endothelial cells with cell surface carbohydrate ligands (left). Close contact with chemokines then triggers an intracellular signal that changes leukocyte integrins from a low-affinity to a high-affinity state and mediates stable arrest through interaction with endothelial immunoglobulin family members (center). Dynamic regulation of integrin adhesion then allows migration of adherent leukocytes across the endothelium and into and through the extravascular space (right).

The next step in leukocyte recruitment, firm adhesion to the vessel wall, is mediated principally by the interaction of integrins on leukocytes with immunoglobulin superfamily members on endothelial cells. The specific integrins involved, and the corresponding immunoglobulin family counter-receptors, vary among different leukocytes. Neutrophils were initially thought to arrest their movement by virtue of adhesion of the integrins αLβ2 and αMβ2 with the endothelial counter-receptor ICAM-1, whereas monocytes, lymphocytes, and eosinophils [16] utilize α4β1, which interacts with endothelial VCAM-1 [17]. However, neutrophils also express the integrin α9β1, which avidly binds VCAM-1 [18], and lymphocytes, monocytes, and eosinophils also utilize αLβ2 and thus arrest on ICAM-1.

As with the initial rolling step, firm arrest is tightly regulated through signaling in both leukocytes and endothelial cells. On the leukocyte side, integrins in circulating leukocytes are generally maintained in a nonadhesive (inactive) conformation. In response to signals generally initiated by the binding of chemokines to their cognate G-protein-coupled receptors (which are presented to integrins on rolling leukoctyes by virtue of their interactions with proteoglycans on the luminal surface of endothelial cells), the short cytoplasmic domains of the integrin-α and -β subunits are separated by interactions with one or more cytoplasmic binding partners, and this change in subunit spacing induces a massive conformational change in the extracellular domains of both subunits that result in a high-affinity conformation that is competent to bind to endothelial ligands. On the endothelial side, the same pro-inflammatory cytokines that induce expression of P and E selectins increase transcription and subsequent surface expression of ICAM-1 and VCAM-1.

ROLES OF INTEGRINS AND THEIR LIGANDS IN LEUKOCYTE MIGRATION AND RETENTION IN THE EXTRAVASCULAR SPACE

Whereas the molecular steps involved in rolling and firm adhesion are relatively well characterized, much less is know about how subsequent steps of extravasation and migration and retention at specific sites are regulated. These steps also require repeated cycles of adhesion and de-adhesion. Although multiple molecular pathways have been shown to modulate these steps in various *in vitro* systems, the relative significance of most of these for specific leukocyte populations *in vivo* remains to be determined. One of the steps that has just begun to receive attention is the retention of leukocytes at specific tissue sites. The potential importance of this step in asthma was demonstrated by mice lacking the metalloprotease, MMP2, which were found to have dramatically increased susceptibility to fatal respiratory failure in response to airway allergen challenge after prior allergic sensitization [19]. Mortality in these mice was shown to be associated with abnormal retention of recruited leukocytes in the pulmonary interstitium, with a marked decrease in the subsequent migration of these cells into the airway lumen.

One of the critical steps involved in retention of inflammatory cells within the airway epithelium is induced expression of ICAM-1 on airway epithelial cells. Another key step is the induction of the integrin αEβ7, which can be induced on lymphocytes and mast cells by the action of the cytokine transforming growth factor β (TGF-β), activated on epithelial cells at sites of airway injury (see below). αEβ7 recognizes an epithelial adhesion receptor, E Cadherin, as its principal ligand [20], and through this interaction increases the residence time of mast cells and lymphocytes within the epithelium. These leukocytes utilize the αEβ7-integrin and presumably E Cadherin for retention within the epithelium.

POTENTIAL THERAPEUTIC RELEVANCE OF LEUKOCYTE CELL ADHESION RECEPTORS AND THEIR ENDOTHELIAL AND EPITHELIAL COUNTER-RECEPTORS

Because airway inflammation is a prominent feature of asthma, and because leukocytes are the principal source of cytokines (e.g., interleukin-13) known to play a major role in inducing airway hyperresponsiveness and mucus metaplasia, leukocyte adhesion molecules and their endothelial and/or epithelial ligands have been extensively studied as potential therapeutic targets. Nearly two decades ago, blockade of the β2-integrin ligand, ICAM-1, was shown to prevent airway hyperresponsiveness in a murine model of allergic asthma [21]. Subsequent studies with mice lacking the integrin β2 have confirmed an important role for this family of integrins in the specific homing of Th2 effector T cells to the lungs [22].

α4-containing integrins are more widely expressed than β2-integrins but are also expressed on a number of leukocytes, including lymphocytes, eosinophils, and mast cells. Blockade of α4-integrins has been shown to protect against induction of airway hyperresponsiveness in models of allergic asthma in mice and sheep [23, 24]. Mice lacking α4 only on leukocytes have also been shown to be protected from induction of airway hyperresponsiveness, suggesting that α4-integrins (α4β1 and α4β7) might be reasonable targets for treatment of asthma. Although this strategy was conceived as a way to prevent the recruitment of leukocytes into the airways, local delivery of blocking antibody into the airways of allergic sheep prevented the development of airways hyperresponsiveness (AHR) without affecting leukocyte recruitment [24]. These results suggest that protection from AHR might be due to inhibition of α4-integrin-mediated signals initiated on cells that are already present in the airways, rather than simply to effects on leukocyte recruitment.

Based on these encouraging pre-clinical studies, both humanized antibodies and small molecule antagonists of α4-integrins have been developed as potential drug treatments for asthma. Humanized antibody has been shown to be dramatically effective for prevention of new brain lesions in another inflammatory disease, multiple sclerosis [25]. However, the development of fatal or disabling progressive multifocal leukoencephalopathy in three patients treated with this antibody, presumably as a consequence of activation of latent JC virus infection, has dramatically slowed efforts to use system blockade of α4-integrins for treatment of asthma [26, 27]. Because of the potential effectiveness of local inhibition of these integrins, inhalational therapy with α4-antagonists is still worthy of consideration.

Selectins have also been targeted in pre-clinical studies in mice and sheep. Inhibition of one of the three selectin family members (L selectin) also inhibited airway inflammation, allergen-induced late airway narrowing, and AHR in allergic sheep [14]. As with α4-blockade, local delivery of blocking antibody to L selectin was as effective as systemic administration, suggesting relevant effects of ligation of selectins on inflammatory cells after their recruitment into the airway wall.

ADHESION RECEPTORS ON AIRWAY SMOOTH MUSCLE

Airway smooth muscle is clearly an important target tissue in asthma. An increase in smooth muscle mass is a common feature in both human asthma and experimental models of chronic allergen challenge, and alterations in *in vivo* smooth muscle responsiveness might be an important determinant of airway hyperresponsiveness. Airway smooth muscle cells can express ICAM-1 and VCAM-1 and a number of integrins, including at least α1β1, α2β1, α3β1, α7β1, α8β1, α9β1, and αvβ3. These integrins are localized at intercalated disks that connect adjacent smooth muscle cells and are thus likely to play some role in regulating the strength of smooth muscle contraction. Airway smooth muscle cells exhibit markedly different proliferative responses and modulate their production of pro-inflammatory cytokines when they are plated on different integrin ligands [28, 29]. Furthermore, integrins in vascular smooth muscle cells have been shown to modulate contraction by regulating the activity of associated L-type voltage-gated calcium channels [30, 31]. The *in vivo* significance of adhesion receptors on airway smooth muscle cells thus seems like a fruitful area for future investigation.

ADHESION RECEPTORS ON PULMONARY EPITHELIAL CELLS

One potential contributing factor to increased responsiveness to environmental stimuli in asthma and COPD would be localized increases in epithelial permeability. The normal impermeable character of the airway epithelium is determined by the function of the adhesion molecules that make up adherens junctions (E Cadherin) and tight junctions (claudins and occludins). In contrast to responses in endothelial cells, where permeability can be rapidly (within minutes) increased in response to a number of physiological stimuli (e.g., VEGF, TGF-β, and thrombin), changes in epithelial permeability usually occur with a time course of several hours and are associated with loss of adhesion molecules (especially E Cadherin) from the cell surface. In the setting of chronic inflammation, for example with exposure to active TGF-β, E Cadherin is internalized and epithelial permeability increases. However, the *in vivo* relevance of these events to chronic airway disease has not been determined.

Airway epithelial cells express several members of the integrin family, including α2β1, α3β1, α6β4 α9β1, αvβ5, αvβ6, and αvβ8 [32]. Epithelial injury and inflammation can dramatically alter the expression of these integrins [33]. The integrin αvβ6, which is normally expressed at low levels in the conducting airway epithelium of healthy individuals, is upregulated in the airways of patients with COPD and asthma [34].

INTEGRIN-MEDIATED TGF-β ACTIVATION

At least two of the integrins expressed on airway epithelial cells, αvβ6 [35] and αvβ8 [36], play a role in activating latent complexes of TGF-β. TGF-β is a pleiotropic cytokine that plays important roles in regulating tissue fibrosis, in induced proliferation of airway smooth muscle, in recruiting and retaining tissue mast cells and in negatively regulating adaptive immunity. It is thus clear that TGF-β could play important roles in several basic processes relevant to asthma and COPD. Mice with a knockin mutation of the TGF-β1 gene, which uniquely prevents integrin-mediated activation, have an identical inflammatory phenotype as mice completely lacking TGF-β1 have [37], suggesting that integrin-mediated activation may be the central *in vivo* mechanism of TGF-β1 activation. Indirect evidence suggests similar importance for activation of TGF-β3.

αvβ6 and αvβ8 have different patterns of cellular expression. Each binds to the same sequence of the latency-associated peptides of TGF-β1 and -3, but these two integrins appear to utilize completely different mechanisms for TGF-β activation. αvβ6 is restricted to epithelial cells and activates TGF-β by mechanically deforming the latent complex (Annes). This process provides access to the active site in TGF-β without releasing free TGF-β for diffusion to distant receptors and thus spatially restricts TGF-β activity to the surface of αvβ6 expressing epithelial cells [35]. In contrast, αvβ8, which is much more widely expressed, presents the latent complex to transmembrane metalloproteinases expressed on the same cell, which cleave the latent complex allowing release of diffusible active TGF-β that could act at a distance [36].

Current evidence suggests that αvβ6 plays a central role in activating TGF-β on the surface of alveolar epithelial cells, where the activated TGF-β can exert effects on adjacent fibroblasts, endothelial cells, and alveolar macrophages which make direct physical contacts with alveolar epithelial cells. This process has been shown to play important roles in regulating pulmonary fibrosis [35] and several models of acute lung injury [38], where genetic deletion or antibody blockade of the αvβ6-integrin dramatically protects against induction of *in vivo* pathology. Microarray studies on the lungs of mice lacking the β6 subunit (and thus the αvβ6-integrin) identified the macrophage metalloprotease, MMP12 as an important target of negative regulation by this integrin, since MMP12 levels were increased 200-fold in the lungs of β6 knockout mice [39]. Induction of MMP12 leads to the development of age-related emphysema in these mice and can be prevented by expression of the integrin or low levels of active TGF-β in lung epithelium [40]. These findings suggest that αvβ6-mediated TGF-β activation plays an important homeostatic role in preventing the development of emphysema in response to normally inconsequential environmental insults (Fig. 21.3).

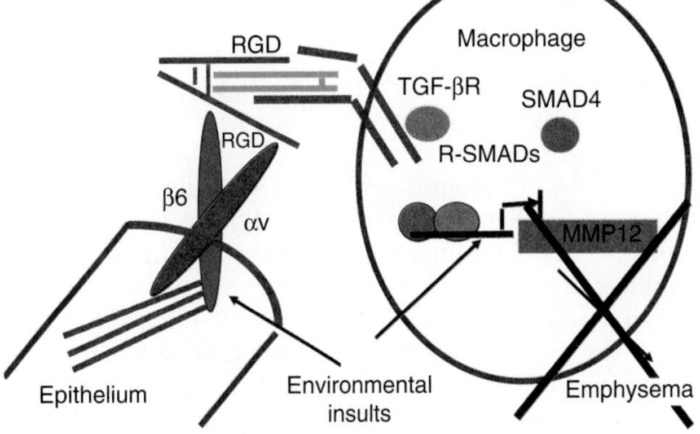

FIG. 21.3 Homeostatic regulation of macrophage MMP12 expression by integrin-mediated activation of TGF-β. Environmental insults (e.g., cigarette smoke) increase macrophage expression of MMP12 (and other proteases that can degrade the extracellular matrix of the lung. In parallel, the same insults activate a pathway through which the epithelial integrin, αvβ6, activates TGF-β. Activated TGF-β binds to TGF-β receptors on macrophages, activating a signaling pathway involving transcription factors (receptor SMADs – R-SMADs) that pair with co-SMADs to shut off transcription of the MMP12 gene, thereby providing a brake for this potentially destructive process.

INTEGRIN-MEDIATED TGF-β ACTIVATION IN THE CONDUCTING AIRWAYS

Both $\alpha v\beta 6$ and $\alpha v\beta 8$ are expressed on epithelial cells in the conducting airways. Under resting conditions, cultured airway epithelial cells express both integrins and secrete both TGF-β1 and TGF-β2 but do not appear to activate either isoform [41]. In response to mechanical wounding, these cells can be induced to activate TGF-β1 (but not TGF-β2) and both integrins appear to contribute to this activation. The net effect of this "autocrine" TGF-β activation is a slowing of the rate of sheet migration and subsequent wound closure. Interestingly, whereas blockade of $\alpha v\beta 8$, like blockade of TGF-β1, accelerates the rate of wound closure, blockade of $\alpha v\beta 6$ has no net effect on closure rates. This appears to be due to opposing effects of this integrin on activating TGF-β (which slows closure rates) and on directly enhancing epithelial migration, as shown by slowing of wound closure when anti-$\alpha v\beta 6$ antibody was added to wounded cells in the presence of inhibitors of TGF-β or $\alpha v\beta 8$.

In contrast to $\alpha v\beta 6$, which is restricted in its expression to airway epithelial cells and is highly induced in response to injury, $\alpha v\beta 8$ is also expressed on airway fibroblasts [42]. Recent work has shown that $\alpha v\beta 8$ can mediate autocrine activation of TGF-β by airway fibroblasts and that this effect inhibits to production of hepatocyte growth factor by adjacent epithelial cells. Through this process, airway fibroblasts can also induce expression of $\alpha v\beta 6$ on epithelial cells (which is highly inducible by TGF-β) generating a positive feedback loop that would be predicted to inhibit airway epithelial cell proliferation and contribute to the development of airway wall fibrosis.

SUMMARY

Adhesion molecules play critical roles in maintaining the structural integrity of all components of the airway wall. In the epithelium and endothelium, occludins, claudins, cadherins, and JAMs regulate paracellular permeability. Airway injury or inflammation, as occurs in patients with asthma and COPD, can perturb epithelial and endothelial permeability by altering the expression and/or function of these proteins. Selectins, integrins, and their immunoglobulin family member counter-receptors, play central roles in the recruitment of leukocytes into the airway wall and in regulating the function of leukocytes that have been recruited. Promising pre-clinical studies suggest that members of each of these classes of adhesion molecules are potential drug targets for treatment of asthma. Integrins also contribute to signaling responses of airway cells to input from other receptor families, including growth factor and cytokine receptors, and can modulate survival, growth, and differentiation of each of the cell types present in the airway wall. Two integrins, $\alpha v\beta 6$ and $\alpha v\beta 8$, are also largely responsible for *in vivo* activation of two of three mammalian isoforms of TGF-β. TGF-β activation by the $\alpha v\beta 6$-integrin is critical for the maintenance of normal alveolar homeostasis, and loss of this function could contribute to the development of emphysema. TGF-β activation by the $\alpha v\beta 8$-integrin plays an important role in homeostasis in the airway wall, and alterations in this function could contribute to airway wall remodeling in COPD. The multiple critical roles played by adhesion molecules in airway health and disease make these proteins attractive candidates for improvements in treatment and monitoring of patients with asthma and COPD.

REFERENCES

1. Takeichi M. Cadherin cell adhesion receptors as a morphogenetic regulator. *Science* 251(5000): 1451–55, 1991.
2. Gumbiner BM. Cell adhesion: The molecular basis of tissue architecture and morphogenesis. *Cell* 84(3): 345–57, 1996.
3. Anderson JM. Molecular structure of tight junctions and their role in epithelial transport. *News Physiol Sci* 16: 126–30, 2001.
4. Wang F, Daugherty B, Keise LL, Wei Z, Foley JP, Savani RC, Koval M. Heterogeneity of claudin expression by alveolar epithelial cells. *Am J Respir Cell Mol Biol* 29(1): 62–70, 2003.
5. Furuse M, Hirase T, Itoh M, Nagafuchi A, Yonemura S, Tsukita S, Tsukita S. Occludin: A novel integral membrane protein localizing at tight junctions. *J Cell Biol* 123(6 Pt 2): 1777–88, 1993.
6. Nusrat A, Chen JA, Foley CS, Liang TW, Tom J, Cromwell M, Quan C, Mrsny RJ. The coiled-coil domain of occludin can act to organize structural and functional elements of the epithelial tight junction. *J Biol Chem* 275(38): 29816–22, 2000.
7. Hynes RO. Integrins: Versatility, modulation, and signaling in cell adhesion. *Cell* 69: 11–25, 1992.
8. Clark EA, Brugge JS. Integrins and signal transduction pathways: The road taken. *Science* 268(5208): 233–39, 1995.
9. McEver RP. Leukocyte-endothelial cell interactions. *Curr Opin Cell Biol* 4(5): 840–49, 1992.
10. Luscinskas FW, Kansas GS, Ding H, Pizcueta P, Schleiffenbaum BE, Tedder TF, Gimbrone MA Jr. Monocyte rolling, arrest and spreading on IL-4-activated vascular endothelium under flow is mediated via sequential action of L-selectin, beta 1-integrins, and beta 2-integrins. *J Cell Biol* 125(6): 1417–27, 1994.
11. McEver RP. Selectins: Lectins that initiate cell adhesion under flow. *Curr Opin Cell Biol* 14(5): 581–86, 2002.
12. Alon R, Kassner PD, Carr MW, Finger EB, Hemler ME, Springer TA. The integrin VLA-4 supports tethering and rolling in flow on VCAM-1. *J Cell Biol* 128(6): 1243–53, 1995.
13. Jones DA, McIntire LV, Smith CW, Picker LJ. A two-step adhesion cascade for T cell/endothelial cell interactions under flow conditions. *J Clin Invest* 94(6): 2443–50, 1994.
14. Abraham WM, Ahmed A, Sabater JR, Lauredo IT, Botvinnikova Y, Bjercke RJ, Hu X, Revelle BM, Kogan TP, Scott IL, Dixon RA, Yeh ET, Beck PJ. Selectin blockade prevents antigen-induced late bronchial responses and airway hyperresponsiveness in allergic sheep. *Am J Respir Crit Care Med* 159(4 Pt 1): 1205–14, 1999.
15. Rosen SD, Tsay D, Singer MS, Hemmerich S, Abraham WM. Therapeutic targeting of endothelial ligands for L-selectin (PNAd) in a sheep model of asthma. *Am J Pathol* 166(3): 935–44, 2005.
16. Sriramarao P, von Andrian UH, Butcher EC, Bourdon MA, Broide DH. L-selectin and very late antigen-4 integrin promote eosinophil rolling at physiological shear rates *in vivo*. *J Immunol* 153(9): 4238–46, 1994.
17. Elices MJ, Osborn L, Takada Y, Crouse C, Luhowskyj S, Hemler ME, Lobb RR. VCAM-1 on activated endothelium interacts with the leukocyte integrin VLA-4 at a site distinct from the VLA-4/fibronectin binding site. *Cell* 60(4): 577–84, 1990.
18. Taooka Y, Chen J, Yednock T, Sheppard D. The integrin alpha9beta1 mediates adhesion to activated endothelial cells and transendothelial

neutrophil migration through interaction with vascular cell adhesion molecule-1. *J Cell Biol* 145(2): 413–20, 1999.
19. Corry DB, Rishi K, Kanellis J, Kiss A, Song Lz LZ, Xu J, Feng L, Werb Z, Kheradmand F. Decreased allergic lung inflammatory cell egression and increased susceptibility to asphyxiation in MMP2-deficiency. *Nat Immunol* 3(4): 347–53, 2002.
20. Cepek KL, Shaw SK, Parker CM, Russell GJ, Morrow JS, Rimm DL, Brenner MB. Adhesion between epithelial cells and T lymphocytes mediated by E-cadherin and the alpha E beta 7 integrin. *Nature* 372(6502): 190–93, 1994.
21. Wegner CD, Gundel RH, Reilly P, Haynes N, Letts LG, Rothlein R. Intercellular adhesion molecule-1 (ICAM-1) in the pathogenesis of asthma. *Science* 247(4941): 456–59, 1990.
22. Lee SH, Prince JE, Rais M, Kheradmand F, Shardonofsky F, Lu H, Beaudet AL, Smith CW, Soong L, Corry DB. Differential requirement for CD18 in T-helper effector homing. *Nat Med* 9(10): 1281–86, 2003.
23. Abraham WM, Gill A, Ahmed A, Sielczak MW, Lauredo IT, Botinnikova Y, Lin KC, Pepinsky B, Leone DR, Lobb RR, Adams SP. A small-molecule, tight-binding inhibitor of the integrin alpha(4)beta(1) blocks antigen-induced airway responses and inflammation in experimental asthma in sheep. *Am J Respir Crit Care Med* 162(2 Pt 1): 603–11, 2000.
24. Abraham WM, Sielczak MW, Ahmed A, Cortes A, Lauredo IT, Kim J, Pepinsky B, Benjamin CD, Leone DR, Lobb RR *et al.* Alpha 4-integrins mediate antigen-induced late bronchial responses and prolonged airway hyperresponsiveness in sheep. *J Clin Invest* 93(2): 776–87, 1994.
25. Miller DH, Khan OA, Sheremata WA, Blumhardt LD, Rice GP, Libonati MA, Willmer-Hulme AJ, Dalton CM, Miszkiel KA, O'Connor PW. A controlled trial of natalizumab for relapsing multiple sclerosis. *N Engl J Med* 348(1): 15–23, 2003.
26. Kleinschmidt-DeMasters BK, Tyler KL. Progressive multifocal leukoencephalopathy complicating treatment with natalizumab and interferon beta-1a for multiple sclerosis. *N Engl J Med* 353(4): 369–74, 2005.
27. Langer-Gould A, Atlas SW, Green AJ, Bollen AW, Pelletier D. Progressive multifocal leukoencephalopathy in a patient treated with natalizumab. *N Engl J Med* 353(4): 375–81, 2005.
28. Nguyen TT, Ward JP, Hirst SJ. beta1-Integrins mediate enhancement of airway smooth muscle proliferation by collagen and fibronectin. *Am J Respir Crit Care Med* 171(3): 217–23, 2005.
29. Peng Q, Lai D, Nguyen TT, Chan V, Matsuda T, Hirst SJ. Multiple beta 1 integrins mediate enhancement of human airway smooth muscle cytokine secretion by fibronectin and type I collagen. *J Immunol* 174(4): 2258–64, 2005.
30. Waitkus-Edwards KR, Martinez-Lemus LA, Wu X, Trzeciakowski JP, Davis MJ, Davis GE, Meininger GA. alpha(4)beta(1) Integrin activation of L-type calcium channels in vascular smooth muscle causes arteriole vasoconstriction. *Circ Res* 90(4): 473–80, 2002.
31. Wu X, Mogford JE, Platts SH, Davis GE, Meininger GA, Davis MJ. Modulation of calcium current in arteriolar smooth muscle by alphav beta3 and alpha5 beta1 integrin ligands. *J Cell Biol* 143(1): 241–52, 1998.
32. Sheppard D. Airway epithelial integrins: Why so many? *Am J Respir Cell Mol Biol* 19(3): 349–51, 1998.
33. Sheppard D. Epithelial integrins. *Bioessays* 18(8): 655–60, 1996.
34. Weinacker A, Ferrando R, Elliott M, Hogg J, Balmes J, Sheppard D. Distribution of integrins alpha v beta 6 and alpha 9 beta 1 and their known ligands, fibronectin and tenascin, in human airways. *Am J Respir Cell Mol Biol* 12: 547–56, 1995.
35. Munger JS, Huang X, Kawakatsu H, Griffiths MJ, Dalton SL, Wu J, Pittet JF, Kaminski N, Garat C, Matthay MA, Rifkin DB, Sheppard D. The integrin alpha v beta 6 binds and activates latent TGF beta 1: A mechanism for regulating pulmonary inflammation and fibrosis. *Cell* 96(3): 319–28, 1999.
36. Mu D, Cambier S, Fjellbirkeland L, Baron JL, Munger JS, Kawakatsu H, Sheppard D, Broaddus VC, Nishimura SL. The integrin alpha(v)beta8 mediates epithelial homeostasis through MT1-MMP-dependent activation of TGF-beta1. *J Cell Biol* 157(3): 493–507, 2002.
37. Yang Z, Mu Z, Dabovic B, Jurukovski V, Yu D, Sung J, Xiong X, Munger JS. Absence of integrin-mediated TGFbeta1 activation *in vivo* recapitulates the phenotype of TGFbeta1-null mice. *J Cell Biol* 176(6): 787–93, 2007.
38. Pittet JF, Griffiths MJ, Geiser T, Kaminski N, Dalton SL, Huang X, Brown LA, Gotwals PJ, Koteliansky VE, Matthay MA, Sheppard D. TGF-beta is a critical mediator of acute lung injury. *J Clin Invest* 107(12): 1537–44, 2001.
39. Kaminski N, Allard JD, Pittet JF, Zuo F, Griffiths MJ, Morris D, Huang X, Sheppard D, Heller RA. Global analysis of gene expression in pulmonary fibrosis reveals distinct programs regulating lung inflammation and fibrosis. *Proc Natl Acad Sci USA* 97(4): 1778–83, 2000.
40. Morris DG, Huang X, Kaminski N, Wang Y, Shapiro SD, Dolganov G, Glick A, Sheppard D. Loss of integrin alpha(v)beta6-mediated TGF-beta activation causes Mmp12-dependent emphysema. *Nature* 422(6928): 169–73, 2003.
41. Neurohr C, Nishimura SL, Sheppard D. Activation of transforming growth factor-beta by the integrin alphavbeta8 delays epithelial wound closure. *Am J Respir Cell Mol Biol* 35(2): 252–59, 2006.
42. Araya J, Cambier S, Morris A, Finkbeiner W, Nishimura SL. Integrin-mediated transforming growth factor-beta activation regulates homeostasis of the pulmonary epithelial-mesenchymal trophic unit. *Am J Pathol* 169(2): 405–15, 2006.

Extracellular Matrix

CHAPTER 22

Maurice Godfrey

Center for Human Molecular Genetics, Munroe Meyer Institute, University of Nebraska Medical Center, Omaha, NE, USA

INTRODUCTION

Until recently the extracellular matrix was thought of solely as a static structural support network. We now know that the extracellular matrix is comprised by a large and varied group of dynamic macromolecules and their regulatory factors [1] which provides structural support and is a physical barrier. However, it also elicits cellular responses and its interactions are involved in development and organ formation [2]. Disruption of normal extracellular matrix during disease processes can lead to an inflammatory response that exacerbate aberrant remodeling of the lung [3, 4]. Much has also been learned about the role of the extracellular matrix in asthma and chronic obstructive pulmonary disease (COPD) using mouse models [5, 6]. These studies tend to support observations of altered function in people with polymorphic variants of extracellular matrix molecules [7–9].

Extracellular matrix molecules are a part of a finely regulated system of development, maintenance, and repair. In addition to the structural macromolecules that are discussed in this chapter, there are regulatory molecules that are essential components of the extracellular matrix [10, 11]. In this chapter we will review the extracellular matrix constituents of the respiratory system.

COMPONENTS OF THE EXTRACELLULAR MATRIX

The biomechanical properties of the lung are primarily determined by the extracellular matrix components that comprise the tissue [2, 12]. In the lung they can be largely categorized into four groups: the collagens, proteoglycans, elastic fibers, and non-collagenous proteins. Connective tissue in the lung imparts the appropriate mechanical sturdiness and elastic resilience that permits the lung to expand and relax repeatedly. In addition to this mechanical function, the extracellular matrix is organized to allow for efficient gas exchange (Table 22.1).

Collagens

By far the most abundant of the extracellular matrix elements are the collagens. The collagens are the largest, in number, and most abundant, in content, of the extracellular matrix elements. There are more than 20 different collagen types and even more genes that encode these collagens [1, 2, 13].

Collagens are characterized by a triple helical structure whereby three polypeptide chains wind around each other. To form this helical configuration, every third amino acid must be a glycine. Glycine, with its very small side chain is the only amino acid that is able to fit in the center of the helix without causing any distortion. The primary structure of the triple helical region of collagens is described by the general pattern of $[\text{Gly-X-Y}]_n$ as a repeating triplet. About 20% of the X and Y positions in this triplet are the imino acids, proline, and hydroxyproline. The hydroxyl group of hydroxyproline is essential for hydrogen bonding which imparts stability to the triple helix. Similarly, lysine and hydroxylysine residues in the helical region are important for covalent cross-links within and between collagen molecules. Hydroxylysines are also subject to glycosylation.

TABLE 22.1 Summary of the major extracellular matrix constituents of the lung and their localization.

Collagens	
Type I	Alveolar, bronchial, vascular walls
Type II	Bronchial and tracheal cartilage
Type III	Co-localizes with Type I collagen
Type IV	Basement membrane
Type V	Basement membrane, interstitium
Elastin	Alveolar septa, blood vessels, bronchial walls, pleura
Microfibrils	Found with elastin, scaffold for elastin deposition
Proteoglycans	Alveolar walls, basement membrane
Laminin	Basement membrane
Integrin (receptors for extracellular matrix molecules)	Mesenchymal and epithelial cells
Fibronectin	Basement membrane, interstitium

The number of glycosylated residues vary with the different collagen type [14, 15].

Collagen α-chains are synthesized as pro-α-chains with non-helical amino and carboxy terminal ends. Cleavage of the propeptides is essential for the proper stacking of triple helical collagen molecules to form an extracellular matrix. Amino acid substitutions, especially the invariant glycine residues, cause a delay (momentary pause, ie. nanoseconds) in formation of the triple helix. This delay leads to an increase in hydroxylated and glycosylated residues. These, so-called, overmodified collagens are characteristic of a number of heritable connective tissue disorders [16].

About 60% of the connective tissue protein mass in the adult lung is collagen. As expected Type I is the predominant type. The lung also contains a significant amount of Type III collagen, a major constituent of skin and vascular tissues. The Type III fibers are thin which allows for a more compliant tissue. Studies have shown that Type III collagen is essential for fibrillogenesis of Type I collagen in normal development [17–20]. Types I and III collagen provide the structural framework for the alveolar wall [2].

Type IV collagen is an important constituent of basement membranes [1, 21, 22]. The basement membrane, as visualized by electron microscopy, consists of two layers, the lamina lucida and lamina densa. In the lung, the basement membrane forms continuously along the bronchial epithelia from before 12 weeks gestation [23]. In addition to Type IV collagen, basement membranes contain other extracellular matrix molecules such as proteoglycans and laminins. The Type IV collagen appears to serve as a scaffold for binding of laminin and proteoglycans leading to the formation of the basement membrane [24]. During lung branching, the epithelium of the elongating bronchial buds and tubules are lined with Type IV collagen. Type III collagen, however, does not appear to play a significant role in lung branching [25]. Type V collagen has also been localized in the lung. While it is found near the basement membrane, it is unclear whether this minor collagen is part of the basement membrane or solely in the nearby interstitial matrix [26]. The basement membrane of the lung also contains Type XVIII collagen, a member of the collagen subfamily called MULTIPLEXINs [27]. These multiplexins have multiple triple helical domains and interruptions, hence their name [1]. Cartilage collagens, Types II, IX, and XI along with cartilage matrix protein (not a collagen) are primary components of the trachea [28].

Despite the biochemical, immunochemical, ultrastructural in addition to molecular characterization of collagens in normal lung, their precise function and roles in pulmonary homeostasis is still not completely elucidated. Studies continue to determine the cellular influence of collagens; their mechanical properties as they relate to lung mechanics, interactions between collagen types and other matrix molecules; kinetics of maturation and turnover; and their self-assembly. Nevertheless, the extracellular matrix components of the lung show that they are part of a dynamic structure. When perturbed by environmental assault (eg. smoking, industrial polution) or disease (genetic mutations), this dynamic balance can be altered and the biomechanical properties of the lung changed [2].

Elastin

Elastin is the main elastic protein in vertebrates and is responsible for the ability of the lung to recoil following each cycle of expansion and contraction. Thus, like collagen, elastin is an integral component of the interstitial matrix of the lung. Given its importance for this elastic resilience, elastin has been localized to the alveoli, pleura, conducting airways, and vascular tissues [29, 30]. Elastin is secreted as a monomer of so-called tropoelastin, generally by cells that are of mesenchymal origin. The exquisite ability of repeated recoil is a unique property of the elastic fiber. The ability to undergo repeated rounds of recoil are believed to be due to the high hydrophobic amino acid content of elastin [31]. These hydrophobic amino acids are arranged as to allow cross-links to form that are believed to be critical for its function. Recent studies have also shown the importance of optimal iron levels for elastic fiber integrity [32]. It is important to note that none of these extracellular matrix elements function in a vacuum. They comprise a characteristic architecture and thus a function in an interrelated and integrated fashion. Therefore, elastin provides its unique function in the context of the collagens, proteoglycans, and microfibrils.

The elastic fiber is among the most long-lasting structures in the body. Studies have shown that humans retain elastin an entire lifetime [33]. This stability appears to be due to very specific cross-links between elastin monomers in the mature elastic fibers. It is in fact these unique cross-links, formed from lysine residues, that produce novel amino acids, desmosine and isodesmosine, that are markers for the presence of elastin in tissues.

Early structural studies tended to describe elastic fibers as having an amorphous appearance. More recent studies, however, have shown a rather complex structure.

Two morphologically and chemically distinct entities have been identified. Elastin, the insoluble polymer composed of tropoelastin monomers and microfibrils. Structure and function of microfibrils are discussed below [34, 35].

Elastin, in its polymerized form, constitutes more than 90% of the mature elastic fiber. The amino acid structure of tropoelastin is highly conserved in evolution [36]. In all species studied, tropoelastin has a modular structure of alternating hydrophobic and cross-linking domains. While sequence variation exists, it is the hydrophobicity that is exquisitely conserved. The conservation appears greater in the cross-linking regions. Unlike the large variety and numerous gene products that encode the collagens, elastin is present as a single gene in a single copy. Interestingly, the various isoforms of elastin that are present are produced by alternative splicing of a common transcript [37–39]. At present, however, the functional roles of these isoforms are not understood [40].

Mouse studies have shown that animals that lack any elastin expression do not undergo normal alveogenesis. Mice with only 50% normal elastin do develop normal lungs while expression of less than 50% elastin leads to abnormal lung development [30]. Significantly, however, is the fact that injurious stimuli make mice with less than optimal elastin levels more prone to severe lung disease [30]. Recent studies have also demonstrated a polymorphic variant in human elastin causing early onset COPD in a family [41]. Severe early onset pulmonary disease has also been documented in smokers whose elastin mutation causes autosomal dominant cutis laxa [42].

Microfibrils

Elastin-associated microfibrils have been classically defined as 10–12 nm diameter fibrils when seen by electron microscopy [43]. They were initially identified as structures that surrounded or were within mature elastic fibers. It is now known that microfibrils can also exist without the presence of elastin [44]. Analyses of developing tissues, including the lung and major blood vessels, have shown that microfibrils are deposited first followed by the deposition of elastin [45]. Thus, the hypothesis that microfibrils act as a scaffold for the deposition of elastin to produce mature elastic fibers has emerged. The complete function and, for the matter, constituents of the microfibrils are presently unknown.

Ultrastructurally, microfibrils display a "beads on a string" structure with a diameter of 8 to 12 nm and are composed of several proteins, the most abundant of which is fibrillin-1 [46]. Fibrillin-1, the product of the FBNI gene, is a cysteine rich-glycoprotein with a molecular weight of about 350 kDa. The extracellular domain structure of fibrillin-1 is divided into five distinct regions. The most abundant of these domains are calcium binding epidermal growth factor-like (EGF) motifs, that occur some 43 times. Four additional EGF motifs that do not bind calcium are also present. Interspersed between their calcium binding domains are seven transforming growth factor β1 binding protein domains each containing eight cysteine residues [47, 48]. These eight cysteine domains are globular in structure and interrupt the multiple stretches of the EGF modules that are believed to form rod-like structures. The middle 8-cysteine domains contain an RGD (arginine–glycine–aspartate) site. RGD motifs interact with cell surface receptors to mediate cell adhesion [49]. These receptors are part of the integrin family, transmembrane proteins that interact with extracellular matrix proteins to anchor cells within the extracellular matrix. While the role of elastin in mediating elastic recoil is well established, microfibrils too appear to have some elasticity. In fact, speculation exists that microfibrils alone performed the function of elastic fibers prior to the evolution of tropoelastin. Closely related to fibrillin-1 is fibrillin-2. Fibrillin-2 shares the domain structure with fibrillin-1 and in the EGF containing regions are about 80% identical at the amino acid level. There are several important differences that may reflect differing functional roles. Fibrillin-2 contains two RGD sites and the domain that is proline rich in fibrillin-1 is glycine rich in fibrillin-2 [47, 48, 50].

Fibrillin-1 and -2 are differentially expressed both temporally and spatially. In most cases, developmental expression of the fibrillin genes displays a diphasic pattern. Thus, expression of fibrillin-2 occurs earlier in development than fibrillin-1. Studies at the level of mRNA have shown FBN2 transcript accumulation prior to tissue differentiation followed by their rapid decrease. FBN1 transcripts then begin to increase gradually. Studies have shown that fibrillin-2 is found preferentially in elastic tissues, such as elastic cartilage, tunica media of the aorta, and along the bronchial tree. The two fibrillins, therefore, may have differing functional roles. It has been hypothesized that fibrillin-2 may have a greater role during early morphogenesis in directing the assembly of elastic fibers, while fibrillin-1 is mainly responsible for load bearing [50–52].

While the fibrillins are regarded as the major constituents of microfibrils, there are additional families of proteins that are part of the microfibrils and the elastic fiber. These additional families include the latent transforming growth factor beta binding proteins (LTBPs); microfibril-associated glycoproteins (MAGPs) and microfibril-associated proteins (MFAPs); fibulins; emilin; Big-H3; and lysyl oxidase. The structural and functional roles of most of these proteins are not clear. We do know that LTBP-2 is evolutionarily the closest of the LTBPs to the fibrillins and has been isolated from tissue rich in microfibrils. MAGP-1 has also been immunolocalized to both elastin-associated and naked microfibrils. MAGP-1 interacts directly with tropoelastin monomers and may play a primary role in elastic fiber formation (see Ref. [53] for a review).

Studies in mice have shown that fibulin-5 null animals have disrupted elastic fibers and pulmonary emphysema among other phenotypes [54, 55]. Fibulin-5 mutations in people appear not to be associated with lung abnormalities [56]. The hypothesis that microfibrils play a role in TGF-β signaling has lead to uncovering of emphysema in Marfan-like mice [57]. In addition, pulmonary abnormalities in several LTBP mouse mutants have also been demonstrated [58, 59]. The effects of TGF-β signaling on lung development and branching are reviewed in Ref. [60].

It is apparent that we are just beginning to understand the roles of the extracellular matrix in forming complex tissues. The myriad of molecular interactions involved with the numerous proteins noted above that are required to build an elastic fiber are just being uncovered. The temporal or tissue specific expression of the various microfibrillar components is also largely unknown. Mouse studies will be critical for the continued progress in elucidating the complex signaling role of the extracellular matrix.

Proteoglycans

Proteoglycans comprise a large group of multidomain core proteins to which glycosaminoglycans are attached. Glycosaminoglycans are unbranched carbohydrate chains of repeating disaccharide units. Since most are negatively charged, they bind to other matrix molecules, cell adhesion molecules, and growth factors [61, 62]. They are integral to maintain normal pulmonary structure and function. Of the species of proteoglycans, heparan sulfate and dermatan sulfate appear to be the most abundant [2, 63]. These proteoglycans are also a reservoir for heparin. Chondroitin sulfates, proteoglycans found in cartilagenous tissues, are present in bronchioles, while heparan sulfate is the major proteoglycan of the gas exchange tissue. Heparin is also found primarily in the gas exchange tissues and pleura, but, as expected, not the cartilagenous bronchioles [63, 64]. Biochemical studies over the past quarter century have helped to elucidate the composition of lung proteoglycans, but their precise functions are still not completely understood.

Heparan sulfate interacts with laminin and appears to be essential for human development and lung branching [65–68]. Studies also demonstrate that proteoglycans are important for the function of growth factors. Much of this modulation by proteoglycans appears to involve binding and signaling of TGFβ and its receptor [69–72]. These molecules may also stimulate production of other matrix constituents or act as receptors for the extracellular matrix [73, 74].

Here again mouse targeting of perlecan, a heparan sulfate proteoglycan, have shown that null mice who survive embryonic life die perinatally due to respiratory failure [75, 76].

Laminin

Laminins are cross-shaped molecules that comprise several different types and are a part of lung development [1, 77–79]. Analysis of the developing lung has shown the expression of some laminin chains as early as 10 weeks of human gestation [80]. Laminins are expressed along the basement membrane, but temporal and spatial expression of different laminin subunits has been documented [25, 81]. Laminin expression may play a role in cell differentiation in the lung as well as other tissues. There is some evidence to suggest that laminin may play a role in alveolar morphogenesis [82]. A laminin–heparan sulfate interaction may be essential to lumen formation and branching morphogenesis [83].

Integrins

Integrins are cell surface glycoproteins that serve as receptors for the extracellular matrix [84–86]. Integrins are composed of many subtypes all arranged as heterodimers [85]. These heterodimers can selectively bind several different matrix constituents [87]. In fact, the RGD sequence of extracellular matrix molecules is the binding site for integrins [88, 89]. Integrins are expressed early in human lung development [90, 91]. Animal model manipulation of some integrins have resulted in reduced bronchial branching suggesting a vital role for some integrin dependent interactions in lung development [86, 88, 92, 93]. Integrins also function in signaling pathways to mediate migration and differentiation of epithelial cells [94–96].

Fibronectin

Fibronectin is a widely distributed glycoprotein found in embryonic tissue that plays a material role in morphogenesis [97, 98]. Fibronectin plays an important role in cell attachment [99]. While it has been localized in several regions of the developing lung, it is primarily seen in regions of airway bifurcation [89]. Fibronectin is co-distributed with collagen and may be required for normal collagen deposition [100, 101]. Its precise role, however, has not yet been elucidated.

EXTRACELLULAR MATRIX AND LUNG FUNCTION

The matrix of the normal lung imparts the strength and resilience for the continuous cycling of inspiration and expiration (Fig 22.1). For proper gas exchange, the components of the extracellular matrix are distributed in a fashion to reduce the boundary between erythrocytes and oxygen [26]. When the exquisite balance of matrix glycoproteins is disturbed, (by structural mutations of matrix components or environmental damage) the consequences on the lung are manifested by pathology. In most cases the assault on the lung leads to fibrosis. Fibrosis leads to the thickening of the alveolar walls, reduction in lung volume, reduced lung elasticity, and fundamentally anomalous gas exchange. There are a number of causes of pulmonary fibrosis in addition to those due to genetic abnormalities of matrix molecules or their modifiers (see below).

In some cases, pulmonary fibrosis is a result of some other primary disease [102]. Examples include infections (viral and fungal) or immune disorders (rheumatoid arthritis, scleroderma). As expected, environmental assaults comprise the greatest number of causes of pulmonary fibrosis. As one might predict, tobacco smoke causing emphysema is among the major causes of COPD [103]. Fibrogenic dusts, such as asbestos, toxins, and chemicals, such as insecticides and herbicides, are additional causes of fibrosis in the lung. Iatrogenic causes, such as pharmacotherapy and therapeutic irradiation may also lead to pulmonary fibrosis.

HERITABLE DISORDERS OF CONNECTIVE TISSUE

Heritable disorders of connective tissue (HDCT) are a series of disorders caused by mutations in structural or modifying components of the extracellular matrix. Since most of these gene products are expressed in multiple tissues, pleiotropic manifestations are observed in most HDCT. It is important to distinguish HDCT from connective tissue disorders that are autoimmune in nature such as rheumatoid arthritis and systemic lupus erythematosus. For a comprehensive review of the HDCT, see Refs [16, 104].

Despite the abundance of collagen and elastic fibers in the lung, the number of HDCT with primary and severe pulmonary disease is rather small. While there are numerous HDCT, for the purposes of this abbreviated review only those disorders caused by defects in genes expressed in the pulmonary system or with pulmonary complications will be discussed. Osteogenesis imperfecta (OI) comprises several disorders of varying severity that are caused by mutations in Type I collagen. In its most severe form, OI Type II, is characterized by sever bone fragility and stillbirth or neonatal death. Due to the small chest in these neonates, pulmonary insufficiency and failure to ventilate can cause death in the few who are not stillborn. In OI Type III, characterized by moderate to severe bone fragility, the thorax is often conical in shape. Due to the softness of the bones in the chest, an often lethal, respiratory failure may occur in neonates. Despite these findings, most individuals with OI, except of course Type II OI, do not have severe pulmonary deficiency even though Type I collagen is abundant in the lung matrix.

The Ehlers–Danlos syndromes (EDS) are another series of HDCT. The classic type, previously called Types I and II are now known to be caused by mutations in Type V collagen. The most severe form of EDS is the, so-called, vascular type, previously called Type IV EDS. This form of EDS is caused by mutations in Type III collagen.

Pulmonary complications are rare in EDS. However, some cases of mediastinal and subcutaneous emphysema and spontaneous pneumothorax have been reported. In addition, dilation of the trachea and bronchi has also been described in some cases of EDS.

Pseudoxanthoma elasticum (PXE) is an HDCT whose molecular pathogenesis has been recently elucidated. PXE is characterized by abnormalities in the skin, eyes, and vasculature. Although the mechanism of the pathogenesis remains unclear, the generalized dystrophy of elastic fibers does occasionally manifest in the lung. Degenerative changes in the walls of alveoli and miliary mottling of the lungs have been reported.

The mucopolysaccharidoses (MPS) are a large group of disorders that are defects in enzymes required in the processing of glycosaminoglycan molecules (see "Proteoglycans"). The large number of different enzymatic defects makes a comprehensive review here impossible. Nevertheless, pulmonary complications have been observed in several MPS. For example, Hurler syndrome (MPS I H) is due to defects in α-l-iduronidase. Hurler syndrome is characterized by diagnosis prior to age 2 early corneal clouding, kyphoscoliosis, mental retardation, and death by age 10. All individuals with Hurler syndrome experience severe respiratory problems. Accumulation of glycosaminoglycans in the oropharyngeal trachea leads to airway obstruction. Radical pharyngoplasty provides some temporary relief, but obstruction recurs. These individuals are also prone to repeated upper respiratory infections. Deformity of the thorax and abnormalities of bronchial cartilage contributes to reduced chest expansion and decreased vital capacity. Bronchopneumonia is a frequent cause of death.

Hunter syndrome (MPS II), caused by a deficiency in iduronate 2 sulfatase, is characterized by diagnosis prior to age 4, mental retardation, and death before age 15. A mild form has also been described with moderate skeletal and respiratory involvement and survival to adulthood without intellectual impairment. Respiratory complications include upper airway obstruction, nasal congestion, and thick rhinorrhea. As children grow older, pharyngeal hypertrophy, tongue enlargement, and supraglottic swelling may result in obstructive sleep apnea and death. Upper respiratory infections are common. Abnormalities of the trachea have been documented. MPS II is an X-linked disorder, thus virtually all effected individuals are male. Pulmonary complications are observed in the majority of individuals with the gamut of MPS types.

The chondrodysplasias are another group of HDCT often classified based on radiographic involvement of the long bones. Severe respiratory difficulty is present in a number of these disorders due to a greatly restricted thorax, which leads to early death. Defects in Type II collagen are well documented in several of these disorders.

The Marfan syndrome (MFS) is a prototypical member of the HDCT. It is now well known that mutations in the gene encoding fibrillin-1 cause MFS. Thus, MFS is due to abnormalities of the elastic fiber system. Clinically, MFS is characterized by defects in the cardiovascular, skeletal, and ocular systems. Given the presence of elastic fibers in the lung it is not surprising that pulmonary complications also occur. The principal respiratory system abnormality in MFS is spontaneous pneumothorax, which occurs in approximately 5% of affected individuals. This indicates that spontaneous pneumothorax is statistically several hundred times more likely in a person with MFS than in the general population. In fact, the diagnosis of MFS has often been made after an initial event of a spontaneous pneumothorax. Spontaneous pneumothorax may be familial even in the absence of MFS. Apical bullae are also known to occur in MFS and may be a predisposing factor for spontaneous pneumothorax. Emphysema and congenital cystic lung have also been documented in MFS.

Pulmonary function too has been studied in MFS but has been interpreted as essentially normal. In some cases of MFS, severe kyphoscoliosis may be a major contributor to pulmonary failure, however, that cannot be attributed to abnormal fibrillin in the lung, but to the generalized skeletal dysplasia.

Therefore, one can conclude that primary defects in extracellular matrix proteins or their modifying enzymes can cause an array of pulmonary disease. However, even when the pulmonary abnormality leads to premature death, the manifestations in other organ systems are often more prominent (Table 22.2).

EXTRACELLULAR MATRIX IN ASTHMA AND COPD

Alterations in connective tissue likely play key roles in the pathogenesis of both asthma and COPD [105–107].

TABLE 22.2 Pulmonary manifestations of primary matrix abnormalities.

Matrix element	Disorder	Pulmonary abnormality
Type I collagen	Osteogenesis imperfecta (various types)	Pulmonary insufficiency; Pulmonary hypertension
Type III collagen	Ehlers–Danlos (vascular type)	Spontaneous pneumothorax; hemoptysis
Elastin	Dominant cutis laxa	Bronchiectasis; emphysema
Fibulin-5	Recessive cutis laxa	Emphysema
Fibrillin-1	Marfan syndrome	Pneumothorax; pulmonary blebs; emphysema

Destruction of elastin is thought to be a major feature in the development of emphysema. Exposure of the lung to enzymes with elastolytic activity results in emphysema in animals [108]. Histologic examination of the emphysematous lung shows disrupted elastic fibers [109]. Urinary excretion of desmosine, a specific marker for degradation of elastin, has been observed in smokers and former smokers with COPD. Smokers without COPD did not excrete desmosine [110, 111]. Because individuals with severe deficiency of α-1-protease inhibitor, an inhibitor of several serine proteases, are at markedly increased risk to develop emphysema [112], these proteases have been thought to have a major role in the destruction of elastin. Recent studies, however, suggest that the matrix metalloproteases contribute to the development of emphysema [113, 114].

In addition to tissue destruction, there is evidence of excess deposition of extracellular matrix (i.e. fibrosis in COPD). In addition to the destruction of elastin in emphysema, there is an increase in deposition of collagen [115–117]. Fibrosis of the small airways is also a regular feature of

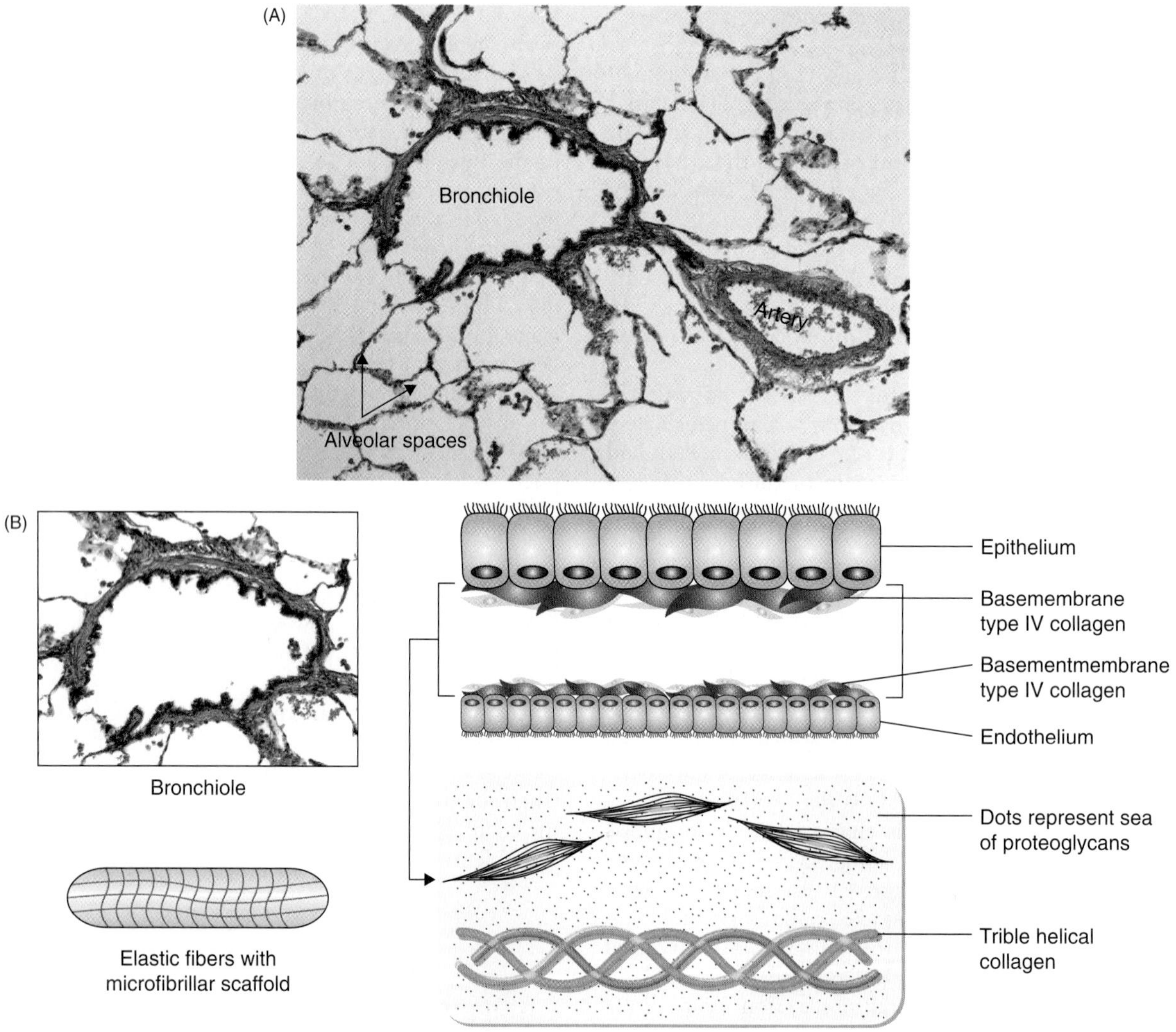

FIG. 22.1 (A) Photomicrograph of normal H&E stained lung (original magnification 100×). Various part are labeled in the figure. (B) An artist's rendition of a bronchiole. The outer (epithelium) and inner (endothelium) cellular layers surround the extracellular matrix containing triple helical collagen, elastic fibers with their associated microfibrils, and proteoglycans.

chronic bronchitis [118] and is related to the development of airflow limitation [119, 120]. The airways in COPD are of smaller than expected diameter; this may be a direct consequence of this fibrotic process [120].

Extracellular matrix alterations have also been documented in asthma [121]. For example, the connective tissue in the large airways, include thickening of the lamina reticularis with increased deposition of collagen types I, III, and fibronectin [122]; tenascin [123]; and fragmentation of the airway elastic fiber network [124, 125]. The thickened lamina reticularis is particularly striking as the epithelium can be seen to rest on a dense nearly acellular layer of apparently homogeneous connective tissue. Whether these changes are altered with therapy of asthma is controversial [126–129], but the connective tissue changes may occur somewhat independently of measures of clinical severity [127, 130]. The functional significance of the altered connective tissue in asthma likely leads to compromised lung function [106]. Similar changes in the small airways may account for increased peripheral airway resistance, reduced airway compliance [131], and progressive airflow limitation in asthma [132, 133]. Importantly, growth factors are critical for the extracellular matrix remodeling in asthma [134, 135].

Recognizing the complexity of the extracellular matrix, it seems likely that many alterations in these structural molecules will contribute to the pathophysiology of asthma and COPD. It seems likely as well that the many genetic variations in connective tissue metabolism, which are beginning to be characterized, will contribute to the heterogeneity observed in asthma and COPD. Understanding the genetic milieu will undoubtedly help to predict those at greatest risk [4, 30, 136].

References

1. Ayad S, Boot-Handford RP, Humphries MJ *et al.* The Extracellular Matrix: Facts Book. . New York: Academic Press, 1998.
2. Suki B, Ito S, Stamenovic D *et al.* Biomechanics of the lung parenchyma: Critical roles of collagen and mechanical forces. *J Appl Physiol* 98: 1892–99, 2005.
3. Chapman HA. Disorders of lung matrix remodeling. *J Clin Invest* 113: 148–57, 2004.
4. Fixman ED, Stewart A, Martin JG. Basic mechanisms of development of airway structural changes in asthma. *Eur Respir J* 29: 379–89, 2007.
5. Aszodi A, Legate KR, Nakchbandi I *et al.* What mouse mutants teach us about extracellular matrix function. *Annu Rev Cell Dev Biol* 22: 591–621, 2006.
6. Warburton D, Gauldie J, Bellusci S *et al.* Lung development and susceptibility to chronic obstructive pulmonary disease. *Proc Am Thorac Soc* 3: 668–72, 2006.
7. Hersh CP, DeMeo DL, Lazarus R *et al.* Genetic association analysis of functional impairment in chronic obstructive pulmonary disease. *Am J Respir Crit Care Med* 173: 977–84, 2006.
8. Joos L, He JQ, Shepherdson MB *et al.* The role of matrix metalloproteinase polymorphisms in the rate of decline in lung function. *Hum Mol Genet* 11: 569–76, 2002.
9. Kaartinen V, Warburton D. Fibrillin controls TGF-beta activation. *Nat Genet* 33: 331–32, 2003.
10. Barnes PJ. Mediators of chronic obstructive pulmonary disease. *Pharmacol Rev* 56: 515–48, 2004.
11. Bourbon J, Boucherat O, Chailley-Heu B *et al.* Control mechanisms of lung alveolar development and their disorders in bronchopulmonary dysplasia. *Pediatr Res* 57: 38R–46R, 2005.
12. Garcia CS, Prota LF, Morales MM *et al.* Understanding the mechanisms of lung mechanical stress. *Braz J Med Biol Res* 39: 697–706, 2006.
13. Ricard-Blum S, Ruggiero F. The collagen superfamily: From the extracellular matrix to the cell membrane. *Pathol Biol (Paris)* 53: 430–42, 2005.
14. Bornstein P, Sage H. Structurally distinct collagen types. *Annu Rev Biochem* 1003: 41003–9957, 1980.
15. Burgeson RE, Morris NP. The collagen family of proteins. In: Uitto J, Perejda AJ (eds). *Connective Tissue Disease: Molecular Pathology of the Extracellular Matrix*, pp. 3–28. New York: Marcel Dekker, Inc, 1987.
16. *McKusick's Heritable Disorders of Connective Tissue*, P. Beighton, ed. 5th edn, pp. 1–748. St. Louis: Mosby, 1993.
17. Bienskowski RS. Interstitial collagens. In: Crystal RG, West JB (eds). *The Lung*, pp. 381–88. New York: Raven Press, 1991.
18. Bradley K, McConnell Breul S, Crystal RG. Lung collagen heterogeneity. *Proc Natl Acad Sci USA* 71: 2828–32, 1974.
19. Bradley KH, McConnell SD, Crystal RG. Lung collagen composition and synthesis: Characterization and changes with age. *J Biol Chem* 249: 2674–83, 1974.
20. Hay ED. Cell Biology of the Extracellular Matrix, 2nd edn. New York: Plenum Press, 1991.
21. Maina JN, West JB. Thin and strong! The bioengineering dilemma in the structural and functional design of the blood–gas barrier. *Physiol Rev* 85: 811–44, 2005.
22. Yurchenco PD, Furthmayr H. Self-assembly of basement membrane collagen. *Biochemistry* 23: 1839–50, 1984.
23. Lallemand AV, Ruocco SM, Gaillard DA. Expression and immunohistochemical localization of laminin and type IV collagen in developing human fetal tracheal glands. *Int J Dev Biol* 37: 491–95, 1993.
24. Leblond CP, Inoue S. Structure, composition, and assembly of basement membrane. *Am J Anat* 185: 367–90, 1989.
25. Virtanen I, Laitinen A, Tani T *et al.* Differential expression of laminins and their integrin receptors in developing and adult human lung. *Am J Respir Cell Mol Biol* 15: 184–96, 1996.
26. Clark JG, Kuhn C3, McDonald JA *et al.* Lung connective tissue. *Int Rev Connect Tissue Res* 331: 10249–331, 1983.
27. Saarela J, Rehn M, Oikarinen A *et al.* The short and long forms of type XVIII collagen show clear tissue specificities in their expression and location in basement membrane zones in humans. *Am J Pathol* 153: 611–26, 1998.
28. Kelley J. Collagen. In: Massaro D (ed.), *Lung Cell Biology*, pp. 821–66. New York: Marcel Dekker, 1989.
29. Foster JA, Rich CB, Curtiss SW *et al.* Elastin. In: Massaro D (ed.), *Lung Cell Biology*, pp. 867–905. New york: Marcel Dekker, 1989.
30. Shifren A, Mecham RP. The stumbling block in lung repair of emphysema: Elastic fiber assembly. *Proc Am Thorac Soc* 3: 428–33, 2006.
31. Sandberg LB, Soskel NT, Leslie JG. Elastin structure, biosynthesis, and relation to disease states. *N Engl J Med* 304: 566–79, 1981.
32. Hill CH, Ashwell CM, Nolin SJ *et al.* Dietary iron deficiency compromises normal development of elastic fibers in the aorta and lungs of chicks. *J Nutr* 137: 1895–900, 2007.
33. Shapiro SD, Endicott SK, Province MA *et al.* Marked longevity of human lung parenchymal elastic fibers deduced from prevalence of D-aspartate and nuclear weapons-related radiocarbon. *J Clin Invest* 87: 1828–34, 1991.
34. Brown PL, Mecham L, Tisdale C *et al.* The cysteine residues in the carboxy terminal domain of tropoelastin form an intrachain disulfide bond that stabilizes a loop structure and positively charged pocket. *Biochem Biophys Res Commun* 186: 549–55, 1992.
35. Mecham RP, Heuser J. Three-dimensional organization of extracellular matrix in elastic cartilage as viewed by quick freeze, deep etch electron microscopy. *Connect Tissue Res* 24: 83–93, 1990.
36. Boyd CD, Christiano AM, Pierce RA *et al.* Mammalian tropoelastin: Multiple domains of the protein define an evolutionarily divergent amino acid sequence. *Matrix* 11: 235–41, 1991.

37. Heim RA, Pierce RA, Deak SB *et al.* Alternative splicing of rat tropoelastin mRNA is tissue-specific and developmentally regulated. *Matrix* 11: 359–66, 1991.
38. Indik Z, Yeh H, Ornstein-Goldstein N *et al.* Structure of the elastin gene and alternative splicing elastin mRNA. In: Sandell L, Boyd CD (eds). *Extracellular Matrix Genes*, pp. 221–50. New York: Academic Press, 1990.
39. Parks WC, Secrist H, Wu LC *et al.* Developmental regulation of tropoelastin isoforms. *J Biol Chem* 263: 4416–23, 1988.
40. Parks WC, Deak SB. Tropoelastin heterogeneity: Implications for protein function and disease. *Am J Respir Cell Mol Biol* 2: 399–406, 1990.
41. Kelleher CM, Silverman EK, Broekelmann T *et al.* A functional mutation in the terminal exon of elastin in severe, early-onset chronic obstructive pulmonary disease. *Am J Respir Cell Mol Biol* 33: 355–62, 2005.
42. Urban Z, Gao J, Pope FM *et al.* Autosomal dominant cutis laxa with severe lung disease: Synthesis and matrix deposition of mutant tropoelastin. *J Invest Dermatol* 124: 1193–99, 2005.
43. Low FN. Microfibrils: Fine filamentous components of the tissue space. *Anat Rec* 142: 131–37, 1962.
44. Streeten BW, Licari PA, Marucci AA *et al.* Immunohistochemical comparison of ocular zonules and the microfibrils of elastic tissue. *Invest Ophthalmol Vis Sci* 21: 130–35, 1981.
45. Cleary EG. The microfibrillar component of the elastic fibers: Morphology and biochemistry. In: Uitto J, Perejda AJ (eds). *Connective Tissue Disease: Molecular Pathology of the Extracellular Matrix*, pp. 55–81. New York: Marcel Dekker, 1987.
46. Sakai LY, Keene DR, Engvall E. Fibrillin, a new 350-kD glycoprotein, is a component of extracellular microfibrils. *J Cell Biol* 103: 2499–509, 1986.
47. Lee B, Godfrey M, Vitale E *et al.* Linkage of Marfan syndrome and a phenotypically related disorder to two different fibrillin genes. *Nature* 352: 330–34, 1991.
48. Pereira L, D'Alessio M, Ramirez F *et al.* Genomic organization of the sequence coding for fibrillin, the defective gene product in Marfan syndrome. *Hum Mol Genet* 2: 961–68, 1993.
49. Sakamoto H, Broekelmann T, Cheresh DA *et al.* Cell-type specific recognition of RGD- and non-RGD-containing cell binding domains in fibrillin-1. *J Biol Chem* 271: 4916–22, 1996.
50. Zhang H, Apfelroth SD, Hu W *et al.* Structure and expression of fibrillin-2 a novel microfibrillar component preferentially located in elastic matrices. *J Cell Biol* 124: 855–63, 1994.
51. Mariencheck MC, Davis EC, Zhang H *et al.* Fibrillin-1 and fibrillin-2 show temporal and tissue-specific expression in developing elastic tissues. *Connect Tissue Res* 31: 87–97, 1995.
52. Zhang H, Hu W, Ramirez F. Developmental expression of fibrillin genes suggests heterogeneity of extracellular microfibrils. *J Cell Biol* 129: 1165–76, 1995.
53. Robinson PN, Godfrey M. The molecular genetics of Marfan syndrome and related microfibrillopathies. *J Med Genet* 37: 9–25, 2000.
54. Nakamura T, Lozano PR, Ikeda Y *et al.* Fibulin-5/DANCE is essential for elastogenesis *in vivo*. *Nature* 415: 171–75, 2002.
55. Yanagisawa H, Davis EC, Starcher BC *et al.* Fibulin-5 is an elastin-binding protein essential for elastic fibre development *in vivo*. *Nature* 415: 168–71, 2002.
56. Hu Q, Loeys BL, Coucke PJ *et al.* Fibulin-5 mutations: Mechanisms of impaired elastic fiber formation in recessive cutis laxa. *Hum Mol Genet* 15: 3379–86, 2006.
57. Neptune ER, Frischmeyer PA, Arking DE *et al.* Dysregulation of TGF-beta activation contributes to pathogenesis in Marfan syndrome. *Nat Genet* 33: 407–11, 2003.
58. Colarossi C, Chen Y, Obata H *et al.* Lung alveolar septation defects in LTBP-3-null mice. *Am J Pathol* 167: 419–28, 2005.
59. Sterner-Kock A, Thorey IS, Koli K *et al.* Disruption of the gene encoding the latent transforming growth factor-beta binding protein 4 (LTBP-4) causes abnormal lung development, cardiomyopathy, and colorectal cancer. *Genes Dev* 16: 2264–73, 2002.
60. Roth-Kleiner M, Post M. Similarities and dissimilarities of branching and septation during lung development. *Pediatr Pulmonol* 40: 113–34, 2005.
61. Ruoslahti E. Structure and biology of proteoglycans. *Annu Rev Cell Biol*: 4229–55, 1988.
62. Ruoslahti E. Proteoglycans in cell regulation. *J Biol Chem* 264: 13369–72, 1989.
63. Radhakrishnamurthy B, Berenson SG. Proteoglycans of the lung. In: Massaro D (ed.), *Lung Cell Biology*, pp. 981–1010. New York: Marcel Dekker, Inc, 1989.
64. Hance AJ, Crystal RG. The connective tissue of lung. *Am Rev Respir Dis* 112: 657–711, 1975.
65. Schuger L, O'Shea KS, Nelson BB *et al.* Organotypic arrangement of mouse embryonic lung cells on a basement membrane extract: Involvement of laminin. *Development* 110: 1091–99, 1990.
66. Schuger L, Skubitz AP, Gilbride K *et al.* Laminin and heparan sulfate proteoglycan mediate epithelial cell polarization in organotypic cultures of embryonic lung cells: Evidence implicating involvement of the inner globular region of laminin beta 1 chain and the heparan sulfate groups of heparan sulfate proteoglycan. *Dev Biol* 179: 264–73, 1996.
67. Schuger L, Skubitz AP, O'Shea KS *et al.* Identification of laminin domains involved in branching morphogenesis: Effects of anti-laminin monoclonal antibodies on mouse embryonic lung development. *Dev Biol* 146: 531–41, 1991.
68. Thompson SM, Connell MG, Fernig DG *et al.* Novel phage display antibodies identify distinct heparan sulfate domains in developing mammalian lung. *Pediatr Surg Int* 23: 411–17, 2007.
69. Border WA, Ruoslahti E. Transforming growth factor-beta in disease: The dark side of tissue repair. *J Clin Invest* 90: 1–7, 1992.
70. Hildebrand A, Romaris M, Rasmussen LM *et al.* Interaction of the small interstitial proteoglycans biglycan, decorin and fibromodulin with transforming growth factor beta. *Biochem J* 302: 34, 1994.
71. Lopez Casillas F, Wrana JL, Massague J. Betaglycan presents ligand to the TGF beta signaling receptor. *Cell* 73: 1435–44, 1993.
72. Yamaguchi Y, Mann DM, Ruoslahti E. Negative regulation of transforming growth factor-beta by the proteoglycan decorin. *Nature* 346: 281–84, 1990.
73. Bernfield M, Kokenyesi R, Kato M *et al.* Biology of the syndecans: A family of transmembrane heparan sulfate proteoglycans. *Annu Rev Cell Biol* 93: 8365–93, 1992.
74. Mast BA, Diegelmann RF, Krummel TM *et al.* Hyaluronic acid modulates proliferation, collagen and protein synthesis of cultured fetal fibroblasts. *Matrix* 13: 441–46, 1993.
75. Arikawa-Hirasawa E, Watanabe H, Takami H *et al.* Perlecan is essential for cartilage and cephalic development. *Nat Genet* 23: 354–58, 1999.
76. Costell M, Gustafsson E, Aszodi A *et al.* Perlecan maintains the integrity of cartilage and some basement membranes. *J Cell Biol* 147: 1109–22, 1999.
77. Burgeson RE, Chiquet M, Deutzmann R *et al.* A new nomenclature for the laminins. *Matrix Biol* 14: 209–11, 1994.
78. Nguyen NM, Senior RM. Laminin isoforms and lung development: All isoforms are not equal. *Dev Biol* 294: 271–79, 2006.
79. Timpl R, Rohde H, Robey PG *et al.* Laminin – a glycoprotein from basement membranes. *J Biol Chem* 254: 9933–37, 1979.
80. Lallemand AV, Ruocco SM, Gaillard DA. Synthesis and expression of laminin during human foetal lung development. *Anat Rec* 242: 233–41, 1995.
81. Uehara Y, Minowa O, Mori C *et al.* Placental defect and embryonic lethality in mice lacking hepatocyte growth factor/scatter factor. *Nature* 373: 702–5, 1995.
82. Kouretas D, Karinch AM, Rishi A *et al.* Conservation analysis of rat and human SP-A gene identifies 5′ flanking sequences of rat SP-A that bind rat lung nuclear proteins. *Exp Lung Res* 19: 485–503, 1993.
83. Schuger L, O'Shea S, Rheinheimer J *et al.* Laminin in lung development: Effects of anti-laminin antibody in murine lung morphogenesis. *Dev Biol* 137: 26–32, 1990.

84. Albelda SM, Buck CA. Integrins and other cell adhesion molecules. *FASEB J* 4: 2868–80, 1990.
85. Hynes RO. Integrins: Versatility, modulation, and signaling in cell adhesion. *Cell* 69: 11–25, 1992.
86. Wu JE, Santoro SA. Differential expression of integrin alpha subunits supports distinct roles during lung branching morphogenesis. *Dev Dyn* 206: 169–81, 1996.
87. Glukhova MA, Koteliansky VE. Integrins, cytoskeletal and extracellular matrix proteins in developing smooth muscle cells of human aorta. In: Schwartz SM, Mecham RP (eds). *The Vascular Smooth Muscle Cell*, pp. 37–79. San Diego: Academic Press, 1995.
88. Roman J, Little CW, McDonald JA. Potential role of RGD-binding integrins in mammalian lung branching morphogenesis. *Development* 112: 551–58, 1991.
89. Roman J, McDonald JA. Expression of fibronectin, the integrin alpha 5, and alpha-smooth muscle actin in heart and lung development. *Am J Respir Cell Mol Biol* 6: 472–80, 1992.
90. Coraux C, Delplanque A, Hinnrasky J *et al.* Distribution of integrins during human fetal lung development. *J Histochem Cytochem* 46: 803–10, 1998.
91. Sheppard D. Functions of pulmonary epithelial integrins: From development to disease. *Physiol Rev* 83: 673–86, 2003.
92. Kreidberg JA, Donovan MJ, Goldstein SL *et al.* Alpha 3 beta 1 integrin has a crucial role in kidney and lung organogenesis. *Development* 122: 3537–47, 1996.
93. Sheppard D. Roles of alphav integrins in vascular biology and pulmonary pathology. *Curr Opin Cell Biol* 16: 552–57, 2004.
94. Caniggia I, Liu J, Han R *et al.* Identification of receptors binding fibronectin and laminin on fetal rat lung cells. *Am J Physiol* 270: L459–68, 1996.
95. Clark EA, Brugge JS. Integrins and signal transduction pathways: The road taken. *Science* 268: 233–39, 1995.
96. Sheppard D. Epithelial integrins. *Bioessays* 18: 655–60, 1996.
97. Hynes RO, Yamada KM. Fibronectins: Multifunctional modular glycoproteins. *J Cell Biol* 95: 369–77, 1982.
98. Murphy-Ullrich JE, Mosher DF. Fibronectin and disease processes. In: Uitto J, Perejda AJ (eds). *Connective Tissue Disease: Molecular Pathology of the Extracellular Matrix*, pp. 455–73. New York: Marcel Dekker, Inc, 1987.
99. Ruoslahti E. Fibronectin and its receptors. *Annu Rev Biochem*: 57375–413, 1988.
100. Furie MB, Frey AB, Rifkin DB. Location of a gelatin-binding region of human plasma fibronectin. *J Biol Chem* 255: 4391–94, 1980.
101. Hahn LH, Yamada KM. Identification and isolation of a collagen-binding fragment of the adhesive glycoprotein fibronectin. *Proc Natl Acad Sci USA* 76: 1160–63, 1979.
102. Clark JG. The molecular pathology of pulmonary fibrosis. In: Uitto J, Perejda AJ (eds). *Connective Tissue Disease: Molecular Pathology of the Extracellular Matrix*. New York: Marcel Dekker, Inc, 1987.
103. Rennard SI, Togo S, Holz O. Cigarette smoke inhibits alveolar repair: A mechanism for the development of emphysema. *Proc Am Thorac Soc* 3: 703–8, 2006.
104. *Connective Tissue and its Heritable Disorders – Molecular, Genetic, and Medical Aspects*, B. Steinmann, ed. pp. 1–709. New York: Wiley-Liss, 1993.
105. Davidson W, Bai TR. Lung structural changes in chronic obstructive pulmonary diseases. *Curr Drug Targets Inflamm Allergy* 4: 643–49, 2005.
106. Pascual RM, Peters SP. Airway remodeling contributes to the progressive loss of lung function in asthma: An overview. *J Allergy Clin Immunol* 116: 477–86, 2005.
107. Postma DS, Timens W. Remodeling in asthma and chronic obstructive pulmonary disease. *Proc Am Thorac Soc* 3: 434–39, 2006.
108. Snider GL, Lucey EC, Stone PJ. Animal models of emphysema. *Am Rev Respir Dis* 133: 149–69, 1986.
109. Fukuda Y, Masuda Y, Ishizaki M *et al.* Morphogenesis of abnormal elastic fibers in lungs of patients with panacinar and centriacinar emphysema. *Hum Pathol* 20: 652–59, 1989.
110. Gottlieb DJ, Stone PJ, Sparrow D *et al.* Urinary desmosine excretion in smokers with and without rapid decline of lung function: The Normative Aging Study. *Am J Respir Crit Care Med.* 154: 1290–95, 1996.
111. Stone PJ, Gottlieb DJ, O'Connor GT *et al.* Elastin and collagen degradation products in urine of smokers with and without chronic obstructive pulmonary disease. *Am J Respir Crit Care Med* 151: 952–59, 1995.
112. Laurell CB, Eriksson S. The electrophoretic alpha 1-globulin pattern of serum in alpha 1-antitrypsin deficiency. *Scand J Clin Lab Invest* 15: 132–40, 1963.
113. D'Armiento J, Dalal SS, Okada Y *et al.* Collagenase expression in the lungs of transgenic mice causes pulmonary emphysema. *Cell* 71: 955–61, 1992.
114. Hautamaki RD, Kobayashi DK, Senior RM *et al.* Requirement for macrophage elastase for cigarette smoke-induced emphysema in mice. *Science* 277: 2002–4, 1997.
115. Finlay GA, O'Donnell MD, O'Connor CM *et al.* Elastin and collagen remodeling in emphysema. A scanning electron microscopy study. *Am J Pathol* 149: 1405–15, 1996.
116. Lang MR, Fiaux GW, Gillooly M *et al.* Collagen content of alveolar wall tissue in emphysematous and non-emphysematous lungs. *Thorax* 49: 319–26, 1994.
117. Pierce JA, Hocott JB, Ebert RV. The collagen and elastin content of the lung in emphysema. *Ann Intern Med* 55: 210–21, 1961.
118. Cosio M, Ghezzo H, Hogg JC *et al.* The relations between structural changes in small airways and pulmonary-function tests. *N Engl J Med* 298: 1277–81, 1978.
119. Finkelstein R, Ma HD, Ghezzo H *et al.* Morphometry of small airways in smokers and its relationship to emphysema type and hyperresponsiveness. *Am J Respir Crit Care Med* 152: 267–76, 1995.
120. Kuwano K, Bosken CH, Pare PD *et al.* Small airways dimensions in asthma and in chronic obstructive pulmonary disease. *Am Rev Respir Dis* 148: 1220–25, 1993.
121. Stenmark KR, Davie N, Frid M *et al.* Role of the adventitia in pulmonary vascular remodeling. *Physiology (Bethesda)* 21: 134–45, 2006.
122. Roche WR, Beasley R, Williams JH *et al.* Subepithelial fibrosis in the bronchi of asthmatics. *Lancet* 1: 520–24, 1989.
123. Laitinen A, Altraja A, Kampe M *et al.* Tenascin is increased in airway basement membrane of asthmatics and decreased by an inhaled steroid. *Am J Respir Crit Care Med* 156: 951–58, 1997.
124. Bousquet J, Lacoste JY, Chanez P *et al.* Bronchial elastic fibers in normal subjects and asthmatic patients. *Am J Respir Crit Care Med* 153: 1648–54, 1996.
125. Mauad T, Xavier AC, Saldiva PH *et al.* Elastosis and fragmentation of fibers of the elastic system in fatal asthma. *Am J Respir Crit Care Med* 160: 968–75, 1999.
126. Hoshino M, Nakamura Y, Sim JJ *et al.* Inhaled corticosteroid reduced lamina reticularis of the basement membrane by modulation of insulin-like growth factor (IGF)-I expression in bronchial asthma. *Clin Exp Allergy* 28: 568–77, 1998.
127. Laitinen LA, Laitinen A, Altraja A *et al.* Bronchial biopsy findings in intermittent or "early" asthma. *J Allergy Clin Immunol* 98: S3–6, 1996.
128. Olivieri D, Chetta A, Del Donno M *et al.* Effect of short-term treatment with low-dose inhaled fluticasone propionate on airway inflammation and remodeling in mild asthma: A placebo-controlled study. *Am J Respir Crit Care Med* 155: 1864–71, 1997.
129. Trigg CJ, Manolitsas ND, Wang J *et al.* Placebo-controlled immunopathologic study of four months of inhaled corticosteroids in asthma. *Am J Respir Crit Care Med* 150: 17–22, 1994.
130. Laitinen LA, Altraja A, Karjalainen E-M *et al.* Early interventions in asthma with inhaled corticosteroids. *J Allergy Clin Immunol* 105: S582–585, 2000.
131. Wagner EM, Liu MC, Weinmann GG *et al.* Peripheral lung resistance in normal and asthmatic subjects. *Am Rev Respir Dis* 141: 584–88, 1990.

132. Lange P, Parner J, Vestbo J *et al.* A 15-year follow-up study of ventilatory function in adults with asthma. *N Engl J Med* 339: 1194–200, 1998.
133. Peat JK, Woolcock AJ, Cullen K. Rate of decline of lung function in subjects with asthma. *Eur J Respir Dis* 70: 171–79, 1987.
134. Boxall C, Holgate ST, Davies DE. The contribution of transforming growth factor-beta and epidermal growth factor signalling to airway remodelling in chronic asthma. *Eur Respir J* 27: 208–29, 2006.
135. Burgess JK. Connective tissue growth factor: A role in airway remodelling in asthma? *Clin Exp Pharmacol Physiol* 32: 988–94, 2005.
136. Brody JS, Spira A. State of the art. Chronic obstructive pulmonary disease, inflammation, and lung cancer. *Proc Am Thorac Soc* 3: 535–37, 2006.

PART 4

Inflammatory Mediators and Pathways

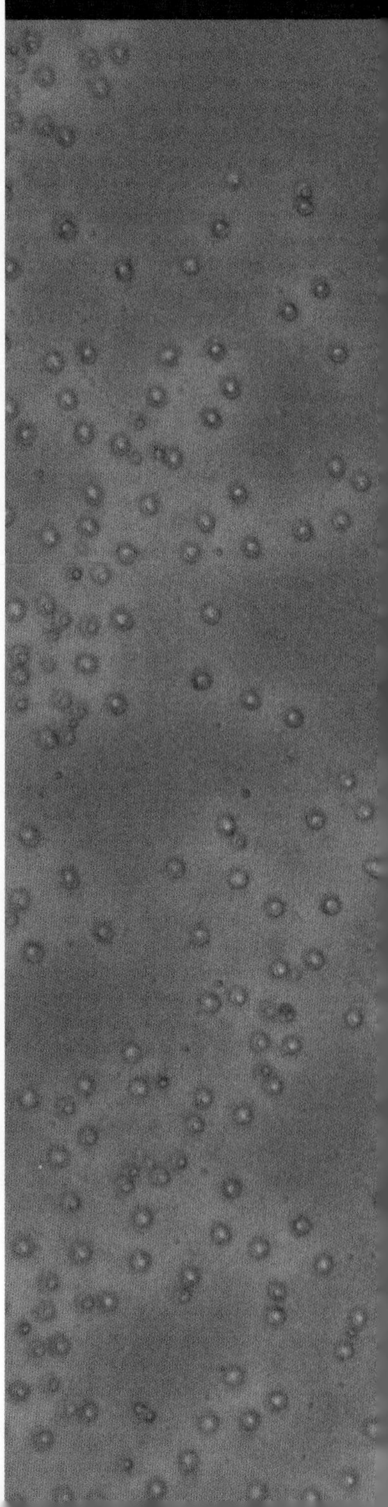

CHAPTER 23

Prostanoids

Paul M. O'Byrne

Firestone Institute for Respiratory Health, St. Joseph's Healthcare and Department of Medicine, McMaster University, Hamilton, ON, Canada

ARACHIDONIC ACID METABOLISM

The release of arachidonic acid from cell membrane phospholipids, through the action of a family of phospholipases, can result in the production of a wide variety of mediators which may be relevant in the pathogenesis of asthma. These lipid mediators have traditionally been considered in two classes:

1. mediators which result from the action of the cycloxygenases on arachidonic acid, which are prostaglandins (PG) or thromboxane (Tx)
2. mediators which result from the action of the enzymes 5-, 12-, or 15-lipoxygenase on arachidonic acid, which are the leukotrienes (LT) and lipoxins (Lx).

Platelet-activating factor (PAF) has also been recognized to be a mediator formed during arachidonic acid metabolism.

The oxidative metabolism of arachidonic acid by cycloxygenases produces the cyclic endoperoxides PGG_2 and PGH_2 (Fig. 23.1). The subsequent action of prostaglandin isomerases produces either PGD_2 or PGE_2, reductive cleavage produces PGF_{2a}, while one of two terminal synthetases on the endoperoxide produces PGI_2 and TxA_2. Cycloxygenase appears to be present in most cells; however, the cycloxygenase metabolite(s) released from a particular cell are quite specific (e.g. TxA_2 from platelets, and PGI_2 from endothelial cells). This suggests that terminal synthetases are cell-specific.

At least two isoforms of cycloxygenase have been identified:

1. cycloxygenase-1 (COX-1), which is constitutively present mainly in the gastric mucosa, kidney and platelets
2. cycloxygenase-2 (COX-2), which is mainly an inducible form, although also to some extent presents constitutively in the CNS, in the juxtaglomerular apparatus of the kidney, in the lung, in platelets, and in the placenta during late gestation.

Both isoforms contribute to the inflammatory process, but COX-2 is induced during acute and chronic inflammation, resulting in an enhanced formation of prostaglandins. Some studies have indicated that associations between COX-2 polymorphisms, prostaglandin production, and the development of asthma and atopy [1], while there is evidence that COX-1 polymorphisms do not appear to play a substantial role in genetic pre-disposition for asthma or asthma severity [2].

Prostaglandins and thromboxane mediate their effects through activation of specific receptors, and there is cross-activation of these receptors by the different agonists. The receptor designation has been accepted as the most potent agonist followed by the term "prostanoid." Thus, the thromboxane receptor is designated the TP receptor, and the PGE receptor is the EP receptor. There are DP, EP, FP, and TP receptors; the EP receptors are subdivided into EP1–EP4 [3].

ROLE IN ASTHMA

All of the cycloxygenase products of arachidonic acid metabolism have been prepared by total chemical synthesis and, with the exception of thromboxane, are readily available for study. Thromboxane has an exceedingly short half-life (about 30 s) and studies with thromboxane have been limited to a few, very circumscribed,

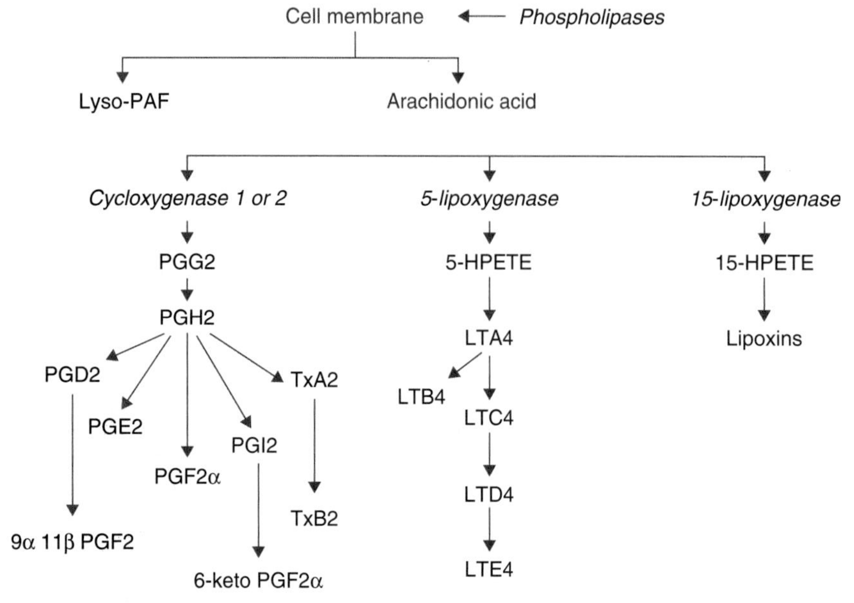

FIG. 23.1 The spectrum of eicosanoids produced as a consequence of arachidonic acid metabolism.

experimental preparations, none of them in the airways. Fortunately, several stable thromboxane mimetics have been developed. These are endoperoxides which activate the thromboxane receptor and mimic the biological actions of thromboxane but have substantially longer half-lives.

Prostaglandins are believed to have a variety of effects on airway function in asthma. The prostaglandins are most easily considered in two classes. There are stimulatory prostaglandins, such as PGD_2, PGF_{2a}, and TxA_2, which are potent bronchoconstrictors, and inhibitory prostaglandins, such as PGE_2, which can reduce bronchoconstrictor responses and attenuate the release of bronchoconstrictor mediators, such as acetylcholine, from airway nerves.

Evidence has been obtained in both animal models of airway hyperresponsiveness [4] and in human subjects with asthma that cycloxygenase metabolites are involved in causing aspirin-induced bronchoconstriction [5], or bronchoconstriction after exercise [6]. There is, however, little convincing evidence that cycloxygenase metabolites are important in causing the ongoing, persisting airway hyperresponsiveness that is characteristic of asthma. This is because several studies have failed to demonstrate any effect of cycloxygenase inhibitors on stable airway hyperresponsiveness in asthmatic subjects.

The initial studies examining the role of cycloxygenase metabolites in the pathogenesis of transient airway hyperresponsiveness after an inflammatory stimulus were carried out using a cycloxygenase inhibitor, indomethacin, in dogs in whom airway hyperresponsiveness had been induced by exposure to inhaled ozone [4]. Indomethacin did not alter baseline airway responsiveness to inhaled acetylcholine, but did prevent the development of airway hyperresponsiveness after inhaled ozone. Despite the absence of airway hyperresponsiveness, the magnitude of the inflammatory response, as measured by the numbers of neutrophils in the airway epithelium, was not altered by indomethacin. This suggested that a cycloxygenase product was not responsible for the chemotaxis of acute inflammatory cells into the airways after inhaled ozone; however, a cycloxygenase product was released during the inflammatory response which caused airway hyperresponsiveness. Subsequently, a reputed combined cycloxygenase and lipoxygenase inhibitor, BW755C, was also demonstrated to prevent the development of airway hyperresponsiveness after inhaled ozone in dogs [7]. Inhibition of cycloxygenase by indomethacin also prevents the development of airway hyperresponsiveness in other species including that which occurs after C_{5a} des Arg exposure in rabbits [8] and following inhaled allergen in sheep [9].

Cycloxygenase products have been implicated in the pathogenesis of allergen-induced early asthmatic as well as late asthmatic responses in human subjects. This has been done by pretreating subjects with several different cycloxygenase inhibitors. One study reported that pretreatment with indomethacin inhibited the late response, without having a major effect on the early response [10]. Another study, however, where subjects were pretreated with indomethacin, could not confirm these observations on allergen-induced either early responses or late responses [11]. However, in this latter study, indomethacin resulted in a significant inhibition of the development of allergen-induced airway hyperresponsiveness, which suggests that a cycloxygenase product is involved in the pathogenesis of this response.

STIMULATORY PROSTAGLANDINS

Prostaglandin D_2

PGD_2 is known to be released from stimulated dispersed human lung cells *in vitro* and from the airways of allergic human subjects which have been stimulated by allergen [12]. PGD_2 synthase is present in mast cells, T helper 2 (T_H2) cells and other leukocytes. PGD_2 is rapidly metabolized (with a half-life of 1.5 min in blood), and the main products that have been detected *in vivo* are PGJ_2 and 9α 11β PGF_2. PGD_2 is a bronchoconstrictor of human airways and is more potent when inhaled by human subjects than $PGF_{2\alpha}$ [13].

The biological effects of PGD_2 occur through stimulation of the G-protein-coupled receptors DP_1, CRTH2 (also known as DP_2), and TP receptors [14], and in part indirectly through a presynaptic action on airway cholinergic nerves to release acetylcholine [15]. The bronchoconstrictor effects of PGD_2 are mediated by stimulation of TP receptors [16]. PGD_2 also caused vasodilatation leading to erythema and edema [17], which occurs through PGD_2 stimulation of the DP_1 receptor [18]. PGD_2 is also known to cause chemotaxis of eosinophils [19] and T_H2 lymphocytes and modulation of T_H2-cytokine production, and these effects are mediated by PGD_2 stimulation of CRTH2 [20]. CRTH2 is a receptor unrelated to DP_1 and which was originally described as an orphan chemoattractant-like receptor expressed by T_H2 cells [21] and which is now known to be selectively expressed by T_H2 lymphocytes, eosinophils, and basophils [22]. PGD_2 can stimulate the production of the cytokines IL-4, IL-5, and IL-13 by human T_H2 cells in the absence of co-stimulation [23]. It has been suggested that DP_1 and CRTH2 cooperate to promote T_H2-dependent allergic responses [14]. Activation of DP_1 promotes an environment in which polarization of T_H2 cells can occur, whereas CRTH2 mediates their recruitment and activation to produce cytokines. Also, stimulation of DP_1 by PGD_2 has been suggested to induce regulatory T-cells and to control the extent of airway inflammation by an IL-10 dependent mechanism [24].

Subthreshold contractile concentrations of PGD_2 have been demonstrated to increase airway responsiveness to inhaled histamine and methacholine in asthmatic subjects [25]. Thus, PGD_2 released in human airways after allergen inhalation has the potential to both cause acute broncho-constriction and increase airway hyperresponsiveness to other constrictor mediators. However, specific receptor antagonists for PGD_2 or inhibitors of its production are not yet available to allow a precise evaluation of the importance of this cycloxygenase metabolite in the spontaneous asthmatic response.

Prostaglandin $PGF_{2\alpha}$

$PGF_{2\alpha}$ may also play a role as a mediator of bronchoconstriction and airway hyperresponsiveness following inhaled allergen in human subjects. $PGF_{2\alpha}$ is a potent bronchoconstrictor in asthmatic airways [26], and when inhaled at subthreshold constrictor concentrations increases airway responsiveness in dogs and human subjects [27]. $PGF_{2\alpha}$ also causes bronchoconstriction in human subjects partially through cholinergic-mediated bronchoconstriction [28]. As with PGD_2, there are no selective $PGF_{2\alpha}$ receptor antagonists available, which would allow identification of the specific importance of these metabolites in mediating theses airway responses. It has been suggested that all contractile prostaglandins act via the TP_1-receptor. Therefore, differentiation of the relative importance of the contractile prostaglandins in causing asthmatic responses may prove to be difficult.

Thromboxane A_2

TxA_2 is a potent constrictor of smooth muscle. It was originally described as being released from platelets, but it is now known to be released from other cells, including macrophages and neutrophils. As noted above, the biological half-life of TxA_2 is very short, so it has been implicated in disease processes through identification of its more stable metabolite thromboxane B_2 (TxB_2) in biological fluids; through the use of the stable TxA_2 analogs U44069 or U46619, which mimic most of the biological effects of TxA_2; and through the use of inhibitors of TxA_2 synthesis and antagonists of the TxA_2 receptor. Using these techniques, TxA_2 has been implicated in

- the pathogenesis of airway hyperresponsiveness in dogs [29]and primates [30];
- in the late cutaneous response to intradermal allergen in humans [31];
- in the late asthmatic response after inhaled allergen in humans [32];
- in airway hyperresponsiveness in asthmatic subjects [33].

Other studies, however, have demonstrated slight, but statistically significant, inhibition of the magnitude of the allergen-induced early, but not the late responses after pretreatment with a thromboxane synthetase inhibitor or receptor agonist [34, 35]. These studies suggest that thromboxane may be released following allergen challenge, and may account for a portion of the airway narrowing observed during the early asthmatic response; however, thromboxane is not important in the airway hyperresponsiveness that occurs following allergen inhalation.

INHIBITORY PROSTAGLANDINS

The differentiation of the prostaglandins into stimulatory or inhibitory mediators is somewhat inappropriate. For example, both PGE_2 and $PGF_{2\alpha}$ can have different effects on the airways depending on the time after inhalation at which the response is measured [36]. However, the main action of PGE_2 and PGI_2 on airway function is to relax airway smooth muscle and to antagonize the contractile responses of other bronchoconstrictor agonists. In addition PGE_2 is extremely potent at inhibiting the release of acetylcholine from airway cholinergic nerves [28].

The evidence that inhibitory prostaglandins play a role in modulating the contractile responses of agonists in asthmatic subjects comes from studies which have demonstrated that tachyphylaxis (a decreased response to repeated stimulation) occurs following repeated challenges with inhaled histamine, when challenges are separated by up to 6 h [37]. Also, repeated exercise challenges in asthmatic subjects, at intervals up to 4 h, also result in less bronchoconstriction occurring after the second when compared to the initial exercise challenge [38]. This has been termed "exercise refractoriness." This inhibitory effect are abolished

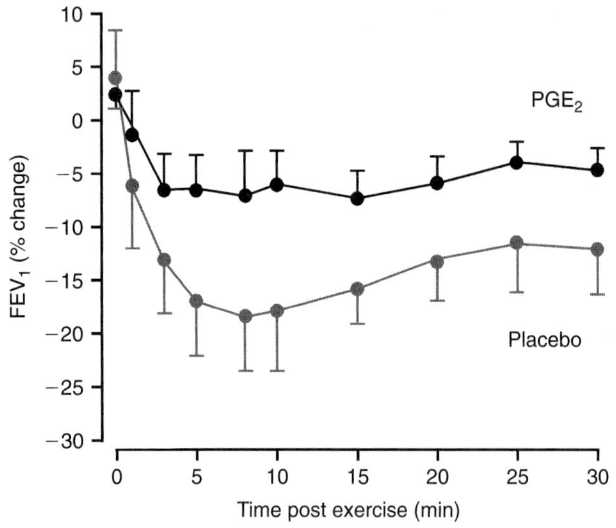

FIG. 23.2 The attenuation of exercise-induced bronchoconstriction by pretreatment with inhaled PGE_2. Reproduced with permission from Ref. [40].

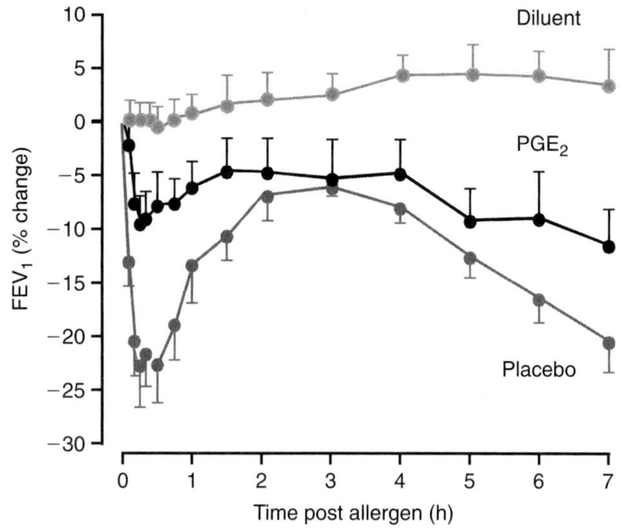

FIG. 23.3 The attenuation of allergen-induced early and late asthmatic responses by pretreatment with inhaled PGE_2. Reproduced with permission from Ref. [51].

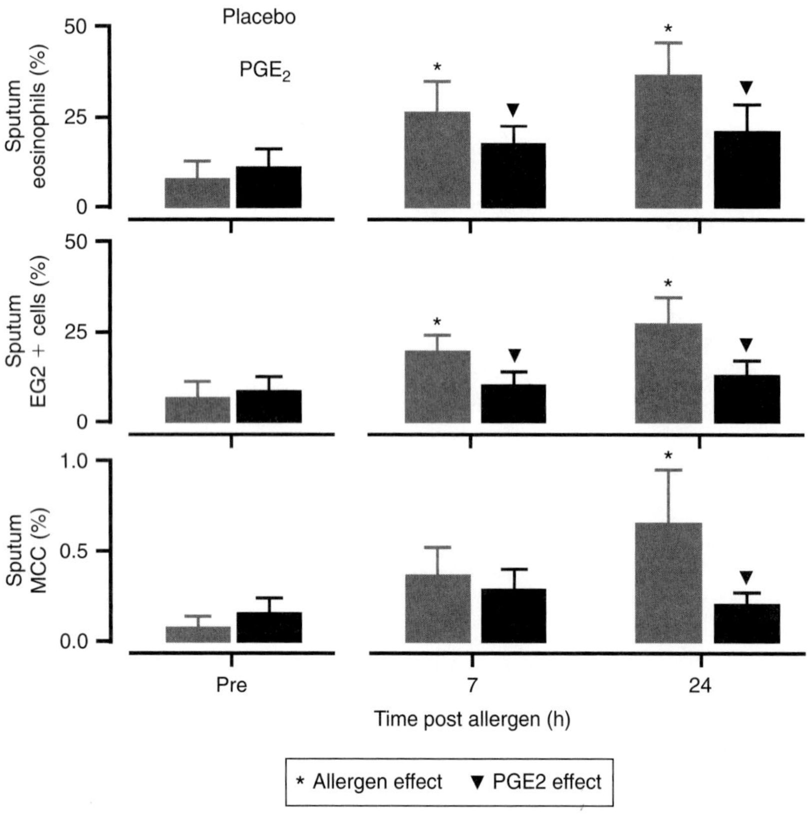

FIG. 23.4 The attenuation of allergen-induced increases in sputum eosinophils, $EG2^+$ cells (activated eosinophils) and metachromatic cells (mast cells and basophils), at 7 h and 24 h after allergen inhalation, by pretreatment with inhaled PGE_2. Reproduced with permission from Ref. [51].

by pretreatment with indomethacin [38]. This suggests that inhibitory prostaglandins, released as a consequence of exercise, could modify bronchoconstrictor responses in asthmatics. This hypothesis is supported by studies which have demonstrated that pretreatment of asthmatic subjects with oral PGE_1, in doses which do not cause bronchodilation, reduce airway responsiveness to both histamine and methacholine [39], and that inhaled PGE_2 largely abolishes exercise-induced bronchoconstriction [40] (Fig. 23.2).

These results are consistent with studies of airway smooth muscle *in vitro*, where histamine tachyphylaxis occurs through inhibitory prostaglandin release [41] and with studies of dogs *in vivo* where histamine tachyphylaxis is inhibited by indomethacin [42]. In addition, histamine tachyphylaxis in asthmatic subjects is blocked by pretreatment with an H_2 receptor antagonist in asthmatics, suggesting that H_2 receptor stimulation is involved with the development of histamine tachyphylaxis [43]. Contraction of asthmatic airways by histamine also reduces airway responsiveness to acetylcholine and exercise [44]. This lack of specificity suggests that either receptor downregulation or an alteration of the contractile properties or airway smooth muscle is occurring.

An initial hypothesis to explain exercise refractoriness in asthmatics was that histamine is released following exercise, causing bronchoconstriction, but this also provides partial protection against subsequent exercise bronchoconstriction, through PGE_2 released by stimulation of histamine H_2 receptors. However, several studies have suggested that this hypothesis is incorrect. First, the marked attenuation of exercise-induced bronchoconstriction by pretreatment with cysLT1, that is leukotriene (LT)D_4-receptor antagonists indicates that LTD_4, rather than histamine, is the main mediator responsible for exercise-induced bronchoconstriction [45]. Second, exercise refractoriness is not prevented by pretreatment with the H_2-receptor antagonists, which effectively prevent histamine tachyphylaxis [46]. Therefore, histamine-stimulated inhibitory prostaglandin release does not appear to be the cause of exercise refractoriness.

These studies raise the possibility that exercise refractoriness is caused by leukotriene-stimulated inhibitory prostaglandin release. This hypothesis is supported by evidence that there is interdependence between the cycloxygenase and lipoxygenase pathways of arachidonate metabolism in causing exercise bronchoconstriction and refractoriness in asthmatic subjects. [47]

The release of cysteinyl leukotrienes in the airways is also important in the pathophysiology of aspirin-intolerant asthma. This is associated with over-expression of the enzyme LTC_4 synthetase in the airways of patients with aspirin-intolerant asthma [48]. In addition, pretreatment with inhaled PGE_2 prevents not only the bronchoconstriction caused by aspirin, but also the increases in urinary LTE_4 associated with it [49]. These results suggest that, in patients with aspirin-intolerant asthma, PGE_2 may play a critical role in inhibiting the overproduction of cysteinyl leukotrienes, as this inhibiting effect is lost when the patients ingest aspirin or other cycloxygenase inhibitors.

Inhaled PGE_2 has also been demonstrated to attenuate allergen-induced bronchoconstrictor responses (Fig. 23.3) and airway hyperresponsiveness [50], and to attenuate allergen-induced airway eosinophilic inflammation [51] (Fig. 23.4). Thus, these findings suggest that endogenous PGE_2, produced in response to inhaled allergen, may play a role in regulating the associated airway inflammation.

SUMMARY

In asthmatic patients, it is likely that PGD_2 and TxA_2 are involved in causing acute bronchoconstriction after stimuli such as inhaled allergen in asthmatic patients. PGD_2 also promotes the chemotaxis of eosinophils and Th2 cells, as well as the release of Th2 cytokines through stimulation of CRTH2. Also, there is evidence that inhibitory prostaglandins are released by asthmatic airways, which reduces bronchoconstrictor responses to stimuli such as exercise, and that this effect is mediated by leukotriene-induced PGE_2 release. PGE_2 also appears to play a role in inhibiting leukotriene overproduction in patients with aspirin-intolerant asthma. Finally, PGE_2 can also attenuate allergen-induced airway responses and eosinophilic inflammation. These studies suggest that endogenous production of PGE_2 does have an important influence on the magnitude of asthmatic responses to stimuli such as exercise or inhaled allergens.

References

1. Chan IH, Tang NL, Leung TF, Ma SL, Zhang YP, Wong GW *et al.* Association of prostaglandin-endoperoxide synthase 2 gene polymorphisms with asthma and atopy in Chinese children. *Allergy* 62(7): 802–9, 2007.
2. Shi J, Misso NL, Duffy DL, Bradley B, Beard R, Thompson PJ *et al.* Cyclooxygenase-1 gene polymorphisms in patients with different asthma phenotypes and atopy. *Eur Respir J* 26(2): 249–56, 2005.
3. Coleman RA, Smith WL, Narumiya S. International Union of Pharmacology classification of prostanoid receptors: properties, distribution, and structure of the receptors and their subtypes. *Pharmacol Rev* 46(2): 205–29, 1994.
4. O'Byrne PM, Walters EH, Aizawa H, Fabbri LM, Holtzman MJ, Nadel JA. Indomethacin inhibits the airway hyperresponsiveness but not the neutrophil influx induced by ozone in dogs. *Am Rev Respir Dis* 130(2): 220–24, 1984.
5. O'Sullivan S, Dahlen B, Roquet A, Larsson L, Dahlen SE, Kumlin M. Urinary 9 alpha, 11 beta-PGF2 as a marker of mast cell activation in allergic and aspirin-intolerant asthma. *Adv Exp Med Biol* 433: 159–62, 1997.
6. O'Sullivan S, Roquet A, Dahlen B, Larsen F, Eklund A, Kumlin M *et al.* Evidence for mast cell activation during exercise-induced bronchoconstriction. *Eur Respir J* 12(2): 345–50, 1998.
7. Fabbri LM, Aizawa H, O'Byrne PM, Bethel RA, Walters EH, Holtzman MJ *et al.* An anti-inflammatory drug (BW755C) inhibits airway hyperresponsiveness induced by ozone in dogs. *J Allergy Clin Immunol* 76(2 Pt 1): 162–66, 1985.
8. Berend N, Armour CL, Black JL. Indomethacin inhibits the increased airway responsiveness to histamine following inhalation of C5a des Arg in rabbits. *Agents Actions* 18(5–6): 468–72, 1986.
9. Lanes S, Stevenson JS, Codias E, Hernandez A, Sielczak MW, Wanner A *et al.* Indomethacin and FPL-57231 inhibit antigen-induced airway hyperresponsiveness in sheep. *J Appl Physiol* 61(3): 864–72, 1986.
10. Fairfax AJ. Inhibition of the late asthmatic response to house dust mite by non-steroidal anti-inflammatory drugs. *Prostaglandins Leukot Med* 8(3): 239–48, 1982.
11. Kirby JG, Hargreave FE, Cockcroft DW, O'Byrne PM. Effect of indomethacin on allergen-induced asthmatic responses. *J Appl Physiol* 66(2): 578–83, 1989.
12. Murray JJ, Tonnel AB, Brash AR, Roberts LJ, Gosset P, Workman R *et al.* Release of prostaglandin D2 into human airways during acute antigen challenge. *N Engl J Med* 315(13): 800–4, 1986.
13. Hardy CC, Robinson C, Tattersfield AE, Holgate ST. The bronchoconstrictor effect of inhaled prostaglandin D2 in normal and asthmatic men. *N Engl J Med* 311(4): 209–13, 1984.
14. Pettipher R, Hansel TT, Armer R. Antagonism of the prostaglandin D2 receptors DP1 and CRTH2 as an approach to treat allergic diseases. *Nat Rev Drug Discov* 6(4): 313–25, 2007.

15. Tamaoki J, Sekizawa K, Graf PD, Nadel JA. Cholinergic neuromodulation by prostaglandin D2 in canine airway smooth muscle. *J Appl Physiol* 63(4): 1396–400, 1987.
16. Coleman RA, Sheldrick RL. Prostanoid-induced contraction of human bronchial smooth muscle is mediated by TP-receptors. *Br J Pharmacol* 96(3): 688–92, 1989.
17. Flower RJ, Harvey EA, Kingston WP. Inflammatory effects of prostaglandin D2 in rat and human skin. *Br J Pharmacol* 56(2): 229–33, 1976.
18. Giles H, Leff P, Bolofo ML, Kelly MG, Robertson AD. The classification of prostaglandin DP-receptors in platelets and vasculature using BW A868C, a novel, selective and potent competitive antagonist. *Br J Pharmacol* 96(2): 291–300, 1989.
19. Shiraishi Y, Asano K, Nakajima T, Oguma T, Suzuki Y, Shiomi T *et al.* Prostaglandin D2-induced eosinophilic airway inflammation is mediated by CRTH2 receptor. *J Pharmacol Exp Ther* 312(3): 954–60, 2005.
20. Hirai H, Tanaka K, Yoshie O, Ogawa K, Kenmotsu K, Takamori Y *et al.* Prostaglandin D2 selectively induces chemotaxis in T helper type 2 cells, eosinophils, and basophils via seven-transmembrane receptor CRTH2. *J Exp Med* 193(2): 255–61, 2001.
21. Abe H, Takeshita T, Nagata K, Arita T, Endo Y, Fujita T *et al.* Molecular cloning, chromosome mapping and characterization of the mouse CRTH2 gene, a putative member of the leukocyte chemoattractant receptor family. *Gene* 227(1): 71–77, 1999.
22. Nagata K, Hirai H, Tanaka K, Ogawa K, Aso T, Sugamura K *et al.* CRTH2, an orphan receptor of T-helper-2-cells, is expressed on basophils and eosinophils and responds to mast cell-derived factor(s). *FEBS Lett* 459(2): 195–99, 1999.
23. Xue L, Gyles SL, Wettey FR, Gazi L, Townsend E, Hunter MG *et al.* Prostaglandin D2 causes preferential induction of proinflammatory Th2 cytokine production through an action on chemoattractant receptor-like molecule expressed on Th2 cells. *J Immunol* 175(10): 6531–36, 2005.
24. Hammad H, Kool M, Soullie T, Narumiya S, Trottein F, Hoogsteden HC *et al.* Activation of the D prostanoid 1 receptor suppresses asthma by modulation of lung dendritic cell function and induction of regulatory T cells. *J Exp Med* 204(2): 357–67, 2007.
25. Fuller RW, Dixon CM, Dollery CT, Barnes PJ. Prostaglandin D2 potentiates airway responsiveness to histamine and methacholine. *Am Rev Respir Dis* 133(2): 252–54, 1986.
26. Thomson NC, Roberts R, Bandouvakis J, Newball H, Hargreave FE. Comparison of bronchial responses to prostaglandin F2 alpha and methacholine. *J Allergy Clin Immunol* 68(5): 392–98, 1981.
27. O'Byrne PM, Aizawa H, Bethel RA, Chung KF, Nadel JA, Holtzman MJ. Prostaglandin F2 alpha increases responsiveness of pulmonary airways in dogs. *Prostaglandins* 28(4): 537–43, 1984.
28. Walters EH, O'Byrne PM, Fabbri LM, Graf PD, Holtzman MJ, Nadel JA. Control of neurotransmission by prostaglandins in canine trachealis smooth muscle. *J Appl Physiol* 57(1): 129–34, 1984.
29. Aizawa H, Chung KF, Leikauf GD, Ueki I, Bethel RA, O'Byrne PM *et al.* Significance of thromboxane generation in ozone-induced airway hyperresponsiveness in dogs. *J Appl Physiol* 59(6): 1918–23, 1985.
30. Letts GL, McFarlane CS. Thromboxane A2 and airway responsiveness to acetylcholine aerosol in the conscious primate. *Prog Clin Biol Res* 263: 91–98, 1988.
31. Dorsch W, Ring J, Melzer H. A selective inhibitor of thromboxane biosynthesis enhances immediate and inhibits late cutaneous allergic reactions in man. *J Allergy Clin Immunol* 72(2): 168–74, 1983.
32. Shephard EG, Malan L, Macfarlane CM, Mouton W, Joubert JR. Lung function and plasma levels of thromboxane B2, 6-ketoprostaglandin F1 alpha and beta-thromboglobulin in antigen-induced asthma before and after indomethacin pretreatment. *Br J Clin Pharmacol* 19(4): 459–70, 1985.
33. Fujimura M, Sakamoto S, Saito M, Miyake Y, Matsuda T. Effect of a thromboxane A2 receptor antagonist (AA-2414) on bronchial hyperresponsiveness to methacholine in subjects with asthma. *J Allergy Clin Immunol* 87(1 Pt 1): 23–27, 1991.
34. Manning PJ, Stevens WH, Cockcroft DW, O'Byrne PM. The role of thromboxane in allergen-induced asthmatic responses. *Eur Respir J* 4(6): 667–72, 1991.
35. Beasley RC, Featherstone RL, Church MK, Rafferty P, Varley JG, Harris A *et al.* Effect of a thromboxane receptor antagonist on PGD2 and allergen-induced bronchoconstriction. *J Appl Physiol* 66(4): 1685–93, 1989.
36. Walters EH. The influence of airways smooth muscle tone upon the response to inhaled PGE2. *Agents Actions Suppl* 13: 185–88, 1983.
37. Manning PJ, Jones GL, O'Byrne PM. Tachyphylaxis to inhaled histamine in asthmatic subjects. *J Appl Physiol* 63(4): 1572–77, 1987.
38. O'Byrne PM, Jones GL. The effect of indomethacin on exercise-induced bronchoconstriction and refractoriness after exercise. *Am Rev Respir Dis* 134(1): 69–72, 1986.
39. Manning PJ, Lane CG, O'Byrne PM. The effect of oral prostaglandin E1 on airway responsiveness in asthmatic subjects. *Pulm Pharmacol* 2(3): 121–24, 1989.
40. Melillo E, Woolley KL, Manning PJ, Watson RM, O'byrne PM. Effect of inhaled PGE2 on exercise-induced bronchoconstriction in asthmatic subjects. *Am J Respir Crit Care Med* 149(5): 1138–41, 1994.
41. Anderson WH, Krzanowski JJ, Polson JB, Szentivanyi A. Prostaglandins as mediators of tachyphylaxis to histamine in canine tracheal smooth muscle. *Adv Prostaglandin Thromboxane Res* 7: 995–1001, 1980.
42. Shore S, Irvin CG, Shenkier T, Martin JG. Mechanisms of histamine-induced contraction of canine airway smooth muscle. *J Appl Physiol* 55(1 Pt 1): 22–26, 1983.
43. Jackson PJ, Manning PJ, O'Byrne PM. A new role for histamine H2-receptors in asthmatic airways. *Am Rev Respir Dis* 138(4): 784–88, 1988.
44. Hamielec CM, Manning PJ, O'Byrne PM. Exercise refractoriness after histamine inhalation in asthmatic subjects. *Am Rev Respir Dis* 138(4): 794–98, 1988.
45. Manning PJ, Watson RM, Margolskee DJ, Williams VC, Schwartz JI, O'Byrne PM. Inhibition of exercise-induced bronchoconstriction by MK-571, a potent leukotriene D4-receptor antagonist. *N Engl J Med* 323(25): 1736–39, 1990.
46. Manning PJ, Watson R, O'Byrne PM. The effects of H2-receptor antagonists on exercise refractoriness in asthma. *J Allergy Clin Immunol* 90(1): 125–26, 1992.
47. Manning PJ, Watson RM, O'Byrne PM. Exercise-induced refractoriness in asthmatic subjects involves leukotriene and prostaglandin interdependent mechanisms. *Am Rev Respir Dis* 148(4 Pt 1): 950–54, 1993.
48. Cowburn AS, Sladek K, Soja J, Adamek L, Nizankowska E, Szczeklik A *et al.* Overexpression of leukotriene C4 synthase in bronchial biopsies from patients with aspirin-intolerant asthma. *J Clin Invest* 101(4): 834–46, 1998.
49. Sestini P, Armetti L, Gambaro G, Pieroni MG, Refini RM, Sala A *et al.* Inhaled PGE2 prevents aspirin-induced bronchoconstriction and urinary LTE4 excretion in aspirin-sensitive asthma. *Am J Respir Crit Care Med* 153(2): 572–75, 1996.
50. Pavord ID, Wong CS, Williams J, Tattersfield AE. Effect of inhaled prostaglandin E2 on allergen-induced asthma. *Am Rev Respir Dis* 148(1): 87–90, 1993.
51. Gauvreau GM, Watson RM, O'Byrne PM. Protective effects of inhaled PGE2 on allergen-induced airway responses and airway inflammation. *Am J Respir Crit Care Med* 159(1): 31–36, 1999.

CHAPTER 24

Leukotrienes and Lipoxins

Bruce D. Levy[1] and Jeffrey M. Drazen[1,2]

[1]Department of Medicine, Pulmonary Division, Brigham and Women's Hospital, Boston, MA, USA
[2]Harvard Medical School, Editorial Office New England Journal of Medicine, Boston, MA, USA

The leukotrienes (LTB_4, LTC_4, LTD_4, and LTE_4) and lipoxins (LXA_4 and LXB_4) are molecules derived by lipoxygenation of arachidonic acid. There is evidence that each of these molecules is potentially important in the pathogenesis of asthma and chronic obstructive pulmonary disease (COPD); this chapter reviews the biology of these molecules as related to these disease entities.

FORMATION AND METABOLISM OF THE LEUKOTRIENES

Arachidonic acid, a normal component of many cell membrane phospholipids, is commonly found esterified to these phospholipids in the middle or *sn2* position. In the presence of an appropriately activated phospholipase A_2, arachidonic acid is cleaved from internal and external cell membranes and serves as the primary substrate for subsequent eicosanoid metabolism (Fig. 24.1) [1]; there is a family of phospholipase A_2 (PLA_2) molecules with this catalytic capacity [2, 3]. The two major forms of interest in leukotriene and lipoxin biosynthesis are cytosolic PLA_2 ($cPLA_2$) and secretory PLA_2 ($sPLA_2$) [4, 5].

Arachidonic acid, released as a result of PLA_2 action, enters into a series of reactions at the perinuclear membrane. The first of these requires a specific 5-lipoxygenase activating protein (FLAP) [6, 7], which allows arachidonate to serve as a substrate for the enzyme 5-lipoxygenase [8]. 5-Lipoxygenase sequentially catalyzes the addition of molecular oxygen to arachidonic acid to form 5-hydroperoxy eicosatetraenoic acid (5-HPETE) and leukotriene A_4 (LTA_4), respectively; LTA_4 is a major branch point in the formation of the leukotrienes [1]. In one major pathway, found in a variety of cells but most notably neutrophilic polymorphonuclear leukocytes (PMN), the addition of water to form LTB_4 is catalyzed by LTA_4 hydrolase. [9].

In addition to LTB_4, the 5-LO product (LTA_4) can also be enzymatically transformed to several other bioactive compounds [1]. Each of the other two major lipoxygenases in human tissues, namely 12-LO and 15-LO, can convert LTA_4 to a distinct class of mediators termed lipoxins for lipoxygenase interaction products [10]. Lipoxins carry 3 alcohol groups on arachidonic acid's 20 carbon backbone with 4 conjugated double bonds. The redundant pathways for lipoxin (LX) biosynthesis indicate the functional importance of these compounds. Because lipoxygenases are most commonly compartmentalized into distinct cell types, LX biosynthesis principally occurs during cell–cell interactions. For example, at sites of vascular injury or inflammation, PMN 5-LO and platelet 12-LO can interact to generate LXA_4 and LXB_4 [11]. This transcellular LX generation is also bidirectional as arachidonic acid can originate from either cell type [12]. Because the airway is lined with epithelial cells that are enriched with 15-LO activity, LX generation also occurs when leukocytes infiltrate airways during inflammatory responses [13]. Of interest, interactions between 5-LO and 12-LO generate an equal amount of LXA_4 and LXB_4, whereas 5-LO and 15-LO interactions strongly favor LXA_4 production [14].

Aspirin inhibits prostaglandin formation and triggers the formation of lipoxin epimers [15] as well. Acetylation of cyclooxygenase-2 at serine 516 blocks formation of PGG_2. When this occurs, enzymatic action results in arachidonic acids being converted to 15(R)-hydroxyeicosatetraenoic acid (15(R)-HETE).

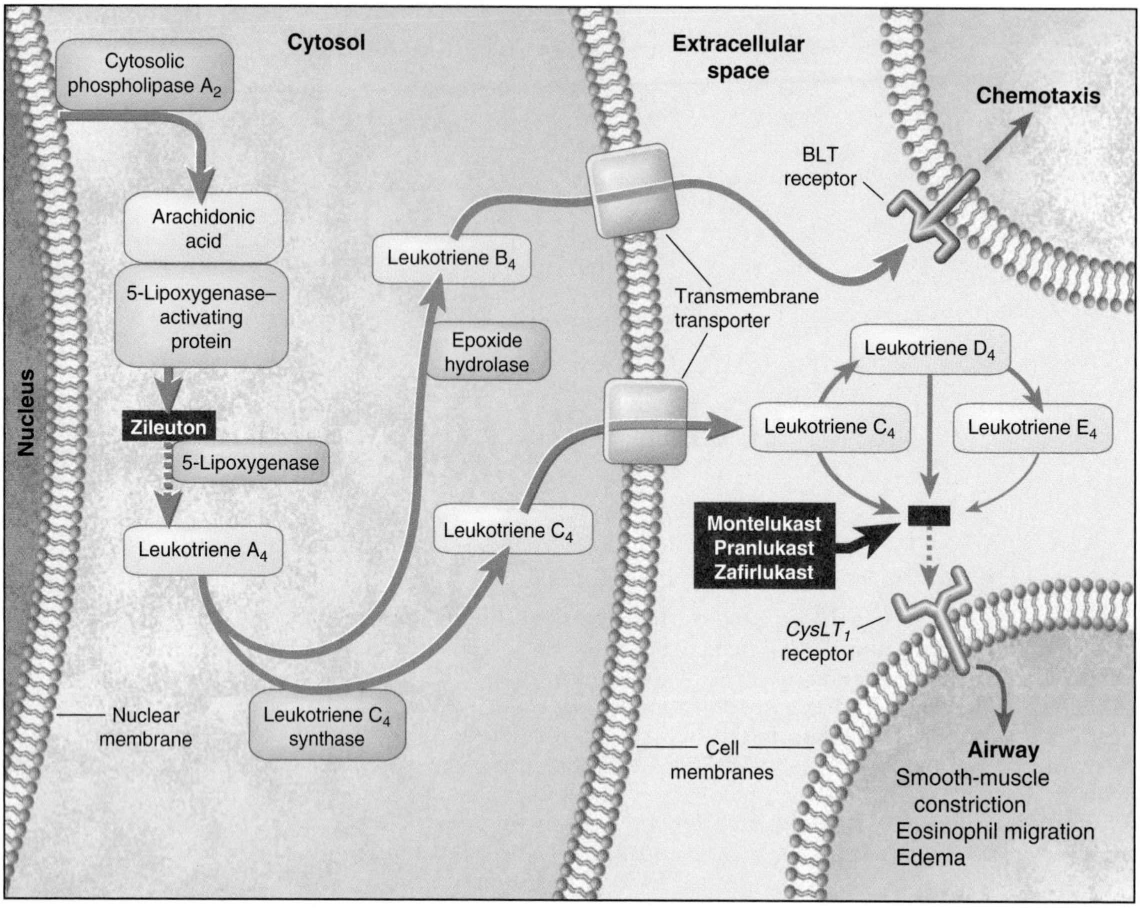

FIG. 24.1 Schematic showing the synthesis of leukotrienes from arachidonic acid and the site of action of various drugs in the pathway. Reproduced from Ref. [75], with permission.

This molecule can be a substrate for conversion of 5-LO to 15-epimer-LXA_4 (15-epi-LXA_4), a compound also known as aspirin-triggered LXA_4. This seemingly subtle switch in stereochemistry at carbon-15 leads to a twofold increase in circulating half-life and potency compared to the 15(S)-LXA_4 product of 15-LO [16]. Because cytochrome p450 enzymes can also catalyze 15(R)-HETE formation from arachidonate, aspirin is not required for 15-epi-LXA_4 production [17].

Similar to LTs, LXs are autacoids that are rapidly formed and rapidly, enzymatically inactivated [18]. The major route for LX inactivation is dehydrogenation by 15-hydroxy/oxo-eicosanoid oxidoreductase (15-PGDH), leading to the conversion of LXA_4 to the biologically inert 15-oxo-LXA_4. This compound is further converted to 13, 14-dihydro-15-oxo-LXA_4 by LXA_4/PGE 13,14-reductase/ LTB_4 12-hydroxydehydrogenase (PGR/LTB_4DH). This compound is subject to additional reduction of the 15-oxo group to 13,14-dihydro-LXA_4 by 15-PGDH. Moreover, LXs also undergo β-oxidation *in vivo* [19].

Other cells of myeloid origin including eosinophils, mast cells, and alveolar macrophages, not only have the capacity to form LTA_4 from arachidonic acid, but also they possess a unique and specific glutathionyl-*S*-transferase, LTC_4 synthase, which catalyzes the conjugation of glutathione to LTA_4, at carbon 6, to form leukotriene C_4 (LTC_4) [20–22]. Once formed, LTC_4 exits the cell via a specific transmembrane transporter [23–25], and when it is outside the cell, LTC_4 serves as a substrate for γ-glutamyl transpeptidase [26, 27] or a specific leukotrienase [28], which cleaves the glutamic acid moiety from its peptide chain to form leukotriene D_4 (LTD_4; LTD_4 is further processed by the removal of the glycine moiety from its peptide chain to form leukotriene E_4, LTE_4) [29, 30]. LTC_4, LTD_4, and LTE_4 make up the material formerly known as slow-reacting substance of anaphylaxis or SRS-A and are collectively known as the cysteinyl leukotrienes.

In the presence of appropriately activated PMNs, the cysteinyl leukotrienes are degraded to their respective sulfoxides and 6-*trans* diastereoisomers of LTB_4 [31]. In the absence of such cells the major degradation and excretion products of the cysteinyl leukotrienes are native LTE_4, *N*-acetyl LTE_4, or the products resulting from ω-oxidation and β-elimination of LTE_4 [32].

LEUKOTRIENES IN ASTHMA

Since the structural identification of the leukotrienes, a number of lines of evidence have accrued indicating a causal role for the cysteinyl leukotrienes in aspects of the asthmatic response. These are reviewed in the following sections.

Cysteinyl leukotrienes in the pathology of chronic mild asthma

The cysteinyl leukotrienes are synthesized and exported into the microenvironment by constitutive and infiltrating cells including mast cells and eosinophils [33, 34]. Cysteinyl leukotrienes produced by these cells have numerous actions but the predominant ones in asthma are their ligation of leukotriene receptors on airway smooth muscle and the bronchial vasculature mediating airway obstruction and microvascular leak [35–37].

Biological effects of cysteinyl leukotrienes relevant to the asthmatic response

The leukotrienes are known to have profound biochemical, immunological, and physiological effects, even in picomolar concentrations, including induction of airway obstruction, tissue edema, and expression of bronchial mucus from submucosal glands [38–40]; these pathobiological effects make them part of the panel of mediators which characterize asthmatic responses.

Although these mediators have pleotropic biological actions, the ability of the cysteinyl leukotrienes to mediate airway narrowing in normal individuals and persons with asthma is critical to the biology of asthma. Inhalation of aerosols of leukotrienes by normal subjects results in airway obstruction as manifested by a decrease in flow rates during a forced exhalation [41–52]. In addition, among normal subjects there is a relationship between responsiveness to the leukotrienes and responsiveness to reference agonists such as histamine or methacholine [46, 47]. Subjects that are more responsive to histamine are those that are more responsive to the leukotrienes.

In subjects with asthma, LTC_4, LTD_4, and LTE_4 are potent bronchoconstrictor agonists when administered by aerosol, as indicated by induced decrements in the FEV_1, in the V_{30P}, or in specific conductance [46–48, 50, 53, 54]. When the latter outcome indicator is used, the nebulizer concentrations required to narrow airways in asthmatics are approximately one-tenth of those required by normal subjects to achieve the same decrement in airflow rates. Since normal subjects are approximately 100-fold less sensitive to histamine or methacholine than asthmatic subjects, whereas they are only 10-fold less responsive to the cysteinyl leukotrienes, this indicates that the relative degree of hyperresponsiveness to the cysteinyl leukotrienes is less than that observed when histamine or methacholine is used as the contractile agonist.

Leukotriene recovery in asthma

Cysteinyl leukotrienes have been recovered after experimental challenges, which elicit clinical symptoms similar to those that occur in spontaneously occurring asthmatic conditions. Leukotrienes have also been recovered in the nasal lavage fluid after intranasal challenge with either antigen or cold air [55, 56]. Leukotrienes are recovered in significantly greater amounts in the bronchoalveolar lavage (BAL) fluid from subjects with symptomatic asthma than from subjects with asymptomatic asthma or normal subjects [57–60], as well as from exhaled breath condensate [61–63], suggesting that the leukotrienes are produced locally in the lungs of patients with asthma.

The profusion of leukotriene production can be estimated from measurements made on urine samples [64–69]. In normal human subjects, after intravenous administration of radiolabeled LTC_4, 12–48% of the counts are recovered in the urine with 4–13% as intact LTE_4 [70, 71], and it is now common to measure leukotriene recovery from the urine as an index of endogenous production of leukotrienes.

Asano *et al.* [72] examined urinary LTE_4 excretion rates in eight patients with mild chronic stable asthma (FEV_1 ~ 70% predicted, inhaled β-agonists as the only asthma treatment) followed on a metabolic ward for 4 days. They demonstrated that, on average, patients with asthma have significantly higher urinary LTE_4 levels than normal subjects. However, among the asthmatics studied there were individuals who were persistent hyperexcretors of urinary LTE_4 and others with urinary LTE_4 levels that were persistently within the normal range. Among the explanations for this finding is the possibility that, among patients with asthma, whose clinical phenotype is similar, there are individuals whose asthma is associated with leukotriene production and others for whom this is not the case. These data clearly indicate that, during spontaneous, induced, or chronic stable asthma, at least in some individuals, there is enhanced urinary LTE_4 excretion and by inference an increased cysteinyl leukotriene production. Among patients with severe asthma, who are receiving treatment with corticosteroids, there are substantially increased levels of LTE_4 in the urine [73].

Leukotriene receptor blockade and synthesis inhibition

Two classes of pharmacological agents have been demonstrated to have a salutary effect on airway obstruction in patients with asthma: leukotriene receptor antagonists and leukotriene synthesis inhibitors. As noted above there are at least two distinct receptors that when ligated transduce the biological activity of the cysteinyl leukotrienes; there are the cysteinyl leukotriene receptors type 1 and type 2 ($CysLT_1$ and $CysLT_2$) [37, 74]. Although many chemically distinct antagonists at the $CysLT_1$ receptor have been recognized, as of the present time, only three have been introduced into medical use; namely, pranlukast, zafirlukast, and montelukast [75].

With regard to leukotriene synthesis inhibitors, agents have been developed which are direct inhibitors of 5-LO, such as zileuton [76], as well as agents that inhibit the interaction between arachidonic acid and FLAP (MK-0591 and Bay x1005) [77, 78] (Fig. 24.2).

Induced asthma

Leukotriene receptor antagonists and synthesis inhibitors have been shown to inhibit various types of laboratory-induced asthma [80–84]. Current data indicate that most

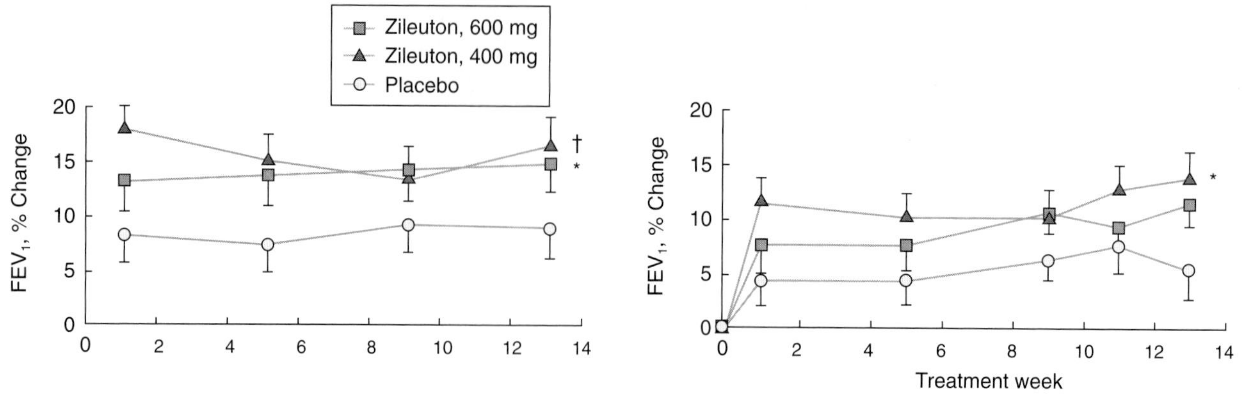

FIG. 24.2 Mean percentage change in forced expiratory volume rate in the first second (FEV_1) (±SE) from pre-randomization baseline during 13 weeks of treatment with 600 mg of zileuton, 400 mg of zileuton, or placebo, each given four times daily. Asterisk indicates $p > 0.05$ versus placebo; and the dagger, $p > 0.01$ versus placebo. Left panel, average FEV_1 2–4 h after drug ingestion (expected time of peak drug levels). Right panel, FEV, before ingestion of morning dose of drug (expected time of low drug levels). Reproduced from Ref. [79], with permission.

TABLE 24.1 Early studies of leukotriene modifiers in patients with chronic persistent asthma[a]

	Zafirlukast	Montelukast	Pranlukast	Zileuton
Number of patients[b]	70	408	45	122
Length of study	6 week	12 week	6 week	6 mo
Dose (oral)	20 mg b.i.d.	10 mg q.d.	337.5 mg b.i.d.	600 mg q.i.d.
Baseline FEV_1 (% of predicted)	66%	66%	66%	62%
Trough improvement in FEV_1 liters[c]	0.23 l[g]	–	~0.31 l[g]	0.34 l
Trough improvement in FEV_1 percent[c,d]	11% (13–14%)[g]	–	~11.5%[g,h]	15(18)%
Peak improvement in FEV_1[d,e]	–	13%	–	20%[i] (23%)[g]
β-adrenergic agonist use reduction[d]	31%	27%[j]	NC	(30%)
Improvement in AM PEFR[d]	6%	6.1%	~5%[g]	7.1% (8.5%)
Treatment failure or glucocorticoid rescue therapy (treatment versus placebo)[f]	2% versus 10%	–	–	8.3% versus 21.5%[g]
Decrease in symptoms: day/night (%)	28/46	20/NC	NC/28	36/33

Reproduced from Ref. [75]. NC: no significant change; PEFR: peak expiratory flow rate.

[a]Compared with pretreatment values. Only data from double-blind, randomized, and placebo-controlled studies are included. Although additional studies with each drug have been reported, the study chosen for display is representative. All values were statistically significant from placebo unless otherwise indicated.

[b]Number of patients receiving active treatment at the dose indicated. Zafirlukast was administered at twice the currently recommended dose.

[c]Trough values: values immediately before next dose or those values not reported as obtained at times of expected peak plasma concentrations or effects.

[d]Figures *not* in parentheses represent means over the study period or end-point analyses. Figures in parentheses represent maximum effect among the study observation intervals reported.

[e]Values recorded at or near time of drug's expected peak plasma concentration or effects.

[f]Treatment failure for zafirlukast; glucocorticoid rescue therapy for zileuton.

[g]Derived from figures or statistics.

[h]225 mg dose.

[i]Endpoint value: NS versus placebo at week 26.

[j]Percentage difference compared with placebo.

but not all of the physiological effects of aspirin-induced asthma derive from the action of the cysteinyl leukotrienes [82, 85]. It is important to note that, although all the physiological effects of aspirin-induced asthma derive from the effects of the leukotrienes, this does not mean that anti-leukotriene treatment can prevent all the clinical manifestations of aspirin ingestion in patients with aspirin-sensitive asthma. Indeed, there have been reports of aspirin-induced asthma in patients receiving treatment with $CysLT_1$ receptor antagonists [86, 87]. These findings are consistent with the known competitive nature of $CysLT_1$ antagonists and the massive release of cysteinyl leukotrienes after aspirin ingestion in patients with aspirin-induced aspirin.

Leukotriene inhibition in chronic stable asthma

Multiple double-blind and placebo-controlled and active compartor clinical trials in both adult and pediatric patients with mild-to-moderate chronic stable asthma reported in the archival literature; these have been the subject of multiple reviews [88–92].

The general design of these trials has been similar. They have recruited and enrolled patients whose asthma was marginally controlled solely by use of inhaled β-agonists, who had FEV_1 values most often between 40% and 80% of predicted, who had moderate asthma symptoms as judged from daily symptom diaries. Each trial incorporated a 1–3 week "run-in" period when all patients had their asthma control monitored while on single-blind oral placebo. During this period baseline lung function, β-agonist use, and symptom data were gathered. Patients then entered a period in which they received randomized treatment with an active agent or placebo. Patients returned on a regular basis to their clinical centers where their lung function and asthma symptom data were recorded; the results from these trials are summarized in Table 24.1.

In the course of the performance of these trials, it has been noted by many investigators that there is a marked variability among individuals in the response to anti-leukotriene treatment. One interpretation of this observation is that there could be genetic differences among individuals that lead to differences in the role of leukotrienes in a given individual's asthmatic responses. Polymorphisms, both with and without known functional consequences have been identified in a number of the genes encoding the proteins that make up the 5-lipoxygenase pathway [94–99]. The gene encoding 5-lipoxygenase has been shown to harbor variability that is associated with the asthma treatment response, but the specific functional variants within that gene have not been identified [97, 100].

LIPOXINS IN ASTHMA

Lipoxins are derived from arachidonic acid, yet structurally and functionally distinct from prostaglandins and leukotrienes [1]. There is a growing body of evidence linking these counter-regulatory eicosanoid mediators to homeostatic airway responses including control of airway inflammation and reactivity [101].

Lipoxins in the pathology of chronic mild asthma

Lipoxins and 15-epi-LXs display several features of mediators inhibiting airway inflammation [10] including inhibition *in vivo* of eosinophil and T-lymphocyte activation and TH2 cytokine release [102, 103]. LXs are generated in respiratory tissues [13] and block airway inflammation and hyperresponsiveness in an experimental model of asthma [102].

Biological effects of lipoxins relevant to the asthmatic response

Lipoxin A_4 displays cell-type selective actions [10]. On the one hand, LXA_4 blocks PMN activation and transmigration across endothelial and epithelial cells [104–107]. Distinct from these stop signals for PMNs, LXA_4 stimulates monocyte locomotion and macrophage phagocytosis of apoptotic PMNs [108, 109]. LXA_4 also reduces eosinophil tissue accumulation [102], dendritic cell (DC) mobilization [110], T-lymphocyte cytokine release [103], and NK cell cytotoxicity [111]. In sum, LXs regulate both innate and acquired immune effector cells to reduce leukocyte entry into tissues and promote clearance of apoptotic cells, important mechanisms in the resolution of acute inflammation.

LXs also regulate airway epithelial cells [112]. LXA_4 blocks proinflammatory cytokine and chemokine release and gene expression via NFκB [113]. Of interest, LX signaling increases expression of bacterial permeability-inducing protein in epithelial cells to enhance mucosal bacterial killing [114]. Thus, in addition to anti-inflammation, LXA_4 signals carry host protective actions.

Lipoxins also potently regulate pulmonary responses. LXA_4 relaxes precontracted pulmonary arteries and bronchi [115]. In addition, a LX stable analog, designed using the 15-epi-LXA_4 structure as a template, markedly inhibits (with as little as 1 μg; ~0.05 mg/kg) allergen-driven airway hyperresponsiveness and inflammation in an experimental model of asthma [102]. Asthma is frequently worsened by aspiration of gastric acid, and LX and 15-epi-LX signaling promote restitution of injured airway epithelium and resolution of acid-initiated acute airway injury [112, 116]. Together, the impact of LXs on both cellular effectors and airway responses indicate protective roles for these counter-regulatory mediators in airway inflammation and hyperresponsiveness, important processes linked to asthma pathogenesis.

As a consequence of LXs' rapid inactivation, there is only limited information on their actions in humans *in vivo*. In one clinical trial, LXA_4 was administered to individuals with asthma, leading to protection from LTC_4-induced bronchoconstriction [117]. The recent development of new LX stable analogs that are topically and orally active should enable further investigation on LX regulation of human illness.

Lipoxin recovery in asthma

Lipoxins have been identified in human subjects with a wide range of respiratory illnesses involving leukocyte activation [13]. Both LXs and 15-epi-LXs are generated during respiratory inflammation [101, 118]. LXA_4 (0.4–2.8 ng/ml) has been detected in BAL and pleural fluid from individuals with inflammatory lung disease [13, 119]. In a murine experimental model of asthma, LX formation at peak airway inflammation is similar in magnitude to the amount of LTB_4 formed, yet one to two log orders less than cysteinyl

LTs and PGE_2 [102]. Of interest, diminished formation of these counter-regulatory mediators has been identified in severe forms of human respiratory illness, including aspirin-intolerant asthma [120] and severe, steroid-dependent asthma [121]. Decreased production of these counter-regulatory substances could predispose the host to more severe, inflammatory responses in the lung.

Lipoxin A_4 receptors

Lipoxins interact with one or more specific receptors to signal their counter-regulatory actions. LXA_4 can serve as a high affinity ligand for the LXA_4 receptor (ALX) [122, 123] and a receptor-level antagonist for a subclass of LTD_4 receptors (i.e. CysLT1) [124]. ALX is a seven-transmembrane spanning, G-protein coupled protein, binding LXA_4 with high affinity ($K_D = 1.7\,nM$) [125]. The human, mouse and rat ALX amino acid sequences share 74% or greater homology with the highest homology found in their second intracellular loop (100% identical) and sixth transmembrane segments (93% identical) [126]. ALX receptors are expressed in several mammalian cells and tissues including human PMNs, monocytes, epithelial cells, and human and murine spleen. Interleukin-13 and interferon-γ dramatically induce ALX expression *in vitro* in airway epithelial cells [124]. Cytokine regulation of ALX is also evident *in vivo* during disease states, as ALX is upregulated in airway epithelial and inflammatory cells during allergic airway inflammation in a murine model of asthma [102]. ALX was the initial receptor identified that binds to both lipid and peptide ligands [123]. Of interest for asthma therapy, corticosteroids induce expression of annexin 1 that is enzymatically cleaved to peptides that can also interact with ALX to initiate anti-inflammatory circuits [127]. In PMNs, ALX signals for LXs' anti-inflammatory effects, in part, by polyisoprenyl phosphate remodeling [107] and by inhibition of leukocyte-specific protein-1 phosphorylation, a downstream regulator of the p38-MAPK cascade [128]. Targeted expression of human LXA_4 receptors to murine leukocytes in transgenic mice also dramatically inhibits allergen sensitization, eosinophil trafficking, and airway inflammation [102].

LEUKOTRIENES AND LIPOXINS IN COPD

Leukotrienes and the biology of the bronchitic airway

There are fewer data concerning the role of leukotrienes in the pathogenesis of COPD than are available about this class of mediators in asthma. Introduction of leukotrienes into the microenvironment of the airway is associated with expression of mucus from glands [129, 130]. The receptor for LTB_4 has been detected in greater profusion on alveolar macrophages from patients who smoke and have airway obstruction than from either normal subjects or patients who smoke and do not have airway obstruction [131]. The leukotrienes have pleomorphic inflammatory actions that are relevant to the microenvironment of the airway in COPD. Although a complete evaluation of these actions is beyond the scope of this chapter, there are excellent review articles that reflect this biology [39, 40, 132–134]. These data indicate that if leukotrienes were to become available in the microenvironment of the airway that they have the potential to contribute to aspects of the bronchotic phenotype; at this time, their exact role is not known.

Recovery of leukotrienes in COPD

Leukotrienes have been recovered in greater quantities from the airways of patients with COPD. For example they are found in the sputum, BAL fluid, or exhaled breath condensate of patients with chronic bronchitis at higher levels than from similar samples obtained from normal subjects [135–140]. In one study the concentration of LTB_4 in sputum was closely correlated with the number of polymorphonuclear leukocytes recovered in the sputum [140]. When patients with COPD were treated with ibuprofen, the amount of LTB_4 recovered in the exhaled breath condensate increased; there was no effect of rofecoxib on this measure [141]. Sputum obtained from patients in the midst of an infective exacerbation of chronic bronchitis contained high levels of LTB_4, which fell with treatment [142]. In these samples LTB_4 constituted about one-third of the total chemotactic activity at the time of disease presentation, and the importance of LTB_4 as a chemotactic factor fell with antibiotic treatment. These data indicate that leukotrienes, both cysteinyl and di-hydroxy moieties, are present and likely active in the airway microenvironment in COPD. While limited information is currently available on LXs and COPD, LXA_4 and 15-epi-LXA_4 are both found in increased amounts in sputum from subjects with COPD [118].

There are few data in the archival literature from studies in which agents active on various components of the 5-lipoxygenase pathway have been studied on lung function in patients with COPD. Small studies, with varying quality of controls, show that treatment of patients with leukotriene receptor antagonists modifies a number of outcome indicators in COPD. In most studies the effects are salutary, but the magnitude of clinical improvement is marginal at best [93, 143–146]. Since these data are not biologically informative with respect to the role the leukotrienes in COPD, larger better-controlled studies are needed to resolve this issue.

CONCLUSIONS

The available data support the following conclusions as to the role of leukotrienes and lipoxins in bronchial asthma and COPD:

- They are produced by constitutive cells (mast cells/macrophages) and infiltrating cells (eosinophils, neutrophils) implicated in asthma and COPD.
- They are potent bronchoconstrictor agonists.

- Their administration results in the expression of mucus from mucous glands.
- Laboratory-induced asthma and spontaneous asthma or COPD is associated with an enhanced recovery of leukotrienes in the urine and exhaled breath condensate of subjects with asthma and sputum, BAL or exhaled breath condensate leukotrienes in COPD.
- Asthma control is enhanced (compared with placebo) by agents capable of interfering with leukotriene action or synthesis. Similar data are not available for COPD.
- The data indicate that the leukotrienes play an important role in the asthmatic response. We are not sure of the importance of the pathogenetic role of the leukotrienes in COPD.
- Lipoxins are generated in respiratory tissues in asthma and COPD and can decrease airway inflammation and hyperresponsiveness in laboratory-induced asthma.

References

1. Samuelsson B, Dahlen SE, Lindgren JA, Rouzer CA, Serhan CN. Leukotrienes and lipoxins: Structures, biosynthesis, and biological effects. *Science* 237: 1171–76, 1987.
2. Dennis EA. The growing phospholipase A2 superfamily of signal transduction enzymes. *Trends Biochem Sci* 22: 1–2, 1999.
3. Murakami M, Kudo I. Phospholipase A(2). *J Biochem* 131(3): 285–92, 2002.
4. DeMarino V, Gentile M, Granata F, Marone G, Triggiani M. Secretory phospholipase A(2): A putative mediator of airway inflammation. *Int Arch Allergy Immunol* 118(2–4): 200–1, 1999.
5. Leslie CC. Properties and regulation of cytosolic phospholipase A2. *J Biol Chem* 272: 16069–72, 1997.
6. Dixon RA, Diehl RE, Opas E *et al.* Requirement of a 5-lipoxygenase-activating protein for leukotriene synthesis. *Nature* 343: 282–84, 1990.
7. Reid GK, Kargman S, Vickers PJ *et al.* Correlation between expression of 5-lipoxygenase-activating protein, 5-lipoxygenase, and cellular leukotriene synthesis. *J Biol Chem* 265: 19818–23, 1990.
8. Petersgolden M, Brock TG. 5-lipoxygenase and FLAP. *Prostagland Leuk Essent Fatty Acids* 69(2–3): 99–109, 2003.
9. Haeggstrom JZ. Leukotriene A(4) hydrolase and the committed step in leukotriene B-4 biosynthesis. *Clin Rev Allergy Immunol* 17(1–2): 111–31, 1999.
10. Serhan CN. Resolution phases of inflammation: Novel endogenous anti-inflammatory and proresolving lipid mediators and pathways. *Annu Rev Immunol*, 2006.
11. Serhan CN, Sheppard KA. Lipoxin formation during human neutrophil-platelet interactions. Evidence for the transformation of leukotriene A4 by platelet 12- lipoxygenase *in vitro*. *J Clin Invest* 85: 772–80, 1990.
12. Fiore S, Serhan CN. Formation of lipoxins and leukotrienes during receptor-mediated interactions of human platelets and recombinant human granulocyte/macrophage colony-stimulating factor-primed neutrophils. *J Exp Med* 172(5): 1451–57, 1990.
13. Lee TH, Crea AE, Gant V *et al.* Identification of lipoxin A4 and its relationship to the sulfidopeptide leukotrienes C4, D4, and E4 in the bronchoalveolar lavage fluids obtained from patients with selected pulmonary diseases. *Am Rev Respir Dis* 141: 1453–58, 1990.
14. Serhan CN. Lipoxins and novel aspirin-triggered 15-epi-lipoxins (ATL): A jungle of cell–cell interactions or a therapeutic opportunity? *Prostaglandins* 53(2): 107–37, 1997.
15. Claria J, Serhan CN. Aspirin triggers previously undescribed bioactive eicosanoids by human endothelial cell-leukocyte interactions. *Proc Natl Acad Sci USA* 92: 9475–79, 1995.
16. Clish CB, O'Brien JA, Gronert K, Stahl GL, Petasis NA, Serhan CN. Local and systemic delivery of a stable aspirin-triggered lipoxin prevents neutrophil recruitment *in vivo*. *Proc Natl Acad Sci USA* 96(14): 8247–52, 1999.
17. Claria J, Lee MH, Serhan CN. Aspirin-triggered lipoxins (15-epi-LX) are generated by the human lung adenocarcinoma cell line (A549)-neutrophil interactions and are potent inhibitors of cell proliferation. *Mol Med* 2(5): 583–96, 1996.
18. Serhan CN, Fiore S, Brezinski DA, Lynch S. Lipoxin-A(4) metabolism by differentiated HL-60 cells and human monocytes – conversion to novel 15-oxo and dihydro products. *Biochem* 32: 6313–19, 1993.
19. Guilford WJ, Parkinson JF. Second-generation beta-oxidation resistant 3-oxa-lipoxin A4 analogs. *Prostaglandins Leukot Essent Fatty Acids* 73(3–4): 245–50, 2005.
20. Lam BK, Penrose JF, Freeman GJ, Austen KF. Expression cloning of a cDNA for human leukotriene C_4 synthase, an integral membrane protein conjugating reduced glutathione to leukotriene A_4. *Proc Natl Acad Sci USA* 91: 7663–67, 1994.
21. Welsch DJ, Creely DP, Hauser SD, Mathis KJ, Krivi GG, Isakson PC. Molecular cloning end expression of human leukotriene-C_4 synthase. *Proc Natl Acad Sci USA* 91: 9745–49, 1994.
22. Lam BK. Leukotriene C-4 synthase. *Prostagland Leuk Essent Fatty Acids* 69(2–3): 111–16, 2003.
23. Loe DW, Almquist KC, Deely RG, Coles SPC. Multidrug resistance protein (MRP)-mediated transport of leukotriene C4 and chemotherapeutic agents in membrane vesicles. *J Biol Chem* 271: 9675–82, 1996.
24. Gao M, Yamazaki M, Loe DW *et al.* Multidrug resistant protein: Identification of regions required for active transport of leukotriene C4. *J Biol Chem* 273: 10733–40, 1998.
25. Qian YM, Qiu W, Gao M, Westlake CJ, Cole SPC, Deeley RG. Characterization of binding of leukotriene C-4 by human multidrug resistance protein – Evidence of differential interactions with NH2- and COOH-proximal halves of the protein. *J Biol Chem* 276(42): 38636–44, 2001.
26. Bernstrom K, Orning L, Hammarstrom S. Gamma-glutamyl transpeptidase, a leukotriene metabolizing enzyme. *Methods Enzymol* 86: 38–45, 1982.
27. Hanigan MH, Frierson HF. Immunohistochemical detection of gamma-glutamyl transpeptidase in normal human tissue. *J Histochem Cytochem* 44(10): 1101–8, 1996.
28. Han B, Luo GY, Shi ZZ *et al.* gamma-Glutamyl leukotrienase, a novel endothelial membrane protein, is specifically responsible for leukotriene D-4 formation *in vivo*. *Amer J Pathol* 161(2): 481–90, 2002.
29. Lewis RA, Drazen JM, Austen KF, Clark DA, Corey EJ. Identification of the C(6)-S-conjugate of leukotriene A with cysteine as a naturally occurring slow reacting substance of anaphylaxis (SRS-A). Importance of the 11-cis-geometry for biological activity. *Biochem Biophys Res Commun* 96: 271–77, 1980.
30. Parker CW, Koch D, Huber MM, Falkenstein SF. Formation of the cysteinyl form of slow-reacting substance of analyphylaxis (leukotriene E) on human plasma. *Biochem Biophys Res Commun* 97: 1038–46, 1980.
31. Murphy RC, Wheelan P. Pathways of Leukotriene Metabolism in Isolated Cell Models and Human Subjects. New York: Macel Dekker, Inc, 1998.
32. Sala A, Voelkel N, Maclouf J, Murphy RC. Leukotriene E4 elimination and metabolism in normal human subjects. *J Biol Chem* 265: 21771–78, 1990.
33. BandeiraMelo C, Weller PF. Eosinophils and cysteinyl leukotrienes. *Prostagland Leuk Essent Fatty Acids* 69(2–3): 135–43, 2003.
34. Boyce JA. The role of mast cells in asthma. *Prostagland Leuk Essent Fatty Acids* 69(2–3): 195–205, 2003.
35. Lynch KR, O'Neill GP, Liu Q *et al.* Characterization of the human cysteinyl leukotriene $CysLT_1$ receptor. *Nature* 399: 789–93, 1999.

36. Heise CE, ODowd BF, Figueroa DJ *et al.* Characterization of the human cysteinyl leukotriene 2 receptor. *J Biol Chem* 275(39): 30531–36, 2000.
37. Evans JF. The cysteinyl leukotriene receptors. *Prostagland Leuk Essent Fatty Acids* 69(2–3): 117–22, 2003.
38. Drazen JM. Leukotrienes as mediators of airway obstruction. *Amer J Respir Crit Care Med* 158(5): S193–S200, 1998.
39. Petersgolden M, Canetti C, Mancuso P, Coffey MJ. Leukotrienes: Underappreciated mediators of innate immune responses. *J Immunol* 174(2): 589–94, 2005.
40. Kanaoka Y, Boyce JA. Cysteinyl leukotrienes and their receptors: Cellular distribution and function in immune and inflammatory responses. *J Immunol* 173: 1503–10, 2004.
41. Holroyde MC, Altounyan RE, Cole M, Dixon M, Elliott EV. Bronchoconstriction produced in man by leukotrienes C and D. *Lancet* 2: 17–18, 1981.
42. Weiss JW, Drazen JM, Coles N *et al.* Bronchoconstrictor effects of leukotriene C in humans. *Science* 216: 196–98, 1982.
43. Bisgaard H, Groth S, Dirksen H. Leukotriene D4 induces bronchoconstriction in man. *Allergy* 38: 441–43, 1983.
44. Weiss JW, Drazen JM, McFadden ER Jr. *et al.* Airway constriction in normal humans produced by inhalation of leukotriene D. Potency, time course, and effect of aspirin therapy. *JAMA* 249: 2814–17, 1983.
45. Barnes NC, Piper PJ, Costello JF. Comparative effects of inhaled leukotriene C4, leukotriene D4, and histamine in normal human subjects. *Thorax* 39: 500–4, 1984.
46. Smith LJ, Greenberger PA, Patterson R, Krell RD, Bernstein PR. The effect of inhaled leukotriene D4 in humans. *Am Rev Respir Dis* 131: 368–72, 1985.
47. Adelroth E, Morris MM, Hargreave FE, O'Byrne PM. Airway responsiveness to leukotrienes C4 and D4 and to methacholine in patients with asthma and normal controls. *N Engl J Med* 315: 480–84, 1986.
48. Roberts JA, Rodger IW, Thomson NC. Effect of verapamil and sodium cromoglycate on leukotriene D4 induced bronchoconstriction in patients with asthma. *Thorax* 41: 753–58, 1986.
49. Bisgaard H, Groth S. Bronchial effects of leukotriene D4 inhalation in normal human lung. *Clin Sci* 72: 585–92, 1987.
50. Davidson AB, Lee TH, Scanlon PD *et al.* Bronchoconstrictor effects of leukotriene E4 in normal and asthmatic subjects. *Am Rev Respir Dis* 135: 333–37, 1987.
51. Smith LJ, Geller S, Ebright L, Glass M, Thyrum PT. Inhibition of leukotriene D4-induced bronchoconstriction in normal subjects by the oral LTD4 receptor antagonist ICI 204, 219. *Am Rev Respir Dis* 141: 988–92, 1990.
52. Smith LJ, Kern R, Patterson R, Krell RD, Bernstein PR. Mechanism of leukotriene D4-induced bronchoconstriction in normal subject. *J Allergy Clin Immunol* 80: 340–47, 1987.
53. Griffin M, Weiss JW, Leitch AG *et al.* Effects of leukotriene D on the airways in asthma. *N Engl J Med* 308: 436–39, 1983.
54. Pichurko BM, Ingram RHJ, Sperling RI *et al.* Localization of the site of the bronchoconstrictor effects of leukotriene C4 compared with that of histamine in asthmatic subjects. *Am Rev Respir Dis* 140: 334–39, 1989.
55. Togias AG, Naclerio RM, Peters SP *et al.* Local generation of sulfidopeptide leukotrienes upon nasal provocation with cold, dry air. *Am Rev Respir Dis* 133: 1133–37, 1986.
56. Silber G, Proud D, Warner J *et al.* *In vivo* release of inflammatory mediators by hyperosmolar solutions. *Am Rev Respir Dis* 137: 606–12, 1988.
57. Lam S, Chan H, LeRiche JC, Chan-Yeung M, Salari H. Release of leukotrienes in patients with bronchial asthma. *J Allergy Clin Immunol* 81: 711–17, 1988.
58. Zehr BB, Casale TB, Wood D, Floerchinger C, Richerson HB, Hunninghake GW. Use of segmental airway lavage to obtain relevant mediators from the lungs of asthmatic and control subjects. *Chest* 95: 1059–63, 1989.
59. Wenzel SE, Larsen GL, Johnston K, Voelkel NF, Westcott JY. Elevated levels of leukotriene C4 in bronchoalveolar lavage fluid from atopic asthmatics after endobronchial allergen challenge. *Am Rev Respir Dis* 142: 112–19, 1990.
60. Kane GC, Tollino M, Pollice M *et al.* Insights into IgE-mediated lung inflammation derived from a study employing a 5-lipoxygenase inhibitor. *Prostaglandins* 50(1): 1–18, 1995.
61. Csoma Z, Kharitonov SA, Balint B, Bush A, Wilson NM, Barnes PJ. Increased leukotrienes in exhaled breath condensate in childhood asthma. *Amer J Respir Crit Care Med* 166(10): 1345–49, 2002.
62. Baraldi E, Carraro S, Alinovi R *et al.* Cysteinyl leukotrienes and 8-isoprostane in exhaled breath condensate of children with asthma exacerbations. *Thorax* 58(6): 505–9, 2003.
63. Carraro S, Corradi M, Zanconato S *et al.* Exhaled breath condensate cysteinyl leukotrienes are increased in children with exercise-induced bronchoconstriction. *J Allergy Clin Immunol* 115(4): 764–70, 2005.
64. Tagari P, Ethier D, Carry M *et al.* Measurement of urinary leukotrienes by reversed-phase liquid chromatography and radioimmunoassay. *Clin Chem* 35: 388–91, 1989.
65. Taylor GW, Taylor I, Black P *et al.* Urinary leukotriene E4 after antigen challenge and in acute asthma and allergic rhinitis. *Lancet* 1: 584–88, 1989.
66. Westcott JY. The measurement of leukotrienes in human fluids. *Clin Rev Allergy Immunol* 17(1–2): 153–77, 1999.
67. Wu YH, Li LYT, Henion JD, Krol GJ. Determination of LTE(4) in human urine by liquid chromatography coupled with ionspray tandem mass spectrometry. *J Mass Spectrom* 31(9): 987–93, 1996.
68. Higashi N, Taniguchi M, Mita H, Osame M, Akiyama K. A comparative study of eicosanoid concentrations in sputum and urine in patients with aspirin-intolerant asthma. *Clin Exp Allergy* 32(10): 1484–90, 2002.
69. Rabinovitch N, Zhang L, Gelfand EW. Urine leukotriene E-4 levels are associated with decreased pulmonary function in children with persistent airway obstruction. *J Allergy Clin Immunol* 118(3): 635–40, 2006.
70. Maltby NH, Taylor GW, Ritter JM, Moore K, Fuller RW, Dollery CT. Leukotriene C4 elimination and metabolism in man. *J Allergy Clin Immunol* 85: 3–9, 1990.
71. Sladek K, Dworski R, Fitzgerald GA *et al.* Allergen-stimulated release of thromboxane A2 and leukotriene E4 in humans Effect of indomethacin. *Am Rev Respir Dis* 141: 1441–45, 1990.
72. Asano K, Lilly CM, Odonnell WJ *et al.* Diurnal variation of urinary leukotriene E_4 and histamine excretion rates in normal subjects and patients with mild-to- moderate asthma. *J Allergy Clin Immunol* 96(5 Part 1): 643–51, 1995.
73. Vachier I, Kumlin M, Dahlen SE, Bousquet J, Godard P, Chanez P. High levels of urinary leukotriene E-4 excretion in steroid treated patients with severe asthma. *Respir Med* 97(11): 1225–29, 2003.
74. Capra V. Molecular and functional aspects of human cysteinyl leukotriene receptors. *Pharmacol Res* 50: 1–11, 2004.
75. Drazen JM, Israel E, Obyrne PM. Treatment of asthma with drugs modifying the leukotriene pathway. *N Engl J Med* 340(3): 197–206, 1999.
76. Carter GW, Young PR, Albert DH *et al.* 5-lipoxygenase inhibitory activity of zileuton. *J Pharmacol Exp Ther* 256: 929–37, 1991.
77. Depre M, Friedman B, Vanhecken A *et al.* Pharmacokinetics and pharmacodynamics of multiple oral doses of MK-0591, a 5-lipoxygenase-activating protein inhibitor. *Clin Pharmacol Ther* 56: 22–30, 1994.
78. Gardiner PJ, Cuthbert NJ, Francis HP *et al.* Inhibition of antigen-induced contraction of guinea-pig airways by a leukotriene synthesis inhibitor, BAY x1005. *Eur J Pharmacol* 258: 95–102, 1994.
79. Israel E, Cohn J, Dube L, Drazen JM. Effect of treatment with zileuton, a 5-lipoxygenase inhibitor, in patients with asthma: A randomized controlled trial. *JAMA* 275(12): 931–36, 1996.

80. Fuller RW, Black PN, Dollery CT. Effect of the oral leukotriene D4 antagonist LY171883 on inhaled and intradermal challenge with antigen and leukotriene D4 in atopic subjects. *J Allergy Clin Immunol* 83: 939–44, 1989.
81. Israel E, Dermarkarian R, Rosenberg M *et al.* The effects of a 5-lipoxygenase inhibitor on asthma induced by cold, dry air. *N Engl J Med* 323: 1740–44, 1990.
82. Israel E, Fischer AR, Rosenberg MA *et al.* The pivotal role of 5-lipoxygenase products in the reaction of aspirin-sensitive asthmatics to aspirin. *Am Rev Respir Dis* 148: 1447–51, 1993.
83. Dahlen B, Kumlin M, Margolskee DJ *et al.* The leukotriene-receptor antagonist MK-0679 blocks airway obstruction induced by inhaled lysine-aspirin in aspirin-sensitive asthmatics. *Eur Respir J* 6: 1018–26, 1993.
84. Dahlen B, Zetterstrom O, Bjorck T, Dahlen SE. The leukotriene-antagonist ICI-204,219 inhibits the early airway reaction to cumulative bronchial challenge with allergen in atopic asthmatics. *Eur Respir J* 7: 324–31, 1994.
85. Dahlen SE, Malmstrom K, Nizankowska E *et al.* Improvement of asprinin-intolerant asthma by montelukast, a leukotriene antagonist – A randomized, double-blink, placebo-controlled trial. *Amer J Respir Crit Care Med* 165(1): 9–14, 2002.
86. Menendez R, Venzor J, Ortiz G. Failure of zafirlukast to prevent ibuprofen-induced anaphylaxis. *Ann Allergy Asthma Immunol* 80(3): 225–26, 1998.
87. Enrique E, GarciaOrtega P, Gaig P, SanMiguel MM. Failure of montelukast to prevent anaphylaxis to diclofenac. *Allergy* 54(5): 529–30, 1999.
88. Barnes NC. Effects of antileukotrienes in the treatment of asthma. *Amer J Respir Crit Care Med* 161(2): S73–76, 2000.
89. Wenzel SE. The role of leukotrienes in asthma. *Prostagland Leuk Essent Fatty Acids* 69(2–3): 145–55, 2003.
90. Busse WW, Mcgill KA, Horwitz RJ. Leukotriene pathway inhibitors in asthma and chronic obstructive pulmonary disease. *Clin Exp Allergy* 29: 110–15, 1999.
91. Obyrne PM, Israel E, Drazen JM. Antileukotrienes in the treatment of asthma. *Ann Intern Med* 127(6): 472–80, 1997.
92. Bisgaard H. Leukotriene modifiers in pediatric asthma management. *Pediatrics* 107(2): 381–90, 2001.
93. Celik P, Sakar A, Havlucu Y, Yuksel H, Turkdogan P, Yorgancioglu A. Short-term effects of montelukast in stable patients with moderate to severe COPD. *Respir Med* 99(4): 444–50, 2005.
94. In KH, Asano K, Beier D *et al.* Naturally occurring mutations in the human 5-lipoxygenase gene promoter that modify transcriptio factor binding and reporter gene transcription. *J Clin Invest* 99: 1130–37, 1997.
95. Silverman ES, Du J, Desanctis GT *et al.* Egr-1 and Sp1 interact functionally with the 5-lipoxygenase promoter and its naturally occurring mutants. *Amer J Respir Cell Mol Biol* 19(2): 316–23, 1998.
96. Sanak M, Simon HU, Szczeklik A. Leukotriene C4 synthase promoter polymorphism and risk of aspirin-induced asthma. *Lancet* 350: 1599–600, 1997.
97. Asano K, Shiomi T, Hasegawa N *et al.* Leukotriene C4 synthase gene A(−444)C polymorphism and clinical response to a CYS-LT1 antagonist, in Japanese patients with moderate asthma. *Pharmacogenetics* 12: 1–6, 2002.
98. Sayers I, Barton S, Rorke S *et al.* Promoter polymorphism in the 5-lipoxygenase (ALOX5) and 5-lipoxygenase-activating protein (ALOX5AP) genes and asthma susceptibility in a Caucasian population. *Clin Exp Allergy* 33(8): 1103–10, 2003.
99. Lima JJ, Zhang S, Grant A *et al.* Influence of leukotriene pathway polymorphisms on response to montelukast in asthma. *Am J Respir Crit Care Med* 173(4): 379–85, 2006.
100. Drazen JM, Yandava C, Dube L *et al.* Pharmacogenetic association between ALOX5 promoter genotype and the response to anti-asthma treatment. *Nat Gen* 22: 170–72, 1999.
101. Levy BD. Lipoxins and lipoxin analogs in asthma. *Prostaglandins Leukot Essent Fatty Acids* 73(3–4): 231–37, 2005.
102. Levy BD, Desanctis GT, Devchand PR *et al.* Multi-pronged inhibition of airway hyper-responsiveness and inflammation by lipoxin A(4). *Nat Med* 8(9): 1018–23, 2002.
103. Ariel A, Chiang N, Arita M, Petasis NA, Serhan CN. Aspirin-triggered lipoxin A4 and B4 analogs block extracellular signal-regulated kinase-dependent TNF-alpha secretion from human T cells. *J Immunol* 170(12): 6266–72, 2003.
104. Colgan SP, Serhan CN, Parkos CA, Delp-Archer C, Madara JL. Lipoxin A4 modulates transmigration of human neutrophils across intestinal epithelial monolayers. *J Clin Invest* 92(1): 75–82, 1993.
105. Serhan CN, Maddox JF, Petasis NA *et al.* Design of lipoxin A4 stable analogs that block transmigration and adhesion of human neutrophils. *Biochemistry* 34(44): 14609–15, 1995.
106. Takano T, Fiore S, Maddox JF, Brady HR, Petasis NA, Serhan CN. Aspirin-triggered 15-epi-lipoxin A4 (LXA4) and LXA4 stable analogues are potent inhibitors of acute inflammation: Evidence for anti-inflammatory receptors. *J Exp Med* 185(9): 1693–704, 1997.
107. Levy BD, Fokin VV, Clark JM, Wakelam MJ, Petasis NA, Serhan CN. Polyisoprenyl phosphate (PIPP) signaling regulates phospholipase D activity: A 'stop' signaling switch for aspirin-triggered lipoxin A4. *FASEB J* 13(8): 903–11, 1999.
108. Maddox JF, Serhan CN. Lipoxin A4 and B4 are potent stimuli for human monocyte migration and adhesion: Selective inactivation by dehydrogenation and reduction. *J Exp Med* 183(1): 137–46, 1996.
109. Godson C, Mitchell S, Harvey K, Petasis NA, Hogg N, Brady HR. Cutting edge: Lipoxins rapidly stimulate nonphlogistic phagocytosis of apoptotic neutrophils by monocyte-derived macrophages. *J Immunol* 164(4): 1663–67, 2000.
110. Aliberti J, Hieny S, Reis eSousa, Serhan CN, Sher A. Lipoxin-mediated inhibition of IL-12 production by DCs: A mechanism for regulation of microbial immunity. *Nat Immunol* 3(1): 76–82, 2002.
111. Ramstedt U, Ng J, Wigzell H, Serhan CN, Samuelsson B. Action of novel eicosanoids lipoxin A and B on human natural killer cell cytotoxicity: Effects on intracellular cAMP and target cell binding. *J Immunol* 135(5): 3434–38, 1985.
112. Bonnans C, Fukunaga K, Levy MA, Levy BD. Lipoxin A(4) regulates bronchial epithelial cell responses to acid injury. *Am J Pathol* 68(4): 1064–72, 2006.
113. Gewirtz AT, Collier-Hyams LS, Young AN *et al.* Lipoxin a4 analogs attenuate induction of intestinal epithelial proinflammatory gene expression and reduce the severity of dextran sodium sulfate-induced colitis. *J Immunol* 168(10): 5260–67, 2002.
114. Canny G, Levy O, Furuta GT *et al.* Lipid mediator-induced expression of bactericidal/permeability-increasing protein (BPI) in human mucosal epithelia. *Proc Natl Acad Sci USA* 99(6): 3902–7, 2002.
115. Dahlen SE, Raud J, Serhan CN, Bjork J, Samuelsson B. Biological activities of lipoxin A include lung strip contraction and dilation of arterioles *in vivo*. *Acta Physiol Scand* 130(4): 643–47, 1987.
116. Fukunaga K, Kohli P, Bonnans C, Fredenburgh LE, Levy BD. Cyclooxygenase 2 plays a pivotal role in the resolution of acute lung injury. *J Immunol* 174(8): 5033–39, 2005.
117. Christie PE, Spur BW, Lee TH. The effects of lipoxin A4 on airway responses in asthmatic subjects. *Am Rev Respir Dis* 145(6): 1281–84, 1992.
118. Vachier I, Bonnans C, Chavis C *et al.* Severe asthma is associated with a loss of LX4, an endogenous anti- inflammatory compound. *J Allergy Clin Immunol* 115(1): 55–60, 2005.
119. Levy BD, Clish CB, Schmidt B, Gronert K, Serhan CN. Lipid mediator class switching during acute inflammation: Signals in resolution. *Nat Immunol* 2(7): 612–19, 2001.
120. Sanak M, Levy BD, Clish CB *et al.* Aspirin-tolerant asthmatics generate more lipoxins than aspirin-intolerant asthmatics. *Eur Respir J* 16(1): 44–49, 2000.
121. Levy BD, Bonnans C, Silverman ES, Palmer LJ, Marigowda G, Israel E. Diminished lipoxin biosynthesis in severe asthma. *Am J Respir Crit Care Med* 172(7): 824–30, 2005.

122. Fiore S, Romano M, Reardon EM, Serhan CN. Induction of functional lipoxin A4 receptors in HL-60 cells. *Blood* 81(12): 3395–403, 1993.
123. Chiang N, Fierro IM, Gronert K, Serhan CN. Activation of lipoxin A(4) receptors by aspirin-triggered lipoxins and select peptides evokes ligand-specific responses in inflammation. *J Exp Med* 191(7): 1197–208, 2000.
124. Gronert K, Martinsson-Niskanen T, Ravasi S, Chiang N, Serhan CN. Selectivity of recombinant human leukotriene D(4), leukotriene B(4), and lipoxin A(4) receptors with aspirin-triggered 15-epi-LXA(4) and regulation of vascular and inflammatory responses. *Am J Pathol* 158(1): 3–9, 2001.
125. Fiore S, Serhan CN. Lipoxin A4 receptor activation is distinct from that of the formyl peptide receptor in myeloid cells: Inhibition of CD11/18 expression by lipoxin A4-lipoxin A4 receptor interaction. *Biochemistry* 34(51): 16678–86, 1995.
126. Chiang N, Serhan CN, Dahlen SE *et al.* The lipoxin receptor ALX: Potent ligand-specific and stereoselective actions *in vivo*. *Pharmacol Rev* 58(3): 463–87, 2006.
127. Perretti M, Chiang N, La M *et al.* Endogenous lipid- and peptide-derived anti-inflammatory pathways generated with glucocorticoid and aspirin treatment activate the lipoxin A4 receptor. *Nat Med* 8(11): 1296–302, 2002.
128. Ohira T, Bannenberg G, Arita M *et al.* A stable aspirin-triggered lipoxin A4 analog blocks phosphorylation of leukocyte-specific protein 1 in human neutrophils. *J Immunol* 173(3): 2091–98, 2004.
129. Coles SJ, Neill KH, Reid LM *et al.* Effects of leukotrienes C4 and D4 on glycoprotein and lysozyme secretion by human bronchial mucosa. *Prostaglandins* 25: 155–70, 1983.
130. Piacentini GL, Kaliner MA. The potential roles of leukotrienes in bronchial asthma. *Am Rev Respir Dis* 143: S96–S99, 1991.
131. Marian E, Baraldo S, Visentin A *et al.* Up-regulated membrane and nuclear leukotriene B4 receptors in COPD. *Chest* 129(6): 1523–30, 2006.
132. Peters-Golden M, Coffey M. Role of leukotrienes in antimicrobial host defense of the lung. *Clin Rev Allergy Immunol* 17(1–2): 261–69, 1999.
133. Hoshino M. Impact of inhaled corticosteroids and leukotriene receptor antagonists on airway remodeling. *Clin Rev Allergy Immunol* 27(1): 59–64, 2004.
134. Busse W, Kraft M. Cysteinyl leukotrienes in allergic inflammation – Strategic target for therapy. *Chest* 127(4): 1312–26, 2005.
135. Kostikas K, Gaga M, Papatheodorou G, Karamanis T, Orphanidou D, Loukides S. Leukotriene B4 in exhaled breath condensate and sputum supernatant in patients with COPD and asthma. *Chest* 27(5): 1553–59, 2005.
136. Wardlaw AJ, Hay H, Cromwell O, Collins JV, Kay AB. Leukotrienes, LTC_4 and LTB_4, in bronchoalveolar lavage in bronchial asthma and other respiratory diseases. *J Allergy Clin Immunol* 84: 19–26, 1989.
137. Efimov W, Blazhko VI, Voeikova LS, Karanysheva SA, Bondar TN. The leukotriene B4 content of the bronchoalveolar lavage fluid and the function of the prostacyclin-thromboxane system in patients with variants of chronic bronchitis. *Terapevticheskii Arkhiv* 62(4): 94–96, 1990.
138. Birring SS, Parker D, Brightling CE, Bradding P, Wardlaw AJ, Pavord ID. Induced sputum inflammatory mediator concentrations in chronic cough. *Am J Respir Crit Care Med* 169(1): 15–19, 2004.
139. Montuschi P. Exhaled breath condensate analysis in patients with COPD. *Clin Chim Acta* 356(1–2): 22–34, 2005.
140. Profita M, Giorgi RD, Sala A *et al.* Muscarinic receptors, leukotriene B4 production and neutrophilic inflammation in COPD patients. *Allergy* 60(11): 1361–69, 2005.
141. Montuschi P, Macagno F, Parente P *et al.* Effects of cyclo-oxygenase inhibition on exhaled eicosanoids in patients with COPD. *Thorax* 60(10): 827–33, 2005.
142. Crooks SW, Bayley DL, Hill SL, Stockley RA. Bronchial inflammation in acute bacterial exacerbations of chronic bronchitis: The role of leukotriene B-4. *Eur Respir J* 15(2): 274–80, 2000.
143. Matsuyama W, Mitsuyama H, Watanabe M *et al.* Effects of omega-3 polyunsaturated fatty acids on inflammatory markers in COPD. *Chest* 128(6): 3817–27, 2005.
144. Rubinstein I, Kumar B, Schriever C. Long-term montelukast therapy in moderate to severe COPD – a preliminary observation. *Respir Med* 98(2): 134–38, 2004.
145. Zuhlke IE, Kanniess F, Richter K *et al.* Montelukast attenuates the airway response to hypertonic saline in moderate-to-severe COPD. *Eur Respir J* 22(6): 926–30, 2003.
146. Nannini LJ, Flores DM. Bronchodilator effect of zafirlukast in subjects with chronic obstructive pulmonary disease. *Pulm Pharmacol Ther* 16(5): 307–11, 2003.

25

CHAPTER

Reactive Oxygen Species and Antioxidant Therapeutic Approaches

Irfan Rahman

Department of Environmental Medicine, Lung Biology and Disease Program, University of Rochester Medical Center, Rochester, NY, USA

INTRODUCTION

The lung is constantly exposed to a high-oxygen environment and, owing to its large surface area and blood supply, is highly susceptible to injury mediated by a phenomenon known as oxidative stress. Oxidative stress is described simply as an increase in the oxidant to antioxidant ratio in the lung. Reactive oxygen species (ROS) such as superoxide anion ($O_2^{\bullet -}$) and the hydroxyl radical ($^{\bullet}OH$) are unstable molecules with unpaired electrons, capable of initiating the oxidation of various molecules in the airway and airspace structures.

Biological systems are continuously exposed to oxidants that are generated either endogenously by metabolic reactions (e.g., from mitochondrial electron transport during respiration or during activation of phagocytes) or exogenously (such as by air pollutants or cigarette smoke). For example, production of ROS has been directly linked to oxidation of proteins, DNA, and lipids which may cause direct lung injury or induce a variety of cellular responses, through the generation of secondary metabolic reactive species. ROS may result in remodeling of extracellular matrix, cause apoptosis and mitochondrial respiration, and regulate cell proliferation [1, 2]. Alveolar repair responses and immune modulation in the lung may also be influenced by ROS [1, 2]. Furthermore, high levels of ROS have been implicated in initiating inflammatory responses in the lungs through the activation of transcription factors such as nuclear factor-κB (NF-κB) and activator protein-1 (AP-1); their activation leads to signal transduction and gene expression of pro-inflammatory mediators [3, 4]. It is proposed that ROS produced by phagocytes that have been recruited to sites of inflammation are a major cause of the cell and tissue damage associated with many chronic inflammatory lung diseases, including asthma and chronic obstructive pulmonary disease (COPD) [5–9] (Fig. 25.1).

The composition of inflammatory cell types that invade tissues varies widely in asthma and COPD; this suggests that there are differences in the characteristics of the ROS produced in these diseases [10–12]. This chapter reviews the evidence for the role of ROS in the pathogenesis of asthma and COPD, and discusses the molecular mechanisms (cell signaling and gene expression) and pathophysiological consequences of increased ROS release in these conditions. Moreover, it also highlights the antioxidant mechanisms in place to protect against the damaging effects of ROS. Finally, it goes on to explore possible therapeutic approaches that involve the use of a variety of antioxidants.

CELL-DERIVED ROS

A common feature of all inflammatory lung diseases is the development of an inflammatory–immune response, characterized by activation of epithelial cells, and resident macrophages, and the recruitment and activation of neutrophils, eosinophils, monocytes, and lymphocytes. The degree to which this occurs and the cell types involved vary widely in asthma and COPD. Inflammatory cells once recruited in the airspace become activated and generate ROS in response to a sufficient stimulus (threshold concentration). The activation of macrophages, neutrophils, and eosinophils generates $O_2^{\bullet -}$, which is rapidly converted to H_2O_2 through

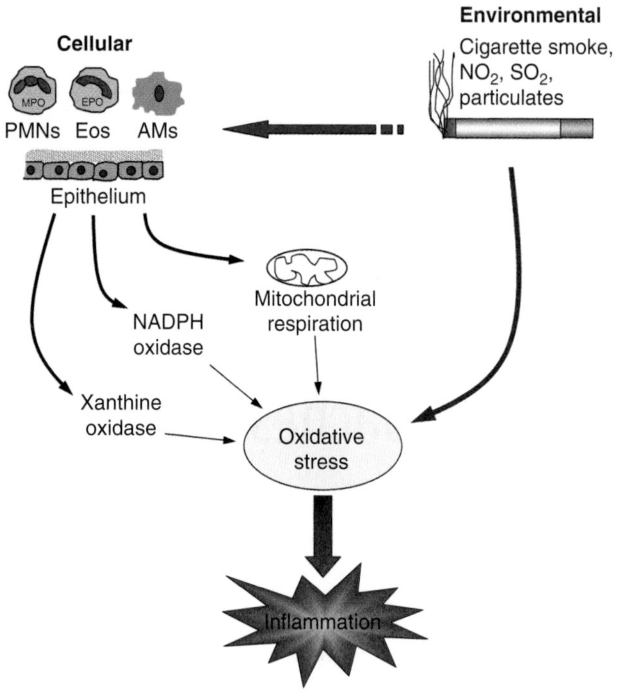

FIG. 25.1 Sources of oxidative stress. Oxidative stress derived from either environmental or cellular origins results in inflammation. Cellular-derived oxidative stress can be produced through mitochondrial respiration, the NADPH oxidase system, or the xanthine oxidase system. Inflammation itself can have a feed-forward effect, triggering inflammatory cells to produce yet more oxidative stress, exacerbating and intensifying the inflammatory response. PMNs: polymorphonucleocytes; Eos: eosinophils; AMs: alveolar macrophages.

the catalytic action of superoxide dismutase (SOD); $^{\bullet}OH$ is formed nonenzymatically in the presence of Fe^{2+} as a secondary reaction. ROS and reactive nitrogen species (RNS) can also be generated intracellularly from several sources such as mitochondrial respiration, the NADPH oxidase system, and xanthine/xanthine oxidase (XO) (Fig. 25.2). However, the primary ROS generating system is NADPH oxidase, a complex enzyme system that is present in phagocytes and epithelial cells.

Activation of this enzyme system involves a complex mechanism with the assembly of various cytosolic and membrane-associated subunits, resulting in the one-electron reduction of oxygen to $O_2^{\bullet-}$ using NADPH as the electron donor. In addition to NADPH oxidase, phagocytes use heme peroxidases such as myeloperoxidase (MPO) or eosinophil peroxidase (EPO) to produce ROS. Activation of EPO results in the formation of the potent oxidants hypochlorous acid (HOCl) and hypobromous acid (HOBr) from H_2O_2 in the presence of chloride (Cl^-) and bromide (Br^-) ions, respectively. It is believed that the oxidant burden produced by eosinophils is substantial because these cells possess several times greater capacity to generate $O_2^{\bullet-}$ and H_2O_2 than do neutrophils, and the content of EPO in eosinophils is 3–10 times higher than the amount of MPO present in neutrophils [13–15] (Fig. 25.3).

The physiological consequences of EPO-dependent formation of brominating oxidants such as HOBr *in vivo* are unknown. HOBr reacts rapidly with a variety of nucleophilic targets such as thiols, thiol ethers, amines, unsaturated groups, and aromatic compounds [16].

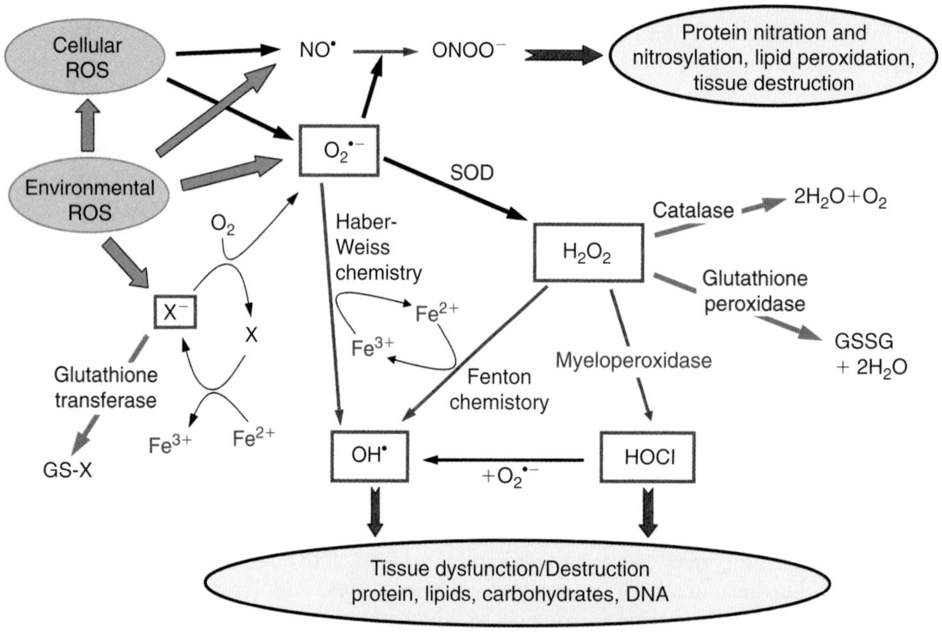

FIG. 25.2 Molecular consequences of oxidative stress. The relatively weak superoxide anion from both cellular and environmental sources can be transformed into more damaging and potent reactive oxygen and reactive nitrogen species, such as HOCl, the hydroxyl radical and peroxynitrite, through a series of enzymatic and nonenzynmatic steps. Xenobiotic radicals from the environment can be long lived and undergo redox cycling, such as the semiquinones, and result in further superoxide anion formation as well as more powerful radical formation in the presence of free metal ions through Haber–Weiss and Fenton chemistry. Endogenous antioxidant defenses glutathione transferase, GPx, SOD, and catalase neutralize and remove these ROS and RNS. ROS: reactive oxygen species; RNS: reactive nitrogen species; X^-: xenobiotic radical; $O_2^{-\bullet}$: superoxide anion; H_2O_2: hydrogen peroxide; $OH^{\bullet}$: hydroxide radical; HOCl: hypochlorous acid; NO: nitric oxide; $ONOO^-$: peroxynitrite; GSH: reduced glutathione; GSSG: oxidized glutathione dimers.

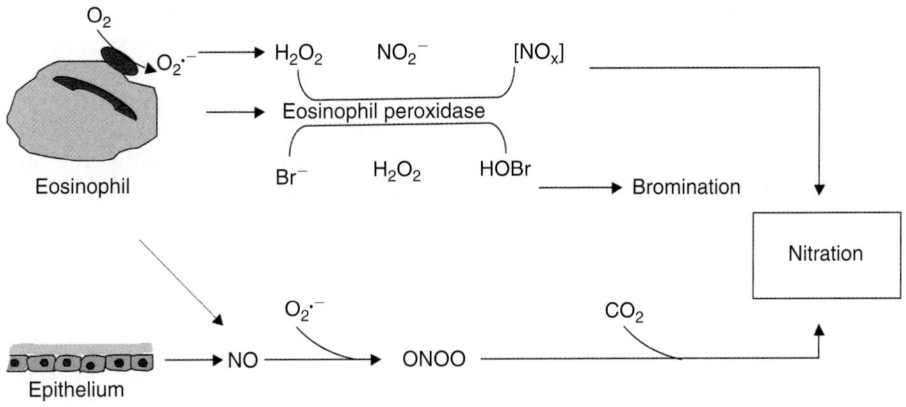

FIG. 25.3 Generation of reactive nitrogen species. Model of potential pathways used by eosinophils and airway epithelial cells for generation of NO-derived reactive nitrogen and halogen species leading to nitration of proteins.

Several transition metal salts react with H_2O_2 to form $^{\bullet}OH$. Most attention *in vivo* for the generation of $^{\bullet}OH$ has focused on the role of iron [17]. Iron is a critical element in many oxidative reactions. Free iron in the ferrous form catalyzes the Fenton reaction and the superoxide-driven Haber–Weiss reaction, which generate the $^{\bullet}OH$, a ROS which damages tissues, particularly cell membranes by peroxidation of their lipid membranes. MPO- and EPO-derived ROS can also interact with nitrite (NO_2^-) and H_2O_2 to promote formation of RNS. ROS may also be released by lung epithelial cells [18] and stimulate inflammatory cells directly, thereby amplifying lung inflammatory and oxidant events. ROS interact with a variety of molecules and donate electrons in biological systems. Reactive oxygen and nitrogen species also act on certain amino acids such as methionine, tyrosine, and cysteine in proteins (e.g. enzymes, kinases), profoundly altering the function of these proteins in inflammatory lung diseases [19].

INHALED OXIDANTS AND CIGARETTE SMOKE

Cigarette smoking, or inhalation of airborne pollutants that may be either oxidant gases [such as ozone, nitrogen dioxide (NO_2), sulfur dioxide (SO_2)] or particulate air pollution, results in direct lung damage as well as the activation of inflammatory responses in the lungs. Cigarette smoke is a complex mixture of over 4700 chemical compounds, including high concentrations of oxidants (10^{14} molecules per puff) [20]. Short-lived oxidants such as $O_2^{\bullet-}$ and nitric oxide (NO) are predominantly found in the gas phase. NO and $O_2^{\bullet-}$ immediately react to form the highly reactive peroxynitrite ($ONOO^-$) molecule. The radicals in the tar phase of cigarette smoke are organic in nature, such as long-lived semiquinone radicals, which can react with $O_2^{\bullet-}$ to form $^{\bullet}OH$ and H_2O_2 [21]. The aqueous phase of cigarette smoke condensate may undergo redox recycling for a considerable period of time in the epithelial lining fluid (ELF) of smokers [22, 23]. The tar phase is also an effective metal chelator and can bind iron to produce tar-semiquinone + tar-Fe^{2+}, which can generate H_2O_2 continuously [22, 23].

Quinone (Q), hydroquinone (QH_2), and semiquinone ($QH^{\bullet}$) in the tar phase are present in equilibrium:

$$Q + QH_2 \rightarrow 2H^+ + 2Q^{\bullet-}.$$

Aqueous extracts of cigarette tar contain the quinone radical ($Q^{\bullet-}$), which can reduce oxygen to form $O_2^{\bullet-}$, which may dismutate in the presence of H^+ to form H_2O_2:

$$Q^{\bullet-} + O_2 \rightarrow Q + O_2^{\bullet-}$$

$$2O_2^{\bullet-} + 2H^+ \rightarrow O_2 + H_2O_2.$$

Furthermore, since both cigarette tar and lung ELF contain metal ions, such as iron, Fenton chemistry will result in the production of the $^{\bullet}OH$ which is a highly reactive and potent ROS. Thus, oxidants and electrophilic compounds may be the initiating or triggering factor in the underlying inflammation seen in the pathogenesis of COPD as detailed below.

ROS AND MEMBRANE LIPID PEROXIDATION

Oxygen species such as $O_2^{\bullet-}$ and $^{\bullet}OH$ are highly reactive, and when generated close to cell membranes they oxidize membrane phospholipids (lipid peroxidation), a process which may continue as a chain reaction. Thus, a single $^{\bullet}OH$ can result in the formation of many molecules of lipid hydroperoxides in the cell membrane [19]. The peroxidative breakdown of polyunsaturated fatty acids impairs membrane function, inactivates membrane-bound receptors and enzymes, and increases tissue permeability. Each of these processes has been implicated in the pathogenesis of many forms of lung injury. There is increasing evidence that aldehydes, generated endogenously during the process of lipid peroxidation, are involved in many of the pathophysiological effects associated with oxidative stress in cells and tissues [19]. Compared with free radicals, lipid peroxidation aldehydes are generally stable, can diffuse within, or even escape from the cell, and attack targets far from the site of the original free radical event. In addition to their cytotoxic properties, lipid peroxides are increasingly recognized as being important in signal transduction for a number of important events in the inflammatory response [24].

Many of the effects of ROS in airways may be mediated by the secondary release of inflammatory lipid mediators such as 4-hydroxy-2-nonenal (4-HNE). 4-HNE, a highly reactive diffusible end-product of lipid peroxidation, is known to induce/regulate various cellular events, such as proliferation, apoptosis, and activation of signaling pathways [24, 25]. 4-HNE has a high affinity toward cysteine, histidine, and lysine residues. It forms adducts with proteins, altering their function. Acrolein is another example of reactive aldehyde. Both acrolein and 4-HNE have a high affinity toward cysteine, histidine, and lysine residues. These reactive aldehydes can form adducts with both intracellular proteins, such as histone deacetylase (HDAC) 2 [26] and extracellular proteins, such as collagen and fibronectin [27] altering their function which in turn can then impact on cell function [27, 28]. In contrast, the isoprostanes, are ROS-catalyzed oxidative products of arachidonic acid metabolism and are stable lipid peroxidation products, which circulate in plasma and are excreted in the urine [29, 30]. It is because of their relative stability *in vivo* that they have been used as markers of oxidative stress in both asthma and COPD [31]. However, one isoprostane member, 8-isoprostane, has been shown to possess a very potent biological activity. It has been demonstrated that 8-isoprostane is a very potent stimulus for smooth muscle contraction through initiation of transduction at the thromboxane A_2 receptor; this could be one of the many causes of small airway contraction in asthma and COPD [32, 33]. ROS-induced degradation of arachidonate-based phospholipids can also produce other bio-active molecules. They are pro-inflammatory affecting both monocytes and neutrophils causing increased endothelial cell interaction [34] as well as increased cytokine release [35, 36], and it is proposed that they may play an important role in various chronic inflammatory diseases [34].

ROLE OF ROS IN SIGNAL TRANSDUCTION

ROS have been implicated in the activation of transcription factors such as NF-κB and AP-1, and in the signal transduction and gene expression involved in cellular pro-inflammatory actions [37]. Both environmental and inflammatory cell-derived ROS can lead to the activation and phosphorylation of the mitogen-activated protein kinase (MAPK) family, including extracellular signal regulated kinase (ERK), c-Jun *N*-terminal kinase (JNK), p38 kinase, and P1-3K, via sensitive cysteine-rich domains; activation of the sphingomyelinase–ceramide pathway also occurs leading to increased gene transcription [3, 4, 37]. Activation of members of the MAPK family leads to the transactivation of transcription factors such as c-Jun, activating factor-2 (ATF2), cyclic AMP response element binding proteins (CREB), CREB-binding protein (CBP), and Elk-1 [37–40]. This eventually results in chromatin remodeling and expression of genes regulating a battery of distinct pro-inflammatory and antioxidant genes involved in several cellular events, including apoptosis, proliferation, transformation, and differentiation. The intracellular molecular mechanisms responsible for these actions of ROS have not been completely characterized.

Redox-sensitive molecular targets usually contain highly conserved cysteine residues; their oxidation, nitration, and the formation of disulfide links are crucial events in oxidant/redox signaling. It is hypothesized that oxidation of sulfide groups in signaling proteins causes structural modifications, resulting in the exposure of active sites and consequent protein activation. Such molecular targets include transcription factors (NF-κB, AP-1), signaling molecules such as ras/rac or JNK, protein tyrosine phosphatases, and $p21^{ras}$. Thiol molecules such as intracellular glutathione (GSH) and thioredoxin are of central importance in regulating such redox signaling pathways, by reducing disulfide bridges or oxidized cysteine residues [37, 41].

In response to tumor necrosis factor (TNF-α) and lipopolysaccharide (LPS), which are relevant stimuli for the inflammatory response in COPD, airway epithelial cells can concurrently produce increased amounts of intracellular ROS and RNS [18]. This intracellular production of oxidants and the subsequent changes in intracellular redox status are important in the molecular events controling the expression of genes for inflammatory mediators [3]. The signaling pathways and activation of transcription factors in response to ROS are the subject of rigorous investigation.

ROS IN ASTHMA AND COPD

In asthma, the airways are the major site of action, characterized by reversible airflow obstruction, airway hyperresponsiveness and hyperreactivity, chronic inflammation characterized by a predominant infusion of lungs with eosinophils, and the influx and activation of inflammatory cells such as macrophages, neutrophils, lymphocytes, and mast cells. Airway smooth muscle contraction, increased airway reactivity and secretions, increased vascular permeability, and increased generation of chemoattractants are the major features of an asthmatic response.

In COPD, the major site of attack is the alveolar spaces. Inflammatory cells become compartmentalized depending on the cell type. For example, neutrophils are mainly located in the airspaces of the lung, and macrophages accumulate in the tissue matrix in COPD. Progression of COPD has been strongly associated with the accumulation of inflammatory mucus exudates in the lumen of the small airways. In other phenotypes of COPD, oxidative stress occurs in the small airways, lung parenchyma, and alveolar regions. Hence the compartmentalization of different inflammatory cells in the lungs may help distinguish the oxidative stress response in asthma and COPD.

OXIDATIVE STRESS IN ASTHMA

It has been shown that inflammation driven by increased oxidative stress occurs in the airways of patients with asthma [8]. Recent evidence indicates that increased oxidative stress occurs in the airways of patients with asthma [8]. Inflammatory and immune cells in the airways, such as macrophages, neutrophils, and eosinophils, release increased amounts of ROS [42–45]. ROS can result in lung injury as a result of direct oxidative damage to epithelial cells and cell shedding [46, 47].

ROS have been shown to be associated with the pathogenesis of asthma by evoking bronchial hyperreactivity [48, 49]. Viral infections, ozone, and cigarette smoke, potential triggers for asthma, may serve as sources of ROS which enhance inflammation and asthmatic symptoms. Animal studies suggest that ROS may contribute to airway hyperresponsiveness by increasing vagal tone due to inhibition of β-adrenergic receptors and by decreasing mucociliary clearance [50, 51]. The actions of ROS can produce many of the pathophysiological features of asthma [51], including changes in the biochemical microenvironment with enhanced arachidonic acid release, increased synthesis of chemoattractants, glucocorticoid resistance, and impaired β-adrenergic responsiveness. These and other effects lead to the physiological changes of airway smooth muscle contraction, increased airway reactivity and secretions, and increased vascular permeability.

ROS-mediated injury to the airway epithelium produces hyperresponsiveness of human peripheral airways, suggesting that ROS may play a role in the pathogenesis of asthma [47]. Much of the evidence for this is indirect, since there are no specific and reliable methods to assess oxidative stress *in vivo.* Neutrophils isolated from peripheral blood of asthmatic patients generate greater amounts of $O_2^{\bullet-}$ and H_2O_2 than do cells from normal subjects, and their ability to produce $O_2^{\bullet-}$ is related directly to the degree of airway hyperresponsiveness to inhaled methacholine [52, 53]. Inflammatory cells obtained from asthmatic patients, particularly eosinophils derived from peripheral blood, produce increased amounts of ROS and RNS such as NO spontaneously and after stimulation *ex vivo* [42–46, 54]. This observation suggests that the inflammatory milieu in asthma contains factors that may prime ROS generation. Other studies have shown that impaired SOD activity is associated with airflow obstruction, airway hyperresposiveness, and remodeling [55, 56]. This inactivation of SOD in asthmatics, which has been reported in other studies [57, 58], was as a result of increased ROS causing tyrosine nitration of SOD.

Eosinophils are thought to play a critical role in the inflammation of asthma. They are present in large numbers in bronchoalveolar lavage (BAL) fluid and blood, and the number of cells correlates with bronchial hyperresponsiveness [59, 60]. BAL fluid eosinophils, alveolar macrophages, and neutrophils from asthmatic patients produce more ROS ($O_2^{\bullet-}$, H_2O_2, hypothalites) than do those from normal subjects [61, 62]. ROS cause direct contraction of airway smooth muscle preparations, and this effect is enhanced when the epithelium is injured or removed. This observation might provide a mechanistic link between epithelial injury arising from a variety of causes and airway hyperresponsiveness [47]. ROS also stimulate histamine release from mast cells and mucus secretion from airway epithelial cells [63].

ROS generation is thought to be a nonspecific process initiated by the concurrent action of numerous inflammatory mediators that have been shown to be present in increased amounts in asthmatics. Several mediators, including lipid mediators, chemokines, adhesion molecules, and eosinophil granule proteins (EPO), are potential stimuli or promoters of ROS production in the airways of asthmatic patients [64, 65].

Numerous surrogate markers of oxidative stress have been measured in exhaled air or breath condensate. The concentration of H_2O_2 in exhaled air condensate is increased in asthmatics [66], and it has been suggested that airway inflammation increases exhaled peroxides. The increased levels of exhaled H_2O_2 may be due to decreased dismutation of $O_2^{\bullet-}$ since SOD activity is reduced in lung cells of patients with asthma [57]. The significance of elevated levels of various ROS markers to the disease pathogenesis has not been studied.

The role of EPO-derived ROS

A specific role for EPO in the generation of oxidants by the phagocytes and in protein oxidation has been described [67]. Eosinophil activation (peroxidase – H_2O_2 system + halides) *in vivo* results in oxidative damage to proteins through bromination of tyrosine residues, as shown by the formation of 3-bromotyrosine in BAL fluid of patients with asthma [16]. The formation of 3-bromotyrosine is a specific response to the release of oxidants from eosinophils. Neutrophil- and monocyte-derived MPO, which are increased in smokers and patients with COPD, produce 3-chlorotyrosine [16]. Thus, distinction between these "footprints" might be useful in assessing the ROS burden in patients with asthma and COPD. A specific marker of protein modification by reactive brominating species 3-bromotyrosine has been shown to be markedly increased in BAL proteins obtained from asthmatics [16]. EPO-generated oxidants can interact with RNS present in asthmatics and promote protein nitration [15]. Thus, oxidative modification of critical biological targets in asthmatic airways may contribute to the pathophysiological features of asthma, such as epithelial cell damage, airway hyperreactivity, bronchoconstriction, β-adrenergic receptor dysfunction, mucus hypersecretion, microvascular leak, and airway edema [68, 69].

Interaction between ROS and RNS

The levels of nitric oxide are elevated in the exhaled air of patients with asthma [70–73]. Increased levels of exhaled NO, together with increased exhaled H_2O_2 in asthmatic patients, are associated with a pro-oxidant activity in airway walls resulting in lipid peroxidation and nitration of proteins [15, 16, 70, 74, 75]. The finding that NO reduces the potency of β-adrenergic signaling pathways may be an important deleterious effect of elevated RNS in asthma [50]. The presence of allergic inflammation, involving the recruitment and activation of eosinophils, may be a contributing factor in these pro-oxidant effects. Eosinophil granule proteins may participate in the formation of nitrating oxidants as well as unique molecules such as brominated products, which have been shown to be elevated in patients with asthma [15, 16]. However, the significance of these inflammatory ROS in the etiology of asthma is not established.

A reaction between NO and $O_2^{\bullet-}$ results in the formation of peroxynitrite anions ($ONOO^-$), a highly reactive oxidant species. $ONOO^-$ adds a nitro group to the 3-position adjacent to the hydroxyl group of tyrosine to produce the stable product nitrotyrosine. $ONOO^-$ induces hyperresponsiveness in airways of guinea pigs, inhibits pulmonary surfactant function, induces membrane lipid peroxidation, results in tyrosine/MAP kinase activation, and damages pulmonary epithelial cells [73, 75–78]. The levels of 3-nitrotyrosine are elevated in the exhaled breath of asthmatic patients [79]. Furthermore, there is strong immunoreactivity for nitrotyrosine in the airway epithelium, lung parenchyma, and inflammatory cells in the airways of patients with asthma [79, 80].

It has been postulated that increased levels of HOBr production as a result of EPO release from eosinophils result in increased peroxynitrite formation by interaction of HOBr with NO. NO itself can be stored in cells as s-nitrosothiols [81], and this may regulate cellular apoptosis through inactivation of caspases by s-nitrosylation of critical cysteine residues [82]. Consequently, increased $ONOO^-$ formation could deplete intracellular stores of NO, liberating active caspases that can then induce epithelial cell apoptosis. This is supported by the evidence showing that s-nitrosothiol levels in asthmatics are significantly depressed [83, 84]. However, there are other data, which show that levels of NO in exhaled breath condensate (EBC) are increased in asthmatics, that appear to conflict with this supposition [71]. This may simply reflect a compartmentalization effect in that different mechanisms are operating in an intracellular versus extracellular environment. The precise mechanistic role that SOD inactivation by $ONOO^-$ plays in asthma is as yet unclear. However, it is postulated that SOD inactivation by $ONOO^-$ could increase the overall redox state by allowing higher levels of H_2O_2 to persist thereby depleting intracellular NO stores through increased $ONOO^-$ formation and lowering s-nitrosothiol levels [85]. Overall, this clearly shows that oxidant stress occurs in asthma which can be reflected/detected in the lungs both systemically and locally.

Generation of ROS-mediated lipid peroxidation products

The measurement of aldehydes in exhaled breath has been proposed as a means to assess lipid peroxidation *in vivo*. The levels of lipid peroxidation products (8-isoprostanes), hydrocarbons (ethane and pentane), and nonspecific products of lipid peroxidation (thiobarbituric acid reactive substances, TBARS) are increased in EBC of patients with asthma [86–89]. The levels of 8-isoprostane are also increased in BAL fluid of patients with asthma [90]. Urinary excretion of 15-F_{2t}-isoprostane (8-isoprostaglandin$_{2\alpha}$, family of F_2-isoprostanes) was increased in mild atopic asthmatics following inhaled allergen provocation, whereas no increase in the urinary excretion of 15-F_{2t}-isoprostane was observed after inhalation of methacholine [91]. The presence of various markers of oxidative stress provide evidence that oxidative stress is present in the airspaces in asthma, but do not provide definitive evidence for a role for oxidative stress in the pathogenesis of asthma.

Measurement of systemic or exhaled isoprostane 8-iso-$PGF_{2\alpha}$ levels may provide a useful tool in monitoring clinical status in asthma [92]. The level of plasma F_2-isoprostanes (an 8-iso$PGF_{2\alpha}$ isomer) is significantly increased in asthmatics and is related to disease severity [92]. Similarly, arachidonic acid is increased in airway inflammatory cells of patients with bronchial asthma [93]. Arachidonic acid may undergo oxidation to produce an end-product of lipid peroxidation such as 4-HNE [24, 25]. The levels of plasma lipid peroxides TBARS have been shown to be elevated in asthma, and are negatively correlated with the Forced Expiratory Volume in one sec (FEV_1), suggesting a role for increased oxidative stress in the pathogenesis of the airway obstruction of asthma [94, 95]. Further studies are needed to define the source, significance, and specificity of these peroxidation products.

OXIDATIVE STRESS IN COPD

More than 90% of patients with COPD are smokers, but not all smokers develop COPD [96–98]. However, 15–20% of cigarette smokers show a rapid decline in FEV_1 over time and develop COPD. An increased oxidant burden in smokers derives from the fact that cigarette smoke contains an estimated 10^{14} oxidants per puff, and many of these are relatively long-lived – such as tar-semiquinone that can generate $^{\bullet}OH$ and H_2O_2 by the Fenton reaction [20–23]. Other factors that may exacerbate COPD, such as air pollutants, infections, and occupational dusts, also have the potential to produce oxidative stress [5, 99]. These include oxidative stress, which has important consequences on lung physiology and for the pathogenesis of COPD including increased sequestration of neutrophils in the pulmonary microvasculature, increased oxidative inactivation of antiproteases, mucus hypersecretion, alteration in mitochondrial respiration, alveolar epithelial injury or permeability, breakdown and remodeling of extracellular matrix, apoptosis of epithelial and endothelial cells, skeletal muscle dysfunction fatigue and exercise limitation, and glucocorticoid resistance.

ROS in the alveolar space

The oxidant burden in the lungs is enhanced in smokers by the release of ROS from macrophages and neutrophils [5]. Oxidants present in cigarette smoke can stimulate alveolar macrophages to produce ROS and to release a host of mediators, some of which attract neutrophils and other inflammatory cells into the lungs. Both neutrophils and macrophages, which are known to migrate in increased numbers into the lungs of cigarette smokers compared with nonsmokers [5], can generate ROS via the NADPH oxidase system. Moreover, the lungs of smokers with airway obstruction have more neutrophils than smokers without airway obstruction [100]. Circulating neutrophils from cigarette smokers and patients with exacerbations of COPD release more $O_2^{\bullet -}$ [95]. Cigarette smoking is associated with increased content of MPO in neutrophils, which correlates with the degree of pulmonary dysfunction [101, 102]. MPO activity is negatively correlated with FEV_1 in patients with COPD, suggesting that neutrophil MPO-mediated oxidative stress may play a role in the pathogenesis of the airway obstruction in COPD [103].

Alveolar macrophages obtained by BAL fluid from the lungs of smokers release increased amounts of ROS such as $O_2^{\bullet -}$ and H_2O_2 in response to stimulation [5, 104, 105]. Exposure to cigarette smoke *in vitro* has also been shown to increase the oxidative metabolism of alveolar macrophages [106]. Subpopulations of alveolar macrophages with a higher granular density appear to be more prevalent in the lungs of smokers and are responsible for the increased $O_2^{\bullet -}$ production of smoker's macrophages [106, 107].

Hydrogen peroxide, measured in exhaled breath, is thought to be a direct measurement of oxidant burden in the airspaces. Smokers and patients with COPD have higher levels of exhaled H_2O_2 than nonsmokers [108–110], and levels are even higher during exacerbations of COPD [110]. The source of the increased H_2O_2 is unknown but may in part derive from increased release of $O_2^{\bullet -}$ from alveolar macrophages in smokers [110], However, in one study smoking did not appear to influence the levels of exhaled H_2O_2 [108]; the levels of exhaled H_2O_2 in this study correlated with the degree of airflow obstruction as measured by the FEV_1. However, the variability of the measurement of exhaled H_2O_2, along with the presence of other confounding factors, such as cigarette smoking and caffeine intake, has led to concerns over the reproducibility of the use of exhaled H_2O_2 as a marker for oxidative stress in smokers and in patients with COPD. The generation of ROS in ELF may be further enhanced by the presence of increased amounts of free iron in the airspaces in smokers [111, 112]. This is relevant to COPD since the intracellular iron content of alveolar macrophages is increased in cigarette smokers and is increased further in those who develop chronic bronchitis, compared with nonsmokers [113]. In addition, macrophages obtained from smokers release more free iron *in vitro* than those from nonsmokers [114].

In some studies, in both stable bronchitis [12] and mild exacerbations [10], eosinophils have been shown to be prominent in the airways. BAL fluid from patients with COPD has also been shown to contain increased eosinophilic cationic protein [102]. Furthermore, peripheral blood eosinophilia is also considered to be a risk factor for the development of airway obstruction in patients with chronic bronchitis and is an adverse prognostic sign [115, 116]. However, despite the presence of an increased number of eosinophils, EPO-mediated generation of 3-bromotyrosine has not been detected in COPD patients [16]. This does not provide support for a role of brominating oxidant in eosinophil-mediated ROS damage in COPD.

Superoxide anion and H_2O_2 can be generated by the xanthine/XO reaction. XO activity has been shown to be increased in cell-free BAL fluid and plasma from COPD patients, compared with normal subjects; and this has been associated with increased $O_2^{\bullet -}$ and lipid peroxide levels [117–119].

ROS in blood

The neutrophil appears to be a critical cell in the pathogenesis of COPD [120]. Previous epidemiological studies have shown a relationship between circulating neutrophil numbers and FEV_1 [121, 122]. A relationship has also been shown between the changes in peripheral blood neutrophil count and in airflow limitation over time [122]. Similarly, a correlation between $O_2^{\bullet -}$ release by peripheral blood neutrophils and bronchial hyperreactivity in patients with COPD has been shown, suggesting a role for systemic ROS in the pathogenesis of the airway abnormalities in COPD [123]. Another study has shown a relationship between peripheral blood neutrophil luminol-enhanced chemiluminescence, as a measure of the release of ROS, and measurements of airflow limitation in young cigarette smokers [124].

Various studies have demonstrated increased production of $O_2^{\bullet -}$ from peripheral blood neutrophils obtained from patients during acute exacerbations of COPD; these levels returned to normal when the patients were clinically stable [95, 125, 126]. Other studies have shown that circulating neutrophils from patients with COPD show upregulation of their surface adhesion molecules, which may also be an oxidant-mediated effect [95, 127]. Activation may be even more pronounced in neutrophils that are sequestered in the pulmonary microcirculation in smokers and in patients with COPD, since neutrophils sequestered in the pulmonary microcirculation in animal models of lung inflammation release more ROS than circulating neutrophils [128]. Thus, neutrophils sequestered in the pulmonary microcirculation may be a source of ROS and may have a role in inducing endothelial adhesion molecule expression in COPD.

Interaction between ROS and RNS

Nitric oxide has been used as a marker of airway inflammation and indirectly as a measure of oxidative stress. There have been reports of increased levels of NO in exhaled breath in patients with COPD, but not so high as the levels reported in asthmatics [129–131]. One study failed to confirm this result [132]. Smoking cessation increases NO levels in exhaled air [133], and the reaction of NO with $O_2^{\bullet -}$ limits the usefulness of this marker in COPD, except perhaps to differentiate from patients with COPD from those with asthma (Table 25.1).

Cigarette smoking increases the formation of RNS and results in nitration and oxidation of plasma proteins. The levels of nitrated proteins (fibrinogen, transferrin, plasminogen, and ceruloplasmin) were higher in smokers than in nonsmokers [142]. Evidence of enhanced NO and

TABLE 25.1 ROS markers in asthma and COPD.

	References	
Biochemical marker	**Asthma**	**COPD**
Elevated hydrogen peroxide level in exhaled breath	[66, 75]	[108–110]
Release of ROS from peripheral blood neutrophils, eosinophils, and macrophages	[42–46, 54, 61]	[95, 121, 123, 126]
Increased release of ROS from alveolar macrophages, eosinophils, and neutrophils	[60, 62]	[104]
Increased MPO and EPO levels	[15]	[101–103]
Increased BAL fluid xanthine/XO activity	–	[117–119]
Elevated plasma and exhaled F_2-isoprostane, ethane, and pentane levels	[86–88, 90, 92]	[134–136]
Elevated plasma and exhaled lipid peroxide (TBARS) levels	[89, 94, 95]	[95, 112, 137, 138]
Increased formation of 4-hydroxy-2-nonenal-protein adducts in lungs	–	[139]
Elevated plasma protein carbonyls	–	[140]
Increased exhaled CO	[141]	[141]
Increased exhaled NO	[70, 71]	[129–132]
Increased 3-nitrotyrosine in plasma, BAL fluid, and exhaled breath	[74, 79, 80]	[142–144]
Increased 3-chlorotyrosine and 3-bromotyrosine	[15, 16]	[16]

$ONOO^-$ activity in plasma has been shown in cigarette smokers [143]. *In vitro*, exposure to gas-phase cigarette smoke results in increased lipid peroxidation and protein carbonyl formation in plasma [140]. It is likely that alpha, beta-unsaturated aldehydes (acrolein, acetaldehyde, and crotonaldehyde) that are abundantly present in cigarette smoke may react with protein-SH and -NH_2 groups leading to the formation of a protein-bound aldehyde functional group and are capable of converting tyrosine to 3-nitrotyrosine and dityrosine [140]. Nitric oxide and $ONOO^-$-mediated formation of 3-nitrotyrosine in plasma and free catalytic iron (Fe^{2+}) levels in ELF are elevated in chronic smokers [144]. Nitration of tyrosine residues or proteins in plasma leads to the production of 3-nitrotyrosine [145]. The levels of nitrotyrosine and inducible NO synthase (iNOS) were higher in airway inflammatory cells obtained by induced sputum from patients with COPD, compared to those with asthma [144]. The levels of nitrotyrosine were negatively correlated with the percentage predicted FEV_1 in patients with COPD. These direct and indirect studies indicate that increased RNS- and ROS-mediated protein nitration and lipid peroxidation respectively may play a role in the pathogenesis of COPD.

Generation of ROS-mediated lipid peroxidation products: Carbonyl stress

Formation of protein carbonyls (aldehyde protein adducts) in response to cigarette smoke-derived lipid peroxides (aldehydes) has been implicated in pathogenesis of COPD. The levels of lipid peroxides, such as 8-isoprostane, and hydrocarbons, such as ethane and pentane, are increased in exhaled air condensate in smokers and in patients with COPD [134–136]. Furthermore, the levels of these lipid peroxidation products have been correlated with airway obstruction [136]. Urinary levels of isoprostane $F_{2\alpha}$-III have been shown to be elevated in patients with COPD compared with control subjects and are even more elevated during exacerbations of COPD [30]. These studies indicate that there is increased lipid peroxidation in patients with COPD. However, it is not known whether the increased level of lipid peroxidation products found in these diseases is the result of primary lung-associated processes such as alveolar macrophage activation, neutrophil activity, or the ongoing lipid peroxidative chain reaction in the alveoli, parenchyma, or airways, which are induced by inhaled oxidants/cigarette smoke [141].

Indirect and nonspecific measurements of lipid peroxidation products, such as TBARS, have also been shown to be elevated in breath condensate and in lungs of patients with stable COPD [112, 137, 138]. The levels of plasma lipid peroxides have been shown to be elevated in COPD and negatively correlated with the FEV_1 [94]. Oxidative stress, measured as lipid peroxidation products in plasma, has also been shown to correlate inversely with the percentage predicted FEV_1 in a population study [146], suggesting that in patients with COPD lipid peroxidation may play a role in the progression of the disease.

4-HNE is a highly reactive and specific diffusible end-product of lipid peroxidation. Increased 4-HNE-modified protein levels (protein carbonyls) are present in airway and alveolar epithelial cells and endothelial cells, and in neutrophils in smokers with airway obstruction compared to subjects without airway obstruction [139]. This demonstrates not only the presence of 4-HNE but that 4-HNE modifies proteins in lung cells to a greater extent in patients with COPD. The increased level of 4-HNE adducts in alveolar epithelium, airway endothelium, and neutrophils was inversely correlated with FEV_1, suggesting a role for 4-HNE in the pathogenesis of COPD. The formation of 4-HNE may have detrimental effect on a variety of signaling molecules, such as Nrf2, HDAC2, and IκB kinase (NF-κB activation). Hence, the carbonyl stress that predominantly occurs in response to cigarette smoke may be a differentiating factor in distinguishing the oxidative stress response in the airways of asthma and COPD.

ALTERATION IN ENDOGENOUS ANTIOXIDANT DEFENSES WITHIN THE LUNG

To combat and neutralize the deleterious effects of ROS, various endogenous antioxidant strategies have evolved which employ both enzymatic and nonenzymatic mechanisms. Within the lung lining fluid, several nonenzymatic antioxidant species exist which include, glutathione, ascorbic acid (vitamin C), uric acid, α-tocopherol (vitamin E), and albumin. The relative abundance of these antioxidants can differ from that observed in blood plasma [147]. Nevertheless, ROS-induced lung injury can increase lung epithelial permeability [148] allowing leakage of plasma constituents, which will also contain antioxidants, into the lung lining fluid providing additional antioxidant protection. Similarly, enzymatic antioxidant defenses can also differ in both anatomical and subcellular localization based on expression. This includes enzymes such as SOD, catalase, thioredoxin, glutathione peroxidase (GPx), and glutathione-S-transferase. Of these, enzymes such as SOD can be found as different isoforms expressed either intracellularly or extracellularly. Moreover, extracellular-SOD is highly expressed in the lungs mainly around blood vessels and airways [149].

Table 25.2 highlights the level of different nonenzymatic antioxidants found in the ELF compared to that in plasma. While differences are clearly evident, both enzymatic and nonenzymatic antioxidant strategies are employed within the lung.

TABLE 25.2 Antioxidant constituents of plasma and lung epithelial lining fluid (ELF).

Antioxidant	Plasma (μM)	ELF (μM)
Ascorbic acid	40	100
Glutathione	1.5	100
Uric acid	300	90
Albumin-SH	500	70
α-tocopherol	25	2.5
β-carotene	0.4	–

Depletion of antioxidant enzymes in asthma and COPD

An important effect of oxidative stress and inflammation is the upregulation of protective antioxidant genes. Amongst antioxidants, glutathione (GSH), and its redox enzymes have an important protective role in the airspaces and intracellularly in lung epithelial cells. Indeed, GSH levels are increased in the ELF of both asthmatics and chronic cigarette smokers [150, 151]. GSH is a tripeptide (L-γ-glutamyl-L-cysteinyl-glycine) that contains a thiol group. It functions as an antioxidant by acting as a sacrificial target for ROS and other products of lipid peroxidation, such as reactive carbonyls. In so doing, GSH becomes oxidized to its dimeric form (glutathione disulfide, GSSG) or forms adducts with reactive carbonyls and other reactive xenobiotics (GS-X). Furthermore, enzymes such as GPx and glutathione transferase can facilitate this process. GSSG can itself be reduced back to GSH by glutathione reductase using NADPH generated from the pentose phosphate pathway. Oxidative stress causes upregulation of glutamate cysteine ligase, GCL (formerly known as γ-glutamylcysteine synthetase) [3, 152], an important enzyme involved in the synthesis of GSH, as an adaptive mechanism against subsequent oxidative stress. We have shown the increased expression of GCL mRNA in lungs of smokers which is even more pronounced in smokers with COPD [153]. This implies that GSH synthesis is upregulated in lungs of smokers with and without COPD. Similarly, bronchial epithelial cells of rats exposed to cigarette smoke have shown increased expression of the antioxidant genes such as manganese superoxide dismutase (MnSOD), metallothionein, and GPx [154]. This would suggest the importance of an adaptive antioxidant gene response against the injurious effects of cigarette smoke [154]. However, Harju and colleagues have found that GCL immunoreactivity was decreased (possibly leading to decreased GSH levels) in the airways of smokers compared to nonsmokers, suggesting that cigarette smoke predisposes lung cells to ongoing oxidant stress [155]. In addition, Neurohr and colleagues showed that decreased GSH levels in BAL fluid cells of chronic smokers were associated with a decreased expression of GCL-light subunit without a change in GCL-heavy subunit expression [156].

Important protective antioxidant genes such as MnSOD, GCL, heme oxygenase-1 (HO-1), GPx, thioredoxin reductase, and metallothionein are similarly induced by various oxidative stresses including hyperoxia and inflammatory mediators, such as TNF-α and LPS in lung cells [37–40]. Indeed, the transcription factor, Nrf2, is redox sensitive (contains –SH groups) and binds to the antioxidant response element (ARE) within DNA regulating a variety of antioxidant genes. The importance of Nrf2 can be gauged from a report by Rangasamy *et al.*, where they have shown that disruption of the Nrf2 gene in mice lead to an early and a more intense emphysema in response to cigarette smoke compared to mice with an intact Nrf2 gene [157]. In the same study they have shown that the expression of nearly 50 antioxidant and cytoprotective genes in the lungs may be transcriptionally controlled by Nrf2 and all genes may work in concert to overcome the effects of cigarette smoke. Moreover, studies examining polymorphisms in these protective antioxidant genes, such as glutathione transferase and HO-1, have suggested a link to onset of COPD and emphysema [158, 159]. Interestingly, Rangasamy and colleagues have recently showed that disruption of Nrf2 also leads to susceptibility to severe allergen-induced asthma in mice. These findings indicate that though Nrf2 is important for the control of antioxidant genes, it is not specific to one inflammatory disease [160]. Recent studies showed depletion of various Nrf-2 dependent Phase II antioxidant genes in lungs of patients with COPD [161]. The compartmentalization and localization of various antioxidant enzymes and their functions/alterations in asthma and COPD are described in Table 25.3.

Depletion of antioxidant small molecules in asthma and COPD

Apart from GSH, several other antioxidants, such as ascorbic acid (vitamin C), uric acid, α-tocopherol (vitamin E), and albumin, are present in the lung. Of these, the major antioxidants in lung lining fluid are GSH, ascorbic acid, and uric acid [147]. Like GSH, uric acid can also be found intracellularly, although at lower concentrations than that seen in plasma. It is a powerful scavenger of both ROS and RNS and can protect proteins against nitration [162] particularly from the nitrogen dioxide radical, nitrous oxide, found in cigarette smoke [163]. Oxidation of uric acid by ROS or RNS results in the formation of allantoin that can then be measured in various body fluids as a marker of oxidative stress [164]. The antioxidant properties of albumin, as well as mucins, come from the presence of exposed −SH groups. Their protective effects are solely extracellular. However, albumin exists in a high concentration and acts as a sacrificial substrate able to react quickly with the very damaging and potent oxidizing $ONOO^-$ and HOCl [165].

The remaining two protective antioxidant molecules, ascorbic acid and α-tocopherol, are derived from the diet. Both antioxidants are decreased in chronic cigarette smokers [5]. Furthermore, the total antioxidant capacity of plasma is decreased in smokers, and in patients with asthma and COPD [95]. It is interesting to note that the decrease in antioxidant capacity in smokers occurs transiently during smoking and resolves rapidly after smoking cessation. Depletion of total antioxidant capacity in smokers is associated with decreased levels of major plasma antioxidants (e.g. ascorbic acid, vitamin E, β-carotene, and selenium). The depletion of antioxidants may thus be a reflection of ongoing oxidative stress due to underlying inflammation in these diseases. There is, however, a limitation of oxidative stress and antioxidant biomarkers at present due to the lack of longitudinal studies, correlation with disease severity or outcome, and the variations observed in the control subjects.

Ascorbic acid, otherwise known as vitamin C, is water soluble. It functions as an antioxidant by accepting free radical electrons, thus forming the ascorbyl radical which is relatively unreactive. However, the ascorbyl radical will undergo a disproportionation reaction to regenerate ascorbate and dehydroascorbate, the latter rapidly breaking down

TABLE 25.3 Antioxidant enzymes of the lungs, their localization, and functions in asthma and COPD.

Enzyme	Lung localization	Function	Expression/Activity
Cu, ZnSOD	Bronchial, alveolar epithelium, macrophages, fibroblasts, pneumocytes	Scavenges $O_2^{\bullet -}$	Decreased in COPD
EC-SOD	Bronchial epithelium, macrophages, neutrophils, vascular walls, pneumocytes	Scavenges $O_2^{\bullet -}$	Decreased in COPD
MnSOD	Bronchial epithelium, macrophages, neutrophils, vascular walls, pneumocytes	Scavenges $O_2^{\bullet -}$	Decreased in Asthma Decreased in Smokers
Catalase	macrophages, fibroblasts, pneumocytes	Hydrogen peroxide to water	Decreased in Asthma
Glutathione peroxidase	ELF cells, epithelium, macrophages, other lung cells	Organic hydroperoxides to organic hydroxides	Decreased in COPD Decreased in Asthma
Heme oxygenase-1	Alveolar, bronchial epithelium, macrophages, inflammatory cells of lungs	Heme to CO and biliverdin	Decreased in Emphysema Decreased in COPD
Thioredoxin	bronchial epithelium, macrophages	Transcriptional modulation, thiol–dithiol exchange, Prot-S-S-Prot to Prot-SH	Decreased in COPD
GCL (catalytic subunit)	Alveolar, bronchial epithelium, macrophages	Synthesis of glutathione (first rate limiting step)	Decreased in COPD Decreased in Smokers

to form oxalic acid and L-threonine [166]. Ascorbic acid has been shown to possess a whole array of antioxidant properties *in vitro*, from scavenging ROS and RNS, preventing lipid peroxidation to regenerating other antioxidants, such as uric acid and α-tocopherol. Like uric acid, ascorbic acid is also a powerful scavenger of nitrous oxide which is a potent promoter of both protein nitration and lipid peroxidation [167, 168]. Indeed, increases in F_2-isoprostane lipid peroxidation products are decreased in smokers who consume more ascorbate [169]. Further evidence of asorbate's antioxidant properties *in vivo* is provided by the evidence highlighting its depletion in clinical conditions associated with a high oxidative stress burden [170]. Intriguingly, there is also a dark side to ascorbic acid in that *in vitro* it can reduce Fe^{3+} to Fe^{2+} thereby facilitating $OH^{\bullet}$ radical formation through Fenton chemistry. However, the relevance of this *in vivo* is still a subject of investigation.

IS THERE A CO-RELATIONSHIP BETWEEN OXIDATIVE STRESS, ANTIOXIDANT DEPLETION AND DECLINE IN LUNG FUNCTION IN ASTHMA AND COPD?

Emerging evidence suggests that oxidative stress is directly correlated with the decline in lung function in asthma and COPD. Epidemiological studies have shown a relationship between circulating increase in neutrophil numbers and decline in FEV_1. Oxidative stress, measured as lipid peroxidation products in plasma, has also been shown to correlate inversely with the FEV_1 percentage in a population study [171].

There is an association between dietary intake of antioxidant vitamins and polyphenols and lung function in the general population as well as in obstructive airway disease, supporting the hypothesis that oxidative stress may have a role in decline in lung function, and chronic cough and breathlessness (but not chronic phlegm) in COPD [146, 172–174]. This is substantiated by a study highlighting the beneficial protective effect of fruit containing polyphenols and vitamin E intake against COPD symptoms in 20-year COPD mortality from three European countries consisting of Finnish, Italian, and Dutch cohorts [175]. It has been shown that there is a correlation between increased dietary antioxidant intake and improved lung function [146]. Moreover, increased dietary antioxidant status, particularly vitamin E (α-tocopherol), correlated with lower levels of lipid peroxidation [176]. α-Tocopherol is a lipid soluble antioxidant and as such is probably one of the most important scavengers and inhibitors of lipid peroxidation. It largely achieves this by reacting faster with the lipid peroxyl radicals than these radicals can react with other lipid molecules. Similarly, another study reported that increased dietary intake of ascorbic acid led to an improvement in lung function in both smokers and asthmatics [177]. These studies support the concept that oxidative stress plays an important role in decline in lung function in smokers, and possibly in asthma patients with COPD.

THERAPEUTIC INTERVENTION WITH ANTIOXIDANTS IN ASTHMA AND COPD

In view of the evidence implicating oxidative stress in the pathogenesis of chronic airway diseases, one rational approach would be to consider antioxidant intervention to neutralize the increased oxidative stress, and the subsequent inflammatory response. Several small molecular weight compounds that target oxidant signaling, or quench oxidants/aldehydes/carbonyls derived from cigarette smoke or boost antioxidant potential in the lung are currently being tested clinically [178]. Antioxidant agents such as thiol molecules (GSH/cysteine and mucolytic drugs, such as *N*-acetyl-L-cysteine (NAC) and nacystelyn (NAL)), dietary polyphenols (curcumin, resveratrol, green tea, and quercetin), erdosteine, fudosteine, and carbocysteine lysine salt, all have been reported to increase intracellular thiol antioxidant levels and induction of GSH biosynthesis genes. The various antioxidant strategies used have been divided into six categories; dietary, thiols, spin traps, enzyme mimetics, and polyphenols as shown in Table 25.4.

Dietary

Reports of clinical benefit in asthma and COPD for increased vitamin C, E, and other dietary antioxidants have been varied. Nevertheless, epidemiological studies have shown that there are decreased vitamin C, E, β-carotene, and selenium levels in cigarette smokers [179–181]. Similarly, other studies have reported a link with decreased lung function [182] and the presence and severity of asthma to dietary antioxidant intake, such as vitamins C and E, β-carotene and selenium [183, 184]. One small trial has shown that there was a clinical improvement in asthmatic symptoms after being given selenium supplementation compared to placebo [185]. In addition, one very early study incorporating selenium supplementation in smokers demonstrated that this resulted in reduced superoxide release from leukocytes [186]. Moreover, two studies by the same group have shown no clinical benefit of supplemental intake of vitamin C or E when compared to current standard therapies for mild-to-moderate asthma [187, 188]. This is in contrast with earlier studies, showing decreased lipid peroxidation and a corresponding improvement in lung function with increased vitamin C intake in smokers [177, 189]. Supplementation with either vitamin E or β-carotene alone resulted in no clinical benefit in COPD [190]. Moreover, β-carotene supplementation may have a detrimental side

TABLE 25.4 Antioxidant therapeutic interventions in asthma and COPD.

Antioxidant compounds
Thiol compounds-*N*-acetyl-L-cysteine (NAC), nacystelyn (NAL), glutathione esters, thioredoxin, Procysteine, erdosteine, fudosteine, carbocysteine, *N*-isobutyrylcysteine, ebselen
Antioxidant vitamins (vitamin A, E, C), β-carotene, CoQ_{10}
Polyphenols (curcumin, resveratrol, quercetin, and green tea-catechins)
Nitrone spin traps
Enzyme mimetics: SOD and GPx mimetics, porphyrins

effect in that it may accelerate the onset of lung cancer in cigarette smokers [189]. Hence, dietary antioxidants intake/supplementation may not be so effective as therapeutic option to control the declining lung function in airway diseases.

Thiols

In the lung, thiols in the form of GSH constitute one of the main antioxidant defenses in both intracellular and extracellular compartments. Attempts to raise GSH levels in the lung through administration of GSH itself have been tried [191]. Unfortunately, aerosolization of GSH into the lung was characterized by a poor half life [192] and led to induction of bronchial hyperreactivity [193]. Intracellular antioxidant protection after GSH oral administration was also limited, due to its inefficient cellular uptake [194]. Alternative methods of raising GSH levels have involved supplementing the GSH precursor cysteine, through the use of NAC. This approach has met with varying success in the past [195]. A Cochrane systematic review and other systematic reviews demonstrated that NAC treatment was associated with a significant reduction of 0.79 exacerbations/patient/year compared with placebo, and a 29% decrease in COPD patients [196–200]. Similarly, randomized, double-blind, placebo-controlled Phase II trials of a 6-12 month oral dose (of 600 mg, twice a day) showed reduction in various plasma and BAL fluid oxidative biomarkers in smokers [201, 202]. Similarly, two further studies showed a clear benefit of NAC in reducing oxidant burden [203, 204]. While there is a body of evidence that the administration of NAC provides a potential benefit for COPD patients, this evidence is not clinically directive. For example, in a recently conducted Phase III multicenter Bronchitis Randomized on NAC Cost-Utility Study (BRONCUS) [205], NAC (600 mg oral, daily) was ineffective in halting the decline in lung function, but did lead to a reduction in hyperinflation in patients with severe COPD and decreased the exacerbation rate in patients who were not treated with inhaled glucocorticoids [205] (Table 25.5). The variability in all the current studies using NAC at 600 mg, oral, daily may simply reflect the fact that the dose was not high enough. Indeed the pharmacokinetics support this explanation. Given orally, 600 mg of NAC is rapidly absorbed by the gut but has a bioavailability of only 10% with a plasma half life of 6.3 h; this led to a transient increase in lung GSH levels after daily dosing for 2 weeks [206, 207]. Though oral intake of NAC (600 mg daily) led to decreased markers of oxidative stress in smokers and patients with COPD, perhaps higher doses of NAC (1200–1800 mg/day) are needed to observe any clinical benefit on lung function.

An alternative to NAC is the lysine salt of NAC, nacystelyn (NAL). It has been used as a mucolytic for cystic fibrosis. However, it also possesses antioxidant properties and can reduce both ROS levels and ROS-mediated inflammatory events *in vitro* [213] and *in vivo* [214]. NAL has several advantages over NAC, firstly it can enhance GSH levels twice as effectively as NAC and secondly it forms a neutral pH when in solution, unlike NAC which is acidic [215]. This has meant that when delivered directly into the lung by aerosol in healthy volunteers, it did not cause any irritation or other side effects [216]. Therefore NAL may be more promising than NAC in reducing the oxidant burden in the lung in chronic airways disease.

The potential of other thiols as antioxidants are also being explored. These include erdosteine, fudosteine, and carbocysteine, which have mucoactive properties as well as reduce bacterial adhesiveness. In the "Equalife" randomized placebo controlled trial, erdosteine was dosed orally 300 mg b.i.d. for a period of 8 months [217]. Patients receiving erdosteine had significantly fewer exacerbations and spent few days in hospital than the placebo group. Moreover, patients receiving erdosteine showed no reduction in lung function over this period and showed a significant improvement in health-related quality of life. Similarly, fudosteine has shown to possess promising potential in treatment of patients with COPD. A clinical trial on the combination of steroids and erdosteine/fudosteine/carbosyteine in patients with COPD is needed. A recent study has shown that long-term use of carbocysteine reduced the rate of exacerbations in patients with COPD, but there was no difference in exacerbation rate between the carbocysteine group and placebo group at an early treatment (3 months) [218]. However, 1-year treatment of carbocysteine (1,500 mg/day) was effective in COPD patients in terms of reductions in numbers of exacerbations and improvements in quality of life. Hence, it was suggested that longer use of carbocisteine is more effective for preventing exacerbations of COPD.

Spin traps

Spin traps are based around a nitrone or nitroxide containing molecule such as isoindole-based nitrones [219] and azulenyl-based nitrones [220]. One such azulenyl nitrone, STANZ, may prove promising for *in vivo* use as it exhibits very potent antioxidant activity. Surprisingly, no studies have been performed in asthma and COPD, looking at the impact of spin traps on clinical endpoints, such as FEV_1. A phenyl-based nitrone spin trap developed by AstraZeneca, NXY-059, is due to enter Phase III clinical trials for use in acute ischemic stroke. The utility of this compound in SAINTII trial was unsuccessul though the drugh was safe. Nevertheless, it remains to be seen whether such compounds could be developed for more long-term use in asthma and COPD.

Enzyme mimetics

Enzyme mimetics are generally small compounds that possess catalytic activity that mimics the activity of larger enzyme-based molecules. In the case of antioxidants, this includes the SOD, catalase, and GPx-like activities. A number of SOD mimetics based around organo–manganese complexes have been developed which retain their antioxidant properties *in vivo*. These include a series of manganese-based macrocyclic ligands, such as M40401, M40403, and M40419 [221–223]. The second class of SOD mimetics are the manganese–metalloporphyrin based compounds as exemplified by AEOL-10113 and AEOL-10150 [224, 225]. The third class of manganese-based SOD mimetic are the "Salens." These are generally aromatic substituted ethylene–diamine metal complexes [226], such as EUK134 [227, 228]. The "Salens" also possess some catalytic activity and can therefore

TABLE 25.5 Clinical trials conducted for the efficacy of antioxidants in asthma and COPD.

Trial name	Antioxidant used	Aim of study	Disease/ Condition	Outcome	References
BRONCUS	*N*-acetyl-L-cysteine	Effect of NAC on FEV_1	COPD	30% reduction in COPD hospitalization obtained without change on decline in FEV_1.	[205]
Systematic Cochrane review of 23 randomized controlled trails	*N*-acetyl-L-cysteine (2 months of oral NAC therapy)	Effect of NAC and antibiotics on number of days of disability	COPD	Significant reduction in days of disability (0.65 day/patient/ month) and 29% reduction in exacerbations. No difference in lung function	[197, 198]
A Cochrane systematic review of randomized, controlled trials of 11 of 39 retrieved trials	*N*-acetyl-L-cysteine	Used a validated score to evaluate the quality of each study	COPD	9 trials showed prevention of exacerbation and 5 of which addressed improvement of symptoms compared with 34.6% of patients receiving placebo	[199]
A meta-analysis of published trials	*N*-acetyl-L-cysteine	To assess the possible prophylactic benefit of prolonged treatment	COPD	23% decrease in number of acute exacerbations	[200]
	N-acetyl-L-cysteine (600 mg once a day for 12 months)	Effect of NAC on H_2O_2 and TBARS in EBC	COPD	No change in TBARS reduce H_2O_2 levels	[202–204]
	β-carotene (20 mg/day) and α-tocopherol (50 mg/day)	Effect on the symptoms (chronic cough, phlegm, or dyspnea)	COPD	No benefit on symptoms	[208]
	β-carotene (20 mg daily for 4 weeks)	Effect on lipid peroxides Levels in exhaled breath	Smokers	Reduce lipid peroxidation (pentane levels) in exhaled breath	[190, 209, 210]
The MORGEN study	Diet rich in polyphenols/ bioflavonoids (catechin, flavonol, and flavone) (58 mg/day)	Effect on FEV_1, chronic cough, breathlessness, and chronic phlegm	COPD	Positively associated with decline in FEV_1 and inversely associated with chronic cough and breathlessness, but not chronic phlegm	[174]
European countries (Finnish, Italian, and Dutch cohorts)	Diet rich in fruits, vegetables and fish intake	Effect on 20-year COPD mortality	COPD	A 24% lower COPD mortality risk	[175]
	Vitamin C (500 mg daily for 4 weeks)	Effect on lipid peroxides Levels in breath and plasma	Smokers	No change in lung function No change in lipid peroxidation (breath pentane and plasma MDA levels)	[209]
	Vitamin C (600 mg) Vitamin E (400 IU) β-carotene (30 mg)	Effect on lipid peroxidation	Smokers	Reduced lipid peroxidation	[211, 212]
	Vitamin E (400 IU b.i.d. For 3 weeks)	Effect on breath ethane levels	Smokers	No effect on breath ethane levels	[208]
	Vitamin C and E, β-carotene, selenium	Effect on lung function and oxidative stress biomarkers	Asthmatics	50% reduction in asthma prevalence and clinical improvement whereas other studies showed no clinical improvement	[185–188]
Peace study	Carbocysteine (Carbocisteine)	Effect on rate of exacerbations in COPD	COPD	Long-term (one year) use of carbocysteine (1,500 mg/day) produced reduction in numbers of exacerbations in patients with COPD	[218]

BONCUS: Bronchitis randomized on NAC cost utility study.

also scavenge hydrogen peroxide, a product of SOD activity. Moreover, they are also able to decompose peroxynitrite [229].

Within the various classes of SOD mimetic, only the metalloporphyrin-based compounds AEOL-10150 and AEOL-10113 (iron-containing porphyrin complexes that decomposes $ONOO^-$ into innocuous nitrate) have been studied in models of airway inflammation [230]. In one study, AEOL-10113 was shown to inhibit both airway inflammation and bronchial hyperreactivity in an ovalbumin challenge model of airway inflammation [224]. This highlighted an important therapeutic benefit in asthma. In another study, AEOL-10150 was demonstrated to inhibit cigarette smoke-induced lung inflammation [225] suggesting a potential therapeutic benefit in COPD.

Another type of catalytic antioxidant is the GPx mimetic ebselen. This is a selenium-based organic complex and has been shown to be a very powerful antioxidant against the highly reactive and destructive $ONOO^-$ [231]. It is able to prevent both NF-κB/AP-1 activation and pro-inflammatory gene expression in human leukocytes exposed to $ONOO^-$. Other studies have shown that ebselen is also active *in vivo* in preventing LPS-induced airway inflammation [23, 232]. However, no studies have been reported so far on the protective effects of ebselen in cigarette smoke-induced lung inflammation. In contrast, the GPx mimetic BXT-51072 (Oxis, USA) and the lipid peroxidation inhibitor BO-653 (Chugai Pharma, Japan) are either currently in Phase I or pre-clinical trials in patients with COPD. A more detailed review of the development and properties of these catalytic antioxidants is described elsewhere [233].

Polyphenols

Several epidemiological studies have been undertaken, which have established a beneficial link between polyphenol intake and lower disease risk with many of the clinical benefits being attributed to both the antioxidant and anti-inflammatory properties of polyphenols [234]. In one Finnish study with over 10,000 participants, a significant inverse correlation was observed between polyphenol intake and the incidence of asthma [235]. Similar beneficial associations were also observed for COPD in a study encompassing over 13,000 adults. This study reported that increased polyphenol intake correlated with improved symptoms, as assessed by cough, phlegm production, and breathlessness and improved lung function as measured by FEV_1 [174]. Two further studies appeared to corroborate these findings. The first study showed a beneficial protective effect against COPD symptoms for increased fruit intake, high in polyphenol and vitamin E content [175]. In the second more recent study, a standardized polyphenol extract administered orally was shown to be effective in reducing oxidant stress and increasing PaO_2, as well as improvements in FEV_1 between the enrollment and the end of the study [236].

While the above studies would appear to demonstrate an epidemiological link between polyphenol intake and clinical benefit in asthma and COPD, other studies have tried to show a direct impact of specific polyphenolic compounds on inflammation *in vitro* and *in vivo*. For example, the flavonoid resveratrol, a constituent of red wine, inhibits inflammatory cytokine release from macrophages isolated from COPD patients [237]. Moreover, Birrell and colleagues have showed that *in vivo*, resveratrol can inhibit inflammatory cytokine expression in response to LPS challenge in rat lungs [238]. Furthermore, in both monocytic U937 and alveolar epithelial A549 cells, resveratrol inhibits NF-κB and AP-1 activation [239, 240]. However, the clinical utility of resveratrol in patients with asthma and COPD is not known.

Another well-studied polyphenol is curcumin. It is the active constituent of *Curcuma longa*, commonly known as turmeric. Like resveratrol, it has also been reported to inhibit NF-κB activation, along with IL-8 release, Cyclooxygenase-2 expression, and neutrophil recruitment in the lungs [241]. Interestingly, one study postulates that curcumin prevents cigarette smoke-induced NF-κB activation through the inhibition of IκBα kinase in human lung epithelial cells [242]. Recently, we have observed that curcumin can inhibit inflammation and restore glucocorticoid efficacy in response to oxidative stress, through the upregulation/restoration of HDAC2 activity in macrophages (U937 and MonoMac6 cells) [243]. This would facilitate steroid-mediated HDAC2 recruitment in attenuating NF-κB-mediated chromatin acetylation and subsequent pro-inflammatory gene expression. As glucocorticoids are the main thrust of anti-inflammatory treatment, any therapeutics that can be used as an add-on to improve steroid responsiveness in COPD and severe asthma would be of significant clinical benefit. Clearly clinical trials by using a combination approach of a steroid with polyphenols are warranted.

SUMMARY

There is now increasing evidence that ROS generation plays a major pathophysiological role in asthma and COPD, and it is important for the severity of these conditions. ROS generation through endogenous mechanisms or exogenous cigarette smoke/environmental oxidants is critical to the inflammatory response through activation of redox-sensitive transcription factors and pro-inflammatory signaling pathways. At the same time, endogenous antioxidant mechanisms are present to attenuate this redox-mediated inflammatory response. It is when these two opposing mechanisms are out of balance that a chronic and more severe inflammatory state becomes apparent. The use of antioxidants or other pharmacological agents to boost the endogenous antioxidant system could be used to redress this imbalance. In so doing, this would provide therapeutic benefit in damping down and curtailing the severity and chronicity of the inflammatory response in asthma and COPD patients.

A variety of oxidants, free radicals, and aldehydes are implicated in the pathogenesis of COPD, therefore it is possible that the therapeutic administration of multiple antioxidants will be effective in the treatment of COPD. An effective wide spectrum antioxidant therapy that has good bioavailability and potency is needed to control the localized oxidative and inflammatory processes that occur in the pathogenesis of asthma and COPD. Furthermore, human clinical trials of newly developed small molecular antioxidants (alone or in combination with anti-inflammatory agents), or higher doses of NAC, are urgently needed to validate these compounds as clinical therapies and to provide the proof of concept that oxidative stress is involved in the pathogenesis of COPD.

Clearly, further studies are required to understand the effect of ROS on basic cellular functions and the differential responses seen in different cell types and how this in turn impacts on the pathology of different inflammatory disease states. At the same time, endeavors into identifying new and more efficacious antioxidants, radical quenchers, inhibitors of lipid peroxidation, and agents that can reverse protein carbonyl protein damage as a therapeutic strategy should continue. Indeed, elucidating the mechanism of action for some of the naturally occurring antioxidants, such as the potent enzyme mimetics and polyphenols, may lead to new therapeutic targets that can be antagonized/agonized through more conventional pharmacological approaches.

References

1. Gutteridge JM, Halliwell B. Free radicals and antioxidants in the year 2000: A historic look to the future. *Ann NY Acad Sci* 899: 136–47, 2000.
2. Richer C, Cogvadze V, Laffranchi R *et al.* Oxidants in mitochondria: From physiology to diseases. *Biochim Biophys Acta* 127: 67–74, 1995.
3. Rahman I, MacNee W. Role of transcription factors in inflammatory lung diseases. *Thorax* 53: 601–12, 1998.
4. Guyton KZ, Liu Y, Gorospe M *et al.* Activation of mitogenactivated protein kinase by H_2O_2. *J Biol Chem* 271: 4138–42, 1996.
5. Rahman I, MacNee W. Role of oxidants/antioxidants in smoking-induced airway diseases. *Free Radic Biol Med* 21: 669–81, 1996.
6. Rahman I, MacNee W. Lung glutathione and oxidative stress: Implications in cigarette smoke-induced airways disease. *Am J Physiol* 277: L1067–L1088, 1999.
7. Rahman I, MacNee W. Oxidative stress and regulation of glutathione synthesis in lung inflammation. *Eur Respir J* 16: 534–54, 2000.
8. Dworski R. Oxidant stress in asthma. *Thorax* 55: S51–53, 2000.
9. Hatch GE. Asthma: Inhaled oxidants and dietary antioxidants. *J Clin Nutr* 61: 625S–630S, 1995.
10. Saetta M. Airway inflammation in chronic obstructive pulmonary disease. *Am J Respir Crit Care Med* 160: S17–20, 1999.
11. Barnes PJ. Mechanisms in COPD: Differences from asthma. *Chest* 117: 10S–14S, 2000.
12. Jeffery PK. Structural and inflammatory changes in COPD: A comparison with asthma. *Thorax* 53: 129–36, 1998.
13. Walsh GM. Advances in the immunobiology of eosinophils and their role in disease. *Crit Rev Clin Lab Sci* 36: 453–96, 1999.
14. Eiserich JP, Hristova M, Cross CE *et al.* Formation of nitric oxide derived inflammatory oxidants by myeloperoxidase in neutrophils. *Nature* 391: 393–97, 1998.
15. MacPherson JC, Comhair SAAA, Eruzurum SC *et al.* Eosinophils are a major source of nitric oxide-derived oxidants in severe asthma: Characterization of pathways available to eosinophils for generating reactive nitrogen species. *J Immunol* 166: 5763–72, 2001.
16. Wu W, Samoszuk MK, Comhair SAAA *et al.* Eosinophils generate brominating oxidants in allergen-induced asthma. *J Clin Invest* 105: 1455–63, 2000.
17. Halliwell B, Gutteridge JMC. Role of free radicals and catalytic metal ions in human disease: An overflow. *Methods Enzymol* 186: 1–85, 1990.
18. Rochelle LG, Fischer BM, Adler KB. Concurrent production of reactive oxygen and nitrogen species by airway epithelial cells *in vitro*. *Free Radic Biol Med* 24: 863–68, 1998.
19. Gutteridge JMC. Lipid peroxidation and antioxidants as biomarkers of tissue damage. *Clin Chem* 41: 1819–28, 1995.
20. Church T, Pryor WA. Free radical chemistry of cigarette smoke and its toxicological implications. *Environ Health Perspect* 64: 111–26, 1985.
21. Pryor WA, Stone K. Oxidants in cigarette smoke: Radicals, hydrogen peroxides, peroxynitrate, and peroxynitrite. *Ann NY Acad Sci* 686: 12–28, 1993.
22. Nakayama T, Church DF, Pryor WA. Quantitative analysis of the hydrogen peroxide formed in aqueous cigarette tar extracts. *Free Radic Biol Med* 7: 9–15, 1989.
23. Zang LY, Stone K, Pryor WA. Detection of free radicals in aqueous extracts of cigarette tar by electron spin resonance. *Free Radic Biol Med* 19: 161–67, 1995.
24. Uchida K, Shiraishi M, Naito Y *et al.* Activation of stress signaling pathways by the end product of lipid peroxidation. *J Biol Chem* 274: 2234–42, 1999.
25. Parola M, Bellomo G, Robino G, Barrera G, Dianzani MU. 4-hydroxynonenal as a biological signal: Molecular basis and pathophysiological implications. *Antioxid Redox Signal* 1: 255–84, 1999.
26. Marwick JA, Kirkham PA, Stevenson CS, Danahay H, Giddings J, Butler K, Donaldson K, MacNee W, Rahman I. Cigarette smoke alters chromatin remodeling and induces proinflammatory genes in rat lungs. *Am J Respir Cell Mol Biol* 31: 633–42, 2004.
27. Kirkham PA, Spooner G, Rahman I, Rossi AG. Macrophage phagocytosis of apoptotic neutrophils is compromised by matrix proteins modified by cigarette smoke and lipid peroxidation products. *Biochem Biophys Res Commun* 318: 32–37, 2004.
28. Kirkham PA, Spooner G, Ffoulkes-Jones C, Calvez R. Cigarette smoke triggers macrophage adhesion and activation: Role of lipid peroxidation products and scavenger receptor. *Free Radic Biol Med* 35: 697–710, 2003.
29. Reilly M, Delanty N, Lawson JA, FitzGerald GA. Modulation of oxidant stress *in vivo* in chronic cigarette smokers. *Circulation* 94: 19–25, 1996.
30. Pratico D, Basili S, Vieri M *et al.* Chronic obstructive pulmonary disease is associated with an increase in urinary levels of isoprostane $F_{2\alpha}$-III, an index of oxidant stress. *Am J Respir Crit Care Med* 158: 1709–14, 1998.
31. Morrow JD, Roberts LJ. The isoprostanes: Unique bioactive products of lipid peroxidation. *Prog Lipid Res* 36: 1–21, 1997.
32. Kinsella BT, O'Mahony DJ, Fitzgerald GA. The human thromboxane A2 receptor alpha isoform (TP alpha) functionally couples to the G proteins Gq and G11 *in vivo* and is activated by the isoprostane 8-epi prostaglandin F2 alpha. *J Pharmacol Exp Ther* 281: 957–64, 1997.
33. Okazawa A, Kawikova I, Cui ZH, Skoogh BE, Lotvall J. 8-Epi-PGF2alpha induces airflow obstruction and airway plasma exudation *in vivo*. *Am J Respir Crit Care Med* 155: 436–41, 1997.
34. Leitinger N, Tyner TR, Oslund L, Rizza C, Subbanagounder G, Lee H, Shih PT, Mackman N, Tigyi G, Territo MC, Berliner JA, Vora DK. Structurally similar oxidized phospholipids differentially regulate endothelial binding of monocytes and neutrophils. *Proc Natl Acad Sci USA* 96: 12010–15, 1999.
35. Lee H, Shi W, Tontonoz P, Wang S, Subbanagounder G, Hedrick CC, Hama S, Borromeo C, Evans RM, Berliner JA, Nagy L. Role for peroxisome proliferator-activated receptor alpha in oxidized phospholipid-induced synthesis of monocyte chemotactic protein-1 and interleukin-8 by endothelial cells. *Circ Res* 87: 516–21, 2000.
36. Yeh M, Leitinger N, de MR, Onai N, Matsushima K, Vora DK, Berliner JA, Reddy ST. Increased transcription of IL-8 in endothelial cells is differentially regulated by TNF-alpha and oxidized phospholipids. *Arterioscler Thromb Vasc Biol* 21: 1585–91, 2001.
37. Sen CK. Redox signaling and the emerging therapeutic potential of thiol antioxidants. *Biochem Pharmacol* 55: 1747–58, 1998.
38. Thannickal VJ, Fanburg BL. Reactive oxygen species in cell signaling. *Am J Physiol Lung Cell Mol Physiol* 279: L1005–28, 2000.
39. Adler V, Yin Z, Tew KD, Ronai Z. Role of redox potential and reactive oxygen species in stress signaling. *Oncogene* 18: 6104–11, 1999.
40. Rahman I, Adcock M. Oxidative stress and redox signaling of lung inflammation in COPD. *Eur Respir J* 28: 219–42, 2006.

41. Rahman I, MacNee W. Regulation of redox glutathione levels and gene transcription in lung inflammation: Therapeutic approaches. *Free Radic Biol Med* 28: 1405–20, 2000.
42. Sedgewick JB, Geiger KM, Busse WW. Superoxide generation by hypodense eosinophils from patients with asthma. *Am Rev Respir Dis* 80: 195–201, 1990.
43. Kanazawa H, Kurihara N, Hirata K, Takeda T. The role of free radicals in airway obstruction in asthmatic patients. *Chest* 100: 1319–22, 1991.
44. Calhoun WJ, Reed HE, Moest DR, Stevens CA. Enhanced superoxide production by alveolar macrophages and air-space cells, airway inflammation, and alveolar macrophage density changes after segmental antigen bronchoprovocation in allergic subjects. *Am Rev Respir Dis* 145: 317–25, 1992.
45. Vachier I, Damon M, Le Doucen C *et al.* Increased oxygen species generation in blood monocytes of asthmatic patients. *Am Rev Respir Dis* 146(1): 161–66, 1992.
46. Cluzel M, Damon M, Chanez P *et al.* Enhanced alveolar cell luminol-dependent chemiluminescence in asthma. *J Allergy Clin Immunol* 80: 195–201, 1987.
47. Hulsmann AR, Raatgeep HR, den Hollander JC *et al.* Oxidative epithelial damage produces hyperresponsiveness of human peripheral airways. *Am J Respir Crit Care Med* 149: 519–25, 1994.
48. Sadeghi-Hashjin G, Folkerts G, Henricks PA *et al.* Peroxynitrite induces airway hyperresponsiveness in guinea pigs *in vitro* and *in vivo*. *Am J Respir Crit Care Med* 153: 1697–701, 1996.
49. Cortijo J, Marti-Cabrera M, de la Asuncion JG *et al.* Contraction of human airways by oxidative stress protection by *N*-acetylcysteine. *Free Radic Biol Med* 27: 392–400, 1999.
50. Adam L, Bouvier M, Jones TLZ. Nitric oxide modulates β_2-adrenergic receptor palmitoylation and signaling. *J Biol Chem* 274: 26337–43, 1999.
51. Owen S, Pearson D, O'Driscoll R. Evidence of free-radical activity in asthma. *N Engl J Med* 325: 586–87, 1991.
52. Seltzer J, Bigby BG, Stulbarg M *et al.* O_3-induced change in bronchial reactivity to methacholine and airway inflammation in humans. *J Appl Physiol* 60: 1321–26, 1986.
53. Hiltermann TJ, Peters EA, Alberts B *et al.* Ozone-induced airway hyperresponsiveness in patients with asthma: Role of neutrophil-derived serine proteinases. *Free Radic Biol Med* 24: 952–58, 1998.
54. Sanders SP, Zweier JL, Harrison SL *et al.* Spontaneous oxygen radical product at sites of antigen challenge in allergic subjects. *Am J Respir Crit Care Med* 151: 1725–33, 1995.
55. Comhair SA, Ricci KS, Arroliga M, Lara AR, Dweik RA, Song W, Hazen SL, Bleecker ER, Busse WW, Chung KF, Gaston B, Hastie A, Hew M, Jarjour N, Moore W, Peters S, Teague WG, Wenzel SE, Erzurum SC. Correlation of systemic superoxide dismutase deficiency to airflow obstruction in asthma. *Am J Respir Crit Care Med* 172: 306–13, 2005a.
56. Comhair SA, Xu W, Ghosh S, Thunnissen FB, Almasan A, Calhoun WJ, Janocha AJ, Zheng L, Hazen SL, Erzurum SC. Superoxide dismutase inactivation in pathophysiology of asthmatic airway remodeling and reactivity. *Am J Pathol* 166: 663–74, 2005b.
57. Smith LJ, Shamsuddin M, Sporn PH, Denenberg M, Anderson J. Reduced superoxide dismutase in lung cells of patients with asthma. *Free Radic Biol Med* 22: 1301–7, 1997.
58. Comhair SA, Bhathena PR, Dweik RA, Kavuru M, Erzurum SC. Rapid loss of superoxide dismutase activity during antigen-induced asthmatic response. *Lancet* 355: 624, 2000.
59. Foreman RC, Mercer PF, Kroegel C, Warner JA. Role of the eosinophil in protein oxidation in asthma: Possible effects on proteinase/antiproteinase balance. *Int Arch Allergy Immunol* 118: 183–86, 1999.
60. Wardlaw AJ, Dunnette S, Gleich GJ, Collins JV, Kay AB. Eosinophils and mast cells in bronchoalveolar lavage in subjects with mild asthma: Relationship to bronchial hyperreactivity. *Am Rev Respir Dis* 137: 62–69, 1988.
61. Teramoto S, Shu CY, Ouchi Y, Fukuchi Y. Increased spontaneous production and generation of superoxide anion by blood neutrophils in patients with asthma. *J Asthma* 33: 149–55, 1996.
62. Shauer U, Leinhaas C, Jager R, Rieger CH. Enhanced superoxide generation by eosinophils from asthmatic children. *Int Arch Allergy Appl Immunol* 96: 317–21, 1991.
63. Krishna MT, Madden J, Teran LM *et al.* Effects of 0.2 ppm ozone on biomarkers of inflammation in bronchoalveolar lavage fluid and bronchial mucosa of healthy subjects. *Eur Respir J* 11: 1294–300, 1998.
64. Agosti JM, Altman LC, Avars GH *et al.* The injurious effect of eosinophil peroxidase, hydrogen peroxide, and halides on pneumocytes *in vitro*. *J Allergy Clin Immunol* 79: 496–504, 1987.
65. McBride DE, Koenig JQ, Luchtel DL, Williams PV, Henderson WR. Inflammatory effects of ozone in the upper airways of subjects with asthma. *Am J Respir Crit Care Med* 149: 1192–97, 1994.
66. Emelyanov A, Fedoseev G, Abulimity A *et al.* Elevated concentrations of exhaled hydrogen peroxide in asthmatic patients. *Chest* 120: 1136–39, 2001.
67. Mitra SN, Slungaard A, Hazen SL. Role of eosinophil peroxidase in the origins of protein oxidation in asthma. *Redox Rep* 5: 215–24, 2000.
68. Motojima S, Fukuda T, Makino S. Effect of eosinophil peroxidase on beta-adrenergic receptor density on guinea pig lung. *Biochem Biophys Res Commun* 189: 1613–19, 1992.
69. Yoshikawa S, Kayes SG, Parker JC. Eosinophils increase lung microvascular permeability via the peroxidase-hydrogen peroxide-halide system: Bronchoconstriction and vasoconstriction unaffected by eosinophil peroxidase inhibition. *Am Rev Respir Dis* 147: 914–20, 1993.
70. Alving K, Weitzberg E, Lundberg JM. Increased amount of nitric oxide in exhaled air of asthmatics. *Eur Respir J* 6: 1368–70, 1993.
71. Silkoff PE, Sylvester JT, Zamel N, Permutt S. Airway nitric oxide diffusion in asthma: Role in pulmonary function and bronchial responsiveness. *Am J Respir Crit Care Med* 161: 1218–28, 2000.
72. Barnes PJ. NO or no NO in asthma? *Thorax* 51: 218–20, 1996.
73. Beckman DL, Mehta P, Hanks V, Rowan WH, Liu L. Effects of peroxynitrite on pulmonary edema and the oxidative state. *Exp Lung Res* 26: 349–59, 2000.
74. Saleh D, Ernst P, Lim S, Barnes PJ, Giaid A. Increased formation of the potent oxidant peroxynitrite in the airways of asthmatic patients is associated with induction of nitric oxide synthase: Effect of inhaled glucocorticoid. *FASEB J* 12: 929–37, 1998.
75. Dohlman AW, Black HR, Royall JA. Expired breath hydrogen peroxide is a marker of acute airway inflammation in paediatric patients with asthma. *Am Rev Respir Dis* 148: 955–60, 1993.
76. Zhang P, Wang YZ, Kagan E, Bonner JC. Peroxynitrite targets the epidermal growth factor receptor, Raf-1, and MEK independently to activate MAPK. *J Biol Chem* 275: 22479–86, 2000.
77. Groves JT. Peroxynitrite: Reactive, invasive and enigmatic. *Curr Opin Chem Biol* 3: 226–35, 1999.
78. Hogg N, Kalyanaraman B. Nitric oxide and lipid peroxidation. *Biochim Biophys Acta* 1411: 378–84, 1999.
79. Hanazawa T, Kharitonov SA, Barnes PJ. Increased nitrotyrosine in exhaled breath condensate of patients with asthma. *Am J Respir Crit Care Med* 162(4 Pt.1): 1273–76, 2000.
80. Kaminsky DA, Mitchell J, Carroll N *et al.* Nitrotyrosine formation in the airways and lung parenchyma of patients with asthma. *J Allergy Clin Immunol* 104: 747–54, 1999.
81. Foster MW, McMahon TJ, Stamler JS. S-Nitrosylation in health and disease. *Trends Mol Med* 9: 160–68, 2003.
82. Liu L, Stamler JS. NO: An inhibitor of cell death. *Cell Death Differ* 6: 937–42, 1999.
83. Gaston B, Sears S, Woods J, Hunt J, Ponaman M, McMahon T, Stamler JS. Bronchodilator S-nitrosothiol deficiency in asthmatic respiratory failure. *Lancet* 351: 1317–19, 1998.
84. Dweik RA, Comhair SA, Gaston B, Thunnissen FB, Farver C, Thomassen MJ, Kavuru M, Hammel J, Abu-Soud HM, Erzurum SC. NO chemical events in the human airway during the immediate and

late antigen-induced asthmatic response. *Proc Natl Acad Sci USA* 98: 2622–27, 2001.

85. Janssen-Heininger Y, Ckless K, Reynaert N, van der Vliet A. SOD inactivation in asthma: Bad news or no news? *Am J Pathol* 166: 649–52, 2005.
86. Montuschi P, Corradi M, Ciabattoni G *et al.* Increased 8-isoprostane, a marker of oxidative stress, in exhaled condensate of asthma patients. *Am J Respir Crit Care Med* 160: 216–20, 1999.
87. Paredi P, Kharitonov SA, Barnes PJ. Elevation of exhaled ethane concentration in asthma. *Am J Respir Crit Care Med* 162: 1450–54, 2000.
88. Olepade CO, Zakkar M, Swedler WI. Exhaled pentane levels in acute asthma. *Chest* 111: 862–65, 1997.
89. Antczak A, Nowak D, Shariati B *et al.* Increased hydrogen peroxide and thiobarbituric acid reactive products in expired breath condensate of asthma patients. *Eur Respir J* 10: 1235–41, 1997.
90. Dworski R, Murry JJ, Roberts LJ *et al.* Allergen-induced synthesis of F2-isoprostanes in atopic asthmatics: Evidence for oxidant stress. *Am J Respir Crit Care Med* 160: 1947–51, 1999.
91. Dworski R, Roberts LJ, Murry JJ *et al.* Assessment of oxidant stress in allergic asthma by measurement of the major urinary metabolite of F2-isoprostane, 15-F2t-IsoP (8-iso-PGF2alpha). *Clin Exp Allergy* 31: 387–90, 2001.
92. Wood LG, Fitzgerald DA, Gibson PG, Cooper DM, Garg ML. Lipid peroxidation as determined by plasma isoprostanes is related to disease severity in mild asthma. *Lipids* 35: 967–74, 2000.
93. Calabrese C, Triggiani M, Marone G, Mazzarella G. Arachidonic acid metabolism in inflammatory cells of patients with bronchial asthma. *Allergy* 55: 27–30, 2000.
94. Tsukagoshi H, Shimizu Y, Iwamae S *et al.* Evidence of oxidative stress in asthma and COPD: Potential inhibitory effect of theophylline. *Respir Med* 94: 584–88, 2000.
95. Rahman I, Morrison D, Donaldson K, MacNee W. Systemic oxidative stress in asthma, COPD, and smokers. *Am J Respir Crit Care Med* 154: 1055–60, 1996.
96. British Thoracic Society guidelines for the management of chronic obstructive pulmonary disease. *Thorax* 52:S1–28, 1997.
97. American Thoracic Society Standards for the diagnosis and care of patients with chronic obstructive pulmonary disease. *Am J Respir Crit Care Med* 152:S77–120, 1995.
98. Snider G. Chronic obstructive pulmonary disease: Risk factors, pathophysiology and pathogenesis. *Ann Rev Med* 40: 411–29, 1989.
99. Repine JE, Bast A, Lankhorst Ithe Oxidative Stress Study Group. Oxidative stress in chronic obstructive pulmonary disease. *Am J Respir Crit Care Med* 156: 341–57, 1997.
100. Bosken CH, Hards J, Gatter K, Hogg JC. Characterization of the inflammatory reaction in the peripheral airways of cigarette smokers using immunocytochemistry. *Am Rev Respir Dis* 145: 911–17, 1992.
101. Aaron SD, Angel JB, Lunau M *et al.* Granulocyte inflammatory markers and airway infection during acute exacerbation of chronic obstructive pulmonary disease. *Am J Respir Crit Care Med* 163: 349–55, 2001.
102. Fiorini G, Crespi S, Rinaldi M *et al.* Serum ECP and MPO are increased during exacerbations of chronic bronchitis with airway obstruction. *Biomed Pharmacother* 54: 274–78, 2000.
103. Gompertz S, Bayley DL, Hill SL, Stockley RA. Relationship between airway inflammation and the frequency of exacerbations in patients with smoking related COPD. *Thorax* 56: 36–41, 2001.
104. Morrison D, Rahman I, Lannan S, MacNee W. Epithelial permeability, inflammation and oxidant stress in the airspaces of smokers. *Am J Respir Crit Care Med* 159: 473–79, 1999.
105. Nakashima H, Ando M, Sugimoto M *et al.* Receptor-mediated O_2^- release by alveolar macrophages and peripheral blood monocytes from smokers and nonsmokers. *Am Rev Respir Dis* 136: 310–15, 1987.
106. Drath DB, Larnovsky ML, Huber GL. The effects of experimental exposure to tobacco smoke on the oxidative metabolism of alveolar macrophages. *J Reticul Soc* 25: 597–604, 1970.
107. Schaberg T, Klein U, Rau M, Eller J, Lode H. Subpopulation of alveolar macrophages in smoker and nonsmokers: Relation to the expression of CD 11/CD 18 molecules and superoxide anion production. *Am J Respir Crit Care Med* 151: 1551–58, 1995.
108. Nowak D, Kasielski M, Pietras T, Bialasiewicz P, Antczak A. Cigarette smoking does not increase hydrogen peroxide levels in expired breath condensate of patients with stable COPD. *Monaldi Arch Chest Dis* 53: 268–73, 1998.
109. Nowak D, Antczak A, Krol M *et al.* Increased content of hydrogen peroxide in expired breath of cigarette smokers. *Eur Respir J* 9: 652–57, 1996.
110. Dekhuijzen PNR, Aben KKH, Dekke I *et al.* Increased exhalation of hydrogen peroxide in patients with stable and unstable chronic obstructive pulmonary disease. *Am J Respir Crit Care Med* 154: 813–16, 1996.
111. Mateos F, Brock JF, Perez-Arellano JL. Iron metabolism in the lower respiratory tract. *Thorax* 53: 594–600, 1998.
112. Lapenna D, Gioia SD, Mezzetti A *et al.* Cigarette smoke, ferritin, and lipid peroxidation. *Am J Respir Crit Care Med* 151: 431–35, 1995.
113. Thompson AB, Bohling T, Heires A, Linder J, Rennard SI. Lower respiratory tract iron burden is increased in association with cigarette smoking. *J Lab Clin Med* 117: 494–99, 1991.
114. Wesselius LJ, Nelson ME, Skikne BS. Increased release of ferritin and iron by iron loaded alveolar macrophages in cigarette smokers. *Am J Respir Crit Care Med* 150: 690–95, 1994.
115. Lacoste JY, Bousquet J, Chanez P *et al.* Eosinophilic and neutrophilic inflammation in asthma, chronic bronchitis, and chronic obstructive pulmonary disease. *J Allergy Clin Immunol* 149: 803–10, 1993.
116. Lebowitz MD, Postma DS. Adverse effects of eosinophilic and smoking on the natural history of newly diagnosed chronic bronchitis. *Chest* 108: 55–61, 1995.
117. Pinamonti S, Muzzuli M, Chicca C *et al.* Xanthine oxidase activity in bronchoalveolar lavage fluid from patients with chronic obstructive lung disease. *Free Radic Biol Med* 21: 147–55, 1996.
118. Heunks LM, Vina J, van Herwaarden CL *et al.* Xanthine oxidase is involved in exercise-induced oxidative stress in chronic obstructive pulmonary disease. *Am J Physiol* 277: R1697–704, 1999.
119. Pinamonti S, Leis M, Barbieri A *et al.* Detection of xanthine oxidase activity products by EPR and HPLC in bronchoalveolar lavage fluid from patients with chronic obstructive pulmonary disease. *Free Radic Biol Med* 25: 771–79, 1998.
120. Chan-Yeung M, Dybuncio A. Leucocyte count, smoking and lung function. *Am J Med* 76: 31–37, 1984.
121. Van Antwerpen VL, Theron AJ, Richards GA *et al.* Vitamin E, pulmonary functions, and phagocyte-mediated oxidative stress in smokers and nonsmokers. *Free Radic Biol Med* 18: 935–43, 1995.
122. Chan-Yeung M, Abboud R, Dybuncio A, Vedal S. Peripheral leucocyte count and longitudinal decline in lung function. *Thorax* 43: 426–68, 1988.
123. Postma DS, Renkema TEJ, Noordhoek JA *et al.* Association between nonspecific bronchial hyperreactivity and superoxide anion production by polymorphonuclear leukocytes in chronic airflow obstruction. *Am Rev Respir Dis* 137: 57–61, 1988.
124. Richards GA, Theron AJ, van der Merwe CA, Anderson R. Spirometric abnormalities in young smokers correlate with increased chemiluminescence responses of activated blood phagocytes. *Am Rev Respir Dis* 139: 181–87, 1989.
125. Rahman I, Skwarska E, MacNee W. Attenuation of oxidant/antioxidant imbalance during treatment of exacerbations of chronic obstructive pulmonary disease. *Thorax* 52: 565–68, 1997.
126. Muns G, Rubinstein I, Bergmann KC. Phagocytosis and oxidative bursts of blood phagocytes in chronic obstructive airway disease. *Scand J Infect Dis* 27: 369–73, 1995.
127. Noguera A, Busquets X, Sauleda J *et al.* Expression of adhesion molecules and G-proteins in circulating neutrophils in COPD. *Am J Respir Crit Care Med* 158: 1664–68, 1998.

128. Brown DM, Drost E, Donaldson K, MacNee W. Deformability and CD 11/CD 18 expression of sequestered neutrophils in normal and inflamed lungs. *Am J Respir Cell Mol Biol* 13: 531–39, 1995.
129. Maziak W, Loukides S, Culpitt S *et al.* Exhaled nitric oxide in chronic obstructive pulmonary disease. *Am J Respir Crit Care Med* 147: 998–1002, 1998.
130. Corradi M, Majori M, Cacciani GC *et al.* Increased exhaled nitric oxide in patients with stable chronic obstructive pulmonary disease. *Thorax* 54: 572–75, 1999.
131. Delen FM, Sippel JM, Osborne ML *et al.* Increased exhaled nitric oxide in chronic bronchitis: Comparison with asthma and COPD. *Chest* 117: 695–701, 2000.
132. Rutgers SR, van der Mark TW, Coers W *et al.* Markers of nitric oxide metabolism in sputum and exhaled air are not increased in chronic obstructive pulmonary disease. *Thorax* 54: 576–680, 1999.
133. Robbins RA, Millatmal T, Lassi K, Rennard S, Daughton D. Smoking cessation is associated with an increase in exhaled nitric oxide. *Chest* 112: 313–18, 1997.
134. Paredi P, Kharitonov SA, Leak D, Ward S, Cramer D, Barnes PJ. Exhaled ethane, a marker of lipid peroxidation, is elevated in chronic obstructive pulmonary disease. *Am J Respir Crit Care Med* 162: 369–73, 2000.
135. Euler DE, Dave SJ, Guo H. Effect of cigarette smoking on pentane excretion in alveolar breath. *Clin Chem* 42: 303–8, 1996.
136. Montuschi P, Collins JV, Ciabattoni G *et al.* Exhaled 8-isoprostane as an *in-vivo* biomarker of lung oxidative stress in patients with COPD and healthy smokers. *Am J Respir Crit Care Med* 162: 1175–77, 2000.
137. Nowak D, Kasielski M, Antczak A, Pietras T, Bialasiewicz P. Increased content of thiobarbiturate reactive acid substances in hydrogen peroxide in the expired breath condensate of patients with stable chronic obstructive pulmonary disease: No significant effect of cigarette smoking. *Respir Med* 93: 389–96, 1999.
138. Fahn H, Wang L, Kao S *et al.* Smoking-associated mitochondrial DNA mutation and lipid peroxidation in human lung tissue. *Am J Respir Cell Mol Biol* 19: 901–9, 1998.
139. Rahman I, Crowther A, de Boer WI *et al.* 4-hydroxy-2-nonenal, a specific lipid peroxidation product is elevated in lungs of patients with chronic obstructive pulmonary disease. *Am J Respir Crit Care Med* 163: A31, 2001.
140. Eiserich JP, van der Vliet A, Handelman GJ, Halliwell B, Cross CE. Dietary antioxidants and cigarette smoke-induced biomolecular damage: A complex interaction. *Am J Clin Nutr* 62: 1490S–500S, 1995.
141. Rahman I, Kelly F. Biomarkers in breath condensate: A promising non-invasive technique in free radical research. *Free Radic Res* 37: 1253–66, 2003.
142. Pignatelli B, Li CG, Boffetta P *et al.* Nitrated and oxidized plasma proteins in smokers and lung cancer patients. *Cancer Res* 61: 778–84, 2001.
143. Petruzzelli S, Puntoni R, Mimotti P *et al.* Plasma 3-nitrotyrosine in cigarette smokers. *Am J Respir Crit Care Med* 156: 1902–7, 1997.
144. Ichinose M, Sugiura H, Yamagata S, Koarai A, Shirato K. Increase in reactive nitrogen species production in chronic obstructive pulmonary disease airways. *Am J Respir Crit Care Med* 162: 701–6, 2000.
145. Van der Vliet A, Smith D, O'Neill CA *et al.* Interactions of peroxynitrite and human plasma and its constituents: Oxidative damage and antioxidant depletion. *Biochem J* 303: 295–301, 1994.
146. Britton JR, Pavord ID, Richards KA, Knox AJ, Wisniewski AF, Lewis SA, Tattersfield AE, Weiss ST. Dietary antioxidant vitamin intake and lung function in the general population. *Am J Respir Crit Care Med* 151: 1383–87, 1995. , 174.
147. Cross CE, van dV, O'Neill CA, Louie S, Halliwell B. Oxidants, antioxidants, and respiratory tract lining fluids. *Environ Health Perspect* 102(Suppl 10): 185–91, 1994.
148. Li XY, Donaldson K, Rahman I, MacNee W. An investigation of the role of glutathione in increased epithelial permeability induced by cigarette smoke *in vivo* and *in vitro*. *Am J Respir Crit Care Med* 149: 1518–25, 1994.
149. Su WY, Folz R, Chen JS, Crapo JD, Chang LY. Extracellular superoxide dismutase mRNA expressions in the human lung by *in situ* hybridization. *Am J Respir Cell Mol Biol* 16: 162–70, 1997.
150. Cantin AM, North SL, Hubbard RC, Crystal RG. Normal alveolar epithelial lining fluid contains high levels of glutathione. *J Appl Physiol* 63: 152–57, 1987.
151. Smith LJ, Houston M, Anderson J. Increased levels of glutathione in bronchoalveolar lavage fluid from patients with asthma. *Am Rev Respir Dis* 147: 1461–64, 1993.
152. Rahman I, Smith CA, Lawson MF, Harrison DJ, MacNee W. Induction of gamma-glutamylcysteine synthetase by cigarette smoke is associated with AP-1 in human alveolar epithelial cells. *FEBS Lett* 396: 21–25, 1996.
153. Rahman I, van Schadewijk AM, Hiemstra PS *et al.* Localisation of γ-glutamylcysteine synthetase messenger RNA expression in lungs of smokers and patients with chronic obstructive pulmonary disease. *Free Radic Biol Med* 28: 920–25, 2000.
154. Gilks CB, Price K, Wright JL, Churg A. Antioxidant gene expression in rat lung after exposure to cigarette smoke. *Am J Pathol* 152: 269–78, 1998.
155. Harju T, Kaarteenaho-Wiik R, Soini Y, Sormunen R, Kinnula VL. Diminished immunoreactivity of gamma-glutamylcysteine synthetase in the airways of smokers' lung. *Am J Respir Crit Care Med* 166: 754–59, 2002.
156. Neurohr C, Lenz AG, Ding I, Leuchte H, Kolbe T, Behr J. Glutamate-cysteine ligase modulatory subunit in BAL alveolar macrophages of healthy smokers. *Eur Respir J* 22: 82–87, 2003.
157. Rangasamy T, Cho CY, Thimmulappa RK, Zhen L, Srisuma SS, Kensler TW, Yamamoto M, Petrache I, Tuder RM, Biswal S. Genetic ablation of Nrf2 enhances susceptibility to cigarette smoke-induced emphysema in mice. *J Clin Invest* 114: 1248–59, 2004.
158. Harrison DJ, Cantlay AM, Rae F, Lamb D, Smith CA. Frequency of glutathione s-transferase M1 deletion in smokers with emphysema and lung cancer. *Hum Exp Toxicol* 16: 356–60, 1997.
159. Exner M, Minar E, Wagner O, Schillinger M. The role of heme oxygenase-1promoter polymorphisms in human disease. *Free Radic Biol Med* 37: 1097–104, 2004.
160. Rangasamy T, Guo J, Mitzner WA, Roman J, Singh A, Fryer AD, Yamamoto M, Kensler TW, Tuder RM, Georas SN, Biswal S. Disruption of Nrf2 enhances susceptibility to severe airway inflammation and asthma in mice. *J Exp Med* 202: 47–59, 2005.
161. Tomaki M, Sugiura H, Koarai A, Komaki Y, Akita T, Matsumoto T, Nakanishi A, Ogawa H, Hattori T, Ichinose M. Decreased expression of antioxidant enzymes and increased expression of chemokines in COPD lung. *Pulm Pharmacol Ther*, 20: 596-605, 2007.
162. Simic, Michael G, Jovanovic , Slobodan V. Antioxidation mechanisms of uric acid. *J Am Chem Soc* 111: 5778–82, 1989.
163. Janoff A, Pryor WA, Bengali ZH. NHLBI Workshop Summary. Effects of tobacco smoke components on cellular and biochemical processes in the lung. *Am Rev Respir Dis* 136: 1058–64, 1987.
164. Grootveld M, Halliwell B. Measurement of allantoin and uric acid in human body fluids. A potential index of free-radical reactions *in vivo*? *Biochem J* 243: 803–8, 1987.
165. Halliwell B, Gutteridge JM. The antioxidants of human extracellular fluids. *Arch Biochem Biophys* 280: 1–8, 1990b.
166. Halliwell B, Gutteridge JM. Free Radicals in Biology and Medicine. , 3rd edn. New York: Oxford University Press Inc, 1999.
167. Spencer JP, Jenner A, Chimel K, Aruoma OI, Cross CE, Wu R, Halliwell B. DNA damage in human respiratory tract epithelial cells: Damage by gas phase cigarette smoke apparently involves attack by reactive nitrogen species in addition to oxygen radicals. *FEBS Lett* 375: 179–82, 1995.
168. Halliwell B. Vitamin C: Antioxidant or pro-oxidant *in vivo*?. *Free Radic Res* 25: 439–54, 1996.

169. Morrow JD, Frei B, Longmire AW, Gaziano JM, Lynch SM, Shyr Y, Strauss WE, Oates JA, Roberts LJ. Increase in circulating products of lipid peroxidation (F2-isoprostanes) in smokers, smoking as a cause of oxidative damage. *N Engl J Med* 332: 1198–203, 1995.
170. Schorah CJ, Downing C, Piripitsi A, Gallivan L, Al-Hazaa AH, Sanderson MJ, Bodenham A. Total vitamin C, ascorbic acid, and dehydroascorbic acid concentrations in plasma of critically ill patients. *Am J Clin Nutr* 6: 760–65, 1996.
171. Schunemann HJ, Muti P, Freudenheim JL *et al.* Oxidative stress and lung function. *Am J Epidemiol* 146: 939–48, 1997.
172. Grievink L, Smit HA, Ocke MC *et al.* Dietary intake of antioxidant (pro)-vitamins, respiratory symptoms and pulmonary function: The MORGEN Study. *Thorax* 53: 166–71, 1998.
173. Sargeant LA, Jaeckel A, Wareham NJ. Interaction of vitamin C with the relation between smoking and obstructive airways disease in EPIC Norfolk. European Prospective Investigation into Cancer and Nutrition. *Eur Respir J* 16: 397–403, 2000.
174. Tabak C, Ilja C, Arts CW *et al.* Chronic obstructive pulmonary disease and intake of catechins, flavonols, and flavones: The MORGEN Study. *Am J Respir Crit Care Med* 164: 61–64, 2001.
175. Walda IC, Tabak C, Smit HA. Diet and 20-year chronic obstructive pulmonary disease mortality in middle-aged men from three European countries. *Eur J Clin Nutr* 56: 638–43, 2002.
176. Mezzetti A, Lapenna D, Pierdomenico SD, Calafiore AM, Costantini F, Riario-Sforza G, Imbastaro T, Neri M, Cuccurullo F. Vitamins E, C and lipid peroxidation in plasma and arterial tissue of smokers and non-smokers. *Atherosclerosis* 112: 91–99, 1995.
177. Schwartz J, Weiss ST. Relationship between dietary vitamin C intake and pulmonary function in the First National Health and Nutrition Examination Survey (NHANES I). *Am J Clin Nutr* 59: 110–14, 1994.
178. Rahman I. Antioxidant therapies in COPD. *Int J COPD* 1: 15–30, 2006.
179. Anderson R. Assessment of the roles of vitamin C, vitamin E, and beta-carotene in the modulation of oxidant stress mediated by cigarette smoke-activated phagocytes. *Am J Clin Nutr* 53: 358S–61S, 1991.
180. Romieu I, Trenga C. Diet and obstructive lung diseases. *Epidemiol Rev* 2: 268–87, 2001.
181. Santos MC, Oliveira AL, Viegas-Crespo AM, Vicente L, Barreiros A, Monteiro P, Pinheiro T, Bugalho De AA. Systemic markers of the redox balance in chronic obstructive pulmonary disease. *Biomarkers* 9: 461–69, 2004.
182. Gilliland FD, Berhane KT, Li YF, Gauderman WJ, McConnell R, Peters J. Children's lung function and antioxidant vitamin, fruit, juice, and vegetable intake. *Am J Epidemiol* 158: 576–84, 2003.
183. Harik-Khan RI, Muller DC, Wise RA. Serum vitamin levels and the risk of asthma in children. *Am J Epidemiol* 159: 351–57, 2004.
184. Rubin RN, Navon L, Cassano PA. Relationship of serum antioxidants to asthma prevalence in youth. *Am J Respir Crit Care Med* 169: 393–98, 2004.
185. Allam MF, Lucane RA. Selenium supplementation for asthma. *Cochrane Database Syst Rev*, 2004, 2: CD003538, 2004
186. Clausen J. The influence of selenium and vitamin E on the enhanced respiratory burst reaction in smokers. *Biol Trace Elem Res* 31: 281–91, 1991.
187. Fogarty A, Lewis SA, Scrivener SL, Antoniak M, Pacey S, Pringle M, Britton J. Oral magnesium and vitamin C supplements in asthma: A parallel group randomized placebo-controlled trial. *Clin Exp Allergy* 33: 1355–59, 2003.
188. Pearson PJ, Lewis SA, Britton J, Fogarty A. Vitamin E supplements in asthma: A parallel group randomised placebo controlled trial. *Thorax* 59: 652–56, 2004.
189. Bucca C, Rolla G, Arossa W, Caria E, Elia C, Nebiolo F, Baldi S. Effect of ascorbic acid on increased bronchial responsiveness during upper airway infection. *Respiration* 55: 214–19, 1989.
190. Rautalahti M, Virtamo J, Haukka J, Heinonen OP, Sundvall J, Albanes D, Huttunen JK. The effect of alpha-tocopherol and beta-carotene supplementation on COPD symptoms. *Am J Respir Crit Care Med* 156: 1447–52, 1997.
191. MacNee W, Rahman I. Oxidants and antioxidants as therapeutic targets in chronic obstructive pulmonary disease. *Am J Respir Crit Care Med* 160: S58–S65, 1999.
192. Borok Z, Buhl R, Grimes GJ, Bokser AD, Hubbard RC, Holroyd KJ, Roum JH, Czerski DB, Cantin AM, Crystal RG. Effect of glutathione aerosol on oxidant-antioxidant imbalance in idiopathic pulmonary fibrosis. *Lancet* 338: 215–16, 1991.
193. Gillissen A, Roum JH, Hoyt RF, Crystal RG. Aerosolization of superoxide dismutase. Augmentation of respiratory epithelial lining fluid antioxidant screen by aerosolization of recombinant human Cu++/Zn++ superoxide dismutase. *Chest* 104: 811–15, 1993.
194. MacNee W, Bridgeman MM, Marsden M, Drost E, Lannan S, Selby C, Donaldson K. The effects of *N*-acetylcysteine and glutathione on smoke-induced changes in lung phagocytes and epithelial cells. *Am J Med* 1(91): 60S–66S, 1991.
195. Dueholm M, Nielson C, Thorshauge H *et al. N*-acetylcysteine by metered dose inhaler in the treatment of chronic bronchitis: A multicentre study. *Respir Med* 86: 89–92, 1992.
196. Dekhuijzen PN. Antioxidant properties of *N*-acetylcysteine: Their relevance in relation to chronic obstructive pulmonary disease. *Eur Respir J* 23: 629–36, 2004.
197. Poole PJ, Black PN. Oral mucolytic drugs for exacerbations of chronic obstructive pulmonary disease: Systematic review. *BMJ* 322: 1271–74, 2001.
198. Poole PJ, Black PN. Preventing exacerbations of chronic bronchitis and COPD: Therapeutic potential of mucolytic agents. *Am J Respir Med* 2: 367–70, 2003.
199. Stey C, Steurer J, Bachmann S, Medici TC, Tramer MR. The effect of oral *N*-acetylcysteine in chronic bronchitis: A quantitative systematic review. *Eur Respir J* 16: 253–62, 2006.
200. Grandjean EM, Berthet P, Ruffmann R, Leuenberger P. Efficacy of oral long-term *N*-acetylcysteine in chronic bronchopulmonary disease: A meta-analysis of published double-blind, placebo-controlled clinical trials. *Clin Ther* 22: 209–21, 2000.
201. Van Schooten FJ, Nia AB, De Flora S *et al.* Effects of oral administration of *N*-acetyl-l-cysteine: A multi-biomarker study in smokers. *Cancer Epidemiol Biomarkers Prev* 11: 167–75, 2002.
202. Kasielski M, Nowak D. Long-term administration of *N*-acetylcysteine decreases hydrogen peroxide exhalation in subjects with chronic obstructive pulmonary disease. *Respir Med* 95: 448–56, 2001, 200.
203. De Benedetto F, Aceto A, Dragani B, Spacone A, Formisano S, Pela R, Donner CF, Sanguinetti CM. Long-term oral *N*-acetylcysteine reduces exhaled hydrogen peroxide in stable COPD. *Pulm Pharmacol Ther* 18: 41–47, 2005.
204. Sadowska AM, van Overveld FJ, Gorecka D, Zdral A, Filewska M, Demkow UA, Luyten C, Saenen E, Zielinski J, De Backer WA. The interrelationship between markers of inflammation and oxidative stress in chronic obstructive pulmonary disease: Modulation by inhaled steroids and antioxidant. *Respir Med* 99: 241–49, 2005.
205. Decramer M, Rutten-van Molken M, Dekhuijzen PN, Troosters T, van Herwaarden C, Pellegrino R, van Schayck CP, Olivieri D, Del Donno M, De Backer W, Lankhorst I, Ardia A. Effects of *N*-acetylcysteine on outcomes in chronic obstructive pulmonary disease (Bronchitis Randomized on NAC Cost-Utility Study, BRONCUS): A randomised placebo-controlled trial. *Lancet* 365: 1552–60, 2005.
206. Bridgeman MM, Marsden M, MacNee W, Flenley DC, Ryle AP. Fluid after treatment with *N*-acetylcysteine. *Thorax* 46: 39–42, 1991.
207. Bridgeman MM, Marsden M, Selby C, Morrison D, MacNee W. Effect of *N*-acetyl cysteine on the concentrations of thiols in plasma, bronchoalveolar lavage fluid, and lung tissue. *Thorax* 49: 670–75, 1994.
208. Habib MP, Tank LJ, Lane LC, Garewal HS. Effect of vitamin E on exhaled ethane in cigarette smokers. *Chest* 115: 684–90, 1999.
209. Allard JP, Royall D, Kurian R, Muggli R, Jeejeebhoy KN. Critical assessment of body-composition measurements in malnourished

subjects with Crohn's disease: The role of bioelectric impedance analysis. *Am J Clin Nutr* 59: 884–90, 1994.
210. Steinberg FM, Chait A. Antioxidant vitamin supplementation and lipid peroxidation in smokers. *Am J Clin Nutr* 68: 319–27, 1998.
211. Aghdassi E, Royall D, Allard JP. Oxidative stress in smokers supplemented with vitamin C. *Int J Vitam Nutr Res* 6(9): 45–51, 1999.
212. Lykkesfeldt J, Christen S, Wallock LM, Chang HH, Jacob RA, Ames BN. Ascorbate is depleted by smoking and repleted by moderate supplementation: A study in male smokers and nonsmokers with matched dietary antioxidant intakes. *Am J Clin Nutr* 71: 530–36, 2000.
213. Antonicelli F, Parmentier M, Drost EM, Hirani N, Rahman I, Donaldson K, MacNee W. Nacystelyn inhibits oxidant-mediated interleukin-8 expression and NF-kappab nuclear binding in alveolar epithelial cells. *Free Radic Biol Med* 32: 492–502, 2002.
214. Antonicelli F, Brown D, Parmentier M, Drost EM, Hirani N, Rahman I, Donaldson K, MacNee W. Regulation of LPS-mediated inflammation *in vivo* and *in vitro* by the thiol antioxidant nacystelyn. *Am J Physiol Lung Cell Mol Physiol* 286: L1319–L1327, 2004.
215. Gillissen A, Jaworska M, Orth M, Coffiner M, Maes P, App EM, Cantin AM, Schultze-Werninghaus G. Nacystelyn, a novel lysine salt of *N*-acetylcysteine, to augment cellular antioxidant defence *in vitro*. *Respir Med* 91: 159–68, 1997.
216. App EM, Baran D, Dab I, Malfroot A, Coffiner M, Vanderbist F, King M. Dose-finding and 24-h monitoring for efficacy and safety of aerosolized nacystelyn in cystic fibrosis. *Eur Respir J* 19: 294–302, 2002.
217. Moretti M, Bottrighi P, Dallari R, Da PR, Dolcetti A, Grandi P, Garuti G, Guffanti E, Roversi P, De GM, Potena A. The effect of long-term treatment with erdosteine on chronic obstructive pulmonary disease: The EQUALIFE Study. *Drugs Exp Clin Res* 30: 143–52, 2004.
218. Zheng JP, Kang J, Huang SG, Chen P, Yao WZ, Yang L, Bai, CX, Wang CZ, Wang C, Chen BY, Shi Y, Liu CT, Chen P, Li Q, Wang ZS, Huang YJ, Luo ZY, Chen FP, Yuan JZ, Yuan BT, Qian HP, Zhi RC, Zhong NS. *Lancet*. 371: 2013-8, 2008.
219. Bottle SE, Micallef AS. Synthesis and EPR spin trapping properties of a new isoindole-based nitrone: 1,1,3-Trimethylisoindole *N*-oxide (TMINO). *Org Biomol Chem* 1: 2581–84, 2003.
220. Becker DA, Ley JJ, Echegoyen L, Alvarado R. Stilbazulenyl nitrone (STAZN): A nitronyl-substituted hydrocarbon with the potency of classical phenolic chain-breaking antioxidants. *J Am Chem Soc* 124: 4678–84, 2002.
221. Salvemini D, Wang ZQ, Zweier JL, Samouilov A, Macarthur H, Misko TP, Currie MG, Cuzzocrea S, Sikorski JA, Riley DP. A non-peptidyl mimic of superoxide dismutase with therapeutic activity in rats. *Science* 286: 304–6, 1999.
222. Tuder RM, Zhen L, Cho CY, Taraseviciene-Stewart L, Kasahara Y, Salvemini D, Voelkel NF, Flores SC. Oxidative stress and apoptosis interact and cause emphysema due to vascular endothelial growth factor receptor blockade. *Am J Respir Cell Mol Biol* 29: 88–97, 2003.
223. Muscoli C, Sacco I, Alecce W, Palma E, Nistico R, Costa N, Clementi F, Rotiroti D, Romeo F, Salvemini D, Mehta JL, Mollace V. The protective effect of superoxide dismutase mimetic M40401 on balloon injury-related neointima formation: Role of the lectin-like oxidized low-density lipoprotein receptor-1. *J Pharmacol Exp Ther* 311: 44–50, 2004.
224. Chang LY, Crapo JD. Inhibition of airway inflammation and hyperreactivity by an antioxidant mimetic. *Free Radic Biol Med* 33: 379–86, 2002.
225. Smith KR, Uyeminami DL, Kodavanti UP, Crapo JD, Chang LY, Pinkerton KE. Inhibition of tobacco smoke-induced lung inflammation by a catalytic antioxidant. *Free Radic Biol Med* 33: 1106–14, 2002.
226. Doctrow SR, Huffman K, Marcus CB, Musleh W, Bruce A, Baudry M, Malfroy B. Salen-manganese complexes: Combined superoxide dismutase/catalase mimics with broad pharmacological efficacy. *Adv Pharmacol* 38: 247–69, 1997.
227. Izumi M, McDonald MC, Sharpe MA, Chatterjee PK, Thiemermann C. Superoxide dismutase mimetics with catalase activity reduce the organ injury in hemorrhagic shock. *Shock* 18: 230–35, 2002.
228. Chatterjee PK, Patel NS, Kvale EO, Brown PA, Stewart KN, Mota-Filipe H, Sharpe MA, Di Paola R, Cuzzocrea S, Thiemermann C. EUK-134 reduces renal dysfunction and injury caused by oxidative and nitrosative stress of the kidney. *Am J Nephrol* 24: 165–77, 2004.
229. Sharpe MA, Ollosson R, Stewart VC, Clark JB. Oxidation of nitric oxide by oxomanganese-salen complexes: A new mechanism for cellular protection by superoxide dismutase/catalase mimetics. *Biochem J* 366: 97–107, 2002.
230. Cuzzocrea S, Thiemermann C, Salvemini D. Potential therapeutic effect of antioxidant therapy in shock and inflammation. *Curr Med Chem* 11: 1147–62, 2004.
231. Jozsef L, Filep JG. Selenium-containing compounds attenuate peroxynitrite-mediated NF-KappaB and AP-1 activation and interleukin-8 gene and protein expression in human leukocytes. *Free Radic Biol Med* 35: 1018–27, 2003.
232. Haddad E, McCluskie K, Birrell MA, Dabrowski D, Pecoraro M, Underwood S, Chen B, De Sanctis GT, Webber SE, Foster ML, Belvisi MG. Differential effects of ebselen on neutrophil recruitment, chemokine, and inflammatory mediator expression in a rat model of lipopolysaccharide-induced pulmonary inflammation. *J Immunol* 169: 974–82, 2002.
233. Day BJ. Catalytic antioxidants: A radical approach to new therapeutics. *Drug Discov Today* 9: 557–66, 2004.
234. Arts IC, Hollman PC. Polyphenols and disease risk in epidemiologic studies. *Am J Clin Nutr* 81: 317S–325S, 2005.
235. Knekt P, Kumpulainen J, Jarvinen R, Rissanen H, Heliovaara M, Reunanen A, Hakulinen T, Aromaa A. Flavonoid intake and risk of chronic diseases. *Am J Clin Nutr* 76: 560–68, 2002.
236. Santus P, Sola A, Carlucci P, Fumagalli F, Di GA, Mondoni M, Carnini C, Centanni S, Sala A. Lipid peroxidation and 5-lipoxygenase activity in chronic obstructive pulmonary disease. *Am J Respir Crit Care Med* 171: 838–43, 2005.
237. Culpitt SV, Rogers DF, Fenwick PS, Shah P, De Matos C, Russell RE, Barnes PJ, Donnelly LE. Inhibition by red wine extract, resveratrol, of cytokine release by alveolar macrophages in COPD. *Thorax* 58: 942–46, 2003.
238. Birrell MA, McCluskie K, Wong S, Donnelly LE, Barnes PJ, Belvisi MG. Resveratrol, an extract of red wine, inhibits lipopolysaccharide induced airway neutrophilia and inflammatory mediators through an NF-KappaB-independent mechanism. *FASEB J* 19: 840–41, 2005.
239. Manna SK, Mukhopadhyay A, Aggarwal BB. Resveratrol suppresses TNF-induced activation of nuclear transcription factors NF-kappa B, activator protein-1, and apoptosis: Potential role of reactive oxygen intermediates and lipid peroxidation. *J Immunol* 164: 6509–19, 2000.
240. Donnelly LE, Newton R, Kennedy GE, Fenwick PS, Leung RH, Ito K, Russell RE, Barnes PJ. Anti-inflammatory effects of resveratrol in lung epithelial cells: Molecular mechanisms. *Am J Physiol Lung Cell Mol Physiol* 287: L774–83, 2004.
241. Biswas SK, McClure D, Jimenez LA, Megson IL, Rahman I. Curcumin induces glutathione biosynthesis and inhibits NF-kappaB activation and interleukin-8 release in alveolar epithelial cells: Mechanism of free radical scavenging activity. *Antioxid Redox Signal* 7: 32–41, 2005.
242. Shishodia S, Potdar P, Gairola CG, Aggarwal BB. Curcumin (diferuloylmethane) Down-regulates cigarette smoke-induced NF-kappaB activation through inhibition of IkappaBalpha kinase in human lung epithelial cells: Correlation with suppression of COX-2, MMP-9 and cyclin D1. *Carcinogenesis* 24: 1269–79, 2003.
243. Meja KK, Rajendrasozhan S, Adenuga D, Biswas SK, Sundar IK, Spooner G, Marwick JA, Chakravarty P, Fletcher D, Whittaker P, Megson IL, Kirkham PA, Rahman I. *Am J Respir Cell Mol Biol*. 2008 (In press).

26

CHAPTER

Chemokines

James E. Pease and Timothy J. Williams

Leukocyte Biology Section, National Heart and Lung Institute, Imperial College London, London, UK

Tissue infection or injury induces a local inflammatory response in which the accumulation of white blood cells from the blood is an essential feature. The type of infection or injury determines the types of leukocyte which are recruited as appropriate for the necessary local host defense and repair response. Inflammatory diseases can result from an inappropriate or overzealous inflammatory response. Such inflammation, when it occurs in the lung, can seriously compromise the normal function of the organ and can be life-threatening. Thus, there is considerable interest in the mechanisms underlying leukocyte recruitment and the endogenous molecules regulating the process, with an ultimate aim of designing targeted therapy to attenuate the trafficking of specific cell types. Leukocyte recruitment represents a refinement of a fundamental property of a motile cell to detect a chemical gradient, orient itself, and move along the gradient.

This process, known as chemotaxis, has been elegantly described and analyzed in primitive single-celled organisms and is an essential property of cells at different stages of their life history in multicellular organisms. In higher animals, this property has been adapted to trap specific leukocyte types on the inner surface of small blood vessels and induce migration through the vessel wall into the tissues. Pivotal to the process of chemoattraction of leukocytes are the chemokines, a large family of chemotactic proteins, typically of 10 kDa; see Rot and Von Andrian [1] for a review. These proteins act upon specific G-protein-coupled receptors (GPCRs) found on the surface of cells and have a prominent role in the immune system, where they are essential for both the localization of different types of leukocytes in tissue compartments under basal conditions and the rapid recruitment of these cells to sites of injury or infection.

The chemokine gradient, which the cells follow, is produced constitutively in specialized locations, or upregulated at sites of inflammation, such as in asthma or chronic obstructive pulmonary disease (COPD). The chemokines produced locally stimulate integrins on the leukocyte to attach to complementary receptors on the microvascular endothelium and cytoskeletal changes necessary for attachment and migration through the barrier. Similar processes mediate the movement of the cell in the tissue toward the inflammatory focus and elsewhere if appropriate (e.g. to lymph nodes).

Until recently, little was known about how the immune system organizes its geography under basal conditions or how specific cell types can be recruited to sites of inflammation. Chemoattractants, such as complement-derived C5a, leukotriene B4, and platelet activity factor, had been discovered to be chemotactic *in vitro* in Boyden chamber chemotaxis assays, but none had the specificity necessary for many of the migratory responses seen in the body. This situation changed dramatically with the discovery of the first chemokines in the late 1980s and the growing number of family members that have been discovered over the last decade.

CHEMOKINES

Chemokines have multiple roles in the organization of the immune system under basal conditions and during infection, and are also involved in angiogenesis. A crucial role of chemokines is the recruitment of different types of leukocytes from the blood to the sites of inflammation. Selective cell recruitment is achieved by means of locally produced chemokines, the particular chemokine secreted depends on the type of stimulus and the tissue involved. Specificity is provided by the expression of different types of receptor on different leukocyte types.

Most of the known chemokines (Table 26.1) belong to two major families defined by the position of four conserved cysteine residues (Fig. 26.1). The largest family is the CC chemokine family (28 in number) which possess two adjacent cysteines in the vicinity of the N-terminus of the mature peptide. The CXC family (16 members) have two cysteine residues in this same region, but with an interposed amino acid. In both families, the first cysteine forms a disulfide bond with the third, and the second with the fourth cysteine. CXC chemokines are characteristically chemotactic for neutrophils and lymphocytes, whereas CC chemokines act upon the majority of leukocytes, but generally have little activity on neutrophils. CXC chemokines can be further subdivided into those possessing a glutamate–leucine–arginine (ELR) motif immediately before the first cysteine, and those lacking the ELR motif. Members of the former, such as CXCL1, CXCL8, and CXCL7, are associated with angiogenic properties, while those lacking the ELR motif are typically inducible by IFN-γ and are angiostatic, such as CXCL9, CXCL10, and CXCL11 [2]. There are some nonconforming chemokines: lymphotactin has only one pair of cysteines and a single disulfide bond [3] and fractalkine has a CXXXC motif at the N-terminus [4]. Fractalkine is also unusual in that it is produced attached to a cell-bound mucin stalk, a property shared by the CXC chemokine CXCL16 [5]. This allows them to also function as adhesion molecules, tethering cells expressing the cognate receptor.

Chemokines are thought to bind to presenting molecules on the luminal surface of the venular endothelium, such as glycosaminoglycans, where the chemokines engage their receptors on the leukocyte surface. Leukocytes typically roll on the endothelium through low affinity intermolecular interactions mediated by selectins. Stimulation of chemokine receptors induces the upregulation of adhesion molecules on the leukocyte surface and cytoskeletal changes necessary for firm attachment, followed by emigration.

Chemokines bind to and transduce their actions via specific GPCRs [7, 8]. To date, 6 CXC and 10 CC chemokine receptors have been characterized, with single receptors identified for fractalkine and for lymphotactin. The majority of receptors bind and signal in response to several chemokine ligands, but such promiscuity is generally class-restricted, with CC chemokines binding only to CC chemokine receptors. One exception to this is the Duffy antigen receptor for chemokines (DARC) which can bind chemokines of both classes and has a role in clearance of chemokines. This receptor is present on erythrocytes and endothelial

TABLE 26.1 CXC Chemokines, and their receptors.

Chemokine	Original name	Chemokine receptor(s)	Chemokine	Original name	Chemokine receptor(s)
CXCL1	GRO-α/MGSA-α	CXCR2 > CXCR1	CCL8	MCP-2	CCR3
CXCL2	GRO-β/MGSA-β	CXCR2	(CCL9/10)	Unknown	Unknown
CXCL3	GRO-γ/MGSA-γ	CXCR2	CCL11	Eotaxin	CCR3
CXCL4	PF4	Unknown	(CCL12)	Unknown	Unknown
CXCL5	ENA-78	CXCR2	CCL13	MCP-4	CCR2, CCR3
CXCL6	GCP-2	CXCR1, CXCR2	CCL14	HCC-1	CCR1
CXCL7	NAP-2	CXCR2	CCL15	HCC-2/Lkn-1/MIP-1δ	CCR1, CCR3
CXCL8	IL-8	CXCR1, CXCR2	CCL16	HCC-4/LEC	CCR1
CXCL9	Mig	CXCR3	CCL17	TARC	CCR4
CXCL10	IP-10	CXCR3	CCL18	DC-CK1/PARC AMAC-1	Unknown
CXCL11	I-TAC	CXCR3	CCL19	MIP-3β/ELC/exodus-3	CCL7
CXCL12	SDF-1α/β	CXCR4	CCL20	MIP-3α/LARC/exodus-1	CCR6
CXCL13	BLC/BCA-1	CXCR5	CCL21	6Ckine/SLC/exodus-2	CCR7
CXCL14	BRAK/bolekine	Unknown	CCL22	MDC/STCP-1	CCR4
(CXCL15)	Unknown	Unknown	CCL23	MPIF-1	CCR1
CXCL16	SexCkine	CXCR6	CCL24	MPIF-2/eotaxin-2	CCR3

The systematic names, together with the colloquial names and agoinst activity of known receptors ae shown. Some chemokines appear to be missing from the list, e.g. CCL6. In these cases please note that whilst a chemokinase of that name has been identified in the mouse, no human orthologue has been documented.

cells, whereas a second receptor, D6 is located primarily on the lymphatic endothelium and is involved in the selective uptake of inflammatory chemokines [9].

The chemokines and their receptors have now been given a nomenclature to alleviate some of the problems associated with multiple colloquial names for the same ligands. Chemokine receptors are designated as CXCRn, CCRn, CX3CR1, and XCR1 with the chemokine ligands that stimulate these receptors now designated CXCLn, CCLn, CX3CLn, and XCLn, respectively (Table 26.1).

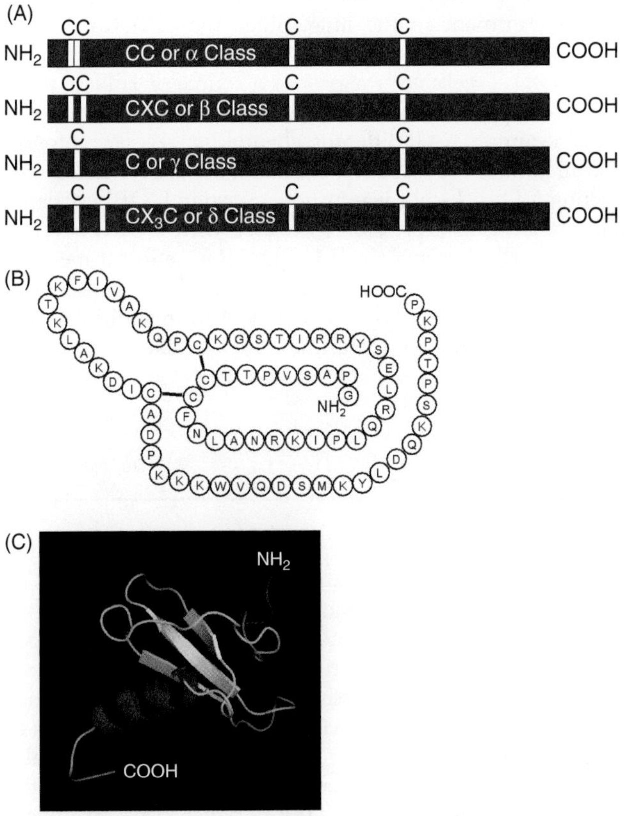

FIG. 26.1 (A) Linear representations of the major chemokine classes. (B) Primary structure of the human homolog of CCL11/Eotaxin showing disulfide bridges. (C) The secondary structure of human CCL11 is represented as a ribbon model, modeled from the coordinates given in Crump *et al.* using the Pymol software [6].

CHEMOKINES IN COPD

Neutrophils have been strongly implicated in the pathogenesis of COPD and are found in high numbers in lung tissue and sputum [10]. It is believed that activation of these cells to release activated oxygen species and proteases, such as elastase, is important in the specific pattern of lung damage and chronic dysfunction that characterize COPD. Consequently, there is considerable interest in the mechanisms involved in the recruitment of these cells to the lung with the aim of identifying opportunities for potential therapeutic intervention (Fig. 26.2).

CXCL8/IL-8 is a CXC chemokine that is a potent chemoattractant for neutrophils and was amongst the first of the chemokines to be discovered and characterized [11–13]. Early studies showed that this chemokine was present in the inflammatory exudate induced in a model of microbial infection *in vivo* [14, 15] and its production was often preceded by a phase of appearance of the complement fragment, C5a, which is also a potent neutrophil attractant but is generated in tissue fluid rather than as a secretory product of a cell [16]. Neutrophils themselves have been shown to produce CXCL8 during phagocytosis [17, 18], these effects can be blocked by antibodies to integrins and platelet-activating factor antagonists [19]. A similar pattern of chemoattractant production, C5a followed by CXCL8, has also been observed in response to myocardial ischemia, and interestingly the CXCL8 production is dependent on the presence of neutrophils in the heart – implying that neutrophils are the source of the chemokine [20]. In a lung model of ischemia, neutralization of CXCL8 by an antibody

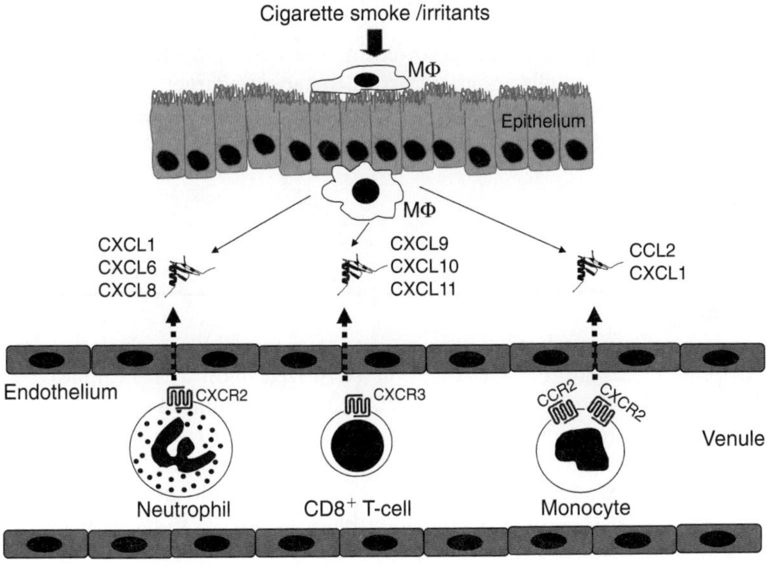

FIG. 26.2 Mechanisms underlying leukocyte recruitment in COPD. The roles of CXC chemokines in recruiting neutrophils, monocytes and T-cells to the COPD lung are illustrated.

suppresses tissue damage, thus providing a link between neutrophils and injury to lung tissues [21].

CXCL8 was first discovered as a secretory product of stimulated macrophages and other cells [11]. It stimulates neutrophils via two different receptors namely CXCR1 and CXCR2 [22, 23]; another related chemokine, CXCL6/GCP-2 can stimulate both receptors [24, 25]. There are several other chemokines that can stimulate CXCR2 only (e.g. CXCL1/GRO-α, CXCL2/GRO-β, CXCL3/GRO-γ, and CXCL7/NAP-2). CXCL8 appears to be an important mediator of neutrophil accumulation in COPD. Elevated levels of CXCL8 have been detected in the bronchoalveolar lavage (BAL) fluid, induced sputum, and in the circulation of patients with the disease compared with smoking and nonsmoking controls [26–29] with the levels of CXCL8 correlating with neutrophil accumulation [30–32]. Increases in CXCL8 levels in the sputum from COPD patients have also been documented during disease exacerbation [33, 34] although there appears to be some debate as to whether there is correlation with acute inflammation and viral/bacterial airway infections. Likewise, elevated levels of CXCL1 [35] and CXCL5 [36] have been detected in the sputum and BAL fluid of COPD patients compared with smoking and nonsmoking controls, although levels of the latter chemokine were not significantly different in normal smokers versus patients with emphysema [36]. The source of CXCL1 in COPD patients appears to be predominantly bronchial epithelial cells under the influence of TNF-α [37]. The chemotactic activity of CXCL1 for monocytes in addition to neutrophils, may explain in part the marked increase in alveolar macrophage numbers observed in COPD patients compared with normal controls [38]. The upregulation of MCP-1/CCL2 in the sputum and BAL of patients with COPD [35, 39] may also play a role in this process as it too is a potent recruiter of monocytes [40]. Since the engagement of chemokines with their cognate receptor usually leads to receptor endocytosis, this is the likely explanation for the finding that, compared with control subjects, the cell surface expression of CXCR2 on peripheral blood neutrophils is lower in patients with COPD than healthy controls [41]. Enhanced chemotaxis of monocytes from COPD patients in response to the CXC chemokines CXCL1/GRO-α and NAP-2/CXCL7, compared with monocytes from smoking and nonsmoking control subjects has also been reported, although this was apparently not due to differences in expression of either CXCR1 or CXCR2 on the cell surface [42].

In addition to the neutrophil influx observed in the lungs of COPD patients, increased numbers of $CD8^+$ T-cells have also been reported [43]. These cells are a major source of IFN-γ which itself, when expressed as a lung-targeted transgene in mice, can induce emphysema with classical symptoms of alveolar enlargement, enhanced lung volumes, enhanced pulmonary compliance, and an inflammatory infiltrate rich in both macrophages and neutrophils [44]. This suggests a primary role for $CD8^+$ T-cells in the orchestration of leukocyte recruitment and subsequent pathology. Supportive of this, a recent study showed that $CD8^+$ T-cell-deficient mice had a blunted inflammatory response and did not develop emphysema when exposed to long-term cigarette smoke [45]. Increased expression of CXCR3 by peripheral blood $CD8^+$ T-cells has been reported to be increased in smokers with COPD compared to controls [46]. Similarly, increased numbers of $CXCR3^+$ $CD8^+$ T-cells have been reported within the epithelium and submucosa of lung biopsies from smokers with COPD, compared with nonsmoking control subjects [47]. This suggests that this receptor and its ligands may play a role in their recruitment to the lung. Since the CXCR3 ligands CXCL9, CXCL10, and CXCL11 are all induced by IFN-γ and $CD8^+$ T-cells are themselves a source of IFN-γ, there is a great potential for positive feedback. A more recent study examining 11 different chemokine receptors, found that in addition to CXCR3, CCR5, and CXCR6 expression by lung $CD8^+$ T-cells was also increased in COPD patients and the levels of expression correlated with COPD severity [48], the cells displaying a Tc1, $CD45RA^+$ effector memory phenotype. Since the LTB4 receptor BLT1 has also been shown to be expressed by this subset [49], an additional role for leukotrienes in cell recruitment may be likely.

CHEMOKINES IN ASTHMA

Considerable evidence underpins the link between the infiltration of inflammatory cells in the lung and the progressive changes in lung pathophysiology that characterize asthma. For example, the late asthmatic response to allergen exposure is associated with activation of Th2-lymphocytes that are believed to regulate the massive eosinophil influx that occurs. Eosinophils are thought to make an important contribution to lung damage and dysfunction by releasing their toxic granular contents which directly induce tissue damage and also by the release of cytokines, such as TGF-β, which plays an important role in tissue remodeling [50]. The recent generation of mice deficient in eosinophils has allowed the role of the cell to be probed. Data from allergen-challenge models with these mice implicate the eosinophil in both airways remodeling and airway hyperresposiveness [51–53], although this is the subject of controversy [54]. For this reason, there is considerable interest in the endogenous chemoattractants involved in the recruitment of eosinophils to the lung. Since eosinophils are believed to have evolved in host defense to parasitic worms, these cells are an attractive therapeutic target in the Western world. In contrast, neutrophils present difficulties as a target, as they are essential for survival against microbial infection.

EOSINOPHIL CHEMOATTRACTION

The selective recruitment of eosinophils during the late response to allergen in sensitized individuals implies the existence of endogenous selective eosinophil chemoattractants (Fig. 26.3). A protein was purified from BAL fluid of allergen-challenged, sensitized guinea pigs that had this property [55]. This protein was sequenced and named "Eotaxin" as it was found to exhibit high specifically in recruiting eosinophils when intradermally injected into naïve recipients [55].

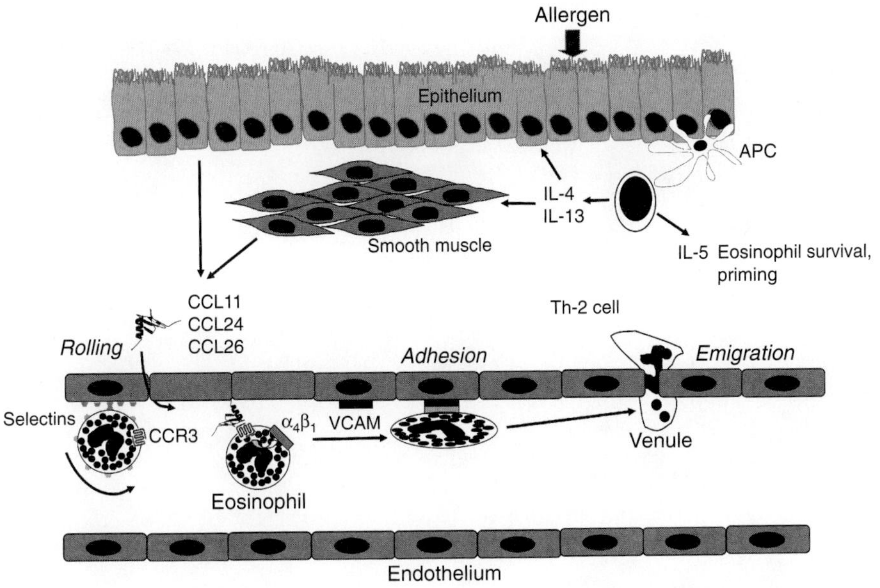

FIG. 26.3 Mechanisms underlying eosinophil recruitment in asthma. An example of the links between the cytokine, chemokine and cellular adhesion systems mediating the recruitment of eosinophils to sites of allergic inflammation .

Eotaxin/CCL11 is expressed by many cells [56–66] and its generation within the lung appears to be dependent upon the presence of T-cells [67]. CCL11 orthologs have been discovered in several species, including human, mouse, rat, and macaque [68–72]. CCL11 is a highly potent chemoattractant for eosinophils acting via a single chemokine receptor, CCR3, that is highly expressed on eosinophils [72], but also on Th2 cells [73], basophils [74], and mast cells [75, 76], supportive of a critical role for this molecule in the process of allergic inflammation. Two additional functional homologs, with low sequence similarity but closely related biological properties, have been described in the human and are named CCL24/Eotaxin-2 and CCL26/Eotaxin-3 [77–81]. Despite sharing the same colloquial name, both CCL24 and CCL26 are only distantly related to CCL11, being encoded in different chromosomal locations and sharing only low identity at the amino acid level. In terms of their ability to recruit eosinophils in assays of chemotaxis *in vitro*, CCL11 and CCL24 exhibit similar potencies, whilst CCL26 has been reported to be an order of magnitude less potent [77]. All three signal specifically via CCR3. Animal studies have highlighted the production of both CCL11 mRNA and CCL11 protein in the lung during the early response to allergen exposure [55, 82–84]. In humans, increased expression of CCL11 and CCL24 at both the mRNA and the protein level has been observed in the allergic lung of both atopic and nonatopic asthmatics [56, 59, 60, 85] and the proteins are also found in induced sputum [60, 86]. In both animal and human studies of allergic airways disease, airway epithelium, microvascular endothelium, tissue macrophages, and infiltrating inflammatory cells have been shown to express eotaxins [56, 57, 59, 60]. CCL11 has also been shown to act in concert with the important eosinophil survival factor IL-5, promoting eosinophil recruitment from the microcirculation [87] and eosinophil mobilization from the bone marrow [84, 88, 89]. Thus, CCL11 along with other cytokines may further contribute to eosinophilia. Indeed, the number of eosinophils in the blood appears to be an important determinant of the number that can be recruited to a site of allergic inflammation [87]. In addition to the eotaxins, an increasing number of CC chemokines, have been found to activate CCR3 with varying potency, including CCL5/RANTES and members of the monocyte chemotactic proteins such as CCL2, CCL7, and CCL13, but as these chemokines also activate other chemokine receptors, they are relatively nonselective (Table 26.2).

Different approaches have been made to determine the role of particular chemokines in eosinophil recruitment *in vivo,* including the use of neutralizing antibodies against both CCR3 and CCL11 which inhibited the migration of eosinophils *in vivo* [90, 91]. Studies with mice harboring a disrupted gene for CCL11 have shown conflicting results; BALB/c mice deficient in CCL11 showed a reduction in ovalbumin-induced lung eosinophilia [92] whilst, in contrast, CCL11-deficient mice of the outbred ICR strain exhibited no difference in allergen-induced eosinophil recruitment.

Use of CCR3-deficient mice has also been informative. Reduced eosinophil numbers in the airways of intraperitoneal-sensitized, aerosol-challenged CCR3-deficient mice have been reported by Humbles *et al.*, with the eosinophils trapped within the vasculature, unable to migrate into the lung parenchyma [93]. Intriguingly, CCR3-deficient mice also exhibited elevated airway hyperresponsiveness (AHR) to methacholine upon ovalbumin challenge which was attributed to increased numbers of intraepithelial mast cells within the trachea. The routes of immunization used for allergen sensitization appear to make an important contribution, as intradermal sensitization of mice with ovalbumin led to reduced AHR responses in CCR3-deficient mice following methacholine and ovalbumin challenge [94]. More recently, a study using *Aspergillus* as the challenge, rather than ovalbumin, found reduced numbers of perivascular and peribronchial eosinophils in both CCR3-deficient mice and double knockout CCL11/CCL24-deficient mice [52]. Deletion of the genes for CCL11 and the Th2 cytokine IL-5 in double knockout mice appears to be particularly effective in suppressing

TABLE 26.2 Cellular specificity of chemokines.

Systematic name	Colloquial name	Receptor usage	Principle cells recruited	Systematic name	Colloquial name	Receptor usage	Principle cells recruited
CCL1	I-309	CCR8	Mo, T, NK	CXCL1	GRO-α/MGSA-α	CXCR2	No, Mo
CCL2	MCP-1/MCAF	CCR2	Mo, DC, T, Bs	CXCL2	GRO-β/MGSA-β	CXCR2	No, Mo
CCL3	MIP-1α/LD78 α	CCR1, 5	Mo, DC, Eo, Bs, T, No, NK	CXCL3	GRO-γ/MGSA-γ	CXCR2	No, Mo
CCL3L1	LD78β	CCR1, 5	Mo, No, DC, Eo, Bs, T, NK	CXCL4	PF-4	Unknown	T, No, Mo
CCL4	MIP-1β	CCR5	Mo, DC, T	CXCL5	ENA-78	CXCR2	No, Mo
CCL4L1	LAG-1	CCR5	Mo, DC, T	CXCL6	GCP-2	CXCR1, 2	No, Mo
CCL5	RANTES	CCR1, 3, 5	Mo, DC, Eo, Bs, T, No, NK	CXCL7	NAP-2	CXCR2	No, Mo
CCL7	MCP-3	CCR1, 2, 3	Mo, DC, Eo, Bs, T, No, NK	CXCL8	IL-8	CXCR1, 2	No, Mo
CCL8	MCP-2	CCR3	Eo, T, Bs, Mc	CXCL9	Mig	CXCR3	T, B
CCL11	Eotaxin	CCR3	Eo, Bs, T	CXCL10	IP-10	CXCR3	T, B
CCL13	MCP-4	CCR2, 3	Eo, T, Bs, Mc Mo, DC,	CXCL11	I-TAC	CXCR3	T, B
CCL14	HCC-1	CCR1	Mo, DC, Eo, Bs, T, No, NK	CXCL12	SDF-1α/β	CXCR4	T, B, DC, Mo
CCL15	HCC-2/Lkn-1/MIP-1δ/MIP-5	CCR1, 3	Eo, T, Bs, Mc	CXCL13	BLC/BCA-1	CXCR5	T, B
CCL16	HCC-4/LEC	CCR1	Mo, DC, Eo, Bs, T, No, NK	CXCL14	BRAK/bolekine	unknown	
CCL17	TARC	CCR4	DC, T, Bs, NK	CXCL16	SR-PSOX	CXCR6	T
CCL18	DC-CK 1/AMAC-1/MIP-5/PARC	unknown	T				
CCL19	MIP-3β/ELC/	CCR7	DC, T, B, NK	XCL1	Lymphotactin/ SCM-1α/ATAC	XCR1	T, NK
CCL20	MIP-3α/LARC	CCR6	DC, T	XCL2	SCM-1β	XCR1	T, NK
CCL21	SLC/6Ckine	CCR7	DC, T, B, NK				
CCL22	MDC/STCP-1	CCR4	DC, T, Bs, NK	CX3CL1	Fractalkine, neurotactin	CX3CR1	T, NK, DC, Mo
CCL23	MPIF-1	CCR1	Mo, DC, Eo, Bs, T, No, NK				
CCL24	MPIF-2/eotaxin-2	CCR3	Eo, Bs, T				
CCL25	TECK	CCR9	T				
CCL26	Eotaxin-3	CCR3	Eo, Bs, T				
CCL27	CTACK/ALP/ILC/ESkine	CCR10	T				
CCL28	MEC	CCR10, 3	Eo, T, Bs, Mc				

B: B-lymphocyte; Bs: basophil; DC: dendritic cell: Eo, eosinophil; Mc: mast cell; Mo: monocyte; NK: natural killer cell; No: neutrophil; T: T-lymphocyte.

airway eosinophilia and hyperresponsiveness [95]; interestingly, IL-13 levels were markedly reduced in these animals.

CHEMOATTRACTION OF DENDRITIC CELLS

Dendritic cells (DCs) are important sentinels of the immune system, and their migration to lymph nodes following activation and encounter with antigens is crucial to the adaptive immune response. In the context of allergic lung inflammation, CCR6 has been shown to be expressed by lung-derived DCs [96] allowing them to be recruited by the chemokine CCL20, produced by lymphocytes and monocytes and markedly upregulated by tumor necrosis factor or lipopolysaccharide [97, 98]. This receptor is downregulated during maturation and is replaced by CCR7, allowing the cells to respond to the chemokines CCL19 [99, 100] and CCL21 produced in the lymphatic system [101]. This maturation is accompanied by the production of chemokines, such as CCL22, which can recruit activated T-cells expressing CCR4 [102]. The importance of CCR7 in animal models of allergy has been demonstrated by the finding that mAb-mediated neutralization of CCR7 in a Severe Combined Immunodeficiency mouse reconstituted with Peripheral blood mononuclear cells from allergic humans, impaired DC homing to lymph nodes following antigen challenge, accompanied by a decrease in both Th2 cytokine production and T-cell recruitment [103]. Likewise, mice deficient in the transcription factor Runx3, which have enhanced expression of CCR7 on alveolar DCs, have an accumulation of activated DCs within the lymph nodes which is associated with features characteristic of allergic asthma, such as increased serum IgE levels and AHR to methacholine [104].

CHEMOATTRACTION OF T-LYMPHOCYTES

T-cell depletion has been shown to prevent allergen-induced CCL11 generation and eosinophil accumulation in the lungs of sensitized mice, emphasizing the pivotal role for T-cells in allergic disease [67]. A link between Th2 cells and eosinophil recruitment is well established, with the Th2 cytokines, IL-4, and IL-13 stimulating CCL11 and CCL26 production by other cells *in vitro* [58, 63, 105–107]. In addition, expression of CXCR4 [108] by naïve T-cells can be upregulated by IL-4 [109]. In murine models of allergic airway disease, neutralizing mAbs to CXCR4 and its with ligand CXCL12 were reported to reduce lung eosinophilia and airway hyperresponsiveness [110]. Likewise, treatment of allergic mice with the CXCR4 antagonist AMD3100 resulted in significant reduction in airway hyperresponsiveness, eosinophilia, and the production of Th2 associated cytokines such as IL-4, IL-5, CCL17, and CCL22 [111] supportive of a role for the CXCR4:CXCL12 axis in allergic airways disease. Th2 cells polarized *in vitro* express the chemokine receptors CCR3, CCR4, and CCR8 [73, 112, 113]. Such polarization has also been observed *in vivo*, with T-cells recovered from the allergic airways of humans shown to preferentially express CCR3, CCR4, and CCR8 [114–116]. Consistent with this, the CCR4 ligands CCL22 and CCL17 are upregulated in the human lung following allergen challenge [117, 118].

In mice, mAb-mediated neutralization of CCL11 and CCL22 has been shown to impede the early recruitment of Th2 cells to the allergic lung with CCL22 responsible for Th2 cell recruitment following repeated antigen stimulation [119]. Studies using mice deficient in CCR4 and CCR8 have failed to conclusively prove a requirement for either receptor in allergic airways disease [120–122] and likewise, mAb-mediated neutralization of CCR4 in guinea pigs or CCL1 in mice had little effect upon T helper cell recruitment following allergen challenge [123, 124].

Regulatory T-cells

An inbalance or defect in the activity of regulatory $CD4^+CD25^+$ T-cells may result in allergic disease [125, 126]. CCR4 and CCR8 are also reported to be co-expressed by $CD4^+CD25^+$ regulatory T-cells, in both murine and human systems. In the former, these cells have been reported to suppress the differentiation of murine $CD4^+$ T-cells toward Th2 cells via a contact-dependent mechanism [127]. Similarly in the human, $CD4^+CD25^+$ regulatory T-cells purified from peripheral blood have been reported to express CCR4 and CCR8 which is likely to facilitate their migration toward antigen-presenting cells (APCs) and activated T-cells, enabling them to inhibit APC function or suppress responding T-cells respectively [128]. T-cell polarization may also be amplified by chemokines acting as antagonists for example, CXCL9, CXCL10, and CXCL11, ligands for the Th1 associated receptor CXCR3, can also act as antagonists of CCR3 both *in vitro* [129] and *in vivo* [130].

CHEMOATTRACTION OF BASOPHILS

Basophils are the least common granulocyte in peripheral blood, but they accumulate in significant numbers in certain types of inflammatory responses [131]. Basophils contribute to the cellular infiltrate in asthmatic lung and are prominent in late-phase cutaneous allergic reactions. Although, relative to eosinophils and neutrophils, basophils contain small amounts of cytotoxic proteins and proteases, they are able to release mediators, such as histamine, in response to cross-linking of their FcεR1 receptors by allergens. Further, basophils are thought to contribute large amounts of the Th2 cytokines, IL-4 and IL-13 [132].

A major observation was that basophils express the CCR3 receptor and CCL11 is highly chemotactic for these cells [74, 133]. However, the preferential recruitment of basophils in certain types of inflammatory response (e.g. the cutaneous response to ectoparasitic ticks) suggests that more selective basophil chemoattractant molecules exist. Despite this, no such molecule has been identified. One possibility is that two or more chemokines acting in combination control basophil recruitment *in vivo.* Using a system for measuring shape change in basophils with flow cytometry, some evidence has been provided for this [134]. Combinations of ligands acting via CCR2 and CCR3 were the most effective in this system. Alternatively, nonchemokine molecules may have potent stimulatory effects on basophils. In this context, an extract of nasal polyps was found to have potent stimulatory effects on basophils. This activity was found to be due to insulin-like growth factor-1 and -2 (IGF-1 and -2) [135]. The IGFs induce increase random migration (i.e. chemokinesis). This action, not observed with eosinophils, had a marked potentiating effect on chemotaxis along a gradient induced by CCL11.

CHEMOATTRACTION OF MAST CELLS

Mast cells pose a particular challenge in the identification of chemoattractant mechanisms as recruitment is from a small circulating pool of progenitors in the blood. Recently, it has been found that mast cell progenitors express the high affinity BLT1 receptor [136]. As the ligand for this receptor, LTB4, is secreted by activated mature mast cells, this may explain part of the mechanism underlying local mast cell hyperplasia associated with allergic reactions. The cells mature and proliferate in the tissue under the influence of local factors, such as stem cell factor. During maturation, the BTL1 receptor is downregulated [136], and it is likely that other chemoattractants and their receptors are involved in moving the cells along local tissue gradients to mediate the localization of the long-lived tissue-resident mature cells. Several chemokine receptors have been identified on mast cells [76, 137, 138], and their ligands may be important in this localization in the tissues. Of particular interest is the finding that mast cells move into airway smooth muscle layers in asthmatic patients [139]. These mast cells have been found to express CXCR3 and one of its ligands, CCL10, is produced by asthmatic smooth muscle cells [140]. These observations may indicate an important mechanism underlying lung dysfunction in asthmatic patients.

CHEMOKINES AND THEIR RECEPTORS AS THERAPEUTIC TARGETS

The observations outlined above provide the basis for the development of low-molecular-weight compounds aimed at selectively preventing the recruitment of a particular leukocyte type to a site of inflammation and thus suppressing the pathogenesis associated with that cell type. High throughput screening of vast libraries comprising several millions of small molecules have identified many molecules that can modify the interaction between GPCRs and their ligands; such compounds, either agonists or antagonists, provide a large proportion of the drugs currently prescribed [141]. A selection of prototypic chemokine receptor antagonists of relevance to this chapter are shown in Fig. 26.4.

The first low-molecular-weight compound to be developed against a chemokine receptor was a CXCR2 antagonist [142]. This compound, SB225002, is a potent antagonist that blocks CXCL8 binding and inhibits neutrophil chemotaxis *in vitro* and neutrophil accumulation *in vivo* . It is therefore plausible that CXCR2 antagonists could be developed to prevent neutrophil accumulation as observed in the lung of COPD patients. However, such effects would have to be monitored carefully because neutrophils are essential for host defense against microbial pathogens, as previously discussed. The SB225002 compound has shown efficacy in several animal models of disease including bronchopulmonary dysplasia [143] and arthritis [144] although some of this efficacy may be due to "off-target" effects as recent profiling of the compound suggests it has potency at several additional targets, including histamine and prostanoid receptors [145].

Because of its prominence on cells associated with Th2 cell-driven allergic reactions, CCR3 has become a prime target for the development of antagonists. The first publication of such an antagonist described the effects of a low-molecular-weight compound (UCB 35625, a structure based on a compound produced by Banyu, Japan) that antagonizes CCR3 [146]. This compound also antagonizes CCR1, a receptor for CCL3 and CCL5. Such a bispecific compound may be advantageous as around 15–20% of individuals also express CCR1 at high levels on their eosinophils [147, 148]. UCB 35625 is unusual amongst chemokine receptor antagonists as, although it blocks the activation of chemokine receptors at nanomolar concentrations, it does not readily displace chemokine from the receptor. It appears that the antagonist binding site is located within the transmembrane helices of the receptor [149], whilst

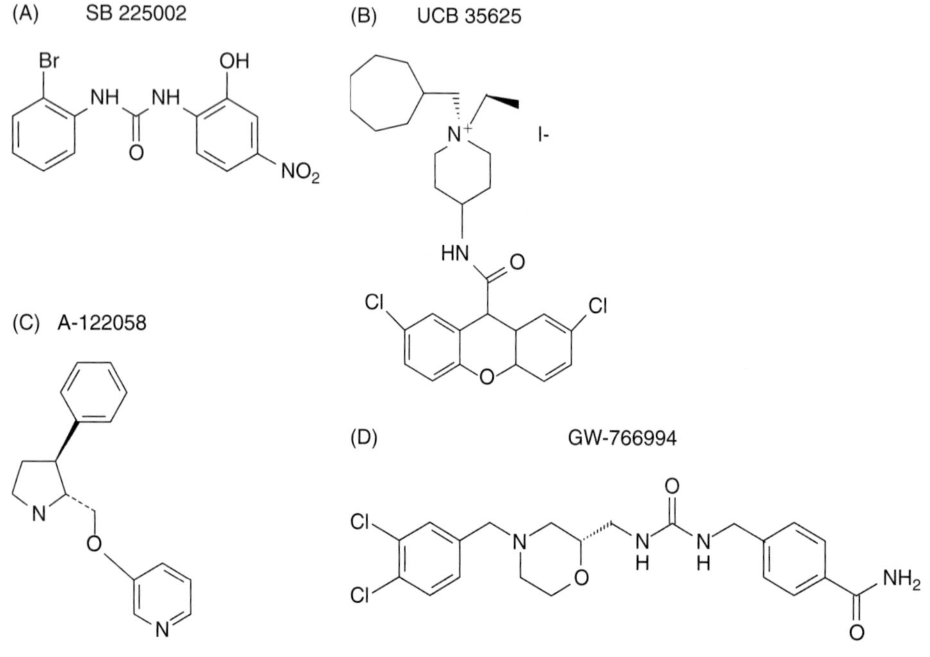

FIG. 26.4 Chemical structures of the CXCR2 antagonist (A) the bispecific CCR1 and CCR3 antagonist UCB 35625 (B), and the CCR3 antagonists A-122058 (C) and GW-701897B (D).

the ligand binding sites are located extracellularly [150]. Despite several publications describing the *in vitro* efficacy of CCR3 specific antagonists [151–154], descriptions of their efficacy *in vivo* lag well behind, as the molecules generally have poor cross-reactivity with the rodent ortholog receptors. The small molecule CCR3 antagonist A-122058 (Abbott Laboratories) has been reported to be efficacious in reducing the number of eosinophils in a mouse peritoneal model of eosinophil recruitment following injection with CCL11 [155]. Likewise, the Yamanouchi Pharmaceutical Company have recently described compounds with efficacy in a murine model of cutaneous inflammation [156] and in a macaque model of eosinophil recruitment to the lung following bronchoprovocation with CCL11 [151].

To date, the only small molecule antagonist reported to have entered Phase II trails is the compound GW-766994 from GlaxoSmithKline, with little apparent efficacy in an allergic rhinitis study [157]. This follows on from the recent reports of the efficacy of the GlaxoSmithKline compound GW-701897B in reducing vagally mediated bronchoconstriction in antigen-challenged guinea pigs [158]. In an alternative approach, neutralization of CCL11 has been achieved by the generation of a human single-chain fragment variable antibody named CAT-213, isolated by phage display methodologies [159]. CAT-213 is selective for CCL11 over closely related chemokines, such as CCL2, with sub-nanomolar IC_{50} values. In a Phase I single-dose clinical study, CAT-213 was reported to be well tolerated, and long-lived, remaining in an active form in serum for over 2 months, following intravenous administration [160]. In subsequent Phase II studies, administration of CAT-213 prior to the allergen challenge of rhinitic subjects was observed to reduce the mucosal eosinophil infiltration [161, 162].

CONCLUSION

A combination of clinical observations, *in vitro* cell biology and *in vivo* animal modeling has delineated potentially important mechanisms underlying lung inflammation in COPD and asthma. These extensive studies support the hypothesis that chemokines have a fundamental role in regulating leukocyte trafficking in inflammatory disease. The chemokine receptors represent novel targets for treatment and therapeutic compounds have been described with efficacy *in vitro* and *in vivo*. Clinical trials will ultimately provide the crucial information that will define the chemokine receptors that can be blocked to provide future therapy in these diseases.

References

1. Rot A, Von Andrian UH. Chemokines in innate and adaptive host defense: Basic chemokinese grammar for immune cells. *Annu Rev Immunol* 22: 891–928, 2004.
2 Strieter RM, Belperio JA, Keane MP. CXC chemokines in angiogenesis related to pulmonary fibrosis. *Chest* 122: 298S–301S, 2002.
3 Kelner GS, Kennedy J, Bacon KB *et al.* Lymphotactin: A cytokine that represents a new class of chemokine. *Science* 266: 1395–99, 1994.
4. Bazan JF, Bacon KB, Hardiman G *et al.* A new class of membrane-bound chemokine with a CX_3C motif. *Nature* 385: 640–44, 1997.
5. Matloubian M, David A, Engel S *et al.* A transmembrane CXC chemokine is a ligand for HIV-coreceptor Bonzo. *Nat Immunol* 1: 298–304, 2000.
6. DeLano WL. The PyMOL Molecular Graphics System on World Wide Web, 2002; Available from: http://www.pymol.org.
7. Murphy PM, Baggiolini M, Charo IF *et al.* International union of pharmacology. XXII. Nomenclature for chemokine receptors. *Pharmacol Rev* 52: 145–76, 2000.
8. Murphy PM. International Union of Pharmacology. XXX. Update on chemokine receptor nomenclature. *Pharmacol Rev* 54: 227–29, 2002.
9. Mantovani A, Bonecchi R, Locati M. Tuning inflammation and immunity by chemokine sequestration: Decoys and more. *Nat Rev Immunol* 6: 907–18, 2006.
10. Barnes PJ, Shapiro SD, Pauwels RA. Chronic obstructive pulmonary disease: Molecular and cellular mechanisms. *Eur Respir J* 22: 672–88, 2003.
11. Yoshimura T, Matsushima K, Tanaka S *et al.* Purification of a human monocyte-derived neutrophil chemotactic factor that has peptide sequence similarity to other host defense cytokines. *Proc Natl Acad Sci USA* 84: 9233–37, 1987.
12. Schröder J-M, Mrowietz U, Christophers E. Purification and partial biological characterization of a human lymphocyte-derived peptide with potent neutrophil-stimulating activity. *J Immunol* 140: 3534–40, 1988.
13. Walz A, Peveri P, Aschauer H, Baggiolini M. Purification and amino acid sequencing of NAF, a novel neutrophil-activating factor produced by monocytes. *Biochem Biophys Res Commun* 149: 755–61, 1987.
14. Beaubien BC, Collins PD, Jose PJ *et al.* A novel neutrophil chemoattractant generated during an inflammatory reaction in the rabbit peritoneal cavity *in vivo*: Purification, partial amino acid sequence and structural relationship to interleukin 8. *Biochem J* 271: 797–801, 1990.
15. Jose PJ, Collins PD, Perkins JA *et al.* Identification of a second neutrophil chemoattractant cytokine generated during an inflammatory reaction in the rabbit peritoneal cavity *in vivo*: Purification, partial amino acid sequence and structural relationship to melanoma growth stimulatory activity. *Biochem J* 278: 493–97, 1991.
16. Collins PD, Jose PJ, Williams TJ. The sequential generation of neutrophil chemoattractant proteins in acute inflammation in the rabbit *in vivo*. Relationship between C5a and proteins with the characteristics of IL-8/neutrophil-activating protein 1. *J Immunol* 146: 677–84, 1991.
17. Bazzoni F, Cassatella MA, Rossi F *et al.* Phagocytosing neutrophils produce and release high amounts of the neutrophil-activating peptide 1/interleukin 8. *J Exp Med* 173: 771–74, 1991.
18. Au B-T, Williams TJ, Collins PD. Zymosan-induced interleukin-8 release from human neutrophils involves activation via the CD11b/CD18 receptor and endogenous platelet activating factor as an autocrine modulator. *J Immunol* 152: 5411–19, 1994.
19. Au B-T, Teixeira MM, Collins PD, Williams TJ. Effect of PDE4 inhibitors on zymosan-induced IL-8 release from human neutrophils: Synergism with prostanoids and salbutamol. *Br J Pharmacol* 123: 1260–66, 1998.
20. Ivey CL, Williams FM, Collins PD *et al.* Neutrophil chemoattractants generated in two phases during reperfusion of ischemic myocardium in the rabbit: Evidence for a role for C5a and interleukin-8. *J Clin Invest* 95: 2720–28, 1995.
21. Sekido N, Mukaida N, Harada A *et al.* Prevention of lung reperfusion injury in rabbits by a monoclonal antibody against interleukin-8. *Nature* 365: 654–57, 1993.
22. Holmes WLJ, Kuang WJ, Rice G, Wood W. Structure and functional expression of a human interleukin08 receptor. *Science* 253: 1278, 1991.
23. Murphy P, Tiffany H. Cloning of complimentary DNA encoding a functional human IL-8 receptor. *Science* 253: 1280, 1991.

24. Proost P, De Wolf-Peeters C, Conings R *et al.* Identification of a novel granulocyte chemotactic protein (GCP-2) from human tumor cells. *J Immunol* 150: 1000–10, 1993.
25 Ahuja SK, Lee JC, Murphy PM. CXC chemokines bind to unique sets of selectivity determinants that can function independently and are broadly distributed on multiple domains of human interleukin-8 receptor B: Determinants of high affinity binding and receptor activation are distinct. *J Biol Chem* 271: 225–32, 1996.
26. Soler N, Ewig S, Torres A *et al.* Airway inflammation and bronchial microbial patterns in patients with stable chronic obstructive pulmonary disease. *Eur Respir J* 14: 1015–22, 1999.
27. Pesci A, Balbi B, Majori M *et al.* Inflammatory cells and mediators in bronchial lavage of patients with chronic obstructive pulmonary disease. *Eur Respir J* 12: 380–86, 1998.
28. Nocker RE, Schoonbrood DF, van de Graaf EA *et al.* Interleukin-8 in airway inflammation in patients with asthma and chronic obstructive pulmonary disease. *Int Arch Allergy Immunol* 109: 183–91, 1996.
29. Keatings VM, Collins PD, Scott DM, Barnes PJ. Differences in interleukin-8 and tumor necrosis factor-alpha in induced sputum from patients with chronic obstructive pulmonary disease or asthma. *Am J Respir Crit Care Med* 153: 530–34, 1996.
30. Keman S, Willemse B, Tollerud DJ *et al.* Blood interleukin-8 production is increased in chemical workers with bronchitic symptoms. *Am J Ind Med* 32: 670–73, 1997.
31. Yamamoto C, Yoneda T, Yoshikawa M *et al.* Airway inflammation in COPD assessed by sputum levels of interleukin-8. *Chest* 112: 505–10, 1997.
32. Woolhouse IS, Bayley DL, Stockley RA. Sputum chemotactic activity in chronic obstructive pulmonary disease: Effect of alpha(1)-antitrypsin deficiency and the role of leukotriene B(4) and interleukin 8. *Thorax* 57: 709–14, 2002.
33. Aaron SD, Angel JB, Lunau M *et al.* Granulocyte inflammatory markers and airway infection during acute exacerbation of chronic obstructive pulmonary disease. *Am J Respir Crit Care Med* 163: 349–55, 2001.
34. Sethi S, Muscarella K, Evans N *et al.* Airway inflammation and etiology of acute exacerbations of chronic bronchitis. *Chest* 118: 1557–65, 2000.
35. Traves SL, Culpitt SV, Russell RE *et al.* Increased levels of the chemokines GROalpha and MCP-1 in sputum samples from patients with COPD. *Thorax* 57: 590–95, 2002.
36. Tanino M, Betsuyaku T, Takeyabu K *et al.* Increased levels of interleukin-8 in BAL fluid from smokers susceptible to pulmonary emphysema. *Thorax* 57: 405–11, 2002.
37. Schulz C, Kratzel K, Wolf K *et al.* Activation of bronchial epithelial cells in smokers without airway obstruction and patients with COPD. *Chest* 125: 1706–13, 2004.
38. Turato G, Zuin R, Miniati M *et al.* Airway inflammation in severe chronic obstructive pulmonary disease: Relationship with lung function and radiologic emphysema. *Am J Respir Crit Care Med* 166: 105–10, 2002.
39. Capelli A, Di Stefano A, Gnemmi I *et al.* Increased MCP-1 and MIP-1beta in bronchoalveolar lavage fluid of chronic bronchitics. *Eur Respir J* 14: 160–65, 1999.
40. Rollins BJ, Walz A, Baggiolini M. Recombinant human MCP-1/JE induces chemotaxis, calcium flux, and the respiratory burst in human monocytes. *Blood* 78: 1112–16, 1991.
41. Pignatti P, Moscato G, Casarini S *et al.* Downmodulation of CXCL8/IL-8 receptors on neutrophils after recruitment in the airways. *J Allergy Clin Immunol* 115: 88–94, 2005.
42. Traves SL, Smith SJ, Barnes PJ, Donnelly LE. Specific CXC but not CC chemokines cause elevated monocyte migration in COPD: A role for CXCR2. *J Leukoc Biol* 76: 441–50, 2004.
43. Saetta M, Di Stefano A, Turato G *et al.* CD8+ T-lymphocytes in peripheral airways of smokers with chronic obstructive pulmonary disease. *Am J Respir Crit Care Med* 157: 822–26, 1998.
44. Wang Z, Zheng T, Zhu Z *et al.* Interferon gamma induction of pulmonary emphysema in the adult murine lung. *J Exp Med* 192: 1587–600, 2000.
45. Maeno T, Houghton AM, Quintero PA *et al.* CD8+ T Cells are required for inflammation and destruction in cigarette smoke-induced emphysema in mice. *J Immunol* 178: 8090–96, 2007.
46. Koch A, Gaczkowski M, Sturton G *et al.* Modification of surface antigens in blood CD8+ T-lymphocytes in COPD: Effects of smoking. *Eur Respir J* 29: 42–50, 2007.
47. Saetta M, Mariani M, Panina-Bordignon P *et al.* Increased expression of the chemokine receptor CXCR3 and its ligand CXCL10 in peripheral airways of smokers with chronic obstructive pulmonary disease. *Am J Respir Crit Care Med* 165: 1404–9, 2002.
48. Freeman CM, Curtis JL, Chensue SW. CC chemokine receptor 5 and CXC chemokine receptor 6 expression by lung CD8+ cells correlates with chronic obstructive pulmonary disease severity. *Am J Pathol* 171: 767–76, 2007.
49. Goodarzi K, Goodarzi M, Tager AM *et al.* Leukotriene B4 and BLT1 control cytotoxic effector T cell recruitment to inflamed tissues. *Nat Immunol* 4: 965–73, 2003.
50. Kay AB, Phipps S, Robinson DS. A role for eosinophils in airway remodelling in asthma. *Trends Immunol* 25: 477–82, 2004.
51. Humbles AA, Lloyd CM, McMillan SJ *et al.* A critical role for eosinophils in allergic airways remodeling. *Science* 305: 1776–79, 2004.
52. Fulkerson PC, Fischetti CA, McBride ML *et al.* A central regulatory role for eosinophils and the eotaxin/CCR3 axis in chronic experimental allergic airway inflammation. *Proc Natl Acad Sci USA* 103: 16418–23, 2006.
53. Lee JJ, Dimina D, Macias MP *et al.* Defining a link with asthma in mice congenitally deficient in eosinophils. *Science* 305: 1773–76, 2004.
54. Wills-Karp M, Karp CL. Biomedicine. Eosinophils in asthma: Remodeling a tangled tale. *Science* 305: 1726–29, 2004.
55. Jose PJ, Griffiths-Johnson DA, Collins PD *et al.* Eotaxin: A potent eosinophil chemoattractant cytokine detected in a guinea-pig model of allergic airways inflammation. *J Exp Med* 179: 881–87, 1994.
56. Ying S, Robinson DS, Meng Q *et al.* Enhanced expression of eotaxin and CCR3 mRNA and protein in atopic asthma. Association with airway hyperresponsiveness and predominant co-localization of eotaxin mRNA to bronchial epithelial and endothelial cells. *Eur J Immunol* 27: 3507–16, 1997.
57. Komiya A, Nagase H, Yamada H *et al.* Concerted expression of eotaxin-1, eotaxin-2, and eotaxin-3 in human bronchial epithelial cells. *Cell Immunol* 225: 91–100, 2003.
58. Kobayashi I, Yamamoto S, Nishi N *et al.* Regulatory mechanisms of Th2 cytokine-induced eotaxin-3 production in bronchial epithelial cells: Possible role of interleukin 4 receptor and nuclear factor-kappaB. *Ann Allergy Asthma Immunol* 93: 390–97, 2004.
59. Ying S, Meng Q, Zeibecoglou K *et al.* Eosinophil chemotactic chemokines (eotaxin, eotaxin-2, RANTES, monocyte chemoattractant protein-3 (MCP-3), and MCP-4), and C-C chemokine receptor 3 expression in bronchial biopsies from atopic and nonatopic (Intrinsic) asthmatics. *J Immunol* 163: 6321–29, 1999.
60. Zeibecoglou K, Macfarlane AJ, Ying S *et al.* Increases in eotaxin-positive cells in induced sputum from atopic asthmatic subjects after inhalational allergen challenge. *Allergy* 54: 730–35, 1999.
61. Abonyo BO, Alexander MS, Heiman AS. Autoregulation of CCL26 synthesis and secretion in A549 cells: A possible mechanism by which alveolar epithelial cells modulate airway inflammation. *Am J Physiol Lung Cell Mol Physiol* 289: L478–L488, 2005.
62. Miyamasu M, Yamaguchi M, Nakajima T *et al.* Th1-derived cytokine IFN-gamma is a potent inhibitor of eotaxin synthesis *in vitro*. *Int Immunol* 11: 1001–4, 1999.
63. Teran LM, Mochizuki M, Bartels J *et al.* Th1- and Th2-type cytokines regulate the expression and production of eotaxin and RANTES by human lung fibroblasts. *Am J Respir Cell Mol Biol* 20: 777–86, 1999.
64. Stellato C, Matsukura S, Fal A *et al.* Differential regulation of epithelial-derived C-C chemokine expression by IL-4 and the glucocorticoid budesonide. *J Immunol* 163: 5624–32, 1999.

65. Ghaffar O, Hamid Q, Renzi PM *et al.* Constitutive and cytokine-stimulated expression of eotaxin by human airway smooth muscle cells. *Am J Respir Crit Care Med* 159: 1933–42, 1999.
66. Fukagawa K, Nakajima T, Saito H *et al.* IL-4 induces eotaxin production in corneal keratocytes but not in epithelial cells. *Int Arch Allergy Immunol* 121: 144–50, 2000.
67. MacLean JA, Ownbey R, Luster AD. T cell-dependent regulation of eotaxin in antigen-induced pulmonary eosinophilia. *J Exp Med* 184: 1461–69, 1996.
68. Ponath PD, Qin S, Ringler DJ *et al.* Cloning of the human eosinophil chemoattractant, eotaxin. Expression, receptor binding and functional properties suggest a mechanism for the selective recruitment of eosinophils. *J Clin Invest* 97: 604–12, 1996.
69. Rothenberg ME, Luster AD, Leder P. Murine eotaxin: An eosinophil chemoattractant inducible in endothelial cells and in interleukin 4-induced tumor suppression. *Proc Natl Acad Sci USA* 92: 8960–64, 1995.
70. Williams CMM, Newton DJ, Wilson SA *et al.* Conserved structure and tissue expression of rat eotaxin. *Immunogenetics* 47: 178–80, 1998.
71. Zhang L, Soares MP, Guan Y *et al.* Functional expression and characterization of macaque C-C chemokine receptor 3 (CCR3) and generation of potent antagonistic anti-macaque CCR3 monoclonal antibodies. *J Biol Chem* 277: 33799–810, 2002.
72. Ponath PD, Qin S, Post TW *et al.* Molecular cloning and characterization of a human eotaxin receptor expressed selectively on eosinophils. *J Exp Med* 183: 2437–2448, 1996.
73. Sallusto F, Mackay CR, Lanzavecchia A. Selective expression of the eotaxin receptor CCR3 by human T helper 2 cells. *Science* 277: 2005–7, 1997.
74. Uguccioni M, Mackay CR, Ochensberger B *et al.* High expression of the chemokine receptor CCR3 in human blood basophils. Role in activation by eotaxin, MCP-4, and other chemokines. *J Clin Invest* 100: 1137–43, 1997.
75. Romagnani P, De Paulis A, Beltrame C *et al.* Tryptase–chymase double-positive human mast cells express the eotaxin receptor CCR3 and are attracted by CCR3-binding chemokines. *Am J Pathol* 155: 1195–204, 1999.
76. Ochi H, Hirani WM, Yuan Q *et al.* T helper cell type 2 cytokine-mediated comitogenic responses and CCR3 expression during differentiation of human mast cells *in vitro*. *J Exp Med* 190: 267–80, 1999.
77. Kitaura M, Suzuki N, Imai T *et al.* Molecular cloning of a novel human CC chemokine (Eotaxin-3) that is a functional ligand of CC chemokine receptor 3. *J Biol Chem* 274: 27975–80, 1999.
78. Patel VP, Kreider BL, Li Y *et al.* Molecular and functional characterization of two novel human C-C chemokines as inhibitors of two distinct classes of myeloid progenitors. *J Exp Med* 185: 1163–72, 1997.
79. White JR, Imburgia C, Dul E *et al.* Cloning and functional characterization of a novel human CC chemokine that binds to the CCR3 receptor and activates human eosinophils. *J Leukoc Biol* 62: 667–75, 1997.
80. Forssmann U, Uguccioni M, Loetscher P *et al.* Eotaxin-2, a novel CC chemokine that is selective for the chemokine receptor CCR3, and acts like eotaxin on human eosinophil and basophil leukocytes. *J Exp Med* 185: 2171–76, 1997.
81. Shinkai A, Yoshisue H, Koike M *et al.* A novel human CC chemokine, eotaxin-3, which is expressed in IL-4- stimulated vascular endothelial cells, exhibits potent activity toward eosinophils. *J Immunol* 163: 1602–10, 1999.
82. Rothenberg ME, Luster AD, Lilly CM *et al.* Constitutive and allergen-induced expression of eotaxin mRNA in the guinea pig lung. *J Exp Med* 181: 1211–16, 1995.
83. Gonzalo J-A, Jia G-Q, Aguirre V *et al.* Mouse eotaxin expression parallels eosinophil accumulation during lung allergic inflammation but it is not restricted to a Th2-type response. *Immunity* 4: 1–14, 1996.
84. Humbles AA, Conroy DM, Marleau S *et al.* Kinetics of eotaxin generation and its relationship to eosinophil accumulation in allergic airways disease: Analysis in a guinea pig model *in vivo*. *J Exp Med* 186: 601–12, 1997.
85. Yamada H, Yamaguchi M, Yamamoto K *et al.* Eotaxin in induced sputum of asthmatics: Relationship with eosinophils and eosinophil cationic protein in sputum. *Allergy* 55: 392–97, 2000.
86. Yamada H, Yamaguchi M, Yamamoto K *et al.* Eotaxin in induced sputum of asthmatics: Relationship with eosinophils and eosinophil cationic proteins in sputum. *Allergy* 55: 392–97, 2000.
87. Collins PD, Marleau S, Griffiths-Johnson DA *et al.* Co-operation between interleukin-5 and the chemokine eotaxin to induce eosinophil accumulation *in vivo*. *J Exp Med* 182: 1169–74, 1995.
88. Palframan RT, Collins PD, Severs NJ *et al.* Mechanisms of acute eosinophil mobilization from the bone marrow stimulated by interleukin 5: The role of specific adhesion molecules and phosphatidylinositol 3-kinase. *J Exp Med* 188: 1621–32, 1998.
89. Palframan RT, Collins PD, Williams TJ, Rankin SM. Eotaxin induces a rapid release of eosinophils and their progenitors from the bone marrow. *Blood* 91: 2240–48, 1998.
90. Sabroe I, Conroy DM, Gerard NP *et al.* Cloning and characterisation of the guinea pig eosinophil eotaxin receptor, CCR3: Blockade using a monoclonal antibody *in vivo*. *J Immunol* 161: 6139–47, 1998.
91. Campbell EM, Kunkel SL, Strieter RM, Lukacs NW. Temporal role of chemokines in a murine model of cockroach allergen-induced airway hyperreactivity and eosinophilia. *J Immunol* 161: 7047–53, 1998.
92. Rothenberg ME, MacLean JA, Pearlman E *et al.* Targeted disruption of the chemokine eotaxin partially reduces antigen-induced tissue eosinophilia. *J Exp Med* 185: 785–90, 1997.
93. Humbles AA, Lu B, Friend DS *et al.* The murine CCR3 receptor regulates both the role of eosinophils and mast cells in allergen-induced airway inflammation and hyperresponsiveness. *Proc Natl Acad Sci USA* 99: 1479–84, 2002.
94. Ma W, Bryce PJ, Humbles AA *et al.* CCR3 is essential for skin eosinophilia and airway hyperresponsiveness in a murine model of allergic skin inflammation. *J Clin Invest* 109: 621–28, 2002.
95. Mattes J, Yang M, Mahalingam S *et al.* Intrinsic defect in T cell production of interleukin (IL)-13 in the absence of both IL-5 and eotaxin precludes the development of eosinophilia and airways hyperreactivity in experimental asthma. *J Exp Med* 195: 1433–44, 2002.
96. Power CA, Church DJ, Meyer A *et al.* Cloning and characterization of a specific receptor for the novel CC chemokine MIP-3alpha from lung dendritic cells. *J Exp Med* 186: 825–35, 1997.
97. Greaves DR, Wang W, Dairaghi DJ *et al.* CCR6, a CC chemokine receptor that interacts with macrophage inflammatory protein 3alpha and is highly expressed in human dendritic cells. *J Exp Med* 186: 837–44, 1997.
98. Hromas R, Gray PW, Chantry D *et al.* Cloning and characterization of exodus, a novel beta-chemokine. *Blood* 89: 3315–22, 1997.
99. Dieu MC, Vanbervliet B, Vicari A *et al.* Selective recruitment of immature and mature dendritic cells by distinct chemokines expressed in different anatomic sites. *J Exp Med* 188: 373–86, 1998.
100. Sozzani S, Allavena P, D'Amico G *et al.* Differential regulation of chemokine receptors during dendritic cell maturation: A model for their trafficking properties. *J Immunol* 161: 1083–86, 1998.
101. Yoshida R, Imai T, Hieshima K *et al.* Molecular cloning of a novel human CC chemokine EBI1-ligand chemokine that is a specific functional ligand for EBI1, CCR7. *J Biol Chem*: 13803–9, 1997.
102. Tang HL, Cyster JG. Chemokine up-regulation and activated T cell attraction by maturing dendritic cells. *Science* 284: 819–22, 1999.
103. Hammad H, Lambrecht BN, Pochard P *et al.* Monocyte-derived dendritic cells induce a house dust mite-specific Th2 allergic inflammation in the lung of humanized SCID mice: Involvement of CCR7. *J Immunol* 169: 1524–34, 2002.
104. Fainaru O, Shseyov D, Hantisteanu S, Groner Y. Accelerated chemokine receptor 7-mediated dendritic cell migration in Runx3 knockout mice and the spontaneous development of asthma-like disease. *Proc Natl Acad Sci USA* 102: 10598–603, 2005.

105. Li L, Xia Y, Nguyen A *et al.* Effects of Th2 cytokines on chemokine expression in the lung: IL-13 potently induces eotaxin expression by airway epithelial cells. *J Immunol* 162: 2477–87, 1999.
106. Wenzel SE, Trudeau JB, Barnes S *et al.* TGF-beta and IL-13 synergistically increase eotaxin-1 production in human airway fibroblasts. *J Immunol JID – 2985117R* 169: 4613–19, 2002.
107. Zuyderduyn S, Hiemstra PS, Rabe KF. TGF-beta differentially regulates TH2 cytokine-induced eotaxin and eotaxin-3 release by human airway smooth muscle cells. *J Allergy Clin Immunol* 114: 791–98, 2004.
108. Bleul CC, Wu L, Hoxie JA *et al.* The HIV coreceptors CXCR4 and CCR5 are differentially expressed and regulated on human T lymphocytes. *Proc Natl Acad Sci USA* 94: 1925–30, 1997.
109. Jourdan P, Abbal C, Noraz N *et al.* IL-4 induces functional cell-surface expression of CXCR4 on human T cells. *J Immunol* 160: 4153–57, 1998.
110. Gonzalo JA, Lloyd CM, Peled A *et al.* Critical involvement of the chemotactic axis CXCR4/stromal cell-derived factor-1 alpha in the inflammatory component of allergic airway disease. *J Immunol* 165: 499–508, 2000.
111. Lukacs NW, Berlin A, Schols D *et al.* AMD3100, a CxCR4 antagonist, attenuates allergic lung inflammation and airway hyperreactivity. *Am J Pathol* 160: 1353–60, 2002.
112. Bonecchi R, Bianchi G, Bordignon PP *et al.* Differential expression of chemokine receptors and chemotactic responsiveness of type 1 T helper cells (Th1s) and Th2s. *J Exp Med* 187: 129–34, 1998.
113. Sallusto F, Lenig D, Mackay CR, Lanzavecchia A. Flexible programs of chemokine receptor expression on human polarised T helper 1 and 2 lymphocytes. *J Exp Med* 187: 875–83, 1998.
114. Morgan AJ, Symon FA, Berry MA *et al.* IL-4-expressing bronchoalveolar T cells from asthmatic and healthy subjects preferentially express CCR 3 and CCR 4. *J Allergy Clin Immunol* 116: 594–600, 2005.
115. Panina-Bordignon P, Papi A, Mariani M *et al.* The C-C chemokine receptors CCR4 and CCR8 identify airway T cells of allergen-challenged atopic asthmatics. *J Clin Invest* 107: 1357–64, 2001.
116. Banwell ME, Robinson DS, Lloyd CM. Adenoid-derived TH2 cells reactive to allergen and recall antigen express CC chemokine receptor 4. *J Allergy Clin Immunol* 112: 1155–61, 2003.
117. Pilette C, Francis JN, Till SJ, Durham SR. CCR4 ligands are up-regulated in the airways of atopic asthmatics after segmental allergen challenge. *Eur Respir J* 23: 876–84, 2004.
118. Bochner BS, Hudson SA, Xiao HQ, Liu MC. Release of both CCR4-active and CXCR3-active chemokines during human allergic pulmonary late-phase reactions. *J Allergy Clin Immunol* 112: 930–34, 2003.
119. Lloyd CM, Delaney T, Nguyen T *et al.* CC Chemokine receptor (CCR)3/eotaxin is followed by CCR4/monocyte-derived chemokine in mediating pulmonary T helper lymphocyte type 2 recruitment after serial antigen challenge *in vivo*. *J Exp Med* 191: 265–73, 2000.
120. Chvatchko Y, Hoogewerf AJ, Meyer A *et al.* A key role for CC chemokine receptor 4 in lipopolysaccharide-induced endotoxic shock. *J Exp Med* 191: 1755–64, 2000.
121. Chung CD, Kuo F, Kumer J *et al.* CCR8 is not essential for the development of inflammation in a mouse model of allergic airway disease. *J Immunol* 170: 581–87, 2003.
122. Goya I, Villares R, Zaballos A *et al.* Absence of CCR8 does not impair the response to ovalbumin-induced allergic airway disease. *J Immunol* 170: 2138–46, 2003.
123. Conroy DM, Jopling LA, Lloyd CM *et al.* CCR4 blockade does not inhibit allergic airways inflammation. *J Leukoc Biol* 74: 558–63, 2003.
124. Bishop B, Lloyd CM. CC chemokine ligand 1 promotes recruitment of eosinophils but not Th2 cells during the development of allergic airways disease. *J Immunol* 170: 4810–17, 2003.
125. Bellinghausen I, Klostermann B, Knop J, Saloga J. Human CD4+ CD25+ T cells derived from the majority of atopic donors are able to suppress TH1 and TH2 cytokine production. *J Allergy Clin Immunol* 111: 862–68, 2003.
126. Ling EM, Smith T, Nguyen XD *et al.* Relation of CD4 + CD25+ regulatory T-cell suppression of allergen-driven T-cell activation to atopic status and expression of allergic disease. *Lancet* 363: 608–15, 2004.
127. Stassen M, Jonuleit H, Muller C *et al.* Differential regulatory capacity of CD25+ T regulatory cells and preactivated CD25+ T regulatory cells on development, functional activation, and proliferation of Th2 cells. *J Immunol* 173: 267–74, 2004.
128. Iellem A, Mariani M, Lang R *et al.* Unique chemotactic response profile and specific expression of chemokine receptors CCR4 and CCR8 by CD4(+)CD25(+) regulatory T cells. *J Exp Med* 194: 847–53, 2001.
129. Xanthou G, Duchesnes CE, Williams TJ, Pease JE. CCR3 functional responses are regulated by both CXCR3 and its ligands CXCL9, CXCL10 and CXCL11. *Eur J Immunol* 33: 2241–50, 2003.
130. Fulkerson PC, Zimmermann N, Brandt EB *et al.* Negative regulation of eosinophil recruitment to the lung by the chemokine monokine induced by IFN-gamma (Mig, CXCL9). *Proc Natl Acad Sci USA* 101: 1987–92, 2004.
131. Falcone FH, Haas H, Gibbs BF. The human basophil: A new appreciation of its role in immune responses. *Blood* 96: 4028–38, 2000.
132. Gibbs BF, Haas H, Falcone FH *et al.* Purified human peripheral blood basophils release interleukin-13 and preformed interleukin-4 following immunological activation. *Eur J Immunol* 26: 2493–98, 1996.
133. Yamada H, Hirai K, Miyamasu M *et al.* Eotaxin is a potent chemotaxin for human basophils. *Biochem Biophys Res Commun* 231: 365–68, 1997.
134. Heinemann A, Hartnell A, Stubbs VE *et al.* Basophil responses to chemokines are regulated by both sequential and cooperative receptor signaling. *J Immunol* 165: 7224–33, 2000.
135. Hartnell A, Heinemann A, Conroy DM *et al.* Identification of selective basophil chemoattractants in human nasal polyps as insulin-like growth factor-1 and insulin-like growth factor-2. *J Immunol* 173: 6448–57, 2004.
136. Weller CL, Collington SJ, Brown JK *et al.* Leukotriene B4, an activation product of mast cells, is a chemoattractant for their progenitors. *J Exp Med* 201: 1961–71, 2005.
137. Oliveira SH, Lukacs NW. Stem cell factor and IgE-stimulated murine mast cells produce chemokines (CCL2, CCL17, CCL22) and express chemokine receptors. *Inflamm Res* 50: 168–74, 2001.
138. Juremalm M, Hjertson M, Olsson N *et al.* The chemokine receptor CXCR4 is expressed within the mast cell lineage and its ligand stromal cell-derived factor-1alpha acts as a mast cell chemotaxin. *Eur J Immunol* 30: 3614–22, 2000.
139. Brightling CE, Bradding P, Symon FA *et al.* Mast-cell infiltration of airway smooth muscle in asthma. *N Engl J Med* 346: 1699–705, 2002.
140. Brightling CE, Ammit AJ, Kaur D *et al.* The CXCL10/CXCR3 axis mediates human lung mast cell migration to asthmatic airway smooth muscle. *Am J Respir Crit Care Med* 171: 1103–8, 2005.
141. Fredriksson R, Lagerstrom MC, Lundin LG, Schioth HB. The G-protein-coupled receptors in the human genome form five main families. Phylogenetic analysis, paralogon groups, and fingerprints. *Mol Pharmacol* 63: 1256–72, 2003.
142. White JR, Lee JM, Young PR *et al.* Identification of a potent, selective non-peptide CXCR2 antagonist that inhibits interleukin-8-induced neutrophil migration. *J Biol Chem* 273: 10095–98, 1998.
143. Auten RL, Richardson RM, White JR *et al.* Nonpeptide CXCR2 antagonist prevents neutrophil accumulation in hyperoxia-exposed newborn rats. *J Pharmacol Exp Ther* 299: 90–95, 2001.
144. Podolin PL, Bolognese BJ, Foley JJ *et al.* A potent and selective non-peptide antagonist of CXCR2 inhibits acute and chronic models of arthritis in the rabbit. *J Immunol* 169: 6435–44, 2002.
145. Validation of different FLIPR calcium no wash dyes for performance through Millipore's GPCRProfilerTM service. Oral presentation SBS 13th Annual Conference & Exhibition, Montreal, Canada April 15th–19th, 2007.

146. Sabroe I, Peck MJ, Jan Van Keulen B *et al.* A small molecule antagonist of the chemokine receptors CCR1 and CCR3 : Potent inhibition of Eosinophil function and CCR3-mediated HIV-1 entry. *J Biol Chem* 275: 25985–92, 2000.
147. Sabroe I, Hartnell A, Jopling LA *et al.* Differential regulation of eosinophil chemokine signaling via CCR3 and non-CCR3 pathways. *J Immunol* 162: 2946–55, 1999.
148. Phillips R, Stubbs VELS, Henson MR, Williams TJ, Pease JE, Sabroe I. Variations in eosinophil chemokine responses: An investigation of CCR1 and CCR3 function, expression in atopy, and identification of a functional CCR1 promoter. *J Immunol* 170: 6190–201, 2003.
149. de Mendonca FL, da Fonseca PC, Phillips RM *et al.* Site-directed mutagenesis of CC chemokine receptor 1 reveals the mechanism of action of UCB 35625, a small molecule chemokine receptor antagonist. *J Biol Chem* 280: 4808–16, 2005.
150. Pease JE, Wang J, Ponath PD, Murphy PM. The N-terminal extracellular segments of the chemokine receptors CCR1 and CCR3 are determinants for MIP-1α and eotaxin binding, respectively, but a second domain is essential for receptor activation. *J Biol Chem* 273: 19972–76, 1998.
151. Morokata T, Suzuki K, Masunaga Y *et al.* A novel, selective, and orally available antagonist for CC chemokine receptor 3. *J Pharmacol Exp Ther* 317: 244–50, 2006.
152. Anderskewitz R, Bauer R, Bodenbach G *et al.* Pyrrolidinohydroquin azolines – a novel class of CCR3 modulators. *Bioorg Med Chem Lett* 15: 669–73, 2005.
153. De Lucca GV, Kim UT, Vargo BJ *et al.* Discovery of CC chemokine receptor-3 (CCR3) antagonists with picomolar potency. *J Med Chem* 48: 2194–211, 2005.
154. Naya A, Kobayashi K, Ishikawa M *et al.* Structure–activity relationships of 2-(benzothiazolylthio)acetamide class of CCR3 selective antagonist. *Chem Pharm Bull (Tokyo)* 51: 697–701, 2003.
155. Warrior U, McKeegan EM, Rottinghaus SM *et al.* Identification and characterization of novel antagonists of the CCR3 receptor. *J Biomol Screen* 8: 324–31, 2003.
156. Suzuki K, Morokata T, Morihira K *et al. In vitro* and *in vivo* characterization of a novel CCR3 antagonist, YM-344031. *Biochem Biophys Res Commun* 339: 1217–23, 2006.
157. Murdoch RD. *The Challenges of Drug Discovery & Development.* Oral presentation, 4th James Black Conference, University of Hertfordshire, September 13, 2006.
158. Fryer AD, Stein LH, Nie Z *et al.* Neuronal eotaxin and the effects of CCR3 antagonist on airway hyperreactivity and M2 receptor dysfunction. *J Clin Invest* 116: 228–36, 2006.
159. Main S, Handy R, Wilton J *et al.* A potent human anti-eotaxin1 antibody, CAT-213: Isolation by phage display and *in vitro* and *in vivo* efficacy. *J Pharmacol Exp Ther* 319: 1395–404, 2006.
160. Brennan N, Case N, Meyers T, Amakye D, Doughty J, Forward JA, Varley P, Powell J, Glover DR. Pharmacokinetics of CAT-123, a human anti-eotaxin-1 monoclonal antibody, following single intravenous administration to healthy volunteers. *Br J Clin Pharmacol* 53: 441P, 2002.
161. Pereira S, Clark T, Darby Y, Salib R, Salagean M, Hewitt L, Powell J, Howarth P, Scadding G. Effects of anti-eotaxin monoclonal antibody CAT-213 on allergen-induced rhinitis. *J Allergy Clin Immunol* 111: S268, 2003.
162. Salib R, Salagean M, Lau L, Di Giovanna I, Brennan N, Scadding G, Howarth P. The anti-inflammatory response of anti-eotaxin monoclonal antibody CAT-213 on nasal allergen-induced cell infiltration and activation. *J Allergy Clin Immunol* 111: S347, 2003.

27

Cytokines

CHAPTER

Kian Fan Chung

Department of Thoracic Medicine, National Heart and Lung Institute, Imperial College, London, UK

INTRODUCTION

Cytokines are signaling proteins, usually less than 80 kDa in size, which regulate a wide range of biological functions including innate and acquired immunity, hematopoiesis, inflammation and repair, and proliferation through mostly extracellular signaling. They are secreted by many cell types at local high concentrations and are involved in cell-to-cell interactions, have an effect on closely adjacent cells, and therefore function in a predominantly paracrine fashion. They may also act at a distance by secretion of soluble products into the circulation (endocrine or systemic effect) and may have effects on the cell of origin itself (autocrine effect). The first group of cytokines described was immune-response factors, labeled as lymphokines that were soluble products released from activated lymphocytes in response to specific or polyclonal antigen. Other groups of cytokines were named according to their actions such as the colony stimulating factors (e.g., granulocyte colony stimulating factor, G-CSF or granulocyte-macrophage colony stimulating factor, GM-CSF), the chemoattractants now grouped under chemokines with a special nomenclature (e.g., CCL11/eotaxin or CXCL1/MIP-1α), or the growth factors that mediate proliferation, differentiation, and survival of cells (e.g. transforming growth factor-β, TGF-β). However, the effect of a particular cytokine is not restricted to only one set of biological functions but often has more than one function extending beyond the function that may be implied in its name. The wide pleiotropy and element of redundancy in the cytokine family, with each cytokine having many overlapping functions, and with each function potentially mediated by more than one cytokine make the classification of cytokines a difficult task. In the context of airways disease, nevertheless, a broad functional classification remains helpful (Table 27.1). The major classes of cytokines include: pro- and anti-inflammatory cytokines, cytokines of neutrophil and eosinophil recruitment and activation, cytokines derived from T-helper (Th) and T-regulatory (Tregs) cells, and cytokines of T-cell recruitment and growth factors.

The receptors for many cytokines have been grouped into superfamilies based on the presence of common homology regions (Table 27.2). Cytokine receptors are linked to multiple signaling pathways in the cytoplasm and nucleus, leading to transcriptional and post-transcriptional activation of many factors including cytokines. Some cytokines may stimulate their own production in an autocrine manner while others stimulate the synthesis of different cytokines that have a stimulatory feedback effect on the first cytokine. Cytokines may also induce the expression of receptors which may change the responsiveness of both source and target cells. This complexity of cytokine interactions is referred to as the "cytokine network", the overall effect of which will depend on the type of cytokines and cytokine receptors involved and the general milieu and context of the response.

The potential contribution of cytokines to disease has been explored in studies using cytokines as agonists (mainly in cell-based studies), by blocking the effects of specific cytokines mainly in animal models, by conditional overexpression and deletion of cytokines in transgenic mice, and by genetic studies. In humans, implication for particular cytokines in disease has rested on the ability to detect their expression in the tissue of interest, and rarely on the effect of selective inhibition of specific cytokines. The sources and effects of cytokines are summarized in Table 27.2. In this chapter, the cytokines will

TABLE 27.1 Classification of cytokines and cytokine receptors.

Cytokines	
Proinflammatory cytokines	IL-1α/β, TNF-α/β, IL-6, IL-11, IL-18, IFN-γ
Anti-inflammatory cytokines	IL-10, TGFβ, IL-1ra
Cytokines of neutrophil recruitment and activation	CXCL8/IL-8, IL-1α/β, TNF-α/β, G-CSF, IL-17A, IL-17F
Cytokines of eosinophil recruitment and activation	IL-2, IL-3, IL-4, IL-5, GM-CSF, CCL5/RANTES, CCL11/eotaxin, CCL7/MCP-3, CCL13/MCP-4
Cytokines from T-helper (Th) cells	*Th1*: IFN-γ, IL-2, IL-12, GM-CSF, TNF-α, TNF-β, IL-18
	Th2: IL-4, IL-5, IL-6, IL-9, IL-10, IL-13, IL-25 (IL-17E)
Cytokines from T-regulatory cells (Tregs)	IL-10, TGFβ
Cytokines of T-cell recruitment	IL-16, CCL5/RANTES, CCL3/MIP-1α, CCL4/MIP-1β, TSLP, CCL17/TARC, CCL22/MDC
Growth factors	PDGF, VEGF, TGF-β, FGF, EGF, SCF
Cytokine receptor superfamilies	
Cytokine receptor superfamily	IL-2R β and γ chains, IL-4R, IL-3R α and β chains, IL-5 α and β chains, IL-6R, gp130, IL-12R, GM-CSFR Soluble forms by alternative splicing (e.g. IL-4R)
Immunoglobulin superfamily	IL-1R, IL-6R, PDGFR, M-CSFR
Protein kinase receptor superfamily	PDGFR, EGFR, FGFR
Interferon receptor superfamily	IFN-α/β receptor, IFN-γ receptor, IL-10 receptor
Nerve growth factor superfamily	NGFR, TNFR-I (p55), TNFR-II (p75)
Seven-transmembrane G-protein-coupled receptor superfamily	Chemokine receptors: CXCR1 to CXCR7, CCR1 to CCR10, CX_3CR1

EGF = epidermal growth factor; FGF = fibroblast growth factor; GM-CSF = granulocyte–macrophage colony stimulating factor; IFN = interferon; IL = interleukin; MCP = monocyte chemotactic protein; MDC = macrophage-derived chemokine; MIP = macrophage inflammatory protein; NGFR = nerve growth factor receptor; PDGF = platelet-derived growth factor; R = receptor; RANTES = regulated on activation, normal T-cell expressed, and secreted; SCF = stem cell factor; TARC = thymus and activation regulated chemokine; TGF = transforming growth factor; TNF = tumor necrosis factor; TSLP = thymic stromal lymphopoietin; VEGF = vascular endothelial growth factor.

TABLE 27.2 Sources and effects of cytokines.

Cytokine	Sources	Important cellular and mediator effects
Th2 cytokines		
IL-3	Th2 cell Mast cells Eosinophils	• Eosinophilia *in vivo* • Pluripotential hematopoietic factor
IL-4	Th2 cell Mast cells Basophils Eosinophils	• ↑ Eosinophil growth • ↑ Th2; ↓ Th1 • ↑ IgE • ↑ Mucin expression and goblet cells
IL-5	Th2 cell Mast cells Eosinophils	• Eosinophil maturation • ↓ Apoptosis • ↓ Th2 cells • BHR
IL-9	Th2 cells Eosinophils	• ↑ Activated T-cells and IgE from B-cells • ↑ Mast cell growth and differentiation • ↑ Mucin expression and goblet cells • Causes eosinophilic inflammation and BHR

(Continued)

TABLE 27.2 (Continued)

Cytokine	Sources	Important cellular and mediator effects
IL-13	Th2 cells Basophils Eosinophils Natural killer cells	● Activates eosinophils ● ↓ Apoptosis ● ↑ IgE ● ↑ Mucin expression and goblet cells
IL-15	T-cells Lung fibroblasts Monocytes	● As for IL-2 ● Growth and differentiation of T-cells
IL-23	Antigen presenting cells	● Induces IL-17A from activated memory $CD4^+$ or $CD8^+$ T-cells ● Induces secretion of IL-17F, IL-6, TNFα in Th17 cells
IL-25 (IL-17E)	Th2 cells Mast cells	● Limits Th1-induced inflammation ● Causes eosinophilic inflammation and bbronhicl hyperresponsiveness ● Increases IL-5, IL-13, CXCL5/ENA-78, CCL11/eotaxin expression ● Increases mucus secretion
Th1 cytokines		
IFN-γ	Th1 cells Natural killer cells	● ↓ Eosinophil influx after allergen ● ↓ Th2 cells ● Activates endothelial cells, epithelial cells, alveolar macrophages/monocytes ● ↓ IgE ● ↓ BHR
IL-2	Th0 & Th1 cells Eosinophils Airway epithelial cells	● Eosinophilia ***in vivo*** ● Growth and differentiation of T-cells
IL-12	B-cells Monocytes/macrophages Dendritic cells Eosinophils	● Regulates Th1 cell differentiation ● ↓ Expansion of Th2 cells ● ↓ IL-4-induced IgE synthesis ● Stimulates NK cells and T-cells to produce IFN-γ
IL-18	Th1 cells Airway epithelium	● Induces IFN-γ release from mitogen-stimulated blood mononuclear cells ● Induces Th1 cell development together with IL-12 ● Causes release of CXCL8/IL-8, CCL1/MIP1α and CCL2/MCP-1 from mononuclear cells
Proinflammatory		
IL-1	Monocytes/macrophages Fibroblasts B-cells Th1 and Th2 cells Neutrophils Endothelial cells Airway epithelial cells Airway smooth muscle cells	● ↑ Adhesion to vascular endothelium; eosinophil accumulation ***in vivo*** ● Growth factor for Th2 cells ● B-cell growth factor; neutrophil chemoattractant; T-cell and epithelial activation ● BHR
TNF-α	Macrophages T-cells Mast cells Airway epithelial cells	● Activates epithelium, endothelium, antigen-presenting cells, monocytes/macrophages ● BHR ● ↑ CXCL8/IL-8 from epithelial cells ● ↑ Matrix metalloproteinases from macrophages
IL-6	Monocytes/macrophages T-cells B-cells Fibroblasts Airway epithelial cells	● T-cell growth factor ● B-cell growth factor ● ↑ IgE

(Continued)

TABLE 27.2 (Continued)

Cytokine	Sources	Important cellular and mediator effects
CXCL-8/IL-8	Monocytes/macrophages Airway epithelial cells Airway smooth muscle cells Eosinophils	• Neutrophil chemoattractant and activator • Chemotactic for $CD8^+$ T-cells • Activates 5-lipoxygenase in neutrophils • Induces release of histamine and cys-leukotrienes from basophils
IL-11	Airway epithelial cells Airway smooth muscle cells Fibroblasts Eosinophils	• B-cell growth factor • Activates fibroblast • BHR
IL-16	$CD8^+$ T-cells Epithelial cells Eosinophils Mast cells	• Eosinophil migration • Growth factor and chemotaxis of $CD4^+$ T-cells
IL-17A	Activated $CD4^+$ and $CD8^+$ memory T-cells Neutrophils Monocytes	• Activates epithelial and endothelial cells, and fibroblasts • Induces release of IL-6, CXCL5 (ENA-78), CXCL8 (IL-8), CCL1 (GROα) and GM-CSF • Neutrophil chemoattractant and activator • Induces mucins, Muc5Ac and 5B from epithelial cells
GM-CSF	T-cells Macrophages Eosinophils Fibroblasts Endothelial cells Airway smooth muscle cells Airway epithelial cells	• Eosinophil apoptosis and activation; induces release of leukotrienes • Proliferation and maturation of hematopoietic cells • Endothelial cell migration • BHR
SCF	Bone marrow stromal cells Fibroblasts Epithelial cells	• ↑ VCAM-1 on eosinophils • Growth factor for mast cells
Chemokines		
CXCL1/GROα	Macrophages/monocytes Neutrophils Airway smooth muscle cells	• Neutrophil chemoattractant through CXCR2 receptor • Angiogenesis
CXCL10/IP10	Epithelial cells Monocytes Endothelial cells Fibroblasts	• Chemoattractant for CXCR3-bearing cells including Th1 cells • Chemoattractant for mast cells and monocytes • Promotes T-cell adhesion to endothelial cells • Inhibits angiogenesis
CCL2/MCP-1	Monocytes/macrophages Vascular and airway smooth muscle	• Activates CCR2 and CCR5 receptors • Monocyte chemoattractant • Induces MMP seceretion from macrophages • Recruits NK T-cells
CCL5/ RANTES	Epithelial cells Fibroblasts $CD8^+$ T-cells Airway smooth muscle cells	• Activates CCR3 receptors, and also CCR1, CCR4 and CCR5 • Chemoattractant for eosinophil, basophils, memory T-cells, NK T-cells and monocytes
CCL11/ eotaxin	Epithelial cells Fibroblasts Vascular endothelium	• Activates CCR3 and CCR5 receptors • Eosinophils and monocytes • Mobilises eosinophils and its precursors from bone marrow • Eosinophil chemoattractant • Adhesion of eosinophils to endothelium

(Continued)

TABLE 27.2 (Continued)

Cytokine	Sources	Important cellular and mediator effects
Inhibitory cytokines		
IL-10	Th2 cells Tregs ($CD4^+$) $CD8^+$ T-cells Monocytes/macrophages	• ↓ Eosinophil survival • ↓ Th1 and Th2 • ↓ Monocyte/macrophage activation; ↑ B-cell; ↓ Mast cell growth and release of pro-inflammatory cytokines • Suppresses allergen-specific IgE; induces allergen-specific IgG4 • ↓ BHR
IL-1ra	Monocytes/macrophages	• ↓ Th2 proliferation • ↓ BHR
IL-18	Dendritic cells Monocytes NK cells	• Enhances Th1 cells via IFN-γ release • Releases IFN-γ from Th1 cells • Activates NK cells, monocytes • ↓ IgE
Growth factors		
PDGF	Macrophages Airway epithelium Fibroblasts	• Fibroblast and airway smooth muscle proliferation • Release of collagen
TGF-β	Macrophages Eosinophils Airway epithelium Tregs	• ↓ allergen-specific IgE, and Th1 and Th2 cells • Blocks IL-2 effects • Down-regulates FceRI expression and co-stimulatory ligand expression • Fibroblast proliferation • Chemoattractant for monocytes, fibroblasts, mast cells • ↓ ASM proliferation
VEGF	Macrophages Airway epithelium Airway and vascular smooth muscle Mesenchymal cells Neutrophils Eosinophils	• Angiogenic factor • Increases vascular permeability • Eosinophil chemotaxis • Apoptosis of lung parenchymal cells

ASM = airway smooth muscle; BHR = bronchial hyperresponsiveness; GM-CSF = granulocyte–macrophage colony stimulating factor; IgE = immunoglobulin E; IFN = interferon; IL = interleukin; IL-1ra = interleukin-1 receptor antagonist; MCP = monocyte chemoattractant protein; MIP = macrophage inflammatory factor; NK = natural killer; TNF = tumor necrosis factor; Treg = $CD4^+CD25^+$ regulatory T-cells; VCAM = vascular adhesion molecule.

be considered separately in terms of asthma and chronic obstructive pulmonary disease (COPD), although there is often a clear overlap in the role of many cytokines in the context of these two airway diseases.

INFLAMMATION AND CYTOKINES IN ASTHMA

T-cell-derived cytokine expression

The chronic airway inflammation of asthma is characterized by an infiltration of T-cells, eosinophils, macrophages/monocytes, and mast cells, and sometimes neutrophils; the latter are a possible marker of severity. Acute-on-chronic inflammation may be observed with acute exacerbations of the disease, with an increase in eosinophils and neutrophils in the airway submucosa and release of mediators such as histamine and cysteinyl leukotrienes from eosinophils and mast cells to induce bronchoconstriction, airway edema, and mucus secretion. Airway wall remodeling changes include increase in the thickness of the airway smooth muscle (ASM) layer result from both hypertrophy and hyperplasia and thickening of the lamina reticularis with increased collagen and tenascin deposition, in blood vessels and goblet cell numbers in the airway epithelium [1].

Cytokines play an integral role in the coordination and persistence of the chronic allergic inflammatory process in asthma (Fig. 27.1). Particular subsets of $CD4^+$ T-cells may be induced as a result of microenvironmental events resulting the preferential secretion of defined patterns of cytokines, leading to initiation and propagation of distinct immune effector mechanisms. Studies in mouse $CD4^+$ T-cell clones,

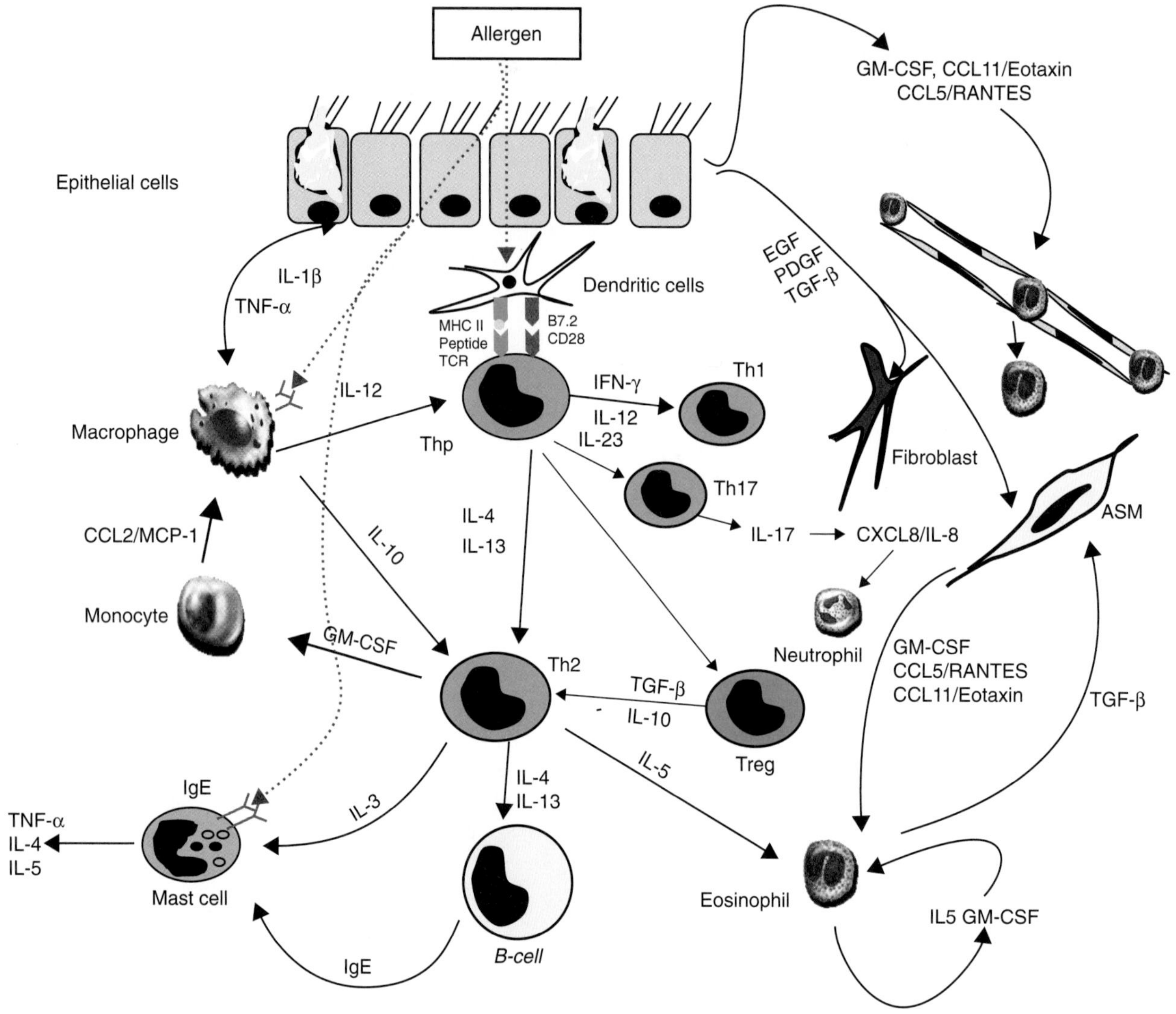

FIG. 27.1 Interactions of cells and cytokines in airway inflammation of asthma. Antigen presentation by dendritic cells to T-cells with the subsequent polarization to Th2 cells appears to be important initial process. The roles of other cells are also crucial, such as airway epithelium, eosinophils, neutrophils, macrophages, and ASM cells. Arrows shown in red with dashed lines indicate potential interactions of allergens through high-affinity IgE receptors with macrophages and mast cells. GM-CSF: granulocyte-macrophage colony stimulating factors; IFN = interferon; IgE = immunoglobulin E; IL = interleukin; MCP-1 = monocyte chemoattractant protein-1; PDGF = platelet-derived growth factor; RANTES = regulated on activation, normal T-cell expressed, and secreted; TCR = T-cell receptor; Th2 = T-cell helper type 2; Th17 = T-cell helper producing IL-17; Treg = $CD4^+$ regulatory T-cell; TGF = transforming growth factor; TNF = tumor necrosis factor.

and later in human $CD4^+$ T-cells, indicate the presence of two basic functional polarized subsets, termed T-helpers (Th1 and Th2). The Th2-associated cytokines are one of the more important classes of cytokines identified in the airway mucosa of patients with asthma. By using *in situ* hybridization techniques, increased proportions of cells in bronchial biopsies and in bronchoalveolar lavage (BAL) fluid from patients with atopic asthma express mRNA for IL-3, IL-4, IL-5, IL-13, and granulocyte-macrophage colony-stimulating factor (GM-CSF) that is the Th2 gene cluster, when compared with those of nonatopic control subjects [2–4]. In a model of inhalational allergen challenge, an increase in mRNA for these Th2 cytokines, but not for IFN-γ, has been demonstrated in bronchial biopsies and BAL cells of atopic asthmatics [5]. Thus, lung T-cells are activated and express and release high levels of Th2-type cytokines.

A subset of $CD4^+$ T-cells expressing CD25, termed T-regulatory cells (Tregs), are capable of suppressing both Th1 and Th2-mediated adaptive immune responses, and these cells may mediate their suppressive actions through release of TGF-β or IL-10 [6]. An imbalance between allergen-specific Tregs and effector Th2 cells has been shown in allergic diseases [7].

Recruitment of Th2 cells to the airways may occur through certain chemokines such as CCL17/TARC and CCL22/MDC which are ligands for CCR4 expressed on Th2 T-cells. Increased expression of these chemokines in the bronchial mucosa of asthma patients has been reported, together with the expression of thymic stromal lymphopoietin that can increase the expression of these Th2 chemoattractants [8].

Although $CD4^+$ T-cells appear to be the main cell expressing these Th2 cytokine transcripts, mast cells and eosinophils also express IL-4 and IL-5 mRNA [9–11]. Mast cells in mucosal biopsies from atopic asthmatics were positive for IL-3, IL-4, IL-5, IL-6, and tumor necrosis factor-α (TNF-α) by immunohistochemistry, while immunoreactive IL-5 and GM-CSF in eosinophils have been detected after endobronchial allergen challenge in asthmatics [10]. Alveolar macrophages from patients with asthma release more proinflammatory cytokines (i.e. IL-1β, TNF-α, GM-CSF, and MIP-1α) than do macrophages from normal subjects [12, 13].

Antigen presentation and release of cytokines

The primary signals that activate Th2 cells are related to the presentation of a restricted panel of antigens in the presence of appropriate cytokines. Allergens are taken up and processed by specialized cells within the mucosa, such as dendritic cells (antigen presenting cells, APC), followed by presentation of peptide fragments to naive T-cells. The activation of naive T-cells occurs firstly via the T-cell receptor on $CD4^+$ cells through the APC-bound antigen to major histocompatibility complex (MHC)-II complex and, secondly, via the costimulatory pathway linked by the B7 family and T-cell-bound CD28 [14]. CD28 itself has two major ligands: B7.1 that inhibits Th2 cell activation and development, and B7.2 which induces T-cell activation and Th2 cell proliferation. Under the influence of IL-12 and IFN-γ, Th cells are polarised toward Th1 which predominantly secrete interleukin (IL-2), interferon (IFN-γ), and tumor necrosis factor (TNF), triggering both cell-mediated immunity and production of opsonizing antibodies. Th2 cells secrete IL-4, IL-5, IL-10, and IL-13, responsible for IgE and IgG_4 antibody production, and activation of mast cells and eosinophils [15]. A third subset of T-helper cells, Th0, shows a composite profile producing both Th1- and Th2-associated cytokines. These soluble factors can also be expressed by macrophages, epithelial cells, and mast cells and may take part in more general inflammatory processes.

Cytokines may play an important role in antigen presentation. Airway macrophages are usually poor at antigen presentation and suppress T-cell proliferative responses (possibly via release of cytokines such as IL-1 receptor antagonist), but in asthma there is reduced suppression after exposure to allergen [16]. Both GM-CSF and IFN-γ increase the ability of macrophages to present allergen and express HLA-DR [17]. IL-1 is important in activating T lymphocytes and is an important costimulator of the expansion of Th2 cells after antigen presentation [18].

Airway macrophages may also be an important source of "first wave" cytokines, such as IL-1, TNF-α, and IL-6, which may be released on exposure to inhaled allergens via transduction mediated by FcεRI receptors. These cytokines may then act on epithelial cells to cause release of a "second wave" of cytokines, including GM-CSF, CCL11/eotaxin, and CCL5/RANTES, which then leads to influx of secondary cells, such as eosinophils, which themselves may release multiple cytokines.

The immunoglobulin-E response and mast cells

IL-4 is the most important cytokine mediating IgE synthesis through isotype switching by B-cells, but IL-13 is also capable of a similar action on B-cells [19]. IL-4 also activates B-cells through increasing the expression of class II MHC molecules, as well as enhancing the expression of CD23, the low-affinity IgE (FcεRII) receptor. CD40 antigen is expressed on B-cells after antigen recognition. IL-4 together with the engagement of CD40 antigen with its ligand, CD40-L on activated T-cells, promotes IgE class switching and B-cell growth. Because IL-4 can also be produced by basophils, mast cells, and eosinophils and also have surface CD40-L, these cells may contribute to the amplification of IgE responses. IL-4 and IL-13 share the α-chain of the IL-4 receptor (IL-4Rα). On engagement of the ligand with IL-4Rα, signal transducer and activator of transcription 6 (STAT 6) translocates to the nucleus, and germline ε mRNA transcription is initiated together with ε-class switching of immunoglobulin genes. Individuals with gain of function mutations in both the extracytoplasmic and intracellular domains of IL-4Rα show an enhanced IgE response and predisposition to atopic disease [20].

IgE produced in asthmatic airways binds to FcεRI receptors ("high affinity" IgE receptors) on mast cells and basophils, priming them for activation by antigen. Cross-linking of FcεRI receptors upregulates its own expression and leads to mast cell degranulation, with the release of mediators such as histamine and cysteinyl leukotrienes. The importance of IgE in asthmatic inflammation has been emphasized by the demonstration that a humanized monoclonal antibody to IgE inhibits eosinophilic inflammation and reduced $CD4^+$ T-cells, $IL\text{-}4^+$ staining cells and B-cells in the airway mucosa of asthmatic patients [21].

The maturation and expansion of mast cells from bone marrow cells involves growth factors and cytokines such as stem cell factor (SCF) and IL-3 derived from structural cells. Bronchoalveolar mast cells from asthmatics show enhanced release of mediators such as histamine. Mast cells also elaborate IL-4 and IL-5 [9]. IL-4 also increases the expression of an inducible form of the low-affinity receptor for IgE (FcεRII or CD23) on B lymphocytes and macrophages [22]. IL-4 drives the differentiation of $CD4^+$ T-helper precursors into Th2-like cells.

Eosinophil-associated cytokines

The differentiation, migration, and pathobiological effects of eosinophils may occur through the actions of GM-CSF, IL-3, IL-5, and certain chemokines such as CCL11/eotaxin [23–25]. IL-5 influences the production, maturation, and activation of eosinophils, acting predominantly at the later stages of eosinophil maturation and activation, and can also prolong the survival of eosinophils. IL-5 causes eosinophils to be released from the bone marrow, while the local release of an eosinophil chemoattractant such as eotaxin may be necessary for the tissue localization of eosinophils [26]. Mature eosinophils may show increased survival in bronchial tissue, secondary to the effects of GM-CSF, IL-3, and

IL-5 [27]. Eosinophils themselves may also generate other cytokines such as IL-3, IL-5, and GM-CSF [11].

Cytokines such as IL-4 may also exert an important regulatory effect on the expression of adhesion molecules such as VCAM-1, both on endothelial cells of bronchial blood vessels and on airway epithelial cells. IL-1 and TNF-α increase the expression of intercellular adhesion molecule 1 (ICAM-1) in both vascular endothelium and airway epithelium [28]. An anti-IL5 monoclonal antibody administered to mild allergic asthmatics has been shown to suppress allergen-induced blood and sputum eosinophilia, without affecting the degree of the late-phase response [29], indicating that IL-5 may not be the only cytokine involved in allergen-induced late-phase responses. Eotaxin is a chemoattractant cytokine (chemokine) selective for eosinophils and acts through the chemokine receptor CCR3 present on eosinophils, basophils, and T-cells. Cooperation between IL-5 and eotaxin appears necessary for the mobilization of eosinophils from the bone marrow during allergic reactions and for the local release of chemokines to induce homing and migration into tissues.

Th17 cytokines: IL-17 family

A separate lineage of effector T-helper cells termed Th-17 cells produced by the action of IL-15 or IL-23 on CD4 or $CD8^+$ memory T-cells has been described. Several members of the IL-17 family have been described, with the founder member, IL-17 or IL-17A, originally reported from $CD4^+$ memory T-cells. Together with IL-17F, these two members are the most studied and have overlapping functions particularly mediating neutrophil mobilization through activation of T-cells (Th17). IL-17A activates neutrophils to cause release of neutrophil elastase and myeloperoxidase activity and regulate the recruitment of neutrophils in response to allergen, while attenuating allergen-induced increase in IL-5 and eosinophils [30]. IL-17A may be necessary for the proliferation of $CD4^+$ T-cells for the production of IL4 and IL5 and for bronchial hyperresponsiveness [31]. IL-17A induces the release of CXCL1/GROα, CXCL8/IL-8 and G-CSF, and GM-CSF from the airway epithelium, which may be a mechanism for neutrophilic chemotaxis and activation, as is found in patients with severe asthma. IL-17A induces the expression of the mucins, MUC-5AC and MUC-5B, in bronchial epithelial cells. An increase in the expression of IL-17A in eosinophils has been shown in patients with asthma [32].

Unlike IL-17A and IL-17F, IL-17E, also known as IL-25, is a Th2-like cytokine produced by Th2 cells and mast cells. IL-17E induces overproduction of Th2 cytokines, IL-4 and IL-13, production of IgA and IgE, mucus, epithelial cell hyperplasia, and eosinophilia. $IL\text{-}17E^+$ cells, probably eosinophils, were detected in the bronchial submucosa of patients with asthma [33].

Airway wall remodeling cytokines

Experimental approaches made possible by genetically modified mice have highlighted the potential role of various T-cell-derived cytokines in aspects of airway wall remodeling. Overexpression of these cytokines in the airway epithelium causes various features of airway wall remodeling. A CC10-driven overexpression of IL-13 in the lungs caused eosinophilic and mononuclear inflammation with goblet cell hyperplasia, subepithelial fibrosis, airway obstruction, and airway hyperresponsiveness [34]. IL-4, IL-5, and IL-9 overexpression lead to substantial mucus metaplasia, while IL-9 and IL-5 overexpression also caused subepithelial fibrosis and airways hyperresponsiveness [35–37]. ASM hyperplasia occurs following IL-11 overexpression [38]. IL-5 and IL-13 have been associated with the production of TGF-β in asthma [39, 40].

Proliferation of myofibroblasts and the hyperplasia of ASM may also occur through the action of several growth factors, such as platelet-derived growth factor (PDGF) and TGF-β. They may be released from inflammatory cells in the airways, such as macrophages and eosinophils, but also by structural cells such as airway epithelium, endothelial cells, and fibroblasts. These growth factors may stimulate fibrogenesis by recruiting and activating fibroblasts or transforming myofibroblasts [41]. Epithelial cells may release growth factors, since collagen deposition occurs underneath the basement membrane of the airway epithelium [42].

Growth factors may also stimulate the proliferation and growth of ASM cells. PDGF, TGF-β, and epidermal growth factor (EGF) are potent stimulants of human ASM proliferation [43, 44]. Pro-angiogenic factors such as angiogenin, vascular endothelial growth factor (VEGF), and CCL2/MCP-1 may be involved in angiogenesis of chronic asthma [45]. There is greater expression of airway mucosal eosinophils expressing TGF-β mRNA and protein, correlating with the severity of asthma and the degree of subepithelial fibrosis [46]. Increased expression of TGF-β is also observed in epithelial cells and airway smooth cells of patients with asthma [47]. EGF expression is increased in bronchial epithelium, smooth muscle, and the bronchial submucosa of asthmatic patients [48, 49].

Airway smooth muscle-associated cytokines

Airway smooth muscle (ASM) cells have the capacity to elaborate a range of cytokines, including IL-4, IL-5, GM-CSF, CXCL8/IL-8, CCL5/RANTES, CCL11/eotaxin, CCL2/MCP-1, and TGF-β and therefore may play a role in the induction of local inflammatory and remodeling responses [50]. In addition, these cytokines may provide the mechanisms for interacting with cells within the ASM bundle of asthmatic patients, such as mast cells [51]. Thus, increased production of CXCL10/IP-10 from ASM cells of asthmatics could be the cause for enhanced chemotactic activity of lung mast cells toward ASM cells in asthma [52].

INFLAMMATION AND CYTOKINES IN COPD

Cytokine profile in COPD

The characteristic pathology of stable COPD includes chronic inflammation of the small bronchi and bronchioles less than

2 mm diameter, together with structural changes to all the airways and lungs with alveolar wall destruction or emphysema. The inflammation is characterized by an accumulation of neutrophils, macrophages, B-cells, lymphoid aggregates, and $CD8^+$ T-cells particularly in the small airways and is proportionate to the severity of disease [53]. Chronic bronchitis characterized by glandular hypertrophy and hyperplasia with dilated ducts of 2–4 mm internal diameter may be present with mucus plugging, goblet cell hyperplasia, increased ASM mass, and distortion due to fibrosis of bronchioles. Neutrophils are localized particularly to the bronchial epithelium, the bronchial glands [54], and also in close apposition to ASM bundles [55]. Some patients with COPD have a preponderance of eosinophils in sputum, which has been associated with improvement of FEV_1 with corticosteroid therapy [56]. CD8+ T-cells expressing IFN-γ and CXCR3, probably of the type 1 cytotoxic (Tc1) T-cells [57], are increased throughout the airways and in lung parenchyma [58].

In COPD the basic abnormality is thought to be the response of airway cells, particularly epithelial cells and macrophages to toxic gases and particulates in cigarette smoke, which generate an inflammatory and immune response with the release of proinflammatory cytokines, chemokines, growth factors, cathepsins, and serine and matrix metalloproteinase (MMP) proteases. In COPD, as contrasted to healthy smokers also exposed to cigarette smoke, the release of cytokines may be amplified by the effect of oxidative stress, perhaps viruses and by unknown genetic factors (Fig. 27.2).

Cytokines in COPD may be classified as (i) proinflammatory cytokines including chemokines involved in the recruitment of neutrophils, macrophages, T-cells, and B-cells (ii) cytokines of airway wall remodeling, including goblet cell, ASM hyperplasia (iii) cyokines of emphysema (iv) systemic cytokines, that is extrapulmonary source of cytokines involved in systemic inflammatory responses.

Proinflammatory cytokines and chemokines

Increased levels of IL-6, IL-1β, TNF-α, CXCL8/IL-8, and CCL2/MCP-1 have been observed in induced sputum of patients with stable COPD [59, 60]. Increased release of proinflammatory cytokines CXCL8/IL-8, CCL1/GROα, CXCL5/ENA-78, IL-1, TNF-α and of the anti-inflammatory cytokine, IL-10 from alveolar macrophages of cigarette smokers and of COPD patients have been reported [61, 62]. Levels of TNF-α, IL-1β, IL-6, CXCL8/IL-8, and

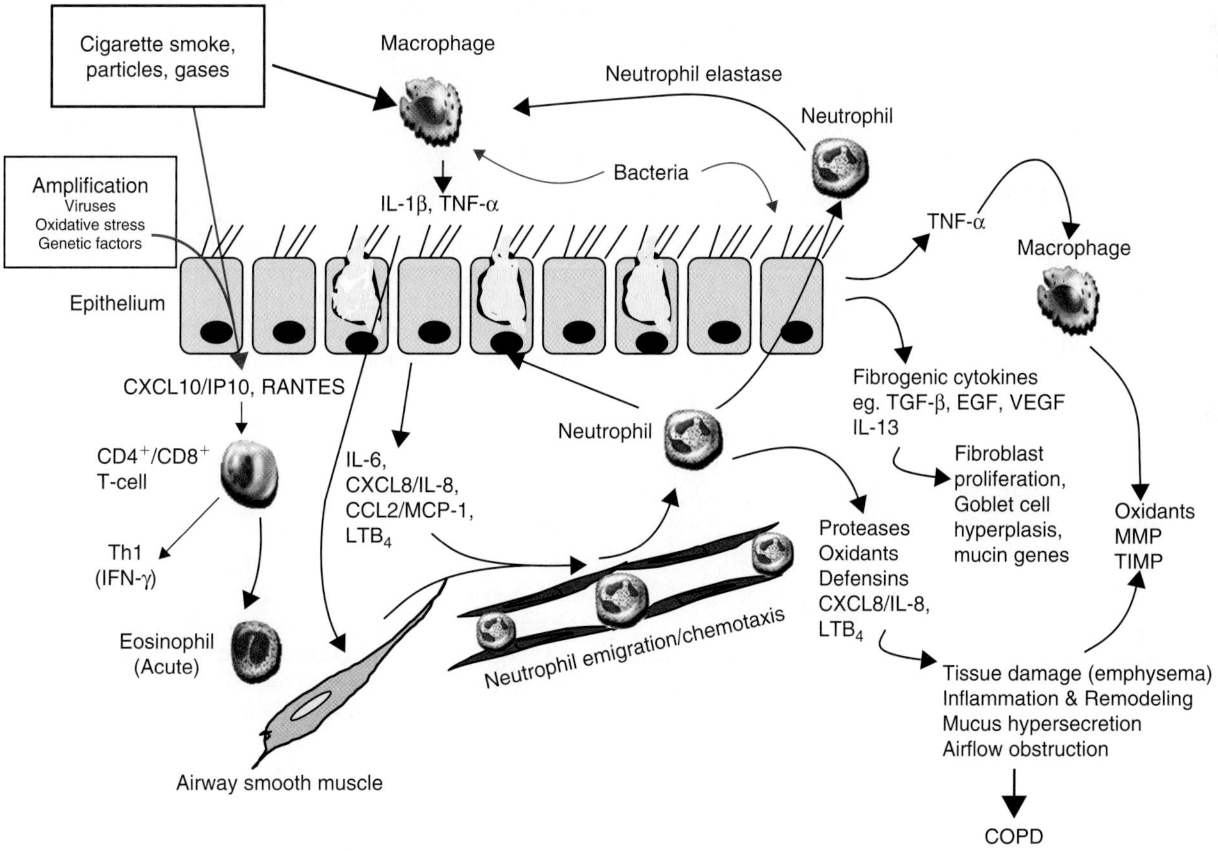

FIG. 27.2 Interaction of cells and cytokines in airway inflammation of COPD. The initiating factors include cigarette smoke with macrophages and airway epithelium, inducing the release of chemotactic factors for neutrophils, which in turn are important effector cells for inflammation, tissue damage, and repair. Potential amplification factors include viruses, oxidative stress, and genetic factors. EGF = epidermal growth factor; IL = interleukin; IP-10 = IFN-γ-induced protein 10; LTB_4 = leukotriene B4; MCP = monocyte chemotactic protein; MMP = matrix metalloproteinase; TGF = transforming growth factor; TIMP = tissue inhibitor of matrix metalloproteinase; TNF = tumor necrosis factor; VEGF = vascular endothelial growth factor. Black arrows indicate interactions with inflammatory cells leading to release of cytokines and mediators; red arrows indicate interactions of bacteria with macrophages and epithelial cells that could amplify the inflammatory process.

CCL2/MCP-1 were increased in BAL fluid of chronic smokers compared to non-smokers [63], and of CXCL8/IL-8 and GM-CSF levels in COPD patients [64, 65]. In general, the expression of cytokines from COPD patients is higher than of asymptomatic smokers. Sputum neutrophil counts and level of CXCL8/IL-8 and circulating levels of TNF-α and C-reactive protein (CRP) have been reported to be the best markers for distinguishing the various levels of severity of COPD [66].

Cigarette smoke increases CXCL8/IL-8 gene expression and release by bronchial epithelial cells, and TNF-α and IL-6 by alveolar macrophages [67, 68], through oxidant mechanisms that include activation of the transcription factor, nuclear factor-κB (NF-κB) [69]. Exposure of lung epithelial cells to smoke extract causes the release of neutrophil and monocytic chemotactic activities with CXCL8/IL-8 and G-CSF accounting for the neutrophilic activity and CCL-2/MCP-1 for the monocytic activity [70].

A higher expression of CCL2/MCP-1, TGF-β1 and CXCL8/IL-8 mRNA and protein has been observed in bronchiolar epithelium in macrophages of smokers with COPD compared with those smokers without COPD [71]; a similar difference has been found for CCR2. Since CCL2/MCP-1 binds to CCR2 and can induce T-cell and monocytic migration, CCL2 [72] may contribute to the recruitment of these cells in COPD.

Different chemokines in induced sputum from COPD patients have been found to be involved in the chemotaxis of monocytes. Thus, there is increased chemotaxis of monocytes in COPD for CCL1/GROα or CXCL7/NAP2 but not for CXCL8/IL-8 or CXCL5/ENA-78, possibly due to differential regulation of CXCR2 receptors in COPD [73]. CD8 T-cells expressing IFN-γ infiltrate the peripheral airways of smokers with COPD and have an increased expression of the chemokine receptor, CXCR3, which is paralleled by a strong epithelial expression of its ligand, CXCL10/IP-10 [57], a chemoattractant for Th1 cells. The chemokine ligand 20, CCL20, is the most potent known chemoattractant for dendritic cells and is upregulated in lung and in induced sputum of patients with COPD [74]. This could account for increased number of dendritic cells observed in the small airways of COPD as these cells express CCR6 which is the receptor for CCL20. Microarray analysis of lung tissues from COPD patients revealed the expression of CX3CL1/fractalkine [75].

Th2 cytokines

Expression of Th2 cytokines is less prominent in COPD as compared to asthma. IL-5 expression is absent in COPD airways associated with eosinophilia, but other eosinophil chemoattractants such as eotaxin or RANTES may be implicated, particularly during exacerbations [76]. Increased IL-4^+ cells in mucus secreting cells and associated plasma cells of the airway mucosa of chronic bronchitis patients have been reported [77].

Overexpression of the Th2 cytokine, IL-13, in lungs of adult mice induces emphysema, mucus goblet cell hyperplasia, and airway inflammation with macrophages, lymphocytes, and eosinophils, and increased MMPs, which are many of the features associated with COPD [78]. The induction of emphysema in these mice was related to the release of metallo- and cysteine-proteases [78]. Interestingly, the proinflammatory cytokine, IL-18, can induce lung inflammation and emphysema through the production of IL-13 but not through IFN-γ [79], despite IFN-γ's capability of inducing emphysema [80]. Although IL-18 expression appears to be increased in pulmonary macrophages of COPD patients, IL-13 expression is decreased in emphysema lung [81]. IL-9-mRNA is overexpressed in CD3+ T-cells in patients with COPD [82].

Systemic cytokines

Evidence of systemic inflammatory processes in COPD can be judged from increased circulating levels of cytokines such as IL-6, CXCL8/IL-8, TNF-α, and TNF receptors 55 and 75, together with CRP levels [83]. The source of these circulating cytokines may represent a spill-over from pulmonary sources but may also originate from other sources such as circulating blood monocytes [84], striated muscle or from concomitant atherosclerotic lesions. In stable COPD patients with respiratory muscle impairment, increased TNF-α and IL-6 gene and protein expression in muscle has been reported [85].

Exacerbations of COPD

Increases in plasma levels of IL-6, CXCL8/IL-8, and LTB4 but not of TNF-α have been reported during COPD exacerbations severe enough to require the patient to be hospitalized [86]. During exacerbations of COPD associated with neutrophilia requiring mechanical ventilatory support, gene expression for CXCL5/ENA78, CXCL8/IL-8, and the receptor CXCR2 is increased in tissue, particularly in the bronchiolar epithelium [87]. Both CXCL5/ENA-78 and CXCL8/IL-8 bind to CXCR2 and are neutrophil chemoattractants.

In acute exacerbations of COPD, eosinophils may be prominent among the cells recovered in sputum or in bronchial biopsies, but there is no increased expression of IL-5 protein in tissues [88]. Higher levels of IL-6 and CXCL8/IL-8 in induced sputum of patients with recurrent exacerbations of COPD (>3 per year) have been reported [89]. In patients with stable COPD, lymphomononuclear cells expressing eotaxin mRNA was increased compared to healthy non-smokers. In patients with an exacerbation of chronic bronchitis, RANTES mRNA expression was upregulated and strongly expressed on the surface epithelium and subepithelial lympho-mononuclear cells, together with increased number of eosinophils [76]. CCL11/eotaxin and its receptor, CCR3, are upregulated during exacerbations of chronic bronchitis, with CCR3 colocalized mainly to eosinophils [90]. The increased expression of CCL11/eotaxin and CCL5/RANTES may underlie the eosinophilia sometimes observed during exacerbations of COPD.

Properties of some specific cytokines in COPD

TNF-α

TNF-α is produced by many cells including macrophages, T-cells, mast cells, and epithelial cells, but the principal source is the macrophage. The secretion of TNF-α by monocytes/macrophages is greatly enhanced by other cytokines such as IL-1, GM-CSF, and IFN-γ. TNF-α activates NFκB that switches on the transcription of the CXCL8/IL-8 gene and increases CXCL8/IL-8 release from the airway epithelium and neutrophils. Bacteria and bacterial products such as lipopolysaccharide can induce CXCL8/IL-8 expression, probably as a result of initial TNF-α production [91, 92]. TNF-α may be important in inducing the expression of CXCL2/MIP-2 and CCL2/MCP-1, and in connective tissue breakdown following cigarette smoke exposure [93]. TNF-α through the activation of EGF receptor leads to the expression of mucin genes, in particular MUC5AC [94].

TNF-α increases the expression of ICAM-1, which is increased in serum of COPD patients [95]. TNF-α may activate macrophages to produce MMPs. This effect is inhibited by the anti-inflammatory cytokine, IL-10, which also enhances the release of tissue inhibitor of metalloproteinases (TIMP) in macrophages from normal volunteers. In smokers IL-10 increases TIMP-1 release, without modifying MMP-9 release, from alveolar macrophages [61]. TNF-α also stimulates bronchial epithelial cells to produce tenascin, an extracellular matrix glycoprotein that may contribute to airway remodeling. In animal models, MMP-12 induces the release of TNF-α from lung macrophages exposed to cigarette smoke, causing endothelial activation and neutrophil influx [96].

Increased serum concentrations of TNF-α have been measured in patients with COPD with weight loss [97, 98], and increased TNF-α production has been noted by peripheral blood monocytes from COPD patients during the catabolic phase of their illness leading to weight loss [99]. TNF-α has direct effects on skeletal muscle cells and reduces adult myosin heavy chain content and induces apoptosis of skeletal muscle via several signaling pathways [100, 101]. Given the potential role for TNF-α in the inflammation of COPD, the effect of TNF-α blockers in this disease is of therapeutic interest. However, a multicenter randomized trial of an anti-TNF-α antibody, infliximab, found no treatment benefit in terms of quality of life, lung function, and the 6-min walking test [102].

IL-1β

IL-1β induces leukocytosis by the release of neutrophils from the bone marrow and induces the release of many other cytokines such as IL-1, IL-2, IL-3, IL-4, IL-5, IL-6, CXCL8/IL-8, CCL5/RANTES, GM-CSF, IFN-γ, TNF, and PDGF from a variety of cells. It induces fibroblasts to proliferate, and increases prostaglandin secretion, and the synthesis of fibronectin and collagen. Together with TNF-α, IL-1β induces ICAM-1 expression on endothelial cells. IL-1β can induce attenuation of β-adrenergic receptor-induced airway relaxation through reduction of β-adrenoreceptors, an increase in the G-protein, Giα subunit, and a defect in adenylate cyclase activity [103].

CXCL8/IL-8

CXCL8/IL-8 is a neutrophil chemoattractant and activator that induces a transient shape change, rise in intracellular calcium, exocytosis with release of enzymes and proteins from intracellular storage organelles, and respiratory burst. It also upregulates the expression of two integrins (CD11b/CD18 and CD11c/CD18) during exocytosis of specific granules. CXCL8/IL-8 activates neutrophil 5-lipoxygenase with the formation of LTB_4 and 5-hydroxy-eicosanotetraenoic acid (5-HETE) and together with LTB_4 contributes to the neutrophil chemotactic activity of sputum from COPD patients [104, 105]. CXCL8/IL-8 also has chemoattractant properties for T-cells. Bacteria for example due to a soluble cytoplasmic factor of nontypeable Haemophilus Influenzae can induce CXCL8/IL-8 expression in epithelial cells [106]. The increased levels of CXCL8/IL-8 found in sputum samples of COPD patients correlate with airway bacterial load and with myeloperoxidase released from activated neutrophils. Elastase released from neutrophils may also stimulate epithelial cells to produce more CXCL8/IL-8 [107]. A study of a blocking anti-IL-8 antibody in moderate COPD patients showed a small effect in improving dyspnea [108].

TGF-β

TGF-β is a multifunctional growth factor with biological effects on fibroblast proliferation, deposition of extracellular matrix, epithelial repair, chemotactic activities, and also as an immune suppressor. TGF-β is secreted as latent TGF-β that is sequestered by many binding proteins including decorin, biglycan, fibronectin, and elastin. TGF-β can be activated *in vivo* by mechanisms including thrombospondin, acid pH, reactive oxygen species, and nitric oxide nitrosylation. TGF-β regulates epithelial repair by inhibiting proliferation but increasing epithelial cell migration [109]. TGF-β1 is involved in the transformation of fibroblasts to myofibroblasts, and enhancing myofibroblast survival. In addition, TGF-β1 and TGF-β2 can convert myofibroblasts into smooth muscle cells [110]. TGF-β induces hypertrophy and hyperplasia of ASM cells [47, 111]. TGF-β induces collagen production in human lung fibroblasts [112] and ASM cells [113]. Therefore, TGF-β is involved in remodeling of the airway wall in both asthma and COPD.

Increased expression of a number of related growth factors, CTGF, TGF-β1, and PDGFRA in the lungs using gene microarray analysis of lungs from smoking COPD patients compared to healthy smokers has been reported [75]. Expression of TGF-β in the bronchiolar and alveolar epithelium and submucosal cells of patients with COPD has been reported [114–116]. In addition, increased release of TGF-β and of fibronectin from alveolar macrophages from patients with chronic bronchitis has been observed [117].

VEGF

Increased levels of VEGF have been observed in sputum of patients with COPD [118], and its augmented expression appears to be localized to pulmonary arteries [119]. Inhibition of VEGF receptor activation in mice led to the induction of enlargement of air spaces, and induced alveolar septal cell apoptosis, indicating that VEGF receptor signaling is required for maintenance of alveolar structures [120]. Oxidative stress and apoptosis may be the down-stream effects of VEGF receptor activation [121]. The finding that VEGF levels are increased in patients with COPD reinforces a potential non-proteolytic mechanism for emphysema through VEGF.

CONCLUSION

Because there has been little direct comparison of tissues between asthma and COPD patients, it is not possible to state categorically the degree of overlap of cytokine profiles or networks that exist between the two diseases. However, analysis of cytokine expression in asthma and COPD from the published literature reveals many similarities in terms of increased expression of many cytokines (Table 27.3) that may reflect similarities in established disease but not necessarily during the initial phase of these diseases. For example, severe asthma may resemble more closely COPD with a neutrophilic inflammatory response, and remodeling aspects of the airways may be similar despite being differently localized in the airways in terms of the cytokines and growth factors involved. Further analysis of cytokine expression in asthma and COPD may help determine whether these conditions have common mechanisms or similar predisposing factors. Analysis of cytokine expression may help determine which ones may be useful to target for therapeutic effects, although the limited experience so far has been disappointing.

TABLE 27.3 Expression of cytokines in asthma and COPD[a].

	Asthma	COPD
IL-3	+	?
IL-4	+ +	+
IL-5	+ +	–
IL-9	+	+
IL-10	+ or –	+
IL-13	+	+
IL-17A	+	?
IL-25 (IL-17E)	+	?
IFNγ	–	+
CCL2/MCP-1	+	+
CCL5/RANTES	+	+*
CCL11/Eotaxin	+	+*
CCL17/TARC	+	?
CCL20	?	+
CCL22/MDC	+	?
CXCL1/GROα	?	+
CXCL8/IL-8	+	+ +
CXCL10/IP10	+	+
IL-1β	+	+
TNF-α	+	+
TGF-β1	+	+
CTGF	?	+
GM-CSF	+	+
EGF	+	+
VEGF	+	+

[a]From review of literature: expression is reported as gene or protein in airway cells or tissues, in addition to measurement of protein release in airway fluids. The degree of expression has been arbitrarily assessed as: – = reduced; + or + + = increased; ? = not known or uncertain; * = associated with exacerbations.

References

1. Bousquet J, Jeffery PK, Busse WW, Johnson M, Vignola AM. Asthma. From bronchoconstriction to airways inflammation and remodeling. *Am J Respir Crit Care Med* 161(5): 1720–45, 2000.
2. Hamid Q, Azzawi M, Ying S, Moqbel R, Wardlaw AJ, Corrigan CJ *et al.* Expression of mRNA for interleukins in mucosal bronchial biopsies from asthma. *J Clin Invest* 87: 1541–46, 1991.
3. Broide DH, Lotz M, Cuomo AJ, Coburn DA, Federman EC, Wasserman SI. Cytokines in symptomatic asthmatic airways. *J Allergy Clin Immunol* 89: 958–67, 1992.
4. Humbert M, Durham SR, Kimmitt P, Powell N, Assoufi B, Pfister R *et al.* Elevated expression of messenger ribonucleic acid encoding IL-13 in the bronchial mucosa of atopic and nonatopic subjects with asthma. *J Allergy Clin Immunol* 99(5): 657–65, 1997.
5. Robinson DS, Hamid Q, Ying S, Tsigopoulos A, Barkans T, Bentley AM *et al.* Predominant Th2-type bronchoalveolar lavage T-lymphocyte population in atopic asthma. *New Engl J Med* 326: 298–304, 1992.
6. Taylor A, Verhagen J, Blaser K, Akdis M, Akdis CA. Mechanisms of immune suppression by interleukin-10 and transforming growth factor-beta: The role of T regulatory cells. *Immunology* 117(4): 433–42, 2006.
7. Ling EM, Smith T, Nguyen XD, Pridgeon C, Dallman M, Arbery J *et al.* Relation of $CD4^+CD25^+$ regulatory T-cell suppression of allergen-driven T-cell activation to atopic status and expression of allergic disease. *Lancet* 363(9409): 608–15, 2004.
8. Ying S, O'Connor B, Ratoff J, Meng Q, Mallett K, Cousins D *et al.* Thymic stromal lymphopoietin expression is increased in asthmatic airways and correlates with expression of Th2-attracting chemokines and disease severity. *J Immunol* 174(12): 8183–90, 2005.
9. Bradding P, Roberts JA, Britten KM, Montefort S, Djukanovic R, Mueller R *et al.* Interleukin-4, -5 and -6 and tumor necrosts factor-α in normal and asthmatic airways: Evidence for the human mast cell as a source of these cytokines. *Am J Respir Cell Mol Biol* 10: 471–80, 1994.
10. Broide D, Paine MM, Firestein GS. Eosinophils express interleukin S and granulocyte-macrophage colony-stimulating factor mRNA at sites of allergic inflammation in asthmatics. *J Clin Invest* 90: 1414–24, 1992.
11. Moqbel R, Hamid Q, Ying S, Barkans J, Hartnell A, Tsicopoulos A *et al.* Expression of mRNA and immunoreactivity for the granulocyte/macrophage colony-stimulating factor in activated human eosinophils. *J Exp Med* 174: 749–52, 1991.
12. John M, Lim S, Seybold J, Jose P, Robichaud A, O'Connor B *et al.* Inhaled corticosteroids increase interleukin-10 but reduce macrophage inflammatory protein-1alpha, granulocyte-macrophage colony-stimulating factor, and interferon-gamma release from alveolar macrophages in asthma. *Am J Respir Crit Care Med* 157(1): 256–62, 1998.
13. Hallsworth MP, Soh CPC, Lane SJ, Arm JP, Lee TH. Selective enhancement of GM-CSF, TNF-α, IL-1β and IL-8 production by monocytes and macrophages of asthmatic subjects. *Eur Respir J* 7: 1096–102, 1994.

14. Green JM. The B7/CD28/CTLA4 T-cell activation pathway. Implications for inflammatory lung disease. *Am J Respir Cell Mol Biol* 22(3): 261–64, 2000.
15. Mossman TR, Coffman RL. TH1 and TH2 cells : Different patterns of lymphokine secretion lead to different functional properties. *Ann Rev Immunol* 7: 145–73, 1989.
16. Spiteri M, Knight RA, Jeremy JY, Barnes PJ, Chung KF. Alveolar macrophage-induced suppression of T-cell hyperresponsiveness in bronchial asthma is reversed by allergen exposure. *Eur Respir J* 7: 1431–38, 1994.
17. Fischer HG, Frosch S, Reske K, Reske-Kunz AB. Granulocyte-macrophage colony-stimulating factor activates macrophages derived from bone marrow cultures to synthesis of MHC class II molecules and to augmented antigen presentation function. *J Immunol* 141: 3882–88, 1988.
18. Chang TL, Shea CH, Urioste S, Thompson RC, Boom WH, Abbas AK. Heterogeneity of helper/inducer T lymphocytes: Lymphokine production and lymphokine responsiveness. *J Immunol* 145: 2803–8, 1990.
19. Minty A, Chalon P, Derocq J-M, Dumont X, Guillemot J-C, Kaghad M *et al.* Interleukin-13 is a new human lymphokine regulating inflammatory and immune responses. *Nature* 362: 248–50, 1993.
20. Khuruna Hershey GK, Friedrich MF, Esswein LA, Thomas ML, Chatila TA. The association of atopy with a gain-of-function mutation in the α-subunit of the IL-4 receptor. *New Eng J Med* 337: 1720–25, 1997.
21. Djukanovic R, Wilson SJ, Kraft M, Jarjour NN, Steel M, Chung KF *et al.* Effects of treatment with anti-immunoglobulin E antibody omalizumab on airway inflammation in allergic asthma. *Am J Respir Crit Care Med* 170(6): 583–93, 2004.
22. Vercelli D, Jabara HH, Lee BW, Woodland N, Geha RS, Leung DY. Human recombinant interleukin-4 induces FC4RII/CD23 on normal human monocytes. *J Exp Med* 167: 1406–16, 1988.
23. Sanderson CJ, Warren DJ, Strath M. Identification of a lumphokine that stimulates eosinophil differentiation *in vitro*. Its relationship to interleukin 3, and functional properties of eosinophils produced in cultures. *J Exp Med* 162: 60–74, 1985.
24. Jose PJ, Griffiths-Johnson DA, Collins PD, Walsh DT, Moqbel R, Totty NF *et al.* Eotaxin: A potent eosinophil chemoattractant cytokine detected in a guinea pig model of allergic airways inflammation. *J Exp Med* 179: 881–87, 1994.
25. Ponath PD, Qin S, Post TW, Wang J, Wu L, Gerard N *et al.* Molecular cloning and characterization of a human eotaxin receptor expressed selectively on eosinophils. *J Exp Med* 183: 2437–48, 1996.
26. Collins PD, Griffiths-Johnson DA, Jose PJ, Williams TJ, Marleau S. Co-operation between interleukin-5 and the chemokine, eotaxin, to induce eosinophil accumulation *in vivo*. *J Exp Med* 182: 1169–74, 1995.
27. Rothenberg ME, Owen WFJ, Siberstein DS. Human eosinophils have prolonged survival, enhanced functional properties and become hypodense when exposed to human interleukin. *J Clin Invest* 81: 1986–92, 1988.
28. Tosi MF, Stark JM, Smith WC, Hamedani A, Gruenert DC, Infeld MD. Induction of ICAM-1 expression on human airway epithelial cells by inflammatory cytokines: Effects on neutrophil-epithelial cell adhesion. *Am J Respir Cell Mol Biol* 7: 214–21, 1992.
29. Leckie MJ, ten BA, Khan J, Diamant Z, O'Connor BJ, Walls CM *et al.* Effects of an interleukin-5 blocking monoclonal antibody on eosinophils, airway hyper-responsiveness, and the late asthmatic response. *Lancet* 356(9248): 2144–48, 2000.
30. Hellings PW, Kasran A, Liu Z, Vandekerckhove P, Wuyts A, Overbergh L *et al.* Interleukin-17 orchestrates the granulocyte influx into airways after allergen inhalation in a mouse model of allergic asthma. *Am J Respir Cell Mol Biol* 28(1): 42–50, 2003.
31. Nakae S, Komiyama Y, Nambu A, Sudo K, Iwase M, Homma I *et al.* Antigen-specific T cell sensitization is impaired in IL-17-deficient mice, causing suppression of allergic cellular and humoral responses. *Immunity* 17(3): 375–87, 2002.
32. Molet S, Hamid Q, Davoine F, Nutku E, Taha R, Page N *et al.* IL-17 is increased in asthmatic airways and induces human bronchial fibroblasts to produce cytokines. *J Allergy Clin Immunol* 108(3): 430–38, 2001.
33. Letuve S, Lajoie-Kadoch S, Audusseau S, Rothenberg ME, Fiset PO, Ludwig MS *et al.* IL-17E upregulates the expression of proinflammatory cytokines in lung fibroblasts. *J Allergy Clin Immunol* 117(3): 590–96, 2006.
34. Zhu Z, Homer RJ, Wang Z, Chen Q, Geba GP, Wang J *et al.* Pulmonary expression of interleukin-13 causes inflammation, mucus hypersecretion, subepithelial fibrosis, physiologic abnormalities, and eotaxin production. *J Clin Invest* 103(6): 779–88, 1999.
35. Temann UA, Geba GP, Rankin JA, Flavell RA. Expression of interleukin 9 in the lungs of transgenic mice causes airway inflammation, mast cell hyperplasia, and bronchial hyperresponsiveness. *J Exp Med* 188: 1307–20, 1998.
36. Rankin JA, Picarella DE, Geba GP, Temann UA, Prasad B, DiCosmo B *et al.* Phenotypic and physiologic characterization of transgenic mice expressing interleukin 4 in the lung: Lymphocytic and eosinophilic inflammation without airway hyperreactivity. *Proc Natl Acad Sci USA* 93(15): 7821–25, 1996.
37. Lee JJ, McGarry MP, Farmer SC, Denzler KL, Larson KA, Carrigan PE *et al.* Interleukin-5 expression in the lung epithelium of transgenic mice leads to pulmonary changes pathognomonic of asthma. *J Exp Med* 185(12): 2143–56, 1997.
38. Einarsson O, Geba GP, Zhu Z, Landry M, Elias JA. Interleukin-11: Stimulation *in vivo* and *in vitro* by respiratory viruses and induction of airways hyperresponsiveness. *J Clin Invest* 97(4): 915–24, 1996.
39. Flood-Page P, Menzies-Gow A, Phipps S, Ying S, Wangoo A, Ludwig MS *et al.* Anti-IL-5 treatment reduces deposition of ECM proteins in the bronchial subepithelial basement membrane of mild atopic asthmatics. *J Clin Invest* 112(7): 1029–36, 2003.
40. Lee CG, Homer RJ, Zhu Z, Lanone S, Wang X, Koteliansky V *et al.* Interleukin-13 induces tissue fibrosis by selectively stimulating and activating transforming growth factor beta(1). *J Exp Med* 194(6): 809–21, 2001.
41. Boxall C, Holgate ST, Davies DE. The contribution of transforming growth factor-beta and epidermal growth factor signalling to airway remodelling in chronic asthma. *Eur Respir J* 27(1): 208–29, 2006.
42. Brewster CEP, Howarth PH, Djukanovic R, Wilson J, Holgate ST, Roche WR. Myofibroblasts and subepithelial fibrosis in bronchial "asthma". *Am J Resp Cell Mol Biol* 3: 507–11, 1990.
43. Hirst SJ, Martin JG, Bonacci JV, Chan V, Fixman ED, Hamid QA *et al.* Proliferative aspects of airway smooth muscle. *J Allergy Clin Immunol* 114(2 Suppl): S2–17, 2004.
44. Leung SY, Niimi A, Noble A, Oates T, Williams AS, Medicherla S *et al.* Effect of transforming growth factor-beta receptor I kinase inhibitor 2,4-disubstituted pteridine (SD-208) in chronic allergic airway inflammation and remodeling. *J Pharmacol Exp Ther* 319(2): 586–94, 2006.
45. Simcock DE, Kanabar V, Clarke GW, O'Connor BJ, Lee TH, Hirst SJ. Proangiogenic activity in bronchoalveolar lavage fluid from patients with asthma. *Am J Respir Crit Care Med* 176(2): 146–53, 2007.
46. Minshall EM, Leung DYM, Martin RJ, Song YL, Cameron L, Ernst P *et al.* Eosinophil-associated TGFβ1 mRNA expression and airways fibrosis in asthma. *Am J Respir Cell Mol Biol* 17: 326–33, 1997.
47. Xie S, Sukkar MB, Issa R, Khorasani NM, Chung KF. Mechanisms of induction of airway smooth muscle hyperplasia by transforming growth factor-beta. *Am J Physiol Lung Cell Mol Physiol* 293(1): L245–53, 2007.
48. Vignola AM, Chanez P, Chiappara G, Merendino A, Pace E, Rizzo A *et al.* Transforming growth factor-beta expression in mucosal biopsies in asthma and chronic bronchitis. *Am J Respir Crit Care Med* 156(2 Pt 1): 591–99, 1997.
49. Amishima M, Munakata M, Nasuhara Y, Sato A, Takahashi T, Homma Y *et al.* Expression of epidermal growth factor and epidermal growth factor receptor immunoreactivity in the asthmatic human airway. *Am J Respir Crit Care Med* 157(6 Pt 1): 1907–12, 1998.
50. Howarth PH, Knox AJ, Amrani Y, Tliba O, Panettieri RA Jr., Johnson M. Synthetic responses in airway smooth muscle. *J Allergy Clin Immunol* 114(2 Suppl): S32–50, 2004.

51. Brightling CE, Bradding P, Symon FA, Holgate ST, Wardlaw AJ, Pavord ID. Mast-cell infiltration of airway smooth muscle in asthma. *New Engl J Med* 346(22): 1699–705, 2002.
52. Brightling CE, Ammit AJ, Kaur D, Black JL, Wardlaw AJ, Hughes JM *et al.* The CXCL10/CXCR3 axis mediates human lung mast cell migration to asthmatic airway smooth muscle. *Am J Respir Crit Care Med* 171(10): 1103–8, 2005.
53. Hogg JC, Chu F, Utokaparch S, Woods R, Elliott WM, Buzatu L *et al.* The nature of small-airway obstruction in chronic obstructive pulmonary disease. *New Engl J Med* 350(26): 2645–53, 2004.
54. Saetta M, Turato G, Facchini FM, Corbino L, Lucchini RE, Casoni G *et al.* Inflammatory cells in the bronchial glands of smokers with chronic bronchitis. *Am J Respir Crit Care Med* 156(5): 1633–39, 1997.
55. Baraldo S, Turato G, Badin C, Bazzan E, Beghe B, Zuin R *et al.* Neutrophilic infiltration within the airway smooth muscle in patients with COPD. *Thorax* 59(4): 308–12, 2004.
56. Chanez P, Vignola AM, O'Shaugnessy T, Enander I, Li D, Jeffery PK *et al.* Corticosteroid reversibility in COPD is related to features of asthma. *Am J Respir Crit Care Med* 155(5): 1529–34, 1997.
57. Saetta M, Mariani M, Panina-Bordignon P, Turato G, Buonsanti C, Baraldo S *et al.* Increased Expression of the Chemokine Receptor CXCR3 and Its Ligand CXCL10 in Peripheral Airways of Smokers with Chronic Obstructive Pulmonary Disease. *Am J Respir Crit Care Med* 165(10): 1404–9, 2002.
58. O'Shaughnessy TC, Ansari TW, Barnes NC, Jeffery PK. Inflammation in bronchial biopsies of subjects with chronic bronchitis: Inverse relationship of CD8+ T lymphocytes with FEV1. *Am J Respir Crit Care Med* 155: 852–57, 1997.
59. Keatings VM, Collins PD, Scott DM, Barnes PJ. Differences in interleukin-8 and tumor necrosis factor-alpha in induced sputum from patients with chronic obstructive pulmonary disease or asthma. *Am J Respir Crit Care Med* 153(2): 530–34, 1996.
60. Traves SL, Culpitt SV, Russell RE, Barnes PJ, Donnelly LE. Increased levels of the chemokines GROalpha and MCP-1 in sputum samples from patients with COPD. *Thorax* 57(7): 590–95, 2002.
61. Lim S, Roche N, Oliver BG, Mattos W, Barnes PJ, Chung KF. Balance of matrix metalloprotease-9 and tissue inhibitor of metalloprotease-1 from alveolar macrophages in cigarette smokers. Regulation by interleukin-10. *Am J Respir Crit Care Med* 162(4 Pt 1): 1355–60, 2000.
62. Morrison D, Strieter RM, Donnelly SC, Burdick MD, Kunkel SL, MacNee W. Neutrophil chemokines in bronchoalveolar lavage fluid and leukocyte-conditioned medium from nonsmokers and smokers. *Eur Respir J* 12(5): 1067–72, 1998.
63. Kuschner WG, D'Alessandro A, Wong H, Blanc PD. Dose-dependent cigarette smoking-related inflammatory responses in healthy adults. *Eur Respir J* 9(10): 1989–94, 1996.
64. Pesci A, Balbi B, Majori M, Cacciani G, Bertacco S, Alciato P *et al.* Inflammatory cells and mediators in bronchial lavage of patients with chronic obstructive pulmonary disease. *Eur Respir J* 12(2): 380–86, 1998.
65. Balbi B, Bason C, Balleari E, Fiasella F, Pesci A, Ghio R *et al.* Increased bronchoalveolar granulocytes and granulocyte/macrophage colony-stimulating factor during exacerbations of chronic bronchitis. *Eur Respir J* 10(4): 846–50, 1997.
66. Franciosi LG, Page CP, Celli BR, Cazzola M, Walker MJ, Danhof M *et al.* Markers of disease severity in chronic obstructive pulmonary disease. *Pulm Pharmacol Ther* 19(3): 189–99, 2006.
67. Mio T, Romberger DJ, Thompson AB, Robbins RA, Heires A, Rennard SI. Cigarette smoke induces interleukin-8 release from human bronchial epithelial cells. *Am J Respir Crit Care Med* 155(5): 1770–76, 1997.
68. Dubar V, Gosset P, Aerts C, Voisin C, Wallaert B, Tonnel AB. *In vitro* acute effects of tobacco smoke on tumor necrosis factor alpha and interleukin-6 production by alveolar macrophages. *Exp Lung Res* 19(3): 345–59, 1993.
69. Nishikawa M, Kakemizu N, Ito T, Kudo M, Kaneko T, Suzuki M *et al.* Superoxide mediates cigarette smoke-induced infiltration of neutrophils into the airways through nuclear factor-kappaB activation and IL-8 mRNA expression in guinea pigs *in vivo*. *Am J Respir Cell Mol Biol* 20(2): 189–98, 1999.
70. Masubuchi T, Koyama S, Sato E, Takamizawa A, Kubo K, Sekiguchi M *et al.* Smoke extract stimulates lung epithelial cells to release neutrophil and monocyte chemotactic activity. *Am J Pathol* 153(6): 1903–12, 1998.
71. de Boer WI, Sont JK, van Schadewijk A, Stolk J, van Krieken JH, Hiemstra PS. Monocyte chemoattractant protein 1, interleukin 8, and chronic airways inflammation in COPD. *J Pathol* 190(5): 619–26, 2000.
72. Premack BA, Schall TJ. Chemokine receptors: Gateways to inflammation and infection. *Nat Med* 2(11): 1174–78, 1996.
73. Traves SL, Smith SJ, Barnes PJ, Donnelly LE. Specific CXC but not CC chemokines cause elevated monocyte migration in COPD: A role for CXCR2. *J Leukoc Biol* 76(2): 441–50, 2004.
74. Demedts IK, Bracke KR, Van Pottelberge G, Testelmans D, Verleden GM, Vermassen FE *et al.* Accumulation of dendritic cells and increased CCL20 levels in the airways of patients with chronic obstructive pulmonary disease. *Am J Respir Crit Care Med* 175(10): 998–1005, 2007.
75. Ning W, Li CJ, Kaminski N, Feghali-Bostwick CA, Alber SM, Di YP *et al.* Comprehensive gene expression profiles reveal pathways related to the pathogenesis of chronic obstructive pulmonary disease. *Proc Natl Acad Sci USA* 101(41): 14895–900, 2004.
76. Zhu J, Qiu YS, Majumdar S, Gamble E, Matin D, Turato G *et al.* Exacerbations of Bronchitis: Bronchial eosinophilia and gene expression for interleukin-4, interleukin-5, and eosinophil chemoattractants. *Am J Respir Crit Care Med* 164(1): 109–16, 2001.
77. Zhu J, Qiu Y, Valobra M, Qiu S, Majumdar S, Matin D *et al.* Plasma cells and IL-4 in chronic bronchitis and chronic obstructive pulmonary disease. *Am J Respir Crit Care Med* 175(11): 1125–33, 2007.
78. Zheng T, Zhu Z, Wang Z, Homer RJ, Ma B, Riese RJ Jr. *et al.* Inducible targeting of IL-13 to the adult lung causes matrix metalloproteinase- and cathepsin-dependent emphysema. *J Clin Invest* 106(9): 1081–93, 2000.
79. Hoshino T, Kato S, Oka N, Imaoka H, Kinoshita T, Takei S *et al.* Pulmonary Inflammation and Emphysema: Role of the Cytokines IL-18 and IL-13. *Am J Respir Crit Care Med* 176(1): 49–62, 2007.
80. Wang Z, Zheng T, Zhu Z, Homer RJ, Riese RJ, Chapman HA Jr. *et al.* Interferon gamma induction of pulmonary emphysema in the adult murine lung. *J Exp Med* 192(11): 1587–600, 2000.
81. Boutten A, Bonay M, Laribe S, Leseche G, Castier Y, Lecon-Malas V *et al.* Decreased expression of interleukin 13 in human lung emphysema. *Thorax* 59(10): 850–54, 2004.
82. Panzner P, Lafitte JJ, Tsicopoulos A, Hamid Q, Tulic MK. Marked up-regulation of T lymphocytes and expression of interleukin-9 in bronchial biopsies from patients with chronic bronchitis with obstruction. *Chest* 124(5): 1909–15, 2003.
83. Gan WQ, Man SF, Senthilselvan A, Sin DD. Association between chronic obstructive pulmonary disease and systemic inflammation: A systematic review and a meta-analysis. *Thorax* 59(7): 574–80, 2004.
84. de GI, Donahoe M, Calhoun WJ, Mancino J, Rogers RM. Elevated TNF-alpha production by peripheral blood monocytes of weight-losing COPD patients. *Am J Respir Crit Care Med* 153(2): 633–37, 1996.
85. Casadevall C, Coronell C, Ramirez-Sarmiento AL, Martinez-Llorens J, Barreiro E, Orozco-Levi M *et al.* Upregulation of proinflammatory cytokines in the intercostal muscles of COPD patients. *Eur Respir J*, 2007.
86. Pinto-Plata VM, Livnat G, Girish M, Cabral H, Masdin P, Linacre P *et al.* Systemic cytokines, clinical and physiological changes in patients hospitalized for exacerbation of COPD. *Chest* 131(1): 37–43, 2007.
87. Qiu Y, Zhu J, Bandi V, Atmar RL, Hattotuwa K, Guntupalli KK *et al.* Biopsy neutrophilia, neutrophil chemokine and receptor gene expression in severe exacerbations of chronic obstructive pulmonary disease. *Am J Respir Crit Care Med* 168(8): 968–75, 2003.
88. Saetta M, Di SA, Maestrelli P, Turato G, Mapp CE, Pieno M *et al.* Airway eosinophilia and expression of interleukin-5 protein in asthma and in exacerbations of chronic bronchitis [see comments]. *Clin Exp Allergy* 26(7): 766–74, 1996.

89. Wedzicha JA, Seemungal TA, MacCallum PK, Paul EA, Donaldson GC, Bhowmik A *et al.* Acute exacerbations of chronic obstructive pulmonary disease are accompanied by elevations of plasma fibrinogen and serum IL-6 levels. *Thromb Haemost* 84(2): 210–15, 2000.
90. Bocchino V, Bertorelli G, Bertrand CP, Ponath PD, Newman W, Franco C *et al.* Eotaxin and CCR3 are up-regulated in exacerbations of chronic bronchitis. *Allergy* 57(1): 17–22, 2002.
91. Khair OA, Devalia JL, Abdelaziz MM, Sapsford RJ, Tarraf H, Davies RJ. Effect of Haemophilus influenzae endotoxin on the synthesis of IL-6, IL-8, TNF-alpha and expression of ICAM-1 in cultured human bronchial epithelial cells. *Eur Respir J* 7(12): 2109–16, 1994.
92. Inoue H, Hara M, Massion PP, Grattan KM, Lausier JA, Chan B *et al.* Role of recruited neutrophils in interleukin-8 production in dog trachea after stimulation with Pseudomonas *in vivo*. *Am J Respir Cell Mol Biol* 13(5): 570–77, 1995.
93. Churg A, Dai J, Tai H, Xie C, Wright JL. Tumor necrosis factor-alpha is central to acute cigarette smoke-induced inflammation and connective tissue breakdown. *Am J Respir Crit Care Med* 166(6): 849–54, 2002.
94. Takeyama K, Dabbagh K, Lee HM, Agusti C, Lausier JA, Ueki IF *et al.* Epidermal growth factor system regulates mucin production in airways. *Proc Natl Acad Sci USA* 96(6): 3081–86, 1999.
95. Riise GC, Larsson S, Lofdahl CG, Andersson BA. Circulating cell adhesion molecules in bronchial lavage and serum in COPD patients with chronic bronchitis. *Eur Respir J* 7(9): 1673–77, 1994.
96. Churg A, Wang RD, Tai H, Wang X, Xie C, Dai J *et al.* Macrophage metalloelastase mediates acute cigarette smoke-induced inflammation via tumor necrosis factor-alpha release. *Am J Respir Crit Care Med* 167(8): 1083–89, 2003.
97. Di FM, Barbier D, Mege JL, Orehek J. Tumor necrosis factor-alpha levels and weight loss in chronic obstructive pulmonary disease. *Am J Respir Crit Care Med* 150(5 Pt 1): 1453–55, 1994.
98. Schols AM, Buurman WA, Staal van den Brekel AJ, Dentener MA, Wouters EF. Evidence for a relation between metabolic derangements and increased levels of inflammatory mediators in a subgroup of patients with chronic obstructive pulmonary disease. *Thorax* 51(8): 819–24, 1996.
99. de GI, Donahoe M, Calhoun WJ, Mancino J, Rogers RM. Elevated TNF-alpha production by peripheral blood monocytes of weight-losing COPD patients. *Am J Respir Crit Care Med* 153(2): 633–37, 1996.
100. Reid MB, Lannergren J, Westerblad H. Respiratory and limb muscle weakness induced by tumor necrosis factor-alpha: Involvement of muscle myofilaments. *Am J Respir Crit Care Med* 166(4): 479–84, 2002.
101. Li YP, Lecker SH, Chen Y, Waddell ID, Goldberg AL, Reid MB. TNF-alpha increases ubiquitin-conjugating activity in skeletal muscle by up-regulating UbcH2/E220k. *FASEB J* 17(9): 1048–57, 2003.
102. Rennard SI, Fogarty C, Kelsen S, Long W, Ramsdell J, Allison J *et al.* The safety and efficacy of infliximab in moderate to severe chronic obstructive pulmonary disease. *Am J Respir Crit Care Med* 175(9): 926–34, 2007.
103. Koto H, Mak JC, Haddad EB, Xu WB, Salmon M, Barnes PJ *et al.* Mechanisms of impaired beta-adrenoceptor-induced airway relaxation by interleukin-1beta *in vivo* in the rat. *J Clin Invest* 98(8): 1780–87, 1996.
104. Beeh KM, Kornmann O, Buhl R, Culpitt SV, Giembycz MA, Barnes PJ. Neutrophil chemotactic activity of sputum from patients with COPD: Role of interleukin 8 and leukotriene B4. *Chest* 123(4): 1240–47, 2003.
105. Woolhouse IS, Bayley DL, Stockley RA. Sputum chemotactic activity in chronic obstructive pulmonary disease: Effect of alpha(1)-antitrypsin deficiency and the role of leukotriene B(4) and interleukin 8. *Thorax* 57(8): 709–14, 2002.
106. Wang B, Cleary PP, Xu H, Li JD. Up-regulation of interleukin-8 by novel small cytoplasmic molecules of nontypeable Haemophilus influenzae via p38 and extracellular signal-regulated kinase pathways. *Infect Immun* 71(10): 5523–30, 2003.
107. Nakamura H, Yoshimura K, McElvaney NG, Crystal RG. Neutrophil elastase in respiratory epithelial lining fluid of individuals with cystic fibrosis induces interleukin-8 gene expression in a human bronchial epithelial cell line. *J Clin Invest* 89: 1478–84, 1992.
108. Mahler DA, Huang S, Tabrizi M, Bell GM. Efficacy and safety of a monoclonal antibody recognizing interleukin-8 in COPD: A pilot study. *Chest* 126(3): 926–34, 2004.
109. Howat WJ, Holgate ST, Lackie PM. TGF-beta isoform release and activation during *in vitro* bronchial epithelial wound repair. *Am J Physiol Lung Cell Mol Physiol* 282(1): L115–23, 2002.
110. Chambers RC, Leoni P, Kaminski N, Laurent GJ, Heller RA. Global expression profiling of fibroblast responses to transforming growth factor-beta1 reveals the induction of inhibitor of differentiation-1 and provides evidence of smooth muscle cell phenotypic switching. *Am J Pathol* 162(2): 533–46, 2003.
111. Goldsmith AM, Bentley JK, Zhou L, Jia Y, Bitar KN, Fingar DC *et al.* Transforming growth factor-beta induces airway smooth muscle hypertrophy. *Am J Respir Cell Mol Biol* 34(2): 247–54, 2006.
112. Eickelberg O, Kohler E, Reichenberger F, Bertschin S, Woodtli T, Erne P *et al.* Extracellular matrix deposition by primary human lung fibroblasts in response to TGF-beta1 and TGF-beta3. *Am J Physiol* 276(5 Pt 1): L814–24, 1999.
113. Coutts A, Chen G, Stephens N, Hirst S, Douglas D, Eichholtz T *et al.* Release of biologically active TGF-beta from airway smooth muscle cells induces autocrine synthesis of collagen. *Am J Physiol Lung Cell Mol Physiol* 280(5): L999–1008, 2001.
114. Takizawa H, Tanaka M, Takami K, Ohtoshi T, Ito K, Satoh M *et al.* Increased expression of transforming growth factor-beta1 in small airway epithelium from tobacco smokers and patients with chronic obstructive pulmonary disease (COPD). *Am J Respir Crit Care Med* 163(6): 1476–83, 2001.
115. Vignola AM, Chanez P, Chiappara G, Merendino A, Pace E, Rizzo A *et al.* Transforming growth factor-beta expression in mucosal biopsies in asthma and chronic bronchitis. *Am J Respir Crit Care Med* 156(2 Pt 1): 591–99, 1997.
116. de Boer WI, van Schadewijk A, Sont JK, Sharma HS, Stolk J, Hiemstra PS *et al.* Transforming growth factor beta1 and recruitment of macrophages and mast cells in airways in chronic obstructive pulmonary disease. *Am J Respir Crit Care Med* 158(6): 1951–57, 1998.
117. Vignola AM, Chanez P, Chiappara G, Merendino A, Zinnanti E, Bousquet J *et al.* Release of transforming growth factor-beta (TGF-beta) and fibronectin by alveolar macrophages in airway diseases. *Clin Exp Immunol* 106(1): 114–19, 1996.
118. Kanazawa H, Asai K, Hirata K, Yoshikawa J. Possible effects of vascular endothelial growth factor in the pathogenesis of chronic obstructive pulmonary disease. *Am J Med* 114(5): 354–58, 2003.
119. Santos S, Peinado VI, Ramirez J, Morales-Blanhir J, Bastos R, Roca J *et al.* Enhanced expression of vascular endothelial growth factor in pulmonary arteries of smokers and patients with moderate chronic obstructive pulmonary disease. *Am J Respir Crit Care Med* 167(9): 1250–56, 2003.
120. Kasahara Y, Tuder RM, Taraseviciene-Stewart L, Le Cras TD, Abman S, Hirth PK *et al.* Inhibition of VEGF receptors causes lung cell apoptosis and emphysema. *J Clin Invest* 106(11): 1311–19, 2000.
121. Tuder RM, Zhen L, Cho CY, Taraseviciene-Stewart L, Kasahara Y, Salvemini D *et al.* Oxidative stress and apoptosis interact and cause emphysema due to vascular endothelial growth factor receptor blockade. *Am J Respir Cell Mol Biol* 29(1): 88–97, 2003.

Matrix Degrading Proteinases in COPD and Asthma

28

CHAPTER

Steven D. Shapiro

Department of Medicine, University of Pittsburgh, Pittsburgh, PA, USA

INTRODUCTION

In chronic obstructive pulmonary disease (COPD), we now appreciate that cigarette smoke sets into motion a complex inflammatory network leading to inflammation, destruction, and when coupled with abnormal repair results in COPD. Elastolytic proteinases are well established in the destructive phase. Recent studies have focused on the roles of proteinases in control of inflammation and repair. With respect to asthma, proteinases are prominently expressed on many airway cells and also appear be involved in controlling inflammation as well as participating in other aspects of remodeling such as subepithelial fibrosis, mucus metaplasia, and smooth muscle hypertophy. This discussion will be limited to host proteinases with potential extracellular matrix degrading capacity, excluding intracellular cysteine proteinases, such as caspases, that regulate cell death, and non-host serine proteinases of dust mites and other allergens that may also play a role in asthma.

PROTEINASES

Serine proteinases

Serine proteinases have diverged evolutionarily from a single gene product; as a result of duplications and mutations, enzymes with diverse biological functions including digestive enzymes of exocrine glands, clotting factors, and leukocyte granule associated proteinases some of which degrade extracellular matrix proteins and are of relevance to emphysema have been derived (Table 28.1). Host serine proteinases associated with COPD and asthma largely belong to the SA clan, S1(trypsin/chymotrypsin) family. S1 serine proteinases are characterized by conserved His, Asp, and Ser residues that form a charge-relay system that functions by transfer of electrons from the carboxyl group of Asp to the oxygen of Ser which then becomes a powerful nucleophile able to attack the carbonyl carbon atom of the peptide bond of the substrate. These enzymes are synthesized as pre-proenzymes in the endoplasmic reticulum and processed by cleavage of the signal peptide (pre-) and removal of a dipeptide (pro-) by cathepsin C, and stored in granules as active packaged proteins. Distinct subsets of serine proteinases are expressed in a lineage-restricted manner in immune and inflammatory cells. Serine proteinases are also expressed in a developmentally specific manner. For example, neutrophil elastase (NE), proteinase 3, and cathepsin G are major components of primary or azurophil granules that are formed during a very specific stage during the development of myeloid cells.

Neutrophil elastase (NE), cathepsin G, and *proteinase 3* are ~30 kDa glycoprotein containing ~20% neutral sugars. NE is a more potent elastase than the other serine proteinases and has received the most attention in COPD. NE's tertiary structure is similar to other chymotrypsin-like serine proteinases with two interacting antiparallel β-barrel cylindrical domains that form a crevice encompassing the catalytic triad [1]. NE prefers substrates with Val > Ala > Ser, Cys at the P1 position. NE has activity against a broad range of extracellular matrix proteins including the highly resistant elastin. NE expression is localized to primary or azurophil granules of neutrophils and a subset of proinflammatory monocytes (Fig. 28.1) [2].

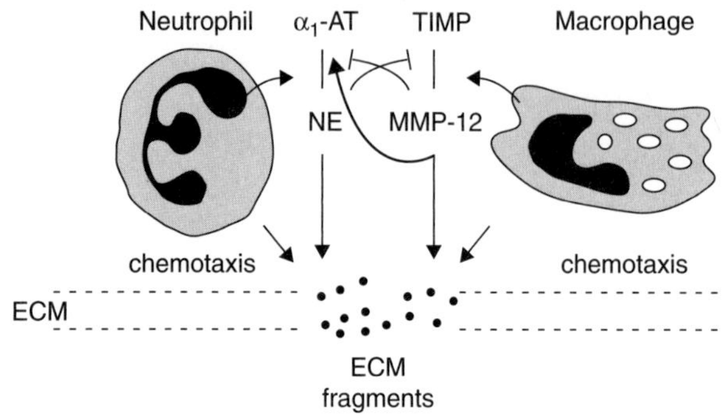

FIG. 28.1 Interactions between neutrophil and macrophage proteinases in emphysema. Neutrophil elastase (NE) originating from azurophil granules directly degrades elastin and indirectly augments elastolysis via degradation of TIMP(s). Macrophage MMP-12 also directly degrades elastin and also indirectly augments NE activity by degrading its inhibitor α1-AT. Extracellular matrix (ECM) fragments released during proteolysis play a role in perpetuating inflammatory cell recruitment.

Studies from several laboratories have shown that NE is responsible for most of the degradative activity of neutrophils toward extracellular matrix structures. However, when optimally primed and stimulated by biologically relevant agonists, neutrophils release less than 2% of their content of NE freely into the extracellular space, and that they are able to translocate as much as 12% of their total NE to the cell membrane, where it is catalytically active and resistant to inhibitors [3]. The high concentration of NE (~5 mM) within azurophil granules transiently overwhelm local NE inhibitors, resulting in a quantum burst of obligate catalytic activity near the cell surface when each granule is released [4]. These mechanisms allow NE to proteolyze extracellular matrix (ECM) in a focused and protected manner while limiting widespread tissue destruction. NE might also be freely released during "frustrated phagocytosis," or by necrotic or apoptotic neutrophils if the latter are not cleared efficiently by macrophages. Regardless of the mechanisms by which active NE reaches ECM, it is relatively protected from inhibitors following binding to insoluble macromolecules, and NE is intimately associated with elastic fibers in emphysematous tissue [5]. During bacterial exacerbations of COPD, overwhelming (micromolar) concentrations of active NE can be found as a result of brisk neutrophil influx, and active NE can often be found within the airways of patients with alpha-1-antitrypsin (α_1-AT) deficiency. NE is also a potent secretagogue and induces mucin gene expression [6], which might aggravate airflow obstruction in COPD. Neutrophil influx and NE is also believed to be important in severe, steroid-resistant asthma.

NE deficient mice have demonstrated a role for NE in killing gram-negative bacteria [7]. NE-mediated bacterial killing is related to proteolytic degradation of Omp proteins on the outer wall of gram-negative bacteria [8]. Application of $NE^{-/-}$ mice to cigarette smoke also demonstrates a role for NE in the development of emphysema as discussed below [9].

Serine proteinase inhibitors

Serine proteinase inhibitors are abundant in the plasma. *Alpha-2 macroglobulin*, a large protein usually restricted to the bloodstream because of its mass, 725,000 kDa, inhibits proteinases of several classes by "entrapping" proteinases following cleavage of susceptible regions of the molecule. At a concentration of 150–350 mg/dl, α_1-AT has the highest concentration of the plasma inhibitors. α_1-AT belongs to a family of serine proteinase inhibitors called the *serpins*. Serpins have considerable sequence homology, particularly around their reactive sites. They are important for homeostasis, since they exert some control over such major proteolytic cascades as the complement system and coagulation. α_1-AT is a 52 kDa glycoprotein synthesized primarily by the liver, consisting of a single polypeptide chain of 394 amino acids. Proteolytic inhibition by α_1-AT involves cleavage of the "strained" reactive open center of α_1-AT between Met^{358} and Ser^{359}, resulting in an altered, "relaxed" α_1-AT conformation in complex with the proteinase. Formation of the complex renders the proteinase inactive and, because the complex is quite stable, inactivation is essentially permanent. The association and inhibition of NE by α_1-AT is much faster than with other serine proteinases including trypsin, yet the name "α_1-antitrypsin" is retained for historical respect. α_1-AT is the major inhibitor of serine proteinases in the lower airspace. As discussed below, inherited deficiency of α_1-AT represents the only genetic abnormality to date associated with COPD.

Additional low molecular weight serine proteinase inhibitors are abundant in airway fluid and hence thought to represent the primary defense against proteinase-mediated airway damage. *Secretory leukoprotease inhibitor (SLPI)* is a12 kDa protein produced by mucus-secreting and epithelial cells in the airway as well as type 2 pneumocytes [10]. SLPI inhibits NE and cathepsin G and many other serine proteinases, but not proteinase 3. *Elafin*, also produced by airway secretory and epithelial cells is released as a 12 kDa precursor which is processed to a 6 kDa form that specifically inhibits NE and proteinase 3 [11]. These inhibitors are able to inhibit NE bound to substrate giving them an added dimension that α_1-AT lacks. An additional serpin, serpin B1 or monocyte/neutrophil elastase inhibitor, also has considerable activity against NE [12]. Airway mucus contains several other substances that partially inhibit NE including polyanionic molecules such as mucins, other glycosaminoglycans, and fatty acids. DNA, released from inflammatory leukocytes, binds to SLPI greatly enhancing its rate of association with NE. The relative contribution of each of these molecules to proteinase inhibition is unknown.

Matrix metalloproteinases

Matrix metalloproteinases (MMPs) [13] comprise a family of 23 human matrix degrading enzymes believed to be essential for development and physiologic tissue remodeling and repair (Table 28.1). Abnormal expression of MMPs has been implicated in many destructive processes, including tumor cell invasion and angiogenesis, arthritis, atherosclerosis, arterial aneurysms, and pulmonary emphysema [14]. MMPs are secreted as inactive proenzymes that are activated at the cell membrane surface or within the extracellular space by proteolytic cleavage of the N-terminal domain. Catalytic activity is dependent upon binding of a zinc ion at the active site and is specifically inhibited by members of another gene family, called tissue inhibitor of metalloproteinases (TIMPs) for tissue inhibitors of MMPs. Currently, four TIMPs have been described. Optimal activity of MMPs is around pH 7.4. MMP family members share 40–50% identity at the amino acid level, and they possess common structural domains. Domains include a proenzyme domain that maintains the enzyme in its latent form, an active domain that coordinates binding of the catalytic zinc molecule, and (except for MMP-7) a C-terminal domain involved in substrate, cell and TIMP binding. The gelatinases A and B (MMP-2 and MMP-9, respectively) have an additional fibronectin-like domain which mediates their high binding affinity to gelatins and elastin. MMP-9 has one more domain with homology to type V collagen. Membrane-type MMPs (MT1-6-MMP or MMP-14 to -19) have an additional C-terminal membrane-spanning domain.

Individual members of the MMP family can be loosely divided into groups based on their matrix degrading capacity. As a whole, they are able to cleave all extracellular matrix components. Those MMPs implicated in COPD and asthma include the *collagenases* (MMPs – -1, -8, -13) that have the unique capacity to cleave native triple helical interstitial collagens but not elastin. MMP-1 is prominently expressed in epithelial cells in patients with asthma and COPD [15]. However, mice do not have MMP-1 but only MMP-8 and -13. Gene targeting of MMP-8 has shown that this neutrophil-derived proteinase is involved in turning off neutrophil recruitment, likely via degradation of neutrophil chemokines [16]. MMP-13$^{-/-}$ mice have developmental bone defects [17].

Gelatinases of 72 kDa (gelatinase A, MMP-2) and 92 kDa *(gelatinase B, MMP-9)* differ in their cellular origin and regulation, but share the capacity to degrade gelatins (denatured collagens), type IV collagen, elastin, and other matrix proteins. The gelatinases are by far the most thoroughly studied members of the MMP family and MMP-9 clearly the most thoroughly investigated in COPD and asthma. Research focus to gelatinases is in part due to their importance in biology, but also reflects the ease of detecting them by the exquisitely sensitive technique of gelatin zymography in which the active moiety is detected by virtue of its cleavage of an artificial substrate.

In both COPD [18] and asthma [19, 20] many studies have found increased expression of MMP-9, often in association with decrease in its inhibitor TIMP-1, in sputum, bronchoalveolar lavage (BAL), lung tissue, and cells obtained from asthmatic patients. For example, airway epithelial cells produce MMP-9 upon injury and proportional to the severity of asthma [21]. Inflammatory and immune cells are a rich source of MMP-9 particularly eosinophils, neutrophils, macrophages [22,23], *and DCs* [24,25].

MMP-9$^{-/-}$ mice are protected from airway subepithelial fibrosis in models of asthma [26]. Yet, they have no role in cigarette smoke-induced airspace enlargement (Shapiro unpublished). Other phenotypes of MMP-9$^{-/-}$ mice include abnormal long bone development (through failed activation of vascular endothelial growth factor, VEGF) [27], and many interesting vascular phenotypes, such as a requirement for MMP-9, for abdominal aortic aneurysms [28].

Macrophage elastase (MMP-12) is characterized by macrophage-specific expression and broad potent matrix degrading capacity for an MMP [29]. MMP-12, like many MMPs also has important non-matrix substrates. MMP-12 is required for TNF-α shedding following smoke exposure [30], and MMP-12 to a greater degree than other MMPs degrades and inactivates α_1-AT [29], thus indirectly enhancing the activity of NE.

MMP-12 is predominantly a macrophage product and it has been found in macrophages in association with cigarette smoke exposure and COPD. In fact, it was the most highly upregulated gene by expression profiling in macrophages obtained from both murine and human smokers [31]. MMP-12 has also been found in asthmatic lungs, and has even been detected in airway epithelial cells [32] and smooth muscle cells as well [33].

Macrophages of MMP-12$^{-/-}$ mice have a markedly diminished capacity to degrade extracellular matrix components and are essentially unable to penetrate reconstituted basement membranes both *in vitro* and *in vivo* [34]. As discussed below unlike wild-type (WT) mice, MMP-12-deficient mice were protected from macrophage accumulation and the development of emphysema despite heavy long-term smoke exposure [35]. MMP-12 appears to be important in inflammatory cell recruitment in several murine models of asthma described below as well (Fig. 28.2).

Membrane-type metalloproteinases (MT-MMPs) represent six MMPs that are localized at the cell surface and at least one MT-MMP, MT1-MMP, activates MMP-2. MT-MMPs also appear to directly degrade ECM proteins, but their catalytic capacities is not well defined at present. MT1-MMP is among the MMPs associated with COPD [36].

ADAMs (a disintegrin and metalloprotease domain) represents a family of related metalloproteases that are bound to the cell surface. ADAM-33 was described as an asthma gene in humans following an extensive genetic linkage analysis [37]. Subsequently, ADAM-33 has been replicated in some but not all asthma populations, as well as in COPD. The role of ADAM-33 is not clear, in fact it does not appear to be catalytically active due to splicing. Expressed by fibroblasts and smooth muscle cells, it might be involved in airway remodeling and hyperresponsiveness [38].

Tissue inhibitors of metalloproteinases

Tissue inhibitors of metalloproteinases (TIMPS) comprise a family of four with molecular masses ranging between 21 (TIMP-2, nonglycosylated) and 27.5 (TIMP-1, glycosylated). Each TIMP inhibits MMPs via tight, non-covalent binding with 1:1 stoichiometry. TIMP-1 binds to the C-terminal domain of MMPs, but how this leads to inhibition of catalysis is unknown. Those MMPs that lack the C-terminal domain, including MMP-7 and fully processed form of MMP-12 are still susceptible to TIMP inhibition although with a lower Ki. TIMP-2 is secreted complexed to MMP-2 in fibroblasts. TIMP-2, not only inhibits MMP-2, but also is involved in docking pro-MMP-2 to the cell surface where the enzyme is activated by MT1-MMP.

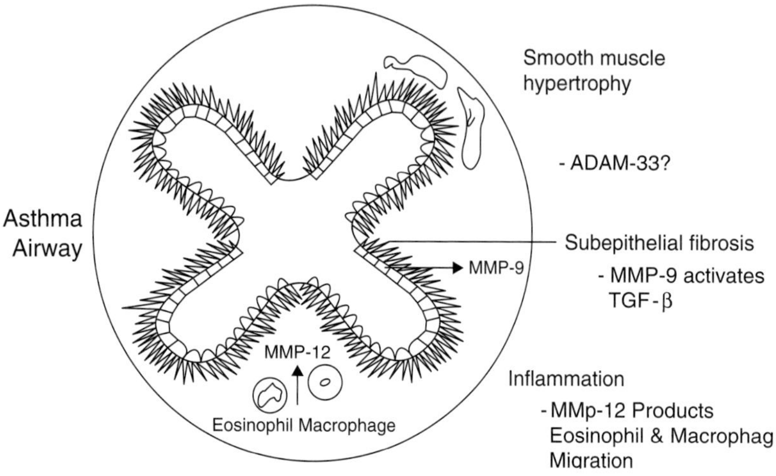

FIG. 28.2 Potential functions of proteinases in asthma. Tryptase mediates bronchoconstriction and fibroblast proliferation, most likely via proteolytic activation of protease activated receptor-2. MMPs have been linked to both fibroblast proliferation (MMP-2) and collagenolysis (MMP-1). MMP-9 might promote eosinophil transvascular migration. MMPs and related ADAMs (a disintegrin and metalloproteinase domain) might regulate cell survival via shedding of surface molecules such as Fas, TNF/TNF receptor.

TIMPs are secreted from many cell types and are abundant in tissues. For example, alveolar macrophages secrete both a variety of metalloproteinases and TIMP-1 and TIMP-2. TIMP-3 is expressed predominantly by epithelial cells and binds to extracellular matrix and thus may be important in preventing emphysema and in allowing excess matrix accumulation in asthmatic airways. Expression of MMPs and TIMPs may depend on the inflammatory stimulus, MMPs, and TIMPs may be coordinately regulated, perhaps to limit tissue injury during normal remodeling associated with inflammation, or regulation may be discoordinate, potentially leading to tissue injury.

Cysteine proteinases

Cysteine proteinases utilize the sulfhydryl group of Cys as a nucleophile. Similar to serine proteinases a proton donor from His form a catalytic dyad (and in some cases triads) required for endopeptidase activity (Table 28.1). Cysteine proteinase represent a large, diverse group of plant and animal enzymes with amino acid homology at the active site only [39]. Human alveolar macrophages produce the lysosomal thiol proteinases, cathepsins B, H, L, and S that have been implicated in COPD [40]. CD4$^+$ cells express cathepsin S which functions in antigen processing [41] and thus might prove to play a role in asthma . These enzymes share similar sizes of 24–32 kDa and high mannose side chains (typical of proteins targeted for lysosomal accumulation). *Cathepsins B and H* have little endopeptidase activity and may function to activate other proteins similar to a distant relative; interleukin converting enzyme. *Cathepsin C* or dipeptidyl peptidase I has limited extracellular matrix degrading activity but is required for activation of nearly all matrix degrading serine proteinase proenzymes to their active form. *Cathepsins L and S* have large active pockets with relatively indiscriminate substrate specificities that include elastin and other matrix components. These enzymes have an acidic pH optima but cathepsin S retains ~25% of its elastolytic capacity at neutral pH (making it approximately equal to NE). *Cathepsin K* is a potent elastase predominantly expressed in osteoclasts but also by macrophages in the vasculature and perhaps other tissues. These enzymes clearly have the capacity to cause lung destruction if targeted to the cell surface or extracellular space particularly in acidic microenvironments.

The role of cysteine proteinases in COPD and asthma remains speculative. However, cathepsin S was shown to be involved in the emphysematous process associated with IFN-γ transgenic mice [42].

Cysteine proteinase inhibitors

Cysteine proteinases are inhibited by *cystatins*. Some cystatins are strictly intracellular, while others, such as cystatin C, possess a signal peptide and are secreted by a variety of cells into the extracellular fluid. Cystatin C, comprised of a single non-glycosylated 120 amino acid peptide chain (13 kDa) forms reversible 1:1 complexes with enzymes in competition with substrates. Cystatin C, the most ubiquitous cystatin, is found in all human tissues and body fluids tested, providing general protection against tissue destruction by intracellular cathepsin enzymes leaking from dying cells. Lack of cystatin C has been associated with destructive lesions in the vasculature [43].

COPD

COPD is currently defined by the Global Initiative for Chronic Obstructive Lung Disease (GOLD) as a disease state characterized by exposure to a noxious agent resulting in airflow limitation that is not fully reversible (http://www.goldcopd.com/) [44]. Airflow limitation may be secondary to airway obstruction or alveolar destruction with loss of elastic recoil. In fact, most patients have an admixture of large airway changes (accounting for symptoms of chronic bronchitis), small airway changes, and parenchymal lung involvement.

Small airway disease

The small airways (diameter <2 mm) are the major sites of airway obstruction in COPD. Characteristic cellular changes leading to luminal narrowing include goblet cell

TABLE 28.1 Main inflammatory cell matrix degrading proteinases in the lung.

Cell (major source)	Proteinase	Class	Molecular mass*(kDa)	Matrix substrates	Other cells expressing proteinase
Parenchyma					
Neutrophil	Neutrophil elastase	Serine	27–31	Elastin, bm components+	Proinflammatory monocyte
	Proteinase 3	Serine	28–34	Elastin, bm components+	Monocyte Mast cell
	Cathepsin G	Serine	27–32	(elastin) bm components+	Monocyte Mast cell
	MMP-8	MMP	#Pro-55	Interstitial collagens	
	MMP-9	MMP	Pro-92–95	Denatured collagens, types IV, V, and VII collagen	Macrophage Endothelial cell
Macrophage	MMP-12	MMP	Pro-54	Elastin, bm components+	
	MMP-1	MMP	Pro-55	Interstitial collagens	Fibroblasts Epithelial cells
	Cathepsin L	Cysteine	29	Elastin (at acidic pH)	
	Cathepsin S	Cysteine	28	Elastin	CD4 T-cell

*denotes (pre)proenzyme forms. basement membrane (bm) components include fibronectin, laminin, entactin, vitronectin, and type IV collagen (non-helical domains). #Pro, molecular mass of proenzyme form.
Note: parentheses denote minor cellular sources.

metaplasia with excess mucus production, smooth muscle hypertrophy, infiltrating mononuclear inflammatory cells, and submucosal fibrosis. Surrounding lung parenchymal elastin provides radial traction on bronchioles at points where alveolar septa attach. Hence, elastolysis may lead to loss of alveolar attachments with airway distortion and narrowing in COPD. Destruction of small airways may directly contribute to an increase in flow resistance through destruction of parallel conducting pathways.

The role of proteinases in small airways disease in COPD is less well appreciated than in the lung parenchyma. Proteinase expression has been well studied in asthma (see below) where there may be some common mechanisms of airway remodeling. One distinction is that neutrophil influx is a hallmark of COPD, while prominent only in severe asthma. Neutrophil influx in COPD is likely related directly to both cigarette smoke exposure and colonization by pathogens later in the disease process. In addition to its elastolytic activity, NE is a potent secretagogue. As in asthma, MMP-9 is expressed by several cell types in the airway might paradoxically lead to increased, rather than decreased, collagen accumulation in the airway. Similarly, despite collagenase-1 (MMP-1) expression by airway epithelial cells [15], collagen accumulates in the airways, highlighting the complexity of matrix turnover.

Emphysema

Emphysema is characterized by abnormal, permanent enlargement of airspaces distal to the terminal bronchiole, accompanied by destruction of their walls with coalescence into larger airspaces. The explanation for this pathology stems from the elastase:antielastase hypothesis, the foundation of which was built upon two seminal observations made over 35 years, one experimental and the other clinical. First, in 1963 Gross instilled papain into the lungs of rodents which resulted in emphysema [45]. Subsequently, investigators have instilled a variety of proteinases into animal lungs, yet only elastases result in emphysema. Second, in 1964 Laurel and Eriksson described five patients with deficiency of α_1-AT, the physiologic inhibitor of NE. Three of these initial subjects had emphysema [46].

Over time, there have been revisions to the elastase: antielastase hypothesis, but the basic concept has stood the test of time remarkably well [47]. A current rendition of the pathogenesis of emphysema includes the following steps. First, cigarette smoke causes inflammatory cell recruitment. Next, these inflammatory cells release matrix degrading proteinases that locally overwhelm or evade inhibitors, causing destruction of lung elastin and other extracellular matrix (ECM) proteins. In addition to loss of ECM, alveolar cells die by apoptosis. Whether this is a primary or secondary event remains a point of controversy. Finally, lung destruction coupled with failure to repair abnormal structures leads to airspace enlargement characteristic of emphysema.

Inflammation and proteinases in human COPD

A major achievement in our understanding of emphysema has been our appreciation of the complex inflammatory milieu and the interrelationships between inflammatory/immune cells leading to lung destruction [48]. Macrophages patrol the lower airspace under normal conditions. Acutely following cigarette smoke exposure, macrophages may become activated and neutrophils quickly arrive. Subacutely, macrophages accumulate in respiratory bronchioles. Chronically, macrophages, neutrophils, and $CD8^+ > CD4^+$ T-cells as well as B-cells accumulate in the lung. Moreover, loss of cilia predisposes to airway infection with a prominent neutrophilic response.

While cigarette smoke initiates the inflammation and destruction in emphysema, in severe disease inflammation is sustained by factors other than cigarette smoke. Indeed intense inflammation with many cell types was observed in lung tissue taken from patients with end-stage lung COPD undergoing volume reduction surgery, who had discontinued smoking on average 9 years earlier [49].

A more careful examination of inflammatory cell interrelationships demonstrates that proteinases are not only involved in lung destruction, but also in regulation of inflammation. Upon initial exposure to cigarette smoke, constitutive macrophages secrete MMP-12 which is responsible for "shedding" of active TNF-α from the cell surface and subsequent recruitment of neutrophils in response to cigarette smoke in mice [50]. In addition, cigarette smoke related oxidant activity, inactivates macrophage histone deacetylase-2 promoting transcription of MMPs and neutrophil chemokines including IL-8 [51]. Both MMP-12 and neutrophil NE contribute to lung destruction in murine models of cigarette smoke-induced emphysema with MMP-$12^{-/-}$ mice being fully protected and $NE^{-/-}$ mice being two-third protected from airspace enlargement [9]. These two proteolytic systems interact to achieve maximal lung damage. NE degrades TIMP-1 augmenting MMP activity [52] and MMPs degrade α_1-AT, indirectly augmenting NE activity [53].

Elastin degradation products generated by these enzymes themselves appear to be a critical chemokine in further macrophage accumulation in COPD [54]. Hence, a positive feedback loop is established between elastolysis, macrophage accumulation, and neutrophil recruitment. Most recently, elastin fragments have also been implicated an auto-immune target in COPD [55]. Interest in B-cells in COPD has grown with the appreciation that the amount of BALT (bronchial associated lymphoid tissue) correlates with the severity of COPD [56]. BALT could simply be associated with airway remodeling and bacterial colonization. However, patients with COPD have been show to develop antibodies to elastin and this might be involved in disease progression [55]. Hence, these new wrinkles continue to strengthen the elastase:antielastase hypothesis demonstrating that not only elastin is critical to the structure and function of the lung, but by-products of its dissolution are both proinflammatory and may induce auto-immune disease.

Currently, focus has shifted to understanding the role of adaptive immunity in COPD, particularly the $CD8^+$ T-cell. $CD8^+$ T-cells, are known to be increased in lungs of patients with COPD [57]. Transgenic mice overexpressing IFN-γ develop inflammation, apoptosis, and consequent emphysema [58]. Humans with COPD have increased amounts of IP-10 and its receptor CXCR3 [59]. Induction of IP-10 in $CD8^+$ T-cells and bronchiolar epithelium signals through CXCR3 expressed on alveolar macrophages to produce MMP-12 [60]. As discussed above, MMP-12 production leads to elastolysis and sets in motion a cycle of further lung destruction. These concepts were confirmed in a mouse model where cigarette exposure to mice deficient in $CD8^+$ T-cells fails to produce IP-10, activate/recruit macrophages, produce MMP-12, or develop airspace enlargement [61].

Emphysema associated with α_1-AT deficiency

The clearest example of the association of proteinase–antiproteinase imbalance and emphysema occurs with inherited deficiency of α_1-AT. Several abnormal α1-AT alleles are associated with very low serum concentrations of α_1-AT and enhanced risk for emphysema. Of these, the Pi Z variant is by far the most common, and greater than 95% of α_1-AT deficient individuals have only the Z variant detectable. A variety of rare deficiency variants, including non-expressing alleles that do not result in detectable α_1-AT in plasma, comprise the remaining individuals. Pi Z individuals have about 15% of the normal serum concentration of α_1-AT. The abnormality leading to the Pi Z variant is a point mutation involving a single nucleotide at codon 342 that results in coding for Lys instead of Glx. This amino acid substitution changes the charge attraction between the amino acids at positions 342 and 290 present in the normal form of α_1-AT and prevents the formation of a fold in the molecule. With this change in tertiary structure, the molecule is susceptible to loop sheet polymerization of α_1-AT in the endoplasmic reticulum that impedes secretion of the protein from the hepatocyte [62]. In addition, its rate of association with NE is slightly but significantly slower than the association rate of normal α_1-AT with NE. The prevalence of the Pi Z phenotype in the United States is about one in 2800 people [63]. The Z allele is not found in Asians and African populations at measurable frequencies.

Most Pi Z individuals eventually become symptomatic with COPD, but there is considerable variation and some individuals reach advanced age with minimal symptoms. Smoking has a marked effect on the age at which shortness of breath appears. On the average Pi Z smokers have symptoms by age 40, about 15 years earlier than Pi Z nonsmokers [63]. Additional genetic modifiers, such as polymorphisms in IL-10 [64], also help explain differences in the clinical course of patients with α_1-AT deficiency.

ASTHMA

Human asthma is characterized by persistent airway hyperresponsiveness (AHR) and episodic airflow obstruction. Asthma is also associated with an immune/inflammatory response characterized by activated T-cells with Th2 cytokines (IL-4, -5, -9, -13), eosinophils, and mast cells that infiltrate the mucosa and submucosa. The AHR may be a distinct trait, separate and preceding the inflammation. While several proteinases, particularly MMPs have been associated with asthma, their role in disease pathogenesis is still emerging. Proteinases might be particularly important in controlling inflammatory cell migration and airway remodeling.

Initiation of asthma

The initiation of asthma is driven, in large part, by stimulation of allergen-specific CD4+ T helper cells by professional antigen presenting cells (APCs), most notably lung dendritic cells (DCs). DCs play a prominent role in the pathogenesis of asthma because they are able to efficiently present major histocompatibility complex (MHC) class II-restricted antigenic determinants to naïve T-cells to direct CD4+ T-cell polarization to promote a Th2-type immune response within the lung.

MMP-9 is expressed in DCs and has been shown to be promote DC trafficking [24] lending support for its potential importance in the initiation of asthma. Cathepsin S, a potent matrix degrading cysteine proteinase in its own right, has been shown to be required for normal MHC class II trafficking and function within DCs [65]. Pharmacological inhibition of cathepsin S attenuated lung eosinophilia and IgE generation in a mouse model of asthma [41].

Asthma progression

Chronic asthma is characterized by airway remodeling, a term that incorporates chronic airway inflammation, marked smooth muscle hyperplasia (with mast cell infiltration), mucus gland and goblet cell hyperplasia, myofibroblast accumulation with subepithelial fibrosis, and angiogenesis.

Inflammation

Although MMP-9 is expressed in a variety of inflammatory cells including eosinophils, macrophages, and neutrophils, the role of MMP-9 deficiency in asthma pathogenesis has been confusing to date. Application of MMP-9$^{-/-}$ mice to acute allergic/inflammatory models have reported both proinflammatory [66] and anti-inflammatory roles [67,68]. It is likely that depending upon the protocol, amount of lipopolysaccharide (LPS) and other conditions, one might suppress or exacerbate inflammation in asthma. Future studies will require control for these factors. Moreover, the role of MMPs in newer models of chronic asthma with repetitive antigen stimulation in susceptible strains will allow us to establish roles for MMPs in models of chronic asthma, with potential human applicability.

Application of MMP-12$^{-/-}$ mice to several models of asthma has established the importance of the enzyme in inflammatory cell migration in asthma. For example, IL-13 inducible transgenic mice, described above, develop both inflammation and emphysema. Inflammation and emphysema were dependent upon MMP-12 [69]. In addition, in mice sensitized and challenged with cockroach antigen (CRA) [70], MMP-12 deficiency resulted in fewer peribronchial leukocytes and correspondingly less protein and fewer cells in BAL compared to WT mice. Eosinophils were reduced to the greatest degree (80% reduction). In addition, there were marked decreases in IL-5, MIP-1α, MCP-1, and TNF-α in MMP-12$^{-/-}$ BAL. AHR did not differ between groups. Another model of acute allergic airway inflammation that involved sensitization by aerosolized OVA $\times$10 days in the context of adenovirus-mediated GMCSF [71], resulted in a Th2 cytokine profile, OVA-specific IgE, and airway eosinophilia. In WT mice, MMP-12 mRNA and protein were upregulated 70-fold in OVA treated versus saline controls. MMP-12$^{-/-}$ mice had an 80% reduction in eosinophil accumulation in BAL compared to WT mice. It is noteworthy that, MMP-12 production was IL-13 dependent, since it was abrogated in similar experiments using IL-13$^{-/-}$ mice.

Airway remodeling

Inflammation is generally thought to drive other aspects of airway remodeling. For example, eosinophils were shown to be required for fibrosis and smooth muscle hypertrophy, but not AHR or mucus production [72]. In turn, smooth muscle changes might also regulate inflammation [73]. Alternatively, many changes related to airway remodeling have been replicated by pathological stretch applied purely to airway epithelial cells (i.e. in the absence of inflammation) [74]. Expression profiling, followed by confirmative studies demonstrated that mechanical deformation of airway epithelial cells induces production of MMP-9 as well as its activator plasmin(ogen) [75].

MMP-9 has been implicated in *subepithelial airway fibrosis*. The IL-13 transgenic mice described above also have features of airway remodeling including *mucus cell hyperplasia* and subepithelial fibrosis [76]. The airway changes were attenuated upon back-crossing to MMP-9$^{-/-}$ mice [26]. This appears to work through the capacity of MMP-9 to activate TGF-β promoting collagen production. NE also has been shown to activate latent TGF-β in a model of bleomycin-induced lung fibrosis. It is noteworthy that, NE also induces epithelial cell production of MUC5 [77] and mucus cell metaplasia in mouse lungs [78]. Hence, in addition to the local concentration of proteinases and their inhibitors, additional non-matrix substrates will influence the effect of a proteinase is turnover.

References

1. Bode W, Meyer EJ, Powers JC. Human leukocyte elastase: X-ray crystal structures, mechanism, substrate specificity, and mechanism-based inhibitors. *Biochemistry* 28: 1951–63, 1989.
2. Owen CA, Campbell MA, Boukedes SS, Stockley RA, Campbell EJ. A discrete subpopulation of human monocytes expresses a neutrophil-like proinflammatory (P) phenotype. *Am J Physiol* 267: 775–85, 1994.
3. Owen CA, Campbell MA, Sannes PL, Boukedes SS, Campbell EJ. Cell surface-bound elastase and cathepsin G on human neutrophils: A novel, non-oxidative mechanism by which neutrophils focus and preserve catalytic activity of serine proteinases. *J Cell Biol* 131(3): 775–89, 1995.
4. Campbell EJ, Campbell MA, Boukedes SS, Owen CA. Quantum proteolysis by neutrophils: Implications for pulmonary emphysema in alpha 1-antitrypsin deficiency. *J Clin Invest* 104(3): 337–44, 1999.
5. Damiano VV, Tsang A, Kucich U, Abrams WR, Rosenbloom J, Kimbel P, Fallahnejed M, Weinbaum G. Immunolocalization of elastase in human emphysematous lungs. *J Clin Invest* 78: 482–93, 1986.
6. Fahy JV, Schuster A, Ueki I, Boushey HA, Nadel JA. Mucus hypersecretion in bronchiectasis. The role of neutrophil proteases. *Am Rev Respir Dis* 146(6): 1430–33, 1992.

7. Belaaouaj A, McCarthy R, Baumann M, Gao Z, Ley TJ, Abraham SN, Shapiro SD. Impaired host defense against bacterial infection in mice lacking neutrophil elastase. *Nat Med*, 1998.
8. Belaaouaj A, Kim KS, Shapiro SD. Degradation of outer membrane protein A in *Escherichia coli* killing by neutrophil elastase. *Science* 289(5482): 1185–88, 2000.
9. Shapiro SD, Goldstein NM, Houghton AM, Kobayashi DK, Kelley D, Belaaouaj A. Neutrophil elastase contributes to cigarette smoke-induced emphysema in mice. *Am J Pathol* 163(6): 2329–35, 2003.
10. Sallenave JM. The role of secretory leukocyte proteinase inhibitor and elafin (elastase-specific inhibitor/skin-derived antileukoprotease) as alarm antiproteinases in inflammatory lung disease. *Respir Res* 1(2): 87–92, 2000.
11. Sallenave JM, Silva A. Characterization and gene sequence of the precursor of elafin, an elastase-specific inhibitor in bronchial secretions. *Am J Respir Cell Mol Biol* 8(4): 439–45, 1993.
12. Cooley J, Takayama TK, Shapiro SD, Schechter NM, Remold-O'Donnell E. The serpin MNEI inhibits elastase-like and chymotrypsin-like serine proteases through efficient reactions at two active sites. *Biochemistry* 40(51): 15762–70, 2001.
13. Parks WC, Shapiro SD. Matrix metalloproteinases in lung biology. *Respir Res* 2(1): 10–19, 2001.
14. Shapiro SD. Matrix metalloproteinase degradation of extracellular matrix: Biological consequences. *Curr Opin Cell Biol* 10(5): 602–8, 1998.
15. Imai K, Dalal SS, Chen ES, Downey R, Schulman LL, Ginsburg M, D'Armiento J. Human collagenase (matrix metalloproteinase-1) expression in the lungs of patients with emphysema. *Am J Respir Crit Care Med* 163(3 Pt 1): 786–91, 2001.
16. Balbin M, Fueyo A, Tester AM, Pendas AM, Pitiot AS, Astudillo A, Overall CM, Shapiro SD, Lopez-Otin C. Loss of collagenase-2 confers increased skin tumor susceptibility to male mice. *Nat Genet* 35(3): 252–57, 2003.
17. Deguchi JO, Aikawa E, Libby P, Vachon JR, Inada M, Krane SM, Whittaker P, Aikawa M. Matrix metalloproteinase-13/collagenase-3 deletion promotes collagen accumulation and organization in mouse atherosclerotic plaques. *Circulation* 112(17): 2708–15, 2005.
18. Atkinson JJ, Senior RM. Matrix metalloproteinase-9 in lung remodeling. *Am J Respir Cell Mol Biol* 28(1): 12–24, 2003.
19. Cundall M, Sun Y, Miranda C, Trudeau JB, Barnes S, Wenzel SE. Neutrophil-derived matrix metalloproteinase-9 is increased in severe asthma and poorly inhibited by glucocorticoids. *J Allergy Clin Immunol* 112(6): 1064–71, 2003.
20. Wenzel SE, Balzar S, Cundall M, Chu HW. Subepithelial basement membrane immunoreactivity for matrix metalloproteinase 9: Association with asthma severity, neutrophilic inflammation, and wound repair. *J Allergy Clin Immunol* 111(6): 1345–52, 2003.
21. Mattos W, Lim S, Russell R, Jatakanon A, Chung KF, Barnes PJ. Matrix metalloproteinase-9 expression in asthma: Effect of asthma severity, allergen challenge, and inhaled corticosteroids. *Chest* 122(5): 1543–52, 2002.
22. Shapiro SD. Biological consequences of extracellular matrix cleavage by matrix metalloproteinases. In Extracellular matrix and cell-to-cell contact. *Curr Opin Cell Biol* 10: 602–8, 1998.
23. Parks W. Matrix metalloproteinases. San Diego, CA: Academic Press, 1998.
24. Ichiyasu H, McCormack JM, McCarthy KM, Dombkowski D, Preffer FI, Schneeberger EE. Matrix metalloproteinase-9-deficient dendritic cells have impaired migration through tracheal epithelial tight junctions. *Am J Respir Cell Mol Biol* 30(6): 761–70, 2004.
25. Vermaelen KY, Cataldo D, Tournoy K, Maes T, Dhulst A, Louis R, Foidart JM, Noel A, Pauwels R. Matrix metalloproteinase-9-mediated dendritic cell recruitment into the airways is a critical step in a mouse model of asthma. *J Immunol* 171(2): 1016–22, 2003.
26. Lee CG, Homer RJ, Zhu Z, Lanone S, Wang X, Koteliansky V, Shipley JM, Gotwals P, Noble P, Chen Q, Senior RM, Elias JA. Interleukin-13 induces tissue fibrosis by selectively stimulating and activating transforming growth factor beta(1). *J Exp Med* 194(6): 809–21, 2001.
27. Vu TH, Shipley JM, Bergers G, Berger JE, Helms JA, Hanahan D, Shapiro SD, Senior RM, Werb Z. MMP-9/gelatinase B is a key regulator of growth plate angiogenesis and apoptosis of hypertrophic chondrocytes. *Cell* 93(3): 411–22, 1998.
28. Pyo R, Lee JK, Shipley JM, Curci JA, Mao D, Ziporin SJ, Ennis TL, Shapiro SD, Senior RM, Thompson RW. Targeted gene disruption of matrix metalloproteinase-9 (gelatinase B) suppresses development of experimental abdominal aortic aneurysms. *J Clin Invest* 105(11): 1641–49, 2000.
29. Gronski TJ, Martin R, Kobayashi DK, Walsh BC, Holman MC, Van Wart E, Shapiro SD. Hydrolysis of a broad spectrum of extracellular matrix proteins by human macrophage elastase. *J Biol Chem* 272: 12189–94, 1997.
30. Churg A, Wang RD, Tai H, Wang X, Xie C, Dai J, Shapiro SD, Wright JL. Macrophage metalloelastase mediates acute cigarette smoke-induced inflammation via tumor necrosis factor-alpha release. *Am J Respir Crit Care Med* 167(8): 1083–89, 2003.
31. Woodruff PG, Koth LL, Yang YH, Rodriguez MW, Favoreto S, Dolganov GM, Paquet AC, Erle DJ. A distinctive alveolar macrophage activation state induced by cigarette smoking. *Am J Respir Crit Care Med* 172(11): 1383–92, 2005.
32. Lavigne MC, Thakker P, Gunn J, Wong A, Miyashiro JS, Wasserman AM, Wei SQ, Pelker JW, Kobayashi M, Eppihimer MJ. Human bronchial epithelial cells express and secrete MMP-12. *Biochem Biophys Res Commun* 324(2): 534–46, 2004.
33. Xie S, Issa R, Sukkar MB, Oltmanns U, Bhavsar PK, Papi A, Caramori G, Adcock I, Chung KF. Induction and regulation of matrix metalloproteinase-12 in human airway smooth muscle cells. *Respir Res* 6: 148, 2005.
34. Shipley JM, Wesselschmidt RL, Kobayashi DK, Ley TJ, Shapiro SD. Metalloelastase is required for macrophage-mediated proteolysis and matrix invasion in mice. *Proc Natl Acad Sci* 93: 3942–46, 1996.
35. Hautamaki RD, Kobayashi DK, Senior RM, Shapiro SD. Macrophage elastase is required for cigarette smoke-induced emphysema in mice. *Science* 277: 2002–4, 1997.
36. Ohnishi K, Takagi M, Kurokawa Y, Satomi S, Konttinen YT. Matrix metalloproteinase-mediated extracellular matrix protein degradation in human pulmonary emphysema. *Lab Invest* 78(9): 1077–87, 1998.
37. Van Eerdewegh P, Little RD, Dupuis J, Del Mastro RG, Falls K, Simon J, Torrey D, Pandit S, McKenny J, Braunschweiger K, Walsh A, Liu Z, Hayward B, Folz C, Manning SP, Bawa A, Saracino L, Thackston M, Benchekroun Y, Capparell N, Wang M, Adair R, Feng Y, Dubois J, FitzGerald MG, Huang H, Gibson R, Allen KM, Pedan A, Danzig MR, Umland SP, Egan RW, Cuss FM, Rorke S, Clough JB, Holloway JW, Holgate ST, Keith TP. Association of the ADAM33 gene with asthma and bronchial hyperresponsiveness. *Nature* 418(6896): 426–30, 2002.
38. Holgate ST, Yang Y, Haitchi HM, Powell RM, Holloway JW, Yoshisue H, Pang YY, Cakebread J, Davies DE. The genetics of asthma: ADAM33 as an example of a susceptibility gene. *Proc Am Thorac Soc* 3(5): 440–43, 2006.
39. Kung TT, Jones H, Adams GK, Umland SP, Kreutner W, Egan RW, Chapman RW, Watnick AS. Characterization of a murine model of allergic pulmonary inflammation. *Int Arch Allergy Immunol* 105: 83–90, 1994.
40. Chapman HA, Riese RJ, Shi GP. Emerging roles for cysteine proteases in human biology. *Annu Rev Physiol* 59: 63–88, 1997.
41. Riese RJ, Mitchell RN, Villadangos JA, Shi GP, Palmer JT, Karp ER, De Sanctis GT, Ploegh HL, Chapman HA. Cathepsin S activity regulates antigen presentation and immunity. *J Clin Invest* 101(11): 2351–63, 1998.
42. Zheng T, Kang MJ, Crothers K, Zhu Z, Liu W, Lee CG, Rabach LA, Chapman HA, Homer RJ, Aldous D, De Sanctis GT, Underwood S, Graupe M, Flavell RA, Schmidt JA, Elias JA. Role of cathepsin

S-dependent epithelial cell apoptosis in IFN-gamma-induced alveolar remodeling and pulmonary emphysema. *J Immunol* 174(12): 8106–15, 2005.
43. Shi GP, Sukhova GK, Grubb A, Ducharme A, Rhode LH, Lee RT, Ridker PM, Libby P, Chapman HA. Cystatin C deficiency in human atherosclerosis and aortic aneurysms. *J Clin Invest* 104(9): 1191–97, 1999.
44. Pauwels R, Buist A, Calverley P, Jenkins C, Hurd S. Global strategy for the diagnosis, management, and prevention of chronic obstructive pulmonary disease. NHLBI/WHO Global Initiative for Chronic Obstructive Lung Disease (GOLD) Workshop summary. *Am J Respir Crit Care Med* 163: 1256–76, 2001.
45. Gross P, Pfitzer E, Tolker E, Babyak M, Kaschak M. Experimental emphysema: its production with papain in normal and silicotic rats. *Arch Environ Health* 11: 50–58, 1965.
46. Laurell. C. B. a. S. E. The electrophoretic alpha-globulin pattern of serum in alpha-antitrypsin deficiency. *Scand J Clin Invest* 15: 132–40, 1963.
47. Shapiro SD, Ingenito EP. The pathogenesis of chronic obstructive pulmonary disease: Advances in the past 100 years. *Am J Respir Cell Mol Biol* 32(5): 367–72, 2005.
48. Shapiro SD. COPD unwound. *N Engl J Med* 352(19): 2016–19, 2005.
49. Retamales I, Elliott WM, Meshi B, Coxson HO, Pare PD, Sciurba FC, Rogers RM, Hayashi S, Hogg JC. Amplification of inflammation in emphysema and its association with latent adenoviral infection. *Am J Respir Crit Care Med* 164(3): 469–73, 2001.
50. Churg A, Dai J, Zay K, Karsan A, Hendricks R, Yee C, Martin R, MacKenzie R, Xie C, Zhang L, Shapiro S, Wright JL. Alpha-1-antitrypsin and a broad spectrum metalloprotease inhibitor, RS113456, have similar acute anti-inflammatory effects. *Lab Invest* 81(8): 1119–31, 2001.
51. Tomita K, Barnes PJ, Adcock IM. The effect of oxidative stress on histone acetylation and IL-8 release. *Biochem Biophys Res Commun* 301(2): 572–77, 2003.
52. Itoh Y, Nagase H. Preferential inactivation of tissue inhibitor of metalloproteinases-1 that is bound to the precursor of matrix metalloproteinase 9 (progelatinase B) by human neutrophil elastase. *J Biol Chem* 270(28): 16518–21, 1995.
53. Desrochers PE, Jeffrey JJ, Weiss SJ. Interstitial collagenase (matrix metalloproteinase-1) expresses serpinase activity. *J Clin Invest* 87(6): 2258–65, 1991.
54. Houghton AM, Quintero PA, Perkins DL, Kobayashi DK, Kelley DG, Marconcini LA, Mecham RP, Senior RM, Shapiro SD. Elastin fragments drive disease progression in a murine model of emphysema. *J Clin Invest* 116(3): 753–59, 2006.
55. Lee SH, Goswami S, Grudo A, Song LZ, Bandi V, Goodnight-White S, Green L, Hacken-Bitar J, Huh J, Bakaeen F, Coxson HO, Cogswell S, Storness-Bliss C, Corry DB, Kheradmand F. Antielastin autoimmunity in tobacco smoking-induced emphysema. *Nat Med* 13(5): 567–69, 2007.
56. Hogg JC, Chu F, Utokaparch S, Woods R, Elliott WM, Buzatu L, Cherniack RM, Rogers RM, Sciurba FC, Coxson HO, Pare PD. The nature of small-airway obstruction in chronic obstructive pulmonary disease. *N Engl J Med* 350(26): 2645–53, 2004.
57. Saetta M, Di Stefano A, Turato G, Facchini FM, Corbino L, Mapp CE, Maestrelli P, Ciaccia A, Fabbri LM. CD8+ T-lymphocytes in peripheral airways of smokers with chronic obstructive pulmonary disease. *Am J Respir Crit Care Med* 157(3 Pt 1): 822–26, 1998.
58. Wang Z, Zheng T, Zhu Z, Homer RJ, Riese RJ, Chapman HA Jr., Shapiro SD, Elias JA. Interferon gamma induction of pulmonary emphysema in the adult murine lung. *J Exp Med* 192(11): 1587–600, 2000.
59. Saetta M, Mariani M, Panina-Bordignon P, Turato G, Buonsanti C, Baraldo S, Bellettato CM, Papi A, Corbetta L, Zuin R, Sinigaglia F, Fabbri LM. Increased expression of the chemokine receptor CXCR3 and its ligand CXCL10 in peripheral airways of smokers with chronic obstructive pulmonary disease. *Am J Respir Crit Care Med* 165(10): 1404–9, 2002.
60. Grumelli S, Corry DB, Song LZ, Song L, Green L, Huh J, Hacken J, Espada R, Bag R, Lewis DE, Kheradmand F. An immune basis for lung parenchymal destruction in chronic obstructive pulmonary disease and emphysema. *PLoS Med* 1(1): e8, 2004.
61. Maeno T, Houghton AM, Quintero PA, Grumelli S, Owen CA, Shapiro SD. CD8+ T Cells are required for inflammation and destruction in cigarette smoke-induced emphysema in mice. *J Immunol* 178(12): 8090–96, 2007.
62. Lomas DA, Evans DL, Stone SR, Chang W-SW, Carre RW. Effect of the Z mutation on the physical and inhibitory properties of alpha-1-antitrypsin. *Biochemistry* 32: 500–8, 1993.
63. Silverman EK, Pierce JA, Province MA, Rao DC, Campbell EJ. Variability of pulmonary function in alpha-1-antitrypsin deficiency: Clinical correlates. *Ann Intern Med* 111(12): 982–91, 1989.
64. Demeo DL, Campbell EJ, Barker AF, Brantly ML, Eden E, McElvaney NG, Rennard SI, Sandhaus RA, Stocks JM, Stoller JK, Strange C, Turino G, Silverman EK. IL10 polymorphisms are associated with airflow obstruction in severe alpha 1-antitrypsin deficiency. *Am J Respir Cell Mol Biol* 38(1): 114–20, 2008.
65. Riese RJ, Wolf PR, Bromme D, Natkin LR, Villadangos JA, Ploegh HL, Chapman HA. Essential role for cathepsin S in MHC class II-associated invariant chain processing and peptide loading. *Immunity* 4(4): 357–66, 1996.
66. Cataldo DD, Tournoy KG, Vermaelen K, Munaut C, Foidart JM, Louis R, Noel A, Pauwels RA. Matrix metalloproteinase-9 deficiency impairs cellular infiltration and bronchial hyperresponsiveness during allergen-induced airway inflammation. *Am J Pathol* 161(2): 491–98, 2002.
67. McMillan SJ, Kearley J, Campbell JD, Zhu XW, Larbi KY, Shipley JM, Senior RM, Nourshargh S, Lloyd CM. Matrix metalloproteinase-9 deficiency results in enhanced allergen-induced airway inflammation. *J Immunol* 172(4): 2586–94, 2004.
68. Greenlee KJ, Corry DB, Engler DA, Matsunami RK, Tessier P, Cook RG, Werb Z, Kheradmand F. Proteomic identification of *in vivo* substrates for matrix metalloproteinases 2 and 9 reveals a mechanism for resolution of inflammation. *J Immunol* 177(10): 7312–21, 2006.
69. Zheng T, Zhu Z, Wang Z, Homer RJ, Ma B, Riese RJ Jr., Chapman HA Jr., Shapiro SD, Elias JA. Inducible targeting of IL-13 to the adult lung causes matrix metalloproteinase- and cathepsin-dependent emphysema. *J Clin Invest* 106(9): 1081–93, 2000.
70. Warner RL, Lukacs NW, Shapiro SD, Bhagarvathula N, Nerusu KC, Varani J, Johnson KJ. Role of metalloelastase in a model of allergic lung responses induced by cockroach allergen. *Am J Pathol* 165(6): 1921–30, 2004.
71. Pouladi MA, Robbins CS, Swirski FK, Cundall M, McKenzie AN, Jordana M, Shapiro SD, Stampfli MR. Interleukin-13-dependent expression of matrix metalloproteinase-12 is required for the development of airway eosinophilia in mice. *Am J Respir Cell Mol Biol* 30(1): 84–90, 2004.
72. Humble A, Lloyd C, McMillan S, Friend D, Zanthou G, McKenna E, Ghiran S, Gerard N, Yu C, Orkin S, Gerard C. A critical role for eosinophils in allergic airways remodeling. *Science* 305: 1776–79, 2004.
73. Lazaar A, Panettieri RJ. Airway smooth muscle: A modulator of airway remodeling in asthma. *J Allergy Clin Immunol* 116: 488–95, 2005.
74. Tschumperlin DJ, Shively JD, Kikuchi T, Drazen JM. Mechanical stress triggers selective release of fibrotic mediators from bronchial epithelium. *Am J Respir Cell Mol Biol* 28(2): 142–49, 2003.
75. Chu EK, Cheng J, Foley JS, Mecham BH, Owen CA, Haley KJ, Mariani TJ, Kohane IS, Tschumperlin DJ, Drazen JM. Induction of the plasminogen activator system by mechanical stimulation of human bronchial epithelial cells. *Am J Respir Cell Mol Biol* 35(6): 628–38, 2006.
76. Cho JY, Miller M, McElwain K, McElwain S, Shim JY, Raz E, Broide DH. Remodeling associated expression of matrix metalloproteinase 9 but not tissue inhibitor of metalloproteinase 1 in airway

epithelium: Modulation by immunostimulatory DNA. *J Allergy Clin Immunol* 117(3): 618–25, 2006.
77. Voynow JA, Young LR, Wang Y, Horger T, Rose MC, Fischer BM. Neutrophil elastase increases MUC5AC mRNA and protein expression in respiratory epithelial cells. *Am J Physiol* 276(5 Pt 1): L835–43, 1999.
78. Voynow JA, Fischer BM, Malarkey DE, Burch LH, Wong T, Longphre M, Ho SB, Foster WM. Neutrophil elastase induces mucus cell metaplasia in mouse lung. *Am J Physiol Lung Cell Mol Physiol* 287(6): L1293–302, 2004.

29

Growth Factors

CHAPTER

Martin Kolb, Zhou Xing, Kjetil Ask and Jack Gauldie

Centre for Gene Therapeutics, Department of Pathology and Molecular Medicine, McMaster University Hamilton, ON, Canada

INTRODUCTION

It has long been recognized that after acute injury to the lung, the tissue repair process is engaged to return the organ to normal function. In chronic tissue injury, it appears the process of repair loses many of the control mechanisms, and continued repair results in remodeling of the tissue with alteration of normal structure and compromise of normal lung function. The remodeling process involves excess matrix synthesis along with distorted deposition of that matrix; in addition there is the appearance of altered tissue cell phenotypes, most notably that of the myofibroblast. These cells are derived from existing fibroblasts under the influence of cytokines and growth factors, such as transforming growth factor-β (TGF-β), or possibly from circulating (stem cell like) precursors that have migrated into the tissue injury site. Given the morphology seen in lung disorders, asthma is mostly associated with minor, primarily peribronchial, matrix deposition, while chronic obstructive pulmonary disease (COPD) has wide evidence of remodeling, primarily at the alveolar level and fibrosis of the small airways [1].

Many cytokines and growth factors can be found at the site of tissue remodeling using immunohistochemistry, *in situ* hybridization, and gene expression assessments, so much so, that it is difficult to determine causative versus secondary presence. However, some, in comparison to others, have recently been shown to play a primary role in induction, and thus represent more likely targets for developing therapeutic interventions. Transgenic and gene knockout models in mice point to possible primary targets in growth factors such as TGF-β, IL-13, or hepatocyte growth factor (HGF), while others implicate factors such as IL-1β or TNF-α. The presence of acute inflammation, such as in ARDS, or more chronic inflammation states such as COPD and usual interstitial pneumonia (UIP) are usually associated with the sequelae of tissue remodeling. However, there may be situations involving structural cell phenotype or extra cellular matrix alterations that can propagate the remodeling process through autocrine and paracrine pathways independent of the state of inflammation that may have preceded the alteration. Thus, identification of growth factors known to modulate the synthesis and deposition of matrix or influence the make-up of the cell phenotype in the tissue may be crucial in defining the therapy that halts remodeling and progression and/or induces the tissue to return to normal structure and function.

GROWTH FACTORS IN HUMAN LUNG AND AIRWAY DISEASE

A large number of studies have investigated the role of growth factors in human lung disease in which tissue remodeling is a prominent feature. Most of these studies are based on immunohistochemical analysis of bronchial and pulmonary tissue, obtained by biopsy, and on analysis of bronchoalveolar lavage (BAL) fluid and cells. Unfortunately there is minimal information available about growth factors in airway disease, but the presence of fibrotic responses and mesenchymal cell proliferation would suggest factors that are prevalent in fibrogenesis. Three different factors are considered to play a prominent role: epidermal growth factor (EGF), GM-CSF, and TGF-β [2, 3]. Increased EGF and EGF receptor are present in the submucosa of asthmatic airways

[4, 5] and it is suggested that the epithelium does not respond adequately to EGF after damage in asthma [3]. GM-CSF expression is increased in asthmatic epithelium and lymphocytes after allergen challenge [4] and GM-CSF can mediate airway remodeling through its survival effect on eosinophils [2] and release of profibrotic cytokines [6]. TGF-β has received particular interest. In one study TGF-β expression in airways is shown to correlate with the degree of subepithelial fibrosis in asthma, although the data are controversial [4, 7]. More telling is the presence of TGF-β in BAL fluid after segmental allergen challenge [5]. Moreover, neutralization of TGF-β in an animal model of repeat allergen exposure showed decreased bronchial fibrotic sequelae [8] and TNF-α and IL-1 are found at elevated levels in acutely inflamed asthmatic airways [4]. Since both acute inflammatory cytokines are able to trigger profibrotic tissue reactions [9–11], this strongly suggests a role in airway remodeling. Other growth factors detected, such as platelet-derived growth factor (PDGF) [2], insulin growth factor-I (IGF-I), and basic fibroblast growth factors (bFGF) [3], could also participate in the repair process in airways.

Most knowledge about growth factors in lung tissue remodeling has been accumulated through studies of parenchymal lung disease, such as asbestosis, sarcoidosis, or idiopathic pulmonary fibrosis (IPF). We include these disorders in our discussion to provide the reader perspective on the potential biological actions of these factors and thus to provide a better perspective on asthma and COPD. The presence of two key profibrotic growth factors, TGF-β and PDGF, has been demonstrated in various fibrosing disorders. In IPF, TGF-β was shown to be elevated in BAL fluid [12, 13] and can be expressed by bronchial and alveolar epithelial cells, alveolar macrophages and fibroblast foci [14–17]. PDGF-mRNA was shown to be upregulated in BAL cells retrieved from patients with IPF and in fibrotic areas of the lung, while lungs affected by asbestosis and silicosis express increased TGF-β mRNA [18]. In human fibrotic disorders TNF-α and IL-1 are present in BAL fluid or BAL cell supernatants [19]. In lungs of patients with progressive pulmonary fibrosis, TNF-α and G-CSF are expressed at higher than normal levels [19, 20], while similar tissue expresses lower than normal levels of HGF [21]. An important consideration is the relationship between growth factors and extracellular matrix (ECM). Many growth factors can induce ECM synthesis, but many are also associated with the matrix through adherence via RGD or related sequences to proteoglycans, collagens, and fibronectins [22] and require activation (release) through interaction with cell integrins and ECM components [23–25] or via proteolysis, as seen with plasminogen-mediated activation and release of HGF from ECM [26]. (Table 29.1 and Fig. 29.1)

SPECIFIC GROWTH FACTORS

Tumor necrosis factor-α

Tumor necrosis factor-α (TNF-α) is a peptide secreted as membrane bound form and released after cleavage by TNF-α-converting enzyme [27]. The cytokine is not constitutively present in the lung but is secreted rapidly upon a variety of stimuli, mainly by alveolar macrophages and type II epithelial cells [18]. TNF-α can bind to two different types of receptors, which are expressed on most cells and signals through various intracellular pathways [27]. It is a potent proinflammatory cytokine and exerts a variety of effects, which may contribute to the process of remodeling and fibrosis: TNF-α induces inflammatory cell migration and adhesion, initiates a cytokine cascade, and regulates apoptosis [27]. Further, it is mitogenic for mesenchymal cells and influences collagen metabolism, being either pro- or antifibrotic [18, 28]. Numerous studies have demonstrated that TNF-α is involved in acute and chronic tissue changes seen after bleomycin, asbestos, silica, and irradiation damage. Animal strains that do not develop fibrosis following exposure to these agents show less TNF-α upregulation. Transient overexpression of TNF-α in the lung induces a limited fibrosis, likely by upregulation of TGF-β [29]. Others have shown that TNF-α can act also through PDGF pathways [18]. TNF-α was detected in BAL fluid and biopsies of patients with IPF, BOOP, and asbestosis, and likely plays a role in many pulmonary diseases [18, 30].

In asthma, TNF-α can amplify the inflammatory process and have indirect influence on airway remodeling through induction of other growth factors [4]. It has been detected in BAL fluid of asthmatics, in alveolar macrophages after allergen challenge and in bronchial mucosa of asthma patients [4] and can be rapidly released from mast cells on degranulation [31].

Interleukin-1β

Two forms of IL-1 (α and β) are known with almost identical biological properties. In the lung, many parenchymal and epithelial cell types are able to produce IL-1β; however, the major sources are activated macrophages [4]. IL-1 is a pleiotropic proinflammatory cytokine, often acting synergistically with TNF-α [32]. In remodeling and fibrosis, several actions of IL-1 are important. IL-1 stimulates fibroblasts to secrete other cytokines including IL-1β, IL-8, MCP-1, PDGF, and TGF-β [33, 34], while the direct effect on fibroblast proliferation and ECM synthesis is controversial [28, 32, 35, 36]. *In vivo*, IL-1 is elevated in BAL fluid and alveolar macrophages of patients with ARDS, but not in IPF [19, 33], while in animal studies IL-1 is involved in the early stage of pulmonary bleomycin injury [33], and IL-1 receptor antagonist ameliorates the fibrotic response following silica and bleomycin administration [37]. Recent studies show that transient transgene expression of IL-1 in the lung causes marked alveolar damage, induction of TGF-β and fibroblast foci with progressive fibrosis [38], and similar findings are seen in mice with inducible IL-1β transgene [11]. Moreover, IL-1β is a potent stimulator of osteopontin expression in lung fibroblasts, one of the most highly expressed genes found in IPF tissues [39, 40] and a cytokine that recently has been shown to play a critical role in allergic airway disease [41].

In asthma, IL-1β was found in BAL fluid and macrophages of asthma patients [4]. Its role in airway disease is probably similar to TNF-α, perpetuating primarily acute

TABLE 29.1 Growth factors in pulmonary tissue remodeling.

	Induce fibroblast accumulation		Induce myofibroblasts		Induce matrix expression	
	In vitro	*In vivo*	*In vitro*	*In vivo*	*In vitro*	*In vivo*
TNF	+/−	++		++	+/−	++
IL-1β	+/−	++++		++++	+/−	++++
GM-CSF		+++		+++		+++
PDGF	+++	+/−	++		++	+/−
IGF	+/−		?		++	
FGFs	+++	?	?	?	++	?
KGF	+++	?	?	?	?	?
HGF	−	−	−	−	−	−
TGF-α	+/−	++	?	++	++	++
TGF-β	+++	++++	++++	++++	++++	++++
CTGF	+++	?	?	?	++++	?

TNF: tumor necrosis factor; IL: interleukin; GM-CSF: granulocyte/macrophage colony stimulating factor; PDGF: platelet-derived growth factor; IGF: insulin-like growth factor; FGF: fibroblast growth factor; TGF: transforming growth factor; CTGF: connective tissue growth factor; KGF: keratinocyte growth factor (FGF7); HGF: hepatocyte growth factor; "−" = not detectable; "+/−" = detectable; "+" = low; "++" = moderate; "+++" = medium; "++++" = strong; "?" = unknown.

Interplay between growth factors and pulmonary parenchymal cells.

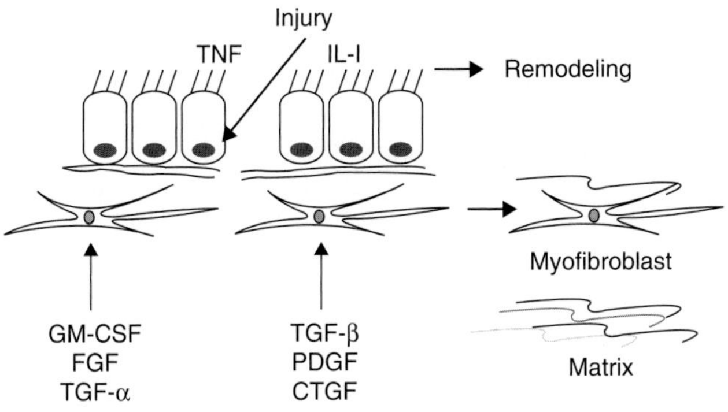

FIG. 29.1 Growth factors released from injured epithelial and parenchymal cells modify the behaviour and phenotype of mesenchymal cells in the parenchyma to induce myofibroblast differentiation and matrix deposition.

inflammatory processes and affecting remodeling directly, through modification of the matrix microenvironment, or indirectly via induction of growth factors, such as PDGF and TGF-β.

Granulocyte macrophage-colony stimulating factor

GM-CSF is a colony-stimulating factor that regulates growth and differentiation of hematopoietic cells. The lung is a major source of GM-CSF and most pulmonary cells are able to synthesize this cytokine in response to various stimuli [4]. In the context of tissue remodeling and fibrosis, the effects of GM-CSF in the local environment of the lung depend on dose and time of expression and on concomitant epithelial cell damage. In the model of transient transgene overexpression, low levels of GM-CSF induce little peribronchial inflammation and facilitate allergic reactions to exogenous allergen [42]. At higher levels, GM-CSF in the airways results in sustained eosinophilia and macrophage accumulation with moderate development of tissue fibrosis mediated by induction of TGF-β[43] and may represent a target for manipulation in diseases such as COPD [44]. In the bleomycin model, early upregulation of GM-CSF in pulmonary cells is present and followed by enhanced TGF-β expression [45], while other studies imply a protective effect of GM-CSF on the development of fibrosis in rats [46] and mice [47]. The profibrotic response may be due to stimulation and enhanced survival of macrophages and eosinophils, both cells major sources for TGF-β[4, 43, 45]. In the setting of damaged bronchial and alveolar epithelium, the proliferative response of epithelium to GM-CSF stimulation might result in a more protective effect [46].

GM-CSF is present in asthmatic airways of humans; there are increased levels in BAL after allergen challenge; significant levels of GM-CSF are found in plasma of patients with severe asthma [4]. These observations indicate that GM-CSF likely contributes to airway remodeling and plays a major role in the pathogenesis of asthma [2].

Platelet-derived growth factor

PDGF is a glycoprotein homo- and heterodimer composed of A and B chains, and is secreted by epithelial and endothelial cells, macrophages, and fibroblasts. PDGF-A chains can bind to both PDGF-receptors α and β, whereas PDGF-B bind only

PDGF-receptor β[48]. Signal transduction proceeds through tyrosine kinases, a major pharmacologic target [49, 50]. PDGF isoforms are chemoattractive not only for fibroblasts, but also for neutrophils and macrophages and they upregulate fibronectin and procollagen gene expression and synthesis. PDGF can induce TGF-β expression, suggesting that some of the long-term effects are partly mediated through this cytokine. However, PDGF and PDGF-receptor expression is stimulated by TGF-β, IL-1β, TNF-α, and bFGF, indicating that parts of the profibrotic activities of these molecules are due to PDGF-dependent pathways [18]. In animals that are exposed to asbestos, both PDGF and its receptors are upregulated in bronchial bifurcations with developing fibroproliferative lesions [51]. PDGF-A knockout mice die from pulmonary emphysema apparently because of impaired alveolar development and lack of myofibroblasts [52]. Interference with PDGF activity, either through transgene overexpression of a truncated receptor or through inhibition of tyrosine kinase results in amelioration of the fibrotic response in the bleomycin model and in the bronchiolitis obliterans model [18]. In human fibrotic disorders of the lung, such as IPF, scleroderma and bronchiolitis obliterans, PDGF genes were shown to be upregulated in BAL cells or in affected tissues [18].

In asthma and airway remodeling, a contribution of PDGF to the pathogenesis is likely, but available data are controversial. Eosinophils in biopsies of asthmatic airways have been shown to produce PDGF-B chain [4]. Bronchial fibroblasts from asthma patients show enhanced responsiveness to the mitogenic effects of PDGF, but increased levels of PDGF in BAL are either not present or do not correlate with airway fibrosis [2].

Insulin-like growth factor

IGF-I and II are single chain peptides with structural homology to pro-insulin [53]. In the lung, the major sources of IGFs are macrophages, but mesenchymal and bronchial epithelial cells are also able to produce IGF [4, 54]. IGFs can bind to two receptors, type I and II, and biological activity is mainly determined by IGF-binding protein that releases IGF [55]. IGFs regulate proliferation and differentiation of a variety of cells [53], being strong mitogens for fibroblasts and causing enhanced collagen synthesis [54]. IGFs are also found in tissue and BAL fluids of various fibroproliferative diseases [12, 56]. However, little information is available about the role of IGF's in asthma or COPD, but IGF's can stimulate the proliferation of airway smooth muscle cells and may be involved in airway fibrosis [4].

Fibroblast growth factors

FGFs are a group of nine heparin-binding peptides, amongst them acidic and basic FGF (now FGF-1 and 2) and keratinocyte growth factor (KGF or FGF-7) [53, 57]. FGF-1 and 2 have been shown to accelerate granulation tissue formation, fibroblast proliferation and collagen synthesis [53]. They are strong mitogens for angiogenesis and endothelial cell migration [4]. FGF-1 and 2 are not present at significant levels in normal lungs but are produced after tissue injury by alveolar macrophages, epithelial cells, and fibroblasts [58]. Major sources for FGF-2 in the lung are mast cells [59]. In humans, FGF-2 was found in BAL and serum of patients with IPF and scleroderma [60] and in human airway smooth muscle cells; FGF-2 increases the expression of PDGF-receptor α and therefore indirectly stimulates proliferation [4].

KGF (FGF-7) is primarily produced by fibroblasts and is mitogenic mainly for alveolar type II epithelial cells through the FGFR2-IIIb receptor, essentially expressed only by epithelial cells [57, 61]. IL-1β appears to be the most potent inducer of KGF from fibroblasts [35] and in the context of epithelial-mesenchymal cell interaction, IL-1β from epithelial cells stimulates the release of KGF from fibroblasts and results in the stimulation of the epithelium in a positive paracrine system [62]. In the bleomycin model of fibrosis instillation of KGF prevented the progression to fibrogenesis [63] and KGF was able to limit tissue damage and enhance epithelial repair in a syngeneic tracheal transplant model in rats [64]. In humans with ARDS, KGF in BAL fluid was taken as a marker of the severity of tissue injury and correlated with poor prognosis [57], while KGF administration limited the decreased lung permeability and inflammation seen on allergen challenge in an OVA-sensitized rat model [65].

Epidermal growth factor

The EGF family is an enlarging group of related proteins and includes EGF and transforming growth factor-α (TGF-α). Both factors have potential roles in wound healing and remodeling. They share 42% homology and signal through the EGF receptor that activates tyrosine kinase [49, 66].

EGFs are important in the repair of epithelial injury. It has been shown in damaged bronchial epithelial cells that EGF receptors become phosphorylated and consequently the defect is repaired and supplementation with exogenous EGF further enhances this process [3]. In asthma, EGF receptor is highly expressed in airway epithelium, but epithelial cell proliferation is still impaired [3], and impaired re-epithelialization induces the formation of granulation tissue [67]. This suggests that an abnormal response of the epithelium to growth factors could be a central factor in the pathogenesis of asthma and airway remodeling. In humans, EGF and EGF receptor are present in the submucosa of asthmatic airways [3, 4].

Similar to EGF, TGF-α is able to promote wound epithelialization [53], while tissue specific overexpression of TGF-α in the lung of mice leads to alveolar enlargement and interstitial and pleural fibrosis. Conversely, elimination of the TGF-α gene significantly decreases the fibrogenic tissue response to bleomycin [66]. In humans, TGF-α protein is elevated in BAL fluid of patients with ARDS and IPF [68], and TGF-α and EGF receptor expression is seen in biopsy material in IPF [69].

Hepatocyte growth factor

HGF was originally identified as a potent stimulator of mature hepatocytes, but more recently is recognized as a participant in many tissue responses to injury and is a potentially important stimulator of alveolar epithelial cells and vascular endothelial cells [70, 71]. In lung development, HGF is shown to be required for alveologenesis in the rodent [72], and recent studies indicate this growth factor may play a role in airway sensitization through modulation of dendritic cell activation [73]. Importantly, there are defects in HGF production from fibroblasts in tissue samples from both emphysema and IPF patients [21, 74]. Most impressive is the experimental attenuation of both elastase-induced emphysema and bleomycin-induced fibrosis by administration of either recombinant HGF or by gene transfer to the lung [71, 75, 76]. These data suggest that HGF has activities that could balance the profibrotic activities of factors such as TGF-β and establishes an interplay between these powerful factors that could distinguish correct tissue repair from disordered pathology (Fig. 29.2).

Transforming growth factor-β

The TGF-β family includes 5 isoforms, of which mammalian cells express three. The isoforms TGF-β 1 to 3 reveal considerable sequence homology, and their biological properties in wound healing and tissue remodeling are similar. Most studies have been performed with TGF-β1, which is the most abundant and best-characterized isoform [77]. TGF-β is profibrotic by various effects on ECM turnover and stromal cell biology. The immunomodulatory activities of TGF-β are crucial and aim mainly at the limitation of inflammatory processes. Many different cells secrete TGF-β, in the lung mainly macrophages, epithelial cells, and fibroblasts. The secreted molecule is inactive and has to be activated by cleavage of the latency associated peptide (LAP) through oxygen radicals or proteases (e.g. thrombospondin or plasmin) [78] or through interaction with cell integrins, most notably αVβ6 and αVβ8 on epithelial cells or fibroblasts [24, 79]. TGF-β binds to two receptors, type I and II, and signals are transduced by serine/threonine kinases to different intracellular pathways [80, 81]. In the context of tissue remodeling, the *Smad*-signaling pathway is the most prominent and a potential target for pharmacologic intervention in fibrotic disorders [82, 83]. TGF-β stimulates ECM production, predominantly collagen and fibronectin, and reduces matrix degradation by changing the balance of collagenases and collagenase inhibitors (TIMP-1) [78]. It is anti-proliferative for epithelial cells, inducing apoptosis and subsequent fibrotic responses in the lung [84, 85]. The mitogenic effect on fibroblasts is not uniform, but it is apparent that TGF-β induces the transformation of fibroblasts into myofibroblasts [86]. Myofibroblasts are contractile cells and synthesize most matrix proteins, both important factors in the final step of wound healing, but potentially a pathogenic process in fibrosis. In the remodeling process, TGF-β acts in concert with many other cytokines and growth factors. On the afferent, it can be upregulated by TNF-α, IL-1, GM-CSF, PDGF, and TGF-β itself. On the effector, TGF-β not only acts through direct interference with collagen metabolism but may induce the expression and secretion of FGF, PDGF, and connective tissue growth factor (CTGF), as well as its own upregulation [9, 18, 86].

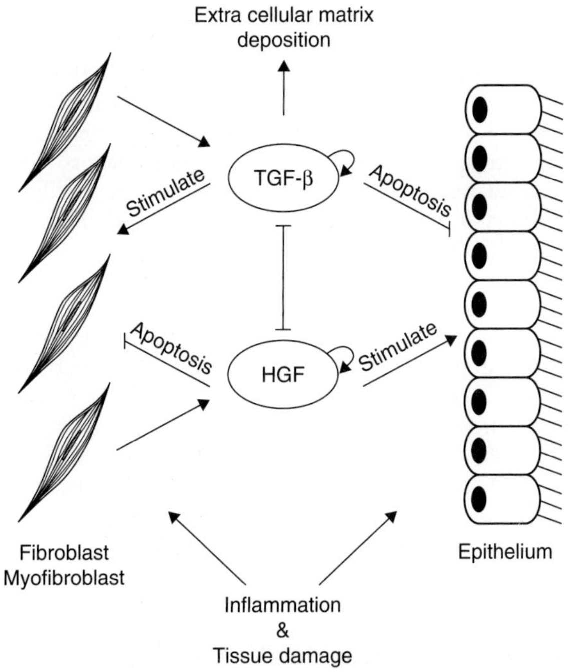

FIG. 29.2 Two main growth factors balance cell survival and tissue remodeling. TGF-β, released from the matrix in the parenchyma during tissue injury, is a major stimulator of myofibroblast and fibroblast survival and ECM production. TGF-β also induces apoptosis in pulmonary epithelial cells. HGF, released from fibroblasts in the parenchyma by molecules such as IL-1β, balance the pro-fibrotic effects of TGF-β by stimulating the survival of epithelial cells and inducing the apoptosis of myofibroblasts, as well as dampening the inflammatory response that accompanies tissue injury. Both are autocrine and paracrine in function, and have an apparent inhibition effect on release of the counterpart.

The role of TGF-β in pulmonary remodeling and fibrosis has been investigated mainly in interstitial lung disease, where the key role of this growth factor is widely accepted, with evidence from a large number of animal models and human studies [18, 53, 83, 85]. In models using chemical agents such as bleomycin, asbestos or silica, or irradiation to induce progressive tissue damage, TGF-β was shown to be upregulated at the sites of developing fibrosis. In our own studies, in which transient overexpression of active TGF-β1 is achieved by adenoviral gene transfer, severe and diffuse pulmonary fibrosis developed in rats and mice [78], mediated through the Smad3-signaling pathway [83]. The importance of TGF-β in tissue remodeling after pulmonary injury is highlighted by several experimental settings that demonstrate the beneficial effect of anti-TGF-β agents on the course of fibrosis. Neutralizing antibodies, dominant negative TGF-β-receptors or natural antagonists such as the proteoglycan decorin have all been able to reduce fibrotic reactions in the lung [87, 88] and TGF-β was shown to be present in a variety of human lung diseases, including asthma and IPF. Also, as COPD could be considered a disorder with inadequate tissue repair [89], the genetic association of the TGF gene with COPD disease is a positive indication of involvement [90].

The importance of TGF-β in pulmonary remodeling is not only restricted to the interstitium, but is also present in airways. TGF-β genes are upregulated in bronchial walls in a model of chronic airway disease [91], and anti-TGF-β is shown to modulate the subepithelial fibrotic response seen in a chronic allergen challenge model in the mouse [8]. Interestingly, TGF-β1 and TGF-β3 may have opposite effects on connective tissue synthesis in cultured airway smooth muscle cells, suggesting a protective role for TGF-β3 in airway remodeling [92]. In asthma patients, TGF-β was found in airway walls and correlated with the degree of subepithelial fibrosis [7], but others have failed to find a correlation to disease severity [93], while TGF-β is released into BAL fluid after endobronchial allergen challenge [3]. There are findings that link the pathogenesis of fibrogenesis to that of emphysema; interference with either the activation of TGF-β (αVβ6 null mouse) or signaling through the receptor (Smad3 null mouse) leads to spontaneous airspace enlargement and aspects of emphysema [89, 94]. These changes are accompanied by enhanced expression of MMP12 by macrophages, normally controlled by exposure of these cells to TGF-β [89], implying tissue repair abnormalities lie at the center of both disorders.

Connective tissue growth factor

Connective tissue growth factor is a growth factor with similar activities on collagen metabolism as TGF-β. It is a cysteine-rich peptide produced by fibroblasts and endothelial cells, but not by leukocytes or epithelial cells [18, 95]. CTGF is constitutively expressed in human lung fibroblasts and is stimulated by TGF-β, but not TNF-α or IL-1β [96] In mouse lungs, TGF-β and CTGF are upregulated following bleomycin injury. It has been suggested that TGF-β exerts its profibrotic effects in part through PDGF pathways and through a PDGF-independent pathway using CTGF [18].

Less information is known about CTGF in airway remodeling. However, the association with TGF-β makes it likely that CTGF might have similar effects in the airway as TGF-β. Indeed, it has been recently shown that TGF-β stimulates expression of CTGF in cultured airway smooth muscle cells [97].

ANIMAL MODELS

Chronic airway challenge

Airway remodeling in human chronic airway disease is a concept not yet clearly defined [2]. The principal morphologic features are subepithelial fibrosis, myofibroblast hyperplasia, airway smooth muscle hypertrophy, mucus gland and goblet cell hyperplasia, and epithelial disruption [98]. The mechanisms leading to persistent airway changes are poorly characterized because of a lack of simple and suitable animal models. An impaired response of asthmatic bronchi to EGF after epithelial injury seems to be important to initiate the process, while other growth factors, such as TGF-β isoforms, PDGF, and bFGF might be responsible for the proliferative response [3]. Recently described animal models resemble certain aspects of human asthma induced by repeat allergen exposure with *Aspergillus*, ovalbumin or house dust mite antigens. These models produce in mice changes similar to human asthmatic airway remodeling, and airway fibrosis, accompanied by upregulation of TGF-β and GM-CSF [8, 17, 91, 99, 100].

Transgene animal models

In transgenic animal models certain genes are deleted, overexpressed, or mutated. Transgenic animals, either developed by gene insertion (or deletion) in embryonic stem cells or developed by transient transgene expression in adult animals by gene transfer (e.g. adenoviral vectors) have proven useful to answer questions about the role of specific growth factors in the pathogenesis of pulmonary disease. Animals transiently overexpressing IL-1 or TGF-β develop progressive fibrotic lesions in the lung, IL-1, likely through induction of TGF-β [10, 29, 43, 78]. Conversely, TNF-α-receptor and TGF-α knockout mice are resistant to bleomycin-induced pulmonary fibrosis [18, 66]. Also implicated by transgenic studies are IL-11 and IL-13; these molecules were not previously known to be factors involved in tissue remodeling and repair, but showing evidence of fibrotic changes, emphysematous lesions, and deranged mucus production [101, 102].

SUMMARY

Of the many cytokines and growth factors that are found within the tissue or surrounding fluids in COPD and asthma, only a few can be shown to have direct impact on the process of tissue remodeling. *In vitro* and *in vivo* studies outlined above indicate that factors such as TGF-β, which induces chronic repair without accompanying tissue injury, and IL-1β, which induces tissue injury and chronic repair, likely through induction of TGF-β, may be considered the most critical targets for intervention. The fact that these growth factors act mainly at a local site in association with matrix helps explain the progressive and tissue-restricted nature of remodeling. Development of potent inhibitors of these growth factors or of genes activated downstream of them could prove beneficial in modifying the altered tissue in asthma and COPD allowing conjoint therapy with anti-inflammatory drugs, preferably delivered to the local remodeled site, to halt the destructive process and return the lung to normal function.

References

1. Hogg JC, Chu F, Utokaparch S *et al.* The nature of small-airway obstruction in chronic obstructive pulmonary disease. *N Engl J Med* 350(26): 2645–53, 2004.
2. Fish JE, Peters SP. Airway remodeling and persistent airway obstruction in asthma. *J Allergy Clin Immunol* 104(3 Pt 1): 509–16, 1999.
3. Holgate ST. Epithelial damage and response. *Clin Exp Allergy* 30 (Suppl 1): 37–41, 2000.

4. Chung KF, Barnes PJ. Cytokines in asthma. *Thorax* 54(9): 825–57, 1999.
5. Redington AE, Madden J, Frew AJ *et al.* Transforming growth factor-beta 1 in asthma. Measurement in bronchoalveolar lavage fluid. *Am J Respir Crit Care Med* 156(2 Pt 1): 642–47, 1997.
6. Xing Z, Ohkawara Y, Jordana M, Graham FL, Gauldie J. Transfer of granulocyte-macrophage colony-stimulating factor gene to rat lung induces eosinophilia, monocytosis, and fibrotic reactions. *J Clin Invest* 97(4): 1102–10, 1996.
7. Minshall EM, Leung DY, Martin RJ *et al.* Eosinophil-associated TGF-beta1 mRNA expression and airways fibrosis in bronchial asthma. *Am J Respir Cell Mol Biol* 17(3): 326–33, 1997.
8. McMillan SJ, Xanthou G, Lloyd CM. Manipulation of allergen-induced airway remodeling by treatment with anti-TGF-{beta} antibody: Effect on the Smad signaling pathway. *J Immunol* 174(9): 5774–80, 2005.
9. Xing Z, Jordana M, Gauldie J, Wang J. Cytokines and pulmonary inflammatory and immune diseases. *Histol Histopathol* 14(1): 185–201, 1999.
10. Kolb M, Margetts PJ, Anthony DC, Pitossi F, Gauldie J. Transient expression of IL-1 beta induces acute lung injury and chronic repair leading to pulmonary fibrosis. *J Clin Invest* 107(12): 1529–36, 2001.
11. Lappalainen U, Whitsett JA, Wert SE, Tichelaar JW, Bry K. Interleukin-1beta causes pulmonary inflammation, emphysema, and airway remodeling in the adult murine lung. *Am J Respir Cell Mol Biol* 32(4): 311–18, 2005.
12. Vanhee D, Gosset P, Wallaert B, Voisin C, Tonnel AB. Mechanisms of fibrosis in coal workers' pneumoconiosis. Increased production of platelet-derived growth factor, insulin-like growth factor type I, and transforming growth factor beta and relationship to disease severity. *Am J Respir Crit Care Med* 150(4): 1049–55, 1994.
13. Hiwatari N, Shimura S, Yamauchi K, Nara M, Hida W, Shirato K. Significance of elevated procollagen-III-peptide and transforming growth factor-beta levels of bronchoalveolar lavage fluids from idiopathic pulmonary fibrosis patients. *Tohoku J Exp Med* 181(2): 285–95, 1997.
14. Broekelmann TJ, Limper AH, Colby TV, McDonald JA. Transforming growth factor beta 1 is present at sites of extracellular matrix gene expression in human pulmonary fibrosis. *Proc Natl Acad Sci USA* 88(15): 6642–46, 1991.
15. Khalil N, Greenberg AH. The role of TGF-beta in pulmonary fibrosis. *Ciba Found Symp* 157: 194–211, 1991.
16. Corrin B, Butcher D, McAnulty BJ *et al.* Immunohistochemical localization of transforming growth factor-beta 1 in the lungs of patients with systemic sclerosis, cryptogenic fibrosing alveolitis and other lung disorders. *Histopathology* 24(2): 145–50, 1994.
17. Kelly MM, Leigh R, Bonniaud P *et al.* Epithelial expression of profibrotic mediators in a model of allergen-induced airway remodeling. *Am J Respir Cell Mol Biol* 32(2): 99–107, 2005.
18. Lasky JA, Brody AR. Interstitial fibrosis and growth factors. *Environ Health Perspect* 108(Suppl 4): 751–62, 2000.
19. Ziegenhagen MW, Schrum S, Zissel G, Zipfel PF, Schlaak M, Muller-Quernheim J. Increased expression of proinflammatory chemokines in bronchoalveolar lavage cells of patients with progressing idiopathic pulmonary fibrosis and sarcoidosis. *J Investig Med* 46(5): 223–31, 1998.
20. Ashitani J, Mukae H, Taniguchi H *et al.* Granulocyte-colony stimulating factor levels in bronchoalveolar lavage fluid from patients with idiopathic pulmonary fibrosis. *Thorax* 54(11): 1015–20, 1999.
21. Marchand-Adam S, Marchal J, Cohen M *et al.* Defect of hepatocyte growth factor secretion by fibroblasts in idiopathic pulmonary fibrosis. *Am J Respir Crit Care Med* 168(10): 1156–61, 2003.
22. Postma DS, Timens W. Remodeling in asthma and chronic obstructive pulmonary disease. *Proc Am Thorac Soc* 3(5): 434–39, 2006.
23. Munger JS, Huang X, Kawakatsu H *et al.* The integrin alpha v beta 6 binds and activates latent TGF beta 1: A mechanism for regulating pulmonary inflammation and fibrosis. *Cell* 96(3): 319–28, 1999.
24. Araya J, Cambier S, Morris A, Finkbeiner W, Nishimura SL. Integrin-mediated transforming growth factor-beta activation regulates homeostasis of the pulmonary epithelial-mesenchymal trophic unit. *Am J Pathol* 169(2): 405–15, 2006.
25. Parameswaran K, Willems-Widyastuti A, Alagappan VK, Radford K, Kranenburg AR, Sharma HS. Role of extracellular matrix and its regulators in human airway smooth muscle biology. *Cell Biochem Biophys* 44(1): 139–46, 2006.
26. Matsuoka H, Sisson TH, Nishiuma T, Simon RH. Plasminogen-mediated activation and release of hepatocyte growth factor from extracellular matrix. *Am J Respir Cell Mol Biol* 35(6): 705–13, 2006.
27. Sporn MB, Roberts AB (eds). *Peptide Growth Factors and Their Receptors*. New York: Springer-Verlag, 1991.
28. Tufvesson E, Westergren-Thorsson G. Alteration of proteoglycan synthesis in human lung fibroblasts induced by interleukin-1beta and tumor necrosis factor-alpha. *J Cell Biochem* 77(2): 298–309, 2000.
29. Sime PJ, Marr RA, Gauldie D *et al.* Transfer of tumor necrosis factor-alpha to rat lung induces severe pulmonary inflammation and patchy interstitial fibrogenesis with induction of transforming growth factor-beta 1 and myofibroblasts. *Am J Pathol* 153(3): 825–32, 1998.
30. Mukhopadhyay S, Hoidal J, Mukherjee T. Role of TNFalpha in pulmonary pathophysiology. *Respir Res* 7(1): 125, 2006.
31. Kendall JC, Li XH, Galli SJ, Gordon JR. Promotion of mouse fibroblast proliferation by IgE-dependent activation of mouse mast cells: Role for mast cell tumor necrosis factor-alpha and transforming growth factor-beta 1. *J Allergy Clin Immunol* 99(1 Pt 1): 113–23, 1997.
32. Rochester CL *et al.* Cytokines and cytokine networking in the pathogenesis in interstitial and fibrotic lung disorders. *Semin Respir Crit Care Med*(14): 389–416, 1994.
33. Sime PJ, Gauldie J. Mechanisms of scarring. In: Tw E, Haslett C (eds). *ARDS: Acute Respiratory Distress in Adults*, pp. 215–31. London: Chapman and Hall Medical, 1996.
34. Boyle JE, Lindroos PM, Rice AB, Zhang L, Zeldin DC, Bonner JC. Prostaglandin-E2 counteracts interleukin-1beta-stimulated upregulation of platelet-derived growth factor alpha-receptor on rat pulmonary myofibroblasts. *Am J Respir Cell Mol Biol* 20(3): 433–40, 1999.
35. Chedid M, Rubin JS, Csaky KG, Aaronson SA. Regulation of keratinocyte growth factor gene expression by interleukin 1. *J Biol Chem* 269(14): 10753–57, 1994.
36. Postlethwaite AE, Raghow R, Stricklin GP, Poppleton H, Seyer JM, Kang AH. Modulation of fibroblast functions by interleukin 1: Increased steady-state accumulation of type I procollagen messenger RNAs and stimulation of other functions but not chemotaxis by human recombinant interleukin 1 alpha and beta. *J Cell Biol* 106(2): 311–18, 1988.
37. Piguet PF, Vesin C, Grau GE, Thompson RC. Interleukin 1 receptor antagonist (IL-1ra) prevents or cures pulmonary fibrosis elicited in mice by bleomycin or silica. *Cytokine* 5(1): 57–61, 1993.
38. Kolb M, Bonniaud P, Galt T *et al.* Differences in the fibrogenic response after transfer of active transforming growth factor-beta1 gene to lungs of "fibrosis-prone" and "fibrosis-resistant" mouse strains. *Am J Respir Cell Mol Biol* 27(2): 141–50, 2002.
39. Pardo A, Gibson K, Cisneros J *et al.* Up-regulation and profibrotic role of osteopontin in human idiopathic pulmonary fibrosis. *PLoS Med* 2(9), 2005.
40. Serlin DM, Kuang PP, Subramanian M *et al.* Interleukin-1beta induces osteopontin expression in pulmonary fibroblasts. *J Cell Biochem*(97): 519–29, 2006.
41. Xanthou G, Alissafi T, Semitekolou M *et al.* Osteopontin has a crucial role in allergic airway disease through regulation of dendritic cell subsets. *Nat Med* 13(5): 570–78, 2007.
42. Stampfli MR, Wiley RE, Neigh GS *et al.* GM-CSF transgene expression in the airway allows aerosolized ovalbumin to induce allergic sensitization in mice. *J Clin Invest* 102(9): 1704–14, 1998.
43. Xing Z, Tremblay GM, Sime PJ, Gauldie J. Overexpression of granulocyte-macrophage colony-stimulating factor induces pulmonary granulation tissue formation and fibrosis by induction of transforming

growth factor-beta 1 and myofibroblast accumulation. *Am J Pathol* 150(1): 59–66, 1997.

44. Vlahos R, Bozinovski S, Hamilton JA, Anderson GP. Therapeutic potential of treating chronic obstructive pulmonary disease (COPD) by neutralising granulocyte macrophage-colony stimulating factor (GM-CSF). *Pharmacol Ther* 112(1): 106–15, 2006.
45. Andreutti D, Gabbiani G, Neuville P. Early granulocyte-macrophage colony-stimulating factor expression by alveolar inflammatory cells during bleomycin-induced rat lung fibrosis. *Lab Invest* 78(12): 1493–502, 1998.
46. Christensen PJ, Bailie MB, Goodman RE, O'Brien AD, Toews GB, Paine R III.. Role of diminished epithelial GM-CSF in the pathogenesis of bleomycin-induced pulmonary fibrosis. *Am J Physiol Lung Cell Mol Physiol* 279(3): L487–IL495, 2000.
47. Piguet PF, Grau GE, de Kossodo S. Role of granulocyte-macrophage colony-stimulating factor in pulmonary fibrosis induced in mice by bleomycin. *Exp Lung Res* 19(5): 579–87, 1993.
48. Fabisiak JK, Kelley JP. Platelet derived growth factor. In: Kelley J (ed.), *Cytokines of the Lung*, pp. 3–39. New York: Marcel Dekker, 1992.
49. Rice AB, Moomaw CR, Morgan DL, Bonner JC. Specific inhibitors of platelet-derived growth factor or epidermal growth factor receptor tyrosine kinase reduce pulmonary fibrosis in rats. *Am J Pathol* 155(1): 213–21, 1999.
50. Ingram JL, Bonner JC. EGF and PDGF receptor tyrosine kinases as therapeutic targets for chronic lung diseases. *Curr Mol Med* 6(4): 409–21, 2006.
51. Lasky JA, Tonthat B, Liu JY, Friedman M, Brody AR. Upregulation of the PDGF-alpha receptor precedes asbestos-induced lung fibrosis in rats. *Am J Respir Crit Care Med* 157(5 Pt 1): 1652–57, 1998.
52. Bostrom H, Willetts K, Pekny M *et al.* PDGF-A signaling is a critical event in lung alveolar myofibroblast development and alveogenesis. *Cell* 85(6): 863–73, 1996.
53. Mutsaers SE, Bishop JE, McGrouther G, Laurent GJ. Mechanisms of tissue repair: From wound healing to fibrosis. *Int J Biochem Cell Biol* 29(1): 5–17, 1997.
54. Uh ST, Inoue Y, King TE Jr., Chan ED, Newman LS, Riches DW. Morphometric analysis of insulin-like growth factor-I localization in lung tissues of patients with idiopathic pulmonary fibrosis. *Am J Respir Crit Care Med* 158(5 Pt 1): 1626–35, 1998.
55. Bayes-Genis A, Conover CA, Schwartz RS. The insulin-like growth factor axis: A review of atherosclerosis and restenosis. *Circ Res* 86(2): 125–30, 2000.
56. Krein PM, Sabatini PJ, Tinmouth W, Green FH, Winston BW. Localization of insulin-like growth factor-I in lung tissues of patients with fibroproliferative acute respiratory distress syndrome. *Am J Respir Crit Care Med* 167(1): 83–90, 2003.
57. Stern JB, Fierobe L, Paugam C *et al.* Keratinocyte growth factor and hepatocyte growth factor in bronchoalveolar lavage fluid in acute respiratory distress syndrome patients. *Crit Care Med* 28(7): 2326–33, 2000.
58. Barrios R, Pardo A, Ramos C, Montano M, Ramirez R, Selman M. Upregulation of acidic fibroblast growth factor during development of experimental lung fibrosis. *Am J Physiol* 273(2 Pt 1): L451–L458, 1997.
59. Liebler JM, Picou MA, Qu Z, Powers MR, Rosenbaum JT. Altered immunohistochemical localization of basic fibroblast growth factor after bleomycin-induced lung injury. *Growth Factors* 14(1): 25–38, 1997.
60. Kadono T, Kikuchi K, Kubo M, Fujimoto M, Tamaki K. Serum concentrations of basic fibroblast growth factor in collagen diseases. *J Am Acad Dermatol* 35(3 Pt 1): 392–97, 1996.
61. Finch PW, Rubin JS. Keratinocyte growth factor/fibroblast growth factor 7, a homeostatic factor with therapeutic potential for epithelial protection and repair. *Adv Cancer Res* 91: 69–136, 2004.
62. Marchand-Adam S, Plantier L, Bernuau D *et al.* Keratinocyte growth factor expression by fibroblasts in pulmonary fibrosis: Poor response to interleukin-1 {beta}. *Am J Respir Cell Mol Biol* 32(5): 470–77, 2005.
63. Sugahara K, Iyama K, Kuroda MJ, Sano K. Double intratracheal instillation of keratinocyte growth factor prevents bleomycin-induced lung fibrosis in rats. *J Pathol* 186(1): 90–98, 1998.
64. Gomperts BN, Belperio JA, Fishbein MC, Keane MP, Burdick MD, Strieter RM. Keratinocyte growth factor improves repair in the injured tracheal epithelium. *Am J Respir Cell Mol Biol* 37(1): 48–56, 2007.
65. Tillie-Leblond I, Gosset P, Le Berre R *et al.* Keratinocyte growth factor improves alterations of lung permeability and bronchial epithelium in allergic rats. *Eur Respir J* 30(1): 1317–25, 2007.
66. Madtes DK, Elston AL, Hackman RC, Dunn AR, Clark JG. Transforming growth factor-alpha deficiency reduces pulmonary fibrosis in transgenic mice. *Am J Respir Cell Mol Biol* 20(5): 924–34, 1999.
67. Khalil N, O'Connor RN, Flanders KC, Shing W, Whitman CI. Regulation of type II alveolar epithelial cell proliferation by TGF-beta during bleomycin-induced lung injury in rats. *Am J Physiol* 267(5 Pt 1): L498–507, 1994.
68. Madtes DK, Rubenfeld G, Klima LD *et al.* Elevated transforming growth factor-alpha levels in bronchoalveolar lavage fluid of patients with acute respiratory distress syndrome. *Am J Respir Crit Care Med* 158(2): 424–30, 1998.
69. Baughman RP, Lower EE, Miller MA, Bejarano PA, Heffelfinger SC. Overexpression of transforming growth factor-alpha and epidermal growth factor-receptor in idiopathic pulmonary fibrosis. *Sarcoidosis Vasc Diffuse Lung Dis* 16(1): 57–61, 1999.
70. Ware LB, Matthay MA. Keratinocyte and hepatocyte growth factors in the lung: Roles in lung development, inflammation, and repair. *Am J Physiol Lung Cell Mol Physiol* 282(5): L924–L940, 2002.
71. Shigemura N, Sawa Y, Mizuno S *et al.* Amelioration of pulmonary emphysema by *in vivo* gene transfection with hepatocyte growth factor in rats. *Circulation* 111(11): 1407–14, 2005.
72. Padela S, Cabacungan J, Shek S *et al.* Hepatocyte growth factor is required for alveologenesis in the neonatal rat. *Am J Respir Crit Care Med* 172(7): 907–14, 2005.
73. Okunishi K, Dohi M, Nakagome K *et al.* A novel role of hepatocyte growth factor as an immune regulator through suppressing dendritic cell function. *J Immunol* 175(7): 4745–53, 2005.
74. Plantier L, Marchand-Adam S, Marchal-Somme J *et al.* Defect of hepatocyte growth factor production by fibroblasts in human pulmonary emphysema. *Am J Physiol Lung Cell Mol Physiol* 288(4): L641–L647, 2005.
75. Mizuno S, Matsumoto K, Li MY, Nakamura T. HGF reduces advancing lung fibrosis in mice: A potential role for MMP-dependent myofibroblast apoptosis. *FASEB J* 19(6): 580–82, 2005.
76. Gazdhar A, Fachinger P, van Leer C *et al.* Gene transfer of hepatocyte growth factor by electroporation reduces bleomycin-induced lung fibrosis. *Am J Physiol Lung Cell Mol Physiol* 292(2): L529–L536, 2007.
77. O'Kane S, Ferguson MW. Transforming growth factor beta s and wound healing. *Int J Biochem Cell Biol* 29(1): 63–78, 1997.
78. Sime PJ, Xing Z, Graham FL, Csaky KG, Gauldie J. Adenovector-mediated gene transfer of active transforming growth factor-beta1 induces prolonged severe fibrosis in rat lung. *J Clin Invest* 100(4): 768–76, 1997.
79. Munger JS, Huang X, Kawakatsu H *et al.* A mechanism for regulating pulmonary inflammation and fibrosis: The integrin [alpha]v[beta]6 binds and activates latent TGF [beta]1. *Cell* 96(3): 319, 1999.
80. Derynck R, Zhang YE. Smad-dependent and Smad-independent pathways in TGF-beta family signalling. *Nature* 425(6958): 577–84, 2003.
81. Feng XH, Derynck R. Specificity and versatility in tgf-beta signaling through Smads. *Annu Rev Cell Dev Biol* 21: 659–93, 2005.
82. Bonniaud P, Margetts PJ, Kolb M *et al.* Progressive transforming growth factor {beta}1-induced lung fibrosis is blocked by an orally active ALK5 kinase inhibitor. *Am J Respir Crit Care Med* 171(8): 889–98, 2005.
83. Gauldie J, Kolb M, Ask K, Martin G, Bonniaud P, Warburton D. Smad3 signaling involved in pulmonary fibrosis and emphysema. *Proc Am Thorac Soc* 3(8): 696–702, 2006.
84. Lee CG, Cho SJ, Kang MJ *et al.* Early growth response gene 1-mediated apoptosis is essential for transforming growth factor beta1-induced pulmonary fibrosis. *J Exp Med* 200(3): 377–89, 2004.

85. Lee CG, Kang HR, Homer RJ, Chupp G, Elias JA. Transgenic modeling of transforming growth factor-beta(1): Role of apoptosis in fibrosis and alveolar remodeling. *Proc Am Thorac Soc* 3(5): 418–23, 2006.
86. Gauldie J *et al.* TGFb gene transfer to the lung induces myofibro-blast presence and pulmonary fibrosis. In: Desmoliere A, Tuchweber B (eds). *Tissue Repair and Fibrosis: Current Topics in Pathology*. Berlin: Springer-Verlag, 1999.
87. Giri SN, Hyde DM, Braun RK, Gaarde W, Harper JR, Pierschbacher MD. Antifibrotic effect of decorin in a bleomycin hamster model of lung fibrosis. *Biochem Pharmacol* 54(11): 1205–16, 1997.
88. Kolb M, Margetts PJ, Galt T *et al.* Transient transgene expression of decorin in the lung reduces the fibrotic response to bleomycin. *Am J Respir Crit Care Med* 163(3): 770–77, 2001.
89. Morris DG, Sheppard D. Pulmonary emphysema: When more is less. *Physiology (Bethesda)* 21: 396–403, 2006.
90. Celedon JC, Lange C, Raby BA *et al.* The transforming growth factor-beta1 (TGFB1) gene is associated with chronic obstructive pulmonary disease (COPD). *Hum Mol Genet* 13(15): 1649–56, 2004.
91. Hogaboam CM, Blease K, Mehrad B *et al.* Chronic airway hyper-reactivity, goblet cell hyperplasia, and peribronchial fibrosis during allergic airway disease induced by *Aspergillus fumigatus*. *Am J Pathol* 156(2): 723–32, 2000.
92. Coutts A *et al.* TGF-beta3 inhibits connective tissue synthesis by airway smooth muscle cells. *Am J Respir Crit Care Med* 161: A699, 2000.
93. Chu HW, Halliday JL, Martin RJ, Leung DY, Szefler SJ, Wenzel SE. Collagen deposition in large airways may not differentiate severe asthma from milder forms of the disease. *Am J Respir Crit Care Med* 158(6): 1936–44, 1998.
94. Bonniaud P, Kolb M, Galt T *et al.* Smad3 null mice develop airspace enlargement and are resistant to TGF-beta-mediated pulmonary fibrosis. *J Immunol* 173(3): 2099–108, 2004.
95. Grotendorst GR. Connective tissue growth factor: A mediator of TGF-beta action on fibroblasts. *Cytokine Growth Factor Rev* 8(3): 171–79, 1997.
96. Lasky JA, Ortiz LA, Tonthat B *et al.* Connective tissue growth factor mRNA expression is upregulated in bleomycin-induced lung fibrosis. *Am J Physiol* 275(2 Pt 1): L365–L371, 1998.
97. Douglas DA. TGFb stimulates the expression of CTGF by human bronchial smooth muscle cells. *Am J Respir Crit Care Med* 161: A699, 2000.
98. Redington AE. Fibrosis and airway remodelling. *Clin Exp Allergy* 30(Suppl 1): 42–45, 2000.
99. Cates EC, Fattouh R, Wattie J *et al.* Intranasal exposure of mice to house dust mite elicits allergic airway inflammation via a GM-CSF-mediated mechanism. *J Immunol* 173(10): 6384–92, 2004.
100. Johnson JR, Wiley RE, Fattouh R *et al.* Continuous exposure to house dust mite elicits chronic airway inflammation and structural remodeling. *Am J Respir Crit Care Med* 169(3): 378–85, 2004.
101. Zhu Z, Homer RJ, Wang Z *et al.* Pulmonary expression of interleukin-13 causes inflammation, mucus hypersecretion, subepithelial fibrosis, physiologic abnormalities, and eotaxin production. *J Clin Invest* 103(6): 779–88, 1999.
102. Lee PJ, Zhang X, Shan P *et al.* ERK1/2 mitogen-activated protein kinase selectively mediates IL-13-induced lung inflammation and remodeling *in vivo*. *J Clin Invest* 116(1): 163–73, 2006.

30

CHAPTER

Nitric Oxide

Sergei A. Kharitonov and Kazuhiro Ito

Section of Airway Disease, National Heart and Lung Institute, Imperial College, London, UK

INTRODUCTION

Nitric oxide (NO) in human biology was mostly considered to be a by-product of the combustion of fossil fuels and therefore as an air pollutant, until the mid- to late 1980s when it was discovered that this free radical was the previously uncharacterized endothelial-derived relaxing factor [1].

NO is now known to play a central role in the physiology and the pathophysiology of many human organ systems. Within the respiratory system, NO promotes vascular and bronchial dilation, is a key mediator of the coordinated beating of ciliated epithelial cells, promotes mucus secretion, and is an important neurotransmitter for nonadrenergic, noncholinergic neurons in the bronchial wall [2, 3]. Additionally, NO is a mediator of inflammatory phenomena within the lung by virtue of its ability to influence the phenotype of inflammatory cells and its contribution to the formation of reactive nitrogen products. Given its wide distribution within the lung and airway, it is not surprising that NO can be detected in exhaled gas in levels that we now know vary in health and disease.

This chapter reviews the formation of NO, novel approaches of its measurements, and its role in the pathobiology and the management of asthma and chronic obstructive pulmonary disease (COPD).

SOURCE OF NO IN ASTHMA AND COPD

The enzyme system responsible for producing NO is NO synthase (NOS), consists of at least three isoforms: (1) constitutive neuronal NOS (NOS I or nNOS); (2) inducible NOS (NOS II or iNOS); and (3) constitutive endothelial NOS (NOS III or eNOS). The existence of a genetic association between a polymorphism in the nNOS gene and the diagnosis of asthma has led to the speculation that nNOS is a candidate gene for asthma [4].

Functionally, constitutive (or c) NOSs (nNOS and eNOS) are Ca^{2+}- and calmodulin-dependent enzymes and rapidly release low concentrations of NO upon receptor stimulation by selective agonists. The activity of iNOS is largely regulated at a pretranslational level and can be induced by proinflammatory cytokines, such as tumor necrosis factor-α (TNF-α), interferon-γ (IFN-γ), and interleukin (IL)-1β [5]. Several hours after exposure to these inflammatory stimuli, iNOS action releases large quantities of NO, which may continue in a sustained manner (hours or days).

All the three NOS isoforms are found in the respiratory system and may co-operatively regulate various functions in the respiratory tract from bronchomotor tone to immunologic/inflammatory responses:

a. eNOS is expressed in human bronchial epithelium [6]and in type II human alveolar epithelial cells [7, 8].

b. nNOS is localized in the airway nerves and in the airway smooth muscle where NO is the major mediator for the neuronal smooth muscle relaxation [9], mediated by the inhibitory nonadrenergic, noncholinergic (iNANC) system [10]. In the human lung nNOS is expressed in capillary endothelial cells of alveolar septa and plays a role as an endothelium-derived regulator of capillary permeability, a modulator of cholinergic neuronal transmission, or an inhibitor of platelet aggregation [11]. These low NO concentrations produced by nNOS are detectable in exhaled air [12], and contributes to bronchial responsiveness [13]. We recently found that nNOS is upregulated with increased disease severity and dominant isoform as a source of NO in COPD (Ito et al., in preparation).

c. iNOS is expressed in alveolar type II endothelial cells, lung fibroblasts, airway and vascular smooth muscle cells, airway epithelial cells, mast cells, endothelial cells, neutrophils, and chondrocytes.
- The increased levels of exhaled NO in asthma, for example, have a predominant lower airway origin and are most likely due to activation of NOS2 in airway epithelial and inflammatory cells [14–18], with a small contribution from NOS [14].
- Exhaled NO levels in stable COPD [19, 20] are lower than in either smoking or nonsmoking asthmatics [21] and are not different from normal subjects. This reduction in exhaled NO is due to the effect of tobacco smoking, which downregulates eNOS [22] and reduces exhaled NO [23], suggesting that this may contribute to the high risk of pulmonary and cardiovascular disease in cigarette smokers. In addition to the effects of cigarette smoking, a relatively low value of exhaled NO in COPD may reflect more peripheral inflammation than in asthma, low NOS2 expression [24], and increased oxidative stress that may consume NO in the formation of peroxynitrite [25].

POTENTIAL MECHANISMS OF NO-RELATED DYSFUNCTION IN ASTHMA AND COPD

Nitration of proteins

NO production and release are directly linked to the formation of reactive nitrogen species (RNS) that modify proteins in asthma and COPD. The magnitude of this modification correlates with the degree of oxidative and nitrosative stresses.

Protein nitration is unique among posttranslational modifications in its dependency on reactivity of tyrosine residues in the protein target that may be achieved:

a. by peroxynitrite (formed from the reaction between NO and $O_2^{\bullet -}$),

b. through the reaction of NO with protein tyrosyl radicals,

c. by the reaction of nitrite with peroxidases [26, 27] including myeloperoxidase and eosinophil peroxidase [28],

d. by combination of all the above pathways.

Nitration of proteins by peroxynitrite is a concentration-dependent process related to formation of two distinctive forms of nitrated proteins: stable 3-nitrotyrosine (nitration) and labile *S*-nitrosocysteine (*S*-nitrosation) [29]. Both of these nitrated proteins can be further enzymatically modified by glutathione *S*-transferase or glutathione peroxidase *via* either of the following:

a. converting NO_2^- to NH_2^- in tyrosine residues

b. denitrating NO_2^- directly/indirectly in tyrosine residues

c. changing SNO to SH^- in cysteine residues, or denitrosation.

Nitration of tyrosine residues by lipopolysaccharide (LPS) [30], for example, induces the formation of hydrophilic negatively charged nitrotyrosine residue which alters the function [31], catalytic activity, structure, susceptibility to proteolytic degradation, and reduces the kinase substrate efficiency of proteins [32–34]. Nitration of mitochondrial proteins may change the activity of several enzymes involved in energy production (glutamate dehydrogenase), or in the electron transport chain (cytochrome oxidase and ATPase), or in energy production (creatine kinase) [30] (Fig. 30.1).

Interestingly, both energy production and apoptosis are affected by NO and related oxides [35] and are different in asthma and COPD.

Nonasthmatic lungs showed little or no nitrotyrosine staining, whereas lungs of patients even with mild asthma [15] and especially who died of status asthmaticus [36] have high presence of nitrotyrosine in both the airways and the lung parenchyma. Nevertheless, nitrotyrosine formation in airway epithelial and inflammatory cells is significantly higher in COPD than in asthma [37] and is related to COPD severity [38]. This exaggerated level of nitration may explain low protective properties of antioxidant enzymes [30] in the central bronchial epithelium in COPD [39].

On the other hand, protein nitration may be also beneficial:

a. Surfactant protein A (SP-A), a product of tyrosine nitration of human pulmonary surfactant, downregulates T-cell-dependent alveolar inflammation and protect against idiopathic pneumonia injury [40].

b. Tyrosine nitration in proteins is also sufficient to induce an accelerated degradation of the modified proteins by the proteasome, which may be critical for the removal of nitrated proteins *in vivo* [41].

c. Nitration of tyrosine residues in tyrosine kinase substrates may prevent phosphorylation and, therefore, inhibit tyrosine kinase function in cellular signaling [27]; this mechanism remains highly speculative.

d. Despite the discovery of enzymatic "nitrotyrosine dinitrase" activity in lung homogenates [42], suggesting the potential role of nitration (and dinitration) as a signaling mechanism, it remains to be established whether nitrotyrosine is merely a biomarker of increased nitrosative stress or whether it actively contributes to cellular dysfunction and development of the airway inflammatory processes in asthma or COPD.

Tyrosine hydroxylase

Tyrosine hydroxylase (TH), the enzyme which catalyses the production of L-dihydroxyphenylalanine (L-DOPA) is also

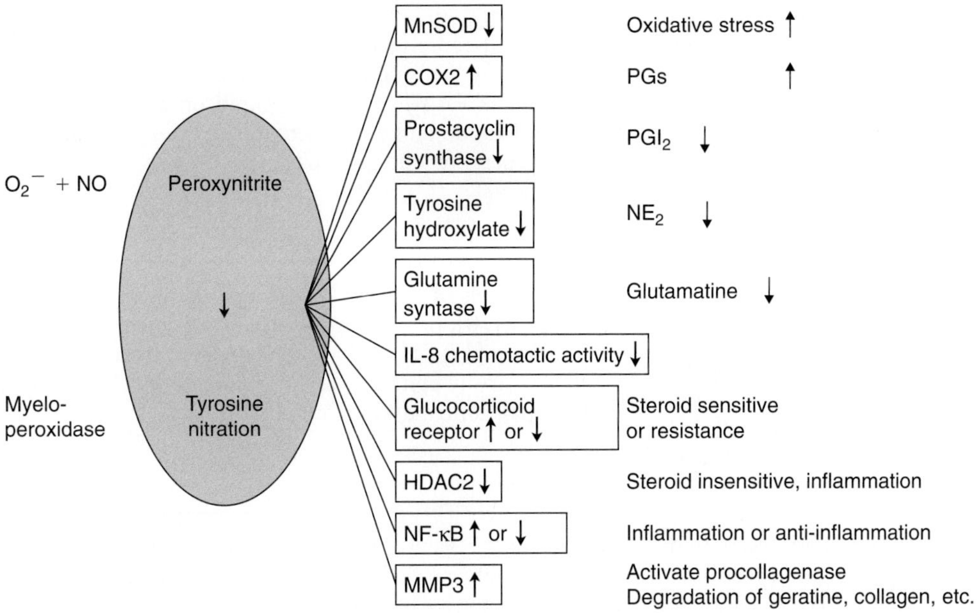

FIG. 30.1 Effect of peroxynitrite or impact of tyrosine nitration on various molecules. PG: prostaglandin, NE: norepinephrine.

reported to be inactivated following tyrosine nitration [43]. The loss in TH activity and expression is thought to contribute to the L-DOPA deficiency observed in certain neurological disorders such as Parkinson's disease [43, 44].

Repeated exposures to cigarette smoke cause a variety of molecular neuroadaptations in the cAMP signaling pathway in postmortem tissue from the brains of human smokers and former smokers [45]. Activity levels of two major components of cAMP signaling, cAMP-dependent protein kinase A (PKA), and adenylate cyclase, have been found abnormally elevated in nucleus accumbens of smokers and in ventral midbrain dopaminergic region of both smokers and former smokers. Although protein levels of other candidate neuroadaptations, including glutamate receptor subunits, TH, and other protein kinases, were within normal range, protein levels of the catalytic subunit of PKA were correspondingly higher in the ventral midbrain dopaminergic region of both smokers and former smokers [45]. These findings suggest that smoking-induced brain neuroadaptations can persist for significant periods in former smokers.

Histones

Histones are appreciably more stable than most cellular proteins and their slower turnover may permit them to accumulate 3-nitrotyrosine more than high turnover proteins [46]. Since nitrotyrosine can readily be detected immunohistochemically, nitrated histones may prove to be useful as a marker of extended exposure of cells or tissues to reactive nitrogen oxygen species.

The fact that only a limited number of the tyrosine residues were nitrated in histones [46] allows us to consider nitrotyrosine as one of the posttranslational modifications of histones that can occur *in vivo*. Nitration of histones using Mutatect cells (contain a relatively low level of nitrated proteins) cultured under standard conditions was not apparent until 3 days of sodium nitroprusside exposure, reaching a maximum at about 6 days [46]. This suggests that presence of nitrated histones in tissues may reflect the long-term exposure to RNS [47].

Histone deacetylase

Inducible NOS and other NF-κB-dependent genes involved in inflammation may be regulated by the specific recruitment of histone deacetylases (HDAC) [48], that may further enhance cytokine induction of both the iNOS and the NF-κB. HDAC activity/expression is reduced in airways of mild asthmatics [49] compared with normal subjects, but may be restored by steroids that activate histone acetyltransferases [49]. It has been shown that in subjects with severe asthma and COPD with reduced HDAC activity, the ability of inhaled steroids to control inflammation may be lost [49].

Molecular mechanisms underlying this reduction in HDAC2 activity is currently believed to be linked to nitrative stress. Peroxynitrite has been reported to alter HDAC2 function and stability via nitration of tyrosine residues on proteins *in vitro* and *in vivo* [50, 51].

Fate of nitrated protein

Protein nitration is reported to be more sensitive to trypsinization, and TH, once nitrated by peroxynitrite, is eliminated by proteasomal degradation [41, 52]. An increase in free nitrotyrosine is also an evidence of nitrated protein degradation. Oxidative/nitrative stress may delay the degradation of nitrated proteins due to the inactivation of the proteasome [53].

Interestingly, protein tyrosine nitration is a reversible process catalysed by an unknown enzyme [54] that removes the nitro (NO_2) group on nitrated proteins and has been referred to as a "denitrase" [42, 55].

We have found that denitrase activity in human cells and lung tissue is reduced in COPD (Ito and Osoata, unpublished), suggesting that reduced denitration activity could be involved in accumulation of *n*-tyrosine in the lung as well as elevated NO production.

MEASUREMENT OF NO IN EXHALED AIR

Several techniques are currently used to measure the fraction of NO in the exhaled air (Fig. 30.2); these are published in internationally recognized guidelines [56, 57]. The major issues are to control the expiratory flow rate [58] and to use an orapharnygeal pressure that prevents contamination of the expirate with high NO gas derived from the nasopharynx [59].

Measurement of exhaled NO from a single expiration at a fixed flow are simple, highly reproducible [60], have been used to monitor FE_{NO} in asthma research [61] and are now moving into clinical practice [62].

Methods for measuring exhaled NO at multiple expiratory flows (ME FE_{NO}) [63, 64] have been used to detect elevated levels of alveolar NO in asthma [65, 66] leading to the refinement of analytical methods to potentially discriminate exhaled NO sources in the lung [67] (Fig. 30.2).

Until recently, no day-to-day and home FE_{NO} monitoring was possible, as portable and simple NO analysers were not available. The arrival of the first hand-held portable NO analyser (NIOX MINO, Aerocrine, Sweden) (Fig. 30.2) that allows FE_{NO} measurements with sufficient accuracy and reproducibility in both children and adults [68, 69], may considerably change current management of asthma, as FE_{NO} can be measured on a daily basis by patients at home and frequently during their regular visits to their general practitioner [62].

Asthma

Increased levels of exhaled NO of a predominant lower airway origin [70], have been widely documented in patients with asthma [17, 71], and are likely related to activation of NOS2 in airway epithelial and inflammatory cells [14–18], with a small contribution from NOS1 [4].

Diagnosis

The diagnostic value of exhaled NO measurements (with 90% specificity and 95% positive predictive value) to differentiate between healthy subjects with or without respiratory symptoms and patients with confirmed asthma has been reported by several groups [72, 73]. This suggests that the simple measurement of exhaled NO can be used as an additional diagnostic tool for the screening of patients with a suspected diagnosis of asthma. Importantly, normal values of FE_{NO} for adults have been established and can be predicted on the basis of age and height [74].

It has been confirmed that the diagnostic utility of exhaled NO and induced sputum are superior to the conventional tests (peak flow measurements, spirometry, and changes in these parameters after a trial of steroid) [75],

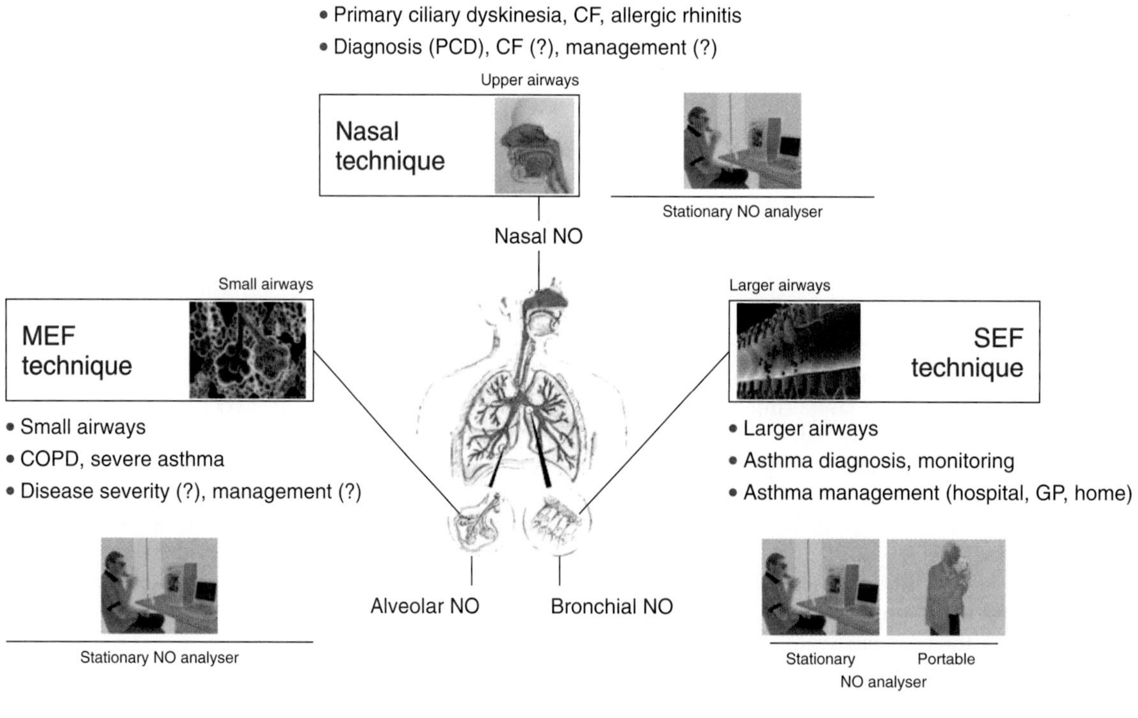

FIG. 30.2 Exhaled NO: A marker of airway (SEF technique) and lung inflammation (MEF technique). MEF: multiple exhalation flow; SEF: single exhalation flow.

with exhaled NO being most advantageous because the test is quick and easy to perform.

Asthma control

Monitoring of asthma may be much more conclusive [76] when repetitive FE_{NO} measurements, especially made at home by a portable NO analyzer, instead of single assessment were used, *changes* in FE_{NO} correlates significantly not only with changes in sputum eosinophils and hyperresponsiveness, but also with lung function and asthma symptoms.

Jones *et al.* [77] have demonstrated that repetitive exhaled NO measurements have a positive predictive value between 80% and 90% for predicting and diagnosing loss of control in asthma and are as useful as induced sputum eosinophils and airway hyperresponsiveness to hypertonic saline, but with the enormous advantage that they are easy to perform.

An advantage of FE_{NO} as a "loss-of-control-marker" [78] is that increase in FE_{NO} and asthma symptoms may be seen before any significant deterioration in airway hyperresponsiveness, sputum eosinophils, or lung function during asthma exacerbation induced by steroid reduction [18, 79]. Exhaled NO levels were (median, interquartile range) 11 ppb (9–21) in children who had good asthma control, 15 ppb (11–26) acceptable asthma control, and 28 ppb (19–33) with insufficiently controlled asthma [80], suggesting NO measurements may be useful for monitoring pediatric asthma in clinic.

Inhaled corticosteroids

Onset and cessation of action

Exhaled NO behaves as a "rapid response" marker that is extremely sensitive to steroid treatment, because it may be significantly reduced even 6 h after a single dose of nebulized budesonide [81], or within 2–3 days [16, 82] after treatment with inhaled corticosteroids (ICS) from a metered-dose or dry-powder inhaler.

We observed that the onset of action of inhaled budesonide on exhaled NO was dose dependent, both within the initial phase (first 3–5 days of treatment) and during treatment weeks 1, 2, and 3 [82]. Cessation of the actions of ICS is a rapid process too and can also be monitored by repetitive NO during the first 3–5 days in patients who stopped inhaling budesonide [82].

Dose dependency

We have shown that the acute reduction in exhaled NO (within the first 3–5 days of treatment) and the chronic reduction (days 7–21) are dose dependent in patients with mild asthma who are treated with low doses of budesonide [82]. Serial exhaled NO measurements, as we recently suggested [61], may therefore be useful in studying the onset and duration of action of ICS, as well as in monitoring patient compliance.

It is unclear, however, whether exhaled NO can guide an asthma treatment strategy. Thus, recent study based on the measurement of exhaled NO did not result in a large reduction in asthma exacerbations, but resulted in more patients on lower doses of ICS over 12 months, when compared with management by current asthma guidelines and no measure of FE_{NO} [83].

iNOS inhibitors

The flux of NO through the airways can be decreased by a single dose of inhaled NOS inhibitors in both asthma and healthy controls subjects with a greater reduction in patients with asthma (80 versus 61%) [84] (Fig. 30.3). This finding supports the hypothesis that iNOS induction may be

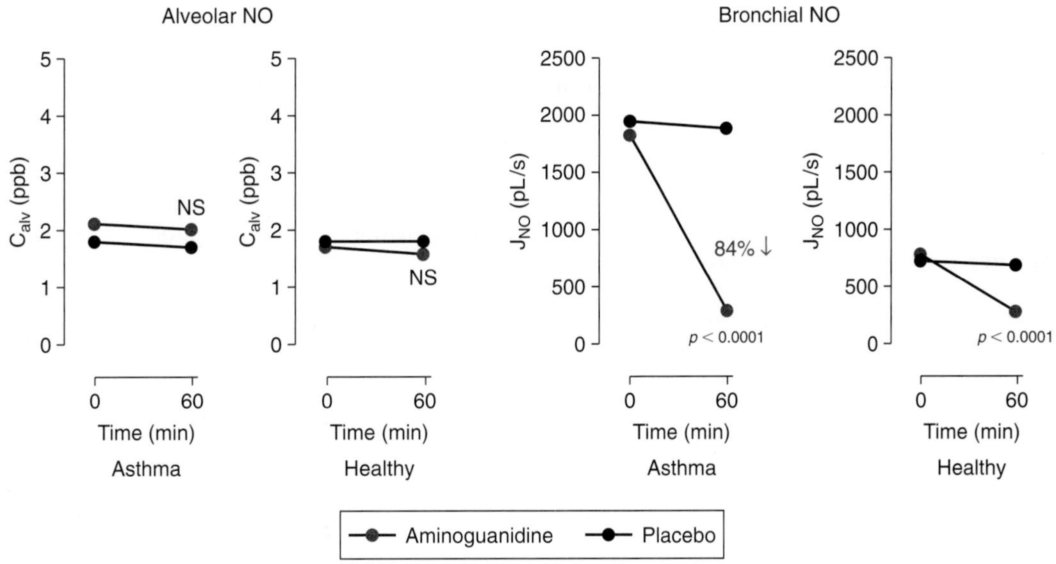

FIG. 30.3 Fast and "differential" effect of inhaled aminoguanidine versus placebo on bronchial NO and alveolar NO in asthma and healthy. From Ref. [84].

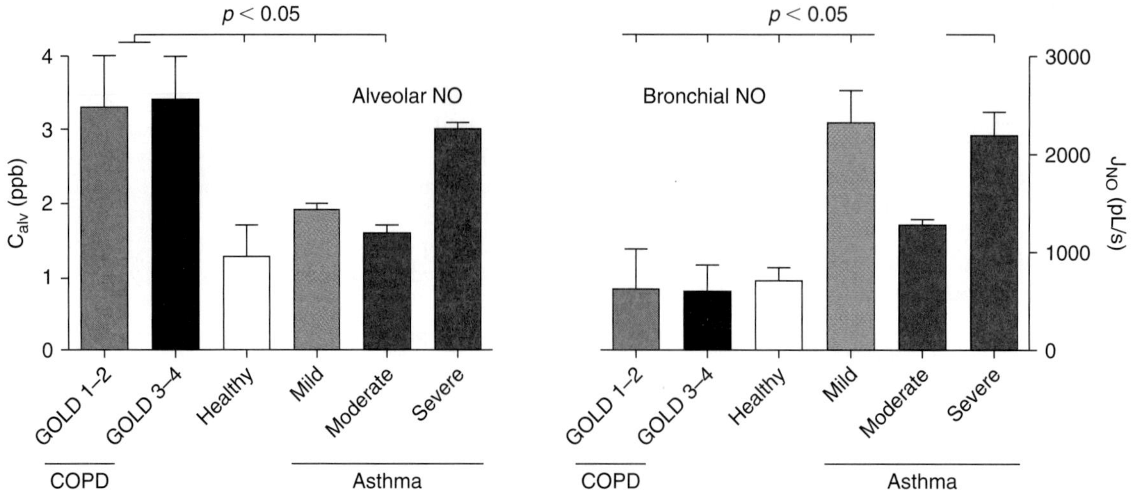

FIG. 30.4 Exhaled NO measured with the MEF technique: Alveolar versus bronchial inflammation is asthma and COPD of different stage and severity. From Ref. [84, 90].

responsible for the raised levels of NO in asthma, whereas the levels seen in normal subjects seem due to cNOS and a low level of iNOS activity.

COPD

Exhaled NO levels in stable COPD [19, 20] are lower than in either smoking or nonsmoking asthmatics [21] and are not different from normal subjects. Patients with unstable COPD, however, have high NO levels compared with stable smokers or ex-smokers with COPD [19], which are still lower than expired NO in nonsteroid-treated asthmatic subjects [85].

Data from experiments using multiple expiratory flows have demonstrated that the peripheral airways/alveolar region is the predominant source of elevated exhaled NO in COPD. In contrast, increased exhaled NO levels in asthma (Fig. 30.4) are mainly of larger airways/bronchial origin [86]. Prevalence of alveolar-derived NO in COPD is possibly related to the iNOS in macrophages, alveolar walls, and bronchial epithelium of COPD patients [87].

Disease severity and progression

Severity-related increase of the inferred concentration of NO in the alveolar zone, C_{alv}, may be due to higher presence of iNOS-positive cells in alveolar walls in more severe COPD patients [88], but not with severe emphysema that shows a lower percentage of iNOS-positive alveolar macrophages than patients with milder disease [89]. Patient with COPD rated as stage Global Initiative for Chronic Obstructive Lung Disease II (GOLD II) have C_{alv} levels [90] similar to those of more severe asthma patients [91] (Fig. 30.4) suggesting similar mechanisms may be responsible.

MONITORING OF SMALL AIRWAY INFLAMMATION IN BOTH COPD AND SEVERE ASTHMA

MEFT FE_{NO} measurements are highly reproducible, free of diurnal variation, unaffected by smoking, bronchodilator, or ICS [90] and therefore could be useful for COPD monitoring (Fig. 30.5).

Interestingly, the elevated levels of alveolar NO are similar in COPD [90] and severe asthma [91], supporting our previous finding of elevated exhaled FE_{NO} levels in patients with severe [92] and difficult steroid-resistant asthma [93].

Therefore, differential flow analysis of exhaled NO provides additional information about the sites of inflammation in asthma and COPD which are clearly different (Figs. 30.3, 30.4, 30.6), may be useful in assessing the response of peripheral inflammation to therapy, including its combination with iNOS inhibitors [94].

iNOS inhibitors

Under conditions of inflammation and oxidative stress, superoxide anion will preferentially react with available NO rather than its endogenous neutralizer superoxide dismutase, thus increasing the formation of peroxynitrite in tissues [95].

Diffusion of peroxynitrite through biomembranes can cause oxidative damage at one to two cell diameters from its site of formation [96] that may be of importance in COPD

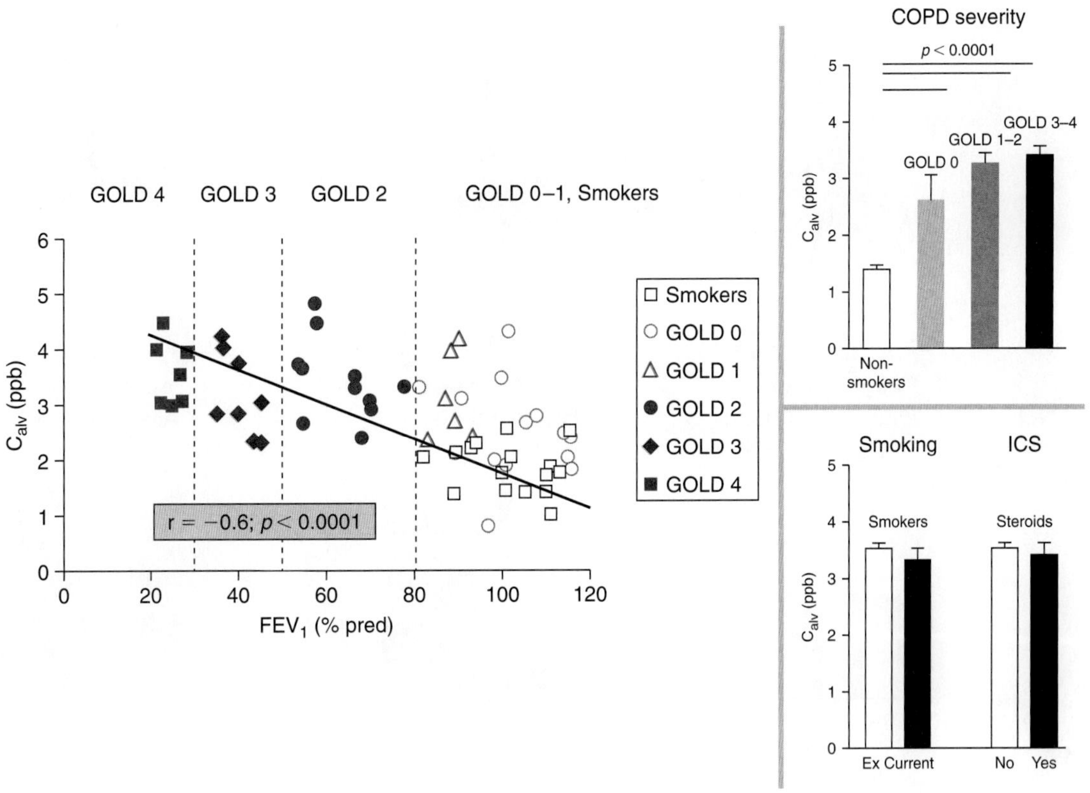

FIG. 30.5 Alveolar NO in COPD: relationship to the disease stage and severity, smoking and treatment with inhaled corticosteroids. ICS: inhaled corticosteroids. From Ref. [90].

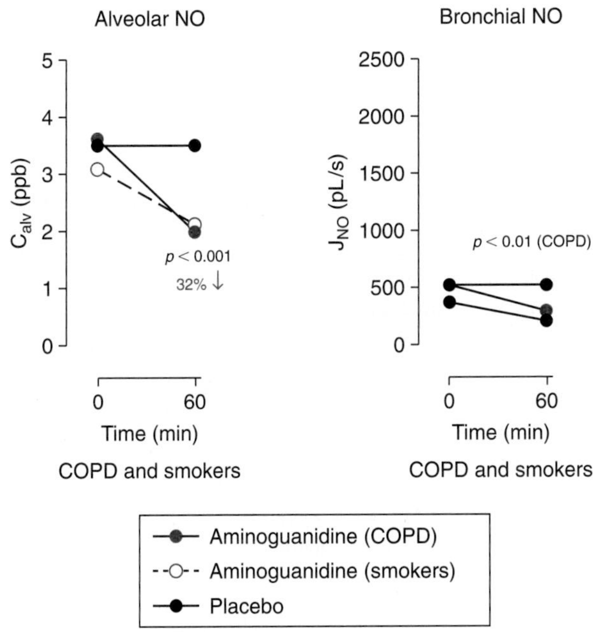

FIG. 30.6 Fast and "differential" effect of inhaled aminoguanidine versus placebo on bronchial and alveolar NO in COPD and healthy smokers. From Ref. [97].

with its high local NO production [90]. Therefore, the use of iNOS inhibitors by reducing "available" NO may restore the preferred pathway of superoxide radicals detoxification, *via* superoxide dismutase.

We have reported this "selective inhibition" by aminoguanidine versus placebo of bronchial NO flux (J_{NO}) in COPD patients but not in smokers or normal volunteers [97] (Fig. 30.6). Aminoguanidine also caused a decrease in alveolar NO (C_{alv}) in smokers and in COPD but not in healthy controls. These findings give potential new insights into the differences between asthma (Fig. 30.3) and COPD (Fig. 30.6).

SUMMARY

The precise inflammatory pathways and cells, NOS enzyme isoforms, and anatomical compartments responsible for production of the NO captured in the expirate are areas of active investigation (Fig. 30.7).

There are two interlinked and equally important future areas of NO research: exploring further potential and prevalent sources of NO and its role in different diseases and the use of NO in clinical medicine, including home measurements and routine asthma monitoring.

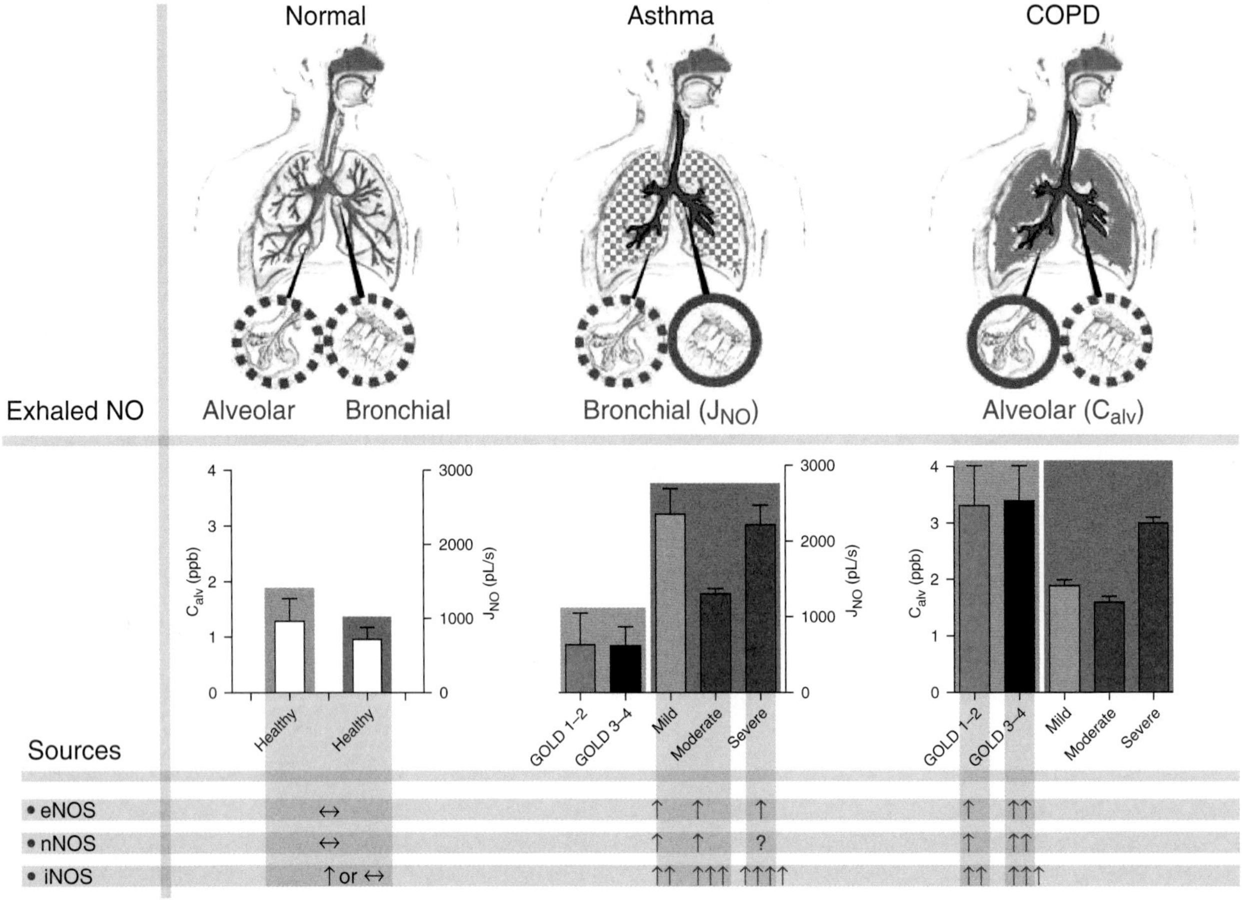

FIG. 30.7 Prevalent anatomical and enzymatic sources of NO in exhaled air of normal subjects and patients with asthma and COPD.

References

1. Palmer RM, Ferrige AG, Moncada S. Nitric oxide release accounts for the biological activity of endothelium-derived relaxing factor. *Nature* 327(6122): 524–26, 1987.
2. Ialenti A, Ianaro A, Moncada S, Di Rosa M. Modulation of acute inflammation by endogenous nitric oxide. *Eur J Pharmacol* 211(2): 177–82, 1992.
3. Guo FH, De Raeve HR, Rice TW, Stuehr DJ, Thunnissen FB, Erzurum SC. Continuous nitric oxide synthesis by inducible nitric oxide synthase in normal human airway epithelium *in vivo*. *Proc Natl Acad Sci USA* 92(17): 7809–13, 1995.
4. Wechsler ME, Grasemann H, Deykin A, Silverman EK, Yandava CN, Israel E, Wand M, Drazen JM. Exhaled nitric oxide in patients with asthma. Association with *NOS1* genotype. *Am J Respir Crit Care Med* 162: 2043–47, 2000.
5. Morris SM Jr, Billiar TR. New insights into the regulation of inducible nitric oxide synthesis. *Am. J Physiol* 266(6 Pt 1): E829–39, 1994.
6. Shaul PW, North AJ, Wu LC, Wells LB, Brannon TS, Lau KS, Michel T, Margraf LR, Star RA. Endothelial nitric oxide synthase is expressed in cultured human bronchiolar epithelium. *J Clin Invest* 94(6): 2231–36, 1994.
7. Pechkovsky DV, Zissel G, Stamme C, Goldmann T, Ari JH, Einhaus M, Taube C, Magnussen H, Schlaak M, Muller-Quernheim J. Human alveolar epithelial cells induce nitric oxide synthase-2 expression in alveolar macrophages. *Eur Respir J* 19(4): 672–83, 2002.
8. Pechkovsky DV, Zissel G, Goldmann T, Einhaus M, Taube C, Magnussen H, Schlaak M, Muller-Quernheim J. Pattern of NOS2 and NOS3 mRNA expression in human A549 cells and primary cultured AEC II. *Am J Physiol Lung Cell Mol Physiol* 282(4): L684–92, 2002.
9. Fischer A, Hoffmann B. Nitric oxide synthase in neurons and nerve fibers of lower airways and in vagal sensory ganglia of man. Correlation with neuropeptides. *Am J Respir Crit Care Med* 154(1): 209–16, 1996.
10. Widdicombe JG. Autonomic regulation. i-NANC/e-NANC. *Am J Respir Crit Care Med* 158(5 Pt 3): S171–75, 1998.
11. Luhrs H, Papadopoulos T, Schmidt HH, Menzel T. Type I nitric oxide synthase in the human lung is predominantly expressed in capillary endothelial cells. *Respir Physiol* 129(3): 367–74, 2002.
12. De Sanctis GT, Mehta S, Kobzik L, Yandava C, Jiao A, Huang PL, Drazen JM. Contribution of type I NOS to expired gas NO and bronchial responsiveness in mice. *Am. J Physiol* 273(4 Pt 1): L883–88, 1997.
13. Jang AS, Choi IS, Lee JU, Park SW, Lee JH, Park CS. Changes in the expression of NO synthase isoforms after ozone: The effects of allergen exposure. *Respir Res* 5: 5, 2004.
14. Hamid Q, Springall DR, Riveros-Moreno V, Chanez P, Howarth PH, Redington A, Bousquet J, Godard P, Holgate S, Polak JM. Induction of nitric oxide synthase in asthma. *Lancet* 342(8886-8887): 1510–13, 1993.
15. Saleh D, Ernst P, Lim S, Barnes PJ, Giaid A. Increased formation of the potent oxidant peroxynitrite in the airways of asthmatic patients is associated with induction of nitric oxide synthase: Effect of inhaled glucocorticoid. *FASEB J* 12(11): 929–37, 1998.

16. Kharitonov SA, Yates DH, Barnes PJ. Inhaled glucocorticoids decrease nitric oxide in exhaled air of asthmatic patients. *Am J Respir Crit Care Med* 153(1): 454–57, 1996.
17. Kharitonov SA, Yates DH, Robbins RA, Logan-Sinclair R, Shinebourne EA, Barnes PJ. Increased nitric oxide in exhaled air of asthmatic patients. *Lancet* 343(8890): 133–35, 1994.
18. Kharitonov SA, Yates DH, Chung KF, Barnes PJ. Changes in the dose of inhaled steroid affect exhaled nitric oxide levels in asthmatic patients. *Eur Respir J* 9: 196–201, 1996.
19. Maziak W, Loukides S, Culpitt SV, Sullivan P, Kharitonov SA, Barnes PJ. Exhaled nitric oxide in chronic obstructive pulmonary disease. *Am J Respir Crit Care Med* 157(3 Pt 1): 998–1002, 1998.
20. Robbins RA, Floreani AA, Von Essen SG, Sisson JH, Hill GE, Rubinstein I, Townley R. Measurement of exhaled nitric oxide by three different techniques. *Am J Respir Crit Care Med* 153: 1631–35, 1996.
21. Verleden GM, Dupont LJ, Verpeut AC, Demedts MG. The effect of cigarette smoking on exhaled nitric oxide in mild steroid- naive asthmatics. *Chest* 116(1): 59–64, 1999.
22. Su Y, Han W, Giraldo C, De Li Y, Block ER. Effect of cigarette smoke extract on nitric oxide synthase in pulmonary artery endothelial cells. *Am J Respir Crit Care Med* 19: 819–25, 1998.
23. Kharitonov SA, Robbins RA, Yates D, Keatings V, Barnes PJ. Acute and chronic effects of cigarette smoking on exhaled nitric oxide. *Am J Respir Crit Care Med* 152(2): 609–12, 1995.
24. Rutgers SR, van der Mark TW, Coers W, Moshage H, Timens W, Kauffman HF, Ko Postma,DS. Markers of nitric oxide metabolism in sputum and exhaled air are not increased in chronic obstructive pulmonary disease. *Thorax* 54(7): 576–80, 1999.
25. Eiserich JP, Hristova M, Cross CE, Jones AD, Freeman BA, Halliwell B, van-der VA. Formation of nitric oxide-derived inflammatory oxidants by myeloperoxidase in neutrophils. *Nature* 391(6665): 393–97, 1998.
26. Gunther MR, Sturgeon BE, Mason RP. Nitric oxide trapping of the tyrosyl radical-chemistry and biochemistry. *Toxicology* 177(1): 1–9, 2002.
27. Davis KL, Martin E, Turko IV, Murad F. Novel effects of nitric oxide. *Ann Rev Pharmacol Toxicol* 41: 203–36, 2001.
28. Eiserich JP, Baldus S, Brennan ML, Ma W, Zhang C, Tousson A, Castro L, Lusis AJ, Nauseef WM, White CR, Freeman BA. Myeloperoxidase, a leukocyte-derived vascular NO oxidase. *Science* 296(5577): 2391–94, 2002.
29. Kuo WN, Kocis JM. Nitration/S-nitrosation of proteins by peroxynitrite-treatment and subsequent modification by glutathione S-transferase and glutathione peroxidase. *Mol. Cell Biochem.* 233(1-2): 57–63, 2002.
30. Aulak KS, Miyagi M, Yan L, West KA, Massillon D, Crabb JW, Stuehr DJ. Proteomic method identifies proteins nitrated *in vivo* during inflammatory challenge. *Proc Natl Acad Sci USA* 98(21): 12056–61, 2001.
31. Turko IV, Murad F. Protein nitration in cardiovascular diseases. *Pharmacol Rev* 54(4): 619–34, 2002.
32. Berlett BS, Friguet B, Yim MB, Chock PB, Stadtman ER. Peroxynitrite-mediated nitration of tyrosine residues in Escherichia coli glutamine synthetase mimics adenylylation: Relevance to signal transduction. *Proc Natl Acad Sci USA* 93(5): 1776–80, 1996.
33. Lennon CW, Cox HD, Hennelly SP, Chelmo SJ, McGuirl MA. Probing structural differences in prion protein isoforms by tyrosine nitration. *Biochemistry* 46(16): 4850–60, 2007.
34. Salvemini D, Doyle TM, Cuzzocrea S. Superoxide, peroxynitrite and oxidative/nitrative stress in inflammation. *Biochem Soc Trans* 34(Pt 5): 965–70, 2006.
35. Brown GC, Borutaite V. Nitric oxide, mitochondria, and cell death. *IUBMB Life* 52(3-5): 189–95, 2001.
36. Kaminsky DA, Mitchell J, Carroll N, James A, Soultanakis R, Janssen Y. Nitrotyrosine formation in the airways and lung parenchyma of patients with asthma. *J Allergy Clin Immunol* 104(4 Pt 1): 747–54, 1999.
37. Ichinose M, Sugiura H, Yamagata S, Koarai A, Shirato K. Increase in reactive nitrogen species production in chronic obstructive pulmonary disease airways. *Am J Respir Crit Care Med* 162(2): 701–6, 2000.
38. Ricciardolo FL, Caramori G, Ito K, Capelli A, Brun P, Abatangelo G, Papi A, Chung KF, Adcock I, Barnes PJ, Donner CF, Rossi A, Di Stefano A. Nitrosative stress in the bronchial mucosa of severe chronic obstructive pulmonary disease. *J Allergy Clin Immunol* 116(5): 1028–35, 2005.
39. Harju T, Kaarteenaho-Wiik R, Soini Y, Sormunen R, Kinnula VL. Diminished immunoreactivity of gamma-glutamylcysteine synthetase in the airways of smokers' lung. *Am J Respir Crit Care Med* 166(5): 754–59, 2002.
40. Yang S, Milla C, Panoskaltsis-Mortari A, Ingbar DH, Blazar BR, Haddad IY. Human surfactant protein a suppresses T cell-dependent inflammation and attenuates the manifestations of idiopathic pneumonia syndrome in mice. *Am J Respir Cell Mol Biol* 24(5): 527–36, 2001.
41. Souza JM, Choi I, Chen Q, Weisse M, Daikhin E, Yudkoff M, Obin M, Ara J, Horwitz J, Ischiropoulos H. Proteolytic degradation of tyrosine nitrated proteins. *Arch Biochem Biophys* 380(2): 360–66, 2000.
42. Kamisaki Y, Wada K, Bian K, Balabanli B, Davis K, Martin E, Behbod F, Lee YC, Murad F. An activity in rat tissues that modifies nitrotyrosine-containing proteins. *Proc Natl Acad Sci USA* 95(20): 11584–89, 1998.
43. Blanchard-Fillion B, Souza JM, Friel T, Jiang GC, Vrana K, Sharov V, Barron L, Schoneich C, Quijano C, Alvarez B, Radi R, Przedborski S, Fernando GS, Horwitz J, Ischiropoulos H. Nitration and inactivation of tyrosine hydroxylase by peroxynitrite. *J Biol Chem* 276(49): 46017–23, 2001.
44. Souza JM, Chen Q, Blanchard-Fillion B, Lorch SA, Hertkorn C, Lightfoot R, Weisse M, Friel T, Paxinou E, Themistocleous M, Chov S, Ischiropoulos H. Reactive nitrogen species and proteins: Biological significance and clinical relevance. *Adv Exp Med Biol* 500: 169–74, 2001.
45. Hope BT, Nagarkar D, Leonard S, Wise RA. Long-term upregulation of protein kinase A and adenylate cyclase levels in human smokers. *J Neurosci* 27(8): 1964–72, 2007.
46. Haqqani AS, Kelly JF, Birnboim HC. Selective nitration of histone tyrosine residues *in vivo* in mutatect tumors. *J Biol Chem* 277(5): 3614–21, 2002.
47. Hanazawa T, Kharitonov SA, Barnes PJ. Increased nitrotyrosine in exhaled breath condensate of patients with asthma. *Am J Respir Crit Care Med* 162(4 Pt 1): 1273–76, 2000.
48. Ito K, Adcock IM. Histone acetylation and histone deacetylation. *Mol Biotechnol* 20(1): 99–106, 2002.
49. Ito K, Caramori G, Lim S, Oates T, Chung KF, Barnes PJ, Adcock IM. Expression and activity of histone deacetylases in human asthmatic airways. *Am J Respir Crit Care Med* 166(3): 392–96, 2002.
50. Ito K, Hanazawa T, Tomita K, Barnes PJ, Adcock IM. Oxidative stress reduces histone deacetylase 2 activity and enhances IL-8 gene expression: Role of tyrosine nitration. *Biochem Biophys Res Commun* 315(1): 240–45, 2004.
51. Marwick JA, Kirkham PA, Stevenson CS, Danahay H, Giddings J, Butler K, Donaldson K, MacNee W, Rahman I. Cigarette smoke alters chromatin remodeling and induces proinflammatory genes in rat lungs. *Am J Respir Cell Mol Biol* 31(6): 633–42, 2004.
52. Gole MD, Souza JM, Choi I, Hertkorn C, Malcolm S, Foust RF III, Finkel B, Lanken PN, Ischiropoulos H. Plasma proteins modified by tyrosine nitration in acute respiratory distress syndrome. *Am J Physiol Lung Cell Mol. Physiol* 278(5): L961–67, 2000.
53. Reinheckel T, Sitte N, Ullrich O, Kuckelkorn U, Davies KJ, Grune T. Comparative resistance of the 20S and 26S proteasome to oxidative stress. *Biochem J* 335(Pt 3): 637–42, 1998.
54. Gow AJ, Duran D, Malcolm S, Ischiropoulos H. Effects of peroxynitrite-induced protein modifications on tyrosine phosphorylation and degradation. *FEBS Lett* 385(1-2): 63–66, 1996.
55. Irie Y, Saeki M, Kamisaki Y, Martin E, Murad F. Histone H1.2 is a substrate for denitrase, an activity that reduces nitrotyrosine immunoreactivity in proteins. *Proc Natl Acad Sci USA* 100(10): 5634–39, 2003.
56. Kharitonov SA, Alving K, Barnes PJ. Exhaled and nasal nitric oxide measurements: Recommendations. *Eur Respir J* 10: 1683–93, 1997.
57. ATS/ERS. Recommendations for Standardized Procedures for the Online and Offline Measurement of Exhaled Lower Respiratory

Nitric Oxide and Nasal Nitric Oxide, 2005. *Am J Respir Crit Care Med* 171(8): 912–30, 2005.
58. Silkoff PE, McClean PA, Slutsky AS, Furlott HG, Hoffstein E, Wakita S, Chapman KR, Szalai JP, Zamel N. Marked flow-dependence of exhaled nitric oxide using a new technique to exclude nasal nitric oxide. *Am J Respir Crit Care Med* 155(1): 260–67, 1997.
59. Kharitonov SA, Barnes PJ. Nasal contribution to exhaled nitric oxide during exhalation against resistance or during breath holding. *Thorax* 52(6): 540–44, 1997.
60. Kharitonov SA, Gonio F, Kelly C, Meah S, Barnes PJ. Reproducibility of exhaled nitric oxide measurements in healthy and asthmatic adults and children. *Eur Respir J* 21(3): 433–38, 2003.
61. Kharitonov SA, Barnes PJ. Exhaled markers of pulmonary disease. *Am J Respir Crit Care Med* 163(7): 1693–722, 2001.
62. Kharitonov SA. Exhaled markers of inflammatory lung diseases: Ready for routine monitoring?. *Swiss Med Wkly* 134(13-14): 175–92, 2004.
63. Lehtimaki L, Turjanmaa V, Kankaanranta H, Saarelainen S, Hahtola P, Moilanen E. Increased bronchial nitric oxide production in patients with asthma measured with a novel method of different exhalation flow rates. *Ann Med* 32(6): 417–23, 2000.
64. Lehtimaki L, Kankaanranta H, Saarelainen S, Hahtola P, Jarvenpaa R, Koivula T, Turjanmaa V, Moilanen E. Extended exhaled NO measurement differentiates between alveolar and bronchial inflammation. *Am J Respir Crit Care Med* 163(7): 1557–61, 2001.
65. Silkoff PE, Sylvester JT, Zamel N, Permutt S. Airway Nitric Oxide Diffusion in Asthma. Role in pulmonary function and bronchial responsiveness. *Am J Respir Crit Care Med* 161(4 Part 1): 1218–28, 2000.
66. Lehtimaki L, Kankaanranta H, Saarelainen S, Turjanmaa V, Moilanen E. Increased alveolar nitric oxide concentration in asthmatic patients with nocturnal symptoms. *Eur Respir J* 20(4): 841–45, 2002.
67. George SC, Hogman M, Permutt S, Silkoff PE. Modeling pulmonary nitric oxide exchange. *J Appl Physiol* 96(3): 831–39, 2004.
68. Vahlkvist S, Sinding M, Skamstrup K, Bisgaard H. Daily home measurements of exhaled nitric oxide in asthmatic children during natural birch pollen exposure. *J Allergy Clin Immunol* 117(6): 1272–76, 2006.
69. Gill M, Graff GR, Adler AJ, Dweik RA. Validation study of fractional exhaled nitric oxide measurements using a handheld monitoring device. *J Asthma* 43(10): 731–34, 2006.
70. Kharitonov SA, Chung FK, Evans DJ, O'Connor BJ, Barnes PJ. The elevated level of exhaled nitric oxide in asthmatic patients is mainly derived from the lower respiratory tract. *Am J Respir Crit Care Med* 153: 1773–80, 1996.
71. Massaro AF, Gaston B, Kita D, Fanta C, Stamler JS, Drazen JM. Expired nitric oxide levels during treatment of acute asthma. *Am J Respir Crit Care Med* 152(2): 800–3, 1995.
72. Dupont LJ, Demedts MG, Verleden GM. Prospective evaluation of the accuracy of exhaled nitric oxide for the diagnosis of asthma. *Am J Respir Crit Care Med* 159: A861, 1999.
73. Dupont LJ, Demedts MG, Verleden GM. Prospective evaluation of the validity of exhaled nitric oxide for the diagnosis of asthma. *Chest* 123(3): 751–56, 2003.
74. Olin AC, Bake B, Toren K. Fraction of exhaled nitric oxide at 50 mL/s: Reference values for adult lifelong never-smokers. *Chest* 131(6): 1852–56, 2007.
75. Smith AD, Cowan JO, Filsell S, McLachlan C, Monti-Sheehan G, Jackson P, Taylor DR. Diagnosing Asthma: Comparisons between Exhaled Nitric Oxide Measurements and Conventional Tests. *Am J Respir Crit Care Med* 169(4): 473–78, 2004.
76. Kharitonov SA, Barnes PJ. Does Exhaled Nitric Oxide Reflect Asthma Control?. Yes, it does!. *Am J Respir Crit Care Med* 164(5): 727–28, 2001.
77. Jones SL, Kittelson J, Cowan JO, Flannery EM, Hancox RJ, McLachlan CR, Taylor DR. The predictive value of exhaled nitric oxide measurements in assessing changes in asthma control. *Am J Respir Crit Care Med* 164(5): 738–43, 2001.
78. Kharitonov SA. Exhaled nitric oxide and carbon monoxide in asthma. *Eur Respir J* 9(68): 212–18, 1999.
79. Jatakanon A, Lim S, Barnes PJ. Changes in sputum eosinophils predict loss of asthma control. *Am J Respir Crit Care Med* 161(1): 64–72, 2000.
80. Meyts I, Proesmans M, De Boeck K. Exhaled nitric oxide corresponds with office evaluation of asthma control. *Pediatr Pulmonol* 36(4): 283–89, 2003.
81. Kharitonov SA, Barnes PJ, O'Connor BJ. Reduction in exhaled nitric oxide after a single dose of nebulised budesonide in patients with asthma. *Am J Respir Crit Care Med* 153: A799, 1996.
82. Kharitonov SA, Donnelly LE, Montuschi P, Corradi M, Collins JV, Barnes PJ. Dose-dependent onset and cessation of action of inhaled budesonide on exhaled nitric oxide and symptoms in mild asthma. *Thorax* 57(10): 889–96, 2002.
83. Shaw DE, Berry MA, Thomas M, Green RH, Brightling CE, Wardlaw AJ, Pavord ID. The use of exhaled nitric oxide to guide asthma management: A randomized controlled trial. *Am J Respir Crit Care Med* 176(3): 231–37, 2007.
84. Brindicci C, Ito K, Barnes PJ, Kharitonov SA. Effect of an inducible nitric oxide synthase inhibitor on differential flow exhaled nitric oxide in asthmatic patients and healthy volunteers. *Chest*, 2007.
85. Massaro AF, Mehta S, Lilly CM, Kobzik L, Reilly JJ, Drazen JM. Elevated nitric oxide concentrations in isolated lower airway gas of asthmatic subjects. *Am J Respir Crit Care Med* 153: 1510–14, 1996.
86. Brindicci C, Cosio B, Gajdocsi R, Collins JV, Bush A, Abdallah S, Barnes PJ, Kharitonov SA. Extended exhaled NO measurements at different exhalation flows may differentiate between bronchial and alveolar inflammation in patients with asthma and COPD. *Eur Respir J* 20: 174s, 2002.
87. Paska C, Maestrelli P, Formichi B, Monti S, Baldi S, Miniati M, Saetta M, Fabbri LM. Increased expression of inducible NOS in peripheral lung of severe COPD patients. *Eur Respir J* 20: 95s, 2002.
88. Maestrelli P, Paska C, Saetta M, Turato G, Nowicki Y, Monti S, Formichi B, Miniati M, Fabbri LM. Decreased haem oxygenase-1 and increased inducible nitric oxide synthase in the lung of severe COPD patients. *Eur Respir J* 21(6): 971–76, 2003.
89. van Straaten JF, Postma DS, Coers W, Noordhoek JA, Kauffman HF, Timens W. Macrophages in lung tissue from patients with pulmonary emphysema express both inducible and endothelial nitric oxide synthase. *Mod Pathol* 11(7): 648–55, 1998.
90. Brindicci C, Ito K, Resta O, Pride NB, Barnes PJ, Kharitonov SA. Exhaled nitric oxide from lung periphery is increased in COPD. *Eur Respir J* 26(1): 52–59, 2005.
91. Brindicci C, Ito K, Barnes PJ, Kharitonov SA. Differential flow analysis of exhaled nitric oxide in patients with asthma of differing severity. *Chest* 131(5): 1353–62, 2007.
92. Jatakanon A, Lim S, Kharitonov SA, Chung KF, Barnes PJ. Correlation between exhaled nitric oxide, sputum eosinophils, and methacholine responsiveness in patients with mild asthma. *Thorax* 53(2): 91–95, 1998.
93. Stirling RG, Kharitonov SA, Campbell D, Robinson D, Durham SR, Chung KF, Barnes PJ. Exhaled NO is elevated in difficult asthma and correlates with symptoms and disease severity despite treatment with oral and inhaled corticosteroids. *Thorax* 53: 1030–34, 1998.
94. Kharitonov SA, Barnes PJ. Nitric oxide, nitrotyrosine, and nitric oxide modulators in asthma and chronic obstructive pulmonary disease. *Curr Allergy Asthma Rep* 3(2): 121–29, 2003.
95. Cooke CL, Davidge ST. Peroxynitrite increases iNOS through NF-kappaB and decreases prostacyclin synthase in endothelial cells. *Am J Physiol Cell Physiol* 282(2): C395–402, 2002.
96. Szabo C, Ischiropoulos H, Radi R. Peroxynitrite: Biochemistry, pathophysiology and development of therapeutics. *Nat Rev Drug Discov* 6: 662–80, 2007.
97. Brindicci C, Torre O, Barnes PJ, Ito K, Kharitonov SA. Effects of aminoguanidine, an inhibitor of inducible nitric oxide synthase, on nitric oxide production and its metabolites in healthy controls, healthy smokers and COPD patients. *Chest*, 2008.

31
CHAPTER

Transcription Factors

Ian M. Adcock and Gaetano Caramori

National Heart and Lung Institute, Imperial College School of Medicine, London, UK

INTRODUCTION

Inflammation is a central feature of asthma and chronic obstructive pulmonary disease (COPD). The specific characteristics of the inflammatory response in each disease and the site of inflammation differ, but both involve the recruitment and activation of inflammatory cells and changes in the structural cells of the lung [1]. Asthma and COPD are characterized by an increased expression of components of the inflammatory cascade. These inflammatory proteins include cytokines, chemokines, growth factors, enzymes, receptors, and adhesion molecules [2]. The increased expression of these proteins seen in asthma and COPD is the result of enhanced gene transcription since many of the genes are not expressed in normal cells but are induced in a cell-specific manner during the inflammatory process [2].

Changes in gene transcription are regulated by transcription factors, which are proteins that bind to DNA and modulate the transcriptional apparatus. Transcription factors regulate the expression of many genes, including inflammatory genes, and may play a key role in the pathogenesis of asthma and COPD since they regulate the increased gene expression that may underlie the acute and chronic inflammatory mechanisms that characterize these diseases [1, 3]. Transcription factors may amplify and perpetuate the inflammatory process, so it is possible that abnormal functioning of transcription factors may determine disease severity and response to treatment. The increased understanding of the role of transcription factors in the pathogenesis of asthma and COPD has also opened an opportunity for the development of new potential anti-inflammatory drugs [4].

Several new compounds based on interacting with specific transcription factors or their activation pathways are now in development for the treatment of asthma and COPD, and some drugs already in clinical use (such as glucocorticoids) are thought to work via transcription factors. Glucocorticoids are effective therapy in the long-term control of asthma. One of their mechanisms of action in asthmatic airways is by inhibiting the action of transcription factors that regulate inflammatory gene expression [5]. Glucocorticoids may also have a beneficial effect in COPD during exacerbations, although this is less marked than in asthma, indicating that different genes and transcription factors are involved and emphasizing the importance of cell-specific transcription factors.

One concern about this approach is the specificity of such drugs, but it is clear that transcription factors have selective effects on the expression of certain genes and this may make it possible to be more selective. In addition, there are cell-specific transcription factors that may be targeted for inhibition, which could provide selectivity of drug action. One such example is GATA3, which has been reported to have a restricted cellular distribution [6]. In asthma and COPD it may be possible to target drugs to the airways by inhalation, as is currently performed for inhaled glucocorticoids, to systemic effects.

Despite the fact that many transcription factors have now been discovered there is still a paucity of data concerning the regulation of transcription factors in the human lungs. This chapter briefly reviews the physiological function of the transcription factors in the normal cells and the role of some transcription factors that may be relevant in the pathogenesis of asthma and COPD.

TRANSCRIPTION FACTORS

Transcription factors are proteins that bind to DNA-regulatory sequences (enhancers and silencers), usually localized in the 5′-upstream region of target genes, to modulate the rate of gene transcription. This may result in increased or decreased gene transcription, protein synthesis, and subsequent altered cellular function. Many transcription factors have now been identified and a large proportion of the human genome appears to code for these proteins.

Several families of transcription factors exist and members of each family may share structural characteristics. These families include

- helix-turn-helix (e.g. Oct-1),
- helix-loop-helix (e.g. E2A),
- zinc finger (e.g. glucocorticoid receptors, GATA proteins),
- basic protein-leucine zipper [cyclic AMP response element-binding factor (CREB), activator protein-1 (AP-1)],
- β-sheet motifs [e.g. nuclear factor-κB (NF-κB)] [4, 7] (Table 31.1).

Many transcription factors are common to several cell types (ubiquitous), such as AP-1 and NF-κB, and may play a general role in the regulation of inflammatory genes, whereas others are cell-specific and may determine the phenotypic characteristics of a cell. Transcription factor activation is complex and may involve multiple intracellular signal transduction pathways, including the kinases PKA, MAPKs, JAKs, and PKCs, stimulated by cell-surface receptors [8, 9]. Transcription factors may also be directly activated by ligands such as glucocorticoids and vitamins A and D [5]. Transcription factors may therefore convert transient environmental signals at the cell surface into long-term changes in gene transcription, thus acting as "nuclear messengers." Transcription factors may be activated within the nucleus, often with the transcription factor already bound to DNA, or within the cytoplasm, resulting in exposure of nuclear localization signals and targeting to the nucleus [5]. It is becoming clear that post-translational modifications of transcription factors by phosphorylation, acetylation, and nitration can have profound effects on either the DNA-binding capacity or the transcriptional activity of a transcription factor [10]. This is exemplified by changes in p65 NF-κB phosphorylation and acetylation which alter its association with transcriptional coactivator and repressor complexes resulting in changes in gene expression profile [11–13].

TABLE 31.1 Transcription factor families involved in the pathogenesis of asthma and COPD.

GR	NFATs
NF-κB	GATA
AP-1	HIF-1
CREB	C/EBP
STATs	SP1

GR: glucocorticoids receptor; NF-κB: nuclear factor-kappa B; AP-1: activator protein 1; CREB: cyclic AMP response element binding protein; STATs: signal transducers and activators of transcription; NFATs: nuclear factors of activated T cells; GATA: GATA binding proteins; HIF-1: hypoxia-inducible factor-1; C/EBP: CAAT/enhancer-binding protein; SP1: specificity protein 1.

Cross-talk

One of the most important concepts to have emerged is the demonstration that transcription factors may physically interact with each other to form homodimers or heterodimers, resulting in inhibition or enhancement of transcriptional activity at a site distinct from the consensus target for a particular transcription factor (Fig. 31.1). This then allows cross-talk between different signal transduction pathways at the level of gene expression. Generally it is necessary to have coincident activation of several transcription factors in order to have maximal gene expression. This may explain how transcription factors that are ubiquitous may regulate particular genes in certain types of cells [4, 10].

The complexity of the activation pathways and their ability to engage in cross-talk enables cells to overcome inhibition of one pathway and retain a capacity to activate specific transcription factors. This cross-talk and redundancy may also hinder the search for novel anti-inflammatory agents targeted toward transcription factor activation. Binding of transcription factors to their specific binding motifs in the promoter region may alter transcription by interacting directly with components of the basal transcription apparatus or via cofactors that link the transcription factor to the basal transcription apparatus [14]. Large proteins that bind to the basal transcription apparatus bind many transcription factors and thus act as integrators of gene transcription. These coactivator molecules include CREB-binding protein (CBP), and the related p300, thus allowing complex interactions between different signaling pathways [14].

Histone acetylation

DNA is wound around histone proteins to form nucleosomes and the chromatin fiber in chromosomes [15]. It has long been recognized at a microscopic level that chromatin may become dense or opaque owing to the winding or unwinding of DNA around the histone core [16]. CBP, p300, and other coactivators have histone acetylase activity (HAT) which is activated by the binding of transcription factors, such as AP-1, NF-κB, and STATs (Fig. 31.2) [14, 15]. Acetylation of histone residues results in unwinding of DNA coiled around the histone core, thus opening up the chromatin structure, allowing increased transcription. Histone deacetylation by specific histone deacetylases (HDACs) reverses this process, leading to gene repression [17].

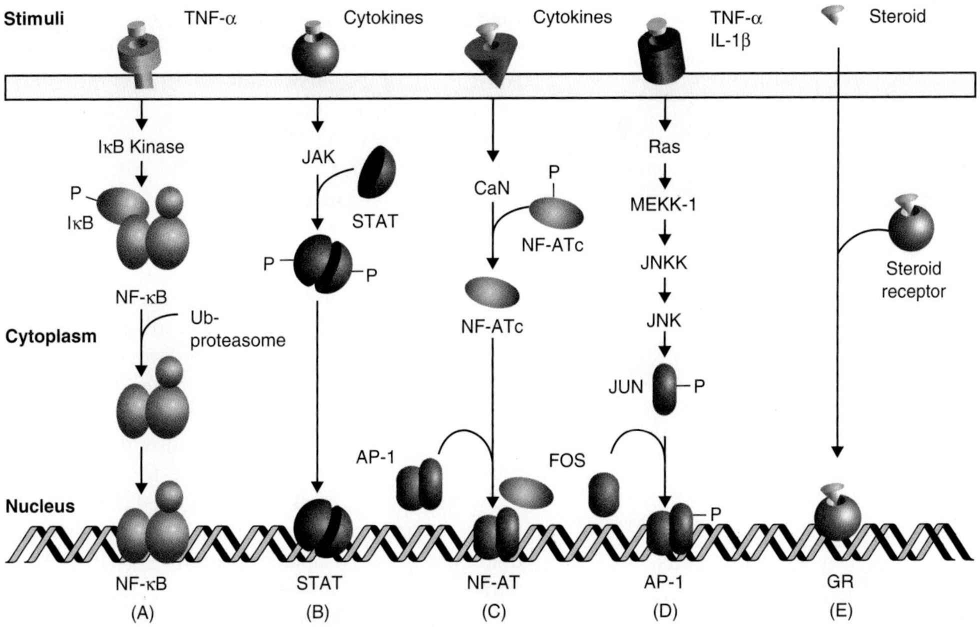

FIG. 31.1 Multiple pathways mediating transcription factor modulation of inflammatory genes, (A) Inflammatory mediator signal transduction activation. The binding of cytokines, growth factors, or chemokines to their respective receptors sets in train the activation of a number of signal transduction pathways, including the receptor tyrosine kinases, mitogen-activated protein kinases (MAPKs including MEKK1 and JNK), Janus kinases (JAKs), and other kinase pathways involved in NF-κB activation. Activation of nuclear factor-κB involves phosphorylation of the inhibitory protein IκB by specific kinase(s), with subsequent ubiquitination and proteolytic degradation by the proteasome. The free NF-κB then translocates to the nucleus, where it binds to κB sites in the promoter regions of inflammatory genes. Activation of the IκB gene results in increased synthesis of IκB to terminate the activation NF-κB. (B) JAK-STAT pathways. Cytokine binding to its receptor results in activation of JAK which phosphorylate intracellular domains of the receptor, resulting in phosphorylation of signal transduction-activated transcription factors (STATs). Activated STATs dimerize and translocate to the nucleus where they bind to recognition elements on certain genes. (C and D) Nuclear factor of activated T-cells (NF-AT) is activated via dephosphorylation by calcineurin (CaN) and translocates to the nucleus where it interacts with AP-1 to induce gene transcription. (E) Classical mechanism of steroid action. Glucocorticoids are lipophilic molecules which diffuse readily through cell membranes to interact with cytoplasmic receptors. Upon ligand binding receptors are activated and translocate into the nucleus where they bind to specific DNA elements. The foregoing pathways can interact so that the final signal may be amplified or altered depending upon the exact combination of stimuli. The final response to each stimulus or combination of stimuli by a particular cell depends upon the receptors present in a particular cell along with the exact intracellular transduction pathway activated.

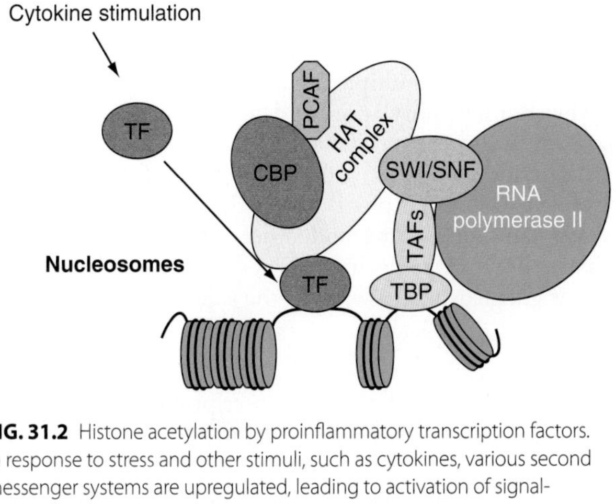

FIG. 31.2 Histone acetylation by proinflammatory transcription factors. In response to stress and other stimuli, such as cytokines, various second messenger systems are upregulated, leading to activation of signal-dependent transcription factors such as CREB, NF-κB, AP-1, and STAT proteins. Binding of these factors leads to recruitment of CBP and/or other coactivators to signal-dependent promoters and acetylation of histones by an intrinsic acetylase activity (HAT). Induction of histone acetylation allows the formation of a more loosely packed nucleosome structure which enables access to TATA box-binding protein (TBP) and associated factors (TAFs) and the recruitment of further remodeling factors including switch/sucrose nonfermentable (SWI/SNF). Remodeling thereby allows RNA polymerase II recruitment and the activation of inflammatory gene transcription.

TRANSCRIPTION FACTORS IN ASTHMA

Asthma is a complex chronic inflammatory disease of the airways that involves the activation of many inflammatory and structural cells, all of which release inflammatory mediators that result in the typical clinicopathological changes of asthma. The chronic airway inflammation of asthma is unique in that the airway wall is infiltrated by T-lymphocytes of the T-helper (Th) type 2 phenotype, eosinophils, macrophages/monocytes, and mast cells. In addition, an "acute-on-chronic" inflammation may be observed during exacerbations, with an increase in eosinophils and sometimes also neutrophils [5].

The role of transcription factors in differentiation of Th1/Th2 cells

$CD4^+$ T-helper (Th) cells can be divided into three major subsets, termed Th1, Th2, and Th0 based on the pattern of cytokines they produce [18]. Th1 cells produce predominantly IL-2 and interferon gamma (IFN-γ) and predominantly promote cell-mediated immune responses. Th2 cells, which

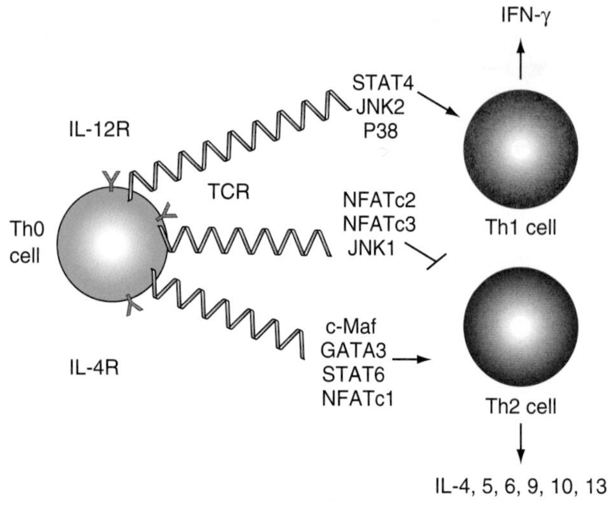

FIG. 31.3 Differentiation of T-cell subtypes depends upon the action of a number of transcription factors. Th0 cells can differentiate into Th1 and Th2 cells following IL-12 or IL-4 stimulation. Stimulation through the IL-12 receptor (IL-12) activates the transcription factor STAT4 and the MAPK pathways JNK2 and p38 and drives Th1 cell differentiation, leading to cells capable of releasing IFN-γ. In contrast, activation of the IL-4R in concert with T-cell receptor (TCR) stimulation leads to activation of the transcription factors STAT6, GATA3, c-Maf, and NFATc1, and drives Th2 cell differentiation. Th2 cells can release a number of cytokines including interleukins 4, 5, 6, 9, 10, and 13. In addition, activation of JNK1, NFATc2, and NFATc3 can prevent Th2 cell differentiation.

produce mainly interleukins (IL-4, IL-5, and IL-13), augment certain B-cell responses. IL-4 in particular is the major inducer of B-cell switching to immunoglobulin E (IgE) production, and therefore plays a crucial role in allergic reactions involving IgE and mast cells. Th cells producing cytokines typical of both Th1 and Th2 clones have also been described and they have been named Th0 [1].

Evidence indicates that Th1 and Th2 cells differentiate from a common precursor, naive T-cells. This appears to be a multistep process, in which naive T-cells pass through an intermediate stage (Th0) at which both Th1 and Th2 cytokines are produced (Fig. 31.3). Transcription factors, including GATA3, T-bet, c-maf, and STAT6, are involved in the molecular mechanisms by which Th1 and Th2 cells differentially express the Th1 and Th2 cytokines [19, 20] and that this may have differing consequences on subepithelial fibrosis, airway smooth muscle hyperplasia and mucus hyperplasia as a consequence of effects on TGF-β and CCL11 expression [21]. However, most of these studies were conducted on murine T-cells, and the situation in human T-cells, in physiological and pathological conditions (e.g. bronchial asthma) remain largely unknown. A key role for regulatory T-cells (e.g. Treg and Th-17) has also been proposed in asthma, particularly in those patients with severe disease [22, 23]. Production of the Treg phenotype is associated with the transcription factor Foxp3 whose DNA binding to gene promoters has recently been mapped [24] and whose activity is also dependent upon NF-AT [25].

Determination of asthma severity

Activation of NF-κB leads to the coordinated induction of multiple genes that are expressed in inflammatory and immune responses. Many of these genes are induced in inflammatory and structural cells and play an important role in the inflammatory process.

NF-κB

While NF-κB is not the only transcription factor involved in regulation of the expression of these genes, it often appears to have a decisive regulatory role. NF-κB often functions in cooperation with other transcription factors, such as AP-1 and C/EBP, which are also involved in regulation of inflammatory and immune genes [10, 26]. Genes induced by NF-κB include those for the proinflammatory cytokines IL-1β, TNF-α, and GM-CSF and the chemokines IL-8, macrophage inflammatory protein-1α (MIP-1α), macrophage chemotactic protein-1 (MCP-1), RANTES, and eotaxins, that are largely responsible for attracting inflammatory cells into sites of inflammation [4, 7]. NF-κB also regulates the expression of inflammatory enzymes, including the inducible form of nitric oxide synthase (NOS_2) that produces large amounts of nitric oxide (NO). NF-κB also plays an important role in regulating expression of adhesion molecules, such as vascular cell adhesion molecule-1 (VCAM-1) and intercellular adhesion molecule-1 (ICAM-1), that are expressed on endothelial and epithelial cells at inflammatory sites and play a key role in the initial recruitment of inflammatory cells [4, 7].

Data also suggest that NF-κB may be activated by many of the stimuli that exacerbate asthmatic inflammation (e.g. rhinovirus infection, allergen exposure, proinflammatory cytokines, and oxidants) [27–29]. There is also evidence for activation of NF-κB in bronchial epithelial and sputum cells of patients with asthma [30]. NF-κB is an amplifying and perpetuating mechanism that exaggerates the disease-specific inflammatory process through the coordinated activation of multiple inflammatory genes. The degree of NF-κB phosphorylation and acetylation in asthma is unknown as is the specific transcriptome associated with each modified NF-κB [10]. Inhibitors of the controlling kinase (Inhibitor of κB kinase 2, IKK2) are in clinical development for asthma and other inflammatory diseases with the hope of achieving as good an anti-inflammatory effect as glucocorticoids without the side effects [31].

AP-1

AP-1 is a collection of related transcription factors belonging to the Fos (c-Fos, FosB) and Jun (c-Jun, JunB, JunD) families which dimerize in various combinations through their leucine zipper region. Fos/Jun form heterodimers with high affinity and are the predominant form of AP-1 in most cells, whereas Jun/Jun homodimers bind with low affinity and are less abundant [32]. AP-1 may be activated by various cytokines, including TNF-α and IL-1β, via several types of protein tyrosine kinase (PTK) and mitogen-activated protein (MAP) kinases in particular JUN-N-terminal kinase (JNK), which themselves activate a cascade of intracellular kinases [9]. AP-1, like NF-κB, regulates many of the inflammatory and immune genes that are overexpressed in asthma. Indeed many of these genes require the simultaneous activation of both transcription factors that work together cooperatively. There is evidence for increased expression of c-Fos in bronchial epithelial cells in asthmatic airways [33] and an even greater expression in patients with severe, treatment-insensitive asthma [34, 35], and many of the stimuli relevant to asthma that activate NF-κB will also activate AP-1. Inhibitors of JNK are being developed [32] and these show good effectiveness against bronchial hyperresponsiveness, bronchoalveolar lavage (BAL) inflammatory cells, and airway remodeling in animal models of asthma [36]. Interestingly, JNKs are also involved in T-cell class switching and their inhibitors may also have a profound immunomodulatory role [32].

STATs and other transcription factors

Signal transduction-activated transcription factor-6 (STAT6) also provides a target as a potential treatment for allergic asthma. STAT6 knockout mice have no response to IL-4, do not develop Th2 cells in response to IL-4, and fail to produce IgE, bronchial hyperresponsiveness, or BAL eosinophilia after allergen sensitization indicating the critical role of STAT6 in allergic responses [37]. STAT6 has been reported to be overexpressed in bronchial biopsies from asthmatic patients [38]. In addition, evidence exists for enhanced expression of STAT1 in the airways of asthmatic, but not of COPD subjects [39]. Indeed, decoy oligonucleotides against STAT1 are in early clinical development for asthma [40].

Cyclosporin has been used as an immunomodulator for many years although its side effect profile limits its utility in asthma [1]. Its target is cyclophilin which controls the dephosphorylation of NF-AT (nuclear factor of activated T-cells) and thereby the expression of key Th2 cytokines including IL-13 [4, 7].

Sp1 is a ubiquitous transcription factor that binds to GC-boxes and related motifs, which are frequently occurring DNA elements present in many promoters and enhancers. Modification of these sites within the IL-10 and IL-4 and the 5-lipoxygenase promoters has been associated with altered expression of these genes in distinct patient groups [41, 42].

Transcription factors and glucocorticoid action

Glucocorticoids belong to the family of nuclear steroid hormone receptors and are important anti-inflammatory agents used in the treatment of chronic inflammatory diseases such as asthma [5]. Functionally they act by suppressing airway hyperresponsiveness, reducing airway edema and the infiltration of inflammatory cells from the blood to the airway and thereby reducing the airway inflammatory response [5]. Glucocorticoid receptors (GR) are predominantly localized to the airway epithelium, alveolar macrophages and endothelium [5], which are, therefore, probably important sites for the anti-inflammatory action of steroids, especially those delivered by the inhaled route. Airway epithelial cells

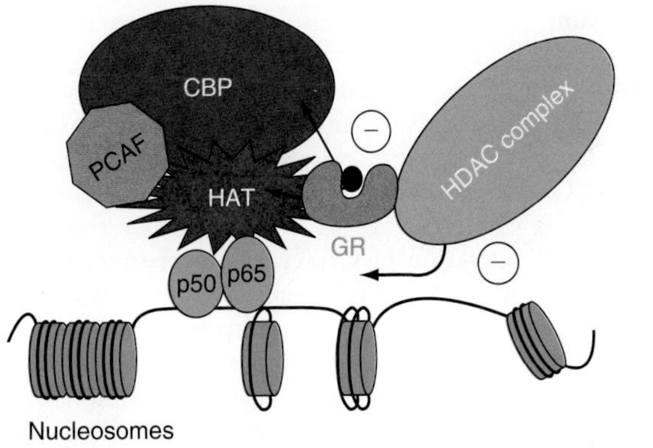

FIG. 31.4 The actions of the dexamethasone/GR complex on inhibition of IL-1β-stimulated histone acetylation. DNA-bound p65 induces histone acetylation via activation of CBP and a CBP-associated HAT complex to occur. This causes local unwinding of DNA and increased gene transcription. GR, acting as a monomer, interacts with CBP causing an inhibition of CBP-mediated HAT activity. In addition, GR also recruits HDAC2 to the p65/CBP complex, further reducing local HAT activity, leading to enhanced nucleosome compaction and repression of transcription.

act as important regulators of the inflammatory reaction, responding to various inflammatory mediators, such as cytokines, by the production of a wide range of cytokines, chemokines, and other inflammatory mediators [5].

Glucocorticoids bind to and activate a cytosolic receptor (GR) which then translocates to the nucleus. Within the nucleus two GR subunits form a dimer and bind to specific DNA elements (glucocorticoid response elements or GREs) in the promoter regions of glucocorticoid-responsive genes, resulting in modulation of transcription [5]. Several genes are upregulated by glucocorticoids, including the β_2-receptor, MAPK phosphatase (MKP1), and serum leukoprotease inhibitor (SLPI). However, in inflammation the major role of glucocorticoids appears to be gene repression. Evidence accumulated over the past few years has suggested that a major mechanism of glucocorticoids is repression of the actions of AP-1 and NF-κB [5].

The major effect of glucocorticoids is on gene transcription although actions on mRNA stability do occur with specific genes under the control of MKP-1 [43]. Several molecular mechanisms for the suppressive actions of the activated GR on NF-κB-mediated gene transcription have been proposed, but it is likely that a combination of inhibition of histone H4 acetylation on lysines 8 and 12 induced by inflammatory mediators, such as IL-1β and TNF-α, lead to changes in the ability of RNA polymerase II to transcribe mRNA [5, 44–46]. This effect on histone acetylation is achieved by a direct inhibition of CBP-associated histone acetylation and by recruitment of HDAC2 to the activated DNA-bound NF-κB complex [44] (Fig. 31.4). Of interest is that oxidative stress that modulates HDAC activity is able to markedly alter the expression of some inflammatory mediators and reduce glucocorticoid responsiveness [47]. We have recently reported that the expression of HDAC2 is reduced in the lung of patients with COPD [48] and that overexpression of HDAC2 restores glucocorticoid sensitivity in BAL macrophages from these patients [49]. The failure of glucocorticoids to regulate proopiomelanocortin (POMC) expression in Cushing's disease has also been related to a lack of HDAC2 [46] and targeting of these enzymes may provide another novel target for future drug discovery.

Glucocorticoids are able to activate the expression of a number of anti-inflammatory genes due to GR binding to specific GREs in the promoter regions of responsive genes [5]. This is often in conjunction with other transcription factors and activation of CAAT/enhancer-binding protein (C/EBPα) by long-acting β_2-agonists in airway smooth muscle cells has been proposed to account, at least in part, for the enhanced effect of combination treatment in asthmatic patients [50]. Loss of C/EBPα expression in some asthmatic patients may account for steroid effectiveness in these subjects [50].

A very small number of asthmatic patients are steroid-resistant and fail to respond to even high doses of oral glucocorticoids (so called steroid-resistant asthma). This defect is also present in their peripheral blood mononuclear cells (PBMCs) and T-lymphocytes [35]. Although ligand binding data show no change in GR number in these patients, immunocytochemistry indicates a reduced GR nuclear translocation which may account for the reduced binding to DNA seen in these patients [51]. Altered nuclear translocation is affected by GR phosphorylation status [52], which may also be a target for long-acting β-agonists [5]. In the same patients there is a reduced inhibitory effect of glucocorticoids on AP-1 activation, but not on NF-κB. Furthermore, there is an increase in the baseline activity of AP-1 which appears to be due to excessive activation of JNK [53, 54]. This resistance will be seen at the site of inflammation where AP-1 is activated but not at uninflamed sites. This may explain why patients with steroid-resistant asthma are not resistant to the endocrine and metabolic effects of glucocorticoids, and thus develop the drug's systemic side effects.

TRANSCRIPTION FACTORS IN COPD

In contrast to the enormous increase in our understanding of the pathogenesis of asthma, less is known about the molecular pathogenesis of COPD [55].

Transcription factors and disease susceptibility

The etiology of COPD appears to be due to interactions between environmental factors (particularly cigarette smoking) and genetic factors. Chronic heavy cigarette smoking is currently the cause of more than 90% of cases of COPD in Westernized countries, so environmental factors are clearly very important. However, it is important to identify the factors that determine why only 15–20% of chronic heavy cigarette smokers develop symptomatic COPD [55]. So far, this is little understood, although it is likely that genetic factors are important. There is convincing evidence that several genes influence the development of COPD. In a complex polygene disease such as COPD, it is likely that multiple genes are operating and that the influence of each gene in isolation may be relatively weak. The susceptibility to develop COPD with smoking is likely to depend on the coincidence of several gene polymorphisms that act together [55].

The only clearly established, but rare, genetic risk factor for COPD is α_1-antitrypsin deficiency (α_1-AT) [56]. Approximately 95% of cases of clinical α_1-AT deficiency are caused by a single amino acid substitution at position 342 (lysine for glutamic acid) in the coding region of the α_1-AT gene (Z allele). This genotype is uncommon, but a Taq1 polymorphism in the 3'-flanking region of the α_1-AT gene has been reported to be present in 18% of COPD patients but in only 5% of the general population in the United Kingdom [56]. Although this mutation is not associated with an abnormal α_1-AT protein expression, this genetic variant is within an enhancer sequence (which contains C/EBP-binding sites) and may impair the acute-phase increase in α_1-AT gene expression in response to IL-6 [57].

Transcription factors and clinical manifestations of COPD

The chronic airflow obstruction in tobacco smoking-related COPD results from a combination of airway disease, which particularly affects small airways, and loss of lung elasticity because of destruction of the lung parenchyma. The modulation of the inflammatory response in COPD by genetic factors associated with transcription factors may, in part, cause of the different clinical phenotypes of patients with COPD [4]; however, it is clear that COPD patients have marked inflammation, which increases with disease stage [58]. Patients with COPD are reported to have increased activation of NF-κB in sputum and lung tissue macrophages and bronchial epithelial cells [59–61]. The activation status of NF-κB and its repression by glucocorticoids may also be affected by altered expression of HDAC2 [48, 49] and of the protein deacetylase sirtuin (SIRT1) [62].

Hogg and colleagues [58] were able to show that the mRNA expression of the transcription factors early growth response (Egr)-1 and c-Fos were increased in bronchial biopsies and that exposure of fibroblasts to cigarette smoke enhanced Egr-1 expression [63]. This results in subsequent induction of MMP2 expression *in vitro* and *in vivo* [64]. In contrast to the tissue destruction seen in the alveoli in COPD patients, the small airways are characterized by fibrosis and increased smooth muscle. Activation of Smad factors by growth factors, such as TGFβ, may be important in this process and Smad3 knockout mice develop emphysema [65]. In addition, expression of the inhibitory Smad7 is reduced in epithelial cells of COPD (Global Initiative for Chronic Obstructive Lung Disease, GOLD stage II) patients [66].

In contrast to asthma, where C/EBPα is abnormally expressed, Borger and colleagues have reported a reduction of C/EBPδ in patients with COPD [67]. Activation of STAT4 is critical for the differentiation of Th1/Tc1 cells T-cells and the production of IFN-γ [68]. We have reported increased expression of activated phospho-STAT4 in bronchial biopsies and BAL lymphocytes of COPD patients and that this correlated with the degree of airflow obstruction and the number of IFN-γ^+ cells. In contrast, T-bet expression was not altered [68].

Recently it has been suggested that in patients with COPD and symptoms of chronic bronchitis there is an increased prevalence of a common polymorphism in the 5'-flanking region of the TNF-α gene, which was previously found to be associated with high baseline and induced TNF-α expression [69]. Unfortunately, recent results indicate that anti-TNFα therapy is ineffective in COPD [70].

CONCLUSION

In the future the role of transcription factors and the genetic regulation of their expression in asthma and COPD may be an increasingly important aspect of research, as this may be one of the critical mechanisms regulating the expression of clinical phenotypes and their responsiveness to therapy. Despite recent advances in the knowledge of the pathogenesis of asthma and COPD, much more research on the molecular mechanisms of asthma and COPD are needed to aid the logical development of new therapies for these common and important diseases, particularly in COPD where no effective treatments currently exist.

References

1. Caramori G, Adcock I. Pharmacology of airway inflammation in asthma and COPD. *Pulm Pharmacol Ther* 16: 247–77, 2003.
2. Barnes PJ. Mechanisms in COPD: Differences from asthma. *Chest* 117: 10S–14S, 2000.
3. Barnes PJ, Chung KF, Page CP. Inflammatory mediators of asthma: An update. *Pharmacol Rev* 50: 515–96, 1998.
4. Barnes PJ. Transcription factors in airway diseases. *Lab Invest* 86: 867–72, 2006.
5. Ito K, Chung KF, Adcock IM. Update on glucocorticoid action and resistance. *J Allergy Clin Immunol* 117: 522–43, 2006.
6. Caramori G, Lim S, Ito K *et al.* Expression of GATA family of transcription factors in T-cells, monocytes and bronchial biopsies. *Eur Respir J* 18: 466–73, 2001.
7. Caramori G, Ito K, Adcock IM. Transcription factors in asthma and COPD. *IDrugs* 7: 764–70, 2004.
8. Gilmore TD. Introduction to NF-kappaB: Players, pathways, perspectives. *Oncogene* 25: 6680–84, 2006.
9. Saklatvala J. The p38 MAP kinase pathway as a therapeutic target in inflammatory disease. *Curr Opin Pharmacol* 4: 372–77, 2004.
10. Perkins ND. Post-translational modifications regulating the activity and function of the nuclear factor kappaB pathway. *Oncogene* 25: 6717–30, 2006.
11. Zhong H, May MJ, Jimi E, Ghosh S. The phosphorylation status of nuclear NF-kappaB determines its association with CBP/p300 or HDAC-1. *Mol Cell* 9: 625–36, 2002.
12. Chen L, Fischle W, Verdin E, Greene WC. Duration of nuclear NF-kappaB action regulated by reversible acetylation. *Science* 293: 1653–57, 2001.
13. Ogawa S, Lozach J, Benner C *et al.* Molecular determinants of crosstalk between nuclear receptors and toll-like receptors. *Cell* 122: 707–21, 2005.
14. Li B, Carey M, Workman JL. The Role of Chromatin during Transcription. *Cell* 128: 707–19, 2007.
15. Kouzarides T. Chromatin modifications and their function. *Cell* 128: 693–705, 2007.
16. Allfrey VG, Faulkner R, Mirsky AE. Acetylation and methylation of histones and their possible role in the regulation of RNA synthesis. *Proc Natl Acad Sci USA* 51: 786–94, 1964.
17. de Ruijter AJ, van Gennip AH, Caron HN, Kemp S, van Kuilenburg AB. Histone deacetylases (HDACs): Characterization of the classical HDAC family. *Biochem J* 370: 737–49, 2003.
18. Neurath MF, Finotto S, Glimcher LH. The role of Th1/Th2 polarization in mucosal immunity. *Nat Med* 8: 567–73, 2002.
19. Zheng W, Flavell RA. The transcription factor GATA-3 is necessary and sufficient for Th2 cytokine gene expression in CD4 T cells. *Cell* 89: 587–96, 1997.
20. Zhang S, Lukacs NW, Lawless VA, Kunkel SL, Kaplan MH. Cutting edge: Differential expression of chemokines in Th1 and Th2 cells is dependent on STAT6 but not STAT4. *J Immunol* 165: 10–14, 2000.
21. Kiwamoto T, Ishii Y, Morishima Y *et al.* Transcription factors T-bet and GATA-3 regulate development of airway remodeling. *Am J Respir Crit Care Med* 174: 142–51, 2006.
22. Xystrakis E, Kusumakar S, Boswell S *et al.* Reversing the defective induction of IL-10-secreting regulatory T cells in glucocorticoid-resistant asthma patients. *J Clin Invest* 116: 146–55, 2006.
23. Bullens DM, Truyen E, Coteur L *et al.* IL-17 mRNA in sputum of asthmatic patients: Linking T cell driven inflammation and granulocytic influx? *Respir Res* 7: 135, 2006.
24. Marson A, Kretschmer K, Frampton GM *et al.* Foxp3 occupancy and regulation of key target genes during T-cell stimulation. *Nature* 445: 931–35, 2007.
25. Wu Y, Borde M, Heissmeyer V *et al.* FOXP3 controls regulatory T cell function through cooperation with NFAT. *Cell* 126: 375–87, 2006.
26. Baldwin AS Jr. Series introduction: The transcription factor NF-kappaB and human disease. *J Clin Invest* 107: 3–6, 2001.
27. Papi A, Johnston SL. Rhinovirus infection induces expression of its own receptor intercellular adhesion molecule 1 (ICAM-1) via increased NF-kappaB-mediated transcription. *J Biol Chem* 274: 9707–20, 1999.
28. Poynter ME, Irvin CG, Janssen-Heininger YM. Rapid activation of nuclear factor-kappaB in airway epithelium in a murine model of allergic airway inflammation. *Am J Pathol* 160: 1325–34, 2002.
29. Rahman I. Redox signaling in the lungs. *Antioxid Redox Signal* 7: 1–5, 2005.
30. Hart LA, Krishnan VL, Adcock IM, Barnes PJ, Chung KF. Activation and localization of transcription factor, nuclear factor-kappaB, in asthma. *Am J Respir Crit Care Med* 158: 1585–92, 1998.
31. Adcock IM, Chung KF, Caramori G, Ito K. Kinase inhibitors and airway inflammation. *Eur J Pharmacol* 533: 118–32, 2006.
32. Manning AM, Davis RJ. Targeting JNK for therapeutic benefit: From junk to gold? *Nat Rev Drug Discov* 2: 554–65, 2003.
33. Demoly P, Basset-Seguin N, Chanez P *et al.* c-fos proto-oncogene expression in bronchial biopsies of asthmatics. *Am J Respir Cell Mol Biol* 7: 128–33, 1992.
34. Loke TK, Mallett KH, Ratoff J *et al.* Systemic glucocorticoid reduces bronchial mucosal activation of activator protein 1 components in glucocorticoid-sensitive but not glucocorticoid-resistant asthmatic patients. *J Allergy Clin Immunol* 118: 368–75, 2006.
35. Adcock IM, Lane SJ. Corticosteroid-insensitive asthma: Molecular mechanisms. *J Endocrinol* 178: 347–55, 2003.
36. Eynott PR, Nath P, Leung SY *et al.* Allergen-induced inflammation and airway epithelial and smooth muscle cell proliferation: Role of JUN N-terminal kinase. *Br J Pharmacol* 140: 1373–80, 2003.
37. Tomkinson A, Kanehiro A, Rabinovitch N *et al.* The failure of STAT6-deficient mice to develop airway eosinophilia and airway hyperresponsiveness is overcome by interleukin-5. *Am J Respir Crit Care Med* 160: 1283–91, 1999.
38. Mullings RE, Wilson SJ, Puddicombe SM *et al.* Signal transducer and activator of transcription 6 (STAT-6) expression and function in asthmatic bronchial epithelium. *J Allergy Clin Immunol* 108: 832–38, 2001.

39. Sampath D, Castro M, Look DC, Holtzman MJ. Constitutive activation of an epithelial signal transducer and activator of transcription (STAT) pathway in asthma. *J Clin Invest* 103: 1353–61, 1999.
40. Quarcoo D, Weixler S, Groneberg D *et al.* Inhibition of signal transducer and activator of transcription 1 attenuates allergen-induced airway inflammation and hyperreactivity. *J Allergy Clin Immunol* 114: 288–95, 2004.
41. Hobbs K, Negri J, Klinnert M, Rosenwasser LJ, Borish L. Interleukin-10 and transforming growth factor-beta promoter polymorphisms in allergies and asthma. *Am J Respir Crit Care Med* 158: 1958–62, 1958.
42. Lim S, Crawley E, Woo P, Barnes PJ. Haplotype associated with low interleukin-10 production in patients with severe asthma. *Lancet* 352: 113, 1998.
43. Abraham SM, Lawrence T, Kleiman A *et al.* Antiinflammatory effects of dexamethasone are partly dependent on induction of dual specificity phosphatase 1. *J Exp Med* 203: 1883–89, 2006.
44. Ito K, Barnes PJ, Adcock IM. Glucocorticoid receptor recruitment of histone deacetylase 2 inhibits interleukin-1beta-induced histone H4 acetylation on lysines 8 and 12. *Mol Cell Biol* 20: 6891–903, 2000.
45. Luecke HF, Yamamoto KR. The glucocorticoid receptor blocks P-TEFb recruitment by NFkappaB to effect promoter-specific transcriptional repression. *Genes Dev* 19: 1116–27, 2005.
46. Bilodeau S, Vallette-Kasic S, Gauthier Y *et al.* Role of Brg1 and HDAC2 in GR trans-repression of the pituitary POMC gene and misexpression in Cushing disease. *Genes Dev* 20: 2871–86, 2006.
47. Ito K, Lim S, Caramori G *et al.* Cigarette smoking reduces histone deacetylase 2 expression, enhances cytokine expression, and inhibits glucocorticoid actions in alveolar macrophages. *FASEB J* 15: 1110–12, 2001.
48. Ito K, Ito M, Elliott WM *et al.* Decreased histone deacetylase activity in chronic obstructive pulmonary disease. *N Engl J Med* 352: 1967–76, 2005.
49. Ito K, Yamamura S, Essilfie-Quaye S *et al.* Histone deacetylase 2-mediated deacetylation of the glucocorticoid receptor enables NF-kappaB suppression. *J Exp Med* 203: 7–13, 2006.
50. Roth M, Johnson PR, Borger P *et al.* Dysfunctional interaction of C/EBPalpha and the glucocorticoid receptor in asthmatic bronchial smooth-muscle cells. *N Engl J Med* 351: 560–74, 2004.
51. Matthews JG, Ito K, Barnes PJ, Adcock IM. Defective glucocorticoid receptor nuclear translocation and altered histone acetylation patterns in glucocorticoid-resistant patients. *J Allergy Clin Immunol* 113: 1100–8, 2004.
52. Irusen E, Matthews JG, Takahashi A *et al.* p38 Mitogen-activated protein kinase-induced glucocorticoid receptor phosphorylation reduces its activity: Role in steroid-insensitive asthma. *J Allergy Clin Immunol* 109: 649–57, 2002.
53. Lane SJ, Adcock IM, Richards D *et al.* Corticosteroid-resistant bronchial asthma is associated with increased c-fos expression in monocytes and T lymphocytes. *J Clin Invest* 102: 2156–64, 1998.
54. Sousa AR, Lane SJ, Soh C, Lee TH. *In vivo* resistance to corticosteroids in bronchial asthma is associated with enhanced phosyphorylation of JUN N-terminal kinase and failure of prednisolone to inhibit JUN N-terminal kinase phosphorylation. *J Allergy Clin Immunol* 104: 565–74, 1999.
55. Barnes PJ, Kleinert S. COPD – a neglected disease. *Lancet* 364: 564–65, 2004.
56. Kalsheker NA, Morgan K. Regulation of the alpha 1-antitrypsin gene and a disease-associated mutation in a related enhancer sequence. *Am J Respir Crit Care Med* 150: S183–S189, 1994.
57. Morgan K, Scobie G, Marsters P, Kalsheker NA. Mutation in an alpha1-antitrypsin enhancer results in an interleukin-6 deficient acute-phase response due to loss of cooperativity between transcription factors. *Biochim Biophys Acta* 1362: 67–76, 1997.
58. Hogg JC, Chu F, Utokaparch S *et al.* The nature of small-airway obstruction in chronic obstructive pulmonary disease. *N Engl J Med* 350: 2645–53, 2004.
59. Caramori G, Romagnoli M, Casolari P *et al.* Nuclear localisation of p65 in sputum macrophages but not in sputum neutrophils during COPD exacerbations. *Thorax* 58: 348–51, 2003.
60. Di Stefano A, Caramori G, Oates T *et al.* Increased expression of nuclear factor-kappaB in bronchial biopsies from smokers and patients with COPD. *Eur Respir J* 20: 556–63, 2002.
61. Szulakowski P, Crowther AJ, Jimenez LA *et al.* The effect of smoking on the transcriptional regulation of lung inflammation in patients with chronic obstructive pulmonary disease. *Am J Respir Crit Care Med* 174: 41–50, 2006.
62. Rajendrasozhan S, Yang SR, Kinnula VL, Rahman I. SIRT1, an anti-inflammatory and antiaging protein, is decreased in lungs of patients with chronic obstructive pulmonary disease. *Am J Respir Crit Care Med* 177: 861–870, 2008.
63. Ning W, Li CJ, Kaminski N *et al.* Comprehensive gene expression profiles reveal pathways related to the pathogenesis of chronic obstructive pulmonary disease. *Proc Natl Acad Sci USA* 101: 14895–900, 2004.
64. Ning W, Dong Y, Sun J *et al.* Cigarette Smoke Stimulates Matrix Metalloproteinase-2 Activity via EGR-1 in Human Lung Fibroblasts. *Am J Respir Cell Mol Biol*, 2006.
65. Chen H, Sun J, Buckley S *et al.* Abnormal mouse lung alveolarization caused by Smad3 deficiency is a developmental antecedent of centrilobular emphysema. *Am J Physiol Lung Cell Mol Physiol* 288: L683–91, 2005.
66. Zandvoort A, Postma DS, Jonker MR *et al.* Altered expression of the Smad signalling pathway: Implications for COPD pathogenesis. *Eur Respir J* 28: 533–41, 2006.
67. Borger P, Matsumoto H, Boustany S *et al.* Disease-specific expression and regulation of CCAAT/enhancer-binding proteins in asthma and chronic obstructive pulmonary disease. *J Allergy Clin Immunol* 119: 98–105, 2007.
68. Di Stefano A, Caramori G, Capelli A *et al.* STAT4 activation in smokers and patients with chronic obstructive pulmonary disease. *Eur Respir J* 24: 78–85, 2004.
69. Huang SL, Su CH, Chang SC. Tumor necrosis factor-alpha gene polymorphism in chronic bronchitis. *Am J Respir Crit Care Med* 156: 1436–39, 1997.
70. Rennard SI, Fogarty C, Kelsen S *et al.* The Safety and Efficacy of Infliximab in Moderate-To-Severe Chronic Obstructive Pulmonary Disease. *Am J Respir Crit Care Med* 156: 926–34, 2007.

Neural and Humoral Control of the Airways

CHAPTER 32

Peter J. Barnes[1] and Neil C. Thomson[2]

[1]National Heart and Lung Institute (NHLI), Clinical Studies Unit, Imperial College, London, UK

[2]Department of Respiratory Medicine, Division of Immunology, Infection and Inflammation, University of Glasgow, Glasgow, UK

INTRODUCTION

Airway nerves and circulating hormones regulate the caliber of the airways and have an influence on airway smooth muscle tone, airway blood flow, and mucus secretion. They may also influence the inflammatory process and play an integral role in host defense. There is increasing evidence that neural and humoral mechanisms may play a role in the pathophysiology of asthma and COPD, and several of the treatments used interact with neural or humoral control.

NEURAL CONTROL

Overview of airway innervation

Neural control of airway function is complex and many neurotransmitters are autonomic receptors are involved. Three types of airway nerve are recognized:

1. parasympathetic nerves which primarily release acetylcholine (ACh)
2. sympathetic nerves which primarily release norepinephrine (noradrenaline)
3. afferent (sensory nerves) whose primary transmitter is glutamate.

In addition to these classical transmitters, multiple neuropeptides have now been localized to airway nerves and may have potent effects on airway function [1]. All of these neurotransmitters act on autonomic receptors that are expressed on the surface of target cells in the airway. It is increasingly recognized that a single transmitter may act on several subtypes of receptor, which may lead to different cellular effects mediated via different second messenger systems.

Several neural mechanisms are involved in the regulation of airway caliber and abnormalities in neural control may contribute to airway narrowing in disease (Fig. 32.1). Neural

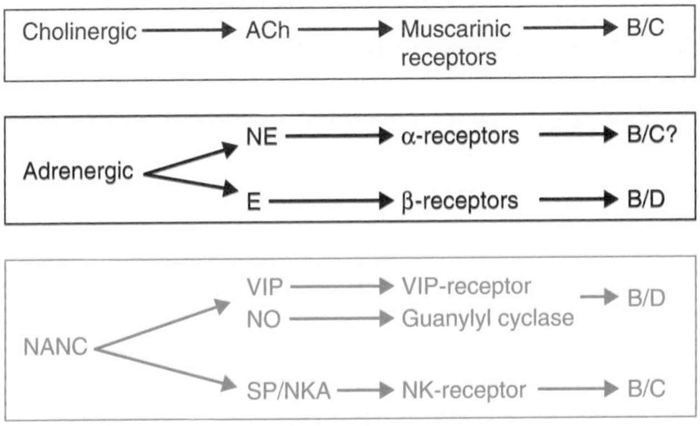

FIG. 32.1 Autonomic control of airway smooth muscle tone. Neural mechanisms resulting in bronchoconstriction (B/C) and bronchodilatation (B/D). Ach: acetylcholine; NE: norepinephrine; E: epinephrine; VIP: vasoactive intestinal peptide; NO: nitric oxide; i-NANC: inhibitory non-adrenergic non-cholinergic nerves; e-NANC: excitatory non-adrenergic non-cholinergic nerves; NK: neurokinin.

mechanisms are involved in the pathophysiology of asthma and COPD, contributing to the symptoms, and possibly to the inflammatory response [2]. There is a close interrelationship between inflammation and neural responses in the airways, since inflammatory mediators may influence the release of neurotransmitters via activation of sensory nerves leading to reflex effects and via stimulation of pre-junctional receptors that influence the release of neurotransmitters [3]. In turn, neural mechanisms may influence the nature of the inflammatory response, either reducing inflammation or exaggerating the inflammatory response. Neuroimmmune interactions in asthma are increasingly recognized, particularly the role of neurotrophins released from inflammatory and immune cells in the airways [4].

Neural interactions

Complex interactions between various components of the autonomic nervous system are now recognized. Adrenergic nerves may modulate cholinergic neurotransmission in the airways and sensory nerves may influence neurotransmission in parasympathetic ganglia and at post-ganglionic nerves. This means that changes in the function of one neural pathway may have effects on other pathways.

Co-transmission

Almost every nerve contains multiple transmitters. Thus airway parasympathetic nerves, in which the primary transmitter is ACh, also contain the neuropeptides vasoactive intestinal polypeptide (VIP), peptide histidine isoleucine/methionine (PHI/M), pituitary adenylyl cyclase activating peptide (PACAP), helodermin, galanin, and neuronal nitric oxide synthase (Fig. 32.2). These co-transmitters may have either facilitatory effects or antagonistic effects on target cells, or may influence the release of the primary transmitter via pre-junctional receptors. Thus VIP modulates the release of ACh from airway cholinergic nerves. Sympathetic nerves, which release noradrenaline, may also release neuropeptide Y (NPY) and enkephalins, whereas afferent nerves may contain a variety of peptides including substance P (SP), neurokinin A (NKA), calcitonin gene-related peptide (CGRP), galanin, VIP, and cholecystokinin.

The physiological role of neurotransmission may be in "fine tuning" of neural control. Neuropeptides may be preferentially released by high frequency firing of nerves, and their effects may therefore only become manifest under condition of excessive nerve stimulation. Neuropeptide neurotransmitters may also act on target cells different from the primary transmitter, resulting in different physiological effects. Thus in airways ACh causes bronchoconstriction, but VIP which is co-released may have its major effect on bronchial vessels, thus increasing blood flow to the airways. In chronic inflammation the role of co-transmitters may be increased by alterations in the expression of their receptors or by increased synthesis of transmitters via increased gene transcription.

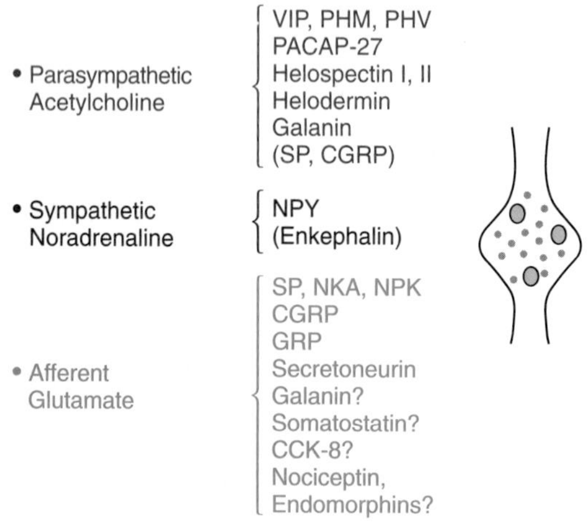

FIG. 32.2 Neurotransmitters and co-transmitters in airway nerves. SP: substance P, NKA: neurokinin A, CGRP: calcitonin gene-related peptide, VIP: vasoactive intestinal peptide, PHI/PHM/PHV: peptide histidine isoleucine/methionine/valine, PACAP: pituitary adenylate cyclase activating peptide, NPY: neuropeptide Y.

SENSORY NERVES

The sensory innervation of the respiratory tract is mainly carried in the vagus nerve. The neuronal cell bodies are localized to the nodose and jugular ganglia and input to the solitary tract nucleus in the brain stem. A few sensory fibers supplying the lower airways enter the spinal cord in the upper thoracic sympathetic trunks, but their contribution to respiratory reflexes is minor and it is uncertain whether they are represented in humans. There is a tonic discharge of sensory nerves that has a regulatory effect on respiratory function and also triggers powerful protective reflex mechanisms in response to inhaled noxious agents, physical stimuli, or certain inflammatory mediators.

At least three types of afferent fiber have been identified in the lower airways [5] (Fig. 32.3). Most of the information on their function has been obtained from studies in anesthetized animals, so it is difficult to know how much of the information obtained in anesthetized animals can be extrapolated to human airways.

Slowly adapting receptors

Myelinated fibers associated with smooth muscle of proximal airways are probably slowly adapting (pulmonary stretch) receptors (SARs) that are involved in reflex control of breathing. Activation of SARs reduces efferent vagal discharge and mediates bronchodilatation. During tracheal constriction the activity of SARs may serve to limit the bronchoconstrictor response. SARs may play a role in the cough reflex since when these receptors are destroyed by high concentration of SO_2 the cough response to mechanical stimulation is lost.

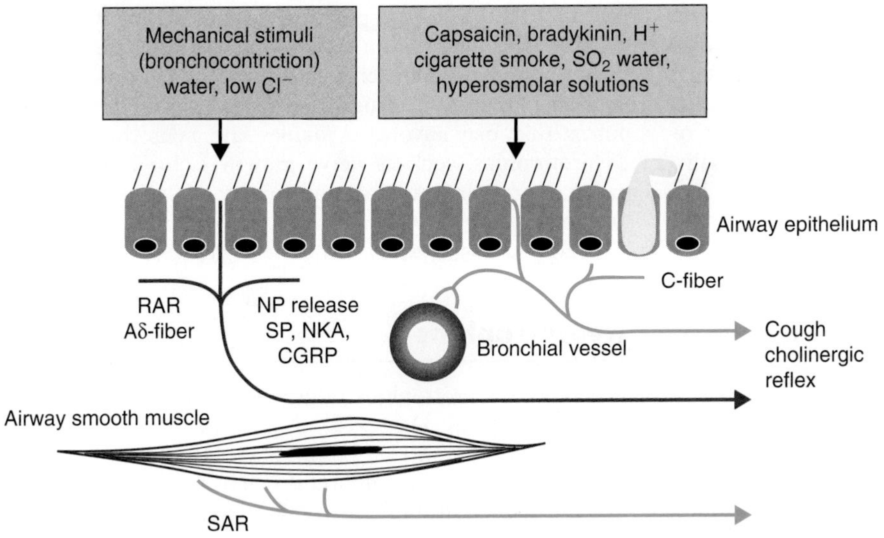

FIG. 32.3 Afferent nerves in airways. Slowly adapting receptors (SARs) are found in airway smooth muscle, whereas rapidly adapting myelinated (RAR) and unmyelinated C-fibers are present in the airway mucosa.

Rapidly adapting receptors

Myelinated fibers in the epithelium, particularly at the branching points of proximal airways, show rapid adaptation. Rapidly adapting receptors (RARs) account for 10–30% of the myelinated nerve endings in the airways. These endings are sensitive to mechanical stimulation and to mediators such as histamine. The response of RAR to histamine is partly due to mechanical distortion consequent on bronchoconstriction, although if this is prevented by pre-treatment with isoprenaline the RAR response is not abolished, indicating a direct stimulatory effect of histamine. It is likely that mechanical distortion of the airway may amplify irritant receptor discharge.

RAR with widespread arborizations are very numerous in the area of the carina, where they have been termed "cough receptors" as cough can be evoked by even the slightest touch in this region. RAR respond to inhaled cigarette smoke, ozone, serotonin, and prostaglandin $F_{2\alpha}$, although it is possible that these responses are secondary to the mechanical distortion produced by the bronchoconstrictor response to these irritants. Neurophysiological studies using an *in vitro* preparation in guinea pig trachea and bronchi show that a majority of afferent fibers are myelinated and belong to the Aδ-fiber group. Although these fibers are activated by mechanical stimulation and low pH, they are not sensitive to capsaicin, histamine, or bradykinin [6, 7].

C-Fibers

There is a high density of unmyelinated (C-fibers) in the airways and they greatly outnumber myelinated fibers. In the bronchi C-fibers account for 80–90% of all afferent fibers in cats. C-fibers play an important role in the defense of the lower respiratory tract. C-fibers contain neuropeptides, including SP, NKA, and CGRP, that confers a motor function on these nerves. Bronchial C-fibers are insensitive to lung inflation and deflation but typically respond to chemical stimulation. *In vivo* studies suggest that bronchial C-fibers in dogs respond to the inflammatory mediators histamine, bradykinin, serotonin, and prostaglandins. They are selectively stimulated by capsaicin given either by intravenously or by inhalation and are also stimulated by SO_2 and cigarette smoke. Since these fibers are relatively unaffected by lung mechanics, it is likely that these agents act directly on the unmyelinated endings in the airway epithelium. In the *in vitro* guinea pig trachea preparation C-fibers are stimulated by capsaicin and by bradykinin, but not by histamine, serotonin, or prostaglandins (with the possible exception of prostacyclin) [6].

RARs and C-fibers are sensitive to water and hyperosmotic solutions, with RARs showing a greater sensitivity to hypotonic and C-fibers to hypertonic saline. *In vitro* guinea pig trachea preparation Aδ fibers and C-fibers are stimulated by water and by hyperosmolar solutions; a small proportion of Aδ fibers are also stimulated by low-chloride solutions, whereas the majority of C-fibers are [8]. C-fibers are activated by capsaicin, which acts through vanilloid or TRPV1 receptors, for which small molecule inhibitors are now in development [9].

Cough

Cough is an important defense reflex, which may be triggered from either laryngeal or lower airway afferents and is an important symptom of asthma and COPD [10]. There is debate about which are the most important afferents for initiation of cough and this may be dependent on the stimulus. Thus RARs are activated by mechanical stimuli (e.g. particulate matter), bronchoconstrictors and hypotonic saline and water, whereas C-fibers are more sensitive to hypertonic solutions, bradykinin, and capsaicin. In normal humans inhaled capsaicin is a potent tussive stimulus and this is associated with a transient bronchoconstrictor reflex that is abolished by an anticholinergic drug. It is not certain whether this is due to stimulation of C-fibers in the larynx, but as these are very sparse it is likely that bronchial C-fibers are also involved. Citric acid is commonly used to stimulate coughing in experimental challenges in human subjects; it is likely that it produces cough by a combination of low pH (which stimulates C-fibers) and low chloride (which may stimulate laryngeal and lower airway afferents). Inhaled bradykinin causes coughing and a raw sensation retrosternally which may be due to stimulation of C-fibers in the lower airways. Bradykinin appears to be a relatively pure stimulant of C-fibers [6]. Prostaglandins E_2 and $F_{2\alpha}$ are potent tussive agents in humans and also sensitize the cough reflex [11, 12].

Afferent nerves in airway disease

Airway afferent nerves may become sensitized in inflammatory airway diseases, resulting in increased symptoms, such as cough and chest tightness (Fig. 32.4). Cough is a prominent symptom of asthma and COPD, and there is evidence that cough sensitivity is increased [13]. This may be due to senitization of afferent nerves in the airways as a result of inflammatory mediators produced during asthma and COPD. PGE_2 is a potent sensitizer of airway sensory nerves and is increased in asthma and COPD. Bradykinin is also a potent afferent nerve sensitizer [7]. Chronic inflammation may lead to neural hyperesthesia through mechanisms that may involve cytokines and neurotrophins [5]. Neurotrophins, such as nerve growth factor (NGF), may result in proliferation of airway sensory nerves and a change in the nerve phenotype, with a reduced threshold of activation and increased expression of neuropeptides [14].

Neurotrophins

The role of neurotrophins in airway diseases is increasingly recognized (Fig. 32.5) [15]. NGF, brain-derived neurotrophic factor, and neurotrophins 3 and 4/5 are nerve-related cytokines that play an important role in the function, proliferation, and survival of autonomic nerves. In sensory nerves

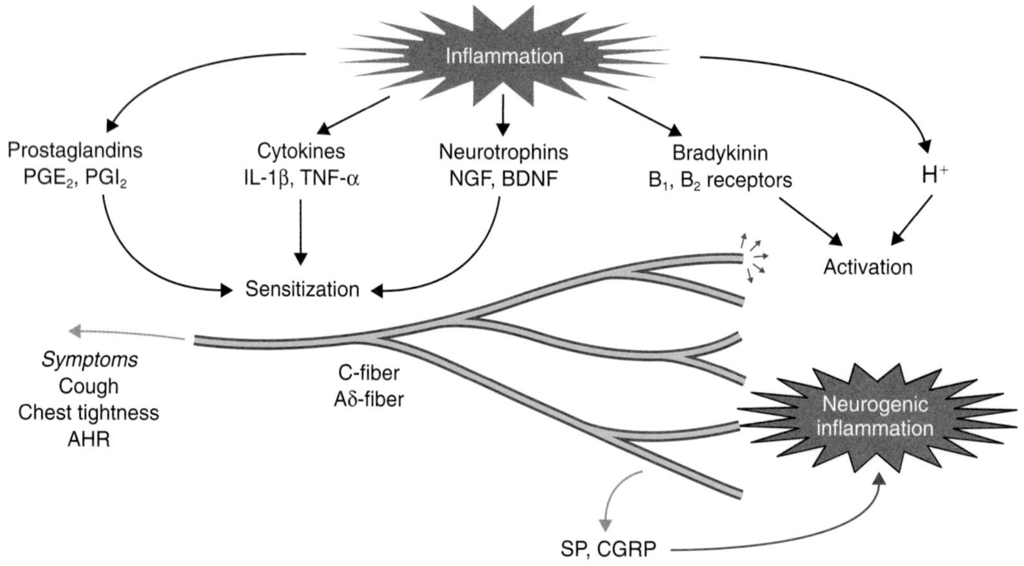

FIG. 32.4 Airway hyperaesthesia. Increased sensitivity of airway nerves induced by inflammation as a result of sensitization and activation of airway sensory nerves, resulting in increased symptoms and possibly neurogenic inflammation through the release of neuropeptides, such as substance P (SP) and calcitonin gene-related peptide (CGRP). AHR: airway hyperresponsiveness.

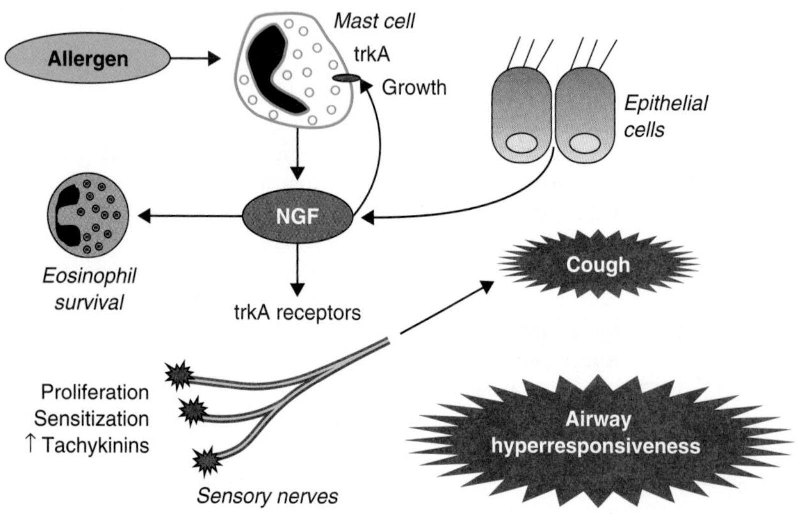

FIG. 32.5 Neurotrophin effects in airways. NGF is released from mast cell and epithelial cells and has effects of airway afferent nerves, which may lead to cough and airway hyperresponsiveness. NGF also promotes the growth of mast cells through its receptor TrkA and is chemotactic for eosinophils.

neurotrophins increase responsiveness and may also lead to expression of tachykinins. Neurotrophins may be produced by inflammatory cells, such as mast cells, lymphocytes, macrophages, and eosinophils, as well as structural cells, such as epithelial cells, fibroblasts, and airway smooth muscle cells [16]. Although neurotrophins have predominant effects on neuronal cells they may also act as growth factors for inflammatory cells such as mast cells, as well as increasing chemotaxis and survival of eosinophils [4]. NGF induces airway hyperresponsiveness in various animal models of asthma, including guinea pig, mice, and rats and this is blocked by antibodies to NGF or its receptor TrkA [15]. In guinea pig NGF enhances airway hyperresponsiveness by increasing the release of SP from sensory nerves [17] and increases sensory nerve activation [4]. The role of neurotrophins in human asthma is uncertain. However, NGF levels are increased in bronchoalveolar lavage fluid of asthmatic patients and there is a further increase after allergen challenge [18]. Blocking NGF or its receptors may therefore have some therapeutic potential. A decoy TrkA receptor which blocks NGF *in vivo* reduces airway hyperresponsiveness in animal models [19].

CHOLINERGIC NERVES

Cholinergic nerves are the major neural bronchoconstrictor mechanism in human airway and are the major determinant of airway caliber in humans.

Cholinergic control of airways

Cholinergic nerve fibers arise in the nucleus ambiguous in the brain stem and travel down the vagus nerve and synapse in parasympathetic ganglia which are located within the airway wall. From these ganglia short post-ganglionic fibers travel to airway smooth muscle and submucosal glands (Fig. 32.6). In animals, electrical stimulation of the vagus nerve causes release of ACh from cholinergic nerve terminals, with activation of muscarinic cholinergic receptors on smooth muscle and gland cells, which results in bronchoconstriction and mucus secretion. Prior administration of a muscarinic receptor antagonist, such as atropine, prevents vagally induced bronchoconstriction.

Non-neuronal cholinergic system

There is increasing evidence that ACh is also synthesized and released from several non-neuronal cells, including airway epithelial cells and fibroblasts [20, 21]. Inflammatory signals may increase the synthesis of ACh and thus enhance cholinergic effects. The contribution of non-neuronal ACh to cholinergic responses in the airways is currently unknown but may play an important role in regulating the tone of small airways where cholinergic innervation is very sparse. Non-neuronal ACh may have effects on inflammatory cells through nicotinic and muscarinic receptors that are expressed on their surface [22].

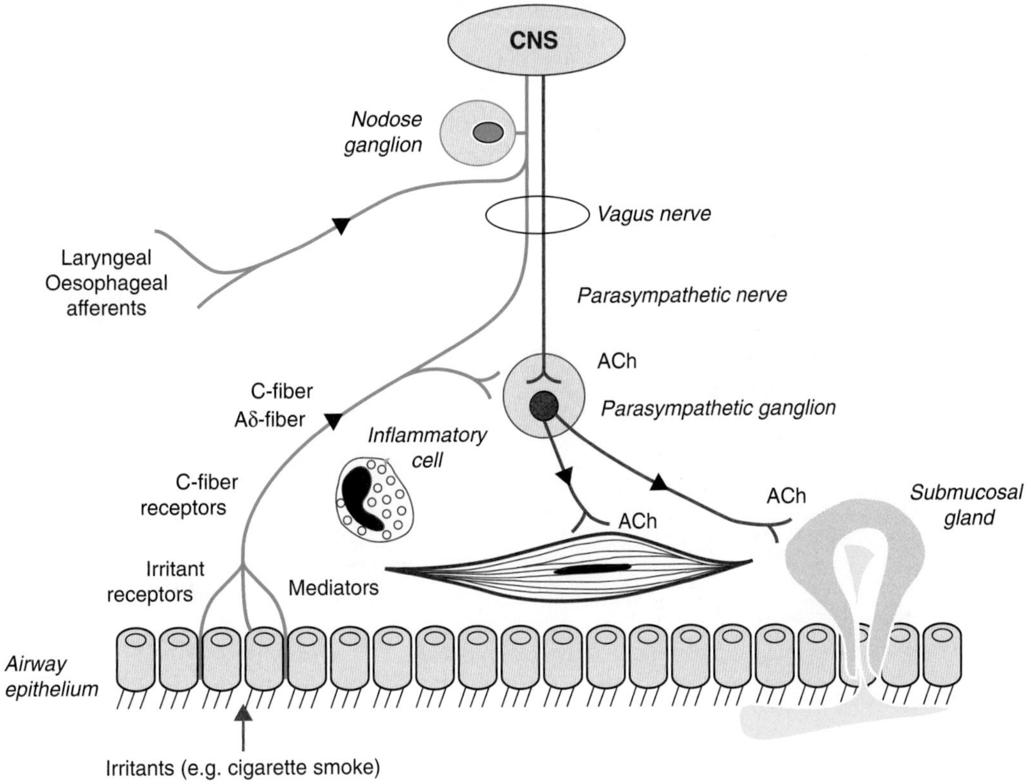

FIG. 32.6 Cholinergic control of airway smooth muscle. Pre-ganglionic and post-ganglionic parasympathetic nerves release acetylcholine (ACh) and can be activated by airway and extra-pulmonary afferent nerves.

Muscarinic receptors

Of the five known subtypes of muscarinic receptor, four have been identified by binding studies and pharmacologically in lung [23]. The muscarinic receptors that mediate bronchoconstriction in human and animal airways belong to the M_3-receptor subtype, whereas mucus secretion appears to be mediated by M_1- and M_3-receptors. M_1-receptors are also localized to parasympathetic ganglia, where they facilitate the neurotransmission mediated via nicotinic receptors (Fig. 32.7).

Inhibitory muscarinic receptors (autoreceptors) have been demonstrated on cholinergic nerves of airways in animals *in vivo*, and in human bronchi *in vitro* [24]. These pre-junctional receptors inhibit ACh release and may serve to limit vagal bronchoconstriction. Autoreceptors in human airways belong to the M_2-receptor subtype, whereas those on airway smooth muscle and glands belong to the M_3-receptor subtype [25]. Drugs such as atropine and ipratropium bromide, which block both pre-junctional M_2-receptors and post-junctional M_3-receptors on smooth muscle with equal efficacy, therefore increase in ACh release may then overcome the post-junctional blockade. This means that such drugs will not be as effective against vagal bronchoconstriction as against cholinergic agonists, and it may be necessary to re-evaluate the contribution of cholinergic nerves when drugs which are selective for the M_3-receptors are developed for clinical use. The presence of muscarinic autoreceptors has been demonstrated in human subjects *in vivo* [26]. A cholinergic agonist, pilocarpine, which selectively activates M_2-receptors, inhibits cholinergic reflex bronchoconstriction induced by sulfur dioxide in normal subjects, but such an inhibitory mechanism does not appear to operate in asthmatic subjects, suggesting that there may be dysfunction of these autoreceptors. Such a defect in muscarinic autoreceptors may then result in exaggerated cholinergic reflexes in asthma, since the normal feedback inhibition of ACh release may be lost. This might also explain the sometimes catastrophic bronchoconstriction which occurs with β-blockers in asthma which, at least in mild asthmatics, appears to be mediated by cholinergic pathways [27]. Antagonism of inhibitory β-receptors on cholinergic nerves would result in increased release of ACh which could not be switched off in the asthmatic patient (Fig. 32.8). This explains why anticholinergics prevent β-blocker-induced asthma. The mechanisms which lead to dysfunction of pre-junctional M_2-receptors in asthmatic airways are not certain, but it is possible that M_2-receptors may be more susceptible to damage by oxidants or other products of the inflammatory response in the airways. Experimental studies have demonstrated that influenza virus infection and eosinophils in guinea pigs may result in a selective loss of M_2-receptors compared with M_3-receptors, resulting in a loss of autoreceptor function and enhanced cholinergic bronchoconstriction.

Cholinergic innervation is greatest in large airways and diminishes peripherally, although in humans muscarinic receptors are localized to airway smooth muscle in all airways [28]. In humans, studies which have tried to distinguish large and small airway effects have shown that cholinergic bronchoconstriction predominantly involves larger airways, whereas β-agonists are equally effective in large and small airways. This relative diminution of cholinergic control in small airways may have important clinical implications, since anticholinergic drugs are likely to be less useful than β-agonists when bronchoconstriction involves small airways. Normal human subjects also have resting bronchomotor tone, since atropine causes bronchodilatation.

Cholinergic reflexes

A wide variety of stimuli are able to elicit reflex cholinergic bronchoconstriction through activation of sensory receptors in the larynx or lower airways. Activation of cholinergic reflexes may result in bronchoconstriction and an increase in

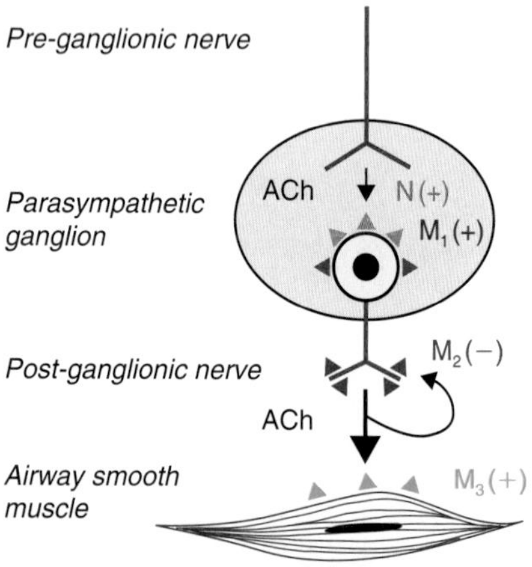

FIG. 32.7 Muscarinic receptor subtypes in the airways. Acetylcholine (ACh) from pre-ganglionic vagal nerves activates nicotinic receptors (N) and may be facilitated by M_1-receptors. ACh release from pre-ganglionic nerves activate M_3-receptors on airway smooth muscle and feeds back to activate pre-junctional M_2 receptors (autoreceptors) which inhibit further ACh release.

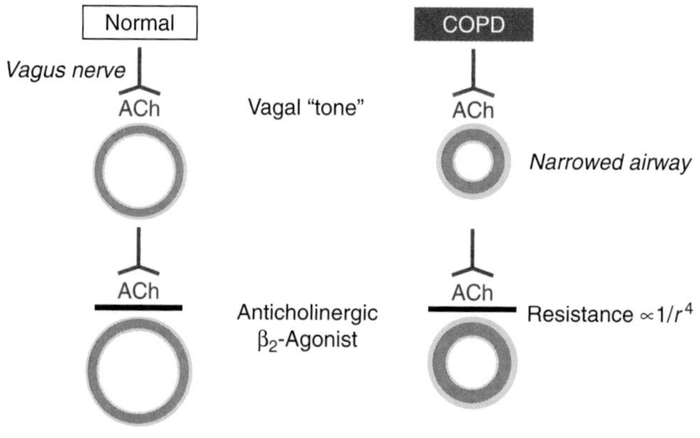

FIG. 32.8 Cholinergic control of airways in COPD. The normal vagal cholinergic tone has a greater effect in COPD airways because of the geometric relationship between airway diameter and resistance, so that anticholinergics are effective bronchodilators in COPD. β_2-Agonists are also effective and act as functional antagonists of cholinergic tone.

airway mucus secretion through the activation of muscarinic receptors on airway smooth muscle cells and submucosal glands. Cholinergic reflexes may also be activated from extrapulmonary afferents and these reflexes may also contribute to airway defenses. Esophageal reflux may be associated with bronchoconstriction in asthmatic patients. In some patients this may be due to aspiration of acid into the airways, in other cases acid reflux into the esophagus activates a reflex cholinergic bronchoconstriction (the "reflux reflex").

Modulation of cholinergic neurotransmission

Many agonists may modulate cholinergic neurotransmission via pre-junctional receptors on post-ganglionic nerves [29]. Some receptors increase (facilitate), whereas others inhibit the release of ACh. Inflammatory mediators may influence cholinergic neurotransmission via pre-junctional receptors. For example, thromboxane and prostaglandin (PG)D_2 facilitate ACh release from post-ganglionic nerves in the airways. Facilitation may also occur at parasympathetic ganglia in the airways; these structures are surrounded by inflammatory cells and have an afferent neural input. Electrophysiological recordings show a prolonged potentiation of neurotransmission in ganglia after allergen exposure in sensitized guinea pigs [30].

Role in asthma

Many of the stimuli which produce bronchospasm in asthma activate sensory nerves and reflex cholinergic bronchoconstriction in animals and there is some evidence that cholinergic tone is increased in asthmatic airways. There are several mechanisms by which cholinergic tone might be increased in asthma:

- Increased afferent stimulation by inflammatory mediators, such as histamine or prostaglandins.
- Increased release of ACh from cholinergic nerve terminals by an action on cholinergic nerve endings themselves, or by an increase in nerve traffic through cholinergic ganglia (local airway reflex).
- Abnormal muscarinic receptor expression, either via an increase in M_3-receptors or via reduction in M_2-receptors. There is no evidence for increased M_1- or M_3-receptor expression in asthmatic lungs [31], but there is functional evidence for a defect in M_2-receptor function that may be secondary to the inflammatory process.
- Decrease in the neuromodulators (VIP, NO) that have a "braking" effect on neurotransmission.

The effect of ACh on asthmatic airways is exaggerated, as a manifestation of non-specific hyperresponsiveness of the airways. Anticholinergic agents will only counteract the cholinergic reflex component of bronchoconstriction, which are less prominent in human airways than animal studies had indicated. By contrast β_2-agonists reverse bronchoconstriction irrespective of the mechanism, since they act as functional antagonists and always have a much greater bronchodilator effect than anticholinergic drugs. Vagal tone increases the airway narrowing further, and for geometric reasons will have a greater effect on airway resistance in narrowed airways. This may explain why anticholinergics are often of greater use in chronic asthmatics with a major element of fixed airway obstruction.

Role in COPD

The structural narrowing of the airways in COPD means that even normal vagal tone will exert a greater effect on airway caliber than in normal airways (Fig. 32.8). This may account for the efficacy of anticholinergics as bronchodilators in COPD, as cholinergic tone is the only reversible element and have similar or even greater effects than β_2-agonists. In addition, cholinergic mechanisms may account for the mucus hypersecretion of chronic bronchitis.

ADRENERGIC CONTROL

The airways are also under adrenergic control, which includes sympathetic nerves (which release noradrenaline), circulating catecholamines (predominantly adrenaline) and α- and β-adrenoceptors (Fig. 32.9). The fact that β-adrenergic antagonists cause bronchoconstriction in asthmatic patients, but not in normal individuals, suggests that adrenergic control of airway smooth muscle may be abnormal in asthma.

Sympathetic innervation

Although sympathetic bronchodilator nerves have been demonstrated in several species, including cats, dogs, and guinea pigs, most evidence suggests that adrenergic nerves do not control human airway smooth muscle directly. However, sympathetic nerves may influence cholinergic tone of airway smooth muscle via adrenoceptors localized to parasympathetic ganglia and pre-junctionally on post-ganglionic nerves [29], and sympathetic nerves may play an important role in the regulation of airway blood flow and in mucus secretion.

β-Adrenoceptors

β-Adrenoceptors regulate many aspects of airway function, including airway smooth muscle tone, mast cell mediator release, and plasma exudation. The possibility that β-receptors are abnormal in asthma has been extensively investigated. The suggestion that there is a primary defect in β-receptor function in asthma has not been substantiated and any defect in β-receptors is likely to be secondary to the disease, perhaps as a result of inflammation or as a consequence of adrenergic therapy. Some studies have demonstrated that airways from asthmatic patients fail to relax normally to isoproterenol, suggesting a possible defect in β-receptor function in airway smooth muscle [32]. Whether this is due to a reduction in β-receptors, a defect in receptor coupling, or some abnormality in the biochemical pathways leading to relaxation, is not yet known, although the density of β-receptors in airway smooth muscle appears to be normal [33] and there is no reduction in the density of β_1- or β_2-receptors in asthmatic lung, either at the receptor or at the mRNA level [31].

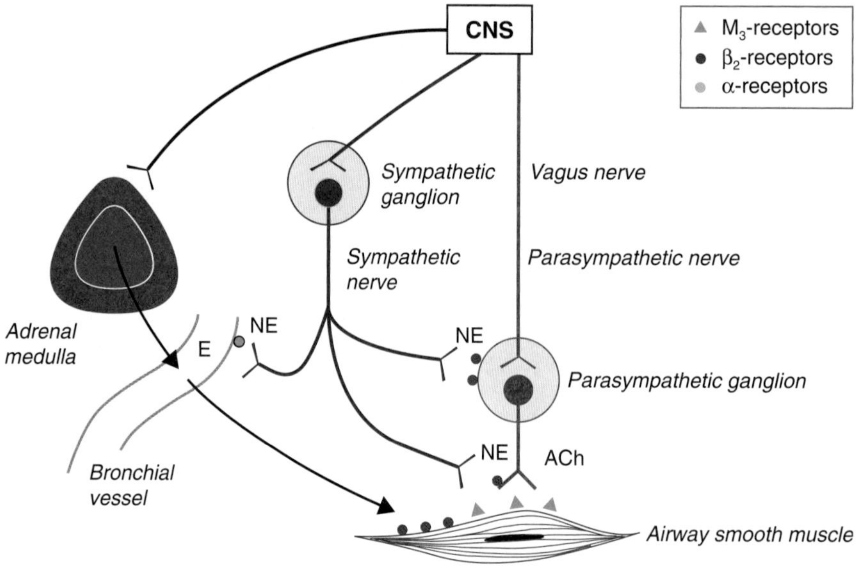

FIG. 32.9 Adrenergic control of airway smooth muscle. Sympathetic nerves release norepinephrine (NE), which may modulate cholinergic nerves at the level of the parasympathetic ganglion or post-ganglionic nerves, rather than directly at smooth muscle in human airways. Circulating epinephrine (E) is more likely to be important in adrenergic control of airway smooth muscle.

There is some evidence that proinflammatory cytokines may affect β_2-receptor function. IL-1β reduces the bronchodilator effect of isoprenaline *in vitro* and *in vivo* and this appears to be due to uncoupling of β_2-receptors due to increased expression of the inhibitory G protein, G_i [34, 35]. However, studies of β_2-receptor expression in asthmatic airways obtained by biopsy have demonstrated only small defects in coupling after local allergen challenge [36].

α-Adrenoceptors

α-Receptors which mediate bronchoconstriction have been demonstrated in airways of several species, and may only be demonstrated under certain experimental conditions. There is now considerable doubt about the role of α-receptors in the regulation of tone in human airways, however, since it has proved difficult to demonstrate their presence functionally or by autoradiography [37]. α-Adrenoceptor antagonists are not effective as bronchodilators, but it is possible that α-receptors may play an important role in regulating airway blood flow, which may indirectly influence airway responsiveness.

NANC NERVES AND NEUROPEPTIDES

Neural responses that are not blocked by a combination of adrenergic and cholinergic antagonists are known as non-adrenergic non-cholinergic (NANC) nerves. These NANC responses appear to be due to the release of neurotransmitters from classical autonomic nerves, which include neuropeptides, nitric oxide (NO), and adenosine triphosphate (ATP). In the airways both inhibitory NANC (bronchodilator) and excitatory NANC (bronchoconstrictor) nerves have been described.

i-NANC nerves

i-NANC nerves which mediate bronchodilatation have been described in may species, including humans, in whom they are of particular importance in the absence of any direct sympathetic innervation of airway smooth muscle [38]. The neurotransmitter for these nerves in some species, including guinea pigs and cats, is VIP and related peptides. The i-NANC bronchodilator response is blocked by α-chymotrypsin, an enzyme, which very efficiently degrades VIP and by antibodies to VIP. However, although VIP is present in human airways and VIP is a potent bronchodilator of human airways *in vitro*, there is no evidence that VIP is involved in neurotransmission of i-NANC responses in human airways, and α-chymotrypsin that completely blocks the response to exogenous VIP has no effect on neural bronchodilator responses [39]. It is likely that VIP and related peptides may be more important in neural vasodilatation responses and may result on increased blood flow to bronchoconstricted airways.

The predominant neurotransmitter of human airways is NO. NO synthase inhibitors, such as N^G-L-arginine methyl ester, virtually abolish the i-NANC response [40]. This effect is more marked in proximal airways, consistent with the demonstration that nitrergic innervation is greatest in proximal airways. NO appears to be a co-transmitter with ACh, and NO acts as a "braking" mechanism for the cholinergic system by acting as a functional antagonist to ACh at airway smooth muscle [41] (Fig. 32.10).

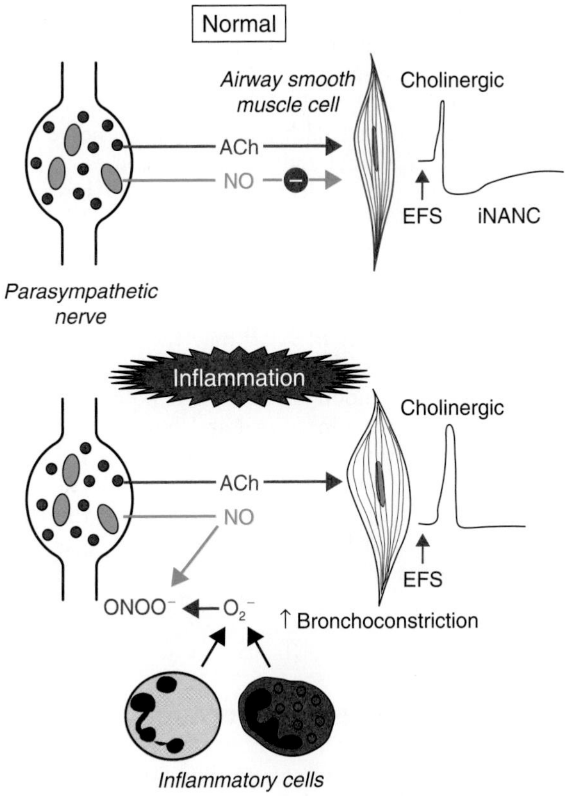

FIG. 32.10 Nitric oxide (NO) and vasoactive intestinal peptide (VIP) may modulate cholinergic neural effects mediated via acetylcholine (ACh). In inflammation NO may be removed by superoxide anions (O_2^-) generated form inflammatory cells and VIP by mast cell tryptase, therefore diminishing their "braking" effects, resulting in exaggerated cholinergic bronchoconstriction.

TABLE 32.1 Neuropeptides in the respiratory tract.

Neuropeptide	Localization
Vasoactive intestinal peptide	Parasympathetic
Peptide histidine isoleucine/methionine	(+Afferent)
Peptide histidine valine-42	
Helodermin	
Helospectins I and II	
PACAP-27	
Galanin	
Substance P	Afferent
Neurokinin A	(C-fibers)
Neuropeptide K	
Calcitonin gene-related peptide	
Gastrin-releasing peptide	
Secretoneurin	
Nociceptin	
Neuropeptide Y	Sympathetic
Opioids?	
Somatostatin	Afferent/uncertain
Enkephalin	
Endomorphins Cholecystokinin octapeptide	

Airway neuropeptides

Many neuropeptides are localized to sensory, parasympathetic, and sympathetic neurones in the human respiratory tract (Table 32.1). These peptides have potent effects on bronchomotor tone, airway secretions, bronchial circulation, and on inflammatory and immune cells [1]. Although the precise physiological roles of each peptide are not yet fully understood, clues are provided by their localization and functional effects. Recently the development of specific neuropeptide receptor antagonists has provided important new insights into the roles of these neurotransmitters. Many of the inflammatory and functional effects of neuropeptides are relevant to asthma and there is compelling evidence for the involvement of neuropeptides in the pathophysiology and symptomatology of asthma and COPD.

Although classically neuropeptides are released from autonomic nerves, there is increasing evidence that these peptides may be synthesized and released from inflammatory and non-neural structural cells, particularly in disease. Inflammatory cytokines may increase the expression of neuropeptide genes in inflammatory cells, so that inflammatory become a major source of the neuropeptide at the inflammatory site. For example, both VIP and SP have been localized to human eosinophils and SP to macrophages [42, 43].

Neuropeptides and airway inflammation

Neuropeptides have multiple inflammatory and immune effects on the airways, thereby intensifying the ongoing inflammation [44]. In turn, inflammatory mediators may amplify or sometimes dampen neuropeptide effects. Inflammatory mediators may increase the release of neuropeptides from sensory and other nerves, may increase the expression of neuropeptide genes in neural and inflammatory cells, may increase the expression of neuropeptide receptors and may decrease the degradation of neuropeptides.

Vasoactive intestinal peptide and related peptides

VIP-immunoreactive nerves are widely distributed throughout the respiratory tract in humans and there is also evidence for the presence of several closely related peptides (peptide histidine methionine, peptide histidine valine, helodermin, helospectins I and II, and pituitary adenylate cyclase activating peptide (PACAP-38) and PACAP-27 which have

TABLE 32.2 Effects of tachykinins on airways.

Effect	Neurokinin receptor
Bronchoconstriction (small > large airways)	NK_2, NK_1
Plasma exudation	NK_1
Bronchial vasodilatation	NK_1
↑ Neurotransmission (ganglia, cholinergic nerves)	NK_1, NK_2, NK_3
Cough (peripheral and central mechanisms)	NK_1, NK_2, NK_3
Mucus secretion (submucosal glands and goblet cells)	NK_1
↑ Adhesion molecules (ICAM-1, E-selectin)	NK_1
Activation of inflammatory cells (macrophages, T-lymphocytes, eosinophils)	NK_1, NK_2
Angiogenesis	NK_1
Fibroblast activation	NK_1, NK_2

similar functional effects. VIP may be localized to parasympathetic and sensory nerves. It is a potent vasodilator, a bronchodilator, increases mucus secretion and may have anti-inflammatory effects. In some species it is a mediator of neurogenic bronchodilatation, but this is not the case in human airways. A defect in VIP has been proposed in asthma, but there is little evidence for this [45].

Tachykinins

SP and NKA, but not NKB, are localized to C-fibers are abundant in rodent airways, but are sparse in human airways [46]. Tachykinins are also be expressed human macrophages, which also express tachykinin receptors. Tachykinins have many different effects on the airways which may be relevant to asthma and these effects are mediated via NK_1-receptors (preferentially activated by SP) and NK_2-receptors (activated by NKA) (Table 32.2). Tachykinins constrict smooth muscle of human airways *in vitro* via NK_2-receptors. Tachykinin receptors are widely distributed in human airways, with localization predominantly to airway smooth muscle, mucus-secreting cells and bronchial cells [47]. NKA causes bronchoconstriction after both intravenous and inhaled administration in asthmatic subjects. Mechanical removal of airway epithelium potentiates the bronchoconstrictor response to tachykinins, largely because the neutral endopeptidase (NEP), which is a key enzyme in the degradation of tachykinins in airways is strongly expressed on epithelial cells [48].

SP stimulates mucus secretion from submucosal glands in human airways *in vitro* and is a potent stimulant to goblet cell secretion in guinea pig airways via activation of NK_1-receptors. NK_1-receptors also mediate the increased plasma exudation and the vasodilator response to tachykinins. Tachykinins may activate alveolar macrophages and monocytes to release inflammatory cytokines, such as IL-6. Tachykinins also enhance cholinergic neurotransmission by facilitating acetylcholine release at cholinergic nerve terminals and by enhancing ganglionic transmission.

Tachykinins are subject to degradation by at least two enzymes, angiotensin-converting enzyme (ACE), and NEP. ACE is predominantly localized to vascular endothelial cells and therefore degrades intravascular peptides, whereas NEP is important for degrading tachykinins in the airways. The activity of NEP may therefore determine tachykinin responsiveness in the airways. Inhibition of NEP by phosphoramidon or thiorphan potentiates bronchoconstriction *in vitro* in animal and human airways and after inhalation *in vivo*. The activity of NEP is reduced by mechanical removal of the epithelium, virus infections, cigarette smoke, and hypertonic saline. Several of the stimuli known to induce bronchoconstrictor responses in asthmatic patients have been found to reduce the activity of airway NEP [49].

Calcitonin gene-related peptide

CGRP-immunoreactive nerves are abundant in the respiratory tract of several species and is co-stored and co-localized with SP in afferent nerves. CGRP is a potent vasodilator, which has long-lasting effects and potently dilates bronchial vessels *in vitro* and *in vivo*. It is possible that CGRP may be the predominant mediator of arterial vasodilatation and increased blood flow in response to sensory nerve stimulation in the bronchi. CGRP may be an important mediator of airway hyperemia in asthma. CGRP has variable effects on airway smooth muscle tone and appears to act indirectly through the relapse of other constrictors, such as endothelin. Like tachykinins, CGRP is chemotactic for eosinophils.

Neurogenic inflammation in airway disease

Sensory nerves may be involved in inflammatory responses through the antidromic release of neuropeptides from nociceptive nerves or C-fibers via a local (axon) reflex [50] (Fig. 32.11). The phenomenon is well documented in several organs, including skin, eye, gastrointestinal tract, and bladder. Neurogenic inflammation occurs in the respiratory tract and may contribute to the inflammatory response in asthma. Neurogenic inflammation is well documented in the airways of rodents, and there is good evidence that tachykinins contribute to the airway hyperresponsiveness in several animal models of asthma, using capsaicin depletion or specific tachykinin antagonists. However, although it was proposed several years ago that neurogenic inflammation and peptides released from sensory nerves might be important as an amplifying mechanism in asthmatic inflammation, there is little evidence to date to support this idea.

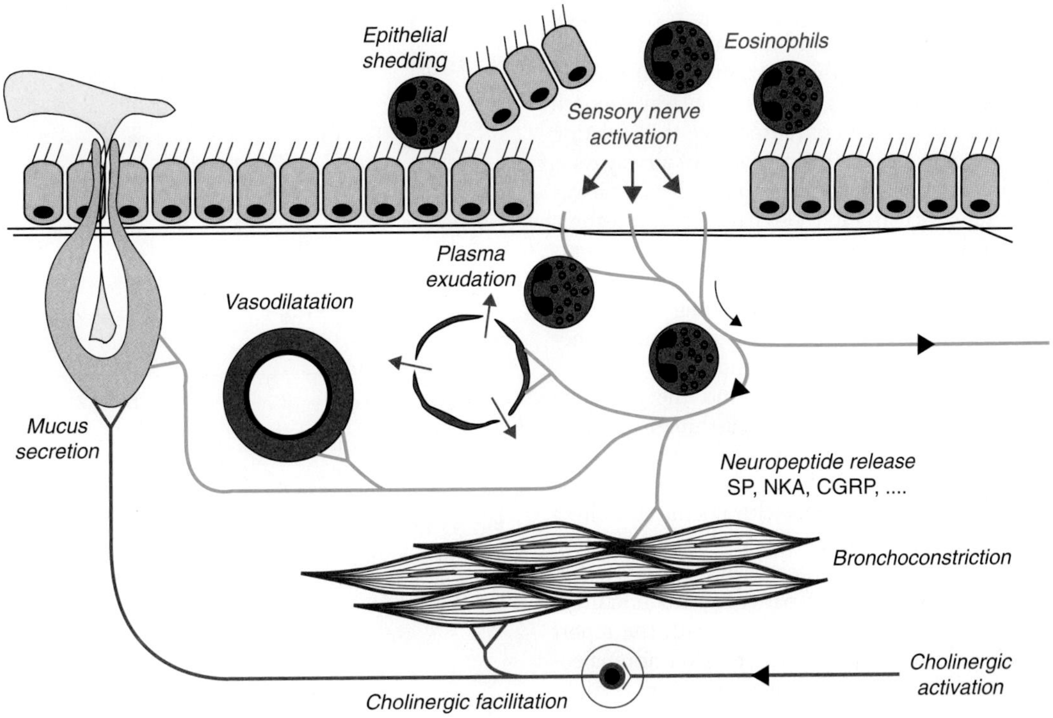

FIG. 32.11 Axon reflex in asthma. Possible neurogenic inflammation in asthmatic airways via retrograde release of peptides from sensory nerves via an axon reflex. Substance P (SP) causes vasodiatation, plasma exudation and mucus secretion, whereas neurokinin A (NKA) causes bronchoconstriction and enhanced cholinergic reflexes and calcitonin gene-related peptide (CGRP) vasodilatation.

There is some evidence in support of a role for tachykinins in asthma:

- An increase in SP-immunoreactive nerves has been described in patients with severe asthma [51].
- SP and NKA levels are increased in bronchoalveolar lavage fluid and induced sputum of asthmatic patients [46].
- There is increased expression of NK_1- and NK_2-receptors in asthmatic lungs and airways [52, 53]. However this is more evidence against a role for tachykinins:
- SP-immunoreactive nerves are sparse in human airways and are not increased in lungs and biopsies from asthmatic patients [45].
- Capsaicin has no effect on human airways *in vitro*, whereas it potently constructs guinea pig airways. Similarly, inhaled capsicum causes cough and transient bronchoconstriction, but not prolonged bronchoconstriction as in rodents.
- NEP inhibitors have no different effects in patients with asthma than normal subjects [54].
- Tachykinin antagonists are so far ineffective in asthma.

Neurogenic inflammation plays a role in COPD. SP levels are elevated in induced sputum of patients with COPD [55]. Cigarette smoke activates C-fibers in airways and may result in mucus hypersecretion and goblet cell discharge [56] and tachykinins are potent stimuli of mucus secretion in human airways [57].

Other neuropeptides

Several other neuropeptides have been identified in human airways (Table 32.1) [58], but their function is even less well defined that the neuropeptides discussed. Secretoneurin is a neuropeptide released by sensory nerves in the nose and is chemotactic for inflammatory cells such as eosinophils [59], but its role in asthma is uncertain.

CIRCULATING VASOACTIVE PEPTIDES AND HORMONES

Airway function can be altered by vasoactive peptides and hormones that reach the lungs from the bloodstream, as well as by neurotransmitters released from nerve endings and by molecules released locally from other cells within the airway [60].

Circulating catecholamines

Circulating epinephrine (adrenaline) is released from the adrenal medulla into the circulation. It may reduce bronchial smooth muscle tone directly by stimulating β_2-adrenergic receptors on airway smooth muscle, or indirectly by reducing acetylcholine release from cholinergic nerves. The lack of a bronchoconstrictor effect of β-antagonists in normal subjects suggests that in this group, basal concentrations of circulating adrenaline are probably not important in the regulation of resting bronchomotor tone. In contrast, β_2-antagonists cause bronchoconstriction in some asthmatics; in the absence of an important sympathetic nerve supply to airway smooth muscle, this suggests that basal concentrations of circulating epinephrine are important in the maintenance of airway tone. The controlling influence of circulating epinephrine on airway tone might operate particularly in those patients with asthma in whom resting airway caliber is already reduced. In patients with non-asthmatic COPD, however, β_2-antagonists do not normally cause bronchoconstriction. This finding would suggest that basal concentrations of circulating adrenaline are not important in the maintenance of airway tone in COPD.

Basal epinephrine concentrations and the circadian variation in epinephrine concentrations in asthmatic patients appear to be similar to those found in normal subjects [61–63]. Although Bates *et al.* reported that plasma epinephrine levels at 10 pm were lower in patients with nocturnal asthma than in a non-nocturnal asthma group [64]; correction of the nocturnal fall in plasma epinephrine does not alter the peak expiratory flow values of patients with nocturnal asthma [65]. These findings, taken together with the report of nocturnal asthma occurring in a patient after adrenalectomy [66], suggest that a fall in plasma epinephrine at night is not a dominant factor in nocturnal asthma.

Allergen- or pharmacological-induced bronchoconstriction do not cause the release of epinephrine, which suggests that epinephrine does not appear to have an important homeostatic role in the regulation of airway caliber during bronchoconstriction to these stimuli [67, 68]. Even during acute exacerbations of asthma there may be no elevation in plasma epinephrine levels [69, 70], although very high-epinephrine concentrations have been found in some patients with acute severe asthma. The elevated epinephrine concentrations achieved after strenuous exercise [71] cause bronchodilation in both normal and asthmatic subjects [62, 63, 72] and may counteract bronchospasm induced by exercise in asthma [73]. Although a blunted catecholamine response to exercise in asthmatic patients has been reported by some investigators, [74] other studies have found no significant difference in either the peak plasma catecholamine level between normal and asthmatic subjects or in the response to increasing levels of exercise (Fig. 32.12) [71, 75, 76].

Norepinephrine, which has β_1- and weak β_2-adrenergic activity in addition to α-adrenergic effects, acts as a neurotransmitter in the sympathetic nervous system but overspills into the circulation. The infusion of noradrenaline, producing circulating concentrations within the physiological and pathophysiological range, has no effect on airway caliber in either normal subjects or asthmatic subjects [62, 63]. The third catecholamine present in the blood, dopamine, has no influence on resting bronchomotor tone in normal or stable asthmatic subjects [77] although it may attenuate induced bronchoconstriction [78] and can decrease the rate of ventilation in animals and man [79].

Natriuretic peptides

Natriuretic peptides are a family of hormones that have an important role in salt and water homeostasis [80, 81].

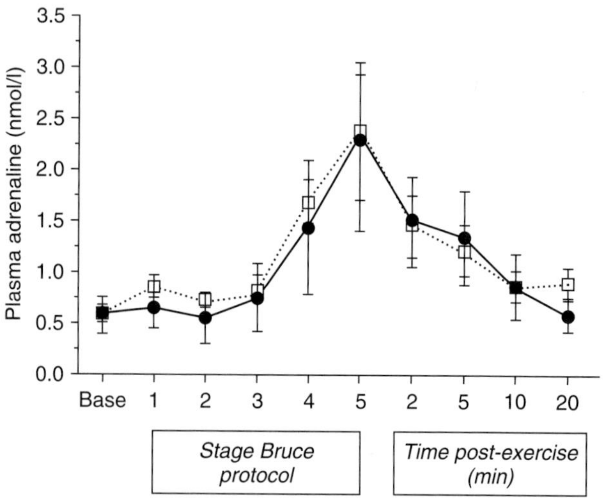

FIG. 32.12 Plasma epinephrine concentrations during the course of and for 20 min after maximal treadmill exercise in normal subjects (broken line) and asthmatic subjects (solid) (no significant difference). Reproduced from Ref. [76] with permission.

The human natriuretic peptides include atrial natriuretic peptide (ANP), brain natriuretic peptide (BNP), C-type natriuretic peptide (CNP), and urodilatin [82]. Most natriuretic peptides are produced primarily in the heart but are released also in other tissues, including the kidneys, lungs, and central nervous system. Specific ANP receptors have been localized to lung (including airway smooth muscle), of which some may be the ANP-$_c$ or clearance receptor subtype [83], although the receptor subtype(s) in human airway smooth muscle is unknown. In isolated human airway tissue, ANP has a direct relaxant effect and confers protection against agonist-induced contraction [84, 85]. Two principal mechanisms have been proposed for the inactivation of ANP: degradation by NEP and binding to a non-guanylyl cyclase clearance receptor (ANP-$_c$ receptor). NEP is widely distributed within the airways and plays a role in modulating the effect of ANP on airway smooth muscle [84, 85].

An intravenous infusion of exogenous ANP causes bronchodilation and attenuates bronchial reactivity to inhaled histamine and to fog challenge [86–90]. The rise in plasma ANP levels during exercise is similar to these obtained during the lowest rates of ANP infusion; these results suggest that the rise in circulating ANP levels during exercise may attenuation exercise-induced bronchospasm [76] (Fig. 32.13). Plasma ANP levels are elevated in patients with cardiac failure [91] and pulmonary hypertension secondary to COPD [92]; under these circumstances ANP may also play a protective role on the airways. Circulating ANP at physiological concentrations, however, appears unlikely to have any influence on bronchomotor tone in healthy subjects [88].

Angiotensin II

The renin-angiotensin system plays an important role in fluid and electrolyte homeostasis through the actions of the

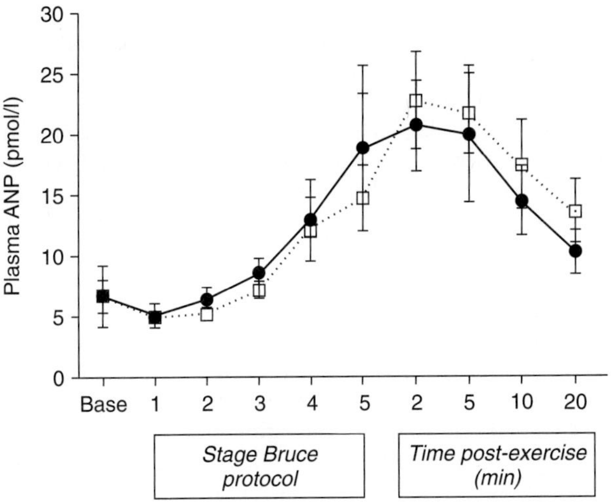

FIG. 32.13 Plasma ANP concentrations during exercise after maximal treadmill exercise in normal subjects (broken line) and asthmatic subjects (solid line) (no significant difference). Reproduced from Ref. [76] with permission.

octapeptide angiotensin II. Angiotensin II is formed from angiotensinogen by the action of renin and then ACE, 60–80% occurring within the pulmonary vascular endothelium. An alternative ACE-independent pathway, possibly mediated by several inflammatory proteases [93], may also cause the formation of angiotensin II.

The effect of physiological concentrations of angiotensin II on basal bronchial tone of normal individuals is not know, whereas infusion of angiotensin II in mild asthmatic patients to plasma levels found in acute asthma causes bronchoconstriction [94] (Fig. 32.14). Angiotensin II, although causing only weak contraction of isolated human and bovine bronchial rings, potentates the effects of methacholine and endothelin-1 *in vitro* [95, 96]. In patients with mild asthma, angiotensin II at subthreshold concentrations potentates methacholine-induced bronchoconstriction [96] but has no effect on endothelin- or histamine-evoked bronchoconstriction *in vivo* [97, 98]. Angiotensin II receptor antagonists reduce allergen-induced airway hyperresponsiveness and eosinophil accumulation in guinea pigs [99] and methacholine-induced bronchoconstriction in asthma [100]. These results suggest a role for angiotensin II as a putative mediator in asthma.

The renin-angiotensin system is activated in acute severe asthma, but not in stable chronic asthma [70, 94]. The mechanism of activation is unclear, but nebulized β_2-agonists cause elevation of renin and angiotensin II in normal and mild asthmatic subjects through an ACE-dependent pathway [101, 102]. This may occur via stimulation of β-adrenoceptors on juxtaglomerular cells, but the levels of angiotensin II seen in acute severe asthma are higher, suggesting the existence of an alternative pathway of angiotensin II formation. Exercise activates the renin-angiotensin system [103, 104], raising the possibility that elevated angiotensin II levels during exercise could contribute to exercise-induced bronchospasm. The renin-angiotensin system is also activated in patients with COPD and edema [105, 106].

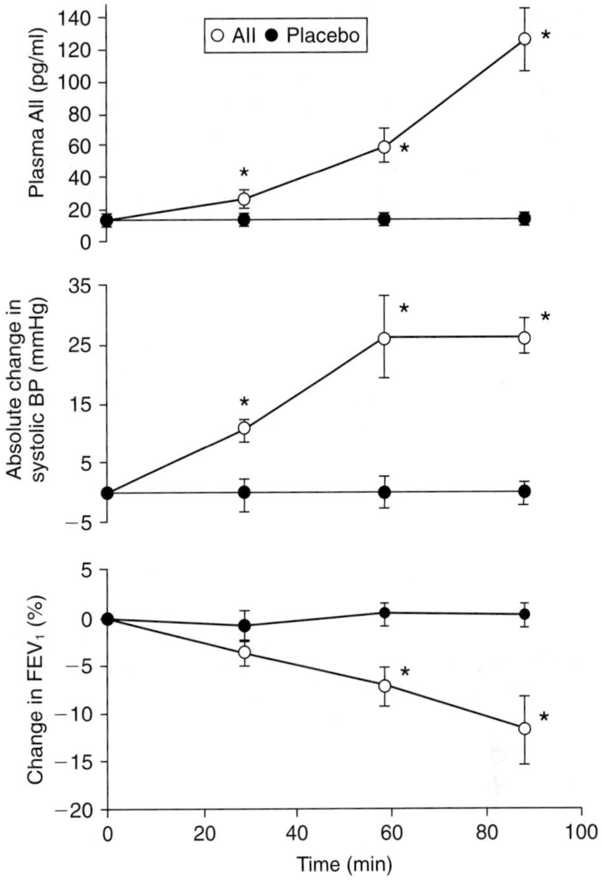

FIG. 32.14 Effect of infused angiotensin II (AII) on plasma levels of AII, changes in systolic blood pressure (BP) and change in FEV_1 from baseline values in eight asthmatic patients; *$p < 0.05$ versus placebo. Reproduced from Ref. [94] with permission.

Adrenomedullin

Adrenomedullin is a 52-amino acid peptide that possesses vasodilator and natriuretic properties [107]. The chemical structure of adrenomedullin is similar to CGRP. Immunoreactivity to adrenomedullin is found in tissues throughout the body including the lung [108]. Adrenomedullin has a long-lasting bronchodilator action in guinea pig trachea [109, 110], inhibits IL-8 secretion from alveolar macrophages [111] and attenuates antigen-induced microvascular leakage and bronchoconstriction in guinea pigs [112]. Plasma levels of adrenomedullin are elevated in hypoxic patients with COPD [113] and in severe chronic asthma [114]. Its role in the control of airway smooth tone and inflammation in asthma or COPD, however, remains uncertain.

Urotensin II

Urotensin II is an 11-amino acid cyclic peptide, which activates specific receptors that are widely distributed in smooth muscle [115]. It is a potent vasoconstrictor and a more potent constrictor of airway smooth muscle than endothelins [116]. Its role in regulating airway smooth muscle and vascular tone in asthma and COPD are currently unknown.

Cortisol

Pharmacological doses of intravenous cortisol have no short-term effect on airway caliber in normal subjects [117]. Although glucocorticoids can potentate the response to catecholamines in isolated bronchial tissue, the effect occurs only at supraphysiological concentrations [118]. These results suggest that endogenous cortisol is unlikely to have an important direct modulating effect on airway tone in normal individuals.

In asthma the role of physiological concentrations of circulating cortisol in airway function is uncertain. In nocturnal asthma, the nadir in the circadian variation in plasma cortisol occurs 4h before maximal bronchoconstriction [119], although the delayed action of cortisol means that it could still have an influence on airway caliber. A reduced nadir of plasma cortisol is found in patients with nocturnal asthma compared to a group without nocturnal asthma, but this finding may have been influenced by previous corticosteroid therapy [120]. Other studies have found no direct association between plasma cortisol concentrations and nocturnal asthma [63]. Furthermore, the infusion of physiological concentrations of hydrocortisone, eliminating the fall in plasma cortisol at night, does not prevent the nocturnal fall in peak flow rate in most asthmatic patients [119], suggesting that the circulating cortisol level is not the only factor in determining nocturnal asthma. Endogenous glucocorticoids rise following allergen challenge in allergic asthmatic subjects and may play a role in the modulation of allergen-induced bronchoconstriction [121].

Thyroid hormones

The relationship between asthma and thyroid disease provides indirect evidence of a role for thyroid hormones in maintaining airway function. The development of hyperthyroidism can be associated with deterioration in asthma control, with subsequent improvement in symptoms following appropriate treatment [122, 123]. Conversely the occurrence of hypothyroidism has been reported to be associated with improvement in asthma control, which relapses following subsequent thyroxine replacement [124]. The possible mechanisms by which thyroid hormones could influence airway smooth muscle tone and responsiveness are unclear [60].

Sex hormones

Progesterone has an important role in reducing the contractility of uterine smooth muscle during pregnancy, and this effect may be due to its influence on gap junction formation between smooth muscle cells. It has been suggested that progesterone might cause similar effects on bronchial smooth muscle. Progesterone could influence airway smooth muscle tone by other mechanisms indirectly either by potentiating the effect of catecholamines [118] or by potentiating airway inflammation [125]. Progesterone levels and airway responsiveness do not show a clear relationship during either the pregnancy or the menstrual cycle, although changes in the levels of other hormones may obscure an effect of progesterone on the airways [126, 127]. However, one report found that female athletes develop more severe exercise-induced asthma during the mid-luteal phase of the menstrual cycle when salivary progesterone levels were at there highest, suggesting that the menstrual cycle phase is an important determinant of the severity of exercise-induced asthma in female athletes with mild atopic asthma [128]. It is of interest that intramuscular progesterone has a beneficial effect in some women with severe menstrual asthma [129].

Estrogen possesses both immunostimulatory and immunosuppressive properties and causes increased acetylcholine activity in the lungs of animals [130], which could result in an increase or decrease in airway tone. Estrogen receptor-α deficient mice exhibit increased airway responsiveness, possibly due to reduced M_2 muscarinic receptor function and resultant increased acetylcholine release, suggesting a possible role for estrogen in regulating airway function [131]. A preliminary report suggested that estrogen treatment may have steroid-sparing effects in women with severe asthma [132].

Glucagon

Glucagon has bronchodilator actions, but whether it has a role in the control of airway smooth tone in humans is not established [133]. Certainly intravenous glucagon has no significant bronchodilation action in patients with acute asthma [134].

ROLE OF HUMORAL MECHANISMS IN ASTHMA AND COPD

Circulating hormones and vasoactive peptides appear to play a minor role in the physiological regulation of airway tone in normal individuals. Epinephrine is the only hormone known to influence bronchomotor tone, and it is only during strenuous exercise that concentrations are elevated sufficiently to cause bronchodilation.

Humoral mechanisms play a more important role in the regulation of airway tone in diseased states of the airways such as asthma and possibly in other disorders such as COPD, cor pulmonale, congestive cardiac failure, respiratory failure, and thyroid disease. Circulating epinephrine has a role in the maintenance of resting airway tone in asthma, perhaps particularly in those patients in whom resting airway caliber is already reduced. The elevated epinephrine and ANP concentrations achieved after vigorous exercise may act to counteract exercise-induced asthma. It has not been established in asthma whether elevated circulating angiotensin II achieved during exercise, or more particularly in acute severe asthma, contributes to bronchospasm.

References

1. Barnes PJ, Baraniuk J, Belvisi MG. Neuropeptides in the respiratory tract. *Am Rev Respir Dis* 144(1187–1198): 1391–99, 1991.
2. Barnes PJ. Autonomic Control of the Respiratory System. London: Harwood, 1997.

3. Barnes PJ. Neurogenic inflammation in the airways. *Respir Physiol* 125: 145–54, 2001.
4. Nassenstein C, Kutschker J, Tumes D, Braun A. Neuro-immune interaction in allergic asthma: Role of neurotrophins. *Biochem Soc Trans* 34: 591–93, 2006.
5. Taylor-Clark T, Undem BJ. Transduction mechanisms in airway sensory nerves. *J Appl Physiol* 101: 950–59, 2006.
6. Fox AJ, Barnes PJ, Urban L, Dray A. An *in vitro* study of the properties of single vagal afferents innervating guinea-pig airways. *J Physiol* 469: 21–35, 1993.
7. Fox AJ, Lalloo UG, Belvisi MG, Bernareggi M, Chung KF, Barnes PJ. Bradykinin-evoked sensitization of airway sensory nerves: A mechanism for ACE-inhibitor cough. *Nature Med* 2: 814–17, 1996.
8. Fox AJ, Barnes PJ, Dray A. Stimulation of afferent fibres in the guinea pig trachea by non-isomotic and low chloride solutions and its modulation by frusemide. *J Physiol* 482: 179–87, 1995.
9. Jia Y, Lee LY. Role of TRPV receptors in respiratory diseases. *Biochim Biophys Acta* 1772: 915–27, 2007.
10. McGarvey LP, Morice AH. Clinical cough and its mechanisms. *Respir Physiol Neurobiol* 152: 363–71, 2006.
11. Choudry NB, Fuller RW, Pride NB. Sensitivity of the human cough reflex: Effect of inflammatory mediators prostaglandin E*2, bradykinin and histamine. Am Rev Respir Dis* 140: 137–41, 1989.
12. Nicholson CD, Challiss RAJ, Shahid M. Differential modulation of tissue function and therapeutic potential of selective inhibitors of cyclic nucleotide phosphodiesterase isoenzymes. *Trends Pharmacol Sci* 12: 19–27, 1991.
13. Doherty MJ, Mister R, Pearson MG, Calverley PM. Capsaicin responsiveness and cough in asthma and chronic obstructive pulmonary disease. *Thorax* 55: 643–49, 2000.
14. Carr MJ, Hunter DD, Undem BJ. Neurotrophins and asthma. *Curr Opin Pulm Med* 7: 1–7, 2001.
15. Freund-Michel V, Frossard N. The nerve growth factor and its receptors in airway inflammatory diseases. *Pharmacol Ther* 117: 52–76, 2008.
16. Fox AJ, Patel HJ, Barnes PJ, Belvisi MG. Release of nerve growth factor by human pulmonary epithelial cells: Role in airway inflammatory diseases. *Eur J Pharmacol* 424: 159–62, 2001.
17. de Vries A, Engels F, Henricks PA, Leusink-Muis T, McGregor GP, Braun A *et al.* Airway hyper-responsiveness in allergic asthma in guinea-pigs is mediated by nerve growth factor via the induction of substance P: A potential role for trkA. *Clin Exp Allergy* 36: 1192–200, 2006.
18. Kassel O, de Blay F, Duvernelle C, Olgart C, Israel-Biet D, Krieger P *et al.* Local increase in the number of mast cells and expression of nerve growth factor in the bronchus of asthmatic patients after repeated inhalation of allergen at low-dose. *Clin Exp Allergy* 31: 1432–40, 2001.
19. Watson JJ, Fahey MS, van den WE, Engels F, Nijkamp FP, Stroemer P *et al.* TrkAd5: A novel therapeutic agent for treatment of inflammatory pain and asthma. *J Pharmacol Exp Ther* 316: 1122–29, 2006.
20. Grando SA, Kawashima K, Kirkpatrick CJ, Wessler I. Recent progress in understanding the non-neuronal cholinergic system in humans. *Life Sci* 80: 2181–85, 2007.
21. Gwilt CR, Donnelly LE, Rogers DF. The non-neuronal cholinergic system in the airways: An unappreciated regulatory role in pulmonary inflammation? *Pharmacol Ther* 115: 208–22, 2007.
22. Neumann S, Razen M, Habermehl P, Meyer CU, Zepp F, Kirkpatrick CJ *et al.* The non-neuronal cholinergic system in peripheral blood cells: Effects of nicotinic and muscarinic receptor antagonists on phagocytosis, respiratory burst and migration. *Life Sci* 80: 2361–64, 2007.
23. Barnes PJ. Muscarinic receptor subtypes in airways. *Life Sci* 52: 521–28, 1993.
24. Fryer AD, Adamko DJ, Yost BL, Jacoby DB. Effects of inflammatory cells on neuronal M_2 muscarinic receptor function in the lung. *Life Sci* 64: 449–55, 1999.
25. Patel HJ, Barnes PJ, Takahashi T, Tadjkarimi S, Yacoub MH, Belvisi MG. Characterization of prejunctional muscarinic autoreceptors in human and guinea-pig trachea *in vitro. Am J Respir Crit Care Med* 152: 872–78, 1995.
26. Minette PAH, Lammers J, Dixon CMS, McCusker MT, Barnes PJ. A muscarinic agonist inhibits reflex bronchoconstriction in normal but not in asthmatic subjects. *J Appl Physiol* 67: 2461–65, 1989.
27. Ind PW, Dixon CMS, Fuller RW, Barnes PJ. Anticholinergic blockade of beta-blocker induced bronchoconstriction. *Am Rev Respir Dis* 139: 1390–94, 1989.
28. Mak JCW, Baraniuk JN, Barnes PJ. Localization of muscarinic receptor subtype mRNAs in human lung. *Am J Respir Cell Mol Biol* 7: 344–48, 1992.
29. Barnes PJ. Modulation of neurotransmission in airways. *Physiol Rev* 72: 699–729, 1992.
30. Undem BJ, Riccio MM, Weinreich D, Ellis JL, Myers AC. Neurophysiology of mast cell–nerve interactions in the airways. *Int Arch Allergy Immunol* 107: 199–201, 1995.
31. Haddad E-B, Mak JCW, Barnes PJ. Expression of β-adrenergic and muscarinic receptors in human lung. *Am J Physiol* 270: L947–53, 1996.
32. Bai TR. Abnormalities in airway smooth muscle in fatal asthma: A comparison between trachea and bronchus. *Am Rev Respir Dis* 143: 441–43, 1991.
33. Spina D, Rigby PJ, Paterson JW, Goldie RG. Autoradiographic localization of beta-adrenoceptors in asthmatic human lung. *Am Rev Respir Dis* 140: 1410–15, 1989.
34. Hakonarson H, Herrick DJ, Serrano PG, Grunstein MM. Mechanism of cytokine-induced modulation of β-adrenoceptor responsiveness in airway smooth muscle. *J Clin Invest* 97: 2593–600, 1996.
35. Koto H, Mak JCW, Haddad EB, Xu WB, Salmon M, Barnes PJ *et al.* Mechanisms of impaired β-adrenergic receptor relaxation by interleukin-1β *in vivo* in rat. *J Clin Invest* 98: 1780–87, 1996.
36. Penn RB, Shaver JR, Zangrilli JG, Pollice M, Fish JE, Peters SP *et al.* Effects of inflammation and acute β-agonist inhalation on $β_2$-AR signaling in human airways. *Am J Physiol* 271: L601–08, 1996.
37. Spina D, Rigby PJ, Paterson JW, Goldie RG. β-Adrenoceptor function and autoradiographic distribution in human asthmatic lung. *Br J Pharmacol* 97: 701–8, 1989.
38. Lammers JWJ, Barnes PJ, Chung KF. Non-adrenergic, non-cholinergic airway inhibitory nerves. *Eur Respir J* 5: 239–46, 1992.
39. Belvisi MG, Stretton CD, Miura M, Verleden GM, Tadjarimi S, Yacoub MH *et al.* Inhibitory NANC nerves in human tracheal smooth muscle: A quest for the neurotransmitter. *J Appl Physiol* 73: 2505–10, 1992.
40. Belvisi MG, Ward JR, Mitchell JA, Barnes PJ. Nitric oxide as a neurotransmitter in human airways. *Arch Int Pharmacodyn Ther* 329: 111–20, 1995.
41. Ward JK, Belvisi MG, Fox AJ, Miura M, Tadjkarimi S, Yacoub MH *et al.* Modulation of cholinergic neural bronchoconstriction by endogenous nitric oxide and vasoactive intestinal peptide in human airways *in vitro. J Clin Invest* 92: 736–43, 1993.
42. Ho WZ, Lai JP, Zhu XH, Uvaydova M, Douglas SD. Human monocytes and macrophages express substance P and neurokinin-1 receptor. *J Immunol* 159: 5654–60, 1997.
43. Germonpre PR, Bullock GR, Lambrecht BN, Van DVV, Luyten WH, Joos GF *et al.* Presence of substance P and neurokinin 1 receptors in human sputum macrophages and U-937 cells. *Eur Respir J* 14: 776–82, 1999.
44. Barnes PJ. Cytokine modulators as novel therapies for airway disease. *Eur Respir J Suppl* 34: 67s–77s, 2001.
45. Lilly CM, Bai TR, Shore SA, Hall AE, Drazen JM. Neuropeptide content of lungs from asthmatic and nonasthmatic patients. *Am J Respir Crit Care Med* 151: 548–53, 1995.
46. De Swert KO, Joos GF. Extending the understanding of sensory neuropeptides. *Eur J Pharmacol* 533: 171–81, 2006.
47. Mapp CE, Miotto D, Braccioni F, Saetta M, Turato G, Maestrelli P *et al.* The distribution of neurokinin-1 and neurokinin-2 receptors in human central airways. *Am J Respir Crit Care Med* 161: 207–15, 2000.
48. Frossard N, Rhoden KJ, Barnes PJ. Influence of epithelium on guinea pig airway responses to tachykinins: Role of endopeptidase and cyclooxygenase. *J Pharmacol Exp Ther* 248: 292–98, 1989.

49. Di Maria GU, Bellofiore S, Geppetti P. Regulation of airway neurogenic inflammation by neutral endopeptidase [In Process Citation]. *Eur Respir J* 12: 1454–62, 1998.
50. Barnes PJ. Asthma as an axon reflex. *Lancet* i: 242–45, 1986.
51. Ollerenshaw SL, Jarvis D, Sullivan CE, Woolcock AJ. Substance P immunoreactive nerves in airways from asthmatics and non-asthmatics. *Eur Respir J* 4: 673–82, 1991.
52. Adcock IM, Peters M, Gelder C, Shirasaki H, Brown CR, Barnes PJ. Increased tachykinin receptor gene expression in asthmatic lung and its modulation by steroids. *J Mol Endocrinol* 11: 1–7, 1993.
53. Bai TR, Zhou D, Weir T, Walker B, Hegele R, Hayashi S *et al.* Substance P (NK_1)- and neurokinin A (NK_2)-receptor gene expression in inflammatory airway diseases. *Am J Physiol* 269: L309–17, 1995.
54. Cheung D, Timmers MC, Zwinderman AH, den Hartigh J, Dijkman JH, Sterk PJ. Neonatal endopeptidase activity and airway hyperesponsiveness to neurokin A in asthmatic subjects *in vivo*. *Am Rev Respir Dis* 148: 1467–73, 1993.
55. Tomaki M, Ichinose M, Miura M, Hirayama Y, Yamauchi H, Nakajima N *et al.* Elevated substance P content in induced sputum from patients with asthma and patients with chronic bronchitis. *Am J Respir Crit Care Med* 151: 613–17, 1995.
56. Kuo H-P, Rohde JAL, Barnes PJ, Rogers DF. Cigarette smoke induced goblet cell secretion: Neural involvement in guinea pig trachea. *Eur Respir J* 3: 1895, 1990.
57. Rogers DF, Aursudkij B, Barnes PJ. Effects of tachykinins on mucus secretion on human bronchi *in vitro*. *Eur J Pharmacol* 174: 283–86, 1989.
58. Uddman R, Hakanson R, Luts A, Sundler F. Distribution of neuropeptides in airways. In: Barnes PJ (ed.), *Autonomic Control of the Respiratory System*, pp. 21–37. London: Harwood Academic, 1997.
59. Korsgren M, Fischer-Colbrie R, Andersson M, Coman WB, Mackay-Sim A, Persson CG *et al.* Secretoneurin is released into human airways by topical histamine but not capsaicin. *Allergy* 60: 459–63, 2005.
60. Thomson N, Dagg K, Ramsay S. Humoral control of airway tone. *Thorax* 51: 461–64, 1996.
61. Berkin K, Inglis G, Ball S, Thomson N. Effect of low dose adrenaline and noradrenaline infusions on airway calibre in asthmatic patients. *Clin Sci* 70: 347–52, 1986.
62. Berkin K, Inglis G, Ball S, Thomson N. Airway responses to low concentrations of adrenaline and noradrenaline in normal subjects. *Q J Exp Physiol* 70: 203–9, 1985.
63. Barnes P, FitzGerald G, Brown M, Dollery C. Nocturnal asthma and changes in circulating epinephrine, histamine, and cortisol. *N Engl J Med* 303: 263–67, 1980.
64. Bates M, Clayton M, Calhoun W, Jarjour N, Schrader L, Geiger K *et al.* Relationship of plasma epinephrine and circulating eosinophils to nocturnal asthma. *Am J Respir Crit Care Med* 149: 667–72, 1994.
65. Morrison J, Teale C, Pearson S, Marshall P, Dwyer N, Jones S *et al.* Adrenaline and nocturnal asthma. *BMJ* 301: 473–76, 1990.
66. Morice A, Sever P, Ind P. Adrenaline, bronchoconstriction, and asthma. *BMJ* 293: 539–40, 1986.
67. Arvidsson P, Larsson S, Löfdahl C, Melander B, Wåhlander L, Svedmyr N. Formoterol, a new long-acting bronchodilator for inhalation. *Eur Respir J* 2: 325–30, 1989.
68. Larsson K, Gronneberg R, Hjemdahl P. Bronchodilatation and inhibition of allergen-induced bronchoconstriction by circulating epinephrine in asthmatic subjects. *J Allergy Clin Immunol* 75: 586–93, 1985.
69. Ind P, Causon R, Brown M, Barnes P. Circulating catecholamines in acute asthma. *BMJ* 290: 267–69, 1985.
70. Ramsay SG, Dagg KD, McKay IC, Lipworth BJ, McSharry C, Thomson NC. Investigations on the renin-angiotensin system in acute severe asthma. *Eur Respir J* 10: 2766–71, 1997.
71. Berkin K, Walker G, Inglis G, Ball S, Thomson N. Circulating adrenaline and noradrenaline concentrations during exercise in patients with exercise induced asthma and normal subjects. *Thorax* 43: 295–99, 1988.
72. Warren J, Dalton N. A comparison of the bronchodilator and vasopressor effects of exercise levels of adrenaline in man. *Clin Sci* 64: 475–79, 1983.
73. Knox A, Campos-Gongora H, Wisniewski A, MacDonald I, Tattersfield A. Modification of bronchial reactivity by physiological concentrations of plasma epinephrine. *J Appl Physiol* 73: 1004–7, 1992.
74. Barnes P, Brown M, Silverman M, Dollery C. Circulating catecholamines in exercise and hyperventilation induced asthma. *Thorax* 36: 435–40, 1981.
75. Gilbert I, Lenner K, McFadden EJ. Sympathoadrenal response to repetitive exercise in normal and asthmatic subjects. *J Appl Physiol* 64: 2667–74, 1988.
76. Hulks G, Mohammed A, Jardine A, Connell J, Thomson NC. Circulating plasma concentrations of atrial natriuretic peptide and catecholamines in response to maximal exercise in normal and asthmatic subjects. *Thorax* 46: 824–28, 1991.
77. Thomson N, Patel K. Effect of dopamine on airways conductance in normals and extrinsic asthmatics. *Br J Clin Pharmacol* 5: 421–24, 1978.
78. Michoud M, Amyot R, Jeanneret-Grosjean A. Dopamine effect on bronchomotor tone *in vivo*. *Am Rev Respir Dis* 130: 755–58, 1984.
79. Saaresranta T, Polo O. Hormones and breathing. *Chest* 122: 2165–82, 2002.
80. Ruskoaho H. Atrial natriuretic peptide: Synthesis, release, and metabolism. *Pharmacol Rev* 44: 479–602, 1992.
81. Levin ER, Gardner DG, Samson WK. Natriuretic peptides. *N Engl J Med* 339: 321–28, 1998.
82. Hamad AM, Clayton A, Islam B, Knox AJ. Guanylyl cyclases, nitric oxide, natriuretic peptides, and airway smooth muscle function. *Am J Physiol Lung Cell Mol Physiol* 285: L973–83, 2003.
83. James S, Burnstock G. Atrial and brain natriuretic peptides share binding sites on cultured cells from the rat trachea. *Cell Tissue Res* 265: 555–65, 1991.
84. Angus R, Nally J, McCall R, Young L, McGrath J, Thomson NC. Modulation of the effect of atrial natriuretic peptide in human and bovine bronchi by phosphoramidon. *Clin Sci* 86: 291–95, 1994.
85. Nally J, Clayton R, Thomson N, McGrath J. The interaction of alpha-human atrial natriuretic peptide (ANP) with salbutamol, sodium nitroprusside and isosorbide dinitrate in human bronchial smooth muscle. *Br J Pharmacol* 113: 1328–32, 1994.
86. Hulks G, Jardine A, Connell J, Thomson N. Bronchodilator effect of atrial natriuretic peptide in asthma. *BMJ* 299: 1081–82, 1989.
87. Chanez P, Mann CJB, Chabrier P, Godard P, Braquet P *et al.* Atrial natriuretic factor (ANF) is a potent bronchodilator in asthma. *J Allergy Clin Immunol* 86: 321–24, 1990.
88. Hulks G, Jardine A, Connell J, Thomson NC. Effect of atrial natriuretic factor on bronchomotor tone in the normal human airway. *Clin Sci* 79: 51–55, 1990.
89. Mcalpine L, Hulks G, Thomson N. Effect of atrial natriuretic peptide given by intravenous infusion on bronchoconstriction induced by ultrasonically nebulized distilled water (fog). *Am Rev Respir Dis* 146: 912–15, 1992.
90. Angus R, McCallum M, Hulks G, Thomson N. Bronchodilator, cardiovascular, and cyclic guanylyl monophosphate response to high-dose infused atrial natriuretic peptide in asthma. *Am Rev Respir Dis* 147: 1122–25, 1993.
91. Raine A, Erne P, Bürgisser E, Müller F, Bolli P, Burkart F *et al.* Atrial natriuretic peptide and atrial pressure in patients with congestive heart failure. *N Engl J Med* 315: 533–37, 1986.
92. Burghuber O, Hartter E, Punzengruber C, Weissel M, Woloszczuk W. Human atrial natriuretic peptide secretion in precapillary pulmonary hypertension. Clinical study in patients with COPD and interstitial fibrosis. *Chest* 93: 31–37, 1988.
93. Husain A. The chymase–angiotensin system in humans. *J Hypertens* 11: 1155–59, 1993.
94. Millar E, Angus R, Hulks G, Morton J, Connell J, Thomson N. Activity of the renin–angiotensin system in acute severe asthma and the effect of angiotensin II on lung function. *Thorax* 49: 492–95, 1994.
95. Nally J, Clayton R, Wakelam M, Thomson N, McGrath J. Angiotensin II enhances responses to endothelin-1 in bovine bronchial smooth muscle. *Pulm Pharmacol* 7: 409–13, 1994.

96. Millar EA, Nally JE, Thomson NC. Angiotensin II potentiates methacholine-induced bronchoconstriction in human airway both *in vitro* and *in vivo*. *Eur Respir J* 8: 1838–41, 1995.
97. Chalmers GW, Millar EA, Little SA, Shepherd MC, Thomson NC. Effect of infused angiotensin II on the bronchoconstrictor activity of inhaled endothelin-1 in asthma. *Chest* 115: 352–56, 1999.
98. Ramsay S, Clayton R, Dagg K, Thomson L, Nally J, Thomson N. Effect of angiotensin II on histamine-induced bronchoconstriction in the human airway both *in vitro* and *in vivo*. *Respir Med* 91: 609–15, 1997.
99. Myou S, Fujimura M, Kurashima K, Tachibana H, Watanabe K, Hirose T. Type 1 angiotensin II receptor antagonism reduces antigen-induced airway reactions. *Am J Respir Crit Care Med* 162: 45–49, 2000.
100. Myou S, Fujimura M, Kamio Y, Ishiura Y, Kurashima K, Tachibana H *et al.* Effect of Losartan, a type 1 angiotensin II receptor antagonist, on bronchial hyperresponsiveness to methacholine in patients with bronchial asthma. *Am J Respir Crit Care Med* 162: 40–44, 2000.
101. Millar E, McInnes G, Thomson N. Investigation of the mechanism of beta 2-agonist-induced activation of the renin-angiotensin system. *Clin Sci* 88: 433–37, 1995.
102. Millar EA, Connell JM, Thomson NC. The effect of nebulized albuterol on the activity of the renin–angiotensin system in asthma. *Chest* 111: 71–74, 1997.
103. Kosunen KJ, Pakarinen AJ, Kuoppasalmi K, Adlercreutz H. Plasma renin activity, angiotensin II, and aldosterone during intense heat stress. *J Appl Physiol* 41: 323–27, 1976.
104. Milledge JS, Catley DM. Renin, aldosterone, and converting enzyme during exercise and acute hypoxia in humans. *J Appl Physiol* 52: 320–23, 1982.
105. Reihman DH, Farber MO, Weinberger MH, Henry DP, Fineberg NS, Dowdeswell IRG *et al.* Effect of hypoxemia on sodium and water excretion in chronic obstructive lung disease. *Am J Med* 78: 87–94, 1985.
106. Anand I, Chandrashekhar Y, Ferrari R, Sarma R, Guleria R, Jindal S *et al.* Pathogenesis of congestive state in chronic obstructive pulmonary disease. Studies of body water and sodium, renal function, hemodynamics, and plasma hormones during edema and after recovery. *Circulation* 86: 12–21, 1992.
107. Jougasaki M, Burnett JC. Adrenomedullin: Potential in physiology and pathophysiology. *Life Sci* 66: 855–72, 2000.
108. Ichiki Y, Kitamura K, Kangawa K, Kawamoto M, Matsuo H, Eto T. Distribution and characterization of immunoreactive adrenomedullin in human tissue and plasma. *FEBS Lett* 338: 6–10, 1994.
109. Kanazawa H, Kurihara N, Hirata K, Kudoh S, Kawaguchi T, Takeda T. Adrenomedullin, a newly discovered hypotensive peptide, is a potent bronchodilator. *Biochem Biophys Res Commun* 205: 251–54, 1994.
110. Nishimura J, Seguchi H, Sakihara C, Kureishi Y, Yoshimura H, Kobayashi S *et al.* The relaxant effect of adrenomedullin on particular smooth muscles despite a general expression of its mRNA in smooth muscle, endothelial and epithelial cells. *Br J Pharmacol* 120: 193–200, 1997.
111. Kamoi H, Kanazawa H, Hirata K, Kurihara N, Yano Y, Otani S. Adrenomedullin inhibits the secretion of cytokine-induced neutrophil chemoattractant, a member of the interleukin-8 family, from rat alveolar macrophages. *Biochem Biophys Res Commun* 211: 1031–35, 1995.
112. Ohbayashi H, Suito H, Yoshida N, Ilto Y, Kume H, Yamaki K. Adrenomedullin inhibits ovalbumin-induced bronchoconstriction and airway microvascular leakage in guinea-pigs. *Eur Respir J* 14: 1076–81, 1999.
113. Cheung B, Leung R. Elevated plasma levels of human adrenomedullin in cardiovascular, respiratory, hepatic and renal disorders. *Clin Sci* 92: 59–62, 1997.
114. Ceyhan B, Karakurt S, Hekim N. Plasma adrenomedullin levels in asthmatic patients. *J Asthma* 38: 221–27, 2001.
115. Ames RS, Sarau HM, Chambers JK, Willette RN, Aiyar NV, Romanic AM *et al.* Human urotensin-II is a potent vasoconstrictor and agonist for the orphan receptor GPR14. *Nature* 401: 282–86, 1999.
116. Hay DWP, Luttmann MA, Douglas SA. Human urotensin-II is a potent spasmogen of primate airway smooth muscle. *Br J Pharmacol* 131: 10–12, 2000.
117. Ramsdell JW, Berry CC, Clausen JL. The immediate effects of cortisol on pulmonary function in normals and asthmatics. *J Allergy Clin Immunol* 72: 69–74, 1983.
118. Foster P, Goldie R, Paterson J. Effect of steroids on beta-adrenoceptor-mediated relaxation of pig bronchus. *Br J Pharmacol* 78: 441–45, 1983.
119. Soutar C, Costello J, Ijaduola O, Turner-Warwick M. Nocturnal and morning asthma, relationship to plasma corticosteroids and response to cortisol infusion. *Thorax* 30: 436–40, 1975.
120. Kallenbach JM, Panz VR, Joffe BI, Jankelow D, Anderson R, Haitas B *et al.* Nocturnal events related to "morning dipping" in bronchial asthma. *Chest* 93: 751–57, 1988.
121. Stokes P, Togias, Bickel, Diemer, Hubbard, Schleimer. Endogenous glucocorticoids and antigen-induced acute and late phase pulmonary responses. *Clin Exp Allergy* 30: 1257–65, 2000.
122. Ayres J, Clark T. Asthma and the thyroid. *Lancet* 2: 1110–11, 1981.
123. Lipworth B, Dhillon D, Clark R, Newton R. Problems with asthma following treatment of thyrotoxicosis. *Br J Dis Chest* 82: 310–14, 1988.
124. Bush RK, Ehrlich EN, Reed CE. Thyroid disease and asthma. *J Allergy Clin Immunol* 59: 398–401, 1977.
125. Hellings PW, Vandekerckhove P, Claeys R, Billen J, Kasran A, Ceuppens JL. Progesterone increases airway eosinophilia and hyper-responsiveness in a murine model of allergic asthma. *Clin Exp Allergy* 33: 1457–63, 2003.
126. Juniper E, Daniel E, Roberts R, Kline P, Hargreave F, Newhouse M. Improvement in airway responsiveness and asthma severity during pregnancy. A prospective study. *Am Rev Respir Dis* 140: 924–31, 1989.
127. Tan KS, Thomson NC. Asthma in pregnancy. *Am J Med* 109: 727–33, 2000.
128. Stanford K, Mickleborough T, Ray S, Lindley M, Koceja D, Stager J. Influence of menstrual cycle phase on pulmonary function in asthmatic athletes. *Eur J Appl Physiol* 96: 703–10, 2006.
129. Beynon H, Garbett N, Barnes P. Severe premenstrual exacerbations of asthma: Effect of intramuscular progesterone. *Lancet* 2: 370–72, 1988.
130. Abdul-Karim R, Marshall L, Nesbitt RJ. Influence of estradiol-17 beta on the acetylcholine content of the lung in the rabbit neonate. *Am J Obstet Gynecol* 107: 641–44, 1970.
131. Carey MA, Card JW, Bradbury JA, Moorman MP, Haykal-Coates N, Gavett SH *et al.* Spontaneous airway hyperresponsiveness in estrogen receptor-{alpha}-deficient mice. *Am J Respir Crit Care Med* 175: 126–35, 2007.
132. Myers JR, Sherman CB. Should supplemental estrogens be used as steroid-sparing agents in asthmatic women? *Chest* 106: 318–19, 1994.
133. Sherman M, Lazar E, Eichacker P. A bronchodilator action of glucagon. *J Allergy Clin Immunol* 81: 908–11, 1988.
134. Wilber ST, Wilson JE, Blanda M, Gerson LW, Meerbaum SO, Janas G. The bronchodilator effect of intravenous glucagon in asthma exacerbation: A randomized, controlled trial. *Ann Emerg Med* 36: 427–31, 2000.

PART 5

Pathogenic Mechanisms in Asthma and COPD

CHAPTER 33

Pathophysiology of Asthma

Peter J. Barnes[1] and Jeffrey M. Drazen[2]

[1]National Heart and Lung Institute (NHLI), Clinical Studies Unit, Imperial College, London, UK
[2]Department of Medicine, Pulmonary Division, Brigham and Women's Hospital, Boston, MA, USA

INTRODUCTION

This chapter aims to provide a brief overview of disease mechanisms to integrate some of the detailed information provided in earlier chapters into a framework with clinical relevance. Asthma and chronic obstructive pulmonary disease (COPD) are both highly complex conditions whose pathobiology remains inadequately defined. In both diseases there are many different and distinct inflammatory cells and multiple mediators with complex acute and chronic effects on the airways and airspaces. We now appreciate that these changes may vary among patients because of genetic variance in susceptibility [1]. Although our understanding of asthma pathogenesis has been enhanced by the application of new molecular, cell biological, and genetic techniques to the condition, and the key test of the acquired knowledge is the success of new specifically targeted therapies. In this area the successes have been few in asthma and are virtually absent in COPD. Thus, even though we have made considerable advances in understanding asthma, there remain many fundamental questions that need to be answered.

Our views on asthma and COPD have continued to evolve. Although it is recognized that chronic inflammation underlies the clinical syndromes, it has been hard to define the precise nature of this inflammation, much less its primary etiology. Nevertheless it is appreciated that the final consequence of this chronic inflammatory response is an abnormal *control* of airway caliber *in vivo* in asthma. In contrast, the evolution of understanding in COPD has been much slower; key insights are sought which will open up the understanding of this complex disorder (see Chapter 34).

ASTHMA AS AN INFLAMMATORY DISEASE

It had been recognized for many years that patients who die of asthma attacks have grossly inflamed airways. The airway lumen is occluded by tenacious mucus plugs composed of plasma proteins exuded from airway vessels and mucus glycoproteins secreted from surface epithelial cells. The airway wall is edematous and infiltrated with inflammatory cells, which are predominantly eosinophils and T-lymphocytes [2]. The airway epithelium is invariably shed in a patchy manner and clumps of epithelial cells are found in the airway lumen. Occasionally there have been opportunities to examine the airways of asthmatic patients who die in accidents thought to be unrelated to their asthma. In this setting inflammatory changes have been observed but they are less marked than those observed in patients with active asthma [3]. These studies also reveal that the inflammation in asthmatic airways involves not only the trachea and bronchi, but also extends to the terminal bronchioles [4]; some investigators using transbronchial biopsies have shown inflammatory cells in the parenchyma [5].

It has also been possible to examine the airways of asthmatic patients by fiberoptic and rigid bronchoscopy, bronchial biopsy, and bronchoalveolar lavage (BAL). Direct bronchoscopic examination reveals that the airways of asthmatic patients are often erythematous and swollen, indicating acute inflammation. Lavage of the airways has revealed an increase in the numbers of lymphocytes, mast cells, and eosinophils and evidence for activation of macrophages in comparison with nonasthmatic controls. Biopsies have provided evidence for increased numbers and activation of mast cells, macrophages, eosinophils,

T-lymphocytes including regulatory T-cells, and mucin secreting cells [6–9]. These changes are found even in patients with mild asthma who have few symptoms [10], suggesting that inflammation may be found in all asthmatic patients who are symptomatic. Similar inflammatory changes have been found in bronchial biopsies of children with asthma, even at the onset of symptoms [11, 12].

AIRWAY HYPERRESPONSIVENESS

The relationship between inflammation and clinical symptoms of asthma is not clear. There is evidence that the degree of inflammation is loosely related to airway hyperresponsiveness (AHR), as measured by histamine or methacholine challenge. However, the degree of inflammation, as measured by the number of various types of inflammatory cells in the lesion, is not closely linked to the clinical severity of asthma or AHR [13]. This suggests that other factors, such as structural changes in the airway wall, are important in mediating the clinical symptoms of asthma. The increased airway responsiveness in asthma is a striking physiological abnormality that is present even when airway function is otherwise normal. It is likely that there are several factors that underlie this increased responsiveness to constrictor agents, particularly those that act indirectly by releasing bronchoconstrictor mediators from airway cells or that modify the mechanical characteristics of the airway wall. AHR may be due to increased release of mediators (such as histamine and leukotrienes from mast cells), abnormal behavior of airway smooth muscle, thickening of the airway wall by reversible (edema), and irreversible (airway smooth muscle thickening, fibrosis) elements. Airway sensory nerves may also contribute importantly to symptoms, such as cough and chest tightness, as the nerves become sensitized by the chronic inflammation in the airways.

Although most attention has been focused on the acute inflammatory changes seen in asthmatic airways (bronchoconstriction, plasma exudation, mucus secretion), asthma is a *chronic* disease, persisting over many years in most patients. Superimposed on this chronic inflammatory state are acute inflammatory episodes, which correspond to exacerbations of asthma. It is clearly important to understand the mechanisms of acute and chronic inflammation in asthmatic airways and to investigate the long-term consequences of this chronic inflammation on airway function.

INFLAMMATORY CELLS

Many different inflammatory cells are involved in asthma, although the precise role of each cell type is not yet certain (Fig. 33.1). It is evident that no single inflammatory cell is

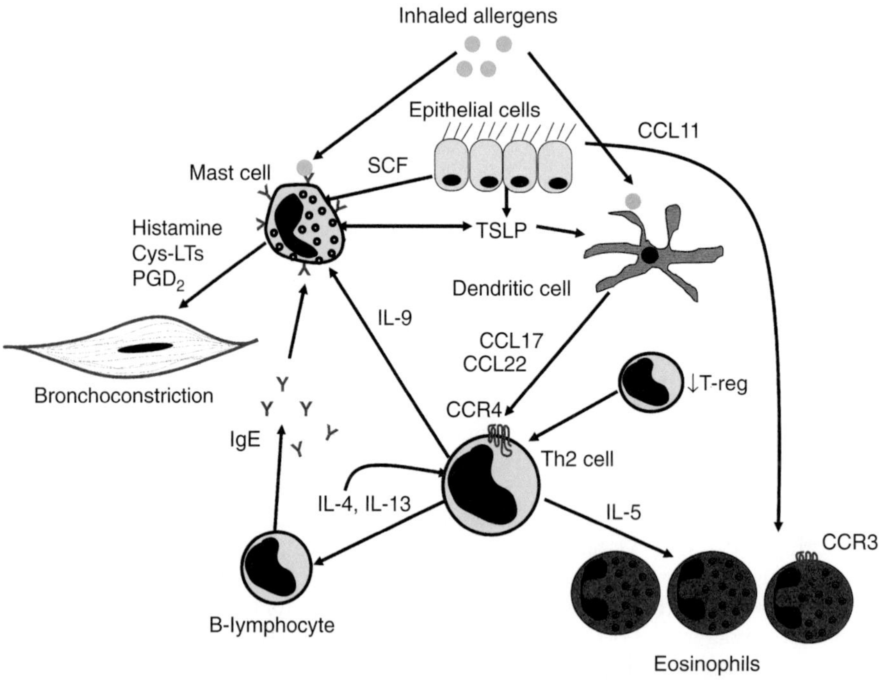

FIG. 33.1 A potential schema of Inflammation in asthma. Inhaled allergens activate sensitized mast cells by crosslinking surface-bound IgE molecules to release several bronchoconstrictor mediators, including cysteinyl-leukotrienes (*cys*-LT) and prostaglandin D_2 (PGD_2). Epithelial cells release stem-cell factor (SCF), which is important for maintaining mucosal mast cells at the airway surface. Allergens are processed by myeloid dendritic cells, which are conditioned by thymic stromal lymphopoietin (TSLP) secreted by epithelial cells and mast cells to release the chemokines CC-chemokine ligand 17 (CCL17) and CCL22, which act on CC-chemokine receptor 4 (CCR4) to attract T helper 2 (Th2) cells. Th2 cells have a central role in orchestrating the inflammatory response in allergy through the release of interleukin-4 (IL-4) and IL-13 (which stimulate B-cells to synthesize IgE), IL-5 (which is necessary for eosinophilic inflammation) and IL-9 (which stimulates mast-cell proliferation). Epithelial cells release CCL11, which recruits eosinophils via CCR3. Patients with asthma may have a defect in regulatory T-cells (T-reg), which may favor further Th2-cell proliferation.

able to account for the complex pathophysiology of asthma but some cells predominate in asthmatic inflammation. The inflammation in asthmatic airways differs strikingly from that observed in COPD, where there is a predominance of macrophages, cytotoxic ($CD8^+$) T-lymphocytes, and neutrophils, although both of these common diseases may coexist in some patients. However, although the differences in inflammation between mild asthma and COPD are well described, it is not recognized that patients with severe asthma have inflammatory changes that are similar to those of COPD, with increased numbers of neutrophils, $CD4^+$ Th1 and $CD8^+$ cells [14].

Mast cells

Mast cells are clearly important in initiating the acute bronchoconstrictor responses to allergen and probably to other indirect stimuli, such as exercise and hyperventilation (via osmolality or thermal changes) and fog. Mast cell activation is likely to play a key role in the symptoms of asthma and during acute exacerbations. There is infiltration of airway smooth muscle by mast cells in patients with asthma that is associated with the characteristic disordered airway function of this condition [15]. Treatment of asthmatic patients with prednisone results in a decrease in the number of tryptase-only positive mast cells [16]. Furthermore, the number of mast cells in induced sputum in patients with seasonal allergic rhinitis is related to the degree of AHR [17]. There appears to be an infiltration of mast cells into airway smooth muscle of asthmatic patients and this is associated with AHR [15]. The key role of mast cells in asthma has been the subject of comprehensive reviews [7, 18]. Mast cells are maintained at the mucosal surface by the secretion of stem cell factor, which acts on c-kit receptors, by epithelial cells [19]. However, it is now clear that mast cells alone cannot account for all the pathobiological changes in asthma and that other cells, such as macrophages, eosinophils, and T-lymphocytes, play key roles in the chronic inflammatory process including induction and maintenance of AHR.

Classically mast cells are activated by allergens through an IgE-dependent mechanism. The importance of IgE in the pathophysiology of asthma has been highlighted by clinical studies with humanized anti-IgE antibodies, which inhibit IgE-mediated effects. Although anti-IgE antibody results in a reduction in circulating IgE to undetectable levels, this treatment results in minimal clinical improvement in patients with severe steroid-dependent asthma [20]. Examination of the patients receiving anti-IgE therapy with omalizumab reveals downregulated expression of IgE receptors (FcεRI) on mast cells and basophils [21]. Biopsy studies show a profound reduction in tissue eosinophilia, together with reductions in T-cell and B-cell numbers in the airway submucosa [22]. These observations suggest that treatment with anti-IgE may be more effective in early stage rather than late stage asthma; however, no clinically directive trials with omalizumab have been conducted in this population. Furthermore, unless the cost of such therapy is dramatically reduced it is unlikely that such treatment would be used in patients with mild disease.

Macrophages

Macrophages, which are derived from blood monocytes, may traffic into the airways in asthma and may be activated by allergen via low-affinity IgE receptors (FcεRII) [23]. The enormous immunological repertoire of macrophages allows these cells to produce many different products, including a large variety of cytokines that may orchestrate the inflammatory response. Macrophages have the capacity to initiate a particular type of inflammatory response via the release of a certain pattern of cytokines. Macrophages may increase or decrease inflammation depending on the stimulus. For example, alveolar macrophages from normal and patients with mild asthma phagocytose apoptic cells resulting from lipopolysaccharide (LPS) exposure *in vitro* and produce the anti-inflammatory mediator PGE_2. In contrast alveolar macrophages from patients with severe asthma are impaired with respect to apoptotic cell uptake and production of PGE_2 [24]. These observations indicate that the severity of asthma and perhaps its chronicity can modulate the role of macrophages in asthma. It is also possible that the opposite is true, that is, patients with severe asthma are the individuals with defective macrophage function. It is established that repeated short term exposure to ozone recruits alveolar macrophages to the airways of patients with mild asthma [25] and that there is an association between the amount of carbon in alveolar macrophages, which had to be activated to ingest this material, and lung function [26]. Thus the role of alveolar macrophages appears to be to modulate the expression and severity of asthma.

Dendritic cells

By contrast, dendritic cells (which are specialized macrophage-like cells in the airway epithelium) are very effective antigen-presenting cells and may therefore play a very important role in the initiation of allergen-induced responses in asthma [27, 28]. Dendritic cells take up allergens though their long finger-like projections, which extend into the airway lumen, process them to peptides, and migrate to local lymph nodes where they present the allergenic peptides to uncommitted T-lymphocytes, to program the production of allergen-specific T-cells. Although there is no established ways to manipulate dendritic cells, it has been shown that during experimental allergen challenge in humans that a significantly greater proportion of $CD33^+$ cells expressed the cys-LT_1-receptor compared with $CD123^+$ cells. Pranlukast, cys-LT_1-receptor antagonist, decreased the allergen-induced decrease in $CD33^+$ dendritic cells at 3h compared with placebo treatment. Based on this finding it is reasonable to speculate that anti-leukotriene treatment could modify the initial responses to allergen exposure [29].

Eosinophils

Eosinophil infiltration is a characteristic feature of asthmatic airways and differentiates asthma from other non-infectious inflammatory conditions of the airway. Indeed, asthma was described as "chronic eosinophilic bronchitis" as early as 1916.

Allergen inhalation results in a marked increase in eosinophils in BAL fluid at the time of the late reaction, and there is a correlation between eosinophil counts in peripheral blood or bronchial lavage and AHR. Eosinophils are linked to the development of AHR through the release of basic proteins and oxygen-derived free radicals [30, 31].

An important area of research is now concerned with the mechanisms involved in recruitment of eosinophils into asthmatic airways. Eosinophils are derived from bone marrow precursors; the Th2 cytokine IL-5 is a unique mediator of eosinophil differentiation and survival in response to allergen provocation. After allergen challenge eosinophils appear in BAL fluid during the late response, and this appearance is associated with a decrease in peripheral eosinophil counts and with the appearance of eosinophil progenitors in the circulation. The signal for increased eosinophil production is likely IL-5 derived from the inflamed airway. Eosinophil recruitment initially involves adhesion of eosinophils to vascular endothelial cells in the airway circulation, their migration into the submucosa, and their subsequent activation. The role of individual adhesion molecules, cytokines, and mediators in orchestrating these responses has been extensively investigated but since there are no pathways that are unique to eosinophils, this aspect of eosinophil biology has not been extensively probed. Adhesion of eosinophils involves the expression of specific glycoprotein molecules on the surface of eosinophils (integrins) and their expression of such molecules as intercellular adhesion molecule-1 (ICAM-1) on vascular endothelial cells. An antibody directed at ICAM-1 markedly inhibits eosinophil accumulation in the airways after allergen exposure and also blocks the accompanying hyperresponsiveness [32]. However, ICAM-1 is not selective for eosinophils and cannot account for the selective recruitment of eosinophils in allergic inflammation. Eosinophil migration may be due to the effects of lipid mediators, such as leukotrienes [33, 34] and possibly platelet-activating factor (PAF), to the effects of cytokines, such as GM-CSF and IL-5 both of which are important for the survival of eosinophils in the airways and "prime" eosinophils to exhibit enhanced responsiveness. Eosinophils from asthmatic patients show exaggerated responses to PAF and phorbol esters, compared with eosinophils from atopic nonasthmatic individuals [35]. This is further increased by allergen challenge [36, 37] suggesting that they may have been primed by exposure to cytokines in the circulation.

There are several mediators involved in the migration of eosinophils from the circulation to the surface of the airway. There are data showing that cysteinyl leukotrienes, through stimulation at the cys-LT_1-receptor, are in part responsible for eosinophilopoiesis in human airways [38], but other mediators such as the eotaxins, a family of chemoattractant cytokines that promote eosinophil recruitment to tissues in response to allergic provocation via CCR3 receptors also play a role [39–42]. Other potent and selective eosinophil chemoattractants include chemokines, such as CCL5 (RANTES) and CCL13 (MCP-4), that are expressed in epithelial cells [40, 43, 44] but data for their involvement in intact humans are minimal. There appears to be a co-operative interaction between IL-5, IL-13, and chemokines, so that more than one cytokine may be necessary for the eosinophilic response in airways [45, 46]. Once recruited to the airways, eosinophils require the presence of various growth factors, of which GM-CSF and IL-5 appear to be the most important [47]. In the absence of these growth factors eosinophils undergo programmed cell death (apoptosis) [48].

A humanized monoclonal antibody to IL-5 has been administered to asthmatic patients [49]; and, as in animal studies, there is a profound and prolonged reduction in circulating eosinophils. Although the infiltration of eosinophils into the airway after inhaled allergen challenge is completely blocked, there is no effect on the response to inhaled allergen and no reduction in AHR. Anti-CCR3, anti-integrin, and antiselectin treatments are currently being considered as anti-eosinophil strategies for treatment of asthma; the results of these studies should better define the role of eosinophils as mediators of asthmatic events. Eosinophils may play a role in airway remodeling rather than AHR in asthma as anti-IL-5 treatment is able to reduce extracellular matrix deposition in asthma [50].

Neutrophils

The eosinophil has been the recipient of current attention as an effector cell in asthma, but attention has been returning to the role of neutrophils [51, 52]. Although neutrophils are not a predominant cell type observed in the airways of most patients with mild-to-moderate chronic asthma, they appear to be a more prominent cell type in airways and induced sputum of patients with more severe asthma [53–57]. The mechanism of neutrophilic inflammation are not yet well understood, but several mediators, including CXCL8 (IL-8) are likely to be involved (Fig. 33.2). Sputum analysis shows that ~20% of asthmatics, even with mild disease, have a predominantly neutrophilic pattern of inflammation [58]. Also in patients who die suddenly of asthma, large numbers of neutrophils are found in their airways [59] although this may reflect the rapid kinetics of neutrophil recruitment compared to eosinophil inflammation. Rapid withdrawal of corticosteroids results in the appearance of neutrophils in the airways of patients with asthma [60]. There are data to suggest that monomeric IgE may mediate the appearance of neutrophils in the airways of asthmatics and may represent the anti-apoptic activity of myeloid cell leukemia-1 protein, decreased caspase-3 activity, diminished availability of Smac from mitochondria, and sequestration of the pro-apoptic protein Bax in the cytoplasm [61]. Thus the presence of neutrophils in the airway may reflect the inflammatory microenvironment rather than indicate a causal role for this cell. Further research into the role of neutrophils especially in severe asthma is ongoing [62].

T-lymphocytes

Th2 cells

T-lymphocytes play a very important role in co-ordinating the inflammatory response in asthma through the release of specific patterns of cytokines, resulting in the recruitment and survival of eosinophils and in the maintenance of mast cells in the airways. T-lymphocytes are coded to express a distinctive pattern of cytokines, which are similar to that described

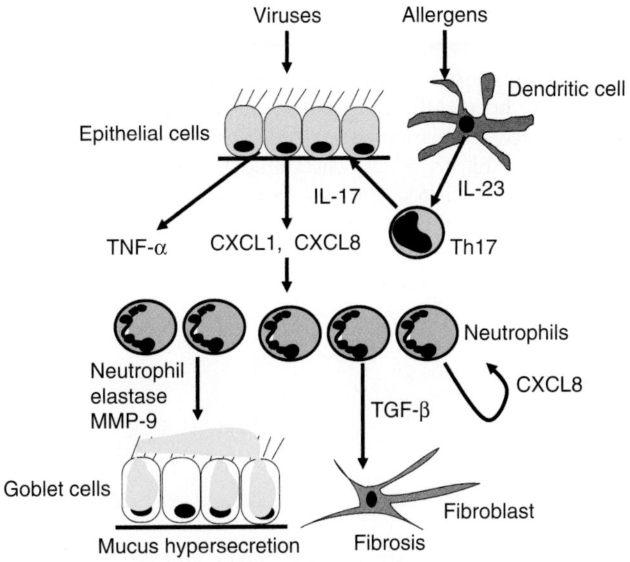

FIG. 33.2 Neutrophilic inflammation in asthma. Viruses, such as rhinovirus, stimulate the release of CXCL8 (IL-8) and CXCL1 (GRO-α) from airway epithelial cells. Allergens activate dendritic cells to release IL-23, which recruits helper T-cells that secrete IL-17 (Th17 cells) to release tumor necrosis factor-α (TNF-α), which amplifies inflammation and CXCL1 and CXCL8, which recruit neutrophils into the airways. Neutrophils release more CXCL8 and also transforming growth factor-β (TGF-β), which activates fibroblasts to cause fibrosis, and neutrophil elastase and matrix metalloproteinase-9 (MMP-9) which stimulate mucus hypersecretion from goblet cells.

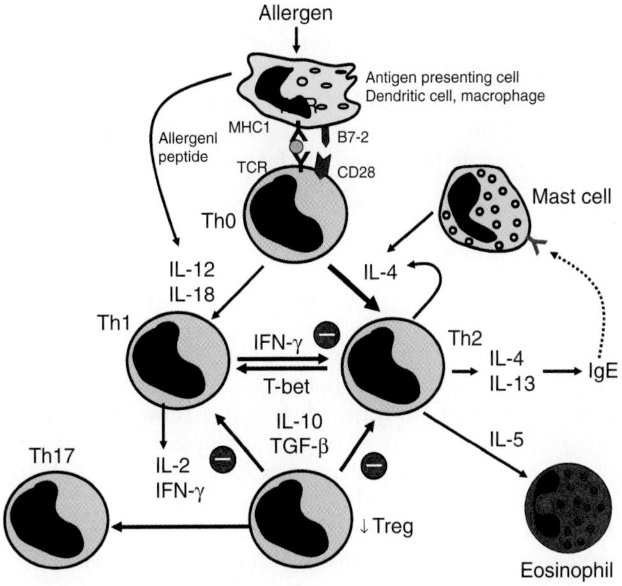

FIG. 33.3 T-lymphocytes in asthma. Asthmatic inflammation is characterized by a preponderance of T helper 2 (Th2) lymphocytes over T helper 1 (Th1) cells. Regulatory T-cells (Treg) have an inhibitory effect, whereas T helper 17 (Th17) cells have a proinflammatory effects. MHC1: Class 1 major histocompatibility complex; IL: interleukin; IFN-γ: interferon gamma; TGF-β: transforming growth factor beta; IgE: immunoglobulin E, Th0: uncommitted T-cell.

in the murine Th2 type of T-lymphocytes, which characteristically express IL-4, IL-5, and IL-13 [6]. This programming of T-lymphocytes is presumably due to antigen-presenting cells such as dendritic cells, which may migrate from the epithelium to regional lymph nodes or which interact with lymphocytes resident in the airway mucosa. The naive immune system is skewed to express the Th2 phenotype; data now indicate that children with atopy are more likely to retain this skewed phenotype than are normal children [63]. There is also evidence that chitinase, a hydrolase that cleaves environmental chitin, an insect protein, plays a critical role in murine Th2 differentiation and has been identified in human tissues from patients with asthma [64, 65]. The recruitment of Th2 cells in the airway is dependent on several chemotactic mediators, including the chemokines CCL17 (TARC) and CCL22 (MDC) which act on CCR4 that are selectively expressed on Th2 cells. These CCR4 ligands show increased expression in airway epithelial cells, macrophages, and mast cells of asthmatic patients [66]. PGD_2, mainly derived from mast cells, is also an important chemoattractant of Th2 cells through the activation of DP_2-receptors (or CRTH2) [67]. There is some evidence that steroid treatment may differentially affect the balance between IL-12 expression and IL-13 expression [68]. Data from murine models of asthma [69–71] have strongly suggested that IL-13 is both necessary and sufficient for induction of the asthmatic phenotype. One of the most important areas of asthma research in the next few years will be to establish the importance of IL-13 in the induction of the Th2 phenotype and asthma in humans.

Regulatory T-cells

A subset of $CD4^+$ T-cells expressing CD25, termed T-regulatory cells (Tregs), are capable of suppressing both Th1 and Th2-mediated adaptive immune responses, and these cells may mediate their suppressive actions through release of TGF-β or IL-10 [72]. An imbalance between allergen-specific Tregs and effector Th2 cells has been shown in allergic diseases [73]. There is some evidence that early infections or exposure to endotoxins might promote Th1-mediated responses to predominate and that a lack of infection or a clean environment in childhood may favor Th2-cell expression and thus atopic diseases [74, 75]. Indeed, the balance between Th1 cells and Th2 cells is thought to be determined by locally released cytokines, such as IL-12, which tip the balance in favor of Th1 cells, or IL-4 or IL-13, which favor the emergence of Th2 cells (Fig. 33.3).

Th17 cells

Another subset of $CD4^+$ T-cells, known as Th17 cells, has recently been described and shown to have an important role in inflammatory and autoimmune diseases [76]. Little is known about the role of Th17 cells in asthma, but increased concentrations of IL-17A (the predominant product of Th17 cells) have been reported in the sputum of asthma patients [77]. IL-17A and the closely related cytokine IL-17F have been linked to neutrophilic inflammation by inducing the release of CXCL1 and CXCL8 from airway epithelial cells [78]. As well as IL-17, Th17 cells also produce IL-21, which is important for the differentiation of these cells and thus acts as a positive autoregulatory mechanism, but it also inhibits FOXP3 expression and regulatory T-cell development [79, 80]. Another cytokine IL-22 is also released by these cells and stimulates the production of IL-10 and acute-phase proteins [81]. However, more work is needed to understand the role and regulation of Th17 cells in asthma, as they may represent important new targets for future therapies.

Natural killer T-cells

A subset of $CD4^+$ T-cells termed invariant natural killer T (*i*NKT) cells, which secrete IL-4 and IL-13, has been shown to account for 60% of all $CD4^+$ T-cells in bronchial biopsies from asthmatic patients [82], but this has been disputed in another study, which failed to show any increase in *i*NKT-cell numbers in bronchial biopsies, BAL or sputum of either asthma patients or COPD patients [83]. The role of *i*NKT cells in asthma is currently uncertain as there appears to be a discrepancy between the data from murine models of asthma and humans with the disease [84].

$CD8^+$ cells

$CD8^+$ (cytotoxic) T-cells are present in patients with more severe disease and irreversible airflow obstruction [85] and these cells may be of either the T_C1 or the T_C2 type, releasing the same patterns of cytokines as $CD4^+$ cells [86].

B-lymphocytes

B-cells have an important role in asthma through the release of allergen-specific IgE which binds to FcεRI on mast cells and basophils and to low-affinity FcεRII on other inflammatory cells, including B-cells, macrophages, and possibly eosinophils [87]. The Th2-type cytokines IL-4 and IL-13 induce B-cells to undergo class switching to produce IgE. In both atopic asthma and non-atopic asthma, IgE may be produced locally by B-cells in the airways [88].

Basophils

The role of basophils in asthma is uncertain, as these cells have previously been difficult to detect by immunocytochemistry [89]; indeed there may be multiple phenotypes of basophils with some able to release histamine on IgE challenge and others not able to do so [90]. Using a basophil-specific marker, a small increase in basophils has been documented in the airways of asthmatic patients, with an increased number after allergen challenge. However, these cells are far outnumbered by eosinophils [91].

STRUCTURAL CELLS AND AIRWAY REMODELING

Structural cells of the airways, including epithelial cells, fibroblasts, and even airway smooth muscle cells are the site of influence of many of the cytokines, growth factors, and inflammatory mediators that characterize the asthmatic diathesis. Interestingly they may also be an important source of these same molecular entities [92, 93]. Although the concept is poorly defined, the idea that there are long-lasting changes in the cells that make up the airway wall and that these cells narrow the lumen and contribute to chronic airflow limitation has been termed airway remodeling. Given the absence of a clean case definition, most investigators agree that certain changes, such as thickening of the collagen layer deposited below the true basement membrane, is a critical component of this aspect of asthma [94–98]. Indeed, because structural cells far outnumber inflammatory cells they may become the major source of mediators driving chronic inflammation in asthmatic airways. For example, epithelial cells may have a key role in translating inhaled environmental signals into an airway inflammatory response and are probably a target cell for inhaled glucocorticoids and anti-leukotrienes (Fig. 33.4) [99].

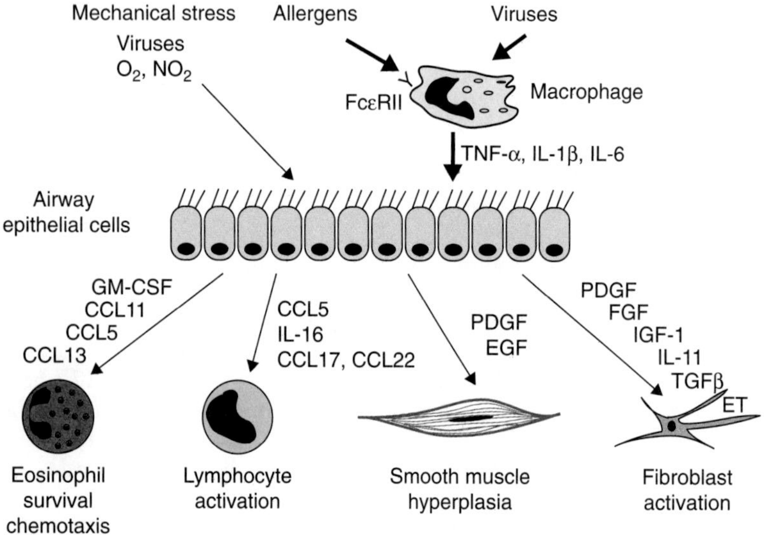

FIG. 33.4 Airway epithelial cells in asthma. Airway epithelial cells may play an active role in asthmatic inflammation through the release of many inflammatory mediators, cytokines, chemokines and growth factors. TNF-α: tumor necrosis factor alpha; IL: interleukin, CCL: C-C chemokine, PDGF: platelet-derived growth factor; EGF: epidermal growth factor; FGF: fibroblast growth factor; IGF: insulin-like growth factor.

Structural changes in the airways may account for the accelerated decline in airway function seen in asthmatic patients over several years [100, 101]. Structural changes in asthmatic airways include

- increased thickness of airway smooth muscle;
- infiltration of the airway smooth muscle with mast cells
- fibrosis (which is predominantly subepithelial);
- increased mucus-secreting cells;
- increased numbers of blood vessels (angiogenesis).

It is not known if these changes are reversible with therapy. These changes may occur in some patients to a greater extent than others and may be increased by other factors such as concomitant cigarette smoking. It is likely that genetic factors will influence the extent of remodeling that occurs in individual patients.

There are good data to support the idea that mechanical deformation of the airway epithelium as a result of airway smooth muscle constriction results in the creation of a microenvironment that promotes airway remodeling. Deformed airway epithelial cells activate signaling via the epidermal growth factor receptor (EGFR) and the urokinase plasminogen activator system. Experimental data show that soluble signals generated by airway epithelial cells can influence the phenotype of fibroblasts to produce collagen similar to that found in remodeled airways [102–104]. These data could help explain why combination treatments with inhaled steroids and long-acting β-agonists are effective asthma therapies.

INFLAMMATORY MEDIATORS

Many different mediators have been implicated in asthma. They may have a variety of effects on the airways, which could account for the pathological features of asthma (Fig. 33.5) [105, 106]. Mediators such as histamine, prostaglandins, and leukotrienes contract airway smooth muscle, increase microvascular leakage, increase airway mucus secretion, and attract

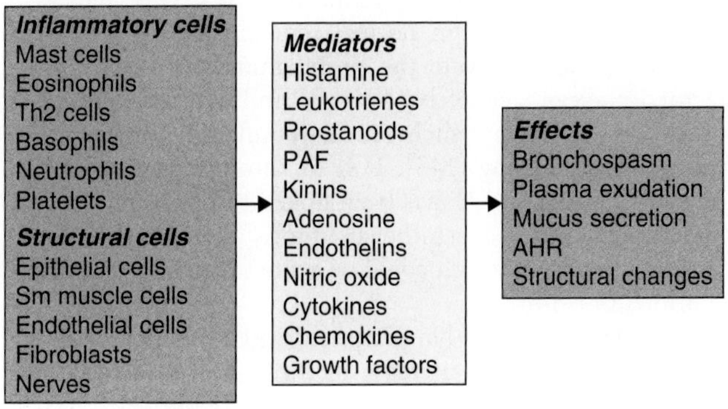

FIG. 33.5 Multiple cells, mediators and effects. Many cells and mediators are involved in asthma and lead to several effects on the airways. Th2: T helper 2 cells; Sm: smooth; PAF: platelet-activating factor; AHR: airway hyperresponsiveness.

other inflammatory cells. Because each mediator has many effects in the role of individual mediators in the pathophysiology of asthma is not yet clear. Although the multiplicity of mediators makes it unlikely that preventing the synthesis or action of a single mediator will have a major impact in clinical asthma, recent clinical studies with anti-leukotrienes suggest that cysteinyl leukotrienes have a clinically important effect.

Leukotrienes

The cysteinyl leukotrienes LTC_4, LTD_4, and LTE_4 are potent constrictors of human airways and have been reported to increase AHR and may play an important role in asthma [107–110] (see Chapter 24). The recent development of potent specific leukotriene antagonists has made it possible to evaluate the role of these mediators in asthma. Potent LTD_4 antagonists protect (by about 50%) against exercise- and allergen-induced bronchoconstriction [111–114], suggesting that leukotrienes contribute to bronchoconstrictor responses. Combined treatment with an anti-histamine and an anti-leukotriene is particularly effective [115, 116] Chronic treatment with anti-leukotrienes improves lung function and symptoms in asthmatic patients, although the degree of lung function improvement is not as great as that seen with an inhaled corticosteroid. It is only through the use of specific antagonists that the role of individual mediators of asthma may be defined. In the future, pharmaceuticals with specific targets of action will be of special value in providing pathobiological insights into the basic mechanisms of asthma; their potential role in the treatment of asthma remains to be determined.

Platelet-activating factor

Platelet-activating factor (PAF) and PAF-acetylhydrolase, the enzyme that degrades PAF, constitute a potent inflammatory mediator system that mimics many of the features of asthma, including eosinophil recruitment and activation and induction of AHR; yet even potent PAF antagonists, such as modipafant, do not control asthma symptoms, at least in chronic asthma [117–120]. However, genetic studies in Japan, where there is a high frequency of a genetic mutation which disables the PAF metabolizing enzyme, PAF-acetylhydrolase, have shown that there is an association between the presence of the mutant form of the enzyme and severe asthma [121, 122]. In addition, there are data showing that patients with a deficiency of the enzyme that degrades PAF is associated with enhanced asthma severity [123]. These data suggest that there may be certain conditions associated with a significant role for PAF in asthma.

Cytokines

Cytokines are increasingly recognized to be important in chronic inflammation and play a critical role in orchestrating the type of inflammatory response (see Chapter 27) (Fig. 33.6) [124]. Many inflammatory cells (macrophages, mast cells, eosinophils, and lymphocytes) are capable of synthesizing

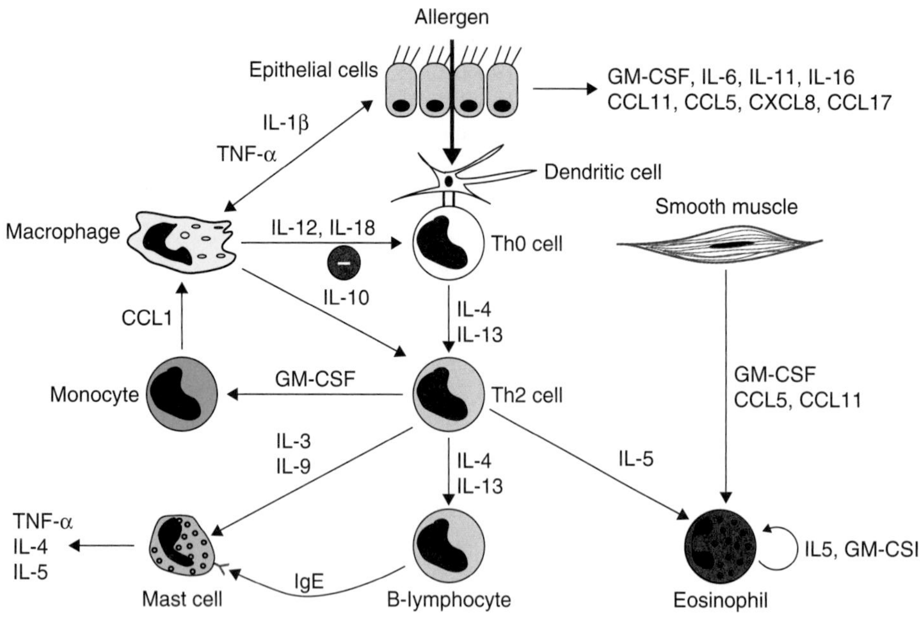

FIG. 33.6 The cytokine network in asthma. Many inflammatory cytokines are released from inflammatory and structural cells in the airway and orchestrate and perpetuate the inflammatory response.

and releasing these proteins, and structural cells such as epithelial cells and endothelial cells may also release a variety of cytokines and may therefore participate in the chronic inflammatory response [125, 126]. There are number of major classes of cytokines. One class includes T-cell derived cytokines, examples of which include interferon (IFN)γ and TNFα from Th1 cells and IL-4 and IL-5 from Th2 cells. Another class includes pro- and anti-inflammatory cytokines with IL-1β and IL-10 being respective examples. Chemoattractant cytokines, such as CCL11 and CXCL8 (IL-8), and growth factors cytokines, including PDGF, TGF-β, and VEGF are other major classes. It is now clear that cytokines play a dominant role in chronic inflammation of asthma.

T-cell-derived cytokines play a key role in creating the microenvironment that lends itself to initiating and perpetuating the chronic inflammatory response of asthma [127]. The key cytokines here are Th2 cytokines, IL-3, IL-4, and IL-5. IL-3 induces eosinophilia *in vivo* and maintains mast cell viability. IL-4 induces differentiation of Th2 cells and creates the microenvironment in which IgE antibody is synthesized [128] and enhances mucin expression in goblet cells [129]. IL-5 enhances eosinophil maturation and prevents apoptosis of inflammatory cells [130]. The recently described IL-17 family of cytokines, produced by cells as a result of the action of IL-15 or IL-23 on CD4 or CD8 memory cells may be important regulators of the asthmatic response [131–133].

A lack of specific small molecule antagonists has made cytokine related research in humans difficult. Although important observations have been made using specific neutralizing antibodies [134], interventions involving immune function can be difficult.

There is increased gene expression of IL-5 in lymphocytes in bronchial biopsies of patients with symptomatic asthma [135, 136]. The role of IL-5 in eosinophil recruitment in asthma has been confirmed in a study in which administration of an anti-IL-5 antibody to asthmatic patients was associated with a decrease in eosinophil counts in the blood and BAL fluid [49]. Interestingly in this study there was no effect on the physiology of the allergen-induced asthmatic response; although this is not the last word on eosinophils in asthma, it provides evidence that the eosinophil, as recruited by IL-5, is not the major pathogenetic cell in asthma. This has been confirmed by a study of the same antibody in symptomatic asthmatic patients, who showed no improvement in asthma control despite a profound reduction in circulating eosinophils [137]. Another Th2 cytokine produced by peripheral blood cells of patients with asthma, IL-9, may play a critical role in airway remodeling, eosinophilopoiesis, and may interact with the IL-4 receptor$_\alpha$ in mediating some of the pathobiological features of asthma [138–141].

Other cytokines, such as IL-1β, IL-6, TNF-α, TGF-β, and GM-CSF, are released from a variety of cells, including macrophages and epithelial cells and may be important in amplifying the inflammatory response. TNF-α may be an amplifying mediator in asthma and is produced in increased amounts in asthmatic airways. Inhalation of TNF-α increased airway responsiveness in normal individuals [142]. TNF-α and IL-1β both activate the proinflammatory transcription factors, nuclear factor-κB (NF-κB) and activator protein-1 (AP-1) which then switch on many inflammatory genes in the asthmatic airway [143]. TGF-β, through ligation of two receptors, type I and II, has been linked to tissue remodeling, and potentially to subepithelial fibrosis through the *Smad*-signaling pathway which could be a site for therapeutic intervention [144–146].

Thymic stromal lymphopoeitin (TSLP) is a cytokine that shows a marked increase in expression in airway epithelium and mast cells of asthmatic patients [66]. TSLP appears to play a key role in programming airway dendritic cells to release CCL17 and CCL22 to attract Th2 cells (Fig. 33.7) [147]. It therefore acts as an upstream cytokine which

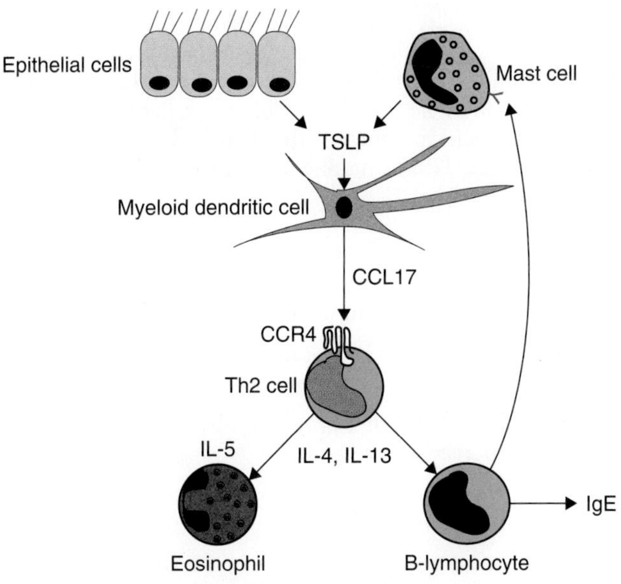

FIG. 33.7 Thymic stromal lymphopoetin in asthma. TSLP is an upstream cytokine produced by airway epithelial cells and mast cells in asthma that acts on immature dendritic cells to mature and release CCL17 (TARC), which attracts Th2 cells via CCR4. Th2: T helper 2; CCL: chemokine; CCR: chemokine receptor.

may play a key role in orchestrating an allergic pattern of inflammation in asthmatic airways.

Complement

A decade ago it was postulated that complement had little role in the biology of asthma. However, studies in mice harboring a targeted deletion of the C5a receptor showed diminished bronchial hyperresponsiveness induced after allergen challenge [148]. In addition, genetic linkage studies in mice have linked regions of the mouse genome containing the gene for C5a to the phenotype of AHR [149]. Furthermore, the complement peptide C5a has been recovered from BAL fluid of patients after allergen challenge [148]. Current thinking is that complement contributes to the airway inflammation, AHR and mucus production observed in human asthma and may also prevent the formation of a Th2-directed immune response to allergens [150–152].

Oxidative stress

In asthma, activation of epithelial cells, and resident macrophages, and the recruitment and activation of neutrophils, eosinophils, monocytes, and lymphocytes leads to the generation of reactive oxygen species (ROS) and reactive nitrogen species (RNS). Action of many systems but primarily NADPH oxidase results in the microenvironmental availability of superoxide anions ($O_2^{\bullet -}$), which are rapidly converted to H_2O_2 through the catalytic action of superoxide dismutase (SOD). Hydroxyl ion ($^{\bullet}OH$) is then formed nonenzymatically; this damages tissues, particularly cell membranes by peroxidation of their lipid membranes. Phagocytes use heme peroxidases (myeloperoxidase, MPO) or eosinophil peroxidase (EPO) to produce ROS; these moieties can also interact with nitrite (NO_2^-) and hydrogen peroxide (H_2O_2) to promote formation of RNS. Eosinophils, a key cell in asthma, have several times greater capacity to generate $O_2^{\bullet -}$ and H_2O_2 than do neutrophils, and the content of EPO in eosinophils is 3–10 times higher than the amount of MPO present in neutrophils [153–155].

Evidence for increased oxidative stress in asthma is provided by the increased concentrations of 8-isoprostane (a product of oxidized arachidonic acid) in exhaled breath condensates [156], and increased ethane (a product of oxidative lipid peroxidation) in exhaled breath of asthmatic patients [156–158]. Airflow obstruction, airway hyperresponsiveness and airway remodeling have been associated with impaired SOD activity [159, 160]. Oxidative stress is increased particularly in patients with severe asthma (perhaps as a result of the neutrophilic infiltration in these patients) and this may reduce the activity of histone deacetylase (HDAC), which may account for the reduced response to corticosteroids in these patients [161, 162].

Endothelins

Endothelins are potent peptide mediators that are vasoconstrictors, bronchoconstrictors and promote fibrosis [163, 164]. Endothelin-1 levels are increased in the sputum, serum, and BAL fluid of patients with asthma; these levels are modulated by allergen exposure and steroid treatment [165–167]. Endothelins also induce airway smooth muscle cell proliferation and promote a profibrotic phenotype and may therefore play a role in the chronic inflammation of asthma [164, 168].

Nitric oxide

Nitric oxide (NO) is produced by several cells in the airway by NO synthases [169, 170] (see Chapter 30). Although the cellular source of NO within the lung is not known, inferences based on mathematical models suggest that it is the large airways which are the source of NO [171]. However, in patients with severe asthma NO is also derived form peripheral airways [172]. Current data indicate that the level of NO in the exhaled air of patients with asthma is higher than the level of NO in the exhaled air of normal subjects (see Chapter 30) [173–177]. The elevated levels of NO in asthma are closely linked to airway pH [178–180] and thus are more likely reflective of inflammatory mechanisms than of a direct pathogenetic role of this gas in asthma [181, 182]. The combination of increased oxidative stress and NO may lead to the formation of the potent radical peroxynitrite that may result in nitrosylation of proteins in the airways [183, 184]. Asthmatics managed by measurements of exhaled nitric oxide achieved equivalent asthma control with less inhaled corticosteroids than those managed using conventional algorithms [185]; the availability of portable devices to measure exhaled nitric oxide may allow management of asthma in manner similar to the self management of diabetics [186].

EFFECTS OF INFLAMMATION ON THE AIRWAYS

The idea that the chronic inflammatory response results in the characteristic pathophysiological changes associated with asthma has been widely promulgated. However, biopsies from children with wheezing show many of the characteristic features of asthma [12]. Thus it is probable that some of the differences between the condition of the airways in normal and asthmatic subjects reflects the genetic makeup of the patient influenced by environmental factors such as airway infections and allergen exposures. The descriptions below present the observations without adequate evidence to support whether changes are markers for or the result of a chronic inflammatory response. The inflammatory response has many effects on structural cells of the airways resulting an amplification and perpetuation of the inflammatory response (Fig. 33.8).

Airway epithelium

Airway epithelial cells play a critical role in asthma and link environmental influences to inflammatory responses in the airway (Fig. 33.4). Airway epithelial shedding may be important in contributing to AHR and may explain how several different mechanisms, such as ozone exposure, certain virus infections, chemical sensitizers and allergen exposure can lead to its development, since all these stimuli may lead to epithelial disruption. Epithelium may be shed because of an inflammatory milieu made up of eosinophil basic proteins, oxygen-derived free radicals, proteases released from inflammatory cells, and activated complement. Epithelial cells shed from the airway wall may be found in clumps in the BAL fluid or sputum (Creola bodies) of asthmatics, suggesting that there has been a loss of attachment to the basal layer or basement membrane. The damaged epithelium may contribute to AHR in a number of ways, including

- loss of its barrier function to allow penetration of allergens and irritants;
- loss of enzymes (such as neutral endopeptidase) which normally degrade inflammatory mediators such as kinins;
- loss of a relaxant factor (so-called epithelial-derived relaxant factor);
- edema and swelling that promotes obstruction by lumenal encroachment;
- loss of surfactant function leading to increased increased recoil of the airway;
- exposure of sensory nerves which may lead to reflex neural effects on the airway.

Fibrosis

The basement membrane in asthma appears on light microscopy to be thickened; on closer inspection by electron microscopy it has been demonstrated that this apparent thickening is due to subepithelial fibrosis with deposition of types III and V collagen below the true basement membrane; however, the ratio of interstitial collagen fibrils to matrix is not different from that found in normal subjects [187–189]. Data from a number of investigative groups show that the thickness of the deposited collagen is related to airway obstruction

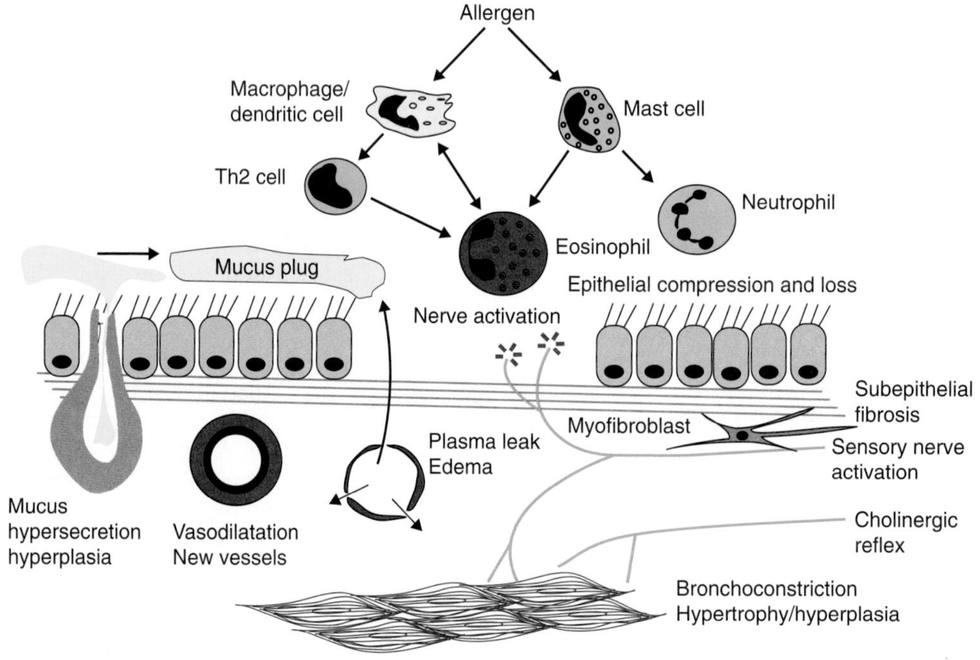

FIG. 33.8 The complexity of asthma. The pathophysiology of asthma is complex, with participation of several interacting inflammatory cells which result in acute and chronic inflammatory effects on the airway.

and airway responsiveness [190–193]. The mechanism of the collagen deposition is not known. However, it is known that several profibrotic cytokines, including TGF-β and PDGF, and mediators such as endothelin-1, can be produced by epithelial cells or macrophages in the inflamed airway [126]. Even simple mechanical manipulations can alter the phenotype of airway epithelial cells in a profibrotic fashion [102–104, 194]. The role of airway fibrosis in asthma is unclear, as subepithelial fibrosis has been observed even in mild asthmatics at the onset of disease; it is not certain whether the collagen deposition has any functional consequences. Subepitheial fibrosis is also seen in patients with chronic eosinophilic bronchitis, who usually present with cough and have an infiltration of eosinophils in the airways but no AHR or airway obstruction [195]. This suggests that the subepithelial fibrosis seen in asthmatics is likely to be secondary to eosinophilic inflammation in the airways and may not have important functional consequences.

Airway smooth muscle

When the concept that asthma was primarily an inflammatory disease was developed in the late 1980s to early 1990s, airway smooth muscle was thought to be predominantly a bystander tissue. That is the inflammatory process resulted in the availability of mediators that led to smooth muscle constriction and expression of an asthma phenotype. There are now substantial data to support the concept that airway smooth muscle in asthma is phenotypically different from normal airway smooth muscle and that these changes reflect the impact of the inflammatory asthmatic airway microenvironment on smooth muscle [15, 196–201]. Although some of the abnormalities identified to date in asthma impact on the contractility of individual units of airway smooth muscle, the more recently identified ones reflect differences in the ability of this tissue to proliferate or to function as an endocrine tissue (Fig. 33.9). For example, airway smooth muscle may increase in mass because of an increased tendency to proliferate due to the absence of an inhibitory transcription factor C/EBP-α [197].

Airway smooth muscle is not only influenced by the inflammatory mediators and cytokines in which it is immersed [202], but also recent studies suggest that airway smooth muscle modulates the remodeling process by secreting multiple cytokines, growth factors, or matrix proteins and by expressing cell adhesion molecules and other potential costimulatory molecules [196]. For example, increased numbers of mast cells have been identified in airway smooth muscle of patients with asthma [15, 198].

The importance of airway smooth muscle in chronic asthma was highlighted by a recent trial in which airway smooth muscle was rendered dysfunctional by bronchial thermoplasty. Although this intervention was limited to relatively large airways there was a measurable, although not overwhelming, clinical benefit [203]. The idea of manipulating the mass or contractile capacity of airway smooth muscle in asthma remains an important and unexplored avenue of asthma therapeutics [199, 204].

Vascular responses

It has been known for decades that the subepithelial connective tissue of the asthmatic airway has many more blood vessels than are found in similar locations in normal subjects [205]. It is now recognized that bronchial vessels play a key role in the pathophysiology of asthma (Fig. 33.10). Despite this anatomic knowledge, little is known about the role of the bronchial circulation in asthma. This is partly because of the difficulties involved in measuring airway blood flow. Recent studies using an inhaled absorbable gas have demonstrated increased airway mucosal blood flow in asthma [206–208].

The bronchial circulation may also play an important role in regulating airway caliber, since an increase in the vascular volume may contribute to airway narrowing. Vascular blanching of the skin is used to measure the potency of topical steroids and a similar mechanism may be involved with the effects of inhaled steroids in asthma [207, 209]. Increased airway blood flow may be important in removing inflammatory mediators from the airway, and may play a role in the development of exercise-induced asthma [210]. Increased shear stress due to high expiratory pressures may lead to gene transduction and enhanced production of nitric oxide by type III (endothelial) NO synthase [211, 212]; there are *in vivo* experimental data in sheep that support the importance of this mechanism [213]. There may also be an

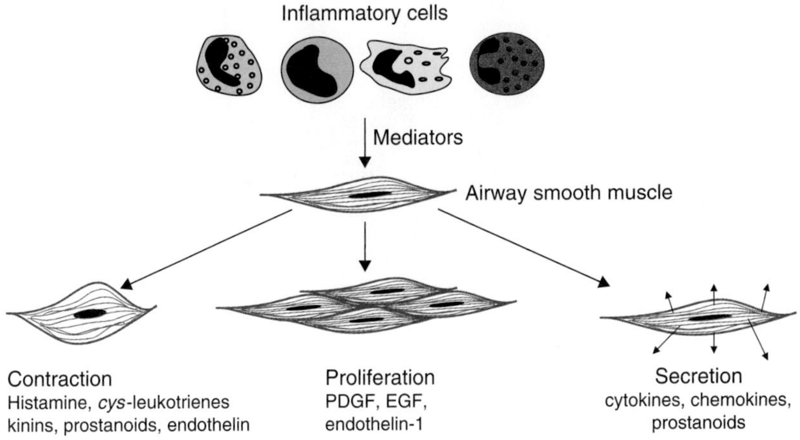

FIG. 33.9 Airway smooth muscle in asthma. Inflammation has several effects on airway smooth muscle cells, resulting in contraction, proliferation and secretion of inflammatory mediators. Cys: Cysteinyl; PDGF: platelet-derived growth factor; EGF: epidermal growth factor.

increase in the number of blood vessels in asthmatic airways as a result of angiogenesis in response to VEGF [214–216].

Microvascular leakage is an essential component of the inflammatory response and many of the inflammatory mediators implicated in asthma produce this leakage [217, 218]. There is a good evidence for microvascular leakage in asthma, and it may have several consequences on airway function, including increased airway secretions, impaired mucociliary clearance, formation of new mediators from plasma precursors (such as kinins), and mucosal edema which may contribute to airway narrowing and increased AHR [219–221].

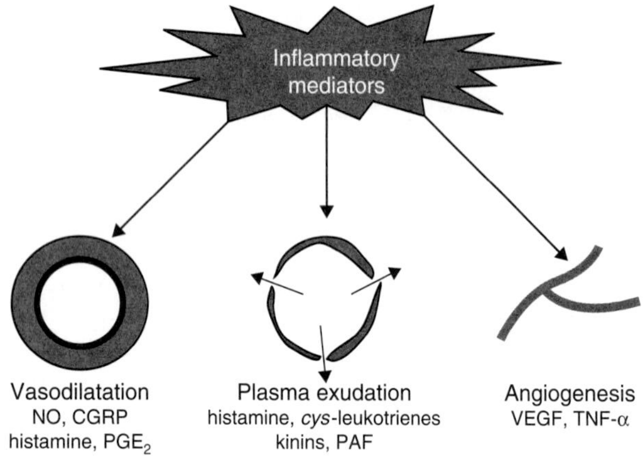

FIG. 33.10 Vascular abnormalities in asthma. Allergic inflammation has several vascular effects, including vasodilatation, plasma exudation from post-capillary venules and new vessel formation (angiogenesis). NO: Nitric oxide; CGRP: calcitonin gene-related peptide; PGE_2: prostaglandin E_2; PAF: platelet-activating factor; VEGF: vascular-endothelial growth factor; TNF-α: tumor necrosis factor-alpha.

Mucus hypersecretion

Mucus hypersecretion is a common inflammatory response in secretory tissues and is an important component of the inflammatory response in asthma (Fig. 33.11). Increased mucus secretion contributes to the viscid mucus plugs which occlude asthmatic airways, particularly in fatal asthma. There is evidence for hyperplasia of submucosal glands which are confined to large airways, and of increased numbers of epithelial goblet cells in patients with asthma, especially those with severe asthma [8]. One of the major mucin genes, MUC5AC is induced by complement factor C3a [222], NO [223], TNF-α [224], EGF family ligands [225], or neutrophil elastase [226]. Since both neutrophils and enhanced expression of mucin genes have been found in great profusion in the airways of patients with severe asthma [52], it is tempting to speculate that the neutrophils may have led to the enhanced mucin gene expression. Thus enhanced mucin expression in epithelial cells can be viewed as an indicator of the inflammatory state of the airways. These findings are not limited to cell culture, since in intact humans with asthma, exercise-induced bronchospasm is associated with enhanced expression of MUC5AC [227]. It is highly likely that methods to manipulate mucin gene expression in response to inflammatory stimuli will be identified and could be of value in the treatment of asthma as well as COPD.

Neural effects

Neural mechanisms play an important role in the pathogenesis in asthma as discussed in Chapter 32 [228]. Autonomic nervous control of the airways is complex; in addition to classical cholinergic and adrenergic mechanisms,

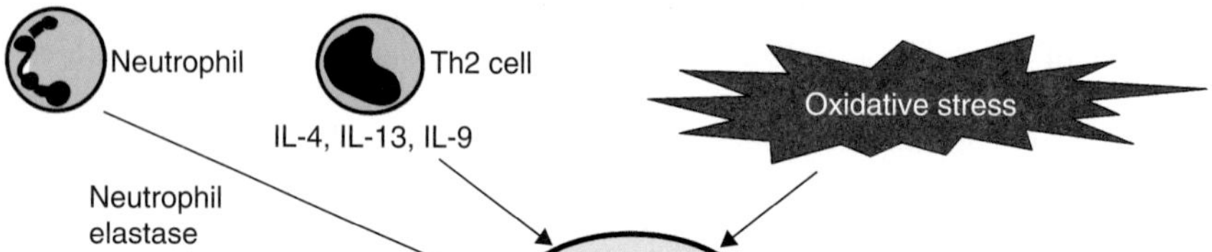

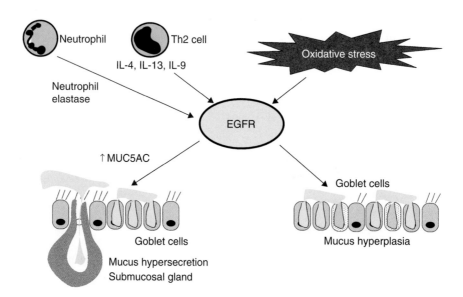

FIG. 33.11 Mucus hypersecretion in asthma. Increased mucus secretion in asthma may be stimulated by neutrophils through the release of neutrophil elastase, T helper 2 (Th2) cytokines and oxidative stress. This may be mediated via epidermal growth factor receptors (EGFR) which may result in mucus hyperplasia and increased expression of the mucin gene MUC5AC. Mucus hypersecretion is also enhanced by neural mechanisms through the release of acetylcholine (ACh) and substance P (SP).

nonadrenergic noncholinergic (NANC) nerves and several neuropeptides have been identified in the respiratory tract [229–231]. Many studies have investigated the possibility that defects in autonomic control may contribute to AHR in asthma, and abnormalities of autonomic function, such as enhanced cholinergic and α-adrenergic responses or reduced β-adrenergic responses, have been proposed. Current thinking suggests that these abnormalities are likely to be secondary to the disease, rather than primary defects [228]. It is possible that airway inflammation may interact with autonomic control by several mechanisms.

There is a close interaction between airway inflammation and airway nerves (Fig. 33.12). Inflammatory mediators may act on various prejunctional receptors on airway nerves to modulate the release of neurotransmitters [232, 233]. Thus thromboxane and PGD_2 facilitate the release of acetylcholine from cholinergic nerves in canine airways, whereas histamine inhibits cholinergic neurotransmission at both parasympathetic ganglia and postganglionic nerves via histamine H_3-receptors. Inflammatory mediators may also activate sensory nerves, resulting in reflex cholinergic bronchoconstriction or release of inflammatory neuropeptides. Inflammatory products may also sensitize sensory nerve endings in the airway epithelium, so that the nerves become hyperalgesic. Hyperalgesia and pain (dolor) are cardinal signs of inflammation, and in the asthmatic airway may mediate cough and chest tightness, which are such characteristic symptoms of asthma. The precise mechanisms of hyperalgesia are not yet certain, but mediators such as prostaglandins, certain cytokines, and neurotrophins may be important. Neurotrophins, which may be released from various cell types in peripheral tissues, may cause proliferation and sensitization of airway sensory nerves [234].

Bronchodilator nerves which are nonadrenergic are prominent in human airways and may be dysfunctional in asthma [235]. In human airways, the bronchodilator neurotransmitter appears to be NO [236].

Airway nerves may also release neurotransmitters which have proinflammatory effects. Thus neuropeptides such as substance P (SP), neurokinin A, and calcitonin-gene related peptide may be released from sensitized inflammatory nerves in the airways which increase and extend the ongoing inflammatory response [237]. There is evidence for an increase in SP-immunoreactive nerves in airways of patients with severe asthma [238], which may be due to proliferation of sensory nerves and increased synthesis of sensory neuropeptides as a result of nerve growth factors released during chronic inflammation – although this has not been confirmed in milder asthmatic patients [239]. There may also be a reduction in the activity of enzymes, such as neutral endopeptidase, which degrade neuropeptides such as SP [240]. Thus chronic asthma may be associated with increased neurogenic inflammation, which may provide a mechanism for perpetuating the inflammatory response even in the absence of initiating inflammatory stimuli.

TRANSCRIPTION FACTORS

The chronic inflammation of asthma is due to increased expression of multiple inflammatory proteins (cytokines, enzymes, receptors, adhesion molecules). In many cases these inflammatory proteins are induced by transcription factors, DNA-binding factors that increase the transcription of selected target genes (Fig. 33.13) [143].

A proinflammatory transcription factor that may play a critical role in asthma is nuclear factor-kappa B (NF-κB) which can be activated by multiple stimuli, including protein kinase C activators, oxidants, and proinflammatory cytokines (such as IL-1β and TNF-α) [241–243]. There is evidence for increased activation of NF-κB in asthmatic airways, particularly in epithelial cells and macrophages [244]. Although many aspects of the asthmatic airway are influenced by the expression of transcription factors, there are data linking the mechanical deformation of the airway to expression of proinflammatory transcription factors; these findings have direct implications for the mechanism of exercise-induced asthma and may explain how this physical stimulus could

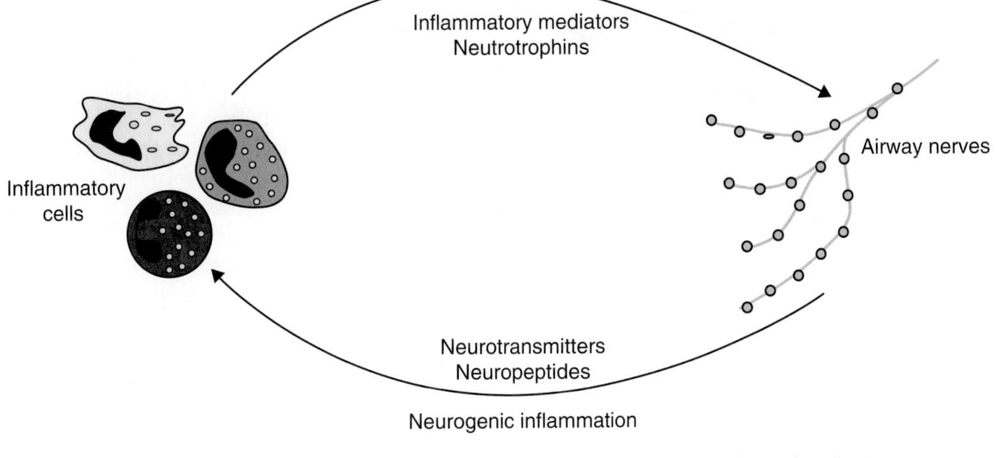

FIG. 33.12 Neural mechanism in asthma. There is a close interaction between asthmatic inflammation and neural mechanisms.

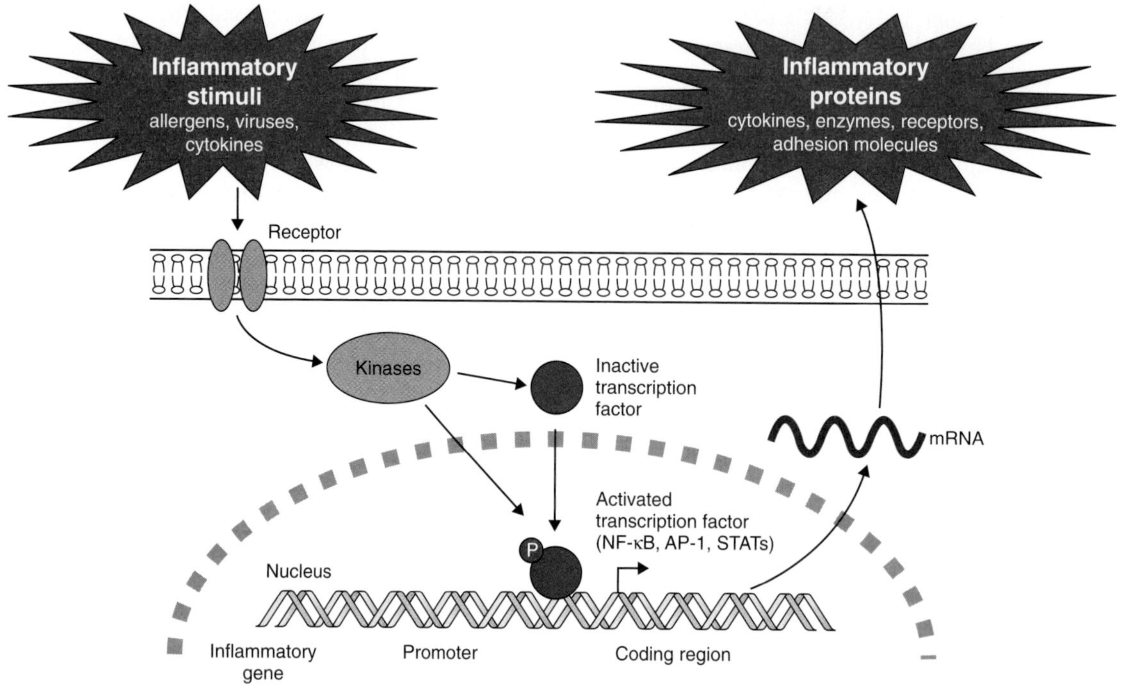

FIG. 33.13 Proinflammatory transcription factors in asthma. Transcription factors play a key role in amplifying and perpetuating the inflammatory response in asthma. Transcription factors, including nuclear factor kappa-B (NF-κB), activator protein-1 (AP-1), and signal transduction-activated transcription factors (STATs), are activated by inflammatory stimuli and increase the expression of multiple inflammatory genes.

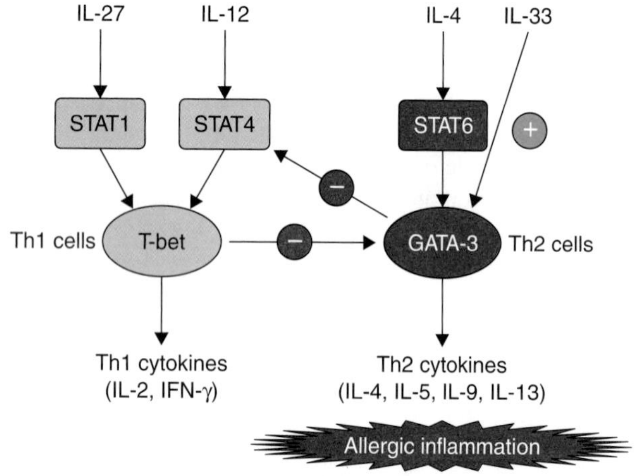

FIG. 33.14 GATA-3 in asthma. The transcription factor GATA-3 (GATA-binding protein-3) is regulated by interleukin-4 (IL-4) via STAT-6 (signal transducer and activator of transcription 6) and regulates the expression of IL-4, IL-5, IL-9, and IL-13 from T helper 2 (Th2) cells and inhibits the expression of T-bet via inhibition of STAT4. IL-33 enhances the actions of GATA-3. T-bet regulates T_H1-cell secretion of IL-2 and interferon-γ (IFN-γ) and has an inhibitory action on GATA-3. T-bet is regulated by IL-12 via STAT4 and by IL-27 via STAT1. This demonstrates the complex interplay of cytokines and transcription factors in asthma.

modify the inflammatory microenvironment of the airway [104, 245–247].

The transcription factor GATA-3 (GATA-binding protein 3) is crucial for the differentiation of uncommitted naive T-cells into Th2 cells and it regulates the secretion of T_H2-type cytokines [248]. There is an increase in the number of GATA-3$^+$ T-cells in the airways of asthmatic subjects compared with normal subjects [249]. Following simultaneous ligation of the T-cell receptor (TCR) and co-receptor CD28 by antigen-presenting cells, GATA-3 is phosphorylated and activated by p38 mitogen-activated protein kinase. Activated GATA-3 then translocates from the cytoplasm to the nucleus, where it activates gene transcription [250]. GATA-3 expression in T-cells is regulated by the transcription factor STAT-6 (signal transducer and activator of transcription 6) via IL-4 receptor activation. Another transcription factor T-bet has the opposite effect and promotes Th1 cells and suppresses Th2 cells. T-bet expression is reduced in T-cells from the airways of asthmatic patients compared with nonasthmatic patients [251]. When phosphorylated, T-bet can associate with and inhibit the function of GATA-3, by preventing it from binding to its DNA target sequences [252]. GATA-3 inhibits the production of Th1-type cytokines by inhibiting STAT-4, the major transcription factor activated by the T-bet-inducing cytokine IL-12 [253] (Fig. 33.14). Nuclear factor of activated T-cells (NFAT) is a T-cell-specific transcription factor and appears to enhance the transcriptional activation of GATA-3 at the IL-4 promoter [254]. FOXP3 is a transcription factor that is expressed in Tregs, whereas the equivalent transcription factor in Th17 cells is RORγt. There appears to be a negative interaction between these transcription factors so that a reduction of Tregs in asthma may be a factor increasing Th17 cells [255].

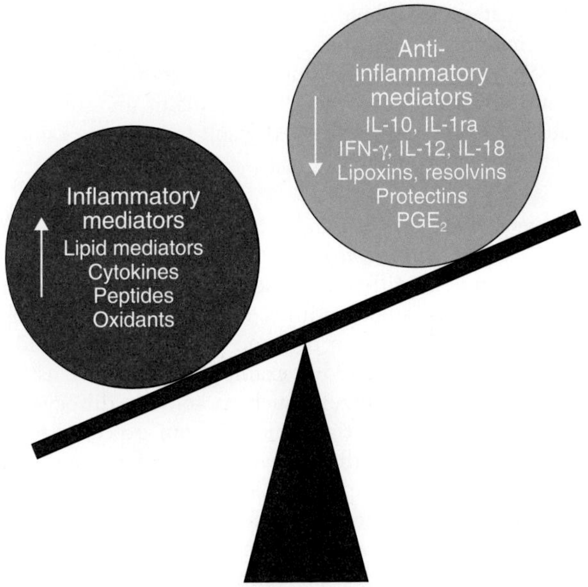

FIG. 33.15 Anti-inflammatory mediators in asthma. There may be an imbalance between increased proinflammatory mediators and a deficiency in anti-inflammatory mediators. IL: interleukin; IL-1ra: interleukin-1 receptor antagonist; IFN-γ: interferon gamma; PGE_2: prostaglandin E_2.

ANTI-INFLAMMATORY MECHANISMS IN ASTHMA

Although most emphasis has been placed on inflammatory mechanisms, there may be important anti-inflammatory mechanisms that may be defective in asthma, resulting in increased inflammatory responses in the airways (Fig. 33.15).

Lipid anti-inflammatory mediators

Several lipid mediators appear to have anti-inflammatory effects, including PGE_2, lipoxins, resolvins, and protectins [256, 257]. There is a strong evidence supporting the role of signaling through the lipoxin A receptor as being deficient in severe asthma in general and aspirin-sensitive asthma in particular [258–263]. In addition it has now been established that presqualene diphosphate (PSDP) is an endogenous regulator of phosphatidylinositol 3-kinase (PI3K). PSDP, but not presqualene monophosphate (PSMP), directly inhibits recombinant human PI3K-γ activity and a new PSDP structural mimetic blocks human neutrophil activation and murine lung PI3K activity and inflammation [264]. Resolvin E1 (RvE1: 5S,12R,18R-trihydroxyei cosapentaenoic acid) is an anti-inflammatory lipid mediator derived from the ω-3 fatty acid eicosapentaenoic acid. RvE1 is involved in the resolution of inflammation and reduces allergic inflammation in mice exposed to inhaled allergen [265].

Another important anti-inflammatory mediator is protectin D1, a natural lipid based chemical mediator that counters leukocyte activation. PD1 has been recovered from exhaled breath condensates from healthy subjects but levels were significantly lower in condensates from patients with asthma exacerbations [266]. PD1 when administered before inhaled allergen challenge, in mice led to decreased airway eosinophil and T-lymphocyte recruitment, decreased airway mucus, and less of an increment in AHR than without the PD1. These data suggest that endogenous PD1 is a novel and important counterregulatory signal in allergic airway inflammation.

IL-10

Airway and alveolar macrophages have a predominantly suppressive effect in asthma and inhibit T-cell proliferation. The mechanism of macrophage-induced immunosuppression is not yet certain, but PGE_2 and IL-10 secretion may contribute [267]. IL-10 inhibits many of the inflammatory mechanisms in asthma and its secretion is defective from macrophages of asthmatic patients [268]. IL-10 is secreted by a subset of Tregs and this is enhanced by specific allergen immunotherapy [267]. Interestingly treatment of asthmatic children with an inhaled corticosteroid or with montelukast increases serum levels of IL-10 [269]; these findings suggest that increasing IL-10 may be an important therapeutic approach in asthma.

GENETIC INFLUENCES

Over the past 15 years there have been multiple studies on the genetics of asthma (see Chapter 4). A number of potential asthma genes have been identified [270], but very few have been replicated in multiple populations [271]. However all the data suggest that there is not a single major asthma susceptibility gene, but rather there are multiple genes each explaining only a small proportion of the total variance in the asthma phenotype. The keys to find the most important asthma genes will be to use comprehensive methods and to replicate the findings in an individual population in multiple distinct populations.

ADAM33 is an asthma gene, discovered by positional cloning, that has been replicated in many but not all populations tested [272]. It was initially identified in Caucasian families from the United States and United Kingdom and was associated with asthma characterized by bronchial hyperresponsiveness. There were many variants within *ADAM33* that were significantly associated with these phenotypes. Subsequently, there have been a large number of case-control and family-based association studies focused on *ADAM33*, with the majority but not all, confirming the original finding [273]. Current data suggest that *ADAM33* may be involved in lung development and remodeling.

In genome wide-association studies microarrays capable of typing 500,000 to 1 million single nucleotide polymorphisms are used to identify associations between a given phenotype and genotype. This technology has been applied to childhood asthma [274]. These investigators identified a strong statistical association between the diagnosis of asthma in childhood in families from the United Kingdom and markers located on chromosome 17q21. The results were replicated in two European populations. They then used the genetic data to examine transcript levels in immortalized lymphocytes from children harboring the implicated variants and showed a strong association between variants in *ORMDL3* and asthma. *ORMDL3* is a member of a family of genes that encode endoplasmic reticulum proteins of unknown function. If these studies are replicated, then the community of asthma researchers will need to determine how variants in the function of this gene could lead to childhood asthma.

An area of importance is the relationship between clinical responses to anti-asthma treatments and genotype at various drug targets [275]. For example, there are known to be a number of SNPs in the β_2-adrenoceptor gene that result in structural changes in the β_2-receptor structure associated with functional changes in isolated cell systems [276]. These are related to the acute response to bronchodilator aerosols acting via this receptor [277, 278] and with deleterious effects of chronic use of albuterol [279, 280]. In a double blind randomized controlled trial, the first in a non-malignant condition where genotype at a specific locus was an entry criterion, those patients harboring the alleles encoding for homozygous expression of arginine rather than glycine at the 16th amino acid in the β_2-adrenergic receptor had diminished morning and evening peak flow and lower FEV_1 after the regularly scheduled use of albuterol than when albuterol was used on an "as needed only" basis [281]. In a retrospective analysis of two other randomized clinical trials similar changes have been observed by some groups [282] but not by others [283]. However, other retrospective analyses of long-acting β_2-agonists have not demonstrated a functional effect of these SNPs [284]. Similarly, there are genetically based functional variants in the regulation of the ALOX-5 gene [285]. Although these variants were originally associated with the response to anti-leukotriene treatment [286], current studies suggest that it is not these variants per se, but variants in the ALOX-5 gene in general that may confer differences in treatment response [287].

UNANSWERED QUESTIONS

As a syndrome much about asthma remains an enigma. Although we have reasonably effective treatments and a general idea of it pathophysiology, there are still many unanswered questions about its pathobiology.

Why is the prevalence of asthma increasing throughout the world as a consequence of Westernization?
It is clear that allergic diseases appear as societies become more Westernized, but what environmental factors are responsible is not known. It is likely that several factors are operating together [288]. Among the factors identified to date are diet (reduced intake of antioxidants, reduced unsaturated fats), lack of early childhood infections (with consequent tendency to develop Th2-driven responses), greater exposure to allergens in the home (tight housing, mattresses, central heating providing more favorable environment), cigarette smoking (pregnancy and early childhood exposure), and possibly air pollution due to road traffic. The role of exposures to pets in early childhood remains unresolved and because of confounding will be difficult to ascertain using an epidemiological approach.

Why does asthma once established become chronic in some individuals but remain intermittent and episodic in others?
Is there a genetic difference among people or an environmental difference among stimuli that results in full resolution of airway obstruction in some patients and the failure for this to occur in others? For example, occupational asthma due to chemical sensitizers, such as toluene diisocyanate, may remit if the patient is removed from exposure to the sensitizer within 6 months of development of asthma symptoms, whereas longer exposure is often associated with persistent asthma even when avoidance of exposure is complete [289, 290]. This suggests that once inflammation is established it may continue independently of a causal mechanism, whereas in others this self perpetuation does not occur.

Are there different types of asthma, each characterized by a common genetic background with its attendant immunological and inflammatory response?
Asthma remains a syndromic diagnosis. How can we get beyond a physiological and descriptive definition to one in which the primary disease pathobiology, for each of the many asthma syndromes, can be developed?

How does inflammation of the airways translate into clinical symptoms of asthma?
Airway thickening, as a consequence of the inflammatory response, may contribute to increased responsiveness to spasmogens [193, 291]. However, there is no obvious relationship between the inflammatory response in airways and asthma severity; patients with mild asthma may have a similar eosinophil response to patients with severe asthma, suggesting that there are other factors that determine clinical severity. The nature of these "factors" remains elusive, but we must identify them if we are to make true scientific progress.

How important are genetic factors (genetic polymorphisms) in determining the phenotype of asthma?
Asthma is a complex genetic disease but likely one in which there is both differences among individuals in the genes that pre-dispose to asthma and a requirement for more than one genetic variant in a given person for asthma to be manifest. Since each "asthma gene" may account for such a small proportion of the total genetic variance in asthma, how can we use genetics to develop better asthma diagnostics? How can we use genetics to help stratify patients to achieve uniform cohorts for clinical trials?

Are asthma and atopy causally related or just closely linked?
Much of our understanding of asthma derives from study of models of allergy. Although these models are of great value in defining the pathobiology of asthma, how closely will they relate to the human disease?

References

1. Bosse Y, Hudson TJ. Toward a comprehensive set of asthma susceptibility genes. *Annu Rev Med* 58: 171–84, 2007.
2. Hogg JC. The pathology of asthma. *APMIS* 105(10): 735–45, 1997.
3. Dunnill MS. The pathology of asthma, with special reference to the changes in the bronchial vasculature. *J Clin Pathol* 13: 27–33, 1960.
4. Hamid Q, Song YL, Kotsimbos TC *et al.* Inflammation of small airways in asthma. *J Allergy Clin Immunol* 100(1): 44–51, 1997.
5. Martin RJ. Small airway and alveolar tissue changes in nocturnal asthma. *Am J Respir Crit Care Med* 157(5 Suppl.): S188–90, 1998.
6. Larche M, Robinson DS, Kay AB. The role of T lymphocytes in the pathogenesis of asthma. *J Allergy Clin Immunol* 111(3): 450–63, 2003.
7. Bradding P, Walls AF, Holgate ST. The role of the mast cell in the pathophysiology of asthma. *J Allergy Clin Immunol* 117(6): 1277–84, 2006.
8. Ordonez CL, Khashayar R, Wong HH *et al.* Mild and moderate asthma is associated with airway goblet cell hyperplasia and abnormalities in mucin gene expression. *Am J Respir Crit Care Med* 163(2): 517–23, 2001.
9. Seroogy CM, Gern JE. The role of T regulatory cells in asthma. *J Allergy Clin Immunol* 116(5): 996–99, 2005.
10. Laitinen LA, Laitinen A, Altraja A *et al.* Bronchial biopsy findings in intermittent or "early" asthma. *J Allergy Clin Immunol*, 1996.
11. Payne DN, Rogers AV, Adelroth E *et al.* Early thickening of the reticular basement membrane in children with difficult asthma. *Am J Respir Crit Care Med* 167(1): 78–82, 2003.
12. Saglani S, Payne DN, Zhu J *et al.* Early detection of airway wall remodelling and eosinophilic inflammation in preschool wheezers. *Am J Respir Crit Care Med*, 2007.
13. Brusasco V, Pellegrino R. Complexity of factors modulating airway narrowing *in vivo*: Relevance to assessment of airway hyperresponsiveness. *J Appl Physiol* 95(3): 1305–13, 2003.
14. Barnes PJ. Immunology of asthma and chronic obstructive pulmonary disease. *Nat Rev Immunol* 8(3): 183–92, 2008.
15. Brightling CE, Bradding P, Symon FA, Holgate ST, Wardlaw AJ, Pavord ID. Mast-cell infiltration of airway smooth muscle in asthma. *N Engl J Med* 346(22): 1699–705, 2002.
16. Bentley AM, Hamid Q, Robinson DS *et al.* Prednisolone treatment in asthma. Reduction in the numbers of eosinophils, T cells, tryptase-only positive mast cells, and modulation of IL-4, IL-5, and interferon-gamma cytokine gene expression within the bronchial mucosa. *Am J Respir Crit Care Med* 153(2): 551–56, 1996.
17. Foresi A, Leone C, Pelucchi A. Eosinophils, mast cells, and basophils in induced sputum from patients with seasonal allergic rhinitis and perennial asthma: Relationship to methacholine responsiveness. *J Allergy Clin Immunol* 100(5): 58–64, 1997.
18. Boyce JA. The role of mast cells in asthma. *Prostaglandins Leukot Essent Fatty Acids* 69(2–3): 195–205, 2003.
19. Reber L, Da Silva CA, Frossard N. Stem cell factor and its receptor c-Kit as targets for inflammatory diseases. *Eur J Pharmacol* 533(1–3): 327–40, 2006.
20. Milgrom H, Fick RB, Su JQ *et al.* Treatment of allergic asthma with monoclonal anti-IgE antibody. *N Engl J Med* 341(26): 1966–73, 1999.
21. Avila PC. Does anti-IgE therapy help in asthma? Efficacy and controversies. *Annu Rev Med* 58: 185–203, 2007.
22. Holgate S, Casale T, Wenzel S, Bousquet J, Deniz Y, Reisner C. The anti-inflammatory effects of omalizumab confirm the central role of IgE in allergic inflammation. *J Allergy Clin Immunol* 115(3): 459–65, 2005.
23. Sabroe I, Parker LC, Dockrell DH, Davies DE, Dower SK, Whyte MK. Targeting the networks that underpin contiguous immunity in asthma and chronic obstructive pulmonary disease. *Am J Respir Crit Care Med* 175(4): 306–11, 2007.
24. Huynh ML, Malcolm KC, Kotaru C *et al.* Defective apoptotic cell phagocytosis attenuates prostaglandin E_2 and 15-hydroxyeicosatetraenoic acid in severe asthma alveolar macrophages. *Am J Respir Crit Care Med* 172(8): 972–79, 2005.
25. Arjomandi M, Witten A, Abbritti E *et al.* Repeated exposure to ozone increases alveolar macrophage recruitment into asthmatic airways. *Am J Respir Crit Care Med* 172(4): 427–32, 2005.
26. Kulkarni N, Pierse N, Rushton L, Grigg J. Carbon in airway macrophages and lung function in children. *N Engl J Med* 355(1): 21–30, 2006.
27. Hammad H, Lambrecht BN. Recent progress in the biology of airway dendritic cells and implications for understanding the regulation of asthmatic inflammation. *J Allergy Clin Immunol* 118(2): 331–36, 2006.
28. Hammad H, Lambrecht BN. Dendritic cells and epithelial cells: Linking innate and adaptive immunity in asthma. *Nat Rev Immunol* 8(3): 193–204, 2008.
29. Parameswaran K, Liang H, Fanat A, Watson R, Snider DP, Obyrne PM. Role for cysteinyl leukotrienes in allergen-induced change in circulating dendritic cell number in asthma. *J Allergy Clin Immunol* 114: 73–79, 2004.
30. Gleich GJ. Mechanisms of eosinophil associated inflammation. *J Allergy Clin Immunol* 195: 651–63, 2000.
31. Rosenberg HF, Phipps S, Foster PS. Eosinophil trafficking in allergy and asthma. *J Allergy Clin Immunol* 119(6): 1303–10, 2007.
32. Wegner CD, Gundel RH, Reilly P, Haynes N, Letts LG, Rothlein R. Intercellular adhesion molecule-1 (ICAM-1) in the pathogenesis of asthma. *Science* 247(4941): 456–59, 1990.
33. Laitinen LA, Laitinen A, Haahtela T, Vilkka V, Spur BW, Lee TH. Leukotriene-E(4) and granulocytic infiltration into asthmatic airways. *Lancet* 341: 989–90, 1993.
34. Laitinen A, Lindqvist A, Halme M, Altraja A, Laitinen LA. Leukotriene E-4-induced persistent eosinophilia and airway obstruction are reversed by zafirlukast in patients with asthma. *J Allergy Clin Immunol* 115(2): 259–65, 2005.
35. Chanez P, Dent G, Yukawa T, Barnes PJ, Chung KF. Generation of oxygen free radicals from blood eosinophils from asthma patients after stimulation with PAF or phorbol ester. *Eur Respir J* 3(9): 1002–7, 1990.
36. Evans DJ, Barnes PJ, Spaethe SM, Vanalstyne EL, Mitchell MI, Oconnor BJ. Effect of a leukotriene B_4 receptor antagonist, LY293111, on allergen induced responses in asthma. *Thorax* 51(12): 1178–84, 1996.
37. Evans DJ, Lindsay MA, O'Connor BJ, Barnes PJ. Priming of circulating human eosinophils following late response to allergen challenge. *Eur Respir J* 9(4): 703–8, 1996.
38. Parameswaran K, Watson R, Gauvreau GM, Sehmi R, O'Byrne PM. The effect of pranlukast on allergen-induced bone marrow eosinophilopoiesis in subjects with asthma. *Am J Respir Crit Care Med* 169(8): 915–20, 2004.
39. Griffithsjohnson DA, Collins PD, Rossi AG, Jose PJ, Williams TJ. The chemokine, eotaxin, activates guinea-pig eosinophils *in vitro* and causes their accumulation into the lung *in vivo*. *Biochem Biophys Res Commun* 197: 1167–72, 1993.
40. Ponath PD, Qin S, Ringler DJ *et al.* Cloning of the human eosinophil chemoattractant, eotaxin. Expression, receptor binding, and functional properties suggest a mechanism for the selective recruitment of eosinophils. *J Clin Invest* 97(3): 604–12, 1996.
41. Luster AD, Rothenberg ME. Role of the monocyte chemoattractant protein and eotaxin subfamily of chemokines in allergic inflammation. *J Leukocyte Biol* 62(5): 620–33, 1997.
42. Ravensberg AJ, Ricciardolo FL, van Schadewijk A *et al.* Eotaxin-2 and eotaxin-3 expression is associated with persistent eosinophilic bronchial inflammation in patients with asthma after allergen challenge. *J Allergy Clin Immunol* 115(4): 779–85, 2005.
43. Berkman N, Krishnan VL, Gilbey T *et al.* Expression of RANTES mRNA and protein in airways of patients with mild asthma. *Am J Respir Crit Care Med* 154(6): 1804–11, 1996.
44. Ponath PD, Qin SX, Post TW *et al.* Molecular cloning and characterization of a human eotaxin receptor expressed selectively on eosinophils. *J Exp Med* 183(6): 2437–48, 1996.
45. Collins PD, Marleau S, Griffithsjohnson DA, Jose PJ, Williams TJ. Cooperation between interleukin-5 and the chemokine eotaxin to induce eosinophil accumulation *in vivo*. *J Exp Med* 182: 1169–74, 1995.

46. Zimmermann N, King NE, Laporte J *et al.* Dissection of experimental asthma with DNA microarray analysis identifies arginase in asthma pathogenesis. *J Clin Invest* 111(12): 1863–74, 2003.
47. Park CS, Choi YS, Ki SY *et al.* Granulocyte macrophage colony-stimulating factor is the main cytokine enhancing survival of eosinophils in asthmatic airways. *Eur Respir J* 12(4): 872–78, 1998.
48. Walsh GM. Mechanisms of human eosinophil survival and apoptosis. *Clin Exp Allergy* 27(5): 482–87, 1997.
49. Leckie MJ, ten Brincke A, khan j *et al.* Effects of an interleukin-5 blocking monoclonal antibody on eosinophils, airway hyperresponsiveness and the late asthmatic response. *Lancet* 356: 2144–48, 2000.
50. FloodPage P, MenziesGow A, Phipps S *et al.* Anti-IL-5 treatment reduces deposition of ECM proteins in the bronchial subepithelial basement membrane of mild atopic asthmatics. *J Clin Invest* 112(7): 1029–36, 2003.
51. Empey DW, Laitinen LA, Jacobs L, Gold WM, Nadel JA. Mechanisms of bronchial hyperreactivity in normal subjects after upper respiratory tract infection. *Am Rev Respir Dis* 113: 131–39, 1976.
52. Wenzel SE. Asthma: Defining of the persistent adult phenotypes. *Lancet* 368: 804–13, 2007.
53. Wenzel S. Severe asthma in adults. *Am J Respir Crit Care Med* 172(2): 149–60, 2005.
54. Comhair SA, Ricci KS, Arroliga M *et al.* Correlation of systemic superoxide dismutase deficiency to airflow obstruction in asthma. *Am J Respir Crit Care Med* 172(3): 306–13, 2005.
55. Chu HW, Balzar S, Seedorf GJ *et al.* Transforming growth factor-beta 2 induces bronchial epithelial mucin expression in asthma. *Am J Pathol* 165(4): 1097–106, 2004.
56. Wenzel SE. The role of leukotrienes in asthma. *Prostaglandins Leukot Essent Fatty Acids* 69(2–3): 145–55, 2003.
57. Wenzel SE, Szefler SJ, Leung DY, Sloan SI, Rex MD, Martin RJ. Bronchoscopic evaluation of severe asthma. Persistent inflammation associated with high dose glucocorticoids. *Am J Respir Crit Care Med* 156(3 Pt 1): 737–43, 1997.
58. Simpson JL, Scott R, Boyle MJ, Gibson PG. Inflammatory subtypes in asthma: Assessment and identification using induced sputum. *Respirology* 11(1): 54–61, 2006.
59. Sur S, Crotty TB, Kephart GM *et al.* Sudden-onset fatal asthma: A distinct entity with few eosinophils and relatively more neutrophils in the airway submucosa. *Am Rev Respir Dis* 148: 713–19, 1993.
60. Maneechotesuwan K, Essilfie-Quaye S, Kharitonov SA, Adcock IM, Barnes PJ. Loss of control of asthma following inhaled corticosteroid withdrawal is associated with increased sputum interleukin-8 and neutrophils. *Chest* 132(1): 98–105, 2007.
61. Saffar AS, Alphonse MP, Shan L, Hayglass KT, Simons FE, Gounni AS. IgE modulates neutrophil survival in asthma: Role of mitochondrial pathway. *J Immunol* 178(4): 2535–41, 2007.
62. Holgate ST, Holloway J, Wilson S *et al.* Understanding the pathophysiology of severe asthma to generate new therapeutic opportunities. *J Allergy Clin Immunol* 117(3): 496–506, 2006.
63. Holt PG, Upham JW, Sly PD. Contemporaneous maturation of immunologic and respiratory functions during early childhood: Implications for development of asthma prevention strategies. *J Allergy Clin Immunol* 116: 16–24, 2005.
64. Zhu Z, Zheng T, Homer RJ *et al.* Acidic mammalian chitinase in asthmatic Th2 inflammation and IL-13 pathway activation. *Science* 304(5677): 1678–82, 2004.
65. Elias JA, Homer RJ, Hamid Q, Lee CG. Chitinases and chitinase-like proteins in T(H)2 inflammation and asthma. *J Allergy Clin Immunol* 116(3): 497–500, 2005.
66. Ying S, O'Connor B, Ratoff J *et al.* Thymic stromal lymphopoietin expression is increased in asthmatic airways and correlates with expression of Th2-attracting chemokines and disease severity. *J Immunol* 174(12): 8183–90, 2005.
67. Kostenis E, Ulven T. Emerging roles of DP and CRTH2 in allergic inflammation. *Trends Mol Med* 12(4): 148–58, 2006.
68. Naseer T, Minshall EM, Leung DY *et al.* Expression of IL-12 and IL-13 mRNA in asthma and their modulation in response to steroid therapy. *Am J Respir Crit Care Med* 155(3): 845–51, 1997.
69. Grunig G, Warnock M, Wakil AE *et al.* Requirement for IL-13 independently of IL-4 in experimental asthma. *Science* 282(5397): 2261–63, 1998.
70. Wills-Karp M, Luyimbazi J, Xu X *et al.* Interleukin-13: Central mediator of allergic asthma. *Science* 282(5397): 2258–61, 1998.
71. Zhou Z, Homer RJ, Wang ZD *et al.* Pulmonary expression of interleukin-13 causes inflammation, mucus hypersecretion, subepithelial fibrosis, physiologic abnormalities, and eotaxin production. *J Clin Invest* 103(6): 779–88, 1999.
72. Taylor A, Verhagen J, Blaser K, Akdis M, Akdis CA. Mechanisms of immune suppression by interleukin-10 and transforming growth factor-beta: The role of T regulatory cells. *Immunology* 117(4): 433–42, 2006.
73. Ling EM, Smith T, Nguyen XD *et al.* Relation of $CD4^+CD25^+$ regulatory T-cell suppression of allergen-driven T-cell activation to atopic status and expression of allergic disease. *Lancet* 363(9409): 608–15, 2004.
74. Ball TM, Castro-Rodriguez JA, Griffith KA, Holberg CJ, Martinez FD, Wright AL. Siblings, day-care attendance, and the risk of asthma and wheezing during childhood. *N Engl J Med* 343(8): 538–43, 2000.
75. Braun-Fahrlander C, Riedler J, Herz U *et al.* Environmental exposure to endotoxin and its relation to asthma in school-age children. *N Engl J Med* 347(12): 869, 2002.
76. Stockinger B, Veldhoen M. Differentiation and function of Th17 T cells. *Curr Opin Immunol* 19(3): 281–86, 2007.
77. Bullens DM, Truyen E, Coteur L *et al.* IL-17 mRNA in sputum of asthmatic patients: Linking T cell driven inflammation and granulocytic influx?. *Respir Res* 7: 135, 2006.
78. Laan M, Lotvall J, Chung KF, Linden A. IL-17-induced cytokine release in human bronchial epithelial cells *in vitro*: Role of mitogen-activated protein (MAP) kinases. *Br J Pharmacol* 133(1): 200–6, 2001.
79. Nurieva R, Yang XO, Martinez G *et al.* Essential autocrine regulation by IL-21 in the generation of inflammatory T cells. *Nature* 448(7152): 480–83, 2007.
80. Spolski R, Leonard WJ. Interleukin-21: Basic biology and implications for cancer and autoimmunity. *Annu Rev Immunol* 26: 57–79, 2008.
81. Wolk K, Sabat R. Interleukin-22: A novel T- and NK-cell derived cytokine that regulates the biology of tissue cells. *Cytokine Growth Factor Rev* 17(5): 367–80, 2006.
82. Akbari O, Faul JL, Hoyte EG *et al.* $CD4^+$ invariant T-cell-receptor$^+$ natural killer T cells in bronchial asthma. *N Engl J Med* 354(11): 1117–29, 2006.
83. Vijayanand P, Seumois G, Pickard C *et al.* Invariant natural killer T cells in asthma and chronic obstructive pulmonary disease. *N Engl J Med* 356(14): 1410–22, 2007.
84. Meyer EH, DeKruyff RH, Umetsu DT. T cells and NKT cells in the pathogenesis of asthma. *Annu Rev Med* 59: 281–92, 2008.
85. van Rensen EL, Sont JK, Evertse CE *et al.* Bronchial CD8 cell infiltrate and lung function decline in asthma. *Am J Respir Crit Care Med* 172(7): 837–41, 2005.
86. Cho SH, Stanciu LA, Holgate ST, Johnston SL. Increased interleukin-4, interleukin-5, and interferon-gamma in airway $CD4^+$ and $CD8^+$ T cells in atopic asthma. *Am J Respir Crit Care Med* 171(3): 224–30, 2005.
87. Gould HJ, Beavil RL, Vercelli D. IgE isotype determination: Epsilon-germline gene transcription, DNA recombination and B-cell differentiation. *Br Med Bull* 56(4): 908–24, 2000.
88. Takhar P, Corrigan CJ, Smurthwaite L *et al.* Class switch recombination to IgE in the bronchial mucosa of atopic and nonatopic patients with asthma. *J Allergy Clin Immunol* 119(1): 213–18, 2007.
89. Costa JJ, Weller PF, Galli SJ. The cells of the allergic response: Mast cells, basophils, and eosinophils. *JAMA* 278(22): 1815–22, 1997.
90. Youssef LA, Schuyler M, Gilmartin L *et al.* Histamine release from the basophils of control and asthmatic subjects and a comparison of gene

expression between "releaser" and "nonreleaser" basophils. *J Immunol* 178(7): 4584–94, 2007.
91. Macfarlane AJ, Kon OM, Smith SJ *et al.* Basophils, eosinophils, and mast cells in atopic and nonatopic asthma and in late-phase allergic reactions in the lung and skin. *J Allergy Clin Immunol* 105(1 Pt 1): 99–107, 2000.
92. Levine SJ. Bronchial epithelial cell-cytokine interactions in airway inflammation. *J Invest Med* 43(3): 241–49, 1995.
93. Boxall C, Holgate ST, Davies DE. The contribution of transforming growth factor-beta and epidermal growth factor signalling to airway remodelling in chronic asthma. *Eur Respir J* 27(1): 208–29, 2006.
94. Bai TR, Roberts CR, Pare PD. *Airway remodelling. Asthma: Basic Mechanisms and Clinical Management*, 3rd edn, pp. 475–486, London; San Diego: Academic press, 1998.
95. Pascual RM, Peters SP. Airway remodeling contributes to the progressive loss of lung function in asthma: An overview. *J Allergy Clin Immunol* 116(3): 477–86, 2005.
96. Homer RJ, Elias JA. Airway remodeling in asthma: Therapeutic implications of mechanisms. *Physiology* 20(1): 28–35, 2005.
97. Bergeron C, Page N, Barbeau B, Chakir J. Interleukin-4 promote's airway remodeling in asthma-regulation of procollagen I (Alpha(1)) gene by interleukin-4. *Chest* 123(3): 424S, 2003.
98. James AL, Wenzel S. Clinical relevance of airway remodelling in airway diseases. *Eur Respir J* 30(1): 134–55, 2007.
99. Hoshino M. Impact of inhaled corticosteroids and leukotriene receptor antagonists on airway remodeling. *Clin Rev Allergy Immunol* 27(1): 59–64, 2004.
100. Lange P, Parner J, Vestbo J, Schnohr P, Jensen G. A 15-year follow-up study of ventilatory function in adults with asthma. *N Engl J Med* 339(17): 1194–200, 1998.
101. Mcfadden ER. Natural history of chronic asthma and its long-term effects on pulmonary function. *J Allergy Clin Immunol* 105(2): S535–39, 2000.
102. Tschumperlin DJ, Dai G, Maly IV *et al.* Mechanotransduction through growth-factor shedding into the extracellular space. *Nature* 429(6987): 83–86, 2004.
103. Chu EK, Foley JS, Cheng J, Patel AS, Drazen JM, Tschumperlin DJ. Bronchial epithelial compression regulates epidermal growth factor receptor family ligand expression in an autocrine manner. *Am J Respir Cell Mol Biol* 32(5): 373–80, 2005.
104. Tschumperlin DJ, Drazen JM. Chronic effects of mechanical force on airways. *Annu Rev Physiol* 68: 563–83, 2006.
105. Barnes PJ, Chung KF, Page CP. Inflammatory mediators of asthma: An update. *Pharmacol Rev* 50(4): 515–96, 1998.
106. Spina D. Asthma mediators: Current views. *J Pharm Pharmacol* 52(2): 125–45, 2000.
107. Samuelsson B, Dahlen SE, Lindgren JA, Rouzer CA, Serhan CN. Leukotrienes and lipoxins: Structures, biosynthesis, and biological effects. *Science* 237: 1171–76, 1987.
108. Drazen JM, Israel E, Obyrne PM. Treatment of asthma with drugs modifying the leukotriene pathway. *N Engl J Med* 340(3): 197–206, 1999.
109. Ogawa Y, Calhoun WJ. The role of leukotrienes in airway inflammation. *J Allergy Clin Immunol* 118(4): 789–98, 2006.
110. Obyrne PM. Leukotriene bronchoconstriction induced by allergen and exercise. *Am J Respir Crit Care Med* 161(2): S68–72, 2000.
111. Leff JA, Busse WW, Pearlman D *et al.* Montelukast, a leukotriene-receptor antagonist, for the treatment of mild asthma and exercise-induced bronchoconstriction. *N Engl J Med* 339(3): 147–52, 1998.
112. Adelroth E, Inman MD, Summers E, Pace D, Modi M, Obyrne PM. Prolonged protection against exercise-induced bronchoconstriction by the leukotriene D_4-receptor antagonist cinalukast. *J Allergy Clin Immunol* 99(2): 210–15, 1997.
113. Taylor IK, O'Shaughnessy KM, Fuller RW, Dollery CT. Effect of cysteinyl-leukotriene receptor antagonist ICI 204.219 on allergen-induced bronchoconstriction and airway hyperreactivity in atopic subjects. *Lancet* 337: 690–94, 1991.
114. Wenzel SE. The role of leukotrienes in asthma. *Prostaglandins Leukot Essent Fatty Acids* 69(2–3): 145–55, 2003.
115. Dahlen B, Roquet A, Inman MD *et al.* Influence of zafirlukast and loratadine on exercise-induced bronchoconstriction. *J Allergy Clin Immunol* 109(5): 789–93, 2002.
116. Roquet A, Dahlen B, Kumlin M *et al.* Combined antagonism of leukotrienes and histamine produces predominant inhibition of allergen-induced early and late phase airway obstruction in asthmatics. *Am J Respir Crit Care Med* 155(6): 1856–63, 1997.
117. Freitag A, Watson RM, Matsos G, Eastwood C, Obyrne PM. Effect of a platelet activating factor antagonist, WEB-2086, on allergen induced asthmatic responses. *Thorax* 48: 594–98, 1993.
118. Karasawa K. Clinical aspects of plasma platelet-activating factor-acetylhydrolase. *Biochim Biophys Acta* 176: 1359–72, 2006.
119. Kuitert LM, Angus RM, Barnes NC *et al.* Effect of a novel potent platelet-activating factor antagonist, modipafant, in clinical asthma. *Am J Respir Crit Care Med* 151(5): 1331–35, 1995.
120. Spence DP, Johnston SL, Calverley PM *et al.* The effect of the orally active platelet-activating factor antagonist WEB 2086 in the treatment of asthma. *Am J Respir Crit Care Med* 149(5): 1142–48, 1994.
121. Stafforini DM, Satoh K, Atkinson DL *et al.* Platelet-activating factor acetylhydrolase deficiency: A missense mutation near the active site of an anti-inflammatory phospholipase. *J Clin Invest* 97(12): 2784–91, 1996.
122. Stafforini DM, Numao T, Tsodikov A *et al.* Deficiency of platelet-activating factor acetylhydrolase is a severity factor for asthma. *J Clin Invest* 103(7): 989–97, 1999.
123. Ito S, Noguchi E, Shibasaki M, YamakawaKobayashi K, Watanabe H, Arinami T. Evidence for an association between plasma platelet-activating factor acetylhydrolase deficiency and increased risk of childhood atopic asthma. *J Hum Genet* 47(2): 99–101, 2002.
124. O'Byrne PM. Cytokines or their antagonists for the treatment of asthma. *Chest* 130(1): 244–50, 2006.
125. Barnes PJ, Lim S. Inhibitory cytokines in asthma. *Mol Med Today* 4(10): 452–58, 1998.
126. Elias JA. Cytokines in inflammation. *Chest*, 2002.
127. Meiler F, Zimmermann M, Blaser K, Akdis CA, Akdis M. T-cell subsets in the pathogenesis of human asthma. *Curr Allergy Asthma Rep* 6(2): 91–96, 2006.
128. Oettgen HC, Geha RS. IgE regulation and roles in asthma pathogenesis. *J Allergy Clin Immunol* 107(3): 429–40, 2001.
129. Karras JG, Crosby JRGM, Tung D *et al.* Anti-inflammatory activity of inhaled IL-4 receptor-alpha antisense oligonucleotide in mice. *Am J Respir Cell Mol Biol* 36: 276–85, 2007.
130. Chung KF, van Rensen E, Stirling R, Barnes PJ. IL-5 in asthma. *Thorax* 57(8): 751, 2002.
131. Chakir J, shannon J, Molet S *et al.* Airway remodeling-associated mediators in moderate to severe asthma: effect of steroids on TGF-beta, IL-11, IL-17, and type I and type III collagen expression. *J Allergy Clin Immunol* 111: 1293–98, 2003.
132. Molet S, Hamid Q, davoine F *et al.* IL-17 is increased in asthmatic airways and induces human bronchial fibroblasts to produce cytokines. *J Allergy Clin Immunol* 108: 430–38, 2001.
133. Letuve S, Lajoie-Kadoch S, Audusseau S *et al.* IL-17E upregulates the expression of proinflammatory cytokines in lung fibroblasts. *J Allergy Clin Immunol* 117(3): 590–96, 2006.
134. Berry MA, Hargadon B, Shelley M *et al.* Evidence of a role of tumor necrosis factor alpha in refractory asthma. *N Engl J Med* 354(7): 697–708, 2006.
135. Hamid Q, Azzawi M, Ying S *et al.* Expression of mRNA for interleukin-5 in mucosal bronchial biopsies from asthma. *J Clin Invest* 87: 1541–46, 1991.
136. Teran LM. Chemokines and IL-5: Major players of eosinophil recruitment in asthma. *Clin Exp Allergy* 29(3): 287–90, 1999.
137. Flood-Page P, Swenson C, Faiferman I *et al.* A study to evaluate safety and efficacy of mepolizumab in patients with moderate persistent asthma. *Am J Respir Crit Care Med* 176(11): 1062–71, 2007.
138. Umezu-Goto M, Kajiyama Y, Kobayashi N, Kaminuma O, Suko M, Mori A. IL-9 production by peripheral blood mononuclear cells of

atopic asthmatics. *Int Arch Allergy Immunol* 143(Suppl 1): 76–79, 2007, Epub@2007 May 1
139. van den BS, Heymans J, Havaux X *et al.* Profibrotic effect of IL-9 overexpression in a model of airway remodeling. *Am J Respir Cell Mol Biol* 37(2): 202–9, 2007.
140. Melen E, Umerkajeff S, Nyberg F *et al.* Interaction between variants in the interleukin-4 receptor alpha and interleukin-9 receptor genes in childhood wheezing: Evidence from a birth cohort study. *Clin Exp Allergy* 36(11): 1391–98, 2006.
141. Sitkauskiene B, Radinger M, Bossios A, Johansson AK, Sakalauskas R, Lotvall J. Airway allergen exposure stimulates bone marrow eosinophilia partly via IL-9. *Respir Res* 6: 33, 2005.
142. Thomas PS, Yates DH, Barnes PJ. Tumor necrosis factor-alpha increases airway responsiveness and sputum neutrophilia in normal human subjects. *Am J Respir Crit Care Med* 152: 76–80, 1995.
143. Barnes PJ. Transcription factors in airway diseases. *Lab Invest* 86(9): 867–72, 2006.
144. Gauldie J, Kolb M, Ask K, Martin G, Bonniaud P, Warburton D. Smad3 signaling involved in pulmonary fibrosis and emphysema. *Proc Am Thorac Soc* 3(8): 696–702, 2006.
145. Kelly MM, Leigh R, Bonniaud P *et al.* Epithelial expression of profibrotic mediators in a model of allergen-induced airway remodeling. *Am J Respir Cell Mol Biol* 32(2): 99–107, 2005.
146. Bonniaud P, Margetts PJ, Ask K, Flanders K, Gauldie J, Kolb M. TGF-beta and Smad3 signaling link inflammation to chronic fibrogenesis. *J Immunol* 175(8): 5390–95, 2005.
147. Liu YJ, Soumelis V, Watanabe N *et al.* TSLP: An epithelial cell cytokine that regulates T cell differentiation by conditioning dendritic cell maturation. *Annu Rev Immunol* 25: 193–219, 2007.
148. Humbles AA, Lu B, Nilsson CA *et al.* A role for the C3a anaphylatoxin receptor in the effector phase of asthma. *Nature* 406(6799): 998–1001, 2000.
149. Karp CL, Grupe A, Schadt E *et al.* Identification of complement factor 5 as a susceptibility locus for experimental allergic asthma. *Nat Immunol* 1(3): 221–26, 2000.
150. Lambrecht BN. An unexpected role for the anaphylatoxin C5a receptor in allergic sensitization. *J Clin Invest* 116(3): 628–32, 2006.
151. Kohl J, Wills-Karp M. A dual role for complement in allergic asthma. *Curr Opin Pharmacol* 7(3): 283–89, 2007.
152. Kohl J, Baelder R, Lewkowich IP *et al.* A regulatory role for the C5a anaphylatoxin in type 2 immunity in asthma. *J Clin Invest* 116(3): 783–96, 2006.
153. Walsh GM. Advances in the immunobiology of eosinophils and their role in disease. *Crit Rev Clin Lab Sci* 36(5): 453–96, 1999.
154. Cross CE, van der Vilet A, Eiserich JP. Peroxidases wheezing their way into asthma. *Am J Respir Crit Care Med* 164(7): 1102–3, 2001.
155. Chambellan A, Cruickshank PJ, McKenzie P *et al.* Gene expression profile of human airway epithelium induced by hyperoxia *in vivo*. *Am J Respir Cell Mol Biol* 35(4): 424–35, 2006.
156. Montuschi P, Corradi M, Ciabattoni G, Nightingale J, Kharitonov SA, Barnes PJ. Increased 8-isoprostane, a marker of oxidative stress, in exhaled condensate of asthma patients. *Am J Respir Crit Care Med* 160(1): 216–20, 1999.
157. Baraldi E, Carraro S, Alinovi R *et al.* Cysteinyl leukotrienes and 8-isoprostane in exhaled breath condensate of children with asthma exacerbations. *Thorax* 58(6): 505–9, 2003.
158. Baraldi E, Ghiro L, Piovan V *et al.* Increased exhaled 8-isoprostane in childhood asthma. *Chest* 124(1): 25–31, 2003.
159. Comhair SA, Ricci KS, Arroliga M *et al.* Correlation of systemic superoxide dismutase deficiency to airflow obstruction in asthma. *Am J Respir Crit Care Med* 172(3): 306–13, 2005.
160. Comhair SA, Xu W, Ghosh S *et al.* Superoxide dismutase inactivation in pathophysiology of asthmatic airway remodeling and reactivity. *Am J Pathol* 166(3): 663–74, 2005.
161. Cosio BG, Mann B, Ito K *et al.* Histone acetylase and deacetylase activity in alveolar macrophages and blood mononocytes in asthma. *Am J Respir Crit Care Med* 170(2): 141–47, 2004.
162. Hew M, Bhavsar P, Torrego A *et al.* Relative corticosteroid insensitivity of peripheral blood mononuclear cells in severe asthma. *Am J Respir Crit Care Med* 174(2): 134–41, 2006.
163. Barnes PJ. Endothelins and pulmonary diseases. *J Appl Physiol* 77: 1051–59, 1994.
164. Dosanjh A, Zuraw B. Endothelin-1 (ET-1) decreases human bronchial epithelial cell migration and proliferation: Implications for airway remodeling in asthma. *J Asthma* 40(8): 883–86, 2003.
165. Redington AE, Springall DR, Meng QH *et al.* Immunoreactive endothelin in bronchial biopsy specimens: Increased expression in asthma and modulation by corticosteroid therapy. *J Allergy Clin Immunol* 100(4): 544–52, 1997.
166. Chalmers GW, Thomson L, Macleod KJ *et al.* Endothelin-1 levels in induced sputum samples from asthmatic and normal subjects. *Thorax* 52(7): 625–27, 1997.
167. Gawlik R, Jastrzebski D, Ziora D, Jarzab J. Concentration of endothelin in plasma and BAL fluid from asthmatic patients. *J Physiol Pharmacol* 57(Suppl 4): 103–10, 2006.
168. Shi-Wen X, Chen Y, Denton CP *et al.* Endothelin-1 promotes myofibroblast induction through the ETA receptor via a rac/phosphoinositide 3-Kinase/Akt-dependent pathway and is essential for the enhanced contractile phenotype of fibrotic fibroblasts. *Mol Biol Cell* 15(6): 2707–19, 2004.
169. Barnes PJ. Nitric oxide and airway disease. *Ann Med* 27(3): 389–93, 1995.
170. De Sanctis G, Maclean JA, Hamada K *et al.* Contribution of nitric oxide synthases 1, 2, and 3 to airway hyperresponsiveness and inflammation in a murine model of asthma. *J Exp Med* 189(10): 1621–30, 1999.
171. Silkoff PE, Sylvester JT, Zamel N, Permutt S. Airway nitric oxide diffusion in asthma: Role in pulmonary function and bronchial responsiveness. *Am J Respir Crit Care Med* 161(4 Pt 1): 1218–28, 2000.
172. Brindicci C, Ito K, Barnes PJ, Kharitonov SA. Differential flow analysis of exhaled nitric oxide in patients with asthma of differing severity. *Chest* 131(5): 1353–62, 2007.
173. Gustafsson LE, Leone AM, Persson MG, Wiklund NP, Moncada S. Endogenous nitric oxide is present in the exhaled air of rabbits, guinea pigs and humans. *Biochem Biophys Res Commun* 181: 852–57, 1991.
174. Kharitonov SA, Yates D, Robbins RA, Logan-Sinclair R, Shinebourne EA, Barnes PJ. Increased nitric oxide in exhaled air of asthmatic patients. *Lancet* 343: 133–35, 1994.
175. Massaro AF, Gaston B, Kita D, Fanta C, Stamler JS, Drazen JM. Expired nitric oxide levels during treatment of acute asthma. *Am J Respir Crit Care Med* 152: 800–3, 1995.
176. Kharitonov SA, Gonio F, Kelly C, Meah S, Barnes PJ. Reproducibility of exhaled nitric oxide measurements in healthy and asthmatic adults and children. *Eur Respir J* 21(3): 433–38, 2003.
177. Covar RA, Szefler SJ, Martin RJ *et al.* Relations between exhaled nitric oxide and measures of disease activity among children with mild-to-moderate asthma. *J Pediatr* 142(5): 469–75, 2003.
178. Hunt JF, Fang KZ, Malik R *et al.* Endogenous airway acidification: Implications for asthma pathophysiology. *Am J Respir Crit Care Med* 161(3): 694–99, 2000.
179. Gaston B, Kelly R, Urban P *et al.* Buffering airway acid decreases exhaled nitric oxide in asthma. *J Allergy Clin Immunol* 118(4): 817–22, 2006.
180. Ricciardolo FL, Gaston B, Hunt J. Acid stress in the pathology of asthma. *J Allergy Clin Immunol* 113(4): 610–19, 2004.
181. Lim S, Jatakanon A, Meah S, Oates T, Chung KF, Barnes PJ. Relationship between exhaled nitric oxide and mucosal eosinophilic inflammation in mild to moderately severe asthma. *Thorax* 55(3): 184–88, 2000.
182. Kharitonov SA, Barnes PJ. Exhaled biomarkers. *Chest* 130(5): 1541–46, 2006.
183. Murphy MP, Packer MA, Scarlett JL, Martin SW. Peroxynitrite: A biologically significant oxidant. *Gen Pharmacol* 31(2): 179–86, 1998.

184. Robbins RA, Hadeli K, Nelson D, Sato E, Hoyt JC. Nitric oxide, peroxynitrite, and lower respiratory tract inflammation. *Immunopharmacology* 48(3): 217–21, 2000.
185. Smith AD, Cowan JO, Brassett KP, Herbison GP, Taylor DR. Use of exhaled nitric oxide measurements to guide treatment in chronic asthma. *N Engl J Med* 352(21): 2163–73, 2005.
186. Menzies D, Nair A, Lipworth BJ. Portable exhaled nitric oxide measurement: Comparison with the "gold standard" technique. *Chest* 131(2): 410–14, 2007.
187. Jeffery PK, Godfrey RW, Adelroth E, Nelson F, Rogers A, Johansson SA. Effects of treatment on airway inflammation and thickening of basement membrane reticular collagen in asthma. A quantitative light and electron microscopic study. *Am Rev Respir Dis* 145(4 Pt 1): 890–99, 1992.
188. Wilson JW, Li X. The measurement of reticular basement membrane and submusocal collagen in the asthmatic airway. *Clin Exp Allergy* 27: 363–71, 1997.
189. Saglani S, Molyneux C, Gong H *et al.* Ultrastructure of the reticular basement membrane in asthmatic adults, children and infants. *Eur Respir J* 28(3): 505–12, 2006.
190. Jeffery PK, Wardlaw AJ, Nelson FC, Collins JV, Kay AB. Bronchial biopsies in asthma. An ultrastructural, quantitative study and correlation with hyperreactivity. *Am Rev Respir Dis* 140(6): 1745–53, 1989.
191. Chetta A, Foresi A, Del Donno M *et al.* Bronchial responsiveness to distilled water and methacholine and its relationship to inflammation and remodeling of the airways in asthma. *Am J Respir Crit Care Med* 153(3): 910–17, 1996.
192. Chetta A, Foresi A, Del Donno M, Bertorelli G, Pesci A, Olivieri D. Airways remodeling is a distinctive feature of asthma and is related to severity of disease. *Chest* 111(4): 852–57, 1997.
193. James AL, Maxwell PS, Pearcepinto G, Elliot JG, Carroll NG. The relationship of reticular basement membrane thickness to airway wall remodeling in asthma. *Am J Respir Crit Care Med* 166(12): 1590–95, 2002.
194. Chu EK, Foley J, Cheng J, Drazen JM, Tschumperlin DJ. Mechanical regulation of the urokinase system as a potential mediator of remodeling in a human airway cell culture model of asthma, *Am J Respir Crit Care Med* 35(6): 628–38, 2006.
195. Birring SS, Berry M, Brightling CE, Pavord ID. Eosinophilic bronchitis: Clinical features, management and pathogenesis. *Am J Respir Med* 2(2): 169–73, 2003.
196. Lazaar AL, Panettieri RA Jr. Airway smooth muscle: A modulator of airway remodeling in asthma. *J Allergy Clin Immunol* 116(3): 488–95, 2005.
197. Roth M, Johnson PRA, Borger P *et al.* Dysfunctional interaction of C/EBPα and the glucocorticoid receptor in asthmatic bronchial smooth-muscle cells. *N Engl J Med* 351(6): 560–74, 2004.
198. Sutcliffe A, Kaur D, Page S *et al.* Mast cell migration to Th2 stimulated airway smooth muscle from asthmatics. *Thorax* 61(8): 657–62, 2006.
199. An SS, Bai TR, Bates JH *et al.* Airway smooth muscle dynamics: A common pathway of airway obstruction in asthma. *Eur Respir J* 29(5): 834–60, 2007.
200. Oliver BG, Black JL. Airway smooth muscle and asthma. *Allergol Int* 55(3): 215–23, 2006.
201. Brightling CE, Ammit AJ, Kaur D *et al.* The CXCL10/CXCR3 axis mediates human lung mast cell migration to asthmatic airway smooth muscle. *Am J Respir Crit Care Med* 171(10): 1103–8, 2005.
202. Fernandes DJ, Mitchell RW, Lakser O, Dowell M, Stewart AG, Solway J. Invited review: Do inflammatory mediators influence the contribution of airway smooth muscle contraction to airway hyperresponsiveness in asthma? *J Appl Physiol* 95(2): 844–53, 2003.
203. Cox G, Thomson NC, Rubin AS *et al.* Asthma control during the year after bronchial thermoplasty. *N Engl J Med* 356(13): 1327–37, 2007.
204. Solway J, Irvin CG. Airway smooth muscle as a target for asthma therapy. *N Engl J Med* 356(13): 1367–69, 2007.
205. Huber HL, Koessler KK. The pathology of bronchial asthma. *Arch Intern Med* 30: 689–760, 1922.
206. Kumar SD, Emery MJ, Atkins ND, Danta I, Wanner A. Airway mucosal blood flow in bronchial asthma. *Am J Respir Crit Care Med* 158(1): 153–56, 1998.
207. Brieva JL, Danta I, Wanner A. Effect of an inhaled glucocorticosteroid on airway mucosal blood flow in mild asthma. *Am J Respir Crit Care Med* 161(1): 293–96, 2000.
208. Paredi P, Kharitonov SA, Barnes PJ. Correlation of exhaled breath temperature with bronchial blood flow in asthma. *Respir Res* 6: 15, 2005.
209. Knox AJ, Deacon K, Clifford R. Blanching the airways: Steroid effects in asthma. *Thorax* 62(4): 283–85, 2007.
210. McFadden ER Jr. Hypothesis: Exercise-induced asthma as a vascular phenomenon. *Lancet* 335(8694): 880–83, 1990.
211. Searles CD. Transcriptional and posttranscriptional regulation of endothelial nitric oxide synthase expression. *Am J Physiol Cell Physiol* 291(5): C803–16, 2006.
212. Boo YC, Jo H. Flow-dependent regulation of endothelial nitric oxide synthase: Role of protein kinases. *Am J Physiol Cell Physiol* 285(3): C499–C508, 2003.
213. Abraham WM, Ahmed A, Serebriakov I *et al.* Whole-body periodic acceleration modifies experimental asthma in sheep. *Am J Respir Crit Care Med* 174(7): 743–52, 2006.
214. Bhandari V, Choo-Wing R, Chapoval SP *et al.* Essential role of nitric oxide in VEGF-induced, asthma-like angiogenic, inflammatory, mucus, and physiologic responses in the lung. *Proc Natl Acad Sci USA* 103(29): 11021–26, 2006.
215. Feltis BN, Wignarajah D, Zheng L *et al.* Increased vascular endothelial growth factor and receptors: Relationship to angiogenesis in asthma. *Am J Respir Crit Care Med* 173(11): 1201–7, 2006.
216. Siddiqui S, Sutcliffe A, Shikotra A *et al.* Vascular remodeling is a feature of asthma and nonasthmatic eosinophilic bronchitis. *J Allergy Clin Immunol* 120(4): 813–19, 2007.
217. Persson CG. Plasma exudation and asthma. *Lung* 166: 1–23, 1988.
218. Chung KF, Rogers DF, Barnes PJ, Evans TW. The role of increased airway microvascular permeability and plasma exudation in asthma. *Eur Respir J* 3(3): 329–37, 1990.
219. Persson CGA, Andersson M, Greiff L *et al.* Airway permeability. *Clin Exp Allergy* 25: 807–14, 1995.
220. Yager D, Shore S, Drazen JM. Airway luminal liquid. Sources and role as an amplifier of bronchoconstriction. *Am Rev Respir Dis* 143: S52–S54, 1991.
221. Yager D, Butler JP, Bastacky J, Israel E, Smith G, Drazen JM. Amplification of airway constriction due to liquid filling of airway interstices. *J Appl Physiol* 66: 2873–84, 1989.
222. Dillard P, Wetsel RA, Drouin SM. Complement C3a regulates Muc5ac expression by airway Clara cells independently of Th2 responses. *Am J Respir Crit Care Med* 175(12): 1250–58, 2007.
223. Song JS, Kang CM, Yoo MB *et al.* Nitric oxide induces MUC5AC mucin in respiratory epithelial cells through PKC and ERK dependent pathways. *Respir Res* 8: 28, 2007.
224. Lora JM, Zhang DM, Liao SM *et al.* Tumor necrosis factor-alpha triggers mucus production in airway epithelium through an IkappaB kinase beta-dependent mechanism. *J Biol Chem* 280(43): 36510–17, 2005.
225. Perrais M, Pigny P, Copin MC, Aubert JP, Van SI. Induction of MUC2 and MUC5AC mucins by factors of the epidermal growth factor (EGF) family is mediated by EGF receptor/Ras/Raf/extracellular signal-regulated kinase cascade and Sp1. *J Biol Chem* 277(35): 32258–67, 2002.
226. Shao MXG, Nadel JA. Neutrophil elastase induces MUC5AC mucin production in human airway epithelial cells via a cascade involving protein kinase C, reactive oxygen species, and TNF-α-converting enzyme. *J Immunol* 175(6): 4009–16, 2005.
227. Hallstrand TS, Debley JS, Farin FM, Henderson WR Jr.. Role of MUC5AC in the pathogenesis of exercise-induced bronchoconstriction. *J Allergy Clin Immunol* 119(5): 1092–98, 2007.
228. Barnes PJ. Is asthma a nervous disease? The Parker B. Francis lectureship. *Chest* 107(3 Suppl): 119S–25S, 1995.

229. Lou YP, Francocereceda A, Lundberg JM. Variable alpha2-adrenoceptor-mediated inhibition of bronchoconstriction and peptide release upon activation of pulmonary afferents. *Eur J Pharmacol* 210: 173–81, 1992.
230. Barnes PJ, Baraniuk JN, Belvisi MG. Neuropeptides in the respiratory tract, part I. *Am Rev Respir Dis* 144: 1187–98, 1991.
231. Joos GF, Germonpre PR, Pauwels R. Role of tachykinins in asthma. *Allergy* 55: 321–37, 2000.
232. Canning BJ, Undem BJ. Evidence that antidromically stimulated vagal afferents activate inhibitory neurones innervating guinea-pig trachealis. *J Physiol (Lond.)* 480: 613–25, 1994.
233. Barnes PJ. Modulation of neurotransmission in airways. *Physiol Rev* 72: 699–729, 1992.
234. Nockher WA, Renz H. Neurotrophins and asthma: Novel insight into neuroimmune interaction. *J Allergy Clin Immunol* 117(1): 67–71, 2006.
235. Lammers JW, Barnes PJ, Chung KF. Nonadrenergic, noncholinergic airway inhibitory nerves. *Eur Respir J* 5(2): 239–46, 1992.
236. Belvisi MG, Stretton CD, Yacoub M, Barnes PJ. Nitric oxide is the endogenous neurotransmitter of bronchodilator nerves in humans. *Eur J Pharmacol* 210: 221–22, 1992.
237. Joos GF, De Swert KO, Schelfhout V, Pauwels RA. The role of neural inflammation in asthma and chronic obstructive pulmonary disease. *Ann NY Acad Sci* 992: 218–30, 2003.
238. Ollerenshaw SL, Jarvis D, Sullivan CE, Woolcock AJ. Substance P immunoreactive nerves in airways from asthmatics and nonasthmatics. *Eur Respir J* 4: 673–82, 1991.
239. Howarth PH, Springall DR, Redington AE, Djukanovic R, Holgate ST, Polak JM. Neuropeptide-containing nerves in endobronchial biopsies from asthmatic and nonasthmatic subjects. *Am J Respir Cell Mol Biol* 13(3): 288–96, 1995.
240. Nadel JA. Neutral endopeptidase modulates neurogenic inflammation. *Eur Respir J* 4(6): 745–54, 1991.
241. Ahn KS, Aggarwal BB. Transcription factor NF-kappaB: A sensor for smoke and stress signals. *Ann NY Acad Sci* 1056: 218–33, 2005.
242. Wright JG, Christman JW. The role of nuclear factor kappa B in the pathogenesis of pulmonary diseases: Implications for therapy. *Am J Respir Med* 2(3): 211–19, 2003.
243. Ulevitch RJ. Therapeutics targeting the innate immune system. *Nat Rev Immunol* 4(7): 512–20, 2004.
244. Hart LA, Krishnan VL, Adcock IM, Barnes PJ, Chung KF. Activation and localization of transcription factor, nuclear Factor-kappa B, in asthma. *Am J Respir Crit Care Med* 158(5): 1585–92, 1998.
245. Swartz MA, Tschumperlin DJ, Kamm RD, Drazen JM. Mechanical stress is communicated between different cell types to elicit matrix remodeling. *Proc Natl Acad Sci USA* 98(11): 6180–85, 2001.
246. Hilberg T. Etiology of exercise-induced asthma: Physical stress-induced transcription. *Curr Allergy Asthma Rep* 7(1): 27–32, 2007.
247. Kumar A, Lnu S, Malya R *et al.* Mechanical stretch activates nuclear factor-kappaB, activator protein-1, and mitogen-activated protein kinases in lung parenchyma: Implications in asthma. *FASEB J* 17(13): 1800–11, 2003.
248. Ho IC, Pai SY. GATA-3-not just for Th2 cells anymore. *Cell Mol Immunol* 4(1): 15–29, 2007.
249. Caramori G, Lim S, Ito K *et al.* Expression of GATA family of transcription factors in T-cells, monocytes and bronchial biopsies. *Eur Respir J* 18(3): 466–73, 2001.
250. Maneechotesuwan K, Xin Y, Ito K *et al.* Regulation of Th2 cytokine genes by p38 MAPK-mediated phosphorylation of GATA-3. *J Immunol* 178(4): 2491–98, 2007.
251. Finotto S, Neurath MF, Glickman JN *et al.* Development of spontaneous airway changes consistent with human asthma in mice lacking T-bet. *Science* 295(5553): 336–38, 2002.
252. Hwang ES, Szabo SJ, Schwartzberg PL, Glimcher LH. T helper cell fate specified by kinase-mediated interaction of T-bet with GATA-3. *Science* 307(5708): 430–33, 2005.
253. Usui T, Preiss JC, Kanno Y *et al.* T-bet regulates Th1 responses through essential effects on GATA-3 function rather than on IFNG gene acetylation and transcription. *J Exp Med* 203(3): 755–66, 2006.
254. Avni O, Lee D, Macian F, Szabo SJ, Glimcher LH, Rao A. T(H) cell differentiation is accompanied by dynamic changes in histone acetylation of cytokine genes. *Nat Immunol* 3(7): 643–51, 2002.
255. Du J, Huang C, Zhou B, Ziegler SF. Isoform-specific inhibition of RORα-mediated transcriptional activation by human FOXP3. *J Immunol* 180(7): 4785–92, 2008.
256. Bonnans C, Levy BD. Lipid mediators as agonists for the resolution of acute lung inflammation and injury. *Am J Respir Cell Mol Biol* 36(2): 201–5, 2007.
257. Schwab JM, Chiang N, Arita M, Serhan CN. Resolvin E1 and protectin D1 activate inflammation-resolution programmes. *Nature* 447(7146): 869–74, 2007.
258. Bonnans C, Fukunaga K, Levy MA, Levy BD. Lipoxin A(4) regulates bronchial epithelial cell responses to acid injury. *Am J Pathol* 168(4): 1064–72, 2006.
259. Levy BD. Lipoxins and lipoxin analogs in asthma. *Prostaglandins Leukot Essent Fatty Acids* 73(3–4): 231–37, 2005.
260. Levy BD, Bonnans C, Silverman ES, Palmer LJ, Marigowda G, Israel E. Diminished lipoxin biosynthesis in severe asthma. *Am J Respir Crit Care Med* 172(7): 824–30, 2005.
261. Levy BD, Desanctis GT, Devchand PR *et al.* Multi-pronged inhibition of airway hyper-responsiveness and inflammation by lipoxin A(4). *Nat Med* 8(9): 1018–23, 2002.
262. Levy BD, Clish CB, Schmidt B, Gronert K, Serhan CN. Lipid mediator class switching during acute inflammation: Signals in resolution. *Nat Immunol* 2(7): 612–19, 2001.
263. Sanak M, Levy BD, Clish CB *et al.* Aspirin-tolerant asthmatics generate more lipoxins than aspirin-intolerant asthmatics. *Eur Respir J* 16(1): 44–49, 2000.
264. Bonnans C, Fukunaga K, Keledjian R, Petasis NA, Levy BD. Regulation of phosphatidylinositol 3-kinase by polyisoprenyl phosphates in neutrophil-mediated tissue injury. *J Exp Med* 203(4): 857–63, 2006.
265. Aoki H, Hisada T, Ishizuka T *et al.* Resolvin E1 dampens airway inflammation and hyperresponsiveness in a murine model of asthma. *Biochem Biophys Res Commun* 367(2): 509–15, 2008.
266. Levy BD, Kohli P, Gotlinger K *et al.* Protectin D1 is generated in asthma and dampens airway inflammation and hyperresponsiveness. *J Immunol* 178(1): 496–502, 2007.
267. Hawrylowicz CM. Regulatory T cells and IL-10 in allergic inflammation. *J Exp Med* 202(11): 1459–63, 2005.
268. John M, Lim S, Seybold J *et al.* Inhaled corticosteroids increase interleukin-10 but reduce macrophage inflammatory protein-1alpha, granulocyte-macrophage colony-stimulating factor, and interferon-gamma release from alveolar macrophages in asthma. *Am J Respir Crit Care Med* 157(1): 256–62, 1998.
269. Stelmach I, Jerzynska J, Kuna P. A randomized, double-blind trial of the effect of glucocorticoid, antileukotriene and beta-agonist treatment on IL-10 serum levels in children with asthma. *Clin Exp Allergy* 32(2): 264–69, 2002.
270. Martinez FD. Genes, environments, development and asthma: A reappraisal. *Eur Respir J* 29(1): 179–84, 2007.
271. Hersh CP, Raby BA, Soto-Quiros ME *et al.* Comprehensive testing of positionally cloned asthma genes in two populations. *Am J Respir Crit Care Med*, 2007.
272. Van Eerdewegh P, Little RD, Dupuis J *et al.* Association of the ADAM33 gene with asthma and bronchial hyperresponsiveness. *Nature*, 2002.
273. Holgate ST, Yang Y, Haitchi HM *et al.* The genetics of asthma: ADAM33 as an example of a susceptibility gene. *Proc Am Thorac Soc* 3(5): 440–43, 2006.
274. Moffatt MF, Kabesch M, Liang L *et al.* Genetic variants regulating ORMDL3 expression contribute to the risk of childhood asthma. *Nature* 448(7152): 470–73, 2007.
275. Weiss ST, Litonjua AA, Lange C *et al.* Overview of the pharmacogenetics of asthma treatment. *Pharmacogenomics J* 6(5): 311–26, 2006.
276. Liggett SB. Polymorphisms of the beta$_2$-adrenergic receptor. *N Engl J Med* 346: 536–38, 2002.

277. Martinez FD, Graves PE, Baldini M, Solomon S, Erickson R. Association between genetic polymorphisms of the beta(2)-adrenoceptor and response to albuterol in children with and without a history of wheezing. *J Clin Invest* 100(12): 3184–88, 1997.
278. Thakkinstian A, McEvoy M, Minelli C *et al.* Systematic review and meta-analysis of the association between beta(2)-adrenoceptor polymorphisms and asthma: A HuGE review. *Am J Epidemiol* 162(3): 201–11, 2005.
279. Israel E, Drazen JM, Liggett SB *et al.* Effect of polymorphism of the beta(2)-adrenergic receptor on response to regular use of albuterol in asthma. *Int Arch Allergy Immunol* 124(1–3): 183–86, 2001.
280. Hall IP. Pharmacogenetics of asthma. *Chest* 130(6): 1873–78, 2006.
281. Israel E, Chinchilli VM, Ford JG *et al.* Use of regularly scheduled albuterol treatment in asthma: Genotype-stratified, randomised, placebo-controlled cross-over trial. *Lancet* 364(9444): 1505–12, 2004.
282. Wechsler ME, Lehman E, Lazarus SC *et al.* beta-Adrenergic receptor polymorphisms and response to salmeterol. *Am J Respir Crit Care Med* 173(5): 519–26, 2006.
283. Bleecker ER, Yancey SW, Baitinger LA *et al.* Salmeterol response is not affected by beta2-adrenergic receptor genotype in subjects with persistent asthma. *J Allergy Clin Immunol* 118(4): 809–16, 2006.
284. Bleecker ER, Postma DS, Lawrance RM, Meyers DA, Ambrose HJ, Goldman M. Effect of ADRB2 polymorphisms on response to longacting beta2-agonist therapy: A pharmacogenetic analysis of two randomised studies. *Lancet* 370(9605): 2118–25, 2007.
285. In KH, Asano K, Beier D *et al.* Naturally occurring mutations in the human 5-lipoxygenase gene promoter that modify transcription factor binding and reporter gene transcription. *J Clin Invest* 99(5): 1130–37, 1997.
286. Drazen JM, Yandava C, Dube LM *et al.* Pharmacogenetic association between *ALOX5* promoter genotype and the response to anti-asthma treatment. *Nat Genet* 22: 168–70, 1999.
287. Lima JJ, Zhang S, Grant A *et al.* Influence of leukotriene pathway polymorphisms on response to montelukast in asthma. *Am J Respir Crit Care Med* 173(4): 379–85, 2006.
288. Eder W, Ege MJ, Vonmutius E. Current concepts: The asthma epidemic. *N Engl J Med* 355(21): 2226–35, 2006.
289. Malo JL, Chanyeung M. Occupational asthma. *J Allergy Clin Immunol* 108(3): 317–28, 2001.
290. Beckett WS. Current concepts: Occupational respiratory diseases. *N Engl J Med* 342(6): 406–13, 2000.
291. Niimi A, Matsumoto H, Takemura M, Ueda T, Chin K, Mishima M. Relationship of airway wall thickness to airway sensitivity and airway reactivity in asthma. *Am J Respir Crit Care Med* 168(8): 983–88, 2003.

34

CHAPTER

Pathophysiology of COPD

Peter J. Barnes[1] and Stephen I. Rennard[2]

[1]National Heart and Lung Institute (NHLI), Clinical Studies Unit, Imperial College, London, UK

[2]Pulmonary and Critical Care Medicine, University of Nebraska Medical Center, Omaha, NE, USA

INTRODUCTION

COPD is a complex inflammatory disease that involves multiple interacting cells and mediators and various tissue destruction and repair mechanisms leading to structural changes that result in progressive airflow limitation with little reversibility. The preceding chapters discuss details of the individual inflammatory cells and mediators and the structural changes that are found in COPD. The links between the cellular and molecular mechanisms, the pathology, the physiological abnormalities, and symptoms are still not well understood. This is complicated by the fact that there is heterogeneity of the disease, with some patients showing a predominant emphysema pattern, whereas in others small airway disease predominates, although many patients have a mixed pattern, as cigarette smoking is a common causal mechanism. There are differences in the degree of mucus hypersecretion between patients and in the frequency of infective exacerbations. Finally it is nowrecognized that COPD affects extrapulmonary organs, such as skeletal muscle, as well as increasing the prevalence of common co-morbid diseases, such as cardiovascular disease and lung cancer, which together are the commonest cause of death in patients with COPD.

The definition of COPD adopted by the Global initiative on Obstructive Lung Disease (GOLD) encompasses the idea that COPD is a chronic inflammatory disease and much of the recent research has focused on the nature of this inflammatory response [1]. COPD is an obstructive disease of the lungs which slowly progresses over many decades leading to death from respiratory failure unless patients die of co-morbidities such as heart disease and lung cancer before this stage. Although the commonest cause of COPD is chronic cigarette smoking, some patients, particularly in developing countries, develop the disease from inhalation of wood smoke from biomass fuels or other inhaled irritants [2]. However, only about 25% of smokers develop COPD of at least moderate severity [3], although as many as half will develop abnormal lung function [4], suggesting that there may be genetic or host factors that predispose to its development, although these have not yet been clearly identified (see Chapter 4). The disease is thought to be relentlessly progressive and, to date, only smoking cessation has been demonstrated to reduce the rate of decline in lung function, although as the disease becomes more severe there is less effect of smoking cessation and lung inflammation persists. However, it is not yet clear that all patients with mild and moderate COPD show progression. Some patients may have small lungs due to abnormalities in lung development due to fetal or early childhood influences and develop symptoms of COPD as a result of the normal decline in lung function with aging but starting from a lower peak (Fig. 34.1). These patients may end up with airflow obstruction without abnormal inflammatory changes.

AIRFLOW LIMITATION

Airflow limitation in COPD may be the result of three different pathological mechanisms (Fig. 34.2):

1. Thickening and fibrosis of small airways (chronic obstructive bronchiolitis) which is presumed to be due to the effect of chronic inflammation.

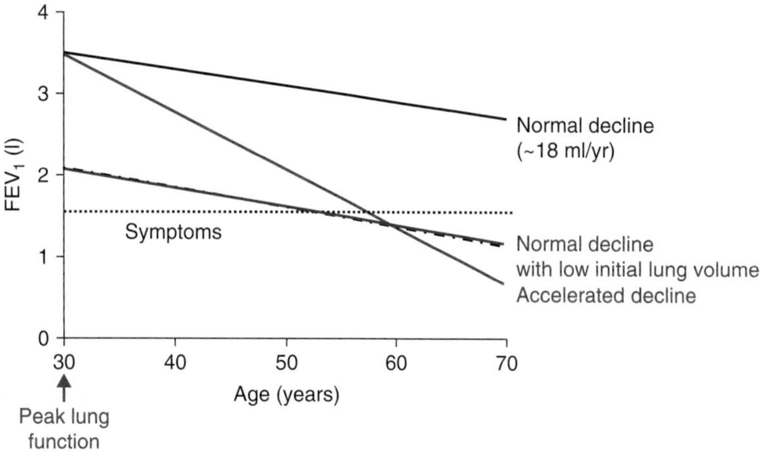

FIG. 34.1 Small airways in COPD patients. The airway wall is thickened and infiltrated with inflammatory cells, predominately macrophages and CD8$^+$ lymphocytes, with increased numbers of fibroblasts. In severe COPD there are also lymphoid follicles, which consist of a central core of B-lymphocytes, surrounded by T-lymphocytes and thought to indicate chronic exposure to antigens (bacterial, viral, or autoantigens). Similar changes are also reported in larger airways. The lumen is often filled with an inflammatory exudate and mucus. There is peribronchial fibrosis, and, resulting in progressive and irreversible narrowing of the airway. Airway smooth muscle may be increased slightly.

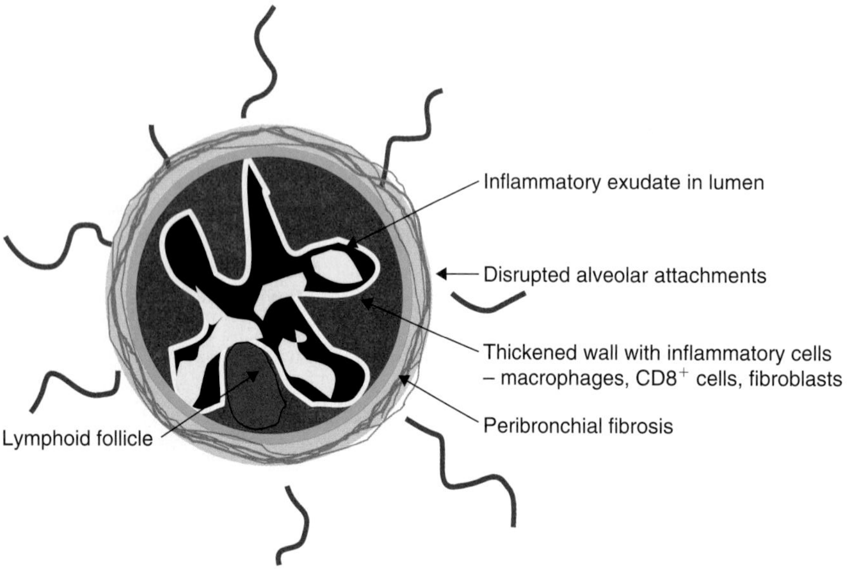

FIG. 34.2 Mechanisms of airflow limitation in COPD. Small airways are obstructed as a result of loss of alveolar attachments, thickening and distortion of the airway wall as a result of inflammation and fibrosis and occlusion of the airway lumen by mucus and inflammatory exudate.

2. Emphysema results in expiratory airway collapse for two reasons: Disruption of alveolar attachments permits small airways to close on expiration due to lack of elastic support; in addition, loss of lung elastic recoil decreases the intralumenal pressure, which also predisposes to small airway collapse, particularly with forced exhalation.
3. Luminal obstruction with mucus and plasma exudate as a result of inflammation.

All three mechanisms may contribute to air trapping and lung hyperinflation, resulting in dyspnea and exercise limitation [5] (Fig. 34.3) (see Chapter 5). The relative roles of emphysema and small airway disease in causing airflow limitation are debated, but recent studies with measurements of emphysema by computerized tomography show that varying severity of emphysema can be present with similar impairment of FEV$_1$, although there may be abnormalities in diffusion capacity and hypoxia [6]. The variable association of emphysema with airflow limitation has focused increased attention on small airways in COPD, but it is difficult to measure small airway function in patients and novel imaging and physiological approaches are needed [7].

ETIOLOGIC FACTORS

Inhalational exposures

The most important etiologic risk factors for the development of COPD are inhalational exposures and cigarette smoke is by far the most important of these. Exposures other than cigarette smoking, however, also contribute to the risk for COPD. Such exposures can both cause COPD independently of cigarette smoking and can increase the risk for COPD in the presence of concurrent cigarette smoking. Exposures leading to COPD include a range of both environmental fumes and dusts. Coal dust, for example, alone can lead to airflow limitation. Use of biomass for home cooking can result in significant inhalational exposure and a marked increased risk for the development of COPD [8]. Such exposure is a currently a major risk factor for COPD in parts of the developing world. Passive cigarette smoke exposure also is related to symptoms of cough and sputum production and is likely a risk factor for the development of airflow limitation [9].

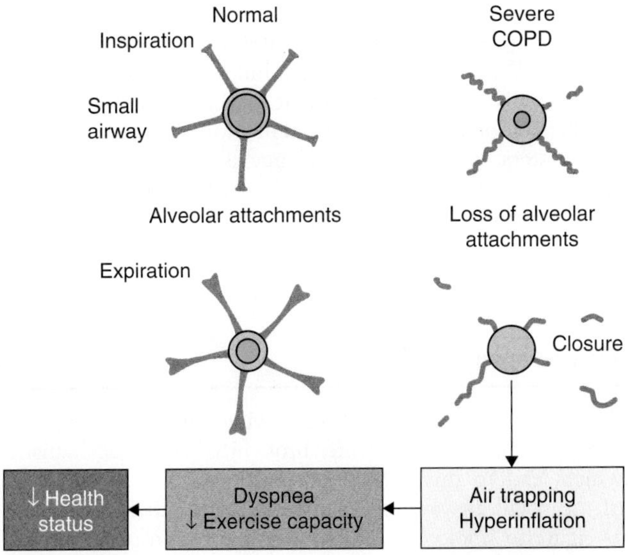

FIG. 34.3 Mechanism of dyspnea in COPD. The airway narrows on expiration but in normal individuals the elasticated alveolar attachments prevent closure so that the alveoli empty. In COPD there is a loss of elastin fibers as a result of elastases, which means that even in mild COPD small airways close to a greater extent. As COPD becomes more severe, the thickness of small airways increases and alveolar attachments may be disrupted so that peripheral airway may close during expiration. This results in air trapping and hyperinflation, leading to dyspnea and reduced exercise capacity.

Some exposures which represent risk factors for the development of COPD, for example intravenous drug abuse and cadmium may act through pathways distinct from the more common inhalational exposures. Intravenous drug abuse may lead to the development of COPD by a vascular mechanism, and cadmium, which can cause emphysema without inflammation, may act by altering the repair response. These rare causes of COPD are particularly instructive as they indicate that a variety of mechanisms may lead to altered lung structure and function. To what degree such mechanisms are involved in the more common etiologies of COPD remains to be determined.

It is likely that the varied inhalational exposures, which lead to COPD, share the ability to initiate some of the same pathogenetic mechanisms. This may account for some of the similarities of the disease induced by cigarette smoke and other environmental exposures. Most exposures, however, are complex and contain many candidate toxins. Importantly, not all exposed individuals will develop COPD. Genetic factors, most of which remain to be defined, can affect risk.

Infections

Infections play a key role in COPD exacerbations (see Chapter 37). Up to two-thirds of exacerbations can be associated with a potential infectious pathogen, about half being bacterial and half being viral. Individuals with frequent exacerbations have been demonstrated to have an inferior quality of life as assessed by the St. George's Respiratory Questionnaire. The suggestion has also been made that individuals with frequent exacerbations experience a more rapid decline in lung function [10]. Damage induced by infections and the ensuing inflammatory process may make the airways more susceptible to subsequent infection [11].

Infections also have been related to increased risk for the development of COPD. A history of severe childhood viral infection has been associated with reduced lung function and with increased respiratory symptoms in some studies. There are several, non-exclusive, possible explanations for these associations. First, there might be an increased diagnosis of severe infections in children who have underlying airways responsiveness which itself may be a risk factor. Second, viral infections may be related to another factor such as socioeconomic status or birth weight which itself is related to COPD. Viral infections, moreover, may directly contribute to the development of COPD. In this regard, viral infections may result in incorporation of viral DNA into airway cells which could alter the response to subsequent exposures increasing the risk for COPD. In contrast, a more recent study failed to show an accelerated rate of decline among adults age 35–45 who had a history of childhood infections. However, another study did not demonstrate an accelerated rate of decline in adults aged 35–45 who had childhood infections [12].

Animal studies suggest the incorporation of adenoviral DNA can amplify the inflammatory response induced when airway epithelial cells are exposed to cigarette smoke and increased levels of adenoviral DNA have been detected in the lungs of patients with COPD compared to control patients [13]. In addition, HIV infection has also been related to the development of emphysema [14]. HIV-induced pulmonary inflammation may play a role, although weight loss associated with HIV infection has also been suggested as a possible mechanism.

Nutrition

Nutritional status, independent of concurrent viral infection, is also likely a risk factor for the development of COPD. The famine and resulting starvation in the Warsaw ghetto in World War II was associated with the development of emphysema. Reduced body weight, often expressed as body mass index, is a poor prognostic sign in individuals with established COPD. Underweight individuals with COPD, moreover, appear to be at risk for the development of emphysema. The association of starvation and anabolic/catabolic status with the development of COPD, particularly with emphysema, is also supported by experimental studies in animals.

Birth weight, lung growth, and development

Low birth weight has been associated with the development of COPD although this association remains controversial. Low birth weight has also been associated with increased risk for asthma and with reduced lung function in both childhood and young adulthood. In this regard, reduced maximal attained lung function may identify individuals who are at increased risk for the development of COPD. While some of these results are controversial it remains possible that developmental problems associated with low birth weight represent a risk factor for the development of COPD.

Aging

COPD is a disease of the elderly. While decrements in lung function can be observed at any age and decline in lung function can begin in young adulthood, clinically significant COPD is unusual before the fifth or sixth decade of life and is most common even later [15]. Lung function also declines with normal aging and compliance increases. Associated with this process are changes in alveolar structure including increased size and number of alveolar pores together with enlargement in alveolar spaces. These changes, which are sometimes termed "senile emphysema," resemble, in at least some features, clinical emphysema. This suggests the concept that aging may lead to emphysema or may accelerate its development. Further supporting a connection between COPD and aging, circulating lymphocytes in smokers have reduced telomere length [16] and chronic exposure to cigarette smoke can induce senescence in cells cultured *in vitro* [17]. In addition, fibroblasts cultured from the lungs of COPD patients proliferate more slowly and also show markers of senescence [18, 19]. Finally, emphysema spontaneously develops in mice with genetic forms of accelerated aging [20, 21]. Taken together, these observations support the concept that COPD may represent a form of accelerated senescence in the lung.

Reactive airways

Asthma and increased airways reactivity have been identified as risk factors for the development of COPD. This relationship was originally proposed by Orie and colleagues and termed the Dutch hypothesis. Asthmatics, as a group, experience accelerated loss of lung function as do smokers with increased airways reactivity. Whether airways reactivity results from an inflammatory process, which, in turn, also leads to loss of lung function in COPD or whether the airways reactivity directly causes COPD remains undetermined.

Genetics

It is likely that many genetic factors interact with and increase (or decrease) the risk for developing COPD. Family studies have demonstrated an increased risk of COPD within families with COPD probands. Some of this risk may be due to shared environmental factors, but several studies in diverse populations also suggest shared genetic risk (see Chapter 4).

To date the only well-defined genetic risk factor is deficiency of alpha-1 protease inhibitor (PI). Severe deficiency in this major circulating inhibitor of serine proteases is associated with development of emphysema in nonsmokers, although not all deficient individuals are affected. In smokers, alpha-1 PI deficiency is associated with the accelerated development of emphysema and mortality. A number of other genetic factors, while not unequivocally established, have been suggested to be related to risk for COPD (see Chapter 4).

Tissue remodeling

Peribronchial fibrosis and narrowing contributes to airflow limitation in chronic bronchitis. The mechanisms which lead to this process are unknown. However, injury of the small airways, either directly by inhaled toxins such as cigarette smoke or indirectly by the action of inflammatory mediators likely initiates repair processes [22]. In this regard, the airway epithelium has considerable capacity to mediate repair. Following mechanical injury, for example, airway epithelial cells from the edge of the wound rapidly flatten and migrate to cover the defect.

In vitro studies suggest that fibronectin and transforming growth factor (TGF)-β produced by the epithelial cells present in the wound may help direct these processes. The newly recruited cells then replicate and undergo an orderly sequence of differentiation. It is likely that these processes often can restore both anatomic structure and airway function. In addition to the epithelial cells, mesenchymal cells (fibroblasts and myofibroblasts) are also activated in the repair response. The wave of accumulation and replication of epithelial cells which occurs 24 h after mechanical injury is followed about 2 days later by the accumulation and proliferation of mesenchymal cells. Under normal circumstances, these cells disappear over the next few weeks.

However, as in many tissues, repair in the airways can result in the excessive deposition of fibrotic extracellular matrix, and like most scars, these contract [22]. If this were to happen circumferentially around an airway, airway narrowing would result. In this context, airway epithelial cells can also produce factors which drive fibroblast recruitment, matrix production and remodeling. Interestingly, the fibrotic process may be driven by some of the same mediators which may also lead to epithelial repair such as TGF-β and fibronectin. Consistent with a role for these mediators, both TGF-β and fibronectin have been reported in the airways and bronchoalveolar lavage (BAL) fluid in asthma and chronic bronchitis. While the processes which regulate airway repair following injury are only partly delineated, it seems likely that disordered repair processes can lead to tissue remodeling with altered airway structure and function.

Altered repair mechanisms may contribute to the development of emphysema [22]. In this context, the net tissue loss which characterizes emphysema may result from inadequate repair in the face of injury. Several lines of evidence support this concept. First, several animal models of emphysema have been developed in which lung destruction is induced either by exposure to large concentrations of neutrophil elastase (NE), to cigarette smoke or to other similar insults. Such injury is characterized by a rapid loss of lung connective tissue consistent with tissue destruction taking place. These models, however, are also associated with a rapid onset of new connective tissue synthesis. In many cases, total tissue matrix macromolecule concentration is restored to normal or increased levels within a few weeks of tissue injury. Similarly, in mild human emphysema, lung collagen content has been reported to be increased consistent with initiation of matrix molecule production.

The concept that repair processes initiated in model systems of emphysema serve to mitigate the severity of the resulting emphysema is supported by studies in which repair processes are disrupted. For example, starvation can greatly potentiate the development of emphysema following elastase exposure in rats [23]. Similarly, inhibition of matrix macromolecule cross-linking can also exacerbate the development of emphysema. In this context, cigarette smoke can interfere with repair processes by a number of mechanisms. Smoke can inhibit parenchymal cell recruitment, proliferation, matrix production, and tissue remodeling [22]. Smoke can also interfere with matrix macromolecule cross-linking [24]. It is possible, therefore, that smoke can lead to the development of emphysema by three interacting mechanisms: (1) by initiating an inflammatory response which causes tissue destruction; (2) by interfering with the defenses which normally protect tissues from injury during inflammation; and (3) by disrupting the repair processes which have the potential for restoring tissue architecture in the face of injury.

Several lines of evidence suggest that tissue repair or maintenance may be abnormal in emphysema. Reduced VEGF and VEGF receptors in the lungs of patients with COPD and, in an animal model, that interruption of VEGF signaling can lead to the development of emphysema through apoptosis [25]. Fibroblasts from patients with COPD proliferate more slowly [18] and are deficient in several *in vitro* measures of tissue repair [26].

Pathogenesis: Integrating concepts

Taken together, the currently available data suggest that the majority of COPD results from exposures to noxious agents. These, in turn, lead to activation of inflammatory processes within the lower respiratory tract. Damage to lung structures results both from the exposures and, more importantly, from the ensuing inflammatory responses. Inflammation and injury also activate repair responses. Some individuals, either on a genetic or on a developmental basis, appear to be particularly sensitive to the injurious effects of exposures and inflammation. Similarly, individuals likely differ in their ability to repair lung injury, differences which reflect both genetic and acquired (e.g. nutritional) heterogeneity. The following sections will review briefly the cells and mediators believed to play a role in COPD.

COPD AS AN INFLAMMATORY DISEASE

Both the small airway remodeling and narrowing and the emphysema are believed to be due to chronic inflammation in the lung periphery. Quantitative studies have shown that the inflammatory response in small airways and lung parenchyma increases as the disease progresses [27] (see Chapter 6). There is a specific pattern of inflammation in COPD airways and lung parenchyma with increased numbers of macrophages, T-lymphocytes, with predominance of $CD8^+$ (cytotoxic) T-cells, and in more severe disease B-lymphocytes with increased numbers of neutrophils in the lumen [28] (Fig. 34.4). The inflammatory response in COPD involves both innate and adaptive immune responses. Multiple inflammatory mediators are increased in COPD and are derived from inflammatory cells and structural cells of the airways and lungs [29]. A similar pattern of inflammation is seen in smokers without airflow limitation, but in COPD this inflammation is amplified and even further amplified during acute exacerbations of the disease which are usually precipitated by bacterial and viral infections (Fig. 34.5). The molecular basis of this amplification of inflammation is not yet understood but may be, at least in part, genetically determined. Cigarette smoke and other irritants in the respiratory tract may activate surface macrophages and airway epithelial cells to release chemotactic factors that then attract circulating leukocytes into the lungs. Amongst chemotactic factors, chemokines predominate and therefore play a key role in orchestrating the chronic inflammation in COPD lungs and its further amplification during acute exacerbations [30]. These might be the initial inflammatory events occurring in all smokers. However in smokers who develop COPD this inflammation progresses into a more complicated inflammatory pattern of adaptive immunity and involves T- and B-lymphocytes and possibly dendritic cells along with a complicated interacting array of cytokines and other mediators [31].

Differences from asthma

Histopathological studies of COPD show a predominant involvement of peripheral airways (bronchioles) and lung parenchyma, whereas asthma involves inflammation in all airways (particularly proximal airways) but usually without involvement of the lung parenchyma [32]. In COPD there is narrowing of bronchioles, with fibrosis and infiltration with macrophages and T-lymphocytes, along with destruction of lung parenchyma and an increased number of macrophages and T-lymphocytes, with a greater increase in $CD8^+$ than $CD4^+$ (helper) cells [27] (Fig. 34.6). Bronchial biopsies show similar changes with an infiltration of macrophages and $CD8^+$ cells and an increased number of neutrophils in patients with severe COPD. BAL fluid and induced sputum demonstrate a marked increase in macrophages and neutrophils. In contrast to asthma, eosinophils are not prominent except during exacerbations or when patients have concomitant asthma [32, 33]. In contrast to the epithelial fragility of asthma, the airway epithelium in patients with COPD may be metaplastic, as a result of the chronic release of epithelial growth factors such as epithelial growth factor (EGF) from these cells. Another difference between COPD and asthma is the pattern of fibrosis, with subepithelial fibrosis (sometimes called basement membrane thickening) as a very characteristic feature of asthma,

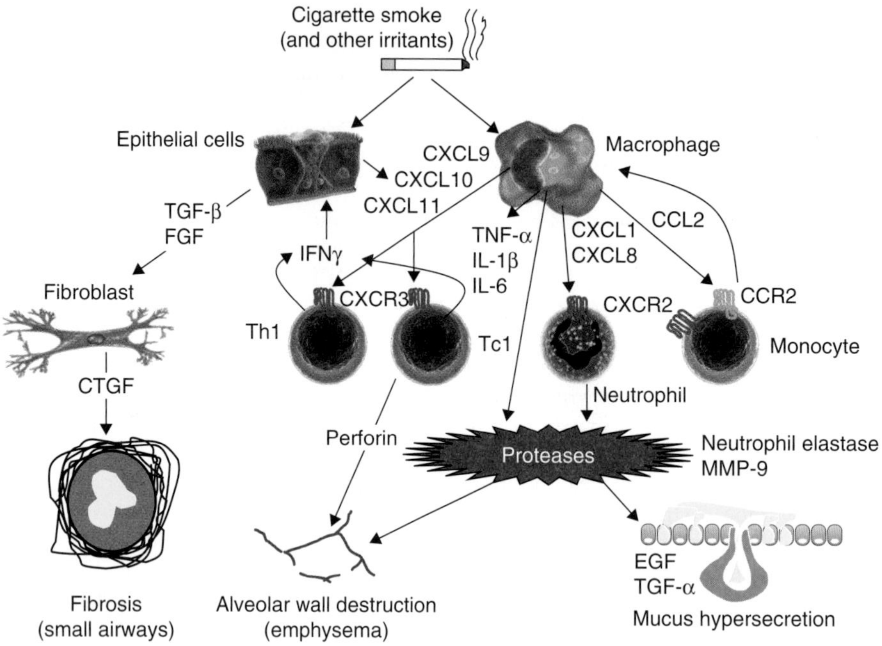

FIG. 34.4 Inflammatory cells and mediators involved in COPD. Inhaled cigarette smoke and other irritants activate epithelial cells and macrophages to release several chemotactic factors that attract inflammatory cells to the lungs, including CC-chemokine ligand 2 (CCL2), which acts on CC-chemokine receptor 2 (CCR2) to attract monocytes, CXC-chemokine ligand 1 (CXCL1) and CXCL8, which act on CXCR2 to attract neutrophils and monocytes (which differentiate into macrophages in the lungs) and CXCL9, CXCL10, and CXCL11, which act on CXCR3 to attract T helper 1 (Th1) cells and type 1 cytotoxic T-cells (Tc1 cells), both of which release interferon-γ (IFN-γ). These inflammatory cells together with macrophages and epithelial cells release proteases, such as matrix metalloproteinase 9 (MMP9), which cause elastin degradation and emphysema. Neutrophil elastase also causes mucus hypersecretion. Epithelial cells and macrophages also release transforming growth factor-β (TGF-β) and fibroblast growth factors (FGF), which stimulate fibroblast proliferation, resulting in fibrosis in the small airways as well as the proinflammatory cytokines tumor necrosis factor-α (TNF-α), IL-1β and IL-6 which amplify inflammation. Mucus hypersecretion is stimulated by epithelial growth factor (EGF) and TGF-α.

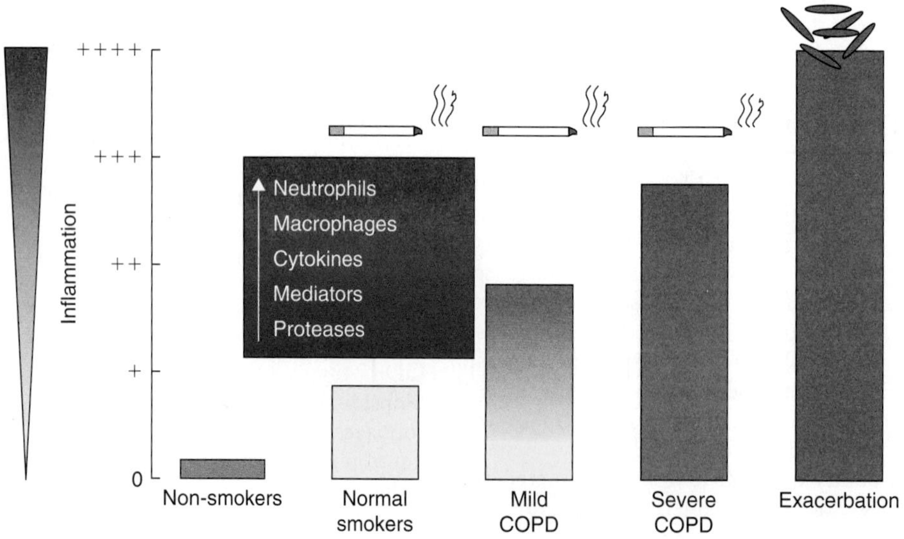

FIG. 34.5 Amplification of lung inflammation in COPD. Normal smokers have a mild inflammatory response, which represents the normal (probably protective) reaction of the respiratory mucosa to chronic inhaled irritants. In COPD this same inflammatory response is markedly amplified, and this amplification increases as the disease progresses. It is further increased during exacerbations triggered by infective organisms. The molecular mechanisms of this amplification are currently unknown, but may be determined by genetic factors or possibly latent viral infection. Oxidative stress is an important amplifying mechanism and may increase the expression of inflammatory genes through impairing the activity of histone deacetylase-2 (HDAC2), which is needed to switch off inflammatory genes.

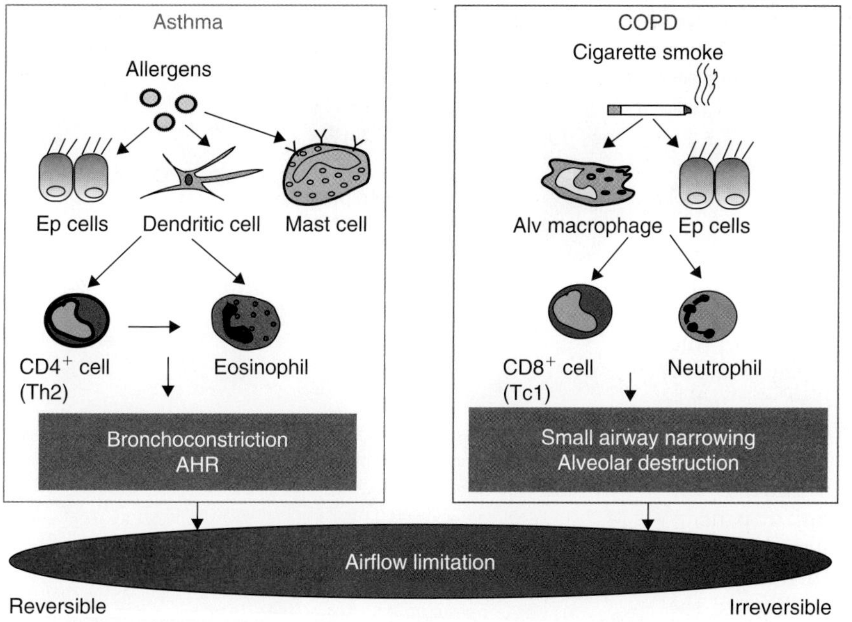

FIG. 34.6 Differences in inflammation between COPD and asthma. Left panel shows that inflammation in asthma is driven by allergens resulting in activation of mast cells, dendritic cells and epithelial cells and leading to the recruitment of T helper 2 (Th2) cells and eosinophils. This results in airway hyperresponsiveness, the defining physiologic defect in asthma. Right panel shows that COPD is driven by inhaled irritants such as cigarette smoke, which results in activation of epithelial cells and macrophages, which recruit neutrophils and T-lymphocytes into the lung, resulting in small airway narrowing (chronic obstructive bronchiolitis) and alveolar destruction (emphysema).

whereas in COPD fibrosis is found mainly around small airways (peribronchiolar fibrosis). The airway smooth muscle layer in asthma may be thickened as a result of hypertrophy and hyperplasia of airway smooth muscle cells, whereas this layer is not very much increased in patients with COPD [34]. Mucus hypersecretion is a feature of asthma and COPD, with hyperplasia of goblet cells and submucosal glands [35] (see Chapter 17).

Although the inflammatory picture in asthma and COPD differs, in patients with severe asthma there are greater similarities between the two diseases, with increased neutrophils, $CD8^+$ cells, and similar patterns of mediators in severe asthma as in COPD [31].

INFLAMMATORY CELLS

For many years, it was believed that the inflammatory reaction in the lungs of smokers consisted of neutrophils and macrophages and those proteinases from these cells were responsible for the lung destruction in COPD. More recently it has been recognized that there is a prominent T-cell infiltration in the lungs of patients with COPD, with a predominance of $CD8^+$ T-cells, although $CD4^+$ (helper) T-cells are also numerous. Although abnormal numbers of inflammatory cells have been documented in COPD, the relationship between these cell types and the sequence of their appearance and their persistence are not yet understood in detail [28]. Most studies have been cross-sectional, based on selection of patients with different stages of the disease, and comparisons have been made between smokers without airflow limitation (normal smokers) and those with COPD who have smoked a similar amount. There are no serial studies; selection biases (such as selecting tissue from patients suitable for lung volume reduction surgery) may give misleading results. Nonetheless, there is a progressive increase in the numbers of inflammatory cells in small airways and lung parenchyma as COPD becomes more severe even though the patients with most severe obstruction have stopped smoking for many years [27], indicating the existence of mechanisms which perpetuate the inflammatory reaction in COPD. This is in contrast to many other chronic inflammatory diseases, such as rheumatoid arthritis, where the inflammation tends to diminish in severe disease. The inflammation of COPD lungs involves both innate immunity (neutrophils, macrophages, eosinophils, mast cells, NK cells, $\gamma\delta$-T-cells, and dendritic cells) and adaptive immunity (T- and B-cells).

The inflammation in the airways of COPD patients does not seem to disappear after smoking cessation, even after a period of several years [36, 37].

Epithelial cells

Epithelial cells are activated by cigarette smoke to produce inflammatory mediators, including tumor necrosis factor (TNF)-α, interleukin (IL)-1β, IL-6, granulocyte-macrophage colony-stimulating factor (GM-CSF) and CXCL8 (IL-8). Epithelial cells in small airways may be an important source of TGF-β, which then induces local fibrosis. Airway epithelial cells are also important in defense of the airways, with mucus production from goblet cells, and secretion of antioxidants, antiproteases, and defensins (see Chapter 16). It is possible that cigarette smoke and other noxious agents impair these innate and adaptive immune responses of the airway epithelium, increasing susceptibility to infection. The airway epithelium in chronic bronchitis and COPD often shows squamous metaplasia, which may result from increased proliferation of basal airway epithelial cells but the nature of the growth factors involved in epithelial cell proliferation, cell cycle, and differentiation in

COPD are not yet known. Epidermal growth factor receptors (EGFR) show increased expression in airway epithelial cells of smokers and may contribute to basal cell proliferation, resulting in squamous metaplasia and an increased risk of bronchial carcinoma [38].

Mesenchymal cells

Mesenchymal cells, including fibroblasts, myofibroblasts, and smooth muscle cells, are the major structural cells responsible for the production and maintenance of the interstitial connective tissue. As such, they play a major role in the tissue remodeling that characterizes the structural alterations of COPD. In addition, mesenchymal cells can also produce inflammatory mediators and may play a role in persistent inflammation in asthma and possibly COPD. Fibroblasts from the lung parenchyma of patients with COPD proliferate more slowly [18] and are less active in chemotaxis and contraction of extracellular matrix [26]. This suggests that deficient repair could contribute to the development of emphysema. Conversely, a "pro-fibrotic" phenotype has been reported in fibroblasts from various fibrotic diseases [39, 40]. Whether such cells play a role in the airways fibrosis in COPD is unknown.

Neutrophils

Increased numbers of activated neutrophils are found in sputum and BAL fluid of patients with COPD [41], yet are relatively little increased in the airways or lung parenchyma. This may reflect their rapid transit through the airways and parenchyma. The role of neutrophils in COPD is not yet clear, however, neutrophil numbers in induced sputum are correlated with COPD disease severity [41] and with the rate of decline in lung function [42]. Smoking has a direct stimulatory effect on granulocyte production and release from the bone marrow and survival in the respiratory tract, possibly mediated by GM-CSF and G-CSF released from lung macrophages [43]. Smoking may also increase intravascular neutrophil retention in the lung. Neutrophil recruitment to the airways and parenchyma involves adhesion to endothelial cells and E-selectin which is up-regulated on endothelial cells in the airways of COPD patients [44]. Adherent neutrophils then migrate into the respiratory tract under the direction of neutrophil chemotactic factors. There are several chemotactic signals that have the potential for neutrophil recruitment in COPD, including leukotriene (LT)B_4, CXCL8, and related CXC chemokines, including CXCL1 (GRO-α) and CXCL5 (ENA-78), which are increased in COPD airways [45] (Fig. 34.7). These mediators may be derived form alveolar macrophages T-cells and epithelial cells, but the neutrophil itself may be a major source of CXCL8. Neutrophils from the circulation marginate in the pulmonary circulation and adhere to endothelial cells in the alveolar wall before passing into the alveolar space. The neutrophils recruited to the airways of COPD patients are activated as there is increased concentration of granule proteins, such as myeloperoxidase (MPO) and human neutrophil lipocalin, in the sputum supernatant [46].

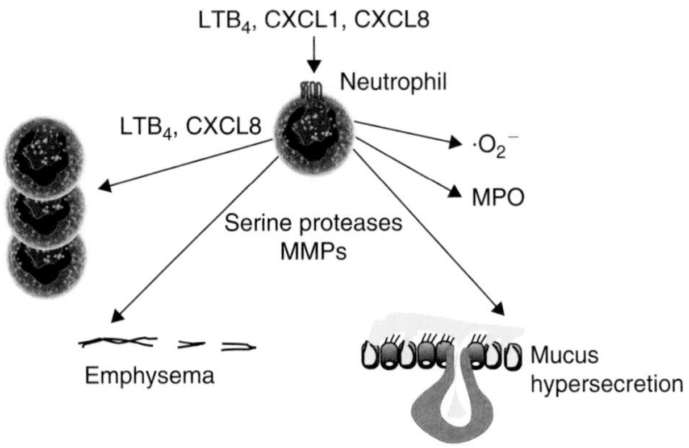

FIG. 34.7 Neutrophils in COPD. Neutrophils are attracted to the lungs of COPD patients by neutrophil chemotactic mediators, such as leukotriene B_4 (LTB_4) and the chemokines CXCL1 (also known as GRO-α) and CXCL8 (also known as IL-8). Activated neutrophils release more LTB_4 and CXCL8 to attract even more neutrophils, reactive oxygen species, such as superoxide anions (O_2^-), and myeloperoxidase (MPO). They also release the serine proteinases NE, cathpsin G, and proteinase-3, as well as the matrix metalloproteinases (MMP)-8 and MMP-9. These proteinases stimulate mucus hypersecretion and elastolysis, resulting in emphysema.

Neutrophils secrete serine proteases, including NE, cathepsin G, and proteinase-3, as well as matrix metalloproteinase (MMP)-8 and MMP-9, which may contribute to alveolar destruction. Neutrophils have the capacity to induce tissue damage through the release of serine proteases and oxidants. However, while neutrophils have the capacity to cause elastolysis, this is not a prominent feature of other pulmonary diseases where chronic airway neutrophilia is even more prominent, including cystic fibrosis and bronchiectasis. This suggests that other factors are involved in the generation of emphysema. Indeed neutrophils are not a prominent feature of parenchymal inflammation in COPD. It is likely that airway neutrophilia is more linked to mucus hypersecretion in chronic bronchitis. Serine proteases from neutrophils, including NE, cathepsin G, and proteinase-3 are all potent stimulants of mucus secretion from submucosal glands and goblet cells in the epithelium. There is a marked increase in neutrophil numbers in the airways in acute exacerbations of COPD accounting for the increased purulence of sputum. This may reflect increased production of neutrophil chemotactic factors including LTB_4 and CXCL8 [47, 48].

Macrophages

Macrophages appear to play a pivotal role in the pathophysiology of COPD and can account for most of the known features of the disease [49] (Fig. 34.8). There is a marked increase (5- to 10-fold) in the numbers of macrophages in airways, lung parenchyma, BAL fluid, and sputum in patients with COPD. A careful morphometric analysis of macrophage numbers in the parenchyma of patients with emphysema showed a 15-fold increase in the numbers of macrophages in the tissue and alveolar space compared with normal smokers [50]. Furthermore, macrophages are localized to sites of alveolar wall destruction in patients with

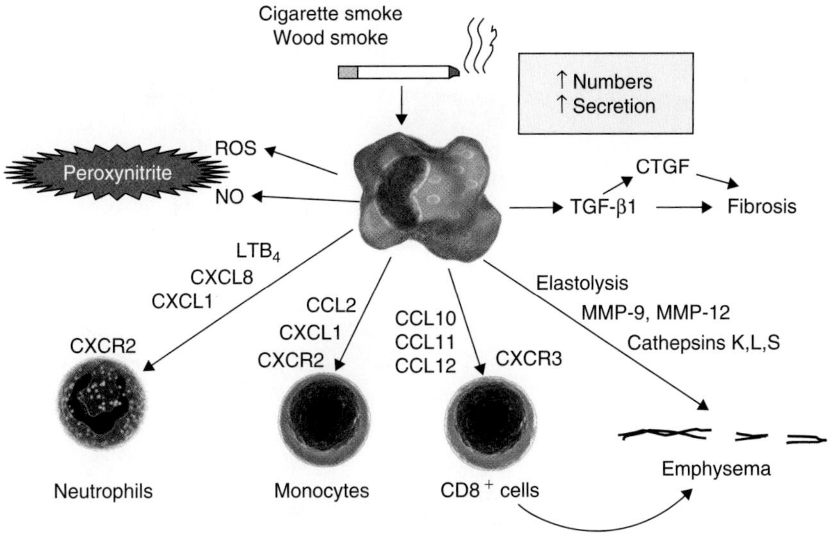

FIG. 34.8 Macrophages in COPD. Macrophages may play a pivotal role in COPD as they are activated by cigarette smoke extract and secrete many inflammatory proteins that may orchestrate the inflammatory process in COPD. Neutrophils may be attracted by CXCL8, CXCL1 and leukotriene B_4 (LTB_4), monocytes by CCL2, and $CD8^+$ lymphocytes by CXCL10 and CXCL11. Release of elastolytic enzymes including matrix metalloproteinases (MMP) and cathepsins cause elastolysis, and release of transforming growth factor (TGF)-β1 and connective tissue growth factor (CTGF). Macrophages also generate reactive oxygen species (ROS) and nitric oxide (NO) which together form peroxynitrite and may contribute to steroid resistance.

emphysema and there is a correlation between macrophage numbers in the parenchyma and severity of emphysema [51]. Macrophages may be activated by cigarette smoke extract to release inflammatory mediators including TNF-α, CXCL8 and other CXC chemokines, CCL2 (MCP-1), LTB_4, and reactive oxygen species (ROS), providing a cellular mechanism that links smoking with inflammation in COPD. Alveolar macrophages also secrete elastolytic enzymes, including MMP-2, MMP-9, MMP-12, cathepsins K, L and S, and NE taken up from neutrophils [52]. Alveolar macrophages from patients with COPD secrete more inflammatory proteins and have a greater elastolytic activity at baseline than those from normal smokers and this is further increased by exposure to cigarette smoke [52]. Macrophages demonstrate this difference even when maintained in culture for 3 days and therefore appear to be intrinsically different from the macrophages of normal smokers and non-smoking normal control subjects [52]. The predominant elastolytic enzyme secreted by alveolar macrophages in COPD patients is MMP-9. Most of the inflammatory proteins that are up-regulated in COPD macrophages are regulated by the transcription factor nuclear factor-κB (NF-κB) which is activated in alveolar macrophages of COPD patients, particularly during exacerbations [53].

The increased numbers of macrophages in smokers and COPD patients may be due to increased recruitment of monocytes from the circulation in response to the monocyte-selective chemokines CCL2 and CXCL1, which are increased in sputum and BAL of patients with COPD [45]. Monocytes from patients with COPD show a greater chemotactic response to GRO-α than cells from normal smokers and non-smokers, but this is not explained by an increase in CXCR2 [54]. Interestingly, while all monocytes express CCR2, the receptor for CCL2, only ~30% of monocytes express CXCR2. It is possible that these CXCR2 expressing monocytes transform into macrophages that are more inflammatory. Macrophages also release the chemokines CXCL9, CXCL10, and CXCL11, which are chemotactic for $CD8^+$ Tc1 and $CD4^+$ Th1 cells, via interaction with the chemokine receptor CXCR3 expressed on these cells [55].

The increased numbers of macrophages in COPD are mainly be due to increased recruitment of monocytes, as macrophages have a very low proliferation rate in the lungs. Macrophages have a long survival time so this is difficult to measure directly. However, in macrophages from smokers, there is markedly increased expression of the anti-apoptotic protein Bcl-X_L and increased expression of $p21^{CIP/WAF1}$ in the cytoplasm [56]. This suggests that macrophages may have a prolonged survival in smokers and patients with COPD. Once activated, macrophages will increase production of ROS, nitric oxide, and lysosomal enzymes and will increase secretion of many cytokines, including TNF-α, IL-1β, IL-6, CXCL8, and IL-18 among others. This activated macrophage response increases the efficiency of killing of organisms but may promote tissue damage and further inflammation.

Corticosteroids are ineffective in suppressing inflammation, including cytokines, chemokines, and proteases, in patients with COPD [57]. *In vitro* the release of CXCL8, TNF-α, and MMP-9 macrophages from normal subjects and normal smokers are inhibited by corticosteroids, whereas corticosteroids are ineffective in macrophages from patients with COPD [58]. The reasons for resistance to corticosteroids in COPD and to a lesser extent macrophages from smokers may be the marked reduction in activity of histone deacetylase-2 (HDAC2) [59], which is recruited to activated inflammatory genes by glucocorticoid receptors (GRs) to switch off inflammatory genes. The reduction in HDAC activity in macrophages is correlated with increased secretion of cytokines like TNF-α and CXCL8 and reduced response to corticosteroids. The reduction of HDAC activity in COPD patients may be mediated through oxidative stress and peroxynitrite formation [60].

Eosinophils

While eosinophils are the predominant leukocyte in asthma, their role in COPD is much less certain. In some studies, increased numbers of eosinophils have been described in the airways and BAL of patients with stable COPD, whereas others have not found increased numbers in airway biopsies, BAL or induced sputum. The presence of eosinophils in patients with COPD predicts a response to corticosteroids and may indicate coexisting asthma [33]. Increased numbers of eosinophils have been reported in bronchial biopsies and BAL fluid during acute exacerbations of chronic bronchitis [61]. Surprisingly the levels of eosinophil basic proteins in induced sputum are as elevated in COPD, as in asthma, despite the absence of eosinophils, suggesting the eosinophils may have degranulated and are no longer recognizable by microscopy [46]. Perhaps this is due to the high levels of NE that have been shown to cause degranulation of eosinophils.

Dendritic cells

Dendritic cells play a central role in the initiation of the innate and adaptive immune response and it is believed that they provide a link between them [62]. The airways and lungs contain a rich network of dendritic cells that are localized near the surface, so that they are ideally located to signal the entry of foreign substances that are inhaled. Dendritic cells can activate a variety of other inflammatory and immune cells, including macrophages, neutrophils, T- and B-lymphocytes so dendritic cells may play an important role in the pulmonary response to cigarette smoke and other inhaled noxious agents. However, there does not appear to be an increase in dendritic cells in the airways of COPD patients, in contrast to asthma patients [63].

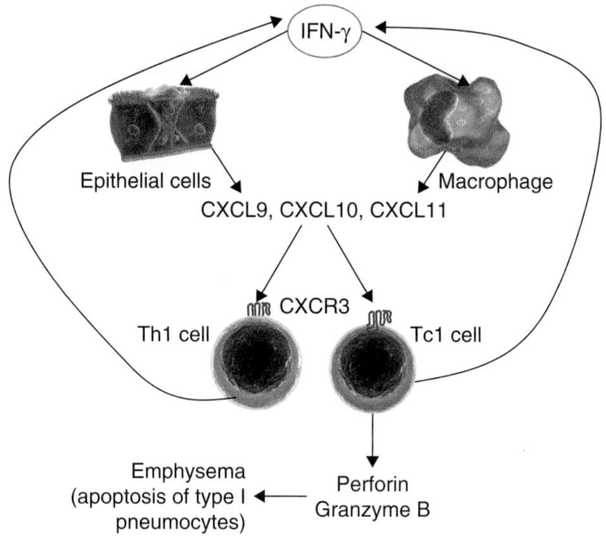

FIG. 34.9 T-lymphocytes in COPD. Epithelial cells and macrophages are stimulated by interferon-γ (IFN-γ) to release the chemokines CXC-chemokine ligand 9 (CXCL9), CXCL10, and CXCL11, which together act on CXC-chemokine receptor 3 (CXCR3) expressed on T helper 1 (Th1) cells and type 1 cytotoxic T (Tc1) cells to attract them into the lungs. Tc1 cells, through the release of perforin and granzyme B, induce apoptosis of type 1 pneumocytes, thereby contributing to emphysema. IFN-γ released by Th1 and Tc1 cells then stimulates further release of CXCR3 ligands, resulting in a persistent inflammatory activation.

T-lymphocytes

There is an increase in the total numbers of T-lymphocytes in lung parenchyma, peripheral and central airways of patients with COPD, with the greater increase in $CD8^+$ than $CD4^+$ cells [27, 55]. There is a correlation between the numbers of T-cells and the amount of alveolar destruction and the severity of airflow obstruction. Furthermore, the only significant difference in the inflammatory cell infiltrate in asymptomatic smokers and smokers with COPD is an increase in T-cells, mainly $CD8^+$, in patients with COPD. There is also an increase in the absolute number of $CD4^+$ T-cells albeit in smaller numbers, in the airways of smokers with COPD and these cells express activated STAT-4, a transcription factor that is essential for activation and commitment of the Th1 lineage and IFN-γ [64].

The ratio of $CD4^+$ and $CD8^+$ cells are reversed in COPD. The majority of T-cells in the lung in COPD are of the Tc1 and Th1 subtypes [55]. There is a marked increase in T-cells in the walls of small airways in patients with severe COPD and the T-cells are formed into lymphoid follicles, surrounding B-lymphocytes [27].

The mechanisms by which $CD8^+$, and to a lesser extent $CD4^+$ cells, accumulate in the airways and parenchyma of patients with COPD is not yet understood [65]. However, homing of T-cells to the lung must depend upon some initial activation (only activated T-cells can home to the organ source of antigenic products), then adhesion and selective chemotaxis. $CD4^+$ and $CD8^+$ T-cells in the lung of COPD patients show increased expression of CXCR3, a receptor activated by the chemokines CXCL9 (Mig), CXCL10 (IP-10), and CXCL11 (I-TAC), all of which are increased in COPD [66]. There is increased expression of CXCL10 by bronchiolar epithelial cells and this could contribute to the accumulation of $CD4^+$ and $CD8^+$ T-cells, which preferentially express CXCR3 [67] (Fig. 34.9). $CD8^+$ cells are typically increased in airway infections and it is possible that the chronic colonization of the lower respiratory tract of COPD patients by bacterial and viral pathogens is responsible for this inflammatory response. It is possible that cigarette-induce lung injury may uncover previously sequestered autoantigens or cigarette smoke itself may damage lung interstitial and structural cells and make them antigenic [68]. The role of increased numbers of $CD4^+$ cells in COPD, particularly in severe disease is also unknown [50]; however, it is now clear that T-cell help is required for the priming of cytotoxic T-cell responses, for maintaining $CD8^+$ T-cell memory and for ensuring $CD8^+$ T-cell survival. It is also possible that $CD4^+$ T-cells have immunological memory and play a role in perpetuating the inflammatory process in the absence of cigarette smoking. In a mouse model of cigarette-induced emphysema there is a predominance of T-cells which are directly related to the severity of emphysema [69].

The role of T-cells in the pathophysiology of COPD is not yet certain, although they have the potential to produce extensive damage in the lung. $CD8^+$ cells have the capacity cause cytolysis and apoptosis of alveolar epithelial cells through release of perforins, granzyme B, and TNF-α [70, 71]. There is an association between $CD8^+$ cells and apoptosis of alveolar cells in emphysema [72]. Apoptotic cells are potential sources of antigenic material that could reach the DC and perpetrate the T-cell response. In addition $CD8^+$ T-cells also produce a number of cytokines of the Tc1 phenotype including TNF-α, lymphotoxin, and IFN-γ, and there is evidence that $CD8^+$ in the lungs of COPD patients expresses IFN-γ [73]. All these cytokines would enhance the inflammatory reaction in the lung besides the direct killing by $CD8^+$ cells. COPD has been considered an autoimmune disease triggered by smoking, as previously suggested [68] and the presence of highly activated oligoclonal T-cells in emphysema patients supports this [74]. There is evidence for anti-elastin antibodies in experimental models of COPD and in COPD patients [75]. In addition to activated Th1 cells there is some evidence for an increase in Th2 cells that express IL-4 in BAL fluid of COPD patients in COPD patients [76].

NK cells are also increased in COPD lungs and may contribute to epithelial cell apoptosis in emphysema [77]. Invariant NK T-cells are $CD4^+$ cells that release IL-4 but are not increased in COPD [78].

MEDIATORS OF INFLAMMATION

Many inflammatory mediators have now been implicated in COPD, including lipids, free radicals, cytokines, chemokines, and growth factors [29]. These mediators are derived from inflammatory and structural cells in the lung and interact with each other in a complex manner.

Lipid mediators

The profile of lipid mediators in exhaled breath condensates of patients with COPD shows an increase in prostaglandins and leukotrienes [79]. There is a significant increase in PGE_2 and $PGF_{2\alpha}$ and an increase in LTB_4 but not cysteinyl leukotrienes. This is a different pattern to that seen in asthma, in which increases in thromboxane and cysteinyl leukotrienes have been shown. The increased production of prostanoids in COPD is likely to be secondary to the induction of cyclo-oxygenase-2 (COX2) by inflammatory cytokines and increased expression of COX2 is found in alveolar macrophages of COPD patients. LTB_4 concentrations are also increased in induced sputum and are further increased in sputum and exhaled breath condensate during acute exacerbations [47]. LTB_4 is a potent chemoattractant for neutrophils, acting through high affinity BLT_1-receptors. A BLT_1-receptor antagonist reduces the neutrophil chemotactic activity of sputum by ~25% [80]. Recently BLT_1-receptors have been identified on T-lymphocytes and there is evidence that LTB_4 is involved in recruitment of T-cells.

Oxidative stress

Oxidative stress occurs when ROS are produced in excess of the antioxidant defense mechanisms and result in harmful effects including damage to lipids, proteins, and DNA. There is increasing evidence that oxidative stress is an important feature in COPD [81] (see Chapter 25). Inflammatory and structural cells that are activated in the airways of patients with COPD produce ROS, including, neutrophils, eosinophils, macrophages, and epithelial cells. Superoxide anions (O_2^-) are generated by NADPH oxidase and converted to hydrogen peroxide (H_2O_2) by superoxide dismutases (SODs). H_2O_2 is then dismuted to water by catalase. O_2^- and H_2O_2 may interact in the presence of free iron to form the highly reactive hydroxyl radical (OH). O_2^- may also combine with NO to form peroxynitrite, which also generates OH. Oxidative stress leads to the oxidation of arachidonic acid and the formation of a new series of prostanoid mediators called isoprostanes, which may exert significant functional effects including bronchoconstriction and plasma exudation [82] (Fig. 34.10).

The normal production of oxidants is counteracted by several antioxidant mechanisms in the human respiratory tract [81]. The major intracellular antioxidants in the airways are catalase, SOD, and glutathione, formed by the enzyme γ-glutamyl cysteine synthetase, and glutathione synthetase. In the lung intracellular antioxidants are expressed at relatively low levels and are not induced by oxidative stress, whereas the major antioxidants are extracellular. A transcription factor Nrf2 plays a key role in switching on several genes for antioxidants in response to oxidative stress and there is evidence that Nrf2 knock-out mice are more susceptible to developing emphysema after cigarette smoke exposure [83, 84]. Extracellular antioxidants, particularly glutathione peroxidase, are markedly up-regulated in response to cigarette smoke and oxidative stress. Extracellular antioxidants also include the dietary antioxidants vitamin C (ascorbic acid) and vitamin E (α-tocopherol), uric acid, lactoferrin, and extracellular superoxide dismutase (SOD3), which is highly expressed in human lung, but its role in COPD is not yet clear.

ROS have several effects on the airways and parenchyma and increase the inflammatory response. ROS activate NF-κB, which switches on multiple inflammatory genes resulting in amplification of the inflammatory response. The molecular pathways by which oxidative stress activates NF-κB have not been fully elucidated, but there are several redox-sensitive steps in the activation pathway. Oxidative stress results in activation of histone acetyltransferase activity which opens up the chromatin structure and is associated with increased transcription of multiple inflammatory genes [85]. Exogenous oxidants may also be important in worsening airway disease. There is considerable evidence for increased oxidative stress in COPD [81]. Cigarette smoke itself contains a high concentration of ROS. Inflammatory cells, such as activated macrophages and neutrophils, also generate ROS, as discussed earlier. There are several markers of oxidative stress that may be detected in the breath and several studies have demonstrated increased production of oxidants, such as H_2O_2, 8-isoprostane, and ethane, in exhaled air or breath condensates, particularly during exacerbations [47].

The increased oxidative stress in the lung epithelium of COPD patient may play an important pathophysiological role in the disease by amplifying the inflammatory response in COPD. This may reflect the activation of

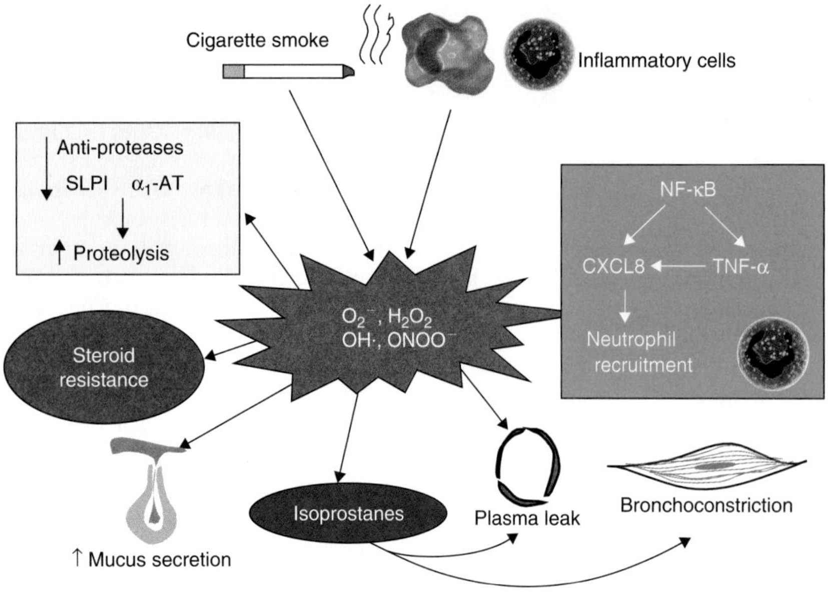

FIG. 34.10 Oxidative stress in COPD. Oxidative stress plays a key role in the pathophysiology of COPD and amplifies the inflammatory and destructive process. Reactive oxygen species from cigarette smoke or from inflammatory cells (particularly macrophages and neutrophils) result in several damaging effects in COPD, including decreased anti-protease defenses, such as α1-antitrypsin (AT) and secretory leukoprotease inhibitor (SLPI), activation of nuclear factor-κB (NF-κB) resulting in increased secretion of the cytokines CXCL8 and tumor necrosis factor-α (TNF-α), increased production of isoprostanes and direct effects on airway function. In addition recent evidence suggests that oxidative stress induces steroid resistance.

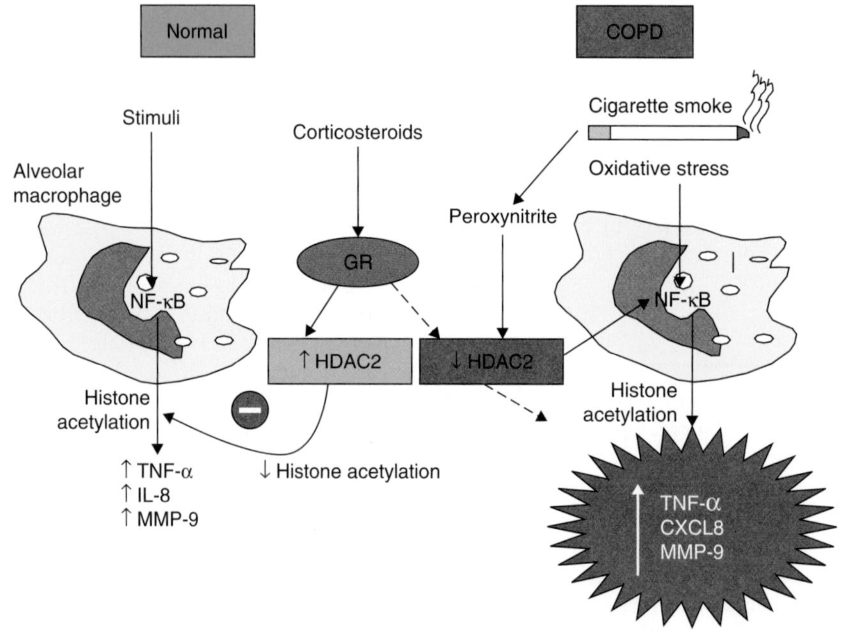

FIG. 34.11 Mechanism of corticosteroid resistance in COPD. Stimulation of normal alveolar macrophages activates nuclear factor-κB (NF-κB) and other transcription factors to switch on histone acetyltransferase leading to histone acetylation and subsequently to transcription of genes encoding inflammatory proteins, such as tumor necrosis factor-α (TNF-α) and CXCL8 (IL-8). Corticosteroids reverse this by binding to glucocorticoid receptors (GRs) and recruiting histone deacetylase-2 (HDAC2). This reverses the histone acetylation induced by NF-κB and switches off the activated inflammatory genes. In COPD patients cigarette smoke activates macrophages, as in normal subjects, but oxidative stress (acting in part through the formation of peroxynitrite) impairs the activity of HDAC2. This amplifies the inflammatory response to NF-κB activation, but also reduces the anti-inflammatory effect of corticosteroids as HDAC2 is now unable to reverse histone acetylation.

NF-κB and AP-1, which then induce a neutrophilic inflammation via increased expression of CXC chemokines, TNF-α, and MMP-9. Oxidative stress may also impair the function of antiproteases such as α_1-antitrypsin and SLPI, and thereby accelerates the breakdown of elastin in lung parenchyma. Corticosteroids are much less effective in COPD than in asthma and do not reduce the progression or mortality of the disease. Alveolar macrophages from patients with COPD show a marked reduction in responsiveness to the anti-inflammatory effects of corticosteroids, compared to cells from normal smokers and non-smokers [58]. In patients with COPD there is a marked reduction in activity of HDAC and reduced expression of HDAC2 in alveolar macrophages and peripheral lung tissue [59], which is correlated with increased expression of inflammatory cytokines and a reduced response to corticosteroids. This may result directly or indirectly from oxidative stress and is mimicked by the effects of H_2O_2 in cell lines [86] (Fig. 34.11).

Nitrative stress

The increase in exhaled nitric oxide (NO) is less marked in COPD than in asthma, partly because cigarette smoking reduces exhaled NO. Recently exhaled NO has been partitioned into central and peripheral portions and this shows reduced NO in the bronchial fraction but increased NO in the peripheral fraction, which includes lung parenchyma and small airways [87] (see Chapter 30). The increased peripheral NO in COPD patients may reflect increased expression of inducible NO synthase in epithelial cells and macrophages of patients with COPD [88]. NO and superoxide anions combine to form peroxynitrite which nitrates certain tyrosine residues in proteins and there is increased expression of 3-nitrotyrosine in peripheral lung and macrophages of COPD patients [88]. There is tyrosine nitration of HDAC2 which may lead to impaired activity and degradation of this enzyme, resulting in steroid resistance [86].

Inflammatory cytokines

Cytokines are the mediators of chronic inflammation and several have now been implicated in COPD [29, 89]. There is an increase in concentration of TNF-α in induced sputum in stable COPD with a further increase during exacerbations [41, 48]. TNF-α production from peripheral blood monocytes is also increased in COPD patients and has been implicated in the cachexia and skeletal muscle apoptosis found in some patients with severe disease. TNF-α is a potent activator of NF-κB and this may amplify the inflammatory response. Unfortunately anti-TNF therapies have not proved to be effective in COPD patients. IL-1β and IL-6 are proinflammatory cytokines that may amplify the inflammation in COPD and may be important for systemic effects.

Chemokines

Chemokines are small chemotactic cytokines that play a key role in the recruitment and activation if inflammatory cells through specific chemokine receptors. Several chemokines have now been implicated in COPD and have been of particular interest since chemokine receptors are G-protein coupled receptors, for which small molecule antagonists have now been developed [30, 89]. CXCL8 concentrations are increased in induced sputum of COPD patients and increase further during exacerbations [41, 48]. CXCL8 is secreted from macrophages, T-cells, epithelial cells, and neutrophils. CXCL8 activates neutrophils via low affinity specific receptors CXCR1, and is chemotactic for neutrophils via high affinity receptors CXCR2, which are also activated by related CXC chemokines such as CXCL1. CXCL1 concentrations are markedly elevated in sputum and BAL fluid of COPD patients and this chemokine may be more important as a chemoattractant than CXCL8, acting via CXCR2 which are expressed on neutrophils and monocytes [45]. CXCL1 induces significantly more chemotaxis of monocytes of COPD patient compared to those of normal smokers and this may reflect increased turnover and recovery of CXCR2 in monocytes of COPD patients [54]. CXCL5 shows a marked increase in expression in airway epithelial cells during exacerbations of COPD and this is accompanied by a marked up-regulation of epithelial CXCR2.

CCL2 is increased in concentration in COPD sputum and BAL fluid [45] and plays a role in monocyte chemotaxis via activation of CCR2. CCL2 appears to cooperate with CXCL1 in recruiting monocytes into the lungs. The chemokine CCL5 (RANTES) is also expressed in airways of COPD patients during exacerbations and activates CCR5 on T-cells and CCR3 on eosinophils, which may account for the increased eosinophils and T-cells in the wall of large airways that have been reported during exacerbations of chronic bronchitis. As discussed earlier, CXCR3 are up-regulated on Tc1 and Th1 cells of COPD patients with increased expression of their ligands CXCL9, CXCL10, and CXCL11.

Growth factors

Several growth factors have been implicated in COPD and mediate the structural changes that are found in the airways (see Chapter 29). TGF-β1 is expressed in alveolar macrophages and airway epithelial cells of COPD patients and is released from epithelial cells of small airways. TGF-β is released in a latent form and is activated by various factors including MMP-9. TGF-β may play an important role in the characteristic peribronchiolar fibrosis of small airways, either directly or through the release of connective tissue growth factor (Fig. 34.12). TGF-β down-regulates β_2-adrenergic receptors by inhibiting gene transcription in human cell lines and may reduce the bronchodilator

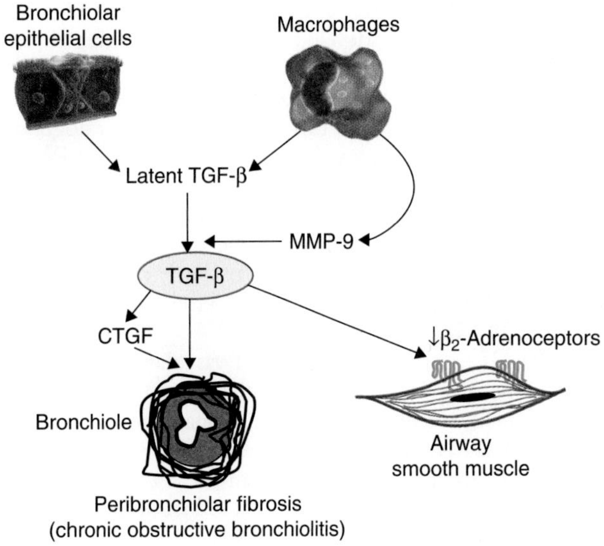

FIG. 34.12 Transforming growth factor (TGF)-β in COPD. TGF-β is released in a latent form that may be activated by matrix metalloproteinase-9 (MMP-9). It may then cause fibrosis directly through effects on fibroblasts or indirectly via the release of connective tissue growth factor (CTGF). TGF-β may also down-regulate β_2-adrenoceptors on cells such as airway smooth muscle to diminish the bronchodilator response to β-agonists.

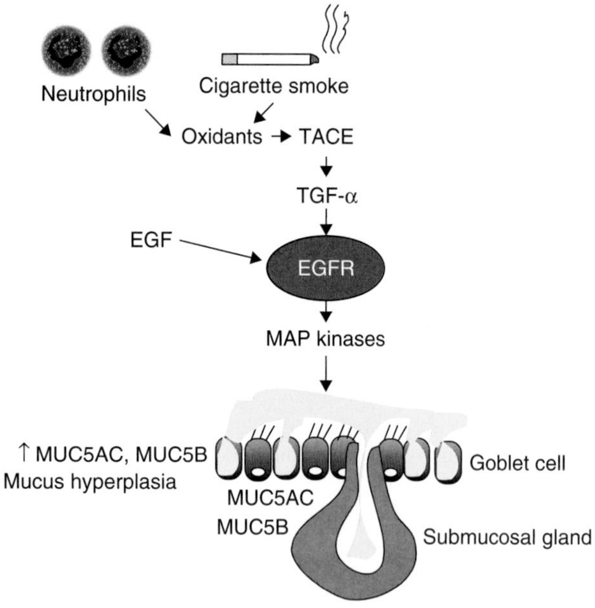

FIG. 34.13 Epidermal growth factor receptors (EGFR) in COPD. EGFR play a key role in the regulation of mucus hypersecretion, with increased expression of mucin genes (MUC5AC, MUCB) and differentiation of goblet cells and hyperplasia of mucus-secreting cells. These effects are mediated via the activation of mitogen-activated protein (MAP) kinases. EGFR are activated by transforming growth factor-α (TGF-α), which is in turn activated by tumor necrosis factor-α converting enzyme (TACE), activated via release of oxidants from cigarette smoke and neutrophils. EGFR may also be activated by EGF.

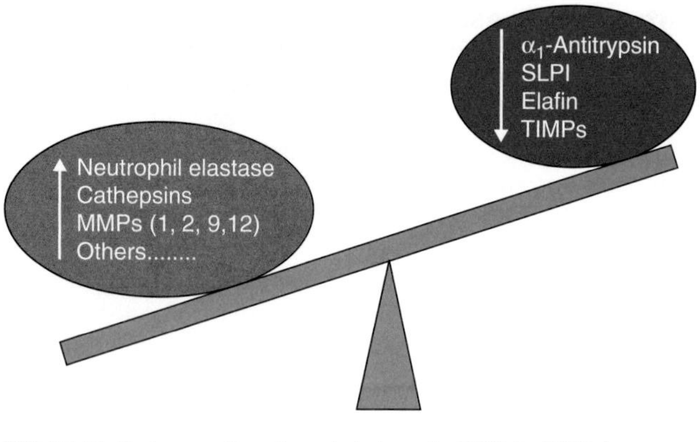

FIG. 34.14 Protease-anti-protease imbalance in COPD. In COPD the balance appears to be tipped in favor of increased proteolysis, either because of an increase in proteases, including NE, cathepsins, and matrix metalloproteinases (MMP), or a deficiency in antiproteases, which may include α_1-antitrypsin, elafin, secretory leukoprotease inhibitor (SLPI), and tissue inhibitors of matrix metalloproteinases (TIMPs).

response to β-agonists in airway smooth muscle. TGF-β is also a potent anti-inflammatory molecule and TGF-β1 deficient mice die early from overwhelming inflammation, which occurs even in germ-free animals [90]. In contrast, mice deficient in Smad3, a TGFβ signal transducing molecule are resistant the development of fibrotic lung disease but spontaneously develop emphysema [91].

Alveolar macrophages produce TGF-α in greater amounts than TGF-β and this may be a major endogenous activator of epithelial growth factor receptors (EGFR) that play a key role in regulating mucus secretion in response to many stimuli, including cigarette smoke. Cigarette smoke activates TNF-α converting enzyme on airway epithelial cells which results in the shedding of TGF-α and the activation of EGFR, resulting in increased mucus secretion [92, 93] (Fig. 34.13).

VEGF is a major regulator of vascular growth and is likely to be involved in the pulmonary vascular remodeling that occurs as a result of hypoxic pulmonary vasoconstriction in severe COPD. There is increased expression of VEGF in pulmonary vascular smooth muscle of patients with mild and moderate COPD but paradoxically a reduction in expression in severe COPD with emphysema [94]. Inhibition of VEGF receptors using a selective inhibitor induces apoptosis of alveolar endothelial cells in rats, resulting in emphysema and this appears to be driven by oxidative stress [25] and may be mediated via the sphingolipid ceramide [95]. In addition, VEGF is also an important proinflammatory cytokine produced by epithelial and endothelial cells, macrophages, and activated T-cells, which acts by increasing endothelial cell permeability, by inducing expression of endothelial adhesion molecules and via its ability to act as a monocyte chemoattractant. VEGF also stimulates the expression of CXCL10 and its receptor CXCR3. Thus VEGF is acts as an intermediary between cell-mediated immune inflammation and the associated angiogenesis reaction.

Proteinases

An imbalance between proteinases that degrade connective tissue and anti-proteinases has long been postulated in COPD (Fig. 34.14). There has been debate about which are the most important proteinases involved in emphysema. While previous attention focused on NE, more recently a role of MMP has been recognized (see Chapter 28). MMP-9 appears to be particularly important in degrading elastin fibers in lung parenchyma and is secreted in large amounts by alveolar macrophages of COPD patients [52, 96]. MMP-9 also activates TGF-β accounting for the simultaneous occurrence of lung destruction and fibrosis in COPD (Fig. 34.15). In the mouse, MMP-12 appears to play a key role in the development of cigarette smoke-induced emphysema as MMP-12 derived elastin fragments are major factors driving macrophage recruitment [97]. A role for MMP-1 has also been suggested in both human and animal studies [98].

SYSTEMIC MANIFESTATIONS

It is now recognized that COPD involves organs outside the lung, but the mechanism for these systemic effects of COPD are not well understood (see Chapter 44). Inflammatory cytokines, such as TNF-α and IL-6, may spill over from the lung periphery into the systemic circulation, where they may have detrimental effects on

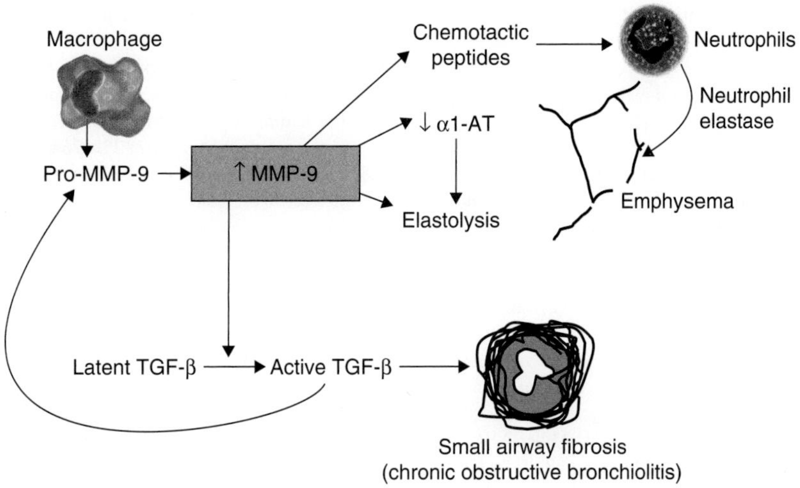

FIG. 34.15 Connective tissue destruction and fibrosis coexist in COPD. Matrix metalloproteinase-9 (MMP-9) is the major elastolytic enzyme released by alveolar macrophages and is generated from an inactive precursor pro-MMP-9. MMP-9 causes elastolysis directly, but also releases chemotactic peptides that attract neutrophils, which release neutrophil elastase. The elastolytic effect of neutrophil elastase is enhanced as MMP-9 inactivates its specific anti-proteinase α_1-antitrypsin, resulting in emphysema. MMP-9 also activates transforming growth factor-β (TGF-β) from a latent precursor and this leads to fibrosis of small airways. TGF-β also plays a role in activating MMP-9, so the interplay between these mediators leads to tissue destruction and fibrosis simultaneously, as observed in COPD lungs.

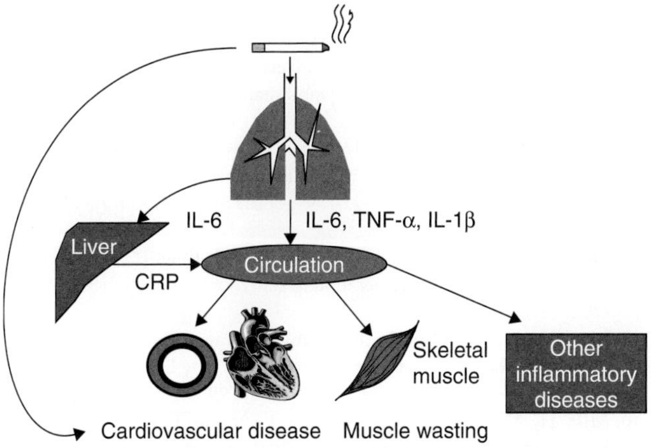

FIG. 34.16 Systemic features of COPD and co-morbidities. Inflammation in the lungs may spill over into the systemic circulation resulting in systemic inflammatory effects and exacerbation of cardiovascular diseases and skeletal muscle wasting. Increased concentrations of interleukin-6 (IL-6) and tumor necrosis factor-α (TNF-α) are found in systemic circulation of patients with severe COPD and these may directly result in skeletal muscle atrophy. IL-6 releases the acute phase protein C-reactive protein (CRP) which is associated with increased cardiovascular disease risk [101]. Cigarette smoke also contributes to increased cardiovascular risk.

skeletal muscle function and endothelial function [99] (Fig. 34.16). IL-6 causes the release of the acute phase protein C-reactive protein, which is increased in the circulation of patients with COPD and correlates with disease severity [100]. Skeletal muscle wasting is a feature of severe COPD and this may in part be due to systemic inflammation but is more likely due to disuse atrophy as a result of impaired exercise performance.

CONCLUSIONS

Cigarette smoke exposure induces a florid inflammatory response in the lung involving structural and inflammatory cells and a large array of inflammatory mediators. The interaction of these complex steps eventually leads to airway remodeling and obstruction and emphysema, albeit variable among individual smokers. Of interest, the main difference between smokers who develop COPD and the ones who do not seems to be the presence of an adaptive immune response with $CD8^+$, $CD4^+$, and B-cells, which express obvious signs of being activated effector cells. It is likely that genetic and epigenetic factors (such as histone acetylation) are involved in determining the progression of the inflammatory cascade, as this is supported by animal models, where different strains appear to have different sensitivities to cigarette smoke. COPD is a complex inflammatory disease and the interactions between different inflammatory cells and mediators are still uncertain. The role played by structural cells in responding to inflammatory cell-mediated injury and the degree to which it is deficient in COPD remains to be fully defined. More research into these mechanisms is needed in order to identify novel targets that may lead to the discovery of more effective therapies that are able to prevent disease progression and reduce the high mortality of this common disease [102].

References

1. Rabe KF, Hurd S, Anzueto A, Barnes PJ, Buist SA, Calverley P *et al.* Global strategy for the diagnosis, management, and prevention of COPD-2006 update. *Am J Respir Crit Care Med* 176: 532–55, 2007.
2. Mannino DM, Buist AS. Global burden of COPD: Risk factors, prevalence, and future trends. *Lancet* 370: 765–73, 2007.

3. Lokke A, Lange P, Scharling H, Fabricius P, Vestbo J. Developing COPD: A 25 year follow up study of the general population. *Thorax* 61: 935–39, 2006.
4. Rennard SI, Vestbo J. COPD: The dangerous underestimate of 15%. *Lancet* 367: 1216–19, 2006.
5. O'Donnell DE, Laveneziana P. Dyspnea and activity limitation in COPD: Mechanical factors. *COPD* 4: 225–36, 2007.
6. Makita H, Nasuhara Y, Nagai K, Ito Y, Hasegawa M, Betsuyaku T *et al.* Characterisation of phenotypes based on severity of emphysema in chronic obstructive pulmonary disease. *Thorax* 62: 932–37, 2007.
7. Sturton G, Persson C, Barnes PJ. Small airways: An important but neglected target in the treatment of obstructive airway diseases. *Trends Pharmacol Sci*, 2008.
8. Zhang JJ, Smith KR. Household air pollution from coal and biomass fuels in China: Measurements, health impacts, and interventions. *Environ Health Perspect* 115: 848–55, 2007.
9. Yin P, Jiang CQ, Cheng KK, Lam TH, Lam KH, Miller MR *et al.* Passive smoking exposure and risk of COPD among adults in China: The Guangzhou Biobank Cohort Study. *Lancet* 370: 751–57, 2007.
10. Donaldson GC, Seemungal TA, Bhowmik A, Wedzicha JA. Relationship between exacerbation frequency and lung function decline in chronic obstructive pulmonary disease. *Thorax* 57: 847–52, 2002.
11. Sethi S, Murphy TF. Bacterial infection in chronic obstructive pulmonary disease in 2000: A state-of-the-art review. *Clin Microbiol Rev* 14: 336–63, 2001.
12. Marossy AE, Strachan DP, Rudnicka AR, Anderson HR. Childhood chest illness and the rate of decline of adult lung function between ages 35 and 45 years. *Am J Respir Crit Care Med* 175: 355–59, 2007.
13. Higashimoto Y, Elliott WM, Behzad AR, Sedgwick EG, Takei T, Hogg JC *et al.* Inflammatory mediator mRNA expression by adenovirus E1A-transfected bronchial epithelial cells. *Am J Respir Crit Care Med* 166: 200–7, 2002.
14. Petrache I, Diab K, Knox KS, Twigg HL III, Stephens RS, Flores S *et al.* HIV associated pulmonary emphysema: A review of the literature and inquiry into its mechanism. *Thorax* 63: 463–69, 2008.
15. Fletcher C, Peto R. The natural history of chronic airflow obstruction. *BMJ* 1: 1645–48, 1977.
16. Morla M, Busquets X, Pons J, Sauleda J, Macnee W, Agusti AG. Telomere shortening in smokers with and without COPD. *Eur Respir J* 27: 525–28, 2006.
17. Nyunoya T, Monick MM, Klingelhutz A, Yarovinsky TO, Cagley JR, Hunninghake GW. Cigarette smoke induces cellular senescence. *Am J Respir Cell Mol Biol* 35: 681–88, 2006.
18. Holz O, Zuhlke I, Jaksztat E, Muller KC, Welker L, Nakashima M *et al.* Lung fibroblasts from patients with emphysema show a reduced proliferation rate in culture. *Eur Respir J* 24: 575–79, 2004.
19. Muller KC, Welker L, Paasch K, Feindt B, Erpenbeck VJ, Hohlfeld JM *et al.* Lung fibroblasts from patients with emphysema show markers of senescence *in vitro*. *Respir Res* 7: 32, 2006.
20. Sato A, Hirai T, Imura A, Kita N, Iwano A, Muro S *et al.* Morphological mechanism of the development of pulmonary emphysema in klotho mice. *Proc Natl Acad Sci USA* 104: 2361–65, 2007.
21. Sato T, Seyama K, Sato Y, Mori H, Souma S, Akiyoshi T *et al.* Senescence marker protein-30 protects mice lungs from oxidative stress, aging and smoking. *Am J Respir Crit Care Med* 174: 530–7, 2006.
22. Rennard SI, Togo S, Holz O. Cigarette smoke inhibits alveolar repair: A mechanism for the development of emphysema. *Proc Am Thorac Soc* 3: 703–8, 2006.
23. Sahebjami H, Domino M. Effects of starvation and refeeding on elastase-induced emphysema. *J Appl Physiol* 66: 2611–16, 1989.
24. Laurent P, Janoff A, Kagan HM. Cigarette smoke blocks cross-linking of elastin *in vitro*. *Chest* 83: 63S–65S, 1983.
25. Kasahara Y, Tuder RM, Taraseviciene-Stewart L, Le Cras TD, Abman S, Hirth PK *et al.* Inhibition of VEGF receptors causes lung cell apoptosis and emphysema. *J Clin Invest* 106: 1311–19, 2006.
26. Togo S, Holz O, Liu X, Sugiura H, Kamio K, Wang X *et al.* Lung fibroblast repair functions in COPD patients are altered by multiple mechanisms. *Am J Respir Crit Care Med*, 2008.
27. Hogg JC, Chu F, Utokaparch S, Woods R, Elliott WM, Buzatu L *et al.* The nature of small-airway obstruction in chronic obstructive pulmonary disease. *N Engl J Med* 350: 2645–53, 2004.
28. Barnes PJ, Shapiro SD, Pauwels RA. Chronic obstructive pulmonary disease: Molecular and cellular mechanisms. *Eur Respir J* 22: 672–88, 2003.
29. Barnes PJ. Mediators of chronic obstructive pulmonary disease. *Pharmacol Rev* 56: 515–48, 2004.
30. Donnelly LE, Barnes PJ. Chemokine receptors as therapeutic targets in chronic obstructive pulmonary disease. *Trends Pharmacol Sci* 27: 546–53, 2006.
31. Barnes PJ. Immunology of asthma and chronic obstructive pulmonary disease. *Nat Rev Immunol* 8: 183–92, 2008.
32. Fabbri LM, Romagnoli M, Corbetta L, Casoni G, Busljetic K, Turato G *et al.* Differences in airway inflammation in patients with fixed airflow obstruction due to asthma or chronic obstructive pulmonary disease. *Am J Respir Crit Care Med* 167: 418–24, 2003.
33. Papi A, Romagnoli M, Baraldo S, Braccioni F, Guzzinati I, Saetta M *et al.* Partial reversibility of airflow limitation and increased exhaled NO and sputum eosinophilia in chronic obstructive pulmonary disease. *Am J Respir Crit Care Med* 162: 1773–77, 2000.
34. Jeffery PK. Comparison of the structural and inflammatory features of COPD and asthma. *Chest* 117: 251S–260S, 2000.
35. Rogers DF. The role of airway secretions in COPD: Pathophysiology, epidemiology and pharmacotherapeutic options. *COPD* 2: 341–53, 2005.
36. Willemse BW, ten Hacken NH, Rutgers B, Lesman-Leegte IG, Postma DS, Timens W. Effect of 1-year smoking cessation on airway inflammation in COPD and asymptomatic smokers. *Eur Respir J* 26: 835–45, 2005.
37. Gamble E, Grootendorst DC, Hattotuwa K, O'Shaughnessy T, Ram FS, Qiu Y *et al.* Airway mucosal inflammation in COPD is similar in smokers and ex-smokers: A pooled analysis. *Eur Respir J* 30: 467–71, 2007.
38. Franklin WA, Veve R, Hirsch FR, Helfrich BA, Bunn PA Jr. Epidermal growth factor receptor family in lung cancer and premalignancy. *Semin Oncol* 29: 3–14, 2002.
39. Mio T, Nagai S, Kitaichi M, Kawatani A, Izumi T. Proliferative characteristics of fibroblast lines derived from open lung biopsy specimens of patients with IPF (UIP). *Chest* 102: 832–37, 1992.
40. Miki H, Mio T, Nagai S, Hoshino Y, Nagao T, Kitaichi M *et al.* Fibroblast contractility: Usual interstitial pneumonia and nonspecific interstitial pneumonia. *Am J Respir Crit Care Med* 162: 2259–64, 2000.
41. Keatings VM, Collins PD, Scott DM, Barnes PJ. Differences in interleukin-8 and tumor necrosis factor-a in induced sputum from patients with chronic obstructive pulmonary disease or asthma. *Am J Respir Crit Care Med* 153: 530–34, 1996.
42. Stanescu D, Sanna A, Veriter C, Kostianev S, Callagni PG, Fabbri LM *et al.* Airways obstruction, chronic expectoration and rapid decline in FEV_1 in smokers are associated with increased levels of sputum neutrophils. *Thorax* 51: 267–71, 1996.
43. Vlahos R, Bozinovski S, Hamilton JA, Anderson GP. Therapeutic potential of treating chronic obstructive pulmonary disease (COPD) by neutralising granulocyte macrophage-colony stimulating factor (GM-CSF). *Pharmacol Ther* 112: 106–15, 2006.
44. Di Stefano A, Maestrelli P, Roggeri A, Turato G, Calabro S, Potena A *et al.* Upregulation of adhesion molecules in the bronchial mucosa of subjects with chronic obstructive bronchitis. *Am J Respir Crit Care Med* 149: 803–10, 1994.
45. Traves SL, Culpitt S, Russell REK, Barnes PJ, Donnelly LE. Elevated levels of the chemokines GRO-a and MCP-1 in sputum samples from COPD patients. *Thorax* 57: 590–95, 2002.
46. Keatings VM, Barnes PJ. Granulocyte activation markers in induced sputum: Comparison between chronic obstructive pulmonary disease,

asthma and normal subjects. *Am J Respir Crit Care Med* 155: 449–53, 1997.
47. Biernacki WA, Kharitonov SA, Barnes PJ. Increased leukotriene B4 and 8-isoprostane in exhaled breath condensate of patients with exacerbations of COPD. *Thorax* 58: 294–98, 2003.
48. Aaron SD, Angel JB, Lunau M, Wright K, Fex C, Le Saux N *et al.* Granulocyte inflammatory markers and airway infection during acute exacerbation of chronic obstructive pulmonary disease. *Am J Respir Crit Care Med* 163: 349–55, 2001.
49. Barnes PJ. Macrophages as orchestrators of COPD. *COPD* 1: 59–70, 2004.
50. Retamales I, Elliott WM, Meshi B, Coxson HO, Pare PD, Sciurba FC *et al.* Amplification of inflammation in emphysema and its association with latent adenoviral infection. *Am J Respir Crit Care Med* 164: 469–73, 2001.
51. Meshi B, Vitalis TZ, Ionescu D, Elliott WM, Liu C, Wang XD *et al.* Emphysematous lung destruction by cigarette smoke. The effects of latent adenoviral infection on the lung inflammatory response. *Am J Respir Cell Mol Biol* 26: 52–57, 2002.
52. Russell RE, Thorley A, Culpitt SV, Dodd S, Donnelly LE, Demattos C *et al.* Alveolar macrophage-mediated elastolysis: Roles of matrix metalloproteinases, cysteine, and serine proteases. *Am J Physiol Lung Cell Mol Physiol* 283: L867–lL873, 2002.
53. Caramori G, Romagnoli M, Casolari P, Bellettato C, Casoni G, Boschetto P *et al.* Nuclear localisation of p65 in sputum macrophages but not in sputum neutrophils during COPD exacerbations. *Thorax* 58: 348–51, 2003.
54. Traves SL, Smith SJ, Barnes PJ, Donnelly LE. Specific CXC but not CC chemokines cause elevated monocyte migration in COPD: A role for CXCR2. *J Leukoc Biol* 76: 441–50, 2004.
55. Grumelli S, Corry DB, Song L-X, Song L, Green L, Huh J *et al.* An immune basis for lung parenchymal destruction in chronic obstructive pulmonary disease and emphysema. *PLoS Med* 1: 75–83, 2004.
56. Tomita K, Caramori G, Lim S, Ito K, Hanazawa T, Oates T *et al.* Increased p21CIP1/WAF1 and B cell lymphoma leukemia-xL expression and reduced apoptosis in alveolar macrophages from smokers. *Am J Respir Crit Care Med* 166: 724–31, 2002.
57. Keatings VM, Jatakanon A, Worsdell YM, Barnes PJ. Effects of inhaled and oral glucocorticoids on inflammatory indices in asthma and COPD. *Am J Respir Crit Care Med* 155: 542–48, 1997.
58. Culpitt SV, Rogers DF, Shah P, de Matos C, Russell RE, Donnelly LE *et al.* Impaired inhibition by dexamethasone of cytokine release by alveolar macrophages from patients with chronic obstructive pulmonary disease. *Am J Respir Crit Care Med* 167: 24–31, 2003.
59. Ito K, Ito M, Elliott WM, Cosio B, Caramori G, Kon OM *et al.* Decreased histone deacetylase activity in chronic obstructive pulmonary disease. *N Engl J Med* 352: 1967–76, 2005.
60. Barnes PJ. Reduced histone deacetylase in COPD: Clinical implications. *Chest* 129: 151–55, 2006.
61. Zhu J, Qiu YS, Majumdar S, Gamble E, Matin D, Turato G *et al.* Exacerbations of Bronchitis: Bronchial eosinophilia and gene expression for interleukin-4, interleukin-5, and eosinophil chemoattractants. *Am J Respir Crit Care Med* 164: 109–16, 2001.
62. Hammad H, Lambrecht BN. Dendritic cells and epithelial cells: Linking innate and adaptive immunity in asthma. *Nat Rev Immunol* 8: 193–204, 2008.
63. Rogers AV, Adelroth E, Hattotuwa K, Dewar A, Jeffery PK. Bronchial mucosal dendritic cells in smokers and ex-smokers with COPD: An electron microscopic study. *Thorax*, 2007.
64. Di Stefano A, Caramori G, Capelli A, Gnemmi I, Ricciardolo F, Oates T *et al.* STAT4 activation in smokers and patients with chronic obstructive pulmonary disease. *Eur Respir J* 24: 78–85, 2004.
65. Barnes PJ, Cosio MG. Characterization of T lymphocytes in chronic obstructive pulmonary disease. *PLoS Med* 1: 25–27, 2004.
66. Costa C, Rufino R, Traves SL, Silva LE, Barnes PJ, Donnelly LE. CXCR3 and CCR5 chemokines in the induced sputum from patients with COPD. *Chest* 133: 26–33, 2008.
67. Saetta M, Mariani M, Panina-Bordignon P, Turato G, Buonsanti C, Baraldo S *et al.* Increased expression of the chemokine receptor CXCR3 and its ligand CXCL10 in peripheral airways of smokers with chronic obstructive pulmonary disease. *Am J Respir Crit Care Med* 165: 1404–9, 2002.
68. Cosio MG, Majo J, Cosio MG. Inflammation of the airways and lung parenchyma in COPD: Role of T cells. *Chest* 121: 160S–65S, 2002.
69. Takubo Y, Guerassimov A, Ghezzo H, Triantafillopoulos A, Bates JH, Hoidal JR *et al.* Alpha1-antitrypsin determines the pattern of emphysema and function in tobacco smoke-exposed mice: Parallels with human disease. *Am J Respir Crit Care Med* 166: 1596–603, 2002.
70. Hashimoto S, Kobayashi A, Kooguchi K, Kitamura Y, Onodera H, Nakajima H. Upregulation of two death pathways of perforin/granzyme and FasL/Fas in septic acute respiratory distress syndrome. *Am J Respir Crit Care Med* 161: 237–43, 2000.
71. Chrysofakis G, Tzanakis N, Kyriakoy D, Tsoumakidou M, Tsiligianni I, Klimathianaki M *et al.* Perforin expression and cytotoxic activity of sputum CD8$^+$ lymphocytes in patients with COPD. *Chest* 125: 71–76, 2004.
72. Majo J, Ghezzo H, Cosio MG. Lymphocyte population and apoptosis in the lungs of smokers and their relation to emphysema. *Eur Respir J* 17: 946–53, 2001.
73. Cosio MG. T-lymphocytes. In: Barnes PJ (ed.), *Chronic Obstructive Pulmonary disease: Cellular and Molecular mechanisms*, pp. 321–25. New York: Taylor & Francis Group, 2005.
74. Sullivan AK, Simonian PL, Falta MT, Mitchell JD, Cosgrove GP, Brown KK *et al.* Oligoclonal CD4$^+$ T cells in the lungs of patients with severe emphysema. *Am J Respir Crit Care Med* 172: 590–96, 2005.
75. Lee SH, Goswami S, Grudo A, Song LZ, Bandi V, Goodnight-White S *et al.* Antielastin autoimmunity in tobacco smoking-induced emphysema. *Nat Med* 13: 567–69, 2007.
76. Barczyk A, Pierzchala W, Kon OM, Cosio B, Adcock IM, Barnes PJ. Cytokine production by bronchoalveolar lavage T lymphocytes in chronic obstructive pulmonary disease. *J Allergy Clin Immunol* 117: 1484–92, 2006.
77. Fairclough L, Urbanowicz RA, Corne J, Lamb JR. Killer cells in chronic obstructive pulmonary disease. *Clin Sci (Lond)* 114: 533–41, 2008.
78. Vijayanand P, Seumois G, Pickard C, Powell RM, Angco G, Sammut D *et al.* Invariant natural killer T cells in asthma and chronic obstructive pulmonary disease. *N Engl J Med* 356: 1410–22, 2007.
79. Montuschi P, Kharitonov SA, Ciabattoni G, Barnes PJ. Exhaled leukotrienes and prostaglandins in COPD. *Thorax* 58: 585–88, 2003.
80. Beeh KM, Kornmann O, Buhl R, Culpitt SV, Giembycz MA, Barnes PJ. Neutrophil chemotactic activity of sputum from patients with COPD: Role of interleukin 8 and leukotriene B4. *Chest* 123: 1240–47, 2003.
81. Rahman I. Oxidative stress in pathogenesis of chronic obstructive pulmonary disease: Cellular and molecular mechanisms. *Cell Biochem Biophys* 43: 167–88, 2005.
82. Montuschi P, Barnes PJ, Roberts LJ. Isoprostanes: Markers and mediators of oxidative stress. *FASEB J* 18: 1791–800, 2004.
83. Rangasamy T, Cho CY, Thimmulappa RK, Zhen L, Srisuma SS, Kensler TW *et al.* Genetic ablation of Nrf2 enhances susceptibility to cigarette smoke-induced emphysema in mice. *J Clin Invest* 114: 1248–59, 2004.
84. Ishii Y, Itoh K, Morishima Y, Kimura T, Kiwamoto T, Iizuka T *et al.* Transcription factor Nrf2 plays a pivotal role in protection against elastase-induced pulmonary inflammation and emphysema. *J Immunol* 175: 6968–75, 2005.
85. Tomita K, Barnes PJ, Adcock IM. The effect of oxidative stress on histone acetylation and IL-8 release. *Biochem Biophys Res Commun* 301: 572–77, 2003.
86. Ito K, Tomita T, Barnes PJ, Adcock IM. Oxidative stress reduces histone deacetylase (HDAC)2 activity and enhances IL-8 gene expression: Role of tyrosine nitration. *Biochem Biophys Res Commun* 315: 240–45, 2004.

87. Brindicci C, Ito K, Resta O, Pride NB, Barnes PJ, Kharitonov SA. Exhaled nitric oxide from lung periphery is increased in COPD. *Eur Respir J* 26: 52–59, 2005.
88. Ricciardolo FL, Caramori G, Ito K, Capelli A, Brun P, Abatangelo G *et al.* Nitrosative stress in the bronchial mucosa of severe chronic obstructive pulmonary disease. *J Allergy Clin Immunol* 116: 1028–35, 2005.
89. Barnes PJ. Cytokine networks in asthma and chronic obstructive pulmonary disease. *J Clin Invest*, 2008.
90. Boivin GP, Ormsby I, Jones-Carson J, O'Toole BA, Doetschman T. Germ-free and barrier-raised TGF beta 1-deficient mice have similar inflammatory lesions. *Transgenic Res* 6: 197–202, 1997.
91. Bonniaud P, Kolb M, Galt T, Robertson J, Robbins C, Stampfli M *et al.* Smad3 null mice develop airspace enlargement and are resistant to TGF-beta-mediated pulmonary fibrosis. *J Immunol* 173: 2099–108, 2004.
92. Shao MX, Nakanaga T, Nadel JA. Cigarette smoke Induces MUC5AC mucin overproduction via tumor necrosis factor-a converting enzyme in human airway epithelial (NCI-H292) cells. *Am J Physiol Lung Cell Mol Physiol* 287: L420–L427, 2004.
93. Burgel PR, Nadel JA. Roles of epidermal growth factor receptor activation in epithelial cell repair and mucin production in airway epithelium. *Thorax* 59: 992–96, 2004.
94. Rothenberg ME. VEGF obstructs the lungs. *Nat Med* 10: 1041–42, 2004.
95. Petrache I, Natarajan V, Zhen L, Medler TR, Richter AT, Cho C *et al.* Ceramide upregulation causes pulmonary cell apoptosis and emphysema-like disease in mice. *Nat Med* 11: 491–98, 2005.
96. Russell RE, Culpitt SV, DeMatos C, Donnelly L, Smith M, Wiggins J *et al.* Release and activity of matrix metalloproteinase-9 and tissue inhibitor of metalloproteinase-1 by alveolar macrophages from patients with chronic obstructive pulmonary disease. *Am J Respir Cell Mol Biol* 26: 602–9, 2002.
97. Hautamaki RD, Kobayashi DK, Senior RM, Shapiro SD. Requirement for macrophage metalloelastase for cigarette smoke-induced emphysema in mice. *Science* 277: 2002–4, 1997.
98. Imai K, Dalal SS, Chen ES, Downey R, Schulman LL, Ginsburg M *et al.* Human collagenase (matrix metalloproteinase-1) expression in the lungs of patients with emphysema. *Am J Respir Crit Care Med* 163: 6–91, 2001.
99. Agusti A. Chronic obstructive pulmonary disease: A systemic disease. *Proc Am Thorac Soc* 3: 478–81, 2006.
100. van Eeden SF, Sin DD. Chronic obstructive pulmonary disease: A chronic systemic inflammatory disease. *Respiration* 75: 224–38, 2008.
102. Barnes PJ, Hansel TT. Prospects for new drugs for chronic obstructive pulmonary disease. *Lancet* 364: 985–96, 2004.
101. Rabe KF, Beghe B, Luppi F, Fabbri LM. Update in chronic obstructive pulmonary disease 2006. *Am J Respir Crit Care Med* 175: 1222–32, 2007.

Triggers of Asthma and COPD

PART 6

35

Allergens

CHAPTER

Donald W. Cockcroft

Division of Respirology, Critical Care and Sleep Medicine, Department of Medicine, University of Saskatchewan, Royal University Hospital, Saskatoon, Saskatchewan, Canada

HISTORY

Introduction

Allergy, that is type I IgE-mediated sensitization and sequelae to predominantly inhaled allergens, has become increasingly recognized as important in the pathogenesis of asthma. The development of this increased recognition is briefly outlined focusing on identification of allergy and its mechanisms, on the relevance of allergen in the pathogenesis of asthma, and on increased prevalence of the condition(s).

Allergy

Although *rose catarrh* (likely actually due to grass pollen) had been known for some time, the first comprehensive clinical description of allergic rhinitis is attributed to John Bostock in 1819 [1]. The disease was given the medical name *catarrhus aestivus* (summer cold) but was recognized in the lay literature as *hay fever*, a label which persists to this day. This disease was felt to be limited to the middle and upper classes, and a major prevailing hypothesis was that it was due to the effects of summer heat and/or sunshine [1]. Grass pollen had been suggested as a potential cause, but this was not confirmed until the elegant experiments of Charles Blackley in 1873 [2]. The term *allergy* (different from normal) was coined by Von Pirquet in 1906 [3]. In 1912, allergy skin testing was first used in the investigation of this type of hypersensitivity [4]. With the mistaken belief that pollen contained a toxin, the concept of pollen immunization was introduced by Dunbar [5] and popularized by Noon in 1911 [6, 7]. Prausnitz and Kustner disproved the toxin hypothesis by the demonstration of passive transfer of allergic sensitivity using the injection of the serum of a sensitized individual into the skin of a nonsensitized subject [8] – the basis of the so-called PK test. Coca demonstrated that a heat labile serum factor that could not be precipitated from serum by the usual methods was the case of this sensitivity and, in 1923, he coined the terms *atopy* (strange disease) and *atopic reagin* as the name for this serum factor [9]. Reaginic antibody was eventually identified as IgE by Ishizaka in 1967 [10]. The complete allergen response, including allergen presenting cells, lymphocytes (T-helper 2 or Th2 and B cells), mast cells, basophils, and mediators, is still being elucidated.

Allergy and asthma

The recognition of allergens as potential causes of asthma parallels the recognition of allergy. Classic asthma symptoms were a major component of Bostock's original description of *catarrhus aestivus* [1]. Pollen, both ragweed [11] and grass [2], was specifically shown to provoke attacks of asthma in the 1870s. Cat, horse, and house-dust mite were identified as relevant asthma producing allergens in the early part of the 20th century [12], and more recently fungal spore sensitivity has been identified as a risk factor for asthma [13]. Asthma became lumped with the atopic diseases by many in the early part of the 20th century. However, the very influential Osler's textbook had stated that asthma was a *neurosis* [14], an opinion that continued to hold weight for many years. The opposing view, namely that asthma was primarily an atopic allergic disease, was argued by Cooke and colleagues [15]. This controversy persisted well into the 20th century. The identification of allergen-induced late asthmatic responses [16], allergen-induced airway hyperresponsiveness[1] [17], and allergen-induced airway inflammation [18], that is features that define clinical asthma, has led to allergens becoming recognized as important position in the pathogenesis of asthma. Although still a subject of some controversy [19], many investigators now regard the majority, that is 75–80%, of nonselected asthmatics as being atopic [20, 21].

Prevalence

There has been a striking increase in the prevalence of allergic rhinitis from the beginning to the end of the 19th century [22]. Within this time, frame allergic disease has gone from rare in the early 1800s to a common disorder by 1900. A case can be made that this is at least in part related to earlier under-recognition of a condition whose symptoms (rhinitis, wheezing) resemble other conditions believed to have been more common, and whose symptoms are often so mild as to potentially go unreported at a time when there were many more serious diseases to be concerned with. As early as 1873, Blackley noted the rising prevalence of allergic disease and felt that this could not adequately be explained by increased recognition which he admitted did play a role [2]. The prevalence of atopic disease and asthma in particular has continued to increase throughout the 20th century with more objective data to support these trends. The reasons for the remarkable increase in the prevalence of atopy and asthma are not completely understood. Genetics play a role in determining the presence and severity of atopy, however, large changes in prevalence over a short period cannot be due to genetics and therefore must be due to one or more environmental factors. One hypothesis is that increased exposure to allergens, particularly indoor allergens in modern air-tight homes, may lead to increased atopy and asthma [23]. This has been demonstrated prospectively in Papua New Guinea where blanket introduction was followed by house-dust mites, house-dust mite sensitivity, and an asthma epidemic [24]. Another intriguing hypothesis is the *hygiene hypothesis* [25]. This hypothesis suggests that early childhood infections drive the immune system toward the T helper 1 (Th1) paradigm that suppresses or reduces the Th2 arm of the immune system. Control/prevention of infections will remove this Th2 suppression and lead to an increased prevalence of Th2-related atopic disease. The evidence for this, including family size studies and epidemiologic studies related to infection, is stronger for atopy than for asthma alone [25]. Other factors related to lifestyle issues including diet and level of activity may be important [23].

[1]The term "airway (hyper)responsiveness" throughout this chapter, unless otherwise stated, refers to the nonallergic (hyper)responsiveness to histamine, cholinergic agonists, exercise, etc., which is a characteristic feature of symptomatic asthma.

ATOPY

Atopy is the tendency to develop IgE antibodies to commonly encountered environmental allergens by natural exposure in which the route of entry of allergen is across intact mucosal surfaces [26]. The recognized familial nature of atopy is due to complex (multiple gene) inheritance (genetic heterogeneity) [21, 27]. The remarkable relatively short-term increase in atopic prevalence indicates that environmental factors are also important [22–25]. The pathophysiologic basis for subjects developing atopy is uncertain and is the topic of a recent symposium [28]. An old hypothesis favoring allergen handling perhaps at the mucosal surface [29] rather than increased capacity to produce IgE has not been excluded.

The prevalence of atopy in random populations, defined as the presence of positive(s) on prick skin testing with a small battery of common relevant allergens, ranges from 30% to almost 50% [30–32]. The peak period of sensitization is in the third decade; thereafter the prevalence falls [26]. Our experience with a random young population would suggest that about 50% of atopic (skin test positive) subjects will have symptoms referable to atopy, which will include asthma in about 50% [32]. Thus, atopy is common, affecting about one in three, with about one in six having symptomatic atopy and about one in twelve having atopic asthma; the prevalences will be proportionately higher in populations with a higher prevalence of atopy.

ALLERGENS

Inhaled complete allergens that provoke asthma by IgE-mediated mechanisms are soluble organic high molecular weight (20,000–40,000 MW) protein or protein-containing molecules, which may be derived from any phylum of either the plant or the animal kingdoms (including bacteria) [33, 34]. The structural characteristics that make a protein allergenic have been recently reviewed [35]. In clinical nonoccupational settings, the important inhalant allergens fall into four groups; pollen, fungal spores, animal danders, and household arthropods (mite and insects) [36].

Pollen allergens that trigger asthma are predominantly from wind-pollinated plants, namely trees, grass, and weeds [36]. The relevant allergens and seasonal fluctuations will vary with geography and climate, with tree pollens predominant in spring months, grasses in summer, and weeds in late summer and autumn [34]. Although whole pollen grains may have limited access to the lower respiratory tract [37], the relationship of pollen to clinical asthma is convincing [36].

Atmospheric fungal spores of many groups of fungi are smaller and more respirable than pollen, and are recognized as causing atopic sensitization [13]. Their role in triggering asthma is less certain than pollen [34, 36]. Fungal spore types and seasons will also vary with geographic and climatic (temperature and humidity) conditions. Fungal spores are associated with decaying vegetation resulting in a late summer and autumn peak for common fungal spores, *Alternaria*, *Cladosporium*, *Aspergillus*, *Sporobolomyces*, etc. [34]. A spring peak for atmospheric fungal spores may be seen in some areas especially where late melting of snow cover leads to so-called snow mold [34]. Thus, atmospheric fungal spores may be responsible for fall or spring–fall asthma symptoms. Fungi may also be present inside living areas in moist basements, food storage areas, and waste receptacles [34]. *Aspergillus* may cause a distinct clinical syndrome, allergic bronchopulmonary aspergillosis, which will be covered separately.

Household animals [34], particularly cats and dogs, but also small mammals (gerbils, hamsters, rabbits, etc.) and birds, may release allergens in secretions (e.g. saliva) or excretions (e.g. urine, feces). Large animals, particularly horses, may also provoke atopic sensitivity. House dust, due to its content of mite antigens from various *Dermatophagoides* species or insect antigens such as cockroach [38], is an important source of atopic sensitization. *Dermatophagoides* spp., in particular, are likely the most important cause of atopic sensitization worldwide. Again, climatic conditions are important, since areas of low indoor relative humidity do not favor growth of house-dust mites [34, 39].

Other allergens are encountered less frequently, often in occupational settings, and include various plant parts [castor bean, cocoa bean, tobacco leaf, psyllium (laxative), vegetable gums, etc.], insect dusts, bacterial enzymes, and in the very highly sensitized even atmospheric molecular levels of foods (e.g. cooking fish) [34].

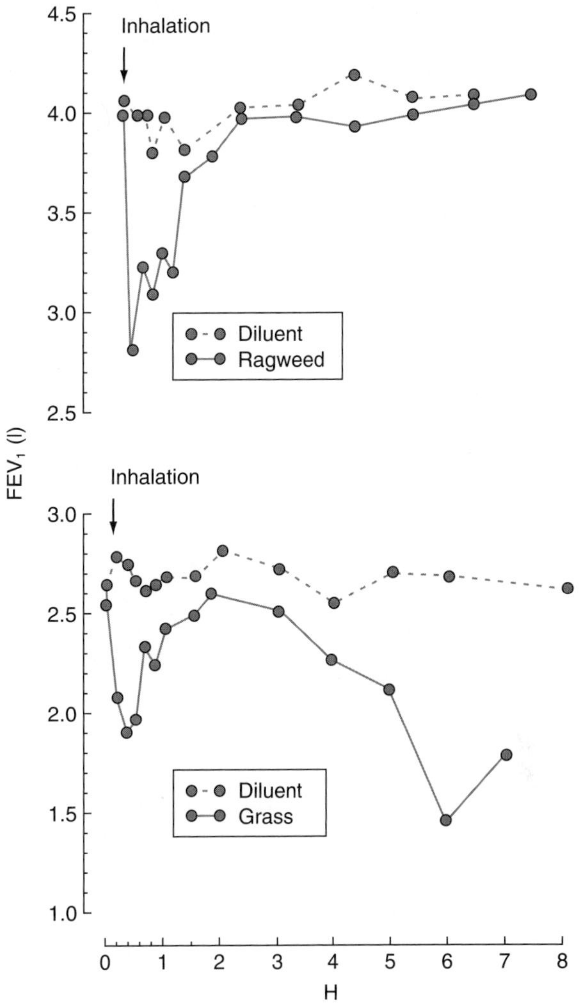

FIG. 35.1 Early and dual asthmatic responses to allergen. The top graph shows an isolated EAR following ragweed pollen inhalation and the bottom graph a dual asthmatic response (in another subject) following grass pollen inhalation. Modified and reproduced from Ref. [49] with permission.

INHALED ALLERGENS

Patterns of airway response

Airway responses to inhaled allergens have been assessed by the somewhat artificial inhalation tests in the laboratory with aqueous allergen extracts [17, 40, 41]. Nevertheless, the results of such challenges, especially the late sequelae, appear to be clinically relevant [42, 43], and allergen inhalation tests allow study of both the pharmacology and the pathophysiology of allergen-induced asthma. Airway responses to allergen can be divided into early and late sequelae. The early asthmatic response (EAR) is an episode of airflow obstruction which is maximal 10–20 min after allergen inhalation and resolves spontaneously in 1–2 h [17, 40, 41, 44]. The late sequelae include the late asthmatic response (LAR) [16, 17, 40, 41, 43–46], airway hyperresponsiveness [17, 41, 42], recurrent nocturnal asthma [46], and airway inflammation [18, 47]. The LAR is an episode of airflow obstruction which develops after spontaneous resolution of the EAR between 3 and 5 h after exposure, occasionally earlier, rarely later [17, 40, 41,43,44,45]. Resolution usually begins by 6–8 h but may require in excess of 12 h [41, 43, 44]. Modest late responses respond well to bronchodilators [48]; unpublished observations suggest bronchodilators may be required often (e.g. up to 2 h). More severe late airway obstruction is not always completely reversible by bronchodilator [44]. Examples of early and late responses are shown in Fig. 35.1. Allergen-induced increase in airway responsiveness (e.g. to histamine/methacholine) occurs following both experimental [17, 41, 42, 47] and natural [42, 50, 51] allergen exposure. This is correlated with the occurrence and severity of the late response, often appearing with small, previously ignored, late responses (5–15% FEV_1 (1-s forced expiratory volume) fall) [17, 41]. Airway responsiveness develops between 2 [52] and 3 h [53–55] after exposure, is present at 7–8 h [17, 41, 52], and may persist for days, occasionally worsening despite the return of airway caliber to baseline [41] (Fig. 35.2). As expected [56], the increased airway responsiveness is associated with symptoms of asthma [17, 41], including recurrent nocturnal asthma [46]. Both the LAR [18, 47] and the increased airway responsiveness [47] are associated with increases in airway inflammation. The occurrence of seasonal increases in airway responsiveness [42, 50, 51], and airway inflammation [57] provides support for both the relevance of the bronchoprovocation model and the importance of the inflammation in the pathogenesis of the late sequelae.

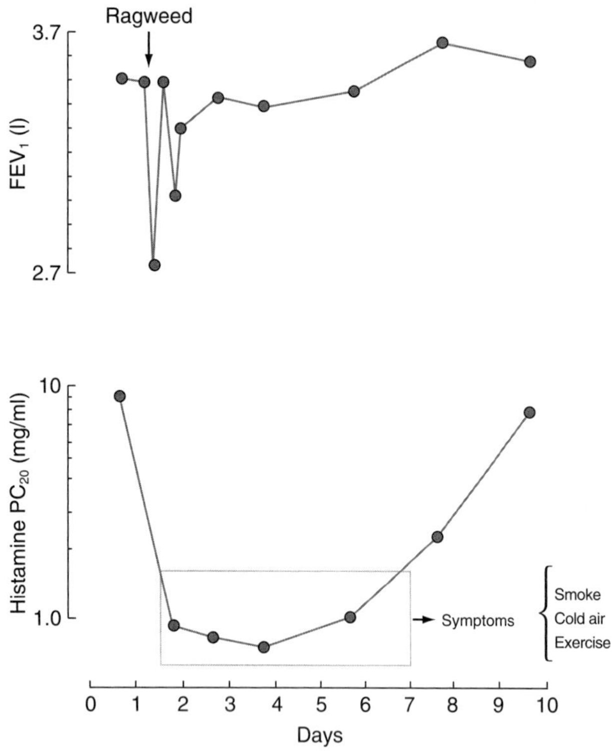

FIG. 35.2 Allergen-induced increase in nonallergic airway responsiveness to inhaled histamine. A dual asthmatic response, with spontaneous recovery, occurred after a single inhalation of ragweed pollen extract. Airway responsiveness to inhaled histamine, expressed as the provocation concentration causing a 20% FEV1 fall (PC20), increased after allergen exposure, and was associated with asthma symptoms on exposure to nonallergic stimuli. Reproduced from Ref. [49] with permission.)

Pharmacology

Pharmacologic inhibition of airway responses to inhaled allergen has been studied for both its therapeutic relevance and further understanding of the pathophysiology of the responses.

Inhaled β_2-agonists are the best inhibitors of the EAR [45, 58–61] due to a combination of effects on smooth muscle and on mediator release [62, 63]. Despite the latter, intermediate acting inhaled β_2-agonists (salbutamol, terbutaline, etc.) do not inhibit the LAR [45, 58, 59, 61] or the increased airway responsiveness [61]. The long acting inhaled β_2-agonists salmeterol and formoterol completely inhibit or mask all aspects of the allergen-induced airway response [64, 65], likely due to functional antagonism rather than anti-inflammatory effect [65]. Regular use of inhaled β_2-agonists for a week or more increases the EAR [66, 68, 69], the LAR [69, 70, 184], mast cell mediator release [69], and allergen-induced airway inflammation [69, 70]. A larger dose of allergen can be administered after an inhaled β_2-agonist and will lead to a larger LAR [71]. These features, failure to inhibit the LAR, enhanced airway responses, and ability to tolerate a larger dose of allergen, may be relevant in β_2-agonist-worsened asthma control [72].

Muscarinic blockers cause variable minor inhibition of the EAR [60, 73–75] and no inhibition of the LAR [73, 75]. Allergen-induced airway hyperresponsiveness appears uninfluenced by anticholinergics [75]. The enhanced airway responsiveness to histamine following allergen inhalation is no more responsive to atropine than it was prior to allergen inhalation [76].

Ingested theophylline offers partial protection against both the EAR and the LAR [77–80] and variable protection against the induced airway hyperresponsiveness [78–80]. It is not clear whether this is a functional antagonist or an anti-inflammatory effect.

Inhaled sodium cromoglycate (SCG) given prior to allergen exposure inhibits both the EAR and LAR [44, 45, 58, 61, 78, 81], as well as the allergen-induced increased responsiveness to both histamine [61] and methacholine [78]. Nedocromil sodium appears to have similar effects on allergen-induced asthmatic responses [82]. SCG given after the EAR will slightly delay but not inhibit the LAR [67].

The leukotriene pathway can be altered in several ways. Leukotriene receptor antagonists (LTRAs) are the most commonly used leukotriene modifiers. LTRAs provide protection against both EAR [83–85] and LAR [83, 84]. In addition, 5-lipoxygenase inhibitors [86] or 5-lipoxygenase activating protein inhibitors [87] provide some protection. Recently, the combination of desloratadine (H_1 blocker) and montelukast (LTRA) has shown a synergistic effect regarding inhibition of the EAR [85]; hence, studies on the LAR are required.

A single dose of inhaled corticosteroid, given prior to allergen, has no influence on the EAR but provides effective and often complete inhibition of the LAR [44, 45, 58, 61, 66, 67, 81, 82, 89, 90, 91, 92]. A single dose given after the EAR will inhibit the LAR [67]. Longer treatment periods with inhaled corticosteroids will partially inhibit the EAR as well [68, 91, 92]. Corticosteroid-induced improvement in airway responsiveness [68, 93, 94] provides only partial explanation [68]. Reduction in mucosal mast cells [95–97] is likely more important.

H_1 blockers partially inhibit the early portion of the EAR [45, 58, 98–100]. Newer H_1 blockers may also show some inhibition of the LAR [100]; hence, further studies are necessary. Ingested antiallergic drugs such as ketotifen and repirinast have produced variable effects on allergen-induced asthma [101–107]. Most studies have failed to show any significant protection [101, 102, 106, 107]. Nonsteroidal anti-inflammatory agents (cyclooxygenase inhibitors), particularly indomethacin, appear to have no effect or perhaps enhance the EAR [108]; there is conflicting evidence regarding the late response [88, 109, 110]. Allergen-induced increase in airway responsiveness appears to be partially inhibited by indomethacin [88]. A thromboxane synthetase inhibitor had no effect on allergen-induced early or late responses or increased airway responsiveness [111]. A platelet-activating factor antagonist proved ineffective against allergen-induced asthma [112]. Inhaled furosemide provides inhibition of both EAR and LAR [113]. Allergen injection therapy has produced variable results in modulating the EAR [114, 115] but may be particularly effective versus the LAR [116]. A novel recombinant anti-IgE molecule directed against the Fc component of IgE is very effective at inhibiting both the EAR [117] and the LAR [118]. Anti-IL-5 [119] and IL-12 [120] suppressed allergen-induced eosinophilia with little effect on allergen-induced airways hyperresponsiveness (AHR) casting some doubt on the causal relationship between inflammation and AHR.

Mechanisms

The mechanisms of allergen-induced asthmatic responses have been studied in humans by indirect means. Animal studies, *in vitro* studies on excised human tracheobronchial smooth muscle, drug-inhibition studies and, more recently, bronchoalveolar lavage (BAL), and induced sputum, have all been used to assess mechanisms. The EAR is due to the allergen-IgE mast cell acute mediator release including histamine [121], prostaglandins [122], and leukotrienes [123] and is primarily bronchospastic. Individual mediator blockers are only partially effective in inhibiting the EAR [83, 84, 85, 98, 99], however, an *in vitro* study on human tracheal smooth muscle demonstrated complete inhibition of the EAR by combined H_1 blocker, cyclooxygenase inhibitor, and lipoxygenase inhibitor [124]. The pathogenesis of the LAR is not so clear. An outdated hypothesis that late responses were type III precipitin-mediated responses [44], controversial at that time [58], has been disproved since both cutaneous [125, 126] and pulmonary [118, 127] late responses are IgE mediated. Animal studies have documented the requirement for inflammatory cells (eosinophils, neutrophils) in the LAR [128, 129] and induced airway hyperresponsiveness [130–132]. This has been confirmed in humans using BAL [18, 133] and induced sputum [47, 84]. The precise role of the chemokines and their cellular origin in the recruitment of inflammatory cells are a topic of current research [134–136].

Allergens as a cause of asthma

The importance of allergens as a *cause* of asthma (i.e. symptomatic airway hyperresponsiveness and airway inflammation [137], which was hypothesized several years ago [49] (Fig. 35.3), is now generally accepted [138]. The lines of reasoning include the high prevalence of atopy amongst asthmatics [20, 21], the correlation of both airway hyperresponsiveness and asthma with atopy in epidemiologic population studies [32,139–144] (Fig. 35.4), the relationship of both seasonal [42, 50, 51, 145, 146] and indoor [23, 24, 39] allergen exposure to symptoms and airway hyperresponsiveness, and their

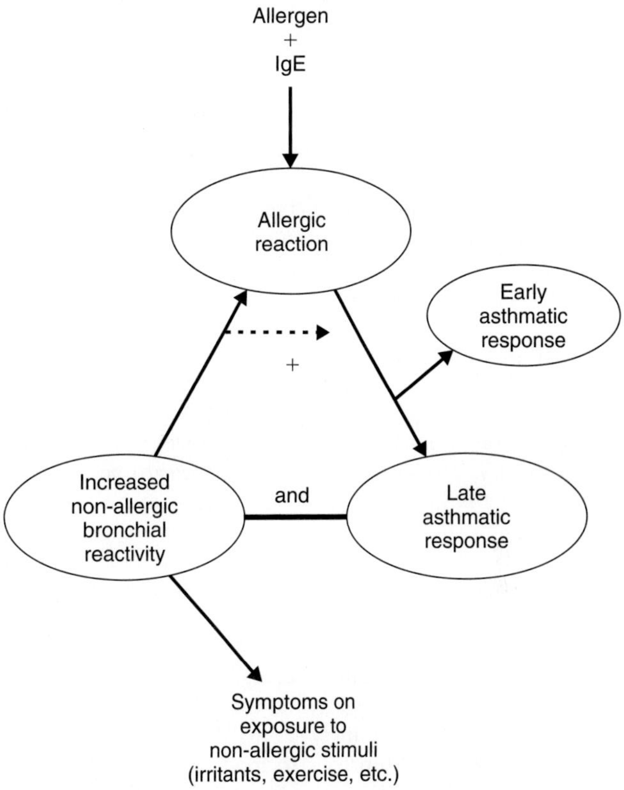

FIG. 35.3 Diagram of hypothesis explaining development and maintenance of perennial allergen-induced asthma. Reproduced from Ref. [49] with permission.

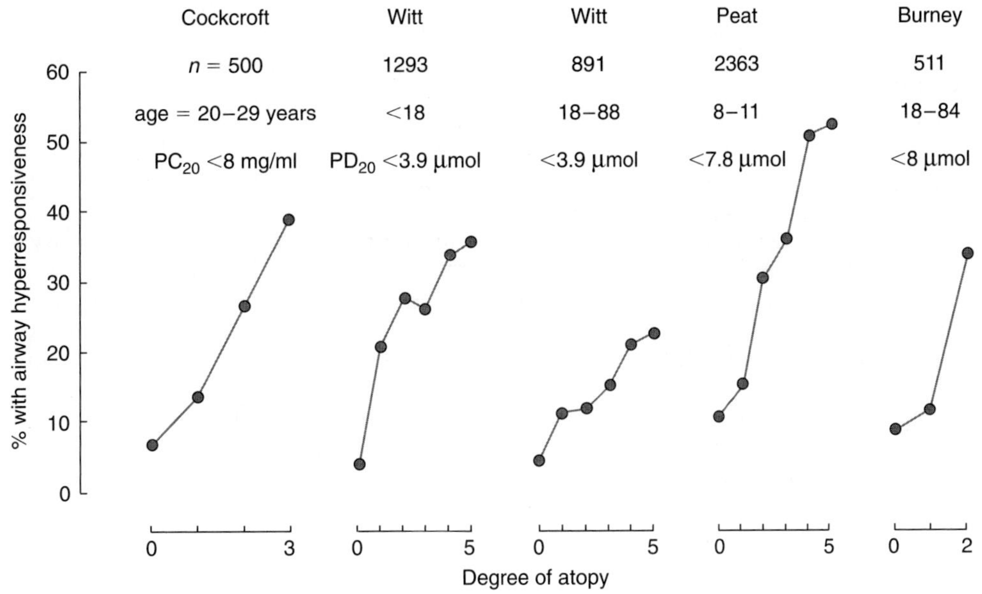

FIG. 35.4 Degree of atopy (various scales) from nonatopic (0) to highly atopic (highest number) on the horizontal axis versus prevalence of airway hyperresponsiveness (%) on the vertical axis from four population studies [20, 140–142]. Reproduced from Ref. [143] with permission.

reduction with allergen avoidance [147, 183]. It is speculated that, in sensitized individuals, the duration and magnitude of airways allergic exposure, if not suppressed pharmacologically, may lead not only to transient but also to persistent airway hyperresponsiveness and clinical asthma [49]. This is supported by parallel observations in animals [148] and in human occupational asthma [149, 150].

INGESTED/INJECTED ALLERGENS

Isolated asthma caused by allergens introduced via routes other than inhalation is uncommon but has been reported [151, 152]. Ingested allergens [152] (foods, drugs) or injected allergens [153] (hyposensitization injections, intradermal allergen tests, drugs, insect bites, and stings) can produce IgE-mediated hypersensitivity reactions. Most often, these produce systemic allergen reactions [151, 153] (violent gastrointestinal upset, urticaria, angioedema, laryngeal edema, anaphylactic shock, with or without bronchospasm). However, occasionally such exposures produce reactions that appear to be centered primarily in the lung [152, 154]. It is likely that these represent systemic allergen reactions in subjects with pre-existing asthma and high levels of airway hyperresponsiveness who develop disproportionately severe bronchospasm. Some cases of food/bite/sting-induced *asthma* may actually represent laryngeal spasm or edema which has been misdiagnosed. Although relatively uncommon, allergen (asthmatic or otherwise) responses to both ingested and injected allergens occur rapidly and can be exceedingly severe. We have seen subjects with sudden severe asthma due to foods (nuts, shellfish) and two subjects with sudden onset status asthmaticus circumstantially linked to unrecognized black fly bites.

CLINICAL FEATURES

Clinical presentation

Allergic asthma usually begins at a young age, between about 2 and 20, but can develop at any time. A positive family history of asthma or atopy is common. Other atopic symptoms are often present and include food sensitivity (infancy), childhood eczema, urticaria, conjunctivitis, and allergic rhinitis. Asthma or asthma exacerbations should correlate with allergen exposure or seasons. Spring, summer, or fall exacerbations suggest fungal spore or pollen sensitivity whereas winter exacerbations are typical of indoor allergen sensitivity but may also occur in nonatopic asthmatics. The polyallergic subject may show little seasonal variability. Because of the gradual onset of airway inflammation and hyperresponsiveness, allergen-induced asthma exacerbations are usually of gradual onset often lagging behind and persisting beyond allergen exposure. An acute exposure–symptom relationship may be lacking. For these reasons, the patient and physician may miss the importance of allergens. When present, nonrespiratory, particularly ocular or cutaneous and, to a lesser extent, nasal symptoms will bear a much closer relationship to exposure and may provide an important historical clue to sensitization. Irritant- and exercise-induced bronchospasm are symptomatic of underlying airway hyperresponsiveness and, consequently, can be due to (often unrecognized) allergen exposure that may not be acutely temporally related.

Sudden severe symptoms can occur following inhaled allergen exposure if the exposure is marked, if the individual is highly sensitized, or if previous asthma symptoms/airflow obstruction have been either under-perceived or ignored. However, genuinely sudden asthma exacerbations raise the possibility of ingested or injected allergen (foods, drugs, bites, stings, allergen shots); this clinical scenario should prompt an aggressive historical review for relevant clues. Non-IgE-mediated responses to ASA and NSAIDs, β-adrenergic blockers, food additives (e.g. metabisulphites), or cholinesterase inhibiting insecticides are included in the differential diagnosis of sudden severe asthma.

Diagnosis

The diagnosis of allergic asthma rests predominantly on the historical features noted above. The inquiry into possible allergens should not stop with the home but must include work, school, social, and recreational exposures. Because of the frequently subtle exposure–symptom relationships, nonrespiratory symptoms related to allergen exposure should be sought. Following completion of a complete history, allergen prick-skin testing with a small series of relevant allergens should be done.

Treatment

The treatment of allergen-induced asthma is the same as for asthma in general and is outlined elsewhere in this book. The pathophysiologic and clinical features outlined above stress the importance of environmental control. The possibility that chronic allergen exposure might lead to permanent pathology suggests that therapeutic strategies that are largely effective by preventing the allergic response and sequelae (allergen avoidance, cromones, anti-IgE) will be less effective if their introduction is delayed.

Allergic bronchopulmonary mycoses

A distinctive clinical syndrome occurs when atopic individuals have organisms, against which they have IgE antibodies, growing in their airways. The prototype, by far the commonest of these "allergic bronchopulmonary mycoses," is allergic bronchopulmonary aspergillosis (ABPA) usually caused by *Aspergillus fumigatus* [155–158]. Other fungi, including *Helminthosporium* [159, 160], *Curvularia* and *Drechslera* [160, 161], *Stemphylium* [162], *Candida* [163], *Fusarium* [164], and rarely bacteria such as *Pseudomonas* [165], may cause a similar syndrome. The syndrome that these organisms can produce involves complex immunologic and mechanical pathogenesis.

The pathogenesis of ABPA initially involves acquisition of IgE-mediated type I hypersensitivity to the fungus. *Aspergillus* fungal spores released from decaying vegetation, particularly in the autumn germinate and grow in the airway provoking reduced airway caliber and mucous hypersecretion leading to mucous plugs containing fungal hyphae producing "cast-like" outlines of the bronchial tree. The chronic high-level allergen exposure leads to very high levels of both allergen-specific and total serum IgE, an intense peripheral and bronchial eosinophilia, and in most, but not all, to the development of IgG precipitating antibodies. Other immunologic responses including cell-mediated immunity may be stimulated. The continuous airways allergic reaction will lead to exacerbation of asthma while the other immunologic reactions may be responsible for bronchial and parenchymal damage that includes (proximal) bronchiectasis and (upper lobe) interstitial pulmonary fibrosis. Fleeting or fixed pulmonary infiltrates may be produced by immunologic reactions within the lung, by the mechanical effect of obstruction of major bronchi by mucous plugs, or both.

ABPA is a fairly common condition in some areas; the precise prevalence is not certain. Amongst asthmatics, the prevalence of type I (IgE) sensitivity to *Aspergillus* may approximate 25% [166–168], and the prevalence of type III (IgG) sensitivity to *Aspergillus* is approximately half of this [169, 170]. By contrast, on the dry Canadian prairies, *Aspergillus* skin sensitivity is uncommon, and we have seen no new cases of ABPA in over 20 years. Other organisms (e.g. *Helminthosporium*) are involved only rarely. The clinical picture is generally that of a subject with pre-existent atopy, and usually previous asthma, presenting with exacerbation of asthma accompanied by the expectoration of characteristic firm brown plugs [155]. Pulmonary infiltrates with eosinophils with or without fever may be seen. Chronic disease may manifest as bronchiectasis with chronic or recurrent pulmonary infection or pulmonary fibrosis with progressive dyspnea or both [155, 156].

The two features that must be present in all subjects with ABPA are type I hypersensitivity to the *Aspergillus* and the presence of *Aspergillus* in the airways. However, it is not always possible to grow the organism [171]. Specific IgG precipitating antibodies are found in about 90% of cases [157]. Other features that are commonly seen include intense peripheral and bronchial eosinophilia [171], marked elevations of total serum IgE [172], and transient pulmonary infiltrates [155–158]; these all tend to correlate with activity of disease. Chronic changes in established or recurrent disease include radiographic demonstration of an unusual and essentially pathognomonic *proximal* bronchiectasis [173] and, in more severe cases, progressive upper lobe interstitial pulmonary fibrosis similar to tuberculosis and other upper lobe scarring conditions [156]. The computerized tomographic scan may be particularly helpful as an adjunct to the diagnosis of ABPA [174, 175]. ABPA is a common complication of cystic fibrosis [176]; the 2% prevalence in one observational cohort was felt to represent under recognition [177]. Atopic skin testing has been suggested as a useful screen in this population [176], however, may not be reliable [178]; specific skin tests, allergen-specific IgE, total IgE, and precipitins are recommended in this population [178].

Allergic bronchopulmonary infestations with *Aspergillus* or other fungal organisms are not true infections; progression to invasive fungal infections or mycetoma formation is uncommon. Treatment is directed against the asthma and the immunologic abnormalities; intensive asthma treatment with systemic corticosteroids in doses sufficient to suppress clinical and laboratory features of the disease is indicated [171, 179, 180]. Total serum IgE may be useful to predict exacerbations [180]. With such treatment, the prognosis is favorable; however, under-treatment can lead to substantial, even severe, permanent bronchopulmonary damage. Itraconazole has been reported to be effective in reducing corticosteroid requirements in ABPA [181] and may be useful in cystic fibrosis [176]. However, a recent Cochrane Database report concludes there is insufficient information available to recommend the use of azoles (itraconazole, ketoconazole) in this condition [182].

SUMMARY

1. The majority of subjects with asthma are atopic.
2. Allergen-induced airway hyperresponsiveness and airway inflammation point to allergens as a *cause* of asthma (i.e. symptomatic airway hyperresponsiveness with inflammation).
3. The remarkable rise in the prevalence of atopy ("hygiene" hypothesis, airtight home/indoor allergen exposure hypothesis, other hypotheses or some combination) over the past 200 years may explain in whole or in part the rising prevalence of asthma.
4. Chronic/recurrent allergen exposure may lead to persistent asthma.
5. This provides a plausible basis to suggest that early prophylactically or actively anti-inflammatory therapeutic strategies should improve long-term outcomes in allergic asthma.
6. This also provides a rationale to speculate re-*primary* prevention of (allergic) asthma.

ACKNOWLEDGEMENT

The author wishes to thank Jacquie Bramley for assisting in the preparation of this manuscript.

References

1. Bostock J. Case of a periodical affection of the eyes and chest. *Med Chir Trans* 10: 161–65, 1819.
2. Blackley CH. Experimental Researches on the Cause and Nature of Catarrhus Aestivus (Hay-Fever or Hay-Asthma). London: Balliere Tindall & Cox, 1873.

3. von Pirquet C. Allergie. Munch Med Wochenschr 1906;53:1457. (Translated from the German by Prausnitz C). In: Gell PGH, Coombs RRA (eds). *Clinical Aspects of Immunology*. Philadelphia, PA: F.A. Davis Company, 1963.
4. Schloss OM. A case of allergen to common foods. *Am J Dis Child* 3: 341–62, 1912.
5. Dunbar WP. The present state of our knowledge of hay fever. *J Hygiene* 13: 105–48, 1913.
6. Noon L. Prophylactic inoculation against hay fever. *Lancet* 1: 1572–73, 1911.
7. Freeman J. Vaccination against hay fever: Report of results during the last three years. *Lancet* 1: 1178–80, 1914.
8. Prausnitz C, Kustner H. Studien uber Uberempfindlichkeit. Centralb Bakteriol 1 Abt Orig 1921;86:160. (Translated by Prausnitz C). In: Gell PGH, Coombs RRA (eds). *Clinical Aspects of Immunology*, pp. 808–16. Oxford: Blackwell Scientific Publications, 1962.
9. Coca AF, Cooke RA. On the classification of the phenomena of hypersensitiveness. *J Immunol* 8: 163–82, 1923.
10. Ishizaka K, Ishizaka T. Identification of gamma-E antibodies as a carrier of reaginic activity. *J Immunol* 99: 1187–98, 1967.
11. Wyman M. Autumnal Catarrah. Cambridge: Hurd and Houston, 1872.
12. Simons FER. Ancestors of Allergy. New York: Global Medical Communications, 1994.
13. Herxheimer H, Hyde HA, Williams DA. Allergic asthma caused by fungal spores. *Lancet* 1: 572–73, 1966.
14. Osler W. Principles and Practice of Medicine. pp. 628–632, 3rd edn. New York: D. Appleton and Company, 1896.
15. Cooke RA. Asthma. In: Musser JH (ed.), *Internal Medicine, Theory and Practice*, pp. 994–1001. Philadelphia, PA: Lea & Febiger, 1932.
16. Herxheimer H. The late bronchial reaction in induced asthma. *Int Arch Allergy Appl Immunol* 3: 323–33, 1952.
17. Cockcroft DW, Ruffin RE, Dolovich J *et al.* Allergen-induced increase in nonallergic bronchial reactivity. *Clin Allergy* 7: 503–13, 1977.
18. de Monchy JGR, Kauffman HF, Venge P *et al.* Bronchoalveolar eosinophilia during allergen-induced late asthmatic reactions. *Am Rev Respir Dis* 131: 373–76, 1985.
19. Pearce N, Pekkanen J, Beasley R. How much asthma is really attributable to atopy?. *Thorax* 51: 268–72, 1999.
20. Cockcroft DW, Berscheid BA, Murdock KY *et al.* Sensitivity and specificity of histamine PC_{20} measurements in a random selection of young college students. *J Allergy Clin Immunol* 89: 23–30, 1992.
21. Postma DS, Koppelman GH, Meyers DA. The genetics of atopy and airway hyperresponsiveness. *Am J Respir Crit Care Med* 162: S118–S123, 2000.
22. Emanuel MB. Hay fever, a post industrial revolution epidemic: A history of its growth during the 19th century. *Clin Allergy* 18: 295–304, 1988.
23. Platts-Mills TA, Blumenthal K, Perzanowski M. Determinants of clinical allergic disease. The relevance of indoor allergens to the increase in asthma [in process citation]. *Am J Respir Crit Care Med* 162: S128–33, 2000.
24. Dowse GK, Turner KJ, Stewart GA *et al.* The association between *Dermatophagoides* mites and the increasing prevalence of asthma in village communities within the Papua New Guinea highlands. *J Allergy Clin Immunol* 75: 75–83, 1985.
25. Strachan DP. Family size, infection and atopy: The first decade of the "hygiene hypothesis". *Thorax* 55: S2–S10, 2000.
26. Pepys J. Atopy. In: Gell PGH, Coombs RRA, Lachman PJ (eds). *Clinical Aspects of Immunology*, pp. 877–902. Oxford: Blackwell Scientific Publications, 1975.
27. Sandford A, Weir T, Pare P. The genetics of asthma. *Am J Respir Crit Care Med* 153: 1749–65, 1996.
28. Aalberse RC, Sterk PJ. The roots of allergic sensitisation: Fourth Lunteren conference. *Am J Respir Crit Care Med* 162: S61–136, 2000.
29. Leskowitz S, Salvaggio JE, Schwartz HJ. An hypothesis for the development of atopic allergy in man. *Clin Allergy* 2: 237–46, 1972.
30. Woolcock AJ, Colman MH, Jones MW. Atopy and bronchial reactivity in Australian and Melanesian populations. *Clin Allergy* 8: 155–64, 1978.
31. Brown WG, Halonen MJ, Kaltenborn WT *et al.* The relationship of respiratory allergy, skin test reactivity, and serum IgE in a community population sample. *J Allergy Clin Immunol* 63: 328–35, 1979.
32. Cockcroft DW, Murdock KY, Berscheid BA. Relationship between atopy and bronchial responsiveness to histamine in a random population. *Ann Allergy* 53: 26–29, 1984.
33. King TP. Immunochemical properties of antigens that cause disease. In: Weis EB, Stein M (eds). *Bronchial Asthma: Mechanisms and Therapies*, pp. 43–49. Boston, MA: Little, Brown and Co., 1993.
34. Platts-Mills TAE, Solomon WR. Aerobiology and inhalant allergens. In: Middleton E, Reed CE, Ellis EF, Adkinson NF, Yuninger JW, Busse WW (eds). *Allergy Principles and Practice*, 4th edn., pp. 469–528. St. Louis, MO: Mosby YearBook Inc., 1993.
35. Aalberse RC. Structural biology of allergens. *J Allergy Clin Immunol* 106: 228–38, 2000.
36. Dolovich J, Zimmerman B, Hargreave FE. Allergy in asthma. In: Clark TJH, Godfrey S (eds). *Asthma*, pp. 132–57. London: Chapman and Hall, 1983.
37. Busse WW, Reed CE, Hoehne JH. Where is the allergic reaction in ragweed asthma?. *J Allergy Clin Immunol* 50: 289–93, 1972.
38. Pollart SM, Chapman MD, Fiocco GP *et al.* Epidemiology of acute asthma: IgE antibodies to common inhalant allergens as a risk factor for emergency room visits. *J Allergy Clin Immunol* 83: 875–82, 1989.
39. Murray AB, Ferguson AC, Morrison B. The seasonal variation of allergic respiratory symptoms induced by house dust mites. *Ann Allergy* 45: 347–50, 1980.
40. Robertson DG, Kerigan AT, Hargreave FE *et al.* Late asthmatic responses induced by ragweed pollen allergen. *J Allergy Clin Immunol* 54: 244–54, 1974.
41. Cartier A, Thomson NC, Frith PA *et al.* Allergen-induced increase in bronchial responsiveness to histamine: Relationship to the late asthmatic response and change in airway caliber. *J Allergy Clin Immunol* 70: 170–77, 1982.
42. Boulet LP, Cartier A, Thomson NC *et al.* Asthma and increases in nonallergic bronchial responsiveness from seasonal pollen exposure. *J Allergy Clin Immunol* 71: 399–406, 1983.
43. O'Byrne PM, Dolovich J, Hargreave FE. State of the art: Late asthmatic responses. *Am Rev Respir Dis* 136: 740–51, 1987.
44. Pepys J. Immunopathology of allergic lung disease. *Clin Allergy* 3: 1–22, 1973.
45. Booij-Noord H, deVries K, Sluiter HJ *et al.* Late bronchial obstructive reaction to experimental inhalation of house dust extract. *Clin Allergy* 2: 43–61, 1972.
46. Newman Taylor AJ, Davies RJ, Hendrick DJ *et al.* Recurrent nocturnal asthmatic reactions to bronchial provocation tests. *Clin Allergy* 9: 213–19, 1979.
47. Pin I, Freitag AP, O'Byrne PM *et al.* Changes in the cellular profile of induced sputum after allergen-induced asthmatic responses. *Am Rev Respir Dis* 145: 1265–69, 1992.
48. Dorsch W, Baur X, Emslander HP *et al.* Zur pathogenese und therapie der allergeninduzierten verzogerten bronchialostruktion. *Prax Klin Pneumol* 34: 461–68, 1980.
49. Cockcroft DW. Mechanism of perennial allergic asthma. *Lancet* 2: 253–56, 1983.
50. Altounyan REC. Changes in histamine and atropine responsiveness as a guide to diagnosis and evaluation of therapy in obstructive airways disease. In: Pepys J, Franklands AW (eds). *Disodium Cromoglycate in Allergic Airways Disease*, pp. 47–53. London: Butterworths, 1970.
51. Lowhagen O, Rak S. Modification of bronchial hyperreactivity after treatment with sodium cromoglycate during pollen season. *J Allergy Clin Immunol* 75: 460–67, 1985.
52. Cockcroft DW, Murdock KY. Changes in bronchial responsiveness to histamine at intervals after allergen challenge. *Thorax* 42: 302–8, 1987.
53. Millilo G. Discussion. In: Nadel JA, Pauwels R (eds). *International Conference on Bronchial Hyperreactivity*, p. 17. Oxford: The Medicine Publishing Foundation, 1982.

54. Durham SR, Graneek BJ, Hawkins R *et al.* The temporal relationship between increases in airway responsiveness to histamine and late asthmatic responses induced by occupational agents. *J Allergy Clin Immunol* 79: 398–406, 1987.
55. Thorpe J, Steinberg D, Bernstein D *et al.* Bronchial hyperreactivity occurs soon after the immediate asthmatic response in dual responders. *Am Rev Respir Dis* 133: A93, 1986.
56. Hargreave FE, Ryan G, Thomson NC *et al.* Bronchial responsiveness to histamine or methacholine in asthma: Measurement and clinical significance. *J Allergy Clin Immunol* 68: 347–55, 1981.
57. Djukanovic R, Feather I, Gratziou C *et al.* Effect of natural allergen exposure during the grass pollen season on airways inflammatory cells and asthma symptoms. *Thorax* 51: 575–81, 1996.
58. Orie NGM, Van Lookeren Campagne JG, Knol K, *et al.* Late reactions in bronchial asthma. In: Pepys J, Yamamura I (eds.), *Intal in Bronchial Asthma, Proceedings of the 8th International Congress of Allergollogy*, pp. 17–29. Tokyo, 1974.
59. Hegardt B, Pauwels R, Van Der Straeten M. Inhibitory effect of KWD 2131, terbutaline, and DSCG on the immediate and late allergen-induced bronchoconstriction. *Allergy* 36: 115–22, 1981.
60. Ruffin RE, Cockcroft DW, Hargreave FE. A comparison of the protective effect of Sch1000 and fenoterol on allergen-induced asthma. *J Allergy Clin Immunol* 61: 42–47, 1978.
61. Cockcroft DW, Murdock KY. Comparative effects of inhaled salbutamol, sodium cromoglycate and beclomethasone dipropionate on allergen-induced early asthmatic response, late asthmatic responses and increased bronchial responsiveness to histamine. *J Allergy Clin Immunol* 79: 734–40, 1987.
62. Church MK, Young KD. The characteristics of inhibition of histamine release from human lung fragments by sodium cromoglycate, salbutamol, and chlorpromazine. *Br J Pharmacol* 78: 671–79, 1983.
63. Howarth PH, Durham SR, Lee TH *et al.* Influence on albuterol, cromolyn sodium, and ipratropium bromide on the airway and circulating mediator responses to allergen bronchial provocation in asthma. *Am Rev Respir Dis* 132: 986–92, 1985.
64. Twentyman OP, Finnerty JP, Harris A *et al.* Protection against allergen-induced asthma by salmeterol. *Lancet* 336: 1338–42, 1990.
65. Wong BJ, Dolovich J, Ramsdale EH *et al.* Formoterol compared with beclomethasone and placebo on allergen-induced asthmatic responses. *Am Rev Respir Dis* 146: 1158–60, 1992.
66. Cockcroft DW, McParland CP, Britto SA *et al.* Regular inhaled salbutamol and airway responsiveness to allergen. *Lancet* 342: 833–37, 1993.
67. Cockcroft DW, McParland CP, O'Byrne PM *et al.* Beclomethasone given after the early asthmatic response inhibits the late response and the increased methacholine responsiveness and cromolyn does not. *J Allergy Clin Immunol* 91: 1163–68, 1993.
68. Cockcroft DW, Swystun VA, Bhagat R. Interaction of inhaled β_2 agonist and inhaled corticosteroid on airway responsiveness to allergen and methacholine. *Am J Respir Crit Care Med* 152: 1485–89, 1995.
69. Swystun VA, Gordon JR, Davis EB *et al.* Mast cell tryptase release and asthmatic responses to allergen increase with regular use of salbutamol. *J Allergy Clin Immunol* 106: 57–64, 2000.
70. Gauvreau GM, Watson RM, Jordana M *et al.* The effect of regular inhaled salbutamol on allergen-induced airway responses and inflammatory cells in blood and induced sputum. *Am J Respir Crit Care Med* 151: A39, 1995.
71. Lai CKW, Twentyman OP, Holgate ST. The effect of an increase in inhaled allergen dose after rimiterol hydrobromide on the occurrence and magnitude of the late asthmatic response and the associated change in nonspecific bronchial responsiveness. *Am Rev Respir Dis* 140: 917–23, 1989.
72. van Schaych CP, Cloosterman SGM, Hofland ID *et al.* How detrimental is chronic use of bronchodilators in asthma and chronic obstructive pulmonary disease?. *Am J Respir Crit Care Med* 151: 1317–19, 1995.
73. Yu DYC, Galant SP, Gold WM. Inhibition of antigen-induced bronchoconstriction by atropine in asthmatic patients. *J Appl Physiol* 32: 823–28, 1972.
74. Orehek J, Gayrard P, Grimaud Ch *et al.* Bronchoconstriction provoquee par inhalation d'allergene dans l'asthme: Effet antagoniste d'un anticholinergique de synthese. *Bull Eur Physiopathol Respir* 11: 193–201, 1975.
75. Cockcroft DW, Ruffin RE, Hargreave FE. Effect of Sch1000 in allergen-induced asthma. *Clin Allergy* 8: 361–72, 1978.
76. Boulet LP, Latimer KM, Roberts RS *et al.* The effect of atropine on allergen-induced increases in bronchial responsiveness to histamine. *Am Rev Respir Dis* 130: 368–72, 1984.
77. Pauwels R, van Renterghem D, Van Der Straeten M *et al.* The effect of theophylline and enprofylline on allergen-induced bronchoconstriction. *J Allergy Clin Immunol* 76: 583–90, 1985.
78. Cockcroft DW, Murdock KY, Gore BP *et al.* Theophylline does not inhibit allergen-induced increase in airway responsiveness to methacholine. *J Allergy Clin Immunol* 83: 913–20, 1989.
79. Crescioli S, Spinazzi A, Plebani M *et al.* Theophylline inhibits early and late asthmatic reactions induced by allergens in asthmatic subjects. *Ann Allergy* 66: 245–51, 1991.
80. Hendeles L, Harman E, Huang D *et al.* Theophylline attenuation of airway responses to allergen: Comparison with cromolyn metered-dose inhaler. *J Allergy Clin Immunol* 95: 505–14, 1995.
81. Pepys J, Chan M, Hargreave FE *et al.* Inhibitory effects of disodium cromoglycate on allergen-inhalation tests. *Lancet* 2: 134–37, 1968.
82. Dahl R, Pedersen B. Influence of nedocromil sodium on the dual asthmatic reaction after allergen challenge: A double-blind, placebo-controlled study. *Eur J Respir Dis* 69(Suppl 147): 263–65, 1986.
83. Taylor IK, O'Shaughnessy KM, Fuller RW *et al.* Effect of cysteinyl-leukotriene receptor antagonist ICI 204.219 on allergen-induced bronchoconstriction and airway hyperreactivity in atopic subjects. *Lancet* 337: 690–94, 1991.
84. Diamant Z, Grootendorst DC, Veselic-Charvat M *et al.* The effect of Montelukast (MK-0476), a cysteinyl leukotriene receptor antagonist, on allergen-induced airway responses and sputum cell counts in asthma. *Clin Exp Allergy* 29: 42–51, 1999.
85. Davis BE, Todd DC, Cockcroft DW. Effect of combined montelukast and desloratadine on the early asthmatic response to inhaled allergen. *J Allergy Clin Immunol* 116: 768–72, 2005.
86. Hui KP, Taylor IK, Taylor GW *et al.* Effect of a 5-lipoxygenase inhibitor on leukotriene generation and airway responses after allergen challenge in asthmatic patients. *Thorax* 46: 184–89, 1991.
87. Diamant Z, Timmers MC, van der Veen H *et al.* The effect of MK-0591, a novel 5-lipoxygenase activating protein inhibitor, on leukotriene biosynthesis and allergen-induced airway response in asthmatic subjects in vivo. *J Allergy Clin Immunol* 95: 42–51, 1995.
88. Kirby JG, Hargreave FE, Cockcroft DW *et al.* Indomethacin inhibits allergen-induced airway hyperresponsiveness but not allergen-induced asthmatic responses. *J Appl Physiol* 66: 578–83, 1989.
89. Booij-Noord H, Orie NGM, deVries K. Immediate and late bronchial obstructive reactions to inhalation of house dust and protective effects of disodium cromoglycate and prednisolone. *J Allergy Clin Immunol* 48: 344–54, 1971.
90. Pepys J, Davies RJ, Breslin ABX *et al.* The effects of inhaled beclomethasone dipropionate (Becotide) and sodium cromoglycate on asthmatic reactions to provocation tests. *Clin Allergy* 4: 13–24, 1974.
91. Van Der Star JG, Berg WC, Steenhuis EJ *et al.* Invloed van beclometason-dipropionaat per aerosol op de obstructieve reactie in de bronchien na huisstofinhalatie. *Ned Tijdschr Geneeskd* 120: 1928–32, 1976.
92. Burge PS, Efthimiou J, Turner-Warwick M *et al.* Double-blind trials of inhaled beclomethasone dipropionate and fluocortin butyl ester in allergen-induced immediate and late reactions. *Clin Allergy* 12: 523–31, 1982.
93. Du Toit JI, Salome CM, Woolcock AJ. Inhaled corticosteroids reduce the severity of bronchial hyperresponsiveness in asthma but oral theophylline does not. *Am Rev Respir Dis* 136: 1174–78, 1987.
94. Woolcock AJ, Yan K, Salome CM. Effect of therapy on bronchial hyperresponsiveness in the long-term management of asthma. *Clin Allergy* 18: 165–76, 1988.

95. Laitinen LA, Laitinen A, Haahtela T. A comparative study of the effects of an inhaled corticosteroid, budesonide, and of a β_2-agonist, terbutaline, on airway inflammation in newly diagnosed asthma. *J Allergy Clin Immunol* 90: 32–42, 1992.
96. Djukanovic R, Wilson JW, Britton YM *et al.* Effect of an inhaled corticosteroid on airway inflammation and symptoms of asthma. *Am Rev Respir Dis* 145: 669–74, 1992.
97. Trigg CJ, Manolitsas ND, Wang J *et al.* Placebo-controlled immunopathologic study of four months of inhaled corticosteroids in asthma. *Am J Respir Crit Care Med* 150: 17–22, 1994.
98. Holgate ST, Emanuel MB, Howarth PH. Astemizole and other H_1-antihistaminic drug treatment of asthma. *J Allergy Clin Immunol* 76: 375–80, 1985.
99. Rafferty P, Beasley R, Holgate S. The contribution of histamine to immediate bronchoconstriction provoked by inhaled allergen and adenosine 5' monophosphate in atopic asthma. *Am Rev Respir Dis* 136: 369–73, 1987.
100. Hamid M, Rafferty P, Holgate ST. The inhibitory effect of terfenadine and flurbiprofen on early and late-phase bronchoconstriction following allergen challenge in atopic asthma. *Clin Exp Allergy* 20: 261–67, 1990.
101. Wells A, Taylor B. A placebo-controlled trial of ketotifen (HC20-511, Sandoz) in allergen induced asthma and comparison with disodium cromoglycate. *Clin Allergy* 9: 237–40, 1979.
102. Pauwels R, Lamont H, Van Der Straeten M. Comparison between ketotifen and DSCG in bronchial challenge. *Clin Allergy* 8: 289–93, 1978.
103. Craps L, Greenwood C, Radielovic P. Clinical investigation of agents with prophylactic anti-allergic effects in bronchial asthma. *Clin Allergy* 8: 373–82, 1978.
104. Klein G, Urbanek R, Matthys H. Long-term study of the protective effect of ketotifen in children with allergic bronchial asthma. The value of a provocation test in assessment of treatment. *Respiration* 41: 128–32, 1981.
105. Adachi M, Kobayashi H, Aoki N *et al.* A comparison of the inhibitory effects of ketotifen and disodium cromoglycate on bronchial responses to house dust, with special reference to the late asthmatic response. *Pharmatherapeutics* 4: 36–42, 1984.
106. Cockcroft DW, Keshmiri M, Murdock KY *et al.* Allergen-induced increase in airway responsiveness is not inhibited by acute treatment with ketotifen or clemastine. *Ann Allergy* 68: 245–50, 1992.
107. Patel PC, Rutherford BC, Lux J *et al.* The effect of repirinast on airway responsiveness to methacholine and allergen. *J Allergy Clin Immunol* 90: 782–88, 1992.
108. Fish JE, Ankin MG, Adkinson NF *et al.* Indomethacin modification of immediate-type immunologic airway responses in allergic asthmatic and non-asthmatic subjects; evidence for altered arachidonic acid metabolism in asthma. *Am Rev Respir Dis* 123: 609–14, 1981.
109. Nakazawa T, Toyoda T, Furukawa M *et al.* Inhibitory effects of various drugs on dual asthmatic responses in wheat flour-sensitive subjects. *J Allergy Clin Immunol* 58: 1–9, 1976.
110. Fairfax AJ. Inhibition of the late asthmatic response to house dust mite by non-steroidal anti-inflammatory drugs. *Prostaglandins Leuko Med* 8: 239–48, 1982.
111. Manning PJ, Stevens WH, Cockcroft DW *et al.* The role of thromboxane in allergen-induced asthmatic responses. *Eur J Respir Dis* 4: 667–72, 1991.
112. Wilkens H, Wilkens JH, Bosse S *et al.* Effects of an inhaled PAF-antagonist (WEB 2086 BS) on allergen-induced early and late asthmatic responses and increased bronchial responsiveness to methacholine. *Am Rev Respir Dis* 143: A812, 1991.
113. Bianco S, Pieroni MG, Refini RM *et al.* Protective effect of inhaled furosemide on allergen-induced early and late asthmatic reactions. *New Engl J Med* 321: 1069–73, 1989.
114. Ortolani C, Pastorello E, Moss RB *et al.* Grass pollen immunotherapy: A single year double-blind, placebo-controlled study in patients with grass pollen-induced asthma and rhinitis. *J Allergy Clin Immunol* 73: 283–90, 1984.
115. Van Metre TE, Marsh DG, Adkinson NF *et al.* Immunotherapy for cat asthma. *J Allergy Clin Immunol* 82: 1055–68, 1988.
116. Warner JD, Soothill JF, Price JF *et al.* Controlled trial of hyposensitization to Dermatographoides pteronyssinus in children with asthma. *Lancet* 2: 912–15, 1978.
117. Boulet L-P, Chapman KR, Cote J *et al.* Inhibitory effects of an anti-IgE antibody E25 on allergen-induced early asthmatic response. *Am J Respir Crit Care Med* 155: 1835–40, 1997.
118. Fahy JV, Fleming HE, Wong HH *et al.* The effect of an anti-IgE monoclonal antibody on the early- and late-phase responses to allergen inhalation in asthmatic subjects. *Am J Respir Crit Care Med* 155: 1828–34, 1997.
119. Leckie MJ, ten Brinke A, Khan J *et al.* Effects of an interleukin-5 blocking monoclonal antibody on eosinophils, airway hyperresponsiveness, and the late asthmatic response. *Lancet* 356: 2144–48, 2000.
120. Bryan SA, O'Connor BJ, Matti S *et al.* Effects of recombinant human interleukin-12 on eosinophils, airway hyperresponsiveness, and the late asthmatic response. *Lancet* 356: 2149–53, 2000.
121. Lee TH, Brown MJ, Nagy L *et al.* Exercise-induced release of histamine and neutrophil chemotactic factor in atopic asthmatics. *J Allergy Clin Immunol* 70: 73–81, 1982.
122. Murray JJ, Tonnel AB, Brash AR *et al.* Release of prostaglandin D_2 into human airways during acute antigen challenge. *New Engl J Med* 315: 800–4, 1986.
123. Lewis RA, Austen KF. The biologically active leukotrienes: Biosynthesis, metabolism, receptors, functions and pharmacology. *J Clin Invest* 73: 889–97, 1984.
124. Schellenberg RR, Duff MJ, Foster A. Human bronchial responses to anti-IgE in vitro. *Clin Invest Med* 8: A41, 1985.
125. Dolovich J, Hargreave FE, Chalmers R *et al.* Late cutaneous allergic responses in isolated IgE-dependent reactions. *J Allergy Clin Immunol* 52: 38–46, 1973.
126. Solley GO, Gleich GJ, Jordon RE *et al.* The late phase of the immediate wheal and flare skin reactions. *J Clin Invest* 58: 408–20, 1976.
127. Kirby JG, Robertson DG, Hargreave FE *et al.* Asthmatic responses to inhalation of anti-human IgE. *Clin Allergy* 16: 191–94, 1986.
128. Schampain MP, Behrens BL, Larsen GL *et al.* An animal model of late pulmonary responses to Alternaria challenge. *Am Rev Respir Dis* 126: 493–98, 1982.
129. Abraham WM, Delehunt JC, Yerger L *et al.* Characterization of a late phase pulmonary response after antigen challenge in allergic sheep. *Am Rev Respir Dis* 128: 839–44, 1983.
130. Chung KF, Becker AB, Lazarus SC *et al.* Antigen-induced hyperresponsiveness and pulmonary inflammation in allergic dogs. *J Appl Physiol* 58: 1347–53, 1985.
131. O'Byrne PM, Walters EH, Gold ED *et al.* Neutrophil depletion inhibits airway hyperresponsiveness induced by ozone exposure. *Am Rev Respir Dis* 130: 214–19, 1984.
132. Murphy KR, Wilson MC, Irvin CG *et al.* The requirement for polymorphonuclear leukocytes in the late asthmatic response and heightened airways reactivity in an animal model. *Am Rev Respir Dis* 134: 62–68, 1986.
133. Metzger WJ, Richerson WB, Worden K *et al.* Bronchoalveolar lavage of allergic asthmatic patients following allergen bronchoprovocation. *Chest* 89: 477–83, 1986.
134. O'Byrne P. Asthma pathogenesis and allergen-induced late responses. *J Allergy Clin Immunol* 102: S85–89, 1998.
135. O'Byrne PM, Gauvreau GM, Wood LJ. Interaction between haemopoietic regulation and airway inflammation. *Clin Exp Allergy* 29(Suppl 2): 27–32, 1999.
136. Varga EM, Wachholz P, Nouri-Aria KT *et al.* T cells from human allergen-induced late asthmatic responses express IL-12 receptor beta 2 subunit mRNA and respond to IL-12 *in vitro*. *J Immunol* 165: 2877–85, 2000.
137. Boulet L-P, Becker A, Berube D, Beveridge R, Ernst P. Canadian Asthma Consensus Group. *CMAJ* 161: S1–61, 1999.

138. Nelson HS. The importance of allergens in the development of asthma and the persistence of symptoms. *J Allergy Clin Immunol* 105: S628–32, 2000.
139. Cookson WOCM, Musk AW, Ryan G. Association between asthma history, atopy, and non-specific bronchial responsiveness in young adults. *Clin Allergy* 16: 425–32, 1984.
140. Witt C, Stuckey MS, Woolcock AJ *et al.* Positive allergy prick skin tests associated with bronchial histamine responsiveness in an unselected population. *J Allergy Clin Immunol* 77: 698–702, 1986.
141. Peat JK, Britton WJ, Salome CM *et al.* Bronchial hyperresponsiveness in two populations of Australian school children: III. Effect of exposure to environmental allergens. *Clin Allergy* 17: 291–300, 1987.
142. Burney PFJ, Britton JR, Chinn S *et al.* Descriptive epidemiology of bronchial reactivity in an adult population: Results from a community study. *Thorax* 42: 38–44, 1987.
143. Cockcroft DW, Hargreave FE. Relationship between atopy and airway responsiveness. In: Sluiter HJ, van der Lende R (eds). *Bronchitis IV*, pp. 23–32. The Netherlands: Royal Vangorcum, Assen, 1988.
144. Peat JK, Salome CM, Woolcock AJ. Longitudinal changes in atopy during a 4-year period: Relation to bronchial hyperresponsiveness and respiratory symptoms in a population sample of Australian school children. *J Allergy Clin Immunol* 85: 65–74, 1990.
145. Sotomayor H, Badier M, Vervloet D *et al.* Seasonal increase of carbachol airway responsiveness in patients allergic to grass pollen. *Am Rev Respir Dis* 130: 56–58, 1984.
146. Dorward AJ, Roberts JA, Thomson NC. Effect of nedocromil sodium on histamine airway responsiveness in grass-pollen sensitive asthmatics during the pollen season. *Clin Allergy* 16: 309–15, 1986.
147. Platts-Mills TAE, Mitchell EB, Nock P *et al.* Reduction of bronchial hyperreactivity during prolonged allergen avoidance. *Lancet* 2: 675–78, 1982.
148. Becker AB, Hershkovich J, Simons FER *et al.* Development of chronic airway hyperresponsiveness in ragweed sensitized dogs. *J Appl Physiol* 66: 2691–97, 1989.
149. Chan-Yeung M, Lam S, Koener S. Clinical features and natural history of occupational asthma due to Western Red Cedar (*Thuja plicata*). *Am J Med* 72: 411–15, 1982.
150. Mapp CE, Corona PC, De Marzo N *et al.* Persistent asthma due to isocyanates: A follow-up study of subjects with occupational asthma due to toluene di-isocyanate (TDI). *Am Rev Respir Dis* 137: 1326–29, 1988.
151. Metcalfe DD. The diagnosis of food allergy: Theory and practice. In: Spector SL (ed.), *Provocative Challenge Procedures: Bronchial, Oral, Nasal, and Exercise*, pp. 119–32. Boca Raton, FL: CRC Press Inc., 1983.
152. Bock SA, Lee W-Y, Remigio LK *et al.* Studies of hypersensitivity reactions to foods in infants and children. *J Allergy Clin Immunol* 62: 327–34, 1978.
153. Orange RP, Donsky GJ. Anaphylaxis. In: Middleton E Jr., Reed CE, Ellis EF (eds). *Allergy Principles and Practice*, pp. 563–73. St.Louis, MO: The CV Mosby Company, 1978.
154. Gluck JC, Pacin MP. Asthma from mosquito bites: A case report. *Ann Allergy* 56: 492–93, 1986.
155. Malo JL, Hawkins R, Pepys J. Studies in chronic allergic bronchopulmonary aspergillosis 1: Clinical and physiological findings. *Thorax* 32: 254–61, 1977.
156. Malo JL, Pepys J, Simon G. Studies in chronic allergic bronchopulmonary aspergillosis 2: Radiological findings. *Thorax* 32: 262–68, 1977.
157. Malo JL, Longbottom J, Mitchell J *et al.* Studies in chronic allergic bronchopulmonary aspergillosis 3: Immunological findings. *Thorax* 32: 269–74, 1977.
158. Malo JL, Inouye T, Hawkins R *et al.* Studies in chronic allergic bronchopulmonary aspergillosis 4: Comparison with a group of asthmatics. *Thorax* 32: 275–80, 1977.
159. Dolan CT, Weed LA, Dines DE. Bronchopulmonary helminthosporiosis. *Am J Clin Pathol* 53: 235–42, 1970.
160. Matthiesson AM. Allergic bronchopulmonary disease caused by fungi other than Aspergillus. *Thorax* 36: 719, 1981.
161. McAleer R, Kroenert DB, Elder JL *et al.* Allergic bronchopulmonary disease caused by *Curvularia lunata* and *Drechslera hawaiiensis*. *Thorax* 36: 338–44, 1981.
162. Benatar SR, Allan B, Hewitson RP *et al.* Allergic bronchopulmonary stemphyliosis. *Thorax* 35: 515–18, 1980.
163. Voisin C, Tonnel AB, Jacob M *et al.* Infiltrats pulminaires avec grande eosinophilie sanguine associes a une candidose bronchique. *Rev Fr Allergie Immunol Clin* 16: 279–81, 1976.
164. Saini SK, Boas SR, Jerath A *et al.* Allergic bronchopulmonary mycosis to *Fusarium vasinfectum* in a child. *Ann Allergy Asthma Immunol* 80: 377–80, 1998.
165. Gordon DS, Hunter RG, O'Reilly RJ *et al. Pseudomonas aeruginosa* allergy and humoral antibody-mediated hypersensitivity pneumonia. *Am Rev Respir Dis* 108: 127–31, 1973.
166. Longbottom JL, Pepys J. Pulmonary aspergillosis: Diagnostic and immunologic significance of antigens and C-substance in *Aspergillus fumigatus*. *J Pathol Bacteriol* 88: 141–51, 1964.
167. Hendrick DJ, Davies RJ, D'Souza MF *et al.* An analysis of prick skin test reactions in 656 asthmatic patients. *Thorax* 30: 2–8, 1975.
168. Malo JL, Paquin R. Incidence of immediate sensitivity to *Aspergillus fumigatus* in a North American asthma population. *Clin Allergy* 9: 377–84, 1979.
169. Hoehne JH, Reed CE, Dickie HA. Allergic bronchopulmonary aspergillosis is not rare. *Chest* 63: 177–81, 1973.
170. Malo JL, Paquin R, Longbottom JL. Prevalence of precipitating antibodies to different extracts of *Aspergillus fumigatus* in a North American asthmatic population. *Clin Allergy* 11: 333–41, 1981.
171. McCarthy DS, Pepys J. Allergic bronchopulmonary aspergillosis. *Clin Allergy* 1: 261–86, 1971.
172. Patterson R, Fink JN, Pruzansky JJ *et al.* Serum immunoglobulin levels in pulmonary allergic aspergillosis and certain other lung disease, with special reference to immunoglobulin E. *Am J Med* 54: 16–22, 1973.
173. Scadding JG. The bronchi in allergic bronchopulmonary aspergillosis. *Scand J Respir Dis* 48: 372–77, 1967.
174. Panchal N, Bhagat R, Pant C, Shah A. Allergic bronchopulmonary aspergillosis: The spectrum of computed tomography appearances. *Respir Med* 91: 213–19, 1997.
175. Johkoh T, Muller NL, Akira M *et al.* Eosinophilic lung diseases: Diagnostic accuracy of thin-section CT in 111 patients. *Radiology* 216: 773–80, 2000.
176. Nepomuceno IB, Esrig S, Moss RB. Allergic bronchopulmonary aspergillosis in cystic fibrosis: Role of atopy and response to itraconazole. *Chest* 115: 364–70, 1999.
177. Geller DE, Kaplowitz H, Light MJ *et al.* Allergic bronchopulmonary aspergillosis in cystic fibrosis: Reported prevalence, regional distribution, and patient characteristics. Scientific Advisory Group, Investigators and Coordinators of the Epidemiologic Study of Cystic Fibrosis. *Chest* 116: 639–46, 1999.
178. Skov M, Koch C, Reimert CM *et al.* Diagnosis of allergic bronchopulmonary aspergillosis (ABPA) in cystic fibrosis. *Allergy* 55: 50–58, 2000.
179. Safirstein BH, D'Souza MF, Simon G *et al.* Five year follow-up of allergic bronchopulmonary aspergillosis. *Am Rev Respir Dis* 108: 450–59, 1973.
180. Wang JLF, Patterson R, Roberts M *et al.* The management of allergic bronchopulmonary aspergillosis. *Am Rev Respir Dis* 120: 87–92, 1979.
181. Salez F, Brichet A, Desurmont S *et al.* Effects of itraconazole therapy in allergic bronchopulmonary aspergillosis. *Chest* 116: 1665–68, 1999.
182. Wark P, Wilson AW, Gibson PG. Azoles for allergic bronchopulmonary aspergillosis. *Cochrane Database Syst Rev* 3: CD001108, 2000.
183. Dorward AJ, Colloff MJ, MacKay NS *et al.* Effect of house dust mite avoidance measures on adult atopic asthma. *Thorax* 43: 98–102, 1988.
184. Cockcroft DW, O'Byrne PM, Swystun VA *et al.* Regular use of inhaled albuterol and the allergen-induced late asthmatic response. *J Allergy Clin Immunol* 96: 44–49, 1995.

36

CHAPTER

Occupational Agents

Jean-Luc Malo[1], Moira Chan-Yeung[2] and Susan Kennedy[3]

[1]Department of Chest Medicine, Sacré-Coeur Hospital, Montreal, Canada
[2]Department of Medicine, Respiratory Division, University of British Columbia, Vancouver, BC, Canada
[3]School of Environmental Health, University of British Columbia, Vancouver, BC, Canada

OCCUPATIONAL ASTHMA

Definition

A proposed definition of occupational asthma (OA) should take into account at least two essential aspects of the condition. First, it should reflect the fact that OA shares the clinical, functional, and pathological features of asthma, that is variable, spontaneously or as a result of treatment, symptomatic airway caliber, hyperresponsiveness, and inflammation of the airways. Second, it should state that OA is caused by the workplace in which, in most cases, an agent has been identified. The definition of OA which has been retained in a textbook on OA reads as follows: "OA is a disease characterized by variable airflow limitation and/or airway hyperresponsiveness and/or inflammation due to causes and conditions attributable to a particular occupational environment and not to stimuli encountered outside the workplace" [1]. Within this definition, two types of OA are distinguished by whether they appear after or without a latency period.

Agents that cause OA and the pathophysiological mechanisms

Over 300 agents in the workplace have been implicated in causing asthma [2]. Table 36.1 shows a list of the more common agents responsible for OA. The agents can be divided into two groups according to the pathogenic mechanisms: those that give rise to asthma by immunological mechanisms and those by nonimmunological mechanisms. Agents in the former group can cause asthma by immunoglobulin (Ig)E-dependent or IgE-independent mechanisms. Irritant-induced asthma (IIA) is a type of OA caused by an apparently nonimmunological mechanism [3].

Agents that induce OA by immunological mechanisms

Agents that induce asthma by IgE-dependent mechanisms

The agents causing OA by IgE-dependent mechanisms include both high molecular weight compounds (>10 kDa) and some low molecular weight compounds (<10 kDa). High molecular weight compounds are usually proteins or polysaccharides and induce specific IgE antibodies. Some low molecular weight agents, such as platinum [4] and acid anhydrides [5], can also induce specific IgE antibodies by acting as haptens, which combine with a body protein to form complete antigens. There is evidence to support the fact that the asthmogenic potency of a compound is determined to a certain extent by its chemical structure [6, 7]. Often, agents such as trimellitic anhydride and isocyanates can modify body proteins, an important step toward becoming a hapten, and thus stimulate the immune system.

These agents often affect atopic subjects. The inhaled occupational sensitizer can bind to specific IgE antibodies on the surface of mast cells, basophils, and possibly macrophages and eosinophils. The mechanism of asthma induction by these sensitizers is similar to nonoccupational allergens [8].

Agents that induce asthma by IgE-independent mechanisms

The agents that give rise to OA through IgE-independent mechanisms are mostly low molecular weight compounds such as isocyanates and

TABLE 36.1 Agents that cause immunologically mediated occupational asthma.

	Common sources of exposure
High molecular weight agents	
Animal-derived material	Animal, poultry and insect work, veterinary medicine, fishing and fish processing, laboratory work
Dander	
Excreta	
Secretions	
Serum	
Plant-derived material	
Flour	Bakery
Grain	Grain elevator and terminal and feed mill
Castor bean	Oil manufacture
Coffee bean	Food processing
Wood dust	Sawmill, carpentry, furniture work
Vegetable gum	Printing
Psyllium	Health care
Latex	Latex
Enzymes	
α-amylase	Bakery
Papain	Food processing
Alcalase	Pharmaceutical industry
Bacillus subtilis-derived enzyme	Detergent enzyme industry
Low molecular weight agents	
Spray paints	
Toluene diisocyanate	Manufacture of plastic, foam
Dimethylphenyl diisocyanate	Insulation
Hexamethylene diisocyanate	Automobile spray paint
Wood dust	
Western red cedar	Sawmill worker, carpenter, furniture maker
Acid anhydride	Users of plastics, epoxy resins
Biocides	
Formaldehyde	Health care workers
Glutaraldehyde	
Chloramine T	
Colophny fluxes	Electronic workers
Irritant agents	
Chlorine	Pulp and paper mills
Acetic acid	Hospital setting
Isocyanates	Spray paint

For further information see Ref. [2].

plicatic acid (the agent responsible for Western red cedar asthma) [8]. In subjects with both isocyanate and Western red cedar asthma, specific IgE antibodies have been found in only a small proportion proven to have the disease [9] although, if present in high concentrations, they appear specific for the disease [10] though, apparently, not causal for the disease [11]. It is still uncertain whether the presence of specific IgG antibodies is a marker of the disease or merely of exposure. In Western red cedar asthma, the significance of the presence of specific IgE antibodies is not clear since anti-IgE antibodies failed to inhibit the release of histamine by plicatic acid from granulocytes of patients with the disease [12].

Studies of both isocyanate and Western red cedar asthma have demonstrated that T-lymphocytes might play a direct role in mediating the inflammatory response in the airways [13, 14]. In patients with isocyanate-induced asthma, the majority of T-cell clones derived from bronchial mucosa of subjects were found to be $CD8^+$ T-lymphocytes that produced IL-5 [13]. Among patients with Western red cedar asthma, peripheral blood lymphocytes released IL-5 and IFN-γ after stimulation with plicatic acid in sensitized subjects [14]. Animal studies have shown that diisocyanate-induced asthma is driven primarily by $CD4^+$ T-cells and is dependent upon the expression of Th2 cytokines. However, animal models are not always reflective of human responses [15].

A more recent study of bronchial biopsies taken after specific inhalation challenge testing from patients suspected of diisocyanate-induced asthma showed a striking absence of C and IL-4 mRNA-positive cells in bronchial biopsy specimens irrespective of whether the response was positive or negative indicating the absence of IgE involvement at least in a proportion of patients with diisocyanate-induced asthma [11].

Toluene diisocyanate was also found to induce production of proinflammatory cytokines and chemokines from bronchial epithelial cells through the epidermal growth factor receptor and p38 mitogen-activated protein kinase pathways [16].

Agents that induce OA by nonimmunological mechanism

Irritant-induced asthma is apparently a nonimmunologically induced type of asthma. There seems to be a dose-dependent relationship between exposure and likelihood of permanent disability/impairment [3, 17–19].

Pathology

Immunologically mediated OA

There is no difference between the pathology of the airways of patients with occupational and those with nonoccupational asthma. Studies of bronchoalveolar lavage (BAL) fluid have shown influx of eosinophils and neutrophils and a marked increase in albumin concentration during late asthmatic reaction after inhalation challenge with isocyanates [20]. In patients with Western red cedar asthma, increases in eosinophils and epithelial cells, as well as in albumin, histamine, and LTE4, have also been found in the BAL fluid [21]. Irrespective of the sensitizing agent, the pathological features in the airways are similar. There is subepithelial fibrosis, hypertrophy of airway smooth muscle, edema of the airway wall, accumulation of inflammatory cells, mostly eosinophils, and obstruction of the airway lumen by exudate and/or mucus in patients with OA [22]. Cessation of exposure to the sensitizing agent is associated with a decrease in the number of inflammatory cells in the airway mucosa [23]. In isocyanate-induced asthma, some reversal of the subepithelial fibrosis has been found [24] but airway eosinophilia and neutrophilia may persist. Studies of induced sputum of patients with OA after cessation of exposure for a mean interval of 8.7 years indeed showed that some subjects still had increased levels of neutrophils and eosinophils in induced sputum [25].

Nonimmunologically induced asthma

The mechanism of IIA is not known. IIA is a "big bang" phenomenon. Exposure to a high level of the irritant leads to acute sloughing of the epithelium [3]. There are some differences between the pathological features of IIA and those of immunologically mediated asthma. In general, subepithelial fibrosis is more evident in IIA [26]. In the acute phase, the airway epithelium is extensively damaged and the submucosa infiltrated by mononuclear cells [27]. One study reported fewer T-lymphocytes in the airway mucosa in IIA than in allergen-induced asthma [28].

Epidemiology: Frequency and determinants

Frequency

OA has become one of the two most common occupational lung diseases in developed countries (the other is mesothelioma) [29]. Estimates of incidence of OA have been made using registers based on mandated or voluntary physician reporting, medicolegal statistics, and various national or disability registers. There are considerable between-country differences in the estimated incidence of OA ranging from 22 per million per year in the United Kingdom to 187 per million per year in Finland. The differences are likely due in part to the methods used to derive these estimates. However, community-based studies on population-attributable risk (PAR) of occupational exposure for asthma carried out in countries involved in the European Community Respiratory Health Survey using similar methodology have shown considerable variation from 5% in Spain to 41% in New Zealand depending on the local industries, options for employment, and the population's susceptibility [30–32]. Studies in the United States of adult patients with new-onset asthma have shown that about 8% reported a history of exposure to sensitizers at work and 13% when irritants were included [33, 34]. The generally accepted PAR of occupational exposure for asthma is around 9–15% [35, 36].

Table 36.2 summarizes the findings from selected cross-sectional studies for various workplaces [37–46]. There is considerable variation in the prevalence of OA in different industries, ranging from 2% in latex-exposed workers to 50% among detergent-enzyme workers [37]. While the between-workforce differences can be due to varying data collection methodology, different definitions used for OA, and intensity of exposure, it is quite possible that the asthmogenic potential of the agents is different.

Few longitudinal studies have been carried out in high-risk apprenticeship programs including baking [47, 48] and students (bakers, animal health technicians and veterinarians, dental hygienists, welders, spray painters) exposed to various high [49, 50] and low [51, 52] molecular weight agents. This model is particularly interesting because it assesses young individuals who have not been previously exposed to the relevant

TABLE 36.2 Prevalence and determinants of work-related asthma: results of selected studies.

Exposure/Industry	No.	Latency (months)	Prevalence		
			Work-related asthma (%)	Smoking (%)	Atopy (%)
High molecular weight compounds					
Enzyme/detergent [37]	98	N/A	50	52	64
Clam/Shrimp [38]	59	N/A	26 (8)	49	21
Snow-crab processor [39]	303	number unknown	21 (16)	67	11
Laboratory animal worker [40]	238	26	6	30	40
Bakery [41]	344	26	6	57	34
Latex/Hospital workers [42]	289	120	2 (3)	22	25
Low molecular weight compounds					
Platinum refinery [43]	91	12–24	54	63	33
Anhydrides, TCPA[44]	329	Up to 24	3.2	50	22
Toluene diisocyanate [45]	241	Up to 36	9.5	51	35
Plicatic acid/Red cedar sawmills [46]	652	N/A	4	38	19

For further information see Ref. [1].

workplace allergens and can best evaluate the natural history of sensitization, symptoms, and disease even after beginning of exposure and even after removal of exposure.

Host determinants

Atopy (defined as positive skin test to one or more common allergens) has been shown to be associated with sensitization to some high molecular weight agents but not others (Table 36.2). The positive predictive value of atopy in various studies of OA is low: 34% in animal laboratory workers [53, 54] and 7% in psyllium workers [55]. Moreover, over 40% of young adults have positive skin-test reactions to common allergens. These findings do not justify routine screening for atopy in high-risk workplaces.

The relationship between smoking and OA is complex and findings of studies are often contradictory [56]. It appears that the effect of smoking is dependent on the type of occupational agent. An interaction between smoking and atopy has been found in OA in animal laboratory handlers [54] and in workers exposed to tetrachlorophthalic anhydride [54]; atopic smokers had the highest prevalence of sensitization and nonatopic nonsmokers, the lowest. Among platinum workers, smoking, not atopy, is the most important risk factor for sensitization [43]. Cigarette smoke probably acts as an adjuvant to promote IgE sensitization [57]. When agents cause OA by IgE-independent mechanisms, as in the cases of isocyanate-induced asthma and Western red cedar asthma, both atopy and smoking are unimportant [58].

Studies of genetics of OA has proven to be difficult because of the small numbers of patients with asthma due to one type of occupational agents and the difficulties of getting appropriate controls. Nevertheless, there have been some studies showing association between HLA class II genes and several types of OA such as isocyanate-[59], trimellitic anhydride-[60], Western red cedar-[61], and platinum-induced asthma [62]. However, such findings should be considered as preliminary [63] and this type of testing cannot be used for screening of susceptible subjects. Exposure to diisocyanates increases reactive oxygen species which are taken up by the glutathione S-transferase enzyme [64]. Individuals with the glutathione S-transferase M1 null gene had a two-fold increase risk of diisocyanate-induced asthma [65]. The combination of GSTM1 null and GTM3 AA phenotype was strongly associated with lack of diisocyanate-specific IgE antibodies and with late asthmatic reaction [65].

Exposure factors

Improved industrial hygiene techniques have resulted in the ability to measure several low molecular weight compounds such as isocyanates, formaldehyde, and amines [66]. Immunochemical techniques for quantitating aeroallergens have been developed and used in a number of workforce-based studies [67, 68].

Several studies have shown that there is a dose-response relationship between the level of exposure to occupational agents and the prevalence of sensitization and/or nonallergic bronchial hyperresponsiveness and/or asthma. These include exposure to high molecular weight allergens such as α-amylase [69], animal laboratory allergen [70], and low molecular weight compounds such as Western red cedar [71], acid anhydride [44], and colophony [72]. As a result, a "permissible exposure limit" has been proposed for a few occupational agents [67]. The minimum concentration of occupational allergen that causes sensitization may be one to two orders of magnitude greater than the concentration for eliciting symptoms. Once an individual is sensitized to an agent, a minute dose can trigger an attack of asthma. Concomitant environmental exposures, such as low levels of irritants and cigarette smoke, may enhance sensitization to some occupational agents. Further studies are required in this area.

Diagnosis

All asthmatic subjects should be questioned regarding possible exposure to causal agents in their current or previous workplaces. It has been estimated that approximately 5% of all asthmatics seen in a specialized asthma clinic may suffer from OA. Also, it has to be remembered that because affected subjects could be left with permanent asthma after removal from the workplace, their current asthma symptoms could result from previous exposure to occupational causal agents. Physicians should be aware that certain workplaces have a high risk of exposing subjects to asthma-causing agents. Information on workplaces at risk and on all agents causing OA is available on the web site ASMANET.COM as prepared and reviewed by Henriette Dhivert-Donnadieu and on ASTHMA-WORKPLACE.COM. Safety data sheets (SDS) on all products used in the workplace can be obtained from employers and/or from safety agencies. However, these SDS do not necessarily report all products present in the workplace. Although flour and isocyanates are still the most frequent causes of OA, many other high and low molecular weight agents emerge every year on the list of agents that often cause OA.

Clinical questionnaires should be sensitive although, in general, they are not very specific tools [73, 74]. Exposure to a known causal agent at work and the presence of asthma should be sufficient to alert the physician to the possibility of OA even though the temporal relationship between exposure and symptoms may seem discordant. The possibility of the presence of nasal or conjunctival symptoms should be addressed. Ocular and nasal symptoms often accompany or even precede, in the case of high molecular weight agents, the occurrence of OA. Moreover, ocular and nasal symptoms are good predictors of OA in the case of high molecular weight agents [74].

Skin-prick tests can be done with extracts derived from specific high molecular weight agents although these extracts are not generally standardized. The presence of immediate skin sensitivity only indicates the presence of sensitization and does not confirm the diagnosis of OA. The target organ, in this case the bronchial, should be shown to be hyperresponsive. This can be done through the assessment of nonallergic airway responsiveness. The combination of positive skin-test reaction to a relevant occupational allergen and nonallergic airway hyperresponsiveness in a subject means that there is an ~80% likelihood that he/she has OA [55]. Nonallergic airway hyperresponsiveness can be shown either by assessing the response to bronchodilators if there is airway obstruction, or by estimating the degree of bronchoconstrictive response to a pharmacological agent. If the worker is still employed, nonallergic airway responsiveness should be evaluated on a working day after a minimum period of 2 weeks at work. The absence of nonallergic airway hyperresponsiveness in a subject when still working and exposed to the agent(s) suspected of causing OA virtually excludes OA.

Serial measurement of peak expiratory flow rates (PEF) in the diagnosis and management of asthma [75, 76] has been used since the late 1970s in the investigation of OA [77]. Workers are asked to measure and register their PEF at least four times a day and to record their medication, symptoms, and whether they are at work or away from work. Although monitoring of PEF has been found to be a useful tool in the investigation of OA [77], there are several pitfalls including generally unsatisfactory compliance [78] and tracing, analysis and interpretation of results, for which computerized methods have been suggested. [79]. In our hands, while positive tracings are most likely associated with confirmed diagnosis, negative graphs cannot exclude OA [80]. Adding markers of airway inflammation (exhaled NO, induced sputum) for a period at work and off work improves the accuracy of diagnosis [81].

Exposing individuals to the potential causal agent(s) in a hospital laboratory, or at the workplace under careful supervision, is a good method to confirm OA. This method was first proposed in the 1970s by Professor Jack Pepys at the Brompton Hospital in London, UK [82]. However, such tests require the expertise of highly trained personnel and can only be done in specialized centers. They should be performed in a dose-response manner, exposing subjects to increasing concentrations but nonirritant levels of the agent with serial monitoring of FEV_1 following exposure for at least 8 h [83].

The investigation of OA is a stepwise procedure, as illustrated in Fig. 36.1.

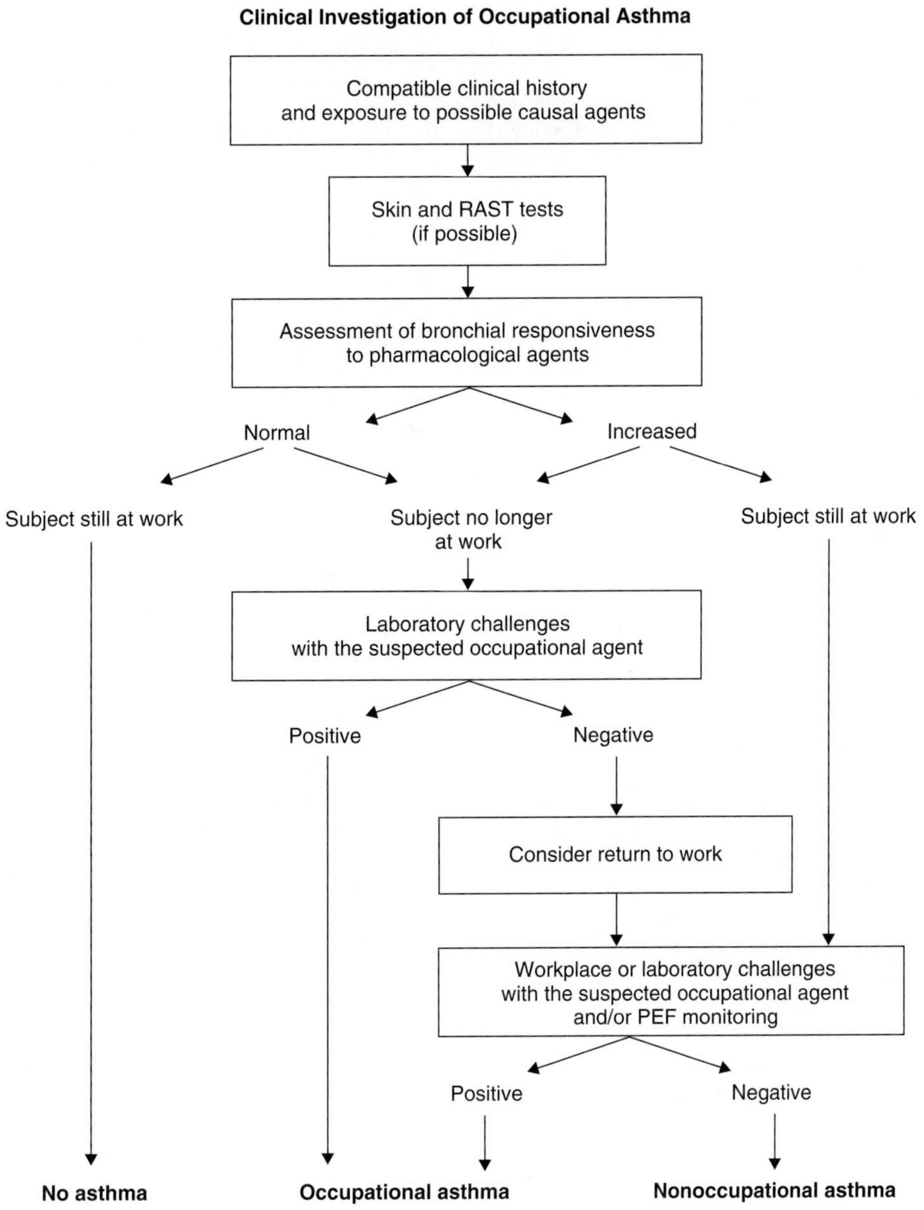

FIG. 36.1 Flow chart for the investigation of OA.

Management and compensation

Physicians may be asked to screen for OA in subjects exposed to known causal agents or in a workplace where cases of OA have been identified. These surveillance programs may include preemployment testing and periodic assessment. A cost-effectiveness examination of such programs needs to be done. Preemployment testing to document the baseline status may include a questionnaire, spirometry with assessment of nonallergic airway responsiveness, and skin testing to work-related allergens in the case of exposure to high molecular weight allergens. Workers with preexisting asthma should not be excluded as there is no reason to believe that asthmatic subjects are more likely to develop OA than anyone else. It is difficult to recommend the frequency for screening assessments. It has been estimated that 40% of subjects with OA due to low molecular weight agents develop their symptoms during the first year of exposure, while the corresponding figure for high molecular weight agents is 20% [84]. After that, there is a progressive reduction in the rate of development of OA.

Prevention is relevant in high-risk workplaces where inhalational accidents leading to IIA can occur. Assessment of airway responsiveness should also be considered as a preemployment test so that if an accident occurs comparisons are possible [85].

Once the diagnosis of OA is confirmed, subjects should be advised to avoid exposure to the causal agent. Whereas wearing cartridge masks does not apparently reduce symptomatology or functional abnormalities [86], helmet masks may eliminate exposure completely and may be considered in some instances. Treatment with inhaled steroids, while keeping the subject at work, improves the asthma but not to the extent of removal from exposure complemented by inhaled steroids [87], especially when steroids are given early after removal [88]. Subjects with OA whose symptoms persist after removal from exposure should be treated in the same way as patients with non-OA.

As for OA with a latency period, there might be functional recovery in the 2 years following the inhalation accident leading to IIA [89]. Steroids may be beneficial [90].

Patients with OA should be offered suitable help in finding another job where there is no further exposure to the causal agent, either with the same employer or another one. Subjects aged 55 or over should be offered early retirement, and young subjects should be retrained for a new job, all with financial compensation. Compensation boards or similar medicolegal agencies should offer these programs to workers and assess their cost-effectiveness. The time needed to make a diagnosis of OA and to implement a program is generally too long, causing hardship to the subjects [91, 92]. It has been estimated that during the 1990s, a single case of OA in Québec cost approximately C$50,000, a figure that is currently C$75,000 [91].

Because OA can lead to permanent impairment/disability, subjects should be reassessed periodically. The first assessment should take place 2 years after removal from exposure, when a plateau of improvement can occur, as in the case of a high molecular weight agent such as snow crab [93], although improvement can also occur at a later stage though at a slower pace [94]. The three main criteria for assessing impairment/disability for asthma are: (1) airway caliber; (2) airway responsiveness to a bronchodilator if airway obstruction is present or to a bronchoconstrictor if it is not the case; (3) the need for medication, which is a reflection of the clinical severity of asthma [95]. Adding evaluation of persistent airway inflammation may help in this assessment [96].

Conclusion

Asthma is a common respiratory occupational ailment. Whereas improvement in diagnostic tools and pathophysiological mechanisms still need to be considered, emphasis should be put on prevention programs through identification and application of permissible respirable levels and of affected subjects at an early stage of sensitization or disease to prevent long-term sequelae of permanent asthma.

COPD

In most patients, COPD is caused by chronic inhalation of polluted air, specifically, air contaminated by proinflammatory agents. The air can be polluted by personal choice (e.g. cigarette smoking), or by exposure sources in the ambient or work environment. Many patients may have been exposed to multiple sources, and the relative contributions of each may be impossible to disentangle. COPD is the symptomatic and functional consequence of chronic exposure to these polluted air sources. Therefore, it seems appropriate to pose a definition for occupational COPD that allows it to coexist with "smoking-induced COPD" or "urban air pollution-induced COPD." For the purpose of this section, we will define occupational COPD as "the existence of COPD in a patient with a history of chronic exposure to proinflammatory agents in workplace air." This definition is useful for clinical management and public health prevention as it points to the potential for eliminating or modifying occupational risk factors for the disease. Additional challenges with defining occupational COPD for legal or compensation purposes will be discussed below.

Epidemiology: Frequency and determinants

The general population prevalence of occupational COPD is difficult to estimate due to the lack of occupational exposure information in most population-based studies. After reviewing all available published evidence from general population epidemiologic studies, an *ad hoc* committee of the American Thoracic Society recently estimated the contribution of occupational exposure to the overall population burden of COPD to be at least 15% [97]. A 2001 study of 517 lifetime nonsmokers referred to a hospital pulmonary function laboratory estimated the PAR among nonsmokers from occupational exposure to be 23.6% for bronchitis and 29.6% for obstructive lung disease [98].

Occupational COPD is associated with chronic exposure to mineral and/or metal particulate matter and fumes, organic particulate matter, combustion products, irritant gases, and various combinations of these exposures. Evidence for this has emerged primarily from epidemiological studies, some of which are listed in Fig. 36.2.

Mineral particulate

Prevalence rates for COPD with airflow obstruction among miners range from 6% to 20% among nonsmokers, and up to 60% among smokers [104–107]. Exposure-response relationships are also seen. Airflow obstruction was seen among 10.5% of nonsmoking British coal miners exposed to low dust levels and 20.6% of those exposed to medium and high dust levels [106]. Similarly, among nonsmoking US underground coal miners the corresponding rates were 7.4% in the lower dust exposure category and 14.3% in the higher dust exposure category [108, 109]. Chronic bronchitis symptoms related to mining exposures are persistent even after cessation of exposure, as was seen among former Swedish iron ore miners after an average of 16 years of not being exposed [110].

Mine dust with greater crystalline silica content appears to produce even higher COPD rates. Among South African gold miners the estimated effect of dust exposure on airflow obstruction was approximately ten times greater than that seen among coal miners [107, 111]. COPD mortality is significantly increased (in a dose-related fashion) among workers exposed to silica, especially in the presence of silicosis [112, 113]. COPD has also been clearly demonstrated in association with asbestos exposure [114–118]. From the evidence to date, airflow obstruction appears more pronounced among workers installing or handling asbestos products than among miners [116]; however, only a few studies have looked specifically for airflow obstruction in these populations. Among miners and millers of wollastonite (another fibrous dust) significant dose-response relationships for cumulative dust and airflow obstruction were seen in both nonsmokers and smokers [119]. COPD is also linked to particulate exposure (with and without crystalline silica) in other industries, such as quarrying, carbon black manufacturing [120–122], and cement manufacturing [123].

Metal fumes, irritant gases, combustion products

Exposure to metal fumes, irritant gases, and combustion products appears to augment the effect of exposure to dust alone on COPD. This has been seen in mining and smelter workers [124, 125], workers in rubber manufacturing [126, 127], welders [128–132], tunnel workers [133], and fire fighters [134–137]. There is some evidence that the risk of airflow obstruction in welders and smelter workers may be higher among atopic workers, raising the possibility that at least some of the excess airflow obstruction seen in these groups may be related to asthma [130]. Indeed, as discussed above in the section on OA, several metal fumes and irritant gases are recognized risk factors for OA.

Organic dusts

Although organic dust exposure is also associated with asthma and hypersensitivity pneumonitis, there is increasing evidence that chronic respiratory symptoms and nonasthmatic airflow obstruction are caused in part by exposure to both "allergenic" and "nonallergenic" organic dusts. For example, a recent study of nonasthmatic cedar sawmill workers found that the annual decline in FVC was significantly related to cumulative

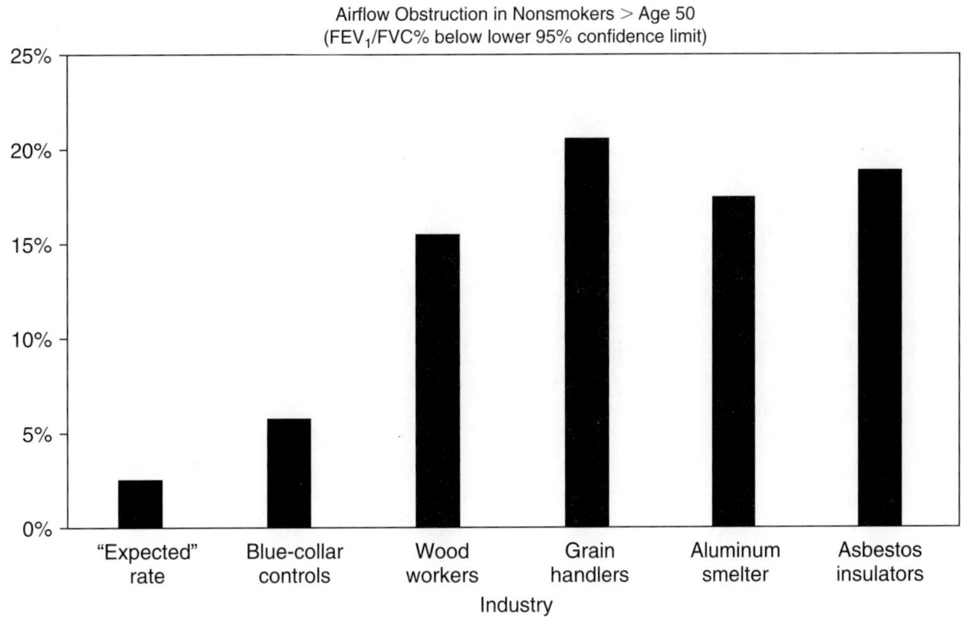

FIG. 36.2 Prevalence rates for airflow obstruction (defined as FEV_1/FVC % below the lower 95% confidence limit, based on age) among nonsmokers, aged 50 and over, exposed to mixed dusts and fumes and organic dusts, compared with a blue-collar control population [100] and to "expected rates" [101]; from studies conducted by the UBC Occupational Lung Diseases Research Unit [99, 102, 103].

dust exposure, even when the average dust exposure was well below the accepted limit [138]. Studies of sawmill and furniture workers exposed to wood dusts from species not currently known to cause asthma also show exposure-related airflow obstruction [139–144]. Generally, in these studies, the excess airflow obstruction linked to dust exposure is similar among nonsmokers and smokers.

There is also considerable evidence confirming a link between grain dust and COPD in both smokers and nonsmokers, with consistent dose-response relationships found [145–150]. Among retired grain workers, moderate to severe airflow obstruction was found in 40% of nonsmokers and 50% of smokers [151].

Mixed exposure to grains, other animal feeds, and animal by-products are found in animal confinement buildings, and there is growing evidence that exposure to these environments leads to COPD. There is clear evidence of excess chronic bronchitis among farmers and workers in poultry and swine confinement buildings, with prevalence rates remarkably consistent across studies at about 25–35% [99, 152–157]. Evidence for nonasthmatic chronic airflow obstruction in this industry is mixed [99, 157] although the majority of studies indicate reductions in airflow rates associated with duration of exposure [155, 158, 159]. COPD was found among 17% of 105 European farmers who had never smoked and who worked in animal confinement buildings, with 10% having moderate or severe disease by GOLD criteria. The highest disease prevalence was seen in farmers with the highest dust and endotoxin exposure levels [160].

Relationship between cigarette smoking and occupational exposures

The research indicates that occupational exposure plays a clinically important role in the development of COPD in exposed workers. Among aluminum smelter workers, the impact of a 30-year working career with particulate exposure at the permitted level was about the same as smoking, 75 g/week [161]. Among tunnel workers exposed to dust and diesel exhaust, the decline in FEV_1 associated with each year of tunnel work was twice that associated with each pack year of cigarette smoking [133]. Among fire fighters, the cumulative impact of "being a fire fighter" on longitudinal decline in FEV_1 was about half as strong as the impact of cigarette smoking [162].

Cigarette smoking and dust exposure appear to exert their effect on airways in a roughly additive fashion regardless of the dust type [106, 111, 138] although there is a suggestion that the effect is more than additive in patients with marked airflow obstruction, and in the presence of high crystalline silica [163].

Natural history

Most research suggests that the natural history of COPD is similar regardless of the source of the polluted air. However, there is growing evidence that there may be a link between early rapid decline in airflow rates due to occupational exposures and later airflow obstruction. This has been seen in British [106], US [108], and Italian [164] coal miners, among workers exposed to asbestos [165, 166], and grain [151, 156].

Assessment of exposure

The relevant exposure for COPD is that to which the patient was exposed in the past. Therefore, the best clinical tool for assessing this is a detailed occupational history, augmented by specific inquiry about exposures to dusts, gases, and fumes. In fact, research has shown that a positive response to the simple question "have you been exposed to dusts, gases, or fumes at work?" is linked to a more rapid decline in FEV_1 and increased prevalence of chronic bronchitis in population studies [167, 168]. Ideally, the occupational history should be reviewed by a professional with knowledge of the typical occupational exposures in the region. However, evidence has shown that the patient is also a reliable source of information about exposure duration and intensity. For each main job, the patient should be asked what year the job began and ended (to estimate duration of exposure); whether or not there was noticeable dust, fumes, or gas exposure; and if so, how often (to estimate intensity of exposure).

Although it is not possible to provide a clear answer to the question, "How much exposure is necessary before one should suspect an occupational contribution to COPD?" most research indicates that the relevant exposure duration is measured in years (or even decades), not months or days. Many patients will have held more than one job, and exposure duration should be summed over all jobs with relevant exposures. Time spent in irregular work environments should also be considered (e.g. in the armed forces, in prison, during extended periods of casual or temporary employment) as hazardous exposures are common in these situations. Dust and fume exposures are seldom present every day, all day, so patients should be asked whether exposure occurred most days, only a few times a month, or only seldom (i.e. a few days a year). Exposure that occurred at least a few times a month should be considered relevant.

Clinicians should be cautious about assuming that workplace exposures within "safe" limits were not high enough to produce disease. Many of the studies discussed above found significant airflow obstruction among workers exposed at or below regulated allowable limits. It is useful to consider the example of grain dust, which in many countries (including Canada and much of the United States) is still regulated with reference to the "nuisance dust" standard, despite overwhelming evidence that this standard is unacceptable. The American Conference of Governmental and Industrial Hygienists has adopted the term "particulate not otherwise classified" rather than nuisance dust, to indicate "that all materials are potentially toxic and to avoid the implication that these materials are harmless" and to emphasize that "although these materials may not cause fibrosis and systemic effects, they are not biologically inert" [169].

Occupational COPD among women

There is growing evidence that COPD mortality and morbidity is rising among women relative to men and it has been hypothesized that this may be linked to an increase in tobacco exposure or susceptibility among women [170, 171]. However, other evidence indicates that smoking rates among adult women have declined in parallel with men since the 1980s, at least in Western countries [172], suggesting that other factors must also be contributing to the apparent COPD increase among women. Environmental (or domestic occupational) exposure to biomass particulate matter (from indoor cooking) has been strongly linked to COPD among women in developing countries. The potential impact on COPD of women's work and changing occupational environments has only been explored in a few very recent studies, but there is some evidence of possible differential susceptibility to biological dusts among women. In a 20-year follow-up study of Chinese cotton textile workers, nonsmoking men demonstrated a small improvement in airflow rates after cessation of exposure (reduced decline in FEV_1), whereas no such improvement was seen among women, despite higher dust and endotoxin exposures among men [173]. Among a random sample of Australians, employment in a job with exposure to biological dust was a strong and a significant risk factor for COPD among women (sevenfold increase in risk); this effect was not seen in men (no increased risk) [174]. Jobs with biological dust exposure held by women in this study included health care professions, food and textile workers, artists, and cleaners. An analysis of data from the US National Health and Nutrition Examination Survey [175] indicated an increased risk for COPD, using GOLD criteria (compared to the population as a whole) among women in personal services, agriculture, textiles, rubber and plastics industries, and sales-related jobs. These very preliminary findings require further investigation to understand and quantify the potential role of occupational exposures in contributing to the growing prevalence of COPD among women. They do point, however, to the importance of considering occupational risks even among female patients with jobs not traditionally considered "dusty" or "industrial."

Management and compensation

When the clinician sees a patient (male or female) with chronic nonspecific airflow limitation (regardless of whether or not that patient is a smoker), it is useful to consider the possibility that occupational exposures may have contributed to the disease. This is important as continued occupational exposure may contribute to a worsening of the disease (in the patient, or in others in that workplace), if no preventive action is taken. For the patient still exposed to particulate or irritant gases or fumes, a recommendation should be made to reduce or eliminate the potential for the exposure. For the patient no longer exposed (or no longer working), it is important to record and report the disease as "potentially occupational" in order that public health officials can assess the role of the occupational exposure and use this information to direct prevention activities in the workplace.

Whether or not a patient with occupational COPD is entitled to compensation or disability, benefits will depend on the specific requirements of the compensation or insurance carrier. These requirements vary broadly. Disability evaluation should be carried out without reference to the question of etiology, as the level of disability or impairment is unaffected by the cause. The decision to attach the label "possibly occupational" to the diagnosis should also be based on the occupational history. Although requirements vary among jurisdictions, it should be the attending physician's role to raise the possibility of occupational etiology, not to make the final determination in law. General guidance can be found in the following statement from the BC Workers' Compensation Board policy manual: "Since workers' compensation ... operates on an enquiry basis rather than on an adversarial basis, there is no onus on the worker to prove his or her case. All that is needed is for the worker to describe his or her experience of the disease and the reasons why they suspect the disease has an occupational basis. Then it is the responsibility of the Board to research the available scientific literature and carry out any other investigations into the origin of the worker's condition which may be necessary."

That said, unfortunately, physicians do find themselves called upon by legal tribunals, to quantify the relative contribution of occupational exposure to COPD. This can be extremely difficult, if not impossible, in the absence of detailed exposure information and the expertise to interpret it. A useful benchmark may be the research among smelter workers, discussed earlier in this section, that found the effect of working for 30 years at today's "accepted" exposure concentration had an impact on airflow of about the same magnitude as a similar duration of cigarette smoking.

Prevention

Some evidence suggests that a rapid decline in FEV_1 (but still in the "normal" range) in young exposed workers may be associated with a worse prognosis. Therefore, although these workers are unlikely to seek treatment, if such a patient is encountered (e.g. as a result of routine screening or for other reasons) the index of concern should rise. The evidence is not strong enough to suggest that such a worker should be removed from exposure, but the patient should be made aware of his or her excess decline in lung function and the potential role of occupational exposure should be explored.

Global prevention of occupational COPD will require increased recognition of the disease (by physicians, by regulators, and by employers), and a willingness by regulators and employers to act to reduce exposure to particulates and irritant gases and fumes in the work environment. Primary care and pulmonary physicians can play a major role in prevention by increasing the recognition of the disease, by reporting it, even if just as "suspect" or "possible" occupational COPD, to the local agency responsible for occupational disease prevention.

References

1. Bernstein IL, Chan-Yeung M, Malo JL, Bernstein DI. Asthma in the Workplace, 3rd edn. New York, NY: Taylor & Francis, 2006.
2. Malo JL, Chan-Yeung M. Agents causing occupational asthma with key references. Appendix. In: Bernstein IL, Chan-Yeung M, Malo JL, Bernstein DI (eds). *Asthma in the Workplace*, 825–66. New York: Taylor & Francis Inc, 2006.

3. Gautrin D, Bernstein IL, Brooks SM, Henneberger PK. Reactive airways dysfunction syndrome and irritant-induced asthma. In: Bernstein IL, Chan-Yeung M, Malo JL, Bernstein DI (eds). *Asthma in the Workplace*, 3rd edn, 579–627. New York, NY: Taylor & Francis, 2006.
4. Patterson R, Zeiss CR, Roberts M, Pruzansky JJ. Human antihapten antibodies in trimellitic anhydride inhalation reactions. *J Clin Invest* 62: 971–78, 1978.
5. Pepys J. Occupational allergy due to platinum complex salts. *Clin Immunol Allergy* 4(1): 131–58, 1984.
6. Graham C, Rosenkranz HS, Karol MH. Structure-activity model of chemicals that cause human respiratory sensitization. *Regul Toxicol Pharmacol* 26: 296–306, 1997.
7. Jarvis J, Seed MJ, Elton R, Sawyer L, Agius R. Relationship between chemical structure and the occupational asthma hazard of low molecular weight organic compounds. *Occup Environ Med* 62: 243–50, 2005.
8. Maestrelli P, Fabbri LM, Mapp CE. Pathophysiology. In: Bernstein IL, Chan-Yeung M, Malo JL, Bernstein DI (eds). *Asthma in the Workplace*, 3rd edn, 109–40. New York: Taylor & Francis, 2006.
9. Cartier A, Grammer L, Malo JL, Lagier F, Ghezzo H, Harris K, Patterson R. Specific serum antibodies against isocyanates: Association with occupational asthma. *J Allergy Clin Immunol* 84: 507–14, 1989.
10. Tee RD, Cullinan P, Welch J, Burge PS, Newman-Taylor AJ. Specific IgE to isocyanates: A useful diagnostic role in occupational asthma. *J Allergy Clin Immunol* 101: 709–15, 1998.
11. Jones MG, Floyd A, Nouri-Aria KT, Jacobson MR, Durham SR, Taylor AN. Is occupational asthma to diisocyanates a non-IgE-mediated disease? *J Allergy Clin Immunol* 117: 663–69, 2006.
12. Frew A, Chan H, Dryden P, Salari H, Lam S, Chan-Yeung M. Immunologic studies of the mechanisms of occupational asthma caused by western red cedar. *J Allergy Clin Immunol* 92: 466–78, 1993.
13. Maestrelli P, DelPrete GF, DeCarli M, Saetta M, DiStefano A, Ricci M, Romagnani S, Fabbri LM. Activated CD8 T lymphocytes producing interferon-gamma (IFN-gamma) and interleukin-5 (IL-5) in bronchial mucosa of subjects sensitized to toluene diisocyanate (TDI). *Scand J Work Environ Health* 20: 376–81, 1999.
14. Frew A, Chang JH, Chan H, Quirce S, Noertjojo K, Keown P, Chan-Yeung M. T-lymphocyte responses to plicatic acid-human serum albumin conjugate in occupational asthma caused by western red cedar. *J Allergy Clin Immunol* 101: 841–47, 1998.
15. Mapp CE, Boschetto P, Miotto D, de RE. Asthma induced by isocyanates: A model of IgE independent asthma. *Acta Biomed* 76(Suppl 2): 15–19, 2005.
16. Ogawa H, Inoue S, Ogushi F, Ogura H, Nakamura Y. Toluene diisocyanate (TDI) induces production of inflammatory cytokines and chemokines by bronchial epithelial cells via the epidermal growth factor receptor and p38 mitogen-activated protein kinase pathways. *Exp Lung Res* 32: 245–62, 2006.
17. Kennedy SM, Enarson DA, Janssen RG, Chan-Yeung M. Lung health consequences of reported accidental chlorine gas exposures among pulpmill workers. *Am Rev Respir Dis* 143: 74–79, 1991.
18. Kern DG. Outbreak of the reactive airways dysfunction syndrome after a spill of glacial acetic acid. *Am Rev Respir Dis* 144: 1058–64, 1991.
19. Bhérer L, Cushman R, Courteau JP, Quévillon M, Côté G, Bourbeau J, L'Archevêque J, Cartier A, Malo JL. Survey of construction workers repeatedly exposed to chlorine over a three to six month period in a pulpmill: II. Follow up of affected workers by questionnaire, spirometry and assessment of bronchial responsiveness 18 to 24 months after exposure. *Occup Environ Med* 51: 225–28, 1994.
20. Saetta M, Di Stefano A, Maestrelli P, De Marzo N, Milani GF, Pivirotto F, Mapp CE, Fabbri LM. Airway mucosal inflammation in occupational asthma induced by toluene diisocyanate. *Am Rev Respir Dis* 145: 160–68, 1992.
21. Lam S, LeRiche J, Phillips D, Chan-Yeung M. Cellular and protein changes in bronchial lavage fluid after late asthmatic reaction in patients with red cedar asthma. *J Allergy Clin Immunol* 80: 44–50, 1987.
22. Boulet LP, Boutet M, Laviolette M, Dugas M, Milot J, Leblanc C, Paquette L, Côté J, Cartier A, Malo JL. Airway inflammation after removal from the causal agent in occupational asthma due to high and low molecular weight agents. *Eur Respir J* 7: 1567–75, 1994.
23. Chan-Yeung M, Leriche J, Maclean L, Lam S. Comparison of cellular and protein changes in bronchial lavage fluid of symptomatic and asymptomatic patients with red cedar asthma on follow-up examination. *Clin Exp Allergy* 18: 359–65, 1988.
24. Saetta M, Maestrelli P, Turato G, Mapp CE, Milani G, Pivirotto F, Fabbri LM, DiStefano A. Airway wall remodeling after cessation of exposure to isocyanates in sensitized asthmatic subjects. *Am J Respir Crit Care Med* 151: 489–94, 1995.
25. Maghni K, Lemière C, Ghezzo H, Yuquan W, Malo JL. Airway inflammation after cessation of exposure to agents causing occupational asthma. *Am J Respir Crit Care Med* 169: 367–72, 2004.
26. Gautrin D, Boulet LP, Boutet M, Dugas M, Bhérer L, L'Archevêque J, Laviolette M, Côté J, Malo JL. Is reactive airways dysfunction syndrome a variant of occupational asthma? *J Allergy Clin Immunol* 93: 12–22, 1994.
27. Lemière C, Malo JL, Boulet LP, Boutet M. Reactive airways dysfunction syndrome induced by exposure to a mixture containing isocyanate: Functional and histopathologic behaviour. *Allergy* 51: 262–65, 1996.
28. Chan-Yeung M, Lam S, Kennedy SM, Frew AJ. Persistent asthma after repeated exposure to high concentrations of gases in pulpmills. *Am J Respir Crit Care Med* 149: 1676–80, 1994.
29. McDonald JC, Keynes HL, Meredith SK. Reported incidence of occupational asthma in the United Kingdom, 1989–97. *Occup Environ Med* 57: 823–29, 2000.
30. Kogevinas M, Anto JM, Soriano JB, Tobias A, Burney P. The risk of asthma attributable to occupational exposures. *Am J Respir Crit Care Med* 154: 137–43, 1996.
31. Kogevinas M, Anto JM, Sunyer J, Tobias A, Kromhout H, Burney P Group and the European Community Respiratory Health Survey Study. Occupational asthma in Europe and other industrialised areas: A population-based study. *Lancet* 353: 1750–54, 1999.
32. Fishwick D, Pearce N, D'Souza W, Lewis S, Town I, Armstrong R, Kogevinas M, Crane J. Occupational asthma in New Zealanders: A population based study. *Occup Environ Med* 54: 301–6, 1997.
33. Blanc PD, Cisternas M, Smith S, Yelin E. Occupational asthma in a community-based survey of adult asthma. *Chest* 109: 56S–57S, 1996.
34. Blanc PD, Eisner MD, Israel L, Yelin EH. The association between occupation and asthma in general medical practice. *Chest* 115: 1259–64, 1999.
35. Blanc PD, Toren K. How much asthma can be attributed to occupational factors? *Am J Med* 107: 580–87, 1999.
36. American Thoracic Society. Occupational contribution to the burden of airway disease. *Am J Respir Crit Care Med* 167: 787–97, 2003.
37. Mitchell CA, Gandevia B. Respiratory symptoms and skin reactivity in workers exposed to proteolytic enzymes in the detergent industry. *Am Rev Respir Dis* 104: 1–12, 1971.
38. Desjardins A, Malo JL, L'Archevêque J, Cartier A, McCants M, Lehrer SB. Occupational IgE-mediated sensitization and asthma due to clam and shrimp. *J Allergy Clin Immunol* 96: 608–17, 1995.
39. Cartier A, Malo JL, Forest F, Lafrance M, Pineau L, St-Aubin JJ, Dubois JY. Occupational asthma in snow crab-processing workers. *J Allergy Clin Immunol* 74: 261–69, 1984.
40. Cullinan P, Lowson D, Nieuwenshuijsen MJ, Sandiford C, Tee RD, Venables KM, McDonald JC, Newman Taylor AJ. Work related symptoms, sensitisation, and estimated exposure in workers not previously exposed to flour. *Occup Environ Med* 51: 579–83, 1994.
41. Cullinan P, Lowson D, Nieuwenhuijsen MJ, Gordon S, Tee RD, Venables KM, McDonald JC, Newman Taylor AJ. Work related symptoms, sensitisation, and estimated exposure in workers not previously exposed to laboratory rats. *Occup Environ Med* 51: 589–92, 1994.
42. Vandenplas O, Delwich JP, Evrard G, Aimont P, van der Brempt X, Jamart J, Delaunois L. Prevalence of occupational asthma due to latex among hospital personnel. *Am J Respir Crit Care Med* 151: 54–60, 1995.

43. Venables KM, Dally MB, Nunn AJ, Stevens JF, Stephens R, Farrer N, Hunter JV, Stewart M, Hughes EG, Newman Taylor AJ. Smoking and occupational allergy in workers in a platinum refinery. *BMJ* 299: 939–42, 1989.
44. Venables KM, Newman Taylor AJ. Exposure-response relationships in asthma caused by tetrachlorophthalic anhydride. *J Allergy Clin Immunol* 85: 55–58, 1990.
45. Venables KM, Dally MB, Burge PS, Pickering CAC, Newman Taylor AJ. Occupational asthma in a steel coating plant. *Br J Ind Med* 42: 517–24, 1985.
46. Chan-Yeung M, Vedal S, Kus J, Maclean L, Enarson D, Tse KS. Symptoms, pulmonary function, and bronchial hyperreactivity in Western Red Cedar workers compared with those in office workers. *Am Rev Respir Dis* 130: 1038–41, 1984.
47. De Zotti R, Bovenzi M. Prospective study of work related respiratory symptoms in trainee bakers. *Occup Environ Med* 57: 58–61, 2000.
48. Walusiak J, Hanke W, Gorski P, Palczynski C. Respiratory allergy in apprentice bakers: Do occupational allergies follow the allergic march? *Allergy* 59: 442–50, 2004.
49. Gautrin D, Ghezzo H, Infante-Rivard C, Malo JL. Incidence and determinants of IgE-mediated sensitization in apprentices: A prospective study. *Am J Respir Crit Care Med* 162: 1222–28, 2000.
50. Malo JL, Chan YM. Occupational asthma. *J Allergy Clin Immunol* 108: 317–28, 2001.
51. Dragos DMC, Malo JL, Gautrin D. Specific IgG and IgE to hexamethylene diisocyanate (HDI) among car-painting apprentices at risk of developing occupational asthma. *Am J Respir Crit Care Med* 169: A647, 2004.
52. El-Zein M, Malo JL, Infante-Rivard C, Gautrin D. Incidence of probable occupational asthma and of changes in airway calibre and responsiveness in apprentice welders. *Eur Respir J* 22: 513–18, 2003.
53. Slovak AJ, Hill RN. Laboratory animal allergy: A clinical survey of an exposed population. *Br J Ind Med* 38: 38–41, 1981.
54. Venables KM, Upton JL, Hawkins ER, Tee RD, Longbottom JL, Newman-Taylor AJ. Smoking, atopy and laboratory animal allergy. *Br J Ind Med* 45: 667–71, 1988.
55. Malo JL, Cartier A, L'Archevêque J, Ghezzo H, Lagier F, Trudeau C, Dolovich J. Prevalence of occupational asthma and immunologic sensitization to psyllium among health personnel in chronic care hospitals. *Am Rev Respir Dis* 142: 1359–66, 1990.
56. Siracusa A, Marabini A, Folletti I, Moscato G. Smoking and occupational asthma. *Clin Exp Allergy* 36: 577–84, 2006.
57. Nielsen GD, Olsen O, Larsen ST, Lovik M, Poulsen LK, Glue C, Brandorff NP, Nielsen PJ. IgE-mediated sensitisation, rhinitis and asthma from occupational exposures. Smoking as a model for airborne adjuvants?. *Toxicology* 216: 87–105, 2005.
58. Mapp CE, Saetta M, Maestrelli P, Ciacia A, Fabbri LM. Low molecular weight pollutants and asthma: Pathogenetic mechanisms and genetic factors. *Eur Respir J* 7: 1559–63, 1994.
59. Bignon JS, Aron Y, Ju LY, Kopferschmitt MC, Garnier R, Mapp C, Fabbri LM, Pauli G, Lockhart A, Charron D, Swierczewski E. HLA class II alleles in isocyanate-induced asthma. *Am J Respir Crit Care Med* 149: 71–75, 1994.
60. Young RP, Barker RD, Pile KD, Cookson WOCM, Newman TAJ. The association of HLA DR3 with specific IgE to inhaled acid anhydrides. *Am J Respir Crit Care Med* 151: 219–21, 1995.
61. Horne C, Quintana PJE, Keown PA, Dimich-Ward H, Chan-Yeung M. Distribution of HLA class II DQB1 alleles in patients with occupational asthma due to Western red cedar. *Eur Respir J* 15: 911–14, 2000.
62. Newman Taylor AJ, Cullinan P, Lympany PA, Harris JM, Dowdeswell RJ, DuBois R. Interaction of HLA phenotype and exposure intensity in sensitization to complex platinum salts. *Am J Respir Crit Care Med* 160: 435–38, 1999.
63. Newman Taylor AJ, Yucesov B. Genetics and occupational asthma. In: Bernstein IL, Chan-Yeung M, Malo JL, Bernstein DI (eds). *Asthma in the Workplace*, 3rd edn, pp. 87–108. New York, NY: Taylor & Francis, 2006.
64. Lange RW, Day BW, Lemus R, Tyurin VA, Kagan VE, Karol MH. Intracellular *S*-glutathionyl adducts in murine lung and human bronchoepithelial cells after exposure to diisocyanatotoluene. *Chem Res Toxicol* 12: 931–36, 1999.
65. Piirila P, Wikman H, Luukkonen R, Kaaria K, Rosenberg C, Nordman H, Norppa H, Vainio H, Hirvonen A. Glutathione *S*-transferase genotypes and allergic responses to diisocyanate exposure. *Pharmacogenetics* 11: 437–45, 2001.
66. Lesage J, Perrault G. Environmental monitoring of chemical agents. In: Bernstein IL, Chan-Yeung M, Malo JL, Bernstein DI (eds). *Asthma in the Workplace*, 3rd edition. 297–318. New York: Taylor & Francis, 2006.
67. Nieuwenhuijsen M, Baur X, Heederik D. Environmental monitoring: General considerations, exposure-response relationships, and risk assessment. In: Bernstein IL, Chan-Yeung M, Malo JL, Bernstein DI (eds). *Asthma and the Workplace*, 3rd edition. 253–74. New York: Taylor & Francis, 2006.
68. Swanson MC, Heederik D. Quantification of bio-active protein aerosols. In: Bernstein IL, Chan-Yeung M, Malo JL, Bernstein DI (eds). *Asthma in the Workplace*, 3rd edition. 275–96. New York: Taylor & Francis, 2006.
69. Houba R, Heederik DJJ, Doekes G, van Run PEM. Exposure-sensitization relationship for α-amylase allergens in the baking industry. *Am J Respir Crit Care Med* 154: 130–36, 1996.
70. Heederik D, Venables KM, Malmberg P, Hollander A, Karlsson AS, Renstrom A, Doekes G, Nieuwenhijsen M, Gordon S. Exposure-response relationships for work-related sensitization in workers exposed to rat urinary allergens: Results from a pooled study. *J Allergy Clin Immunol* 103: 678–84, 1999.
71. Vedal S, Chan-Yeung M, Enarson D, Fera T, Maclean L, Tse KS, Langille R. Symptoms and pulmonary function in western red cedar workers related to duration of employment and dust exposure. *Arch Environ Health* 41: 179–83, 1986.
72. Burge PS, Edge G, Hawkins R, White V, Taylor AN. Occupational asthma in a factory making flux-cored solder containing colophony. *Thorax* 36: 828–34, 1981.
73. Malo JL, Ghezzo H, L'Archevêque J, Lagier F, Perrin B, Cartier A. Is the clinical history a satisfactory means of diagnosing occupational asthma? *Am Rev Respir Dis* 143: 528–32, 1991.
74. Vandenplas O, Ghezzo H, Munoz X, Moscato G, Perfetti L, Lemière C, Labrecque M, L'Archevêque J, Malo JL. What are the questionnaire items most useful in identifying subjects with occupational asthma? *Eur Respir J* 26: 1056–63, 2005.
75. Epstein SW, Fletcher CM, Oppenheimer EA. Daily peak flow measurements in the assessment of steroid therapy for airway obstruction. *BMJ* 1: 223–25, 1969.
76. Turner-Warwick M. On observing patterns of airflow obstruction in chronic asthma. *Br J Dis Chest* 71: 73–86, 1977.
77. Burge PS, Moscato G, Johnson A, Chan-Yeung M. Physiological assessment: Serial measurements of lung function and bronchial responsiveness. In: Bernstein IL, Chan-Yeung M, Malo JL, Bernstein DI (eds). *Asthma in the Workplace*, 3rd edn, 199–226. New York, NY: Francis & Taylor, 2006.
78. Malo JL, Trudeau C, Ghezzo H, L'Archevêque J, Cartier A. Do subjects investigated for occupational asthma through serial PEF measurements falsify their results? *J Allergy Clin Immunol* 96: 601–7, 1995.
79. Burge PS, Pantin CFA, Newton DT, Gannon PFG, Bright P, Belcher J, McCoach J, Baldwin DR, Burge CBSG. Development of an expert system for the interpretation of serial peak expiratory flow measurements in the diagnosis of occupational asthma. *Occup Environ Med* 56: 758–64, 1999.
80. Leroyer C, Perfetti L, Trudeau C, L'Archevêque J, Chan-Yeung M, Malo JL. Comparison of serial monitoring of peak expiratory flow and FEV_1 in the diagnosis of occupational asthma. *Am J Respir Crit Care Med* 158: 827–32, 1998.
81. Girard F, Chaboillez S, Cartier A, Côté J, Hargreave FE, Labrecque M, Malo JL, Tarlo S, Lemière C. An effective strategy for diagnosing occupational asthma. Induced sputum. *Am J Respir Crit Care Med* 170: 845–50, 2004.

82. Pepys J, Hutchcroft BJ. Bronchial provocation tests in etiologic diagnosis and analysis of asthma. *Am Rev Respir Dis* 112: 829–59, 1975.
83. Vandenplas O, Malo JL. Inhalation challenges with agents causing occupational asthma. *Eur Respir J* 10: 2612–29, 1997.
84. Malo JL, Ghezzo H, D'Aquino C, L'Archevêque J, Cartier A, Chan-Yeung M. Natural history of occupational asthma: Relevance of type of agent and other factors in the rate of development of symptoms in affected subjects. *J Allergy Clin Immunol* 90: 937–44, 1992.
85. Leroyer C, Malo JL, Infante-Rivard C, Dufour JG, Gautrin D. Changes in airway function and bronchial responsiveness after acute occupational exposure to chlorine leading to treatment in a first-aid unit. *Occup Environ Med* 55: 356–59, 1998.
86. Côté J, Kennedy S, Chan-Yeung M. Outcome of patients with cedar asthma with continuous exposure. *Am Rev Respir Dis* 141: 373–76, 1990.
87. Moscato G, Dellabianca A, Perfetti L, Brame B, Galdi E, Niniano R, Paggiaro P. Occupational asthma. A longitudinal study on the clinical and socioeconomic outcome after diagnosis. *Chest* 115: 249–56, 1999.
88. Malo JL, Cartier A, Côté J, Milot J, Leblanc C, Paquette L, Ghezzo H, Boulet LP. Influence of inhaled steroids on the recovery of occupational asthma after cessation of exposure: An 18-month double-blind cross-over study. *Am J Respir Crit Care Med* 153: 953–60, 1996.
89. Malo JL, Cartier A, Boulet LP, L'Archevêque J, Saint-Denis F, Bhérer L, Courteau JP. Bronchial hyperresponsiveness can improve while spirometry plateaus two to three years after repeated exposure to chlorine causing respiratory symptoms. *Am J Respir Crit Care Med* 150: 1142–45, 1994.
90. Demnati R, Fraser R, Martin JG, Plaa G, Malo JL. Effects of dexamethasone on functional and pathological changes in rat bronchi caused by high acute exposure to chlorine. *Toxicol Sci* 45: 242–46, 1998.
91. Dewitte JD, Chan-Yeung M, Malo JL. Medicolegal and compensation aspects of occupational asthma. *Eur Respir J* 7: 969–80, 1994.
92. Ameille J, Parion JC, Bayeux MC, Brochard P, Choudat D, Conso F, Devienne A, Garnier R, Iwatsubo Y. Consequences of occupational asthma on employment and financial status: A follow-up study. *Eur Respir J* 10: 55–58, 1997.
93. Malo JL, Cartier A, Ghezzo H, Lafrance M, Mccants M, Lehrer SB. Patterns of improvement of spirometry, bronchial hyperresponsiveness, and specific IgE antibody levels after cessation of exposure in occupational asthma caused by snow-crab processing. *Am Rev Respir Dis* 138: 807–12, 1988.
94. Malo JL, Ghezzo H. Recovery of methacholine responsiveness after end of exposure in occupational asthma. *Am J Respir Crit Care Med* 169: 1304–7, 2004.
95. American Medical Association. *Guides to the Evaluation of Permanent Impairment*, 5th edn, Cocchiarella L, Andersson GBJ (eds). Chicago, IL, 2001.
96. Yacoub MR, Lavoie K, Lacoste G, Daigle S, L'Archevêque J, Ghezzo H, Lemière C, Malo JL. Assessment of impairment/disability due to occupational asthma through a multidimensional approach. *Eur Respir J* 29:889–96, 2007.
97. Balmes J, Becklake M, Blanc P *et al.* American thoracic society statement on occupational contribution to the burden of airway disease. *Am J Respir Crit Care Med* 167: 787–97, 2003.
98. Mak GK, Gould MK, Kuschner WG. Occupational inhalant exposure and respiratory disorders among never-smokers referred to a hospital pulmonary function laboratory. *Am J Med Sci* 322: 121–26, 2001.
99. Holness DL, O'Blenis EL, Sass-Kortsak AM, Pilger CW, Nethercott JR. Respiratory effects and dust exposures in hog confinement farming. *Am J Ind Med* 11: 571–80, 1987.
100. Kennedy SM, Chan-Yeung M, Marion S, Lea J, Teschke K. Maintenance of stellite and tungsten carbide saw tips: Respiratory health and exposure-response evaluations. *Occup Environ Med* 52: 185–91, 1995.
101. Crapo RO, Morris AH, Gardner RM. Reference spirometric values using techniques and equipment that meet ATS recommendations. *Am Rev Respir Dis* 123: 659–64, 1981.
102. Chan-Yeung M, Enarson DA, MacLean L, Irving D. Longitudinal study of workers in an aluminum smelter. *Arch Environ Health* 44: 134–39, 1989.
103. Kennedy SM, Vedal S, Muller N, Kassam A, Chan-Yeung M. Lung function and chest radiograph abnormalities among construction insulators. *Am J Ind Med* 20: 673–84, 1991.
104. Fairman RP, O'Brien RJ, Swecker S, Amandus HE, Shoub EP. Respiratory status of surface coal miners in the United States. *Arch Environ Health* 32: 211–15, 1977.
105. Kibelstis JA, Morgan EJ, Reger R, Lapp NL, Seaton A, Morgan WKC. Prevalence of bronchitis and airway obstruction in American bituminous coal miners. *Am Rev Respir Dis* 108: 886–93, 1973.
106. Marine WM, Gurr D, Jacobsen M. Clinically important respiratory effects of dust exposure and smoking in British coal miners. *Am Rev Respir Dis* 137: 106–12, 1988.
107. Oxman AD, Muir DC, Shannon HS, Stock SR, Hnizdo E, Lange HJ. Occupational dust exposure and chronic obstructive pulmonary disease. A systematic overview of the evidence (see comments). *Am Rev Respir Dis* 148: 38–48, 1993.
108. Seixas NS, Robins TG, Attfield MD, Moulton LH. Exposure-response relationships for coal mine dust and obstructive lung disease following enactment of the Federal Coal Mine Health and Safety Act of 1969. *Am J Ind Med* 21: 715–34, 1992.
109. Seixas NS, Robins TG, Attfield MD, Moulton LH. Longitudinal and cross sectional analyses of exposure to coal mine dust and pulmonary function in new miners. *Br J Ind Med* 50: 929–37, 1993.
110. Hedlund U, Jarvholm B, Lundback B. Respiratory symptoms and obstructive lung diseases in iron ore miners: Report from the obstructive lung disease in northern Sweden studies. *Eur J Epidemiol* 19: 953–58, 2004.
111. Hnizdo E. Combined effect of silica dust and tobacco smoking on mortality from chronic obstructive lung disease in gold miners. *Br J Ind Med* 47: 656–64, 1990.
112. Calvert GM, Rice FL, Boiano JM, Sheehy JW, Sanderson WT. Occupational silica exposure and risk of various diseases: An analysis using death certificates from 27 states of the United States. *Occup Environ Med* 60: 122–29, 2003.
113. Tse LA, Yu IT, Leung CC, Tam W, Wong TW. Mortality from non-malignant respiratory diseases among people with silicosis in Hong Kong: Exposure-response analyses for exposure to silica dust. *Occup Environ Med* 64: 87–92, 2007.
114. Rodriquez-Roisin R, Merchant JA, Cochrane GE, Hickey BPH. Maximal expiratory flow volume curves in workers exposed to asbestos. *Respiration* 39: 158–65, 1980.
115. Kilburn KH, Warshaw RH, Einstein K, Bernstein J. Airway disease in non-smoking asbestos workers. *Arch Environ Health* 40: 293–95, 1985.
116. Bégin R, Boileau R, Peloquin S. Asbestos exposure, cigarette smoking, and airflow limitation in long-term Canadian chrysotile miners and millers. *Am J Ind Med* 11: 55–66, 1987.
117. Kilburn KH, Warshaw RH. Abnormal pulmonary function associated with diaphragmatic pleural plaques due to exposure to asbestos. *Br J Ind Med* 47: 611–14, 1990.
118. Demers RY, Neale AV, Robins T, Herman SC. Asbestos-related disease in boilermakers. *Am J Ind Med* 17: 327–39, 1990.
119. Hanke W, Sepulveda MJ, Watson A, Jankovic J. Respiratory morbidity in wollastonite workers. *Br J Ind Med* 41: 474–79, 1984.
120. Malmberg P, Hendenstöm H, Sundblad BM. Changes in lung function in granite crushers exposed to moderately high silica concentrations: A 12 year follow-up. *Br J Ind Med* 50: 725–31, 1993.
121. Gardiner K, Thethowan NW, Harrington JM, Rossiter CE, Glass DC. Occupational exposure to carbon monoxide and sulphur dioxide during the manufacture of carbon black. *Ann Occup Hyg* 36: 363–72, 1992.
122. Gardiner K, Thethowan NW, Harrington JM, Rossiter CE, Calvert IA. Respiratory health effects of carbon black: A survey of European carbon black workers. *Br J Ind Med* 50: 1082–96, 1993.

123. Mwaiselage J, Bratveit M, Moen BE, Mashalla Y. Respiratory symptoms and chronic obstructive pulmonary disease among cement factory workers. *Scand J Work Environ Health* 31: 316–23, 2005.
124. Manfreda J, Cheang M, Warren CP. Chronic respiratory disorders related to farming and exposure to grain dust in a rural adult community. *Am J Ind Med* 15: 7–19, 1989.
125. Kennedy SM, Wright JL, Mullen JB, Pare PD, Hogg JC. Pulmonary function and peripheral airway disease in patients with mineral dust or fume exposure. *Am Rev Respir Dis* 132: 1294–99, 1985.
126. Fine LJ, Peters JM. Respiratory morbidity in rubber workers: II. Pulmonary function in curing workers. *Arch Environ Health* 31: 6–10, 1976.
127. Fine LJ, Peters JM. Respiratory morbidity in rubber workers: III. Respiratory morbidity in processing workers. *Arch Environ Health* 31: 136–40, 1976.
128. Hayden SP, Pincock AC, Hayden J, Tyler LE, Cross KW, Bishop JM. Respiratory symptoms and pulmonary function of welders in the engineering industry. *Thorax* 39: 442–47, 1984.
129. Cotes JE, Feinmann EL, Male VJ, Rennie FS, Wickham CAC. Respiratory symptoms and impairment in shipyard welders and caulker/burners. *Br J Ind Med* 46: 292–301, 1989.
130. Chinn DJ, Stevenson IC, Cotes JE. Longitudinal respiratory survey of shipyard welders: Effects of trade and atopic status. *Br J Ind Med* 47: 83–90, 1990.
131. Hjortsberg U, Orbaek P, Arborelius M. Small airways dysfunction among non-smoking shipyard arc welders. *Br J Ind Med* 49: 441–44, 1992.
132. Luo JC, Hsu KH, Shen WS. Pulmonary function abnormalities and airway irritation symptoms of metal fumes exposure on automobile spot welders. *Am J Ind Med* 49: 407–16, 2006.
133. Ulvestad B, Bakke B, Melbostad E, Fuglerud P, Kongerud J, Lund MB. Increased risk of obstructive pulmonary disease in tunnel workers. *Thorax* 55: 277–82, 2000.
134. Musk AW, Peters JM, Bernstein L, Rubin C, Monroe CB. Pulmonary function in firefighters: A six-year follow-up in the Boston fire department. *Am J Ind Med* 3: 3–9, 1982.
135. Minty BD, Royston D, Jones JG, Smith DJ, Searing CSM, Beeley M. Changes in permeability of the alveolar-capillary barrier in firefighters. *Br J Ind Med* 42: 631–34, 1985.
136. Rosen G, Lundstrom S. Concurrent video filming and measuring for visualization of exposure. *Am Ind Hyg Assoc J* 48: 688–92, 1987.
137. Horsfield K, Guyatt AR, Cooper FM, Buckman M, Cumming G. Lung function in West Sussex firemen: A four year study. *Br J Ind Med* 45: 116–21, 1988.
138. Noertjojo HK, Dimich-Ward H, Peelen S, Dittrick M, Kennedy SM, Chan-Yeung M. Western red cedar dust exposure and lung function: A dose-response relationship. *Am J Respir Crit Care Med* 154: 968–73, 1996.
139. Chan-Yeung M, Wong R, MacLean L et al. Respiratory survey of workers in a pulp and paper mill in Powell River, British Columbia. *Am Rev Respir Dis* 122: 249–57, 1980.
140. Brooks SM, Edwards JJ, Apol A, Edwards FH. An epidemiologic study of workers exposed to western red cedar and other wood dusts. *Chest* 81(Suppl 79): 81S–85S, 1981.
141. Hessel PA, Herbert FA, Melenka LS, Yoshida K, Michaelchuk D, Nakaza M. Lung health in sawmill workers exposed to pine and spruce. *Chest* 108: 642–46, 1995.
142. Whitehead LW, Ashkaga T, Vacek P. Pulmonary function status of workers exposed to hardwood or pine dust. *Am Ind Hyg Assoc J* 41: 178–86, 1981.
143. Whitehead LW. Health effects of wood dust – Relevance for an occupational standard. *Am Ind Hyg Assoc J* 43: 674–78, 1982.
144. Goldsmith DF, Shy CM. Respiratory health effects from occupational exposure to wood dusts. *Scand J Work Environ Health* 14: 1–15, 1988.
145. Corey P, Hutcheon M, Broder I, Mintz S. Grain elevator workers show work-related pulmonary function changes and dose-effect relationships with dust exposure. *Br J Ind Med* 39: 330–37, 1982.
146. Enarson DA, Vedal S, Chan-Yeung M. Rapid decline in FEV_1 in grain handlers. Relation to level of dust exposure. *Am Rev Respir Dis* 132: 814–17, 1985.
147. Huy T, De Schipper K, Chan-Yeung M, Kennedy SM. Grain dust and lung function. Dose-response relationships. *Am Rev Respir Dis* 144: 1314–21, 1991.
148. Smid T, Heederik D, Houba R, Quanjer PH. Dust- and endotoxin-related respiratory effects in the animal feed industry. *Am Rev Respir Dis* 146: 1474–79, 1992.
149. Jorna THJM, Borm PJA, Valds J, Houba R, Wouters EFM. Respiratory symptoms and lung function in animal feed workers. *Chest* 106: 1050–55, 1994.
150. Peelen SJ, Heederik D, Dimich-Ward HD, Chan-Yeung M, Kennedy SM. Comparison of dust related respiratory effects in Dutch and Canadian grain handling industries: A pooled analysis. *Occup Environ Med* 53: 559–66, 1996.
151. Kennedy SM, Dimich-Ward H, Desjardins A, Kassam A, Vedal S, Chan-Yeung M. Respiratory health among retired grain elevator workers. *Am J Respir Crit Care Med* 150: 59–65, 1994.
152. Iversen M, Dahl R, Korsgaard J, Hallas T, Jensen EJ. Respiratory symptoms in Danish farmers: An epidemiological study of risk factors. *Thorax* 43: 872–77, 1988.
153. Leistikow B, Pettit W, Donham K, Merchant J, Popendorf W. Respiratory risks in poultry farmers. In: Dosman J, Cockcroft D (eds). *Principles of Health and Safety in Agriculture*, pp. 62–65. New York: Academic Press, 1989.
154. Donham KJ. Health effects from work in swine confinement buildings. *Am J Ind Med* 17: 17–25, 1990.
155. Morris PD, Lenhart SW, Service WS. Respiratory symptoms and pulmonary function in chicken catchers in poultry confinement units. *Am J Ind Med* 19: 195–204, 1991.
156. Zejda JE, Hurst TS, Rhodes CS, Barber EM, McDuffie HH, Dosman JA. Respiratory health of swine producers. Focus on young workers. *Chest* 103: 702–9, 1993.
157. Choudat D, Goehen M, Korobaeff M, Boulet A, Dewitte JD, Martin MH. Respiratory symptoms and bronchial reactivity among pig and dairy farmers. *Scand J Work Environ Health* 20: 48–54, 1994.
158. Zejda JE, Pahwa P, Dosman JA. Decline in spirometric variables in grain workers from start of employment: Differential effect of duration of follow up. *Br J Ind Med* 49: 576–80, 1992.
159. Schwartz DA, Donham KJ, Olenchock SA *et al.* Determinants of longitudinal changes in spirometric function among swine confinement operators and farmers. *Am J Respir Crit Care Med* 151: 47–53, 1995.
160. Monso E, Riu E, Radon K, Magarolas R, Danuser B, Iversen M *et al.* Chronic obstructive pulmonary disease in never-smoking animal farmers working inside confinement buildings. *Am J Ind Med* 46: 357–62, 2004.
161. Soyseth V, Boe J, Kongerud J. Relation between decline in FEV_1 and exposure to dust and tobacco smoke in aluminium potroom workers. *Occup Environ Med* 54: 27–31, 1997.
162. Sparrow D, Bosse R, Rosner B, Weiss ST. The effect of occupational exposure on pulmonary function. *Am Rev Respir Dis* 125: 319–22, 1982.
163. Holman CD, Psaila-Savona P, Roberts M, McNulty JC. Determinants of chronic bronchitis and lung dysfunction in Western Australian gold miners. *Br J Ind Med* 44: 810–18, 1987.
164. Carta P, Aru G, Barbieri MT, Avataneo G, Casula D. Dust exposure, respiratory symptoms, and longitudinal decline of lung function in young coal miners. *Occup Environ Med* 53: 312–19, 1996.
165. Hall SK, Cissik JH. Effects of cigarette smoking on pulmonary function in asymptomatic asbestos workers with normal chest radiographs. *Am Ind Hyg Assoc J* 43: 381–86, 1982.
166. Copes R, Thomas D, Becklake MR. Temporal patterns of exposure and non-malignant pulmonary abnormality in Quebec chrysotile workers. *Arch Environ Health* 40: 80–87, 1985.
167. Kauffmann F, Drouet D, Lellouch J, Brille D. Occupational exposure and 12-year spirometric changes among Paris area workers. *Br J Ind Med* 39: 221–32, 1982.

168. Le Moual N, Bakke P, Orlowski E *et al.* Performance of population specific job exposure matrices (JEMs): European collaborative analyses on occupational risk factors for chronic obstructive pulmonary disease with job exposure matrices (ECOJEM). *Occup Environ Med* 157: 126–32, 2000.
169. American Conference of Governmental and Industrial Hygienists. Threshold Limit Values for Chemical Substances and Physical Agents and Biological Exposure Indices. Cincinnati, OH: ACGIH, 1998.
170. Mannino D, Homa D, Akinbami L, Ford E, Redd S. Chronic obstructive pulmonary disease surveillance – United States, 1971–2000. *MMWR CDC Surveill Summ* 51: 1–16, 2002.
171. Prescott E, Bjerg AM, Andersen PK, Lange P, Vestbo J. Gender difference in smoking effects on lung function and risk of hospitalization for COPD: Results from a Danish longitudinal population study. *Eur Respir J* 10: 822–27, 1997.
172. U.S. Department of Health and Human Services. *Women and smoking: A report of the Surgeon General: U.S. Department of Health and Human Services*, Centers for Disease Control and Prevention, Office on Smoking and Health; 2001.
173. Wang XR, Zhang HX, Sun BX, Dai HL, Hang JQ, Eisen EA *et al.* A 20-year follow-up study on chronic respiratory effects of exposure to cotton dust. *Eur Respir J* 26: 881–86, 2005.
174. Matheson MC, Benke G, Raven J, Sim MR, Kromhout H, Vermeulen R *et al.* Biological dust exposure in the workplace is a risk factor for chronic obstructive pulmonary disease. *Thorax* 60: 645–51, 2005.
175. Hnizdo E, Sullivan PA, Bang KM, Wagner G. Association between chronic obstructive pulmonary disease and employment by industry and occupation in the US population: A study of data from the Third National Health and Nutrition Examination Survey. *Am J Epidemiol* 156: 738–46, 2002.

37 CHAPTER

Infections

Simon D. Message and Sebastian L. Johnston

Department of Respiratory Medicine, National Heart and Lung Institute, Imperial College London, London, UK

INTRODUCTION

Asthma and chronic obstructive pulmonary disease (COPD) are common diseases and both result in significant morbidity and mortality. Although they share some clinical features and although they may coexist in the same individual, they are distinct disease syndromes with different pathogenetic mechanisms. In each case much of the morbidity and mortality is associated with exacerbations of disease, in response to a variety of trigger factors. A common feature of asthma and COPD is the important role of infection in triggering exacerbations. Infections have also been implicated in the etiology of the two diseases. This chapter will review the epidemiological evidence implicating infectious pathogens as triggers and will discuss the mechanisms of interaction between the host–pathogen response and preexisting airway pathology that result in an exacerbation.

ASTHMA

Asthma affects 20–33% of children in the United Kingdom [1]. It is a multifaceted syndrome involving atopy, bronchial hyperreactivity, and IgE and non-IgE-mediated acute and chronic immune responses. The asthmatic airway is characterized by an infiltrate of eosinophils and of T-lymphocytes expressing the type 2 cytokines IL-4, IL-5, and IL-13. Trigger factors associated with acute exacerbations of asthma include exposure to environmental allergens, especially animals, molds, pollens and mites, cold, exercise, and drugs. The link between respiratory infection and asthma exacerbations is well established although incompletely understood. In the 1950s this association was attributed to bacterial allergy [2] but it is now clear that the majority of exacerbations are due to viral rather than bacterial infection.

Epidemiology

Viral respiratory tract infections are a major cause of wheezing in infants and in adult patients with asthma. Their role may have been underestimated in early epidemiological studies because of difficulties in isolation and identification [3]. The introduction of PCR to such studies has implicated viral infection in the majority of asthma exacerbations.

Indirect evidence from population studies has established a significant correlation between the seasonal variation in wheezing episodes in young children and peaks of virus identification [4]. Seasonal patterns of identification of respiratory viruses are associated with peaks in hospital admissions for both children and adults with asthma indicating a role for such infections in severe asthma attacks [5]. Direct evidence implicating viral infection in asthma exacerbations has been provided by studies showing an increased rate of virus detection in individuals suffering asthma attacks. Viruses have been detected in 80–85% of asthma exacerbations in children [4–10] and in 75–80% in adults [11–14]. The highest rates of identification are in those studies where subjects were followed prospectively allowing collection of clinical specimens early in the course of the illness, where PCR-based methods of diagnosis were used instead of or in addition to serology and culture, and where the methodology used allowed for detection of rhinoviruses. The rate of detection of viruses between exacerbations when individuals are asymptomatic is only of the order of 3–12%. In contrast a study of transtracheal aspirates in adult asthmatics during exacerbations [15] yielded sparse bacterial cultures with no correlation to clinical illness and no difference from those of normal subjects.

In almost all studies of asthmatics, the predominant viruses are rhinoviruses (RV), influenza, RSV, and parainfluenza viruses. RV alone are detected in around 50% of virus-induced asthma attacks. Adenoviruses, enteroviruses, metapneumoviruses, bocaviruses, and coronaviruses are also detected but less frequently. Influenza is only found during annual epidemics.

Experimental virus infection

The effects of respiratory virus infection in the nasal mucosa and upper respiratory tract have been extensively investigated. The effects of such viruses in the lower respiratory tract have been studied but detailed knowledge of the pathogenetic mechanisms involved in asthma exacerbations remains limited. Experimental respiratory virus infection in human volunteers is limited to mild disease by concerns of safety [16]. Most such studies have therefore focused on the experimental innoculation of rhinovirus in allergic rhinitic or mild asthmatic individuals and normal control subjects [17–28]. Such studies provide a useful model of natural virus infection in asthma and offer the advantages of patient selection and monitoring, under controlled conditions before, during, and after infection, of administration of active and placebo medication, of ability to sample the lower airway with timing from onset of infection accurately defined and the study of RV-induced effects including asthma symptomatology, lung function, and airway pathology/immunology.

Recent epidemiological evidence confirms a synergistic interaction between virus infection and allergen exposure in precipitating hospital admissions for asthma [29, 30]. Other trigger factors that may interact with infection include air pollution. A study of asthmatic children demonstrated an increased risk of developing an asthmatic episode within 7 days of an upper respiratory tract infection if the nitrogen dioxide level was greater than $28\,\mu g/m^3$ [9, 31].

Most studies of experimental virus infection in allergic subjects are performed outside the relevant season for allergen exposure. One attempt to provide a model combining allergen exposure and virus infection utilized RV infection in subjects with allergic rhinitis. Individuals received three high dose allergen challenges in the week prior to innoculation to try to mimic combined allergen exposure and virus infection [32]. Interestingly, prior allergen challenge in this model, somewhat unexpectedly, appeared to protect against an RV cold with delayed nasal leukocytosis, increased generation of the proinflammatory cytokines IL-6 and IL-8 and a delayed, less severe clinical course. There was an inverse correlation between nasal lavage eosinophilia and the severity of cold symptoms. The explanation proposed by the authors of this study is that limited high dose allergen challenge may not reproduce the effects of chronic low dose allergen exposure and may stimulate the production of anti-inflammatory mediators such as IL-10 or antiviral cytokines such as IFN-γ or TNF-α. In work by de Kluijver *et al.* the effects of a 10-day period of low dose allergen exposure in house-dust mite sensitive and/or experimental RV16 infection were studied. No synergistic or additive effects were observed as regards lung function parameters [33]. Further development of models of experimental combined allergen exposure and virus infection is clearly required.

We have recently adopted the approach of infecting asthmatic volunteers with RV and then sending them home to continue their normal allergen exposure in the natural environment [34]. We investigated physiologic, virologic, and immunopathologic responses to experimental rhinovirus infection in blood, induced sputum, and bronchial lavage in 10 atopic mild asthmatic and 15 nonatopic normal volunteers. Rhinovirus infection induced significantly greater lower respiratory symptoms, lung function impairment, increases in bronchial hyperreactivity, and eosinophilic lower airway inflammation in asthmatic compared to normal subjects. We also saw trends to increased neutrophils and lymphocytes in the lower airway in asthmatics, and coincident reductions in blood lymphocytes, suggesting trafficking to the airway. In asthmatic, but not normal subjects, virus load was significantly related to lower respiratory symptoms, bronchial hyperreactivity, and reductions in blood total and $CD8^+$ lymphocytes and lung function impairment was significantly related to neutrophilic and eosinophilic lower airway inflammation. This study demonstrated increased rhinovirus-induced clinical illness severity in asthmatic compared to normal subjects, provided evidence of strong relationships between virus load, lower airway virus-induced inflammation, and asthma exacerbation severity and suggests that this approach could provide a very good model in which to examine asthma exacerbation pathogenesis as well as treatment interventions.

Rhinovirus infection of the lower airway

Whereas other respiratory viruses such as influenza, parainfluenza, RSV, and adenovirus are well recognized causes of lower airway syndromes such as pneumonia and bronchiolitis and are capable of replication in the lower airway, until recently there was uncertainty as to whether RV infection occurred in the lower airway or solely in the upper respiratory tract. Although the possibility of nasopharyngeal contamination cannot be ruled out, RV has been detected in lower airway clinical specimens such as sputum [35], tracheal brushings [26], and BAL [36] by both RT-PCR and culture. RV has been cultured in cell lines of bronchial epithelial cell origin [37] and replication has been demonstrated in primary cultures of bronchial epithelial cells [38–40]. The preference of RV for culture at 33°C rather than 37°C has been used as an argument against lower airway infection but there is now evidence that replication does occur at lower airway temperatures [41]. Finally the use of *in situ* hybridization has conclusively demonstrated RV replication in bronchial biopsies of subjects following experimental infection [38] and recent immunochemistry data suggests a preference for basal cells [42]. These data confirm that RV infection of the lower airway does occur and directly implicate lower airway infection in the pathogenesis of asthma exacerbations.

A mouse model of rhinovirus-induced asthma exacerbation

Investigation into the pathogenesis of rhinovirus infections and rhinovirus-induced asthma exacerbations has been severely hampered for the $\sim$50 years since their discovery, as it has been believed that rhinoviruses only infect humans and chimpanzees. However, a mouse model of rhinovirus infection has recently been successfully developed for the first time. New methods of purification and concentration of rhinoviruses, were used to show that for the minor group of rhinoviruses (the $\sim$10% that use the LDL receptor as their mode of entry into cells), wild-type BALB/c mice can be successfully infected and that most of the disease-related outcomes observed in humans were reproduced in this unique new model. These outcomes include induction of both innate and acquired immune responses, induction of mucin synthesis and secretion, induction of both acute neutrophilic and prolonged lymphocytic airway inflammation, and induction of chemokines responsible for chemoattraction of neutrophils, lymphocytes and dendritic cells as well as a range of proinflammatory cytokines. Mice transgenic for a chimera of ICAM-1, the receptor for the major group ($\sim$90%) of rhinoviruses, in which the rhinovirus-binding domains were human, but the remainder of the molecule mouse were then developed. This transgenic mouse was then able to be infected by major group strains, thus generating mouse models capable of being infected by all rhinovirus serotypes. Finally an established mouse model of allergic airway inflammation was used to demonstrate that rhinovirus infection of this model resulted in rhinovirus-induced exacerbation of allergic airway inflammation. The asthma-related outcomes exacerbated by infection in this model include exacerbation of airway hyperresponsiveness, exacerbation of mucin synthesis and secretion (MUC5AC and MUC5B), exacerbation of neutrophilic, eosinophilic, and lymphocytic airway inflammation, and augmented induction of both Th1 (IFN-γ) and Th2 (IL-4 and -13) cytokines. The development of this novel mouse model of rhinovirus-induced asthma exacerbations, should allow mechanisms of disease to be investigated *in vivo* and true causation be established *in vivo*. [43].

Physiological effects of experimental rhinovirus infection

Subjects with asthma and/or allergic rhinitis exhibit increased pathophysiological effects as a result of RV infection as compared to nonatopic, nonasthmatic controls. With detailed monitoring, it is possible to detect reductions in both peak flow [44] and home recordings of FEV_1 [24] in atopic asthmatic patients in the acute phase of experimental RV16 infection. There is an enhanced sensitivity to histamine and allergen challenge after RV16 innoculation in nonasthmatic atopic rhinitic subjects [19, 45]. RV16 increases asthma symptoms, coinciding with an increase in the maximal bronchoconstrictive response to methacholine up to 15 days after infection [20]. There is also a significant increase in sensitivity to histamine in asthmatic subjects after RV16 infection, most pronounced in those with severe cold symptoms [25] and our recent study confirmed that these reductions in lung function and increases in symptoms and airway hyperresponsiveness were observed only in asthmatic, but not in normal subjects [34].

Components of the antiviral immune response

Current concepts of a typical antiviral immune response, as reviewed in detail elsewhere [46, 47], result from research in human volunteers and patients but also in experimental animals, especially inbred mice. Results of animal studies may not be directly applicable to the outbreed human population but ethical considerations often limit direct investigation of the human immune system. All immune responses are a combination of nonspecific (innate) and specific (adaptive) immunity.

Nonspecific or innate [48] elements include: phagocytes such as neutrophils and macrophages that engulf and destroy viruses; natural killer (NK) cells that recognize and destroy virus-infected cells on the basis of reduced HLA class I expression; cells including NK cells, neutrophils, macrophages, mast cells, basophils, epithelial cells that release cytokines, such as interferons, with immunoregulatory or antiviral actions; components of body fluids such as complement, defensins, and surfactant proteins that are capable of neutralizing viruses independently of, or in combination with, antibodies.

Complement

Some viruses may also cause complement-mediated damage. Complement components bind to epithelial cells both *in vitro* and *in vivo* during RSV infections. C3a and C5a are increased in human volunteers infected with influenza A virus [49]. There is little information on the role of complement in immunity to RV. Recent data suggests that the RV 3C protease cleaves the complement factors C3 and C5 which may interfere with the destruction of virus-infected cells [50]. For other viruses, for example influenza, the complement system forms an important link between the innate and specific immune systems. Mice deficient for the third component of complement are highly susceptible to primary influenza, showing reduced priming of T-helper cells and cytotoxic T-cells in lung draining lymph nodes and severely impaired recruitment into the lung of virus-specific $CD4^+$ and $CD8^+$ effector T-cells producing IFN-γ [51]. Activation of the complement cascade may be necessary for the function of other innate antiviral proteins such as serum mannose-binding protein [52]

Defensins

The α and β defensins are small cationic antimicrobial peptides which have the capacity to kill bacteria, fungi, and enveloped viruses by disruption of the microbial membrane.

In vivo they are probably most important in phagocytic vacuoles and on the surface of skin and mucosal epithelia. In addition to their direct antibiotic role, defensins are increasingly being found to have immunomodulatory actions [53] and to play a role in cell recruitment through activation of certain chemokine receptors, for example hBD3 and CCR6 on dendritic cell (DC).

Specific immunity involves production of antibody by B-lymphocytes and the activities of cytotoxic T-cells following processing and presentation of viral antigens by additional cells of the immune system, the most important of which are probably dendritic cells. Immunological memory modifies the overall response to reinfection by previously encountered virus and alters the timing and magnitude of contributions due to different components.

Time course of innate and adaptive immunity in primary and secondary infections

In primary infection, viruses replicate in the respiratory tract reaching peak levels at around days 2–4. At this time type I interferons are first detected, peaking around days 2–3 and falling to become undetectable once active replication has ceased. Interferons activate NK cells, first detectable around day 3 and peaking around day 4. In addition to destruction of virally infected cells NK cells release cytokines including IFN-γ that activate additional inflammatory cells in the airway including macrophages. Such nonspecific immune mechanisms are essential in early defense against virus in the first few days. In addition, the innate immune system plays a role in stimulating specific immunity and may influence the nature of the specific response, for example whether this is characterized by type 1 or type 2 cytokines.

Meanwhile, viral antigens are processed locally and in regional lymph nodes by dendritic cells and presented to T-cells. $CD4^+$ and $CD8^+$ T-cells are detectable from around day 4 then generally decline as infection resolves to become undetectable by day 14. However, memory $CD4^+$ and $CD8^+$ responses may persist for life. T-cell recruitment is dependent on the production of chemokines and on alterations in the expression of adhesion molecules on the endothelium of inflamed tissues. Time is also required to generate B-cell responses. Mucosal IgA may be detected around day 3, serum IgM from days 5–6 and IgG days 7–8, increasing in amount and avidity over the next 2–3 weeks. IgA falls normally to low or undetectable levels over 3–6 months. Serum IgG may remain detectable for life. Specific immune mechanisms such as $CD8^+$ T-cells and immunoglobulin are responsible for the eradication of infectious virus usually by 7–10 days after infection.

Secondary infection with the same virus results in rapid mobilization of B- and T-cell specific immunity with an earlier T-cell peak coinciding with the NK cell peak around days 3–4. If reinfection is with the same serotype a rapid increase in levels of preexisting neutralizing antibodies may limit viral replication to such an extent that infection is clinically silent. Because this results in fewer infected cells there is relatively less activation of nonspecific immunity and it may be difficult to detect a $CD8^+$ T-cell response.

Following experimental infection of seronegative subjects with RV2 [54] serum-specific antibodies are detectable at 1–2 weeks, reach a maximum at 5 weeks, persist for at least a year and may remain elevated many years after infection. Local specific antibody levels may be lost more rapidly. High levels of serum neutralizing antibody or specific IgA protect against reinfection with the same rhinovirus serotype. However, since it appears relatively late, recovery from illness for seronegative hosts which usually occurs at 7–10 days must be due to other components of the immune response. In seropositive subjects preexisting serum neutralizing antibodies to RV39 and to RV-Hanks modify experimental infections in human subjects [55, 56]. Local IgA and IgG passing from the vasculature into the pulmonary interstitium contribute to viral clearance. However, the 100+ RV serotypes mean that repeated infection with RV to which an individual lacks appropriate antibodies is common.

T-cell responses to RV demonstrate MHC class I restricted cross-reactivity between serotypes due to specificity for conserved epitopes within the capsid proteins VP 1–3 [57]. RV16- and RV49-specific T-cell clones from human peripheral blood demonstrate recognition of both serotype specific and shared viral epitopes [58]. Vigorous proliferation of and IFN-γ production by PBMC in response to RV16 in seronegative subjects is associated with reduced viral shedding after inoculation [59], thus T-cells responses also appear protective.

Interactions between virus infection and asthmatic airway inflammation

The interaction of respiratory virus infection and chronic asthmatic airway inflammation results in respiratory symptoms that are more severe than those suffered by nonasthmatic individuals [34, 60] and case-control studies have demonstrated clear synergistic interactions between virus infection and allergen exposure in increasing risk of exacerbation [29, 30]. The detailed immunological mechanisms underlying this interaction are currently being investigated, but recent data suggest deficient production of type I (β), type II (γ), and type III (λ) IFNs, as well as Th1 cytokines (IL-12) and anti-inflammatory cytokines (IL-10), are likely to increase virus-induced lower airway inflammation [34, 61, 62]. These deficiencies are also accompanied by augmented production of Th2 cytokines suggesting that perhaps allergen-induced inflammation is also increased in the pathogenesis of virus-induced asthma exacerbations.

Bronchial inflammation is likely therefore a central event for virus-induced asthma exacerbations. The processes involved include interacting cascades from the complement, coagulation, fibrinolytic, and kinin systems of the plasma as well as cell-derived cytokines, chemokines, and arachidonic acid metabolites. Our understanding of the interaction of viruses with these cascades in asthma is incomplete and it is likely that different viruses interact with each system to different extents. However, it is reasonable to believe that in all cases the initial trigger of the inflammatory reactions is epithelial cell–virus interaction.

TABLE 37.1 Current hypotheses for the pathogenesis of virus-induced asthma exacerbations.

Epithelial disruption	Reduced ciliary clearance Increased permeability Loss of protective functions Kinins
Mediator production	Complement Arachidonic acid metabolites Nitric oxide Reactive oxygen products Cytokines
Induction of inflammation	Chemokines Immune cell activation Adhesion molecule induction Impaired innate IFN production
Immune dysregulation	Impaired apoptosis Impaired Th1 immunity Impaired IL-10 production Augmented Th2 immunity Increased total IgE
IgE dysregulation	Antiviral IgE production Airway smooth muscle
Airway remodeling	Fibroblasts Myofibroblasts Growth factors Increased cholinergic sensitivity
Alterations of neural responses	Neuropeptide metabolism modulation β-adrenergic receptor dysfunction

Table 37.1 summarizes some of the current hypotheses proposed to explain the mechanisms of exacerbation of asthma following respiratory virus infection. The evidence supporting these hypotheses is reviewed in detail below.

The role of the airway epithelial cell

The airway epithelium is an important component of antiviral defense. In addition to its function as a physical barrier to the entry of viruses, the responses of epithelial cells (EC) following viral infection, whether or not this results in destruction of the cell, contribute to both innate and adaptive antiviral immune responses. Information regarding the effects of RV on EC comes from *in vivo* studies and from *in vitro* models using either cultured primary airway EC or cell lines of epithelial cell origin such as A549, BEAS-2B, and H292.

EC contribute to the immune response following virus infection through the production of cytokines and chemokines (Fig. 37.1). They may also act as antigen presenting cells particularly during secondary respiratory viral infections. Epithelial cells express MHC class I and the costimulatory molecules B7-1 and B7-2 and this expression is upregulated *in vitro* by RV16 [63] (Fig. 37.2).

The extent of epithelial cell destruction observed in the airway varies according to virus type. Influenza typically causes extensive necrosis [64], whereas RV causes little or only patchy damage. Destruction of epithelial cells results in both an increase in epithelial permeability, and increased penetration of irritants and allergens, and exposure of the extensive network of afferent nerve fibers. Both effects may contribute to increased bronchial hyperresponsiveness.

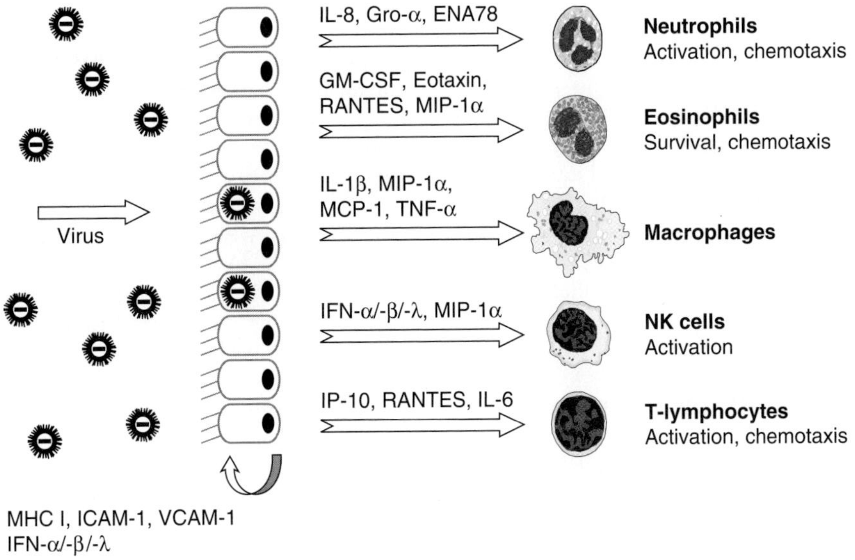

FIG. 37.1 Airway epithelial cells participate in the immune response to respiratory virus, producing a variety of cytokines and chemokines with actions on other cells. In addition the migration of inflammatory cells is aided by the upregulation of adhesion molecules and interferons help to establish an antiviral state in neighboring epithelial cells. Upregulation of MHC class I may facilitate presentation of viral antigens.

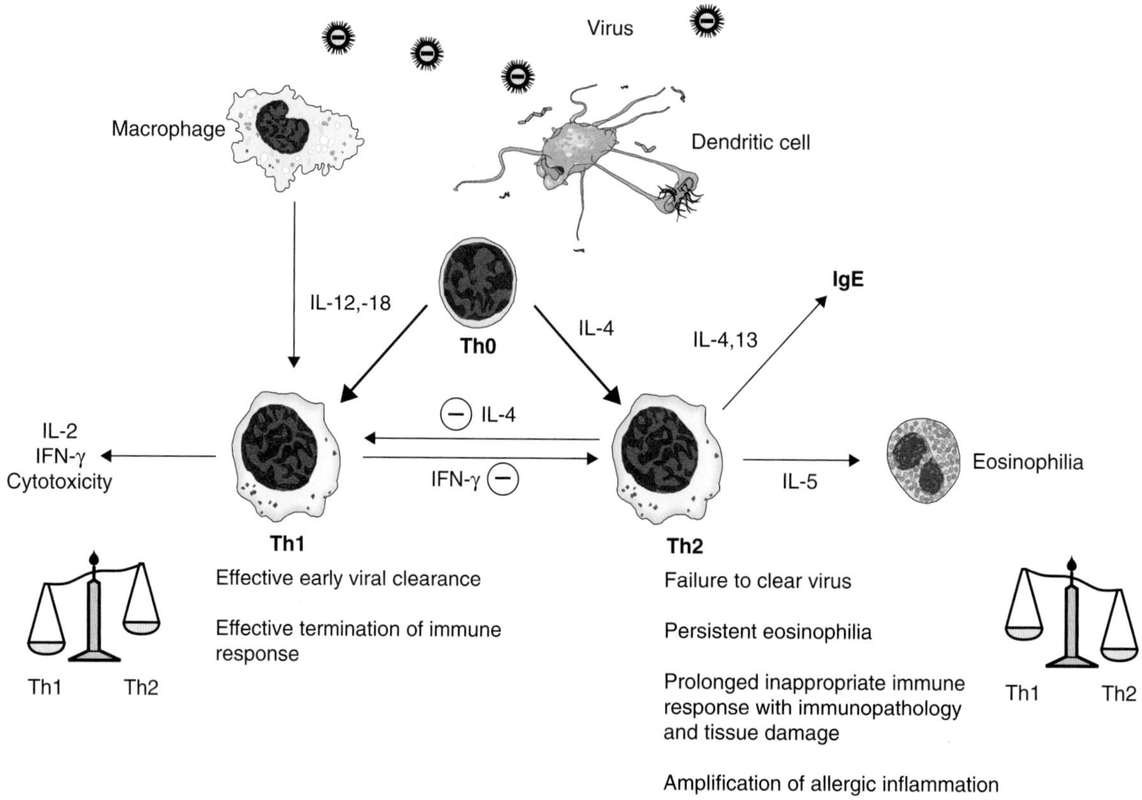

FIG. 37.2 Preexisting asthmatic airway inflammation may modify a predominantly Th1 antiviral immune response, favoring a Th2 or mixed response which may provide less efficient viral clearance and result in prolonged virus-induced inflammation, increased associated immunopathology and increased tissue damage.

In vivo RV causes some shedding of infected ciliated EC [65], but the extent of viral infection of the epithelium may be incomplete even in the nose [66]. *In vitro* studies exposing monolayer cultures of nasal epithelial cells to respiratory viruses at 10^3–10^4 $TCID_{50}$/ml demonstrate no detectable CPE (carboxypeptidase E) with RV or coronavirus in contrast to the extensive destruction with influenza and adenovirus [67]. *Ex vivo* infection of cells from both the upper and the lower respiratory tract suggest that less than 10% of cells in the epithelium are infected by RV [68], however the extent of virus-induced epithelial damage may be considerably greater in asthmatic than in normal subjects, as asthmatic epithelium has been shown to be much more susceptible *in vitro* [61]. *In vivo* studies of degrees of epithelial damage would be technically challenging, but could generate interesting findings.

Receptors for entry of RV into host cells

Viruses enter into and replicate within airway EC. Entry is dependent on the interaction with host cell surface proteins which function as receptors. In the case of the major group RV this is ICAM-1 [69] and infection can be blocked by antibodies to ICAM-1 or with soluble ICAM-1 [37]. There is relatively limited expression of ICAM-1 in airway epithelium prior to RV infection [70] and this may explain the patchy nature of infection. The upregulation of ICAM-1 in the asthmatic airway is one possible explanation for the increased severity of RV infection in asthma. RV upregulates expression of its own receptor ICAM-1 both *in vitro* and *in vivo*. Following experimental infection with RV, ICAM-1 expression is upregulated in nasal epithelium within 24h, declining by day 5 [71]. RV has similar effects on EC from the lower airway. RV has been shown to upregulate ICAM-1 in primary bronchial EC *in vitro* [72] and ICAM-1 is upregulated in bronchial biopsies following experimental infection of asthmatic subjects with RV16 [73].

There are two forms of ICAM-1: membrane bound (mICAM-1) which favors viral infection by acting as a virus receptor and soluble (sICAM-1) which binds virus outside the cell and can thereby inhibit virus infection. RV infection of EC is reported to alter the balance in favor of further infection by inducing mRNA for mICAM-1 whilst suppressing that of sICAM-1 [74]. The LDL receptor is the receptor for the minor group RV. RV2 infection of primary human tracheal EC (PHTEC) is blocked by an antibody to the LDL receptor and is also reported to upregulate LDL-R expression [48].

ICAM-1 expression by human nasal EC is upregulated *in vitro* by exposure to a number of inflammatory cytokines and mediators including IL-1β, IL-8, IFN-γ, TNF-λ, and the eosinophil-derived proteins MBP and ECP [75]. IL-1β in particular may be important in RV-induced induction of

ICAM-1. Antibodies to IL-1β but not TNF-α decreased viral replication and ICAM-1 expression by PHTEC [76]. Not all respiratory epithelial cell lines behave in the same way as primary EC – for example A549 cells express ICAM-1 at lower levels constitutively and show upregulation by IFN-γ and TNF-α but not by ECP or MBP. The effect of IFN-γ is complex. Whilst IFN-γ upregulates ICAM-1 in uninfected cells this cytokine inhibits ICAM-1 upregulation by RV14 in H292 cells and its presence results in reduced viral titers [72, 77].

A preexisting elevation of ICAM-1 expression in the asthmatic airway may contribute to increased symptom severity of RV infection. Type 2 cytokines (IL-4, IL-5, IL-13) upregulate ICAM-1 in H-292 cells [78]. Allergen challenge results in upregulation of ICAM-1 on conjunctival and nasal EC in atopics [79]. In nasal brushing EC from atopics, basal ICAM-1 levels were increased relative to nonatopics and elevated in the relevant allergen season. Nasal EC from atopics showed further upregulation after *in vitro* culture with allergen. The highest basal ICAM-1 was found on nasal polyp EC and this was increased further after RV14 infection. Viral titers after RV14 infection were significantly higher for polyp EC than for nonatopic and atopic nonpolyp EC [80].

Modification of EC ICAM-1 expression is therefore of possible therapeutic benefit. *In vitro* RV increases expression of ICAM-1 and VCAM-1 in primary bronchial EC (PBEC) cultures and in A549 cells via a mechanism involving NF-κB [5, 81, 82]. One of the actions of corticosteroids is inhibition of NF-κB [83]. In both A549 cells and in PBEC pretreatment with three corticosteroids, hydrocortisone, dexamethasone, and mometasone furoate inhibits RV16-induced increases in ICAM-1 surface expression, mRNA, and promoter activation without alteration of virus infectivity or replication. Dexamethasone suppresses ICAM-1 in PHTEC and inhibits RV infections [84]. Dexamethasone does not inhibit infection of PHTEC by minor group RV2 [85]. Disappointingly, a study of inhaled corticosteroids in asthmatics prior to experimental RV infection failed to show reduced virus-induced ICAM-1 expression in bronchial biopsies [73] but it is possible that a longer course and/or a higher dose of inhaled steroid or administration of oral steroids might have demonstrated a significant effect.

Other drugs which affect EC ICAM-1 include reducing agents [86], the H_1 receptor antagonists desloratidine/loratidine which inhibit RV-induced ICAM-1 upregulation in HPBEC and in A549 cells [87] and erythromycin which inhibits infection of PHTEC by both major group RV14 and minor group RV2 through effects including ICAM-1 reduction, blockage of RV RNA entry into endosomes and small reductions in LDL receptor expression [88].

RV induction of EC production of cytokines and chemokines

EC can activate and recruit a variety of other cell types such as lymphocytes, eosinophils, and neutrophils through the production of chemokines and cytokines (Fig. 37.1). Such cells are important components of the antiviral response but may also contribute to airway inflammation and dysfunction in asthma.

Type 1 interferons

Interferons (IFN) play an important role in innate resistance to viruses [89], acting on virus-infected cells and surrounding cells to produce an antiviral state characterized by the expression and antiviral activity of IFN-stimulated genes (ISGs). There are three main types of IFN, type 1 (IFN-α, IFN-β, IFN-ω, IFN-τ), type 2 (IFN-γ) and the recently discovered type 3 (IFN-λs [90, 91]). EC can produce both type 1 and type 3 IFNs. There are 14 IFN-α genes but only 1 IFN-β gene. IFN-β synthesis involves NF-κB, ATF/JUN and the interferon regulatory factors (IRFs) (up to 10 of which are currently identified), activation of which occurs in response to virus-specific signals including dsRNA, a product of the replication of ssRNA viruses such as RV, RSV, and influenza.

IFN-β and IFN-α4 are expressed early through the action of IRF3. Activation of the IFN intracellular signaling pathway is required for induction of IRF7 which is required for transcription of the full range of IFNs. DNA microarray analysis has shown that following binding to their receptors on target cells IFNs trigger a complex signaling pathway (mainly JAK-STAT) resulting in the transcription of hundreds of ISGs [92].

Several ISGs have been well studied. These include the dsRNA-activated serine/threonine protein kinase (PKR) which reduces cellular mRNA translation and transcriptional events, two enzymes involved in mRNA degradation, 2′5′oligoadenylate synthetase (OAS) and RNase L, the myxovirus resistance (Mx) proteins and RNA-specific adenosine deaminase (ADAR) which is involved in RNA editing. These ISGs inhibit virus replication at a number of levels and not surprisingly, viruses have evolved mechanisms to resist the actions of IFNs, for example blocking of PKR by the influenza NS1 protein [93]. IFNs also upregulate cellular expression of MHC class I and II molecules therefore increasing antigen presentation to $CD8^+$ and $CD4^+$ T-cells and enhancing cellular immune responses.

One recent study has examined type 1 interferon production by primary bronchial epithelial cells from normal and asthmatic subjects infected *ex vivo* with RV [40]. Asthmatic EC following infection released a higher titer of virus into culture supernatant and exhibited impaired apoptosis and a greater degree of necrotic cell death, favoring release of virus from dying cells. This was accompanied by lower concentrations of IFN-β after infection. Addition of IFN-β to asthmatic ECs inhibited virus replication to levels observed in ECs from normal subjects. This study suggests that the production of IFN-β is deficient in asthmatic ECs and that replacement/augmentation of IFN-β to boost the innate immune response could be a novel approach to treatment of virus-induced asthma exacerbations [94].

Type III interferons

A new family of interferons, called type III IFN-λs, and characterized by three elements: λ1, λ2, and λ3, also termed IL-29, IL-28A, and IL-28B, has recently been described

[90, 91]. The three highly homologous IFN-λ proteins demonstrate limited (about 20%) homology to type I IFNs [95]. Human IFN-λs bind to a unique heterodimeric receptor (IFN-λR), composed of CRF2-12 (also designated IFN-λR1), and CRF2-4 (also designated IL-10R2) shared with other class II cytokine-receptor ligands including IL-10, IL-22, and IL-26 [90].

Viral infection induces upregulation of IFN-λ mRNA in epithelial cells, peripheral blood mononuclear cells (PBMCs), and dendritic cells [90, 91, 96–99]. IFN-λs exhibit some similar biological properties to type I IFNs: they induce Jak/STAT pathways that lead to the upregulation of several antiviral proteins and enzymes including 2′,5′-OAS and MxA, have antiviral activity *in vitro* [90, 91, 97] and have also exhibited antviral activity in an *in vivo* model of vaccinia virus-infected mice [100]. Based on current knowledge, it thus appears that both IFN-α/-β and IFN-λ ligand-receptor systems can independently induce an antiviral state by engaging similar participants of the antiviral response, though the signaling pathways involved in IFN-λ production are currently largely unknown. Recent data suggest that IFN-λ may be involved in antiviral responses against RV. *In vitro* RV infection of a bronchial epithelial cell line (BEAS-2B) led to IFN-λ production and this cytokine demonstrated a dose-dependent antiviral effect against RV [62]. Moreover, IFN-λ production occurred after *in vitro* infection of primary bronchial epithelial cells, macrophages, and BAL cells from healthy volunteers.

We have recently investigated the production of IFN-λs in response to RV in primary bronchial ECs and in BAL cells (90% macrophages) from normal and asthmatic subjects [62]. Production of IFN-λ1 and IFN-λ2–3 was deficient in ECs and BAL cells from asthmatic subjects after *in vitro* RV infection and induction of IFN-λs by RV infection of ECs was strongly inversely related to RV replication. To determine whether IFN-λ production was important in determining responses to RV infection *in vivo*, the same volunteers were then experimentally infected with RV16, the severity of symptoms and reductions in lung function were monitored and the virus load was determined in BAL. *In vitro* production of IFN-λs by RV infection of BAL cells was strongly inversely correlated with both common cold symptoms and *in vivo* virus load and strongly positively correlated with severity of falls in lung function, in asthmatic and normal volunteers experimentally infected with RV16. Asthmatic patients, in whom *in vitro* IFN-λ production in BAL cells was significantly lower than in normal subjects, exhibited increased common cold symptoms and reductions in lung function and virus load after *in vivo* RV16 infection. In marked contrast normal subjects had robust IFN-λ responses, less severe cold symptoms, lower virus load, and no significant changes in lung function. These results document the importance of IFN-λ in the host defense against RV infection *in vitro* and *in vivo* and indicate that deficient IFN-λ production is likely to be important in the pathogenesis of virus-induced asthma exacerbations.

Proinflammatory cytokines and chemokines

Viral infection of the respiratory tract results in significant changes in the pattern of cytokine expression by a number of cell types, by both cells of the immune system, which may be increased in number and activation status, and by cells often considered to be structural but which in fact contribute significantly to the immune response such as EC. Efficient orchestration of the immune response by cytokines is essential for eradication of virus. Modification of cytokine expression in the airway may contribute to the increased severity of virus infection in asthma.

In vitro studies of bronchial EC lines or macrophages have demonstrated the production of a wide range of proinflammatory cytokines such as IL-1, IL-6, IL-11, IFN-α, IFN-γ, TNF-α, and granulocyte-macrophage colony stimulating factor (GM-CSF) and the chemokines IL-8, ENA-78, RANTES, and IP-10 and macrophage inflammatory protein (MIP)-1α in response to RV and RSV [37, 101–103]. *In vivo* these cytokines can be found in nasal lavage in association with RV infection [104].

The specific roles of individual cytokines in the human lower airway during viral infection are not well understood, but increasing information is becoming available. Such cytokines and chemokines activate and recruit a variety of other cells including lymphocytes, eosinophils, and neutrophils. IL-1, TNF-α, and IL-6 share proinflammatory properties such as the induction of the acute phase response and the activation of both T- and B-lymphocytes. IL-1 enhances the adhesion of inflammatory cells to endothelium, facilitating chemotaxis [105]. TNF-α is a potent antiviral cytokine but *in vitro* increases the susceptibility of cultured epithelial cells to infection by RV14 through upregulation of ICAM-1 [37]. IL-6 has been shown to stimulate IgA-mediated immune responses. IL-11 may also be important in virus-induced asthma [106]. It appears to cause bronchoconstriction by a direct effect on bronchial smooth muscle [102]. Production of this cytokine by human stromal cells *in vitro* is increased by RV14, RSV, and parainfluenza type 3 but not by cytomegalovirus (CMV), herpes simplex virus (HSV)-2 or adenovirus. *In vivo* IL-11 is elevated in nasal aspirates from children with colds, levels correlating with the presence of wheezing.

Similarly the chemokine MIP-1α is increased in nasal secretions during natural viral exacerbations of asthma [107]. Studies in MIP-1α knock-out mice suggest that it mediates pneumonitis due to influenza [108]. The other chemokines IL-8 and ENA-78, will recruit and activate neutrophils, while RANTES and IP-10 will do the same for lymphocytes [103].

Viral upregulation of cytokines and chemokines may be mediated through certain key transcription factors. Increases in IL-6 and IL-8 production by cultured epithelial cells due to RV was dependent on NF-κB [82, 107, 109] and further upstream, protein kinase R (PKR)-mediated RV-induced RANTES, IL-8, and IL-6 [110]. Rhinovirus induction of IL-8 was shown to require IkappaB kinase-beta (IKKβ) and the transcription factor NF-IL-6 as well as NF-κB. Similar observations have been made with regard to the induction of IL-1, -6, -8, -11, and TNF-α by RSV [111, 112], thus the potential role of inhibition of NF-κB in this context has generated considerable interest.

In addition to the induction of IL-1α and IL-1β RV infection results in substantial increases in IL-1ra both *in vivo* and *in vitro*. This is a relatively late effect, occurring

48–72 h after infection and may contribute to symptom resolution [113].

Kinins and nitric oxide

A multitude of inflammatory mediators are generated or act on the epithelial surface. Bradykinin, a-nine-amino acid peptide generated from plasma precursors as part of the inflammatory process has been shown to be present in nasal secretions of RV-infected individuals [114]. Bradykinin given intranasally is able to reproduce some of the symptoms of the common cold such as sore throat and rhinitis [115]. Although the presence of kinins in the lungs of virus-infected individuals has not been reported they are present in both the upper and lower airways in allergic reactions [114–118].

Nitric oxide (NO) is produced by diverse sources including epithelial, endothelial, and smooth muscle cells. In human airways NO appears to be important in relaxation of the human airway smooth muscle [119]. Nitric oxide (NO) may be important in a range of respiratory diseases [120] including asthma [121] and in virus infection [122]. NO is produced both by the constitutive enzymes, nitric oxide synthase (NOS)1, and NOS3 and by the inducible, calcium-independent NOS2 expressed by airway EC [123] and macrophages.

In asthma there is increased NOS2 expression and an elevated level of exhaled NO [124] that falls with corticosteroid therapy; the level of exhaled NO correlates with sputum eosinophilia and methacholine responsiveness [125]. In contrast in stable COPD, exhaled NO levels are not different from normal subjects [126] although there may be an increase during exacerbations [127]. In fact *in vitro* cigarette smoke reduces cytokine-induced NOS2 mRNA expression in the LA-4 murine cell line, in A549 cells and in PHBEC [128].

The relative importance of the beneficial antimicrobial activity of NO versus the potentially disadvantageous suppression of IFN-γ may be dependent on the specific pathogen. NOS2 knock-out mice show an increased susceptibility to infections [129], perhaps because release of NO may be important for NK cell-mediated target cell killing [130]. However NO may also possess antiviral activity. *In vitro* RV induces NOS expression in HPBEC [131]. There is increased expression of NOS2 mRNA in cultured HPBEC after RV16 infection [132]. NO inhibits RV-induced production of IL-6, IL-8, and GM-CSF and viral replication in a human respiratory epithelial cell line [132, 133]. RSV also induces NOS2 and increases nitrite levels in supernatant from A549 cells, from HPBEC culture and in BAL fluid from RSV-infected BALB/c mice, effects opposed by IL-4 and dexamethasone but unaffected by IL-13 or IFN-γ [134]. Replication of RSV in Hep2 cells is inhibited following transfection with a retroviral construct containing NOS and this inhibition is abolished by the NOS inhibitor, NG-methyl-l-arginine [135]. Replication of influenza A and B in Mabin Darby kidney cells is severely impaired by the NO donor, *S*-nitroso-*N*-acetylpenicillamine [136].

Overall there is evidence that *in vivo* increased lower airway NO production may be of benefit in virus-induced asthma exacerbations. Work in a guinea pig model suggests that one mechanism for increased airway hyperresponsiveness during respiratory virus infection is through inhibition of NOS enzymes and a loss of NO-related relaxation of airway smooth muscle [137]. Studies of human asthmatics would also suggest that NO has a protective role in virus-induced exacerbations. Following experimental RV16 infection patients with the greatest increase in exhaled NO had smaller increases in histamine airway responsiveness [131].

In experimental animals parainfluenza virus-induced hyperreactivity correlates with a deficiency in constitutive NO production [137]. Increased levels of exhaled NO are found in nonasthmatic volunteers following natural colds [138] as well as in asthmatic patients after experimental RV infection [131]. In the latter study, an inverse association between NO increase and worsening of airway hyperresponsiveness was demonstrated arguing in favor of a protective role for this substance. This is further supported by the observation that NO reduces cytokine production and viral replication in an *in vitro* model of RV infection [133]. Interestingly, studies of viral upper respiratory tract infections have failed to demonstrate an increase in nasal NO after experimental RSV, RV, and influenza infections [139]. In normal subjects experimental influenza infection increased oral NO 8 days postinfection but had no effect on nasal NO [140]. This raises the possibility that during respiratory virus infection induction of NO is selective for the lower respiratory tract.

Signaling pathways

The responses of airway EC to virus infection are consequences of the interactions between virus and the intracellular signaling pathways of the host cell [141]. Knowledge of the mechanisms involved for rhinoviruses is currently very limited. Activation of signaling pathways may be dependent on cell surface receptor (ICAM-1, LDL-R) binding or may occur during viral replication within the cell. The need for replicative virus is demonstrated by the inhibition of RV induction of EC cytokines after UV inactivation. One product of replication, common to ssRNA viruses such as RV and also RSV and influenza, is dsRNA, which has been shown to activate components of signaling pathways including dsRNA-dependent protein kinase PKR, IKKβ, NF-κB, and p38 mitogen-activated protein kinase with resultant induction of IL-6, IL-8, and RANTES [110, 142]. Activation of EC by dsRNA may be direct or indirect through the interferon system as discussed above. It has also been reported that dsRNA and virus infections activate EC through binding to TLR3 [143, 144].

Effects of viruses on airway smooth muscle cells

Studies utilizing isolated rabbit tissues and human cultured airway smooth muscle cells suggest that, for RV16, exposure to the virus may have a direct effect on smooth muscle cells, resulting in increased contractility to acetylcholine and impaired relaxation to isoproterenol. This effect is dependent on ICAM-1 and appears to involve an autocrine signaling mechanism including upregulation of production of

IL-5 and IL-1β by the airway smooth muscle itself [145]. A more recent study demonstrated that RV induction of IL-6 and IL-8 was increased in smooth muscle cells from asthmatic compared to normal subjects [146]. Whether rhinovirus reaches airway smooth muscle cells in sufficient quantity to produce a significant effect by this mechanism *in vivo* is as yet unknown. The effects of other respiratory viruses on smooth muscle require further investigation.

The cellular immune response to virus infection in the lower airway

A variety of leukocytes show changes in number, site of accumulation, and activation state in response to virus infection. Since these cells are also implicated in asthmatic inflammation of the lower airway they provide potential sites of interaction between the immunopathologies of virus infection and asthma.

Monocytes/macrophages

Alveolar macrophages are present in large numbers in the lower airway. They make up around 90% of the cells seen in BAL from normal volunteers [28]. They are ideally placed for early phagocytosis of virus particles and are likely to play an important role in the immune response through antigen presentation to T-cells and through the production of cytokines and other mediators. RV has been shown to enter human monocytes and macrophages which express high levels of the major RV receptor ICAM-1. It has not been possible to demonstrate RV replication within alveolar macrophages although low grade productive infection has been shown in the monocyte cell line THP-1 [147]. Replication also occurred in THP-1-derived macrophages but was limited in monocyte-derived macrophages which are relatively resistant to viral replication at least partly because of higher levels of type I interferon production [148]. RV entry into monocytes results in activation and the production of both IL-8 [147] and TNF-α [149]. In monocyte and THP-1-derived macrophages RV induction of TNF-α is NF-κB dependent [148].

A recent study reported that infectious but not UV-inactivated RV-increased TNF-α and IL-8 release by macrophages derived from resected lung tissue. Interestingly, infectious rhinovirus-impaired LPS and lipoteichoic acid-induced TNF-α and IL-8 secretion by macrophages as well as the macrophage phagocytic response to labeled bacterial particles [150]. This RV-induced impairment of cytokine responses to bacterial LPS and lipoteichoic acid and of phagocytosis in alveolar macrophages could lead to impairment of antibacterial host defense may have important implications in the pathogenesis of exacerbations of respiratory diseases including both asthma [151] and COPD [152]. In contrast, infection of human monocytes *in vitro* with influenza A causes alterations in structure and activation status and the production of IL-1β IL-6, TNF-α IFN-α, and IFN-β [153], effects dramatically potentiated by subsequent exposure to bacterial LPS.

Dendritic cells

Dendritic cells are key cells in IFN production, as well as in antigen presentation both of allergens and pathogens with a capacity to induce both primary and secondary immune responses. They may also play a role in the regulation of the type of T-cell-mediated immune response [154]. RV infection has been shown to induce production of the dendritic cell attracting chemokine MIP-3α [43] and to increase in number in the lung during RSV infection [155], suggesting they are recruited to the lung during respiratory virus infections. However, they have also been shown to be produced from local precursors during RSV infection [156]. Plasmacytoid dendritic cells are likely protective against infection as they have been shown to limit virus replication in RSV infections, as well as reducing airway inflammation and airway hyperresponsiveness [157]. In contrast, others have reported induction of the high affinity IgE receptor on dendritic cells during Sendai virus infection, and linked this with induction of mucus cell metaplasia and airway hyperreactivity [158]. There is thus increasing knowledge of the immunobiology of these cells during respiratory infections but their role in the context of viral exacerbations of asthma remains unclear and further studies are needed.

Lymphocytes

Bronchial biopsies demonstrate increases in cells positive for CD3, CD4, and CD8 within the epithelium and submucosa of both normal and asthmatic subjects following experimental RV infection [21] and we have recently demonstrated a trend toward increased numbers of lymphocytes in BAL from asthmatic compared to normal subjects ($p = 0.06$)[34]. Such increases coincided with peripheral blood lymphopenia, and reductions in blood total lymphocytes and $CD8^+$ T-cells correlated strongly with virus load only in asthmatic subjects[34] suggesting increased recruitment of T-cells to the asthmatic airway may be important in the context of asthma exacerbations. Since T-cells are believed to be key cells in the pathogenesis of asthma the effects of viruses on T-cells are of particular importance.

T-cell recruitment into the airway is at least partly under the influence of chemokines, including those whose production by EC is upregulated by viruses. The nature and the effectiveness of the specific immune response may be influenced by the balance of chemokine production by airway EC. This balance may in turn be influenced by preexisting chronic inflammation as found in asthma.

Studies of cloned T-cells suggest that Th1 and Th2 cells show differential expression of chemokine receptors. There is increased expression of CXCR3 (receptor for IP-10, I-TAC, and Mig) and CCR5 (MIP-1β) in human Th1 cells and increased expression of CCR4 (TARC and MDC) and to a lesser extent CCR3 (eotaxin and MCP-3) in Th2 cells, with selective migration of cells in response to the appropriate chemokines. CCR1 (RANTES, MIP-1α, MCP-3) and CCR2 (MCP-1, -2 ,-3, -4) were found on both Th1 and Th2 cells [159]. Bronchial biopsies from asthmatics show high levels of expression of CCR4 and significant levels of CCR8 by T-cells [160].

Increased recruitment of T-cells to the airway as a result of virus-induced chemokine production by EC could amplify preexisting allergic inflammation. If the asthmatic airway microenvironment influences the pattern of chemokine expression following virus infection then this could alter the Th1/Th2 balance of the antiviral immune response.

$CD4^+$ T-cells

The $CD4^+$ T-cell response to virus infection is thought to be of the T-helper 1 (Th1) type. It is thought that an effective antiviral immune response is characterized by the production of type 1 cytokines such as IFN-γ. IFN-γ, in addition to IFN-α, IFN-β, and IFN-λ from monocytes and macrophages, plays a role in establishing an "antiviral state" in neighboring cells. IFN-γ has a complex role in the pathogenesis of asthma. It appears to increase basophil and mast cell histamine release [161] but on the other hand inhibits the expression of type 2 cytokines. Production of IFN-γ is increased in PBMC [162] and in nasal secretions [104] during RV colds and in human and animal models of influenza, parainfluenza, and RSV infection [119, 163, 164]. There are exceptions where the antiviral response exhibits a Th2 character or a mixture of Th1/Th2. In animal models of RSV, different proteins of the virus may induce either Th1 or Th2 type responses and priming with such proteins prior to infection with whole virus can influence the character, effectiveness, and associated immunopathology of the immune response [165].

Asthma is believed to be characterized by type 2 inflammation. Many studies have demonstrated mutual inhibition of Th1 and Th2 cells [166, 167]. It is therefore possible within an airway with a preexisting type 2 allergic asthmatic microenvironment that there may be inhibition of the normal effective type 1 antiviral immune response or that the system may be skewed toward type 2 responses.

Papadopoulos *et al.* have shown that type 1 responses to RV are deficient in individuals with asthma [168]. PBMC taken from asthmatics and exposed *in vitro* to RV show lower levels of IFN-γ and IL-12 and higher levels of IL-4 and IL-10 in culture supernatants than cells from normal subjects. The IFN-γ/IL-4 ratio was three times lower in the asthmatic group [168].

In a study by Gern *et al.* of experimental RV16 infection in subjects with allergic rhinitis or asthma, the balance of airway Th1 and Th2 cytokines in induced sputum induced by viral infection was found to be related to clinical symptoms and viral clearance. Although protein could not be detected in sputum due to the presence of inhibitors of the ELISA assay used, there were increases in mRNA, as determined by semi-quantitative RT-PCR, for both IL-5 and IFN-γ. An inverse correlation was demonstrated between the ratio of IFN-γ mRNA to IL-5 mRNA and peak cold symptoms. In addition subjects with RV16 still detectable 14 days after inoculation had lower IFN-γ/IL-5 ratios during the acute phase of the cold than those subjects who had cleared the virus [169].

We have recently investigated the production of type 1 and type 2 cytokines from BAL cells in asthmatic and normal subjects. We found that production of the type 1 cytokines IL-12 and IFN-γ were suppressed in the asthmatics, while production of the type 2 cytokines IL-4, -5 and, -13 were all increased [34]. Importantly $CD4^+$ T-cell production of IFN-γ was strongly inversely correlated with virus load and reductions in lung function in the asthmatic subjects when they then underwent RV experimental infection, suggesting that $CD4^+$ T-cell production of IFN-γ is protective in the context of RV-induced asthma [34]. Conversely, $CD4^+$ T-cell production of each of IL-4, -5, and -13 was positively correlated with lower respiratory symptom severity, suggesting $CD4^+$ T-cell production of each of IL-4, -5, and -13 are associated with more severe exacerbations. These data are novel and important, but causal roles cannot be established in such human challenge studies. Investigation of the possible causal role of each these cytokines *in vivo* is now required using the newly developed mouse model [43].

$CD8^+$ T-cells

$CD8^+$ T-cells are important effector cells in specific cell-mediated antiviral immunity. They also demonstrate polarization of cytokine production, the major Tc1 cytokine again being IFN-γ and are believed to regulate CD4 Th1/Th2 balance [170]. In a murine asthma model induction of bystander $CD4^+$ Th2 responses to ovalbumin resulted in a switch of virus-peptide specific lung $CD8^+$ T-cells to production of Tc2 cytokines including IL-5 with, after virus peptide challenge, induction of airway eosinophilia [171]. If this occurs in man it suggests a means whereby CD8 antiviral function could be inhibited at the same time as CD8 amplification of allergic inflammation through IL-5 induction of airway eosinophilia. The role of $CD8^+$ T-cell production of type 1 and type 2 cytokines in virus-induced asthma exacerbations requires investigation.

$\gamma\delta$-TCR^+ T-cells

$\gamma\delta$-TCR^+ T-cells are a minor subset of T-cells expressing receptors distinct from the $\alpha\beta$ receptors found on the majority of T-cells involved in adaptive immunity. There appear to be at least two types of $\gamma\delta$-TCR^+ T-cells. The first type is found in the lymphoid tissue of all vertebrates and displays highly diversified receptors. The second type, intraepithelial $\gamma\delta$-TCR^+ T-cells, display receptors of limited diversity. It has been suggested that this second subset recognize molecules expressed only by nearby infected cells. Candidate ligands are heat-shock proteins, MHC class IB molecules and unorthodox nucleotides and phospholipids. Antigen is recognized directly rather than as processed peptide presented by MHC. Recognition of molecules expressed as a consequence of infection rather than pathogen-specific molecules themselves would place $\gamma\delta$-TCR^+ T-cells at the intersection of innate and adaptive immunity [172].

However, exaggerated responses to various pathogens and self tissues have been found in studies of mice deficient in $\gamma\delta$-TCR^++ T-cells rather than deficiencies in control of pathogens. Such work has suggested that at least some $\gamma\delta$-TCR^++ T-cells have a regulatory role in modulating immune responses [173], a function consistent with their demonstrated ability to secrete regulatory cytokines when activated.

It has been reported that $\gamma\delta$-TCR$^+$ T-cells are more numerous in the asthmatic airway [174]. A recent study found a greater capacity for production of IL-5 and IL-13 in bronchoalveolar lavage $\gamma\delta$-TCR$^+$ T-cells from asthmatic subjects [175]. If virus infection results in the release of molecules from epithelial cells that activate $\gamma\delta$-TCR$^+$ T-cells in the respiratory mucosa, such cells could provide a source of type 2 cytokines that influence the nature of the subsequent immune response. The role of $\gamma\delta$-TCR$^+$ T-cells in virus-induced asthma exacerbations requires investigation.

Eosinophils

Eosinophils are increased in bronchial epithelium in biopsies taken from normal and asthmatic volunteers following experimental RV infection; in a small study eosinophilic inflammation persisted for up to 6 weeks in asthmatic subjects [21] and in our recent study eosinophil numbers in the BAL were significantly (threefold) increased in asthmatic compared to normal subjects during the acute RV infection and correlated significantly with reductions in lung function only in the asthmatic subjects [34]. In allergic rhinitis experimental RV infection increases BAL eosinophils following segmental allergen challenge, again persisting for 6 weeks [28], and increased levels of ECP are found in the sputum of RV-infected subjects [23] and during naturally occurring acute exacerbations of asthma [13]. Eosinophils accumulate in the airway under the influence of IL-5, GM-CSF, IL-8, RANTES, and eotaxin [176]. Of these only IL-5 has not been shown to be produced by airway EC *in vitro* after infection by RV. Expression of RANTES is increased in nasal secretions of children with natural virus-induced asthma [107]. RANTES is upregulated in primary nasal EC cultures by RSV [177] and RV [178]. GM-CSF is important in bone marrow eosinophil production and in eosinophil survival [176] but levels are not increased during viral upper respiratory tract infections [107, 179, 180]. Levels of eotaxin in nasal lavage rise after experimental RV16 infection [181]. These data suggest a pathogenic role for eosinophils in virus-induced asthma. However, a protective role is also possible. In allergic rhinitic subjects, infected with RV after high dose allergen challenge, the severity and duration of cold symptoms were inversely related to the NL eosinophil count prior to infection [32]. Eosinophils may contribute to viral antigen presentation. Eosinophils pretreated with GM-CSF bind RV16 via ICAM-1 and present viral antigen to RV16-specific T-cells, inducing proliferation and secretion of IFN-γ [182]. Eosinophils have antiviral actions in parainfluenza-infected guinea pigs [183]. EDN and ECP have ribonuclease activity and reduce RSV infectivity [184]. The role of the eosinophil in the antiviral immune response thus requires further evaluation.

Mast cells/basophils

These cells are important sources of inflammatory mediators, characteristic of allergic inflammation in asthma. Mast cell basal and stimulated histamine release increases after virus infection [185]. Airway mast cell numbers are upregulated in a rat model of parainfluenza infection. Several viruses can enhance basophil IgE-mediated histamine release, but the role of this cell in human asthma is controversial.

Mast cells are also important sources of inflammatory mediators. Their function and localization suggest an early interaction with viruses. Leukotriene (LT) C_4 is among the mediators responsible for the late phase of bronchospasm in asthma. During RSV infection increased levels of LTC_4 were found in the nasopharyngeal secretions of infants [186]. Levels correlated well with the symptoms of the disease with concentrations in infants presenting with bronchiolitis being fivefold higher than in those with only upper respiratory tract symptomatology. Cultured alveolar macrophages can be infected with parainfluenza virus and respond with an increase in arachidonic acid metabolism. Several of the products of this pathway are known inducers of airway constriction, including LTC_4, LTD_4, $PGF_{2\alpha}$, and thromboxanes and/or stimulants of mucous secretion such as $PGF_{2\alpha}$, LTB_4, and 5-hydroxyeicosatetraenoic acid [187]. RV infection has been shown to induce prostaglandin and LT synthetic enzymes in bronchial biopsies in normal subjects, as well as trends for increased numbers of mast cells ($p = 0.07$) bronchoalveolar lavage fluid cysteinyl-leukotriene levels ($p = 0.13$), but these outcomes have not been studied in asthmatic subjects[188].

Neutrophils

Neutrophils are recruited early during respiratory viral infection in response to the production of IL-8, Gro-α, and ENA-78 by EC and activated neutrophils are a prominent feature of severe asthma. Induced sputum IS in asthmatics and nonasthmatics demonstrates a significant increase in neutrophils at day 4 of a natural cold, correlating with sputum IL-8 [189]. Similar results were obtained in IS taken 2 and 9 days after experimental RV16 infection in asthmatics. Intracellular staining demonstrated an increase in cells positive for IL-8 at day 2 attributable to increased IL-8 positive neutrophils [25]. The chemokine IL-8 is a potent chemoattractant for neutrophils but also acts on lymphocytes, basophils, and primed eosinophils. Increased IL-8 has been found in NL from children with natural colds [104]. Experimental RV16 infection of asthmatics resulted in elevated NL IL-8, correlating with cold/asthma symptom scores and histamine PC20 [25]. IS from asthmatics with exacerbations has both elevated IL-8 and neutrophilia [23, 190]. A study of experimental infection in asthmatic children also demonstrated elevated IL-8 and neutrophilia in NL during the acute infection and levels of neutrophil myeloperoxidase correlated with symptom severity [191]. In asthma, exacerbations in asthmatic adults [13] those with virus infection had increased sputum neutrophils and increased neutrophil elastase and more severe clinical disease. Such studies suggest a prominent role for the neutrophil in tissue damage during virus-induced asthma.

Natural killer cells

Natural killer (NK) cells are an important part of the innate immune response, their function being the elimination of a variety of target cells including virus-infected cells and the modulation of adaptive immunity toward viruses [192]. Cell killing by NK cells may occur through natural killing, antibody-dependent cellular cytotoxicity (ADCC), or apoptotic killing of Fas-positive target cells via membrane bound FasL. The ability to directly kill virus-infected cells is regulated by a balance between inhibitory and activating receptors [193]. Killer inhibitory receptors (KIRs), Ig-like receptors that recognize HLA-A, -B, or -C molecules, and the lectin-like CD94/NKG2A receptor that interacts with HLA-E allow NK cells to recognize cells expressing normal self MHC class I [194]. Loss of inhibition occurs if potential target cells have lost class I expression following virus infection or if they display abnormal class I/peptide complexes.

NK cells are rapid and efficient producers of cytokines such as IFN-γ, important both in early viral infection in the antigen-independent activation of antigen presenting cells such as macrophages, dendritic cells, and epithelial cells, and for biasing the development of $CD4^+$ Th1 and $CD8^+$ Tc1 cells. Cytokines and chemokines shown to enhance the activities of NK cells *in vitro* and *in vivo* include IFN-α/-β, IFN-γ, TNF-α, IL-2, IL-12, IL-15, IL-18, MIP-1α MIP-1β, MCP-1, 2, 3, and RANTES. Transforming growth factor (TGF)-β and IL-10 inhibit NK cell activity [195]. Type 2 cytokines may also modulate NK function, increasing NK type 2 activities and decreasing NK type 1 activities. Human NK cells cultured in medium supplemented with IL-4 differentiated into NK type 2 cells, secreting IL-5 and IL-13 and when cultured in the presence of IL-12, differentiated into NK type 1 cells secreting IL-10 and IFN-γ. IL-4 and IL-13 have also been shown to suppress IL-2-induced cytolytic and proliferative activities and IFN-γ production of human NK cells [196]. NK cell production of IL-5 is enhanced by IL-4 and reduced by IL-10 and IL-12 [197]. In a mouse model of asthma, intracellular staining of NK cells has demonstrated IL-5 production and depletion of NK cells resulted in reduced airway eosinophilia [198].

The function of NK cells in the asthmatic airway is as yet unexplored. It may be that, in an airway environment rich in type 2 cytokines, that NK type 1 function and effective antiviral activity are inhibited. If this is the case then a key component of the early immune response would be deficient and viral clearance would be impaired. In addition, if NK type 2 function is favored by the asthmatic microenvironment, production of type 2 cytokines by NK cells in response to virus infection might be one mechanism for amplification of allergic inflammation. These hypotheses are as yet untested in human studies of experimental virus infection.

B-lymphocytes and interaction of viruses with IgE-dependent mechanisms

An elevated serum total and allergen-specific IgE are features of "extrinsic" or atopic asthma. IgE-mediated mechanisms are certainly important in the pathophysiology of extrinsic asthma. Recent studies suggest a similar airway pathology in both extrinsic and "intrinsic" nonatopic asthma [199] where there is an absence of specific serum IgE and negative skin prick tests to aeroallergens. It has been suggested that there may be the production of local IgE to as yet unknown environmental allergens in intrinsic asthma.

Upregulation of total IgE or virus-/allergen-specific IgE locally or systemically during respiratory virus infection would be expected to contribute to the duration and severity of symptoms of an asthma exacerbation.

Intranasal challenge with RV39 results in an increase in total serum IgE in allergic rhinitic subjects but no increase in preexisting allergen-specific IgE [200]. In children with asthma, during infection with influenza A there was no change in total IgE but increases were observed in specific serum IgE to house dust mite and in *ex vivo* proliferative and IL-2 responses of lymphocytes challenged with house dust mite allergen [201]. In a study of RSV infection in infants the development of serum RSV-specific IgE occurred more frequently in atopics and correlated with clinical wheezing, histamine levels in nasal secretions, and hypoxia [202]. There is no information as yet on the presence of local virus-specific IgE in the airway during asthma exacerbations.

COPD

Increasing interest in the clinical features and pathogenesis of COPD reflects the worldwide importance of the disease. More than 14 million patients are affected in the United States alone. It is predicted to become the third leading cause of death worldwide by 2020 [203]. National and global initiatives have been launched and management guidelines have been published [204, 205].

The frequency of exacerbations is a major factor in the quality of life of patients with COPD [206]. The typical clinical features of an exacerbation include increased dyspnea, wheezing, cough, sputum production, and worsened gas exchange. Although noninfectious causes of exacerbations such as allergy, air pollution, or inhaled irritants including cigarette smoke may be important, acute airway infections are the major precipitants [207]. The infection and consequent host inflammatory response result in increased airway obstruction.

Epidemiology

It is likely that two-thirds to three-quarters of COPD exacerbations may be caused by viral infections. In a study of 186 patients rhinoviruses, influenza virus, parainfluenza virus, and coronavirus were significantly associated with COPD exacerbations [208]. Between 60% and 70% of exacerbations are associated with preceding symptoms of a common cold. The frequency of exacerbations requiring hospitalization is higher in the winter. One explanation for this could be the increased frequency of respiratory viruses at this time of the year. A recent study of 321 exacerbations in 83 patients with moderate to severe COPD using new diagnostic methods including RT-PCR shows a high incidence of viral infection [209]. Viruses were detected in nasal aspirates at exacerbation in almost 40% of cases. Rhinovirus was the most common, occurring in 58% of cases where a virus was present. The presence of virus was associated with increased dyspnea, cold symptoms, and sore throat and with prolonged recovery from exacerbation. Earlier studies relying on serology and virus culture quote lower virus detection rates of 15–20% [208, 210–212]. Other studies of more severe exacerbations using more comprehensive PCR methods confirmed the importance of virus infection, with viruses being detected in around 50% of exacerbations [152, 213]. Because these studies all involve sampling relatively late in the course of illness, it is likely that these detection rates underestimate the true importance of virus infections.

The role of bacteria in precipitating exacerbations is also somewhat controversial. Bacteria may have a primary role in the development of an exacerbation and/or represent a secondary superinfection of an initial viral process. Various bacterial species are present in the airways of 25–40% of patients, even when the COPD is stable but increased frequency of recovery of bacteria during exacerbations (~55%) as well as higher bacterial loads during exacerbations both suggest that they play an important role in a significant number [152, 214, 215]. Significant bacterial infection has been suggested when there is an abundance of neutrophils in the sputum [216] and when the sputum is purulent and green (due to neutrophil myeloperoxidase) [217]. Bacteria may contribute to the pathogenesis of an exacerbation due to increased bacterial loads of bacteria already colonizing diseased airways, however in addition to this, acquisition of new bacterial strains has also been shown to be important, increasing the risk of exacerbation over twofold [218].

The major bacterial organisms associated with COPD exacerbations are nontypable *Haemophilis influenzae*, *Streptococcus pneumoniae*, and *Moraxella (Branhamella) catarrhalis* [219, 220]. *Mycoplasma pneumoniae* and *Chlamydia pneumoniae* may play a part [221, 222]. Evidence also suggests that in more severe patients with a baseline FEV_1 of 35% predicted or less, gram-negative bacteria especially *enterobacteriaceae* and *pseudomonas* play an important part in acute exacerbations [223].

Recent studies have addressed the role of coinfection with both bacteria and viruses – one study showed this to occur in 25% of exacerbations, and that patients with dual infection had more marked lung function impairment and longer hospitalizations [152]. Another reported that exacerbations with both cold symptoms (a marker of putative viral infection) and a bacterial pathogen, the FEV_1 fall was greater and symptom count was higher than those with a bacterial pathogen alone [224]. Thus even in exacerbations in which viruses are detected, bacteria can also contribute to exacerbation severity.

Although the results of placebo-controlled trials show conflicting results, overall the effects of antibiotic treatment also support an etiological role for bacteria in exacerbations in some patients. A meta-analysis of nine studies showed a small overall benefit when antibiotics were used for COPD exacerbations [225]. The largest study included 362 exacerbations in 173 outpatients [216]. Compared with placebo, the rate of symptom resolution and improvement of peak expiratory flow during exacerbations was slightly but significantly faster when patients were treated with co-trimoxazole, amoxicillin, or doxycycline. More importantly, treatment failures as defined by respiratory deterioration were nearly twice as likely in the placebo group. Benefit from antibiotics was most evident for patients with most symptoms (dyspnea, increased sputum volume, and sputum purulence).

Guidelines for the use of antibiotics in acute exacerbations of COPD are unclear because of the difficulties in defining the role of bacterial infection in an individual case. The American Thoracic Society statement on COPD [205] suggests using antibiotics if there is evidence of infection (fever, leukocytosis, CXR changes) but not all patients with bacterial bronchial infection have fever (this is more common in viral infection or pneumonia) and few have CXR changes. The European Respiratory Society recommends antibiotics if the sputum is purulent, using standard antibiotics as first line, and sputum culture if these fail [226].

Evidence for a role for bacterial infection in pathogenesis/progression of COPD

Bacterial infection has a definite role in the pathogenesis of other chronic lung diseases such as cystic fibrosis and bronchiectasis where bacterial infection is chronic, causing not only acute exacerbations but also influencing long-term prognosis [207].

In these diseases chronic bacterial infection occurs as the host immune response is unable to clear the bacteria, the continuous infection leads to continuous inflammatory responses and continuous tissue damage [207]. Host and bacterial factors attract and activate neutrophils, which produce proteinases and reactive oxygen species. Lung antiproteinase defenses are overwhelmed. Both proteinase enzymes and reactive oxygen species cause damage to the epithelium, stimulating mucus production and impairing mucociliary clearance. Neutrophil elastase stimulates epithelial cell production of the chemokine IL-8 which attracts further neutrophils and in addition impairs phagocytosis by destroying antibody and cleaving complement receptors from neutrophils and complement components from bacteria. Neutrophils are also stimulated by cigarette smoke.

Identification of bacteria during exacerbation is also associated with increased levels of inflammatory mediators in BAL and/or sputum. These include reactive oxidant species, IL-8, TNF-α, neutrophil elastase, LTB_4, and myeloperoxidase and many others. These clearly have potential to cause considerable tissue damage, as well as further recruitment and activation of inflammatory cells. COPD patients, particularly those at the more severe end of the disease spectrum, may also be chronically colonized by bacteria between exacerbations, bacterial numbers then increasing during exacerbations. In a study using bronchoscopic protected brush specimens [214] 10 of 40 COPD patients were colonized with bacteria when stable. During exacerbations 50% had bacteria present and when present, bacterial numbers were greater. When protected brush specimens were taken during severe acute exacerbations of COPD requiring ventilation [227] bacteria were detected in 50% but it was not possible to distinguish patients more likely to have bacteria on the basis of clinical features or other investigations.

The major bacterial pathogens isolated during bronchial infections all form part of the commensal flora in the nasopharynx. Bronchial infections occur in patients with abnormal airways with reduced host defenses. Persistence of bacteria within the bronchial tree may come about through toxins that impair mucociliary clearance, enzymes that breakdown local immunoglobulin, products that alter immune effector cell function, adherence to mucus and damaged epithelium, or other mechanisms of avoiding immune surveillance [207, 228].

Bacterial colonization in the stable state represents an equilibrium in which the number of bacteria present in the bronchial tree is contained by the host defenses but not eliminated. During an exacerbation this equilibrium is upset and bacterial numbers increase, inciting an inflammatory response. Change will usually occur because of a change in the host rather than altered virulence of the bacteria, for example as a result of viral infection.

Evidence for a role for viruses in pathogenesis/progression of COPD

Exacerbations associated with viral infections also have increased levels of many inflammatory mediators also found in bacterial exacerbations including TNF-α, IL-8, neutrophil elastase, myeloperoxidase, and LTB_4. In contrast to bacterial exacerbations where neutrophils and neutrophil products predominate, during a viral exacerbation both neutrophils and eosinophils are present and eosinophil products such as ECP are also increased [152]. Other mediators implicated include ENA-78, RANTES, and endothelin–1 [229].

It has also been suggested that persistent virus infection contributes to the progression of COPD. In particular, adenovirus appears to persist in a latent form in which viral proteins are produced without replication of complete virus. Such latent infection may amplify lung inflammation due to cigarette smoke [230]. Adenoviral E1A DNA persists in human lungs from patients with COPD compared with patients of similar age, sex, and smoking history who do not have COPD [231]. The E1A protein has been demonstrated in airway epithelial cells from smokers [232]. It is able to amplify many host genes through attachment to the DNA-binding sites of transcription factors [233]. Airway epithelial cells transfected with E1A produce excess inflammatory cytokines such as IL-8 [234] and surface adhesion molecules such as ICAM-1 [235] after *in vitro* challenge by an NF-κB-dependent mechanism [236].

RSV has been identified in induced sputum from patients with stable COPD. These individuals have a higher plasma fibrinogen and serum IL-6, a higher pCO_2 and increased frequency of exacerbations [209]. This suggests either that low grade persistent RSV infection contributes to COPD severity or that patients with more severe COPD are less able to clear RSV from the airway.

The immunology of virus infection in COPD is not well understood. Less data is available than for virus infection in asthma since this has not been a major subject of human experimental infection studies. In a small safety study of four patients, inoculation with low dose RV16 resulted in symptomatic colds, viral replication, significant increases in lower respiratory tract symptoms, and reductions in PEF and FEV_1 typical of an acute exacerbation of COPD [237]. Further studies are clearly needed in view of the increasing evidence for a major role for viruses in causing COPD exacerbations.

Therapy for infective exacerbations of asthma and COPD

Currently much of the treatment of infective exacerbations of asthma and COPD is symptomatic, consisting of increased bronchodilators, either short-acting β_2-agonists in inhaled or intravenous form or anticholinergics or theophyllines, or supportive in the form of oxygen and in severe cases noninvasive or invasive ventilatory measures. Corticosteroids are widely used in inhaled or oral form for their anti-inflammatory actions. The effects of corticosteroids are the result of actions at many points in various inflammatory cascades. Whilst this undoubtedly contributes to their beneficial effects it also results in significant local and systemic side effects, in particular if oral steroid treatment is prolonged or frequent. In addition systemic steroids may interfere with the antiviral immune response resulting in reduced viral clearance [238].

In persistent asthma, control of disease is achieved predominantly with inhaled corticosteroids. There is a role for additional drugs such as long-acting β_2-agonists and leukotriene antagonists. The long-acting β_2-agonists in particular appear to increase the effectiveness of inhaled corticosteroids allowing the dose needed to achieve control to be reduced [239]. There is also evidence that these drugs in combination with inhaled corticosteroids may further reduce exacerbation frequency [240]. The leukotriene antagonists appear to be most effective in treating or preventing exacerbations in children [241, 242].

Regular corticosteroid treatment is however only partially effective at preventing exacerbations. In adult asthma inhaled steroids reduce exacerbation frequency by only 40% [243]. In school age children inhaled steroids are ineffective at reducing exacerbation frequency, duration, or severity [244]. In preschool age children with virus-induced wheeze oral steroids are ineffective even in those with primed eosinophils [245].

Specific antibiotic therapy is available for bacterial infections and is indicated where there is good evidence of such infection or when the exacerbation is severe and bacterial involvement is a possibility.

However, as discussed above the majority of infective asthma exacerbations are of viral rather than bacterial origin and viruses are also common in exacerbations of COPD.

Vaccination

The success of vaccination to prevent respiratory virus infections has been limited by significant variation within the major virus types causing disease. There are 102 serotyped strains of rhinovirus and several more that have not been serotyped and no effective vaccine has been introduced. A decavalent vaccine [246] developed in the 1970s was ultimately of limited efficacy. The influenza viruses display antigenic shift and drift. Vaccines must be modified every 2–3 years to cover the strains prevalent at the time. Vaccination against RSV experienced a major setback in the 1960s when the use of formalin-inactivated virus in young babies resulted in increased disease severity following subsequent virus infection [247]. Eighty percent of vaccinated children required hospitalization when subsequently infected with RSV, as compared to 5% of controls. The lungs of two vaccinated children who died contained eosinophilic infiltrates. It has been suggested that formalin inactivation may have modified epitopes within the RSV G and F surface glycoproteins, resulting in a modified immune response to subsequent infection with enhanced immunopathology [248]. Vaccinated individuals demonstrate a number of differences from individuals who have suffered natural RSV infection including a lack of specific mucosal antibodies and deficient neutralizing and fusion-inhibiting serum antibodies [249]. There are also differences in the cell-mediated immune response with some vaccinated individuals demonstrating peripheral eosinophilia and exaggerated lymphocytic proliferative responses to RSV [250]. To protect against RSV infection a successful vaccine would need to provide more effective protection than natural infection, which is itself frequently followed by reinfection [251], and would have to be administered early in infancy to have an effect on infant bronchiolitis.

Treatment for virus-induced asthma exacerbations

Simple nonspecific treatments for the common cold do exist although their efficacy is debated. Vitamin C and zinc gluconate [252] both may shorten the duration of a cold by 1–2 days. The inhalation of humidified hot air provides symptomatic relief [253]. Nasal IFN-α is an effective treatment for the common cold [254] but must be given either prior to or shortly after exposure to the virus. It is also expensive and is associated with significant local side effects such as bleeding and discharge. These problems have limited its clinical use. However IFN therapy for virus-induced asthma exacerbations may be more useful in view of the deficiencies identified by recent studies. Further because of the large number of viruses producing similar clinical syndromes, the general antiviral properties of IFNs would provide significant advantage over the use of specific antiviral drugs.

Antivirals

Specific antiviral agents exist for influenza. Amantidine and rimantidine are effective against influenza A. The use of amantidine has been limited by CNS side effects such as dizziness and insomnia; fewer such side effects are seen with rimantidine. Both drugs are indicated during epidemics for treatment and prophylaxis in high risk groups including asthmatics. Neither is active against influenza B. Two neuraminidase inhibitors, zanamivir and oseltamivir, are active against both influenza A and B [255–257]. These agents are effective in preventing infection when used as prophylaxis during the influenza season and, as treatment, they reduce the duration of illness if started within 36–48 h of the onset of illness. Zanamivir must be given by inhalation whereas oseltamivir can be given orally. Ribavirin is a nucleoside analog active against RSV *in vivo* and also against influenza *in vitro*. Nebulized ribavirin therapy is licensed for use in hospitalized infants and children in the first 3 days of RSV bronchiolitis. It is however expensive and of unproven benefit on clinical outcome. Because of its toxicity it is not appropriate for asthma. RSV enriched immunoglobulin was effective as prophylaxis for infants at high risk of RSV bronchiolitis [258] but has been superseded by RSV neutralizing monoclonal antibodies [259].

Antirhinoviral agents

RV are a major target for drug treatment. It has been estimated that rhinoviruses result in 6–10 colds per year in young children [260]. As yet no effective agent is available for clinical use. Capsid-binding/canyon inhibitors block RV binding to host cell receptor (ICAM-1 in the case of the major group). One example in phase 3 clinical trials is pleconaril (Picovir). These drugs can be extremely potent but their clinical usefulness is often limited by toxicity, the need for rapid initiation of therapy and the possible development of resistance. Alternative targets include soluble ICAM-1 which inhibits major rhinovirus infection and conserved viral enzymes such as protein 3D, the RNA-dependent RNA transcriptase, protein 2C, the associated ATP-helicase, and the cysteine protease 3C.

New approaches

Alternative approaches to direct antiviral therapy are suppression of virus-induced inflammation, or strategies that promote innate or type 1 immune responses in individuals with excessive type 2 responses. Understanding the complexities of the antiviral immune response, in particular how it may be altered in the context of preexisting chronic airway diseases such as asthma is an essential first step. Further work is needed to elucidate the important sites of interaction between the immunological networks of asthma and of virus infection. Greater knowledge is required if we are to identify key targets for therapeutic intervention, the aim of which will be to minimize immunopathology whilst maintaining or enhancing the host anti-viral immune response.

References

1. Asher MI, Montefort S, Bjorksten B, Lai CK, Strachan DP, Weiland SK, Williams H, and ISAAC Phase Three Study Group. Worldwide time trends in the prevalence of symptoms of asthma, allergic rhinoconjunctivitis, and eczema in childhood: ISAAC Phases One and Three repeat multicountry cross-sectional surveys. *Lancet* 368: 733–43, 2006.
2. Stevens FA. Acute asthmatic episodes in children caused by upper respiratory bacteria during colds, with and without bacterial sensitization. J Allergy 24: 221–26, 1953.

3. Bardin PG, Johnston SL, Pattemore PK. Viruses as precipitants of asthma symptoms. II. Physiology and mechanisms. *Clin Exp Allergy* 22: 809–22, 1992.
4. McIntosh K, Ellis EF, Hoffman LS, Lybass TG, Eller JJ, Fulginiti VA. The association of viral and bacterial respiratory infections with exacerbations of wheezing in young asthmatic children. *J Pediatr* 82: 578–90, 1973.
5. Johnston SL, Pattemore PK, Sanderson G, Smith S, Campbell MJ, Josephs LK, Cunningham A, Robinson BS, Myint SH, Ward ME et al. The relationship between upper respiratory infections and hospital admissions for asthma: A time-trend analysis. *Am J Respir Crit Care Med* 154: 654–60, 1996.
6. Minor TE, Dick EC, DeMeo AN, Ouellette JJ, Cohen M, Reed CE. Viruses as precipitants of asthmatic attacks in children. *JAMA* 227: 292–98, 1974.
7. Horn ME, Brain EA, Gregg I, Inglis JM, Yealland SJ, Taylor P. Respiratory viral infection and wheezy bronchitis in childhood. *Thorax* 34: 23–28, 1979.
8. Johnston SL, Pattemore PK, Sanderson G, Smith S, Lampe F, Josephs L, Symington P, O'Toole S, Myint SH, Tyrrell DA. Community study of role of viral infections in exacerbations of asthma in 9–11 year old children. *BMJ* 310: 1225–29, 1995.
9. Chauhan AJ, Inskip HM, Linaker CH, Smith S, Schreiber J, Johnston SL, Holgate ST. Personal exposure to nitrogen dioxide (NO_2) and the severity of virus-induced asthma in children. *Lancet* 361: 1939–44, 2003.
10. Freymuth F, Vabret A, Brouard J, Toutain F, Verdon R, Petitjean J, Gouarin S, Duhamel JF, Guillois B. Detection of viral, *Chlamydia pneumoniae* and *Mycoplasma pneumoniae* infections in exacerbations of asthma in children. *J Clin Virol* 13: 131–39, 1999.
11. Nicholson KG, Kent J, Ireland DC. Respiratory viruses and exacerbations of asthma in adults. *BMJ* 307: 982–86, 1993.
12. Beasley R, Coleman ED, Hermon Y, Holst PE, O'Donnell TV, Tobias M. Viral respiratory tract infection and exacerbations of asthma in adult patients. *Thorax* 43: 679–83, 1988.
13. Wark PA, Johnston SL, Moric I, Simpson JL, Hensley MJ, Gibson PG. Neutrophil degranulation and cell lysis is associated with clinical severity in virus-induced asthma. *Eur Respir J* 19: 68–75, 2002.
14. Grissell TV, Powell H, Shafren DR, Boyle MJ, Hensley MJ, Jones PD, Whitehead BF, Gibson PG. Interleukin-10 gene expression in acute virus-induced asthma. *Am J Respir Crit Care Med* 172: 433–39, 2005.
15. Berman SZ, Mathison DA, Stevenson DD, Tan EM, Vaughan JH. Transtracheal aspiration studies in asthmatic patients in relapse with "infective" asthma and in subjects without respiratory disease. *J Allergy Clin Immunol* 56: 206–14, 1975.
16. Gwaltney JMJ, Hendley O, Hayden FG, McIntosh K, Hollinger FB, Melnick JL, Turner RB. Updated recommendations for safety-testing of viral inocula used in volunteer experiments on rhinovirus colds. *Prog Med Virol* 39: 256–63, 1992.
17. Halperin SA, Eggleston PA, Beasley P, Suratt P, Hendley JO, Groschel DH, Gwaltney JM. Exacerbations of asthma in adults during experimental rhinovirus infection. *Am Rev Respir Dis* 132: 976–80, 1985.
18. Bardin PG, Fraenkel DJ, Sanderson G, Dorward M, Lau LC, Johnston SL, Holgate ST. Amplified rhinovirus colds in atopic subjects. *Clin Exp Allergy* 24: 457–64, 1994.
19. Lemanske RFJ, Dick EC, Swenson CA, Vrtis RF, Busse WW. Rhinovirus upper respiratory infection increases airway hyperreactivity and late asthmatic reactions. *J Clin Invest* 83: 1–10, 1989.
20. Cheung D, Dick EC, Timmers MC, de Klerk EP, Spaan WJ, Sterk PJ. Rhinovirus inhalation causes long-lasting excessive airway narrowing in response to methacholine in asthmatic subjects *in vivo*. *Am J Respir Crit Care Med* 152: 1490–96, 1995.
21. Fraenkel DJ, Bardin PG, Sanderson G, Lampe F, Johnston SL, Holgate ST. Lower airways inflammation during rhinovirus colds in normal and in asthmatic subjects. *Am J Respir Crit Care Med* 151: 879–86, 1995.
22. Grunberg K, Kuijpers EA, de Klerk EP, de Gouw HW, Kroes AC, Dick EC, Sterk PJ. Effects of experimental rhinovirus 16 infection on airway hyperresponsiveness to bradykinin in asthmatic subjects *in vivo*. *Am J Respir Crit Care Med* 155: 833–38, 1997.
23. Grunberg K, Smits HH, Timmers MC, de Klerk EP, Dolhain RJ, Dick EC, Hiemstra PS, Sterk PJ. Experimental rhinovirus 16 infection. Effects on cell differentials and soluble markers in sputum in asthmatic subjects. *Am J Respir Crit Care Med* 156: 609–16, 1997.
24. Grunberg K, Timmers MC, de Klerk EP, Dick EC, Sterk PJ. Experimental rhinovirus 16 infection causes variable airway obstruction in subjects with atopic asthma. *Am J Respir Crit Care Med* 160: 1375–80, 1999.
25. Grunberg K, Timmers MC, Smits HH, de Klerk EP, Dick EC, Spaan WJ, Hiemstra PS, Sterk PJ. Effect of experimental rhinovirus 16 colds on airway hyperresponsiveness to histamine and interleukin-8 in nasal lavage in asthmatic subjects *in vivo* [see comments]. *Clin Exp Allergy* 27: 36–45, 1997.
26. Halperin SA, Eggleston PA, Hendley JO, Suratt PM, Groschel DH, Gwaltney JM. Pathogenesis of lower respiratory tract symptoms in experimental rhinovirus infection. *Am Rev Respir Dis* 128: 806–10, 1983.
27. Calhoun WJ, Swenson CA, Dick EC, Schwartz LB, Lemanske RFJ, Busse WW. Experimental rhinovirus 16 infection potentiates histamine release after antigen bronchoprovocation in allergic subjects. *Am Rev Respir Dis* 144: 1267–73, 1991.
28. Calhoun WJ, Dick EC, Schwartz LB, Busse WW. A common cold virus, rhinovirus 16, potentiates airway inflammation after segmental antigen bronchoprovocation in allergic subjects. *J Clin Invest* 94: 2200–8, 1994.
29. Green RM, Custovic A, Sanderson G, Hunter J, Johnston SL, Woodcock A. Synergism between allergens and viruses and risk of hospital admission with asthma: Case-control study. *BMJ* 324: 763, 2002.
30. Murray CS, Poletti G, Kebadze T, Morris J, Woodcock A, Johnston SL, Custovic A. Study of modifiable risk factors for asthma exacerbations: Virus infection and allergen exposure increase the risk of asthma hospital admissions in children. *Thorax* 61: 376–82, 2006.
31. Linaker CH, Coggon D, Holgate ST, Clough J, Josephs L, Chauhan AJ, Inskip HM. Personal exposure to nitrogen dioxide and risk of airflow obstruction in asthmatic children with upper respiratory infection. *Thorax* 55: 930–33, 2000.
32. Avila PC, Abisheganaden JA, Wong H, Liu J, Yagi S, Schnurr D, Kishiyama JL, Boushey HA. Effects of allergic inflammation of the nasal mucosa on the severity of rhinovirus 16 cold. *J Allergy Clin Immunol* 105: 923–32, 2000.
33. de Kluijver J, Evertse CE, Sont JK, Schrumpf JA, van Zeijl-van der Ham CJ, Dick CR, Rabe KF, Hiemstra PS, Sterk PJ. Are rhinovirus-induced airway responses in asthma aggravated by chronic allergen exposure? [see comment]. *Am J Respir Crit Care Med* 168: 1174–80, 2003.
34. Message SD, Laza-Stanca V, Mallia P, Parker HL, Zhu J, Kebadze T, Contoli M, Sanderson G, Kon OM, Papi A, Jeffery PK, Stanciu L, Johnston SL. Rhinovirus induced lower respiratory illness is increased in asthma and related to viral load and Th1/2 cytokine and IL-10 production. *ProcNatl AcadSci USA* 24, 2008, (in press).
35. Horn ME, Reed SE, Taylor P. Role of viruses and bacteria in acute wheezy bronchitis in childhood: A study of sputum. *Arch Dis Childhood* 54: 587–92, 1979.
36. Gern JE, Galagan DM, Jarjour NN, Dick EC, Busse WW. Detection of rhinovirus RNA in lower airway cells during experimentally induced infection. *Am J Respir Crit Care Med* 155: 1159–61, 1997.
37. Subauste MC, Jacoby DB, Richards SM, Proud D. Infection of a human respiratory epithelial cell line with rhinovirus. Induction of cytokine release and modulation of susceptibility to infection by cytokine exposure. *J Clin Invest* 96: 549–57, 1995.
38. Papadopoulos NG, Bates PJ, Bardin PG, Papi A, Leir SH, Fraenkel DJ, Meyer J, Lackie PM, Sanderson G, Holgate ST et al. Rhinoviruses infect the lower airways. *J Infect Dis* 181: 1875–84, 2000.

39. Schroth MK, Grimm E, Frindt P, Galagan DM, Konno SI, Love R, Gern JE. Rhinovirus replication causes RANTES production in primary bronchial epithelial cells. *Am J Respir Cell Mol Biol* 20: 1220–28, 1999.
40. Wark PA, Johnston SL, Bucchieri F, Powell R, Puddicombe S, Laza-Stanca V, Holgate ST, Davies DE. Asthmatic bronchial epithelial cells have a deficient innate immune response to infection with rhinovirus. *J Exp Med* 201: 937–47, 2005.
41. Papadopoulos NG, Sanderson G, Hunter J, Johnston SL. Rhinoviruses replicate effectively at lower airway temperatures. *J Med Virol* 58: 100–4, 1999.
42. Jakiela B, Brockman-Schneider R, Amineva S, Lee WM, Gern JE. Basal cells of differentiated bronchial epithelium are more susceptible to rhinovirus infection. *Am J Respir Cell Mol Biol* 38(5): 517–23, 2008.
43. Bartlett NW, Walton RP, Edwards MR, Aniscenko J, Caramori G, Zhu J, Glanville N, Choy KJ, Jourdan P, Burnet J et al. Mouse models of rhinovirus-induced disease and exacerbation of allergic airway inflammation. *Nat Med* 14: 199–204, 2008.
44. Bardin PG, Fraenkel DJ, Sanderson G, van Schalkwyk EM, Holgate ST, Johnston SL. Peak expiratory flow changes during experimental rhinovirus infection. *Eur Respir J* 16: 980–85, 2000.
45. Gern JE, Calhoun W, Swenson C, Shen G, Busse WW. Rhinovirus infection preferentially increases lower airway responsiveness in allergic subjects. *Am J Respir Crit Care Med* 155: 1872–76, 1997.
46. Whitton JL, Oldstone MBA. Immune response to viruses. In: Fields BN, Knipe DN, Howley PM. *Fields Virology*, 345–74. Philadelphia, PA: Lippincott-Raven, 1996.
47. Yewdell JW, Bennink JR. Immune responses to viruses. In: Richman DR, Whiteley RJ, Hayden FG. *Clinical Virology*, 271–306. New York: Churchill Livingstone, 1997.
48. Suzuki T, Yamaya M, Kamanaka M, Jia YX, Nakayama K, Hosoda M, Yamada N, Nishimura H, Sekizawa K, Sasaki H. Type 2 rhinovirus infection of cultured human tracheal epithelial cells: Role of LDL receptor. *Am J Physiol Lung Cell Mol Physiol* 280: L409–20, 2001.
49. Bjornson AB, Mellencamp MA, Schiff GM. Complement is activated in the upper respiratory tract during influenza virus infection. *Am Rev Respir Dis* 143: 1062–66, 1991.
50. Amineva SP, Gern JE. Rhinovirus 3C protease cleaves the C3 and C5 complement factors. *Am J Respir Crit Care Med* 167: A212, 2003.
51. Kopf M, Abel B, Gallimore A, Carroll M, Bachmann MF. Complement component C3 promotes T-cell priming and lung migration to control acute influenza virus infection. *Nat Med* 8: 373–78, 2002.
52. Anders EM, Hartley CA, Reading PC, Ezekowitz RA. Complement-dependent neutralization of influenza virus by a serum mannose-binding lectin. *J Gen Virol* 75: 615–22, 1994.
53. Yang D, Biragyn A, Kwak LW, Oppenheim JJ. Mammalian defensins in immunity: more than just microbicidal. *Trends Immunol* 23: 291–96, 2002.
54. Barclay WS, al-Nakib W, Higgins PG, Tyrrell DA. The time course of the humoral immune response to rhinovirus infection. *Epidemiol Infect* 103: 659–69, 1989.
55. Alper CM, Doyle WJ, Skoner DP, Buchman CA, Seroky JT, Gwaltney JM, Cohen SA. Prechallenge antibodies: Moderators of infection rate, signs, and symptoms in adults experimentally challenged with rhinovirus type 39. *Laryngoscope* 106: 1298–305, 1996.
56. Alper CM, Doyle WJ, Skoner DP, Buchman CA, Cohen S, Gwaltney JM. Prechallenge antibodies moderate disease expression in adults experimentally exposed to rhinovirus strain hanks. *Clin Infect Dis* 27: 119–28, 1998.
57. Hastings GZ, Francis MJ, Rowlands DJ, Chain BM. Epitope analysis of the T cell response to a complex antigen: Proliferative responses to human rhinovirus capsids. *Eur J Immunol* 23: 2300–5, 1993.
58. Gern JE, Dick EC, Kelly EA, Vrtis R, Klein B. Rhinovirus-specific T cells recognize both shared and serotype-restricted viral epitopes. *J Infect Dis* 175: 1108–14, 1997.
59. Parry DE, Busse WW, Sukow KA, Dick CR, Swenson C, Gern JE. Rhinovirus-induced PBMC responses and outcome of experimental infection in allergic subjects. *J Allergy Clin Immunol* 105: 692–98, 2000.
60. Corne JM, Marshall C, Smith S, Schreiber J, Sanderson G, Holgate ST, Johnston SL. Frequency, severity, and duration of rhinovirus infections in asthmatic and non-asthmatic individuals: A longitudinal cohort study. *Lancet* 359: 831–34, 2002.
61. Wark PA, Johnston SL, Bucchieri F, Powell R, Puddicombe S, Laza-Stanca V, Holgate ST, Davies DE. Asthmatic bronchial epithelial cells have a deficient innate immune response to infection with rhinovirus. *J Exp Med* 201: 937–47, 2005.
62. Contoli M, Message SD, Laza-Stanca V, Edwards MR, Wark PA, Bartlett NW, Kebadze T, Mallia P, Stanciu LA, Parker HL et al. Role of deficient type III interferon-lambda production in asthma exacerbations. *Nat Med* 12: 1023–26, 2006.
63. Papi A, Stanciu LA, Papadopoulos NG, Teran LM, Holgate ST, Johnston SL. Rhinovirus infection induces major histocompatibility complex class I and costimulatory molecule upregulation on respiratory epithelial cells. *J Infect Dis* 181: 1780–84, 2000.
64. Hers JF. Disturbances of the ciliated epithelium due to influenza virus. *Am Rev Respir Dis* 93 (Suppl 77), 1966.
65. Turner RB, Hendley JO, Gwaltney JM. Shedding of infected ciliated epithelial cells in rhinovirus colds. *J Infect Dis* 145: 849–53, 1982.
66. Turner RB, Winther B, Hendley JO, Mygind N, Gwaltney JM. Sites of virus recovery and antigen detection in epithelial cells during experimental rhinovirus infection. *Acta Otolaryngol Suppl* 413: 9–14, 1984.
67. Winther B, Gwaltney JM, Hendley JO. Respiratory virus infection of monolayer cultures of human nasal epithelial cells. *Am Rev Respir Dis* 141: 839–45, 1990.
68. Mosser AG, Brockman-Schneider R, Amineva S, Burchell L, Sedgwick JB, Busse WW, Gern JE. Similar frequency of rhinovirus-infectible cells in upper and lower airway epithelium. *J Infect Dis* 185: 734–43, 2002.
69. Greve JM, Davis G, Meyer AM, Forte CP, Yost SC, Marlor CW, Kamarck ME, McClelland A. The major human rhinovirus receptor is ICAM-1. *Cell* 56: 839–47, 1989.
70. Winther B, Greve JM, Gwaltney JMJ, Innes DJ, Eastham JR, McClelland A, Hendley JO. Surface expression of intercellular adhesion molecule-1 on epithelial cells in the human adenoid. *J Infect Dis* 176: 523–25, 1997.
71. Winther B, Arruda E, Witek TJ, Marlin SD, Tsianco MM, Innes DJ, Hayden FG. Expression of ICAM-1 in nasal epithelium and levels of soluble ICAM-1 in nasal lavage fluid during human experimental rhinovirus infection. *Arch OtolaryngolHead Neck Surg* 128: 131–36, 2002.
72. Bianco A, Spiteri MA. A biological model to explain the association between human rhinovirus respiratory infections and bronchial asthma. *Monaldi Arch Chest Dis* 53: 83–87, 1998.
73. Grunberg K, Sharon RF, Hiltermann TJ, Brahim JJ, Dick EC, Sterk PJ, van Krieken JH. Experimental rhinovirus 16 infection increases intercellular adhesion molecule-1 expression in bronchial epithelium of asthmatics regardless of inhaled steroid treatment. *Clin Exp Allergy* 30: 1015–23, 2000.
74. Whiteman SC, Bianco A, Knight RA, Spiteri MA. Human rhinovirus selectively modulates membranous and soluble forms of its intercellular adhesion molecule-1 (ICAM-1) receptor to promote epithelial cell infectivity. *J Biol Chem* 278: 11954–61, 2003.
75. Altman LC, Ayars GH, Baker C, Luchtel DL. Cytokines and eosinophil-derived cationic proteins upregulate intercellular adhesion molecule-1 on human nasal epithelial cells. *J Allergy Clin Immunol* 92: 527–36, 1993.
76. Terajima M, Yamaya M, Sekizawa K, Okinaga S, Suzuki T, Yamada N, Nakayama K, Ohrui T, Oshima T, Numazaki Y et al. Rhinovirus infection of primary cultures of human tracheal epithelium: Role of ICAM-1 and IL-1beta. *Am J Physiol* 273: L749–59, 1997.
77. Sethi SK, Bianco A, Allen JT, Knight RA, Spiteri MA. Interferon-gamma (IFN-gamma) down-regulates the rhinovirus-induced expression of intercellular adhesion molecule-1 (ICAM-1) on human airway epithelial cells. *Clin Exp Immunol* 110: 362–69, 1997.

78. Bianco A, Sethi SK, Allen JT, Knight RA, Spiteri MA. Th2 cytokines exert a dominant influence on epithelial cell expression of the major group human rhinovirus receptor, ICAM-1. *Eur Respir J* 12: 619–26, 1998.
79. Canonica GW, Ciprandi G, Pesce GP, Buscaglia S, Paolieri F, Bagnasco M. ICAM-1 on epithelial cells in allergic subjects: A hallmark of allergic inflammation. *Int Arch Allergy Immunol* 107: 99–102, 1995.
80. Bianco A, Whiteman SC, Sethi SK, Allen JT, Knight RA, Spiteri MA. Expression of intercellular adhesion molecule-1 (ICAM-1) in nasal epithelial cells of atopic subjects: A mechanism for increased rhinovirus infection?. *Clin Exp Immunol* 121: 339–45, 2000.
81. Papi A, Johnston SL. Respiratory epithelial cell expression of vascular cell adhesion molecule-1 and its up-regulation by rhinovirus infection via NF-kappaB and GATA transcription factors. *J Biol Chem* 274: 30041–51, 1999a.
82. Papi A, Johnston SL. Rhinovirus infection induces expression of its own receptor intercellular adhesion molecule 1 (ICAM-1) via increased NF-kappaB-mediated transcription. *J Biol Chem* 274: 9707–20, 1999b.
83. Barnes PJ, Adcock IM. Transcription factors and asthma. *Eur Respir J* 12: 221–34, 1998.
84. Papi A, Papadopoulos NG, Degitz K, Holgate ST, Johnston SL. Corticosteroids inhibit rhinovirus-induced intercellular adhesion molecule-1 up-regulation and promoter activation on respiratory epithelial cells. *J Allergy Clin Immunol* 105: 318–26, 2000.
85. Suzuki T, Yamaya M, Sekizawa K, Yamada N, Nakayama K, Ishizuka S, Kamanaka M, Morimoto T, Numazaki Y, Sasaki H. Effects of dexamethasone on rhinovirus infection in cultured human tracheal epithelial cells. *Am J Physiol Lung Cell Mol Physiol* 278: L560–71, 2000.
86. Papi A, Papadopoulos NG, Stanciu LA, Bellettato CM, Pinamonti S, Degitz K, Holgate ST, Johnston SL. Reducing agents inhibit rhinovirus-induced up-regulation of the rhinovirus receptor intercellular adhesion molecule-1 (ICAM-1) in respiratory epithelial cells. *FASEB J* 16: 1934–36, 2002.
87. Papi A, Papadopoulos NG, Stanciu LA, Degitz K, Holgate ST, Johnston SL. Effect of desloratadine and loratadine on rhinovirus-induced intercellular adhesion molecule 1 upregulation and promoter activation in respiratory epithelial cells. *J Allergy Clin Immunol* 108: 221–28, 2001.
88. Suzuki T, Yamaya M, Sekizawa K, Hosoda M, Yamada N, Ishizuka S, Yoshino A, Yasuda H, Takahashi H, Nishimura H et al. Erythromycin inhibits rhinovirus infection in cultured human tracheal epithelial cells. *Am J Respir Crit Care Med* 165: 1113–18, 2002.
89. Samuel CE. Antiviral actions of interferons. *Clinical Microbiology Reviews* 14:778–809, 2001.
90. Kotenko SV, Gallagher G, Baurin VV, Lewis-Antes A, Shen M, Shah NK, Langer JA, Sheikh F, Dickensheets H, Donnelly RP. IFN-lambdas mediate antiviral protection through a distinct class II cytokine receptor complex. *Nat Immunol* 4: 69–77, 2003.
91. Sheppard P, Kindsvogel W, Xu W, Henderson K, Schlutsmeyer S, Whitmore TE, Kuestner R, Garrigues U, Birks C, Roraback J et al. IL-28, IL-29 and their class II cytokine receptor IL-28R. *Nat Immunol* 4: 63–68, 2003.
92. Katze MG, He Y, Gale M. Viruses and interferon: A fight for supremacy. *Nat Rev Immunol* 2: 675–87, 2002.
93. Bergmann M, Garcia-Sastre A, Carnero E, Pehamberger H, Wolff K, Palese P, Muster T. Influenza virus NS1 protein counteracts PKR-mediated inhibition of replication. *J Virol* 74: 6203–6, 2000.
94. Contoli M, Stanciu L, Message SD, Papi A, Johnston SL. Susceptibility to asthma exacerbations: Antiviral immunity and protection against asthma exacerbations. In: Johnston SL, O'Byrne PM. *Exacerbations of asthma*, 167–85. UK: Informa, 2007.
95. Kotenko SV. The family of IL-10-related cytokines and their receptors: Related but to what extent?. *Cytokine Growth Factor Rev* 13: 223–40, 2002.
96. Robek MD, Boyd BS, Chisari FV. Lambda interferon inhibits hepatitis B and C virus replication. *J Virol* 79: 3851–54, 2005.
97. Osterlund P, Veckman V, Siren J, Klucher KM, Hiscott J, Matikainen S, Julkunen I. Gene expression and antiviral activity of alpha/beta interferons and interleukin-29 in virus-infected human myeloid dendritic cells. *J Virol* 79: 9608–17, 2005.
98. Coccia EM, Severa M, Giacomini E, Monneron D, Remoli ME, Julkunen I, Cella M, Lande R, Uze G. Viral infection and Toll-like receptor agonists induce a differential expression of type I and lambda interferons in human plasmacytoid and monocyte-derived dendritic cells. *Eur J Immunol* 34: 796–805, 2004.
99. Spann KM, Tran KC, Chi B, Rabin RL, Collins PL. Suppression of the induction of alpha, beta, and lambda interferons by the NS1 and NS2 proteins of human respiratory syncytial virus in human epithelial cells and macrophages. *J Virol* 78: 4363–69, 2004.
100. Bartlett NW, Buttigieg K, Kotenko SV, Smith GL. Murine interferon lambdas (type III interferons) exhibit potent antiviral activity *in vivo* in a poxvirus infection model. *J GenVirol* 86: 1589–96, 2005.
101. Becker S, Quay J, Soukup J. Cytokine (tumor necrosis factor, IL-6, and IL-8) production by respiratory syncytial virus-infected human alveolar macrophages. *J Immunol* 147: 4307–12, 1991.
102. Einarsson O, Geba GP, Zhu Z, Landry M, Elias JA. Interleukin-11: Stimulation *in vivo* and *in vitro* by respiratory viruses and induction of airways hyperresponsiveness. *J Clin Invest* 97: 915–24, 1996.
103. Edwards MR, Johnson MW, Johnston SL. Combination therapy: Synergistic suppression of virus-induced chemokines in airway epithelial cells. *Am J Respir Cell Mol Biol* 34: 616–24, 2006.
104. Corne JM, Lau L, Scott SJ, Davies R, Johnston SL, Howarth PH. The relationship between atopic status and IL-10 nasal lavage levels in the acute and persistent inflammatory response to upper respiratory tract infection. *Am J Respir Crit Care Med* 163: 1101–7, 2001.
105. Proud D, Gwaltney JMJ, Hendley JO, Dinarello CA, Gillis S, Schleimer RP. Increased levels of interleukin-1 are detected in nasal secretions of volunteers during experimental rhinovirus colds. *J Infect Dis* 169: 1007–13, 1994.
106. Einarsson O, Geba GP, Zhou Z, Landry ML, Panettieri RAJ, Tristram D, Welliver R, Metinko A, Elias JA. Interleukin-11 in respiratory inflammation. *AnnNY Acad Sci* 762: 89–100, 1995.
107. Teran LM, Seminario MC, Shute JK, Papi A, Compton SJ, Low JL, Gleich GJ, Johnston SL. RANTES, macrophage-inhibitory protein 1alpha, and the eosinophil product major basic protein are released into upper respiratory secretions during virus-induced asthma exacerbations in children. *J Infect Dis* 179: 677–81, 1999.
108. Cook DN, Beck MA, Coffman TM, Kirby SL, Sheridan JF, Pragnell IB, Smithies O. Requirement of MIP-1 alpha for an inflammatory response to viral infection. *Science* 269: 1583–85, 1995.
109. Johnston SL, Papi A, Bates PJ, Mastronarde JG, Monick MM, Hunninghake GW. Low grade rhinovirus infection induces a prolonged release of IL-8 in pulmonary epithelium. *J Immunol* 160: 6172–81, 1998.
110. Edwards MR, Hewson CA, Laza-Stanca V, Lau HT, Mukaida N, Hershenson MB, Johnston SL. Protein kinase R, IkappaB kinase-beta and NF-kappaB are required for human rhinovirus induced pro-inflammatory cytokine production in bronchial epithelial cells. *Mol Immunol* 44: 1587–97, 2007.
111. Mastronarde JG, He B, Monick MM, Mukaida N, Matsushima K, Hunninghake GW. Induction of interleukin (IL)-8 gene expression by respiratory syncytial virus involves activation of nuclear factor (NF)-kappa B and NF-IL-6. *J Infect Dis* 174: 262–67, 1996.
112. Bitko V, Velazquez A, Yang L, Yang YC, Barik S. Transcriptional induction of multiple cytokines by human respiratory syncytial virus requires activation of NF-kappa B and is inhibited by sodium salicylate and aspirin. *Virology* 232: 369–78, 1997.
113. Yoon HJ, Zhu Z, Gwaltney JM, Elias JA. Rhinovirus regulation of IL-1 receptor antagonist *in vivo* and *in vitro*: A potential mechanism of symptom resolution. *J Immunol* 162: 7461–69, 1999.
114. Proud D, Naclerio RM, Gwaltney JM, Hendley JO. Kinins are generated in nasal secretions during natural rhinovirus colds. *J Infect Dis* 161: 120–23, 1990.

115. Proud D, Reynolds CJ, Lacapra S, Kagey-Sobotka A, Lichtenstein LM, Naclerio RM. Nasal provocation with bradykinin induces symptoms of rhinitis and a sore throat. *Am Rev Respir Dis* 137: 613–16, 1988.
116. Christiansen SC, Proud D, Cochrane CG. Detection of tissue kallikrein in the bronchoalveolar lavage fluid of asthmatic subjects. *J Clin Invest* 79: 188–97, 1987.
117. Christiansen SC, Zuraw BL, Proud D, Cochrane CG. Inhibition of human bronchial kallikrein in asthma. *Am Rev Respir Dis* 13: 1125–31, 1989.
118. Christiansen SC, Proud D, Sarnoff RB, Juergens U, Cochrane CG, Zuraw BL. Elevation of tissue kallikrein and kinin in the airways of asthmatic subjects after endobronchial allergen challenge. *Am Rev Respir Dis* 145: 900–5, 1992.
119. Nijkamp FP, Folkerts G. Nitric oxide and bronchial reactivity. *Clin Exp Allergy* 24: 905–14, 1994.
120. Nevin BJ, Broadley KJ. Nitric oxide in respiratory diseases. *Pharmacol Ther* 95: 259–93, 2002.
121. Fischer A, Folkerts G, Geppetti P, Groneberg DA. Mediators of asthma: Nitric oxide. *Pulm Pharmacol Ther* 15: 73–81, 2002.
122. Akaike T, Maeda H. Nitric oxide and virus infection. *Immunology* 101: 300–8, 2000.
123. Donnelly LE, Barnes PJ. Expression and regulation of inducible nitric oxide synthase from human primary airway epithelial cells. *Am J Respir Cell Mol Biol* 26: 144–51, 2002.
124. Kharitonov SA, Yates D, Robbins RA, Logan-Sinclair R, Shinebourne EA, Barnes PJ. Increased nitric oxide in exhaled air of asthmatic patients. *Lancet* 343: 133–35, 1994.
125. Jatakanon A, Lim S, Kharitonov SA, Chung KF, Barnes PJ. Correlation between exhaled nitric oxide, sputum eosinophils, and methacholine responsiveness in patients with mild asthma. *Thorax* 53: 91–95, 1998.
126. Rutgers SR, van der Mark TW, Coers W, Moshage H, Timens W, Kauffman HF, Koeter GH, Postma DS. Markers of nitric oxide metabolism in sputum and exhaled air are not increased in chronic obstructive pulmonary disease. *Thorax* 54: 576–80, 1999.
127. Maziak W, Loukides S, Culpitt S, Sullivan P, Kharitonov SA, Barnes PJ. Exhaled nitric oxide in chronic obstructive pulmonary disease. *Am J Respir Crit Care Med* 157: 998–1002, 1998.
128. Hoyt JC, Robbins RA, Habib M, Springall DR, Buttery LD, Polak JM, Barnes PJ. Cigarette smoke decreases inducible nitric oxide synthase in lung epithelial cells. *Exp Lung Res* 29: 17–28, 2003.
129. Wei XQ, Charles IG, Smith A, Ure J, Feng GJ, Huang FP, Xu D, Muller W, Moncada S, Liew FY. Altered immune responses in mice lacking inducible nitric oxide synthase. *Nature* 375: 408–11, 1995.
130. Cifone MG, Ulisse S, Santoni A. Natural killer cells and nitric oxide. *Int Immunopharmacol* 1: 1513–24, 2001, [Review] [103 refs].
131. de Gouw HW, Grunberg K, Schot R, Kroes AC, Dick EC, Sterk PJ. Relationship between exhaled nitric oxide and airway hyperresponsiveness following experimental rhinovirus infection in asthmatic subjects. *Eur Respir J* 11: 126–32, 1998.
132. Sanders SP, Kim J, Connolly KR, Porter JD, Siekierski ES, Proud D. Nitric oxide inhibits rhinovirus-induced granulocyte macrophage colony-stimulating factor production in bronchial epithelial cells. *Am J Respir Cell Mol Biol* 24: 317–25, 2001.
133. Sanders SP, Siekierski ES, Porter JD, Richards SM, Proud D. Nitric oxide inhibits rhinovirus-induced cytokine production and viral replication in a human respiratory epithelial cell line. *J Virol* 72: 934–42, 1998.
134. Kao YJ, Piedra PA, Larsen GL, Colasurdo GN. Induction and regulation of nitric oxide synthase in airway epithelial cells by respiratory syncytial virus. *Am J Respir Crit Care Med* 163: 532–39, 2001.
135. Ali-Ahmad D, Bonville CA, Rosenberg HF, Domachowske JB. Replication of respiratory syncytial virus is inhibited in target cells generating nitric oxide *in situ*. *Front Biosci* 8: A48–53, 2003.
136. Rimmelzwaan GF, Baars MM, de Lijster P, Fouchier RA, Osterhaus AD. Inhibition of influenza virus replication by nitric oxide. *J Virol* 73: 8880–83, 1999.
137. Folkerts G, van der Linde HJ, Nijkamp FP. Virus-induced airway hyperresponsiveness in guinea pigs is related to a deficiency in nitric oxide. *J Clin Invest* 95: 26–30, 1995.
138. Kharitonov SA, Yates D, Barnes PJ. Increased nitric oxide in exhaled air of normal human subjects with upper respiratory tract infections. *Eur Respir J* 8: 295–97, 1995.
139. Gentile DA, Doyle WJ, Belenky S, Ranck H, Angelini B, Skoner DP. Nasal and oral nitric oxide levels during experimental respiratory syncytial virus infection of adults. *Acta Otolaryngol* 122: 61–66, 2002.
140. Murphy AW, Platts-Mills TA, Lobo M, Hayden F. Respiratory nitric oxide levels in experimental human influenza. *Chest* 114: 452–56, 1998.
141. Mogensen TH, Paludan SR. Molecular pathways in virus-induced cytokine production. *Microbiol Mol Biol Rev* 65: 131–50, 2001.
142. Gern JE, French DA, Grindle KA, Brockman-Schneider RA, Konno S, Busse WW. Double-stranded RNA induces the synthesis of specific chemokines by bronchial epithelial cells. *Am J Respir Cell Mol Biol* 28: 731–37, 2003.
143. Alexopoulou L, Holt AC, Medzhitov R, Flavell RA. Recognition of double-stranded RNA and activation of NF-kappaB by Toll-like receptor 3. *Nature* 413: 732–38, 2001.
144. Hewson CA, Jardine A, Edwards MR, Laza-Stanca V, Johnston SL. Toll-like receptor 3 is induced by and mediates antiviral activity against rhinovirus infection of human bronchial epithelial cells. *J Virol* 79: 12273–79, 2005.
145. Grunstein MM, Hakonarson H, Maskeri N, Chuang S. Autocrine cytokine signaling mediates effects of rhinovirus on airway responsiveness. *Am J Physiol Lung Cell Mol Physiol* 278: L1146–53, 2000.
146. Oliver BG, Johnston SL, Baraket M, Burgess JK, King NJ, Roth M, Lim S, Black JL. Increased proinflammatory responses from asthmatic human airway smooth muscle cells in response to rhinovirus infection. *Respir Res* 7: 71, 2006.
147. Johnston SL, Papi A, Monick MM, Hunninghake GW. Rhinoviruses induce interleukin-8 mRNA and protein production in human monocytes. *J Infect Dis* 175: 323–29, 1997.
148. Laza-Stanca V, Stanciu L, Message SD, Edwards MR, Gern JE, Johnston SL. Rhinovirus replication in human macrophages induces NFkB dependent tumour necrosis factor alpha production. *J Virol* 80(16): 8248–58, 2006.
149. Gern JE, Dick EC, Lee WM, Murray S, Meyer K, Handzel ZT, Busse WW. Rhinovirus enters but does not replicate inside monocytes and airway macrophages. *J Immunol* 156: 621–27, 1996.
150. Oliver BG, Lim S, Wark P, Laza-Stanca V, King NJ, Black JL, Burgess JK, Roth M, Johnston SL. Rhinovirus exposure impairs immune responses to bacterial products in human alveolar macrophages. *Thorax* 43, 2008.
151. Johnston SL. Macrolide antibiotics and asthma treatment. *J Allergy Clin Immunol* 117: 1233–36, 2006.
152. Papi A, Bellettato CM, Braccioni F, Romagnoli M, Casolari P, Caramori G, Fabbri LM, Johnston SL. Infections and airway inflammation in chronic obstructive pulmonary disease severe exacerbations. *Am J Respir Crit Care Med* 173: 1114–21, 2006.
153. Peschke T, Bender A, Nain M, Gemsa D. Role of macrophage cytokines in influenza A virus infections.. *Immunobiology* 189: 340–55, 1993, [Review] [33 refs].
154. Banchereau J, Briere F, Caux C, Davoust J, Lebecque S, Liu YJ, Pulendran B, Palucka K. Immunobiology of dendritic cells. [Review] [289 refs]. *Ann Rev Immunol* 18: 767–811, 2000.
155. Beyer M, Bartz H, Horner K, Doths S, Koerner-Rettberg C, Schwarze J. Sustained increases in numbers of pulmonary dendritic cells after respiratory syncytial virus infection. *J Allergy Clin Immunol* 113(1): 127–33, 2004.
156. Wang H, Peters N, Laza-Stanca V, Nawroly N, Johnston SL, Schwarze J. Local CD11c+ MHC class II- precursors generate lung dendritic cells during respiratory viral infection, but are depleted in the process. *J Immunol* 177: 2536–42, 2006.

157. Wang H, Peters N, Schwarze J. Plasmacytoid dendritic cells limit viral replication, pulmonary inflammation, and airway hyperresponsiveness in respiratory syncytial virus infection. *J Immunol* 177(9): 6263–70, 2006.
158. Grayson MH, Cheung D, Rohlfing MM, Kitchens R, Spiegel DE, Tucker J, Battaile JT, Alevy Y, Yan L, Agapov E, Kim EY, Holtzman MJ. Induction of high-affinity IgE receptor on lung dendritic cells during viral infection leads to mucous cell metaplasia. *J Exp Med* 204(11): 2759–69, 2007.
159. Bonecchi R, Bianchi G, Bordignon PP, D'Ambrosio D, Lang R, Borsatti A, Sozzani S, Allavena P, Gray PA, Mantovani A et al. Differential expression of chemokine receptors and chemotactic responsiveness of type 1 T helper cells (Th1s) and Th2s. *J Exp Med* 187: 129–34, 1998.
160. Panina-Bordignon P, Papi A, Mariani A, Di Lucia P, Casoni G, Bellettato C, Buonsanti C, Miotto D, Mapp C, Villa A, Arrigoni G, Fabbri L, Sinigaglia F. The C-C chemokine receptors CCR4 and CCR8 identify airway T cells of allergen-challenged atopic asthmatics. *J Clin Invest* 107(11): 1357–64, 2001.
161. Huftel MA, Swensen CA, Borcherding WR, Dick EC, Hong R, Kita H, Gleich GJ, Busse WW. The effect of T-cell depletion on enhanced basophil histamine release after *in vitro* incubation with live influenza A virus. *Am J Respir Cell Mol Biol* 7: 434–40, 1992.
162. Hsia J, Goldstein AL, Simon GL, Sztein M, Hayden FG. Peripheral blood mononuclear cell interleukin-2 and interferon-gamma production, cytotoxicity, and antigen-stimulated blastogenesis during experimental rhinovirus infection. *J Infect Dis* 162: 591–97, 1990.
163. Corne JM, Holgate ST. Mechanisms of virus induced exacerbations of asthma. *Thorax* 52: 380–89, 1997.
164. Folkerts G, Nijkamp FP. Virus-induced airway hyperresponsiveness. Role of inflammatory cells and mediators. *Am J Respir Crit Care Med* 151: 1666–73, 1995.
165. Alwan WH, Kozlowska WJ, Openshaw PJ. Distinct types of lung disease caused by functional subsets of antiviral T cells. *J Exp Med* 179: 81–89, 1994.
166. Mosmann TR, Sad S. The expanding universe of T-cell subsets: Th1, Th2 and more. *Immunol Today* 17: 138–46, 1996.
167. Romagnani S. The Th1/Th2 paradigm. *Immunol Today* 18: 263–66, 1997.
168. Papadopoulos NG, Stanciu LA, Papi A, Holgate ST, Johnston SL. A defective type 1 response to rhinovirus in atopic asthma. *Thorax* 57: 328–32, 2002.
169. Gern JE, Vrtis R, Grindle KA, Swenson C, Busse WW. Relationship of upper and lower airway cytokines to outcome of experimental rhinovirus infection. *Am J Respir Crit Care Med* 162: 2226–31, 2000.
170. Hussell T, Spender LC, Georgiou A, O'Garra A, Openshaw PJ. Th1 and Th2 cytokine induction in pulmonary T cells during infection with respiratory syncytial virus. *J Gen Virol* 77: 2447–55, 1996.
171. Coyle AJ, Erard F, Bertrand C, Walti S, Pircher H, Le Gros G. Virus-specific CD8+ cells can switch to interleukin 5 production and induce airway eosinophilia. *J Exp Med* 181: 1229–33, 1995.
172. Holtmeier W, Kabelitz D. gammadelta T cells link innate and adaptive immune responses. *Chem Immunol Allergy* 86: 151–83, 2005.
173. Holtmeier W. Compartmentalization gamma/delta T cells and their putative role in mucosal immunity. *Crit Rev Immunol* 23: 473–88, 2003.
174. Spinozzi F, Agea E, Bistoni O, Forenza N, Bertotto A. gamma delta T cells, allergen recognition and airway inflammation. *Immunol Today* 19: 22–26, 1998.
175. Krug N, Erpenbeck VJ, Balke K, Petschallies J, Tschernig T, Hohlfeld JM, Fabel H. Cytokine profile of bronchoalveolar lavage-derived CD4(+), CD8(+), and gammadelta T cells in people with asthma after segmental allergen challenge. *Am J Respir Cell Mol Biol* 25: 125–31, 2001.
176. Gleich GJ. Mechanisms of eosinophil-associated inflammation. *J Allergy Clin Immunol* 105: 651–63, 2000.
177. Saito T, Deskin RW, Casola A, Haeberle H, Olszewska B, Ernst PB, Alam R, Ogra PL, Garofalo R. Respiratory syncytial virus induces selective production of the chemokine RANTES by upper airway epithelial cells. *J Infect Dis* 175: 497–504, 1997.
178. Schroth MK, Grimm E, Frindt P, Galagan DM, Konno SI, Love R, Gern JE. Rhinovirus replication causes RANTES production in primary bronchial epithelial cells. *Am J Respir Cell Mol Biol* 20: 1220–28, 1999.
179. Noah TL, Henderson FW, Henry MM, Peden DB, Devlin RB. Nasal lavage cytokines in normal, allergic, and asthmatic school-age children. *Am J Respir Crit Care Med* 152: 1290–96, 1995a.
180. Noah TL, Henderson FW, Wortman IA, Devlin RB, Handy J, Koren HS, Becker S. Nasal cytokine production in viral acute upper respiratory infection of childhood. *J Infect Dis* 171: 584–92, 1995b.
181. Greiff L, Andersson M, Andersson E, Linden M, Myint S, Svensson C, Persson CG. Experimental common cold increases mucosal output of eotaxin in atopic individuals. *Allergy* 54: 1204–8, 1999.
182. Handzel ZT, Busse WW, Sedgwick JB, Vrtis R, Lee WM, Kelly EA, Gern JE. Eosinophils bind rhinovirus and activate virus-specific T cells. *J Immunol* 160: 1279–84, 1998.
183. Adamko DJ, Yost BL, Gleich GJ, Fryer AD, Jacoby DB. Ovalbumin sensitization changes the inflammatory response to subsequent parainfluenza infection. Eosinophils mediate airway hyperresponsiveness, m(2) muscarinic receptor dysfunction, and antiviral effects. *J Exp Med* 190: 1465–78, 1999.
184. Domachowske JB, Dyer KD, Adams AG, Leto TL, Rosenberg HF. Eosinophil cationic protein/RNase 3 is another RNase A-family ribonuclease with direct antiviral activity. *Nucleic Acids Res* 26: 3358–63, 1998.
185. Folkerts G, Busse WW, Nijkamp FP, Sorkness R, Gern JE. Virus-induced airway hyperresponsiveness and asthma. *Am J Respir Crit Care Med* 157: 1708–20, 1998.
186. Volovitz B, Faden H, Ogra PL. Release of leukotriene C4 in respiratory tract during acute viral infection. *J Pediatr* 112: 218–22, 1988.
187. Barnes PJ, Chung KF, Page CP. Inflammatory mediators of asthma: An update. *Pharmacol Rev* 50: 515–96, 1998.
188. Seymour ML, Gilby N, Bardin PG, Fraenkel DJ, Sanderson G, Penrose JF, Holgate ST, Johnston SL, Sampson AP. Rhinovirus infection increases 5-lipoxygenase and cyclooxygenase-2 in bronchial biopsy specimens from nonatopic subjects. *J Infect Dis* 185: 540–44, 2002.
189. Pizzichini MM, Pizzichini E, Efthimiadis A, Chauhan AJ, Johnston SL, Hussack P, Mahony J, Dolovich J, Hargreave FE. Asthma and natural colds. Inflammatory indices in induced sputum: A feasibility study. *Am J Respir Crit Care Med* 158: 1178–84, 1998.
190. Fahy JV, Kim KW, Liu J, Boushey HA. Prominent neutrophilic inflammation in sputum from subjects with asthma exacerbation. *J Allergy Clin Immunol* 95: 843–52, 1995.
191. Teran LM, Johnston SL, Schroder JM, Church MK, Holgate ST. Role of nasal interleukin-8 in neutrophil recruitment and activation in children with virus-induced asthma. *Am J Respir Crit Care Med* 155: 1362–66, 1997.
192. Biron CA, Nguyen KB, Pien GC, Cousens LP, Salazar-Mather TP. Natural killer cells in antiviral defense: Function and regulation by innate cytokines. *Ann Rev Immunol* 17: 189–220, 1999, [Review] [253 refs].
193. Moretta L, Biassoni R, Bottino C, Mingari MC, Moretta A. Human NK-cell receptors. *Immunol Today* 21: 420–22, 2000.
194. Mingari MC, Ponte M, Bertone S, Schiavetti F, Vitale C, Bellomo R, Moretta A, Moretta L. HLA class I-specific inhibitory receptors in human T lymphocytes: Interleukin 15-induced expression of CD94/NKG2A in superantigen- or alloantigen-activated CD8+ T cells. *Proc Natl Acad Sci USA* 95: 1172–77, 1998.
195. Biron CA. Role of early cytokines, including alpha and beta interferons (IFN-alpha/beta), in innate and adaptive immune responses to viral infections. *Semin Immunol* 10: 383–90, 1998, [Review] [76 refs].
196. Peritt D, Robertson S, Gri G, Showe L, Aste-Amezaga M, Trinchieri G. Differentiation of human NK cells into NK1 and NK2 subsets. *J Immunol* 161: 5821–24, 1998.

197. Warren HS, Kinnear BF, Phillips JH, Lanier LL. Production of IL-5 by human NK cells and regulation of IL-5 secretion by IL-4, IL-10, and IL-12. *J Immunol* 154: 5144–52, 1995.
198. Walker C, Checkel J, Cammisuli S, Leibson PJ, Gleich GJ. IL-5 production by NK cells contributes to eosinophil infiltration in a mouse model of allergic inflammation. *J Immunol* 161: 1962–69, 1998.
199. Humbert M, Menz G, Ying S, Corrigan CJ, Robinson DS, Durham SR, Kay AB. The immunopathology of extrinsic (atopic) and intrinsic (non-atopic) asthma: More similarities than differences. *Immunol Today* 20: 528–33, 1999.
200. Skoner DP, Doyle WJ, Tanner EP, Kiss J, Fireman P. Effect of rhinovirus 39 (RV-39) infection on immune and inflammatory parameters in allergic and non-allergic subjects. *Clin Exp Allergy* 25: 561–67, 1995.
201. Lin CY, Kuo YC, Liu WT, Lin CC. Immunomodulation of influenza virus infection in the precipitating asthma attack. *Chest* 93: 1234–38, 1988.
202. Welliver RC, Wong DT, Sun M, Middleton EJ, Vaughan RS, Ogra PL. The development of respiratory syncytial virus-specific IgE and the release of histamine in nasopharyngeal secretions after infection. *New Engl J Med* 305: 841–46, 1981.
203. Murray CJ, Lopez AD. Alternative projections of mortality and disability by cause 1990–2020: Global burden of disease study. *Lancet* 349: 1498–504, 1997.
204. Anonymous. BTS guidelines for the management of chronic obstructive pulmonary disease. The COPD Guidelines Group of the Standards of Care Committee of the BTS. *Thorax* 52(Suppl 5): S1–28, 1997.
205. Anonymous. Standards for the diagnosis and care of patients with chronic obstructive pulmonary disease. American Thoracic Society. *Am J Respir Crit Care Med* 152: S77–121, 1995.
206. Seemungal TA, Donaldson GC, Paul EA, Bestall JC, Jeffries DJ, Wedzicha JA. Effect of exacerbation on quality of life in patients with chronic obstructive pulmonary disease. *Am J Respir Crit Care Med* 157: 1418–22, 1998.
207. Wilson R. The role of infection in COPD. *Chest* 113: 242S–48S, 1998.
208. Smith CB, Golden CA, Kanner RE, Renzetti AD. Association of viral and *Mycoplasma pneumoniae* infections with acute respiratory illness in patients with chronic obstructive pulmonary diseases. *Am Rev Respir Dis* 121: 225–32, 1980.
209. Seemungal TA, Harper-Owen R, Bhowmik A, Moric I, Sanderson G, Message SD, MacCallum P, Meade TW, Jeffries DJ, Johnston SL, Wedzicha JA. The role of respiratory viral infections in COPD. *Am J Respir Crit Care Med*, 164(9): 1618–23, 2001.
210. Gump DW, Phillips CA, Forsyth BR, McIntosh K, Lamborn KR, Stouch WH. Role of infection in chronic bronchitis. *Am Rev Respir Dis* 113: 465–74, 1976.
211. Tager I, Speizer FE. Role of infection in chronic bronchitis. *New Engl J Med* 292: 563–71, 1975.
212. Greenberg SB, Allen M, Wilson J, Atmar RL. Respiratory viral infections in adults with and without chronic obstructive pulmonary disease. *Am J Respir Crit Care Med* 162: 167–73, 2000.
213. Rohde G, Wiethege A, Borg I, Kauth M, Bauer TT, Gillissen A, Bufe A, Schultze-Werninghaus G. Respiratory viruses in exacerbations of chronic obstructive pulmonary disease requiring hospitalisation: A case-control study. *Thorax* 58: 37–42, 2003.
214. Monso E, Ruiz J, Rosell A, Manterola J, Fiz J, Morera J, Ausina V. Bacterial infection in chronic obstructive pulmonary disease. A study of stable and exacerbated outpatients using the protected specimen brush. *Am J Respir Crit Care Med* 152: 1316–20, 1995.
215. Irwin RS, Erickson AD, Pratter MR, Corrao WM, Garrity FL, Myers JR, Kaemmerlen JT. Prediction of tracheobronchial colonization in current cigarette smokers with chronic obstructive bronchitis. *J Infect Dis* 145: 234–41, 1982.
216. Anthonisen NR, Manfreda J, Warren CP, Hershfield ES, Harding GK, Nelson NA. Antibiotic therapy in exacerbations of chronic obstructive pulmonary disease. *Ann Int Med* 106: 196–204, 1987.
217. Stockley RA, O'Brien C, Pye A, Hill SL. Relationship of sputum color to nature and outpatient management of acute exacerbations of COPD. *Chest* 117: 1638–45, 2000.
218. Sethi S, Evans N, Grant BJ, Murphy TF. New strains of bacteria and exacerbations of chronic obstructive pulmonary disease. *N Engl J Med* 347(7): 465–71, 2002.
219. Murphy TF, Sethi S. Bacterial infection in chronic obstructive pulmonary disease. *Am Rev Respir Dis* 146: 1067–83, 1992.
220. Ball P, Tillotson G, Wilson R. Chemotherapy for chronic bronchitis. *Controversies Presse Med* 24: 189–94, 1995.
221. Blasi F, Cosentini R, Schoeller MC, Lupo A, Allegra L. *Chlamydia pneumoniae* seroprevalence in immunocompetent and immunocompromised populations in Milan. *Thorax* 48: 1261–63, 1993.
222. Blasi F, Legnani D, Lombardo VM, Negretto GG, Magliano E, Pozzoli R, Chiodo F, Fasoli A, Allegra L. *Chlamydia pneumoniae* infection in acute exacerbations of COPD. *Eur Res J* 6: 19–22, 1993.
223. Eller J, Ede A, Schaberg T, Niederman MS, Mauch H, Lode H. Infective exacerbations of chronic bronchitis: Relation between bacteriologic etiology and lung function. *Chest* 113: 1542–48, 1998.
224. Wilkinson TM, Hurst JR, Perera WR, Wilks M, Donaldson GC, Wedzicha JA. Effect of interactions between lower airway bacterial and rhinoviral infection in exacerbations of COPD. *Chest* 129(2): 317–24, 2006.
225. Saint S, Bent S, Vittinghoff E, Grady D. Antibiotics in chronic obstructive pulmonary disease exacerbations. A meta-analysis. *JAMA* 273: 957–60, 1995.
226. Siafakas NM, Vermeire P, Pride NB, Paoletti P, Gibson J, Howard P, Yernault JC, Decramer M, Higenbottam T, Postma DS. Optimal assessment and management of chronic obstructive pulmonary disease (COPD). The European Respiratory Society Task Force. *Eur Respir J* 8: 1398–420, 1995.
227. Fagon JY, Chastre J, Trouillet JL, Domart Y, Dombret MC, Bornet M, Gibert C. Characterization of distal bronchial microflora during acute exacerbation of chronic bronchitis. Use of the protected specimen brush technique in 54 mechanically ventilated patients. *Am Rev Respir Dis* 142: 1004–8, 1990.
228. Wilson R, Dowling RB, Jackson AD. The biology of bacterial colonization and invasion of the respiratory mucosa. *Eur Respir J* 9: 1523–30, 1996.
229. Sykes A, Mallia P, Johnston SL. Diagnosis of pathogens in exacerbations of chronic obstructive pulmonary disease. *Proc Am Thorac Soc* 4: 642–46, 2007.
230. Hogg JC. Childhood viral infection and the pathogenesis of asthma and chronic obstructive lung disease. *Am J Respir Critl Care Med* 160: S26–28, 1999.
231. Matsuse T, Hayashi S, Kuwano K, Keunecke H, Jefferies WA, Hogg JC. Latent adenoviral infection in the pathogenesis of chronic airways obstruction. *Am Rev Respir Dis* 146: 177–84, 1992.
232. Elliott WM, Hayashi S, Hogg JC. Immunodetection of adenoviral E1A proteins in human lung tissue. *Am J Respir Cell Mol Biol* 12: 642–48, 1995.
233. Liu F, Green MR. Promoter targeting by adenovirus E1a through interaction with different cellular DNA-binding domains. *Nature* 368: 520–25, 1994.
234. Keicho N, Elliott WM, Hogg JC, Hayashi S. Adenovirus E1A upregulates interleukin-8 expression induced by endotoxin in pulmonary epithelial cells. *Am J Physiol* 272: L1046–52, 1997.
235. Keicho N, Elliott WM, Hogg JC, Hayashi S. Adenovirus E1A gene dysregulates ICAM-1 expression in transformed pulmonary epithelial cells. *Am J Respir Cell Mol Biol* 16: 23–30, 1997.
236. Keicho N, Higashimoto Y, Bondy GP, Elliott WM, Hogg JC, Hayashi S. Endotoxin-specific NF-kappaB activation in pulmonary epithelial cells harboring adenovirus E1A. *Am J Physiol* 277: L523–32, 1999.
237. Mallia P, Message SD, Kebadze T, Parker HL, Kon OM, Johnston SL. An experimental model of rhinovirus induced chronic obstructive pulmonary disease exacerbations: A pilot study. *Respir Res* 7(116), 2006.

238. Gustafson LM, Proud D, Hendley JO, Hayden FG, Gwaltney JM. Oral prednisone therapy in experimental rhinovirus infections. *J Allergy Clin Immunol* 97: 1009–14, 1996.
239. Eickelberg O, Roth M, Lorx R. Ligand-independent activation of the glucocorticoid receptor by β2-adrenergic receptor agonists in primary human lung fibroblasts and vascular smooth muscle cells. *J Biol Chem* 274: 1005–10, 1999.
240. Pauwels RA, Lofdahl CG, Postma DS, Tattersfield AE, O'Byrne P, Barnes PJ, Ullman A. Effect of inhaled formoterol and budesonide on exacerbations of asthma. Formoterol and Corticosteroids Establishing Therapy (FACET) International Study Group. *New Engl J Med* 337: 1405–11, 1997.
241. Robertson CF, Price D, Henry R, Mellis C, Glasgow N, Fitzgerald D, Lee AJ, Turner J, Sant M. Short-course montelukast for intermittent asthma in children: A randomised controlled trial. *Am J Respir Crit Care Med* 175(4): 323–29, 2007.
242. Bisgaard H, Zielen S, Garcia-Garcia ML, Johnston SL, Gilles L, Menten J, Tozzi CA, Polos P. Montelukast reduces asthma exacerbations in 2- to 5-year-old children with intermittent asthma. *Am J Respir Crit Care Med* 171: 315–22, 2005.
243. Pauwels RA, Pedersen S, Busse WW, Tan WC, Chen YZ, Ohlsson SV, Ullman A, Lamm CJ, O'Byrne PM START Investigators Group. Early intervention with budesonide in mild persistent asthma: A randomised, double-blind trial. *Lancet* 361: 1071–76, 2003.
244. Doull IJ, Lampe FC, Smith S, Schreiber J, Freezer NJ, Holgate ST. Effect of inhaled corticosteroids on episodes of wheezing associated with viral infection in school age children: Randomised double blind placebo controlled trial. *BMJ* 315: 858–62, 1997.
245. Oommen A, Lambert PC, Grigg J. Efficacy of a short course of parent-initiated oral prednisolone for viral wheeze in children aged 1–5 years: Randomised controlled trial. *Lancet* 362: 1433–38, 2003.
246. Hamory BH, Hamparian VV, Conant RM, Gwaltney JM. Human responses to two decavalent rhinovirus vaccines. *J Infect Dis* 132: 623–29, 1975.
247. Kim HW, Canchola JG, Brandt CD, Pyles G, Chanock RM, Jensen K, Parrott RH. Respiratory syncytial virus disease in infants despite prior administration of antigenic inactivated vaccine. *Am J Epidemiol* 89: 422–34, 1969.
248. Hall CB. Respiratory syncytial virus and parainfluenza virus. *New Engl J Med* 344: 1917–28, 2001.
249. Murphy BR, Prince GA, Walsh EE, Kim HW, Parrott RH, Hemming VG, Rodriguez WJ, Chanock RM. Dissociation between serum neutralizing and glycoprotein antibody responses of infants and children who received inactivated respiratory syncytial virus vaccine. *J Clin Microbiol* 24: 197–202, 1986.
250. Kim HW, Leikin SL, Arrobio J, Brandt CD, Chanock RM, Parrott RH. Cell-mediated immunity to respiratory syncytial virus induced by inactivated vaccine or by infection. *Pediatr Res* 10: 75–78, 1976.
251. Glezen WP, Taber LH, Frank AL, Kasel JA. Risk of primary infection and reinfection with respiratory syncytial virus. *Am J Dis Child* 140: 543–46, 1986.
252. Marshall S. Zinc gluconate and the common cold. Review of randomized controlled trials. *Can Fam Physician* 44: 1037–42, 1998.
253. Singh M. Heated, humidified air for the common cold. *Cochrane Database of Syst Rev*4: CD001728, 2004.
254. Hayden FG, Gwaltney JM. Intranasal interferon-alpha 2 treatment of experimental rhinoviral colds. *J Infect Dis* 150: 174–80, 1984.
255. Khare MD, Sharland M. Influenza. *Expert Opinion on Pharmacotherapy* 1: 367–75, 2000.
256. Lalezari J, Campion K, Keene O, Silagy C. Zanamivir for the treatment of influenza A and B infection in high-risk patients: a pooled analysis of randomized controlled trials. *Archives of Internal Medicine* 161: 212–17, 2001.
257. Monto AS, Robinson DP, Herlocher ML, Hinson JMJ, Elliott MJ, Crisp A. Zanamivir in the prevention of influenza among healthy adults: a randomized controlled trial. *JAMA* 282: 31–35, 1999.
258. Rodriguez WJ, Gruber WC, Welliver RC, Groothuis JR, Simoes EA, Meissner HC, Hemming VG, Hall CB, Lepow ML, Rosas AJ et al. Respiratory syncytial virus (RSV) immune globulin intravenous therapy for RSV lower respiratory tract infection in infants and young children at high risk for severe RSV infections: Respiratory Syncytial Virus Immune Globulin Study Group. *Pediatrics* 99: 454–61, 1997, [see comments].
259. Wu H, Pfarr DS, Johnson S, Brewah YA, Woods RM, Patel NK, White WI, Young JF, Kiener PA. Development of motavizumab, an ultra-potent antibody for the prevention of respiratory syncytial virus infection in the upper and lower respiratory tract. *J Mol.Biol* 368(3): 652–65, 2008.
260. Pattemore PK, Johnston SL, Bardin PG. Viruses as precipitants of asthma symptoms. I. Epidemiology. *Clinical & Experimental Allergy* 22: 325–36, 1992.

CHAPTER 38

Exercise as a Stimulus

Sandra D. Anderson[1] and Jennifer A. Alison[2]

[1]Department of Respiratory Medicine, Royal Prince Alfred Hospital, Camperdown, NSW, Australia
[2]Discipline of Physiotherapy, Faculty of Health Sciences, University of Sydney, Lidcombe, NSW, Australia

OVERVIEW

There are substantial differences in the pulmonary response between those with asthma and those with chronic obstructive pulmonary disease (COPD) when exercise is used as a stimulus. For the person with asthma, exercise is a stimulus for bronchodilatation and for bronchoconstriction and the airway response varies according to resting lung function, duration and intensity of exercise, and the temperature and water content of the air inspired. The stimulus for exercise-induced bronchoconstriction (EIB) relates to the thermal and osmotic consequences of water loss by evaporation from the airway surface when large volumes of air need to be conditioned in a short time. For most asthmatics however, EIB can be well controlled with treatment and the emphasis has moved from exercise as a stimulus for provoking attacks of asthma to exercise for the potential benefits to health and lifestyle. For those with COPD, exercise does not usually provide a dehydrating stimulus sufficient to provoke EIB simply because the subjects cannot ventilate at a rate sufficient to achieve significant loss of water from the airway surface. Some people with COPD however do demonstrate the airway hyperresponsiveness of asthma in response to inhaling aerosols of hyperosmolar agents such as saline or mannitol. The more classic response exercise provides to the patient with COPD is dynamic hyperinflation that increases end-expiratory lung volume (EELV) and decreases inspiratory capacity (IC). This strategy serves to reduce expiratory flow limitation and increase tidal volume but at a cost of greater respiratory effort that contributes to exercise limitation. As with treatment of asthma there has been a marked improvement in treatment of COPD in recent years. Although treatment has not allowed patients with COPD to achieve normal exercise performance, as asthmatics have done, exercise has become an important intervention in the management of patients with COPD.

EXERCISE AND THE PERSON WITH ASTHMA

The cardiopulmonary response to exercise in asthmatics is extremely variable. When expiratory flow rates and airways resistance are normal, at rest and during exercise, the cardiopulmonary response to exercise is not significantly different to healthy subjects [1]. Asthmatics however commonly have a higher ventilatory equivalent (ventilation per unit of oxygen consumption) compared with non-asthmatics although this can normalize following physical training [2]. Asthmatic children with moderate to severe asthma have reduced aerobic capacity, endurance time, and cardiac function [3], particularly at high altitude [4]. Expiratory flow limitation and hyperinflation are likely contributors to the increased work of breathing and shortness of breath in some people with asthma.

Physical inactivity is thought to contribute to the development of "asthma" because bronchial smooth muscle can become stiff and hyperresponsive without the benefit of periodic stretch from the increased tidal volume of exercise [5]. There are many reports of asthmatic children being physically inactive and unfit, particularly those who are obese or overweight, but being unfit is not a universal finding in asthmatic children [1]. Neither body mass index nor baseline lung function predicted exercise limitation in children with asthma [6].

LUNG FUNCTION AND GAS EXCHANGE DURING AND AFTER EXERCISE

It is the propensity for airway resistance and arterial oxygen tension to change rapidly in response to exercise that is of primary physiological importance to a person with asthma

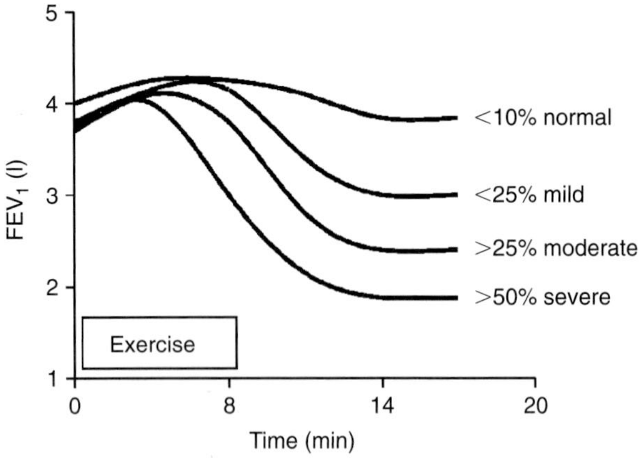

FIG. 38.1 Changes in forced expiratory volume during and after exercise. The severity of the response is assessed by the fall in FEV_1 after exercise expressed as a percentage of the pre-exercise value. A normal response is usually regarded as a fall less than 10% of baseline. A value less than 10–25% is mild, 25.1–49.9% as moderate, and 50% or more as severe. The values refer to the severity of EIB in those not taking an ICS. In those taking ICS a value greater than 30% would be generally accepted as severe.

([7–9], Figs 38.1 and 38.2). There are three phases of changes in the airway resistance (Raw) to exercise. During exercise of variable intensity, Raw decreases when exercise intensity is increased and increases when exercise intensity is reduced [10]. During intense steady-state exercise there is an early phase in which Raw falls rapidly. This is not due to release of nitric oxide or due to prostaglandins (PGs) [11, 12] but maybe due to tidal volume stretch reducing bronchial smooth muscle tone [13]. This early fall in Raw during exercise is followed by a slow and progressive rise in Raw after the first few minutes, then on the cessation of 6–8 min of vigorous exercise there is a marked increase in Raw. It is common for the increase in Raw to be associated with a fall in peak expiratory flow (PEF) or FEV_1 from pre-exercise level of 10% or more within the first 5 min. The terms commonly used to describe these changes in lung function following exercise are EIB or exercise-induced asthma (Fig. 38.1).

In "untrained" asthmatics with moderate airflow limitation at rest (% predicted $FEV_1 > 75\%$), the fall in Raw or bronchodilatation of the first few minutes of exercise is associated with an increase in arterial oxygen tension (PaO_2) [7, 8, 14] (Fig. 38.2). After exercise ceases the PaO_2 falls in concert with the reduction in lung function and as a result of ventilation–perfusion inequality [7, 15, 16]. Carbon dioxide levels in asthmatics usually, but not always, remain within normal limits (<40 mmHg) both during exercise and after exercise [8, 15]. Recovery to pre-exercise lung function occurs spontaneously over 30–60 min, or usually within 10 min in response to a bronchodilator, with PaO_2 increasing with the improvement in lung function. The lung function and gas exchange changes

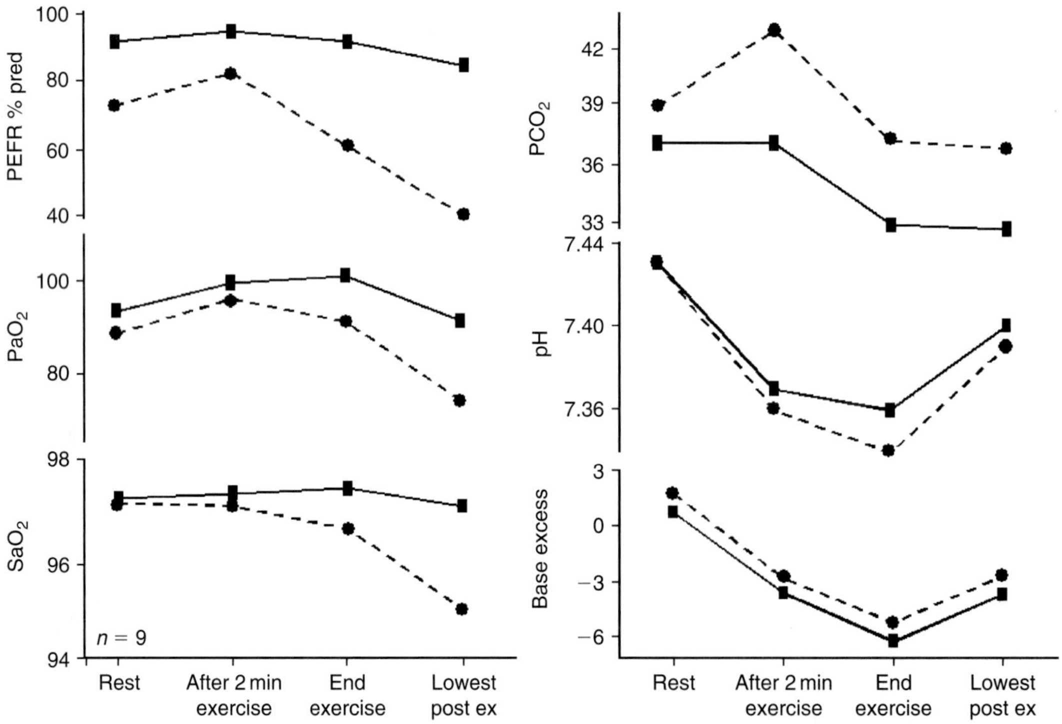

FIG. 38.2 Mean values for PEF expresses as a percentage of predicted, arterial oxygen tension (PaO_2 mmHg), arterial saturation (SaO_2 %), arterial carbon dioxide tension ($PaCO_2$ mmHg), pH, and base excess (BE) at baseline, during, and after exercise after placebo (broken line) and 70 min later after pre-medication with terbutaline sulfate (1250 μg – 20 min after exercise, and another 1250 μg – 50 min later). The time between exercise tests was 90 min. The data are from the study of Ref. [8].

associated with EIB can sometimes be extreme, and deaths in young athletes, though uncommon, have been reported [17].

Differences from this classic pattern of gas exchange described above have been reported in some "trained" asthmatics with good lung function (i.e. FEV_1 94% predicted) exercising at higher workloads for longer periods than 8 min. In one study, 8 of 21 asthmatic subjects were unable to sustain their SaO_2 above 94% during 11 min of strenuous treadmill exercise ($\dot{V}O_2$ maximum range from 39.9 to 61.8 ml/min/kg). The low SaO_2 was attributed to a widened alveolar-arterial oxygen gradient ($A\text{-}aO_2$) and ventilatory insufficiency that resulted in a significant increase in $PaCO_2$ in some subjects [9]. It was not possible to predict from pre-exercise lung function or blood gases which asthmatic subject would desaturate in response to exercise. In these "trained" asthmatics, who were not taking inhaled corticosteroids (ICS), the arterial hypoxemia, expiratory flow limitation, and dynamic hyperinflation during exercise occurred at lower metabolic rates than would be sometimes documented in highly "trained" healthy subjects [9]. The findings have significant implications for untreated asthmatics wishing to participate in competitive high endurance sports and support the policy that athletes with asthma would benefit from having objective tests [18].

PHYSICAL TRAINING AND EIB

The aim of exercise training in an asthmatic who is "unfit" is to increase the exercise intensity and thus the threshold of ventilation that will provoke EIB [2]. Exercise training in an unfit asthmatic results in an increase in maximum oxygen uptake per kilogram and a reduction in ventilatory equivalent and a modest improvement in maximum heart rate [19]. Since ventilation determines severity of EIB for an individual, physical fitness should be encouraged, and at an early age.

It is important to know fitness does not eliminate potential for EIB [20], but becoming fit will increase the workload to provoke EIB. Physical training does not improve any parameters of lung function including PEF, FEV_1, FVC, or maximum ventilation [19]. It is for this reason that treatment of the airway inflammation is recommended so that an asthmatic has the best opportunity to perform normally at any level of competition [21].

FACTORS WITH POTENTIAL TO INFLUENCE THE ASTHMATIC RESPONSE TO EXERCISE

There are many endogenous factors that have the potential to impact on the airway response to exercise in asthmatics. These include severity of the underlying airway inflammation, the associated bronchial smooth muscle responsiveness and increased vascular permeability, the degree of atopy, and the pre-exercise lung function in relation to predicted normal. Other factors include age and fitness of the subject, and body mass index, particularly for its potential to impact on functional residual capacity.

The intervention that has the greatest influence on these factors is the medication being taken by the asthmatic subject, either daily as with ICS, long acting β_2-agonist (LABA), or leukotriene receptor antagonists (LTRA), or prophylactically before exercise as with short acting β_2 agonists (SABA), sodium cromoglycate (SCG), or nedocromil sodium [8, 22–26]. Treatment of airway inflammation and prevention of bronchoconstriction during and after exercise can lead to improved gas exchange (Fig. 38.2) and normal exercise performance [8]. Diets that include reduction in sodium intake [27], and supplementation with omega 3 free fatty acids [28] have been reported to reduce severity of EIB but their effect on gas exchange has not been reported.

Many young people with self-reported exercise symptoms are diagnosed clinically as having EIB but fail to demonstrate EIB on objective testing and may thus be over treated [29, 30]. In contrast, many young fit and clinically "well controlled" asthmatics with normal lung function can have a significant reduction in lung function and gas exchange abnormalities after exercise. Some may have become tolerant to the protective effects of β_2-agonists [23], and others may be receiving too low a dose of inhaled steroids [25] (Fig. 38.3). These are reasons why exercise is and should remain a concern for people with asthma.

DETERMINANTS OF EIB

The most important external determinants of the severity of the airway response to exercise in asthmatics are the level

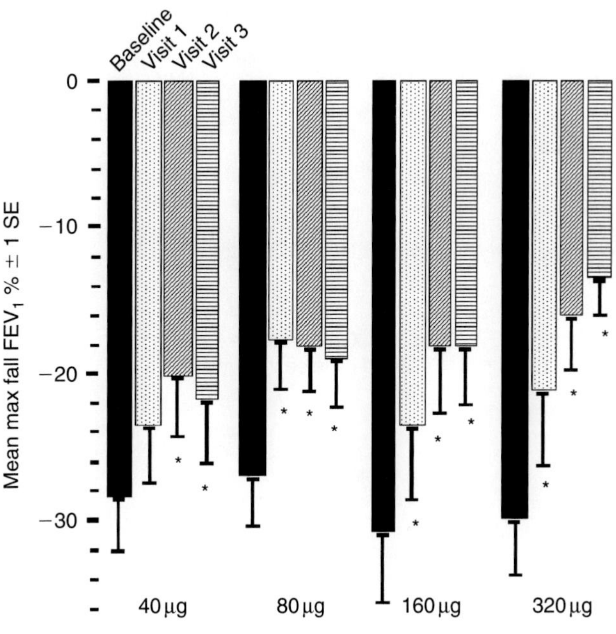

FIG. 38.3 Maximum percent fall in FEV_1 from baseline after 8 min of running exercise breathing dry air exercise in two groups of young adults with repeatable EIB, that is, 15% or more fall in FEV_1 at baseline. The fall is shown at baseline and after treatment with an ICS in a dose of 40 and 160 μg in one group and 80 and 320 μg in another for a duration of 1, 2, and 3 weeks. Reproduced from the study of Ref. [25].

of ventilation reached and sustained during exercise and the water content of the inspired air [31]. Thus exercise load is very important [32]. Whilst temperature of the air inspired during exercise can influence severity of EIB, it is of less importance than the absolute water content. Inspiring air close to body temperature and fully saturated with water prevents EIB.

That hyperpnea of dry air at high ventilation rates is required to provoke EIB suggests that severity of EIB depends on the rate at which generations of airways are recruited into the air conditioning process [33]. This concept explains why, for an individual, EIB occurs more often and more severely the drier and cooler the inspired air and the more rapid the increase in ventilation. It may also explain why EIB may be reported for the first time in athletes when they have reached their full ventilatory capacity in their early twenties and can achieve ventilation rates of 100 l/min or more [34]. It also explains why patients with COPD are unlikely to have EIB provoked by exercise, that is, they cannot reach high enough ventilation during exercise.

At the very high ventilation chronically achieved by cross-country skiers, rowers, and cyclists, a large surface area of airways including very small airways (<1 mm in diameter) will be used to condition the air. If very small airways become dehydrated and sustain injury then the restorative process of plasma exudation [35] that follows may explain pathogenesis of EIB [34]. If exposure to plasma-derived products occurs regularly then the contractile properties of the bronchial smooth muscle may be altered over time [36]. In allergic subjects repeated exposure to plasma may induce passive sensitization of the bronchial smooth muscle *in vivo* [34], increasing its sensitivity to mast cell mediators [37]. In healthy subjects mast cells are more abundant in the smaller airways compared with the larger airway [38]. Thus they are well positioned to release histamine and other mediators that increase the plasma exudation from post-capillary venules.

STIMULUS AND MECHANISM OF EIB

The stimulus for exercise to provoke the airways of asthmatics to narrow is the loss of water from the airway surface that occurs in conditioning large volumes of air in a short time (Fig. 38.4). That is why EIB occurs more often when the exercise is intense [32, 39] and performed in a dry environment [40]. It is also the reason that eucapnic voluntary hyperpnea of dry air (EVH) is an excellent surrogate for exercise to provoke EIB [41]. The mechanism whereby the loss of water causes airway narrowing relates to the thermal (cooling) and osmotic consequences of dehydration of the airway surface and the underlying mucosa [42].

The thermal hypothesis proposes that EIB is a vascular event. Thus cooling of the airway surface from conditioning inspired air causes vasoconstriction of the bronchial vessels and a reactive hyperemia follows immediately after

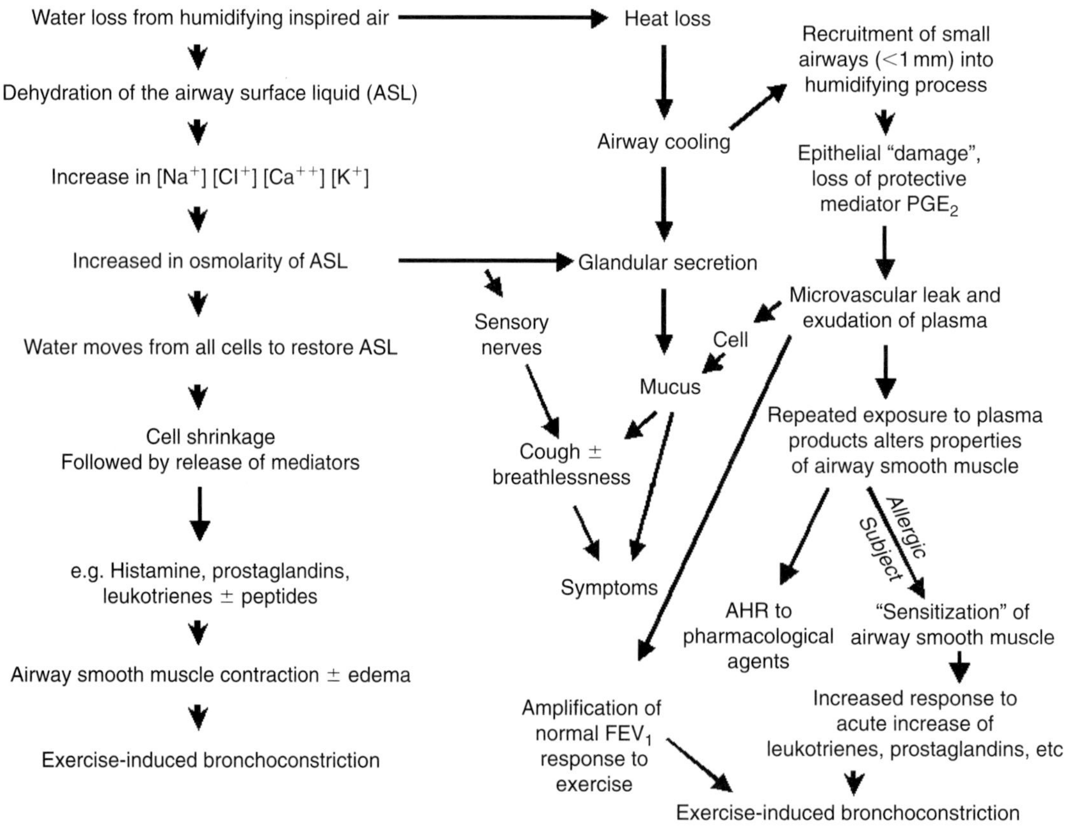

FIG. 38.4 The events that lead to EIB in people with asthma and events proposed to contribute to the pathogenesis of EIB and airway hyperresponsiveness in people exercising strenuously. Reproduced from Ref. [34].

exercise when ventilation falls and the airways re-warm rapidly [43]. The mechanical events of vascular engorgement, amplified by edema, serve to narrow the airways.

The osmotic hypothesis of EIB proposes that a transient increase in osmolarity of the airway surface liquid occurs in response to evaporative water loss, and that the cooler and drier the air inspired the greater the surface area that will become dehydrated and hyperosmotic [31, 44]. It is the hyperosmolarity of the airway surface liquid and the underlying mucosa that provides the appropriate microenvironment for release of mediators [45, 46]. In contrast to the thermal hypothesis, the osmotic hypothesis proposes airway narrowing occurs as a result of the contraction of bronchial smooth muscle by mediators [31]. The same mediators may increase vascular permeability and exudation of plasma, an event that would serve to normalize osmolarity.

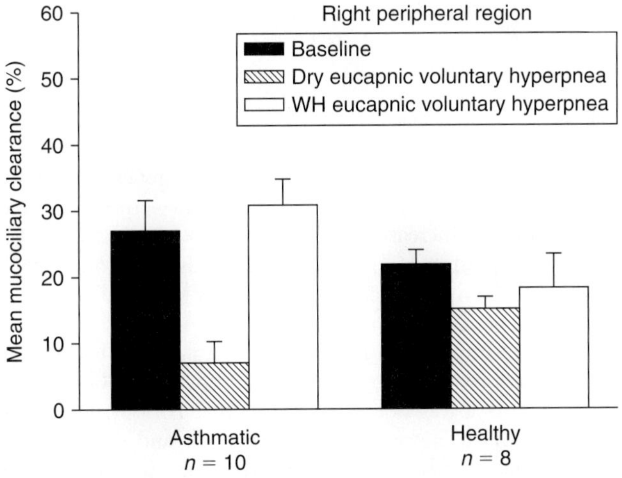

FIG. 38.5 Mean values for the percentage (%) mucociliary clearance of a radionuclide from the airways at baseline and following 6 min of hyperpnea in asthmatic and healthy subjects. Hyperpnea with dry air but not humid air caused a significant reduction in mean mucociliary clearance in the asthmatics suggesting the loss of water from the airway surface exceeded the rate of replacement and dehydration occurred. Redrawn data from Ref. [47].

Healthy subjects have the same burden to condition the air inspired during exercise as asthmatics. That water replacement to the airway surface is compromised at a high ventilation (at least for the first 5–6 min) in both healthy and asthmatic subjects has been demonstrated by an acute reduction in mucociliary clearance during hyperpnea of dry, but not warm humid air ([47], Fig. 38.5).

The difference in the lung function response to exercise between healthy subjects and asthmatic subjects is likely due to the bronchial smooth muscle of asthmatics being hyperresponsive and having a greater abundance of mast cells and their mediators close by [48]. Thus treatment that reduces the source (e.g. ICS), or production (fish oil diet), of mediators or their release (β_2-agonists, SCG, nedocromil sodium), or their contractile action on smooth muscle (β_2-agonists, leukotriene antagonists, histamine antagonists), are effective in inhibiting or preventing EIB [22-28] (Fig. 38.6).

OPTIMIZING PROTOCOLS TO IDENTIFY EIB

The benefit of optimizing an exercise protocol as a provocative stimulus for inducing an attack of asthma is that it provides a simple and inexpensive method to monitor treatment of asthma and can be utilized by both the patient and their doctor [49].

An exercise load that increases heart rate to 95% predicted maximum [32], and ventilation to $>17.5 \times FEV_1$ in l/min within a few minutes, and one that can be sustained for at least 6–8 min whilst breathing dry air, is more likely to provoke EIB than protocols using a lower intensity of

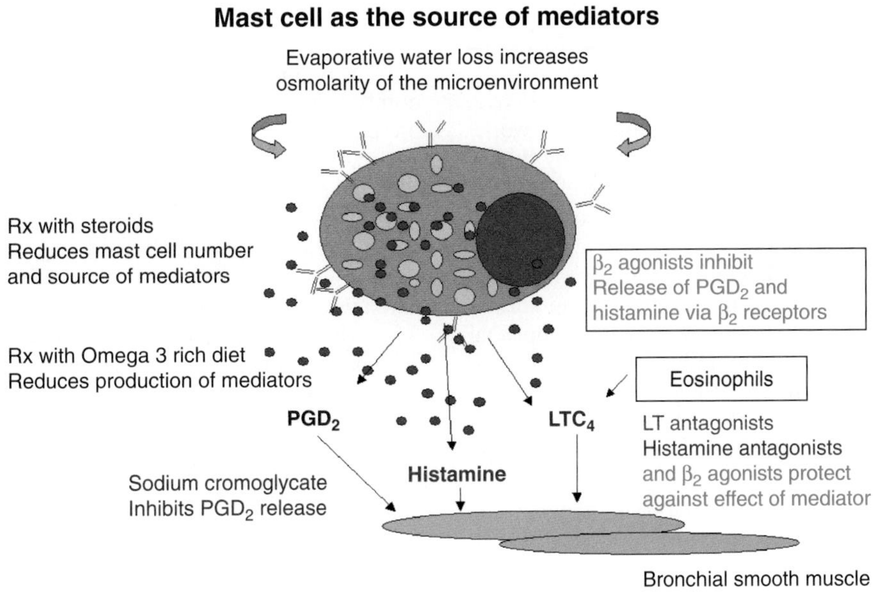

FIG. 38.6 Schematic diagram of the mast cell showing the stimulus for the release of mediators such as prostaglandin D_2 (PGD_2), histamine and LTC_4 and the mode of action of various pharmaceutical agents in preventing or inhibiting EIB.

TABLE 38.1 Challenge tests that can be used as a surrogate for exercise to identify EIB.

The surrogate stimuli (eucapnic voluntary hyperpnea with dry air, hyperosmolar aerosols of salt or mannitol) are easy to perform.

- These stimuli can be easily repeated for monitoring the response to treatment.
- These stimuli are more potent than exercise so false negative results are less likely.
- Some tests (e.g. mannitol) are portable and provide a common operating standard and can be performed at the point of care.
- Less expensive in manpower than field testing.

exercise for a longer duration [50]. There are, however, difficulties in achieving optimal conditions to provoke EIB, particularly in fit young adults with normal lung function. For this reason many laboratories have abandoned exercise in preference to using EVH or other surrogates to identify EIB (Table 38.1).

The protocol for EVH requires the subject to ventilate dry air containing 5% CO_2 for 6 min [51]. An "untrained" person can easily hyperventilate at $21 \times FEV_1$ l/min for 6 min without respiratory muscle fatigue and a "trained subject" can usually achieve and maintain a ventilation of $30 \times FEV_1$ l/min [52].

Other surrogates used to identify EIB include inhalation of aerosols of 4.5% saline or dry powder mannitol that increase osmolarity and cause release of mediators [53–56]. Some patients with COPD have been shown to have asthmatic responses to hyperosmolar agents [57, 58] even though they occasionally asthmatic responses to dry air challenge [59].

REFRACTORINESS TO THE BRONCHO-CONSTRICTING EFFECTS OF EXERCISE

Protocols that use a progressive increase in workload (e.g. 10 or 20 W increments each minute) to maximum working capacity are inappropriate to identify EIB likely due to refractoriness developing at the lower workloads. This was recognized many years ago when these protocols became popular in laboratories for assessing both heart disease and lung disease in children and adults. Similarly submaximal exercise for 20 min [60], or multiple warm up sprints of 30 s duration [61], induce refractoriness to an optimal stimulus. Even when an optimal stimulus is used to provoke EIB, 50% of subjects become refractory for about 2 h [62]. Refractoriness also develops to the effects of EVH when ventilation is increased progressively [63] so single stage protocols are recommended [51].

Prostaglandins and leukotrienes (LTs) are thought to be important in protecting the airways and for inducing a state of refractoriness [64]. The evidence for this is the absence of refractoriness following treatment for a few days with clinically recommended doses of non-steroidal anti-inflammatory agents, such as indomethacin [65]. The role of these mediators might be to increase the rate of return of water to the airway surface, probably by increasing bronchial blood flow. In this way the osmotic threshold for mediator release from mast cells may not be reached either during warm up exercise or later when an optimal exercise stimulus is used. The capacity of the epithelium to produce and release protective mediators may also be related to the extent to which the epithelium is injured in response to the hyperpnea of dry air [66]. It is also likely that the increase in tidal volume causes stretch on the bronchial smooth muscle reducing the "latch state" and making it less responsive to a subsequent stimulus [13].

INFLAMMATORY MARKERS AND EXERCISE AND EIB

Exercise can enhance transcription of two genes coding for 5-lipoxygenase (ALOX5) and ALOX5-activating protein leading to increased levels of LTB_4 and leukotriene C_4 (LTC_4) in plasma [67]. Leukotrienes sustain bronchoconstriction following exercise in asthmatics [68] and they may contribute to EIB in elite athletes [34]. A mast cell-derived mediator of importance in EIB is prostaglandin D_2 (PGD_2), usually reported as its metabolite 9α, 11β-PGF_2 [69]. Histamine is a less potent mediator of EIB, but unlike the PGs and LTs it is preformed in mast cells and thus is likely to contribute to the reduction in lung function that occurs immediately after exercise [9, 70]. It is however, the LTs and PGDs that sustain the fall in lung function after exercise and this is the reason that antihistamines are not useful as a single therapy for controlling EIB in most asthmatics [71].

The role of neuropeptides in EIB has not been well studied although the data are supportive and fluctuations in osmolarity of the airway surface could stimulate their release [72]. Recent data suggest an association between increase in neurokinin A levels in induced sputum and cysteinyl LTs in sputum following exercise in subjects with EIB [73].

Eosinophils are a potential source of LTs and increased numbers in sputum [74] and blood [75] are associated with presence of EIB independently of the presence of atopy. An acute increase in the percentage of eosinophils has been reported after exercise [76, 77].

The percentage of columnar epithelial cells in sputum relates to severity of EIB [66], and the concentration of these cells is higher in asthmatic subjects with EIB compared with asthmatics without EIB [78].

The severity of EIB is related to the degree of atopy. This is unsurprising in that atopy is an indirect marker of IgE on mast cells and it is the mediators from these cells that are implicated in EIB. The presence of IgE on mast cells renders them less stable to hyperosmotic stress [45].

The severity of EIB is also related to concentration of expired nitric oxide (eNO) although this appears to be true only for atopic subjects and thus may be an epiphenomenon [79]. A value for eNO $\leq$25 ppb has a negative predictive value of 100% for EIB but a positive predictive value of only 28% [80]. Severity of EIB is related to markers of vascular permeability and vascular endothelial growth factor [81].

EXERCISE AND THE PERSON WITH COPD

COPD is characterized by airway inflammation and airflow limitation that is not fully reversible. Despite optimal medical management, people with COPD have significant abnormalities in exercise capacity and ability to participate in everyday life. Often, the presenting symptom is breathlessness on exertion. Breathlessness during activity is evident in everyday life and varies with disease severity, with those with mild disease only experiencing breathlessness at high levels of exertion, whereas those with moderate to severe COPD becoming breathless during everyday tasks. In terms of the World Health Organization descriptors [82], the "impairments" (i.e. the physiological and anatomical changes) resulting from COPD lead to "activity limitation" (inability to perform daily tasks or activities), and "participation restriction" (inability to contribute fully in societal roles such as work and recreation).

The impairments that lead to exercise intolerance and activity limitation are related to the effects of COPD on the ventilatory, cardiovascular, and muscular systems.

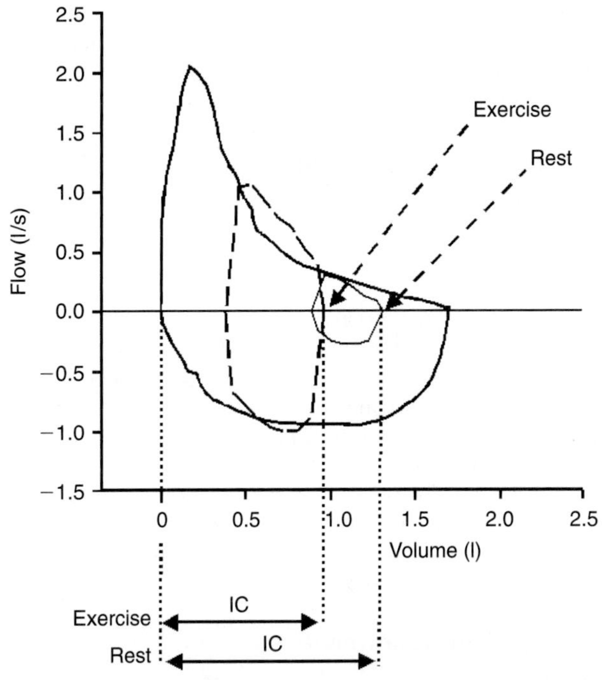

FIG. 38.7 Tidal flow-volume loops at rest and at peak exercise in COPD. Dynamic hyperinflation is evident by the shift of the exercise flow-volume loop towards TLC. The reduction in IC indicates an increase in EELV.

Ventilatory limitations to exercise

Flow limitation and altered mechanics of breathing

The increased airway resistance and expiratory flow limitation that occurs in COPD results in hyperinflation at rest. This alters the length–tension relationship of the respiratory muscles (i.e. muscles are in a more shortened position) which places them at a mechanical disadvantage. During exercise, ventilation increases (via increases in tidal volume and inspiratory and expiratory flow rates) to meet the metabolic demands. The need for increased flow rates in the presence of expiratory flow limitation results in further hyperinflation during exercise (i.e. dynamic hyperinflation) which is measured as a decrease in IC, the corollary of which is an increase in end-expiratory lung volume EELV (Fig. 38.7). Such dynamic hyperinflation further impacts on respiratory muscle length–tension relationships as well as limiting the tidal volume response to exercise, resulting in a greater respiratory effort to achieve a given tidal volume (V_T) (Fig. 38.8).

At low levels of exercise, dynamic hyperinflation may enable higher flow rates and greater tidal volume than would have been possible without dynamic hyperinflation, allowing exercise to continue. With increasing intensity of exercise the inability to further increase ventilation due to expiratory flow constraints becomes a limiting factor. A measure of this ventilatory limitation is the ratio of peak exercise ventilation to maximum voluntary ventilation (V_E/MVV). A V_E/MVV ratio above 0.9, while heart rate and other physiological functions are below maximal capacity, suggests severe ventilatory constraints to exercise [83]. However, since the estimate of MVV is usually obtained from resting measures this approach to the determination of ventilatory limitation may be imprecise.

Altered mechanics of breathing in COPD may also impact on the capacity for arm exercise, especially in those

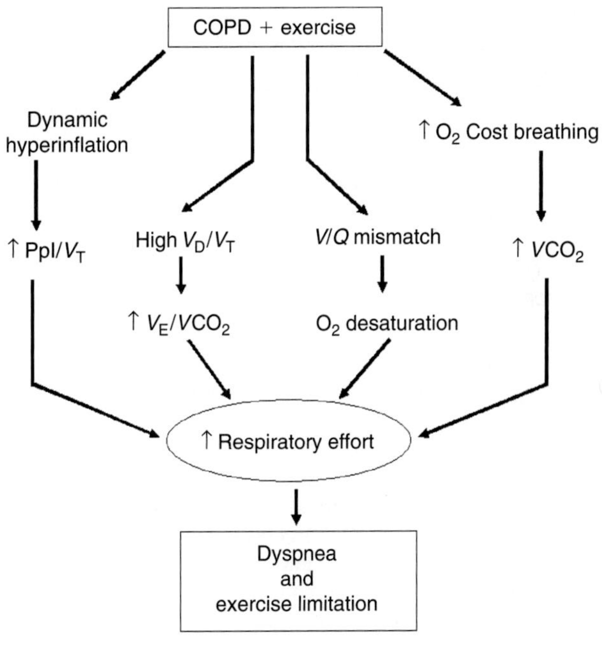

FIG. 38.8 Ventilatory factors leading to dyspnea and exercise limitation in COPD. Ppl: pleural pressure as a measure of inspiratory effort; V_T: tidal volume; VCO_2: carbon dioxide production; O_2: oxygen; V/Q: ventilation/perfusion; V_E: ventilation; V_D: dead space.

with severe hyperinflation. Arm elevation reduces the ability of the accessory muscles to contribute to breathing due to involvement in the arm task [84]. This shifts more of the respiratory work to the diaphragm which is already shortened due to hyperinflation, and thus is less effective in responding to the added respiratory load. Dynamic hyperinflation during arm exercise [85–87] and a higher EELV with arms elevated [86] may also contribute to exercise limitation, although this has not been examined in detail.

Increased work of breathing

Compared to the healthy population, respiratory muscle work during exercise is greater in people with COPD to overcome the increased airway resistance. In addition, due to dynamic hyperinflation, elastic work increases as tidal volume (V_T) nears total lung capacity (TLC). It has been shown that the oxygen consumption of the respiratory system in COPD during exercise may be 35–40% of total body oxygen consumption ($\dot{V}O_2$) [88]. Such a large oxygen demand by the respiratory muscles may lead to a higher carbon dioxide production ($\dot{V}CO_2$) for a given level of lower limb exercise, further increasing ventilation, as well as possible diversion of blood flow from the lower limb muscles to meet respiratory muscle requirements. Indirect evidence for blood flow diversion comes from studies in healthy subjects [89] and in COPD [90].

Gas exchange abnormalities

Gas exchange abnormalities arise from alveolar and capillary bed destruction leading to increased physiological dead space (i.e. a high dead space to tidal volume ratio (V_D/V_T) and ventilation–perfusion ($\dot{V}/\dot{Q}$) mismatch. The high V_D/V_T means that there is an increased requirement for ventilation for a given level of $\dot{V}CO_2$, while the $\dot{V}/\dot{Q}$ mismatch results in oxygen desaturation which may further increase ventilatory demand [83]. An added consequence of hypoxia is increased ventilatory drive resulting from early onset of lactic acidosis.

The above anatomical and physiological changes (i.e. the impairments) in the ventilatory system that result in dyspnea and exercise limitation are depicted in Fig. 38.8.

Cardiovascular constraints to exercise

Besides ventilatory constraints to exercise, people with COPD commonly have coexisting cardiovascular constraints. Cardiac output may be reduced due to a reduction in right ventricular stroke volume related to (1) increased pulmonary vascular resistance from hypoxic vasoconstriction and/or loss of capillary bed and vascular restructuring; (2) a reduction in right heart pre-load from decreased venous return consequent to hyperinflation [91]. Such reductions in cardiac output may limit blood supply, and thus oxygen availability, to the exercising muscles, with consequent early onset of lactic acidosis [92] adding to the constraints to exercise in COPD.

Skeletal muscle dysfunction and exercise

Skeletal muscle dysfunction in COPD may occur as a consequence of deconditioning related to inactivity caused by exertional dyspnea. In approximately 40% of patients with COPD the main symptom limiting exercise is leg fatigue. Besides deconditioning, other factors such as malnutrition, skeletal muscle myopathy, low levels of anabolic hormones, systemic inflammation, corticosteroid use, aging, and hypoxia may contribute to skeletal muscle dysfunction [93]. The physiological abnormalities evident in skeletal muscle of people with COPD are reduced oxidative enzyme capacity [94], reduced muscle fiber capillarization [95], atrophy of Type I [96] and Type IIa muscle fibers. Combined, these abnormalities result in lactic acidosis at lower levels of exercise and a reduced exercise endurance capacity compared to healthy people [97, 98]. Muscle mass and strength are also reduced in people with COPD. Importantly, some of these abnormalities can be partially reversed by exercise training.

Interestingly, skeletal muscle dysfunction is less evident in arm muscles compared to leg muscles in COPD [99] and therefore limitation of upper limb exercise capacity is more likely due to alterations in mechanics of breathing, described previously, than to skeletal muscle dysfunction. Maintenance of arm skeletal muscle function has been attributed to the requirements to continue performing upper limb tasks for daily functional activities, while lower limb activity can be more easily reduced, leading to decondition. However, further research is required to validate this.

INTERVENTIONS TO IMPROVE EXERCISE CAPACITY IN COPD

Interventions that can either decrease ventilatory demand or that can increase ventilatory capacity may improve the ability of people with COPD to exercise.

Interventions that reduce ventilatory demand

Exercise training

Lower limb endurance exercise training is effective in improving exercise capacity and quality of life and reducing exertional dyspnea in people with COPD [100], largely due to physiological changes that decrease ventilatory demand at a given level of exercise. These physiological changes include increases in muscle fiber capillarization [101], mitochondrial density, and muscle oxidative/metabolic capacity [102–104]. Such training adaptations improve the aerobic capacity of the muscle, delaying the onset of lactic acidosis and, as such, reduce the ventilatory requirements for exercise at equivalent pre-training work rates [92]. This is reflected in the ability of people with COPD to perform equivalent work rates for longer after training as well as achieving higher peak work rates [92] and translates into a reduction in "activity limitation" and "participation restriction" and improved quality of life.

The endurance exercise prescription in terms of the mode of training (cycle, treadmill, etc.) the intensity (high or low), duration of a training session, frequency of training, and the length of the program has generated a great deal of research, much of which is outlined in the major recent guidelines of on Pulmonary Rehabilitation [105–109] and Chapter 58. Information on the practical application of exercise training in COPD patients can be accessed through the Pulmonary Rehabilitation Toolkit (http://www.pulmonaryrehab.com.au).

It should be noted that while exercise training is considered the key component to achieve changes in exercise capacity and quality of life, exercise training often occurs as part of a comprehensive pulmonary rehabilitation program which includes education, anxiety and dyspnea management, smoking cessation support, and nutritional advice. These additional components of pulmonary rehabilitation may enhance the outcomes of training regimen.

In addition to lower limb training, endurance training of the upper limb using either supported arm exercise (arm cranking) or unsupported arm exercise (free weights) has been shown to reduce ventilatory demand and improve arm exercise capacity [110, 111]. Such improvements are task specific [112], therefore exercise mimicking daily activities, which are mostly unsupported arm tasks, are recommended for inclusion in exercise training programs [108, 109].

Supplemental oxygen

During exercise gas exchange abnormalities result is oxygen desaturation in COPD patients with severe disease, with consequent increases in ventilation (through earlier production of lactate) and higher dyspnea levels [83]. Acute oxygen supplementation during exercise in COPD reduces ventilatory demand for an equivalent work rate as well as reducing EELV [113] and exercise-induced pulmonary hypertension [114], and improving leg blood flow [90]. All these factors contribute to an increased exercise capacity and reduced dyspnea. However, no significant relationship has been found between exercise desaturation on room air and improvement in exercise capacity with supplemental oxygen [114, 115], suggesting that responses to oxygen supplementation need to be assessed on an individual basis.

The effect of oxygen supplementation during exercise training in COPD subjects who desaturate during exercise has been equivocal, with most studies showing no greater improvement in exercise capacity compared with exercise training on room air [109]. However, one study of oxygen supplementation during exercise training in nonhypoxaemic COPD subjects, demonstrated significantly greater improvements in endurance exercise time in the oxygen trained group, most likely related to the higher training intensities achieved [116]. Further studies are required to validate the effectiveness of similar exercise training protocols using supplemental oxygen in COPD subjects who desaturate during exercise.

Strength training

Although strength training does not results in reduced ventilatory demand per se it had been included in this section, since the short bouts of this form of exercise induce lower ventilatory demand and dyspnea compared to endurance training and are usually well tolerated by people with COPD [117].

Reductions in skeletal muscle strength are evident in people with COPD and may affect their ability to perform functional activities as well as increase the risk of falls; however, the latter has not been specifically studied in the COPD population. Whether improvements in muscle strength gained by resistance training programs [118] translate into improved exercise capacity is not clear. Significant improvements in cycle endurance capacity were demonstrated in studies comparing strength training with no intervention [119, 120]. However, the gains in endurance capacity from strength training were small compared to those that could be elicited from endurance training [119, 120].

Interventions that increase ventilatory capacity

Bronchodilator therapy

In general, lung function is poorly correlated with exercise capacity in COPD. However, as one of the main factors limiting exercise in COPD is expiratory flow limitation, medications such as inhaled anticholinergic and β_2-sympathomimetic agents, that reduce airway resistance, may improve exercise capacity in COPD. Bronchodilator medication delivered prior to exercise has been shown to decrease resting hyperinflation and dynamic hyperinflation during exercise, which correlated strongly with improved exercise capacity in severe COPD despite minimal change in spirometry [121]. A recent study of bronchodilator therapy prior to exercise combined with supplemental oxygen during exercise has shown an additive effect on exercise endurance, due to the combined physiological effects of reduced hyperinflation and reduced ventilatory drive [122].

Non-invasive ventilation

In people with severe COPD, positive pressure non-invasive ventilation reduces respiratory muscle load and prolongs the tolerance to lactic acid accumulation resulting in a reduction in dyspnea and improved exercise duration [123]. Exercise training using non-invasive ventilation in such patient groups, especially those with a ventilatory limitation to exercise, enabled higher training intensities to be achieved resulting in a reduction in lactate at equivalent workloads and greater improvements in exercise capacity in the intervention group [124–126]. The clinical significance of this form of intervention in terms of improvements in functional capacity and quality of life, as well as the cost-effectiveness has yet to be determined.

Lung volume reduction surgery

Lung volume reduction surgery (LVRS), the removal of areas of emphysematous lung tissue, increases static lung recoil, thus reducing flow limitation and dynamic hyperinflation [127]. Combined with post-operative exercise training, LVRS improved exercise capacity for both lower limbs [128] and upper limbs [129]. The improvements in exercise capacity were additional to those gained from pre-surgery exercise training. LVRS is mostly effective in selected patients with upper-lobe emphysema and persistent low exercise capacity after an exercise training program.

Body positioning

The lean forward position with shoulder girdle fixation is often adopted by people with COPD when breathless. Leaning forward has been shown to reduce accessory muscle activity, increase transdiaphragmatic pressure, increase maximum inspiratory pressures, and reduce dyspnea at rest in COPD subjects with severe disease [130, 131], presumably by increased abdominal pressure improving the length–tension relationship of the diaphragm in people who are hyperinflated. Shoulder girdle fixation allows the accessory muscles to function with reverse origin and insertion (i.e. rather than acting to move the arm on a fixed torso these muscles can act to move the ribs in an inspiratory direction). Shoulder girdle fixation has been shown to improve MVV, potentially reducing the ventilatory constraint to exercise [132]. Some evidence that the lean forward position with shoulder fixation can enhance exercise capacity in people with COPD comes from studies of rollator walking which have shown improved functional exercise capacity and a reduction in dyspnea [132, 133].

Breathing strategies

Pursed-lip breathing is another strategy adopted spontaneously by people with severe COPD to prolong active expiration and limit airway collapse. Pursed-lip breathing at rest has been shown to reduce EELV [134] which could potentially increase ventilatory capacity. However, volitional pursed-lip breathing during exercise has not been shown to increase exercise capacity in people with COPD [135]. The variable response to volitional pursed-lip breathing during exercise suggests that benefit may only occur in those who spontaneously adopt pursed-lip breathing.

Respiratory muscle training

Respiratory muscle weakness is evident in a subgroup of people with COPD. The impact of inspiratory muscle weakness on exercise capacity is unclear and may only be important if such weakness reduces ventilatory capacity such that it is a factor limiting exercise. Improvements in exercise capacity gained with inspiratory muscle training are significantly less than those gained with lower limb endurance training [136]. However, a number of studies have shown that the addition of inspiratory muscle training to a lower limb training program improved functional exercise capacity above that gained from lower limb training alone [137, 138]. Inspiratory muscle training for people with COPD may only be of benefit in those with suspected or proven respiratory muscle weakness [108] and current guidelines do not support routine inclusion of this type of training for people with COPD [109].

References

1. Welsh L *et al.* Fitness and physical activity in children with asthma. *Sports Med* 34: 861–70, 2004.
2. Hallstrand TS *et al.* Aerobic conditioning in mild asthma decreases the hyperpnea of exercise and improves exercise and ventilatory capacity. *Chest* 118: 1460–69, 2000.
3. Alioglu B *et al.* Cardiopulmonary responses of asthmatic children to exercise: Analysis of systolic and diastolic cardiac function. *Pediatr Pulmonol* 42: 283–89, 2007.
4. Varray A *et al.* Cardiac role in exercise limitation in asthmatic subjects with special reference to disease severity. *Eur Respir J* 6: 1011–17, 1993.
5. Lucas SR, Platts-Mills TA. Physical activity and exercise in asthma: Relevance to etiology and treatment. *J Allergy Clin Immunol* 115: 928–34, 2005.
6. Joyner BL *et al.* Cardiopulmonary exercise testing in children and adolescents with asthma who report symptoms of exercise-induced bronchoconstriction. *J Asthma* 43: 675–78, 2006.
7. Anderson SD *et al.* Metabolic and ventilatory changes in asthmatic patients during and after exercise. *Thorax* 27: 718–25, 1972.
8. Bye PTP *et al.* Plasma cyclic AMP levels in response to exercise and terbutaline sulphate aerosol in normal and asthmatic patients. *Eur J Respir Dis* 61: 287–97, 1980.
9. Haverkamp HC *et al.* Gas exchange during exercise in habitually active asthmatic subjects. *J Appl Physiol* 99: 1938–50, 2005.
10. Beck KC *et al.* Bronchoconstriction occurring during exercise in asthmatic subjects. *Am J Respir Crit Care Med* 149: 352–57, 1994.
11. Suman OE *et al.* Airway function after cyclooxygenase inhibition during hyperpnea-induced bronchoconstriction in guinea pigs. *J Appl Physiol* 89: 1971–78, 2000.
12. Suman OE, Beck KC. Role of airway endogenous nitric oxide on lung function during and after exercise in mild asthma. *J Appl Physiol* 93: 1932–38, 2002.
13. Fredberg JJ *et al.* Airway smooth muscle, tidal stretches, and dynamically determined contractile states. *Am J Respir Crit Care Med* 156: 1752–59, 1997.
14. Silverman M *et al.* Metabolic changes preceding exercise-induced bronchoconstriction. *BMJ* 1: 207–9, 1972.
15. Munoz P. Pulmonary gas exchange response to exercise and mannitol challenges in exercise-induced asthma patients. *Am J Respir Crit Care Med* ASM 3(15): A721, 2006.
16. Young IH *et al.* Pattern and time course of ventilation-perfusion inequality in exercise-induced asthma. *Am Rev Respir Dis* 125: 304–11, 1982.
17. Becker JM *et al.* Asthma deaths during sports: Report of a 7-year experience. *J Allergy Clin Immunol* 113: 264–67, 2004.
18. Anderson SD *et al.* Responses to bronchial challenge submitted for approval to use inhaled β_2 agonists prior to an event at the 2002 Winter Olympics. *J Allergy Clin Immunol* 111: 44–49, 2003.
19. Ram FSF *et al.* Physical training for asthma. *Cochrane Database Syst Rev* 19: 2005, CD001116.
20. Thio BJ *et al.* Exercise-induced asthma and cardiovascular fitness in asthmatic children. *Thorax* 51: 207–9, 1996.
21. Anderson SD *et al.* Bronchial challenges in athletes applying to inhale a β_2-agonist at the 2004 Summer Olympics. *J Allergy Clin Immunol* 117: 767–73, 2006.
22. Anderson SD. Single dose agents in the prevention of exercise-induced asthma: A descriptive review. *Treat Respir Med* 3: 365–79, 2004.
23. Anderson SD *et al.* β_2-Agonists and exercise-induced asthma. *Clin Rev Allergy Immunol* 31: 163–80, 2006.
24. Storms W *et al.* A comparison of the effects of oral montelukast and inhaled salmeterol on response to rescue bronchodilation after challenge. *Respir Med* 98: 1051–62, 2004.
25. Subbarao P *et al.* Effect of ciclesonide dose and duration of therapy on exercise-induced bronchoconstriction in patients with asthma. *J Allergy Clin Immunol* 117: 1008–13, 2006.
26. Weiler JM *et al.* Effect of fluticasone/salmeterol administered via a single device on exercise-induced bronchospasm in patients with persistent asthma. *Ann Allergy Asthma Immunol* 94: 65–72, 2005.
27. Mickleborough T, Gotshall R. Dietary components with demonstrated effectiveness in decreasing the severity of exercise-induced asthma. *Sports Med* 33: 671–81, 2003.
28. Mickleborough TD *et al.* Protective effect of fish oil supplementation on exercise-induced bronchoconstriction in asthma. *Chest* 129: 39–49, 2006.
29. Hallstrand TS *et al.* Effectiveness of screening examinations to detect unrecognised exercise-induced bronchoconstriction. *J Pediatr* 141: 343–49, 2002.
30. Rundell KW *et al.* Self-reported symptoms and exercise-induced asthma in the elite athlete. *Med Sci Sports Exerc* 33: 208–13, 2001.
31. Anderson SD, Daviskas E. The mechanism of exercise-induced asthma is .... *J Allergy Clin Immunol* 106: 453–59, 2000.
32. Carlsen KH *et al.* Exercise induced bronchoconstriction depends on exercise load. *Respir Med* 94: 750–55, 2000.
33. Anderson SD, Holzer K. Exercise-induced asthma: Is it the right diagnosis in elite athletes?. *J Allergy Clin Immunol* 106: 419–28, 2000.

34. Anderson SD, Kippelen P. Exercise-induced bronchoconstriction: Pathogenesis. *Curr Allergy Asthma Rep* 5: 116–22, 2005.
35. Erjefalt JS *et al.* Epithelial pathways for luminal entry of bulk plasma. *Clin Exp Allergy* 25: 187–95, 1995.
36. Johnson PR, Burgess JK. Airway smooth muscle and fibroblasts in the pathogenesis of asthma. *Curr Allergy Asthma Rep* 4: 102–8, 2004.
37. Schmidt D *et al.* Passive sensitization of human airway increases responsiveness to leukotriene C_4. *Eur Respir J* 14: 315–19, 1999.
38. Carroll NG *et al.* Distribution and degranulation of airway mast cells in normal and asthmatic subjects. *Eur Respir J* 19: 879–85, 2002.
39. Silverman M, Anderson SD. Standardization of exercise tests in asthmatic children. *Arch Dis Child* 47: 882–89, 1972.
40. Anderson SD *et al.* Sensitivity to heat and water loss at rest and during exercise in asthmatic patients. *Eur J Respir Dis* 63: 459–71, 1982.
41. Rundell KW *et al.* Field exercise vs laboratory eucapnic voluntary hyperventilation to identify airway hyperresponsiveness in elite cold weather athletes. *Chest* 125: 909–15, 2004.
42. Anderson SD, Daviskas E. The airway microvasculature and exercise-induced asthma. *Thorax* 47: 748–52, 1992.
43. McFadden ER *et al.* Postexertional airway rewarming and thermally induced asthma. *J Clin Invest* 78: 18–25, 1986.
44. Anderson SD. Is there a unifying hypothesis for exercise-induced asthma? *J Allergy Clin Immunol* 73: 660–65, 1984.
45. Gulliksson M *et al.* Release of prostaglandin D2 and leukotriene C in response to hyperosmolar stimulation of mast cells. *Allergy* 61: 1473–79, 2006.
46. Moloney ED *et al.* Release of inflammatory mediators from eosinophils following a hyperosmolar stimulus. *Respir Med* 97: 1–5, 2003.
47. Daviskas E *et al.* Changes in mucociliary clearance during and after isocapnic hyperventilation in asthmatic and healthy subjects. *Eur Respir J* 8: 742–51, 1995.
48. Brightling CE *et al.* Mast-cell infiltration of airway smooth muscle in asthma. *N Engl J Med* 346: 1699–705, 2002.
49. Haby MM *et al.* An exercise challenge for epidemiological studies of childhood asthma: Validity and repeatability. *Eur Respir J* 8: 729–36, 1995.
50. Anderson SD *et al.* Laboratory protocol for exercise asthma to evaluate salbutamol given by two devices. *Med Sci Sports Exerc* 33: 893–900, 2001.
51. Anderson SD *et al.* Provocation by eucapnic voluntary hyperpnoea to identify exercise induced bronchoconstriction. *Br J Sports Med* 35: 344–47, 2001.
52. Spiering BA *et al.* Standardized estimated ventilation for eucapnic voluntary hyperventilation: An evaluation. *Med Sci Sports Exerc* 34: S52, 2002.
53. Brannan JD *et al.* Responsiveness to mannitol in asthmatic subjects with exercise- and hyperventilation-induced asthma. *Am J Respir Crit Care Med* 158: 1120–26, 1998.
54. Brannan JD *et al.* Inhibition of mast cell PGD_2 release protects against mannitol-induced airway narrowing. *Eur Respir J* 27: 944–50, 2006.
55. Holzer K *et al.* Mannitol as a challenge test to identify exercise-induced bronchoconstriction in elite athletes. *Am J Respir Crit Care Med* 167: 534–47, 2003.
56. Smith CM, Anderson SD. Inhalational challenge using hypertonic saline in asthmatic subjects: A comparison with responses to hyperpnoea, methacholine and water. *Eur Respir J* 3: 144–51, 1990.
57. Leuppi JD *et al.* Prediction of treatment-response to inhaled corticosteroids by mannitol-challenge test in COPD. A proof of concept. *Pulm Pharmacol Ther* 18: 83–88, 2005.
58. Taube C *et al.* Airway response to inhaled hypertonic saline in patients with moderate to severe chronic obstructive pulmonary disease. *Am J Respir Crit Care Med* 164: 1810–15, 2001.
59. Ramsdale EH *et al.* Bronchial responsiveness to methacholine in chronic bronchitis: Relationship to airflow limitation and cold air responsiveness. *Thorax* 39: 912–18, 1984.
60. Reiff DB *et al.* The effect of prolonged submaximal warm-up exercise on exercise-induced asthma. *Am Rev Respir Dis* 139: 479–84, 1989.
61. Schnall RP, Landau LI. Protective effects of repeated short sprints in exercise-induced asthma. *Thorax* 35: 828–32, 1980.
62. Anderson SD, Schoeffel RE. Respiratory heat and water loss during exercise in patients with asthma: Effect of repeated exercise challenge. *Eur J Respir Dis* 63: 472–80, 1982.
63. Argyros GJ *et al.* The refractory period after eucapnic voluntary hyperventilation challenge and its effect on challenge technique. *Chest* 108: 419–24, 1995.
64. Manning PJ *et al.* Exercise-induced refractoriness in asthmatic subjects involves leukotriene and prostaglandin interdependent mechanisms. *Am Rev Respir Dis* 148: 950–54, 1993.
65. Wilson BA *et al.* The effects of indomethacin on refractoriness following exercise both with and without bronchoconstriction. *Eur Respir J* 12: 2174–78, 1994.
66. Hallstrand TS *et al.* Inflammatory basis of exercise-induced bronchoconstriction. *Am J Respir Crit Care Med* 172: 679–86, 2005.
67. Hilberg T *et al.* Transcription in response to physical stress-clues to the molecular mechanisms of exercise-induced asthma. *FASEB J* 19: 1492–94, 2005.
68. Reiss TF *et al.* Increased urinary excretion of LTE_4 after exercise and attenuation of exercise-induced bronchospasm by montelukast, a cysteinyl leukotriene receptor antagonist. *Thorax* 52: 1030–35, 1997.
69. O'Sullivan S *et al.* Evidence for mast cell activation during exercise-induced bronchoconstriction. *Eur Respir J* 12: 345–50, 1998.
70. Anderson SD *et al.* Arterial plasma histamine levels at rest, during and after exercise in patients with asthma: Effects of terbutaline aerosol. *Thorax* 36: 259–67, 1981.
71. Dahlén B *et al.* Influence of zafirlukast and loratadine on exercise-induced bronchoconstriction. *J Allergy Clin Immunol* 109: 789–93, 2002.
72. Ichinose M *et al.* A neurokinin 1-receptor antagonist improves exercise-induced airway narrowing in asthmatic patients. *Am J Respir Crit Care Med* 153: 936–41, 1996.
73. Hallstrand TS *et al.* Role of MUC5AC in the pathogenesis of exercise-induced bronchoconstriction. *J Allergy Clin Immunol*, 119: 1092-1098, 2007.
74. Yoshikawa T *et al.* Severity of exercise-induced bronchoconstriction is related to airway eosinophilic inflammation in patients with asthma. *Eur Respir J* 12: 879–84, 1998.
75. Lee SY *et al.* Eosinophils play a major role in the severity of exercise-induced bronchoconstriction in children with asthma. *Pediatr Pulmonol* 41: 1161–66, 2006.
76. Haverkamp HC *et al.* Repeat exercise normalizes the gas-exchange impairment induced by a previous exercise bout in asthmatic subjects. *J Appl Physiol* 99: 1843–52, 2005.
77. Kivity S *et al.* Eosinophil influx into the airways in patients with exercise-induced asthma. *Respir Med* 94: 1200–5, 2000.
78. Hallstrand TS *et al.* Airway immunopathology of asthma with exercise-induced bronchoconstriction. *J Allergy Clin Immunol* 116: 586–93, 2005.
79. Franklin PJ *et al.* Measuring exhaled nitric oxide levels in adults: The importance of atopy and airway responsiveness. *Chest* 126: 1540–45, 2004.
80. Lex C *et al.* Value of surrogate tests to predict exercise-induced bronchoconstriction in atopic childhood asthma. *Pediatr Pulmonol* 42: 225–30, 2007.
81. Kanazawa H *et al.* Vascular involvement in exercise-induced airway narrowing in patients with bronchial asthma. *Chest* 122: 166–70, 2002.
82. World Health Organization. Towards a Common Language for Functioning, Disability and Health. Geneva: ICF, 2002.
83. O'Donnell DE. Ventilatory limitations in chronic obstructive pulmonary disease. *Med Sci Sports Exerc* 33: S647–655, 2001.
84. Epstein SK *et al.* Ventilatory response to arm elevation, its determinants and use in patients with chronic obstructive pulmonary disease. *Am J Respir Crit Care Med* 152: 211–16, 1995.
85. Gigliotti F *et al.* Arm exercise and hyperinflation in patients with COPD: Effect of arm training. *Chest* 128: 1225–32, 2005.
86. McKeough ZJ *et al.* Arm positioning alters lung volumes in subjects with COPD and healthy subjects. *Aust J Physiother* 49: 133–37, 2003.
87. McKeough ZJ *et al.* Arm exercise capacity and dyspnea ratings in subjects with chronic obstructive pulmonary disease. *J Cardiopulm Rehabil* 23: 218–25, 2003.
88. Levison H, Cherniack RM. Ventilatory cost of exercise in chronic obstructive pulmonary disease. *J Appl Physiol* 25: 21–27, 1968.

89. Harms CA *et al.* Respiratory muscle work compromises leg blood flow during maximal exercise. *J Appl Physiol* 82: 1573–83, 1997.
90. Maltais F *et al.* Effects of oxygen on lower limb blood flow and O_2 uptake during exercise in COPD. *Med Sci Sports Exerc* 33: 916–22, 2001.
91. Sietsema K. Cardiovascular limitations in chronic pulmonary disease. *Med Sci Sports Exerc* 33: S656–61, 2001.
92. Casaburi R *et al.* Reductions in exercise lactic acidosis and ventilation as a result of exercise training in patients with obstructive lung disease. *Am Rev Respir Dis* 143: 9–18, 1991.
93. A statement of the American Thoracic Society and European Respiratory Society. Skeletal muscle dysfunction in chronic obstructive pulmonary disease. *Am J Respir Crit Care Med* 159: S1–40, 1999.
94. Maltais F *et al.* Oxidative capacity of the skeletal muscle and lactic acid kinetics during exercise in normal subjects and in patients with COPD. *Am J Respir Crit Care Med* 153: 288–93, 1996.
95. Jobin J *et al.* Chronic obstructive pulmonary disease: Capillarity and fiber-type characteristics of skeletal muscle. *J Cardiopulm Rehabil* 18: 432–37, 1998.
96. Jakobsson P *et al.* Skeletal muscle metabolites and fibre types in patients with advanced chronic obstructive pulmonary disease (COPD), with and without chronic respiratory failure. *Eur Respir J* 3: 192–96, 1990.
97. Casaburi R. Skeletal muscle dysfunction in chronic obstructive pulmonary disease. *Med Sci Sports Exerc* 33: S662–70, 2001.
98. Polkey MI. Muscle metabolism and exercise tolerance in COPD. *Chest* 121: 131S–135S, 2002.
99. Gea JG *et al.* Metabolic characteristics of the deltoid muscle in patients with chronic obstructive pulmonary disease. *Eur Respir J* 17: 939–45, 2001.
100. Lacasse Y *et al.* Pulmonary rehabilitation for chronic obstructive pulmonary disease. *Cochrane Database Syst Rev*: CD003793, 2006.
101. Whittom F *et al.* Histochemical and morphological characteristics of the vastus lateralis muscle in patients with chronic obstructive pulmonary disease. *Med Sci Sports Exerc* 30: 1467–74, 1998.
102. Maltais F *et al.* Skeletal muscle adaptation to endurance training in patients with chronic obstructive pulmonary disease. *Am J Respir Crit Care Med* 154: 442–47, 1996.
103. McKeough ZJ *et al.* Exercise capacity and quadriceps muscle metabolism following training in subjects with COPD. *Respir Med* 100: 1817–25, 2006.
104. Sala E *et al.* Effects of endurance training on skeletal muscle bioenergetics in chronic obstructive pulmonary disease. *Am J Respir Crit Care Med* 159: 1726–34, 1999.
105. ACCP/AACVPR Pulmonary Rehabilitation Guidelines Panel. Pulmonary rehabilitation: Joint ACCP/AACVPR evidence-based guidelines. American College of Chest Physicians. American Association of Cardiovascular and Pulmonary Rehabilitation. *Chest* 112: 1363–96, 1997.
106. American Thoracic Society. Pulmonary rehabilitation – 1999. *Am J Respir Crit Care Med* 159: 1666–82, 1999.
107. British Thoracic Society Standards of Care Subcommittee on Pulmonary Rehabilitation. Pulmonary rehabilitation. *Thorax* 56: 827–34, 2001.
108. Nici L *et al.* American Thoracic Society/European Respiratory Society statement on pulmonary rehabilitation. *Am J Respir Crit Care Med* 173: 1390–413, 2006.
109. Ries AL *et al.* Pulmonary rehabilitation: Joint ACCP/AACVPR evidence-based clinical practice guidelines. *Chest* 131: 4S–42S, 2007.
110. Couser JI Jr. *et al.* Pulmonary rehabilitation that includes arm exercise reduces metabolic and ventilatory requirements for simple arm elevation. *Chest* 103: 37–41, 1993.
111. Epstein SK *et al.* Arm training reduces the VO_2 and V_E cost of unsupported arm exercise and elevation in chronic obstructive pulmonary disease. *J Cardiopulm Rehabil* 17: 171–77, 1997.
112. Martinez FJ *et al.* Supported arm exercise vs unsupported arm exercise in the rehabilitation of patients with severe chronic airflow obstruction. *Chest* 103: 1397–402, 1993.
113. Somfay A *et al.* Dose-response effect of oxygen on hyperinflation and exercise endurance in nonhypoxaemic COPD patients. *Eur Respir J* 18: 77–84, 2001.
114. Fujimoto K *et al.* Benefits of oxygen on exercise performance and pulmonary hemodynamics in patients with COPD with mild hypoxemia. *Chest* 122: 457–63, 2002.
115. Jolly EC *et al.* Effects of supplemental oxygen during activity in patients with advanced COPD without severe resting hypoxemia. *Chest* 120: 437–43, 2001.
116. Emtner M *et al.* Benefits of supplemental oxygen in exercise training in nonhypoxemic chronic obstructive pulmonary disease patients. *Am J Respir Crit Care Med* 168: 1034–42, 2003.
117. Probst VS *et al.* Cardiopulmonary stress during exercise training in patients with COPD. *Eur Respir J* 27: 1110–18, 2006.
118. O'Shea SD *et al.* Peripheral muscle strength training in COPD: A systematic review. *Chest* 126: 903–14, 2004.
119. Ortega F *et al.* Comparison of effects of strength and endurance training in patients with chronic obstructive pulmonary disease. *Am J Respir Crit Care Med* 166: 669–74, 2002.
120. Spruit MA *et al.* Resistance versus endurance training in patients with COPD and peripheral muscle weakness. *Eur Respir J* 19: 1072–78, 2002.
121. O'Donnell DE *et al.* Spirometric correlates of improvement in exercise performance after anticholinergic therapy in chronic obstructive pulmonary disease. *Am J Respir Crit Care Med* 160: 542–49, 1999.
122. Peters MM *et al.* Combined physiological effects of bronchodilators and hyperoxia on exertional dyspnoea in normoxic COPD. *Thorax* 61: 559–67, 2006.
123. van't Hul A *et al.* The acute effects of noninvasive ventilatory support during exercise on exercise endurance and dyspnea in patients with chronic obstructive pulmonary disease: A systematic review. *J Cardiopulm Rehabil* 22: 290–97, 2002.
124. Costes F *et al.* Noninvasive ventilation during exercise training improves exercise tolerance in patients with chronic obstructive pulmonary disease. *J Cardiopulm Rehabil* 23: 307–13, 2003.
125. Hawkins P *et al.* Proportional assist ventilation as an aid to exercise training in severe chronic obstructive pulmonary disease. *Thorax* 57: 853–59, 2002.
126. van't Hul A *et al.* Training with inspiratory pressure support in patients with severe COPD. *Eur Respir J* 27: 65–72, 2006.
127. Martinez FJ *et al.* Lung-volume reduction improves dyspnea, dynamic hyperinflation, and respiratory muscle function. *Am J Respir Crit Care Med* 155: 1984–90, 1997.
128. Fishman A *et al.* A randomized trial comparing lung-volume-reduction surgery with medical therapy for severe emphysema. *N Engl J Med* 348: 2059–73, 2003.
129. McKeough ZJ *et al.* Supported and usnsupported arm exercise capacity following lung volume reduction surgery: A pilot study. *Chron Respir Dis* 2: 59–65, 2005.
130. O'Neill S, McCarthy DS. Postural relief of dyspnoea in severe chronic airflow limitation: Relationship to respiratory muscle strength. *Thorax* 38: 595–600, 1983.
131. Sharp JT *et al.* Postural relief of dyspnea in severe chronic obstructive pulmonary disease. *Am Rev Respir Dis* 122: 201–11, 1980.
132. Probst VS *et al.* Mechanisms of improvement in exercise capacity using a rollator in patients with COPD. *Chest* 126: 1102–7, 2004.
133. Solway S *et al.* The short-term effect of a rollator on functional exercise capacity among individuals with severe COPD. *Chest* 122: 56–65, 2002.
134. Bianchi R *et al.* Chest wall kinematics and breathlessness during pursed-lip breathing in patients with COPD. *Chest* 125: 459–65, 2004.
135. Garrod R *et al.* An evaluation of the acute impact of pursed lips breathing on walking distance in nonspontaneous pursed lips breathing chronic obstructive pulmonary disease patients. *Chron Respir Dis* 2: 67–72, 2005.
136. Crowe J *et al.* Inspiratory muscle training compared with other rehabilitation interventions in adults with chronic obstructive pulmonary disease: A systematic literature review and meta-analysis. *COPD* 2: 319–29, 2005.
137. Dekhuijzen PN *et al.* Target-flow inspiratory muscle training during pulmonary rehabilitation in patients with COPD. *Chest* 99: 128–33, 1991.
138. Wanke T *et al.* Effects of combined inspiratory muscle and cycle ergometer training on exercise performance in patients with COPD. *Eur Respir J* 7: 2205–11, 1994.

Atmospheric Pollutants

CHAPTER 39

Jon Ayres

Department of Environmental and Occupational Medicine, Liberty Safe Work Research Centre, University of Aberdeen, Aberdeen, UK

In the 1940s and 1950s, air pollution was a fact of life in industrialized cities, but the true extent of its health risks was not appreciated until the London Fog Incident of 1952 [1], which caused 4000 excess deaths largely from cardiac disease, bronchitis, or pneumonia. At the time, it was believed that the combination of particles and SO_2 was the cause of such effects, but subsequent reanalysis of these historical data suggested that the acidity of the aerosol could have been the major contributor to mortality [2]. As the Clean Air Act of 1956 began to reduce black smoke levels, day-to-day changes in pollution had progressively less impact on symptoms in chronic bronchitis [3] and the advice given to Government was that air pollution would never again represent a risk to human health, even though studies of urban compared with rural areas showed a higher prevalence of productive cough in urban areas [4]. However, by the late 1970s such reassurance was being questioned as vehicle-generated air pollution was being shown to impact on human health.

CURRENT POLLUTANT EXPOSURES

The air pollutants relevant to asthma and chronic obstructive pulmonary disease (COPD) are:

- sulfur dioxide (SO_2)
- nitrogen dioxide (NO_2)
- ozone
- particles

Levels of these pollutants vary worldwide and health-based air quality standards have been produced by many individual countries or agencies (Table 39.1). These give an indication of the levels to which populations are currently exposed, although levels exceeding these are commonplace in large cities. The time frame for these standards varies between pollutants. For example, the UK air quality standard for SO_2 is 100 ppb over a 15-min sampling time frame, because of its ability to cause acute bronchoconstriction in asthma, whereas the standard for ozone is based on 8 h, being the timescale in which ozone is formed during sunlight.

Sources of gaseous pollutants

The main source of SO_2 is the burning of fossil fuel in coal-fired power stations, domestic fires, or industrial processes with a variable contribution from sulfur-containing engine fuels. NO_2 is largely derived from vehicle emissions, although the highest individual exposures are from gas appliances indoors. Ozone is a secondary pollutant formed by the action of ultraviolet light on components of vehicle exhaust.

Sources of particulate pollutants

The main sources of particles are vehicle emissions. Particles are often measured by a reflectance method, black smoke (BS) which, while simple and cheap has been superceded by a mass measure of particles either using a high volume filter system or a TEOM (tapered element oscillating microbalance). This produces measures such as PM_{10} (particles broadly $<10\,\mu m$ in diameter) or $PM_{2.5}$ ($<2.5\,\mu m$ in diameter), depending on the size of the air inlet. There is some evidence that at least part of the toxic fraction of particles resides in the ultrafine

TABLE 39.1 Air quality standards.

	WHO (2005 revision)	EC	UK
Sulfur dioxide			
10 min	500 μg/m³		
15 min	–		100 ppb (286 μg/m³)
24 h	125 μg/m³ (interim target 1) 20 μg/m³ (AQS)		
Annual	Now not needed as adherence to the 24-h standard will control annual exposures	48.8 ppb if smoke >60 μg/m³	
		67.5 ppb if smoke <60 μg/m³	
Nitrogen dioxide			
1 h	200 μg/m³	70.6 ppb (133 μg/m³)[a]	150 ppb (282 μg/m³)
Annual	40 μg/m³	26.2 ppb (49 μg/m³)[b]	
Ozone			
8 h moving average	100 μg/m³	55 ppb (110 μg/m³)	50 ppb (100 μg/m³)
Particles			
24-h mean	PM_{10} 50 μg/m³ $PM_{2.5}$ 25 μg/m³	PM_{10} 80 μg/m³	PM_{10} 50 μg/m³ [c]
Annual mean	PM_{10} 20 μg/m³ $PM_{2.5}$ 10 μg/m³		

EC: European Community; UK: United Kingdom; WHO: World Health Organization.
[a] 98 percentile.
[b] 50 percentile.
[c] Currently under revision to a $PM_{2.5}$ standard.

fraction (<100 nm). The relevance of the surface chemistry of particles in terms of health effects is now recognized and the need for a measure of particle surface area has arisen. This is difficult, but particle numbers can be used as a surrogate measure, and there is some evidence that this metric relates to health effects.

Exposures

Exposure of an individual to pollutants is only very crudely estimated by levels obtained from sentinel monitoring sites, which do not take into account indoor levels of pollutants and factors, such as personal activity and preexisting lung disease, which will affect the amount of a given pollutant inhaled.

Particles, once emitted, agglomerate to produce secondary particles of varying size and shape and undergo chemical reactions depending on the emission source and the atmosphere into which they are emitted. There are marked differences in the deposition of particles by size both regionally in the lung and with respect to the airway level at which deposition occurs. Larger particles (>10 μm in diameter) mostly impact in the upper airways with those in the respirable range (<7.2 μm) penetrating deeper (Fig. 39.1). The belief that ultrafine particles (<1 μm) do not remain in the lung and are simply breathed out again leaving none in the lung is now not held. Once penetrating to the distal lung, some particles will deposit and will either be taken up by alveolar macrophages or become interstitialized.

Deposition of gases will depend on, amongst other things, gas solubility, minute ventilation, and the presence of airflow obstruction. The endogenous antioxidant systems (largely urate in the nose and glutathione nitrate/nitrite in the lower respiratory tract) will, up to certain concentrations, counteract the proinflammatory effects of oxidant gases (ozone and NO_2) thus also altering the exposure of the epithelium to the inhaled gas.

ASSESSING THE HEALTH EFFECTS OF AIR POLLUTANTS

Health effects of air pollution

Exposure to air pollutants can result in:

- acute, or day-to-day effects;
- chronic effects;
- latent effects.

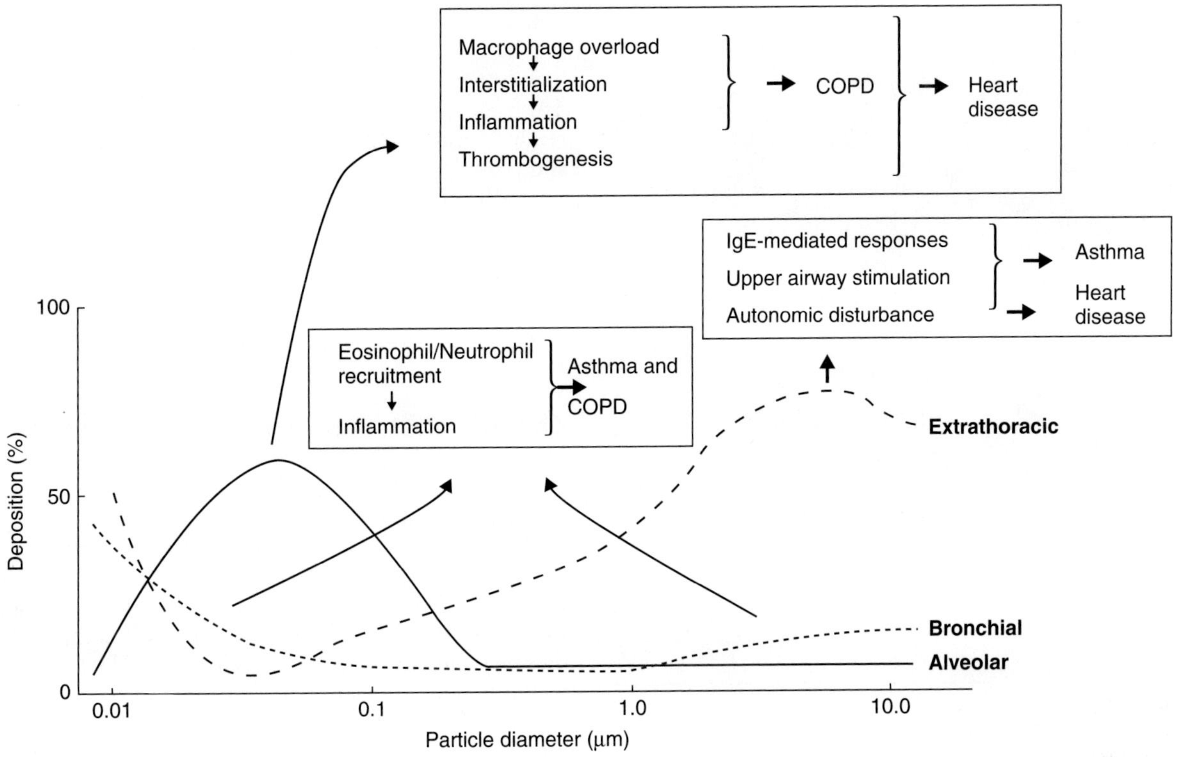

FIG. 39.1 Deposition of particles at various levels of the respiratory tract by particle size and likely mechanisms of health effect.

Acute effects occur in patients with preexisting disease, while chronic effects may impact on patients who are apparently disease-free prior to exposure. The major at-risk groups for acute effects are patients with asthma or COPD, although both deaths and hospital admissions for patients with cardiac disease increase on days of higher particulate levels [5]. While the cardiac effects will not be covered here, the frequent coincidence of coronary heart disease and COPD means that patients with COPD may be affected by air pollution more through their cardiac than their pulmonary disease.

Methods of assessment of health effects

- Epidemiological studies provided the first evidence of and quantified, to an extent, the adverse health effects of air pollution. Time series studies of routinely collected data (mortality, hospital admissions) and panel studies of putative susceptible groups (asthma, COPD) have provided information across a range of severities.
- Experimental studies in man have relied on controlled laboratory exposure of volunteers (almost exclusively normal subjects or patients with mild asthma) to specific pollutants, on *in vitro* work on cells obtained from nasal or bronchoalveolar lavage (BAL) and studies of cultured cell monolayers.
- Animal studies can be used for the study of more prolonged exposures than is possible with man. Experimental and animal studies both have the benefit of being able to study specific, controlled exposures of single or combined pollutant exposures.
- Computer modeling can predict outcomes providing the specific exposure characteristics of the pollutant(s) and recognized coexposures (e.g. meteorological variables) and host factors (e.g. exercise) are known.

Most health effects of air pollutants are respiratory although in COPD these are limited to mortality and hospital admissions as there are no contemporary panel studies of well-characterized patients with COPD and little human challenge evidence. In asthma, in contrast, there is a substantial body of evidence ranging from hospital admissions to challenge studies.

Responses to an air pollutant

Individual response to an air pollutant is affected by many factors (Table 39.2). Where possible these need to be allowed for when assessing the effects of a specific pollutant, usually by complex statistical manipulation of the data [6]. The observation of an effect will depend on whether there is a threshold for an effect which will always be present at an individual level but which may be more difficult to determine for populations.

TABLE 39.2 Factors affecting response to air pollutants.

Pollutant	Degree of exposure (concentration) Duration of exposure
Coexposure	Other pollutants (both indoor and outdoor) Allergens Viral infections Meteorological conditions (e.g. cold air) Degree of physical activity Cigarette smoke (both active and passive)
Host factors	Atopy Age (infants, perhaps the elderly) The fetus Preexisting disease state Use of treatment (e.g. bronchodilators) Airflow obstruction Bronchial hyperresponsiveness

CHRONIC EFFECTS OF AIR POLLUTION

The importance of a possible chronic effect of exposure to air pollution is that it brings more members of the population into the subpopulation susceptible to triggering by air pollutant exposure. At present, the evidence for air pollution contributing to the prevalence of both COPD and asthma is not compelling. However, it is generally believed that the chronic effects of air pollution may prove to be, at least in public health terms, more important than the acute effects. A chronic effect can best be identified either from prospective, longitudinal studies such as the US Six Cities Study [7] and the American Cancer Society Study [8], from cross-sectional studies comparing prevalence between areas of differing pollutant exposure or from life-table analysis [9]. All types of study are open to the criticism that early life pollutant exposures could have contributed to later life symptoms and morbidity, but are the best currently available methods for assessing chronic effects.

Asthma

In some cases, serial cross-sectional studies have provided important insights into changes in respiratory disease in the face of alterations in pollutant exposures. The German pre- and post-reunification studies [10, 11] have demonstrated that, with marked reductions in black smoke and SO_2 levels (although no change in NO_2 levels) in what was previously East Germany, the prevalence of episodes of bronchitis in children has fallen. Asthma prevalence has, however, remained unchanged at a lower level than in the old West Germany, although both hay fever and eczema have begun to increase. The implications of these findings are still unclear. It is likely that the higher levels of SO_2 and black smoke experienced in East Germany before reunification contributed to sputum production and episodes of bronchitis, in parallel with the pattern of symptoms seen in industrialized cities in the Western world up to the 1960s. Whether these exposures protected the airway against asthmagenic stimuli is debatable. Any such effect may take longer to be expressed than we have had time to observe to date, but the change in hay fever rates does suggest that allergic conditions may be emerging as a result of this differing exposure, although indoor exposures and possible changes in diet may also be playing a part.

The studies of Seventh Day Adventists [12] (a non-smoking population) suggest that chronic exposure to different components of the pollutant mix may result in an increased incidence of new cases of diagnosed asthma over time, but these are isolated findings in an unusual population and it is difficult to extrapolate these findings more widely.

Chronic obstructive pulmonary disease

The evidence for pollution contributing to the prevalence of COPD is limited to studies from the early 1960s in the United Kingdom [4]. The overwhelming effect of cigarette smoking as the main cause of this condition means that contributions from other environmental exposures to initiation of COPD are difficult to quantify. The recent findings that coal-dust exposure may contribute to chronic productive bronchitis and airflow obstruction independent of cigarette smoke [13] adds some supportive evidence to the possibility that current air pollutant exposure may contribute to the prevalence of COPD, although the character of the particulate exposures is clearly greatly different both in amount and in type. There are more recent data from Germany which suggests that in women living near busy roads, COPD is more prevalent.

ACUTE EFFECTS OF AIR POLLUTION

Asthma

Panel studies

Panel studies of children with asthma, mostly from North America, have shown small day-to-day changes in peak flow and symptoms [14], the most frequently reported pollutants associated with effects being NO_2, ozone, and particles. On average, inhaler use and symptoms increase by around 3% for every $10\,\mu g/m^3$ rise in PM_{10} with a less than 1% fall in peak flow for the same change [14]. However, not all studies show such effects and the PEACE study from Europe was essentially negative [15, 16]. Where they occur, effects are most marked in children either with more preexisting symptoms or who are atopic or both. It is not clear whether greater disease severity predicts greater effects [17]. There is some evidence to suggest that personal exposure to NO_2 (most of which is likely to be from indoor sources) may predispose asthmatic subjects to exacerbations in association with respiratory tract infections [18].

Hospital admissions

Hospital admissions for asthma have been shown to be related to air pollutant levels on a day-to-day basis both in adults and in children [19, 20], although not consistently. This may appear to conflict with the small effects seen in the panel studies, although it can be argued that some individuals may be more likely to be admitted on days when other asthmagenic factors combine to produce an attack if they are concurrently exposed to higher levels of pollution.

Mortality

No relationship has been shown between death from asthma and day-to-day changes in air pollution although the numbers of asthma deaths occurring in a city on a given day are so few that a small effect might yet exist.

Episodes

Episodes of air pollution might in theory result in asthma deaths or hospitalization. The London smog of 1952 [1] resulted in many deaths from bronchitis, but no clear evidence of an effect in asthma. The more recent 1991 [21] episode in the same city (where the sources of pollutants were quite different) showed no effect on any index of asthma, suggesting that any effect that might have occurred would have been in terms of small changes in symptoms rather than severe attacks.

Chronic obstructive pulmonary disease

Panel studies

There is limited information on day-to-day changes in clinical state in patients with COPD. There is some evidence for changes in symptoms in relation to particles and treatment use in relation to NO_2 and for changes in lung function [22]. These changes were modest (the lung function change amounted to a 0.2% fall in FEV_1 for a 10 μg/m^3 rise in PM_{10}) and changes were either same day effects or lagged by 24 h.

Hospital admissions

There is a consistent effect of air pollutant exposure and hospital admissions for COPD from studies across a wide range of countries. In the United States, hospital admissions increase between 1% and 3% for a 10 μg/m^3 rise in PM_{10} [14], while in Europe, the APHEA studies have shown smaller overall effects of around 1% [23, 24]. In most instances particles seem to be the most important pollutant, although the London study [25] showed no effect of particles, but an effect of ozone in the warmer months only. There is a range of effects however, across countries, with effects being less marked in eastern Europe and more marked in western countries (Fig. 39.2).

Mortality

Deaths from COPD are similarly associated with pollution and the effects are usually most strongly seen at lags of around 3 days [14] mediated largely through particles, although SO_2 also appears to contribute in Europe. On average the increase in respiratory mortality for a 10 μg/m^3 rise in PM_{10} is 3.4% on a day-to-day basis [14], although in those parts of Europe where particle exposures are generally higher, this effect is less marked suggesting a tolerance effect [26, 27].

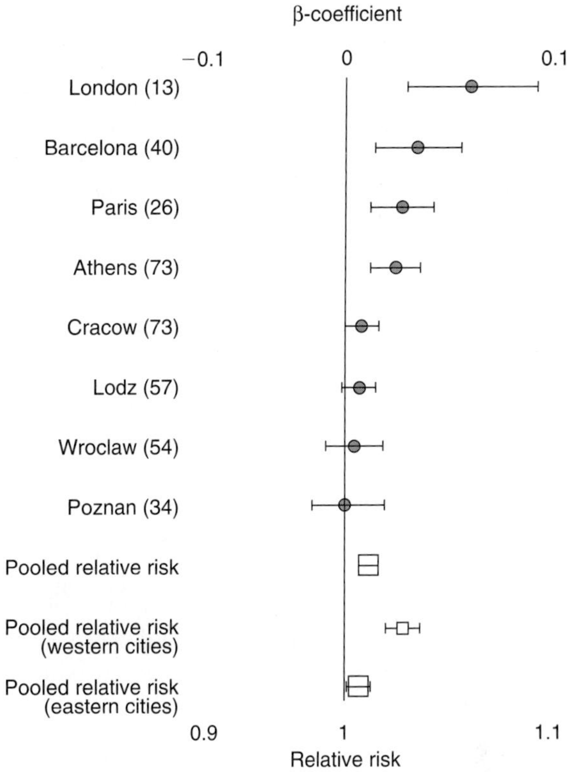

FIG. 39.2 Relative risks for death by city from respiratory disease for a 50 μg/m^3 rise in black smoke for a range of European cities (the APHEA project [26]).

MECHANISMS OF THE EFFECTS OF AIR POLLUTANTS

Mechanisms of damage by particles

Particles are likely to exert health effects via a number of mechanisms. One hypothesis is that macrophages become overloaded allowing particles to penetrate through the alveolar/bronchiolar wall and become interstitialized [28]. This initiates an inflammatory response which could both affect gas exchange in COPD and also release prothrombogenic cytokines, which, in an individual with a compromised coronary circulation, could lead to an acute cardiac event [29]. The inflammation may be caused through release of free radicals, perhaps because of transition metals on the particles [30, 31] although just the very small size of the particles themselves may be responsible [32]. Patients with COPD have a condition which is almost exclusively caused by cigarette smoking and many will thus have coexistent atheromatous coronary disease. It is possible, therefore, that increase in hospital admissions and deaths seen in patients with COPD associated with days of higher air pollution might not simply be due to enhancement of airway inflammation or facilitation of microbial pathogenicity, but due to cardiac decompensation.

There have been few studies of particulate challenge in asthma and none in COPD. One series of freshly generated diesel challenges in healthy normal subjects [33] has revealed infiltration of inflammatory cells and increases in some proinflammatory cytokines, notably IL-8, albeit at high pollutant concentrations which may be of relevance in both the asthmatic and COPD settings. For asthma, there is also evidence that diesel exhaust can potentiate the local production of specific IgE in the nose with an associated shift in T-cell patterns compatible with this [34, 35].

Mechanisms of damage by gases

Sulfur dioxide

SO_2 is a highly soluble, irritant gas quickly absorbed by the respiratory tract. It is a potent bronchoconstrictor in asthma particularly with exercise, but there is marked variability in the response between individuals, individuals responding to between 200 ppb and over 1500 ppb [36]. An airway effect is only seen in normal subjects at exposures of 4–5 ppm. In asthma SO_2 is thought to activate rapidly adapting receptors (RAR) [37], thus leading directly to bronchoconstriction. Lower doses inhaled at rest (200 ppb) can alter autonomic balance in asthma invoking the possibility that the gas initially activates neurogenic inflammation through a RAR initiated, vagally mediated reflex which subsequently results in an autonomic modulating response which differs from that seen in normal subjects [38].

Nitrogen dioxide

Nitrogen dioxide is an oxidant gas which at high concentrations can cause acute pulmonary edema (e.g. silo filler's disease). At levels usually experienced in ambient air (15–30 ppb as urban background levels) effects are difficult to find. At high concentrations (1000–3000 ppb usually with exercise) some changes in spirometric indices and bronchial hyperresponsiveness (BHR) have been recorded in asthmatic subjects, but these are of modest degree [39]. At these high concentrations some inflammatory markers are raised in BAL fluid after challenge in normal subjects, although there is considerable inconsistency between studies [40, 41].

At lower exposures (400 ppb) NO_2 has been shown to enhance the acute bronchoconstrictor effect of house dust mite allergen exposure, both at high dose [42] or as repeated smaller doses of allergen [43]. This suggests that exposure to gaseous air pollutants may exert more subtle effects than can be identified by epidemiological studies and which consequently may prove very difficult to quantify at a population level. There is no evidence that any such potentiating effects are seen in COPD, for instance with respect to viral infections.

Ozone

Ozone is a highly reactive, oxidant gas and can cause inflammatory changes in the bronchial mucosa (largely in the terminal bronchioles) in animals at concentrations similar to those which can be experienced in ambient air (80–120 ppb) [41, 44]. With exercise, it can produce similar degrees of bronchoconstriction in both normal and asthmatic subjects although there are large interindividual variations in response [45]. Ozone can also increase nonspecific airway responsiveness in both normal and asthmatic subjects in a dose-related manner [46]. Repeated daily exposures of asthmatic subjects to ozone lead to progressive loss of the response suggesting tolerance to its effects [47]. Ozone also potentiates the bronchoconstrictor effect of allergen in asthma either alone [48] or in combination with NO_2 [49]. The inflammatory response to ozone is very variable. Although most studies have shown an increase in neutrophils in BAL some, but not all, have shown increases in certain adhesion molecules and cytokines (notably IL-8) [41, 50, 51]. There are, however, marked interindividual differences and any changes in airflow so induced do not necessarily match changes in inflammatory markers which are often present at elevated levels some hours after exposure even though the cellular response has settled. There is suggestive evidence that patients with mild COPD exposed to ozone (along with exercise) at 200–300 ppb may show small decrements in both lung function and oxygen saturations although this needs confirmation with adequately powered studies [52, 53].

SUMMARY

Air pollution can affect some patients with asthma, in terms of symptoms and hospital admissions although the picture varies between countries and perhaps according to severity. There is evidence for all components of the pollutant mix to be capable of playing a role, although the mechanisms are only just being unraveled. For patients with COPD, air pollution increases the risks of hospital admission and death, particles being the most important pollutant. Particles may exacerbate inflammation by generation of free radicals, perhaps through transition metals. Interstitialization of particles at bronchioloalveolar level may exacerbate COPD itself or induce cardiac events by release of prothrombotic cytokines. Figure 39.1 summarizes the possible mechanisms of the health effects of particles in both COPD and asthma.

At present there is no specific approach for dealing with any air pollution-induced changes in clinical state in either asthma or COPD, although increasing inhaled anti-inflammatory treatment makes intuitive sense, while avoiding exertion and other known triggers during periods of higher air pollution may also be sensible.

The overall impact on public health of these triggering events of air pollution on patients with airway diseases is significant although may prove to be swamped by the size of the chronic effects once these are adequately quantified [54, 55].

References

1. Ministry of Health. Mortality and Morbidity During the London Fog of December 1952. London: HMSO, 1954.
2. Lippmann M, Ito K. Separating the effects of temperature and season on daily mortality from those of air pollution in London: 1965–1972. *Inhal Toxicol* 7: 85–97, 1995.
3. Lawther PJ, Waller RE, Henderson M. Air pollution and exacerbations of bronchitis. *Thorax* 25: 525–39, 1970.

4. Holland WW, Reid D. The urban factor in chronic bronchitis. *Lancet* i: 445–48, 1965.
5. Poloniecki JD, Atkinson RW, Ponce de Leon A, Anderson HR. Daily time series for cardiovascular hospital admissions and previous day's air pollution in London. *Occup Environ Med* 54: 534–40, 1997.
6. Schwartz J. Air pollution and hospital admissions for respiratory disease. *Epidemiology* 7: 20–28, 1996.
7. Dockery DW, Pope CA, Xu X *et al.* An association between air pollution and mortality in six US cities. *N Engl J Med* 329: 1753–59, 1993.
8. Pope CA, Thun MJ, Namboodiri MM *et al.* Particulate air pollution as a predictor of mortality in a prospective study of US adults. *Am J Respir Crit Care Med* 151: 669–74, 1995.
9. Hurley JF, Holland MR, Markandya A *et al.* Towards assessing the health impacts of ambient particulate air pollution in the UK. Final report for UK Department of Health, 2000.
10. von Mutius E, Fritzsch C, Weiland SK, Roll G, Magnussen H. Prevalence of asthma and allergic disorders among children in united Germany. *BMJ* 305: 1395–99, 1992.
11. Weiland SK, von Mutius E, Hirsch T *et al.* Prevalence of respiratory and atopic disorders among children in the East and West of Germany five years after unification. *Eur Respir J* 14: 862–70, 1999.
12. Abbey DE, Nishino N, McDonnell WF *et al.* Long-term inhalable particles and other air pollutants related to mortality in non-smokers. *Am J Respir Crit Care Med* 159: 373–82, 1999.
13. Henneberger PK, Attfield MD. Respiratory symptoms and spirometry in experienced coal miners: Effects of both distant and recent coal mine dust exposures. *Am J Ind Med* 32: 268–74, 1997.
14. Dockery DW, Pope CA. Acute respiratory effects of particulate air pollution. *Ann Rev Public Health* 15: 107–32, 1994.
15. Roemer W, Hoek G, Brunekreef B, Haluszka J, Kalandidi A, Pekkanen J. Daily variations in air pollution and respiratory health in a multicentre study: The PEACE project. *Eur Respir J* 12: 1354–61, 1998.
16. Roemer W, Hoek G, Brunekreef B *et al.* The PEACE project: General discussion. *Eur Respir Rev* 8: 125–30, 1998.
17. Brunekreef B, Kinney PL, Ware JH *et al.* Sensitive subgroups and normal variation in pulmonary function response to air pollution episodes. *Environ Health Perspect* 90: 189–93, 1991.
18. Linaker CH, Coggon D, Holgate ST *et al.* Personal exposure to nitrogen dioxide and risk of airflow obstruction in asthmatic children with upper respiratory infection. *Thorax* 55: 930–33, 2000.
19. Walters S, Griffiths RK, Ayres JG. Temporal association between hospital admissions for asthma in Birmingham and ambient levels of sulphur dioxide and smoke. *Thorax* 49: 133–40, 1994.
20. Sunyer J, Spix C, Quenel P *et al.* Urban air pollution and emergency admissions for asthma in four European cities: The APHEA project. *Thorax* 52: 760–65, 1997.
21. Anderson HR, Limb ES, Bland JM, Ponce de Leon A, Strachan DP, Bower JS. Health effects of an air pollution episode in London, December 1991. *Thorax* 50: 1188–93, 1995.
22. Harre ESM, Price PD, Ayrey RB, Toop LJ, Martin IR, Town GI. Respiratory effects of air pollution in chronic obstructive pulmonary disease: A three month prospective study. *Thorax* 52: 1040–44, 1997.
23. Spix C, Anderson HR, Schwartz J *et al.* Short-term effects of air pollution on hospital admissions of respiratory diseases in Europe: A quantitative summary of APHEA study results. Air pollution and health: A European approach. *Arch Environ Health* 53: 54–64, 1998.
24. Anderson HR, Spix C, Medina S *et al.* Air pollution and daily admissions for chronic obstructive pulmonary disease in 6 European cities: Results from the APHEA project. *Eur Respir J* 10: 1064–71, 1997.
25. Ponce de Leon A, Anderson HR, Bland JM, Strachan DP, Bower J. Effects of air pollution on daily hospital admissions for respiratory disease in London between 1987–88 and 1991–92. *J Epidemiol Comm Health* 50(Suppl. 1): S63–70, 1996.
26. Katsouyanni K, Touloumi G, Spix C *et al.* Short term effects of ambient sulphur dioxide and particulate matter on mortality in 12 European cities: Results from time series data from the APHEA project. *BMJ* 314: 1658–63, 1997.
27. Sunyer J, Castellsagué J, Sáez M, Tobias A, Antó JM. Air pollution and mortality in Barcelona. *J Epidemiol Comm Health* 50(Suppl 1): S76–80, 1996.
28. Oberdörster G, Ferin J, Lehnert BE. Correlation between particle size, *in vivo* particle persistence and lung injury. *Environ Health Perspect* 102(Suppl 5): 173–79, 1994.
29. Seaton A, MacNee W, Donaldson K, Godden D. Particulate air pollution and acute health effects. *Lancet* 345: 176–78, 1995.
30. Carter JD, Ghio AJ, Samet JM, Devlin RB. Cytokine production by human airway epithelial cells after exposure to an air pollution particle is metal dependant. *Toxicol Appl Pharmacol* 146: 180–88, 1997.
31. Costa DL, Dreher K-L. Bioavailable transition metals in particulate matter mediate cardiopulmonary injury in health and compromised animal models. *Environ Health Perspect* 105(S5): 1053–60, 1997.
32. Donaldson K, Li XY, MacNee W. Ultrafine (nanometer) particle-mediated lung injury. *J Aerosol Sci* 29: 553–60, 1998.
33. Salvi SS, Nordenhall C, Blomberg A *et al.* Acute exposure to diesel exhaust increases IL-8 and GRO-alpha production in healthy human airways. *Am J Respir Crit Care Med* 161: 550–57, 2000.
34. Diaz-Sanchez D, Dotson AR, Takenaka H, Saxon A. Diesel exhaust particles induce local IgE production *in vivo* and alter the pattern of IgE messenger RNA isoform. *J Clin Invest* 94: 1417–25, 1994.
35. Diaz-Sanchez D, Tsien A, Fleming J, Saxon A. Combined diesel exhaust particulate and ragweed allergen challenge markedly enhances human *in vivo* nasal ragweed specific IgE and skews cytokine production to a Th2 type phenotype. *J Immunol* 158: 2406–13, 1997.
36. Sheppard D, Saisho A, Nadel JA, Boushey HA. Exercise increases sulfur dioxide-induced bronchoconstriction in asthmatic subjects. *Am Rev Respir Dis* 123: 486–91, 1981.
37. Atzori L, Bannenberg G, Corriga AM *et al.* Sulfur dioxide-induced bronchoconstriction via ruthenium red-sensitive activation of sensory nerves. *Respiration* 59: 272–78, 1992.
38. Tunnicliffe WS, Mark D, Harrison RM, Ayres JG. Effect of particle and sulphur dioxide challenge on heart rate variability in normal and asthmatic subjects. *Eur Respir J* 17: 604–8, 2001.
39. Folinsbee L. Does nitrogen dioxide exposure increase airways responsiveness? *Toxicol Ind Health* 8: 273–83, 1992.
40. Blomberg A, Krishna MT, Bocchino V *et al.* The inflammatory effects of 2 ppm NO_2 on the airways of healthy subjects. *Am J Respir Crit Care Med* 156: 418–24, 1997.
41. Blomberg A. Airway inflammatory and antioxidant responses to oxidative and particulate air pollutants – experimental exposure studies in humans. *Clin Exp Allergy* 30: 310–17, 2000.
42. Tunnicliffe WS, Burge PS, Ayres JG. Effect of domestic concentrations in nitrogen dioxide on airway responses to inhaled allergen in asthmatic patients. *Lancet* 344: 1733–36, 1994.
43. Strand V, Svartengren M, Rak S, Barck C, Bylin G. Repeated exposures to an ambient level of NO_2 enhances asthmatic response to a non-symptomatic allergen dose. *Eur Respir J* 12: 6–12, 1998.
44. Aris RM, Christian D, Hearne PQ, Kerr K, Finkbeiner WE, Balmes JR. Ozone-induced airway inflammation in human subjects as determined by airway lavage and biopsy. *Am Rev Respir Dis* 148: 1363–72, 1993.
45. Hazucha MJ. Relationship between ozone exposure and pulmonary function changes. *J Appl Physiol* 62: 1671–80, 1987.
46. Horstman DH, Folinsbee LJ, Ives PJ, Abdul-Saleem S, McDonnell WF. Ozone concentration and pulmonary response relationships for 6.6 hour exposures with five hours of moderate exercise to 0.08, 0.10 and 0.12 ppm. *Am Rev Respir Dis* 142: 1158–63, 1990.
47. Horvath SM, Gliner JA, Folinsbee LJ. Adaptation to ozone: Duration of effect. *Am Rev Respir Dis* 123: 496–99, 1981.
48. Jörres R, Nowak D, Magnussen H. The effect of ozone exposure on allergen responsiveness in subjects with asthma or rhinitis. *Am J Respir Crit Care Med* 153: 56–64, 1996.
49. Jenkins HS, Devalia JL, Mister RL, Bevan AM, Rusznak C, Davies RJ. The effect of exposure to ozone and nitrogen dioxide on the airway

response of atopic asthmatics to inhaled allergen. *Am J Respir Crit Care Med* 160: 33–39, 1999.
50. Holz O, Jörres R, Timm P *et al.* Ozone-induced airway inflammatory changes differ between individuals and are reproducible. *Am J Respir Crit Care Med* 159: 776–84, 1999.
51. Jörres RA, Holz O, Zachgo W *et al.* The effect of repeated ozone exposures on inflammatory markers in bronchoalveolar lavage fluid and mucosal biopsies. *Am J Respir Crit Care Med* 161: 1855–61, 2000.
52. Solic JJ, Hazucha MJ, Bromberg PA. The acute effects of 0.2 ppm ozone in patients with chronic obstructive pulmonary disease. *Am Rev Respir Dis* 125: 664–69, 1982.
53. Kehrl HR, Hazucha MJ, Solic JJ, Bromberg PA. Responses of subjects with chronic obstructive pulmonary disease after exposure to 0.3 ppm ozone. *Am Rev Respir Dis* 131: 719–24, 1985.
54. Committee on the Medical Effects of Air Pollution. Quantification of the health effects of air pollution in the UK. Department of Health: The Stationery Office, 1997.
55. Künzli N, Kaiser R, Medina S *et al.* Public health impact of outdoor and traffic-related air pollution: A European assessment. *Lancet* 356: 795–801, 2000.

40

CHAPTER

Drugs

Neil C. Thomson[1] and Peter J. Barnes[2]

[1]Department of Respiratory Medicine, Division of Immunology, Infection and Inflammation, University of Glasgow, Glasgow, UK

[2]National Heart and Lung Institute (NHLI), Clinical Studies Unit, Imperial College, London, UK

NONSTEROIDAL ANTI-INFLAMMATORY DRUGS

Introduction

In the early part of the 20th century it was first recognized that aspirin could precipitate asthma in susceptible individuals [1]. Similar asthmatic reactions were later shown to occur with other nonsteroidal anti-inflammatory drugs (NSAIDs) [2] and patients with aspirin-induced asthma were found to be sensitive to other cyclooxygenase (COX)-1 inhibitors. Various terms have been used to describe the sensitivity to NSAIDs, including aspirin-induced asthma, aspirin-sensitive asthma, aspirin hypersensitivity, idiosyncratic reaction to aspirin, aspirin intolerance, and non-allergic hypersensitivity reactions to aspirin and NSAIDs [3].

Prevalence

The prevalence of hypersensitivity reactions to aspirin and NSAIDs in individuals with asthma is unclear, with estimates ranging from 4.3% to 21% [4–6]. The prevalence figures are higher when determined by oral provocation testing (e.g. 21% in adults and 5% in children [6]). Oral aspirin challenges undertaken in subgroups of patients with asthma, nasal polyps, and chronic sinusitis attending specialized allergy centers provide high positive rates of 30% or greater.

Clinical features

Aspirin-induced asthma is characterized by the development of bronchoconstriction within minutes to several hours after the ingestion of aspirin or other NSAIDs [2, 3]. The asthmatic reaction can be associated with other symptoms, including rhinorrhea, flushing, and loss of consciousness, and very rarely the attack may be fatal.

In the typical case, symptoms of chronic rhinitis are present for many years before asthma develops. The rhinitis starts as intermittent watery rhinorrhea, which develops during the second or third decade of life. The rhinorrhea becomes progressively more severe and is complicated by nasal polyp formation and sinusitis. On average 2 years later symptoms of asthma appear associated with the development of acute asthma after the ingestion of COX-1 inhibitors. Although in this group drugs may precipitate asthma, these patients continue to have nasal and asthmatic symptoms in the absence of ingesting aspirin or other NSAIDs. Respiratory symptoms are often chronic, severe, and perennial in nature. Individuals with aspirin-induced asthma are more commonly female. Skin tests to common external allergens are positive in a third to over a half of subjects. A small percentage of patients with aspirin-induced asthma also develop associated symptoms of urticaria and angioedema after ingestion of aspirin.

Diagnosis

In most clinical circumstances, the diagnosis is based on a history of bronchoconstriction following the ingestion of a COX-1 inhibitor. It should be appreciated, however, that a history of either the presence or the absence of aspirin-induced asthma may result in both a false positive and a false negative diagnosis. For example, patients with a history of aspirin-induced asthma have a positive oral aspirin challenge in approximately 60% to over 90% of cases. A negative oral challenge may occur in patients with a history of aspirin-induced asthma due to changing responsiveness to aspirin, a high threshold to aspirin, or a misleading history.

The only reliable method of establishing a diagnosis of aspirin-induced asthma is by aspirin challenge. Oral aspirin challenge is indicated in patients with adult asthma who require treatment with an NSAID. The oral challenge procedure has been reported to be safe, if a standardized protocol is used [3, 7]. An alternative but less widely used method of detecting aspirin-induced asthma involves administering aspirin-lysine by inhalation [7]. It is claimed that this method is safe and when compared to oral challenge the procedure is shorter and the reaction is localized to the respiratory tract. Aspirin-lysine administered as a nasal challenge has also been used to diagnose aspirin-sensitive asthma and this technique does not induce bronchoconstriction [8]. No *in vitro* test is currently available to detect aspirin-induced asthma.

Tartrazine is a yellow dye used for coloring foods and drugs such as confectionery, soft drinks, antibiotics, and antihistamines. A number of earlier reports suggested that tartrazine caused bronchospasm in some patients with aspirin-induced asthma, but further research has failed to support any cross-reactivity between tartrazine and aspirin in patients with aspirin-induced asthma [9].

Mechanisms

Alterations in the arachidonic acid pathway

A characteristic feature of patients with aspirin-induced asthma is the occurrence of airway obstruction following the ingestion of drugs which inhibit COX-1, the enzyme that converts arachidonic acid into prostaglandins (PG) and thromboxane. An abnormality in the arachidonic acid pathway in the respiratory tract of patients with aspirin-induced asthma seems to be central to the development of an asthmatic attack after the ingestion of NSAIDs (Fig. 40.1). In support of this hypothesis, the severity of bronchoconstriction induced by aspirin or other NSAIDs, such as indomethacin, flufenamic acid, and naproxen, is directly proportional to their ability to inhibit COX *in vitro*. Furthermore, specimens of nasal polyps removed from patients with aspirin-induced asthma are more sensitive to the inhibition effects of aspirin on the COX-1 pathway.

The influence of PG on bronchomotor tone may differ between patients with or without aspirin-induced asthma. The control of airway caliber in the latter group could be more dependent on the effects of PG, such as PGE_2, either due to its direct bronchodilator activity or indirectly by suppression of the release of bronchoconstrictor mediators. Alternately, the production of PGE_2 may be impaired in these individuals. In support of this later mechanism, peripheral blood mononuclear cells release reduced amounts of PGE_2 in patients with aspirin-induced asthma [10], although a recent clinical challenge study with aspirin found that urinary metabolites of PGE_2 were not altered in patients with aspirin-induced asthma [11]. The expression of the E-prostanoid (EP) 2 receptor for PGE_2 is reduced in nasal tissue of patients with aspirin-induced asthma [12] and polymorphisms in the promoter region of the gene coding for the EP2 gene are associated with sensitivity to aspirin [13].

The lipoxygenase pathway also appears to be abnormal in these individuals, since the synthesis of leukotriene (LT) E4 is frequently increased in aspirin-induced asthma [14], the expression of cys LT_1 receptors is increased in nasal inflammatory cells [15], and leukotriene C_4 synthase

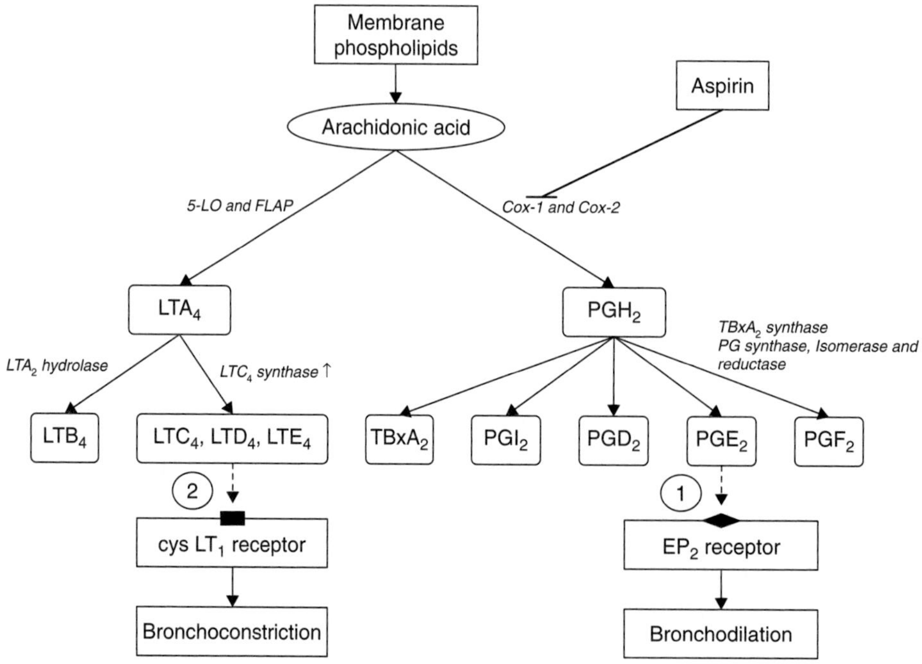

FIG. 40.1 Biosynthetic pathways leading to the production of LT and PG. Possible mechanisms by which aspirin and NSAIDs inhibit COX-1 to induce bronchoconstriction: (1) Impaired PGE2 synthesis and/or reduced PGE2 receptor function. (2) Increased cys LT synthesis and/or increased cys LT_1 receptor function. 5-LO: 5-lipoxygenase; FLAP: 5-lipoxygenase-activating protein; $TBxA_2$: thromboxane A_2; PGI_2: prostacyclin.

(LTC_4S), an enzyme involved in the synthesis of LTs, is often overexpressed in eosinophils and in bronchial tissue [16]. Furthermore, genetic polymorphisms of LTC_4S promoter region have been found in some patients with aspirin-sensitive asthma [17], but this has not been confirmed in other populations [18].

Taken together, these findings point to abnormalities in the arachidonic acid pathway in patients sensitive to NSAIDs in particular impaired PGE_2 synthesis and actions and excessive LT synthesis and effects. There is, however, an overlap in some of the defects in the arachidonic acid pathway found in patients with aspirin-induced asthma when compared with individuals without sensitivity to NSAIDs, suggesting that the exact mechanisms underlying aspirin-induced asthma are not yet fully elucidated.

Other mechanisms

There is no convincing evidence that aspirin-induced asthma is due to IgE-mediated mechanisms. For example, specific anti-aspiryl IgE antibodies are generally absent and skin tests to aspiryl-polylysine are negative in these patients. Basal and post-aspirin challenge complement levels are not altered in patients with aspirin-induced asthma. Inflammatory mediators are released following aspirin challenge, such as neutrophil chemotactic factor, but this is thought to be secondary to the primary biochemical events.

Management

Avoidance

Patients with a history of acute bronchoconstrictor response after the ingestion of NSAIDs must be warned to avoid these drugs. If they require a simple analgesic, then acetaminophen (paracetamol) is usually safe in doses up to 500 mg. Cross-reactivity can occur with high doses of acetaminophen (paracetamol), which is a weak inhibitor of COX-1, although the respiratory reaction tends to be mild. There is considerable data to indicate that COX-2 inhibitors do not induce bronchoconstriction in individuals sensitive to aspirin [19], although a few case reports of reactions have been published.

If a patient with aspirin-induced asthma should require an NSAID for the treatment of another condition, such as arthritis, then the patient can be desensitized to aspirin. This procedure should be undertaken by doctors with experience in aspirin challenge.

Treatment of respiratory disease

These patients often have severe chronic asthma and nasal disease, which should be treated along similar lines to that of patients with chronic asthma who are not sensitive to NSAIDs. Pretreatment with a number of drugs, including sodium cromoglycate, the H_1 receptor antagonist clemastine, and ketotifen, can inhibit the bronchoconstrictor response to aspirin ingestion in patients with aspirin-induced asthma. A small percentage of patients with aspirin-induced asthma develop bronchoconstriction after an intravenous injection of hydrocortisone, and occasionally the bronchoconstriction can be severe. These patients do not react adversely to other intravenous steroids such as methylprednisolone, dexamethasone, or betamethasone. Although cys LTs have been implicated in the pathogenesis of aspirin-induced asthma, LT receptor antagonists and a 5′-lipoxygenase' inhibitor have not been found to be any more effective than in nonsensitive asthmatic patients [20].

Desensitization

Following the administration of oral or inhaled aspirin to patients with aspirin-induced asthma there is a refractory period to further aspirin challenge, which lasts for 2–5 days. Continued administration of aspirin on a daily basis will maintain this refractory state, and this effect has been termed desensitization. The process of outpatient desensitization is safe [21] and results in desensitization not only to aspirin but also to other NSAIDs. Those patients who react to small doses of aspirin require several aspirin challenges before desensitization is accomplished, whereas less aspirin-sensitive patients are more quickly desensitized. The process by which aspirin ingestion causes desensitization is unknown.

β-BLOCKERS

Nonselective and cardioselective β-blockers

Nonselective β-adrenergic receptor antagonists precipitate acute bronchoconstriction in asthma, even in individuals with mild disease [22], and attacks can be fatal. The dose of β-blockers that causes bronchospasm may be low and there are case reports of severe asthma attacks induced by eye drops of timolol, a nonselective β-blocker used to treat glaucoma [23]. Compared to individuals with asthma, patients with chronic obstructive airways disease (COPD) are less likely to develop deterioration in lung function after a nonselective β-adrenergic receptor antagonist [24], although there are a few case reports of acute bronchospasm induced by nonselective β-blockers [25]. In patients with COPD who have acute exacerbations there is no evidence that β-blockers are detrimental and may even reduce mortality [26].

A meta-analysis of 19 studies on single-dose treatment and 10 studies on continued treatment with cardioselective β_1-blockers concluded that this group of drugs does not precipitate bronchospasm in patients with mild to moderate reactive airway disease [27]. Nevertheless there are case reports of individuals with asthma who have developed bronchospasm due to cardioselective β-blockers [28]. Thus despite the recommendation that cardioselective β-blockers should not be withheld from patients with mild to moderate asthma [27] it seems prudent to administer these drugs with great care to patients with moderate persistent asthma and to avoid their use in individuals with severe persistent asthma [28]. A recent systematic review of 20 studies of cardioselective β-blockers given to patients with COPD found no evidence of adverse effects on lung function or on the response to β_2-agonists [25, 29].

Acute bronchospasm induced by a nonselective β-blocker should be treated with an inhaled anticholinergic bronchodilator such as ipratropium or oxitropium [30].

Any fall in lung function due to a cardioselective β_1-blocker can be reversed by an inhaled short-acting β_2-agonist.

Possible mechanisms

β-blocker-induced bronchoconstriction occurs in individuals with asthma, but not in healthy subjects, which suggests that endogenous activation of β-receptors is important in maintaining airway tone in asthma against neural and inflammatory bronchoconstrictor stimuli.

Circulating catecholamines

β-blockers could antagonize the bronchodilator effect of circulating catecholamines in patients with asthma but not in normal subjects. However, against this suggestion, circulating catecholamines are not elevated in asthmatic subjects, even in those subjects who have demonstrable bronchoconstriction after propranolol and the concentrations of adrenaline in plasma (<0.3 nmol/l) are too low to have a direct effect on human airway smooth muscle tone [31, 32]. β-Blockers may inhibit the action of catecholamines on some other target cell, such as airway mast cells or cholinergic nerves. Mediator release from human lung mast cells is potently inhibited by β-agonists [33]. The effect of β-blockers may, therefore, be an increase in mediator release, which may be more marked in the "leaky" mast cells of asthmatic individuals. This idea is supported by the observation that cromolyn sodium, a mast cell "stabilizer," prevents the bronchoconstriction produced by inhaled propranolol. However after intravenous propranolol, no increase in plasma histamine has been detected [34].

Neural pathways

A more likely explanation for β-blocker-induced asthma is that there is an increase in neural bronchoconstrictor mechanisms. β_2-adrenergic receptors on cholinergic nerves in human airways may be tonically activated by adrenaline to modulate acetylcholine (ACh) release [35] and therefore to dampen cholinergic tone. Blockage of these receptors would therefore increase the amount of ACh released tonically, but this would be compensated for by the increased stimulation of prejunctional M_2-autoreceptors, which would act homeostatically to inhibit any increase in ACh release and therefore no increase in airway tone would occur, even with high doses of a β-blocker. By contrast, in patients with asthma β-blockers inhibit prejunctional β-receptors in the same way, increasing the release of ACh [36]; however, there may be a defect in M_2-receptor function in asthmatic airways, so that the increased release in ACh cannot be compensated. Thus increased ACh reaches M_3 receptors on airway smooth muscle. In addition, bronchoconstrictor responses to ACh are exaggerated in asthma, a manifestation of airway hyperresponsiveness, and thus two interacting amplifying mechanisms may lead to marked bronchoconstriction (Fig. 40.2). Evidence to support this hypothesis is provided by the inhibitory effect of an inhaled anticholinergic drug oxitropium bromide on β-blocker-induced asthma [30].

In patients with more severe asthma there may be an additional neural mechanism by which β-blockers may

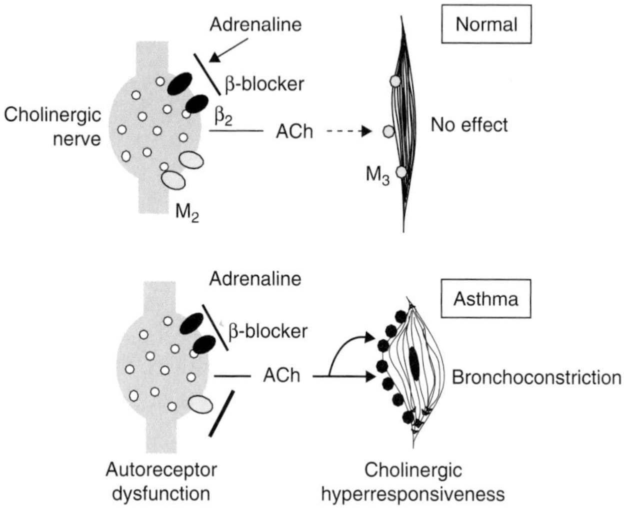

FIG. 40.2 Possible mechanisms of β-blocker-induced asthma. Blockade of prejunctional β_2-receptors on cholinergic nerves in normal individuals results in increased release of acetylcholine (ACh), but this is compensated by stimulation of prejunctional muscarinic M_2 receptors to inhibit any increase in ACh. In patients with asthma, prejunctional M_2 receptors are dysfunctional, so that there is a net release of ACh; ACh also has a greater bronchoconstrictor effect on the airways due to airway hyperresponsiveness.

cause bronchoconstriction. β_2-adrenergic receptors inhibit the release of tachykinins from airway sensory nerves [37], thus β-blockers may increase the release of these neuropeptides, thereby increasing bronchoconstriction and airway inflammation. While this mechanism may not be relevant in patients with mild asthma, in whom cholinergic mechanisms appear to account for the bronchoconstrictor response to β-blockers [30], it may be relevant in more severely affected asthmatic patients in whom cholinergic mechanisms do not appear to be as important.

Inverse agonism

Another possible mechanism may be related to the recently recognized phenomenon of inverse agonism [38]. It has been found that some mutants of the β-receptor have constitutive activity and activate the coupling protein Gs, even in the absence of occupation by agonist. In this situation β-blockers function as inverse agonists and have an inhibitory effect on baseline function. It is possible that in asthmatic patients β_2-receptors are constitutively active, so that β-blockers result in adverse effects. Different β-blockers have differing potencies as inverse agonists that are unrelated to their β-blocking potency. Thus, propranolol is a potent inverse agonist whereas pindolol is not and this may relate to the different tendency of these two agents to induce asthma.

It has recently been suggested that some β-blockers with inverse agonist activity may even be beneficial in asthma when used chronically [39]. In murine models β-blockers with inverse agonist activity (nadolol and carvedilol) increased β_2-receptor expression and caused bronchodilation [40]. In a pilot open study nadolol has been reported to be safe in asthmatic patients and to reduce airway hyperresponsiveness [41]. Further controlled studies may therefore be indicated.

ANGIOTENSIN CONVERTING ENZYME INHIBITORS

A retrospective cohort study suggested that bronchospasm was twice as common in patients treated with angiotensin converting enzyme (ACE) inhibitors (5.5%) compared to the reference group treated with lipid-lowering drugs (2.3%) [42]. However, the prevalence of a past history of bronchospasm in patients reporting ACE inhibitor-induced bronchospasm was not significantly different from the prevalence in patients on ACE inhibitors without an adverse reaction [42]. In a controlled trial in asthmatic and hypersensitive patients (with and without cough), there was no change in lung function following administration of captopril and no increase in reactivity to histamine or bradykinin [43]. Similar findings were obtained in a group of 21 patients with asthma given ACE inhibitors for 3 weeks, although one subject developed a slight wheeze [44]. Administration of a potent ACE inhibitor (ramipril) to a group of individuals with mild asthma showed no change in lung function or bronchial reactivity to inhaled histamine, nor was there any increase in bronchoconstrictor response to inhaled bradykinin [45]. Taken together these findings suggest that ACE inhibitors are unlikely to worsen asthma in the majority of patients, although there may be occasional patients in whom this occurs. When there is a possibility that asthma has been worsened or precipitated by an ACE inhibitor, the drug should be withdrawn and an alternative agent selected. As many as 20% of hypertensive patients treated with ACE inhibitors may develop an irritant cough, although this is unrelated to the presence of underlying airway disease or atopic status.

ADDITIVES

Several chemicals used as additives in drug preparations and food have been associated with worsening of asthma and should, where possible, be avoided. Bisulphites and metabisulphites (E220, E221, E222, E226, and E227) are antioxidants used as preservatives in several foods, including wines (especially sparkling wines), beer, fruit juices, salads, and medications. Characteristically, they produce bronchoconstriction within 30 min of ingestion and this may account for several cases of "food allergy." The mechanism of metabisulphite-induced asthma is probably explained by release of sulfur dioxide (SO_2) after ingestion that is then inhaled, since nebulized metabisulphite solutions generate SO_2 in sufficient quantities to provoke bronchoconstriction in asthmatic subjects [46]. Tartrazine (E102), a yellow dye, is used as a coloring in many foods, beverages (such as orange squash), and pharmaceutical preparations. Tartrazine sensitivity is relatively common and may affect 4% of asthmatic individuals, especially children [9]. Ingestion of tartrazine may result in urticarial rashes and bronchoconstriction. The mechanism may depend upon mediator release from mast cells. Monosodium glutamate (MSG, E621) is added to food as a flavor enhancer. It is found in soy sauce, spices, stock cubes, hamburgers, and in Chinese restaurant food. Some people react with sweating, flushing, and numbness of the chest; in patients with asthma this may be accompanied by wheezing, which may begin several hours after the ingestion ("Chinese restaurant asthma syndrome") [47] Precipitation of asthma symptoms by MSG is uncommon, however.

LOCAL ANESTHETICS

Aerosols of the local anesthetics, such as bupivacaine and lignocaine (lidocaine), cause bronchoconstriction in a proportion of asthmatic patients [48, 49]. The degree of bronchial reactivity to histamine does not predict the development or extent of bronchoconstriction following lignocaine inhalation [49]. The mechanism of local anesthetic-induced bronchoconstriction is unclear. Pretreatment with anticholinergic drugs partially attenuates the bronchoconstrictor response to aerosols of local anesthetics, suggesting that they may be acting in part via a vagal reflex pathway. Inhaled local anesthetics may selectively inhibit nonadrenergic noncholinergic bronchodilator nerves and so allow unopposed vagal tone. Some evidence for this is provided by the demonstration that lignocaine inhalation blocks nonadrenergic noncholinergic reflex bronchodilation in human subjects, leading to a reflex bronchoconstrictor response [50]. It is important to be aware that some patients with asthma may develop bronchoconstriction with topical local anesthetics during fiber optic bronchoscopy. All patients with asthma should receive premedication with a bronchodilator prior to bronchoscopy.

OTHER DRUGS

Many other drugs have been reported to lead to exacerbation of asthma in occasional patients. Bronchoconstriction may constitute part of an anaphylactic reaction to a drug, such as penicillin or to intravenous dextran or to contrast media. Other drugs, such as opiates, may cause direct degranulation of mast cells. Bronchodilator aerosols may occasionally cause a paradoxical bronchoconstriction. This is presumed to be due to the propellant (Freon) or other additives (such as oleic acid, which is used as a surfactant). The mechanism of bronchoconstriction may be via a cholinergic reflex. The treatment of paracetamol poisoning with intravenous *N*-acetylcysteine has been reported to exacerbate asthma. Cholinergic agents such as pilocarpine, used for the treatment of dry mouth and pyridostigmine, used to treat myasthenia gravis can induce attacks of asthma.

References

1. Cooke R. Allergy in drug idiosyncrasy. *JAMA* 73: 759–60, 1919.
2. Smith A. Response to aspirin-allergic patients to challenge by some analgesics in common use. *Br Med J* 2: 494–96, 1977.
3. Stevenson DD, Szczeklik A. Clinical and pathologic perspectives on aspirin sensitivity and asthma. *J Allergy Clin Immunol* 118(4): 773–86, 2006.
4. Kasper L, Sladek K, Duplaga M, Bochenek G, Liebhart J, Gladysz U *et al.* Prevalence of asthma with aspirin hypersensitivity in the adult population of Poland. *Allergy* 58(10): 1064–66, 2003.
5. Vally H, Taylor ML, Thompson PJ. The prevalence of aspirin intolerant asthma (AIA) in Australian asthmatic patients. *Thorax* 57(7): 569–74, 2002.
6. Jenkins C, Costello J, Hodge L. Systematic review of prevalence of aspirin induced asthma and its implications for clinical practice. *Br Med J* 328(7437): 434, 2004.
7. Nizankowska E, Bestynska-Krypel A, Cmiel A, Szczeklik A. Oral and bronchial provocation tests with aspirin for diagnosis of aspirin-induced asthma. *Eur Respir J* 15(5): 863–69, 2000.
8. Alonso-Llamazares A, Martinez-Cocera C, Dominguez-Ortega J, Robledo-Echarren T, Cimarra-Alvarez M, Mesa del Castillo M. Nasal provocation test (NPT) with aspirin: A sensitive and safe method to diagnose aspirin-induced asthma (AIA). *Allergy* 57(7): 632–35, 2002.
9. Stevenson D, Simon R, Lumry W, Mathison D. Adverse reactions to tartrazine. *J Allergy Clin Immunol* 78: 182–91, 1986.
10. Schafer D, Schmid M, Gode UC, Baenkler HW. Dynamics of eicosanoids in peripheral blood cells during bronchial provocation in aspirin-intolerant asthmatics. *Eur Respir J* 13(3): 638–46, 1999.
11. Mastalerz L, Sanak M, Gawlewicz-Mroczka A, Gielicz A, Cmiel A, Szczeklik A. Prostaglandin E2 systemic production in patients with asthma with and without aspirin hypersensitivity. *Thorax* 63(1): 27–34, January 1, 2008.
12. Ying S, Meng Q, Scadding G, Parikh A, Corrigan CJ, Lee TH. Aspirin-sensitive rhinosinusitis is associated with reduced E-prostanoid 2 receptor expression on nasal mucosal inflammatory cells. *J Allergy Clin Immunol* 117(2): 312–18, 2006.
13. Jinnai N, Sakagami T, Sekigawa T, Kakihara M, Nakajima T, Yoshida K *et al.* Polymorphisms in the prostaglandin E2 receptor subtype 2 gene confer susceptibility to aspirin-intolerant asthma: A candidate gene approach. *Hum Mol Genet* 13(24): 3203–17, 2004.
14. Daffern P, Muileburg D, Hugli T, Stevenson D. Association of urinary leukotriene E4 excretion during aspirin challenges with severity of respiratory responses. *J Allergy Clin Immunol* 104: 559–64, 1999.
15. Sousa AR, Parikh A, Scadding G, Corrigan CJ, Lee TH. Leukotriene-receptor expression on nasal mucosal inflammatory cells in aspirin-sensitive rhinosinusitis. *N Eng J Med* 347(19): 1493–99, 2002.
16. Cowburn AS, Sladek K, Soja J, Adamek L, Nizankowska E, Szczeklik A *et al.* Overexpression of leukotriene C4 synthase in bronchial biopsies from patients with aspirin-intolerant asthma. *J Clin Invest* 101(4): 834–46, 1998.
17. Sanak M, Simon H-U. Leukotriene C4 synthase promoter polymorphism and risk of aspirin-induced asthma. *Lancet* 350(9091): 1599, 1997.
18. Kedda M-A, Shi J, Duffy D, Phelps S, Yang I, O'Hara K *et al.* Characterization of two polymorphisms in the leukotriene C4 synthase gene in an Australian population of subjects with mild, moderate, and severe asthma. *J Allergy Clin Immunol* 113(5): 889–95, 2004.
19. Stevenson DD, Simon RA. Lack of cross-reactivity between rofecoxib and aspirin in aspirin-sensitive patients with asthma. *J Allergy Clin Immunol* 108(1): 47–51, 2001.
20. Dahlen S-E, Malmstrom K, Nizankowska EWA, Dahlen B, Kuna P, Kowalski M *et al.* Improvement of aspirin-intolerant asthma by Montelukast, a leukotriene antagonist. A randomized, double-blind, placebo-controlled trial. *Am J Respir Crit Care Med* 165(1): 9–14, January 1, 2002.
21. Williams AN, Simon RA, Woessner KM, Stevenson DD. The relationship between historical aspirin-induced asthma and severity of asthma induced during oral aspirin challenges. *J Allergy Clin Immunol* 120(2): 273–77, 2007.
22. McNeill R. Effect of a beta-adrenergic-blocking agent, propranol, on asthmatics. *Lancet* 2: 1101–2, 1964.
23. Schoene R, Martin T, Charan N, French C. Timolol-induced bronchospasm in asthmatic ronchitis. *JAMA* 245(14): 1460–61, 1981.
24. Lammers J, Folgering H, van Herwaarden C. Ventilatory effects of long-term treatment with pindolol and metoprolol in hypertensive patients with chronic obstructive lung disease. *Br J Clin Pharmacol* 20(3): 205–10, 1985.
25. Salpeter S, Buckley N. Use of beta-blockers and beta-agonists in COPD: A review of clinical outcomes. *COPD Update* 2: 133–39, 2007.
26. Dransfield MT, Rowe SM, Johnson JE, Bailey WC, Gerald LB. Beta-blocker use and the risk of death in hospitalized patients with acute exacerbations of COPD. *Thorax*. 2008 63: 301–05.
27. Salpeter SR, Ormiston TM, Salpeter EE. Cardioselective {beta}-blockers in patients with reactive airway disease: A meta-analysis. *Ann Intern Med* 137(9): 715–25, 2002.
28. Self T, Soberman JE, Bubla JM, Chafin CC. Cardioselective beta-blockers in patients with asthma and concomitant heart failure or history of myocardial infarction: When do benefits outweigh risks?. *J Asthma* 40(8): 839–45, 2003.
29. Salpeter S, Ormiston T, Salpeter E. Cardioselective beta-blockers for chronic obstructive pulmonary disease. *Cochrane Database Syst Rev*. 2005 (Issue 4): CD003566. DOI: 10.1002/14651858.CD003566.pub2.
30. Ind P, Dixon C, Fuller R, Barnes P. Anticholinergic blockade of beta-blocker-induced bronchoconstriction. *Am Rev Respir Dis* 139(6): 1390–94, 1989.
31. Barnes P. Neural control of human airways in health and disease. *Am Rev Respir Dis* 134(6): 1289–314, 1986.
32. Thomson N, Dagg K, Ramsay S. Humoral control of airway tone. *Thorax* 51: 461–64, 1996.
33. Church M, Hiroi J. Inhibition of IgE-dependent histamine release from human dispersed lung mast cells by anti-allergic drugs and salbutamol. *Br J Pharmacol* 90(2): 421–29, 1987.
34. Ind P, Barnes P, Brown M, Dollery C. Plasma histamine concentration during propranolol induced bronchoconstriction. *Thorax* 40(12): 903–9, 1985.
35. Barnes P. Muscarinic receptor subtypes: Implications for lung disease. *Thorax* 44(3): 161–67, 1989.
36. Barnes P. Modulation of neurotransmission in airways. *Physiol Rev* 72(3): 699–729, 1992.
37. Verleden G, Belvisi M, Rabe K, Miura M, Barnes P. Beta 2-adrenoceptor agonists inhibit NANC neural bronchoconstrictor responses in vitro. *J Appl Physiol* 73(3): 1195–99, 1993.
38. Leff P. Inverse agonism: Theory and practice. *Trends Pharmacol Sci* 16(8): 256–59, 1995.
39. Bond RA, Spina D, Parra S, Page CP. Getting to the heart of asthma: Can "[beta] blockers" be useful to treat asthma?. *Pharmacol Ther* 115(3): 360–74, 2007.
40. Callaerts-Vegh Z, SinghEvans KLJ, Dudekula N, Cuba D, Knoll BJ, Callaerts PFK et al. Effects of acute and chronic administration of {beta}-adrenoceptor ligands on airway function in a murine model of asthma. *Proc Nat Acad Sci* 101(14): 4948–53, January 1, 2004.
41. Hanania NA, Singh S, El-Wali R, Flashner M, Franklin AE, Garner WJ et al. The safety and effects of the beta-blocker, nadolol, in mild asthma: An open-label pilot study. *Pulm Pharmacol Therapeut* 21(1): 134–41, 2008.
42. Wood R. Bronchospasm and cough as adverse reactions to the ACE inhibitors captopril, enalapril and lisinopril. A controlled retrospective cohort study. *Br J Clin Pharmacol* 39(3): 265–70, 1995.
43. Overlack A, Müller B, Schmidt L, Scheid M, Müller M, Stumpe K. Airway responsiveness and cough induced by angiotensin converting enzyme inhibition. *J Hum Hypertens* 6(5): 387–92, 1992.

44. Kaufman J, Schmitt S, Barnard J, Busse W. Angiotensin-converting enzyme inhibitors in patients with bronchial responsiveness and asthma. *Chest* 101(4): 922–25, 1992.
45. Dixon C, Fuller RPJB. The effect of an angiotensin converting enzyme inhibitor, ramipril, on bronchial responses to inhaled histamine and bradykinin in asthmatic subjects. *Br J Clin Pharmacol* 23(1): 91–93, 1987.
46. Stevenson D, Simon R. Sulfites and asthma. *J Allergy Clin Immunol* 74: 469–72, 1984.
47. Allen D, Baker G. Chinese-restaurant asthma. *N Eng J Med* 305(19): 1154–55, 1981.
48. Thomson N. The effect of different pharmacological agents on respiratory reflexes in normal and asthmatic subjects. *Clin Sci* 56(3): 235–41, 1979.
49. McAlpine LG, Thomson NC. Lidocaine-induced bronchoconstriction in asthmatic patients. Relation to histamine airway responsiveness and effect of preservative. *Chest* 96(5): 1012–15, 1989.
50. Lammers J, Barnes P, Chung K. Nonadrenergic, noncholinergic airway inhibitory nerves. *Eur Respir J* 5(2): 239–46, 1992.

PART 7

Clinical Assessment of Asthma and COPD

41

CHAPTER

Diagnosis of Asthma and COPD

Fabrizio Luppi[1], Bianca Beghè[1], Lorenzo Corbetta[2] and Leonardo M. Fabbri[1]

[1]Department of Respiratory Diseases, University of Modena and Reggio Emilia, Modena, Italy
[2]Department of Respiratory Diseases, University of Firenze, Firenze, Italy

INTRODUCTION

Asthma and chronic obstructive pulmonary disease (COPD) are different chronic inflammatory respiratory disorders that may share a common functional abnormality, that is, poorly reversible airflow limitation [1–4]. According to current guidelines, airflow limitation in asthma is reversible or partly reversible [1], whereas airflow limitation in COPD is poorly reversible or not reversible at all [2].

In the pathogenesis of both asthma and COPD, individual genetic susceptibility and environmental exposures are relevant for disease expression. Cigarette smoking is the major cause of COPD [2]. The causes of asthma are largely unknown, although atopy and allergen exposure have major roles [1, 2]. Asthma is a phenotypically heterogeneous disorder that, over the years, has been divided into many different clinical subtypes. In particular, asthma starting in adulthood, asthma in smokers, non-eosinophilic asthma, and asthma in obese subjects are important subtypes in the adult asthma population that are still poorly characterized and that may overlap with COPD [5–7].

The differential diagnosis between asthma and COPD is quite simple when the typical clinical and functional features of either disease are present. It is easy to recognize asthma in a young, atopic, nonsmoking subject with recurrent dyspnea, wheezing, or chest tightness and fully reversible airflow limitation. Similarly, it is easy to diagnose COPD in a subject older than 40, a smoker, who presents with chronic dyspnea, cough, sputum, and fixed airflow limitation and no history of asthma or allergic diseases.

The difficulty comes when trying to make a diagnosis of asthma or COPD in a middle-aged or elderly patient, a smoker, who may be atopic or have a history of asthma, who complains of chronic dyspnea but not wheezing, chronic cough, or sputum, and who presents with poorly reversible airflow limitation. It is also difficult to make a diagnosis of asthma or COPD in a middle-aged or elderly patient who has a clear history of atopy and asthma, bronchodilator reversibility, and recurrent wheezing, but who also smokes and has chronic cough and sputum and dyspnea that are not suppressed by inhaled steroids. In such patients, differential diagnosis might become important from a clinical and therapeutic point of view. Inhaled glucocorticosteroids are the first choice of regular medication in asthma but not in COPD, whereas regular long-acting bronchodilators are the first choice of regular medication in COPD but not in asthma. Thus, in patients with overlapping features, the differential diagnosis between asthma and COPD is important in making the decision to prescribe regular treatment with either steroids or bronchodilators.

DEFINITIONS OF ASTHMA AND COPD

Asthma is a chronic inflammatory disease of the airways clinically characterized by recurrent respiratory symptoms as follows: dyspnea, wheezing, chest tightness, or cough associated with reversible airflow limitation. Other important features of asthma are an exaggerated responsiveness of the airways to various stimuli, and a specific chronic inflammation of the airways characterized by an increased number of $CD4^+$ Th2 lymphocytes, eosinophils, and methacromatic cells in the airway mucosa, and increased thickness of the reticular layer of the epithelial basement membrane. Familial predisposition, atopy, and exposure to allergens and sensitizing agents are important risk factors for asthma, even though the causes of asthma – the factors responsible for the development of asthma rather than its exacerbations – remain largely undetermined [1].

COPD is a syndrome characterized by poorly reversible airflow limitation, usually progressive, and often associated with chronic respiratory symptoms, such as dyspnea and/or chronic cough and sputum [2]. COPD is associated with chronic inflammation of the airways that is remarkably different from asthmatic inflammation, and that is characterized by an increased number of $CD8^+$ Th1/Tc1 lymphocytes in the airway mucosa and neutrophils in the lumen, with no increased thickness of the reticular layer of the epithelial basement membrane [8–13]. Even though genetic and familial predisposition as well as occupational exposure are considered risk factors, cigarette smoking is by far the most important risk factor for COPD [2].

TABLE 41.1 Differential diagnosis of asthma.

Localized pathology	Inhaled foreign body Endobronchial tumor Vocal cord dysfunction
Diffuse airway pathology	COPD Eosinophilic bronchitis Postinfectious airway hyperresponsiveness Cystic fibrosis Bronchiectasis Left ventricular failure
Other pathologies	Gastroesophageal reflux Pulmonary embolism Pulmonary eosinophilia Drug-induced airway hyperresponsiveness

MINIMUM REQUIREMENTS FOR THE DIAGNOSIS OF ASTHMA OR COPD

The diagnosis of asthma or COPD is based on clinical history and lung function tests, particularly peak expiratory flow (PEF) and spirometry, with assessment of spontaneous or postbronchodilator reversibility of airflow limitation. Allergy tests are also usually performed for the diagnosis of asthma, but not of COPD patients, to identify allergens responsible for asthma exacerbations and to consider the opportunity to treat the patient with immunotherapy.

Symptoms and medical history

Asthma

Most patients who are diagnosed with asthma seek medical attention because of respiratory symptoms. A typical feature of asthma symptoms is their variability. One or more of the following symptoms – wheezing, chest tightness, cough, and episodic shortness of breath – are reported by more than 90% of patients with asthma [14]. However, the simple presence of these symptoms is not diagnostic, because identical symptoms may be triggered by different stimuli in nonasthmatics, such as in children by acute viral infections [15]. In some asthmatics, wheezing and chest tightness are absent, and the only symptom the patient complains of may be chronic cough or cough after exercise [16]. This clinical entity is also called "cough-variant asthma;" it is particularly common in children and is often more problematic at night [17, 18]. Symptoms of asthma may be triggered or worsened by several factors, such as exercise, exposure to allergens, viral infections, and emotions. Recurrent exacerbations of respiratory symptoms, worsening of lung function requiring change of treatment, unscheduled requests for medical assistance, and sometimes hospitalization are also among the characteristic clinical features of asthma.

Asthma clusters in families and its genetic determinants appear to be linked to those of other allergic IgE-mediated diseases [19–21]. Thus, a personal or family history of asthma and/or allergic rhinitis, atopic dermatitis, or eczema increases the likelihood of a diagnosis of asthma.

Physical activity is an important cause of symptoms (wheezing and/or cough) for most asthma patients, particularly in children, and for some it is the only cause [15, 16]. Exercise-induced asthma usually develops not during exercise but 5–10 min afterward, and it resolves spontaneously within 30–45 min. Prompt relief of symptoms after the use of inhaled beta2-agonist, or their prevention by pretreatment with an inhaled beta2-agonist before exercise, supports a diagnosis of asthma. Important aspects of personal history are exposure to agents known to worsen asthma in the home (heating system, cooking system, house-dust mites), workplace conditions, air-conditioning, pets, cockroaches, environmental tobacco smoke [22–26], or even the general environment (e.g. diesel fumes in traffic [27]).

Since respiratory symptoms of asthma are nonspecific, the differential diagnosis is quite extensive, and the main goal for the physician is to consider and exclude other possible diagnoses (Table 41.1). This is even more important if the response to a trial of therapy (i.e. bronchodilators) has been negative.

Asthma is often classified by severity, but asthma severity changes over time. It also depends not only on the severity of the underlying disease but also on its responsiveness to treatment, which becomes the most important criterion in treated subjects. An asthmatic patient might be completely asymptomatic, either because he or she has mild intermittent asthma and long periods without symptoms even without treatment, or because he or she has severe asthma and is receiving full anti-asthmatic treatment, including systemic steroids.

While respiratory symptoms suggest asthma, the *sine qua non* for the objective diagnosis of asthma is the presence of reversible airflow limitation in subjects with persistent airways obstruction, and/or airway hyperresponsiveness or increased PEF variability in subjects without airways obstruction [1].

COPD

A clinical diagnosis of COPD should be considered in any patient who has dyspnea, chronic cough, or sputum production, and/or a history of exposure to risk factors for the disease [2]. Most patients who are diagnosed with COPD seek medical attention because of respiratory symptoms, particularly dyspnea [28]. Since the early stages COPD may manifest as chronic cough and sputum production, they may be present even in smokers without airflow limitation. Cough and sputum may precede the development of airflow limitation: in fact, respiratory symptoms may be an important risk factor for the development of COPD [29, 30]. Regular production of sputum for 3 or more months in 2 consecutive years is defined as chronic bronchitis [31]. In some subjects, chronic cough may be unproductive [32–35], and airflow limitation may develop in the absence of cough. In COPD, dyspnea is characteristically persistent, unlike in asthma where it is variable and progressive [36]. In the early stages of the disease, dyspnea is noted only during the patient's usual effort; as lung function decreases, dyspnea becomes more serious and is present during everyday activities or at rest. Dyspnea is not closely correlated with arterial blood gases; for example, the typical "blue bloater" with peripheral edema, hypoxemia, and hypercapnia has generally less dyspnea than the "pink puffer," who generally does not have these blood gas abnormalities but is much more dyspneic. Wheezing and chest tightness are nonspecific symptoms of COPD and may vary on different days or over the course of a single day. Recurrent exacerbations of respiratory symptoms requiring change of treatment, unscheduled requests for medical assistance, and sometimes hospitalization are also among the characteristic clinical features of COPD.

A detailed medical history of a patient with symptoms suggestive of COPD should include exposure to risk factors (e.g. smoking and occupational or environmental exposures), family history of COPD or other chronic respiratory disease, pattern of symptoms, history of exacerbations, presence of comorbidities, medical treatment, and the patient's quality of life.

Diagnosis and assessment of the severity of COPD are mainly based on the degree of airflow limitation during spirometric measurement. Thus, according to current guidelines, the *sine qua non* for the diagnosis of COPD is the presence of poorly reversible airflow limitation, that is, the presence of a postbronchodilator forced expiratory volume in 1 s/forced vital capacity ratio (FEV_1/FVC) <0.70 and FEV_1 $<80\%$ predicted. These values confirm the presence of airflow limitation that is not fully reversible [2].

Airflow limitation in COPD is due to both small airways disease (obstructive bronchiolitis) and parenchymal destruction (emphysema), the relative contributions of which vary among patients [1, 2]. In contrast, airflow limitation in asthma is almost exclusively due to airways disease [10]. Imaging (see below), particularly thin-section computed tomography (CT), has been used to quantify emphysema by detecting areas of low attenuation. However, airflow limitation as assessed by FEV_1 correlates poorly with the severity of emphysema as evaluated by CT [4], possibly because small airways disease contributes significantly to airflow limitation [37]. Recent progress in CT technology has made it possible to detect and quantify airway abnormalities [38, 39], but more work is required before the technique can be used in clinical practice.

TABLE 41.2 Differential diagnosis of COPD.

Other airway inflammatory diseases	Asthma Bronchiectasis Diffuse panbronchiolitis
Infectious airway diseases	Tuberculosis
Cardiac diseases	Congestive heart failure

Studies in COPD patients recruited by experienced pulmonary specialists in hospitals [40] or by general practitioners in outpatient clinics [41] report that up to 40% of smokers/ex-smokers with chronic respiratory symptoms and clinical findings compatible with COPD do not fit the spirometric definition of COPD reported above. Thus, even when there is a clear history of COPD – chronic respiratory symptoms, exacerbations, smoking, and age >50 – a large proportion of such patients either have normal spirometric values or present with reduced lung volumes (a restrictive pattern), and thus the diagnosis of COPD cannot be confirmed [40, 41]. Clinical features of COPD correlate poorly with airflow limitation. Thus, although not yet recommended by current guidelines, proper diagnosis and assessment of the severity of COPD require more than a comprehensive approach that includes imaging [38, 39] assessment of exercise tolerance [42], body mass index [42, 43], and chronic comorbidities (e.g. chronic heart failure (CHF), arterial hypertension, metabolic syndrome) that are often associated with COPD [44, 45].

Because chronic respiratory symptoms (particularly dyspnea), clinical features, and poorly reversible airflow limitation may also be present in other pathological conditions, a careful differential diagnosis between COPD and these conditions should always be performed (Table 41.2).

Physical examination

In mild asthma, physical examination is usually normal under stable conditions but becomes characteristically abnormal during asthma attacks. Typical physical signs of asthma attacks are wheezing on auscultation, cough, expiratory ronchi throughout the chest, and signs of acute hyperinflation (e.g. poor diaphragmatic excursion at percussion, use of accessory muscles of respiration). Some patients, particularly children, may present with a predominant nonproductive cough (cough-variant asthma). In some asthmatics, wheezing – which usually reflects airflow limitation – may be absent or detectable only on forced expiration, even in the presence of significant airflow limitation; this may be due to hyperinflation or to very marked airflow limitation. In these patients, however, the severity of asthma is mostly indicated by other signs, such as cyanosis, drowsiness, difficulty in speaking, tachycardia, hyperinflated chest, use of accessory muscles, and intercostal recession.

In the early stages of COPD, physical examination is usually normal. Current smokers may have signs of active smoking, including an odor of smoke or nicotine staining of fingernails. In more severe COPD, prolonged expiration and wheezing and signs of hyperinflation (e.g. barrel chest and poor diaphragmatic excursion at percussion, use of accessory muscles of respiration) are usually present. Cyanosis of the lips and nail beds and signs of cor pulmonale, such as edema of the ankle or lower leg, are often present in patients with reduced oxyhemoglobin percentage. Even though clubbing of the digits may be present in patients with severe COPD and is considered a sign of COPD, it is nonspecific; in fact, its presence should alert the physician to the possible presence of other diseases, particularly lung cancer. Patients with COPD often have reduced breath sounds and wheezing during quiet breathing or after forced expiration.

Physical examination is usually not very useful in making the differential diagnosis between asthma and COPD, but it can be useful in assessing the severity of exacerbations of both asthma and COPD [46].

Lung function tests

Spirometry

Lung function tests play a crucial role in the diagnosis and follow-up of asthma and COPD. Spirometric measurements – FEV_1 and slow vital capacity (VC) or FVC – are the standard means for assessing airflow limitation. Spirometry is recommended at the time of diagnosis and for the assessment of the severity of both asthma and COPD [1, 2, 47, 48] it should be repeated to monitor the disease and when there is a need for reassessment, such as during exacerbations [3].

Poorly reversible airflow limitation is indicated by the absolute reduction of postbronchodilator FEV_1/VC or FEV_1/FVC ratios <0.7 but it should be confirmed with postbronchodilator FEV_1/VC values below the lower limit of normal [4, 49]. Measurements of residual volume and total lung capacity may also be useful in determining the degree of hyperinflation and/or enlargement of airspaces [1, 2, 47, 48].

In asthma, airflow limitation is usually reversible, either spontaneously or after treatment, except for moderate/severe asthma with fixed airway obstruction [1, 3, 10]. In COPD, airflow limitation is by definition not reversible (i.e. FEV_1/FVC does not reach 0.7 even after inhalation of a bronchodilator or a short course of long-acting beta2-agonist (LABA) and inhaled steroids). However, up to one-third of COPD patients show a significant increase in FEV_1 ($>15\%$) after receiving inhaled beta-adrenergic agonists [50–54], which simulate the reversible airflow limitation observed in asthmatics.

In conclusion, while the best spirometric values are useful to define whether airflow obstruction is reversible or not (i.e. does not return within normal values), the degree of reversibility after treatment does not help to make the differential diagnosis between asthma and COPD (see below).

Peak expiratory flow

An important tool for the diagnosis and subsequent treatment of asthma is the PEF meter [55]. PEF is the highest expiratory flow obtained during a forced expiration starting immediately after a deep inspiration from total lung capacity. PEF is a simple, reproducible index and can be measured with inexpensive and portable meters [56]. If spirometry does not reveal airflow limitation, the home monitoring of PEF for 2–4 weeks may help to detect an increased variability of airway caliber, and thus help to diagnose asthma [57]. For most asthmatic patients, PEF correlates well with FEV_1 [58]. Daily monitoring of PEF (at least in the morning at awakening and in the evening hours, preferably after bronchodilator inhalation) [1, 3] is also useful to assess the severity of asthma and its response to treatment, and it can help patients to detect early signs of asthma deterioration [59]. However, PEF measurements have some limitations. PEF is effort dependent and mainly reflects the caliber of large airways and may therefore underestimate the degree of airflow limitation present in peripheral airways [60]. Diurnal variability is calculated as follows:

$$\frac{PEF_{max} - PEF_{min} \times 100}{PEF_{max} + PEF_{min}/2}$$

A diurnal variability of PEF of more than 20% is diagnostic of asthma, and the magnitude of the variability is broadly proportional to disease severity. PEF monitoring may be of use not only in establishing a diagnosis of asthma and assessing its severity, but also in uncovering an occupational cause for asthma. When used in this way, PEF should be measured more frequently than twice daily, and special attention should be paid to changes occurring in and out of the workplace [55, 61–63].

Even though PEF is at least as important to prognosis as FEV_1 in moderate to severe COPD [64], PEF monitoring is not frequently used in COPD for various reasons. First, PEF reflects the patency of central airways, and airflow limitation in COPD starts from peripheral airways. Thus, the PEF value may underestimate airflow limitation, particularly if it occurs in peripheral airways. Second, by definition, airflow limitation is poorly reversible in COPD, and thus PEF usually does not vary significantly. Finally, there is only limited evidence to support a role for PEF in detecting COPD exacerbations [65, 66].

Reversibility to bronchodilators

The reversibility of airflow limitation following bronchodilator therapy is no longer an accepted criterion in support of the diagnosis of asthma, or to establish the differential diagnosis between asthma and COPD [1–3]; even though there is a large increase in FEV1, and particularly its return is above the lower limit of normal values, normal expiratory flows after anti-asthma treatment strongly suggests asthma. Subjects with moderate to severe asthma may develop poorly reversible airflow limitation and have a response to treatment but not a return to normal values. Similarly, COPD patients may show a significant response to treatment, even without a return to normal expiratory flows.

In subjects with airflow limitation, an improvement in FEV_1 of $>12–15\%$ predicted and more than 200 ml after administration of a bronchodilator (e.g. 200 μg of inhaled salbutamol from a metered dose inhaler) is no longer considered a pathognomonic hallmark of asthma [1], or a criterion for differential diagnosis between asthma and COPD [3, 67–69]. In fact, an incomplete response to a single administration of a bronchodilator does not exclude the possibility of reversibility to longer treatment with bronchodilators or steroids [70, 71]. Thus, more attention should be paid to the response to long-term treatment.

In COPD patients, bronchodilator reversibility testing should generally be performed at least once. Airflow limitation in COPD is usually not reversible, but the one-third of COPD patients who show a significant response to bronchodilator agents [50, 51, 53, 54, 72] are likely to benefit from treatment with glucocorticosteroids [73, 74]. The absence of a response to a bronchodilator should never be a reason to withhold bronchodilator therapy, as the response to bronchodilators in COPD is mainly symptomatic rather than functional [75–79]. Bronchodilator responsiveness is a continuous and poorly reproducible variable; thus, classifying patients as responders or nonresponders can be misleading and does not predict disease progression [68].

As recommended by recent guidelines [1], the terms "reversibility" and "variability" should refer to changes in symptoms accompanied by changes in airflow limitation, which occur spontaneously or in response to treatment. At present, however, these terms often refer only to rapid improvements in FEV_1 measured within minutes after inhalation of a rapid-acting bronchodilator, for example, after 200–400 μg salbutamol or salbutamol and ipratropium bromide [54, 80]. In contrast, a history of symptoms and/or functional reversibility (spontaneous or after any kind of treatment) is the essential component in the diagnosis of asthma.

The assessment of reversibility of both clinical features and functional abnormalities may be useful in obtaining the best level of asthma control achievable and/or the best lung function for individual patients [81]. Achieving and maintaining lung function at the best possible level is one of the objectives of both asthma and COPD management [1, 81, 82].

In summary, while the best spirometric values are useful to define whether airflow obstruction is reversible or not (i.e. does not return within normal values), the degree of reversibility after treatment does not help to make the differential diagnosis between asthma and COPD (see below).

Arterial blood gases

In severe asthma and COPD and, more importantly, during acute exacerbations of both asthma and COPD, the measurement of arterial blood gases while the patient is breathing air and/or after oxygen administration is essential for the diagnosis of chronic and/or acute respiratory failure. This test should be performed in all patients with clinical signs of acute or chronic respiratory and/or heart failure [2].

Allergy tests

The presence of allergic disorders in a patient's family history should be investigated in all patients in whom symptoms are suggestive of asthma [83]. A history provides important information about the patient's lifestyle and occupation, both of which influence exposure to allergens and the time and factors possibly involved in onset and in exacerbations of asthma [84, 85]. In asthmatics, the relationship between exposure to one or more allergens and the occurrence of asthma and/or ocular and nasal symptoms should be established [86]. Also, the relationship of symptoms to the time of the year (seasonal pollen asthma) and to the presence of pets in the home should be assessed, together with a description of the patient's living environment with special attention to carpets, pillows, and other dust collectors [87]. Identifying the presence of an allergic component in asthma adds little to the diagnosis, but it can help in identifying potential triggers and directing allergen immunotherapy [1].

Skin tests with all relevant allergens are present in the geographic area in which the patient lives are the primary diagnostic tool in determining allergic status. Deliberate provocation of the airways with a suspected allergen or sensitizing agent may also be helpful in establishing causality, especially in the workplace [63, 88]. Measurement of specific IgE is not usually more informative than a skin test, and is more expensive. Measurement of total IgE in serum has no value as a diagnostic test for atopy. The main limitation of the allergy test is that a positive test does not necessarily mean that the disease is allergic in nature or that it is causing asthma, as some individuals have specific IgE antibodies without any symptoms. The cost–benefit ratio of performing inhalation tests with allergens or other sensitizing agents should be carefully examined for each patient because of the high cost and the potential risk involved [69].

The assessment of atopy is not useful in COPD. Even though atopy may be a risk factor for both asthma and COPD [89, 90], the demonstration of atopy in COPD patients does not help in the identification of potential triggers as in allergic asthma. Allergen immunotherapy has no role in COPD.

ADDITIONAL TESTS

While the diagnosis and assessment of severity of asthma and COPD can be fully established on the basis of clinical history and lung function tests (including arterial blood gases, see below), additional tests might be helpful to better characterize individual patients.

Reversibility to corticosteroids

In patients with airflow limitation that is not reversed by a single dose of a short-acting bronchodilator, a 2-week treatment with oral or inhaled glucocorticosteroids and bronchodilators might be considered. Glucocorticosteroids can be administered orally (e.g. 40 mg daily prednisone) by aerosol (e.g. 2 mg daily beclomethasone, or equivalent) or both [91–93] for at least 14 days [94]. Unfortunately, patients with COPD cannot be separated into discrete groups of glucocorticosteroid responders and nonresponders, and thus glucocorticosteroid testing is an unreliable predictor of the benefit from inhaled glucocorticosteroids in individual patients.

Because of their efficacy and infrequent adverse effects, inhaled glucocorticosteroids alone or in combination with long-acting bronchodilators are increasingly used in practice as first-choice therapy to investigate the reversibility of airflow limitation [74, 95, 96].

In most patients with a clear history of asthma or COPD, the reversibility to glucocorticosteroids confirms the diagnosis, even if significant overlap exists [10]. Asthma is usually responsive to bronchodilators and/or glucocorticosteroids, whereas COPD is usually less responsive or not responsive at all. In asthma, a combination of sputum eosinophilia and increased nitric oxide (NO) levels may be useful in predicting the response to a trial of oral steroids [97, 98], and sputum eosinophilia may also predict the response to steroids in COPD [99, 100]. However, some COPD patients may show a significant improvement in function after glucocorticosteroid treatment [73], particularly if they present with pathological abnormalities similar to those in asthma.

The simplest and potentially the safest way of identifying these COPD patients is by an *ex juvantibus* treatment trial with inhaled glucocorticosteroids [92] in combination with long-acting bronchodilators for 6 weeks to 3 months, using the same criteria for reversibility as in the bronchodilator trial (FEV_1 increase of 200 ml and 12%) [2, 73, 92, 101]. The response to glucocorticosteroids alone or in combination with long-acting bronchodilators should be evaluated with respect to the postbronchodilator FEV_1 [2, 4].

COPD patients with a response to glucocorticosteroids present some pathological features of asthma, such as a significantly higher number of eosinophils and higher levels of eosinophil cationic protein in their bronchoalveolar lavage fluid, and a thicker reticular basement membrane [73, 101].

All long-term studies in COPD have demonstrated the lack of any effect of inhaled glucocorticosteroids on the natural history of COPD, as evaluated by the decline in FEV_1 [2, 102–106]. Therefore, given the documented risks of chronic glucocorticosteroid therapy in both asthma [107] and COPD, such as osteoporosis [102, 108], the decision to start long-term treatment with inhaled glucocorticosteroids must be made very carefully.

In patients with poorly reversible airflow limitation due to asthma, the beneficial effects of inhaled glucocorticosteroids are likely to overcome the risks of negative systemic effects [74]. However, in patients with poorly reversible airflow limitation due to smoking, this may not be the case, particularly considering that inhaled glucocorticosteroid alone may be associated with increased mortality [109], and that inhaled glucocorticosteroid both alone and in combination with a long-acting beta2-adrenergic agonist may slightly increase the risk of pneumonia in COPD patients [109, 110].

Exercise testing

Exercise testing is useful for assessing the degree of disability, the role of comorbidities, prognosis for survival, presence of exercise-induced hypoxemia, and response to treatment in individuals with COPD. Simple walking tests are increasingly used in the assessment of COPD patients [111]. Severe COPD might be better assessed by a composite score such as BODE (body mass index, degree of airway obstruction, severity of dyspnea, and exercise tolerance), which has been shown to be a better predictor of subsequent survival and is increasingly used in clinical assessment of patients.

Diffusion capacity

Measurement of the diffusing capacity of the lung for carbon monoxide (Dlco) has been recommended for distinguishing asthma from COPD [112]. In asthma, Dlco is usually normal or increased [10, 113, 114]. In contrast, Dlco is usually reduced in COPD, possibly due to emphysema [10, 114, 115], but it may also be reduced in smokers without airflow limitation [116]. Dlco is lower in COPD patients than in asthmatics with incomplete reversible airflow limitation [10, 117, 118]. However, patients with severe alpha-1 antitrypsin deficiency may present with normal Dlco, despite having a significant component of fixed airway obstruction and prominent panacinar emphysema on a high-resolution CT (HRCT) scan, suggesting the limitations of measuring Dlco in these patients [119, 120].

Airway hyperresponsiveness

In patients who have symptoms consistent with asthma but have normal lung function, bronchial provocation tests (methacholine, histamine, adenosine 5′-monophosphate, mannitol, and exercise) are helpful in measuring airway hyperresponsiveness and thereby confirming or excluding the diagnosis of active asthma [121, 122]. Methacholine is mainly used to identify bronchial hyperresponsiveness and to guide treatment. Exercise is used as a bronchial provocation test because demonstrating prevention of exercise-induced asthma is an indication for use of a drug [121].

These measurements are sensitive for a diagnosis of asthma, but they have low specificity [123]. This means that a negative test can be used to exclude a diagnosis of active asthma, but a positive test does not always mean that a patient has asthma [123]. Airway hyperresponsiveness has been described in workers who are acutely exposed to irritants [124, 125], allergic rhinitis [126–128], and other diseases with airflow limitation, such as cystic fibrosis [129–131] and COPD [132–134]. Indeed COPD, especially in current smokers, is often accompanied by airway hyperresponsiveness [135] that is no different from that in asthmatics with a similar degree of airflow limitation [10, 136]. In patients with fixed airflow limitation, a similar degree of airway hyperresponsiveness was observed in those with a history of COPD and those with a history of asthma [10]. In these patients, hyperresponsiveness might be largely due to the airflow limitation itself.

In conclusion, the measurement of airway hyperresponsiveness may be useful to confirm asthma in subjects with normal baseline lung function, but it is not useful in the differential diagnosis between asthma and COPD, particularly when patients have a similar degree of poorly reversible airflow limitation.

Imaging

While chest radiography may be useful to exclude diseases that may mimic asthma and COPD, it is not required in the confirmation of the diagnosis and management of asthma. The utility of chest radiography is to exclude other conditions that may imitate or complicate asthma, particularly acute asthma. Examples include pneumonia, cardiogenic pulmonary edema, pulmonary thromboembolism, tumors (especially those that result in airway obstruction with resulting peripheral atelectasis), and pneumothorax [137].

A number of novel imaging methods for assessing airway pathology in asthmatic patients have been proposed [39, 138]. Both direct and indirect signs of airway pathology have been described using HRCT. Direct signs are obtained by measuring airway or bronchial wall thickness, evaluating the ratio of bronchial diameters to adjacent pulmonary arteries, and identifying a lack of bronchial tapering. Indirect signs include foci of mucoid impaction (including the finding of a tree-in-bud configuration of small peripheral lung nodules indicative of bronchiolitis) and mosaic attenuation [139]. Mosaic attenuation is the presence of geographic zones of decreased lung density adjacent to areas of apparent increased lung density in the absence of architectural distortion or honeycombing. This finding may also be seen in patients with diffuse infiltrative lung disease or, more rarely, chronic embolic pulmonary hypertension. In most cases, reliable identification of air trapping resulting in mosaic lung attenuation requires expiratory imaging. During exhalation, areas of air trapping become accentuated, simplifying the differential diagnosis. Each of these methods has serious limitations, and in clinical practice they have not yet proved sufficiently accurate to warrant their use in diagnosing most cases in which extensive airway remodeling has occurred [140].

Airflow limitation in COPD is due to both small airways disease (obstructive bronchiolitis) and parenchymal destruction (emphysema), the relative contributions of which vary among patients [141, 142]. Thin-section CT has been used to quantify emphysema by detecting low-attenuation areas, and the role of CT in diagnosing emphysema is well established. However, airflow limitation evaluated by FEV_1 does not show a good correlation with the severity of emphysema as evaluated by CT [143, 144], because small airways disease appears to contribute significantly to airflow limitation [145]. Recent progress in CT technology has made it possible to detect and quantify airway abnormalities [146]. Theoretically, thin-section CT can depict the dimensions of airways as small as ~1 to 2 mm in inner diameter, suggesting that CT can be used to evaluate airway dimensions in a variety of diseases [141, 142]. Hasegawa and colleagues [38], who developed new software for measuring airway dimension using curved multiplanar reconstruction, demonstrated that airway luminal area and wall area significantly correlated with FEV_1 (% predicted). The correlation coefficients improved as the airways became smaller.

TABLE 41.3 HRCT features in asthma and COPD.

	Mild-to-persistent asthma	Severe asthma	COPD	Healthy subjects
Bronchial wall thickening	++	+++	+	−
Emphysema	−	+	+++	−
Bronchiectasis	+/−	++	−	+/−

Scintigraphic approaches may be used to assess COPD or emphysema and to provide functional imaging. Ultrafine ^{133}Xe gas particles are being used for ventilation scintigraphy, including single photon emission CT (SPECT). SPECT imaging has been shown to be more useful than morphologic HRCT in the evaluation of small airways disease, including pulmonary emphysema [147]. Diffusion-weighted, hyperpolarized ^{3}He magnetic resonance imaging has been shown to correlate with pulmonary function tests, particularly Dlco [148]. Also, dynamic contrast-enhanced magnetic resonance imaging may detect abnormalities of the pulmonary peripheral microvasculature [149]. These techniques might be useful in the assessment of pulmonary emphysema.

The clinical application of HRCT is mandatory inCOPD patients who are candidates for volume reduction surgery and in whom the regional distribution of emphysema, particularly upper lobe emphysema, is critical for the outcome of the intervention [150–152]. HRCT features of asthma and COPD [39, 153–156] are reported in Table 41.3.

Laboratory examinations

Circulating eosinophils: circulating eosinophilia is a feature of many different lung diseases. In some conditions, eosinophils are increased in the blood but not in the lung tissue; in other diseases, there may be significant eosinophilia in the lung tissue but not in the peripheral blood. In others, there may be lung eosinophilia without any radiographic evidence of disease, as in asthma. Churg–Strauss syndrome (CSS) is characterized by peripheral and pulmonary eosinophilia with infiltrates on chest radiograph. However, the primary features that distinguish CSS from other pulmonary eosinophilic syndromes are the presence of eosinophilic vasculitis in the setting of asthma and the involvement of multiple end organs. Therefore, in these cases it is important to establish a differential diagnosis between these diseases, which may occur without any radiographic evidence of disease. Although perceived to be quite rare, the incidence of this disease seems to have increased in the last few years, particularly in association with various asthma therapies.

Clinical presentation of COPD exacerbation includes worsening of dyspnea, increased cough and sputum, and changes in the aspect of expectorations.

Biomarkers of respiratory bacterial infections

Although clinical criteria are still used to determine which patient should be treated with antibiotics [157], these criteria are neither sensitive nor specific enough to exclude other causes of exacerbation of respiratory symptoms in these patients. Novel biomarkers (e.g. pro-calcitonin) have been recommended for guiding antibiotic treatment both in exacerbations of COPD and pneumonia, but these studies need confirmation [158–160].

Other frequent clinical conditions may mimic the symptoms of COPD exacerbation, including congestive heart failure, pneumonia, pneumothorax, pleural effusion and pulmonary embolism [2]. However, COPD is often cited among the risk factors for acute venous thromboembolism and is an independent predictor of pulmonary embolism [161]. In a small series of patients, the prevalence of deep vein thrombosis in patients admitted with acute exacerbation of COPD was 31% [162, 163]. Similarly, on the basis of ventilation–perfusion lung scintigraphy, the prevalence of pulmonary embolism in patients admitted with acute exacerbation of COPD was as high as 20% [164]. More recently, Tillie-Leblond *et al.* [165] explored the prevalence of pulmonary embolism in a cohort of patients with COPD with unexplained dyspnea and found a rate of 25% in this population. D-dimer is a product of lysis of stabilized fibrin-clot that is considered an indirect marker of coagulation activation. Measurement of plasma D-dimer has a well established diagnostic role in acute pulmonary embolism because of its high negative predictive value [166]. The usefulness of D-dimer testing remains controversial in inpatients, in part due to a high percentage of "positive" D-dimer values among them, which is consequent to a broad spectrum of diseases (other than pulmonary embolism) and procedures related to the hospitalization. Moreover, for inpatients, a negative D-dimer reduces suspicion but its sensitivity is only 89%, unsatisfactory to exclude pulmonary embolism [167]. If the helical contrast CT angiogram is negative and ultrasound is negative, there is still a 5% false-negative rate for inpatients [168]. There are no convincing data regarding a negative CT alone for inpatients. Accordingly, CT and D-dimer evidence may add information but conventional pulmonary arteriography may still be required to make a secure diagnosis. In its absence, a "clinical" decision to treat (and suspend treatment, if contraindications supervene) may be required [169].

Troponin and/or N-BNP

COPD and CHF are common conditions. The diagnosis of CHF can remain unsuspected in patients with COPD, because shortness of breath is attributed to COPD. Measurement of plasma B-type natriuretic peptide (BNP) levels helps to uncover unsuspected CHF in patients with COPD and clinical deterioration [170]. Amino-terminal pro-B-type natriuretic peptides (NT-proBNP) are strong and independent prognostic indicators, representing a particularly strong predictor of heart failure or death [171]. This risk is independent of all other variables, including renal function or troponin, and is proportional to the magnitude of NT-proBNP release, with higher risk observed among those with a more marked elevation of the marker. An elevated initial NT-proBNP concentration should prompt consideration of an early invasive management approach. Consideration should be given to repeating the NT-proBNP measurement after 24–72 h and again at 3–6 months because these follow-up measurements provide more long-term prognostic information than single measures at presentation. In acute ischemic heart disease, an NT-proBNP value >250 ng/l is associated with an adverse prognosis. In patients with stable coronary artery disease, measurement may be performed for prognostication purposes at 6- to 18-month intervals. In the case of clinical suspicion of disease progression, a new sample may be warranted.

Alpha-1 antitrypsin

Severe hereditary deficiency of alpha-1 antitrypsin is usually associated with early-onset panacinar emphysema [172]. Thus, in patients who develop COPD before the age of 45 and/or who have a strong family history of COPD, alpha-1 antitrypsin should be measured; if the serum concentration is <15–20% of the normal value, the patient should be considered for alpha-1 antitrypsin augmentation therapy. Like other genetic tests, this test has no clinical value in asthma [21].

Assessment of airway inflammation

Bronchopulmonary inflammation is markedly different in asthma and COPD [9, 12, 173–182] (Fig. 41.1).

While airway biopsies and bronchoalveolar lavage clearly distinguish between asthma and COPD in subjects with overlapping features that may provide useful information in research protocols, they are considered too invasive for the diagnosis or staging of either asthma or COPD [179, 183, 184]. In contrast, noninvasive markers of airway inflammation have been increasingly used in research protocols to differentiate asthma from COPD [183, 185].

Sputum

Sputum induction has been widely used in the study of airway inflammation in asthma and COPD because it is a safe, reproducible, and noninvasive technique that can be used repeatedly, even during exacerbations [186–188]. Sputum findings mainly represent the bronchial compartment. Induced sputum from asthmatic patients during stable conditions is usually characterized by a higher percentage of eosinophils and metachromatic cells than that found in samples from healthy subjects [189, 190]. Sputum neutrophilia may also be present across the range of disease severity; its identification is important, as it is associated with a poor response to glucocorticosteroids [191–193].

In stable conditions, ex- or current smokers with COPD characteristically show an increased total cell number in spontaneous or induced sputum, with a predominance of neutrophils and a small percentage of eosinophils in some subjects [133, 194, 195]. In some smokers with chronic bronchitis, with or without chronic airflow limitation,

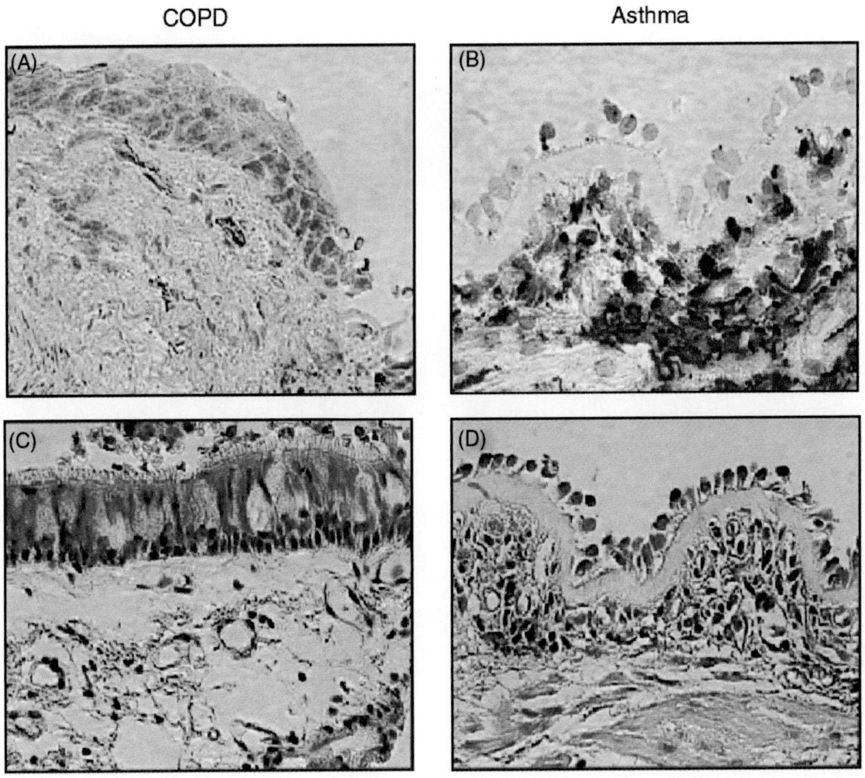

FIG. 41.1 Photomicrographs showing bronchial biopsy specimens immunostained with anti-EG-2 (eosinophil cationic protein) from a patient with fixed airflow obstruction and a history of COPD (A) and from a patient with fixed airflow obstruction and a history of asthma (B). The two patients had a similar degree of fixed airflow obstruction. In (B), there is prominent eosinophilia beneath the destroyed epithelium that is not present in (A). Photomicrographs showing bronchial biopsy specimens stained with H&E from a patient with fixed airflow obstruction and a history of COPD (C) and from a patient with fixed airflow obstruction and a history of asthma (D). The two patients had a similar degree of fixed airflow obstruction. In (D), there is a thicker reticular layer of the epithelial basement membrane compared with (C) (from Ref. [10]).

an excess proportion of eosinophils (>3%) in lower respiratory secretions (called "eosinophilic bronchitis") can also occur [100, 186, 187]. A recent study has elegantly shown that in patients with COPD the main cells in sputum are neutrophils and macrophages, with more macrophages and eosinophils than in a sub-phenotype of patients with chronic bronchitis [196].

Analysis of sputum from patients with exacerbations of asthma or COPD has provided interesting new information. Mild exacerbations of asthma induced by tapering the dose of inhaled steroids are associated with sputum eosinophilia [197, 198]. In contrast, mild exacerbations that are spontaneous are associated with eosinophilia in about 50% of subjects, but the other 50% do not have sputum eosinophilia [199]. In children, eosinophilic airway inflammation is associated with deteriorating asthma over time. This is consistent with the hypothesis that airway inflammation has an adverse effect on the prognosis of childhood asthma and suggests a role for monitoring inflammation in asthma management [200]. Severe asthma exacerbations are associated with more prominent sputum neutrophilia [201]. Bronchial neutrophilia has also been observed in bronchial lavage fluid from asthmatics during status asthmaticus [202].

Interestingly, exacerbations of chronic bronchitis or COPD are associated with quite similar changes in sputum cell count. Mild exacerbations of chronic bronchitis or COPD are associated with eosinophilia in sputum and in biopsy specimens [203, 204], whereas severe exacerbations of COPD are associated with sputum neutrophilia [205].

Thus, at least in sputum, the changes in inflammatory cells during exacerbations may be no different in asthma and COPD. Once again, this evidence underlines the similarities between the two diseases and the difficulty in making a differential diagnosis in those few cases in which clinical findings are not definitive.

TABLE 41.4 History, symptoms, and results of pulmonary function tests in the differential diagnosis between asthma and COPD.

	Asthma	COPD
Onset	Mainly in childhood	In mid to late adult life
Smoking	Usually non-smokers	Almost invariably smokers
Chronic cough and sputum	Absent	Frequent (chronic bronchitis)
Dyspnoea on effort	Variable and reversible to treatment	Constant, poorly reversible and progressive
Nocturnal symptoms	Relatively common	Uncommon
Airflow limitation	Increased diurnal variability	Normal diurnal variability
Response to bronchodilator	Good	Poor
Airway hyperresponsiveness	In most patients, with or without airflow limitation	In most patients with airflow limitation

Several biochemical markers have been studied in induced sputum from both asthma and COPD patients [206–210] (Table 41.4). Although some markers are markedly different in asthma and COPD, and studying

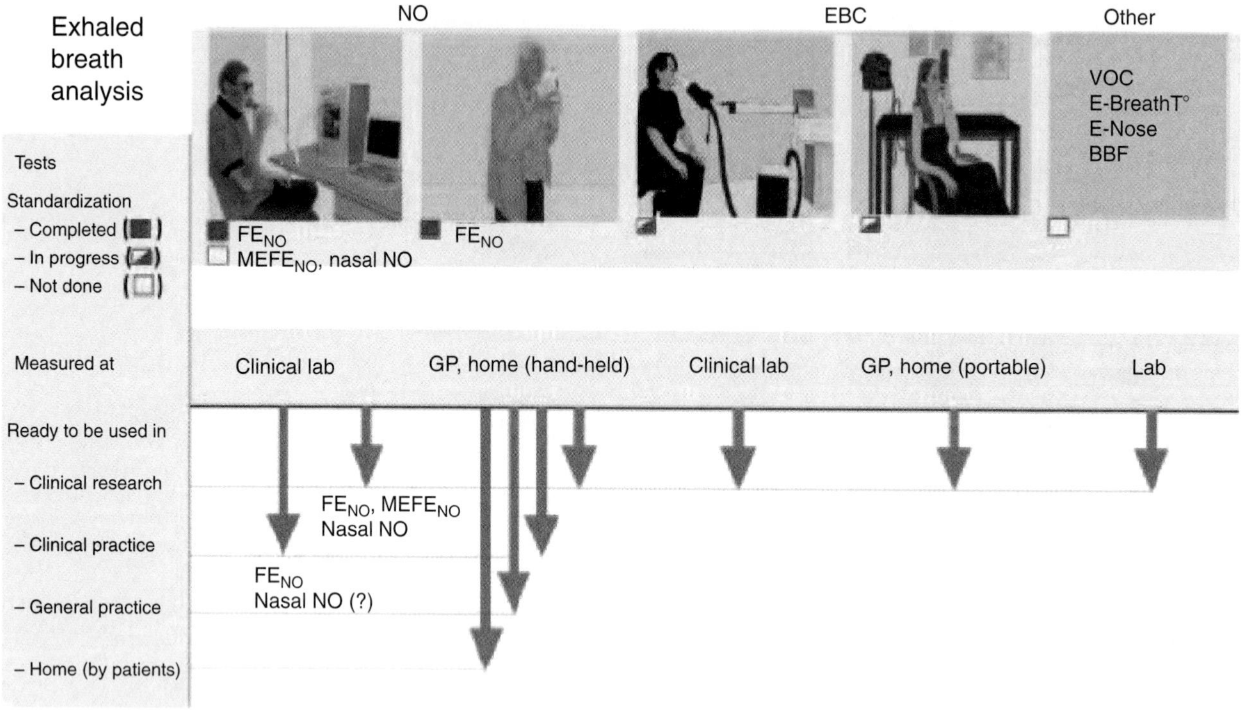

FIG. 41.2 Exhaled breath analysis: current state of standardization, research, and clinical use (from Ref. [212]).

these markers may provide useful information in research protocols, their use in clinical practice has not been shown to be superior to simple cell counts.

Exhaled NO

While endogenous NO may be involved in the pathophysiology of asthma and COPD [211, 212] (Fig. 41.2), exhaled NO is increased in atopic asthma [213–217] but less at lesser extent so in nonatopic asthma [218, 219]. Furthermore, exhaled NO is reduced by glucocorticosteroids [220] but not by bronchodilators [221]. Conflicting results have been obtained in COPD [222–228].

In patients with stable COPD, a partial bronchodilator response to inhaled salbutamol is associated with increased exhaled NO and sputum eosinophilia [101]. Taken together with previous findings [73], this suggests that there is a subset of patients with COPD who share some characteristics of asthmatic inflammation and who may be responsive to steroids [100].

TABLE 41.5 Ancillary tests in the differential diagnosis between stable asthma and COPD.

Ancillary test	Asthma	COPD
Reversibility to bronchodilator and/or glucocorticosteroids	Usually present	Usually absent
Lung volumes		
Residual volume, total lung capacity	Usually normal or, if increased, reversible	Usually irreversibly increased
Diffusion capacity	Normal	Decreased
Airway hyperresponsiveness	Increased	Usually not measurable due to airflow limitation
Allergy tests	Often positive	Often negative
Imaging of the chest	Usually normal	Usually abnormal
Sputum	Eosinophilia	Neutrophilia
Exhaled NO	Increased	Usually normal

DIFFERENTIAL DIAGNOSIS BETWEEN ASTHMA AND COPD

In most patients, the clinical presentation and particularly the history provide the strongest diagnostic criteria to distinguish asthma from COPD (Table 41.5). Results of pulmonary function tests, particularly spirometry, that show a nearly complete reversibility of airflow limitation may help to confirm a diagnosis of asthma, and those that show poorly reversible airflow limitation may help to confirm the diagnosis of COPD (Table 41.5).

Differential diagnosis between asthma and COPD becomes more difficult in elderly patients, in whom some features may overlap, such as smoking and atopy, and, more importantly, when the patient develops poorly reversible airflow limitation that responds only partially to treatment. In these cases, symptoms, lung function, airway responsiveness, imaging, and even pathological findings may overlap and thus may not provide solid information for the differential diagnosis. Because the differential diagnosis mainly aims to provide better treatment, it is important in these cases to undertake an individual approach and to perform additional tests. Reversibility to corticosteroids alone or in combination with long-acting bronchodilators, measurements of lung volumes and diffusion capacity, analysis of sputum and exhaled NO, and imaging of the chest may demonstrate whether asthma or COPD is the predominant cause of airflow limitation (Table 41.5). In contrast, reversibility to bronchodilator and assessment of airway hyperresponsiveness or skin testing may not be useful in these patients.

COMORBIDITIES OF ASTHMA AND COPD

The coexistence of chronic rhinitis, nasal polyposis, and sinusitis may contribute to the severity of asthma [229, 230]. There is broad evidence to show that adequate treatment of these upper airway diseases is beneficial to asthma by mechanisms that are not clearly understood. The "one airway" concept developed by the WHO ARIA Group [231] has drawn attention to the importance of treating the whole respiratory tract while managing asthma. Gastroesophageal reflux is also occasionally associated with asthma, both in adults and in children [232], but treatment of reflux usually has little overall effect on mild to moderate asthma [233]. Two recent studies [234, 235] indicate that proton-pump inhibitors in patients with symptomatic reflux improve asthma control in severe disease. A frequent and quite important comorbidity of asthma in adults is COPD, most likely due to smoking, which is quite common in asthmatics. Smoking modifies the airway pathology of asthmatics to a COPD-like pattern and reduces the response to treatment [7]. Comorbidities may become important in severe asthmatics, whereas overall they play a less important role overall in the clinical manifestations of mild to moderate asthma [236, 237].

Recent research suggests that inflammation in COPD is not confined to the lungs, because the main risk factor, smoking, may simultaneously cause pulmonary and systemic inflammation. This may account for the observation that patients with COPD often present with one or more comorbid conditions. The most common comorbidities that have been described in association with COPD are *hypertension*, *diabetes*, coronary artery disease [238, 239], *chronic heart failure* [240], pulmonary infections, *cancer* [241], and *pulmonary vascular disease* [242, 243].

Comorbidities are highly likely to affect health outcomes in COPD. Progressive respiratory failure accounts for only about one-third of COPD-related deaths: COPD patients are more likely to die of *cardiovascular* complications or *cancer* than of respiratory failure [242]. Therefore, factors other than the progression of lung disease must play a substantial role. The number of preexisting comorbidities in patients with COPD is associated with increased in-hospital mortality [239]. Comorbid conditions that have been associated in particular with an increased mortality risk in COPD patients include chronic renal failure, cor pulmonale [244], and *pulmonary vascular disease* [245]. Underlying heart diseases have not been consistently associated with a higher mortality risk. However, since COPD is frequently underreported, it is difficult to make an accurate estimate of how comorbid conditions influence COPD mortality or, conversely, how COPD affects the outcome of other diagnoses [242].

The complexity of chronic comorbidities applies to acute exacerbations of asthma and COPD to a similar extent. Acute exacerbations of respiratory symptoms may be present in several other acute conditions that should always be carefully considered and excluded, such as acute left ventricular failure, pulmonary thromboembolism, pneumonia, metabolic acidosis, and anemia.

ACKNOWLEDGMENTS

We thank M. McKenney for scientific assistance with the manuscript and E. Veratelli for her scientific secretarial assistance.

Supported by MURST (Grants 60% and 40%), Consorzio Ferrara Ricerche (CFR), Associazione per la Cura e la Ricerca dell'Asma (ARCA), Associazione per lo Studio dei Tumori e delle Malattie Polmonari (ASTMP).

References

1. Global Strategy for Asthma Management and Prevention. Global Initiative for Asthma (GINA), 2007; Available from: http://www.ginasthma.org
2. Global Strategy for the Diagnosis, Management and Prevention of COPD. Global Initiative for Chronic Obstructive Lung Disease (GOLD), 2007; Available from: http://www.goldcopd.org
3. National Heart, Lung, and Blood Institute. National Asthma Education and Prevention Program. Expert Panel Report 3. *Guidelines for the Diagnosis and Management of Asthma*, 2007, Publication No. 07-4051.
4. Celli BR, MacNee W. Standards for the diagnosis and treatment of patients with COPD: A summary of the ATS/ERS position paper. *Eur Respir J* 23(6): 932–46, 2004.
5. Bel EH. Clinical phenotypes of asthma. *Curr Opin Pulm Med* 10(1): 44–50, 2004.
6. Beuther DA, Weiss ST, Sutherland ER. Obesity and asthma. *Am J Respir Crit Care Med* 174(2): 112–19, 2006.
7. Thomson NC. Smokers with asthma: What are the management options? *Am J Respir Crit Care Med* 175(8): 749–50, 2007.
8. Baraldo S, Lokar Oliani K, Turato G, Zuin R, Saetta M. The role of lymphocytes in the pathogenesis of asthma and COPD. *Curr Med Chem* 14(21): 2250–56, 2007.
9. Curtis JL, Freeman CM, Hogg JC. The immunopathogenesis of chronic obstructive pulmonary disease: Insights from recent research. *Proc Am Thorac Soc* 4(7): 512–21, 2007.
10. Fabbri LM, Romagnoli M, Corbetta L, Casoni G, Busljetic K, Turato G, Ligabue G, Ciaccia A, Saetta M, Papi A. Differences in airway inflammation in patients with fixed airflow obstruction due to asthma or chronic obstructive pulmonary disease. *Am J Respir Crit Care Med* 167(3): 418–24, 2003.

11. Hogg JC. Pathophysiology of airflow limitation in chronic obstructive pulmonary disease. *Lancet* 364(9435): 709–21, 2004.
12. Hogg JC, Chu F, Utokaparch S, Woods R, Elliott WM, Buzatu L, Cherniack RM, Rogers RM, Sciurba FC, Coxson HO, Pare PD. The nature of small-airway obstruction in chronic obstructive pulmonary disease. *N Engl J Med* 350(26): 2645–53, 2004.
13. MacNee W. Pathogenesis of chronic obstructive pulmonary disease. *Proc Am Thorac Soc* 2(4): 258–66, discussion 290–91, 2005.
14. Elliott MW, Adams L, Cockcroft A, MacRae KD, Murphy K, Guz A. The language of breathlessness. Use of verbal descriptors by patients with cardiopulmonary disease. *Am Rev Respir Dis* 144(4): 826–32, 1991.
15. Townshend J, Hails S, McKean M. Diagnosis of asthma in children. *BMJ* 335(7612): 198–202, 2007.
16. McFadden ER. Exertional dyspnea and cough as preludes to acute attacks of bronchial asthma. *N Engl J Med* 292(11): 555–59, 1975.
17. Dicpinigaitis PV. Chronic cough due to asthma: ACCP evidence-based clinical practice guidelines. *Chest* 129(1 Suppl): 75S–79S, 2006.
18. Pavord ID. Cough and asthma. *Pulm Pharmacol Ther* 17(6): 399–402, 2004.
19. Burrows B, Martinez FD, Cline MG, Lebowitz MD. The relationship between parental and children's serum IgE and asthma. *Am J Respir Crit Care Med* 152(5 Pt 1): 1497–500, 1995.
20. Burrows B, Martinez FD, Halonen M, Barbee RA, Cline MG. Association of asthma with serum IgE levels and skin-test reactivity to allergens. *N Engl J Med* 320(5): 271–77, 1989.
21. Zhang J, Pare PD, Sandford AJ. Recent advances in asthma genetics. *Respir Res* 9: 4, 2008.
22. Chauhan AJ, Krishna MT, Frew AJ, Holgate ST. Exposure to nitrogen dioxide (NO_2) and respiratory disease risk. *Rev Environ Health* 13(1–2): 73–90, 1998.
23. Garrett MH, Hooper BM, Hooper MA. Indoor environmental factors associated with house-dust-mite allergen (Der p 1) levels in south-eastern Australian houses. *Allergy* 53(11): 1060–65, 1998.
24. Garrett MH, Hooper MA, Hooper BM, Abramson MJ. Respiratory symptoms in children and indoor exposure to nitrogen dioxide and gas stoves. *Am J Respir Crit Care Med* 158(3): 891–95, 1998.
25. Garrett MH, Rayment PR, Hooper MA, Abramson MJ, Hooper BM. Indoor airborne fungal spores, house dampness and associations with environmental factors and respiratory health in children. *Clin Exp Allergy* 28(4): 459–67, 1998.
26. Jarvis D, Chinn S, Luczynska C, Burney P. Association of respiratory symptoms and lung function in young adults with use of domestic gas appliances. *Lancet* 347(8999): 426–31, 1996.
27. McCreanor J, Cullinan P, Nieuwenhuijsen MJ, Stewart-Evans J, Malliarou E, Jarup L, Harrington R, Svartengren M, Han IK, Ohman-Strickland P, Chung KF, Zhang J. Respiratory effects of exposure to diesel traffic in persons with asthma. *N Engl J Med* 357(23): 2348–58, 2007.
28. O'Donnell DE, Banzett RB, Carrieri-Kohlman V, Casaburi R, Davenport PW, Gandevia SC, Gelb AF, Mahler DA, Webb KA. Pathophysiology of dyspnea in chronic obstructive pulmonary disease: A roundtable. *Proc Am Thorac Soc* 4(2): 145–68, 2007.
29. de Marco R, Accordini S, Cerveri I, Corsico A, Anto JM, Kunzli N, Janson C, Sunyer J, Jarvis D, Chinn S, Vermeire P, Svanes C, Ackermann-Liebrich U, Gislason T, Heinrich J, Leynaert B, Neukirch F, Schouten JP, Wjst M, Burney P. Incidence of chronic obstructive pulmonary disease in a cohort of young adults according to the presence of chronic cough and phlegm. *Am J Respir Crit Care Med* 175(1): 32–39, 2007.
30. Viegi G, Pistelli F, Sherrill DL, Maio S, Baldacci S, Carrozzi L. Definition, epidemiology and natural history of COPD. *Eur Respir J* 30(5): 993–1013, 2007.
31. Definition and classification of chronic bronchitis for clinical and epidemiological purposes. A report to the Medical Research Council by their Committee on the Aetiology of Chronic Bronchitis. *Lancet* 1(7389): 775–9, 1965.
32. Burrows B, Knudson RJ, Cline MG, Lebowitz MD. Quantitative relationships between cigarette smoking and ventilatory function. *Am Rev Respir Dis* 115(2): 195–205, 1977.
33. Choudry NB, Fuller RW. Sensitivity of the cough reflex in patients with chronic cough. *Eur Respir J* 5(3): 296–300, 1992.
34. Clark KD, Wardrobe-Wong N, Elliott JJ, Gill PT, Tait NP, Snashall PD. Cigarette smoke inhalation and lung damage in smoking volunteers. *Eur Respir J* 12(2): 395–99, 1998.
35. Smith J, Woodcock A. Cough and its importance in COPD. *Int J Chron Obstruct Pulmon Dis* 1(3): 305–14, 2006.
36. Buist AS. Similarities and differences between asthma and chronic obstructive pulmonary disease: Treatment and early outcomes. *Eur Respir J Suppl* 39: 30s–35s, 2003.
37. Rabe KF, Hurd S, Anzueto A, Barnes PJ, Buist SA, Calverley P, Fukuchi Y, Jenkins C, Rodriguez-Roisin R, van Weel C, Zielinski J. Global strategy for the diagnosis, management, and prevention of chronic obstructive pulmonary disease: GOLD executive summary. *Am J Respir Crit Care Med* 176(6): 532–55, 2007.
38. Hasegawa M, Nasuhara Y, Onodera Y, Makita H, Nagai K, Fuke S, Ito Y, Betsuyaku T, Nishimura M. Airflow limitation and airway dimensions in chronic obstructive pulmonary disease. *Am J Respir Crit Care Med* 173(12): 1309–15, 2006.
39. Mitsunobu F, Tanizaki Y. The use of computed tomography to assess asthma severity. *Curr Opin Allergy Clin Immunol* 5(1): 85–90, 2005.
40. Kohler D, Fischer J, Raschke F, Schonhofer B. Usefulness of GOLD classification of COPD severity. *Thorax* 58(9): 825, 2003.
41. Rutten FH, Moons KG, Cramer MJ, Grobbee DE, Zuithoff NP, Lammers JW, Hoes AW. Recognising heart failure in elderly patients with stable chronic obstructive pulmonary disease in primary care: Cross sectional diagnostic study. *BMJ* 331(7529): 1379, 2005.
42. Celli BR, Cote CG, Marin JM, Casanova C, Montes de Oca M, Mendez RA, Pinto Plata V, Cabral HJ. The body-mass index, airflow obstruction, dyspnea, and exercise capacity index in chronic obstructive pulmonary disease. *N Engl J Med* 350(10): 1005–12, 2004.
43. Ischaki E, Papatheodorou G, Gaki E, Papa I, Koulouris N, Loukides S. Body mass and fat-free mass indices in COPD: Relation with variables expressing disease severity. *Chest* 132(1): 164–69, 2007.
44. Fabbri LM, Luppi F, Beghe B, Rabe KF. Complex chronic comorbidities of COPD. *Eur Respir J* 31(1): 204–12, 2008.
45. Fabbri LM, Rabe KF. From COPD to chronic systemic inflammatory syndrome? *Lancet* 370(9589): 797–99, 2007.
46. Flaherty KR, Kazerooni EA, Martinez FJ. Differential diagnosis of chronic airflow obstruction. *J Asthma* 37(3): 201–23, 2000.
47. Miller MR, Hankinson J, Brusasco V, Burgos F, Casaburi R, Coates A, Crapo R, Enright P, van der Grinten CP, Gustafsson P, Jensen R, Johnson DC, MacIntyre N, McKay R, Navajas D, Pedersen OF, Pellegrino R, Viegi G, Wanger J. Standardisation of spirometry. *Eur Respir J* 26(2): 319–38, 2005.
48. Wanger J, Clausen JL, Coates A, Pedersen OF, Brusasco V, Burgos F, Casaburi R, Crapo R, Enright P, van der Grinten CP, Gustafsson P, Hankinson J, Jensen R, Johnson D, Macintyre N, McKay R, Miller MR, Navajas D, Pellegrino R, Viegi G. Standardisation of the measurement of lung volumes. *Eur Respir J* 26(3): 511–22, 2005.
49. Pellegrino R, Viegi G, Brusasco V, Crapo RO, Burgos F, Casaburi R, Coates A, van der Grinten CP, Gustafsson P, Hankinson J, Jensen R, Johnson DC, MacIntyre N, McKay R, Miller MR, Navajas D, Pedersen OF, Wanger J. Interpretative strategies for lung function tests. *Eur Respir J* 26(5): 948–68, 2005.
50. Anthonisen NR, Wright EC. Bronchodilator response in chronic obstructive pulmonary disease. *Am Rev Respir Dis* 133(5): 814–19, 1986.
51. Calverley PM, Rennard SI. What have we learned from large drug treatment trials in COPD? *Lancet* 370(9589): 774–85, 2007.
52. Celli BR. The importance of spirometry in COPD and asthma: Effect on approach to management. *Chest* 117(Suppl 2): 15S–19S, 2000.
53. Soriano JB, Mannino DM. Reversing concepts on COPD irreversibility. *Eur Respir J* 31(4): 695–96, 2008.

54. Tashkin DP, Celli B, Decramer M, Liu D, Burkhart D, Cassino C, Kesten S. Bronchodilator responsiveness in patients with COPD. *Eur Respir J* 31(4): 742–50, 2008.
55. Chiry S, Cartier A, Malo JL, Tarlo SM, Lemiere C. Comparison of peak expiratory flow variability between workers with work-exacerbated asthma and occupational asthma. *Chest* 132(2): 483–88, 2007.
56. Reddel HK, Marks GB, Jenkins CR. When can personal best peak flow be determined for asthma action plans? *Thorax* 59(11): 922–24, 2004.
57. Wensley D, Silverman M. Peak flow monitoring for guided self-management in childhood asthma: A randomized controlled trial. *Am J Respir Crit Care Med* 170(6): 606–12, 2004.
58. Stahl E. Correlation between objective measures of airway calibre and clinical symptoms in asthma: A systematic review of clinical studies. *Respir Med* 94(8): 735–41, 2000.
59. Harrison TW, Oborne J, Wilding PJ, Tattersfield AE. Randomised placebo controlled trial of beta agonist dose reduction in asthma. *Thorax* 54(2): 98–102, 1999.
60. Sawyer G, Miles J, Lewis S, Fitzharris P, Pearce N, Beasley R. Classification of asthma severity: Should the international guidelines be changed? *Clin Exp Allergy* 28(12): 1565–70, 1998.
61. Kennedy WA, Girard F, Chaboillez S, Cartier A, Cote J, Hargreave F, Labrecque M, Malo JL, Tarlo SM, Redlich CA, Lemiere C. Cost-effectiveness of various diagnostic approaches for occupational asthma. *Can Respir J* 14(5): 276–80, 2007.
62. Mapp CE, Boschetto P, Maestrelli P, Fabbri LM. Occupational asthma. *Am J Respir Crit Care Med* 172(3): 280–305, 2005.
63. Venables KM, Chan-Yeung M. Occupational asthma. *Lancet* 349(9063): 1465–69, 1997.
64. Hansen EF, Vestbo J, Phanareth K, Kok-Jensen A, Dirksen A. Peak flow as predictor of overall mortality in asthma and chronic obstructive pulmonary disease. *Am J Respir Crit Care Med* 163(3 Pt 1): 690–93, 2001.
65. Mallia P, Message SD, Kebadze T, Parker HL, Kon OM, Johnston SL. An experimental model of rhinovirus induced chronic obstructive pulmonary disease exacerbations: A pilot study. *Respir Res* 7: 116, 2006.
66. Wilkinson TM, Donaldson GC, Johnston SL, Openshaw PJ, Wedzicha JA. Respiratory syncytial virus, airway inflammation, and FEV1 decline in patients with chronic obstructive pulmonary disease. *Am J Respir Crit Care Med* 173(8): 871–76, 2006.
67. Burge PS, Calverley PM, Jones PW, Spencer S, Anderson JA. Prednisolone response in patients with chronic obstructive pulmonary disease: Results from the ISOLDE study. *Thorax* 58(8): 654–58, 2003.
68. Calverley PM, Burge PS, Spencer S, Anderson JA, Jones PW. Bronchodilator reversibility testing in chronic obstructive pulmonary disease. *Thorax* 58(8): 659–64, 2003.
69. Sterk PJ, Fabbri LM, Quanjer PH, Cockcroft DW, O'Byrne PM, Anderson SD, Juniper EF, Malo JL. Airway responsiveness. Standardized challenge testing with pharmacological, physical and sensitizing stimuli in adults. Report Working Party Standardization of Lung Function Tests, European Community for Steel and Coal. Official Statement of the European Respiratory Society. *Eur Respir J Suppl* 16: 53–83, 1993.
70. Hansen EF, Phanareth K, Laursen LC, Kok-Jensen A, Dirksen A. Reversible and irreversible airflow obstruction as predictor of overall mortality in asthma and chronic obstructive pulmonary disease. *Am J Respir Crit Care Med* 159(4 Pt 1): 1267–71, 1999.
71. Ulrik CS, Backer V. Nonreversible airflow obstruction in life-long non-smokers with moderate to severe asthma. *Eur Respir J* 14(4): 892–96, 1999.
72. Gross NJ. COPD: A disease of reversible air-flow obstruction. *Am Rev Respir Dis* 133(5): 725–26, 1986.
73. Chanez P, Vignola AM, O'Shaugnessy T, Enander I, Li D, Jeffery PK, Bousquet J. Corticosteroid reversibility in COPD is related to features of asthma. *Am J Respir Crit Care Med* 155(5): 1529–34, 1997.
74. Kerstjens HA, Brand PL, Hughes MD, Robinson NJ, Postma DS, Sluiter HJ, Bleecker ER, Dekhuijzen PN, de Jong PM, Mengelers HJ *et al.* A comparison of bronchodilator therapy with or without inhaled corticosteroid therapy for obstructive airways disease. Dutch Chronic Non-Specific Lung Disease Study Group [see comments]. *N Engl J Med* 327(20): 1413–19, 1992.
75. Chrystyn H, Mulley BA, Peake MD. Dose response relation to oral theophylline in severe chronic obstructive airways disease. *BMJ* 297(6662): 1506–10, 1988.
76. Gross NJ, Petty TL, Friedman M, Skorodin MS, Silvers GW, Donohue JF. Dose response to ipratropium as a nebulized solution in patients with chronic obstructive pulmonary disease. A three-center study. *Am Rev Respir Dis* 139(5): 1188–91, 1989.
77. Higgins BG, Powell RM, Cooper S, Tattersfield AE. Effect of salbutamol and ipratropium bromide on airway calibre and bronchial reactivity in asthma and chronic bronchitis. *Eur Respir J* 4(4): 415–20, 1991.
78. Vathenen AS, Britton JR, Ebden P, Cookson JB, Wharrad HJ, Tattersfield AE. High-dose inhaled albuterol in severe chronic airflow limitation. *Am Rev Respir Dis* 138(4): 850–55, 1988.
79. Tashkin D, Kesten S. Long-term treatment benefits with tiotropium in COPD patients with and without short-term bronchodilator responses. *Chest* 123(5): 1441–49, 2003.
80. Decramer M, Celli B, Tashkin DP, Pauwels RA, Burkhart D, Cassino C, Kesten S. Clinical trial design considerations in assessing long-term functional impacts of tiotropium in COPD: The UPLIFT trial. *COPD* 1(2): 303–12, 2004.
81. Bateman ED, Boushey HA, Bousquet J, Busse WW, Clark TJ, Pauwels RA, Pedersen SE. Can guideline-defined asthma control be achieved? The Gaining Optimal Asthma ControL study. *Am J Respir Crit Care Med* 170(8): 836–44, 2004.
82. Spencer S, Mayer B, Bendall KL, Bateman ED. Validation of a guideline-based composite outcome assessment tool for asthma control. *Respir Res* 8: 26, 2007.
83. Hemminki K, Li X, Sundquist K, Sundquist J. Familial risks for asthma among twins and other siblings based on hospitalizations in Sweden. *Clin Exp Allergy* 37(9): 1320–25, 2007.
84. Berger Z, Rom WN, Reibman J, Kim M, Zhang S, Luo L, Friedman-Jimenez G. Prevalence of workplace exacerbation of asthma symptoms in an urban working population of asthmatics. *J Occup Environ Med* 48(8): 833–39, 2006.
85. Kogevinas M, Zock JP, Jarvis D, Kromhout H, Lillienberg L, Plana E, Radon K, Toren K, Alliksoo A, Benke G, Blanc PD, Dahlman-Hoglund A, D'Errico A, Hery M, Kennedy S, Kunzli N, Leynaert B, Mirabelli MC, Muniozguren N, Norback D, Olivieri M, Payo F, Villani S, van Sprundel M, Urrutia I, Wieslander G, Sunyer J, Anto JM. Exposure to substances in the workplace and new-onset asthma: An international prospective population-based study (ECRHS-II). *Lancet* 370(9584): 336–41, 2007.
86. Corren J. The connection between allergic rhinitis and bronchial asthma. *Curr Opin Pulm Med* 13(1): 13–18, 2007.
87. Apter AJ, Szefler SJ. Advances in adult and pediatric asthma. *J Allergy Clin Immunol* 117(3): 512–18, 2006.
88. Busse WW, Wanner A, Adams K, Reynolds HY, Castro M, Chowdhury B, Kraft M, Levine RJ, Peters SP, Sullivan EJ. Investigative bronchoprovocation and bronchoscopy in airway diseases. *Am J Respir Crit Care Med* 172(7): 807–16, 2005.
89. Pearce N, Pekkanen J, Beasley R. How much asthma is really attributable to atopy. *Thorax* 54(3): 268–72, 1999.
90. Weiss ST. Atopy as a risk factor for chronic obstructive pulmonary disease: Epidemiological evidence. *Am J Respir Crit Care Med* 162(3 Pt 2): S134–36, 2000.
91. Weir DC, Burge PS. Assessment of reversibility of airway obstruction in patients with chronic obstructive airways disease. *Thorax* 45(11): 907–8, 1990.
92. Weir DC, Burge PS. Effects of high dose inhaled beclomethasone dipropionate, 750 micrograms and 1500 micrograms twice daily, and 40 mg per day oral prednisolone on lung function, symptoms, and bronchial hyperresponsiveness in patients with non-asthmatic chronic airflow obstruction. *Thorax* 48(4): 309–16, 1993.

93. Weir DC, Gove RI, Robertson AS, Burge PS. Corticosteroid trials in non-asthmatic chronic airflow obstruction: A comparison of oral prednisolone and inhaled beclomethasone dipropionate. *Thorax* 45(2): 112–17, 1990.
94. Weir DC, Robertson AS, Gove RI, Burge PS. Time course of response to oral and inhaled corticosteroids in non-asthmatic chronic airflow obstruction. *Thorax* 45(2): 118–21, 1990.
95. Pauwels RA, Lofdahl CG, Postma DS, Tattersfield AE, O'Byrne P, Barnes PJ, Ullman A. Effect of inhaled formoterol and budesonide on exacerbations of asthma. Formoterol and Corticosteroids Establishing Therapy (FACET) International Study Group [see comments] [published erratum appears in *N Engl J Med* 338(2): 139, January 1998]. *N Engl J Med* 337(20): 1405–11, 1997.
96. van Grunsven PM, van Schayck CP, Derenne JP, Kerstjens HA, Renkema TE, Postma DS, Similowski T, Akkermans RP, Pasker-de Jong PC, Dekhuijzen PN, van Herwaarden CL, van Weel C. Long term effects of inhaled corticosteroids in chronic obstructive pulmonary disease: A meta-analysis [see comments]. *Thorax* 54(1): 7–14, 1999.
97. Little SA, Chalmers GW, MacLeod KJ, McSharry C, Thomson NC. Non-invasive markers of airway inflammation as predictors of oral steroid responsiveness in asthma. *Thorax* 55(3): 232–34, 2000.
98. Smith AD, Taylor DR. Is exhaled nitric oxide measurement a useful clinical test in asthma? *Curr Opin Allergy Clin Immunol* 5(1): 49–56, 2005.
99. Brightling CE, McKenna S, Hargadon B, Birring S, Green R, Siva R, Berry M, Parker D, Monteiro W, Pavord ID, Bradding P. Sputum eosinophilia and the short term response to inhaled mometasone in chronic obstructive pulmonary disease. *Thorax* 60(3): 193–98, 2005.
100. Brightling CE, Monteiro W, Ward R, Parker D, Morgan MD, Wardlaw AJ, Pavord ID. Sputum eosinophilia and short-term response to prednisolone in chronic obstructive pulmonary disease: A randomised controlled trial. *Lancet* 356(9240): 1480–85, 2000.
101. Papi A, Romagnoli M, Baraldo S, Braccioni F, Guzzinati I, Saetta M, Ciaccia A, Fabbri LM. Partial reversibility of airflow limitation and increased exhaled NO and sputum eosinophilia in chronic obstructive pulmonary disease. *Am J Respir Crit Care Med* 162(5): 1773–77, 2000.
102. The Lung Health Study Research Group. Effect of inhaled triamcinolone on the decline in pulmonary function in chronic obstructive pulmonary disease. *N Engl J Med* 343(26): 1902–9, 2000.
103. Burge PS, Calverley PM, Jones PW, Spencer S, Anderson JA, Maslen TK. Randomised, double blind, placebo controlled study of fluticasone propionate in patients with moderate to severe chronic obstructive pulmonary disease: The ISOLDE trial. *BMJ* 320(7245): 1297–303, 2000.
104. Paggiaro PL, Dahle R, Bakran I, Frith L, Hollingworth K, Efthimiou J. Multicentre randomised placebo-controlled trial of inhaled fluticasone propionate in patients with chronic obstructive pulmonary disease. International COPD Study Group. *Lancet* 351(9105): 773–80, 1998.
105. Pauwels RA, Lofdahl CG, Laitinen LA, Schouten JP, Postma DS, Pride NB, Ohlsson SV. Long-term treatment with inhaled budesonide in persons with mild chronic obstructive pulmonary disease who continue smoking. European Respiratory Society Study on Chronic Obstructive Pulmonary Disease. *N Engl J Med* 340(25): 1948–53, 1999.
106. Vestbo J, Sorensen T, Lange P, Brix A, Torre P, Viskum K. Long-term effect of inhaled budesonide in mild and moderate chronic obstructive pulmonary disease: A randomised controlled trial. *Lancet* 353(9167): 1819–23, 1999.
107. Wong CA, Walsh LJ, Smith CJ, Wisniewski AF, Lewis SA, Hubbard R, Cawte S, Green DJ, Pringle M, Tattersfield AE. Inhaled corticosteroid use and bone-mineral density in patients with asthma. *Lancet* 355(9213): 1399–403, 2000.
108. Mapp CE. Inhaled glucocorticoids in chronic obstructive pulmonary disease. *N Engl J Med* 343(26): 1960–61, 2000.
109. Calverley PM, Anderson JA, Celli B, Ferguson GT, Jenkins C, Jones PW, Yates JC, Vestbo J. Salmeterol and fluticasone propionate and survival in chronic obstructive pulmonary disease. *N Engl J Med* 356(8): 775–89, 2007.
110. Nannini LJ, Cates CJ, Lasserson TJ, Poole P. Combined corticosteroid and long-acting beta-agonist in one inhaler versus inhaled steroids for chronic obstructive pulmonary disease. *Cochrane Database Syst Rev* (4): CD006826, 2007.
111. Brown CD, Wise RA. Field tests of exercise in COPD: The six-minute walk test and the shuttle walk test. *COPD* 4(3): 217–23, 2007.
112. Macintyre N, Crapo RO, Viegi G, Johnson DC, van der Grinten CP, Brusasco V, Burgos F, Casaburi R, Coates A, Enright P, Gustafsson P, Hankinson J, Jensen R, McKay R, Miller MR, Navajas D, Pedersen OF, Pellegrino R, Wanger J. Standardisation of the single-breath determination of carbon monoxide uptake in the lung. *Eur Respir J* 26(4): 720–35, 2005.
113. Collard P, Njinou B, Nejadnik B, Keyeux A, Frans A. Single breath diffusing capacity for carbon monoxide in stable asthma. *Chest* 105(5): 1426–29, 1994.
114. Sin BA, Akkoca O, Saryal S, Oner F, Misirligil Z. Differences between asthma and COPD in the elderly. *J Investig Allergol Clin Immunol* 16(1): 44–50, 2006.
115. Clausen JL. The diagnosis of emphysema, chronic bronchitis, and asthma. *Clin Chest Med* 11(3): 405–16, 1990.
116. Sansores RH, Pare P, Abboud RT. Effect of smoking cessation on pulmonary carbon monoxide diffusing capacity and capillary blood volume. *Am Rev Respir Dis* 146(4): 959–64, 1992.
117. Boulet LP, Turcotte H, Hudon C, Carrier G, Maltais F. Clinical, physiological and radiological features of asthma with incomplete reversibility of airflow obstruction compared with those of COPD. *Can Respir J* 5(4): 270–77, 1998.
118. Papi A, Romagnoli M, Bellettato CM, Cogo AL, Zanin A, Turato G, Saetta M, Ciaccia A, Fabbri LM. Pulmonary function and pathology in asthma and chronic obstructive pulmonary disease (COPD) with similar degree of irreversible airflow limitation. *Eur Respir J* 16(Suppl 31): 551s, 2000.
119. Wilson JS, Galvin JR. Normal diffusing capacity in patients with PiZ alpha(1)-antitrypsin deficiency, severe airflow obstruction, and significant radiographic emphysema. *Chest* 118(3): 867–71, 2000.
120. Holme J, Stockley RA. Radiologic and clinical features of COPD patients with discordant pulmonary physiology: Lessons from alpha1-antitrypsin deficiency. *Chest* 132(3): 909–15, 2007.
121. Anderson SD. Provocative challenges to help diagnose and monitor asthma: Exercise, methacholine, adenosine, and mannitol. *Curr Opin Pulm Med* 14(1): 39–45, 2008.
122. Cockcroft DW, Hargreave FE. Airway hyperresponsiveness. Relevance of random population data to clinical usefulness. *Am Rev Respir Dis* 142(3): 497–500, 1990.
123. Birnbaum S, Barreiro TJ. Methacholine challenge testing: Identifying its diagnostic role, testing, coding, and reimbursement. *Chest* 131(6): 1932–35, 2007.
124. Miedinger D, Chhajed PN, Tamm M, Stolz D, Surber C, Leuppi JD. Diagnostic tests for asthma in firefighters. *Chest* 131(6): 1760–67, 2007.
125. Padoan M, Pozzato V, Simoni M, Zedda L, Milan G, Bononi I, Piola C, Maestrelli P, Boschetto P, Mapp CE. Long-term follow-up of toluene diisocyanate-induced asthma. *Eur Respir J* 21(4): 637–40, 2003.
126. Allergic Rhinitis and Its Impact on Asthma (ARIA). World Health Organization Initiative. Geneva, 2000.
127. Polosa R, Ciamarra I, Mangano G, Prosperini G, Pistorio MP, Vancheri C, Crimi N. Bronchial hyperresponsiveness and airway inflammation markers in nonasthmatics with allergic rhinitis. *Eur Respir J* 15(1): 30–35, 2000.
128. Ramsdale EH, Morris MM, Roberts RS, Hargreave FE. Asymptomatic bronchial hyperresponsiveness in rhinitis. *J Allergy Clin Immunol* 75(5): 573–77, 1985.

129. Ip M, Lauder IJ, Wong WY, Lam WK, So SY. Multivariate analysis of factors affecting pulmonary function in bronchiectasis. *Respiration* 60(1): 45–50, 1993.
130. Nielsen KG, Pressler T, Klug B, Koch C, Bisgaard H. Serial lung function and responsiveness in cystic fibrosis during early childhood. *Am J Respir Crit Care Med* 169(11): 1209–16, 2004.
131. van Haren EH, Lammers JW, Festen J, van Herwaarden CL. Bronchial vagal tone and responsiveness to histamine, exercise and bronchodilators in adult patients with cystic fibrosis. *Eur Respir J* 5(9): 1083–88, 1992.
132. Grootendorst DC, Rabe KF. Mechanisms of bronchial hyperreactivity in asthma and chronic obstructive pulmonary disease. *Proc Am Thorac Soc* 1(2): 77–87, 2004.
133. Rutgers SR, Timens W, Tzanakis N, Kauffman HF, van der Mark TW, Koeter GH, Postma DS. Airway inflammation and hyperresponsiveness to adenosine 5′-monophosphate in chronic obstructive pulmonary disease. *Clin Exp Allergy* 30(5): 657–62, 2000.
134. Schafroth Torok S, Leuppi JD. Bronchial hyper-responsiveness and exhaled nitric oxide in chronic obstructive pulmonary disease. *Swiss Med Wkly* 137(27–28): 385–91, 2007.
135. Scichilone N, Battaglia S, La Sala A, Bellia V. Clinical implications of airway hyperresponsiveness in COPD. *Int J Chron Obstruct Pulmon Dis* 1(1): 49–60, 2006.
136. James AL, Wenzel S. Clinical relevance of airway remodelling in airway diseases. *Eur Respir J* 30(1): 134–55, 2007.
137. Sung A, Naidich D, Belinskaya I, Raoof S. The role of chest radiography and computed tomography in the diagnosis and management of asthma. *Curr Opin Pulm Med* 13(1): 31–36, 2007.
138. Nakano Y, Muller NL, King GG, Niimi A, Kalloger SE, Mishima M, Pare PD. Quantitative assessment of airway remodeling using high-resolution CT. *Chest* 122(Suppl 6): 271S–75S, 2002.
139. Sharma V, Shaaban AM, Berges G, Gosselin M. The radiological spectrum of small-airway diseases. *Semin Ultrasound CT MR* 23(4): 339–51, 2002.
140. Niimi A, Matsumoto H, Takemura M, Ueda T, Nakano Y, Mishima M. Clinical assessment of airway remodeling in asthma: Utility of computed tomography. *Clin Rev Allergy Immunol* 27(1): 45–58, 2004.
141. Coxson HO, Rogers RM. Quantitative computed tomography of chronic obstructive pulmonary disease. *Acad Radiol* 12(11): 1457–63, 2005.
142. Coxson HO, Rogers RM. New concepts in the radiological assessment of COPD. *Semin Respir Crit Care Med* 26(2): 211–20, 2005.
143. Baldi S, Miniati M, Bellina CR, Battolla L, Catapano G, Begliomini E, Giustini D, Giuntini C. Relationship between extent of pulmonary emphysema by high-resolution computed tomography and lung elastic recoil in patients with chronic obstructive pulmonary disease. *Am J Respir Crit Care Med* 164(4): 585–89, 2001.
144. Omori H, Fujimoto K, Katoh T. Computed-tomography findings of emphysema: Correlation with spirometric values. *Curr Opin Pulm Med* 14(2): 110–14, 2008.
145. Hogg JC. State of the art. Bronchiolitis in chronic obstructive pulmonary disease. *Proc Am Thorac Soc* 3(6): 489–93, 2006.
146. de Jong PA, Muller NL, Pare PD, Coxson HO. Computed tomographic imaging of the airways: Relationship to structure and function. *Eur Respir J* 26(1): 140–52, 2005.
147. Magnant J, Vecellio L, de Monte M, Grimbert D, Valat C, Boissinot E, Guilloteau D, Lemarie E, Diot P. Comparative analysis of different scintigraphic approaches to assess pulmonary ventilation. *J Aerosol Med* 19(2): 148–59, 2006.
148. Fain SB, Panth SR, Evans MD, Wentland AL, Holmes JH, Korosec FR, O'Brien MJ, Fountaine H, Grist TM. Early emphysematous changes in asymptomatic smokers: Detection with 3He MR imaging. *Radiology* 239(3): 875–83, 2006.
149. Morino S, Toba T, Araki M, Azuma T, Tsutsumi S, Tao H, Nakamura T, Nagayasu T, Tagawa T. Noninvasive assessment of pulmonary emphysema using dynamic contrast-enhanced magnetic resonance imaging. *Exp Lung Res* 32(1–2): 55–67, 2006.
150. Fishman A, Martinez F, Naunheim K, Piantadosi S, Wise R, Ries A, Weinmann G, Wood DE. A randomized trial comparing lung-volume-reduction surgery with medical therapy for severe emphysema. *N Engl J Med* 348(21): 2059–73, 2003.
151. Martinez FJ, Chang A. Surgical therapy for chronic obstructive pulmonary disease. *Semin Respir Crit Care Med* 26(2): 167–91, 2005.
152. Lenfant C. Will lung volume reduction surgery be widely applied? *Ann Thorac Surg* 82(2): 385–87, 2006.
153. McLean AN, Sproule MW, Cowan MD, Thomson NC. High resolution computed tomography in asthma. *Thorax* 53(4): 308–14, 1998.
154. Webb WR. Radiology of obstructive pulmonary disease. *AJR Am J Roentgenol* 169(3): 637–47, 1997.
155. Boschetto P, Quintavalle S, Zeni E, Leprotti S, Potena A, Ballerin L, Papi A, Palladini G, Luisetti M, Annovazzi L, Iadarola P, De Rosa E, Fabbri LM, Mapp CE. Association between markers of emphysema and more severe chronic obstructive pulmonary disease. *Thorax* 61(12): 1037–42, 2006.
156. Kurashima K, Takayanagi N, Sato N, Kanauchi T, Hoshi T, Tokunaga D, Ubukata M, Yanagisawa T, Sugita Y, Kanazawa M. High resolution CT and bronchial reversibility test for diagnosing COPD. *Respirology* 10(3): 316–22, 2005.
157. Anthonisen NR, Manfreda J, Warren CP, Hershfield ES, Harding GK, Nelson NA. Antibiotic therapy in exacerbations of chronic obstructive pulmonary disease. *Ann Intern Med* 106(2): 196–204, 1987.
158. Stolz D, Christ-Crain M, Morgenthaler NG, Leuppi J, Miedinger D, Bingisser R, Muller C, Struck J, Muller B, Tamm M. Copeptin, C-reactive protein, and procalcitonin as prognostic biomarkers in acute exacerbation of COPD. *Chest* 131(4): 1058–67, 2007.
159. Stolz D, Christ-Crain M, Bingisser R, Leuppi J, Miedinger D, Muller C, Huber P, Muller B, Tamm M. Antibiotic treatment of exacerbations of COPD: A randomized, controlled trial comparing procalcitonin-guidance with standard therapy. *Chest* 131(1): 9–19, 2007.
160. Stolz D, Christ-Crain M, Gencay MM, Bingisser R, Huber PR, Muller B, Tamm M. Diagnostic value of signs, symptoms and laboratory values in lower respiratory tract infection. *Swiss Med Wkly* 136(27–28): 434–40, 2006.
161. Poulsen SH, Noer I, Moller JE, Knudsen TE, Frandsen JL. Clinical outcome of patients with suspected pulmonary embolism. A follow-up study of 588 consecutive patients. *J Intern Med* 250(2): 137–43, 2001.
162. Schonhofer B, Kohler D. Prevalence of deep-vein thrombosis of the leg in patients with acute exacerbation of chronic obstructive pulmonary disease. *Respiration* 65(3): 173–77, 1998.
163. Winter JH, Buckler PW, Bautista AP, Smith FW, Sharp PF, Bennett B, Douglas AS. Frequency of venous thrombosis in patients with an exacerbation of chronic obstructive lung disease. *Thorax* 38(8): 605–8, 1983.
164. Erelel M, Cuhadaroglu C, Ece T, Arseven O. The frequency of deep venous thrombosis and pulmonary embolus in acute exacerbation of chronic obstructive pulmonary disease. *Respir Med* 96(7): 515–18, 2002.
165. Tillie-Leblond I, Marquette CH, Perez T, Scherpereel A, Zanetti C, Tonnel AB, Remy-Jardin M. Pulmonary embolism in patients with unexplained exacerbation of chronic obstructive pulmonary disease: Prevalence and risk factors. *Ann Intern Med* 144(6): 390–96, 2006.
166. Palareti G, Cosmi B, Legnani C. Diagnosis of deep vein thrombosis. *Semin Thromb Hemost* 32(7): 659–72, 2006.
167. Kuruvilla J, Wells PS, Morrow B, MacKinnon K, Keeney M, Kovacs MJ. Prospective assessment of the natural history of positive D-dimer results in persons with acute venous thromboembolism (DVT or PE). *Thromb Haemost* 89(2): 284–87, 2003.
168. Musset D, Parent F, Meyer G, Maitre S, Girard P, Leroyer C, Revel MP, Carette MF, Laurent M, Charbonnier B, Laurent F, Mal H, Nonent M, Lancar R, Grenier P, Simonneau G. Diagnostic strategy for patients with suspected pulmonary embolism: A prospective multicentre outcome study. *Lancet* 360(9349): 1914–20, 2002.

169. Davidson BL, Tomkowski WZ. Management of pulmonary embolism in 2005. *Dis Mon* 51(2–3): 116–23, 2005.
170. Le Jemtel TH, Padeletti M, Jelic S. Diagnostic and therapeutic challenges in patients with coexistent chronic obstructive pulmonary disease and chronic heart failure. *J Am Coll Cardiol* 49(2): 171–80, 2007.
171. Omland T, de Lemos JA. Amino-terminal pro-B-type natriuretic peptides in stable and unstable ischemic heart disease. *Am J Cardiol* 101(3A): 61–66, 2008.
172. Blanco I, de Serres FJ, Fernandez-Bustillo E, Lara B, Miravitlles M. Estimated numbers and prevalence of PI*S and PI*Z alleles of alpha1-antitrypsin deficiency in European countries. *Eur Respir J* 27(1): 77–84, 2006.
173. Mauad T, Dolhnikoff M. Pathologic similarities and differences between asthma and chronic obstructive pulmonary disease. *Curr Opin Pulm Med* 14(1): 31–38, 2008.
174. Bochner BS, Busse WW. Allergy and asthma. *J Allergy Clin Immunol* 115(5): 953–59, 2005.
175. Busse WW, Lemanske RF Jr. Asthma. *N Engl J Med* 344(5): 350–62, 2001.
176. Fabbri LM, Romagnoli M, Luppi F, Corbetta L. Asthma versus COPD: Cellular and molecular differences. In: Barnes PJ (ed.), *COPD Cellular and Molecular Mechanisms. Lung Biology in Health and Disease Series*, pp. 55–81. Marcel Dekker, Inc, New York – Basel, 2004.
177. Jeffery PK. Structural and inflammatory changes in COPD: A comparison with asthma. *Thorax* 53(2): 129–36, 1998.
178. Fabbri L, Beghe B, Caramori G, Papi A, Saetta M. Similarities and discrepancies between exacerbations of asthma and chronic obstructive pulmonary disease. *Thorax* 53(9): 803–8, 1998.
179. Fabbri LM, Durham S, Holgate ST, O'Byrne PM, Postma DS. Assessment of airway inflammation: An overview. *Eur Respir J*(Suppl 26): 6S–8S, 1998.
180. Saetta M. Airway pathology of COPD compared with asthma. *Eur Respir Rev* 7: 29–33, 1997.
181. Saetta M. Airway inflammation in chronic obstructive pulmonary disease. *Am J Respir Crit Care Med* 160(5 Pt 2): S17–20, 1999.
182. Saetta M, Timens W, Jeffrey PK. Pathology. In: Postma DS, Siafakas NM (eds.), *Management of Chronic Obstructive Pulmonary Disease.* Sheffield, UK: Eur Respir Mon ERS Journals Ltd; 92–111,1998.
183. Barnes PJ, Chowdhury B, Kharitonov SA, Magnussen H, Page CP, Postma D, Saetta M. Pulmonary biomarkers in chronic obstructive pulmonary disease. *Am J Respir Crit Care Med* 174(1): 6–14, 2006.
184. Zhang JY, Wenzel SE. Tissue and BAL based biomarkers in asthma. *Immunol Allergy Clin North Am* 27(4): 623–32, vi, 2007.
185. Menzies D, Nair A, Lipworth BJ. Non-invasive measurement of airway inflammation in asthma. *J Asthma* 43(6): 407–15, 2006.
186. Brightling CE. Clinical applications of induced sputum. *Chest* 129(5): 1344–48, 2006.
187. Hargreave FE, Leigh R. Induced sputum, eosinophilic bronchitis, and chronic obstructive pulmonary disease. *Am J Respir Crit Care Med* 160(5 Pt 2): S53–57, 1999.
188. Pizzichini MM, Pizzichini E, Clelland L, Efthimiadis A, Mahony J, Dolovich J, Hargreave FE. Sputum in severe exacerbations of asthma: Kinetics of inflammatory indices after prednisone treatment. *Am J Respir Crit Care Med* 155(5): 1501–8, 1997.
189. Kips JC, Fahy JV, Hargreave FE, Ind PW, in't Veen JC. Methods for sputum induction and analysis of induced sputum: A method for assessing airway inflammation in asthma. *Eur Respir J Suppl* 26: 9S–12S, 1998.
190. Pin I, Gibson PG, Kolendowicz R, Girgis-Gabardo A, Denburg JA, Hargreave FE, Dolovich J. Use of induced sputum cell counts to investigate airway inflammation in asthma. *Thorax* 47(1): 25–29, 1992.
191. Green RH, Brightling CE, Woltmann G, Parker D, Wardlaw AJ, Pavord ID. Analysis of induced sputum in adults with asthma: Identification of subgroup with isolated sputum neutrophilia and poor response to inhaled corticosteroids. *Thorax* 57(10): 875–79, 2002.
192. Kikuchi S, Nagata M, Kikuchi I, Hagiwara K, Kanazawa M. Association between neutrophilic and eosinophilic inflammation in patients with severe persistent asthma. *Int Arch Allergy Immunol* 137(Suppl 1): 7–11, 2005.
193. Pavord ID, Brightling CE, Woltmann G, Wardlaw AJ. Non-eosinophilic corticosteroid unresponsive asthma. *Lancet* 353(9171): 2213–14, 1999.
194. Grootendorst DC, Gauw SA, Verhoosel RM, Sterk PJ, Hospers JJ, Bredenbroker D, Bethke TD, Hiemstra PS, Rabe KF. Reduction in sputum neutrophil and eosinophil numbers by the PDE4 inhibitor roflumilast in patients with COPD. *Thorax* 62(12): 1081–87, 2007.
195. Pizzichini E, Pizzichini MM, Gibson P, Parameswaran K, Gleich GJ, Berman L, Dolovich J, Hargreave FE. Sputum eosinophilia predicts benefit from prednisone in smokers with chronic obstructive bronchitis. *Am J Respir Crit Care Med* 158(5 Pt 1): 1511–17, 1998.
196. Snoeck-Stroband JB, Lapperre TS, Gosman MM, Boezen HM, Timens W, ten Hacken NH, Sont JK, Sterk PJ, Hiemstra PS. Chronic bronchitis sub-phenotype within COPD: Inflammation in sputum and biopsies. *Eur Respir J* 31(1): 70–77, 2008.
197. in't Veen JC, Smits HH, Hiemstra PS, Zwinderman AE, Sterk PJ, Bel EH. Lung function and sputum characteristics of patients with severe asthma during an induced exacerbation by double-blind steroid withdrawal. *Am J Respir Crit Care Med* 160(1): 93–99, 1999.
198. Jatakanon A, Lim S, Barnes PJ. Changes in sputum eosinophils predict loss of asthma control. *Am J Respir Crit Care Med* 161(1): 64–72, 2000.
199. Turner MO, Hussack P, Sears MR, Dolovich J, Hargreave FE. Exacerbations of asthma without sputum eosinophilia. *Thorax* 50(10): 1057–61, 1995.
200. Lovett CJ, Whitehead BF, Gibson PG. Eosinophilic airway inflammation and the prognosis of childhood asthma. *Clin Exp Allergy* 37(11): 1594–601, 2007.
201. Fahy JV, Boushey HA, Lazarus SC, Mauger EA, Cherniack RM, Chinchilli VM, Craig TJ, Drazen JM, Ford JG, Fish JE, Israel E, Kraft M, Lemanske RF, Martin RJ, McLean D, Peters SP, Sorkness C, Szefler SJ. Safety and reproducibility of sputum induction in asthmatic subjects in a multicenter study. *Am J Respir Crit Care Med* 163(6): 1470–75, 2001.
202. Lamblin C, Gosset P, Tillie-Leblond I, Saulnier F, Marquette CH, Wallaert B, Tonnel AB. Bronchial neutrophilia in patients with noninfectious status asthmaticus. *Am J Respir Crit Care Med* 157(2): 394–402, 1998.
203. Saetta M, Di Stefano A, Maestrelli P, Turato G, Ruggieri MP, Roggeri A, Calcagni P, Mapp CE, Ciaccia A, Fabbri LM. Airway eosinophilia in chronic bronchitis during exacerbations. *Am J Respir Crit Care Med* 150(6 Pt 1): 1646–52, 1994.
204. Siva R, Green RH, Brightling CE, Shelley M, Hargadon B, McKenna S, Monteiro W, Berry M, Parker D, Wardlaw AJ, Pavord ID. Eosinophilic airway inflammation and exacerbations of COPD: A randomised controlled trial. *Eur Respir J* 29(5): 906–13, 2007.
205. Piattella M, Maestrelli P, Saetta M, Mapp C, Caramori G, Ciaccia A, Fabbri LM. Sputum eosinophilia during mild exacerbations and sputum neutrophilia during severe exacerbations of COPD. *Am J Respir Crit Care Med* 153(Suppl): A822, 1996.
206. Hill AT, Bayley D, Stockley RA. The interrelationship of sputum inflammatory markers in patients with chronic bronchitis. *Am J Respir Crit Care Med* 160(3): 893–98, 1999.
207. Keatings VM, Barnes PJ. Granulocyte activation markers in induced sputum: Comparison between chronic obstructive pulmonary disease, asthma, and normal subjects. *Am J Respir Crit Care Med* 155(2): 449–53, 1997.
208. Keatings VM, Barnes PJ. Comparison of inflammatory cytokines in chronic obstructive pulmonary disease, asthma and controls. *Eur Respir Rev* 7: 146–50, 1997.
209. O'Driscoll BR, Cromwell O, Kay AB. Sputum leukotrienes in obstructive airways diseases. *Clin Exp Immunol* 55(2): 397–404, 1984.

210. Pizzichini E, Pizzichini MM, Efthimiadis A, Evans S, Morris MM, Squillace D, Gleich GJ, Dolovich J, Hargreave FE. Indices of airway inflammation in induced sputum: Reproducibility and validity of cell and fluid-phase measurements. *Am J Respir Crit Care Med* 154(2 Pt 1): 308–17, 1996.
211. Barnes PJ. Nitric oxide and airway disease. *Ann Med* 27(3): 389–93, 1995.
212. Kharitonov SA, Barnes PJ. Exhaled biomarkers. *Chest* 130(5): 1541–46, 2006.
213. Alving K, Weitzberg E, Lundberg JM. Increased amount of nitric oxide in exhaled air of asthmatics. *Eur Respir J* 6(9): 1368–70, 1993.
214. Kharitonov SA, Yates D, Robbins RA, Logan-Sinclair R, Shinebourne EA, Barnes PJ. Increased nitric oxide in exhaled air of asthmatic patients. *Lancet* 343(8890): 133–35, 1994.
215. Massaro AF, Gaston B, Kita D, Fanta C, Stamler JS, Drazen JM. Expired nitric oxide levels during treatment of acute asthma. *Am J Respir Crit Care Med* 152(2): 800–3, 1995.
216. Persson MG, Zetterstrom O, Agrenius V, Ihre E, Gustafsson LE. Single-breath nitric oxide measurements in asthmatic patients and smokers. *Lancet* 343(8890): 146–47, 1994.
217. Turner S. Exhaled nitric oxide in the diagnosis and management of asthma. *Curr Opin Allergy Clin Immunol* 8(1): 70–76, 2008.
218. Gratziou C, Lignos M, Dassiou M, Roussos C. Influence of atopy on exhaled nitric oxide in patients with stable asthma and rhinitis. *Eur Respir J* 14(4): 897–901, 1999.
219. Henriksen AH, Lingaas-Holmen T, Sue-Chu M, Bjermer L. Combined use of exhaled nitric oxide and airway hyperresponsiveness in characterizing asthma in a large population survey. *Eur Respir J* 15(5): 849–55, 2000.
220. van Rensen EL, Straathof KC, Veselic-Charvat MA, Zwinderman AH, Bel EH, Sterk PJ. Effect of inhaled steroids on airway hyperresponsiveness, sputum eosinophils, and exhaled nitric oxide levels in patients with asthma. *Thorax* 54(5): 403–8, 1999.
221. Yates DH, Kharitonov SA, Barnes PJ. Effect of short- and long-acting inhaled beta2-agonists on exhaled nitric oxide in asthmatic patients. *Eur Respir J* 10(7): 1483–88, 1997.
222. Corradi M, Majori M, Cacciani GC, Consigli GF, de'Munari E, Pesci A. Increased exhaled nitric oxide in patients with stable chronic obstructive pulmonary disease. *Thorax* 54(7): 572–75, 1999.
223. Kanazawa H, Shoji S, Yoshikawa T, Hirata K, Yoshikawa J. Increased production of endogenous nitric oxide in patients with bronchial asthma and chronic obstructive pulmonary disease. *Clin Exp Allergy* 28(10): 1244–50, 1998.
224. Maziak W, Loukides S, Culpitt S, Sullivan P, Kharitonov SA, Barnes PJ. Exhaled nitric oxide in chronic obstructive pulmonary disease. *Am J Respir Crit Care Med* 157(3 Pt 1): 998–1002, 1998.
225. Robbins RA, Floreani AA, Von Essen SG, Sisson JH, Hill GE, Rubinstein I, Townley RG. Measurement of exhaled nitric oxide by three different techniques. *Am J Respir Crit Care Med* 153(5): 1631–35, 1996.
226. Rutgers SR, Meijer RJ, Kerstjens HA, van der Mark TW, Koeter GH, Postma DS. Nitric oxide measured with single-breath and tidal-breathing methods in asthma and COPD. *Eur Respir J* 12(4): 816–19, 1998.
227. Rutgers SR, van der Mark TW, Coers W, Moshage H, Timens W, Kauffman HF, Koeter GH, Postma DS. Markers of nitric oxide metabolism in sputum and exhaled air are not increased in chronic obstructive pulmonary disease. *Thorax* 54(7): 576–80, 1999.
228. Taylor DR, Pijnenburg MW, Smith AD, De Jongste JC. Exhaled nitric oxide measurements: Clinical application and interpretation. *Thorax* 61(9): 817–27, 2006.
229. Bresciani M, Paradis L, Des Roches A, Vernhet H, Vachier I, Godard P, Bousquet J, Chanez P. Rhinosinusitis in severe asthma. *J Allergy Clin Immunol* 107(1): 73–80, 2001.
230. ten Brinke A, Zwinderman AH, Sterk PJ, Rabe KF, Bel EH. Factors associated with persistent airflow limitation in severe asthma. *Am J Respir Crit Care Med* 164(5): 744–48, 2001.
231. Bachert C, van Cauwenberge P, Khaltaev N. Allergic rhinitis and its impact on asthma. In collaboration with the World Health Organization. Executive summary of the workshop report. December 7–10, 1999, Geneva, Switzerland. *Allergy* 57(9): 841–55, 2002.
232. Gibson PG, Henry RL, Coughlan JL. Gastro-oesophageal reflux treatment for asthma in adults and children. *Cochrane Database Syst Rev* (2): CD001496, 2003.
233. Coughlan JL, Gibson PG, Henry RL. Medical treatment for reflux oesophagitis does not consistently improve asthma control: A systematic review. *Thorax* 56(3): 198–204, 2001.
234. Kiljander TO, Harding SM, Field SK, Stein MR, Nelson HS, Ekelund J, Illueca M, Beckman O, Sostek MB. Effects of esomeprazole 40 mg twice daily on asthma: A randomized placebo-controlled trial. *Am J Respir Crit Care Med* 173(10): 1091–97, 2006.
235. Littner MR, Leung FW, Ballard ED II, Huang B, Samra NK. Effects of 24 weeks of lansoprazole therapy on asthma symptoms, exacerbations, quality of life, and pulmonary function in adult asthmatic patients with acid reflux symptoms. *Chest* 128(3): 1128–35, 2005.
236. Slavin RG. Medical management of rhinosinusitis comorbidities-asthma, aspirin sensitivity, gastroesophageal reflux, immune deficiencies. *Clin Allergy Immunol* 20: 273–86, 2007.
237. Peroni DG, Piacentini GL, Ceravolo R, Boner AL. Difficult asthma: Possible association with rhinosinusitis. *Pediatr Allergy Immunol* 18(Suppl 18): 25–27, 2007.
238. Holguin F, Folch E, Redd SC, Mannino DM. Comorbidity and mortality in COPD-related hospitalizations in the United States, 1979 to 2001. *Chest* 128(4): 2005–11, 2005.
239. Sidney S, Sorel M, Quesenberry CP Jr, DeLuise C, Lanes S, Eisner MD. COPD and incident cardiovascular disease hospitalizations and mortality: Kaiser Permanente Medical Care Program. *Chest* 128(4): 2068–75, 2005.
240. Sin DD, Wu L, Anderson JA, Anthonisen NR, Buist AS, Burge PS, Calverley PM, Connett JE, Lindmark B, Pauwels RA, Postma DS, Soriano JB, Szafranski W, Vestbo J. Inhaled corticosteroids and mortality in chronic obstructive pulmonary disease. *Thorax* 60(12): 992–97, 2005.
241. Balkwill F, Mantovani A. Inflammation and cancer: Back to Virchow? *Lancet* 357(9255): 539–45, 2001.
242. Mannino DM, Watt G, Hole D, Gillis C, Hart C, McConnachie A, Davey Smith G, Upton M, Hawthorne V, Sin DD, Man SF, Van Eeden S, Mapel DW, Vestbo J. The natural history of chronic obstructive pulmonary disease. *Eur Respir J* 27(3): 627–43, 2006.
243. Sevenoaks MJ, Stockley RA. Chronic Obstructive Pulmonary Disease, inflammation and co-morbidity – a common inflammatory phenotype? *Respir Res* 7: 70, 2006.
244. Naeije R. Pulmonary hypertension and right heart failure in chronic obstructive pulmonary disease. *Proc Am Thorac Soc* 2(1): 20–22, 2005.
245. Naeije R, Vizza D. Current perspectives modern hemodynamic evaluation of the pulmonary circulation. Application to pulmonary arterial hypertension and embolic pulmonary hypertension. *Ital Heart J* 6(10): 784–88, 2005.

Non-invasive Assessment of Airway Inflammation

42

CHAPTER

Ian D. Pavord[1] and Sergei A. Kharitonov[2]

[1]Department of Respiratory Medicine and Thoracic Surgery, Institute for Lung Health, Glenfield Hospital, Leicester, UK
[2]Section of Airway Disease, National Heart and Lung Institute, Imperial College, London, UK

INTRODUCTION

Airway inflammation is thought to be an important component of asthma, chronic obstructive pulmonary disease (COPD), and the various cough syndromes. However, it is not routinely assessed in clinical practice and much of what we know about the nature and importance of airway inflammation is derived from bronchoscopy studies. These studies are limited by the invasive nature of bronchoscopy, bronchial biopsy, and bronchoalveolar lavage so it has not been possible to study large heterogeneous populations nor has not been standard practice to assess airway inflammation in clinical practice or clinical trials.

Over the last 15–20 years there has been an explosion of interest in the assessment of airway inflammation using non-invasive means and there are now a large number of different techniques to assess airway inflammation, each with its own strengths and weaknesses (Table 42.1). Two techniques are particularly well developed and are already widely used in clinical trials and impacting on clinical practice: induced sputum, where a sputum differential and total cell count is used to determine the characteristics and intensity of the lower airway inflammatory response; and exhaled nitric oxide (FE_{NO}), where the concentration of nitric oxide (NO) in exhaled air is used to provide information about the presence of eosinophilic, corticosteroid responsive airway inflammation. Induced sputum has the advantage of providing measurements of the type of airway inflammation (eosinophilic versus neutrophilic) as well as its severity. The main limitation is that results are not available immediately, limiting the application of the technique in asthma monitoring. FE_{NO} has the advantage of being simple to measure and provides an immediate result but it provides only limited information on the nature of the inflammatory response. The different strengths and weaknesses of the techniques suggests that they may find different roles, with FE_{NO} being used mainly in primary care to facilitate diagnosis and to titrate corticosteroid therapy and induced sputum used in secondary and tertiary care, where more detailed information on the type of lower airway inflammation is necessary.

In this chapter we will review the methodology and validation of induced sputum and FE_{NO} and outline other more experimental approaches to the assessment of airway inflammation. We will finish by discussing how the widespread application of induced sputum to large and heterogeneous populations of patients with airway disease has furthered our understanding of the complex relationship between airway inflammation and the clinical expression of airway disease and opened the way to a new approach to the management of airway disease based on assessment of airway inflammation (inflammometry).

INDUCED SPUTUM

Methodology

Sputum induction using nebulized hypertonic saline is used to collect respiratory secretions from the airways of patients who do not expectorate spontaneously. The precise mechanism leading to production of increased secretions is not known but increased production of mucous by the submucosal glands, increased vascular permeability of the bronchial mucosa and release of pro-inflammatory mediators may all contribute. Eighty to ninety percent of patients with airway disease and a similar proportion of healthy controls are able to produce a satisfactory sputum sample following hypertonic saline inhalation.

TABLE 42.1 Comparison of methods measuring airway inflammation.

	Ease of performing technique	Ease of analyzing result	Time to result	Cost	Influence on outcome proved	Potential use
Induced sputum	Simple sputum sample	Simple, but requires experience	3–4 h	Moderate	Yes	Secondary care
Blood eosinophil count	Simple blood test	Simple	30 min	Inexpensive	Suggests eosinophilic airway inflammation when raised in correct context	Secondary care
Eosinophil cationic protein	Urine test/sputum test/ blood test	Simple	3–4 h	Moderate	Not proven conclusively	Research
Exhaled nitric oxide	Very easy breathing test	Easy	Immediate	Expensive though could become cheaper	Not proven	Research Possibly primary care
Exhaled breath temperature	Thermometer. Complex currently but could be simplified.	Easy	Immediate	Inexpensive	Studies awaited	Research
Breath condensate	Cooling/freezing of exhaled air	Simple	Moderate	Inexpensive	Studies awaited	Research
BAL, bronchial wash, and biopsy	Biopsy- invasive	Complex	2 days	Moderate	Yes	Secondary care

BOX 42.1PROTOCOL FOR SPUTUM INDUCTION USING NEBULIZED HYPERTONIC SALINE.

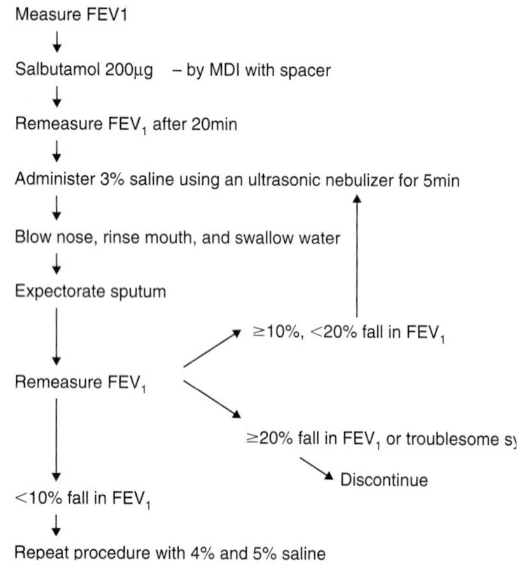
Measure FEV1
Salbutamol 200μg – by MDI with spacer
Remeasure FEV1 after 20min
Administer 3% saline using an ultrasonic nebulizer for 5min
Blow nose, rinse mouth, and swallow water
Expectorate sputum
≥10%, <20% fall in FEV1
Remeasure FEV1
≥20% fall in FEV1 or troublesome sy
Discontinue
<10% fall in FEV1
Repeat procedure with 4% and 5% saline

A variety of protocols for sputum induction have been published and shown to be safe provided patients are pre-treated with bronchodilators and monitored carefully [1, 2]. Risk factors for bronchoconstriction include a low baseline FEV_1% predicted [3], overuse of short-acting β_2-agonists [4] and poor asthma control [4]. Symptomatic bronchoconstriction is unusual in patients with COPD, even those with severe airflow obstruction [5]; sputum induction has also been carried out repeatedly and safely in patients with severe asthma [6]. It is likely that higher nebulizer output, higher concentration of inhaled saline, a longer duration of saline inhalation and reduced frequency and timing of safety assessment by forced expiratory volume in 1 s or peak expiratory flow reduce safety. Higher output nebulizers have also been associated with the development of a sputum neutrophilia 24 h after sputum induction [7]. Whether this is seen with the low output nebulizers is unknown. A common and well-validated protocol is to use a relatively low output ultrasonic nebulizer (output 0.7–0.9 ml/min) delivering increasing concentrations of hypertonic saline [5, 6, 8]. The method is summarized in Box 42.1.

Once expectorated sputum should be processed within 2 h. There is evidence that sputum can be stored for up to 9 h in a refrigerator at 4°C or that sputum can be snap frozen for longer without affecting cell counts [9], although experience with these techniques is limited. The whole expectorate or selected sputum plugs can be processed. The latter approach has the advantage of producing better quality cytospins and more repeatable differential cell counts [10]. A common method used to process induced sputum is outlined in Box 42.2. The total cell count, cell viability and squamous cell contamination are assessed using a hemocytometer. Differential cell counts are determined by counting 400 leukocytes on an appropriately stained cytospin. Other biomarkers can also be measured in the sputum supernatant. Some of the molecular markers of airway inflammation that have been successfully measured in sputum are shown in Table 42.2.

BOX 42.2 PROTOCOL FOR PROCESSING OF SPUTUM.

Process at 4°C within 2 h of expectoration

Select sputum (if necessary using inverted microscope)

↓

Weight and incubate with 4 × volume 0.1% dithiothreitol (DTT)

↓

Gently aspirate with pasteur pipette, vortex for 15 s

↓

Rock on bench rocker for 15 min on ice

↓

Mix with equal volume (to DTT) of Dulbecco's phosphate buffered saline (DPBS)

↓

Vortex for 15 s

↓

Filter through 48 μm nylon gauze (pre-wet with DPBS)

↓

Centrifuge 790 g 10 min → Store supernatant at −70°C → ↓ Measure fluid phase

↓

Resuspend cell pellet in DPBS

↓

Perform total cell count and viability by trypan blue exclusion method in Neubauer hemocytometer

↓

Adjust cell suspension to 0.5–0.75 × 10^6 cells/ml with DPBS

↓

Prepare cytospins by placing 50 or 75 μl cell suspension in cups of cytocentrifuge and centrifuge at 450 rpm for 6 min

↓

Air dry and stain

TABLE 42.2 Cell types and molecular markers that have been successfully measured in induced sputum.

Cells	Effector mediators	Cellular markers	Cytokines/ chemokines
Eosinophils Neutrophils Macrophages Lymphocytes Epithelial Mast cells	Leukotrienes C/D/E_4 Prostaglandin D_2 Histamine	Eosinophilic cationic protein Neutrophil elastase	Interleukin-8

It is generally agreed that the central airways are sampled with induced sputum. This view is supported by studies showing a greater proportion of granulocytes in both sputum and bronchial samples compared with bronchoalveolar lavage [11–13] and by the demonstration that sputum induction results in greater clearance of radiolabeled aerosol from the central airways than the peripheral airway [14]. There is evidence that increasing the duration of sputum induction leads to sampling of more distal airways although, as yet, the clinical utility of this technique has not been explored [15].

Validation

Sputum differential cell counts have been shown to correlate closely with bronchial wash cell counts [16], less closely with bronchoalveolar lavage counts [13, 16] and not at all with biopsy cell counts [17]. This may reflect sampling of different airway compartments by the different techniques and the fact that some cell types (notably eosinophils and neutrophils) are not tissue dwelling.

Normal ranges have been published for a large adult population [18–20] (Table 42.3). Age has been shown to influence differential sputum neutrophil counts, with the higher values occurring in the older age groups [20]. Sputum differential eosinophils, macrophage and neutrophil counts and the sputum supernatant concentration of eosinophilic cationic protein (ECP), cysteinyl–leukotrienes, prostanoids, and IL-8 can be measured repeatably in asthma [21–23] and in COPD [5] (Table 42.3). The differential lymphocyte and epithelial cell count and the total cell count are less repeatable [22]. Spontaneous and induced sputum have similar cell and molecular characteristics and they can be used interchangeably in asthma and COPD; there is no evidence that pre-treatment with salbutamol or a prior methacholine inhalation test influences sputum cell counts.

The sputum eosinophil count is responsive in that it increases when asthma worsens (e.g. after allergen challenge and following relevant occupational exposures [24, 25]), and decreases when asthma improves with inhaled corticosteroid (ICS) treatment [26]. Sputum and bronchoscopy studies with corticosteroids [26, 27] and anti-IL-5 [28, 29] suggest that the sputum eosinophil count is more responsive than tissue eosinophil counts. The sputum differential neutrophil count is reduced following treatment with the fixed dose combination inhaler seretide in patients with COPD [30]. There are theoretical reasons to suggest that the total neutrophil count may be a more responsive measure than the differential count since the relationship between them becomes relatively flat over a differential neutrophil count of 80% [31].

Based on what is known about the responsiveness and repeatability of the sputum eosinophil count, a 2-fold change is regarded as clinically significant [32]. A reduction of the sputum neutrophil differential cell count of 13% or more has been regarded as significant since longitudinal, observational studies suggest that a difference of this magnitude is associated with a clinically important reduction in the rate of decline in lung function in patients with COPD [33].

Findings in disease

Clinically relevant patterns of airway inflammation include neutrophilic, eosinophilic, and mixed eosinophilic and

TABLE 42.3 Normal ranges and within subject repeatability of induced sputum cell counts in adults.

	Normal ranges			Repeatability		
Author	Belda [18]	Spanevello [19]	Thomas [20]	Pizzichini [22]	in't Veen [21]	Spanavello [23]
Number (male)	96 (54)	96 (46)	66 (24)	39 (20)	21 (10)	88 (44)
Characteristics	Healthy	Healthy	Healthy	Asthma (19) Healthy (10), Smokers (10)	Asthma	Asthma (53) Healthy (19) Rhinitis (16)
Mean age	Not recorded	Not recorded	46 (25)	39	24 (4)	38
TCC ($\times 10^6$/g)	4.1 (4.8)	2.7 (2.5)	2.1 (2.4)	0.35		0.44
Eosinophils (%)	0.4 (0.9)	0.6 (0.8)	0.3 (0.6)	0.94 (0.75^a)^b	0.85 (6.2)	0.84 (0.17^a)
Neutrophils (%)	37.5 (20.1)	27.3 (13.0)	47.0 (27.0)	0.81 (14.0)	0.57 (15.5)	0.75
Macrophages (%)	58.8 (21.0)	69.2 (13.0)	49.0 (25.2)	0.71	0.64 (7.7)	0.76
Lymphocytes (%)	1.0 (1.1)	1.0 (1.2)	1.0 (1.4)	0.25	0.76 (3.9)	0.39
Epithelial cells (%)	1.6 (3.9)	1.5 (1.8)	2.5 (3.2)		0.64 (7.7)	0.56

Figures represent mean (SD). Repeatability expressed as intra-class correlation coefficient (within subject SD or [a]log within subject SD). Repeatability was assessed over 2-days (in't Veen *et al.*), 6-days (Pizzichini *et al.*) and 1-week (Spanavello *et al.*). [b]Within subject log standard deviation of sputum eosinophil count overly influenced by small absolute differences in normals; in subjects with asthma it was 0.25.

	Normal eosinophil count (<1.9%)	Raised eosinophil count
Normal neutrophil count (<61%)	*Paucigranulocytic* Well controlled or intermittent asthma Consider alternative diagnosis	*Eosinophilic* Asthma Eosinophilic bronchitis
Raised neutrophil count	*Neutrophilic* Acute infection (viral or bacterial) Chronic infection (chlamydia, adenovirus) Smoking and COPD Environmental pollutants (ozone, NO_2) Occupational antigens Endotoxin exposure Obesity Chronic cough syndromes	*Mixed granulocytic* Exacerbations Refractory asthma COPD

FIG. 42.1 Classification based on induced sputum patterns of cellular inflammation in airway diseases.

neutrophilic (Fig. 42.1). The latter pattern is most commonly seen in COPD, refractory asthma, and in exacerbations of airway disease. A sputum neutrophilia with a raised total cell count suggests infective bronchitis, COPD, and bronchiectasis; a neutrophilia with a normal total cell count is more non-specific but is commonly seen in chronic cough and non-eosinophilic asthma. A significant minority of patients with airway disease have paucigranulocytic sputum characterized by normal sputum eosinophil and neutrophil counts (Fig. 42.1).

A raised sputum eosinophil count is the most characteristic finding in patients with asthma. 60–80% of patients with symptomatic untreated asthma and around 50% of patients taking ICS have a sputum eosinophil count above the normal range. A minority of patients with asthma of all severity grades have consistently non-eosinophilic sputum; many have a sputum neutrophilia. A raised sputum eosinophil count is not exclusive to asthma: 30–40% of patients with cough and a similar proportion of patients with COPD have a sputum eosinophilia (Fig. 42.2). Roughly a half of patients with cough and a sputum eosinophilia have evidence of variable airflow obstruction and/or airway hyperresponsiveness and meet diagnostic criteria for cough variant asthma. The remainder consistently do not and are classified as having eosinophilic bronchitis. Patients with COPD and a sputum eosinophilia may have more asthma like features such as bronchodilator reversibility, atopy, and airway hyperresponsiveness but there is too much overlap in the presence of these features to make it possible to reliably predict pathology on the basis of these tests.

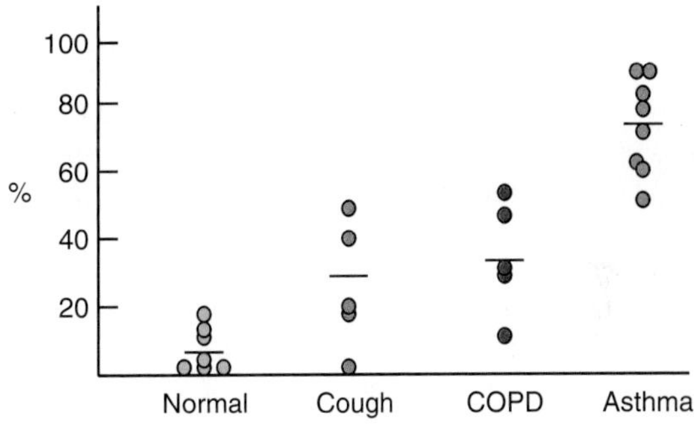

FIG. 42.2 Approximate prevalence of eosinophilic airway inflammation (defined as a sputum eosinophil count >1.9%) in different patient populations. Each dot represents an estimate of prevalence from a single study. Adapted from Gibson *et al.* [75].

EXHALED NITRIC OXIDE

Methodology

FE_{NO} is usually measured by chemiluminescence. This uses ozone to react with NO and produce nitrogen dioxide which then emits photon in a stoichiometric relationship with the amount of NO present in the gas sample. This technique is sensitive, allowing measurements down to 1 part per billion (ppb) [34]. FE_{NO} can be measured either "offline" or "online". Online measurement involves the inhalation of NO free air immediately followed by exhalation at a steady flow directly into the measuring apparatus, whereas in offline measurements the exhaled air is collected in a Mylar® balloon and then transported to the NO analyzer.

There is now a consensus that FE_{NO} is best measured before other spirometric maneuvers, at an exhaled rate of 50 ml/s maintained within 10% for >6 s, and with an oral pressure of 5–20 cmH_2O to ensure velum closure [35]. The technique is simple and well within the scope of most patients in primary care, including children as young as 5 years of age [36]. Results are expressed as the NO concentration in ppb (equivalent to nanoliters/liter) based on the mean of two or three values within 10% [37]. NO output can be calculated as the product of the exhaled NO concentration in nl/l and the exhalation flow in l/min. Alveolar NO concentration can be estimated from measurement of NO output at multiple exhalation flows [38]. Alveolar NO may reflect distal lung inflammation; it is increased in patients with refractory asthma and is reduced by oral, but not inhaled, corticosteroids [39]. The clinical role of this derived measurement has not been firmly established and it will not be considered further.

Validation

Robust estimates of FE_{NO} measured using the above methods are available in children [40] and, to a lesser extent, adults [37]. FE_{NO} tends to increase in healthy children to the age of 17; thereafter the values are similar to adults. There is no age effect in adults. There is conflicting data on the effects of gender [37, 40]. One study suggested fluctuations in FE_{NO} in healthy pre-menopausal women related to the menstrual cycle, with peak values mid-cycle [41]. A reasonable estimate of the normal ranges of FE_{NO} in adults is <30 ppb [37, 40]; this is reduced to <24 ppb when outliers are removed from the analysis and <22.4 ppb when outliers and atopics are removed. Various factors have been shown to influence FE_{NO} and some do so independently of the presence of eosinophilic airway inflammation. These factors and the likely mechanism of the effect on FE_{NO} are summarized in Table 42.4. Genetic factors are known to affect the production of NO: van's Gravesande *et al.* demonstrated a strong relationship between a known functional NOS3 missense sequence variant in the endothelial nitric oxide gene (G894T) and FE_{NO} level in a cohort of subjects with asthma [42], and the number of alpha-1 antitrypsin (AAT) repeats in intron 20 of this gene correlate with NO levels, with a higher number of repeats associated with lower NO levels [43].

The between subject standard deviation of FE_{NO} is around 25 ppb in asthma and 8 ppb in normal controls [37]. The within subject standard deviation has been estimated at 1.6–2 ppb in a range of subjects [37, 40]. FE_{NO} is reduced 2- to 4-fold by corticosteroids in patients with asthma [44–46] and is increased (by about 60%) by during the late response to allergen in subjects with atopic asthma [47]. There is a dose-related effect of low dose ICS on FE_{NO} but no additional benefit above a dose of budesonide of 400 μg/day [44, 48, 49]. Based on what is known about the effect of ICS on FE_{NO} and the predictive value of a change in FE_{NO} [50], a reasonable minimally important difference is a 60% change either way.

Findings in disease

Any form of airway inflammation increases FE_{NO} levels, including that seen in bronchiectasis, viral infection, fibrosing alveolitis, allergic rhinitis, pulmonary tuberculosis, COPD, and pulmonary sarcoidosis, although the relationship between inflammation and FE_{NO} is complex as pneumonia [51], ciliary dyskinesia, and cystic fibrosis [52] have been shown to be associated with reduced FE_{NO}. Cross-sectionally, the relationship between FE_{NO} levels and eosinophilic airway inflammation is much closer than the relationship with other markers of airway inflammation [53] and it is not clear to what extent the finding of a raised FE_{NO} are a true reflection of the presence of eosinophilic airway inflammation in these conditions.

OTHER METHODS TO MEASURE AIRWAY INFLAMMATION

Exhaled condensate

The collection of exhaled breath condensate (EBC) and assay of inflammatory markers is the most recent development in non-invasive asthma monitoring technologies. Methodological aspects of EBC collection and analysis have been reviewed by a European Respiratory Society (ERS) working party [54]. Water vapor and respiratory droplets and particles are cooled and collected and used for conventional assay or inflammatory markers such as leukotrienes [55], 8-isoprostane [56], *pH* [57], and hydrogen peroxide [58–60]. The current methods for the collection of EBC vary primarily in the type of condensers that are employed. The physical surface properties of each condenser system may influence the condensate that is collected. Therefore it is possible that there may be variation between systems in terms of the particles

TABLE 42.4 Factors associated with increased and decreased FE_{NO}.

Factor	Effect on FE_{NO}	Likely mechanism
Age	Increase until age 17, no effect thereafter	Increased airway size
Gender	Mixed evidence. If anything lower in females independent of height but not a major effect	? Increased airway size in males
Height	Small increase with increasing height	Increased airway size
Atopy	Increased in atopic patients	Underlying eosinophilic airway inflammation, which may be asymptomatic in some
Smoking	Increased in non-smokers	Unclear. May depend on time of last cigarette. Relationship between FE_{NO} and sputum eosinophil count lost in smokers
Spirometry	Transient reduction after spirometry	Unclear. FE_{NO} should be done before other tests.
Rhinitis	Increased, independent of atopy	? Increased eosinophilic airway inflammation

collected. The influence of salivary contamination on EBC values may also be a problem as many of the mediators assayed are found in high concentration in saliva [61]. The measurement characteristics of some of the mediators detectable in EBC are summarized in Table 42.5.

Assessment of airway inflammation using EBC is non-invasive and feasible in almost all patients. It has the potential to provide information about different aspects of the airway inflammatory response and this information may be clinically important. However, results are not available immediately, limiting the clinical utility of the technique. Future technological advances may overcome this problem. There is evidence of significant variability in the concentration of some EBC markers and the reported relationship between the concentration of EBC markers and other inflammatory or clinical measures is variable between studies raising the possibility of unresolved methodological issues and a need for further validation work.

Exhaled breath temperature

Exhaled breath temperature is a potential marker of airway inflammation as it is likely to reflect vascular hyperperfusion secondary to tissue inflammation and airway remodeling. Measurement of exhaled breath temperature and bronchial blood flow [62] is feasible in asthma [63] and COPD [64]. Asthma is associated with increased exhaled breath temperature presumable reflecting increased blood flow due to new vessel formation and vasodilatation. Exhaled breath temperature is not consistently increased in COPD, suggesting differences in bronchial blood flow and tissue remodeling. Whether exhaled breath temperature provides clinically important information over and above other techniques has not been established but if it does, then the simplicity of the technique makes it an attractive option as a non-invasive marker of airway inflammation although the lack of specific information on the pattern of airway inflammation may limit its clinical value.

Serum ECP

ECP is a product of eosinophils and there has been interest in the use of serum, sputum or urinary ECP as a non-invasive marker of eosinophilic airway inflammation. Serum ECP concentrations are higher than EDTA plasma concentration, probably because blood eosinophils continue to produce ECP *ex-vivo* in the absence of additives. Serum ECP concentrations are thought to be better at discriminating health from disease and are preferred. A standardized collection, processing, and testing method has been described [65].

Serum ECP increases with spontaneous allergen exposure or after a laboratory allergen challenge and decreases following allergen avoidance and ICS therapy. However, serum ECP is a less responsive measure than the sputum eosinophil count and FE_{NO} [66, 67]. Compared with eosinophil counts, ECP measurements in either induced sputum or serum failed to reflect treatment-related changes in chronic asthma [68], supporting the view that serum ECP is not a sensitive or reliable means of evaluating eosinophilic airway inflammation. Moreover, serum ECP was found to be insensitive marker in titrating and monitoring therapy with ICS over a wide dose range in childhood asthma [69] and it does not help predict a response to corticosteroid therapy [70]. Finally, a randomized trial which compared a serum ECP based algorithm with a conventional algorithm for managing asthma found no improvement in symptom scores, in spite of increased doses of ICS [71]. These findings suggest that serum ECP is an imperfect and insensitive measure of eosinophilic airway inflammation, which provides no particular advantages over other techniques.

NEW INSIGHTS INTO THE IMPORTANCE OF AIRWAY INFLAMMATION IN AIRWAY DISEASE

Our view of the importance of inflammation in the pathogenesis of asthma and other airway diseases has been heavily influenced by bronchoscopy studies performed over the

TABLE 42.5 Measurement characteristics of markers assayed in exhaled breath condensate.

	8-isoprostane	Cysteinyl leukotrienes	Hydrogen Peroxide	pH
Methodology	EIA, GCMS	EIA		De-aerated sample, pH meter
Normal range	?	?	?	6.8–8.1
Repeatability	?	?	Acceptable	Acceptable
Result in asthma	Increased 3-fold	Increased 3-fold	Increased	Reduced
Effect of corticosteroid	Inhaled: nil Oral: reduced	Nil	Reduced	pH normalises
Correlation with other markers	FEV_1: NS	FEV_1: NS	Correlates with eosinophils	Correlated with eosinophils, lung function

Notes: EIA: enzyme immunoassay; GCMS: gas chromatography–mass spectrometry; NS: not significant.

last 20 years [72]. These, by necessity, were largely limited to young volunteers with mild disease. The development of a non-invasive technique to assess airway inflammation has made it possible to relate the presence of airway inflammation to objective measures of disordered airway function in larger and more heterogeneous populations than was possible with bronchoscopy studies. These studies have suggested a complex relationship between airway inflammation and the pattern and severity of airway dysfunction [8, 73, 74]. Clinicians and researchers interested in using measures of airway inflammation need to be aware of this relationship in order to fully appreciate the added value such measures provide. A number of key observations are particularly important. All have been primarily made using induced sputum as the measure of airway inflammation but they are potentially relevant to the use of FE_{NO} given the close relationship between FE_{NO} and eosinophilic airway inflammation.

Firstly, the presence of pattern of airway inflammation is not closely related to either the pattern or the severity of the airway dysfunction or symptoms. A raised sputum eosinophil count is seen in 60–80% of corticosteroid naive patients with asthma, 50% of corticosteroid treated patients with symptomatic asthma [75], 30–40% of patients with cough [76], and up to 40% of patients with COPD [77]. Within diagnostic groups there is a weak correlation between the presence of eosinophilic airway inflammation and the severity of symptoms or disordered airway function [73]. Thus, little can be deduced about the presence, nature and severity of airway inflammation from a standard clinical assessment.

This observation is important as it strongly implies that there is not a close causal link between eosinophilic airway inflammation and the airway dysfunction that underlies many of the day-to-day clinical manifestations of asthma. Further support for this view is provided by studies showing that treatment with anti-IL-5 markedly reduced eosinophilic airway inflammation but does not affect asthma symptoms, lung function or airway responsiveness [29, 78]. A comparative bronchial biopsy study of asthma and eosinophilic bronchitis (a condition where one sees eosinophilic airway inflammation in the absence of the abnormalities of airway function that characterize asthma) has shown that infiltration of the airway smooth muscle with mast cell is only seen in asthma [79]. Moreover, the extent of infiltration is closely related to airway hyperresponsiveness implying that it is the interaction between mast cells and airway smooth muscle that is of fundamental importance in the genesis of airway hyperresponsiveness.

Secondly, the presence of eosinophilic airway inflammation is more closely associated with a positive response to corticosteroids than other clinical measures. Moreover, a positive response to corticosteroids is seen irrespective of the pattern of airway disease in which eosinophilic airway inflammation occurs (Fig. 42.3) [26, 77, 80–82]. Thus, if the clinical questions were whether a patient with symptoms suggesting airway disease should receive corticosteroid treatment (as it often is), then the identification of eosinophilic airway inflammation would be a better basis for making this decision than the findings of other tests.

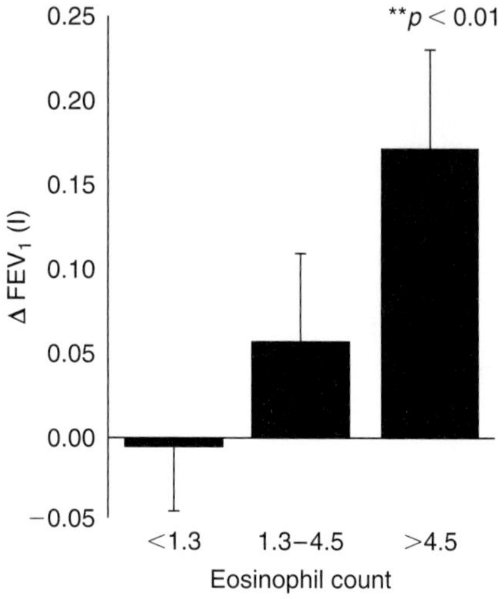

FIG. 42.3 Mean net change in FEV_1 with after 2 weeks treatment with prednisone 30 mg od in 69 patients with moderate and severe COPD participating in a randomized, double blind placebo controlled trial. Patients are stratified into tertiles based on baseline sputum eosinophil count. Reproduced with permission from Brightling *et al.* [77].

Thirdly, the sputum eosinophil count is a better marker for titrating corticosteroid therapy than standard clinical measures. Studies in asthma [6, 83] and COPD [84] have shown that management strategies where decisions about corticosteroid use and dose are guided by the sputum eosinophil count results in a lower frequency of exacerbations and more economical use of corticosteroids than management guided by traditional clinical measures (Fig. 42.4). The findings of these studies, together with evidence that a sputum eosinophilia is an important and independent predictor of the occurrence of an asthma exacerbation following corticosteroid withdrawal [85, 86], suggest that eosinophilic airway inflammation is a more valid surrogate marker of preventable exacerbation risk than other currently available markers.

Fourthly, there is a clear and consistent positive correlation between a raised sputum neutrophil count and the presence of fixed airflow obstruction as reflected by a reduced post-bronchodilator FEV_1 and FEV_1/FVC. This is seen in patients with asthma [87] and COPD [88] and in COPD a raised sputum neutrophil count is associated with longitudinal decline in FEV_1 [33]. These findings raise the possibility of a common mechanism for the development of fixed airflow obstruction involving neutrophilic airway inflammation. Much more information is required before we can be sure that there is a causal link between neutrophilic airway inflammation and fixed airflow obstruction. In particular, there is a need for longitudinal studies evaluating the effect of reducing neutrophilic airway inflammation. Nevertheless, there is sufficient evidence to speculate that a raised sputum neutrophil count is a useful biomarker of risk of subsequent decline in lung function.

The new understanding of the importance of airway inflammation in airway disease has opened the way for a new approach to the management of airways disease in clinical practice where assessment of airway inflammation

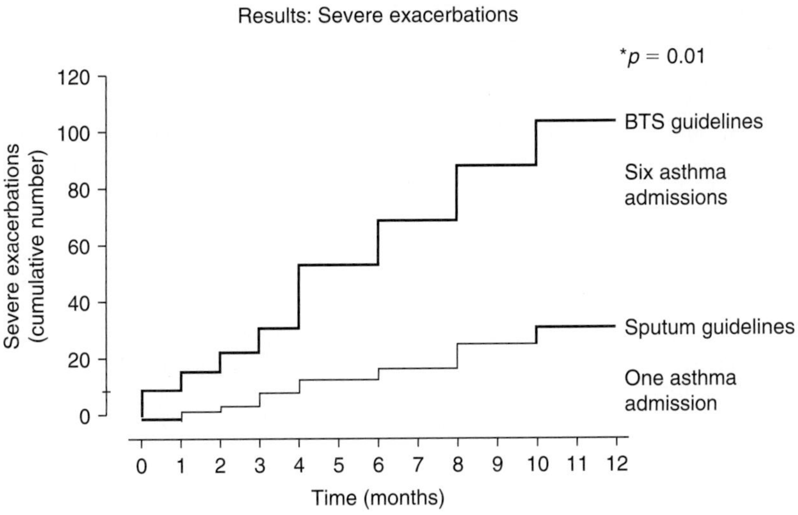

FIG. 42.4 Comparison of effects of two treatment strategies on rates of severe exacerbations of asthma. One strategy (British Thoracic Society (BTS) guidelines) utilized standard guidelines of the BTS and the other (sputum guidelines) adjusted the anti-inflammatory treatment with corticosteroids based on the eosinophil counts. Reproduced from Green *et al.* [6].

is used to identify risk of future adverse outcome and likely response to treatment.

FE_{NO} AS A MEANS OF MONITORING AIRWAY INFLAMMATION

The main limitations of induced sputum are that the technique is labor intensive, it requires a significant amount of training to get reliable results and results are not available immediately, limiting the application of the technique in disease monitoring. FE_{NO} has the advantages of being simple to measure and providing an immediate result making it much more suitable for a monitoring tool.

FE_{NO} is a reasonably robust measure of the presence of eosinophilic airway inflammation across a wide range of patients differing in atopic status, diagnosis, and medication [53]. The association is lost in current smokers [53]. As for sputum eosinophils, it has been shown that a raised FE_{NO} level is a reliable indicator of a positive response to corticosteroids in a heterogeneous population of patients with symptoms suggesting airway disease [89]. This finding was independent of the clinical diagnosis at presentation, in particular the label of asthma. One corticosteroid reduction study has shown that a raised F_ENO is predictive of loss of asthma control [50] but two others have shown less convincing evidence for such an effect [86, 90]. The relationship between airway eosinophilia, FE_{NO} levels, steroid responsiveness, and exacerbation risk raises the possibility that FE_{NO} can be used as a guide to corticosteroid dose requirements. This has been investigated in a number of small studies in different populations [91–93]. The findings show a consistent trend to reduction in asthma exacerbation of around 25%, in one study this was achieved with a 40% lower mean daily dose of ICS [91]. More definitively, studies have shown that FE_{NO} levels may be used to identify which patients do and do not require long-term maintenance therapy with inhaled steroids: the predictive values for either high (>50 ppb) or low (<25 ppb) FE_{NO} levels are highly significant and may be used with confidence to guide decision-making [94, 95]. More needs to be learnt about the optimum cut-points for FE_{NO} as a marker of corticosteroid responsive airway disease and as a tool for asthma monitoring, but the evidence to date suggests that the use of regular FE_{NO} measurements is a promising way to achieve more economical and effective use of ICS [96].

POTENTIAL CLINICAL ROLE OF INFLAMMOMETRY

Role in diagnosis

None of the currently available diagnostic tests are sufficiently sensitive to rule out asthma [97, 98] with the result that treatment trials are often instigated without good evidence variable airflow obstruction, airway hyperresponsiveness or airway inflammation. One study has shown that out of 263 subjects referred to a tertiary referral center with suspected asthma, 160 received an alternative diagnosis [99]. Many of these had received prolonged treatment with potentially toxic therapy before the correct diagnosis was reached. Even in tertiary referral centers the diagnosis of refractory asthma can be difficult to make with certainty [100]. The presence of a sputum eosinophilia or raised FE_{NO} in asthma is sufficiently common to suggest that these findings may have a role in the diagnosis of asthma. Hunter *et al.* [98] showed that the sensitivity of a sputum eosinophil count outside the normal range in identifying asthma (defined as consistent symptoms with objective evidence of abnormal variable airflow obstruction) was around 80%, significantly better than PEF amplitude % mean and the acute bronchodilator response and approached the sensitivity and specificity of measurement of airway responsiveness. Smith *et al.* [97] have reported similar findings; in this study a high FE_{NO} concentration achieved a similarly high diagnostic accuracy. Similar results have been reported by Berkman *et al.* [101] and Dupont *et al.* [102]. FE_{NO} performed less well in a community study [103], possibly because the diagnosis of asthma was less rigorous and because subjects were tested whilst taking ICS.

Arguably it matters more to patients what can be done to help them than the diagnostic label attached to them. Thus, testing strategies that identify patients who are going to respond well to corticosteroid and provide guidance on the dosing of corticosteroids might be particularly helpful. There is now compelling evidence that the presence of a sputum eosinophilia or a raised FE_{NO} predicts corticosteroid responsiveness. This was first clearly demonstrated by Morrow-Brown in the 1950s who showed that patients with airway disease and a sputum eosinophilia responded to corticosteroid treatment whereas those without a sputum eosinophilia did not [104]. It has since been shown that patients with non-eosinophilic asthma respond less well to inhaled budesonide than a group with more typical sputum features [26, 105]. This is also the case with longer-term corticosteroid treatment in patients with more severe asthma [6]. A sputum eosinophilia or raised FE_{NO} [91] is a predictor of a steroid response irrespective of the clinical context: patients with chronic cough respond well to inhaled corticosteroids if there is a sputum eosinophilia [80, 106] and patients with COPD with a sputum eosinophilia respond better to systemic and ICS than those without [77, 81, 107].

Role in monitoring asthma

Once a decision is made to start corticosteroid treatment the next question is how best to titrate this therapy. Traditionally dose titration is done by assessing the clinical response to treatment and attempting to define the lowest dose of ICS that maintains this. Many patients do very well with this approach [108] but there is evidence that the use of the induced sputum eosinophil count to titrate therapy results in a lower exacerbation frequency with no overall increase in treatment, particularly in patients with more severe asthma (Fig. 42.4) [6, 83]. A number of researchers have investigated whether a similar effect is seen using FE_{NO} as the monitoring tool (Table 42.6). Smith *et al.* [91] used FE_{NO} measurements at a flow of 250ml/s to titrate ICS in a single blind parallel group trial involving 92 subjects with asthma. Subjects were treated according to their FE_{NO} measurements or Global Initiative for Asthma (GINA) guidelines. Following a run in period the steroid dose was reduced in the FE_{NO} group if the FE_{NO} was <15 ppb (equivalent to an FE_{NO} of about 45 ppb using an expiratory flow of 50 ml/s). In the control group, steroid reduction was based on current GINA guidelines and only occurred when the subjects had, over the course of the previous week, achieved all of the following: less than 2 night-time awakenings, a mean peak flow amplitude of <20%, bronchodilator use <4 times on 1 or 2 days, minimal asthma symptoms, and a FEV_1 >90% predicted.

TABLE 42.6 Summary of studies investigation FE_{NO} as a tool for monitoring corticosteroid treatment in asthma.

Study	Design	Population	Duration	Intervention	Reduction in exacerbation frequency	Change in ICS dose
Smith *et al.* [91]	Single blind, parallel group	97 adults with mild-moderate asthma	12 months, after an optimization phase of up to 12 months	Control: conventional measures of asthma control Intervention: FE_{NO} < 15* ppb	0.49 moderate and severe exacerbation/ patient/year in intervention and 0.9 in control ($p = 0.27$)	40% lower in intervention group ($p = 0.003$)
Pijnenburg *et al.* [92]	Single blind, parallel group	85 children with atopic asthma	12 months	Control: control of symptoms Intervention: Control of symptoms and FE_{NO} < 30 ppb	Eight courses of prednisolone in intervention and 18 in control; $p = 0.6$	Cumulative dose over 5 visits of 4407 µg in intervention, 4332 µg in control
Shaw *et al.* [93]	Single blind, parallel group	118 adults with a primary care diagnosis of asthma	12 months	Control: control of symptoms Intervention: FE_{NO} < 26 ppb	0.42 severe exacerbation/ patient/year in control and 0.33 in intervention ($p = 0.4$)	12% higher in intervention group ($p = 0.4$)

Note: ICS: inhaled corticosteroid.
*Measured at 250 ml/s.

In the FE_{NO} group the optimal dose was one dose above the dose at which the subjects FE_{NO} was >15 ppb. In the control group the optimal dose was one dose above the dose at which a loss of control had occurred. These became the optimal doses at which subjects entered the final year of the study. During this stage the steroid dose was increased in the FE_{NO} group if the FE_{NO} was >15 ppb and in the control group the optimal dose was increased if a loss of control occurred. Although there were fewer exacerbations in the FE_{NO} group, this did not reach statistical significance. However, over the year of the study the FE_{NO} managed group used 40% less ICS.

A similar study in a pediatric population showed a significant improvement in airway responsiveness in children whose FE_{NO} was maintained below 30 ppb when compared to a population whose corticosteroid dose was titrated using traditional measures [92]. Few exacerbations were seen with either management strategy and there was no overall reduction in corticosteroid dose.

Shaw *et al.* [93] took a different approach in a single blind parallel group trial of 118 adults with asthma managed in primary care. In the control group the goal of treatment was to maintain the Juniper asthma control score below 1.57 [109]; in the intervention group the aim was to achieve this and to maintain the FE_{NO} between 16 and 26 ppb. Exacerbation frequency assessed over 12 months was 21% lower ($p = 0.4$) in the intervention group and there was no overall difference in corticosteroid dose over the year. The study was underpowered to show a clinically important effect on exacerbation frequency although the effect on severe exacerbation frequency was similar to that seen by Smith *et al.*

The results of corticosteroid dose–titration studies with FE_{NO} are therefore mixed. There is a fairly consistent finding of a small reduction in severe asthma exacerbations in patients managed with reference to FE_{NO} (Table 42.6) but a much larger study will be needed to determine whether such an effect is real, and whether it can be achieved cost effectively and with no overall increase in corticosteroid therapy.

The induced sputum studies showing a reduction in exacerbation frequency have generally included a more severe patient population than the FE_{NO} studies and it is possible that more clear evidence of benefit of FE_{NO} monitoring would be seen in this population. In support of this, Jayaram *et al.* [83] noted that the benefit of sputum eosinophil directed management was much more clearly seen in patients with more severe asthma. Patients taking long-acting β_2-agonists did particularly well and it is possible that the increased dissociation between eosinophilic airway inflammation and symptoms seen after treatment with long-acting β_2-agonists [110] is an important determinate of this. FE_{NO} is measurable in most patients with severe asthma and it appears to be as good a marker of eosinophilic airway inflammation as it is in less severe asthma [111], so future studies should evaluate its use in severe asthma. In a

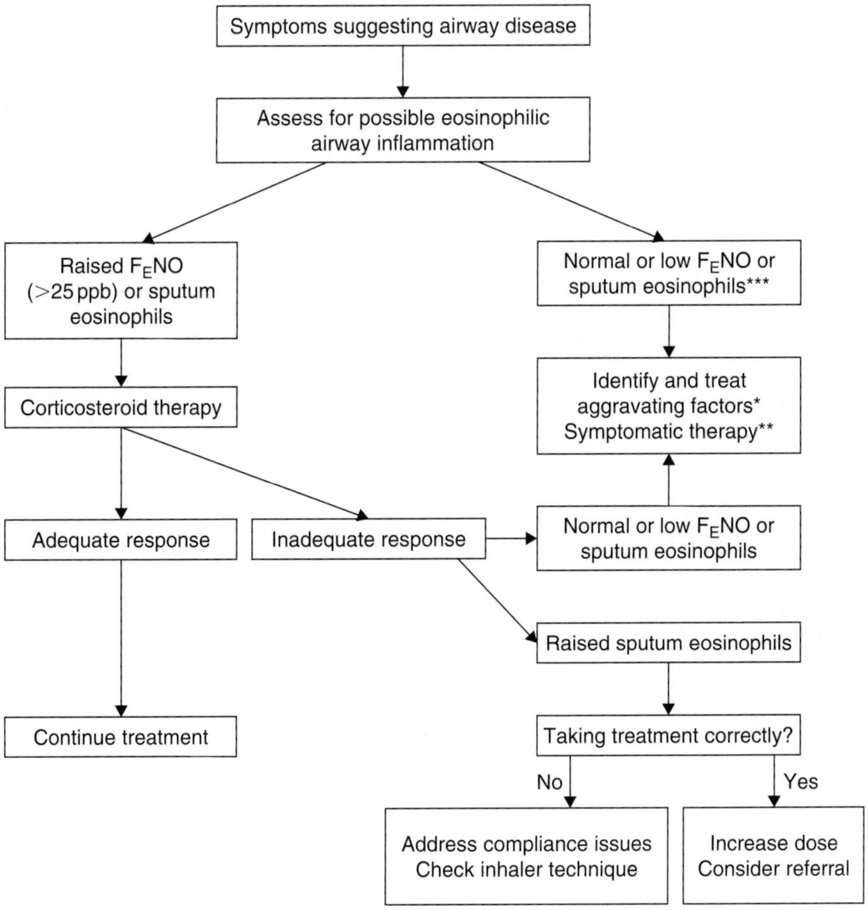

FIG. 42.5 Suggested algorithm for assessment of patients presenting with untreated airway disease. Eosinophilic airway inflammation can be assessed using induced sputum eosinophils or FE_{NO}. *Potentially treatable aggravating factors include rhinitis, anxiety–hyperventilation syndrome, vocal cord dysfunction, bronchiectasis, and gastroesophageal reflux disease. **Symptomatic therapy includes short- and long-acting bronchodilator therapy, oral theophylline, and mucolytics as well as specific treatments for the aggravating factors. ***The suggested normal range for sputum eosinophils is <2% and FE_{NO} in adults is <25 ppb. FE_{NO} > 50 ppb is very predictive of a response to corticosteroid and a relapse when treatment is withdrawn; the interpretation of FE_{NO} in the range 25–50 ppb varies between individual patients, and in some individuals different cut-points may apply. The induced sputum eosinophil count is likely to be a more reliable indicator of corticosteroid response and relapse risk when FE_{NO} is in this range.

TABLE 42.7 An approach to monitoring ICS treatment in patients with airway disease using FE_{NO}.

FE_{NO}	Low	Normal	Intermediate	High
Ppb	<5	5–25	25–50	>50
Symptoms controlled	Decrease/stop ICS	Decrease/stop ICS	No change	Increase ICS/refer
Symptoms uncontrolled	Decrease ICS, increase bronchodilators. Investigate for PCD and CF	Decrease/stop ICS, increase bronchodilators	Assess compliance. Increase ICS	Assess compliance. Increase ICS/refer

Notes: ppb: parts per billion; PCD: primary ciliary dyskinesia; CF: cystic fibrosis.

less severe population the main impact of inflammometry might be a reduction in corticosteroid dose without loss of efficacy.

Trials evaluating FE_{NO} as a monitoring tool for asthma have used a wide variety of target ranges for FE_{NO} (Table 42.6). Since than there has been an increase in our understanding of the relationship between FE_{NO} and eosinophilic airway inflammation an there is now a consensus that an FE_{NO} of <25 ppb is a sensitive indicator of the absence of eosinophilic airway inflammation [93, 96] and is therefore an appropriate cut off for a reduction in ICS [112]. Conversely, an FE_{NO} above 50 ppb is a specific marker for the presence of eosinophilic airway inflammation [93, 96] and a good response to corticosteroids is consistently seen above this value [89]. These therefore would appear to be the most appropriate values for a reduction and an increase in corticosteroid dose. Values lying between 25 and 50 ppb are indeterminate and clinical criteria should be used to guide corticosteroid treatment. A suggested approach to interpretation of FE_{NO} values in the diagnosis and monitoring of asthma is outlined in Fig. 42.5 and Table 42.7.

FUTURE DIRECTIONS

Recently it has become possible to investigate the contribution of the distal lung in the inflammatory process in asthma and other airway diseases by measurement of FE_{NO} at different expiratory flows [38]. Lehtimaki *et al.* have used this method to demonstrate that patients with nocturnal symptoms have elevated alveolar NO concentrations [113] and that alveolar NO concentration does not fall in response to treatment with ICS [114]. Patients with refractory asthma have a raised alveolar NO concentration which is reduced by oral but not ICS [39] suggesting that it reflects corticosteroid responsive inflammation in a site that can't be accessed by ICS. Induced sputum may also have a role in investigating proximal and distal airway inflammation as there is evidence that distal airways are sampled with increasing output and duration of nebulized hypertonic saline [15]. The clinical utility of these methods has not been explored. They could provide useful information on the need for a systemic rather than an inhaled approach to anti-inflammatory treatment with agents such as prednisolone, anti-IgE, and perhaps anti-IL-5.

There is an urgent need for a non-invasive technique capable of providing accurate information on the characteristics of the lower airway inflammatory response from an easily accessible sample. The technique should be capable of providing an immediate result and be useable in acute as well as outpatient settings. Whether FE_{NO} or EBC analysis as it currently stands are ideal for this purpose is doubtful. Refinements of breath analysis, or perhaps identification of novel biomarkers in urine or blood are perhaps most likely to advance this area. Work already done with induced sputum provide a strong scientific rationale for the use of these markers as a means of defining risk and likely treatment response.

References

1. Paggiaro PL, Chanez P, Holz O, Ind PW, Djukanovic R, Maestrelli P *et al.* Sputum induction. *Eur Respir J Suppl* 37: 3s–s8s, 2002.
2. Hunter CJ, Ward R, Woltmann G, Wardlaw AJ, Pavord ID. The safety and success rate of sputum induction using a low output ultrasonic nebuliser. *Respir Med* 93(5): 345–48, 1999.
3. de la Fuente PT, Romagnoli M, Godard P, Bousquet J, Chanez P. Safety of inducing sputum in patients with asthma of varying severity. *Am J Respir Crit Care Med* 157(4 Pt 1): 1127–30, 1998.
4. ten Brinke A, de Lange C, Zwinderman AH, Rabe KF, Sterk PJ, Bel EH. Sputum induction in severe asthma by a standardized protocol: Predictors of excessive bronchoconstriction. *Am J Respir Crit Care Med* 164(5): 749–53, 2001 September 1.
5. Brightling CE, Monterio W, Green RH, Parker D, Morgan MD, Wardlaw AJ *et al.* Induced sputum and other outcome measures in chronic obstructive pulmonary disease: Safety and repeatability. *Respir Med* 95(12): 999–1002, 2001.
6. Green RH, Brightling CE, McKenna S, Hargadon B, Parker D, Bradding P *et al.* Asthma exacerbations and sputum eosinophil counts: A randomised controlled trial. *Lancet* 360(9347): 1715–21, 2002 November 30.
7. Pavord ID. Sputum induction to assess airway inflammation: Is it an inflammatory stimulus?. *Thorax* 53(2): 79–80, 1998 February 1.
8. Green RH, Brightling CE, Woltmann G, Parker D, Wardlaw AJ, Pavord ID. Analysis of induced sputum in adults with asthma: Identification of subgroup with isolated sputum neutrophilia and poor response to inhaled corticosteroids. *Thorax* 57(10): 875–79, 2002.
9. Efthimiadis A, Jayaram L, Weston S, Carruthers S, Hargreave FE. Induced sputum: Time from expectoration to processing. *Eur Respir J* 19(4): 706–8, 2002.
10. Ward R, Woltmann G, Wardlaw AJ, Pavord ID. Between-observer repeatability of sputum differential cell counts. Influence of cell viability and squamous cell contamination. *Clin Exp Allergy* 29(2): 248–52, 1999.

11. Alexis N, Soukup J, Ghio A, Becker S. Sputum phagocytes from healthy individuals are functional and activated: A flow cytometric comparison with cells in bronchoalveolar lavage and peripheral blood. *Clin Immunol* 97(1): 21–32, 2000.
12. Moodley YP, Krishnan V, Lalloo UG. Neutrophils in induced sputum arise from central airways. *Eur Respir J* 15(1): 36–40, 2000.
13. Pizzichini E, Pizzichini MM, Kidney JC, Efthimiadis A, Hussack P, Popov T *et al.* Induced sputum, bronchoalveolar lavage and blood from mild asthmatics: Inflammatory cells, lymphocyte subsets and soluble markers compared. *Eur Respir J* 11(4): 828–34, 1998.
14. Alexis NE, Hu SC, Zeman K, Alter T, Bennett WD. Induced sputum derives from the central airways: Confirmation using a radiolabeled aerosol bolus delivery technique. *Am J Respir Crit Care Med* 164 (10 Pt 1): 1964–70, 2001.
15. Gershman NH, Liu H, Wong HH, Liu JT, Fahy JV. Fractional analysis of sequential induced sputum samples during sputum induction: Evidence that different lung compartments are sampled at different time points. *J Allergy Clin Immunol* 104(2 Pt 1): 322–28, 1999.
16. Keatings VM, Evans DJ, O'Connor BJ, Barnes PJ. Cellular profiles in asthmatic airways: A comparison of induced sputum, bronchial washings, and bronchoalveolar lavage fluid. *Thorax* 52(4): 372–74, 1997.
17. Maestrelli P, Saetta M, Di SA, Calcagni PG, Turato G, Ruggieri MP *et al.* Comparison of leukocyte counts in sputum, bronchial biopsies, and bronchoalveolar lavage. *Am J Respir Crit Care Med* 152(6 Pt 1): 1926–31, 1995.
18. Belda Jose, Leigh Rich, Parameswaran Kris, O'Byrne PM, Sears MR, Hargreave FE. Induced sputum cell counts in healthy adults. *Am J Respir Crit Care Med* 161(2): 475–78, 2000 February 1.
19. Spanevello A, Confalonieri M, Sulotto F, Romano F, Balzano G, Migliori GB *et al.* Induced sputum cellularity. Reference values and distribution in normal volunteers. *Am J Respir Crit Care Med* 162 (3 Pt 1): 1172–74, 2000.
20. Thomas RA, Green RH, Brightling CE, Birring SS, Parker D, Wardlaw AJ *et al.* The influence of age on induced sputum differential cell counts in normal subjects. *Chest* 126(6): 1811–14, 2004.
21. in't Veen JC, de Gouw HW, Smits HH, Sont JK, Hiemstra PS, Sterk PJ *et al.* Repeatability of cellular and soluble markers of inflammation in induced sputum from patients with asthma. *Eur Respir J* 9(12): 2441–47, 1996.
22. Pizzichini E, Pizzichini MM, Efthimiadis A, Evans S, Morris MM, Squillace D *et al.* Indices of airway inflammation in induced sputum: Reproducibility and validity of cell and fluid-phase measurements. *Am J Respir Crit Care Med* 154(2 Pt 1): 1138–44, 1996.
23. Spanevello A, Migliori GB, Sharara A, Ballardini L, Bridge P, Pisati P *et al.* Induced sputum to assess airway inflammation: a study of reproducibility. *Clin Exp Allergy* 10: 1138–44, 1997.
24. Lemiere C, Pizzichini MM, Balkissoon R, Clelland L, Efthimiadis A, O'Shaughnessy D *et al.* Diagnosing occupational asthma: Use of induced sputum. *Eur Respir J* 13(3): 482–88, 1999.
25. Pizzichini MM, Kidney JC, Wong BJ, Morris MM, Efthimiadis A, Dolovich J *et al.* Effect of salmeterol compared with beclomethasone on allergen-induced asthmatic and inflammatory responses. *Eur Respir J* 9(3): 449–55, 1996.
26. Pavord ID, Brightling CE, Woltmann G, Wardlaw AJ. Non-eosinophilic corticosteroid unresponsive asthma. *Lancet* 353(9171): 2213–14, 1999 June 26.
27. Bentley AM, Hamid Q, Robinson DS, Schotman E, Meng Q, Assoufi B *et al.* Prednisolone treatment in asthma. Reduction in the numbers of eosinophils, T cells, tryptase-only positive mast cells, and modulation of IL-4, IL-5, and interferon-gamma cytokine gene expression within the bronchial mucosa. *Am J Respir Crit Care Med* 153(2): 551–56, 1996.
28. Flood-Page P, Menzies-Gow A, Phipps S, Ying S, Wangoo A, Ludwig MS *et al.* Anti-IL-5 treatment reduces deposition of ECM proteins in the bronchial subepithelial basement membrane of mild atopic asthmatics. *J Clin Invest* 112(7): 1029–36, 2003.
29. Leckie MJ, ten BA, Khan J, Diamant Z, O'Connor BJ, Walls CM *et al.* Effects of an interleukin-5 blocking monoclonal antibody on eosinophils, airway hyper-responsiveness, and the late asthmatic response. *Lancet* 356(9248): 2144–48, 2000 December 23.
30. Barnes NC, Qiu YS, Pavord ID, Parker D, Davis PA, Zhu J *et al.* Antiinflammatory effects of salmeterol/fluticasone propionate in chronic obstructive lung disease. *Am J Respir Crit Care Med* 173(7): 736–43, 2006 April 1.
31. Neale N, Parker D, Barlow S, Green RH, Brightling CE, Pavord ID. The relationship between total and differential cell counts in induced sputum. *Eur Respir J* 20: 274s, 2002.
32. Pavord ID, Sterk PJ, Hargreave FE, Kips JC, Inman MD, Louis R *et al.* Clinical applications of assessment of airway inflammation using induced sputum. *Eur Respir J* 37(Suppl.): 40s–43s, 2002.
33. Stanescu D, Sanna A, Veriter C, Kostianev S, Calcagni PG, Fabbri LM *et al.* Airways obstruction, chronic expectoration, and rapid decline of FEV1 in smokers are associated with increased levels of sputum neutrophils. *Thorax* 51(3): 267–71, 1996.
34. Lundberg JO, Weitzberg E, Lundberg JM, Alving K. Nitric oxide in exhaled air. *Eur Respir J* 9(12): 2671–80, 1996.
35. ATS/ERS Recommendations for Standardized Procedures for the Online and Offline Measurement of Exhaled Lower Respiratory Nitric Oxide and Nasal Nitric Oxide, 2005. *Am J Respir Crit Care Med* 171(8): 912–30, 15 April 2005.
36. Gruffyd-Jones K, Ward S, Stonham C, Macfarlane TV, Thomas M. The use of exhaled nitric oxide monitoring in primary care asthma clinics: A pilot study. *Prim Care Respir J* 16: 349–56, 2007.
37. Kharitonov SA, Gonio F, Kelly C, Meah S, Barnes PJ. Reproducibility of exhaled nitric oxide measurements in healthy and asthmatic adults and children. *Eur Respir J* 21(3): 433–38, 2003.
38. Tsoukias NM, George SC. A two-compartment model of pulmonary nitric oxide exchange dynamics. *J Appl Physiol* 85(2): 653–66, 1998.
39. Berry M, Hargadon B, Morgan A, Shelley M, Richter J, Shaw D *et al.* Alveolar nitric oxide in adults with asthma: Evidence of distal lung inflammation in refractory asthma. *Eur Respir J* 25(6): 986–91, 2005.
40. Buchvald F, Baraldi E, Carraro S, Gaston B, De JJ, Pijnenburg MW *et al.* Measurements of exhaled nitric oxide in healthy subjects age 4 to 17 years. *J Allergy Clin Immunol* 115(6): 1130–36, 2005.
41. Kharitonov SA, Logan-Sinclair RB, Busset CM, Shinebourne EA. Peak expiratory nitric oxide differences in men and women: Relation to the menstrual cycle. *Br Heart J* 72(3): 243–45, 1994.
42. van's Gravesande KS, Wechsler ME, Grasemann H, Silverman ES, Le L, Palmer LJ *et al.* Association of a missense mutation in the NOS3 gene with exhaled nitric oxide levels. *Am J Respir Crit Care Med* 168(2): 228–31, 2003 July 15.
43. Henriksen AH, Sue-Chu M, Lingaas HT, Langhammer A, Bjermer L. Exhaled and nasal NO levels in allergic rhinitis: Relation to sensitization, pollen season and bronchial hyperresponsiveness. *Eur Respir J* 13(2): 301–6, 1999.
44. Jatakanon A, Kharitonov S, Lim S, Barnes PJ. Effect of differing doses of inhaled budesonide on markers of airway inflammation in patients with mild asthma. *Thorax* 54(2): 108–14, 1999.
45. Massaro AF, Gaston B, Kita D, Fanta C, Stamler JS, Drazen JM. Expired nitric oxide levels during treatment of acute asthma. *Am J Respir Crit Care Med* 152(2): 800–3, 1995.
46. Kharitonov SA, Yates D, Robbins RA, Logan-Sinclair R, Shinebourne EA, Barnes PJ. Increased nitric oxide in exhaled air of asthmatic patients. *Lancet* 343(8890): 133–35, 1994 January 15.
47. Kharitonov SA, O'Connor BJ, Evans DJ, Barnes PJ. Allergen-induced late asthmatic reactions are associated with elevation of exhaled nitric oxide. *Am J Respir Crit Care Med* 151(6): 1894–99, 1995.
48. Jones SL, Herbison P, Cowan JO, Flannery EM, Hancox RJ, McLachlan CR *et al.* Exhaled NO and assessment of anti-inflammatory effects of inhaled steroid: Dose–response relationship. *Eur Respir J* 20(3): 601–8, 2002.

49. Kharitonov SA, Donnelly LE, Montuschi P, Corradi M, Collins JV, Barnes PJ. Dose-dependent onset and cessation of action of inhaled budesonide on exhaled nitric oxide and symptoms in mild asthma. *Thorax* 57(10): 889–96, 2002.
50. Jones SL, Kittelson J, Cowan JO, Flannery EM, Hancox RJ, McLachlan CR *et al.* The predictive value of exhaled nitric oxide measurements in assessing changes in asthma control. *Am J Respir Crit Care Med* 164(5): 738–43, 2001 September 1.
51. Adrie C, Monchi M, Dinh-Xuan AT, Dall'Ava-Santucci J, Dhainaut JF, Pinsky MR. Exhaled and nasal nitric oxide as a marker of pneumonia in ventilated patients. *Am J Respir Crit Care Med* 163(5): 1143–49, 2001.
52. Balfour-Lynn IM, Laverty A, Dinwiddie R. Reduced upper airway nitric oxide in cystic fibrosis. *Arch Dis Child* 75(4): 319–22, 1996.
53. Berry MA, Shaw DE, Green RH, Brightling CE, Wardlaw AJ, Pavord ID. The use of exhaled nitric oxide concentration to identify eosinophilic airway inflammation: An observational study in adults with asthma. *Clin Exp Allergy* 35(9): 1175–79, 2005.
54. Horvath I, Hunt J, Barnes PJ, Alving K, Antczak A, Baraldi E *et al.* Exhaled breath condensate: Methodological recommendations and unresolved questions. *Eur Respir J* 26(3): 523–48, 2005.
55. Montuschi P, Barnes PJ. Exhaled leukotrienes and prostaglandins in asthma. *J Allergy Clin Immunol* 109(4): 615–20, 2002.
56. Mondino C, Ciabattoni G, Koch P, Pistelli R, Trove A, Barnes PJ *et al.* Effects of inhaled corticosteroids on exhaled leukotrienes and prostanoids in asthmatic children. *J Allergy Clin Immunol* 114(4): 761–67, 2004.
57. Jobsis Q, Raatgeep HC, Schellekens SL, Hop WC, Hermans PW, de Jongste JC. Hydrogen peroxide in exhaled air of healthy children: Reference values. *Eur Respir J* 12(2): 483–85, 1998.
58. Antczak A, Nowak D, Shariati B, Krol M, Piasecka G, Kurmanowska Z. Increased hydrogen peroxide and thiobarbituric acid-reactive products in expired breath condensate of asthmatic patients. *Eur Respir J* 10(6): 1235–41, 1997.
59. Horvath I, Donnelly LE, Kiss A, Balint B, Kharitonov SA, Barnes PJ. Exhaled nitric oxide and hydrogen peroxide concentrations in asthmatic smokers. *Respiration* 71(5): 463–68, 2004.
60. Dohlman AW, Black HR, Royall JA. Expired breath hydrogen peroxide is a marker of acute airway inflammation in pediatric patients with asthma. *Am Rev Respir Dis* 148(4 Pt 1): 955–60, 1993.
61. Montuschi P, Corradi M, Ciabattoni G, Nightingale J, Kharitonov SA, Barnes PJ. Increased 8-isoprostane, a marker of oxidative stress, in exhaled condensate of asthma patients. *Am J Respir Crit Care Med* 160(1): 216–20, 1999.
62. Paredi P, Ward S, Cramer D, Barnes PJ, Kharitonov SA. A new method for the non-invasive measurement of bronchial blood flow. *Am J Respir Crit Care Med* 167: A448, 2003.
63. Paredi P, Kharitonov SA, Barnes PJ. Faster rise of exhaled breath temperature in asthma. A novel marker of airway inflammation?. *Am J Respir Crit Care Med* 165(2): 181–84, 2002 January 15.
64. Paredi P, Caramori G, Cramer D, Ward S, Ciaccia A, Papi A *et al.* Slower rise of exhaled breath temperature in chronic obstructive pulmonary disease. *Eur Respir J* 21(3): 439–43, 2003.
65. Venge P, Bystrom J, Carlson M, Hakansson L, Karawacjzyk M, Peterson C *et al.* Eosinophil cationic protein (ECP): Molecular and biological properties and the use of ECP as a marker of eosinophil activation in disease. *Clin Exp Allergy* 29(9): 1172–86, 1999.
66. Kips JC, Pauwels RA. Serum eosinophil cationic protein in asthma: What does it mean?. *Clin Exp Allergy* 28(1): 1–3, 1998.
67. Currie GP, Syme-Grant NJ, McFarlane LC, Carey FA, Lipworth BJ. Effects of low dose fluticasone/salmeterol combination on surrogate inflammatory markers in moderate persistent asthma. *Allergy* 58(7): 602–7, 2003.
68. Aldridge RE, Hancox RJ, Cowant JO, Frampton CM, Town GI, Taylor DR. Eosinophils and eosinophilic cationic protein in induced sputum and blood: Effects of budesonide and terbutaline treatment. *Ann Allergy Asthma Immunol* 89(5): 492–97, 2002.
69. Visser MJ, Postma DS, Brand PL, Arends LR, Duiverman EJ, Kauffman HF. Influence of different dosage schedules of inhaled fluticasone propionate on peripheral blood cytokine concentrations in childhood asthma. *Clin Exp Allergy* 32(10): 1497–503, 2002.
70. Meijer RJ, Postma DS, Kauffman HF, Arends LR, Koeter GH, Kerstjens HA. Accuracy of eosinophils and eosinophil cationic protein to predict steroid improvement in asthma. *Clin Exp Allergy* 32(7): 1096–103, 2002.
71. Lowhagen O, Wever AM, Lusuardi M, Moscato G, De Backer WA, Gandola L *et al.* The inflammatory marker serum eosinophil cationic protein (ECP) compared with PEF as a tool to decide inhaled corticosteroid dose in asthmatic patients. *Respir Med* 96(2): 95–101, 2002.
72. Djukanovic R, Roche WR, Wilson JW, Beasley CR, Twentyman OP, Howarth RH *et al.* Mucosal inflammation in asthma. *Am Rev Respir Dis* 142(2): 434–57, 1990.
73. Rosi E, Ronchi MC, Grazzini M, Duranti R, Scano G. Sputum analysis, bronchial hyperresponsiveness, and airway function in asthma: Results of a factor analysis. *J Allergy Clin Immunol* 103(2 Pt 1): 232–37, 1999.
74. Crimi E, Spanevello A, Neri M, Ind PW, Rossi GA, Brusasco V. Dissociation between airway inflammation and airway hyperresponsiveness in allergic asthma. *Am J Respir Crit Care Med* 157(1): 4–9, 1998.
75. Gibson PG, Fujimura M, Niimi A. Eosinophilic bronchitis: Clinical manifestations and implications for treatment. *Thorax* 57(2): 178–82, 2002.
76. Brightling CE, Ward R, Goh KL, Wardlaw AJ, Pavord ID. Eosinophilic bronchitis is an important cause of chronic cough. *Am J Respir Crit Care Med* 160(2): 406–10, 1999.
77. Brightling CE, Monteiro W, Ward R, Parker D, Morgan MD, Wardlaw AJ *et al.* Sputum eosinophilia and short-term response to prednisolone in chronic obstructive pulmonary disease: A rando mised controlled trial. *Lancet* 356(9240): 1480–85, 2000.
78. Kips JC, O'Connor BJ, Langley SJ, Woodcock A, Kerstjens HA, Postma DS *et al.* Effect of SCH55700, a humanized anti-human interleukin-5 antibody, in severe persistent asthma: A pilot study. *Am J Respir Crit Care Med* 167(12): 1655–59, 2003.
79. Brightling CE, Bradding P, Symon FA, Holgate ST, Wardlaw AJ, Pavord ID. Mast-cell infiltration of airway smooth muscle in asthma. *N Engl J Med* 346(22): 1699–705, 2002.
80. Brightling CE, Ward R, Wardlaw AJ, Pavord ID. Airway inflammation, airway responsiveness and cough before and after inhaled budesonide in patients with eosinophilic bronchitis. *Eur Respir J* 15(4): 682–86, 2000.
81. Pizzichini E, Pizzichini MM, Gibson P, Parameswaran K, Gleich GJ, Berman L *et al.* Sputum eosinophilia predicts benefit from prednisone in smokers with chronic obstructive bronchitis. *Am J Respir Crit Care Med* 158(5 Pt 1): 1511–17, 1998.
82. Berry MA, Morgan A, Shaw DE, Parker D, Bradding P, Green RH *et al.* Pathological features and inhaled corticosteroid response of eosinophilic and non-eosinophilic asthma. *Thorax*, in press 2007.
83. Jayaram L, Pizzichini MM, Cook RJ, Boulet LP, Lemiere C, Pizzichini E *et al.* Determining asthma treatment by monitoring sputum cell counts: Effect on exacerbations. *Eur Respir J* 27(3): 483–94, 2006.
84. Siva R, Green RH, Brightling CE, Shelley M, Hargadon B, McKenna S *et al.* Eosinophilic airway inflammation and exacerbations of COPD: A randomised controlled trial. *Eur Respir J* 29(5): 906–13, 2007.
85. Jatakanon A, Lim S, Barnes PJ. Changes in sputum eosinophils predict loss of asthma control. *Am J Respir Crit Care Med* 161(1): 64–72, 2000.
86. Leuppi JD, Salome CM, Jenkins CR, Anderson SD, Xuan W, Marks GB *et al.* Predictive markers of asthma exacerbation during stepwise dose reduction of inhaled corticosteroids. *Am J Respir Crit Care Med* 163(2): 406–12, 2001.
87. Shaw DE, Berry MA, Hargadon B, McKenna S, Shelley MJ, Green RH *et al.* Association between neutrophilic airway inflammation and airflow limitation in adults with asthma. *Chest* 132(6): 1871–75, 2007.
88. Keatings VM, Collins PD, Scott DM, Barnes PJ. Differences in interleukin-8 and tumor necrosis factor-alpha in induced sputum from

patients with chronic obstructive pulmonary disease or asthma. *Am J Respir Crit Care Med* 153(2): 530–34, 1996.

89. Smith AD, Cowan JO, Brassett KP, Filsell S, McLachlan C, Monti-Sheehan G *et al.* Exhaled nitric oxide: A predictor of steroid response. *Am J Respir Crit Care Med* 172(4): 453–59, 2005.
90. Deykin A, Lazarus SC, Fahy JV, Wechsler ME, Boushey HA, Chinchilli VM *et al.* Sputum eosinophil counts predict asthma control after discontinuation of inhaled corticosteroids. *J Allergy Clin Immunol* 115(4): 720–27, 2005.
91. Smith AD, Cowan JO, Brassett KP, Herbison GP, Taylor DR. Use of exhaled nitric oxide measurements to guide treatment in chronic asthma. *N Engl J Med* 352(21): 2163–73, 2005.
92. Pijnenburg MW, Bakker EM, Hop WC, de Jongste JC. Titrating steroids on exhaled nitric oxide in children with asthma: A randomized controlled trial. *Am J Respir Crit Care Med* 172(7): 831–36, 2005.
93. Shaw DE, Berry MA, Thomas M, Green RH, Brightling CE, Wardlaw AJ *et al.* The use of exhaled nitric oxide to guide asthma management: A randomized controlled trial. *Am J Respir Crit Care Med* 176(3): 231–37, 2007.
94. Pijnenburg MW, Hofhuis W, Hop WC, de Jongste JC. Exhaled nitric oxide predicts asthma relapse in children with clinical asthma remission. *Thorax* 60(3): 215–18, 2005.
95. Zacharasiewicz A, Wilson N, Lex C, Erin EM, Li AM, Hansel T *et al.* Clinical use of non-invasive measurements of airway inflammation in steriod reduction in children. *Am J Respir Crit Care Med*, 2005.
96. Taylor DR, Pijnenburg MW, Smith AD, de Jongste JC. Exhaled nitric oxide measurements: Clinical application and interpretation. *Thorax* 61(9): 817–27, 2006.
97. Smith AD, Cowan JO, Filsell S, McLachlan C, Monti-Sheehan G, Jackson P *et al.* Diagnosing asthma: Comparisons between exhaled nitric oxide measurements and conventional tests. *Am J Respir Crit Care Med* 169(4): 473–78, 2004.
98. Hunter CJ, Brightling CE, Woltmann G, Wardlaw AJ, Pavord ID. A comparison of the validity of different diagnostic tests in adults with asthma. *Chest* 121(4): 1051–57, 2002.
99. Joyce DP, Chapman KR, Kesten S. Prior diagnosis and treatment of patients with normal results of methacholine challenge and unexplained respiratory symptoms. *Chest* 109(3): 697–701, 1996.
100. Robinson DS, Campbell DA, Durham SR, Pfeffer J, Barnes PJ, Chung KF. Systematic assessment of difficult-to-treat asthma. *Eur Respir J* 22(3): 478–83, 2003.
101. Berkman N, Avital A, Breuer R, Bardach E, Springer C, Godfrey S. Exhaled nitric oxide in the diagnosis of asthma: Comparison with bronchial provocation tests. *Thorax* 60(5): 383–88, 2005.
102. Dupont LJ, Demedts MG, Verleden GM. Prospective evaluation of the validity of exhaled nitric oxide for the diagnosis of asthma. *Chest* 123(3): 751–56, 2003.
103. Travers J, Marsh S, Aldington S, Williams M, Shirtcliffe P, Pritchard A *et al.* Reference ranges for exhaled nitric oxide derived from a random community survey of adults. *Am J Respir Crit Care Med* 176(3): 238–42, 2007.
104. Brown HM. Treatment of chronic asthma with prednisolone; significance of eosinophils in the sputum. *Lancet* 2(7059): 1245–47, 1958.
105. Berry M, Morgan A, Shaw DE, Parker D, Green R, Brightling C *et al.* Pathological features and inhaled corticosteroid response of eosinophilic and non-eosinophilic asthma. *Thorax* 62(12): 1043–49, 2007.
106. Pizzichini MM, Pizzichini E, Parameswaran K, Clelland L, Efthimiadis A, Dolovich J *et al.* Nonasthmatic chronic cough: No effect of treatment with an inhaled corticosteroid in patients without sputum eosinophilia. *Can Respir J* 6(4): 323–30, 1999.
107. Brightling CE, McKenna S, Hargadon B, Birring S, Green R, Siva R *et al.* Sputum eosinophilia and the short term response to inhaled mometasone in chronic obstructive pulmonary disease. *Thorax* 60(3): 193–98, 2005.
108. Bateman ED, Boushey HA, Bousquet J, Busse WW, Clark TJ, Pauwels RA *et al.* Can guideline-defined asthma control be achieved? The Gaining Optimal Asthma Control study. *Am J Respir Crit Care Med* 170(8): 836–44, 2004.
109. Juniper EF, Bousquet J, Abetz L, Bateman ED. Identifying 'well-controlled' and 'not well-controlled' asthma using the Asthma Control Questionnaire. *Respir Med* 100(4): 616–21, 2006.
110. Mcivor RA, Pizzichini E, Turner MO, Hussack P, Hargreave FE, Sears MR. Potential masking effects of salmeterol on airway inflammation in asthma. *Am J Respir Crit Care Med* 158(3): 924–30, 1998.
111. Haldar P, Shaw D, Siva R, Berry M, Brightling C, Green R *et al.* The correlation between exhaled nitric oxide and sputum eosinophil counts in patients with airway disease. *Thorax* 62: A128, 2008.
112. Zacharasiewicz A, Wilson N, Lex C, Erin EM, Li AM, Hansel T *et al.* Clinical use of noninvasive measurements of airway inflammation in steroid reduction in children. *Am J Respir Crit Care Med* 171(10): 1077–82, 2005.
113. Lehtimaki L, Kankaanranta H, Saarelainen S, Turjanmaa V, Moilanen E. Increased alveolar nitric oxide concentration in asthmatic patients with nocturnal symptoms. *Eur Respir J* 20(4): 841–45, 2002.
114. Lehtimaki L, Kankaanranta H, Saarelainen S, Turjanmaa V, Moilanen E. Inhaled fluticasone decreases bronchial but not alveolar nitric oxide output in asthma. *Eur Respir J* 18(4): 635–39, 2001.

43

CHAPTER

Quantitative Imaging of the Lung

Harvey O. Coxson

Department of Radiology, James Hogg iCAPTURE Centre for Cardiovascular and Pulmonary Research, University of British Columbia, Vancouver, BC, Canada

INTRODUCTION

Imaging of the lung has always played a very important role in pulmonary medicine. The advent of quantitative imaging allowed lung researchers to move beyond descriptive studies in the lung and actually apply measurements that could be applied to humans generally and disease conditions specifically. Weibel was one of the first researchers to meticulously measure the lung and open up a whole new field of pathologic research known as quantitative stereology [1–6]. These techniques have undergone some modification over the last 40 years, with the proliferation of computers, but the basic principles remain unchanged today. Unfortunately, pathologic techniques have some considerable drawbacks most notably that they require the tissue from either resected or autopsy specimens. The second limitation is that they are histologic examinations and because of their depth of scale require an extensive amount of work to cover a small region of the lung. However, pathology and histology are still considered the gold standard by which all other measuring devices are judged and remains the cornerstone of quantitative analysis.

Today the search for quantitative imaging techniques that are not as invasive as pathology and allow the investigator to obtain multiple measurements over time is becoming very important. With the explosion worldwide of chronic lung diseases and, therefore, our need to accurately measure the structure of the lung and possibly any changes due to the pathogenic mechanism or interventions is key. It is for these reasons that medical imaging has become so popular and the introduction of the computed tomography (CT) scanner by Hounsfield has changed the way that we look at organs today. As the CT scanner developed and images became easier in terms of time and financial considerations, it has become one of the first lines of clinical medicine. The CT technology is now so pervasive that nearly every hospital has at least one CT scanner and the technology behind CT has changed so much that now images of entire organs can be obtained in less time than it took to acquire one image in the original CT scanner. Of course many other techniques have also started to proliferate including magnetic resonance imaging, which provides not only anatomic information, but also some functional information as well. This chapter will focus on these non-invasive quantitative measurements of lung structure specifically as they apply to lung structure and the changes involved in chronic obstructive pulmonary disease (COPD). This chapter will start with an examination of CT since it is the most popular both in terms of market penetration of the CT scanner, and quantitative methodologies. The other major focus of this chapter will be hyperpolarized magnetic resonance (MR) imaging since it is a growing field and potentially has some valuable insights. This chapter will also divide COPD into two separate sections based on widely accepted phenotypic expressions of the disease, emphysema and small airways disease because the exact pathogenesis of COPD is still unknown and it varies widely between individuals.

It has been established that the root cause of COPD is an exaggerated inflammatory response that, in susceptible people, results in either a destruction of lung tissue (emphysema) or a remodeling of the airway wall structure. Both of these processes result in changes to the function of the lung that termed COPD, but the exact role and contribution of these two processes varies between individuals and is thought to be under genetic control. Therefore, studies of COPD must take into account the differences in these pathogenic pathways particularly as therapeutic interventions are developed because it is very likely that a specific intervention for the inflammatory response in airway walls may be the exact wrong intervention for the destructive components of emphysema.

QUANTITATIVE MEASUREMENTS OF THE LUNG IN COPD

Emphysema analysis

In brief, CT images are obtained by passing a beam of X-rays through the body and measuring their absorbance which, in biological tissues, is directly proportional to the density of tissue. This apparent absorbance of X-rays is converted to a relative scale known as the Hounsfield Unit (HU) scale which is a linear scale based on the attenuation water. In the Hounsfield scale the apparent attenuation of water is set at 0 HU while the absorbance of air is −1000 HU. The CT scan is a transaxial image so each of the pixels, or picture elements, in the image has an X and a Y dimension. However, the X-ray beam and detector system of the scanner also has at width which gives the CT scan a "slice thickness" making the pixel of the CT image more appropriately known as a voxel or volume element. In a classic conventional CT scan the slice thickness was 10 mm which allowed adequate coverage of the lung using the shortest breath hold maneuver. The X-ray detectors then became thinner so that images could be acquired down to 1 mm in thickness, which combined with a high spatial frequency reconstruction algorithm, became known as the high-resolution CT scan (HRCT). CT scanners have seen immense growth over the last 10–15 years as the introduction of slip ring technology allowed continuous acquisition of images while the patient moved through the gantry, the helical or spiral CT scan, to the sub-division of the X-ray detector into multiple elements allowing for multiple slice acquisitions simultaneously, the multi-detector row or multi-slice CT scanner.

The normal human lung is approximately 80 % airspace and 20% tissue and blood [1] and CT densitometry studies have shown that the normal human lung has a symmetrical distribution of X-ray attenuation values around a mean of approximately −800 HU [7–10]. This distribution of X-ray attenuation values was the first quantitative feature of the CT scan that was exploited to quantify structural changes in disease states because changes in the quantity of tissue and gas in the lung causes changes in the shape of this distribution. Independently, investigators led by Hayhurst [11] in Edinburgh and Müller [12] in Vancouver found that the extent of emphysema quantified pathologically was directly correlated to either the percentage of lung voxels with attenuation values less than a HU specific threshold value, "density mask" [12, 13], or the actual HU value at the lowest 5th percentile of the frequency distribution curve [14]. While both of these techniques are similar in idea they are different in application and have led to numerous disagreements in the literature concerning their applicability.

Since these initial studies CT scanners have undergone remarkable changes as described above. The initial description of the "density mask" technique found that for CT scans that were 10 mm thick and reconstructed using a low spatial frequency reconstruction algorithm, emphysematous holes greater than 5 mm in diameter correlated well with a threshold value of −910 HU (Fig. 43.1). This means that the percentage of lung voxels with attenuation values less than −910 HU was correlated with the percentage of the lung occupied by holes greater than 5 mm in diameter [12, 13]. Coxson and colleagues took this analysis a step further and found that by estimating the average maximal lung inflation for a subject (based on a 6 l TLC and 1000 g lung mass) they found that a cut-off value of 6.0 ml gas per gram of tissue (−856 HU) estimated the extent of emphysematous holes between 2 and 5 mm in size [15].

As CT scanner technology changed threshold values have also had to change. Gevenois and co-workers re-examined the "density mask" technique using 1 mm thick CT slices

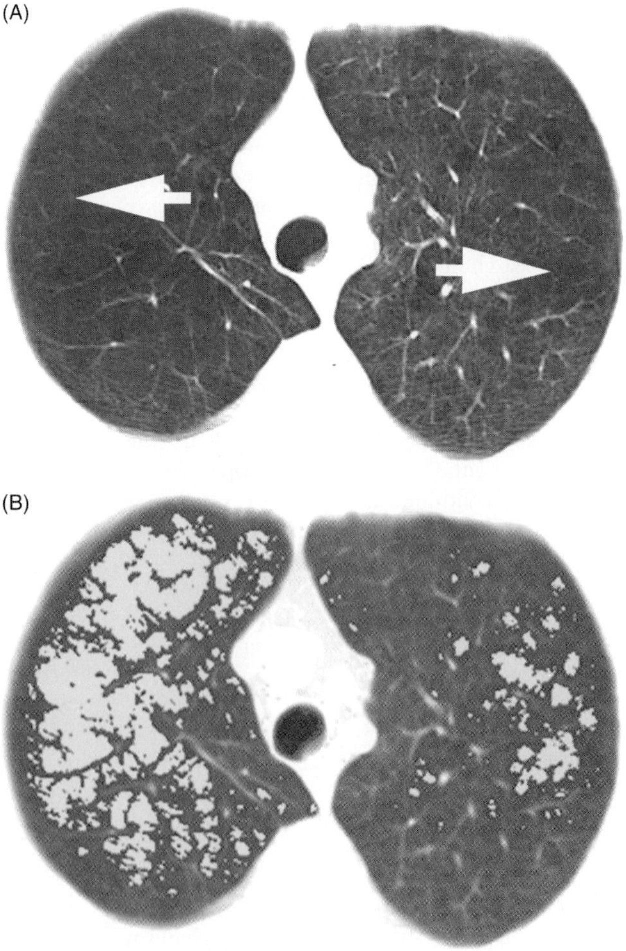

FIG. 43.1 (A) This figure shows a CT scan from a subject with relatively severe emphysema indicated by the arrows. (B) The voxels with attenuation values less than −950 HU or below the threshold or "density mask" are highlighted on the CT scan.

reconstructed using a high spatial frequency reconstruction algorithm and found that the best correlation between lung pathology and CT was obtained using the −950 HU threshold value [16, 17]. Furthermore, the lowest 5th percentile was also redefined to be the lowest 15th percentile [18, 19]. Finally with the advent of multi-detector row CT scanners that obtain multiple CT images at the same time and acquire these images helically while the subject is moving through the CT gantry. Madani and colleagues found that the best correlations with pathology were obtained using the −960 HU threshold and all percentile points below the lowest 18th percentile correlated with macroscopic and microscopic measures of emphysema, although the strongest correlation with the lowest first percentile [20].

Uses of the "density mask" and percentile method have proliferated in the literature in recent years. The "density mask" technique has been used to measure the extent of emphysema in very severe subjects undergoing lung volume reduction surgery (LVRS) [15, 21–23] or other therapeutic interventions such as retinoic acid treatment [24]. It has also been used to describe differences in the extent of disease in males and females and has reported that for a given level of obstruction males have more "emphysema" than females [25]. It has also even been used to investigate emphysema like changes in lung structure in subjects with Anorexia Nervosa and found that the subjects with the lowest body mass index had the lowest diffusing capacity for carbon monoxide and also had X-ray attenuation values that were similar to heavy smokers and normally attributed to emphysema [26].

The percentile technique has also been applied to LVRS studies [27] and numerous studies of α_1-antitrypsin deficiency [18, 19, 27–30]. These investigators have shown that the lowest 15th percentile point is a robust threshold and allows subjects to be compared over time and is less susceptible to minor changes that may be induced by the CT scanner or the size of the breath the subject takes during the scan [18, 28, 30].

Having a robust metric of lung change is an extremely important tool because with the proliferation of CT scanners and the continuous changing of the devices by manufacturers it makes it very difficult come up with a metric that changes with the disease process and not just the with features of the scanner. Therefore, when conducting longitudinal studies all aspects of the scan must be considered, including patient parameters such as breath size and motion during the scan, but also calibration of the scanner, slice thickness, reconstruction algorithm, type of scanner and radiation dose. In a longitudinal study of subjects with α_1 deficiency, Parr *et al.* noticed a drift in density mask measurements which they managed to trace back to calibration of the CT scanner [31]. They found that while the water calibration of the scanner was performed routinely, there was no attempt to calibrate the scanner to air and therefore the change they were observing in lung density was due entirely to the scanner and not the disease. For this reason the authors had to develop a correction factor for their CT scans to take into account this shift in X-ray attenuation values [31]. In another study, Boedeker *et al.* found that even when all other aspects, such as scanner type, size of breath, etc. were carefully controlled, the reconstruction algorithm could make up to a 15% (average 9.4%) difference in the extent of emphysema measured using a "density mask" technique [32]. Their data showed that some reconstruction algorithms, described as "over-enhancing," improve spatial resolution by over-enhancing the difference between lung voxels; thereby changing the frequency distribution of the apparent X-ray attenuation values within the CT image and ultimately the number of voxels below a certain cut-off value [32]. Other investigators have looked at the effect of dose, and scanner manufacturer. Using a phantom, Stoel found that there were significant differences between single and multi-slice scanners [33]. In this study they reported that there were more errors in the volume measurements of single slice scanners than multi-slice and the higher degree of apparent random error in the density measurements led them to conclude that only multi-slice CT scanners should be used in clinical multi-center studies. Madanai *et al.* recently examined the effect of slice thickness and radiation dose on measurements of emphysema. In this study the authors used one multi-slice CT scanner and acquired images using different radiation doses (mA) and then reconstructed the images using different slice thicknesses [34]. They found that there were significant correlations between their metrics of emphysema, either threshold cut-offs or percentile technique, however X-ray dose had an effect on the threshold cut-off measurements but not percentile method while slice thickness had a significant effect of both threshold and percentile techniques. In a similar study, Yuan and colleagues compared two multi-slice CT scanners as well as changes in radiation dose and showed that there was no difference between the techniques in terms of volume measurements if scanner or dose was changed [35]. However, there were differences in mean lung density between scanners and, most importantly, large differences in the extent of emphysema measured using either the threshold or the percentile technique when the X-ray dose was changed [35]. Therefore, the conclusions of all of these studies are that if you intend to use CT as a measurement of emphysema in a longitudinal and/or multi-center study you must pay careful attention to the details. A multi-slice scanner should be use for these studies and if you are to use a Toshiba scanner a correction factor must be applied to the X-ray attenuation values. Furthermore, the radiation dose must be carefully monitored because this factor alone can have large effects on the emphysema measurements. In conclusion, the CT scanner can be used to measure emphysema, but it must be treated like all other measuring tools and all of the parameters must be carefully monitored.

An aspect of emphysema measurements that has been largely overlooked in the literature is some type of emphysema hole size measurement. While it has been shown that there are decent correlations between the threshold cut-off and the percentile techniques and pathologic measurements of emphysema, these techniques do not provide any information on the size or location of the emphysematous hole (Fig. 43.2). This is an area that needs more research because the number of publications using these types of techniques is very small. One of the first attempt to quantify hole size used a technique known as the "fractal" or "cluster" analysis [36, 37] whereby the size of low attenuation areas, as defined by the number of connected voxels below a given

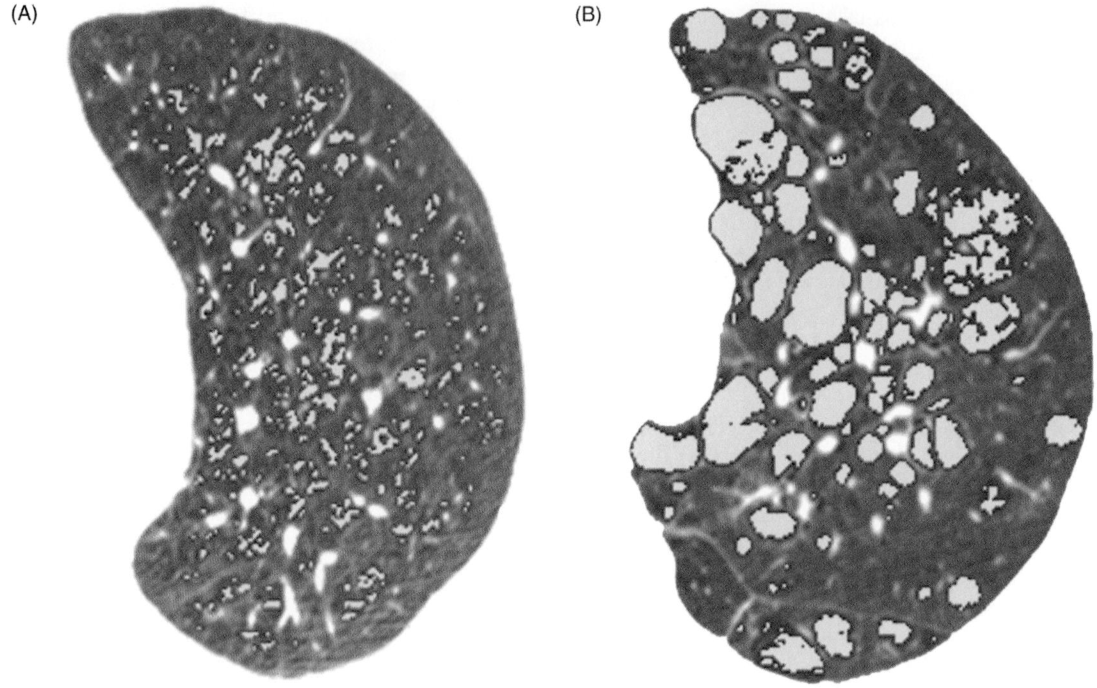

FIG. 43.2 This figure shows the voxels identified as below −950 HU in two different subjects, one with predominately smaller holes (A) and one with predominately large holes (B). A cluster analysis can be used to separate the subjects with large holes from the one with smaller holes.

threshold, is correlated to the number of these low attenuating areas. Mishima validated this technique using an "elastic spring" model and showed that as the emphysema become more extensive the holes become larger and they decrease in number [36]. They also found that in subjects with COPD but similar percentage emphysema as control subjects the "fractal" measurement was lower in the COPD subjects suggesting that while they had the same "extent" of emphysema the emphysematous lesions tended to be larger in subjects with COPD. This study was follow-up by another study that examined the size and location of the lesion in subjects undergoing lung volume reduction surgery. This study showed that that subjects with large upper lung region lesions have the best outcome following LVRS [37]. The National Emphysema Treatment Trial (NETT) applied a similar technique to their study subjects and found that there was a good correlation between the radiologist assessed hole size and the computer assessed metric [22]. Furthermore, the size and location of these lesions, large lesions located in the apical regions of the lung, seemed to have a predictive effect on survival following surgery [22]. More complex techniques that quantify multiple features within the CT scan, such as the adaptive multiple feature method, have reported decent results in disease stratification including the ability to separate smokers from non-smokers in the absence of other disease [38]. However, the disadvantage of these new "texture" analysis techniques is that they often use complex mathematical terminology to describe the lung rather than anatomic or physiologic terminology that are more commonly used in medicine and have not been validated using pathologic assessment of the underlying lung structure. Finally, the location of the emphysema seems to be important as well as indicated by the NETT study and the emphysematous hole size studies above [22, 37]. Nakano first showed that emphysema located in the upper-outer regions of the lung predicted better outcome following surgery, presumably because it is more accessible to surgical resection than those with more centrally located or diffuse disease [39]. The studies of location of emphysema and the effect that it has on lung function are sparse in the literature, but it is hoped that with the widespread use of multi-slice CT scanners that these types of studies will shed more light on the pathogenesis of emphysema (Fig. 43.3).

Finally, there is great interest in using CT scans in multi-center studies as an end point for a therapeutic intervention. This has already been accomplished in studies such as the NETT trial but the work of Dirksen really adds importance to this approach. In a small study of α_1-antitrypsin deficient subjects, Dirksen found that there was no significant change in the functional or structural (assessed using CT) characteristics over the course of the study [28]. However, when the authors performed a power calculation of their data they found that to show a difference in FEV_1 the authors would have had to study 550 subjects but to show a difference using CT the authors would only have had to study 130 [28]. These data add further credence to the important role that CT scans can play in clinical intervention studies.

In conclusion, these studies all show that CT measurements of emphysema are correlated with pathologic measurements of emphysema. Furthermore, these measurements can be used to follow disease progression, possibly with more sensitivity than the traditional metrics of FEV_1, and that size and location of the disease is important. Also, these techniques can be used to measure early or small changes in the lungs

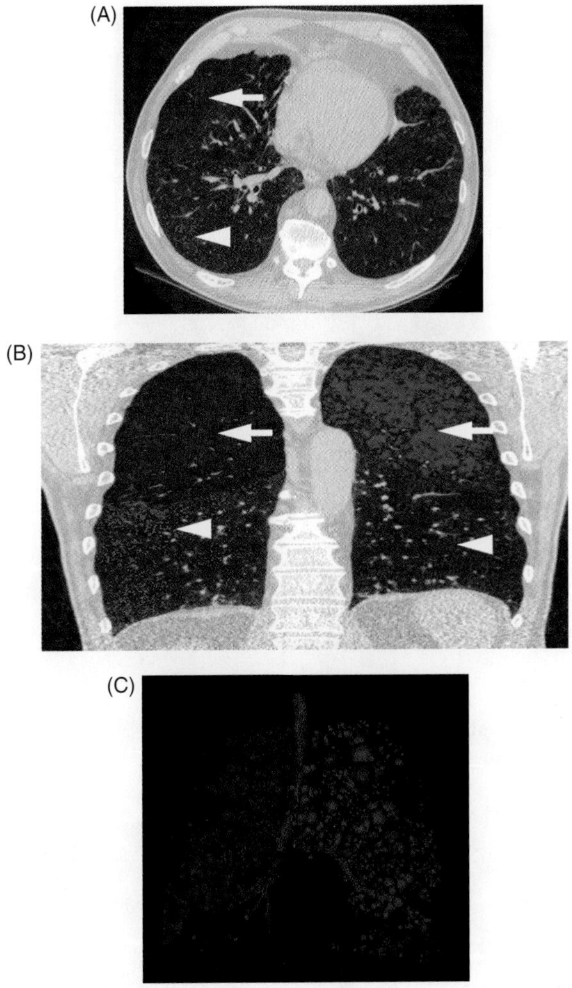

FIG. 43.3 This figure shows a three dimension analysis of emphysema in a subject with predominately upper lobe emphysema. The trans-axial images (A) show more highlighted regions (arrows) in the upper lobes while the coronal images (B) show quite clearly that there are more highlighted voxels in the upper lobes (arrows) than the lower lobes (arrow heads). A cluster analysis (C) that highlights the connected regions clearly shows larger clusters in the upper lobes. Images were created using the Pulmonary Workstation 2.0 software (VIDA Diagnostics Inc.).

of early disease subjects and can be used to compare differences in disease between genders. Overall, the assessment of emphysema using CT has a long and well developed history and is now starting to be applied in more and more longitudinal studies. However, it is important to keep in mind that CT, like any measuring device, must be used properly. The CT scanners must be comparable and the CT image acquisition techniques must be comparable. It has been shown that low dose studies can be performed in these groups of subjects, but because dose effects the measurements of emphysema the dose must be kept constant throughout the study.

Airway analysis

Of course assessing the emphysema is only one part of the analysis of the structural changes that cause the chronic airflow limitation because it has long been recognized that the site of major airflow limitation is the small airways [40]. There have been numerous investigations of small airway dimensions and COPD that have confirmed the importance of the remodeling in these structures, but once again, these rely on histologic specimens [40–45]. The introduction of CT scanning peaked investigators interest in measuring airways non-invasively so several methods were implemented to manually trace the airway walls using the CT images printed on film, similar to how it was accomplished using histology. However, one thing that became immediately apparent was that optimal measurement of airway lumen using this technique required very specific display settings. For example, it was found that an accurate measurement of wall area required the use of a window level of −450 HU and a window width of 1000–1400 HU [46–49]. Unfortunately, these display settings produce images that are too dark for optimal visualization of the lung parenchyma and so are never used clinically. Furthermore, because the technique is cumbersome, time consuming, and associated with considerable intra- and interobserver variability, this approach was quickly abandoned [47–49].

Since the CT scan is a digital image and contains gray scale information attention quickly turned to using automated airway analysis techniques. McNitt-Gray *et al.* found that the lumen area could be accurately measured using a threshold cut off of −500 HU [50] which was further validated using excised formalin fixed pig lungs by King and co-workers who reported that a threshold of −577 HU produced the least error in the measurements [51]. However, these techniques just measured the airway lumen so attention turned to using the X-ray attenuation values of the CT scan to measure both the airway lumen and wall. One of the more common techniques is the "Full-Width-At-Half-Maximum" ("half-max") method (Fig. 43.4) where the distribution of apparent X-ray attenuation values along a ray projected from a central point of the lumen to the parenchyma is measured. The distance between the point at which the attenuation is half way between the local minimum in the lumen or parenchyma and the maximum within the wall is considered to be wall thickness [52, 53]. Unfortunately, the shape of this curve is dependant on various parameters including: the reconstruction algorithm used to create the image, partial volume averaging due to field of view and orientation of the airway within the CT image and the inevitable blurring of edges that occurs due to the point spread function of the CT scanner. Validation studies show that the CT scans over-estimate airway wall area and underestimate lumen area and that these errors become very large in small airways [52, 54]. Furthermore, this algorithm requires that the airway being analyzed on CT is cut in cross section. While it is relatively easy for observers to find airways cut in cross section, because the CT slice has a thickness, typically on the order of 1 mm, the airway observed on CT may or may not have actually been cut in cross section (Fig. 43.5). This is another source of error within the airway wall measurements and makes them especially unreliable as the airways approach the resolution of the CT scanner. For these reasons investigators have developed numerous other algorithms to measure these airways such as the "maximum-likelihood method" whereby

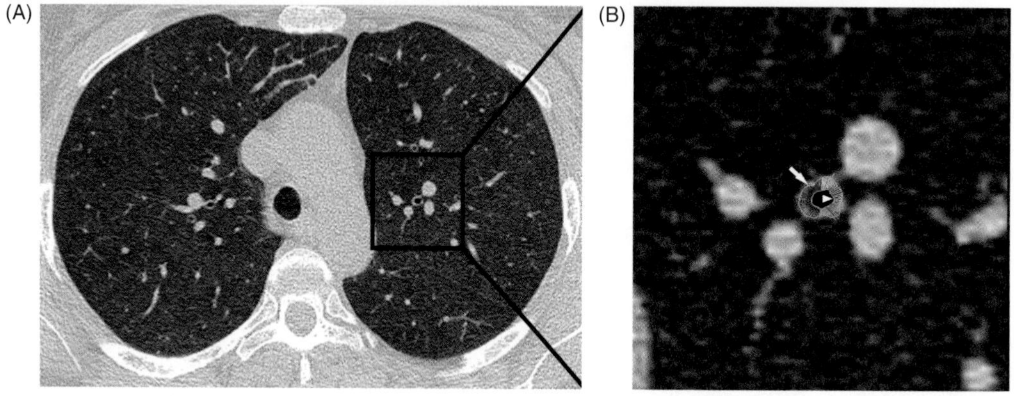

FIG. 43.4 This figure shows a CT scan (A) from which an airway is chosen and displayed using a magnified view (B). The measurement of the airway wall dimensions is performed by using the Full-Width-At-Half-Maximum method whereby the length of the rays is determined by choosing the halfway point between the minimum X-ray attenuation value in the lumen or lung parenchyma and the maximum value within the airway wall. The lumen area and the internal perimeter of the airway wall are measured using the internal boundary of the rays (shown by arrowhead) and the wall area and the outer perimeter are measured by the external boundary of the rays (arrow).

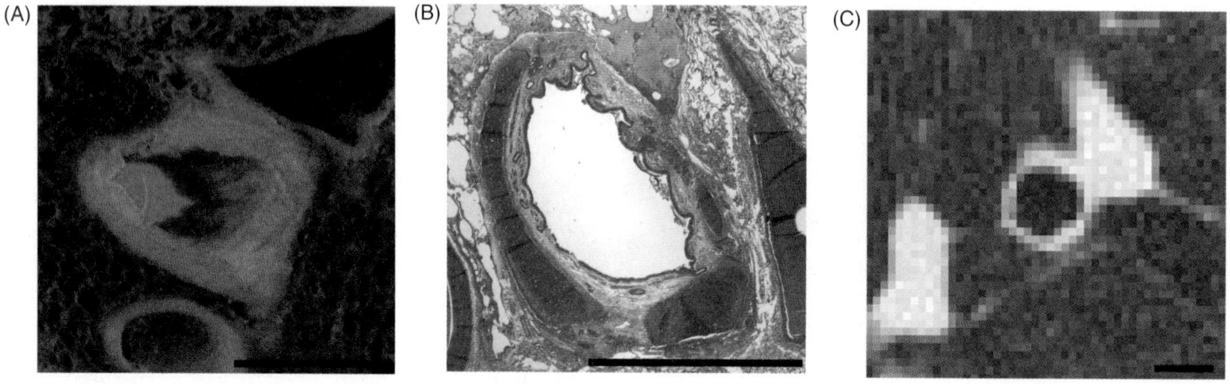

FIG. 43.5 This figure shows the same airway from gross pathology (A), histology (B) and from the pre-operative CT scan (C). From this image some of the measuring artifacts can be seen including the fact that the airway may not actually be cut in cross section in the CT scan because the airway is slightly different shape in the CT and can be seen to not be in perfect cross section in the gross pathology section. Secondly, the resolution issues can clearly be seen as the actual pixels are visible in the CT scan and the image appears very pixilated. All of the scale bars are 5 mm in length.

the attenuation threshold along each ray is matched to an ideal calculated ray [54], the "score-guided erosion algorithm" [51] where airway wall edges are found using an edge finding algorithm that assumes that airways are circular and have a relatively high density compared to the surrounding parenchyma, and an algorithm where ellipses are fit to the airway lumen and wall [55].

Because it is well recognized that airway wall remodeling plays a very important role in COPD there is great interest in non-invasive measurement of airway wall dimensions. However, the applications of these techniques have so far been very limited. One of the first papers published in this field was a study by Nakano and co-workers [53] who evaluated the right apical segmental bronchus of asymptomatic smokers and COPD subjects. The authors chose this airway to measure because it is routinely cut in cross section on transaxial CT scans, and they could compare the same airway across all subjects. The authors also measured the extent of emphysema using a threshold technique and correlated the emphysema and airway wall dimension data with pulmonary function data. The data from this study showed that the thickening in this large airway correlated with FEV_1 percent predicted, FVC percent predicted and RV/TLC while the emphysema correlated with FEV_1, FEV/FVC ratio and the DL_{CO}. Furthermore, a multiple regression analysis suggested that, for a given FEV_1, subjects with more extensive emphysema had less airway wall thickening than those with less extensive emphysema [53]. Obviously a segmental airway is unlikely to be the major source of airflow limitation in subjects with COPD. Furthermore, it is questionable whether any of the airways that can be observed using CT are the airways responsible for the airflow limitation. Since histology is considered the gold standard for small airway wall remodeling, Nakano compared the measurements of wall area obtained using CT (average diameter of 3.2 mm) to those obtained using histology of small airways (1.27 mm diameter) in resected specimens [56]. These data show a significant association ($R^2 = 0.57$, $p = 0.001$) between the

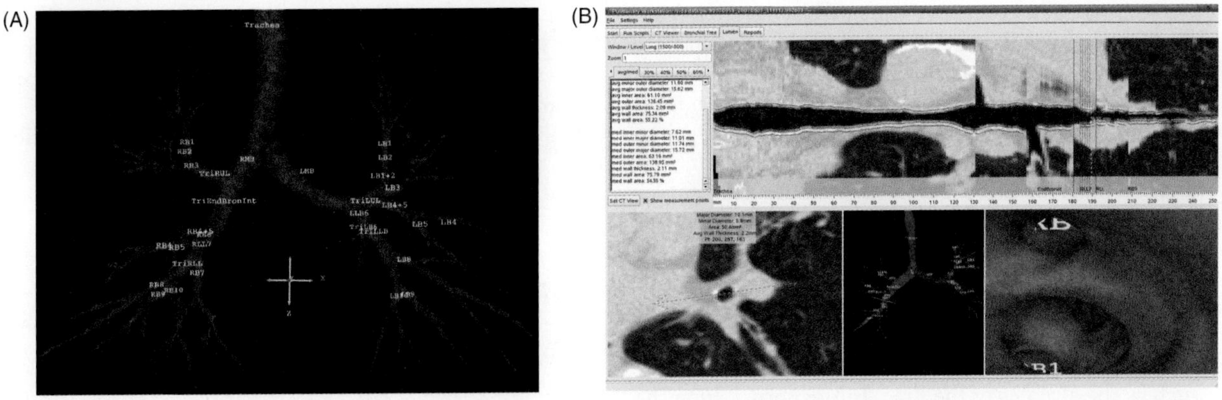

FIG. 43.6 This figure shows multi-planar reformatted reconstructions of the airway tree (A). Using these new software tools it is now possible to segment the airway tree from the lung parenchyma and, using a pre-defined pathway, reconstruct it as a long straight tube and obtain measurements at a cross section to the center line of the airway at any point along the pathway (B). Images were created using the Pulmonary Workstation 2.0 software (VIDA Diagnostics Inc.).

dimensions of the smaller and larger airways. The authors conclude that, at least for COPD, measuring airway dimensions in the larger bronchi, which are more accurately assessed by CT, can provide an estimate of small airway remodeling. It is likely that the same pathophysiologic process, which causes small airway obstruction, also takes place in larger airways where it has less functional effect.

Recently focus has turned away from measurements of airways cut in cross section on transaxial CT images to what is thought to be a true cross section of the airway obtained by multi-planar reformats of three-dimensional airway data. Multi-slice scanners can now obtain CT images with resolution in the Z dimension that approaches that of the X–Y dimensions, although most studies still use CT slices with 1–1.25 mm slice thickness. Investigators can now segment the airway tree in three dimensions starting in the trachea and projecting out to the fifth or sixth generation (Fig. 43.6). Also, this technique allows the investigator to section the airway at any place and any angle that they choose. Of course this technique assumes that the voxel dimension in the Z dimension is small, that is 1 mm or less, or the assumptions inherent to the re-format become too large. Another important advance to this technique is that now investigators can know exactly where they are in the airway tree and if they wish to compare the airway size of a specific airway between individuals and over time they have the landmarks to do so and no longer have to count on random selections of airways. While random airway selections have their advantages in that you have an overall picture of airways within a subject they also have their drawbacks. For example, investigators often normalize airway wall dimensions between individuals by comparing the percent of the total airway (lumen + wall) that is occupied by wall, the so called Wall Area Percent (WA%). This is a very good estimate of wall area between subjects but it assumes that you are comparing equal sized airways because in large airways the WA% is always small because the wall area is such a small fraction of the total airway. On the other hand, in small airways, the WA% is always large because now the lumen is very small and the wall starts to occupy a much larger fraction of the total airway. Therefore, you can potentially measure a statistical difference in WA% between individuals, because there is an inflammatory process occurring in the airway wall that is making the wall area bigger, or you could simply be comparing big airways to small airways due to improper sampling. The new three-dimensional reconstruction technique can avoid that trap by allowing investigators to compare the exact airway between individuals and over time. That was one of the original goals of the Nakano paper where they measured the right apical segmental bronchus only, but it is now better achieved and in smaller airways using current CT and image analysis technology.

This approach was recently put to the test by Hasegawa and co-workers [57] who used multi-slice CT scans to measure the right apical and basal segmental bronchus at different branch points along its pathway. These authors were not able to verify the findings of Nakano by comparing the right apical segmental bronchus to FEV_1 however, they did show that as they moved more distally, to the sixth generation of the airway tree, the correlation with FEV_1 improved. The authors concluded that the airflow limitations are more strongly associated with the dimensions of the distal airways than the proximal ones and demonstrate that these new three-dimensional techniques can be used to obtain measurements from a specific location within a specific airway [57]. Also, it once again shows the importance of airway measurements in understanding COPD.

Hyperpolarized noble gas imaging of the lung

Other techniques such as MR imaging of the lung are becoming very attractive techniques particularly because they do not expose the subject to any ionizing radiation. This has always been a strong point of MR, but factors such as the small nuclear spin polarization at thermal equilibrium and the field inhomogeneity problems associated with the lung have greatly limited the application of this technique [58]. However, the last few years have seen resurgence in

MR imaging research because of the introduction of hyperpolarized noble gases. Research has shown that by exposing noble gases such as ^{129}Xenon or ^{3}Helium to a laser you can hyperpolarize the gas by aligning more of their magnetic moments into the same orientation. This allows the gas to be used as a contrast agent within the lungs during an MR scan and has allowed investigators to measure the ventilation of the lung [59, 60], the size of the alveolus airspace [61] and airway dimensions [62, 63]. The most well established of these techniques is the measurement of alveolar size by a technique known as the "Apparent Diffusion Coefficient" or ADC. This technique works on the principle that gas molecules will move, or diffuse, within a given volume such as the alveolus of the lung. In the normal lung the airspaces are much smaller so that the movement of Helium is restricted but in emphysema where the airspaces become much larger. Several studies have correlated the ADC with spirometry measurements in animal models, normal volunteers and in subjects with airflow limitations emphysema [60–62, 64–69]. In a recent study Woods *et al.* showed that there was a very good correlation between the ADC and alveolar size measured using either the surface area to volume ratio of the lung parenchyma or the mean linear intercept and that the ADC was able to separate emphysematous lung from normal lung with greater sensitivity than histology [61].

Other investigators have shown that it is possible to create three-dimensional reconstruction of the airway lumen to the seventh-generation of airways [62, 63]. While this is an interesting technique it does not provide measurements of airway wall thickness and so the applicability is still uncertain.

Finally, while it appears that hyperpolarized gas MR has some very powerful applications in the study of COPD, the technique its self is very limited by availability of a hyperpolarized helium or xenon source and the extremely technical nature and cost of the procedure. While it is an exciting and innovative technique, it is very likely that hyperpolarized gas MR will remain a research tool for the immediate future.

CONCLUSION

In conclusion, the last 20 years have seen a large amount of growth in both the interest and the application of non-invasive quantitative imaging of the lung. As long as the scanners are treated properly it is now possible to obtain reliable quantitative information on the extent of emphysema, the size of the emphysematous lesions and airway wall dimensions. The advent of multi-slice CT scanners has opened up new avenues of research into regional distribution of disease and specific tagging and measurements of airways. On the downside CT imaging has some inherent limitations: particularly concerning radiation exposure to the subject and the standardization and quality control of the CT scan parameters. Hyperpolarized gas MR imaging has come a long ways in the last several years and has made some important steps to provide both quantitative structure and functional information. However, because of technical and financial reasons CT is currently the only readily accessible, relatively non-invasive technique that provides quantitative structural data *in vivo*. The proper application of these techniques should produce valuable information on the pathogenesis of COPD and the effects of new therapeutic interventions.

References

1. Weibel ER. *Morphometry of the Human Lung*. Berlin: Springer-Verlag, 1963.
2. Weibel ER. *Stereological Methods, Volume 1: Practical Methods for Biological Morphometry*, London: Academic Press, 1980.
3. Weibel ER. *Stereological Methods, Volume 2: Theoretical Foundations*, London: Academic Press, 1980.
4. Weibel ER. Functional morphology of lung parenchyma. In: Macklem PT, Mead J, (eds). *Handbook of Physiology*, Section 3: The Respiratory System Volume III, Part I, pp. 89–111. Bethesda: American Physiological Society, 1986.
5. Weibel ER. Measuring through the microscope: Development and evolution of stereological methods. *J Microsc* 155: 393–403, 1989.
6. Weibel ER. The structural basis of lung function. In: West JB (ed.). *Respiratory Physiology People and Ideas*, pp. 3–46. New York: Oxford University Press, 1996.
7. Wegener OH, Koeppe P, Oeser H. Measurement of lung density by computed tomography. *J Comput Assist Tomogr* 2: 263–73, 1978.
8. Robinson PJ, Kreel L. Pulmonary tissue attenuation with computed tomography: Comparison if Inspiration and expiration scans. *J Comput Assist Tomogr* 3: 740–48, 1979.
9. Wollmer P, Albrechtsson U, Brauer K, Eriksson L, Jonson B, Tylèn U. Measurement of pulmonary density by means of X-ray computerized tomography. *Chest* 90: 387–91, 1986.
10. Coxson HO, Mayo JR, Behzad H, Moore BJ, Verburgt LM, Staples CA, Pare PD, Hogg JC. Measurement of lung expansion with computed tomography and comparison with quantitative histology. *J Appl Physiol* 79: 1525–30, 1995.
11. Hayhurst MD, Flenley DC, McLean A, Wightman AJA, MacNee W, Wright D, Lamb D, Best J. Diagnosis of pulmonary emphysema by computerized tomography. *Lancet* 2: 320–22, 1984.
12. Müller NL, Staples CA, Miller RR, Abboud RT. "Density mask". An objective method to quantitate emphysema using computed tomography. *Chest* 94: 782–87, 1988.
13. Miller RR, Müller NL, Vedal S, Morrison NJ, Staples CA. Limitations of computed tomography in the assessment of emphysema. *Am Rev Respir Dis* 139: 980–83, 1989.
14. Gould GA, MacNee W, McLean A, Warren PM, Redpath A, Best JJ, Lamb D. CT measurements of lung density in life can quantitate distal airspace enlargement – an essential defining feature of human emphysema. *Am Rev Respir Dis* 137: 380–92, 1988.
15. Coxson HO, Rogers RM, Whittall KP, D'Yachkova Y, Pare PD, Sciurba FC, Hogg JC. A quantification of the lung surface area in emphysema using computed tomography. *Am J Respir Crit Care Med* 159: 851–56, 1999.
16. Gevenois PA, de Maertelaer V, De Vuyst P, Zanen J, Yernault JC. Comparison of computed density and macroscopic morphometry in pulmonary emphysema. *Am J Respir Crit Care Med* 152: 653–57, 1995.
17. Gevenois PA, De Vuyst P, de Maertelaer V, Zanen J, Jacobovitz D, Cosio MG, Yernault JC. Comparison of computed density and microscopic morphometry in pulmonary emphysema. *Am J Respir Crit Care Med* 154: 187–92, 1996.
18. Dirksen A, Friis M, Olesen KP, Skovgaard LT, Sorensen K. Progress of emphysema in severe α_1-antitrypsin deficiency as assessed by annual CT. *Acta Radiol* 38: 826–32, 1997.

19. Stolk J, Ng WH, Bakker ME, Reiber JH, Rabe KF, Putter H, Stoel BC. Correlation between annual change in health status and computer tomography derived lung density in subjects with alpha1-antitrypsin deficiency. *Thorax* 58: 1027–30, 2003.
20. Madani A, Zanen J, de Maertelaer V, Gevenois PA. Pulmonary emphysema: Objective quantification at multi-detector row CT – Comparison with macroscopic and microscopic morphometry. *Radiology* 238: 1036–43, 2006.
21. Bae KT, Slone RM, Gierada DS, Yusen RD, Cooper JD. Patients with emphysema: Quantitative CT analysis before and after lung volume reduction surgery. Work in progress. *Radiology* 203: 705–14, 1997.
22. Martinez FJ, Foster G, Curtis JL, Criner G, Weinmann G, Fishman A, DeCamp MM, Benditt J, Sciurba F, Make B *et al.* Predictors of mortality in patients with emphysema and severe airflow obstruction. *Am J Respir Crit Care Med* 173: 1326–34, 2006.
23. Rogers RM, Coxson HO, Sciurba FC, Keenan RJ, Whittall KP, Hogg JC. Preoperative severity of emphysema predictive of improvement after lung volume reduction surgery: Use of CT morphometry. *Chest* 118: 1240–47, 2000.
24. Mao JT, Goldin JG, Dermand J, Ibrahim G, Brown MS, Emerick A, McNitt-Gray MF, Gjertson DW, Estrada F, Tashkin DP *et al.* A pilot study of all-trans-retinoic acid for the treatment of human emphysema. *Am J Respir Crit Care Med* 165: 718–23, 2002.
25. Dransfield MT, Washko GR, Foreman MG, Estepar RS, Reilly J, Bailey WC. Gender differences in the severity of CT emphysema in COPD. *Chest* 132: 464–70, 2007.
26. Coxson HO, Chan IH, Mayo JR, Hlynsky J, Nakano Y, Birmingham CL. Early emphysema in patients with anorexia nervosa. *Am J Respir Crit Care Med* 170: 748–52, 2004.
27. Stolk J, Versteegh MIM, Montenij LJ, Bakker ME, Grebski E, Tutic M, Wildermuth S, Weder W, el Bardiji M, Reiber JHC *et al.* Densitometry for assessment of effect of lung volume reduction surgery for emphysema. *Eur Respir J* 29: 1075–77, 2007.
28. Dirksen A, Dijkman JH, Madsen IF, Stoel B, Hutchison DCS, Skovgaard LT, Kok-Jensen A, Rudolphus A, Seersholm N *et al.* A randomized clinical trial of a-1 antitrypsin augmentation therapy. *Am J Respir Crit Care Med* 160: 1468–72, 1999.
29. Parr DG, Stoel BC, Stolk J, Stockley RA. Pattern of emphysema distribution in alpha1-antitrypsin deficiency influences lung function impairment. *Am J Respir Crit Care Med* 170: 1172–78, 2004.
30. Stolk J, Dirksen A, van der Lugt AA, Hutsebaut J, Mathieu J, de Ree J, Reiber JH, Stoel BC. Repeatability of lung density measurements with low-dose computed tomography in subjects with alpha-1-antitrypsin deficiency-associated emphysema. *Invest Radiol* 36: 648–51, 2001.
31. Parr DG, Stoel BC, Stolk J, Nightingale PG, Stockley RA. Influence of calibration on densitometric studies of emphysema progression using computed tomography. *Am J Respir Crit Care Med* 170: 883–90, 2004.
32. Boedeker KL, McNitt-Gray MF, Rogers SR, Truong DA, Brown MS, Gjertson DW, Goldin JG. Emphysema: Effect of reconstruction algorithm on CT imaging measures. *Radiology* 232: 295–301, 2004.
33. Stoel BC, Bakker ME, Stolk J, Dirksen A, Stockley RA, Piitulainen E, Russi EW, Reiber JH. Comparison of the sensitivities of 5 different computed tomography scanners for the assessment of the progression of pulmonary emphysema: A phantom study. *Invest Radiol* 39: 1–7, 2004.
34. Madani A, De Maertelaer V, Zanen J, Gevenois PA. Pulmonary emphysema: Radiation dose and section thickness at multidetector CT quantification – Comparison with macroscopic and microscopic morphometry. *Radiology* 243: 250–57, 2007.
35. Yuan R, Mayo JR, Hogg JC, Pare PD, McWilliams AM, Lam S, Coxson HO. The effects of radiation dose and CT manufacturer on measurements of lung densitometry. *Chest* 132: 617–23, 2007.
36. Mishima M, Hirai T, Itoh H, Nakano Y, Sakai H, Muro S, Nishimura K, Oku Y, Chin K, Ohi M *et al.* Complexity of terminal airspace geometry assessed by lung computed tomography in normal subjects and patients with chronic obstructive pulmonary disease. *Proc Natl Acad Sci USA* 96: 8829–34, 1999.
37. Coxson HO, Whittall KP, Nakano Y, Rogers RM, Sciurba FC, Keenan RJ, Hogg JC. Selection of patients for lung volume reduction surgery using a power law analysis of the computed tomographic scan. *Thorax* 58: 510–14, 2003.
38. Uppaluri R, Mitsa T, Sonka M, Hoffman EA, McLennan G. Quantification of pulmonary emphysema from lung computed tomography images. *Am J Respir Crit Care Med* 156: 248–54, 1997.
39. Nakano Y, Coxson HO, Bosan S, Rogers RM, Sciurba FC, Keenan RJ, Walley KR, Pare PD, Hogg JC. Core to rind distribution of severe emphysema predicts outcome of lung volume reduction surgery. *Am J Respir Crit Care Med* 164: 2195–99, 2001.
40. Hogg JC, Macklem PT, Thurlbeck WM. Site and nature of airway obstruction in chronic obstructive lung disease. *N Engl J Med* 278: 1355–60, 1968.
41. Bosken CH, Hards J, Gatter K, Hogg JC. Characterization of the inflammatory reaction in the peripheral airways of cigarette smokers using immunocytochemistry. *Am Rev Respir Dis* 145: 911–17, 1992.
42. Bosken CH, Wiggs BR, Pare PD, Hogg JC. Small airway dimensions in smokers with obstruction to airflow. *Am Rev Respir Dis* 142: 563–70, 1990.
43. Hogg JC, Chu F, Utokaparch S, Woods R, Elliott WM, Buzatu L, Cherniack RM, Rogers RM, Sciurba FC, Coxson HO *et al.* The nature of small-airway obstruction in chronic obstructive pulmonary disease. *N Engl J Med* 350: 2645–53, 2004.
44. Hogg JC, Chu FS, Tan WC, Sin DD, Patel SA, Pare PD, Martinez FJ, Rogers RM, Make BJ, Criner GJ *et al.* Survival after lung volume reduction in chronic obstructive pulmonary disease: Insights from small airway pathology. *Am J Respir Crit Care Med* 176: 454–59, 2007.
45. Kim WD, Ling SH, Coxson HO, English JC, Yee J, Levy RD, Pare PD, Hogg JC. The association between small airway obstruction and emphysema phenotypes in COPD. *Chest* 131: 1372–78, 2007.
46. Bankier AA, Fleischmann D, Mallek R, Windisch A, Winkelbauer FW, Kontrus M, Havelec L, Herold CJ, Hubsch P. Bronchial wall thickness: Appropriate window settings for thin-section CT and radiologic-anatomic correlation. *Radiology* 199: 831–36, 1996.
47. McNamara AE, Muller NL, Okazawa M, Arntorp J, Wiggs BR, Pare PD. Airway narrowing in excised canine lung measured by high-resolution computed tomography. *J Appl Physiol* 73: 307–16, 1992.
48. Okazawa M, Muller NL, McNamara AE, Child S, Verburgt L, Pare PD. Human airway narrowing measured using high resolution computed tomography. *Am J Respir Crit Care Med* 154: 1557–62, 1996.
49. Webb WR, Gamsu G, Wall SD, Cann CE, Proctor E. CT of a bronchial phantom: Factors affecting appearance and size measurements. *Invest Radiol* 19: 394–98, 1984.
50. McNitt-Gray MF, Goldin JG, Johnson TD, Tashkin DP, Aberle DR. Development and testing of image-processing methods for the quantitative assessment of airway hyperresponsiveness from high-resolution CT images. *J Comput Assist Tomogr* 21: 939–47, 1997.
51. King GG, Muller NL, Whittall KP, Xiang QS, Pare PD. An analysis algorithm for measuring airway lumen and wall areas from high-resolution computed tomographic data. *Am J Respir Crit Care Med* 161: 574–80, 2000.
52. Nakano Y, Whittall KP, Kalloger SE, Coxson HO, Pare PD. Development and validation of human airway analysis algorithm using multidetector row CT. *Proc SPIE* 4683: 460–69, 2002.
53. Nakano Y, Muro S, Sakai H, Hirai T, Chin K, Tsukino M, Nishimura K, Itoh H, Pare PD, Hogg JC *et al.* Computed tomographic measurements of airway dimensions and emphysema in smokers. Correlation with lung function. *Am J Respir Crit Care Med* 162: 1102–8, 2000.
54. Reinhardt JM, D'Souza ND, Hoffman EA. Accurate measurement of intrathoracic airways. *IEEE Trans Med Imaging* 16: 820–27, 1997.
55. Saba OI, Hoffman EA, Reinhardt JM. Maximizing quantitative accuracy of lung airway lumen and wall measures obtained from X-ray CT imaging. *J Appl Physiol* 95: 1063–75, 2003.
56. Nakano Y, Wong JC, de Jong PA, Buzatu L, Nagao T, Coxson HO, Elliott WM, Hogg JC, Paré PD. The prediction of small airway dimensions using computed tomography. *Am J Respir Crit Care Med* 171: 142–46, 2005.

57. Hasegawa M, Nasuhara Y, Onodera Y, Makita H, Nagai K, Fuke S, Ito Y, Betsuyaku T, Nishimura M. Airflow limitation and airway dimensions in chronic obstructive pulmonary disease. *Am J Respir Crit Care Med* 173: 1309–15, 2006.
58. Mayo JR. Magnetic resonance imaging of the chest. Where we stand. *Radiol Clin North Am* 32: 795–809, 1994.
59. Salerno M, Altes TA, Brookeman JR, de Lange EE, Mugler JP. Dynamic spiral MRI of pulmonary gas flow using hyperpolarized (3)He: Preliminary studies in healthy and diseased lungs. *Magn Reson Med* 46: 667–77, 2001.
60. Swift AJ, Woodhouse N, Fichele S, Siedel J, Mills GH, van Beek EJ, Wild JM. Rapid lung volumetry using ultrafast dynamic magnetic resonance imaging during forced vital capacity maneuver: Correlation with spirometry. *Invest Radiol* 42: 37–41, 2007.
61. Woods JC, Choong CK, Yablonskiy DA, Bentley J, Wong J, Pierce JA, Cooper JD, Macklem PT, Conradi MS, Hogg JC. Hyperpolarized 3He diffusion MRI and histology in pulmonary emphysema. *Magn Reson Med* 56: 1293–300, 2006.
62. Tooker AC, Hong KS, McKinstry EL, Costello P, Jolesz FA, Albert MS. Distal airways in humans: Dynamic hyperpolarized 3He MR imaging – feasibility. *Radiology* 227: 575–79, 2003.
63. Lewis TA, Tzeng YS, McKinstry EL, Tooker AC, Hong K, Sun Y, Mansour J, Handler Z, Albert MS. Quantification of airway diameters and 3D airway tree rendering from dynamic hyperpolarized 3He magnetic resonance imaging. *Magn Reson Med* 53: 474–78, 2005.
64. Fain SB, Panth SR, Evans MD, Wentland AL, Holmes JH, Korosec FR, O'Brien MJ, Fountaine H, Grist TM. Early emphysematous changes in asymptomatic smokers: Detection with 3He MR imaging. *Radiology* 239: 875–83, 2006.
65. Fichele S, Woodhouse N, Swift AJ, Said Z, Paley MN, Kasuboski L, Mills GH, van Beek EJ, Wild JM. MRI of helium-3 gas in healthy lungs: Posture related variations of alveolar size. *J Magn Reson Imaging* 20: 331–35, 2004.
66. Ley S, Zaporozhan J, Morbach A, Eberle B, Gast KK, Heussel CP, Biedermann A, Mayer E, Schmiedeskamp J, Stepniak A *et al.* Functional evaluation of emphysema using diffusion-weighted 3Helium-magnetic resonance imaging, high-resolution computed tomography, and lung function tests. *Invest Radiol* 39: 427–34, 2004.
67. Salerno M, de Lange EE, Altes TA, Truwit JD, Brookeman JR, Mugler JP III. Emphysema: Hyperpolarized helium 3 diffusion MR imaging of the lungs compared with spirometric indexes – initial experience. *Radiology* 222: 252–60, 2002.
68. Swift AJ, Wild JM, Fichele S, Woodhouse N, Fleming S, Waterhouse J, Lawson RA, Paley MN, Van Beek EJ. Emphysematous changes and normal variation in smokers and COPD patients using diffusion 3He MRI. *Eur J Radiol* 54: 352–58, 2005.
69. Mata JF, Altes TA, Cai J, Ruppert K, Mitzner W, Hagspiel KD, Patel B, Salerno M, Brookeman JR, de Lange EE *et al.* Evaluation of emphysema severity and progression in a rabbit model: Comparison of hyperpolarized 3He and 129Xe diffusion MRI with lung morphometry. *J Appl Physiol* 102: 1273–80, 2007.

44

CHAPTER

Systemic Manifestations

Alvar Agusti

Hospital University Son Dureta, Servi Pneumologia, Palma de Majorca, Spain
Fundacion Caubet-CIMERA
CIBER Enfermedades Respiratorias

INTRODUCTION

Single organ diseases can have systemic manifestations. Acute diseases, such as urinary tract infections or pneumonia can cause septic shock, multiple organ failure, and eventually, death. Likewise, chronic diseases like cancer, heart failure, or chronic obstructive pulmonary disease (COPD), to name only a few, are associated with systemic manifestations as well. In a much more relentless way than in acute conditions, systemic manifestations in chronic diseases influence the course of the disease, and eventually, contribute to death. This chapter reviews the systemic manifestations associated to two chronic respiratory diseases, COPD, and asthma, and discusses their potential implications for the management and treatment of these patients.

SYSTEMIC MANIFESTATIONS OF COPD

The COPD guidelines published jointly by the American Thoracic Society (ATS) and European Respiratory Society (ERS) in 2004 were the first to explicitly recognize that "*although chronic obstructive pulmonary disease (COPD) affects the lungs, it also produces significant systemic consequences*" [1]. More recently, the last update of the Global Obstructive Lund Disease (GOLD) initiative published in 2006 incorporates and extends this concept by defining COPD as "*a preventable and treatable disease with some significant extra-pulmonary effects that may contribute to the severity in individual patients*" [2]. It is interesting to note that the systemic manifestations of the disease are quoted in this update even before stating that "*its pulmonary component is characterized by airflow limitation that is not fully reversible*" [2]. This highlights the clinical relevance of these extra-pulmonary effects because: (1) they contribute to disease progression independently of airflow obstruction. In fact, it is likely that important clinical outcomes, such as mortality or health status, among others, eventually result from the interplay between the intra-pulmonary and extra-pulmonary effects that occur in COPD (Fig. 44.1) [3]; and (2) they constitute novel therapeutic targets [4]. Table 44.1 presents the systemic manifestations of COPD identified so far.

Nutritional abnormalities

Unexplained weight loss occurs in about 50% of patients with severe COPD, but it can also be seen in patients with mild to moderate disease (10–15%) [5]. It is a poor prognostic indicator [6, 7] and interestingly, it is independent of other traditional indices, such as the volume of air exhaled in the first second of a forced spirometry maneuver (FEV_1) or the arterial partial pressure of oxygen [6]. Therefore, weight loss identifies a new systemic domain of COPD that needs to be taken into consideration in their clinical management. In this context, Celli *et al.* have recently proposed a composite index (the BODE index) that includes body weight (assessed by the body mass index, BMI), the degree of airflow obstruction (assessed by the FEV_1 value expressed as percentage of the reference value), the level of dyspnea experienced by the patient (assessed by the modified Medical Research Council scale), and their exercise capacity (assessed by the 6-min walking test), that predicts survival much more accurately than FEV_1 alone [8]. This approach has the

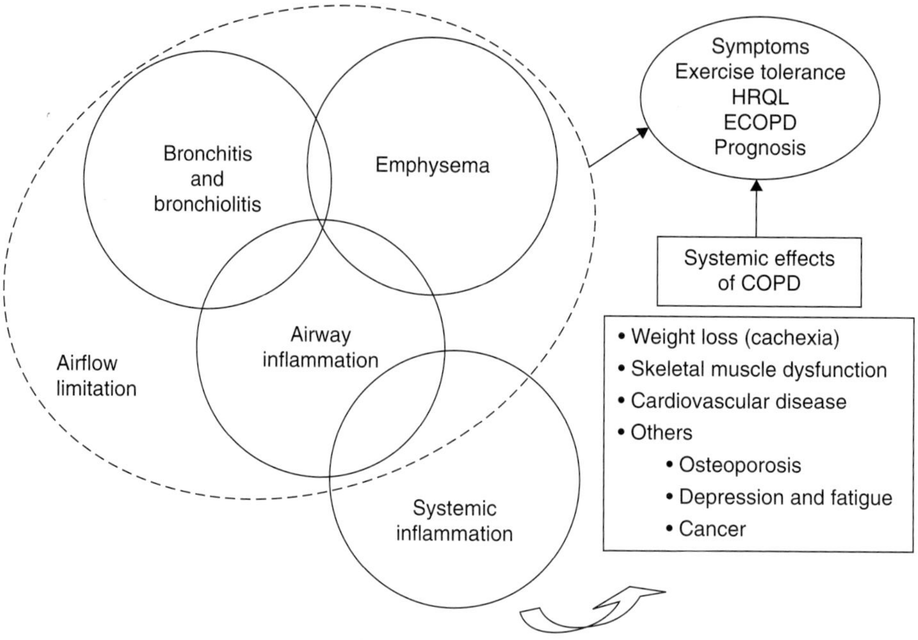

FIG. 44.1 The interaction between the pulmonary and extra-pulmonary components of COPD contributes the clinical presentation of the disease. Reproduced from Ref. [3].

TABLE 44.1 Systemic effects of COPD.

Nutritional abnormalities
Skeletal muscle dysfunction
Cardiovascular disease
Depression
Osteoporosis
Metabolic syndrome
Systemic inflammation

historical merit of being the first to assess COPD multidimensionally, but it is likely that the BODE index may be modified in the future by the inclusion of other relevant domains of the disease, such as the degree of inflammation [9], the number and severity of exacerbations episodes [10], and/or the severity of lung hyperinflation [11], to name only a few.

Skeletal muscle dysfunction

Unexplained weight loss in COPD is mostly due to skeletal muscle atrophy [12, 13]. Further, the remaining muscle mass is often dysfunctional [14]. This combination contributes significantly to reduce the exercise capacity (thus, health status) of patients with COPD [14] and is likely to contribute to the BODE index discussed earlier. The mechanisms underlying skeletal muscle abnormalities in COPD are not precisely defined, but they are probably multiple and interdependent, including sedentarism, malnutrition, systemic inflammation, and tissue hypoxia among others [13].

In this context, systemic inflammation is probably a key contributor [15]. Several molecular abnormalities have been described. On the one hand, a number of cytokines, particularly TNF-α, activate the transcription factor NF-κB, upregulate the inducible form of the nitric oxide synthase (iNOS) and facilitate the degradation of myosin heavy chains through the ubiquitin–proteasome complex [16]. On the other hand, these same cytokines have the potential to promote apoptosis (programmed cell death) in skeletal muscle cells, and this has been shown to occur in patients with COPD and low body weight [17]. Finally, abnormalities of the sarcoplasmic–endoplasmic reticulum calcium adenosine triphosphatase 2 (SERCA2) have been described by Morlá *et al.* [18]. These investigators found that the expression of SERCA2 was significantly lower in COPD patients with low BMI (another domain captured by the BODE index discussed earlier) and, interestingly, related to iNOS [18]. All these molecular abnormalities are likely to contribute to skeletal muscle atrophy and dysfunction in COPD [19, 20].

Cardiovascular disease

Cardiovascular disease (CVD) is another clinically relevant systemic manifestation of COPD [21]. Despite the fact that tobacco smoking is a major risk factor for both COPD and CVD, recent investigations have shown that the risk of CVD is higher among those smokers who develop COPD as compared to those who do not develop it [21]. For instance, Sin *et al.* reported that when the lowest quintile of FEV_1 is compared with the highest quintile, the risk of cardiovascular mortality increases by approximately 75% both in men and in women, and that the presence of symptoms of chronic bronchitis increases the risk of coronary deaths by 50% [22]. Further, the presence of ventricular arrhythmias increases the risk of coronary events two-fold, suggesting that the cardiovascular effects of COPD are amplified in those who have underlying cardiac rhythm disturbances [22]. Sin *et al.* calculated that, for every 10% decrease in FEV_1, all-cause mortality increases by 14%, CVD mortality by 28%, and nonfatal coronary event increases by 20%, and concluded that COPD is a powerful, independent risk factor for CVD morbidity and mortality [22]. These findings were later confirmed in a large ($n = 11{,}493$), retrospective cohort study in health-care databases maintained by the government of Saskatchewan (Canada). Curkendall *et al.* found that persons diagnosed and treated for COPD have a significantly increased risk for hospitalizations and deaths due to CVD [23]. The mechanism(s) for the synergy between COPD and CVD are still under investigation, but the low-grade persistent systemic inflammation that characterizes COPD is likely to contribute significantly [21, 24] by causing endothelial dysfunction [25, 26].

Depression

In patients with severe COPD (FEV_1 <50% reference value), the prevalence of depression was 25.0% whereas it was 19.6% in patients with mild to moderate COPD and 17.5% in controls [27]. When results were adjusted for demographic variables and comorbidity, the risk for depression was 2.5 times greater for patients with severe COPD than for controls [27].

Osteoporosis

Osteoporosis in COPD might develop possibly due to a number of factors related or not related to the disease, including age, sedentarism, smoking, steroid use, and systemic inflammation [28, 29]. In a study by Jorgensen *et al.*, 68% of patients with severe COPD (FEV_1 33% of reference value) had osteoporosis or osteopenia and the use steroids could not explain this high prevalence [30]. The occurrence of fractures, as a consequence of osteoporosis, can contribute to the disability and mortality of patients with COPD and add to the economic burden of the disease [31]. Randomized placebo-controlled trials are required to assess the effect of intervention, such as bisphosphonates, hormone replacement, calcium supplementation, on the prevention and treatment of osteoporosis and fractures in these patients [31]. Interestingly, very recently Sabit and coworkers have described a link between CVD and osteoporosis in patients with COPD [32]. These authors found that arterial stiffness and serum inflammatory mediators were particularly increased in patients with osteoporosis suggesting that age-related bone and vascular changes occur prematurely in COPD [32].

Metabolic syndrome

The metabolic syndrome is characterized by the presence of abdominal obesity, atherogenic dyslipidemia, raised blood pressure, presence of insulin resistance, and prothrombotic and inflammatory states that predispose to CVD [33]. It is frequently present in patients with COPD [34]. For instance, its prevalence in patients with moderate to severe COPD (FEV_1 43% of reference value) who participated in a cardiopulmonary rehabilitation program was 47%, whereas it was 21% in controls [33].

Insulin resistance and diabetes mellitus are a key components of the metabolic syndrome [33]. In the Nurses' Health Study (a prospective cohort that included 103,614 female nurses) during 8 years of follow-up, a total of 2,959 new cases of type 2 diabetes were identified [35]. The risk of type 2 diabetes was significantly higher (multivariate relative risk 1.8) for patients with COPD, suggesting that COPD may be a risk factor for developing type 2 diabetes [35]. Authors speculated that systemic inflammation in these patients might contribute to the increased risk of type 2 diabetes [35].

Systemic inflammation

The systemic manifestations of COPD reviewed earlier have been often related to the presence of systemic inflammation in these patients [34]. Many different studies have now provided convincing evidence that COPD is associated with increased levels of several pro-inflammatory cytokines (such as TNF-α and its soluble receptors soluble (sTNF-R55 and sTNF-R75)), interleukins 6 (IL-6) and 8 (IL-8), acute phase reactants (such as C-reactive protein, CRP), oxidative stress and activations of several inflammatory cells (such as neutrophils, monocytes, and lymphocytes) [13, 36]. Gan and associates conducted a meta-analysis encompassing all these previous studies and showed beyond doubt that COPD is associated with systemic inflammation [15]. Interestingly, other chronic conditions such as chronic heart failure, asthma (reviewed later), obesity, diabetes, and the normal process of aging are also associated with a similar low-grade systemic inflammatory process [37–39].

Many questions related to systemic inflammation in COPD remain unanswered [3]. For instance, we do not know if systemic inflammation occurs in all patients with COPD or only in a subgroup of them. Also, because all studies carried out so far are cross-sectional, we do not know the quantitative and qualitative longitudinal variation of systemic inflammation in a given patient, although recent data by Pinto Plata *et al.* indicate that, in the absence of an episode of exacerbation, CRP values tend to be relatively stable [40, 41]. In contrast, systemic inflammation (like pulmonary inflammation) appears to burst during the episodes of exacerbation of the disease [42–46].

The origin of systemic inflammation in COPD is unclear. It is possible that the pulmonary inflammation that characterizes COPD "spills-over" into the systemic circulation and/or contribute to activate the inflammatory cells during their transit through the pulmonary circulation [13]. Yet, several observations appear to contradict this hypothesis. For instance, Vernooy *et al.* could not find any relationship between the levels of soluble TNF-α (and its receptors) and IL-8 measured in induced sputum and in plasma of 18 COPD patients [47]. More recently, Hurst *et al.* were also unable to find such relationships for individual cytokines, albeit overall, global pulmonary inflammation correlated with global systemic inflammation [48]. Another mechanism of systemic inflammation in COPD is smoking.

It can cause by itself, that is, in the absence of COPD, systemic inflammation [49]. However, Vernooy *et al.* found that former smoker patients with COPD also had systemic inflammation [47]. Interestingly, the persistence of inflammation after smoking cessation occurs also in the lungs of patients with COPD [9, 50] and has actually raised the possibility that the pathogenesis of COPD may include an auto-immune component [51]. Another potential mechanism of systemic inflammation in COPD is the lung hyperinflation that occurs as a consequence of the chronic airflow obstruction that characterizes the disease [52], because it can stimulate the production of cytokines from the lungs [53–55]. Importantly, if this was the case, systemic inflammation in COPD should be responsive to bronchodilator therapy. This hypothesis is testable. In fact, in support of this possibility is the very recent observation that patients undergoing lung volume reduction surgery (a surgical procedure aimed at, basically, reduce lung hyperinflation) present a reduced incidence of episodes of exacerbation of COPD, which are characterized by a burst of inflammation [56]. Other potential origin of systemic inflammation in COPD is the skeletal muscle, which, as discussed above, is often abnormal and dysfunctional in these patients. In support of this possibility, Rabinovitch *et al.* have shown that, contrary to healthy subjects, systemic inflammation increases after muscular exercise in patients with COPD [57]. It has to be considered also that the bone marrow is the site of production of inflammatory cells and that smoking, as well as air pollution, can stimulate their release from the bone marrow [58–60]. Finally, the possibility that bone marrow abnormalities can contribute to systemic inflammation in COPD has not been studied systematically, but a recent study by Palange *et al.* showed that circulating hemopoietic progenitors ($CD34^+$ cells) were significantly reduced in COPD patients as compared to controls and that CD34[+] cell counts correlated with exercise capacity and severity of airflow obstruction [61]. Two final considerations deserve comment when considering the origin of systemic inflammation in COPD. First, the normal aging process is also associated with low-grade systemic inflammation [62, 63], and COPD is an age-related disease [39]. Further, smoking has been shown to enhance telomeric loss (a marker of cell aging) (Fig. 44.2) [64], and evidence for cell senescence has been recently identified in the lung parenchyma of patients with emphysema [65, 66]. Thus, a better understanding of the mechanisms of lung aging and their relationship with the pathogenesis of COPD in general, and to systemic inflammation in particular, may be very valuable in this setting. Second, even in subjects with lung function values within the normal range (FEV_1 values above 80% predicted), an inverse linear relationship exists between CRP concentrations and FEV_1 [67]. This relationship occurs also in never-smoker individuals, suggesting that systemic inflammation may be linked to early perturbations of pulmonary function [67]. Whether systemic inflammation precedes or follows lung function changes requires longitudinal studies.

Irrespective of its origin, an intriguing possibility about systemic inflammation in COPD is that it may "feed back" to the lungs and contribute to enhance the chronic destructive process that characterizes the disease. In other words, a systemic manifestation of COPD may contribute to the core pulmonary domain of the disease. It is possible,

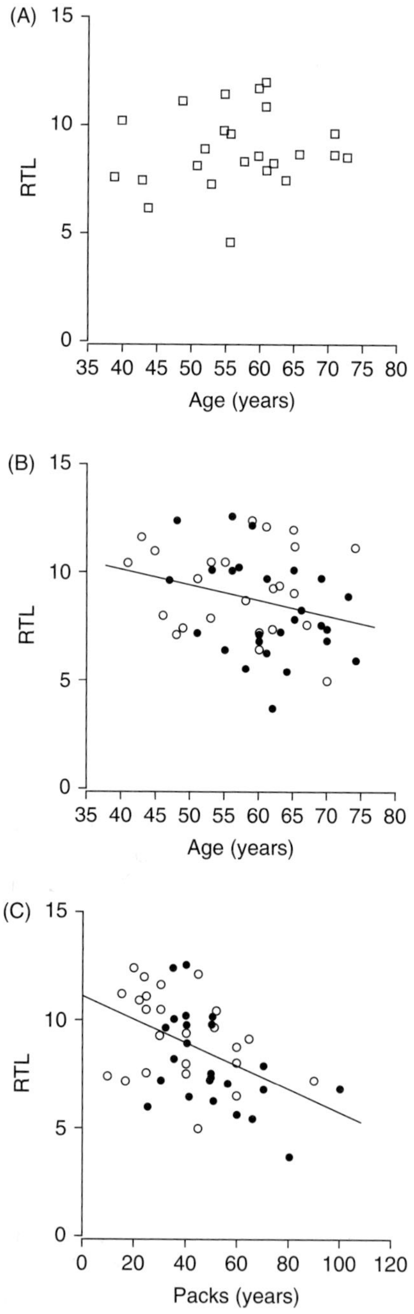

FIG. 44.2 Relationship between relative telomeric length (RTL), a marker of biological ageing, and chronological age in never smokers (*panel A*) and smokers with (closed symbols) and without COPD (open symbols) (*panel B*). *Panel C* shows the relationship between RTL and FEV_1 in smokers with (closed symbols) and without COPD (open symbols). For further explanations, see text. Reproduced from Ref. [64].

although not proven, that systemic oxidative stress [68] and the increased plasma levels of pro-inflammatory cytokines (most notably TNF-α [69]) can contribute to the pathogenesis of lung damage in COPD because, first, the inhibition of the vascular endothelial growth factor (VEGF) receptors cause lung cell apoptosis and emphysema [70], second, oxidative stress and apoptosis interact and cause emphysema due to VEGF receptor blockade [71], and third, endothelial cell death and decreased expression of VEGF and the VEGF receptor KDR/FLK-1 occur in patients with smoking-induced emphysema [72].

Therapeutic implications

The systemic manifestations of COPD discussed earlier offer new opportunities for therapy because they constitute novel targets beyond the traditional therapies in COPD (broncho-dilators, inhaled steroids), which aimed at improving lung function [2]. For instance, malnutrition and poor skeletal muscle performance can be improved by diet supplementation and adequate rehabilitation programs [2, 73]. Tissue hypoxia and the associated systemic manifestations are potentially treatable by domiciliary oxygen therapy [74, 75]. Likewise, those systemic manifestations of COPD related to systemic inflammation may be responsive to anti-inflammatory therapy. Sin and coworkers showed that withdrawal of inhaled corticosteroids in patients with COPD increased the plasma levels of CRP by about 30% and that 2 weeks of treatment with inhaled fluticasone (or oral prednisolone) reduced them by about 50% [76]. If low-grade chronic systemic inflammation is really a relevant mechanism in the pathogenesis of many of the systemic effects of COPD described earlier, then it is likely that the biological effects described by Sin *et al.* may well be clinically relevant. In fact, a retrospective study has suggested that the risk of acute myocardial infarction in COPD patients was reduced by 32% in those receiving low doses of inhaled steroids [77]. The recently published, study, TORCH has explored the potential beneficial effect on survival of the combination of salmeterol and fluticasone propionate (versus placebo) in patients with moderate and severe COPD [78]. Strictly speaking, the *p* value of the study did not reach the predetermined level of statistical significance, but this occurred only after adjusting for the potential effect of the interim analysis performed. In absolute terms, there was a reduction in the risk of death of 17.5% in the patients receiving the combination of a salmeterol and fluticasone propionate [78], a reduction that is in the same order of magnitude as that provided by statins. Unfortunately, no measurement of systemic inflammation was obtained in these subjects, so the interpretation of these results is still open [78]. Other therapeutic alternatives, such as the use of the antioxidant N-acetyl-cysteine [79, 80], anti-TNF-α or other cytokine antibodies, NF-κB blockers and i-NOS inhibitors, among others, require large and rigorously conducted clinical trials before entering clinical practice but, as recently reviewed [81, 82], they may be interesting alternatives. Finally, other systemic manifestations of COPD may deserve specific therapy. For instance, because patients with COPD are at increased risk of developing depression, particularly those with severe airflow obstruction, living alone, with airflow reversibility, respiratory symptoms, and physical impairment [27], it is important to reduce symptoms and improve physical functioning both (rehabilitation) as well as use anti-depressants drugs if needed. Likewise, as the prevalence of osteoporosis and osteopenia is increased in severe COPD, it is necessary to select the individuals at risk and to initiate prophylaxis or treatment for the disease [30].

TABLE 44.2 Systemic manifestations of asthma.

Systemic inflammation
Obesity
Diabetes mellitus
Cardiovascular disease
Skeletal muscle abnormalities
Osteoporosis
Depression and panic disorders
Inflammatory bowel disease

SYSTEMIC MANIFESTATIONS OF ASTHMA

Although often overlooked, asthma is frequently associated with systemic manifestations. The link between asthma and rhinitis, sinusitis, or allergic dermatitis is a good examples [83]. However, other potential systemic manifestations of asthma (Table 44.2) have been less extensively studied than those of COPD discussed earlier. Several features can contribute to explain this. First, asthma tends to be a disease of young (otherwise healthy) individuals [84], whereas COPD is rarely diagnosed before the sixth decade of age in individuals who often have other comorbidities [2, 85]. Second, lung disease is reversible in asthma but mostly irreversible in COPD [86]. It is plausible that the chronicity of the disease may favor the development of abnormalities in other distant organs. Finally, available therapies (mostly inhaled steroids with or without long-acting bronchodilators) are very effective in the vast majority of asthmatics [84] whereas, interestingly, the same therapy is considerably less effective in patients with COPD [2].

Systemic inflammation

Airway inflammation is a key pathologic mechanism in asthma [84]. As in COPD, there may be cross-communication between the airways and the bone marrow in patients with asthma through inflammatory mediators that can eventually lead to systemic inflammation [83]. Sauleda *et al.* were the first to report that the activity of cytochrome oxidase (CytOx), the terminal enzyme of the mitochondrial respiratory chain, was abnormally increased in circulating lymphocytes of patients with asthma (as compared to those with COPD, arthritis and/or healthy subjects) [87] (Fig. 44.3). This was a totally unexpected and serendipitous observation because the goal of that study was to investigate if the observation made previously by these same investigators of increased CytOx activity in skeletal muscle of patients with COPD [88] could be reproduced in other cell types and other chronic inflammatory diseases. As shown in Fig. 44.2, CytOx activity was higher in COPD than in healthy subjects (as previously observed in skeletal muscle [88]), but its activity was one order of magnitude greater in patients

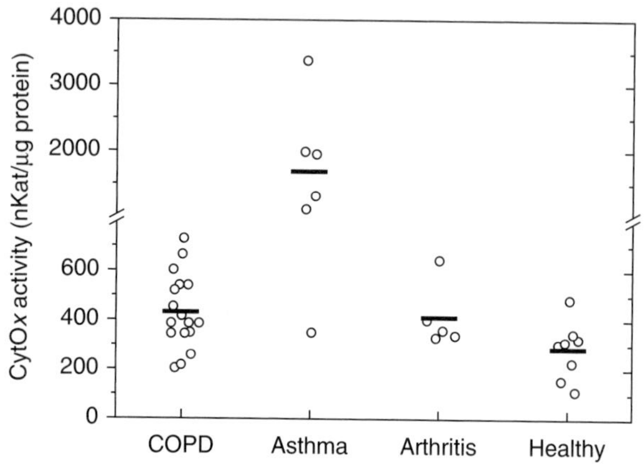

FIG. 44.3 Cytochrome oxidase activity in circulating lymphocytes of healthy subjects and patients with COPD, asthma and rheumatoid arthritis. Reproduced from Ref. [87].

with asthma [87] (Fig. 44.3). The mechanisms and implications of these findings were unclear, but authors speculated that they were likely to reflect activation of circulating lymphocytes (thus systemic inflammation) in these conditions [87]. Other studies that were published later, provided further evidence of systemic inflammation in asthma. Jousilahti *et al.* found that the age-adjusted odds ratios of asthma increased in proportion to the levels of several inflammatory markers in plasma (CRP, serum amyloid-A, and fibrinogen), and that this association was independent of smoking, the waist-to-hip ratio (a marker of central obesity), and symptoms of chronic bronchitis [89]. Silvestri *et al.* reported that patients with severe asthma had higher circulating neutrophil and eosinophil number, as well as increased serum levels of IL-8 and TNF-α, than controls and that there was a positive correlation between circulating neutrophil counts, plasma TNF-α levels, and exhaled nitric oxide (a well-established marker of pulmonary inflammation in asthma) as well as between TNF-α and IL-8 plasma levels and the degree of airflow obstruction (FEV_1, % of reference) [90]. Higashimoto *et al.* found that the systemic inflammatory profile of patients with asthma was similar to that of patients with COPD, including increases in serum CRP, fibrinogen, TNF-α, IL-6, and the tissue inhibitor of matrix metalloproteinase TIMP-1 [91]; interestingly, however, serum TGF-β1 levels were higher in asthma patients than in COPD patients [91]. Finally, the latest US National Health and Nutrition Examination Survey in 8020 adults found that those with asthma have higher levels of CRP in plasma, suggesting that this may be a clinically useful marker for asthma severity and control [92]. From all these studies, therefore, it is clear that asthma is also associated with systemic inflammation. Like in COPD, thus, it is possible that systemic inflammation in asthma contributes to some (or all) the systemic manifestations reviewed later.

Obesity

At variance with COPD, where unexplained weight loss is frequent (see above), in patients with asthma, a relationship with obesity has been suggested [93, 94]. For instance, the First National Health and Nutrition Examination Survey studied 14,407 subjects, aged 25–74, followed-up during a mean of 10 years and found that, both at baseline and during the follow-up, increasing BMI was associated with increased prevalence of asthma [95]. Other studies have shown that obesity precedes and predicts the onset of asthma, increased obesity leads to more severe asthma, weight reduction improves asthmatic symptoms, and obesity co-occurs with intermediate asthma phenotypes [93, 94]. These observations suggest that obesity and asthma may be causally related through immunologic and inflammatory, hormonal, genetic, and/or nutritional mechanisms [93].

Diabetes mellitus

It is unclear whether or not asthma and type 1 diabetes mellitus are related (and the direction of such potential relationship). Contrary to findings in COPD, the Nurses' Health Study could not detect an increased risk of type 2 diabetes among asthmatic patients [35]. Yet, other authors have suggested an inverse relationship between asthma and type 1 diabetes mellitus (DM1) [96]. It has been hypothesized that because, according to the "hygiene hypothesis," the risk of asthma decreases with infections early in childhood by shifting the Th2 profile, common at birth, to the Th1 phenotype, the latter may favor the development of DM1 [96]. Thus further research is required to find out whether or not asthma and DM1 are related.

Cardiovascular disease

Several pieces of evidences indicate that the risk of CVD is increased in asthmatics [97]. Schanen *et al.* reported that asthma was an independent risk factor for incident stroke (but not coronary heart disease) in a cohort of 13,501 adults followed during 14 years [98]. In an even larger cohort (74,342 participants in the Canadian Community Health Survey), Dogra *et al.* found that asthmatics were 43% more likely to have heart disease and 36% more likely to have high blood pressure than non-asthmatics [97]. Interestingly, the age of onset of asthma did not influence this relationship significantly [97]. Finally, Wu *et al.* reported increases in multiple inflammatory markers of atherosclerosis in patients with asthma [86]. Thus, available evidence supports the concept that CVD risk is increased in patients with asthma.

Skeletal muscle abnormalities

Skeletal muscle abnormalities are uncommon (or at least, seldom studied) in patients with asthma, at variance with COPD (see above). In one of the rare studies that have addressed this issue, Picado *et al.* could not find significant differences between oral steroid-dependent asthmatics and age- and sex-matched steroid naïve asthmatics in respiratory and skeletal muscle strength, endurance time, or fiber-type composition [99].

Osteoporosis

A recent systematic review of the literature by Kearney *et al.* found that patients with asthma (and COPD, as discussed earlier) are at risk of osteoporosis, particularly those treated with high doses of corticosteroids [100].

Depression and panic disorders

Depression and panic disorders are frequent in patients with asthma. Schneider *et al.* found that about 4% of asthmatics visited in primary care suffered from major depressive disorder, 23% from minor depressive disorder, and 8% from panic disorder [101]. Further, the presence of depression predicted hospitalization within the subsequent year (odds ratios, OR 6.1) and panic disorder predicted unscheduled emergency home visits (OR 4.8). Likewise, McCauley *et al.* found that 125 out of 767 adolescents with asthma (16.2%) reported significant anxiety and depressive disorders in the last 12 months, and this was related to functional status [102]. Overall, these results indicate that the impact of psychiatric comorbidity on health outcomes for patients with asthma is substantial [103]. Interestingly, the frequency of symptoms of anxiety and depression is greater among asthma patients (particularly those with uncontrolled asthma) than among patients with COPD [104].

Inflammatory bowel disease

The prevalence of asthma is particularly increased in patients with inflammatory bowel disease. Berstein *et al.* found that patients with ulcerative colitis (n = 3879) and Crohn's disease (n = 4193) had a significantly greater likelihood of having arthritis, asthma, bronchitis, psoriasis, and pericarditis than population controls, and that asthma was the most common comorbidity in Crohn's disease patients [105]. The reasons for these associations are unknown but shared inflammatory mechanisms have been postulated [83].

Therapeutic implications

Inhaled steroids (with or without long-acting bronchodilators) are very effective in patients with asthma [84]. They reduce local inflammation and improve pulmonary function but their effects upon the systemic manifestations of asthma are less well studied [83]. In the context of the systemic manifestations of asthma, oral medications, such as leukotriene receptor antagonists, can block the action of cysteinyl leukotrienes and thus improve both asthma and rhinitis and other conditions systemically linked with asthma [83]. Other potential treatments include receptor-blocking molecules and synthesis inhibitors related to eicosanoid inflammation [83]. Further, as depression and panic disorders are frequent in patients with asthma and both are predictive of health-care resources utilization, specific psychiatric interventions may reduce inappropriate health-care utilization and avoid adverse outcomes [101].

References

1. Celli BR, MacNee W, Agusti AG, Anzueto A, Berg BR, Buist AS, Calverley P, Chavannes N, Dillard T, Fahy B, Fein A, Heffner J, Lareau SC, Meek P, Martinez F, McNicholas WT, Muris J, Austergard E, Pauwels R, Rennard S, Rossi A, Siafakas N, Tiep B, Vestbo J, Wouters E, ZuWallack R. Standards for the diagnosis and treatment of patients with COPD: A summary of the ATS/ERS position paper. *Eur Respir J* 23: 932–46, 2004.
2. Rabe KF, Hurd S, Anzueto A, Barnes PJ, Buist SA, Calverley P, Fukuchi Y, Jenkins C, Rodriguez-Roisin R, van WC, Zielinski J. Global strategy for the diagnosis, management, and prevention of chronic obstructive pulmonary disease: GOLD executive summary. *Am J Respir Crit Care Med* 176: 532–55, 2007.
3. Agusti A. Systemic effects of chronic obstructive pulmonary disease: What we know and what we don't know (but should). *Proceedings of the American Thoracic Society*, 4:522–25, 2007.
4. Agusti A, Thomas A. Neff lecture. Chronic obstructive pulmonary disease: A systemic disease. *Proc Am Thorac Soc* 3: 478–81, 2006.
5. Schols AM, Soeters PB, Dingemans AM, Mostert R, Frantzen PJ, Wouters EF. Prevalence and characteristics of nutritional depletion in patients with stable COPD eligible for pulmonary rehabilitation. *Am Rev Respir Dis* 147: 1151–56, 1993.
6. Schols AM, Slangen J, Volovics L, Wouters EF. Weight loss is a reversible factor in the prognosis of chronic obstructive pulmonary disease. *Am J Respir Crit Care Med* 157: 1791–97, 1998.
7. Landbo C, Prescott E, Lange P, Vestbo J, Almdal TP. Prognostic value of nutritional status in chronic obstructive pulmonary disease. *Am J Respir Crit Care Med* 160: 1856–61, 1999.
8. Celli BR, Cote CG, Marin JM, Casanova C, Montes de Oca M, Mendez RA, Pinto Plata V, Cabral HJ. The Body-mass index, airflow obstruction, dyspnea, and exercise capacity index in chronic obstructive pulmonary disease. *N Engl J Med* 350: 1005–12, 2004.
9. Hogg JC, Chu F, Utokaparch S, Woods R, Elliott WM, Buzatu L, Cherniack RM, Rogers RM, Sciurba FC, Coxson HO, Pare PD. The nature of small-airway obstruction in chronic obstructive pulmonary disease. *N Engl J Med* 350: 2645–53, 2004.
10. Soler-Cataluna JJ, Martinez-Garcia MA, Roman Sanchez P, Salcedo E, Navarro M, Ochando R. Severe acute exacerbations and mortality in patients with chronic obstructive pulmonary disease. *Thorax* 60: 925–31, 2005.
11. Casanova C, Cote C, de Torres JP, guirre-Jaime A, Marin JM, Pinto-Plata V, Celli BR. Inspiratory-to-total lung capacity ratio predicts mortality in patients with chronic obstructive pulmonary disease. *Am J Respir Crit Care Med* 171: 591–97, 2005.
12. Wouters EFM. A wasting disease. In: Voelkel NF, MacNee W (eds). *Chronic Obstructive Lung Disease*, pp. 364–76. Hamilton, ON, Canada: B.C. Dekker, 2002.
13. Agusti AG, Noguera A, Sauleda J, Sala E, Pons J, Busquets X. Systemic effects of chronic obstructive pulmonary disease. *Eur Respir J* 21: 347–60, 2003.
14. American Thoracic Society and European Respiratory Society. Skeletal muscle dysfunction in chronic obstructive pulmonary disease. *Am J Respir Crit Care Med* 159: S1–S40, 1999.
15. Gan WQ, Man SF, Senthilselvan A, Sin DD. Association between chronic obstructive pulmonary disease and systemic inflammation: A systematic review and a meta-analysis. *Thorax* 59: 574–80, 2004.
16. Agusti A, Morla M, Sauleda J, Saus C, Busquets X. NF-KB activation and iNOS upregulation in skeletal muscle of patients with COPD and low body weight. *Thorax* 59: 483–87, 2004.
17. Agusti AGN, Sauleda J, Miralles C, Gomez C, Togores B, Sala E, Batle S, Busquets X. Skeletal muscle apoptosis and weight loss in chronic obstructive pulmonary disease. *Am J Respir Crit Care Med* 166: 485–89, 2002.
18. Morla M, Iglesias A, Sauleda J, Cosio B, Agusti A, Busquets X. [Reduced expression of the sarcoplasmic calcium pump SERCA2 in

skeletal muscle from patients with chronic obstructive pulmonary disease and low body weight]. *Arch Bronconeumol* 43: 4–8, 2007.
19. Montes de Oca M, Loeb E, Torres SH, De Sanctis J, Hernandez N, Talamo C. Peripheral muscle alterations in non-COPD smokers. *Chest* 133: 13–18, 2008.
20. Barreiro EA, Schols MWJ, Polkey MI, Galdiz JB, Gosker HR, Swallow EB, Coronell C, Gea J, on behalf of the ENIGMA in COPD project. Cytokine profile in quadriceps muscles of patients with severe COPD. *Thorax* 63: 100–107, 2008.
21. Sin DD, Man SF. Why are patients with chronic obstructive pulmonary disease at increased risk of cardiovascular diseases? The potential role of systemic inflammation in chronic obstructive pulmonary disease. *Circulation* 107: 1514–19, 2003.
22. Sin DD, Man SF. Chronic obstructive pulmonary disease as a risk factor for cardiovascular morbidity and mortality. *Proc Am Thorac Soc* 2: 8–11, 2005.
23. Curkendall SM, DeLuise C, Jones JK, Lanes S, Stang MR, Goehring E Jr, She D. Cardiovascular disease in patients with chronic obstructive pulmonary disease, Saskatchewan Canada cardiovascular disease in COPD patients. *Ann Epidemiol* 16: 63–70, 2006.
24. McAllister DA, Maclay JD, Mills NL, Mair G, Miller J, Anderson D, Newby DE, Murchison JT, MacNee W. Arterial Stiffness is Independently associated with emphysema severity in patients with COPD. *Am J Respir Crit Care Med* 176: 1208–14, 2007.
25. Dinh-Xuan AT, Higenbottam TW, Clelland CA, Pepke-Zaba J, Cremona G, Butt Y, Large SR, Wells FC, Wallwork J. Impairment of endothelium-dependent pulmonary-artery relaxation in chronic obstructive lung disease. *N Engl J Med* 324: 1539–47, 1991.
26. Saadjian A, Philip-Joet F, Levy S, Arnaud A. Vascular and cardiac reactivity in pulmonary hypertension due to chronic obstructive lung disease: Assessment with varios oxygen concentrations. *Eur Respir J* 5: 525–30, 1992.
27. van Manen JG, Bindels PJ, Dekker FW, IJzermans CJ, van der Zee JS, Schade E. Risk of depression in patients with chronic obstructive pulmonary disease and its determinants. *Thorax* 57: 412–16, 2002.
28. Incalzi RA, Caradonna P, Ranieri P, Basso S, Fuso L, Pagano F, Ciappi G, Pistelli R. Correlates of osteoporosis in chronic obstructive pulmonary disease. *Respir Med* 94: 1079–84, 2000.
29. Gross NJ. Extrapulmonary effects of chronic obstructive pulmonary disease. *Curr Opin Pulm Med* 7: 84–92, 2001.
30. Jorgensen NR, Schwarz P, Holme I, Henriksen BM, Petersen LJ, Backer V. The prevalence of osteoporosis in patients with chronic obstructive pulmonary disease: A cross sectional study. *Respir Med* 101: 177–85, 2007.
31. Ionescu AA, Schoon E. Osteoporosis in chronic obstructive pulmonary disease. *Eur Respir J Suppl* 46: 64s–75s, 2003.
32. Sabit R, Bolton CE, Edwards PH, Pettit RJ, Evans WD, McEniery CM, Wilkinson IB, Cockcroft JR, Shale DJ. Arterial stiffness and osteoporosis in chronic obstructive pulmonary disease. *Am J Respir Crit Care Med* 175: 1259–65, 2007.
33. Marquis K, Maltais F, Duguay V, Bezeau AM, Leblanc P, Jobin J, Poirier P. The metabolic syndrome in patients with chronic obstructive pulmonary disease. *J Cardiopulm Rehabil* 25: 226–32, 2005.
34. Fabbri LM, Rabe KF. From COPD to chronic systemic inflammatory syndrome?. *Lancet* 370: 797–99, 2007.
35. Rana JS, Mittleman MA, Sheikh J, Hu FB, Manson JE, Colditz GA, Speizer FE, Barr RG, Camargo CA Jr. Chronic obstructive pulmonary disease, asthma, and risk of type 2 diabetes in women. *Diabetes Care* 27: 2478–84, 2004.
36. Wouters EF, Creutzberg EC, Schols AM. Systemic effects in COPD. *Chest* 121: 127S–130S, 2002.
37. Sevenoaks MJ, Stockley RA. Chronic Obstructive Pulmonary Disease, inflammation and co-morbidity—a common inflammatory phenotype? *Respir Res* 7: 70, 2006.
38. De MM, Franceschi C, Monti D, Ginaldi L. Inflamm-ageing and lifelong antigenic load as major determinants of ageing rate and longevity. *FEBS Lett* 579: 2035–39, 2005.
39. Fabbri LM, Ferrari R. Chronic disease in the elderly: Back to the future of internal medicine. *Breathe* 3: 40–49, 2006.
40. Pinto-Plata V, Toso J, Lee K, Parks D, Bilello J, Mullerova H, Desouza M, Vessey RS, Celli B. Profiling serum biomarkers in patients with COPD: Associations with Clinical Parameters. *Thorax* 62: 595–601, 2007.
41. Pinto-Plata VM, Mullerova H, Toso JF, Feudjo-Tepie M, Soriano JB, Vessey RS, Celli BR. C-reactive protein in patients with COPD, control smokers and non-smokers. *Thorax* 61: 23–28, 2006.
42. Malo O, Sauleda J, Busquets X, Miralles C, Agusti AG, Noguera A. Inflamación sistémica durante las agudizaciones de la enfermedad pulmonar obstructiva crónica. *Arch Bronconeumol* 38: 172–76, 2002.
43. Dentener MA, Creutzberg EC, Schols AM, Mantovani A, van't Veer C, Buurman WA, Wouters EF. Systemic anti-inflammatory mediators in COPD: Increase in soluble interleukin 1 receptor II during treatment of exacerbations. *Thorax* 56: 721–26, 2001.
44. Drost EM, Skwarski KM, Sauleda J, Soler N, Roca J, Agusti A, MacNee W. Oxidative stress and airway inflammation in severe exacerbations of COPD. *Thorax* 60: 293–300, 2005.
45. Wedzicha JA, Seemungal TA, MacCallum PK, Paul EA, Donaldson GC, Bhowmik A, Jeffries DJ, Meade TW. Acute exacerbations of chronic obstructive pulmonary disease are accompanied by elevations of plasma fibrinogen and serum IL-6 levels. *Thromb Haemost* 84: 210–15, 2000.
46. Pinto-Plata VM, Livnat G, Girish M, Cabral H, Masdin P, Linacre P, Dew R, Kenney L, Celli BR. Systemic cytokines, clinical and physiological changes in patients hospitalized for exacerbation of COPD. *Chest* 131: 37–43, 2007.
47. Vernooy JH, Kucukaycan M, Jacobs JA, Chavannes NH, Buurman WA, Dentener MA, Wouters EF. Local and systemic inflammation in patients with chronic obstructive pulmonary disease: Soluble tumor necrosis factor receptors are increased in sputum. *Am J Respir Crit Care Med* 166: 1218–24, 2002.
48. Hurst JR, Wilkinson TM, Perera WR, Donaldson GC, Wedzicha JA. Relationships among bacteria, upper airway, lower airway, and systemic inflammation in COPD. *Chest* 127: 1219–26, 2005.
49. Yanbaeva DG, Dentener MA, Creutzberg EC, Wesseling G, Wouters EF. Systemic effects of smoking. *Chest* 131: 1557–66, 2007.
50. Retamales I, Elliot MW, Meshi B, Coxson HO, Pare PD, Sciurba FC, Rogers RM, Hayashi S, Hogg JC. Amplification of inflammation in emphysema and its association with latent adenoviral infection. *Am J Respir Crit Care Med* 164: 469–73, 2001.
51. Agusti A, MacNee W, Donaldson K, Cosio M. Hypothesis: Does COPD have an autoimmune component? *Thorax* 58: 832–34, 2003.
52. Parker CM, O'Donnell DE. Dynamic hyperinflation, dyspnoea and exercise intolerance in chronic obstructive pulmonary disease. In: Barnes PJ, Postma DS (eds). *Essentials of COPD Van Zuiden Communications*, pp. 2–30. The Netherlands: Van Zuiden Communications BV, 2005.
53. Vassilakopoulos T, Roussos C, Zakynthinos S. The immune response to resistive breathing. *Eur Respir J* 24: 1033–43, 2004.
54. Vassilakopoulos T, Katsaounou P, Karatza MH, Kollintza A, Zakynthinos S, Roussos C. Strenuous resistive breathing induces plasma cytokines: Role of antioxidants and monocytes. *Am J Respir Crit Care Med* 166: 1572–78, 2002.
55. Vassilakopoulos T, Zakynthinos S, Roussos C. Strenuous resistive breathing induces proinflammatory cytokines and stimulates the HPA axis in humans. *Am J Physiol* 277: R1013–19, 1999.
56. Washko GR, Fan VS, Ramsey SD, Mohsenifar Z, Martinez F, Make BJ, Sciurba FC, Criner GJ, Minai O, DeCamp MM, Reilly JJ, for the National Emphysema Treatment Trial Research Group. The Effect of Lung Volume Reduction Surgery on Chronic Obstructive Pulmonary Disease Exacerbations. *Am J Respir Crit Care Med* 177: 164–69, 2008.
57. Rabinovich RA, Figueras M, Ardite E, Carbo N, Troosters T, Filella X, Barbera JA, Fernandez-Checa JC, Argiles JM, Roca J. Increased

tumour necrosis factor-alpha plasma levels during moderate-intensity exercise in COPD patients. *Eur Respir J* 21: 789–94, 2003.
58. van Eeden SF, Yeung AC, Quinlam K, Hogg JC. Systemic response to ambient particulate matter. Relevance to chronic obstructive pulmonary disease. *Proc Am Thorac Soc* 2: 61–67, 2005.
59. Mukae H, Vincent R, Quinlan K, English D, Hards J, Hogg JC, van Eeden SF. The effect of repeated exposure to particulate air pollution (pm(10)) on the bone marrow. *Am J Respir Crit Care Med* 163: 201–9, 2001.
60. van Eeden SF, Hogg JC. The response of human bone marrow to chronic cigarette smoking. *Eur Respir J* 15: 915–21, 2000.
61. Palange P, Testa U, Huertas A, Calabro L, Antonucci R, Petrucci E, Pelosi E, Pasquini L, Satta A, Morici G, Vignola MA, Bonsignore MR. Circulating haemopoietic and endothelial progenitor cells are decreased in COPD. *Eur Respir J* 27: 529–41, 2006.
62. De Martinis M, Franceschi C, Monti D, Ginaldi L. Inflammation markers predicting frailty and mortality in the elderly. *Exp Mol Pathol* 80: 219–27, 2006.
63. Ginaldi L, De MM, Monti D, Franceschi C. Chronic antigenic load and apoptosis in immunosenescence. *Trends Immunol* 26: 79–84, 2005.
64. Morla M, Busquets X, Pons J, Sauleda J, MacNee W, Agusti AG. Telomere shortening in smokers with and without COPD. *Eur Respir J* 27: 525–28, 2006.
65. Muller KC, Welker L, Paasch K, Feindt B, Erpenbeck VJ, Hohlfeld JM, Krug N, Nakashima M, Branscheid D, Magnussen H, Jorres RA, Holz O. Lung fibroblasts from patients with emphysema show markers of senescence in vitro. *Respir Res* 7: 32, 2006.
66. Tsuji T, Aoshiba K, Nagai A. Alveolar cell senescence in patients with pulmonary emphysema. *Am J Respir Crit Care Med* 174: 886–93, 2006.
67. Aronson D, Roterman I, Yigla M, Kerner A, Avizohar O, Sella R, Bartha P, Levy Y, Markiewicz W. Inverse association between pulmonary function and C-reactive protein in apparently Healthy Subjects. *Am J Respir Crit Care Med* 174: 626–32, 2006.
68. Rahman I, Morrison D, Donaldson K, MacNee W. Systemic oxidative stress in asthma, COPD, and smokers. *Am J Respir Crit Care Med* 154: 1055–60, 1996.
69. Reid MB, Li YP. Cytokines and oxidative signalling in skeletal muscle. *Acta Physiol Scand* 171: 225–32, 2001.
70. Kasahara Y, Tuder RM, Taraseviciene-Stewart L, Le Cras TD, Abman S, Hirth PK, Waltenberger J, Voelkel NF. Inhibition of VEGF receptors causes lung cell apoptosis and emphysema. *J Clin Invest* 106: 1311–19, 2000.
71. Tuder RM, Zhen L, Cho CY, Taraseviciene-Stewart L, Kasahara Y, Salvemini D, Voelkel NF, Flores SC. Oxidative stress and apoptosis interact and cause emphysema due to vascular endothelial growth factor receptor blockade. *Am J Respir Cell Mol Biol* 29: 88–97, 2003.
72. Kasahara Y, Tuder RM, Cool CD, Lynch DA, Flores SC, Voelkel NF. Endotelial cell death and decreased expression of vegf and KDR/FLK-1 in smoking-induced emphysema. *Am J Respir Crit Care Med* 161: A583, 2000.
73. Wilt TJ, Niewoehner D, MacDonald R, Kane RL. Management of stable chronic obstructive pulmonary disease: A Systematic Review for a Clinical Practice Guideline. *Ann Intern Med* 147: 639–53, 2007.
74. Nocturnal Oxygen Therapy Trial Group. Continuous or nocturnal oxygen therapy in hypoxemic chronic obstructive lung disease. A clinical trial. *Ann Intern Med* 93: 391–98, 1980.
75. Report of the Medical Research Council Working Party. Long term domiciliary oxygen therapy in chronic hypoxic cor pulmonale complicating chronic bronchitis and emphysema. *Lancet* 1: 681–85, 1981.
76. Sin DD, Lacy P, York E, Man SFP. Effects of fluticasone on systemic markers of inflammation in chronic obstructive pulmonary disease. *Am J Respir Crit Care Med* 170: 760–65, 2004.
77. Huiart L, Ernst P, Ranouil X, Suissa S. Low-dose inhaled corticosteroids and the risk of acute myocardial infarction in COPD. *Eur Respir J* 25: 634–39, 2005.
78. Vestbo J. The TORCH (towards a revolution in COPD health) survival study protocol. *Eur Respir J* 24: 206–10, 2004.
79. Koechlin C, Couillard A, Simar D, Cristol JP, Bellet H, Hayot M, Prefaut C. Does oxidative stress alter quadriceps endurance in chronic obstructive pulmonary disease? *Am J Respir Crit Care Med* 169: 1022–27, 2004.
80. Decramer M, Rutten-Van MM, Dekhuijzen PN, Troosters T, van HC, Pellegrino R, Van Schayck CP, Olivieri D, Del DM, De BW, Lankhorst I, Ardia A. Effects of N-acetylcysteine on outcomes in chronic obstructive pulmonary disease (Bronchitis Randomized on NAC Cost-Utility Study, BRONCUS): A randomised placebo-controlled trial. *Lancet* 365: 1552–60, 2005.
81. Agusti AG. COPD, a multicomponent disease: Implications for management. *Respir Med* 99: 670–82, 2005.
82. Barnes PJ, Stockley RA. COPD: Current therapeutic interventions and future approaches. *Eur Respir J* 25: 1084–106, 2005.
83. Bjermer L. Time for a paradigm shift in asthma treatment: From relieving bronchospasm to controlling systemic inflammation. *J Allergy Clin Immunol* 120: 1269–75, 2007.
84. Bateman ED, Hurd SS, Barnes PJ, Bousquet J, Drazen JM, FitzGerald M, Gibson P, Ohta K, O'Byrne P, Pedersen SE, Pizzichini E, Sullivan SD, Wenzel SE, Zar HJ. Global strategy for asthma management and prevention: GINA executive summary. *Eur Respir J* 31: 143–78, 2008.
85. Fabbri LM, Luppi F, Beghe B, Rabe KF. Complex chronic comorbidities of COPD. *Eur Respir J* 31: 204–12, 2008.
86. Wu TL, Chang PY, Tsao KC, Sun CF, Wu LL, Wu JT. A panel of multiple markers associated with chronic systemic inflammation and the risk of atherogenesis is detectable in asthma and chronic obstructive pulmonary disease. *J Clin Lab Anal* 21: 367–71, 2007.
87. Sauleda J, Garcia-Palmer FJ, Gonzalez G, Palou A, Agusti AG. The activity of cytochrome oxidase is increased in circulating lymphocytes of patients with chronic obstructive pulmonary disease, asthma, and chronic arthritis. *Am J Respir Crit Care Med* 161: 32–35, 2000.
88. Sauleda J, García-Palmer FJ, Wiesner R, Tarraga S, Harting I, Tomas P, Gomez C, Saus C, Palou A, Agustí AGN. Cytochrome oxidase activity and mitochondrial gene expression in skeletal muscle of patients with chronic obstructive pulmonary disease. *Am J Respir Crit Care Med* 157: 1413–17, 1998.
89. Jousilahti P, Salomaa V, Hakala K, Rasi V, Vahtera E, Palosuo T. The association of sensitive systemic inflammation markers with bronchial asthma. *Ann Allergy Asthma Immunol* 89: 381–85, 2002.
90. Silvestri M, Bontempelli M, Giacomelli M, Malerba M, Rossi GA, Di SA, Rossi A, Ricciardolo FL. High serum levels of tumour necrosis factor-alpha and interleukin-8 in severe asthma: Markers of systemic inflammation? *Clin Exp Allergy* 36: 1373–81, 2006.
91. Higashimoto Y, Yamagata Y, Taya S, Iwata T, Okada M, Ishiguchi T, Sato H, Itoh H. Systemic inflammation in chronic obstructive pulmonary disease and asthma: Similarities and differences. *Respirology* 13: 128–33, 2008.
92. Arif AA, Delclos GL, Colmer-Hamood J. Association between asthma, asthma symptoms and C-reactive protein in US adults: Data from the National Health and Nutrition Examination Survey, 1999–2002. *Respirology* 12: 675–82, 2007.
93. Castro-Rodriguez JA. Relationship between obesity and asthma. *Arch Bronconeumol* 43: 171–75, 2007.
94. Brand C, Sundararajan V, Jones C, Hutchinson A, Campbell D. Readmission patterns in patients with chronic obstructive pulmonary disease, chronic heart failure and diabetes mellitus: An administrative dataset analysis. *Intern Med J* 35: 296–99, 2005.
95. Stanley AH, Demissie K, Rhoads GG. Asthma development with obesity exposure: Observations from the cohort of the National Health and Nutrition Evaluation Survey Epidemiologic Follow-up Study (NHEFS). *J Asthma* 42: 97–99, 2005.
96. Alves C, Diniz AB, Souza MB, Ponte EV, Araujo MI. Controversies in the association between type 1 diabetes and asthma. *Arq Bras Endocrinol Metabol* 51: 930–37, 2007.
97. Dogra S, Ardern CI, Baker J. The relationship between age of asthma onset and cardiovascular disease in Canadians. *J Asthma* 44: 849–54, 2007.

98. Schanen JG, Iribarren C, Shahar E, Punjabi NM, Rich SS, Sorlie PD, Folsom AR. Asthma and incident cardiovascular disease: The Atherosclerosis Risk in Communities Study. *Thorax* 60: 633–38, 2005.
99. Picado C, Fiz JA, Montserrat JM, Grau JM, Fernandez-Sola J, Luengo MT, Casademont J, gusti-Vidal A. Respiratory and skeletal muscle function in steroid-dependent bronchial asthma. *Am Rev Respir Dis* 141: 14–20, 1990.
100. Kearney DM, Lockey RF. Osteoporosis and asthma. *Ann Allergy Asthma Immunol* 96: 769–74, 2006.
101. Schneider A, Lowe B, Meyer FJ, Biessecker K, Joos S, Szecsenyi J. Depression and panic disorder as predictors of health outcomes for patients with asthma in primary care. *Respir Med* 102(3): 359–66, 2007.
102. McCauley E, Katon W, Russo J, Richardson L, Lozano P. Impact of anxiety and depression on functional impairment in adolescents with asthma. *Gen Hosp Psychiatry* 29: 214–22, 2007.
103. Kullowatz A, Kanniess F, Dahme B, Magnussen H, Ritz T. Association of depression and anxiety with health care use and quality of life in asthma patients. *Respir Med* 101: 638–44, 2007.
104. Carvalho NS, Ribeiro PR, Ribeiro M, Nunes MP, Cukier A, Stelmach R. Comparing asthma and chronic obstructive pulmonary disease in terms of symptoms of anxiety and depression. *J Bras Pneumol* 33: 1–6, 2007.
105. Bernstein CN, Wajda A, Blanchard JF. The clustering of other chronic inflammatory diseases in inflammatory bowel disease: A population-based study. *Gastroenterology* 129: 827–36, 2005.

Therapies for Asthma and COPD

PART 8

45

CHAPTER

Cardiovascular Effects

Don D. Sin

The James Hogg iCAPTURE Center for Cardiovascular and Pulmonary Research (St. Paul's Hospital), Vancouver, BC, Canada

INTRODUCTION

Chronic obstructive pulmonary disease (COPD) and asthma are inflammatory disorders of the lung, characterized by airflow limitation and symptoms of dyspnea and cough [1]. In both of these conditions spirometric parameters are usually abnormal with a forced expiratory volume in one second (FEV_1) to forced vital capacity (FVC) ratio of less than 80%. In asthma, this ratio may normalize following administration of a bronchodilator, whereas in COPD, there is little or no reversibility of lung function [1]. Although these disorders are distinct and have different pathophysiologies, in both conditions, cardiovascular complications are frequently observed. For instance, in patients with COPD, the risk of cardiovascular hospitalizations and mortality including those related to ischemic heart disease, stroke, arrhythmias, heart failure, and sudden death is two to three times higher than those without COPD [2]. In mild to moderate cases of COPD, the leading causes of hospitalization are cardiovascular events, accounting for approximately 50% of all cases [3]. In this chapter, we will review the epidemiological evidence and explore the potential mechanistic link(s) between obstructive airways diseases and cardiovascular comorbidities.

EPIDEMIOLOGY

Reduced lung function and CVD

One of the defining features of asthma and COPD is reduced FEV_1 in addition to a reduced FEV_1 to FVC ratio. There have been many studies that have evaluated the relationship between FEV_1 and cardiovascular diseases (CVD) in various populations. Almost universally, these studies have shown that reduced FEV_1, regardless of the cause, is a significant risk factor for CVD. As an example, in the First National Health and Examination Survey Epidemiologic Follow-up Study ($n = 1861$), Sin and colleagues found that individuals in the lowest quintile of FEV_1 (i.e. reduced lung function) had the highest risk for cardiovascular mortality (relative risk, RR, 3.36; 95% confidence interval, CI, 1.54–7.34), while individuals in the highest FEV_1 quintile had the lowest risk. The mortality risk was particularly notable for deaths related to ischemic heart disease (RR, 5.65; 95% CI, 2.26–14.13). The CVD risk was slightly lower in FEV_1 quintiles 2 and 3 but still significantly higher than that for the lowest quintile group (RR, 2.00, 95% CI, 1.03–3.89 for quintile 2; and RR, 2.22; 95% CI, 1.23–4.01 for quintile 3) suggesting that the relationship between reduced FEV_1 and CVD is linear [2]. Similar findings were observed when the analysis was limited to nonsmokers suggesting that the relationship between reduced FEV_1 and CVD is independent of the effects of cigarette smoking.

In another population-based study, the SALIA Cohort (Study on the influence of Air pollution on Lung function, Inflammation and Aging) investigators found that the RR of cardiovascular mortality of women with FEV_1 less than 80% of predicted was 3.79 (95% CI, 1.64–8.74) at 5 years and 1.35 (95% CI, 0.66 to 2.77) at 12 years of follow-up compared with women who had normal lung function [4]. In a Norway study, the investigators found that the risk of CVD mortality increased by 7% with every 10% decrease in FEV_1 (as percent

TABLE 45.1 Baseline characteristics of included studies and their reported association between FEV_1 and cardiovascular mortality (from Ref. [2]).

Author	Publication year	Study population	Sample size	Age (year)	Male (%)	Mean FEV_1 (liter or % of predicted)	Current smokers (%)	FEV_1 categorization (% predicted or liter)	Follow-up (years)	RR of cardiovascular mortality (95% CI)
Marcus [11]	1989	Honolulu Heart Program, US (Japanese American)	5924	54	100	2.71 (94%)	48	Quintiles (2.10 l versus 3.28 l)	15–18	1.93 (1.46, 2.54)
Higgins [7]	1970	Tecumseh, US	5140	16–75 (range)	47.7	NA	NA	<2.0 l (1.4[b]) versus ≥2.0 l (1.4[b])	2–6	5.03(3.07, 8.22)
Hole [10]	1996	Renfrew & Paisley, UK	15,411	45–64 (range)	45.8	2.83[a] 1.99[b]	36	Quintiles (≤73–75% versus ≥108–113%)	15	1.56 (1.26, 1.92)[a] 1.88 (1.44, 2.47)[b]
Beaty [12]	1985	Baltimore Longitudinal Study of Aging, US	874	50.7	100	95.6%	21	≤80% versus ≥80%	24	1.58 (0.96, 2.60)
Schunemann [9]	2000	Buffalo/Erie County, US	1195	46.8	46.4	2.8	58.3	Quintiles (<80% versus ≥109–114%)	29	2.11 (1.20, 3.71)[a] 1.96 (0.99, 3.88)[b]
Ebi-Kryston [13]	1988	Whitehall Civil Servants, UK	17,717	40–64 (range)	100	NA	NA	<65% versus ≥65%	10	1.49 (1.24, 1.80)
Lange [14]	1991	Copenhagen City Study, Denmark	12,511	53.1	NA	NA	NA	<60% versus >80%	6.5	1.8 (1.4, 2.4)
Tockman [15]	1989	Washington County, Maryland	884	NA	100	NA	32.3	Quartiles (<65% versus >100%)	10	3.66 (1.76, 7.61)
Krzyzanowski [16]	1986	Cracow, Poland	3047	19–70 (range)	59	NA	38.4	<65% versus 76–100%	13	2.56 (1.35, 4.85)[a] 1.99 (0.95, 4.18)[b]
Hospers [17]	1999	Vlagtwedde-Vlaardingen, Netherlands	5382	36	54	98%	55	<80% versus ≥100%	~25	1.82 (1.42, 2.34)
Kuller [18]	1989	Multiple Risk Factor Intervention Trial	7368[c]	46	100	3.38	64	Quintiles (<2.8–3.0 l versus ≥3.8–4.0 l)	7	2.33 (1.35, 4.03)[d]
Speizer [8]	1989	Harvard Six Cities Study	8427	49	45	2.85	39.9	Quartiles (2.0–2.6L versus 2.9–4.1 l)	12	1.42 (1.07, 1.90)[a] 2.74 (1.93, 3.90)[b]
Pooled summary			83,880							1.99 (1.71, 2.29)

Test for heterogeneity, $p = 0.001$.

NA: not available; FEV_1: forced expiratory volume in 1 s;RR: relative risk of cardiovascular mortality; CI: confidence interval.

[a]Male values.

[b]Female values.

[c]Although 12,866 participants were originally enrolled in the study, only 7368 participants had an acceptable pulmonary function testing.

[d]Although the smoking-adjusted RR was reported, its 95% CI was not provided; thus, the adjusted data could not be used for the meta-analysis.

predicted) over a 26-year follow-up [5]. In the Honolulu Heart Program, which prospectively followed 5924 healthy, middle-aged men for up to 18 years, Curb and colleagues found that the individuals in the lowest FEV_1 quintile had a 1.93-fold increase in the risk of cardiovascular mortality compared to those in the best FEV_1 quintile (RR, 1.93; 95% CI, 1.46–2.54) [6]. In the Tecumseh Cohort Study, Higgins and colleagues observed that individuals with FEV_1 values less than 2.01 had a fivefold increase in the risk for CVD mortality compared with those with FEV_1 greater than this value (RR, 5.03; 95% CI, 3.07–8.22) [7]. In the Harvard Six Cities Study, Speizer and colleagues reported a RR of 2.74 (95% CI, 1.93–3.90) in women and 1.42 (95% CI, 1.07–1.90) in men, comparing the lowest FEV_1 quartile to the highest quartile [8]. Schunemann and colleagues reported a RR of 1.96 (95% CI, 0.99–3.88) in women and 2.11 (95% CI, 1.20–3.71) in men, once again comparing the lowest FEV_1 quintile to the highest quintile [9]. Hole and colleagues reported a RR of 1.88 (95% CI, 1.44–2.47) in women and 1.56 (95% CI, 1.26–1.92) in men, comparing the lowest to the highest quintile of FEV_1 [10]. Overall, these large population-based studies showed that reduced FEV_1 is an important, independent risk factor for CVD morbidity and mortality (pooled RR, 1.99; 95% CI, 1.71–2.29) even among nonsmokers where the RR is 1.67 (95% CI, 1.35–2.01) [2]. The relationship between reduced FEV_1 and cardiovascular mortality appears to be similar between men (RR, 1.64; 95% CI, 1.48–1.84) and women (RR, 2.14; 95% CI, 1.75–2.59). In general, when the lowest quintile of FEV_1 is compared with the highest quintile, the risk of cardiovascular mortality increases by ~75% in both men and women [2]. Taken together, these population-based studies clearly support the notion that CVD risk increases as FEV_1 decreases (Table 45.1).

The population attributable risk of ischemic cardiac deaths imposed by reduced FEV_1 was estimated by Hole *et al.* [10]. They showed that when the lowest quintile of FEV_1 was compared with the highest quintile, the population attributable risk for deaths related to ischemic heart disease was 26% (95% CI, 19–34%) in men and 24% (95% CI, 14–34%) in women, independent of the effects of cigarette smoking. The magnitude of the mortality burden attributed to reduced FEV_1 in this model was similar to that imposed by hypercholesterolemia. In the same study, comparison of total serum cholesterol (between the lowest to the highest quintile levels) produced a population attributable risk of 21% in men and 25% in women for deaths related to ischemic heart disease [10] (Fig. 45.1). In certain populations, reduced FEV_1 may be responsible for 20–30% of deaths related to ischemic heart disease [10].

The risk of CVD in those with reduced FEV_1 is amplified by the presence of another risk factor for CVD such as hypertension. In the "Malmo Born in 1914" Study, the risk of cardiac events defined as fatal or nonfatal myocardial infarction, was only 10% higher in individuals who had reduced FEV_1 but without hypertension [19]. The risk nearly tripled (RR, 2.9; 95% CI, 1.8–4.5) when subjects had both reduced FEV_1 and hypertension together. The incidence of stroke was nearly fourfold higher (RR, 3.8; 95% CI, 1.9–7.6) when subjects had both reduced FEV_1 and hypertension. In contrast, reduced FEV_1 alone was not associated with increased risk of stroke (RR, 1.2; 95% CI, 0.70–2.2) [19]. In the same cohort, Engstrom and colleagues found that individuals who had reduced FEV_1 and ventricular dysrhythmia on the baseline electrocardiogram (ECG) experienced a twofold increase in the risk of coronary events (RR, 2.31; 95% CI, 1.28–4.20) [20]. In contrast, reduced FEV_1 alone in the absence of ventricular ectopy was not significantly associated with coronary events (RR, 1.24; 95% CI, 0.67–2.27). Similarly, ventricular ectopy by itself was not associated with coronary events (RR, 1.16; 95% CI, 0.58–2.33).

The relationship between rate of FEV_1 decline and CVD

Another important physiologic phenotype associated with COPD is accelerated decline in lung function over time [21]. Individuals who experience rapid decline in FEV_1 are two times more likely to experience COPD-related hospitalizations and mortality [22]. They are also more likely to experience CVD events. In the Malmo "Men Born in 1914" Study, the cardiovascular event rate among smokers in the high, middle, and low thirds with regard to the decline in FEV_1 was 56.0, 41.0, and 22.7 events per 1000 person-years, respectively (p for trend = 0.01) [23]. The Baltimore Longitudinal Study of Aging showed that individuals who experienced the most rapid decline in FEV_1 over 16 years were three to five times more likely to die from a cardiac

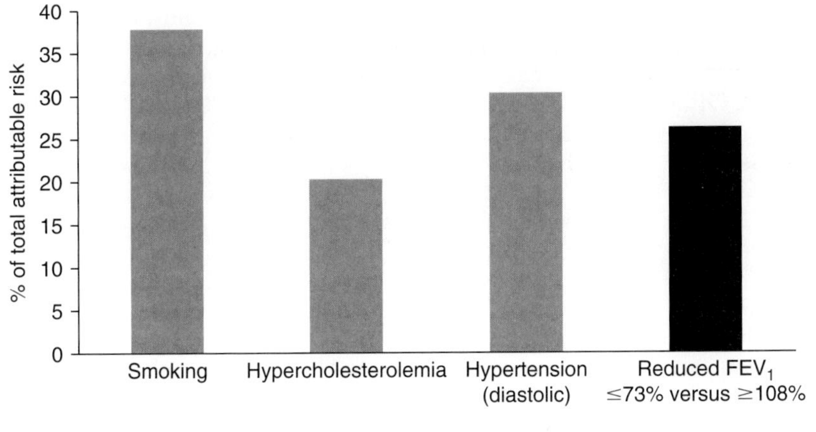

FIG. 45.1 Population attributable risk of reduced FEV_1 on mortality from ischemic heart disease in men [10].

cause of death than those who had the slowest decline in FEV_1 [24]. In lifetime nonsmokers, accelerated decline in FEV_1 was associated with a 5- to 10-fold increase in the risk for cardiac deaths. In the US National Study of Coal Workers' Pneumoconiosis, a longitudinal health survey of male underground coal miners, Beeckman and colleagues found that workers who had an accelerated decline in FEV_1 (defined as FEV_1 decline of 60 ml/year or greater) were more likely to develop incident asthma or COPD compared to individuals with a slower FEV_1 decline. Importantly, the former group had nearly 50% excess mortality than the latter group (RR of all-cause mortality, 1.45), driven mostly by excess CVD deaths (RR, 1.90) and pulmonary causes of mortality (RR, 3.20) [25].

The relationship between FEV_1 to FVC ratio and CVD

Reduced FEV_1/FVC ratio is a more specific indicator of airways disease than is reduced FEV_1 alone in the general population. Using data from the Third National Health and Nutrition Examination Survey in the United States [26], Sin and Man found that individuals with severe airflow obstruction (defined as FEV_1 >50% of predicted and FEV_1/FVC ratio ≥70%) were twice as likely to have ECG evidence of probable or possible myocardial infarction than individuals with normal lung function. In the "Men Born in 1914 Study," Engstrom *et al.* found that compared to subjects in the highest FEV_1/FVC quintile (ratio ≥77.3%), those in the lowest FEV_1/FVC (≤66.3%) quintile were 73% more likely to experience coronary events ($p = 0.01$) [20]. The risk for frequent or complex ventricular arrhythmia was 83% higher in the lowest FEV_1/FVC quintile compared with the highest quintile. As with FEV_1, reduced FEV_1 to FVC ratio may have a modifying effect on CVD. In the study by Engstrom *et al.*, reduced FEV_1/FVC ratio by itself was a modest independent risk factor for coronary events (RR, 1.30). Presence of arrhythmias in those with normal FEV_1/FVC was not associated with coronary events (RR, 1.01). However, the combination of reduced FEV_1/FVC ratio and presence of arrhythmias increased the risk of coronary events by over twofold (RR, 2.43; 95% CI, 1.36–4.32) [20]. These data suggest that airflow obstruction impacts synergistically on the diseased heart to make it more vulnerable to acute coronary events.

COPD and CVD

The Lung Health Study (LHS) is one of the largest COPD cohorts that studied the effects of smoking cessation and inhaled anticholinergic drugs on disease progression in patients with mild COPD (Global Initiative for Chronic Obstructive Lung Disease, GOLD, Stages 1 and 2) [27]. Although the initial study was designed as a 5-year trial, the follow-up was subsequently extended to 14.5 years. The latter was called Lung Health Study-3 (LHS-3) [28]. In this study, the LHS investigators confirmed the benefits of smoking cessation in retarding disease progression and attenuating morbidity and mortality [28]. Notably in this study, the leading cause of hospitalization was CVD [3] and the leading cause of mortality was lung cancer [28]. CVDs accounted for nearly half of all hospitalizations and 22% of all deaths, while lung cancer accounted for 33% of all deaths. In contrast, only 8% of the cohort died from respiratory failure [28]. These data are a sobering reminder of the importance of CVD and cancer in patients with mild COPD.

Even in more advanced disease (GOLD Stages 3 and 4), CVD and lung cancer are important causes of morbidity and mortality. In TORCH (Towards a Revolution in COPD Health) study, a 3-year randomized controlled trial that assessed the effects of salmeterol/fluticasone in severe COPD (FEV_1 <60% of predicted), CVDs accounted for 26% and cancer accounted for 21% of all deaths [29]. In an analysis of data from the Kaiser Permanente Medical Group, Sidney and coworkers found that having COPD increased the risk of hospitalization for cardiac arrest and ventricular fibrillation by 2.8 fold, atrial fibrillation by 2.1 fold, angina by 2 fold, myocardial infarction by 87%, congestive heart failure by nearly 4 fold, stroke by 39%, pulmonary embolism by 2.7 fold, and other CVD by 86%, independent of age, sex, and history of hypertension, hyperlipidemia, and diabetes [30]. Curkendall and colleagues reported similar findings using a completely different cohort [31]. On average, for every 10% decrease in FEV_1, all-cause mortality increases by 14%, cardiovascular mortality increases by 28%, and nonfatal coronary event increased by almost 20%, after adjustments for relevant confounders such as age, sex, and smoking status [3].

In COPD, there are two major phenotypes: emphysema and small airways disease. Emphysema can be determined noninvasively using quantitative computerized tomography (CT) scanning [32]. McAllister and coworkers used this method and found a significant relationship between emphysema severity as assessed by quantitative CT scanning of the lungs and arterial stiffness as measured by pulse wave velocity [33]. Arterial stiffness is a good surrogate of atherosclerosis and a robust predictor of future CVD risk [34].

The relationship between asthma and CVD

Although most of the attention has been directed at the relationship between COPD and CVD, there is emerging literature supporting the notion that asthma is also a risk factor for CVD. In one study from Sweden, Toren and colleagues estimated the risk of mortality in patients with asthma. Compared to the expected mortality, the mortality rate among patients with severe asthma, defined as one who needed daily treatment of oral corticosteroids for at least a year, was twofold higher (RR, 2.1, 95% CI, 1.8–2.5) [35]. The RR of ischemic cardiac deaths was also nearly twofold higher (RR, 1.9, 95% CI, 1.4–2.4) with women having a particularly high RR at 2.5 (95% CI, 1.7–3.3) [35]. In another population-based study, Musk and colleagues found that patients with asthma had excess mortality compared to the rest of the population (standardized mortality ratio, SMR, for all causes of death 1.6 for men ($p < 001$) and 1.7 for women ($p < 0.001$)). The risk of ischemic cardiac death was also elevated (SMR, 1.3) but only in the older age group [36]. A Canadian study found that patients with self-reported asthma were 43% more likely to have heart disease and 36% more likely to have hypertension compared to nonasthmatics [37]. Using the Northern California Kaiser Permanente database, a large Health Maintenance Organization providing to over 3 million people, Iribarren and colleagues found that patients with asthma defined based on self-report or hospitalization for asthma had a 24% increase in the risk for coronary heart disease hospitalization or death adjusted for confounders such as age, race/ethnicity, education level, smoking status, white blood cell count, hypertension and diabetes, and occupational exposures. The relationship was particularly strong in women [38].

However, there are some dissenting studies. For instance, Belli and coworkers failed to find a significant association between asthma and mortality in an elderly group of individuals. In this study, the risk of cardiovascular mortality was similar between those with and without asthma [39]. Similarly, in the Atherosclerosis Risk In Communities (ARIC) Study, Zureik and colleagues found that compared with individuals without asthma, those with (ever) asthma (defined by self-report) had similar RR of coronary heart disease (RR, 0.87; 95% CI, 0.66–1.14). The RR using a time-dependent analysis was 0.88 (95% CI, 0.69–1.11). However, they found that individuals with asthma had increased risk of stroke (RR, 1.43; 95% CI, 1.03–1.98 using a time-dependent analysis) [40].

One of the characteristic physiologic signs in asthma is bronchial hyperresponsiveness (BHR). Zureik and colleagues used data from the European Community Respiratory Health Survey to determine the relationship between BHR and carotid intima-media thickness, which is a marker of atherosclerosis [41] and a risk factor for stroke [42]. They found that in men but not in women who demonstrated BHR (defined as a 20% fall in baseline FEV_1 with methacholine exposure of 4 mg or less) the mean carotid intima-media thickness was greater compared to men who did not have BHR (0.68 ± 0.11 in men with BHR versus 0.62 ± 0.09 mm in men without BHR, $p = 0.002$) [43]. Although the mechanism for this observation is not completely known, a recent study indicates that BHR is associated with systemic inflammation, as measured by circulating C-reactive protein (CRP) [44].

POSSIBLE MECHANISMS

While the epidemiological link between COPD (and to lesser extent with asthma) and cardiovascular events is solid, the mechanism(s) that underpin this relationship are far from certain.

The inflammation theory

It is well recognized and accepted that both asthma and COPD are inflammatory conditions associated with both lung and systemic inflammation, though the content and the nature of the inflammatory process differs between the two conditions [1]. It is postulated that the inflammatory process starts in the lungs and then "spills" into the systemic circulation through the rich network of capillary beds in the pulmonary microcirculation [45]. The rate of spillage is heavily influenced by inflammation (which generally increases permeability of the vasculature) and by underlying parenchymal disease. It is believed (though not proven) that in asthma and in COPD, proinflammatory cytokines and chemokines, such as tumor necrosis factor-α (TNF-α), interleukin (IL)-1β, IL-6, and others, escape into the systemic circulation from the lungs inducing a systemic inflammatory response in the liver, bone marrow, and other organs [46]. In asthma, the inflammatory process in the lungs is thought to reflect an "allergic" or "atopic" response to environmental allergens, while in COPD, it is thought to be an abnormal (exaggerated) response to cigarette smoke and other irritants. Systemic inflammation is believed to contribute to plaque build-up in coronary and cerebral vasculature and to their rupture during periods of stress [47].

Consistent with the systemic inflammation theory, CRP levels in COPD patients have been associated with CVD and all-cause morbidity and mortality [48]. In LHS, for instance, the risk of all-cause and CVD mortality over 7 years of follow-up increased as a function of CRP levels. Compared to the lowest CRP quintile group (mean level 0.21 mg/l), the highest CRP quintile (mean level 7.1 mg/l) had a mortality risk that was 1.8 times higher, adjusted for sex, race, age, body mass index, pack-years of smoking, biochemically validated smoking status (i.e. continued smokers, sustained quitters, or intermittent quitters), rate of descent in FEV_1 (in quintiles), and predicted FEV_1 (in quintiles). The risk of fatal and nonfatal CVD also increased along the CRP gradient. Compared to the lowest quintile, the highest CRP quintile group experienced a 1.5-fold increase in fatal and nonfatal cardiovascular events [49].

DISTURBANCE OF NEUROHUMORAL TONE

Patients with severe COPD have a perturbed neurohumoral regulatory system, leading to excess sympathetic nervous activity and reduced vagal tone [50], which is inversely related to their oxyhemoglobin saturation. Accordingly, COPD patients have raised resting heart rate and have an increased risk of rhythm disturbances and ectopic beats [19], which may increase their risk for CVD events. Metaiodobenzylguanidine (MIBG), an analog of guanethidine, has similar metabolism to norepinephrine in systemic nervous tissues [51]. MIBG has been used to image the heart for assessment of cardiac sympathetic activity. Increased sympathetic nervous activity leads to reduced cardiac to mediastinal activity ratio in the delayed images [51]. In a group of 28 COPD patients and 7 control subjects, COPD patients demonstrated reduced cardiac accumulation of MIBG and a higher washout rate from the heart than control subjects, indicating excess activity of the sympathetic nervous system with increased norepinephrine turnover [52]. Interestingly, the MIBG accumulation rate correlated significantly with the intensity of dyspnea as reported by these patients ($p < 0.05$). Consistent with the MIBG data, COPD patients had higher plasma norepinephrine levels than did the control subjects (449 versus 69 pg/ml; $p < 0.01$). There is little information on whether or not patients with mild or moderate COPD have disturbed neurohumoral regulatory system. However, there is emerging data to indicate that in mild disease, patients are frequently exercise limited by an abnormal cardiovascular response to exercise (e.g. reduced anaerobic threshold). In contrast, subjects without COPD are limited mainly from leg muscle fatigue [53]. These data suggest that even in mild COPD, cardiac performance may be perturbed.

Whether certain inhaled drugs commonly used in COPD and asthma worsen cardiac status is very controversial. Pharmacoepidemiological studies have linked the use of short- or long-acting bronchodilators with increased risk of cardiovascular events in COPD and asthma. In one study, the use of ipratropium bromide, a short-acting anticholinergic medication, was associated with a 60% increase in adjusted mortality compared with nonuse among 827 patients with COPD [54]. In the LHS in which one-third of the participants were assigned in a random fashion to usual care; another one-third to special intervention (for smoking cessation) in addition to ipratropium bromide; and the remaining subjects to special intervention (for smoking cessation) in addition to placebo puffers, ipratropium bromide was associated with a 26% increase in the risk of cardiovascular events compared with placebo. The risk of fatal cardiovascular events was even higher (relative risk, 2.6). Although this study was underpowered to detect cardiovascular events, these data, nonetheless, raised concerns that ipratropium bromide may increase cardiovascular morbidity and mortality in COPD [3].

There is a scarcity of cardiovascular safety data on tiotropium. In the 1-year clinical trials, the risk of heart rate and rhythm disturbance disorders, such as arrhythmias, atrial fibrillation, and tachycardia, were about twofold higher in those assigned to tiotropium ($n = 550$ patients) compared with those assigned to placebo ($n = 371$). However, these events were rare occurring in only 4.4% of this carefully selected group of COPD patients [55]. Nevertheless, tiotropium should be used with caution in patients with COPD who also have coexisting rhythm disorders.

A systematic review and meta-analysis showed that a single dose of a β_2-agonist caused an increase in heart rate of 9 beats/min and reduced serum potassium by 0.36 mmol/l compared with placebo [56]. The use of short-acting β_2-agonist has been associated with increased cardiovascular events. Au and colleagues, for example, studied 630 patients with unstable angina or myocardial infarction and 10,486 control subjects enrolled in 7 Veterans Administration Medical Centers, and found that compared with subjects who did not fill a short-acting β_2-agonist, patients who had filled one β_2-agonist prescription in the 3 months prior to their index date had ~70% increase in the risk for an acute coronary event [57]. Importantly, the excess risk was limited to those patients who had a prior history of CVD; their risk was over threefold higher than those who did not use β_2-agonists. Additionally, new users of β_2-agonists had a sevenfold increase in the risk of cardiovascular events. Overall, treatment with short-acting β_2-adrenoceptor agonists was associated with a 2.5-fold increase in the risk of adverse cardiovascular events, which included ventricular tachycardia, myocardial infarction, and sudden deaths [56]. The exact mechanisms by which these medications may potentially increase cardiovascular morbidity and mortality are unknown. Future studies are needed to validate these initial epidemiologic observations and to determine the potential mechanisms by which these medications may increase cardiovascular risk in susceptible COPD and asthmatic patients.

CAN HEART-PROTECTIVE DRUGS REDUCE COPD MORBIDITY?

In view of the intrinsic link between COPD and CVD, some have suggested that cardioprotective drugs may have beneficial effects in COPD. There have been no randomized controlled studies that have carefully and prospectively tested this hypothesis. There are, however, several epidemiological studies that have generated promising data. For instance, Mancini and coworkers conducted a time-matched nested case-control study using administrative health data in Quebec to determine whether or not statins (hydroxymethylglutaryl CoA reductase inhibitors), angiotensin-converting enzyme (ACE) inhibitors, and/or angiotensin receptor blockers (ARB) could improve survival in COPD. They found that the combined use of statins and ACE or ARBs was associated with a 34% reduction in hospitalization for COPD, a 61% reduction in myocardial infarction, and a 58% reduction in all-cause mortality [58]. Similar findings were noted by an independent group in Japan [59]. One of the major limitations of the previous two studies was that the diagnosis of COPD was not confirmed by spirometry. A Norweigan group using a retrospective cohort of 854 patients with spirometry confirmed COPD (mean FEV_1, 48% of predicted) demonstrated that treatment with statins following hospitalization for COPD was associated with a 43% reduction in all-cause mortality. Interestingly, in the same study, inhaled corticosteroids, which are anti-inflammatory drugs, were also associated with reduced mortality [60], a finding that has been observed in other epidemiological studies [61, 62] but not in a randomized controlled trial [63]. Low-dose inhaled corticosteroid therapy has been associated with reduced risk of myocardial infarction, arrhythmias, and CVD mortality [64–66]. Other groups have shown that beta-blockers may also reduce mortality in COPD patients but the overall magnitude of the benefit appears to be small [67]. Moreover, they (especially noncardioselective beta-blockers) are associated with a slight increase in the risk for COPD hospitalizations [68]. Thus, for safety reasons, in general, only cardioselective beta-blockers in low doses should be employed in COPD [69].

SUMMARY

Cardiovascular events are common complications in COPD and asthma. While the mechanisms linking these disorders have not been well established, the epidemiological evidence is strong. Reduced lung function (regardless of the cause) increases the risk of cardiovascular hospitalization by twofold may contribute up to 25% of all ischemic cardiac deaths. In mild to moderate COPD, cardiovascular events are the leading cause of hospitalization and the second leading cause of mortality (trailing only behind lung cancer). System inflammation and neurohumoral disturbances are the leading putative mechanisms for the association. There are emerging data that cardioprotective drugs may also have salutary effects in COPD. However, these promising observations must be validated in clinical trials before they can be routinely recommended in patients with obstructive airways disease.

References

1. O'Byrne PM, Postma DS. The many faces of airway inflammation. Asthma and chronic obstructive pulmonary disease. Asthma Research Group. *Am J Respir Crit Care Med* 159(5 Pt 2): S41–63, 1999.
2. Sin DD, Wu L, Man SF. The relationship between reduced lung function and cardiovascular mortality: A population-based study and a systematic review of the literature. *Chest* 127(6): 1952–59, 2005.
3. Anthonisen NR, Connett JE, Enright PL, Manfreda J. Hospitalizations and mortality in the Lung Health Study. *Am J Respir Crit Care Med* 166(3): 333–39, 2002.
4. Stavem K, Aaser E, Sandvik L, Bjornholt JV, Erikssen G, Thaulow E *et al.* Lung function, smoking and mortality in a 26-year follow-up of healthy middle-aged males. *Eur Respir J* 25(4): 618–25, 2005.
5. Schikowski T, Sugiri D, Ranft U, Gehring U, Heinrich J, Wichmann HE *et al.* Does respiratory health contribute to the effects of long-term air pollution exposure on cardiovascular mortality? *Respir Res* 8(1): 20, 2007.
6. Curb JD, Marcus EB, Reed DM, MacLean C, Yano K. Smoking, pulmonary function, and mortality. *Ann Epidemiol* 1(1): 25–32, 1990.
7. Higgins MW, Keller JB. Predictors of mortality in the adult population of Tecumseh. *Arch Environ Health* 21(3): 418–24, 1970.
8. Speizer FE, Fay ME, Dockery DW, Ferris BG Jr. Chronic obstructive pulmonary disease mortality in six U.S. cities. *Am Rev Respir Dis* 140(3 Pt 2): S49–55, 1989.
9. Schunemann HJ, Dorn J, Grant BJ, Winkelstein W Jr, Trevisan M. Pulmonary function is a long-term predictor of mortality in the general population: 29-year follow-up of the Buffalo Health Study. *Chest* 118(3): 656–64, 2000.
10. Hole DJ, Watt GC, Davey-Smith G, Hart CL, Gillis CR, Hawthorne VM. Impaired lung function and mortality risk in men and women: Findings from the Renfrew and Paisley prospective population study. *BMJ* 313(7059): 711–15, discussion 5–6, 1996.
11. Marcus EB, Curb JD, MacLean CJ, Reed DM, Yano K. Pulmonary function as a predictor of coronary heart disease. *Am J Epidemiol* 129(1): 97–104, 1989.
12. Beaty TH, Newill CA, Cohen BH, Tockman MS, Bryant SH, Spurgeon HA. Effects of pulmonary function on mortality. *J Chronic Dis* 38(8): 703–10, 1985.
13. Ebi-Kryston KL. Respiratory symptoms and pulmonary function as predictors of 10-year mortality from respiratory disease, cardiovascular disease, and all causes in the Whitehall Study. *J Clin Epidemiol* 41(3): 251–60, 1988.
14. Lange P, Nyboe J, Jensen G, Schnohr P, Appleyard M. Ventilatory function impairment and risk of cardiovascular death and of fatal or non-fatal myocardial infarction. *Eur Respir J* 4(9): 1080–87, 1991.
15. Tockman MS, Comstock GW. Respiratory risk factors and mortality: Longitudinal studies in Washington County, Maryland. *Am Rev Respir Dis* 140(3 Pt 2): S56–63, 1989.
16. Krzyzanowski M, Wysocki M. The relation of thirteen-year mortality to ventilatory impairment and other respiratory symptoms: The Cracow Study. *Int J Epidemiol* 15(1): 56–64, 1986.
17. Hospers JJ, Postma DS, Rijcken B, Weiss ST, Schouten JP. Histamine airway hyper-responsiveness and mortality from chronic obstructive pulmonary disease: A cohort study. *Lancet* 356(9238): 1313–17, 2000.
18. Kuller LH, Ockene JK, Townsend M, Browner W, Meilahn E, Wentworth DN. The epidemiology of pulmonary function and COPD mortality in the multiple risk factor intervention trial. *Am Rev Respir Dis* 140(3 Pt 2): S76–81, 1989.
19. Engstrom G, Hedblad B, Valind S, Janzon L. Increased incidence of myocardial infarction and stroke in hypertensive men with reduced lung function. *J Hypertens* 19(2): 295–301, 2001.
20. Engstrom G, Wollmer P, Hedblad B, Juul-Moller S, Valind S, Janzon L. Occurrence and prognostic significance of ventricular arrhythmia is related to pulmonary function: A study from "men born in 1914," Malmo, Sweden. *Circulation* 103(25): 3086–91, 2001.
21. Rabe KF, Hurd S, Anzueto A, Barnes PJ, Buist SA, Calverley P *et al.* Global strategy for the diagnosis, management, and prevention of chronic obstructive pulmonary disease: GOLD executive summary. *Am J Respir Crit Care Med* 176(6): 532–55, 2007.
22. Mannino DM, Davis KJ. Lung function decline and outcomes in an elderly population. *Thorax* 61(6): 472–77, 2006.
23. Engstrom G, Hedblad B, Janzon L, Valind S. Respiratory decline in smokers and ex-smokers – an independent risk factor for cardiovascular disease and death. *J Cardiovasc Risk* 7(4): 267–72, 2000.
24. Tockman MS, Pearson JD, Fleg JL, Metter EJ, Kao SY, Rampal KG *et al.* Rapid decline in FEV1. A new risk factor for coronary heart disease mortalityIVI. *Am J Respir Crit Care Med* 151(2 Pt 1): 390–98, 1995.
25. Beeckman LA, Wang ML, Petsonk EL, Wagner GR. Rapid declines in FEV1 and subsequent respiratory symptoms, illnesses, and mortality in coal miners in the United States. *Am J Respir Crit Care Med* 163176(3 Pt 112): 6331208–91214, 2001.
26. Sin DD, Man SF. Why are patients with chronic obstructive pulmonary disease at increased risk of cardiovascular diseases? The potential role of systemic inflammation in chronic obstructive pulmonary disease. *Circulation* 107(11): 1514–19, 2003.
27. Anthonisen NR, Connett JE, Kiley JP, Altose MD, Bailey WC, Buist AS *et al.* Effects of smoking intervention and the use of an inhaled anticholinergic bronchodilator on the rate of decline of FEV1. The Lung Health Study. *JAMA* 272(19): 1497–505, 1994.
28. Anthonisen NR, Skeans MA, Wise RA, Manfreda J, Kanner RE, Connett JE. The effects of a smoking cessation intervention on 14.5-year mortality: A randomized clinical trial. *Ann Intern Med* 142(4): 233–39, 2005.
29. McGarvey LP, John M, Anderson JA, Zvarich M, Wise RA. Ascertainment of cause-specific mortality in COPD: Operations of the TORCH Clinical Endpoint Committee. *Thorax* 62(5): 411–15, 2007.
30. Sidney S, Sorel M, Quesenberry CP Jr, DeLuise C, Lanes S, Eisner MD. COPD and incident cardiovascular disease hospitalizations and mortality: Kaiser Permanente Medical Care Program. *Chest* 128(4): 2068–75, 2005.
31. Curkendall SM, DeLuise C, Jones JK, Lanes S, Stang MR, Goehring E Jr *et al.* Cardiovascular disease in patients with chronic obstructive pulmonary disease, Saskatchewan Canada cardiovascular disease in COPD patientsM. *Ann Epidemiol* 16(1): 63–70, 2006.
32. Muller NL, Coxson H. Chronic obstructive pulmonary disease. IV: Imaging the lungsT in patients with chronic obstructive pulmonary disease. *Thorax* 57(11): 982–85, 2002.
33. McAllister DA, Maclay JD, Mills NL, Mair G, Miller J, Anderson D *et al.* Arterial stiffness is independently associated with emphysema severity in patients with COPD. *Am J Respir Crit Care Med* 176(12): 1208–14, 2007.
34. Willum-Hansen T, Staessen JA, Torp-Pedersen C, Rasmussen S, Thijs L, Ibsen H *et al.* Prognostic value of aortic pulse wave velocity as index of arterial stiffness in the general population. *Circulation* 113(5): 664–70, 2006.

35. Toren K, Lindholm NB. Do patients with severe asthma run an increased risk from ischaemic heart disease? *Int J Epidemiol* 25(3): 617–20, 1996.
36. Musk AW, Ryan GF, Perera DM, D'Souza BP, Hockey RL, Hobbs MS. Mortality from asthma in Western Australia. *Med J Aust* 147(9): 423–27, 1987.
37. Dogra S, Ardern CI, Baker J. The relationship between age of asthma onset and cardiovascular disease in Canadians. *J Asthma* 44(10): 849–54, 2007.
38. Iribarren C, Tolstykh IV, Eisner MD. Are patients with asthma at increased risk of coronary heart disease? *Int J Epidemiol* 33(4): 743–48, 2004.
39. Bellia V, Pedone C, Catalano F, Zito A, Davi E, Palange S *et al.* Asthma in the elderly: Mortality rate and associated risk factors for mortality. *Chest* 132(4): 1175–82, 2007.
40. Schanen JG, Iribarren C, Shahar E, Punjabi NM, Rich SS, Sorlie PD *et al.* Asthma and incident cardiovascular disease: The Atherosclerosis Risk in Communities Study. *Thorax* 60(8): 633–38, 2005.
41. Touboul PJ, Labreuche J, Vicaut E, Amarenco P. Carotid intima-media thickness, plaques, and Framingham risk score as independent determinants of stroke risk. *Stroke* 36(8): 1741–45, 2005.
42. Tsivgoulis G, Vemmos K, Papamichael C, Spengos K, Manios E, Stamatelopoulos K *et al.* Common carotid artery intima-media thickness and the risk of stroke recurrence. *Stroke* 37(7): 1913–16, 2006.
43. Zureik M, Kony S, Neukirch C, Courbon D, Leynaert B, Vervloet D *et al.* Bronchial hyperresponsiveness to methacholine is associated with increased common carotid intima-media thickness in men. *Arterioscler Thromb Vasc Biol* 24(6): 1098–103, 2004.
44. Kony S, Zureik M, Driss F, Neukirch C, Leynaert B, Neukirch F. Association of bronchial hyperresponsiveness and lung function with C-reactive protein (CRP): A population based study. *Thorax* 59(10): 892–96, 2004.
45. Sin DD, Man SF. Systemic inflammation and mortality in chronic obstructive pulmonary disease. *Can J Physiol Pharmacol* 85177(1): 141622–7629, 2007.
46. Mannino DM, Watt G, Hole D, Gillis C, Hart C, McConnachie A *et al.* The natural history of chronic obstructive pulmonary disease. *Eur Respir J* 27(3): 627–43, 2006.
47. Ross R. Atherosclerosis – an inflammatory disease. *N Engl J Med* 340(2): 115–26, 1999.
48. Stork S, Feelders RA, van den Beld AW, Steyerberg EW, Savelkoul HF, Lamberts SW *et al.* Prediction of mortality risk in the elderlyA. *Am J Med* 119(6): 519–25, 2006.
49. Man SF, Connett JE, Anthonisen NR, Wise RA, Tashkin DP, Sin DD. C-reactive protein and mortality in mild to moderate chronic obstructive pulmonary disease. *Thorax* 61(10): 849–53, 2006.
50. Heindl S, Lehnert M, Criee CP, Hasenfuss G, Andreas S. Marked sympathetic activation in patients with chronic respiratory failure. *Am J Respir Crit Care Med* 164(4): 597–601, 2001.
51. Wieland DM, Wu J, Brown LE, Mangner TJ, Swanson DP, Beierwaltes WH. Radiolabeled adrenergi neuron-blocking agents: Adrenomedullary imaging with [131I]iodobenzylguanidine. *J Nucl Med* 21(4): 349–53, 1980.
52. Sakamaki F, Satoh T, Nagaya N, Kyotani S, Nakanishi N, Ishida Y. Abnormality of left ventricular sympathetic nervous function assessed by (123)I-metaiodobenzylguanidine imaging in patients with COPD. *Chest* 116(6): 1575–81, 1999.
53. Ofir D, Laveneziana P, Webb KA, Lam YM, O'Donnell DE. Mechanisms of dyspnea during cycle exercise in symptomatic patients with GOLD Stage I COPD. *Am J Respir Crit Care Med* 177: 622–29, 2007.
54. Ringbaek T, Viskum K. Is there any association between inhaled ipratropium and mortality in patients with COPD and asthma? *Respir Med* 97(3): 264–72, 2003.
55. Committee FaDAP-ADA. Spiriva® (tiotropium bromide) Inhalation Powder, NDA 21-395, Boehringer Ingelheim Pharmaceuticals. September 6, 2002.
56. Salpeter SR, Ormiston TM, Salpeter EE. Cardiovascular effects of beta-agonists in patients with asthma and COPD: A meta-analysis. *Chest* 125(6): 2309–21, 2004.
57. Au DH, Curtis JR, Every NR, McDonell MB, Fihn SD. Association between inhaled beta-agonists and the risk of unstable angina and myocardial infarction. *Chest* 121(3): 846–51, 2002.
58. Mancini GB, Etminan M, Zhang B, Levesque LE, FitzGerald JM, Brophy JM. Reduction of morbidity and mortality by statins, angiotensin-converting enzyme inhibitors, and angiotensin receptor blockers in patients with chronic obstructive pulmonary disease. *J Am Coll Cardiol* 47(12): 2554–60, 2006.
59. Ishida W, Kajiwara T, Ishii M, Fujiwara F, Taneichi H, Takebe N *et al.* Decrease in mortality rate of chronic obstructive pulmonary disease (COPD) with statin use: A population-based analysis in Japan. *Tohoku J Exp Med* 212(3): 265–73, 2007.
60. Soyseth V, Brekke PH, Smith P, Omland T. Statin use is associated with reduced mortality in COPD. *Eur Respir J* 29(2): 279–83, 2007.
61. Sin DD, Tu JV. Inhaled corticosteroids and the risk of mortality and readmission in elderly patients with chronic obstructive pulmonary disease. *Am J Respir Crit Care Med* 164(4): 580–84, 2001.
62. Soriano JB, Kiri VA, Pride NB, Vestbo J. Inhaled corticosteroids with/without long-acting beta-agonists reduce the risk of rehospitalization and death in COPD patients. *Am J Respir Med* 2(1): 67–74, 2003.
63. Calverley PM, Anderson JA, Celli B, Ferguson GT, Jenkins C, Jones PW *et al.* Salmeterol and fluticasone propionate and survival in chronic obstructive pulmonary disease. *N Engl J Med* 356(8): 775–89, 2007.
64. Huiart L, Ernst P, Ranouil X, Suissa S. Low-dose inhaled corticosteroids and the risk of acute myocardial infarction in COPD. *Eur Respir J* 25(4): 634–39, 2005.
65. Lofdahl CG, Postma DS, Pride NB, Boe J, Thoren A. Possible protection by inhaled budesonide against ischaemic cardiac events in mild COPD. *Eur Respir J* 29(6): 1115–19, 2007.
66. Macie C, Wooldrage K, Manfreda J, Anthonisen NR. Inhaled corticosteroids and mortalityL in COPD. *Chest* 130(3): 640–46, 2006.
67. Chen J, Radford MJ, Wang Y, Marciniak TA, Krumholz HM. Effectiveness of beta-blocker therapy after acute myocardial infarction in elderly patients with chronic obstructive pulmonary disease or asthma. *J Am Coll Cardiol* 37(7): 1950–56, 2001.
68. Brooks TW, Creekmore FM, Young DC, Asche CV, Oberg B, Samuelson WM. Rates of hospitalizations and emergency department visits in patients with asthma and chronic obstructive pulmonary disease taking beta-blockersA. *Pharmacotherapy* 27(5): 684–90, 2007.
69. Salpeter S, Ormiston T, Salpeter E. Cardioselective beta-blockers for chronic obstructive pulmonary disease. *Cochrane Database Syst Rev*(4), 2005. CD003566

46

CHAPTER

Allergen Avoidance

Ashley Woodcock and Adnan Custovic

University of Manchester, Manchester, UK

INTRODUCTION

Senitization to inhalant allergens is a major risk factor for asthma, rhinitis, and eczema [1, 2]. Exposure of individuals with established allergic disease to high levels of sensitizing allergens causes exacerbation of symptoms and worsening of the underlying inflammatory process [3–10]. Complete cessation of exposure leads to the improvement in disease control. For example, patients with seasonal allergic rhinitis have no symptoms in the absence of exposure to pollen, and moving atopic asthmatics into low-allergen environments such as hospitals [11] or high-altitude sanatoria [12–14] improves markers of asthma severity.

In occupational asthma, cessation of exposure to an allergen may be associated with a dramatic improvement in symptoms, and sometimes cure, but only *early* in the natural history of the disease [15]. However, exposure over a longer period can result in asthma becoming persistent in spite of removal from exposure. This early "window of opportunity" appears variable between individuals and may explain why environmental control is ineffective in some circumstances. It suggests that any allergen avoidance strategy must achieve an early and major reduction in exposure and be targeted to individuals likely to benefit.

ALLERGEN AVOIDANCE MEASURES

Allergen avoidance measures are usually not specific to any given allergen, or even environmental contaminant. For example, measures against house dust mite allergen will also impact on levels of microbial products such as endotoxin and glucan. This may have unanticipated consequences and may alter the effectiveness of allergen avoidance on allergic disease.

When designing effective methods to reduce personal exposure, the aerodynamic characteristics and distribution of allergens have to be taken into account [16]. Mite and cockroach allergens are contained within particles >10 μm diameter and become airborne after vigorous cleaning, whereas substantial cat and dog allergen circulates in the air on small particles (<5 μm diameter) [17–19]. Consequently, pet allergic patients often develop wheezing, sneezing, or itchy eyes immediately on entering a home with a pet, but mite and cockroach-sensitized asthmatics are unaware of the relationship between exposure and symptoms. Air filtration units have little effect on personal exposure to mite or cockroach but may be useful in removing airborne pet allergens.

Mite allergen avoidance

Reduction of mites and mite allergens in the home can be achieved by a number of measures (Table 46.1; for review see Ref. [16]).

Bed and bedding

The greatest exposure to mite allergen is in bed at night, and synthetic quilts provide much greater risk than feather cover does. The most effective mite allergen avoidance measure is to cover the mattress, duvet, and pillows with impermeable encasings (or get rid of the quilt if it is not covered). Finely woven fabrics are the most effective covers [20]. Sheets should be washed weekly, and blankets should be washed regularly in a hot cycle (above 55°C) to kill the mites [21].

TABLE 46.1 Practical measures for reducing house dust mite, pet, fungal, and cockroach allergen levels.

House dust mite
Bed and bedding ● Encase mattress, pillow, and quilt in allergen impermeable covers (preferably finely woven fabric) ● Wash all bedding weekly. Use hot cycle (55–60°C) if possible Replace carpets with hard flooring (e.g., linoleum or wood) Minimize upholstered furniture/replace with leather furniture Replace curtains with blinds Minimize dust accumulating objects; keep in closed cupboards Remove soft toys (if impossible, hot wash/freeze soft toys) Reduce indoor humidity if possible
Cat/dog
Remove cat/dog from the home
Fungi
Reduce indoor humidity if possible Use HEPA air filters in main living areas and bedrooms Use Fungicides on heavily contaminated surfaces Minimize upholstered furniture Replace carpets with hard flooring (e.g., linoleum or wood) Ensure regular inspection of heating and air conditioning units to prevent contamination
Cockroaches
Remove food and water sources Use suitable pesticide in bait form Remove all dead carcasses and frass Wash down all surfaces, floors, and walls with detergent Seal cracks in walls and plaster work to reduce further access Wash all bedding, clothing, and curtains

Carpets and upholstered furnishings

Fitted carpets should be removed and replaced with hard floor coverings (e.g., wooden or linoleum floor) since they are impossible to clean (vacuuming provides substantial personal exposure). Exposing carpets to direct sunlight, steam cleaning, use of acaricides or tannic acid, freezing with liquid nitrogen, etc. are only partially effective [21]. Allergen reduction in upholstered furniture is impossible, without impermeable (leather or PVC) covers. Fabric curtains can be replaced with venetian blinds. Soft toys can be frozen then washed to kill mites and remove allergens.

Controlling humidity

Mites live in beds because of the high levels of humidity; but reducing relative humidity of indoor air may be insufficient to reduce humidity in the middle of a mattress [22, 23]. This depends on the local climate, housing design, and effectiveness of dehumidification. Central mechanical ventilation heat recovery units are effective in areas where outdoor humidity is low and home insulation is good, but not where outdoor humidity is high and homes are poorly insulated. Portable dehumidifiers are generally inadequate [24].

Major reductions in personal exposure are possible, but single measures are unlikely to work. A stringent environmental control regime combining several measures can achieve and maintain a low-allergen environment [25], but this will be expensive and require long-term discipline – this may be beyond many patients.

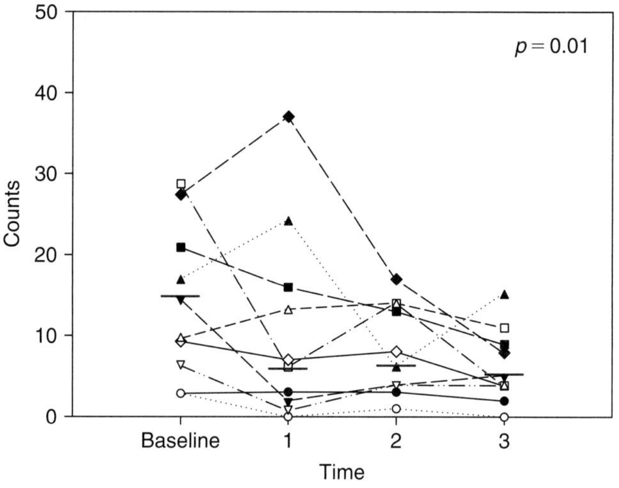

FIG. 46.1 Air filtration with HEPA units results in only moderate reduction in personal exposure to cat allergen (assessed by intranasal air filters). From Ref. [28] with permission.

Pet allergen avoidance

Cat and dog allergen levels are low in homes without a pet [16]. After permanent removal of a pet, allergen levels remain high for months [26]. Pet-sensitized children exposed to pets may lose lung function in early life which may never be recovered [3]. Almost all pet owning and pet-sensitized asthmatics continue keeping their pets, in spite of asthma which is hard to control.

Air cleaning units with high-efficiency particulate arrest (HEPA) filters can reduce the airborne concentration of cat and dog allergens in homes with pets [27], but field studies demonstrated that the magnitude of reduction in personal exposure is small (Fig. 46.1) [28]. Washing pets produces a modest reduction in airborne allergen [29, 30] but needs to be done more than twice a week [31] and is unlikely to translate into clinical benefit.

In experimental chambers, vacuum cleaners with built-in HEPA filters and double thickness bags do not leak pet allergens. However, in real life (actually vacuuming a carpet for which they were designed!) intranasal air samplers have shown a fivefold increase in the amount of cat allergen inhaled while using a range of brand new high-efficiency vacuum cleaners (Fig. 46.2) [32]. Data from experimental chamber studies are not a relevant index of personal exposure in a real-life situation and thus cannot be used to recommend high-efficiency vacuum cleaners to allergy sufferers.

Pet removal is the only effective advice to patients with pet allergy who experience symptoms on exposure.

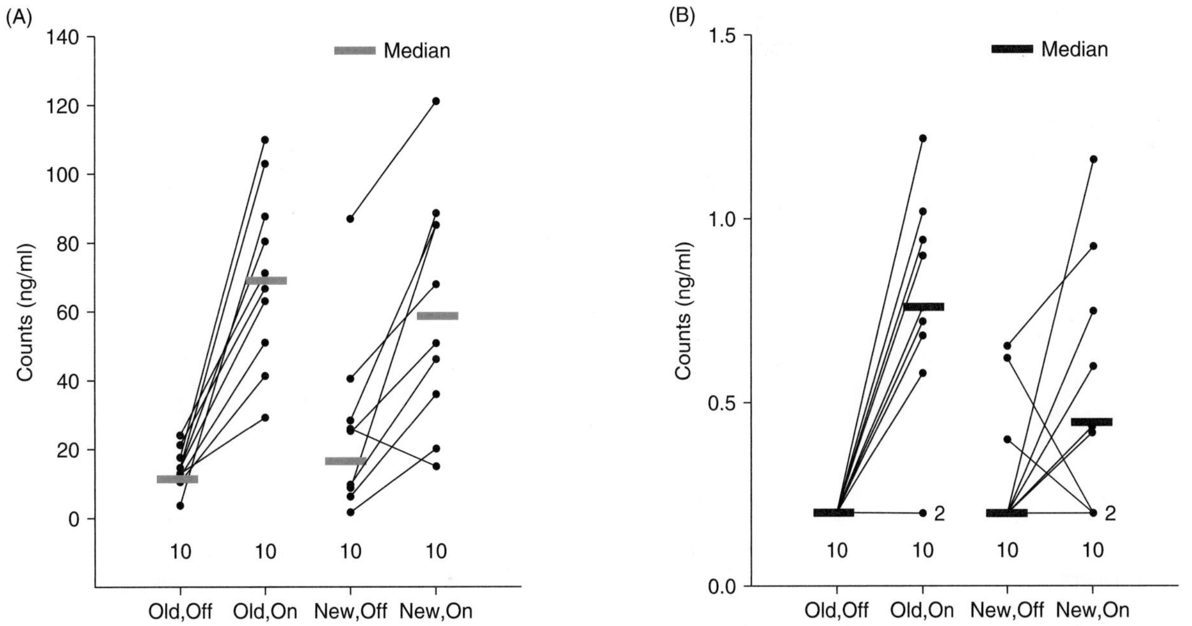

FIG. 46.2 Vacuum cleaning and cat allergen exposure: Fel d 1 bearing particle counts (halo units; (A) and Fel d 1 level (ELISA; (B). Vacuum cleaners are grouped into either control (old) or unused (new). Each value is the median (range) of the observations in each cell (2 observers × 10 homes). There is a fivefold increase in the amount of cat allergen inhaled while using a range of brand new high-efficiency vacuum cleaners. From Ref. [32] with permission.

Cockroach allergen reduction

Physical and chemical procedures can control cockroach populations in infested houses [33–35]. Several pesticides are available, in either a gel or a bait form. Elimination of food and water sources, sealing holes in plasterwork and floors, and household cleaning (of contaminated floors, surfaces, and bedding) are essential. Successful treatments reduce exposure in ~2 weeks, have maximal effect within 1 month, and reduce populations for up to 6 months. Discipline in cleaning and repeat treatments are critical.

ALLERGEN AVOIDANCE IN THE TREATMENT OF ALLERGIC DISEASE

Vulnerable patients are bombarded with advertisements for allergen avoidance, often with spurious or exaggerated claims for extremely expensive vacuum cleaners, dehumidifiers, etc. We have a duty to only recommend clinically effective controls, which are relevant to specific patients, often relevant to their geographical location. The questions that remain controversial are whether home environmental changes are adequate, can be sustained in real life sufficient to impact disease, and how to identify those patients who will benefit [36].

Mite allergen avoidance

Recent updates of the Cochrane meta-analysis reported no effect of the interventions and concluded that current methods of mite allergen avoidance should not be recommended to mite-sensitive asthmatics [37, 38]. The authors concluded that they were unable to demonstrate any clinical benefit to mite-sensitive patients with asthma of measures designed to reduce mite exposure (Figs 46.3 and 46.4). The most recent meta-analysis [37] included 3002 patients included into 54 trials, which is more than doubling the number of patients since the first such review [39]. The authors suggested that allergen avoidance methods did not adequately reduce mite antigen levels but also stressed that mite-sensitive asthmatic patients are usually sensitized to other allergens, so that successful elimination of only one allergen may have limited benefit, whatever its success [37]. Recent Cochrane review of dehumidification showed no satisfactory studies.

In *adults* with asthma, three older studies (98 study participants in total) suggested improvements in bronchial reactivity with mite allergen avoidance in mite-sensitive patients [40–42]. In contrast, the largest randomized double-blind, placebo-controlled SMAC trial assessed the effectiveness of mite-impermeable bed covers as a single intervention, in over 1000 adult asthmatics. There was no improvement in any of the outcome measures [morning peak expiratory flow rate, PEFR, during the first 6 months or 12 months (Fig. 46.5), the proportion of patients who are able to discontinue inhaled steroids during the second 6 months of the study, symptoms scores, quality of life, etc.] [43]. Over two-thirds had positive mite-specific IgE and one in four beds had mite allergen levels >10 μg/g Der p 1. In a *post-hoc* analysis of the subgroup of 130 patients with high mite-specific IgE (≥10 kU_A/l) and high mite allergen exposure (>10 μg/g Der p 1 in mattress dust), there was no difference between active and placebo groups. This was confirmed by a study in 55 adult patients with asthma who

Review: House dust mite control measures for asthma
Comparison: 01 House dust mite reduction versus control
Outcome: 02 Asthma symptoms score

Study	Treatment *N*	Treatment Mean (SD)	Control *N*	Control Mean (SD)	Weight (%)	Standardized mean difference (fixed) 95% CI
01 Chemical methods						
Bahir (1997)	13	1.60 (1.50)	17	1.40 (1.50)	2.2	0.13 (−0.59, 0.85)
Chang (1996)	12	1.10 (1.70)	14	0.40 (0.50)	1.8	0.56 (−0.23, 1.35)
Dietemann (1993)	11	1.40 (1.24)	12	1.18 (0.36)	1.7	0.24 (−0.58, 1.06)
Reiser (1990)	23	5.50 (4.30)	23	3.30 (3.50)	3.3	0.55 (−0.04, 1.14)
Subtotal (95% CI)	59		66		9.0	0.39 (0.04, 0.75)
Test for heterogeneity chi-square = 1.10 df = 3 p = 0.78 I^2 = 0.0%						
Test for overall effect z = 2.16 p = 0.03						
02 Physical methods – parallel group studies						
Chen (1996)	20	0.50 (0.66)	15	0.82 (1.01)	2.5	−0.38 (−1.05, 0.30)
Cinti (1996)	10	0.30 (0.68)	10	1.00 (1.15)	1.4	−0.71 (−1.62, 0.20)
Dharmage (2006)	15	−0.02 (0.15)	15	−0.04 (0.17)	2.2	0.12 (−0.60, 0.84)
Fang (2001)	22	10.00 (6.80)	21	17.30 (10.30)	2.9	−0.83 (−1.45, −0.20)
Huss (1992)	26	8.80 (10.70)	26	13.10 (11.20)	3.8	−0.39 (−0.94, 0.16)
Rijssenbeek (2002)	16	2.25 (2.24)	14	2.37 (3.17)	2.2	−0.04 (−0.76, 0.67)
Sheikh (2002)	23	−3.40 (29.50)	20	−18.10 (27.80)	3.1	0.50 (−0.11, 1.11)
Thiam (1999)	18	0.80 (0.50)	6	1.80 (0.18)	0.9	−2.16 (−3.30, −1.02)
Woodcock (2003)	315	1.03 (0.70)	310	1.03 (0.73)	46.2	0.0 (−0.16, 0.16)
de Vries (2007)	48	1.23 (0.86)	48	1.13 (0.83)	7.1	0.12 (−0.28, 0.52)
Subtotal (95% CI)	513		485		72.2	−0.07 (−0.20, 0.05)
Test for heterogeneity chi-square = 27.72 df = 9 p = 0.001 I^2 = 67.5%						
Test for overall effect z = 1.11 p = 0.3						
03 Pyhsical methods – crossover studies						
Antonicelli (1991)	9	0.16 (0.32)	9	0.26 (0.34)	1.3	−0.29 (−1.22, 0.64)
Warner (1993)	14	0.20 (0.26)	14	0.19 (0.34)	2.1	0.03 (−0.71, 0.77)
Zwemer (1973)	12	0.70 (0.51)	12	1.40 (0.43)	1.4	−1.43 (−2.35, −0.52)
Subtotal (95% CI)	35		35		4.7	−0.48 (−0.97, 0.01)
Test for heterogeneity chi-square = 6.17 df = 2 p = 0.05 I^2 = 67.6%						
Test for overall effect z = 1.91 p = 0.06						
04 Combination methods						
Cloosterman (1999)	76	5.50 (6.10)	81	6.30 (6.70)	11.6	−0.12 (−0.44, 0.19)
Marks (1994)	17	0.98 (0.57)	18	0.67 (0.55)	2.5	0.54 (−0.14, 1.22)
Subtotal (95% CI)	93		99		14.1	−0.01 (−0.29, 0.28)
Test for heterogeneity chi-square = 3.06 df = 1 p = 0.08 I^2 = 67.3%						
Test for overall effect z = 0.05 p = 1						
Total (95% CI)	700		685		100.0	−0.04 (−0.15, 0.07)
Test for heterogeneity chi-square = 47.02 df = 18 p = 0.0002 I^2 = 61.7%						
Test for overall effect z = 0.73 p = 0.5						

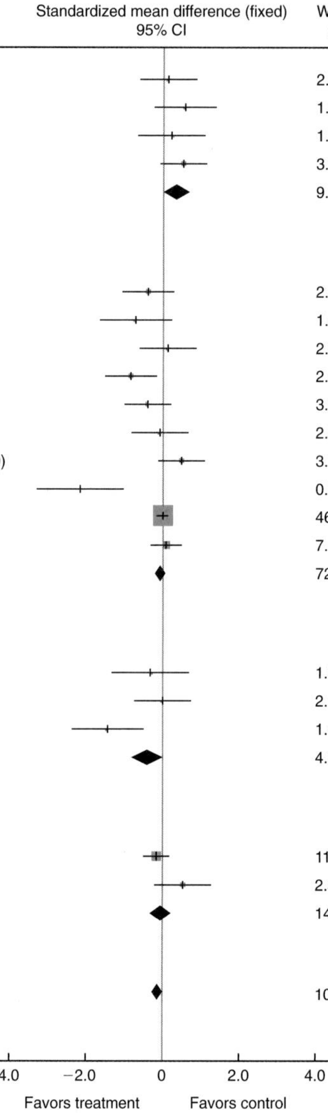

FIG. 46.3 Cochrane meta-analysis reported no effect of the interventions. House dust mite reduction versus control, outcome: asthma symptoms score.

were sensitized to mite (mite-specific IgE > 0.7 kU_A/l) and exposed to high levels of dust mite allergen in their mattresses (Der p 1 > 2 μg/g) [44]. Recent Dutch studies have given opposing results [45]. In asthma in adults, a single intervention with allergen-impermeable encasings for the mattress, duvet, and pillows is ineffective in long-term management, even in individuals who are highly allergic to dust mite and exposed to high levels of mite allergens.

In contrast, several studies in *children* with asthma have indicated that environmental interventions may improve airway reactivity [46], lung function [47], reduce acute Emergency Room visits [48], and substantially reduce maintenance inhaled steroids [49]. The Inner-City Asthma Study assessed individually tailored environmental interventions [50]. A total of 937 children aged 5–11 years with uncontrolled moderate/severe asthma and at least one positive skin test were recruited from seven inner city areas of the USA. Mattress and pillow encasings and a high-filtration vacuum cleaner were supplied to all homes, and products required for specific interventions (e.g., a HEPA air filter for the reduction in passive smoke exposure) were supplied free of charge (the control group had no placebo interventions). Children in the intervention group had significantly fewer days with asthma symptoms, and fewer emergency room visits [50] (Fig. 46.5). An additional period of 34 symptom-free days over 2 years was obtained at a cost of $2000/child (Fig. 46.6) [51].

Rhinitis

A Cochrane systematic review of mite avoidance measures in the management of perennial allergic rhinitis published in 2001 reported that the quality of studies was poor and found no evidence of benefit [52]. In the very large SMAC

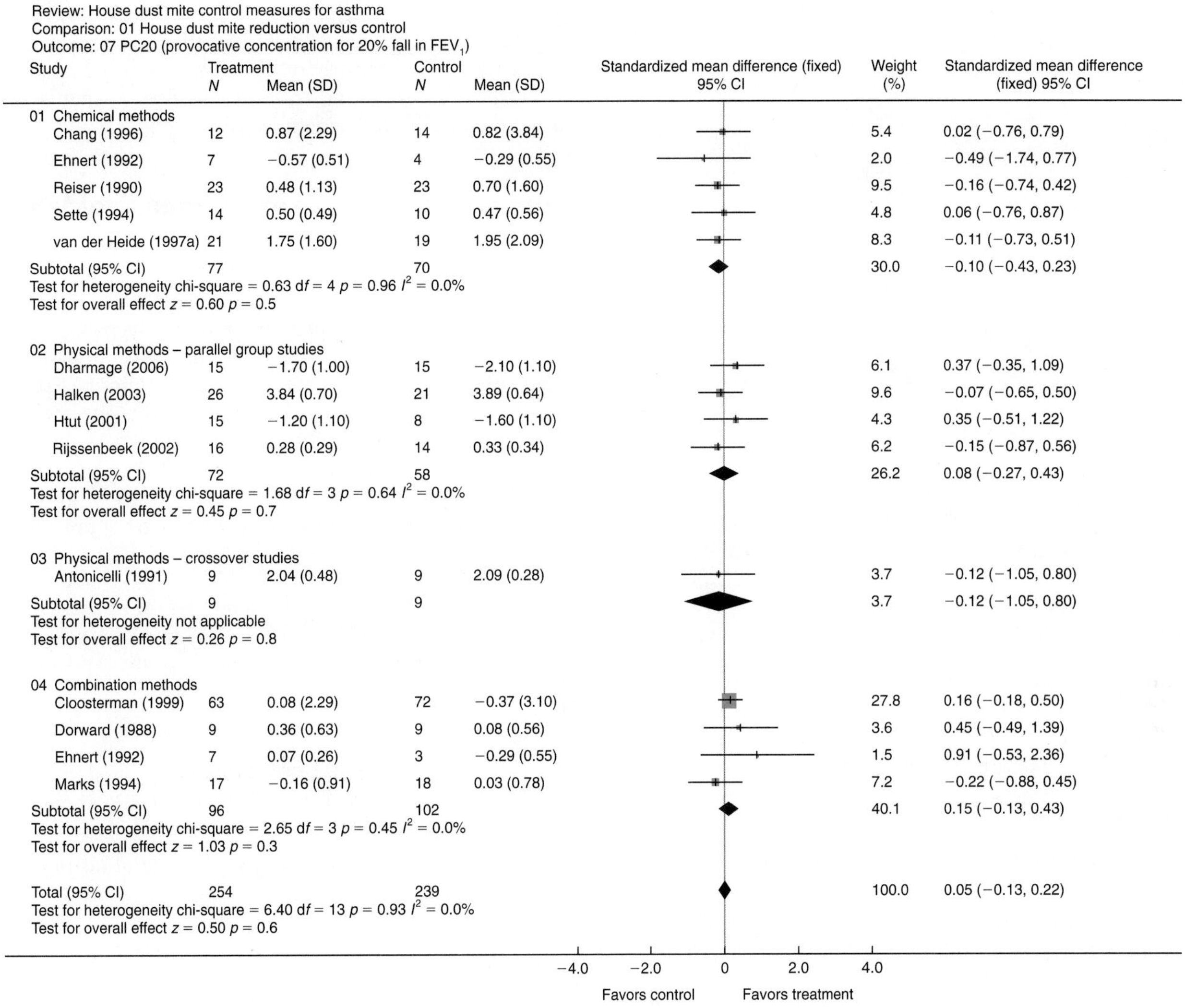

FIG. 46.4 Cochrane meta-analysis reported no effect of the interventions. House dust mite reduction versus control, outcome PC20 (provocative concentration for 20% fall in FEV_1).

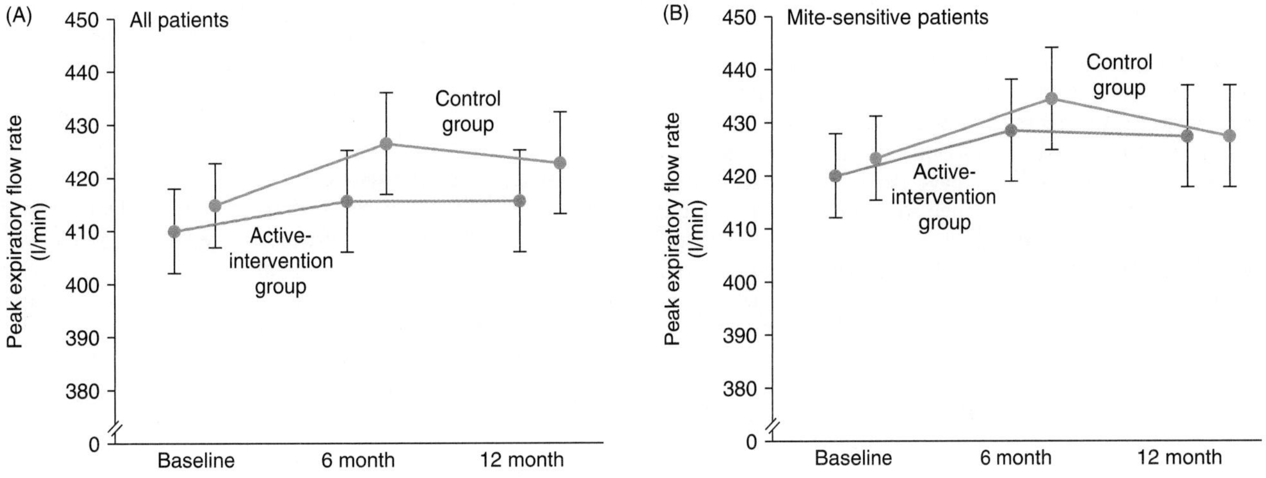

FIG. 46.5 No effect of covering mattress, pillow, and quilt with allergen-impermeable covers amongst adults with asthma. Mean morning peak expiratory flow rate in the active–intervention and control groups at base line, 6 months, and 12 months among all patients (A) and among mite-sensitive patients (B). From Ref. [43] with permission.

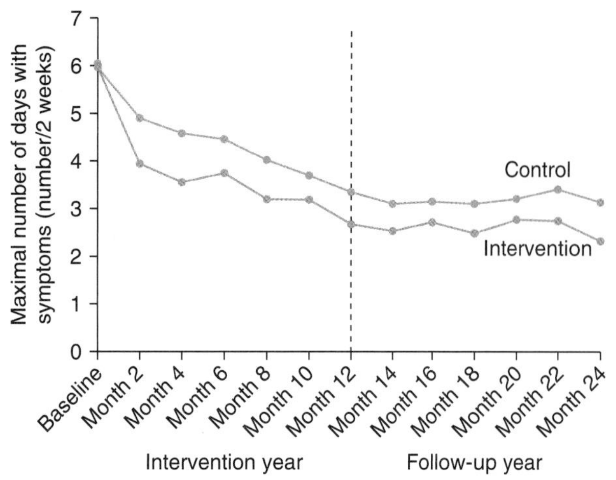

FIG. 46.6 Environmental control is effective amongst children with asthma. Mean maximal number of days with symptoms for every 2-week period before a follow-up assessment during the 2 years of the study. The difference between the environmental intervention and control group was significant in both the intervention year ($p < 0.001$) and the follow-up year ($p < 0.001$). From Ref. [50] with permission.

study [40], there was no difference in rhinitis symptoms between active bedding and control groups (data unpublished). A further large randomized, double-blind, placebo-controlled study has since confirmed this [53].

Eczema

There are as yet no systematic reviews assessing the effect of mite avoidance measures in patients with eczema. Two small studies [54, 55] compared mite-impermeable encasings +/− acaricides with controls and failed to show any improvement in eczema. In contrast, an earlier study in which half of the study population were children, with a more comprehensive environmental intervention which included a combination of bed covers, acaricides, and high-filtration vacuum cleaners for 6 months, demonstrated a significantly greater improvement in the severity score and area affected by eczema in the active compared to the control group [56].

Pet allergen avoidance

While experience suggests that clinical improvement can occur when a pet to whom a patient is sensitized is removed [57], it rarely happens. There have been three small studies of cat and dog allergen avoidance with the pet still in home with contrasting outcomes [58, 59], A Cochrane Airways review emphasized the lack of evidence on the clinical effectiveness of pet allergen avoidance [59]. Larger studies are needed before definitive recommendations can be made.

Food allergen avoidance

It is difficult to get patients to comply with food allergen avoidance. The beneficial effects of food avoidance in food-allergic patients without other allergic disorders are unambiguous, this approach being indeed the mainstay of their treatment. Some controversy still exists, however, regarding the role of exclusion diets in patients with combined eczema and food sensitivity. The few small studies on this issue, most investigating the impact of an elemental diet, showed no or limited benefit [60, 61].

Summary on the use of allergen avoidance in the treatment of asthma

There is no evidence to support the use of simple single allergen avoidance methods amongst adults with asthma; with respect to the effects of a multifaceted intervention, there are as yet no adequately powered studies based upon which to make an informed decision in this age group. In contrast, the majority of studies in children suggest that allergen avoidance may be of benefit.

Until unequivocal evidence for all age groups is obtained, we suggest the following pragmatic approach:

- Single avoidance measures ineffective
  - use a comprehensive environmental control regime aiming to achieve a complete cessation of exposure, or as great a reduction in personal exposure as possible
- Tailor the intervention to the patient's sensitization and exposure status
  - if unable to assess the exposure, use the level of allergen-specific IgE antibodies or the size of skin test wheal as an indicator
- Start the intervention as early in the natural history of the disease as possible.

ALLERGEN AVOIDANCE IN THE PREVENTION OF ALLERGIC DISEASE

The consistent finding of an association between allergic sensitization and childhood asthma raises the question as to whether reducing exposure to allergens early in life can reduce the risk of subsequent development of sensitization and/or symptoms of allergic disease.

Secondary prevention

Secondary preventive measures are designed to halt the progression of disease in individuals who are at high risk for the development of allergic disease, but who have not yet developed specific symptoms (e.g., prevention of asthma in individuals with eczema, rhinitis, or evidence of allergen senitization). The evidence for effective secondary prevention measures is scarce and only a small number of secondary prevention trials using allergen avoidance measures have been conducted (most research to date has focused on primary and tertiary prevention).

The multicentre Study on the Prevention of Allergy in Children in Europe (SPACE) secondary prevention study investigated the protective effect of mite avoidance in high-risk allergic children (family history of atopy asthma, rhinitis, eczema, but not sensitized to mite). In one study, 636 young children aged 1.5–5 years were randomized to mite-impermeable mattress covers and specific mite avoidance advice versus general advice only [62]. The rate of senitization to mite after 1 year was reduced in the active group. A similar study and intervention including 242 children aged 5–7 years was reported by another group [63]. After 1 year, fewer children were newly sensitized to mite in the active group, with a trend for fewer wheezing symptoms.

These two studies indicate that senitization to mite can be prevented by allergen avoidance in the short term; however, further follow-up of these children is essential to show whether this effect can be sustained, as well as to ascertain its impact on the natural history of allergic disease.

Primary prevention

So, does successful reduction in exposure to allergens early in life prevent the development of sensitization and allergic disease? This topic has been a subject of several recent review articles [64, 65]. Seven "primary prevention" studies are ongoing, all focussed on children at high risk of developing allergic disease (e.g., both parents atopic, single parent atopic, mother with asthma, etc.). The studies used different environmental control approaches (e.g., four combined dietary intervention with environmental control). Different outcomes were assessed at different ages. While the results are not directly comparable, they do provide interesting and complementary data.

Isle of Wight was the first primary prevention study [66–69]. A complex intervention involved avoidance of both food and dust mite allergens until age 9 months. Mothers who were breast-feeding their children were asked to avoid allergenic foods, and these foods were not introduced into the infant's diet until after 9 months of age. Infants who were not breast-fed were given a hydrolyzed formula. Environmental control comprised of the application of acaricide to carpets and upholstered furniture. Infants in both groups slept on polyvinyl-covered mattresses with vented head area. Mite allergen levels at age 9 months were lower in the active than in the control group but were still higher than levels seen in most studies.

By age 8 years in the active group, sensitization to mite was reduced by more than 50%, and current wheeze, nocturnal cough, wheeze with bronchial hyper-responsiveness, and atopy were lower in the multivariate analysis. [69]. However, due to the study design it is impossible to determine which intervention is responsible.

The Canadian Primary Prevention Study (CaPPS) exemplifies the complexity of adherence to multifaceted interventions [70, 71]. Compliance with mattress encasings and acaricides was excellent. Mite allergens were significantly reduced in the parental bed throughout the study, while child's mattress levels were low in both groups. However, despite advice, there was no reduction in the use of carpets and in pet ownership in the active group compared to controls. There was a significant reduction in "probable asthma" and "rhinitis without colds" at age 2 in the active group. At age 7, physician-diagnosed asthma was reduced in the intervention group compared to the control group (14.9% versus 23.0%) [71], but there was no difference in allergic rhinitis, eczema, atopy, and bronchial hyper-responsiveness [71].

In the Study on the Prevention of Allergy in Children in Europe (SPACE), the multifaceted intervention included both inhalant and food allergens, studied in new born children from Austria, Germany, and UK [72, 73]. Mattress covers were applied to the infants' bed. Mothers were advised to breast-feed, and hypoallergenic formula was recommended as an alternative (Austria and Germany, but not UK). Parents were advised to delay the introduction of allergenic foods. At age 2, no differences were found between the control and the intervention groups in the prevalence of mite sensitization or respiratory symptoms.

The Childhood Asthma Prevention Study (CAPS) is a multicentre, parallel group, randomized controlled trial in Sydney, Australia [74–77]. Children from at-risk families were randomized into four groups (mite avoidance with placebo dietary intervention, mite avoidance with active dietary intervention, active dietary intervention alone, no mite avoidance; placebo dietary intervention, no mite avoidance). A significant reduction in the very high baseline mite levels was achieved in the mite intervention group; the dietary intervention consisted of omega 3 fatty acids supplementation. At age 3 and 5 years, there was a significant reduction in sensitization to house dust mite in the active mite allergen avoidance group, but no significant differences in wheeze were found with either intervention [75]. At 5, the prevalence of eczema was higher in the active mite avoidance group (26% versus 19%) [76].

In the Primary Prevention of Asthma in Children Study (PREVACS) in the Netherlands, children were recruited during the prenatal period and randomized to either a control group (receiving usual care) or an intervention group in which families received instruction from nurses on how to reduce exposure of newborns to mite, pet and food allergens, and passive smoking. This study was not effective in reducing asthma-like symptoms in high-risk children during the first 2 years of life, and although a modest effect was observed at age 2 years, no significant differences in the total and specific IgE were found between the groups [78].

The Prevention and Incidence of Asthma and Mite Allergy (PIAMA) is a multi-centre, population-based cohort study of >4000 children in the Netherlands [79, 80]. Nested within this cohort is an intervention study amongst 810 high-risk infants who were randomly allocated to receive active or placebo mite-proof encasings for the mattress and pillows of the parental and infant beds. Advice was given to the active group to wash bedding regularly at more than 60°C. Baseline mite allergens levels were low in both groups, and even lower in the active group at 1 year, there was little change in the absolute value [79]. At age 2 and 4 years, sensitization and allergic symptoms were similar in both groups [79, 80].

The Manchester Asthma and Allergy Study (MAAS) is a whole population birth cohort study of >1000 children, with a nested environmental intervention study in the high-risk group. Children with two atopic parents who had no pets in their home were randomly allocated before birth to an extremely stringent environmental control (n = 145) or normal regime (n = 146) [81, 82]. This included mite proof encasings to the parental mattress, duvet and pillows by 16th week of pregnancy; mite proof cot and carry cot mattresses; advice to hot wash bedding weekly at over 55°C, supply of a high-filtration vacuum cleaner, carpets removed from the nursery and a vinyl cushion floor fitted, acaricide for carpets with high mite allergen levels, hot washable toy. No attempt was made to introduce placebo devices into the control homes. A substantial and sustained reduction in exposure to mite, cat and dog allergens was observed in the homes of children in the active group.

At age 3, children in the intervention group were significantly and *more* frequently sensitized to dust mite compared to controls (risk ratio 2.85) [82]. However, lung function (assessed by the measurement of specific airway resistance) was markedly and significantly better amongst children in the intervention group. Since there was no difference between the groups in infant lung function at age 4 weeks [82], the difference had occurred between 4 weeks and 3 years of age. Thus, in MAAS stringent environmental control was associated with *increased* risk of mite senitization but better lung function at age 3 years.

CONCLUSIONS

- Established adult asthma: there is little evidence to support the use of simple physical or chemical methods as single interventions to control dust mite or pet allergen levels (e.g., mattress encasings, acaricides, or HEPA filters). While it remains possible that a multifaceted intervention in carefully selected patients could have some benefit, this has not as yet been addressed in an adequately designed study in adults. There are serious concerns about the effectiveness and sustainability outside the context of a clinical trial. There is inadequate evidence to advise allergen avoidance in adults.
- In childhood asthma, several smaller trials of allergen-impermeable bed encasings as a single intervention in asthmatic children reported benefits, as did a large study of a more comprehensive environmental intervention in children living in poor quality housing.
- The reasons for the apparent differences in response between adults and children are not clear, but the situation may be analogous to that observed in the occupational asthma, where early intervention is associated with a better long-term outcome.
- In the prevention of allergic disease, the clinical outcomes reported from different intervention studies appear inconsistent and often confusing. Much longer follow-up is required before we can be sure that the interventions do not cause harm. There is very unlikely to be a "one-size-fits-all" strategy for primary prevention which will be applicable to the whole population; Specific environmental interventions will be designed and targeted to genetically susceptible individuals [83, 84].

REFERENCES

1. Illi S, von Mutius E, Lau S, Niggemann B, Gruber C, Wahn U. Perennial allergen sensitisation early in life and chronic asthma in children: A birth cohort study. *Lancet* 368(9537): 763–70, 2006.
2. Simpson BM, Custovic A, Simpson A, Hallam CL, Walsh D, Marolia H *et al.* NAC Manchester Asthma and Allergy Study (NACMAAS): Risk factors for asthma and allergic disorders in adults. *Clin Exp Allergy* 31(3): 391–99, 2001.
3. Lowe LA, Woodcock A, Murray CS, Morris J, Simpson A, Custovic A. Lung function at age 3 years: Effect of pet ownership and exposure to indoor allergens. *Arch Pediatr Adolesc Med* 158(10): 996–1001, 2004.
4. Langley SJ, Goldthorpe S, Craven M, Morris J, Woodcock A, Custovic A. Exposure and sensitization to indoor allergens: Association with lung function, bronchial reactivity, and exhaled nitric oxide measures in asthma. *J Allergy Clin Immunol* 112(2): 362–68, 2003.
5. Langley SJ, Goldthorpe S, Craven M, Woodcock A, Custovic A. Relationship between exposure to domestic allergens and bronchial hyperresponsiveness in non-sensitised, atopic asthmatic subjects. *Thorax* 60(1): 17–21, 2005.
6. Rosenstreich DL, Eggleston P, Kattan M, Baker D, Slavin RG, Gergen P *et al.* The role of cockroach allergy and exposure to cockroach allergen in causing morbidity among inner-city children with asthma. *N Engl J Med* 336(19): 1356–63, 1997.
7. Tunnicliffe WS, Fletcher TJ, Hammond K, Roberts K, Custovic A, Simpson A *et al.* Sensitivity and exposure to indoor allergens in adults with differing asthma severity. *Eur Respir J* 13(3): 654–59, 1999.
8. Murray CS, Poletti G, Kebadze T, Morris J, Woodcock A, Johnston SL *et al.* Study of modifiable risk factors for asthma exacerbations: Virus infection and allergen exposure increase the risk of asthma hospital admissions in children. *Thorax* 61(5): 376–82, 2006.
9. Simpson A, Custovic A, Pipis S, Adisesh A, Faragher B, Woodcock A. Exhaled nitric oxide, sensitization, and exposure to allergens in patients with asthma who are not taking inhaled steroids. *Am J Respir Crit Care Med* 160(1): 45–49, 1999.
10. Custovic A, Taggart SC, Francis HC, Chapman MD, Woodcock A. Exposure to house dust mite allergens and the clinical activity of asthma. *J Allergy Clin Immunol* 98(1): 64–72, 1996.
11. Platts-Mills TA, Tovey ER, Mitchell EB, Moszoro H, Nock P, Wilkins SR. Reduction of bronchial hyperreactivity during prolonged allergen avoidance. *Lancet* 2(8300): 675–78, 1982.
12. Peroni DG, Piacentini GL, Costella S, Pietrobelli A, Bodini A, Loiacono A *et al.* Mite avoidance can reduce air trapping and airway inflammation in allergic asthmatic children. *Clin Exp Allergy* 32(6): 850–55, 2002.
13. Milanese M, Peroni D, Costella S, Aralla R, Loiacono A, Barp C *et al.* Improved bronchodilator effect of deep inhalation after allergen avoidance in asthmatic children. *J Allergy Clin Immunol* 114(3): 505–11, 2004.
14. Sensi LG, Piacentini GL, Nobile E, Ghebregzabher M, Brunori R, Zanolla L *et al.* Changes in nasal specific IgE to mites after periods of allergen exposure-avoidance: A comparison with serum levels. *Clin Exp Allergy* 24(4): 377–82, 1994.
15. Chan-Yeung M, Malo JL. Occupational asthma. *N Engl J Med* 333(2): 107–12, 1995.
16. Custovic A, Murray CS, Gore RB, Woodcock A. Controlling indoor allergens. *Ann Allergy Asthma Immunol* 88(5): 432–41, 2002. quiz 442–43, 529
17. Custovic A, Green R, Fletcher A, Smith A, Pickering CA, Chapman MD *et al.* Aerodynamic properties of the major dog allergen Can f 1: Distribution in homes, concentration, and particle size of allergen in the air. *Am J Respir Crit Care Med* 155(1): 94–98, 1997.
18. Custovic A, Simpson A, Pahdi H, Green RM, Chapman MD, Woodcock A. Distribution, aerodynamic characteristics, and removal of the major cat allergen Fel d 1 in British homes. *Thorax* 53(1): 33–38, 1998.

19. Custovic A, Simpson B, Simpson A, Hallam C, Craven M, Woodcock A. Relationship between mite, cat, and dog allergens in reservoir dust and ambient air. *Allergy* 54(6): 612–16, 1999.
20. Mahakittikun V, Boitano JJ, Tovey E, Bunnag C, Ninsanit P, Matsumoto T *et al.* Mite penetration of different types of material claimed as mite proof by the Siriraj chamber method. *J Allergy Clin Immunol* 118(5): 1164–68, 2006.
21. Tovey E, Marks G. Methods and effectiveness of environmental control. *J Allergy Clin Immunol* 103(2 Pt 1): 179–91, 1999.
22. Simpson A, Woodcock A, Custovic A. Housing characteristics and mite allergen levels: To humidity and beyond. *Clin Exp Allergy* 31(6): 803–5, 2001.
23. Fletcher AM, Pickering CA, Custovic A, Simpson J, Kennaugh J, Woodcock A. Reduction in humidity as a method of controlling mites and mite allergens: The use of mechanical ventilation in British domestic dwellings. *Clin Exp Allergy* 26(9): 1051–56, 1996.
24. Custovic A, Taggart SC, Kennaugh JH, Woodcock A. Portable dehumidifiers in the control of house dust mites and mite allergens. *Clin Exp Allergy* 25(4): 312–16, 1995.
25. Custovic A, Simpson BM, Simpson A, Hallam C, Craven M, Brutsche M *et al.* Manchester Asthma and Allergy Study: Low-allergen environment can be achieved and maintained during pregnancy and in early life. *J Allergy Clin Immunol* 105(2 Pt 1): 252–58, 2000.
26. Wood RA, Chapman MD, Adkinson NF Jr, Eggleston PA. The effect of cat removal on allergen content in household-dust samples. *J Allergy Clin Immunol* 83(4): 730–34, 1989.
27. Green R, Simpson A, Custovic A, Faragher B, Chapman M, Woodcock A. The effect of air filtration on airborne dog allergen. *Allergy* 54(5): 484–88, 1999.
28. Gore RB, Bishop S, Durrell B, Curbishley L, Woodcock A, Custovic A. Air filtration units in homes with cats: Can they reduce personal exposure to cat allergen?. *Clin Exp Allergy* 33(6): 765–69, 2003.
29. de Blay F, Chapman MD, Platts-Mills TA. Airborne cat allergen (Fel d I). Environmental control with the cat in situ. *Am Rev Respir Dis* 43(6): 1334–39, 1991.
30. Avner DB, Perzanowski MS, Platts-Mills TA, Woodfolk JA. Evaluation of different techniques for washing cats: Quantitation of allergen removed from the cat and the effect on airborne Fel d 1. *J Allergy Clin Immunol* 100(3): 307–12, 1997.
31. Hodson T, Custovic A, Simpson A, Chapman M, Woodcock A, Green R. Washing the dog reduces dog allergen levels, but the dog needs to be washed twice a week. *J Allergy Clin Immunol* 103(4): 581–85, 1999.
32. Gore RB, Durrell B, Bishop S, Curbishley L, Woodcock A, Custovic A. High-efficiency particulate arrest-filter vacuum cleaners increase personal cat allergen exposure in homes with cats. *J Allergy Clin Immunol* 111(4): 784–87, 2003.
33. Eggleston PA. Cockroach allergen abatement in inner-city homes. *Ann Allergy Asthma Immunol* 91(6): 512–14, 2003.
34. Eggleston PA, Arruda LK. Ecology and elimination of cockroaches and allergens in the home. *J Allergy Clin Immunol* 107(3 Suppl): S422–29, 2001.
35. Eggleston PA, Butz A, Rand C, Curtin-Brosnan J, Kanchanaraksa S, Swartz L *et al.* Home environmental intervention in inner-city asthma: A randomized controlled clinical trial. *Ann Allergy Asthma Immunol* 95(6): 518–24, 2005.
36. Custovic A, Wijk RG. The effectiveness of measures to change the indoor environment in the treatment of allergic rhinitis and asthma: ARIA update (Collaboration with GA(2)LEN). *Allergy* 60(9): 1112–15, 2005.
37. Gotzsche P, Johansen H. House dust mite control measures for asthma. *Cochrane Database Syst Rev* 2: 2008, CD001187.
38. Gotzsche PC, Johansen HK, Schmidt LM, Burr ML. House dust mite control measures for asthma. *Cochrane Database Syst Rev* 4: 2004, CD001187.
39. Gotzsche PC, Johansen HK, Hammarquist C, Burr ML. House dust mite control measures for asthma. *Cochrane Database Syst Rev* 2: 2001, CD001187.
40. Walshaw MJ, Evans CC. Allergen avoidance in house dust mite sensitive adult asthma. *Q J Med* 58(226): 199–215, 1986.
41. van der Heide S, Kauffman HF, Dubois AE, de Monchy JG. Allergen reduction measures in houses of allergic asthmatic patients: Effects of air-cleaners and allergen-impermeable mattress covers. *Eur Respir J* 10(6): 1217–23, 1997.
42. Htut T, Higenbottam TW, Gill GW, Darwin R, Anderson PB, Syed N. Eradication of house dust mite from homes of atopic asthmatic subjects: A double-blind trial. *J Allergy Clin Immunol* 107(1): 55–60, 2001.
43. Woodcock A, Forster L, Matthews E, Martin J, Letley L, Vickers M *et al.* Control of exposure to mite allergen and allergen-impermeable bed covers for adults with asthma. *N Engl J Med* 349(3): 225–36, 2003.
44. Luczynska C, Tredwell E, Smeeton N, Burney P. A randomized controlled trial of mite allergen-impermeable bed covers in adult mite-sensitized asthmatics. *Clin Exp Allergy* 33(12): 1648–53, 2003.
45. van den Bemt L, van Knapen L, de Vries MP, Jansen M, Cloosterman S, van Schayck CP. Clinical effectiveness of a mite allergen-impermeable bed-covering system in asthmatic mite-sensitive patients. *J Allergy Clin Immunol* 114(4): 858–62, 2004.
46. Ehnert B, Lau-Schadendorf S, Weber A, Buettner P, Schou C, Wahn U. Reducing domestic exposure to dust mite allergen reduces bronchial hyperreactivity in sensitive children with asthma. *J Allergy Clin Immunol* 90(1): 135–38, 1992.
47. Carswell F, Birmingham K, Oliver J, Crewes A, Weeks J. The respiratory effects of reduction of mite allergen in the bedrooms of asthmatic children – a double-blind controlled trial. *Clin Exp Allergy* 26(4): 386–96, 1996.
48. Carter MC, Perzanowski MS, Raymond A, Platts-Mills TA. Home intervention in the treatment of asthma among inner-city children. *J Allergy Clin Immunol* 108(5): 732–37, 2001.
49. Halken S, Host A, Niklassen U, Hansen LG, Nielsen F, Pedersen S *et al.* Effect of mattress and pillow encasings on children with asthma and house dust mite allergy. *J Allergy Clin Immunol* 111(1): 169–76, 2003.
50. Morgan WJ, Crain EF, Gruchalla RS, O'Connor GT, Kattan M, Evans R *et al.* Results of a home-based environmental intervention among urban children with asthma. *N Engl J Med* 351(11): 1068–80, 2004.
51. Kattan M, Stearns SC, Crain EF, Stout JW, Gergen PJ, Evans R 3rd *et al.* Cost-effectiveness of a home-based environmental intervention for inner-city children with asthma. *J Allergy Clin Immunol* 116(5): 1058–63, 2005.
52. Sheikh A, Hurwitz B. House dust mite avoidance measures for perennial allergic rhinitis. *Cochrane Database Syst Rev* 4: 2001, CD001563.
53. Terreehorst I, Hak E, Oosting AJ, Tempels-Pavlica Z, de Monchy JG, Bruijnzeel-Koomen CA *et al.* Evaluation of impermeable covers for bedding in patients with allergic rhinitis. *N Engl J Med* 349(3): 237–46, 2003.
54. Oosting AJ, de Bruin-Weller MS, Terreehorst I, Tempels-Pavlica Z, Aalberse RC, de Monchy JG *et al.* Effect of mattress encasings on atopic dermatitis outcome measures in a double-blind, placebo-controlled study: The Dutch mite avoidance study. *J Allergy Clin Immunol* 110(3): 500–6, 2002.
55. Gutgesell C, Heise S, Seubert S, Seubert A, Domhof S, Brunner E *et al.* Double-blind placebo-controlled house dust mite control measures in adult patients with atopic dermatitis. *Br J Dermatol* 145(1): 70–74, 2001.
56. Tan BB, Weald D, Strickland I, Friedmann PS. Double-blind controlled trial of effect of housedust-mite allergen avoidance on atopic dermatitis. *Lancet* 347(8993): 15–18, 1996.
57. Shirai T, Matsui T, Suzuki K, Chida K. Effect of pet removal on pet allergic asthma. *Chest* 127(5): 1565–71, 2005.
58. Francis H, Fletcher G, Anthony C, Pickering C, Oldham L, Hadley E *et al.* Clinical effects of air filters in homes of asthmatic adults sensitized and exposed to pet allergens. *Clin Exp Allergy* 33(1): 101–5, 2003.
59. van der Heide S, van Aalderen WM, Kauffman HF, Dubois AE, de Monchy JG. Clinical effects of air cleaners in homes of asthmatic children sensitized to pet allergens. *J Allergy Clin Immunol* 104(2 Pt 1): 447–51, 1999.

60. Lever R, MacDonald C, Waugh P, Aitchison T. Randomised controlled trial of advice on an egg exclusion diet in young children with atopic eczema and sensitivity to eggs. *Pediatr Allergy Immunol* 9(1): 13–19, 1998.
61. Mabin DC, Sykes AE, David TJ. Controlled trial of a few foods diet in severe atopic dermatitis. *Arch Dis Child* 73(3): 202–7, 1995.
62. Arshad SH, Bojarskas J, Tsitoura S, Matthews S, Mealy B, Dean T *et al.* Prevention of sensitization to house dust mite by allergen avoidance in school age children: A randomized controlled study. *Clin Exp Allergy* 32(6): 843–49, 2002.
63. Tsitoura S, Nestoridou K, Botis P, Karmaus W, Botezan C, Bojarskas J *et al.* Randomized trial to prevent sensitization to mite allergens in toddlers and preschoolers by allergen reduction and education: One-year results. *Arch Pediatr Adolesc Med* 156(10): 1021–27, 2002.
64. Simpson A, Custovic A. Allergen avoidance in the primary prevention of asthma. *Curr Opin Allergy Clin Immunol* 4(1): 45–51, 2004.
65. Simpson A, Custovic A. The role of allergen avoidance in primary and secondary prevention. *Pediatr Pulmonol Suppl* 26: 225–28, 2004.
66. Arshad SH, Matthews S, Gant C, Hide DW. Effect of allergen avoidance on development of allergic disorders in infancy. *Lancet* 339(8808): 1493–97, 1992.
67. Hide DW, Matthews S, Matthews L, Stevens M, Ridout S, Twiselton R *et al.* Effect of allergen avoidance in infancy on allergic manifestations at age two years. *J Allergy Clin Immunol* 93(5): 842–46, 1994.
68. Hide DW, Matthews S, Tariq S, Arshad SH. Allergen avoidance in infancy and allergy at 4 years of age. *Allergy* 51(2): 89–93, 1996.
69. Arshad SH, Bateman B, Matthews SM. Primary prevention of asthma and atopy during childhood by allergen avoidance in infancy: A randomised controlled study. *Thorax* 58(6): 489–93, 2003.
70. Becker A, Watson W, Ferguson A, Dimich-Ward H, Chan-Yeung M. The Canadian Asthma Primary Prevention Study: Outcomes at 2 years of age. *J Allergy Clin Immunol* 113(4): 650–56, 2004.
71. Chan-Yeung M, Ferguson A, Watson W, Dimich-Ward H, Rousseau R, Lilley M *et al.* The Canadian Childhood Asthma Primary Prevention Study: Outcomes at 7 years of age. *J Allergy Clin Immunol* 116(1): 49–55, 2005.
72. Halmerbauer G, Gartner C, Schierl M, Arshad H, Dean T, Koller DY *et al.* Study on the Prevention of Allergy in Children in Europe (SPACE): Allergic sensitization at 1 year of age in a controlled trial of allergen avoidance from birth. *Pediatr Allergy Immunol* 14(1): 10–17, 2003.
73. Horak F Jr, Matthews S, Ihorst G, Arshad SH, Frischer T, Kuehr J *et al.* Effect of mite-impermeable mattress encasings and an educational package on the development of allergies in a multinational randomized, controlled birth-cohort study – 24 months results of the Study of Prevention of Allergy in Children in Europe. *Clin Exp Allergy* 34(8): 1220–25, 2004.
74. Mihrshahi S, Peat JK, Marks GB, Mellis CM, Tovey ER, Webb K *et al.* Eighteen-month outcomes of house dust mite avoidance and dietary fatty acid modification in the Childhood Asthma Prevention Study (CAPS). *J Allergy Clin Immunol* 111(1): 162–68, 2003.
75. Peat JK, Mihrshahi S, Kemp AS, Marks GB, Tovey ER, Webb K *et al.* Three-year outcomes of dietary fatty acid modification and house dust mite reduction in the Childhood Asthma Prevention Study. *J Allergy Clin Immunol* 114(4): 807–13, 2004.
76. Marks GB, Mihrshahi S, Kemp AS, Tovey ER, Webb K, Almqvist C *et al.* Prevention of asthma during the first 5 years of life: A randomized controlled trial. *J Allergy Clin Immunol* 118(1): 53–61, 2006.
77. Almqvist C, Li Q, Britton WJ, Kemp AS, Xuan W, Tovey ER *et al.* Early predictors for developing allergic disease and asthma: Examining separate steps in the "allergic march". *Clin Exp Allergy* 37(9): 1296–302, 2007.
78. Schonberger HJ, Dompeling E, Knottnerus JA, Maas T, Muris JW, van Weel C *et al.* The PREVACS study: The clinical effect of a multifaceted educational intervention to prevent childhood asthma. *Eur Respir J* 25(4): 660–70, 2005.
79. Koopman LP, van Strien RT, Kerkhof M, Wijga A, Smit HA, de Jongste JC *et al.* Placebo-controlled trial of house dust mite-impermeable mattress covers: Effect on symptoms in early childhood. *Am J Respir Crit Care Med* 166(3): 307–13, 2002.
80. Corver K, Kerkhof M, Brussee JE, Brunekreef B, van Strien RT, Vos AP *et al.* House dust mite allergen reduction and allergy at 4 yr: Follow up of the PIAMA-study. *Pediatr Allergy Immunol* 17(5): 329–36, 2006.
81. Custovic A, Simpson BM, Simpson A, Kissen P, Woodcock A. Effect of environmental manipulation in pregnancy and early life on respiratory symptoms and atopy during first year of life: A randomised trial. *Lancet* 358(9277): 188–93, 2001.
82. Woodcock A, Lowe LA, Murray CS, Simpson BM, Pipis SD, Kissen P *et al.* Early life environmental control: Effect on symptoms, sensitization, and lung function at age 3 years. *Am J Respir Crit Care Med* 170(4): 433–39, 2004.
83. Simpson A, John SL, Jury F, Niven R, Woodcock A, Ollier WE *et al.* Endotoxin exposure, CD14, and allergic disease: An interaction between genes and the environment. *Am J Respir Crit Care Med* 174(4): 386–92, 2006.
84. Custovic A, Simpson A. Environmental allergen exposure, sensitisation and asthma: From whole populations to individuals at risk. *Thorax* 59(10): 825–27, 2004.

47

CHAPTER

Smoking Cessation

Michael A. Chandler and Stephen I. Rennard

University of Nebraska Medical Center, Omaha, NE, USA

INTRODUCTION

Cigarette smoking is a chronic relapsing disorder with secondary complications including atherosclerotic cardiovascular disease, chronic obstructive pulmonary disease (COPD), and lung cancer.

Smoking remains the leading cause of preventable death in the world [1]. Tobacco control policies and smoking cessation therapies have been shown to be effective deterrents and treatments (Fig. 47.1). This chapter will outline the epidemiology, pathophysiology, and treatment of cigarette smoking with particular emphasis on current pharmacotherapies including nortriptyline, clonidine, nicotine replacement, bupropion, and varenicline.

THE GLOBAL TOBACCO EPIDEMIC

Cigarette smoking led to five million premature deaths worldwide in 2000 [2] (Fig. 47.2). Due to declines in smoking prevalence, smoking-related mortality is trending downward in developed countries like the United States and United Kingdom; however, other developing countries are experiencing dramatic increases in mortality [3]. In China, smoking is attributed to a million deaths annually, a figure that is still on the rise [4]. India is expected to achieve this dubious milestone by 2010 [5]. Barring alterations in present tobacco use, the tobacco epidemic will lead to an estimated one billion deaths this century alone [1].

A HETEROGENEOUS DISEASE

Heterogeneity in the susceptibility to and persistence of cigarette smoking is governed by genes, environment, and gene by environment interactions [6]. Genes that modulate neurotransmitter signaling, best characterized for dopamine and opiod signaling, have been implicated [7, 8]. Polymorphisms in the enzymes that metabolize nicotine into inactive metabolites like cotinine

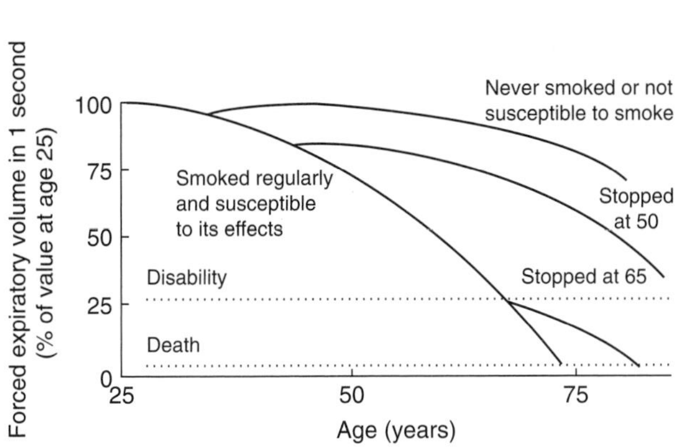

FIG. 47.1 The natural history of chronic bronchitis and emphysema. Adapted from Fletcher M, Peto R. *BMJ.* 1, 1645–48, 1977.

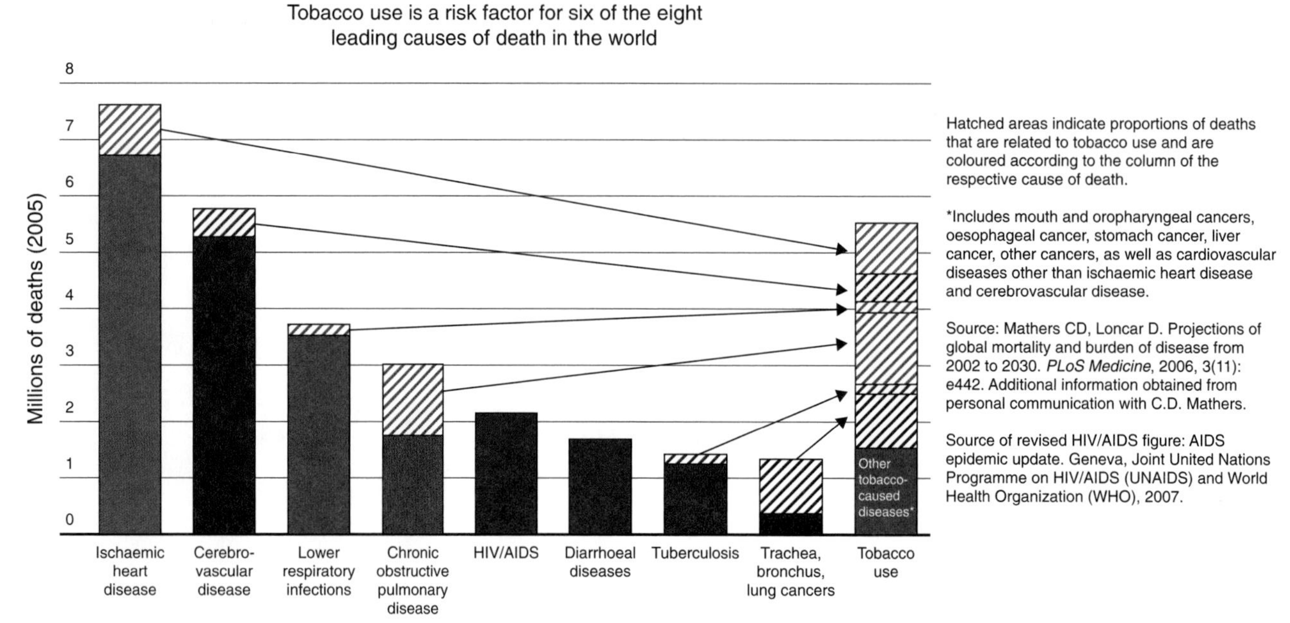

FIG. 47.2 Individual and Composite Impact of Tobacco Use on Death Rate of the Eight Leading Causes of Death in the World. From World Health Organization (WHO), report on the global tobacco epidemic, 2008: The MPOWER package, Geneva.

and 3-hydroxycotinine also play a role. For example, slow metabolizers are less likely to become smokers. Nevertheless, they comprise about 20% of total smokers but are likely to smoke fewer cigarettes, perhaps because they "overdose" experiencing higher levels of nicotine for greater duration. The opposite is true for fast metabolizers [9]. Environmental cues, such as social prohibition or acceptance of smoking, also impact smoking prevalence. These factors and their interplay lead to a broad spectrum of smoking behaviors.

NICOTINE ADDICTION

Nicotine as an addictive euphoriant is comparable to amphetamines, cocaine, or opiates [10]. Nicotinic acetylcholine receptors in the central nervous system regulate downstream neurotransmitter release. For example, nicotine addiction appears to be mediated by nicotine-stimulated dopamine release in the mesolimbic system, a pathway involved in endogenous reward and behavioral consolidation [11, 12]. Additionally, smokers habitually smoke during common situations, for example, when driving or socializing or after eating or during stressful conversations. This leads to classic or "Pavlovian" conditioning, which is potentiated by nicotine [13].

The interplay between these neurophysiologic and psychologic forces leads to addiction for roughly 85% of smokers [14] (Fig. 47.3). The presence of addiction is further evidenced by a well-defined withdrawal syndrome upon smoking cessation. Features of this syndrome include dysphoria, insomnia, irritability, anxiety, diminished concentration, restlessness, increased appetite, decreased heart rate, and weight gain [15]. Craving for smoking often persists years following cessation and may wane in frequency but not in intensity, a feature that

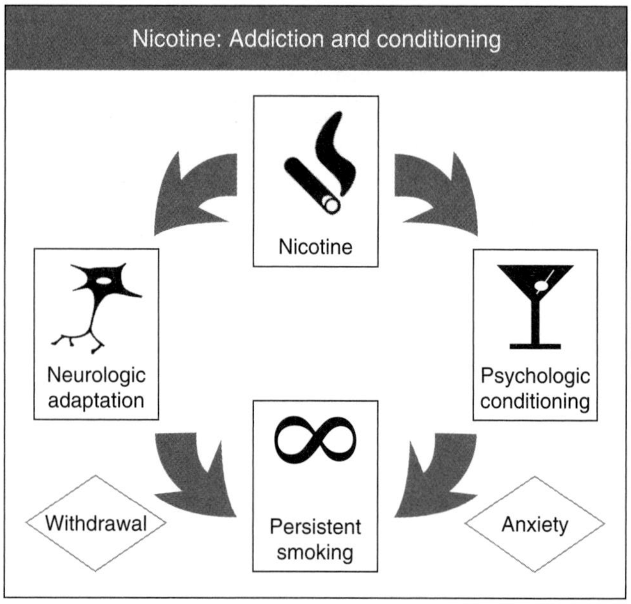

FIG. 47.3 Biopsychosocial model of nicotine addiction.

resembles a grief response. Insights such as these have led to our current understanding of tobacco dependence as a chronic relapsing disorder requiring recurrent therapy.

SPECTRUM OF COMPLICATIONS

Smoking leads to about one-third of deaths from coronary artery disease (CAD), the most common cause of morbidity

TABLE 47.1 The 5 A's: brief smoking cessation strategies for routine practice. The 5 R's: generating or reinforcing motivation to quit smoking.

Step	Action
Five A's	
Ask about smoking	Systematically assess smoking status Smoking should be a vital sign
Advise all smokers to stop	Give a clear, strong, and personalized message to stop smoking
Assess willingness to stop	Determine whether the smoker is ready to stop currently or soon. Provide assistance to motivated smokers; motivate those unwilling to stop
Assist motivated smokers to stop	Help smoker with a quit plan, counseling, pharmacotherapy and additional materials
Arrange follow-up	Follow up soon after the quit date. Assess success and difficulties. Consider referring for a more intensive intervention if there is relapse
Five R's	
Relevance	Encourage the patient to indicate why quitting is personally relevant
Risks	Ask the patient to identify potential negative consequences of tobacco use
Rewards	Ask the patient to identify potential benefits of stopping tobacco use
Roadblocks	Ask the patient to identify barriers to quitting
Repetition	Apply 5 R's whenever interacting with unmotivated or relapsed smokers

Adapted from Ref. [45] and *http:ahrq.gov/clinic/tobacco/5rs.htm.*

and mortality in the developed world [8, 16, 17]. Active and passive smoke exposure leads to increases in oxidative stress, vascular inflammation, blood coagulability, platelet aggregation, and thrombus formation as well as reduced oxygen delivery and coronary vasoconstriction [16, 18]. Heart disease risk drops off significantly after 1–2 years of cessation and normalizes at 3–5 years [18, 19].

COPD and cigarette smoking are closely linked. Smoking directly damages the lung and triggers a cascade of secondary inflammation and impaired repair that leads to ongoing damage [20]. These effects lead to COPD in roughly 50% of active smokers and 7% of passive smoke exposed individuals [21–23]. Exposure *in utero* or in adolescence compromises subsequent lung growth and predisposes to COPD [24–27].

In countries where smoking is common, the risk of lung cancer increases 20-fold leading to 90% of lung cancers [28]. Smoking has been clearly linked to genetic and epigenetic alterations that promote carcinogenesis [16, 29, 30].

PUBLIC HEALTH RESPONSE

Smoking prevention and cessation should reduce the prevalence of smoking and, thus, prevent millions of premature deaths [3]. Public health measures (advertisement bans, indoor smoking bans and measures to increase cigarette purchase price) increase smoking cessation and reduce tobacco use and passive smoke exposure [31–33]. While smoking trends have been decreasing in the United States, recent estimates suggest this decrease has slowed and nearly stopped with a current smoker rate of 20.8% [34]. Public health initiatives and tobacco control programs in the United States remain underfunded and susceptible to having monies withdrawn and allocated to support other programs or to cover budget deficits. Despite ample evidence supporting the achievements and returns to society of such programs, they are transient fixtures in public health [35]. Nevertheless, many states have continued to push for total bans on smoking and 24 states have been successful. Internationally, several countries now too have similar laws prohibiting smoking nationwide.

SMOKING CESSATION: BEHAVIORAL INTERVENTION

Smoking abstinence rates at 6–12 months in motivated quitters without behavioral or pharmacologic therapy approach 3–5% [36]. Even minimal behavioral therapy delivered by a medical practitioner increases this abstinence rate to 5.5–7.5% [37].

The "5A's" provides general guidelines to smoking cessation in clinical practice (Table 47.1) [38]. The "stages of change" model predicts smoking abstinence at 1 year and has become a useful tool in tailoring behavioral tobacco cessation efforts [40] (Fig. 47.4). In this model, smokers are regarded as "precontemplators" who are not thinking of

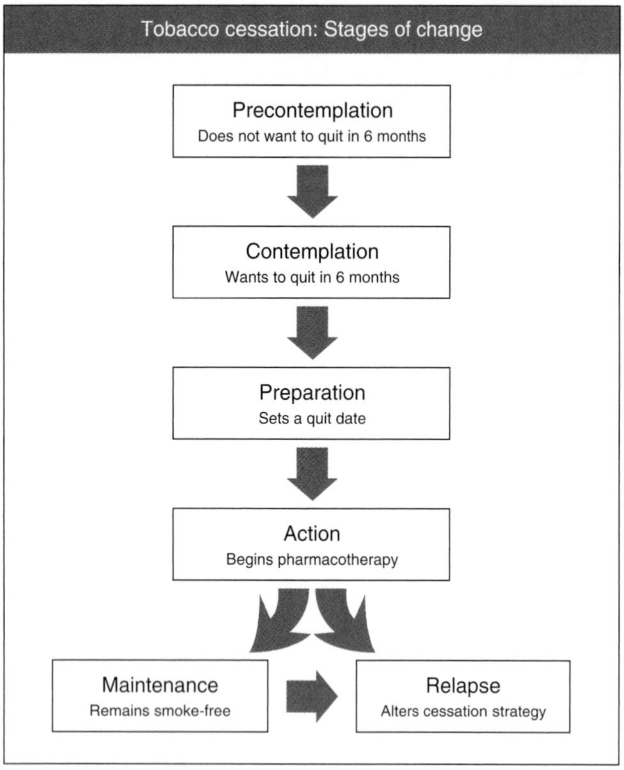

FIG. 47.4 Transtheoretical model of change. Adapted from Ref. [39].

quitting, "contemplators" who are considering quitting, or "preparers" who are ready to trial cessation. Precontemplators and contemplators may be advanced by discussing the "5R's" (Table 47.1) [38]. With physician guidance, preparers create a smoking cessation plan including a quit date. At this stage, smokers should be advised to inform others of the impending quit attempt and counseled to avoid places or situations with close associations to smoking. The quit date marks their graduation into the action then maintenance stages, marked by initial and chronic tobacco cessation, respectively. Attempts to bypass stages generally decrease the likelihood of success. Inpatient or outpatient episodes of acute illness are an exception to this and represent windows of opportunity when smokers are more liable to quit [41, 42]. The chronic, relapsing nature of cigarette smoking means patients often will cycle through stages several times before achieving stable abstinence [39]. Additional counseling advice may be reviewed elsewhere [38, 39].

More intensive behavioral therapies like group or individual counseling sessions improve the odds of cessation when compared to minimal advice (odds ratio 1.44, 95% CI = 1.24–1.67) [37]. Many smokers, however, are unwilling to participate in more intensive programs. Augmenting behavioral support with additional healthcare personnel, follow-up visits, or even a brief phone call increases quit rates but is not consistently covered by private or public health insurance [37, 38].

The majority of patients will relapse in the first 6–12 months. Reasons for relapse should be reviewed and accounted for in future cessation trials. All tobacco cessation attempts should supplement behavioral interventions with pharmacotherapy as described below.

SMOKING CESSATION: PHARMACOTHERAPY

Because the subject characteristics and the support offered differ across clinical trials, the most conventional means to assess the effect of pharmacotherapy is the quit rate relative to placebo.

When compared to placebo, pharmacotherapy doubles or triples quit rates achieved by nonpharmacological means (Table 47.2). The relatively high placebo quit rates seen in many clinical trials, which are higher than the rates noted above for minimal advice, are likely due to selection of highly motivated individuals for entry.

Nicotine replacement therapy

Smokers who quit not only lose the euphoric effects of nicotine but frequently experience symptoms related to nicotine withdrawal [4]. Nicotine replacement therapy (NRT), the most widely used tobacco cessation agent, may partially ameliorate these ill effects and is central to American and British guidelines [1, 5–7]. NRT is given at doses sufficient to temper withdrawal without providing the reinforcing effects of bolus nicotine. NRT almost doubles the untreated quit rate regardless of whether behavioral counseling is applied [8, 9]. Smoking abstinence rates at 6–12 months with combined NRT and limited behavioral therapy approach 17% (range 12–34%) [43, 47–50]. However, a single round of NRT does not prevent relapse. About one-third of quitters will relapse by 4 years whether NRT is employed or not – a result consistent with our current understanding of tobacco dependence as a chronic relapsing disorder requiring recurrent therapy [7, 8].

Nicotine gums, lozenges, patches, nasal sprays, and inhalers are equally efficacious whereas heavy smokers may benefit from higher doses [9, 15–19, 51]. The patch and nasal spray are beneficial for up to 2 and 3 months, respectively. Other forms of NRT are used for 3–6 month courses [9]. There is no benefit to tapering of NRT versus abrupt withdrawal [9]. Physician-prescribed NRT appears to be more effective than voluntary over-the-counter NRT [9]. The cost of 6 weeks of NRT is comparable to that of smoking one pack per day and diminishes as the course progresses.

Nicotine is metabolized by the liver so that NRT with gum or lozenge is dependent on absorption through the buccal mucosa. Moreover, nicotine is poorly absorbed at low oral pH; thus, gum or lozenge use should be deferred during and 15 min after snacks or meals and should not be used with acidic beverages [52–54]. Oral irritation, indigestion, and diarrhea, which likely result from swallowed drug, are the most common side effects [52, 53].

Transdermal nicotine patches require no special technique for application, have no restrictions regarding dosage

TABLE 47.2 Pharmacotherapies for smoking cessation.

Product	Time peak venous nicotine blood level	Common side effects	Dosage	Duration	Odds ratio 1-year smoking abstinence versus placebo
Nicotine gum or lozenge	25–30 min	Oral irritation, indigestion, diarrhea	1 piece every 1–2 h as needed, then taper Daily maximum: 24 pieces gum or 20 lozenges	3 months	1.80
Nicotine nasal spray	5–15 min	Nasal irritation	1 inhalation, each nostril, every 1–2 h as needed, then taper Daily maximum: 40 doses (80 sprays)	3 months	2.03
Nicotine inhaler	15 min	Oropharyngeal irritation	Multiple inhalations over 20 min every 1–2 h as needed, then taper Daily maximum: 16 cartridges	3 months	2.14
Nicotine patch	6–12 h	Vivid nightmares	1 patch daily for 16 or 24 h	2 months	2.16
Bupropion	NA	Insomnia, lowered seizure threshold	150 mg daily for 3 d, then 150 mg twice daily	3 months*	1.83
Varenicline	NA	Nausea	0.5 mg daily increasing to 1 mg twice daily	3 months*	3.22

Derived from manufacturers' product data sheets and Refs [43–46].
*Efficacy may be increased by extending duration of therapy beyond 3 months.

timing with oral intake, and are associated with fewer gastrointestinal complaints. They also have the lowest addiction potential, likely due to their slow pharmacokinetics [55]. Patches are often used at night to treat morning withdrawal symptoms; however, this may lead sleep disturbances, particularly vivid dreams or nightmares [56]. Skin irritation occurs in up to 54% of patch users but is generally mild and self-limited [57]. Rotating the site of patch application is recommended to minimize irritation.

Nicotine nasal spray delivers an aqueous solution of nicotine to the nasal mucosa and produces the most rapid rise in plasma nicotine concentration amongst various forms of NRT [58]. This may lead to prolonged nicotine dependence. For example, the rate of continued use after 12 months in clinical trials ranges from 3% to 13% for nicotine nasal spray versus 1% and 8% for nicotine gum [9, 59]. The most common side effect is nasal irritation that persists in up to 81% of patients after 3 weeks of use [60].

The nicotine inhaler is composed of a mouthpiece and a plastic, nicotine-containing cartridge that releases nicotine upon inhalation. Mild mouth or throat irritation and cough occurs with initial use in one-third of cases and wanes with time [61]. While named an "inhaler" nicotine does not effectively reach the lower respiratory tract with this device and absorption is mostly through the mouth and pharynx. Nevertheless, patients with severe airway reactivity should consider alternate forms of NRT as inhaled nicotine may cause bronchospasm.

Serious side effects are extremely rare despite short- and long-term NRT use by millions of smokers over the past two decades [9, 59]. Patients suffering acute cardiovascular events should defer NRT use; otherwise, NRT appears safe in patients with known cardiovascular disease [9, 58–65]. Though not recommended, concurrent smoking and NRT use does not appear to enhance cardiac risk [66]. Nicotine crosses the placenta and is excreted in breast milk. While animal studies link various fetal and reproductive toxicities to nicotine exposure, NRT is devoid of the hazardous compounds present in tobacco smoke [67]. Ongoing tobacco smoking is the single largest modifiable risk for pregnancy-related morbidity and mortality and pregnant smokers have additional motivation to quit [68]. Thus, NRT is a reasonable alternative to ongoing smoking, should behavioral intervention fail. Nicotine gum and lozenge are considered safer in this population than other forms of NRT by the US Food and Drug Administration (USFDA) [28, 29, 60, 61, 67–70].

Bupropion

Bupropion is an atypical antidepressant that has been used as a smoking cessation agent for over a decade [71]. It is believed to block the reuptake of dopamine, serotonin, and norepinephrine and thereby treat occult depression, withdrawal dysphoria, and addiction [44, 72]. Smoking cessation rates with bupropion (19% versus placebo) are comparable to rates with NRT (17% versus placebo) by indirect comparison [68].

The standard dose of sustained-released bupropion used for tobacco cessation is 150 mg twice daily. A meta-analysis comparing standard-dose to low-dose bupropion (150 mg once daily) did not demonstrate a benefit to standard therapy but was underpowered to establish equivalency [71, 73, 74]. Nonetheless, low-dose therapy remains a reasonable alternative to discontinuation in patients who experience side effects at standard doses. Bupropion is taken for up to 12 weeks though longer courses may be beneficial. One randomized, placebo-controlled trial found that ongoing use of bupropion decreased weight gain (3.8 versus 5.6 kg) and relapse rates in quitters at 1 year of treatment [75]. This benefit was lost 1 year after drug discontinuation. Two subsequent studies found no benefit in relapse rates in quitters 1 year after cessation with limited duration bupropion (3–6 months) at standard or low doses [76, 77].

Insomnia, the most common side effect, occurs in roughly one-third of patients. Bupropion may cause seizures in 1 patient per 1000 at standard doses leading to its contraindication in persons predisposed to or suffering from seizure disorders [44, 72, 78]. Bupropion, like other antidepressants, increases the risk of suicidality in patients under age 24 with psychiatric disorders, though this does not preclude its use in carefully monitored settings if the benefits are felt to outweigh the risks [79]. The USFDA considers bupropion in the same risk category as nicotine gum and lozenge when used by pregnant women [28, 29, 78]. A higher rate of spontaneous abortions has been identified in prospective trials of bupropion users who were pregnant or planning a pregnancy [78, 80].

Other antidepressants

Nortriptyline, which is not approved by the USFDA for smoking cessation, is suggested for consideration for off-label use in United States and United Kingdom guidelines [38]. Quit rates with nortriptyline are comparable to those of NRT and bupropion therapy per indirect meta-analysis and direct head-to-head comparison [44]. It is worth noting that the most recent studies in the aforementioned meta-analysis failed to show benefit. Side effects include nausea, sedation, and anticholinergic symptoms. No benefit has been found with selective serotonin reuptake inhibitors including fluoxetine, sertraline, and paroxetine, nor with serotonin-norepinephrine uptake inhibitors including venlafaxine, in either individual studies or meta-analyses [44].

Clonidine

Clonidine, an α-2 adrenergic and imidazoline agonist used for the treatment of essential hypertension, has been suggested for off-label use for smoking cessation [1]. Clonidine reduces withdrawal symptoms; however, only one of six randomized, placebo-controlled studies selected for recent meta-analysis found a statistically significant benefit [81]. Inclusion of this study raised the pooled odds ratio at 3 months up to 1.89 versus placebo. This second- or third-line drug may be useful in quitters with intense withdrawal symptoms, but clear benefit beyond 3 months of smoking cessation has not been established. Dry mouth and sedation, the most common side effects, occur in two-thirds of cases [81]. Postural hypotension is also common. Clonidine should be tapered off to avoid rebound hypertension.

Varenicline

Varenicline is the first nicotine receptor partial agonist approved for smoking cessation by the USFDA [82]. As mentioned above, neuronal nicotinic acetylcholine receptors stimulate the mesolimbic dopamine release that leads to subsequent nicotine addiction. Varenicline is a partial agonist of one such nicotinic receptor, the $\alpha 4\beta 2$ nicotinic receptor. As a partial agonist, varenicline elicits less dopamine release than nicotine and blunts subsequent nicotine-induced dopamine release [83]. This leads to less intense craving in the absence of nicotine and less intense pleasure in the presence of nicotine [84–87]. By eroding the reinforcing properties of smoking, varenicline continues to capture quitters weeks after the onset of therapy, an effect that seems to differ from other pharmacotherapies [85, 86]. Cessation rates up to 44% have been documented in clinical trials [88, 89]. In a meta-analysis of four randomized, placebo-controlled trials encompassing 3300 participants, the number-needed-to-treat for ongoing smoking abstinence with varenicline was 8 versus 15 and 20 for bupropion and NRT, respectively [88]. Trials subsequent to this analysis continue to demonstrate the efficacy and safety of varenicline [86, 87, 90, 91].

Nausea occurs in up to one-third of patients [88]. This is limited by the initial dose titration and in clinical trials led to discontinuation rates of 2.5–7.6% [88]. Dose should be adjusted for renal disease (creatinine clearance <30 ml/min) [89]. At doses 36–50 times the recommended daily maximum, low birthweight and subsequent reproductive toxicity, but not teratogenicity, were demonstrated. Thus, varenicline is USFDA pregnancy category C, the same risk category as nicotine gum, nicotine lozenge, and bupropion.

In 2007, the FDA informed health care providers about suicidal thoughts and aggressive and erratic behavior in patients treated with varenicline, even non-quitters [92]. This has lead to recommendations that physicians and caregivers monitor for neuropsychiatric symptoms such as agitation, behavior change, depressed mood, suicidal ideation, and suicidal behavior.

Combination therapy

Supplementing *trans*-dermal nicotine with short-acting NRT products, such as nicotine gum, nasal spray, or inhaler, increases the likelihood of cessation by 40% compared to the use of the same products independently [43]. Using bupropion or nortriptyline in combination with NRT has not been shown to be beneficial by meta-analysis [44]. The efficacy of combination therapy with varenicline and either NRT or bupropion is unknown at present.

Cost-effectiveness

Consumer expense, lost productivity, and direct and indirect medical costs compound the burden of morbidity and mortality inherent with cigarette smoking. Despite this understanding, many practitioners reserve pharmacotherapy for refractory cases in an effort to minimize costs. Smoking cessation is cost-effective, regardless of cessation method used, statistical analysis utilized, and geographic or socioeconomic strata examined [93–95]. Thus, current guidelines suggest that all individuals who are willing to make a serious quit attempt should be given the best opportunity to achieve success, including pharmacological support [96].

In the United States in 2000, the cost per life-year saved with NRT and bupropion was cost $1914 to $11,217, a relative bargain. During this time period, the cost per quality-adjusted life-year saved with lovastatin, a popular drug used for the primary prevention of cardiovascular disease, was $56,415 to $438,780 [93]. Subsequent studies support the cost-effectiveness of varenicline [94]. In the near future, gene testing to predict treatment responses may offer additional cost reduction [97].

HARM REDUCTION

Harm reduction is an alternative strategy for the smoker who cannot, or will not, quit. It is designed to mitigate the medical complications of smoking without achieving full abstinence. In theory, this could be achieved by pharmacotherapy with nicotine or other agents or by substitution of less toxic tobacco products [98]. A similar strategy may also reduce the exposure to secondhand smoke among children of smokers. To date, however, there are no demonstrated health benefits of a harm reduction approach. Even if a harm reduction approach were to benefit individual smokers, the effect of such interventions in a population is difficult to predict. Some individuals may use reduction as a bridge to cessation. Others may opt not to quit but to reduce. These uncertainties have led to considerable debate and controversy.

FUTURE DIRECTIONS

A number of novel smoking cessation pharmacotherapies have followed in the wake of recent insights into nicotine addiction. Several cannabinoid receptor antagonists are in development. One such cannabinoid receptor antagonist, rimonobant, demonstrated efficacy but did not lead to drug approval due to concerns of toxicity [45]. A nicotine vaccine that stimulates the production of antibodies against nicotine has demonstrated efficacy in Phase II trials [99]. Other potential targets for smoking cessation drugs under investigation include the γ-amino butyric acid β receptor, the μ-opioid receptor, and the $\beta 2$ nicotinic receptor.

Finally, pharmacogenomics studies continue to cast light on the mechanisms behind variations in individual response to specific pharmacological agents. Research in this area will help tailor maximally effective smoking cessation pharmacotherapy.

References

1. World Health Organization. WHO report on the global tobacco epidemic. Geneva: The mpower package, 2008.
2. Ezzati M, Lopez AD. Estimates of global mortality attributable to smoking in 2000. *Lancet* 362(9387): 847–52, 2003.
3. MMWR. Annual smoking-attributable mortality, years of potential life lost and productivity losses – United States 1997–2001. *MMWR Morb Mortal Wkly Rep* 54(25): 625–628.
4. Liu BQ, Peto R, Chen ZM, Boreham J, Wu YP, Li JY, Campbell TC, Chen JS. Emerging tobacco hazards in China: 1. Retrospective proportional mortality study of one million deaths. *BMJ* 317(7170): 1411–22, 1998.
5. Jha P, Jacob B, Gajalakshmi V, Gupta PC, Dhingra N, Kumar R, Sinha DN, Dikshit RP, Parida DK, Kamadod R, Boreham J, Peto R. A nationally representative case-control study of smoking and death in India. *New Engl J Med* 358(11): 1137–47, 2008.
6. Hardie TL, Moss HB, Lynch KG. Genetic correlations between smoking initiation and smoking behaviors in a twin sample. *Addict Behav* 31(11): 2030–37, 2006.
7. Morley KI, Medland SE, Ferreira MA, Lynskey MT, Montgomery GW, Heath AC, Madden PA, Martin NG. A possible smoking susceptibility locus on chromosome 11p12: Evidence from sex-limitation linkage analyses in a sample of Australian twin families. *Behav Genet* 36(1): 87–99, 2006.
8. Arinami T, Ishiguro H, Onaivi ES. Polymorphisms in genes involved in neurotransmission in relation to smoking. *Eur J Pharmacol* 410(2–3): 215–26, 2000.
9. Lerman CE, Schnoll RA, Munafo MR. Genetics and smoking cessation improving outcomes in smokers at risk. *Am J Prev Med* 33(6 Suppl): S398–405, 2007.
10. Henningfield JE, Miyasato K, Jasinski DR. Abuse liability and pharmacodynamic characteristics of intravenous and inhaled nicotine. *J Pharmacol Exp Ther* 234: 1–12, 1985.
11. Picciotto MR, Zoli M, Rimondini R, Lena C, Marubio LM, Pich EM, Fuxe K, Changeux JP. Acetylcholine receptors containing the beta2 subunit are involved in the reinforcing properties of nicotine. *Nature* 391(6663): 173–77, 1998.
12. Tapper AR, McKinney SL, Nashmi R, Schwarz J, Deshpande P, Labarca C, Whiteaker P, Marks MJ, Collins AC, Lester HA. Nicotine activation of $\alpha 4^*$ receptors: Sufficient for reward, tolerance, and sensitization. *Science* 306(5698): 1029–32, 2004.
13. Olausson P, Jentsch JD, Taylor JR. Repeated nicotine exposure enhances responding with conditioned reinforcement. *Psychopharmacology (Berl)* 173(1–2): 98–104, 2004.
14. Shiffman S. Tobacco "chippers" – individual differences in tobacco dependence. *Psychopharmacology (Berl)* 97(4): 539–47, 1989.
15. *Diagnostic and Statistical Manual of Mental Disorders*. American Psychiatric Association, Washington, D.C., 1994.
16. The 2004 United States Surgeon General's Report. The health consequences of smoking. *NSW Public Health Bull* 15(5–6): 107, 2004.
17. Yusuf S, Reddy S, Ounpuu S, Anand S. Global burden of cardiovascular diseases: part I: General considerations, the epidemiologic transition, risk factors, and impact of urbanization. *Circulation* 104(22): 2746–53, 2001.
18. Reid RD, Quinlan B, Riley DL, Pipe AL. Smoking cessation: Lessons learned from clinical trial evidence. *Curr Opin Cardiol* 22(4): 280–85, 2007.
19. White WB. Smoking-related morbidity and mortality in the cardiovascular setting. *Prev Cardiol* 10(2 Suppl 1): 1–4, 2007.

20. Berlin AA, Hogaboam CM, Lukacs NW. Inhibition of SCF attenuates peribronchial remodeling in chronic cockroach allergen-induced asthma. *Lab Invest* 86(6): 557–65, 2006.
21. Lokke A, Lange P, Scharling H, Fabricius P, Vestbo J. Developing COPD: A 25 year follow up study of the general population. *Thorax* 61(11): 935–39, 2006.
22. Yin P, Jiang CQ, Cheng KK, Lam TH, Lam KH, Miller MR, Zhang WS, Thomas GN, Adab P. Passive smoking exposure and risk of COPD among adults in China: The Guangzhou Biobank Cohort Study. *Lancet* 370(9589): 751–57, 2007.
23. Rennard SI, Vestbo J. COPD: The dangerous underestimate of 15%. *Lancet* 367: 1216–19, 2006.
24. DiFranza JR, Aligne CA, Weitzman M. Prenatal and postnatal environmental tobacco smoke exposure and children's health. *Pediatrics* 113(4 Suppl): 1007–15, 2004.
25. Gold DR, Wang X, Wypij D, Speizer FE, Ware JH, Dockery DW. Effects of cigarette smoking on lung function in adolescent boys and girls. *New Engl J Med* 335: 931–37, 1996.
26. Wang X, Mensinga TT, Schouten JP, Rijcken B, Weiss ST. Determinants of maximally attained level of pulmonary function. *Am J Respir Crit Care Med* 169(8): 941–49, 2004.
27. Gilliland FD, Berhane K, McConnell R, Gauderman WJ, Vora H, Rappaport EB, Avol E, Peters JM. Maternal smoking during pregnancy, environmental tobacco smoke exposure and childhood lung function. *Thorax* 55(4): 271–76, 2000.
28. Alberg AJ, Ford JG, Samet JM. Epidemiology of lung cancer: ACCP evidence-based clinical practice guidelines (2nd edition). *Chest* 132(3 Suppl): 29S–55S, 2007.
29. Peebles KA, Lee JM, Mao JT, Hazra S, Reckamp KL, Krysan K, Dohadwala M, Heinrich EL, Walser TC, Cui X, Baratelli FE, Garon E, Sharma S, Dubinett SM. Inflammation and lung carcinogenesis: Applying findings in prevention and treatment. *Expert Rev Anticancer Ther* 7(10): 1405–21, 2007.
30. Wu X, Zhao H, Suk R, Christiani DC. Genetic susceptibility to tobacco-related cancer. *Oncogene* 23(38): 6500–23, 2004.
31. MMWR. Reducing tobacco use. A report of the Surgeon General. Executive Summary. *MMWR Recomm Rep* 49: 1–27, 2000.
32. Gorini G, Chellini E, Galeone D. What happened in Italy? A brief summary of studies conducted in Italy to evaluate the impact of the smoking ban. *Ann Oncol* 18(10): 1620–22, 2007.
33. Quentin W, Neubauer S, Leidl R, Konig HH. Advertising bans as a means of tobacco control policy: A systematic literature review of time-series analyses. *Int J Public Health* 52(5): 295–307, 2007.
34. MMWR. Cigarette smoking among adults – United States, 2006. *MMWR Morb Mortal Wkly Rep* 56(44): 1157–61, 2006.
35. Farrelly MC, Pechacek TF, Thomas KY, Nelson D. The impact of tobacco control programs on adult smoking. *Am J Public Health* 98(2): 304–9, 2008.
36. Hughes JR, Keely J, Naud S. Shape of the relapse curve and long-term abstinence among untreated smokers. *Addiction* 99(1): 29–38, 2004.
37. Lancaster T, Stead L. Physician advice for smoking cessation. *Cochrane Database Syst Rev*: CD000165, 2004.
38. Fiore MC. Treating tobacco use and dependence: An introduction to the US public health service clinical practice guideline. *Respir Care* 45(10): 1196–99, 2000.
39. Mallin R. Smoking cessation: Integration of behavioral and drug therapies. *Am Fam Physician* 65(6): 1107–14, 2002.
40. DiClemente CC, Prochaska JO, Fairhurst SK, Velicer WF, Velasquez MM, Rossi JS. The process of smoking cessation: An analysis of precontemplation, contemplation, and preparation stages of change. *J Consult Clin Psychol* 59(2): 295–304, 1991.
41. Rigotti NA, Munafo MR, Murphy MF, Stead LF. Interventions for smoking cessation in hospitalised patients. *Cochrane Database Syst Rev*(1): CD001837, 2003.
42. Daughton DM, Heatley SA, Prendergast JJ, Causey D, Knowles M, Rolf CN, Cheney RA, Hatelid K, Thompson AB, Rennard SI. Effect of transdermal nicotine delivery as an adjunct to low-intervention smoking cessation therapy. *Arch Intern Med* 151: 749–52, 1991.
43. Silagy C, Lancaster T, Stead L, Mant D, Fowler G. Nicotine replacement therapy for smoking cessation. *Cochrane Database Syst Rev*(3): 2004, CD000146.
44. Hughes JR, Stead LF, Lancaster T. Antidepressants for smoking cessation. *Cochrane Database Syst Rev*: 2007, CD000031.
45. Cahilll K, Ussher M. Cannabinoid type 1 receptor antagonists (rimonabant) for smoking cessation. *Cochrane Database Syst Rev*: 2007, CD005353.
46. Etter JF, Stapleton JA. Nicotine replacement therapy for long-term smoking cessation: A meta-analysis. *Tob Control* 15(4): 280–85, 2006.
47. Ahluwalia JS, McNagny SE, Clark WS. Smoking cessation among inner-city African Americans using the nicotine transdermal patch. *J Gen Intern Med* 13(1): 1–8, 1998.
48. Croghan GA, Sloan JA, Croghan IT, Novotny P, Hurt RD, DeKrey WL, Mailliard JA, Ebbert LP, Swan DK, Walsh DJ, Wiesenfeld M, Levitt R, Stella P, Johnson PA, Tschetter LK, Loprinzi C. Comparison of nicotine patch alone versus nicotine nasal spray alone versus a combination for treating smokers: A minimal intervention, randomized multicenter trial in a nonspecialized setting. *Nicotine Tob Res* 5(2): 181–87, 2003.
49. Fiore MC, Kenford SL, Jorenby DE, Wetter DW, Smith SS, Baker TB. Two studies of the clinical effectiveness of the nicotine patch with different counseling treatments. *Chest* 105(2): 524–33, 1994.
50. Tonnesen P, Paoletti P, Gustavsson G, Russell MA, Saracci R, Gulsvik A, Rijcken B, Sawe U. Higher dosage nicotine patches increase one-year smoking cessation rates: Results from the European CEASE trial. Collaborative European Anti-Smoking Evaluation. European Respiratory Society. *Eur Respir J* 13(2): 238–46, 1999.
51. Yusuf S, Reddy S, Ounpuu S, Anand S. Global burden of cardiovascular diseases: Part II: Variations in cardiovascular disease by specific ethnic groups and geographic regions and prevention strategies. *Circulation* 104(23): 2855–64, 2001.
52. GlaxoSmithKline Healthcare. *Product Information: Commit Oral Lozenge, Nicotine Polacrilex Oral Lozenge*. Pittsburgh, PA, 2003.
53. GlaxoSmithKline Healthcare. *Product Information: Nicorette Gum, Nicotine Polacrilex Gum*. Moon Township, PA, 2005.
54. Henningfield JE, Radzius A, Cooper TM, Clayton RR. Drinking coffee and carbonated beverages blocks absorption of nicotine from nicotine polacrilex gum. *JAMA* 264(12): 1560–64, 1990.
55. Mulligan SC, Masterson JG, Devane JG, Kelly JG. Clinical and pharmacokinetic properties of a transdermal nicotine patch. *Clin Pharmacol Ther* 47(3): 331–37, 1990.
56. Page F, Coleman G, Conduit R. The effect of transdermal nicotine patches on sleep and dreams. *Physiol Behav* 88(4–5): 425–32, 2006.
57. Greenland S, Satterfield MH, Lanes SF. A meta-analysis to assess the incidence of adverse effects associated with the transdermal nicotine patch. *Drug Saf* 18(4): 297–308, 1998.
58. Hughes JR, Goldstein MG, Hurt RD, Shiffman S. Recent advances in the pharmacotherapy of smoking [see comments]. *JAMA* 281(1): 72–76, 1999.
59. Hajek P, McRobbie H, Gillison F. Dependence potential of nicotine replacement treatments: Effects of product type, patient characteristics, and cost to user. *Prev Med* 44(3): 230–34, 2007.
60. Pfizer Consumer Healthcare. *Product Information: Nicotrol vs Nicotine Nasal Spray*. Morris Plains, NJ, 2005.
61. Pfizer Consumer Healthcare. Product Information: Nicotrol Inhalation Solution, Nicotine Inhalation Solution. Morris Plains, NJ, 2005.
62. Joseph AM, Fu SS. Safety issues in pharmacotherapy for smoking in patients with cardiovascular disease. *Prog Cardiovasc Dis* 45(6): 429–41, 2003.
63. Joseph AM, Norma SM, Ferry LH. The safety of transdermal nicotine as an aid to smoking cessation in patients with cardiac disease. *New Engl J Med* 335: 1792–98, 1996.
64. Tzivoni D, Keren A, Meyler S, Khoury Z, Lerer T, Brunel P. Cardiovascular safety of transdermal nicotine patches in patients with

coronary artery disease who try to quit smoking. *Cardiovasc Drugs Ther* 12(3): 239–44, 1998.

65. Working Group for the Study of Transdermal Nicotine in Patients with Coronary Artery Disease. Nicotine replacement therapy for patients with coronary artery disease. *Arch Intern Med* 154: 989–95, 1994.
66. Ford CL, Zlabek JA. Nicotine replacement therapy and cardiovascular disease. *Mayo Clin Proc* 80(5): 652–56, 2005.
67. Dempsey DA, Benowitz NL. Risks and benefits of nicotine to aid smoking cessation in pregnancy. *Drug Saf* 24(4): 277–322, 2001.
68. Le Houezec J. What smoking cessation interventions are effective in pregnant women? *J Gynecol Obstet Biol Reprod* 1: 182–93, 2005.
69. Drugdex evaluations retrieved January 18 from online micromedex database, Available from: URL: www.micromedex.com. Nicotine (2007).
70. Dempsey D, Jacob P. III, Benowitz NL. Accelerated metabolism of nicotine and cotinine in pregnant smokers. *J Pharmacol Exp Ther* 301(2): 594–98, 2002.
71. National Institute on Drug Abuse. *Buproprion Helps People with Schizophrenia Quit Smoking*. Research Findings 20(5), 2006.
72. Stack NM. Smoking cessation: An overview of treatment options with a focus on varenicline. *Pharmacotherapy* 27(11): 1550–57, 2007.
73. Hurt RD, Sachs DP, Glover ED, Offord KP, Johnston JA, Dale LC, Khayrallah MA, Schroeder DR, Glover PN, Sullivan CR, Croghan IT, Sullivan PM. A comparison of sustained-release bupropion and placebo for smoking cessation. *New Engl J Med* 337: 1195–202, 1997.
74. Swan GE, McAfee T, Curry SJ, Jack LM, Javitz H, Dacey S, Bergman K. Effectiveness of bupropion sustained release for smoking cessation in a health care setting: A randomized trial. *Arch Intern Med* 163(19): 2337–44, 2003.
75. Hays JT, Hurt RD, Rigotti NA, Niaura R, Gonzales D, Durcan MJ, Sachs DP, Wolter TD, Buist AS, Johnston JA, White JD. Sustained-release bupropion for pharmacologic relapse prevention after smoking cessation. A randomized, controlled trial. *Ann Intern Med* 135(6): 423–33, 2001.
76. Killen JD, Fortmann SP, Murphy GM Jr, Hayward C, Arredondo C, Cromp D, Celio M, Abe L, Wang Y, Schatzberg AF. Extended treatment with bupropion SR for cigarette smoking cessation. *J Consult Clin Psychol* 74(2): 286–94, 2006.
77. Hurt RD, Krook JE, Croghan IT, Loprinzi CL, Sloan JA, Novotny PJ, Kardinal CG, Knost JA, Tirona MT, Addo F, Morton RF, Michalak JC, Schaefer PL, Porter PA, Stella PJ. Nicotine patch therapy based on smoking rate followed by bupropion for prevention of relapse to smoking. *J Clin Oncol* 21(5): 914–20, 2003.
78. GlaxoSmithKline Healthcare. *Product Information: Zyban Buproprion Hydrochloride Sustained-release Tablets*. Research Triangle Park, 2007.
79. Center for Drug Evaluation and Research US Food and Drug Administration, 2008. Background information on the suicidality classification project and buproprion; Available from: URL: http://www.Fda.Gov/cder/drug/infopage/buproprion/default.Htm.
80. Chun-Fai-Chan B, Koren G, Fayez I, Kalra S, Voyer-Lavigne S, Boshier A, Shakir S, Einarson A. Pregnancy outcome of women exposed to bupropion during pregnancy: A prospective comparative study. *Am J Obstet Gynecol* 192(3): 932–36, 2005.
81. Gourlay SG, Stead LF, Benowitz NL. Clonidine for smoking cessation. *Cochrane Database Syst Rev*(3): 2004, CD000058.
82. The FDA approves new drug for smoking cessation. *FDA Consum* 40:29, 2006.
83. Coe JW, Brooks PR, Vetelino MG, Wirtz MC, Arnold EP, Huang J, Sands SB, Davis TI, Lebel LA, Fox CB, Shrikhande A, Heym JH, Schaeffer E, Rollema H, Lu Y, Mansbach RS, Chambers LK, Rovetti CC, Schulz DW, Tingley FD III, O'Neill BT. Varenicline: An alpha4beta2 nicotinic receptor partial agonist for smoking cessation. *J Med Chem* 48(10): 3474–77, 2005.
84. Jorenby DE, Hays JT, Rigotti NA, Azoulay S, Watsky EJ, Williams KE, Billing CB, Gong J, Reeves KR. Efficacy of varenicline, an alpha4beta2 nicotinic acetylcholine receptor partial agonist, vs placebo or sustained-release bupropion for smoking cessation: A randomized controlled trial. *JAMA* 296(1): 56–63, 2006.
85. Gonzales D, Rennard SI, Nides M, Oncken C, Azoulay S, Billing CB, Watsky EJ, Gong J, Williams KE, Reeves KR. Varenicline, an alpha4beta2 nicotinic acetylcholine receptor partial agonist, vs sustained-release bupropion and placebo for smoking cessation: A randomized controlled trial. *JAMA* 296(1): 47–55, 2006.
86. Nakamura M, Oshima A, Fujimoto Y, Maruyama N, Ishibashi T, Reeves KR. Efficacy and tolerability of varenicline, an alpha4beta2 nicotinic acetylcholine receptor partial agonist, in a 12-week, randomized, placebo-controlled, dose–response study with 40-week follow-up for smoking cessation in Japanese smokers. *Clin Ther* 29(6): 1040–56, 2007.
87. Tsai ST, Cho HJ, Cheng HS, Kim CH, Hsueh KC, Billing CB Jr, Williams KE. A randomized, placebo-controlled trial of varenicline, a selective alpha4beta2 nicotinic acetylcholine receptor partial agonist, as a new therapy for smoking cessation in Asian smokers. *Clin Ther* 29(6): 1027–39, 2007.
88. Cahill K, Stead LF, Lancaster T. Nicotine receptor partial agonists for smoking cessation. *Cochrane Database Syst Rev*(1): 2007, CD006103.
89. Pfizer Labs. *Product Information: Chantix Varenicline Tablets*. New York, 2008.
90. Aubin HJ, Bobak A, Britton JR, Oncken C, Billing CB Jr, Gong J, Williams KE, Reeves KR. Varenicline versus transdermal nicotine patch for smoking cessation: Results from a randomised, open-label trial. *Thorax*, 2008, doi:10.1136/thx.2007.090647
91. Williams KE, Reeves KR, Billing CB Jr, Pennington AM, Gong J. A double-blind study evaluating the long-term safety of varenicline for smoking cessation. *Curr Med Res Opin* 23(4): 793–801, 2007.
92. Pfier Product Center Database, accessed 2/3/2008. Safety update to chantix package insert; (2008). Available from: URL: https://www.Pfizerpro.Com/product_info/chantix_isi_about_letter.Jsp
93. Cornuz J, Gilbert A, Pinget C, McDonald P, Slama K, Salto E, Paccaud F. Cost-effectiveness of pharmacotherapies for nicotine dependence in primary care settings: A multinational comparison. *Tob Control* 15(3): 152–59, 2006.
94. Hoogendoorn M, Welsing P, Rutten-van Molken MP. Cost-effectiveness of varenicline compared with bupropion, NRT, and nortriptyline for smoking cessation in the Netherlands. *Curr Med Res Opin* 24(1): 51–61, 2008.
95. Faulkner MA, Lenz TL, Stading JA. Cost-effectiveness of smoking cessation and the implications for COPD. *Int J Chron Obstruct Pulmon Dis* 1(3): 279–87, 2006.
96. Fiore M, Bailey W, Cohen S, Dorfman S, Goldstein M, Gritz E, Heyman R, Jaen C, Kottke T, Lando H, Mecklenburg R, Mullen P, Nett L, Robinson L, Sistzer M, Tommasello A, Villejo L, Wewers M. Treating Tobacco Use and Dependence. Rockville, MD: U.S. Department Of Health and Human Services, 2000.
97. Welton NJ, Johnstone EC, David SP, Munafo MR. A cost-effectiveness analysis of genetic testing of the DRD2 Taq1A polymorphism to aid treatment choice for smoking cessation. *Nicotine Tob Res* 10(1): 231–40, 2008.
98. *Clearing the Smoke*. National Academy Press, Washington, D.C., 2001.
99. Hatsukami DK, Rennard SI, Jorenby D, Fiore M, Koopmeiners J, de Vos A, Horwith G, Pentel PR. Safety and immunogenicity of a nicotine conjugate vaccine in current smokers. *Clin Pharmacol Ther* 78(5): 456–67, 2005.

48

CHAPTER

β_2-Adrenoceptor Agonists

Ian P. Hall

Division of Therapeutics and Molecular Medicine, University Hospital of Nottingham, Nottingham, UK

INTRODUCTION

β_2-Agonists have been the mainstay bronchodilator agents used for the treatment of asthma and chronic obstructive pulmonary disease (COPD) since the development of inhaled isoprenaline (isoproterenol) preparations in the 1960s. While the initial preparations were marketed at relatively high doses and had little β_2-selectivity, the side-effect profile of these agents was markedly improved by the development of short-acting β_2-selective agents such as salbutamol (albuterol) and terbutaline. More recently, long-acting β_2-selective agents have assumed an increasingly important role in the management of asthma and COPD.

MECHANISM OF ACTION

β_2-Agonists bind to the β_2-adrenoceptor which is present in the cell membrane of a number of airway cells including airway smooth muscle, airway epithelial cells, inflammatory cells including mast cells, vascular and endothelium and vascular smooth muscle [1]. However, the major site of action of β_2-agonists in the airways is the airway smooth muscle cell. Following binding of β_2-agonist to the β_2-adrenoceptor on airway smooth muscle, a signaling cascade is triggered which results in a number of events, all of which contribute to relaxation of airway smooth muscle (Table 48.1) [2, 3].

The majority of these events are dependent on elevation of cell cyclic adenosine monophosphate (AMP) content, which is brought about following binding of β_2-agonist to the β_2-adrenoceptor by stimulation of adenylyl cyclase as a result of activation of the G protein coupled to the β_2-adrenoceptor, Gs [4]. This exists as a heterotrimeric complex but following stimulation of the β_2-adrenoceptor Gs dissociates releasing free α-subunits that are able to stimulate adenylyl cyclase. Adenylyl cyclase exists in a number of different isoforms although there is at least some evidence suggesting that adenylyl cyclase VI is important in airway smooth muscle; however, most of the other adenylyl cyclase isoforms are also present in this tissue [5]. Adenylyl cyclase catalyzes the formation of cyclic AMP from adenosine triphosphate (ATP). Cyclic AMP is able to convert protein kinase A from an inactive form to the active form in which the catalytic and regulatory subunits dissociate. The catalytic subunit of protein kinase A then phosphorylates key targets within the cell bringing about the majority of the physiological effects of β_2-adrenoceptor stimulation. However, there is at least some evidence to suggest that cyclic AMP-independent actions may also occur; for example, direct stimulation by Gsα of the BKCa channel present in the airway smooth muscle cell membrane has

TABLE 48.1 Mechanisms underlying airway smooth muscle relaxation by β_2 adrenoceptor agonists.

- Inhibition of spasmogen-induced inositol 1,4,5 trisphosphate production
- Inhibition of spasmogen-induced rises in intracellular free calcium
- Activation of calcium-activated K^+ channels
- Alteration of sensitivity of contractile apparatus
- Increased extrusion/re-uptake of calcium from cytoplasm
- Hyperpolarization of cell membrane

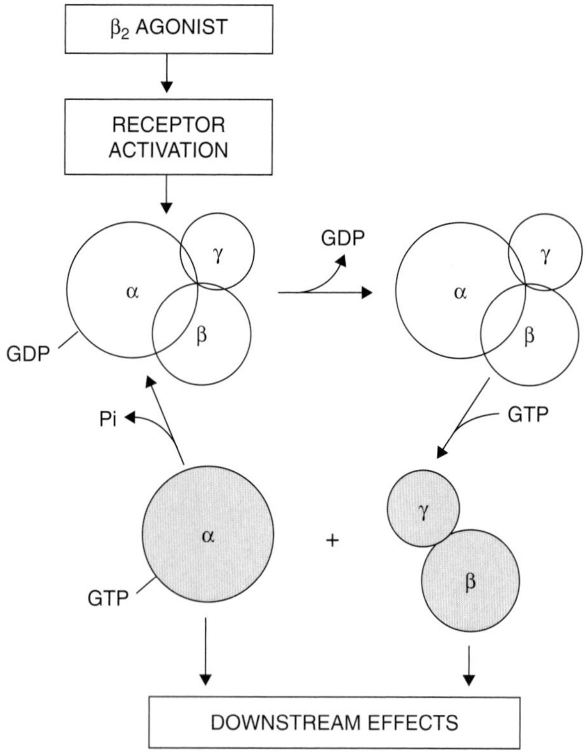

FIG. 48.1 G-protein regulatory cycle of activation and deactivation for transmission of the signal from receptor to effector. When GDP is bound, the heterotrimeric G-protein is inactive, receptor stimulation causes conformational change in both the receptor and the G-protein which decreases GDP binding affinity. GTP is abundant in the cell and replaces GDP; the active conformation of the Gα subunit dissociates from βγ. This activated state remains until GTP is hydrolysed to GDP whereby the subunits reassociate.

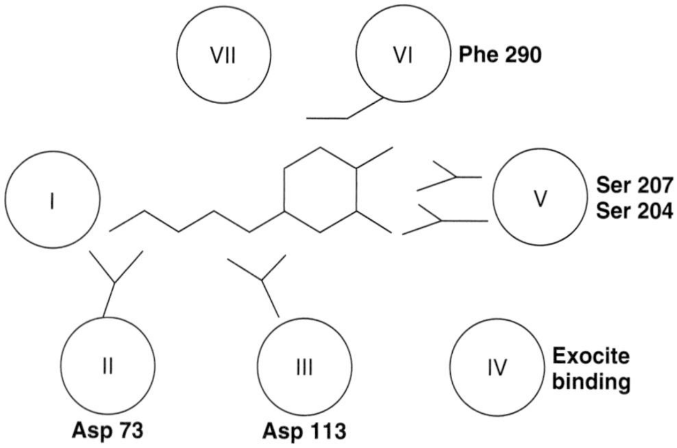

FIG. 48.2 Cross-sectional view of transmembrane spanning domains of the β_2-adrenoreceptor core. The β_2 agonist ligand is shown sitting in the binding pocket in the receptor. Key amino acid residues for binding of ligand are shown. Amino acids in transmembrane domain IV are implicated in salmeterol exocite binding.

TABLE 48.2 Frequently used β_2-adrenoceptor agonists.

Short-acting β_2-adrenoceptor agonists
Salbutamol (albuterol)*
Terbutaline
Fenoterol
Long-acting β_2-adrenoceptor agonists
Salmeterol
Formoterol

*USA name.

been described [2]. This is relevant because the BKCa channel (the calcium activated potassium channel) is thought to be important in modulating changes in cell membrane potential following stimulation with β_2-agonist, and thus can contribute to the relaxant response of β_2-adrenoceptor stimulation.

The intracellular effects of β_2-adrenoceptor stimulation are relatively short lived. Cyclic AMP is broken down by phosphodiesterase isoenzymes present in the cell, with type 3 and type 4 phosphodiesterase activities believed to be the most important in regulating cyclic AMP content in airway smooth muscle and type 4 phosphodiesterase being physiologically important in many inflammatory cells including eosinophils. Continued stimulation by Gsα is prevented by the free Gsα rapidly reassociating with βγ to reconstitute the heterotrimeric Gs complex (Fig. 48.1).

STRUCTURE AND FUNCTIONAL RELATIONSHIPS OF β_2-AGONISTS AND THE β_2-ADRENOCEPTOR

The β_2-adrenoceptor is a member of the G protein coupled receptor superfamily with the typical seven transmembrane spanning domains (Fig. 48.2) [6, 7]. The binding site for β_2-agonists consists of residues in at least three of the α-helices that pass through the cell membrane. The prolonged duration of action of salmeterol is believed to be due to the binding of the lipophilic tail to residues deep in the fourth transmembrane domain [8, 9]. This process is essentially irreversible. The explanation for the prolonged duration of action of formoterol is less clear although it has been proposed that because of its lipophilicity formoterol partitions in the cell membrane which forms a reservoir allowing prolonged interaction with the receptor [10]. The most commonly used β_2-agonists in clinical practice are listed in Table 48.2.

All the clinically important β_2-agonists consist of a benzene ring with a chain of two carbon atoms and either an amine head or a substituted amine head. If a hydroxyl (OH) group is present at positions 3 or 4 on the benzene ring, the structure is a catechol nucleus and hence the agent a catecholamine. If these hydroxyl groups are substituted or repositioned, the drug is generally less potent than the synthetic catecholamine isoprenaline, which is a full agonist at β_2-adrenoceptors. This potential disadvantage may be outweighed by the relative resistance of substituted catecholamines to metabolic degradation by the enzyme catechol-*O*-methyltransferase (COMT). Examples of such agents are salbutamol and terbutaline with salbutamol only being a partial agonist. As mentioned above, the prolonged duration of action of salmeterol is due to a long side chain substitution which is believed to bind to an additional site in the fourth transmembrane spanning domain of the receptor. Substitutions on the α carbon atom help block oxidation by monoamine oxidase (MAO).

The effects of catecholamines such as adrenaline (epinephrine), noradrenaline (norepinephrine), and isoprenaline are terminated by uptake into either sympathetic nerve endings (uptake 1) or other innervated tissues such as smooth muscle (uptake 2). The dominant enzyme present in innervated tissues is COMT whereas the dominant metabolic degradation route in sympathetic nerve endings is through oxidation by MAO. In addition to degradation, exogenously administered β_2-agonists can be conjugated to sulfates or glucuronides in the liver, or the lung. Following ingestion the drugs are partially conjugated during first past metabolism which accounts for roughly 50% of the metabolism of the short-acting drug salbutamol.

CLINICAL PHARMACOLOGY OF β_2-AGONISTS

General pharmacology

β_2-Adrenoceptor agonists are predominantly used in the treatment of airflow obstruction because of their bronchodilator properties. These differ markedly between asthma and COPD (Table 48.3); indeed, reversibility of airflow obstruction with inhaled β_2-agonist is often used as a diagnostic marker of asthma and helps distinguish asthma from COPD in patients where the distinction is in doubt (e.g., chronic "asthmatics" who have smoked, or previous smokers who develop symptoms of wheeze in later life). However, in addition to reversing airflow obstruction in asthmatic individuals β_2-agonists also protect against bronchoconstrictor challenge [11]. Because of this latter effect, β_2-agonists have been considered to have potential anti-inflammatory actions (see below). *In vitro* β_2-agonists prevent mediator release from inflammatory cells including mast cells, an effect which if present in human airways would be expected to reduce airway inflammation [12, 13]. However, the concentrations of these agents required to demonstrate these effects are in general much higher than those seen in the lungs *in vivo*.

SHORT-TERM EFFECTS OF β_2-AGONISTS IN ASTHMA AND COPD

As mentioned above, the ability of β_2-adrenoceptor agonists to reverse airflow obstruction is a hallmark of asthma, with reversibility >15% being considered diagnostic. β_2-Agonists do produce a measurable bronchodilator effect in normal individuals and in patients with other diseases characterized by airflow obstruction such as COPD and bronchiectasis, although the magnitude of these effects is generally much smaller. In normal individuals bronchodilator responses are generally only observable by measuring specific airway conductance. This contrast between the marked effects of β_2-agonists in asthmatic patients and the minimal effects in normal individuals led early investigators to hypothesize that a primary defect in the β_2-adrenoceptor signaling pathway was the actual cause of asthma. This hypothesis would not explain many of the inflammatory features present in the disease, but interest in the potential for a primary abnormality of β_2-adrenoceptors to contribute in part to the pathophysiology of asthma resurfaced following the description of polymorphic variation within the β_2-adrenoceptor (see below).

TABLE 48.3 β_2-agonists in asthma and COPD.

Asthma	COPD
Bronchodilator response greater than 15% change in FEV_1	Bronchodilator response less than 15% change in FEV_1
Bronchodilator hyperreactivity present	Bronchial hyperreactivity usually absent
β_2-agonist protects against nonspecific airway challenge	
Marked symptomatic benefit with β_2-agonist	Small or moderate symptomatic benefit with β_2-agonist
Effective in long-term management	No disease-modifying effects

In addition to having a bronchodilator action, β_2-agonists in asthmatics protect against bronchoconstrictor stimuli [14–17]. One feature demonstrated by most (but not all) asthmatics is nonspecific airway hyperreactivity to inhaled irritant challenge including allergen. The most frequently used stimuli are histamine and methacholine although abnormal responses are also seen to exercise, allergen, and other challenges. The inhaled dose of histamine and other agents which required to provoke a 20% fall in FEV_1 (PC_{20}) is increased markedly (usually by about three doubling doses) by previous treatment with a short-acting β_2-agonist. While a degree of such bronchial hyperreactivity can also be demonstrated in other diseases including COPD, bronchiectasis, cystic fibrosis, and left ventricular failure, this is usually far less marked than in asthmatic subjects.

LONG-TERM EFFECTS OF β_2-AGONISTS IN ASTHMA AND COPD

While regular treatment of inhaled or nebulized β_2-agonist is frequently used in the management of moderate or severe COPD, despite relatively small improvements in symptoms, controversy has reigned over the long-term use of regular β_2-agonist in the management of asthma [18, 19]. This controversy originated from discussions following epidemics of asthma deaths in New Zealand in the late 1960s and late 1970s which were linked to the prescribing of high-dose isoprenaline and fenoterol, respectively. Several studies subsequently concentrated on the possibility that tachyphylaxis may develop as a result of inhaled β_2-agonists in asthma. In general, studies attempting to demonstrate tachyphylaxis to the bronchodilator effects of β_2-agonists have failed to identify clinically important loss of responsiveness although small effects have been observed when looked for carefully (see for example, Ref. [20]). However, in contrast, the bronchoprotective effects of β_2-agonists against nonspecific airway challenge show tachyphylaxis that generally develops within 24h [21, 22]. Thus, the magnitude of the protective effect against exercise, histamine challenge, or methacholine challenge is reduced in magnitude compared with the level of initial protection following administration of β_2-agonist for periods of over 24h. However, there is still overall protection against bronchoconstrictor challenge even if this is less than the protection seen in the initial hours after treatment. Thus, while it seems clear that tachyphylaxis develops to the bronchoprotective effects of β_2-agonists against nonspecific airway challenge, this is only partial, and one would presume that patients would still be better despite this tachyphylaxis than if they were not taking β_2-agonist at all. The main concern therefore has revolved around patients taking intermittent treatment or in those discontinuing treatment. Small rebound increases in airway reactivity have been demonstrated following the cessation of β_2-agonist therapy [23]. Clinical studies comparing regular (e.g., four times a day) short-acting β_2-agonist versus "as required" β_2-agonist have not shown clinically important differences despite earlier reports that asthma control deteriorated following regular treatment [24, 25]. None the less, there is no reason to suppose that regular treatment is better than "as required" usage which remains the preferred way of prescribing β_2-agonists in asthma.

The controversy regarding β_2-agonists resurfaced with the introduction into the marketplace of long-acting β_2-agonists (LABAs). There were initial concerns that LABAs might worsen asthma control in asthmatics in general, but most asthmatics respond well to LABAs. At least for salmeterol, this might initially seems surprising, given the fact that salmeterol is a partial agonist and essentially binds irreversibly with the β_2-adrenoceptor. The issue of LABA safety has recently resurfaced, following a large study examining efficacy of salmeterol in asthma (the SMART study [26]). In this study, there were a small number of deaths and severe exacerbations in the salmeterol treatment group which led to premature termination of the study and a subsequent black box warning from the FDA which was later extended to all LABAs. The explanation for these findings is unclear, although it is possible that reduced compliance with inhaled corticosteroids may have played a role. The other commonly used LABA, formoterol, has undergone recent studies which has led to a license in the UK for use as an "as required" agent (in a combination inhaler with inhaled budesonide).

Although mild COPD is often treated with "as required" β_2-agonist, regular high-dose inhaled or nebulized β_2-agonists have been much more widely used in COPD than in asthma. While small improvements in lung function and symptomatic improvement have been demonstrated in severe COPD, the overall effect on lung function has generally been small, which is hardly surprising given the fixed nature of the airflow obstruction in the majority of patients. There have been no data suggesting deterioration in lung function following chronic administration of high doses of β_2-agonist in COPD (these would be difficult to observe in short-term studies in any case) although there have been concerns about other effects of high doses of β_2-agonist in this setting. In particular, high-dose nebulized β_2-agonists are known to cause hypokalaemia and both supraventricular and ventricular arrhythmias [27], and there have been concerns that these may occur in patients on regular high-dose nebulized bronchodilator therapy for either asthma or COPD at home.

β_2-ADRENOCEPTOR POLYMORPHISM IN ASTHMA AND COPD

As mentioned above, the identification of polymorphic variation within the gene for the β_2-adrenoceptor reawakened interest in the possibility that primary abnormalities of β_2-adrenoceptor signaling pathways may be involved in the pathogenesis of these airway diseases. The gene for the β_2-adrenoceptor is situated on chromosome 5q31-33 in a region showing linkage to intermediate phenotypes for asthma and/or atopy [28]. The β_2-adrenoceptor gene and its immediate controlling regions show a high degree of polymorphic variation with nine single nucleotide type polymorphisms (SNPs) having been identified within the coding region of the gene and a further eight in the immediate 5 prime untranslated region (reviewed in Ref. [29]). Of the nine coding region polymorphisms, five are degenerate (i.e. do not alter the amino acid code of the receptor) [30]. However, the other four all result in single amino acid substitutions. While the polymorphism at codon 34 (Val34Met) is rare and appears to have no functional effects, the other three nondegenerate polymorphisms appear to produce functional alteration in receptor behavior. Thus, the rare Thr164Ile polymorphism results in reduced affinity for catechol ligands and an altered receptor sequestration profile [31]. Interestingly, Thr164 is very close to the salmeterol binding site within the fourth transmembrane spanning domain of the receptor, and it seems likely that the isoleucine 164 substitution may alter the binding characteristics of salmeterol to the β_2-adrenoceptor. However, the allelic frequency of this polymorphism is only around 2–3% in Caucasian populations, hence homozygous individuals are very rare and to date have not been adequately studied.

In contrast, the two N-terminal polymorphisms at codon 16 (Arg16Gly) and 27 (Gln27Glu) are common. While neither alter agonist binding properties of the receptor, both result in altered downregulation profiles following long-term agonist exposure. Thus, the Gly16 and Gln27 forms of the receptor show increased receptor downregulation following agonist exposure while the Glu 27 form of the receptor appears to be partially protected from downregulation

[9]. These effects have been shown both in transformed cell systems and in primary cultures of human airways smooth muscle. Of the 5 prime untranslated region polymorphisms, the strongest evidence for possible functional effects is for the −47 T-C SNP which alters the terminal amino acid in a short open reading frame that codes for the β-upstream peptide (also known as the β_2-adrenoceptor 5 prime leader cistron); this peptide is believed to be important in maintaining a "brake" on receptor expression, and the Cys19Arg polymorphism in this peptide may possibly increase translational inhibition of the β_2-adrenoceptor [32, 33].

The role of β_2-adrenoceptor polymorphism has been extensively studied in asthma but less so in COPD. A recent major study in UK 1958 birth cohort [34] failed to demonstrate an association between β_2-adrenoceptor polymorphism and asthma *per se*, although weak association with IgE and the degree of bronchial hyperresponsiveness have previously been described in some (but not in all) studies [35]. These effects may, in part, be due to linkage disequilibrium with other important genes on chromosome 5q such as the nearby Th2 cytokine locus. The possibility that these polymorphisms may contribute to the development of COPD has not been reported upon to date, although studies in this area are currently in progress.

The other potential importance of β_2-adrenoceptor polymorphism is in pharmacogenetic studies. The possibility that β_2-adrenoceptor polymorphism may predict treatment response, particularly following long-term exposure to agonists due to the altered downregulation profile of individuals carrying particular genotypes (e.g., Gly16, Gln27), has been studied by a number of groups. Initial studies demonstrated an association between Gly16 and subsensitivity to the bronchodilator effects of formoterol following chronic dosing and also a reduced bronchodilator response to salbutamol [20, 36]. Gly16 has also been shown to be associated with nocturnal asthma. Two relatively large studies, one retrospective and one prospective, suggested that treatment with regular (four times a day) salbutamol (albuterol) resulted in poorer outcome in Arg16Arg asthmatics (15% of the Caucasian population) [37, 38]. However, salbutamol is not currently used in most countries in this way, and the asthmatics studied were not on inhaled corticoseroids, so the clinical relevance of these observations is unclear. More recently, a small retrospective study suggested that these effects may extend to LABAs [39], although subsequently a larger retrospective analysis of pharma-funded studies suggests that this does not extend to patients on combined LABA/inhaled steroid preparations [40]. In general, therefore, it appears that the effects of these polymorphisms upon treatment response to these drugs used as recommended in national guidelines are likely to be relatively small and of doubtful clinical significance overall. It is also possible that the combination of groups of polymorphisms around this region (i.e. the haplotype of an individual at this locus) may be the main determinant of functional effects rather than single polymorphisms in isolation [41]. It will be interesting to determine whether these polymorphisms are relevant to treatment response in COPD; intuitively, one might imagine that this could be the case given the higher doses of agents used in COPD although, to date, no published data are available on this issue.

References

1. Davis C, Conolly ME, Greenacre JK. Beta-adrenoceptors in human lung, bronchus and lymphocytes. *Br J Clin Pharm* 10: 425, 1980.
2. Kume H, Hall IP, Washabow RJ, Takagi K, Kotlikoff MI. β-Adrenergic agonists regulate KCa channels in airway smooth muscle cAMP dependent mechanisms. *J Clin Invest* 93: 371, 1994.
3. Torphy TJ, Hall IP. Cyclic AMP and the control of airways smooth muscle tone. In: Raeburn D, Giembycz MA (eds). *Airways Smooth Muscle: Biochemical Control of Contraction and Relaxation*, p. 215. Basel: Birkhauser Verlag, 1994.
4. Johnson M. The β-adrenoceptor. *Am J Respir Crit Care Med* 158: S146, 1999.
5. Billington CK, Hall IP, Stuart J, Mundell JLP *et al.* Inflammatory and contractile agents sensitize specific adenylyl cyclase isoforms in human airway smooth muscle. *Am J Respir Cell Mol Biol* 21: 597–606, 1999.
6. Dohlman HG, Bouvier M, Benovic JL, Caron MG, Lefkowitz RJ. The multiple membrane spanning topography of the β_2-adrenergic receptor. *J Biol Chem* 262: 14282, 1987.
7. Emorine J, Marullo S, Delavier-Klutchko C, Kaveri SV, Duriev-Trautmann O, Strosberg AD. Structure of the gene for human β_2-adrenergic receptor: Expression and promoter characterisation. *Proc Natl Acad Sci USA* 84: 6995, 1987.
8. Green SA, Spasoff AP, Coleman RA, Johnson M, Liggett SB. Sustained activation of a G protein-coupled receptor via "anchored" agonist binding: Molecular localisation of the salmeterol exocite within the β_2-adrenergic receptor. *J Biol Chem* 271: 24029, 1996.
9. Green SA, Turki J, Bejarano P, Hall IP, Liggett SB. Influence of β_2-adrenergic receptor genotypes on signal transduction in human airway smooth muscle cells. *Am J Respir Cell Mol Biol* 13: 25, 1995.
10. Lofdahl CG, Svedmyr N. Formoterol fumarate, a new β_2-adrenoceptor agonist. *Allergy* 44: 264, 1989.
11. Tattersfield AE. Effect of beta agonists and anticholinergic drugs on bronchial reactivity. *Am Rev Respir Dis* 136: S64, 1987.
12. Assem ESK, Schild HO. Inhibition by sympathomimetic amines of histamine release induced by antigen in passively sensitized human lung. *Nature* 224: 1028, 1969.
13. Howarth PH, Durham SR, Lee TH, Kay AB, Church MK, Holgate ST. Influence of albuterol, cromolyn sodium and ipratropium bromide on the airway and circulating mediator responses to allergen bronchial provocation in asthma. *Am Rev Respir Dis* 132: 986, 1985.
14. Cheung D, Timmers MC, Zwinderman AH, Bel EH, Dijkman JH, Sterk PJ. Long-term effects of a long-acting β_2-adrenoceptor agonist, salmeterol, on airway hyperresponsiveness with mild asthma. *N Engl J Med* 327: 1198, 1992.
15. Twentyman OP, Finnerty JP, Harris A, Palmer J, Holgate ST. Protection against allergen-induced asthma by salmeterol. *Lancet* 336: 1338, 1990.
16. Twentyman OP, Finnerty JP, Holgate ST. The inhibitory effect of nebulized albuterol on the early and late asthmatic reactions and increase in airway responsiveness provoked by inhaled allergen in asthma. *Am Rev Respir Dis* 144: 782, 1991.
17. Wong BJ, Dolovich J, Ramsdale EH *et al.* Formoterol compared with beclomethasone and placebo on allergen-induced asthmatic responses. *Am Rev Respir Dis* 146: 1156, 1992.
18. Sears MR, Taylor DR, Print CG *et al.* Regular inhaled β-agonist treatment in bronchial asthma. *Lancet* 336: 1391, 1990.
19. Taylor DR, Sears MR, Herbison GP *et al.* Regular inhaled β-agonist in asthma: Effects on exacerbations and lung function. *Thorax* 48: 134, 1993.
20. Tan S, Hall IP, Dewar J, Dow E, Lipworth B. Association between β_2-adrenoceptor polymorphism and susceptibility to bronchdilator desensitisation in moderately severe stable asthmatics. *Lancet* 350: 995, 1997.
21. Connor BJ, Aikman SL, Barnes BJ. Tolerance to the nonbronchodilator effects of inhaled β_2-agonists in asthma. *N Engl J Med* 327: 1204, 1992.

22. Ramage L, Lipworth BJ, Ingram CG, Cree IA, Dhillon DP. Reduced protection against exercise-induced bronchoconstriction after chronic dosing with salmeterol. *Respir Med* 88: 363, 1994.
23. Wahedna I, Wong CS, Wisniewski AFZ, Pavord ID, Tattersfield AE. Asthma control during and after cessation of regular β_2-agonist treatment. *Am Rev Respir Dis* 148: 707, 1993.
24. Drazen JM, Israel E, Boushey HA *et al.* The National Heart, Lung, and Blood Institute's Asthma Clinical Research Network Comparison of regularly scheduled with as-needed use of albuterol in mild asthma. *N Engl J Med* 335: 841, 1996.
25. van Schayck CP, Dompeling E, van Herwaarden CLA *et al.* Bronchodilator treatment in moderate asthma or chronic bronchitis: Continuous or on demand? A randomised controlled study. *BMJ* 303: 1426, 1991.
26. Nelson HS, Weiss ST, Bleecker ER, Yancey SW, Dorinsky PM SMART Study Group. The Salmeterol Multicenter Asthma Research Trial: A comparison of usual pharmacotherapy for asthma or usual pharmacotherapy plus salmeterol. *Chest* 129: 15–26, 2006.
27. Wong CS, Pavord ID, Williams J, Britton JR, Tattersfield AE. Bronchodilator, cardiovascular, and hypokalaemic effects of fenoterol, salbutamol, and terbutaline in asthma. *Lancet* 336: 1396, 1990.
28. Kobilka BK, Dixon RAF, Frielle T *et al.* cDNA for the human β_2-adrenergic receptor: A protein with multiple membrane-spanning domains and encoded by a gene whose chromosomal location is shared with that of the receptor for platelet-derived growth factor. *Proc Natl Acad Sci USA* 84: 46, 1987.
29. Fenech A, Hall IP. Pharmacogenetics of asthma. *Br J Clin Pharmacol* 53: 3–15, 2005.
30. Reihsaus E, Innis M, MacIntyre N, Ligghett SB. Mutations in the gene encoding for β_2-adrenergic receptor in normal and asthmatic subjects. *Am J Respir Cell Mol Biol* 8: 334, 1993.
31. Green SA, Cole G, Jacinto M, Innis M, Liggett SB. A polymorphism of the human β_2-adrenergic receptor within the fourth transmembrane domain alters ligand binding and functional properties of the receptor. *J Biol Chem* 268: 23116, 1993.
32. McGraw DW, Forbes SL, Kramer LA, Liggett SB. Polymorphisms of the 5′ leader cistron of the human β_2-adrenergic receptor regulate receptor expression. *J Clin Invest* 102: 1927, 1998.
33. Scott MG, Swan C, Wheatley AP, Hall IP. Identification of novel polymorphisms within the promoter region of the human β_2-adrenergic receptor gene. *Br J Pharmacol* 126: 841–44, 1999.
34. Hall IP, Blakey JD, Al Balushi KA, Wheatley A, Sayers I, Pembrey ME, Ring SM, McArdle WL, Strachan DP. Beta2-adrenoceptor polymorphisms and asthma from childhood to middle age in the British 1958 birth cohort: A genetic association study. *Lancet* 368: 771–79, 2006.
35. Dewar JC, Wilkinson J, Wheatley A *et al.* The glutamine 27 β_2-adrenoceptor polymorphism is associated with elevated IgE levels in asthmatic families. *J Allergy Clin Immunol* 100: 261, 1997.
36. Martinez FD, Graves PE, Baldini M, Solomon S, Erickson R. Association between genetic polymorphisms of the β_2-adrenoceptor and response to albuterol in children with and without a history of wheezing. *J Clin Invest* 100: 3184, 1997.
37. Israel E, Drazen JM, Liggett SB, Boushey HA, Cherniack RM, Chinchilli VM, Cooper DM, Fahy JV, Fish JE, Ford JG, Kraft M, Kunselman S, Lazarus SC, Lemanske RF, Martin RJ, McLean DE, Peters SP, Silverman EK, Sorkness CA, Szefler SJ, Weiss ST, Yandava CN. The effect of polymorphisms of the beta(2)-adrenergic receptor on the response to regular use of albuterol in asthma. *Am J Respir Crit Care Med* 162: 75–80, 2000.
38. Israel E, Chinchilli VM, Ford JG, Boushey HA, Cherniack R, Craig TJ, Deykin A, Fagan JK, Fahy JV, Fish J, Kraft M, Kunselman SJ, Lazarus SC, Lemanske RF Jr., Liggett SB, Martin RJ, Mitra N, Peters SP, Silverman E, Sorkness CA, Szefler SJ, Wechsler ME, Weiss ST, Drazen JM National Heart, Lung, and Blood Institute's Asthma Clinical Research Network. Use of regularly scheduled albuterol treatment in asthma: Genotype-stratified, randomised, placebo-controlled cross-over trial. *Lancet* 364: 1505–12, 2004.
39. Wechsler ME, Lehman E, Lazarus SC, Lemanske RF Jr., Boushey HA, Deykin A *et al.* Beta-Adrenergic receptor polymorphisms and response to salmeterol. *Am J Respir Crit Care Med* 173: 519–26, 2006.
40. Bleecker ER, Yancey SW, Baitinger LA, Edwards LD, Klotsman M, Anderson WH, Dorinsky PM. Salmeterol response is not affected by beta2-adrenergic receptor genotype in subjects with persistent asthma. *J Allergy Clin Immunol* 118: 809–16, 2006.
41. Drysdale CM, McGraw DW, Stack CB *et al.* Complex promoter and coding region β_2-adrenergic receptor haplotypes alter receptor expressions and predict *in vivo* responsiveness. *Proc Natl Acad Sci USA* 97: 10483–88, 2000.

49

CHAPTER

Anticholinergic Bronchodilators

Trevor T. Hansel, Andrew J. Tan, Peter J. Barnes and Onn Min Kon

National Heart and Lung Institute (NHLI), Clinical Studies Unit, Imperial College, London, UK

INTRODUCTION

The major prescribed long-acting muscarinic antagonist (LAMA) is now tiotropium bromide (Spiriva) that is a once daily inhaled therapy for symptomatic management of chronic obstructive pulmonary disease (COPD). Indeed, tiotropium is recommended on a regular daily basis by the Global Obstructive Lung Disease (GOLD) guidelines (2007) for moderate (Stage II), severe (Stage III), and very severe (Stage IV) COPD to prevent or reduce symptoms, improve exercise performance, and decrease exacerbations [1, 2] (Fig. 49.1). Tiotropium has been the subject of a number of recent reviews that address the clinical development process from research to clinical practice, and their

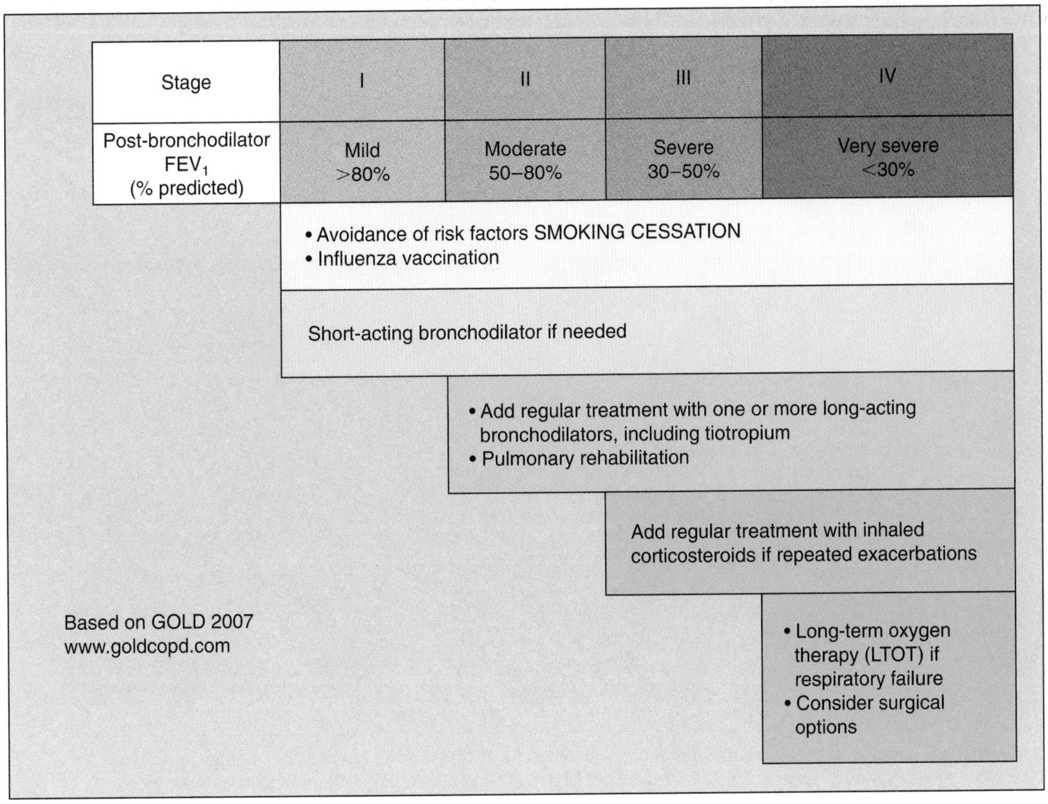
Stage
I
II
III
IV
Post-bronchodilator FEV1 (% predicted)
Mild >80%
Moderate 50–80%
Severe 30–50%
Very severe <30%
• Avoidance of risk factors SMOKING CESSATION
• Influenza vaccination
Short-acting bronchodilator if needed
• Add regular treatment with one or more long-acting bronchodilators, including tiotropium
• Pulmonary rehabilitation
Add regular treatment with inhaled corticosteroids if repeated exacerbations
• Long-term oxygen therapy (LTOT) if respiratory failure
• Consider surgical options
Based on GOLD 2007
www.goldcopd.com

FIG. 49.1 Treatment of COPD at different GOLD stages.

place in the management of patients with COPD [3–7]. Certain articles have centered on review and meta-analyses of experience in clinical studies [8–10]. This chapter will consider the pharmacology, biochemistry, and potential anti-inflammatory actions of anticholinergic bronchodilators, including possible anti-inflammatory actions, as well as clinical trial evidence for the safety and efficacy of these therapies. Important outcome measures for LAMAs involve considerations of effects on symptoms, lung function, dynamic hyperinflation and exercise responses, quality of life, health economics, exacerbations, survival, and the natural history of COPD. We shall then discuss clinical trials that have assessed combined inhaled LAMAs and LABAs, as well as combinations of LAMAs, LABAs, and ICSs. Finally, we shall look to the future with novel LAMAs such as glycopyrrolate and aclidinium bromide (Almirall).

PLANT ORIGINS OF ANTICHOLINERGICS

Herbal remedies have been used since ancient times to treat human disease; and the leaves, roots, and seeds of many plants contain anticholinergic agents [11–16] (Fig. 49.2). Ayurvedic medicine sources from India at around 450 AD describe the use of *Datura stramonium* (thorn-apple or jimson weed) extract for the relief of asthma. Atropine and scopolamine represent parent anticholinergic agents, and are alkaloids found in *Datura* species as well as *Atropa belladonna* (deadly nightshade). After serving in India, General Gent in 1802 recommended the smoking of *Datura* to relieve bronchial complaints.

In the last century, plant extracts have largely been replaced by synthetic chemicals, and in the late 1970s

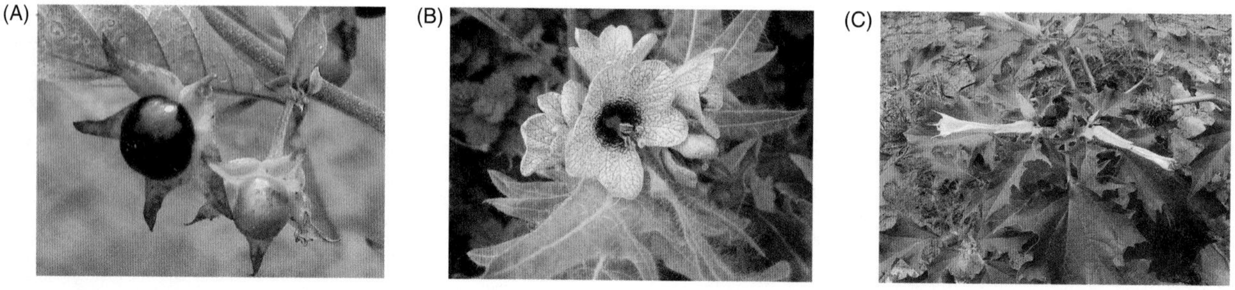

FIG. 49.2 Botanical origin of anticholinergics. (A) The black berries of *Atropa belladonna* (deadly nightshade) contain the deadly plant alkaloid, atropine. Atropine is a competitive antagonist to acetylcholine at muscarinic receptor sites, M_1, M_2, and M_3. Atropine, when taken orally, causes blocked vagal innervation of the heart (tachycardia), decreased mucus secretion and bronchial smooth muscle relaxation in the respiratory system, mydriasis (pupillary dilation), decreased salivation with dry mouth, decreased gut motility and acid secretion, bladder wall relaxation, and decreased sweating. (B) *Hyosyamus niger* is the plant source of scopolamine (hyosine) that was employed by the notorious Dr Crippen to murder his wife in London in 1910. (C) *Datura stramonium* (thorn-apple or jimsonweed) is the source of stramonium which was inhaled to alleviate respiratory disorders. These images are taken from the Wikimedia Commons, free public domain.

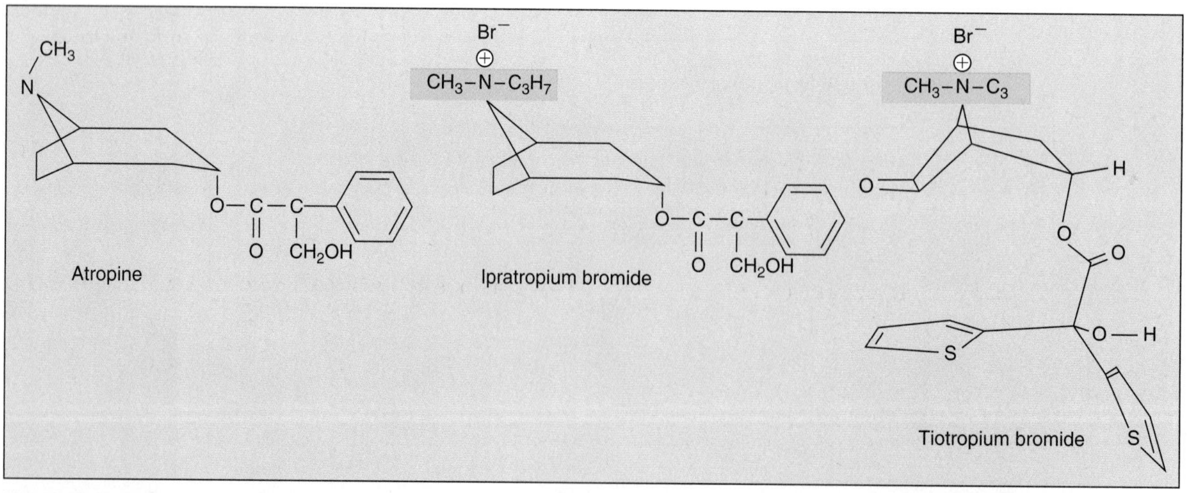

The plant alkaloids are represented by atropine. Ipratropium and tiotropium are synthetic anticholinergic agents that contain a positively charged quaternary nitrogen moiety (highlighted in green). This ensures local action following inhalation, without systemic absorption.

FIG. 49.3 Structure of anticholinergics (see text on figure).

short-acting synthetic anticholinergic drugs were introduced: ipratropium bromide (4 times a day) and oxitropium bromide (3 times a day). These have largely been superseded by tiotropium bromide (Spiriva®) that is administered once daily from the HandiHaler and was licensed in most of the European Union in 2002, while a license in the USA was obtained in 2003.

MOLECULAR STRUCTURES

The parent compounds for anticholinergic therapy are the plant alkaloids atropine and scopolamine (Fig. 49.3). Atropine methyl nitrate, ipratropium bromide, oxitropium bromide, and tiotropium bromide share a quaternary ammonium structure [17–19]. Quaternary ammonium compounds are poorly absorbed from respiratory or pharyngeal mucosal surfaces, and they have minimal oral bioavailability, ensuring that systemic anticholinergic effects are kept to a minimum.

MUSCARINIC RECEPTOR PHARMACOLOGY

Anticholinergics are muscarinic receptor antagonists, inhibiting cholinergic reflex bronchoconstriction and reducing vagal cholinergic tone, which is the main reversible component in COPD. Normal airways have a small degree of vagal cholinergic tone, but because the airways are patent this has no perceptible effect and does not reduce airflow. When the airways are irreversibly narrowed in COPD, vagal cholinergic tone has a much greater effect on airway resistance for geometric reasons, so when this increased constriction is relieved by an anticholinergic drug, there is a perceptible improvement in airflow. In addition, anticholinergics may reduce mucus hypersecretion [20] (Fig. 49.4). Anticholinergics have no apparent effect on pulmonary vessels, and therefore do not cause a fall in Pao_2, as may sometimes be seen with β_2-agonists and theophylline.

The existence of several subtypes of muscarinic receptor in human airways has suggested that more selective muscarinic antagonists may have advantages over non-selective agents, such as ipratropium bromide and oxitropium bromide [21, 22] (Fig. 49.4).

- M_1-receptors are localized to parasympathetic ganglia, and their blockade results in reduced reflex bronchoconstriction.
- M_2-receptors located at cholinergic nerve terminals inhibit the release of acetylcholine, thus acting as autoreceptors.
- M_3-receptors mediate the bronchoconstrictor and mucus secretion action of acetylcholine in human airways. The most important effects of anticholinergics appear to be mediated through blockage of M_3-receptors.

Non-selective anticholinergics, such as atropine and ipratropium bromide, act on M_1-, M_2-, and M_3-receptors [17]. Blockage of M_1- and M_3-receptors leads to bronchodilatation. However, by blocking pre-junctional M_2-receptors this leads to an increase in acetylcholine release and this may work against the post-junctional blockade of M_3-receptors, making these antagonists less efficient. It has been difficult to develop M_3-selective antagonists [23].

Tiotropium bromide is a quaternary ammonium derivative like ipratropium bromide and is a long-acting anticholinergic drug that has unique kinetic selectivity for M_1 and M_3 muscarinic receptors [17, 24, 25]. Tiotropium has similar affinity for the five types of muscarinic receptors, M_1 to M_5. In the airways, tiotropium has a high affinity and dissociates very slowly from M_1- and M_3-receptors in human lung and produces long-term blockade of cholinergic neural bronchoconstriction in human airway smooth muscle, but much more rapidly from M_2-receptors giving a unique kinetic selectivity for M_1- and M_3-receptors [26, 27].

ANTI-INFLAMMATORY ACTIONS

There has been major recent progress in understanding cholinergic systems mediated by muscarinic acetylcholine receptors [28–30] as well as nicotinic pathways [31] (Fig. 49.4). In particular, lymphocytes have been found to express a variety of cholinergic receptors [32], and acetylcholine/muscarinic receptor signaling involving epithelial and inflammatory cells has been documented in both asthma and COPD. Interestingly, anticholinergic may benefit asthmatics with COPD-like inflammation in asthma, viral hyperreactivity [33], and emergency asthma [34]. In guinea pigs tiotropium inhibits allergen-induced airway remodeling [35, 36], while muscarinic receptors mediate lung fibroblasts proliferation [37]. Tiotropium has been shown to suppress acetylcholine-induced release of chemotactic mediators *in vitro* with a special activity on LTB_4 [38]. A recent clinical study did not show an anti-inflammatory effect on sputum inflammatory markers during exacerbations of COPD [39], but this study was flawed by the methodology of detection of sputum IL-6 and IL-8. This was because the supernatants were formed after liquefaction of sputum with dithiothreitol, and this agent can denature cytokines and thus lower detectable levels on immunoassay.

TIOTROPIUM DELIVERY

Tiotropium is licensed in Europe as a bronchodilator for the maintenance treatment of COPD but may have a role in the chronic treatment of severe asthma, as well as in acute severe emergency room asthma. Tiotropium is present in capsules containing 22.5 μg tiotropium bromide monohydrate equivalent to 18 μg tiotropium. These capsules are administered in the HandiHaler, a specialized dry powder inhalation system [40]. As with all inhaled therapy, attention to effective drug delivery and training in inhaler technique is essential. COPD patients may have difficulty in coordinating a Metered Dose Inhaler (MDI), and breath-activated Dry Powder Inhalers

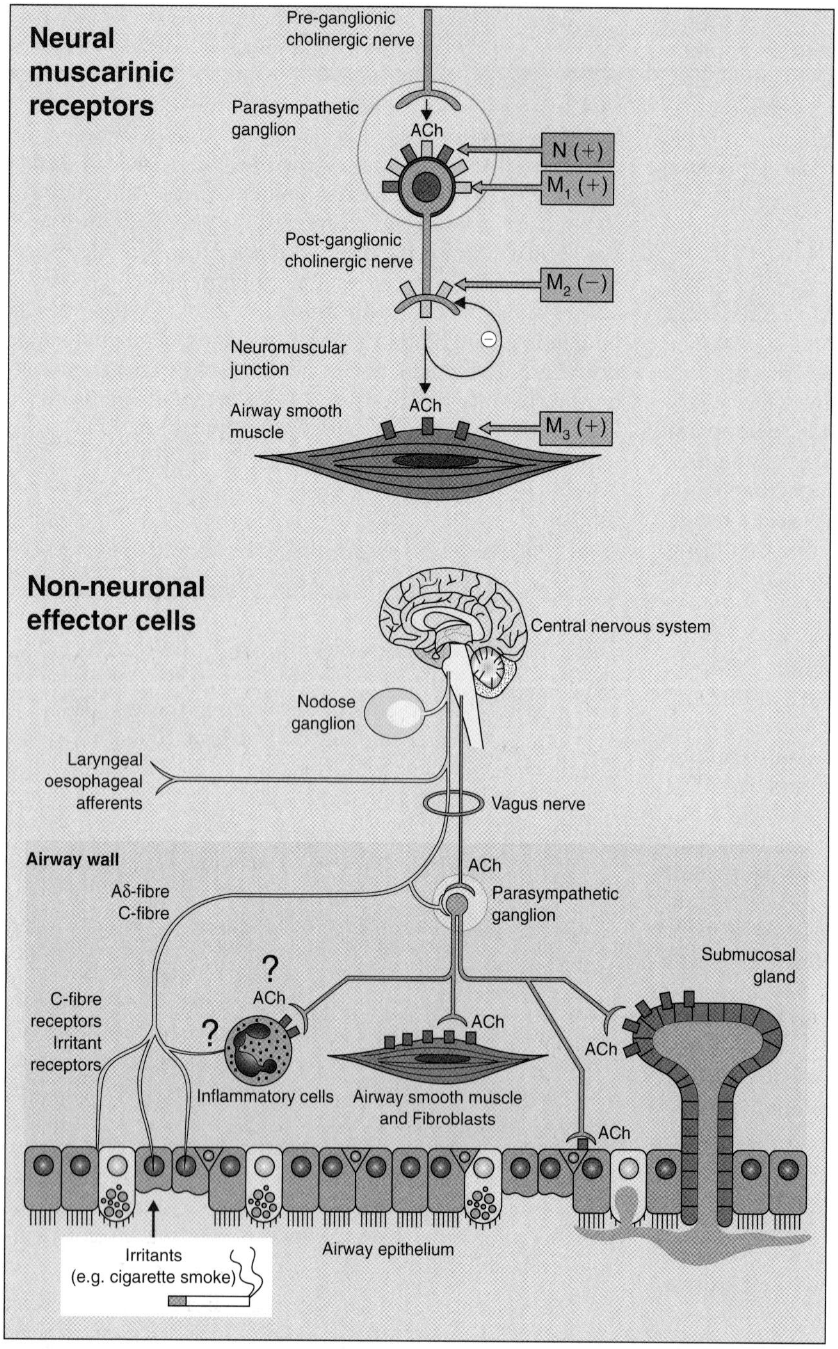

The mechanism of vagally mediated reflex bronchoconstriction is induced by irritants and non-specific stimuli acting on irritant receptors. Pre- and postganglionic parasympathetic nerves release acetylcholine (ACh). Preganglionic nervefibers from the vagus nerve relay in parasympathetic ganglia in the airway wall, and short postganglionic pathways release ACh to cause bronchoconstriction and mucus secretion. Cholinergic muscarinic receptors are present on inflammatory cells, airway smooth muscle, fibroblasts, submucosal gland cells, and epithelial cells. The role of cholinergic receptors on inflammatory cells and inflammatory mediators remains controversial (?).
For details see: Gosens R *et al.* Respir Res 2006; 7:73 doi: 10.1186/1465-9921-7-73

FIG. 49.4 Cholinergic control of airways (see text on figure).

(DPI) have advantages in elderly subjects. HandiHaler requires low inspiratory flow rates to evacuate the capsule and so can be used by patients with severe COPD. The capsule begins to vibrate at 15 l/min and evacuates with concomitant clinical effect at 20 l/min. The delivered dose that leaves the HandiHaler is 10 μg. The contents of one capsule of tiotropium bromide should be inhaled once daily with the HandiHaler at the same time of the day.

Patient performance with the HandiHaler was found in a clinical study to be superior to that with a MDI [41].

PHARMACOKINETICS

Tiotropium bromide provides a bronchodilator effect for over 24h with a plasma concentration of ~2pg/ml. After inhalation of a 10-μg dose there is a rapid absorption with a peak plasma concentration within 5min of 6pg/ml, followed by a rapid fall within 1h to the steady state level of 2pg/ml and a terminal half-life of 5–6 days that is independent of the dose [24]. It has been calculated that this concentration would occupy <5% muscarinic receptors, perhaps accounting for the very low incidence of systemic side effects. There is no evidence for accumulation of tiotropium after repeated administration. The pharmacokinetics and tissue distribution of tiotropium in the rat and dog have also been described [42]. Following dry powder inhalation, the absolute bioavailability of 19.5% suggests that the lung fraction is highly bioavailable. Since tiotropium is a quaternary ammonium compound, it is poorly absorbed from the gastrointestinal tract (10–15%). Food is not expected to influence absorption.

Tiotropium at a level of 72% to plasma proteins shows a volume of distribution of 32l/kg. At steady state, tiotropium plasma levels in COPD patients at peak were 17–19pg/ml when measured 5min after inhalation of an 18-μg dose, and steady state trough plasma concentrations were 3–4pg/ml. The extent of biotransformation is small, with urinary excretion of 74% of unchanged substance after an intravenous dose to young healthy volunteers. Tiotropium should be used with care in patients with moderate-to-severe renal insufficiency [43]. The termination half-life of tiotropium is between 5 and 6 days following inhalation. Tiotropium has linear pharmacokinetics in the therapeutic range after both intravenous administration and dry powder inhalation.

ONSET AND DURATION OF ACTION

Tiotropium bromide is given as 18μg once daily, usually in the morning or at noon. In studies of patients with COPD, single doses of tiotropium bromide give an almost immediate bronchodilator response that is preserved for over 24h [44]. Approximately 70% of the improvement in FEV_1 is seen after 48h, which is judged to be the steady state, while continued improvements can be expected beyond the first week of therapy [45], while dyspnea and health-related quality of life improve over the longer term. Interestingly in patients with and without short-term bronchodilator responses, long-term benefits of treatment with tiotropium are found in COPD patients [46]. Tiotropium was found to improve significantly in lung functions in African-American patients with COPD [47].

In contrast, ipratropium and oxitropium bromide have a slower onset of action than β_2-agonists, with a peak effect being obtained in between 30 and 90min. Ipratropium bromide has a duration of 6–8h, while oxitropium is claimed to maintain bronchodilation for 8h, although this could be due to a higher dose of oxitropium [17]. On the basis of the duration of action, ipratropium bromide is given four times daily, while oxitropium bromide is given three times daily.

CLINICAL STUDIES

A comprehensive range of clinical studies have been performed with tiotropium up to registration with the drug regulatory authorities, and other studies that investigate the effects of tiotropium on exacerbations and the natural history of COPD are ongoing (Table 49.1).

Single-dose relationships of tiotropium bromide have been determined with nebulized doses of up to 160μg in patients with COPD [23, 44], as well as doses of up to 80μg from a DPI in asthma [72] (Fig. 49.5). In the open label first study in patients prolonged bronchodilation without significant adverse events was noted in COPD [44], and in a randomized placebo-controlled crossover design a clear dose response could be demonstrated between for tiotropium at single doses of between 10 and 80μg [44]. In addition, in the single dose study in patients with asthma, tiotropium mediated bronchoprotection against the constrictor effects of methacholine [72].

A 1-week study in patients with COPD demonstrated that ~70% of the bronchodilator effect of tiotropium are obtained after two doses, and that the FEV_1 steady state is reached within 48h [45]. A 4 week study of tiotropium at doses of up to 36μg given daily at noon found comparable responses at doses from 9 to 36μg, representing a flat dose response curve, with little dose dependency in terms of spirometric outcomes [49]. On this basis, a dose of 18μg was selected for use in long-term studies, since the 36μg dose increased the incidence of dry mouth without significantly increasing FEV_1.

The Dutch Tiotropium Study Group has undertaken a 13-week study with 288 patients with COPD, in which tiotropium at 18μg from the HandiHaler administered once daily was significantly more effective than ipratropium administered four times daily [61]. The US Tiotropium Study Group also undertook a 13-week multicenter study, but compared tiotropium with placebo, and found significant improvements in lung function (FEV_1 and PEFR), symptoms, as-needed albuterol/salbutamol usage and physician global assessment [48].

Long-term studies with tiotropium bromide in moderate-to-severe COPD have demonstrated significant improvement in lung function, symptoms, and health-related quality of life, as well as a reduction in exacerbations [50, 62] (Fig. 49.6). These studies confirm that the positive effects seen in 4- and 13-week studies are extended to 1 year and suggest that tolerance does not occur in this period [73]. In the study from the Dutch/Belgian Tiotropium Study Group,

TABLE 49.1 Summary of selected randomized controlled clinical studies on tiotropium bromide.

Reference	n	Duration (weeks)	Mean FEV_1 at baseline (% predicted)	FEV_1	Dyspnea/ symptoms	Exacerbation frequency	HRQOL score
Tiotropium versus Placebo							
Casaburi [48]	470	13	39	↑	↓	NA	NA
Littner [49]	169	4	39	↑	NA	NA	NA
Casburi [50]	921	52	39	↑	↓	↓	↑
Donohue [51]	623	26	40	↑	↓	↔	↑
Calverley [52]	121	6	41	↑	NA	↔	NA
Brusasco [53]	1207	26	40	↑	↓	↓	↑
Celli [54]	81	4	43	↑	NA	NA	NA
Hasani [55]	34	3	44	↑	NA	NA	NA
O'Donnell [56]	187	6	42	↑	↓	↔	NA
McNicholas [57]	95	4	32	↑	NA	NA	NA
Dusser [58]	1010	52	48	↑	↓	↓	NA
Niehwoehner [59]	1829	26	36	↑	NA	↓	NA
Verkindre [60]	100	12	36	↑	↓	NA	↑
Tiotropium versus Ipratropium							
Van Noord [61]	288	13	41	↑	NA	↔	NA
Vincken [62]	535	52	41	↑	↓	↓	↑
Tiotropium versus Salmeterol							
Donohue [51]	623	26	40	↑	↓	↔	↔
Brusasco [53]	1207	26	40	↑	↔	↔	NA
Briggs [63]	653	12	38	↑	NA	↔	NA
Tiotropium versus Formoterol							
Van Noord [64]	71	20	37	↑	NA	NA	NA
Cazzola [65]	20	Single doses	35	↑	NA	NA	NA
Di Marco [66]	21	Single doses	38	↑	NA	NA	NA
Tiotropium (TIO) combinations with Salmeterol (SALM)/Fluticasone (FLU)/Formoterol (FORM)							
Cazzola [67] • TIO • SALM–FLU • TIO + SALM–FLU	90	12	37–39	↑	↑	NA	NA
Aaron [68] • TIO • TIO + SALM • TIO + SALM–FLU	449	52	38–39	↑	↑	↔	↑
Singh [69] • TIO • SALM–FLU • TIO + SALM–FLU	41	2	47	↑	↑	NA	NA
Van Noord [70] • TIO • TIO + FORM	95	2	38	↑	↑	NA	NA
Di Marco [66] • TIO • TIO + FORM	21	Single doses	38	↑	NA	NA	NA
Tashkin [71] Nebulized • TIO • TIO + FORM	129	6	38	↑	↑	NA	↔

This table is adapted with permission from a review by Ref. [8].

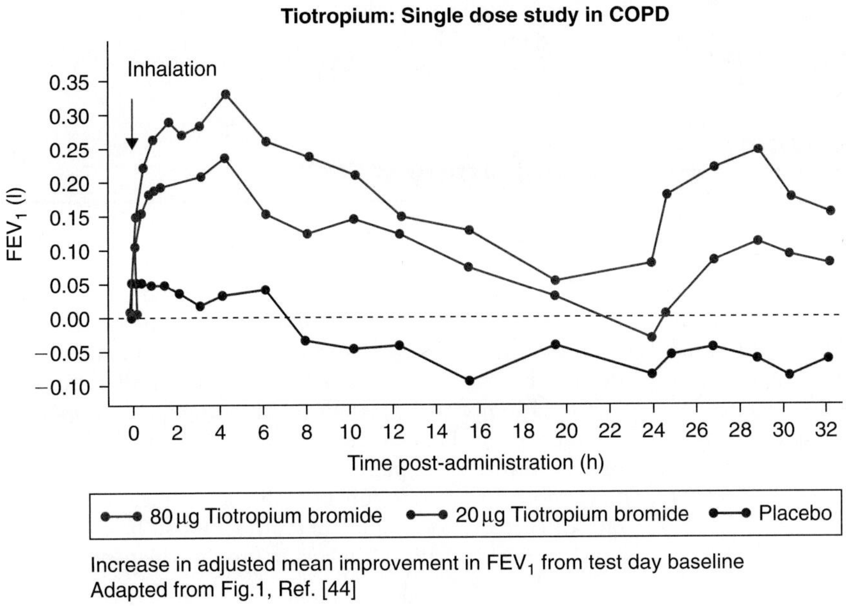

FIG. 49.5 Single-dose study of tiotropium in COPD (see text on figure).

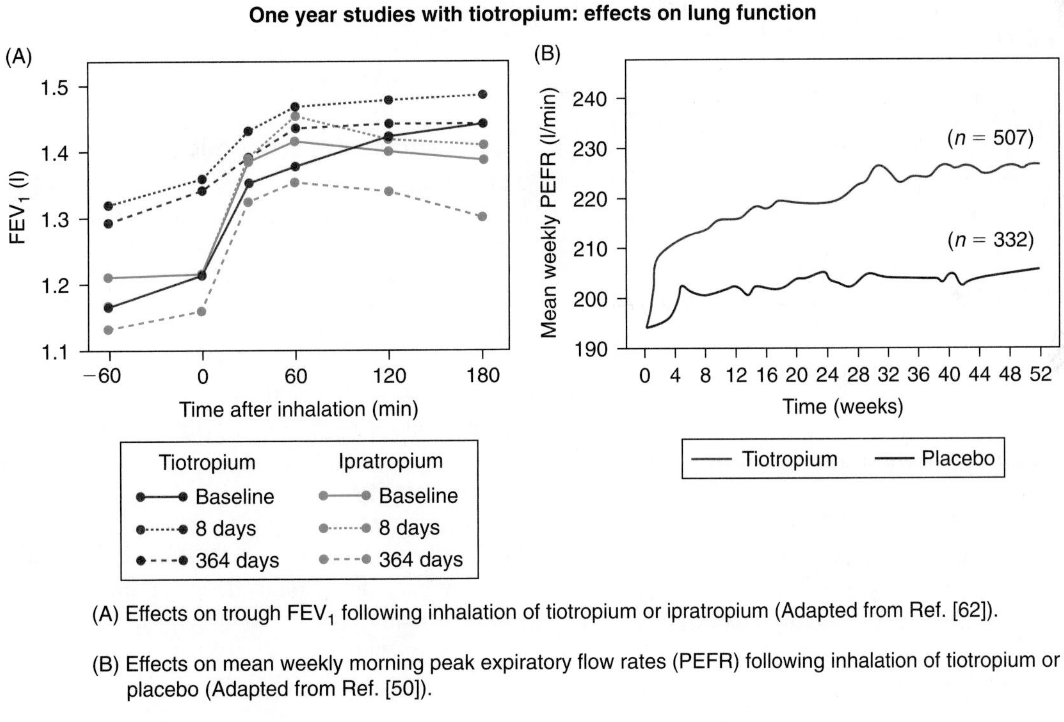

FIG. 49.6 One year studies with tiotropium (see text on Figure).

tiotropium was effective in improving lung function, dyspnea, exacerbations and health-related quality of life, and improved trough FEV_1 at 1 year [62]. In this study tiotropium was found to be more effective than ipratropium inhaled four times a day. In a separate placebo-controlled multicenter study from Casaburi and colleagues, tiotropium also showed significant effects on the Transition Dyspnea Index (TDI) as well as health status assessed by the disease-specific St. George's Respiratory Questionnaire (SGRQ) and the generic Short Form (SF) 36 [50].

EFFECTS OF TIOTROPIUM ON OUTCOMES IN COPD

Bronchodilatation, symptoms of dyspnea

Bronchodilators lessen airway smooth muscle tone and reduce dynamic hyperinflation, to cause relatively rapid relief of symptoms and improved exercise tolerance [74–76]. Relaxation of airway smooth muscle tone causes an increase in FEV_1, but in COPD these changes are generally small (<10%). Bronchodilator efficacy in COPD is generally assessed as the FEV_1 measured by spirometry. Symptomatic improvement may be greater than changes in FEV_1, especially in moderate-to-severe COPD [62], and tiotropium causes improvement in dyspnea [50, 77, 78].

Dynamic hyperinflation

Improved exercise performance may be a conspicuous feature of bronchodilator usage in COPD [54]. Reduced dynamic hyperinflation with a reduction in residual volume (RV) and functional residual capacity (FRC) but improved resting inspiratory capacity (IC) [54] contributes markedly to symptomatic benefits, since this makes breathing more comfortable and dyspnea is lessened on exertion [56,79]. Improvement of exercise tolerance is aided by the combination of tiotropium and pulmonary rehabilitation [80]. Other effects of anticholinergics are to increase mucociliary clearance and decrease mucous production. However, improved respiratory muscle function is unlikely at the dose of tiotropium used clinically.

Exacerbations

The reader is recommended to consult a recent review on the effect of tiotropium on exacerbations and hospitalizations [81].

- The 1-year tiotropium registration studies demonstrated that tiotropium causes a decreased incidence of exacerbations when compared with placebo [46, 50, 62].
- Following this, Niewoehner and colleagues prospectively demonstrated significant tiotropium effects on reducing COPD exacerbations, and may reduce related healthcare utilization [59].
- The MISTRAL (*Mesure de L'Influence de Spiriva sur les Troubles Aigus à Long Terme*) study studied 500 patients with COPD treated with tiotropium for a 1-year period in comparison with a placebo group [58].
- The INSPIRE (Investigating New Standards for Prophylaxis in Reduction of Exacerbations) study compares salmeterol in addition to fluticasone (SF) in a combination inhaler compared with tiotropium [82]. This study found no difference in exacerbation rates between the two arms, with more patients failing to complete the study while receiving tiotropium, yet more cases of pneumonia were reported in the salmeterol/fluticasone arm. In addition, a small significant beneficial effect was found on health status with an unexpected finding in lower death in SF-treated patients [83].

Natural history of COPD

There is little evidence that bronchodilators, including tiotropium, modify the rate of decline in lung function (FEV_1) in COPD [84]. It is very important to assess whether chronic therapy with tiotropium can affect the natural history of COPD, which can be defined by the annual rate of decline in post-bronchodilatory FEV_1. A rather unsatisfactory attempt to assess effects of tiotropium on longitudinal effects of lung function was made by pooled analysis of two 1-year tiotropium studies [85]. These studies have the problem that lung function should be measured at the end of chronic tiotropium therapy, with a sufficiently large time interval to have no lingering bronchodilatory or other effects of tiotropium.

The UPLIFT (Understanding the Potential Long-Term Impacts on Function with Tiotropium) study is currently evaluating the long-term effects of 4 years of therapy with tiotropium on the rate of decline in lung function, health status, and frequency of exacerbations [86]. The UPLIFT study measures trough FEV_1, as well as FEV_1, 90 min after maximal bronchodilator administration and includes a spectrum of patient-centered outcomes [87].

Costs and health economics

Health outcomes have been found to be positive following treatment of COPD patients with once daily tiotropium for 6 months [53], and similar results have been found after 1 year of treatment with tiotropium [88], and in comparison with treatment with ipratropium over a year [89]. In a probabilistic Markov model to assess the cost-effectiveness of bronchodilator therapy in COPD, tiotropium was associated in terms of exacerbations with maximum expected net benefit for plausible values of the ceiling ratio [90].

Sleep quality

Sustained LAMA therapy with tiotropium improves sleeping oxygen saturation in COPD without affecting sleep quality [57]. Tiotropium does not appear to abolish circadian variation in airway caliber [52].

COMPARISON OF TIOTROPIUM WITH SALMETEROL AND FORMOTEROL

A 6-month clinical trial in 623 patients with COPD has demonstrated superior effects of tiotropium on bronchodilation, dyspnea, and health-related quality of life compared to salmeterol [51]. In other studies with tiotropium being given once daily, while salmeterol was given twice daily, tiotropium has been shown to be superior in terms of daytime spirometric efficacy [63] and exacerbations and health outcomes [53].

In comparison with formoterol, tiotropium was better in terms of lung function, although a combination of both drugs was most effective [64]. The benefits of tiotropium over formoterol have also been demonstrated in single-dose studies [65].

COMBINATION STUDIES WITH TIOTROPIUM

Combination bronchodilator products have already been demonstrated to have advantages over a single class of bronchodilator, so that both short-acting and long-acting β_2-agonists may be combined with anticholinergics, and additional oral theophylline may have benefit. Combination bronchodilator inhalers with an anticholinergic and a short-acting β_2-agonist, such as Combivent (ipratropium bromide + salbutamol), are a convenient a and more effective way of giving bronchodilators [91–93]. Tiotropium bromide has additive effects with both short- and long-acting β_2-agonists (LABAs) [94].

An important study demonstrated that adding tiotropium to formoterol causes improvements in airflow obstruction, resting hyperinflation, and use of rescue salbutamol [70] (Fig. 49.7). Concomitant treatment with nebulized formoterol and tiotropium provided benefits in patients with COPD [71], and the combination of formoterol and tiotropium is complementary in treating acute exacerbations of COPD [66].

Combined LABAs and ICS in a single inhaler are licensed for asthma and COPD, with the combination of salmeterol and fluticasone available as Seretide or Advair (GlaxoSmithKline) [95], with formoterol and budesonide as Symbicort (AstraZeneca) [96]. Recently, there have been studies to show the benefits of triple therapy with the combination of salmeterol/fluticasone and tiotropium [67, 69]. However, addition of salmeterol/fluticasone to tiotropium did not statistically influence rates of COPD exacerbation in a study of 449 patients over 1 year of treatment [68].

TIOTROPIUM IN ASTHMA

Anticholinergics are less useful in asthma than in COPD, as inhaled β_2-agonists are generally more effective [5]. However, with regard to allergen-induced airway remodeling in guinea pigs, tiotropium and budesonide have similar inhibitory efficacy [35]. Patients with concomitant asthma and COPD have been shown to benefit from tiotropium [97], and in patients with severe asthma adding salmeterol/tiotropium to patients with fluticasone has benefits [98]. Tiotropium could reduce chronic viral inflammation and may be helpful in acute severe asthma [34]. Hence, tiotropium is likely to be effective as an additional bronchodilator in patients with severe disease, especially when there is an element of fixed airflow limitation. It is likely that patients with severe asthma will benefit from the addition of tiotropium to combination inhalers containing long-acting β_2-agonists and corticosteroids, but clinical trials are needed to document this.

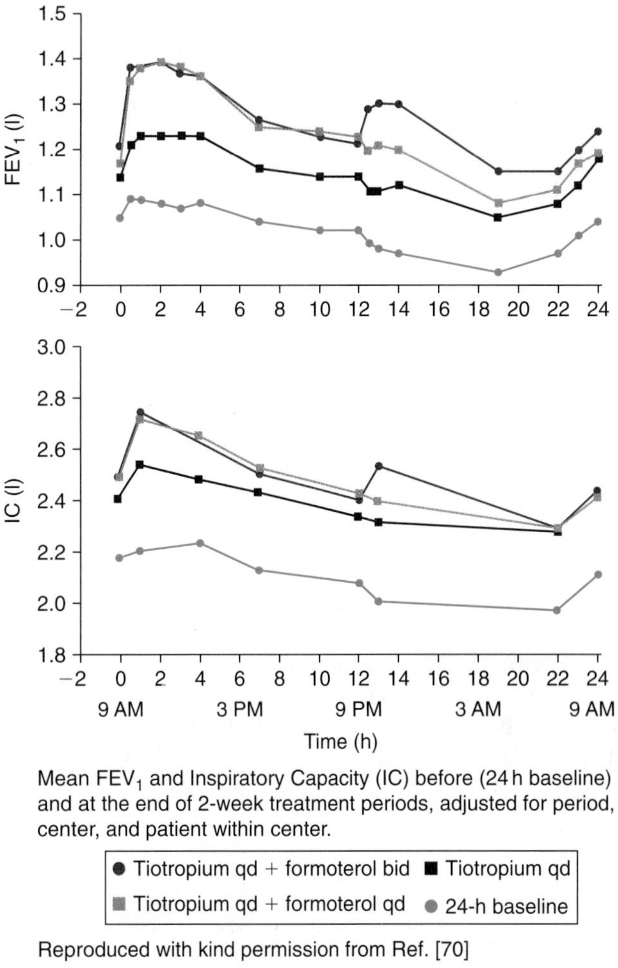

FIG. 49.7 Bronchodilator effect of tiotropium with and without formoterol in COPD.

SAFETY OF TIOTROPIUM

A pooled clinical trial analysis (meta-analysis) of tiotropium safety has been carried out on a population of 4435 patients receiving inhaled tiotropium [99]. Dyspnea, dry mouth, COPD exacerbation, and upper respiratory tract infections were the most commonly reported adverse events. A careful study demonstrated that tiotropium does not retard mucus clearance from the lungs [55].

A cohort study using UK primary care databases found that tiotropium and LABAs had similar risks of total mortality and cardiovascular endpoints [100], while the SPRUCE primary care study showed benefits and not adverse effects when adding tiotropium to COPD care [101]. Detailed electrocardiography (ECG) monitoring has been performed in COPD patients receiving tiotropium with ECGs performed up to six times during 12 months of treatment [102], and tiotropium was not associated with any cardiac safety concerns in a separate ECG analysis in COPD patients [103].

Side effects of anticholinergics are pharmacologically predictable and dose dependent, and inhaled delivery means that systemic adverse effects are less likely. However,

COPD patients are older and more likely to have comorbidities than patients with asthma, so they have a greater risk of developing adverse reactions to therapy. Side effects of tiotropium bromide are very uncommon as tiotropium acts topically and there is virtually no systemic absorption. This limits the type of adverse anticholinergic effects seen with atropine; and extensive experience has shown inhaled anticholinergics to be a remarkably safe class of drug.

- Dry mouth may be associated with dental caries
- Cough, pharyngitis, upper respiratory infections
- Ocular effects: narrow angle glaucoma is a recognized adverse effect of nebulized anticholinergics and tiotropium may cause eye pain, blurred vision, visual halos, and conjunctival and corneal congestion (redness).
- Taste disturbance: a bitter metallic taste has been reported by some patients.
- Immediate hypersensitivity reactions may occur.
- CNS: headache, nervousness, irritation, dizziness.
- Theoretical systemic anticholinergic effects: prostatic hypertrophy, urinary retention, bladder disorders, constipation, tachycardia, cardiac arrhythmias, palpitations.
- Renal and hepatic impairment: patients can use tiotropium bromide at the recommended dose.

NEW ANTICHOLINERGICS

There have been considerable efforts from the pharmaceutical industry to develop novel bronchodilators, and especially LABAs and LAMAs, for COPD [104, 105]. Glycopyrrolate has been used extensively systemically for decades to control salivation as a pre-anesthetic agent and causes bronchodilation on isolated guinea pig trachea [106, 107]. Interestingly, glycopyrrolate is a quaternary ammonium derivative which has minimal mucosal absorption and systemic toxicity when inhaled, causing prolonged bronchodilation in patients with asthma [106] and a family of anticholinergics based on glycopyrrolate have been synthesized [108]. A variant on this agent is racemically pure R,R-glycopyrrolate as a LAMA [109], while an inhaled glycopyrrolate derivative, NVA-237, is currently being developed by Novartis (license from Arakis and Ventura). An oral selective M3-selective M3-cholinergic receptor antagonist that was studied in patients with COPD (OrM3, Merck) has apparently been discontinued in clinical development [110]. However, Phase 3 studies have been successfully carried with the inhaled LAMA, aclidinium bromide (LAS-34273, Almirall), that is likely to be licensed in the near future. Recently there have been efforts to develop a combined muscarinic antagonist and beta-agonist in a single molecule: Theravance describes this as a MABA. In the future we can expect LABAs and LAMAs to be combined in single inhalers (eg. indacaterol-glycopyrrolate) and there may be the possibility of 3 agents in a single inhaler (LABA, LAMA, ICS).

References

1. Rabe KF, Hurd S, Anzueto A *et al.* Global strategy for the diagnosis, management, and prevention of chronic obstructive pulmonary disease: GOLD executive summary. *Am J Respir Crit Care Med* 176(6): 532–55, 2007.
2. National Institutes of Health (NIH), National Heart Lung and Blood Institute (NHLBI), World Health Organisation (WHO). *Global Initiative for Chronic Obstructive Lung Disease (GOLD): Global Strategy for the Diagnosis, Management, and Prevention of Chronic Obstructive Pulmonary Disease.* Updated 2007.
3. Tashkin DP, Cooper CB. The role of long-acting bronchodilators in the management of stable COPD. *Chest* 125(1): 249–59, 2004.
4. Decramer M. Tiotropium as essential maintenance therapy in COPD. *Eur Respir Rev* 15: 51–57, 2006.
5. Gross NJ. Anticholinergic agents in asthma and COPD. *Eur J Pharmacol* 533(1–3): 36–39, 2006.
6. Koumis T, Samuel S. Tiotropium bromide: A new long-acting bronchodilator for the treatment of chronic obstructive pulmonary disease. *Clin Ther* 27(4): 377–92, 2005.
7. Restrepo RD. Use of inhaled anticholinergic agents in obstructive airway disease. *Respir Care* 52(7): 833–51, 2007.
8. Currie GP, Rossiter C, Miles SA, Lee DK, Dempsey OJ. Effects of tiotropium and other long acting bronchodilators in chronic obstructive pulmonary disease. *Pulm Pharmacol Ther* 19(2): 112–19, 2006.
9. Voduc N. Tiotropium in the treatment of chronic pulmonary disease: Review of clinical trials. *Today's Therapeutic Trends*: 37–47, 2006.
10. Barr RG, Bourbeau J, Camargo CA, Ram FS. Tiotropium for stable chronic obstructive pulmonary disease: A meta-analysis. *Thorax* 61(10): 854–62, 2006.
11. Sakula A. A history of asthma: The FitzPatrick lecture 1987. *J R Coll Physicians Lond* 22: 36–44, 1988.
12. Rau JL. *Respiratory Care Pharmacology*, 6th edn. St Louis, USA: Mosby, 2003.
13. Gross NJ. Ipratropium bromide. *N Engl J Med* 319(8): 486–94, 1988.
14. Chapman KR. History of anticholinergic treatment in airways disease. In: Gross NJ (ed.), *Anticholinergic Therapy in Obstructive Airways Disease*, pp. 9–17. London: Franklin Scientific Publications, 1993.
15. Herxheimer H. Atropine cigarettes in asthma emphysema. *BMJ* 2: 167, 1959.
16. Gandevia B. Historical review of the use of parasympatholytic agents in the treatment of respiratory disorders. *Postgrad Med J* 51: S13–20, 1975.
17. Barnes PJ. Tiotropium bromide. *Expert Opin Investig Drugs* 10(4): 733–40, 2001.
18. Hvizdos KM, Goa KL. Tiotropium bromide. *Drugs* 62: 1195–203, 2002.
19. Hansel TT, Barnes PJ. Tiotropium bromide: A novel once-daily anticholinergic bronchodilator for the treatment of COPD. *Drugs Today* 38(9): 585–600, 2002.
20. Tamaoki J, Chiyotani A, Tagaya E, Sakai N, Konno K. Effect of long term treatment with oxitropium bromide on airway secretion in chronic bronchitis and diffuse panbronchiolitis. *Thorax* 49(6): 545–48, 1994.
21. Barnes PJ. Muscarinic receptor subtypes in airways. *Life Sci* 52: 521–28, 1993.
22. Eglen RM, Choppin A, Watson N. Therapeutic opportunities from muscarinic receptor research. *Trends Pharmacol Sci* 22: 409–14, 2001.
23. Maesen FPV, Smeets JJ, Costongs MAL, Wald FDM, Comelissen PJG. Ba 679 Br, a new long-acting antimuscarinic bronchodilator: A pilot dose-escalation study in COPD. *Eur Respir J* 6: 1031–36, 1993.
24. Disse B, Speck GA, Rominger KL, Witek TJ, Hammer R. Tiotropium (Spiriva) mechanistical considerations and clinical profile in obstructive lung disease. *Life Sci* 64: 457–64, 1999.
25. Barnes PJ. The pharmacological properties of tiotropium. *Chest* 117(2 Suppl): 63S–66S, 2000.
26. Disse B, Reichl R, Speck G, Traunecker W, Rominger KL, Hammer R. A novel long-acting anticholinergic bronchodilator. *Life Sci* 52: 537–44, 1993.

27. Haddad EB, Mak JC, Barnes PJ. Characterization of [3H]Ba 679 BR, a slowly dissociating muscarinic antagonist, in human lung: Radioligand binding and autoradiographic mapping. *Mol Pharmacol* 45(5): 899–907, 1994.
28. Wess J, Eglen RM, Gautam D. Muscarinic acetylcholine receptors: Mutant mice provide new insights for drug development. *Nat Rev Drug Discov* 6(9): 721–33, 2007.
29. Grando SA, Kawashima K, Kirkpatrick CJ, Wessler I. Recent progress in understanding the non-neuronal cholinergic system in humans. *Life Sci* 80(24–25): 2181–85, 2007.
30. Wessler I, Kilbinger H, Bittinger F, Unger R, Kirkpatrick CJ. The non-neuronal cholinergic system in humans: Expression, function and pathophysiology. *Life Sci* 72(18–19): 2055–61, 2003.
31. Ulloa L. The vagus nerve and the nicotinic anti-inflammatory pathway. *Nat Rev Drug Discov* 4(8): 673–84, 2005.
32. Kawashima K, Fujii T. The lymphocytic cholinergic system and its contribution to the regulation of immune activity. *Life Sci* 74(6): 675–96, 2003.
33. Adamko DJ, Fryer AD, Bochner BS, Jacoby DB. $CD8^+$ T lymphocytes in viral hyperreactivity and M_2 muscarinic receptor dysfunction. *Am J Respir Crit Care Med* 167(4): 550–56, 2003.
34. Kanazawa H. Anticholinergic agents in asthma: Chronic bronchodilator therapy, relief of acute severe asthma, reduction of chronic viral inflammation and prevention of airway remodeling. *Curr Opin Pulm Med* 12(1): 60–67, 2006.
35. Bos IS, Gosens R, Zuidhof AB *et al.* Inhibition of allergen-induced airway remodelling by tiotropium and budesonide: A comparison. *Eur Respir J* 30(4): 653–61, 2007.
36. Gosens R, Bos IS, Zaagsma J, Meurs H. Protective effects of tiotropium bromide in the progression of airway smooth muscle remodeling. *Am J Respir Crit Care Med* 171(10): 1096–102, 2005.
37. Matthiesen S, Bahulayan A, Kempkens S *et al.* Muscarinic receptors mediate stimulation of human lung fibroblast proliferation. *Am J Respir Cell Mol Biol*, 2006.
38. Buhling F, Lieder N, Kuhlmann UC, Waldburg N, Welte T. Tiotropium suppresses acetylcholine-induced release of chemotactic mediators *in vitro*. *Respir Med* 101(11): 2386–94, 2007.
39. Powrie DJ, Wilkinson TM, Donaldson GC *et al.* Effect of tiotropium on sputum and serum inflammatory markers and exacerbations in COPD. *Eur Respir J* 30(3): 472–78, 2007.
40. Chodosh S, Flanders JS, Kesten S, Serby CW, Hochrainer D, Witek TJ. Effective delivery of particles with the HandiHaler dry powder inhalation system over a range of chronic obstructive pulmonary disease severity. *J Aerosol Med* 14: 309–15, 2001.
41. Dahl R, Backer V, Ollgaard B, Gerken F, Kesten S. Assessment of patient performance of the HandiHaler compared with the metered dose inhaler four weeks after instruction. *Respir Med* 97(10): 1126–33, 2003.
42. Leusch A, Eichhorn B, Muller G, Rominger KL. Pharmacokinetics and tissue distribution of the anticholinergics tiotropium and ipratropium in the rat and dog. *Biopharm Drug Dispos* 22(5): 199–212, 2001.
43. Turck D, Weber W, Sigmund R *et al.* Pharmacokinetics of intravenous, single-dose tiotropium in subjects with different degrees of renal impairment. *J Clin Pharmacol* 44(2): 163–72, 2004.
44. Maesen FPV, Smeets JJ, Sledsens TJH, Wald FDM, Cornelissen PJG. Tiotropium bromide, a new long-acting antimuscarinic bronchodilator: A pharmacodynamic study in patients with chronic obstructive disease (COPD). *Eur Respir J* 8: 1506–13, 1995.
45. van Noord JA, Smeets JJ, Custers FL, Korducki L, Cornelissen PJ. Pharmacodynamic steady state of tiotropium in patients with chronic obstructive pulmonary disease. *Eur Respir J* 19(4): 639–44, 2002.
46. Tashkin D, Kesten S. Long-term treatment benefits with tiotropium in COPD patients with and without short-term bronchodilator responses. *Chest* 123: 1441–49, 2003.
47. Criner GJ, Sharafkhaneh A, Player R *et al.* Efficacy of tiotropium inhalation powder in african-american patients with chronic obstructive pulmonary disease. *COPD* 5(1): 35–41, 2008.
48. Casaburi R, Serby W, Menjoge SS, Witek TJ. The spirometric efficacy of once daily dosing with tiotropium in stable COPD. *Am J Respir Crit Care Med* 159: A524, 1999.
49. Littner MR, Ilowite JS, Tashkin DP *et al.* Long-acting bronchodilation with once-daily dosing of tiotropium (Spiriva) in stable chronic obstructive pulmonary disease. *Am J Respir Crit Care Med* 161(4 Pt 1): 1136–42, 2000.
50. Casaburi R, Mahler DA, Jones PW *et al.* A long-term evaluation of once-daily inhaled tiotropium in chronic obstructive pulmonary disease. *Eur Respir J* 19(2): 217–24, 2002.
51. Donohue JF, van Noord JA, Bateman ED *et al.* A 6-month, placebo-controlled study comparing lung function and health status changes in COPD patients treated with tiotropium or salmeterol. *Chest* 122(1): 47–55, 2002.
52. Calverley PM, Lee A, Towse L, van Noord J, Witek TJ, Kelsen S. Effect of tiotropium bromide on circadian variation in airflow limitation in chronic obstructive pulmonary disease. *Thorax* 58(10): 855–60, 2003.
53. Brusasco V, Hodder R, Miravitlles M, Korducki L, Towse L, Kesten S. Health outcomes following treatment for six months with once daily tiotropium compared with twice daily salmeterol in patients with COPD. *Thorax* 58(5): 399–404, 2003.
54. Celli B, ZuWallack R, Wang S, Kesten S. Improvement in resting inspiratory capacity and hyperinflation with tiotropium in COPD patients with increased static lung volumes. *Chest* 124(5): 1743–48, 2003.
55. Hasani A, Toms N, Agnew JE, Sarno M, Harrison AJ, Dilworth P. The effect of inhaled tiotropium bromide on lung mucociliary clearance in patients with COPD. *Chest* 125(5): 1726–34, 2004.
56. O'Donnell DE, Fluge T, Gerken F *et al.* Effects of tiotropium on lung hyperinflation, dyspnoea and exercise tolerance in COPD. *Eur Respir J* 23(6): 832–40, 2004.
57. McNicholas WT, Calverley PM, Lee A, Edwards JC. Long-acting inhaled anticholinergic therapy improves sleeping oxygen saturation in COPD. *Eur Respir J* 23(6): 825–31, 2004.
58. Dusser D, Bravo ML, Iacono P. The effect of tiotropium on exacerbations and airflow in patients with COPD. *Eur Respir J* 27(3): 547–55, 2006.
59. Niewoehner DE, Rice K, Cote C *et al.* Prevention of exacerbations of chronic obstructive pulmonary disease with tiotropium, a once-daily inhaled anticholinergic bronchodilator: A randomized trial. *Ann Intern Med* 143(5): 317–26, 2005.
60. Verkindre C, Bart F, Aguilaniu B *et al.* The effect of tiotropium on hyperinflation and exercise capacity in chronic obstructive pulmonary disease. *Respiration* 73(4): 420–27, 2006.
61. van Noord JA, Bantje TA, Eland ME, Korducki L, Cornelissen PJ. A randomised controlled comparison of tiotropium and ipratropium in the treatment of chronic obstructive pulmonary disease. The Dutch Tiotropium Study Group. *Thorax* 55(4): 289–94, 2000.
62. Vincken W, van Noord JA, Greefhorst AP *et al.* Improved health outcomes in patients with COPD during 1 yr's treatment with tiotropium. *Eur Respir J* 19(2): 209–16, 2002.
63. Briggs DD Jr, Covelli H, Lapidus R, Bhattycharya S, Kesten S, Cassino C. Improved daytime spirometric efficacy of tiotropium compared with salmeterol in patients with COPD. *Pulm Pharmacol Ther* 18(6): 397–404, 2005.
64. van Noord JA, Aumann JL, Janssens E *et al.* Comparison of tiotropium once daily, formoterol twice daily and both combined once daily in patients with COPD. *Eur Respir J* 26(2): 214–22, 2005.
65. Cazzola M, Di MF, Santus P *et al.* The pharmacodynamic effects of single inhaled doses of formoterol, tiotropium and their combination in patients with COPDA. *Pulm Pharmacol Ther* 17(1): 35–39, 2004.
66. Di MF, Verga M, Santus P, Morelli N, Cazzola M, Centanni S. Effect of formoterol, tiotropium, and their combination in patients with acute exacerbation of chronic obstructive pulmonary disease: A pilot study. *Respir Med* 100(11): 1925–32, 2006.
67. Cazzola M, Ando F, Santus P *et al.* A pilot study to assess the effects of combining fluticasone propionate/salmeterol and tiotropium on

the airflow obstruction of patients with severe-to-very severe COPD. *Pulm Pharmacol Ther* 20(5): 556–61, 2007.
68. Aaron SD, Vandemheen KL, Fergusson D *et al.* Tiotropium in combination with placebo, salmeterol, or fluticasone–salmeterol for treatment of chronic obstructive pulmonary disease: A randomized trial. *Ann Intern Med* 146(8): 545–55, 2007.
69. Singh D, Brooks J, Hagan G, Cahn T, O'Connor B. Triple therapy with salmeterol/fluticasone propionate and tiotropium bromide versus individual components in moderate to severe COPD. *Thorax*, 2008.
70. van Noord JA, Aumann JL, Janssens E *et al.* Effects of tiotropium with and without formoterol on airflow obstruction and resting hyperinflation in patients with COPD. *Chest* 129(3): 509–17, 2006.
71. Tashkin DP, Littner M, Andrews CP, Tomlinson L, Rinehart M, Denis-Mize K. Concomitant treatment with nebulized formoterol and tiotropium in subjects with COPD: A placebo-controlled trial. *Respir Med*, 2008. doi:10.1016/j.rmed.2007.12.019
72. O'Connor BJ, Towse LJ, Barnes PJ. Prolonged effect of tiotropium bromide on methacholine-induced bronchoconstriction in asthma. *Am J Respir Crit Care Med* 154: 876–80, 1996.
73. Rees PJ. Tiotropium in the management of chronic obstructive pulmonary disease. *Eur Respir J* 19(2): 205–6, 2002.
74. O'Donnell DE. Impacting patient-centred outcomes in COPD: Breathlessness and exercise tolerance. *Eur Respir Rev* 15: 37–44, 2006.
75. O'Donnell DE. Lung mechanics in COPD: The role of tiotropium. *Eur Respir Rev* 13(89): 40–44, 2004.
76. Ferguson GT. The ins and outs of breathing: An overview of lung mechanics. *Eur Respir Rev* 13: 30–34, 2004.
77. Magnussen H. Exercise limitation in COPD: Mechanisms, assessment and treatment. *Eur Respir Rev* 13: 35–39, 2004.
78. Maltais F, Hamilton A, Marciniuk D *et al.* Improvements in symptom-limited exercise performance over 8 h with once-daily tiotropium in patients with COPD. *Chest* 128(3): 1168–78, 2005.
79. O'Donnell DE, Lam M, Webb KA. Measurement of symptoms, lung hyperinflation, and endurance during exercise in chronic obstructive pulmonary disease. *Am J Respir Crit Care Med* 158(5 Pt 1): 1557–65, 1998.
80. Casaburi R, Kukafka D, Cooper CB, Witek TJ Jr, Kesten S. Improvement in exercise tolerance with the combination of tiotropium and pulmonary rehabilitation in patients with COPD. *Chest* 127(3): 809–17, 2005.
81. Rodriguez-Roisin R. Impacting patient-centred outcomes in COPD: Exacerbations and hospitalisations. *Eur Respir Rev* 15: 47–50, 2006.
82. Seemungal T, Stockley R, Calverley P, Hagan G, Wedzicha JA. Investigating new standards for prophylaxis in reduction of exacerbations – the INSPIRE study methodology. *COPD* 4(3): 177–83, 2007.
83. Wedzicha JA, Calverley PM, Seemungal TA, Hagan GW, Ansari Z, Stockley RA. The prevention of COPD exacerbations by salmeterol/fluticasone propionate or tiotropium bromide. *Am J Respir Crit Care Med* 177: 19–26, 2007.
84. Calverley PM, Rennard SI. What have we learned from large drug treatment trials in COPD? *Lancet* 370(9589): 774–85, 2007.
85. Anzueto A, Tashkin D, Menjoge S, Kesten S. One-year analysis of longitudinal changes in spirometry in patients with COPD receiving tiotropium. *Pulm Pharmacol Ther* 18(2): 75–81, 2005.
86. Decramer M *et al.* Clinical trial design considerations in assessing long-term functional impacts of tiotropium in COPD: The Uplift trial. *J Chronic Obstr Pulm Dis* 1: 303–12, 2004.
87. Tashkin DP. The role of patient-centered outcomes in the course of chronic obstructive pulmonary disease: How long-term studies contribute to our understanding. *Am J Med* 119(10 Suppl 1): 63–72, 2006.
88. Friedman M, Menjoge SS, Anton SF, Kesten S. Healthcare costs with tiotropium plus usual care versus usual care alone following 1 year of treatment in patients with chronic obstructive pulmonary disorder (COPD). *Pharmacoeconomics* 22(11): 741–49, 2004.
89. Oostenbrink JB, Rutten-van Molken MP, Al MJ, van Noord JA, Vincken W. One-year cost-effectiveness of tiotropium versus ipratropium to treat chronic obstructive pulmonary disease. *Eur Respir J* 23(2): 241–49, 2004.
90. Oostenbrink JB, Rutten-van Molken MP, Monz BU, Fitzgerald JM. Probabilistic Markov model to assess the cost-effectiveness of bronchodilator therapy in COPD patients in different countries. *Value Health* 8(1): 32–46, 2005.
91. Fitzgerald JM, Grunfield A, Pare PD *et al.* The clinical efficacy of combination nebulized anticholinergic and adrenergic bronchodilators vs nebulized adrenergic bronchodilator alone in acute asthma. *Chest* 111: 311–15, 1997.
92. Chin JE, Hatfield CA, Winterrowd GE *et al.* Preclinical evaluation of anti-inflammatory activities of the novel pyrrolopyrimidine PNU-142731A, a potential treatment for asthma. *J Pharmacol Exp Ther* 290: 188–95, 1999.
93. Wilson JD, Serby CW, Menjoge SS, Witek TJ. The efficacy and safety of combination bronchodilator therapy. *Eur Respir Rev* 6: 286–89, 1996.
94. Tennant RC, Erin EM, Barnes PJ, Hansel TT. Long-acting beta(2)-adrenoceptor agonists or tiotropium bromide for patients with COPD: Is combination therapy justified? *Curr Opin Pharmacol* 3(3): 270–76, 2003.
95. Nelson HS. Advair: Combination treatment with fluticasone propionate/salmeterol in the treatment of asthma. *J Allergy Clin Immunol* 107: 397–416, 2001.
96. McGavin JK, Goa KL, Jarvis B. Inhaled budesonide/formoterol combination. *Drugs* 61(1): 71–78, 2001.
97. Magnussen H, Bugnas B, van NJ, Schmidt P, Gerken F, Kesten S. Improvements with tiotropium in COPD patients with concomitant asthma. *Respir Med* 102(1): 50–56, 2008.
98. Fardon T, Haggart K, Lee DK, Lipworth BJ. A proof of concept study to evaluate stepping down the dose of fluticasone in combination with salmeterol and tiotropium in severe persistent asthma. *Respir Med* 101(6): 1218–28, 2007.
99. Kesten S, Jara M, Wentworth C, Lanes S. Pooled clinical trial analysis of tiotropium safety. *Chest* 130(6): 1695–703, 2006.
100. Jara M, Lanes SF, Wentworth C, May C, Kesten S. Comparative safety of long-acting inhaled bronchodilators: A cohort study Using the UK THIN Primary Care Database. *Drug Saf* 30(12): 1151–60, 2007.
101. Freeman D, Lee A, Price D. Efficacy and safety of tiotropium in COPD patients in primary care – the SPiRiva Usual CarE (SPRUCE) study. *Respir Res* 8: 45, 2007.
102. Morganroth J, Golisch W, Kesten S. Electrocardiographic monitoring in COPD patients receiving tiotropium. *COPD* 1(2): 181–90, 2004.
103. Covelli H, Bhattacharya S, Cassino C, Conoscenti C, Kesten S. Absence of electrocardiographic findings and improved function with once-daily tiotropium in patients with chronic obstructive pulmonary disease. *Pharmacotherapy* 25(12): 1708–18, 2005.
104. Bailey WC, Tashkin DP. Pharmacologic therapy: Novel approaches for chronic obstructive pulmonary disease. *Proc Am Thorac Soc* 4(7): 543–48, 2007.
105. Hanania NA, Donohue JF. Pharmacologic interventions in chronic obstructive pulmonary disease: Bronchodilators. *Proc Am Thorac Soc* 4(7): 526–34, 2007.
106. Hansel TT, Neighbour H, Erin EM *et al.* Glycopyrrolate causes prolonged bronchoprotection and bronchodilatation in patients with asthma. *Chest* 128(4): 1974–79, 2005.
107. Villetti G, Bergamaschi M, Bassani F *et al.* Pharmacological assessment of the duration of action of glycopyrrolate vs tiotropium and ipratropium in guinea-pig and human airways. *Br J Pharmacol* 148(3): 291–98, 2006.
108. Ji F, Wu W, Dai X *et al.* Synthesis and pharmacological effects of new, *N*-substituted soft anticholinergics based on glycopyrrolate. *J Pharm Pharmacol* 57: 1427–35, 2005.
109. Pahl A, Bauhofer A, Petzold U *et al.* Synergistic effects of the anticholinergic R,R-glycopyrrolate with anti-inflammatory drugs. *Biochem Pharmacol*, 2006.
110. Lu S, Parekh DD, Kuznetsova O, Green SA, Tozzi CA, Reiss TF. An oral selective M3 cholinergic receptor antagonist in COPD. *Eur Respir J* 28(4): 772–80, 2006.

CHAPTER

50

Theophylline

Peter J. Barnes

National Heart and Lung Institute (NHLI), Clinical Studies Unit, Imperial College, London, UK

INTRODUCTION

Theophylline remains one of the most widely prescribed drugs for the treatment of asthma and chronic obstructive pulmonary disease (COPD) worldwide, since it is inexpensive and widely available. In many industrialized countries, however, theophylline has become a third-line treatment that is only used in poorly controlled patients. This has been reinforced by various national and international guidelines on asthma therapy. Some have even questioned whether theophylline is indicated in any patients with asthma [1], although others have emphasized the special beneficial effects of theophylline which still give it an important place in management of asthma [2]. The frequency of side-effects at the previously recommended doses and the relatively low efficacy of theophylline have recently led to reduced usage, since inhaled β_2-agonists are far more effective as bronchodilators and inhaled corticosteroids have a greater anti-inflammatory effect. Despite the fact that theophylline has been used in asthma therapy for over 70 years, there is still considerable uncertainty about its molecular mode of action in asthma and its logical place in therapy. Recently, novel mechanisms of action that may account for the effectiveness of theophylline in severe asthma have been elucidated [3]. Because of problems with side effects, there have been attempts to improve on theophylline, and recently there has been increasing interest in selective phosphodiesterase (PDE) inhibitors, which have the possibility of improving the beneficial and reducing the adverse effects of theophylline.

CHEMISTRY

Theophylline is a methylxanthine similar in structure to the common dietary xanthines caffeine and theobromine. Several substituted derivatives have been synthesized, but none has any advantage over theophylline [4]. The 3-propyl derivative, enprofylline, is more potent as a bronchodilator and may have fewer toxic effects; however, its clinical development was halted because of hepatic toxicity problems [5]. Many salts of theophylline have also been marketed, the most common being aminophylline, the ethylene diamine salt used to increase solubility at neutral pH, so that intravenous administration is possible. Other salts, such as choline theophyllinate, do not have any advantage and others, such as acepifylline, are virtually inactive [4].

MOLECULAR MECHANISMS OF ACTION

Although theophylline has been in clinical use for more than 70 years, its mechanism of action at a molecular level and its site of action remain uncertain, although there have been important recent advances. Several molecular mechanisms of action have been proposed, many of which appear to occur only at higher concentrations of theophylline than are effective clinically (Table 50.1).

TABLE 50.1 Proposed mechanisms of action of theophylline.

PDE inhibition (non-selective)
Adenosine receptor antagonism (A_1-, A_{2A}-, A_{2B}-receptors)
Inhibition of nuclear factor-κB (↓ nuclear translocation)
Inhibition of phosphoinositide 3 kinase-δ
↑ IL-10 secretion
↑ Apoptosis of inflammatory cells
↓ Poly(ADP-ribose)polymerase-1 (inhibits cell death)
↑ Histone deacetylase activity (↑ efficacy of corticosteroids)

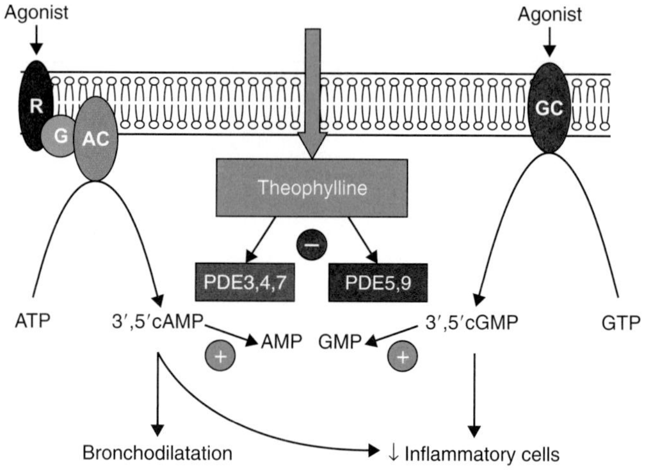

FIG. 50.1 Effect of PDE inhibitors in the breakdown of cyclic nucleotides in airway smooth muscle and inflammatory cells. AC: adenylyl cyclase; ATP: adenosine triphosphate; cAMP: cyclic adenosine monophosphate; cGMP: cyclic guanosine monophosphate; GC: guanylyl cyclase; Gs: stimulatory G-protein; GTP: guanosine triphosphate; PDE: phosphodiesterase; R: receptor.

PDE inhibition

Theophylline is a weak and nonselective inhibitor of PDEs, which break down cyclic nucleotides in the cell, thereby leading to an increase in intracellular cyclic 3′5′ adenosine monophosphate (AMP) and cyclic 3′,5′ guanosine monophosphate (GMP) concentrations (Fig. 50.1); however, the degree of inhibition is small at concentrations of theophylline which are therapeutically relevant. Thus, total PDE activity in human lung extracts is inhibited by only 5–10% by therapeutic concentrations of theophylline [6]. There is convincing *in vitro* evidence that theophylline relaxes airway smooth muscle by inhibition of PDE activity, but relatively high concentrations are needed for maximal relaxation [7]. Similarly, the inhibitory effect of theophylline on mediator release from alveolar macrophages appears to be mediated by inhibition of PDE activity in these cells [8]. There is no evidence that airway smooth muscle or inflammatory cells concentrate theophylline to achieve higher intracellular than circulating concentrations. Inhibition of PDE should lead to synergistic interaction with β-agonists, but this has not been convincingly demonstrated *in vivo* or in clinical studies; however, this might be because the relaxation of airway smooth muscle by β-agonists may involve direct coupling of β-receptors via a stimulatory G-protein to the opening of potassium channels, without the involvement of cyclic AMP [9].

At least 11 isoenzyme families of PDE have now been recognized, and some (PDE3, PDE4, PDE5) are important in smooth muscle relaxation [10]; however, there is no convincing evidence that theophylline has any greater inhibitory effect on the PDE isoenzymes involved in smooth muscle relaxation. It is possible that PDE isoenzymes may have an increased expression in asthmatic airways, as a result of either the chronic inflammatory process or the therapy. Elevation of cyclic AMP by β-agonists may result in increased PDE activity, thus limiting the effect of β-agonists. Indeed, alveolar macrophages from asthmatic patients appear to have increased PDE activity [11]. This would mean that theophylline might have a greater inhibitory effect on PDE in asthmatic airways than in normal airways. Support for this is provided by the lack of bronchodilator effect of theophylline in normal subjects, compared to a bronchodilator effect in asthmatic patients [12].

Inhibition of PDEs is likely to account for some of the moist frequent side effects of theophylline, including nausea and vomiting (PDE4), palpitations and cardiac arrhythmias (PDE3), and headaches (PDE4).

Adenosine receptor antagonism

Theophylline is a potent inhibitor of adenosine receptors at therapeutic concentrations. Both A_1- and A_2-receptors are inhibited, but theophylline is less effective against A_3-receptors, suggesting that this could be the basis for its bronchodilator effects [13]. Although adenosine has little effect on normal human airway smooth muscle *in vitro*, it constricts airways of asthmatic patients via the release of histamine and leukotrienes, suggesting that adenosine releases mediators from mast cells [14]. The receptor involved appears to be an A_3-receptor in rat mast cells [15, 16], but in humans there is evidence for the involvement of an A_{2B}-receptor [17, 18]. Adenosine causes bronchoconstriction in asthmatic subjects when given by inhalation [19]. The mechanism of bronchoconstriction is indirect and involves release of histamine from airway mast cells [14, 20]. The bronchoconstrictor effect of adenosine is prevented by therapeutic concentrations of theophylline [19]; however, this only confirms that theophylline is capable of antagonizing the effects of adenosine at therapeutic concentrations and does not necessarily indicate that this is important for its anti-asthma effect. Adenosine antagonism is likely to account for some of the side effects of theophylline, such as central nervous system stimulation, cardiac arrhythmias (both via blockade of A_1-receptors), gastric hypersecretion, gastroesophageal reflux, and diuresis. A novel AMP receptor, $P2Y_{15}$, has been identified which is more potently inhibited by theophylline [21], although the function of these receptors has been questioned.

IL-10 release

IL-10 has a broad spectrum of anti-inflammatory effects, and there is evidence that its secretion is reduced in COPD [22]. IL-10 release is increased by theophylline, and this effect may be mediated via PDE inhibition [23], although this has not been seen at the low doses that are effective in asthma [24].

Effect on gene transcription

Theophylline prevents the translocation of the proinflammatory transcription factor, nuclear factor-κB (NF-κB), into the nucleus, thus potentially reducing the expression of inflammatory genes in asthma and COPD [25]. Inhibition of NF-κB appears to be due to a protective effect against the degradation of the inhibitory protein I-κBα, so that nuclear translocation of activated NF-κB is prevented [26]. However, these effects are seen at high concentrations and may be mediated by inhibition of PDE.

Effect on kinases

Theophylline directly inhibits phosphoinositie-3-kinases, with greatest potency for the PI3K (p110)-δ subtype (IC_{50} 75 μM) [27], a subtype of the enzyme that has been implicated in responses to oxidative stress [28]. However, it is a relatively weak effect against the PI3K-γ subtype (IC_{50} 800 μM), which is involved in chemotactic responses of neutrophils and monocytes. The inhibitory effect of theophylline on PI3K-δ may account for the ability of theophylline to reverse corticosteroid resistance, which may be of critical importance for its clinical effects in severe asthma and COPD.

Effects on apoptosis

Prolonged survival of granulocytes due to a reduction in apoptosis may be important in perpetuating chronic inflammation in COPD. Theophylline promotes inhibits apoptosis in neutrophils *in vitro* [29]. This is associated with a reduction in the anti-apoptotic protein Bcl-2 [30]. This effect is not mediated via PDE inhibition but in neutrophils may be mediated by antagonism of adenosine A_{2A}-receptors [31]. Theophylline also induces apoptosis of T-lymphocytes thus reducing their survival, and this effect appears to be mediated via PDE inhibition [32]. Theophylline also inhibits the enzyme poly(ADP-ribose)polymerase-1 (PARP-1), which is activated by oxidative stress and leads to a reduction in nicotine adenine diamine levels resulting in an energy crisis that leads to cell death [33].

Other effects

Several other effects of theophylline have been described, including an increase in circulating catecholamines, inhibition of calcium influx into inflammatory cells, inhibition of prostaglandin effects, and antagonism of tumor necrosis factor-α. These effects are generally seen only at high concentrations of theophylline which are above the therapeutic range and are therefore unlikely to contribute to the anti-inflammatory actions of theophylline. Despite intense efforts, it has been difficult to find any molecular mechanisms that can account for the anti-inflammatory effects of theophylline in asthma and COPD. Although several potential mechanisms have been demonstrated *in vitro*, there is little evidence that these occur at plasma concentrations of 5–10 mg/l where clinical benefit and anti-inflammatory effects are seen. A novel mechanism of action involving activation of histone deacetylases (HDAC) has been described which, in contrast to the proposed molecular mechanisms discussed above, is seen at therapeutically relevant concentrations [3].

HDAC activation

Expression of inflammatory genes is regulated by the balance between histone acetylation and deacetylation [34]. In asthma multiple inflammatory genes are activated through proinflammatory transcription factors, such as NF-κB, leading to histone acetylation and increased transcription. This process is reversed by the recruitment of HDAC to the activated inflammatory gene promoter site within the nucleus. Corticosteroids suppress inflammation by recruiting HDAC2 to activated inflammatory genes, thus switching off their expression [35] (see Chapter 53). This molecular mechanism is defective in COPD patients as HDAC2 activity and expression are markedly reduced, thus accounting for the steroid resistance of COPD [36]. There is also a defect in HDAC2 function in patients with severe asthma and in asthmatic patients who smoke [37, 38]. We have shown that theophylline is an activator of HDACs and enhances the anti-inflammatory effect of corticosteroids, as well as reversing steroid resistance in cells from COPD patients [39, 40] (Fig. 50.2). This action of theophylline is seen at low plasma concentrations (optimally 5 mg/l) and is completely independent of PDE inhibition and adenosine antagonism. The effect of theophylline is completely blocked by an HDAC inhibitor called trichostatin A and by knocking out HDAC2 using interference RNA [41]. The reason why theophylline selective activates HDAC activity is not yet known but appears to be indirect through the activation of kinase and phosphatase pathways in the cell that are activated by oxidative stress and involved in the regulation of HDAC2 activity. This effect of theophylline is seen particularly in the presence of oxidative and nitrative stress, and this accounts for why theophylline is effective particularly in severe asthma, where oxidative and nitrative stress are greatest. Increased reactive oxygen species and nitric oxide from increased expression of inducible nitric oxide synthase result in the formation of peroxynitrite radicals. Peroxynitrite is unstable and nitrates tyrosine residues in proteins, which may result in altered protein function. 3-Nitrotyrosine adducts are increased in asthmatic airways [42]. Peroxynitrite is also increased in COPD lungs [43] and is associated with tyrosine nitration and inactivation of HDAC2 [44]. Theophylline also appears to reduce the formation of peroxynitrite, and this provides

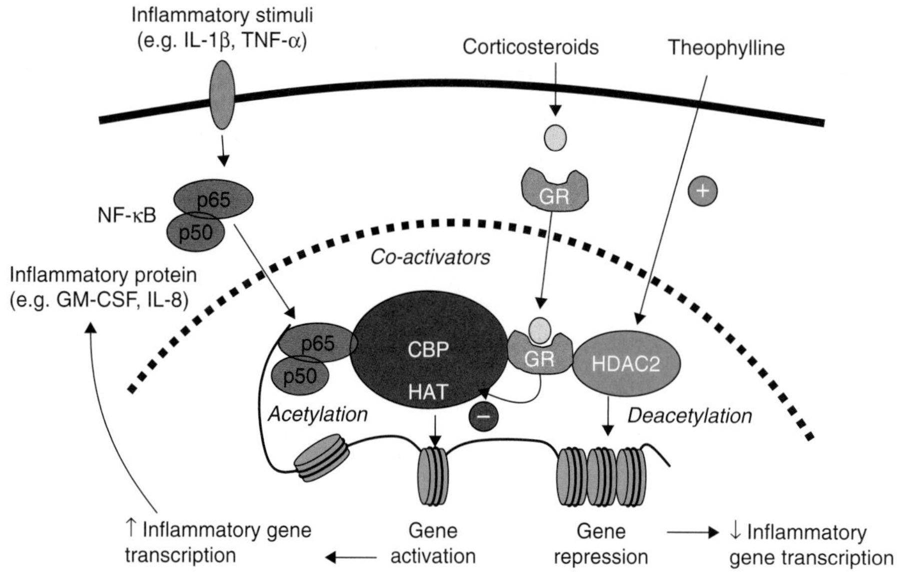

FIG. 50.2 Theophylline directly activates HDACs which deacetylate core histones that have been acetylated by the histone acetyltransferase (HAT) activity of co-activators, such as CREB binding protein (CBP). This results in suppression of inflammatory genes and proteins, such as granulocyte-macrophage colony stimulating factor (GM-CSF) and IL-8 that have been switched on by proinflammatory transcription factors, such as nuclear factor-κB (NF-κB). Corticosteroids also activate HDACs, but through a different mechanism resulting in the recruitment of HDACs to the activated transcriptional complex via activation of the glucocorticoid receptors (GR) which function as a molecular bridge. This predicts that theophylline and corticosteroids may have a synergistic effect in repressing inflammatory gene expression.

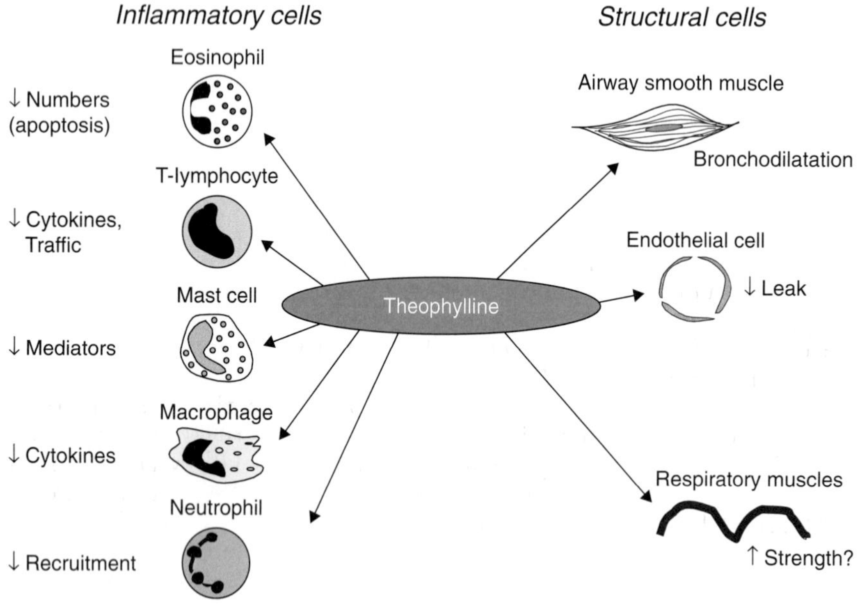

FIG. 50.3 Multiple cellular effects of theophylline.

a further mechanism for increasing HDAC2 function in asthma [45].

CELLULAR EFFECTS

Theophylline has several cellular effects that may contribute to its clinical efficacy in the treatment of asthma (Fig. 50.3).

Airway smooth muscle effects

The primary effect of theophylline is assumed to be the relaxation of airway smooth muscle; as *in vitro* studies have shown, it is equally effective in large and small airways [46]. In airways obtained at lung surgery ~25% of preparations fail to relax with a β-agonist but all relax with theophylline [47]. The molecular mechanism of bronchodilatation is almost certainly related to PDE inhibition, resulting in an increase in cyclic AMP [7]. The bronchodilator effect of theophylline is reduced in human airways by the toxin charybdotoxin, which inhibits large conductance Ca^{2+}-activated K^+ channels (maxi-K channels), suggesting that theophylline opens maxi-K channels via an increase in cyclic AMP [48]. Theophylline acts as a functional antagonist and inhibits the contractile response of multiple spasmogens. In airways obtained at post-mortem from patients who have died from asthma, the relaxant response to β-agonists is reduced, whereas the bronchodilator response to theophylline is no different from that seen in normal airways [49]. There is evidence that β-adrenoceptors in airway smooth muscle of patients with fatal asthma become uncoupled [50], and theophylline may therefore have a theoretical advantage over β-agonists in severe asthma exacerbations. However, theophylline is a weak bronchodilator at therapeutically relevant concentrations, suggesting that some other target cell may be more relevant for its anti-asthma effect. In human airways the EC_{50} for airway smooth muscle relaxation by theophylline is $\sim 1.5 \times 10^{-4}$M, which is equivalent to 67 mg/l assuming 60% protein binding [47], that is above the therapeutic range. As discussed above, it is important to consider the possibility that PDE activity may be increased in asthmatic airways, so that theophylline may have a greater than expected effect.

In vivo intravenous aminophylline has an acute bronchodilator effect in asthmatic patients, which is most likely to be due to a relaxant effect on airway smooth muscle [51]. The bronchodilator effect of theophylline in chronic asthma is small in comparison with β_2-agonists, however. Several studies have demonstrated a small protective effect of theophylline on histamine, methacholine, or exercise challenge [52–55]. This protective effect does not correlate well with any bronchodilator effect, and it is interesting that in some studies the protective effect of theophylline is observed at plasma concentrations of $<$10 mg/l. These clinical studies suggest that theophylline may have anti-asthma effects that are unrelated to any bronchodilator action.

Anti-inflammatory effects

There is increasing evidence that theophylline has anti-inflammatory effects in asthma and that these are seen at lower plasma concentrations than are needed for bronchodilatation. Theophylline inhibits mediator release from chopped human lung [56], although high concentrations are necessary and it is likely that this effect involves an increase in cyclic AMP concentration due to PDE inhibition. Theophylline also has an inhibitory effect on superoxide anion release from human neutrophils [57] and inhibits the feedback stimulatory effect of adenosine on neutrophils *in vivo* [58]. At therapeutic concentrations *in vitro*, theophylline may *increase* superoxide release via an inhibitory effect on adenosine receptors, since endogenous adenosine may normally exert an inhibitory action on these cells [59]. Similar results are also seen in guinea pig and human eosinophils [60]. At therapeutic concentrations there is an increased release of superoxide anions from eosinophils, which appears to be mediated via inhibition of adenosine A_2-receptors and is mimicked by the adenosine antagonist 8-phenyltheophylline. Inhibition of eosinophil superoxide generation occurs only at high concentrations of theophylline ($>10^{-4}$M) which are likely to inhibit PDE. Similar results have also been obtained in human alveolar macrophages [8]. Macrophages in bronchoalveolar lavage fluid from patients taking theophylline have been found to have a reduced oxidative burst response [61], but there is no reduction in the release of the proinflammatory cytokines TNF-α or GM-CSF [62]. Theophylline inhibits neutrophil chemotaxis via inhibition of adenosine A_{2A}-receptors, and this may be relevant in severe asthma [63].

In vivo theophylline inhibits mediator-induced airway microvascular leakage in rodents when given in high doses [64], although this is not seen at therapeutically relevant concentrations [65]. Theophylline has an inhibitory effect on plasma exudation in nasal secretions induced by allergen in patients with allergic rhinitis, although this could be secondary to inhibition of mediator release [66].

In allergen challenge studies, chronic oral treatment with theophylline reduces the late response to allergen [67]. This has been interpreted as an effect on the chronic inflammatory response and is supported by a reduced infiltration of eosinophils into the airways after allergen challenge following low doses of theophylline [68]. In patients with nocturnal asthma low-dose theophylline inhibits the influx of neutrophils and, to a lesser extent, eosinophils seen in the early morning [69]. Chronic treatment with low-dose theophylline reduces the numbers of eosinophils in bronchial biopsies, bronchoalveolar lavage and induced sputum of patients with mild asthma [62]. However, these effects are less than those expected with low doses of inhaled corticosteroids, and there is no reduction of exhaled nitric oxide, indicating a lesser effect on suppression of inflammation than corticosteroids.

In patients with COPD, theophylline reduces the proportion of neutrophils in induced sputum and reduces the concentration of IL-8, suggesting that it may have an anti-inflammatory effect unlike corticosteroids [70, 71]. Since patients with severe asthma may have increased neutrophils in the airways [72, 73], this may provide a mechanism whereby theophylline is effective as an add-on therapy to high does of inhaled corticosteroids in these patients.

Immunomodulatory effects

T-lymphocytes are now believed to play a central role in coordinating the chronic inflammatory response in asthma. Theophylline has been shown to have several actions on T-lymphocyte function, suggesting that it might have an immunomodulatory effect in asthma. Theophylline has a stimulatory effect on suppressor ($CD8^+$) T-lymphocytes which may be relevant to the control of chronic airway inflammation [74, 75], and has an inhibitory effect on graft rejection [76]. *In vitro* theophylline inhibits IL-2 synthesis in human T-lymphocytes, an effect that is secondary to a rise in intracellular cyclic AMP concentration [77]. At high concentrations theophylline inhibits proliferation in $CD4^+$ and $CD8^+$ cells, an effect that is mediated via inhibition of PDE4 [78]. Theophylline also inhibits the chemotactic response of T-lymphocytes, an effect that is also mediated through PDE inhibition [79]. In allergen-induced airway inflammation in guinea pigs, theophylline has a significant inhibitory effect on eosinophil infiltration [80], suggesting that it may inhibit the T-cell-derived cytokines responsible for this eosinophilic response. Theophylline has been reported to decrease circulating concentrations of IL-4 and IL-5 in asthmatic patients [81]. In asthmatic patients low-dose theophylline treatment results in an increase in activated circulating $CD4^+$ and $CD8^+$ T-cells, but a decrease in these cells in the airways, suggesting that it may reduce the trafficking of activated T-cells into the airways [82]. This is supported by studies in allergen challenge, where low-dose theophylline decreases the number of activated $CD4^+$ and $CD8^+$ T-cells in bronchoalveolar lavage fluid after allergen challenge, and this is mirrored by an increase in these cells in peripheral blood [83]. These effects are seen even in patients treated with high does of inhaled corticosteroids, indicating that the molecular effects of theophylline are likely to be different from those of corticosteroids. Theophylline induces apoptosis of T-lymphocytes, thus reducing their survival [32]. This effect may be mediated via PDE4 inhibition so may not be relevant to clinical doses of theophylline. The therapeutic range of theophylline was based on measurement of immediate bronchodilatation in response to the acute administration of theophylline [51]. However, it is possible that the nonbronchodilator effects of theophylline, which may reflect some anti-inflammatory or immunomodulatory effect, may be exerted at lower plasma concentrations and that different molecular mechanisms may be involved [84].

Extrapulmonary effects

It has been suggested that theophylline may exert its effects in asthma via some action outside the airways. It may be relevant that theophylline is ineffective when given by inhalation until therapeutic plasma concentrations are achieved [85]. This may indicate that theophylline has effects on cells other than those in the airway.

An effect of theophylline which remains controversial is its action on respiratory muscles. Aminophylline increases diaphragmatic contractility and reverses diaphragm fatigue [86]. This effect has not been observed by all investigators, and there are now doubts about the relevance of these observations to the clinical benefit provided by theophylline [87].

PHARMACOKINETICS

There is a close relationship between the acute improvement in airway function and the serum theophylline concentration. Below 10 mg/l therapeutic effects (at least in terms of rapid improvement in airway function) are small, and above 25 mg/l additional benefits are outweighed by side effects, so that the therapeutic range was usually taken as 10–20 mg/l (55–110 μM) [4]. It is now apparent that nonbronchodilator effects of theophylline may be seen at plasma concentrations of <10 mg/l and that clinical benefit may be derived from these lower concentrations of theophylline. This suggests that it may be necessary to redefine the therapeutic range of theophylline based on anti-asthma effect, rather than the acute bronchodilator response that requires a higher plasma concentration. The dose of theophylline required to give therapeutic concentrations varies among patients, largely because of differences in clearance. In addition, there may be differences in bronchodilator response to theophylline and, with acute bronchoconstriction, higher concentrations may be required to produce bronchodilatation [88].

Theophylline is rapidly and completely absorbed, but there are large inter-individual variations in clearance, due to differences in hepatic metabolism (Table 50.2). Theophylline is metabolized in the liver by the cytochrome P450 microsomal enzyme system, and a large number of factors may influence hepatic metabolism. Theophylline is predominantly metabolized by the CYP1A2 enzyme, while at higher plasma concentrations CYP2E1 is also involved [89].

TABLE 50.2 Factors affecting clearance of theophylline.

Increased clearance
Enzyme induction (rifampicin, phenobarbitone, ethanol)
Smoking (tobacco, marijuana)
High protein, low carbohydrate diet
Barbecued meat
Childhood
Decreased clearance
Enzyme inhibition (cimetidine, erythromycin, ciprofloxacin, allopurinol, zileuton)
Congestive heart failure
Liver disease
Pneumonia
Viral infection
Vaccination (immunization)
High carbohydrate diet
Old age

Increased clearance

Increased clearance is seen in children (1–16 years), and in cigarette and marijuana smokers. Concurrent administration of phenytoin and phenobarbitone increases activity of P450, resulting in increased metabolic breakdown, so that higher doses may be required.

Reduced clearance

Reduced clearance is found in liver disease, pneumonia, and heart failure and doses need to be reduced to half and plasma levels monitored carefully [90]. Decreased clearance is also seen with certain drugs, including erythromycin, certain quinolone antibiotics (ciprofloxacin, but not ofloxacin), allopurinol, cimetidine (but not ranitidine), serotonin uptake inhibitors (fluvoxamine), and the 5-lipoxygenase inhibitor zileuton, which interfere with CYP1A2 function. Thus, if a patient on maintenance theophylline requires a course of erythromycin, the dose of theophylline should be halved. Viral infections and vaccinations (immunizations) may also reduce clearance, and this may be particularly important in children. Because of these variations in clearance, individualization of theophylline dosage is required and plasma concentrations should be measured 4h after the last dose with slow-release preparations, when steady state has usually been achieved. There is no significant circadian variation in theophylline metabolism [91], although there may be delayed absorption at night, which may relate to the supine posture [92].

ROUTES OF ADMINISTRATION

Intravenous

Intravenous aminophylline has been used for many years in the treatment of acute severe asthma. The recommended dose is now 6mg/kg given intravenously over 20–30min, followed by a maintenance dose of 0.5mg/kg/h. If the patient is already taking theophylline, or there are any factors that decrease clearance, these doses should be halved and the plasma level checked more frequently.

Oral

Plain theophylline tablets or elixir, which are rapidly absorbed, give wide fluctuations in plasma levels and are not recommended. Several effective sustained-release preparations now available are absorbed at a constant rate and provide steady plasma concentrations over a 12–24h period [93]. Although there are differences between preparations, these are relatively minor and of no clinical significance. Both slow-release aminophylline and theophylline are available and are equally effective (although the ethylene diamine component of aminophylline has very occasionally been implicated in allergic reactions). For continuous treatment, twice daily therapy (~8mg/kg b.i.d.) is needed, although some preparations are designed for once daily administration. For nocturnal asthma, a single dose of slow-release theophylline at night is effective [94, 95], and more effective than an oral slow-release β-agonist preparation. Once optimal doses have been determined, plasma concentrations usually remain stable, providing no factors which alter clearance are introduced.

Other routes

Aminophylline may be given as a suppository, but rectal absorption is unreliable and proctitis may occur, so this route should be avoided. Inhalation of theophylline is irritating and ineffective [85]. Intramuscular injections of theophylline are very painful and should never be given.

CLINICAL USE

Acute exacerbations

Intravenous aminophylline has been used in the management of acute severe asthma for over 50 years, but this use has been questioned in view of the risk of adverse effects compared with nebulized β_2-agonists. In patients with acute asthma, intravenous aminophylline is less effective than nebulized β_2-agonists [96] and should therefore be reserved for those who fail to respond to β-agonists. In a meta-analysis of 27 studies which looked at addition of intravenous aminophylline to nebulized β_2-agonists, there is no evidence for significant benefit in adults [97] or children [98]. This indicates that aminophylline should not be added routinely to nebulized β-agonists. Indeed, addition of aminophylline only increases adverse effects. Several deaths have been reported after intravenous aminophylline. In one study of 43 asthma deaths in southern England, there was a significantly greater frequency of toxic theophylline concentrations (21%) compared with matched controls (7%) [99]. These concerns have lead to the view that intravenous aminophylline should be reserved for the few patients with acute severe asthma who fail to show a satisfactory response to nebulized β_2-agonists. When intravenous aminophylline is used, it should be given as a slow intravenous infusion with careful monitoring of vital signs, and plasma theophylline concentrations should be measured prior to and after infusion.

Aminophylline similarly has no place in the routine management of COPD exacerbations [100, 101].

Chronic asthma

In most guidelines for asthma management theophylline is recommended as an additional bronchodilator if asthma remains difficult to control after high doses of inhaled corticosteroids. The introduction of long-acting inhaled β_2-agonists has further threatened the position of theophylline, since the side effects of these agents may be less frequent than those associated with theophylline, and long-acting inhaled β_2-agonists are more effective controllers than theophylline [102]. Whether theophylline has some additional benefit over its bronchodilator action is now an important consideration. In chronic asthma oral theophylline appears to be as effective as cromolyn sodium in controlling young allergic asthmatics [103] and provides additional control of asthma symptoms even in patients talking regular inhaled steroids [104]. In one study a group of adolescent patients with severe asthma who were controlled with oral and inhaled steroids, nebulized β_2-agonists, inhaled anticholinergics, and cromolyn, in addition to regular oral theophylline, withdrawal of the oral theophylline resulted in a marked deterioration of asthma control which could not be controlled by further increase in steroids and only responded to reintroduction of theophylline [105]. This suggests that there may be a group of severe asthmatic patients who particularly benefit from theophylline. In a controlled trial of theophylline withdrawal in patients with severe asthma controlled only on high doses of inhaled corticosteroids, there was a significant deterioration in symptoms and lung function when placebo was substituted for the relatively low maintenance dose of theophylline [82]. There is also evidence that addition of theophylline improves asthma control to a greater extent than β_2-agonists in patients with severe asthma treated with high-dose inhaled steroids [106].

This suggests that theophylline may have a useful place in the optimal management of moderate-to-severe asthma and appears to provide additional control above that provided by high-dose inhaled steroids [107, 108].

Theophylline may be a useful treatment for nocturnal asthma, and a single dose of a slow-release theophylline preparation given at night may provide effective control of nocturnal asthma symptoms [94, 109]. There is evidence that slow-release theophylline preparations are more effective than slow-release oral β-agonists and inhaled β-agonists in controlling nocturnal asthma [95, 110, 111]. Theophylline has equal efficacy to salmeterol in controlling nocturnal asthma, but the quality of sleep is better with salmeterol compared to theophylline [112]. The mechanism of action of theophylline in nocturnal asthma may involve more than long-lasting bronchodilatation and could involve inhibition of some components of the inflammatory response, which may increase at night [69].

Add-on therapy

Several studies have demonstrated that adding low-dose theophylline to inhaled corticosteroids in patients who are not controlled gives better asthma control than doubling the dose of inhaled corticosteroids. This has been demonstrated in patients with moderate-to-severe and mild asthma [113–115]. Interestingly, there is a greater degree of improvement in forced vital capacity than in FEV_1, possibly indicating an effect on peripheral airways. Since the improvement in lung function was relatively slow, this suggests that the effect of the added theophylline may be having an anti-inflammatory rather than a bronchodilator effect, particularly as the plasma concentration of theophylline in these studies was <10 mg/ml. These studies suggest that low-dose theophylline may be preferable to increase the dose of inhaled steroids when asthma is not controlled on moderate doses of inhaled steroids; such a therapeutic approach would be much less expensive than adding long-acting inhaled β_2-agonists. However, theophylline is a less effective option than adding a long-acting inhaled β_2-agonist [102].

Chronic obstructive pulmonary disease

Theophylline may also benefit patients with COPD and increases exercise tolerance [116, 117]. Theophylline reduces air trapping, suggesting an effect on peripheral airways, and this may explain why some patients with COPD may obtain considerable symptomatic improvement without any increase in spirometric values [118]. Although the effect of theophylline on respiratory muscle weakness was believed to be important in contributing to symptomatic improvement in patients with COPD [86], this seems unlikely as several investigators have failed to confirm any effect on respiratory muscle function at therapeutic concentrations of theophylline [87]. The demonstration that low doses of theophylline reduce neutrophils in induced sputum of patients with COPD suggests that theophylline may have some anti-inflammatory effect [71, 119]. In COPD macrophages *in vitro* theophylline restores HDAC activity to normal and thus reverses corticosteroid resistance [40]. It also reduces nitrative stress in macrophages from patients with COPD, whereas high doses of an inhaled corticosteroid are without effect [45]. This suggests that corticosteroids may be useful in reversing corticosteroids resistance in patients with COPD, but long-term clinical trials are now needed to confirm this [120, 121].

Interaction with β_2-agonists

If theophylline exerts its effects by PDE inhibition, a synergistic interaction with β-agonists would be expected. Many studies have investigated this possibility, but while there is good evidence that theophylline and β-agonists have additive effects, true synergy is not seen [122]. This can now be understood in terms of the molecular mechanisms of action of β-agonists and theophylline. β-Agonists may cause relaxation of airway smooth muscle via several mechanisms. Classically, they increase intracellular cyclic AMP concentrations, which were believed to be an essential event in the relaxation response. It has recently become clear that β-agonists may cause bronchodilatation, at least in part, by opening maxi-K channels in airway smooth muscle cells [9, 48]. Maxi-K channels are opened by low concentrations of β_2-agonists which are likely to be therapeutically relevant. There is now evidence that β-receptors may be coupled directly to maxi-K channels via the α-subunit of G_s [123], and therefore may induce relaxation without any increase in cyclic AMP, thus accounting for a lack of synergy.

Repeated administration of β_2-agonists may result in tolerance; however, this may be explained by down-regulation of β_2-receptors, an additional mechanism may involve upregulation of PDE enzymes (especially PDE4D) which then break down cyclic AMP more readily [124]. Theophylline may therefore theoretically prevent the development of tolerance. However, in a clinical study theophylline failed to prevent the development of tolerance to the bronchoprotective effect of salmeterol in asthmatic patients [125].

SIDE EFFECTS

There is no doubt that theophylline provides clinical benefit in obstructive airway disease, but the main limitation to its use is the frequency of adverse effects [126]. Unwanted effects of theophylline are usually related to plasma concentration and tend to occur when plasma levels exceed 20 mg/l; however, some patients develop side-effects even at low plasma concentrations. To some extent side effects may be reduced by gradually increasing the dose until therapeutic concentrations are achieved.

The commonest side effects are headache, nausea and vomiting, abdominal discomfort, and restlessness. There may also be increased acid secretion, gastroesophageal reflux, and diuresis. There has recently been concern that theophylline, even at therapeutic concentrations, may lead to behavioral disturbance and learning difficulties in school children [127], although it is difficult to design adequate controls for such studies. At high concentrations convulsions and cardiac arrhythmias may occur and, as stated previously, there is concern that intravenous aminophylline administered in the emergency room may be a contributory factor to the deaths of some patients with severe asthma [99].

Some of the side-effects of theophylline (central stimulation, gastric secretion, diuresis, and arrhythmias) may be due to adenosine receptor antagonism (A_{1A} receptors), and these may therefore be avoided by PDE inhibitors, as discussed above. The commonest side effects of theophylline are nausea and headaches, which may be due to inhibition of certain PDEs (e.g. PDE4 in the vomiting center) and cardiac arrhythmias due to inhibition of PDE3 [128].

FUTURE OF THEOPHYLLINE

Although theophylline has recently been used much less in developed countries, there are reasons for thinking that it may come back in fashion for the treatment of chronic asthma and COPD, with the recognition that it may have an anti-inflammatory and immunomodulatory effect when given in low doses (plasma concentration 5–10 mg/l) [3]. At these low doses the drug is easier to use, side effects are uncommon, and the problems of drug interaction are less of a problem, thus making the clinical use of theophylline less complicated. Theophylline appears to have an effect that is different from those of corticosteroids, and it may therefore be a useful drug to combine with low-dose inhaled steroids. The molecular mechanism of anti-inflammatory effects of theophylline is now becoming clearer, and it seems likely that there is a synergistic interaction with the anti-inflammatory mechanism of corticosteroids through restoration of HDAC activity. This interaction may underlie the beneficial effects of theophylline when added to inhaled corticosteroids. This may be particularly appropriate in patients with more severe asthma in whom corticosteroids are less effective as there may be a reduction in HDAC activity in these patients [37] as well as in smoking asthmatics patients and patients with COPD. As slow-release theophylline preparations are cheaper than long-acting inhaled β_2-agonists and leukotriene modifiers, this may justify the choice of low-dose theophylline as the add-on therapy for asthma control. In addition, compliance with oral therapy is likely to be greater than with inhaled therapies [129]. This suggests that low-dose theophylline may find an important place in modern asthma management in patients with moderate asthma as well as with severe asthma.

References

1. Lam A, Newhouse MT. Management of asthma and chronic airflow limitation. Are methylxanthines obsolete? *Chest* 98: 44–52, 1990.
2. Weinberger M, Hendeles L. Theophylline in asthma. *N Engl J Med* 334: 1380–88, 1996.
3. Barnes PJ. Theophylline: New perspectives on an old drug. *Am J Respir Crit Care Med* 167: 813–18, 2003.
4. Weinburger M. The pharmacology and therapeutic use of theophylline. *J Allergy Clin Immunol* 73: 525–40, 1984.
5. Persson CGA. Development of safer xanthine drugs for the treatment of obstructive airways disease. *J Allergy Clin Immunol* 78: 817–24, 1986.
6. Poolson JB, Kazanowski JJ, Goldman AL, Szentivanyi A. Inhibition of human pulmonary phosphodiesterase activity by therapeutic levels of theophylline. *Clin Exp Pharmacol Physiol* 5: 535–39, 1978.
7. Rabe KF, Magnussen H, Dent G. Theophylline and selective PDE inhibitors as bronchodilators and smooth muscle relaxants. *Eur Respir J* 8: 637–42, 1995.
8. Dent G, Giembycz MA, Rabe KF, Wolf B, Barnes PJ, Magnussen H. Theophylline suppresses human alveolar macrophage respiratory burst through phosphodiesterase inhibition. *Am J Respir Cell Mol Biol* 10: 565–72, 1994.
9. Kume H, Hall IP, Washabau RJ, Takagi K, Kotlikoff MI. Adrenergic agonists regulate K_{Ca} channels in airway smooth muscle by cAMP-dependent and -independent mechanisms. *J Clin Invest* 93: 371–79, 1994.
10. Beavo JA. Cyclic nucleotide phosphodiesterases: Functional implications of multiple isoforms. *Physiol Rev* 75: 725–48, 1995.
11. Bachelet M, Vincent D, Havet N, Marrash-Chahla R, Pradalier A, Dry J *et al.* Reduced responsiveness of adenylate cyclase in alveolar macrophages from patients with asthma. *J Allergy Clin Immunol* 88: 322–28, 1991.
12. Estenne M, Yernault J, De Troyer A. Effects of parenteral aminophylline on lung mechanics in normal humans. *Am Rev Respir Dis* 121: 967–71, 1980.
13. Pauwels RA, Joos GF. Characterization of the adenosine receptors in the airways. *Arch Int Pharmacodyn Ther* 329: 151–56, 1995.
14. Björk T, Gustafsson LE, Dahlén S-E. Isolated bronchi from asthmatics are hyperresponsive to adenosine, which apparently acts indirectly by liberation of leukotrienes and histamine. *Am Rev Respir Dis* 145: 1087–91, 1992.
15. Fozard JR, Pfannkuche HJ, Schuurman HJ. Mast cell degranulation following adenosine A3 receptor activation in rats. *Eur J Pharmacol* 298: 293–97, 1996.
16. Hannon JP, Tigani B, Williams I, Mazzoni L, Fozard JR. Mechanism of airway hyperresponsiveness to adenosine induced by allergen challenge in actively sensitized Brown Norway rats. *Br J Pharmacol* 132: 1509–23, 2001.
17. Feoktistov I, Biaggioni I. Pharmacological characterization of adenosine A_{2B} receptors: Studies in human mast cells co-expressing A_{2A} and A_{2B} adenosine receptor subtypes. *Biochem Pharmacol* 55: 627–33, 1998.
18. Holgate ST. The identification of the adenosine A_{2B} receptor as a novel therapeutic target in asthma. *Br J Pharmacol* 145: 1009–15, 2005.
19. Cushley MJ, Tattersfield AE, Holgate ST. Adenosine-induced bronchoconstriction in asthma: Antagonism by inhaled theophylline. *Am Rev Respir Dis* 129: 380–84, 1984.
20. Cushley MJ, Holgate ST. Adenosine induced bronchoconstriction in asthma: Role of mast cell mediator release. *J Allergy Clin Immunol* 75: 272–78, 1985.
21. Inbe H, Watanabe S, Miyawaki M, Tanabe E, Encinas JA. Identification and characterization of a cell-surface receptor, P2Y15, for AMP and adenosine. *J Biol Chem* 279: 19790–99, 2004.
22. Takanashi S, Hasegawa Y, Kanehira Y, Yamamoto K, Fujimoto K, Satoh K *et al.* Interleukin-10 level in sputum is reduced in bronchial asthma, COPD and in smokers. *Eur Respir J* 14: 309–14, 1999.
23. Mascali JJ, Cvietusa P, Negri J, Borish L. Anti-inflammatory effects of theophylline: Modulation of cytokine production. *Ann Allergy Asthma Immunol* 77: 34–38, 1996.
24. Oliver B, Tomita K, Keller A, Caramori G, Adcock I, Chung KF *et al.* Low-dose theophylline does not exert its anti-inflammatory effects in mild asthma through upregulation of interleukin-10 in alveolar macrophages. *Allergy* 56: 1087–90, 2001.
25. Tomita K, Chikumi H, Tokuyasu H, Yajima H, Hitsuda Y, Matsumoto Y *et al.* Functional assay of NF-kappaB translocation into nuclei by laser scanning cytometry: Inhibitory effect by dexamethasone or theophylline. *Naunyn Schmiedebergs Arch Pharmacol* 359: 249–55, 1999.

26. Ichiyama T, Hasegawa S, Matsubara T, Hayashi T, Furukawa S. Theophylline inhibits NF-kB activation and IkBa degradation in human pulmonary epithelial cells. *Naunyn Schmiedebergs Arch Pharmacol* 364: 558–61, 2001.
27. Foukas LC, Daniele N, Ktori C, Anderson KE, Jensen J, Shepherd PR. Direct effects of caffeine and theophylline on p110 delta and other phosphoinositide 3-kinases. Differential effects on lipid kinase and protein kinase activities. *J Biol Chem* 277: 37124–30, 2002.
28. Yamamori T, Inanami O, Nagahata H, Kuwabara M. Phosphoinositide 3-kinase regulates the phosphorylation of NADPH oxidase component p47(phox) by controlling cPKC/PKCdelta but not Akt. *Biochem Biophys Res Commun* 316: 720–30, 2004.
29. Yasui K, Hu B, Nakazawa T, Agematsu K, Komiyama A. Theophylline accelerates human granulocyte apoptosis not via phosphodiesterase inhibition. *J Clin Invest* 100: 1677–84, 1997.
30. Chung IY, Nam-Kung EK, Lee NM, Chang HS, Kim DJ, Kim YH *et al.* The downregulation of bcl-2 expression is necessary for theophylline-induced apoptosis of eosinophil. *Cell Immunol* 203: 95–102, 2000.
31. Yasui K, Agematsu K, Shinozaki K, Hokibara S, Nagumo H, Nakazawa T *et al.* Theophylline induces neutrophil apoptosis through adenosine A_{2A} receptor antagonism. *J Leukoc Biol* 67: 529–35, 2000.
32. Ohta K, Yamashita N. Apoptosis of eosinophils and lymphocytes in allergic inflammation. *J Allergy Clin Immunol* 104: 14–21, 1999.
33. Moonen HJ, Geraets L, Vaarhorst A, Bast A, Wouters EF, Hageman GJ. Theophylline prevents NAD+ depletion via PARP-1 inhibition in human pulmonary epithelial cells. *Biochem Biophys Res Commun* 338: 1805–10, 2005.
34. Barnes PJ, Adcock IM, Ito K. Histone acetylation and deacetylation: Importance in inflammatory lung diseases. *Eur Respir J* 25: 552–63, 2005.
35. Barnes PJ. How corticosteroids control inflammation. *Br J Pharmacol* 148: 245–54, 2006.
36. Ito K, Ito M, Elliott WM, Cosio B, Caramori G, Kon OM *et al.* Decreased histone deacetylase activity in chronic obstructive pulmonary disease. *N Engl J Med* 352: 1967–76, 2005.
37. Hew M, Bhavsar P, Torrego A, Meah S, Khorasani N, Barnes PJ *et al.* Relative corticosteroid insensitivity of peripheral blood mononuclear cells in severe asthma. *Am J Respir Crit Care Med* 174: 134–41, 2006.
38. Murahidy A, Ito M, Adcock IM, Barnes PJ, Ito K. Reduction is histone deacetylase expression and activity in smoking asthmatics: A mechanism of steroid resistance. *Proc Am Thorac Soc* 2: A889, 2005.
39. Ito K, Lim S, Caramori G, Cosio B, Chung KF, Adcock IM *et al.* A molecular mechanism of action of theophylline: Induction of histone deacetylase activity to decrease inflammatory gene expression. *Proc Natl Acad Sci USA* 99: 8921–26, 2002.
40. Cosio BG, Tsaprouni L, Ito K, Jazrawi E, Adcock IM, Barnes PJ. Theophylline restores histone deacetylase activity and steroid responses in COPD macrophages. *J Exp Med* 200: 689–95, 2004.
41. Ito K, Yamamura S, Essilfie-Quaye S, Cosio B, Ito M, Barnes PJ *et al.* Histone deacetylase 2-mediated deacetylation of the glucocorticoid receptor enables NF-kB suppression. *J Exp Med* 203: 7–13, 2006.
42. Saleh D, Ernst P, Lim S, Barnes PJ, Giaid A. Increased formation of the potent oxidant peroxynitrite in the airways of asthmatic patients is associated with induction of nitric oxide synthase: Effect of inhaled glucocorticoid. *FASEB J* 12: 929–37, 1998.
43. Ricciardolo FL, Caramori G, Ito K, Capelli A, Brun P, Abatangelo G *et al.* Nitrosative stress in the bronchial mucosa of severe chronic obstructive pulmonary disease. *J Allergy Clin Immunol* 116: 1028–35, 2005.
44. Ito K, Tomita T, Barnes PJ, Adcock IM. Oxidative stress reduces histone deacetylase (HDAC)2 activity and enhances IL-8 gene expression: Role of tyrosine nitration. *Biochem Biophys Res Commun* 315: 240–45, 2004.
45. Hirano T, Yamagata T, Gohda M, Yamagata Y, Ichgikawa T, Yanagisawa S *et al.* Inhibition of reactive nitrogen species production in COPD airways: Comparison between inhaled corticosteroid and oral theophylline. *Thorax* 61: 761–66, 2006.
46. Finney MJB, Karlson JA, Persson CGA. Effects of bronchoconstriction and bronchodilation on a novel human small airway preparation. *Br J Pharmacol* 85: 29–36, 1985.
47. Guillot C, Fornaris M, Badger M, Orehek J. Spontaneous and provoked resistance to isoproterenol in isolated human bronchi. *J Allergy Clin Immunol* 74: 713–18, 1984.
48. Miura M, Belvisi MG, Stretton CD, Yacoub MH, Barnes PJ. Role of potassium channels in bronchodilator responses in human airways. *Am Rev Respir Dis* 146: 132–36, 1992.
49. Goldie RG, Spina D, Henry PJ, Lulich KM, Paterson JW. *In vitro* responsiveness of human asthmatic bronchus to carbachol, histamine, beta-adrenoceptor agonists and theophylline. *Br J Clin Pharmacol* 22: 669–76, 1986.
50. Bai TR, Mak JCW, Barnes PJ. A comparison of beta-adrenergic receptors and *in vitro* relaxant responses to isoproterenol in asthmatic airway smooth muscle. *Am J Respir Cell Mol Biol* 6: 647–51, 1992.
51. Mitenko PA, Ogilvie RI. Rational intravenous doses of theophylline. *N Engl J Med* 289: 600–3, 1973.
52. McWilliams BC, Menendez R, Kelly WH, Howick J. Effects of theophylline on inhaled methacholine and histamine in asthmatic children. *Am Rev Respir Dis* 130: 193–97, 1984.
53. Cartier A, Lemire I, L'Archeveque J. Theophylline partially inhibits bronchoconstriction caused by inhaled histamine in subjects with asthma. *J Allergy Clin Immunol* 77: 570–75, 1986.
54. Magnusson H, Reuss G, Jorres R. Theophylline has a dose-related effect on the airway response to inhaled histamine and methacholine in asthmatics. *Am Rev Respir Dis* 136: 1163–67, 1987.
55. Magnussen H, Reuss G, Jörres R. Methylxanthines inhibit exercise-induced bronchoconstriction at low serum theophylline concentrations and in a dose-dependent fashion. *J Allergy Clin Immunol* 81: 531–37, 1988.
56. Orange RP, Kaliner MA, Laraia PJ, Austen KF. Immunological release of histamine and slow reacting substance of anaphylaxis from human lung. II. Influence of cellular levels of cyclic AMP. *Fed Proc* 30: 1725–29, 1971.
57. Nielson CP, Crawley JJ, Morgan ME, Vestal RE. Polymorphonuclear leukocyte inhibition by therapeutic concentrations of theophylline is mediated by cyclic 3′,5′ adenosine aminophosphate. *Am Rev Respir Dis* 137: 25–30, 1988.
58. Kraft M, Pak J, Borish L, Martin RJ. Theophylline's effect on neutrophil function and the late asthmatic response. *J Allergy Clin Immunol* 98: 251–57, 1996.
59. Schrier DJ, Imre RM. The effects of adenosine antagonists on human neutrophil function. *J Immunol* 137: 3284–89, 1986.
60. Yukawa T, Kroegel C, Dent G, Chanez P, Ukena D, Barnes PJ. Effect of theophylline and adenosine on eosinophil function. *Am Rev Respir Dis* 140: 327–33, 1989.
61. O'Neill SJ, Sitar DS, Kilass DJ. The pulmonary disposition of theophylline and its influences on human alveolar macrophage bactericidal function. *Am Rev Respir Dis* 134: 1225–28, 1988.
62. Lim S, Tomita K, Carramori G, Jatakanon A, Oliver B, Keller A *et al.* Low-dose theophylline reduces eosinophilic inflammation but not exhaled nitric oxide in mild asthma. *Am J Respir Crit Care Med* 164: 273–76, 2001.
63. Yasui K, Agematsu K, Shinozaki K, Hokibara S, Nagumo H, Yamada S *et al.* Effects of theophylline on human eosinophil functions: Comparative study with neutrophil functions. *J Leukoc Biol* 68: 194–200, 2000.
64. Erjefalt I, Persson CGA. Pharmacologic control of plasma exudation into tracheobronchial airways. *Am Rev Respir Dis* 143: 1008–14, 1991.
65. Boschetto P, Roberts NM, Rogers DF, Barnes PJ. The effect of anti-asthma drugs on microvascular leak in guinea pig airways. *Am Rev Respir Dis* 139: 416–21, 1989.
66. Naclerio RM, Bartenfelder D, Proud D, Togias AG, Meyers DA, Kagey Sobotka A *et al.* Theophylline reduces histamine release during pollen-induced rhinitis. *J Allergy Clin Immunol* 78: 874–76, 1986.

67. Ward AJM, McKenniff M, Evans JM, Page CP, Costello JF. Theophylline – an immunomodulatory role in asthma? *Am Rev Respir Dis* 147: 518–23, 1993.
68. Sullivan P, Bekir S, Jaffar Z, Page C, Jeffery P, Costello J. Anti-inflammatory effects of low-dose oral theophylline in atopic asthma. *Lancet* 343: 1006–8, 1994.
69. Kraft M, Torvik JA, Trudeau JB, Wenzel SE, Martin RJ. Theophylline: Potential antiinflammatory effects in nocturnal asthma. *J Allergy Clin Immunol* 97: 1242–46, 1996.
70. Culpitt S, Maziak W, Loukides S, Keller A, Barnes PJ. Effect of theophylline on induced sputum inflammatory indices in COPD patients. *Am J Respir Crit Care Med* 157: A797, 1997.
71. Kobayashi M, Nasuhara Y, Betsuyaku T, Shibuya E, Tanino Y, Tanino M *et al.* Effect of low-dose theophylline on airway inflammation in COPD. *Respirology* 9: 249–54, 2004.
72. Wenzel SE, Szefler SJ, Leung DY, Sloan SI, Rex MD, Martin RJ. Bronchoscopic evaluation of severe asthma. Persistent inflammation associated with high dose glucocorticoids. *Am J Respir Crit Care Med* 156: 737–43, 1997.
73. Jatakanon A, Uasaf C, Maziak W, Lim S, Chung KF, Barnes PJ. Neutrophilic inflammation in severe persistent asthma. *Am J Respir Crit Care Med* 160: 1532–39, 1999.
74. Shohat B, Volovitz B, Varsano I. Induction of suppressor T cells in asthmatic children by theophylline treatment. *Clin Allergy* 13: 487–93, 1983.
75. Fink G, Mittelman M, Shohat B, Spitzer SA. Theophylline-induced alterations in cellular immunity in asthmatic patients. *Clin Allergy* 17: 313–16, 1987.
76. Guillou PJ, Ramsden C, Kerr M, Davison AM, Giles GR. A prospective controlled clinical trial of aminophylline as an adjunct immunosuppressive agent. *Transplant Proc* 16: 1218–20, 1984.
77. Didier M, Aussel C, Ferrua B, Fehlman M. Regulation of interleukin 2 synthesis by cAMP in human T cells. *J Immunol* 139: 1179–84, 1987.
78. Giembycz MA, Corrigan CJ, Seybold J, Newton R, Barnes PJ. Identification of cyclic AMP phosphodiesterases 3, 4 and 7 in human $CD4^+$ and $CD8^+$ T-lymphocytes. *Br J Pharmacol* 118: 1945–58, 1996.
79. Hidi R, Timmermans S, Liu E, Schudt C, Dent G, Holgate ST *et al.* Phosphodiesterase and cyclic adenosine monophosphate-dependent inhibition of T-lymphocyte chemotaxis. *Eur Respir J* 15: 342–49, 2000.
80. Sanjar S, Aoki S, Kristersson A, Smith D, Morley J. Antigen challenge induces pulmonary eosinophil accumulation and airway hyperreactivity in sensitized guinea pigs: The effect of anti-asthma drugs. *Br J Pharmacol* 99: 679–86, 1990.
81. Kosmas EN, Michaelides SA, Polychronaki A, Roussou T, Toukmatzi S, Polychronopoulos V *et al.* Theophylline induces a reduction in circulating interleukin-4 and interleukin-5 in atopic asthmatics. *Eur Respir J* 13: 53–58, 1999.
82. Kidney J, Dominguez M, Taylor PM, Rose M, Chung KF, Barnes PJ. Immunomodulation by theophylline in asthma: Demonstration by withdrawal of therapy. *Am J Respir Crit Care Med* 151: 1907–14, 1995.
83. Jaffar ZH, Sullivan P, Page C, Costello J. Low-dose theophylline modulates T-lymphocyte activation in allergen-challenged asthmatics. *Eur Respir J* 9: 456–62, 1996.
84. Barnes PJ, Pauwels RA. Theophylline in asthma: Time for reappraisal?. *Eur Respir J* 7: 579–91, 1994.
85. Cushley MJ, Holgate ST. Bronchodilator actions of xanthine derivatives administered by inhalation in asthma. *Thorax* 40: 176–79, 1985.
86. Aubier M, De Troyer A, Sampson M, Macklem PT, Roussos C. Aminophylline improves diaphragmatic contractility. *N Engl J Med* 305: 249–52, 1981.
87. Moxham J. Aminophylline and the respiratory muscles: An alternative view. In: Respiratory Muscles: Function in health and disease. *Clin Chest Med* 2: 325–40, 1988.
88. Vozeh S, Kewitz G, Perruchoud A, Tschan M, Koppe C, Heitz M *et al.* Theophylline serum concentration and therapeutic effect in severe acute bronchial obstruction: The optimal use of intravenously administered aminophylline. *Am Rev Respir Dis* 125: 181–84, 1982.
89. Zhang ZY, Kaminsky LS. Characterization of human cytochromes P450 involved in theophylline 8-hydroxylation. *Biochem Pharmacol* 50: 205–11, 1995.
90. Jusko WJ, Gardner MJ, Mangiore A, Schentag JJ, Kopp JR, Vance JW. Factors affecting aminophylline clearance: Age, tobacco, marijuana, cirrhosis, congestive heart failure, obesity, oral contraceptives, benzodiazepines, barbiturates and ethanol. *J Pharm Sci* 68: 1358–66, 1979.
91. Taylor DR, Ruffin D, Kinney CD, McDevitt DG. Investigation of diurnal changes in the disposition of theophylline. *Br J Clin Pharmacol* 16: 413–16, 1983.
92. Warren JB, Cuss F, Barnes PJ. Posture and theophylline kinetics. *Br J Clin Pharmacol* 19: 707–9, 1985.
93. Weinberger M, Hendeles L. Slow-release theophylline: Rationale and basis for product selection. *N Engl J Med* 308: 760–63, 1983.
94. Barnes PJ, Greening AP, Neville L, Timmers J, Poole GW. Single dose slow-release aminophylline at night prevents nocturnal asthma. *Lancet* i: 299–301, 1982.
95. Heins M, Kurtin L, Oellerich M, Maes R, Sybrecht GW. Nocturnal asthma: Slow-release terbutaline versus slow-release theophylline therapy. *Eur Respir J* 1: 306–10, 1988.
96. Bowler SD, Mitchell CA, Armstrong JG. Nebulised fenoterol and i.v. aminophylline in acute severe asthma. *Eur Resp J* 70: 280–83, 1987.
97. Parameswaran K, Belda J, Rowe BH. Addition of intravenous aminophylline to beta2-agonists in adults with acute asthma. *Cochrane. Database Syst Rev*: 2000, CD002742.
98. Mitra A, Bassler D, Goodman K, Lasserson TJ, Ducharme FM. Intravenous aminophylline for acute severe asthma in children over two years receiving inhaled bronchodilators. *Cochrane Database Syst Rev*: 2005, CD001276.
99. Eason J, Makowe HLJ. Aminophylline toxicity – how many hospital asthma deaths does it cause?. *Respir Med* 83: 219–26, 1989.
100. Duffy N, Walker P, Diamantea F, Calverley PM, Davies L. Intravenous aminophylline in patients admitted to hospital with non-acidotic exacerbations of chronic obstructive pulmonary disease: A prospective randomised controlled trial. *Thorax* 60: 713–17, 2005.
101. Barr RG, Rowe BH, Camargo CA Jr. Methylxanthines for exacerbations of chronic obstructive pulmonary disease: Meta-analysis of randomised trials. *BMJ* 327: 643, 2003.
102. Wilson AJ, Gibson PG, Coughlan J. Long acting beta-agonists versus theophylline for maintenance treatment of asthma. *Cochrane Database Syst Rev*(2), 2000, CD001281.
103. Furukawa CT, Shapiro SG, Bierman CW. A double-blind study comparing the effectiveness of cromolyn sodium and sustained release theophylline in childhood asthma. *Pediatrics* 74: 435–39, 1984.
104. Nassif EG, Weinburger M, Thompson R, Huntley W. The value of maintenance theophylline in steroid-dependent asthma. *N Engl J Med* 304: 71–75, 1981.
105. Brenner MR, Berkowitz R, Marshall N, Strunk RC. Need for theophylline in severe steroid-requiring asthmatics. *Clin Allergy* 18: 143–50, 1988.
106. Rivington RN, Boulet LP, Cote J, Kreisman H, Small DI, Alexander M *et al.* Efficacy of slow-release theophylline, inhaled salbutamol and their combination in asthmatic patients on high-dose inhaled steroids. *Am J Respir Crit Care Med* 151: 325–32, 1995.
107. Barnes PJ. The role of theophylline in severe asthma. *Eur Respir Rev* 6(rev 34): 154S–59S, 1996.
108. Markham A, Faulds D. Theophylline. A review of its potential steroid sparing effects in asthma. *Drugs* 56: 1081–91, 1998.
109. Martin RJ, Pak J. Overnight theophylline concentrations and effects on sleep and lung function in chronic obstructive pulmonary disease. *Am Rev Respir Dis* 145: 540–44, 1992.
110. Zwilli CW, Neagey SR, Cicutto L, White DP, Martin RJ. Nocturnal asthma therapy: Inhaled bitolterol versus sustained-release theophylline. *Am Rev Respir Dis* 139: 470–74, 1989.
111. Fairfax AJ, Clarke R, Chatterjee SS, Connolly CK, Higenbottam T, Holgate ST *et al.* Controlled release theophylline in the treatment of nocturnal asthma. *J Int Med Res* 18: 273–81, 1990.

112. Selby C, Engleman HM, Fitzpatrick MF, Sime PM, Mackay TW, Douglas NJ. Inhaled salmeterol or oral theophylline in nocturnal asthma? *Am J Respir Crit Care Med* 155: 104–8, 1997.
113. Evans DJ, Taylor DA, Zetterstrom O, Chung KF, O'Connor BJ, Barnes PJ. A comparison of low-dose inhaled budesonide plus theophylline and high-dose inhaled budesonide for moderate asthma. *N Engl J Med* 337: 1412–18, 1997.
114. Ukena D, Harnest U, Sakalauskas R, Magyar P, Vetter N, Steffen H *et al.* Comparison of addition of theophylline to inhaled steroid with doubling of the dose of inhaled steroid in asthma. *Eur Respir J* 10: 2754–60, 1997.
115. Lim S, Groneberg D, Fischer A, Oates T, Caramori G, Mattos W *et al.* Expression of heme oxygenase isoenzymes 1 and 2 in normal and asthmatic airways: effect of inhaled corticosteroids. *Am J Respir Crit Care Med* 162: 1912–18, 2000.
116. Taylor DR, Buick B, Kinney C, Lowry RC, McDevitt DG. The efficacy of orally administered theophylline, inhaled salbutamol, and a combination of the two as chronic therapy in the management of chronic bronchitis with reversible airflow obstruction. *Am Rev Respir Dis* 131: 747–51, 1985.
117. Murciano D, Avclair M-H, Parievte R, Aubier M. A randomized controlled trial of theophylline in patients with severe chronic obstructive pulmonary disease. *N Engl J Med* 320: 1521–25, 1989.
118. Chrystyn H, Mulley BA, Peake MD. Dose response relation to oral theophylline in severe chronic obstructive airway disease. *BMJ* 297: 1506–10, 1988.
119. Culpitt SV, de Matos C, Russell RE, Donnelly LE, Rogers DF, Barnes PJ. Effect of theophylline on induced sputum inflammatory indices and neutrophil chemotaxis in COPD. *Am J Respir Crit Care Med* 165: 1371–76, 2002.
120. Barnes PJ. Theophylline in chronic obstructive pulmonary disease: New horizons. *Proc Am Thorac Soc* 2: 334–39, 2005.
121. Barnes PJ. Theophylline for COPD. *Thorax* 61: 742–43, 2006.
122. Jenne JW. Theophylline as a bronchodilator in COPD and its combination with inhaled beta-adrenergic drugs. *Chest* 92: 7S–14S, 1987.
123. Kume H, Graziano MP, Kotlikoff MI. Stimulatory and inhibitory regulation of calcium-activated potassium channels by guanine nucleotide binding proteins. *Proc Natl Acad Sci USA* 89: 11051–55, 1992.
124. Giembycz MA. Phosphodiesterase 4 and tolerance to beta 2-adrenoceptor agonists in asthma. *Trends Pharmacol Sci* 17: 331–36, 1996.
125. Cheung D, Wever AM, de GOEIJ JA, de GRAAFF CS, Steen H, Sterk PJ. Effects of theophylline on tolerance to the bronchoprotective actions of salmeterol in asthmatics *in vivo*. *Am J Respir Crit Care Med* 158: 792–96, 1998.
126. Williamson BH, Milligan C, Griffiths K, Sparta S, Tribe AC, Thompson PJ. An assessment of major and minor side effects of theophylline. *Aust NZ J Med* 19: 539, 1988.
127. Rachelefsky WOJ, Adelson J, Mickey MR, Spector SL, Katz RM, Siegel SC *et al.* Behaviour abnormalities and poor school performance due to oral theophylline use. *Pediatrics* 78: 1113–38, 1986.
128. Nicholson CD, Challiss RAJ, Shahid M. Differential modulation of tissue function and therapeutic potential of selective inhibitors of cyclic nucleotide phosphodiesterase isoenzymes. *Trends Pharmacol Sci* 12: 19–27, 1991.
129. Kelloway JS, Wyatt RA, Adlis SA. Comparison of patients' compliance with prescribed oral and inhaled asthma medications. *Arch Intern Med* 154: 1349–52, 1994.

51

CHAPTER

Corticosteroids

Peter J. Barnes

National Heart and Lung Institute (NHLI), Clinical Studies Unit, Imperial College, London, UK

Corticosteroids (also known as glucocorticosteroids, glucocorticoids, steroids) are by far the most effective controllers used in the treatment of asthma and the only drugs that can effectively suppress the characteristic inflammation in asthmatic airways. By contrast, they are ineffective in suppressing pulmonary inflammation in COPD. After the discussion of the mechanism of action and pharmacology of corticosteroids their current use in the treatment of asthma and COPD will be described.

MECHANISMS OF ACTION

There have been major advances in understanding the molecular mechanisms whereby corticosteroids suppress inflammation, based on recent developments in understanding the fundamental mechanisms of gene transcription [1, 2]. Corticosteroids activate and suppress many genes relevant to understanding their action in asthma and in other allergic diseases (Table 51.1).

Cellular effects

At a cellular level, corticosteroids reduce the numbers of inflammatory cells in the airways, including eosinophils, T-lymphocytes, mast cells, and dendritic cells (Fig. 51.1). These effects of corticosteroids are produced through inhibiting the recruitment of inflammatory cells into the airway by suppressing the production of chemotactic mediators and adhesion molecules and by inhibiting the survival in the airways of inflammatory cells, such as eosinophils, T-lymphocytes, and mast cells. Epithelial cells may be the major cellular target for inhaled corticosteroids, which are the mainstay of modern asthma management. Inhaled corticosteroids suppress many activated inflammatory genes in airway epithelial cells (Fig. 51.2). Epithelial integrity is restored by regular inhaled corticosteroids. The suppression of mucosal inflammation is relatively rapid with a significant reduction in eosinophils detectable within 6 h and associated with reduced airway hyperresponsiveness [3, 4]. Reversal of airway hyperresponsiveness

TABLE 51.1 Effect of corticosteroids on gene transcription.

Increased transcription
• Lipocortin-1
• β_2-Adrenergic receptors
• Secretory leukocyte inhibitory protein
• IκB-α (inhibitor of NF-κB)
• Anti-inflammatory or inhibitory cytokines
• IL-10, IL-12, IL-1 receptor antagonist
• Mitogen-activated protein kinase phosphatase-1 (MKP-1, inhibits MAP kinase pathways)
Decreased transcription
• Inflammatory cytokines
IL-2, IL-3, IL-4, IL-5, IL-6, IL-11, IL-13, IL-15, TNF-α, GM-CSF, SCF
• Chemokines
IL-8, RANTES, MIP-1α, eotaxin
• Inflammatory enzymes
Inducible nitric oxide synthase (iNOS), inducible cyclo-oxygenase (COX-2) inducible phospholipase A_2 ($cPLA_2$)
• Inflammatory peptides
Endothelin-1
• Mediator receptors
Neurokinin (NK_1)-, bradykinin (B_2)-receptors
• Adhesion molecules
ICAM-1,VCAM-1

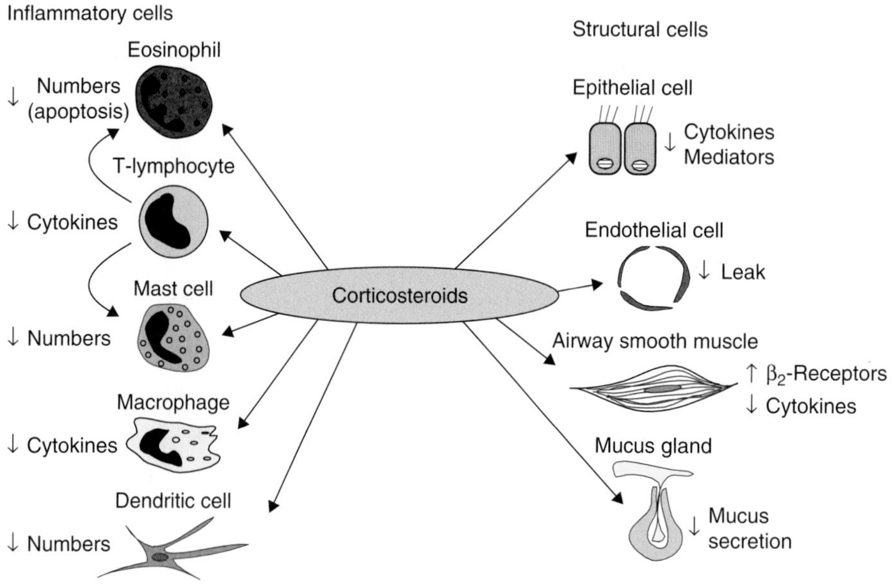

FIG. 51.1 Cellular effects of corticosteroids.

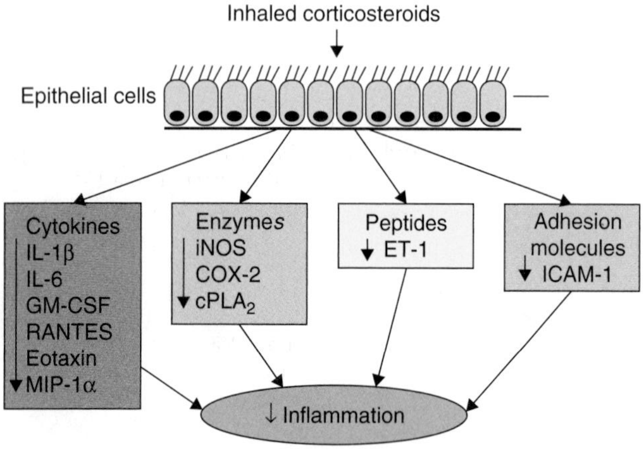

FIG. 51.2 Inhaled corticosteroids may inhibit the transcription of several "inflammatory" genes in airway epithelial cells and thus reduce inflammation in the airway wall. NF-κB: nuclear factor κB; AP-1: activator protein-1; GM-CSF: granulocyte-macrophage colony stimulating factor; IL-1: interleukin-1; iNOS: inducible nitric oxide synthase; NO: nitric oxide; COX-2: inducible cyclooxygenase; $cPLA_2$: cytoplasmic phospholipase A_2; PG: prostaglandin; ET: endothelin; ICAM: intercellular adhesion molecule.

may take several months to reach a plateau, probably reflecting recovery of structural changes in the airway [5].

Glucocorticoid receptors

Corticosteroids diffuse across the cell membrane and bind to glucocorticoid receptors (GR) in the cytoplasm [2]. There is only one form of GR that binds corticosteroids termed GR-α. GR-β is an alternatively spliced form of GR that interacts with DNA but not with corticosteroids, so may act as a dominant negative inhibitor of glucocorticoid action by interfering with the binding of GR to DNA [6]. Whether GR-β is involved in steroid resistance in asthma is controversial [7]. Activated GRs rapidly translocate to the nucleus where they produce their molecular effects. A pair of GRs (GR homodimer) bind to glucocorticoid response elements in the promoter region of steroid-responsive genes and this interaction switches on (and sometimes switches off) gene transcription (Fig. 51.3). Examples of genes that are activated by corticosteroids include genes encoding β_2-adrenergic receptors and the anti-inflammatory proteins secretory leukoprotease inhibitor and mitogen-activated protein kinase phosphatase-1 (MKP-1), which inhibits MAP kinase pathways. These effects may contribute to the anti-inflammatory actions of corticosteroids [8, 9]. GR interaction with negative GREs may suppress gene transcription and it is thought that this may be important in mediating many of the side effects of corticosteroids. For example, corticosteroids inhibit the expression of osteocalcin that is involved in bone synthesis [10].

Switching off inflammation

The major action of corticosteroids is to switch off multiple activated inflammatory genes that encode for cytokines, chemokines, adhesion molecules, inflammatory enzymes, and receptors [11]. These genes are switched on in the airways by proinflammatory transcription factors, such as nuclear factor-κB (NF-κB) and activator protein-1, both of which are activated in asthmatic airways and switch on inflammatory genes by interacting with coactivator molecules, such as CREB-binding protein, that have intrinsic histone acetyltransferase (HAT) activity, resulting in acetylation of core histones, which opens up the chromatin structure so that gene transcription is facilitated [12]. In artificial overexpression systems, activated GR may directly interact with NF-κB and AP-1 to inhibit their activity, but this does not appear to occur in asthmatic patients treated with inhaled corticosteroids [13]. Glucocorticoid-activated GR also interact with coactivator molecules and this inhibits the interaction of NF-κB with coactivators, thus reducing histone acetylation [1, 14]. Reduction of histone

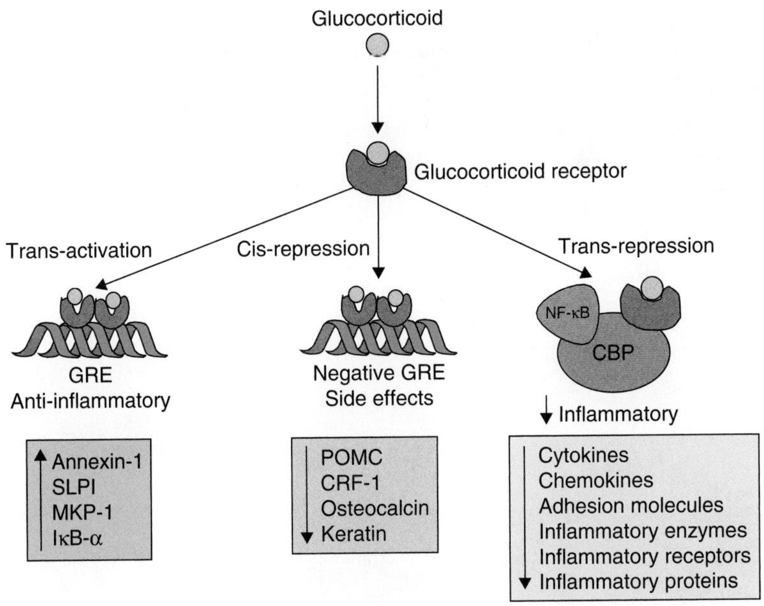

FIG. 51.3 Corticosteroids may regulate gene expression in several ways. Glucocorticoids enter the cell to bind to glucocorticoid receptors in the cytoplasm that translocate to the nucleus. GR homodimers bind to glucocorticoid-response elements (GRE) in the promoter region of steroid-sensitive genes, which may encode anti-inflammatory proteins. Less commonly, GR homodimers interact with negative GREs to suppress genes, particularly those linked to side effects of corticosteroids. Nuclear GR also interact with coactivator molecules, such as CREB-binding protein (CBP), which is activated by proinflammatory transcription factors, such as nuclear factor-κB (NF-κB), thus switching off the inflammatory genes that are activated by these transcription factors. SLPI: secretory leukoprotease inhibitor; MKP-1: mitogen-activated kinase phosphatase-1; IκB-α: inhibitor of NF-κB; GILZ: glucocorticoid-induced leucine zipper protein; POMC: proopiomelanocortin; CRH: corticotrophin releasing factor.

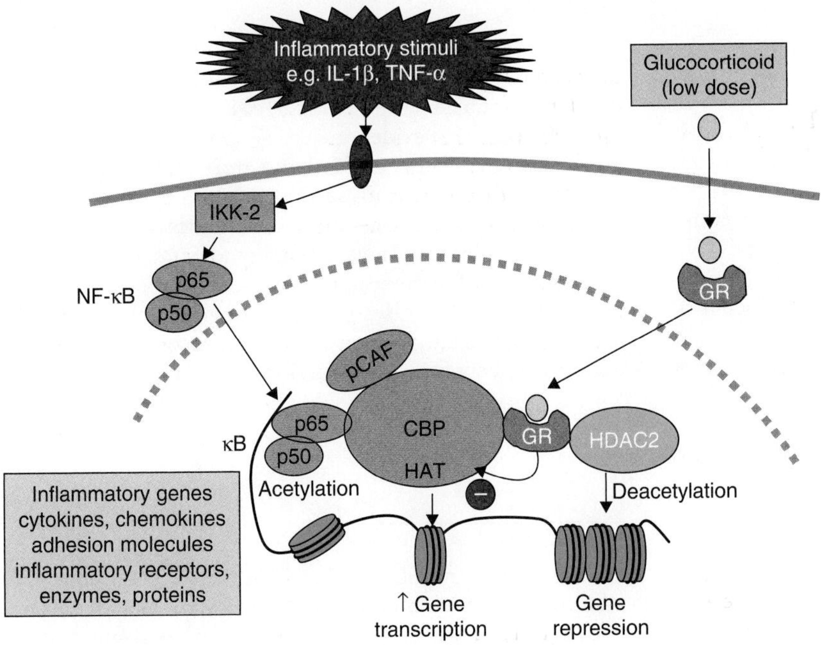

FIG. 51.4 Corticosteroid suppression of activated inflammatory genes. Inflammatory genes are activated by inflammatory stimuli, such as interleukin-1β (IL-1β) or tumor necrosis factor-α (TNF-α), resulting in activation of IKK2 (inhibitor of I-κB kinase-2), which activates the transcription factor, nuclear factor κB (NF-κB). A dimer of p50 and p65 NF-κB proteins translocates to the nucleus and binds to specific κB recognition sites and also to coactivators, such as CREB-binding protein (CBP) or p300/CBP-activating factor (pCAF), which have intrinsic histone acetyltransferase (HAT) activity. This results in acetylation of core histone H4, resulting in increased expression of genes encoding multiple inflammatory proteins. Glucocorticoid receptors (GR) after activation by glucocorticoids translocate to the nucleus and bind to coactivators to inhibit HAT activity directly and recruiting histone deacetylase-2 (HDAC2), which reverses histone acetylation leading in suppression of these activated inflammatory genes.

acetylation also occurs through the recruitment of histone deacetylase-2 (HDAC2) to the activated inflammatory gene complex by activated GR, thereby resulting in effective suppression of all activated inflammatory genes within the nucleus (Fig. 51.4). This accounts for why corticosteroids are so effective in the control of asthmatic inflammation, but also why they are safe, since other activated genes are not affected.

There may be additional mechanisms that are also important in the anti-inflammatory actions of corticosteroids. Corticosteroids have potent inhibitory effects on mitogen-activated kinase signaling pathways through the induction of MKP-1 and this may inhibit the expression of multiple inflammatory genes [8, 9]. Some inflammatory genes, for example granulocyte-macrophage colony stimulating factor, have an unstable mRNA that is rapidly degraded by certain RNAses but stabilized when cells are stimulated by inflammatory mediators. Corticosteroids reverse this effect, resulting in rapid degradation of mRNA and reduced inflammatory protein secretion [15]. This may be through the inhibition of proteins that stabilize mRNAs of inflammatory proteins, such as tristetraprolin [16].

Corticosteroid resistance

Patients with severe asthma have a poor response to corticosteroids, which necessitates the need for high doses and a few patients are completely resistant. All patients with COPD show corticosteroid resistance. Asthmatics who smoke are also

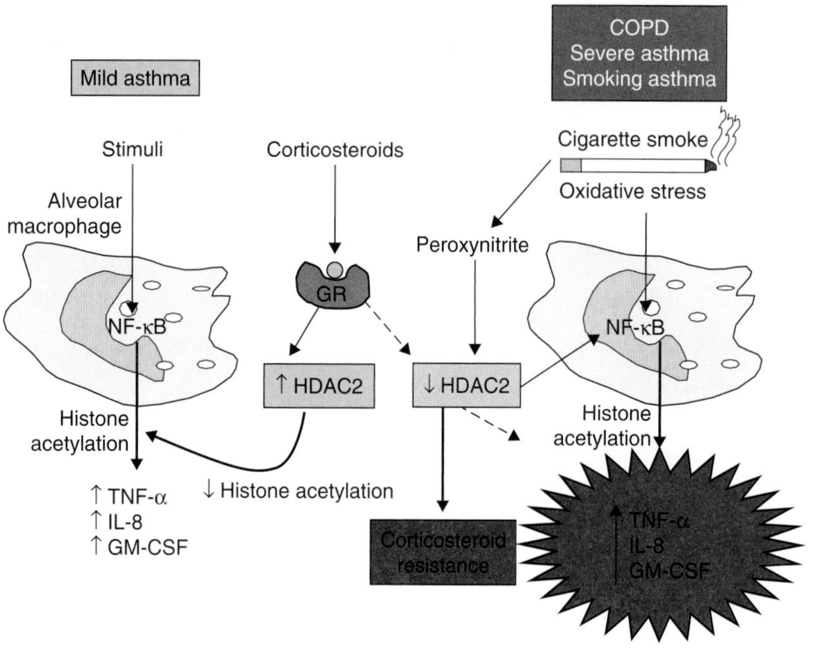

FIG. 51.5 Mechanism of corticosteroid resistance in COPD, smoking asthma, and severe asthma. Stimulation of mild asthmatic alveolar macrophages activates nuclear factor-κB (NF-κB) and other transcription factors to switch on histone acetyltransferase leading to histone acetylation and subsequently to transcription of genes encoding inflammatory proteins, such as tumor necrosis factor-α (TNF-α), interleukin-8 (IL-8), and granulocyte-macrophage colony stimulating factor (GM-CSF). Corticosteroids reverse this by binding to GR and recruiting HDAC2. This reverses the histone acetylation induced by NF-κB and switches off the activated inflammatory genes. In COPD patients and smoking asthmatics cigarette smoke generates oxidative stress (acting through the formation of peroxynitrite) and in severe asthma and COPD intense inflammation generates oxidative stress to impair the activity of HDAC2. This amplifies the inflammatory response to NF-κB activation, but also reduces the anti-inflammatory effect of corticosteroids, as HDAC2 is now unable to reverse histone acetylation.

relatively corticosteroid-resistant and require increased doses of corticosteroids for asthma control [17]. Several molecular mechanisms have now been identified to account for corticosteroid resistance in severe asthma and COPD [18]. In patients with COPD, smoking asthmatics, and severe asthma there is a reduction in HDAC2 activity and expression, which prevents corticosteroids switching off activated inflammatory genes (Fig. 51.5) [19–21]. In steroid-resistant asthma other mechanisms may also contribute to corticosteroid insensitivity, including reduced translocation of GR as a result of phosphorylation by p38 MAP kinase [22] and abnormal histone acetylation patterns [23]. A proposed mechanism is an increase in GR-β, which prevents GR binding to DNA, but there is little evidence that this would be sufficient to account for corticosteroid insensitivity as the amounts of GR-β are too low [7].

INTERACTION WITH β_2-ADRENERGIC RECEPTORS

Inhaled β_2-agonists and corticosteroids are frequently used together in the control of asthma and it is now recognized that there are important molecular interactions between these two classes of drugs (Fig. 51.6) [24]. As discussed above, corticosteroids increase the gene transcription of β_2-receptors, resulting in increased expression of cell surface receptors. This has been demonstrated in human lung *in vitro* [25] and nasal mucosa *in vivo* after topical application of a glucocorticoid [26]. In this way corticosteroids protect against the down-regulation of β_2-receptors after long-term administration [27]. This may be important for the non-bronchodilator effects of β_2-agonists, such as mast cell stabilization. Corticosteroids may also enhance the coupling of β_2-receptors to G-proteins, thus enhancing β_2-agonist effects and reversing the uncoupling of β_2-receptors that may occur in response to inflammatory mediators, such as IL-1β through a stimulatory effect on a G-protein-coupled receptor kinase [28].

There is also evidence that β_2-agonists may affect GR and thus enhance the anti-inflammatory effects of corticosteroids. β_2-Agonists increase the translocation of GR from cytoplasm to the nucleus after activation by corticosteroids [29]. This effect has now been demonstrated in sputum macrophages of asthmatic patients after an inhaled glucocorticoid and inhaled long-acting β_2-agonist (LABA) [30]. This suggests that β_2-agonists and glucocorticoid enhance each others beneficial effects in asthma therapy.

PHARMACOKINETICS

Prednisolone is readily and consistently absorbed after oral administration with little interindividual variation. Prednisone is converted in the liver to the active prednisolosne. Enteric coatings to reduce the incidence of dyspepsia delay absorption but not the total amount of drug absorbed. Prednisolone is metabolized in the liver and drugs such as rifampicin, phenobarbitone, or phenytoin, which induce CYP450 enzymes, lower the plasma half-life of prednisolone [31]. The plasma half-life is 2–3 h, although its biological half-life is approximately 24 h, so that it is suitable for daily dosing. There is no evidence that previous exposure to steroids changes their subsequent metabolism. Prednisolone is approximately 92% protein bound, the majority to a specific protein transcortin and the remainder to albumin; it is the unbound fraction which is biologically active. Corticosteroid-resistant asthma is not explained by impaired absorption or metabolism of steroids, but is due to reduced anti-inflammatory actions of corticosteroids, as discussed above. Measurement of plasma concentrations of prednisolone are useful in monitoring compliance with inhaled corticosteroids and in assessing whether a poor

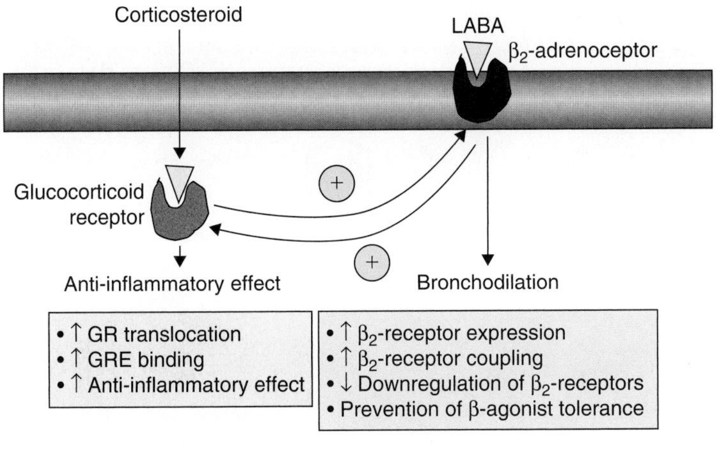

FIG. 51.6 Interaction between corticosteroids and LABA. Corticosteroids have anti-inflammatory effects but also increase the numbers of β_2-receptors, whereas β_2-agonists, as well as inducing direct bronchodilation, act on GR to increase the anti-inflammatory effects of corticosteroids.

therapeutic response to corticosteroids is due to poor absorption or increased metabolism.

Inhaled delivery

The pharmacokinetics of inhaled corticosteroids is important in relation to systemic effects [32, 34]. The fraction of steroid which is inhaled into the lungs acts locally on the airway mucosa, but may be absorbed from the airway and alveolar surface. This fraction therefore reaches the systemic circulation (Fig 51.7). The fraction of inhaled steroid which is deposited in the oropharynx is swallowed and absorbed from the gut. The absorbed fraction may be metabolized in the liver before reaching the systemic circulation (first-pass metabolism). Budesonide and fluticasone propionate (FP) have a greater first-pass metabolism than beclomethasone dipropionate (BDP) and are therefore less likely to produce systemic effects at high inhaled doses. The use of a large volume spacer chamber reduces oropharyngeal deposition and therefore reduces systemic absorption of corticosteroids, although this effect is minimal in corticosteroids with a high first-pass metabolism [35]. Mouth rinsing and discarding the rinse has a similar effect and this procedure should be used with high dose dry powder steroid inhalers, since spacer chambers cannot be used with these devices. The ideal inhaled corticosteroid with optimal therapeutic index should have high lung bioavailability, negligible oral bioavailability, low systemic absorption, high systemic clearance, and high protein binding [36].

A recently introduced corticosteroid ciclesonide is an inactive prodrug that is activated by esterases in the lung to the active metabolite des-ciclesonide [37]. This may reduce oropharyngeal side effects as esterases appear to be less active in this site than in the lower airways. Ciclesonide is also claimed to be effective as a once daily therapy.

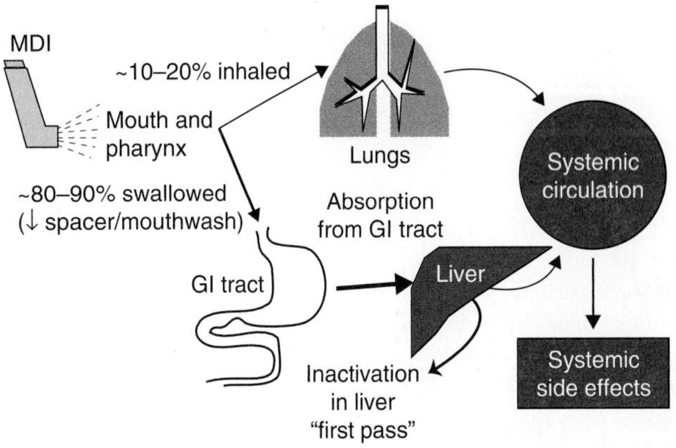

FIG. 51.7 Pharmacokinetics of inhaled glucocorticoids.GI: gastrointestinal.

SYSTEMIC STEROIDS

Hydrocortisone is given intravenously in acute severe asthma. While the value of corticosteroids in acute severe asthma has been questioned, others have found that they speed the resolution of attacks. There is no apparent advantage in giving very high doses of intravenous steroids (such as methylprednisolone 1 g) as this only increases the risk of side effects, such as hyperglycemia and increased susceptibility to infections. Intravenous steroids are indicated in acute asthma if lung function is < 30% predicted and in whom there is no significant improvement with a nebulized β_2-agonist. Intravenous therapy is usually given until a satisfactory response is obtained and then oral prednisolone may be substituted. Oral prednisolone (40–60 mg) has a similar effect to intravenous hydrocortisone and is easier to administer [38, 39]. High doses of inhaled corticosteroids may also substitute for a course of oral steroids in controlling acute exacerbations of asthma. High dose FP (2000 μg daily) was as effective as a course of oral prednisolone in controlling acute exacerbations of asthma in a family practice setting and in children in an emergency room setting, although this route of delivery is more expensive [40, 41]. Although doubling the dose of inhaled corticosteroids was recommended for mild exacerbations of asthma, this does not appear to be useful [42, 43], but a fourfold increase in dose appears to be effective [44]. Inhaled steroids have no proven effect in the management of severe acute asthma in a hospital setting [45], but trials with nebulized steroids, which can deliver large doses, are underway.

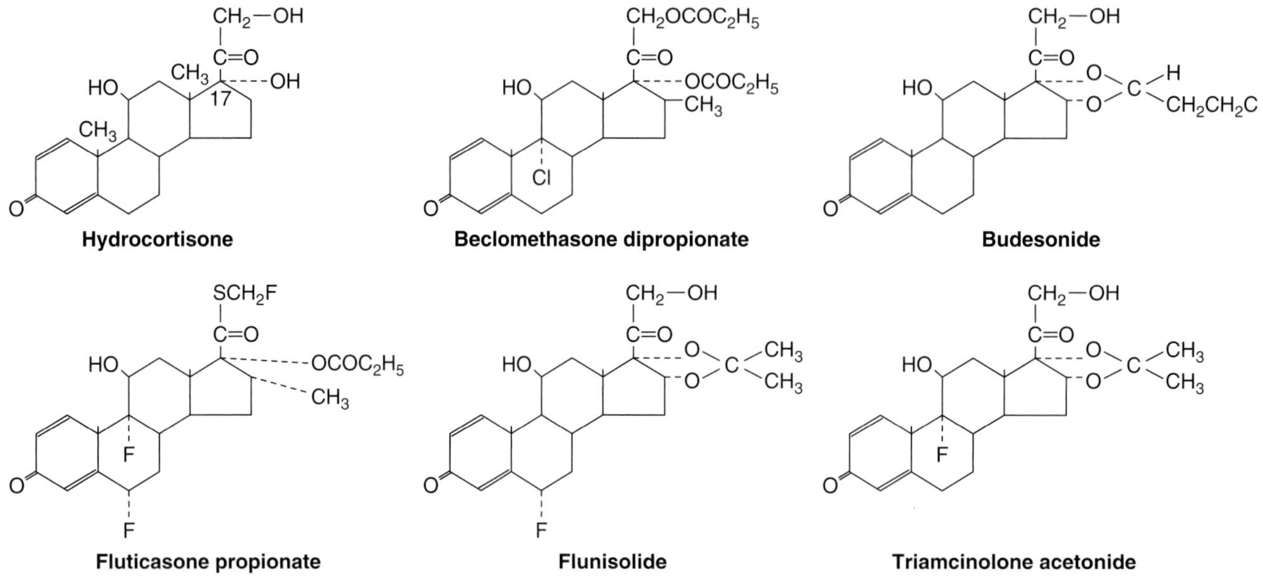

FIG. 51.8 Chemical structures of inhaled glucocorticoids.

Maintenance treatment with oral steroids are reserved for patients who cannot be controlled on maximum doses of other therapy, the dose being titrated to the lowest which provides acceptable control of symptoms. For any patient taking regular oral steroids objective evidence of steroid responsiveness should be obtained before maintenance therapy is instituted. Short courses of oral steroids (30–40 mg prednisolone daily for 1–2 weeks) are indicated for exacerbations of asthma, and the dose may be tailed off over 1 week once the exacerbation is resolved (although the tail-off period is not strictly necessary, patients often find it reassuring).

INHALED CORTICOSTEROIDS

There is no doubt that the early use of inhaled corticosteroids has revolutionized the management of asthma, with marked reductions in asthma morbidity and improvement in health status. Inhaled steroids are now recommended as first-line therapy for all patients with persistent asthma [46]. Several topically acting corticosteroids are now available for inhalation (Fig. 51.8). Inhaled corticosteroids are very effective in controlling asthma symptoms in asthmatic patients of all ages and severity. Inhaled corticosteroids improve the quality of life of patients with asthma and allow many patients to lead normal lives, improve lung function, reduce the frequency of exacerbations, and may prevent irreversible airway changes. They were first introduced to reduce the requirement for oral corticosteroids in patients with severe asthma and many studies have confirmed that the majority of patients can be weaned off oral corticosteroids [32].

USE IN ASTHMA

Studies in adults

As experience has been gained with inhaled corticosteroids they have been introduced in patients with milder asthma, with the recognition that inflammation is present even in patients with mild asthma. Inhaled anti-inflammatory drugs have now become first-line therapy in any patient who needs to use a β_2-agonist inhaler more than two to three times a week and this is reflected in national and international guidelines for the management of chronic asthma. In patients with newly diagnosed asthma an inhaled corticosteroid (budesonide 600 μg twice daily) reduced symptoms and β_2-agonist inhaler usage and improved lung function. These effects persisted over the 2 years of the study, whereas in a parallel group treated with inhaled β_2-agonists alone there was no significant change in symptoms or lung function [47]. In another study, patients with mild asthma treated with a low dose of inhaled corticosteroid (budesonide 400 μg daily) showed less symptoms and a progressive improvement in lung function over several months and many patients became completely asymptomatic [48]. There was also a significant reduction in the number of exacerbations; in patients with mild asthma a low dose of corticosteroids (budesonide 400 μg daily) significantly reduces exacerbation by around 40% over a 3-year period [49]. Although the effects of inhaled corticosteroids on AHR may take several months to reach a plateau, the reduction in asthma symptoms occurs much more rapidly and reduced inflammation is seen within hours [3, 4].

High dose inhaled corticosteroids may be used for the control of more severe asthma. This markedly reduces the need for maintenance oral corticosteroids [50]. With the use of add-on therapies, particularly LABA, most patients can now be controlled on much lower doses of inhaled corticosteroids so that high doses are needed in only a few patients with severe disease. Inhaled corticosteroids are the treatment of choice in nocturnal asthma, which is a manifestation of inflamed airways, reducing nocturnal awakening and reducing the diurnal variation in airway function.

Inhaled corticosteroids effectively control asthmatic inflammation but must be taken regularly. When inhaled corticosteroids are discontinued there is usually a gradual increase in symptoms and airway responsiveness back to pretreatment values [51]. Reduction in the dose of inhaled corticosteroids is associated with an increase in symptoms and this is preceded by an increase in exhaled NO and sputum eosinophils [52, 53].

Studies in children

Inhaled corticosteroids are equally effective in children. In an extensive study of children aged 7–17 years there was a significant improvement in symptoms, peak flow variability, and lung function compared to a regular inhaled β_2-agonist which was maintained over the 22 months of the study [54], but asthma deteriorated when the inhaled corticosteroids were withdrawn [55]. There was a high proportion of dropouts (45%) in the group treated with inhaled β_2-agonist alone. Inhaled corticosteroids are also effective in younger children. Nebulized budesonide reduces the need for oral corticosteroids and also improved lung function in children under the age of three [56]. Inhaled corticosteroids given via a large volume spacer improve asthma symptoms and reduce the number of exacerbations in preschool children and in infants.

Dose-response studies

Surprisingly, the dose-response curve for the clinical efficacy of inhaled corticosteroids is relatively flat and, while all studies have demonstrated a clinical benefit of inhaled corticosteroids, it has been difficult to demonstrate differences between doses, with most benefit obtained at the lowest doses used [57, 58]. This is in contrast to the steeper dose response for systemic effects, implying that while there is little clinical benefit from increasing doses of inhaled corticosteroids the risk of adverse effects is increased. However, the dose-response effect of inhaled corticosteroids may depend on the parameters measured and, while it is difficult to discern a dose response when traditional lung function parameters are measured, there may be a dose-response effect in prevention of asthma exacerbations. Thus, there is a significantly greater effect of budesonide 800 μg daily compared to 200 μg daily in preventing severe and mild asthma exacerbations [59]. Normally, a fourfold or greater difference in dose has been required to detect a statistically significant (but often small) difference in effect on commonly measured outcomes such as symptoms, PEF, use of rescue β_2-agonist and lung function and even such large differences in dose are not always associated with significant differences in response. These findings suggest that pulmonary function tests or symptoms may have a rather low sensitivity in the assessment of the effects of inhaled corticosteroids. This is obviously important for the interpretation of clinical comparisons between different inhaled corticosteroids or inhalers. It is also important to consider the type of patient included in clinical studies. Patients with relatively mild asthma may have relatively little room for improvement with inhaled corticosteroids, so that maximal improvement is obtained with relatively low doses. Patients with more severe asthma or with unstable asthma may have more room for improvement and may therefore show a greater response to increasing doses, but it is often difficult to include such patients in controlled clinical trials.

More studies are needed to assess whether other outcome measures such as AHR or more direct measurements of inflammation, such as sputum eosinophils or exhaled NO, may be more sensitive than traditional outcome measures such as symptoms or lung function tests [60–63]. Higher doses of inhaled corticosteroids are needed to control AHR than to improve symptoms and lung function, and this may have a better long-term outcome in terms of reduction in structural changes of the airways [64]. Measurement of sputum eosinophils to adjust the dose of inhaled corticosteroids may reduce the overall dose requirement for inhaled corticosteroids and exacerbations [65, 66]. Monitoring of exhaled nitric oxide also reduces the requirement for corticosteroids but is not yet practical in clinical practice [67].

Prevention of irreversible airway changes

Some patients with asthma develop an element of irreversible airflow obstruction, but the pathophysiological basis of these changes is not yet understood. It is likely that they are the result of chronic airway inflammation and that they may be prevented by treatment with inhaled corticosteroids. There is some evidence that the annual decline in lung function may be slowed by the introduction of inhaled corticosteroids [68] and this is supported by a 5-year study of low dose budesonide in patients with mild asthma [69, 70]. Increasing evidence also suggests that delay in starting inhaled corticosteroids may result in less overall improvement in lung function in both adults and children [71, 73]. These studies suggest that introduction of inhaled corticosteroids at the time of diagnosis is likely to have the greatest impact [72, 73]. So far there is no evidence that early use of inhaled corticosteroids is curative and even when inhaled corticosteroids are introduced at the time of diagnosis, symptoms and lung function reverts to pretreatment levels when corticosteroids are withdrawn [71].

Reduction in mortality

In a retrospective review of the risk of mortality and prescribed antiasthma medication, there was a significant protection provided by regular inhaled corticosteroid therapy [74]. By contrast, asthma mortality appears to increase with increasing usage of short-acting β_2-agonists, reflecting the fact that increased rescue therapy is a marker of poor asthma control [75]. The increase in use of rescue therapy should result in an increase in the maintenance dose of inhaled corticosteroids. The long-acting inhaled β_2-agonist salmeterol is associated with a small increase in asthma mortality, but the excess deaths appear to be related to underuse of inhaled corticosteroids [76].

Comparison between inhaled corticosteroids

Several inhaled corticosteroids are currently prescribable in asthma, although their availability varies between countries. There have been relatively few studies comparing efficacy of the different inhaled corticosteroids, and it is important to take into account the delivery system and the type of patient under investigation when such comparisons are made. Because of the relatively flat dose-response curve for the clinical parameters normally used in comparing doses of inhaled corticosteroids, it may be difficult to see differences in efficacy of inhaled corticosteroids. Most comparisons have concentrated on differences in systemic effects at equally efficacious doses, although it has often proved difficult to establish dose equivalence. There are few studies comparing different doses of inhaled corticosteroids in asthmatic patients. Budesonide has been compared with BDP and in adults and children it appears to have comparable antiasthma effects at equal doses, whereas FP appears to be approximately twice as potent as BDP and budesonide [58]. Studies have consistently shown that FP and budesonide have less systemic effects than BDP, triamcinolone, and flunisolide [33]. The new inhaled corticosteroids mometasone and ciclesonide are claimed to have less systemic effects [77, 78].

Clinical application

Inhaled corticosteroids are now recommended as first-line therapy for all patients with persistent symptoms. Inhaled corticosteroids should be started in any patient who needs to use a β_2-agonist inhaler for symptom control more than three times weekly. It is conventional to start with a low dose of inhaled corticosteroid and to increase the dose until asthma control is achieved. However, this may take time and a preferable approach is to start with a dose of corticosteroids in the middle of the dose range (400 μg twice daily) to establish asthma control. Once control is achieved (defined as normal or best possible lung function and infrequent need to use an inhaled β_2-agonist) the dose of inhaled corticosteroid should be reduced in a stepwise manner to the lowest dose needed for optimal control. It may take as long as 3 months to reach a plateau in response and any changes in dose should be made at intervals of 3 months or more. When daily doses of ≥800 μg daily are needed, a large volume spacer device should be used with a metered dose inhaler (MDI) and mouth washing with a dry powder inhaler in order to reduce local and systemic side effects. Inhaled corticosteroids are usually given as a twice daily dose in order to increase compliance. When asthma is unstable four times daily dosage is preferable [79].

The dose of inhaled corticosteroid should be increased to 2000 μg daily if necessary, but higher doses may result in systemic effects. It may be preferable to add a low dose of oral corticosteroid, since higher doses of inhaled corticosteroids are expensive and have a high incidence of local side effects. Nebulized budesonide has been advocated in order to give an increased dose of inhaled corticosteroid and to reduce the requirement for oral corticosteroids [80], but this treatment is expensive and may achieve its effects largely via systemic absorption. The dose of inhaled corticosteroid should be the minimal dose that controls asthma and once control is achieved the dose should be slowly reduced [81].

USE IN COPD

Patients with COPD have a poor response to corticosteroids in comparison to asthma with little improvement in lung function. High doses of inhaled corticosteroids have consistently been shown a reduction (20–25%) in exacerbations in patients with severe disease and this is the main clinical indication for their use [82]. However, several large studies have shown that corticosteroids fail to reduce the progression in COPD (measured by annual fall in FEV_1) [83] and they have not been found to reduce mortality in a large study [84]. These results are likely to reflect the resistance of pulmonary inflammation to corticosteroids in COPD patients as a result of the reduction in HDAC2 [21].

ADD-ON THERAPY

Previously it was recommended to increase the dose of inhaled corticosteroids if asthma was not controlled, on the assumption that there was residual inflammation of the airways. However the dose-response effect of inhaled corticosteroids is relatively flat, so that there is little improvement in lung function after increasing the dose of inhaled corticosteroids. An alternative strategy is to add some other class of controller drug and this is more effective than increasing the dose of inhaled corticosteroids for most patients [85].

LABA

In patients who are not controlled on BDP 200 μg twice daily, addition of salmeterol 50 μg twice daily was more effective than increasing the dose of inhaled corticosteroid to 500 μg twice daily, in terms of lung function improvement, use of rescue β_2-agonist and symptom control [86]. This has been confirmed in several other studies [87]. Similar results have been found with another long-acting inhaled β_2-agonist formoterol, which in addition reduced the frequency of mild and severe asthma exacerbations in patients with mild, moderate, and severe persistent asthma [59, 88]. Analysis of several studies clearly show that adding a LABA is more effective than increasing the dose of inhaled corticosteroids in terms of improving asthma control and reducing exacerbations [89]. These studies showing the great efficacy of combined corticosteroids and LABA compared to increased doses of LABA have led to the development of fixed combinations of corticosteroids and

LABA, such as FP/salmeterol and budesonide/formoterol, which may be more convenient for patients. These fixed combination inhalers also ensure that patients do not discontinue their inhaled corticosteroids when a long-acting bronchodilator is used. For patients with mild persistent asthma, combination inhalers are no more effective than the inhaled corticosteroids alone in controlled trials [90], but may have an advantage in the real world where adherence to regular inhaled corticosteroids is very low.

Recently, studies have demonstrated that when formoterol combined with budesonide is used as a reliever therapy this gives better control of asthma compared to the normally used short-acting β_2-agonist as a rescue therapy with either the same dose of combination inhaler or a high dose of inhaled corticosteroids as maintenance treatment [91, 92]. This advantage is particularly striking in terms of reducing the number of severe exacerbations. When formoterol was used as the reliever therapy this reduce exacerbations to a greater extent than the short-acting β_2-agonist terbutaline but the combination was even more effective [93]. This suggests that the "as required" use of inhaled corticosteroids contributes to the marked reduction in acute exacerbations. The mechanisms by which corticosteroids as required improve asthma control and reduce exacerbations are not completely understood, but exacerbations of asthma evolve over several days when patients take increasing amounts of rescue medication [94]. During this time there is increasing inflammation of the airways, as may be measured by exhaled nitric oxide and sputum eosinophils [52]. Taking the inhaled corticosteroid at the same time as the formoterol to relieve symptoms may suppress this evolving inflammation, particularly since corticosteroids appear to have a relatively rapid onset of effect in suppressing airway inflammation [95].

LABA/corticosteroid inhalers are also more effective in COPD patients than either treatment alone and reduce exacerbations and improve symptoms [96, 97]. They are more effective than LABA alone in reducing exacerbations [98]. This may be explained by molecular interactions between LABA and corticosteroids as discussed above. However, there is a reduction in COPD mortality although this fails to reach significance in one large study (TORCH) [84], although this has been seen in another smaller study (INSPIRE) [99].

Theophylline

Addition of low doses of theophylline (giving plasma concentrations of <10 mg/l) are more effective than doubling the dose of inhaled budesonide, either in mild or in severe asthma [100–102]. However, this is less effective than using a long-acting inhaled β_2-agonist as an add-on therapy [103]. Theophylline has not been examined as an add-on therapy in COPD patients but there are theoretical reasons to believe that low dose theophylline may reverse corticosteroid resistance in COPD (see Chapter 52).

Antileukotrienes

Antileukotrienes have also been used as an add-on therapy in asthma [104, 105], although this is less effective than addition of LABA [106–108]. There is no role of antileukotrienes in COPD patients.

SIDE EFFECTS

The efficacy of inhaled corticosteroids is now established in short- and long-term studies in adults and in children, but there are still concerns about side effects, particularly in children and when high inhaled doses are used. Several side effects have been recognized (Table 51.2).

Local side effects

Side effects due to the local deposition of the inhaled corticosteroid in the oropharynx may occur with inhaled corticosteroids, but the frequency of complaints depends on the dose and frequency of administration and on the delivery system used.

The commonest complaint is of hoarseness of the voice (dysphonia) and may occur in over 50% of patients using MDI. Dysphonia is not appreciably reduced by using spacers, but may be less with dry powder devices. Dysphonia may be due to myopathy of laryngeal muscles and is reversible when treatment is withdrawn [109]. For most patients it is not troublesome but may be disabling in singers and lecturers.

Oropharyngeal candidiasis (thrush) may be a problem in some patients, particularly in the elderly, with concomitant oral corticosteroids and more than twice daily administration [110]. Large volume spacer devices protect against this local side effect by reducing the dose of inhaled corticosteroid that deposits in the oropharynx.

Infections

There is no evidence that inhaled corticosteroids, even in high doses, increase the frequency of infections, including tuberculosis, in the lower respiratory tract in asthmatic

TABLE 51.2 Side effects of inhaled corticosteroids.

Local side effects
Dysphonia
Oropharyngeal candidiasis
Cough
Pneumonia (COPD patients)
Systemic side effects
Adrenal suppression
Growth suppression
Bruising
Osteoporosis
Cataracts
Glaucoma
Metabolic abnormalities (glucose, insulin, triglycerides)
Psychiatric disturbances

patients. Recently several large controlled studies have shown that high dose ICS increase physician-diagnosed pneumonias, either used alone or in combination with a LABA [83, 98] and this has been confirmed in an epidemiological study of hospital admissions for pneumonia amongst COPD patients [111].

Systemic side effects

The efficacy of inhaled corticosteroids in the control of asthma is undisputed, but there are concerns about systemic effects of inhaled corticosteroids, particularly as they are likely to be used over long periods and in children of all ages [33, 112]. The safety of inhaled corticosteroids has been extensively investigated since their introduction 30 years ago [32]. One of the major problems is to decide whether a measurable systemic effect has any significant clinical consequence and this necessitates careful long-term follow-up studies. As biochemical markers of systemic corticosteroid effects become more sensitive, then systemic effects may be seen more often, but this does not mean that these effects are clinically relevant. There are several case reports of adverse systemic effects of inhaled corticosteroids, and these may be idiosyncratic reactions, which may be due to abnormal pharmacokinetic handing of the inhaled corticosteroid. The systemic effect of an inhaled corticosteroid will depend on several factors, including the dose delivered to the patient, the site of delivery (gastrointestinal tract and lung), the delivery system used, and the individual differences in the patient's response to the corticosteroid. Recent studies suggest that systemic effects of inhaled corticosteroid are less in patients with more severe asthma, presumably as less drug reaches the lung periphery [113, 114].

The systemic effect of an inhaled corticosteroid is dependent on the amount of drug absorbed into the systemic circulation. Approximately 80–90% of the inhaled dose from an MDI deposits in the oropharynx and is swallowed and subsequently absorbed from the gastrointestinal tract. Use of a large volume spacer device markedly reduces the oropharyngeal deposition, and therefore the systemic effects of inhaled corticosteroids, although this is less important when oral bioavailability is minimal, as with FP. For dry powder inhalers similar reductions in systemic effects may be achieved with mouth washing and discarding the fluid. All patients using a daily dose of $\geq$800 μg of an inhaled corticosteroid should therefore use either a spacer or a mouthwash to reduce systemic absorption. Approximately 10% of an MDI enters the lung and this fraction (which presumably exerts the therapeutic effect) may be absorbed into the systemic circulation. As the fraction of inhaled corticosteroid deposited in the oropharynx is reduced, the proportion of the inhaled dose entering the lungs is increased. More efficient delivery to the lungs is therefore accompanied by increased systemic absorption, but this is offset by a reduction in the dose needed for optimal control of airway inflammation. For example, a multiple dry powder delivery system, the Turbuhaler, delivers approximately twice as much corticosteroid to the lungs as other devices, and therefore has increased systemic effects. However this is compensated for by the fact that only half the dose is required.

Adrenal suppression

Corticosteroids may cause hypothalamic–pituitary–adrenal (HPA) axis suppression by reducing corticotrophin (ACTH) production, which reduces cortisol secretion by the adrenal gland. The degree of HPA suppression is dependent on dose, duration, frequency, and timing of corticosteroid administration. Measurement of HPA axis function provides evidence for systemic effects of an inhaled corticosteroid. Basal adrenal cortisol secretion may be measured by a morning plasma cortisol, 24 h urinary cortisol, or by plasma cortisol profile over 24 h. Other tests measure the HPA response following stimulation with tetracosactrin (which measures adrenal reserve) or stimulation with metyrapone and insulin (which measure the response to stress). There are many studies of HPA axis function in asthmatic patients with inhaled corticosteroids, but the results are inconsistent as they have often been uncontrolled and patients have also been taking courses of oral corticosteroids (which may affect the HPA axis for weeks) [32]. BDP, budesonide, and FP at high doses by conventional MDI (>1600 μg daily) give a dose-related decrease in morning serum cortisol levels and 24 h urinary cortisol, although values still lie well within the normal range. However, when a large volume spacer is used doses of 2000 μg daily of BDP or budesonide have little effect on 24 h urinary cortisol excretion. Stimulation tests of HPA axis function similarly show no consistent effects of doses of 1500 μg or less of inhaled corticosteroid. At high doses (>1500 μg daily) budesonide, and FP have less effect than BDP on HPA axis function. In children no suppression of urinary cortisol is seen with doses of BDP of 800 μg or less. In studies where plasma cortisol has been measured at frequent intervals there was a significant reduction in cortisol peaks with doses of inhaled BDP as low as 400 μg daily, although this does not appear to be dose related in the range 400–1000 μg. The clinical significance of these effects is not certain, however.

Bone metabolism

Corticosteroids lead to a reduction in bone mass by direct effects on bone formation and resorption and indirectly by suppression of the pituitary–gonadal and HPA axes, effects on intestinal calcium absorption, renal tubular calcium reabsorption, and secondary hyperparathyroidism [115]. The effects of oral corticosteroids on osteoporosis and increased risk of vertebral and rib fractures are well known, but there are no reports suggesting that long-term treatment with inhaled corticosteroids is associated with an increased risk of fractures. Bone densitometry has been used to assess the effect of inhaled corticosteroids on bone mass. Although there is evidence that bone density is less in patients taking high dose inhaled corticosteroids, interpretation is confounded by the fact that these patients are also taking intermittent courses of oral corticosteroids. Changes in bone mass occur very slowly and several biochemical indices have been used to assess the short-term effects of inhaled corticosteroids on bone metabolism. Bone formation has been measured by plasma concentrations of bone-specific alkaline phosphatase, serum osteocalcin, or procollagen peptides. Bone resorption may be assessed by urinary hydroxyproline after a 12-h fast, urinary calcium excretion, and pyridinium cross-link excretion. Inhaled corticosteroids, even at doses up to 2000 μg daily, have no significant effect on calcium excretion, but acute and reversible dose-related suppression of serum osteocalcin has been reported with BDP and budesonide when given by conventional MDI in several studies. Budesonide consistently has less effect than BDP at equivalent doses and only BDP increases urinary hydroxyproline at high doses. With a large volume spacer even doses of 2000 μg daily of either BDP or budesonide are without effect on plasma osteocalcin concentrations, however. Urinary pyridinium and deoxypyridinoline cross-links, which are a more accurate and stable measurement of bone and collagen degradation, are not increased with inhaled corticosteroids (BDP > 1000 μg daily), even with intermittent courses of oral corticosteroids. It is important to monitor changes in markers of bone formation as well as bone degradation, as the net effect on bone turnover is important. There is no evidence that inhaled corticosteroids increase the frequency of fractures. Long-term treatment with high dose inhaled corticosteroids has not been associated with any consistent change in bone density. Indeed, in elderly patients there may be an increase in bone density due to increased mobility.

Connective tissue effects

Oral and topical corticosteroids cause thinning of the skin, telangiectasiae, and easy bruising, probably as a result of loss of extracellular ground substance within the dermis, due to an inhibitory effect on dermal fibroblasts. There are reports of increased skin bruising and purpura in patients using high doses of inhaled BDP, but the amount of intermittent oral corticosteroids in these patients is not known. Easy bruising in association with inhaled corticosteroids is more frequent in elderly patients [116] and there are no reports of this problem in children. Long-term prospective studies with objective measurements of skin thickness are needed with different inhaled corticosteroids.

Cataracts

Long-term treatment with oral corticosteroids increases the risk of posterior subcapsular cataracts and there are several case reports describing cataracts in individual patients taking inhaled corticosteroids [32]. In a recent cross-sectional study in patients aged 5–25 years taking either inhaled BDP or budesonide no cataracts were found on slit-lamp examination, even in patients taking 2000 μg daily for over 10 years [117]. A slight increase in the risk of glaucoma in patients taking very high doses of inhaled corticosteroids has also been identified [118].

Growth

There has been particular concern that inhaled corticosteroids may cause stunting of growth and several studies have addressed this issue. Asthma itself (as with other chronic diseases) may have an effect on the growth pattern and has been associated with delayed onset of puberty and deceleration of growth velocity that is more pronounced with more severe disease [119]. However, asthmatic children appear to grow for longer, so that their final height is normal. The effect of asthma on growth make it difficult to assess the effects of inhaled corticosteroids on growth in cross-sectional studies, particularly as courses of oral corticosteroids is a confounding factor. Longitudinal studies have demonstrated that there is no significant effect of inhaled corticosteroids on statural growth in doses of up to 800 μg daily and for up to 5 years of treatment [32]. A meta-analysis of 21 studies, including over 800 children, showed no effect of inhaled BDP on statural height, even with higher doses and long duration of therapy [120] and in a large study of asthmatics treated with inhaled corticosteroids during childhood there was no difference in statural height compared to normal children [121]. Another long-term follow-up study showed no effect of corticosteroids on final height in children treated over several years [122]. Short-term growth measurements (knemometry) have demonstrated that even a low dose of an oral corticosteroid (prednisolone 2.5 mg) is sufficient to give complete suppression of lower leg growth. However inhaled budesonide up to 400 μg is without effect, although some suppression is seen with 800 μg and with 400 μg BDP. The relationship between knemometry measurements and final height are uncertain since low doses of oral corticosteroid that have no effect on final height cause profound suppression.

Metabolic effects

Several metabolic effects have been reported after inhaled corticosteroids, but there is no evidence that these are clinically relevant at therapeutic doses. In adults, fasting glucose and insulin are unchanged after doses of BDP up to 2000 μg daily and in children with inhaled budesonide up to 800 μg daily. In normal individuals high dose inhaled BDP may slightly increase resistance to insulin. However, in patients with poorly controlled asthma high doses of BDP and budesonide paradoxically decrease insulin resistance and improve glucose tolerance, suggesting that the disease itself may lead to abnormalities in carbohydrate metabolism. Neither BDP 2000 μg daily in adults nor budesonide 800 μg daily in children have any effect on plasma cholesterol or triglycerides.

Psychiatric effects

There are various reports of psychiatric disturbance, including emotional lability, euphoria, depression, aggressiveness and insomnia, after inhaled corticosteroids. Only eight such patients have so far been reported, suggesting that this is very infrequent and a causal link with inhaled corticosteroids has usually not been established.

Pregnancy

Based on extensive clinical experience inhaled corticosteroids appear to be safe in pregnancy, although no controlled studies have been performed. There is no evidence for any adverse effects of inhaled corticosteroids on the pregnancy, the delivery, or on the fetus [123]. It is important to recognize that poorly controlled asthma may increase the incidence of perinatal mortality and retard intrauterine growth, so that more effective control of asthma with inhaled corticosteroids may reduce these problems.

Side effects in COPD

Patients with COPD are elderly and are likely to have increased systemic side effects from inhaled corticosteroids as they have several additional risk factors. There have been fewer studies of systemic side effects in COPD patients. However, a systematic review found no reduction in bone mineral density or increase in fractures in COPD patients treated for up to 3 years with inhaled corticosteroids [83]. An epidemiological study showed an increase in cataracts which are more common in an elderly population [124]. Many patients with COPD suffer from comorbidities, including hypertension, metabolic syndrome and diabetes, and may therefore have a worsening of these conditions, but this has not yet been systematically investigated.

SYSTEMIC CORTICOSTEROIDS

Oral or intravenous corticosteroids may be indicated in several situations. Prednisone is converted in the liver to the active prednisolone. In pregnant patients prednisone is preferable as it is not converted to prednisolone in the fetal liver, thus diminishing the exposure of the fetus to corticosteroids. Enteric-coated preparations of prednisolone are used to reduce side effects (particularly gastric side effects) and give delayed and reduced peak plasma concentrations, although the bioavailability and therapeutic efficacy of these preparations is similar to uncoated tablets. Prednisolone and prednisone are preferable to dexamethasone, betamethasone, or triamcinolone, which have longer plasma half-lives and therefore an increased frequency of adverse effects.

Short courses

Short courses of oral corticosteroids (30–40 mg prednisolone daily for 1–2 weeks or until the peak flow values return to best attainable) are indicated for exacerbations of asthma, and the dose may be tailed off over 1 week once the exacerbation is resolved. The tail-off period is not strictly necessary, but some patients find it reassuring.

Maintenance corticosteroids

Maintenance oral corticosteroids are only needed in a small proportion of asthmatic patients (approximately 1%) with the most severe asthma that cannot be controlled with maximal doses of inhaled corticosteroids (2000 μg daily) and additional bronchodilators. The minimal dose of oral corticosteroid needed for control should be used and reductions in the dose should be made slowly in patients who have been on oral corticosteroids for long periods (e.g. by 2.5 mg per month for doses down to 10 mg daily and thereafter by 1 mg per month). Oral corticosteroids are usually given as a single morning dose, as this reduces the risk of adverse effects since it coincides with the peak diurnal concentrations. There is some evidence that administration in the afternoon may be optimal for some patients who have severe nocturnal asthma [125]. Alternate day administration may also reduce adverse effects, but control of asthma may not be as good on the day when the oral dose is omitted in some patients.

Intramuscular triamcinolone acetonide (80 mg monthly) has been advocated in patients with severe asthma as an alternative to oral corticosteroids [126, 127]. This may be considered in patients in whom compliance is a particular problem, but the major concern is the high frequency of proximal myopathy associated with this fluorinated corticosteroid. Some patients who do not respond well to prednisolone are reported to respond to oral betamethasone, presumably because of pharmacokinetic handling problems with prednisolone.

Acute severe asthma

Intravenous hydrocortisone is given in acute severe asthma, with a recommended dose of 200 mg intravenously. While the value of corticosteroids in acute severe asthma has been questioned, others have found that they speed the resolution of attacks [128]. There is no apparent advantage in giving very high doses of intravenous corticosteroids (such as methylprednisolone 1 g). Indeed, intravenous corticosteroids have occasionally been associated with an acute severe myopathy [129]. No difference in recovery from acute severe asthma was seen whether intravenous hydrocortisone in doses of 50, 200, or 500 mg 6 hourly were used [130], and another placebo-controlled study showed no beneficial effect of intravenous corticosteroids [131]. Intravenous corticosteroids are indicated in acute asthma if lung function is <30% predicted and in whom there is no significant improvement with nebulized β_2-agonist. Intravenous therapy is usually given until a satisfactory response is obtained and then oral prednisolone may be substituted. Oral prednisolone (40–60 mg) has a similar effect to intravenous hydrocortisone and is easier to administer [38, 128]. Oral prednisolone is the preferred treatment for acute severe asthma, providing there are no contraindications to oral therapy. There is some evidence that high doses of nebulized corticosteroids may also be effective in acute exacerbations of asthma, with a more rapid onset of action [132].

References

1. Barnes PJ. How corticosteroids control inflammation. *Br J Pharmacol* 148: 245–54, 2006.
2. Rhen T, Cidlowski JA. Antiinflammatory action of glucocorticoids – New mechanisms for old drugs. *N Engl J Med* 353: 1711–23, 2005.
3. Gibson PG, Saltos N, Fakes K. Acute anti-inflammatory effects of inhaled budesonide in asthma: A randomized controlled trial. *Am J Respir Crit Care Med* 163: 32–36, 2001.
4. Ketchell RI, Jensen MW, Lumley P, Wright AM, Allenby MI, O'Connor BJ. Rapid effect of inhaled fluticasone propionate on airway responsiveness to adenosine 5'-monophosphate in mild asthma. *J Allergy Clin Immunol* 110: 603–6, 2002.
5. Juniper EF, Kline PA, Yan Zieleshem MA, Ramsdale EH, O'Byrne PM, Hargreave FE. Long-term effects of budesonide on airway responsiveness and clinical asthma severity in inhaled steroid-dependent asthmatics. *Eur Respir J* 3: 1122–27, 1990.
6. Lewis-Tuffin LJ, Cidlowski JA. The physiology of human glucocorticoid receptor beta (hGRbeta) and glucocorticoid resistance. *Ann NY Acad Sci* 1069: 1–9, 2006.
7. Pujols L, Mullol J, Picado C. Alpha and beta glucocorticoid receptors: Relevance in airway diseases. *Curr Allergy Asthma Rep* 7: 93–99, 2007.
8. Barnes PJ. Corticosteroid effects on cell signalling. *Eur Respir J* 27: 413–26, 2006.

9. Clark AR. MAP kinase phosphatase 1: A novel mediator of biological effects of glucocorticoids?. *J Endocrinol* 178: 5–12, 2003.
10. Dostert A, Heinzel T. Negative glucocorticoid receptor response elements and their role in glucocorticoid action. *Curr Pharm Des* 10: 2807–16, 2004.
11. Barnes PJ, Adcock IM. How do corticosteroids work in asthma?. *Ann Intern Med* 139: 359–70, 2003.
12. Barnes PJ, Adcock IM, Ito K. Histone acetylation and deacetylation: Importance in inflammatory lung diseases. *Eur Respir J* 25: 552–63, 2005.
13. Hart L, Lim S, Adcock I, Barnes PJ, Chung KF. Effects of inhaled corticosteroid therapy on expression and DNA-binding activity of nuclear factor-kB in asthma. *Am J Respir Crit Care Med* 161: 224–31, 2000.
14. Ito K, Barnes PJ, Adcock IM. Glucocorticoid receptor recruitment of histone deacetylase 2 inhibits IL-1b-induced histone H4 acetylation on lysines 8 and 12. *Mol Cell Biol* 20: 6891–903, 2000.
15. Bergmann MW, Staples KJ, Smith SJ, Barnes PJ, Newton R. Glucocorticoid inhibition of GM-CSF from T cells is independent of control by NF-κB and CLE0. *Am J Respir Cell Mol. Biol* 30: 555–63, 2004.
16. Brook M, Tchen CR, Santalucia T, McIlrath J, Arthur JS, Saklatvala J *et al.* Posttranslational regulation of tristetraprolin subcellular localization and protein stability by p38 mitogen-activated protein kinase and extracellular signal-regulated kinase pathways. *Mol Cell Biol* 26: 2408–18, 2006.
17. Thomson NC, Spears M. The influence of smoking on the treatment response in patients with asthma. *Curr Opin Allergy Clin Immunol* 5: 57–63, 2005.
18. Adcock IM, Lane SJ. Corticosteroid-insensitive asthma: Molecular mechanisms. *J Endocrinol* 178: 347–55, 2003.
19. Ito K, Ito M, Elliott WM, Cosio B, Caramori G, Kon OM *et al.* Decreased histone deacetylase activity in chronic obstructive pulmonary disease. *N Engl J Med* 352: 1967–76, 2005.
20. Hew M, Bhavsar P, Torrego A, Meah S, Khorasani N, Barnes PJ *et al.* Relative corticosteroid insensitivity of peripheral blood mononuclear cells in severe asthma. *Am J Respir Crit Care Med* 174: 134–41, 2006.
21. Barnes PJ. Reduced histone deacetylase in COPD: Clinical implications. *Chest* 129: 151–55, 2006.
22. Irusen E, Matthews JG, Takahashi A, Barnes PJ, Chung KF, Adcock IM. p38 Mitogen-activated protein kinase-induced glucocorticoid receptor phosphorylation reduces its activity: Role in steroid-insensitive asthma. *J Allergy Clin Immunol* 109: 649–57, 2002.
23. Matthews JG, Ito K, Barnes PJ, Adcock IM. Defective glucocorticoid receptor nuclear translocation and altered histone acetylation patterns in glucocorticoid-resistant patients. *J Allergy Clin Immunol* 113: 1100–8, 2004.
24. Barnes PJ. Scientific rationale for combination inhalers with a long-acting β2-agonists and corticosteroids. *Eur Respir J* 19: 182–91, 2002.
25. Mak JCW, Nishikawa M, Barnes PJ. Glucocorticosteroids increase β_2-adrenergic receptor transcription in human lung. *Am J Physiol* 12: L41–46, 1995.
26. Baraniuk JN, Ali M, Brody D, Maniscalco J, Gaumond E, Fitzgerald T *et al.* Glucocorticoids induce β_2-adrenergic receptor function in human nasal mucosa. *Am J Respir Crit Care Med* 155: 704–10, 1997.
27. Mak JCW, Nishikawa M, Shirasaki H, Miyayasu K, Barnes PJ. Protective effects of a glucocorticoid on down-regulation of pulmonary β_2-adrenergic receptors *in vivo*. *J Clin Invest* 96: 99–106, 1995.
28. Mak JC, Chuang TT, Harris CA, Barnes PJ. Increased expression of G protein-coupled receptor kinases in cystic fibrosis lung. *Eur J Pharmacol* 436: 165–72, 2002.
29. Roth M, Johnson PR, Rudiger JJ, King GG, Ge Q, Burgess JK *et al.* Interaction between glucocorticoids and β2 agonists on bronchial airway smooth muscle cells through synchronised cellular signalling. *Lancet* 360: 1293–99, 2002.
30. Usmani OS, Ito K, Maneechotesuwan K, Ito M, Johnson M, Barnes PJ *et al.* Glucocorticoid receptor nuclear translocation in airway cells following inhaled combination therapy. *Am J Respir Crit Care Med* 172: 704–12, 2005.
31. Gambertoglio JG, Amend WJC, Benet LZ. Pharmacokinetics and bioavailability of prednisone and prednisolone in healthy volunteers and patients: A review. *J Pharmacokin Biopharm* 8: 1–52, 1980.
32. Barnes PJ, Pedersen S, Busse WW. Efficacy and safety of inhaled corticosteroids: An update. *Am J Respir Crit Care Med* 157: S1–53, 1998.
33. Lipworth BJ. Systemic adverse effects of inhaled corticosteroid therapy: A systematic review and meta-analysis [see comments]. *Arch Intern Med* 159: 941–55, 1999.
34. Derendorf H, Hochhaus G, Meibohm B, Mollmann H, Barth J. Pharmacokinetics and pharmacodynamics of inhaled corticosteroids. *J Allergy Clin Immunol* 101: S440–46, 1998.
35. Brown PH, Greening AP, Crompton GK. Large volume spacer devices and the influence of high dose beclomethasone diproprionate on hypothalamo-pituitary-adrenal axis function. *Thorax* 48: 233–38, 1993.
36. Derendorf H. Corticosteroid pharmacokinetic/pharmacodynamic parameters and their relationship to safety and efficacy. *Allergy Asthma Proc* 26: 327–35, 2005.
37. Derendorf H. Pharmacokinetic and pharmacodynamic properties of inhaled ciclesonide. *J Clin Pharmacol* 47: 782–89, 2007.
38. Harrison BDN, Stokes TC, Hart GJ, Vaughan DA, Ali NJ, Robinson AA. Need for intravenous hydrocortisone in addition to oral prednisolone in patients admitted to hospital with severe asthma without ventilatory failure. *Lancet* i: 181–84, 1986.
39. Storr J, Barrell E, Barry W, Lenney W, Hatcher G. Effect of a single oral dose of prednisolone in acute childhood asthma. *Lancet* 1: 879–82, 1987.
40. Levy ML, Stevenson C, Maslen T. Comparison of short courses of oral prednisolone and fluticasone propionate in the treatment of adults with acute exacerbations of asthma in primary care. *Thorax* 51: 1087–92, 1996.
41. Manjra AI, Price J, Lenney W, Hughes S, Barnacle H. Efficacy of nebulized fluticasone propionate compared with oral prednisolone in children with an acute exacerbation of asthma. *Respir Med* 94: 1206–14, 2000.
42. Harrison TW, Oborne J, Newton S, Tattersfield AE. Doubling the dose of inhaled corticosteroid to prevent asthma exacerbations: Randomised controlled trial. *Lancet* 363: 271–75, 2004.
43. FitzGerald JM, Becker A, Sears MR, Mink S, Chung K, Lee J. Doubling the dose of budesonide versus maintenance treatment in asthma exacerbations. *Thorax* 59: 550–56, 2004.
44. Foresi A, Morelli MC, Catena E. Low-dose budesonide with the addition of an increased dose during exacerbations is effective in long-term asthma control. On behalf of the Italian Study Group. *Chest* 117: 440–46, 2000.
45. Edmonds ML, Camargo CA Jr., Pollack CV Jr., Rowe BH. Early use of inhaled corticosteroids in the emergency department treatment of acute asthma. *Cochrane Database Syst Rev*, 2003, CD002308.
46. Global Initiative for Asthma. Global strategy for asthma management and prevention. NHLBI/WHO Workshop Report. http://*www.ginasthma.com* 2006.
47. Haahtela T, Jarvinen M, Kava T, Kiviranta K, Koskinen S, Lehtonen K *et al.* Comparison of a β2-agonist terbutaline with an inhaled steroid in newly detected asthma. *N Engl J Med* 325: 388–92, 1991.
48. Juniper EF, Kline PA, Vanzieleghem MA, Ramsdale EH, O'Byrne PM, Hargreave FE. Effect of long-term treatment with an inhaled corticosteroid (budesonide) on airway hyperresponsiveness and clinical asthma in nonsteroid-dependent asthmatics. *Am Rev Respir Dis* 142: 832–36, 1990.
49. Pauwels RA, Pedersen S, Busse WW, Tan WC, Chen YZ, Ohlsson SV *et al.* Early intervention with budesonide in mild persistent asthma: A randomised, double-blind trial. *Lancet* 361: 1071–76, 2003.
50. Mash B, Bheekie A, Jones PW. Inhaled vs oral steroids for adults with chronic asthma. *Cochrane Database Syst Rev* 2, 2002, CD002160: CD002160.
51. Vathenen AS, Knox AJ, Wisniewski A, Tattersfield AE. Time course of change in bronchial reactivity with an inhaled corticosteroid in asthma. *Am Rev Respir Dis* 143: 1317–21, 1991.
52. Jatakanon A, Lim S, Barnes PJ. Changes in sputum eosinophils predict loss of asthma control. *Am J Respir Crit Care Med* 161: 64–72, 2000.
53. Leuppi JD, Salome CM, Jenkins CR, Anderson SD, Xuan W, Marks GB *et al.* Predictive markers of asthma exacerbation during stepwise

dose reduction of inhaled corticosteroids. *Am J Respir Crit Care Med.* 163: 406–12, 2001.

54. van Essen-Zandvliet EE, Hughes MD, Waalkens HJ, Duiverman EJ, Pocock SJ, Kerrebijn KF. Effects of 22 months of treatment with inhaled corticosteroids and/or β_2-agonists on lung function, airway responsiveness and symptoms in children with asthma. *Am Rev Respir Dis* 146: 547–54, 1992.
55. Waalkens HJ, van Essen-Zandvliet EE, Hughes MD, Gerritsen J, Duiverman EJ, Knol K *et al.* Cessation of long-term treatment with inhaled corticosteroids (budesonide) in children with asthma results in deterioration. *Am Rev Respir Dis* 148: 1252–57, 1993.
56. Berger WE. Budesonide inhalation suspension for the treatment of asthma in infants and children. *Drugs* 65: 1973–89, 1973.
57. Busse WW, Chervinsky P, Condemi J, Lumry WR, Petty TL, Rennard S *et al.* Budesonide delivered by Turbuhaler is effective in a dose-dependent fashion when used in the treatment of adult patients with chronic asthma. *J Allergy Clin Immunol* 101: 457–63, 1998.
58. Adams NP, Jones PW. The dose-response characteristics of inhaled corticosteroids when used to treat asthma: An overview of Cochrane systematic reviews. *Respir Med* 100: 1297–306, 2006.
59. Pauwels RA, Lofdahl C-G, Postma DS, Tattersfield AE, O'Byrne PM, Barnes PJ *et al.* Effect of inhaled formoterol and budesonide on exacerbations of asthma. *N Engl J Med* 337: 1412–18, 1997.
60. Lim S, Jatakanon A, John M, Gilbey T, O'Connor BJ, Chung KF *et al.* Effect of inhaled budesonide on lung function and airway inflammation. Assessment by various inflammatory markers in mild asthma. *Am J Respir Crit Care Med* 159: 22–30, 1999.
61. Jatakanon A, Lim S, Chung KF, Barnes PJ. An inhaled steroid improves markers of inflammation in asymptomatic steroid-naive asthmatic patients. *Eur Respir J* 12: 1084–88, 1998.
62. Jatakanon A, Kharitonov S, Lim S, Barnes PJ. Effect of differing doses of inhaled budesonide on markers of airway inflammation in patients with mild asthma. *Thorax* 54: 108–14, 1999.
63. Kharitonov SA, Donnelly LE, Montuschi P, Corradi M, Collins JV, Barnes PJ. Dose-dependent onset and cessation of action of inhaled budesonide on exhaled nitric oxide and symptoms in mild asthma. *Thorax* 57: 889–96, 2002.
64. Sont JK, Willems LN, Bel EH, van Krieken JH, Vandenbroucke JP, Sterk PJ. Clinical control and histopathologic outcome of asthma when using airway hyperresponsiveness as an additional guide to long-term treatment. The AMPUL Study Group. *Am J Respir Crit Care Med* 159: 1043–51, 1999.
65. Green RH, Brightling CE, McKenna S, Hargadon B, Parker D, Bradding P *et al.* Asthma exacerbations and sputum eosinophil counts: A randomised controlled trial. *Lancet* 360: 1715–21, 2002.
66. Jayaram L, Pizzichini MM, Cook RJ, Boulet LP, Lemiere C, Pizzichini E *et al.* Determining asthma treatment by monitoring sputum cell counts: Effect on exacerbations. *Eur Respir J* 27: 483–94, 2006.
67. Smith AD, Cowan JO, Brassett KP, Herbison GP, Taylor DR. Use of exhaled nitric oxide measurements to guide treatment in chronic asthma. *N Engl J Med* 352: 2163–73, 2005.
68. Dompeling E, Van Schayck CP, Molema J, Folgering H, van Grusven PM, van Weel C. Inhaled beclomethasone improves the course of asthma and COPD. *Eur Resp J* 5: 945–52, 1992.
69. O'Byrne PM, Pedersen S, Busse WW, Tan WC, Chen YZ, Ohlsson SV *et al.* Effects of early intervention with inhaled budesonide on lung function in newly diagnosed asthma. *Chest* 129: 1478–85, 2006.
70. Lange P, Scharling H, Ulrik CS, Vestbo J. Inhaled corticosteroids and decline of lung function in community residents with asthma. *Thorax* 61: 100–4, 2006.
71. Haahtela T, Järvinsen M, Kava T, Kiviranta K, Koskinen S, Lemtonen K *et al.* Effects of reducing or discontinuing inhaled budesonide in patients with mild asthma. *N Engl J Med* 331: 700–5, 1994.
72. Agertoft L, Pedersen S. Effects of long-term treatment with an inhaled corticosteroid on growth and pulmonary function in asthmatic children. *Resp Med* 5: 369–72, 1994.
73. Selroos O, Pietinalcho A, Lofroos A-B, Riska A. Effect of early and late intervention with inhaled corticosteroids in asthma. *Chest* 108: 1228–34, 1995.
74. Suissa S, Ernst P, Benayoun S, Baltzan M, Cai B. Low-dose inhaled corticosteroids and the prevention of death from asthma. *N Engl J Med* 343: 332–36, 2000.
75. Ernst P, Habbick B, Suissa S, Hemmelgarn B, Cockcroft D, Buist AS *et al.* Is the association between inhaled beta-agonist use and life-threatening asthma because of confounding by severity? *Am Rev Respir Dis* 148: 75–79, 1993.
76. Nelson HS, Weiss ST, Bleecker ER, Yancey SW, Dorinsky PM. The Salmeterol Multicenter Asthma Research Trial: A comparison of usual pharmacotherapy for asthma or usual pharmacotherapy plus salmeterol. *Chest* 129: 15–26, 2006.
77. Nathan RA, Nayak AS, Graft DF, Lawrence M, Picone FJ, Ahmed T *et al.* Mometasone furoate: Efficacy and safety in moderate asthma compared with beclomethasone dipropionate. *Ann Allergy Asthma Immunol* 86: 203–10, 2001.
78. Reynolds NA, Scott LJ. Ciclesonide. *Drugs* 64: 511–19, 2004.
79. Malo J-L, Cartier A, Merland N, Ghezzo H, Burke A, Morris J *et al.* Four-times-a-day dosing frequency is better than twice-a-day regimen in subjects requiring a high-dose inhaled steroid, budesonide, to control moderate to severe asthma. *Am Rev Respir Dis* 140: 624–28, 1989.
80. Otulana BA, Varma N, Bullock A, Higenbottam T. High dose nebulized steroid in the treatment of chronic steroid-dependent asthma. *Resp Med* 86: 105–8, 1992.
81. Hawkins G, McMahon AD, Twaddle S, Wood SF, Ford I, Thomson NC. Stepping down inhaled corticosteroids in asthma: Randomised controlled trial. *BMJ* 326: 1115, 2003.
82. Rabe KF, Hurd S, Anzueto A, Barnes PJ, Buist SA, Calverley P *et al.* Global strategy for the diagnosis, management, and prevention of COPD-2006 Update. *Am J Respir Crit Care Med* 176: 532–55, 2007.
83. Yang IA, Fong KM, Sim EH, Black PN, Lasserson TJ. Inhaled corticosteroids for stable chronic obstructive pulmonary disease. *Cochrane Database.Syst Rev*, 2007, CD002991.
84. Calverley PM, Anderson JA, Celli B, Ferguson GT, Jenkins C, Jones PW *et al.* Salmeterol and fluticasone propionate and survival in chronic obstructive pulmonary disease. *N Engl J Med* 356: 775–89, 2007.
85. Kankaanranta H, Lahdensuo A, Moilanen E, Barnes PJ. Add-on therapy options in asthma not adequately controlled by inhaled corticosteroids: A comprehensive review. *Respir Res* 5: 17, 2004.
86. Greening AP, Ind PW, Northfield M, Shaw G. Added salmeterol versus higher-dose corticosteroid in asthma patients with symptoms on existing inhaled corticosteroid. *Lancet* 344: 219–24, 1994.
87. Shrewsbury S, Pyke S, Britton M. Meta-analysis of increased dose of inhaled steroid or addition of salmeterol in symptomatic asthma (MIASMA). *BMJ* 320: 1368–73, 2000.
88. O'Byrne PM, Barnes PJ, Rodriguez-Roisin R, Runnerstrom E, Sandstrom T, Svensson K *et al.* Low dose inhaled budesonide and formoterol in mild persistent asthma: The OPTIMA randomized trial. *Am J Respir Crit Care Med* 164: 1392–97, 2001.
89. Gibson PG, Powell H, Ducharme FM. Differential effects of maintenance long-acting beta-agonist and inhaled corticosteroid on asthma control and asthma exacerbations. *J Allergy Clin Immunol* 119: 344–50, 2007.
90. Ni CM, Greenstone IR, Ducharme FM. Addition of inhaled long-acting beta2-agonists to inhaled steroids as first line therapy for persistent asthma in steroid-naive adults. *Cochrane Database Syst.Rev*, 2005, CD005307.
91. O'Byrne PM, Bisgaard H, Godard PP, Pistolesi M, Palmqvist M, Zhu Y *et al.* Budesonide/formoterol combination therapy as both maintenance and reliever medication in asthma. *Am J Resp Crit Care Med* 171: 129–36, 2005.
92. Rabe KF, Pizzichini E, Stallberg B, Romero S, Balanzat AM, Atienza T *et al.* Budesonide/formoterol in a single inhaler for maintenance and relief in mild-to-moderate asthma: A randomized, double-blind trial. *Chest* 129: 246–56, 2006.

93. Rabe KF, Atienza T, Magyar P, Larsson P, Jorup C, Lalloo UG. Effect of budesonide in combination with formoterol for reliever therapy in asthma exacerbations: A randomised controlled, double-blind study. *Lancet* 368: 744–53, 2006.
94. Tattersfield AE, Postma DS, Barnes PJ, Svensson K, Bauer CA, O'Byrne PM *et al.* Exacerbations of asthma. A descriptive study of 425 severe exacerbations. *Am J Respir Crit Care Med* 160: 594–99, 1999.
95. Barnes PJ. New therapies for asthma. *Trends Mol Med* 12: 515–20, 2006.
96. Calverley P, Pauwels R, Vestbo J, Jones P, Pride N, Gulsvik A *et al.* Combined salmeterol and fluticasone in the treatment of chronic obstructive pulmonary disease: A randomoised controlled trial. *Lancet* 361: 449–56, 2003.
97. Szafranski W, Cukier A, Ramirez A, Menga G, Sansores R, Nahabedian S *et al.* Efficacy and safety of budesonide/formoterol in the management of chronic obstructive pulmonary disease. *Eur Respir J* 21: 74–81, 2003.
98. Nannini L, Cates C, Lasserson T, Poole P. Combined corticosteroid and long-acting beta-agonist in one inhaler versus long-acting beta-agonists for chronic obstructive pulmonary disease. *Cochrane Database Syst Rev*, 2007, CD006829.
99. Wedzicha JA, Calverley PM, Seemungal TA, Hagan G, Ansari Z, Stockley RA. The prevention of chronic obstructive pulmonary disease exacerbations by salmeterol/fluticasone propionate or tiotropium bromide. *Am J Respir Crit Care Med* 177: 19–26, 2008.
100. Evans DJ, Taylor DA, Zetterstrom O, Chung KF, O'Connor BJ, Barnes PJ. A comparison of low-dose inhaled budesonide plus theophylline and high-dose inhaled budesonide for moderate asthma. *N Engl J Med* 337: 1412–18, 1997.
101. Ukena D, Harnest U, Sakalauskas R, Magyar P, Vetter N, Steffen H *et al.* Comparison of addition of theophylline to inhaled steroid with doubling of the dose of inhaled steroid in asthma. *Eur Respir J* 10: 2754–60, 1997.
102. Lim S, Jatakanon A, Gordon D, Macdonald C, Chung KF, Barnes PJ. Comparison of high dose inhaled steroids, low dose inhaled steroids plus low dose theophylline, and low dose inhaled steroids alone in chronic asthma in general practice. *Thorax* 55: 837–41, 2000.
103. Wilson AJ, Gibson PG, Coughlan J. Long acting beta-agonists versus theophylline for maintenance treatment of asthma. *Cochrane Database Syst Rev*, 2000. , CD001281
104. Laviolette M, Malmstrom K, Lu S, Chervinsky P, Pujet JC, Peszek I *et al.* Montelukast added to inhaled beclomethasone in treatment of asthma. *Am J Respir Crit Care Med* 160: 1862–68, 1999.
105. Price DB, Hernandez D, Magyar P, Fiterman J, Beeh KM, James IG *et al.* Randomised controlled trial of montelukast plus inhaled budesonide versus double dose inhaled budesonide in adult patients with asthma. *Thorax* 58: 211–16, 2003.
106. Nelson HS, Busse WW, Kerwin E, Church N, Emmett A, Rickard K *et al.* Fluticasone propionate/salmeterol combination provides more effective asthma control than low-dose inhaled corticosteroid plus montelukast. *J Allergy Clin Immunol* 106: 1088–95, 2000.
107. Ducharme F, Schwartz Z, Hicks G, Kakuma R. Addition of anti-leukotriene agents to inhaled corticosteroids for chronic asthma. *Cochrane Database Syst Rev*, 2004, CD003133.
108. Ducharme FM, Lasserson TJ, Cates CJ. Long-acting beta2-agonists versus anti-leukotrienes as add-on therapy to inhaled corticosteroids for chronic asthma. *Cochrane Database Syst Rev*, 2006., CD003137
109. Williamson IJ, Matusiewicz SP, Brown PH, Greening AP, Crompton GK. Frequency of voice problems and cough in patients using pressurised aerosol inhaled steroid preparations. *Eur Resp J* 8: 590–92, 1995.
110. Toogood JA, Jennings B, Greenway RW, Chung L. Candidiasis and dysphonia complicating beclomethasone treatment of asthma. *J Allergy Clin Immunol* 65: 145–53, 1980.
111. Ernst P, Gonzalez AV, Brassard P, Suissa S. Inhaled corticosteroid use in chronic obstructive pulmonary disease and the risk of hospitalization for pneumonia. *Am J Respir Crit Care Med* 176: 162–66, 2007.
112. Kamada AK, Szefler SJ, Martin RJ, Boushey HA, Chinchilli VM, Drazen JM *et al.* Issues in the use of inhaled steroids. *Am J Respir Crit Care Med* 153: 1739–48, 1996.
113. Brutsche MH, Brutsche IC, Munawar M, Langley SJ, Masterson CM, Daley-Yates PT *et al.* Comparison of pharmacokinetics and systemic effects of inhaled fluticasone propionate in patients with asthma and healthy volunteers: A randomised crossover study. *Lancet* 356: 556–61, 2000.
114. Harrison TW, Wisniewski A, Honour J, Tattersfield AE. Comparison of the systemic effects of fluticasone propionate and budesonide given by dry powder inhaler in healthy and asthmatic subjects. *Thorax* 56: 186–91, 2001.
115. Efthimou J, Barnes PJ. Effect of inhaled corticosteroids on bone and growth. *Eur Respir J* 11: 1167–77, 1998.
116. Roy A, Leblanc C, Paquette L, Ghezzo H, Cote J, Cartier A *et al.* Skin bruising in asthmatic subjects treated with high doses of inhaled steroids: Frequency and association with adrenal function. *Eur Respir J* 9: 226–31, 1996.
117. Simons FER, Persaud MP, Gillespie CA, Cheang M, Shuckett EP. Absence of posterior subcapsular cataracts in young patients treated with inhaled glucocorticoids. *Lancet* 342: 736–38, 1993.
118. Garbe E, LeLorier J, Boivin J-F, Suissa S. Inhaled and nasal glucocorticoids and the risks of ocular hypertension or open-angle glaucoma. *JAMA* 227: 722–27, 1997.
119. Pedersen S. Do inhaled corticosteroids inhibit growth in children? *Am J Respir Crit Care Med* 164: 521–35, 2001.
120. Allen DB, Mullen M, Mullen B. A meta-analysis of the effects of oral and inhaled corticosteroids on growth. *J Allergy Clin Immunol* 93: 967–76, 1994.
121. Silverstein MD, Yunginger JW, Reed CE, Petterson T, Zimmerman D, Li JT *et al.* Attained adult height after childhood asthma: Effect of glucocorticoid therapy. *J Allergy Clin Immunol* 99: 466–74, 1997.
122. Agertoft L, Pedersen S. Effect of long-term treatment with inhaled budesonide on adult height in children with asthma. *N Engl J Med* 343: 1064–69, 2000.
123. Schatz M. Asthma and pregnancy. *Lancet* 353: 1202–4, 1999.
124. Ernst P, Baltzan M, Deschenes J, Suissa S. Low-dose inhaled and nasal corticosteroid use and the risk of cataracts. *Eur Respir J* 27: 1168–74, 2006.
125. Beam WR, Ballard RD, Martin RJ. Spectrum of corticosteroid sensitivity in nocturnal asthma. *Am Rev Respir Dis* 145: 1082–86, 1992.
126. McLeod DT, Capewell SJ, Law J, MacLaren W, Seaton A. Intramuscular triamcinolone acetamide in chronic severe asthma. *Thorax* 40: 840–45, 1985.
127. Ogirala RG, Aldrich TK, Prezant DJ, Sinnett MJ, Enden JB, Williams MH. High dose intramuscular triamcinolone in severe life-threatening asthma. *N Engl J Med* 329: 585–89, 1991.
128. Engel T, Heinig JH. Glucocorticoid therapy in acute severe asthma—a critical review. *Eur Respir J* 4: 881–89, 1991.
129. Decramer M, Lacquet LM, Fagard R, Rogiers P. Corticosteroids contribute to muscle weakness in chronic airflow obstruction. *Am J Respir Crit Care Med* 150: 11–16, 1995.
130. Bowler SD, Mitchell CA, Armstrong JG. Corticosteroids in acute severe asthma: Effectiveness of low doses. *Thorax* 47: 584–87, 1992.
131. Morell F, Orkiols R, de Gracia J, Curul V, Pujol A. Controlled trial of intravenous corticosteroids in severe acute asthma. *Thorax* 47: 588–91, 1992.
132. Devidayal, Singhi S, Kumar L, Jayshree M. Efficacy of nebulized budesonide compared to oral prednisolone in acute bronchial asthma. *Acta Paediatr* 88: 835–40, 1999.

52

CHAPTER

Mediator Antagonists

Kian Fan Chung and Peter J. Barnes

National Heart and Lung Institute (NHLI), Clinical Studies Unit, Imperial College, London, UK

INTRODUCTION

Inflammatory mediators contribute to the pathophysiology of asthma and chronic obstructive pulmonary disease (COPD) (see Chapters 23–27) suggesting that antagonists of mediator receptors or inhibitors of their synthesis would be beneficial in treatment. However, because a large number of mediators are involved and many mediators share similar effects on the airways, inhibitors of single mediators may have had little or no clinical benefit to date. This may underlie the fact that despite many years of continuing efforts by pharmaceutical companies, only one class of mediator antagonists, the anti-leukotrienes, has become established in the treatment of asthma, representing the first new class of therapy for asthma introduced in more than 40 years. So far, there has been no mediator antagonists introduced for the treatment of COPD. In addition to reviewing the anti-leukotrienes, we will also review other classes of mediator antagonists that have failed to show beneficial effects or those currently under-investigation but whose potential has yet to be tested.

ANTI-LEUKOTRIENES

Anti-leukotrienes can be divided into cysteinyl-leukotriene (cys-LT) receptor antagonists that antagonize the effects of cys-LTs, such as LTD_4, and leukotriene synthesis inhibitors that are inhibitors of 5′-lipoxygenase (5-LO) enzyme that generates cys-LTs and LTB_4 (Fig. 52.1). Anti-leukotrienes are mainly indicated for the treatment of asthma, and LTB_4 inhibitors, such as LTB_4 receptor antagonists, have no effect in asthma and are being considered for the treatment of COPD.

5′-Lipoxygenase inhibitors

5-LO is a critical enzyme involved in the generation of leukotrienes. Inhibitors of 5-LO may be classified as direct inhibitors of the enzyme and indirect inhibitors that interfere with a nuclear membrane docking protein, 5-LO activating protein (FLAP), that is necessary for enzyme activation. Many hydroxamates and *N*-hydroxyureas are 5-LO inhibitors and act by interfering with the redox state of the active binding site. Zileuton is the most extensively investigated and is the only 5-LO inhibitor available for prescription in asthma (but only in the USA) [1]. The effect of zileuton is similar to that of leukotriene receptor antagonists, and zileuton inhibits allergen- and exercise-induced asthma, as well as aspirin-induced asthma. It decreases airway hyperresponsiveness and inflammatory cells in nocturnal exacerbations of asthma [2]. In addition, zileuton inhibits eosinophil influx induced by allergen challenge [3].

Zileuton has a short duration of action and has to be taken four times daily; a twice daily controlled-release preparation is also available [4]. Its side effects are mainly on the liver, with frequent abnormalities of liver function tests. Other redox 5-LO inhibitors have been developed but have not reached the market. Non-redox 5-LO inhibitors and inhibitors of FLAP (e.g. MK-886, MK-591, Bay-x1005) have been developed and are currently being developed further for clinical use in asthma and possibly COPD. One theoretical advantage of the 5-LO inhibitors on the receptor antagonists

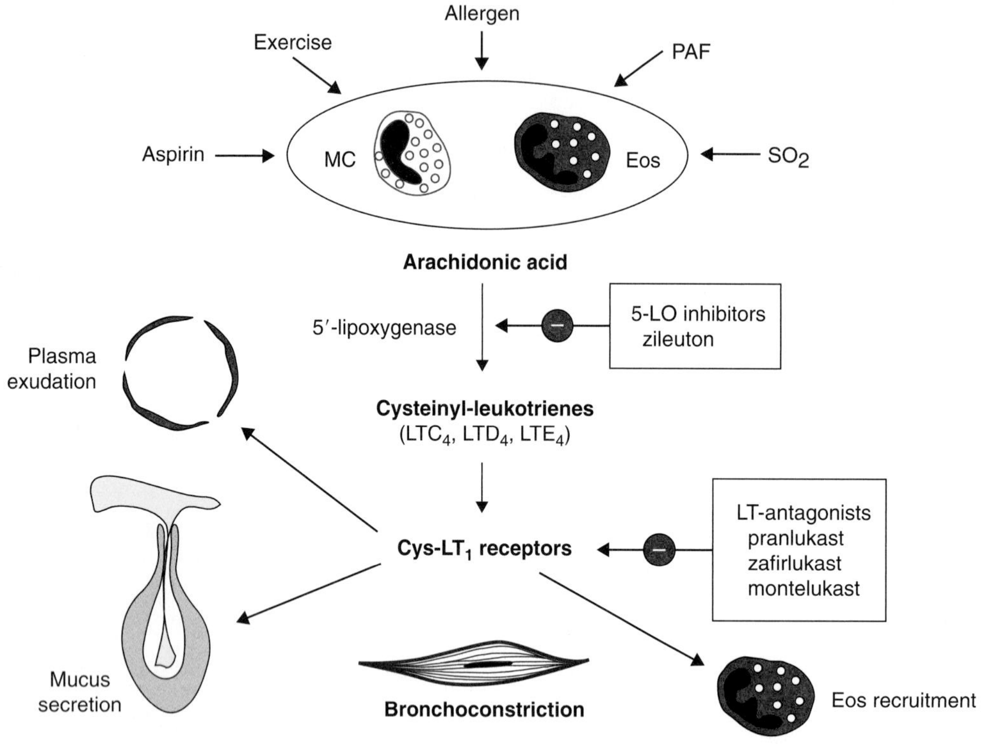

FIG. 52.1 Inhibition of effects of cysteinyl-leukotrienes (cys-LT) in asthma.

is that they inhibit the formation of LTB_4, and other 5-LO products such as eosxins, as well as cys-LTs. This may make the drugs more applicable to other airway diseases where LTB_4 may be involved, such as COPD and cystic fibrosis. There is anecdotal evidence that zileuton was more effective in controlling severe asthma than cys-LT_1 antagonists, supporting the view that other 5-LO products may be important, particularly in severe asthma.

cys-LT antagonists

The pathophysiological role of cys-LTs is discussed in Chapter 24. Cys-LTs cause airway obstruction through the stimulation of specific receptors termed the cys-LT receptor type 1 (Cyst-LT_1) present on airway smooth muscle. This a seven-transmembrane spanning, G-protein-coupled receptor, where the gene is mapped to the X chromosome [5]. Signaling through cyst-LT_1 occurs through stimulation of phosphoinositide hydrolysis [6]. A second cys-LT receptor, cyst-LT_2, is also expressed on vascular endothelium, blood eosinophils, mast cells, and lung macrophages, and could mediate activation of these cells [7] Zafirlukast and montelukast are potent leukotriene receptor antagonists available in most countries, while pranlukast is currently available only in Japan and Korea. There is only one anti-leukotriene a 5-lipoxygenase inhibitor, zileuton, which is available in the USA.

Clinical studies in asthma

Leukotriene receptor antagonists in clinical use inhibit the bronchoconstrictor effects of inhaled cys-LTs. For example, a single 40 mg dose of zafirlukast produces a 100-fold shift of the LTD_4 dose–response curve, and significant protection is present for 24 h [8]. Oral administration of leukotriene receptor antagonist inhibits both the early and late responses to allergen, and exercise-induced asthma [9, 10]. Leukotriene receptor antagonists are able to cause bronchodilation, and their effect is additive to that of short-acting β_2-agonists [11, 12].

In many studies of anti-leukotrienes such as zileuton, zafirlukast, pranlukast, and montelukast, their effectiveness has been compared with that of placebo in short-term studies of 4–6 week duration [13–16]. A greater increase in FEV_1, a reduction in asthma medication use and in asthma symptoms, with an increase in morning peak flow has been demonstrated in mild-to-moderate severe asthma with some degree of airflow obstruction and usually in cohorts not on inhaled corticosteroid therapy. These findings have been confirmed and extended in longer studies in patients with mild-to-moderate chronic stable asthma with zileuton, zafirlukast, montelukast, all demonstrating clinical benefit [17–19]. Their additive effect to the bronchodilation achieved with high doses of inhaled β-agonist indicates that they may have a place in the treatment of acute severe asthma. Intravenous administration of montelukast in acute severe asthma provided additional benefit to the receovery when given together with standard therapy [20]. Clinical benefit has also been demonstrated with the addition of anti-leukotrienes to the patients with poor asthma control who are already taking high doses of inhaled corticosteroids [21]. Anti-leukotrienes may also reduce to a small extent the doses of inhaled corticosteroids required for asthma control [22]. Anti-leukotrienes may reduce the risk of acute severe asthma exacerbations [23, 24].

The effects of anti-leukotrienes have been compared with these inhaled corticosteroids [25]. In a recent meta-analysis, inhaled corticosteroids (250–400 μg beclomethasone dipropionate equivalent per day) was found to provide better improvements in lung function and quality of life, as well as reduction in symptoms, night awakenings, and need for rescue β-agonist. The rate of asthma exacerbations was similar when the anti-leukotrienes were compared with inhaled corticosteroids. The benefit of adding leukotriene-receptor antagonist with inhaled corticosteroids has been evaluated. Addition of pranlukast to half the usual dose of inhaled corticosteroids in patients with moderate-to-severe persistent asthma led to maintained control, while the placebo group demonstrated less improved asthma control [26]. Montelukast maintained control of asthma in patients in whom removal of inhaled corticosteroids caused worsening of asthma [26]. When compared to long-acting β-agonists, montelukast was less superior for preventing exacerbations and for improving lung function, symptoms, and the use of rescue β-agonists [27, 28]. Combination of a leukotriene receptor antagonists with long-acting β-agonists is not advisable for the treatment of moderate asthma [29].

Anti-leukotrienes are effective in blocking aspirin-induced asthmatic responses [30] and may be particularly indicated in patients with aspirin-sensitive asthma. In addition, anti-leukotrienes are particularly effective in inhibiting exercise-induced asthma, [10, 31, 32] without the loss of protection with prolonged usage. In addition, leukotriene inhibitors improve concomitant symptoms of allergic rhinitis [33, 34].

In United States, zileuton, zafirlukast, and montelukast have been approved as a first-line choice for the trteatment of asthma, while in Europe, montelukast and zafirlukast have been approved as an add-on therapy when treatment with inhaled corticosteroids has failed, and for the prophylaxis of exercise-induced asthma. In the GINA 2006 guidelines, anti-leukotrienes are listed as a less preferred initial treatment compared to inhaled corticosteroids and a less preferred option as an add-on therapy compared to long-acting β_2-agonists [35].

Anti-inflammatory effects

Cys-LTs can induce airway eosinophilia in patients with asthma, and conversely, leukotriene-receptor antagonists or synthesis inhibitors reduce blood and airway eosinophilia associated with poorly controlled asthma, [23] and airway inflammation associated with allergen-induced airway responses [36]. Therefore, anti-leukotrienes can be considered as an anti-inflammatory therapy for asthma, but this remains to be confirmed in the clinical situation. In a study of the combination therapy of montelukast with inhaled corticosteroid, there was no effect of montelukast on sputum eosinophil counts [37]. Although there is cellular and animal data to support a role for cys-LTs in airway smooth muscle hyperplasia and airway fibrosis [38, 39], there is as yet no data of the effect of leukotriene inhibitors on these parameters in asthma.

Safety

At the recommended doses, all leukotriene receptor antagonists have not resulted in nonrespiratory symptoms nor laboratory abnormalities when compared with placebo-treated groups. However, with zileuton, asymptomatic 3-fold or greater increases in serum alanine-aminotransferase levels were found in 4.6% of patients receiving zileuton at the standard dose of 600 mg four times per day, compared with 1% of patients receiving standard asthma treatment together with placebo. Controlled-release zileuton at a dose of 1200 mg twice daily was shown to have efficacy within the range of zileuton 600 mg four times daily, with similar side effect profile [40]. These elevations of liver enzymes usually occurred during the first 3 months of therapy, with sometimes normalization of the values despite continuation of treatment.

A rare syndrome of Churg–Strauss, marked by circulating eosinophilia and evidence of tissue or organ infiltration and vasculitis by eosinophils in association with heart failure, cutaneous or gastrointestinal involvement, and peripheral neuropathy, has been associated with treatment with zafirlukast and montelukast. Most patients developing Churg–Strauss syndrome have previously received oral glucocorticoid therapy or high-dose inhaled corticosteroid therapy to control their asthma. This may be due to unmasking of vasculitis of Churg–Strauss syndrome, as corticosteroids are tapered with the introduction of leukotriene receptor antagonist therapy [41].

Role in COPD

Small bronchodilator effects of zafirlukast has been observed in some patients with COPD [42, 43] Because LTB_4 may play a role in the recruitment of neutrophils in COPD, the use of 5-LO inhibitors may bring some additional effects. In a small study of the leukotriene synthesis inhibitor, a 5-lipoxygenase activating protein inhibitor, BAYx1005, a small reduction in sputum LTB4 was observed with no clinical benefit [44]. No studies of zileuton in COPD have been reported yet.

PROSTAGLANDIN AND CRTH2 INHIBITORS

Prostaglandin D_2 (PGD_2) is produced by activated mast cells and is a relatively selective marker for mast cells, and two G-protein-coupled receptors, DP_1 and DP_2 (also known as CRTH2), mediate many of its effects. However, it is quite likely that the PGD_2-bronchoconstrictor effect is mediated via the TP receptor, since the activation of DP_1 receptor in airway smooth muscle cells leads to bronchodilation. DP_2-receptors play an important role in Th2 lymphocyte recruitment and activation [45] (Fig. 52.2). Ramatroban is an effective TP antagonist which has been approved as a treatment for perennial rhinitis. It reduces bronchial hyperresponsiveness to methacholine in asthmatics [46]. Ramatroban also has a partial action as an

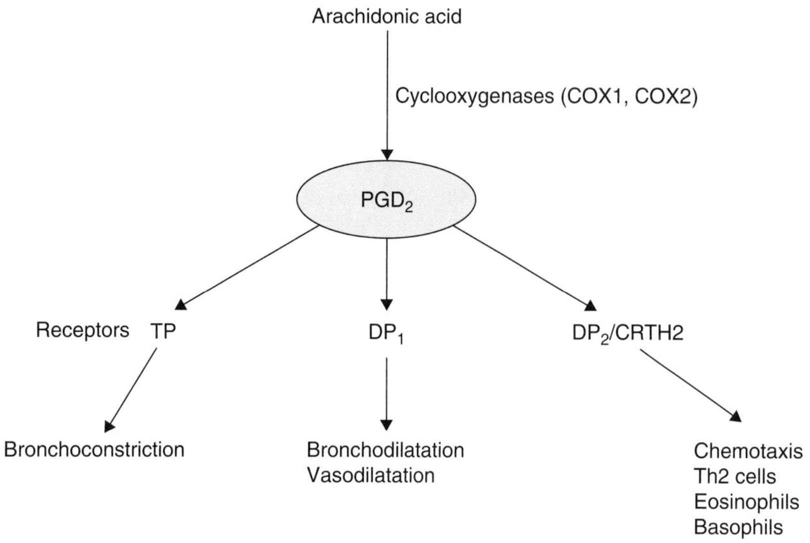

FIG. 52.2 Prostaglandin D_2 (PGD_2) effects. PGD_2 acts on three distinct prostanoid receptiors: the thromboxane receptor (TP) mediates bronchonstriction, DP_1-receptor mediates vasodilatation, and bronchodilatation and DP_2-receptor (also know as CRTH2) is chemotactic for Th2 cells, eosinophils, and basophils.

antagonist at the DP_2 receptor as shown by its inhibitory effect on PGD_2-stimulated eosinophil chemotaxis [47]. More selective and potent DP_2 antagonists have now been developed and are undergoing trials in asthma and allergic rhinitis [48].

THROMBOXANE INHIBITORS

Although thromboxane production is increased in asthma and thromboxane analogs are potent bronchoconstrictors in asthmatic patients [49], there is no convincing evidence that thromboxane receptor (TP) antagonists or thromboxane synthase inhibitors are effective in asthma [50]. A selective TP antagonist, GR32191, caused a modest inhibition of early phase bronchoconstriction due to allergen but was not clinically effective in improving lung function or reducing symptoms in mild asthma [51]. However, a thromboxane synthase inhibitor (ozagrel) and a receptor antagonist (seratrodast) are used in the treatment of asthma in Japan, although their effects in asthma are small [52–54].

There is no evidence for a role for TP or DP receptors in COPD.

ANTI-HISTAMINES

Histamine mediates most of its effects on airway function via H_1-receptors. Non-sedating potent H_1-receptor antagonists, such as terfenadine, fexafenadine, loratadine, desloratadine, ebastine, and astemizole, have useful clinical effects in allergic rhinitis, but they are far from effective in asthmatic patients [55]. The effects of anti-histamines are small and clinically insignificant. Terfenadine causes about 50% inhibition of the immediate response to allergen, but has no effect on the late response [56]. Anti-histamines cause a small degree of bronchodilatation in asthmatic patients, indicating a certain degree of histamine "tone," presumably due to the basal release of histamine from activated mast cells [57]. Chronic administration of terfenadine has a small clinical effect in mild allergic asthmatic patients, [58] but is far less effective that other anti-asthma therapies. H_1-receptor antagonists have not been found to be useful in more severe asthmatic patients [48]. The new generation anti-histamines, cetirizine and astemizole, have some beneficial effects in asthma, [59, 60] that may be unrelated to their H_1-antagonist effects [61].

H_2-antagonists, such as cimetidine and ranitidine, may be contraindicated in asthma on theoretical grounds, if H_2-receptors are important in counteracting the bronchoconstrictor effect of histamine. In clinical practice, however, there is no evidence that H_2-antagonists have any deleterious effect in asthma. H_3-receptor agonists may have some theoretical benefit in asthma, since they may modulate cholinergic bronchoconstriction and inhibit neurogenic inflammation. Although (R)-α-methylhistamine relaxes rodent peripheral airways *in vitro,* it has no effect when given by inhalation on airway caliber or metabisufite-induced bronchoconstriction in asthmatic patients, indicating that a useful clinical effect is unlikely [62].

Histamine H_4-receptors are expressed on eosinophils, T- cells, dendritic cells, basophils and mast cells, mediate mast cell, eosinophil and dendritic cell chemotaxis, and modulate cytokine production from dendritic cells and T-cells, indicating that blockade of histamine H_4-receptors may lead to anti-allergic and anti-inflammatory effects. Several histamine H_4-receptor antagonists are now available but remain to be tested in allergic asthma or rhinitis [63]. Antagonists that block both histamine H_1- and H_4-receptors may be an effective combination. Anti-histamines have a useful effect in the treatment of rhinitis, and particularly the rhinorrhea. As a large proportion of patients with asthma have concomitant rhinitis, an H_1-antagonist may help the overall management of asthma [64]. While H_1-receptor antagonists alone may be ineffective, some studies suggest that they may have some efficacy in combination with other antagonists. Thus, an H_1-receptor antagonist when added to an anti-leukotriene was able to inhibit the early and late responses to allergen more effectively than the anti-leukotriene alone [65, 66], but as yet there has been no studies of combination mediator antagonists in asthma.

There is no evidence that anti-histamines have any role in the treatment of COPD.

SEROTONIN ANTAGONISTS

The evidence for involvement of serotonin in asthma is weak. There is no evidence that serotonin is a direct constrictor in human airways, and it is not stored in and released from human mast cells, as in rodents. Serotonin receptor antagonists have been studied experimentally in asthmatic patients. Ketanserin, which antagonizes $5HT_2$ receptors and blocks the bronchoconstrictor effects of serotonin in animals, has no effect on airway function in asthmatic patients, but there is a small inhibitory effect on methacholine-induced bronchoconstriction [67]. Inhaled ketanserin has no effect on histamine-induced bronchoconstriction but has a small inhibitory effect on adenosine-induced bronchoconstriction, indicating a possible action on mast cells [68]. Tianeptine, which enhances serotonin uptake by platelets, lowers the elevated plasma serotonin levels reported in patients with asthma and is associated with a reduction in asthmatic symptoms [69].

PAF ANTAGONISTS

Although PAF mimics many of the pathophysiological features of asthma including induction of airway hyperresponsiveness, PAF antagonists have proved to be very disappointing in asthma therapy.

Apafant (WEB 2086) potently inhibits the effects of inhaled PAF on airway function [70], but had no effect on the early or late responses to allergen or on airway hyperresponsiveness in patients with mild asthma [71]. A 3-month study of oral apafant in patients with symptomatic moderate asthma failed to show any effect on lung function, symptoms or on rescue β_2-agonist use [72]. Similarly, another PAF antagonist, modipafant (UK80067) had no beneficial effects in moderately severe asthmatics [73]. Another potent and long-acting PAF receptor antagonist, foropafant (SR27417A), was effective in inhibiting systemic, cellular, and pulmonary effects of PAF challenge in patients with mild bronchial asthma [74]. This antagonist produced a modest but significant reduction in the magnitude of the allergen-induced late response, but no effect on the early response, allergen-induced hyperresponsiveness, or on baseline lung function [75]. These clinical studies indicate that PAF plays little or no part in human allergic asthma, despite convincing data in animal models. There is no evidence that PAF is involved in COPD, and no studies with PAF antagonists in this disease have been reported.

BRADYKININ ANTAGONISTS

There is evidence for the involvement of bradykinin in asthma, in particular a role in sensitizing and activating airway sensory nerves. However, a peptide bradykinin antagonist [D-Arg^0,Hyp^3,D-Phe^7]bradykinin (NPC567) was unable to inhibit the effect of bradykinin on nasal secretions, even when given at the same time as bradykinin, [76] presumably because of rapid local metabolism. There has been other more stable antagonists developed [77, 78]. Icatibant (HOE 140, D-Arg[Hyp^3,Thi^6,D-Tic^7,Oic^8] bradykinin is a potent selective B_2-receptor antagonist, [79] which has a long duration of action *in vivo* since it is resistant to enzymatic degradation. This antagonist is potent in inhibiting the bronchoconstrictor and microvascular leakage response to bradykinin [80] and the effect of bradykinin on airway sensory nerves [81]. Nasal application of icatibant reduces the nasal blockage induced by allergen in patients with allergic rhinitis [82]. In a clinical study of nebulized icatibant in asthma, there was a small improvement in airway function after 4 weeks without an improvement in asthma symptoms [83]. WIN 64338 is a nonpeptide B_2-receptor antagonist that blocked the bronchoconstrictor action of bradykinin in airway smooth muscle *in vitro* [84] and more potent nonpeptide antagonists, such as FR167344 have now been developed that have clinical potential [85].

TACHYKININ ANTAGONISTS

Tachykinins interacts with the tachykinin NK_1-, NK_2-, and NK_3-receptors which exist in the airways, particularly on human bronchial glands, bronchial vessels, and bronchial smooth muscle. Inflammatory cells also express NK_1-receptors. NK_1- and NK_2-receptor antagonists have been shown to inhibit allergen-induced bronchial hyperresponsiveness and inflammation in various animal models. Potent NK_1- and NK_2-receptor antagonists have been recently available but have not shown any effects in asthma [86]. Thus, an NK_2-receptor antagonist, SR48968, that caused a significant inhibition of neurokinin-A bronchoconstriction by 3–5 fold shift did not affect baseline airway caliber or bronchial responsiveness to adenosine in asthma [87]. Dual NK1/NK2 and triple NK1/NK2/NK3 receptor antagonists are available with greater potencies, but it is unlikely that there will be trials in asthma. There has been no studies of tachykinin antagonists in COPD.

ADENOSINE AGONISTS AND ANTAGONISTS

Adenosine is released following allergen exposure and exercise of sensitized asthmatics and is a bronchoconstrictor in asthmatics through the activation of mast cells. Adenosine may also be involved in allergic inflammation and remodeling. It interacts with G-protein-coupled receptors, adenosine A_1-, A_{2A}-, A_{2B}- and A_3-receptors that are present on inflammatory and structural cells of the asthmatic airway. Various agonists and antagonists of these receptors have been developed as potential therapies for a variety of cardiac and pulmonary conditions, including asthma and COPD [88]. Activation of A_{2A}-receptors has been shown to have anti-inflammatory effects in animal models, and a selective A_{2A}- agonist has been shown to reduce the number of eosinophils and neutrophils in allergen-induced inflammation. However, two agonists of the A_{2A}-receptor, GW328267 and UK-432,097, have been tested in asthma. GW328267 had no effect on allergen-induced early or late phase response, and UK-432097 had no effect on lung function or bronchial hyperreactivity of mild asthma [89].

Adenosine has been shown to act through the A_{2B}-receptors to enhance various inflammatory responses seen in asthma. An antagonist of A_{2B}-receptors, CVT-6883, has been shown to inhibit the proinflammatory cytokines and profibrotic effects found in adenosine deaminase-deficient mice [90]. It is under development as an oral treatment for asthma.

References

1. Berger W, De Chandt MT, Cairns CB. Zileuton: Clinical implications of 5-Lipoxygenase inhibition in severe airway disease. *Int J Clin Pract* 61: 663–76, 2007.
2. Wenzel SE, Trudeau JB, Kaminsky DA, Cohn J, Martin RJ, Westcott JY. Effect of 5-lipoxygenase inhibition on bronchoconstriction and airway inflammation in nocturnal asthma. *Am J Respir Crit Care Med* 152: 897–905, 1995.
3. Kane GC, Pollice M, Kim CJ, Cohn J, Dworski RT, Murray JJ *et al.* A controlled trial of the effect of the 5-lipoxygenase inhibitor, zileuton, on lung inflammation produced by segmental antigen challenge in human beings. *J Allergy Clin Immunol* 97: 646–54, 1996.
4. Wenzel S, Busse W, Calhoun W, Panettieri R Jr., Peters-Golden M, Dube L *et al.* The safety and efficacy of zileuton controlled-release tablets as adjunctive therapy to usual care in the treatment of moderate persistent asthma: A 6-month randomized controlled study. *J Asthma* 44: 305–10, 2007.
5. Lynch KR, O'Neill GP, Liu Q, Im DS, Sawyer N, Metters KM *et al.* Characterization of the human cysteinyl leukotriene CysLT1 receptor. *Nature* 399: 789–93, 1999.
6. Crooke ST, Sarau HM, Saussy D, Winkler J, Foley J. Signal transduction processes for the LTD4 receptor. *Adv Prost Thromb Leuko Res* 20: 127–37, 1990.
7. Kanaoka Y, Boyce JA. Cysteinyl leukotrienes and their receptors: Cellular distribution and function in immune and inflammatory responses. *J Immunol* 173: 1503–10, 2004.
8. Smith LJ, Geller S, Ebright L, Glass M, Thyrum PT. Inhibition of leukotriene D4-induced bronchoconstriction in normal subjects by the oral LTD4 receptor antagonist ICI 204,2. *Am Rev Respir Dis* 141: 988–92, 1990.
9. Taylor IK, O'Shaughnessy KM, Fuller RW, Dollery CT. Effect of cysteinyl-leukotriene receptor antagonist ICI 204-219 on allergen-induced bronchoconstriction and airway hyperreactivitiy in atopic subjects. *Lancet* 337: 690–94, 1991.
10. Finnerty JP, Wood Baker R, Thomson H, Holgate ST. Role of leukotrienes in exercise-induced asthma. Inhibitory effect of ICI 204219, a potent leukotriene D4 receptor antagonist. *Am Rev Respir Dis* 145: 746–49, 1992.
11. Hui KP, Barnes NC. Lung function improvement in asthma with a cystemyl-leukotriene receptor antagonist. *Lancet* 337: 1062–63, 1991.
12. Reiss TF, Sorkness CA, Stricker W, Botto A, Busse WW, Kundu S *et al.* Effects of montelukast (MK-0476); a potent cysteinyl leukotriene receptor antagonist, on bronchodilation in asthmatic subjects treated with and without inhaled corticosteroids. *Thorax* 52: 45–48, 1997.
13. Spector SL, Smith LJ, Glass M. Effects of six weeks of therapy with oral doses of ICI 204,219, a leukotriene D4 receptor antagonist, in subjects with bronchial asthma. *Am J Respir Crit Care Med* 150: 618–23, 1994.
14. Altman LC, Munk Z, Seltzer J, Noonan N, Shingo S, Zhang J *et al.* A placebo-controlled, dose-ranging study of montelukast, a cysteinyl leukotriene-receptor antagonist. Montelukast Asthma Study Group. *J Allergy Clin Immunol* 102: 50–56, 1998.
15. Barnes NC, Pujet JC. Pranlukast, a novel leukotriene receptor antagonist: Results of the first European, placebo controlled, multicentre clinical study in asthma. *Thorax* 52: 523–27, 1997.
16. Israel E, Rubin P, Kemp JP, Grossman J, Pierson W, Siegel SC *et al.* The effect of inhibition of 5-lipoxygenase by zileuton in mild- to-moderate asthma. *Ann Intern Med* 119: 1059–66, 1993.
17. Liu MC, Dube L, Lancaster J. and the Zileuton Study Group. Acute and chronic effects of a 5-lipoxygenase inhibitor in asthma: A 6 month randomized multicentre trial. *J Allergy Clin Immunol* 98: 859–71, 1996.
18. Calhoun WJ. Summary of clinical trials with zafirlukast. *Am J Respir Crit Care Med* 157: S238–45, 1998.
19. Blake KV. Montelukast: Data from clinical trials in the management of asthma. *Ann Pharmacother* 33: 1299–314, 1999.
20. Camargo CA Jr., Smithline HA, Malice MP, Green SA, Reiss TF. A randomized controlled trial of intravenous montelukast in acute asthma. *Am J Respir Crit Care Med* 167: 528–33, 2003.
21. Virchow CJ, Prasse A, Naya I, Summerton L, harris A. Zafirlukast improves asthma control in patients receiving high-dose inhaled corticosteroids. *Am J Respir Crit Care Med* 162: 578–85, 2000.
22. Lofdahl C-G, Reiss T, Leff JA, Israel E, Noonan M, Finn AF *et al.* Randomized, placebo controlled trial of effect of a leukotriene receptor antagonist, montelukast, on tapering inhaled corticosteroids in asthmatic patients. *B MJ* 319: 87–90, 1999.
23. Laviolette M, Malmstrom K, Lu S, Chervinsky P, Pujet JC, Peszek I *et al.* Montelukast added to inhaled beclomethasone in treatment of asthma. Montelukast/Beclomethasone Additivity Group. *Am J Respir Crit Care Med* 160: 1862–68, 1999.
24. Barnes NC, Miller CJ. Effect of leukotriene receptor antagonist therapy on the risk of asthma exacerbations in patients with mild to moderate asthma: An integrated analysis of zafirlukast trials. *Thorax* 55: 478–83, 2000.
25. Ducharme FM, Hicks GC. Anti-leukotriene agents compared to inhaled corticosteroids in the management of recurrent and/or chronic asthma (Cochrane review) [In Process Citation]. *Cochrane Database Syst Rev*(3), 2000., *CD002314.*CD002314.
26. Tamaoki J, Kondo M, Sakai N, Nakata J, Takemura H, Nagai A *et al.* Leukotriene antagonist prevents exacerbations of asthma during reduction of high-dose inhaled corticosteroid. *Am J Respir Crit Care Med* 155: 1235–40, 1997.
27. Ducharme FM, Lasserson TJ, Cates CJ. Long-acting beta2-agonists versus anti-leukotrienes as add-on therapy to inhaled corticosteroids for chronic asthma. *Cochrane Database Syst Rev* 43, 2006, CD003137.
28. Peters SP, Anthonisen N, Castro M, Holbrook JT, Irvin CG, Smith LJ *et al.* Randomized comparison of strategies for reducing treatment in mild persistent asthma. *N Engl J Med* 356: 2027–39, 2007.
29. Deykin A, Wechsler ME, Boushey HA, Chinchilli VM, Kunselman SJ, Craig TJ *et al.* Combination therapy with a long-acting beta-agonist and a leukotriene antagonist in moderate asthma. *Am J Respir Crit Care Med* 175: 228–34, 2007.
30. Israel E, Fischer AR, Rosenberg MA, Lilly CM, Callery JC, Shapiro J *et al.* The pivotal role of 5-lipoxygenase products in the reaction of aspirin-sensitive asthmatics to aspirin. *Am Rev Respir Dis* 148: 1447–51, 1993.

31. Leff JA, Busse WW, Pearlman D, Bronsky EA, Kemp J, Hendeles L *et al.* Montelukast, a leukotriene-receptor antagonist, for the treatment of mild asthma and exercise-induced bronchoconstriction. *N Engl J Med* 339: 147–52, 1998, [see comments]
32. Manning PJ, Watson RM, Margolskee DJ, Williams VC, Schwartz JI, O'Byrne PM. Inhibition of exercise-induced bronchoconstriction by MK-751, a potent leukotriene-D4 receptor antagonist. *N Engl J Med* 323: 1736–39, 1990.
33. Donnelly AL, Glass M, Minkwitz MC, Casale TB. The leukotriene D4-receptor antagonist, ICI 204,219, relieves symptoms of acute seasonal allergic rhinitis. *Am J Respir Crit Care Med* 151: 1734–39, 1995.
34. Patel P, Philip G, Yang W, Call R, Horak F, Laforce C *et al.* Randomized, double-blind, placebo-controlled study of montelukast for treating perennial allergic rhinitis. *Ann Allergy Asthma Immunol* 95: 551–57, 2005.
35. Bateman ED, Hurd SS, Barnes PJ, Bousquet J, Drazen JM, Fitzgerald M *et al.* Global strategy for asthma management and prevention: GINA executive summary. *Eur Respir J* 31: 143–78, 2008.
36. Calhoun WJ, Lavins BJ, Minkwitz MC, Evans R, Gleich GJ, Cohn J. Effect of zafirlukast (Accolate) on cellular mediators of inflammation: Bronchoalveolar lavage fluid findings after segmental antigen challenge. *Am J Respir Crit Care Med* 157: 1381–89, 1998.
37. Barnes N, Laviolette M, Allen D, Flood-Page P, Hargreave F, Corris P *et al.* Effects of montelukast compared to double dose budesonide on airway inflammation and asthma control. *Respir Med* 101: 1652–58, 2007.
38. Salmon M, Walsh DA, Huang TJ, Barnes PJ, Leonard TB, Hay DW *et al.* Involvement of cysteinyl leukotrienes in airway smooth muscle cell DNA synthesis after repeated allergen exposure in sensitized Brown Norway rats. *Br J Pharmacol* 127: 1151–58, 1999.
39. Henderson WR, Chiang GK, Tien YT, Chi EY. Reversal of allergen-induced airway remodeling by CysLT1 receptor blockade. *Am J Respir Crit Care Med* 173: 718–28, 2006.
40. Wenzel S, Busse W, Calhoun W, Panettieri R, Peters-Golden M, Dube L *et al.* The safety and efficacy of zileuton controlled-release tablets as adjunctive therapy to usual care in the treatment of moderate persistent asthma: A 6-month randomized controlled study. *J Asthma* 44: 305–10, 2007.
41. Wechsler ME, Garpestad E, Flier SR, Kocher O, Weiland DA, Polito AJ *et al.* Pulmonary infiltrates, eosinophilia, and cardiomyopathy following corticosteroid withdrawal in patients with asthma receiving zafirlukast. *JAMA* 279: 455–57, 1998.
42. Cazzola M, Santus P, Di Marco F, Boveri B, Castagna F, Carlucci P *et al.* Bronchodilator effect of an inhaled combination therapy with salmeterol + fluticasone and formoterol + budesonide in patients with COPD. *Respir Med* 97: 453–57, 2003.
43. Nannini LJ, Flores DM. Bronchodilator effect of zafirlukast in subjects with chronic obstructive pulmonary disease. *Pulm Pharmacol Ther* 16: 307–11, 2003.
44. Gompertz S, Stockley RA. A randomized, placebo-controlled trial of a leukotriene synthesis inhibitor in patients with COPD. *Chest* 122: 289–94, 2002.
45. Kostenis E, Ulven T. Emerging roles of DP and CRTH2 in allergic inflammation. *Trends Mol Med* 12: 148–58, 2006.
46. Aizawa H, Shigyo M, Nogami H, Hirose T, Hara N. BAY u3405, a thromboxane A2 antagonist, reduces bronchial hyperresponsiveness in asthmatics. *Chest* 109: 338–42, 1996.
47. Mimura H, Ikemura T, Kotera O, Sawada M, Tashiro S, Fuse E *et al.* Inhibitory effect of the 4-aminotetrahydroquinoline derivatives, selective chemoattractant receptor-homologous molecule expressed on T helper 2 cell antagonists, on eosinophil migration induced by prostaglandin D2. *J Pharmacol Exp Ther* 314: 244–51, 2005.
48. Pettipher R, Hansel TT, Armer R. Antagonism of the prostaglandin D2 receptors DP1 and CRTH2 as an approach to treat allergic diseases. *Nat Rev Drug Discov* 6: 313–25, 2007.
49. Saroea HG, Inman MD, O'Byrne PM. U46619-induced bronchoconstriction in asthmatic subjects is mediated by acetylcholine release. *Am J Respir Crit Care Med* 151: 321–24, 1995.
50. O'Byrne PM, Fuller RW. The role of thromboxane A2 in the pathogenesis of airway hyperresponsiveness. *Eur Respir J* 2: 782–86, 1989.
51. Coleman RA. GR32191 and the role of thromboxane A2 in asthma–preclinical and clinical findings. *Agents Actions* 34(Suppl): 211–20, 1991.
52. Manning PJ, Stevens WH, Cockcroft DW, O'Byrne PM. The role of thromboxane in allergen-induced asthmatic responses. *Eur Respir J* 4: 667–72, 1991.
53. Obase Y, Shimoda T, Matsuo N, Matsuse H, Asai S, Kohno S. Effects of cysteinyl-leukotriene receptor antagonist, thromboxane A2 receptor antagonist, and thromboxane A2 synthetase inhibitor on antigen-induced bronchoconstriction in patients with asthma. *Chest* 114: 1028–32, 1998.
54. Fukuoka T, Miyake S, Umino T, Inase N, Tojo N, Yoshizawa Y. The effect of seratrodast on eosinophil cationic protein and symptoms in asthmatics. *J Asthma* 40: 257–64, 2003.
55. Van Ganse E, Kaufman L, Derde MP, Yernault JC, Delaunois L, Vincken W. Effects of antihistamines in adult asthma: A meta-analysis of clinical trials. *Eur Respir J* 10: 2216–24, 1997.
56. Hamid M, Rafferty P, Holgate ST. The inhibitory effect of terfenadine and flurbiprofen on early and late-phase bronchoconstriction following allergen challenge in atopic asthma. *Clin Exp Allergy* 20: 261–67, 1990.
57. Eiser N, Mills J, Snashall P, Guz A. The role of histamine receptors in asthma. *Clin Sci* 60: 363–70, 1981.
58. Taytard A, Beaumont D, Pujet JC, Sapene M, Lewis PJ. Treatment of bronchial asthma with terfenadine: A randomized controlled trial. *Br J Clin Pharmacol* 24: 743–46, 1987.
59. Spector SL, Nicodemus CF, Corren J, Schanker HM, Rachelefsky GS, Katz RM *et al.* Comparison of the bronchodilatory effects of cetirizine, albuterol, and both together versus placebo in patients with mild-to-moderate asthma. *J Allergy Clin Immunol* 96: 174–81, 1995.
60. Busse WW, Middleton E, Storms W, Dockhorn RJ, Chu TJ, Grossman J *et al.* Corticosteroid-sparing effect of azelastine in the management of bronchial asthma. *Am J Respir Crit Care Med* 153: 122–27, 1996.
61. Walsh GM. The anti-inflammatory effects of cetirizine. *Clin Exp Allergy* 24: 81–85, 1994.
62. O'Connor BJ, Lecomte JM, Barnes PJ. Effect of an inhaled histamine H3-receptor agonist on airway responses to sodium metabisulphite in asthma. *Br J Clin Pharmacol* 35: 55–57, 1993.
63. Fung-Leung WP, Thurmond RL, Ling P, Karlsson L. Histamine H4 receptor antagonists: The new antihistamines?. *Curr Opin Investig Drugs* 5: 1174–83, 2004.
64. Simons FE. Is antihistamine (H1-receptor antagonist) therapy useful in clinical asthma? *Clin Exp Allergy* 29(Suppl 3): 98–104, 1999.
65. Roquet A, Dahlen B, Kumlin M, Ihre E, Anstren G, Binks S *et al.* Combined antagonism of leukotrienes and histamine produces predominant inhibition of allergen-induced early and late phase airway obstruction in asthmatics. *Am J Respir Crit Care Med* 155: 1856–63, 1997.
66. Meltzer EO, Malmstrom K, Lu S, Prenner BM, Wei LX, Weinstein SF *et al.* Concomitant montelukast and loratadine as treatment for seasonal allergic rhinitis: A randomized, placebo-controlled clinical trial. *J Allergy Clin Immunol* 105: 917–22, 2000.
67. Cazzola M, Assogna G, Lucchetti G, Cicchitto G, D'Amato G. Effect of ketanserin, a new blocking agent of the 5-HT2 receptor, on airway responsiveness in asthma. *Allergy* 45: 151–53, 1990.
68. Cazzola M, Matera MG, Santangelo G, Assogna G, D'Amato G, Rossi F *et al.* Effect of the selective 5-HT2 antagonist ketanserin on adenosine-induced bronchoconstriction in asthmatic subjects. *Immunopharmacology* 23: 21–28, 1992.
69. Lechin F, van der DB, Orozco B, Jara H, Rada I, Lechin ME *et al.* The serotonin uptake-enhancing drug tianeptine suppresses asthmatic symptoms in children: A double-blind, crossover, placebo-controlled study. *J Clin Pharmacol* 38: 918–25, 1998.

70. Adamus WS, Heuer H, Meade CJ, Frey G, Brecht HM. Inhibitory effect of oral WEB 2086, a novel selective PAF-acether antagonist, on ex vivo platelet aggregation. *Eur J Clin Pharmacol* 35: 237–40, 1988.
71. Freitag A, Watson RM, Matsos G, Eastwood C, O'Byrne PM. The effect of an oral platelet activating factor antagonist, WEB 2086, on allergen induced asthmatic responses. *Thorax* 48: 594–98, 1993.
72. Spence DP, Johnston SL, Calverley PM, Dhillon P, Higgins C, Ramhamadany E *et al.* The effect of the orally active platelet-activating factor antagonist WEB 2086 in the treatment of asthma. *Am J RespirCrit Care Med* 149: 1142–48, 1994.
73. Kuitert LM, Hui KP, Uthayarkumar S, Burke W, Newland AC, Uden S *et al.* Effect of a platelet activating factor (PAF) antagonist UK-74, 505 on allergen-induced early and late response. *Am Rev Respir Dis* 147: 82–86, 1993.
74. Gomez FP, Marrades RM, Iglesia R, Roca J, Barbera JA, Chung KF *et al.* Gas exchange response to a PAF receptor antagonist, SR 27417A, in acute asthma: A pilot study. *Eur Respir J* 14: 622–26, 1999.
75. Evans DJ, Barnes PJ, Cluzel M, O'Connor BJ. Effects of a potent platelet-activating factor antagonist, SR27417A, on allergen-induced asthmatic responses. *Am J Respir Crit Care Med* 156: 11–16, 1997.
76. Pongracic JA, Naclerio RM, Reynolds CJ, Proud D. A competitive kinin receptor antagonist, [DArg0, Hyp3, DPhe7]- bradykinin, does not affect the response to nasal provocation with bradykinin. *Br J Clin Pharmacol* 31: 287–94, 1991.
77. Abraham WM, Scuri M, Farmer SG. Peptide and non-peptide bradykinin receptor antagonists: Role in allergic airway disease. *Eur J Pharmacol* 533: 215–21, 2006.
78. Abraham WM, Scuri M, Farmer SG. Peptide and non-peptide bradykinin receptor antagonists: Role in allergic airway disease. *Eur J Pharmacol* 533: 215–21, 2006.
79. Wirth K, Hock FJ, Albus V *et al.* HOE 140, a new potent and long-acting bradykinin antagonist; in vivo studies. *Br J Pharmacol* 102: 774–77, 1991.
80. Wirth KJ, Gehring D, Scholkens BA. Effect of Hoe 140 on bradykinin-induced bronchoconstriction in anesthetized guinea pigs. *Am Rev Resp Dis* 148: 702–6, 1993.
81. Miura M, Belvisi MG, Barnes PJ. Effect of bradykinin on airway neural responses in vitro. *J Appl Physiol* 73: 1537–41, 1992.
82. Austin CA, Foreman JC, Scadding GK. Reduction by HOE140, the B2 kinin receptor antagonist, of antigen-induced nasal blockage. *Br J Pharmacol* 111: 969–71, 1994.
83. Akbary AM, Wirth KJ, Scholkens BA. Efficacy and tolerability of Icatibant (Hoe 140) in patients with moderately severe chronic bronchial asthma. *Immunopharmacology* 33: 238–42, 1996.
84. Scherrer D, Daeffler L, Trifilieff A, Gies JP. Effects of WIN 64338, a nonpeptide bradykinin B2 receptor antagonist, on guinea-pig trachea. *Br J Pharmacol* 115: 1127–28, 1995.
85. Inamura N, Asano M, Hatori C, Sawai H, Hirosumi J, Fujiwara T *et al.* Pharmacological characterization of a novel, orally active, non-peptide bradykinin B2 receptor antagonist, FR1673. *Eur J Pharmacol* 333: 79–86, 1997.
86. De Swert KO, Joos GF. Extending the understanding of sensory neuropeptides. *Eur J Pharmacol* 533: 171–81, 2006.
87. Kraan J, Vink-Klooster H, Postma DS. The NK-2 receptor antagonist SR 48968C does not improve adenosine hyperresponsiveness and airway obstruction in allergic asthma. *Clin Exp Allergy* 31: 274–78, 2001.
88. Jacobson KA, Gao ZG. Adenosine receptors as therapeutic targets. *Nat Rev Drug Discov* 5: 247–64, 2006.
89. Polosa R, Holgate ST. Adenosine receptors as promising therapeutic targets for drug development in chronic airway inflammation. *Curr Drug Targets* 7: 699–706, 2006.
90. Sun CX, Zhong H, Mohsenin A, Morschl E, Chunn JL, Molina JG *et al.* Role of A2B adenosine receptor signaling in adenosine-dependent pulmonary inflammation and injury. *J Clin Invest* 116: 2173–82, 2006.

CHAPTER 53

Antibiotics

Sanjay Sethi

Division of Pulmonary, Critical Care and Sleep Medicine, Department of Medicine, University of Buffalo, State University of New York, and Veterans Affairs Western New York Health Care System, Buffalo, New York, USA

INTRODUCTION

Exacerbations of chronic obstructive pulmonary disease (COPD) are a frequent cause for office and emergency room visits as well as hospitalizations. Exacerbations are intermittent episodes of increased respiratory symptoms and worse pulmonary function that may be accompanied by fever and other constitutional symptoms. These episodes contribute significantly to the morbidity associated with COPD, and in advanced disease, they are also the most frequent cause of death [1–3]. The frequency of exacerbations varies widely between patients. Though severity of airflow obstruction is correlated with the frequency of exacerbations, other yet poorly understood factors predispose some patients to have more frequent exacerbations.

Several investigators have reported that exacerbations are important determinants of the decline in health-related quality of life in COPD [4, 5]. Following an exacerbation, acute symptoms usually subside over 2–3 weeks, however, quality of life indices takes several months to return to baseline. Furthermore, recurrent exacerbations are clearly associated with a more rapid decline in quality of life in COPD [6]. Based on studies performed by Fletcher and Peto in the 1960s it was widely believed that exacerbations do not contribute to the decline in lung function (measured as FEV_1). Recent data has disputed these results and shown that the frequency of exacerbations is associated with accelerated long-term decline in lung function, in both mild COPD and more advanced diseases [7, 8].

Undisputedly, exacerbations are a major contributor to the morbidity, costs, and mortality associated with COPD. Substantial progress has been made in understanding their etiology and pathogenesis of exacerbations. In contrast, such richness of data is unfortunately missing in the field of clinical management of exacerbations. Major modalities used to treat exacerbations include bronchodilators, systemic corticosteroids, and antibiotics. Only for systemic corticosteroid use in severe exacerbations requiring emergency room care or hospitalization is the adequate clinical trial data available. Adequately powered, well-designed clinical trials studies for the other modalities used to treat exacerbations are relatively few. Consequently, whether the use of antibiotics for exacerbations is appropriate is still debated and whether antibiotic choice makes a difference is even more controversial [9, 10].

Principles that should guide appropriate use of antibiotics in exacerbations are listed in Table 53.1. The comments in Table 53.1 illustrate the several barriers that currently exist in fully applying these principles in everyday practice. This chapter will describe the optimal approach to antibiotic treatment of exacerbations recognizing these limitations. Such an approach relies upon an accurate diagnosis of an exacerbation, including judicious application of diagnostic tests. This is followed by determining the severity of an exacerbation, the probability that it is bacterial and whether antibiotics are indicated. If antibiotics are indicated, then a risk stratification approach is described to choose an appropriate antibiotic.

DIAGNOSIS OF ACUTE EXACERBATION

Clear, objective, universally accepted definition of a disease or syndrome is a pre-requisite for its accurate diagnosis. Unfortunately, for

TABLE 53.1 Principles of appropriate antibiotic use in exacerbations of COPD.

Principle	Comment
Bacteria cause significant proportion of exacerbations	40–50% of exacerbations are of bacterial origin
Bacterial exacerbations can be reliably distinguished from nonbacterial episodes	Sputum purulence is a useful marker but not always easily assessed or reliable
Placebo-controlled antibiotic trials have shown benefit of antibiotics	This is true for moderate-to-severe exacerbations, but not mild episodes
Antibiotics used are appropriate for the causative pathogen	Sputum cultures are inaccurate and usually not performed, therefore therapy is usually empiric
Antibiotic dose and duration are appropriate	Shorter durations of 3–7 days appear to be as good as 10–14 days
Antibiotic choice makes a difference in outcome	Inadequate study methods have limited their ability to show differences among antibiotics
Risk stratification is useful in choosing antibiotics	Though not prospectively validated, it is widely advocated

exacerbations of COPD, the current definitions are imprecise, variable, and lack objective measures [11]. There are two widely used definitions of exacerbation. In 1987, in a large placebo-controlled trial in exacerbations of COPD, Anthonisen and colleagues defined exacerbations based on the presence of one or more of the three cardinal symptoms, including an increase or new onset of dyspnea, sputum production, and sputum purulence [12]. When only one cardinal symptom is present, then one or more supporting symptoms or signs are required to make the diagnosis, including an upper respiratory tract infection in the past 5 days, wheezing, cough, fever without an obvious source, or a 20% increase in the respiratory rate or heart rate above baseline. Though simple and clinically useful, this definition does have several limitations. It is narrow in its scope, and several important symptoms of exacerbation such as cough, chest congestion, chest tightness, fatigue, and sleep disturbance are not included. It is subjective, however, that limitation applies to all definitions of exacerbations as reliable objective measures of exacerbations are currently unavailable.

A more recent and also commonly used definition came from a consensus panel that defined an exacerbation as an acute sustained worsening of the patients' condition from stable state, beyond day-to-day variability and which requires additional treatment [11]. This definition though more inclusive of symptoms of an exacerbation, is not specific with regard to the nature and duration of symptoms. It lacks objective measures. Longitudinal cohort studies with daily recording of symptoms have revealed that a significant proportion, up to 50%, of episodes of increased symptoms are not reported by patients to their health care providers and therefore are not associated with additional treatment. Such episodes, which likely represent mild exacerbations would not meet this definition of exacerbation.

Missing in both definitions is the clinical exclusion of entities that could lead to increased respiratory symptoms in a manner similar to exacerbation, such as pneumonia, congestive heart failure, upper respiratory infection, noncompliance with medications etc. These clinical entities have distinct etiology, pathogenesis and treatment, and therefore should be in the differential diagnosis of an exacerbation rather than be included under the definition. In our clinical studies, we suspect an exacerbation when a patient with COPD reports a minor increase (or new onset) of two or a major increase (or new onset) of one of the following respiratory symptoms: dyspnea, cough, sputum production, sputum tenacity, sputum purulence [13]. The increase in symptoms should be of at least 24 h duration and should be of greater intensity than their normal day-to-day variability. Furthermore, as described above, clinical evaluation should exclude other clinical entities that could present in a similar manner.

DETERMINATION OF SEVERITY

The decision to treat exacerbations is often based on severity, with antibiotic treatment recommended for moderate to severe exacerbations. However, currently our determination of severity of exacerbations is imprecise, and like the diagnosis, is determined in a variable manner among studies. The severity of an exacerbation is a complicated concept, determined by at least two factors, the severity of the underlying COPD and the acute change induced by the exacerbation. A severe exacerbation may therefore be assessed when a patient with very severe underlying COPD has a relatively small change from his baseline, or when a patient with mild COPD has much larger acute change in his symptoms and lung function. The exacerbation etiology and pathophysiology may differ considerably in these circumstances and may warrant distinct treatment approaches.

Exacerbation severity has been variously defined among studies. Ideally, changes in lung function should be used to define severity of exacerbations. However, spirometry and lung volumes are difficult to measure accurately during exacerbations, especially severe episodes. With simpler

TABLE 53.2 Anthonisen classification of COPD exacerbations based on cardinal symptoms.

Severity of exacerbation	Type of exacerbation	Characteristics
Severe	Type 1	Increased dyspnea, sputum volume, and sputum purulence
Moderate	Type 2	Any two of the above three cardinal symptoms
Mild	Type 3	Any one of the above three cardinal symptoms and one or more of the following minor symptoms or signs - Cough - Wheezing - Fever without an obvious source - Upper respiratory tract infection in the past 5 days - Respiratory rate increase >20% over baseline - Heart rate increase >20% over baseline

Based on data from Ref. [12].

measures, such as peak flow, often the change with an exacerbation is of the same magnitude as its day-to-day variability. Severity has been also measured by site of care, with hospitalized exacerbations regarded as severe, outpatient exacerbations regarded as moderate and self medicated exacerbations as mild [14]. Site of care is unreliable as a measure of exacerbation severity. Though the major factor determining site of care is undoubtedly illness severity, it is also dependent on differences in hospital admission practices among countries and health care systems, patient reporting of exacerbations, physician preferences, etc. The intensity of recommended treatment has also been used as a measure of severity of exacerbations, with treatment with bronchodilators only indicating mild exacerbations, while treatment with antibiotics and steroids in addition to bronchodilators regarded as indicating moderate or severe exacerbations. Such measurement of severity is beset with the same problems of preferences and practice approach as discussed above.

Another widely used determination of severity of exacerbations is the Anthonisen classification [12]. This classification relies on the number of cardinal symptoms and the presence of some supporting symptoms (Table 53.2). Though not developed as a severity classification, it has become so over time. Advantages of this determination of severity are its simplicity and that it correlates with benefit with antibiotics, with such benefit seen only in Type 1 and 2 exacerbations. Limitations include its lack of validation against objective measures of severity and that its ability to predict benefit with antibiotics has not been consistent in other studies. Another limitation is the lack of gradation of severity within each symptom, such that a Type 2 exacerbation with mild changes in two cardinal symptoms would be regarded as the same severity in this classification as an exacerbation with a marked increase in both symptoms.

It is evident that we need a better definition and objective measures of severity of exacerbations. Recent work in the development of patient reported outcomes to measure exacerbations has demonstrated that the experience of an exacerbation from the patient perspective includes not only respiratory symptoms but extra-respiratory manifestations including fatigue, anxiety, sleep disturbance etc. Future definitions of exacerbation should therefore include such symptoms. An expectation in the future is that a properly developed patient reported outcome measure would become universally applied to define exacerbations, where a certain change from baseline in this measure would constitute an exacerbation. Furthermore, the degree of change in such a measure from baseline would represent an objective measurement of severity.

Biomarkers in exacerbations are being vigorously explored, as they hold the promise of being objective measures to define an exacerbation and determine its severity. A recent study explored 36 plasma biomarkers in 90 patients with exacerbations and found that none of them alone or in combination were adequate to define an exacerbation [15]. In another study of multiple serum biomarkers in 20 hospitalized patients with exacerbation, reduction in interleukin-6 (IL-6) and interleukin-8 (IL-8) correlated with decrease in dyspnea during recovery from exacerbation, while decreases in IL-6 and tumor necrosis factor (TNF-α) were proportional to recovery in FEV_1 [16]. Changes from baseline in sputum IL-8 and TNF-α as well as sputum neutrophil elastase (NE) and serum C-reactive protein (CRP) correlate with clinical severity of an exacerbation as assessed by a clinical score based on symptoms and signs [17, 18]. It appears unlikely that a single biomarker will be capable of defining an exacerbation, because of the heterogeneity of these episodes. However, biomarkers do hold promise in objectively determining severity of exacerbations and defining etiology to guide appropriate treatment.

PATHOGENESIS OF EXACERBATIONS

An increase in airway inflammation from the baseline level in a patient appears central to the pathogenesis of most acute exacerbations [19, 20]. Airway inflammation measured in induced or expectorated sputum, bronchoalveolar lavage or bronchial biopsy has revealed that increased inflammation accompanies exacerbations and resolves with treatment [21–25]. Both neutrophilic and eosinophilic inflammations have been described. This acute increase in airway inflammation leads to increased bronchial tone, edema in the bronchial wall and mucus production. In a diseased lung, these processes worsen ventilation-perfusion mismatch and expiratory flow limitation. Clinically, these pathophysiologic changes present as an increase in or new onset of dyspnea, cough, sputum production, tenacity and purulence along with worsening gas exchange, which are the cardinal manifestations of an exacerbation. Inflammation in exacerbations extends beyond the lung, and increased plasma fibrinogen, interleukin 6 (IL-6) and CRP have been described during exacerbations [18, 26, 27]. These and other mediators likely cause the systemic manifestations of exacerbations, including fatigue and in some instances fever.

The etiology of exacerbations appears to determine the nature and degree of inflammation in exacerbations. Neutrophilic inflammation is characteristic of bacterial exacerbations, while both neutrophilic and eosinophilic inflammations have been described with viral infection [28]. The intensity of neutrophilic inflammation, when measured as associated cytokines/chemokines (IL-8, TNF-α) and products of neutrophil degranulation (NE and myeloperoxidase), is much greater in well-characterized bacterial exacerbations than exacerbations of nonbacterial etiology [18]. Systemic inflammation, measured as serum CRP is also more intense in bacterial exacerbations [18]. These findings have important implications. They can form the basis for biomarkers to distinguish etiology of exacerbations in a reliable and rapid manner, facilitating appropriate therapy. The heightened airway and systemic inflammation with bacterial exacerbations can be potentially damaging to the lungs. Effective antibiotic therapy to eradicate the bacteria responsible for exacerbations and reduce the inflammation to baseline levels becomes desirable and could have potential long-term benefits in COPD.

A variety of noninfectious and infectious stimuli can induce an acute increase in airway inflammation in COPD, thereby causing an exacerbation. Increased respiratory symptoms and respiratory mortality among patients with COPD during periods of increased air pollution have been described [29–31]. Environmental pollutants, both particulate matter, such as PM-10 and diesel exhaust particles, and nonparticulate gases, such as ozone, nitrogen dioxide, sulfur dioxide, are capable of inducing inflammation *in vitro* and *in vivo* [32–34]. Infectious agents, including bacteria, viruses, and atypical pathogens are implicated as causes of up to 80% of acute exacerbations [35].

MICROBIAL PATHOGENS IN COPD EXACERBATIONS

Potential pathogens in COPD exacerbations includes typical respiratory bacterial pathogens, respiratory viruses, and atypical bacteria (Table 53.3). *Pneumocystis jiroveci*, a fungus, appears to cause chronic infection in COPD. Whether it induces exacerbations is being investigated. Among the typical bacteria, Nontypeable *Haemophilus influenzae* (NTHI) is the most common pathogen in COPD and is the best understood [36]. Among the viruses, *Rhinovirus* and *Respiratory syncytial virus* (RSV) have received considerable attention in recent years and their importance in COPD is now better appreciated [37, 38].

The predilection of these pathogens for causing infections in COPD may be related to certain shared characteristics. NTHI, *Streptococcus pneumoniae* and *Moraxella catarrhalis*

TABLE 53.3 Microbial pathogens in exacerbations of COPD.

Proportion of exacerbations (%)	Specific species	Proportion of class of pathogens (%)
Bacteria		
40–50	Nontypeable *Haemophilus influenzae*	30–50
	Streptococcus pneumoniae	15–20
	Moraxella catarrhalis	15–20
	Pseudomonas spp. and *Enterobacteriaceae*	Isolated in very severe COPD, concomitant bronchiectasis, and recurrent exacerbations
	Haemophilus parainfluenzae	Isolated frequently, pathogenic significance undefined
	Haemophilus hemolyticus	Isolated frequently, pathogenic significance undefined
	Staphylococcus aureus	Isolated infrequently, pathogenic significance undefined
Viruses		
30–40	*Rhinovirus*	40–50
	Parainfluenzae	10–20
	Influenza	10–20
	RSV	10–20
	Coronavirus	10–20
	Adenovirus	5–10
Atypical bacteria		
5–10	*Chlamydia pneumoniae*	90–95
	Mycoplasma pneumoniae	5–10

are the predominant bacterial causes of two other common respiratory mucosal infections, acute otitis media in children and acute sinusitis in children and adults. *Pseudomonas aeruginosa* is the dominant mucosal pathogen in cystic fibrosis and noncystic fibrosis bronchiectasis [39]. These mucosal infections have been related to anatomical abnormalities with impaired drainage of secretions, antecedent viral infections and defects in innate and adaptive immunity. All these predisposing factors likely exist in COPD. NTHI, *S. pneumoniae* and *M. catarrhalis* are exclusively human pathogens. Their usual environmental niche in healthy humans is confined to the upper airway which they usually colonize without any clinical manifestations. Acquisition of these pathogens in COPD, because of compromised lung defense, allows establishment of infection in the upper as well as the lower respiratory tract (below the vocal cords). Infection of the lower respiratory tract in COPD can be with or without overt clinical manifestations, the former being addressed as exacerbation while the latter as colonization.

Respiratory viruses implicated in COPD cause acute tracheobronchial infections in healthy hosts, clinically referred to as acute bronchitis. In the setting of COPD, with diminished respiratory reserve, this acute bronchitis has more profound manifestations and serious clinical consequences.

PATHOGENESIS OF INFECTIOUS EXACERBATIONS

Significant progress has been made in our understanding of acute exacerbation pathogenesis, especially in relation to bacterial infection over the last few years. The current model of bacterial exacerbation pathogenesis involves both host and pathogen factors (Fig. 53.1). Acquisition of strains of bacterial pathogens that are new to the host from the environment is the primary event that puts the patient with COPD at risk for an exacerbation [13]. Variation among strains of a species in the surface antigenic structure, as is seen with NTHI, *S. pneumoniae*, *M. catarrhalis*, and *P. aeruginosa*, is crucial to the development of recurrent exacerbations with these pathogens. This variation allows these newly acquired strains to escape the pre-existing host immune response that had developed following prior exposure to other strains of the same species. These newly acquired strains can therefore proliferate in the lower airways and induce acute inflammation. The virulence of the strain and as yet unidentified host factors may determine if the acute inflammatory response to the pathogen reaches the threshold to cause symptoms that present as an exacerbation [40]. In the majority of instances, mucosal and systemic antibodies develop to the pathogen [41, 42]. This immune response, in combination with appropriate antibiotics, eliminates or controls proliferation of the infecting bacteria. However, because of antigenic variability among strains of these bacterial species, these antibodies directed at the infecting strain are usually strain-specific, and do not protect the host from antigenically distinct strains of the same species. This allows recurrent bacterial infection and exacerbations in these patients.

The pathogenesis of acute viral exacerbations is less well understood, but may be similar to bacterial infections. A common cause of exacerbations, the rhinovirus, demonstrates considerable antigenic variation among its more than 100 serotypes, allowing for recurrent infections. The influenza virus demonstrates drift in the antigenic makeup of its major surface proteins, thereby leading to recurrent

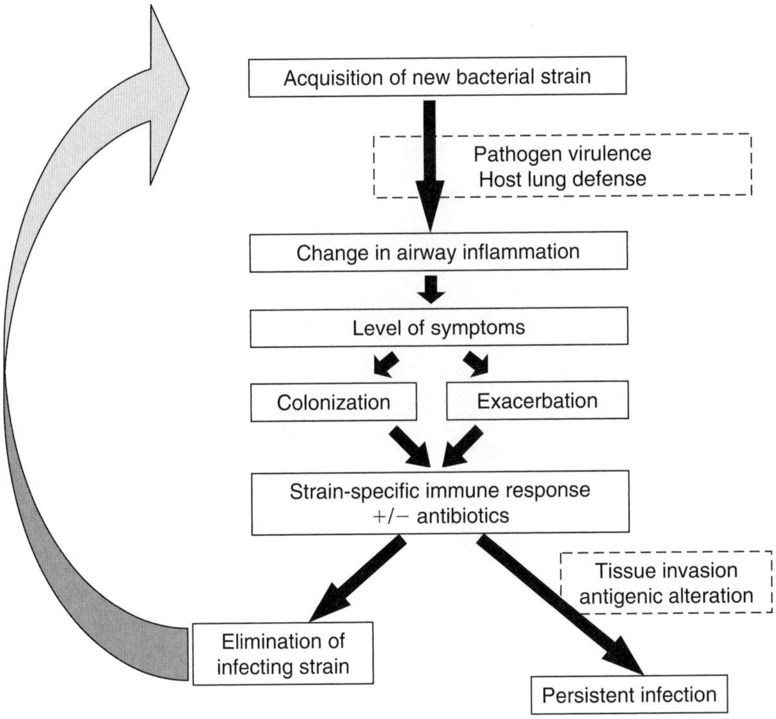

FIG. 53.1 Proposed model of bacterial exacerbation pathogenesis in COPD. Reproduced with permission from Ref. [43].

infections. *In vitro*, viruses can damage airway epithelium, stimulate muscarinic receptors, and induce eosinophil and neutrophil influx [44]. Whether these pro-inflammatory actions are enhanced in epithelial cells from patients with COPD is not known. Bronchial epithelial cells obtained from patients with asthma have diminished production of interferons and increased ICAM-1 expression, changes which could result in increased inflammation, cell lysis, and viral replication on viral infection. Increased ICAM-1 expression in COPD bronchial epithelium has been seen, however the other changes have not been described.

GOALS OF TREATMENT OF EXACERBATIONS

The traditional aims of treatment of an exacerbation are improvement in clinical status and the prevention of complications. Though undoubtedly important, several new observations question the adequacy of these goals. These include the importance of exacerbations in the course of COPD, the role of infection in exacerbations, the high rates of relapse with an adequate initial clinical response, and the potential damaging effects of chronic colonization in COPD. To draw an analogy, confining our goal in the treatment of COPD exacerbations to short-term resolution of symptoms would be the equivalent of treating acute myocardial infarction with the only aim being resolution of chest pain. Several other goals of treatment, both clinical and biological, should therefore be considered (Table 53.4).

A good example of an inadequate goal in the treatment of exacerbations is "clinical success," which is defined as resolution or improvement of symptoms to a degree that no further treatment is required in the opinion of the treating physician. Recent observations have shown that symptoms of an exacerbation are correlated with exaggerated airway and systemic inflammation. Hence, acceptance of clinical improvement as adequate rather than clinical resolution to baseline has important implications. Clinical improvement likely reflects inadequate treatment, permitting the inflammatory process accompanying the exacerbation to persist for prolonged periods of time, causing progressive airway damage [45]. Therefore, clinical resolution of symptoms to baseline is a more appropriate goal of treatment of exacerbations.

Additional important clinical goals of treatment include delaying the next exacerbation, prevention of early relapse and more rapid resolution of symptoms [46–48]. Lengthening the inter-exacerbation interval and prevention of early relapse ultimately translate to a decrease in the frequency of exacerbations, which is now a major focus of COPD treatment. Though most patients and physicians would accept faster recovery to baseline as a desirable goal of treatment, development of this parameter has been hampered by lack of well-validated instruments to reliably measure exacerbation resolution. Patient reported outcomes in development will address this need in the near future.

Biological goals of treatment are either still in their infancy or, in the case of bacteriologic eradication, inadequately assessed in clinical studies to satisfy regulatory requirements for approval of new antibiotics. Most exacerbations are inflammatory events, therefore it is logical that resolution of inflammation to baseline should be an important goal of treatment. Similarly, exacerbations are in many instances induced by infection, therefore eradication of the offending infectious pathogen should be a goal of treatment. Practical application of these biological goals of treatment of exacerbations awaits the development of biomarkers that provide simple, rapid and reliable measurements of inflammation and infection.

A multi-modality approach to treatment of exacerbations is common, that utilizes several modalities simultaneously, to relieve symptoms, to treat the underlying cause or to provide support till recovery occurs [14, 49]. These therapies include bronchodilators, corticosteroids, antimicrobials,

TABLE 53.4 Goals of treatment of COPD exacerbation.

Goals	Comments
Clinical	
Faster resolution of symptoms	Needs validated symptom assessment tools
Clinical resolution to baseline	Needs baseline assessment prior to exacerbation onset for comparison
Prevention of relapse	Relapse within 30 days is quite frequent
Increasing exacerbation-free interval	Needs long-term follow-up after treatment
Preservation of health-related quality of life	Sustained decrements seen after exacerbations
Biological	
Bacterial eradication	Often presumed in usual antibiotic comparison studies
Resolution of airway inflammation	Shown to be incomplete if bacteria persist
Resolution of systemic inflammation	Persistence of systemic inflammation predicts early relapse
Restoration of lung function to baseline	Incomplete recovery is seen in significant proportion
Preservation of lung function	Needs long-term studies

mucolytics and expectorants and, in the more severe cases, oxygen supplementation and mechanical ventilation for acute respiratory failure.

ANTIBIOTICS IN THE TREATMENT OF EXACERBATIONS

The role of antibiotics in the treatment of COPD exacerbations has been a matter of controversy. Even more contentious has been the issue whether antibiotic choice is relevant to clinical outcome of exacerbations. Recommendations for antibiotic use among published guidelines are inconsistent [14, 50–52]. There is a paucity of well-designed, large randomized controlled trials with adequate goals of treatment comparing antibiotics to placebo or among antibiotic classes. This paucity of evidence upon which to base solid recommendations has undoubtedly contributed to the controversy and inconsistency of recommendations regarding antibiotic use [53].

Recently, a few well-designed placebo controlled and antibiotic comparison trials have been reported. Furthermore, epidemiologic studies have consistently identified certain "clinical risk factors," which in the setting of an exacerbation are predictive for failure of treatment or early relapse. The clinical outcomes of exacerbation in observational real-life studies are clearly sub-optimal, with as many as 25–33% of patients experiencing treatment failure or early relapse. Considering the heterogeneity of COPD and of exacerbations, it is clear that the "one size fits all" approach of using the same antibiotic in all episodes is sub-optimal. It is likely that a proportion of treatment failures in exacerbations are related to ineffective antibiotics. Patients "at risk" for poor outcome are the logical candidates for aggressive initial antibiotic treatment, with the expectation that such an approach would improve overall exacerbation outcomes. This "risk stratification" approach has also been advocated for other community-acquired infections such as pneumonia and acute sinusitis [23, 54]. Though improved outcomes with such risk stratification has not yet been demonstrated in prospective controlled trials, this approach takes into account concerns of disease heterogeneity, antibiotic resistance and judicious antibiotic use.

Placebo-controlled antibiotic trials

Exacerbations of COPD result in significant antibiotic consumption, however, there are only a handful of placebo controlled trials in this disease. Two meta-analyses of placebo controlled trials in exacerbations have been published. The first such analysis published in 1995 included nine trials and found a small but significant beneficial effect of antibiotics over placebo [55]. In the second analysis published in 2006, 11 trials were included, and a much larger beneficial effect on mortality and prevention of clinical failure was demonstrated, especially in moderate to severe exacerbations [56]. In this analysis, the number needed to treat in severe exacerbations in hospitalized patients to prevent one death was only three patients and the number needed to treat to prevent one clinical failure was six patients. Diarrhea was the most frequently related adverse effect, with one episode per seven patients treated. Antibiotic treatment was also beneficial in resolving sputum purulence. A benefit on lung function and gas exchange was not observed, however, the data examining this end point was scanty.

The reason that the two meta-analyses came up with different results is in large part inclusion in the later analysis of a study performed by Nouira *et al.* published in 2003. In this randomized double blind study, 93 patients with exacerbations of severe underlying COPD requiring ventilator support in an intensive care unit were randomly assigned to receive a fluoroquinolone antibiotic, ofloxacin, or placebo [57]. No systemic corticosteroids were administered. Bacterial pathogens were isolated in tracheobronchial aspirates in 61% of patients. Ofloxacin administration was associated with dramatic benefits compared to placebo, reducing mortality (4% versus 22%) and the need for additional antibiotics (6% versus 35%) by 17.5-fold and 28.4-fold, respectively.

Another important study, which for some reason has not been included in either meta-analysis, was performed in Italy by Allegra *et al.* and published in 1991. In this double blind randomized trial, amoxicillin/clavulanate was compared with placebo in 414 exacerbations in 369 patients with varying severity of underlying COPD [58]. A unique feature of this study was the measurement of primary outcome at 5 days after the start of treatment, instead of the traditional 2–3 weeks. This earlier timing is clinically of greater relevance than the usual later one, as in clinical practice if patients are not improved within 3–5 days, they are reassessed and their therapy altered. Clinical success (including resolution and improvement) was significantly better with the antibiotic, seen in 86.4% of patients, compared with 50.6% in the placebo arm. In addition, with increasing severity of underlying COPD, the benefit with antibiotics as compared to placebo was larger.

Results of the meta-analyses, the Allegra study, and of the previous classic large placebo-controlled trial conducted by Anthonisen *et al.*, clearly demonstrate that antibiotics are beneficial in the treatment of moderate to severe exacerbations [12, 56–58]. Furthermore, the benefit with antibiotics is more marked early in the course of the exacerbation, suggesting that antibiotics hasten resolution of symptoms [58, 59]. The benefit with antibiotics is also greater as the severity of underlying airflow obstruction increases. This could be related to more frequent bacterial infections and/or a decreased ability of the host in dealing with infections, therefore requiring "help" with antibiotics to resolve them.

Important questions regarding the role of antibiotics in exacerbations still remain. The benefit of antibiotics in mild exacerbations in the context of mild underlying COPD is unproven and warrants a placebo controlled trial. The effect of concomitant treatment with systemic corticosteroids on the benefits of antibiotic therapy in exacerbations is not known. Placebo controlled trials of antibiotics have not systematically regulated concomitant therapy, and all the placebo controlled trials of steroids had antibiotics administered to all patients. Because inflammation and infection are linked, it is likely that there would be additive benefits when both treatments are used over either treatment alone [56, 60].

Antibiotic comparison trials

Antibiotics are clearly useful in moderate to severe exacerbations of COPD. However, there remains considerable controversy as to antibiotic choice, especially for initial empiric therapy of exacerbations [14, 50–52, 61]. Most exacerbations nowadays are treated without obtaining sputum bacteriology and with the trend to short course antibiotic therapy, this initial empiric choice often becomes the only choice made of antibiotics in exacerbations. Results of antibiotic comparison trials should guide the recommendations for appropriate empiric antibiotics in exacerbations. However, though the literature is replete with such trials, in the vast majority, antibiotic choice does not apparently affect the clinical outcome. However, differences in bacteriological eradication rates among antibiotics are seen, with a dissociation between clinical and bacteriological outcomes [62]. These results are contrary to expectations that antibiotics with better *in vitro* and *in vivo* antimicrobial efficacy and better pharmacodynamic and pharmacokinetic characteristics should show superior clinical outcomes. A closer examination of the trial design of these studies reveals several shortcomings that offer potential explanations for this paradox (Table 53.5) [53]. Many of these deficiencies are related to the fact that these trials are performed for regulatory approval of the drugs, therefore are designed for demonstrating noninferiority rather than differences between the two antibiotics. In the face of this large body of data showing clinical equivalence, it is not surprising that several guidelines do not differentiate between antibiotics for therapy of exacerbations.

Most antibiotic comparison trials are underpowered to detect differences among antibiotics. However, because of regulatory requirements, these studies are conducted in a very similar manner and in similar patient populations. This makes these trials very amenable to a meta-analytic approach. Dimopoulos *et al.* used such an approach to determine whether there was any difference in clinical outcomes among first-line antibiotics (amoxicillin, ampicillin, pivampicillin, trimethoprim/sulfamethoxazole, and doxycycline) and second-line antibiotics (amoxicillin/clavulanate, macrolides, second-generation or third-generation cephalosporins, and fluoroquinolones) in the treatment of exacerbations of chronic bronchitis [63]. They identified 12 randomized controlled trials that had enrolled 2261 patients, 10 of these trials included the penicillins as the first-line antibiotic. Only a single trial each with trimethoprim/sulfamethoxazole and doxycycline was included. In the clinically evaluable patients, first-line antibiotics were only half as effective as second-line antibiotics with an odds ratio for clinical treatment success of 0.51 (95% CI, 0.34–0.75). This result was consistent in several sensitivity analyses, with the exception of trials published before 1991, where the difference between first line and second line antibiotics was not seen. There was no difference between the first-line and second-line antibiotics in adverse effects.

This meta-analysis provides additional evidence that antibiotic choice does make a difference in the treatment of exacerbations. Similar treatment success for first line and second line antibiotics in trials before 1991 but not after 1991, suggests that resistance emergence in causative bacterial pathogens (*H. influenzae, M. catarrhalis, S. pneumoniae*) is responsible for the findings of this meta-analysis. Because of the limited number of studies in which the first line antibiotics were not penicillins, the results of this meta-analysis mainly applies to the penicillins. Based on this meta-analysis, recommendations to use amoxicillin and ampicillin in the treatment of exacerbations cannot be supported.

These investigators also performed a similar meta-analysis where they compared second-line antibiotics,

TABLE 53.5 Limitations of published placebo-controlled antibiotic trials in acute exacerbations of COPD.

Limitation of study design	Potential consequences
Small number of subjects	Type 2 error
Subjects with mild or no underlying COPD included	Diminished overall perceived efficacy of antibiotics
Nonbacterial exacerbations included	Type 2 error
Endpoints compared at 3 weeks after onset	– Spontaneous resolution mitigates differences between arms – Clinically irrelevant as most decisions about antibiotic efficacy are made earlier
Speed of resolution not measured	Clinically relevant endpoint not assessed
Lack of long-term follow-up	Time to next exacerbation not assessed
Antibiotic resistance to agents with limited *in vitro* antimicrobial efficacy	Diminished overall perceived efficacy of antibiotics
Poor penetration of antibiotics used in to respiratory tissues	Diminished overall perceived efficacy of antibiotics
Concurrent therapy not controlled	Undetected bias in use of concurrent therapy

Reproduced with permission from Ref. [53].

the macrolides, the fluoroquinolones and amoxicillin/clavulanate in a similar manner [64]. In this analysis of 19 randomized controlled trials that had enrolled 7405 patients, no differences were found among these agents in clinical treatment success defined in the conventional manner.

In addition to these meta-analyses, welcome additions to the literature on antibiotic treatment of exacerbations are two recent antibiotic comparison trials that were designed as superiority studies. They also measured some unconventional but clinically relevant end-points. The GLOBE (Gemifloxacin and Long term Outcome of Bronchitis Exacerbations) trial, a double blind, randomized trial, compared a fluoroquinolone, gemifloxacin, with a macrolide, clarithromycin [65]. End of therapy and long-term outcome assessments were made at the conventional 10–14 day and 28-day time intervals. These assessments, in line with most antibiotic comparison trials, did not demonstrate statistically significant differences in the two arms, with clinical success rates of 85.4% and 84.6% for gemifloxacin and clarithromycin respectively. Also in line with similar studies, bacteriological success, measured as eradication and presumed eradication, was significantly higher with gemifloxacin (86.7%) compared to clarithromycin (73.1%).

Patients with a successful clinical outcome at 28 days were enrolled in a follow-up period for a total of 26 weeks of observation. In this time period, the primary outcomes were the rate of repeat exacerbations, hospitalizations for respiratory disease and health-related quality of life measures. A significantly lower rate of repeat exacerbations was observed with gemifloxacin, with 71% of the patients remaining exacerbation free at 26 weeks compared to 58.5% in the clarithromycin arm. The relative risk reduction for recurrence of exacerbation was 30%. The rate of hospitalization for respiratory tract illness in the 26 weeks was also lower in the gemifloxacin treated than in the clarithromycin treated patients (2.3% versus 6.3%, $p = 0.059$) [65]. Patients who remained free of recurrence in the 26-week period regained more of their health-related quality of life than those who had a recurrent exacerbation [6]. This trial clearly demonstrates that conventional medium-term clinical outcomes are unsuitable for measuring differences among antibiotics in exacerbations. If the 26-week follow-up period had not been included in the GLOBE study, significant differences in the two treatment arms in clinically relevant outcomes of recurrence of exacerbations and respiratory related hospitalization would have been missed.

The MOSAIC trial is another recent landmark antibiotic comparison trial in exacerbations of COPD. Patients in this study were randomized to a fluoroquinolone, moxifloxacin or to standard therapy (which could be one of the following: amoxicillin, cefuroxime or clarithromycin) [66]. Several unique design features of this trial are noteworthy, which relate to observations made in this study and set the standard for future antibiotic comparison trials in this disease. The number of patients enrolled was much larger than previous studies, in order to provide adequate power to demonstrate superiority. Patients were enrolled when stable to establish a baseline as a comparison to reliably distinguish between clinical improvement (enough improvement that no additional antibiotic treatment is required) from clinical cure (improvement of symptoms to baseline) following treatment. A substantial proportion of the patients enrolled had one or more risk factors that would predispose to a poor outcome as discussed below. Patients were followed up to 9 months after randomization to provide an estimate of recurrence of exacerbation.

In line with usual antibiotic comparison trials, moxifloxacin and standard therapy were equivalent (88% versus 83%) for clinical success (resolution and improvement) at 7–10 days after the end of therapy. However, moxifloxacin therapy was associated with a superior clinical cure rate (defined as resolution of symptoms to baseline, rather than simply improvement) than standard therapy (71% versus 63%), as well as with superior bacteriologic response (91.5% versus 81%). Several other *a priori* unconventional end-points were examined. Moxifloxacin treatment resulted in significantly fewer courses of additional antibiotic therapy (8% versus 14%) and an extended time to the next exacerbation (131 versus 104 days) [66]. A composite end-point of clinical failure, requirement of additional antibiotics and recurrence of exacerbation demonstrated a clear difference between the two arms, with moxifloxacin being statistically superior to standard therapy for up to 5 months of follow-up. As with the GLOBE trial, if conventional clinical success would have been measured solely in this study, all the other significant differences in the two arms would have not been discovered.

The GLOBE and MOSAIC trials demonstrate that *in vitro* microbiological superiority as well as the enhanced pharmacokinetic/pharmacodynamic properties in the respiratory tract of the fluoroquinolones does translate to greater *in vivo* effectiveness in the treatment of exacerbation. Antibiotics for exacerbation have very similar results for the standard regulatory end-point of clinical success at 7–14 days after the end of therapy. This standard end-point not only lacks discriminatory power, it also has little clinical relevance. Most decisions about antibiotic benefit in the clinical setting are made within the 1st week of therapy. Differences among antibiotics are perceptible when clinically relevant end-points such as speed of resolution, clinical cure, need for additional antimicrobials and time to next exacerbation are considered [65, 66].

RISK STRATIFICATION OF PATIENTS

Fluoroquinolones are excellent antimicrobials for exacerbations, and based on the MOSAIC and GLOBE studies, it is tempting to prescribe them for all moderate to severe exacerbations. Such a strategy, though likely to be successful in the short-term, would foster antimicrobial resistance to these valuable antibiotics in the long term. Therefore, it would be judicious to make an effort to identify those patients and exacerbations that are most likely to benefit from these antibiotics and use them in those circumstances.

Observational real life studies of the outcome of exacerbations in the community have clearly demonstrated that our current treatment approach is sub-optimal. In these studies, treatment failure rates, either defined as failure to improve or relapse within 30 days of 25–33% are seen

[67, 68]. These studies have also demonstrated that certain patient characteristics that antedate the onset of the exacerbation impact the outcome of the exacerbation [48, 67–71]. Interestingly, several of these characteristics, such as co-morbid cardiac disease and frequent exacerbations, were found to be relevant to outcome in more than one study. These risk factors for poor outcome should be considered in the decision regarding choice of empiric antibiotics when treating exacerbations. In theory, patients at risk for poor outcome would have the greatest benefit from early aggressive antibiotic therapy, such as with the fluoroquinolones. These are the patients in whom the consequences of treatment with an antibiotic ineffective against the pathogen causing the exacerbation are likely to be significant, with clinical failures, hospitalizations and early recurrences likely. These at risk patients contribute substantially to the overall poor clinical outcomes of exacerbations. Therefore, targeting these patients with potentially more effective treatment is likely to have a significant impact on the overall outcomes of exacerbations.

Risk factors for poor outcome identified in various studies are increasing age, severity of underlying airway obstruction, presence of co-morbid illnesses (especially cardiac disease), a history of recurrent exacerbations, use of home oxygen, use of chronic steroids, hypercapnia and acute bronchodilator use [48, 67–71]. It is likely that home oxygen use, hypercapnia and chronic steroid use mainly reflect increasing severity of underlying COPD. Acute bronchodilator use could either be related to the severity of underlying COPD or reflect the wheezy phenotype of exacerbation that would be less responsive to antibiotic treatment. Many of these risk factors are continuous in severity, however, certain thresholds have been defined in studies that are clinically useful and predictive of poor outcome and easier to use clinically. These include an age of more than 65 years, forced expiratory volume in 1 s (FEV_1) <50%, and >3 exacerbations in the previous 12 months.

Another important consideration in choosing antibiotics for exacerbation is recent antibiotic use, especially within the past 3 months. Because of the recurrent nature of exacerbation and the high prevalence of co-morbid conditions in COPD, such antibiotic exposure is likely to be prevalent in COPD patients. In other respiratory infections such as pneumonia, recent antibiotic use leads to increased risk for harboring antibiotic resistant pathogens and therefore having a poor outcome following treatment. This phenomenon has been best described for *S. pneumoniae* among patients with community acquired pneumococcal pneumonia and recently also described for this pathogen among patients with COPD [72, 73]. Whether such selection for antibiotic resistant strains occurs among NTHI and *M. catarrhalis* after antibiotic exposure is not known, but is possible.

RISK STRATIFICATION APPROACH TO ANTIBIOTIC THERAPY IN ACUTE EXACERBATION

A risk stratification approach has been advocated by several experts for the initial empiric antibiotic treatment of exacerbations based on the risk factors discussed above as well as the *in vitro* and *in vivo* efficacy of antibiotics. Our current treatment algorithm is shown in Fig. 53.2 [14, 51, 61]. Once an exacerbation is diagnosed, the initial step in the algorithm is determination of the severity of the exacerbation. We use the Anthonisen criteria by defining single cardinal symptom exacerbations as mild, while the presence of two or all three of the cardinal symptoms defines moderate and severe exacerbations.

Mild exacerbations are initially managed with symptomatic treatment, including bronchodilators, antitussives and expectorants and antibiotics are with held. However, patients are counseled regarding the cardinal symptoms and if additional symptoms appear, then antibiotics are prescribed. If a moderate to severe exacerbation is diagnosed, the next important step is the differentiation of "uncomplicated" patients from the "complicated" patients. Uncomplicated patients do not have any of the risk factors for poor outcome. Complicated patients have one or more of the following risk factors for poor outcome: Age >65 years, $FEV_1 < 50\%$, co-morbid cardiac disease and three or more exacerbations in the previous 12 months [14, 51, 61]. A threshold of four or more exacerbations in the previous 12 months has been used to define frequent exacerbations in previous studies and guidelines. However, current therapy of COPD with long acting bronchodilators and inhaled steroids has reduced overall frequency of exacerbations in this disease by about 25%. Therefore, we use a threshold of three or more exacerbations to define the complicated patient on contemporary COPD treatment.

In uncomplicated patients, antibiotic choices include an advanced macrolide (azithromycin, clarithromycin), a cephalosporin (cefuroxime, cefpodoxime, or cefdinir), doxycycline or trimethoprim/sulfamethoxazole. Amoxicillin (or ampicillin) is not an appropriate choice because of prevalent antibiotic resistance and the results of the meta-analysis discussed above. In complicated patients, our usual antibiotic choice is a respiratory fluoroquinolone (moxifloxacin, gemifloxacin, levofloxacin) with amoxicillin/clavulanate as an alternative.

Other important considerations in choosing antibiotics include exposure to antibiotics within the past 3 months. This exposure history should be elicited not only for respiratory infections, but includes antibiotics prescribed for any indication. The antibiotic chosen to treat the exacerbation should be from a different class of agents from the one prescribed within the past 3 months. For example, exposure to a macrolide in the past 3 months should lead to use of a cephalosporin in an uncomplicated patient. Similarly, prior use of a fluoroquinolone in a complicated patient should lead to use of amoxicillin/clavulanate.

Another important consideration in the complicated patients is a sub-group of these patients who are at risk for infection by *P. aeruginosa* and *Enterobacteriaceae* or have a documented infection by these pathogens [74]. These patients usually have very severe underlying COPD ($FEV_1 < 35\%$), have developed bronchiectasis, are hospitalized (often requiring intensive care), or have been recently hospitalized or have received multiple courses of antibiotics. In such patients, empiric treatment with ciprofloxacin is appropriate. However, resistance among *P. aeruginosa* and

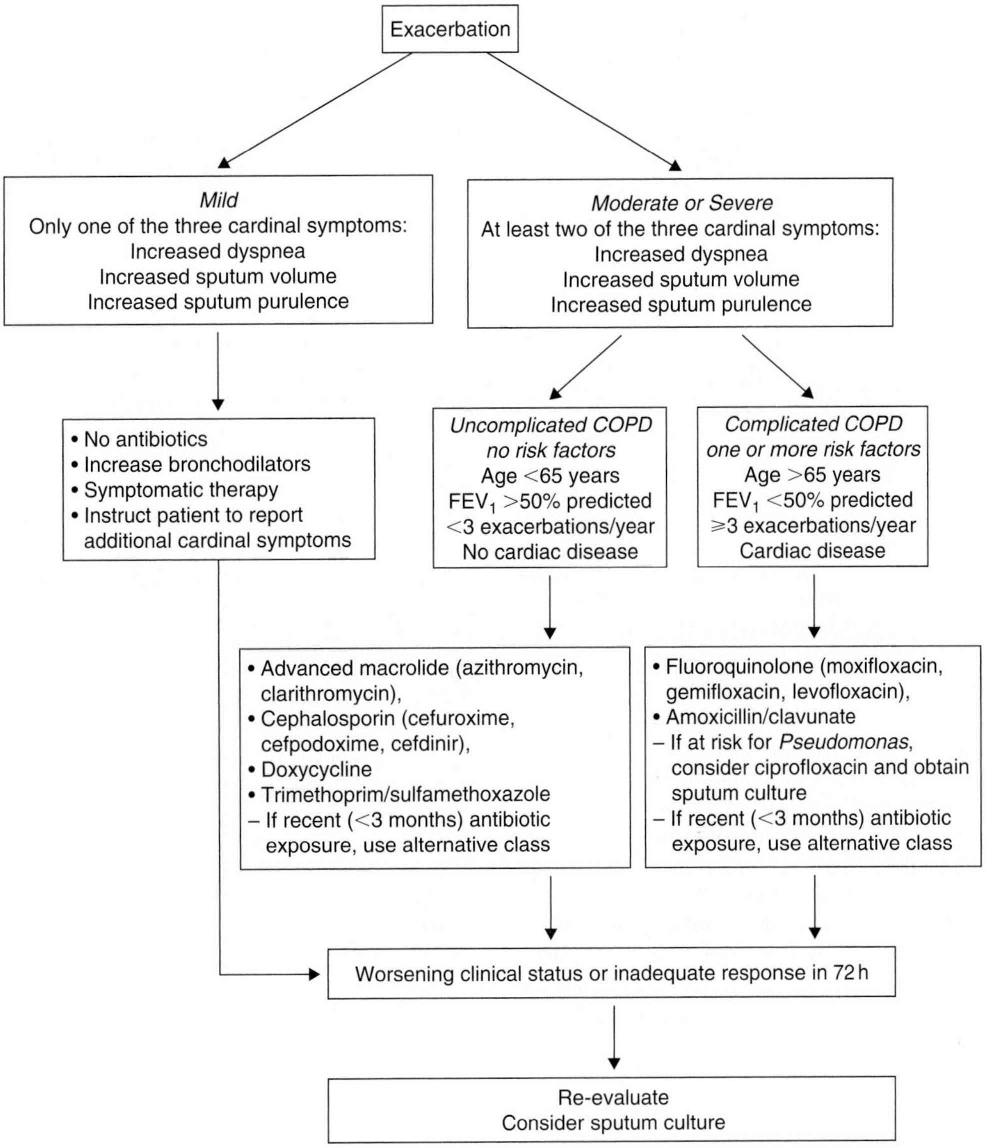

FIG. 53.2 Algorithm for antibiotic treatment of acute exacerbations of COPD.

the *Enterobacteriaceae* to the fluoroquinolones may compromise their efficacy. Therefore, in this sub-group of patients, we obtain a sputum (or tracheobronchial aspirate if intubated) culture to allow adjustment of antibiotics based on the *in vitro* susceptibility of pathogens isolated. Unless dictated by poor clinical response and *in vitro* antimicrobial susceptibility, we do not use combination or parenteral antibiotic therapy for *P. aeruginosa* as this approach has never been systematically examined and is of unproven benefit in exacerbations.

We instruct our patients to report to us any deterioration or lack of improvement at 48–72 h, because in this time frame clinical improvement should be apparent. In these patients who are failing initial empiric antimicrobial therapy, we re-examine the patient to confirm the diagnosis, consider sputum studies to ascertain for resistant or difficult to treat pathogens and treat with an alternative agent with better *in vitro* microbiological efficacy.

ALTERNATIVE APPROACHES TO ANTIBIOTIC THERAPY IN ACUTE EXACERBATION

Patients with COPD exacerbation often experience a change in the color of sputum from white or gray (mucoid) to yellow, green or brown (purulent). Purulence of sputum is related to the presence of myeloperoxidase, a product of neutrophil degranulation. Neutrophil degranulation is associated with bacterial infection, therefore sputum purulence

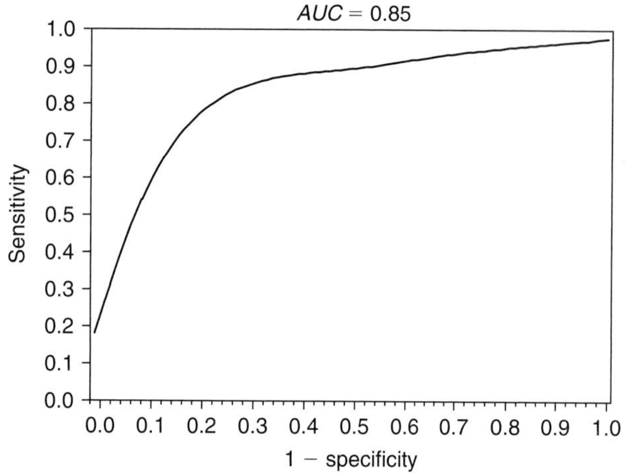

FIG. 53.3 Receiver operating characteristic (ROC) curve for distinguishing new bacterial strain exacerbations from exacerbations without new strains with levels of sputum TNF-α, sputum NE, and serum CRP at exacerbation included in the prediction model. The predictive values for each of the mediators in this model were sputum TNF-α: 0.32 ng/ml; sputum NE: 0.76nM; and serum CRP: 2.37 mg/l. AUC: Area under the ROC. Reproduced with permission from Ref. [18].

at exacerbation is a marker of bacterial infection, defined by quantitative cultures of sputum and bronchoscopic protected brush specimens [75, 76]. Presence of sputum purulence has been advocated as the sole determinant for antibiotic treatment of exacerbations. However, its accuracy and reproducibility as an indicator of bacterial infection is limited, and is likely to be even more so in clinical practice than in research studies [75, 76]. Often, patients have purulent sputum production even when stable, have intermittent purulent sputum production during exacerbation, or have not have observed the sputum color. Sputum purulence is one of the cardinal Anthonisen criteria, and we recommend it should be used in conjunction with other symptoms and other measures of risk stratification in antibiotic treatment of exacerbations (Fig 53.2).

Utilization of biomarkers to rapidly identify bacterial exacerbations and to guide antibiotic use is another potential approach in exacerbations. Several potential biomarkers have been identified in sputum and serum. The best studied biomarker of bacterial infection is serum procalcitonin level. In a recent study in patients hospitalized for exacerbations, antibiotic treatment was only recommended if the procalcitonin level was above a certain threshold. There was no difference in outcomes in spite of reduction in antibiotic use from 72% to 40% with procalcitonin guidance [77]. However, this approach needs to be validated in multi-center trials with varied populations before widespread application [78]. Furthermore, it needs to be tested in outpatients which represent the majority of exacerbations treated, where patients are not as closely supervised and other supportive care is less rigorous. Only a minority of patients received a fluoroquinolone in this study, which should have been the antibiotics of choice in these complicated patients. Therefore, the lack of effect of withholding antibiotics could have been related to their relative inefficacy in the first place. In this study, short-term goals as well as biologic goals of bacterial eradication and inflammation reduction as discussed above were also not recorded.

Other biomarkers that have been explored in COPD exacerbations include sputum TNF-α and NE and CRP (Fig. 53.3). Though individually they do not appear to discriminate between exacerbations, a combination of these parameters reliably distinguishes well-characterized bacterial exacerbations from others [18]. Additional studies are required to confirm and extend these observations.

Other important considerations in antibiotic prescribing are safety and tolerability of the agent, drug interactions and cost of treatment. It is important that cost of the antibiotic not be considered in isolation. Miravittles *et al.* have shown that exacerbations in which initial empiric treatment fails cost 10 times more than clinical successes [79]. The investigators estimated that the overall cost of care could be reduced by half with reduction in clinical failure rates by a third. Though not shown as yet in properly designed studies, it is likely that an appropriate and logical approach to antibiotic choice, as discussed above, would reduce clinical failure rates in exacerbations.

References

1. Burrows B, Earle RH. Course and prognosis of chronic obstructive lung disease. A prospective study of 200 patients. *N Engl J Med* 280(8): 397–404, 1969.
2. Calverley PM *et al.* Salmeterol and fluticasone propionate and survival in chronic obstructive pulmonary disease. *N Engl J Med* 356(8): 775–89, 2007.
3. Soler-Cataluna JJ *et al.* Severe acute exacerbations and mortality in patients with chronic obstructive pulmonary disease. *Thorax* 60(11): 925–31, 2005.
4. Seemungal TAR *et al.* Effect of exacerbation on quality of life in patients with chronic obstructive pulmonary disease. *Am J Respir Crit Care Med* 157: 1418–22, 1998.
5. Spencer S *et al.* Health status deterioration in patients with chronic obstructive pulmonary disease. *Am J Respir Crit Care Med* 163: 122–28, 2001.
6. Spencer S, Jones PW. Time course of recovery of health status following an infective exacerbation of chronic bronchitis. *Thorax* 58(7): 589–93, 2003.
7. Kanner R *et al.* Lower respiratory illnesses promote FEV(1) decline in current smokers but not ex-smokers with mild chronic obstructive pulmonary disease: Results from the lung health study. *Am J Respir Crit Care Med* 164(3): 358–64, 2001.
8. Donaldson GC *et al.* Relationship between exacerbation frequency and lung function decline in chronic obstructive pulmonary disease. *Thorax* 57(10): 847–52, 2002.
9. Hirschmann JV. Do bacteria cause exacerbations of COPD? *Chest* 118: 193–203, 2000.
10. Murphy TF, Sethi S, Niederman MS. The role of bacteria in exacerbations of COPD. A constructive view. *Chest* 118(1): 204–9, 2000.
11. Rodriguez-Roisin R. Toward a consensus definition for COPD exacerbations. *Chest* 117(5 Suppl 2): 398S–401S, 2000.
12. Anthonisen NR *et al.* Antibiotic therapy in exacerbations of chronic obstructive pulmonary disease. *Ann Intern Med* 106: 196–204, 1987.
13. Sethi S *et al.* Acquisition of a new bacterial strain and occurrence of exacerbations of chronic obstructive pulmonary disease. *N Engl J Med* 347(7): 465–71, 2002.
14. Globial Initiative for Obstructive Lung Disease; Available from: www.goldcopd.com [cited].

15. Hurst JR *et al.* Use of plasma biomarkers at exacerbation of chronic obstructive pulmonary disease. *Am J Respir Crit Care Med* 174(8): 867–74, 2006.
16. Pinto-Plata VM *et al.* Systemic cytokines, clinical and physiological changes in patients hospitalized for exacerbation of COPD. *Chest* 131(1): 37–43, 2007.
17. Sethi S *et al.* Airway inflammation and etiology of acute exacerbations of chronic bronchitis. *Chest* 118(6): 1557–65, 2000.
18. Sethi S *et al.* Inflammatory profile of new bacterial strain exacerbations of chronic obstructive pulmonary disease. *Am J Respir Crit Care Med.* 177(5): 491-7 (2008).
19. White AJ, Gompertz S, Stockley RA. Chronic obstructive pulmonary disease. 6: The aetiology of exacerbations of chronic obstructive pulmonary disease. *Thorax* 58(1): 73–80, 2003.
20. Sethi S. New developments in the pathogenesis of acute exacerbations of chronic obstructive pulmonary disease. *Curr Opin Infect Dis* 17(2): 113–19, 2004.
21. Heckerling PS. The need for chest roentgenograms in adults with acute respiratory illness. *Arch Intern Med* 146: 1321–24, 1986.
22. Bhushan R *et al.* Antigenic characterization and analysis of the human immune response to outer membrane protein E of *Branhamella catarrhalis. Infect Immun* 65(7): 2668–75, 1997.
23. Mandell LA *et al.* Update of practice guidelines for the management of community-acquired pneumonia in immunocompetent adults. *Clin Infect Dis* 37(11): 1405–33, 2003.
24. Kjaergard LL *et al.* Basophil-bound IgE and serum IgE directed against *Haemophilus influenzae* and *Streptococcus pneumoniae* in patients with chronic bronchitis during acute exacerbations. *APMIS* 104: 61–67, 1996.
25. Hiltke TJ, Sethi S, Murphy TF. Sequence stability of the gene encoding outer membrane protein P2 of nontypeable *Haemophilus influenzae* in the human respiratory tract. *J Infect Dis* 185(5): 627–31, 2002.
26. Wedzicha JA *et al.* Acute exacerbations of chronic obstructive pulmonary disease are accompanied by elevations of plasma fibrinogen and serum IL-6 levels. *Thromb Haemost* 84(2): 210–15, 2000.
27. Dev D *et al.* Value of C-reactive protein measurements in exacerbations of chronic obstructive pulmonary disease. *Respir Med* 92(4): 664–67, 1998.
28. Papi A *et al.* Infections and airway inflammation in chronic obstructive pulmonary disease severe exacerbations. *Am J Respir Crit Care Med* 173(10): 1114–21, 2006.
29. Garcia-Aymerich J *et al.* Air pollution and mortality in a cohort of patients with chronic obstructive pulmonary disease: A time series analysis. *J Epidemiol Community Health* 54(1): 73–74, 2000.
30. Sunyer J *et al.* Air pollution and emergency room admissions for chronic obstructive pulmonary disease: A 5-year study. *Am J Epidemiol* 137(7): 701–5, 1993.
31. Sunyer J *et al.* Patients with chronic obstructive pulmonary disease are at increased risk of death associated with urban particle air pollution: A case-crossover analysis. *Am J Epidemiol* 151(1): 50–56, 2000.
32. Ohtoshi T *et al.* Diesel exhaust particles stimulate human airway epithelial cells to produce cytokines relevant to airway inflammation *in vitro*. *J Allergy Clin Immunol* 101(6 Pt 1): 778–85, 1998.
33. Devalia JL *et al.* Effect of nitrogen dioxide and sulphur dioxide on airway response of mild asthmatic patients to allergen inhalation. *Lancet* 344(8938): 1668–71, 1994.
34. Rudell B *et al.* Bronchoalveolar inflammation after exposure to diesel exhaust: Comparison between unfiltered and particle trap filtered exhaust. *Occup Environ Med* 56: 527–34, 1999.
35. Sethi S. Infectious etiology of acute exacerbations of chronic bronchitis. *Chest* 117: 380S–85S, 2000.
36. Eldika N, Sethi S. Role of Nontypeable *Haemophilus influenzae* in exacerbations and progression of chronic obstructive pulmonary disease. *Curr Opin Pulm Med* 12(2): 118–24, 2006.
37. Seemungal T *et al.* Respiratory viruses, symptoms, and inflammatory markers in acute exacerbations and stable chronic obstructive pulmonary disease. *Am J Respir Crit Care Med* 164(9): 1618–23, 2001.
38. Falsey AR *et al. Respiratory syncytial virus* infection in elderly and high-risk adults. *N Engl J Med.* 352(17): 1749–59 (2005).
39. Aaron SD *et al.* Adult cystic fibrosis exacerbations and new strains of *Pseudomonas aeruginosa. Am J Respir Crit Care Med* 169(7): 811–15, 2004.
40. Chin CL *et al. Haemophilus influenzae* from patients with chronic obstructive pulmonary disease exacerbation induce more inflammation than colonizers. *Am J Respir Crit Care Med* 172(1): 85–91, 2005.
41. Sethi S *et al.* Strain specific immune response to *Haemophilus influenzae* in chronic obstructive pulmonary disease. *Am J Respir Crit Care Med* 169: 448–53, 2004.
42. Murphy TF *et al. Moraxella catarrhalis* in chronic obstructive pulmonary disease: Burden of disease and immune response. *Am J Respir Crit Care Med* 172(2): 195–99, 2005.
43. Veeramachaneni SB, Sethi S. Pathogenesis of bacterial exacerbations of COPD. *COPD: J Chronic Obstructive Pul Dis* 3: 109–15, 2006.
44. Johnston SL *et al.* Low grade rhinovirus infection induces a prolonged release of IL-8 in pulmonary epithelium. *J Immunol* 160(12): 6172–81, 1998.
45. White AJ *et al.* Resolution of bronchial inflammation is related to bacterial eradication following treatment of exacerbations of chronic bronchitis. *Thorax* 58(8): 680–85, 2003.
46. Aaron SD *et al.* Outpatient oral prednisone after emergency treatment of chronic obstructive pulmonary disease. *N Engl J Med* 348(26): 2618–25, 2003.
47. Anzueto A, Rizzo JA, Grossman RF. The infection-free interval: Its use in evaluating antimicrobial treatment of acute exacerbation of chronic bronchitis. *Clin Infect Dis* 28(6): 1344–45, 1999.
48. Miravitlles M, Murio C, Guerrero T. Factors associated with relapse after ambulatory treatment of acute exacerbations of chronic bronchitis. DAFNE Study Group. *Eur Respir J* 17(5): 928–33, 2001.
49. Sethi S. Acute exacerbations of COPD: A "multipronged" approach. *J Respir Dis* 23(4): 217–55, 2002.
50. Bach PB *et al.* Management of acute exacerbations of chronic obstructive pulmonary disease: A summary and appraisal of published evidence. *Ann Intern Med* 134(7): 600–20, 2001.
51. Balter MS *et al.* Canadian guidelines for the management of acute exacerbations of chronic bronchitis. *Can Respir J* 10(Suppl B): 3B–32B, 2003.
52. Celli BR, MacNee W. Standards for the diagnosis and treatment of patients with COPD: A summary of the ATS/ERS position paper. *Eur Respir J* 23(6): 932–46, 2004.
53. Sethi S. Bacteria in exacerbations of chronic obstructive pulmonary disease. Phenomenon or epiphenomenon?. *Proc Am Thorac Soc* 1: 109–14, 2004.
54. Sinus A, Allergy Health P. Antimicrobial treatment guidelines for acute bacterial rhinosinusitis 2004. *Otolaryngol Head Neck Surg* 130(1): 1–45, 2004.
55. Saint S *et al.* Antibiotics in chronic obstructive pulmonary disease exacerbations. A meta-analysis. *JAMA* 273(12): 957–96, 1995.
56. Ram FS *et al.* Antibiotics for exacerbations of chronic obstructive pulmonary disease. *Cochrane Database Syst Rev*(2), 2006. , CD004403
57. Nouira S *et al.* Once daily oral ofloxacin in chronic obstructive pulmonary disease exacerbation requiring mechanical ventilation: A randomised placebo-controlled trial. *Lancet* 358(9298): 2020–25, 2001.
58. Allegra L *et al.* Antibiotic treatment and baseline severity of disease in acute exacerbations of chronic bronchitis: A re-evaluation of previously published data of a placebo-controlled randomized study. *Pulm Pharmacol Ther* 14(2): 149–55, 2001.
59. Miravitlles M *et al.* The efficacy of moxifloxacin in acute exacerbations of chronic bronchitis: A Spanish physician and patient experience. *Int J Clin Pract* 55(7): 437–41, 2001.
60. Wood-Baker RR *et al.* Systemic corticosteroids for acute exacerbations of chronic obstructive pulmonary disease. *Cochrane Database Syst Rev* (1), 2005, CD001288.
61. Sethi S, Murphy TF. Acute exacerbations of chronic bronchitis: new developments concerning microbiology and pathophysiology – impact

on approaches to risk stratification and therapy. *Infect Dis Clin North Am* 18(4): 861–82, 2004, ix.
62. Obaji A, Sethi S. Acute exacerbations of chronic bronchitis. What role for the new fluoroquinolones?. *Drugs Aging* 18(1): 1–11, 2001.
63. Dimopoulos G *et al.* Comparison of first-line with second-line antibiotics for acute exacerbations of chronic bronchitis: A meta-analysis of randomized controlled trials. *Chest* 132(2): 447–55, 2007.
64. Siempos II *et al.* Macrolides, quinolones and amoxicillin/clavulanate for chronic bronchitis: A meta-analysis. *Eur Respir J* 29(6): 1127–37, 2007.
65. Wilson R *et al.* A comparison of gemifloxacin and clarithromycin in acute exacerbations of chronic bronchitis and long-term clinical outcomes. *Clin Ther* 24(4): 639–52, 2002.
66. Wilson R *et al.* Short-term and long-term outcomes of moxifloxacin compared to standard antibiotic treatment in acute exacerbations of chronic bronchitis. *Chest* 125(3): 953–64, 2004.
67. Adams SG *et al.* Antibiotics are associated with lower relapse rates in outpatients with acute exacerbations of COPD. *Chest* 117: 1345–52, 2000.
68. Dewan NA *et al.* Acute exacerbation of COPD: Factors associated with poor treatment outcome. *Chest* 117(3): 662–71, 2000.
69. Wilson R *et al.* Antibiotic treatment and factors influencing short and long term outcomes of acute exacerbations of chronic bronchitis. *Thorax* 61(4): 337–42, 2006.
70. Ball P *et al.* Acute infective exacerbations of chronic bronchitis. *Q J Med* 88: 61–68, 1995.
71. Groenewegen KH, Schols AM, Wouters EF. Mortality and mortality-related factors after hospitalization for acute exacerbation of COPD. *Chest* 124(2): 459–67, 2003.
72. Vanderkooi OG *et al.* Predicting antimicrobial resistance in invasive pneumococcal infections. *Clin Infect Dis* 40(9): 1288–97, 2005.
73. Sethi S *et al.* Antibiotic exposure in COPD and the development of penicillin and erythromycin resistance in *Streptococcus pneumoniae*. San Francisco, CA: ICAAC, 2006.
74. Soler N *et al.* Bronchial microbial patterns in severe exacerbations of chronic obstructive pulmonary disease (COPD) requiring mechanical ventilation. *Am J Respir Crit Care Med* 157: 1498–505, 1998.
75. Stockley RA *et al.* Relationship of sputum color to nature and outpatient managment of acute exacerbations of COPD. *Chest* 117: 1638–45, 2000.
76. Soler N *et al.* Bronchoscopic validation of the significance of sputum purulence in severe exacerbations of chronic obstructive pulmonary disease. *Thorax* 62(1): 29–35, 2007.
77. Stolz D *et al.* Antibiotic treatment of exacerbations of COPD: A randomized, controlled trial comparing procalcitonin-guidance with standard therapy. *Chest* 131(1): 9–19, 2007.
78. Martinez FJ, Curtis JL. Procalcitonin-guided antibiotic therapy in COPD exacerbations: Closer but not quite there. *Chest* 131(1): 1–2, 2007.
79. Miravitlles M *et al.* Pharmacoeconomic evaluation of acute exacerbations of chronic bronchitis and COPD. *Chest* 121(5): 1449–55, 2002.

54

CHAPTER

Long-Term Oxygen Therapy

Bartolome R. Celli

Division of Pulmonary and Critical Care, Caritas St. Elizabeth's Medical Center, Tufts University, Boston, MA, USA

The use of oxygen as a therapeutic agent is generally thought to have began in the 1920s [1]. Since then, much has been learned about the effects of oxygen and many methods of delivery have been developed. In this chapter, we will review the known effects of chronic oxygen therapy, its indications, and the various delivery systems now available.

PATHOPHYSIOLOGY OF OXYGENATION

To understand the scientific rationale and guidelines concerning oxygen therapy, it is necessary to know about the physiology of gas exchange, transportation of oxygen to tissues, and the consequences of tissue hypoxia. Hypoxemia is defined as an abnormally low arterial oxygen tension P_aO_2. The physiologic causes of hypoxemia include a low inspired partial pressure of oxygen (F_iO_2), abnormal ventilation–perfusion relationship ($\dot{V}/\dot{Q}$ mismatch), decreased diffusion capacity, alveolar hypoventilation, and right to left shunt. Oxygen therapy increases the F_iO_2 and is the primary treatment for hypoxemia resulting from the first three causes. The hypoxemia of alveolar hypoventilation is best treated by increased ventilation while a true shunt is by definition unresponsive to hyperoxia.

Oxygen delivery (DO_2) to the tissues is dependent upon the arterial oxygen content (C_aO_2) and the cardiac output ($\dot{Q}_t$) as illustrated by the following equation:

$$DO_2 = C_aO_2 \times \dot{Q}_t \qquad (54.1)$$

Arterial oxygen content is further determined by

$$C_aO_2 = (\text{Hgb} \times 1.39 \times \%\ \text{sat}/100) + (0.003 \times P_aO_2)$$

where Hgb is blood hemoglobin concentration in g/100 ml, 1.39 is the maximum amount of oxygen (ml) that can combine with 1 g of hemoglobin, % sat is the amount of oxygen combined with hemoglobin divided by the maximal amount possible, 0.003 is the amount of oxygen (ml) soluble in 100 ml of blood per torr P_aO_2 in the absence of Hgb, and P_aO_2 is the partial pressure of oxygen in arterial blood. Tissue hypoxia occurs when the amount of oxygen present cannot meet the metabolic needs. Inadequate oxygen transport is dependent upon any of the above factors and is the most common cause of tissue hypoxia. However, decreased oxygen utilization is another possible cause of hypoxia. This is demonstrated by shunting at the capillary level in sepsis or more dramatically by mitochondrial poisoning secondary to certain substances such as cyanide.

The physiologic response to acute hypoxemia is to maintain oxygen transport (Fig. 54.1). At a P_aO_2 below 55, ventilatory drive increases rapidly leading to a higher arterial P_aO_2 and hypocapnia. Vascular beds supplying hypoxic tissue vasodilate, resulting in tachycardia, increased cardiac output, and oxygen delivery. The lung responds to hypoxemia by vasoconstriction to increase ventilation–perfusion matching and the P_aO_2. Subsequently, secretion of erythropoietin results in erythrocytosis and improved oxygen carrying capacity. These adaptations improve oxygen delivery. Unfortunately, the short-term benefit of these responses may have a detrimental long-term effect. Prolonged pulmonary

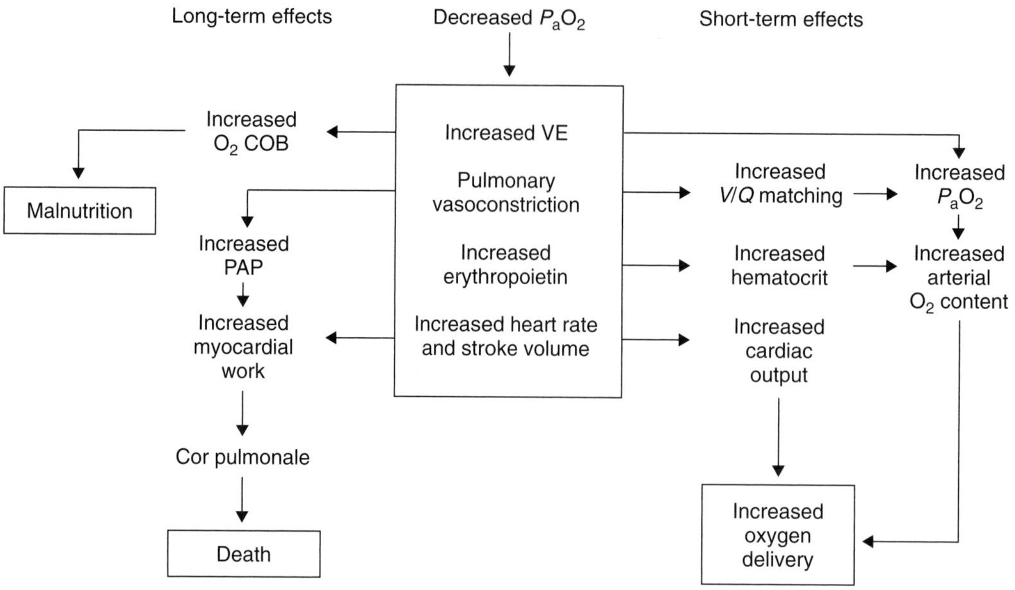

FIG. 54.1 The short- and long-term physiologic consequences of hypoxemia.

TABLE 54.1 Known effects of chronic oxygen therapy.

Survival
Increased in hypoxemic patients with COPD after receiving continuous O_2.
Pulmonary hemodynamics
Decreases pulmonary artery pressure and pulmonary vascular resistance.
Increases stroke volume.
Exercise
Improves exercise endurance.
Improves ventilatory muscle function.
Decreases minute ventilation.
Oxygen cost of ventilation
Decreases resting oxygen cost of breathing.
Reduces oxygen cost of ventilation during carbon dioxide induced hyperventilation.
Neuropsychology
Improves neuropsychologic performance.
Sleep
Improves sleep quality.

vasoconstriction and increased cardiac output can result in pulmonary hypertension (PH), right ventricular failure, and decreased survival [2–5]. In addition, the increased minute ventilation and increased oxygen cost of breathing (O_2 COB) may contribute to chronic malnutrition [6]. To interrupt this sequence of events, oxygen therapy has become important in hypoxemic patients [4, 5, 7, 8]. We shall review what is known about the long-term effects of oxygen therapy with particular attention to COPD because this is the only disease in which chronic oxygen therapy has been studied in detail (Table 54.1).

EFFECTS OF CHRONIC OXYGEN THERAPY

Survival

Although the cause of death in patients with hypoxic cor pulmonale is uncertain, the factors predicting survival in patients on long-term oxygen have been investigated. In general, variables reflecting worse severity of COPD, such as reduction of P_aO_2 or increased PCO_2, [6–9] lower FEV_1 [6, 9], and elevated mean pulmonary artery pressure (PAP) [9], correlate with survival. On the other hand, the effect of transient hypoxemia measured by oxygen desaturation appears to have limited impact on survival [10]. In the most hypoxemic patients as determined by arterial hypoxemia, the most important evidence supporting the use of chronic oxygen has been derived from studies investigating its effect on survival [11].

In 1970, Neff and Petty [12] found a 30–40% decrease in mortality in their severely hypoxic patients on continuous oxygen when they were compared with those in the literature. Subsequently, two large controlled trials studied the effect of oxygen on survival in COPD [4, 5]. The British Medical Research Council Study randomized patients to 15 h of continuous oxygen each day compared with no oxygen. During the 5-year follow-up, 19 of 42 oxygen therapy patients compared with 30 of 45 controls died. The Nocturnal Oxygen Therapy Trial (NOTT) randomly assigned patients to either 12 or 24 h of oxygen per day. After 26 months, the mortality of the continuous group (mean 19 h/day) was one-half that of the 12-h group. Taken together, these studies found survival to improve proportionally to the number of hours of supplemental oxygen per day. As a result, the current recommendation for hypoxemic patients (P_aO_2 < 55 mm Hg or S_aO_2 $< 88\%$) is continuous 24 h/day oxygen therapy. In addition, patients with a P_aO_2 of 55–59 mm Hg or a S_aO_2 $< 89\%$ in the presence of cor pulmonale or polycythemia also qualify for long-term continuous oxygen. The findings of these trials have been confirmed in the routine clinical setting [13].

Because of the lifestyle changes and economic implications of chronic oxygen therapy, patients must meet certain criteria for longer-term prescription. First, patients must be stable, on optimal medical therapy and not smoking. In addition, patients must be observed for at least 3 months to document persistent hypoxemia requiring long-term oxygen. During both the NOTT study [4] and a multicenter French study [14], approximately 40% of patients had improved enough after 1 month to no longer qualify for chronic oxygen. Consequently, patients should be reassessed 1–3 months after initiation of oxygen therapy to determine continued need for its use.

Pulmonary hemodynamics

Although the reason for increased survival with oxygen is not clear, there is evidence that O_2 can improve pulmonary hemodynamics and lead to reduced cardiac work and greater oxygen delivery. Initially, uncontrolled studies in hypoxemic patients suggested that continuous oxygen for 6–8 weeks could reduce pulmonary artery hypertension (PAH) [15, 16]. Subsequently, however, the controlled NOTT and MRC studies revealed no definite hemodynamic improvement [4, 5]. In the MRC trial, the mean PAP remained stable in patients receiving oxygen 15 h/day, but increased significantly in the control group. In NOTT, there was only a slight decrease in pulmonary vascular resistance (PVR) after 6 months on continuous oxygen while the nocturnal oxygen group had a minimal increase in PVR. However, the results of both these studies may have been flawed by the fact that each was devoted mainly to prognosis, follow-up was short, and not all patients were evaluated.

Some studies have specifically examined the effect of oxygen therapy on PAH. Weitzenblum and colleagues [3] performed right heart catheterizations on an average of 41 months prior to oxygen, just before starting oxygen, and 31 months after oxygen in 16 severely hypoxemic COPD patients on oxygen therapy. Before oxygen therapy, there was a mean yearly increase in PAP of 1.47 ± 2.3 mm Hg. Following oxygen, PH improved in 12 of 16 patients as demonstrated by an annual decrease in PAP of 2.15 ± 4.4 mm Hg. Complete normalization of PAP rarely occurred, but the changes in PAP were related to differences in PVR.

Further analysis of the NOTT data showed that the continuous oxygen therapy group had significant improvements in resting and exercise mean PAP, PVR, and stroke volume index after 6 months [2]. In addition, the nocturnal oxygen therapy group showed stable hemodynamic variables. However, the absolute differences between the two groups were not significant. Despite this, the authors suggested that further follow-up may show hemodynamic improvement as a factor altering the longevity between the two groups. One investigator has suggested that the reason continuous oxygen therapy improves survival may be due to the detrimental effect of oxygen withdrawal from patients on long-term therapy [16]. Selinger and colleagues [17] showed that removal of oxygen caused a 31% increase in PVR index coupled with a decrease in stroke volume index and oxygen delivery. They propose that the effect of recurrent oxygen withdrawal may be responsible for the worse prognosis of intermittent oxygen therapy.

In the longest of all the observational therapeutical oxygen trials, Zelinski and coworkers [7] evaluated pulmonary hemodynamics at 2-year intervals for 6 years in a group of 12 patients with hypoxemic COPD. The PAP which was 27 ± 7 torr at entry, decreased to 21 ± 4 at 2 years, to then return to stable values of 26 ± 7 and 26 ± 6 torr at 4 and 6 years, respectively. This was observed in spite of further deterioration of gas exchange with worsening of room air P_aO_2 and P_aCO_2. This study lends support to the concept that oxygen supplementation to hypoxemic patients halts progression of PAH.

Many patients continue to worsen despite long-term oxygen. Early identification of those patients who are unlikely to benefit from oxygen would avoid the cost and inconvenience of oxygen in these patients. As a result, various investigators have attempted to correlate an acute hemodynamic response to oxygen with long-term survival. In most studies, acute oxygen delivery (30 min to 24 h) decreases the PAP in a minority of patients [2, 18, 19]. Conversely, others have shown 28% oxygen to reduce PAP >5 mm Hg in greater than half of their patients [20, 21] and this acute response to oxygen predicted 2-year survival in patients with COPD and cor pulmonale [22]. However, Sliwinski *et al.* [19] have found little correlation between acute response to oxygen and long-term survival. They propose that the difference in results of these studies may be attributable to the lack of a pre-entry period in Ashutosh's study, which may have allowed inclusion of patients who were not stable. Consequently, these patients may have had acute hypoxic vasoconstriction superimposed upon the more irreversible anatomic changes. In their study, responders had lower mean PAP (26 mm Hg) after oxygen compared with the nonresponders (33 mm Hg) suggesting they could have had less severe baseline PH which might have accounted for their better prognosis.

Patients who have an acute hemodynamic response to oxygen may have structural differences in their pulmonary vasculature as compared with those who do not improve. However, an autopsy study examining the pulmonary arteries of patients who died during the NOTT study found no difference in the vascular structure of those who responded to oxygen and those who did not [23]. In addition, these authors found structural alterations of the muscular pulmonary arteries in patients with PAH secondary to COPD, but these changes did not correlate with the degree of PAH.

Because determination of PAP is invasive, noninvasive methods have been assessed to predict response to oxygen therapy. Ashutosh and Dunsky [21] studied exercise $\dot{V}O_2$ max and found a $\dot{V}O_2$ max >6.5 ml/min/kg to predict a fall in PAP >5 mm Hg and 3-year survival on oxygen. Attempts to show an acute or chronic benefit of oxygen on RVEF (right ventricular ejection fraction) [24, 25], and a change in RVEF on oxygen to predict a change in PAP [21, 25] or survival [21] have been disappointing. Therefore, the response of RVEF to oxygen does not appear to predict a beneficial effect of this therapy.

The possibility that long-term oxygen therapy could benefit patients with a lesser degree of resting hypoxemia

(P_aO_2 between 56 and 69 torr) was explored by Gorecka *et al.* [26]. They prospectively evaluated patients with moderate hypoxemia and observed no difference in mortality between treated and untreated patients. This lack of beneficial effect is further highlighted by the multicenter trial reported by Chaouet and coworkers [27] who after 2 years of therapy observed no effect on survival or pulmonary artery hemodynamics in a cohort of moderate hypoxemic patients randomized to supplemental oxygen or routine therapy.

Most of the studies evaluating oxygenation and survival have centered on the actual value of P_aO_2 and to some degree on the oxygen saturation. Recent evidence shows that the level of hemoglobin may be as important in predicting outcome in patients with COPD. The long held belief that most patients with COPD develop compensatory polycythemia has proven to not be the case as those large series show that anemia is highly prevalent in those patients and is associated not only with decreased exercise capacity but also with impaired survival [28, 29].

In summary, long-term oxygen has beneficial effects on survival and PH in patients with COPD and significant hypoxemia. Currently, most studies do not support the ability of acute hemodynamic response to oxygen measured either invasively or noninvasively to predict improvements in PAP, pulmonary vascular changes, or survival. Therefore, long-term oxygen is currently administered to all significantly hypoxemic patients ($P_aO_2 \leq 55$ torr), since some benefit cannot be ruled out and other therapeutic options are limited. There is no evidence to support the administration of long-term oxygen therapy to patients with mild or transitory hypoxemia. The value of correcting the hemoglobin in patients with COPD and anemia has not been explored.

Exercise

Ventilatory, rather than circulatory, factors limit maximum exercise performance in most patients with COPD. Performance is most closely related to airflow limitation [30]. Other investigators have suggested that the ventilatory muscles may contribute to exercise limitation in COPD [31, 32]. The process involves hyperinflation secondary to inability to exhale trapped air which in turn leads to more inspiratory muscle dysfunction at an unfavorable length. This muscles have to generate pressures, which are close to their maximum capacity and this may lead to fatigue as well as a lack of appropriate mechanical response to the increased central drive [33]. The resulting dyspnea leads to premature termination of exercise.

Oxygen therapy has been shown to improve exercise endurance in many studies. Various investigators have found oxygen to increase with distance walked [34, 35], time on a treadmill [36, 37], and time on a cycle ergometer [38, 39]. The precise mechanisms resulting in improvement are unknown [40], but a number have been suggested. In hypoxemic patients and those who desaturate with exercise, oxygen therapy results in greater oxygen delivery and utilization by exercising muscles [41–45]. However, increased oxygen saturation is not always predictive of improved exercise performance [37, 38] and other factors may contribute to this. Some studies have shown oxygen to reduce minute ventilation and respiratory rate for a given workload [35, 38]. Indeed, van Helvoort and colleagues have shown that administration of supplemental oxygen to patients with COPD even if the levels of oxygen are not critical may ameliorate the generation of oxidative stress and inflammatory burst observed during exercise in these patients [46].

Oxygen has been found to improve ventilatory muscle function during exercise, by postponing the onset of fatigue as demonstrated by a delay in the appearance of abdominal paradox and a slower fall in the high to low frequency of the diaphragm electromyogram power spectral density [38]. Criner and Celli [47] found that exercising COPD patients with oxygen resulted in the diaphragm performing more ventilatory work which may prevent overloading of the accessory muscles and contribute to the decrease in dyspnea. Finally, others have suggested that supplemental oxygen may decrease dyspnea and improve endurance by directly reducing chemoreceptor activity from the carotid body [39, 48]. Currently, qualification for oxygen during exercise requires documentation of a P_aO_2 less than or equal to 55 mm Hg or an $S_aO_2 < 88$ [49, 50]. A helpful clue to predicting which patients will desaturate with exercise is the diffusing capacity. Owens and colleagues [51] found a diffusing capacity above 55% predicted to be 100% specific in excluding desaturation.

Oxygen cost of ventilation

Beyond the effect of oxygen on arterial P_aO_2, there is evidence that oxygen may improve dyspnea by decreasing airway resistance and work of breathing (WOB). Astin and Penman [52] studied 18 COPD patients and found a significant association between airway resistance and P_aO_2. Administration of 30% oxygen for 20 min to these patients resulted in an average reduction of airway resistance of 20%. Subsequently, it was also found that 30% oxygen improved flow rates in mildly hypoxemic patients (mean $P_aO_2 = 61$ torr) [53]. Furthermore, atropine caused a similar increase in flow rates that aborted the effect of oxygen. This led the authors to conclude that hypoxemia-related bronchoconstriction may be mediated by vagal tone as had been previously suggested [52, 54].

In addition to its effects on airways resistance, oxygen via a face mask or transtracheal flow in patients with COPD has also been found to decrease minute ventilation [55, 56]. As a consequence of these effects, oxygen may lead to a decrease in the WOB or the O_2 COB and thereby improve dyspnea. Astin [57] studied the effect of 30% oxygen on 15 patients with COPD and found a decrease in WOB in 10 patients and an increase in 5 patients. Mannix *et al.* [58] calculated the COB induced by breathing 7% carbon dioxide. They found 30% oxygen to decrease the O_2 COB by 42% in both COPD patients and controls. In the COPD patients, the decrease in O_2 COB was only due to a smaller increment in VO_2 for a given level of ventilation, while controls had both a lower VO_2 and a VE. They concluded that supplemental oxygen decreases the O_2 demand of the respiratory muscles. Benditt and coworkers [59] observed a decrease in the oxygen cost of breathing with both transtracheal air and oxygen, but the effect was greater with oxygen. Because most patients receive long-term oxygen by a nasal cannula, the effect of nasal flow oxygen on the oxygen cost of breathing was recently assessed [60]. At high flows (O_2 5 l/min), oxygen significantly decreased the O_2 COB while compressed air did not. These initial studies have been recently confirmed with a study that shows that in a dose-dependent ways, supplemental oxygen decreases breathing frequency and dynamic hyperinflation thereby decreasing WOB [61]. The same group has shown that the administration of oxygen may help better train patients undergoing pulmonary rehabilitation [62].

In summary, there is overwhelming evidence that supplemental oxygen increases exercise capacity in hypoxemic patients with COPD and the effect may occur even in patients with mild hypoxemia. The long-term impact of this effect is less clear.

Neuropsychologic effects

Beginning in the 1930s and 1940s, aviation research demonstrated the acute effects of mild to severe hypoxia on healthy young men [63]. These studies found that even mild hypoxia (P_aO_2 45–60) could impair judgment, learning, and short-term memory in normal subjects. To investigate the effects of chronic hypoxemia, research has focused on the neurobehavioral effects of hypoxemia in patients with COPD.

Early work by Krop and colleagues [64] demonstrated poorer neuropsychologic performance in a group of COPD patients with a $P_aO_2 < 55$ mm Hg. Administration of continuous oxygen at 2 l/min by nasal cannula for 4 weeks resulted in significant improvement in 8 of 10 neuropsychologic tests. Because this study and another [65] only looked at a small group of patients without a stabilization period, two large multicenter studies in the United States and Canada examined the neuropsychologic consequences of hypoxemia [66, 67]. The US NOTT studied 203 patients with a mean age of 64 years and P_aO_2 of 51 torr and found that 42% of patients had moderate to severe impairment of cerebral function [66]. The Canadian IPPB Trial examined 100 patients with less hypoxemia (mean P_aO_2 66 torr) which resulted in more selective impairments [67]. The results of both these studies were combined, demonstrating an increase in the rate of deficits from 27% in mild hypoxemia ($P_aO_2 > 60$) to 61% in severe hypoxemia ($P_aO_2 < 50$) [68].

The NOTT study has evaluated the effect of continuous (19 h/day) and nocturnal oxygen on neuropsychologic functioning and life quality in 150 COPD patients. The results show an improvement after 6 months in both groups in general alertness, motor speed, and hand grip [69]. Despite these benefits, patients had no change in emotional status or quality of life. Although improvements after 6 months were no greater on continuous as compared with nocturnal oxygen, a subgroup of patients followed for 12 months showed significantly better neuropsychologic function on continuous oxygen.

The poor neuropsychologic performance in hypoxemic COPD patients has been explained by weakness, fatigue, and depression. However, there is increasing evidence that brain hypoxia itself is responsible for the impaired function. The conventional belief that mild cerebral hypoxia could impair ATP generation and neuron function has been disproved [70]. Some studies suggest that hypoxia can result in reduced synthesis of acetylcholine from labeled precursors [71], which is interesting, since there is growing evidence to support the role of acetylcholine in memory and learning [72].

Sleep

In the original description of rapid eye movement (REM) sleep in 1953, it was observed that breathing became more variable during this stage of sleep [73]. Subsequently, it was found that decreased chest movement and oxygen saturation occurred in this sleep stage [74]. Other groups also demonstrated worsening hypoxemia and carbon dioxide retention during sleep [75, 76]. Koo and colleagues [77] correlated arterial blood gas tension with sleep stage in COPD patients and found that P_aO_2 decreased shortly after sleep from 64 to 58 mm Hg during non-REM sleep. Upon entering REM sleep, P_aO_2 fell further to 50 and the P_aCO_2 rose from 48 to 58 mm Hg. These findings were later confirmed by many others [78–82].

The mechanisms of hypoxemia during sleep in COPD include hypoventilation, a reduction in functional residual capacity (FRC), and alterations in ventilation perfusion matching [83] The major cause of hypoxemia during REM sleep is hypoventilation, which appears related to rapid shallow breathing [83, 84] and long episodes of hypopneas, but not actual apneas [81]. In addition, both the hypoxic [85, 86] and hypercapnic [87, 88] ventilatory responses are diminished. The cause of REM sleep hypoventilation has not been fully determined, but appears related to altered brain stem function during phasic neuronal activity [89]. During REM sleep, the ribcage contribution to ventilation decreases by 18–34% [90, 91], as a consequence of hypotonia of the intercostal muscles and decreased activity of accessory muscles [92]. This may be particularly important in patients with COPD who greatly depend on accessory muscles for breathing [93, 94]. Though it is generally believed that hypoventilation during REM sleep will result in ventilation–perfusion mismatch [83], its relative importance with regard to REM hypoxemia is unknown. Although some patients with COPD have sleep apnea, there is no evidence that this occurrence is more common than would be expected by chance alone [83].

A few studies have measured pulmonary hemodynamics during sleep in COPD and found an increase in PAP with acute desaturation [95–99]. In addition, some studies support an association between nocturnal desaturation and PH in COPD patients who have a daytime $P_aO_2 > 60$ mm Hg [98, 99]. Both studies found significantly higher PAP in the patients who desaturated as compared with the non-desaturators.

Patients with COPD have poor sleep quality as compared with age-matched controls [100–101] and arousals are frequent during periods of desaturation [102]. Whether oxygen therapy improves sleep quality in hypoxemic patients is debatable, since various studies have shown conflicting results [100, 103]. Some studies have found fewer hypoxemic episodes as well as increased total sleep and REM sleep with oxygen [100, 104].

Although mean nocturnal oxygen saturation (S_aO_2) and lowest nocturnal S_aO_2 have been found to correlate with survival, one study found these parameters to have no prognostic value over vital capacity or awake S_aO_2 [104, 105]. Furthermore, patients with greater nocturnal hypoxemia than would be predicted, had no difference in survival. In addition, the clinical characteristics of patients with COPD who had desaturation during 24 h monitoring were not different from that of patients who did not desaturate during the same period [106]. Fletcher's group retrospectively compared survival in COPD patients with a daytime $P_aO_2 > 60$ mm Hg with and without O_2 desaturation at night [107]. They found an improved survival in those patients without nocturnal desaturation. There was also a trend toward improved survival with nocturnal oxygen in those patients who did desaturate at night. These authors attribute the difference in survival in their study and one by Connaughton *et al.* [105] to the worse daytime P_aO_2 of the patients in that study (mean P_aO_2 53±10 mm Hg). They proposed that as the disease becomes more advanced, the daytime S_aO_2 may become a strong predictor, while sleep S_aO_2 correlates less well with survival. However, the recent report by Chaouet *et al.* [10] suggests that oxygen supplementation at night to mildly hypoxemic patients has a limited, if any, beneficial effect in pulmonary hemodynamics or survival.

Presently, patients who are hypoxemic while awake should also be prescribed oxygen during sleep. In addition, patients who are well saturated during wakefulness, but desaturate at night can qualify for oxygen if they have complications attributable to sleep hypoxemia, such as PH, daytime somnolence, or cardiac arrhythmias [50, 108].

Although it seems intuitive that patients who manifest oxygen desaturation during sleep should have a worse prognosis, this has not been confirmed in the few prospectives studies that have been conducted. However, it has been accepted that patients with prolonged destauration at night without sleep apnea as the cause, be provided with supplemental oxygen.

OXYGEN PRESCRIPTION

Currently, there are strict guidelines for oxygen prescription based upon resting hypoxemia ($P_aO_2 < 55$ mm Hg or $S_aO_2 < 88$) and hypoxemia induced by sleep or exercise (Table 54.2).

Once a patient qualifies for long-term oxygen, prescription requires completion of the Certificate of Medical Necessity for Home Oxygen Therapy form in the United States. This form should be completed before a patient leaves the hospital to ensure inclusion of all information such as patient data, reason for oxygen, blood gases, type of system, and liter flow at rest, exercise, and sleep.

Long-term oxygen administration can be accomplished with:

- oxygen concentrator
- compressed gas
- liquid oxygen.

TABLE 54.2 Indications for chronic oxygen therapy.

Continuous oxygen
Resting $P_aO_2 < 55$ mm Hg or $S_aO_2 \leq 88\%$
Resting P_aO_2 56–59 mm Hg or $S_aO_2 \leq 89\%$ in the presence of the following:
1 Dependent edema suggesting congestive heart failure
2 "P" pulmonale on EKG (P wave greater than 3 mm in standard leads II, III, or AVF)
3 Erythrocythemia (hematocrit ≥56%)
Resting $P_aO_2 \geq 59$ mm Hg or $S_aO_2 \geq 89\%$
Only reimbursable with additional documentation justifying the oxygen prescription and a summary of more conservative therapy which has failed
Noncontinuous oxygen
Must specify the oxygen flow rate and the number of hours/day
During exercise, $P_aO_2 \leq 55$ mm Hg or an $S_aO_2 \leq 88\%$ with low level exertion
During sleep, $P_aO_2 \leq 55$ mm Hg or an S_aO_2 88%, with associated complications such as pulmonary hypertension, daytime somnolence, and cardiac arrhythmias

TABLE 54.3 Modes of oxygen delivery.

System	Advantages	Disadvantages
Gas	Low cost	Heavy weight
	Wide availability	Difficult to refill
	Fair portability	Short supply time
Liquid	Light weight	High cost
	Very portable	Vendor incompatibility
	Easy refill	Pressure venting
Concentrator	Low cost	Heavy weight
	Good availability	Least portable

Table 54.3 summarizes the advantages and disadvantages of each mode of therapy comparing weight, cost, portability, ease of refill, and availability. Because home oxygen is supplied under a fixed reimbursement policy regardless of the system used [104], oxygen vendors will attempt to provide the least expensive system.

Oxygen concentrator

Most patients will require a stationary source of oxygen which is usually provided by an oxygen concentrator. Since concentrators are relatively inexpensive and require less frequent home visits than liquid oxygen, they have become the system of choice for suppliers. These electrically powered devices utilize a molecular sieve to separate oxygen from air resulting in delivery of oxygen to the patient, while nitrogen is returned to the atmosphere. The typical sieve achieves oxygen purity of 97% at low flows and 94% at higher flows. However, due to their voltage requirement and their weight, they are primarily a fixed source of oxygen. Consequently, patients need either compressed gas or liquid oxygen as an ambulatory source of oxygen. New, more portable devices should improve availability of these units.

Compressed gas oxygen

Compressed gas oxygen has been provided for many years in high pressure metal or aluminum cylinders. These containers vary in size weighing 90.8, 7.2, 4, and 1.8 kg and provide O_2 at 2 l/min flow for 2.4 days, 5.2 h, 2 h and 1.2 h, respectively. The smaller cylinders can be refilled but the process is inefficient and potentially hazardous. The major advantages of compressed gas are its low cost and wide availability. Its disadvantages are the weight of the cylinders, the difficulty in refilling and the short oxygen supply time.

Liquid oxygen

Liquid oxygen is stored at very cold temperatures which reduces its volume to less than 1% of the volume of atmospheric oxygen. Stationary units weigh 140 lbs and can provide several days of continuous oxygen at 2 l/min flow. There are portable containers weighing as low as 4 lbs that will last for up to 12 h at 2 l/min flow. Compared with compressed gas, an equivalent weight container of liquid oxygen is more portable and easier to refill, its liabilities include a higher cost, manufacturer incompatibility, and the need for pressure relief venting which wastes unused oxygen.

Deciding on which portable system to prescribe depends upon the activity level of the patient. For patients who are more active, liquid oxygen is the ambulatory system of choice since the oxygen lasts longer and the canister is easier to refill and to carry. Lock *et al.* [109] compared liquid and gaseous oxygen and found patients would use liquid oxygen more hours per week (23.5 versus 10 h) and leave their house more hours per week (19.5 versus 15.5 h).

A recently conducted study to evaluate whether portable oxygen improves quality of life has been disappointing with no benefit shown in the largest of them [110].

OXYGEN ADMINISTRATION DEVICES

Nasal cannula

Patients with COPD and other chronic lung diseases usually receive continuous flow of oxygen via a nasal cannula. Delivery at a flow of 2 l/min increases the F_iO_2 to approximately 27%, which provides an adequate O_2 saturation in the majority of patients. Although this method is effective, it is quite inefficient. During the respiratory cycle, alveolar gas exchange is limited to early inhalation which accounts for only one-sixth of the cycle. Conversely, alveolar ventilation does not occur during late or dead space inspiration and exhalation [111]. Consequently, only oxygen flowing during early inspiration is available to the patient, while the remainder is wasted into the environment. To improve efficiency and mobility, oxygen-conserving devices have been designed to deliver oxygen during early inhalation. These devices include reservoir nasal cannulas, transtracheal catheters, and electronic oxygen demand devices. In general, these three devices each result in oxygen savings of two to four times a conventional nasal cannula. For a more detailed review the reader may refer to Ref. [111].

Reservoir cannula

The reservoir cannula was the first method of oxygen conservation to be used [111, 112]. This device stores 20 ml of oxygen in a reservoir during expiration and delivers an oxygen bolus at the onset of inspiration. The two available reservoirs are shaped as either a mustache attached to nasal prongs or a pendant placed over the anterior chest wall. The mustache reservoir was developed first, but because of dissatisfaction with its high visible location, the pendant was designed to be hidden by clothing [113]. Although this device has better cosmesis, it is still more noticeable than the standard nasal cannula.

Electronic demand devices

Electronic demand devices sense the beginning of inspiration and deliver a pulse of oxygen during early inhalation [114, 115]. There are several units available with many similarities and differences. Most devices are relatively inexpensive and have built-in rechargeable batteries and alarms [116]. Some systems can be switched manually from demand to continuous flow [116], while others automatically change to continuous flow if an inspiratory signal is not sensed [117]. Another demand valve is able to maintain constant oxygen delivery by adjusting the pulse size depending on the respiratory rate. A recent comparison of standard nasal cannula, reservoir nasal cannula, and a demand flow device in 15 patients with COPD found no difference in exercise tolerance among the three systems. However, oxygen saturation tended to be lower during the exercise using the demand flow device [118].

Transtracheal catheters

Transtracheal catheters improve oxygen delivery by bypassing anatomic dead space and utilizing the upper airways as a reservoir during end expiration [119]. Heimlich [120] developed the first catheter after experiments in dogs and demonstrated an increased P_aO_2 as a catheter was placed more distally. In addition to increased oxygen savings, other advantages of TTO include its relative inconspicuousness, the lack of nasal, auricular, or facial skin irritation, its infrequent displacement during exercise and sleep [120]. In addition, there are patients with refractory hypoxemia on a nasal cannula that have been successfully oxygenated on TTO [121–124]. Patient acceptance of transtracheal catheters has been variable with rates ranging from 50% to 96% during 9- to 20-month periods [120–125].

Complications

Complications of TTO vary depending on the system implemented and whether the acute or late phase of treatment is examined. The most common procedure-related problems include subcutaneous emphysema, bronchospasm, and paroxysmal coughing [123, 126]. Frequently encountered late complications and sequelae include dislodged catheters, stomal infections, and symptomatic mucous balls. Mucous balls result from oxygen drying the sputum and causing it to adhere to the catheter. The greater frequency of mucous balls seen in 10–25% patients with the Scoop catheter (Transtracheal Systems, Denver, CO, USA) appears to be the result of its larger surface area and the presence of side holes which allow accumulation of secretions. Some reports show life threatening [127, 128] and even fatal airway obstruction due to mucous balls [129]. Many other uncommon complications have been described including broken catheter tips, hemoptysis, keloid formation, hoarseness, and cardiac arrhythmias. After an initial burst of enthusiasm, the number of transtracheal earlier initiations has decreased primarily due to the need for continuous supervision and relatively poor long-term compliance.

OXYGEN AND AIR TRAVEL

Commercial airline travel exposes passengers to hypobaric hypoxia since aircraft cabins are not routinely pressurized to sea level [130]. In patients who have compensated COPD at sea level, lowering the partial pressure of oxygen in the aircraft cabin can produce severe hypoxemia. Physical exertion during the flight can increase the risk of an exacerbation of symptoms. The proportion of patients suffering from complications during air travel is not well known.

Aircraft are usually pressurized to between 5000 and 7000 ft (1500–2100 m). For the preflight evaluation of most patients, clinicians should consider 8000 ft (2438 m) of altitude above sea levels as realistic "worse case scenario."

Preflight assessment can be accomplished with the following elements:

- Estimation of the expected degree of hypoxemia at altitude.
- Identification of comorbid disease conditions.
- Provision of an oxygen prescription if necessary.

Documentation of the recent clinical condition and laboratory tests, particularly if the patient is traveling abroad, and counseling are also desirable elements of the preflight patient care [50, 118].

The two means of estimating the degree of hypoxia at altitude are:

- Hypoxia inhalation test (HIT), which is not performed in many clinical laboratories in the United States and is not recommended for routine use.
- Use of regression formulae. Regression equations offer the opportunity to compare a patient with a group of patients with similar clinical characteristics who have been previously studied during exposure to hypoxia. While regression equations may provide a more physiological basis for the effects of high altitude than the HIT, the regression approach does not assess the individual's susceptibility to the development of symptoms or electrocardiographic changes during hypoxia.

The A-aO_2 gradient and the a-AO_2 ratio generally have no advantages over regression equations. It is currently recommended that the P_aO_2 during air travel should be maintained above 50 mm Hg [50]. While 2–3 l of oxygen by nasal cannula will replace the inspired oxygen partial pressure lost at 8000 ft compared with sea level, lesser increments of oxygen will maintain the P_aO_2 above 50 mm Hg in many patients.

COPD patients receiving continuous oxygen at home will require supplementation during air flight. Such patients should receive greater oxygen supplementation during the flight than at sea level. Increments equivalent to 1–2 ls of oxygen by nasal cannula during flight should suffice for most patients. Patients will also require additional oxygen supplementation if the elevation at the destination is significantly greater than at home. The Federal Aviation Administration requires a physician's statement of oxygen need in order for a patient to receive continuous oxygen during flight. There is no uniform airline request form, so each airline must be contacted by the patient to determine what is required. As the airlines do not provide oxygen for ground use in the airline terminal, patients who require continuous oxygen should be advised to make plans for such locations. The American Lung Association provides patient education materials for individuals who travel with oxygen entitled "Airline Travel with Oxygen."

Many ambulatory COPD patients not receiving oxygen at home can tolerate PO_2 values below 50 mm Hg for brief periods of time without serious consequences. Stable COPD patients without comorbid disease who have previously traveled without incident and who are currently clinically stable compared with their previous air travel, may be advised to travel by air as there is little risk to them.

SUMMARY

Oxygen therapy has become very important in the treatment of severe COPD. Because long-term oxygen improves survival in hypoxemic COPD, patients with a daytime $P_aO_2 < 55$ mm Hg or $S_aO_2 < 88\%$ qualify for continuous oxygen therapy. Other benefits of oxygen include better exercise tolerance, decreased dyspnea, and improvements in neuropsychological performance. In addition, there appear to be beneficial effects on pulmonary hemodynamics, sleep quality, reduced minute ventilation, and WOB. The consequences of exercise induced and nocturnal desaturation are not absolutely known. Clinicians should be aware of air travel-induced hypoxemia in COPD and be able to identify those who need oxygen in flight. Once a patient meets the criteria for oxygen prescription, the physician must complete the certificate of Medical Necessity Form specifying the indication for oxygen, type of system and liter flow at rest, exercise and sleep. Patients should be reevaluated at 1 to 3 months to document a persistent need for oxygen therapy. Occasionally, patients should be considered for oxygen-conserving devices to decrease oxygen flow, cost and to improve mobility.

References

1. Barach AL. The therapeutic uses of oxygen. *JAMA* 79: 693–98, 1922.
2. Timms RM, Khaja FU, Williams GW. Nocturnal oxygen therapy trial group. Hemodynamic response to oxygen therapy in chronic obstructive pulmonary disease. *Ann Intern Med* 102: 29–36, 1985.
3. Weitzenblum E, Sautegeau A, Ehrhart M, Mammosser M, Pelletier A. Long-term oxygen therapy can reverse the progression of pulmonary hypertension in patients with chronic obstructive pulmonary disease. *Am Rev Respir Dis* 131: 493–98, 1985.
4. Nocturnal Oxygen Therapy Trial Group. Continuous or nocturnal oxygen therapy in hypoxemic chronic obstructive lung disease. *Ann Intern Med* 93: 391–98, 1980.
5. Medical Research Council Working Party. Long-term domiciliary oxygen therapy in chronic hypoxic cor pulmonale complicating chronic bronchitis and emphysema. *Lancet* 1: 681–86, 1981.
6. Donahue M, Rogers RM, Wilson DO, Pennock BE. Oxygen consumption of the respiratory muscles in normal and in malnourished patients with chronic obstructive pulmonary disease. *Am Rev Respir Dis* 140: 383–91, 1989.
7. Zelinski J, Tobiasz M, Hawrylkiewicz I, Sliwnski P, Palasiewicz G. Hemodynamics in COPD patients: A 6-year prospective study. *Chest* 113: 65–70, 1998.
8. Cooper C, Howard P. An analysis of sequential physiologic changes in hypoxic cor pulmonale during long-term oxygen therapy. *Chest* 100: 76–80, 1991.
9. Skwarski K, Macnee W, Wraith PK, Sliwinski P, Zielinski J. Predictors of survival in patients with chronic obstructive pulmonary disease treated with long-term oxygen therapy. *Chest* 100: 1522–27, 1991.
10. Chaouat A, Weitzenblum E, Kessler R, Schott R, Charpentier C, Levi-Valensi P, Zielinski J, Delaunois L, Cornudella R, Moutinho dos Santos J. Outcome of COPD patients with mild daytime hypoxaemia with or without sleep-related oxygen desaturation. *Eur Respir J* 17: 848–55, 2001.
11. Cooper CB, Waterhouse J, Howard P. Twelve-year clinical study of patients with hypoxic cor pulmonale given long-term domiciliary oxygen therapy. *Thorax* 42: 105–10, 1987.
12. Neff TA, Petty TL. Long-term continuous oxygen therapy in chronic airway obstruction. *Ann Intern Med* 72: 621–25, 1970.
13. Crockett AJ, Cranston JM, Moss JR, Alpers JH. Survival on long-term oxygen therapy in chronic airflow limitation: From evidence to outcomes in the routine clinical setting. *Intern Med J* 31: 448–54, 2001.
14. Levi-Valensi P, Duwoos H, Racineaux JL. Multicenter study of oxygen therapy: GEMOS preliminary results. *Med Thorac* 5: 502–6, 1983.
15. Levine BE, Bigelow DB, Hamstra RD *et al.* The role of long-term continuous oxygen administration in patients with chronic airway obstruction with hypoxemia. *Ann Intern Med* 66: 639–50, 1967.
16. Abraham AS, Cole RB, Bishop JM. Reversal of pulmonary hypertension by prolonged oxygen administration to patients with chronic bronchitis. *Circ Res* 24: 147–57, 1968.
17. Selinger SR, Kennedy TP, Buescher P *et al.* Effects of removing oxygen from patients with chronic obstructive pulmonary disease. *Am Rev Respir Dis* 136: 85–91, 1987.
18. Aber GM, Harris AM, Bishop JM. The effect of acute changes in inspired oxygen concentration on cardiac, respiratory and renal function in patients with chronic obstructive airways disease. *Clin Sci* 26: 133–43, 1967.
19. Sliwinski P, Hawrylkiewicz I, Gorecka D, Zielinski J. Acute effects of oxygen on pulmonary arterial pressure does not predict survival on long-term oxygen therapy in patients with chronic obstructive pulmonary disease. *Am Rev Respir Dis* 146: 665–69, 1992.
20. Anthonisen NR. Home oxygen therapy in chronic obstructive pulmonary disease. *Clin Chest Med* 7: 673–78, 1986.
21. Ashutosh K, Dunsky M. Noninvasive tests for the responsiveness of pulmonary hypertension to oxygen. Prediction of survival in patients with chronic obstructive lung disease and cor pulmonale. *Chest* 92: 393–99, 1987.
22. Ashutosh K, Mead G, Dunsky M. Early effects of oxygen administration and prognosis in chronic obstructive pulmonary disease and cor pulmonale. *Am Rev Respir Dis* 127: 399–404, 1983.
23. Wright JL, Petty T, Thurlbeck WM. Analysis of the structure of the muscular pulmonary arteries in patients with pulmonary hypertension and COPD: National Institutes of Health Nocturnal Oxygen Therapy Trial. *Lung* 170: 109–24, 1992.
24. Macnee W, Wathen CG, Flenley DC, Muir AD. The effects of controlled oxygen therapy on ventricular function in patients with stable and decompensated cor pulmonale. *Am Rev Respir Dis* 137: 1289–95, 1988.
25. Morrison DA, Henry R, Goldman S. Preliminary study of the effects of low flow oxygen on oxygen delivery and right ventricular function in chronic lung disease. *Am Rev Respir Dis* 133: 390–95, 1986.
26. Gorecka D, Gorzelak K, Siliwinski P, Tobiasz M, Zielinski J. Effects of long-term oxygen therapy on survival in patients with chronic obstructive pulmonary disease with moderate hypoxemia. *Thorax* 52: 674–79, 1997.
27. Chouet A, Weitzenblum E, Kessler R *et al.* A randomized trial of nocturnal oxygen therapy in chronic obstructive pulmonary disease patients. *Eur Respir J* 14: 1002–8, 1999.
28. Chambellan A, Chailleux E, Similowski T. Prognostic value of the hematocrit in patients with severe COPD receiving long-term oxygen therapy. *Chest* 128: 1201–8, 2005.
29. Cote C, Zilberberg MD, Mody SH, Dordelly LJ, Celli B. Haemoglobin level and its clinical impact in a cohort of patients with COPD. *Eur Respir J* 29: 923–29, 2007.
30. Jones NL, Jones G, Edwards RHT. Exercise tolerance in chronic airway obstruction. *Am Rev Respir Dis* 103: 477–91, 1971.
31. Dodd DS, Brancatisano T, Engel LA. Chest wall mechanics during exercise in patients with severe chronic air-flow obstruction. *Am Rev Respir Dis* 129: 33–38, 1984.
32. Montes de Oca M, Rassulo J, Celli B. Respiratory muscle function and cardiopulmonary response to exercise in severe COPD. *Am J Respir Crit Care Med* 154: 1284–89, 1996.
33. O'Donnell D, Flugre T, Gerken F, Hamilton A, Webb K, Aguilaniu B *et al.* Effects of tiotropium on lung hyperinflation, dyspnea and exercise tolerance in COPD. *Eur Respir J* 23: 832–40, 2004.
34. Woodcock AA, Gross ER, Geddes DM. Oxygen relieves breathlessness in "pink puffers". *Lancet* 1: 907–9, 1981.
35. Leggett RJE, Flenley DC. Portable oxygen and exercise tolerance in patients with chronic hypoxic cor pulmonale. *BMJ* 2: 84–86, 1977.
36. Cotes JE, Gilson JC. Effect of oxygen on exercise ability in chronic respiratory insufficiency. *Lancet* 1: 872–76, 1956.
37. Bradley BL, Garner AE, Billiu D, Mestas JM, Forman J. Oxygen-assisted exercise in chronic obstructive lung disease. The effect on exercise capacity and arterial blood gas tension. *Am Rev Respir Dis* 118: 239–43, 1978.
38. Bye PTP, Esau SA, Levy RD *et al.* Ventilatory muscle function during exercise in air and oxygen in patients with chronic air-flow limitation. *Am Rev Respir Dis* 132: 236–40, 1985.
39. Dean NC, Brown JK, Himelman RB, Doherty JJ, Gold WM, Stulbarg MS. Oxygen may improve dyspnea and endurance in patients with chronic obstructive pulmonary disease and only mild hypoxemia. *Am Rev Respir Dis* 146: 941–45, 1992.

40. Stevenson NJ, Calverley PM. Effect of oxygen on recovery from maximal exercise in patients with chronic obstructive pulmonary disease. *Thorax* 59: 668–72, 2004.
41. Stanek KA, Nagle FJ, Bisgard GE, Burnes WC. Effect of hyperoxia on oxygen consumption in exercising ponies. *J Appl Physiol* 46: 1115–18, 1979.
42. Vyas MN, Banister EW, Morton JW, Grzybowski S. Response to exercise in patients with chronic airway obstruction. II. Effects of breathing 40 percent oxygen. *Am Rev Respir Dis* 103: 401–12, 1971.
43. Morrison DA, Stovall JR. Increased exercise capacity in hypoxemic patients after long-term oxygen therapy. *Chest* 102: 542–50, 1992.
44. Gosselin N, Durand F, Poulain M, Lambert K, Ceugniet F, Prefaut C, Varray A. Effect of acute hyperoxia during exercise on quadriceps electrical activity in active COPD patients. *Acta Physiol Scand* 181: 333–43, 2004.
45. Stein DA, Bradley BL, Miller WC. Mechanisms of oxygen effects on exercise in patients with chronic obstructive pulmonary disease. *Chest* 81: 6–10, 1982.
46. van Helvoort HA, Heijdra YF, Heunks LM, Meijer PL, Ruitenbeek W, Thijs HM, Dekhuijzen PN. Supplemental oxygen prevents exercise-induced oxidative stress in muscle-wasted patients with chronic obstructive pulmonary disease. *Am J Respir Crit Care Med* 176: 1122–29, 2006.
47. Criner GJ, Celli BR. Ventilatory muscle recruitment in exercise with O_2 in obstructed patients with mild hypoxemia. *J Appl Physiol* 63: 195–200, 1987.
48. Guz A, Adams L, Minty K, Murphy K. Breathlessness and the ventilatory drives of exercise, hypercapnoea and hypoxia. *Clin Sci* 60: 17–18, 1981.
49. Tiep BL. Long-term home oxygen therapy. *Clin Chest Med* 11: 505–21, 1990.
50. Celli BR, MacNee W. Standards for the diagnosis and treatment of COPD. *Eur Respir J* 23: 932–46, 2004.
51. Owens GR, Rogers RM, Pennock BE, Levin D. The diffusing capacity as a predictor of arterial oxygen desaturation during exercise in patients with chronic obstructive pulmonary disease. *N Engl J Med* 310: 1218–21, 1984.
52. Astin TW, Penman RWB. Airway obstruction due to hypoxemia in patients with chronic lung disease. *Am Rev Respir Dis* 95: 567–75, 1967.
53. Libby DM, Briscoe WA, King TKC. Relief of hypoxia-related bronchoconstriction by breathing 30 percent oxygen. *Am Rev Respir Dis* 123: 171–75, 1981.
54. Nadel JA, Widdicombe JG. Effect of changes in blood gas tensions and carotid sinus pressure on tracheal volume and total lung resistance to airflow. *J Physiol* 163: 13–33, 1962.
55. Swinburn CR, Mould H, Stone TN, Corris PA, Gibson GJ. Symptomatic benefit of supplemental oxygen in hypoxemic patients with chronic lung disease. *Am Rev Respir Dis* 143: 913–15, 1991.
56. Couser JI, Make BJ. Transtracheal oxygen decreases inspired minute ventilation. *Am Rev Respir Dis* 139: 627–31, 1989.
57. Astin TW. The effect of oxygen inhalation on the work of breathing in patients with chronic obstructive bronchitis. *Respiration* 27: 51–62, 1970.
58. Mannix ET, Manfredi F, Palange P, Dowdeswell IRG, Farber MO. Oxygen may lower the O_2 cost of ventilation in chronic obstructive lung disease. *Chest* 101: 910–15, 1992.
59. Benditt JO, Rassulo J, Celli BR. Work of breathing during direct tracheal O_2 administration in patients with severe chronic lung disease. *Am Rev Respir Dis* 14: A883, 1990.
60. Tarpy S, Epstein S, Gottlieb D, Celli B. The effect of oxygen and air via nasal cannula on the oxygen cost of breathing in chronic airflow obstruction. *Am Rev Respir Dis* 145: A646, 1992.
61. Somfay A, Porszasz J, Lee SM, Casaburi R. Dose-response effect of oxygen on hyperinflation and exercise endurance in nonhypoxaemic COPD patients. *Eur Respir J* 18: 77–84, 2001.
62. Emtner M, Porszasz J, Burns M, Somfay A, Casaburi R. Benefits of supplemental oxygen in exercise training in nonhypoxemic chronic obstructive pulmonary disease patients. *Am J Respir Crit Care Med* 168: 1034–42, 2003.
63. Luft U. Aviation physiology – the effects of altitude. In: Fenn W, Rahn K (eds). *Handbook of Physiology Respiration,* 2, p. 1099. Washington, DC: American Physiol. Soc, 1965.
64. Krop HD, Block AJ, Cohen E. Neuropsychologic effects of continuous oxygen therapy in chronic obstructive pulmonary disease. *Chest* 64: 317–22, 1973.
65. Block AJ, Castle JR, Keitt AS. Chronic oxygen therapy. Treatment of chronic obstructive pulmonary disease at sea level. *Chest* 65: 279–88, 1974.
66. Grant I, Heaton RK, Mcsweeny AJ, Adams KM, Timms RM. Neuropsychologic findings in hypoxemic chronic obstructive pulmonary disease. *Arch Intern Med* 142: 1470–76, 1982.
67. Prigatano GP, Parsons OA, Wright E, Levin DC, Hawryluk G. Neuropsychologic test performance in mildly hypoxemic patients with chronic obstructive pulmonary disease. *J Consult Clin Psychol* 51: 108–16, 1983.
68. Grant I, Prigatano GP, Heaton RK, McSweeny AJ, Wright EC, Adams KM. Progressive neuropsychologic impairment and hypoxemia. Relationship in chronic obstructive pulmonary disease. *Arch Gen Psychiatry* 44: 999–1006, 1987.
69. Heaton RK, Grant I, McSweeny AJ, Adams KM, Petty TL. Psychological effects of continuous and nocturnal oxygen therapy in hypoxemic chronic obstructive pulmonary disease. *Arch Intern Med* 143: 1941–47, 1983.
70. Siesjo B, Johannsson H, Ljunggren B, Norberg K. Brain dysfunction in mild to moderate hypoxia. *Am J Med* 70: 1247–53, 1981.
71. Gibson GE, Shimada M, Blass JP. Alterations in acetylcholine synthesis and in cyclic nucleotides in mild cerebral hypoxia. *J Neurochem* 31: 757–60, 1978.
72. Bartus RT. Effects of cholinergic agents on learning and memory in animal models of aging. *Aging* 19: 271–80, 1982.
73. Aserinsky E, Kleitman N. Regularly occurring periods of eye motility and concomitant phenomena during sleep. *Science* 118: 273–74, 1953.
74. Aserinsky E. Periodic respiratory pattern occurring in conjunction with eye movements during sleep. *Science* 150: 763–66, 1965.
75. Robin ED, Whaley RD, Crump CH, Travis DM. Alveolar gas tensions, pulmonary ventilation and blood pH during physiological sleep in normal subjects. *J Clin Invest* 37: 981–89, 1958.
76. Trask CH, Cree EM. Oximeter studies on patients with chronic obstructive emphysema, awake and during sleep. *N Engl J Med* 266: 639–42, 1962.
77. Koo KW, Sax DS, Snider GL. Arterial blood gases and pH during sleep in chronic obstructive pulmonary disease. *Am J Med* 58: 663–70, 1975.
78. Leitch AG, Clancy LJ, Leggett RJE, Tweeddale P, Dawson P, Evans JI. Arterial blood gas tensions, hydrogen ion, and electroencephalogram during sleep in patients with chronic ventilatory failure. *Thorax* 31: 730–35, 1976.
79. Wynne JW, Block AJ, Hemenway J, Hunt LA, Flick MR. Disordered breathing and oxygen desaturation during sleep in patients with chronic obstructive lung disease (COLD). *Am J Med* 66: 573–79, 1979.
80. Catterall JR, Douglas NJ, Calverley PMA *et al.* Transient hypoxemia during sleep in chronic obstructive pulmonary disease is not a sleep apnea syndrome. *Am Rev Respir Dis* 128: 24–29, 1983.
81. Hudgel DW, Martin RJ, Capehart M, Johnson B, Hill P. Contribution of hypoventilation to sleep oxygen desaturation in chronic obstructive pulmonary disease. *J Appl Physiol* 55: 669–77, 1983.
82. Ballard RD, Clover CW, Suh BY. Influence of sleep on respiratory function in emphysema. *Am J Respir Crit Care Med* 151: 945–51, 1995.
83. Douglas NJ. Sleep in patients with chronic obstructive lung disease. *Clin Chest Med* 19: 115–25, 1998.
84. Douglas NJ, White DP, Pickett CK, Weil JV, Zwillich CW. Respiration during sleep in normal man. *Thorax* 37: 840–44, 1982.
85. Douglas NJ, White DP, Weil JV *et al.* Hypoxic ventilatory response decreases during sleep in normal men. *Am Rev Respir Dis* 125: 286–89, 1982.

86. Berthon-Jones M, Sullivan CE. Ventilatory and arousal responses to hypoxia in sleeping humans. *Am Rev Respir Dis* 126: 758–62, 1982.
87. Douglas NJ, White DP, Weil JV, Pickett CK, Zwillich CW. Hypercapnic ventilatory response in sleeping adults. *Am Rev Respir Dis* 126: 758–62, 1982.
88. Berthon-Jones M, Sullivan CE. Ventilation and arousal responses to hypercapnia in normal sleeping adults. *J Appl Physiol* 57: 59–67, 1984.
89. Orem J. Medullary respiratory neuron activity: Relationship to tonic and phasic REM sleep. *J Appl Physiol* 48: 54–65, 1980.
90. Millman RP, Knight H, Kline LR, Shore ET, Chung D-CC, Pack AI. Changes in compartmental ventilation in association with eye movements during REM sleep. *J Appl Physiol* 65: 1196–202, 1988.
91. Tabachnik E, Muller NL, Bryan AC, Levison H. Changes in ventilation and chest wall mechanics during sleep in normal adolescents. *J Appl Physiol* 51: 557–64, 1981.
92. Tusiewicz K, Moldofsky H, Bryan AC, Bryan MH. Mechanisms of the rib cage and diaphragm during sleep. *J Appl Physiol* 43: 600–2, 1977.
93. Johnson MW, Remmers JE. Accessory muscle activity during sleep in chronic obstructive pulmonary disease. *J Appl Physiol* 57: 1011–17, 1984.
94. Martinez F, Couser J, Celli B. Factors influencing ventilatory muscle recruitment in patients with chronic airflow obstruction. *Am Rev Respir Dis* 142: 276–82, 1990.
95. Fletcher EC, Gray BA, Levin DC. Nonapneic mechanisms of arterial oxygen desaturation during rapid-eye-movement sleep. *J Appl Physiol* 54: 632–39, 1983.
96. Coccagna G, Lugaresi E. Arterial blood gases and pulmonary and systemic arterial pressure during sleep in chronic obstructive pulmonary disease. *Sleep* 1: 117–24, 1978.
97. Boysen PG, Block AJ, Wynne JW, Hunt LA, Flick MR. Nocturnal pulmonary hypertension in patients with chronic obstructive pulmonary disease. *Chest* 76: 536–42, 1979.
98. Douglas NJ, Calverley PMA, Leggett RJE, Brash HM. Transient hypoxaemia during sleep in chronic bronchitis and emphysema. *Lancet* 1: 1–4, 1979.
99. Fletcher EC, Luckett RA, Miller T, Castrangos C, Kutka N, Fletcher JG. Pulmonary vascular hemodynamics in chronic lung disease patients with and without oxyhemoglobin desaturation sleep. *Chest* 95: 757–64, 1989.
100. Levi-Valensi P, Weitzenblum E, Rida Z *et al.* Sleep-related oxygen desaturation and daytime pulmonary haemodynamics in COPD patients. *Eur. Respir. J.* 5: 301–7, 1992.
101. Calverley PMA, Brezinova V, Douglas NJ, Catterall JR, Flenley DC. The effect of oxygenation on sleep quality in chronic bronchitis and emphysema. *Am Rev Respir Dis* 126: 206–10, 1982.
102. Fleetham J, West P, Mezon B, Conway W, Roth T, Kryger M. Sleep arousals, and oxygen desaturation in chronic obstructive pulmonary disease: the effect of oxygen therapy. *Am Rev Respir Dis* 126: 429–33, 1982.
103. Tiep BL, Lewis MI. Oxygen conservation and oxygen-conserving devices in chronic lung disease: A review. *Chest* 92: 263–72, 1987.
104. Goldstein RS, Ramcharan V, Bowes G, McNicholas WT, Bradley D, Phillipson EA. Effect of supplemental nocturnal oxygen on gas exchange in patients with severe obstructive lung disease. *N Engl J Med* 310: 425–29, 1984.
105. Connaughton JJ, Catterall JR, Elton RA, Stradling JR, Douglas NJ. Do sleep studies contribute to the management of patients with chronic obstructive pulmonary disease?. *Am Rev Respir Dis* 138: 341–44, 1988.
106. Casanova C, Hernandez MC, Sanchez A, Garcia-Talavera I, de Torres JP, Abreu J, Valencia JM, Aguirre-Jaime A, Celli BR. Twenty-four-hour ambulatory oximetry monitoring in COPD patients with moderate hypoxemia. *Respir Care* 51: 1416–23, 2006.
107. Fletcher EC, Donner CF, Midgren B *et al.* Survival in COPD patients with a daytime $P_aO_2 > 60$ mm Hg with and without nocturnal oxyhemoglobin desaturation. *Chest* 101: 649–55, 1992.
108. Conference report. New problems in supply, reimbursement, and certification of medical necessity for long-term oxygen therapy. *Am Rev Respir Dis* 142: 721–74, 1990.
109. Lock SH, Blower G, Prynne M, Wedzicha JA. Comparison of liquid and gaseous oxygen for domiciliary portable use. *Thorax* 47: 98–100, 1992.
110. Lacasse Y, Lecours R, Pelletier C, Begin R, Maltais F. Randomised trial of ambulatory oxygen in oxygen-dependent COPD. *Eur Respir J* 25: 1032–38, 2005.
111. Tiep BL, Belman MJ, Mittman C, Phillips RE, Otsap B. A new oxygen-saving nasal cannula. *Am Rev Respir Dis* 130: 500–2, 1983.
112. Soffer M, Tashkin DP, Shapiro BJ, Littner M, Harvey E, Farr S. Conservation of oxygen supply using a reservoir nasal cannula in hypoxemic patients at rest and during exercise. *Chest* 88: 663–68, 1985.
113. Tiep BL, Belman MJ, Mittman C, Phillips RE, Otsap B. A new pendant storage oxygen-conserving nasal cannula. *Chest* 87: 381–83, 1985.
114. Franco MA, Llompart JA, Teague R, Bloom K, Wilson R. Pulse dose oxygen delivery system. *Respir Care* 29: 1034–41, 1984.
115. Tiep BL, Nicotra B, Carter R, Phillips R, Otsap B. Low-concentration oxygen therapy via a demand oxygen delivery system. *Chest* 87: 636–38, 1985.
116. Brook CJ, Bower JS, Davis DM, Zimmer AK. Performance of a demand cannula system during rest, exercise, and sleep in hypoxemic patients. *Am Rev Respir Dis* 133: 209A, 1986.
117. Shigeoka JW. Oxygen conservers, home oxygen prescriptions, and the role of the respiratory care practitioner. *Respir Care* 36: 178–83, 1991.
118. Hagarty E, Skorodin M, Langbein L, Hultman C, Jessen J, Maki K. Comparison of three oxygen delivery systems during exercise in hypoxemic patients with chronic obstructive pulmonary disease. *Am J Respir Crit Care Med* 155: 893–98, 1997.
119. Bower JS, Brook CJ, Zimmer K, Davis D. Performance of a demand oxygen saver system during rest, exercise, and sleep in hypoxemic patients. *Chest* 94: 77–80, 1988.
120. Heimlich HJ. Respiratory rehabilitation with transtracheal oxygen system. *Ann Otol Rhinol Laryngol* 91: 643–47, 1982.
121. Christopher KL, Spofford BT, Petrun MD, McCarty DC, Goodman JR, Petty TL. A program for transtracheal oxygen delivery: Assessment of safety and efficacy. *Ann Intern Med* 107: 802–8, 1987.
122. Hoffman LA, Dauber JD, Ferson PF, Openbrier DR, Zullo TG. Patient response to transtracheal oxygen delivery. *Am Rev Respir Dis* 135: 153–56, 1987.
123. Hoffman LA, Johnson JT, Wesmiller SW *et al.* Transtracheal delivery of oxygen: Efficacy and safety for long-term continuous therapy. *Ann Otol Rhinol Laryngol* 100: 108–15, 1991.
124. Christopher KL, Spofford BT, Brannin PK, Petty TL. Transtracheal oxygen therapy for refractory hypoxemia. *JAMA* 256: 494–97, 1986.
125. Adamo JP, Mehta AC, Stelmach K, Meeker D, Rice T, Stoller JK. The Cleveland Clinic's initial experience with transtracheal oxygen therapy. *Respir Care* 35: 153–60, 1990.
126. Heimlich HJ, Carr GC. The micro-trach: A seven-year experience with transtracheal oxygen therapy. *Chest* 95: 1008–12, 1989.
127. Heimlich HJ, Carr GC. Transtracheal catheter technique for pulmonary rehabilitation. *Ann Otol Rhinol Laryngol* 94: 502–4, 1985.
128. Fletcher EC, Nickeson D, Costrarangos-Galaraza C. Endotracheal mass resulting from a transtracheal oxygen catheter. *Chest* 93: 438–39, 1988.
129. Burton GG, Wagshul FA, Henderson D, Kime SW. Fatal airway obstruction caused by a mucous ball from a transtracheal oxygen catheter. *Chest* 99: 1520–23, 1991.
130. Gong HJ. Airtravel and oxygen therapy in cardiopulmonary patients. *Chest* 101: 1104–13, 1992.

55

CHAPTER

Immunomodulators

Christopher J. Corrigan

Department of Asthma, Allergy and Respiratory Science, Guy's Hospital, London, UK
MRC and Asthma UK Centre in Allergic Mechanisms of Asthma King's College London, UK

INTRODUCTION: THE CLINICAL NEED

Asthma affects 10% of children and 4% of adults, with an overall prevalence of 6% (3 million patients) in the United Kingdom. Mortality from asthma is relatively low (1680 deaths in the United Kingdom in 1994), but considerable morbidity arises from disease in that minority of patients whose symptoms are inadequately controlled by conventional therapy (inhaled corticosteroid, long-acting β_2-agonist, leukotriene receptor antagonist) even when optimal delivery has been assured, compliance has been verified and the effects of other exacerbating factors minimized. In these patients oral corticosteroid is often employed, but even then patients may remain symptomatic and become vulnerable to unwanted effects of the therapy. No existing therapy for asthma is preventative, curative, or clearly disease modifying. Estimates from studies of prescribing practice in the United Kingdom [1] suggest that approximately 6% of asthmatics are receiving therapy that would place them at steps 4 or 5 of the British Thoracic Society asthma treatment guidelines. Although in a minority, these patients account for the majority of the costs of asthma care because of their frequent need for physician consultations and emergency and inpatient care. For such patients, new approaches to therapy are urgently required.

In contrast with asthma, chronic obstructive pulmonary disease (COPD), a disease caused largely by cigarette smoking, has a lower prevalence (1% or 600,000 patients, usually 65 or more years of age in the United Kingdom), but a much higher mortality (23,550 deaths in 1994) and considerable morbidity. Once established, the clinical progression of airways obstruction and destruction is relentless. Yet, no existing therapy has been shown to be preventative, curative, or disease modifying.

IMMUNOMODULATORY THERAPY IN ASTHMA

Asthma is associated with chronic, cell-mediated inflammation of the bronchial mucosa in which eosinophil-active cytokine products of activated T-cells play a prominent role (Fig. 55.1). Evidence suggests that glucocorticoids ameliorate asthma at least partly by inhibiting T-cell activation and cytokine production [2–4]. There has been a great deal of research directed at why T-cells in some asthmatics are relatively resistant to glucocorticoid inhibition, but it will be some time before this research evolves toward new approaches to therapy. In the meantime, other T-cell immunomodulatory agents have been investigated for their possible therapeutic effects in asthma. As many of these agents have potentially serious unwanted effects, attention has generally been focused on those asthmatics who continue to have severe disease despite maximal therapy (i.e., in those patients in whom the benefit/risk ratio of therapy is most likely to be acceptable).

Gold salts

Evaluation of these drugs in asthma has been based on empirical observation of their "antiinflammatory" effects in diseases such as rheumatoid arthritis, rather than cogent hypothesis regarding their possible mechanisms of action. Actions of gold salts which may be relevant to a glucocorticoid-sparing effect in asthma are ill-defined, but may include inhibition of T-cell proliferation, IL-5-mediated prolongation of eosinophil survival [5], IgE-mediated degranulation of mast cells and basophils [6], and leukotriene production by granulocytes [7]. Some of these effects may be secondary to inhibition by gold salts of pro-inflammatory

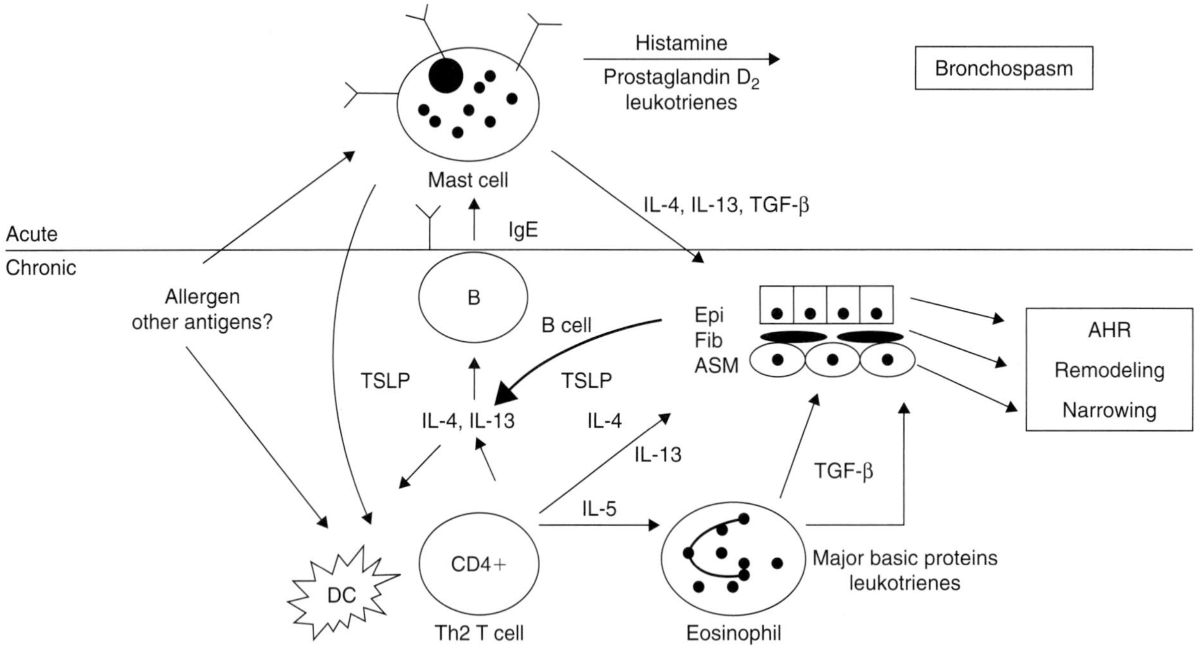

FIG. 55.1 Working hypothesis of asthma pathogenesis. Mast cells sensitized by surface, allergen-specific IgE release acute mediators of bronchoconstriction on allergen exposure, which may serve as a mechanism for acute exacerbation of asthma in atopic patients. Th2 T-cell differentiation in the airways is driven by dendritic cells (DC) which present allergens and possibly other antigens to T-cells in an MHC Class II restricted fashion, possibly under the influence of thymic stromal lymphopoietin (TSLP) produced by mast cells, epithelial cells and other cells. Th2 cytokines control IgE synthesis by B cells (IL-4, IL-13) and encourage recruitment and activation of eosinophils (IL-5) which may damage the mucosa through release of major basic protein, and are also a source of remodeling cytokines such as transforming growth factor-β (TGF-β). Th2 cytokines such as IL-4 and IL-13 also induce release of growth factors by structural cells in the airways such as epithelial cells (Epi), myofibroblasts (Fib) and airways smooth muscle cells (ASM) causing remodeling changes (fibrous protein lay down, mucous hyperplasia, smooth muscle hypertrophy/hyperplasia). Mast cells are also a source of remodeling cytokines.

transcriptional regulatory proteins, in particular NF-κB [8]. Two double-blind, placebo-controlled parallel group studies [9, 10] in which oral gold salts were administered for 6 months showed a modest but significant glucocorticoid-sparing effect of the therapy as compared with placebo. Lung function was not improved, and additional antiasthma therapy was not reduced. Not all of the patients showed a significant response. The unwanted effects of oral gold salts include dermatitis, hepatic dysfunction, proteinuria, interstitial pneumonitis, and, rarely, blood dyscrasias [11]. Nevertheless, it has been concluded in one study [12] comparing the glucocorticoid-sparing effects of gold salts, methotrexate, and cyclosporin A (CsA) in severe asthma that gold salts provide the optimal risk/benefit ratio in terms of efficacy and tolerability.

Methotrexate

Methotrexate is a folic acid analog used at low dosage for its anti-inflammatory activity in an increasing variety of chronic diseases. It exerts a delayed but sustained therapeutic effect with weekly dosage regimens. This probably reflects its accumulation in cells as polyglutamate complexes, resulting in accumulation of S-adenosyl methionine and adenosine, both of which are inhibitory to T-cell function [13]. Methotrexate therapy also appears to increase T-cell sensitivity to glucocorticoid inhibition [14]. Meta-analysis of placebo-controlled trials of methotrexate therapy in oral glucocorticoid-dependent asthma [15, 16] suggest that methotrexate therapy for a minimum of 3–6 months allows significant (overall 20%, but only in about 60% of responding patients) reduction in oral glucocorticoid requirements, with no significant improvement in lung function. In a more exacting analysis of these trials by the Cochrane reviewers [17], however, it was concluded that existing trial data do not justify the premise that methotrexate is glucocorticoid-sparing. The most serious potential unwanted effect of methotrexate therapy is cumulative hepatic toxicity and hepatic fibrosis, with isolated reports of deaths from opportunistic infections and pneumonitis.

Cyclosporin A

CsA is a lipophilic, cyclic undecapeptide derived from the fungus *Tolypocladium inflatum*. In a complex with the cytoplasmic binding protein cyclophilin, it inhibits calcineurin-mediated dephosphorylation and nuclear translocation of the cytoplasmic subunit of the T-cell transcriptional activator NF-AT, thus inhibiting T-cell proliferation and cytokine production relatively specifically [18].

Two blinded, placebo-controlled trials in severe, glucocorticoid-dependent asthmatics [19, 20] together showed that concomitant CsA therapy improved lung function while reducing oral prednisolone requirements. As with methotrexate, not all patients showed a significant response. Unwanted effects of low-dosage CsA therapy include hypertension and renal impairment. Regular monitoring of renal function, blood pressure, and whole blood trough concentrations of CsA is necessary. Lymphoproliferative disorders and serious opportunistic infections appear to be very uncommon.

Intravenous immunoglobulin therapy

Therapy with pooled intravenous immunoglobulin (IVIG), originally designed to restore immune deficiency, also appears to have immunomodulatory effects in diseases involving immune-effector mechanisms. In a blinded, parallel group study of oral glucocorticoid-dependent asthmatics [21], patients were randomized to receive IVIG 2 g/kg, 1 g/kg, or albumin (placebo) 2 g intravenously monthly for a total of 6 months (7 infusions). During the therapy, oral prednisone requirements were reduced to an almost identical extent (33–39%) in all three groups, with no significant differences in changes in symptoms, lung function, and the frequency of disease exacerbations. In a second, similar blinded trial [22], treatment of patients with both IVIG and placebo allowed significant reduction in oral prednisone dosages, which was significantly greater in the IVIG-treated group when only those asthmatics on relatively high-initial dosages of oral prednisone were included in the analysis. Another double-blind, placebo-controlled study of IVIG therapy in severe childhood asthma [23] failed to show any benefit of IVIG over placebo in terms of changes in symptoms and lung function. The possible mechanisms by which IVIG therapy could exert a beneficial effect in asthma are not clear. Some of the benefits may result from immunoglobulin replacement itself, since a proportion of chronic asthmatics have depressed serum IgG concentrations, although there is little evidence that this results in defective humoral immunity [24]. Additionally, IVIG has been shown to increase T-cell susceptibility to glucocorticoid inhibition *in vitro* [25], to abrogate IgE synthesis by B-cells *in vitro* [26], and to contain other potential immunomodulatory products including soluble CD4, CD8, and HLA molecules and cytokines.

In summary, the evidence that IVIG is of any benefit in glucocorticoid-dependent asthma is at present equivocal. In addition, the therapy is very expensive and is associated with a high incidence of unpleasant urticarial and anaphylactic reactions, as well as fever and aseptic meningitis.

Newer immunomodulatory drugs

Tacrolimus (FK506) is a macrolide derived from the soil organism *Streptomyces tsukudaiensis* which, despite having a different structure to CsA, similarly inhibits NF-AT activation. Its increased potency may be outweighed by similarly increased toxicity.

Sirolimus (rapamycin) is another macrolide derived from *Streptomyces hygroscopius*. It has a fundamentally different inhibitory action on T-cells [27], since it inhibits IL-2 signaling at least partly by inhibiting phosphorylation/activation of the kinase p70 S6 ($p70^{S6k}$), and by inhibiting the enzymatic activity of the cyclin-dependent kinase cdk2-cyclin E complex.

Other new immunomodulatory drugs, the inhibitory actions of which are relatively specific for T-cells continue to appear, including brequinar sodium and mycophenolate mofetil (inhibitors of *de novo* synthesis of pyrimidines and purines respectively, particularly in T-cells), leflunomide and the napthopyrans. Whether or not these drugs will offer opportunities for the therapy of severe asthma with a more favorable benefit/risk ratio remains to be seen. One new immunosuppressive agent, suplatast tosilate, appears to inhibit the production of asthma-relevant cytokines relatively selectively, and early studies suggest clinical benefit in asthma.

CYTOKINE-DIRECTED THERAPY IN ASTHMA

T helper 2 (Th2) lymphocytes are thought to play a key role in asthma pathogenesis as they are a principal source of asthma-relevant cytokines. Therapeutic strategies directed toward inhibition of Th2 cytokines would thus seem to offer an attractive immunomodulatory strategy for asthma (Table 55.1). Alternatively, because a range of Th1 cytokines are inhibitory to the development of Th2 T-cells, or antagonize the actions of Th2 cytokines on target pro-inflammatory leukocytes, therapy with key Th1 cytokines may also be effective in asthma (Table 55.1).

Inhibition of Th2 and other cytokines

Of the Th2 cytokines, IL-5 has been regarded as of particular importance because its expression most closely correlates with asthma severity and local eosinophil infiltration (one of the most constant features of asthma). Humanized monoclonal antibodies against human IL-5 have been manufactured. A single intravenous infusion of one of these antibodies (mepolizumab) markedly reduced circulating blood eosinophils and greatly reduced eosinophil recruitment to the airways following allergen bronchial challenge of mild, atopic asthmatics [28]. Unfortunately, however, this had no effect on early or late-phase bronchoconstriction or associated changes in bronchial hyperresponsiveness. A more prolonged study [29] on patients with moderate to severe asthma was also disappointing in terms of improvements in asthma symptoms or lung function. A biopsy study [30] suggested that, while anti-IL-5 antibody profoundly reduced circulating eosinophils, it was less effective in removing eosinophils from the bronchial mucosa but

TABLE 55.1 Cytokine-directed therapeutic strategies in asthma.

Inhibition of pro-inflammatory cytokines	Therapy with inhibitory cytokines
IL-5	IL-10
IL-4	IL-12
IL-13	IFN-α/γ
IL-9	IL-18
IL-25	IL-23
TNF-α	

IL: interleukin; IFN-α/γ: interferon-α/γ; TNF: tumor necrosis factor.

did reduce extracellular matrix deposition [31]. These studies have cast doubt upon the role of infiltrating cells such as eosinophils in the causation of the clinical abnormalities associated with asthma, although since current strategies have not been effective in removing eosinophils entirely from the bronchial mucosa, this conclusion must be guarded. Alternatively, cells such as eosinophils may participate in longer term remodeling changes associated with asthma. Whatever the case be, these observations have at least temporarily quenched enthusiasm for further evaluation of anti-IL-5 strategies.

IL-4, another Th2 cytokine, is one of the only two cytokines (the other being IL-13) which causes switching of B-lymphocytes to IgE synthesis and is therefore implicated in the pathogenesis of atopy. IL-4 is also involved in recruitment of eosinophils to the airways in asthma and acts in an autocrine fashion further to promote the differentiation of Th2 lymphocytes. Soluble, humanized IL-4 receptors (altrakincept), administered by nebulization, have been tested in clinical trials and have been shown to prevent deterioration in lung function induced by withdrawal of corticosteroids in patients with moderately severe asthma [32]. In addition, weekly nebulization of soluble IL-4 receptor improved asthma control over a 12-week period [33]. Further, more prolonged studies were more disappointing and this strategy also is currently not being pursued. IL-4 and the closely related IL-13 signal through the transcriptional activator STAT-6 and endogenous inhibitors of STAT-6 may prove to be useful therapeutic agents [34].

Further on the theme of cytokines causing airways changes in asthma independently of cellular infiltration, IL-13, which shares a surface receptor, IL-4Rα, with IL-4, when selectively over-expressed in the airways of mice causes many of the features associated with airways "remodeling" [35]. Deletion of the IL-13 gene in mice prevents development of airways hyperresponsiveness following allergen sensitization and challenge despite infiltration of eosinophils (another effect of IL-13 is to increase the production of eosinophil attracting chemokines such as eotaxin by airways epithelial cells) [36]. Apart from IL-4Rα, IL-13 has two other specific receptors named IL-13Rα1 and IL-13Rα2. The latter receptor also exists naturally in a soluble form and may act as a "decoy" receptor, scavenging and thus inhibiting the effects of free IL-13. Soluble IL-13Rα2 is currently in clinical development as a possible new therapeutic approach for asthma. In addition, a neutralizing, anti-IL-13 monoclonal antibody [37] and a soluble anti-IL-13 fusion protein [38] were shown to prevent airways hyperresponsiveness produced by allergen challenge in mice. Recently, a recombinant human IL-4 mutant protein called pitrakinra has been developed which competitively inhibits the IL-4Rα complex and thus blocks the actions of IL-4 and IL-13 simultaneously. When given by inhalation or subcutaneous injection it very convincingly reduced the late-phase response to allergen bronchial challenge of atopic asthmatics [39]. The mechanism of this effect is unknown but could reflect blockade of the actions of IL-4 and/or IL-13 signaling. In addition to possible anti-inflammatory actions of IL-4Rα blockade, pitrakinra also accelerated recovery from the acute-phase fall in FEV_1 immediately after allergen challenge, which may also suggest an effect of IL-4/13 blockade on acute mast-cell releasability [40].

Other cytokines are of similar interest as possible therapeutic targets in asthma. The cytokines IL-9 and IL-25 both amplify the production of Th2 cytokines. In addition, IL-9 promotes mast-cell differentiation and mediator release and augments IgE production, and also promotes some of the changes associated with asthmatic airways remodeling when over-expressed experimentally in animals. In an animal model, a blocking anti-IL-9 antibody inhibited airways inflammation [41]. Humanized, blocking anti-IL-9 antibodies are currently in development [42].

The cytokine tumor necrosis factor (TNF)-α is expressed in asthmatic airways and may play a pivotal role in amplifying a wide range of inflammatory responses. A blocking, humanized anti-TNF-α antibody (infliximab) and soluble TNF-α "decoy" receptors (etanercept) have shown remarkable (not to mention unprecedented and unexpected) clinical efficacy in treating other inflammatory diseases such as rheumatoid arthritis and inflammatory bowel disease, although there are concerns about the immunosuppressive actions of this therapy on resistance to microbial infections (recrudescence of tuberculosis, e.g., is a real problem) and possibly even tumor surveillance. In one study, treatment of moderate to severe asthmatics with etanercept 25 mg subcutaneously twice weekly for 12 weeks produced impressive improvements in symptoms and reductions in bronchial hyperresponsiveness [43] to a degree approaching that expected with systemic corticosteroid therapy. A further study [44] replicated these findings, while a third [45] preliminary study was negative, suggesting that the therapy might be effective only in a subgroup of asthmatics in whom TNF-α plays a relatively prominent role. Anti-TNF-α therapies thus seem to hold considerable promise, which may however be outweighed by their unwanted effects. Because these therapies have to be administered by injection, a search is on for small molecule TNF-α inhibitors. TNF-α-converting enzyme (TACE) is a matrix metalloproteinase-related enzyme required for the release of TNF-α from cell surfaces where it is stored in an extracellular matrix "reservoir." Small molecule oral TACE inhibitors are currently in development [46].

Therapy with potential antiasthma cytokines

Some cytokines inhibit the Th2 cytokine responses of T-cells which characterize asthma with varying degrees of specificity. While it may not be possible to administer such cytokines to patients in the long term, it may be possible to develop strategies that increase their release or drugs which activate their particular signaling receptors and pathways.

IL-10 has a number of anti-inflammatory properties which may be relevant to asthma therapy. It is one of the few cytokines the expression of which is enhanced, rather than inhibited by corticosteroids. It inhibits the function of antigen-presenting cells (APC) and impairs their ability to prime T-cells for Th2 cytokine release. It also reduces Th2 cytokine production by differentiated Th2 T-cells and is thought to be one of the principal effector cytokines which mediate the activity of T regulatory cells. In addition, it exerts inhibitory actions on a range of inflammatory leukocytes such as mast cells and eosinophils. Recombinant human IL-10 has proven effective in controlling other inflammatory diseases such as inflammatory bowel disease and psoriasis, where it is given as a weekly injection [47]. It is reasonably well tolerated, although hematological unwanted effects have been noted. With regard to asthma, peripheral blood T-cells from clinically corticosteroid resistant, as compared with sensitive severe asthmatics showed a marked defect in corticosteroid-induced IL-10 production [48]. There is also evidence that IL-10 production plays a role in the well-known synergistic clinical effects of inhaled glucocorticoids and long-acting β-agonists in asthma [49]. These data strongly support a role for IL-10 in regulating asthma severity. Recent studies suggest that the defect in IL-10 production observed in T-cells from corticosteroid resistant asthmatics may be overcome by other agents such as vitamin D3, which are known to increase IL-10 production, both *in vitro* and *ex vivo* [50]. Manipulation of IL-10 production by such relatively innocuous agents, as distinct from therapy with IL-10 itself, offers great promise for future asthma therapy.

IL-12 is a cytokine produced largely by APC which is a key regulator of Th1 T-cell development. Th1 cytokines, particularly interferon-γ strongly inhibit Th2 responses. Recombinant human IL-12 has been administered to humans but has a range of toxic effects which are somewhat minimized by slow escalation of the dosage. In a single study on atopic asthmatics [51], weekly infusions of IL-12 over 4 weeks at escalating dosage caused a progressive fall in circulating eosinophils and also reduced ingress of eosinophils into induced sputum following allergen bronchial challenge. Unfortunately, as with anti-IL-5, there was no evidence of reduction of the early or late-phase bronchoconstrictor responses to allergen challenge or alteration of the associated increased bronchial hyperresponsiveness. Most of the patients suffered with malaise and one with cardiac dysrhythmia. Consequently, IL-12 is not likely to be used alone as a therapeutic agent for asthma.

Cytokines such as IL-18, which synergizes with IL-12 to enhance the release of Th1 cytokines by T-cells, and IL-23, which is related to, and shares some of the effects of IL-12, are also potential therapeutic agents for asthma waiting to be explored, although problems with unwanted effects similar to those of IL-12 might be anticipated. Interferon-γ, a signature Th1 cytokine has been administered by nebulization to asthmatics [52]. It did not seem to reduce inflammation, at least as judged by eosinophil infiltration, although it is not certain whether sufficient concentrations reached the airways. Preliminary reports suggest that interferon-α may be useful for the treatment of severe, corticosteroid refractory asthmatics [53].

Inhibition of chemokines

Chemokines are particularly involved in the regulation of transit of inflammatory cells out of the bone marrow and into target tissues. Over 50 different chemokines, acting on over 20 different surface receptors have been described, allowing for much possible redundancy of action. Chemokine receptors belong to the seven transmembrane receptor superfamily of G-protein-coupled receptors, for which it is possible to manufacture small molecule inhibitors. Many such inhibitors have been developed but only few, if any, have progressed into clinical trials in asthma.

Several chemokines, including eotaxin, eotaxin-2, eotaxin-3, regulated on activation, normal T-cell expressed and secreted (RANTES) and monocyte chemoattractant protein-4 (MCP-4) activate a common receptor on Th2 T-cells, mast cells and eosinophils called CCR3. All of these chemokines are over expressed in the asthmatic airways mucosa [54, 55]. Several small molecule inhibitors of CCR3 block eosinophil recruitment in animal models of asthma.

The chemokine receptor CCR4 is selectively expressed on Th2 T-cells, while CCR8 is expressed on a subset of these cells. Ligands for CCR4 include thymus and activation dependent cytokine (TARC) and monocyte-derived chemokine (MDC), while a ligand for CCR8 is I-309. In animal models of asthma, antibody blockade of all of these molecules attenuated eosinophil influx, although only MDC and TARC blockade attenuated bronchial hyperresponsiveness [56–58]. On the other hand, blockade of the CCR4 receptor [59] and gene deletion of the CCR8 receptor [60] in animal models of asthma failed to reduce cellular influx, perhaps reflecting the extreme redundancy of these pathways. The chemokine receptor CXCR4 is also selectively expressed on Th2 T-cells, and a small molecule inhibitor of this receptor inhibited allergen-induced airways inflammation in a murine model [61].

THE WORTH OF CURRENT IMMUNOMODULATORY THERAPY IN ASTHMA

Immunosuppressive therapy

In view of these clinical observations, it will be clear that many reservations remain about the use of currently available immunosuppressive therapy for the treatment of severe, corticosteroid-dependent asthma since:

- Not all patients respond, and the response cannot be predicted *a priori*;
- The high incidence of unwanted effects makes it difficult to assess overall benefit/risk ratios even in asthmatics who are able to reduce oral corticosteroids;
- There is a risk of opportunistic infection and (at least theoretically) neoplasia;
- There are many relative or absolute contraindications to therapy, such as pregnancy;
- There is lack of knowledge about the long-term effects, beneficial or otherwise, of therapy.

Consequently, it is clear that any further investigation of immunosuppressive therapy for asthma should be performed within the confines of a controlled trial. There is an urgent need to produce a global definition of precisely which patients are suitable for such trials, and what constitutes an appropriate trial of therapy. None of the immunosuppressive strategies discussed above has made a significant impact on the management of severe, oral corticosteroid-dependent asthma, although isolated patients do respond. A large gap in our knowledge is whether or not these drugs do what they are assumed to do in asthma, which is to reduce

T-cell activation and cytokine production. Few studies have addressed possible effects of immunosuppressive therapy on the cellular and molecular immunopathology of the disease.

Cytokine-directed therapy

The similar general lack of success of cytokine-directed therapy so far in asthma has also been disappointing. The apparent failure of anti-IL-5 sent a wave of shock and controversy through the asthma research community, because the eosinophil/IL-5 axis was thought to be fundamental to asthma pathogenesis. These problems have, however, served little more than to highlight the chasm of ignorance about how the cellular, immunological and structural changes one can observe in asthma are related, if at all, to the clinical manifestations of disease. There is a pressing need to delineate the precise functions of cells and cytokines implicated in asthma pathogenesis, otherwise the success of cytokine-directed strategies (as with TNF-α, perhaps the least "asthma related" cytokine) will remain serendipitous. It is also essential to understand what mechanisms cause irreversible airways obstruction in asthma so that therapy can be directed to these.

IMMUNOMODULATORY THERAPY FOR COPD

Key elements of the pathogenesis of COPD include the following:

- Chronic exposure to cigarette smoke (a causal factor in more than 90% of cases in the United Kingdom);
- Individual (possibly inherited) susceptibility, since only 10–20% of heavy smokers develop symptomatic disease.
- A smoking-induced inflammatory cellular infiltrate in the airways, more marked peripherally, comprising of neutrophils, CD8 T-cells and variable numbers of eosinophils [62, 63]. The amount of this infiltrate is closely related to total smoke (pack years) exposure, but not so closely related to the presence/absence of COPD, although some studies [64]seem to demonstrate a COPD-specific, elevated influx of CD8 T–cells.
- Excessive production of proteinases, which digest the lung parenchyma, and overwhelm endogenous antiproteinases which normally protect against this. The principal proteinases are elastase and other serine proteinases from neutrophils, and matrix metalloproteinases (a group of over 20 related neutral endopeptidases) from macrophages and epithelial cells.

A working hypothesis for the pathogenesis of COPD is shown in Fig. 55.2. At present this remains largely theoretical; in particular, there is only circumstantial evidence for a role for T-cells, although in transgenic animal "models" of COPD where cytokines are artificially expressed in the respiratory mucosa, the cytokines TNF-α, IFN-γ, and IL-13 can produce changes reminiscent of emphysema and airways remodeling changes which characterize the human disease, probably through distinct mechanisms [65]. As with asthma (Fig. 55.1), the scheme takes no account of why the process persists in some individuals but not others. It may be inferred that therapeutic approaches to COPD may involve the use of inhibitors of macrophage and neutrophil function, leukotriene inhibitors, and inhibitors of cytokines or chemokines and/or their actions on target cells, and measures which increase proteinase inhibition. It should not be forgotten that, at present, the cheapest and only effective therapeutic maneuver is smoking cessation.

Glucocorticoids as immunomodulators in COPD

In contrast with asthma, glucocorticoids have no significant effect on progressive loss of lung function which characterizes

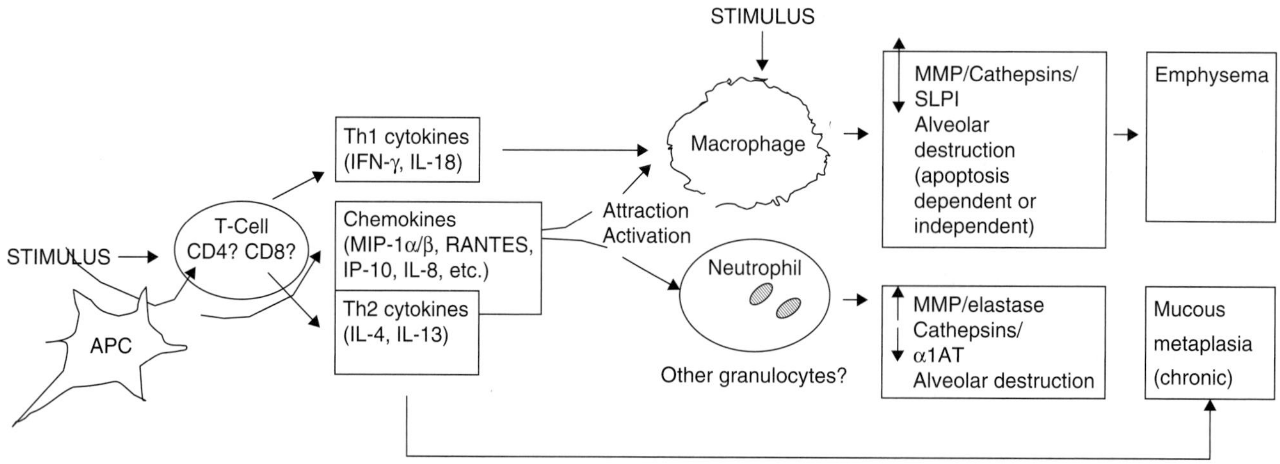

FIG. 55.2 Theoretical scheme of COPD. Both Th1/Tc1 and Th2/Tc2 cytokines from CD4 or CD8 T-cells cause emphysema through secondary release of chemokines, which attract monocyte/macrophages and granulocytes which in turn are a source of proteases such as matrix metalloproteinase (MMP), and also by direct induction of apoptosis and down regulation of anti-proteases such as α1-anti-trypsin (α1AT) and SLPI. CD8+ cytotoxic T-cells may also attack alveolar epithelial cells directly. Although COPD is much more common in cigarette smokers, the precise stimulus to T-cell and monocyte/macrophage activation is unknown, as is whether T-cell activation is in any way antigen-specific involving antigen-presenting cells (APC) (which are also a potential source of chemokines).

COPD [66] and little effect in reducing inflammatory-cell infiltration into induced sputum [67] or the bronchial mucosa [68], although when administered systemically they reduce hospital stay and re-admission rates in patients with exacerbations [69, 70]. Inhaled glucocorticoids also reduce exacerbation rates in "stable" disease [71] and this is likely the principal (and possibly the only) action of glucocorticoids which accounts for their improvement in the quality of life of COPD sufferers.

It is not clear why this disparity exists in the anti-inflammatory action of glucocorticoids in asthma and COPD. It is possible to hypothesize that the environment of the mucosa of chronic smokers with COPD renders cells insensitive to glucocorticoid inhibition [72]. The possibility that T-cells, particularly of the CD8+ phenotype, produce cytokines which have been implicated in causing COPD in animal models provides the only precedent for the hypothesis that immunomodulatory therapy will ameliorate symptoms of COPD or alter the natural history of the disease. On the other hand, this approach might be particularly troublesome in view of the fact that these patients are much more susceptible to debilitating infection than asthmatics.

Inhibition of cytokines

Concentrations of TNF-α and its soluble receptor are increased in the induced sputum of patients with COPD [73] and this mediator has been implicated in the pathogenesis of the weight loss sometimes associated with COPD [74]. A recent multi-center, placebo-controlled trial of therapy with the humanized antiTNF monoclonal antibody infliximab for 6 months (with evaluation for nearly 1 year) however showed no impact on quality of life, exercise tolerance, or disease exacerbations in patients with COPD [75].

Furthermore, more patients on active treatment developed more severe exacerbations and there was a trend toward more cases of cancer and pneumonia. It is not understood why antiTNF- α therapy is apparently more effective, at least in the short term, in asthma as compared with COPD, and again this reflects our ignorance of the true role of cytokines and indeed any mediators in disease causation. Longer term blockade might be more effective, but not without serious hazards. Other cytokines which have been implicated in COPD either from animal models (as discussed earlier) or circumstantially because their concentrations are elevated in the target organ include IFN-γ, IL-13, IL-1β, IL-6, IL-11, and IL-17 [76], but it is fair to say that in our current state of knowledge any further advances with cytokine blockade in COPD are most likely to be serendipitous.

Inhibition of chemokines

The CXC chemokine IL-8 is increased in the sputum of COPD patients and may contribute to the bronchial mucosal and sputum neutrophil influx that characterizes COPD, although there is little direct evidence for its involvement in the causation of COPD. A humanized, monoclonal neutralizing antibody to IL-8 has been tested in COPD, and while it had a small effect in reducing dyspnea scores there was no other clinical improvement [77]. IL-8 interacts with the CXCR1 and CXCR2 chemokine receptors. Since other ligands for CXCR2 such as growth-related oncoprotein (GRO)-α are also circumstantially implicated in COPD [78], CXCR2 inhibitors may be more effective than blockade of single chemokines [79], although there have been no clinical trials of such agents as yet in COPD.

PHOSPHODIESTERASE INHIBITORS IN ASTHMA AND COPD

Phosphodiesterase (PDE) inhibitors may be regarded as honorary immunomodulatory drugs since they modify the functions of a wide range of immune as well as structural cells. PDE hydrolyzes cyclic AMP, and inhibitors elevate intracellular cyclic AMP concentrations, which in turn inhibits the pro-inflammatory actions of many immune leukocytes including T-cells, macrophages, and neutrophils and relaxes airways smooth muscle cells. There are many sub-types of PDE and some of these sub-types also exist in multiple forms [80]. PDE4 is the predominant enzyme in immune cells. Two PDE4 inhibitors, cilomilast, and roflumilast have been evaluated for their possible therapeutic benefits in asthma and COPD. Roflumilast improved lung function in patients with mild/moderate asthma to a degree comparable to that achieved with inhaled glucocorticoid [81]. In a 6-month trial in patients with moderate COPD, treatment with roflumilast 0.5 mg daily resulted in a small but significant improvement in post-bronchodilator FEV_1 (0.0971 compared with placebo) and a reduction in the numbers of exacerbations (0.75/patient compared to 1.13/patient with placebo) [82]. Quality of life was also improved in the patients taking active drug (−3.5 points on the St. George's Hospital respiratory questionnaire). Cilomilast at a dose of 15 mg twice daily for 12 weeks reduced the numbers of T-cells, macrophages, and neutrophils in bronchial biopsies of patients with COPD [83]. Unfortunately, dose-limiting adverse events (nausea, vomiting, and mesenteric arteritis) are still a cause for concern with these drugs and neither has yet been licensed for the treatment of asthma or COPD [84].

CONCLUDING REMARKS

New drugs for the prevention and treatment of asthma and COPD are still greatly needed. In order to develop them, it seems of paramount importance not just to describe the pathophysiological features of each disease but to understand exactly how these features cause clinical symptoms and compromise lung function. Despite three decades of research, evidence for a role for T-cells and cytokines in asthma and COPD, which would mandate continuing the development of drugs which block their production or actions, remains largely circumstantial, and the fundamental questions as to why asthma develops in some individuals but not others, and why COPD develops only in a minority of heavy smokers [85] remain unanswered. Most of our current information comes from static, cross-sectional studies of diseased subjects and there is very little known about what role cytokines and other mediators play in acute, short-term exacerbations of disease or in progressive decline in lung function. Hence there is a need to develop novel short- and long-term outcome measures and surrogate biomarkers of disease in the short term, and better measures of lung functions decline in the longer term, such as computerized tomography [86]. Powerful new techniques such as positional genomics may in future allow analysis of heritable susceptibility factors in a manner unbiased by our limited understanding of disease pathophysiology, and allow new approaches to disease prevention which may not depend on the inhibition of immunological processes. This will be the future challenge to counter these common and significant diseases for which no effective, preventative therapies currently exist.

References

1. Neville RG, Pearson MG, Richards N, Patience J, Sondhi S, Wagstaff B, Wells N. A cost analysis of asthma prescribing in the UK. *Eur Respir J* 14: 605–9, 1999.
2. Corrigan CJ, Haczku A, Gemou-Engesaeth V, Doi S, Kikuchi Y, Takatsu K, Durham SR, Kay AB. CD4T lymphocyte activation in asthma is accompanied by increased serum concentrations of interleukin-5: Effect of glucocorticoid therapy. *Am Rev Respir Dis* 147: 540–47, 1993.
3. Doi S, Gemou-Engesaeth V, Kay AB, Corrigan CJ. Polymerase chain reaction quantification of cytokine messenger RNA expression in peripheral blood mononuclear cells of patients with severe asthma: Effect of glucocorticoid therapy. *Clin Exp Allergy* 24: 854–67, 1994.
4. Corrigan CJ, Hamid Q, North J, Barkans J, Moqbel R, Durham S, Kay AB. Peripheral blood CD4, but not CD8T lymphocytes in patients with exacerbation of asthma transcribe and translate messenger RNA encoding cytokines which prolong eosinophil survival in the context of a Th2-type pattern: effect of glucocorticoid therapy. *Am J Respir Cell Mol Biol* 12: 567–78, 1995.
5. Suzuki S, Okubo M, Kaise S, Ohara M, Kasukawa R. Gold sodium thiomalate selectively inhibits interleukin-5 mediated eosinophil survival. *J Allergy Clin Immunol* 96: 251–56, 1995.
6. Marone G, Columbo M, Galeone D, Guidi G, Kagey-Sobotka A, Lichtenstein LM. Modulation of the release of histamine and arachidonic acid metabolites from human basophils and mast cells by auranofin. *Agents Actions* 18: 100–2, 1986.
7. Parente JE, Wong K, Davis P, Burka JF, Percy JS. Effects of gold compounds on leukotriene B_4, leukotriene C_4 and prostaglandin E_2 production by polymorphonuclear leukocytes. *J Rheumatol* 13: 47–51, 1986.
8. Bratt J, Belcher J, Vercellotti GM, Palmblad J. Effects of anti-rheumatic gold salts on NF-kappa B mobilisation and tumour necrosis factor-alpha-induced neutrophil-dependent cytotoxicity for human endothelial cells. *Clin Exp Immunol* 120: 79–84, 2000.
9. Nierop G, Gijzel WP, Bel EH, Zwinderman AH, Dijkman JH. Auranofin in the treatment of steroid dependent asthma: A double blind study. *Thorax* 47: 349–54, 1992.
10. Bernstein IL, Bernstein DI, Dubb JW, Faiferman I, Wallin B. A placebo-controlled multicenter study of auranofin in the treatment of patients with corticosteroid-dependent asthma. Auranofin Multicenter Drug Trial. *J Allergy Clin Immunol* 98: 317–24, 1996.
11. Tomioka H, King TE. Gold-induced pulmonary disease: Clinical features, outcome, and differentiation from rheumatoid lung disease. *Am J Respir Crit Care Med* 155: 1011–20, 1997.
12. Bernstein IL, Bernstein DI, Bernstein JA. How does auranofin compare with methotrexate and cyclosporin as a corticosteroid-sparing agent in severe asthma? *Biodrugs* 8: 205–15, 1997.
13. Cronstein BN. Methotrexate and its mechanism of action. *Arthritis Rheum* 39: 160–95, 1996.
14. Corrigan CJ, Shiner R, Shakur BH, Ind PW. Methotrexate therapy in asthma increases T cell susceptibility to corticosteroid inhibition. *Clin Exp Allergy* 33: 1090–96, 2003.
15. Aaron SD, Dales RE, Pham B. Management of steroid-dependent asthma with methotrexate: A meta-analysis of randomised clinical trials. *Respir Med* 92: 1059–65, 1998.
16. Marin MG. Low-dose methotrexate spares steroid usage in steroid-dependent chronic asthmatic patients: A meta-analysis. *Chest* 112: 29–33, 1997.
17. Davies H, Olson L, Gibson P. Methotrexate as a steroid sparing agent in adult asthma (Cochrane review). *In The Cochrane Library*. Oxford: Update Software, 2000. Issue 3.
18. Schreiber SL, Crabtree GR. The mechanism of action of cyclosporin A and FK506. *Immunol Today* 13: 136–42, 1992.
19. Alexander AG, Barnes NC, Kay AB. Trial of cyclosporin in corticosteroid-dependent chronic severe asthma. *Lancet* 339: 324–28, 1992.
20. Lock SH, Kay AB, Barnes NC. Double blind, placebo-controlled study of cyclosporin A as a corticosteroid-sparing agent in corticosteroid-dependent asthma. *Am J Respir Crit Care Med* 153: 509–14, 1993.
21. Kishiyama JL, Valacer D, Cunningham-Rundles C *et al.* A multicentre, randomised, double blind, placebo-controlled trial of high-dose intravenous immunoglobulin for oral corticosteroid-dependent asthma. *Clin Immunol* 91: 126–33, 1999.
22. Salmun LM, Barlan I, Wolf HM, Eibl M, Twarog FJ, Geha RS, Schneider LC. Effect of intravenous immunoglobulin on steroid consumption in patients with severe asthma: A double blind, placebo-controlled, randomised trial. *J Allergy Clin Immunol* 103: 810–15, 1999.
23. Niggemann B, Leupold W, Schuster A, Schuster R, Berg AV, Hardt HVD, Eibl MM, Wahn U. Prospective, double-blind, placebo-controlled, multicentre study on the effect of high-dose intravenous immunoglobulin in children and adolescents with severe bronchial asthma. *Clin Exp Allergy* 28: 205–10, 1998.
24. Lack G, Ochs HD, Gelfand EW. Humoral immunity in steroid-dependent children with asthma and hypogammaglobulinaemia. *J Pediatr* 129: 898–903, 1996.
25. Spahn JD, Leung DYM, Chan MTS, Szefler SJ, Gelfand EW. Mechanisms of glucocorticoid reduction in subjects treated with intravenous immunoglobulin. *J Allergy Clin Immunol* 103: 421–26, 1999.
26. Sigman K, Ghibu F, Sommerville W, Toledano BJ, Bastein Y, Cameron L, Hamid QA, Mazer B. Intravenous immunoglobulin inhibits IgE production in human B lymphocytes. *J Allergy Clin Immunol* 102: 421–27, 1998.
27. Dumont FJ, Su Q. Mechanism of action of the immunosuppressant rapamycin. *Life Sci* 58: 373–95, 1995.
28. Leckie MJ, ten Brinke A, Khan J *et al.* Effects of an interleukin-5 blocking monoclonal antibody on eosinophils, airway hyperresponsiveness and the late asthmatic response. *Lancet* 356: 2144–48, 2000.
29. Kips JC, O'Connor BJ, Langely SJ, Woodcock A, Kerstjens HA *et al.* Effect of SCH55700, a humanised anti-human interleukin-5 antibody, in severe persistent asthma: A pilot study. *Am J Respir Crit Care Med* 167: 1655–59, 2003.
30. Flood-Page P, Menzies-Gow AN, Kay AB, Robinson DS. Eosinophil's role remains uncertain as anti-interleukin-5 only partially depletes numbers in asthmatic airway. *Am J Respir Crit Care Med* 167: 199–204, 2003.
31. Flood-Page P, Menzies-Gow A, Phipps S, Ying S, Wangoo A *et al.* Anti-IL-5 treatment reduces deposition of ECM proteins in the bronchial subepithelial basement membrane of mild atopic asthmatics. *J Clin Invest* 112: 1029–36, 2003.
32. Borish LC, Nelson HS, Lanz MJ, Claussen L, Whitmore JB *et al.* Interleukin-4 receptor in moderate atopic asthma. A phase I/II randomized, placebo-controlled trial. *Am J Respir Crit Care Med.* 160: 1816–23, 1999.
33. Borish LC, Nelson HS, Corren J, Bensch G, Busse WW *et al.* Efficacy of soluble IL-4 receptor for the treatment of adults with asthma. *J Allergy Clin Immunol* 107: 963–70, 2001.

34. Jiang H, Harris MB, Rothman P. IL-4/IL-13 signalling beyond JAK/STAT. *J Allergy Clin Immunol* 105: 1063–70, 2000.
35. Wills-Karp M, Chiaramonte M. Interleukin-13 in asthma. *Curr Opin Pulm Med* 9: 21–27, 2003.
36. Walter DM, McIntire JJ, Berry G, McKenzie AN, Donaldson DD *et al.* Critical role for IL-13 in the development of allergen-induced hyperreactivity. *J Immunol* 167: 4668–75, 2001.
37. Yang G, Volk A, Petley T, Emmell E, Giles-Komar J *et al.* Anti-IL-13 monoclonal antibody inhibits airway hyperresponsiveness, inflammation and airway remodelling. *Cytokine* 28: 224–32, 2004.
38. Leigh R, Ellis R, Wattie J, Donaldson DD, Inman MD. Is interleukin-13 critical in maintaining airway hyperresponsiveness in allergen-challenged mice?. *Am J Respir Crit Care Med* 170: 851–56, 2004.
39. Wenzel S, Wilbraham D, Fuller R, Burmeister Getz E, Longphre M. Effect of an interleukin-4 variant on late phase response to allergen challenge in asthmatic patients: Results of two phase 2a studies. *Lancet* 370: 1422–31, 2007.
40. Bischoff SC. Role of mast cells in allergic and non-allergic immune responses: Comparison of human and murine data. *Nat Rev Immunol* 7: 93–104, 2007.
41. Cheng G, Arima M, Honda K, Hirata H, Eda F, Yoshida N *et al.* Anti-interleukin-9 antibody treatment inhibits airway inflammation and hyperreactivity in mouse asthma model. *Am J Respir Crit Care Med* 166: 409–16, 2002.
42. Zhou Y, McLane M, Levitt RC. Th2 cytokines and asthma. Interleukin-9 as a therapeutic target for asthma. *Respir Res* 2: 80–84, 2001.
43. Howarth PH, Babu KS, Arshad HS, Lau L, Buckley M *et al.* Tumour necrosis factor (TNF-alpha) as a novel therapeutic target in symptomatic corticosteroid dependent asthma. *Thorax* 61: 1012–18, 2006.
44. Berry MA, Hargadon B, Shelley M, Parker D, Shaw SE *et al.* Evidence of a role of tumour necrosis factor alpha in refractory asthma. *N Engl J Med* 354: 697–708, 2006.
45. Morjaria JB, Chauhan AJ, Bobu KS, Mehta Rl, Smith S *et al.* Assessment of a soluble TNF-alpha receptor (etanercept) as a novel therapeutic agent for severe refractory asthma (abstract). *Proc Am Thorac Soc* 3: A16, 2006.
46. Baraam B, Bird TG, Lambert-Van Der Brempt C, Campbell D, Foster SJ, Maciewicz R. New alpha-substituted succinate-based hydroxamic acids as TNF-alpha convertase inhibitors. *J Med Chem* 42: 4890–908, 1999.
47. Fedorak RN, Ganl A, Elson CO, Rutgeerts P, Schreiber S *et al.* Recombinant human interleukin 10 in the treatment of patients with mild to moderately active Crohn's disease. The Interleukin 10 Inflammatory Bowel Disease Cooperative Study Group. *Gastroenterology* 119: 1473–82, 2000.
48. Hawrylowicz C, Richards D, Loke TK, Corrigan C, Lee T. A defect in corticosteroid-induced IL-10 production in T lymphocytes from corticosteroid-resistant asthmatic patients. *J Allergy Clin Immunol* 109: 369–70, 2002.
49. Peek EJ, Richards DF, Faith A, Lavender P, Lee TH, Corrigan CJ, Hawrylowicz CM. Interleukin-10 secreting "regulatory" T cells induced by glucocorticoids and β2-agonists. *Am J Respir Cell Mol Biol* 33: 105–11, 2005.
50. Xystrakis E, Kusumakar S, Boswell S, Peek E, Urry Z *et al.* Reversing the defective induction of IL-10-secreting regulatory T cells in glucocorticoid-resistant asthma patients. *J Clin Invest* 116: 146–55, 2006.
51. Bryan SA, O'Connor BJ, Matti S, Leckie MJ, Kanabar V *et al.* Effects of recombinant interleukin-12 on eosinophils, airway hyperresponsiveness and the late asthmatic response. *Lancet* 356: 2149–53, 2000.
52. Boguniewicz M, Martin RJ, Martin D, Gibson U, Celniker A, Williams M, Leung DY. The effects of nebulised recombinant interferon-gamma in asthmatic airways. *J Allergy Clin Immunol* 95: 133–35, 1995.
53. Gratzl S, Palca A, Schmitz M, Simon HU. Treatment with IFN-alpha in corticosteroid-unresponsive asthma. *J Allergy Clin Immunol* 105: 1035–36, 2000.
54. Ying S, Robinson DS, Meng Q, Rottman J, Kennedy R *et al.* Enhanced expression of eotaxin and CCR3 mRNA and protein in atopic asthma. Association with airway hyperresponsiveness and predominant co-localization of eotaxin mRNA to bronchial epithelial and endothelial cells. *Eur J Immunol* 27: 3507–16, 1997.
55. Ying S, Meng Q, Zeibecoglou K, Robinson DS, Macfarlane A *et al.* Eosinophil chemotactic chemokines (eotaxin, eotaxin-2, RANTES, monocyte chemoattractant protein-3 (MCP-3), and MCP-4), and C-C chemokine receptor 3 expression in bronchial biopsies from atopic and non-atopic (intrinsic) asthmatics. *J Immunol* 163: 6321–29, 1999.
56. Gonzalo JA, Pan Y, Lloyd CM, Jia GQ, Yu G *et al.* Mouse monocyte-derived chemokine is involved in airway hyperreactivity and lung inflammation. *J Immunol* 163: 403–11, 1999.
57. Kawasaki S, Takizawa H, Yoneyama H, Nakayama T, Fujisawa R *et al.* Intervention of thymus and activation-regulated chemokine attenuates the development of allergic airway inflammation and hyperresponsiveness in mice. *J Immunol* 166: 2055–62, 2001.
58. Bishop B, Lloyd CM. CC chemokine ligand 1 promotes recruitment of eosinophils but not Th2 cells during the development of allergic airways disease. *J Immunol* 170: 4810–17, 2003.
59. Conroy DM, Jopling LA, Lloyd CM, Hodge MR, Andrew DP *et al.* CCR4 blockade does not inhibit allergic airways inflammation. *J Leukocyte Biol* 74: 558–63, 2003.
60. Chung CD, Kuo F, Kumer J, Motani AS, Lawrence CE *et al.* CCR8 is not essential for the development of inflammation in a mouse model of allergic airway disease. *J Immunol* 170: 581–87, 2003.
61. Lukacs NW, Berlin A, Schols D, Skerlj RT, Bridger GJ. AMD3100, a CXCR4 antagonist, attenuates allergic lung inflammation and airway hyperreactivity. *Am J Pathol* 160: 1353–60, 2002.
62. Lams BEA, Sousa AR, Rees PJ, Lee TH. Immunopathology of the small airways submucosa in smokers with and without chronic obstructive pulmonary disease. *Am J Respir Crit Care Med* 158: 1518–23, 1998.
63. Lams BEA, Sousa AR, Rees PJ, Lee TH. Immunopathology of the large airways submucosa in smokers with and without chronic obstructive pulmonary disease. *Eur Respir J* 15: 512–16, 2000.
64. Saetta M, Di Stefano A, Turato G, Facchini FM, Corbino L *et al.* CD8+ T lymphocytes in peripheral airways of smokers with chronic obstructive pulmonary disease. *Am J Respir Crit Care Med* 157: 822–26, 1998.
65. Elias JA, Kang MJ, Crouthers K, Homer R, Lee CG. State of the art. Mechanistic heterogeneity in chronic obstructive pulmonary disease: Insights from transgenic mice. *Proc Am Thorac Soc* 3: 494–98, 2006.
66. Alsaeedi A, Sin DD, McAlister FA. The effects of inhaled corticosteroids in chronic obstructive pulmonary disease: A systematic review of randomized placebo-controlled trials. *Am J Med* 113: 59–65, 2002.
67. Culpitt SV, Nightingale JA, Barnes PJ. Effect of high dose inhaled steroid on cells, cytokines and proteases in induced sputum in chronic obstructive pulmonary disease. *Am J Respir Crit Care Med* 160: 1635–39, 1999.
68. Hattotuwa Kl, Gizycki MJ, Ansari TW, Jeffery PK, Barnes NC. The effects of inhaled fluticasone on airway inflammation in chronic obstructive pulmonary disease: A double-blind, placebos-controlled biopsy study. *Am J Respir Crit Care Med* 165: 1592–96, 2002.
69. Niewoehner DE. The role of systemic corticosteroids in acute exacerbation of chronic obstructive pulmonary disease. *Am J Respir Med* 1: 243–48, 2002.
70. Aaron SD, Vandemheen KL, Hebert P *et al.* Outpatient oral prednisone after emergency treatment of chronic obstructive pulmonary disease. *N Engl J Med* 348: 2618–25, 2003.
71. Burge PS, Calverley PMA, Jones PW, Spencer S, Anderson JA, Maslen T. Randomised, double-blind, placebo-controlled study of fluticasone propionate in patients with moderate to severe chronic obstructive pulmonary disease: The ISOLDE trial. *BMJ* 320: 1297–303, 2000.
72. Barnes PJ, Ito K, Adcock IM. A mechanism of corticosteroid resistance in COPD: Inactivation of histone deactylase. *Lancet* 363: 731–33, 2004.
73. Keatings VM, Collins PD, Scott DM, Barnes PJ. Differences in interleikin-8 and tumour necrosis factor-α in induced sputum from patients with chronic obstructive pulmonary disease or asthma. *Am J Respir Crit Care Med* 153: 530–34, 1996.
74. Pitsiou G, Kyriazia G, Hatzizisi O, Argyropoulou P, Mavrofridis E, Patakas D. Tumour necrosis factor-alpha serum levels, weight loss and tissue oxygenation in chronic obstructive pulmonary disease. *Respir Med* 96: 594–98, 2002.

75. Rennard SI, Fogarty C, Kelsen S *et al.* The safety and efficacy of infliximab in moderate to severe chronic obstructive pulmonary disease. *Am J Respir Crit Care Med* 175: 926–34, 2007.
76. Barnes PJ, Stockley RA. COPD: current therapeutic interventions and future approaches. *Eur Respir J* 25: 1084–106, 2005.
77. Mahler DA, Huang S, Tabrizi M, Bell GM. Efficacy and safety of a monoclonal antibody recognizing interleukin-8 in COPD: a pilot study. *Chest* 126: 926–34, 2004.
78. Travesm SL, Culpitt S, Russell REK, Barnes PJ, Donnelly LE. Elevated levels of the chemokines GRO-α and MCP-1 in sputum samples from COPD patients. *Thorax* 57: 590–95, 2002.
79. Donnelly LE, Barnes PJ. Chemokine receptors as therapeutic targets in chronic obstructive pulmonary disease. *Trends Pharmacol Sci* 27: 546–53, 2006.
80. Chung KF. Phosphodiesterase inhibitors in airways disease. *Eur J Pharmacol* 533: 110–17, 2006.
81. Bousquet J, Aubier M, Sastre J, Izquierdo JL, Adler LM *et al.* Comparison of roflumilast, an oral anti-inflammatory, with beclomethasone dipropionate in the treatment of persistent asthma. *Allergy* 61: 72–78, 2006.
82. Rabe KF, Bateman ED, O'Donnell D, Witte S, Bredenbroeker D, Bethke TD. Roflumilast – an oral anti-inflammatory treatment for chronic obstructive pulmonary disease: A randomised controlled trial. *Lancet* 366: 563–71, 2005.
83. Gamble E, Grootendorst DC, Brightling CE *et al.* Anti-inflammatory effects of the phosphodiesterase-4 inhibitor cilomilast (Ariflo) in chronic obstructive pulmonary disease. *Am J Respir Crit Care Med* 168: 976–82, 2003.
84. Giembycz MA. An update and appraisal of the cilomilast phase III clinical development programme for chronic obstructive pulmonary disease. *Br J Clin Pharmacol* 62: 138–52, 2006.
85. Sandford AJ, Silverman EK. Chronic obstructive pulmonary disease. 1: Susceptibility factors for COPD: The genotype-environment interaction. *Thorax* 57: 736–41, 2002.
86. Dowson LJ, Guest PJ, Stockley RA. Longitudinal changes in physiological, radiological and health status measurements in alpha(1)-antitrypsin deficiency and factors associated with decline. *Am J Respir Crit Care Med* 164: 1805–9, 2001.

Pulmonary Vasodilators in COPD

56

CHAPTER

Richard N. Channick and Lewis J. Rubin

Division of Pulmonary and Critical Care Medicine, University of California, San Diego School of Medicine, La Jolla, CA, USA

INTRODUCTION

Pulmonary hypertension is a common complication of severe chronic obstructive pulmonary disease (COPD) and contributes substantively to both its morbidity and mortality [1]. Although generally mild, pulmonary hypertension in the setting of COPD is a marker of poor prognosis. Furthermore, acute exacerbations of COPD which result in worsening gas exchange are likely to produce acute deteriorations in pulmonary hemodynamics, which further compromise oxygen transport to peripheral tissues. Accordingly, considerable effort has been given to understanding the pathogenesis of pulmonary hypertension in COPD and in attempting to develop strategies for its treatment.

PATHOGENESIS OF PULMONARY HYPERTENSION IN COPD

Several factors contribute to the development of pulmonary hypertension in COPD.

Loss of cross-sectional vascular surface area

As a result of the widespread destruction of airways and lung parenchyma typical of emphysema, there is also concomitant loss of vasculature, leading to an increase in perfusion pressure to accommodate pulmonary blood flow. In addition, increases in pulmonary blood flow, such as those that occur with physical activity, are associated with further increases in pulmonary arterial pressure owing to the loss of vascular distensibility and the inability to recruit unused vasculature. This process, by its very nature, is irreversible [2]. However, there has not been found a clear relationship between the severity of pulmonary parenchymal destruction and the presence or severity of pulmonary hypertension. Several recent investigations have found more "primary" vascular changes at the cellular and the molecular levels. In one study, Kasahara and coworkers found significant abnormalities in vascular endothelial growth factor expression in emphysema [3]. In addition, abnormalities in endothelial mediators, such as endothelin 1 [4], and increases in proinflammatory cytokines, such as IL-6 and C-reactive protein (CRP), have been found in COPD patients with pulmonary hypertension [5].

Dynamic pulmonary vasoconstriction

The muscular pulmonary arteries respond with constriction to a variety of stimuli that are common in COPD. Of these, hypoxia is the most potent and the frequent stimulus for pulmonary vasoconstriction; hypoxic pulmonary vasoconstriction (HPV) is unique to pulmonary artery smooth muscle cells and is due to inhibition of an oxygen (or redox)-sensitive membrane-bound K^+ channel, leading to cell depolarization and increased intracellular Ca^{++} concentrations [6]. Impaired endothelial function may also be present and further contributes to the vasoconstriction resulting from diminished production of endothelial-derived relaxing factors such as nitric oxide (NO) and prostacyclin (PGI_2).

As a result of vasoconstriction and altered endothelial function, severe COPD results in the elaboration of a variety of promoters of vascular growth and remodeling. Among these, endothelin, platelet-derived growth factor, and transforming growth factor have been suggested to play major roles in the evolution of the chronically hypertensive pulmonary vascular bed. In concert, these and other mitogens promote endothelial and smooth muscle proliferation and extension of smooth muscle cells into the smaller, peripheral vessels [7]. Although this remodeling process may be modified by restoration of normoxic conditions, the amelioration is usually incomplete.

PATHOPHYSIOLOGY OF PULMONARY HYPERTENSION IN COPD

As a result of the aforementioned processes, the narrowing of the vascular lumen produces elevations in pulmonary arterial pressure and pulmonary vascular resistance. Furthermore, pressures increase further with exercise or with acute exacerbations associated with deteriorations in gas exchange. The increased right ventricular afterload may eventually lead to impaired right ventricular output and, if severe and progressive, right ventricular failure ensues. Although the term *cor pulmonale* is frequently used to describe right heart failure, it is more appropriately used to define pulmonary vascular disease in the setting of respiratory disease; right heart failure is a late and an unusual complication of *cor pulmonale,* particularly in the present era of widespread availability of supplemental oxygen therapy for chronic hypoxemia.

How severe is the PH caused by COPD? The answer to this question is both clinically relevant and debatable. What is known is that in several large series of COPD patients with varying severity of disease, the average level of resting pulmonary arterial pressure is either normal or mildly elevated. However, within these series, there are between 4% and 7% of patients who have what could be termed "disproportionate" pulmonary hypertension. Thabut and coworkers found 7.4% of patients in this subgroup, defined as a mean pulmonary arterial pressure of at least 40 mmHg with relatively modest severity of COPD (Fig. 56.1) [8]. Although the mechanistic reasons for this "disproportionate" PH are unclear, it seems plausible that an excessive vascular response to a stimulus, such as alveolar hypoxia, leads to an exaggerated degree of pulmonary hypertension in these patients. Another possibility is that these patients, in fact, have idiopathic pulmonary arterial hypertension (PAH), completely distinct from the COPD.

In addition, it has been shown that some patients with COPD manifest an excessive rise in pulmonary arterial pressure during exercise. In one study, COPD patients with polymorphisms in the angiotensin converting enzyme (ACE) gene had greater exercise rises in pulmonary arterial pressure (Fig. 56.2) [9], again pointing to an intrinsic "susceptibility" to excessive pulmonary vascular disease in some COPD subjects.

Because it appears that some patients with COPD have more severe PH, it would seem likely that these patients would be limited, not by the underlying lung disease, but by the pulmonary vascular disease. It would then follow that benefit from therapy directed at the PH could be realized. What data exist to support this hypothesis?

VASODILATOR THERAPY FOR PULMONARY HYPERTENSION IN COPD

Rationale

As mentioned, the rationale for the use of PAH-directed therapy in COPD is based on the premise that pulmonary vasculopathy and, possibly pulmonary vasoconstriction, are key elements in its pathogenesis. In the older era, before the development of true PAH-targeted therapy, numerous systemic vasodilator agents were evaluated. Indeed, a variety of systemic antihypertensive agents have been shown to reduce pulmonary arterial pressure in experimental animals with pulmonary hypertension produced by acute hypoxic

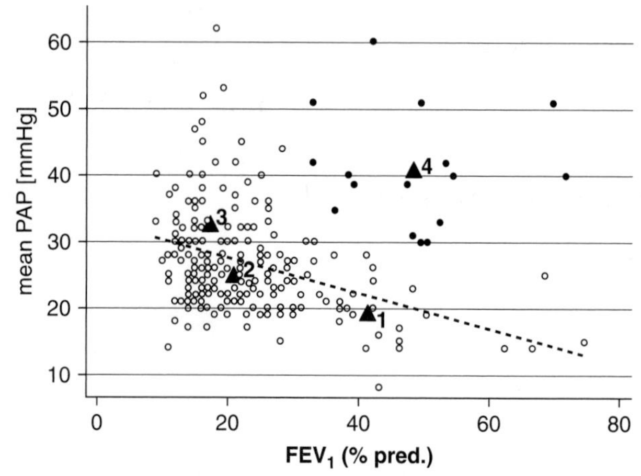

FIG. 56.1 Spectrum of pulmonary arterial pressures in patients with COPD (from Ref. [8]).

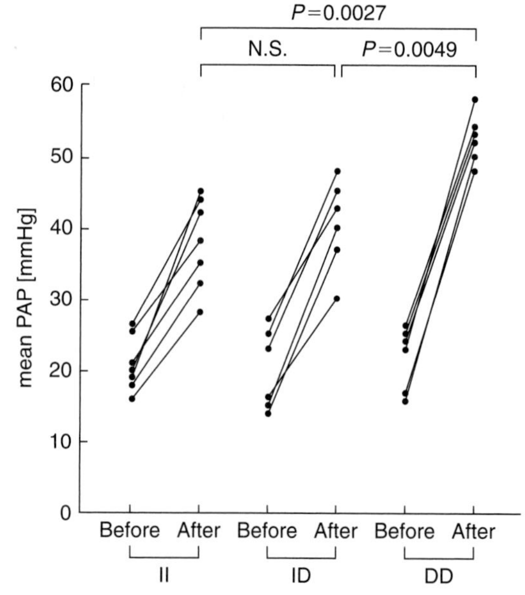

FIG. 56.2 Deletion polymorphisms in the ACE gene are associated with greater rise in mean pulmonary arterial pressures during exercise (from Ref. [9]).

ventilation or the administration of pulmonary vasoconstrictor agents [10]. Unfortunately, as described above, the pulmonary hypertensive state in COPD is the result of more than simple vasoconstriction, with remodeling playing a pivotal role. Thus, it would be conceptually unlikely that simple vasodilators would achieve much efficacy.

Oxygen

It is important to remember that supplemental oxygen, a selective pulmonary vasodilator, is the only pulmonary "vasodilator" which has been demonstrated to improve survival in COPD, although the hemodynamic effects are variable [11–13]. It is likely that the favorable impact of oxygen therapy on survival in COPD is due to a variety of effects, including optimizing myocardial and other peripheral tissue oxygen delivery.

Older vasodilators

Several older studies have evaluated other vasodilators in patients with COPD. Studies, until recently focused on oral vasodilators, including hydralazine [14], oral nitrates, and calcium channel antagonists [15–22]. Several, small reports consistently demonstrate a small, but significant acute pulmonary vasodilating effect from the administration of calcium channel blockers, including nifedipine [15–18, 22], felodipine [19, 20], verapamil [21], and diltiazem [21]. The magnitude of this effect is quite variable, ranging from 10% to 30%, depending on the study. In addition, nifedipine has been shown to reduce PaO_2, presumably by inhibition of HPV and worsening V/Q matching, an effect which is magnified during exercise and with supplemental oxygen administration [15]. Possibly mitigating the worsening of arterial oxygenation by nifedipine, however, is data demonstrating improvement in cardiac output and oxygen delivery during nifedipine administration [15, 22].

Given the modest hemodynamic effects, at least at rest, of oral vasodilators, and the relatively mild nature of the pulmonary hypertension, at least at rest, observed in COPD, it reasonable to question whether oral vasodilators have any long-term clinical utility in these patients. To date, there have been no large placebo-controlled trials addressing this issue. There are several controlled studies, all yielding unimpressive results. Vestri and coworkers found that, compared to a well-matched control group, no significant objective improvement occurred with 1 year of nifedipine [23]. Interestingly, a significant improvement in dyspnea was seen in the nifedipine group. Similarly, Domenighetti *et al.* found that, even in a subgroup of COPD patients with marked acute pulmonary vasodilation in response to nifedipine (43% decrease in PAP mean), "typical" clinical deterioration over 12 months occurred [17].

Sajkov *et al.*, in a 12-week study of another calcium channel antagonist, felodipine, found reductions in pulmonary arterial pressure and in pulmonary vascular resistance that were felt to be significant (22% and 30%, respectively) [19]. However, exercise capacity was not improved.

One theoretical limitation of pulmonary vasodilator therapy, in general, is that although the presence of pulmonary hypertension is an ominous prognostic sign in COPD, it may, in the majority of patients, merely serve as a marker of disease severity and not contribute substantively, *per se*, to symptom limitation or survival. Unlike patients with other forms of pulmonary vascular disease, such as IPAH, CTEPH, and connective tissue disease, patients with COPD are primarily symptom-limited not by the cardiovascular process but by the airway and the parenchymal lung disease [24]. Right ventricular function is not normal, but is usually not severely depressed, although the compensatory mechanisms may fall short during acute exacerbations of airway disease. Accordingly, improving the pulmonary hemodynamic state may not result in improved activity tolerance or survival in this disease.

Another concern regarding vasodilator use is that COPD is typically a disease of older patients, a population that is also prone to have coexistent coronary artery disease. Vasodilators can increase cardiac work if tachycardia is produced, while at the same time they can also "steal" blood flow away from areas of anatomic obstruction, leading to worsening cardiac ischemia.

Finally, systemically administered agents exert nonselective cardiovascular effects, including systemic hypotension, tachycardia, and potentially negative inotropic myocardial effects. While these effects may be dose related, they can occur in patients even at low doses and can limit the utility of vasodilators.

In summary, presently available data suggest that oral calcium channel blockers do cause acute reductions in pulmonary arterial pressure and pulmonary vascular resistance, and modest increases in cardiac output in COPD patients. However, there is no convincing evidence of any improvement in exercise capacity or outcome in these patients.

New vasodilators

Recently, interest has begun to be focused on the utility of the "modern" PAH therapies in COPD patients with, what is felt to be, more clinically significant pulmonary hypertension, the so-called "outliers." These modern therapies include prostacyclin analogs, endothelin receptor antagonists, and phosphodiesterase type 5 inhibitors (Table 56.1). Unique features of these newer agents that might make them beneficial in COPD patients with more advanced PH include their relative lack of systemic hemodynamic effects and their targeting of the pulmonary vasculopathy through direct effects on the endothelial dysfunction, which is a prominent feature of pulmonary hypertension.

Although the data on the efficacy and safety of these therapies for patients with IPAH and PAH associated with some other conditions, such as connective tissue disease, have been clearly proven in several randomized, placebo-controlled trials [25–30], there have been no such systematic trials in the COPD/PH subgroup. It should, however, be noted that patients with modest degrees of COPD (FEV_1 between 60% and 80% predicted) were not necessarily excluded from PAH trials, although the number of these patients was too small to allow subgroup analysis.

There are, on the other hand, uncontrolled reports and case series reporting the effects of various PAH therapies in COPD patients. Madden and coworkers treated 4 severe

TABLE 56.1 FDA-approved therapies for pulmonary arterial hypertension.

Prostacyclin analogs
Epoprostenol (intravenous)
Treprostinil (intravenous, subcutaneous)
Iloprost (inhaled)
Endothelin receptor antagonists
Bosentan
Ambrisentan
Phosphodiesterase type 5 inhibitors
Sildenafil

COPD patients with sildenafil citrate for 8 weeks in an open label trial [31]. These authors reported consistent but non-statistically significant improvements in pulmonary vascular resistance and cardiac output as well as significant increase in 6 minute walk distance.

In a larger, uncontrolled study of 18 COPD patients (GOLD II–IV) with mild resting pulmonary hypertension or exercise-induced pulmonary hypertension, a single dose of 50 mg of sildenafil attenuated the rise in mean pulmonary arterial pressure during exercise [32]. However, this effect was not accompanied by an augmentation in cardiac output or an increase in maximal exercise capacity. In a recently published study by the same group, 15 stable COPD patients were treated with 3 months of sildenafil 50 mg three times daily [33] No improvement in stroke volume as assessed by cardiac MRI or exercise capacity were noted in this group. Although these small trials do not exclude an effect of sildenafil in COPD patients with PH, without a larger, controlled study, this agent should not be routinely used in this group.

Another group of drugs commonly used to treat PAH are the prostacyclin analogs, including epoprostenol, treprostinil, and iloprost. A rationale exists for use of these agents in COPD/PH patients, as one study has demonstrated reduction in prostacyclin synthase expression in the small pulmonary arteries of patients with emphysema [34]. However, to date no clinical data has been published on the effects of prostacyclin analogs in COPD patients. Moreover, it has been the authors' experience that intravenous epoprostenol may lead significant worsening of arterial hypoxemia, presumably due to increased intrapulmonary shunting in patients with underlying ventilation/perfusion inequality.

Endothelin receptor antagonists, such as bosentan, also have theoretical rationale in COPD/PH. Increase in endothelin expression has consistently been found in COPD patients with PH [4]. In addition, these levels have been found to correlate with exercise increases in pulmonary arterial pressures in COPD [35]. However, no clinical studies have yet been published examining the effects of endothelin receptor antagonists in COPD patients.

Inhaled pulmonary vasodilators, such as iloprost or nitric oxide, theoretically are attractive candidates for treating COPD-related PH, as these agents often improve shunt fraction and have minimal systemic hemodynamic effects. However, no data have yet been published on inhaled iloprost in this group. Inhaled nitric oxide, although not available for chronic therapy has shown some promise in patients with underlying lung disease. Acutely, in ARDS [36], marked increases in PaO_2 occur in over two-third of patients given inhaled nitric oxide. The data in COPD, however, is conflicting. Both improvement [37–39] and deterioration [40, 41] in resting arterial oxygenation have been reported in COPD patients given inhaled NO, despite comparable degrees of pulmonary vasodilation. These "conflicting" results may, in fact, reflect the heterogeneity of the V/Q disturbances seen in COPD. For instance, if more low V/Q areas are present, inhaled NO may get into some of these lung units and actually worsen V/Q matching, a phenomenon reproduced in an animal study by Hopkins *et al.* [42]. On the other hand, if the predominant disturbance is shunt, inhaled NO would tend to improve oxygenation by shifting perfusion away from the shunt.

Additionally, the effects of inhaled NO on gas exchange during exercise may be even more important. Roger and colleagues found, in fact, that inhaled NO, despite worsening oxygenation at rest, actually prevented exercise-induced O_2 desaturation in patients with COPD, possibly due to preferential distribution of NO to well-ventilated regions during exercise [38].

However, despite potential beneficial effects of inhaled NO during exercise in COPD patients, exercise limitation is generally not due to either O_2 desaturation or pulmonary hypertension. Thus, it is uncertain whether or not chronic NO administration would benefit these patients. Again, no controlled trials have been done.

SUGGESTED GUIDELINES FOR THE USE OF VASODILATORS IN COPD

Current experience does not support the widespread use of PAH therapy in COPD-related PH, owing to the hazards described above and the dearth of studies demonstrating sustained clinical benefit. Nevertheless, there are selected patients who appear to have alterations in pulmonary hemodynamics which are well out of proportion to the severity of chronic respiratory disease and in whom conventional approaches to therapy are only partially successful. These patients might be considered as having true PAH, idiopathic or associated, and might be candidates for PAH therapy. However if used, the effects of PAH therapy should be monitored using some clinically meaningful parameter, such as objective testing of cardiopulmonary exercise tolerance, noninvasive evaluation of right heart size and function using echocardiography, or invasive measurements of cardiopulmonary hemodynamics.

Theoretically, the use of pulmonary vasodilators in the setting of acute *cor pulmonale* may be more appealing than their use in the chronic state, owing to the acute and, therefore potentially reversible component and the impact of this alteration on morbidity and mortality. Unfortunately, no clinical trials have addressed the role of vasodilators in this population; furthermore, the deleterious effects of vasodilators may be particularly problematic in this setting. The use of more selective agents, such as nitric oxide by inhalation, would be preferred in the acutely decompensated patient. The impact of this therapy, however, remains unproven.

SUMMARY AND CONCLUSIONS

While pulmonary hypertension remains a serious complication of COPD, effective therapy has been elusive. It is likely that pharmacological agents that exert both vasodilator and antiproliferative effects are most likely to produce sustained improvement in cardiopulmonary hemodynamics, although the impact of this therapy on symptoms and the survival in COPD remain unknown. In addition, delivery of these agents in a manner that facilitates selectivity of their effects to the pulmonary vasculature, such as by the inhaled route, would be preferable in order to minimize the adverse effects of systemic vasodilation. As our understanding of the molecular basis for the pulmonary vascular remodeling in COPD unfold, targeted therapy should be feasible and could be combined with other modalities in order to both prevent and treat the vascular complications of end-stage chronic obstructive pulmonary disease.

References

1. Traver G, Kline M, Burrows B. Predictors of mortality in chronic obstructive pulmonary disease. *Am Rev Respir Dis* 119: 895–902, 1979.
2. Rubin LJ, Tod ML, Yoshimura K. Effects of nitrendipine and hypoxia on pulmonary vascular resistance in emphysema. *Am Rev Respir Dis* 142: 625–30, 1990.
3. Kasahara Y, Tuder RM, Cool CD, Lynch DA, Flores SC, Voelkel NF. Endothelial cell death and decreased expression of vascular endothelial growth factor and vascular endothelial growth factor receptor 2 in emphysema. *Am J Respir Crit Care Med* 163: 737–44, 2001.
4. Bacakoglu F, Atasever A, Ozhan MH, Gurgun C, Ozkilic H, Guzelant A. Plasma and bronchoalveolar lavage fluid levels of endothelin-1 in patients with chronic obstructive pulmonary disease and pulmonary hypertension. *Respiration* 70(6): 594–99, 2003 November–December.
5. Joppa P, Petrasova D, Stancak B, Tkacova R. Systemic inflammation in patients with COPD and pulmonary hypertension. *Chest*: 326–33, 2006.
6. Wang J, Juhaszova M, Rubin LJ *et al.* Hypoxia inhibits gene expression of voltage-gated K channel alpha subunits in pulmonary artery smooth muscle cells. *J Clin Invest* 100: 2347–53, 1997.
7. Voelkel NF, Tuder RM. Hypoxia-induced pulmonary vascular remodeling: A model for what human disease? *J Clin Invest* 106: 733–38, 2000.
8. Thabut G, Dauriat G, Stern JB, Logeart D, Levy A, Marrash-Chahla R, Mal H. Pulmonary hemodynamics in advanced COPD candidates for lung volume reduction surgery or lung transplantation. *Chest* 127(5): 1531–36, 2005.
9. Kanazawa H, Okamoto T, Hirata K, Yoshikawa J. Deletion polymorphisms in the angiotensin converting enzyme gene are associated with pulmonary hypertension evoked by exercise challenge in patients with chronic obstructive pulmonary disease. *Am J Respir Crit Care Med* 162: 1235–38, 2000.
10. Young T, Lundquist L, Chesler E, Weir E. Comparative effects of nifedipine, verapamil and diltiazem on experimental pulmonary hypertension. *Am J Cardiol* 51: 195–200, 1983.
11. Long-term domiciliary oxygen therapy in chronic hypoxic cor pulmonale complicating chronic bronchitis and emphysema. Report of the Medical Research Council Working Party. *Lancet* 1(8222): 681–686, 1981.
12. Tims R, Khaja F, Williams G. Hemodynamic response to oxygen therapy in chronic obstructive pulmonary disease. *Ann Intern Med* 102: 29–36, 1985.
13. Weitzenblum E, Oswald M, Mirhom R, Kessler R, Apprill M. Evolution of pulmonary haemodynamics in COLD patients under long-term oxygen therapy. *Eur Respir J* 2(Suppl 7): 669S–673S, 1989.
14. Dal Nogare A, Rubin LJ. Effects of hydralazine on exercise capacity in pulmonary hypertension secondary to chronic obstructive pulmonary disease. *Am Rev Respir Dis* 133: 385–89, 1986.
15. Kennedy TP, Michael JR, Huang CK, Kallman CH, Zahka K, Schlott W, Summer W. Nifedipine inhibits hypoxic pulmonary vasoconstriction during rest and exercise in patients with chronic obstructive pulmonary disease. A controlled double-blind study. *Am Rev Respir Dis* 129: 544–51, 1984.
16. Agostoni P, Doria E, Galli C, Tamborini G, Guazzi MD. Nifedipine reduces pulmonary presure and vascular tone during short- but not long-term treatment of pulmonary hypertension in patients with chronic obstructive pulmonary disease. *Am Rev Respir Dis* 139: 120–25, 1989.
17. Domenighetti GM, Saglini VG. Short- and long-term hemodynamic efects of oral nifedipine in patients with pulmonary hypertension secondary to COPD and lung fibrosis. Deleterious effects in patients with restrictive disease. *Chest* 102: 708–14, 1992.
18. Burghuber OC. Nifedipine attenuates acute hypoxic pulmonary vasoconstriction in patients with chronic obstructive pulmonary disease. *Respiration* 52: 86–93, 1987.
19. Sajkov D, McEvoy RD, Cowie RJ, Bradley JA, Antic R, Morris RG, Frith PA. Felodipine improves pulmonary hemodynamics in chronic obstructive pulmonary disease. *Chest* 103: 1354–61, 1993.
20. Bratel T, Hedenstierna G, Nyquist O, Ripe E. The use of a vasodilator, felodipine as an adjuvant to long-term oxygen treatment in COLD patients. *Eur Respir J* 3: 46–54, 1990.
21. Gassner A, Sommer G, Fridrich L, Magometschnigg D, Priol A. Differential therapy with calcium antagonists in pulmonary hypertension secondary to COPD. Hemodynamic effects of nifedipine, diltiazem, and verapamil. *Chest* 98: 829–34, 1990.
22. Saadjian AY, Philip-Joet FF, Vestri R, Arnaud AG. Long-term treatment of chronic obstructive lung disease by Nifedipine: An 18-month haemodynamic study. *Eur Respir J* 1: 716–20, 1988.
23. Vestri R, Philip-Joet F, Surpas P, Arnaud A, Saadjian A. One-year clinical study on nifedipine in the treatment of pulmonary hypertension in chronic obstructive lung disease. *Respiration* 54: 139–44, 1988.
24. Rubin LJ. Primary pulmonary hypertension. *N Engl J Med* 336: 111–17, 1997.
25. Barst RJ, Rubin LJ, Long WA, McGoon MD, Rich S, Badesch DB, Groves BM, Tapson VF, Bourge RC, Brundage BH. A comparison of continuous intravenous epoprostenol (prostacyclin) with conventional therapy for primary pulmonary hypertension. *N Engl J Med* 334: 296–301, 1996.
26. Simonneau G, Barst RJ, Galie N, Naeije R, Rich S, Bourge RC, Keogh A, Oudiz R, Frost A, Blackburn SD, Crow JW, Rubin LJ Treprostinil Study Group. Continuous subcutaneous infusion of treprostinil, a prostacyclin analogue, in patients with pulmonary arterial hypertension: A double-blind, randomized, placebo-controlled trial. *Am J Respir Crit Care Med* 15;165: 800–4, 2002 March.
27. Channick RN, Simonneau G, Sitbon O, Robbins IM, Frost A, Tapson VF, Badesch DB, Roux S, Rainisio M, Bodin F, Rubin LJ. Effects of the dual endothelin-receptor antagonist bosentan in patients with pulmonary hypertension: A randomised placebo-controlled study. *Lancet* 358(9288): 1119–23, 2001 October 6.
28. Rubin LJ, Badesch DB, Barst RJ, Galie N, Black CM, Keogh A, Pulido T, Frost A, Roux S, Leconte I, Landzberg M, Simonneau G. Bosentan therapy for pulmonary arterial hypertension. *N Engl J Med* 346(12): 896–903, 2002 March 21.
29. Galiè N, Ghofrani HA, Torbicki A, Barst RJ, Rubin LJ, Badesch D, Fleming T, Parpia T, Burgess G, Branzi A, Grimminger F, Kurzyna M, Simonneau G Sildenafil Use in Pulmonary Arterial Hypertension (SUPER) Study Group. Sildenafil citrate therapy for pulmonary arterial hypertension. *N Engl J Med* 353(20): 2148–57, 2005 November 17.
30. Olschewski H, Simonneau G, Galiè N, Higenbottam T, Naeije R, Rubin LJ, Nikkho S, Speich R, Hoeper MM, Behr J, Winkler J, Sitbon O, Popov W, Ghofrani HA, Manes A, Kiely DG, Ewert R, Meyer A, Corris PA, Delcroix M, Gomez-Sanchez M, Siedentop H, Seeger W. Aerosolized Iloprost Randomized Study Group. Inhaled iloprost for severe pulmonary hypertension. *N Engl J Med* 347(5): 322–29, 2002 August 1.
31. Madden BP, Allenby M, Loke TK, Sheth A. A potential role for sildenafil in the management of pulmonary hypertension in patients with parenchymal lung disease. *Vascul Pharmacol* 44: 372–76, 2006.

32. Holverda S, Rietema H, Bogaard HJ, Westerhof N, Postmus PE, Boonstra A, Vonk-Noordegraaf A. Acute effects of sildenafil on exercise pulmonary hemodynamics and capacity in patients with COPD. *Pulm Pharmacol Ther*, 2008., epub ahead of print
33. Rietema H, Holverda S, Bogaard HJ, Marcus JT, Smit HJ, Westerhof N, Postmus PE, Boonstra A, Vonk-Noordegraaf A. Sildenafil treatment in COPD does not affect stroke volume or exercise capacity. *Eur Respir J* 31(4): 759–64, 2008 April.
34. Lee JD, Taraseviciene-Stewart L, Keith R, Geraci MW, Voelkel NF. The expression of prostacyclin synthase is decreased in the small pulmonary arteries from patients with emphysema. *Chest* 128(Suppl 6): 575S, 2005 December.
35. Yamakami T, Taguchi O, Gabazza EC, Yoshida M, Kobayashi T, Kobayashi H, Yasui H, Ibata H, Adachi Y. Arterial endothelin-1 level in pulmonary emphysema and interstitial lung disease. Relation with pulmonary hypertension during exercise. *Eur Respir J* 10(9): 2055–60, 1997 September.
36. Rossaint R, Falke KJ, Lopez F, Slama K, Pison U, Zapol WM. Inhaled nitric oxide for the adult respiratory distress syndrome. *N Engl J Med* 328: 399–405, 1993.
37. Adnot S, Kouyoumdjian C, Defouilloy C, Andrivet P, Sediame S, Herigault R, Fragacci MD. Hemodynamic and gas exchange responses to infusion of acetylcholine and inhalation of nitric oxide in patients with chronic obstructive lung disease and pulmonary hypertension. *Am Rev Respir Dis* 148: 310–16, 1993.
38. Roger N, Barbera JA, Roca J, Rovira I, Gomez FP, Rodriguez-Roisin R. Nitric oxide inhalation during exercise in chronic obstructive pulmonary disease. *Am J Respir Crit Care Med* 156: 800–6, 1997.
39. Yoshida M, Taguchi O, Gabazza EC, Kobayashi T, Yamakami T, Kobayashi H, Maruyama K, Shima T. Combined inhalation of nitric oxide and oxygen in chronic obstructive pulmonary disease. *Am J Respir Crit Care Med* 155: 526–29, 1997.
40. Moinard J, Manier G, Pillet O, Castaing Y. Effect of inhaled nitric oxide on hemodynamics and VA/Q inequalities in patients with chronic obstructive pulmonary disease. *Am J Respir Crit Care Med* 149: 1482–87, 1994.
41. Barbera JA, Roger N, Roca J, Rovira I, Higenbottam TW, Rodriguez-Roisin R. Worsening of pulmonary gas exchange with nitric oxide inhalation in chronic obstructive pulmonary disease. *Lancet* 347: 436–40, 1996.
42. Hopkins SR, Johnson EC, Richardson RS, Wagner H, De Rosa M, Wagner PD. Effects of inhaled nitric oxide on gas exchange in lungs with shunt or poorly ventilated areas. *Am J Respir Crit Care Med* 156: 484–89, 1997.

57

CHAPTER

Ventilator Support

Samuel L. Krachman[1] and Martin J. Tobin[2]

[1]Section of Pulmonary and Critical Care Medicine, Temple University School of Medicine, Philadelphia, PA, USA
[2]Loyola University of Chicago, Stritch School of Medicine and Hines Veterans Administration Hospital, Maywood, IL, USA

In patients with obstructive lung disease, the development of acute respiratory failure is a common cause for admission to the intensive care unit (ICU) [1–8]. Between 2% and 61% of patients [1, 2, 4–9] require mechanical ventilation, with an associated mortality of 6–30% [6–12]. The high mortality may be a consequence of both disease severity and complications of mechanical ventilation.

INDICATIONS FOR VENTILATOR SUPPORT

When severe hypoxemia develops in patients with asthma [13] and chronic obstructive pulmonary disease (COPD) [14], intubation and mechanical ventilation may be necessary to ensure the delivery of a sufficiently high fractional inspired oxygen concentration (F_1O_2). The development of an acute respiratory acidosis is another major indication for mechanical ventilation [15], although simpler measures can sometimes reverse the process [16]. An increase in work of breathing, secondary to an increase in airway resistance [17] or an inspiratory threshold load associated with auto-positive end-expiratory pressure or intrinsic positive end-expiratory pressure ($PEEP_i$), often requires mechanical ventilation to rest the respiratory muscles and decrease the oxygen cost of breathing.

INVASIVE POSITIVE PRESSURE VENTILATION

Modes of mechanical ventilation

The term *mode* refers to the relationship among various breath types (mandatory, assisted, supported, spontaneous), as well as inspiratory phase variables.

Controlled ventilation

With controlled mechanical ventilation, the ventilator delivers all breaths at a preset rate and the patient cannot trigger the machine. With volume-controlled ventilation, breaths have a preset volume (volume targeted), whereas with pressure-controlled ventilation breaths are pressure limited and time cycled [18]. Use of volume-controlled ventilation is largely restricted to patients who are apneic as a result of brain damage, sedation, or paralysis.

Assist-control ventilation

In the assist-control (AC) mode, the ventilator delivers a breath either when triggered by the patient's inspiratory effort or independently if such an effort does not occur within a preselected period. All breaths are delivered under positive pressure by the machine, but unlike controlled ventilation the patient's triggering efforts commonly exceed the preset rate. The amount of active work performed by a patient is critically dependent on the trigger sensitivity and inspiratory flow settings. Even when these settings are optimized, patients actively perform about one-third of the work performed by the ventilator during passive conditions [19, 20].

Intermittent mandatory ventilation

With intermittent mandatory ventilation (IMV), the patient receives periodic positive-pressure breaths from the ventilator at a preset volume and rate, but the patient can also breathe spontaneously between these ventilator breaths [21]. With synchronized IMV, the ventilator waits until the patient begins to inhale and then synchronizes the machine breath with the patient's inspiratory effort. If a patient does not make an effort within a preset time, the ventilator delivers a positive-pressure breath.

With IMV, patients frequently have difficulty in adapting to the intermittent nature of assistance. It was previously assumed that respiratory muscle rest was proportional to the number of machine-assisted breaths. Recent studies, however, indicate that inspiratory effort is equivalent for spontaneous and assisted breaths [22–24]. At a moderate level of machine assistance, that is, where the ventilator accounted for 20–50% of the total ventilation, electromyographic activities of the diaphragm and sternomastoid muscles were equivalent for assisted and spontaneous breaths [23]. These findings suggest that respiratory center output is preprogrammed and is unable to adjust to breath-to-breath changes in unloading. As a result, IMV may contribute to the development of respiratory muscle fatigue or prevent its recovery.

Pressure support ventilation

Pressure support (PS) ventilation is patient-triggered like AC and IMV, but differs in being pressure targeted and flow cycled [25]. The physician sets a level of pressure that augments every spontaneous effort, and the patient can alter respiratory frequency, inspiratory time, and tidal volume. A fall in airway pressure triggers the ventilator, which in turn increases pressure to a preset value. Accordingly, the pressure to inflate the lungs and chest wall is provided jointly by the ventilator and the patient. Unlike the guaranteed volume achieved by AC and IMV, tidal volume varies depending on the set pressure, patient effort, and pulmonary mechanics. The lack of guaranteed assistance in the absence of patient effort can cause apnea in patients with an unstable respiratory center output. Cycling to exhalation is triggered by a decrease in inspiratory flow to a preset level, such as 5 l/min or 25% of the peak inspiratory flow. Inspiratory assistance can also be terminated by a small increase in pressure (1–3 cm H_2O) above the preset level, resulting from expiratory effort [25].

PS is very effective in decreasing the work of inspiration. The degree of inspiratory muscle unloading, however, is variable, with a coefficient of variation of up to 96% among patients [26]. In patients with COPD, the variability in the work of breathing can be explained by the inspiratory resistance, minute ventilation, and degree of $PEEP_i$ [26]. It is noteworthy that, PS does not decrease $PEEP_i$ in patients with COPD [26, 27]. Instead, increasing levels of PS increase the contribution from $PEEP_i$ to the overall work of breathing, accounting for approximately two-thirds of the inspiratory effort [26]. The algorithm used to terminate inspiratory assistance during PS can also affect the work of breathing in patients with COPD. Patients with a prolonged time constant require more time for flow to fall to this threshold, and consequently, mechanical inflation may persist into neural expiration. To counteract such neural–mechanical dysynchrony, patients may activate their expiratory muscles at a time when the ventilator is still inflating the thorax, causing the patient to fight the ventilator.

There is no consensus for the appropriate level of PS for an individual patient. In one study, the level minimizing activity of the sternomastoid muscles also reversed electromyographic evidence of excessive diaphragmatic stress [25]. Commonly, PS is titrated to achieve a decrease in respiratory frequency and an increase in tidal volume. Considerable discrepancy exists among studies as to the preferred target or "optimal" level of PS. Achieving a decrease in respiratory rate to ≤30/min was previously found to be a better predictor of a decrease in the work of breathing than achieving a tidal volume of >0.6 l [26]. Increasing PS to achieve a low respiratory frequency and inspiratory muscle unloading may be accompanied by an increase in expiratory muscle activity, causing the patient to fight the ventilator [26]. Thus, selecting the optimal level of PS can be quite complex.

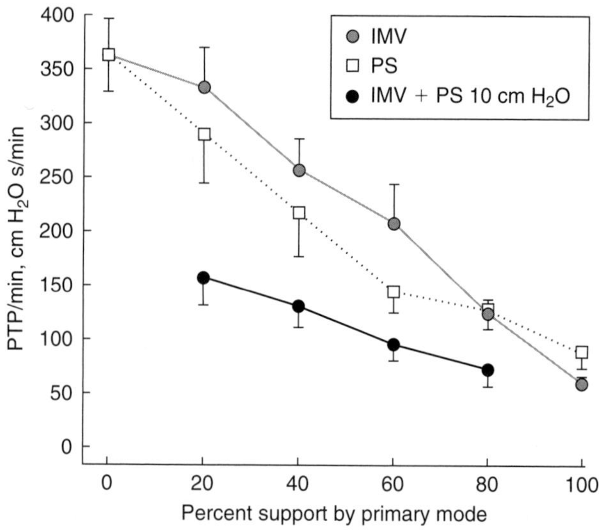

FIG. 57.1 At proportional levels of ventilator assistance, work, as measured by the pressure–time product (PTP/min), was significantly lower for the combination of IMV and PS of 10 cm H_2O than for either PS or IMV alone (from Ref. [20] with permission).

Comparison between modes of mechanical ventilation

A head-to-head comparison of the common modes of assisted ventilation was previously conducted, with most of the patients having COPD [24]. Compared with spontaneous breathing, AC achieved the largest decrease in the work of breathing. The decrease in the work was greater for PS than for IMV at lower levels of support. Addition of PS of 10 cm H_2O to a given level of IMV caused a greater reduction in the work of breathing (Fig. 57.1), not only during the intervening PS breaths, but also during the mandatory IMV breaths. The decrease in work during the mandatory breaths was proportional to the decrease in respiratory drive during the intervening breaths.

VENTILATOR SETTINGS

Fractional inspired oxygen concentration

The lowest F_IO_2 that achieves satisfactory arterial oxygenation should be selected. With arterial blood samples, a SaO_2 target of 90% is appropriate, but with pulse oximetry the same target can be associated with PaO_2 values as low as 41 torr in black patients [28].

Trigger sensitivity

Most ventilators employ pressure triggering, whereby a decrease in circuit pressure is required to initiate the ventilator. With flow triggering (sometimes termed "flow-by"), a base flow of gas (usually set at 5 to 20 l/min) is delivered during both the expiratory and the inspiratory phases of the respiratory cycle [29]. With patient effort, gas enters the patient's lungs and is diverted from the exhalation port. The difference between inspiratory and expiratory base flow is sensed, causing the ventilator to switch phase; sensitivity is usually set at 2 l/min. A decrease in breathing effort is noted with flow triggering in the PS, but not the AC, mode [30].

Triggering the ventilator is more difficult in patients with dynamic hyperinflation. In this situation, the patient must first generate sufficient pressure to offset the elastic recoil associated with hyperinflation, and thereafter overcome the sensitivity threshold. With all modes of ventilation, ineffective triggering increases in proportion to the level of assistance, secondary to a decrease in respiratory drive [24]. Breaths preceding the non-triggered effort have a higher volume, a shorter expiratory time, and a higher $PEEP_i$.

Tidal volume

To prevent alveolar overdistension and lung injury, tidal volumes of 5–7 ml/kg (or less) have become increasingly popular [31]. This ventilator strategy is termed permissive hypercapnia or controlled hypoventilation [32], because it commonly produces an increase in $PaCO_2$. If pH falls below 7.20, some physicians administer intravenous bicarbonate, although its benefit is unproven [32]. In patients with severe asthma requiring mechanical ventilation, most studies suggest that permissive hypercapnia decreases mortality [33, 34]. In a controlled study [4] of patients with asthma and acute respiratory failure, pulmonary barotrauma and hypotension were significantly lower in patients who were electively hypoventilated.

Respiratory rate

Setting the ventilator rate depends on the mode being employed. With AC ventilation, the ventilator supplies a breath in response to each patient effort; the backup rate should be set at approximately four breaths below the patient's spontaneous rate. With IMV, the mandatory rate is initially set high and then gradually decreased according to patient tolerance. As discussed above, this common approach does not ensure adequate rest. With pressure-support ventilation, the ventilator rate is not set.

Inspiratory flow rate

In patients with COPD, increasing inspiratory flow to 100 l/min produces better gas exchange, probably because the resulting increase in expiratory time allows more complete emptying of gas-trapped regions [35]. Studies in healthy subjects [36], as well as in intubated patients [37], have demonstrated that increasing the inspiratory flow setting causes an immediate increase in respiratory frequency and respiratory drive. Yet, it has been demonstrated that imposed ventilator inspiratory time, independent of delivered inspiratory flow and tidal volume, can determine respiratory frequency [38]. More recently, in patients with COPD, it was demonstrated that the increase in respiratory rate noted as inspiratory flow was increased, was associated with an increase in expiratory time and a decrease in $PEEP_i$ [39] (Fig. 57.2). A shortened inspiratory time combined with time-constant inhomogeneity of COPD will cause overinflation of some lung units to persist into neural expiration. Continued inflation during neural expiration causes stimulation of the vagus, which prolongs expiratory time [40].

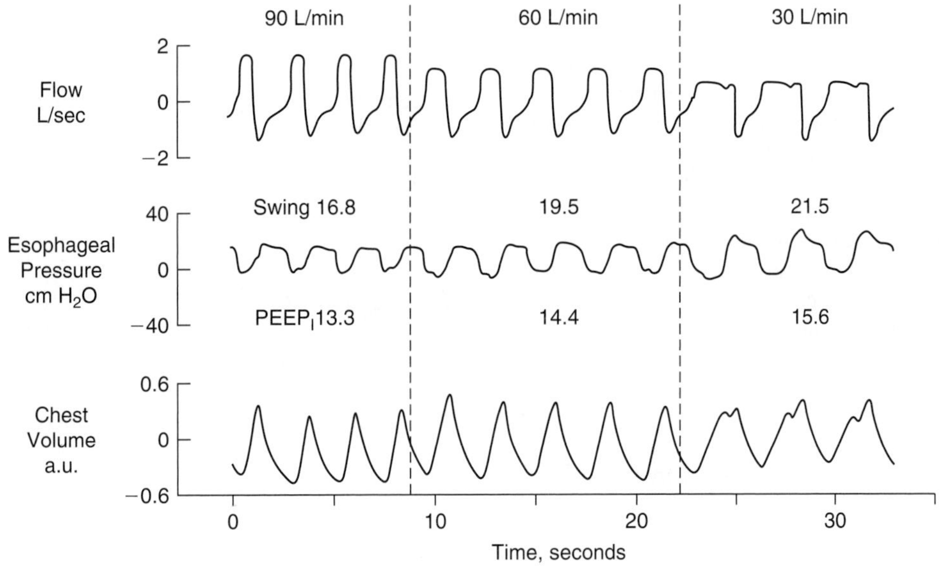

FIG. 57.2 Continuous recordings of flow, esophageal pressure (Pes), and the sum of ribcage and abdominal motion, in a patient with COPD receiving AC ventilation at a constant tidal volume. As flow increased from 30 to 60 and 90 l/min (from right to left), frequency increased (from 18 to 23 and 26 breaths/min, respectively), $PEEP_i$ decreased (from 15.6 to 14.4 and 13.3 cm H_2O, respectively), and end-expiratory lung volume also fell. Increases in flow from 30 to 60 and 90 l/min also led to decreases in the swings in Pes from 21.5 to 19.5 and 16.8 cm H_2O, respectively (from Ref. [39] with permission).

POSITIVE END-EXPIRATORY PRESSURE

In patients with COPD, dynamic hyperinflation results from dynamic airway collapse. The accompanying increase in alveolar pressure is termed $PEEP_i$. To trigger the ventilator in this situation, the patient must first generate a negative inspiratory pressure equal in magnitude to $PEEP_i$ and then overcome the trigger sensitivity setting. Provided the increase in $PEEP_i$ is secondary to increased elastic recoil, rather than expiratory effort, the addition of external PEEP can decrease the inspiratory threshold load [41,42].

Deciding the appropriate amount of external PEEP in patients with airflow limitation can be difficult because of regional inhomogeneities among lung units, each with its own critical closing pressure [43]. If application of external PEEP causes an increase in end-expiratory lung volume, decreases in cardiac output and blood pressure are likely to follow [44–47]. In general, a level of external PEEP, approximately 70% of the $PEEP_i$ value, is used in patients with COPD [44–47]. In contrast to patients with airflow limitation secondary to COPD, PEEP should be avoided in patients with asthma, because it is likely to increase lung volume and decrease cardiac output [48].

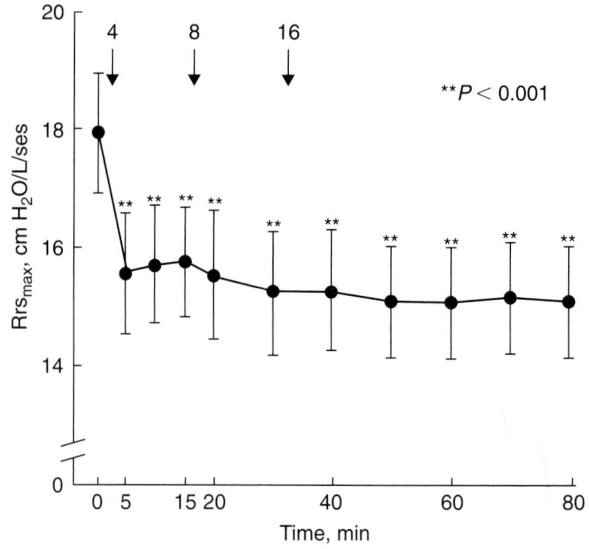

FIG. 57.3 In mechanically ventilated patients with COPD, four puffs of albuterol with an MDI resulted in a significant decrease in maximal inspiratory airway resistance (Rrs_{max}) within 5 min, with no further improvement after a total of 28 puffs (from Ref. [41] with permission).

ANCILLARY THERAPY

Studies have established that it is possible to achieve satisfactory bronchodilation using a metered-dose inhaler (MDI) despite the presence of an endotracheal tube in mechanically ventilated patients. It is essential to follow a specified protocol, use an inline chamber device [49], and actuate the MDI at the onset of inspiratory airflow. Maximal bronchodilation can be achieved in patients with COPD with as few as four puffs of a sympathomimetic aerosol (Fig. 57.3) [50].

Analgesic, anxiolytic, and neuromuscular blocking agents are frequently used in mechanically ventilated patients. When selecting an analgesic in a patient with asthma, use of morphine should be avoided because of the risk of histamine release and worsening bronchospasm [51]. Neuromuscular blocking agents can result in prolonged paralysis and the development of an acute myopathy [52, 53]. The concurrent use of corticosteroids appears to increase the risk. The inadvertent discontinuation of mechanical ventilation in patients receiving a neuromuscular blocking agent will cause complete apnea, leading to cardiopulmonary arrest. Train-of-four monitoring is recommended while patients receive neuromuscular blocking agents, although clinical assessment at the bedside is equally important [54].

Heliox is a mixture of either 70% or 80% helium with oxygen, that has a density one-third that of room air. It has been theorized that heliox may decrease the work of breathing and improve gas exchange during an acute exacerbation in patients with asthma and COPD. Controlled studies involving noninvasive ventilation or intubated patients have demonstrated mixed results [55–58]; with only small changes in $PaCO_2$ [56] and hemodynamics [55], and a slight decrease in the work of breathing and $PEEP_i$ [55, 57, 58]. In patients receiving noninvasive ventilation, heliox does not decrease the intubation rate or the length of stay in the ICU [59].

While previously investigated in patients with acute lung injury [60], prone positioning has more recently been examined in intubated patients with COPD during an acute exacerbation. While improvements in oxygenation [61, 62], as well as chest wall and lung mechanics [62] have been reported, these studies were uncontrolled and involved small numbers of patients.

COMPLICATIONS

Patients receiving mechanical ventilation are at risk of numerous complications, including oxygen toxicity, air leaks, infection, endotracheal tube complications, and decreased cardiac output. These problems occur frequently and they can be life threatening if not promptly detected and treated.

Barotrauma

The development of extraalveolar air, in the form of pneumomediastinum, subpleural air cysts, subcutaneous emphysema, pneumothorax, or pneumoperitoneum is termed pulmonary barotrauma. These findings have been found in 14–27% of patients with obstructive lung disease receiving mechanical ventilation [4, 11, 12].

Infection

Pneumonia develops in 21% of patients receiving mechanical ventilation for over 48 h [63]. In mechanically ventilated patients with COPD, more than 40% of the patients have potentially pathogenic microorganisms present in their upper and lower airways, and have an increased risk of life-threatening pneumonia [64]. The diagnosis of pneumonia is often difficult in patients receiving mechanical ventilation, because clinical criteria are unreliable and airway colonization confounds the identification of causative organisms. Studies comparing the use of quantitative cultures of specimens obtained by bronchoalveolar lavage or with a protected specimen brush passed through the bronchoscope to nonquantitative culture of endotracheal aspirates have demonstrated conflicting results [65–67]. Therefore, the optimal approach to diagnosis ventilator-associated pneumonia is yet to be determined.

Effects on cardiac output

Both PEEP and $PEEP_i$ can cause similar decreases in cardiac output, primarily by decreasing venous return. In addition, increases in pulmonary vascular resistance, secondary to alveolar distension and stretching of adjacent vessels, increases right ventricular afterload. As a result, the interventricular septum may shift to the left and decrease left ventricular compliance. Distended lung parenchyma may also increase juxtacardiac pressure and decrease left ventricular compliance. $PEEP_i$ has been noted in patients with COPD experiencing hemodynamic embarrassment [68]. On discontinuation of mechanical ventilation, hemodynamics improved as the amount of $PEEP_i$ decreased.

WEANING

About 70% of patients tolerate the first attempt to discontinue mechanical ventilation [69, 70]. Nevertheless, about 40–60% of total ventilator time is devoted to the process of weaning.

Pathophysiology of weaning failure

In patients with COPD, a decrease in respiratory motor output appears to be a rare cause for failure to wean [17]. The major reason that patients with COPD fail weaning trials is because of a progressive increase in work of breathing secondary to increases in resistance, elastance, and $PEEP_i$. In contrast, pulmonary mechanics during passive ventilation are virtually indistinguishable in weaning success and weaning failure patients before a weaning trial [71]. Respiratory muscle weakness is commonly seen in patients that fail a weaning trial [72]. However, the development of respiratory muscle fatigue does not appear to be responsible for weaning failure [72]. One reason may be that failure patients become progressively distressed during a weaning trial, leading clinicians to reinstate ventilator support before patients have breathed long enough to develop fatigue [72, 73]. That is, monitoring clinical signs of distress provides sufficient warning to avoid respiratory muscle fatigue. Only a fraction of patients develop abnormal gas exchange, with the development of hypercapnia and/or hypoxemia [74]. An increase in right and left ventricular afterload during a weaning trial may affect how the cardiovascular system meets oxygen demand in weaning failure patients, and contributes to the development of hypoxemia [75]. Rigorous studies of the pathophysiology of weaning failure in patients with asthma have not been conducted.

Timing of weaning process

Previous studies emphasize that clinical bedside assessment and physician judgment are insufficient in determining the appropriate time to initiate weaning [76]. Weaning involves the use of three diagnostic tests in sequence: measurement of predictors, a weaning trial, and a trial of extubation. The fundamental job of a weaning-predictor test is screening; with an ideal screening test having a very high sensitivity [77]. The best predictor of weaning outcome is the measurement of the frequency-to-tidal volume ratio, a measure of rapid shallow breathing. A frequency-to-tidal volume ratio above 100 breaths/min/l suggests that weaning is not likely to be successful [78].

Weaning techniques

When a screening test is positive, the clinician proceeds to a confirmatory test; an ideal confirmatory test has a very low rate of false-positive results. That is, a high specificity. Four weaning techniques are generally used. With IMV and PS, the level of ventilator assistance is gradually reduced. PS became popular as a means of overcoming the resistance of the endotracheal tube. This line of thinking, however, ignored the fact that the upper airway becomes swollen when an endotracheal tube has been in place, and after extubation patients will experience increased upper airway resistance. Indeed, the work of breathing following extubation is similar to that while breathing on a T-piece [79]. With IMV, at respiratory rates of 14 breaths/min or less, patient inspiratory efforts are increased to a level likely to cause respiratory muscle fatigue [22, 23].

Several randomized trials have been performed comparing weaning techniques [69, 80]. IMV has been shown to delay the weaning process. A once-a-day trial of spontaneous breathing resulted in a threefold increase in the rate of successful weaning compared with IMV, and a twofold increase in successful weaning compared with PS. Performing trials of spontaneous breathing once a day is as effective as performing such trials several times a day [69], but much simpler. In patients not expecting to pose any particular difficulty with weaning, a half-hour trial of spontaneous breathing is as effective as a 2-h trial [70]. The use of protocol-driven algorithms for weaning has not shown superiority in shortening the time to successful extubation [81–86].

Before removing the endotracheal tube, the clinician must judge whether or not the patient will be able to maintain a patent upper airway after extubation. Mortality among patients that require reintubation is more than six times as high as mortality among patents who can tolerate extubation [70]. The reason is unknown, but may reflect a greater severity of the underlying illness [87].

NONINVASIVE POSITIVE PRESSURE VENTILATION

In patients with COPD and acute respiratory failure, the use of noninvasive positive pressure ventilation results in less frequent intubation, decreased complications, and a shorter hospital stay. More recently, noninvasive positive pressure ventilation was shown to be equally effective in those with more severe as compared to mild acute respiratory acidosis, both in regards to normalization of pH and $PaCO_2$, as well as duration of use and hospital length of stay [88]. In two of these studies, noninvasive positive pressure ventilation caused a decrease in mortality [89–91].

CONCLUSION

Patients with obstructive lung disease commonly develop acute respiratory failure, requiring the institution of mechanical ventilation. Abnormalities in pulmonary gas exchange, lung mechanics, and respiratory muscle function contribute to the development of respiratory failure. Adapting ventilator strategies that account for these abnormalities, such as controlled hypoventilation in acute asthma or the application of external PEEP in patients with COPD, may decrease morbidity and possibly improve patient outcome.

References

1. Seneff MG, Wagner DP, Wagner RP *et al.* Hospital and 1-year survival of patients admitted to intensive care units with acute exacerbations of chronic obstructive pulmonary disease. *JAMA* 274: 1852–57, 1995.
2. Connors AF, Dawson NV, Thomas C *et al.* Outcomes following acute exacerbation of severe chronic obstructive lung disease. *Am J Respir Crit Care Med* 154: 959–67, 1996.
3. Derenne JP, Fleury B, Pariente R. Acute respiratory failure of chronic obstructive pulmonary disease. *Am Rev Respir Dis* 138: 1006–33, 1988.
4. Williams TJ, Tuxen DV, Scheinkestel CD *et al.* Risk factors for morbidity in mechanically ventilated patients with acute severe asthma. *Am Rev Respir Dis* 146: 607–15, 1992.
5. Pacht ER, Lingo S, St John RC. Clinical features, management, and outcome of patients with severe asthma admitted to the intensive care unit. *J Asthma* 32: 373–77, 1995.
6. Afessa B, Morales IJ, Scanlon PD *et al.* Prognostic factors, clinical course, and hospital outcome of patients with chronic obstructive pulmonary disease admitted to an intensive care unit for acute respiratory failure. *Crit Care Med* 30: 1610–15, 2002.
7. Breen D, Churches T, Hawker F *et al.* Acute respiratory failure secondary to chronic obstructive pulmonary disease treated in the intensive care unit: Along term follow up study. *Thorax* 57: 29–33, 2002.
8. Esteban A, Anzueto A, Frutos F *et al.* Mechanical ventilation international study group. *JAMA* 287: 345–55, 2002.
9. Pendergraft TB, Standord RH, Beasley R *et al.* Rates and characteristics of intensive care unit admissions and intubations among asthma-relate hospitalizations. *Ann Allergy Asthma Immunol* 93: 29–35, 2004.
10. Zimmerman JL, Dellinger RP, Shah AN *et al.* Endotracheal intubation and mechanical ventilation in severe asthma. *Crit Care Med* 21: 1727–30, 1993.
11. Luksza AR, Smith P, Coakley J *et al.* Acute severe asthma treated by mechanical ventilation: 10 year's experience from a district general hospital. *Thorax* 41: 459–63, 1986.
12. Mansel JK, Stogner SW, Petrini MF *et al.* Mechanical ventilation in patients with acute severe asthma. *Am J Med* 89: 42–48, 1990.
13. Rodriguez-Roisin R, Ballester E, Roca J *et al.* Mechanisms of hypoxemia in patients with status asthmaticus requiring mechanical ventilation. *Am Rev Respir Dis* 139: 732–39, 1989.
14. Wagner PD, Dantzker DR, Dueck R *et al.* Ventilation-perfusion inequality in chronic obstructive pulmonary disease. *J Clin Invest* 59: 203–16, 1977.
15. Torres A, Reyes A, Roca J *et al.* Ventilation-perfusion mismatching in chronic obstructive pulmonary disease during ventilator weaning. *Am Rev Respir Dis* 140: 1246–50, 1989.
16. Mountain RD, Sahn SA. Clinical features and outcome in patients with acute asthma presenting with hypercapnia. *Am Rev Respir Dis* 138: 535–39, 1988.
17. Jubran A, Tobin MJ. Pathophysiologic basis of acute respiratory distress in patients who fail a trial of weaning from mechanical ventilation. *Am J Respir Crit Care Med* 155: 906–15, 1997.
18. Marini JJ. Pressure-controlled and inverse ratio ventilation. In: Tobin MJ (ed.), *Principles and Practice of Mechanical Ventilation*, pp. 251–72. New York: McGraw-Hill, 2006.
19. Marini JJ, Capps JS, Culver BH. The inspiratory work of breathing during assisted mechanical ventilation. *Chest* 87: 612–18, 1985.
20. Ward ME, Corbeil C, Gibbons W *et al.* Optimization of respiratory muscle relaxation during mechanical ventilation. *Anesthesiology* 69: 29–35, 1988.
21. Sassoon CSH. Intermittent mandatory ventilation. In: Tobin MJ (ed.), *Principles and Practice of Mechanical Ventilation*, pp. 201–20. New York: McGraw-Hill, 2006.
22. Marini JJ, Smith TC, Lamb VJ. External work output and force generation during synchronized intermittent mechanical ventilation. *Am Rev Respir Dis* 138: 1169–79, 1988.
23. Imsand C, Feihl F, Perret C *et al.* Regulation of inspiratory neuromuscular output during synchronized intermittent mechanical ventilation. *Anesthesiology* 80: 13–22, 1994.
24. Leung P, Jubran A, Tobin MJ. Comparison of assisted ventilator modes on triggering, patient effort and dyspnea. *Am J Respir Crit Care Med* 155: 1940–48, 1997.
25. Brochard L, Lellouche F. Pressure support ventilation. In: Tobin MJ (ed.), *Principles and Practice of Mechanical Ventilation*, pp. 221–50. New York: McGraw-Hill, 2006.
26. Jubran A, Van de Graaff WB, Tobin MJ. Variability of patient–ventilator interaction with pressure support ventilation in patients with chronic obstructive pulmonary disease. *Am J Respir Crit Care Med* 152: 129–36, 1995.
27. Appendini L, Patessio A, Zanaboni S *et al.* Physiologic effects of positive end-expiratory pressure and mask pressure support during exacerbations of chronic obstructive pulmonary disease. *Am J Respir Crit Care Med* 149: 1009–78, 1994.
28. Jubran A, Tobin MJ. Reliability of pulse oximetry in titrating supplemental oxygen therapy in ventilator-dependent patients. *Chest* 90: 1420–25, 1990.
29. Holets S, Hubmayr RD. Setting the ventilator. In: Tobin MJ (ed.), *Principles and Practice of Mechanical Ventilation*, pp. 163–81. New York: McGraw-Hill, 2006.
30. Aslanian P, El Atrous S, Isabey D *et al.* Effects of flow triggering on breathing effort during partial ventilatory support. *Am J Respir Crit Care Med* 157: 135–43, 1998.
31. Dreyfuss D, Ricard JD, Saumon G. Ventilator-induced injury. In: Tobin MJ (ed.), *Principles and Practice of Mechanical Ventilation*, pp. 903–30. New York: McGraw-Hill, 1994, 1994.
32. Laffey JG, Kavanagh BP. Permissive hypercapnia. In: Tobin MJ (ed.), *Principles and Practice of Mechanical Ventilation*, pp. 373–92. New York: McGraw-Hill, 2006.
33. Darioli R, Perret C. Mechanical controlled hypoventilation in status asthmaticus. *Am Rev Respir Dis* 129: 385–87, 1984.
34. Tuxen DV, Williams TJ, Scheinkestel CD *et al.* Use of a measurement of pulmonary hyperinflation to control the level of mechanical ventilation

in patients with acute severe asthma. *Am Rev Respir Dis* 146: 1136–42, 1992.
35. Connors AF Jr., McCaffree DR, Gray BA. Effect of inspiratory flow rate on gas exchange during mechanical ventilation. *Am Rev Respir Dis* 124: 537–43, 1981.
36. Puddy A, Younes M. Effect on inspiratory flow rate on respiratory output in normal subjects. *Am Rev Respir Dis* 146: 787–89, 1992.
37. Corne S, Gillespie D, Roberts D *et al.* Effect of inspiratory flow rate on respiratory rate in intubated patients. *Am J Respir Crit Care Med* 156: 304–8, 1997.
38. Laghi F, Karamchandani K, Tobin M. Influence of ventilator settings in determining respiratory frequency during mechanical ventilation. *Am J Respir Crit Care Med* 160: 1766–70, 1999.
39. Laghi F, Segal J, Choe WK *et al.* Effect of imposed inflation time on respiratory frequency and hyperinflation in patients with chronic obstructive pulmonary disease. *Am J Respir Crit Care Med* 163: 1365–70, 2001.
40. Tobin MJ, Jubran A, Laghi F. Patient–ventilator interaction. *Am J Respir Crit Care Med* 163: 1059–63, 2001.
41. Rossi A, Gottfried SB, Zocchi L *et al.* Measurement of static compliance of the total respiratory system in patients with acute respiratory failure: The effect of intrinsic positive end-expiratory pressure. *Am Rev Respir Dis* 131: 672–77, 1985.
42. Smith TC, Marini JJ. Impact of PEEP on lung mechanics and work of breathing in severe airflow obstruction. *J Appl Physiol* 65: 1488–99, 1988.
43. Schnader J. Estimation of auto-PEEP (letter). *Chest* 99: 520, 1991.
44. Petrof BJ, Legare M, Goldberg P *et al.* Continuous positive airway pressure reduces work of breathing and dyspnea during weaning from mechanical ventilation in severe chronic obstructive pulmonary disease. *Am Rev Respir Dis* 141: 281–89, 1990.
45. Georgopoulos D, Giannouli E, Patakas D. Effects of extrinsic positive end-expiratory pressure on mechanically ventilated patients with chronic obstructive pulmonary disease and dynamic hyperinflation. *Intens Care Med* 19: 197–203, 1993.
46. Ranieri VM, Guiliani R, Cinnella G. Physiologic effects of positive end-expiratory pressure in patients with chronic obstructive pulmonary disease during acute ventilatory failure and controlled mechanical ventilation. *Am Rev Respir Dis* 147: 5–13, 1993.
47. Rossi A, Santos C, Roca J *et al.* Effects of PEEP on V/Q mismatching in ventilated patients with chronic airflow obstruction. *Am J Respir Crit Care Med* 149: 1077–84, 1994.
48. Tuxen DV. Detrimental effects of positive end-expiratory pressure during controlled mechanical ventilation of patients with severe airflow obstruction. *Am Rev Respir Dis* 140: 5–9, 1989.
49. Dhand R, Tobin MJ. Bronchodilator delivery with metered-dose inhalers in mechanically ventilated patients. *Eur Respir J* 9: 585–95, 1996.
50. Dhand R, Duarte AG, Jubran A *et al.* Dose response to bronchodilator delivered by metered-dose inhaler in ventilator-supported patients. *Am J Respir Crit Care Med* 154: 388–93, 1996.
51. Wheeler AP. Sedation, analgesia, and paralysis in the intensive care unit. *Chest* 104: 566–77, 1993.
52. Manthous CA, Chatila W. Prolonged weakness after the withdrawal of atracurium. *Am J Respir Crit Care Med* 150: 1441–43, 1994.
53. Leatherman JW, Fluegel WL, David WS *et al.* Muscle weakness in mechanically ventilated patients with severe asthma. *Am J Respir Crit Care Med* 153: 1686–90, 1996.
54. Strange C, Vaughan L, Franklin C *et al.* Comparison of train-of-four and best clinical assessment during continuous paralysis. *Am J Respir Crit Care Med* 156: 1556–61, 1997.
55. Lee DL, Lee H, Chang HW *et al.* Heliox improves hemodynamics in mechanically ventilated patients with chronic obstructive pulmonary disease with systolic pressure variations. *Crit Care Med* 33: 968–73, 2005.
56. Jolliet P, Tassaux D, Thouret JM *et al.* Beneficial effects of helium:oxygen versus air:oxygen noninvasive pressure support in patients with decompensated chronic obstructive pulmonary disease. *Crit Care Med* 27: 2422–29, 1999.
57. Jaber S, Fodil R, Carlucci A *et al.* Noninvasive ventilation with helium-oxygen in acute exacerbation of chronic obstructive pulmonary disease. *Am J Respir Crit Care Med* 161: 1191–200, 2000.
58. Diehl JL, Mercat A, Guerot E *et al.* Helium/oxygen mixture reduces the work of breathing at the end of the weaning process in patient with severe chronic obstructive pulmonary disease. *Crit Care Med* 31: 1415–20, 2003.
59. Jolliet P, Tassaux D, Roeseler J *et al.* Helium–oxygen versus air-oxygen noninvasive pressure support in decompensated chronic obstructive disease: A prospective, multicenter study. *Crit Care Med* 31: 878–84, 2003.
60. Gattinoni L, Gianni T, Pesenti A *et al.* Effect of prone positioning on the survival of patients with acute respiratory failure. *N Engl J Med* 345: 568–73, 2001.
61. Reisgnier J, Lejeune O, Renard B *et al.* Short-term effects of prone position in chronic obstructive pulmonary disease patients with severe acute hypoxemic and hypercapnic respiratory failure. *Intens Care Med* 31: 1128–31, 2005.
62. Mentzelopoulos SD, Zakynthinos SG, Roussos C *et al.* Prone position improves lung mechanical behavior and enhances gas exchange efficiency in mechanically ventilated chronic obstructive pulmonary disease patients. *Anesth Analg* 96: 1756–67, 2003.
63. Fagon JY, Chastre J, Hance A *et al.* Nosocomial pneumonia in ventilated patients: A cohort study evaluating attributable mortality and hospital stay. *Am J Med* 94: 281–88, 1993.
64. Soler N, Torres A, Ewig S *et al.* Bronchial microbial patterns in severe exacerbations of chronic obstructive pulmonary disease (COPD) requiring mechanical ventilation. *Am J Respir Crit Care Med* 157: 1498–505, 1998.
65. Chastre J, Viau F, Brun P *et al.* Prospective evaluation of the protected specimen brush for the diagnosis of pulmonary infections in ventilated patients. *Am Rev Respir Dis* 130: 924–29, 1984.
66. Fagon JY, Chastre J, Wolff M *et al.* Invasive and noninvasive strategies for management of suspected ventilator-associated pneumonia. *Ann Intern Med* 132: 621–30, 2000.
67. Chastre J, Fagon JY. Ventilator-associated pneumonia. *Am J Respir Crit Care Med* 165: 867–903, 2002.
68. Pepe PE, Marini JJ. Occult positive end-expiratory pressure in mechanically ventilated patients with airflow obstruction. *Am Rev Respir Dis* 126: 166–70, 1982.
69. Esteban A, Frutos F, Tobin MJ *et al.* A comparison of four methods of weaning patients from mechanical ventilation. Spanish Lung failure Collaborative Group. *N Engl J Med* 332: 345–50, 1995.
70. Esteban A, Alia I, Tobin MJ. Effect of spontaneous breathing trial duration on outcome of attempts to discontinue mechanical ventilation. *Am J Respir Crit Care Med* 159: 512–18, 1999.
71. Jubran A, Tobin MJ. Passive mechanics of lung and chest wall in patients who failed or succeeded in trials of weaning. *Am J Respir Crit Care Med* 155: 916–21, 1997.
72. Laghi F, Cattapan SE, Jubran A *et al.* Is weaning failure caused by low-frequency fatigue of the diaphragm? *Am J Respir Crit Care Med* 167: 120–27, 2003.
73. Laghi F, Tobin MJ. Disorders of the respiratory muscles. *Am J Respir Crit Care Med* 168: 10–48, 2003.
74. Tobin MJ, Perez W, Guenther SM *et al.* The pattern of breathing during successful and unsuccessful trials of weaning from mechanical ventilation. *Am Rev Respir Dis* 134: 1111–18, 1986.
75. Jubran A, Mathru M, Dries D *et al.* Continuous recordings of mixed venous oxygen saturation during weaning from mechanical ventilation and the ramifications thereof. *Am J Respir Crit Care Med* 158: 1763–69, 1998.
76. Stroetz RW, Hubmayr R. Tidal volume maintenance during weaning with pressure support. *Am J Respir Crit Care Med* 152: 1034–40, 1995.
77. Tobin MJ, Jubran A. Weaning from mechanical ventilation. In: Tobin MJ (ed.), *Principles and Practice of Mechanical Ventilation*, 2nd edn, 1185–220. New York: McGraw-Hill Inc., 2006.
78. Yang KL, Tobin MJ. A prospective study of indexes predicting the outcome of trials of weaning from mechanical ventilation. *N Engl J Med* 324: 1445–50, 1991.

79. Strauss C, Louis B, Isabey D *et al.* Contribution of the endotracheal tube and the upper airway to breathing workload. *Am J Respir Crit Care Med* 157: 23–30, 1998.
80. Brochard L, Rauss A, Benito S *et al.* Comparison of three methods of gradual withdrawal from ventilatory support during weaning from mechanical ventilation. *Am J Respir Crit Care Med* 150: 896–903, 1994.
81. Ely EW, Baker AM, Dunagan DP *et al.* Effect on the duration of mechanical ventilation of identifying patients capable of breathing spontaneously. *N Engl J Med* 335: 1864–69, 1996.
82. Namen AM, Ely EW, Tatter SB *et al.* Predictors of successful extubation in neurosurgical patients. *Am J Respir Crit Care Med* 163: 658–64, 2001.
83. Randolph AG, Wypij D, Venkataraman ST *et al.* Effect of mechanical ventilator weaning protocols on respiratory outcomes in infants and children: A randomized controlled trial. *JAMA* 288: 2561–68, 2002.
84. Krishnan JA, Moore D, Robeson C *et al.* A prospective, controlled trial of a protocol-based strategy to discontinue mechanical ventilation. *Am J Respir Crit Care Med* 169: 673–78, 2004.
85. Kollef MH, Shapiro SD, Silver P *et al.* A randomized, controlled trial of protocol-directed versus physician-directed weaning from mechanical ventilation. *Crit Care Med* 25: 567–74, 1997.
86. Marelich GP, Murin S, BAttistella F *et al.* Protocol weaning of mechanical ventilation in medical and surgical patients by respiratory care practitioners and nurses: Effect on weaning time and incidence of ventilator-associated pneumonia. *Chest* 118: 459–67, 2000.
87. Tobin MJ, Laghi F. Extubation. In: Tobin MJ (ed.), *Principles and Practice of Mechanical Ventilation*, 2nd edn, 1221–38. New York: McGraw-Hill, 2006.
88. Crummy F, Buchan C, Miller B *et al.* The use of noninvasive mechanical ventilation in COPD and severe hypercapnic acidosis. *Respir Med* 101: 53–61, 2007.
89. Bott J, Carroll MP, Conway JH *et al.* Randomized controlled trial of nasal ventilation in acute ventilatory failure due to chronic obstructive airways disease. *Lancet* 341: 1555–57, 1993.
90. Kramer N, Meyer TJ, Meharg J *et al.* Randomized, prospective trial of noninvasive positive pressure ventilation in acute respiratory failure. *Am J Respir Crit Care Med* 151: 1799–806, 1995.
91. Brochard L, Mancebo J, Wysocki M *et al.* Noninvasive ventilation for acute exacerbations of chronic obstructive pulmonary disease. *N Engl J Med* 333: 817–22, 1995.

58

CHAPTER

Pulmonary Rehabilitation

Thierry Troosters[1,2], Wim Janssens[1] and Marc Decramer[1,2]

[1]Respiratory Rehabilitation and Respiratory Division, University Hospital, Leuven, Belgium
[2]Faculty of Kinesiology and Rehabilitation Sciences, Department of Rehabilitation Sciences, Katholieke Universiteit Leuven, Leuven, Belgium

INTRODUCTION

Pulmonary rehabilitation is nowadays a recognized evidence-based therapy which can be applied to patients with lung diseases who are symptomatic and have reduced activities of daily living, despite optimal medical therapy [1, 2]. Unlike most drugs, pulmonary rehabilitation does not target the lungs directly, but aims at reversing or stabilizing the extra-pulmonary effects of lung diseases [1].

Lung disease, especially chronic obstructive pulmonary disease (COPD), is a poorly diagnosed, but a very prevalent problem in our society. More than half the patients with mild to moderate COPD remain undiagnosed until late in the disease [3]. Physical inactivity is a natural defense strategy to avoid the distressing symptom of exercise induced dyspnea. Unfortunately, inactivity inevitably leads to many of the systemic consequences associated with COPD, such as skeletal muscle weakness, osteoporosis, reduced insulin sensitivity, cardiovascular morbidity, and mood disturbance. Further catalyzed by smoking and perhaps systemic inflammation, oxidative stress, and exacerbations, these systemic consequences may develop along with the progressing airflow obstruction. In fact, particularly patients with comorbidity or systemic consequences were characterized by an inactive lifestyle even in milder stages of COPD [4]. Figure 58.1 shows that the vast majority of patients referred to our rehabilitation program had developed skeletal muscle weakness to some degree, independently of the lung function impairment. Undoubtedly, skeletal muscle dysfunction is an important extra-pulmonary consequence. In COPD, skeletal muscle weakness has been shown to be related to impaired survival [5, 6]. In addition, skeletal muscle weakness is an important driver of utilization of health-care recourses [7]. In the context

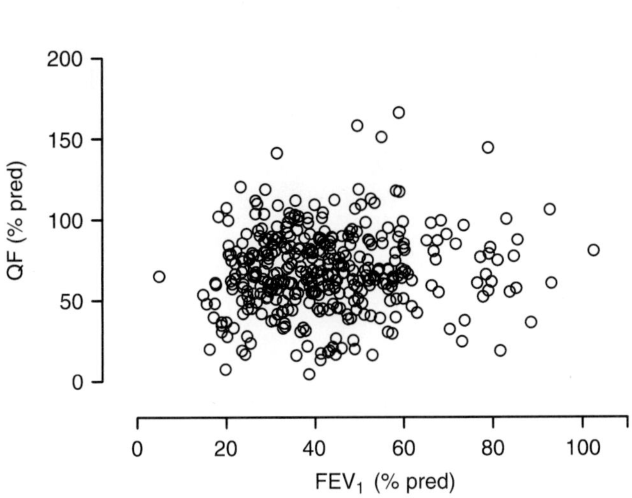

FIG. 58.1 Relation between lung function impairment (FEV_1, expressed as a percentage of the predicted normal value) and Quadriceps force (QF, expressed as a percentage of the predicted value) in 380 consecutive patients with COPD (age 65 ± 8 years, FEV_1 42 ± 16), referred to the authors' outpatient rehabilitation program. Pearsons correlation coefficient was 0.094, *p*-value 0.07.

TABLE 58.1 An overview of the effects of pulmonary rehabilitation as reported in evidence-based practice guidelines [11], a state of the art review [12] and a Cochrane meta-analysis [13].

Effect	Evidence grade [11]	Size of effect
Improvement of dyspnea	1A	CRDQ-Dys: 1.06 (0.85 to 1.26) points [13]
Improvement of HRQoL	1A	SGRQ total: −6.11 (−8.98 to 3.24)% [13]
Reduction in hospital days and utilization of health care recourses	2B	
Survival	None provided	
Psychosocial benefits (self-efficacy with exercise, cognitive function, anxiety, depression)	2B	
Exercise tolerance		
Peak work rate		8W (3–13W) [13] + 18 (IQR 13–24) % baseline [12]
6MWD		48 min (32–65 min) [13] + 34 min if <28 s, 50 min if >28 s [12]
Whole body endurance		+ 87% of baseline [12]

Evidence grading 1A: Strong recommendation; 2B: Weak recommendation; CRDQ-Dys: Dyspnea subscale of the chronic respiratory disease questionnaire; SGRQ: saint Georges Respiratory Questionnaire; 6MWD: six minute walking distance.

of pulmonary rehabilitation, it is also important to recognize the non-physiological extra-pulmonary consequences of chronic lung disease. Depression, for example, is very prevalent in stable patients [8], and is significantly related to the adverse outcome of COPD, particularly after acute exacerbations [9]. In the recent Global Initiative for Obstructive Lung Disease (GOLD), the systemic consequences of COPD were specifically noted as an important marker of morbidity, complicating the management of obstructive lung disease [10].

As mentioned above, pulmonary rehabilitation targets the systemic consequences of lung diseases. It is now generally recognized as an evidence-based and effective therapy for patients with lung diseases, who are symptomatic and have reduced participation the in activities of daily life. The evidence on the effectiveness of pulmonary rehabilitation is impressive, and has been reviewed in several recent documents. Table 58.1 summarizes the most important effects of rehabilitation and where applicable the strength of the evidence, as provided in the different evidence-based reviews. To be successful, pulmonary rehabilitation is typically designed as a comprehensive intervention offered by a team of health-care providers over a substantial period of time. Typically, rehabilitation programs are carried out for 6 weeks to 6 months, with longer programs yielding more substantial effects [14]. Patients participate in an individualized tailored program which takes into account the complexity of the presenting patient. Exercise training is the cornerstone of such a program. Several other interventions may complement the program to maximize its effectiveness. Pulmonary rehabilitation may serve many goals, which depend on the perspective one takes. Examples of the goals from the different perspectives are as follows:

Patient's perspective:

- Reduced symptoms of breathlessness or fatigue
- Enhanced exercise tolerance and skeletal muscle strength
- Improved physiological and emotional function
- Increased health-related quality of life.

The health-care provider's perspective:

- Optimal self-management by the patient
- Optimal utilization of available recourses.

The societal perspective:

- Increased productivity and active participation of patients in society
- Reduction of cost associated with the management of the disease
- Enhanced coping of the family when a family member is suffering from COPD.

It is important to realize that patients suffering from other respiratory diseases, such as pulmonary sarcoidosis [15], cystic fibrosis [16], or after lung transplantation [17], equally present systemic consequences of their disease. Consequently, pulmonary rehabilitation programs should not be restricted to patients with COPD, but should be open to any patient suffering from extra-pulmonary consequences of lung diseases in general.

THE INDICATION FOR REHABILITATION AND SETTING

According to the definition of the American Thoracic Society and the European Respiratory Society, pulmonary rehabilitation is aiming at reducing symptoms, optimizing function, increasing participation, and reducing health-care cost through stabilizing or reversing the systemic consequences of the disease [1]. Hence, theoretically, patients with an indication for the need of rehabilitation are those suffering from extra-pulmonary consequences. This definition is, however, not very practical as the systemic consequences that can be targeted by pulmonary rehabilitation are not defined. Clearly, not all "systemic consequences" of the disease are amendable to pulmonary rehabilitation. For example, osteoporosis and anemia are the recognized systemic consequences of COPD [18], but it is unlikely that pulmonary rehabilitation would have a significant impact on these outcomes. In addition, the spin-off of pulmonary rehabilitation may well be an enhanced self-management, which leads to more efficient utilization of health care recourses and cost savings [19], without altering the "systemic consequences of the disease."

A more practical approach to identifying candidates for rehabilitation and guide the assessment of patients before enrollment in rehabilitation could be to select patients who are optimally pharmacologically treated, but still present with one of the following:

- Disabling symptoms due to deconditioning
- Skeletal or respiratory muscle weakness
- Poor health-related quality of life
- Repeated exacerbations or inefficient use of available recourses
- Depressed mood status
- Malnutrition (obesity of cachexia)
- Poor coping with the symptoms of their disease.

This allows to set up active screening for rehabilitation programs, which typically includes assessment of exercise tolerance, skeletal and respiratory muscle force, nutritional status, symptoms, health-related quality of life, and capability of self-management.

Depending on the complexity of the individual problem, the program can be designed across a range of complexities. Programs can be as simple as an intervention consisting of exercise training in the home setting of patients with uncomplicated COPD [20] or as complex as an in-patient program in a mechanically ventilated patient [21]. On the basis of proper assessment the individual program, its setting and components can be designed, taking into account the available recourses in a given region, current best practice, and evidence. Several guidelines may help those setting up pulmonary rehabilitation facilities to structure their program [1, 11, 22, 23].

Exercise training has now the status of the cornerstone of each program, and will be dealt with below. Exercise training has a strong evidence base. A program of exercise training of the muscles of ambulation is recommended and is a mandatory component of any pulmonary rehabilitation program [11]. For several other components of a rehabilitation program, however, evidence is also building up, particularly for specific subgroups of patients.

In depressed patients (some 40% (95% CI: 36–44%) of patients with COPD [24]), for example, *psychological counseling* may be of benefit in order to help reducing symptoms of depression or anxiety. Admittedly, a large trial to confirm the point is currently still missing [25, 26], but since depressive symptoms do significantly impact on health-related quality of life [27] and even impact readmission [9] in these patients, psychological counseling may well be worth the effort in patients who suffer from symptoms of depression, anxiety, or poor coping. It should be stressed, however, that exposure to exercise therapy may in itself have an anti-depressant effect [28]. *Nutritional interventions* have been shown to be unsuccessful in poorly selected COPD patients [29]. However, when nutritional interventions are successful in improving body mass in cachectic patients, they do spin-off in an important survival benefit [30]. Besides patients suffering from pulmonary cachectia, a less recognized role, but likely of equal importance, is the care for obese COPD patients. Clearly obesity is linked to increased pulmonary ventilation to carry out activities of daily living. Consequently, weight loss in obese patients might yield important functional benefits in activities carried out against gravity (e.g. stair climbing, walking). The authors believe that weight loss, particularly loss of fat mass, might be an important target in overweight COPD patients referred to pulmonary rehabilitation programs. *Occupational therapists* may be consulted within the context of a rehabilitation program [31]. Occupational therapists may advise patients on the mode and pace of carrying out activities of daily life. Oftentimes occupational therapists might advise on the use of wheeled walking aids (rollators). Although many patients show poor compliance with the daily use of a rollator [32], those who use it may substantially and suddenly improve their exercise tolerance [33, 34]. Other potentially cost-effective components of a rehabilitation program are interventions aiming at enhancing self-management. Generally, these interventions are supervised by *specially trained advanced practice nurses*, who work integrated in the rehabilitation team. Interventions aiming at enhancing self-management have shown variable success [35]. Studies showing cost-effectiveness of these interventions focused at a subgroup of patients with at least one hospital admission [19]. Hence it seems reasonable to direct efforts specifically to this subgroup of patients.

Another point to consider when a patient is referred to a pulmonary rehabilitation program is the setting of such a program. Programs have been successfully set up in a primary care (home) setting [20, 36] as an outpatient program in a rehabilitation center [37] or in the community [38, 38], or as an in-patient program [39, 40]. Clearly each of these programs has advantages and disadvantages. One of the most important problems of home-based programs is the limited staff and equipment, which makes it difficult to deal with more severe patients [41]. The most important problem of outpatient programs is the transportation to the center [42]. In-patient programs are costly and should be restricted to those patients with very limited mobility. Ideally, a reference rehabilitation center should have access to all modalities of rehabilitation. This can be done by establishing strong links between the different lines of health care.

Screening of patients for rehabilitation

From the above it is clear that proper design of a rehabilitation program depends on a thorough screening of patients. This screening should answer the following questions:

1. Is there an indication for pulmonary rehabilitation in a given patient and what results can be expected?
2. Which components should be included in the program for a given patient and what will be the optimal setting of the program?

An example of the intake procedure in the authors' institute is given in Table 58.2.

Indication and expected results

To establish the indication for a rehabilitation program, a formal exercise test should be carried out. Exercise intolerance is indeed an important reason to refer a patient for rehabilitation. We found that patients with skeletal and respiratory muscle weakness were most responsive to exercise training [43], justifying the assessment of peripheral and respiratory muscle strength. Rehabilitation programs are oriented toward patients who are symptomatic and have reduced activities of daily life, encouraging to assess symptoms and health-related quality of life, and physical activity. With evolving technology it becomes possible to assess the latter, rather than estimate physical activity levels with questionnaires [44, 45]. In patients admitted to the hospital with a severe exacerbation of COPD, pulmonary rehabilitation

TABLE 58.2 An example of the intake procedure for pulmonary rehabilitation at the authors' center.

Test	Reason or consequence
Lung function assessment	Disease severity
Exercise test	
Incremental cycle ergometry	Exercise intolerance/factors limiting exercise
6MWD	Exercise intolerance, treatment effect
Constant work rate test	Treatment effect
Muscle function	
Quadriceps force, biceps, triceps	Muscle weakness? Need for resistance training
Respiratory muscle force	Need for inspiratory muscle training
Symptoms, Function	
MRC	Severity of impairement in daily life
PFSDQ	Symptoms in daily life (occupational therapists)
HRQoL	
CRDQ	Indication for rehabilitation, identification of problems in four domains
Activity monitoring	
1 week of monitoring	Actual physical activity level
Interview psychologist	
HADS	Psychological burden
Social anamnesis	Social support
Nutritional anamnesis	Calorie intake and nutritional + protein balance
Bioelectrical impedance	Assessment of body composition

6MWD: six minute walking distance; MRC: medical council dyspnea scale; PFSDQ: Pulmonary functional status and dyspnea scale; HRQoL: Health related quality of life with the Chronic Respiratory Disease Questionnaire (CRDQ); HADS: Hospital anxiety and depression scale.

should be part of the proper after care. In these patients rehabilitation does reduce readmission rate by 75% and may even enhance survival [46]. In addition substantial and clinically relevant improvements in exercise tolerance and health-related quality of life were reported [47].

Components of the rehabilitation program

To establish the components of the rehabilitation program and its optimal setting, a multidisciplinary assessment is needed. This consists of a nutritional assessment, including analysis of the fat free mass and a nutritional interview to estimate the caloric balance [48]. A thorough social interview to gain insight in the social network and available social support is helpful since patients lacking social support are at risk to drop out from a program [49]. An intake by a psychologist may allow to identify the large proportion of patients suffering from significant depressive symptoms or other psychiatric morbidity [50]. An important and often forgotten aspect may be the burden on the partner of the patient [51]. A first screening for psychological morbidity can be done using questionnaires such as the Beck depression inventory or the Hospital Anxiety and Depression Scale. Lastly physical activity levels of patients and symptoms during specific activities of daily living can be assessed using questionnaires such as the Pulmonary Functional Status and Dyspnea Questionnaire (PFSDQ) [52, 53].

Depending on the complexity of the individual case one or more disciplines can be engaged in the rehabilitation process of the individual patient. Ideally the patient is referred to a program which optimally meets his needs and takes into account the motivation and social circumstances. Needless to say that a program should also be moduled along the possibilities of the health-care system in a given region.

Exercise training

Exercise training is undoubtedly the cornerstone of a pulmonary rehabilitation program. Programs aim at enhancing exercise tolerance and skeletal muscle force. Skeletal muscle force can be enhanced by specific resistance training (weightlifting) exercise [54–56]. Improvement in exercise tolerance is essentially obtained through two main pathways. First, the oxidative capacity of the skeletal muscle is enhanced [57], leading to reductions in lactate production at identical work [58]. A second pathway involves an improvement in movement efficiency, yielding less oxygen consumption to cover identical work rate. Both pathways are not mutually exclusive, and both lead to reductions in pulmonary ventilation, and dynamic hyperinflation during exercise. The reduction in pulmonary ventilation, the improved dynamic hyperinflation, and the desensitization (better tolerance to high ventilatory levels) are crucial to the improved exertional dyspnea, which is seen after exercise training [13]. To enhance the oxidative capacity of the skeletal muscle, exercise training should be conducted at relatively high training intensity [58] (see below). Improvement in movement efficiency is obtained by practicing specific exercises (e.g. walking training will only improve walking efficiency). For this pathway, the modality of the exercise is more crucial than the intensity at which the exercise is carried out.

In general, exercise training in patients with COPD follows the principles of exercise training in the healthy elderly [59]. Programs generally consist of a brief warm-up, a core program lasting at least 30 min of exercise and a cooling down. An example of an exercise training program is given in Table 58.3. Close supervision and proper monitoring of the patients ensure the safety during the program [12]. In fact, very few exercise-related events and, as far as the authors are aware, no fatal events have been reported after pulmonary rehabilitation in the published literature.

TABLE 58.3 An overview of an outpatient rehabilitation session in the rehabilitation center of the authors.

Time	Duration (min)	Item	Intensity
13:00	5	Brief contact with health care provider to check the health status. Resting saturation, peak flow, and heart rate are recorded	
13:05	10	Low intensity warm-up exercise on bicycle	20–30 W
13:15	32	Cycling exercises e.g. 8 blocks of 2 min interval training with 2 min of rest between the intervals	75% Wmax
13:47	15	Resistance training exercises for the upper and lower limb. Typical muscle groups are quadriceps, triceps, biceps and pectoral muscle	3 series of 8 repetitions
14:02	6	2 blocks of 2 min stair climbing separated by 2 min of rest	Dyspnea 4–6/10
14:08	6	2 blocks of 2 min of arm ergometry separated by 2 min of rest	Dyspnea 4–6/10
14:14	20	2 blocks of 8 min of treadmill walking	80% 6 MWs
14:34	30	Individual contact with another health care worker (psychologist, social worker, dietician)	
15:04		End of session	

Whole body exercise

Exercise training has been included in virtually all studies investigating the benefits of pulmonary rehabilitation. To successfully increase skeletal muscle properties and render measurable physiological benefits, it is important that patients do exercise at relative high work loads. To do so, the exercise training intervention can be adapted to the individual exercise limitations of the patient. The conventionally used form to deliver exercise training to COPD patients is endurance training. In COPD patients with primarily moderate disease, exercise training conducted at approximately 75% of the peak work rate (60% of the difference between the lactate threshold and peak oxygen uptake) resulted in significant physiological effects [58]. A similar training strategy was shown to be effective in patients with severe disease [60]. Others have confirmed that high training intensity is required to elicit physiologic training effects [61, 62].

It is important to adjust and increase the training load in every training session. Trained personnel should be available to ensure close supervision on the training intensity. Training intensity can be monitored using 10-point Borg symptom scales. A score around 4–6 is generally advised as an appropriate training intensity, provided the patients are familiar with the scale. Interestingly, a given Borg symptom score is generally chosen by a patient at an identical relative work rate. Hence as patients improve during training, the same Borg rating will be achieved at higher absolute work rates [63]. Since most patients are not limited by the cardiovascular system, using heart frequency to guide exercise training is not to be advised.

In patients who reach a ventilatory limitation during exercise, it is difficult to obtain a high training intensity for sustained periods of time. Several strategies may help clinicians to ensure high intensity training in more severe COPD. Interval exercise training has been used and showed to result in physiological benefits, comparable to those of endurance training [64–67]. The advantage of interval training is that the ventilatory requirements remain relatively limited [68]. Interval training is achieved by breaking down the long exercise bouts into smaller bouts of exercise separated by periods of active recovery (low work rate) or rest.

An intervention that minimizes the ventilatory burden during exercise training is the use of supplemental oxygen. Oxygen, dose dependently, reduces the ventilation for a given exercise intensity [69, 70]. The application of oxygen supplements hence allows training at higher training intensity [71]. Non-invasive mechanical ventilation reduces the work of breathing and has been used successfully in severe COPD as an adjunct to exercise training [72–74]. In less severe COPD the impact of using non-invasive mechanical ventilation was not significant [75]. Lastly, the required ventilation can be reduced simply by reducing the amount of muscles set at work. If exercise is confined to one leg, ventilation is considerably reduced, allowing a significant increase in training load [76]. A recent trial confirmed that in severe patients, single leg exercise training resulted in 20% more increase in peak oxygen consumption and peak work rate compared to conventional cycling training (two legs) [77]. In the latter study, however, no patient centered outcomes were assessed. Hence, although interesting from a physiological perspective, it is too early to judge whether single leg exercise training is ready for wide clinical use [78]. Similarly, during cycling less muscles are recruited compared to walking exercises; hence it is not surprising that for a given oxygen consumption, cycling is more fatiguing for the quadriceps muscle [79]. It would be interesting to conduct a head-to-head comparison study on the physiological effects of cycling versus walking exercises. On the basis of the larger potential to elicit muscle fatigue, cycling would be a form of exercise training that may result in larger physiological effects, compared to walking. However, such a study is currently unavailable.

All the above techniques reduce the ventilatory burden. Another complementary option is to enhance the ventilatory capacity. Obviously optimal bronchodilator therapy does also allow for larger pulmonary ventilation and less dynamic hyperinflation during exercise. In one study, a potent long-term anti-cholinergic drug showed to enhance exercise training effects compared to the use of short acting bronchodilators only [80]. Breathing gases with light density (e.g. helium and oxygen, HELIOX) also enhance the ventilatory capacity. One study suggests that maximal ventilation is enhanced by 16% when breathing 21% oxygen and 79% helium [81]. Although this would technically allow training at higher intensity, it is doubtful whether this intervention is currently ready for clinical routine. A last intervention which enhances peak ventilatory capacity is bracing the arms. During walking, this can be applied using a wheeled walking aid (rollator). Our group showed that bracing the arms on a rollator enhanced maximal voluntary ventilation by 8 l/min, which partly explained the acute beneficial effect of a rollator [82]. During treadmill walking, bracing of the arms can also easily be achieved.

Resistance training

Another form of conventional training is resistance training. This form of exercise, generally consisting of weight lifting, can be used as the only form of training [55, 56, 83], or in combination with whole body exercises [54, 55, 84]. In all the latter studies, muscle strength was significantly more increased when resistance training was added to the exercise regimen. Increased muscle strength is an important treatment objective in patients with COPD suffering from muscle weakness, as activities of daily life do require strength apart from muscle endurance. As mentioned above, muscle weakness is an important factor related to morbidity and even mortality in COPD. It follows that patients suffering from muscle weakness may be particularly good candidates to a resistance training program.

Resistance training is easy to apply in clinical practice. Patients are instructed to lift weights (generally on a multi-gym device). The weight imposed and the number of repetitions ensure overload of the skeletal muscle. In patients with COPD and several other chronic diseases, resistance training is started at approximately 70% of the weight a patient can lift once (i.e. the one repetition maximum). The effects of resistance training programs may be enhanced in male hypogonadal patients by testosterone replacement therapy. In one study, [85] weekly intramuscular injections with testosterone, aiming at restoring testosterone levels to normal values in initially hypogonadal men, did enhance skeletal muscle force more than either of the interventions alone. Further studies are required to investigate the long-term safety of this intervention. However, since skeletal muscle dysfunction is in itself a negative prognostic factor, a short-term use of testosterone may be beneficial to result in a rapid restoration of this potentially harmful situation.

Neuromuscular electrical stimulation

Another intervention used to specifically stimulate the peripheral muscles is medium to high-frequency neuromuscular electrical stimulation (NMES). Four studies have investigated successfully the effects of this intervention [86–89]. All studies in COPD used transcutaneous electrical stimulation with stimulation frequencies of 35 [89] to 50 Hz [86, 87, 90]. In patients with congestive heart failure, low-frequency (10–15 Hz) transcutaneous electrical stimulation was successfully applied [91, 92]. Studies showed that there was more strength gain in the skeletal muscles treated with electrical stimulation, either as mono-therapy or in combination with general exercise training [89]. This intervention may prove to be attractive in patients who have difficulties to take part in regular rehabilitation, such as patients admitted to hospital with acute exacerbations.

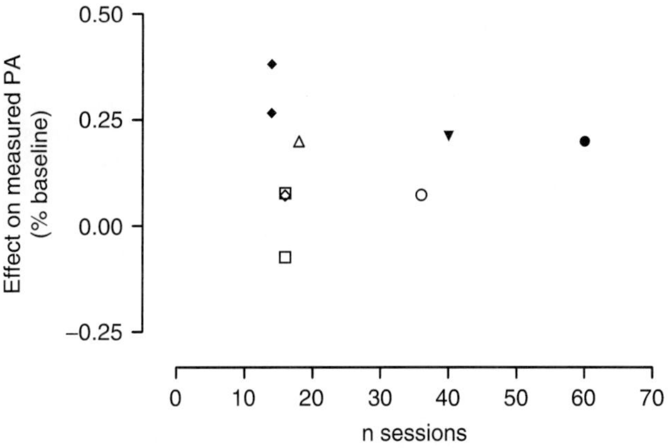

FIG. 58.2 The effect of different pulmonary rehabilitation on assessed activities of daily living. Different activity monitors were used, hence data are expressed as a % of the baseline measure. Open symbols represent studies in which the effect was reported as non-significant, closed symbols are studies that reported a statistically significant effect. ◆ Sewell, *Chest* 2006; ○ Pitta, *Chest* 2008 (after 3 months); ● Pitta, *Chest* 2008 (after 6 months); Δ Corronado, *J Cardiopulm Rehabil* 2003; ▼ Mercken, *Am J Respir Crit Care Med* 2005; □ Steele, *Arch Phys Med Rehabil* 2008; ◇ Steele, *J Rehabil Res Dev* 2003.

Enhancing physical activity

In the last few years, more attention has been given to the effect of pulmonary rehabilitation on physical activity levels. A minimum of 30 min of physical activity of moderate intensity is critical to maintain health [93]. A recent study also showed the importance of maintaining a healthy and physically active lifestyle, which has been shown to be crucial for survival in COPD [94]. Only a handful of studies have investigated the effect of pulmonary rehabilitation on physical activity [95–100]. The results of these studies are summarized in Fig. 58.2. None of these studies were randomized controlled trials, so the differences presented represent pre- to post-program differences. Interestingly not all of them showed significant improvements of the amount and intensity of physical activities in daily life after pulmonary rehabilitation. One trial [95] suggests that the effect of rehabilitation on physical activity is larger after longer programs. However, the relation between the length of the programs and their effectiveness on physical activity can surely not be seen across the different published trials. From the available evidence, it can be appreciated that the transfer of improvements in exercise tolerance toward enhanced activities of daily living should not be taken for granted. Indeed, physical activity is a behavior. The amount and intensity of activities in which patients engage depends, besides their exercise capacity, on intrinsic factors such as motivation, perceived self-efficacy, mood status, and health beliefs. Other factors related to physical activity are extrinsic factors such as social and cultural role, external barriers (e.g. environmental), and climate. Clearly more research is needed to enhance the impact of pulmonary rehabilitation beyond the immediate effects on health-related quality of life, symptoms, exercise tolerance and utilization of health care recourses and also enhance activities of daily living. The missing link toward more endurable effects of pulmonary rehabilitation may very well be the failure to successfully enhance daily physical activity levels of patients toward levels of moderate intense physical activity, sufficient to maintain health, and perhaps, the achieved benefits of the previous rehabilitation program.

References

1. Nici L, Donner C, Wouters E, ZuWallack R, Ambrosino N, Bourbeau J, Carone M, Celli B, Engelen M, Fahy B, Garvey C, Goldstein R, Gosselink R, Lareau S, Macintyre N, Maltais F, Morgan M, O'Donnell D, Prefault C, Reardon J, Rochester C, Schols A, Singh S, Troosters T. American Thoracic Society/European Respiratory Society statement on pulmonary rehabilitation. *Am J Respir Crit Care Med* 173: 1390–413, 2006.
2. Buist AS, McBurnie MA, Vollmer WM, Gillespie S, Burney P, Mannino DM, Menezes AM, Sullivan SD, Lee TA, Weiss KB, Jensen RL, Marks GB, Gulsvik A, Nizankowska-Mogilnicka E. International variation in the prevalence of COPD (the BOLD Study): A population-based prevalence study. *Lancet* 370: 741–50, 2007.
3. Schirnhofer L, Lamprecht B, Vollmer WM, Allison MJ, Studnicka M, Jensen RL, Buist AS. COPD prevalence in Salzburg, Austria: Results from the Burden of Obstructive Lung Disease (BOLD) Study. *Chest* 131: 29–36, 2007.
4. Watz H, Waschki B, Boehme C, Claussen M, Meyer T, Magnussen H. Extrapulmonary effects of chronic obstructive pulmonary disease on physical activity. *Am J Respir Crit Care Med*, 2007.
5. Swallow EB, Reyes D, Hopkinson NS, Man WD, Porcher R, Cetti EJ, Moore AJ, Moxham J, Polkey MI. Quadriceps strength predicts mortality in patients with moderate to severe chronic obstructive pulmonary disease. *Thorax* 62: 115–20, 2007.
6. Marquis K, Debigare R, Lacasse Y, LeBlanc P, Jobin J, Carrier G, Maltais F. Midthigh muscle cross-sectional area is a better predictor of mortality than body mass index in patients with chronic obstructive pulmonary disease. *Am J Respir Crit Care Med* 166: 809–13, 2002.

7. Decramer M, Gosselink R, Troosters T, Verschueren M, Evers G. Muscle weakness is related to utilization of health care resources in COPD patients. *Eur Respir J* 10: 417–23, 1997.
8. van Manen JG, Bindels PJ, Dekker FW, IJzermans CJ, van der Zee JS, Schade E. Risk of depression in patients with chronic obstructive pulmonary disease and its determinants. *Thorax* 57: 412–16, 2002.
9. Ng TP, Niti M, Tan WC, Cao Z, Ong KC, Eng P. Depressive symptoms and chronic obstructive pulmonary disease: Effect on mortality, hospital readmission, symptom burden, functional status, and quality of life. *Arch Intern Med* 167: 60–67, 2007.
10. Rabe KF, Hurd S, Anzueto A, Barnes PJ, Buist SA, Calverley P, Fukuchi Y, Jenkins C, Rodriguez-Roisin R, van WC, Zielinski J. Global strategy for the diagnosis, management, and prevention of chronic obstructive pulmonary disease: GOLD executive summary. *Am J Respir Crit Care Med* 176: 532–55, 2007.
11. Ries AL, Bauldoff GS, Carlin BW, Casaburi R, Emery CF, Mahler DA, Make B, Rochester CL, ZuWallack R, Herrerias C. Pulmonary rehabilitation: Joint ACCP/AACVPR evidence-based clinical practice guidelines. *Chest* 131: 4S–42S, 2007.
12. Troosters T, Casaburi R, Gosselink R, Decramer M. Pulmonary rehabilitation in chronic obstructive pulmonary disease. *Am J Respir Crit Care Med* 172: 19–38, 2005.
13. Lacasse Y, Goldstein R, Lasserson TJ, Martin S. Pulmonary rehabilitation for chronic obstructive pulmonary disease. *Cochrane Database Syst Rev*, 2006, CD003793.
14. Salman GF, Mosier MC, Beasley BW, Calkins DR. Rehabilitation for patients with chronic obstructive pulmonary disease. *J Gen Intern Med* 18: 213–21, 2003.
15. Spruit M, Thomeer M, Gosselink R, Troosters T, Kasran A, Debrock A, Demedts M, Decramer M. Skeletal muscle weakness in patients with sarcoidosis and its relationship with exercise intolerance and reduced health status. *Thorax* 60: 32–38, 2005.
16. Troosters T, Langer D, Vrijsen B, Segers J, Wouters K, Janssens W, Gosselink R, Decramer M, Dupont LJ. Skeletal muscle weakness, exercise tolerance and physical activity in adults with cystic fibrosis. *Eur Respir J*, 2007, Revised.
17. Maury G, Langer D, Verleden G, Dupont L, Gosselink R, Decramer M, Troosters T. Skeletal muscle force and functional exercise tolerance before and after lung transplantation, a cohort study. *Am J Transplant*, 2008, in press.
18. Fabbri LM, Rabe KF. From COPD to chronic systemic inflammatory syndrome? *Lancet* 370: 797–99, 2007.
19. Bourbeau J, Julien M, Maltais F, Rouleau M, Beaupre A, Begin R, Renzi P, Nault D, Borycki E, Schwartzman K, Singh R, Collet JP. Reduction of hospital utilization in patients with chronic obstructive pulmonary disease: A disease-specific self-management intervention. *Arch Intern Med* 163: 585–91, 2003.
20. Strijbos JH, Postma DS, van Altena R, Gimeno F, Koeter GH. A comparison between an outpatient hospital-based pulmonary rehabilitation program and a home-care pulmonary rehabilitation program in patients with COPD. A follow-up of 18 months. *Chest* 109: 366–72, 1996.
21. Nava S. Rehabilitation of patients admitted to a respiratory intensive care unit. *Arch Phys Med Rehabil* 79: 849–54, 1998.
22. National Institute for Clinical Excellence (NICE). Chronic obstructive pulmonary disease: National clinical guidline for the management of COPD in adults in primary and secondary care. *Thorax* 59(Suppl 1), 2004.
23. Société de Pneumologie de Langue Française de Pneumologie. Recommandations de la Société de Pneumologie de Langue Française sur la réhabilitation du malade atteint de BPCO. *Rev Mal Respir* 22: 7S8–7S14, 2005.
24. Yohannes AM, Baldwin RC, Connolly MJ. Depression and anxiety in elderly patients with chronic obstructive pulmonary disease. *Age Ageing* 35: 457–59, 2006.
25. Coventry PA, Hind D. Comprehensive pulmonary rehabilitation for anxiety and depression in adults with chronic obstructive pulmonary disease: Systematic review and meta-analysis. *J Psychosom Res* 63: 551–65, 2007.
26. Coventry PA, Gellatly JL. Improving outcomes for COPD patients with mild-to-moderate anxiety and depression: A systematic review of cognitive behavioural therapy. *Br J Health Psychol*, 2007.
27. Oga T, Nishimura K, Tsukino M, Sato S, Hajiro T, Mishima M. Longitudinal deteriorations in patient reported outcomes in patients with COPD. *Respir Med*, 2006.
28. Craft LL, Perna FM. The benefits of exercise for the clinically depressed. *Prim Care Companion J Clin Psychiatry* 6: 104–11, 2004.
29. Ferreira IM, Brooks D, Lacasse Y, Goldstein RS. Nutritional support for individuals with COPD: A meta-analysis. *Chest* 117: 672–78, 2000.
30. Schols AM, Slangen J, Volovics L, Wouters EF. Weight loss is a reversible factor in the prognosis of chronic obstructive pulmonary disease. *Am J Respir Crit Care Med* 157: 1791–97, 1998.
31. Lorenzi CM, Cilione C, Rizzardi R, Furino V, Bellantone T, Lugli D, Clini E. Occupational therapy and pulmonary rehabilitation of disabled COPD patients. *Respiration* 71: 246–51, 2004.
32. Gupta RB, Brooks D, Lacasse Y, Goldstein RS. Effect of rollator use on health-related quality of life in individuals with COPD. *Chest* 130: 1089–95, 2006.
33. Gupta R, Goldstein R, Brooks D. The acute effects of a rollator in individuals with COPD. *J Cardiopulm Rehabil* 26: 107–11, 2006.
34. Probst V, Heyvaert H, Coosemans I, Pitta F, Spruit M, Troosters T, Gosselink R, Decramer M. Effects of a rollator on exercise capacity, gas exchange and ventilation in COPD patients. *Am J Respir Crit Care Med* 167: A669, 2003, Abstract.
35. Effing T, Monninkhof EM, van der Valk PD, van der Palen J, van Herwaarden CL, Partidge MR, Walters EH, Zielhuis GA. Self-management education for patients with chronic obstructive pulmonary disease. *Cochrane Database Syst Rev*, 2007, CD002990.
36. Wijkstra PJ, Ten Vergert EM, van Altena R, Otten V, Kraan J, Postma DS, Koeter GH. Long term benefits of rehabilitation at home on quality of life and exercise tolerance in patients with chronic obstructive pulmonary disease. *Thorax* 50: 824–28, 1995.
37. Troosters T, Gosselink R, Decramer M. Short- and long-term effects of outpatient rehabilitation in patients with chronic obstructive pulmonary disease: A randomized trial. *Am J Med* 109: 207–12, 2000.
38. Griffiths TL, Burr ML, Campbell IA, Lewis-Jenkins V, Mullins J, Shiels K, Turner-Lawlor PJ, Payne N, Newcombe RG, Ionescu AA, Thomas J, Tunbridge J, Lonescu AA. Results at 1 year of outpatient multidisciplinary pulmonary rehabilitation: A randomised controlled trial. *Lancet* 355: 362–68, 2000.
39. Goldstein RS, Gort EH, Stubbing D, Avendano MA, Guyatt GH. Randomised controlled trial of respiratory rehabilitation. *Lancet* 344: 1394–97, 1994.
40. Franssen FM, Broekhuizen R, Janssen PP, Wouters EF, Schols AM. Effects of whole-body exercise training on body composition and functional capacity in normal-weight patients with COPD. *Chest* 125: 2021–28, 2004.
41. Wedzicha JA, Bestall JC, Garrod R, Garnham R, Paul EA, Jones PW. Randomized controlled trial of pulmonary rehabilitation in severe chronic obstructive pulmonary disease patients, stratified with the MRC dyspnoea scale. *Eur Respir J* 12: 363–69, 1998.
42. Taylor R, Dawson S, Roberts N, Sridhar M, Partridge MR. Why do patients decline to take part in a research project involving pulmonary rehabilitation? *Respir Med* 101: 1942–46, 2007.
43. Troosters T, Gosselink R, Decramer M. Exercise training in COPD: How to distinguish responders from nonresponders. *J Cardiopulm Rehabil* 21: 10–17, 2001.
44. Pitta F, Troosters T, Spruit M, Decramer M, Gosselink R. Activity monitoring for assessment of physical activities of daily life in patients with COPD. *Arch Phys Med Rehabil* 86: 1979–85, 2005.
45. Pitta F, Troosters T, Probst VS, Spruit MA, Decramer M, Gosselink R. Quantifying physical activity in daily life with questionnaires and motion sensors in COPD. *Eur Respir J* 27: 1040–55, 2006.
46. Puhan MA, Scharplatz M, Troosters T, Steurer J. Respiratory rehabilitation after acute exacerbation of COPD may reduce risk for readmission and mortality – a systematic review. *Respir Res* 6: 54, 2005.

47. Man WD, Polkey MI, Donaldson N, Gray BJ, Moxham J. Community pulmonary rehabilitation after hospitalisation for acute exacerbations of chronic obstructive pulmonary disease: Randomised controlled study. *BMJ* 329: 1209, 2004.
48. Brug J, Schols A, Mesters I. Dietary change, nutrition education and chronic obstructive pulmonary disease. *Patient Educ Couns* 52: 249–57, 2004.
49. Young P, Dewse M, Fergusson W, Kolbe J. Respiratory rehabilitation in chronic obstructive pulmonary disease: Predictors of nonadherence. *Eur Respir J* 13: 855–59, 1999.
50. Wagena EJ, Arrindell WA, Wouters EF, van Schayck CP. Are patients with COPD psychologically distressed? *Eur Respir J* 26: 242–48, 2005.
51. Unger DG, Jacobs SB. Couples and chronic obstructive airway diseases: The role of gender in coping and depression. *Womens Health* 1: 237–55, 1995.
52. Lareau SC, Breslin EH, Meek PM. Functional status instruments: Outcome measure in the evaluation of patients with chronic obstructive pulmonary disease. *Heart Lung* 25: 212–24, 1996.
53. Lareau SC, Meek PM, Roos PJ. Development and testing of the modified version of the pulmonary functional status and dyspnea questionnaire (PFSDQ-M). *Heart Lung* 27: 159–68, 1998.
54. Bernard S, Whittom F, LeBlanc P, Jobin J, Belleau R, Berube C, Carrier G, Maltais F. Aerobic and strength training in patients with chronic obstructive pulmonary disease. *Am J Respir Crit Care Med* 159: 896–901, 1999.
55. Ortega F, Toral J, Cejudo P, Villagomez R, Sanchez H, Castillo J, Montemayor T. Comparison of effects of strength and endurance training in patients with chronic obstructive pulmonary disease. *Am J Respir Crit Care Med* 166: 669–74, 2002.
56. Spruit MA, Gosselink R, Troosters T, De Paepe C, Decramer M. Resistance versus endurance training in patients with COPD and skeletal muscle weakness. *Eur Respir J* 19: 1072–78, 2002.
57. Maltais F, LeBlanc P, Simard C, Jobin J, Berube C, Bruneau J, Carrier L, Belleau R. Skeletal muscle adaptation to endurance training in patients with chronic obstructive pulmonary disease. *Am J Respir Crit Care Med* 154: 442–47, 1996.
58. Casaburi R, Patessio A, Ioli F, Zanaboni S, Donner CF, Wasserman K. Reductions in exercise lactic acidosis and ventilation as a result of exercise training in patients with obstructive lung disease. *Am Rev Respir Dis* 143: 9–18, 1991.
59. American College of Sports Medicine Position Stand. Exercise and physical activity for older adults. *Med Sci Sports Exerc* 30: 992–1008, 1998.
60. Casaburi R, Porszasz J, Burns MR, Carithers ER, Chang RS, Cooper CB. Physiologic benefits of exercise training in rehabilitation of patients with severe chronic obstructive pulmonary disease. *Am J Respir Crit Care Med* 155: 1541–51, 1997.
61. Puente-Maestu L, Sanz ML, Sanz P, Ruiz de Ona JM, Rodriguez-Hermosa JL, Whipp BJ. Effects of two types of training on pulmonary and cardiac responses to moderate exercise in patients with COPD. *Eur Respir J* 15: 1026–32, 2000.
62. Gimenez M, Servera E, Vergara P, Bach JR, Polu JM. Endurance training in patients with chronic obstructive pulmonary disease: A comparison of high versus moderate intensity. *Arch Phys Med Rehabil* 81: 102–9, 2000.
63. Mahler DA, Ward J, Mejia-Alfaro R. Stability of dyspnea ratings after exercise training in patients with COPD. *Med Sci Sports Exerc* 35: 1083–87, 2003.
64. Vogiatzis I, Nanas S, Roussos C. Interval training as an alternative modality to continuous exercise in patients with COPD. *Eur Respir J* 20: 12–19, 2002.
65. Vogiatzis I, Terzis G, Nanas S, Stratakos G, Simoes DC, Georgiadou O, Zakynthinos S, Roussos C. Skeletal muscle adaptations to interval training in patients with advanced COPD. *Chest* 128: 3838–45, 2005.
66. Coppoolse R, Schols AM, Baarends EM, Mostert R, Akkermans MA, Janssen PP, Wouters EF. Interval versus continuous training in patients with severe COPD: A randomized clinical trial. *Eur Respir J* 14: 258–63, 1999.
67. Puhan MA, Busching G, Schunemann HJ, VanOort E, Zaugg C, Frey M. Interval versus continuous high-intensity exercise in chronic obstructive pulmonary disease: A randomized trial. *Ann Intern Med* 145: 816–25, 2006.
68. Sabapathy S, Kingsley RA, Schneider DA, Adams L, Morris NR. Continuous and intermittent exercise responses in individuals with chronic obstructive pulmonary disease. *Thorax* 59: 1026–31, 2004.
69. Porszasz J, Emtner M, Goto S, Somfay A, Whipp BJ, Casaburi R. Exercise training decreases ventilatory requirements and exercise-induced hyperinflation at submaximal intensities in patients with COPD. *Chest* 128: 2025–34, 2005.
70. Somfay A, Porszasz J, Lee SM, Casaburi R. Dose–response effect of oxygen on hyperinflation and exercise endurance in nonhypoxaemic COPD patients. *Eur RespirJ* 18: 77–84, 2001.
71. Emtner M, Porszasz J, Burns M, Somfay A, Casaburi R. Benefits of supplemental oxygen in exercise training in non-hypoxemic COPD patients. *Am J Respir Crit Care Med* 168: 1034–42, 2003.
72. Hawkins P, Johnson LC, Nikoletou D, Hamnegard CH, Sherwood R, Polkey MI, Moxham J. Proportional assist ventilation as an aid to exercise training in severe chronic obstructive pulmonary disease. *Thorax* 57: 853–59, 2002.
73. Costes F, Agresti A, Court-Fortune M, Roche F, Vergnon JM, Barthelemy JC. Noninvasive ventilation during exercise training improves exercise tolerance in patients with chronic obstructive pulmonary disease. *J Cardiopulm Rehabil* 23: 307–13, 2003.
74. van't HA, Gosselink R, Hollander P, Postmus P, Kwakkel G. Training with inspiratory pressure support in patients with severe COPD. *Eur Respir J* 27: 65–72, 2006.
75. Bianchi L, Foglio K, Porta R, Baiardi R, Vitacca M, Ambrosino N. Lack of additional effect of adjunct of assisted ventilation to pulmonary rehabilitation in mild COPD patients. *Respir Med* 96: 359–67, 2002.
76. Dolmage TE, Goldstein RS. Response to one-legged cycling in patients with COPD. *Chest* 129: 325–32, 2006.
77. Dolmage TE, Goldstein RS. Effects of one-legged exercise training of patients with COPD. *Chest* 133: 370–76, 2008.
78. Mador MJ. Exercise training in patients with COPD: One leg is better than two? *Chest* 133: 337–39, 2008.
79. Pepin V, Saey D, Whittom F, LeBlanc P, Maltais F. Walking versus cycling: Sensitivity to bronchodilation in chronic obstructive pulmonary disease. *Am J Respir Crit Care Med*, 2005.
80. Casaburi R, Kukafka D, Cooper CB, Witek TJ Jr., Kesten S. Improvement in exercise tolerance with the combination of tiotropium and pulmonary rehabilitation in patients with COPD. *Chest* 127: 809–17, 2005.
81. Eves ND, Petersen SR, Haykowsky MJ, Wong EY, Jones RL. Helium-hyperoxia, exercise, and respiratory mechanics in chronic obstructive pulmonary disease. *Am J Respir Crit Care Med* 174: 763–71, 2006.
82. Probst V, Troosters T, Coosemans I, Spruit M, Pitta F, Decramer M, Gosselink R. Mechanisms of improvement in exercise capacity using a rollator in COPD. *Chest* 126: 1102–7, 2004.
83. Simpson K, Killian K, McCartney N, Stubbing DG, Jones NL. Randomised controlled trial of weightlifting exercise in patients with chronic airflow limitation. *Thorax* 47: 70–75, 1992.
84. Mador MJ, Bozkanat E, Aggarwal A, Shaffer M, Kufel TJ. Endurance and strength training in patients with COPD. *Chest* 125: 2036–45, 2004.
85. Casaburi R, Bhasin S, Cosentino L, Porszasz J, Somfay A, Lewis M, Fournier M, Storer T. Anabolic effects of testosterone replacement and strength training in men with COPD. *Am J Respir Crit Care Med* 170: 870–78, 2004.
86. Neder JA, Sword D, Ward SA, Mackay E, Cochrane LM, Clark CJ. Home based neuromuscular electrical stimulation as a new rehabilitative strategy for severely disabled patients with chronic obstructive pulmonary disease (COPD). *Thorax* 57: 333–37, 2002.
87. Bourjeily-Habr G, Rochester C, Palermo F, Snyder P, Mohsenin V. Randomised controlled trial of transcutaneous electrical muscle stimulation of the lower extremities in patients with chronic obstructive pulmonary disease. *Thorax* 57: 1045–49, 2002.

88. Zanotti E, Felicetti G, Maini M, Fracchia C. Peripheral muscle strength training in bed-bound patients with COPD receiving mechanical ventilation: Effect of electrical stimulation. *Chest* 124: 292–96, 2003.
89. Vivodtzev I, Pepin JL, Vottero G, Mayer V, Porsin B, Levy P, Wuyam B. Improvement in quadriceps strength and dyspnea in daily tasks after 1 month of electrical stimulation in severely deconditioned and malnourished COPD. *Chest* 129: 1540–48, 2006.
90. Dal CS, Napolis L, Malaguti C, Gimenes AC, Albuquerque A, Nogueira CR, De Fuccio MB, Pereira RD, Bulle A, McFarlane N, Nery LE, Neder JA. Skeletal muscle structure and function in response to electrical stimulation in moderately impaired COPD patients. *Respir Med*, 2006.
91. Maillefert JF, Eicher JC, Walker P, Dulieu V, Rouhier-Marcer I, Branly F, Cohen M, Brunotte F, Wolf JE, Casillas JM, Didier JP. Effects of low-frequency electrical stimulation of quadriceps and calf muscles in patients with chronic heart failure. *J Cardiopulm Rehabil* 18: 277–82, 1998.
92. Harris S, LeMaitre JP, Mackenzie G, Fox KA, Denvir MA. A randomised study of home-based electrical stimulation of the legs and conventional bicycle exercise training for patients with chronic heart failure. *Eur Heart J* 24: 871–78, 2003.
93. Haskell WL, Lee IM, Pate RR, Powell KE, Blair SN, Franklin BA, Macera CA, Heath GW, Thompson PD, Bauman A. Physical activity and public health. Updated recommendation for adults from the American College of Sports Medicine and the American Heart Association. *Circulation*, 2007.
94. Garcia-Aymerich J, Lange P, Benet M, Schnohr P, Anto JM. Regular physical activity reduces hospital admission and mortality in chronic obstructive pulmonary disease: A population-based cohort study. *Thorax*, 2006.
95. Pitta F, Troosters T, Probst V, Langer D, Decramer M, Gosselink R. Are patients with COPD more active after pulmonary rehabilitation. *Chest*, 2008, E-pub ahead of Print.
96. Sewell L, Singh SJ, Williams JE, Collier R, Morgan MD. Can individualized rehabilitation improve functional independence in elderly patients with COPD? *Chest* 128: 1194–200, 2005.
97. Mercken EM, Hageman GJ, Schols AM, Akkermans MA, Bast A, Wouters EF. Rehabilitation decreases exercise-induced oxidative stress in chronic obstructive pulmonary disease. *Am J Respir Crit Care Med* 172: 994–1001, 2005.
98. Steele BG, Belza B, Cain KC, Coppersmith J, Lakshminarayan S, Howard J, Haselkorn JK. A randomized clinical trial of an activity and exercise adherence intervention in chronic pulmonary disease. *Arch Phys Med Rehabil* 89: 404–12, 2008.
99. Coronado M, Janssens JP, de Muralt B, Terrier P, Schutz Y, Fitting JW. Walking activity measured by accelerometry during respiratory rehabilitation. *J Cardiopulm Rehabil* 23: 357–64, 2003.
100. Steele BG, Belza B, Hunziker J, Holt L, Legro M, Coppersmith J, Buchner D, Lakshminaryan S. Monitoring daily activity during pulmonary rehabilitation using a triaxial accelerometer. *J Cardiopulm Rehabil* 23: 139–42, 2003.

Surgical and Other Mechanical Procedures

59

CHAPTER

Michael I. Polkey and Pallav L. Shah

Royal Brompton Hospital and National Heart and Lung Institute, London, UK

DECLARATIONS OF INTEREST

Our institutions have received research grants for our group from Asthmalx, Emphasys, Broncus and PortAero who have patents and/or technologies relevant to this article.

INTRODUCTION

Although drugs, together with pulmonary rehabilitation, will form the mainstay of treatment for most patients with chronic obstructive pulmonary disease (COPD) as the lung destruction progresses, these therapies become insufficiently effective for a minority of patients who may then be prepared to consider a surgical treatment. As a general rule, these therapies should not be offered to patients who have not completed pulmonary rehabilitation and should not be offered outside specialist centers where a truly multidisciplinary approach is available. Asthma, being a condition characterized by variable airflow obstruction, is seldom suitable to a surgical approach although the experimental approach of thermoplasty will be briefly discussed.

CHRONIC OBSTRUCTIVE PULMONARY DISEASE

Surgical procedures

In the 1950s Owen Brantigan formed the concept that a surgical treatment of emphysema would be to "reduce lung volume to fit the expiratory phase of the respiratory cavity [by] removing lung tissue that has no respiratory function" [1]. In fact, perhaps due to difficulty with patient selection and the poorer standard of anesthetic and postoperative care at the time the technique was never widespread and except for occasional patients who could benefit from a bullectomy [2, 3] or a Monaldi procedure [4], the only surgical therapy available until recently was single or double lung transplantation. Transplantation remains a moderately dangerous procedure with a mortality in the first year of approximately 15%. More importantly there are insufficient donor organs for the number of patients willing to undergo the procedure.

For this reason surgical lung volume reduction was revived in the 1990s by Joel Cooper and colleagues [5]; their original data was confined to patients with marked hyperinflation and contained no control group but showed, at 6 months, a 51% improvement in FEV_1 and a reduction in oxygen usage such that 70% of patients were able to discontinue domiciliary oxygen therapy.

The technique was later submitted to randomized controlled trials of which the most important was the National Emphysema Treatment Trial (NETT) study [6] (*vide infra*) which accounted for 73% of all presently randomized patients. The results of the smaller studies are summarized in the table and have been usefully summarized in a Cochrane review [7] (Table 59.1). Broadly speaking, the review found that at 90 days there were significant benefits conferred by surgery with regard to quality of life whether measured with generic- or disease-specific instruments but it must be borne in mind that surgery itself has a significant placebo benefit. Other benefits from surgery included improved

TABLE 59.1 Results of lung volume reduction surgery.

Author	ΔFEV_1 (ml)	ΔFEV_1 (%)	Δ Field walking test (m)	Δ Field walking test (%)
Criner, 1999 [8]	200	31	33 (p = ns)	12
Geddes, 2000 [9]	70	10	50	25
Goldstein, 2003 [10]	300	38	42	11
Hillerdal, 2005 [11]	230	35	99	43
Miller, 2006 [12]	265	30	29	10

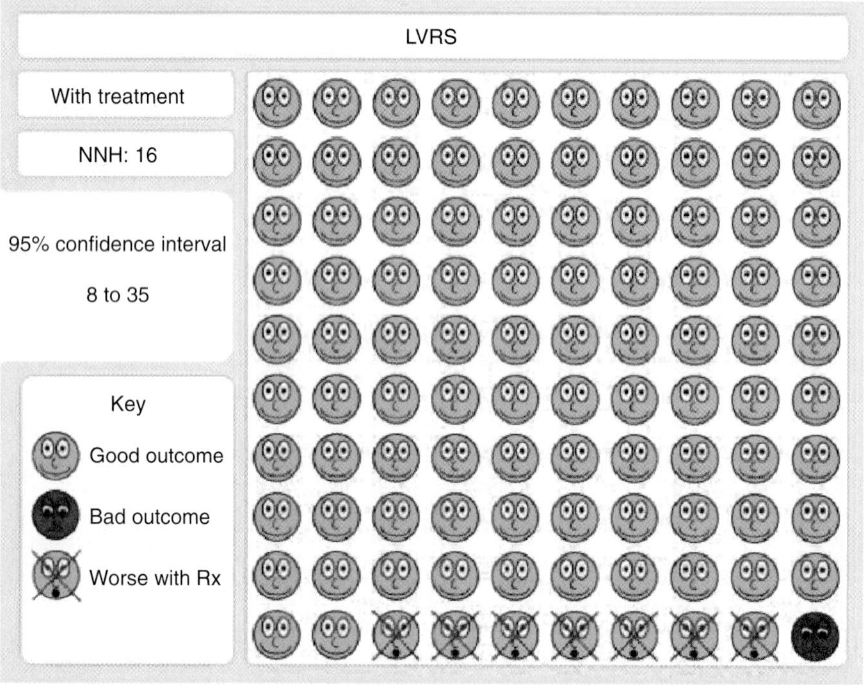

FIG. 59.1 Ninety-day outcome of LVRS reproduced from the meta-analysis by Tiong *et al.* [7]. A second figure may be found in the paper showing the results when the high risk patients from NETT were excluded [13]. Interestingly although this exclusion reduces the number of patients who are worse after surgery but not the mortality; however the number needed to treat rises.

exercise performance, reduced emergency room attendance in postoperative year 2 (but not 1).

The NETT study is by far the biggest study to assess lung volume reduction surgery (LVRS) [6]; briefly after pulmonary rehabilitation 1218 were randomized to receive either LVRS or continued medical care (Fig. 59.1). An early analysis showed that patients with a homogenous distribution of disease and either a $FEV_1 < 20\%$ predicted or a carbon monoxide gas transfer (TL_{CO}) $< 20\%$ predicted had a poor outcome and were excluded from the remainder of the study [13]. The study showed that in the surgical group (compared with medical therapy) the exercise capacity improved on an incremental test by more than 10W in 28% and 15% of operated patients but only 4% and 3% of controls at 6 and 24 months respectively ($p < 0.001$). Surgical patients were more likely to achieve an improved 6-min walk distance and exhibited an approximately 30% reduction in exacerbations [14]. An accompanying paper [15] found the difference in total cost at 3 years between the two groups was $36,392.

Long-term follow up of the randomized controlled trials conducted is confined to data from the Brompton study [9]; Lim and colleagues [16] found that improvement in residual volume (RV)/total lung capacity (TLC) ratio and gas transfer tended to be more sustained than those in FEV_1. Overall the data showed that the surgical group was better than the control group for at least 3 years.

The current position for surgical LVRS is that it may best be described as useful only in patients with the most favorable radiological appearances (heterogenous upper zone disease), who are a minority of emphysema patients. Even so the operation is confined to those patients willing to accept the perioperative mortality (typically 5%) but who meet the safety criteria. These will vary slightly from institution to institution but, for example, we require an $FEV_1 > 20\%$, a $TL_{CO} > 30\%$, and a shuttle walk distance (after rehabilitation) of >150 m. A further problem is that a minority of patients experience no benefit but these patients cannot be reliably predicted prior to surgery. As a result the number of procedures presently performed is small.

Minimally invasive surgical therapies

Since LVRS was demonstrated to work in carefully selected patients the question arose as to whether lung volume reduction could be achieved more safely and cheaply using a nonsurgical mechanical result. In principle, lung volume reduction could be achieved by the introduction of a blocker (subsequently modified to a one-way valve) into the airway. Sabaratnam Sabanathan, an English surgeon, first suggested endobronchial blockade as an alternative approach. Sadly he died whilst developing the concept but his preliminary data, from procedures performed between November 1996 and April 1997, [17] were published posthumously and showed that significant improvement was possible using this approach. However, the blockers initially used were not designed for endobronchial occlusion and later patients in the series had devices inserted that were custom built in Sabanathan's hospital. Significant problems were observed with migration of the first type of blockers and in the case of the custom built devices one episode of blocker disintegration occurred.

Biotechnology companies subsequently took up the challenge and two devices are commercially available. The Emphasys system (Emphasys Medical, Inc; Redwood City, CA) uses a one-way duckbill valve that is inserted in such a way that air is permitted to leave the subtended segment but not enter it. The valve is secured and supported by a collapsible metal framework. The initial valves (Fig. 59.2A) were inserted over a guide wire but later models (Fig. 59.2B) can be inserted through the working channel of a flexible bronchoscope. Five human studies have been reported with the valve [18–22], and the company has recently reported the combined data [23].

The Spiration (Spiration, Inc; Redmond, WA) device is an umbrella arrangement (Fig. 59.2C) arranged so that the convex surface is positioned distally with the aim again of allowing expiratory but not inspiratory flow; so far human data are reported from an uncontrolled study of 30 patients [24]. These patients were subjected to bilateral upper lobe placement. The reported safety data were satisfactory but less than a third of patients showed improvements in FEV_1 or 6 min walking test greater than 15%.

Similarly, although animal data exist for the tissue engineering approach [25]; human data so far are confined to subsegments of a single lung in six subjects [26]. Again the safety profile was satisfactory.

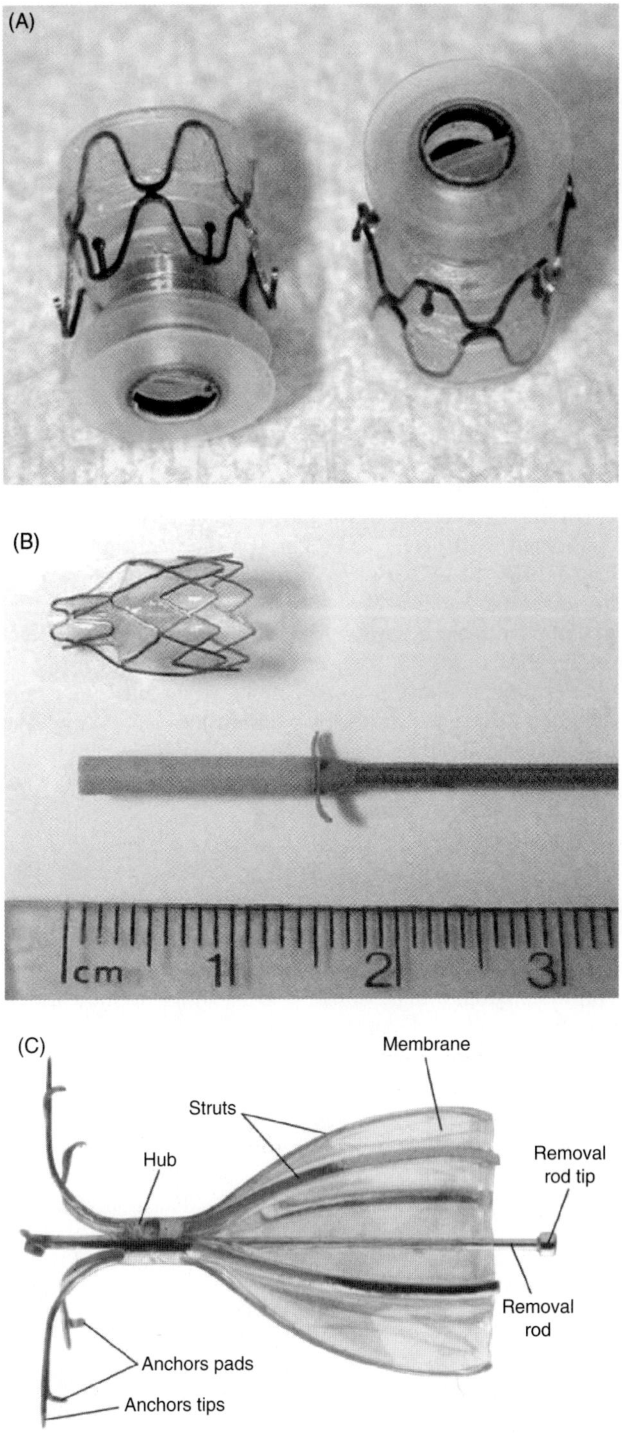

FIG. 59.2 Endobronchial valves (A) Initial Emphasys endobronchial valve. (B) Modified Emphasys endobronchial valve, suitable for transcopic insertion. (C) Intrabronchial valve (Spiration Inc).

Experience with endobronchial therapy

Snell *et al.* [18] used an early version of the valve and aimed for bilateral lung volume reduction. No improvement was observed in any parameter except TL_{CO}, although they also found valve insertion to be safe and well tolerated. Toma *et al.* reported the use of the Emphasys valve for bronchoscopic lung volume reduction [22]. Data from eight patients was presented; despite the small sample size a statistically significant improvement in FEV_1 was observed with an increase from 0.791 (range 0.61–1.07) to 1.061 [0.75–1.22] (difference 34%, $p = 0.028$). Other investigators in uncontrolled series found improvements in FEV_1 either alone [19] or with significant improvements in plethysmographic lung volumes and mean TL_{CO} [20].

Hopkinson *et al.* reported the only study of BLVR in which dynamic hyperinflation was directly measured [21]. We studied 19 patients treated with unilateral endobronchial valve insertion and made key measurements 4 weeks after the procedure. As in previous studies, we observed a statistically significant improvement in mean FEV_1 (28–31% predicted, $p < 0.05$) and in mean TL_{CO} (36–41% predicted, $p = 0.016$). A statistically significant improvement in TLC and functional residual capacity (FRC) was observed though this failed to reach significance in the case of RV.

A priori we divided patients into improvers and nonimprovers. Improvers were defined as patients who were able to show an increase of both 60 s and 30% in endurance time on the constant rate cycle endurance test. Nine of nineteen patients (47%) met this criterion. In stepwise regression analysis only change in resting inspiratory capacity and ΔTL_{CO} were retained as independent predictors, producing an equation that explained 81% of the variation in endurance time ($p < 0.0001$). If patients with atelectasis were excluded, again, only the same two variables were retained in the model ($r^2 = 0.61$; $p = 0.002$).

The combined Emphasys data set was recently reported [23]: 98 patients were studied with an average number of 4 valves per patient, the majority had complete unilateral (48%) or bilateral (21%) lobar occlusion. On average significant improvements were observed in FEV_1 (mean 60 ml, 10%) and in walking distance (37 m, 23%). Clinically the mean improvements are at the lower range of what might be considered worthwhile. The safety profile compared favorably to surgical LVRS. *Post hoc* subset analysis showed that the biggest improvements were seen in those patients submitted to complete lobar occlusion, the lowest FEV_1 and the greatest hyperinflation.

Emphasys have recently concluded a multicenter randomized controlled study (for design see Ref. [27]). The results were presented orally at the Annual Congress of the European Respiratory Society in September 2007 but are not yet available in print form.

Extrapulmonary bypass

An alternative approach to emphysema was proposed by Macklem in 1978 [28]. Based on the observation that emphysematous lungs, unlike normal lungs, demonstrated significant collateral ventilation he proposed that a physical connection between the lung parenchyma and the atmosphere could reduce hyperinflation.

So far, one biotechnology company has created a product which attempts to exploit this property. Broncus Inc (Mountain View CA 94043) has developed a stent which is placed in a bronchial segment to permit exhaled air to bypass the point of flow limitation. So far the device has been demonstrated to be feasible in dogs with patency enhanced by the use of drug eluting stents [29]. The stents increase expiratory flow in explanted human lungs [30, 31]. An uncontrolled study has been reported in 35 patients with emphysema [32]: for the group as a whole a 7.3% reduction in FEV_1 and a 37% increase in 6-min walk were observed. However *post hoc* analysis showed that the greatest benefit was observed in patients with the most preprocedure hyperinflation as one might expect. 69% of stents examined bronchscopically were patent at 6 months after the procedure. An ongoing randomized controlled trial (RCT) is presently underway which is expected to report in 2009.

Overview of options for emphysema

At present the only evidence-based therapy is LVRS but this is unattractive to purchasers, patients, and surgeons because of the cost, risk, and the relatively high safety criteria which are necessary to avoid an unacceptable mortality. Two main noninvasive approaches are feasible. Endobronchial blockade is likely to be the most suitable for the minority of patients with no collateral ventilation and heterogenous upper zone disease. Extrapulmonary bypass is likely to be most suitable for the majority of patients with homogenous disease, resting hyperventilation, and collateral ventilation. In both cases pivotal studies have yet to be reported and neither treatment can be recommended (except on a compassionate basis) until they are shown to be both safe and effective. Collateral ventilation is frequently but not universally present in patients with COPD [33] and can be measured in man [34, 35] and we speculate that these techniques may become part of the work up when assessing patients for these therapies.

ASTHMA

Bronchial thermoplasty is a novel approach in the treatment of asthma, which targets smooth muscle within the airways. In asthma there is frequently an increase in the number and the size of smooth muscle within the airway and this in turn causes bronchoconstriction, hyperresponsiveness and inflammation. Bronchial thermoplasty uses low energy radio frequency to ablate the smooth muscle. The epithelium, mucous glands, and nerves appear to regenerate but there is permanent reduction in the smooth muscle bulk.

A feasibility study was performed in patients with lung cancer who were due to undergo surgical resection. Bronchial thermoplasty was performed in these individuals up to 3 weeks prior to their surgery and segmental bronchi within the lobe that was to be resected were treated with bronchial thermoplasty. This study demonstrated that bronchial thermoplasty was well tolerated and resulted in a significant reduction in smooth muscle mass in the treated airways [36].

A safety study in 16 subjects with mild to moderate asthma was subsequently performed [37]. The patients underwent bronchial thermoplasty in three separate treatments, initially one of the lower lobes was treated, followed by the contralateral lower lobe 3 weeks later, and finally both upper lobes were treated on the last treatment setting a further 3 weeks later. The patients all underwent detailed monitoring with lung function tests, methacholine challenge, and an assessment of their symptom-free days was made. The study demonstrated a significant improvement in morning and evening peak flows and also an increase in symptom-free days following treatment with bronchial thermoplasty. There was a reduction in bronchial hyperresponsiveness in the patients treated, which was maintained at 1 and 2 years following treatment with bronchial thermoplasty.

A further randomized study in 112 subjects with moderate or severe persistent asthma has also shown significant improvements in quality of life, symptom scores, a significant increase in symptom-free days, and morning peak expiratory flow measurements [38]. An increase in adverse events immediately following bronchial thermoplasty was observed. These events were usually mild and well tolerated. They comprised of an increase in breathlessness, cough, wheezing, and chest discomfort. The majority of these events occurred within 1 day of the procedure and resolved within 7 days of onset. There was no increase in adverse events 6 weeks after the last treatment and following this period the adverse event rates were similar in both treatment and controlled groups.

In this study baseline measurements were conducted during a 2-week period when the patients were withdrawn from their long active bronchodilators and similar analyses were performed on bronchodilators at 3, 6, and 12 months. Measurements at this time point demonstrated significant improvement in the number of symptom-free days and on average treatment subjects had 86 additional symptom-free days per year compared to the control group. There was also an improvement in symptom scores, asthma quality of life questionnaire (AQLQ), and asthma control questionnaire (ACQ) scores. There was a significant reduction in β_2-agonists used that would equate to approximately 2 inhalers per patient over the 1-year treatment period. Finally, there was a reduction in mild and severe exacerbations and in the treatment group there were approximately 10 fewer mild exacerbations than in the control group. The main limitation of this large study was that it was an open-label design and hence, subject to bias.

A small study was performed in patients with severe refractory asthma who were on more than 750 μg fluticasone propionate per day in addition to long-acting bronchodilators [39]. A significant proportion of patients were also on maintenance of oral steroids. In this study, 15 patients were randomized to bronchial thermoplasty and there were 17 controlled subjects. The study design was to randomize the patients to treatment or control followed by a 6-week stabilization period. The patients were maintained on their regular treatment for the following 14 weeks, and then the patients entered a steroid weaning phase where the dose of oral steroids or inhaled steroids was reduced by 20–25% in 2-week steps. Following the 14-week steroid-wean phase, patients were maintained on a reduced dose for a further 16 weeks and assessments performed over the two study periods. Overall, bronchial thermoplasty was well tolerated although there was an increase in adverse events over the treatment period. During this treatment period there were seven hospitalizations for adverse events in four subjects in the bronchial thermoplasty group and none in the control group. In the posttreatment period the number of hospitalizations and adverse events were similar in both treatment groups. There were significant improvements in a number of parameters including lung function and quality of life scores in the treatment group compared to the control group both prior to reduction in steroids and also following the steroid-wean phase. Four of the eight patients in the bronchial thermoplasty group were completely weaned off their oral steroids in contrast to only one of the seven patients in the control group who was weaned off oral steroids. Prebronchodilator FEV_1 was improved by around 15% in the treatment group compared to the control group, and there was an over 26% reduction in rescue medication use in the treatment group. There were significant improvements in asthma control scores and quality of life both during the steroid weaning phase and following steroid reduction in the bronchial thermoplasty group compared to the control group.

The clinical studies thus far have demonstrated that bronchial thermoplasty is safe and appears to have significant improvements in quality of life, peak flow measurements, and the number of symptom-free days in asthma patients. FEV_1 only appears to improve in the severe refractory group. The clinical studies that have thus far been published are open-label randomized studies but a pivotal double-blind sham controlled study is nearing completion with the results due in late 2008. This is a uniquely designed interventional study which attempts to reduce the obvious biases of an open-label study and will more critically evaluate the safety and efficacy of bronchial thermoplasty in moderate and severe asthma patients; until this study is formally reported it seems appropriate to reserve thermoplasty for a research setting.

References

1. Noehren TH, Barach AL, Brantigan OC, Smart RH. Pulmonary emphysema – prevention and care. *Dis Chest* 45: 492–502, 1964.
2. Soni R, McKeough ZJ, Dobbin CJ, Young IH. Gas exchange and exercise tolerance following bullectomy. *Respirology* 10(1): 120–23, 2005.
3. O'Donnell DE, Webb KA, Bertley JC, Chau LK, Conlan AA. Mechanisms of relief of exertional breathlessness following unilateral bullectomy and lung volume reduction surgery in emphysema. *Chest* 110(1): 18–27, 1996.
4. Venn GE, Williams PR, Goldstraw P. Intracavity drainage for bullous, emphysematous lung disease: Experience with the Brompton technique. *Thorax* 43(12): 998–1002, 1988.
5. Cooper J, Patterson G, Sundaresan R, Trulock E, Yusen R, Pohl M *et al.* Results of 150 consecutive bilateral lung volume reduction procedures in patients with severe emphysema. *J Thorac Cardiovasc Surg* 112(5): 1319–30, 1996. The. 1996 1996/11//
6. Fishman A, Martinez F, Naunheim K, Piantadosi S, Wise R, Ries A *et al.* A randomized trial comparing lung-volume-reduction surgery with medical therapy for severe emphysema. *N Engl J Med* 348(21): 2059–73, 2003.
7. Tiong LU, Davies R, Gibson PG, Hensley MJ, Hepworth R, Lasserson TJ *et al.* Lung volume reduction surgery for diffuse emphysema. *Cochrane Database Syst Rev* (4), 2006, CD001001.
8. Criner GJ, Cordova FC, Furukawa S, Kuzma AM, Travaline JM, Leyenson V *et al.* Prospective randomized trial comparing bilateral lung volume reduction surgery to pulmonary rehabilitation in severe chronic obstructive pulmonary disease. *Am J Respir Crit Care Med* 160(6): 2018–27, 1999.
9. Geddes D, Davies M, Koyama H, Hansell D, Pastorino U, Pepper J *et al.* Effect of lung-volume-reduction surgery in patients with severe emphysema. *N Engl J Med* 343(4): 239–45, 2000.
10. Goldstein RS, Todd TR, Guyatt G, Keshavjee S, Dolmage TE, van Rooy S *et al.* Influence of lung volume reduction surgery (LVRS) on health related quality of life in patients with chronic obstructive pulmonary disease. *Thorax* 58(5): 405–10, 2003.
11. Hillerdal G, Lofdahl CG, Strom K, Skoogh BE, Jorfeldt L, Nilsson F *et al.* Comparison of lung volume reduction surgery and physical training on health status and physiologic outcomes: A randomized controlled clinical trial. *Chest* 128(5): 3489–99, 2005.
12. Miller JD, Malthaner RA, Goldsmith CH, Goeree R, Higgins D, Cox PG *et al.* A randomized clinical trial of lung volume reduction surgery versus best medical care for patients with advanced emphysema: A two-year study from Canada. *Ann Thorac Surg* 81(1): 314–20, discussion 20–21, 2006.
13. National Emphysema Treatment Trial Research Group. Patients at high risk of death after lung-volume-reduction surgery. *N Engl J Med* 345(15): 1075–83, 2001.
14. Washko GR, Fan VS, Ramsey SD, Mohsenifar Z, Martinez F, Make BJ *et al.* The effect of lung volume reduction surgery on chronic obstructive pulmonary disease exacerbations. *Am J Respir Crit Care Med* 177(2): 164–69, 2008.
15. Ramsey SD, Berry K, Etzioni R, Kaplan RM, Sullivan SD, Wood DE. Cost effectiveness of lung-volume-reduction surgery for patients with severe emphysema. *N Engl J Med* 348(21): 2092–102, 2003.

16. Lim E, Ali A, Cartwright N, Sousa I, Chetwynd A, Polkey M *et al.* Effect and duration of lung volume reduction surgery: Mid-term results of the Brompton trial. *Thorac Cardiovasc Surg* 54(3): 188–92, 2006.
17. Sabanathan S, Richardson J, Pieri-Davies S. Bronchoscopic lung volume reduction. *J Cardiovasc Surg (Torino)* 44(1): 101–8, 2003.
18. Snell G, Holsworth L, Borrill ZL, Thomson KR, Kalff V, Smith JA *et al.* The potential for bronchoscopic lung volume reduction using bronchial prostheses. *Chest* 124: 1073–80, 2003.
19. Yim AP, Hwong TM, Lee TW, Li WW, Lam S, Yeung TK *et al.* Early results of endoscopic lung volume reduction for emphysema. *J Thorac Cardiovasc Surg* 127(6): 1564–73, 2004.
20. Venuta F, de Giacomo T, Rendina EA, Ciccone AM, Diso D, Perrone A *et al.* Bronchoscopic lung-volume reduction with one-way valves in patients with heterogenous emphysema. *Ann Thorac Surg* 79(2): 411–16, discussion 6–7, 2005.
21. Hopkinson NS, Toma TP, Hansell DM, Goldstraw P, Moxham J, Geddes DM *et al.* Effect of bronchoscopic lung volume reduction on dynamic hyperinflation and exercise in emphysema. *Am J Respir Crit Care Med* 171(5): 453–60, 2005.
22. Toma TP, Hopkinson N, Hillier J, Hansell DM, Morgan C, Goldstraw PG *et al.* Bronchoscopic volume reduction with valve implants in patients with severe emphysema. *Lancet* 361: 931–33, 2003.
23. Wan IY, Toma TP, Geddes DM, Snell G, Williams T, Venuta F *et al.* Bronchoscopic lung volume reduction for end-stage emphysema: Report on the first 98 patients. *Chest* 129(3): 518–26, 2006.
24. Wood DE, McKenna RJ Jr., Yusen RD, Sterman DH, Ost DE, Springmeyer SC *et al.* A multicenter trial of an intrabronchial valve for treatment of severe emphysema. *J Thorac Cardiovasc Surg* 133(1): 65–73, 2007.
25. Ingenito EP, Berger RL, Henderson AC, Reilly JJ, Tsai L, Hoffman A. Bronchoscopic lung volume reduction using tissue engineering principles. *Am J Respir Crit Care Med* 167(5): 771–78, 2003.
26. Reilly J, Washko G, Pinto-Plata V, Velez E, Kenney L, Berger R *et al.* Biological lung volume reduction: A new bronchoscopic therapy for advanced emphysema. *Chest* 131(4): 1108–13, 2007.
27. Strange C, Herth FJ, Kovitz KL, McLennan G, Ernst A, Goldin J *et al.* Design of the Endobronchial Valve for Emphysema Palliation Trial (VENT): A non-surgical method of lung volume reduction. *BMC Pulm Med* 7: 10, 2007.
28. Macklem PT. Collateral ventilation. *N Engl J Med* 298(1): 49–50, 1978.
29. Choong CK, Phan L, Massetti P, Haddad FJ, Martinez C, Roschak E *et al.* Prolongation of patency of airway bypass stents with use of drug-eluting stents. *J Thorac Cardiovasc Surg* 131(1): 60–64, 2006.
30. Lausberg HF, Chino K, Patterson GA, Meyers BF, Toeniskoetter PD, Cooper JD. Bronchial fenestration improves expiratory flow in emphysematous human lungs. *Ann Thorac Surg* 75(2): 393–97, discussion 8, 2003.
31. Choong CK, Macklem PT, Pierce JA, Lefrak SS, Woods JC, Conradi MS *et al.* Transpleural ventilation of explanted human lungs. *Thorax* 62(7): 623–30, 2007.
32. Cardoso PF, Snell GI, Hopkins P, Sybrecht GW, Stamatis G, Ng AW *et al.* Clinical application of airway bypass with paclitaxel-eluting stents: Early results. *J Thorac Cardiovasc Surg* 134(4): 974–81, 2007.
33. Higuchi T, Reed A, Oto T, Holsworth L, Ellis S, Bailey MJ *et al.* Interlobar collaterals are related to radiological heterogeneity in severe emphysema. *Thorax*, 2006.
34. Morrell NW, Wignall BK, Biggs T, Seed WA. Collateral ventilation and gas exchange in emphysema. *Am J Respir Crit Care Med* 150(3): 635–41, 1994.
35. Terry PB, Traystman RJ, Newball HH, Batra G, Menkes HA. Collateral ventilation in man. *N Engl J Med* 298(1): 10–15, 1978.
36. Miller JD, Cox G, Vincic L, Lombard CM, Loomas BE, Danek CJ. A prospective feasibility study of bronchial thermoplasty in the human airway. *Chest* 127(6): 1999–2006, 2005.
37. Cox G, Miller JD, McWilliams A, Fitzgerald JM, Lam S. Bronchial thermoplasty for asthma. *Am J Respir Crit Care Med* 173(9): 965–69, 2006.
38. Cox G, Thomson NC, Rubin AS, Niven RM, Corris PA, Siersted HC *et al.* Asthma control during the year after bronchial thermoplasty. *N Engl J Med* 356(13): 1327–37, 2007.
39. Pavord ID, Cox G, Thomson NC, Rubin AS, Corris PA, Niven RM *et al.* Safety and efficacy of bronchial thermoplasty in symptomatic, severe asthma. *Am J Respir Crit Care Med* 176(12): 1185–91, 2007.

60

Other Therapies

CHAPTER

Neil C. Thomson

Department of Respiratory Medicine, Division of Immunology, Infection and Inflammation, University of Glasgow, Glasgow, UK

VACCINES AND ANTIVIRAL AGENTS

Influenza vaccines

Efficacy

A recent systematic review of trials undertaken to assess the efficacy of influenza vaccination in chronic obstructive pulmonary disease (COPD) patients concluded that inactivated vaccine reduces the rate of exacerbations [1]. In asthma the benefits of inactivated vaccination has not been proven [2]. Administration of live-attenuated influenza vaccine by intranasal spray is effective and safe in both healthy adults and in children with asthma. The cold adapted live vaccine produces superior protection compared to the inactive vaccine in children with asthma [3, 4]. The addition of the live-attenuated vaccine to inactivated vaccine offers no additional protection in COPD [1].

Adverse effects

Vaccination is contraindicated in those individuals hypersensitive to eggs, although the risk of reactions is low. The inactivated influenza vaccines can be safely administered to adults and children with asthma and to patients with COPD without any increase in exacerbation rates in either condition [2, 4, 5]. The live-attenuated influenza vaccine can be safely used in children with asthma [4].

Recommendations

Health authorities in the United States and in many European countries recommend annual inactivated influenza vaccination for adults and children with chronic pulmonary disease including asthma [6]. The Global Initiative for Asthma (GINA) guidelines recommend that patients with moderate to severe asthma should receive influenza vaccination every year [7]. The Global Initiative for Chronic Obstructive Lung Disease (GOLD) guidelines emphasize the importance of targeting elderly individuals with COPD [8]. The live-attenuated influenza vaccine has been licensed recently in the United States for use in healthy persons aged 5–49 years and it has been advocated as the preferred vaccine for use in children with asthma [4].

Antiviral agents against influenza

Neuraminidase inhibitors are a class of antiviral agents that are active against both influenza A and influenza B. The neuraminidase inhibitors, inhaled zanamivir (Relenza) and oral oseltamivir (Tamiflu) have received regulatory approval in some countries for the treatment of acute influenza and for chemoprophylaxis [6]. If these agents are started within the first 2 days after the onset of illness, then treatment will shorten the duration of symptoms by 1 to 1 1/2 days and in some studies the need for antibiotics is reduced. Oseltamivir administered to influenza-infected children with asthma aged 6–12 years is well tolerated and produces improvements in pulmonary function and fewer exacerbation of asthma up to day 7 [9]. Antiviral drugs amantadine and rimantadine are not now recommended for treating the influenza A because of concerns about drug resistance [6]. Antiviral drugs should be considered for patients with asthma and COPD, if started within 2 days of infection, particularly in those who have not been vaccinated. It has also been recommended that chemoprophylaxis should be considered in vaccinated members of high-risk groups, including patients with asthma and COPD, because the combination of vaccination and antiviral drug increases protection against influenza.

Pneumococcal vaccines

Efficacy

In elderly individuals with chronic lung disease including COPD, pneumococcal vaccination has been reported to result in fewer outpatient visits, fewer hospitalizations, and fewer deaths [10]. A Cochrane review of four studies of patients with COPD, however, found no evidence for a reduction in exacerbation rates, morbidity, or mortality following pneumococcal vaccination [11]. Routine revaccination is not generally recommended because of the increased incidence of adverse reactions. A Cochrane review found limited evidence of effectiveness in children and adults with asthma [12].

Adverse effects

Transient erythema at the site of injection is common. Hypersensitivity reactions may occur.

Recommendations

Health authorities in some European countries and the United States recommend polyvalent pneumococcal vaccination for all patients with chronic lung disease [13]. The GOLD guidelines recommend vaccination for patients aged 65 years and older with COPD [8].

Haemophilus influenza vaccine

A systematic review of six small studies of oral immunization with monobacterial whole-cell, killed, nontypeable *H. infuenzae* vaccine in patients with recurrent acute exacerbations of chronic bronchitis in the autumn found that vaccination reduced the number and severity of exacerbations over the following winter [14]. The authors suggested that one episode of bronchitis may be prevented for every five individuals vaccinated. The review concluded however that a large clinical trial into the long-term effects of *H. infuenzae* vaccine was needed. Its routine use is not recommended in the management of asthma or COPD.

Multi-component oral bacterial vaccines

OM-85 BV (Broncho–Vaxom) is a mixture of eight bacterial products thought to be important in respiratory infections. It is administered orally and is thought to act as a nonspecific immunostimulant. A randomized controlled study in 381 patients with COPD treated with OM-85 BV did not reduce the risk of having at least one episode of an acute exacerbation (primary outcome) over a 6-month period compared with placebo. Treatment with OM-85, however, reduced the risk and total number of days of hospitalization for a respiratory problem by 30% and 55% respectively [15]. A systematic review of 13 studies, most of which were assessed to be of low quality, found that oral purified bacterial extracts improve symptoms in patients with chronic bronchitis and COPD, but that there was insufficient evidence to suggest that they prevent exacerbations [16]. Skin and urological adverse effects were commonly reported. Further studies are required before oral vaccines are considered for use in COPD.

ANTIOXIDANTS

Antioxidant enzymes form the first line of defense in the lungs. Antioxidants may reduce chronic airway inflammation by neutralizing the damage produced by reactive oxygen species and also possibly by preserving the inhibitory action of antiproteases. Antioxidant genes, however, may be upregulated by oxidative stress [17] and the dose of administered exogenous antioxidant may need to be titrated to avoid suppression of endogenous antioxidant activity.

Asthma

Dietary antioxidants

Dietary antioxidants such as vitamin C, vitamin E, caratenoids, and selenium are derived mainly from fruit and vegetables. Reduced circulating levels of dietary antioxidants have been found in asthma [18, 19] raising the hope that clinical benefits would result from dietary supplementation with these compounds. However, clinical trials of dietary supplementation with vitamin C, vitamin E, or selenium have shown no benefit in adults with mild-to-moderate asthma [20–22].

Inhaled glutathione

The antioxidant glutathione administered to adults with mild asthma-induced bronchoconstriction [23].

COPD

Oxidative stress is increased in smokers and reactive oxygen species may be involved in the pathogenesis of COPD [24]. There is evidence that endogenous antioxidant capacity is reduced in COPD and so antioxidants may be useful in the treatment of COPD.

Dietary antioxidants

There is some evidence linking dietary deficiency of antioxidants and COPD. There are no controlled studies of vitamins C or E supplementation in COPD. β-Carotene and α-tocopherol supplements had no benefit on the symptoms of COPD [25].

N-Acetylcysteine

N-Acetylcysteine, which is a mucolytic, may also act as an antioxidant by providing cysteine intracellularly for increased production of glutathione [26]. *In vitro* *N*-acetylcysteine reduces neutrophil chemotaxis and in smokers it reduces the number and activity of bronchoalveolar neutrophils and alveolar macrophages. *N*-acetylcysteine preserves the inhibitory action of antiproteases from oxidative inactivation. It has not been established, however, whether any clinical benefits of *N*-acetylcysteine and other mucolytics can be attributed to the antioxidant properties of these drugs (see "Mucolytic Drugs").

MUCOLYTIC DRUGS

Mucolytic drugs such as *N*-acetylcysteine, ambroxol, erdosteine, iodinated glycerol, methylcysteine, and carbocysteine reduce the viscosity of mucous *in vitro* and have been assessed in several clinical trials in COPD. These drugs have not been tested in asthma.

COPD

Efficacy

A Cochrane systematic review of 26 randomized controlled trials of mucolytic drugs including *N*-acetylcysteine in the treatment of chronic bronchitis or COPD found that these drugs produced a small reduction in the frequency of acute exacerbations and the total days of disability compared with placebo [27]. The efficacy and side-effect profile was similar between *N*-acetylcysteine and the other mucolytic drugs.

A further systematic review of 11 randomized trials comparing oral *N*-acetylcysteine with placebo administered from 12 to 24 weeks in patients with chronic bronchitis concluded that active treatment reduced the risk of exacerbations and improved symptoms without an increased risk of side effects [28]. Stey and colleagues [28] estimated that of 100 patients with chronic bronchitis taking *N*-acetylcysteine for 12–24 weeks, 17 would be prevented from having an exacerbation and 26 would note an improvement in symptoms compared with placebo. However, a recent large multicentre Bronchitis Randomized on NAC Cost-Utility Study (BRONCUS) in which patients with COPD were followed for 3 years found that *N*-acetylcysteine 600 mg daily was ineffective at preventing exacerbations and rate of decline in lung function [29]. Subgroup analysis suggested that exacerbation rates were reduced in patients not treated with inhaled corticosteroids.

Adverse effects

The main side effects of mucolytic drugs are gastrointestinal in nature.

Recommendations

The GOLD guidelines do not recommend the use of mucolytic drugs in the management of COPD [8].

CROMONES

Mechanism of action

The cromones, sodium cromoglycate and nedocromil sodium, were first introduced as prophylactic drugs for asthma over 40 and 25 years ago respectively. The mechanism of action of the cromones has not been clearly established. Both drugs act as nonspecific chloride channel blockers in a large range of cell types and through this action, these compounds may reduce alterations in cell volume and function [30]. *In vitro* studies of human inflammatory cells and *in vivo* studies in experimental animals have shown that both drugs inhibit functions of a variety of inflammatory cells including mast cells, eosinophils, neutrophils, platelets, and alveolar macrophages [30–32]. Nedocromil sodium has either similar or slightly greater potency to that of sodium cromoglycate. Both sodium cromoglycate and nedocromil sodium may also have inhibitory effects on sensory nerve endings in the lung, thus preventing the release of tachykinins.

Asthma

Anti-inflammatory effects

There is limited and conflicting evidence for an anti-inflammatory effect of cromones in asthma [33–35].

Clinical studies

Bronchial challenge

Sodium cromoglycate and nedocromil sodium are equally effective in attenuating exercise-induced asthma [36]. The early and late response to allergen and the allergen-induced seasonal increase in nonallergic bronchial reactivity can be prevented by both drugs [31]. Nedocromil sodium can also produce small reductions in nonseasonal bronchial reactivity.

Clinical asthma

Comparison with placebo: In both pediatric and adult asthma, treatment with inhaled sodium cromoglycate has been reported to improve asthma control. A systematic review of 24 randomized controlled trials, however, questioned the efficacy of sodium cromoglycate in children with asthma and concluded that it was no longer justified to recommend sodium cromoglycate as a first-line prophylactic agent in chronic childhood asthma [37]. Some studies comparing nedocromil sodium with placebo have demonstrated benefit in both children and adults with asthma. However, efficacy was not confirmed in a large randomized controlled trial of 1041 children with mild-to-moderate asthma [38]. A small number of clinical trials have compared the two cromones in adults with asthma and in most studies the therapeutic effects were found to be comparable.

Comparison with inhaled steroids: The results of most short-to-medium term studies suggest that the improvement in asthma control produced by the cromones is commonly slightly less than that produced by 400 μg of inhaled beclomethasone daily. A large long-term randomized controlled trial in children aged 5–12 years with mild-to-moderate asthma found that inhaled budesonide 200 μg daily provided better control of asthma and improved airway responsiveness when compared with placebo or nedocromil 8 mg daily [38]. Neither drug was better than placebo in terms of improvements in lung function.

Add-on therapy: The addition of inhaled nedocromil sodium to asthmatic patients with symptoms poorly controlled by high-dose (>1000 μg daily) inhaled corticosteroids produces small improvements in symptoms, peak flow readings, and bronchodilator use [39]. Nedocromil sodium does not have a clinically relevant oral corticosteroids sparing effect [40].

Adverse effects: Sodium cromoglycate and nedocromil sodium are safe drugs and uncommonly cause adverse effects. The main side effects reported include a distinctive bitter taste, headache, and nausea.

Place in management

The cromones have a very limited role in the management of adults and children with chronic asthma. The GINA guidelines recommend that the cromones should be considered as an alternative to an inhaled short acting β_2-agonists to prevent exercise-induced asthma but should not be normally used as controller therapies at step 2 because of their poor efficacy, but should be considered as third-line alternative options to inhaled corticosteroids, leukotriene modifiers, or oral sustained release theophylline [7]. There is no role for the use of the cromones as additional therapy for patients already receiving inhaled or systemic corticosteroids.

COPD

A randomized controlled study in patients with COPD reported that nedocromil sodium treatment for 10 weeks reduced the number of dropouts because of exacerbations but had no effect on symptoms, lung function, or airway responsiveness to histamine and adenosine [41]. A further clinical trial of 12-weeks duration noted limited clinical efficacy of nedocromil sodium in patients with obstructive airways disease and sputum production. Nedocromil sodium treatment was associated with a reduction of plasma protein exudation as measured in sputum sol phase suggesting some anti-inflammatory properties in COPD [42]. The GOLD guidelines do not recommend either sodium cromoglycate or nedocromil sodium for the treatment of COPD due to inadequate data [8].

BRONCHIAL THERMOPLASTY

Bronchial thermoplasty to the airways is a new treatment technique that involves the delivery of radio frequency energy to the airways with the aim of reducing airway responsiveness in asthma. Pre-clinical studies showed that bronchial thermoplasty reduces airway smooth muscle, increases airway size, and produces a long-lasting decrease in airway responsiveness to methacholine [43–46]. In the canine model the reduction in airway smooth muscle is directly related to the decrease in airway responsiveness suggesting that the benificial effects of bronchial thermoplasty may be due to a decrease in airway smooth muscle although other mechanisms may be involved, for example altered contractility of airway smooth muscle or stiffening of the airway wall [43]. In an observational study to determine the safety of bronchial thermoplasty in subjects with mild-to-moderate asthma ($n = 16$), there were no severe adverse events in the 2-year-study period [47]. Although the study was not powered to evaluate efficacy, bronchial thermoplasty resulted in improvements in symptom-free days, morning peak expiratory flow (PEF) at 3 months, and a reduction in airway hyperresponsiveness to methacholine that lasted for at least 2 years [47]. In randomized controlled study, the Asthma Intervention Research (AIR) Trial, the effects of bronchial thermoplasty were evaluated in subjects with moderate or severe asthma ($n = 112$) over12 months [48]. Treatment significantly reduced the mean rate of mild exacerbations and use of rescue medication [48] (Fig. 60.1). Bronchial thermoplasty also improved morning PEF, Asthma Quality of Life Questionnaire (AQLQ) scores, Asthma Control Questionnaire (ACQ) scores, and symptom scores compared with the control group. A *post hoc* analysis suggested that the benefits of treatment were particularly marked in subjects requiring high-maintenance doses of inhaled corticosteroids (>1000 μg of beclomethasone

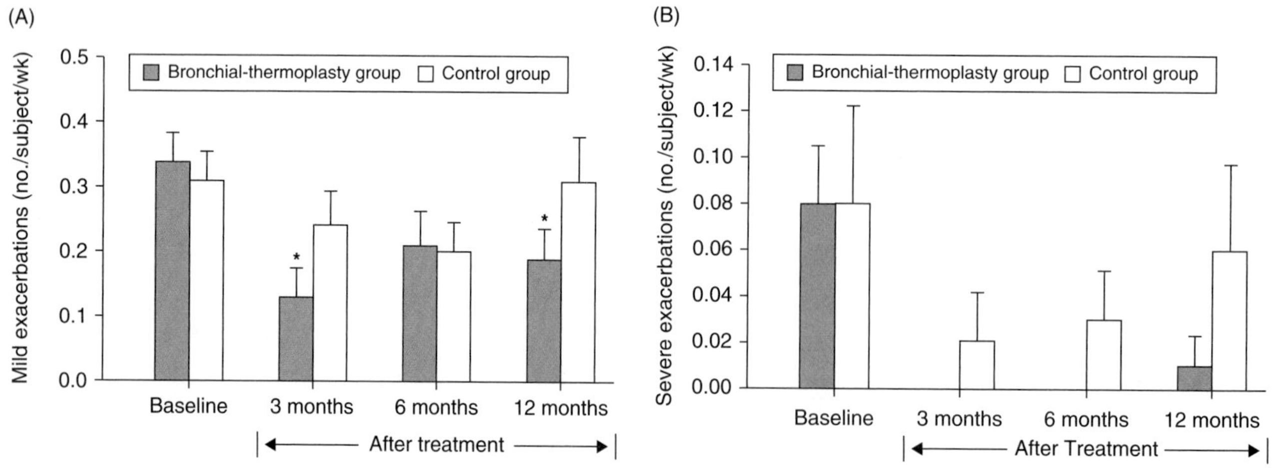

FIG. 60.1 Rates of mild and severe exacerbations per subject per week. Mean values are shown for all subjects receiving inhaled corticosteroids alone for whom data were available at the given time points. Asterisks indicate a statistically significant difference in the mean change from baseline between the two groups, and I bars represent the standard errors. $p = 0.03$ for the comparison between subjects in the two groups treated with inhaled corticosteroids alone at 3 months and at 12 months. Reproduced with permission from Ref. [48].

equivalent daily) [48]. The Research in Severe Asthma (RISA) Trial was designed to determine the safety and efficacy of bronchial thermoplasty in subjects with symptomatic, severe asthma ($n = 32$) [49]. Bronchial thermoplasty was associated with a short-term increase in asthma-related morbidity including increased hospitalizations for asthma. At 22 weeks, bronchial thermoplasty subjects had significant improvements compared to control subjects in rescue medication use, pre-bronchodilator FEV_1% predicted and ACQ scores. Improvements in rescue medication use and ACQ scores remained significantly improved compared to the control group at 52 weeks. The efficacy and safety of bronchial thermoplasty is currently being assessed in a placebo-controlled trial involving the use of sham bronchial thermoplasty.

Taken together, these finding suggest the potential clinical benefits of targeting airway smooth muscle in the treatment of asthma [50]. Future studies need to determine the risk benefit ratio of bronchial thermoplasty in different severities of asthma, to identify factors that predict a beneficial clinical response to bronchial thermoplasty and to investigate whether other less invasive techniques can be used to target airway smooth muscle.

MANAGEMENT OF MALNUTRITION

COPD

Malnutrition (defined as <90% of ideal body weight) is found in 25% of outpatients with COPD and in over 50% of patients admitted to hospital because of COPD. The mechanisms of weight loss are unclear, although changes in energy balance and systemic inflammation are likely to play a part. The mortality rate is increased in malnourished patients with COPD [51]. Malnourished patients with COPD exhibit reduced exercise capacity and respiratory muscle function. No simple recommendation can be given regarding the "best" test for nutritional assessment.

These patients are often unable to regain weight despite receiving nutritional support. High-carbohydrate diets should be avoided to prevent excess carbon dioxide production. A Cochrane review of fourteen nine randomized controlled trials, of which two were double blind, concluded that nutritional support (caloric supplementation for at least 2 weeks) had no effect on improving anthropometric measures, lung function, or exercise capacity among patients with stable COPD [52]. The GOLD guidelines recommend that increased calorie intake is best undertaken with an exercise program [8].

MANAGEMENT OF OBESITY

Asthma

An elevated body mass index has been associated with several features of asthma including an increased prevalence and severity as well as an enhance airway inflammatory response [53–56]. Weight reduction has been shown to improve lung function, symptoms, and health status in obese people with asthma [57, 58] (Fig. 60.2).

COPD

Obesity is also a common feature of COPD and is associated with a poor prognosis, independent of FEV_1. The influence of weight reduction on the morbidity and mortality from COPD has not been systematically studied.

COMPLEMENTARY THERAPIES

Complementary and alternative medicines are frequently used by adults and children with asthma [59, 60]. The

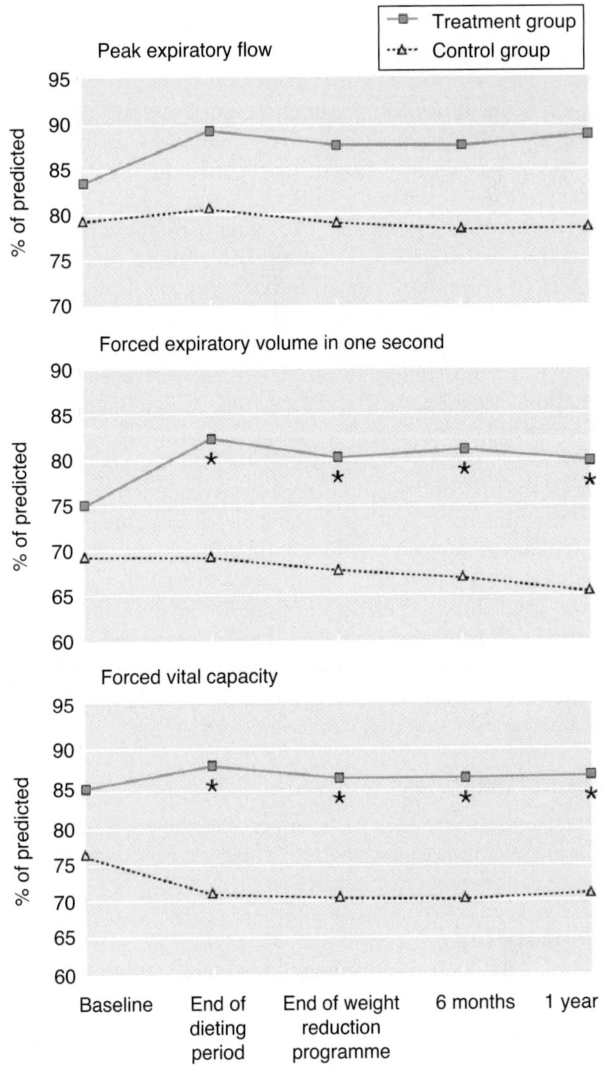

FIG. 60.2 Obesity, weight reduction and asthma: mean morning pre-medication values for PEF, FEV_1, and FVC (% predicted) at different stages during study. Vertical bars show standard errors of the mean. Changes from baseline show significant ($p < 0.05$) difference between groups. Reproduced with permission from Ref. [57].

therapies most frequently used are breathing techniques, homeopathy, herbal medicines, and acupuncture [59, 60]. Complementary therapy use often reflects patients and parents wish to achieve greater self management of their condition [61]. Thirty percent of the general populations in the United States currently use complementary medicine and two-thirds had used at least one complementary and alternative therapy during their lifetime [62].

Asthma

Breathing techniques and inspiratory muscle training

Breathing techniques and inspiratory muscle training have been used to treat asthma. The Buteyko breathing technique is based on the hypothesis that all patients with asthma hyperventilate at rest and that rectifies this problem, by reducing the rate of breathing, results in an improvement in asthma control. The Buteyko breathing technique was compared with a placebo breathing technique in 39 adult asthmatic patients over a 3-month period [63]. At the end of the trial the Buteyko breathing technique group compared with the placebo breathing technique showed reduction in median β-agonist usage. Unfortunately the follow-up arrangements were not identical between the two groups and there was the possibility of greater patient contact in the Buteyko-treated group. Measurements of lung function and quality of life were unaltered in both groups. A larger randomized clinical trial of the Buteyko technique for 6 months reported improved symptoms and reduced bronchodilator use, but no change in bronchial responsiveness or lung function in patients with asthma [64]. A systematic review of breathing exercises for asthma found that although there were trends of improvement in a number of outcomes no firm conclusion can be drawn concerning the use of breathing exercises for asthma in clinical practice [65]. A Cochrane review concluded that there was insufficient evidence to suggest that inspiratory muscle training provides any clinical benefit to patients with asthma [66]. A recent clinical study reported that an integrated breathing and relaxation technique known as the Papworth Method reduced respiratory symptoms, dysfunctional breathing, and adverse mood compared with usual care [67].

Manual therapies

Manual therapies such as massage, chiropractic, and osteopathic techniques are used to treat asthma, but there is insufficient evidence of efficacy [68].

Homeopathy

Homeopathy involves identifying factors that precipitate asthma in an individual and then prescribing a specific homeopathic remedy for that person. In some cases, the remedy may include very small quantities of the allergen to which the patient is allergic, for example. house dust mite. A small randomized controlled parallel group study in 28 patients with allergic asthma compared oral homeopathic immunotherapy to their principal allergen or identical placebo for a 4-week period [69]. All patients continued their unaltered conventional therapy. A daily visual analog scale of overall symptom intensity showed improvements in favor of homeopathic immunotherapy within 1 week of starting treatment which persisted through the treatment period. There were no significant changes in measurement of baseline lung function and bronchial reactivity. A systematic review of placebo-controlled trials of homeopathy have concluded that there is insufficient evidence of efficacy [70].

Acupuncture

Acupuncture involves the insertion of needles at specific points of the body. The technique is based on Chinese theories of the bodies' balance of energies. The effects of acupuncture were assessed in a randomized, double-blind controlled, cross-over study in 22 patients with asthma [71]. There was a statistically significant improvement in assessments of quality of life and reductions in β-agonist use after both the active and sham interventions, but no change in PEF recordings. A Cochrane review of acupuncture for asthma concluded that there was insufficient evidence to make any recommendations about its use, but highlighted the need for well-controlled studies to assess its effectiveness in asthma [72].

Herbal medicines

A large range of herbal medicines including traditional Chinese medicine are used for the treatment of asthma. The efficacy and safety of some of these remedies are now being assessed in controlled trials [73].

Yoga

Yoga is an ancient Hindu discipline that promotes increased mental and physical control of the body. One of the eight steps of yoga, pranayama, deals with control of breathing to promote relaxation and improve fitness. The effects of two pranayama yoga breathing exercises were assessed in a randomized, controlled, cross-over trial in 18 patients with mild asthma [74]. There was a statistically significant reduction in bronchial reactivity to histamine after the active intervention compared with placebo, but no change in baseline lung function tests, symptom scores, or β-agonist use.

Safety of complementary medicine

The safety of complementary therapies is rarely assessed. Although patients often consider these treatments as harmless, this may not always be the case. The health of a patient starting a complementary therapy may be put at risk for several reasons. The patient may abruptly stop their conventional medicines for asthma. Some herbal remedies can cause severe toxic effects due to hepatotoxicity, heavy metal poisoning, or to microorganisms. Herbal medicines may contain conventional pharmaceutical drugs, for example, corticosteroids or nonsteroidal anti-inflammatory drugs. Finally, it is known that acupuncture can cause pneumothoraces and transmit hepatitis.

References

1. Poole P, Chacko E, Wood-Baker R, Cates C. Influenza vaccine for patients with chronic obstructive pulmonary disease. *Cochrane Database Syst Rev*(1), 2006, Art. No.: CD002733. DOI: 10.1002/14651858. CD002733.pub2.

2. Cates C, Jefferson T, Bara A, Rowe B. Vaccines for preventing influenza in people with asthma. *Cochrane Database Syst Rev*(4), 2003, Art. No.: CD000364. DOI: 10.1002/14651858.CD000364.pub2.
3. Fleming D, Crovari P, Wahn U, Klemola T, Schlesinger Y, Langussis A *et al.* Comparison of the efficacy and safety of live attenuated cold-adapted influenza vaccine, trivalent, with trivalent inactivated influenza virus vaccine in children and adolescents with asthma. *Pediatr Infect Dis J* 10: 860–69, 2006.
4. Glezen WP. Asthma, influenza, and vaccination. *J Allergy Clin Immunol* 118(6): 1199–206, 2006.
5. The American Lung Association Asthma Clinical Research Centers. The safety of inactivated influenza vaccine in adults and children with asthma. *N Engl J Med* 345(21): 1529–36, 2001.
6. Fiore A, Shay D, Haber P, Iskander J, Uyeki T, Mootrey G *et al.* Prevention and control of influenza: Recommendations of the advisory committee on immunization practices (ACIP). *Morb Mortal Wkly Rep* 56(RR06): 1–54, 2007.
7. GINA Report, Global Strategy for Asthma Management and Prevention. 2006 [cited December 12, 2007]; Available from: http://www.ginasthma.com/Guideline
8. Global Initiative for Chronic Obstructive Lung Diseases. Workshop Report: Guidelines global strategy for the diagnosis, management and prevention of chronic obstructive pulmonary disease. 2006 [cited May 25, 2007]; Available from: http://www.goldcopd.com
9. Johnston S, Ferrero F, Garcia M, Dutkowski R. Oral oseltamivir improves pulmonary function and reduces exacerbation frequency for influenza-infected children with asthma. *Pediatr Infect Dis J* 24(3): 225–32, 2005.
10. Nichol KL, Baken L, Wuorenma J, Nelson A. The health and economic benefits associated with pneumococcal vaccination of elderly persons with chronic lung disease. *Arch Intern Med* 159(20): 2437–42, 1999.
11. Granger R, Walters J, Poole P, Lasserson T, Mangtani P, Cates C *et al.* Injectable vaccines for preventing pneumococcal infection in patients with chronic obstructive pulmonary disease. *Cochrane Database Syst Rev* (4), 2006, Art. No.: CD001390. DOI: 10.1002/14651858.CD001390.pub2
12. Sheikh A, Alves B, Dhami S. Pneumococcal vaccine for asthma. *Cochrane Database Syst Rev* (3), 2001, CD002165.
13. Prevention of pneumococcal disease: Recommendations of the Advisory Committee on Immunization Practices (ACIP). *Morb Mortal Wkly Rep* 46: 1–24, 1997.
14. Foxwell A, Cripps A, Dear K. *Haemophilus influenzae* oral whole cell vaccination for preventing acute exacerbations of chronic bronchitis. *Cochrane Database Syst Rev* (4), 2006, DOI: 10.1002/14651858.CD001958.pub2.
15. Collet JP, Shapiro S, Ernst P, Renzi P, Ducruet T, Robinson A *et al.* Effects of an immunostimulating agent on acute exacerbations and hospitalizations in patients with chronic obstructive pulmonary disease. *Am J Respir Crit Care Med* 156(6): 1719–24, 1997.
16. Steurer-Stey C, Bachmann LM, Steurer J, Tramer MR. Oral purified bacterial extracts in chronic bronchitis and COPD: Systematic review. *Chest* 126(5): 1645–55, 2004.
17. Rahman I, MacNee W. Oxidative stress and regulation of glutathione in lung inflammation. *Eur Respir J* 16(3): 534–54, 2000.
18. Devereux G, Seaton A. Diet as a risk factor for atopy and asthma. *J Allergy Clin Immunol* 115(6): 1109–17, 2005.
19. Patel BD, Welch AA, Bingham SA, Luben RN, Day NE, Khaw KT *et al.* Dietary antioxidants and asthma in adults. *Thorax* 61(5): 388–93, 2006.
20. Pearson PJK, Lewis SA, Britton J, Fogarty A. Vitamin E supplements in asthma: A parallel group randomised placebo controlled trial. *Thorax* 59(8): 652–56, 2004.
21. Shaheen SO, Newson RB, Rayman MP, Wong APL, Tumilty MK, Phillips JM *et al.* Randomised, double blind, placebo-controlled trial of selenium supplementation in adult asthma. *Thorax* 62(6): 483–90, 2007.
22. Ram F, Rowe B, Kaur B. Vitamin C supplementation for asthma. *Cochrane Database Syst Rev*, 2004, DOI: 10.1002/14651858.CD000993.pub2.
23. Marrades Ramon M, Roca J, Barbera Joan A, de Jover L, Macnee W, Rodriguez-Roisin R. Nebulized glutathione induces bronchoconstriction in patients with mild asthma. *Am J Respir Crit Care Med* 156(2): 425–30, 1997.
24. Kirkham P, Rahman I. Oxidative stress in asthma and COPD: Antioxidants as a therapeutic strategy. *Pharmacol Therap* 111(2): 476–94, 2006.
25. Rautalahti M, Virtamo J, Haukka J, Heinonen O, Sundvall J, Albanes D *et al.* The effect of alpha-tocopherol and beta-carotene supplementation on COPD symptoms. *Am J Respir Crit Care Med* 156(5): 1447–52, 1997.
26. Aitio M-L. *N*-Acetylcysteine-passe-partout or much ado about nothing?. *Br J Clin Pharmacol* 61(1): 5–15, 2006.
27. Poole P, Black P. Mucolytic agents for chronic bronchitis or chronic obstructive pulmonary disease. *Cochrane Database Syst Rev*, 2006, DOI: 10.1002/14651858.CD001287.pub2.
28. Stey C, Steurer J, Bachmann S, Medici TC, Tramer MR. The effect of oral *N*-acetylcysteine in chronic bronchitis: A quantitative systematic review. *Eur Respir J* 16(2): 253–62, 2000.
29. Decramer M, Rutten-van Molken M, Dekhuijzen PNR, Troosters T, van Herwaarden C, Pellegrino R *et al.* Effects of *N*-acetylcysteine on outcomes in chronic obstructive pulmonary disease (Bronchitis Randomized on NAC Cost-Utility Study, BRONCUS): A randomised placebo-controlled trial. *Lancet* 365(9470): 1552–60, 2005.
30. Norris A, Alton E. Chloride transport and the action of sodium cromoglycate and nedocromil sodium in asthma. *Clin Exp Allergy* 26(3): 250–53, 1996.
31. Thomson N. Nedocromil sodium: An overview. *Respir Med* 83(4): 269–76, 1989.
32. Corin RE. Nedocromil sodium: A review of the evidence for a dual mechanism of action. *Clin Exp Allergy* 30(4): 461–68, 2000.
33. Diaz P, Galleguillos FR, Gonzalez MC, Pantin CFA, Kay AB. Bronchoalveolar lavage in asthma: The effect of disodium cromoglycate (cromolyn) on leukocyte counts, immunoglobulins, and complement. *J Allergy Clin Immunol* 74(1): 41–48, 1984.
34. Hoshino M, Nakamura Y. The effect of inhaled sodium cromoglycate on cellular infiltration into the bronchial mucosa and the expression of adhesion molecules in asthmatics. *J Allergy Clin Immunol* 10(4): 858–65, 1997.
35. Manolitsas N, Wang J, Devalia J, Trigg C, McAulay A, Davies RJ. Regular albuterol, nedocromil sodium, and bronchial inflammation in asthma. *Am J Respir Crit Care Med* 151(6): 1925–30, 1995.
36. Kelly KD, Spooner CH, Rowe BH. Nedocromil sodium versus sodium cromoglycate in treatment of exercise-induced bronchoconstriction: A systematic review. *Eur Respir J* 17(1): 39–45, 2001.
37. Tasche MJA, Uijen JHJM, Bernsen RMD, de Jongste JC, van der Wouden JC. Inhaled disodium cromoglycate (DSCG) as maintenance therapy in children with asthma: A systematic review. *Thorax* 55(11): 913–20, 2000.
38. The Childhood Asthma Management Program Research Group. Long-term effects of budesonide or nedocromil in children with asthma. *N Engl J Med* 343(15): 1504–63, 2000.
39. Svendsen U, Jørgensen H. Inhaled nedocromil sodium as additional treatment to high dose inhaled corticosteroids in the management of bronchial asthma. *Eur Respir J* 4(8): 992–99, 1991.
40. Goldin J, Bateman E. Does nedocromil sodium have a steroid sparing effect in adult asthmatic patients requiring maintenance oral corticosteroids? *Thorax* 43(12): 982–86, 1988.
41. de Jong J, Postma D, van der Mark T, Koëter G. Effects of nedocromil sodium in the treatment of non-allergic subjects with chronic obstructive pulmonary disease. *Thorax* 49(10): 1022–24, 1994.
42. Schoonbrood DF, Out TA, Hart AA, Habets FJ, Roos CM, Jansen HM. Nedocromil sodium in obstructive airways disease: Effect on symptoms and plasma protein leakage in sputum. *Eur Respir J* 10(7): 1500–6, 1997.
43. Danek CJ, Lombard CM, Dungworth DL, Cox PG, Miller JD, Biggs MJ *et al.* Reduction in airway hyperresponsiveness to methacholine by the application of RF energy in dogs. *J Appl Physiol* 97(5): 1946–53, 2004.

44. Miller JD, Cox G, Vincic L, Lombard CM, Loomas BE, Danek CJ. A prospective feasibility study of bronchial thermoplasty in the human airway. *Chest* 127(6): 1999–2006, 2005.
45. Brown RH, Wizeman W, Danek C, Mitzner W. *In vivo* evaluation of the effectiveness of bronchial thermoplasty with computed tomography. *J Appl Physiol* 98(5): 1603–6, 2005.
46. Brown RH, Wizeman W, Danek C, Mitzner W. Effect of bronchial thermoplasty on airway distensibility. *Eur Respir J* 26(2): 277–82, 2005.
47. Cox G, Miller JD, McWilliams A, FitzGerald JM, Lam S. Bronchial thermoplasty for asthma. *Am J Respir Crit Care Med* 173(9): 965–69, 2006.
48. Cox G, Thomson NC, Rubin AS, Niven RM, Corris PA, Siersted HC *et al.* Asthma control during the year after bronchial thermoplasty. *N Engl J Med* 356(13): 1327–37, 2007.
49. Pavord ID, Cox G, Thomson NC, Rubin AS, Corris PA, Niven RM *et al.* Safety and efficacy of bronchial thermoplasty in symptomatic, severe asthma. *Am J Respir Crit Care Med* 176(12): 1185–91, 2007.
50. Solway J, Irvin CG. Airway smooth muscle as a target for asthma therapy. *N Engl J Med* 356(13): 1367–69, 2007.
51. Landbo C, Prescott E, Lange P, Vestbo J, Almda IT. Prognostic value of nutritional status in chronic obstructive pulmonary disease. *Am J Respir Crit Care Med* 160(6): 1856–61, 1999.
52. Ferreira I, Brooks D, Lacasse Y, Goldstein R, White J. Nutritional supplementation for stable chronic obstructive pulmonary disease. *Cochrane Database Syst Rev*(2), 2005, 10.1002/14651858.CD000998.pub2.
53. Beuther DA, Weiss ST, Sutherland ER. Obesity and asthma. *Am J Respir Crit Care Med* 174(2): 112–19, 2006.
54. Peters-Golden M, Swern A, Bird SS, Hustad CM, Grant E, Edelman JM. Influence of body mass index on the response to asthma controller agents. *Eur Respir J* 27(3): 495–503, 2006.
55. Shore S. Obesity and asthma: Implications for treatment. *Curr Opin Pulm Med* 13(1): 56–62, 2007.
56. Taylor B, Mannino D, Brown C, Crocker D, Twum-Baah N, Holguin F. Body mass index and asthma severity in the national asthma survey. *Thorax* 63(1): 14–20, 2008.
57. Stenius-Aarniala B, Poussa T, Kvarnstrom J, Gronlund E-L, Ylikahri M, Mustajoki P. Immediate and long term effects of weight reduction in obese people with asthma: Randomised controlled study. *BMJ* 320(7238): 827–32, 2000.
58. Aaron SD, Fergusson D, Dent R, Chen Y, Vandemheen KL, Dales RE. Effect of weight reduction on respiratory function and airway reactivity in obese women. *Chest* 125(6): 2046–52, 2004.
59. Ernst E. Complementary therapies for asthma: What patients use. *J Asthma* 38(8): 667–71, 1998.
60. Slader CA, Reddel HK, Jenkins CR, Armour CL, Bosnic-Anticevich SZ. Complementary and alternative medicine use in asthma: Who is using what? *Respirology* 11(4): 373–87, 2006.
61. Shaw A, Thompson E, Sharp D. Complementary therapy use by patients and parents of children with asthma and the implications for NHS care: A qualitative study. *BMC Health Serv Res* 6: 76, 2006.
62. Kessler RC, Davis RB, Foster DF, Van Rompay MI, Walters EE, Wilkey SA *et al.* Long-term trends in the use of complementary and alternative medical therapies in the United States. *Ann Intern Med* 135(4): 262–68, 2001.
63. Bowler SDGA, Mitchell CA. Buteyko breathing techniques in asthma: A blinded randomised controlled trial. *Med J* 169: 575–78, 1998.
64. Cooper S, Oborne J, Newton S, Harrison V, Thompson Coon J, Lewis S *et al.* Effect of two breathing exercises (Buteyko and pranayama) in asthma: A randomised controlled trial. *Thorax* 58(8): 674–79, 2003.
65. Holloway E, Ram F. Breathing exercises for asthma. *Cochrane Database Syst Rev*(1), 2004, DOI: 10.1002/14651858.CD001277.pub2.
66. Ram F, Wellington S, Barnes NC. Inspiratory muscle training for asthma. *Cochrane Database Syst Rev*(3), 2003, DOI: 10.1002/14651858.CD003792.
67. Holloway EA, West RJ. Integrated breathing and relaxation training (the Papworth method) for adults with asthma in primary care: A randomised controlled trial. *Thorax* 62(12): 1039–42, 2007.
68. Hondras M, Linde K, Jones A. Manual therapy for asthma. *Cochrane Database Syst Rev*(2), 2005. , DOI: 10.1002/14651858.CD001002.pub2
69. Reilly D, Taylor M, Beattie N, Campbell J, McSharry C, Aitchison T *et al.* Is evidence for homoeopathy reproducible? *Lancet* 344: 1601–6, 1994.
70. Altunc U, Pittler MH, Ernst E. Homeopathy for childhood and adolescence ailments: Systematic review of randomized clinical trials. *Mayo Clinic Proceed* 82(1): 69–75, 2007.
71. Biernacki W, Peake MD. Acupuncture in treatment of stable asthma. *Respir Med* 92(9): 1143–45, 1998.
72. McCarney R, Brinkhaus B, Lasserson T, Linde K. Acupuncture for chronic asthma. *Cochrane Database Syst Rev*(3), 2003, CD000008. DOI: 10.1002/14651858.CD000008.pub2.
73. Li X-M. Traditional Chinese herbal remedies for asthma and food allergy. *J Allergy Clin Immunol* 120(1): 25–31, 2007.
74. Singh V, Wisniewski A, Britton J, Tattersfield A. Effect of yoga breathing exercises (pranayama) on airway reactivity in subjects with asthma. *Lancet* 335: 1381–83, 1990.

61

CHAPTER

Future Therapies

Peter J. Barnes

National Heart and Lung Institute (NHLI), Clinical Studies Unit, Imperial College, London, UK

INTRODUCTION

As indicated in previous chapters, asthma and chronic obstructive pulmonary disease (COPD) are different diseases that involve different cells, mediators and inflammatory effects so that different treatments are needed. However, it is increasingly recognized that there are any similarities in the inflammatory process between severe asthma and COPD [1]. Many new classes of drugs are now in development for asthma and COPD and there are several new drugs that are being developed for both diseases [2–4].

THE NEED FOR NEW TREATMENTS

Asthma

Current asthma therapy is highly effective and majority of patients can be well controlled with inhaled corticosteroids and short- and long-acting β_2-agonists. These treatments are not only effective, but safe and relatively inexpensive. This poses a challenge for the development of new treatments, since they will need to be safer or more effective than existing treatments, or offer some other advantage in long-term disease management. However, there are several problems with existing therapies.

- Existing therapies have side effects as they are non-specific. Inhaled β_2-agonists may have side effects and there is some evidence for the development of tolerance, especially to their bronchoprotective effects. Inhaled corticosteroids may also have local and systemic side effect at high doses and there is still a fear of using long-term steroid treatment in many patients. Other treatments, such as theophylline, anticholinergics and anti-leukotrienes are less effective and are largely used as add-on therapies.
- There is still a major problem with poor compliance in the long-term management of asthma, particularly as symptoms come under control with effective therapies [5]. It is likely that a once daily tablet or even an infrequent injection may give improved compliance. However, oral therapy is associated with a much greater risk of systemic side effects and therefore needs to be specific for the abnormality in asthma or COPD.
- Patients with severe asthma are often not controlled on maximal doses of inhaled therapies or may have serious side effects from therapy, especially oral corticosteroids. These patients are relatively resistant to the anti-inflammatory actions of corticosteroids and require some other class of therapy to control the asthmatic process. Corticosteroid resistance is also a major barrier to effective therapy in COPD.
- None of the existing treatments for asthma is disease modifying, which means that the disease recurs as soon as treatment is discontinued.
- None of the existing treatments is curative, although it is possible that therapies which prevent the immune aberration of allergy may have the prospects for a cure in the future.

COPD

In sharp contrast to asthma there are few effective therapies in COPD, despite the fact that it is a common disease that is increasing worldwide [6].

The relative neglect of COPD by pharmaceutical companies is probably as a result of several factors:

- COPD is regarded as largely irreversible and is treated as poorly responsive asthma.
- COPD is self-inflicted and therefore does not deserve substantial investment.
- Animal models do not mimic many of the key aspects of the disease, such as small airway disease and exacerbations.
- Relatively little is understood about the cell and molecular biology of this disease or even about the relative role of small airways disease and parenchymal destruction.

None of the treatments currently available prevents the progression of the disease, and yet the disease is associated with an active inflammatory process that results in progressive obstruction of small airways and destruction of lung parenchyma. Increased understanding of COPD will identify novel targets for future therapy (Fig. 61.1) [7].

There are also uncertainties about how to test drugs for COPD, which may require long-term studies (3 years) in relatively large numbers of patients at an enormous cost. For example, a recent study looking at the effects of drug intervention on mortality costs several hundred million dollars [8]. Many patients with COPD may have co-morbidities, such as ischaemic heart disease and diabetes, which may exclude them from clinical trials of new therapies. There is little information about surrogate markers, for example biomarkers in blood, sputum or breath, to monitor the short-term efficacy and predict the long-term potential of new treatments [9]. Finally, it is difficult to accurately measure small airway function in patients with COPD so there is a need to develop better tests of small airway function that are not affected by emphysema or abnormalities of large airway function [10].

Although asthma and COPD involve different patterns of inflammation, there is increasing evidence that patients with severe asthma have inflammatory features that are very similar to those seen in COPD [1]. This suggests that drugs that are effective in suppressing the inflammation of COPD might also be effective in treating severe asthma.

DEVELOPMENT OF NEW THERAPIES

Several strategies have been adopted in the search for new therapies:

- *Improvement of an existing class of drug*: This is well exemplified by the increased duration of β_2-agonists with salmeterol and formoterol and of anticholinergics with tiotropium bromide, and with the improved

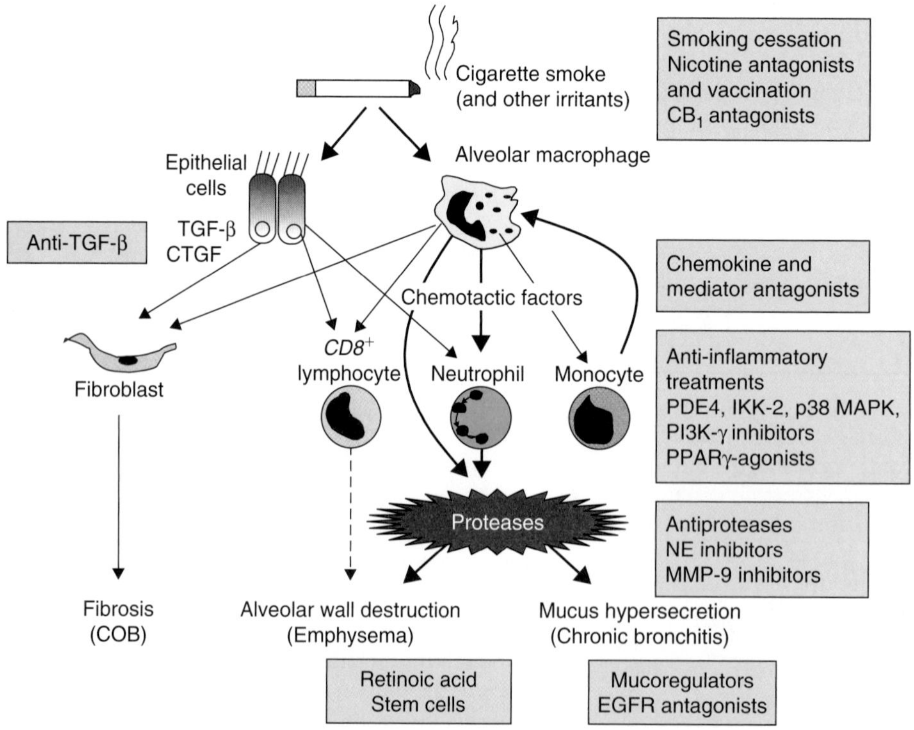

FIG. 61.1 Targets for COPD therapy based on current understanding of the inflammatory mechanisms. Cigarette smoke (and other irritants) activate macrophages in the respiratory tract that release multiple chemotactic factors that attract neutrophils, monocytes, and T-lymphocytes (particularly CD8$^+$ cells). Several cells also release proteases, such as neutrophil elastase (NE) and matrix metalloproteinase-9 (MMP-9) which the break down connective tissue in the lung parenchyma (emphysema) and also stimulate mucus hypersecretion (chronic bronchitis). CD8$^+$ may also be involved in alveolar wall destruction. This inflammatory process may be inhibited at several stages (shown in the boxes). PDE: phosphodiesterase; IKK: inhibitor of nuclear factor-κB kinase; MAPK: mitogen-activated protein kinase; PI3K: phosphoinositide-3-kinase; PPAR: peroxisome proliferator activated receptor; COB: chronic obstructive bronchitis; TGF: transforming growth factor; CB: cannabinoid; EGFR: epithelial growth factor receptor.

pharmacokinetic of the inhaled corticosteroids fluticasone propionate, mometasone, ciclesonide and budesonide, with increased first-pass (hepatic) metabolism and therefore reduced systemic absorption.

- *Development of novel therapies through better understanding of the disease process*: Examples are anti-interleukin(IL)-5 as a potential treatment of asthma and phosphodiesterase (PDE)-4 inhibitors as an anti-inflammatory therapy for COPD.
- *Serendipitous observations, often made in other therapeutic areas*: Examples are tumor necrosis factor-α (TNF-α) antagonists for airway diseases, derived from observations in other chronic inflammatory diseases.
- *Identification of novel targets through gene and protein profiling*: This approach will be increasingly used to identify the abnormal expression of genes (molecular genomics) and proteins (proteomics) from diseased cells, and through identification of single nucleotide polymorphisms (SNPs) that contribute to the disease process [11]. So far there have been no new drugs developed from this approach in any area of medicine.

NEW BRONCHODILATORS

Bronchodilators act predominantly by relaxation of airway smooth muscle cells, although additional effects, such as inhibition of bronchoconstrictor mediator and neurotransmitter release may contribute. β_2-Agonists are by far the most effective bronchodilators in asthma and act as functional antagonists, blocking the effect of all bronchoconstrictors. The increased bronchodilator duration of long-acting inhaled β_2-agonists have been an important advance and there are now several once daily inhaled β_2-agonists in development. By contrast, in COPD anticholinergics appear to be the most effective bronchodilators and there has been a search for selective anticholinergics and drugs with a longer duration. Novel classes of bronchodilator have also been developed (Table 61.1.)

Ultra-LABAs

Several once daily inhaled β_2-agonists, such as indacaterol and carmoterol, are now in clinical development [12].

TABLE 61.1 New bronchodilators.

Drug class	Examples
Ultra-long acting β_2-agonists	Indacaterol, carmoterol
Long-acting muscarinic antagonists	Glycopyrrolate, aclidinium
Combination inhalers	LABA + LAMA, MABA
Rho kinase inhibitors	In development
Myosin light chain kinase inhibitors	In development

Indacaterol is an effective once daily bronchodilator in asthma and appears to be well tolerated with no evidence for the development of tolerance [13]. Indacaterol is a very effective dilator of small human airways measured by videomicroscopy in a precision-cut lung slice preparation, indicating that it will be useful in severe asthma and COPD where small airway obstruction is a problem [14].

New anticholinergics

The once daily inhaled anticholinergic tiotropium bromide has been an important advancement in therapy of COPD and several other long-acting inhaled muscarinic antagonists (LAMA), such as aclidinium bromide and glycopyrrolate are now in development [15, 16]. Combination inhalers with a long-acting β_2-agonist (LABA) with a LAMA are also in development as there is an additive effect between these two bronchodilator classes [17]. Single molecules that link a muscarinic antagonist to a β_2-agonist (MABA) are also in development.

New classes of bronchodilator

It has proved difficult to discover novel classes of bronchodilator drug. Potassium channel openers, while effective in human airways *in vitro*, were not effective in asthma as they were more potent as vasodilators and this limited the dose that could be administered. There has been interest in developing drugs that inhibit the contractile machinery in airway smooth muscle, including Rho kinase inhibitors, inhibitors of myosin light chain kinase and direct smooth muscle myosin inhibitors. As these agents also cause vasodilatation it will be necessary to administer them by inhalation.

MEDIATOR ANTAGONISTS

Blocking the receptors or synthesis of inflammatory mediators is a logical approach to the development of new treatments for asthma and COPD. Several mediator antagonists have now been tested in asthma and COPD (see Chapter 52). However, in both diseases many different mediators are involved and therefore blocking a single mediator may not be very effective, unless it plays a key role in the disease process [18, 19]. Several specific mediator antagonists have been found to be ineffective in asthma, including antagonists/inhibitors of histamine, thromboxane, platelet-activating factor, bradykinin and tachykinins. There is now increasing evidence that this approach is equally ineffective in COPD, with no clinical evidence of benefit for inhibitors of TNF-α, IL-8 or leukotriene B_4. However, it is possible that there may be an upstream mediator that is playing a disproportionate role in the pathogenesis of asthma and COPD so that targeting this mediator may give greater than expected clinical benefits, as is the case for antihistamines in rhinitis and TNF-α inhibitors in rheumatoid arthritis. In asthma there is hope for blocking IL-13 as this cytokine appears to mimic many of the pathophysiological features of asthma. A mutated protein that blocks the common receptor for IL-4 and IL-13 pitrakinra has shown some inhibitory effects on the late response to allergen challenge in patients with mild asthma [20]. The cytokine thymic stomal lymphoietin (TSLP), which is secreted by epithelial cells and mast cells in asthmatic patients, appears to play a critical role in the programing of dendritic cells to orchestrate Th2 cells and allergic inflammation and appears to be a promising target for inhibition in asthma [21].

Chemokines play a key role in attracting inflammatory cells into the lung and act on G-protein-coupled receptors expressed on these cells. Small molecule inhibitors of chemokines receptors are a promising approach in asthma and COPD [22, 23]. Front-runners for asthma are CCR3 expressed predominantly on eosinophils and CCR4 expressed on Th2 cells and for COPD are CXCR2 expressed on neutrophils and monocytes and CXCR3 expressed on T-lymphocytes. Chemokine antagonists have the advantage of oral administration but so far there have been unexpected toxicological problems with many of the small molecule antagonists developed. A dual CXCR1/2 antagonist given orally has been shown to reduce neutrophil and macrophage influx into sputum after endotoxin challenge in normal patients after oral administration [24]. A CCR5 antagonist maraviroc has been approved for use in HIV/AIDS but may also be useful in COPD and these receptors are expressed in T-cells and their agonist CCL5 (RANTES) is increased in COPD sputum [25] (Fig. 61.2).

Anti-oxidants

Oxidative stress is important in severe asthma and COPD, particularly during exacerbations [26]. Oxidative stress activates the proinflammatory transcription factors NF-κB and activator protein-1 (AP-1), resulting in enhanced inflammation. Oxidative stress is increased in patients with COPD, particularly during exacerbations, and reactive oxygen species contribute to its pathophysiology. Oxidative stress reduces steroid responsiveness via a reduction in histone deacetylase-2 (HDAC2) activity and expression [27]. This suggests that antioxidants may reverse corticosteroid resistance and also reduce inflammation. Unfortunately currently available antioxidants based on glutathione are relatively weak and are inactivated by oxidative stress, so new more potent and stable antioxidants are needed, such as superoxide dismutase mimics and NADPH oxidase inhibitors [28].

PROTEASE INHIBITORS

Proteases are involved in the pathophysiology of asthma and COPD and are therefore a logical target for inhibition. However, it has proved very difficult to find drugs that effectively block specific proteases without side effects.

Tryptase inhibitors

Mast cell tryptase has several effects on airways, including increasing responsiveness of airway smooth muscle to constrictors, increasing plasma exudation, potentiating eosinophil recruitment, and stimulating fibroblast proliferation [29]. Some of these effects are mediated by activation of the proteinase-activated receptor PAR2. A tryptase inhibitor APC366 was effective in a sheep model of allergen-induced asthma [30], but was only poorly effective in asthmatic patients in a preliminary study [31] and ineffective in allergic inflammation [32]. More potent tryptase inhibitors and PAR2 antagonists are in development.

Neutrophil elastase inhibitors

Neutrophil elastase (NE), a neutral serine protease, is a major constituent of lung elastolytic activity. In addition it potently stimulates mucus secretion and induces CXCL8 release from epithelial cells and may therefore perpetuate the inflammatory state. This has lead to a search for NE inhibitors. Peptide NE inhibitors, such as ICI 200355, and non-peptide inhibitors, such as ONO-5046, have been developed for use in COPD, but so far all of these drugs have been withdrawn due to toxicological problems or side effects. The NE inhibitor MR889 administered for 4 weeks

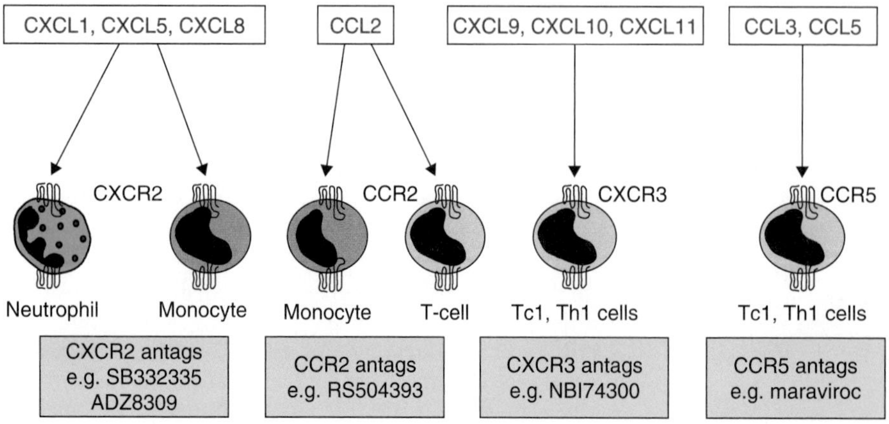

FIG. 61.2 Several chemokines and chemokine receptors are involved in the inflammation of COPD. Chemokines released from epithelial cells and macrophages in the lung recruit inflammatory cells (Tc1 CD8$^+$ T-lymphocytes, neutrophils, and monocytes) from the circulation. Small molecule chemokine receptor antagonists are now in development (shown in boxes).

showed no overall effect on plasma elastic-derived peptides or urinary desmosine (markers of elastolytic activity) [33]. Although NE is likely to be the major mechanism mediating elastolysis in patients with α_1-antitrypsin (α_1-AT) deficiency, it may well not be the major elastolytic enzyme in smoking-related COPD, and it is important to consider other enzymes as targets for inhibition.

Matrix metalloproteinase inhibitors

Matrix metalloproteinases (MMPs) with elastolytic activity appear to play a more important role in the pathogenesis of emphysema than NE so has become a target for drug development. MMP-9 appears to be the predominant enzyme, which is released from macrophages, neutrophils and epithelial cells. Non-selective MMP inhibitors, such as marimastat, appear to have major side effects [34], suggesting that isoenzyme-selective inhibitors or inhaled delivery may be needed. A dual MMP9/MMP12 inhibitor, AZ11557272, has been shown to prevent emphysema and small airway thickening in guinea pigs exposed to cigarette smoke over 6 months, but the clinical development of this inhibitor has been stopped for toxicological reasons [35]. Several other selective MMP inhibitors are now in development [34].

NEW ANTI-INFLAMMATORY DRUGS

Inhaled corticosteroids are by far the most effective therapy for asthma, yet are ineffective in COPD. Thus for asthma, one strategy has been to develop safer inhaled corticosteroids or drugs that mimic their effects, whereas in COPD new non-steroidal anti-inflammatory treatments are needed (Table 61.2). Several novel anti-inflammatory treatments inhibit enzymes that are involved in the signal transduction pathways involved in transcription of multiple inflammatory genes and therefore may act in a similar way to corticosteroids (Fig. 61.3).

New corticosteroids

Inhaled corticosteroids are by far the most effective anti-inflammatory therapy for asthma and work in almost every patient [36]. However, all currently available inhaled corticosteroids are absorbed from the lungs and therefore have the potential for systemic effects. This has led to a concerted effort to find safer corticosteroids, with reduced oral bioavailability, reduced absorption from the lungs or inactivation in the circulation. Ciclesonide, a newly introduced steroid, is a prodrug that becomes activated to the active des-ciclesonide by esterases in the lung. This corticosteroid appears to have less systemic effects than currently available corticosteroids and this may be due to long-term retention in the lung, no oral bioavailability, and a high degree of binding to circulating proteins [37]. Another approach is to develop dissociated steroids that separate the side effect mechanisms from the anti-inflammatory mechanisms. This is theoretically possible as side effects are largely mediated via genomic effects and the binding of glucocorticoid receptors to DNA, whereas anti-inflammatory effects are largely

TABLE 61.2 New anti-inflammatory drugs for asthma and COPD.

Drug class	Example
Phosphodiesterase-4 inhibitors	Roflumilast
NF-κB inhibitors	PS1145
p38 MAP kinase inhibitors	SB203580, SD-208
Phosphoinositide 3-kinase-γ/-δ inhibitors	In development
PPAR-γ agonists	Rosiglitazone, pioglitazone
Syk kinase inhibitors	R112
Adhesion molecule inhibitors	anti-ICAM-1, E-selectin inhibitors, VLA-4 inhibitors

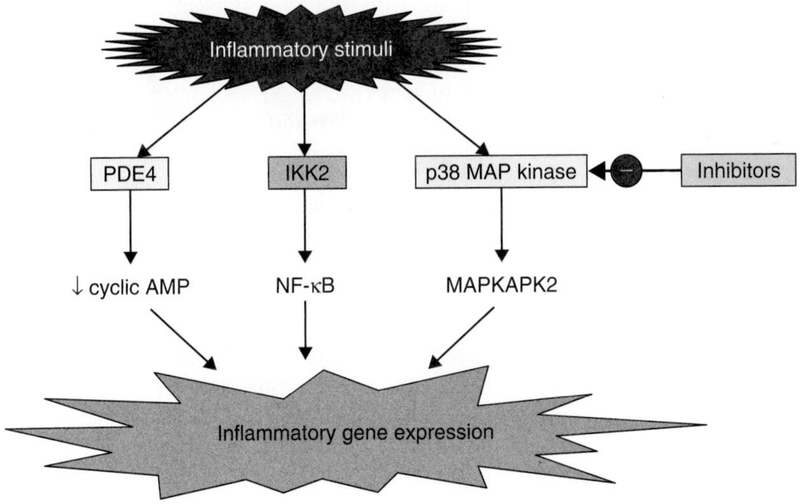

FIG. 61.3 Inhibition of signal transduction pathways. Selective inhibitors have been developed for phosphodieatease-4 (PDE4), which degrades cyclic adenosine monophosphate (AMP), inhibitor of inhibitor of NF-κB kinase (IKK2), which activates NF-κB, and p39 mitogen-activated protein (MAP) kinase, which activates MAP kinase activated protein kinase 2 (MAPKAPK2). Selective inhibitors have now been developed for these enzyme targets.

mediated via inhibition of transcription factors through a non-genomic effect [38]. Some novel corticosteroids have a greater effect on the non-genomic than genomic effect (dissociated steroids) and thus may have a better therapeutic ratio and might even be suitable for oral administration [39]. Non-steroidal glucocorticoid receptor activators, such as AL-438, have now been discovered and are in clinical development [40]. Corticosteroids switch off inflammatory genes by recruiting the nuclear enzyme HDAC2 to the activated inflammatory gene initiation site so that activators of this enzyme may also have anti-inflammatory effects or may enhance the anti-inflammatory effects of corticosteroids [38]. There may be additional mechanisms for the anti-inflammatory effects of corticosteroids which may also be targeted in the future.

Phosphodiesterase inhibitors

The most advanced of the new anti-inflammatory therapies are phosphodiesterase(PDE)-4 inhibitors which have been in development for asthma and COPD as they have a broad spectrum of anti-inflammatory effects. PDE4 inhibitors inhibit T-lymphocytes ($CD4^+$ and $CD8^+$ cells), neutrophils, eosinophils, mast cells, airway smooth muscle, epithelial cells and airway nerves and have been shown to be highly effective in animal models of asthma and COPD [41, 42]. An oral PDE4 inhibitor, roflumilast, has an inhibitory effect on allergen-induced responses in asthma and also reduces symptoms and lung function in a comparable way to low doses of inhaled steroids [43]. In COPD patients oral roflumilast given over 4 weeks significantly reduces the numbers of neutrophils (by 36%) and CXCL8 concentrations in sputum [44]. In clinical trials roflumilast given over 6 months or 12 months improves lung function in COPD patients to a small extent but has no significant effect in reducing exacerbations or improving quality of life [45]. These disappointing results are likely to reflect the fact that side effects, particularly nausea, diarrhea and headaches, limit the dose that can be tolerated. This indicates that it may not be possible to reach an oral dose that is effective and acceptable to patients.

This could be overcome by inhaled delivery, but to date two inhaled PDE4 inhibitors have been found to be ineffective, although well tolerated. Another approach is to develop isoenzyme-selective inhibitors. PDE4D inhibition appears to account for nausea and vomiting, whereas PDE4B inhibition may account for the anti-inflammatory effects, so that PDE4B selective inhibitors may be better tolerated. However, PDE4D inhibition may also have some anti-inflammatory effects, for example in T-lymphocytes, so that PDE4B selective inhibitors may not be as effective as pan-PDE4 inhibitors [46]. PDE7A is also expressed in the same inflammatory cells as PDE4 so inhibition of PDE7 may be beneficial. However, a selective PDE7 inhibitor had only a small anti-inflammatory effect, but potentiated anti-inflammatory effects of a PDE4 inhibitor, suggesting that a combined inhibitor may be useful as it should not increase side effects [47].

NF-κB inhibitors

NF-κB regulates the expression of CXCL8 and other chemokines, TNF-α and other inflammatory cytokines, as well as MMP9. NF-κB is activated in macrophages and epithelial cells of asthma and COPD patients, particularly during exacerbations. Although there are several possible approaches to inhibition of NF-κB, small molecule inhibitors of NF-κB kinase (IKK)2 are the most promising approach. An IKK2 inhibitor is effective in some animal models of COPD (LPS exposure) but not in others (neutrophil elastase instillation), indicating that the effects may be complex [48]. Although several IKK2 inhibitors are now in development, so far none have been tested in asthma or COPD patients. IKK2 inhibitors not only block the activation of NF-κB-activated genes, but also have some unexpected beneficial effects such as inhibition of CXCR3 chemokines, indicating complex interactions between signal transduction pathways [49]. The hope is that IKK2 inhibitors will be effective in suppressing the corticosteroid-resistance inflammation of COPD and severe asthma. One concern about long-term inhibition of NF-κB is that effective inhibitors may result in immune suppression and impair host defenses, since mice which lack NF-κB genes succumb to septicemia. As it is highly likely that there will be systemic toxicity inhaled delivery is likely to be necessary.

MAPK inhibitors

Mitogen-activated protein kinases (MAPK) play a key role in chronic inflammation and several complex enzyme cascades have now been defined. One of these, the p38 MAPK pathway, is activated by cellular stress and regulates the expression of inflammatory cytokines, including CXCL8, TNF-α and MMPs. p38 MAPK (measured by phosphorylated p38 MAPK) is activated in alveolar macrophages of COPD lungs indicating the activation of this pathway in COPD [50]. There is also evidence that this pathway is also activated in severe asthma. p38 MAPK is involved in the activation of GATA3 and allergic inflammation [51]. Furthermore, p38 MAPK inhibitors have also been shown to decrease eosinophil survival by activating apoptotic pathways [52] and an antisense oligonucleotide that blocks p38 MAP kinase demonstrated marked efficacy in inhibiting eosinophilic inflammation in mice [53]. p38 MAPK is also involved in corticosteroid-resistance through phosphorylation of the glucocorticoid receptor in response to lymphokines [54]. Several small molecule inhibitors of the α- and β-isoforms of p38 MAPK (but not the γ- and δ-isoforms) have now been developed. A potent inhibitor of p38-α isoform, SD-282, is effective in inhibiting TNF-α release from human lung macrophages in vitro [55] and the same inhibitor is also effective in suppressing inflammation in a smoking model of COPD in mice in which corticosteroids are ineffective [56]. Several p38 MAPK inhibitors have now entered clinical trials, but there have been major problems of side effects and toxicity, indicating that it will probably be necessary to deliver these drugs by inhalation to reduce systemic exposure. All of these p38 MAPK inhibitors target p38-α and/or -β isoforms of the enzyme. Little is known about the function of the -γ and δ isoforms of the enzymes as there are no selective inhibitors currently available. In human alveolar macrophages there is a high level of expression of the p38-δ isoform, but its function is currently unknown [55].

Recent studies indicate that other MAPK pathways, particularly extracellular signal-regulated kinase (ERK1/2) may also play an important role in regulating the expression of proinflammatory cytokines in alveolar macrophages, in contrast to its lack of effect in blood monocytes [57].

Phosphoinositide 3-kinase inhibitors

PI3Ks are a family of enzymes that lead to the generation of lipid second messengers that regulate a number of cellular events, including innate and adaptive immune responses. A particular isoform, PI3K-γ, is involved in neutrophil recruitment and activation [58]. Knock-out of the PI-3K-γ gene results in inhibition of neutrophil migration and activation, as well as impaired T-lymphocyte and macrophage function, so PI3K-γ inhibitors may be potential anti-inflammatory therapy for COPD [59]. PI3K-δ is also involved in expression of inflammatory genes and several PI3K-δ or mixed PI3K-γ/-δ inhibitors are now in development [60]. Pan-isoform inhibitors of PI3K are likely to be associated with side effects as these enzymes appear to serve a number of key cell function, but the -γ and -δ isoforms have a more restricted distribution of leukocytes and may therefore be better tolerated, especially if delivered by inhalation.

PPAR activators

Peroxisome proliferator-activated receptors (PPARs) are a family of ligand-activated nuclear hormone receptors belonging to the steroid receptor superfamily, and the three recognized subtypes PPAR-α, -γ, and -δ are widely expressed. There is evidence that activation of PPAR-α and PPAR-δ may have anti-inflammatory and immunomodulatory effects [61]. For example PPAR-γ agonists, such as troglitazone and rosiglitazone, inhibit the release of inflammatory cytokines from monocytes and induce apoptosis of T-lymphocytes, suggesting that they may have anti-inflammatory effects in COPD and severe asthma [62]. There is a reduction in PPAR-α expression in skeletal muscle of COPD patients that correlates with muscular weakness, indicating that PPAR-α agonists, such as clofibrate, may be useful in treating muscle weakness in severe disease [63].

Tyrosine kinase inhibitors

Syk kinase (spleen tyrosine kinase) is a protein tyrosine kinase that plays a pivotal role in signaling of the high affinity IgE receptor (FcεRI) in mast cells and in *syk*-deficient mice mast cell degranulation is inhibited, suggesting that this might be an important potential target for the development of mast cell stabilizing drugs [64]. *Syk* is also involved in antigen receptor signaling of B- and T-lymphocytes and in eosinophil survival in response to IL-5 and GM-CSF [65], so that *Syk* inhibitors might have several useful beneficial effects in atopic diseases. Inhibitors of the enzyme *Syk* kinase are currently in development for asthma [66]. An antisense inhibitor of *Syk* kinase is effective in an animal model of asthma [67] and a *Syk* inhibitor reduces mast cell activation and airway hyperresponsiveness in an allergic murine model [68]. The small molecule inhibitor R112 given nasally reduces nasal symptoms in hay fever patients [69]. As with other kinase inhibitors there may be side effects with systemic administration so that inhalation may be the preferred route of delivery. Another tyrosine kinase *Lyn* is upstream of *Syk* and an inhibitor of *lyn* kinase, PP1, has an inhibitory effect on inflammatory and mast cell activation [70].

Cell adhesion blockers

Infiltration of inflammatory cells into pulmonary tissues is dependent on adhesion of blood borne inflammatory cells to endothelial cells prior to migration to the inflammatory site [71]. This depends upon specific glycoprotein adhesion molecules, including integrins and selectins, on both leukocytes and on endothelial cells, which may be up-regulated or show increased binding affinity in response to various inflammatory stimuli, such as cytokines or lipid mediators. Drugs which inhibit these adhesion molecules therefore may prevent inflammatory cell infiltration. Thus a monoclonal antibody to ICAM-1 on endothelial cells prevents the eosinophil infiltration into airways and the increase in bronchial reactivity after allergen exposure in sensitized primates [72]. The interaction between VLA-4 and VCAM-1 is important for eosinophil inflammation and humanized antibodies to VLA-4 ($\alpha_4\beta_1$) have been developed [73]. Small molecule peptide inhibitors of VLA-4 have subsequently been developed which are effective in inhibiting allergen-induced responses in sensitized sheep and in various models of eosinophilic inflammation [74, 75]. Several orally active VLA-4 inhibitors have now been tested in asthma without any clear clinical benefit, although these negative studies so far remain unpublished.

Inhibitors of selectins, particularly L-selectin and E-selectin, based on the structure of sialyl-Lewis^x inhibit the influx of inflammatory cells in response to inhaled allergen in sensitized sheep [76]. These glycoprotein inhibitors, which may inhibit neutrophilic and eosinophilic inflammation, are now under investigation for asthma and COPD [77].

IMMUNOSUPPRESSANTS

T-lymphocytes may play a critical role in initiating and maintaining the inflammatory process in asthma via the release of cytokines that result in eosinophilic inflammation, suggesting that T-cell inhibitors may be useful in controlling asthmatic inflammation. The non-specific calcineurin inhibitor cyclosporin A reduces the dose of oral steroids needed to control asthma in patients with severe asthma [78], but its efficacy is limited. Side effects, particularly nephrotoxicity, also limit its clinical use. The possibility of using inhaled cyclosporin A has been explored, since in animal studies the inhaled drug is effective in inhibiting the inflammatory response in experimental asthma [79]. Immunomodulators, such as tacrolimus (FK506) and rapamycin, appear to be more potent but are also toxic and may offer no real advantage. Aerosolized tacrolimus is effective against allergen challenge in guinea pigs [80] and topical tacrolimus is effective against atopic dermatitis, but the effects of inhaled tacrolimus have not been reported in asthmatic patients. Novel immunomodulators that inhibit purine or pyrimidine pathways, such as mycophenolate mofetil, leflunomide and brequinar sodium, may be less toxic and therefore of greater potential value in asthma therapy. One problem with these non-specific immunomodulators is that they inhibit both Th1 and Th2 cells, and therefore do not restore the imbalance between these Th1 and Th2 cells in atopy. They also inhibit suppresser T-cells (Tc) that may modulate the inflammatory response. Selective inhibition of Th2 cells may be more effective and better tolerated and there is now a search for such drugs.

The role of immunomodulators in COPD is even less certain. There is an increase in Tc1 and Th1 cells in patients with COPD [81], but the role of these cells is uncertain and the usefulness of immunomodulators in COPD has not yet been assessed.

JAK-3 inhibitors

The Janus-kinase (JAK)-3 appears to be important in signal transduction pathways in T-cells and mediates the secretion of IL-2, IL-4 and IL-9, all of which have been implicated in allergic inflammation. A small molecule inhibitor of JAK-3 CP-690550 is effective in inhibiting eosinophilic inflammation in a murine model [82].

Costimulatory receptor inhibitors

Costimulatory molecules may play a critical role in augmenting the interaction between antigen presenting cells and $CD4^+$ T-lymphocytes. The interaction between B7 and CD28 may determine whether a Th2 type cell response develops, and there is some evidence that B7-2 (CD86) skews towards a Th2 response. Blocking antibodies to B7-2 inhibit the development of specific IgE, pulmonary eosinophilia, and AHR in mice, whereas antibodies to B7-1 (CD80) are ineffective [83]. A molecule on activated T-cells CTL4 appears to act as an endogenous inhibitor of T-cell activation and a soluble fusion protein construct CTLA4-Ig is also effective in blocking AHR in a murine model of asthma [84]. Anti-CD28, anti-B7-2 and CTLA4-Ig also block the proliferative response of T-cells to allergen [85], indicating that these are potential targets for novel therapies that should be effective in all atopic diseases.

Th2 cell inhibitors

Non-selective T-cell suppressants, such as cyclosporin A and tacrolimus, may be relatively ineffective in asthma as they inhibit all types of T-cell. $CD4^+$ T-cells have been implicated in asthma and a chimeric antibody directed against $CD4^+$ (keliximab) which reduces circulating $CD4^+$ cells appears to have some beneficial effect in asthma [86], although long term safety of such a treatment might be a problem. Furthermore, there is increasing evidence that $CD8^+$ cells (Tc2 cells), through release of IL-5 and other cytokines, might also be involved in atopic diseases, particularly in response to infections with certain viruses [87]. There has been a search for selective inhibitors of Th2 cells by identifying features that differentiate Th1 and Th2 cells. It is claimed that suplatast tosylate selectively inhibits Th2 cells and is in clinical use in the treatment of asthma in Japan [88]. However, its mechanism of action is uncertain and it does not appear to be very effective in controlling asthma. The transcription factor GATA-3, which regulates Th2 cell differentiation and the expression of Th2 specific cytokines, may be targeted by antisense oligonuleotides, but more easily by p38 MAPK inhibitors, as discussed earlier [51]. The calcium antagonist nicardipine appears to selectively inhibit Th2 cells in mice suggesting that dihydropyrridines may act as immunomodulators in asthma but this has not yet been demonstrated in humans [89]. Indoleamine-2,3-dioxygenase (IDO), a tryptophan-degrading enzyme, plays a key role in the regulation of T-lymphocyte function. IDO inhibits eosinophilic inflammation in a murine asthma model, and concentrations of IDO are reduced in asthmatic patients and restored by corticosteroids [90]. Other drugs which restore IDO activity may therefore be of value in suppressing eosinophilic inflammation in asthma.

Regulatory T-cell activators

Regulatory T-cells (Tregs) play a key role in orchestrating immunity and there is evidence for a defect in Treg function in patients with asthma and COPD that may be linked to increased numbers of T-cells [91, 92]. There is convincing evidence that specific immunotherapy increased IL-10 production from Tregs (Tr-1 cells) [93]. Vaccines which enhance Tr-1 cell function and increase IL-10 release are now in development for asthma. T-cell peptides are also in development as a safer form of immunotherapy [94].

B-lymphocyte depletion

An anti-CD20 blocking monoclonal antibody (rituximab) selectively depletes B-cells and has been found to be very effective in controlling resistant rheumatoid arthritis and other autoimmune diseases [95]. As there is an increase in B-cells in COPD and as B-cells produce IgE in asthma, this approach may also be useful in severe asthma and COPD. However, the potential disadvantages are an increased risk of pulmonary infections, particularly in patients with COPD.

Anti-IgE

Blocking IgE with a chimeric antibody omalizumab is useful in controlling asthma in some patients with severe asthma, although there are no clinical indicators to show how responders can be predicted, so a prolonged trial of therapy is needed [96]. Omalizumab is very expensive as a maintenance therapy, especially in patients with high titers of IgE who require high doses. There is a quest for antibodies with higher affinity, so that patients with high IgE can also be treated. There is also a search for small molecule inhibitors of the IgE response through targeting the signal transduction pathways initiated by IgE interaction with its receptors.

Allergens bind to a low-affinity IgE receptor (FcεRII or CD23) as well as the high affinity receptor FcεRI on several immune cells, including T- and B-lymphocytes [97]. An anti-CD23 antibody (lumiliximab) was well tolerated and reduced IgE concentrations in patients with mild asthma, but did not appear to have clinical efficacy [97].

PREVENTIVE STRATEGIES

The obvious preventive strategy for preventing COPD is stopping smoking, which is covered in Chapter 47. Preventive strategies for asthma are targeted to preventing the Th2 cell preponderance in atopy.

Vaccination

None of the currently available treatments for asthma have long-term effects on airway inflammation or remodeling and therefore are not disease-modifying or curative. Inhaled corticosteroids are usually very effective in controlling asthma, but symptoms usually rapidly return when they are discontinued. The prospects for a cure are remote until the molecular and genetic causes of asthma are better understood, but there is a possibility that vaccination approaches may reverse the abnormal immune regulation found in asthma, restoring Th1 predominance or restoring Treg function [93, 98]. Various approaches, including non-pathogenic bacterial products, such as CpG oligodeoxynucleotides, which target toll-like receptor (TLR)-9, are currently being explored as potential therapies for asthma [99]. A CpG vaccine has been tested in asthma, and despite evidence for an increase in interferons, there was no effect on the response to allergen, in contrast to several positive studies in rodent models [100]. However, the long-term consequences of these approaches need to be carefully evaluated, particularly as they would probably need to be applied in children at the onset of disease. The role of other TLRs in asthma is uncertain, with beneficial and detrimental effects reported in different animal models [101].

REVERSAL OF CORTICOSTEROID RESISTANCE

Even high doses of corticosteroids have minimal effects on the progression of COPD and no effects on mortality [102] and are poorly effective in severe asthma. This may reflect the resistance of inflammation in severe asthma and COPD to the anti-inflammatory effects of corticosteroids. There is increasing evidence that this may be due to a reduction in HDAC2 as a result of oxidative and nitrative stress [27, 103]. This results in increased acetylation of the glucocorticoid receptor which prevents it inhibiting NF-κB driven inflammation [104]. A novel therapeutic strategy is therefore reversal of this corticosteroid resistance by increasing the expression and activity of HDAC2 and this may be achieved in several ways.

Theophylline-like drugs

Low doses of oral theophylline increase HDAC2 expression in alveolar macrophages from COPD patients and thereby restore steroid responsiveness [105, 106]. This has also been demonstrated in mice exposed to cigarette smoke, which develop a steroid-resistant inflammation in the lungs with increased neutrophils and macrophages. This inflammation is not reversed by high doses of corticosteroids or by theophylline alone but is reversed by low dose oral or inhaled theophylline combined with a corticosteroid via an increase in HDAC2 activity [107]. Understanding the molecular mechanisms of action of theophylline, which appear to be independent of PDE inhibition, may lead to novel therapeutic approaches to restoration of corticosteroid responsiveness which avoid the side effects and drug interaction problems of theophylline itself.

Antioxidants

As discussed above, antioxidants may be valuable in reversing corticosteroid resistance since increased oxidative stress appears to be the mechanisms by which severe inflammation and cigarette smoke lead to reduced HDAC2 activity and corticosteroid resistance. Several more potent and stable antioxidants are needed, such as superoxide dismutase mimics and NADPH oxidase inhibitors [28].

Macrolides

It has long been recognised that macrolides have anti-inflammatory effects that may be independent of their antibiotic effects [108]. Macrolides appear to inhibit inflammation by inhibiting NF-κB and other transcription factors. Recently it has been shown that a non-antibiotic macrolide (EM-703) reverses corticosteroid resistance due to oxidative stress by increasing HDAC2 activity [109]. Several non-antibiotic macrolides are now in development as anti-inflammatory therapies.

REVERSAL OF STRUCTURAL CHANGES

In both asthma and COPD structural changes in the airways may lead to reversible narrowing as a result of structural changes. These may be a consequence of persistent inflammation and the generation of growth factors that cause cell proliferation and fibrosis. However, it is possible that the structural changes may become independent of inflammation and may progress in the absence of inflammation. Therapeutic strategies to prevent these structural changes or ideally to reverse them in order to restore lung function are therefore of future interest. In COPR there is also destruction of connective tissue as a result of elastolysis and apoptosis of alveolar epithelial and endothelial cells.

Fibrosis inhibitors

Transforming growth factor-β (TGF-β) appears to play a key role in the fibrosis of asthma and COPD and small molecule inhibitors of TGF-β tyrosine kinase have been developed [110]. However, TGF-β plays a critical role in wound healing and in Treg function, so its inhibition may have deleterious effects. It may be safer to inhibit a downstream cytokine connective tissue growth factor (CTGF), which appears to mediate many of the fibrotic responses to TGF-β.

Reversal of emphysema

Retinoic acid increases alveolar septation during lung development and in adult rats and mice reverses the histological and physiological changes induced by elastase treatment [111]. This has not been seen in several other species and there are doubts whether emphysema is reversible in humans as alveolar formation ceases about the age of 6 years. A clinical trial of all-*trans*-retinoic and 9-*cis*-retinoic acid in patients with emphysema failed to show any improvement in clinical parameters, health status or CT density after 6 months of therapy [112].

Another possible approach to repairing damaged lung in emphysema is the use of stem cells to seed the lung combined with drugs that stimulate their homing and proliferation in the lung. Human embryonic stem cells have been transformed into alveolar type II pneumocytes which have the capacity to repair alveolar damage [113]. Adult bone marrow-derived stem cells may also be suitable for populating the lung, particularly if enhanced by retinoic acid or granulocyte-macrophage colony stimulating factor. However, there are several concerns about the use of stem cells for lung repair as there may be a problem engrafting these cells in the alveoli and there is always a risk of cancer or teratoma development [114]. The lung is a complex organ and it would probably be necessary to grow both endothelial and alveolar cells to repair emphysema.

GENE THERAPY

Since asthma and COPD are polygenic, it is unlikely that gene therapy will be of value in long-term therapy. However, understanding the genes involved in asthma and COPD and in disease severity may identify new molecular targets [115] and may also predict the response to different therapies [116]. Transfer of anti-inflammatory genes may provide specific anti-inflammatory or inhibitory proteins in a convenient manner and gene transfer has been shown to be feasible in animals using viral vectors [117]. Anti-inflammatory proteins relevant to asthma and COPD include IL-10 and IκB. Anti-sense oligonucleotides may switch off specific genes, but there are considerable problems in getting these molecules into cells. An inhaled antisense oligonucleotide directed against the adenosine A_1-receptor has been shown to reduce AHR in a rabbit model of asthma, demonstrating the potential of this approach in treating asthma [118]. Suitable target genes may be IL-4 or IL-5 in asthma and MMP-9 in COPD. Considering the practical problems encountered by gene therapy this approach is unlikely in the foreseeable future, other than for proof of concept studies.

FUTURE DIRECTIONS

New drugs for the treatment of severe asthma and COPD are greatly needed and there has been an enormous effort now invested by the pharmaceutical industry to find such treatments. Both diseases are highly complex so it is unlikely that therapies that target a single cell or mediator will be very effective. While preventing and quitting smoking is the obvious preferred approach, this has proved to be very difficult in the majority of patients. It may be important to more carefully phenotype patients with asthma and COPD in clinical trials. For example, it may also be important to more accurately define the presence of emphysema versus small airway obstruction in COPD patients by using CT scans, as some drugs may be more useful for preventing emphysema, whereas others may be more effective against the small airway inflammatory-fibrotic process. More research on the basic cellular and molecular mechanisms of severe asthma and COPD and on more useful animal models is urgently needed to aid the logical development of new therapies these important diseases, for which no effective preventative drugs currently exist.

Of the drugs currently in development PDE4 inhibitors, p38 MAP kinase inhibitors, and IKK-2 inhibitors appear to be promising for severe asthma and COPD, but there are concerns about side effects so that inhaled administration is likely to be needed. There are also concerns about their long-term safety in increasing lung infection and cancer, particularly in COPD patients. For these reasons alternative strategies, such as reversal of corticosteroid-resistance may be more successful and there have been important advances in understanding the molecular mechanisms of corticosteroid resistance that will identify novel targets for therapy.

Drug delivery is an important issue. Most of the effective therapies used now are given by inhalation in both diseases, but inhaled therapies are more difficult to develop as there is only a poor understanding of inhalation pharmacokinetics and toxicology studies are more difficult to conduct. In addition, the inhaled route may not deliver sufficient drug to peripheral airways, particularly in COPD and severe asthma. Oral administration has the advantage of easier drug development and the treatment of concomitant allergic diseases in the case of asthma and treating systemic complications in the case of COPD. However, oral administration has a high risk of side effects which may outweigh these advantages. Another approach is targeting of drugs to specific cell types, for example to macrophages which take up particles, as this may allow an enhanced therapeutic ratio.

References

1. Barnes PJ. Immunology of asthma and chronic obstructive pulmonary disease. *Nat Immunol Rev* 8: 183–92, 2008.
2. Barnes PJ. New therapies for asthma. *Trends Mol Med* 12: 515–20, 2006.
3. Barnes PJ, Hansel TT. Prospects for new drugs for chronic obstructive pulmonary disease. *Lancet* 364: 985–96, 2004.
4. Holgate ST, Polosa R. Treatment strategies for allergy and asthma. *Nat Rev Immunol* 8: 218–30, 2008.
5. Horne R. Compliance, adherence, and concordance: Implications for asthma treatment. *Chest* 130: 65S–72S, 2006.
6. Barnes PJ. Chronic obstructive pulmonary disease: A growing but neglected epidemic. *PLoS Med* 4: e112, 2007.
7. Barnes PJ, Stockley RA. COPD: Current therapeutic interventions and future approaches. *Eur Respir J* 25: 1084–106, 2005.
8. Calverley PM, Anderson JA, Celli B, Ferguson GT, Jenkins C, Jones PW *et al.* Salmeterol and fluticasone propionate and survival in chronic obstructive pulmonary disease. *N Engl J Med* 356: 775–89, 2007.
9. Barnes PJ, Chowdhury B, Kharitonov SA, Magnussen H, Page CP, Postma D *et al.* Pulmonary biomarkers in chronic obstructive pulmonary disease. *Am J Respir Crit Care Med* 174: 6–14, 2006.
10. Sturton G, Persson CGA, Barnes PJ. Small airways: An important but neglected target in obstructive airway diseases. *Trends in Pharmacol Sci*, 2008, (in press).
11. Roses AD. Pharmacogenetics and future drug development and delivery. *Lancet* 355: 1358–61, 2000.
12. Matera MG, Cazzola M. Ultra-long-acting beta2-adrenoceptor agonists: An emerging therapeutic option for asthma and COPD? *Drugs* 67: 503–15, 2007.
13. Laforce C, Alexander M, Deckelmann R, Fabbri LM, Aisanov Z, Cameron R *et al.* Indacaterol provides sustained 24 h bronchodilation on once-daily dosing in asthma: A 7-day dose-ranging study. *Allergy* 63: 103–11, 2008.
14. Sturton RG, Nicholson AG, Trifilieff A, Barnes PJ. Pharmacological characterisation of indacaterol, a novel once-daily inhaled b2-adrenoceptor agonist, on small airways in human and rat precision-cut lung slices. *J Pharmacol Exp Ther* 324: 270–75, 2008.
15. Hansel TT, Neighbour H, Erin EM, Tan AJ, Tennant RC, Maus JG *et al.* Glycopyrrolate causes prolonged bronchoprotection and bronchodilatation in patients with asthma. *Chest* 128: 1974–79, 2005.
16. Joos GF, Schelfhout VJ, Kanniess F. Bronchodilator effects of aclidinium bromide, a novel long-acting anticholinergic, in COPD patients: A phase II study. *Eur Resp J*, 2007.
17. van Noord JA, Aumann JL, Janssens E, Verhaert J, Smeets JJ, Mueller A *et al.* Effects of tiotropium with and without formoterol on airflow obstruction and resting hyperinflation in patients with COPD. *Chest* 129: 509–17, 2006.
18. Barnes PJ, Chung KF, Page CP. Inflammatory mediators of asthma: An update. *Pharmacol Rev* 50: 515–96, 1998.
19. Barnes PJ. Mediators of chronic obstructive pulmonary disease. *Pharm Rev* 56: 515–48, 2004.
20. Wenzel S, Wilbraham D, Fuller R, Getz EB, Longphre M. Effect of an interleukin-4 variant on late phase asthmatic response to allergen challenge in asthmatic patients: Results of two phase 2a studies. *Lancet* 370: 1422–31, 2007.
21. Liu YJ, Soumelis V, Watanabe N, Ito T, Wang YH, Malefyt RW *et al.* TSLP: An epithelial cell cytokine that regulates T cell differentiation by conditioning dendritic cell maturation. *Annu Rev Immunol* 25: 193–219, 2007.
22. Panina-Bordignon P, D'Ambrosio D. Chemokines and their receptors in asthma and chronic obstructive pulmonary disease. *Curr Opin Pulm Med* 9: 104–10, 2003.
23. Donnelly LE, Barnes PJ. Chemokine receptors as therapeutic targets in chronic obstructive pulmonary disease. *Trends Pharmacol Sci* 27: 546–53, 2006.
24. O'Connor BJ, Leaker BR, Barnes PJ, Nicholson G, Grahames C, Larsson B *et al.* Inhibition of LPS-induced neutrophilic inflammation in healthy volunteers. *Eur Resp J* 30(Suppl 51): 1294, 2007.
25. Costa C, Rufino R, Traves SL, Lapa E, Silva JR, Barnes PJ, Donnelly LE. CXCR3 and CCR5 chemokines in the induced sputum from patients with COPD. *Chest* 133: 26–33, 2008.
26. Rahman I, Biswas SK, Kode A. Oxidant and antioxidant balance in the airways and airway diseases. *Eur J Pharmacol* 533: 222–39, 2006.
27. Barnes PJ. Reduced histone deacetylase in COPD: Clinical implications. *Chest* 129: 151–55, 2006.
28. Kirkham P, Rahman I. Oxidative stress in asthma and COPD: Antioxidants as a therapeutic strategy. *Pharmacol Ther* 111: 476–94, 2006.
29. Reed CE, Kita H. The role of protease activation of inflammation in allergic respiratory diseases. *J Allergy Clin Immunol* 114: 997–1008, 2004.
30. Clark JM, Abraham WM, Fishman CE, Forteza R, Ahmed A, Cortes A *et al.* Tryptase inhibitors block allergen-induced airway and inflammatory responses in allergic sheep. *Am J Respir Crit Care Med* 152: 2076–83, 1995.
31. Krishna MT, Chauhan A, Little L, Sampson K, Hawksworth R, Mant T *et al.* Inhibition of mast cell tryptase by inhaled APC 366 attenuates allergen-induced late-phase airway obstruction in asthma. *J Allergy Clin Immunol* 107: 1039–45, 2001.
32. Erin EM, Leaker BR, Zacharasiewicz A, Higgins LA, Nicholson GC, Boyce MJ *et al.* Effects of a reversible b-tryptase and trypsin inhibitor (RWJ-58643) on nasal allergic responses. *Clin Exp Allergy* 36: 458–64, 2006.
33. Luisetti M, Sturani C, Sella D, Madonini E, Galavotti V, Bruno G *et al.* MR889, a neutrophil elastase inhibitor, in patients with chronic obstructive pulmonary disease: A double-blind, randomized, placebo-controlled clinical trial. *Eur Respir J* 9: 1482–86, 1996.
34. Hu J, Van Den Steen PE, Sang QX, Opdenakker G. Matrix metalloproteinase inhibitors as therapy for inflammatory and vascular diseases. *Nat Rev Drug Discov* 6: 480–98, 2007.
35. Churg A, Wang R, Wang X, Onnervik PO, Thim K, Wright JL. Effect of an MMP-9/MMP-12 inhibitor on smoke-induced emphysema and airway remodelling in guinea pigs. *Thorax* 62: 706–13, 2007.
36. Barnes PJ, Adcock IM. How do corticosteroids work in asthma? *Ann Intern Med* 139: 359–70, 2003.
37. Reynolds NA, Scott LJ. Ciclesonide. *Drugs* 64: 511–19, 2004.
38. Barnes PJ. How corticosteroids control inflammation. *Br J Pharmacol* 148: 245–54, 2006.
39. Schacke H, Schottelius A, Docke WD, Strehlke P, Jaroch S, Schmees N *et al.* Dissociation of transactivation from transrepression by a selective glucocorticoid receptor agonist leads to separation of therapeutic effects from side effects. *Proc Natl Acad Sci USA* 101: 227–32, 2004.
40. Rosen J, Miner JN. The search for safer glucocorticoid receptor ligands. *Endocr Rev* 26: 452–64, 2005.
41. Lipworth BJ. Phosphodiesterase-4 inhibitors for asthma and chronic obstructive pulmonary disease. *Lancet* 365: 167–75, 2005.

42. Martorana PA, Beume R, Lucattelli M, Wollin L, Lungarella G. Roflumilast fully prevents emphysema in mice chronically exposed to cigarette smoke. *Am J Respir Crit Care Med* 172: 848–53, 2005.
43. Bousquet J, Aubier M, Sastre J, Izquierdo JL, Adler LM, Hofbauer P *et al.* Comparison of roflumilast, an oral anti-inflammatory, with beclomethasone dipropionate in the treatment of persistent asthma. *Allergy* 61: 72–78, 2006.
44. Grootendorst DC, Gauw SA, Verhoosel RM, Sterk PJ, Hospers JJ, Bredenbroker D *et al.* Reduction in sputum neutrophil and eosinophil numbers by the PDE4 inhibitor roflumilast in patients with COPD. *Thorax* 62: 1081–87, 2007.
45. Calverley PM, Sanchez-Toril F, McIvor A, Teichmann P, Bredenbroeker D, Fabbri LM. Effect of 1-year treatment with roflumilast in severe chronic obstructive pulmonary disease. *Am J Respir Crit Care Med* 176: 154–61, 2007.
46. Peter D, Jin SL, Conti M, Hatzelmann A, Zitt C. Differential expression and function of phosphodiesterase 4 (PDE4) subtypes in human primary CD4+ T cells: Predominant role of PDE4D. *J Immunol* 178: 4820–31, 2007.
47. Smith SJ, Cieslinski LB, Newton R, Donnelly LE, Fenwick PS, Nicholson AG *et al.* Discovery of BRL 50481, a selective inhibitor of phosphodiesterase 7: In vitro studies in human monocytes, lung macrophages and CD8+ T-lymphocytes. *Mol Pharmacol* 66: 1679–89, 2004.
48. Birrell MA, Wong S, Hardaker EL, Catley MC, McCluskie K, Collins M *et al.* IkappaB kinase-2-independent and -dependent inflammation in airway disease models: Relevance of IKK-2 inhibition to the clinic. *Mol Pharmacol* 69: 1791–800, 2006.
49. Tudhope SJ, Catley MC, Fenwick PS, Russell RE, Rumsey WL, Newton R *et al.* The role of IkB kinase 2, but not activation of NF-kB, in the release of CXCR3 ligands from IFN-g-stimulated human bronchial epithelial cells. *J Immunol* 179: 6237–45, 2007.
50. Renda T, Baraldo S, Pelaia G, Bazzan E, Turato G, Papi A *et al.* Increased activation of p38 MAPK in COPD. *Eur Respir J* 31: 62–69, 2008.
51. Maneechotesuwan K, Xin Y, Ito K, Jazrawi E, Lee KY, Usmani OS *et al.* Regulation of Th2 cytokine genes by p38 MAPK-mediated phosphorylation of GATA-3. *J Immunol* 178: 2491–98, 2007.
52. Kankaanranta H, Giembycz MA, Barnes PJ, Lindsay DA. SB203580, an inhibitor of p38 mitogen-activated protein kinase, enhances constitutive apoptosis of cytokine-deprived human eosinophils. *J Pharmacol Exp Ther* 290: 621–28, 1999.
53. Duan W, Chan JH, McKay K, Crosby JR, Choo HH, Leung BP *et al.* Inhaled p38alpha mitogen-activated protein kinase antisense oligonucleotide attenuates asthma in mice. *Am J Respir Crit Care Med* 171: 571–78, 2005.
54. Matthews JG, Ito K, Barnes PJ, Adcock IM. Defective glucocorticoid receptor nuclear translocation and altered histone acetylation patterns in glucocorticoid-resistant patients. *J Allergy Clin Immunol* 113: 1100–8, 2004.
55. Smith SJ, Fenwick PS, Nicholson AG, Kirschenbaum F, Finney-Hayward TK, Higgins LS *et al.* Inhibitory effect of p38 mitogen-activated protein kinase inhibitors on cytokine release from human macrophages. *Br J Pharmacol* 149: 393–404, 2006.
56. Medicherla S, Fitzgerald M, Spicer D, Woodman P, Ma JY, Kapoun AM *et al.* p38a selective MAP kinase inhibitor, SD-282, reduces inflammation in a sub-chronic model of tobacco smoke-induced airway inflammation. *J Pharmacol Exp Ther*, 2007.
57. Tudhope SJ, Finney-Hayward TK, Nicholson AG, Mayer RJ, Barnette MS, Barnes PJ *et al.* Different mitogen-activated protein kinase-dependent cytokine responses in cells of the monocyte lineage. *J Pharmacol Exp Ther* 324: 306–12, 2008.
58. Sasaki T, Irie-Sasaki J, Jones RG, Oliveira dSA, Stanford WL, Bolon B *et al.* Function of PI3Kgamma in thymocyte development, T cell activation, and neutrophil migration. *Science* 287: 1040–46, 2000.
59. Medina-Tato DA, Ward SG, Watson ML. Phosphoinositide 3-kinase signalling in lung disease: Leucocytes and beyond. *Immunology* 121: 448–61, 2007.
60. Ward S, Sotsios Y, Dowden J, Bruce I, Finan P. Therapeutic potential of phosphoinositide 3-kinase inhibitors. *Chem Biol* 10: 207–13, 2003.
61. Spears M, McSharry C, Thomson NC. Peroxisome proliferator-activated receptor-gamma agonists as potential anti-inflammatory agents in asthma and chronic obstructive pulmonary disease. *Clin Exp Allergy* 36: 1494–504, 2006.
62. Belvisi MG, Hele DJ, Birrell MA. Peroxisome proliferator-activated receptor gamma agonists as therapy for chronic airway inflammation. *Eur J Pharmacol* 533: 101–9, 2006.
63. Remels AH, Schrauwen P, Broekhuizen R, Willems J, Kersten S, Gosker HR *et al.* Peroxisome proliferator-activated receptor expression is reduced in skeletal muscle in COPD. *Eur Respir J* 30: 245–52, 2007.
64. Costello PS, Turner M, Walters AE, Cunningham CN, Bauer PH, Downward J *et al.* Critical role for the tyrosine kinase Syk in signalling through the high affinity IgE receptor of mast cells. *Oncogene* 13: 2595–605, 1996.
65. Yousefi S, Hoessli DC, Blaser K, Mills GB, Simon HU. Requirement of Lyn and Syk tyrosine kinases for the prevention of apoptosis by cytokines in human eosinophils. *J Exp Med* 183: 1407–14, 1996.
66. Wong BR, Grossbard EB, Payan DG, Masuda ES. Targeting Syk as a treatment for allergic and autoimmune disorders. *Expert Opin Investig Drugs* 13: 743–62, 2004.
67. Stenton GR, Ulanova M, Dery RE, Merani S, Kim MK, Gilchrist M *et al.* Inhibition of allergic inflammation in the airways using aerosolized antisense to Syk kinase. *J Immunol* 169: 1028–36, 2002.
68. Matsubara S, Li G, Takeda K, Loader JE, Pine P, Masuda ES *et al.* Inhibition of spleen tyrosine kinase prevents mast cell activation and airway hyperresponsiveness. *Am J Respir Crit Care Med* 173: 56–63, 2006.
69. Meltzer EO, Berkowitz RB, Grossbard EB. An intranasal Syk-kinase inhibitor (R112) improves the symptoms of seasonal allergic rhinitis in a park environment. *J Allergy Clin Immunol* 115: 791–96, 2005.
70. Amoui M, Draber P, Draberova L. Src family-selective tyrosine kinase inhibitor, PP1, inhibits both Fc epsilonRI- and Thy-1-mediated activation of rat basophilic leukemia cells. *Eur J Immunol* 27: 1881–86, 1997.
71. Pilewski JM, Albelda SM. Cell adhesion molecules in asthma: homing activation and airway remodelling. *Am J Respir Cell Mol Biol* 12: 1–3, 1995.
72. Weg VB, Williams TJ, Lobb RR, Noorshargh S. A monoclonal antibody recognizing very late activation antigen-4 inhibits eosinophil accumulation in vivo. *J Exp Med* 177: 561–66, 1993.
73. Yuan Q, Strauch KL, Lobb RR, Hemler ME. Intracellular single-chain antibody inhibits integrin VLA-4 maturation and function. *Biochem J* 318: 591–96, 1996.
74. Lin Kc, Ateeq HS, Hsiung SH, Chong LT, Zimmerman CN, Castro A *et al.* Selective, tight-binding inhibitors of integrin alpha4beta1 that inhibit allergic airway responses. *J Med Chem* 42: 920–34, 1999.
75. Okigami H, Takeshita K, Tajimi M, Komura H, Albers M, Lehmann TE *et al.* Inhibition of eosinophilia in vivo by a small molecule inhibitor of very late antigen (VLA)-4. *Eur J Pharmacol* 559: 202–9, 2007.
76. Abraham WM, Ahmed A, Sabater JR, Lauredo IT, Botvinnikova Y, Bjercke RJ *et al.* Selectin blockade prevents antigen-induced late bronchial responses and airway hyperresponsiveness in allergic sheep. *Am J Respir Crit Care Med* 159: 1205–14, 1999.
77. Kogan TP, Dupre B, Bui H, McAbee KL, Kassir JM, Scott IL *et al.* Novel synthetic inhibitors of selectin-mediated cell adhesion: synthesis of 1,6-bis[3-(3-carboxymethylphenyl)-4-(2-alpha-D- mannopyranosyl oxy)phenyl]hexane (TBC1269). *J MedChem* 41: 1099–111, 1998.
78. Lock SH, Kay AB, Barnes NC. Double-blind, placebo-controlled study of cyclosporin A as a corticosteroid-sparing agent in corticosteroid-dependent asthma. *Am J Respir Crit Care Med* 153: 509–14, 1996.
79. Morley J. Cyclosporin A in asthma therapy: A pharmacological rationale. *J Autoimmun* 5(Suppl A): 265–69, 1992.

80. Morishita Y, Hirayama Y, Miyayasu K, Tabata K, Kawamura A, Ohkubo Y *et al.* FK506 aerosol locally inhibits antigen-induced airway inflammation in Guinea pigs. *Int Arch Allergy Immunol* 136: 372–78, 2005.
81. Grumelli S, Corry DB, Song L-X, Song L, Green L, Huh J *et al.* An immune basis for lung parenchymal destruction in chronic obstructive pulmonary disease and emphysema. *PLoS Med* 1: 75–83, 2004.
82. Kudlacz E, Conklyn M, Andresen C, Whitney-Pickett C, Changelian P. The JAK-3 inhibitor CP-690550 is a potent anti-inflammatory agent in a murine model of pulmonary eosinophilia. *Eur J Pharmacol* 582: 154–61, 2008.
83. Haczku A, Takeda K, Redai I, Hamelmann E, Cieslewicz G, Joetham A *et al.* Anti-CD86 (B7.2) treatment abolishes allergic airway hyperresponsiveness in mice [In Process Citation]. *Am J Respir Crit Care Med* 159: 1638–43, 1999.
84. Van Oosterhout AJ, Hofstra CL, Shields R, Chan B, van Ark I, Jardieu PM *et al.* Murine CTLA4-IgG treatment inhibits airway eosinophilia and hyperresponsiveness and attenuates IgE upregulation in a murine model of allergic asthma. *Am J Respir Cell Mol Biol* 17: 386–92, 1997.
85. van Neerven RJ, Van de Pol MM, van der Zee JS, Stiekema FE, De Boer M, Kapsenberg ML. Requirement of CD28-CD86 costimulation for allergen-specific T cell proliferation and cytokine expression [see comments]. *Clin Exp Allergy* 28: 808–16, 1998.
86. Kon OM, Compton CH, Kay AB, Barnes NC. A dopuble-blind placebo-controlled trial of an anti-CD4 monoclonal antibody SB210396. *Am J Respir Crit Care Med* 155: A203, 1997.
87. Schwarze J, Cieslewicz G, Joetham A, Ikemura T, Hamelmann E, Gelfand EW. CD8λT cells are essential in the development of respiratory syncytial virus-induced lung eosinophilia and airway hyperresponsiveness. *J Immunol* 162: 4207–11, 1999.
88. Tamaoki J, Kondo M, Sakai N, Aoshiba K, Tagaya E, Nakata J *et al.* Effect of suplatast tosilate, a Th2 cytokine inhibitor, on steroid-dependent asthma: A double-blind randomised study. *Lancet* 356: 273–78, 2000.
89. Gomes B, Cabral MD, Gallard A, Savignac M, Paulet P, Druet P *et al.* Calcium channel blocker prevents T helper type 2 cell-mediated airway inflammation. *Am J Respir Crit Care Med* 175: 1117–24, 2007.
90. Maneechotesuwan K, Supawita S, Kasetsinsombat K, Wongkajornsilp A, Barnes PJ. Sputum indoleamine-2, 3-dioxygenase activity is increased in asthmatic airways by using inhaled corticosteroids. *J Allergy Clin Immunol* 121: 43–50, 2008.
91. Larche M. Regulatory T cells in allergy and asthma. *Chest* 132: 1007–14, 2007.
92. Smyth LJ, Starkey C, Vestbo J, Singh D. CD4-regulatory cells in COPD patients. *Chest* 132: 156–63, 2007.
93. Akdis M, Blaser K, Akdis CA. T regulatory cells in allergy: novel concepts in the pathogenesis, prevention, and treatment of allergic diseases. *J Allergy Clin Immunol* 116: 961–68, 2005.
94. Larche M. Update on the current status of peptide immunotherapy. *J Allergy Clin Immunol* 119: 906–9, 2007.
95. Pers JO, Daridon C, Bendaoud B, Devauchelle V, Berthou C, Saraux A *et al.* B-cell depletion and repopulation in autoimmune diseases. *Clin. Rev Allergy Immunol* 34: 50–55, 2008.
96. Humbert M, Beasley R, Ayres J, Slavin R, Hebert J, Bousquet J *et al.* Benefits of omalizumab as add-on therapy in patients with severe persistent asthma who are inadequately controlled despite best available therapy (GINA 2002 step 4 treatment): INNOVATE. *Allergy* 60: 309–16, 2005.
97. Rosenwasser LJ, Meng J. Anti-CD23. *Clin Rev Allergy Immunol* 29: 61–72, 2005.
98. Wohlleben G, Erb KJ. Atopic disorders: A vaccine around the corner? *Trends Immunol* 22: 618–26, 2001.
99. Krieg AM. Therapeutic potential of Toll-like receptor 9 activation. *Nat Rev Drug Discov* 5: 471–84, 2006.
100. Gauvreau GM, Hessel EM, Boulet LP, Coffman RL, O'Byrne PM. Immunostimulatory sequences regulate interferon-inducible genes but not allergic airway responses. *Am J Respir Crit Care Med* 174: 15–20, 2006.
101. Feleszko W, Jaworska J, Hamelmann E. Toll-like receptors–novel targets in allergic airway disease (probiotics, friends and relatives). *Eur J Pharmacol* 533: 308–18, 2006.
102. Yang IA, Fong KM, Sim EH, Black PN, Lasserson TJ. Inhaled corticosteroids for stable chronic obstructive pulmonary disease. *Cochrane Database Syst Rev*: CD002991, 2007.
103. Hew M, Bhavsar P, Torrego A, Meah S, Khorasani N, Barnes PJ *et al.* Relative corticosteroid insensitivity of peripheral blood mononuclear cells in severe asthma. *Am J Respir Crit Care Med* 174: 134–41, 2006.
104. Ito K, Yamamura S, Essilfie-Quaye S, Cosio B, Ito M, Barnes PJ *et al.* Histone deacetylase 2-mediated deacetylation of the glucocorticoid receptor enables NF-kB suppression. *J Exp Med* 203: 7–13, 2006.
105. Cosio BG, Tsaprouni L, Ito K, Jazrawi E, Adcock IM, Barnes PJ. Theophylline restores histone deacetylase activity and steroid responses in COPD macrophages. *J Exp Med* 200: 689–95, 2004.
106. Barnes PJ. Theophylline for COPD. *Thorax* 61: 742–43, 2006.
107. Fox JC, Spicer D, Ito K, Barnes PJ, Fitzgerald MF. Oral or inhaled corticosteroid combination therapy with low dose theophylline reverses corticosteroid insensitivity in a smoking mouse model. *Proc Am Thorac Soc*: A637, 2007.
108. Rubin BK, Henke MO. Immunomodulatory activity and effectiveness of macrolides in chronic airway disease. *Chest* 125: 70S–78S, 2004.
109. Charron C, Sumakuza T, Oomura S, Ito K. EM-703, a non-antibacterial erythromycin derivative, restores HDAC2 activition diminished by hypoxia and oxidative stress. *Proc Am Thorac Soc*, 2007.
110. Prud'homme GJ. Pathobiology of transforming growth factor beta in cancer, fibrosis and immunologic disease, and therapeutic considerations. *Lab Invest* 87: 1077–91, 2007.
111. Maden M, Hind M. Retinoic acid in alveolar development, maintenance and regeneration. *Philos Trans R Soc Lond B Biol Sci* 359: 799–808, 2004.
112. Roth MD, Connett JE, D'Armiento JM, Foronjy RF, Friedman PJ, Goldin JG *et al.* Feasibility of retinoids for the treatment of emphysema study. *Chest* 130: 1334–45, 2006.
113. Wang D, Haviland DL, Burns AR, Zsigmond E, Wetsel RA. A pure population of lung alveolar epithelial type II cells derived from human embryonic stem cells. *Proc Natl Acad Sci USA* 104: 4449–54, 2007.
114. Loebinger MR, Janes SM. Stem cells for lung disease. *Chest* 132: 279–85, 2007.
115. Cookson WO. State of the art, Genetics and genomics of chronic obstructive pulmonary disease. *Proc Am Thorac Soc* 3: 473–75, 2006.
116. Hall IP, Sayers I. Pharmacogenetics and asthma: False hope or new dawn?. *Eur Respir J* 29: 1239–45, 2007.
117. Xing Z, Ohkawara Y, Jordana M, Grahern FL, Gauldie J. Transfer of granulocyte-macrophage colony-stinulating factor gene to rat induces eosinophilia, monocytosis and fibrotic lesions. *J Clin Invest* 97: 1102–10, 1996.
118. Metzger WJ, Nyce JW. Oligonucleotide therapy of allergic asthma. *J Allergy Clin Immunol* 104: 260–66, 1999.

Health Economics in Asthma and COPD

62

CHAPTER

Andrew Briggs, Helen Starkie and Olivia Wu

Section of Public Health and Health Policy, University of Glasgow, Glasgow, Scotland, UK

INTRODUCTION

The aim of this chapter is to introduce the fundamental concepts underlying the use of health economic evaluation methods, with particular reference to treatments for asthma and COPD. The chapter begins with a section that introduces the general framework of cost-effectiveness and cost-utility analyses, the predominant forms of health economic evaluation that have been used to assess value for money in the respiratory field, and indeed in health care evaluation more generally. As part of this background, a distinction is drawn between cost-of-illness studies, which examine only the financial burden that a disease places a health system or society, and full economic evaluations that assess the change in both financial burden and health outcomes that can be expected from implementing a treatment intervention. The following two sections then review the published literature on cost-of-illness studies and health economic evaluations for asthma and COPD treatments separately. A final section summarizes the chapter.

HEALTH ECONOMIC EVALUATION

Health economic evaluations are used to assist decision-makers in allocating scarce resources. Twenty three countries around the world currently have some form of guidelines on pharmacoeconomics, primarily European countries but there are also others, including Australia, Canada, and the United States [1].

In England and Wales, the relevant decision-making agency is the National Institute for Health and Clinical Excellence (NICE), an organization who are independent from Government and responsible for providing guidance on the use of health technologies and the implementation of public health programs. In the process of developing guidance on use of technologies, NICE brings together all available clinical and economic evidence in order to decide whether the adoption of the technology (drug, device or treatment) represents good value for the NHS (National Institute of Health and Clinical Excellence. A guide to NICE. London: National Institute of Health and Clinical Excellence, 2005). The National Institute for Health and Clinical Excellence (NICE) also produces clinical guidelines and published their own guidelines for the management of COPD in 2004 [2].

Within this chapter, two different types of economic evaluations are presented. The first is cost-effectiveness analysis and the second, cost-utility analysis. The main difference between the two relates to how health outcomes are measured. Cost-effectiveness studies present results in terms of natural units such as the cost per exacerbation avoided and improvement in health status [3], whereas a cost-utility evaluation identifies the cost per unit of utility (usually the Quality Adjusted Life Year, QALY). Utilities and the QALY concept are described in detail below.

The outcome measure(s) used should be based upon the aim and intended use of the evaluation. The use of natural units as outcome measures used within a cost-effectiveness analysis, constrains the end use of such a study. For example a study that investigates and reports the cost of an improvement in lung function can only be logically compared with another study where same outcome is also reported.

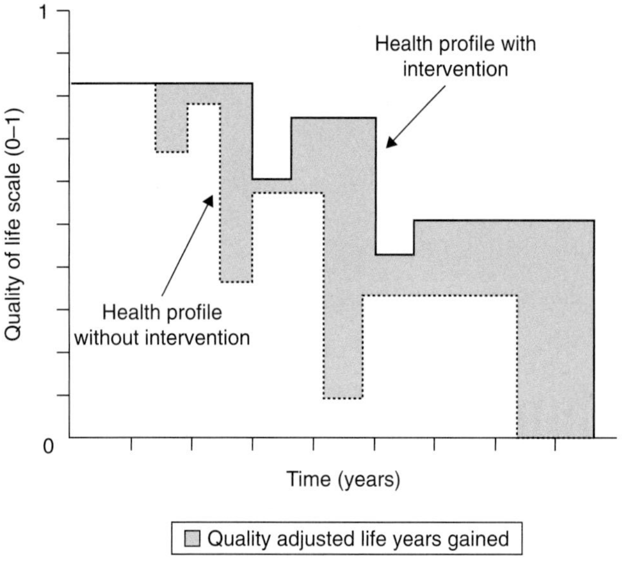

FIG. 62.1 Quality Adjusted Life-Year profiles for a hypothetical patient with and without a treatment intervention. Profiles weight length of life by quality of life on a zero-one scale where one represents perfect health and zero death. In a chronic disease such as COPD, exacerbations might result in reduced quality of life for a period of time with incomplete recovery.

One of the problems found in both asthma and COPD is that there are a number of different outcome measures that could easily be chosen to investigate, for example, cost per exacerbation avoided or cost per symptom-free day. As a result there are few studies that report the same outcome measure and so comparison is difficult. Even if cost per exacerbation avoided was reported in every study, cost-effectiveness studies can only inform the efficient use of resources within the disease itself (or closely linked conditions). It is not possible to directly compare an exacerbation avoided in COPD to an increase in survival time for cancer patients.

A more useful approach for the decision-maker is to use a generic measure that captures the effect of a disease and its treatment on both quantity and quality of life, and which can be compared across disease areas. NICE recommends these cost-utility analyses as the appropriate way to make comparative assessments of value for money within a health system [4]. Although not universally accepted, the QALY is nevertheless acknowledged as the most popular measure of utilities for making comparative assessment of the value for money of alternative treatment choices. The QALY is by far the most accepted health-related utility measure and is the preferred outcome measure in many countries including Canada, New Zealand, Sweden, England and Wales, the Netherlands, and the United States [1].

The QALY quantifies changes in utility over the life of the patient and has two components, quality and quantity of life. This concept can be visualized as shown in Fig. 62.1.

As has previously been described, the use of the QALY within an economic evaluation enables treatments to be compared to one another both across and within disease areas using a cost per QALY ratio. This ratio represses the incremental costs (C) and incremental benefits (B) of a new therapy (x) are compared to the main comparator (y). The ICER is worked out using the simple formula:

$$\text{ICER} = \frac{Cx - Cy}{Bx - By}$$

The reimbursement of one treatment will displace health care resources spent on other treatments within the health system (so called opportunity cost). It is therefore important to ensure that the treatment reimbursed provides sufficient value for money. For this a "guide price" or a threshold from which it can be decided whether or not any one treatment should be reimbursed by the health care system is necessary. In the United Kingdom, NICE's documented threshold is said to be between £20,000 and £30,000 per QALY [5], such that below £20,000 per QALY there is a high probability of the technology being accepted and that above £30,000 per QALY there is less chance of the technology being accepted. In theory, this threshold should reflect the value of those treatments displaced at the margin of the health care budget [6].

Two main designs of economic evaluation exist: either based upon individual patient data from a clinical trial or observational study data, or aggregated using a decision analytic model. Model-based evaluations typically incorporate data from a wider range of sources than clinical trial based studies and can include data from: clinical and observational trials, burden of disease studies, epidemiology, and natural history studies.

The perspective that is adopted by the authors of the study depends upon the target audience [7] and can either be a specific provider/provider institution (i.e. the National Health Service, NHS), the patient/patient group, a third party payer (i.e. an insurer), or the perspective of society [8]. The perspective used will determine the costs employed. For example, a health perspective will only include costs that fall directly within the health care budget, but a broader societal perspective can include cost to individual patients and carers as well as other sectors, such as public sector services or employers.

Particularly important is the choice of comparator used. The use of an inappropriate comparator can bias an analysis and render little value for decision-making. Most published guidelines for economic evaluations assert that the comparator of interest is current treatment [9].

A potentially fundamental problem of conducting economic analyses alongside clinical trials, [10] is that many trials often only assess effectiveness relative to placebo.

Methods of handling uncertainty need to be explored, particularly if the study uses a decision analytical modeling framework. Sensitivity analysis, preferably probabilistic should be employed [11]; as this allows the combined uncertainty of all the parameters in the model to be included [12], by using the full probability distributions of each input into the model, rather than just the point estimates.

An important distinction exists between "full health economic evaluations" that involve a comparison of the resource use costs and health outcome consequences of two or more alternative treatment interventions [8] and so called cost-of-illness studies. Cost-of-illness studies are descriptive, providing information about the financial burden of disease on the health service and on patients and society more broadly. However, while a high financial burden of disease may imply the *potential* for a given treatment to offer value for money (by reducing that burden and thereby offsetting disease costs), value for money can only be proven in a full *incremental* cost-effectiveness analysis that compares the cost and health outcomes of the new treatment with the existing pattern of care. Therefore, in the two sections that follow, the literature on cost-of-illness and cost-effectiveness of treatments in asthma and COPD are presented in separate sections. More space is given to the cost-effectiveness of treatments, since this represents the ability of the health system to positively affect the burden of disease.

COST-OF-ILLNESS STUDIES

Asthma

The costs associated with the management of asthma in adults and in children, from both the health care payer and the societal perspective, are substantial worldwide. A number of cost-of-illness studies have attempted to estimate the annual cost of asthma and had reported a wide range of estimates. This is, in part, due to the variations in study design (e.g. the individual cost components that were included), perspective, and population of interest.

In the United States, the total annual costs for pediatric patients with mild to moderate persistent disease has been estimated to be US$564 [13], while the costs for adult patients ranged from US$1250 [14] to US$7106 [15]. In particular, one study estimated the total annual direct cost of asthma in the United States to be $1.48 billion [16], representing the most costly occupational illness in the United States. In Europe, the estimated average cost for a child with asthma ranged from €883 [17] to €2202 [18] per year, and the estimated average cost for an adult with asthma ranged from €632 [17] to €2745 [18] per year. In Asia, despite a lower prevalence of asthma than in the West, the cost of disease management is a significant and an increasing economic burden. In Japan, the annual direct medical cost associated with managing asthma has been estimated to be US$8.2 billion. This accounted for 1.9% [19] of the annual cost of medical care and 18.9% of the annual cost of medical care for respiratory diseases.

Pharmaceutical expenditures represent an increasing share of total direct medical costs. In the United Kingdom, the estimated annual cost of asthma was £752.6 million, of which, 8% was hospital admissions, 13% general practice consultations, and 79% community prescriptions [20]. Indirect costs represent 40–50% of total societal burden of asthma. Loss of school and working days for children with asthma and their parents are the largest component of indirect costs. The overall cost to the family, health service, and society associated with caring for children aged 1–5 years in the United Kingdom in the 12 months following hospitalization for wheeze or asthma has been estimated to be £14.53 million [21].

Studies that assessed the relationship between costs and disease severity have all showed that costs significantly increase with increasing asthma severity. In particular, one Italian study showed that when stratified by disease severity, the total annual cost-per-patient according to increasing severity of asthma was €720, €1046, €1535, and €3328 for patients with intermittent, mild persistent, moderate persistent, and severe persistent asthma, respectively [22]. Similarly, based on the data from the TENOR database, the health care expenditures for an average uncontrolled patient in the United States, has been shown to be more than double that of an average control patient [15].

COPD

The World Health Organisation (2000) reported that deaths attributable to COPD are the fifth largest killer globally; accounting for 4.5% of deaths worldwide [23]. The proportion of deaths for COPD varies significantly between regions of the world. Of particular concern is the Western Pacific region where COPD accounts for 13.8% of all deaths and where COPD is ranked as the second leading cause of death.

Data from both the United States and the United Kingdom show that the prevalence of COPD is comparatively small among the under 45s but increases markedly throughout later years [24]: approximately 1% of the UK general population and increasing with age to around 5% of men between 65 and 74, rising to 10% in men over 75 [25].

The true rate of COPD within the general population is likely to be greater than that recorded. There is much evidence that suggests COPD is heavily under diagnosed worldwide [25–27], and this is likely to be the situation, particularly among mild cases of COPD.

Total costs to the UK NHS for COPD have been estimated somewhere between £486 million [28, 29] and £848 million [30, 31] per year. The major drivers of this cost are disease severity and exacerbations.

Costs increase most as disease severity moves from moderate to severe (with a much smaller increase between mild and moderate) [32–38]. As FEV_1 deteriorates, a general shift from outpatient care to hospitalization, an increase in the use of oxygen therapy and a subsequent increase in total costs occurs, especially in the most advanced stages of the disease [39].

Exacerbations are the leading drivers of cost in COPD. A serious exacerbation will lead to hospitalization; indeed an exacerbation is the main reason why a COPD patient would attend hospital. COPD patients take up around 1 million bed days per year in the United Kingdom alone.

Reducing or preventing disease progression and/or an exacerbation, (particularly severe exacerbations), will have a direct effect on the total cost of COPD. In England and Wales, McGuire calculated that for every exacerbation-related hospital admission avoided, a total saving of approximately £1200 would be made [40].

REVIEW OF ECONOMIC EVALUATIONS

Asthma

Several cost-effectiveness analyses have been undertaken, alongside small and short-term clinical trials. These studies evaluated the cost-effectiveness of fluticasone propionate–salmeterol compared with fluticasone propionate alone, montelukast, or a combination of the two. In two of these studies FSC was found to dominate its comparators. One study evaluated the cost-effectiveness between three treatments: fluticasone propionate–salmeterol 100/50 μg BID, fluticasone propionate 100 μg BID, and montelukast 10 mg QID [41]. Based on data from four clinical trials assessing patients with symptomatic asthma previously treated with inhaled SABA, FSC showed significantly greater increase in FEV_1 and proportion of symptom-free days and reduced the number of exacerbations. Although daily drug costs were higher for FSC, daily mean exacerbation costs were much lower, making it more effective and less costly. The mean daily costs per patient exacerbation for fluticasone propionate–salmeterol, fluticasone propionate and montelukast were $29, $128, and $154, respectively.

In another study, the cost-effectiveness of fluticasone propionate-salmeterol 100/50μg was compared to montelukast as initial maintenance therapy in patients with persistent asthma uncontrolled with a SABA alone [42]. The economic analysis used symptom-free days and 12% or greater improvement in FEV_1 from a 12-week randomized control trial along with direct medical costs to determine the incremental cost-effectiveness ratio. Considering the SFD endpoint the ICER for FSC versus montelukast was $2.87 indicating that it cost on average an extra $2.87 per day for an additional SFD when FSC is used instead of montelukast. The ICER for FEV_1 was $1.79, indicating that it costs on average an extra $1.79 per day for an additional patient to achieve a lung function improvement of 12% or greater.

The cost-effectiveness of fluticasone propionate–salmeterol 100/50 μg BID was compared to fluticasone propionate 100 μg BID plus montelukast 10 mg QID, based on data from a 12-week randomized control trial [41, 43]. FP/salmeterol was associated with a higher proportion of patients with a 12% improvement in FEV_1 and similar efficacy in terms of symptom-free days at a lower cost than FP plus montelukast. The mean daily cost per successfully treated patient was significantly lower with FP/salmeterol ($6.77) than with FP + montelukast ($14.59). The cost per symptom-free day for FP/salmeterol was $11.96 and $17.10 for FP plus montelukast.

Retrospective database analyses have increased in the past few years. Single treatment regimes (fluticasone propionate, montelukast, and salmeterol) have been shown to incur similar total costs over 1 year [44]. Based on retrospective data from the North Carolina Medicaid program, asthmatic patients starting either fluticasone propionate 44 μg were compared with those using montelukast 5 and 10 mg in a cost-minimization study. No significant differences were found in the adjusted asthma-related health care costs following 1 year. A secondary analysis comparing montelukast to salmeterol also showed no difference between health care service use and costs.

Brüggenjürgen *et al.* [45] evaluated the cost-effectiveness of adjustable maintenance dosing with budesonide 160 μg/formoterol 4.5 μg twice daily in a single inhaler ($n = 1679$) compared with fixed dosing ($n = 1618$) alongside a 12-week randomized-controlled trial [46]. In general the results of the trial showed no significant differences in the primary efficacy outcome – health-related quality of life between the two groups. Despite the differences in fewer daily study drug inhalations (mean inhalations: 2.63 adjustable dosing versus 3.82 fixed dosing; $p < 0.001$) a cost-minimization analysis was conducted. The direct asthma-related cost associated with adjustable dosing was significantly lower than that with fixed dosing (mean costs: €221 versus €292; $p < 0.001$).

Pieters et al [47] suggested greater cost effectiveness when adding salmeterol to fluticasone propionate compared with montelukast/fluticasone in patients with asthma uncontrolled with inhaled corticosteroid alone. In the Netherlands salmeterol 50 μg/fluticasone 100 μg twice daily ($n = 356$) was compared with fluticasone 100 μg twice daily plus oral montelukast 10 mg daily ($n = 369$) in a 12-week randomized-controlled trial [48] that measured the proportion of successfully treated weeks (based on mean morning PEF); in addition, episode-free days and symptom-free days and nights were also recorded. Incremental cost-effectiveness ratios were calculated and sensitivity analysis was carried out. Salmeterol/fluticasone combination was associated with significantly better outcomes compared with fluticasone plus montelukast (median difference 25%; $p < 0.001$). The mean total direct health service costs per patient in the fluticasone plus montelukast group was 16% greater than that in the salmeterol/fluticasone group (€1.98 versus €2.25). Salmeterol/fluticasone showed dominance – clinically more effective in all outcome measures and less costly than fluticasone plus montelukast.

Jönsson *et al.* [49] assessed the cost-effectiveness of increasing the dose of inhaled corticosteroid, the addition of a long-acting β_2-agonist, and both strategies in combination, in patients with mild to moderate persistent asthma. In Sweden four treatment scenarios were compared in a randomized-controlled trial: budesonide 100 μg ($n = 313$); budesonide 200 μg ($n = 302$); budesonide 100 μg plus formoterol 4.5 μg ($n = 310$); and budesonide 200 μg plus formoterol 4.5 μg ($n = 308$) [50]. The primary clinical outcome evaluated was the number of symptom-free days per year. In addition, the number of severe exacerbations per patient per year was also recorded. Health care resource data and medication use were recorded during the trial. Sensitivity analysis was carried out by applying UK and Spanish costs to the analysis. Budesonide 100 μg plus formoterol 4.5 μg was shown to be dominated. The incremental cost-effectiveness ratio for budesonide 200 μg plus formoterol 4.5 μg compared with budesonide 200 μg alone was SEK 21 per symptom-free day gained.

Lindgren *et al.* [51] evaluated the cost-effectiveness for formoterol 4.5 μg ($n = 57$) and salbutamol 200 μg ($n = 58$) as reliever medication in Sweden and Spain, based on data from a 6-month trial. Resource use was recorded in the trial. Costs associated with resource use were calculated and effectiveness was expressed as exacerbations rate per year; the percentage of symptom-free days was also recorded. For Sweden, the total health care cost per severe exacerbation avoided was €504 per year (95%CI 95–2239) and the health care cost per any exacerbation avoided was €140 per year (95%CI 40–573). For Spain, the total health care cost per severe exacerbation avoided was €1088 per year (95%CI 537–3559) and the health care cost per any exacerbation avoided was €489 per year (95%CI 281–924).

Price *et al.* [52] compared the cost-effectiveness of adjustable dosing of budesonide 80–160 μg/formoterol 4.5 μg ($n = 782$) with fixed maintenance dosing ($n = 771$), alongside a pragmatic randomized-controlled trial [53]. From the UK perspective the primary outcome measure was the net proportion of patients experiencing clinically significant improvement in quality of life, assessed by the mini-Asthma Quality of Life Questionnaire (AQLQ); symptom-free days without the use of rescue medication was used as a secondary outcome measure. Data on resource use were collected prospectively during the trial. The trial reported no significant difference in effectiveness between the two groups; therefore, a cost-minimization analysis was carried out. The total per patient daily cost was £1.13 (95%CI £1.08–£1.18) in the adjustable dosing group compared with £1.31 (95%CI £1.27–£1.34) in the fixed dosing group ($p < 0.001$).

COPD

The aim of current treatment for COPD, in the absence of a disease cure, is to prevent and control symptoms, reduce the frequency and severity of exacerbations, improve health status, and improve exercise tolerance [54].

"Smoking cessation is the single most ... cost effective way to reduce exposure to COPD risk factors" [54].

Smoking cessation is recommended at all stages of the disease. Interventions to assist individuals in quitting smoking include: counseling; nicotine replacement products: gum, patches, spray, lozenges; and the drug therapies: Bupropion, Nortriptyline and Varenicline.

A review of smoking cessation on interventions (e.g. advice and nicotine gum) for different subgroups of the US population [55] found that societal cost per life year ranged from \$0 to \$13,000 (1993 prices) (€0–8944). A more recent review of the literature [56], carried out from a UK societal perspective, found cost-effectiveness estimates of between £212 (€304) for advice and £873 (€1253) for advice, self help, nicotine replacement therapy, and a specialist cessation service, per life year saved. Given this evidence, smoking cessation interventions seems to represent good value for money.

There are numerous options available for treating COPD. These are set out in Table 62.1, along with the most

TABLE 62.1 Treatment guidelines for COPD by disease severity [47].

Disease stage	Mild	Moderate	Severe	Very severe
Classification	$FEV_1/FVC < 0.70$ $FEV_1 \geq 80\%$ predicted	$FEV_1/FVC < 0.70$ $50\% \leq FEV_1 < 80\%$ predicted	$FEV_1/FVC < 0.70$ $30\% \leq FEV_1 < 50\%$ predicted	$FEV_1/FVC < 0.70$ $FEV_1 < 30\%$ predicted or $FEV_1 < 50\%$ predicted plus chronic respiratory failure
Treatment	Influenza vaccination Short-acting broncholdilator added when necessary: short-acting β_2-agonists/short acting anticholinergic			
		Add rehabilitation Add regular treatment with one or more long-acting broncholdilators if needed: long-acting β_2-agonists/long acting anticholinergic		
			Add inhaled corticosteroids if exacerbations are repeated	
				Add long-term oxygen if chronic respiratory failure. Consider surgical treatments.

established and up to date treatment guidelines for the management of the disease. Treatment is dependent upon disease severity; classified here in terms of the GOLD guidelines, with four states from mild through to very severe.

Disease management is additive; as the disease progresses through the four stages, more treatments are added. In respect of this, this section has four parts: mild, moderate, severe, and very severe. Under each and where available, the economic evidence for each treatment is presented at the stage where this treatment is first recommended.

Mild COPD

Influenza vaccination

Influenza vaccines are recommended to treat all COPD patients, regardless of severity. The cost for the vaccine for high risk people was found to be \$570 (€392) per life year saved. The Pnuemococcol vaccine is recommended for patients aged 65 years or over. For this subgroup, the cost-effectiveness ratio has been found to be between \$1800 to \$2200 (€1238–1513) per life year [55]. Both vaccines represent good value for money.

Use a short-acting bronchodilator, short-acting anticholinergic when necessary

One study modeled the costs and cost-effectiveness of theophylline compared to ipratropium in moderate to severe COPD. In this case no Incremental Cost-Effectiveness Ratio (ICER) was calculated since ipratropium was found to be both less costly and more cost-effective than theophylline [57].

Sometimes these products are combined and evidence exists to support one such combination. One study examined the efficacy of ipratropium combined with albuterol, compared to each monocomponent in moderate to severe COPD patients. The ICER for the combination product was not calculated: combining ipratropium with albuterol compared to albuterol alone was found to strictly dominate [58].

Moderate COPD

Add rehabilitation

Rehabilitation is a multifaceted approach which incorporates a wide range of programs to improve quality of life and functional independence and to reduce symptoms and disability for the patient. Areas within the rehabilitation program can include exercise and physical training, psychological, social interactions, education around the disease, and about nutrition [59].

A recent US study found a marked reduction in cost following a rehabilitation program in terms of a reduction in health care utilization after 1 year of treatment compared to the previous year of no treatment [60]. A 2001 UK paper reports that the cost-effectiveness of a rehabilitation program leads to overall gains in utility in the patients receiving rehabilitation and also a drop in total costs, compared to the control group [61]. The probability that rehabilitation was cost-effective was 90% at £10,000 and 95% at £17,000 (€ = 24,405) [62]. Given the data, rehabilitation seems to represent a cost-effective treatment.

Add regular treatment with one or more long-acting bronchodilators/anticholinergic

One clinical trial based study compared formoterol to the short-acting anticholinergic, ipratropium and also to placebo in moderate to severe patients. In terms of the ICER for HRQoL, formoterol dominated ipratropium. The ICER for formoterol compared to placebo was found to be \$25.20 over the 12 weeks [63].

A UK study that examined salmeterol compared to placebo in moderate patients found that hospital costs and costs for GP contracts and COPD-related medications were lower in the salmeterol arm of the trial compared to placebo. In addition significant benefits were observed in the measures of health outcome that were used, including improved health status (as measured by the St George's Respiratory Questionnaire, SGRQ). These improvements were found to be made at a moderate cost that is for a four point increase in SGRQ score, was found to have an ICER of £4.44 per patient per day (£1622 per year) [64].

In 2002, the long-acting anticholinergic, tiotropium was launched onto the UK market and later, in 2004 to the United States. Several related economic evaluations have centered on tiotropium.

The earliest cost-effectiveness analysis for tiotropium was carried out for the Netherlands and Belgium and compared tiotropium to ipratropium. The study found that tiotropium led to improved health outcomes, including an improvement in SGRQ score and an improvement in FEV_1 at a small additional cost [65].

Oostenbrink *et al.* then designed a 1-year Markov model around disease severity states (moderate, severe and very severe) and exacerbations (severe) in order to examine the cost-effectiveness of ipratropium, salmeterol, and tiotropium in the Netherlands and in Canada. In the Netherlands and in Canada, tiotropium was found to be associated with maximum expected net benefit for plausible values of the ceiling ratio [66].

Two additional models, adapted from Ref. [66], and applied to different countries have recently been published. The first, to a Greek setting, examined the cost-effectiveness of tiotropium compared to salmeterol, from the perspective of the Greek NHS [67]. Tiotropium was concluded to be cost-effective, but there was no statistically significant difference found between the treatments.

The second study applied the original model to a Spanish setting and sought to compare the cost-effectiveness of tiotropium, salmeterol, ipratropium, and placebo. The ICER per QALY for tiotropium compared to salmeterol was found to be €4118 and for salmeterol compared to ipratropium, €38,931. At low values of the ceiling ratio (below €11,000 per QALY) ipratropium was found to have the highest probability of being cost-effective. Above €11,000, tiotropium had the highest probability of being cost-effective [68].

Severe COPD

Add inhaled corticosteroids if exacerbations are repeated

An economic model examined the effects of adding inhaled corticosteroids to treatment for three groups; all COPD patients, patients with stage 2 or 3 disease, and stage 3 disease.

The model was 3 years in duration with 12 cycles; each cycle being 3 months in length. The model was split into three states according to disease severity, one being least severe and three most severe. Treatment was found to be cost-effective when given to patients with stage 2 or 3 disease [69].

Three clinical trial based studies have compared fluticasone to placebo in COPD patients. The first was by van den Boom *et al.* in a group of patients who, prior to the study, were undiagnosed but had signs of obstructive airway disease. The incremental cost-effectiveness for early treatment was found to be $13,016 per QALY [70].

The second study looked at moderate to severe COPD patients and found that fluticasone produced significant improvements in a range of clinical outcomes including improvement in FEV_1 and in the proportion of patients remaining exacerbation free compared to placebo. Fluticasone was found to be cost-effective in the study population [71].

The third study investigated QALY's over 3 years and also life expectancy between the treatment arms in moderate and severe COPD patients. Cost per QALY was found to be £9500 for treatment with fluticasone compared to placebo and a life expectancy increase was found of 23 days [72].

The combination product Salmeterol and Fluticasone (Seretide) was introduced into the EU market in 2000. Since then it has experienced substantial growth in patient market share. Spencer and colleagues' model [73] employ a Markov model in order to compare the cost-effectiveness of this combination product in relation to usual care. The model has four mutually exclusive states: mild, moderate, severe, and death. Each cycle of the model was 3 months in duration over a time horizon of 25 years. Baseline values for the model were sourced from GSK clinical trial data (TRISTAN), published medical literature, and from expert opinion. The paper concludes that adding a long-acting β_2-agonist to an inhaled corticosteroid may represent a cost-effective treatment in those patients who have a history of frequent exacerbations and poorly reversible COPD.

The combination drug, Budesonide and Formoterol (Symbicort) was introduced into the EU market in 2001. This product has also obtained a significant foothold in the market. Lofdahl *et al.* investigated the cost-effectiveness of combining budesonide with formoterol compared to either monocomponent alone or placebo in severe and very severe patients. Improvements were observed in the combination arm compared to the other arms in terms of fewer accident and emergency admissions, hospitalizations, and specialist visits. The combination was found to be cost-effective if the decision-maker is willing to pay about €2 per day per avoided exacerbation [74].

Two studies have investigated the cost-effectiveness of combining a β_2-agonist with an inhaled corticosteroid. The first examines three treatments arms to one another. The first arm was a combined β_2-agonist and corticosteroid, the second, a β_2-agonist with an anticholinergic, and the third, a β_2-agonist with a placebo. The study was conducted along the side of a two and a half year clinical trial of moderate and severe patients. The addition of an inhaled corticosteroid to a β2 agonist led to significant benefits in respiratory function and reduced the number of restricted activity days. On the other hand, the addition of an anticholinergic was found to be both expensive and of no long-term value. The ICER of the corticosteroid versus the placebo was found to be $200 per 10% improvement in FEV_1 and $5.35 per symptom-free day gained [75].

The second study by Gagnon *et al.* was an observational study and compared four treatment arms. The first, a combination of a long-acting β_2-agonist and an inhaled corticosteroid, the second, just the inhaled corticosteroid, the third, just the long-acting β_2-agonist, and the fourth, placebo. Both the placebo and the inhaled corticosteroid arms were dominated by the long-acting β_2-agonist and the combination. The ICER between the inhaled corticosteroid/long-acting β_2-agonist and the long-acting β_2-agonist was found to be $91,430 [76].

Very severe COPD

Add long-term oxygen if chronic respiratory failure

Oxygen therapy is recommended for use in some very severe patients in order to increase the partial pressure of oxygen in the arterial blood (PaO_2) and has three different uses: to relieve dyspnea, to aid oxygen intake during exercise and for long term, continuous treatment [54]. Whilst it is clear that oxygen therapy forms a high proportion of outpatient costs for COPD patients [77], little is known about the cost-effectiveness of this treatment, outside of findings that concentrators for the delivery of oxygen therapy may be more cost-effective than cylinders [78, 79].

Consider surgical treatments

Lung volume reduction surgery (LVRS) involves cutting away around 30% of the diseased lung tissue in order to increase the effectiveness and efficiency of the remaining lung and surrounding muscle. It is only suitable for some patients with very severe COPD.

Estimates of cost-effectiveness ratios for LVRS, from a societal perspective, compared to medical therapy were found to be $190,000 (€ = 1386) per QALY for 3 years and $53,000 (€ = 36,459) per QALY for 10 years [80]. Because this procedure is very expensive, patients undergoing LVRS must be carefully selected [54] in order to select those patients who are most likely to benefit and least likely to have associated complications. The National Emphysema treatment trial identified the profile of this subset of patients: those with an upper lobe emphysema and a low exercise capacity. The cost-effectiveness ratio for this subgroup, showed a more favorable ICER than for the study average at $98,000 (€ = 67,414) per QALY for 3 years and $21,000 (€ = 14,446) per QALY for 10 years [80].

The cost of surgery, coupled with lifetime follow-up costs, makes lung transplantation very expensive. In addition the supply of suitable organs is very limited. Like with LVRS, lung transplantation is a viable option for some patients who are in the most severe COPD stratification, however it is important to identify those patients who are most likely to benefit and least likely to have adverse reactions to the surgery.

A small observational study yielded a lifetime incremental cost of $176,817 (€ = 121,635) per QALY compared to those not receiving surgery [81]. A larger study which used a simulation model to calculate a cost-effectiveness ratio over 40 years yielded a cost of 167,000 Netherlands Guilders per QALY (€ = 75,781) compared to non surgery [82]. A UK study, extrapolated to 15 years found that the costs per QALY gained were $48,241 (€ = 33,186) for single lung transplantation and $32,803 (€ = 22,566) for double lung transplantation [83]. All three of these studies found substantial gains in quality of life following lung transplantation; however evidence of an effect on survival was mixed.

Bullectomy may be an option when a large air filled bulla exists that fills half of the thoracic volume and compresses the relatively normal adjacent parenchyma [84]. There is no cost-effectiveness information on this surgery option.

SUMMARY

Treatment patterns have changed remarkably in both asthma and COPD in recent years. In both disease areas, the use of pharmacotherapy as part of the armamentarium of treatment options for patient management has increased, with a trend toward using alternative products in combination, especially for the most severe patients. In general terms, the value for money of treatments, including pharmacotherapies, in both disease areas has been demonstrated. Nevertheless, in looking to the future, the challenges of demonstrating value for money of increasingly novel approaches to patient treatment and management in both disease areas are likely to increase. In asthma, the potential for biologic treatments, at high levels of cost-per-patient year, are likely to bring issues of cost-effectiveness to the fore in any regions of the world. Similarly, in COPD, the increasing use of combination therapy (inhaled corticosteroid and long-acting β-agonist) and triple therapy (ICS/LABA combination plus anticholinergic) are likely to put pressure on pharmaceutical budgets in this patient group. Consequently, issues of cost-effectiveness and value for money are likely to become ever more important as all health systems grapple with providing the maximum health gain possible for their population from limited resources.

References

1. ISPOR. Pharmacoeconomic guidelines around the world, 2007; http://www.ispor.org/PEguidelines/index.asp.
2. National Collaborating Centre for Chronic Conditions. Chronic obstructive pulmonary disease: National clinical guideline on management of chronic obstructive pulmonary disease in adults in primary and secondary care. *Thorax* 58(Suppl 1), 2004, No. 1.
3. Jones PW, Wilson KS. Cost-effectiveness of salmeterol in patients with chronic obstructive pulmonary disease: An economic evaluation. *Respir Med* 97(1): 20–26, 2003.
4. National Institute for Clinical Excellence. Guide to the Methods of Technology Appraisal. London: National Institute for Clinical Excellence, 2004.
5. Rawlins MD, Culyer AJ. National institute for clinical excellence and its value judgments. *BMJ* 329: 224–27, 2004.
6. Culyer AJ, McCabe C, Briggs A, Claxton K, Buxton M, Akehurst R, Sculpher M, Brazier J. Searching for a threshold, not setting one: The role of the national institute for health and clinical excellence. *J Health Serv Res and Pol* 12: 56–58, 2007.
7. Philips Z, Bojke L, Sculpher M, Claxton K, Golder S. Good practice guidelines for decision-analytic modelling in health technology assessment: A review and consolidation of quality assessment. *Pharmacoeconomics* 24(4): 355–71, 2006.
8. Drummond MF, Sculpher MJ, Torrance GW, O'Brien BJ, Stoddart GL. Methods for the Economic Evaluation of Health Care Programmes, 3rd edn. Oxford University Press, 2005.
9. Drummond M, Sculpher M. Common methodological flaws in economic evaluations. *Med Care* 43(7(S)): II-5–II-14, 2005.
10. Sculpher M, Claxton K, Drummond M, McCabe C. Whither trial-based economic evaluation for health care decision making?. *Health Econ*(15): 677–87, 2006.
11. Claxton K, Sculpher M, McCabe C, Briggs AH, Akehurst R, Buxton M, Brazier J, O'Hagan T. Probabilistic sensitivity analysis for NICE technology assessment: Not an optional extra. *Health Econ* 14: 339–47, 2005.
12. Briggs AH. Probabilistic analysis of cost-effectiveness models: Choosing between treatment strategies for gastroesophageal reflux disease. *Med Deci Making* 22: 290–308, 2002.
13. Gendo K, Sullivan SD, Lozano P, Finklestein JA, Fuhlbrigge A, Weiss KB. Resource costs for asthma-related care among pediatric patients in managed care. *Ann Allergy Asthma Immunol* 91(3): 251–57, 2003.
14. Sapra S, Nielson K, Martin BC. The net cost of asthma to North Carolina medicaid and the influence of comorbidities that drive asthma costs. *J Asthma* 42: 469–77, 2005.
15. Sullivan SD. The burden of uncontrolled asthma on the U.S. health care system. *Manag Care* 14(Suppl 8): 25–27, 2005.
16. Leigh JP, Yasmeen S, Miller TR. Medical costs of fourteen occupational illnesses in the United States in 1999. *Scand J Work Environ Health* 29(4): 304–13, 2003.
17. Herjavecz I, Nagy GB, Gyurkovits K *et al.* Cost, morbidity, and control of asthma in Hungary: The Hunair Study. *J Asthma* 40(6): 673–81, 2003.
18. Schramm, B., Ehlken, B., Smala, A., Quednau, K., Berger, K., Nowak, D. Cost of illness of atopic asthma and seasonal allergic rhinitis in Germany: 1-yr retrospective study. *Eur Respir J 2003*; 21(1): 116–22..
19. Izumi T. Effects of medication selections on medical expenditures for asthma. *Respir Med* 2: 323–30, 2002.
20. Gupta R, Sheikh A, Strachan DP, Anderson HR. Burdens of allergic disease in the UK: Secondary analyses of national databases. *Clin Exp Allergy* 34(4): 520–26, 2004.
21. Stevens CA, Turner D, Kuehni CE, Couriel JM, Silverman M. The economic impact of preschool asthma and wheeze. *Eur Respir J* 21(6): 1000–6, 2003.
22. Antonicelli I, Bucca C, Neri M *et al.* Asthma severity and medical resource utilisation. *Eur Respir J* 23(5): 723–29, 2004.
23. Murray Christopher JL, Lopez AD, Mathers CD, Stein C. The global burden of disease 2000 project: Aims, methods and data sources. World Health Organization, 2001.
24. US Department of Health and Human Services 2004, *Summary Health Statistics for US Adults: National Health Interview Survey, 2004.*
25. Calverley PM, Bellamy D. The challenge of providing better care for patients with chronic obstructive pulmonary disease: The poor relation of airways obstruction?. *Thorax* 55: 78–82, 2000.
26. Fukuchi Y, Nishimura M, Ichinose M, Adachi M, Nagai A, Kuriyama T, Takahashi K, Nishimura K, Ishioka S, Aizawa H, Zaher C. COPD in Japan: The Nippon COPD epidemiology study. *Respiratory* 9(4), 2004.
27. Zielinski J, Bednarek M, Gorecka D, Viegi G, Hurd SS, Fukuchi Y, Lai CKW, Ran PX. Increasing COPD awareness. *Eur Res J* 27: 833–52, 2006.
28. Britton M, Authors FN. The burden of COPD in the U.K.: Results from the Confronting COPD survey. *Respir Med* 97(Suppl C): S71–79, 2003a.
29. Calverley PM. The burden of obstructive lung disease in the UK-COPD and asthma. *Thorax* 53(Supp 4), 1998.
30. Guest JF. The annual cost of chronic obstructive pulmonary disease to the UK's National Health Service. *Dis Manag Health Outcome* 5(2), 1999.
31. Sullivan SD, Ramsey SD, Lee TA. The economic burden of COPD. *Chest* 117(Suppl 2): 5–9, 2000.

32. Britton M, Authors FN. The burden of COPD in the U.K.: Results from the confronting COPD survey. *Respir Med* 97(Suppl C): S71–79, 2003b.
33. Chapman KR, Bourbeau J, Rance L, Authors FN, Bourbeau J, Rance L. The burden of COPD in Canada: Results from the confronting COPD survey. *Respir Med* 97(Suppl C): S23–31, 2003.
34. Dal Negro R, Rossi A, Cerveri I, Authors FN, Rossi A, Cerveri I. The burden of COPD in Italy: Results from the Confronting COPD survey. *Respir Med* 97(Suppl C): S43–50, 2003.
35. Halpern MT, Stanford RH, Borker R, Authors FN, Stanford RH, Borker R. The burden of COPD in the U.S.A.: Results from the confronting COPD survey. *Respir Med* 97(Suppl C): S81–89, 2003.
36. Izquierdo JL, Authors FN. The burden of COPD in Spain: Results from the confronting COPD survey. *Respir Med* 97(Suppl C): S61–69, 2003.
37. Piperno D, Huchon G, Pribil C, Boucot I, Similowski T. The burden of COPD in France: Results from the confronting COPD survey. *Respir Med* 97(Supplement C): S33–42, 2003.
38. Wouters EF, Authors FN. The burden of COPD in The Netherlands: Results from the confronting COPD survey. *Respir Med* 97(Suppl C): S51–59, 2003.
39. Jansson SA, Andersson F, Borg S, Ericsson A, Jonsson E, Lundback B, Authors FN, Andersson F, Borg S, Ericsson A, Jonsson E, Lundback B. Costs of COPD in Sweden according to disease severity. *Chest* 122(6): 1994–2002, 2002.
40. McGuire A, Irwin DE, Fenn P, Gray A, Anderson P, Lovering A, MacGowan A. The excess cost of acute exacerbations of chronic bronchitis in patients aged 45 and older in England and Wales. *Value Health* 4(5), 2001.
41. O'Connor RD, Stanford R, Crim C *et al.* Effect of fluticasone propionate and salmeterol in a single device, fluticasone propionate and montelukast on overall asthma control, exacerbations, and costs. *Ann Allergy Asthma Immunol* 93(6): 581–88, 2004.
42. Borker r, Emmett A, Jhingran P, Rickard K, Dorinsky P. Determining economic feasibility of fluticasone propionate-salmeterol vs. montelukast in the treatment of persistent asthma using a net benefit approach and cost-effectiveness acceptability curves. *Ann Allergy Asthma Immunol* 95(2): 181–89, 2005.
43. O'Connor RD, Nelson H, Borker R *et al.* Cost effectiveness of fluticasone propionate plus salmeterol versus fluticasone propionate plus montelukast in the treatment of persistent asthma. *Pharmacoeconomics* 22: 815–25, 2004a.
44. Balkrishnan R, Nelsen LM, Kulkarni AS, Pleasants RA, Whitmire JT, Schechter MS. Outcomes associated with initiation of different controller therapies in a medicaid asthmatic population: A retrospective data analysis. *Journal of Asthma* 42(1): 35–40, 2005.
45. Bruggenjurgen B, Selim D, Kardos P *et al.* Economic assessment of adjustable maintenance treatment with budesonide/formoterol in a single inhaler versus fixed treatment in asthma. *Pharmacoeconomics* 23(7): 723–31, 2005.
46. Buhl R, Kardos P, Richter K *et al.* The effect of adjustable dosing with budesonide/formoterol on health-related quality of life and asthma control compared with fixed dosing. *Curr Med Respiry Opin* 20(8): 1209–20, 2004.
47. Pieters WR, Wilson KK, Smith HC, Tamminga JJ, Sondhi S. Salmeterol/fluticasone propionate versus fluticasone propionate plus montelukast: A cost-effective comparison for asthma. *Treat Respir Med* 4(2): 129–38, 2005.
48. Ringdal N, Eliraz A, Pruzinec R *et al.* The salmeterol/fluticasone combination is more effective than fluticasone plus oral montelukast in asthma. *Respir Med* 97(3): 234–41, 2003.
49. Jonsson B, Berggren F, Svensson K, O'Byrne PM. An economic evaluation of combination treatment with budesonide and formoterol in patients with mild-to-moderate persistent asthma. *Respir Med* 98(11): 1146–54, 2004.
50. O'Byrne PM, Barnes PJ, Rodriguez-Roisin R *et al.* Low dose inhaled budesonide and formoterol in mild persistent asthma: The OPTIMA randomized trial. *Am J Respir Crit Care Med* 8(Part 1): 1392–97, 2001 Oct 15.
51. Lindgren B, Sears MR, Campbell M. Cost-effectiveness of formoterol and salbutamol as asthma reliever medication in Sweden and in Spain. *Int J Clin Pract* 59(1): 62–68, 2005.
52. Price D, Haughney J, Lloydm A, Hutchinson J, Plumb J. An economic evaluation of adjustable and fixed dosing with budesonide/formoterol via a single inhaler in asthma patients: the ASSURE study. *Curr Med Respir Opin* 20(10): 1671–79, 2004.
53. Ind PW, Haughney J, Price D, Rosen JP, Kennelly J. Adjustable and fixed dosing with budesonide/formoterol via a single inhaler in asthma patients: The ASSURE study. *Respir Med* 98(5): 464–75, 2004.
54. GOLD 2006, *Global initiative for chronic Obstructive Lung Disease.*
55. Tengs TO, Adams ME, Pliskin JS, Gelb Safran D, Siegel JE, Weinstein MC, Graham JD. Five-hundred life-saving interventions and their cost-effectiveness. *Risk Anal* 15(3): 369–90, 1995.
56. Parrott S, Godfrey C, Raw M, West R, McNeil A. Guidance for commissioners on the cost effectiveness of smoking cessation interventions. *Thorax* 53(Suppl 5, Part 2): S2–37, 1998.
57. Jubran A, Gross N, Ramsdell J, Simonian R, Schuttenhelm K. Comparative cost-effectiveness analysis of theophylline and ipratropium bromide in chronic obstructive pulmonary disease. A three-centre study. *Chest* 103: 678–84, 1993.
58. Friedman M, Serby CW, Menjoge SS, Wilson JD, Hilleman DE, Witek TJJ. Pharmacoeconomic evaluation of a combination of ipratropium plus albuterol compared with ipratropium alone and albuterol alone in COPD. *Chest* 115(3): 635–41, 1999.
59. British Thoracic Society Standards of Care Subcommittee on Pulmonary Rehabilitation. Pulmonary rehabilitation. *Thorax* 56(827): 834, 2001.
60. Raskin J, Spiegler P, McCusker C, ZuWallack R, Bernstein M, Busby J, DiLauro P, Griffiths K, Haggerty M, Hovey L, McEvoy D, Reardon JZ, Stavrolakes K, Stockdale-Woolley R, Thompson P, Trimmer G, Youngson L. The effect of pulmonary rehabilitation on healthcare utilization in chronic obstructive pulmonary disease: The northeast pulmonary rehabilitation consortium. *J Cardiopulm Rehabil* 26(4), 2006.
61. Griffiths TL, Phillips CJ, Davies S, Burr ML, Campbell IA. Cost effectiveness of an outpatient multidisciplinary pulmonary rehabilitation programme. *Thorax* 56(10): 779–84, 2001a.
62. Griffiths TL, Phillips CJ, Davies S, Burr ML, Campbell IA. Cost effectiveness of an outpatient multidisciplinary pulmonary rehabilitation programme. *Thorax* 56(10): 779–84, 2001b.
63. Hogan TJ, Geddes R, Gonzalez E. An economic assessment of inhaled formoterol dry powder versus ipratropium bromide pressurized metered dose inhaler in the treatment of chronic obstructive pulmonary disease. *Clin Ther* 25(1): 285–97, 2003.
64. Jones PW, Wilson K, Sondhi S. Cost-effectiveness of salmeterol in patients with chronic obstructive pulmonary disease: An economic evaluation. *Respir Med* 97(1): 20–26, 2003.
65. Oostenbrink JB, Rutten-van Molken MPMH, Al MJ, Van Noord JA, Vincken W. One-year cost-effectiveness of tiotropium versus ipratopium to treat chronic obstructive pulmonary disease. *Eur Respir J* 23(2), 2004.
66. Oostenbrink JB, Rutten-van Molken MP, Monz BU, FitzGerald JM. Probabilistic Markov model to assess the cost-effectiveness of bronchodilator therapy in COPD patients in different countries. [see comment]. *Value Health* 8(1): 32–46, 2005.
67. Maniadakis N, Tzanakis N, Fragoulakis V, Hatzikou M, Siafakas N. Economic evaluation of tiotropium and salmeterol in the treatment of chronic obstructive pulmonary disease (COPD) in Greece. *Curr Med Res Opin* 22(8), 2006.
68. Rutten-van Molken MP, Oostenbrink JB, Miravitlles M, Monz BU. Modelling the 5-year cost effectiveness of tiotropium, salmeterol and ipratropium for the treatment of chronic obstructive pulmonary disease in Spain. *Eur J Health Econ* 8: 123–35, 2007.
69. Sin DD, Golmohammadi K, Jacobs P. Cost-effectiveness of inhaled corticosteroids for chronic obstructive pulmonary disease according to disease severity. *Am J Med* 116(5): 325–31, 2004.
70. van den Boom G, Rutten-van Molken MP. The cost effectiveness of early treatment with fluticasone propionate 250 microg twice a day in

subjects with obstructive airway disease. Results of the DIMCA program. *Am J Respir Crit Care Med* 164(11): 2057–66, 2001.
71. Ayres JG, Price MJ, Efthimiou J. Cost-effectiveness of fluticasone propionate in the treatment of chronic obstructive pulmonary disease: A double-blind randomized, placebo-controlled trial. *Respir Med* 97(3): 212–20, 2003.
72. Briggs AH, Lozano-Ortega G, Spencer S, Bale G, Spencer MD, Burge PS. Estimating the cost-effectiveness of fluticasone proportionate for treating chronic obstructive pulmonary disease in the presence of missing data. *Value Health* 9(4): 227–35, 2006.
73. Spencer M, Briggs AH, Grossman RF, Rance L. Development of an economic model to assess the cost-effectiveness of treatment interventions for chronic obstructive pulmonary disease. *Pharmacoeconomics* 23(6): 619–37, 2005.
74. Lofdahl CG, Ericsson A, Svensson K, Andreasson E. Cost effectiveness of budesonide/formoterol in a single inhaler for COPD compared with each monocomponent used alone. *Pharmacoeconomics* 23(4): 365–75, 2005.
75. Rutten-van Molken M, Van Doorslaer EK, Jansen MC, Kerstjens HA, Rutten FF. Costs and effects of inhaled corticosteroids and bronchodilators in asthma and chronic obstructive pulmonary disease. *Am J Respir Crit Care Med* 151(4): 975–82, 1995.
76. Gagnon YM, Levy AR, Spencer MD, Hurley JS, Frost FJ, Mapel DW, Briggs AH. Economic evaluation of treating chronic obstructive pulmonary disease with inhaled corticosteroids and long-acting beta2-agonists in a health maintenance organization. *Respir Med* 99(12): 1534–45, 2005.
77. Pelletier-Fleury N, Lanoe JL, Fleury B, Fardeau M. The cost of treating COPD patients with long-term oxygen therapy in a French population. *Chest* 110: 411–16, 1996.
78. O'Neill B, Bradley JM, Heaney L, O'Neill C, MacMahon J. Short burst oxygen therapy in chronic obstructive pulmonary disease: A patient survey and cost analysis. *Int J Clin Pract* 59(7): 751–53, 2005.
79. Heaney L, McAllister D, MacMahon J. Cost minimisation analysis of provision of oxygen at home: Are the drug tariff guidelines cost effective? *BMJ* 319: 19–23, 1999.
80. National Emphysema Treatment Trial Research Group. Cost effectiveness of lung-volume-reduction surgery for patients with severe emphysema. *N Engl J Med* 348(21): 2092–102, 2003.
81. Ramsey SD, Patrick DL, Albert RK, Larson EB, Wood DE. The cost-effectiveness of lung transplantation. A pilot study. University of Washington medical centre lung transplantation study group. *Chest* 108: 1594–601, 1995.
82. Al MJ, Koopmanschap NA, van Enckevort PJ, Geertsma A, Van der Bij W, de Boer WJ, TenVergert EM. Cost-effectiveness of lung transplantation in the Netherlands. *Chest* 113: 124–30, 1998.
83. Anyanwu AC, McGuire A, Rogers CA, Murday AJ. An economic evaluation of lung transplantation. *J Thor Cardiovasc Surg* 123(3): 411–20, 2002.
84. Meyers BF, Patterson GA. Chronic obstructive pulmonary disease v 10: Bullectomy, lung volume reduction surgery, and transplantation for patients with chronic obstructive pulmonary disease. *Thorax* 58: 634–38, 2003.

9

PART

Clinical Management of Asthma and COPD

CHAPTER 63

Management of Chronic Asthma: In Adults

Rodolfo M. Pascual and Stephen P. Peters

Section on Pulmonary, Critical Care, Allergy and Immunologic Diseases, Department of Internal Medicine, Wake Forest University School of Medicine, Medical Center Boulevard, Winston-Salem, NC, USA

INTRODUCTION

Knowledge gained from research into the pathogenesis of asthma combined with numerous randomized clinical trials (RCTs) has revolutionized our understanding of this prevalent disease and its treatment. However, although mortality from asthma appears to be declining in developed nations, the World Health Organization estimates that asthma mortality worldwide will increase in the next 10 years. The reduction in asthma mortality in developed nations appears to coincide with the development of management guidelines [1, 2] that emphasize disease control using anti-inflammatory medications. Unfortunately, asthma morbidity in developed nations remains a significant problem. A recent survey assessing asthma control in the United States demonstrated that the majority of patients surveyed (55%) characterized their asthma as being "uncontrolled." This was despite the fact that for the most part these patients were receiving care in accordance with widely accepted asthma guidelines. These authors also found that most patients do not utilize asthma action plans [3]. Additionally, patients with poorly controlled asthma use a disproportionate amount of healthcare resources. Despite the recognition that patient education strategies and the use of management guidelines improve asthma outcomes, these tools remain underutilized by practitioners, patients often remain ignorant about their asthma and outcomes achieved in the real world fall short of what can be achieved in clinical studies.

Thus, one key overarching theme is that to improve outcomes for individual patients one must improve the process of care. Even though, where guidelines are rigorously implemented, in some patients asthma continues to remain poorly controlled, it has been shown that control can be achieved in a majority of patients [4]. Asthma management is a complex process for a number of reasons. First, asthma is best characterized as a syndrome that exhibits heterogeneity with respect to its etiology, clinical presentation, severity, natural history, and response to therapy. As such, asthma has been termed a complex genetic disorder in that multiple genes and gene–environment interactions lead to the characteristic symptoms and signs in a given patient. One might say that there are many asthma phenotypes so that a single-management approach will not work for all patients, and therapy must be tailored to the individual patient. Second, within each patient, the severity of symptoms and lung function varies over time. Some patients experience prolonged remissions of symptoms whereas others rarely achieve adequate symptom relief. Asthma is also characterized by the occurrence of exacerbations, events where an environmental factor triggers increased airway inflammation, bronchial hyperresponsiveness, and resultant increased symptoms. Such exacerbations may be brief, lasting for minutes and readily ablated by bronchodilators or prolonged lasting days-to-weeks and requiring intensive pharmacotherapy. In earlier studies assessing the efficacy of drugs used to treat asthma changes in lung function is the major outcome of interest; however, more recent studies assess for more patient-centric outcomes like reductions in exacerbations or symptoms and improvements in quality of life. Third, unlike many other diseases where treatment decisions are made predominantly by the healthcare provider during a scheduled encounter, the minute-to-minute variability of asthma requires that the patient frequently makes

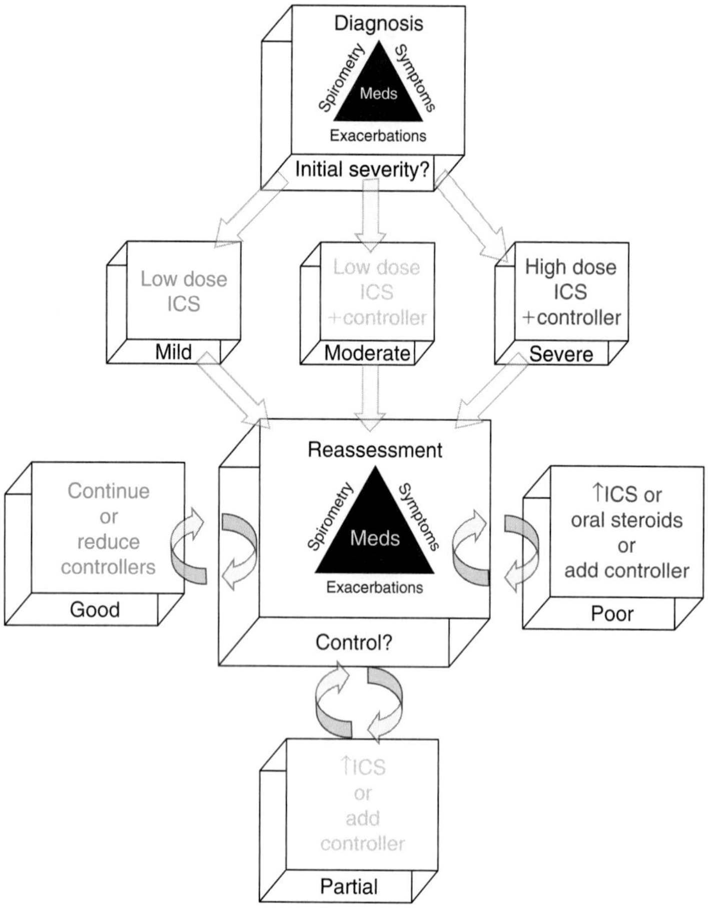

FIG. 63.1 Managing asthma. Once the diagnosis is established, symptoms, medication use, exacerbation history, and spirometry are used to assess initial asthma severity. After treatment is initiated, the patient then undergoes reassessment at an appropriate interval based on severity. From that point on, the emphasis is placed on disease control and control is constantly reassessed at an interval dictated by the degree of recent asthma control. Importantly, the same set of disease measurement tools such as symptoms, medication use, recent exacerbation history, and spirometry are also used to subsequently assess asthma control.

treatment decisions. The quality of this decision-making is dependent on the patient's understanding of their disease and the tools provided to the patient for their disease assessment.

MANAGING ASTHMA NOT IN EXACERBATION

Once the diagnosis of asthma has been established and co-morbidities identified, a management process should be implemented (Fig. 63.1). Though as stated previously the management of asthma should be individualized, guidelines are quite useful in that they outline a process of care. First, the initial severity should be determined; it is especially important to identify those subjects with severe disease. Because patients with severe asthma have a disproportionate morbidity, mortality, and healthcare utilization compared to nonsevere patients, hence facing a higher risk from their disease, it is important to target them for intensive treatment and frequent follow-up. Previous intubation for asthma, ICU care for asthma, hospitalization for asthma, excessive use of asthma medications, psychiatric disease, and low lung function are risk factors for future life-threatening episodes of asthma. Another variable that is not accounted for by merely assessing medication use, symptoms, and lung function is the exacerbation history. Recent severe asthma exacerbations (in the prior 3 months) or the recent use of a steroid burst appear to be powerful indicators of future severe asthma exacerbations [5]. Thus, we recommend that the recent exacerbation history and history of life-threatening episodes be included as part of the assessment of initial asthma severity and subsequently as a measure of asthma control on reassessment. It is our opinion that this initial step of determining severity be performed in any patient with asthma who is new to a practitioner even if the diagnosis had been previously established by another healthcare provider, because asthma frequently is not appropriately treated or misdiagnosed. As shown in Fig. 63.1 and Table 63.1 and consistent with the latest guidelines [1, 2], the assessment of severity integrates the impairment domain that includes frequency of daytime and nocturnal symptoms, the amount of medication used, and lung function with the risk domain that accounts for exacerbation history, medication side effects, and potential loss of lung function. Impairment as it applies to daily activities, exercise, and the performance of job-specific tasks should also be determined. Second, initial therapy should be selected based on severity using increasingly intensive anti-inflammatory regimens for increasingly severe disease as

TABLE 63.1 Initial asthma severity* and treatment steps in adults.*

Components of severity		Intermittent	Persistent: Mild	Persistent: Moderate	Persistent: Severe
Impairment domain	Symptoms	≤2 days/week	>2 days/week, but not daily	Daily	Throughout the day
	Nocturnal symptoms	≤2 ×/month	3–4 ×/month	>1 ×/week but not every night	Often every night
	Short-acting β-agonist use for rescue	≤2 days/week	>2 days/week, but less than daily	Daily	Several times per day
	Interference with normal activity	None	Minor	Some	Extreme
	Lung function when not in exacerbation	FEV_1 > 80% predicted	FEV_1 ≥ 80% predicted	FEV_1 60–79% predicted	FEV_1 < 60% predicted
Risk domain	Exacerbations requiring systemic corticosteroids	0–1/year	≥2/year		
Treatment step		Step 1	Step 2	Step 3 or 4	Step 5 or 6

Adapted from Figures 3–4c in Ref. [1].
*Severity is determined by the most severe component of each domain.

shown in Table 63.2. Third, reassessment should be performed at return visits scheduled at an interval commensurate with either initial asthma severity or later, current asthma control. As shown in Table 63.3, medications are then adjusted in a stepwise manner based on current control and drug-related side effects. When control is achieved for a reasonable period of time, consideration is given to stepping down therapy but if the patient remains uncontrolled, therapy is intensified. In a new patient, a less adherent patient, or in a patient with recent poor asthma control, more frequent reassessment is indicated. Asthma management is thus a dynamic process requiring flexibility from both the patient and the healthcare provider.

Implementing initial therapy

As noted in Table 63.2, inhaled corticosteroids (ICSs) should be used as the initial controller in all patients with persistent asthma of any severity. Of all pharmacological agents ICS have most consistently been shown to improve lung function, quality of life and prevent exacerbations while having a good side effect profile. For *initial* therapy in severe asthma we additionally recommend a short course of oral corticosteroids (0.25–0.5 mg/kg daily). Lung function and asthma control should then be reassessed after 2–4 weeks. Systemic steroid therapy serves two purposes as it improves the chances of quickly achieving asthma control in those patients who will be responsive while at the same time identifying those patients who may be steroid-resistant. In the steroid-phobic patient with mild persistent asthma (Step 2), a leukotriene receptor antagonist (LTRA) can be used as an alternative to ICS, however, it should be understood that for a significant majority of patients, ICS are superior to LTRAs for all relevant asthma treatment outcomes [6, 7]. The benefits of ICS have been known for over two decades [8, 9], and the regular daily use of ICS for asthma control has been the standard of care for persistent asthma. However, it is known that many patients do not use these agents on a regular basis, some studies suggesting that patients refill medications less than half of the time that they are prescribed. Importantly, the use of combination inhalers improves adherence when compared to the use of multiple single-drug devices [10, 11]. Patients with only occasional symptoms (cough, wheezing, dyspnea), no more than one nocturnal awakening/month and normal lung function (Step 1), do not need controller therapy and may be given a reliever to be used as needed. Additionally, such patients that do not need a controller and suffer from exercise-induced bronchospasm can be advised to take a β-agonist with a rapid onset of action whether it be short- or long-acting in duration [12], or an inhaled cromone [13], about 30 min prior to exercise, or a daily LTRA in a prophylactic manner [14]. Patients already taking controller therapy with otherwise well-controlled asthma may also utilize an additional agent (β-agonist, inhaled cromone) which they are not using as a controller, as an add-on for exercise prophylaxis if they experience bothersome symptoms with exercise.

Interestingly, several recent studies of novel treatment protocols using ICS have perhaps emulated the

TABLE 63.2 Treatment for asthma in adults.

Treatment step	Intermittent	Persistent				
		Mild	Moderate		Severe	
	Step 1	Step 2	Step 3	Step 4	Step 5	Step 6
Preferred controller	SABA prn	Low-dose ICS	Low-dose ICS + LABA or Medium dose ICS	Medium-dose ICS + LABA	High-dose ICS + LABA and Consider Omalizumab Consider other additional controllers	High-dose ICS + LABA + low-dose oral corticosteroids and Consider Omalizumab Consider other additional controllers
Alternative controller		Cromolyn, LTRA, Nedocromil, or Theophylline	Low-dose ICS + LTRA, Theophylline, or Zileuton	Medium-dose ICS + LTRA, Theophylline, or Zileuton		
Allergen immunotherapy		Consider	Consider	Consider	No	No
Written action plan and self-monitoring		Consider	Yes	Yes	Yes	Yes
Return visit	Annual	6 months	3–4 months	3–4 months	1–2 months	2–4 weeks

Partially adapted from Figures 4–5 in Ref. [1].
For treatment step see Table 63.1 (initial severity) or Table 63.3 (current control).
ICS, inhaled corticosteroid; SABA, short-acting β_2-agonist; LTRA, leukotriene receptor antagonist; sustained-release theophylline is preferred.

de facto use of these agents by many patients. Since ICS have side effects and as discussed previously adherence to daily use appears to be poor, one obvious question is whether controller agents can be used as needed rather than a regular basis. In one such study, the level of asthma control was examined in patients with mild asthma randomized to an intermittent treatment plan, which instituted the use of ICS or oral steroids only when indicated using a protocol similar to many asthma action plans, or randomized to regular use of ICS, or randomized to regular use of a leukotriene-receptor antagonist. Intermittent treatment appeared to be equivalent to regular use of controllers for rates of asthma exacerbation, even though the overall steroid consumption by the intermittent group was quite low [15]. In another study with a different design, symptom-driven use of an inhaler continuing both an ICS and a short-acting β-agonist was as effective as regular use of ICS and as-needed use of the short-acting β-agonist; steroid consumption again was lower in the group using the controller agent as needed [16]. In yet another trial with a similar theme a long-acting bronchodilator that happens to have an onset of action similar to many relief inhalers was combined with ICS and used as both a regularly taken controller and an as-needed relief inhaler. This strategy proved to be superior in terms of preventing asthma exacerbations when compared to regular use of the ICS/long-acting β-agonist (LABA) combination inhaler as a controller with a short-acting β-agonist used as-needed as a reliever. Total steroid use was actually reduced when the ICS/LABA was used as both a controller and a reliever [17].

All patients should be provided with and instructed on the proper usage of a quick relief medication, preferably a short-acting inhaled β-agonist. Regular use of short-acting bronchodilators is clinically not better than as-needed use, and regular use leads to more medication use and side effects, so short-acting bronchodilators should only be used as-needed for relief of symptoms.

Unfortunately, the proper use of inhaled medication is often not straightforward. The patient should be showed the proper technique for using a device when it is prescribed, and this technique should be reinforced frequently especially if there are excessive side effects or a poor therapeutic response. Quick relief medications should be used as-needed whereas controller medications should be used on a scheduled basis, and this concept should be reinforced frequently with the patient.

Asthma self-management education

Studies suggest that patient education and environmental control programs are effective in reducing asthma morbidity. Asthma self-management education reduces hospitalizations, emergency room visits, unscheduled physician visits, days-off from work or school, and nocturnal asthma symptoms and improves quality of life [18–21]. The goals of asthma education are to improve patient understanding of their disease, medication adherence, self-management and efficiency of healthcare utilization. Educated patients are more likely to recognize and modify environmental factors and avoid triggers that worsen asthma control. In many patients as a prolonged exacerbation is developing, peak

TABLE 63.3 Assessing asthma control in adults.

Components of control		Well controlled (good)	Not well controlled (partial)	Very poorly controlled (poor)
Impairment	Symptoms	≤2 days/week	>2 days/week	Throughout the day
	Nighttime awakenings	≤2 ×/month	1–3 ×/week	≥4 ×/week
	Interference with normal activity	None	Some limitation	Extremely limited
	SABA use for rescue	≤2 days/week	>2 days/week	Several times per day
	[a]Lung function	>80% personal best/predicted	60–80% personal best/predicted	<60% personal best/predicted
	Asthma control questionnaires			
	ATAQ	0	1–2	3–4
	ACQ	≤ 0.75	≥ 1.5	N/A
	ACT	≥ 20	16–19	≤ 15
Risk	Exacerbations requiring systemic steroids	0–1/year	≥2/year	
	[a]Progressive loss of lung function			
	[b]Treatment-related adverse effects			
Recommended action for treatment		1. Maintain current step 2. Regular follow-up 3. Consider step down if well controlled at least 3 months	1. Step up one step 2. Reevaluate in 2–6 weeks	1. Step up 1–2 steps 2. Reevaluate in 2 weeks 3. Consider short course of systemic corticosteroids

Adapted from Figures 4–7 in Ref. [1].
Personal best or predicted FEV_1 or PEF values may be used but the authors recommend using personal best values.
ACQ, Asthma Control Questionnaire©; ACT, Asthma Control Test™; ATAQ, Asthma Therapy Assessment Questionnaire©.
[a]Loss of lung function can be determined by serial measurement during follow-up care.
[b]Treatment-related adverse effects should be considered when assessing risk and making changes in therapy.

expiratory flow (PEF) will begin to deteriorate prior to symptom worsening. Moreover, there appears to be a subset of asthma patients that do not sense worsening airflow obstruction. So we recommend that the action plan include elements of symptom monitoring and PEF measurements and that it be introduced to the patient at the earliest opportunity. An example of an asthma action plan is provided in Fig. 63.2.

Environmental control is an important, often overlooked part of the management plan for all asthmatics, especially those with severe asthma and it consists of measures to control environmental triggers. Avoidance of aeroallergens, viral respiratory pathogens, air pollution, and certain drugs can prevent exacerbations, reduce the need for drug treatment, and decrease utilization of emergency facilities. The relative importance of different indoor allergens may vary among populations, and there is substantial variability among patients. For moderate-to-severe asthma, we recommend an assessment for allergy performed by a subspecialty trained allergist. For mild disease a simpler assessment for other atopic disease (rhinitis, eczema), triggers, seasonal patterns, and home and work environment is often sufficient. The possibility of occupational asthma should be determined during the initial assessment and is discussed in Chapter 36. Most patients with chronic asthma have numerous triggers, and avoidance strategies should be discussed at every opportunity. Allergen immunotherapy can benefit highly selected patients with defined allergic triggers [22]. Though serious complications from immunotherapy are rare they occur more frequently in patients with asthma [23], so immunotherapy should not be used in patients with severe asthma and should only be administered by physicians specifically trained in it use.

Strategies to avoid respiratory infections particularly viral upper respiratory tract infections should be discussed. Inactivated influenza vaccine may be safely administered to patients with asthma [24]. Though it has not conclusively been shown that influenza vaccination decreases the incidence

Asthma Action Plan for ____________________ Last Reviewed:____________________

Instructions:

1. Measure your PEF both upon awakening and at 4 PM each day.
2. Each morning answer the following questions: Did you wake up last night? How many times did you use albuterol in the last 24 hours? Did you have any trouble with exercise or any activities?
3. Based on both items #1 and #2 above find your asthma zone.
4. Follow the instructions in the action box for your current asthma zone
5. If you experience severe or frightening symptoms call EMS (dial 911)
6. If you do not need immediate care but have concerns call our office at__________, if you do not get through proceed to the nearest emergency room.

Green Zone	Action
Your PEF is >_____ (80% personal best)	Continue to monitor PEF twice daily
You are not waking up at night because of your asthma	Continue usual asthma medications
You use your albuterol inhaler less than once daily	ICS____________ ______μg two puffs twice a day
	Albuterol MDI 2 puffs every six hours as needed

Yellow Zone	Action
Your PEF <_____ but >______ for more than 12 hours (50–80% of personal best) OR You woke up last night because of your asthma OR You are using your albuterol inhaler more than twice a day OR You are more short of breath with exertion or exercise	Increase __________ to 4 puffs twice a day AND Monitor your PEF, and if it does not improve to greater than ______within 24 hours take Prednisone 30 mg daily for 4 days AND USE Albuterol MDI 2 puffs every four hours as needed AND *Contact your physician at* __________

Red Zone	Action
Your PEF is <_____ (50% of personal best) at any time OR You are using your albuterol inhaler more than 4 times a day OR Relief from albuterol lasts less than 4 hours OR You are so short of breath that daily activities are difficult	Immediately go to the emergency room or call EMS (dial 911) AND Take prednisone 60 mg once AND USE Albuterol MDI 4 puffs up to every twenty minutes until you obtain medical attention

FIG. 63.2 Asthma action plan. Action plans should be simple, empowering the patient to self-monitor their asthma while providing clear, specific instructions to deal with exacerbations. A combination of symptoms, medication use, and lung function is used by the patient to assess the current control of their asthma. It is important to emphasize that patients should not attempt to manage severe exacerbations without medical assistance.

of exacerbations of asthma [25], patients with asthma should receive inactivated influenza vaccine if there are no other contraindications and especially if they have other conditions that increase the risk of death from influenza infection [26].

Pharmacologic therapy

The categorization of asthma by clinical severity is a key tenet of the major clinical guidelines. The severity indices use symptoms, lung function, and rescue medication use as variables [27, 2]. Asthma drugs currently available are classified as long-term control medications or "controllers" and quick-relief medications or "relievers" based on their principle pharmacodynamic and clinical effects. Short-acting agents that primarily work by inducing rapid bronchodilation such as inhaled β-agonists or anticholinergics are considered quick-relief medications; they should be taken on an as-needed basis. Corticosteroids, long-acting β-agonists, leukotriene pathway inhibitors, cromolyn sodium, nedocromil sodium, sustained-release theophylline, and omalizumab act by promoting sustained bronchodilation, have anti-inflammatory effects or both and are considered as long-term control medications; they should be taken on a scheduled basis. Though the controller versus reliever concept is straightforward to the healthcare provider, it is often not grasped by patients who sometimes will not take controllers when they are symptom-free or may take relief medications on a regular basis. It is important to provide patients with clear instructions regarding the proper use of all of their asthma medications.

Inhaled β_2-adrenergic agonists are the drugs of choice for relief of symptoms due to acute airway obstruction. Short-acting β-agonists have a rapid onset of action and 3–6 h duration of activity. At recommended doses, inhaled β-agonists have few adverse effects. Since regular use of short-acting β-agonists has not been shown to be superior to as needed use of these agents and several studies have associated excessive use of short-acting β-agonists with asthma-related morbidity and mortality, we recommend that these agents be used only as needed. Inhaled anticholinergic drugs that are also effective bronchodilators, though generally less so than β-agonists in asthma, are also indicated for the relief of acute bronchospasm.

Glucocorticoid steroids are the most effective controller agents for treating asthma. Inhaled steroids improve lung function when compared to placebo and reduce exacerbation rates. Patients with asthma stabilized on inhaled steroids experience more exacerbations when the steroids are withdrawn [28]. Unique to this drug class, retrospective data suggests that the consistent use of inhaled steroids reduces asthma mortality [29, 30]. However, as a drug class, the inhaled glucocorticoids suffer from important limitations. First, the dose–response curve of inhaled steroids is relatively flat-higher doses are only incrementally better than low-medium doses. Second, there appears to be substantial variability in both acute [31], and chronic efficacy of glucocorticoids particularly with respect to improvements in bronchial hyperresponsiveness and FEV_1. Finally, some patients appear to be relatively insensitive to the effects of glucocorticoids, some being so insensitive that their asthma may be termed steroid-resistant. The steroid-insensitive phenotype is associated with severe asthma, persistent symptoms, frequent nocturnal symptoms, low and partially irreversible lung function, need for systemic steroids at a young age, higher maintenance doses of oral steroids and African-American race [32]. Fortunately, additional efficacy can often be obtained, using combinations of steroids and other controllers rather than just increasing the inhaled steroid dose. Combinations of ICS with long-acting β-agonists [4, 33–35], LTRAs [36], or theophylline [37], have been shown to provide superior bronchodilation and improve other outcomes when compared to single controller agents alone.

LABAs have at least 12-h duration of action and are not currently recommended for the short-term relief of acute symptoms. When used alone LABAs are clearly inferior to ICS with regards to exacerbations and *there is current consensus that LABAs should not be used as monotherapy for the control of asthma of any severity.* As discussed previously LABA and ICS combinations provide clinical synergy and promote better adherence; the combination of formoterol and budesonide was shown to be effective when used as both a controller, and relief agent and thus provides the advantage of a single device used for both purposes [17].

Theophylline now is primarily used as an adjunct and has steroid sparing effects. Theophylline has a narrow therapeutic index, interacts with many drugs and its use requires monitoring of serum drug levels. The steroid sparing effects of theophylline seem to occur at levels below the traditional therapeutic range of 10-20 mg/l [37]. We recommend that sustained release preparations of theophylline be used and titrated to steady state serum concentrations of 5–10 mg/l or peak concentrations of no higher than 15 mg/l.

Anticholinergic agents by blocking muscarinic receptors induce airway smooth-muscle relaxation. In general, the short-acting anticholinergic agent ipratropium is not so effective as β-agonists as a bronchodilator in asthma. The use of anticholinergic agents in asthma has been reviewed [38]. Inhaled ipratropium can be used as a reliever in chronic asthma; the utility of ipratropium in an individual patient should be determined empirically. Ipratropium, when combined with β-agonists appears to provide faster and greater bronchodilation in acute severe asthma when compared to β-agonists alone and can be added to β-agonists for this purpose when the response to β-agonists is insufficient [2]. One small study showed that the long-acting, inhaled anticholinergic drug tiotropium can provide modest bronchodilation and as would be expected provided prolonged protection against methacholine-induced bronchoconstriction in subjects with asthma [39]. Clinical trials are underway to determine the utility of tiotropium in asthma.

Cromolyn sodium and nedocromil sodium (cromones) are classified as controller agents, and inhibit both antigen-mediated and nonimmunologically (i.e. exercise)-mediated bronchoconstriction in asthmatics. Nedocromil generally has displayed greater potency in protecting patients against exercise-induced bronchospasm [40]. These agents are safe, are considered first-line agents in the treatment of children with asthma but are inferior to inhaled steroids with respect to most relevant asthma outcomes [41, 42].

Leukotriene pathway inhibitors are a group of drugs that mitigate the effects of leukotrienes derived from the 5-lipoxygenation of arachidonic acid. There are two classes of agents: 5-lipoxygenase (5-LO) enzyme inhibitors and cysteinyl-leukotriene receptor type 1 antagonists. When compared head-to-head corticosteroids are superior asthma controllers. [6, 7, 43] Leukotriene pathway inhibitors may be substituted for ICS in selected patients with mild disease and are especially useful when steroids are poorly tolerated, when steroid use is not desired by the patient, or when there is significant, concomitant rhinosinusitis. LTRAs have been shown to prevent exercise-induced bronchospasm [44] and can be used as monotherapy for this indication in selected patients with mild asthma. Finally, leukotriene pathway inhibitors are quite useful as add-on therapy to inhaled steroids in moderate-to-severe asthma and have steroid sparing effects [36, 41, 45–47].

Reassessment

Since asthma is characterized by substantial variability in disease severity over time and response to therapy, reassessment and adjustment of therapy is the cornerstone of management. More severe disease is associated with more frequent exacerbations, more risk of severe events like hospitalization and death and probably with more rapid loss of lung function over time, hence, more severe disease requires frequent reassessment (every few weeks to every few months). On the other hand some patients with mild asthma may be stable clinically for long periods of time and annual reassessment can be appropriate.

Disease monitoring tools

When one is reassessing asthma the key parameter to be ascertained is disease control. When asthma is well controlled, lung function is stable or not declining faster than usual, there is minimal rescue medication use, sleep is not disturbed by asthma symptoms, daytime symptoms are absent or minimal, and there is no limitation to activity. Moreover, there is little day-to-day variability in these clinical parameters; well-controlled asthma has little to no impact on quality of life. In contrast, poorly controlled disease is characterized by frequent symptoms, exacerbations, rescue medication usage and accelerated loss of lung function over time; these patients suffer significant disability. Though well-controlled asthma can be achieved in a majority of patients at least over short periods of time, unfortunately, a significant minority of patients do not achieve good control [4]. If symptoms are not adequately addressed, quality of life is not improved and patient satisfaction is likely to be poor. Additionally, a subset of patients may have few symptoms but poor lung function. Still other patients have controlled symptoms and good lung function but at the cost of heavy medication usage. Generally, such patients are considered to have more severe disease than those otherwise clinically similar patients, but maintained on less medication. For this reason many validated instruments like the Asthma Control Test™ (ACT) (www.asthmacontrol.com) integrate an assessment of symptoms and medication use. We encourage the use of this and similar instruments along with the use of office spirometry to assess for asthma control. These instruments often have important limitations, for example the ACT™ only assesses short-term control using a 1-month recall; though this is valid given the fact that longer periods of recall are associated with diminished accuracy. We suggest that one ask additional questions especially probing for exacerbations that have occurred since the last assessment like asthma-related unscheduled physician visits, emergency department visits, or the use of systemic glucocorticoids. Medication side effects should be noted and when therapy has changed spirometry should be measured. Thus, we recommend the use of a combination of objective tools and subjective assessment to reach an integrated determination as to whether asthma is well controlled, partially controlled, or poorly controlled.

As shown in Fig. 63.1 asthma control can be put into one of three categories: poorly controlled, partially (not well) controlled, or well controlled. Well-controlled asthma is straightforward and one can either keep the medication regimen the same or consider a step down in intensity. In any case the patient should be reminded that control can be lost if adherence to medications or environmental control lapses as many studies have shown that withdrawal of controllers leads to increased rates of exacerbation [28, 34]. Before one considers removing a medication it is important to consider if current control will be lost, that is if that medication was the one that helped to achieve control. When there is only partial control or poor control, generally another agent needs to be added or substituted for one currently in use but that is not working. Unfortunately, there is a tendency to add medications sequentially until control is achieved rather than assessing if an added medication had been helpful. The heterogeneity of response to a given medication is likely due to a combination of genetic and environmental factors. Pharmacogenomics is a growing field investigating the genetic factors responsible for the between-patient variability of response to medications. Unfortunately, we do not currently have the tools to identify ahead of time which drug is most likely to yield a good response in a given patient, so therapeutic challenge is needed. For example, a newly diagnosed moderate asthmatic is started on a low-dose ICS/LABA combination inhaler and asthma is then reassessed and felt to be only partially controlled so a LTRA is added and the patient is then reassessed after a short interval. The LTRA has no effect on control, so SR theophylline is added and then control is achieved. The common error at this point would be to continue the patient on three medications rather than stopping the add-on controller that had no effect.

If asthma is not well controlled, environmental control needs to be reassessed. The patient should be reminded to control those factors known to worsen their asthma and questioned about potential new exacerbating factors or triggers. At this point one also needs to evaluate for co-morbid conditions like sinusitis, gastroesophageal reflux disease (GERD), or other cardiopulmonary disease. If identified and felt to be contributing to poor control these should be treated but this should be done in concert with asthma medication escalation, because one cannot know with certainty if modifying a suspected environmental factor or condition by itself will yield a good therapeutic response. Depending on the disease severity and prior history of severe exacerbations, one should consider using systemic glucocorticoids judiciously to achieve control. Despite the side effects, there should be a low threshold to use systemic steroids in a poorly controlled patient with severe disease especially if there have been previous life-threatening exacerbations. The poorly controlled patient should be reassessed frequently until control improves.

Treatment of Associated Conditions

A clinical assessment for these conditions consisting of focused questions and physical examination should be performed as part of the initial assessment. If treatment for asthma does not easily yield good control a more careful search for these conditions may be indicated.

Cough, breathlessness, and wheeze, also classic asthma symptoms, have been associated with GERD for more than 40 years [48]. Until recently, evidence that GERD can aggravate asthma had been largely based on empiric observations that anti-reflux therapy sometimes improves asthma control. Moreover, GERD has been identified as a risk factor in patients with frequent exacerbations [49], and in asthma patients, esophageal acid perfusion increases bronchial hyperresponsiveness [50]. Interestingly, oral steroids, often used in severe asthma increase esophageal acid contact time in patients with asthma [51]. Studies examining the effect of medical or surgical therapy for GERD on asthma-related outcomes typically have been small and have suffered from poor study design. A recent study [52] showing that surgical anti-reflux therapy but not intensive medical therapy using an H_2-antagonist improved asthma outcomes provides further evidence that treating reflux may be helpful, but this result is tempered by the fact that anti-reflux surgery has been associated with increased mortality when compared to medical therapy for GERD [53]. In a recent RCT in which a high dose proton-pump inhibitor was compared to placebo in moderate-to-severe asthma, efficacy was not universal, patients with nocturnal asthma symptoms and GERD seemed to benefit whereas patients without nocturnal symptoms did not benefit. Moreover, the primary outcome was only a modest improvement in morning peak flow, and there was no effect on symptoms or quality of life making the results of doubtful clinical relevance [54]. If GERD-specific symptoms or nocturnal asthma symptoms are prominent, we recommend ambulatory intraesophageal pH monitoring as a sensitive and specific diagnostic test to verify the diagnosis of GERD. A negative study is valuable in excluding GERD as a cause of asthma symptoms, but a positive study does not prove that GERD is the cause of asthma symptoms; a symptom diary should be used with the study. Empiric treatment without objective confirmation should especially be avoided when patients do not have classic reflux complaints and because chronic GERD may have serious consequences, we suggest specialty consultation. If the study supports a role for GERD in asthma in a patient, we suggest an empiric course of high dose proton-pump inhibitor treatment that should only be continued if it significantly relieves asthma symptoms or is otherwise indicated for GERD itself. We also find that weight loss when appropriate, elevation of the head of the bed, and avoidance of large meals can be helpful.

The relationship between asthma and chronic sinusitis is well established [55], although the underlying mechanisms are not clear. In general, there is a better association between asthma and chronic sinusitis when there is extensive sinus disease. There is little published supportive evidence to support the assertion that aggressive treatment of chronic sinusitis improves asthma control. Therapy for sinusitis includes antibiotics, decongestants, and intranasal topical glucocorticoids. Patients who fail to respond to medical therapy may benefit from endoscopic sphenoethmoidectomy, but the results of endoscopic sinus surgery tend to be poorest in patients with asthma, especially those with aspirin sensitivity and polyposis. Specific allergen immunotherapy is particularly useful in asthmatics with concomitant allergic rhinitis and sinusitis [22].

There is epidemiological evidence that obesity is associated with asthma [56]. Prospective studies showing an association between the body mass index (BMI) and the subsequent development of asthma are generally limited by the fact that they rely on patient-reported asthma. For example, in NHANES III, obese patients reported more asthma, wheezing, and bronchodilator use but were less likely to demonstrate airflow obstruction than non-obese patients [57]. Hence, though obesity may be associated with asthma, obese patients are more likely to be misdiagnosed with asthma and to be incorrectly treated. In one prospective study of physician-diagnosed asthma, a BMI ≥ 28 was associated with asthma largely driven by obese female subjects [58]. Finally, intervention studies have shown that weight loss can favorably affect asthma outcomes [59, 60], so it is prudent to aggressively treat obesity in the asthmatic patient.

Add-on pharmacological therapy

A monoclonal antibody to IgE (omalizumab) reduces serum IgE and should be considered as a controller for atopic asthmatic patients dependent on oral corticosteroids or for use in patients with poorly controlled disease despite high dose ICS also using multiple controllers. Studies in patients with moderate-to-severe corticosteroid-dependent asthma have shown a significant steroid-sparing effect of omalizumab and a reduction in exacerbation frequency [61–64]. Omalizumab is most useful when patients require $>800\,\mu g$ (medium doses) of inhaled steroids daily, take daily oral steroids, have an FEV_1; $< 65\%$ predicted, or have required emergency room treatment within the prior year [65]. Post-marketing data reported to the Food and Drug Administration suggests that omalizumab may cause anaphylaxis with an estimated incidence $>0.1\%$; anaphylaxis after omalizumab administration usually occurs within 2 h but can occur even after 24 h, and omalizumab-associated anaphylaxis can occur even after several treatments (www.fda.gov). The Providers administering this agent should be prepared to deal promptly with this infrequent but important complication.

If the patient remains poorly controlled despite trials of the first-line agents, ICS, LABAs, LTRA, SR theophylline, and omalizumab, there are few therapeutic options besides for further escalation of oral steroids. Studies using macrolide antibiotics in asthma have yielded discordant results [66]. In one clinical trial, the administration of clarithromycin to subjects with asthma resulted in improvements in lung function but *post hoc* analysis showed that improvement only occurred in patients with demonstrable evidence of *mycoplasma* or *chlamydia* infection [67]. Why macrolides may be effective remains under study; macrolides may treat chronic, subclinical airway infection from *mycoplasma* and *chlamydia* bacteria or may have other anti-inflammatory effects [68]. More studies of the utility of macrolide antibiotics in asthma are currently underway.

Several other targeted and nontargeted drugs have been tested in clinical trials and others are under development. Non-targeted drugs useful in other conditions such as gold salts, methotrexate, colchicine, and cyclosporine have been studied in asthma with disappointing results [69–73]. Targeted agents like those designed to inhibit the effects of cytokine products, such as IL-4, and IL-5 [74] and TNF-α [75–77], have been studied in clinical trials with disappointing results to date. Because of the specificity of newer agents, it is likely that some agents will prove more effective in certain patients than others, and this will be likely related to pharmacogenomic and environmental factors.

MANAGING ASTHMA IN EXACERBATION

Asthma is characterized by disease exacerbations that result in substantial morbidity, occasional mortality and considerable medical and economic costs. Most asthma exacerbations occur in a minority of asthma patients. When deaths from asthma have been analyzed, most decedents had experienced worsening symptoms over a period of several hours to several days highlighting the importance of identifying and educating the at-risk patient. Life-threatening episodes can develop in patients whose asthma appears to be mild at baseline, but patients who have at baseline, severe or poorly controlled disease, are those who frequently access the ER, are frequently hospitalized, and are at the highest risk [78]. These patients should be identified and targeted for intensive disease management. There is a spectrum of exacerbation severity, and it is important that a proper action plan based on severity be in place to be rapidly utilized. Patients need to be instructed to promptly seek medical attention for serious symptoms. Less serious exacerbations can be managed in the outpatient setting and are discussed later.

The management of acute severe asthma that requires hospitalization is discussed in Chapter 64.

The cornerstone of therapy for an exacerbation of asthma involves the escalation of both glucocorticoids and quick-relief medications, usually inhaled β_2-adrenergic agonists, with frequent reassessment of symptoms and the degree of airflow obstruction. Initial management in the emergency department or in an urgent care setting involves assessment for hypoxia, severity of symptoms, severity of airflow obstruction, and rapid triage coupled with the administration of intensive glucocorticoid therapy and bronchodilators. Patients who do not immediately need critical care or hospital admission are reassessed after intensive therapy. Those with a good response to treatment are candidates for discharge and continuing outpatient treatment.

Surprising little is known about the optimal dosing of systemic glucocorticoids in acute asthma. It is generally held that inflammation in acute asthma is relatively resistant to glucocorticoids, so relatively high doses of daily oral steroids (0.25–1 mg/kg) are initially used and the steroids tapered at a rate determined by the patient's response. Sometimes a few days of therapy are required before a noticeable improvement in lung function occurs. Steroids may be tapered once there is a response but it is the authors' opinion that systemic steroids should be tapered over a 7–14 day period in cases of severe asthma exacerbation. The patient's prior history of response to steroids often informs that decision. Mild exacerbations in asthmatics with mild persistent disease may be managed in some cases by escalating the dose of or initiating inhaled steroids in cases where patients are taking low dose steroids or no steroids respectively.

More than the usual dose of bronchodilators is often needed to relieve symptoms during an exacerbation: *Exacerbations of asthma should never be treated by escalating bronchodilators alone.* Fatalities from asthma typically occur when patients fail to promptly seek medical attention; instead self-medicating with escalating doses of quick-relief medications or long-acting bronchodilators. Inhaled β_2-adrenergic agonists are safer and equally efficacious to oral preparations. Inhaled β_2-adrenergic agonists may be used safely in quite high doses with close monitoring. Inhaled ipratropium appears to provide further bronchodilation in acute severe asthma. [79, 80] Patients suffering from an exacerbation that do not get relief from reasonable doses of bronchodilators for example 4 puffs from an MDI or one wet nebulizer treatment every 3 h should be hospitalized. Antibiotics are not indicated for acute asthma exacerbations unless there is objective evidence of bacterial pneumonia or co-existing bacterial sinusitis.

References

1. National Asthma Education and Prevention Program, Expert Panel Report 3, Guidelines for the Diagnosis and Management of Asthma. 2007; Department of Health and Human Services, National Institutes of Health, National Heart, Lung, and Blood Institute, Bethesda, MD.
2. Global Strategy for Asthma Management and Prevention, 2005; Available from: www.ginasthma.org
3. Peters SP, Jones CA, Haselkorn T, Mink DR, Valacer DJ, Weiss ST. Real-world evaluation of asthma control and treatment (REACT): Findings from a national web-based survey. *J Allergy Clin Immunol*, 2007.
4. Bateman ED, Boushey HA, Bousquet J, Busse WW, Clark TJ, Pauwels RA, Pedersen SE. Can guideline-defined asthma control be achieved? The Gaining Optimal Asthma Control Study. *Am J Respir. Crit Care Med*: 836–44, 2004.
5. Miller MK, Lee JH, Miller DP, Wenzel SE. Recent asthma exacerbations: A key predictor of future exacerbations. *Respir Med*, 2006.
6. Deykin A, Wechsler ME, Boushey HA, Chinchilli VM, Kunselman SJ, Craig TJ, Dimango E, Fahy JV, Kraft M, Leone F, Lazarus SC, Lemanske RF Jr, Martin RJ, Pesola GR, Peters SP, Sorkness CA, Szefler SJ, Israel E. Combination therapy with a long-acting beta-agonist and a leukotriene antagonist in moderate asthma. *Am J Respir Crit Care Med*: 228–34, 2007.
7. Peters SP, Anthonisen N, Castro M, Holbrook JT, Irvin CG, Smith LJ, Wise RA. Randomized comparison of strategies for reducing treatment in mild persistent asthma. *N Engl J Med*: 2027–39, 2007.
8. Haahtela T, Jarvinen M, Kava T, Kiviranta K, Koskinen S, Lehtonen K, Nikander K, Persson T, Reinikainen K, Selroos O. Comparison of a beta 2-agonist, terbutaline, with an inhaled corticosteroid, budesonide, in newly detected asthma. *N Engl J Med*: 388–92, 1991.
9. Rafferty P, Tucker LG, Frame MH, Fergusson RJ, Biggs BA, Crompton GK. Comparison of budesonide and beclomethasone dipropionate in patients with severe chronic asthma: Assessment of relative prednisolone-sparing effects. *Br J Dis Chest*: 244–50, 1985.
10. Stempel DA, Stoloff SW, Carranza Rosenzweig JR, Stanford RH, Ryskina KL, Legorreta AP. Adherence to asthma controller medication regimens. *Respir Med*: 1263–67, 2005.
11. Stoloff SW, Stempel DA, Meyer J, Stanford RH, Carranza Rosenzweig JR. Improved refill persistence with fluticasone propionate and salmeterol in a single inhaler compared with other controller therapies. *J Allergy Clin Immunol*: 245–51, 2004.
12. Richter K, Janicki S, Jorres RA, Magnussen H. Acute protection against exercise-induced bronchoconstriction by formoterol, salmeterol and terbutaline. *Eur Respir J*: 865–71, 2002.
13. Spooner CH, Saunders LD, Rowe BH. Nedocromil sodium for preventing exercise-induced bronchoconstriction. *Cochrane Database Syst Rev*: 2002, CD001183.

14. Leff JA, Busse WW, Pearlman D, Bronsky EA, Kemp J, Hendeles L, Dockhorn R, Kundu S, Zhang J, Seidenberg BC, Reiss TF. Montelukast, a leukotriene-receptor antagonist, for the treatment of mild asthma and exercise-induced bronchoconstriction. *N Engl J Med*: 147–52, 1998.
15. Boushey HA, Sorkness CA, King TS, Sullivan SD, Fahy JV, Lazarus SC, Chinchilli VM, Craig TJ, Dimango EA, Deykin A, Fagan JK, Fish JE, Ford JG, Kraft M, Lemanske RF, Leone FT Jr., Martin RJ, Mauger EA, Pesola GR, Peters SP, Rollings NJ, Szefler SJ, Wechsler ME, Israel E. Daily versus as-needed corticosteroids for mild persistent asthma. *N Engl J Med*: 1519–28, 2005.
16. Papi A, Canonica GW, Maestrelli P, Paggiaro P, Olivieri D, Pozzi E, Crimi N, Vignola AM, Morelli P, Nicolini G, Fabbri LM. Rescue use of beclomethasone and albuterol in a single inhaler for mild asthma. *N Engl J Med*: 2040–52, 2007.
17. O'Byrne PM, Bisgaard H, Godard PP, Pistolesi M, Palmqvist M, Zhu Y, Ekstrom T, Bateman ED. Budesonide/formoterol combination therapy as both maintenance and reliever medication in asthma. *Am J Respir Crit Care Med*: 129–36, 2005.
18. Gibson PG, Powell H, Coughlan J, Wilson AJ, Abramson M, Haywood P, Bauman A, Hensley MJ, Walters EH. Self-management education and regular practitioner review for adults with asthma. *Cochrane Database Syst Rev*: 2003, CD001117.
19. Gibson PG, Powell H, Coughlan J, Wilson AJ, Hensley MJ, Abramson M, Bauman A, Walters EH. Limited (information only) patient education programs for adults with asthma. *Cochrane Database Syst Rev*: CD001005, 2002.
20. Bailey WC, Richards JM Jr., Brooks CM, Soong SJ, Windsor RA, Manzella BA. A randomized trial to improve self-management practices of adults with asthma. *Arch Intern Med*: 1664–68, 1990.
21. Wilson SR, Scamagas P, German DF, Hughes GW, Lulla S, Coss S, Chardon L, Thomas RG, Starr-Schneidkraut N, Stancavage FB. A controlled trial of two forms of self-management education for adults with asthma. *Am J Med*: 564–76, 1993.
22. Abramson MJ, Puy RM, Weiner JM. Is allergen immunotherapy effective in asthma? A meta-analysis of randomized controlled trials. *Am J Respir Crit Care Med*: 969–74, 1995.
23. Reid MJ, Lockey RF, Turkeltaub PC, Platts-Mills TA. Survey of fatalities from skin testing and immunotherapy 1985–1989. *J Allergy Clin Immunol*: 6–15, 1993.
24. Castro M, Dozor A, Fish J, Irvin C, Scharf S, Scheipeter ME, Holbrook J, Tonascia J, and Wise R for the American Lung Association Asthma Clinical Research Centers. The safety of inactivated influenza vaccine in adults and children with asthma. *N Engl J Med* 1529–36, 2001.
25. Cates CJ, Jefferson TO, Bara AI, Rowe BH. Vaccines for preventing influenza in people with asthma. *Cochrane Database Syst Rev*: CD000364, 2004.
26. Smith NM, Bresee JS, Shay DK, Uyeki TM, Cox NJ, Strikas RA. Prevention and control of influenza: Recommendations of the Advisory Committee on Immunization Practices (ACIP). *MMWR Recomm Rep*: 1–42, 2006.
27. National Asthma Education Program, Expert Panel Report 2, Guidelines for the Diagnosis and Management of Asthma. 1997.
28. Lazarus SC, Boushey HA, Fahy JV, Chinchilli VM, Lemanske RF Jr., Sorkness CA, Kraft M, Fish JE, Peters SP, Craig T, Drazen JM, Ford JG, Israel E, Martin RJ, Mauger EA, Nachman SA, Spahn JD, Szefler SJ. Long-acting beta2-agonist monotherapy vs continued therapy with inhaled corticosteroids in patients with persistent asthma: A randomized controlled trial. *JAMA*: 2583–93, 2001.
29. Suissa S, Ernst P, Benayoun S, Baltzan M, Cai B. Low-dose inhaled corticosteroids and the prevention of death from asthma. *N Engl J Med*: 332–36, 2000.
30. Lanes SF, Garcia Rodriguez LA, Huerta C. Respiratory medications and risk of asthma death. *Thorax*: 683–86, 2002.
31. Szefler SJ, Martin RJ, King TS, Boushey HA, Cherniack RM, Chinchilli VM, Craig TJ, Dolovich M, Drazen JM, Fagan JK, Fahy JV, Fish JE, Ford JG, Israel E, Kiley J, Kraft M, Lazarus SC, Lemanske RF Jr., Mauger E, Peters SP, Sorkness CA. Significant variability in response to inhaled corticosteroids for persistent asthma. *J Allergy Clin Immunol*: 410–18, 2002.
32. Mjaanes CM, Whelan GJ, Szefler SJ. Corticosteroid therapy in asthma: Predictors of responsiveness. *Clin Chest Med*: 119–32, 2006, vii.
33. Greening AP, Ind PW, Northfield M, Shaw G. Added salmeterol versus higher-dose corticosteroid in asthma patients with symptoms on existing inhaled corticosteroid. Allen and Hanburys Limited UK Study Group. *Lancet*: 219–24, 1994.
34. Lemanske RF Jr., Sorkness CA, Mauger EA, Lazarus SC, Boushey HA, Fahy JV, Drazen JM, Chinchilli VM, Craig T, Fish JE, Ford JG, Israel E, Kraft M, Martin RJ, Nachman SA, Peters SP, Spahn JD, Szefler SJ. Inhaled corticosteroid reduction and elimination in patients with persistent asthma receiving salmeterol: A randomized controlled trial. *JAMA*: 2594–603, 2001.
35. Woolcock A, Lundback B, Ringdal N, Jacques LA. Comparison of addition of salmeterol to inhaled steroids with doubling of the dose of inhaled steroids. *Am J Respir. Crit Care Med*: 1481–88, 1996.
36. Laviolette M, Malmstrom K, Lu S, Chervinsky P, Pujet JC, Peszek I, Zhang J, Reiss TF. Montelukast added to inhaled beclomethasone in treatment of asthma. Montelukast/Beclomethasone Additivity Group. *Am J Respir Crit Care Med*: 1862–68, 1999.
37. Evans DJ, Taylor DA, Zetterstrom O, Chung KF, O'Connor BJ, Barnes PJA. comparison of low-dose inhaled budesonide plus theophylline and high-dose inhaled budesonide for moderate asthma. *N Engl J Med*: 1412–18, 1997.
38. Gross NJ. Anticholinergic agents in asthma and COPD. *Eur J Pharmacol*: 36–39, 2006.
39. O'Connor BJ, Towse LJ, Barnes PJ. Prolonged effect of tiotropium bromide on methacholine-induced bronchoconstriction in asthma. *Am J Respir Crit Care Med*: 876–80, 1996.
40. Spooner CH, Saunders LD, Rowe BH. Nedocromil sodium for preventing exercise-induced bronchoconstriction. *Cochrane Database Syst Rev*: 2002, CD001183.
41. Szefler S, Weiss S, and Tonascia J, for The Childhood Asthma Management Program Research Group. Long-term effects of budesonide or nedocromil in children with asthma. The Childhood Asthma Management Program Research Group. *N EnglJ Med* 1054–63, 2000.
42. Donahue JG, Weiss ST, Livingston JM, Goetsch MA, Greineder DK, Platt R. Inhaled steroids and the risk of hospitalization for asthma. *JAMA*: 887–91, 1997.
43. Malmstrom K, Rodriguez-Gomez G, Guerra J, Villaran C, Pineiro A, Wei LX, Seidenberg BC, Reiss TF. Oral montelukast, inhaled beclomethasone, and placebo for chronic asthma. A randomized, controlled trial. Montelukast/Beclomethasone Study Group. *Ann Intern Med*: 487–95, 1999.
44. Edelman JM, Turpin JA, Bronsky EA, Grossman J, Kemp JP, Ghannam AF, DeLucca PT, Gormley GJ, Pearlman DS. Oral montelukast compared with inhaled salmeterol to prevent exercise-induced bronchoconstriction. A randomized, double-blind trial. Exercise Study Group. *Ann Intern Med*: 97–104, 2000.
45. Lofdahl CG, Reiss TF, Leff JA, Israel E, Noonan MJ, Finn AF, Seidenberg BC, Capizzi T, Kundu S, Godard P. Randomised, placebo controlled trial of effect of a leukotriene receptor antagonist, montelukast, on tapering inhaled corticosteroids in asthmatic patients. *BMJ*: 87–90, 1999.
46. Robinson DS, Campbell D, Barnes PJ. Addition of leukotriene antagonists to therapy in chronic persistent asthma: A randomised double-blind placebo-controlled trial. *Lancet*: 2007–11, 2001.
47. Vaquerizo MJ, Casan P, Castillo J, Perpina M, Sanchis J, Sobradillo V, Valencia A, Verea H, Viejo JL, Villasante C, Gonzalez-Esteban J, Picado C. Effect of montelukast added to inhaled budesonide on control of mild to moderate asthma. *Thorax*: 204–10, 2003.
48. Beasley R. The pulmonary complications of oesophageal disease. *Br J Dis Chest*: 342–48, 1960.
49. ten Brinke A, Sterk PJ, Masclee AAM, Spinhoven P, Schmidt JT, Zwinderman AH, Rabe KF, Bel EH. Risk factors of frequent exacerbations in difficult-to-treat asthma. *Eur Respir J*: 812–18, 2005.

50. Wu De Nan, Tanifuji Yukio, Kobayashi Hitoshi, Yamauchi Kohei, Kato Chieko, Suzuki Kazuyuki, Inoue Hiroshi. Effects of esophageal acid perfusion on airway hyperresponsiveness in patients with bronchial asthma. *Chest*: 1553–56, 2000.
51. Lazenby JP, Guzzo MR, Harding SM, Patterson PE, Johnson LF, Bradley LA. Oral corticosteroids increase esophageal acid contact times in patients with stable asthma. *Chest*: 625–34, 2002.
52. Sontag SJ, O'Connell S, Khandelwal S, Greenlee H, Schnell T, Nemchausky B, Chejfec G, Miller T, Seidel J, Sonnenberg A. Asthmatics with gastroesophageal reflux. Long term results of a randomized trial of medical and surgical antireflux therapies. *Am J Gastroenterol*: 987–99, 2003.
53. Spechler SJ, Lee E, Ahnen D, Goyal RK, Hirano I, Ramirez F, Raufman JP, Sampliner R, Schnell T, Sontag S, Vlahcevic ZR, Young R, Williford W. Long-term outcome of medical and surgical therapies for gastroesophageal reflux disease: Follow-up of a randomized controlled trial. *JAMA*: 2331–38, 2001.
54. Kiljander TO, Harding SM, Field SK, Stein MR, Nelson HS, Ekelund J, Illueca M, Beckman O, Sostek MB. Effects of esomeprazole 40 mg twice daily on asthma: A randomized placebo-controlled trial. *Am J Respir Crit Care Med*: 1091–97, 2006.
55. Newman LJ, Platts-Mills TA, Phillips CD, Hazen KC, Gross CW. Chronic sinusitis. Relationship of computed tomographic findings to allergy, asthma, and eosinophilia. *JAMA*: 363–67, 1994.
56. Ford ES. The epidemiology of obesity and asthma. *J Allergy Clin Immunol*: 897–909, 2002.
57. Sin DD, Jones RL, Man SF. Obesity is a risk factor for dyspnea but not for airflow obstruction. *Arch Intern Med*: 1477–81, 2002.
58. Guerra S, Sherrill DL, Bobadilla A, Martinez FD, Barbee RA. The relation of body mass index to asthma, chronic bronchitis, and emphysema. *Chest*: 1256–63, 2002.
59. Hakala K, Stenius-Aarniala B, Sovijarvi A. Effects of weight loss on peak flow variability, airways obstruction, and lung volumes in obese patients with asthma. *Chest*: 1315–21, 2000.
60. Stenius-Aarniala B, Poussa T, Kvarnstrom J, Gronlund EL, Ylikahri M, Mustajoki P. Immediate and long term effects of weight reduction in obese people with asthma: randomised controlled study. *BMJ*: 827–32, 2000.
61. Bousquet J, Cabrera P, Berkman N, Buhl R, Holgate S, Wenzel S, Fox H, Hedgecock S, Blogg M, Cioppa GD. The effect of treatment with omalizumab, an anti-IgE antibody, on asthma exacerbations and emergency medical visits in patients with severe persistent asthma. *Allergy*: 302–8, 2005.
62. Busse W, Corren J, Lanier BQ, McAlary M, Fowler-Taylor A, Cioppa GD, van As A, Gupta N. Omalizumab, anti-IgE recombinant humanized monoclonal antibody, for the treatment of severe allergic asthma. *J Allergy Clin Immunol*: 184–90, 2001.
63. Milgrom H, Fick RB Jr., Su JQ, Reimann JD, Bush RK, Watrous ML, Metzger WJ. Treatment of allergic asthma with monoclonal anti-IgE antibody. RhuMAb-E25 Study Group. *N Engl J Med*: 1966–73, 1999.
64. Soler M, Matz J, Townley R, Buhl R, O'Brien J, Fox H, Thirlwell J, Gupta N, Della CG. The anti-IgE antibody omalizumab reduces exacerbations and steroid requirement in allergic asthmatics. *Eur Respir J*: 254–61, 2001.
65. Bousquet J, Wenzel S, Holgate S, Lumry W, Freeman P, Fox H. Predicting response to omalizumab, an anti-IgE antibody, in patients with allergic asthma. *Chest*: 1378–86, 2004.
66. Richeldi L, Ferrara G, Fabbri L, Lasserson T, Gibson P, Richeldi L. Macrolides for chronic asthma. *Cochrane Database Syst Rev*: 2005, CD002997.
67. Kraft M, Cassell GH, Pak J, Martin RJ. Mycoplasma pneumoniae and Chlamydia pneumoniae in asthma: Effect of clarithromycin. *Chest*: 1782–88, 2002.
68. Shinkai Masaharu Rubin, Bruce K. Macrolides and airway inflammation in children. *Paediat Respir Rev*: 227–35, 2005.
69. Evans DJ, Cullinan P, Geddes DM. Gold as an oral corticosteroid sparing agent in stable asthma. *Cochrane Database Syst Rev*: 2001, CD002985.
70. Alexander AG, Barnes NC, Kay AB. Trial of cyclosporin in corticosteroid-dependent chronic severe asthma. *Lancet*: 324–28, 1992.
71. Evans DJ, Cullinan P, Geddes DM. Cyclosporin as an oral corticosteroid sparing agent in stable asthma. *Cochrane Database Syst Rev*: 2001, CD002993.
72. Schwarz YA, Kivity S, Ilfeld DN, Schlesinger M, Greif J, Topilsky M, Garty MSA. clinical and immunologic study of colchicine in asthma. *J Allergy Clin Immunol*: 578–82, 1990.
73. Fish JE, Peters SP, Chambers CV, McGeady SJ, Epstein KR, Boushey HA, Cherniack RM, Chinchilli VM, Drazen JM, Fahy JV, Hurd SS, Israel E, Lazarus SC, Lemanske RF, Martin RJ, Mauger EA, Sorkness C, Szefler SJ. An evaluation of colchicine as an alternative to inhaled corticosteriods in moderate asthma. National Heart, Lung, and Blood Institute's Asthma Clinical Research Network. *Am J Respir Crit Care Med*: 1165–71, 1997.
74. Leckie MJ, ten Brinke A, Khan J, Diamant Z, O'Connor BJ, Walls CM, Mathur AK, Cowley HC, Chung KF, Djukanovic R, Hansel TT, Holgate ST, Sterk PJ, Barnes PJ. Effects of an interleukin-5 blocking monoclonal antibody on eosinophils, airway hyper-responsiveness, and the late asthmatic response. *Lancet*: 2144–48, 2000.
75. Howarth PH, Babu KS, Arshad HS, Lau L, Buckley M, McConnell W, Beckett P, Al Ali M, Chauhan A, Wilson SJ, Reynolds A, Davies DE, Holgate ST. Tumour necrosis factor (TNF{alpha}) as a novel therapeutic target in symptomatic corticosteroid dependent asthma. *Thorax*: 1012–18, 2005.
76. Berry MA, Hargadon B, Shelley M, Parker D, Shaw DE, Green RH, Bradding P, Brightling CE, Wardlaw AJ, Pavord I. Evidence of a role of tumor necrosis factor {alpha} in refractory asthma. *N Engl J Med*: 697–708, 2006.
77. Erin EM, Leaker BR, Nicholson GC, Tan AJ, Green LM, Neighbour H, Zacharasiewicz AS, Turner J, Barnathan ES, Kon OM, Barnes PJ, Hansel TT. The effects of a monoclonal antibody directed against tumor necrosis factor-{alpha} in asthma. *Am J Respir Crit Care Med*: 753–62, 2006.
78. McFadden ER Jr., Warren EL. Observations on asthma mortality. *Ann Intern Med*: 142–47, 1997.
79. Rodrigo GJ, Rodrigo C. Triple inhaled drug protocol for the treatment of acute severe asthma. *Chest*: 1908–15, 2003.
80. Rodrigo GJ, Rodrigo C. First-line therapy for adult patients with acute asthma receiving a multiple-dose protocol of ipratropium bromide plus albuterol in the emergency department. *Am J Respir Crit Care Med*: 1862–68, 2000.

64

CHAPTER

Asthma Exacerbations

Carlos A. Camargo Jr[1] and Brian H. Rowe[2]

[1]Department of Emergency Medicine, Massachusetts General Hospital, Harvard Medical School, Boston, MA, USA

[2]Department of Emergency Medicine, University of Alberta, and Capital Health, Edmonton, AB, Canada

INTRODUCTION

Asthma is a global health problem of particular importance in developed nations. In the United States, for example, approximately 32 million people have been told by a health professional that they have asthma, while approximately 22 million *currently* have asthma [1]. Children ages 5–17 years have the highest prevalence rates for ever asthma and current asthma, with both rates decreasing with age. Asthma currently affects approximately 8.9% of US children and 7.2% of US adults.

A defining characteristic of asthma is the episodic worsening of the disease, with a more severe episode described as an "asthma attack" or "asthma exacerbation". These episodes also are referred to as "acute asthma", a term that emphasizes the distinction between an acute flare-up of asthma and the day-to-day fluctuations of this chronic, inflammatory disease. In the United States, approximately 12 million Americans have an asthma attack each year [1]. This represents just over half of the 22 million individual with current asthma. As with the prevalence of current (or lifetime) asthma, the highest attack prevalence is among children ages 5–17 years.

Although most exacerbations are mild and can be managed without difficulty outside the acute health care system, or through an urgent (unscheduled) office visit, more severe exacerbations prompt emergency department (ED) visits, and overnight hospital stays. ED visits and hospitalizations together account for approximately 25–50% of the US $14.7 billion in direct asthma-related health care costs [1, 2]. Indirect costs (lost productivity) add another $5 billion for a grand total of $19.7 billion. Asthma exacerbations obviously contribute to these indirect costs as well. For example, fatal exacerbations accounts for $1.9 billion of the $5 billion of indirect costs.

As suggested by these financial data, presentations with acute asthma to the ED or other acute care settings are common. In the United States, acute asthma accounts for approximately 1.8 million ED visits each year and approximately 488,000 hospitalizations [1]. These acute presentations are precipitated by many possible factors, but the most common are an associated upper respiratory tract infection or environmental allergies. For individuals without regular asthma care, an undefined portion of ED visits is likely due to poor control of chronic asthma with or without associated common asthma triggers such as upper respiratory tract infection.

ED visits are important events for asthmatic individuals and their families, as they represent a vulnerable point in the illness and are associated with significant morbidity and occasional mortality. Moreover, in spite of adequate care, patients who experience an exacerbation may suffer from that episode for many days and they remain at risk of subsequent relapses for weeks after the initial ED visit [3, 4].

For all of these reasons, the management of acute asthma has been the focus of an increasing number of studies worldwide. The Cochrane Collaboration has played a vital role in organizing the results of published and non-published randomized trials, and this chapter is based heavily (and appropriately) on these systematic reviews. Moreover, this growing evidence base has led to much greater focus on the management of acute asthma in clinical practice guidelines, such as the recently released US guidelines, *Expert Panel Report 3 (EPR3): Guidelines for the Diagnosis and Management of Asthma* of the National Asthma Education and Prevention

Program (NAEPP) [5]. The new EPR3 guidelines provide an excellent foundation for this chapter.

Despite this growing evidence base, and the increasing attempts to standardize and improve acute asthma care, there remain important gaps in diagnostic and therapeutic knowledge. We will address several of these controversial areas in this chapter. There also remain wide gaps between what is known and what is practiced. Although data are lacking, we believe that most of the latter gap is related to physician variation in the emergency treatment and approach to the patient with acute asthma. There are, however, encouraging signs that this asthma management gap is narrowing in emergency medicine [6].

EVALUATION

The emergency management of acute asthma requires the health care provider to perform a brief history and physical examination [5]. Key elements in the history include the details of the current exacerbation (e.g. time of onset and any potential causes); severity of symptoms (especially compared to previous exacerbations) and response to any treatment given before presentation; all current medications and time of last dose (especially of asthma medications); the patient's previous asthma history (e.g. number of previous unscheduled office visits, ED visits, and hospitalizations for asthma, particularly within the past year; history of intubation for asthma); and other co-morbid conditions (e.g. other pulmonary or cardiac diseases or disease that may be aggravated by systemic corticosteroid therapy).

The key elements of the initial brief physical examination are the assessment of overall patient status (e.g. alertness, fluid status, respiratory distress); vital signs (including pulse oximetry); and chest findings (e.g. use of accessory muscles, wheezing). The examination also should focus on the identification of possible complications (e.g. pneumonia, pneumothorax, or pneumomediastinum); although rare, these complications have the potential to affect patient management. Particularly in children, the examination also should try to rule out upper airway obstruction (e.g. upper airway foreign bodies).

As noted in the EPR3 guidelines, "Most patients who have an asthma exacerbation do not require any initial studies. If laboratory studies are ordered, they must not delay initiation of asthma treatment" [5]. The most important objective of laboratory studies, such as arterial blood gas testing, is the detection of actual or impending respiratory failure. Another objective is to detect conditions that potentially complicate emergency asthma management (e.g. performing an EKG to rule out cardiac ischemia, or performing a chest film to rule out pneumonia). As noted earlier, since most patients have an uncomplicated, single-system disease presentation, they do not require *any* initial studies. Thus, low-yield tests should not be part of routine assessment; they should be reserved for patients that the health care provider believes are at increased risk of a specific complication or co-morbid disease.

By contrast, an important part of the initial assessment of most ED patients with acute asthma is an objective measure of pulmonary function. Although forced expiratory volume in one second (FEV_1) is preferred, serial peak expiratory flow (PEF) measurement can provide a rough, but still helpful, estimate of severity level and be used successfully to guide emergency management. All three editions of the NAEPP guidelines [5, 7, 8] have recommended that emergency management of acute asthma should be based on the FEV_1 or PEF severity level. The 2007 EPR3 guidelines [5] clarify that pulmonary function testing is not necessary in patients in extreme respiratory distress where the severity level is obvious. Thus, one should not use 100% initial PEF testing as a quality measure in the ED; for some patients, it would be highly inappropriate to perform such respiratory maneuvers at presentation or even in the first hour of their ED treatment. Nevertheless, when feeling better, physicians treating these same patients would likely benefit from reviewing objective measurements to assess lung function and monitor improvement.

The severity of asthma exacerbation drives treatment, as outlined in the 2007 EPR3 algorithms for management at home (Fig. 64.1) and in the ED/hospital setting (Fig. 64.2) [5]. An important change in the 2007 EPR3 guidelines, as compared to the 1997 EPR2 guideline [7] or 2002 update [9], is that the 2007 guidelines have reinstated the FEV_1 or PEF cutoffs used in the original 1991 guidelines [8]. Briefly, the 1991 and 2007 percent predicted FEV_1 or PEF cutoffs for asthma exacerbation severity are 40% for severe and 70% for mild. The "new" cutoff for severe exacerbations ($<$40% predicted FEV_1 or PEF) is useful because, as discussed later in this chapter, it is only at that severity level that several adjunct therapies have likely benefit (e.g. continuous nebulization of a β_2- agonist, intravenous (IV) magnesium sulfate, heliox). The "new" cutoff for mild exacerbations ($\geq$70% predicted FEV_1 or PEF) represents the goal for discharge from the urgent/emergent care setting. Although it is certainly possible (indeed common) to discharge ED patients with lower PEF values [10], the 70% cutoff provides an ideal to guide ED and inpatient care. These cutoffs differ from those used to determine long-term asthma control and treatments (50% and 80% predicted FEV1 or PEF), thus underscoring the distinction between acute and chronic asthma. These revised cutoffs will likely be of greater use to ED clinicians than the 1997–2002 cutoff for "severe" exacerbations (i.e. $<$50% predicted FEV_1 or PEF), or the even higher cutoffs used in other asthma guidelines (e.g. $<$60% predicted PEF [11]).

TREATMENT

Although currently available medications for acute asthma are not *always* effective, they have an impressive beneficial impact in most patients. For example, in North American EDs, approximately 80–90% of acute asthma patients will improve enough to be discharged to home [12]. The major therapeutic options are inhaled short-acting β_2-agonists (SABA), systemic (injected) β_2-agonists, anticholinergics, and systemic

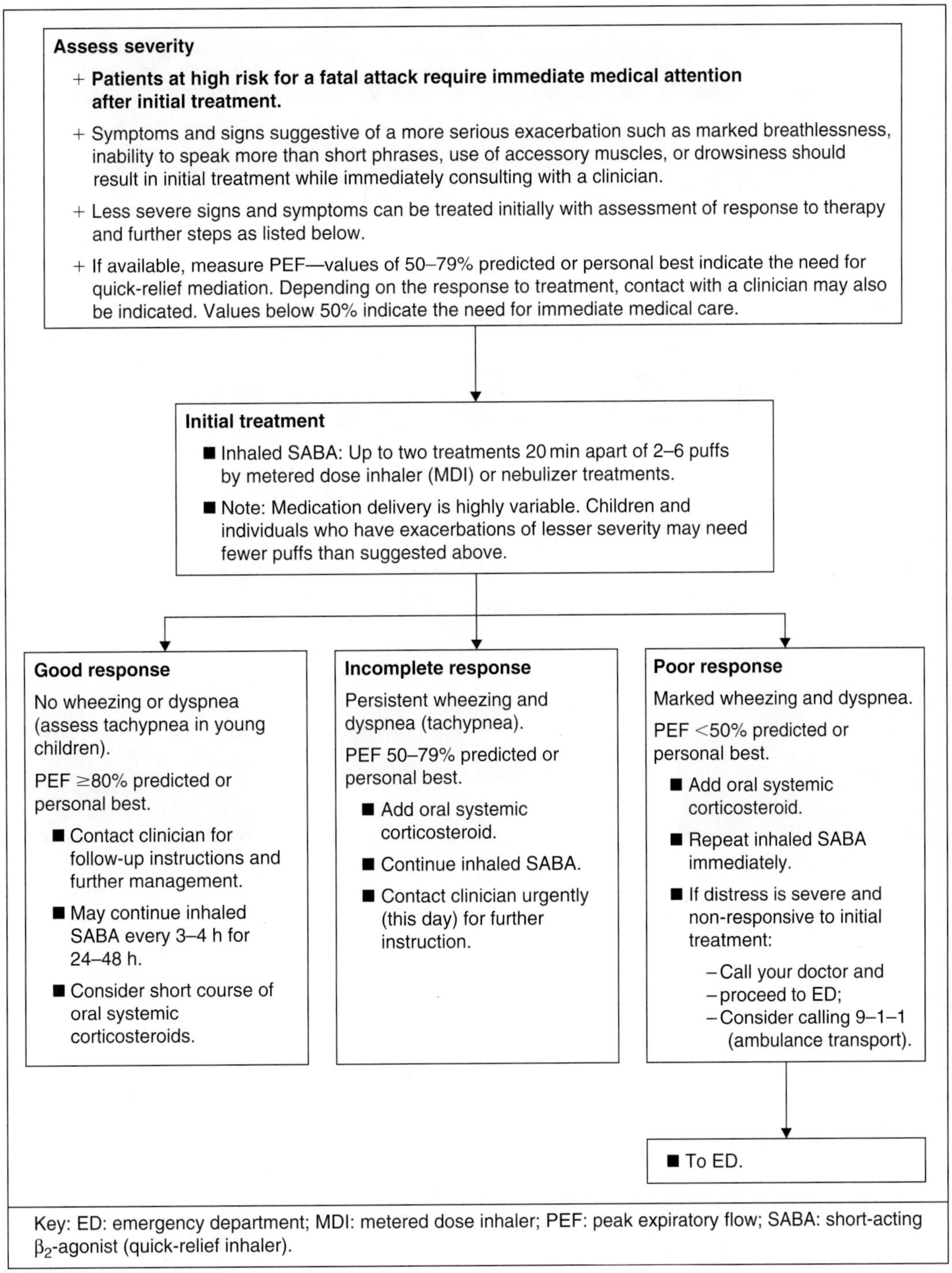

FIG. 64.1 Management of asthma exacerbations: Home treatment [5].

corticosteroids. Although we provide some information about these key "rescue" medications, we devote disproportionate space to more controversial aspects of emergency asthma management. This includes both pharmacological and non-pharmacological approaches. In doing so, we rely heavily on systematic review evidence from the Cochrane Collaboration [13], and the 2007 EPR3 guidelines [5]. As noted earlier, the Cochrane reviews use standardized and rigorous methods to summarize and combine the best evidence from individual randomized controlled trials on a specific topic area, and they were used in the development of the 2007 EPR3 guidelines.

In-ED care

The goals of in-ED are to assure adequate oxygenation, to safely and effectively reverse the acute obstruction to airflow that led to the presentation to the ED, and to initiate

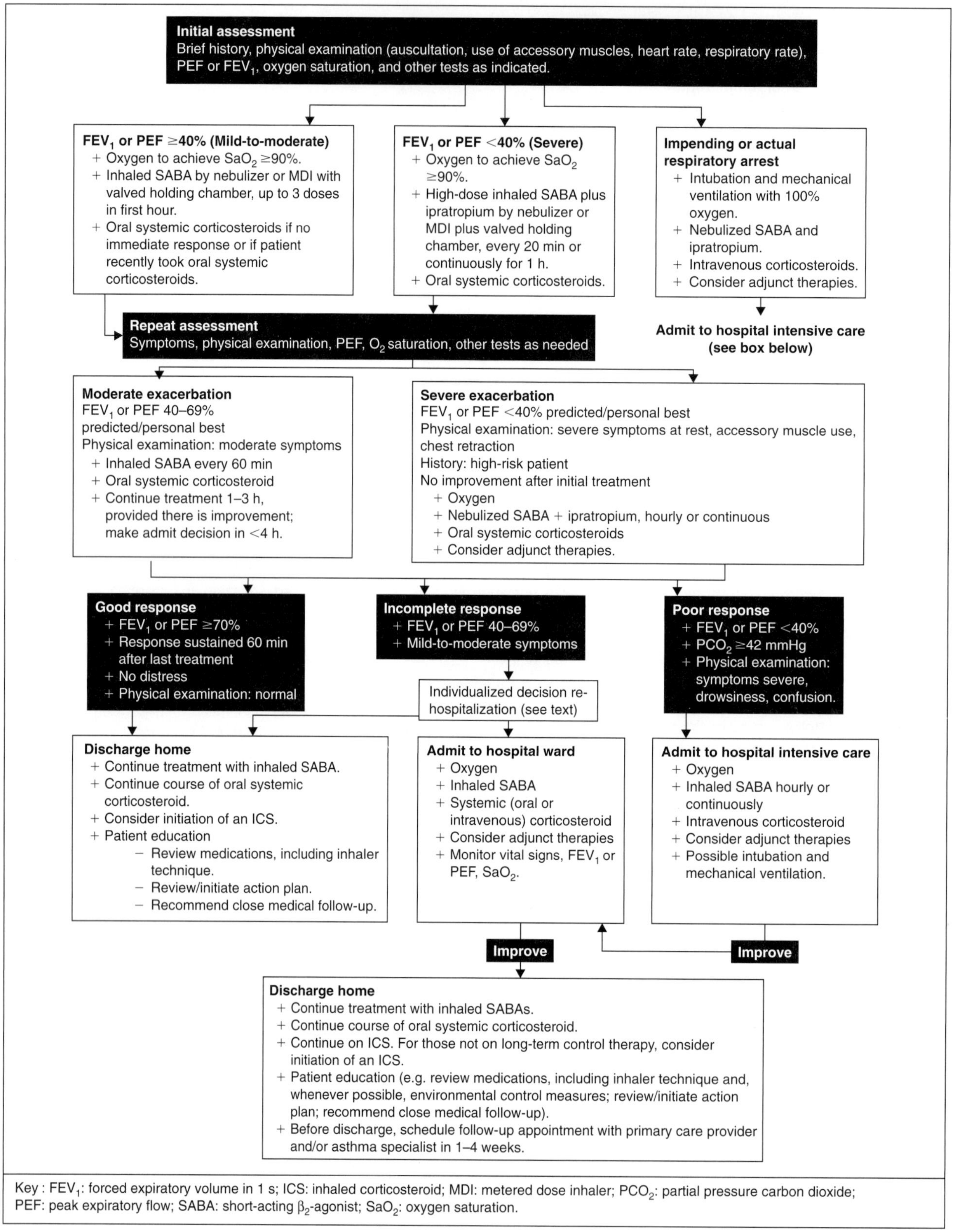

FIG. 64.2 Management of asthma exacerbations: ED and hospital-based care [5].

anti-inflammatory therapy to control the exacerbation and reduce the likelihood of subsequent relapse. The approaches used should reduce the need for hospitalization, and ensure that patients receive the appropriate therapy required to prevent or reduce the potential of an adverse outcome (e.g. intubation, death).

Oxygen

Supplemental oxygen is recommended for the initial treatment of most patients with acute asthma in the ED or inpatient setting [5]. The oxygen can be administered by nasal cannulae or mask, whichever is better tolerated, to maintain an $SaO_2 \geq 90\%$ (or $\geq 95\%$ in pregnant women or in patients with cardiac disease). SaO_2 should be monitored until the patient demonstrates a clear response to bronchodilator therapy. Although the evidence for this recommendation is weak, the risks of oxygen treatment are minimal.

Inhaled short-acting β_2-agonists

In addition to oxygen, early treatment of patients with acute asthma has generally focused on the use of inhaled SABAs because of their undisputed, generally rapid, effect on relieving bronchospasm, and associated breathlessness. Because of the risk of cardiotoxicity, especially in older asthmatic patients, EPR3 guidelines recommend that only selective SABAs (i.e. albuterol (salbutamol), levalbuterol, pirbuterol) be used in high doses [5]. Whether the drug is most effective when delivered via a nebulizer or a metered dose inhaler (MDI) with holding/spacer device has been an area of intense research. A systematic review in the Cochrane Library by Cates *et al.* [14] examined 25 trials from ED and community settings, with participation of 2,066 children and 614 adults. The authors found that the use of either delivery method for inhaled SABAs produces similar outcomes [14]. For example, the use of an MDI was associated with identical admission rates among adults (odds ratio (OR) 0.97; 95% confidence interval (CI), 0.63 to 1.49) and a non-significant difference among children (OR 0.65; 95%CI, 0.4 to 1.06). There were no important differences in FEV_1 or PEF. In children, but not adults, MDI with holding chamber/spacer appeared to offer advantages in terms of ED length-of-stay and fewer adverse effects (e.g. tachycardia). Although economic evaluations of these approaches also tend to favor the use of MDI with spacer over nebulized treatments, these analyses are problematic [15].

Nebulizer therapy may still be preferred, however, for patients who are unable to cooperate effectively in using an MDI because of their age, agitation, or severity of acute asthma. Indeed, the clinical trial data have rarely included patients with severe asthma exacerbations, let alone those with the subset of life-threatening exacerbations. Accordingly, SABA delivery via MDI-spacer may be best confined to patients with mild to moderate asthma exacerbations, which includes many patients in the ED setting (particularly now that severe is defined as <40% predicted). For patients with truly severe asthma exacerbations, the documented benefits of continuous nebulization (reduced admissions, improved pulmonary function) [16] make nebulization a more attractive alternative. Future research will need to resolve the advantages and disadvantages of MDI-spacer versus nebulizer for SABA delivery in severe acute asthma.

Numerous attempts to identify optimal doses or treatment intervals to achieve maximal bronchodilation or symptom relief have not identified a single "best" regimen. For example, several studies have found that higher doses appear to be equivalent to lower doses with regard to affecting bronchodilation or clinical outcome [17–20]. There are a substantial number of patients who achieve a bronchodilation "plateau" and additional β_2-agonist therapy only seems to cause more side effects. For that reason, some guidelines have recommended that β-agonists be titrated to plateau using objective assessment of airway obstruction with pulmonary function measures [20]. The EPR3 recommendations are summarized in Figs 64.1 and 64.2.

Inhaled anticholinergic agents

There is increasing support, particularly in children, for *adding* the anticholinergic agent ipratropium bromide (IB) to β_2-agonist therapy in more severe asthma exacerbations [21, 22]. A systematic review involving children compared IB plus β_2-agonists to β_2-agonists alone [22]. Although a single does of IB was not effective for the treatment of mild and moderate exacerbations, summary statistics indicate that *multiple* doses of IB result in a clinically significant improvement in FEV_1 at 60 (16%; 95%CI, 6 to 27%) and 120 (18%; 95%CI, 4% to 31%) minutes. Combined therapy, with multiple doses, reduced the risk of hospital admission by 25% (relative risk (RR) 0.75; 95%CI, 0.62 to 0.89) in children with moderate and severe exacerbations. The number needed to treat (NNT) to prevent one admission was 12 (95%CI, 8 to 32); the NNT decreased to 7 in severe cases. Neither therapy group was associated with significant adverse effects.

A similar meta-analysis in adults also demonstrated a modest beneficial effect of adding IB to inhaled β_2-agonists [21]. The absolute increase in FEV_1 and PEF, respectively, was 7% (95%CI, 4 to 11%) and 22% (95%CI, 11 to 33%) over 45 to 90 min, respectively. The risk of admission was decreased by 27% (RR 0.73; 95%CI, 0.53 to 0.99). After publication of this systematic review, a large randomized controlled trial confirmed these findings [23]. The trial observed a 21% (95%CI, 3 to 38%) improvement in PEF and a 48% (95%CI, 20 to 76%) improvement in FEV_1 over the control group. High doses of IB reduced the risk of hospital admission by 49% (RR 0.51; 95%CI, 0.31 to 0.83); the NNT to prevent a single admission was 5 (95%CI, 3 to 17). Although questions still need to be resolved regarding IB therapy in the ED setting (such as the dose–response relationship), it seems prudent to use this agent with inhaled β_2-agonists for patients with severe exacerbations [5].

Systemic corticosteroids

The airway edema and secretions associated with acute asthma are most effectively treated with anti-inflammatory agents such as corticosteroids. The early use of systemic corticosteroids (i.e. within 1 h of ED presentation) delivered by oral or intravenous (IV) routes continues to be a principal treatment choice in all evidence-based asthma guidelines [5, 11]. A Cochrane systematic review of this topic identified 12 trials, involving 863 patients [24]. In brief, the early use

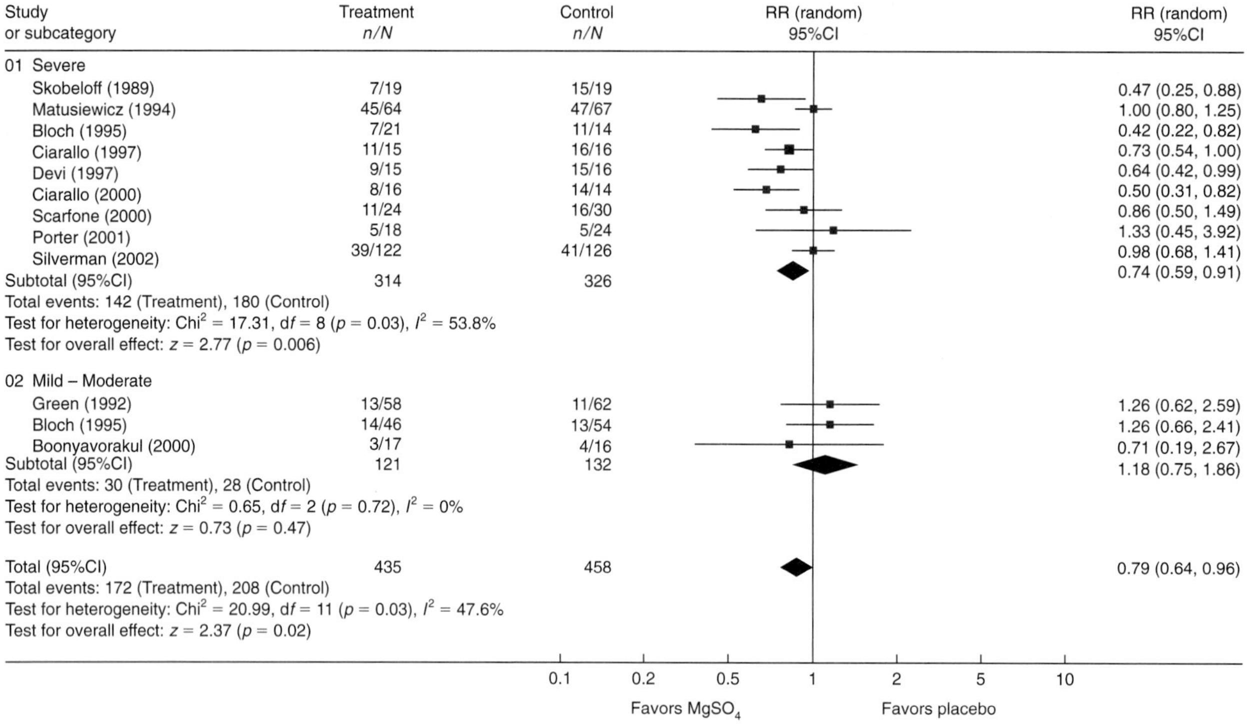

FIG. 64.3 Cochrane review on the effect of IV magnesium sulfate versus placebo on risk of hospital admission in acute asthma [28].

of corticosteroids for acute asthma in the ED significantly reduced admission rates (OR 0.40; 95%CI, 0.21 to 0.78), with an NNT of 8 (95%CI, 5 to 21). This benefit was more pronounced for those not already receiving systemic corticosteroids (OR 0.37; 95%CI, 0.19 to 0.70) and for those experiencing a severe exacerbation (OR 0.35; 95%CI, 0.21 to 0.59). Adverse effect profiles were similar between all systemic corticosteroids treatment routes (e.g. by mouth, IV) and placebo.

The debate about the use of IV versus oral corticosteroids in the ED is one that may never end; however, it now seems to be focused on identifying which patients require the IV route as opposed to the oral route. There is no clear evidence from controlled trials or meta-analyses that suggest the advantage offered by corticosteroids in acute asthma is related to the route of administration. Further, systematic evidence on dosing suggests that high-dose corticosteroids, at least in hospitalized patients, are no more effective than moderate doses in the acute setting [25].

Applying this information to practice requires a clear understanding that not all levels of severity have been assessed with sufficient rigor to confirm equivalency of systemic routes in all situations. For this reason, IV corticosteroids should be reserved for those who are too dyspneic to swallow, obtunded or intubated, or unable to tolerate oral medications (e.g. vomiting). Until further evidence is available, it seems reasonable to select oral agents as the first-line choice unless circumstances suggest otherwise.

Magnesium sulfate

The use of magnesium sulfate ($MgSO_4$) in unresponsive acute asthma has gained support recently. This agent has both immediate bronchodilator effects, as well as mild anti-inflammatory effects. Several systematic reviews have concluded that IV magnesium is not only safe but effective in patients with severe exacerbations [26, 27]. In the original Cochrane review, the *addition* of IV $MgSO_4$ to β-agonist and IV systemic corticosteroid treatment among individuals with severe exacerbations reduced hospitalizations (OR 0.1; 95%CI, 0.04 to 0.27) and improved pulmonary function (mean increase in PEF: 52 l/min; 95%CI, 27 to 77) [27]. Overall, there were no changes in the systolic blood pressure with this treatment and adverse effects were rare or minor. It is important to note that the *routine* use of $MgSO_4$ in all patients with asthma did not produce a statistically significant benefit (OR for admission: 0.31; 95%CI, 0.09 to 1.04). A recently updated analysis (Fig. 64.3) yielded very similar results [28].

Although uncertainties remain, the 2007 EPR3 guidelines recommend consideration of IV $MgSO_4$ in patients who have life-threatening exacerbations and in those whose exacerbations remains in the severe category after 1 h of intensive conventional therapy [5]. Since this agent has been shown to be easy to use, extremely safe and inexpensive, we strongly support consideration of its early use in severe acute asthma. Currently, the recommended dose is 2 g IV over 20 min in adults and 25–100 mg/kg in children (total maximum dose of 2 g).

Given the evidence supporting the use of IV $MgSO_4$ in severe exacerbations, and the rapid uptake of this treatment into emergency asthma management [6], it was not surprising that clinical researchers have tried to treat acute asthma with inhaled magnesium. A 2005 Cochrane review identified six trials involving only 296 patients [29]; four of the trials

Study	Treatment N	Treatment Mean (SD)	Control N	Control Mean (SD)	Weight (%)	Standardized mean difference (random) 95%CI
Carter (1996)	11	56.00 (20.00)	11	50.00 (16.00)	8.1	0.32 (−0.52, 1.16)
Dorfman (2000)	20	63.00 (23.00)	19	68.00 (14.00)	11.9	−0.26 (−0.89, 0.38)
Henderson (1999)	84	310.00 (115.00)	91	302.00 (113.00)	22.2	0.07 (−0.23, 0.37)
Kass (1999)	11	41.50 (32.80)	12	33.70 (50.20)	8.4	0.18 (−0.64, 1.00)
Kress (2002)	23	1.93 (0.84)	22	1.49 (0.83)	12.8	0.52 (−0.08, 1.11)
Lin Lee (2005a)	40	61.00 (15.00)	40	48.00 (13.00)	16.5	0.92 (−0.46, 1.38)
Rose (2002)	18	220.00. (69.00)	18	205.00 (63.00)	11.4	0.22 (−0.43, 0.88)
Xie (2003a)	13	1.86 (0.25)	11	1.84 (0.23)	8.7	0.08 (−0.72, 0.88)
Total (95%CI)	220		224		100.0	0.28 (−0.01, 0.56)

Test for heterogeneity chi-square = 12.91 df = 7 p = 0.07 I^2 = 45.8%
Test for overall effect z = 1.91 p = 0.06

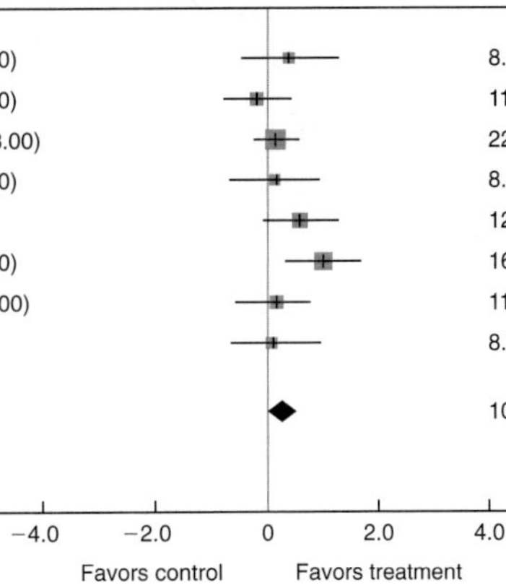

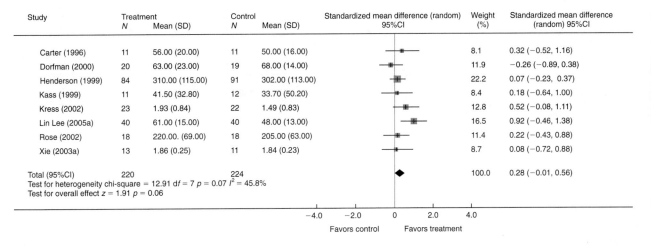

FIG. 64.4 Cochrane review on the effect of heliox therapy versus control on pulmonary function in acute asthma [30, 31].

compared nebulized $MgSO_4$ with β_2-agonist to β_2-agonist alone, and two studies compared $MgSO_4$ alone to β_2-agonist alone. Overall, there was a lack of benefit across studies; however, the heterogeneity of the trials precluded definitive conclusions. For example, in subjects with severe exacerbations, the lung function differences were significant. The area remains controversial [30, 31] and will require further investigation. Currently, there is no evidence to suggest *replacing* inhaled SABAs with magnesium and insufficient evidence to support *combining* inhaled SABAs with magnesium.

Heliox

Another controversial area in acute asthma is the role of heliox. Although this helium–oxygen mixture has been used sporadically since 1930s, there are relatively sparse data supporting its use. A recent Cochrane review identified 10 trials involving 544 patients [30, 31]. Pooling of the eight trials with pulmonary function data yielded a borderline significant group difference (Fig. 64.4). The authors note that there was significant heterogeneity among the studies but that heliox use did improve pulmonary function in the subgroup of patients with the most severe baseline pulmonary function impairment. The heterogeneity may result from small sample sizes and distinct patient populations. More importantly, however, some studies have neglected to account for the different effect of heliox versus oxygen (or room air) on the respirable mass of nebulized albuterol [32]. The technical differences among studies greatly complicate interpretation and synthesis of the literature. A large multicenter trial with optimal heliox delivery would help to resolve lingering concerns. Although uncertainties remain, the 2007 EPR3 guidelines recommend consideration of heliox in patients who have life-threatening exacerbations and in those whose exacerbations remains in the severe category after 1 h of intensive conventional therapy [5].

Inhaled corticosteroids in the ED

Inhaled corticosteroids (ICS) are preferred therapy for chronic asthma, where beneficial effects are seen over a prolonged period of use. In addition, there is now emerging evidence supporting their use in the ED setting for acute asthma. A Cochrane systematic review suggests that ICS may reduce admissions to hospital and improve pulmonary function when administered early in the ED protocol [33]. In acute asthma, ICS reduced admissions (Fig. 64.5; OR 0.33; 95%CI, 0.17 to 0.64) and improved pulmonary function (mean difference PEF 8.0%; 95%CI, 3 to 13) compared to placebo. However, several of the studies included in this review did not administer systemic corticosteroids to either the treatment or control group, which limits the power of the review to determine the *additive* benefit of ICS when systemic corticosteroids are being used.

A large study demonstrated that there may be drawbacks to *replacing* oral corticosteroids with ICS in the treatment of acute asthma [37]. In this randomized double-blind study, patients received ICS versus oral corticosteroids in addition to standard β_2-agonist therapy. The oral corticosteroid-treated subjects performed better and were admitted less frequently than the ICS-treated subjects. Combined with the previous meta-analysis data, these results suggest that ICS may be useful as an adjunct to systemic corticosteroids, but not as a replacement choice [33, 35]. Additional research is needed in this area to determine the optimal dose, frequency, and drug to be used, and to clarify the magnitude of the additional benefit when oral corticosteroids are being administered concurrent with ICS. Until such time, the role of ICS in the treatment of acute asthma is not clear, but we support consideration of its early use in severe acute asthma.

Leukotriene modifiers

Leukotriene modifiers may provide another approach to the management of acute asthma. For example, a randomized trial of IV montelukast in moderate–severe exacerbations demonstrated significant improvement in FEV_1 within 10 min of administration [36]. For rapid bronchodilation, the IV route is required since an oral formulation of montelukast would not begin to provide benefit until approximately

Study	Treatment n/N	Control n/N	Weight (%)	Odds ratio (fixed) 95%CI
01 ICS + systemic corticosteroids versus systemic corticosteroids				
Guttman (1997)	8/30	12/30	17.1	0.55 (0.18, 1.62)
Sung (1998)	2/24	5/20	9.7	0.27 (0.05, 1.60)
Subtotal (95%CI)	54	50	26.8	0.45 (0.18, 1.12)
Total events: 10 (Treatment), 17 (Control)				
Test for heterogeneity chi-square = 0.43 df = 1 p = 0.51 I^2 = 0.0%				
Test for overall effect z = 1.72 p = 0.09				
02 ICS versus placebo				
Afilalo (1999)	2/28	5/26	9.4	0.32 (0.06, 1.84)
Milani (2004)	0/17	1/17	2.8	0.31 (0.01, 8.27)
Rodrigo (1998)	4/47	12/47	21.3	0.27 (0.08, 0.92)
Rodrigo (2003)	6/58	15/62	25.3	0.36 (0.13, 1.01)
Singhi (1999)	0/30	7/30	14.3	0.05 (0.00, 0.95)
Subtotal (95%CI)	180	182	73.2	0.27 (0.14, 0.52)
Total events: 12 (Treatment), 40 (Control)				
Test for heterogeneity chi-square = 1.62 df = 4 p = 0.81 I^2 = 0.0%				
Test for overall effect z = 3.88 p = 0.0001				
Total (95%CI)	234	232	100.0	0.32 (0.18, 0.54)
Total events: 22 (Treatment), 57 (Control)				
Test for heterogeneity chi-square = 2.61 df = 6 p = 0.86 I^2 = 0.0%				
Test for overall effect z = 4.22 p = 0.00002				

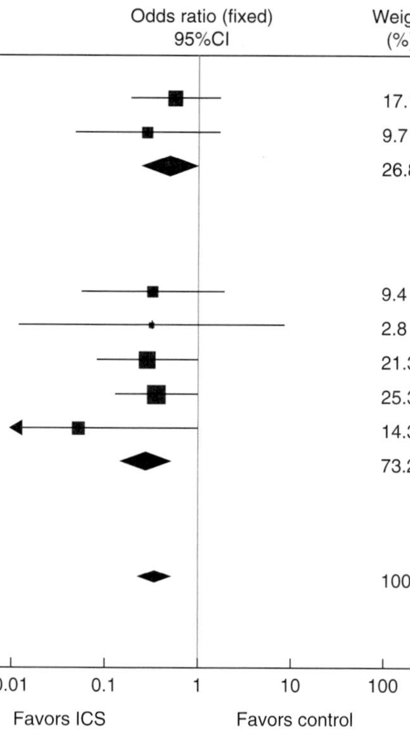

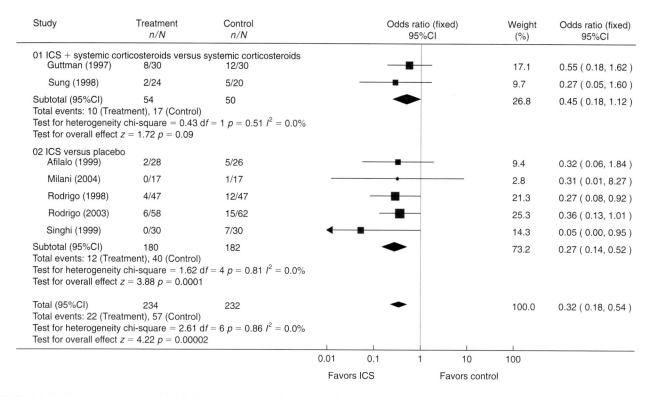

FIG. 64.5 Cochrane review on the effect of inhaled corticosteroids in the ED on risk of hospital admission in acute asthma [33].

90 min after administration [37]. Nonetheless, the observed bronchodilation suggests another novel approach: to give an oral formulation to all patients with moderate-to-severe exacerbations at the time of ED presentation (e.g. as an adjunct to primary therapy with β_2-agonists, anticholinergic agents, and systemic corticosteroids). Early studies on the role of oral leukotriene modifiers in the ED suggest likely benefits [38, 39] but this concept requires further study.

Other therapies

The therapeutic benefits of other agents, such as IV SABAs [40] and aminophylline [41] are limited. The cited Cochrane systematic reviews provide little, if any, support for their use in emergency asthma care; for both agents, the side effects clearly outweigh the benefits. It is important to note, however, that the SABA trials addressed a different question than the one that we would raise today. For example, even if IV SABA provides no benefit over inhaled SABAs in head-to-head comparisons, what is the role of IV SABA in addition to inhaled SABA for severe exacerbations? Few studies have directly examined this more relevant question to practitioners today; the few small trials that have examined this issue provide contradictory results [42–44]. Future, and much larger, trials will need to resolve this important issue. In the meantime, we encourage the use of other, more established therapies in the emergency management of acute asthma.

Likewise, systematic reviews have repeatedly shown that antibiotics should not be used routinely for acute asthma [9, 45]. Antibiotics should be limited to the treatment of co-morbid bacterial infections, such as community-acquired pneumonia or bacterial sinusitis. Despite this evidence-based recommendation, many clinicians continue to prescribe antibiotics for the viral upper respiratory tract infections that trigger acute asthma. For example, national data suggest that 22% of ED patients with acute asthma receive antibiotics as part of their asthma care. (Carlos Camargo MD, unpublished data). The continued prescription of antibiotics for acute asthma represents a potential focus of future quality improvement efforts to decrease in appropriate antibiotic use.

Non-invasive ventilation is another experimental approach for the treatment of respiratory failure due to severe asthma exacerbation, but data are sparse. The Cochrane review on this topic identified only one trial involving 30 patients [46]. Although this small trial [47] showed benefit with non-invasive positive pressure ventilation when compared to usual medical care alone, with significant improvements in hospitalization rate and pulmonary function, there are inadequate data to draw firm conclusions. Future multicenter studies should examine the role of heliox, non-invasive ventilation, or both in patients with severe exacerbations. Until that time, there is insufficient evidence to support use of NIV in the management of acute asthma.

Post-ED care

The goal of post-ED care is to safely return each patient to a level of functioning commensurate with good asthma "control". Per EPR3, this includes reductions in both impairment and risk [5]. With regard to risk, this includes a reduction in asthma exacerbation recurrences, with good control defined as at most one corticosteroid-requiring exacerbation in a year.

The more immediate concern, however, is to successfully treat the current exacerbation and prevent an acute asthma relapse. Please note that the term "relapse" differs from "recurrence", with relapse referring to the acute deterioration of a patient during an exacerbation despite increased medical care (e.g. over a shorter period of time of 1–4 weeks) whereas recurrence refers to repeated exacerbations over longer periods of time (e.g. 6–12 months).

Oral corticosteroids

Approximately 10–20% of ED patients treated for acute asthma and sent home will relapse within 2 weeks of discharge [3, 4], many because of unresolved inflammation that leaves the airways sensitive to inhaled irritants. Guidelines strongly encourage treatment with systemic corticosteroids following ED discharge for an acute exacerbation [5, 11]. Compelling evidence for this approach is found in a recently updated Cochrane systematic review comparing post-ED corticosteroid therapy to placebo [48]. The review identified 6 trials involving 374 people. Surprisingly, despite the small number of total patients involved in this review, it showed that significantly fewer patients in the corticosteroids group relapsed in the first week (OR 0.38; 95%CI, 0.2 to 0.74). This reduced risk was maintained over the first 21 days (OR 0.47; 95%CI, 0.25 to 0.89). The corticosteroid group had less need for β_2-agonists (mean difference: −3 activations/day; 95%CI, −5.6 to −1.0). Changes in pulmonary function tests and adverse effects, while rarely reported, showed no differences between the treatment groups. A subgroup analysis indicated that intra-muscular (IM) corticosteroids and a 7- to 10-day tapering course of corticosteroids were similarly efficacious. IM therapy should be reserved for those patients with questionable compliance, inability to pay for an oral prescription, or those who are otherwise unreliable (cognitive impairment, intoxication, etc.). From these results, the authors conclude that as few as 10 patients need to be treated to prevent one relapse to additional care.

Small sample sizes in the randomized clinical trials (RCTs) performed to date did not permit an examination of the relative effectiveness of various regimens and definitive recommendation concerning dose or dosing protocol(s) cannot be provided. However, given the enhanced compliance associated with once daily dosing and the availability of 40 or 50 mg tablets in North America, the use of oral corticosteroids for a short period (e.g. 5–7 days) seems appropriate for most patients discharged with an acute asthma episode. The need to "taper" these oral corticosteroids over that period appears unwarranted [49, 50], especially when ICS are being employed concurrently [51].

Inhaled corticosteroids

In North America, the majority of patients with acute asthma treated in an emergency setting are discharged and prescribed a short course (5–7 days) of oral corticosteroids. Less information exists regarding the use of ICS; however, the data that exist indicate impressive practice variation with respect to this treatment. For example, in US sites associated with a large North American emergency medicine research network (www.emnet-usa.org), only 11% of discharged patients were prescribed an ICS if they were not regularly taking one, whereas in Canadian sites, 63% of similar patients were treated with an ICS at discharge [12, 48, 52].

There is evidence to support using a combination of inhaled and oral corticosteroids after discharge from the ED [51]. There are three randomized controlled trials [51, 53, 54] which provide somewhat conflicting evidence for the addition of ICS to prevent relapse, but when combined in a systematic review the introduction of ICS appears to be worthwhile [55]. The pooled effect demonstrates a trend in favor of the ICS plus oral corticosteroid group which had fewer relapses after discharge than the group administered oral corticosteroids alone (OR 0.68; 95%CI, 0.46 to 1.02) [55].

The results of this review indicate that patients already taking ICS should be counseled by the ED staff about compliance-enhancing interventions [56–58]. It is important to remember that many of the patients with acute asthma who present to the ED exhibit features associated with poorly controlled chronic asthma [58]. They represent vulnerable patients who are the ideal candidates for ICS. Consequently, it seems prudent to recommend those patients not on ICS agents should be considered for short- or long-term ICS therapy in conjunction with oral prednisone after discharge. The dose and duration of inhaled and oral corticosteroids should be based on recent history of symptom control, health care utilization, and quality of life indicators. For those patients with more severe illness, this clearly would be the optimal treatment strategy. The 2007 EPR3 guidelines encourage clinicians to consider initiation of ICS given the potential for ICS to reduce subsequent ED visits, strong evidence that long-term control ICS therapy reduces exacerbations in patients with persistent asthma, and the opinion of the entire Expert Panel that the initiation (and continuation) of ICS therapy at ED discharge can be an important effort to bridge the gap between emergency and primary care for asthma [5].

There are several recent publications examining the effect of *replacing* oral corticosteroids with a high-dose ICS These generally compare oral prednisone to very high doses of ICS in mild exacerbations after discharge. While the systematic review failed to demonstrate a significant difference in asthma relapse between the two treatments (OR 1.0; 95%CI, 0.48 to 1.42), these results need to be interpreted cautiously [55]. Although the evidence implies equivalence, these results are not conclusive due to the width of the confidence intervals and the inclusion of only patients with mild asthma exacerbations. Given the limited data on this issue to date, use of ICS alone should be reserved for those patients with very mild asthma and those who refuse or cannot take oral corticosteroids (e.g. brittle diabetes, steroid psychosis). Compared to the traditional short course of prednisone, ICS are expensive and more difficult for patients and families to use.

If ICS are used in place of systemic corticosteroids in these subgroups, it is important to not just double the maintenance dose, as recommended in past guidelines, because several recent trials have not found that this approach is not effective [58–62]. In this exceptional situation, and on the basis of admittedly sparse data [63], the current recommendation is to quadruple the ICS dose.

Other therapies

Over the past decade, there have been many exciting advances in chronic asthma care and some of these new approaches are now being tried in patients with acute asthma. A surprisingly under-studied area is the potential use of combination therapies at the time of ED discharge. Recently, Rowe and colleagues reported a small randomized, controlled, double-blind trial of fluticasone/salmeterol versus fluticasone alone at discharge from the ED; all patients were discharged on 7 days of prednisone [52]. Although relapse rates were numerically lower in the combination group, the difference was not statistically significant. Likewise, quality of life measures did not differ between groups but the authors noted that in the subset of patients who already were on ICS at the time of ED presentation, the fluticasone/salmeterol group had significantly improved quality of life ($p < 0.05$). Larger studies are needed to determine the role of inhaled combination therapies, especially those with formoterol, in the post-ED management of acute asthma. This question will become increasingly important as treatment of chronic asthma improves. For example, in the past, emergency clinicians were faced with many asthma patients who were not taking anti-inflammatory medications, such as ICS. In recent years, more patients presenting to the ED are already taking ICS and, increasingly, the combination agents (fluticasone/salmeterol or budesonide/formoterol). Understanding the relapse risks and the best management for patients already receiving anti-inflammatory agents will be an important contribution to evidence-based acute asthma care in the future.

Non-pharmacological approaches

Asthma education

Education is a key recommendation in all asthma guidelines [5, 11]. Emergency physicians, when surveyed, also felt it was an important component of asthma care and rated many educational items as necessary [58]. Despite widespread enthusiasm for asthma education, the overall results of individual trials and systematic reviews have been somewhat disappointing; this is especially true for the acute setting.

A systematic review of RCTs in children who attended the ED was conducted to identify whether asthma education leads to improved health outcomes [64]. The analyses found that structured education, as compared to usual care or lower-intensity education, did not reduce subsequent ED visits (RR 0.87; 95%CI, 0.37 to 2.08), hospital admissions (RR 0.74; 95%CI, 0.38 to 1.46), or unscheduled doctor visits (RR 0.74; 95%CI, 0.49 to 1.12). However, there was a significant heterogeneity among these pooled results ($p < 0.01$) with potential for uncontrolled confounding factors. The authors concluded that there was no firm evidence to support asthma education for children to improve asthma control and that more rigorous research was necessary.

Another Cochrane review examined the impact of limited asthma education programs (information only) for adults with asthma [65]. There was no effect on hospital admissions, doctor visits, lung function, medication use, or normal activity days lost. However, perceived asthma symptoms did improve in the education group (OR 0.40; 95%CI, 0.18 to 0.86). One study found that limited asthma education programs were associated with reduced ED visits (mean decrease visits/person/year: 2.8 visits; 95%CI, −4.3 to 1.2). The reviewers concluded that use of limited asthma education, as it has been practiced, does not appear to improve health outcomes in adults. The usefulness of education in the ED setting is encouraging, but further research is required to confirm these findings.

A second review by the same group assessed the effects of adult asthma self-management programs coupled with regular medical review [66]. Self-management education reduced hospitalizations (OR 0.57; 95%CI, 0.38 to 0.88), ED visits (OR 0.71; 95%CI, 0.57 to 0.90), unscheduled doctor visits (OR 0.57; 95%CI, 0.40 to 0.82), missed work/school days (OR 0.55; 95%CI, 0.38 to 0.79), and nocturnal asthma (OR 0.53; 95%CI, 0.39 to 0.72). Pulmonary function measures were minimally affected by these combined interventions. If a written action plan was added, there was an even greater reduction in hospitalization (OR 0.35; 95%CI, 0.18 to 0.68). Patients who could self-adjust their medications using an individualized written plan had better lung function than those whose medications were adjusted by a doctor.

In the latest Cochrane review [67], the authors examined the effectiveness of educational interventions administered following an ED visit for acute asthma. The review included 12 studies involving 1954 adults. Although education significantly reduced subsequent admission to hospital (Fig. 64.6; RR 0.50; 95%CI, 0.27 to 0.91), it did not significantly reduce the risk of re-presentation at ED during follow-up (RR 0.69; 95%CI, 0.40 to 1.21). The authors identified priority areas for further research but were reluctant to draw further conclusions from these data.

The clinical implications of these reviews are that isolated asthma education provides limited benefit. However, self-management programs associated with regular review by a health care provider and the addition of a written action plans appear to be effective in reducing asthma symptoms, health care use, and improving asthma quality of life. Moreover, educational interventions administered after an ED visit for acute asthma also appear to have tangible benefits (decreased risk of subsequent admission to the hospital). This conclusion receives strong support from a recent single-center randomized trial by Teach and colleagues [68], where the investigators describe striking benefits for children enrolled in an ED-based education program that emphasized patient self-monitoring and management, environmental modification and trigger control, and linkages and referrals to ongoing care.

In light of these findings, clinicians might be well advised to consider training staff, patients, and families in asthma self-management that involves monitoring of asthma (by either PEF or symptoms), regular medical review, and provision of a simple written action plan. Programs that enable people to adjust their medication using a written action plan appear to be more effective than other forms of asthma self-management and should be encouraged. To that end, Camargo and colleagues developed a simple "asthma discharge plan" for use in ED patients with acute asthma (Fig. 64.7) [69].

A conclusion that we draw from the reviewed evidence is the importance of seeing the ED as part of the continuum of asthma care. Given the chronicity of this disease – and notwithstanding the fact that these exacerbations represent teachable moments – it is likely that the main role of the ED will continue to be *reinforcement* of other educational efforts and *facilitation* of follow-up with the primary care provider (assuming one exists). Evidence suggests that only 25–50% of patients see a primary care provider for a routine asthma reassessment within *1 month* of the ED visit [70, 71]. Despite this recognition, efforts to increase this primary care follow-up following an exacerbation have been surprisingly difficult. In one RCT, investigators used increasingly costly and complicated efforts to increase PCP follow-up following an exacerbation of asthma, only to find no impact on any clinically meaningful outcome [71]. The role of facilitated referral of ED patients to asthma specialists is not yet clear and merits rigorous investigation.

EMERGENCY DEPARTMENT – ASTHMA DISCHARGE PLAN

Name: ______________________ was seen by **Dr.** ______________ on ___/___/___

- Take your prescribed medications as directed—do not delay!
- Asthma attacks like this one can be prevented with a long-term treatment plan.
- Even when you feel well, you may need daily medicine to keep your asthma in good control and prevent attacks.
- Visit your doctor or other health care provider as soon as you can to discuss how to control your asthma and to develop *your own* action plan.

Your follow-up appointment with ______________ is on: ___/___/___ **Tel:** ___________

YOUR MEDICINE FOR THIS ASTHMA ATTACK IS:

Medication	Amount	Doses per day, for # days
Prednisone/prednisolone (oral corticosteroid)		_____ a day for _____ days Take the entire prescription, even when you start to feel better.
Inhaled albuterol		_____ puffs every 4–6 h if you have symptoms, for _____ days

YOUR DAILY MEDICINE FOR LONG-TERM CONTROL AND PREVENTING ATTACKS IS:

Medication	Amount	Doses per day
Inhaled corticosteroids		

YOUR QUICK-RELIEF MEDICINE WHEN YOU HAVE SYMPTOMS IS:

Medication	Amount	Number of doses/day
Inhaled albuterol		

ASK YOURSELF 2 TO 3 TIMES PER DAY, EVERY DAY, FOR AT LEAST 1 WEEK:

"How good is my asthma compared to when I left the hospital?"

If you feel much better	**If you feel better, but still need your quick-relief inhaler often:**	**If you feel about the same:**	**If you feel worse:**
• Take your daily long-term control medicine.	• Take your daily long-term control medicine. • See your doctor as soon as possible.	• Use your quick-relief inhaler. • Take your daily long-term control medicine. • See your doctor as soon as possible – don't delay.	• Use your quick-relief inhaler. • Take your daily long-term control medicine. • Immediately go to ED or call 9–1–1.

YOUR ASTHMA IS UNDER CONTROL WHEN YOU:

① Can be active daily and sleep through the night.	② Need fewer than 4 doses of quick-relief medicine in a week.	③ Are free of shortness of breath, wheeze, and cough.	④ Achieve an acceptable "peak flow" (discuss with your health care provider).

FIG. 64.6 ED – Asthma discharge plan [69]. Reprinted by permission from Carlos Camargo, M.D., Principal Investigator of Agency for Health Care Research and Quality. Grant No. R13H31094.

Using an inhaler seems simple, but most patients do not use it the right way. When you use your inhaler the wrong way, less medicine gets to your lungs.

For the next few days, read these steps aloud as you do them or ask someone to read them to you. Ask your doctor, nurse, other health care provider, or pharmacist to check how well you are using your inhaler.

Use your inhaler in one of the three ways pictured below. A or B are best, but C can be used if you have trouble with A and B. Your doctor may give you other types of inhalers.

Steps for using your inhaler

Getting ready

1. Take off the cap and shake the inhaler.
2. Breathe out all the way.
3. Hold your inhaler the way your doctor said (A, B, or C below).

Breathe in slowly

4. As you start breathing in slowly through your mouth, press down on the inhaler one time. (If you use a holding chamber, first press down on the inhaler. Within 5 s, begin to breathe in slowly.)
5. Keep breathing in slowly, as deeply as you can.

Hold your breath

6. Hold your breath as you count to 10 slowly, if you can.
7. For inhaled quick-relief medicine (short-acting β_2-agonists), wait about 15–30 s between puffs. There is no need to wait between puffs for other medicines.

(A) Hold inhaler 1–2 in. in front of your mouth (about the width of two fingers).

(B) Use a spacer/holding chamber. These come in many shapes and can be useful to any patient.

(C) Put the inhaler in your mouth. Do not use for steroids.

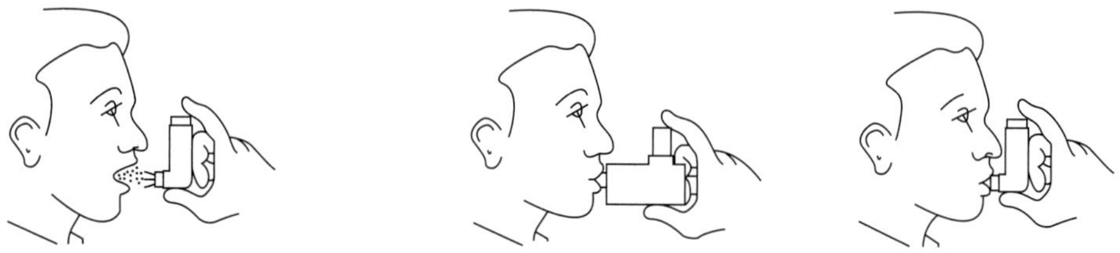

Clean your inhaler as needed, and know when to replace your inhaler. For instructions, read the package insert or talk to your doctor, other health care provider, or pharmacist.

FIG. 64.6 (Continued).

Study	Education n/N	Control n/N	Weight (%)	Relative risk (random) 95%CI
George (1999)	3/30	12/20	17.0	0.17 (0.05, 0.52)
Morice (2001)	10/40	11/40	26.1	0.91 (0.44, 1.90)
Osman (2002)	22/131	38/140	34.0	0.62 (0.39, 0.99)
Pemeger (2002)	4/57	6/58	15.6	0.68 (0.20, 2.28)
Yoon (1993)	1/28	7/28	7.3	0.14 (0.02, 1.09)
Total (95%CI)	286	286	100.0	0.50 (0.27, 0.91)

Total events: 40 (Education), 74 (Control)
Test for heterogeneity chi-square = 8.22 $df = 4$ $p = 0.08$ $I^2 = 51.3\%$
Test for overall effect $z = 2.27$ $p = 0.02$

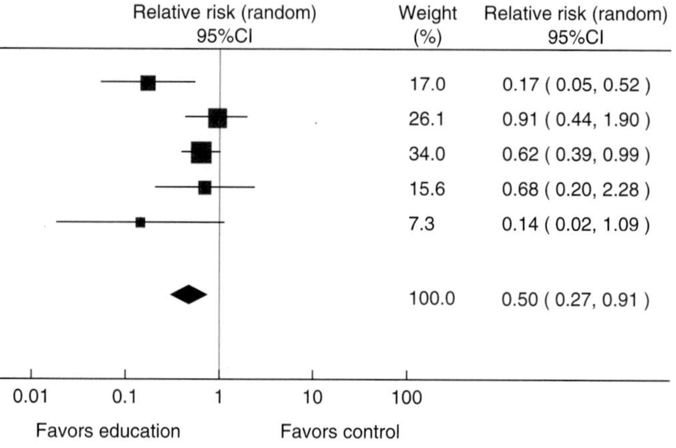

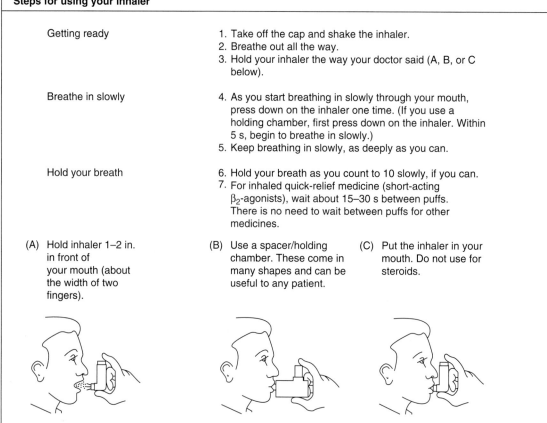

FIG. 64.7 Cochrane review on the effect of education versus usual care on risk of hospital admission or re-admission in acute asthma [67].

Other issues

Studies have documented a large percentage of patients who are repeatedly exposed to environmental irritants. A significant number of asthma patients are either exposed to cigarette smoke or persist with this habit despite the presence of asthma [72]. Smoking cessation counseling and avoidance strategies should be provided to these patients along with their families with every exacerbation in an attempt to motivate smokers to overcome this addiction. Although it is unlikely that a detailed intervention can be undertaken in a busy ED setting, simple approaches are possible and under investigation [73, 74]. At a minimum, ED staff are strongly encouraged to learn about referral options in their hospital (e.g. to refer ED patients to smoking cessation programs).

Acute exacerbations also represent an opportunity for ED physicians and staff to review pharmacological and non-pharmacological management of asthma. Adhering to a long-term plan to keep asthma under control is often problematic, especially in the areas of regular medication use and avoiding environmental triggers. The opportunity to introduce or reinforce helpful behaviors should not be lost, particularly during the inpatient stay.

Other interventions during the inpatient stay could include immunizations against influenza [75], and less orthodox interventions such as the use of acupuncture [76], homeopathy [77], and dietary changes (such as increased fatty acid intake [78]; all of these interventions have been promoted in the past. Despite this, there is little (if any) evidence of benefit from these interventions in the literature and one should be cautious about recommending these approaches – particularly at the expense of other better-established interventions.

All that being said, one unorthodox intervention probably merits further consideration. Almost a decade ago, Camargo and colleagues reported the first prospective data linking obesity and risk of asthma [79]. During the interim, a large number of studies have confirmed this association and linked obesity with poor asthma control, including an increased risk of exacerbations [80]. Although high-quality data are sparse, a recent systematic review suggests that weight loss among obese asthmatics probably leads to improved asthma control [81]. Indeed, the EPR3 guidelines encourage clinicians to start their obese asthmatic patients on a regular exercise program, and to make weight control a part of any long-term asthma management plan [5]. Although this is, of course, a difficult intervention to implement in a busy ED, it is a program that could be started in the inpatient setting – a time of heightened awareness and interest in health matters. Such an approach has been used for cardiovascular disease prevention for decades and we encourage asthma specialists to learn from this example and consider such initiating interventions during asthma hospitalizations.

SUMMARY

Asthma is a common, chronic, and often debilitating disease. The treatment approaches summarized in this chapter provide hope for reduced symptoms, an early return to activities, and improved quality of life in the sub-acute period after an exacerbation. In addition, combining medication and education interventions provides an opportunity for prevention of future exacerbations. Thus, proper asthma management in all clinical settings, including the ED and inpatient ward, emphasizes the reduction of both impairment and risk [5]. After successful treatment of an acutely ill patient, emergency care should increasingly transition toward the optimization of sub-acute and sometimes chronic asthma care issues. Close collaboration between emergency specialists and chronic care specialists will ensure that asthmatic patients receive the best care possible.

ACKNOWLEDGMENTS

Dr. Camargo is supported by grant HL-84401 from the National Institutes of Health (Bethesda, USA). Dr. Rowe is supported through the 21st Century Canada Research Chair Program from the Government of Canada (Ottawa, Canada).

References

1. American Lung Association. Trends in Asthma Morbidity and Mortality (August 2007); Retrieved 10/1/07, from http://www.lungusa.org/site/pp.asp?c = dvLUK9O0E&b = 33347.
2. Weiss KB, Gergen PJ, Hodgson TA. An economic evaluation of asthma in the United States. *N Engl J Med* 326: 862–66, 1992.
3. Emerman CL, Woodruff PG, Cydulka RK, Gibbs MA, Pollack CV Jr., Camargo CA Jr. Prospective multicenter study of relapse following treatment for acute asthma among adults presenting to the emergency department. *Chest* 115: 919–27, 1999.
4. Emerman CE, Cydulka RK, Crain EF, Rowe BH, Radeos MS, Camargo CAJ. Prospective multicenter study of relapse following treatment for acute asthma among children presenting to the emergency department. *J Pediatr* 138: 318–24, 2001.
5. National Asthma Education and Prevention Program. *Expert Panel Report 3: Guidelines for the Diagnosis and Management of Asthma (Full Report).* Bethesda, MD, US: Department of Health and Human Services, National Institutes of Health, 2007.
6. Rowe BH, Camargo CA Jr. The use of magnesium sulfate in acute asthma: Rapid uptake of evidence in North American emergency departments. *J Allergy Clin Immunol* 117: 53–58, 2006.
7. National Asthma Education and Prevention Program. *Expert Panel Report 2: Guidelines for the Diagnosis and Management of Asthma.* Bethesda, MD, US: Department of Health and Human Services, National Institutes of Health, 1997.
8. National Asthma Education Program. *Expert Panel Report: Guidelines for the Diagnosis and Management of Asthma.* Bethesda, MD, US: National Institutes of Health, Publication No. 91-3642, 1991.
9. National Asthma Education and Prevention Program. *Expert Panel Report: Guidelines for the Diagnosis and Management of Asthma—Update on Selected Topics 2002.* Bethesda, MD, US: Department of Health and Human Services, National Institutes of Health, 2002.
10. Weber EJ, Silverman RA, Callaham ML, Pollack CVJ, Woodruff PG, Clark S, Camargo CA Jr. A prospective multicenter study of factors associated with hospital admission among adults with acute asthma. *Am J Med* 113: 371–78, 2002.
11. Global Initiative for Asthma (GINA) Website. Retrieved 10/1/07, from: http://www.ginasthma.com/.

12. Rowe BH, Bota GW, Clark S, Camargo CA Jr. Comparison of Canadian versus American emergency department visits for acute asthma. *Can Respir J* 14: 331–37, 2007.
13. Cochrane Collaboration Website. Retrieved 10/1/07, from http://www.cochrane.org/index.htm.
14. Cates CJ, Crilly JA, Rowe BH. Holding chambers (spacers) versus nebulisers for beta-agonist treatment of acute asthma. *Cochrane Database System Rev*(2): CD000052, 2006.
15. Camargo CA Jr, Kenney PA. Assessing costs of aerosol therapy. *Resp Care* 45: 756–63, 2000.
16. Camargo CA Jr., Spooner CH, Rowe BH. Continuous versus intermittent beta-agonists for acute asthma. *Cochrane Database Syst Rev*(4): CD001115, 2003.
17. Strauss L, Hejal R, Galan G, Dixon L, McFadden ER Jr. Observations on the effects of aerosolized albuterol in acute asthma. *Am J Respir Crit Care Med* 155: 454–58, 1997.
18. Cydulka RK, McFadden ER, Sarver JH, Emerman CL. Comparison of single 7 5-mg dose treatment vs sequential multidose 2.5-mg treatments with nebulized albuterol in the treatment of acute asthma. *Chest* 122: 1982–87, 2002.
19. Stein J, Levitt MA. A randomized, controlled double-blind trial of usual-dose versus high-dose albuterol via continuous nebulization in patients with acute bronchospasm. *Acad Emerg Med* 10: 31–36, 2003.
20. Boulet LP, Becker A, Berube D, Beveridge RC, Ernst P. Canadian asthma consensus report. *Can Med Assoc J* 161: S1–S61, 1999.
21. Stoodley RG, Aaron SD, Dales RE. The role of ipratropium bromide in the emergency management of acute asthma exacerbation: a metaanalysis of randomized clinical trials. *Ann Emerg Med* 34(1): 8–18, 1999.
22. Plotnick LH, Ducharme FM. Combined inhaled anticholinergics and beta2-agonists for initial treatment of acute asthma in children. *Cochrane Database System Rev*(3): CD000060, 2000.
23. Rodrigo GJ, Rodrigo C. First-line therapy for adult patients with acute asthma receiving a mulitple-dose protocol of ipratropium bromide plus albuterol in the emergency department. *Am J Respir Crit Care Med* 161: 1862–68, 2000.
24. Rowe BH, Spooner C, Ducharme FM, Bretzlaff JA, Bota GW. Early emergency department treatment of acute asthma with systemic corticosteroids. *Cochrane Database Syst Rev*(1): CD002178, 2001.
25. Manser R, Reid D, Abramson M. Corticosteroids for acute severe asthma in hospitalized patients. *Cochrane Database Syst Rev*(3): CD001740, 2001.
26. Alter HJ, Koepsell TD, Hilty WM. Intravenous magnesium as an adjunct in acute bronchospasm: A meta-analysis. *Ann Emerg Med* 36: 191–97, 2000.
27. Rowe BH, Bretzlaff JA, Bourdon C, Bota GW, Camargo CAJ. Magnesium sulfate for treating exacerbations of acute asthma in the emergency department. *Cochrane Database Syst Rev*(1): CD001490, 2000.
28. Rowe BH, Camargo CA Jr. The role of magnesium sulfate in the acute and chronic management of asthma. *Curr Opin Pulm Med* 14: 70–76, 2008.
29. Blitz M, Blitz S, Beasely R, Diner BM, Hughes R, Knopp JA, Rowe BH. Inhaled magnesium sulfate in the treatment of acute asthma. *Cochrane Database Syst Rev*(4): CD003898, 2005.
30. Rodrigo G, Pollack C, Rodrigo C, Rowe BH. Heliox for nonintubated acute asthma patients. *Cochrane Database Syst Rev*(4): CD002884, 2006.
31. Rodrigo GJ, Rowe BH, Blitz M, Blitz, S. (2006). There is no evidence to support the use of aerosolized magnesium for acute asthma (letter). *Chest*, 130, 304–5; author reply 305–6.
32. Hess DR, Acosta FL, Ritz RH, Kacmarek RM, Camargo CA Jr. The effect of heliox on nebulizer function using a beta-agonist bronchodilator. *Chest* 115: 184–89, 1999.
33. Edmonds ML, Camargo CA Jr., Pollack CV Jr., Rowe BH. Early use of inhaled corticosteroids in the emergency department treatment of acute asthma. *Cochrane Database Syst Rev*(3): CD002308, 2003.
34. Schuh S, Reisman J, Alshehri M, Dupuis A, Corey M, Arseneault R. A comparison of inhaled fluticasone and oral prednisone for children with severe acute asthma. *N Engl J Med* 343: 689–94, 2000.
35. Anonymous. *Global Initiative for Asthma: Global Strategy for Asthma Management and Prevention*, NHLBI/WHO Workshop Report. Bethesda, MD US: National Institutes of Health; Publication No. 95-3659, 1995.
36. Camargo CA Jr., Smithline HA, Malice MP, Green SA, Reiss TF. A randomized controlled trial of intravenous montelukast in acute asthma. *Am J Respir Crit Care Med* 167(4): 528–33, 2003.
37. Dockhorn RJ, Baumgartner RA, Leff JA, Noonan M, Vandormael K, Stricker W, Weinland DE, Reiss TF. Comparison of the effects of intravenous and oral montelukast on airway function: A double blind, placebo controlled, three period crossover study in asthmatic patients. *Thorax* 55: 260–65, 2000.
38. Silverman RA, Nowak RM, Korenblat PE, Skobeloff E, Chen Y, Bonuccelli CM, Miller CJ, Simonson SG. Zafirlukast treatment for acute asthma: Evaluation in a randomized, double-blind, multicenter trial. *Chest* 126: 1480–89, 2004.
39. Harmanci K, Bakirtas A, Turktas I, Degim T. Oral montelukast treatment of preschool-aged children with acute asthma. *Ann Allergy Asthma Immunol* 96: 731–35, 2006.
40. Travers A, Jones AP, Kelly K, Barker SJ, Camargo CAJ, Rowe BH. Intravenous beta2-agonists for acute asthma in the emergency department. *Cochrane Database Syst Rev*(1): CD002988, 2001.
41. Parameswaran K, Belda J, Rowe BH. Addition of intravenous aminophylline to beta 2-agonists in adults with acute asthma. *Cochrane Database Syst Rev*(4): CD002742, 2001.
42. Browne GJ, Penna AS, Phung X, Soo M. Randomised trial of intravenous salbutamol in early management of acute severe asthma in children. *Lancet* 349: 301–5, 1997.
43. Bogie AL, Towne D, Luckett PM, Abramo TJ, Wiebe RA. Comparison of intravenous terbutaline versus normal saline in pediatric patients on continuous high-dose nebulized albuterol for status asthmaticus. *Pediatr Emerg Care* 23: 355–61, 2007.
44. Browne GJ, Lam LT. Single-dose intravenous salbutamol bolus for managing children with acute severe asthma in the emergency department: Reanalysis of data. *Pediatr Crit Care Med* 3: 117–23, 2002.
45. Graham VA, Lasserson T, Rowe BH. Antibiotics for acute asthma. *Cochrane Database Syst Rev* 3: CD002741, 2001.
46. Ram FSF, Wellington SR, Rowe B, Wedzicha JA. Non-invasive positive pressure ventilation for treatment of respiratory failure due to severe acute exacerbations of asthma. *Cochrane Database System Rev*(3): CD004360, 2005.
47. Soroksky A, Stav D, Shpirer I. A pilot prospective, randomized, placebo-controlled trial of bilevel positive airway pressure in acute asthmatic attack. *Chest* 123: 1018–25, 2003.
48. Rowe BH, Spooner CH, Ducharme FM, Bretzlaff JA, Bota GW. Corticosteroids for preventing relapse following acute exacerbations of asthma. *Cochrane Database System Rev*(3): CD000195, 2007.
49. O'Driscoll BR, Kalra S, Wilson M, Pickering CAC, Carroll KB, Woodcock AA. Double-blind trial of steroid tapering in acute asthma. *Lancet* 341: 324–27, 1993.
50. Verbeek PR, Gerts WH. Nontapering versus tapering prednisone in acute exacerbations of asthma: A pilot study. *J Emerg Med* 13: 15–19, 1995.
51. Rowe BH, Bota GW, Fabris L, Therrien SA, Milner RA, Jacono J. Inhaled budesonide in addition to oral corticosteroids to prevent asthma relapse following discharge from the emergency department. A randomized controlled trial. *JAMA* 281: 2119–26, 1999.
52. Rowe BH, Wong E, Blitz S, Dinner B, Mackey D, Ross S, Senthilselvan A. Adding long-acting beta-agonists to inhaled corticosteroids after discharge from the emergency department for acute asthma: A randomized controlled trial. *Acad Emerg Med* 14: 833–40, 2007.
53. Camargo CA Jr. Randomized trial of medium-dose fluticasone vs placebo after an emergency department visit for acute asthma. *J Allergy Clin Immunol* 105(1 pt 2): S262, 2000.
54. Brenner BE, Chavda KK, Camargo CA Jr. Randomized trial of inhaled flunisolide versus placebo among asthmatics discharged from the emergency department. *Ann Emerg Med* 36: 417–26, 2000.

55. Edmonds ML, Camargo CA Jr., Brenner BE, Rowe BH. Inhaled steroids for acute asthma following emergency department discharge. *Cochrane Database Syst Rev*(3): CD002316, 2000.
56. Stevenson BJ, Rowe BH, Haynes RB, Macharia WM, Leon G. Is this patient taking the treatment as prescribed. *JAMA* 269: 2779–81, 1993.
57. Emond SD, Camargo CA Jr., Nowak RM. 1997 national asthma education and prevention program guidelines: A practical summary for emergency physicians. *Ann Emerg Med* 31: 579–89, 1998.
58. Emond SD, Reed CR, Graff LI, Clark S, Camargo CA Jr. Asthma education in the emergency department. On behalf of the MARC investigators. *Ann Emerg Med* 36(3): 204–11, 2000.
59. Garrett J, Williams S, Wong C, Holdaway D. Treatment of acute asthmatic exacerbations with an increased dose of inhaled steroid. *Arch Dis Child* 79: 12–17, 1998.
60. FitzGerald JM, Becker A, Sears MR, Mink S, Chung K, Lee J. Doubling the dose of budesonide versus maintenance treatment in asthma exacerbations. *Thorax* 59: 550–56, 2004.
61. Harrison TW, Oborne J, Newton S, Tattersfield AE. Doubling the dose of inhaled corticosteroid to prevent asthma exacerbations: Randomised controlled trial. *Lancet* 363: 271–75, 2004.
62. Rice-McDonald G, Bowler S, Staines G, Mitchell C. Doubling daily inhaled corticosteroid dose is ineffective in mild to moderate severe attacks of asthma in adults. *Intern Med J* 35: 693–98, 2005.
63. Foresi A, Morelli MC, Catena E. Low-dose budesonide with the addition of an increased dose during exacerbations is effective in long-term asthma control. *Chest* 117: 440–46, 2000.
64. Haby MM, Waters E, Roberston CF, Gibson G, Ducharme FM. Interventions for educating children who have attended the emergency room for asthma. *Cochrane Database Syst Rev* 1: CD001290, 2001.
65. Gibson PG, Powell H, Coughlan J, Wilson AJ, Hensley MJ, Abramson M, Bauman A, Walters EH. Limited (information only) patient education programs for adults with asthma. *Cochrane Database of Systematic Reviews* 2002, Issue 1. Art. No.: CD001005. DOI: 10.1002/14651858.CD001005.
66. Gibson PG, Powell H, Coughlan J, Wilson AJ, Abramson M, Haywood P, Bauman A, Hensley MJ, Walters EH. (2003). Self-management education and regular practitioner review for adults with asthma. The Cochrane Library 1: CD001117.
67. Tapp S, Lasserson TJ, Rowe BH. Education interventions for adults who attend the emergency room for acute asthma. *Cochrane Database System Rev*(3): CD003000, 2007.
68. Teach SJ, Crain EF, Quint DM, Hylan ML, Joseph JG. Improved asthma outcomes in a high-morbidity pediatric population: Results of an emergency department-based randomized clinical trial. *Arch Pediatr Adolesc Med* 160: 535–41, 2006.
69. Camargo CA Jr, Emond SD, Boulet L, Gibson PG, Kolbe, J. Wagner CW, Brenner BE. (2001). Emergency Department – Asthma Discharge Plan. *Developed at Asthma Education in the Adult Emergency Department: A Multidisciplinary Consensus Conference*, 2 pp. New York Academy of Medicine, New York, NY; Massachusetts General Hospital, Boston, MA, 2001 April 1–5.
70. Baren JM, Shofer FS, Ivey B, Reinhard S, Hornig S, DeGeus J, Stahmer SA, Pantetierri R. A randomized controlled trial of a simple emergency department intervention to improve the rate of primary care follow-up for patients with acute asthma exacerbations. *Ann Emerg Med* 38: 115–22, 2001.
71. Baren JM, Boudreaux ED, Brenner BE, Cydulka RK, Rowe BH, Clark S, Camargo CA Jr. Randomized controlled trial of emergency department interventions to improve primary care follow-up for patients with acute asthma. *Chest* 129: 257–65, 2006.
72. Silverman RA, Boudreaux ED, Woodruff PG, Clark S, Camargo CA Jr. Cigarette smoking among adults presenting to the emergency department with acute asthma. *Chest* 123: 1472–79, 2003.
73. Boudreaux ED, Kim S, Hohrmann JL, Clark S, Camargo CA Jr. Interest in smoking cessation among emergency department patients. *Health Psychol* 24: 220–24, 2005.
74. Bernstein SL, Boudreaux ED, Cydulka RK, Rhodes KV, Lettman NA, Almeida SL, McCullough LB, Mizouni S, Kellermann AL. Tobacco control interventions in the emergency department: A joint statement of emergency medicine organizations. *Ann Emerg Med* 48: e417–e426, 2006.
75. Cates CJ, Jefferson TO, Bara AI, Rowe BH. Vaccines for preventing influenza in people with asthma. *Cochrane Database Syst Rev* 2: CD000364, 2001.
76. Linde K, Jobst K, Panton J. Acupuncture for chronic asthma. *Cochrane Libr* 2: CD000008, 2001.
77. Linde K, Jobst K. Homeopathy for chronic asthma. *Cochrane Database Syst Rev 2*: CD000353, 2001.
78. Woods RK, Thien FCK, Abramson MJ. Dietary marine fatty acids (fish oil) for asthma. *Cochrane Database Syst Rev* 4: CD001283, 2001.
79. Camargo CA Jr., Weiss ST, Zhang S, Willett WC, Speizer FE. Prospective study of body mass index, weight change, and risk of adult-onset asthma in women. *Arch Intern Med* 159: 2582–88, 1999.
80. Ford ES. The epidemiology of obesity and asthma. *J Allergy Clin Immunol* 115: 897–909, 2005.
81. Eneli IU. Skybo T, Camargo CA Jr. (2007). Weight loss and asthma: A systematic review. Thorax 63: 671–76, 2008.

Pediatric Asthma: How Early Life Events Cause Lifelong Respiratory Disease

65

CHAPTER

Andrew Bush

Department of Respiratory Pediatrics, Imperial College of Science, Technology and Medicine, Royal Brompton Hospital and National Heart and Lung Institute, London, UK

INTRODUCTION

Asthma in childhood is a huge subject, and this chapter is therefore perforce selective in the coverage of the subject. It is assumed that all readers of this book are familiar with the British Thoracic Society (BTS): SIGN guidelines [1], and that most will be in adult respiratory practice, so there will be particular emphasis placed on pediatric issues relevant to adult physicians. The most important precept is not to be tempted to extrapolate from adult studies (which are admittedly often very large, very well conducted, and conclusive *for the relevant population*) into childhood, and still less into the preschool years. An obvious example is long-acting β_2-agonists (LABAs), dramatically effective in adults [2–4], but the evidence is much less convincing in children [5–7]. It is increasingly recognized that medicines used in children must be tested in children [8]; the disease may be different, pharmacokinetics may be different, and the risk profile is different (e.g. interference with linear growth).

If therapeutic studies in adults may be poorly relevant to children, are pediatric insights of any relevance to adults and their physicians? We are not born into the world as fully formed adults, from the head of our father, as was the Greek Goddess Pallas Athene, but in fact events prior to birth and during infancy are pivotal in determining long-term lung health, a fact under-recognized in many adult respiratory text books. This chapter will attempt to correct this perspective, as well as covering the basics of pediatric practice. More comprehensive accounts of developmental issues, and the management of childhood asthma, have been published elsewhere [1, 9–11].

DEFINITION OF ASTHMA

It is clear that, in children, asthma is not one disease with a discrete pathology, but many. The best definition, modified from Pediatric Consensus statements [12, 13], is therefore "asthma is most often a diagnosis of exclusion, particularly in preschool children, comprising wheeze and/or excessive cough, accompanied by breathlessness and respiratory distress, which is usually intermittent with periods of remission; other respiratory conditions have to be excluded by clinical history, physical examination, and any indicated special tests." In older children, evidence in support of the diagnosis will be expected to come from demonstration of variable airflow obstruction over time, with treatment, or with a bronchial challenge test; and sometimes with the demonstration of an appropriate phenotype of airway inflammation.

The inclusion of respiratory distress and dyspnea is new to the definition and would preclude the diagnosis of asthma being applied to chronic dry nocturnal cough. Although rarely this can be due to asthma, in most cases it is not (below). Furthermore, since we have no disease-modifying therapies at least in pediatrics (below), there is nothing to be gained by diagnosing asthma at a very early, pre-symptomatic stage, when hypothetically cough might be the first sign of airway inflammation (and in pediatric asthma, but never shown in the general population, there is a relationship between airway inflammation as measured by exhaled nitric oxide [FE_{NO50}] [14]), but we cannot modify the progression of the disease.

If it is accepted that asthma is not one disease but many, then it follows that there are different syndromes or phenotypes. This is particularly important in preschool wheeze. A phenotype may be considered as a cluster of either clinical features or pathological features, resulting from interactions between genes and the environment, which tend to be associated, and which are useful in some way, such as in managing the child or in understanding the mechanisms of disease [15]. There may be overlap between phenotypes, and they may change over time. It may be helpful to consider components of the asthma syndromes or phenotypes to be following [16, 17]:

- Extent and nature of any airway inflammation
- Degree of bronchial hyperreactivity (BHR)
- Extent of persistent, apparently irreversible, airflow obstruction (PAL)

Attempting to dissect out these components may allow a rational treatment plan to be achieved in those children particularly with therapy unresponsive asthma at any age, but also in the preschool years. In the future, more sophisticated factorial analyses [18, 19] may refine these relatively crude phenotypes.

Airway inflammation

Traditionally, asthma, at least in older children, has been considered as a T-helper $(TH)_2$-lymphocyte mediated, eosinophil driven disease of airway inflammation [20, 21]. However, although undoubtedly well founded in many asthmatic children, the concept that this is the sole inflammatory mechanism has been challenged. Studies in particular utilizing bronchoalveolar lavage (BAL) and bronchial biopsy, and induced sputum, have characterized neutrophilic, mixed cellularity, and paucicellular forms of asthma [22–25].

Recent work in children with severe asthma has shown that, unlike in adult asthma, these cellular inflammatory phenotypes are not stable over time, and, over a 1-year period, fewer than 60% retain the same phenotype [26]. Other unexplored issues in pediatric airway inflammation include the relative importance of luminal and intramural inflammation [27], which correlate poorly, and, completely unexplored, the relationship between proximal inflammation and distal (alveolar) inflammation [28–30]. This last currently requires a transbronchial biopsy; the complication rate of this procedure would make it unacceptable for research in children [31].

Extent and nature of BHR

This has traditionally been considered as one of the hallmarks of asthma, and the concepts of direct and indirect airway are well established [32]. The traditional model has been that inflammation causes BHR, and that BHR causes symptoms of asthma. This paradigm has been challenged. First, although there is a good correlation between BHR and severity of asthma for groups, there is much overlap between individuals within the groups, and thus for an individual, only the poorest correlation between level of BHR and asthma severity [33]. There is a very poor correlation between eosinophilic airway inflammation and BHR [34]. Furthermore, therapies can improve inflammation without changing BHR (e.g. anti-IgE monoclonal antibody) [35] and improve BHR without changing airway inflammation (etanercept, the anti-TNF-α monoclonal) [36]. The significance of BHR in infants is discussed later.

Extent of PAL

This is an important practical consideration; there is no point in escalating treatment if there is no further capacity to dilate the airway. PAL is defined pragmatically as a first second forced expired volume (FEV_1) which remains more than two Z- (standard deviation, SD) scores less than normal despite a trial of intramuscular triamcinolone (below) and high-dose bronchodilators; not necessary for the diagnosis, but likely to be a feature, is the absence of any airway inflammation which may be reversible. PAL may be due to anatomical reduction in airway caliber, either antenatal or postnatal; and alteration in airway wall compliance, either as a primary effect on wall structure or due to loss of the alveolar tethering effect (below).

IMPORTANT METHODOLOGICAL ISSUES IN PEDIATRICS

Pediatric lung function tests

The study of infant and preschool lung function is a whole chapter in itself, and will not be reviewed here. The numerous techniques available (including rapid thoraco-abdominal compression, the more sensitive raised volume rapid thoraco-abdominal compression, lung clearance index, plethysmography, the interrupter technique, tidal breathing indices and forced oscillation) all have their proponents. They measure different aspects of airway function, with differing sensitivity to abnormality, and it is important to be clear what the chosen technique is actually measuring, and to ensure appropriate normal ranges are used. The developmental perspective is important; FEV_t, for example, may not represent the same airway generations at age 15 as it does at age 2 [37, 38]. Conventionally, lung function is expressed as percent predicted. In pediatric practice at least, this should be replaced by Z (SD) scores, so that mean Z score at any age for any parameter is 0, with a range of ± 1.96. This is more informative than percent predicted; 80% predicted may be in the normal range, depending on the spread of the data, but -2.5 Z scores is always abnormal.

Significance of BHR

BHR can be measured in small infants, but this is technically very difficult, and data are far scantier than for conventional lung function. Furthermore, just as FEV_t may mean different things at different ages, so does BHR. In older children, BHR probably has the same significance as in adults. However, it is likely but unproven that BHR in babies may predominantly be determined by distal airway caliber rather than inflammation, in other words, "reactive" airways are in fact smaller ones, the same degree of constriction causing a bigger signal in terms of airflow obstruction in an airway that started smaller than one of normal starting caliber. Unlike in asthmatic adults, BHR is not necessarily a feature of wheezy infants [39], underscoring that BHR has different meanings at different ages. Supportive of the "anatomical" view of BHR in the newborn period is that neonatal bronchial responsiveness predicted FEV_1 at age 10 years ($p = 0.03$) but had no relationship with wheeze or diagnosed asthma at this age [40].

The nomenclature of preschool wheeze

The epidemiological phenotypes most commonly used are described below and in Table 65.1. There is a common confusion in nomenclature and thinking between temporal pattern and epidemiological time course of symptoms. Temporal pattern should be described as intermittent (viral) wheeze and multi-trigger wheeze. Either pattern may be transient (disappears by third birthday) or persistent (still present at age 6 years). It must not be assumed that all episodic (viral) wheeze is transient, nor that all multi-trigger wheeze in the first 3 years of life will be persistent. It is likely that these phenotypes, although they have lead to new insights, will prove in the future to be an oversimplification.

Gene: Environment interactions

Genes do not exist in isolation, but within their environment, which may influence gene expression. There are important gene–environment interactions in the pathogenesis of early onset wheeze, which illustrate an important principle. CD14 is involved in the responses to lipopolysaccharide, and as such, a likely candidate "asthma gene." Previous studies have implicated CD14 in the process of immune switching from T_H1 to T_H2, and in responses to viral infections [41, 42]. However, results from other previous epidemiological studies were inconclusive, as was a meta-analysis [43]. However, the CD14/ − 260 C→T polymorphism in the CD14 promoter showed no association with any manifestation of atopic senitization in the total population, but this masked significant findings in a hypothesis driven subgroup analysis. The promotor polymorphism was associated with higher levels of both total IgE and specific IgE to aeroallergens in children in regular contact with domestic pets, but the opposite relationship, not explained by endotoxin levels, in children in contact with stable animals [44]. Hence, a gene could produce exactly opposite effects depending on the environment, so-called phenotypic plasticity, well described in lower organisms. Thus, it is essential to embark on studies with focused hypotheses involving genes and the environment, if important observations are not to be missed.

The role of parental genotype

Maternal smoking illustrates another important principle of gene:environment interactions. The glutathione-S-transferase gene is one of those responsible for protecting the organism against oxidative damage. Polymorphisms of this gene in the mother, but not the father, are important determinant of the effects of maternal smoking during pregnancy on fetal airway development [45, 46]; the null polymorphism exposes the fetus to the threat of oxidant damage, and is associated with airflow obstruction in the newborn period. Maternal smoking negatively affects neonatal birth weight and lung function, but the adverse effect of maternal exposure on neonatal birth weight could be modified by the maternal metabolic genotypes, GSTM1 and GSTT1 [47]. Maternal smoke exposure also resulted in reduced birth weight in the offspring of mothers who had any of the aryl hydrocarbon receptor wild-type genotype, the CYP1A1 variant genotype or the GSTM1 null genotype [48].

Thus if as is explained below, lung function at birth is an important determinant of adult lung function and chronic obstructive pulmonary disease (COPD) risk, at least some of the roots of COPD may lie in the proband's mother's genotype, including genes determining fetal susceptibility to the effects of maternal smoking in pregnancy. This area is fraught with difficulty, and at the moment much remains conjectural, but the collection of DNA from the mothers of COPD sufferers, in particular if the proband has

TABLE 65.1 Wheezing phenotypes in the Tucson study [69, 70].

		Findings at birth	Findings at age 3		Findings at age 6	
	Number (%)	Lung function	Wheeze	Lung function	Wheeze	Lung function
Normals	425 (51)	Normal	−	Not done	−	Normal
Transient wheeze	164 (20)	Reduced	+	Not done	−	Normal
Persistent wheeze	124 (15)	Normal	+	Not done	+	Reduced
Late onset wheeze	113 (14)	Normal	−	Not done	+	Reduced

+ = Present; − = Absent.

only a relatively minor smoking history, could be a fruitful avenue for future research [49].

NORMAL AND ABNORMAL LUNG GROWTH AND DEVELOPMENT IN UTERO

Normal lung growth

This large topic has been reviewed in detail elsewhere [50–52]. Of relevance to the present subject is that the airway branching pattern and all airway generations are determined by 16 weeks of gestation, and it is in the second half of pregnancy that neonatal airway caliber is determined. Alveolar development is largely a postnatal phenomenon; there is a rapid growth phase in the first 6 months of life, probably a slower phase for the next 18 months, and thereafter alveoli increase in size, rather than number. However, since alveolar tethering to the airway is an important mechanism for the maintenance of normal airway caliber, failure of normal alveolarisation may also cause airflow obstruction [53]. The formation of the airway vasculature follows that of the airways and acinar vessels that of the alveoli. Thus efforts at improving long-term airway caliber, for example smoking cessation, are still relevant in the second half of pregnancy, and preservation of alveolar numbers in the first 2 years of life is also of particular relevance, in particular in the context of preterm delivery.

Antenatal adverse influences on lung growth

The major issues are the effects of maternal smoking; maternal atopy; increasingly recognized, the effects of environmental pollution; and obstetric factors which lead to being small for gestational age (SGA), including maternal hypertension of pregnancy. Some, but not all, are amenable to intervention. In considering studies on both ante- and postnatal effects, it is important to consider whether the magnitude is very small, but of interest in generating hypotheses as to mechanisms, or large enough to be the focus of a public health intervention that will be clinically useful in that context.

Maternal smoking

Many studies have demonstrated the adverse effects of smoking during pregnancy. The Boston cohort [54] showed that there was early airflow obstruction in the offspring of smoking mothers, more marked in girls rather than boys, and with diminishing effect over the next 18 months. Other studies have also confirmed an adverse effect of maternal smoking [55–57]. Maternal smoking may cause BHR in the newborn [58], especially in those with a maternal history of asthma. It is important to note that smoking cessation in the second half of pregnancy may improve the outcome for the fetus [59].

There are some pathological correlates of the effects of cigarette smoke on the developing lungs. The offspring of smoke exposed guinea pigs have reduced numbers of alveolar attachment points to the airway [60]. This is likely to result in a reduction in airway stability, and increased collapsibility. A study of the lungs of sudden infant death syndrome victims, whose mothers smoked in pregnancy, revealed increased inner airway wall thickness and increased airway smooth muscle [61, 62]. Whether this is a direct effect of smoke, or related to another factor which caused the sudden death, is still conjectural.

Nicotine has been shown to exert antenatal effects on airway wall structure. Nicotine given by subcutaneous infusion to pregnant baboons caused airway remodeling in the pups [63]. There was increased airway wall area per millimeter of epithelium, and upregulation of mRNA for Types 1 and 111 collagen, confirmed by immunostaining, in the airway and alveolar walls. There were also increases in elastin mRNA, but if anything, elastin protein was decreased, but not increased; this counter-intuitive result did not reach statistical significance. The mechanism is probably via the reaction of nicotine with the $\alpha7$ nicotinic acetylcholine receptor.

Maternal smoking may have immunological as well as physiological and structural effects. There are data showing that maternal smoking leads to lower cord blood IL-4 and IFN-γ [64], and increased cord mononuclear cell proliferation to house dust mite [65]. Other cord blood studies showed that maternal smoking was associated with increased IL-13, and reduced IFN-γ mRNA responses by stimulated cord blood cells [66]. The Perth group [67] has investigated the effects of maternal smoking on fetal Toll-like receptors (TLRs) and their signaling. Smoking during pregnancy was associated with reduced TLR2 mediated IL-6, IL-10, and TNF-α production. TLR3 and TLR4 mediated signaling of TNF-α, but not IL-6, IL-10, and IL-12 was reduced in the infants of mothers who smoked. In terms of TLR9 responses, there were attenuated IL-6 and increased IFN-γ responses in the infants of smoking mothers. Thus, smoking may be the link between (a) reduced airway caliber; (b) reduced alveolar tethering of airways; and (c) altered cord blood immune function, which subsequently interacts with effects of respiratory viral infections in infancy. Interestingly, birth order has an effect on cord blood immune responses. The T_H2 driven cord blood proliferative response of mononuclear cells to timothy grass, PPD, Concanavalin A and house dust mite was lower with successive pregnancies (including miscarriages, placing the effect likely with events in the first trimester of pregnancy) [65]. In other words, immunologically, the effects of a first pregnancy are more like that of pregnancy in a smoker, rather than a fourth pregnancy. This may be an alternative explanation (rather than the occurrence of multiple viral infections) for the occurrence of less atopy if the child has older siblings, the hygiene hypothesis.

Maternal atopy

Infants of asthmatic mothers also have airflow obstruction [56, 57, 68]. Other studies have confirmed an adverse effect of maternal atopy. These findings are difficult to reconcile with the reports of normal lung function at birth in those babies who will become persistent wheezers, as reported from Tucson [69, 70], many of whom would be expected to have an atopic mother. Whether the differences between the studies are real or a manifestation of population or methodological differences is unknown. However, it is clearly not possible to prevent atopic mothers having babies!

SGA effects

Two factors complicate the assessment of being SGA *per se* on lung function. The first is the choice of an appropriate control group, since forced expired flows (FEF) are usually normalized to body length. The second is the confounding effects of smoking, probably one of the commonest causes of SGA delivery. Impaired lung function soon after birth has been reported in SGA infants [71], but it seemed that in the subgroup of non-smoking mothers, the changes related only to reduce body length. The subsequent evolution of lung function changes suggested that they were independent of somatic growth [72]. A follow-up study in SGA infants born to non-smoking mothers showed that there was a consistent 9% reduction in FEFs and forced vital capacity (FVC), with no evidence of "catch-up" growth. Although no longitudinal study has followed these infants into adult life, cross-sectional studies have reported reduced lung function in adults and children who were born SGA, suggestive that the effect of SGA delivery, independent of smoking, persists long term [73–75].

Effects of environmental pollution on the fetus

This is an area of recent and increasing interest. Proof of concept, that environmental pollutants could affect the fetus came with the demonstration that umbilical cord carbon monoxide (CO) correlated with environmental levels [76]. The readout from antenatal pollution studies has been birth weight, not lung function, but as discussed, the two are related. Maternal exposure to particulates ($PM_{2.5}$ and PM_{10}), CO, sulfur dioxide (SO_2), and nitrogen dioxide (NO_2) have all been implicated in reduced birth weight [77, 78]. The insectiside dichlorodiphenyldichloroethylene (DDE) was detectable in cord blood and high levels associated with a later asthma diagnosis [79]. This is an obvious focus for preventive strategies

Other possible antenatal adverse effects

First trimester invasive procedures (chorionic villus sampling, amniocentesis) may be associated with hyperinflation suggestive of airflow obstruction in the infant [80], although another study suggested no influence of these procedures on wheeze [81]. Maternal hypertension of pregnancy is a risk factor for airflow obstruction [57, 81]. Maternal diabetes is a risk factor for persistent wheeze, antibiotics in pregnancy for maternal urinary infection is associated with transient early wheeze, and antibiotics at delivery for both phenotypes [81] (see Table 65.1 for the definition of these phenotypes). It should be noted that antibiotic administration around delivery might be a marker of maternal chorioamnionitis, which accelerates lung maturation and may thus have long-term consequences on lung function independent of antibiotic use [82]. Another maternal medication effect is that paracetamol, but not aspirin, intake during pregnancy is associated with asthma, wheezing, and elevated IgE in school age children [83]. The role of diet is unclear; early work has surprisingly suggested that high maternal vitamin C intake in pregnancy may actually increase the prevalence of wheeze in the second year, but not the first year of life. High vitamin E intake [84], and eating apples and fish, in pregnancy may protect against allergic asthma and atopic disease [85], although since no neonatal lung function data were recorded, it is not possible to determine if this is structurally mediated, or by an effect on neonatal immune responses. No intervention study has determined that adjusting maternal diet affects outcomes in the baby.

Summary: Maternal adverse effects

Maternal smoking is the most important and obvious effect amenable to intervention, followed by air pollution. Paracetamol and antibiotics should be discouraged in pregnancy. Whether dietary manipulation in pregnancy will be beneficial is contentious.

NORMAL AND ABNORMAL LUNG GROWTH AND DEVELOPMENT EX UTERO

Epidemiological studies

The Tucson group [69, 70] recruited over 1200 babies and divided wheezing phenotypes in the first 6 years of life into three categories (Table 65.1). They found that transient wheezers had reduced lung function at birth, and although there was catch-up, they still had evidence of airflow obstruction age 6 years. The persistent wheezers had normal lung function at birth, and by age 6 had evidence of airflow obstruction. The late onset wheeze group had no impairment of lung function. These lung function patterns established at age 6 track to age 16 at least [70]. Recently, they have shown that poor lung function at birth tracks into the early twenties irrespective of confounding factors such as wheeze, smoke exposure, or atopy [86], confirming that the first 6 years of life, and especially antenatally, are a pivotal time for the prevention of COPD.

The Perth birth cohort was smaller than Tucson (n = 243 versus 1246), but more detailed physiological measurements were made, including BHR in the newborn period [87]. Their detailed conclusions, based on smaller numbers, were different, but the importance of early life events again emerges. The discrepancies between the two cohorts have yet to be resolved. The Perth group found that the group wheezing between one year and 3 years of life (n = 17, transient wheeze, in the Tucson nomenclature; compared with more than 100 transient wheezers in the Tucson cohort) had *normal* lung function at birth (mainly using a tidal breathing, indirect measure of airway obstruction), but FEF_{25-75} was reduced at age 11. The persistent wheezers (n = 12) had abnormal lung function shortly after birth, and it remained persistently lower than the never wheeze group. Both studies showed that children who wheezed after age three had no impairment of lung function. In a study of more than 11,000 Sydney schoolchildren, those with a history of wheeze before 2 years of age had small decrements in lung function, supportive of the Tucson findings [88]. Children who were flow limited in the neonatal period (parts of the tidal breathing expiratory flow curve approximated to the forced flow-volume curve) had a trend toward a reduced FEF_{25-75} at age 11 years, again supporting the concept of the tracking of lung function [68, 89].

The Manchester Birth Cohort [90] used the same subdivisions of wheezing phenotypes as Tucson, and measured airway resistance (sRAW) at age 3 and 5. By age 3, persistent wheezers had reduced lung function, and this tracked to age 5. Transient wheezers had normal lung function. sRAW at age 3 could not predict which infants would develop new wheeze in the next 2 years of life. The conclusions were the same irrespective of whether pre- or post-bronchodilator sRAW was studied. It should be noted that sRAW is not a particularly sensitive test to minor degrees of obstruction [91, 92]. These findings have been confirmed in a Dutch cohort, where airway resistance measured by the interrupter technique was higher in persistent wheezers than in normals [93]. They found no effect of transient wheeze on lung function; again, this may reflect the relative insensitivity of the interrupter technique.

Most studies have focused on airway obstruction, with the implication that this is due to airway maldevelopment. However, alternative mechanisms may also be relevant. It is well known that there is interdependence between the airways and the lung parenchyma, with alveolar tethering to the airways being responsible in part for maintaining airway caliber. Alveolar attachment points have been shown to be reduced in number by maternal smoking [60]. This could lead to airflow obstruction without there being a direct effect on the airway itself. Another group, using measurements of input impedance, have suggested that airway wall compliance may be abnormal in wheezing infants. They suggested that compliance, rather than actual caliber, may be the reason for the described lung function abnormalities [94].

Birth weight is an important determinant of lung function [95]. A Norwegian cohort study showed that babies with early tidal breathing lung function below the median were more likely to have a history of asthma and severe airway hyperresponsiveness (AHR) at age 10. There was some evidence of tracking, with early tidal volume indices weakly predicting mid-expiratory flows at age 10. The PIAMA birth cohort study [96] reported on the interactions between birth weight and respiratory symptoms in the first 7 years of life. In accordance with other studies [97–99], they found an increasing risk of wheeze in the first 5 years of life, which declined thereafter, being no longer significant at age 7 years. The novel finding was that smoking in the infant's home increased the effect of low birth weight, such that a 2.5 kg birth weight child had a nearly twofold risk of symptoms compared to a 4.5 kg birth weight child, and in the former, smoke exposure leads to a 12% increase in the prevalence of symptoms.

Studies in adults have extended these observations. A longitudinal birth cohort study of more than 5000 adults has shown that at age 31, each 500 gm increment of birth weight resulted in a 53.1 ml increase in FEV_1 and a 52.5 ml increment in FVC [100]. Another study showed a relationship between lung function age 45–50 years and birth weight [73].

There have been two major studies in which BHR was measured in the neonatal period. The Perth cohort of babies had BHR to histamine measured shortly after birth, and increased responsiveness was associated with reduced spirometry, a physician diagnosis of asthma, and increased respiratory symptoms at age 6 [101]. In the second (United Kingdom) study [39, 40], wheezing was categorized as occurring (1) only before the fourth birthday, (2) after the fourth birthday, or (3) never. No relation was seen between neonatal lung function and later lung function, but neonatal bronchial responsiveness predicted subsequent FEV_1. Increased neonatal BHR was associated with transient wheeze <4 years, but not with later wheeze. Wheeze after 4 years of age was typical of classical asthma, as it was strongly related to atopy and BHR at age 10.

The Groningen [102, 103] group studied 119 allergic asthmatics for the first time between ages 5 and 19, and followed them up into their 40s. They showed that, at the first visit, low FEV_1 and BHR were independent risk factors for low lung function in early adult life. Lower lung function in childhood, and less good evolution of lung function over time, was associated with worse BHR in adult life. Unsurprisingly, young adult asthmatics who smoked had a higher rate of decline in lung function.

The early childhood cohorts overlap a number of others spanning childhood and early adult life. The Dunedin cohort [104] followed over 600 people between the ages of 9 and 26. Their data clearly showed that lung function continued to track over this period, perhaps more in males than females. Pulmonary function was consistently lower in those who wheezed compared with those who did not, and the level of function was clearly set by events prior to recruitment, as the authors themselves recognized. The primary purpose of the Childhood Asthma Management Program (CAMP) study was the treatment of childhood asthma, not lung function tracking, but nonetheless, their data is supportive of this concept [105]. The Melbourne group recruited children age 7 who had either no wheeze, "wheezy bronchitis" (WB; what we would now call episodic (viral) wheeze), or persistent asthma. At age 10, the cohort was enriched by a group of really severe asthmatics. They showed that from recruitment until just below 50 years of age, lung function tracked in all groups, and that the level of lung function was set by early life events prior to recruitment [106]

The next consideration is the rate of lung aging. It has long been known that in the mid-40s, lung function starts to decline from a plateau. The plateau (highest lung function ever) is lower in asthmatics, and the slope of rate of decline is greater in smokers [107]. Recently, patients from the Aberdeen cohort, recruited in 1964 were restudied [108]. They were divided into normals, asthmatics and "wheezy bronchitics," who would now more likely be described as episodic (viral) wheezers. In middle age, the plateau FEV_1 is the same in normals and WB, and lower in asthmatics, as expected. However, when rate of decline of FEV_1 was studied, both the asthmatics and WB groups were declining at the same accelerated rate. This finding persisted after controlling for smoking habit and socioeconomic class.

By contrast, another group found that childhood chest illness (pneumonia, pertussis, wheeze by age 7) did not appear to affect rate of lung function decline between age 35 and 45 [109]. Crucially, the earliest age of study was 7 years, and by the time of this study, perhaps unsurprisingly, >90% of participants had dropped out, leaving data on 1156. Their findings are supported by other groups [106, 110], but crucial methodological factors are the drop out rate and likely recall bias at the first interview, when many will have become symptom free (see e.g. [96]). They had previously showed that [111] parents who gave a history of previous pertussis in their child had forgotten the entire episode at the time of the next questionnaire, emphasizing the importance of recall bias.

The most long-term study relied on indirect measurements but is still enormously informative and challenging. Barker's group [112], in a series of seminal observations, analyzed death certificate data in a large number of individuals in widely differing communities, which were classified as county or London boroughs, or urban and rural areas. We know that death certificate data are often unreliable, but if a parameter of poor individual accuracy is measured thousands of times, inaccuracies will cancel out and important correlations which are robust, may emerge, as here. Barker's group found that, across all these communities, there was an extraordinarily close correlation between infant mortality from pneumonia and other respiratory disease, and standardized mortality rate (SMR) for COPD 50 years later. An area with high infant mortality rate would also have a high SMR for COPD 50 years later. Thus, *any* hypothesis about COPD has to take in the importance of early life events if it is to hold up.

Postnatal adverse influences on lung growth

Postnatal smoke exposure

The adverse effects of tobacco smoke continue postnatally. The adverse effects have been well previously summarized [113]. Environmental smoke exposure in childhood has effects extending well into adulthood [114]. Living with a smoker before the age of 18 increased the odds of chronic dry cough and, to a lesser extent, phlegm, after the age of 18. The associations were strengthened with higher numbers of smokers in childhood.

There are important gene–environment interactions with regard to smoking, which are still incompletely understood. The postnatal interactions between environmental smoke exposure and GSTM1 and GSTT1 polymorphisms are controversial. In a German cohort [115], only the combination of null polymorphisms in one of these genes, and postnatal exposure to environmental tobacco smoke, resulted in evidence of airway obstruction, whereas in a study from the United States [46], antenatal smoke exposure was thought to be more important. Tumor necrosis factor (TNF) polymorphisms are important; exposure to second hand smoke doubled the risk of respiratory related school absences in children with the TNF-308 A compared with the G variant, leading to a fourfold increased risk of lower respiratory tract infections causing school absence [116].

Respiratory viral infections

The observation upon which the "Hygeine hypothesis" is based incontestable, namely that higher birth order is associated with less atopic disease [117]. The interpretation that this relates to early viral infection is controversial. Evidence of the importance of early viral infections comes from studies in which infants were placed in childcare facilities at an early age. Firstborn infants in this environment wheeze more in the first year of life, but less at 6 years of age and older [118]. There is an important gene–environmental interaction, in that the infants of mothers with asthma do not show a subsequent decline in the prevalence of wheeze [119]. Whether viruses mediate all these effects or there are other factors in the environment of the crèche is not clear.

An important question is whether severe viral or other respiratory infections cause airway obstruction or are a manifestation of it. This is best resolved by the antenatal recruitment of babies into prospective, longitudinal studies. The Perth study showed that decrements in lung function preceded bronchiolitis [120]. A study from Tucson also showed impaired lung function preceded pneumonia [121]. However, there is no doubt that, for example, adenovirus infection leads to chronic severe airway obstruction due to obliterative bronchiolitis in putatively previously normal babies or at least babies with at worst only mild airflow obstruction [122]. The ill-studied nonatopic late onset wheeze phenotype may be related to viral bronchiolitis [123], but in this chapter, there were no premorbid data, and it is not possible to determine whether the severe early infections reported were a marker of preprogramming of airway or immune function, or causative of later problems. Furthermore, the diagnosis of "bronchiolitis" was very loosely based.

Much early work has been on respiratory syncytial virus (RSV) bronchiolitis. One study, with no premorbid measurements, but excellent retention during prolonged follow-up, suggested that RSV bronchiolitis causes asthma and atopic sensitization [124–126]. The conclusions have been criticized, partly on the grounds of choice of control group [127], and other work suggested that post-RSV symptoms in fact gradually improve with time [128]. The COAST study has highlighted that rhinovirus (RV) infection may be more strongly associated with wheeze outcomes in later childhood than RV [129]. The authors rightly point out, though, that they cannot distinguish between at least two important mechanisms (1) recurrent infections lead to airway damage and thus to asthma and (2) infants were already predisposed to asthma because of abnormal cord blood IFN-γ or other cytokine responses, and RV infection is just a marker of this preprogramming.

It is obvious that severe viral (particularly adenoviral) infection may cause *de novo* airflow obstruction, but it would be a mistake to forget the antecedent insults which may increase the child's susceptibility to respiratory infection.

Overall, the best evidence is that much of the morbidity of early childhood viral infection is determined antenatally.

Diet

If oxidative stress is important in lung growth and development, then it might be predicted that increasing dietary antioxidants might be beneficial. In one study, vitamin C, fruit, and magnesium intake was associated with better lung function in mid-childhood [130]. The traditional Mediterranean diet consists of a high intake of fruits and vegetables, predominantly whole grain bread and cereals, legumes and nuts, and olive oil as the main source of fats. In those consuming such a diet, atopy is not uncommon, but wheeze and rhinitis are rare [131]. A cross-sectional study showed that high adherence to this diet had a beneficial effect on the symptoms of asthma and rhinitis. Another group [132] confirmed a protective effect of the Mediterranean diet on chronic severe asthma in girls, and that exercise showed a dose response protection against occasional asthma and rhinoconjunctivitis in both sexes. Another group showed that eating bananas and drinking apple juice from concentrate, but not eating apples, may protect against wheezing [133]. The whole subject of dietary manipulation in the ante- and postnatal periods has recently been reviewed [134]. However, whether intervening to alter diet would benefit lung function has not been shown. A healthy diet is desirable in itself but is perhaps not likely to be hugely beneficial to lung development.

Medication administration

This is still controversial. One study showed that frequent paracetamol [135] administration to children lead to increased wheezing, and possibly rhinitis and eczema. Studies of the effects of antibiotic administration for respiratory infection are dogged by reverse causation, that is, the antibiotics may have been erroneously given for asthmatic symptoms. Studies of antibiotics for non-respiratory infections are inconclusive, but recent evidence suggests that any association between antibiotic use and subsequent wheeze is weak at best [136]. On general principles, antibiotics should be avoided in childhood unless really necessary, but I doubt that such a policy will have much effect on population lung health.

Outdoor air pollution

There are adverse effects of outdoor pollution on lung growth in children. NO_2, $PM_{2.5}$, acid vapor, and elemental carbon exposure all caused clinically significant decrements in FEV_1 at age 18 [137]. This group also showed that regional air pollution, and exposure to traffic fumes, had independent and detrimental effects on lung growth [138]. The adverse effects of ozone, NO_2, and particulates on lung growth have been reported from Mexico City. In a study from Leicester, United Kindom, there was a tight relationship between environmental PM_{10}, carbon content of alveolar macrophages, and spirometry [139]. However, the evidence that changing the flow of traffic can improve outcomes for childhood respiratory health is at best contentious [140, 141].

Indoor air pollution: Allergens

There is challenging evidence for the role of allergens in airway development. The Manchester group showed that high house dust mite exposure was associated with airflow obstruction in infancy and stringent mite avoidance was beneficial [142]. The same group showed in adults that house dust mite and dog, but not cat, allergen levels, if high, had adverse effects on airway function and inflammation even if the patient is not sensitized to the allergen, as least on IgE mediated criteria [143]. The implication is that allergens have effects other than by stimulating an allergic response, and there is increasing epidemiological evidence for this [144]. Atopic patients, non-sensitized to cat allergen, still exhibited a dose response worsening BHR with increasing cat allergen levels. One possible mechanism is that allergens cause proteolytic damage to the airway epithelium; and indeed, many allergens are proteolytic enzymes [145]. The effects of allergen reduction are sometimes unpredictable, and high levels may be protective in some circumstances. A detailed description of the important interactions between aeroallergens and environmental endotoxin levels is beyond the scope of this chapter. However, it should be noted that sensitization to allergens in the first 3 years of life is an important marker of wheeze persistence [146]. It is not clear from the literature whether allergic sensitization is best prevented by removing furry pets from the baby's environment, or buying many more. The different effects of pets and farm animals on outlook have been briefly discussed above [44], and the details are beyond the scope of this chapter.

Indoor air pollution: Others

A study from Poland [147] of indoor air quality based on reported environmental tobacco smoke and type of household heating, and the number of winter months that occurred during the first 6 months of life suggested that a lower level of lung function in preadolescent children can be related to postnatal exposure to indoor emissions in the winter. The findings must be extrapolated cautiously to communities with different indoor heating exposures. Multiple indoor factors, including mode of heating and cooking, may adversely affect the developing lung [148]. Housing issues may be important – moisture damage and mold growth in the living quarters were associated with recent onset asthma in a recent study [149]; however, the role of household molds in the initiation or exacerbation of asthma is still controversial [150, 151].

The aging lung

It could be argued that the antenatal adverse effects described above are clinically trivial. However, their effects are magnified when lung aging commences. The effects may be twofold. First, the plateau of lung function reached at the end of development may be lower than normal; second, there may be an accelerated rate of decline in lung function. Both may coexist. The effects of childhood events on lung aging have been little studied, but there is sufficient evidence that aging superposed on an abnormal developmental phenotype or early postnatal wheeze may cause accelerated decline, which presumably will predispose to COPD.

CLINICAL IMPLICATIONS OF THE PHENOTYPES OF PRESCHOOL WHEEZE

The nature of wheezing phenotypes has been discussed above and in Table 65.1. The obvious clinical implication is that these phenotypes, although useful epidemiologically to determine mechanisms, are of no use to the clinician in practice; only time, not clinical acumen, will determine whether an infant wheezing at 2 years of age will have persistent symptoms. A number of predictive indices have been developed, based on severity of symptoms, and atopic manifestations [152–154]. In general, they are good at ruling out persistent symptoms, but less good at predicting those who will have ongoing problems (in fact hardly better than tossing a coin). In general, severe symptoms by age two, but not one, were the best predictor of asthma at age 10 [154], and were superior to stringent and loose indices based on severity but also atopic manifestations in infancy [152, 153]. At this stage, we do not have disease-modifying interventions, and these indices are of most use for giving prognostic advice to parents.

ASTHMA AND COPD FROM ANTENATALLY TO OLD AGE

Introduction

The ideal method of sorting out the effects of early influences on adult disease would be to recruit a large cohort of babies antenatally, preferably preconception, and follow them through until death, making measurements of lung function, airway pathology, infective agents and environmental exposures, both indoor and outdoor. Obviously, no such study exists. The epidemiological data have been summarized previously; this section reviews important pathological and genetic studies.

Pathological studies

Ethical and practical constraints have limited the number of airway biopsy studies, which can be performed in young children. However, recent pathological data have confirmed the importance of the first 3 years of life. We and others have confirmed the safety of endobronchial biopsy in infants and children [155–158], so if a clinically indicated bronchoscopy is being formed, an airway biopsy for research is legitimate, subject to the approval of the institutional review board, and the agreement of the family [159]. What is not legitimate is the performance of a bronchoscopy just for research purposes [160]. Reticular basement membrane (rbm) thickening is one of the pathological hallmarks of asthma. We have shown that in children with severe asthma, rbm thickening is greater than adult and child controls, and equal to that seen in adults with both mild and severe asthma [161]. Until recently, it was not known at what age the changes developed. We have recently shown in a cross-sectional study that even infants (median age 1 year) who are atopic and have bronchodilator responsiveness have no evidence of eosinophilic inflammation and rbm thickening [162]. However, in a further cross-sectional study in a different group of infants, median age 3, we have shown that the presence of wheeze confirmed on a videotape is associated with both eosinophilic inflammation and rbm thickening [163]. The changes were less marked than those we have shown in later childhood [161]. This pathological data confirms epidemiology in showing the likely long-term importance of events in the first 3 years of life.

Genetic studies

The fetal genotype

The possible role of maternal and fetal genotype has been mentioned in a previous section; in summary, both may lead to important interactions with environmental influences. It is important to remember when trying to determine the effects of polymorphisms in antioxidant genes in COPD that the effects may have been modulated early on, as well as when the subject starts to smoke.

COPD genes: Are important clues to be found in antenatal lung development?

The importance of early life events has implications for the hunt for COPD susceptibility genes. It is likely that early genetic effects will be increasingly masked by environmental influences as the child grows up, only to resurface at the time of lung aging. The obvious targets for COPD susceptibility have long been genes related to protection against oxidative stress, and those related to inflammatory pathways. The hypothesis advanced here would open up a completely different avenue to explore, namely genetic polymorphisms related to lung growth and development. There are some confirmatory data connecting genes important in lung growth with those affecting the susceptibility to COPD. This will be illustrated by reviewing the ADAM33, the beta-receptor, the fibroblast growth factor (FGF) and GSTM genes.

ADAM33 was identified as a potential gene for asthma by positional cloning, and has numerous putative effects [164]. ADAM33 is expressed in the developing lung during branching morphogenesis, increases throughout gestation [165], and may have a role in the developmental regulation of airway caliber. Polymorphisms in the ADAM33 have been associated with impaired early life lung function [166]. The Manchester group showed that the SNP F+1 was a determinant of lung function age 3 years. Carriers of the A allele had poorer sRAW at 3 years of age, and AA homozygotes had worse sRAW at 5 years of age, independent of sex, smoking, and allergen exposure. Polymorphisms in ADAM33 are known to be associated with asthma and bronchial hyperresponsiveness in different adult populations [167]. Two studies have demonstrated that ADAM33 polymorphisms are important in COPD. The first demonstrated an association between polymorphisms that were known to convey susceptibility to COPD and rate of lung function decline in the general population [168]. The same group showed that SNPs in ADAM33 also were associated with the degree of inflammation and airway reactivity in COPD [169]. These data suggest a common genetic factor at the start and end of life, suggesting that further probing of this theme might be rewarding.

Another gene linking early lung function, asthma, and COPD is that for the beta-receptor. Although there have been many studies of beta-receptor function in asthma, much less is known about any role in early life. In a UK cohort in whom VmaxFRC and BHR had been measured in the first month of life, the possession of any Glyn 27 allele was predictive of airflow obstruction [40]. The same result was found with any Arg16 allele. In both cases, the association was independent of maternal smoking or atopy. There was no effect of genotype on BHR in either the newborn period or when restudied at age 11 years nor was there any effect of beta-receptor genotype and lung function at age 11; this may reflect the small numbers studied. This was a high-risk population (>80% with a parental history of asthma), and may not be generally applicable. The Perth group [170] found an association between the haplotype arg16gln27 and the prevalence of positive BHR at age 6. The gly16gln27 haplotype was associated with better spirometry at age 6 and age 11. In contrast, arg16gln27 was associated with worse spirometry at age 11. Children with the gly16gln27 haplotype were less likely to have asthma-ever or doctor-diagnosed asthma at age 11. They concluded that haplotypes of beta-receptor polymorphisms were associated with lung function, BHR, and asthma susceptibility in childhood. Another group genotyped the polymorphisms at positions 16 and 27 in >1100 people with a variety of respiratory problems, including asthma and COPD, as well as a control population and found associations with beta-receptor haplotypes [171]. The Arg 16 homozygotes had an increased risk of COPD, asthma, and wheeze. The Glyn 27 homozygotes had a greater risk of asthma and AHR. The Arg16/Glyn27 haplotype was predictive of asthma and COPD, and more respiratory symptoms. Beta-receptor polymorphisms have also been found to be a predictor of persistence of asthma symptoms from childhood into adult life [172] but predict only a small component of long-term prognosis. This is not to say, however, that they may not be of interest in the study of biological mechanisms.

Another illustration of an association between genes important in lung growth and later susceptibility to COPD concerns FGF-1, FGF-2, and their receptor FGF1-R. FGFs are important in lung growth (reviewed in Ref. [173]), and bind to one of four transmembrane, tyrosine kinase receptors FGFR-1 to 4. In an *in vitro* study, human airway smooth muscle cells were shown to proliferate when exposed to FGF1 and 2. *In vivo*, there was cytoplasmic expression of FGF-2 in epithelium, and nuclear localization in airway smooth muscle cells, in COPD patients when compared with controls. There were also elevated levels of FGF-1 in airway smooth muscle cells, and FGF-1, and FGFR-1 in bronchial epithelial cells, in COPD patients as compared with controls [174]. There were inverse associations between FEV_1/FVC ratio and both FGF-2 and FGFR-1 expression in airway smooth muscle. Epithelial FGF-1 correlated with smoking history.

The effects of GSTM polymorphisms in mother and child, and early lung function and symptoms, have been discussed earlier. A study in mid-childhood showed that [175] GSTM1 and GSTP1 polymorphisms are associated with defects in lung growth, particularly in children with asthma. GSTT1 deficiency is also associated with an accelerated decline in lung function, a known risk factor for COPD, in middle aged men, and in another study, a combination of polymorphisms in all three GST genes were associated with accelerated decline in lung function [176]. There is a significant interaction with active smoking. However, the effect of the polymorphisms is independent of smoking history, and thus presumably reflects not later smoke exposure but the antenatal and immediate postnatal effects of this gene on airway growth and development.

In summary, the geneticist looking for COPD susceptibility genes might be well advised to consider genes important in early life events.

PREVENTION OF ASTHMA AND LATER COPD

Smoking in the young

The most obviously beneficial measure would be the prevention of smoking in young people. The public health measures of many governments have been ineffectual, although the increasing momentum to ban smoking in the work place is to be welcomed [177, 178]. If we are not to continue to jeopardize the future of our children, we must build on this:

- Have a total ban on cigarette smoking in ALL public places; this has been shown not to increase the exposure to passive smoke in children at home [179].
- Totally ban the display of tobacco products in any commercial site.
- Have a total ban on advertising tobacco, and the depiction of smoking under any circumstances.
- Purchase of tobacco requiring the production of photo-identity.
- Dramatically increase taxes on tobacco, so each cigarette costs at least £25.
- Introduce really severe penalties, including closing down of businesses and imprisonment for those who sell cigarettes to children.
- Test saliva and urine of high-risk children for cotinine, and take evidence of exposure very seriously.

In this way, hopefully we can prevent young women in particular taking up smoking, and smoking throughout their pregnancies.

In addition, a campaign needs to be targeted in antenatal clinics:

- If you smoke during pregnancy, your unborn baby is smoking too.
- This will affect the baby throughout their life.
- Even if you became pregnant without having given up smoking it is not too late; if you give up early in pregnancy, you can avoid harming the baby.
- (Preferably) link maternity benefits to quitting smoking, with objective checks on urine or salivary cotinine levels.

The benefits of a really effective and hard-hitting campaign now will be felt for generations to come. It is now not sufficient to focus just on quitting smoking, because irreparable damage may in the meantime be done to the next generation; preventing the uptake of this addictive drug is needed to prevent future cases of COPD.

Viral infections?

It is likely that severe RSV infection is a marker for previous damage, not a cause of subsequent problems (above). Admittedly there is some evidence for latent RSV infection, analogous to latent adenovirus infection [180, 181], but whether this is clinically significant awaits further study. At the moment, the likelihood that RSV immunization (even if it could safely be performed) would prevent later COPD is not great. Since RSV infection is virtually universal in the first 2 years of life, and COPD is not universal even in adults who smoke, one would have to postulate other factors in a model whereby RSV was causative of COPD. However, there may be genetic interactions with RSV (polymorphisms in proinflammatory genes for example [182, 183]), and it is possible that there are subgroups of children with anatomical damage, who also have polymorphisms in particular genes, in whom RSV infection may cause an increased likelihood of later COPD, over and above the effects of antenatal factors and the polymorphisms themselves.

Severe adenoviral infections are uncommon in childhood, but the prevention of more mild infections, which may possibly become latent with long-term deleterious effects, may contribute to a reduction in subsequent COPD. The role of other childhood viruses, and mycoplasma and chlamydia, also needs further study. In summary, although there may be a role for viral infection over and above that of airflow obstruction due to antenatal factors, at the moment it seems likely to be minor. We need studies with a clearly focused hypothesis to delineate any possible interactions between genes, environment, antenatal airway damage, and viral infections to move the field forward.

Nutrition?

Obesity is the current childhood epidemic. The relationship between obesity and respiratory disease is not clear – if you are fat, you will become breathless on exertion, while not necessarily having a lung problem. However, obesity may adversely affect BHR [184], and, in combination with obstructive sleep apnea, cause neutrophilic airway inflammation. There have also been reports on adverse effects of eating "fast food" and the benefits of other dietary manipulations [185, 186]. Clearly, healthy eating is a good thing in its own right, but it is likely only to make a minor contribution to the prevention of COPD.

Ensuring optimal obstetric outcomes

It may seem even more bizarre to urge that COPD is an obstetric as well as a pediatric disease, but as discussed above, maternal obstetric problems may have long-term effects. Optimal obstetric care is clearly desirable in its own right, but it may have the spin-off of also preventing prematurity and its consequences, and promoting good lung health in term infants.

DIAGNOSIS OF ASTHMA

All children cough, many are said sometimes to "wheeze" and get breathless, but most children are normal. "Normal" causes of sometimes quite prolonged respiratory symptoms include pertussis and similar syndromes, characterized by paroxysmal coughing, sometimes with a whoop, post-tussive vomiting, or color change; "Nursery School Syndrome", usually in firstborn children who are placed early in a child care facility, and get a succession of viral colds, which merge into each other; and prolonged post-viral cough. None of these respond to asthma therapy; in my practice, I now spend more time telling parents their children do not have asthma than actually making a new diagnosis. Importantly, the vast majority of children with asthma are diagnosed most appropriately on the basis of history and physical examination, supplemented in the older child with simple physiological tests. More detailed investigation is unnecessary for the vast majority of asthmatic children; if they are contemplated they should be carefully directed.

History

As discussed in the definition, the hallmark of asthma is breathlessness with either or both of wheeze and excessive cough. The first point to determine is what the family actually means by the word "wheeze." Wheezing due to airway narrowing sounds like a high-pitched, musical whistle, akin to organ music or the wind whistling in chimneys. However, many parents use the same word to describe many other different noises, for example a palpable crackling in the chest, a noise as if the child needs to clear his throat, or even nasal snuffling [187–189]. The studies using a video-questionnaire [190], or objective recording of lung sounds [191], have demonstrated the unreliability of parental recognition of wheeze. Differentiating stridor from wheeze in the tachypnoeic child may be difficult for parents. The significance of these sounds is very different, and time must be spent on the history to determine exactly what is meant. The evaluation of chronic cough is also notoriously difficult. Coughing is universal in childhood at least at the time of viral upper respiratory infections. There is only poor correlation between objective measures of cough such as diary cards or tape recorders and perception of severity by observers [192, 193]. Ambulatory cough monitoring has been used predominantly in older children [194–196] to document how much coughing is normal, but is not routinely available in clinical practice. Finally, it is important to determine who has the problem; if the *child* makes respiratory noises, but does not have any breathlessness or impairment of quality of life, does the *child* have a problem, or indeed asthma? Having established whether the child truly wheezes, and as far as possible whether there is excessive cough, the next step is to identify the pattern and severity of symptoms. The key distinction in the pattern of symptoms is to determine whether the child has symptoms solely at the time of a viral cold (episodic [viral] wheeze), or additional symptoms in between colds. If the latter, then symptom frequency and triggers should be determined. Specific triggers may include exercise, excited emotional behavior including laughing or crying, dust, exposure to furry pets (the English disease), weather or environmental temperature change, strong perfumes or aerosol sprays, and smoke from cigarettes or open fires. The therapeutic approach to episodic (viral) wheeze is completely different to that to the child with chronic symptoms in between viral colds.

The severity of symptoms should next be determined, both in terms of the disruption to the child and also to the family, in order to ensure that treatment is appropriately focused. The family of a child who coughs intermittently but is not particularly breathless may merely be seeking reassurance that there is no serious underlying disease, rather than seeking a prescription for regular inhaled medication. Conversely, the family of a child who is a so-called fat, happy wheezer may be well aware that their child is not in danger of death, but are very eager for some treatment to try to ensure a good nights sleep. Other factors, which may influence treatment decisions, are a history of atopy in the child or first-degree relatives, which would probably make one more likely to trial prophylactic treatment.

Particularly in the child with symptoms between colds, specific questions which should be asked are summarized in Table 65.2. The upper airway can be the forgotten area of pediatric respirology [197]. Much the commonest cause of chronic cough in the community is the catarrhal child with postnasal drip. Symptoms suggestive of obstructive sleep apnea should be sought, including snoring, apnoeic pauses, restlessness, daytime somnolence, and poor concentration. Adenotonsillectomy may be completely curative of the chronic cough, and prevent the (rare) dangers of nighttime respiratory failure. In general, the earlier the onset of symptoms, the more likely that an important diagnosis will be found. Symptoms from the first day of life should always be investigated; they must be distinguished from symptoms starting at a few weeks of age, which may be due to an asthma syndrome. The mother should be asked whether the problem started literally from day 1 of life. If this is the case, structural abnormalities of the airway should be excluded. If there is prominent and persistent rhinitis from birth (almost inevitably and fatuously diagnosed as the baby being born with a viral cold), then primary ciliary dyskinesia (PCD, Kartagener's syndrome) should be considered [198]. The very sudden onset of symptoms is strongly suggestive of endobronchial foreign body. Parents may not volunteer the history, and should be asked specifically whether choking on a foreign body is a possibility [199]. Note that even babies too young to bring their own hands to their mouth may have older siblings who may have pressed small objects onto their face. Possible endobronchial foreign body should be referred by telephone for immediate investigation.

TABLE 65.2 Points to seek in the history suggesting an underlying serious diagnosis. A detailed history, targeted towards other respiratory conditions is an essential first step in evaluating the child with non-specific respiratory symptoms.

- Are the child/family really describing wheeze or some other noise?
- Upper airway symptoms – snoring, rhinitis, sinusitis
- Symptoms from the first day of life
- Very sudden onset of symptoms
- Chronic moist cough/sputum production
- Worse wheeze or irritable after feed, worse lying down, vomiting
- Choking on feeds
- Any feature of a systemic immunodeficiency
- Continuous, unremitting or worsening symptoms

Chronic sputum production or a moist cough when the child does not have a viral cold should always be a cause for concern. It is helpful to distinguish recurrent bouts of cough, usually with viral colds and with cough-free periods between bouts, from chronic continuous wet cough with no periods of remission. There is good agreement between parental reports of a wet cough and the presence of lower airway secretions at fibreoptic bronchoscopy [200]. *A child who has had more than 6–8 consecutive weeks of a productive cough merits further investigation.* Two series [201, 202] have shown that a proportion of such children have chronic bacterial airway infection, with a neutrophilic BAL and a positive bacterial culture, usually with *Haemophilus influenza*. The (yet unproven) assumption is that such children will go on to develop bronchiectasis if not aggressively treated. Causes of chronic pulmonary sepsis such as cystic fibrosis (CF), PCD, and agammaglobulinaemia may need to be excluded. In areas where there is universal newborn screening for CF, it might be thought that this diagnosis will no longer need to be considered. However, mild atypical cases may still be missed by screening; the child may have been born in regions where screening is not performed; and the child may have missed out on screening due to administrative or other issues. Thus, a new diagnosis of CF will be rare, but not non-existent, and diagnostic alertness will need to be increased, not reduced with the advent of screening.

Gastro-oesophageal reflux (GER) is suspected in an infant who is worse after feeds, is an irritable feeder (often arching away from the breast or bottle) and who vomits or possets easily. A therapeutic trial of thickening of feeds, acid reduction (proton pump inhibitor or H_2 antagonist) and prokinetic therapy (low dose erythromycin, domperidone) is reasonable on clinical suspicion without further investigation. Choking on feeds, particularly in a child with known neurodevelopmental handicap or neuromuscular disease suggests that incoordinate swallowing due to bulbar or pseudo-bulbar palsy may be the cause of symptoms. Laryngeal cleft or H-type tracheo-oesophageal fistula may present with symptoms at the time of feeding.

Another pointer to the need to refer is whether there are any periods of remission. Although symptom-free periods do not exclude the possibility of a serious underlying disease, the child who has no days free of symptoms certainly merits critical consideration of alternative diagnoses. Finally, a history of systemic infections or poor weight gain in the context of chronic respiratory disease should never be dismissed lightly.

Does cough-variant asthma exist?

There is considerable controversy as to whether cough *on its own* can be a manifestation of asthma. There is no doubt that "cough-variant asthma" is greatly over-diagnosed [203]; large epidemiological studies show that *in a community setting*, where by definition the vast majority of children are well, isolated cough is rarely due to asthma and rarely responds to asthma medications [204, 205]. Chronic non-specific cough frequently improves with time and without treatment [206, 207]. However, in a specialist clinic, where a highly selected group of children are seen, children who cough in response to typical asthma triggers, and improve when treated with asthma medications are occasionally seen [208]. My diagnostic criteria are

- Abnormally increased cough, preferably associated with breathlessness and respiratory distress, with no evidence of any nonasthma diagnosis.
- Clear-cut response to asthma medications (measurements of lung function and its variability are mandatory in the child old enough to perform the maneuvers).
- Relapse on stopping medications with second response to recommencing them.

Many children with chronic cough, in fact, have only a non-specific problem and have been shown on bronchoscopic and blind lavage studies to have no evidence of eosinophilic airway inflammation [209, 210]. Follow-up studies show that most will get better over 1–2 years. A few will show evidence of deterioration of BHR over time, wheeze, and develop the picture of classical asthma [211]. If coughing is troublesome and the precautions outlined above are followed, then there is little to be lost attempting a brief therapeutic trial. The only danger is that ineffectual and potentially harmful medication may be continued long term unless a trial off therapy is rigorous. In older children who can perform lung function, there is no justification for a therapeutic trial without making every attempt to document variable airflow obstruction.

Physical signs

Most often there will be no physical signs. Digital clubbing is an obvious and important sign, but will not be found if not actively sought. My experience has been that children are not uncommonly referred with obvious chronic clubbing which has never been noticed. The upper airway should be inspected for rhinitis, and also nasal polyps the latter being virtually pathognomonic of CF in this age group. The nature and severity of any chest deformity should be noted; although a severe Harrison's sulcus and pectus carinatum can be due to uncontrolled asthma, the more severe the deformity, the greater the likelihood of another diagnosis. The child should be asked to cough as part of the respiratory examination; a wet cough is usually a sign of the presence of lower airway secretions. Palpation of the chest with the palms of the hands during quiet breathing or, in an older child, during blowing or huffing may be a better way of detecting airway secretions than auscultation. Careful auscultation may however elicit unexpected findings such as crackles, fixed monophonic wheeze, asymmetric signs, or stridor, all of which mandate a further diagnostic work up. Finally, signs of cardiac and systemic disease should be sought. Key features to be sought on physical examination are given in Table 65.3.

TABLE 65.3 Points to seek on examination suggesting an underlying serious diagnosis. Most children will have no physical signs; however, none will be found unless they are actively sought.

- Digital clubbing, signs of weight loss, failure to thrive
- Nasal polyps
- Really severe chronic secretory otitis media, otorrhea
- Moist sounding cough
- Enlarged tonsils and adenoids, prominent rhinitis
- Unusually severe chest deformity (Harrison's sulcus, barrel chest)
- Fixed monophonic wheeze
- Stridor (monophasic or biphasic)
- Asymmetric wheeze or other auscultatory signs
- Crackles, particularly if coarse in a "well" child
- Palpable rattles
- Signs of cardiac or systemic disease

Physiology

Although there are many techniques for the measurement of lung function in infants and preschool children, these are largely the provenance of the research laboratory. In the older child, a number of techniques are available to demonstrate variable airflow obstruction. The utility of these tests will depend on the context in which they are applied; a negative result of most of these tests cannot exclude asthma, but the harder it is to demonstrate any variability in airway caliber, the less likely it is that of asthma diagnosis is correct. Possible tests include

- Peak flow or spirometry at rest, with a significant increase after inhalation of a β_2-agonist. In most asthmatic children, baseline lung function is normal, and so no bronchodilator response could be expected.
- A drop in lung function in response to an exercise test. Field exercise tests are often not very specific, and that the use of controlled laboratory testing is more reliable.
- Home peak flow monitoring for a short period may reveal diurnal variation. The accuracy of these records is often not high when used for monitoring over prolonged

periods, but the accuracy when the purpose is diagnostic may be different.

- A histamine or methacholine challenge may occasionally be useful. If there is no response to the highest dose of challenge agent, then the child's symptoms are most unlikely to be due to asthma. A positive challenge test is not diagnostic of asthma.

Allergy testing

Allergy testing (usually either skin prick testing, total IgE, or specific RAST tests) can be used (a) to determine if the child is atopic, because if in particular the asthma is thought to be severe, but the child is nonatopic, the diagnosis may need to be revisited; (b) to guide environmental manipulations (discussed later); and (c) possibly to predict prognosis in the wheezy child. Both skin prick tests and RAST testing have the disadvantage of a high false negative rate particularly under age 3. However, if positive, they are likely to be significant.

Atopic status may be clinically obvious in a child with typical flexural eczema. If there is no skin disease, then testing to determine the presence of atopy may be a guide to treatment; most would be more ready to treat an atopic wheezy child with inhaled steroids, although many atopic children will not have airway eosinophilia and will outgrow their symptoms within 2 years even in the absence of anti-inflammatory treatment [212]. Furthermore, even atopic preschool children often respond poorly to inhaled steroids.

A more novel use of allergy testing is to try to predict long-term outlook. One study suggested that a positive RAST to egg was predictive of long-term wheeze [213]. Further prospective studies are needed to clarify whether this or any other allergy test is predictive of prognosis.

Inflammometry

There have been no formalized studies of the measurement of inflammatory markers in the diagnostic process as there have been in adults. Anecdotally, an elevation in exhaled nitric oxide $FeNO_{50}$ to twice the upper limit of normal [214], or the demonstration of an eosinophil count of $\geq 3\%$ on induced sputum would be considered supportive of a diagnosis of asthma, but there are insufficient data to know the likely positive and negative predictive values of these tests. The role of $FeNO_{50}$ measurements in preschool children is even less clear, and they remain largely a research technique.

Response to a therapeutic trial

Especially in preschool children, the only way to settle or otherwise the diagnosis of asthma is a therapeutic trial (below). In a child, old enough to perform lung function tests, these should be incorporated as an endpoint, but in general, the diagnosis should not require a trial of therapy. The second indication is in a child known to have asthma to determine the extent of fixed airflow obstruction (discussed in section "Really Severe Asthma").

TABLE 65.4 Disease which present as recurrent cough and wheeze. These conditions need to be considered and excluded prior to escalating therapy. Most will require referral if suspected in general practice.

- Upper airway disease – adenotonsillar hypertrophy, rhinosinusitis, postnasal drip
- Congenital structural bronchial disease – complete cartilage rings, cysts, webs
- Bronchial/tracheal compression – vascular rings, pulmonary arterial sling, enlarged cardiac chamber or great vessel, lymph nodes enlarged by tuberculosis or lymphoma
- Endobronchial disease – foreign body, tumor
- Esophageal/swallowing problems – reflux, incoordinate swallow, laryngeal cleft or H-type tracheo-esophageal fistula
- Causes of pulmonary suppuration – cystic fibrosis, primary ciliary dyskinesia, persistent bacterial bronchitis [201, 202], any systemic immunodeficiency including agammaglobulinaemia, severe combined immunodeficiency
- Miscellaneous – bronchopulmonary dysplasia, congenital or acquired tracheomalacia, pulmonary edema secondry to left to right shunting or cardiomyopathy

TABLE 65.5 Investigations to be considered in the child with recurrent cough and wheeze. A selective approach is necessary, depending on what clues have been elicited from history, examination and simple investigations. Most of these tests will be carried out only after referral to a specialist.

- Suspected upper airway disease – polysomnography, RAST or skin prick tests (radiograph of postnasal space is rarely useful), MRI or CT of sinuses
- Known or suspected neuromuscular disease with dysfunctional swallow – speech and language therapy assessment, which may be combined with videofluoroscopy
- Suspected aspiration with normal neurology and no reflux – rigid bronchoscopy to exclude laryngeal cleft and H-type fistula
- Suspected esophageal disease – pH probe, barium swallow, tube oesophagram, oesophagoscopy
- Suspected cystic fibrosis – sweat test, nasal potentials, genotype, stool elastase, three day fecal fat collection
- Suspected primary ciliary dyskinesia – saccharine test, nasal ciliary motility, electron microscopy including orientation studies, nasal and exhaled nitric oxide, culture of ciliary brush biopsy, genetic studies becoming available
- Suspected systemic immunodeficiency – immunoglobulins and subclasses; vaccine antibodies; lymphocyte subsets; lymphocyte and neutrophil function tests; HIV test; referral to Pediatric immunologist
- Suspected structural airway disease – fibreoptic bronchoscopy
- Suspected tuberculosis – Heaf test, fibreoptic bronchoscopy and/or gastric lavage, combined with culture and PCR; ELISPOT
- Suspected cardiovascular disease – echocardiogram, barium swallow to exclude a vascular ring or pulmonary artery sling, angiography
- Suspected bronchiectasis – high resolution CT scan, investigations for local or systemic immunodeficiency

DIFFERENTIAL DIAGNOSIS OF ASTHMA

This encompasses most of pediatric respiratory medicine, and will not be considered in detail. A summary guide to differential diagnosis is provided in Table 65.4. Confirmatory diagnostic tests, usually performed after hospital referral, for some of these conditions are listed in Table 65.5. These should not be considered as a catalog to be applied uncritically, but only as guided by history and physical examination. More details of individual conditions can be found in standard texts.

PHARMACOLOGICAL MANAGEMENT OF ASTHMA IN THE PRESCHOOL CHILD

There are two indications for treatment, the immediate relief of symptoms and the prevention of disease progression, in the case of the preschool child from intermittent to continuous symptoms. Prediction of those at high risk for such progression has been discussed above. Detailed evidence based guidelines have been published for all ages, which should be read in conjunction with the commentary below. However, for preschool children, there seems to be a disconnect between knowledge gained from epidemiological and pathological studies, and treatment recommendations.

General use of medications

Critical in the choice of medication is an understanding of the biological basis for the different phenotypes of preschool wheeze (Table 65.6). These are not mutually exclusive, and indeed, if they coexist in a single individual, the relative contributions may be difficult to unravel. What syndrome, for example, would be diagnosed in an infant who was born prematurely and ventilated for a few weeks, has atopic eczema, developed RSV positive bronchiolitis, and now wheezes only with viral colds? The majority of children either have viral associated wheeze or will eventually turn out to have persistent asthma. These can only be diagnosed in retrospect. However, intermediate phenotypes often confuse the picture, and ultimately, a therapeutic trial may be needed to determine best treatment in such a child.

Chronic lung disease of prematurity

This should be obvious from the history. Summarizing the studies of airway function, in mid-childhood, survivors of chronic lung disease (CLD) of prematurity exhibit reduced indices of expiratory airway function (FEV_1 and indices of flow at low lung volumes); increased bronchodilator responsiveness; and BHR [215]. There is limited evidence for airway inflammation, and little evidence that anti-inflammatory medication is helpful. The long-term importance of this group is discussed later.

TABLE 65.6 Features of the different pre-school 'asthma syndromes'. It should be noted that overlap syndromes are very common.

Preschool asthma syndrome	Inflammatory component	Extent of BHR	Extent of PAL
1. Chronic lung disease of prematurity	? (probably none)	+ (Probably related to distal airway narrowing)	+ (antenatal onset)
2. Post-bronchiolitis (usually RSV)	? (probably none) (Some evidence for neurogenic inflammation)	+	+ (antenatal onset)
3. Episodic (viral) wheeze	–	–	+ (antenatal onset)
4. Atopy associated wheeze	+ (probably often eosinophilic in later child, not at age 12 months)	+	+ (probably antenatal and postnatal onset)
5. Obliterative bronchiolitis (e.g. post adenoviral infection)	–	–	+ (postnatal onsetl)
6. Non-atopy associated, later onset wheeze	?present, ?type	Probably present	+ (probably at least postnatal onset)

BHR: Bronchial hyper-reactivity; PAL: persistent airflow limitation; RSV: respiratory syncytial virus. ? = Unknown; + = Present; – = Absent.

Atopy associated wheeze

It should not be assumed that an atopic baby who wheezes has eosinophilic inflammation [162], although, in particular after the age of 3 years, this may be the case. This group will contain some with airway pathology similar to atopic asthma in older children and adults, but some will have a noninflammatory phenotype.

Post-bronchiolitic cough and wheeze

RSV is ubiquitous worldwide, and by the age of around 2 years, virtually all infants have serological evidence of infection. Prolonged respiratory symptoms are common after RSV bronchiolitis. It is unclear whether RSV infection causes asthma, merely unmasks an asthmatic tendency, or is not related to asthma at all. There is a lack of proper prospective studies, with characterization of the atopic status of the children before their RSV infection. For example, atopic children may be more likely to be admitted to hospital with RSV bronchiolitis [124, 216]. In studies of lung function before and after bronchiolitis, it would appear that severe bronchiolitis was seen in those with pre-existing airflow limitation akin to the situation in virus-associated wheeze [120].

Airway lavage studies [217, 218] and autopsy data [219] show that RSV bronchiolitis is a neutrophilic, not an eosinophilic airway disease. In humans invasive studies are not easily performed in the acute phase of RSV bronchiolitis, so it is difficult to know how important eosinophilic inflammation actually is in mild disease. An autopsy study, which by definition would only contain the really severe cases, showed a low number of eosinophils in the bronchial wall [219]. It is also clear from randomised controlled studies that treatment with inhaled or nebulized steroids at the time of RSV bronchiolitis or for subsequent symptoms is ineffective [220]. There is increasing animal and human evidence that neurogenic inflammation may be important after RSV infection [221, 222].

Episodic (viral) wheeze

Isolated wheezing with colds is not due to an inflammatory airway phenotype, but due to reduction in baseline airway caliber, almost certainly on a developmental basis, discussed in detail above. Postnatal physiological studies have demonstrated that infants who wheeze with colds have evidence of airway obstruction *before the first wheezing episode*. Wheezing with viral colds may not be associated with either BHR [39] or airway inflammation [223]. Those who present in infancy with recurrent, viral episodic wheeze have little likelihood of persistent asthma in later childhood [224, 225].

Obliterative bronchiolitis

This syndrome is characterized by PAL, absence of BHR and absence of inflammation, usually the result of adenovirus bronchiolitis, or severe reflux and aspiration. Other causes (lung transplant rejection, drug reactions, and collagen vascular disease) are rare or diagnostically obvious.

Nonatopic, later onset wheeze

Although atopy is a major factor in pediatric asthma, there are children with onset of wheeze in the preschool period or later, who are not atopic by any conventional criterion. This may be more common in developing countries. Little is known of the pathological basis of this group.

Medications to treat symptoms

The first approach is intermittent therapy at the time of symptom episodes. First line therapy would be with either or both of an inhaled anticholinergic (ipratropium bromide) and a β_2-agonist (salbutamol, terbutaline). Although by tradition younger infants, respond better to ipratropium, in practice it is worth trying both medications at any age.

If this approach does not work, a trial of a leukotriene receptor antagonist should be considered. A number of studies have shown that viral infection is associated with transiently elevated levels of the cysteneyl leukotrienes [226, 227]. Traditionally, leukotriene receptor antagonists are prescribed continuously, but a recent study has suggested that they are equally efficacious when given intermittently, for at least 7 days with viral colds [228]. Another approach is to use relatively high doses of inhaled steroids just at the time of viral exacerbations, for example budesonide 1 mg twice daily for 5 days [229]. There are no data on combining this approach with a leukotriene receptor antagonist, although this would seem a reasonable approach for resistant cases. However, although there may be benefits from this approach, a recent study also highlighted that high dose even intermittent fluticasone may cause side effects [230]. The final option for intermittent therapy is parent initiated prednisolone. However, this was not shown to be effective in preventing admission to hospital in a large randomised controlled trial [231].

If intermittent therapy is unavailing, what is the role of continuous therapy? If the child is having really severe exacerbations or symptoms on at least a near daily basis even between viral colds, a therapeutic trial may be justified. This will usually be with inhaled corticosteroids; cromones are not effective, and there is no evidence that continuous leukotriene receptor antagonists are any better in this context than when given intermittently. A therapeutic trial may also be justified if there is doubt as to whether the child has a steroid sensitive asthma phenotype; if this is contemplated, it is important to note that many childhood symptoms improve spontaneously, and it is crucial not to confuse changes with therapy with changes over time. Thus a three stage protocol is advised.

- *Step one*: Commence inhaled corticosteroids at a moderate dose, for example budesonide 400 mcg twice daily
- *Step two*: Reassess at 8 weeks. If there has been no response, the child does not have a steroid sensitive phenotype, so treatment is stopped and the trial terminated. If symptoms have disappeared, the treatment is also stopped.
- *Step three*: If the symptoms which previously regressed return, then treatment is restarted. If the symptoms respond a second time, then the child is considered to have a steroid sensitive phenotype, and the dose of inhaled steroids tapered to the lowest which keeps the child symptom free.

It should be noted that, for the vast majority of preschool children, especially under two years of age long term inhaled steroids bring little benefit. Even in a high-risk group, the increased number of symptom-free days/year was not high despite a dose of inhaled steroids sufficient to cause reduction in linear growth [232].

It might be thought that an oral steroid trial, as in older children and adults, might be more appropriate. However, prednisolone will treat upper airway disease, for example adenoidal hypertrophy, which may confuse the issue, so inhaled steroids are to be preferred.

Medications to prevent progression of airway changes (disease modification)

Clinical experience is that most children with atopic asthma in mid-childhood initially present with intermittent wheeze. Four studies have evaluated whether the early institution of inhaled corticosteroids prevents the progression from intermittent to continuous wheeze. Whether these are administered continuously [232–235] or intermittently just with viral colds [236], they have no effect on disease progression, and are not disease modifying. An initial study suggested that the use of cetirizine may prevent subgroups of children with atopic dermatitis progressing to asthma [237], but this has not been confirmed subsequently (Warner J.O., personal communication). There is an urgent need for discovering new strategies to intervene in the putative window between 12 of age, when there are symptoms but no airway changes, and 30 months, when the first signs of airway inflammation and remodeling are present, in order to prevent long-term morbidity.

Delivery devices

In the preschool years, the choice lies between a spacer, with a mask in children under age three, or a nebulizer. For all but the most severe attacks of acute asthma, spacers are at least as effective as nebulizers, cheaper, and easier to use [238]. Nebulized, as against inhaled, corticosteroids, are rarely if ever indicated. The proper use of inhalers and spacers is summarized in Table 65.7. If a nebulized therapy must be used, it is essential that a tight-fitting mask is used; if the nebulizer cup is hopefully waved near the infants face, lower airway drug deposition will be exactly zero.

TABLE 65.7 Proper use of spacers in pre-school children.

Instruction	Comment
Shake inhaler between each activation	Multiple activations reduce drug delivery
No delay between activating inhaler and applying mask to face	Delay leads to medication being adsorbed onto sides of chamber and not inhaled
Do not give to crying infants	There will be no drug delivered to the airways, even though the infant appears to be inhaling deeply
Wash (non-metal) spacers weekly with washing up liquid, do not rinse or rub dry	Minimizes medication adsorption onto sides of chamber

PHARMACOLOGICAL MANAGEMENT OF ASTHMA IN THE SCHOOL AGE CHILD

General use of medications

The stepwise approach has been well summarized in evidence based guidelines [1], which are underpinned by the reasonable assumption that the pathology of asthma is similar in adults and children over five [239, 240]. The dangers of extrapolation from adult studies to the pediatric age group have been discussed, and there are, however, important differences in the adult and pediatric literature with regard to response to medication.

Medications to treat symptoms

The least difficult aspect is that very mild, intermittent symptoms are treated with as required β_2-agonists. Thereafter there are a large number of unanswered questions:

- How frequently can intermittent β_2-agonists be given before prophylactic inhaled corticosteroids are prescribed?
- Should inhaled corticosteroids be started at a high dose, and weaned down (previous set of BTS guidelines [241]) or started low and worked up (current BTS guidelines [1])?
- At what level of inhaled corticosteroids should a second prophylactic agent be added?
- What prophylactic medication should be added to inhale corticosteroids? Should it be long-acting β_2-agonists, leukotriene receptor antagonists, or low dose theophyllines? Or, should the dose of inhaled corticosteroids be increased further in some children?

It may be that the right approach is to ask, why the asthma has not responded to low dose inhaled corticosteroids? In some, there may be residual BHR, and the best option may be a LABA; in others, it may be relatively steroid resistant eosinophilic inflammation, mandating more inhaled steroids, or leukotriene receptor antagonists; or even, neutrophilic inflammation, which might respond to theophyllines or even macrolide therapy. These phenotypes could be differentiated by simple measurements, such as bronchodilator response, induced sputum, or $FeNO_{50}$. However, this is currently speculative.

A new issue relates to the philosophy of therapy. Should the aim be to suppress every possible symptom of asthma [242], or use flexible, single inhaler treatment with a combination of budesonide and eformoterol [243]? It is not clear in children that big doses of inhaled steroids can suppress all symptoms, even in combination with LABAs, and the risks of side effects are well known. It may be that it is better in the long run to accept that intermittent mild symptoms and a lower dose of baseline medication may be preferable. However, another study has cast doubt on the use of combination therapy in mild asthma in children, suggesting fluticasone alone is superior to either combination therapy or single agent montelukast in at least some aspects of asthma control [244]. The converse approach, single inhaler treatment of symptoms, relies on the ability of the child to detect symptoms and act on them, which may be questionable. There is a real need for a long-term study comparing these two approaches in children.

Medications to prevent progression of airway changes (disease modification)

A retrospective, observational study in children suggested that delay of more than >2 years in initiating inhaled corticosteroid therapy may lead to a worse outcome in terms of pulmonary function [245], similar to the findings in a prospective randomised controlled study in adults [246]. However, even in adults, the concept that even mild asthmatics should be treated with inhaled corticosteroids is losing ground [247], and in pediatrics, the CAMP study has also suggested that inhaled corticosteroids do not help preserve lung function. In this study, inhaled corticosteroids, nedocromil and placebo were compared in mild asthmatics over several years. Overall, 75% maintained their lung function, and 25% showed a deterioration. The proportion of deteriorating was the same in each treatment group, in other words, inhaled corticosteroids were not effective in preventing deterioration in lung function [248]. Thus, as with preschool children, we do not have available a truly disease-modifying therapy in school age children.

Delivery devices

These are covered in standard guidelines. Whatever device is chosen, it is essential that regular checks of technique are carried out in the asthma clinic. If high doses of inhaled corticosteroids are used, spacers are preferable. The special issues of adolescence are covered below.

NON-PHARMACOLOGICAL MANAGEMENT OF ASTHMA IN CHILDHOOD

Environmental manipulation may be undertaken as part of a program of primary prevention, either in high risk or all populations, or as a secondary attempt to ameliorate the severity of established asthma. Despite research from a large number of cohort studies, it is currently not possible to design a prospective primary cohort intervention study. Specifically, it is unclear whether tolerance should be induced by high allergen exposure, or allergens should be avoided. The dose response effects may be different for different allergens, with, in at least one study, the suggestion that there is a threshold effect, above which there is very little evidence of a dose response [249]. The rest of this section details the measures which might be considered in secondary prevention.

Environment

Environmental tobacco smoke

Discouragement of smoke exposure is mandatory. The role of tobacco smoke causing steroid resistance is unequivocal; it is likely that passive smoke exposure will have the same effects.

Allergens

Some evidence has been reviewed above. It is clear that *in vitro* high allergen exposure can lead to a state of steroid resistance via an IL-2 and IL-4 mediated mechanism, the exact pathways being controversial [250, 251]. Controlled low level inhaled allergen exposure in sensitized adults, at a dose too low to cause acute bronchoconstriction, still caused worsening airway eosinophilia and deterioration in BHR [252]. Low dose allergen exposure may come from the school; cat allergic children with asthma develop a pattern of symptoms similar to occupational asthma if more than 20% of their classmates are cat owners [253].

Hence, reduction of allergen exposure should be attempted in the asthmatic child, and, the more troublesome the asthma, the more it should be pursued; this irrespective of IgE mediated senitization (above).

Diet

The search for the food allergy which is the root cause of all symptoms is often something for which parents wish. In practice, the blind use of exclusion diets is rarely helpful, although often attempted by families; occasionally, implementation of a dairy-free diet in a preschool wheezy child may be beneficial. Anecdotally, most successes are in children who have other issues, such as eczema and GER. It is always worth inquiring not only after food allergies, but also whether the child will actually consume milk, eggs, and nuts (including peanuts); allergy may be manifest by an aversion to the food. Food allergy in asthma is managed in the standard way, but erring on the side of caution. Likewise, asthma in the food-allergic is also treated conventionally, but erring on the side of caution. This caution is because there appears to be a higher risk of acute really severe asthma in the food-allergic child, and in particular peanut and milk allergy may be a marker for increased morbidity [254].

MONITORING ASTHMA

Physiology

Testing can be used to make the diagnosis, monitor the effects of therapy, and as part of an asthma treatment plan. Diagnostic testing has already been covered.

Monitoring therapy

Home peak flow monitoring may be useful to try to document the response to therapy and the course of the disease. However, compliance is notoriously poor [255, 256].

Asthma treatment plans

Changes in therapy may be guided by asthma treatment plans. These, although excellent in theory, may be of rather less practical value than the enthusiasts would have us believe. Since perception of dyspnea may be a real issue in asthmatics [257, 258], the idea of monitoring health objectively is fine in theory, but in pediatric practice, few comply. It is doubtful whether giving every asthmatic a treatment plan based on peak flow is a useful exercise, but in selected children, families may find this extremely helpful.

Adverse effects of disease and treatment

All children with any chronic illness should have their height and weight measured accurately and regularly and plotted on a centile chart. Growth failure in asthmatic children may be due to the effects of uncontrolled asthma, treatment with inhaled or oral corticosteroids, or a coincident diagnosis. Recommendations for the monitoring of adrenal function are more difficult. There is no doubt that high dose inhaled corticosteroids may cause adrenal failure and acute hypoglycemic. There is evidence at least from adults that overtreatment, rather than the absolute level of therapy is what is important; in one study, plasma fluticasone levels were much higher for the same dose of inhaled fluticasone in normal adults compared with asthmatics [259]. Theoretically, regular low dose short synacthen tests should be performed in all children prescribed high dose inhaled steroids [260], although the short-term repeatability of this test has been questioned [261]. In practice, the first priority must be to control asthma, with the lowest dose possible of inhaled corticosteroid. Every effort must be made to reduce the dose of inhaled steroid, especially using adjuncts such as long-acting β_2-agonists, leukotriene receptor antagonists, and low dose theophylline. Families must be warned about possible adrenal failure, children should carry a steroid card, and emergency doctors should consider the possibility of acute hypoglycemic [262, 263].

Other potential side effects of high-dose steroids include hypertension, cataract, diabetes, and osteopenia. In most clinical practice, regular ophthalmic review and DEXA scanning are restricted to children taking regular or very frequent intermittent courses of oral steroids. It is arguable that blood pressure measurements and urinalysis should be part of routine pediatric care in any case.

Inflammometry

Studies in adults with asthma have established proof of concept that managing asthma simply by history, physical examination, and simple physiology is no longer optimal [264]. The situation in children is less clear-cut; it is highly likely that inflammometry will have a place in selected groups of children, but what should be measured, and in whom, is more debateable.

Potential choices include measurement of exhaled breath ($FeNO_{50}$), with which there is most experience; exhaled breath condensate (EBC), which many consider a technique looking for an application; induced sputum; measurement of BHR; and more invasive techniques such as bronchoscopy, which will never be suitable for clinic monitoring. The characteristics of the ideal inflammometer are shown in Table 65.8, and of course, there is no such device. $FeNO_{50}$ is the most potentially attractive, because it is quick and easy to measure, is even available for home monitoring, and the answer is immediately available. However, the relationship between $FeNO_{50}$ and sputum eosinophils is not a close one, at least in children [27], and the two cannot be used interchangeably. Furthermore, atopy alone, in the absence of asthma, may elevate $FeNO_{50}$. More work is needed on biomarkers of pediatric asthma, but there is sufficient evidence of proof of concept to encourage the researcher to press on with these studies.

TABLE 65.8 The characteristics of the perfect 'inflammometer.'

Cheap
Easy to maintain and calibrate
Completely non-invasive
Easy to use, no co-operation needed
Direct measurement of all relevant aspects of inflammation
Rapid availability of answers
Evidence of beneficial clinical outcomes

Tailor therapy

An important study comparing standard monitoring with a $FeNO_{50}$ based strategy showed that after a year, the $FeNO_{50}$ group had significantly improved BHR [265]. Another study using BHR as the monitoring tool, rather than an endpoint, also showed improved outcomes [266]. Measurement of BHR as a routine test in a busy clinic is not likely feasible outside a research context. However, these studies confirm that history and simple physiological testing is no longer state of the art.

Predict natural relapse

It is unlikely that $FeNO_{50}$ measurements will predict a viral infection, although conceivably they may give advance warning in the pre-symptomatic phase. The most likely successful model is loss of control in mono-allergic, pollen sensitized children during the pollen season. If $FeNO_{50}$ measurements were standardized to the child's value when well, $FeNO_{50}$ could give an indication of exacerbation within no >5 days of measurement [267]. However, prospective testing of these hypothesis-generating results in the subsequent season were disappointing (unpublished data). It is likely but still not proven that daily measurements, which are now feasible using portable equipment [268, 269], will be more useful than single clinic recordings. More work is needed before these devices can be recommended as a routine in clinical practice.

Predict relapse on stopping therapy

Guidelines rightly stress that after a period of stability, pediatric asthma treatment should be scaled down. Two studies have explored whether inflammometry may be useful in the decision making process. In the first, children who on clinical grounds had a reduction in therapy had prior measurement of inflammatory markers. EBC was not useful. Those without detectable sputum eosinophilia always had a successful dose reduction. Receiver operating curves for FeNO and sputum eosinophils were constructed, and both were useful albeit imperfect adjuncts to clinical decision making [270]. In a second paper, a rising FeNO gave an early warning of relapse in children whose therapy had been stopped altogether [271]. Taken together, these studies suggest a role for inflammometry during treatment reduction.

Inflammometry – what is the role now?

It is clear that inflammometry is rightly here to stay in pediatric asthma. $FeNO_{50}$ seems to be the most promising, but we are a long way from being able to treat an elevation in FeNO with no other indications of any problem. A raised $FeNO_{50}$ may be found apparently constitutionally in completely well individuals [272]. However an acute elevation in $FeNO_{50}$ should prompt a review of management. It may be a sign of an imbalance between pro-asthmatic stimuli (e.g. allergens) and anti-inflammatory therapy (either dose too low or drug delivery device incorrect, or medication not being taken) [273]. The role of home monitoring of $FeNO_{50}$ in pediatric asthma is unclear, and before any widespread introduction, we need more evidence of utility than we had before home peak flow monitoring was widely embraced. The chief needs for the future are more reliable biomarkers of asthma control and impending exacerbation; biomarkers of distal and mucosal inflammation, as well as of airway remodeling; and biomarkers, which can be used in infants and preschool children.

ASTHMA PLUS

As well as being mimicked by conditions such as GER, asthma may coexist with them, and these comorbidities may worsen asthma control. Thus, part of the management of troublesome asthma should be the search for these conditions.

GER

The relationship between asthma and GER is controversial. Opinions are polarized between the surgical approach ("all asthma can be cured with a Nissen's fundoplication") to the medical ("GER is irrelevant, and inhaled steroids cure everything"). The following possibilities may be difficult to distinguish

1. Asthma causes GER including by effects on intrapleural pressure and integrity of the lower esophageal sphincter.
2. GER causes aspiration and symptoms that are mistaken for asthma.
3. GER causes lower esophageal acidification without aspiration, which worsens BHR reflexly.
4. GER and asthma are coincident findings.

GER is more likely to be significant in infants, in whom an empirical trial of therapy may have gratifying results. Anecdotally, children with very severe asthma frequently are found to have GER, but treatment has been almost universally disappointing. The only way to sort out the above options above is probably by carrying out a therapeutic trial.

Rhinosinusitis

The relationship between the upper airway and lower airway is another source of controversy. Whether treatment of allergic rhinitis will improve, asthma is disputed. A large community study suggested that children with BHR and allergic rhinitis were more likely to improve their BHR if the rhinitis was treated with topical corticosteroids [274]. In practice, most would treat any upper airway symptoms on their own merits; any improvement in asthma would be seen as an added bonus. It should be noted that extrapulmonary topical steroid treatments (nose drops, skin creams) may add significantly to the total steroid dose to which the child is exposed [275], and growth failure has been reported just with the use of potent steroid nosedrops [276], so treatment of the upper airway may introduce its own problems.

Psychological issues

Thankfully, the days when asthma was thought to be due to low moral fiber have long gone. However, there is no doubt that there are complex interactions between the brain and the airway, possibly neurally mediated. In one study, high school students were studied at time of stress (sitting examinations) and during student idleness (mid-term). The same dose of inhaled allergen to which they were known to be sensitive produced a more marked airway eosinophilia during the time of stress [277]. This mechanistic, proof of concept study has been confirmed by a number of field studies, which have documented the associations of poor asthma control, and acute exacerbations of the disease, with psychological stresses of various kinds [278, 279]. In severe asthma in particular, anyone who treats a child as merely a pair of lungs is doomed to therapeutic failure. It is essential to explore issues of stress at home and school, in a sensitive manner, in all but the most straightforward cases of pediatric asthma.

A related and important issues are what may rather simplistically be lumped together as dysfunctional breathing disorders, a spectrum which includes hyperventilation and panic attacks, and vocal cord dysfunction (VCD), in which condition there is adduction of the vocal cords in either or both of inspiration or expiration [280, 281]. These frequently coexist with indubitable asthma. Symptoms of hyperventilation such as peri-oral paresthesia will be missed if not actively sought. VCD is characterised by noisy breathing, which careful auscultation establishes comes from the larynx, but not from the lower airway. The child may identify that they cannot breathe in, and actually identify the upper airway as the seat of the problem.

Related to this is the "habit" or "honk" cough, a bizarre and very irritating noise, which is stereotyped, disruptive and repetitive. Those who despite the evidence persist in the unquenchable belief that isolated cough is always due to asthma (it very rarely is) may escalate steroid therapy to try to control the problem.

All these disorders have in common that (1) they do not respond to any asthma therapy and (2) the symptoms invariably disappear when the child is asleep (although they may still be present while the child is lying awake in bed, trying to go to sleep). There may be obvious gains for the child, or more subtle advantages. Management first involves identifying the problem, and ensuring the child and family know that the genuineness and significance of the symptoms is not in doubt, but that they are not due to asthma. Sometimes, symptoms may respond to relaxation and breathing exercises administered by a skilled physiotherapist; if the roots of the problem are more deep seated, then the help of a clinical psychologist may be needed. As a general point, if stress and asthma coexist, it is better to treat their both on independent merits, rather than try to unpick which is chicken and which is egg.

Adherence to therapy is a related issue. Only the minority of children actually even collect enough prescriptions to cover their prophylactic medications [282]. Indeed, complete obedience to everything the pediatrician asks may be considered to be pathological! We know that non-adherence is frequent, even when there are really severe problems with asthma. Detecting it is much more difficult. It is safe to assume that children will not admit to non-adherence. A check on the number of prescriptions dispensed, or a home visit may be useful. If the child has been prescribed prednisolone or theophylline, measurement of plasma levels (including serum cortisol in those on prednisolone) may be informative. Finally, I do not diagnose steroid resistant asthma until I have satisfied myself that there is no response to an intramuscular injection of triamcinolone.

If non-adherence is found to be an issue, sensitive exploration of the underlying problems is indicated, rather than a "blame" culture of mutual recrimination. Non-adherence remains one of the important unsolved problems in pediatric asthma; detection and management are exceedingly difficult. Finally, pediatricians should remember that their judgement about adherence is exactly as good as flipping a coin.

ACUTE ASTHMA IN CHILDHOOD

The management of acute severe childhood asthma has been covered in standard guidelines [1]. Asthma deaths are fortunately rare in childhood, less than 40/year in the United Kingdom. In some, avoidable factors can be identified, such as failure of the child and family to seek medical attention while giving ever larger and more frequent doses of short-acting β_2-agonists, or failure of the pediatrician to recognize the severity of the situation or give systemic steroids. In others, there appears to be a sudden unexpected acute deterioration and even in retrospect, there seems little which could have altered the outcome. All children who have survived a really severe attack of asthma should have a detailed re-evaluation of all aspects of their therapy. It is probably this group in whom treatment plans are most useful, although definitive proof is lacking.

Two particular circumstances should be considered in more detail. Those who have survived a sudden and severe attack may be considered for pre-packed, injectable adrenaline (e.g. the Epipen™).Of course, most children have sufficient time to use inhaled therapy, but for those with really sudden deteriorations may anecdotally benefit from this approach. The other scenario is the child with deteriorating asthma who is unable to recognize that things are getting worse, until a late, catastrophic decline ensues. Failure of perception is therefore a mimic of acute rapid deterioration. These children, if they can be persuaded to use it, should have a home peak flow meter and monitor their status objectively and regularly.

ASTHMA IN SPECIAL CIRCUMSTANCES

Really severe asthma in childhood

The definition of severe asthma has been the subject of international working groups [283, 284]. There are several different patterns of severe asthma in childhood including chronic symptoms despite high-dose therapy; and brittle asthma, either chronic wild swings of lung function or sudden really acute deteriorations on the background of apparent good control. It is likely that the underlying pathology is very different in these groups.

The commonest causes of therapy resistant asthma are non-adherence to treatment and incorrect diagnosis. Really severe, therapy resistant asthma is rare. We should distinguish "difficult" asthma (in which, when the basics are got right, asthma control is achieved) from "really severe, therapy resistant asthma," in which poor asthma control persists despite optimizing all conventional therapies. We now use a three-visit protocol to assess these children (Table 65.9). Obviously, prior to any such approach, a full diagnostic work up is undertaken, as appropriate. This is particularly important in nonatopic children, in whom an alternative or exacerbating diagnosis is found more frequently.

Visit 1: The aim of the first visit is to weed out those asthmatics who are not truly therapy resistant. The child undergoes a full, multi-disciplinary assessment. This includes a visit to the home by the respiratory nurse, a psychological assessment by questionnaire, and a referral to the clinical psychologist if necessary; and contact is made with the school. If not already performed, skin prick tests

TABLE 65.9 The difficult asthma protocol.

	Visit 1	Visit 2 (if no improvement)	Visit 3 (4 weeks later)
1. Clinical Assessments	• Asthma control test • Nurse lead home visit • School visit • Access GP records • Psychological assessment as appropriate	• Asthma control test • Assess symptoms, new peak flow diary	• Asthma control test • Assess symptoms, new peak flow diary • Allocate as steroid responder, partial responder, or non-responder on bases of all results
2. Physiological measurements	• Spirometry including response to β_2-agonist	• Spirometry, including response to β_2-agonist	• Spirometry, including response to β_2-agonist
3. Non-invasive inflammatory and other markers	• Induced sputum • FeNO (variable flow) • RAST or skin prick tests as appropriate • Measure prednisolone and theophylline levels if appropriate	• Induced sputum • FeNO (variable flow)	• Induced sputum • FeNO (variable flow)
4. Invasive studies		• Bronchoscopy, BAL, and bronchial biopsy • Intramuscular triamcinolone • pH study	

are carried out. Adherence is assessed at the home visit, by checking the availability of medications; GP prescriptions are checked; and use of drug delivery devices is assessed. We found that around a third of children with severe asthma could not produce a complete set of in-date medications during a nurse-lead home visit. Environmental measures are instituted as far as possible. As a result, more than 50% of referrals need no further action.

Visit 2: Children who continue to have problematic asthma despite this detailed protocol are further evaluated. They undergo non-invasive measurements of airway inflammation (Table 65.9), and spirometry and bronchodilator reversibility. A fibreoptic bronchoscopy is performed under general anesthetic, with bronchoalveolar lavage and endobronchial biopsy, and an injection of depot triamcinolone administered while the child is still anaesthetized.

Visit 3: this takes place 4 weeks later. Lung function and airway inflammation is measured, and the child phenotyped (Table 65.10); an individual treatment plan is developed on the basis of the phenotype. It is clear that therapy resistant asthma is not one disease, but many, but it also has to be said that the utility of the above approach needs further testing.

Exercise induced symptoms

In most cases, exercise induced asthma is a manifestation of poorly controlled asthma, and should be managed in accord with standard guidelines. Occasionally, exercise symptoms are seen in elite athletes, as the sole problem [285]. They report breathlessness and wheeze at extremes of exercise, just when they want to reach peak performance, and at no other time. The differential diagnosis lies between exercise induced asthma, and exercise induced VCD; the differentiation may be difficult. True exercise induced asthma in the elite athlete may be a different phenotype from standard poorly controlled asthma. Although one study has suggested that fish-oil may be helpful, in practice this is not much used [285]. These athletes usually come to specialist attention because presport, high-dose β_2-agonists have failed. Anecdotally, high-dose nedocromil sodium may be useful, and other strategies include montelukast or Symbicort™ prior to exercise. It is absolutely essential that due care is paid to the requirements of the international drug abuse regulations when treating elite athletes. VCD may also be an issue, particularly if the athlete is being pushed beyond capacity. This may be very difficult to identify – sometimes a Bruce protocol exercise test to exhaustion, with measurement of flow-volume loops, may identify the problem. The standard 6 min exercise induced asthma test is useless in these athletes, because for them, it is insufficiently taxing. Anecdotally, both physiotherapy and the help of a clinical psychologist may be useful.

Adolescent asthma

Normal adolescence is a taxing time for all concerned, and is complicated if there is a comorbidity such as asthma. This big subject has been reviewed in detail recently [286]. The tasks of normal adolescence include establishing increasing independence and autonomy, and, in the context of asthma, taking more responsibility for treatment (or lack of it). Asthma

TABLE 65.10 Summary of proposed difficult asthma management, at the conclusion of the protocol studies.

Clinical scenario	Presumptive diagnosis	Suggested action
1. Continued airflow obstruction, no inflammation, no reversibility to β_2-agonists	Presumed obliterative bronchiolitis, or remodeling secondary to chronic inflammation etc.	• Inspiratory and expiratory CT scan if not already performed • Consider viral and autoimmune studies • Use minimum treatment which maintains lung function
2. Continued airflow obstruction, no inflammation, but with reversibility to β_2-agonists	Presumed steroid resistant, non-inflammatory bronchial reactivity	• Continuous subcutaneous terbutaline treatment • High dose eformoterol by inhalation
3. Persistent eosinophilic inflammation, with either or both of airflow obstruction and symptoms	Presumed steroid partial or complete resistance	• Look for causes of secondary steroid resistance • Treat with either prolonged high dose steroids or steroid sparing agent
4. Persistent eosinophilic inflammation, with no airflow obstruction or symptoms	?Lagging of clearance of inflammation ?Risk of ongoing remodelling despite no symptoms	• Observe closely with repeated spirometry and non-invasive measures of inflammation
5. Presumed inflammation completely resolved with steroids (normal lung function, no symptoms)	Steroid sensitive asthma, but requiring high dose treatment	• Look for causes of secondary steroid resistance • Taper steroids to level at which symptoms are controlled without side-effects • Steroid sparing agent (often less effective in this phenotype)
6. Persistent non-eosinophilic inflammation	Presumed other inflammatory mechanisms (other cells e.g. neutrophilic inflammation; neurogenic mechanisms)	• Reduce steroid treatment to minimum level needed to control eosinophilic inflammation • Consider macrolide therapy, 5-lipoxygenase inhibitor, or theophylline if neutrophilic inflammation
7. Apparently normal lung function, no inflammation, but ongoing symptoms	Poor symptom perception Psychological problems Not asthma at all	• Exercise test with Borg scale • Review by Psychologist

? = Possible.

may remit, return or worsen in adolescence, or present for the first time; epidemiological data are conflicting. Diagnosis of a new presenting asthma may be difficult, with atypical presentations such as chest pain being not uncommon. Management may be very difficult. Risk-taking behaviors by adolescents may be considered almost normal, but smoking and substance abuse do not help asthma control. Compliance is almost always an issue; adolescents often do not want to be different from their peers, and will not take medications for this reason. They may worry about adverse effects of therapy; or probably most usually, be too idle and disorganised to be bothered. Control is frequently poor. Management is very difficult, with many theoretically good ideas being trial, which are often very time-consuming, and with very little objective evidence of benefit. Specialist adolescent clinics are advocated, in which teenagers are seen on their own, at least initially without their parents, in an age-appropriate environment. It is believed that a non-directive, non-threatening approach is optimal, with negotiation over what treatment is reasonable. Others like this author merely hope to weather the storm until adulthood is reached, whereupon the young person is referred to the calm waters of the adult clinic!

THE "NEW" COPD: SURVIVORS OF PREMATURE BIRTH

The skills of neonatal intensive care mean that ever more premature and smaller babies are surviving to go home, and many can be presumed will survive into adult life. The nature of CLD (bronchopulmonary dysplasia, BPD) is also changing, with probably a greater component of alveolar hypoplasia in more recent, very preterm survivors who have received very different ventilatory strategies [287, 288]. Follow-up studies have recently been reviewed in detail [215]. In summary, survivors of CLD are likely to have combinations of airflow obstruction and alveolar-capillary hypoplasia. There is a tendency for improvement of the symptoms and physiological abnormalities in later childhood. The issue that needs attention is what will happen during lung aging, and what follows is surmise, because yet we have no data. By analogy with the episodic (viral) wheezers, who also have early airflow obstruction, an accelerated rate of decline is to be anticipated, and these children are likely to present to adult clinics perhaps as young as in their thirties. The first to arrive will be the "old" BPD patients (relatively big babies, ventilated at slow rates with high pressures, who did not receive surfactant in the newborn period); next, or perhaps simultaneously because they may be younger, "new" BPD survivors, who may have less airflow obstruction and more alveolar-capillary hypoplasia. They will present likely as "COPD" but the significance of their dramatic early life events will not be appreciated unless a good history is taken. These may well provide novel challenges in the COPD clinic of the future.

References

1. British Thoracic Society. Scottish Intercollegiate Guidelines Network. British guideline on the management of asthma. *Thorax* 63(Suppl 4): iv1–iv121, 2008.
2. Greening AP, Ind PW, Northfield M, Shaw G. Added salmeterol versus higher-dose corticosteroid in asthma patients with symptoms on existing inhaled corticosteroid. *Lancet* 344: 219–24, 1994.
3. Pauwels RA, Löfdahl CG, Postma DS, Tattersfield AE, O'Byrne P, Barnes PJ, Ullman A. Effect of inhaled formoterol and budesonide on exacerbations of asthma. Formoterol and Corticosteroids Establishing Therapy (FACET) International Study Group. *N Engl J Med* 337: 1405–11, 1997.
4. Walters EH, Gibson PG, Lasserson TJ, Walters JA. Long-acting beta2-agonists for chronic asthma in adults and children where background therapy contains varied or no inhaled corticosteroid. *Cochrane Database Syst Rev* 1: 2007, CD001385.
5. Meijer GG, Postma DS, Mulder PGH, van Aalderen WM. Long-term circadian effects of salmeterol in asthmatic children treated with inhaled corticosteroids. *Am J Respir Crit Care Med* 152: 1887–92, 1995.
6. Verberne AAPH, Frost C, Roorda RJ, van der Laag H, Kerrebijn KF. One year treatment with salmeterol compared with beclomethasone in children with asthma. *Am J Respir Crit Care Med* 156: 688–95, 1997.
7. Bisgaard H. Long-acting beta(2)-agonists in management of childhood asthma: A critical review of the literature. *Pediatr Pulmonol* 29: 221–34, 2000.
8. Bush A. Evidence-based medicines for children: Important implications for new therapies at all ages. *Eur Respir J* 28: 1069–72, 2006.
9. Silverman M (ed.), *Childhood Asthma and Other Wheezing Disorders*, 2nd edn. London: Publ Arnold, 2002.
10. Couriel J. Asthma in adolescence. *Paediatr Respir Rev* 4: 47–54, 2003.
11. Kaditis AG, Winnie G, Syrogiannopoulos GA. Anti-inflammatory pharmacotherapy for wheezing in preschool children. *Pediatr Pulmonol* 42: 407–20, 2007.
12. Warner JO, Götz M, Landau LI, Levison H, Milner AD, Pedersen S, Silverman M. Management of asthma: a consensus statement. *Arch Dis Child* 64: 1065–79, 1989.
13. Warner JO, Naspitz CK. Third International Pediatric Consensus statement on the management of childhood asthma. International Pediatric Asthma Consensus Group. *Pediatr Pulmonol* 25: 1–17, 1998.
14. Li A, Lex C, Zacharasiewiez A, Wong E, Erin E, Hansel T, Wilson N, Bush A. Cough frequency in children with stable asthma: Correlation with lung function, exhaled nitric oxide, and sputum eosinophil count. *Thorax* 58: 974–78, 2003.
15. Silverman M, Wilson NM. Wheezing phenotypes in childhood. *Thorax* 52: 936–37, 1997.
16. Payne D, Saglani S, Suri R, Hall P, Wilson N, Bush A. Asthma: Beyond the guidelines. *Curr Paediatr* 14: 336–46, 2004.
17. Payne D, Saglani S, Bush A. Are there different phenotypes of childhood asthma? *Clin Pulm Med* 11: 287–97, 2004.
18. Rosi E, Ronchi MC, Grazzini M, Duranti R, Scano G. Sputum analysis, bronchial hyperresponsiveness, and airway function in asthma. Results of a factor analysis. *J Allergy Clin Immunol* 103: 232–37, 1999.
19. Leung TF, Wong GW, Ko FW, Lam CW, Fok TF. Clinical and atopic parameters and airway inflammatory markers in childhood asthma: A factor analysis. *Thorax* 60: 822–26, 2005.
20. Azzawi M, Bradley B, Jeffery PK, Frew AJ, Wardlaw AJ, Knowles G, Assoufi B, Collins JV, Durham S, Kay AB. Identification of activated T lymphocytes and eosinophils in bronchial biopsies in stable atopic asthma. *Am Rev Respir Dis* 142: 1407–13, 1990.
21. Bentley AM, Maestrelli P, Saetta M, Fabbri LM, Robinson DS, Bradley BL, Jeffery PK, Durham SR, Kay AB. Activated T-lymphocytes and eosinophils in the bronchial mucosa in isocyanate-induced asthma. *J Allergy Clin Immunol* 89: 821–29, 1992.
22. Gibson PG, Norzila MZ, Fakes K, Simpson J, Henry RL. Pattern of airway inflammation and its determinants in children with acute severe asthma. *Pediatr Pulmonol* 28: 261–70, 1999.
23. Pavord ID, Brightling CE, Woltmann G, Wardlaw AJ. Non-eosinophilic corticosteroid unresponsive asthma. *Lancet* 353: 2213–14, 1999.
24. Green RH, Brightling CE, Woltmann G, Parker D, Wardlaw AJ, Pavord ID. Analysis of induced sputum in adults with asthma: Identification of subgroup with isolated sputum neutrophilia and poor response to inhaled corticosteroids. *Thorax* 57: 875–79, 2002.
25. Simpson JL, Scott R, Boyle MJ, Gibson PG. Inflammatory subtypes in asthma: Assessment and identification using induced sputum. *Respirology* 11: 54–61, 2006.
26. Fleming L, Wilson N, Regamey N, Bush A. Are inflammatory phenotypes in children with severe asthma stable? *Eur Respir J* 30(Suppl 51): 483S, 2007.
27. Lex C, Ferreira F, Zacharasiewicz A, Nicholson AG, Haslam PL, Wilson NM, Hansel TT, Payne DN, Bush A. Airway eosinophilia in children with severe asthma: Predictive values of non-invasive tests. *Am J Respir Crit Care Med* 174: 1286–91, 2006.
28. Kraft M, Djukanovic R, Wilson S, Holgate ST, Martin RJ. Alveolar tissue inflammation in asthma. *Am J Respir Crit Care Med* 154: 1505–10, 1996.
29. Sutherland ER, Martin RJ, Bowler RP, Zhang Y, Rex MD, Kraft M. Physiologic correlates of distal lung inflammation in asthma. *J Allergy Clin Immunol* 113: 1046–50, 2004.
30. Kraft M, Martin RJ, Wilson S, Djukanovic R, Holgate ST. Lymphocyte and eosinophil influx into alveolar tissue in nocturnal asthma. *Am J Respir Crit Care Med* 159: 228–34, 1999.
31. Whitehead B, Scott JP, Helms P, Malone M, Macrae D, Higenbottam TW, Smyth RL, Wallwork J, Elliott M, de Leval M. Technique and use of transbronchial biopsy in children and adolescents. *Pediatr Pulmonol* 12: 240–46, 1992.
32. Joos GF, O'Connor B, Anderson SD, Chung F, Cockcroft DW, Dahlén B, DiMaria G, Foresi A, Hargreave FE, Holgate ST, Inman M, Lötvall J, Magnussen H, Polosa R, Postma DS, Riedler J. ERS task force. Indirect airway challenges. *Eur Respir J* 21: 1050–68, 2003.
33. Wilson NM, James A, Uasuf C, Payne DNR, Hablas H, Agrofioti C, Bush A. Asthma severity and inflammation markers in children. *Pediatr Allergy Immunol* 12: 125–32, 2001.
34. Crimi E, Spanevello A, Neri M, Ind PW, Rossi GA, Brusasco V. Dissociation between airway inflammation and airway hyperresponsiveness in allergic asthma. *Am J Respir Crit Care Med* 157: 4–9, 1998.
35. Djukanović R, Wilson SJ, Kraft M, Jarjour NN, Steel M, Chung KF, Bao W, Fowler-Taylor A, Matthews J, Busse WW, Holgate ST, Fahy JV. Effects of treatment with anti-immunoglobulin E antibody omalizumab on airway inflammation in allergic asthma. *Am J Respir Crit Care Med* 170: 583–93, 2004.
36. Berry MA, Hargadon B, Shelley M, Parker D, Shaw DE, Green RH, Bradding P, Brightling CE, Wardlaw AJ, Pavord ID. Evidence of a role of tumor necrosis factor alpha in refractory asthma. *N Engl J Med* 354: 697–708, 2006.
37. Ranganathan SC, Stocks J, Dezateux C, Bush A, Wade A, Carr S, Castle R, Dinwiddie R, Hoo AF, Lum S, Price J, Stroobant J, Wallis C. The evolution of airway function in early childhood following clinical diagnosis of cystic fibrosis. *Am J Respir Crit Care Med* 169: 928–33, 2004.

38. Beydon N, Davis SD, Lombardi E *et al.* An official American Thoracic Society/European Respiratory Society statement: Pulmonary function testing in preschool children. *Am J Respir Crit Care Med* 175: 1304–45, 2007.
39. Clarke JR, Salmon B, Silverman M. Bronchial responsiveness in the neonatal period as a risk factor for wheezing in infancy. *Am J Respir Crit Care Med* 151: 1434–40, 1995.
40. Wilson NM, Lamprill JR, Mak JC, Clarke JR, Bush A, Silverman M. Symptoms, lung function, and beta2-adrenoceptor polymorphisms in a birth cohort followed for 10 years. *Pediatr Pulmonol* 38: 75–81, 2004.
41. O'Donnell AR, Toelle BG, Marks GB, Hayden CM, Laing IA, Peat JK, Goldblatt J, Le Souëf PN. Age-specific relationship between CD14 and atopy in a cohort assessed from age 8 to 25 years. *Am J Respir Crit Care Med* 169: 615–22, 2004.
42. Guerra S, Lohman IC, Halonen M, Martinez FD, Wright AL. Reduced interferon production and soluble CD14 levels in early life predict recurrent wheezing by 1 year of age. *Am J Respir Crit Care Med* 169: 70–76, 2004.
43. Kedda M, Lose F, Duffy D, Bell E, Thompson PJ, Upham J. The CD14 C-159T polymorphism is not associated with asthma or asthma severity in an Australian study population. *Thorax* 60: 211–14, 2005.
44. Eder W, Klimecki W, Yu L *et al.* Opposite effects of CD 14/-260 on serum IgE levels in children raised in different environments. *J Allergy Clin Immunol* 116: 601–7, 2005.
45. Flamant C, Henrion-Caude A, Boelle PY *et al.* Glutathione-S-transferase M1, M3, P1 and T1 polymorphisms and severity of lung disease in children with cystic fibrosis. *Pharmacogenetics* 14: 295–301, 2004.
46. Gilliland FD, Li YF, Dubeau L *et al.* Effects of glutathione S-transferase M1, maternal smoking during pregnancy, and environmental tobacco smoke on asthma and wheezing in children. *Am J Respir Crit Care Med* 166: 457–63, 2002.
47. Hong YC, Lee KH, Son BK, Ha EH, Moon HS, Ha M. Effects of the GSTM1 and GSTT1 polymorphisms on the relationship between maternal exposure to environmental tobacco smoke and neonatal birth weight. *J Occup Environ Med* 45: 492–98, 2003.
48. Sasaki S, Kondo T, Sata F *et al.* Maternal smoking during pregnancy and genetic polymorphisms in the Ah receptor, CYP1A1 and GSTM1 affect infant birth size in Japanese subjects. *Mol Hum Reprod* 12: 77–83, 2006.
49. Bush A. COPD: A pediatric disease. *J COPD* 5: 53–67, 2008.
50. Groenman F, Unger S, Post M. The molecular basis for abnormal human lung development. *Biol Neonate* 87: 164–77, 2005.
51. Roth-Kleiner M, Post M. Similarities and dissimilarities of branching and septation during lung development. *Pediatr Pulmonol* 40: 113–34, 2005.
52. Hislop A. Developmental biology of the pulmonary circulation. *Paediatr Respir Rev* 6: 35–43, 2005.
53. Silverman M, Kuehni CE. Early lung development and COPD. *Lancet* 370: 717–19, 2007.
54. Tager IB, Ngo L, Hanrahan JP. Maternal smoking during pregnancy: Effects on lung function during the first 18 months of life. *Am J Respir Crit Care Med* 152: 977–83, 1995.
55. Martinez FD, Wright AL, Taussig LM *et al.* Asthma and wheezing in the first six years of life. *N Engl J Med* 132: 133–38, 1995.
56. Lodrup-Carlsen KC, Jaakkola JJ, Nafstad P, Carlsen KH. In utero exposure to cigarette smoking influences lung function at birth. *Eur Respir J* 10: 1774–79, 1997.
57. Stick SM, Burton PR, Gurrin L, Sly PD, LeSouef PN. Effects of maternal smoking during pregnancy and a family history of asthma on respiratory function in newborn infants. *Lancet* 348: 1060–64, 1996.
58. Young S, LeSouef PN, Geelhoed GC *et al.* The influence of a family history of asthma and parental smoking on airway responsiveness in early infancy. *N Engl J Med* 324: 1166–73, 1991.
59. Dezateux C, Stocks J. Lung development and early origins of childhood respiratory illness. *Br Med Bull* 53: 40–57, 1997.
60. Elliot J, Carroll N, Bosco M, McCrohan M, Robinson P. Increased airway responsiveness and decreased alveolar attachment points following in utero smoke exposure in the guinea pig. *Am J Respir Crit Care Med* 163: 140–44, 2001.
61. Elliot J, Vullermin P, Carroll N, James A, Robinson P. Increased airway smooth muscle in sudden infant death syndrome. *Am J Respir Crit Care Med* 160: 313–16, 1999.
62. Elliot J, Vullermin P, Robinson P. Maternal cigarette smoking is associated with increased inner airway wall thickness in children who die from sudden infant death syndrome. *Am J Respir Crit Care Med* 158: 802–6, 1998.
63. Sekhon HS, Keller JA, Proskocil BJ, Martin EL, Spindel ER. Maternal nicotine exposure upregulates collagen gene expression in fetal monkey lung. Association with alpha7 nicotinic acetylcholine receptors. *Am J Respir Cell Mol Biol* 26: 31–41, 2002.
64. Macaubas C, de Klerk NH, Holt BJ *et al.* Association between antenatal cytokine production and the development of atopy and asthma at age 6 years. *Lancet* 362: 1192–97, 2003.
65. Devereux G, Barker RN, Seaton A. Antenatal determinants of neonatal immune response to allergens. *Clin Exp Allergy* 32: 43–50, 2002.
66. Noakes PS, Holt PG, Prescott SL. Maternal smoking in pregnancy alters neonatal cytokine responses. *Allergy* 58: 1053–58, 2003.
67. Noakes PS, Hale J, Thomas R, Lane C, Devadason SG, Prescott SL. Maternal smoking is associated with impaired neonatal toll-like-receptor-mediated immune responses. *Eur Respir J* 28: 721–29, 2006.
68. Young S, Arnott J, Le Souef PN, Landau LI. Flow limitation during tidal expiration in symptom-free infants and the subsequent development of asthma. *J Pediatr* 124: 681–88, 1994.
69. Martinez FD, Morgan WJ, Wright AL *et al.* Diminished lung function as a predisposing factor for wheezing respiratory illness in infants. *N Engl J Med* 319: 1112–17, 1988.
70. Morgan WJ, Stern DA, Sherrill DL, Guerra S, Holberg CJ, Guilbert TW, Taussig LM, Wright AL, Martinez FD. Outcome of asthma and wheezing in the first 6 years of life: Follow-up through adolescence. *Am J Respir Crit Care Med* 172: 1253–58, 2005.
71. Lum S, Hoo AF, Dezateux C *et al.* The association between birthweight, sex and lung function in the infants of nonsmoking mothers. *Am Rev Respir Crit Care Med* 164: 2078–84, 2001.
72. Hoo AF, Stocks J, Lum S *et al.* Development of lung function in early life. Influence of birthweight in infants of nonsmokers. *Am Rev Respir Crit Care Med* 170: 527–33, 2004.
73. Edwards CA, Osman LM, Godden DJ, Campbell DM, Douglas JG. Relationship between birth weight and adult lung function: Controlling for maternal factors. *Thorax* 58: 1061–65, 2003.
74. Chan KN, Noble-Jamieson CM, Elliman A, Bryan EM, Silverman M. Lung function in children of low birth weight. *Arch Dis Child* 64: 1284–93, 1989.
75. Rona RJ, Gulliford MC, Chinn S. Effects of prematurity and intrauterine growth on respiratory health and lung function in childhood. *BMJ* 306: 817–20, 1993.
76. Gouveia N, Bremner SA, Novaes HM. Association between ambient air pollution and birth weight in São Paulo, Brazil. *J Epidemiol Community Health* 58: 11–17, 2004.
77. Ritz B, Wilhelm M, Hoggatt KJ, Ghosh JK. Ambient air pollution and preterm birth in the environment and pregnancy outcomes study at the University of California, Los Angeles. *Am J Epidemiol* 166: 1045–52, 2007.
78. Dejmek J, Selevan SG, Benes I, Solanský I, Srám RJ. Fetal growth and maternal exposure to particulate matter during pregnancy. *Environ Health Perspect* 107: 475–80, 1999.
79. Sunyer J, Torrent M, Muñoz-Ortiz L, Ribas-Fitó N, Carrizo D, Grimalt J, Antó JM, Cullinan P. Prenatal dichlorodiphenyldichloroethylene (DDE) and asthma in children. *Environ Health Perspect* 113: 1787–90, 2005.
80. Yuksel B, Greenough A, Naik S, Cheeseman P, Nicolaides KH. Perinatal lung function and invasive antenatal procedures. *Thorax* 52: 81–84, 1997.

81. Rusconi F, Galassi C, Forastiere F *et al.* Maternal complications and procedures in pregnancy and at birth and wheezing phenotypes in children. *Am J Respir Crit Care Med* 175: 16–21, 2007.
82. Bush A, Annesi-Maesano I. Beam me up, Scotty. *Am J Respir Crit Care Med* 175: 1–2, 2007.
83. Shaheen SO, Newson RB, Henderson AJ *et al.* Prenatal paracetamol exposure and risk of asthma and elevated immunoglobulin E in childhood. *Clin Exp Allergy* 35: 18–25, 2005.
84. Martindale S, McNeill G, Devereux G, Campbell D, Russell G, Seaton A. Antioxidant intake in pregnancy in relation to wheeze and eczema in the first two years of life. *Am J Respir Crit Care Med* 171: 121–28, 2005.
85. Willers SM, Devereux G, Craig LC, McNeill G, Wijga AH, Abou El-Magd W, Turner SW, Helms PJ, Seaton A. Maternal food consumption during pregnancy and asthma, respiratory and atopic symptoms in 5-year-old children. *Thorax* 62: 773–79, 2007.
86. Stern DA, Morgan WJ, Wright AL, Guerra S, Martinez FD. Poor airway function in early infancy and lung function by age 22 years: A non-selective longitudinal cohort study. *Lancet* 370: 758–64, 2007.
87. Turner SW, Palmer LJ, Rye PJ, Gibson NA, Judge PK, Cox M, Young S, Goldblatt J, Landau LI, Le Souef PN. The relationship between infant airway function, childhood airway responsiveness, and asthma. *Am J Respir Crit Care Med* 169: 921–27, 2004.
88. Woolcock AJ, Leeder SR, Peat JK, Blackburn CR. The influence of lower respiratory illness in infancy and childhood and subsequent cigarette smoking on lung function in Sydney schoolchildren. *Am Rev Respir Dis* 120: 5–14, 1979.
89. Turner SW, Palmer LJ, Rye PJ, Gibson NA, Judge PK, Young S, Landau LI, Le Souef PN. Infants with flow limitation at 4 weeks: outcome at 6 and 11 years. *Am J Respir Crit Care Med* 165: 1294–98, 2002.
90. Lowe LA, Simpson A, Woodcock A, Morris J, Murray CS, Custovic A. NAC Manchester Asthma and Allergy Study Group. Wheeze phenotypes and lung function in preschool children. *Am J Respir Crit Care Med* 171: 231–37, 2005.
91. Nielsen KG, Pressler T, Klug B, Koch C, Bisgaard H. Serial lung function and responsiveness in cystic fibrosis during early childhood. *Am J Respir Crit Care Med* 169: 1209–16, 2004.
92. Ranganathan SC, Bush A, Dezateux C, Carr SB, Hoo AF, Lum S, Madge S, Price J, Stroobant J, Wade A, Wallis C, Wyatt H, Stocks J. Relative ability of full and partial forced expiratory maneuvers to identify diminished airway function in infants with cystic fibrosis. *Am J Respir Crit Care Med* 166: 1350–57, 2002.
93. Brussee JE, Smit HA, Koopman LP *et al.* Interrupter resistance and wheezing phenotypes at 4 years of age. *Am J Respir Crit Care Med* 169: 209–13, 2004.
94. Frey U, Makkonen K, Wellman T, Beardsmore C, Silverman M. Alterations in airway wall properties in infants with a history of wheezing disorders. *Am J Respir Crit Care Med* 161: 1825–29, 2000.
95. Haland G, Lodrup-Carlsen KC, Sandvik L *et al.* Reduced lung function at birth and the risk of asthma at 10 years of age. *N Engl J Med* 355: 1682–89, 2006.
96. Caudri D, Wijga A, Gehring U, Smit HA, Brunekreef B *et al.* Respiratory symptoms in the first 7 years of life and birth weight at term: The PIAMA birth cohort. *Am J Respir Crit Care Med* 175: 1078–85, 2007.
97. Sherriff A, Peters TJ, Henderson J, Strachan D. Risk factors associations with wheezing patterns in children followed longitudinally from birth to 3(1/2) years. *Int J Epidemiol* 30: 1473–84, 2001.
98. Gold DR, Burge HA, Carey V, Milton DK, Platts-Mills T, Weiss ST. Predictors of repeated wheeze in the first year of life: the relative roles of cockroach, birth weight, acute lower respiratory illness, and maternal smoking. *Am J Respir Crit Care Med* 160: 227–36, 1999.
99. Raby BA, Celedon JC, Litonjua AA, Phipatanakul W, Sredl D, Oken E *et al.* Low-normal gestational age as a predictor of asthma at 6 years of age. *Pediatrics* 114: e327–32, 2004.
100. Canoy D, Pekkanen J, Elliott P, Pouta A, Laitinen J, Hartikainen A-L *et al.* Early growth and adult respiratory function in men and women followed from the fetal period to adulthood. *Thorax* 62: 396–402, 2007.
101. Palmer LJ, Rye PJ, Gibson NA, Burton PR, Landau LI, LeSouef PN. Airway responsiveness in infancy predicts asthma, lung function, and respiratory symptoms by school age. *Am J Respir Crit Care Med* 163: 37–42, 2001.
102. Grol MH, Gerritsen J, Vonk JM *et al.* Risk factors for growth and decline of lung function in asthmatic individuals up to 42 years. *Am J Respir Crit Care Med* 160: 1830–37, 1999.
103. Grol MH, Postma D, Vonk JM *et al.* Risk factors from childhood to adulthood for bronchial responsiveness aged 32–42 year. *Am J Respir Crit Care Med* 160: 150–56, 1999.
104. Sears MR, Greene JM, Willan AR, Wiecek EM, Taylor DR, Flannery EM, Cowan JO, Herbison GP, Silva PA, Poulton R. A longitudinal, population-based, cohort study of childhood asthma followed to adulthood. *N Engl J Med* 349: 1414–22, 2003.
105. The Childhood Asthma Management Program Research Group. CAMP study. Long-term effects of budesonide or nedocromil in children with asthma. *N Engl J Med* 343: 1054–63, 2000.
106. Oswald H, Phelan PD, Lanigan A *et al.* Childhood asthma and lung function in mid-adult life. *Pediatr Pulmonol* 23: 14–20, 1997.
107. Fletcher C, Peto R. The natural history of chronic airflow obstruction. *BMJ* 1: 1645–48, 1977.
108. Edwards CA, Osman LM, Godden DJ, Douglas JG. Wheezy bronchitis in childhood: A distinct clinical entity with lifelong significance. *Chest* 124: 18–24, 2003.
109. Marossy AE, Strachan DP, Rudnicka AR, Anderson HR. Childhood chest illness and the rate of decline of adult lung function between ages 35 and 45 Years. *Am J Respir Crit Care Med* 175: 355–59, 2007.
110. Paoletti P, Prediletto R, Carozzi G, Viegi G, Di Pede F, Carmignani G *et al.* Effects of childhood and adolescence-adulthood respiratory infections in a general population. *Eur Respir J* 2: 428–36, 1989.
111. Johnston IDA, Strachan DP, Anderson HR. Effect of pneumonia and whooping cough in childhood on adult lung function. *N Engl J Med* 338: 581–87, 1998.
112. Barker DJP (Ed.), *Fetal and Neonatal Origins of Adult Disease*. London: BMJ, 1992.
113. Cook DG, Strachan DP. Health effects of passive smoking-10: Summary of effects of parental smoking on the respiratory health of children and implications for research. *Thorax* 54: 357–66, 1999.
114. David GL, Koh WP, Lee HP, Yu MC, London SJ. Childhood exposure to environmental tobacco smoke and chronic respiratory symptoms in non-smoking adults: the Singapore Chinese Health Study. *Thorax* 60: 1052–58, 2005.
115. Kabesch M, Hoefler C, Carr D, Leupold W, Weiland SK, von Mutius E. Glutathione S Transferase defiency and maternal smoking increase childhood asthma. *Thorax* 59: 59–73, 2004.
116. Wenten M, Berhane K, Rappaport EB, Avol E, Tsai WW, Gauderman WJ, McConnell R, Dubeau L, Gilliland FD. TNF-308 modifies the effect of second-hand smoke on respiratory illness-related school absences. *Am J Respir Crit Care Med* 172: 1563–68, 2005.
117. Strachan DP. Hay fever, hygiene, and household size. *BMJ* 299: 1259–60, 1989.
118. Ball TM, Castro-Rodriguez JA, Griffith KA, Holberg CJ, Martinez FD, Wright AL. Siblings, day-care attendance, and the risk of asthma and wheezing during childhood. *N Engl J Med* 343: 538–43, 2000.
119. Celedon JC, Wright RJ, Litonjua AA, Sredl D, Ryan L, Weiss ST, Gold DR. Day care attendance in early life, maternal history of asthma, and asthma at the age of 6 years. *Am J Respir Crit Care Med* 167: 1239–43, 2003.
120. Turner SW, Young S, Landau LI, Le Souef PN. Reduced lung function both before bronchiolitis and at 11 years. *Arch Dis Child* 87: 417–20, 2002.
121. Castro-Rodríguez JA, Holberg CJ, Wright AL, Halonen M, Taussig LM, Morgan WJ, Martinez FD. Association of radiologically ascertained pneumonia before age 3 yr with asthma like symptoms and pulmonary

function during childhood: A prospective study. *Am J Respir Crit Care Med* 159: 1891–97, 1999.
122. Castro-Rodriguez JA, Daszenies C, Garcia M, Meyer R, Gonzales R. Adenovirus pneumonia in infants and factors for developing bronchiolitis obliterans: A 5-year follow-up. *Pediatr Pulmonol* 41: 947–53, 2006.
123. Pereira MU, Sly PD, Pitrez PM *et al.* Nonatopic asthma is associated with helminth infections and bronchiolitis in poor children. *Eur Respir J* 29: 1154–60, 2007.
124. Sigurs N, Bjarnason R, Sigurbergsson F, Kjellman B, Bjorksten B. Asthma and immunoglobulin E antibodies after respiratory syncytial virus bronchiolitis: A prospective cohort study with matched controls. *Pediatrics* 95: 500–5, 1995.
125. Sigurs N, Bjarnason R, Sigurbergsson F, Kjellman B. Respiratory syncytial virus bronchiolitis in infancy is an important risk factor for asthma and allergy at age 7. *Am J Respir Crit Care Med* 161: 1501–7, 2000.
126. Sigurs N, Gustafsson PM, Bjarnason R, Lundberg F, Schmidt S, Sigurbergsson F, Kjellman B. Severe respiratory syncytial virus bronchiolitis in infancy and asthma and allergy at age 13. *Am J Respir Crit Care Med* 171: 137–41, 2005.
127. Wilson NM. Virus infections, wheeze and asthma. *Paediatr Respir Rev* 4: 184–92, 2003.
128. Stein RT, Sherrill D, Morgan WJ, Holberg CJ, Halonen M, Taussig LM, Wright AL, Martinez FD. Respiratory syncytial virus in early life and risk of wheeze and allergy by age 13 years. *Lancet* 354: 541–45, 1999.
129. Lemanske R, Jackson DJ, Gangnon RE *et al.* Rhinovirus illnesses during infancy predict subsequent childhood wheezing. *J Allergy Clin Immunol* 116: 571–77, 2005.
130. Tujague J, Bastaki M, Holland N, Balmes JR, Tager IB. Antioxidant intake, GSTM1 polymorphism and pulmonary function in healthy young adults. *Eur Respir J* 27: 282–88, 2006.
131. Chatzi L, Apostolaki G, Bibakis I, Skypala I, Bibaki-Liakou V, Tzanakis N *et al.* Protective effect of fruits, vegetables and the Mediterranean diet on asthma and allergies among children in Crete. *Thorax* 62: 677–83, 2007.
132. Garcia-Marcos L, Canflanca IM, Garrido JB, Varela AL, Garcia-Hernandez G, Grima FG *et al.* Relationship of asthma and rhinoconjunctivitis with obesity, exercise and Mediterranean diet in Spanish schoolchildren. *Thorax* 62: 503–8, 2007.
133. Okoko BJ, Burney PG, Newson RB, Potts JF, Shaheen SO. Childhood asthma and fruit consumption. *Eur Respir J* 29: 1161–68, 2007.
134. Devereux G. Early life events in asthma – diet. *Pediatr Pulmonol* 42: 945–51, 2007.
135. Barragan-Meijueiro MM, Morfin-Maciel B, Nava-Ocampo AA. A Mexican population-based study on exposure to paracetamol and the risk of wheezing, rhinitis, and eczema in childhood. *J Investig Allergol Clin Immunol* 16: 247–52, 2006.
136. Harris JM, Mills P, White C, Moffat S, Newman Taylor AJ, Cullinan P. Recorded infections and antibiotics in early life; associations with allergy in UK children and their parents. *Thorax* 62: 631–7, 2007.
137. Gauderman WJ, Avol E, Gilliland F, Vora H, Thomas D, Berhane K, McConnell R, Kuenzli N, Lurmann F, Rappaport E, Margolis H, Bates D, Peters J. The effect of air pollution on lung development from 10 to 18 years of age. *N Engl J Med* 351: 1057–67, 2004.
138. Gauderman WJ, Vora H, McConnell R *et al.* Effect of exposure to traffic on lung development from 10 to 18 years of age: A cohort study. *Lancet* 369: 571–77, 2007.
139. Kulkarni N, Pierse N, Rushton L, Grigg J. Carbon in airway macrophages and lung function in children. *N Engl J Med* 355: 21–30, 2006.
140. Friedman MS, Powell KE, Hutwagner L, Graham LM, Teague WG. Impact of changes in transportation and commuting behaviors during the 1996 Summer Olympic Games in Atlanta on air quality and childhood asthma. *JAMA* 285: 897–905, 2001.
141. Burr ML, Karani G, Davies B, Holmes BA, Williams KL. Effects on respiratory health of a reduction in air pollution from vehicle exhaust emissions. *Occup Environ Med* 61: 212–18, 2004.
142. Woodcock A, Lowe LA, Murray CS, Simpson BM, Pipis SD, Kissen P, Simpson A, Custovic A. NAC Manchester Asthma and Allergy Study Group. Early life environmental control: Effect on symptoms, sensitization, and lung function at age 3 years. *Am J Respir Crit Care Med* 170: 433–39, 2004.
143. Langley SJ, Goldthorpe S, Craven M, Woodcock A, Custovic A. Relationship between exposure to domestic allergens and bronchial hyperresponsiveness in non-sensitised, atopic asthmatic subjects. *Thorax* 60: 17–21, 2005.
144. Chinn S, Heinrich J, Antó JM, Janson C, Norbäck D, Olivieri M, Svanes C, Sunyer J, Verlato G, Wjst M, Zock JP, Burney PG, Jarvis DL. Bronchial responsiveness in atopic adults increases with exposure to cat allergen. *Am J Respir Crit Care Med* 176: 20–26, 2007.
145. Bush A. Coughs and wheezes spread diseases: But what about the environment? *Thorax* 61: 367–69, 2006.
146. Illi S, von Mutius E, Lau S, Niggemann B, Grüber C, Wahn U. Multicentre Allergy Study (MAS) group. Perennial allergen sensitisation early in life and chronic asthma in children: A birth cohort study. *Lancet* 368: 763–70, 2006.
147. Jedrychowski W, Maugeri U, Jedrychowska-Bianchi I, Flak E. Effect of indoor air quality in the postnatal period on lung function in pre-adolescent children: A retrospective cohort study in Poland. *Public Health* 119: 535–41, 2005.
148. Qian Z, He Q, Kong L, Xu F, Wei F, Chapman RS, Chen W, Edwards RD, Bascom R. Respiratory responses to diverse indoor combustion air pollution sources. *Indoor Air* 17: 135–42, 2007.
149. Pekkanen J, Hyvärinen A, Haverinen-Shaughnessy U, Korppi M, Putus T, Nevalainen A. Moisture damage and childhood asthma: A population-based incident case-control study. *Eur Respir J* 29: 509–15, 2007.
150. McKenzie S. Can I have a letter for the housing, doctor? *Arch Dis Child* 78: 505–6, 1998.
151. Paton J. Can I have a letter for the housing, doctor? *Arch Dis Child* 78: 506–7, 1998.
152. Castro-Rodríguez JA, Holberg CJ, Wright AL, Martinez FD. A clinical index to define risk of asthma in young children with recurrent wheezing. *Am J Respir Crit Care Med* 162: 1403–6, 2000.
153. Guilbert TW, Morgan WJ, Zeiger RS, Bacharier LB, Boehmer SJ, Krawiec M, Larsen G, Lemanske RF, Liu A, Mauger DT, Sorkness C, Szefler SJ, Strunk RC, Taussig LM, Martinez FD. Atopic characteristics of children with recurrent wheezing at high risk for the development of childhood asthma. *J Allergy Clin Immunol* 114: 1282–87, 2004.
154. Devulapalli CS, Carlsen KC, Håland G, Munthe-Kaas MC, Pettersen M, Mowinckel P, Carlsen KH. Severity of obstructive airways disease by age 2 years predicts asthma at 10 years of age. *Thorax* 63: 8–13, 2008.
155. Payne DNR, McKenzie SA, Stacey S, Misra D, Haxby E, Bush A. Safety and ethics of bronchoscopy and endobronchial biopsy in difficult asthma. *Arch Dis Child* 84: 423–26, 2001.
156. Saglani S, Payne DN, Nicholson AG, Scallan M, Haxby E, Bush A. The safety and quality of endobronchial biopsy in children under five years old. *Thorax* 58: 1053–57, 2003.
157. Salva PS, Theroux C, Schwartz D. Safety of endobronchial biopsy in 170 children with chronic respiratory symptoms. *Thorax* 58: 1058–60, 2003.
158. Molina-Teran A, Hilliard TN, Saglani S, Haxby E, Scallan M, Bush A, Davies JC. Safety of endobronchial biopsy in children with cystic fibrosis. *Pediatr Pulmonol* 41: 1021–24, 2006.
159. Bush A, Davies JC. Rebuttal: You are wrong, Dr. Mallory. *Pediatr Pulmonol* 41: 1017–20, 2006.
160. McIntosh N, Bates P, Brykczynska G *et al.* Guidelines for the ethical conduct of medical research involving children. *Arch Dis Child* 82: 177–82, 2000.
161. Payne DN, Rogers AV, Adelroth E, Bandi V, Guntupalli KK, Bush A, Jeffery PK. Early thickening of the reticular basement membrane in children with difficult asthma. *Am J Respir Crit Care Med* 167: 78–82, 2003.
162. Saglani S, Malmstrom K, Pelkonen AS, Malmberg LP, Lindahl H, Kajosaari M, Turpeinen M, Rogers AV, Payne DN, Bush A *et al.*

Airway remodeling and inflammation in symptomatic infants with reversible airflow obstruction. *Am J Respir Crit Care Med* 171: 722–27, 2005.
163. Saglani S, Payne DN, Zhu J, Wang Z, Nicholson AG, Bush A, Jeffery PK. Early detection of airway wall remodelling and eosinophilic inflammation in preschool wheezers. *Am J Respir Crit Care Med* 176: 858–64, 2007.
164. van Eerdewigh P, Little RD, Dupuis J *et al.* Association of the ADAM33 gene with asthma and bronchial hyperresponsiveness. *Nature* 418: 426–30, 2002.
165. Haitchi HM, Powell RM, Shaw TJ *et al.* ADAM33 expression in human lungs and asthmatic airways. *Am J Respir Crit Care Med* 171: 958–65, 2005.
166. Simpson A, Maniatis M, Jury F *et al.* Polymorphisms in a disintegrin and metalloproteinase 33 (ADAM33) predict impaired early lung function. *Am J Respir Crit Care Med* 172: 55–60, 2005.
167. Lee J-Y, Park S-W, Chang HK *et al.* A disintegrin and metalloproteinase 33 in patients with asthma. Relationship to airflow limitation. *Am J Respir Crit Care Med* 173: 729–35, 2006.
168. van Diemen CC, Postma DS, Vonk JM, Bruinenberg M, Schouten JP, Boezen HM. A disintegrin and metalloprotease 33 polymorphisms and lung function decline in the general population. *Am J Respir Crit Care Med* 172: 329–33, 2005.
169. Gosman MM, Boezen HM, van Diemen CC, Snoeck-Stroband JB, Lapperre TS, Hiemstra PS, Ten Hacken NH, Stolk J, Postma DS. A disintegrin and metalloprotease 33 and chronic obstructive pulmonary disease pathophysiology. *Thorax* 62: 242–47, 2007.
170. Zhang G, Hayden CM, Khoo SK *et al.* Association of haplotypes of beta2-adrenoceptor polymorphisms with lung function and airway responsiveness in a pediatric cohort. *Pediatr Pulmonol* 41: 1233–41, 2006.
171. Matheson MC, Ellis JA, Raven J, Johns DP, Walters EH, Abramson MJ. Beta2-adrenergic receptor polymorphisms are associated with asthma and COPD in adults. *J Hum Genet* 51: 943–51, 2006.
172. Hall IP, Blakey JD, Al Balushi KA *et al.* β2-adrenoceptor polymorphisms and asthma from childhood to middle age in the British 1958 birth cohort: A genetic association study. *Lancet* 368: 771–79, 2006.
173. Warburton D, Gauldie J, Bellusci S, Shi W. Lung development and susceptibility to chronic obstructive pulmonary disease. *Proc Am Thorac Soc* 3: 668–72, 2006.
174. Kranenburg AR, Willems-Widyastuti A, Mooi WJ *et al.* Chronic obstructive pulmonary disease is associated with enhanced bronchial expression of FGF-1, FGF-2 and FGFR-1. *J Pathol* 206: 28–38, 2005.
175. Gilliland FD, Gauderman J, Vora H, Rappaport E, Dubeau L. Effects of glutathione-S-transferase M1, P1 and T1 on childhood lung function growth. *Am J Respir Crit Care Med* 166: 710–16, 2002.
176. He JQ, Ruan J, Connett JE, Anthonisen NR, Pare PD, Sandford AJ. Antioxidant gene polymorphisms and susceptibility to a rapid decline in lung function in smokers. *Am J Respir Crit Care Med* 166: 323–28, 2002.
177. Bush A. Update in pediatrics 2005. *Am J Respir Crit Care Med* 173: 585–92, 2006.
178. Bush A. Update in pediatric lung disease 2006. *Am J Respir Crit Care Med* 175: 532–40, 2007.
179. Akhtar PC, Currie DB, Currie CE, Haw SJ. Changes in child exposure to environmental tobacco smoke (CHETS) study after implementation of smoke-free legislation in Scotland: National cross sectional survey. *BMJ* 335: 545, 2007.
180. Hogg JC. Role of latent viral infections in chronic obstructive pulmonary disease and asthma. *Am J Respir Crit Care Med* 164: S71–75, 2001.
181. Schwarze J, O'Donnell DR, Rohwedder A, Openshaw PJ. Latency and persistence of respiratory syncytial virus despite T cell immunity. *Am J Respir Crit Care Med* 169: 801–5, 2004.
182. Wilson J, Rowlands K, Rockett K, Moore C, Lockhart E, Sharland M, Kwiatkowski D, Hull J. Genetic variation at the IL10 gene locus is associated with severity of respiratory syncytial virus bronchiolitis. *J Infect Dis* 191: 1705–9, 2005.
183. Tal G, Mandelberg A, Dalal I, Cesar K, Somekh E, Tal A, Oron A, Itskovich S, Ballin A, Houri S, Beigelman A, Lider O, Rechavi G, Amariglio N. Association between common Toll-like receptor 4 mutations and severe respiratory syncytial virus disease. *Infect Dis* 189: 2057–63, 2004.
184. Guerra S, Wright AL, Morgan WJ, Sherrill DL, Holberg CJ, Martinez FD. Persistence of asthma symptoms during adolescence: Role of obesity and age at the onset of puberty. *Am J Respir Crit Care Med* 170: 78–85, 2004.
185. Hijazi N, Abalkhail B, Seaton A. Diet and childhood asthma in a society in transition: A study in urban and rural Saudi Arabia. *Thorax* 55: 775–79, 2000.
186. Farchi S, Forastiere F, Agabiti N, Corbo G, Pistelli R, Fortes C, Dell'Orco V, Perucci CA. Dietary factors associated with wheezing and allergic rhinitis in children. *Eur Respir J* 22: 772–80, 2003.
187. Cane RS, Raganathan SC, McKenzie SA. What do parents of wheezy children understand by "wheeze"? *Arch Dis Child* 82: 327–32, 2000.
188. Cane RS, McKenzie SA. Parents interpretation of children's respiratory symptoms on video. *Arch Dis Child* 84: 31–34, 2001.
189. Elphick HE, Sherlock P, Foxall G *et al.* Survey of respiratory sounds in infants. *Arch Dis Child* 84: 35–39, 2001.
190. Saglani S, McKenzie SA, Bush A, Payne DN. A video questionnaire identifies upper airway abnormalities in pre-school children with reported wheeze. *Arch Dis Child* 90: 961–64, 2005.
191. Levy ML, Godfrey S, Irving CS, Sheikh A, Hanekom W, Ambulatory Care Nurses. Bush A, Lachman P. Wheeze detection in infants and pre-school children: Recordings versus assessment of physician and parent. *J Asthma* 41: 845–53, 2004.
192. Archer LNJ, Simpson H. Night cough counts and diary card scores in asthma. *Arch Dis Child* 60: 473–74, 1985.
193. Falconer A, Oldman C, Helms P. Poor agreement between reported and recorded nocturnal cough in asthma. *Pediatr Pulmonol* 15: 209–11, 1993.
194. Munyard P, Busst C, Logan-Sinclair R, Bush A. A new device for ambulatory cough recording. *Pediatr Pulmonol* 18: 178–86, 1994.
195. Munyard P, Bush A. How much coughing is normal? *Arch Dis Child* 74: 531–34, 1996.
196. Chang AB, Newman R, Phelan PD, Robertson CF. 24-hour continuous ambulatory cough meter: A new use for an old Holter monitor. *Am J Respir Crit Care Med* 153: A501, 1996.
197. de Benedictis FM, Bush A. Hypothesis paper: Rhinosinusitis and asthma-epiphenomenon or causal association? *Chest* 115: 550–56, 1999.
198. Bush A, Chodhari R, Collins N, et al. Primary ciliary dyskinesia: current state of the art. *Arch. Dis. Child* 92: 1136 – 40, 2007.
199. Puterman M, Gorodischer R, Lieberman A. Tracheobronchial foreign bodies: the impact of a postgraduate educational program on diagnosis, morbidity and treatment. *Pediatrics* 70: 96–98, 1982.
200. Chang AB, Gaffney JT, Eastburn MM, Faoagali J, Cox NC, Masters IB. Cough quality in children: a comparison of subjective vs. bronchoscopic findings. *Respir Res* 6: 3, 2005.
201. Marchant JM, Masters IB, Taylor SM, Cox NC, Seymour GJ, Chang AB. Evaluation and outcome of young children with chronic cough. *Chest* 129: 1132–41, 2006.
202. Saglani S, Nicholson A, Scallan M, Balfour-Lynn I, Rosenthal M, Payne DN, Bush A. Investigation of young children with severe recurrent wheeze. Any clinical benefit? *Eur Respir J* 27: 29–35, 2006.
203. Kelly YJ, Brabin BJ, Milligan PJM, Reid JA, Heaf D, Pearson MG. Clinical significance of cough and wheeze in the diagnosis of asthma. *Arch Dis Child* 75: 489–93, 1996.
204. McKenzie S. Cough-but is it asthma? *Arch Dis Child* 70: 1–3, 1994.
205. Chang AB. Isolated cough-probably not asthma? *Arch Dis Child* 80: 211–13, 1999.
206. Powell CVE, Primhak RA. Stability of respiratory symptoms in unlabelled wheezy illness and nocturnal cough. *Arch Dis Child* 75: 549–54, 1996.

207. Brooke AM, Lambert PC, Burton PR, Clarke C, Luyt DK, Simpson H. Night cough in a population-based sample of children: Characteristics, relation to symptoms and associations with measures of asthma severity. *Eur Respir J* 9: 65–71, 1996.
208. Cloutier MM, Loughlin GM. Chronic cough in children: a manifestation of airway hyperreactivity. *Pediatrics* 67: 6–12, 1981.
209. Marguet C, Jouen-Bodes F, Dean TP, Warner JO. Bronchoalveolar cell profiles in children with asthma, infantile wheeze, chronic cough, or cystic fibrosis. *Am J Respir Crit Care Med* 159: 1533–40, 1999.
210. Forsythe P, McGarvey PA, Heaney LG, MacMahon J, Elborn JS. Neurotrophin levels in BAL fluid from patients with asthma and non-asthmatic cough. *Eur Respir J* 14(Suppl 30): 470s, 1999.
211. Koh YY, Jeong JY, Park Y, Kim CK. Development of wheezing in patients with cough variant asthma during an increase in airway responsiveness. *Eur Respir J* 14: 302–8, 1999.
212. Brooke AM, Lambert PC, Burton PR, Clarke C, Luyt DK, Simpson H. The natural history of respiratory symptoms in preschool children. *Am J Respir Crit Care Med* 152: 1872–78, 1995.
213. Burr ML, Merrett TG, Dunstan FD, Maguire MJ. The development of allergy in high-risk children. *Clin Exp Allergy* 27: 1247–53, 1997.
214. Narang I, Ersu R, Wilson NM, Bush A. Nitric oxide in chronic airway inflammation in children: Diagnostic use and pathophysiological significance. *Thorax* 57: 586–89, 2002.
215. Narang I, Baraldi E, Silverman M, Bush A. Airway function measurements and the long-term follow-up of survivors of preterm birth with and without chronic lung disease. *Pediatr Pulmonol* 41: 497–508, 2006.
216. Henderson FW, Stewart PW, Burchinal MR *et al.* Respiratory allergy and the relationship between early childhood lower respiratory illness and subsequent lung function. *Am Rev Respir Dis* 145: 283–90, 1992.
217. McNamara PS, Ritson P, Selby A, Hart CA, Smyth RL. Bronchoalveolar lavage cellularity in infants with severe respiratory syncytial virus bronchiolitis. *Arch Dis Child* 88: 922–26, 2003.
218. Jones A, Qui JM, Bataki E, Elphick H, Ritson S, Evans GS, Everard ML. Neutrophil survival is prolonged in the airways of healthy infants and infants with RSV bronchiolitis. *Eur Respir J* 20: 651–57, 2002.
219. Neilson K, Yunis E. Demonstration of respiratory syncytial virus in an autopsy series. *Pediatr Pathol* 10: 491–502, 1990.
220. Cade A, Brownlee KG, Conway SP, Haigh D, Short A, Brown J, Dassu D, Mason SA, Phillips A, Eglin R, Graham M, Chetcuti A, Chatrath M, Hudson N, Thomas A, Chetcuti PA. Randomised placebo controlled trial of nebulised corticosteroids in acute respiratory syncytial viral bronchiolitis. *Arch Dis Child* 82: 126–30, 2002.
221. Piedimonte G, Hegele RG, Auais A. Persistent airway inflammation after resolution of respiratory syncytial virus infection in rats. *Pediatr Res* 55: 657–65, 2004.
222. Tortorolo L, Langer A, Polidori G, Vento G, Stampachiacchere B, Aloe L, Piedimonte G. Neurotrophin overexpression in lower airways of infants with respiratory syncytial virus infection. *Am J Respir Crit Care Med* 172: 233–37, 2005.
223. Stevenson EC, Turner G, Heaney LG, Schock BC, Taylor R, Gallagher T, Ennis M, Shields MD. Bronchoalveolar lavage findings suggest two different forms of childhood asthma. *Clin Exp Allergy* 27: 1027–35, 1997.
224. Poder G, Nagy A, Kelemen J, Mezei G. Prognostic data for the second follow-up in childhood wheezy bronchitis. *Acta Paediatr Hung* 32: 43–51, 1992.
225. Sporik R, Holgate ST, Cogswell JJ. Natural history of asthma in childhood — a birth cohort study. *Arch Dis Child* 66: 1050, 1991.
226. Oommen A, Grigg J. Urinary leukotriene E4 in preschool children with acute clinical viral wheeze. *Eur Respir J* 21: 149–54, 2003.
227. Da Dalt L, Callegaro S, Carraro S, Andreola B, Corradi M, Baraldi E. Nasal lavage leukotrienes in infants with RSV bronchiolitis. *Pediatr Allergy Immunol* 18: 100–4, 2007.
228. Robertson CF, Price D, Henry R, Mellis C, Glasgow N, Fitzgerald D, Lee AJ, Turner J, Sant M. Short Course Montelukast for Intermittent Asthma in Children: a Randomised Controlled Trial. *Am J Respir Crit Care Med* 175: 323–29, 2007.
229. Ducharme FM, Lemire C, Noya FJ, Davis GM, Alos N, Leblond H, Savdie C, Collet JP, Rivard G, Platt RW. Randomized Controlled Trial of Intermittent High-Dose Fluticasone Versus Placebo in Young Children with Viral-Induced Asthma. *Am J Respir Crit Care Med* 175: A958, 2007.
230. Ducharme FM, Lemire C, Noya FJ, Davis GM, Alos N, Leblond H, Savdie C, Collet JP, Rivard G, Platt RW. Safety of intermittent high-dose fluticasone vs. placebo in young children with viral-induced Asthma: A multicenter randomized controlled trial. *Am J Respir Crit Care Med* 175: A958, 2007.
231. Oommen A, Lambert P, Grigg J. Efficacy of a short course of patient initiated oral prednisolone for viral wheeze in children aged 1-5 years: Randomised controlled trial. *Lancet* 362: 1433–38, 2003.
232. Guilbert TW, Morgan WJ, Zeiger RS *et al.* Long-term inhaled corticosteroids in preschool children at high risk for asthma. *N Engl J Med* 354: 1985–97, 2006.
233. van Essen-Zandvliet EE, Hughes MD, Waalkens HJ *et al.* Effects of 22 months of treatment with inhaled corticosteroids and/or beta-2-agonists on lung function, airway responsiveness, and symptoms in asthma. The Dutch Chronic Non-specific Lung Disease Study Group. *Am Rev Respir Dis* 146: 547–54, 1992.
234. Waalkens HJ, van Essen-Zandvliet EE, Hughes MD *et al.* Cessation of long-term treatment with inhaled corticosteroid (budesonide) in children with asthma results in deterioration. The Dutch CNSLD Study Group. *Am Rev Respir Dis* 148: 1252–57, 1993.
235. Murray CS, Woodcock A, Langley SJ *et al.* Secondary prevention of asthma by the use of inhaled fluticasone dipropionate in wheezy Infants (IWWIN): Double-blind, randomised controlled study. *Lancet* 368: 754–62, 2006.
236. Bisgaard H, Hermansen MN, Loland L *et al.* Intermittent inhaled corticosteroids in infants with episodic wheezing. *N Engl J Med* 354: 1998–2005, 2006.
237. Warner JO. ETAC Study Group. Early treatment of the atopic child. A double-blinded, randomized, placebo-controlled trial of cetirizine in preventing the onset of asthma in children with atopic dermatitis: 18 months' treatment and 18 months' post-treatment follow-up. *J Allergy Clin Immunol* 108: 929–37, 2001.
238. Cates CJ, Crilly JA, Rowe BH. Holding chambers (spacers) versus nebulisers for beta-agonist treatment of acute asthma. *Cochrane Database Syst Rev* 2: CD000052, 2006.
239. Barbato A, Turato G, Baraldo S, Bazzan E, Calabrese F, Tura M, Zuin R, Beghe B, Maestrelli P, Fabbri LM, Saetta M. Airway inflammation in childhood asthma. *Am J Respir Crit Care Med* 168: 798–803, 2003.
240. Barbato A, Turato G, Baraldo S, Bazzan E, Calabrese F, Panizzolo C, Zanin ME, Zuin R, Maestrelli P, Fabbri LM, Saetta M. Epithelial damage and angiogenesis in the airways of children with asthma. *Am J Respir Crit Care Med* 174: 975–81, 2006.
241. Thoracic Society, British Paediatric Association, Royal College of Physicians of London, et al. Guidelines on the management of asthma. *Thorax* 48 (Suppl 2): S1–24, 1993.
242. Bateman ED, Boushey HA, Bousquet J, Busse WW, Clark TJ, Pauwels RA, Pedersen SE. GOAL Investigators Group. Can guideline-defined asthma control be achieved? The Gaining Optimal Asthma ControL study. *Am J Respir Crit Care Med* 170: 836–44, 2004.
243. O'Byrne PM, Bisgaard H, Godard PP, Pistolesi M, Palmqvist M, Zhu Y, Ekström T, Bateman ED. Budesonide/formoterol combination therapy as both maintenance and reliever medication in asthma. *Am J Respir Crit Care Med* 171: 129–36, 2005.
244. Sorkness CA, Lemanske RF Jr, Mauger DT, Boehmer SJ, Chinchilli VM, Martinez FD, Strunk RC, Szefler SJ, Zeiger RS, Bacharier LB, Bloomberg GR, Covar RA, Guilbert TW, Heldt G, Larsen G, Mellon MH, Morgan WJ, Moss MH, Spahn JD, Taussig LM. Childhood Asthma Research and Education Network of the National Heart, Lung, and Blood Institute. Long-term comparison of 3 controller regimens for mild-moderate persistent childhood asthma: The Pediatric Asthma Controller Trial. *J Allergy Clin Immunol* 119: 64–72, 2007.

245. Agertoft L, Pedersen S. Effects of long-term treatment with an inhaled corticosteroid on growth and pulmonary function in asthmatic children. *Respir Med* 88: 373–81, 1994.
246. Haahtela T, Järvinen M, Kava T, Kiviranta K, Koskinen S, Lehtonen K, Nikander K, Persson T, Reinikainen K, Selroos O *et al.* Comparison of a beta 2-agonist, terbutaline, with an inhaled corticosteroid, budesonide, in newly detected asthma. *N Engl J Med* 325: 388–92, 1991.
247. Boushey HA, Sorkness CA, King TS, Sullivan SD, Fahy JV, Lazarus SC, Chinchilli VM, Craig TJ, Dimango EA, Deykin A, Fagan JK, Fish JE, Ford JG, Kraft M, Lemanske RF Jr, Leone FT, Martin RJ, Mauger EA, Pesola GR, Peters SP, Rollings NJ, Szefler SJ, Wechsler ME, Israel E. National Heart, Lung, and Blood Institute's Asthma Clinical Research Network. Daily versus as-needed corticosteroids for mild persistent asthma. *N Engl J Med* 352: 1519–28, 2005.
248. Covar RA, Spahn JD, Murphy JR, Szefler SJ. Childhood Asthma Management Program Research Group. Progression of asthma measured by lung function in the childhood asthma management program. *Am J Respir Crit Care Med* 170: 234–41, 2004.
249. Cullinan P, MacNeill SJ, Harris JM, Moffat S, White C, Mills P, Newman Taylor AJ. Early allergen exposure, skin prick responses, and atopic wheeze at age 5 in English children: A cohort study. *Thorax* 59: 855–61, 2004.
250. Nimmagadda SR, Szefler SJ, Spahn JD, Surs W, Leung DYM. Allergen exposure decreases glucocorticoid receptor binding affinity and steroid responsiveness in atopic asthmatics. *Am J Respir Crit Care Med* 155: 87–93, 1997.
251. Torrego A, Pujols L, Roca-Ferrer J, Mullol J, Xaubet A, Picado C. Glucocorticoid receptor isoforms alpha and beta in *in vitro* cytokine-induced glucocorticoid insensitivity. *Am J Respir Crit Care Med* 170: 420–28, 2004.
252. Sulakvelidze I, Inman MD, Rerecich T, O'Byrne PM. Increases in airway eosinophils and interleukin-5 with minimal bronchoconstriction during repeated low-dose allergen challenge in atopic asthmatics. *Eur Respir J* 11: 821–27, 1998.
253. Almqvist C, Wickman M, Perfetti L, Berglind N, Renström A, Hedrén M, Larsson K, Hedlin G, Malmberg P. Worsening of asthma in children allergic to cats, after indirect exposure to cat at school. *Am J Respir Crit Care Med* 163: 694–98, 2001.
254. Simpson AB, Glutting J, Yousef E. Food allergy and asthma morbidity in children. *Pediatr Pulmonol* 42: 489–95, 2007.
255. Redline S, Wright EC, Kattan M, Kercsmar C, Weiss K. Short-term compliance with peak flow monitoring: Results from a study of inner city children with asthma. *Pediatr Pulmonol* 21: 203–10, 1996.
256. Verschelden P, Cartier A, L'Archevêque J, Trudeau C, Malo JL. Compliance with and accuracy of daily self-assessment of peak expiratory flows (PEF) in asthmatic subjects over a three month period. *Eur Respir J* 9: 880–85, 1996.
257. Salome CM, Leuppi JD, Freed R, Marks GB. Perception of airway narrowing during reduction of inhaled corticosteroids and asthma exacerbation. *Thorax* 58: 1042–47, 2003.
258. van Gent R, van Essen-Zandvliet LE, Rovers MM, Kimpen JL, de Meer G, van der Ent CK. Poor perception of dyspnoea in children with undiagnosed asthma. *Eur Respir J* 30: 887–91, 2007.
259. Brutsche MH, Brutsche IC, Munawar M, Langley SJ, Masterson CM, Daley-Yates PT, Brown R, Custovic A, Woodcock A. Comparison of pharmacokinetics and systemic effects of inhaled fluticasone propionate in patients with asthma and healthy volunteers: a randomised crossover study. *Lancet* 356: 556–61, 2000.
260. Paton J, Jardine E, McNeill E, Beaton S, Galloway P, Young D, Donaldson M. Adrenal responses to low dose synthetic ACTH (Synacthen) in children receiving high dose inhaled fluticasone. *Arch Dis Child* 91: 808–13, 2006.
261. Shyam R, Ullmann D, Spencer D. Limited repeatability of the low dose synacthen test in children with asthma maintained on inhaled corticosteroids. *Arch Dis Child* 92(Suppl 1): A1, 2006.
262. Drake AJ, Howells RJ, Shield JPH *et al.* Sympptomatic adrenal insufficiency presenting with hypoglycaemia in children with asthma receiving high dose inhaled fluticasone propionate. *BMJ* 324: 1081–83, 2002.
263. Todd GRD, Acerini CL, Ross-Russell R, Zahra S, Warner JT, McCance D. Survey of adrenal crises associated with inhaled corticosteroids in the United Kingdom. *Arch Dis Child* 87: 457–61, 2002.
264. Green RH, Brightling CE, McKenna S, Hargadon B, Parker D, Bradding P, Wardlaw AJ, Pavord ID. Asthma exacerbations and sputum eosinophil counts: a randomised controlled trial. *Lancet* 360: 1715–21, 2002.
265. Pijnenburg MW, Bakker EM, Hop WC, De Jongste JC. Titrating steroids on exhaled nitric oxide in children with asthma: A randomized controlled trial. *Am J Respir Crit Care Med* 172: 831–36, 2005.
266. Nuijsink M, Hop WC, Sterk PJ, Duiverman EJ, de Jongste JC. Long-term asthma treatment guided by airway hyperresponsiveness in children: A randomised controlled trial. *Eur Respir J* 30: 457–66, 2007.
267. Roberts G, Hurley C, Bush A, Lack G. Longitudinal study of grass pollen exposure, symptoms, and exhaled nitric oxide in childhood seasonal allergic asthma. *Thorax* 59: 752–56, 2004.
268. Pijnenburg MW, Floor SE, Hop WC, De Jongste JC. Daily ambulatory exhaled nitric oxide measurements in asthma. *Pediatr Allergy Immunol* 17: 189–93, 2006.
269. Alving K, Janson C, Nordvall L. Performance of a new hand-held device for exhaled nitric oxide measurement in adults and children. *Respir Res* 7: 67, 2006.
270. Zacharasiewicz A, Wilson N, Lex C, Erin EM, Li AM, Hansel T, Khan M, Bush A. Clinical use of non-invasive measurements of airway inflammation in steroid reduction in children. *Am J Respir Crit Care Med* 177: 1077–82, 2005.
271. Pijnenburg MW, Hofhuis W, Hop WC, De Jongste JC. Exhaled nitric oxide predicts asthma relapse in children with clinical asthma remission. *Thorax* 60: 215–18, 2005.
272. Pijnenburg MW, Bakker EM, Lever S, Hop WC, De Jongste JC. High fractional concentration of nitric oxide in exhaled air despite steroid treatment in asthmatic children. *Clin Exp Allergy* 35: 920–25, 2005.
273. Fleming L, Wilson N, Bush A. NO the evidence: What have measurements of exhaled nitric oxide got to tell us? *J Pediatr* 149: 156–58, 2006.
274. Shaaban R, Zureik M, Soussan D, Antó JM, Heinrich J, Janson C, Künzli N, Sunyer J, Wjst M, Burney PG, Neukirch F, Leynaert B. Allergic rhinitis and onset of bronchial hyperresponsiveness: A population-based study. *Am J Respir Crit Care Med* 176: 659–66, 2007.
275. Ekins-Daukes S, Simpson CR, Helms PJ, Taylor PJ, McLay JS. Burden of corticosteroids in children with asthma in primary care: Retrospective observational study. *BMJ* 324: 1374, 2002.
276. Daman Willems CE, Dinwiddie R, Grant DB, Rivers RP, Zahir M. Temporary inhibition of growth and adrenal suppression associated with the use of steroid nose drops. *Eur J Pediatr* 153: 632–34, 1994.
277. Liu LY, Coe CL, Swenson CA, Kelly EA, Kita H, Busse WW. School examinations enhance airway inflammation to antigen challenge. *Am J Respir Crit Care Med* 165: 1062–67, 2002.
278. Chen E, Chim LS, Strunk RC, Miller GE. The role of the social environment in children and adolescents with asthma. *Am J Respir Crit Care Med* 176: 644–49, 2007.
279. Kozyrskyj AL, Mai XM, McGrath P, Hayglass KT, Becker AB, Macneil B. Continued exposure to maternal distress in early life is associated with an increased risk of childhood asthma. *Am J Respir Crit Care Med* 177: 142–47, 2008.
280. Doshi DR, Weinberger MM. Long-term outcome of vocal cord dysfunction. *Ann Allergy Asthma Immunol* 96: 794–99, 2006.
281. Noyes BE, Kemp JS. Vocal cord dysfunction in children. *Paediatr Respir Rev* 8: 155–63, 2007.
282. Warner JO. Review of prescribed treatment for children with asthma in 1990. *BMJ* 311: 663–66, 1995.

283. Chung KF, Godard P, Adelroth E, Ayres J, Barnes N, Barnes P *et al.* Difficult/therapy-resistant asthma: The need for an integrated approach to define clinical phenotypes, evaluate risk factors, understand pathophysiology and find novel therapies. ERS Task Force on Difficult/Therapy-Resistant Asthma. European Respiratory Society. *Eur Respir J* 13: 1198–208, 1999.
284. Wenzel SE, Fahy JV, Irvin CG, Peters SP, Spector S, Szefler SJ. Proceedings of the ATS Workshop on Refractory Asthma: Current understanding, recommendations and unanswered questions. *Am J Respir Crit Care Med* 162: 2341–51, 2000.
285. Mickleborough TD, Murray RL, Ionescu AA, Lindley MR. Fish oil supplementation reduces severity of exercise-induced bronchoconstriction in elite athletes. *Am J Respir Crit Care Med* 168: 1181–89, 2003.
286. de Benedictis D, Bush A. The challenge of asthma in adolescence. *Pediatr Pulmonol* 42: 683–92, 2007.
287. Bland RD. Neonatal chronic lung disease in the post-surfactant era. *Biol Neonate* 88: 181–91, 2005.
288. Chess PR, D'Angio CT, Pryhuber GS, Maniscalco WM. Pathogenesis of bronchopulmonary dysplasia. *Semin Perinatol* 30: 171–78, 2006.

66

CHAPTER

Treatment for Stable COPD

Stephen I. Rennard

Pulmonary and Critical Care Medicine, University of Nebraska Medical Center, Omaha, NE, USA

INTRODUCTION

Chronic obstructive pulmonary disease (COPD) is, for most individuals, a relentlessly progressive disorder, a feature which is recognized in the definition of COPD included in the Guidelines prepared by the Global Initiative for COPD (GOLD) as well as the American Thoracic Society and The European Respiratory Society [1, 2][1]. "Stable" COPD, therefore, is an incorrect term, which is generally applied to individuals who are in their usual state of deteriorating health, that is, not experiencing an acute exacerbation. While the current therapies available to treat so-called stable COPD only partially meet therapeutic goals, these therapies can often offer considerable benefit. It is likely that many patients with COPD, however, are undiagnosed and even those with a diagnosis are inappropriately under treated. The current chapter will outline the current therapeutic strategy for the management of COPD patients who are "stable," that is in their usual state of health.

The medications used to treat patients with COPD are, in many cases, identical to those used to treat patients with asthma. The strategy for the use of these medicines, however, differs importantly between COPD and asthma. Effective patient management, therefore, requires that the clinician make an appropriate diagnosis, accurately stage the severity of disease in an individual patient, and define clearly therapeutic goals. The pharmacology relating to individual classes of drugs is provided in detail in accompanying chapters as is information relating to non-pharmacological treatments. The current chapter will describe the overall strategy for the use of these treatments in various stages of COPD.

A number of guidelines are available for COPD management. Global guidelines for the diagnosis and management of COPD resulted from the collaboration of the World Health Organization and the National Heart Lung and Blood Institute, USA, the Global Initiative for Chronic Obstructive Lung Disease (GOLD Guidelines) [2]. A number of professional societies including the American Thoracic Society and the European Respiratory Society [1] as well as a number of national organizations have also prepared guidelines for the diagnosis and the management of COPD. These guidelines are generally consistent with the GOLD guidelines, but often emphasize issues relevant for specific populations or health care systems.

DIAGNOSIS AND STAGING

Airflow limitation is a characteristic feature of both asthma and COPD. In asthma, the airflow limitation is generally reversible either spontaneously or with treatment. In COPD, in contrast, there may be some degree of reversibility, but expiratory airflow limitation, to some extent, is always present (see Chapter 41). Some definitions of COPD have specifically excluded patients with airflow reversibility [3]. This has led to considerable confusion as bronchodilators are first-line therapy for patients with COPD (see below).

In clinical practice, the distinction between COPD and asthma is often a difficult one for several reasons. Some patients with

[1]The full text of the GOLD Panel Report is available at http://www.goldcopd.com.

TABLE 66.1 Therapeutic goals in COPD.

Prevent disease progression
Relieve symptoms
Improve exercise tolerance
Improve health status (quality of life)
Prevent exacerbations
Recognize and treat co-morbidities
Prolong life
Anticipate end of life issues

COPD may have a considerable degree of reversibility. Such individuals have *both* COPD and asthma, and asthma can progress to the development of fixed airflow limitation [4, 5]. Nevertheless, defining whether asthma and COPD are each present is more than semantic; the therapeutic plans for asthma and COPD differ in important respects and each needs to be addressed and maintained properly.

There are multiple therapeutic goals for the stable COPD patient (Table 66.1). As COPD progresses, the relative importance of the various therapeutic goals change. Appropriate management of the COPD patient, therefore, requires an accurate assessment of disease stage and, as is increasingly recognized, thorough diagnosis of comorbidities. Current staging of COPD depends on quantitative assessment of expiratory airflow to grade disease severity. Since airflow limitation can result from several distinct physiologic processes (see Chapters 5 and 34), this parameter represents an integration of several distinct processes. In addition, many of the features of COPD are very weakly related to the FEV_1 (see below). An accurate staging of the COPD patient using FEV_1, however, can serve as an initial guide to therapeutic intervention and can help the clinician determine appropriate therapeutic goals.

The GOLD staging system

The GOLD Guidelines represent an attempt to develop generally applicable guidelines for COPD and, of necessity, balance a number of factors [2]. The objectives of GOLD include increasing the awareness of COPD among health care professionals, health care authorities, and the general public, stimulating research and improving the diagnosis, management, and prevention of COPD.

GOLD staging

Both the diagnosis and staging of severity in the GOLD system are based on spirometric assessment of FEV_1.

Stage 1. Mild: $FEV_1/FVC < 0.7$, $FEV_1 > 80\%$ predicted. While population differences are well recognized, expiratory airflow is generally considered to be normal if greater than 80% predicted. An individual starting at 100%, therefore, can lose up to one-fifth of their expiratory airflow and remain within the normal range. Early in the development of COPD, however, lung volumes frequently increase due to loss of lung elastic recoil [6]. As a result, the vital capacity is relatively well preserved. This makes the ratio of FEV_1/FVC, termed the Tiffeneau Index, a more sensitive measure of early COPD. Stage 1, therefore, recognizes the earliest physiologic abnormalities as a reduction in the FEV_1/FVC ratio, while the FEV_1 remains within the normal range (that is greater than 80% predicted).

Because the FEV_1 declines with age more rapidly than the FVC, however, the use of the fixed ratio will result in an increased diagnosis of individuals with COPD with increasing age [7, 8]. This potential for overdiagnosis of mild COPD that would occur in the elderly was not felt to be a practical problem for the clinician, particularly as therapy is largely driven by symptoms.

Stage 2. Moderate: $FEV_1/FVC < 0.7$, 50% predicted $< FEV_1 < 80\%$ predicted. This stage includes patients with reductions in expiratory airflow beyond the normal range but generally excludes those likely to have respiratory failure.

Stage 3. Severe: $FEV_1/FVC < 0.7$, 30% predicted $< FEV_1 < 50\%$ predicted. Severe COPD is intended to indicate patients who are likely to require a high degree of support, may have increasingly frequent exacerbations. Severe patients are appropriate for increasingly aggressive therapeutic interventions. The care of these patients is a major determinant of overall health care expenditures for COPD.

Stage 4. Very severe: $FEV_1/FVC < 0.7$, $FEV_1 < 30\%$ predicted, or $FEV_1 < 50\%$ predicted with concurrent respiratory failure. These individuals are very likely to need ICU management and are candidates for aggressive treatment including surgical options, if appropriate.

The GOLD Guidelines emphasize early diagnosis and early physiologic staging. While many patients in the early stages of COPD may not present to the physician with complaints, often these individuals may be experiencing alterations in their lifestyle suggesting that therapeutic interventions designed to improve performance and reduce symptoms might be appropriate. This approach is supported by data from the National Health and Nutrition Examination Survey (NHANES) Study suggesting that nearly two-thirds of adults in the United States with airflow limitation have never been diagnosed [9]. More than half of these individuals, moreover, report some symptoms suggesting a significant burden results from undiagnosed disease.

Current guidelines do not recommend screening of asymptomatic individuals for COPD [2, 10]. However, many COPD patients will be "asymptomatic" despite severe physiologic limitation [11]. A careful and aggressive history, therefore, is needed in order to determine which patients should be further evaluated with spirometry to establish a diagnosis and to stage severity by FEV_1.

Limitations in current COPD staging

It is widely recognized that many features of COPD correlate very poorly with the FEV_1 [12, 13]. Several investigators have suggested multidimensional staging systems for COPD which could independently stage features in addition to FEV_1 [14, 15]. The BODE index, which combines a dyspnea measure, body mass index, and walking distance with FEV_1 is a better predictor of survival than is FEV_1 alone, and has been found to be an effective measure of outcome for clinical interventions [16–18]. One analysis using a dataset from a clinical trial [14] suggests that as many as six dimensions may be relatively independent clinical features in COPD (Table 66.2). Several large trials with less homogeneous groups of COPD patients are being conducted with the intent of characterizing the heterogeneity of the disease [19, 20]. Thus the clinician must recognize that the GOLD COPD staging, while useful as a guide to therapy (Fig. 66.1), is not a complete clinical description. GOLD staging, therefore, must be supplemented with careful clinical assessment. The following therapeutic strategies are based on this approach.

TABLE 66.2 Potential independent dimensions for assessment of COPD patients.

Lung function (FEV_1)
Cough and sputum
Dyspnea
Health status
Bronchodilator reversibility
Body mass index

Comorbidities

It has become clear that COPD is a multisystem disorder (see Chapters 44 and 45). Often disease outside the lung is a major clinical problem. Appropriate clinical suspicion is required for appropriate diagnosis of these conditions (Table 66.3).

TABLE 66.3 System conditions associated with COPD.

Cardiovascular disease
Arrythmias
Abdominal aortic aneurysm
Osteoporosis
Skin wrinkling
Skeletal muscle weakness
Anemia
Increased coagulability/thrombosis
Osteoporosis
Fluid retention
Depression
Cachexia

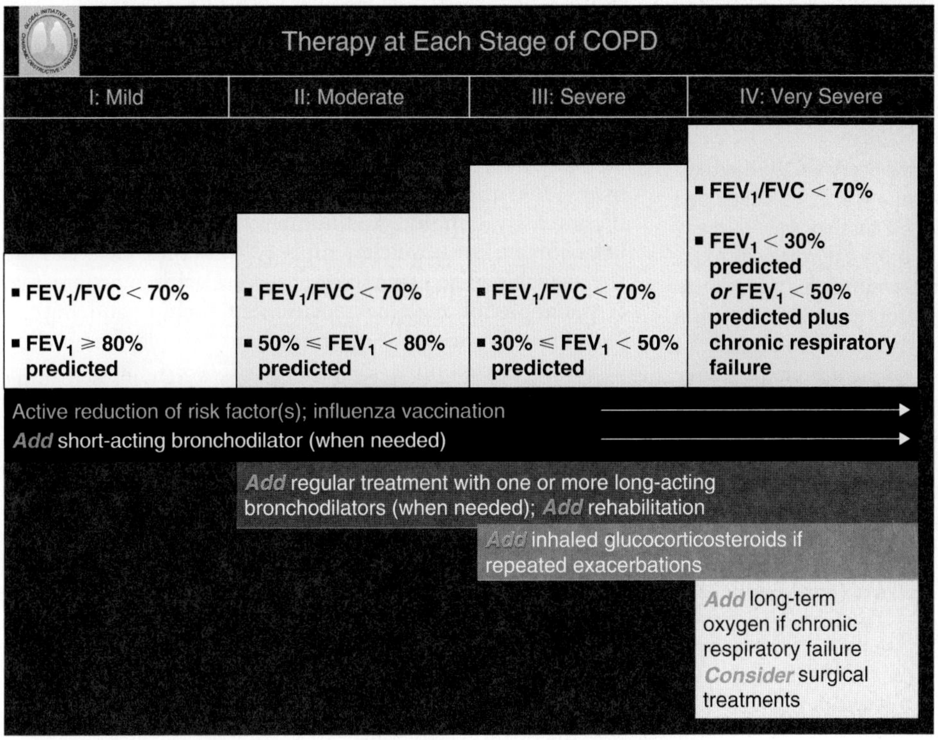

FIG. 66.1 COPD treatment suggestions based on the GOLD staging system. Reproduced from the GOLD Committee Report (www.goldcopd.com). Individual patient management should be guided by clinical response. See text for details.

THERAPEUTIC GOALS IN COPD

Prevent disease progression

The risk factors which contribute to the development of COPD are discussed in detail in Chapter 34. Exogenous risk factors for the development of COPD include cigarette smoking, air pollution, viral and bacterial infections, nutritional deficiencies, and diseases that can affect lung development. Airways hyperreactivity is also a risk factor, which may depend on an interaction of exposures and genetic factors. It is likely that a number of genetic factors will influence the development of COPD (see Chapter 4). The best characterized of these is severe deficiency of α-1 protease inhibitor.

Cigarette smoking

The major means to prevent disease progression in COPD is to eliminate relevant risk factors. Far and away the most important risk factor is cigarette smoking [21]. Addressing cigarette smoking, therefore, is the most important measure to prevent disease progression. The problem of smoking is best addressed within the context of a comprehensive program designed to prevent smoking initiation and to encourage and facilitate smoking cessation [22, 23] (see Chapter 47). Other approaches, for example harm reduction strategies, may play a role but cannot at present be advocated due to lack of supporting data [24].

While social- and community-based approaches to the problem of smoking are essential, it is equally important for the clinician to recognize and appropriately treat the medical aspects of cigarette smoking. The vast majority of smokers are addicted, and nicotine is the major addicting component in cigarette smoke. Smoking, therefore, is most correctly regarded as a chronic disease, which is characterized by frequent remissions (quit attempts that succeed for varying lengths of time) and relapses. In this context, COPD can be regarded as one of the many secondary consequences of the primary disease: smoking.

Smoking should be addressed with every COPD patient on a regular basis. Nonsmoking patients who have developed COPD must be counseled not to begin as they have already demonstrated their unusual susceptibility. Former smokers should be regularly interviewed and counseled in order to anticipate and prevent relapse. Relapse, for example, is common at times of stress and is often associated with concurrent use of alcohol [25]. Finally, COPD patients who continue to smoke should be counseled and encouraged to make quit attempts which can then be properly supported.

A defeatist attitude toward smoking cessation in COPD patients is clearly unwarranted. Smoking cessation can be achieved in COPD patients as documented in several studies [26, 27]. When COPD patients quit, moreover, it is likely that benefits ensue. In one study, symptoms of cough and sputum production were greatly reduced among smokers who quit [28]. The frequently reported complaint that smoking cessation is associated with an increase in cough, therefore, does not appear to be supported among the majority of smokers. The Lung Health Study [26], moreover, in a large randomized prospective trial, clearly demonstrated that when smokers with mild COPD quit, the accelerated rate of loss of lung function, which characterizes these individuals normalizes (Fig. 66.2). In the first year after quitting, lung function improves slightly among quitters. After a year of stability, lung function decline resumes, but the rate of decline resembles much more closely that of nonsmokers rather than that of smokers with mild COPD. This study, therefore, clearly demonstrates that progressive lung function loss can be slowed in COPD by smoking cessation. Whether such benefits would accrue in more severe disease where the pathophysiologic processes may be different remains to be determined. Nevertheless, smoking cessation must be considered a therapeutic goal of prime importance for all patients with COPD.

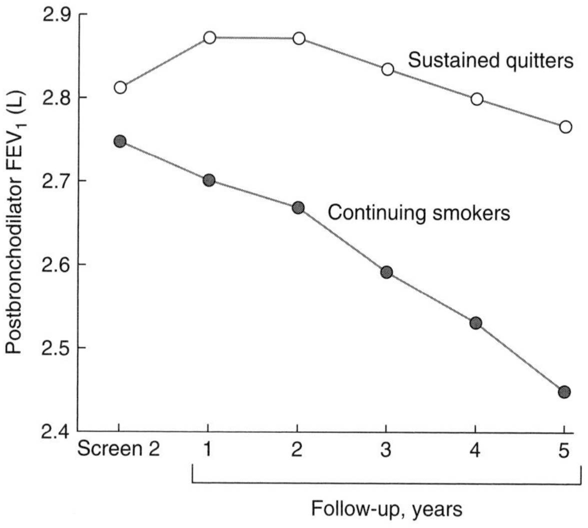

FIG. 66.2 Benefits of smoking cessation. Individuals with mild COPD who quit smoking initially improve lung function then decline at a rate similar to that estimated for nonsmokers. In contrast, continuing smokers decline at an accelerated rate. Reproduced from Ref. [26] with permission.

Several therapeutic strategies are effective in helping smokers quit. These are presented in detail in Chapter 47. Behavioral interventions which can be as brief as a few minutes of personalized practical advice can increase quit rates several fold above the spontaneous quit rate [22]. Increasing the intensity, duration, and number of sessions can increase behaviorally based quit attempts [29]. Several medications are available that can increase success with quitting: five formulations of nicotine replacement therapy and bupropion increase success of quitting about twofold. Varenicline increases success more than threefold. (see Chapter 47 for details in the use of these medications). Current guidelines recommend that each smoker be given combination treatment with non-pharmacological support and pharmacotherapy to assure that each quit attempt has the highest possible likelihood of success [23].

Pharmacologic interventions in continuing smokers

Current concepts suggest that smoke-induced inflammation leads to tissue damage and structural alterations causing airflow limitation in smokers. Agents that modify these pathophysiologic processes are plausible candidates for reducing lung function loss. Unfortunately, no therapeutic intervention based on these concepts has yet been found to alter lung function loss in COPD sufficiently to warrant a clinical recommendation that it be used for this purpose.

Several studies have assessed the effect of inhaled glucocorticoids lung function loss based on the rationale that, as anti-inflammatory agents, they may mitigate inflammation-induced lung damage [30–32]. None of the studies was able to show a statistically significant effect on lung function loss. Meta-analyses have suggested a small effect, although whether it was statistically significant depended on the type of analysis [33, 34]. The TORCH trial, which was a 3-year study of more than 6000 subjects designed to assess the effect of intervention on COPD mortality, also assessed serial FEV_1 [35]. This large trial was able to demonstrate a statistically significant effect on the rate of lung function decline of inhaled fluticasone (42 ml/year), inhaled salmeterol (42 ml/year), and of the combination (39 ml/year) compared to placebo (55 ml/year) [36]. Whether these improvements in the rate of lung function loss are clinically important is not established. However, if sustained for a number of years, this difference in rate would represent a substantial clinical benefit.

The Lung Health Study [26], in addition to evaluating smoking cessation, also evaluated the anticholinergic bronchodilator ipratropium based on the concept that airways reactivity is a risk factor for the development of COPD. While ipratropium had a bronchodilator effect and, therefore, acutely improved airflow, it was completely without effect in altering the progressive loss of lung function. The dose of ipratropium used, however, may have been suboptimal. A 1-year trial with the long-acting anticholinergic bronchodilator tiotropium suggested that lung function might be preserved by this drug [37]. This has led to the UPLIFT trial, a very large 4-year trial evaluating the effect of tiotropium on lung function loss [38]. At the present time, this study has yet to report its results.

Pollution

Indoor and outdoor air pollution as well as exposures to dust and fumes in the workplace are recognized risk factors for the development of COPD [39–42]. It is reasonable to counsel COPD patients regarding these risks, although specific therapeutic interventions to mitigate the risks associated with air pollution have not been assessed. In particular, episodes of worsening pollution, particularly those associated with particulates, are associated with increased exacerbation rates and increased mortality [43], especially the risk of death from cardiac events [44]. Cautioning COPD patients about acute exposures to traffic pollution would seem reasonable.

Infections

Infections in childhood have been suggested to contribute to COPD pathogenesis [45, 46]. To date, no specific antiviral strategies have been assessed to alter COPD natural history. Colonization of the airways with bacteria and acute infections with viruses in adulthood, however, are believed to contribute both to acute exacerbations and, possibly, to progressive lung function loss [47]. It is reasonable, therefore, to advocate vaccination for influenza and for pneumococcus. Whether these will affect progressive lung function loss, however, is not known.

Exacerbations of COPD (see Chapter 67) have been associated with a modest increase in the rate of lung function decline. Both long-acting bronchodilators and inhaled glucocorticoids can reduce exacerbation rates. Whether this accounts for the modest beneficial effect on lung function loss noted above remains to be determined.

Alpha-1 protease inhibitor

Severe congenital deficiency of alpha-1 protease inhibitor (A1PI, formerly termed alpha-1 antitrypsin) is a major risk factor for the development of COPD (see Chapter 4). Deficient individuals are at increased risk for developing emphysema at an early age, particularly if they smoke. Deficient individuals can also develop airways reactivity and chronic bronchitis [48]. Purified A1PI is available for intravenous infusion. Definitive, controlled, randomized trials with A1PI have not yet been done although two are currently in progress. Two registry studies, however, suggest that alpha-1 antitrypsin replacement may slow lung function loss [49, 50]. One smaller prospective-controlled trial also showed a trend in favor of replacement based on CT scan [51]. Replacement is now recommended by the Canadian Thoracic Society [52]. As only a subset of alpha-1 deficient patients will develop COPD, current recommendations are that therapy be offered only to individuals with evidence of compromise of lung function. Therapy is not recommended for continuing smokers as cigarette smoke can inactivate A1PI. The currently available products are purified from pooled human plasma. The product is relatively expensive and, while it is tested for HIV and hepatitis, has the potential for transmission of disease. Therapy requires intravenous infusions generally given at weekly intervals. Interestingly, replacement therapy may also be associated with a reduction in acute exacerbations in deficient individuals [53].

Comorbidities

Awareness of the comorbidities that may be associated with COPD (Table 66.3) is essential to assure that appropriate preventive therapies are initiated (see Chapters 44 and 45). In particular, interventions can successfully reduce the risk of acute cardiac events, osteoporotic fractures, venous thrombosis, and regular exercise can help prevent loss of muscle strength.

RELIEF OF SYMPTOMS

Dyspnea

The primary strategy to improve symptomatic dyspnea in COPD patients is based on the concept that symptoms are secondary to compromised lung function. Improvement in lung function, therefore, is the proximate goal, with improvement in symptoms the desired consequence. To this end, bronchodilators represent the first-line symptom-based pharmacologic intervention in COPD. The connection between bronchodilation and dyspnea relief, however, is likely indirect. There are several factors which contribute to dyspnea [54, 55] (see Chapter 5). It is likely, however, that in COPD the major determinant of dyspnea is increased inspiratory work [56, 57]. This work depends in part on airway caliber and bronchodilators, by improving airflow might have some benefit. A more important determinant of inspiratory work, however, is likely dynamic hyperinflation which develops in many COPD patients with increasing respiratory rate [58, 59]. This mechanism likely accounts for the marked increase in dyspnea on exertion experienced by COPD patients who may be relatively asymptomatic at rest. This also likely accounts for why many COPD patients have greatly restricted activity levels, that is inactivity has been adopted to reduce the risk of dyspnea.

Pulmonary rehabilitation, which does not improve lung function, likely improves both exercise tolerance and dyspnea, in part, because it permits COPD patients to exercise with less tachypnea and, therefore, with less dynamic hyperinflation (see Chapter 5).

Bronchodilators also may reduce the dynamic hyperinflation [60] and thus contribute to reduced dyspnea by a mechanism not directly reflected by the measurement of airflow, for example FEV_1, at rest [61]. The strategic use of bronchodilators in COPD differs fundamentally from the strategy with which they are used in asthma. In asthma, the basic strategy is to prevent episodes of bronchoconstriction with the use of anti-inflammatory agents. Bronchodilators are used only when this approach fails. Dyspneic COPD patients in contrast, will always have significantly impaired lung function. Bronchodilators, even if they are of limited effectiveness, by virtue of improving lung function, have the potential for reducing dyspnea. For this reason, bronchodilators are first-line therapy in the treatment of symptomatic dyspnea in COPD patients, and they are most often used on a regular basis. Paradoxically, therefore, bronchodilators are of higher priority in COPD than in asthma even though COPD patients will, in general, have a more modest response to bronchodilators. It is likely that the optimal use of bronchodilators is when they are combined with a rehabilitation program [62] (see below).

Bronchodilators: Choice of drug and formulation

Several classes of bronchodilators are available and, within each class, there are several agents and formulations available. The pharmacology of these agents is discussed in detail in Chapters 48–50.

The magnitude of the bronchodilator response achieved in most COPD patients is relatively modest and is generally in the range of 100–300 ml improvement in the FEV_1. This response is similar to the improvement in airflow achieved in normal individuals given bronchodilators and may represent inhibition of "normal" airway tone. The modest improvement achieved, however, is relatively more important the more severe the airflow limitation. That is, a 300 ml improvement for an individual with a 4 liter FEV_1 represents a 7.5% improvement. This would generally not be regarded as a positive response. The same improvement, however, for an individual with a 1 liter FEV_1 would represent a 30% improvement and would be a very gratifying clinical response. A number of definitions have been used in order to characterize responders to bronchodilators among COPD patients. Despite the definition used, however, the number of patients showing a response increases as the disease becomes more severe likely because the modest improvements are more meaningful when they are added to a relatively more severe baseline lung function [63].

Bronchodilator response among most COPD patients is distributed as a single normal distribution [64]. The response of a given patient, moreover, is, to some degree, variable [63]. Most likely this reflects varying airway tone, as the pre-bronchodilator FEV_1 is more variable than the post-bronchodilator value [65]. This means that it is not strictly correct to classify patients as bronchodilator-responsive or nonresponsive. Patients who have a marked bronchodilator response and who normalize their lung function should be considered as likely having asthma. Patients who have a very modest response, however, may still derive considerable clinical benefit from bronchodilators, and bronchodilator therapy should not be denied based on pulmonary function testing in a laboratory setting. This is particularly true as the effect of bronchodilators in reducing hyperinflation is likely of more importance in reducing dyspnea than the direct effect on airflow.

Current guidelines do not recommend any specific bronchodilator as being superior as an initial therapeutic choice. Clinical studies suggest that response to individual drugs can be highly variable. In the set of clinical trials of over 800 COPD patients tested with the β-agonist albuterol and the anticholinergic ipratropium, 25% responded (with at least a 200 ml or 12% improvement in FEV_1) only to albuterol, 25% responded only to ipratropium, 30% responded to both, and 25% responded to neither [66, 67]. The choice of an agent, therefore, should depend on a number of factors including local availability and cost, the ability to provide adequate support for training in the use of devices required and, importantly, on individual patient response both with regard to efficacy and side effects (Table 66.4). The GOLD Guidelines recommend that, when possible, bronchodilators be administered via an inhaled route [2]. This offers an increase in the therapeutic index, thus maximizing benefit compared to side effects. Inhaled medications, however, may not be appropriate for some individuals. Oral agents, however, are generally more convenient and, in some populations, have a higher degree of acceptability and such agents should be considered by the clinician for selected patients.

When utilizing bronchodilators, the most appropriate strategy is for the clinician to accurately diagnose and stage each patient, initiate therapy, and then to gauge response. Spirometric assessment of airflow will provide objective measures to complement clinical assessments of the individual patient. More sophisticated physiologic assessment such as exercise testing or lung volume measurement is not

TABLE 66.4 Considerations for choosing bronchodilators.

- Cost
- Availability
- Individual patient response
  - Benefit
  - Side effects
- Route
  - Inhaled
    Better therapeutic index
    Education required (particularly if multiple medications and devices are used)
  - Oral
    Convenience
    Acceptability
    Compliance
- Duration of action
- Combination of agents in different classes

recommended as a routine. It is appealing to quantify the response of hyperinflation by measuring the inspiratory capacity before and after an exercise challenge, for example, with a 6-min walk test, however, this is not currently standard of care, and the degree to which such a strategy will impact clinical care is not established [68].

For patients in whom the clinical response is suboptimal, several issues should be addressed. First, for individuals using inhalation devices, assurance that the device is being used correctly is essential. Several studies suggest that with both metered dose and dry powder inhalers, patient compliance with proper technique is poor and deteriorates significantly with time [69–72], although dry powder inhalers are generally regarded as easier to use. Continual patient education in the use of these devices, therefore, is required. Wet nebulizers are not believed to have any benefits over hand-held devices for most patients [73, 74]. These devices require maintenance and cleaning and do not offer the convenience of portability. Nevertheless, many patients prefer them [75, 76], and they may be particularly beneficial for subjects with altered mental status, in those whose inspiratory flow rate is too low to permit effective inhalation, or in individuals who, in other ways are unable to use hand-held devices [77]. Some patients may prefer wet nebulizers as they are often better reimbursed by health care providers than other formulations.

If the medication is being taken correctly and clinical response is unsatisfactory, it is recommended that bronchodilators of several classes be combined. Combinations of anticholinergic and short-acting β-agonist bronchodilators can be taken as separate inhalers, or these medications can be administered together in the same device [78–80]. The combination device results in better bronchodilation and a slightly prolonged duration of effect [81]. While administration as a combination decreases the ability to regulate the dose of individual components, combination inhalers have achieved a high degree of patient acceptance likely due to both their increased convenience and the improved efficacy that likely accompanies both improved compliance and dual pharmacological benefits. A variety of anticholinergic/β-agonist combinations have been evaluated, and the combination always appears to offer benefits [82, 83]. Interestingly, when ipratropium was combined with salmeterol, the benefits of a single combined inhalation persisted for the entire 12 h of monitoring, despite the fact that the ipratropium, which is generally believed to have a 4–6-h duration of action was not re-administered [84]. Similarly when tiotropium was administered with once daily formoterol benefits were observed that exceeded the 12-h duration expected of the formoterol [83]. These observations are suggestive of potential synergies between anticholinergic and β-agonist bronchodilators. In addition, both anticholinergics and β-agonists can be combined with theophylline [85, 86]. Finally, inhaled corticosteroids (see below and Chapter 51) can also be combined with bronchodilators with increased clinical benefit.

Most COPD patients will have progressive disease. It is likely, therefore, that as disease worsens, bronchodilator therapy will have to be intensified. This suggests that most COPD patients will, eventually, be treated with combination therapy. It also suggests that empirical clinical trials assessing the therapeutic benefit of bronchodilators should be repeated in an organized fashion on a regular basis as disease progression occurs.

Frequency of use/duration of action

Short-acting bronchodilators can be used on an as-needed basis for episodic dyspnea in COPD patients. This "rescue" use, however, is more appropriate for the treatment of asthmatic patients who experience episodes of severe bronchospasm. Patients with COPD are always airflow limited, and episodic dyspnea is more likely related to episodes of increased exertion. Episodic treatment after the fact with a short-acting bronchodilator, therefore, is less likely to be of benefit than is regular maintenance therapy with bronchodilators [87]. Short-acting bronchodilators, even when taken on a regular basis, result in lung function which will be increasing and decreasing throughout the day. In practice, regular use of short-acting agents was not better than PRN use [88]. For this reason, there may be significant advantages in the use of long-acting bronchodilators in patients with COPD. Not only are these agents more convenient, but they also avoid periods of bronchodilation interspersed with relatively poor airflow (Fig. 66.3). Long-acting bronchodilators are recommended for regular use in the GOLD Guidelines.

Perhaps more importantly, long-acting broncodilators reduce the frequency of acute exacerbations (see below).

Glucocorticoids

Glucocorticoids are not bronchodilators. Nevertheless, they may result in a modest improvement in lung function and can, therefore, be considered for patients who do not have an adequate response to aggressive combined bronchodilator treatment. The recommendation made in former guidelines suggesting a 2-week trial with oral glucocorticoids to determine responsiveness [89] was not supported in a clinical trial [32]. Oral challenge is no longer recommended.

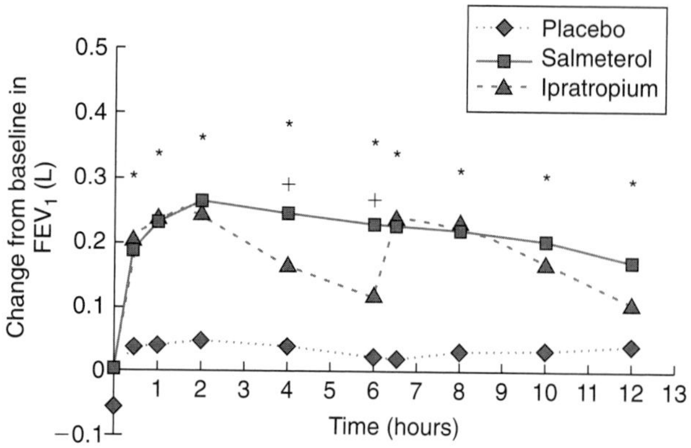

FIG. 66.3 Comparison of bronchodilator response to a short-acting bronchodilator (ipratropium) given every 6 h with a long-acting bronchodilator (salmeterol) given every 12 h. While the maximal bronchodilator effects are similar, the long-acting agent provides more consistent bronchodilation throughout the day. Reproduced from Ref. [66] with permission.

If glucocorticoids are to be given on a trial basis to improve lung function in COPD, inhaled medication is the preferred modality. A clinical trial of 3–6 months may be required and care must be taken to assure adequate compliance before the trial is deemed unsuccessful [32]. The response to glucocorticoids, in terms of FEV_1 improvement, however, is likely to be modest. The large trials assessing the effect of glucocorticoids on COPD natural history noted an improvement compared to placebo of about 50 ml [90, 91]. Caution should be taken if the decision is made to discontinue inhaled glucocorticoid therapy, as discontinuation of the medications may be associated with precipitation of COPD exacerbation [92]. As noted above, inhaled glucocorticoids have a modest effect in reducing the rate at which lung function is lost; the use of inhaled glucocorticoids to reduce the frequency of acute exacerbations is discussed below.

Pulmonary rehabilitation

Pulmonary rehabilitation can also help control symptoms of dyspnea (see Chapter 58). It is likely that rehabilitation can help control symptoms of dyspnea by several mechanisms [93]. First, regular conditioning may decrease the psychological anxiety, which may amplify the subjective perception of the symptoms. In as much as anxiety contributes to tachypnea, this may exacerbate dynamic hyperinflation. Second, improved training may decrease oxygen requirements and, therefore, the increased ventilatory requirements associated with exercise. Finally, it is likely that conditioning allows patients to exercise with a reduced respiratory rate and, therefore, with less dynamic hyperinflation which in turn. Whatever the mechanisms, rehabilitation can contribute greatly to improved symptom control and functionality in COPD patients [93].

In a highly informative study, subjects who received the bronchodilator tiotropium improved exercise tolerance more than those who received placebo [62]. All subjects then received rehabilitation, which was much more effective than tiotropium alone. However, the effect of rehabilitation was greater in subjects treated with tiotropium than in those treated with placebo, suggesting a true synergy between these modalities. This is consistent with the concept that the bronchodilator permitted exercise to greater degrees of intensity, due to reduced dynamic hyperinflation. The exercise at higher levels of intensity, in turn, permitted more effective rehabilitation [62]. Independent of the mechanism, this study supports the approach of optimal bronchodilator therapy combined with rehabilitation to achieve the best clinical results.

While rehabilitation is widely recognized as an important therapeutic intervention for subjects with severe COPD, exercise training is likely important for individuals with milder disease as well [93, 94]. As noted above, many COPD patients deal with symptoms of mild dyspnea on exertion during the early phases of their disease by decreasing their level of activity. This "strategy," however, likely contributes to the severe deconditioning which characterizes COPD patients and likely contributes to their functional compromise. Maintaining a high level of exercise activity in mildly affected patients, therefore, should be an important therapeutic goal. Appropriate use of bronchodilators in order to optimize lung function and thus permit regular sustained activity at a high level can, therefore, be appropriate in patients with relatively mild disease. Long-acting bronchodilators may be particularly important in order to permit a sustained high level of activity throughout the day. As many patients may be "asymptomatic," the clinician should specifically inquire about changes in activity levels, which may be very insidious over time.

Narcotics

As COPD progresses, symptoms of dyspnea can become severe despite maximal functional support. Opiates, likely through a central action, can decrease the subjective perception of dyspnea. These drugs carry significant hazard as they can also significantly depress ventilation and lead to CO_2 retention. They may also suppress cough and may lead to retained secretions. Any subjective effect on dyspnea is not accompanied by an improvement in exercise tolerance [95–99]. The use of opiate narcotics, however, in end-stage COPD may be considered as a potential palliative measure for individuals suffering from severe dyspnea [2]. In this context, the use is analogous to the use of these medications in the control of severe pain for patients with other terminal illnesses recognizing that the predictable adverse effects of the medications may be acceptable in order to relieve suffering. If used for this purpose, systemic administration is generally recommended as administration via inhalation has been demonstrated to have no significant advantages [95, 97, 100].

Surgical interventions in COPD patients

Two surgical options have been suggested for COPD patients: pneumoreductive surgery and lung transplantation (see Chapter 59). Pneumoreductive surgery, by removing areas of relatively nonfunctioning lung in individuals with severe emphysema and hyperinflation can restore the ability of the chest and diaphragm to function as a bellows thus improving function [101–103]. The therapy is most effective in individuals with localized disease, without severe limitation in diffusion capacity and who do not have a beneficial response to maximal medical treatment [104]. In these selected individuals, significant improvement in symptoms and survival have been observed in a controlled randomized trial [104]. This surgery should be distinguished from bullectomy in which isolated nonfunctional bullae are resected. Recent attention to pneumoreductive surgery has re-emphasized the advantages of bullectomy as well, and the availability of CT scanning to define the presence of such lesions has created surgical options for an increasingly large number of COPD patients.

Transplantation can be offered to COPD patients when appropriate. As COPD is, however, generally a disease which progresses with age and as many patients are or have been smokers who have concurrent comorbidities, most patients with COPD are not appropriate candidates for pulmonary transplantation at the present time. For selected individuals such as those with A1PI deficiency, this may represent an important therapeutic option, although whether transplantation improves survival for COPD patients in general, at present is unclear [105, 106].

Cough and sputum production

Cough and sputum production are often major complaints disturbing patients with COPD. As noted above, these symptoms correlate poorly with FEV_1 and may be present in individuals with normal lung function where they can serve as an indicator of disease risk. Management of cough and sputum, however, represent important clinical goals for the COPD patient. Despite the very long history of drugs designed to treat cough and sputum [107], no therapeutic interventions are currently available which have been demonstrated to have sufficient evidence to recommend their routine use [2].

Despite the lack of evidence for their benefit, a number of mucoactive agents in use in various countries are often popular. One large trial evaluated *N*-acetyl cysteine, which has both antioxidant and mucoactive actions for its effect on COPD exacerbations and FEV_1. No clinical benefits were observed [108].

A number of preparations are able to suppress cough likely through a central mechanism [109]. Opiates are particularly potent in this regard. Regular use of these medications, however, is not recommended due to concerns that suppression of cough can lead to retention of secretions and increased risk for infection [2].

Treatment of weakness

COPD patients are frequently bothered by symptoms in addition to cough, sputum, and breathlessness. Weakness is a major feature of COPD. This may, in part, be due to the deconditioning, which characterizes COPD patients. In addition, weakness may result from alterations in skeletal muscle which are secondary to the disease process itself [110, 111]. Weakness in COPD patients, moreover, is a better correlate of exercise performances than is FEV_1 [13, 112]. Therapeutic agents which increase muscle mass are available, but have not yet demonstrated clinical benefit in COPD patients and are not currently recommended [2, 93, 113]. Nevertheless, there are several important considerations for the clinician. First, drugs that could contribute to weakness, particularly systemic glucocorticoids, should be avoided. Second, adequate nutrition should be maintained, particularly at times of intercurrent illness when catabolism may be high and lean muscle mass may be at hazard. Finally, activity levels should be kept high, with or without the benefit of an exercise training program.

Antidepressants

COPD patients are frequently depressed. Whether depression is a consequence of the chronic illness or represents a manifestation of the disease process is undetermined. It is interesting, however, that endogenous depression can increase the risk of an individual to become a smoker and to remain a persistent smoker. Such individuals, of course, would be at risk for developing COPD. As the disease progresses, they may manifest their underlying depression. As depression can sometimes be exacerbated by cigarette smoking cessation, smoking cessation attempts may also unmask underlying depression in these individuals. Close clinical observation should be maintained. If depression develops, an accurate diagnosis and treatment with an appropriate antidepression regimen should be initiated. This may be essential not only in optimizing patient function, but also in preventing smoking relapse.

Depression may have other, indirect effects complicating the management of COPD [114]. Depressed patients are less likely to be compliant with medications, particularly those such as inhalers which require attention to detail to assure proper use. Depressed patients, moreover, are less likely to comply with an exercise program. Mortality is higher among depressed patients [115]. Treatment of depression, therefore, may improve compliance with other aspects of disease management. Conversely, a successfully implemented exercise program can greatly reduce depression [114].

Other management issues

Health status

With the availability of standardized instruments to assess health status (sometimes termed "quality of life"), it has become clear that COPD patients experience symptomtology in a number of quantifiable domains [116, 117]. While groups of COPD patients show a reasonable relationship between FEV_1 and health status, for individual patients, the relationship is very weak [12]. It is reasonable, therefore, for health status to be regarded as a separate therapeutic goal in COPD. Several bronchodilators including ipratropium [64, 66], salmeterol [66, 118], and formoterol [119] have shown improvements in health-related quality of life using these instruments. To date, no drug has regulatory approval for this specific indication, however. Improvement in quality of life, however, should not be regarded as a surrogate for physiologic improvement. Health status measures reflect a variety of inputs, and it is likely that information gained from the use of these measures can help the clinician more effectively develop a strategy to utilize medications. Health status instruments are probably not helpful for clinical management of an individual patient. Recognizing that a number of domains can improve and that they may improve independently, however, can encourage the clinician to continue therapy when spirometric improvements are modest, but a patient relates individual benefit.

Several studies have suggested that therapeutic interventions may slow the rate of decline in health status [32, 35, 93]. Whether this is due to a reduction in exacerbation frequency or severity, an improvement in functional capacity or to another mechanism remains to be determined.

Prevention of exacerbations

As noted above, inhaled glucocorticoids and long-acting bronchodilators reduce exacerbation frequency in COPD [91, 120]. The mechanisms for these effects remain undetermined, but short-acting bronchodilators do not appear to have the same benefits. Although, reduction in exacerbations with therapy with the antioxidant *N*-acetylcysteine was supported by meta-analyses [121, 122], a large prospective randomized trial failed to demonstrate an effect [108]. The immuno-stimulator OM-65 also has been shown to have an effect in reducing the severity of exacerbations [123].

Prevention of exacerbations represents a therapeutic goal distinct from improvement in lung function. This has important clinical implications. If therapy is initiated in order to prevent exacerbations, it will, generally speaking, be impossible for the clinician to gauge whether therapy in an individual patient has been beneficial. That is, patients are likely to experience a small number of exacerbations on an irregular basis. Whether exacerbation frequency is decreased in response to therapy, therefore, may be extraordinarily difficult to gauge in an individual patient. For this reason, when therapies are initiated in COPD patients, clear therapeutic intent should be defined by the clinician. Initiation of bronchodilator therapy to improve symptoms should be reassessed and therapy modified based on functional and clinical (symptomatic) response. Initiation of therapy for exacerbation prevention, in contrast, is likely to be continued life long, barring the onset of adverse side effects.

Prolongation of life

It is frequently stated that oxygen therapy is the only treatment in COPD which has clearly been demonstrated to prolong life. The incorrect implication of this often-quoted (and correct) statement is that other treatments for COPD do not prolong life. Fortunately, several interventions have shown promise in this regard.

Several studies demonstrate the life-prolonging effect of oxygen (see Chapter 54). The first, the MRC trial, compared no oxygen with nocturnal oxygen in hypoxic COPD patients with evidence of right-heart failure [124]. Oxygen supplementation at night time significantly prolonged life. A study conducted under the auspices of the NIH in the United States compared continuous oxygen therapy (actually administered about 19 h/day on average) and found superior survival compared to oxygen administered at night time only [125]. These two studies, which have been supported by several subsequent studies, have led to our current practice of oxygen administration.

Several subsequent studies have also evaluated survival in hypoxic COPD patients [126–128]. These studies, interestingly, show a progressive improvement in the survival of hypoxic COPD patients. While there are a number of potential reasons to explain this historic trend of improving survival including changes in diagnosis (with enrollment of milder patients with a better prognosis), with better concurrent care of non-COPD comorbidities or with more effective utilization of oxygen therapy with current devices, it is also possible, even likely, that some improvement in survival is due to current management of COPD compared to that available 20 years ago [129].

Several studies have utilized mortality as an endpoint for interventions in the management of stable COPD. The National Emphysema Treatment Trial (NETT) (see Chapter 59) demonstrated a survival benefit for selected patients who received volume reduction surgery compared to medical therapy and, conversely, a survival benefit for other patients who received medical therapy [104]. The TORCH trial was designed to prospectively assess survival of a glucocorticoid/long-acting β-agonist combination versus placebo. A 17.5% reduction in the risk of death was observed, although this did not reach the prespecified level of statistical significance ($p = 0.052$) [35]. The UPLIFT trial, which is currently in progress, will assess the effect of tiotropium on mortality. A problem with these large trials is that, over time, there is significant drop-out of the placebo group, many of whom subsequently take the study drug, which is generally available. As a result, these studies have a design bias against demonstrating a benefit of the intervention [91]. In addition, a number of retrospective analyses have suggested improved survival in patients treated with inhaled glucocorticoids [130, 131]. Nevertheless, in the absence of a prospective randomized trial demonstrating a benefit, current guidelines do not indicate a survival advantage for pharmacotherapy [2].

Comorbidities

COPD is associated with a number of comorbidities (Table 66.3, and Chapters 44 and 45). While many of these are common problems that may be associated by chance, there is increased prevalence in COPD. Further, while these conditions share etiologic factors such as smoking, the associations appear to be present even when "adjusted" for smoking status. The association of COPD and cardiac disease, moreover, is present even among nonsmokers [132].

Comorbidities need to be properly diagnosed and treated as this represents reasonable medical care. Interestingly, however, there may be other benefits. Previous recommendations were that β-blockers be avoided in COPD. Perhaps because of the cardiac comorbidity, the use of β1 selective agents has been found to be of significant benefit in COPD patients [133]. Further, retrospective database analyses suggest a reduction in COPD-related complications among patients treated with statins [134–136], and possibly a reduction in the rate of lung function decline [137]. The association of comorbidities raises the interesting possibility of shared mechanisms that may be addressed by specific therapies. The evaluation of comorbidities and the reasons for their association with COPD will be an active area of research in coming years.

End of life issues

COPD is a relentlessly progressive disease. Patients, therefore, will deteriorate and, as the disease progresses, will be increasingly likely to experience episodes of respiratory failure. These episodes may require implementation of invasive and heroic measures such as mechanical ventilation (see Chapter 57).

In communities where such treatments are available, they can frequently be life saving, at least over the short term. Prognosis for COPD patients, however, is limited once respiratory failure has ensued. In one study, patients hospitalized with a PCO_2 of greater than 50 experienced a 2-year mortality of 49%. Thus, while many individuals will survive for extended periods following episodes of respiratory failure many will not (see Chapters 57 and 67). As disease worsens, the concerns of patients and caregivers, which include functional limitations and concerns about oxygen, are somewhat different from those of health care providers [138]. Finally, whether heroic measures should be initiated and to what degree such measures should be extended are issues which should be discussed with COPD patients in advance. Obviously, such decisions are always subject to reconsideration. Nevertheless, careful attention toward preparing advance directives can frequently expedite the delivery of appropriate care as COPD progresses.

References

1. Celli BR, MacNee W. Standards for the diagnosis and treatment of patients with COPD: A summary of the ATS/ERS position paper. *Eur Respir J* 23(6): 932–46, 2004.
2. GOLD Executive and Science Committees, a, G.E., S. Global strategy for the diagnosis, management, and prevention of chronic obstructive pulmonary disease, 2007.
3. Siafakas NM, Vermeire P, Pride NP, Paoletti P, Gibson J, Howard P, Yernault JC, Decramer M, Higenbottam T, Postma DS, Rees J. Optimal assessment and management of chronic obstructive pulmonary disease (COPD). *Eur Respir J* 8: 1398–420, 1995.
4. ten Brinke A. Risk factors associated with irreversible airflow limitation in asthma. *Curr Opin Allergy Clin Immunol* 8(1): 63–69, 2008.
5. Bai TR, Vonk JM, Postma DS, Boezen HM. Severe exacerbations predict excess lung function decline in asthma. *Eur Respir J* 30(3): 452–56, 2007.
6. Coxson HO, Rogers RM, Whittall KP, D'Yachkova Y, Pare PD, Sciurba FC, Hogg JC. A quantification of the lung surface area in emphysema using computed tomography. *Am J Respir Crit Care Med* 159(3): 851–56, 1999.
7. Hardie JA, Buist AS, Vollmer WM, Ellingsen I, Bakke PS, Morkve O. Risk of over-diagnosis of COPD in asymptomatic elderly never-smokers. *Eur Respir J* 20(5): 1117–22, 2002.
8. Hnizdo E, Glindmeyer HW, Petsonk EL, Enright P, Buist AS. Case definitions for chronic obstructive pulmonary disease. *COPD* 3(2): 95–100, 2006.
9. Mannino DM, Gagnon RC, Petty TL, Lydick E. Obstructive lung disease and low lung function in adults in the United States: Data from the National Health and Nutrition Examination Survey, 1988–1994. *Arch Intern Med* 160(11): 1683–89, 2000.
10. Qaseem A, Snow V, Shekelle P, Sherif K, Wilt TJ, Weinberger S, Owens DK. Diagnosis and management of stable chronic obstructive pulmonary disease: A clinical practice guideline from the American College of Physicians. *Ann Intern Med* 147(9): 633–38, 2007.
11. Rennard S, Decramer M, Calverley PM, Pride NB, Soriano JB, Vermeire PA, Vestbo J. Impact of COPD in North America and Europe in 2000: Subjects' perspective of Confronting COPD International Survey. *Eur Respir J* 20(4): 799–805, 2002.
12. Jones PW. Issues concerning health-related quality of life in COPD. *Chest* 107(5 Suppl): 187S–93S, 1995.
13. Schols AM, Mostert R, Soeters PB, Wouters EF. Body composition and exercise performance in patients with chronic obstructive pulmonary disease. *Thorax* 46(10): 695–99, 1991.
14. Celli BR, Calverley PMA, Rennard SI, Wouters MEF. Submitted. Proposal for a new staging system for chronic obstructive pulmonary disease 31: 869–873.
15. Wegner RE, Jorres RA, Kirsten DK, Magnussen H. Factor analysis of exercise capacity, dyspnoea ratings and lung function in patients with severe COPD. *Eur Respir J* 7(4): 725–29, 1994.
16. Celli BR, Cote CG, Marin JM, Casanova C, Montes de Oca M, Mendez RA, Pinto Plata V, Cabral HJ. The body-mass index, airflow obstruction, dyspnea, and exercise capacity index in chronic obstructive pulmonary disease. *N Engl J Med* 350(10): 1005–12, 2004.
17. Cote CG, Celli BR. Pulmonary rehabilitation and the BODE index in COPD. *Eur Respir J* 26(4): 630–36, 2005.
18. Celli BR. Change in the BODE index reflects disease modification in COPD: Lessons from lung volume reduction surgery. *Chest* 129(4): 835–36, 2006.
19. Vestbo J, Anderson W, Coxson HO, Crim C, Dawber F, Edwards L, Hagan G, Knobil K, Lomas DA, Macnee W, Silverman EK, Tal-Singer R. Evaluation of COPD longitudinally to identify predictive surrogate endpoints (ECLIPSE). *Eur Respir J*, 31: 869–73, 2008.
20. SPIROMICS, 2007.
21. Shapiro SD, Snider GL, Rennard SI. Chronic bronchitis and emphysema. In: Mason RJ, Broadus VC, Murray JF, Nadel JA (eds). *Textbook of Respiratory Medicine*, 4th edn, pp. 1115–67. Philadelphia, PA: Elsevier, 2005.
22. Fiore M, Bailey W, Cohen S, Dorfman S, Goldstein M, Gritz E, Heyman R, Jaen C, Kottke T, Lando H, Mecklenburg R, Mullen P, Nett L, Robinson L, Sistzer M, Tommasello A, Villejo L, Wewers M. Treating Tobacco Use and Dependence. Rockville, MD: U.S. Department Of Health and Human Services, 2000.
23. Fiore MC. *Treating Tobacco Use and Dependence*.Department of Health and Human Services, 2008.
24. K. Stratton, P. Shetty, R. Wallace, S. Bondurant. *Clearing the Smoke*. Washington DC: National Academy Press, 2001.
25. Shiffman S, Hickcox M, Paty JA, Gnys M, Kassel JD, Richards TJ. Progression from a smoking lapse to relapse: Prediction from abstinence violation effects, nicotine dependence, and lapse characteristics. *J Consult Clin Psychol* 64(5): 993–1002, 1996.
26. Anthonisen NR, Connett JE, Kiley JP, Altose MD, Bailey WC, Buist AS, Conway WA, Enright PL, Kanner RE, O'Hara P, Owens GR, Scanlon PD, Tashkin DP, Wise RA. Effects of smoking intervention and the use of an inhaled anticholinergic bronchodilator on the rate of decline of FEV1. *JAMA* 272: 1497–505, 1994.
27. Tashkin D, Kanner R, Bailey W, Buist S, Anderson P, Nides M, Gonzales D, Dozier G, Patel KM, Jamerson B. Smoking cessation in patients with chronic obstructive pulmonary disease: A double-blind, placebo-controlled, randomised trial. *Lancet* 357(9268): 1571–75, 2001.
28. Buist AS, Sexton GJ, Nagy JM, Ross BB. The effect of smoking cessation and modification on lung function. *Am Rev Respir Dis* 114: 115–22, 1976.
29. Fiore MC, Bailey WC, Cohen SJ. Smoking cessation. Guideline technical report no. 18. Rockville, MD: U.S. Department of Health and Human Services, Public Health Service, Agency for Health Care Policy and Research. Publication No. AHCPR 97-Noo4, October 1997.

30. Pauwels RA, Lofdahl CG, Laitinen LA, Schouten JP, Postma DS, Pride NB, Ohlsson SV. Long-term treatment with inhaled budesonide in persons with mild chronic obstructive pulmonary disease who continue smoking. *N Engl J Med* 340: 1948–53, 1999.
31. Vestbo J, Sorensen T, Lange P, Brix A, Torre P, Viskum K. Long-term effect of inhaled budesonide in mild and moderate chronic obstructive pulmonary disease: A randomised controlled trial. *Lancet* 353(9167): 1819–23, 1999.
32. Burge PS, Calverley PM, Jones PW, Spencer S, Anderson JA, Maslen TK. Randomised, double blind, placebo controlled study of fluticasone propionate in patients with moderate to severe chronic obstructive pulmonary disease: The ISOLDE trial. *BMJ* 320(7245): 1297–303, 2000.
33. Sutherland ER, Allmers H, Ayas NT, Venn AJ, Martin RJ. Inhaled corticosteroids reduce the progression of airflow limitation in chronic obstructive pulmonary disease: A meta-analysis. *Thorax* 58(11): 937–41, 2003.
34. Highland KB, Strange C, Heffner JE. Long-term effects of inhaled corticosteroids on FEV_1 in patients with chronic obstructive pulmonary disease. A meta-analysis. *Ann Intern Med* 138(12): 969–73, 2003.
35. Calverley PM, Anderson JA, Celli B, Ferguson GT, Jenkins C, Jones PW, Yates JC, Vestbo J. Salmeterol and fluticasone propionate and survival in chronic obstructive pulmonary disease. *N Engl J Med* 356(8): 775–89, 2007.
36. Celli BR, Thomas NE, Anderson JA, Ferguson GT, Jenkins C, Jones PW, Vestbo J, Knobil K, Yates JC, Calverley PMA. Effect of pharmacotherapy on rate of decline of lung function in COPD: Results from the TORCH study [published ahead of print on May 29, 2008]. *Am J Respir Crit Care Med*, doo:10.1164/rccm.200712-1869OC, 2008.
37. Anzueto A, Tashkin D, Menjoge S, Kesten S. One-year analysis of longitudinal changes in spirometry in patients with COPD receiving tiotropium. *Pulm Pharmacol Ther* 18(2): 75–81, 2005.
38. Decramer M, Celli B, Tashkin DP, Pauwels RA, Burkhart D, Cassino C, Kesten S. Clinical trial design considerations in assessing long-term functional impacts of tiotropium in COPD: The UPLIFT trial. *COPD* 1(2): 303–12, 2004.
39. Tashkin DP, Detels R, Simmons M, Liu H, Coulson AH, Sayre J, Rokaw S. The UCLA population studies of chronic obstructive respiratory disease: XI. Impact of air pollution and smoking on annual change in forced expiratory volume in one second. *Am J Respir Crit Care Med* 149(5): 1209–17, 1994.
40. Kauffmann F, Drouet D, Lelouch J. Occupational exposure and 12-year spirometric changes among Paris area workers. *Brit J Industr Med* 39: 221–32, 1982.
41. Samet JM, Marbury MC, Spengler JD. Health effects and sources of indoor air pollution. Part II. *Am Rev Respir Dis* 137(1): 221–42, 1988.
42. Samet JM, Marbury MC, Spengler JD. Health effects and sources of indoor air pollution Part I. *Am Rev Respir Dis* 136(6): 1486–508, 1987.
43. Samet JM, Dominici F, Curriero FC, Coursac I, Zeger SL. Fine particulate air pollution and mortality in 20 U.S. cities, 1987–1994. *N Engl J Med* 343(24): 1742–49, 2000.
44. Peters A, Liu E, Verrier RL, Schwartz J, Gold DR, Mittleman M, Baliff J, Oh JA, Allen G, Monahan K, Dockery DW. Air pollution and incidence of cardiac arrhythmia. *Epidemiology* 11(1): 11–17, 2000, [see comments].
45. Samet J, Tager I. The relationship between respiratory illness in childhood and chronic air-flow obstruction in adulthood. *Am Rev Respir Dis* 127: 508–23, 1983.
46. Hogg JC. Latent adenoviral infection in the pathogenesis of emphysema: The Parker B. Francis Lectureship. *Chest* 117(5 Suppl 1): 282S–285S, 2000.
47. Vestbo J, Prescott E, Lange P. Association of chronic mucus hypersecretion with FEV_1 decline and chronic obstructive pulmonary disease morbidity. *Am J Respir Crit Care Med* 153: 1530–35, 1996.
48. McElvaney NG, Crystal RG. Proteases and lung injury. *The Lung: Scientific Foundations*. Philadelphia, PA: Lippincott-Raven, 1997.
49. Schluchter MD, Stoller JM, Wiedemann HP, Williams GW, Barrett DM, Beck GJ, McCarthy K, Midcalf V, Moore B, Sartori P, Sherer SG, Zhang R. Survival and FEV_1 decline in individuals with severe deficiency of α_1-antitrypsin. *Am J Respir Crit Care Med* 158: 49–59, 1998.
50. Seersholm N, Wencker M, Banik N, Viskum K, Dirksen A, Kok-Jensen A, Konietzko N. Does alpha1-antitrypsin augmentation therapy slow the annual decline in FEV1 in patients with severe hereditary alpha1-antitrypsin deficiency? Wissenschaftliche Arbeitsgemeinschaft zur Therapie von Lungenerkrankungen (WATL) alpha1-AT study group. *Eur Respir J* 10(10): 2260–63, 1997.
51. Dirksen A, Dijkman JH, Madsen F, Stoel B, Hutchison DC, Ulrik CS, Skovgaard LT, Kok-Jensen A, Rudolphus A, Seersholm N, Vrooman HA, Reiber JH, Hansen NC, Heckscher T, Viskum K, Stolk J. A randomized clinical trial of alpha(1)-antitrypsin augmentation therapy. *Am J Respir Crit Care Med* 160(5 Pt 1): 1468–72, 1999.
52. Abboud RT, Ford GT, Chapman KR. Alpha1-antitrypsin deficiency: A position statement of the Canadian Thoracic Society. *Can Respir J* 8(2): 81–88, 2001.
53. Lieberman J. Augmentation therapy reduces frequency of lung infections in antitrypsin deficiency: A new hypothesis with supporting data. *Chest* 118(5): 1480–85, 2000.
54. Mahler DA. Ramirez-Venegas A. 1Dyspnoea. In: *Principles and Practice of Geriatric Medicine* 3rd edn, pp. 655–662. J. Wiley, Chichester: UK.
55. O'Donnell DE. Dyspnea in advanced chronic obstructive pulmonary disease. *J Heart Lung Transplant* 17(6): 544–54, 1998.
56. Mahler DA. Dyspnoea in chronic obstructive pulmonary disease. *Monaldi Arch Chest Dis* 53(6): 669–71, 1998.
57. Gorini M, Misuri G, Corrado A, Duranti R, Iandelli I, De Paola E, Scano G. Breathing pattern and carbon dioxide retention in severe chronic obstructive pulmonary disease. *Thorax* 51: 677–83, 1996.
58. O'Donnell DE, Lam M, Webb KA. Measurement of symptoms, lung hyperinflation, endurance during exercise in chronic obstructive pulmonary disease. *Am J Respir Crit Care Med* 158(5 Pt 1): 1557–65, 1998.
59. O'Donnell DE, Bertley JC, Chau LK, Webb KA. Qualitative aspects of exertional breathlessness in chronic airflow limitation. *Am J Respir Crit Care Med* 1555: 109–15, 1997.
60. O'Donnell DE, Lam M, Webb KA. Spirometric correlates of improvement in exercise performance after anticholinergic therapy in chronic obstructive pulmonary disease. *Am J Respir Crit Care Med* 160(2): 542–49, 1999.
61. O'Donnell DE. Assessment of bronchodilator efficacy in symptomatic COPD: Is spirometry useful?. *Chest* 117(2 Suppl): 42S–47S, 2000.
62. Casaburi R, Kukafka D, Cooper CB, Witek TJ Jr., Kesten S. Improvement in exercise tolerance with the combination of tiotropium and pulmonary rehabilitation in patients with COPD. *Chest* 127(3): 809–17, 2005.
63. Anthonisen NR, Wright E. Bronchodilator response in chronic obstructive pulmonary disease. *Am Rev Respir Dis* 133: 814–19, 1986.
64. Rennard SI, Serby CW, Ghafouri M, Johnson PA, Friedman M. Extended therapy with ipratropium is associated with improved lung function in COPD: A retrospective analysis of data from seven clinical trials. *Chest* 110: 62–70, 1996.
65. Calverley PM, Burge PS, Spencer S, Anderson JA, Jones PW. Bronchodilator reversibility testing in chronic obstructive pulmonary disease. *Thorax* 58(8): 659–64, 2003.
66. Mahler DA, Donohue JF, Barbee RA, Goldman MD, Gross NJ, Wisniewski ME, Yancey SW, Zakes BA, Rickard KA, Anderson WH. Efficacy of salmeterol xinafoate in the treatment of COPD. *Chest* 115(4): 957–65, 1999.
67. Rennard SI, Anderson W, Zu WR, Broughton J, Bailey W, Friedman M, Wisniewski M, Rickard K. Use of a long-acting inhaled beta(2)-adrenergic agonist, salmeterol xinafoate, in patients with chronic obstructive pulmonary disease. *Am J Respir Crit Care Med* 163(5): 1087–92, 2001.
68. Calverley PM. Dynamic hyperinflation: Is it worth measuring?. *Proc Am Thorac Soc* 3(3): 239–44, 2006.
69. van Beerendonk I, Mesters I, Mudde AN, Tan TD. Assessment of the inhalation technique in outpatients with asthma or chronic obstructive pulmonary disease using a metered-dose inhaler or dry powder device. *J Asthma* 35(3): 273–79, 1998.

70. Dompeling E, Van Grunsven PM, Van Schayck CP, Folgering H, Molema J, Van Weel C. Treatment with inhaled steroids in asthma and chronic bronchitis: Long-term compliance and inhaler technique. *Fam Pract* 9(2): 161–66, 1992.
71. Gray SL, Williams DM, Pulliam CC, Sirgo MA, Bishop AL, Donohue JF. Characteristics predicting incorrect metered-dose inhaler technique in older subjects. *Arch Intern Med* 156(9): 984–88, 1996.
72. Tan NC, Ng CJ, Goh S, Lee CE. Assessment of metered dose inhaler technique in family health service patients in Singapore. *Singapore Med J* 40(7): 465–67, 1999.
73. Guthrie SJ, Hill KM, Muers ME. Living with severe COPD. A qualitative exploration of the experience of patients in Leeds. *Respir Med* 95(3): 196–204, 2001.
74. Ikeda A, Nishimura K, Koyama H, Tsukino M, Hajiro T, Mishima M, Izumi T. Comparison of the bronchodilator effects of salbutamol delivered via a metered-dose inhaler with spacer, a dry-powder inhaler, a jet nebulizer in patients with chronic obstructive pulmonary disease. *Respiration* 66(2): 119–23, 1999.
75. Balzano G, Battiloro R, Biraghi M, Stefanelli F, Fuschillo S, Gaudiosi C, De Angelis E. Effectiveness and acceptability of a domiciliary multidrug inhalation treatment in elderly patients with chronic airflow obstruction: Metered dose inhaler versus jet nebulizer. *J Aerosol Med* 13(1): 25–33, 2000.
76. O'Driscoll BR, Bernstein A. A long-term study of symptoms, spirometry and survival amongst home nebulizer users. *Respir Med* 90(9): 561–66, 1996.
77. Tenholder MG, Bryson MJ, Waller RF *et al.* Can MDIs be used effectively by extubated ICU patients? *Am J Med* 77: 834–38, 1992.
78. Group, CIAS. In chronic obstructive pulmonary disease, a combination of ipratropium and albuterol is more effective than either agent alone. *Chest* 105: 1411–19, 1994.
79. Group, CISS. Routine nebulized ipratropium and albuterol together are better than either alone in COPD. *Chest* 112: 1514–21, 1997.
80. Rennard SI. Anticholinergics in combination bronchodilator therapy in COPD. In: Spector SL (ed.), *Anticholinergic Agents in the Upper and Lower Airways*, pp. 119–36. New York: Marcel Dekker, Inc, 1999.
81. Rabe KF, Dent G. Theophylline. In: Barnes PJ, Grunstein MM, Leff AR, Woolcock AJ (eds). *Asthma*, pp. 1535–54. Philadelphia, PA: Lippincott-Raven, 1997.
82. Rennard SI. Anticholinergic bronchodilators. In: *Combination Therapy for Asthma and Chronic Obstructive Pulmonary Disease*. R.J. Martin and M. Kraft eds., New York, Marcel Dekker, 2000; 159–180.
83. van Noord JA, Aumann JL, Janssens E, Smeets JJ, Verhaert J, Disse B, Mueller A, Cornelissen PJ. Comparison of tiotropium once daily, formoterol twice daily and both combined once daily in patients with COPD. *Eur Respir J* 26(2): 214–22, 2005.
84. van Noord JA, de Munck DR, Bantje TA, Hop WC, Akveld ML, Bommer AM. Long-term treatment of chronic obstructive pulmonary disease with salmeterol and the additive effect of ipratropium. *Eur Respir J* 15(5): 878–85, 2000.
85. Roberts AB, Lamb LC, Newton DL, Sporn MB, De Larco JE, Todaro GJ. Transforming growth factors: Isolation of polypeptides from virally and chemically transformed cells by acid/ethanol extraction. *Proc Natl Acad Sci USA* 77: 3494–98, 1980.
86. Nishimura K, Koyama H, Ikeda A, Sugiura N, Kawakatsu K, Izumi T. The additive effect of theophylline on a high-dose combination of inhaled salbutamol and ipratropium bromide in stable COPD. *Chest* 107: 718–23, 1995.
87. Rennard SI, Calverley P. Rescue! Therapy and the paradox of the Barcalounger. *Eur Respir J* 21(6): 916–17, 2003.
88. Cook D, Guyatt G, Wong E, Goldstein R, Bedard M, Austin P, Ramsdale H, Jaeschke R, Sears M. Regular versus as-needed short-acting inhaled beta-agonist therapy for chronic obstructive pulmonary disease. *Am J Respir Crit Care Med* 163(1): 85–90, 2001.
89. Celli BR, Snider GL, Heffner J, Tiep B, Ziment I, Make B, Braman S, Olsen G, Phillips Y. Standards for the diagnosis and care of patients with chronic obstructive pulmonary disease. *Am J Respir Crit Care Med* 152: S77–120, 1995.
90. Rennard SI. Treatment of stable chronic obstructive pulmonary disease. *Lancet* 364(9436): 791–802, 2004.
91. Calverley PM, Rennard SI. What have we learned from large drug treatment trials in COPD?. *Lancet* 370(9589): 774–85, 2007.
92. Jarad NA, Wedzicha JA, Burge PS, Calverley PM. An observational study of inhaled corticosteroid withdrawal in stable chronic obstructive pulmonary disease. ISOLDE Study Group. *Respir Med* 93(3): 161–66, 1999.
93. Ries AL, Bauldoff GS, Carlin BW, Casaburi R, Emery CF, Mahler DA, Make B, Rochester CL, Zuwallack R, Herrerias. C. Pulmonary rehabilitation: Joint ACCP/AACVPR evidence-based clinical practice guidelines. *Chest* 131(5 Suppl): 4S–42S, 2007.
94. Berry MJ, Rejeski WJ, Adair NE, Zaccaro D. Exercise rehabilitation and chronic obstructive pulmonary disease stage. *Am J Respir Crit Care Med* 160(4): 1248–53, 1999.
95. Leung R, Hill P, Burdon J. Effect of inhaled morphine on the development of breathlessness during exercise in patients with chronic lung disease. *Thorax* 51: 596–600, 1996.
96. Light RW, Stansbury DW, Webster JS. Effect of 30 mg of morphine alone or with promethazine or prochlorperazine on the exercise capacity of patients with COPD. *Chest* 109: 975–81, 1996.
97. Masood AR, Reed JW, Thomas SHL. Lack of effect of inhaled morphine on exercise-induced breathlessness in chronic obstructive pulmonary disease. *Thorax* 50: 629–34, 1995.
98. Poole PJ, Veale AG, Black PN. The effect of sustained-release morphine on breathlessness and quality of life in severe chronic obstructive pulmonary disease. *Am J Respir Crit Care Med* 157(6 Pt 1): 1877–80, 1998.
99. Eiser N, Denman WT, West C, Luce P. Oral diamorphine: Lack of effect on dyspnoea and exercise tolerance in the "pink puffer" syndrome. *Eur Respir J* 4(8): 926–31, 1991.
100. Young IH, Daviskas E, Keena VA. Effect of low dose nebulised morphine on exercise endurance in patients with chronic lung disease. *Thorax* 44(5): 387–90, 1989.
101. Cooper JD, Trulock EP, Triantafillou AN, Patterson GA, Pohl MS, Deloney PA, Sundaresan RS, Roper CL. Bilateral pneumectomy (volume reduction) for chronic obstructive pulmonary disease. *J Thorac Cardiovasc Surg* 109: 106–19, 1995.
102. Brenner M, McKenna RJ, Gelb AF, Fischel RJ, Wilson AF. Rate of FEV change following lung volume reduction surgery. *Chest* 113: 652–59, 1998.
103. Lando Y, Boiselle P, Shade D, Travaline JM, Furukawa S, Criner GJ. Effect of lung volume reduction surgery on bony thorax configuration in severe COPD. *Chest* 116(1): 30–39, 1999.
104. Fishman A, Martinez F, Naunheim K, Piantadosi S, Wise R, Ries A, Weinmann G, Wood DE. A randomized trial comparing lung-volume-reduction surgery with medical therapy for severe emphysema. *N Engl J Med* 348(21): 2059–73, 2003.
105. Hosenpud JD, Bennett LE, Keck BM, Fiol B, Novick RJ. The registry of the International Society for Heart and Lung Transplantation: Fourteenth official report – 1997. *J Heart Lung Transplant* 16(7): 691–712, 1997.
106. Hosenpud JD, Bennett LE, Keck BM, Fiol B, Boucek MM, Novick RJ. The Registry of the International Society for Heart and Lung Transplantation: Fifteenth official report – 1998. *J Heart Lung Transplant* 17(7): 656–68, 1998.
107. Ziment I. Historic overview of mucoactive drugs. In: Braga PC, Allegra L (eds). *Drugs in Bronchial Mucology*, pp. 1–33. New York: Raven Press, 1989.
108. Decramer M, Rutten-van Molken M, Dekhuijzen PN, Troosters T, van Herwaarden C, Pellegrino R, van Schayck CP, Olivieri D, Del Donno M, De Backer W, Lankhorst I, Ardia A. Effects of *N*-acetylcysteine on outcomes in chronic obstructive pulmonary disease (Bronchitis Randomized on NAC Cost-Utility Study, BRONCUS): A randomised placebo-controlled trial. *Lancet* 365(9470): 1552–60, 2005.
109. Irwin RS, Curley FJ, Pratter MR. The effects of drugs on cough. *Eur J Respir Dis Suppl* 153: 173–81, 1987.

110. Engelen MP, Schols AM, Lamers RJ, Wouters EF. Different patterns of chronic tissue wasting among patients with chronic obstructive pulmonary disease. *Clin Nutr* 18(5): 275–80, 1999.
111. Casaburi R. Skeletal muscle function in COPD. *Chest* 117(5 Suppl 1): 267S–71S, 2000.
112. Nici L. Mechanisms and measures of exercise intolerance in chronic obstructive pulmonary disease. *Clin Chest Med* 21(4): 693–704, 2000.
113. Casaburi R. Rationale for anabolic therapy to facilitate rehabilitation in chronic obstructive pulmonary disease. *Baillieres Clin Endocrinol Metab* 12(3): 407–18, 1998.
114. Borson S, Claypoole K, McDonald GJ. Depression and chronic obstructive pulmonary disease: Treatment trials. *Semin Clin Neuropsychiatry* 3(2): 115–30, 1998.
115. Fried TR, Pollack DM, Tinetti ME. Factors associated with six-month mortality in recipients of community-based long-term care. *J Am Geriatr Soc* 46(2): 193–97, 1998.
116. Jones PW, Quirk FH, Baveystock CM, Littlejohns P. A self-complete measure of health status for chronic airflow limitation. The St. George's Respiratory Questionnaire. *Am Rev Respir Dis* 145(6): 1321–27, 1992.
117. Guyatt GH, Berman LB, Townsend M, Pugsley SO, Chambers LW. A measure of quality of life for clinical trials in chronic lung disease. *Thorax* 42: 773–78, 1987.
118. Jones PW, Bosh TK. Quality of life changes in COPD patients treated with salmeterol. *Am J Respir Crit Care Med* 155: 1283–89, 1997.
119. Appleton S, Smith B, Veale A, Bara A. Long-acting beta2-agonists for chronic obstructive pulmonary disease. *Cochrane Database Syst Rev* 2, 2000.
120. Sin DD, McAlister FA, Man SF, Anthonisen NR. Contemporary management of chronic obstructive pulmonary disease: Scientific review. *JAMA* 290(17): 2301–12, 2003.
121. Grandjean EM, Berthet P, Ruffmann R, Leuenberger P. Efficacy of oral long-term *N*-acetylcysteine in chronic bronchopulmonary disease: A meta-analysis of published double-blind, placebo-controlled clinical trials. *Clin Ther* 22(2): 209–21, 2000.
122. Stey C, Steurer J, Bachmann S, Medici TC, Tramer MR. The effect of oral *N*-acetylcysteine in chronic bronchitis: A quantitative systemitic review. *Eur Respir J* 16: 253–62, 2000.
123. Collet JP, Shapiro P, Ernst P, Renzi T, Ducruet T, Robinson A. Effects of an immunostimulating agent on acute exacerbations and hospitalizations in patients with chronic obstructive pulmonary disease, The PARI_IS Study steering committee and research group. Prevention of acute respiratory infection by an immunostimulant. *Am J Respir Crit Care Med* 156: 1719–24, 1997.
124. Stuart-Harris C, Bishop JM, Clark TJH, Dornhorst AC, Cotes JE, Flenley DC, Howard P, Oldham PD. Long term domiciliary oxygen therapy in chronic hypoxic cor pulmonale complicating chronic bronchitis and emphysema. *Lancet* 1: 681–86, 1981.
125. Kvale PA, Cugell DW, Anthonisen NR, Timms RM, Petty TL, Boylen CT. Continuous or nocturnal oxygen therapy in hypoxemic chronic obstructive lung disease. *Ann Intern Med* 93: 391–98, 1980.
126. Cooper CB, Waterhouse J, Howard P. Twelve year clinical study of patients with hypoxic cor pulmonale. Thorax 1987; 42: 105–110.
127. Strom K. Survival of patients with chronic obstructive pulmonary disease receiving long-term domiciliary oxygen therapy. *Am Rev Respir Dis* 147(3): 585–91, 1993.
128. Carrera M, Sauleda J, Bauza F, Bosch M, Togores B, Barbe F, Agusti AG. The results of the operation of a monitoring unit for home oxygen therapy. *Arch Bronconeumol* 35(1): 33–38, 1999.
129. Rennard S, Carrera M, Agusti AGN. Management of chronic obstructive pulmonary disease: Are we going anywhere? *Eur Respir J* 16: 1035–36, 2000.
130. Sin DD, Man SF. Pharmacotherapy for mortality reduction in chronic obstructive pulmonary disease. *Proc Am Thorac Soc* 3(7): 624–29, 2006.
131. Sin DD, Wu L, Anderson JA, Anthonisen NR, Buist AS, Burge PS, Calverley PM, Connett JE, Lindmark B, Pauwels RA, Postma DS, Soriano JB, Szafranski W, Vestbo J. Inhaled corticosteroids and mortality in chronic obstructive pulmonary disease. *Thorax* 60(12): 992–97, 2005.
132. Sin DD, Wu L, Man SF. The relationship between reduced lung function and cardiovascular mortality: A population-based study and a systematic review of the literature. *Chest* 127(6): 1952–59, 2005.
133. Salpeter S, Ormiston T, Salpeter E. Cardioselective beta-blockers for chronic obstructive pulmonary disease. *Cochrane Database Syst Rev*(4), 2005, CD003566.
134. Mancini GB, Etminan M, Zhang B, Levesque LE, FitzGerald JM, Brophy JM. Reduction of morbidity and mortality by statins, angiotensin-converting enzyme inhibitors, angiotensin receptor blockers in patients with chronic obstructive pulmonary disease. *J Am Coll Cardiol* 47(12): 2554–60, 2006.
135. Ishida W, Kajiwara T, Ishii M, Fujiwara F, Taneichi H, Takebe N, Takahashi K, Kaneko Y, Segawa I, Inoue H, Satoh J. Decrease in mortality rate of chronic obstructive pulmonary disease (COPD) with statin use: A population-based analysis in Japan. *Tohoku J Exp Med* 212(3): 265–73, 2007.
136. Soyseth V, Brekke PH, Smith P, Omland T. Statin use is associated with reduced mortality in COPD. *Eur Respir J* 29(2): 279–83, 2007.
137. Keddissi JI, Younis WG, Chbeir EA, Daher NN, Dernaika TA, Kinasewitz GT. The use of statins and lung function in current and former smokers. *Chest* 132(6): 1764–71, 2007.
138. Reinke LF, Engelberg RA, Shannon SE, Wenrich MD, Vig EK, Back AL, Curtis JR. Transitions regarding palliative and end-of-life care in severe chronic obstructive pulmonary disease or advanced cancer: Themes identified by patients, families, clinicians. *J Palliat Med* 11(4): 601–9, 2008.

67

CHAPTER

Acute Exacerbations of COPD

Jadwiga A. Wedzicha

Department of Respiratory Medicine,
University College of London,
London, UK

EPIDEMIOLOGY OF COPD EXACERBATION

There has been considerable recent interest into the causes and mechanisms of exacerbations of chronic obstructive pulmonary disease (COPD) as COPD exacerbations are an important cause of the considerable morbidity and mortality found in COPD [1]. COPD exacerbations increase with increasing severity of COPD. Some patients are prone to frequent exacerbations that are an important cause of hospital admission and readmission, and these frequent exacerbations may have considerable impact on quality of life, disease progression, and mortality [2] (Fig. 67.1). COPD exacerbations are also associated with considerable physiological deterioration and increased airway inflammatory changes [3] that are caused by a variety of factors such as viruses, bacteria, and possibly common pollutants. COPD exacerbations are commoner in the winter months and there may be important interactions between cold temperatures and exacerbations caused by viruses or pollutants [4].

Earlier descriptions of COPD exacerbations had concentrated mainly on studies of hospital admission, though most COPD exacerbations are treated in the community and not associated with hospital admission. A cohort of moderate-to-severe COPD patients was followed in East London, UK (East London COPD study) with daily diary cards and peak flow readings, who were asked to report exacerbations as soon as possible after symptomatic onset [2]. The diagnosis of COPD exacerbation was based on criteria modified from those described by Anthonisen and colleagues [5], which require two symptoms for diagnosis, one of which must be a major symptom of increased dyspnoea, sputum volume, or sputum purulence. Minor exacerbation symptoms included cough, wheeze, sore throat, nasal discharge, or fever (Table 67.1). The study found that about 50% of exacerbations were unreported to the research team, despite considerable encouragement provided and only diagnosed from diary cards. But there were no differences in major symptoms or physiological parameters between reported and unreported exacerbations [2]. Patients with COPD are accustomed to frequent symptom changes and thus may tend to underreport exacerbations to physicians. These patients have high levels of anxiety and depression and may accept their situation [6, 7]. The tendency of patients to underreport exacerbations may explain the higher total rate of exacerbation at 2.7 per patient per year, which is higher than that of previously reported by Anthonisen and co-workers at 1.1 per patient per year [5]. However in the latter study, exacerbations were unreported and diagnosed from patients' recall of symptoms.

Using the median number of exacerbations as a cutoff point, COPD patients in the East London Study were classified as frequent and infrequent exacerbators. Quality of life scores measured using a validated disease-specific scale, the St. George's Respiratory Questionnaire (SGRQ), was significantly worse in all of its three component scores (symptoms, activities, and impacts) in the frequent, compared to the infrequent exacerbators. This suggests that exacerbation frequency is an important determinant of health status in COPD and is thus one of the important outcome measures in COPD. Factors predictive of frequent exacerbations included daily cough and sputum and frequent exacerbations in the previous year. A previous

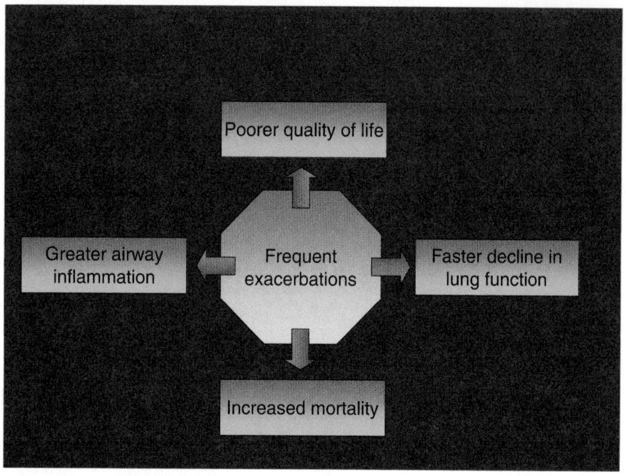

FIG. 67.1 Impact of COPD exacerbations.

TABLE 67.1 Causes of COPD exacerbations.

Viruses
Rhinovirus (common cold)
Influenza
Parainfluenza
Coronavirus
Adenovirus
RSV
Chlamydia pneumoniae
Bacteria
Haemophilus influenzae
Streptococcus penumoniae
Branhamella cattarhalis
Staphylococcus aureus
Pseudomonas aeruginosa
Common pollutants
Nitrogen dioxide
Particulates
Sulphur dioxide
Ozone

study of acute infective exacerbations of chronic bronchitis found that one of the factors predicting exacerbation was also the number in the previous year [8], though this study was limited to exacerbations presenting with purulent sputum and no physiological data was available during the study.

In a further prospective analysis of 504 exacerbations, where daily monitoring was performed, there was some deterioration in symptoms, though no there were significant peak expiratory flow changes [9]. Falls in peak expiratory flow and FEV_1 at exacerbation were generally small and not useful in predicting exacerbations, but larger falls in peak expiratory flow were associated with symptoms of dyspnoea, presence of colds, and longer recovery time from exacerbations. Symptoms of dyspnoea, common colds, sore throat, and cough increased significantly during the prodromal phase, and this suggests that respiratory viruses may have early effects at exacerbations. The median time to recovery of peak expiratory flow was 6 days and 7 days for symptoms, but at 35 days peak expiratory flow had returned to normal in only 75% of exacerbations, while at 91 days, 7.1% of exacerbations had not returned to baseline lung function. Recovery was longer in the presence of increased dyspnoea or symptoms of a common cold at exacerbation. The changes observed in lung function at exacerbation were smaller than those observed at asthmatic exacerbations, though the average duration of an asthmatic exacerbation was longer at 9.6 days [10, 11].

The reasons for the incomplete recovery of symptoms and lung function are not clear, but may involve inadequate treatment or persistence of the causative agent. The incomplete physiological recovery after an exacerbation could contribute to the decline in lung function with time in patients with COPD. However to date, there is no evidence that patients with incomplete recovery of their exacerbation have a greater decline in lung function, and further studies on the natural history of COPD exacerbations are required. A recent audit performed by the Royal College of Physicians, London, showed that ~30% of patients seen at hospital with an index exacerbation will be seen again and possibly readmitted with a recurrent exacerbation within 8 weeks [12]. In a cohort of moderate-to-severe COPD patients 22% of patients had a recurrent exacerbation within 50 days of the first (index) exacerbation, and this event can be separated discretely from the index exacerbation [13]. Thus, exacerbations are complex events and careful follow-up is essential to ensure complete recovery.

The association of the symptoms of increased dyspnoea and of the common cold at exacerbation with a prolonged recovery suggests that viral infections may lead to more prolonged exacerbations. As cold is associated with longer exacerbations, COPD patients who develop cold may be prone to more severe exacerbations and should be considered for therapy early at onset of symptoms. COPD exacerbations are also prone to recurrence in that one exacerbation is more likely to be followed by another one.

INFLAMMATORY CHANGES AT EXACERBATION

COPD exacerbations are associated with rises in airway (upper and lower airway) and systemic inflammation [2, 14]. Increases in systemic markers seen at exacerbations are most likely driven by increases in airway inflammation with exacerbation as the changes in airway and systemic inflammation at exacerbation are directly related [14]. Obviously, biopsy studies are difficult to perform at exacerbation in COPD patients. However in one study, where biopsies were performed at exacerbation in patients with chronic bronchitis, increased airway eosinophilia was found, though the patients studied had only mild COPD [14]. With exacerbation, there were more modest increases observed in neutrophils, T-lymphocytes (CD3), and TNF-α^+ cells, while there were no changes in CD4 or CD8 T-cells, macrophages, or mast cells. Qiu and co-workers have studied biopsies from COPD patients who were intubated and showed that there was considerable airway neutrophilia, neutrophil elastase expression with upregulation of neutrophil chemokine expression [15]. However, intubated COPD patients may have secondary airway infection and thus results may be difficult to interpret. Oxidative stress also plays an important role in the development of airway inflammation at COPD exacerbation. Markers of oxidative stress have been shown to rise in the airways during exacerbations such as hydrogen peroxide and 8-isoprostane, and these markers may take some time to recover to baseline stable levels [16]. Patients with severe exacerbations associated with hospitalization-assisted ventilation showed evidence of increased oxidative stress [17].

Most studies on airway inflammatory markers at exacerbation have been performed using sputum samples, either spontaneous or induced. Sputum inflammatory markers such as IL-6, IL-8, and myeloperoxidase (MPO) rise at the start of the exacerbation and usually recover to normal by 14 days, though in some cases higher airway inflammatory markers may persist for some time, suggesting incomplete recovery of exacerbations. Perera and colleagues also showed that systemic inflammation may persist after the exacerbation and those patients with an elevated C-reactive protein (CRP), 2 weeks after the onset of an exacerbation were more likely to develop an early recurrent exacerbation [13]. Patients with a history of frequent exacerbations have also increased airway and systemic inflammation in the stable state, compared to patients with infrequent exacerbations [2, 18].

ETIOLOGY of COPD exacerbation

COPD exacerbations have been associated with a number of aetiological factors, including infection and pollution episodes (Table 67.1). COPD exacerbations are frequently triggered by upper respiratory tract infections [19], and these are commoner in the winter months, when there are more respiratory viral infections in the community. Patients may also be more prone to exacerbations in the winter months as lung function in COPD patients shows small but significant falls with reduction in outdoor temperature during the winter months [4]. COPD patients have been found to have increased hospital admissions, suggesting increased exacerbation when increasing environmental pollution occurs. During the December 1991 pollution episode in the UK, COPD mortality was increased together with an increase in hospital admission in elderly COPD patients [20]. However, common pollutants especially oxides of nitrogen and particulates may interact with viral infection to precipitate exacerbation rather than acting alone [21].

Viral infections

Viral infections are an important trigger for COPD exacerbations [19, 22, 23]. Studies have shown that at least one-third of COPD exacerbations were associated with viral infections, and that the majority of these were due to human rhinovirus, the cause of the common cold [19, 22, 23]. Viral exacerbations were associated with symptomatic colds and prolonged recovery of the exacerbation [9]. Using molecular techniques, Seemungal and colleagues also showed that rhinovirus can be recovered from induced sputum more frequently than from nasal aspirates at exacerbation, suggesting that wild-type rhinovirus can infect the lower airway and contribute to inflammatory changes at exacerbation [22]. They also found that exacerbations associated with the presence of rhinovirus in induced sputum had larger increases in airway IL-6 levels [22], suggesting that viruses increase the severity of airway inflammation at exacerbation. This finding is in agreement with the data that respiratory viruses produce longer and more severe exacerbations and have a major impact on health care utilization [9, 24]. Other viruses may trigger COPD exacerbation, though coronavirus was associated with only a small proportion of asthmatic exacerbations and is unlikely to play a major role in COPD [25]. RSV (respiratory syncytial virus), influenza, parainfluenza, and adenovirus can all trigger exacerbations. Influenza has become a less prominent cause of exacerbation with the introduction of immunization, though this is still likely to be an important factor at times of influenza epidemics. RSV infection has been found at COPD exacerbation [26], but it is not clear if RSV is a cause of COPD exacerbation as RSV can be frequently detected in the airways of COPD patients when stable [27].

Bacterial infection

Over the past years, the role of bacterial infection at COPD exacerbation has been somewhat controversial as airway bacterial colonization is found when patients are stable state and the same organisms are isolated exacerbations. These include *Haemophilus Influenzae, Streptococcus Pneumoniae, Branhamella cattarhalis, Staphylococcus aureus*, and *Pseudomonas aeruginosa.* [28]. In a study in patients with moderate-to-severe COPD, bacteria were found in 48.2% of patients in the stable state and at exacerbation, bacterial detection rose to 69.6%, with an associated rise in airway bacterial load [29]. The case for involvement of bacteria has come from the studies of antibiotic therapy as exacerbations often present with increased sputum purulence and volume and antibiotics have traditionally been used as first-line therapy in such exacerbations. Anthonisen and colleagues in a classical paper investigating the benefit of antibiotics in over 300 acute exacerbations demonstrated a greater treatment success rate in patients treated with antibiotics, especially if their initial presentation was with the symptoms of increased dyspnoea, sputum volume, and purulence [5]. Patients with mild COPD obtained less benefit from antibiotic therapy. A meta-analysis of trials of antibiotic therapy in COPD has concluded that antibiotic therapy offered a small but significant benefit in outcome in exacerbations [30]. Sethi and colleagues have suggested that isolation of a new bacterial strain in COPD patients who were regularly sampled was associated with an increased risk of exacerbation [31], though this also does not conclusively prove that bacteria are direct causes of exacerbations as not all exacerbations were associated with strain change, and not all strain changes resulted in exacerbation.

At COPD exacerbations both respiratory viruses and bacteria may be isolated. A greater systemic inflammatory response has been reported in those exacerbations associated with both *H. influenzae* and rhinovirus isolations, and if the isolation of *Haemophilus* was associated with new or worsening coryzal symptoms (a surrogate of viral infection) such infections were more severe as assessed by changes in symptoms and lung function at exacerbation onset [29]. This has been confirmed in a further study demonstrating greater lung function impairment and longer hospitalizatons in exacerbations associated with viral and bacterial co-infection [32]. It has also been suggested that atypical micro-organisms such as chlamydia and mycoplasma may cause COPD exacerbations, though evidence on their role is conflicting and these infective agents may interact with other bacteria and viruses in the airways [33, 34].

PATHOPHYSIOLOGICAL CHANGES AT COPD EXACERBATION

In patients with moderate and severe COPD, the mechanical performance of the respiratory muscles is reduced. The airflow obstruction leads to hyperinflation, with the respiratory muscles acting at a mechanical disadvantage and generating reduced inspiratory pressures. The load on the respiratory muscles is also increased in patients with airflow obstruction by the presence of intrinsic positive end-expiratory pressure (PEEP). With an exacerbation of COPD, the increase in airflow obstruction will further increase the load on the respiratory muscles and increase the work of breathing, precipitating respiratory failure in more severe cases. The minute ventilation may be normal, but the respiratory pattern will be irregular with increased frequency and decreased tidal volume. The resultant hypercapnia and acidosis will then reduce inspiratory muscle function, contributing to further deterioration of the respiratory failure.

Hypoxaemia in COPD usually occurs due to a combination of ventilation–perfusion mismatch and hypoventilation, although arterio-venous shunting can also contribute in the acute setting. This causes increase in pulmonary artery pressure, which can lead to salt and water retention and the development of edema. The degree of the ventilation perfusion abnormalities increases during acute exacerbations and then resolves over the following few weeks. Acidosis is an important prognostic factor in survival from respiratory failure during COPD exacerbation, and thus early correction of acidosis is an essential goal of therapy.

TREATMENT

Inhaled bronchodilator therapy

Beta$_2$ agonists and anti-cholinergic agents are the inhaled bronchodilators most frequently used in the treatment of acute exacerbations of COPD. In patients with stable COPD, symptomatic benefit can be obtained with bronchodilator therapy in COPD, even without significant changes in spirometry. This is probably due to a reduction in dynamic hyperinflation that is characteristic of COPD and hence leads to a decrease in the sensation of dyspnoea especially during exertion [35]. In stable COPD greater bronchodilatation has been demonstrated with anti-cholinergic agents than with β_2 agonists, which may be due to the excessive cholinergic neuronal bronchoconstrictor tone [36]. However, studies investigating bronchodilator responses in acute exacerbations of COPD have shown no differences between agents used and no significant additive effect of the combination therapy, even though combination of anti-cholinergic and bronchodilator has benefits in the stable state [37, 38]. This difference in effect between the acute and stable states may be due to the fact that the larger doses of drug delivered in the acute setting produce maximal bronchodilatation, whereas the smaller doses administered in the stable condition may be having a sub-maximal effect.

Methylxanthines such as theophylline are sometimes used in the management of acute exacerbations of COPD. There is some evidence that theophyllines are useful in COPD, though the main limiting factor is the frequency of toxic side effects. The therapeutic action of theophylline is thought to be due to its inhibition of phosphodiesterase that breaks down cyclic 3'5' adenosine monophosphate (AMP), an intracellular messenger, thus facilitating bronchodilatation. However studies of intravenous aminophylline therapy in acute exacerbations of COPD have shown no significant beneficial effect over and above conventional therapy [39, 40]. There are some reports of beneficial effects of methylxanthines upon diaphragmatic and cardiac function, though these mechanisms require further study in patients with COPD exacerbations.

Corticosteroids

Only about 10% to 15% of patients with stable COPD show a spirometric response to oral corticosteroids [41] and, unlike the situation in asthma, steroids have little effect on airway inflammatory markers in patients with COPD [42, 43]. A number of early studies have investigated the effects of corticosteroid therapy at COPD exacerbation. In an early controlled trial in patients with COPD exacerbations and acute respiratory failure, Albert and co-workers found that there were larger improvements in pre- and post-bronchodilator FEV$_1$ when patients were treated for the first 3 days of the hospital admission with intravenous methylprednislone than those treated with placebo [44]. Another trial found that a single dose of methylprednisolone given within 30 min of arrival in the accident and emergency department produced no improvement after 5 h in spirometry, and also had no effect on hospital admission, though another study reduced readmission [45, 46]. A retrospective study comparing patients treated with steroids at exacerbation compared to those not treated showed that the steroid group had a reduced chance of relapse after therapy [47].

Thompson and colleagues gave a 9 day course of prednisolone or placebo in a randomized manner to out-patients presenting with acute exacerbations of COPD [48]. Unlike the previous studies, these patients were either recruited from out-patients or from a group that were pre-enrolled and self reported the exacerbation to the study team. In this study patients with exacerbations associated with acidosis or pneumonia were excluded, so exacerbations of moderate severity were generally included. Patients in the steroid-treated group showed a more rapid improvement in PaO_2, alveolar–arterial oxygen gradient, FEV_1, peak expiratory flow rate, and a trend toward a more rapid improvement in dyspnoea in the steroid-treated group.

In a recent cohort study by Seemungal and colleagues, the effect of therapy with prednisolone on COPD exacerbations diagnosed and treated in the community was studied [9]. Exacerbations treated with steroids were more severe and associated with larger falls in peak expiratory flow. The treated exacerbations also had a longer recovery time to baseline for symptoms and peak expiratory flow. However, the rate of peak expiratory flow recovery was faster in the prednisolone-treated group, though not the rate of symptom score recovery. An interesting finding in this study was that steroids significantly prolonged the median time from the day of onset of the initial exacerbation to the next exacerbation from 60 days in the group not treated with prednisolone to 84 days in the patients treated with prednisolone. In contrast, antibiotic therapy had no effect on the time to the next exacerbation. If short course oral steroid therapy at exacerbation does prolong the time to the next exacerbation, then this could be an important way to reduce exacerbation frequency in COPD patients, which is an important determinant of health status [2].

Davies and colleagues randomized patients admitted to hospital with COPD exacerbations to prednisolone or placebo [49]. In the prednisolone group, the FEV_1 rose faster until day 5, when a plateau was observed in the steroid-treated group. Changes in the pre-bronchodilator and post-bronchodilator FEV_1 were similar suggesting that this is not just an effect on bronchomotor tone, but involves faster resolution of airway inflammatory changes or airway wall edema with exacerbation. Length of hospital stay analysis showed that patients treated with prednisolone had a significantly shorter length of stay. Six weeks later, there were no differences in spirometry between the patient groups, and health status was similar to that measured at 5 days after admission. Thus, the benefits of steroid therapy at exacerbation are most obvious in the early course of the exacerbation. A similar proportion of the patients, ~32% in both study groups required further treatment for exacerbations within 6 weeks of follow-up, emphasizing the high exacerbation frequency in these patients.

Niewoehner and colleagues performed a randomized controlled trial of either a 2-week or an 8-week prednisolone course at exacerbation compared to placebo, in addition to other exacerbation therapy [50]. The primary end point was a first treatment failure, including death, need for intubation, readmission, or intensification of therapy. There was no difference in the results using the 2 or 8 week treatment protocol. The rates of treatment failure were higher in the placebo group at 30 days, compared to the combined 2 and 8 week prednislone groups. As in the study by Davies and colleagues, the FEV_1 improved faster in the prednisolone-treated group, though there were no differences by 2 weeks. In contrast, Niewoehner and colleagues performed a detailed evaluation of steroid complications and found considerable evidence of hyperglycaemia in the steroid-treated patients. Thus, steroids should be used at COPD exacerbation in short courses of no more than 2 weeks duration to avoid risk of complications.

Antibiotics

Acute exacerbations of COPD often present with increased sputum purulence and volume, and antibiotics have traditionally been used as first-line therapy in such exacerbations. However, viral infections may be the triggers in a significant proportion of acute infective exacerbations in COPD and antibiotics used for the consequences of secondary infection. As discussed previously, antibiotic therapy at exacerbations is most useful if patients present with symptoms of increased dyspnoea, sputum volume, and purulence [30]. A randomized placebo-controlled study investigating the value of antibiotics in patients with mild obstructive lung disease in the community concluded that antibiotic therapy did not accelerate recovery or reduce the number of relapses, though patients had mixed pathologies [51].

Management of respiratory failure

Hypoxaemia occurs with more severe exacerbations and usually requires hospital admission. Caution should always be taken in providing supplemental oxygen to patients with COPD, particularly during acute exacerbations, when respiratory drive and muscle strength can be impaired leading to significant increases in carbon dioxide tension at relatively modest oxygen flow rates. However, in the vast majority of cases, the administration of supplemental oxygen increases arterial oxygen tension sufficiently without clinically significant rises in carbon dioxide. It is suggested that supplemental oxygen is delivered at an initial flow rate of 1–2 l/min via nasal cannulae or 24–28% inspired oxygen via Venturi mask, with repeat blood gas analysis after 30–45 min of oxygen therapy.

Hypercapnia during COPD exacerbations may be managed initially with the use of respiratory stimulants. The most commonly used is doxapram, which acts centrally to increase respiratory drive and respiratory muscle activity. The effect is probably only appreciable for 24 to 48 h; the main factor limiting use being side effects which can lead to agitation and are often not tolerated by the patient. There are only a few studies of the clinical efficacy of doxapram and short-term investigations suggest that improvements in acidosis and arterial carbon dioxide tension can be attained [52]. A small study comparing doxapram with non-invasive ventilation (NPPV) in acute exacerbations of COPD suggested that NPPV was superior with regard to correction of blood gases during the initial treatment phase [53]. Increases in pulmonary artery pressure during acute exacerbations of COPD can result in right-sided cardiac dysfunction and development of peripheral edema. Diuretic therapy may thus be necessary if there is edema or a rise in jugular venous pressure.

Ventilatory support

Non-invasive ventilation

The introduction of noninvasive positive pressure ventilation (NPPV) using nasal or face masks has had a major impact on the management of acute exacerbations and has enabled acidosis to be corrected at an early stage. Studies have shown that NIPPV can produce improvements in pH relatively rapidly, at 1h after instituting ventilation [54, 55]. This will allow time for other conventional therapy to work, such as oxygen therapy, bronchodilators, steroids, and antibiotics and thus reverse the progression of respiratory failure and reduce mortality. With NIPPV, there are improvements in minute ventilation, reductions in respiratory rate and in transdiaphragmatic activity. Thus, NIPPV can improve gas exchange and allows respiratory muscle rest in respiratory failure.

With the use of NIPPV patient comfort is improved; there is also no requirement for sedation with preservation of speech and swallowing. The technique can be applied in a general ward, though a high-dependency area is preferable and intensive care is unnecessary. Patient cooperation is important in application of NIPPV. The main advantage of the use of NIPPV is the avoidance of tracheal intubation and the ability to offer ventilatory support to patients with respiratory failure due to severe COPD, who would be considered unsuitable for intubation. A lower incidence of nosocomial penumonia has also been reported with the use of NPPV compared with conventional intubation and ventilation.

Following a number of uncontrolled studies, randomized controlled trials have shown benefit of NIPPV in acute COPD exacerbations. A UK study showed that with the use of NIPPV in exacerbations of respiratory failure, earlier correction of pH can be achieved, together with reduction in breathlessness over the initial 3 days of ventilation, compared with a control standard therapy group [54]. A study from the United States showed a significant reduction in intubation rates with NIPPV from 67% in a group receiving conventional therapy to 9% in the NIPPV group [55]. A third study showed convincingly that in patients with exacerbations of respiratory failure, the use of NIPPV with pressure supports ventilation, reduces the need for intubation and mortality is significantly reduced from 29% in the conventionally treated group to 9% in the NIPPV group [56]. Complications, which were specifically associated with the use of mechanical ventilation, were also reduced. The difference in mortality disappeared after adjustment for intubation, suggesting that the benefits with NIPPV are due to fewer patients requiring intubation. This was also the first study to show that hospital length of stay can be reduced with use of NIPPV. A recent study showed that NPPV can be applied on general wards, though patients with more severe acidosis had a worse outcome [57].

These studies have treated patients where the pH was below 7.35, rather than just below 7.26, when the prognosis of COPD worsens. A number of these patients may have improved without NIPPV, though it seems that the major effect of NIPPV is the earlier correction of acidosis and thus avoidance of tracheal intubation, with all its associated complications. Studies have shown that NIPPV can be successfully implemented in up to 80% of cases [58, 59]. NIPPV is less successful in patients who have worse blood gases at baseline before ventilation, are underweight, have a higher incidence of pneumonia, have a greater level of neurological deterioration, and where compliance with the ventilation is poor [58]. Moretti and colleagues have recently shown that "late treatment failure" (after an initial 48h of therapy with NPPV) is up to 20% and that patients with late failure were more likely to have severe functional and clinical disease with more complications at the time of admission [60]. Identification of patients with a potentially poor outcome is important as delay in intubation can have serious consequences for the patient.

Indications for invasive ventilation

If NIPPV fails, or is unavailable in the hospital, invasive ventilation may be required in the presence of increasing acidosis. It may be considered in any patient when the pH falls below 7.26. Decisions to ventilate these patients may be difficult, though with improved modes of invasive ventilatory support and better weaning techniques, the outlook for the COPD patient is better.

Patients will be suitable for tracheal intubation if this is the first presentation of COPD exacerbation or respiratory failure, or there is a treatable cause of respiratory failure, such as pneumonia. Information will be required on the history and quality of life, especially the ability to perform daily activities. Patients with severe disabling and progressive COPD may be less suitable, but it is important that adequate and appropriate therapy has been used in these patients, with documented disease progression. The patient's wishes and those of any close relatives should be considered in any decision to institute or withhold life supporting therapy.

Supported discharge

Many hospital admissions are related to exacerbations of COPD and thus reductions of admissions especially during the winter months when they are most frequent is particularly desirable. Over the last few years a number of different models of supported discharge have been developed and some evaluated [61–63]. Patients have been discharged early with an appropriate package of care organized, including domiciliary visits made to these patients after discharge by trained respiratory nurses.

Cotton and colleagues randomized patients to discharge on the next day or usual management and found that there were no differences in mortality or readmission rates between the two groups [61]. There was a reduction in hospital stay from a mean of 6.1 days to 3.2 days. In another larger study by Skwarska and colleagues, patients were randomized to discharge on the day of assessment or conventional management [62]. Again there were no differences in readmission rates, no differences in visits to primary care physicians and health status measured 8 weeks after discharge was similar in the two groups. The authors also demonstrated that there were significant cost savings of around 50% for the home support group, compared to the admitted group. However, other considerations need to be taken into account in organizing an assisted discharge service in that resources have to be released for the nurses to follow the patients and the benefits may be seasonal, as COPD admissions are a particular problem in the winter. Further work is required on the different models of supported discharge available and the cost effectiveness of these programmes.

PREVENTION OF COPD EXACERBATION

There has been much recent emphasis on prevention of exacerbations in patients with COPD. As respiratory tract infections are common factors in causing exacerbation, influenza, and pneumococcal vaccinations are recommended for all patients with significant COPD. A study that reviewed the outcome of influenza vaccination in a cohort of elderly patients with chronic lung disease found that influenza vaccinaiton is associated with significant health benefits with fewer outpatient visits, fewer hospitalizations, and a reduced mortality [64].

Long-term antibiotic therapy has been used in the past in patients with very frequent exacerbations, though the evidence was not strong for benefit. However with the advent of novel and more specific antibiotics against airway organisms, the topic of long-term antibiotic therapy in COPD is currently being revisited and results of the first trials are awaited. Recently there has been a report of the effects of an immunostimulatory agent in patients with COPD exacerbations, with reduction in severe complications and hospital admissions in the actively treated group [65]. However, the mechanisms of benefit are not clear, and further studies on the effects of these agents in the prevention of COPD exacerbation are required.

Long-acting bronchodilators (LABA) have been shown to reduce exacerbations. In the recently reported TORCH (Towards a Revolution in COPD Health) study, salmeterol, a long-acting beta agonist, reduced the frequency of exacerbations [66], while a number of other studies have shown that the long-acting anticholinergic tiotropium reduces the exacerbation rate and also a trend to reduction in hospital admission [67–69]. However, there is no good evidence at present that long-acting anticholinergic agents possess anti-inflammatory activity [70], and it is likely that tiotropium reduces exacerbations by reducing dynamic hyperinflation and thus dyspnoea. Combinations of long-acting beta agonists and inhaled corticosteroids have been also evaluated and reduced exacerbations more than the individual components [66]. A direct comparison of inhaled tiotropium with the salmeterol/fluticasone (SFC) combination in the recently reported INSPIRE study in moderate-to-severe COPD patients showed that both interventions had an equal effect on exacerbation rates [71]. However, patients taking tiotropium required more courses of oral corticosteroids with exacerbations whereas patients on the SFC combination required more courses of antibiotics [71]. This for the first time, it has been shown that different interventions have different effects on exacerbations.

The Optimal study recently evaluated the combination of tiotropium with inhaled LABA (salmeterol) and inhaled steroids (fluticasone) [72]. The triple combination reduced hospitalization as a result of exacerbation, but not the total number of exacerbations. In addition a trend was observed in the reduction of the number of exacerbations with the triple combination, which did not reach statistical significance due to the relatively small size of the study and the high dropout rate. Triple therapy may be more effective than other therapies and further studies of these combinations are now required with adequately powered studies.

References

1. Wedzicha JA, Seemungal TAR. COPD exacerbations: Defining their cause and prevention. *Lancet* 370: 786–96, 2007.
2. Seemungal TAR, Donaldson GC, Paul EA, Bestall JC, Jeffries DJ, Wedzicha JA. Effect of exacerbation on quality of life in patients with chronic obstructive pulmonary disease. *Am J Respir Crit Care Med* 151: 1418–22, 1998.
3. Bhowmik A, Seemungal TAR, Sapsford RJ, Wedzicha JA. Relation of sputum inflammatory markers to symptoms and physiological changes at COPD exacerbations. *Thorax* 55: 114–200, 2000.
4. Donaldson GC, Seemungal T, Jeffries DJ, Wedzicha JA. Effect of environmental temperature on symptoms, lung function and mortality in COPD patients. *Eur Respir J* 13: 844–49, 1999.
5. Anthonisen NRJ, Manfreda CPW, Warren ES, Hershfield GKM, Harding , Nelson NA. Antibiotic therapy in exacerbations of chronic obstructive pulmonary disease. *Ann Intern Med* 106: 120–96, 1987.
6. Okubadejo AA, Jones PW, Wedzicha JA. Quality of life in patients with COPD and severe hypoxaemia. *Thorax* 51: 44–47, 1996.
7. Okubadejo AA, O'Shea L, Jones PW, Wedzicha JA. Home assessment of activities of daily living in patients with severe chronic obstructive pulmonary disease on long term oxygen therapy. *Eur Respir J* 10: 1555–72, 1997.
8. Ball P, Harris JM, Lowson D, Tillotson G, Wilson R. Acute infective exacerbations of chronic bronchitis. *Q J Med* 88: 61–68, 1995.
9. Seemungal TAR, Donaldson GC, Bhowmik A, Jeffries DJ, Wedzicha JA. Time course and recovery of exacerbations in patients with chronic obstructive pulmonary disease. *Am J Respir Crit Care Med* 161: 1608–13, 2000.
10. Reddel HS, Ware S, Marks G, Salome C, Jenkins C, Woolcock A. Differences between asthma exacerbations and poor asthma control. *Lancet* 353: 364–69, 1999.
11. Tattersfield AE, Postma DS, Barnes PJ *et al.* Exacerbations of asthma. *Am J Respir Crit Care Med* 160: 594–99, 1999.
12. Roberts CM, Lowe D, Bucknall CE, Ryland I, Kelly Y, Pearson MG. Clinical audit indicators of outcome following admission to hospital with acute exacerbation of chronic obstructive pulmonary disease. *Thorax* 57: 137–41, 2002.
13. Perera WR, Hurst JR, Wilkinson TMA, Sapsford RJ, Müllerova H , Donaldson GC, Wedzicha JA. Inflammatory changes and recurrence at COPD exacerbations. *Eur Respir J* 29: 527–34, 2007.
14. Hurst JR, Perera WR, Wilkinson TMA, Donaldson GC, Wedzicha JA. Systemic and upper and lower airway inflammation at exacerbation of chronic obstructive pulmonary disease. *Am J Respir Crit Care Med* 173: 71–78, 2006.
15. Saetta M, Di Stefano A, Maestrelli P *et al.* Airway eosinophilia in chronic bronchitis during exacerbations. *Am J Respir Crit Care Med* 150: 1646–52, 1994.
16. Qiu Y, Zhu J, Bandi V *et al.* Biopsy neutrophilia, neutrophil chemokin and receptor gene expression in severe exacerbations of chronic obstructive pulmonary disease. *Am J Respir Crit Care Med* 168: 968–75, 2003.
17. Biernacki W, Kharitonov SA, Barnes PJ. Increased leukotriene B4 and 8-isprostane in exhaled breath condensate of patients with exacerbations of COPD. *Thorax* 58: 294–98, 2003.

18. Donaldson GC, Seemungal TAR, Patel IS, Lloyd-Owen SS, Bhowmik A, Wilkinson TMA, Hurst JR, Wilks M, MacCallum PK, Wedzicha JA. Airway and systemic inflammation and decline in lung function, in chronic obstructive pulmonary disease. *Chest* 128: 1995–2004, 2005.
19. Seemungal TAR, Harper-Owen R, Bhowmik A, Moric I, Sanderson G, Message S, MacCallum P, Meade TW, Jeffries DJ, Johnston SL, Wedzicha JA. Respiratory viruses, symptoms and inflammatory markers in acute exacerbations and stable chronic obstructive pulmonary disease. *Am J Respir Crit Care Med* 164: 1618–23, 2001.
20. Anderson HR, Limb ES, Bland JM, Ponce de Leon A, Strachan DP, Bower JS. Health effects of an air pollution episode in London, December 1991. *Thorax* 50: 1188–93, 1995.
21. Linaker CH, Coggon D, Holgate ST *et al.* Personal exposure to nitrogen dioxide and risk of airflow obstruction in asthmatic children with upper respiratory infection. *Thorax* 55: 930–33, 2000.
22. Seemungal TAR, Harper–Owen R, Bhowmik A, Jeffries DJ, Wedzicha JA. Detection of rhinovirus in induced sputum at exacerbation of chronic obstructive pulmonary disease. *Eur Respir J* 16: 677–83, 2000.
23. Rohde G, Wiethege A, Borg I, Kauth M, Bauer TT, Gillissen A, Bufe A, Schultze-Werninghaus G. Respiratory viruses in exacerbations of chronic obstructive pulmonary disease requiring hospitalisation: A case-control study. *Thorax* 58: 37–42, 2003.
24. Greenberg SB, Allen M, Wilson J, Atmar RL. Respiratory viral infections in adults with and without chronic obstructive pulmonary disease. *Am J Respir Crit Care Med* 162: 167–73, 2000.
25. Johnston SL, Pattemore PK, Sanderson G *et al.* Community study of the role of viral infections in exacerbations of asthma in 9-11 year old children. *BMJ* 310: 1225–29, 1995.
26. Falsey AR, Formica MA, Hennessey PA *et al.* Detection of respiratory syncytial virus in adults with chronic obstructive pulmonary disease. *Am J Respir Crit Care Med* 173: 639–43, 2006.
27. Wilkinson TM, Donaldson GC, Johnston SL, Openshaw PJ, Wedzicha JA. Respiratory syncytial virus, airway inflammation and FEV1 decline in patients with COPD. *Am J Respir Crit Care Med* 173: 871–76, 2006.
28. Sapey E, Stockley RA. COPD exacerbations 2: Aetiology. *Thorax* 61: 250–58, 2006.
29. Wilkinson TMA, Hurst JR, Perera WR, Wilks M, Donaldson GC, Wedzicha JA. Interactions between lower airway bacterial and rhinoviral infection at exacerbations of chronic obstructive pulmonary disease. *Chest* 129: 317–24, 2006.
30. Antibiotics for exacerbations of chronic obstructive pulmonary disease. *Cochrane Database Syst Rev* 2006, Issue 2. Art. No.: CD004403. DOI: 10.1002/14651858.CD004403.pub2.
31. Sethi S, Evans N, Grant BJ, Murphy TF. New strains of bacteria and exacerbations of chronic obstructive pulmonary disease. *N Engl J Med* 347: 465–71, 2002.
32. Papi A, Bellettato CM, Braccioni F, Romagnoli M, Casolari P, Caramori G, Fabbri LM, Johnston SL. Infections and airway inflammation in chronic obstructive pulmonary disease severe exacerbations. *Am J Respir Crit Care Med* 173: 1114–21, 2006.
33. Blasi F, Damato S, Consentini R *et al. C. Pneumoniae* and chronic bronchitis: Association with severity and bacterial clearance following treatment. *Thorax* 57: 672, 2002.
34. Seemungal TAR, Wedzicha JA, MacCallum PK, Johnston SL, Lambert PA. *C. Pneumoniae* and COPD exacerbation. *Thorax* 57: 1087–88, 2002.
35. Belman MJ, Botnick WC, Shin JW. Inhaled bronchodilators reduce dynamic hyperinflation during exercise in patients with chronic obstructive pulmonary disease. *Am J Respir Crit Care Med* 153: 967–75, 1996.
36. Braun SR, McKenzie WN, Copeland C, Knight L, Ellersieck M. A comparison of the effect of ipratropium and albuterol in the treatment of chronic obstructive airway disease. *Arch Intern Med* 149: 544–47, 1989.
37. Combivent Inhalation Aerosol Study Group. In chronic obstructive pulmonary disease, a combination of ipratropium and albuterol is more effective than either agent alone. *Chest* 105: 1411–19, 1994.
38. Rebuck AS, Chapman KR, Abboud R, Pare PD, Kreisman H, Wolkove N, Vickerson F. Nebulized anticholinergic and sympathomimetic treatment of asthma and chronic obstructive airways disease in the emergency room. *Am J Med* 82: 59–64, 1987.
39. Rice KL, Leatherman JW, Duane PG, Snyder LS, Harmon KR, Abel J, Niewoehner DE. Aminophylline for acute exacerbations of chronic obstructive pulmonary disease. A controlled trial. *Ann Intern Med* 107: 305–9, 1987.
40. Duffy N, Walker P, Diamantea F, Calverley PMA, Davies L. Intravenous aminophylline in patients admitted to hospital with non-acidotic exacerbations of chronic obstructive pulmonary disease: A prospective randomised controlled trial. *Thorax* 60: 713–17, 2005.
41. Callahan CM, Cittus RS, Katz BP. Oral corticosteroid therapy for patients with stable chronic obstructive pulmonary disease: A meta-analysis. *Ann Intern Med* 114: 216–23, 1991.
42. Keatings VM, Jatakanon A, Worsdell Y, Barnes PJ. Effects of inhaled and oral glucocorticoids on inflammatory indices in asthma and COPD. *Am J Respir Crit Care Med* 155: 542–48, 1997.
43. Culpitt SV, Maziak W, Loukidis S *et al.* Effects of high dose inhaled steroids on cells, cytokines and proteases in induced sputum in chronic obstructive pulmonary disease. *Am J Respir Crit Care Med* 160: 1635–39, 1999.
44. Albert RK, Martin TR, Lewis SW. Controlled clinical trial of methylprednisolone in patients with chronic bronchitis and acute respiratory insufficiency. *Ann Intern Med* 92: 753–58, 1980.
45. Emerman CL, Connors AF, Lukens TW, May ME, Effron D. A randomised controlled trial of methylprednisolone in the emergency treatment of acute exacerbations of chronic obstructive pulmonary disease. *Chest* 95: 563–67, 1989.
46. Bullard MJ, Liaw SJ, Tsai YH, Min HP. Early corticosteroid use in acute exacerbations of chronic airflow limitation. *Am J Emerg Med* 14: 139–43, 1996.
47. Murata GH, Gorby MS, Chick TW, Halperin AK. Intravenous and oral corticosteroids for the prevention of relapse after treatment of decompensated COPD. *Chest* 98: 845–49, 1990.
48. Thompson WH, Nielson CP, Carvalho P *et al.* Controlled trial of oral prednisolone in outpatients with acute COPD exacerbation. *Am J Respir Crit Care Med* 154: 407–12, 1996.
49. Davies L, Angus RM, Calverley PMA. Oral corticosteroids in patients admitted to hospital with exacerbations of chronnic obstructive pulmonary disease: A prospective randomised controlled trial. *Lancet* 354: 456–60, 1999.
50. Niewoehner DE, Erbland ML, Deupree RH *et al.* Effect of systemic glucocorticoids on exacerbations of chronic obstuctive pulmonary disease. *N Engl J Med* 340: 1941–47, 1999.
51. Sachs APE, Koeter GH, Groenier KH, Van der Waaij D, Schiphuis J, Meyboom-de Jong B. Changes in symptoms, peak expiratory flow and sputum flora during treatment with antibiotics of exacerbations in patients with chronic obstructive pulmonary disease in general practice. *Thorax* 50: 758–63, 1995.
52. Moser KM, Luchsinger PC, Adamson JS, McMahon SM, Schlueter DP, Spivack M, Weg JG. Respiratory stimulation with intravenous doxapram in respiratory failure. *N Engl J Med* 288: 427–31, 1973.
53. Angus RM, Ahmed AA, Fenwick LJ, Peacock AJ. Comparison of the acute effects on gas exchange of nasal ventilation and doxapram in exacerbations of chronic obstructive pulmonary disease. *Thorax* 51: 1048–50, 1996.
54. Bott J, Carroll MP, Conway JH, Keilty SEJ, Ward EM, Brown AM *et al.* Randomised controlled trial of nasal ventilation in acute ventilatory failure due to chronic obstructive airways disease. *Lancet* 341: 1555–57, 1993.
55. Kramer N, Meyer TJ, Meharg J, Cece RD, Hill NS. Randomized prospective trial of noninvasive positive pressure ventilation in acute respiratory failure. *Am J Respir Crit Care Med* 151: 1799–806, 1995.
56. Brochard L, Mancebo J, Wysocki M, Lofaso M, Conti G, Rauss A *et al.* Noninvasive ventilation for acute exacerbations of chronic obstructive pulmonary disease. *N Engl J Med* 333: 817–22, 1995.

57. Plant PK, Owen JL, Elliott MW. A multicentre randomised controlled trial of the early use of non-invasive ventilation foracute exacerbations of chronic obstructive pulmonary disease on general respiratory wards. *Lancet* 355: 1931–35, 2000.
58. Ambrosino N, Foglio K, Rubini F, Clini E, Nava S, Vitacca M. Non-invasive mechanical ventilation in acute respiratory failure due to chronic obstructive pulmonary disease: Correlates for success. *Thorax* 50: 755–57, 1995.
59. Brown JS, Meecham Jones DJ, Mikelsons C, Paul EA, Wedzicha JA. Outcome of nasal intermittent positive pressure ventilation when used for acute-on-chronic respiratory failure on a general respiratory ward. *J R Coll Physicians Lond* 32: 219–24, 1998.
60. Morretti M, Cilione C, Tampieri A *et al.* Incidence and causes of non-invasive mechanical ventilation failure after initial success. *Thorax* 55: 819–25, 2000.
61. Gravil JH, Al-Rawas OA, Cotton MM *et al.* Home treatment of exacerbations of COPD by an acute respiratory assessment service. *Lancet* 351: 1853–55, 1998.
62. Cotton MM, Bucknall CE, Dagg KD *et al.* Early discharge for patients with exacerbations of COPD: A randomised controlled trial. *Thorax* 55: 902–6, 2000.
63. Skwarska E, Cohen G, Skwarski KM *et al.* A randomised controlled trial of supported discharge in patients with exacerbations of COPD. *Thorax* 55: 907–12, 2000.
64. Nichol KL, Baken L, Nelson A. Relation between influenza vaccination and out patient visits, hospitalisation and mortality in elderly patients with chronic lung disease. *Ann Intern Med* 130: 397–403, 1999.
65. Collet JP, Shapiro S, Ernst P, Renzi P, Ducruet T, Robinson A *et al.* Effect of an immunostimulating agent on acute exacerbations and hospitalization in COPD patients. *Am J Respir Crit Care Med* 156: 1719–24, 1997.
66. Calverley PM, Anderson JA, Celli B *et al.* Salmeterol and fluticasone propionate and survival in chronic obstructive pulmonary disease. *N Engl J Med* 356: 775–89, 2007.
67. Vincken W, van Noord JA, Greefhorst APM. Improved health outcomes in patients with COPD during 1 yr's treatment with tiotropium. *Eur Respir J* 19: 209–16, 2002.
68. Casaburi R, Mahler DA, Jones PA *et al.* A long-term evaluation of once-daily inhaled tiotropium in chronic obstructive pulmonary disease. *Eur Respir J* 19: 217–24, 2002.
69. Niewoehner DE, Rice K, Cote C *et al.* Prevention of exacerbations of chronic obstructive pulmonary disease with tiotropium, a once daily inhaled anticholinergic bronchodilator: A randomised trial. *Ann Intern Med* 143: 317–26, 2005.
70. Powrie DJ, Wilkinson TMA, Donaldson GC, Jones P, Scrine K, Viel K, Kesten S, Wedzicha JA. Effect of tiotropium on inflammation and exacerbations in chronic obstructive pulmonary disease. *Eur Respir J* 30: 472–78, 2007.
71. Wedzicha JA, Calverley PMA, Seemungal TA, Hagan G, Ansari Z, Stockley RA for the INSPIRE Investigators. The Prevention of Chronic Obstructive Pulmonary Disease Exacerbations by Salmeterol/Fluticasone Propionate or Tiotropium Bromide. *Am J Respir Crit Care Med* 177: 19–26, 2008.
72. Aaron SD, Vandemheen KL, Fergusson D, Maltais F, Bourbeau J, Goldstein R, Balter M, O'donnell D, McIvor A, Sharma S, Bishop G, Anthony J, Cowie R, Field S, Hirsch A, Hernandez P, Rivington R, Rad J, Hoffstein V, Hodder R, Marciniuk D, McCormack D, Fox G, Cox G, Prins HB, Ford G, Bleskie D, Doucette S, Mayers I, Chapman K, Zamel N, Fitzgerald M. Tiotropium in combination with placebo, salmeterol, or fluticasone-salmeterol for treatment of chronic obstructive pulmonary disease: A randomized trial. *Ann Intern Med* 146: 545–55, 2007.

68

CHAPTER

Education and Self-Management

Martyn R. Partridge

The Faculty of Medicine, Imperial College, London, UK

Globally, twice as many people have asthma as the total population of the Russian Federation. In the United States of America alone, the number of people with chronic obstructive pulmonary disease is equivalent to the combined populations of Denmark, Sweden, Finland, Norway, and the Baltic States. For some of these people, we have therapies that can dramatically influence their diseases. Others have persisting and regular symptoms, and all have to live with a long-term condition. It is likely that the majority have only 30–60 min of contact on average with a health professional in any 1 year and this means that for the other 364 days and 23 h, these patients are self managing their own condition. As health professionals we have a responsibility to ensure that these patients are equipped with the tools, knowledge, and skills necessary to self manage and self treat their own condition.

This chapter is concerned with patient education and self-management and will review the evidence in favor of those interventions and highlight similarities and differences between self-management education in asthma and chronic obstructive pulmonary disease (COPD).

WHAT IS PATIENT EDUCATION?

The term patient education has an unpleasant inference suggesting some inadequacy on the patient's behalf which needs rectifying. It seems preferable to make the concept more positive and to list the constituent parts. Van den Borne [1] has defined patient education as "a systematic learning experience in which a combination of methods is generally used, such as the provision of information and advice and behavior modification techniques, which influence the way the patient experiences their illness and/or their knowledge and health behavior, aimed at improving or maintaining health or learning to cope with a condition, usually a chronic one."

For the health care professional an essential prerequisite to patient education is an understanding of how the patient feels about their long-term condition and its management.

HOW DOES IT FEEL TO HAVE COPD?

COPD is largely a smoking-related disease. In many countries the prevalence of smoking is low amongst health care professionals, especially amongst doctors. Personal experience of the condition is therefore likely to be uncommon. Surveys of the views, opinions, and concerns of those with COPD are less numerous than are those amongst people with asthma, and the population affected less diverse – being mainly a disease of the fifth, sixth, and seventh decades. Currently more men than women are affected, but this is rapidly changing.

Those with COPD frequently suffer feelings of guilt about having caused the disease by smoking, and they experience great sadness when they see others around them smoking and proceeding along their path. They frequently report lack of support on social issues, financial issues, and from health care professionals. Professionals imbued with a culture of curing and obsessed with the writing of prescriptions may not actually say "There is nothing I can do for you," but they can convey such negativity by body language or attitude. The late Trevor Clay, a nurse who died from lung disease associated with an inherited condition wrote "There is no cure, no magic, but there is always something that can be done!" He also wrote "Having a long-term condition is not about dying – that only takes a few minutes or less – but I've been struggling to breathe for over 20 years and I've been living a lot and suffering as little as possible"[2].

Others may find it less easy to be positive. Anxiety, frustration, and depression are frequently experienced by sufferers and they may become socially isolated [3].

Psychiatric disorders may reach a prevalence of 50% in those with COPD, [4] and Dudley *et al.* [5] observed those with severe COPD to be in "emotional strait jackets" – no longer able to become angry, depressed, or even happy, as any significant emotional changes triggered distressing symptoms. The mind/body interaction can thus become a vicious cycle with symptoms leading to anxiety, and emotional distress aggravating the symptoms. Psychological impairment may also follow from derangement of blood gases leading to diminished levels of alertness, irritability, restlessness, headaches, and confusion and loved ones may be unaware of the reasons for these factors in their partners. Psychological and personality profiles of those with severe COPD may also influence survival. One study of males with severe COPD showed a significant difference in personality and psychological profile between those who died and those who were alive at the end of 4 years of follow-up, irrespective of the degree of impairment of pulmonary function or oxygenation [6].

In those at the severe end of the spectrum of COPD, a common therapeutic intervention is the use of supplementary oxygen. When this is recommended to be used long term, it could impact significantly upon the patient's life and may enhance feelings of social isolation [7]. In another survey of those on long-term oxygen from oxygen concentrators, the results were more positive. Eighty-three percent of those surveyed reported marked improvement in general well-being on oxygen, 82% reported improvement in breathing and 62% mobility, and 52% reported improvement in sleep pattern. A third thought that the concentrator was too noisy. Sadly, a third of those prescribed long term oxygen continued to smoke [8].

It is also important to be aware of patients' fears, concerns and expectations. Studies of those with COPD which have used rigorous mathematical techniques such as discrete choice modeling, have shown that patients most fear being hospitalized, housebound or bedridden and they fear more attacks in the future [9, 10]. These fears and concerns are greater than those related to the symptoms of the disease, such as breathlessness or a productive cough. Indeed exacerbations of COPD have the most negative impact upon our patients and 50% need additional help with simple tasks during exacerbations and nearly half of all patients have to stop all activities during such exacerbations. It should also be noted that the word "exacerbation" whilst used between health professionals, is not understood by over 60% of patients who are more likely to talk of "a crisis," "an attack," breathing difficulty or "a chest infection."

HOW DOES IT FEEL TO HAVE ASTHMA?

The ages of those suffering from asthma is more diverse than those suffering from COPD. Furthermore, whilst long-term disability and daily symptoms are common in asthma, the symptoms may be more variable. Such variability can induce additional stresses, and uncertainty invokes an unpleasant emotion with fears of holidays, celebrations, or important work events being interrupted by unexpected exacerbations. Denial of the diagnosis or its implication is common [11] and feelings of stigma frequent, but there is no easy comparison to suggest whether it is higher or lower than in COPD. Fears and concerns regarding the medication are common, although steroid phobia amongst patients and parents may be perceived by health professionals to be more common than it really is. Dissatisfaction with dependency on long-term medication may be one of the most common reactions and patients frequently stop medication just to confirm continued need. What is unclear is why there is such a large difference between the goals for asthma management as outlined in guidelines, and the control as discovered by surveys of patients. In one large UK study, 44% of respondents reported at least one activity was "totally or very limited" by their asthma and 20% reported three or more activities to be so limited [12]. Another survey conducted 400 interviews with current asthma patients in seven countries. Over one-third of children and half of the adults reported daytime symptoms at least once a week. Sleep disruption every night was reported by 6.7% of children and 5.3% of adults. A total of 36% of children and 27.9% of adults required an unscheduled urgent care visit in the past 12 months. One or more emergency room visits due to asthma were reported for 18% of children and 11% of adults in the past year [13]. In that survey, only 25% of the patients were taking regular maintenance therapy and it is not therefore clear whether the ongoing morbidity reflected undertreatment or an inadequacy of treatment. A more recent study has looked only at those using regular maintenance therapy and studied over 3400 adults with asthma in several European countries and also in North America and Australia. No significant differences were seen between countries but despite the use of regular medication, over three quarters of these patients needed rescue therapy each day and using the Juniper asthma control questionnaire, only 28% of these patients were shown to be well-controlled. Furthermore, many of these patients perceived their asthma to be better controlled than the ACQ suggested [14]. Some of this morbidity may genuinely reflect the fact that we do not have treatments to control the asthma symptoms of all our patients, but in other cases the patients may be undertreated or may not be reviewed by health care professionals with the result that treatment is not optimized. Only half of those surveyed in the UK [12] had their peak flow measured by a doctor or a nurse in the past year, and 45% of respondents said they had neither had, nor wanted, regular asthma review. Without such review there is a danger that adaptation leads to acceptance of ongoing morbidity. Even with review, there is a danger that the patient may not "offer" symptoms for fear that it may lead to further prescriptions, and the doctor may falsely conclude that all is well.

For these reasons it has been recommended that at each consultation every patient with asthma is asked three questions: [15]

In the last week or month

- Have you had difficulty sleeping because of your asthma symptoms (including cough)?

- Have you had your asthma symptoms during the day (cough, wheeze, chest tightness, or breathlessness)?
- Has there been any limitation of activities (time off work or school or inability to undertake hobbies) because of asthma. Has your asthma interfered with your usual activities (e.g. housework, work/school, etc.)?

Other validated measures of control also exist.

The willingness of patients with asthma to take control of their condition is probably greater than doctors recognize. In one study, whilst patients varied in their desire to be involved in treatment decisions, 55% were less involved than they wished to be [16] and in another study, only 20% of patients had actually been offered a written asthma action plan but when questioned regarding their possession of asthma action plans, over 60% of patients wished to have one [17]. It has also been shown that those with asthma willingly admit to using medication as and when necessary, and express a desire to adjust their dose of maintenance therapy to the changes of their asthma and express concern at being on too much medication when they are well [14]. This suggests an intrinsic willingness on behalf of our patients to take control of their own condition if instructed in the appropriate manner.

COMPLIANCE IN ASTHMA AND COPD

The term noncompliance is used to describe a situation where, for whatever reason, the patient does not take treatment or other actions in a manner as previously discussed with their health professional. Recent trends have been to replace the word compliance with "adherence" or "concordance," but this seems unnecessary if it is emphasized from the outset that the term is not being used in any way in a pejorative sense. Noncompliance may involve noncompliance with lifestyle advice (e.g. continued smoking), failure to attend for follow-up, failure to undertake recommended monitoring, or failure to take therapy. The size of the problem is likely to be large and underestimated.

Noncompliance with medication is either inferred from the presence of poorly controlled disease or confirmed by monitoring drug taking, which may involve measurement of drug levels in urine, plasma or saliva, or by prescription monitoring. More modern methods involve the fitting of microprocessors to the lids of bottles or inhaler devices [18–20]. Such methods are inappropriate at a clinical level and it is preferable to accept that noncompliance is common, and instead make efforts to consider the factors involved and work with the patient to tackle the underlying causes. It is likely that the size of the problem is of similar magnitude in asthma as in COPD, and these two diseases probably do not differ from other long-term conditions such as hypertension or arthritis. Some of the causes may however be disease-specific and possibilities are listed in Table 68.1.

Perhaps the most essential is to understand the importance of good communication between patient and health care professional. In one UK study, [21] only 22% of those with asthma reported having had a good discussion with their doctor, and in another study a median dissatisfaction rate of 38% with medical communication was reported [22]. The effect of this upon compliance may be considerable. In one study, 50 adults with moderate to severe asthma were studied and compliance with inhaled steroid therapy electronically monitored. Mean adherence was 63%. Factors associated with poor adherence included less than 12 years of formal education, and a low household income, but poor patient/clinician communication was independently associated with poor adherence. Those with at least 70% adherence scored the patient/clinician communication significantly better than those with less than 70% adherence [23]. Key elements of good communication are shown in Table 68.2.

TABLE 68.1 Factors which may be involved in noncompliance with the taking of medication in asthma and COPD.

Factors associated with medication
Use of the word "drug"
Difficulties with inhaler devices
Regimens involving multiple medications
Awkward four times daily dosing regimens
Side-effects
Cost of medication
Difficulty getting to the doctor (for a prescription) or the pharmacy (for it to be dispensed)
Perceived lack of effectiveness of medication
Nonmedication factors
Denial of diagnosis
Fears of side-effects
Unexpressed/unanswered concerns
Dissatisfaction with health care professionals
Misunderstanding or lack of instruction
Anger, stigma, or depression
Underestimation of severity
Forgetfulness or complacency
Cultural issues

TABLE 68.2 General guidelines on ways to improve communication with patients and their families.

Be attentive
Elicit underlying concerns
Offer reassuring messages that alleviate fears
Immediately address any concerns that are mentioned
Use interactive dialog (open-ended questions, analogies)
Tailor the therapeutic regimen to lifestyle
Provide a written management plan
Use appropriate nonverbal engagement
Use praise when patient has undertaken correct management strategies
Elicit goals; share goals
Help the patient plan longer-term care and self-management

Adapted from Ref. [47].

WHAT ARE THE CONSTITUENT PARTS NECESSARY FOR SUCCESSFUL EDUCATION AND SELF-MANAGEMENT IN COPD: WHO SHOULD PROVIDE IT AND WHERE?

Clear, evidence-based advice regarding education and self-management is far harder to offer to those with COPD than it is for asthma. In both disease groups there is a problem in published reports that often give too little information about the intervention offered and do not describe the self-management advice given. In studies involving COPD, this is further compounded by a difficulty in separating the educational components from the support and the physical exercise components of pulmonary rehabilitation programs. It is likely that education and self-management programs for COPD should cover the ground outlined in Table 68.3. Self-management involves both alterations in lifestyle and alterations in treatment, and it is immediately apparent that the balance between these two is different for COPD than it is for asthma (as shown in Table 68.4).

One study of such interventions allocated 56 subjects with COPD to either usual care or to receiving a booklet (outlining advice presumed to be similar to that in Table 68.3), an action plan, and a reserve supply of steroids and antibiotics. After 6 months there were no differences in quality of life scores or pulmonary function, but those given the self-treatment advice were significantly more likely to have started steroids or antibiotics in response to deteriorating symptoms [24]. Numbers were too small to look at effects upon hospitalization rates and larger studies would need to include cost-effectiveness evaluation. Some pretest/posttest nonrandomized studies have suggested that "education" may reduce hospitalization rates, [25] but a 2003 systematic review of eight previous studies which looked at self-management education in COPD *alone* (i.e. excluding studies which included multiple interventions such as pulmonary rehabilitation) concluded that there was too little evidence to yet say that self-management in COPD was associated with positive outcomes [26]. The reasons why the benefits of self-management education in COPD appears equivocal compared to the definite advantages in asthma are unclear but the systematic review did highlight the fact that, whereas action plans have been

TABLE 68.3 Self-management advice for those with COPD.

Lifestyle changes	Treatment changes
1. Stop smoking (and avoid smoky environments)	**1.** Continue regular bronchodilators – usually a combination of anticholinergic agents and β-agonists
2. Use nicotine replacement therapies as appropriate as advised	**2.** At times of worsening symptoms increase dose, frequency, and possibly route of administration for example, spacer or nebulizer
3. Use effective breathing methods	**3.** If sputum changes color, consider starting reserve course of antibiotics
4. Use effective coughing methods	**4.** If much more breathless, and lessening response to bronchodilators, consider a course of steroid tablets according to doctor's advice
5. Undertake your exercise program as advised during your pulmonary rehabilitation course	**5.** Use oxygen as advised – either long-term or supplementary during exertion – know when to increase this and be aware of the importance of early morning confusion or headaches which might suggest CO_2 retention.
6. Eat a balanced diet: include plenty of fresh fruits and vegetables and drink plenty of fluids to help keep mucus thin. Avoid gas-forming foods such as broccoli, cabbage, onions, beans, and sauerkraut. If eating makes you breathless, use supplementary oxygen whilst chewing, or liquidize solids	
7. Adjust daily activities of living. Sit down to do personal tasks such as washing or shaving or doing household tasks such as washing up or preparing meals. Use a stool in the shower and use a hairdryer to dry feet or back.	

TABLE 68.4 Self-management advice for those with asthma.

Lifestyle changes	Treatment changes
1. Allergen avoidance	**1.** If you have no symptoms and your peak flow is better than 80–85% of your best peak flow, continue your regular preventative treatment, or talk to your doctor or nurse about taking less treatment
2. Avoiding smoking and smokey environments	**2.** If you get a cold, or have your asthma symptoms during the day or at night, or if your peak flow is less than 80–85% of your best peak flow, increase your preventer treatment according to your written personal action plan
3. Avoid exercising outdoors, especially on the outskirts of cities, at times of high pollution levels	**3.** If you are increasingly breathless, and your reliever therapy is less effective and your peak flow is less than 60–70% of your usual best peak flow, start steroid tablets according to your written personal action plan and contact your doctor
4. Eat a balanced diet	**4.** If you are too breathless to speak, or your peak flow is less than 40–50% of your usual best peak flow, continue to use your reliever, take 8 of the 5 mg prednisolone tablets, and call your doctor or an ambulance urgently

shown to be an important part of asthma self-management, in only two of the studies included in the COPD systematic review were action plans part of the intervention. In one of those studies, the action plan which had been issued was in fact an asthma action plan and did not include the patient self-treating with antibiotics and in the other an appropriate COPD action plan was given but the study was not powered to look at harder outcomes. A subsequent systematic review of the use of action plans in COPD was still only able to find three studies and whilst this demonstrated that use of an action plan in COPD increased the patients' ability to recognize a severe exacerbation and increased their use of antibiotics and steroid tablets, it was not able to demonstrate any effect on health care utilization, quality of life, or symptom scores [27]. A study of the use of self-management plans in the primary care of patients with COPD in New Zealand showed that the use of self-management plans within a structured education program was associated with higher levels of self-management knowledge but actually had no effect on quality of life, health care utilization, or self-reported outcomes [28]. It seems likely that the magnitude of any benefit from self-management education in COPD alone is limited, despite the fact that such tools can alter patient behavior. This may reflect the lesser magnitude of benefit of, for example, steroid tablets in the treatment of exacerbations of COPD compared with that in asthma. We should not preclude the use of action plans on these grounds alone, for we know, for example, that the quicker an exacerbation of COPD is treated, the more effective the intervention and self-treatment is always likely to be quicker than that associated with the inevitable delay in contacting health professionals. Such self-management advice probably begins to contribute to improved outcomes when it is combined with other interventions, such as case management and pulmonary rehabilitation. A multicenter French-Canadian study suggested that amongst those who had one hospitalization for an exacerbation of COPD, such multiple interventions may lead to up to 40% reduction in further emergency department attendances or admissions to hospitals [29].

Most reports of educational and self-management activities in COPD have involved respiratory nurse specialists, physiotherapists, or respiratory therapists. Most studies have been outpatient based. Where outreach home-based programs have been assessed, they have shown no significant reductions in hospitalization rates, but there may be some health-related quality of life gains [30].

PATIENT EDUCATION AND SELF-MANAGEMENT IN ASTHMA

In contrast to the situation in COPD, studies of this area involving those with asthma are numerous and the results in adults and older children are almost always positive, especially involving those who have attended hospital-based programs [31–33]. The key constituents of such interventions are shown in Table 68.4.

A Cochrane review of the subject of self-management education for adults with asthma was undertaken by Gibson and colleagues in 2002 and compared self-management education and follow up with usual care in 36 studies [34]. Self-management education was associated with reduction in hospitalization rates, emergency room visits, unscheduled visits to the doctor, days off work or school, and nighttime asthma. Self-management programs that involved a written action plan showed greater reduction in hospitalization rates than those that did not, and people who managed their asthma by self-adjustment of their asthma treatment using an individualized written plan had better lung function than those whose medications were adjusted by a doctor.

Fewer studies have been undertaken of the value of self-management in children, but good randomized controlled trials involving self-management interventions in those who have been hospitalized, have shown significant reduction in readmission rates [35–36]. Studies of the cost-effectiveness of the teaching of self-management skills in those with asthma have shown significantly beneficial cost-benefit ratios [37–38].

This wealth of evidence in favor of self-management education, regular follow up, and the use of written action plans has not unfortunately led to widespread implementation. In the first British Asthma Guidelines, published in 1990 [39], it said: "As far as possible, patients should be trained to manage their own treatment rather than being required to consult their doctor before making changes." Unfortunately, if we use the possession of a written asthma action plan as a marker of how satisfactorily this process has been undertaken, a UK study 10 years after those Guidelines showed that only 3% of patients had such a plan [40]. However, that figure did rise to 28% of 378 individuals who were surveyed 1 week after need for unscheduled health care because of out-of-control asthma and this could imply some targeting of advice to those most at risk [41]. A more recent study again showed 20% of patients having such a plan [17]. Even in Australia, where with national publicity campaigns they had once reached 48% of patients having action plans, the latest figures show that has fallen to 18.5% of patients owning an asthma plan [42]. The reasons why doctors are not implementing a core part of self-management education which is of proven benefit are not clear, but probably revolve around a perceived lack of time, lack of suitable templates, and possibly a lack of understanding as to what is involved in this process and what needs to be written down for patients. As computer decision support systems evolve, and it becomes easier to link consultation records with information materials, it's likely that ease of issuing of personalized action plans will improve.

HEALTH LITERACY

When offering education and information materials to our patients, it is important to ensure that the patients have the capacity to process and understand the information which we have given them. This entails ensuring not only that spoken information is provided at an appropriate level but that

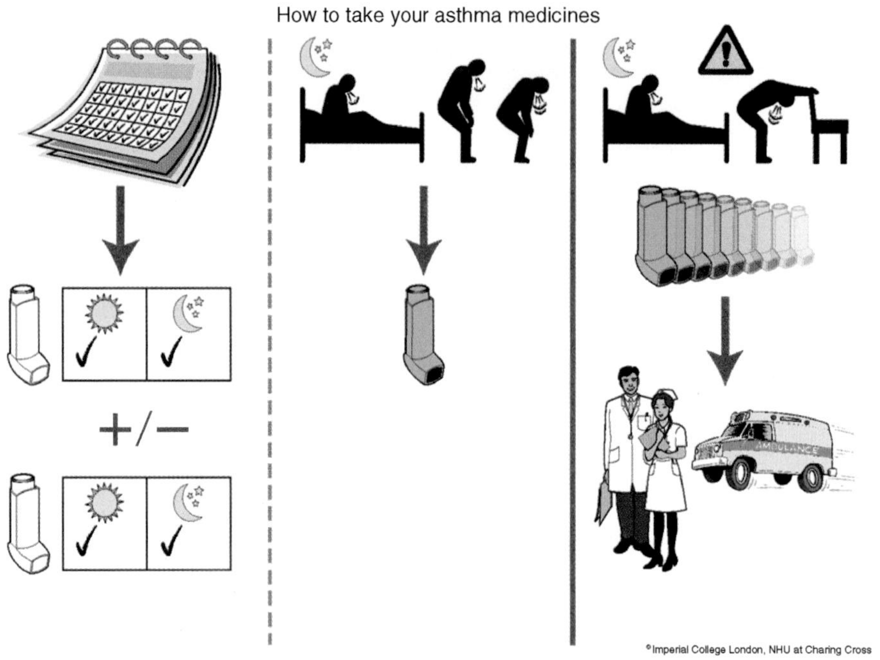

FIG. 68.1 An example of a pictorial asthma action plan.

written materials are readable and reliable. It is possible to formally test patients' ability to use materials using tests such as the Rapid Estimate of Adult Literacy in Medicine (REALM) [43] or the Test of Functional Health Literacy in Adults (TOFHLA) [44]. When such tests have been utilized in studies, 15% of those with rheumatoid arthritis attending a hospital rheumatology clinic in Glasgow had been shown to be functionally illiterate [45] and 13% of patients with asthma attending a US hospital were similarly unable to use the written word in a meaningful manner [46]. Such formal testing is neither possible nor appropriate in clinical practice but we should be aware of the likely magnitude of this problem and consideration should be given not only to producing written materials with an appropriate reading age but also to the possibility of reinforcing the spoken word, not with the written word but for example by the use of pictograms. Such tools have been shown to enhance retention of information, even amongst the fully literate. An example of a pictorial asthma action plan is shown in the Fig. 68.1.

SUMMARY

- Living with a long-term illness can be difficult.
- Health care professionals need an understanding of how their patients feel, and good communication is essential if we are to understand fully what our patients want of us, and how best we can provide optimal care.
- It is important to offer the information the patient wants, rather than what we perceive they need.
- Personalized self-management advice which helps them to adjust their lifestyle or treatment to keep themselves well, is necessary. Such advice needs to be produced in a suitable format to ensure it is usable by those with varying literacy skills.

Current evidence is that self-management education is of significant value in those with asthma but more marginal value in COPD, where it may only be beneficial if combined with other interventions such as pulmonary rehabilitation and case management.

References

1. van den Borne HW. The patient: From receiver of information to informed decision maker. *Patient Edu Couns* 34: 89–102, 1998.
2. Clay T. How to keep the customer satisfied. *Thorax* 49: 279–80, 1994.
3. Dudley DL, Glaser EM, Jorgenson BN, Logan DL. Psychological concomitants to rehabilitation in chronic obstructive pulmonary disease. Part I: Psychosocial and psychological considerations. *Chest* 77: 413–20, 1980.
4. Rutter BM. Some psychological concomitants of chronic bronchitis. *Psych Med* 7: 459–64, 1977.
5. Dudley DL, Wermuth C, Hague W. Psychological aspects of care in chronic obstructive pulmonary disease. *Heart Lung* 2: 289–303, 1973.
6. Ashutosh K, Haldipur C, Boucher ML. Clinical and personality profiles and survival in patients with COPD. *Chest* 111: 95–98, 1997.
7. Ring L, Danielson E. Patients experiences of long term oxygen therapy. *J Adv Nurs* 26: 337–44, 1997.
8. Dilworth JP, Higgs CMB, Jones PA, White RJ. Acceptability of oxygen concentrators: The patient's view. *Br J Gen Pract* 40: 415–17, 1990.
9. Kessler R, Stahl E, Vogelmeier C, Haughney J, Trudeau E, Lofdahl C-G, Partridge MR. Patient understanding, detection and experience of COPD exacerbations: An observational, interview-based study. *Chest* 130: 133–42, 2006.
10. Haughney J, Partridge MR, Vogelmeier C, Larsson T, Kessler R, Stahl E, Brice R, Lofdahl C-G. Exacerbations of COPD: Quantifying the patient's perspective using discrete choice modeling. *Eur Respir J* 26: 623–29, 2005.

11. Adams S, Pill R, Jones A. Medication, chronic illness and identity. The perspective of people with asthma. *Soc Sci Med* 45: 189–201, 1997.
12. Price D, Wolfe S. Delivery of asthma care: Patients use of and views on healthcare services as determined from a nationwide interview survey. *Asthma J* 5: 141–44, 2000.
13. Rabe KF, Vermeire PA, Soriano JB, Maier WC. Clinical management of asthma in 1999: The asthma insights and reality in Europe (AIRE) study. *Eur Respir J* 16: 802–7, 2000.
14. Partridge MR, van der Molen T, Myrseth S-E, Busse WW. Attitudes and actions of asthma patients on regular maintenance therapy: The INSPIRE Study. *BMC Pulm Med* 6: 13, 2006.
15. Clinical Effectiveness and Evaluation Unit. Measuring Clinical Outcome in Asthma. A Patient-Focused Approach. London: Royal College of Physicians of London, 1999.
16. Caress AL, Beaver K, Luker K, Campbell M, Woodcock A. Involvement in treatment decisions: What do adults with asthma want and what do they get? Results of a cross sectional survey. *Thorax* 60(3): 177–78, 2005.
17. Haughney J, Barnes G, Partridge MR, Cleland J. The Living and Breathing Study: A study of patients views of asthma and its treatment. *Prim Care Respir J* 13: 28–35, 2004.
18. Cranmer JA, Mattson RH, Prevey MC, Scheyer RD, Ovelette VL. How often is medication taken as prescribed. *JAMA* 261: 3273–77, 1989.
19. Rand CS, Wise RA, Nide S *et al.* Metered dose inhaler adherence in a clinical trial. *Am Rev Respir Dis* 146: 1559–64, 1992.
20. Cochrane GM. Compliance in asthma: A European perspective. *Eur Respir Rev* 5: 116–19, 1995.
21. Partridge MR. Asthma: Lessons from patient education. *Patient Edu Couns* 26: 81–86, 1995.
22. Ley P. Communicating with Patients. London: Croon Helm, 1988.
23. Apter AT, Reising ST, Affleck G, Barrows C, ZuWallack RL. Adherence with twice daily dosing of inhaled steroids: Socio-economic and health belief difference. *Am J Respir Crit Care Med* 157: 1810–17, 1998.
24. Watson PB, Town GI, Holbrook N, Dwan C, Toop LJ, Drennan CJ. Evaluation of a self-management plan for chronic obstructive pulmonary disease. *Eur Respir J* 10: 1267–71, 1997.
25. Howard JE, Davies JL, Roghmann KJ. Respiratory teaching of patients: How effective is it? *J Adv Nurs* 12: 207–14, 1987.
26. Monninghkhof E, van der Valk P, van der Palen J, van Herwaarden C, Partridge MR, Zielhuis G. Self management education for patients with chronic obstructive pulmonary disease: A systematic review. *Thorax* 58(5): 394–98, 2003.
27. Turnock AC, Walters EH, Walters JAE, Wood-Baker R. Action plans for chronic obstructive pulmonary disease. *Cochrane Database Syst Rev* 4: 2005, Art. No.: CD005074. DOI:10.1002/14651858.CD005074.pub2.
28. McGeoch GRB, Willsman KJ, Dowson CA, Town GI, Frampton CM, McCartin FJ, Cook JM, Epton MJ. Self management plans in the primary care of patients with chronic obstructive pulmonary disease. *Respirology* 11: 611–18, 2006.
29. Bourbeau J, Julien M, Maltais F, Rouleau M, Beaupre A, Begin R, Renzi P, Nault D, Borycki E, Schwartzman K, Singh R, Collet J-P. Reduction of hospital utilization in patients with chronic obstructive pulmonary disease: A disease-specific self-management intervention. *Arch Intern Med* 163: 585–91, 2003.
30. Smith B, Appleton S, Adams R, Southcott A, Ruffin R. Home care by outreach nursing for COPD (Cochrane Review). *Cochrane Libr* 4: Oxford: Update Software, 2000.
31. Ignacio-Garcia JM, Gonalez-Santos P. Asthma self-management education program by home monitoring of peak expiratory flow. *Am J Respir Crit Care Med* 151: 353–59, 1995.
32. Lahdensuo A, Haahtela T, Herrala J *et al.* Randomised comparison of self management. *BMJ* 312: 748–52, 1996.
33. Adams RJ, Smith BJ, Ruffin RC. Factors associated with hospital admissions and repeat emergency department visits for adults with asthma. *Thorax* 55: 566–73, 2000.
34. Gibson PG, Powell H, Coughlan J, Wilson AJ, Abramson M, Haywood P, Bauman A, Hensley MJ, Walters EH. Self-management education and regular practitioner review for adults with asthma. *Cochrane Database of Syst Rev* 3: 2002, Art. No.: CD001117. DOI: 10.1002/14651858.CD001117.
35. Madge P, McColl J, Paton J. Impact of a nurse led home management training programme in children admitted to hospital with acute asthma: A randomised controlled study. *Thorax* 52: 223–28, 1997.
36. Wesseldine L, McCarthy P, Silverman MA. structured discharge procedure for children admitted to hospital with acute asthma: a randomised controlled trial of nursing practice. *Arch Dis Child* 80: 110–14, 1999.
37. Lahdensuo A, Haahtela T, Herrala J. Randomised comparison of cost-effectiveness of guided self-management and traditional treatments of asthma in Finland. *BMJ* 316: 1138–39, 1998.
38. Clark NM, Feldman CH, Evans D, Levison MJ, Wasilewski Y, Mellins RB. The impact of health education and cost of health care use by low income children with asthma. *J Allergy Clin Immunol* 78: 108–15, 1986.
39. Guidelines for Management of Asthma in Adults: I – Chronic Persistent Asthma. Statement by the British Thoracic Society, Research Unit of the Royal College of Physicians of London, Kings Fund Centre, National Asthma Campaign. *BMJ* 301: 651–3, 1990.
40. National Asthma Campaign. Needs of People with Asthma Survey. *Asthma J* 5: 141–44, 2000.
41. Pinnock H. *Asthma J* 5: 130–32, 2000.
42. Ruffin RE, Wilson D, Adams R. *ERJ* 22(2): 295s, 2003.
43. Davis TC, Long SW, Jackson RH *et al.* Rapid estimate of adult literacy in medicine: A shortened screening instrument. *Fam Med* 25: 391–95, 1993.
44. Parker RM, Baker DW, Williams MV *et al.* The test of functional health literacy in adults: A new instrument for measing patients literacy skills. *J Gen Intern Med* 10: 537–41, 1995.
45. Gordon M-M *et al. Rheumatology* 41: 750–54, 2002.
46. Williams MV *et al. Chest* 114: 1008–15, 1998.
47. Partridge MR, Hill SR. Enhancing care for people with asthma: The role of communication, education, training and self-management. *Eur Respir J* 16: 333–48, 2000.

Index

Page numbers in **bold** refer to major discussions
Page numbers in *italics* refer to pages on which figures and/or tables appear.
Since the major subjects of this book are asthma and COPD, entries have been kept to an absolute minimum under these keywords. Readers are advised to seek more specific index entries